ENCYCLOPEDIA OF

AQUACULTURE

ENCYCLOPEDIA OF AQUACULTURE

ENCYCLOPEDIA OF

AQUACULTURE

Robert R. Stickney
Texas Sea Grant College Program
Bryan, Texas

A Wiley-Interscience Publication
John Wiley & Sons, Inc.
New York / Chichester / Weinheim / Brisbane / Singapore / Toronto

This book is printed on acid-free paper. ♾

For ordering and customer service, call 1-800-CALL-WILEY.

Library of Congress Cataloging in Publication Data:

Encyclopedia of aquaculture / [edited by] Robert R. Stickney.
p. cm.
Includes bibliographical references and index.
ISBN 0-471-29101-3 (alk. paper)
1. Aquaculture Encyclopedias. I. Stickney, Robert R.
SH20.3.E53 2000
639.8′03—dc21 99-34744

Printed in the United States of America.

10 9 8 7 6 5 4 3 2 1

FOREWORD

There are many definitions of the word aquaculture. Those concerned with the collation of statistical data concerning food production through aquaculture tend to be very specific; they embody the concept of stock ownership as well as its management, to distinguish between the harvest from capture fisheries and from farming. One simpler definition[1] of aquaculture is the "cultivation of plants or breeding of animals in water." Many different activities fall within this definition. The farming of aquatic animals and plants for direct or indirect human consumption is the field with which I am most familiar but it is clear that this definition of aquaculture would encompass many other activities, including the rearing of aquatic animals and plants for and within public and private aquariums and research facilities, the production of bait fish, and the hatchery and nursery rearing of stock intended for fisheries enhancement or restocking programs. In aquatic food production the word aquaculture has sometimes erroneously been used to imply culture in freshwater, while the word mariculture has been used to refer to culture in seawater. In fact, the word aquaculture embraces culture in all salinities, ranging from freshwater through brackishwater and full-strength seawater to hypersaline water.

The production of an aquaculture encyclopedia at this moment in history is particularly appropriate, since the positive and negative impacts of food production through aquaculture are frequently discussed by scientists not working in this specific field, by the media, and by the public. Often, such discussions are marred by misunderstandings about the various terms utilized. The public image of aquaculture is not always good. While some ventures have undoubtedly caused environmental and/or socioeconomic harm in the past, the emphasis now is on sustainable aquaculture, which implies responsibility. The FAO Code of Conduct for Responsible Fisheries includes many Articles which are specific or related to aquaculture. Many other attempts are being made to enhance the responsibility of aquaculture producers, which range from large commercial enterprises providing products for domestic and export markets to small-scale rural farmers seeking to produce family food and income. Attempts to mollify consumer concern for the environment through the "eco-labelling" of aquaculture products produced under responsible conditions are on-going.

[1] J.B. Sykes (Editor), 1982. *The Concise Oxford Dictionary*, Oxford University Press, Oxford, Seventh Edition 1982, Reprinted 1989.

The scale and importance of food production through aquaculture can be illustrated by a few examples:

- By 1996, more than nine out of every ten oysters, Atlantic salmon, and cyprinids consumed were products of aquaculture. Four out of every five mussels and three out of four scallops were cultured; 27% of all shrimp originate from aquaculture;
- In 1997 (the most recent year for which international statistics are available), global aquaculture production totalled 28.8 million tons of finfish, crustaceans, and molluscs for direct human consumption, worth US$45.5 billion; 7.2 million tons of seaweed (worth US$4.9 billion) were also produced;
- A considerable proportion of the harvest from capture fisheries is destined for the production of fish meal and fish oil, which are primarily used by the feedstuff industry. Capture fisheries production available for human consumption has been on a plateau or increased only slowly for many years;
- Aquaculture thus remains the major means of maintaining current per capita "fish" availability. It has been estimated that global aquaculture production will need to expand to 62 million tons by 2035 to maintain 1993 global average per capita consumption levels.

The *Encyclopedia of Aquaculture* will assist the many scientists, economists, sociologists, administrators, and politicians who are either directly involved in aquaculture itself or are concerned with resource use and environmental matters. The book will also be useful for those concerned with development and planning issues. In addition, this book provides information of relevance to those in the general public who consume aquaculture products, engage in recreational fisheries or keep aquariums, as well as those who belong to organizations concerned with animal welfare and environmental conservation.

The *Encyclopedia of Aquaculture* will thus serve as an essential handy reference book for a very wide audience, and its Editor-in-Chief and Editorial Board are to be congratulated on undertaking the task of producing this unique document. I hope all its readers will find it as useful as I shall.

Michael B. New
Past President, World Aquaculture Society
Board Member, European Aquaculture Society

PREFACE

Aquaculture is the production of aquatic plants and animals under controlled or semicontrolled conditions, or as is sometimes said, aquaculture is equivalent to underwater agriculture (1). The term mariculture refers to the production of marine organisms; thus, it is less inclusive than aquaculture, which relates to both marine and freshwater culture activities.

A primary goal of aquaculturists has been to produce food for human consumption. Various species of carp top the list in terms of aquacultural production. Most of that production is in China, though India and certain European nations also produce significant amounts of carp. In North America, channel catfish farming is the largest aquaculture industry. Others of importance include trout, crawfish, and various species of shellfishes. Seaweed culture as human food is a major industry, particularly in Japan and other Asian nations.

Supplementing the human food supply is not the only goal of aquaculturists. Many of the species taken by recreational anglers are produced in hatcheries and reared to a size where they can be expected to have a good chance of survival before being released into the natural environment. Continuous stocking may be necessary in some bodies of water, while in others resident breeding populations may become established. Examples in North America are largemouth bass, northern pike, muskellunge, red drum, various species of trout, Atlantic salmon and Pacific salmon. Many of the fish produced for stocking purposes are reared in public (state or federal) hatcheries, but increasingly, private hatcheries are becoming a source of fish, particularly in conjunction with stocking farm ponds and private lakes.

The ornamental fish industry depends on animals caught in the wild and on those produced by aquaculturists. Most of the bait minnows available in the marketplace come from fish farms. Seaweeds are not only consumed by people as food, they are also a source of such chemicals as carrageenan and agar, which are utilized in everything from toothpaste and cosmetics to automobile tires. Squid and cuttlefish are not being produced to any extent as human food, but they are reared as a source of giant axons for use in biomedical research. An increasing number of potential pharmaceuticals are being identified from marine organisms. Culture of various species from a number of phyla, many of which have held little or no interest for aquaculture in the past, show promise as one means of meeting the demand for cancer-fighting and other types of drugs. A new, and potentially large aquaculture enterprise could be founded upon such species.

The roots of aquaculture can be traced back to China, perhaps as much as 4,000 years ago. Many nations have had some form of aquaculture in place for one or more centuries, but it is only since about the 1960s that scientists began to conduct research that brought the discipline to its current level of development. Since 1960, typical annual pond production rates have jumped from a few hundred kg/ha (one kg/ha is approximately equivalent to one pound/acre) to several thousand kg/ha. Much higher rates of production are possible in such water systems as raceways and marine net-pens, which are known as intensive culture systems. Ponds are generally considered to be extensive culture systems.

Improvements in production over the past few decades have been associated with the development of sound management techniques that include water quality and disease control, provision of nutritionally complete feeds, and the development of improved stocks through selective breeding, hybridization, and the application of molecular genetics technology. Many species that could not be spawned or reared a few decades ago are now being produced, because of technological breakthroughs, in large quantities by aquaculturists around the world.

Predicted peaking of the world's wild capture fishery at 90 million metric tons (about 99 million short tons) occurred in 1989 (2). Since that time global wild capture landings have been relatively stable. Given increasing demand for seafood, including freshwater aquatic species, and a stable to declining wild catch, the shortfall must come from aquaculture. As of 1992, about 18.5% of global fisheries output was attributable to aquaculture (3), and while aquaculture production is increasing, there is some question as to whether the growth of aquaculture can keep pace with demand.

In 1992, 88.5% of the world's aquaculture production came from Asia (3). Because of suitable growing conditions year round, the vast majority of aquaculture production comes from low temperate and tropical regions. Relatively inexpensive land and labor, accompanied by large expenses of undeveloped coastline with abundant supplies of water and few environmental regulations have contributed to the establishment of much of the industry in developing nations. Conditions are changing however. Many of the best areas for aquaculture have been taken, environmental stewardship is beginning to receive the attention of governments in many developing countries, and the once abundant supplies of high quality water are being fully utilized in many areas. Thus, the face of the industry is changing. Closed system technology, which includes continuous water treatment with little or no effluent, and the development of culture systems located in the open ocean are seen as technologies that will provide opportunities for virtually unlimited expansion of aquaculture. Much of the technology for closed and offshore systems has been developed, but in many instances employment of that technology has not translated into economic feasibility. As greater efficiencies in production are achieved, new species which have higher market prices are developed, and demand increases, the economic picture can be expected to improve.

The field of aquaculture encompasses many technical disciplines and trade as well as business management and economics. Knowledge of plant and/or animal breeding, animal nutrition, water and soils analysis, surveying, computer science, pathology, carpentry, plumbing, electrical wiring, welding, and bookkeeping are among the skills that are required on a working aquaculture facility.

A good aquaculturist is involved in every aspect of the activity, from reproduction of the parent organisms through rearing of the young, to final disposition, whether that involves direct sales to the public, sales to a processor, or stocking of public or private waters. The job of the aquaculturist is not completed until the consumer, whether a patron at a restaurant, a home fish hobbyist, or the angler who is using bait minnows, has received the produce of the aquaculture facility in acceptable condition.

The *Encyclopedia of Aquaculture* has been designed for use by both those who have some knowledge of the field or may even be aquaculture professionals, as well as for individuals who are interested in learning more about aquaculture, perhaps with the idea of becoming involved. Our intent is to provide information that is readily understandable by people who have at least some science background, without insulting professionals in the field. Some topics are mentioned or briefly summarized in several entries, but when a topic is only given cursory treatment the reader is referred to one or more additional contributions that provide more detailed information on the same topic.

The *Encyclopedia of Aquaculture* was written by experts from academia and government agencies and by practicing aquaculturists in the private sector. Entries are followed by bibliographies designed to document the information present, as well as provide readers with an opportunity to further explore each topic in more depth.

References

1. R.R. Stickney, *Principles of Aquaculture*, John Wiley & Sons, New York, 1994.
2. Food and Agriculture Organization of the United Nations, Fisheries Department, Rome, Italy, 1995.
3. Anonymous, Aquaculture Magazine Buyer's Guide '95 pp. 11–22, 1995.

ROBERT R. STICKNEY
Bryan, Texas

CONTRIBUTORS

Geoff Allan, *Port Stephens Research Centre, Taylors Beach, Australia,* Barramundi Culture; Silver Perch Culture

Robert D. Armstrong, *Schering-Plough Animal Health, Forestville, California,* Drugs

C.R. Arnold, *Marine Science Institute, Port Aransas, Texas,* Snapper (Family Lutjanidae) Culture

Dan D. Baliao, *Southeast Asian Fisheries Development Center, Tigbauan, Philippines,* Mud Crab Culture

Frederic T. Barrows, *USFWS, Fish Technology Center, Bozeman, Montana,* Feed Additives; Feed Manufacturing Technology; Larval Feeding—Fish, and more

Bruce A. Barton, *University of South Dakota, Vermillion, South Dakota,* Stress

Daniel D. Benetti, *University of Miami, Miami, Florida,* Grouper Culture

David A. Bengtson, *University of Rhode Island, Kingston, Rhode Island,* Summer Flounder Culture

K.L. Bootes, *Marine Science Institute, Port Aransas, Texas,* Snapper (Family Lutjanidae) Culture

Yolanda J. Brady, *Auburn University, Auburn, Alabama,* Viral Diseases of Fish and Shellfish

Ernest L. Brannon, *University of Idaho, Moscow, Idaho,* Rainbow Trout Culture

Niall Bromage, *University of Stirling, Stirling, Scotland,* Halibut Culture

Nick Brown, *University of Stirling, Stirling, Scotland,* Halibut Culture

Mike Bruce, *University of Stirling, Stirling, Scotland,* Halibut Culture

Martin W. Brunson, *Mississippi State University, Mississippi State, Mississippi,* Fertilization of Fish Ponds; Sunfish Culture

Lucy Bunkley-Williams, *University of Puerto Rico, Mayagüez, Puerto Rico,* Multicellular Parasite (Macroparasite) Problems in Aquaculture

Charles W. Caillouet, Jr., *National Marine Fisheries Service, Galveston, Texas,* Sea Turtle Culture: Kemp's Ridley and Loggerhead Turtles

Newton Castagnolli, *Independent Consultant, San Paulo, Brazil,* Brazil Fish Culture

Joseph J. Cech, Jr, *University of California, Davis, Davis, California,* Osmoregulation in Bony Fishes

Frank A. Chapman, *University of Florida, Gainesville, Florida,* Ornamental Fish Culture, Freshwater

Shulin Chen, *Washington State University, Pullman, Washington,* Effluents: Dissolved Compounds; Effluents: Sludge; Filtration: Mechanical

K.K. Chew, *University of Washington, Seattle, Washington,* Molluscan Culture

W. Craig Clarke, *Pacific Biological Station, Nanaimo, Canada,* Smolting

Angelo Colorni, *National Center for Mariculture, Elat, Israel,* Gilthead Sea Bream Culture; Sea Bass Culture

John Colt, *Northwest Fisheries Science Center, Seattle, Washington,* Aeration Systems; Blowers and Compressors; Degassing Systems, and more

Steven R. Craig, *Texas A&M University, College Station, Texas,* Pompano Culture

R. Leroy Creswell, *Harbor Branch Oceanographic Institution, Inc., Fort Pierce, Florida,* Crab Culture: West Indian Red Spider Crab

Edwin Cryer, *Montgomery Watson, Boise, Idaho,* Ozone

D.A. Davis, *Marine Science Institute, Port Aransas, Texas,* Ingredient and Feed Evaluation; Snapper (Family Lutjanidae) Culture

Gad Degani, *Galilee Technological Center, Qiryat Shemona, Israel,* Eel Culture

M. Richard DeVoe, *South Carolina Sea Grant Consortium, Charleston, South Carolina,* Regulation and Permitting

Beverly A. Dixon, *California State University, Hayward, California,* Antibiotics

Edward M. Donaldson, *Aquaculture and Fisheries Consultant, West Vancouver, Canada,* Hormones in Finfish Aquaculture

Faye M. Dong, *University of Washington, Seattle, Washington,* Antinutritional Factors; Feed Evaluation, Chemical; Lipids and Fatty Acids

Abigail Elizur, *National Center for Mariculture, Elat, Israel,* Gilthead Sea Bream Culture; Sea Bass Culture

Douglas H. Ernst, *Oregon State University, Corvallis, Oregon,* Performance Engineering

Arnold G. Eversole, *Clemson University, Clemson, South Carolina,* Crawfish Culture

William T. Fairgrieve, *Northwest Fisheries Science Center, Seattle, Washington,* Net Pen Culture

Thomas A. Flagg, *National Marine Fisheries Service, Manchester, Washington,* Conservation Hatcheries; Endangered Species Recovery: Captive Broodstocks to Aid Recovery of Endangered Salmon Stocks

Gary C.G. Fornshell, *University of Idaho Cooperative Extension System, Twin Falls, Idaho,* Effluents: Dissolved Compounds; Rainbow Trout Culture

Ian Forster, *The Oceanic Institute, Waimanalo, Hawaii,* Energy; Nutrient Requirements

Joe Fox, *Texas A&M University at Corpus Christi, Corpus Christi, Texas,* Eyestalk Ablation

J. Gabaudan, *Research Centre for Animal Nutrition and Health, Saint-Louis Cedex, France,* Vitamin Requirements; Vitamins Sources for Fish Feeds

Margie Lee Gallagher Ph.D., *East Carolina University, Greenville, North Carolina,* Eel Culture

Delbert M. Gatlin, III, *Texas A&M University, College Station, Texas,* Minerals; Red Drum Culture

Wade L. Griffin, *Texas A&M University, College Station, Texas,* Economics, Business Plans

Nils T. Hagen, *Bodø College, Bodø, Norway,* Echinoderm Culture

Larry A. Hanson, *Mississippi State University, Mississippi State, Mississippi,* Vaccines

Terry Hanson, *Auburn University, Auburn, Alabama,* Market Issues in the United States Aquaculture Industry

Ronald W. Hardy, *Hagerman Fish Culture Experiment Station, Hagerman, Idaho,* Antinutritional Factors; Dietary Protein Requirements; Energy, and more

John Hargreaves, *Mississippi State University, Mississippi State, Mississippi,* Fertilization of Fish Ponds

Upton Hatch, *Auburn University, Auburn, Alabama,* Market Issues in the United States Aquaculture Industry

John P. Hawke, *Louisiana State University, Baton Rouge, Louisiana,* Bacterial Disease Agents

Roy Heidinger, *Southern Illinois University, Carbondale, Illinois,* Black Bass/Largemouth Bass Culture

William K. Hershberger, *National Center for Cool and Cold Water Aquaculture, Lectown, West Virginia,* Reproduction, Fertilization, and Selection

Dave A. Higgs, *West Vancouver Laboratory, West Vancouver, Canada,* Antinutritional Factors; Lipids and Fatty Acids

G. Joan Holt, *University of Texas, Port Aransas, Texas,* Ornamental Fish Culture, Marine

B.R. Howell, *Centre for Environment, Fisheries and Aquaculture Research, Weymouth, United Kingdom,* Sole Culture

W. Huntting Howell, *University of New Hampshire, Durham, New Hampshire,* Winter Flounder Culture

S.K. Johnson, *Texas Veterinary Medical Diagnostic Laboratory, College Station, Texas,* Disinfection and Sterilization; Live Transport; Protozoans as Disease Agents

Walter R. Keithly, *Louisiana State University, Baton Rouge, Louisiana,* Economics: Contrast with Wild Catch Fisheries

T.L. King, *University of Washington, Seattle, Washington,* Molluscan Culture

George Wm. Kissil, *National Center for Mariculture, Elat, Israel,* Gilthead Sea Bream Culture; Sea Bass Culture

Danny Klinefelter, *Texas A&M University, College Station, Texas,* Financing

Christopher C. Kohler, *Southern Illinois University, Carbondale, Illinois,* Striped Bass and Hybrid Striped Bass Culture

Chris Langdon, *Oregon State University, Newport, Oregon,* Microparticulate Feeds, Complex Microparticles; Microparticulate Feeds, Micro Encapsulated Particles

J.P. Lazo, *Marine Science Institute, Port Aransas, Texas,* Ingredient and Feed Evaluation

Cheng-Sheng Lee, *The Oceanic Institute, Waimanalo, Hawaii,* Mullet Culture

William A Lellis, *USGS, Research and Development Laboratory, Wellsboro, Pennsylvania,* Microbound Feeds

Matthew K. Litvak, *University of New Brunswick, St. John, Canada,* Winter Flounder Culture

Meng H. Li, *Mississippi State University, Stoneville, Mississippi,* Dietary Protein Requirements; Protein Sources for Feeds

R.T. Lovell, *Auburn University, Auburn, Alabama,* Mycotoxins

Conrad V.W. Mahnken, *National Marine Fisheries Service, Manchester, Washington,* Conservation Hatcheries; Endangered Species Recovery: Captive Broodstocks to Aid Recovery of Endangered Salmon Stocks

Michael P. Masser, *Texas A&M University, College Station, Texas,* Alligator Aquaculture; Aquatic Vegetation Control; Predators and Pests

Desmond J. Maynard, *National Marine Fisheries Service, Manchester, Washington,* Conservation Hatcheries

Carlos Mazorra, *University of Stirling, Stirling, Scotland,* Halibut Culture

Susan McBride, *University of California Sea Grant Extension, Eureka, California,* Abalone Culture

W. Ray McClain, *Rice Research Station, Crowley, Louisiana,* Crawfish Culture

Joe McElwee, *Galway, Ireland,* Turbot Culture

Russell Miget, *Texas A&M University, Corpus Christi, Texas,* Processing

Makoto Nakada, *Nisshin Feed Co., Tokyo, Japan,* Yellowtail and Related Species Culture

Heisuke Nakagawa, *Hiroshima University, Higashi-hiroshima, Japan,* Ayu Culture

George Nardi, *GreatBay Aquafarms, Portsmouth, New Hampshire,* Summer Flounder Culture

Gianluigi Negroni, *Alveo Co-operative Society, Bologna, Italy,* Frog Culture

Edward J. Noga, *North Carolina State University, Raleigh, North Carolina,* Fungal Diseases

Timothy O'Keefe, *Aqua-Food Technologies, Inc., Buhl, Idaho,* Feed Handling and Storage

Paul Olin, *University of California Sea Grant Extension, Santa Rose, California,* Abalone Culture; Lobster Culture

Anthony C. Ostrowski, *The Oceanic Institute, Waimanalo, Hawaii,* Dolphin (Mahimahi) Culture

David E. Owsley, *Dworshak Fisheries Complex, Ahsahka, Idaho,* Biochemical Oxygen Demand; Chemical Oxygen Demand; Water Management: Hatchery Water and Wastewater Treatment Systems

Nick C. Parker, *U.S. Geological Survey, Lubbock, Texas,* Fisheries Management and Aquaculture

C.O. Patterson, *Texas A&M University, College Station, Texas,* Algae: Toxic Algae and Algal Toxins

Kenneth J. Roberts, *Louisiana State University, Baton Rouge, Louisiana,* Economics: Contrast with Wild Catch Fisheries

Ronald J. Roberts, *Hagerman Fish Culture Experiment Station, Hagerman, Idaho,* Salmon Culture

H. Randall Robinette, *Mississippi State University, Mississippi State, Mississippi,* Sunfish Culture

Edwin H. Robinson, *Mississippi State University, Stoneville, Mississippi,* Dietary Protein Requirements; Protein Sources for Feeds

D.D. Roley, *Bio-Oregon, Inc., Warrenton, Oregon,* Lipid Oxidation and Antioxidants

David B. Rouse, *Auburn University, Auburn, Alabama,* Australian Red Claw Crayfish; Crab Culture

Michael B. Rust, *Northwest Fisheries Science Center, Seattle, Washington,* Larval Feeding—Fish; Recirculation Systems: Process Engineering; Water Sources

John H. Schachte, *New York State Department of Environmental Conservation, Rome, New York,* Disease Treatments

Wendy M. Sealey, *Texas A&M University, College Station, Texas,* Probiotics and Immunostimulants

Tadahisa Seikai, *Fukui Prefectural University, Fukui, Japan,* Flounder Culture, Japanese

William L. Shelton, *University of Oklahoma, Norman, Oklahoma,* Exotic Introductions

Robin Shields, *Sea Fish Industry Authority, Argyll, Scotland,* Halibut Culture

Robert R. Stickney, *Texas Sea Grant College Program, Bryan, Texas,* Barramundi Culture; Cage Culture; Carp Culture, and more

Nathan Stone, *University of Arkansas at Pine Bluff, Pine Bluff, Arkansas,* Baitfish Culture; Fertilization of Fish Ponds

Shozo H. Sugiura, *Hagerman Fish Culture Experiment Station, Hagerman, Idaho,* Digestibility; Environmentally Friendly Feeds

Robert C. Summerfelt, *Iowa State University, Ames, Iowa,* Walleye Culture

Steven T. Summerfelt, *The Conservation Fund's Freshwater Institute, Shepherdstown, West Virginia,* Carbon Dioxide; Tank and Raceway Culture

Amos Tandler, *National Center for Mariculture, Elat, Israel,* Gilthead Sea Bream Culture; Sea Bass Culture

Michael B. Timmons, *Cornell University, Ithaca, New York,* Tank and Raceway Culture

Granvil D. Treece, *Texas Sea Grant College Program, Bryan, Texas,* Brine Shrimp Culture; Eyestalk Ablation; Pollution, and more

Craig S. Tucker, *Mississippi State University, Stoneville, Mississippi,* Channel Catfish Culture

John W. Tucker, Jr., *Harbor Branch Oceanographic Institution, Fort Pierce, Florida,* Grouper Culture

Bjorn Tunberg, *Kristineberg Marine Biological Station, Fiskebäckskil, Sweden,* Crab Culture: West Indian Red Spider Crab

Patricia W. Varner, *Texas Veterinary Medical Diagnostic Lab, College Station, Texas,* Anesthetics

Arietta Venizelos, *National Marine Fisheries Service, NOAA, Virginia Key, Florida,* Grouper Culture

Richard K. Wallace, *Auburn University, Auburn, Alabama,* Crab Culture

Wade O. Watanabe, *The University of North Carolina at Wilmington, Wilmington, North Carolina,* Salinity

Barnaby J. Watten, *U.S. Geological Survey, Kearneysville, West Virginia,* Tank and Raceway Culture

Gary A. Wedemeyer, *Western Fisheries Research Center, Seattle, Washington,* Alkalinity; Buffer Systems; Chlorination/Dechlorination, and more

Gary H. Wikfors, *Northeast Fisheries Science Center, Milford, Connecticut,* Microalgal Culture

Ernest H. Williams, Jr., *University of Puerto Rico, Lajas, Puerto Rico,* Multicellular Parasite (Macroparasite) Problems in Aquaculture

Yonathan Zohar, *University of Maryland Biotechnology Institute, Baltimore, Maryland,* Gilthead Sea Bream Culture

CONVERSION FACTORS, ABBREVIATIONS, AND UNIT SYMBOLS

SI UNITS (Adopted 1960)

The International System of Units (abbreviated SI) is being implemented throughout the world. This measurement system is a modernized version of the MKSA (meter, kilogram, second, ampere) system, and its details are published and controlled by an international treaty organization (The International Bureau of Weights and Measures).

SI units are divided into three classes:

BASE UNITS	
length	meter[†] (m)
mass	solid angle
time	second (s)
electric current	ampere (A)
thermodynamic temperature[‡]	kelvin (K)
amount of substance	mole (mol)
luminous intensity	candela (cd)

SUPPLEMENTARY UNITS	
plane angle	radian (rad)
steradian (sr)	kilogram (kg)

Quantity	Unit	Symbol	Acceptable equivalent
volume	cubic meter	m^3	
	cubic diameter	dm^3	L (liter) (5)
	cubic centimeter	cm^3	mL
wave number	1 per meter	m^{-1}	
	1 per centimeter	cm^{-1}	

In addition, there are 16 prefixes used to indicate order of magnitude, as follows:

Multiplication factor	Prefix	Symbol
10^{18}	exa	E
10^{15}	peta	P
10^{12}	tera	T
10^{9}	giga	G
10^{6}	mega	M
10^{3}	kilo	k
10^{2}	hecto	h[a]
10	deka	da[a]
10^{-1}	deci	d[a]
10^{-2}	centi	c[a]
10^{-3}	milli	m
10^{-6}	micro	μ
10^{-9}	nano	n
10^{-12}	pico	p
10^{-15}	femto	f
10^{-18}	atto	a

[a] Although hecto, deka, deci, and centi are SI prefixes, their use should be avoided except for SI unit-multiples for area and volume and nontechnical use of centimeter, as for body and clothing measurement.

For a complete description of SI and its use the reader is referred to ASTM E380.

A representative list of conversion factors from non-SI to SI units is presented herewith. Factors are given to four significant figures. Exact relationships are followed by a dagger. A more complete list is given in the latest editions of ASTM E380 and ANSI Z210.1.

[†] The spellings "metre" and "litre" are preferred by ASTM; however, "-er" is used in the *Encyclopedia*.

[‡] Wide use is made of Celsius temperature (t) defined by

$$t = T - T_0$$

where T is the thermodynamic temperature, expressed in kelvin, and $T_0 = 273.15$ K by definition. A temperature interval may be expressed in degrees Celsius as well as in kelvin.

CONVERSION FACTORS TO SI UNITS

To convert from	To	Multiply by
acre	square meter (m^2)	4.047×10^3
angstrom	meter (m)	$1.0 \times 10^{-10\dagger}$
are	square meter (m^2)	$1.0 \times 10^{2\dagger}$
astronomical unit	meter (m)	1.496×10^{11}
atmosphere, standard	pascal (Pa)	1.013×10^5
bar	pascal (Pa)	$1.0 \times 10^{5\dagger}$
barn	square meter (m^2)	$1.0 \times 10^{-28\dagger}$
barrel (42 U.S. liquid gallons)	cubic meter (m^3)	0.1590
Bohr magneton (μ_B)	J/T	9.274×10^{-24}
Btu (International Table)	joule (J)	1.055×10^3
Btu (mean)	joule (J)	1.056×10^3
Btu (thermochemical)	joule (J)	1.054×10^3
bushel	cubic meter (m^3)	3.524×10^{-2}
calorie (International Table)	joule (J)	4.187
calorie (mean)	joule (J)	4.190
calorie (thermochemical)	joule (J)	$4.184^\dagger$
centipoise	pascal second (Pa · s)	$1.0 \times 10^{-3\dagger}$
centistokes	square millimeter per second (mm^2/s)	$1.0^\dagger$
cfm (cubic foot per minute)	cubic meter per second (m^3/s)	4.72×10^{-4}
cubic inch	cubic meter (m^3)	1.639×10^{-5}
cubic foot	cubic meter (m^3)	2.832×10^{-2}
cubic yard	cubic meter (m^3)	0.7646
curie	becquerel (Bq)	$3.70 \times 10^{10\dagger}$
debye	coulomb meter (C m)	3.336×10^{-30}
degree (angle)	radian (rad)	1.745×10^{-2}
denier (international)	kilogram per meter (kg/m)	1.111×10^{-7}
	tex‡	0.1111
dram (apothecaries')	kilogram (kg)	3.888×10^{-3}
dram (avoirdupois)	kilogram (kg)	1.772×10^{-3}
dram (U.S. fluid)	cubic meter (m^3)	3.697×10^{-6}
dyne	newton (N)	$1.0 \times 10^{-5\dagger}$
dyne/cm	newton per meter (N/m)	$1.0 \times 10^{-3\dagger}$
electronvolt	joule (J)	1.602×10^{-19}
erg	joule (J)	$1.0 \times 10^{-7\dagger}$
fathom	meter (m)	1.829
fluid ounce (U.S.)	cubic meter (m^3)	2.957×10^{-5}
foot	meter (m)	$0.3048^\dagger$
footcandle	lux (lx)	10.76
furlong	meter (m)	2.012×10^{-2}
gal	meter per second squared (m/s^2)	$1.0 \times 10^{-2\dagger}$
gallon (U.S. dry)	cubic meter (m^3)	4.405×10^{-3}
gallon (U.S. liquid)	cubic meter (m^3)	3.785×10^{-3}
gallon per minute (gpm)	cubic meter per second (m^3/s)	6.309×10^{-5}
	cubic meter per hour (m^3/h)	0.2271
gauss	tesla (T)	1.0×10^{-4}
gilbert	ampere (A)	0.7958
gill (U.S.)	cubic meter (m^3)	1.183×10^{-4}
grade	radian	1.571×10^{-2}
grain	kilogram (kg)	6.480×10^{-5}
gram force per denier	newton per tex (N/tex)	8.826×10^{-2}
hectare	square meter (m^2)	$1.0 \times 10^{4\dagger}$
horsepower (550 ft · lbf/s)	watt (W)	7.457×10^2
horsepower (boiler)	watt (W)	9.810×10^3
horsepower (electric)	watt (W)	$7.46 \times 10^{2\dagger}$
hundredweight (long)	kilogram (kg)	50.80
hundredweight (short)	kilogram (kg)	45.36
inch	meter (m)	$2.54 \times 10^{-2\dagger}$
inch of mercury (32 °F)	pascal (Pa)	3.386×10^3
inch of water (39.2 °F)	pascal (Pa)	2.491×10^2
kilogram-force	newton (N)	9.807
kilowatt hour	megajoule (MJ)	$3.6^\dagger$

CONVERSION FACTORS TO SI UNITS

To convert from	To	Multiply by
kip	newton (N)	4.448×10^{3}
knot (international)	meter per second (m/S)	0.5144
lambert	candela per square meter (cd/m^3)	3.183×10^{3}
league (British nautical)	meter (m)	5.559×10^{3}
league (statute)	meter (m)	4.828×10^{3}
light year	meter (m)	9.461×10^{15}
liter (for fluids only)	cubic meter (m^3)	$1.0 \times 10^{-3\dagger}$
maxwell	weber (Wb)	$1.0 \times 10^{-8\dagger}$
micron	meter (m)	$1.0 \times 10^{-6\dagger}$
mil	meter (m)	$2.54 \times 10^{-5\dagger}$
mile (statute)	meter (m)	1.609×10^{3}
mile (U.S. nautical)	meter (m)	$1.852 \times 10^{3\dagger}$
mile per hour	meter per second (m/s)	0.4470
millibar	pascal (Pa)	1.0×10^{2}
millimeter of mercury (0 °C)	pascal (Pa)	$1.333 \times 10^{2\dagger}$
minute (angular)	radian	2.909×10^{-4}
myriagram	kilogram (kg)	10
myriameter	kilometer (km)	10
oersted	ampere per meter (A/m)	79.58
ounce (avoirdupois)	kilogram (kg)	2.835×10^{-2}
ounce (troy)	kilogram (kg)	3.110×10^{-2}
ounce (U.S. fluid)	cubic meter (m^3)	2.957×10^{-5}
ounce-force	newton (N)	0.2780
peck (U.S.)	cubic meter (m^3)	8.810×10^{-3}
pennyweight	kilogram (kg)	1.555×10^{-3}
pint (U.S. dry)	cubic meter (m^3)	5.506×10^{-4}
pint (U.S. liquid)	cubic meter (m^3)	4.732×10^{-4}
poise (absolute viscosity)	pascal second (Pa · s)	$0.10^{\dagger}$
pound (avoirdupois)	kilogram (kg)	0.4536
pound (troy)	kilogram (kg)	0.3732
poundal	newton (N)	0.1383
pound-force	newton (N)	4.448
pound force per square inch (psi)	pascal (Pa)	6.895×10^{3}
quart (U.S. dry)	cubic meter (m^3)	1.101×10^{-3}
quart (U.S. liquid)	cubic meter (m^3)	9.464×10^{-4}
quintal	kilogram (kg)	$1.0 \times 10^{2\dagger}$
rad	gray (Gy)	$1.0 \times 10^{-2\dagger}$
rod	meter (m)	5.029
roentgen	coulomb per kilogram (C/kg)	2.58×10^{-4}
second (angle)	radian (rad)	$4.848 \times 10^{-6\dagger}$
section	square meter (m^2)	2.590×10^{6}
slug	kilogram (kg)	14.59
spherical candle power	lumen (lm)	12.57
square inch	square meter (m^2)	6.452×10^{-4}
square foot	square meter (m^2)	9.290×10^{-2}
square mile	square meter (m^2)	2.590×10^{6}
square yard	square meter (m^2)	0.8361
stere	cubic meter (m^3)	$1.0^{\dagger}$
stokes (kinematic viscosity)	square meter per second (m^2/s)	$1.0 \times 10^{-4\dagger}$
tex	kilogram per meter (kg/m)	$1.0 \times 10^{-6\dagger}$
ton (long, 2240 pounds)	kilogram (kg)	1.016×10^{3}
ton (metric) (tonne)	kilogram (kg)	$1.0 \times 10^{3\dagger}$
ton (short, 2000 pounds)	kilogram (kg)	9.072×10^{2}
torr	pascal (Pa)	1.333×10^{2}
unit pole	weber (Wb)	1.257×10^{-7}
yard	meter (m)	$0.9144^{\dagger}$

† Exact.
‡ This non-SI unit is recognized by the CIPM as having to be retained because of practical importance or use in specialized fields.

ENCYCLOPEDIA OF

AQUACULTURE

ABALONE CULTURE

PAUL OLIN
University of California Sea Grant Extension
Santa Rose, California
SUSAN MCBRIDE
University of California Sea Grant Extension
Eureka, California

OUTLINE

INTRODUCTION

Abalone are herbivorous marine gastropods represented throughout the world's oceans by about 70 species. Abalone have traditionally been a highly prized seafood item, and they were used more than 5,000 years ago by Native Americans along the Pacific coast of North America for food and for the manufacture of shell implements and mother-of-pearl decorations. The earliest fisheries occurred in China and Japan around 1,500 years ago; and within the past 50 years, fisheries have developed in every country with an exploitable abalone resource. Efforts to manage these fisheries have often been unsuccessful due to a lack of knowledge of population dynamics and the upswing in poaching as harvests declined and the abalone became increasingly valuable. Today, most wild abalone populations are being harvested at or above maximum sustainable yields. This situation provides an excellent opportunity for abalone farming, and considerable effort is underway throughout the world to establish abalone farms.

ABALONE CLASSIFICATION AND BIOLOGY

Abalone are in the phylum Mollusca, a predominantly marine phylum that includes other cultured species such as clams, oysters, and scallops. They are members of the class gastropoda which includes snails. All members of the class are univalves, having one shell, unlike the bivalve oysters and clams which have two. All abalone belong in the family Haliotidae and are members of the genus *Haliotis*.

The prominent shell of the abalone encases the animal and the large centrally located muscular foot that is used to clamp tightly to hard surfaces (Fig. 1). In attached abalone, water enters under the shell, passing through the mantle cavity and over the paired gills before exiting through respiratory pores in the dorsal surface. In the head region, the eyes are located on extended eyestalks; and two enlarged cephalic tentacles extend anteriorly (Fig. 2). The mouth is at the base of the head region and houses a rasp-like radula which is used to scrape food from hard surfaces or consume macroscopic algae. The mouth leads to the esophagus and connects to the gut located between the muscular foot and the shell. The gut wraps around the foot opposite the gonad and terminates in the mantle cavity where waste material is released to exit through the respiratory pores.

A thin mantle and epipodium circle the foot. In a resting animal, small sensory tentacles on the epipodium are visible protruding from the shell periphery. Prominent gonads are visible arcing around about one third of the foot toward the rear of the abalone. The gonads release mature gametes into the mantle cavity where they are broadcast out through the respiratory pores.

The abalone heart is located near the mantle cavity and pumps oxygenated blood from the gills into the foot via two arteries. From there it is distributed via smaller and smaller arterioles to the organs. Returning blood is collected in a sinus located in the muscular foot and flows in veins back to the gills. The location of these arteries, veins, and the blood sinus in the foot make abalone extremely vulnerable to even small cuts in the foot muscle, which can cause them to bleed to death.

Cultured Species

Of the approximately 70 species of abalone in the world there are only 10 that support large commercial fisheries.

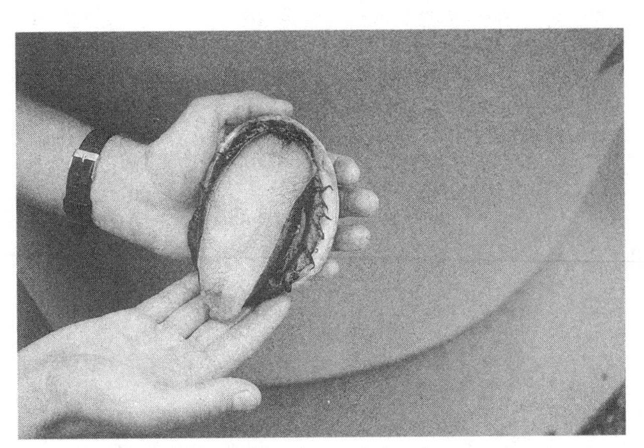

Figure 1. Red abalone (*H. rufescens*) showing prominent muscular foot (35 mm color side).

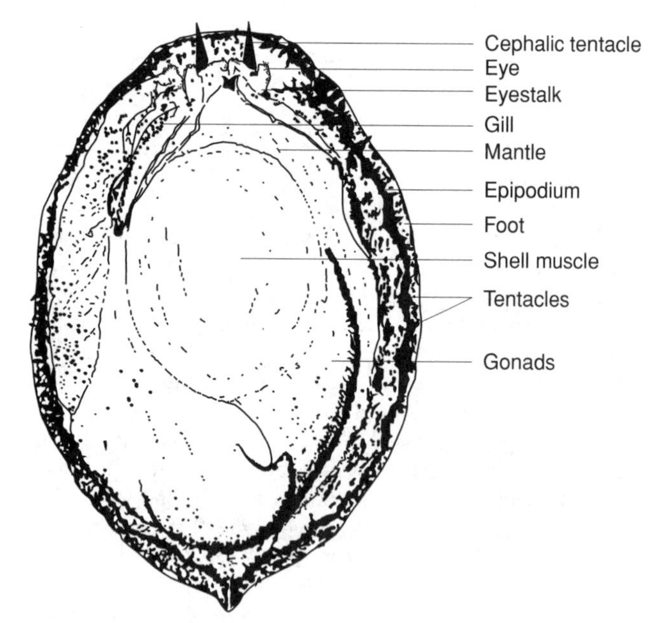

Figure 2. Dorsal view of abalone internal organs with shell removed (California Department of Fish and Game).

Six of those are currently grown in significant numbers. The primary regions where culture is underway are indicated in Table 1 (1). The shell lengths listed are for large wild animals while most cultured product is marketed at 5 to 20 cm (2 to 4 in.). Many of those listed and other species have been transported around the world for research and small-scale growout trials.

Water Quality

Abalone require excellent water quality, which is not surprising given the clean ocean waters in which they evolved. Those coastal waters are saturated with dissolved oxygen and experience fairly stable levels of pH, ammonia, salinity, and temperature. Optimal growth is temperature-dependent and varies between life stages and species. The pH should be around 8.0 and the salinity kept stable between 32 and 35 ppt. Abalone are poor osmoregulators and culture tanks should be shielded from excessive rainfall.

Abalone are very sensitive to hydrogen sulfide, which is produced by the anaerobic breakdown of dead animals, uneaten feed, and feces. Reduced growth has been observed at levels as low as 0.05 ppm H_2S. Ammonia is produced as a metabolite by many aquatic organisms and is toxic to most at levels above 1.0 ppm (un-ionized form). Abalone are especially sensitive to ammonia, showing reduced oxygen consumption at levels as low as 10 μg/L and feeding inhibition at 70 μg/L.

Reproduction and Broodstock

Abalone are dioecious, having separate males and females. The gametes are fertilized externally (Fig. 3). In mature abalone, the sexes are easily distinguished with the male gonads having a cream or pale yellow color, while the mature female gonad has a green coloration. The reproductive cycles of abalone are seasonal and related primarily to water temperature. In temperate species, gonadal development and gamete production increase with temperature; while in tropical species, gonadal development is reduced but not absent at the warmest times of year (1–3). Spawning induction is generally easier in smaller animals, and many hatcheries have also noted that first generation hatchery-reared abalone spawn more readily than wild broodstock. High fertilization rates (above 85%) and larval survival (normally greater than 70%) have resulted in relatively moderate broodstock management requirements.

Abalone are highly fecund molluscs and in temperate species, female abalone measuring 75 to 100 mm (3 to 4 in.) in shell length and weighing 120 to 150 g (4 to 5 oz) routinely release 3 to 6 million eggs per spawn. Large female abalone from temperate waters measuring 20 cm (8 in.) can release over 11 million eggs. Males release

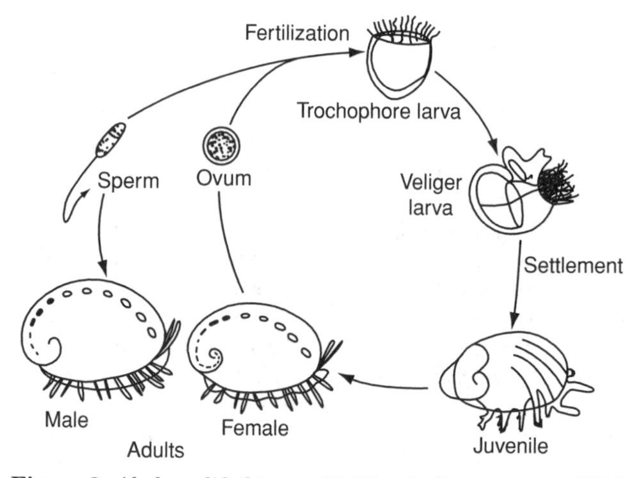

Figure 3. Abalone life history (California Department of Fish and Game).

Table 1. Principal Cultured Abalone (1)

Species	Common Name	Shell Length, [mm (in.)]	Primary Region
H. discus hannai	Ezo awabi	190 (~7.5)	Japan
H. diversicolor supertexta	Tokobushi, small	50 (~2)	China, Taiwan, Japan
H. iris	Paua or black	170 (~7)	New Zealand
H. midae	Perlemoen	90 (~4.5)	South Africa
H. rubra	Black lip	130 (~5)	Australia
H. rufescens	Red	250 (~10)	Mexico, Chile, United States

Table 2. *Haliotis rufescens*, Cultured and Wild Broodstock Induced to Spawn from Four Farms in North America in 1996[a]

Total Number of Abalone Induced to Spawn				Percentage of Abalone that Spawned			
Wild		Cultured		Wild		Cultured	
Male	Female	Male	Female	Male	Female	Male	Female
180	450	540	765	51% ±10	27% ±9	43% ±8	38% ±5

[a]All farms had been in production for a minimum of two years (mean ñ s.d., $n = 4$) (8).

copious amounts of sperm, usually more than is required for breeding purposes in cultured populations. Mature temperate species can become gravid in the hatchery in three to four months, while less time is required with tropical species (1,2,4). The size at first sexual maturity is about 35 to 40 mm (1.5 to 4 in.) in shell length for temperate species.

The goal of broodstock management is to provide sexually mature animals in spawning condition throughout the year and not be reliant on wild broodstock; however, many farms still utilize some wild broodstock to maintain production and genetic diversity (Table 2). Currently about one-half of the abalone production broodstock are from wild populations. Maintaining genetic diversity is an important issue as high fecundity allows relatively small numbers of broodstock to meet hatchery needs, and this could result in inbreeding and genetic drift (9–11). The control of abalone reproduction and spawning induction techniques were developed during the 1970s (5,6). Reproductively mature abalone are induced to spawn by using ultraviolet irradiated seawater at low seawater flows, 180 to 200 ml (6 to 7 oz), or by introducing hydrogen peroxide (5 M) to seawater. Other treatments that may be utilized to induce spawning are desiccation or temperature changes (2). Mature eggs are extruded from the gonads and released through the respiratory pores. Extruded eggs are collected and rinsed prior to being resuspended in seawater and the addition of sperm to achieve a concentration between 10^5 and 10^6 sperm/mL.

Broodstock abalone are maintained separately by sex at low density with a continual supply of feed. Many farms also supply broodstock with a diverse algal diet in addition to the kelp provided to production tanks. In Australia, China, Japan, and New Zealand, abalone reared and maintained on prepared diets have been successfully induced to spawn. The control of broodstock conditioning varies among species, but the primary needs are a balanced diet, good quality seawater, and appropriate temperature and photoperiod. In addition, every effort is made to avoid exposure to potential pathogens and reduce stress related to handling and tank maintenance.

In North America, red abalone (*H. rufescens*) broodstock measuring 8 to 12 cm (3 to 5 in.) are held at low density in tanks supplied with seawater at ambient temperature and photoperiod. They are fed a variety of red and brown algae and are able to digest both (2,7). Broodstock are tagged to maintain records of individual spawnings, fecundity, and performance of progeny. Monitoring this early performance is especially important in an animal with a four-year production cycle. Correlating early growth with overall performance is an important management tool in some shellfish farming enterprises (13). In South Africa, *H. midae* broodstock are conditioned in a similar system but sexually mature individuals become gravid approximately 20 months after spawning, suggesting a two-year reproductive cycle for that species.

In Japan, hatcheries hold *H. discus hannai* in tanks with a controlled photoperiod of 12 hours of light and 12 hours of dark. The water temperature is maintained at 20 °C (68 °F) and seawater flow rate is around 800 L/hr. Japanese hatchery managers routinely hold broodstock at elevated temperatures for maturation and utilize lower temperatures to inhibit spawning in gravid adults until fertilized eggs are required in the hatchery (2).

The difference in the two systems results largely from the natural reproductive cycles of the two species. In wild populations of *H. rufescens*, gravid, sexually mature individuals are found year round, while *H. discus hannai* from Japanese waters exhibit a seasonal reproductive cycle. Gravid *H. discus hannai* are found in summer months, and the elevated broodstock holding temperatures maintain animals in spawning condition throughout the year. As a general rule, broodstock abalone should be maintained in conditions similar to those in the ocean where sexually mature gravid individuals are found. The spawning areas used by northern and southern abalone species are well documented (2).

Larval Rearing and Settlement

Eggs are collected after spawning and fertilization and subsequently held for 24 to 36 hours in a static system. During that time, the microscopic larvae develop to the trochophore stage and hatch out of the egg membrane. The remainder of larval rearing is done by using either flow-through or static systems (Fig. 4). Flow-through systems typically incorporate tanks from 20 to 500 L made of plastic or fiberglass for ease of cleaning. They are supplied with UV-treated 1 μm filtered seawater. Banjo screens of 90 or 100 μm are placed at the seawater outflow to retain the larvae which are generally 220 to 260 μm in diameter (14). The design was first developed in New Zealand and consists of a piece of large diameter plastic pipe attached to the overflow drain (Fig. 4) with 90 to 100 μm screen glued to both ends. The screens provide enough surface area so that the current does not impinge larvae. Flow rates are also low to prevent the weakly swimming larvae from damage in the tanks and on the screens. Gentle aeration is sometimes provided to the larval rearing containers. In static systems, water changes are done one to three

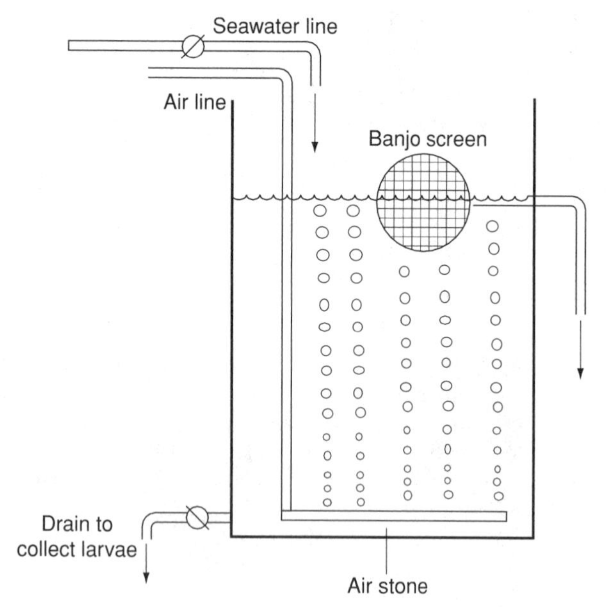

Figure 4. Abalone larval rearing tank with banjo screen.

times per day by flushing the tank or gently collecting the larvae on screens and transferring them to clean rearing containers.

Abalone have swimming planktonic larvae for five to seven days depending on temperature. Temperate and tropical abalone larvae are reared at 13 to 15 °C and 23 to 26 °C, respectively. If excessive bacterial growth develops, larvae are collected on screens and transferred to a clean rearing tank. Healthy larvae swim in a spiraling fashion upward, then drift down, then swim up again. Larval survival is usually around 70% and production ranges from 500,000 to 35 million, depending on the size of the facility.

At the end of the larval rearing period, "competent" planktonic veliger larvae settle on a hard substrate and metamorphose into a crawling benthic juvenile form. Morphological changes include development of the radula, protrusion of the sensory cephalic tentacles, and loss of the swimming organ known as the velum. Behavioral changes signalling the onset of settlement include intermittent swimming and crawling behavior and settlement of some animals at the water line of the rearing container. When these changes are observed, larvae are ready to be placed into the settlement tanks used for their early growth and feeding (2).

Larval Settlement and Nursery Rearing

Indoor and outdoor settlement tanks are used for abalone larvae. Settlement may be enhanced using γ-aminobutyric acid (GABA), diatoms, a diatom/bacterial film or mucous trails of adult abalone (2). When GABA is used, the tanks are cleaned, 1 μm filtered seawater is introduced, and GABA is added to achieve a 10^{-6} M concentration (15). Abalone larvae are then introduced at 2 to 5 larvae/cm^2 of tank surface area. Settlement tanks utilize vertically placed plastic or fiberglass sheeting to increase settlement surface area. Tanks are left static for 12 to 24 hours and then a low flow rate, usually about 1 L/min (0.26 gal/min), is started. The young abalone are very active crawlers and begin feeding during the first ten days after settlement (16). They require microalgae of 10 μm or less in size for the initial feeding. The radula width of *H. rubra* at six weeks of age is 9 to 11 μm, suggesting that this may be a good indicator of optimum diatom or feed size (17).

Diatom cultures are maintained at some commercial farms, while others use only coarse filtration when filling settlement tanks allowing natural diatom populations to settle and grow on tank walls. A light diatom film is desirable and sunscreen covers are used to manage sunlight intensity as a means of regulating diatom growth rates. The correct species and thickness of the diatom film on the tank walls are critical to early survival and growth (18,19). Newly settled animals can become entangled in heavy diatom films that are conducive to the growth of bacteria and protozoans.

Artificial diets are sometimes used after three months when the abalone are about 2 mm in shell length and have formed their first respiratory pore. A thin bladed red algae, *Palmaria mollis*, is also cultured at some farms for young abalone (20,21).

Nursery tanks are rectangular or round and vary from 180 to 1,000 L (47 to 264 gal). Nursery systems also use vertical substrates similar to those used in settlement tanks, but of different dimensions to accommodate the tank and increased abalone size. In Japan, a series of corrugated fiberglass sheets are held in a plastic-coated rigid metal frame suspended in large tanks (2). Abalone are maintained for approximately four to six months in the nursery area of the farm. Tanks are drained and rinsed every one to three weeks depending on abalone grazing rates and diatom growth. The tanks are gently rinsed with seawater, and dislodged abalone are collected on screens at the outflow.

At the end of the nursery period the 6 to 10 mm abalone have a strong radula capable of scraping large macroalgae or prepared diets. They are transitioned to those diets in the nursery system or in 0.2 m^3 (6.2 ft^3) plastic mesh baskets suspended in large production tanks. Abalone are stocked at high densities of around 2,500 per basket, and the baskets are packed with macroalgae, placing the abalone in close proximity to the new food source.

Production of abalone from nursery systems has greatly increased over the past ten years. Some hatcheries are vertically integrated with growout facilities, while other farms lacking hatcheries purchase larvae for growout. Production of abalone is generally not a constraint to commercial production.

GROWOUT SYSTEMS

Growout systems are located on land with tanks and seawater pumping systems, or they are in-water facilities that use long lines or rafts to support cage structures. Land-based growout facilities utilize concrete or fiberglass tanks. As in the nursery systems, vertical panels placed in the tanks slightly above the tank floor provide additional surface area. Some growout tanks have "V" shaped bottoms or false bottoms to allow rapid removal of feces. In-water cages constructed of heavy extruded plastic mesh and screened plastic barrels are also used and contain added vertical substrate. A unique abalone farm is located in South Australia, where cages hold abalone on the seafloor and trap the drifting algae transported by currents. Growout systems involve substantial amounts of labor, power, and feed whether they are land-based or in-water facilities (22). Abalone are held one to three years in production systems depending on the species, growth rate, and market size preference. Farmed abalone are usually transported and sold live in the marketplace (23).

In North and South America, the giant kelp *Macrocystis pyrifera* is the main feed used on abalone farms. The northernmost farms use the bull kelp, *Nereocystis luetkeana*. Other kelp and mixed macrophyte species provide most of the feed used in other abalone producing regions throughout the world. Abalone generally prefer red and brown macrophytes, and they have developed enzymes that lyse the cells walls of their preferred algal species (1,26,31–33). Prepared diets are also used in many production systems, most often for small nursery animals.

In Australia, Japan, New Zealand, and South Africa, some producers rely exclusively on prepared feeds, although this generally increases production costs compared with using harvested kelp (7). While prepared feeds are more expensive, there are considerable labor savings because those feeds are consumed at 2 to 7% of body weight compared with 10 to 30% for natural algal diets. Manufactured diets also have a more reliable composition and provide for more consistent growth (1,7).

Abalone are fed kelp once or twice a week based on seawater temperature, season, and consumption rates. Prepared diets are fed in small amounts daily or every few days. As feed quality deteriorates, it is important to replace it with fresh feed. Toxic hydrogen sulfide and ammonia levels can result if kelp is left to decompose in culture tanks, especially at high water temperatures. Growout tanks are drained and rinsed as needed to remove excess fecal material.

Land-based nursery and growout systems use aeration as do some ocean barrel culture operations. Maintaining near saturation levels of dissolved oxygen is essential as abalone will crawl out of tanks with reduced oxygen. Vigorous aeration also serves to distribute feed to all the animals.

NUTRITIONAL REQUIREMENTS OF ABALONE

Abalone require different foods at different life stages for optimal growth and development. Larval abalone do not feed, although they may absorb some nutrients from seawater (25). Rapidly growing young abalone require higher levels of protein and energy than adults (26). Young abalone actively graze surfaces, removing algal and biofilm nutrients from the substrate. Adult wild abalone primarily consume drift kelp, feeding opportunistically as food drifts their way (27–29).

Development of cost-effective artificial diets will foster continued expansion of abalone aquaculture, and researchers have made significant progress in developing diets that are water stable, nutritionally complete, palatable, and accessible. Knowledge of the specific nutritional requirements of abalone has resulted in diets containing about 30% protein in the form of defatted soybean meal, casein or fish meal. Lipids range from 3 to 5% of prepared diets and the source is usually fish or vegetable oil. Lipids must be stabilized with an antioxidant such as Vitamin E. Abalone require eicosapentanoic (20:5n-3) and docosahexanoic (22:6n-3) fatty acids in their diet. Carbohydrates comprise 30 to 60% of prepared diets and also act as a binder to maintain stability and retard nutrient leaching into the water. Corn and wheat are common sources of carbohydrate. Crude fiber is not readily digested by abalone and generally comprises from 0 to 3% of formulated diets. (7).

Cues that initiate feeding in abalone are not well understood. Abalone feed primarily at night and not all animals will feed on a given night. Because of our poor understanding of feeding behavior and to ensure constant access to food, growers often provide about twice the amount of food the abalone will consume and must be diligent in removing decomposing uneaten feed (30).

Growth

Abalone growth rates are generally slow and variable among individuals and species. From metamorphosis to development of the first respiratory pore at 60 to 90 days, growth is approximately 1 to 1.5 mm/month and the shells are 2 to 2.5 mm long. Growth is rapid for the next 8 to 10 months at 1.5 to 3 mm/month. During this growth phase, the young abalone graze microalgae and mixed biofilms from tank surfaces. Prepared diets with particle sizes between 50 and 300 μm are used during the latter part of that growth phase (1,7,14,34–37).

Some exceptional growth rates have been observed in the tropical species *H. assinina*. Growth of up to 4 to 5 cm the first year are common (2,38). The subtropical abalone, *H. diversicolor supertexta*, average 6–7 cm at two years of age in aquaculture systems in Taiwan. After the first year of accelerated growth, abalone growth rates decrease to 1 to 1.6 mm/month (1,39). Some wild *H. rufescens* populations have shown 5.5 mm/month growth in the summer and 2.5 mm/month in winter, an annual average of 3.5 mm/month for large abalone. *H. discus hannai* reaches 3 cm in 18 months in aquaculture systems and 9 cm in four years in nature.

Variable growth rates are common in abalone but the underlying causes are often unknown (40). There appears to be a genetic component as slow and fast growing abalone maintained their respective growth rates when reared under identical conditions (41). Other factors that influence abalone growth in aquaculture systems are temperature, density, photoperiod, salinity, oxygen, and food intake.

A constant or elevated seawater temperature increases growth in some abalone species (1,42). *H. tuberculata* grew 18 mm/yr when reared in 20 °C (68 °F) seawater (43). *H. fulgens* shows enhanced growth when reared in seawater above 20 °C (68 °F) and *H. discus hannai* will double normal growth rates to reach 6 cm in 2 years when raised at elevated temperatures. Warm water reduces the time for production of 1 to 2 cm *H. rufescens* from one year to six months when the young abalone are raised in 23 °C (73 °F) compared to 16 °C (61 °F). Low temperature reduces growth by affecting feeding rate and duration as well as food absorption (44,45).

Abalone form localized high density aggregations in culture systems that may affect their growth and nutritional status. High stocking densities reduce abalone growth in aquaculture systems (33,37). *H. diversicolor supertexta* measuring 15 mm exhibited reduced growth at densities greater than 500/m^2. Other detrimental effects included shell erosion and splits in the shell where the respiratory pores are normally located.

HUSBANDRY AND HEALTH MANAGEMENT

A hard rule in animal husbandry is to minimize stress when it is economically feasible to do so. To accomplish this with abalone means providing clean, well-oxygenated seawater at a stable temperature that is free of ammonia and other contaminants. Larvae and young juvenile abalone in settlement tanks are very sensitive to temperature changes, and as little as a 2 °C (36 °F) temperature change may cause mortality of *H. rufescens* larvae (35).

Good-quality feeds in sufficient quantity should be regularly available and decomposing feed should be removed. Often production and broodstock tanks require cleaning to prevent populations of copepods, nematodes, or harmful bacteria from flourishing. Animals should be kept moist and handled as little as possible during this maintenance. When handling abalone, always lift them from the substrate with a lifter that does not cut the foot as small cuts can result in animals bleeding to death. Lifting abalone from the posterior end of the shell is essential to avoid damage to delicate tissues and organs in the gill and head region (23).

Since abalone are active and feed primarily at night, it is necessary to walk around the growout tanks each morning to check for animals that may have inadvertently crawled out and fallen to the ground. This is also an ideal time to check all tank flows and aeration systems. Cage culture systems have similar husbandry requirements and must also be kept free of excessive fouling organisms that restrict water circulation.

Every effort should be made to prevent the introduction and spread of disease. Broodstock, hatchery, and production growout systems should be isolated from one another. Care should be taken not to mix populations and animals should not be moved between tanks. Similarly, equipment should be disinfected if used in different areas of a farm. If importing broodstock or seed abalone for growout, a shellfish pathologist should examine the population to determine with reasonable certainty that infectious agents or parasites are absent.

FUTURE CONSTRAINTS AND OPPORTUNITIES

The primary constraints to the continued growth and success of the abalone industry are the need for an economical manufactured feed and management of pathogens and parasites. The industry also needs to domesticate broodstock and begin a genetic selection program to enhance growth and other production traits while eliminating the need for wild broodstock.

Recently, prepared diets have been used throughout the entire culture cycle in Australia where restrictions on algal harvest and phenolic compounds unpalatable to abalone prevent the use of brown macroalgae (24). There are a number of companies now producing manufactured feeds, and as more is learned about nutritional requirements and feed formulation, it is hoped these diets will become more cost effective. When this happens, many new production areas distant from major kelp resources will open up. This will include tropical regions that can benefit from the fast growth rates observed in *H. assinina*.

Pathogens such as sabellid worms and rickettsial bacteria are associated with withering syndrome in California black and red abalone and must be managed so they do not significantly impact growers. Proper management and husbandry techniques using specific pathogen-free broodstock and seed abalone should allow the industry to avoid significant losses from these and other potential pathogens.

Genetic improvements achieved by selection, ploidy manipulations, and transgenic technologies have the potential to improve the performance of abalone in culture systems. Some of this work has begun but is not yet near commercial application (12).

The abalone industry will continue to expand as new technologies develop and are incorporated into farming systems. The high demand coupled with stable or declining fisheries ensures that the market will continue to grow and provide opportunities for abalone growers throughout the world.

BIBLIOGRAPHY

1. P. Jarayabhand and N. Paphavasit, *Aquaculture* **140**, 159–168 (1996).
2. K.O. Hahn, *Handbook of Culture of Abalone and Other Marine Gastropods*, CRC Press, Boca Raton, FL, 1989.
3. N.H.F. Wilson and D.R. Schiel, *Mar. and Freshwater Res.* **46**(3), 629–638 (1995).
4. T. Tutschulte and J.H. Connell, *Veliger.* **23**(3), 195–206 (1981).
5. S. Kikuchi and N. Uki, *Bull. Tohoku Reg. Fish. Res. Lab.* **33**, 79–84 (1974).
6. D.E. Morse, *Science* **196**, 198–200 (1977).
7. A.E. Flemming, R.J. Van Barneveld, and P.W. Hone, *Aquaculture* **140**, 5–54 (1996).
8. S.C. McBride, unpublished data.
9. P.M. Gaffney, V. Powell Rubin, D. Hedgecock, D.A. Powers, G. Morris, and L. Hereford, *Aquaculture* **143**, 257–266.
10. P.J. Smith and A.M. Conroy, *N. Z. J. of Mar. and Freshwater Res.* **26**, 81–85 (1992).
11. Y.D. Mgaya, E.M. Gosling, J.P. Mercer, and J. Donlon, *Aquaculture* **136**, 71–80 (1995).
12. D. Powers, V. Kirby, T. Cole, and L. Hereford, *Mole. Mr. Biol. and Biotech.* **4**(4), 369–375.
13. N.P. Wilkins, *Aquaculture* **22**, 209–228 (1981).

14. L.J. Tong and G.A. Moss, in S.A. Shepherd, M.J. Tegner, and S.A. Guzman del Preo, eds., *Abalone of the World, Fishing News Books*, 1992, pp. 583–591.
15. D.E. Morse, N. Hooker, H. Duncan, and L. Jensen, *Science* **204**, 407–410 (1979).
16. C.L. Kitting and D.E. Morse, *Moll. Res.* **18**, 183–196 (1997).
17. C.D. Garland, S.L. Cooke, J.F. Grant, and T.A. McMeekin, *J. Exp. Mar. Biol. Ecol.* **91**, 137–149 (1985).
18. H. Suzuki, T. Ioriya, T. Seki, and Y. Aruga, *Nippon Suisan Gakkaishi* **53**, 2163–2167 (1987).
19. I. Matthews and P.A. Cook, *Mar. Freshwater Res.* **46**(3), 545–548 (1995).
20. J.E. Levin, M.A. Buchal, and C.J. Langdon, in *Book of Abstracts*, World Aquaculture Society Meeting, Feb. 1–4, San Diego, 1995, pp. 73–74.
21. F. Evans and C.J. Langdon, *Abstract of the 3rd International Abalone Symposium on Abalone Biology*, Fisheries and Culture, Monterey, CA.
22. S.C. McBride, *J. Shellfish Res.* (1998).
23. S.C. McBride, in B. Paust and J.B. Peters, eds., *Marketing and Shipping Live Aquatic Products*, Northeast Regional Agricultural Engineering Service, Ithaca, NY, 1997, pp. 51–59.
24. S.A. Shepherd and P.D. Steinberg, in S.A. Shepherd, M.J. Tegner, and S.A. Guzman del Preo, eds., *Abalone of the World*, Fishing News Books, London, 1992, pp. 169–181.
25. W.B. Jaeckle and D.T. Manahan, *Mar. Biol.* **103**, 87–94 (1989).
26. J.P. Mercer, K.S. Mai, and J. Donlon, *Invert. Rep. Dev.* **23**, 75–88.
27. G. Poore, *N. Z. J. Mar. Freshwater Res.* **6**(1&2), 11–22 (1972).
28. R.W. Day and A.E. Flemming, in S.A. Shepherd, M.J. Tegner, and S.A. Guzman del Preo, eds., *Abalone of the World*, Fishing News Books, London, 1992, pp. 141–168.
29. S. Daume, S. Brand, and W.J. Woelkering, *Moll. Res.* **18**, 119–130 (1997).
30. N. Uki and T. Watanabe, in S.A. Shepherd, M.J. Tegner, and S.A. Guzman del Preo, eds., *Abalone of the World*, Fishing News Books, London, 1992, pp. 504–517.
31. C. Boyen, B. Kloareg, M. Polne-Fuller, and A. Gibor, *Phycologia* **29**, 173–181.
32. Y. Mizukami, M. Okauchi, and H. Kito, *Aquaculture* **108**, 191–205.
33. A.G. Clark and D.A. Jowett, *N. Z. J. Mar. Freshwater Res.* **12**, 221–222.
34. K. Yamaguchi, T. Araki, T. Aoki, C.H. Tseng, and M. Kitamikado, *Bull. Jap. Soc. Sci. Fish.* **55**, 105–110.
35. E.E. Ebert and J.L. Houk, *Aquaculture* **39**, 375–392 (1984).
36. H. Takami, T. Kawamura, and Y. Yamashita, *Moll. Res.* **18**, 143–151 (1997).
37. G.A. Moss, *Moll. Res.* **18**, 153–159 (1997).
38. D.C. McNamara and C.R. Johnson, *Mar. Freshwater Res.* **46**(3), 571–574 (1995).
39. S.A. Shepherd and W.S. Hearn, *Austr. J. Mar. Freshwater Res.* **34**, 461–475 (1983).
40. H. Momma, *Aquaculture* **28**, 142–155 (1980).
41. Z.Q. Nie, M.F. Ji, and J.P. Yan, *Aquaculture* **140**, 177–186 (1996).
42. Y. Koike, J. Flassch, and J. Mazurier, *La Mer.* **17**, 43–52.
43. N. Uki, *Bull. Tohoku Reg. Fish. Res. Lab.* **43**, 861–871 (1981).
44. L.S. Peck, M.B. Culley, and M.M. Helm, *J. Exp. Mar. Biol. Ecol.* **106**, 103–123 (1987).
45. H.C. Chen, *Aquaculture* **39**(1–4), 11–27 (1984).
46. R. Searcy-Bernal, *Aquaculture* **140**, 129–137 (1997).

See also MOLLUSCAN CULTURE.

AERATION SYSTEMS

JOHN COLT
Northwest Fisheries Science Center
Seattle, Washington

OUTLINE

In aquatic culture systems, many of the important water-quality parameters are the levels of dissolved gases, such as oxygen, carbon dioxide, hydrogen sulfide, ammonia, and nitrogen. Aeration, or the addition of dissolved oxygen (DO), is one of the processes most commonly used in aquaculture. The maintenance of environmental quality requires control of levels of dissolved gas. The "best" aeration system for a given application depends on site conditions, production schedules, the layout of the rearing units, and operational procedures. The design of an aeration system must consider the potential impacts on all the dissolved gases in solution.

TYPES AND CONFIGURATION OF AERATORS

Each aeration device can be classified as either surface, subsurface, or gravity. If the source of oxygen is enriched or pure oxygen gas rather than air, the units are called "pure-oxygen aerators"; these units are covered in a separate entry, pure oxygen systems.

Types of Aerators

Surface aerators spray or splash water into the air and thus transfer oxygen from the air into the water. The major types of surface aerators are shown in Figure 1. Subsurface aerators mix water and air together in an aeration basin and transfer oxygen from air bubbles into the water. The major types of subsurface aerators are shown in Figure 2. Gravity aerators are a special type of surface aerator that use gravity rather than mechanical power to transfer oxygen. This type of aerator is commonly used in flow-through systems where adequate head is available. The major types of gravity aerator are shown in Figure 3.

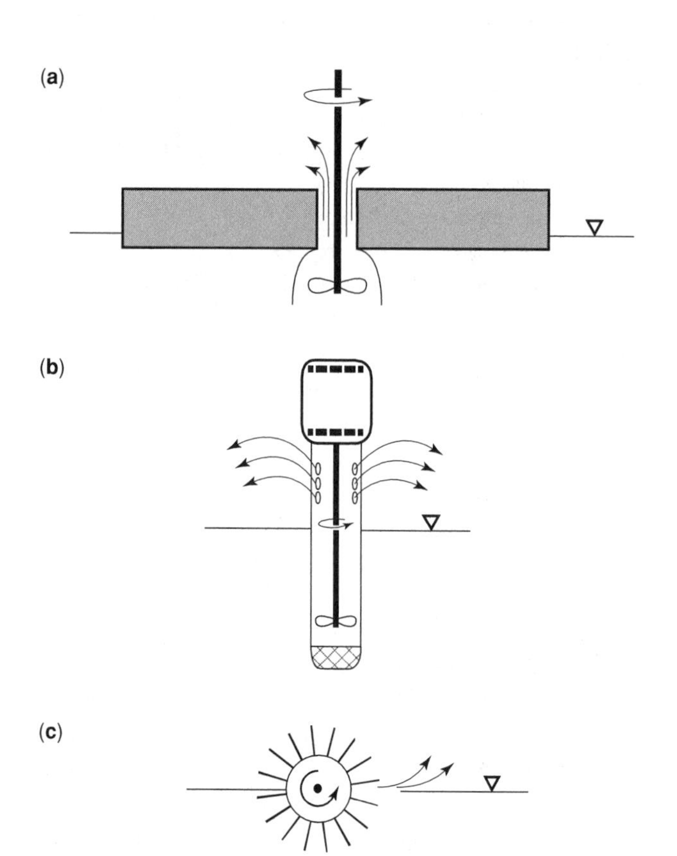

Figure 1. Typical surface aerators: (**a**) floating aerator; (**b**) surface aerator with draft tube; (**c**) brush, rotor, or paddlewheel aerator.

Aerator Configuration and Location

Depending on system configuration and on operational limitations, aerators may be placed at different locations (Fig. 4). The aerators may be located in the influent stream (Figs. 4a, 4b, and 4c), the recycle stream (Figs. 4e and 4f), or the rearing unit (Fig. 4d). In the recycle configuration (Figs. 4e and 4f), the recycle flow may be several times larger than the influent flow. In a side-stream system (Figs. 4b and 4c), only part of the water flow passes through the aeration system.

SOLUBILITY OF GASES IN WATER

The solubility of a gas in water depends on its temperature and composition, the salinity of the water, and the total pressure. The air solubility of oxygen at 1 atm (C^*_{760}), as a function of temperature, is listed in standard references (1,2); it may be adjusted to other barometric pressures by the following equation:

$$C^* = C^*_{760} \frac{(\mathrm{BP} - P_w)}{(760 - P_w)} \tag{1}$$

where

C^* = air-solubility of oxygen (mg/L),
BP = local barometric pressure (mm Hg),
C^*_{760} = air-solubility of oxygen at 760 mm Hg (standard conditions), and
P_w = vapor pressure of water (mm Hg; Reference 1).

Equation 1 is limited to the computation of saturation concentration for air; information for the computation of the saturation concentration of gases other than air can be found in standard references (2). In many aeration systems, hydrostatic head is used to increase the pressure at which gas transfer occurs. A depth of approximately 10 m (30 ft) will double the solubility of a gas.

DISSOLVED-OXYGEN CRITERIA IN INTENSIVE CULTURE

Low-Dissolved-Oxygen Criteria

The growth of fish is not affected until the dissolved oxygen (DO) drops below a critical concentration. This critical concentration is influenced by temperature and by feeding level; it ranges from 5 to 6 mg/L (ppm) for salmon and trout (Salmonidae) and from 3 to 4 mg/L (ppm) for warm-water fish such as channel catfish (*Ictalurus punctatus*) (3,4). The use of oxygen supplementation can also increase survival and improve fish health and quality. Some of the beneficial effects of oxygen supplementation may be due to the stripping of chronic levels of gas supersaturation.

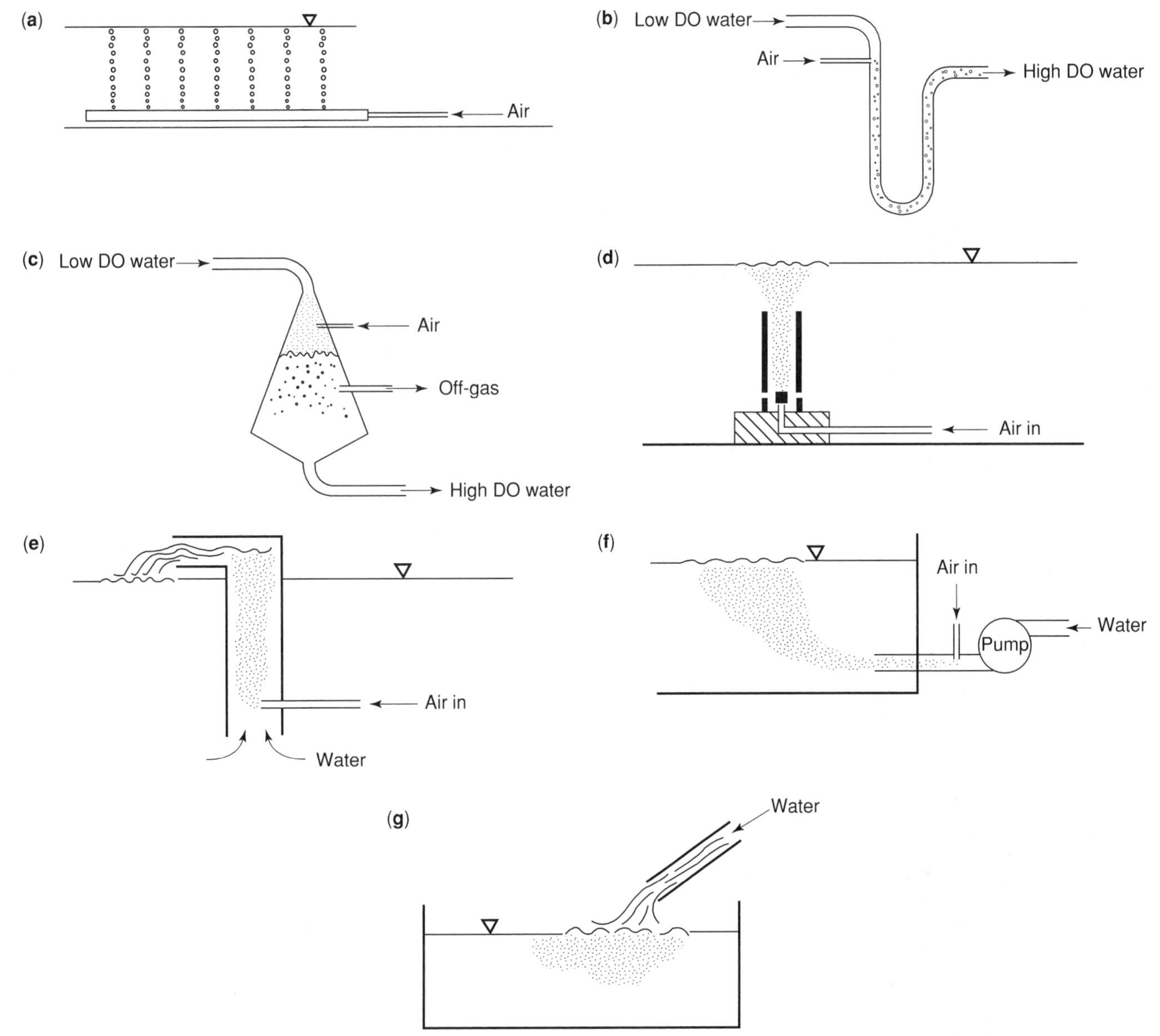

Figure 2. Typical submerged aerators: (**a**) diffused, (**b**) U-tube, (**c**) aerator cone, (**d**) static tube, (**e**) air-lift, (**f**) venturi, (**g**) nozzle.

High-Dissolved-Oxygen Criteria

The maximum allowable DO level depends on several factors, including oxygen toxicity, physiological dysfunctions, and developmental problems. While oxygen is required for the survival of aerobic organisms such as fish, some of the by-products of oxygen metabolism are highly toxic and can overwhelm biochemical defense mechanisms. On the basis of oxygen toxicity considerations, a preliminary maximum oxygen partial pressure of 300 mm Hg has been suggested (5). This corresponds to a dissolved-oxygen concentration equal to 21 mg/L (ppm) at 12 °C (54 °F) and 16 mg/L at 25 °C (77 °F).

The addition of high concentrations of dissolved oxygen can increase the total gas pressure (6). The amount of increase in the total gas pressure depends strongly on the type of aeration unit used and its operating conditions.

Limitations on Maximum Oxygen Consumption

In high-intensity flow-through systems, cumulative oxygen consumption (7) is an important measure of system intensity. The cumulative oxygen consumption (COC) rate for a single rearing unit is equal to the amount of oxygen consumed ($DO_{in} - DO_{out}$). For a serial reuse system, the cumulative oxygen consumption for the overall system is equal to the sum of the oxygen consumed in all of the units.

The utilization of oxygen produces both carbon dioxide and ammonia. The depletion of oxygen may not always be the most severely limiting parameter; when ammonia

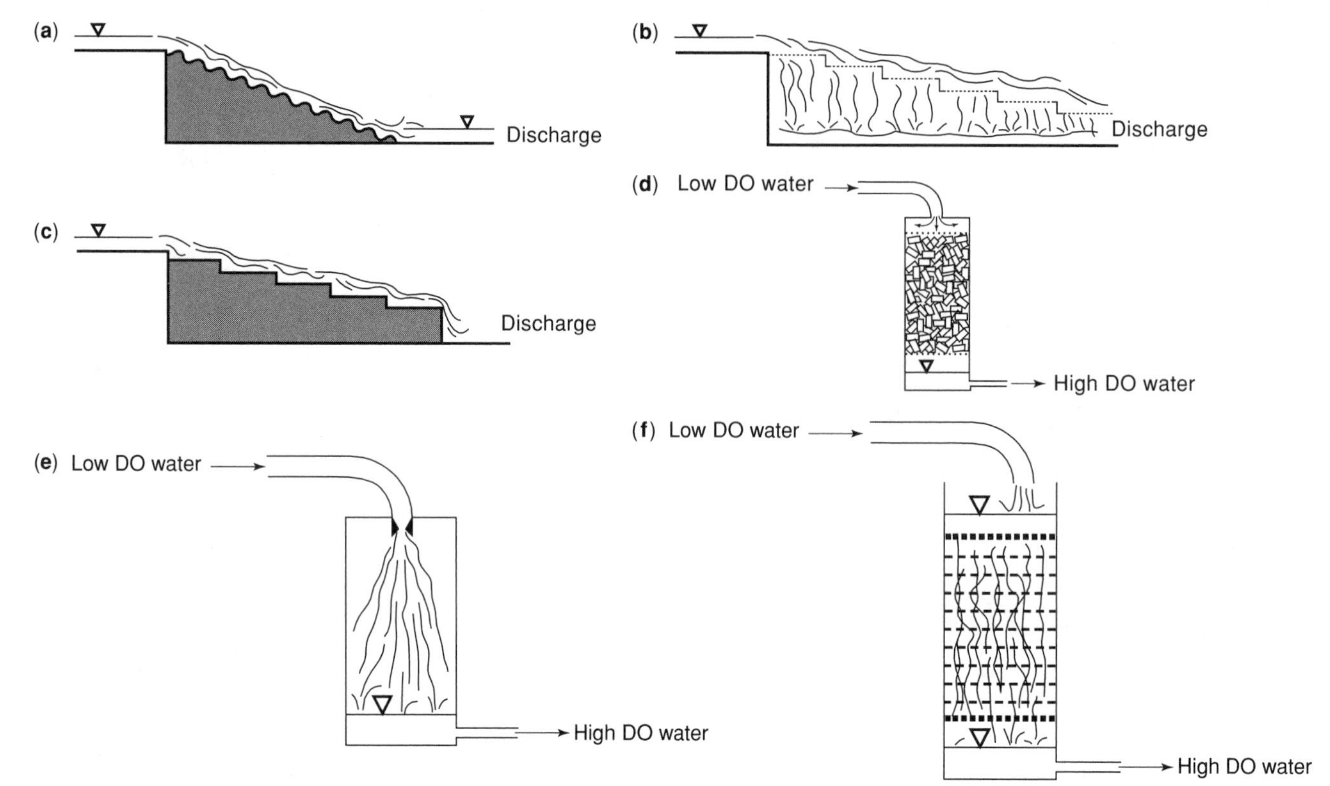

Figure 3. Typical gravity aerators: (**a**) corrugated inclined plane, (**b**) lattice, (**c**) cascade, (**d**) packed column, (**e**) spray column, (**f**) tray or screen.

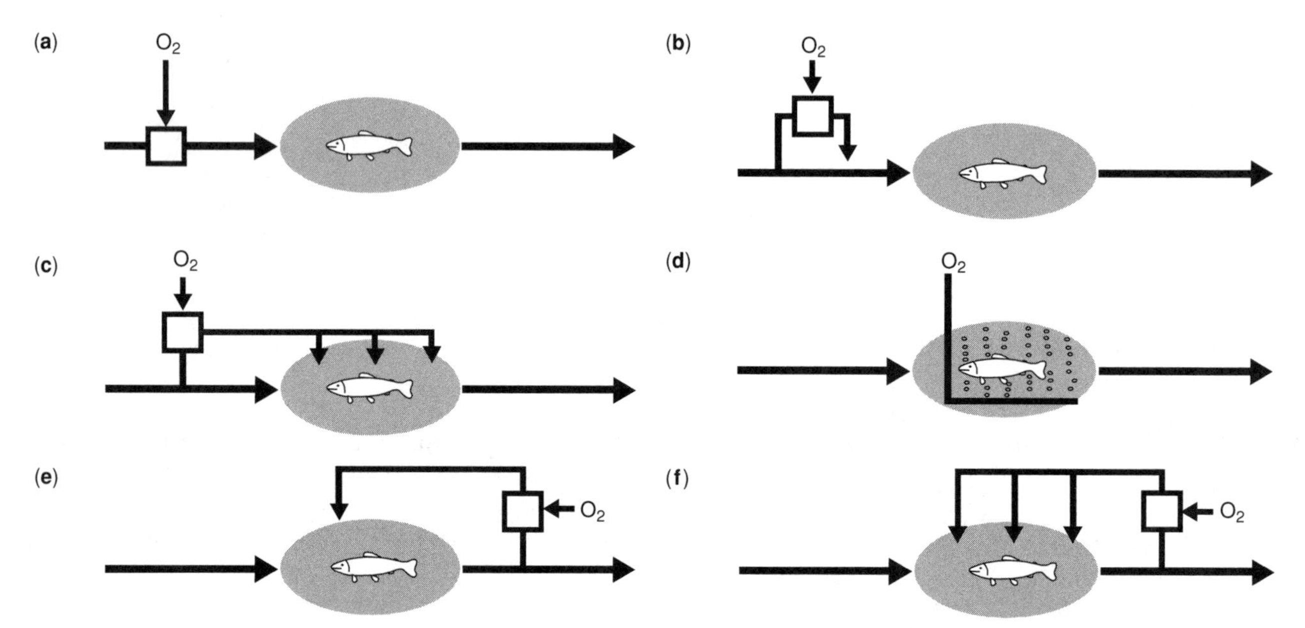

Figure 4. Location of aerators: (**a**) influent; (**b**) influent, side-stream mode with single point of return; (**c**) influent, side-stream mode with multipoint return; (**d**) in-unit; (**e**) recycle, single point of return; (**f**) recycle, multipoint return.

or carbon dioxide is more limiting, aeration will have little effect on carrying capacity (8). Maximum cumulative oxygen consumption (COC), based on limitations due to pH, dissolved oxygen, and un-ionized ammonia, is presented in Figure 5 for water-quality criteria typical in salmon and trout culture.

GAS TRANSFER

The rate at which a slightly soluble gas such as oxygen is transferred into water is proportional to the area of the gas-liquid interface and the difference between the saturation concentration and the existing concentration of

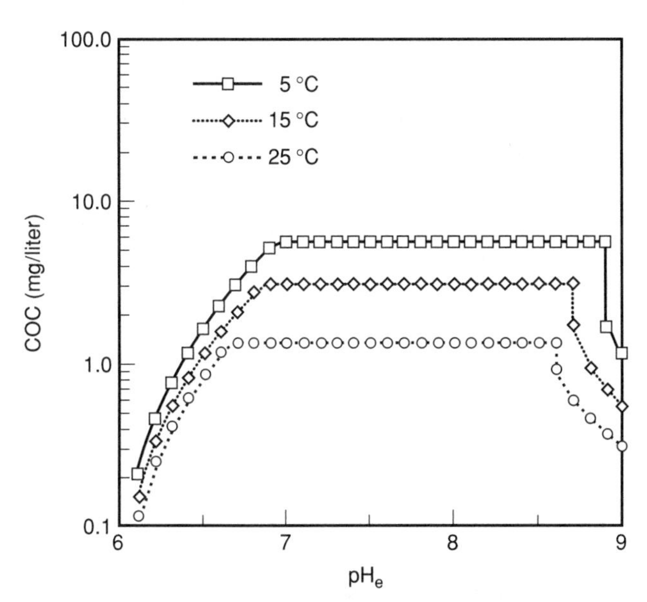

Figure 5. Maximum cumulative oxygen consumption (mg/L) as a function of equilibrium pH_e (pH of a solution in equilibrium with the atmosphere). At low pH_e, COC is limited by the pH criteria; at intermediate pH_e, by the dissolved-oxygen criteria; at high pH_e, by the un-ionized ammonia criteria (8).

the gas in the water (9):

$$\frac{dC}{dt} = K_L \cdot a(C^* - C) \quad (2)$$

where

$\frac{dC}{dt}$ = rate of mass transfer (mass/time),
K_L = overall liquid-phase mass-transfer coefficient (length/time),
a = area of interfacial contact between gas and liquid ($length^2/length^3$), and
C^* = saturation dissolved-gas concentration at a given temperature, pressure, and mole fraction (mass/volume),
C = current dissolved oxygen concentration (mass/volume).

A positive gradient $(C^* - C)$ transfers oxygen into the liquid phase; conversely, a negative gradient transfers oxygen into the gas phase. The transfer rate can be increased by increasing the value of K_L, a, or C^*. In many systems, it is not possible to determine K_L and a independently, and these two variables are combined to a single term (K_La).

STANDARDIZED AERATOR TESTING UNDER CLEAN WATER CONDITIONS

Standardized testing and rating procedures for in-basin aerators have been developed for wastewater applications. These standards are helpful, but their use in aquaculture systems may not always be the most accurate or valid means of rating aerators.

Unsteady-State Testing

Unsteady-state testing procedures (10) are conducted under standard conditions in an experimental test basin. Some test basins are as large as 3,000 to 6,000 m^3 (4,000 to 8,000 yd^3) and can be used to test aerators as powerful as 50 to 100 kW (70 to 140 hb). Typically, a solution of sodium sulfite and cobalt chloride is used to deoxygenate the water by chemical oxidation. The aerator is then started, and the dissolved-oxygen is measured periodically until the saturation concentration is approached. This testing procedure is termed the unsteady-state test, as the amount of oxygen transferred and the dissolved-oxygen concentration are changing during the test.

Fundamental to the rating of in-basin aerators is the experimental determination of K_La (Eq. 2) and the computation of the standardized oxygen transfer rate (SOTR). The SOTR is the maximum rate of transfer into water having a dissolved-oxygen concentration of zero mg/L (ppm), at 20 °C (68 °F), 760 mm Hg; it is expressed in kg/hr (lb/hr). In older literature, this parameter is referred to as N_o. The standardized aeration efficiency is expressed as follows:

$$\mathrm{SAE} = \frac{\mathrm{SOTR}}{P_{\mathrm{in}}} \quad (3)$$

where

SAE = standardized aeration efficiency (kg O_2/kW·hr, lb O_2/hp·hr),
SOTR = standardized oxygen transfer rate (kg/hr, lb/hr), and
P_{in} = power input (kw, hp).

The power input should be the measured wire power, that is, the total power actually used by the entire system: (a) motor, drive, and blower; or (b) motor, coupling, and gearbox (11). The standardized oxygen transfer efficiency is equal to the oxygen transferred into the water divided by the mass flow rate of oxygen supplied to the aerator:

$$\mathrm{OTE}_o = \frac{\mathrm{SOTR}}{\dot{m}} \quad (4)$$

where

OTE_o = standardized oxygen transfer efficiency (%),
SOTR = standardized oxygen transfer rate (kg/hr, lb/hr), and
$\dot{m}$ = mass flow rate of oxygen (kg/hr, lb/hr).

SOTR, OTE_o, and SAE are reported for aerators and can be used to compare different types or brands of aerators. The standardized procedure for the determination of K_La reduces the uncertainty in aerator rating and allows more meaningful comparisons between units. Unfortunately, standards for the rating of aerators change from country to country. In Europe, for example, the standard temperature used in the computation of SOTR and SAE is 10 °C, rather than the 20 °C used in the United States.

For design, it is necessary to estimate the performance of a specific aerator under field conditions. The values of SOTR, OTE_o, and SAE cannot be used directly for design.

Steady-State Testing

For a number of gravity aerators (such as packed columns), both the input and the effluent oxygen concentrations can be directly measured. For these types of aerators, the field oxygen transfer rate, field aeration efficiency, and field oxygen transfer efficiency can be computed directly from the following three equations. The field oxygen transfer rate is calculated as follows:

$$OTR_f = 3.6Q_w\ (DO_{out} - DO_{in}) \tag{5}$$

where

OTR_f = oxygen transfer rate under field conditions (kg/hr),
Q_w = water flow (m^3/s),
DO_{out} = effluent DO concentration (mg/L), and
DO_{in} = influent DO concentration (mg/L).

The field aeration efficiency (FAE) is calculated as follows:

$$FAE = \frac{OTR_f}{P_{in}} \tag{6}$$

where

FAE = field aeration efficiency (kg O_2/kW • hr, lb O_2/hp • hr),
OTR_f = oxygen transfer rate under field conditions (kg/hr, lb/hr), and
P_{in} = Power input (kw, hp).

The field oxygen transfer efficiency (OTE_f) is calculated as follows:

$$OTE_f = \frac{OTR_f}{\dot{m}} \tag{7}$$

where

OTE_f = oxygen transfer efficiency under field conditions (%),
OTR_f = oxygen transfer rate under field conditions (kg O_2/hr, lb O_2/hr), and
$\dot{m}$ = mass flow rate of oxygen (kg/hr, lb/hr).

For some types of atmospheric gravity aerators, $\dot{m}$ can not be measured, and the aeration effectiveness (12) has been used as a rating parameter, calculated from the following equation:

$$AF = \left[\frac{C_{out} - C_{in}}{C^* - C_{in}}\right] \times 100 \tag{8}$$

where

AF = aeration effectiveness under field conditions (%),
C_{out} = effluent dissolved oxygen concentration (mg/L, ppm),
C_{in} = influent dissolved oxygen concentration (mg/L, ppm), and
C^* = saturation dissolved oxygen concentration (mg/L, ppm).

In the unsteady-state tests, the performance of an aerator was rated under standard conditions. In steady-state tests, unless the actual test conditions are equal to the standard conditions, these test results must reduced to standard conditions for comparison of performance values. The interrelationship of standard and field conditions will be presented in the next section.

PERFORMANCE AND RATING OF AERATION SYSTEMS UNDER FIELD CONDITIONS

Values of SOTR, SAE, and OTE_o are computed for water at 20 °C (68 °F), 760 mm Hg, and having zero dissolved oxygen; therefore, they cannot be used directly for the design of aquatic culture systems. Actual performance under field conditions depends primarily on the required dissolved oxygen concentration (C) and to a lesser extent on temperature, on pressure, and on water characteristics. The rating of aerators under field conditions requires the computation of the oxygen transfer rate under field conditions (OTR_f), the field aeration efficiency (FAE), and the oxygen transfer efficiency under field conditions (OTE_f). The computation of these parameters assumes that the field installation is identical to the unit tested under standard conditions. The interrelationship between the rating and the field parameters is presented in the following table (10):

Standard Parameter	Field Parameter	Units
$K_La(20\,°C)$	$K_La(t)$	1/hr
SOTR	OTR_f	kg O_2/hr (lb O_2/hr)
SAE	FAE	kg O_2/kW • hr (lb O_2/hp • hr)
OTE_o	OTE_f	%

Characteristics of Culture Water

Alpha (α). The effects of water characteristics on oxygen transfer are corrected for by the alpha factor, which can be calculated as follows:

$$\alpha = \frac{K_La \text{—field conditions}}{K_La \text{—standard conditions}} \tag{9}$$

Here K_La = volumetric mass transfer coefficient (1/hr).

The value of α depends primarily on the concentration of surfactants in the water. In production catfish ponds, α ranged from 0.66 to 1.07 and averaged 0.94 (12). In recycle systems, α values as low as 0.36 have been measured following feeding (13). The depression of α appears to be caused by the leaching of soluble compounds from feed or from compounds produced by algae.

Beta (β). The effects of water characteristics on oxygen solubility are corrected for by the beta factor, which is computed as follows:

$$\beta = \frac{C^* \text{—field conditions}}{C^* \text{—standard conditions}} \quad (10)$$

Here C^* = saturation DO concentration.

The beta factor is influenced primarily by dissolved solids and to a lesser extent by dissolved organics and suspended solids. In wastewater, beta values typically range from 0.95 to 1.00 (14,15). Beta values for aquaculture conditions are unavailable.

Theta (Θ). The theta factor is used to correct K_La for changes in viscosity, surface tension, and diffusion constants all as a function of temperature (15). The temperature variation of K_La is compensated for as follows:

$$K_la(t) = K_La(20\,^\circ\text{C})\Theta^{(t-20\,^\circ\text{C})} \quad (11)$$

where

20 °C (68 °F) is the standard temperature and t is the field temperature (°C)

A value of 1.024 is recommended (15).

Computation of Field Oxygen Transfer Rate (OTR_f)

The field oxygen transfer rate (OTR_f) is the rate of oxygen transfer under field conditions. It is derived as follows:

$$\text{OTR}_f = \text{SOTR}\left\{\frac{\alpha(1.024^{(t-20)})(\beta C^* - C)}{9.092}\right\} \quad (12)$$

where

OTR_f = field oxygen transfer rate (kg/hr),
C = minimum dissolved oxygen concentration (mg/L, ppm), and
9.092 = dissolved-oxygen concentration at standard conditions (mg/L, ppm).

The computation of the field oxygen transfer rate (OTR_f) requires (*1*) SOTR, (*2*) α and β for the particular water condition, (*3*) local water temperature, and (*4*) C.

Computation of Field Aeration Efficiency (FAE)

The field aeration efficiency (FAE) is the oxygen transfer/unit power input under field conditions. It is derived as follows:

$$\text{FAE} = \text{SAE}\left\{\frac{\alpha(1.024^{(t-20)})(\beta C^* - C)}{9.092}\right\} \quad (13)$$

Here the symbols are as defined for Equations 6, 9, 10, 11, and 12. The impact of hydrostatic pressure on the saturation concentration C^* have been ignored in Equations 12 and 13. These corrections may be significant for aerators submerged in deep aeration basins (10).

PROCESS SELECTION AND DESIGN

A wide range of aeration devices are available for aquaculture. The actual type selected will depend on a variety of factors related to the characteristics of the aerator, the culture system, the site conditions, and system operations. It should be noted that aerator selection may require serious trade-offs between some of these parameters. Many of these tradeoffs may be difficult to quantify, especially if the long-term objectives of the culture system are not well-defined or the oxygen demand of the system changes significantly over the production cycle.

Field Aeration Efficiency

Standardized aeration efficiencies (SAE) for some aerators commonly used in aquatic systems are listed in Table 1.

Table 1. Typical Standardized Aerator Efficiency (SAE) for Aerators Used in Aquaculture (adapted from 16)

Type	SAE (kg O_2/kW·hr)[a]
Surface Aerators	
Low-speed surface	1.2–2.4
Low-speed surface with draft tube	1.2–2.4
High-speed surface	1.2–2.4
Paddlewheel	
Triangular blades	2.7–2.9
PVC pipe blades	1.2–1.9
Tractor powered	1.3–2.0
Gravity Aerators	
Cascade weir (45°)	1.5–1.8
Corrugated inclined-plane (20°)	1.0–1.9
Horizontal screens	1.2–2.6
Lattice aerator	1.8–2.6
Packed column	
Zero head	1.2–2.4
0.5–1.0 m head	10–80[b]
Aeration cone	2.5
Submerged Aerators	
Air-lift pump	2.0–2.1
Diffused air	
Fine bubble	1.2–2.0
Medium bubble	1.0–1.6
Coarse bubble	0.6–1.2
Nozzle aerator	1.3–2.6
Propeller aspirator pump	1.7–1.9
Static tube	1.8–2.4
U-tube	
Zero head	0.72–2.3
0.5–1.0 m head	10–40[a]
Venturi aerator	2.0–3.3

[a] lb O_2/(hp·hr) = kg O_2/(kW·hr × 1.6440).
[b] Does not include pumping power.

The SAE will typically range from 1.0 to 2.6 kg O_2/kW • hr. The SAE of some types of subsurface aerator may range as high as 3.2 to 3.5 kg O_2/kW • hr. If 0.5 to 1.0 m of head is available, the SAE of the U-tube and packed-column aerators can range as high as 40–80 kg O_2/kW • hr. The only power required is for injection of air into the U-tube or for low-pressure fans in the packed columns.

The FAE values for aquaculture systems will be significantly less than the listed SAE values, primarily because of the necessity of maintaining a dissolved oxygen concentration of 5 to 7 mg/L. For example, at 30 °C (86 °F) $\alpha = \beta = 1.0$, and $C = 5$ mg/L (ppm), FAE is equal to only 36% of the SAE value. At high temperatures and C values, the value of FAE is significantly reduced.

Field Aeration Effectiveness

Most gravity aerators can be designed to operate with no power input if 1.0 m or more head is available. Typical values of the aeration effectiveness (AF) are presented in Table 2. Many gravity aerators have very low SAE values; one may, nonetheless, be useful in some applications due to its simplicity of construction and operation. Information on the computation of standard aerator performance under steady-state testing is presented in (16).

Table 2. Aeration Effectiveness (AF) of Typical Gravity Aerators

Type	Height or Head (cm)[a]	AF (%)
Low-Head Types		
Cascade (45°)	25	22–26
	50	36–38
Corrugated inclined-plane (20°)	30	18–29
	60	30–50
Horizontal perforated trays	110	95–100
Lattice aerator	30	29–37
	60	48–61
Simple weir	30	7–10
Splash board	30	23–25
	60	36–41
Packed column	30	94–96
	60	96–98
High-Head Types		
Alfalfa gate	990 to 1,700	61
Ell aspirator	990 to 1,700	83
Gate valve (Half-open)	990 to 1,700	76
Screen	990 to 1,700	59
Screen covered with rocks	990 to 1,700	63
Screen cover	990 to 1,700	52
Screen extension	990 to 1,700	51
Slotted cap	990 to 1,700	65
Splashboard	990 to 1,700	51
Splashboard with holes	990 to 1,700	53
Straight pipe	990 to 1,700	25
Tee aspirator	990 to 1,700	72

[a]Inches = cm/2.54.

Field Oxygen Transfer Rate

In a number of systems, oxygen transfer rate is more important than efficiency. Tractor-powered paddlewheel aerators have been used widely in catfish ponds for emergency aeration (17) and can easily be moved from pond to pond when needed. Diffused aeration with pure oxygen is widely used in transportation systems (18) and in emergency systems for high-intensity systems, because of its ability to transfer large amounts of oxygen without any power input.

Dissolved-Gas Concentrations and Pressures

Dissolved-gas concentrations (or pressures) in the effluent from aeration must be considered in aeration design and operation (6). Lethal dissolved-gas pressures may be produced by some types of submerged aerators (19).

Computation of Oxygen Demand and Supplemental Requirements

The sizing of aeration systems requires estimates of the total oxygen demand by aquatic animals and other organisms, of the available oxygen supplied by water flow (if any), and of the consequent requirement for supplemental oxygen. Because the total oxygen demand of the animals depends on their number, their size, and the water temperature, it is necessary to estimate these parameters over the whole production cycle on a weekly or monthly basis. The temperatures allowed for should include average, extreme maximum, and extreme minimum values.

Average Daily Oxygen Demand

The average daily oxygen demand (20,21) is proportional to the total daily ration:

$$\mathrm{DOD_{aver}} = (\mathrm{OFR}) \times R \tag{14}$$

where

$\mathrm{DOD_{ave}}$ = average daily oxygen demand (kg/d, lb/d),
OFR = Ratio of average daily oxygen demand to daily feed consumption (kg/kg, lb/lb), and
R = daily feed consumption (kg/d, lb/d).

The oxygen requirement to process a given mass of feed depends on animal size, the feeding rate, the composition of the ration, the digestibility of the feed components, and moisture content; it can be characterized by the oxygen:feed ratio (OFR).

In production salmon and trout systems, OFR ranging from 0.20 to 0.22 kg oxygen/kg wet feed have been reported (22,21). In commercial high-density warm-water fish culture, a value for OFR of 1.00 kg oxygen/kg wet feed is commonly used (Anthonie Schuur, Aquaculture Management Services, personal communication). The higher value of OFR for warm-water fish may be due to higher levels of metabolizable energy in the feed, to lower moisture levels in the feed, to lower re-aeration across the water surface, to higher bacteria oxygen demand (from oxidation of organics and ammonia), or to differences in activity and feeding behavior. Limited data is available for OFR in recycle systems. The oxygen demand from bacterial oxidation of organic compounds, ammonia, and solids strongly depends on the unit processes and their operation. The upper bound for OFR is the ultimate Biochemical Oxygen Demand of the feed, which for channel-catfish feed is equal to 1.1 kg O_2/kg dry feed (23). Careful feeding, followed by rapid removal of solids from the system, can significantly reduce the OFR. Due to the minor impact of culture animals on a whole pond's respiration, computation of OFR under pond conditions may not be particularly important. Variation of DO and of aeration demand in ponds can be computed by a variety of techniques (24–26).

Maximum Daily Oxygen Demand

On a daily basis, in a flow-through system, the maximum oxygen consumption occurs at about 4 to 6 hours after feeding. A peaking factor of 1.44, to account for the maximum daily oxygen-consumption rate, has been suggested (22):

$$\mathrm{OD}_{\max} = 1.44(\mathrm{DOD}_{\mathrm{aver}}) \tag{15}$$

where

$\mathrm{OD}_{\max}$ = maximum daily oxygen demand (kg/d, lb/d) and
$\mathrm{DOD}_{\mathrm{aver}}$ = average daily oxygen demand (kg/d, lb/d).

Supplemental Oxygen Requirement

The amount of available oxygen supplied by the flow (kg/d, lb/d) is calculated as follows:

$$\text{Oxygen supplied by flow} = (A)(Q_w)(\mathrm{DO}_{\mathrm{out}} - \mathrm{DO}_{\min}) \tag{16}$$

where

A = constant (84.4, in kms 5.443×10^{-3} in English units),
Q_w = water flow (m^3/s, gpm),
$\mathrm{DO}_{\mathrm{out}}$ = effluent DO concentration (mg/L, ppm), and
$\mathrm{DO}_{\mathrm{in}}$ = influent DO concentration (mg/L, ppm).

The amount of supplemental oxygen (kg/d, lb/d) is calculated by combining Equations 14, 16, and 17, as follows:

$$\text{Supplemental oxygen} = 1.44(\mathrm{OFR})(R) - A(Q_w) \times (\mathrm{DO}_{\mathrm{out}} - \mathrm{DO}_{\min}) \tag{17}$$

where

OFR = ratio of average daily oxygen demand to daily feed consumption (kg/kg, lb/lb),
R = ration (kg/d, lb/d),
A = constant (84.4, in kms 5.443×10^{-3} in English units), and
Q_w = water flow (m^3/s, gpm).

For design purposes, the supplemental oxygen requirement should be based on a weekly (or monthly) biomass and feeding level. Depending on the harvest schedule and temperature, the maximum supplemental oxygen requirement may occur prior to the end of the production cycle. If there is a large variation in biomass between the various rearing units, it may be necessary to compute the supplemental oxygen requirement for each rearing unit.

Number of Units and Power Requirement

The number of units and the power requirement depend on the amount of supplemental oxygen needed (Eq. 17), OTR_f, and FAE.

$$\text{Number of units needed} = \frac{\text{Supplemental oxygen}}{\mathrm{OTR}_f} \tag{18}$$

$$\text{Power requirements (kW)} = \frac{\text{Supplemental oxygen}}{\mathrm{FAE}} \tag{19}$$

System Characteristics

The selection of aerators will also be based on the physical characteristics of (a) the site, (b) the number, size, and configuration of the rearing units, (c) the hydraulics of the rearing units, and (d) the mode of operation. In many cases, the system may be completed before the need for an aeration system is realized. Therefore, it is commonly necessary to design or retrofit an aeration system around a given system, rather than to design a complete culture system from scratch. Some of the most common site considerations are presented in Table 3.

The number, size, and configuration of the rearing units is important in the selection process. In large tanks or ponds, individual mechanical or floating aerators may be used. In the aquarium trade, where a large number of small tanks must be aerated, diffused aeration is commonly used. Although the amount of air available is relatively fixed, the air can be distributed to a large number of individual units inexpensively, and flow to an individual unit can easily be varied.

Aerators that interfere with the normal operations of a culture system or require extensive maintenance will probably not be used long. Operational personnel may lack the time, the knowledge, or the tools to operate and repair some types of aerators. Aerators such as gravity aerators or paddlewheels, which can be constructed and repaired on-site, may be better choices for some operations, even if their overall efficiency is lower than that of other types of aerators. The operational characteristics of different types of aerator systems are presented in Table 3.

Control

The oxygen demand of a production system has a significant diurnal and seasonal variation (Fig. 6). In addition, the oxygen demand from a single raceway or

Table 3. Common Design Considerations

Item	Considerations
	Site
Head	If enough head is available, operation of a gravity aerator may be possible without the use of external power.
Power	If electrical power is unavailable or unreliable, it may be necessary to use a motor-generator system or an engine-powered aerator.
Location	In remote locations, the lack of spare parts and trained personnel may favor simple systems.
Subsurface	In areas with rocky or unstable soils, excavation may be expensive.
Layout	Retro-fitting existing hatcheries may involve careful consideration of problems associated with installing additional electrical lines, piping, and pumps between or around existing structures and utilities. A side-stream pure oxygen system may be easier to retro-fit at some sites.
	Operational
Fouling of diffusers	Diffusers (airstones) may foul from the growth of algae or bacteria. Fine-bubble diffusers may require special air filters and non-metal air lines to prevent clogging due to rust and scale. Diffusers may foul rapidly if not operated continuously.
Icing	Surface aerators (and some types of gravity aerators) produce enough spray to cause ice on walkways and roads. This situation may present a safety hazard to personnel.
Safety	Electrical lines, fuel tanks, or rotating shafts may present a safety hazard to personnel. Diffused aeration systems may be safer than other systems, because electrical lines are required only for the central blower unit. Electrical safety is a major concern in marine systems, because of the high conductivity of seawater.
Harvesting/feeding	In ponds, static-tube or surface aerators may need to be removed prior to harvesting. Aerators should not interfere with the daily operations of the facility.
Repair	The ease of repair may be an important consideration in remote locations. This consideration includes both the skills and tools required and the local availability of spare parts.
Reliability	A simple and highly reliable aerator is desirable. When adequate head is available, gravity aerators will operate during power failures.

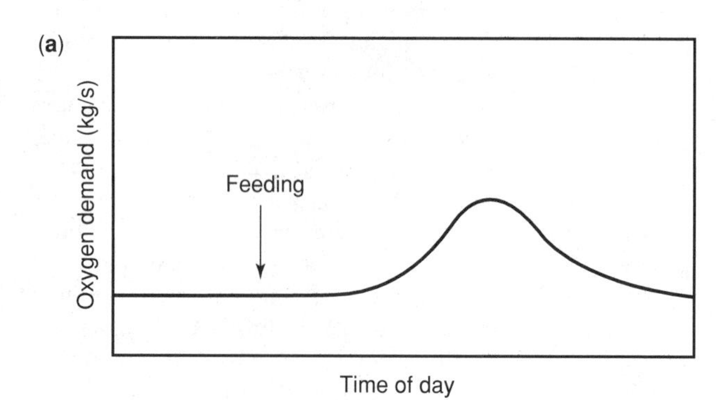

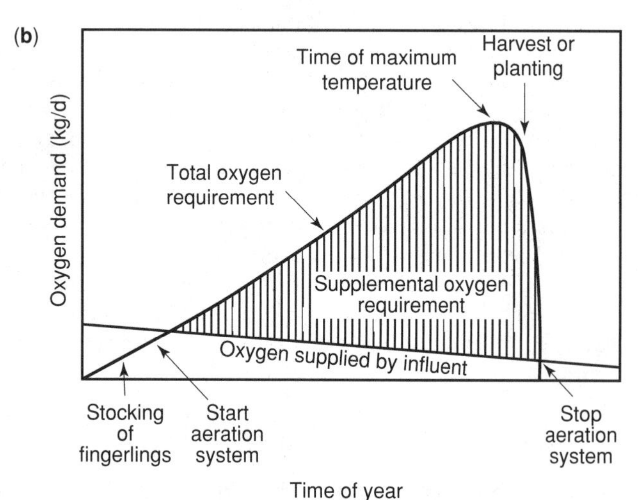

Figure 6. Variation of oxygen demand in a flow-through system with (**a**) time of day (**b**) season of year.

raceway series can change during the transferring or harvesting of fish. The amount of oxygen transfer can be adjusted by turning on another pump or blower. The degree of control depends on the total number of aeration units, the operational characteristics of the aerators, and the layout of the rearing units.

The simplest control strategy for diurnal changes in oxygen demand is to design for the maximum oxygen demand (Fig. 7a). This strategy may result in low FAE values over much of the day.

Step control (Fig. 7b) uses one system to provide base capacity and a second system to provide peak capacity. In raceways, surface aerators are commonly used to provide additional aeration for times at which biomass is high and the output of gravity aerators is insufficient. Surface aerators can also be used to increase the oxygen-transfer rate following feeding, when the oxygen consumption of culture animals increases. In subsurface aerators, oxygen transfer can be changed by changing the air flow to the unit. A system consisting of number of smaller units, each of which can be turned on when needed, may be more efficient than one running a single large blower continuously.

Total required oxygen capacity may be minimized by staggering the feeding times within each raceway series to reduce the peak oxygen demand following feeding (Fig. 7c). This strategy may have the additional benefit of eliminating the need for continuous DO monitoring and for on-line control.

Aerators in ponds are generally run during the night-time period (Fig. 7d). Because of the high oxygen demand

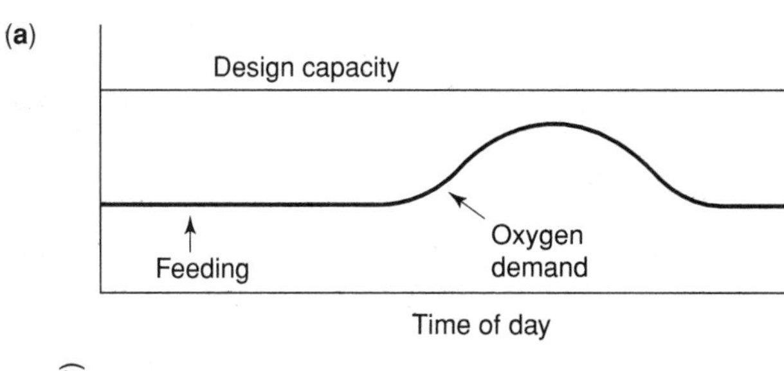

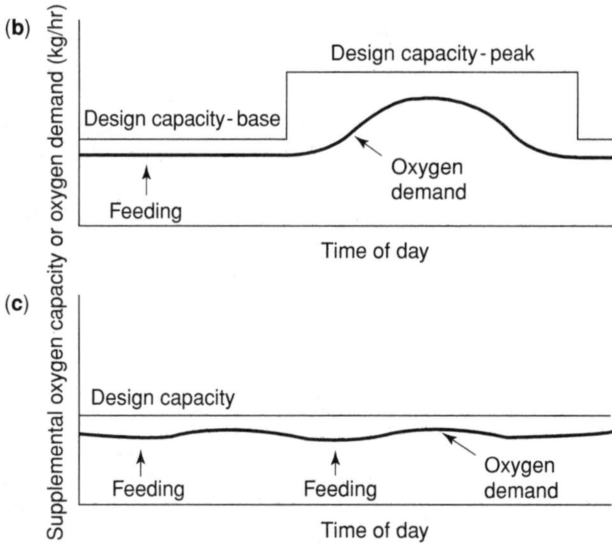

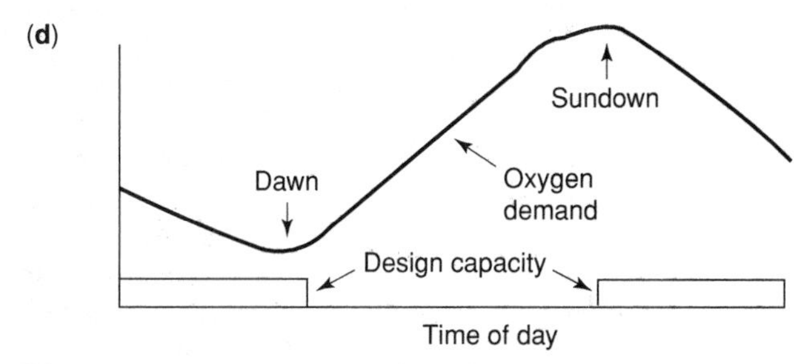

Figure 7. Control strategies: (**a**) peak demand, (**b**) step control, (**c**) reduced peak demand, (**d**) pond systems.

from algae and bacteria, it is generally impossible to maintain adequate dissolved oxygen concentrations in the entire pond.

BIBLIOGRAPHY

1. A.E. Greenberg, L.S. Clesceri, and A.D. Eaton, eds., *Standard Methods for the Examination of Water and Wastewater*, American Public Health Association, Washington, DC, 1992.
2. J. Colt, *Computation of Dissolved Gas Concentrations in Water as Functions of Temperature, Salinity, and Pressure*, Spec. Pub. No. 14, Am. Fish. Soc., Bethesda, MD, 1984.
3. J.W. Andrews, T. Murai, and G. Gibbons, *Trans. Am. Fish. Soc.* **102**, 835–838 (1973).
4. J.R. Brett, in W.S. Hoar, D.J. Randall, and J.R. Brett, eds., *Fish Physiology*, Vol. 8, Academic Press, New York, 1979, pp. 599–675.
5. J. Colt, K.J. Orwicz, and G.R. Bouck, Fisheries Bioengineering Symposium, Am. Fish. Soc. Symp. 10, Am. Fish. Soc., Bethesda, MD, 1991, pp. 372–385.
6. B. Watten, J. Colt, and C. Boyd, Fisheries Bioengineering Symposium, Am. Fish. Soc. Symp. 10, MD, 1991, pp. 474–481.
7. J.W. Meade, *Aquacult. Eng.* **7**, 139–146 (1988).
8. J. Colt and K. Orwicz, *Aquacult. Eng.* **10**, 1–29 (1991).
9. W.K. Lewis and W.C. Whitman, *J. Ind. Eng.* **16**, 1215–1220 (1924).
10. American Society of Civil Engineers, *A Standard for the Measurement of Oxygen Transfer in Clean Water*, New York, 1984.
11. F.W.W. Yunt, in A. Boyle, ed., *Proceedings: Workshop Towards an Oxygen Transfer Standard*, Cincinnati, OH, 1979, pp. 105–127.
12. J.L. Shelton Jr. and C.E. Boyd, *Trans. Am. Fish. Soc.* **112**, 120–122 (1983).
13. D.E. Weaver, Presented at 12th Annual Meeting of the World Mariculture Society, March 8–10, 1981, Seattle, WA (unpublished), 1981.
14. Metcalf and Eddy, Inc., *Wastewater Engineering: Treatment, Disposal, Reuse.*, McGraw-Hill, New York, 1979.
15. M.K. Stenstrom and R.G. Gilbert, *Wat. Res.* **15**, 643–654 (1981).
16. J. Colt and K. Orwicz, in D.A. Brune and J.R. Tomasso, eds., *Water Quality and Aquaculture*, World Aquaculture Society, Baton Rouge, LA, 1992, pp. 198–271.
17. C.E. Boyd and C.S. Tucker, *Trans. Am. Fish. Soc.* **108**, 299–306 (1979).
18. G.J. Carmichael and J.R. Tomasso, *Prog. Fish-Cult.* **50**, 155–159 (1988).
19. J. Colt and H. Westers, *Trans. Am. Fish. Soc.* **111**, 342–360 (1982).
20. D.C. Haskell, R.O. Davies, and J. Reckahn, *NY Fish and Game J.* **7**, 112–129 (1960).
21. H. Willoughby, *Prog. Fish-Cult.* **30**, 173–174 (1968).
22. H. Westers, *Fish Culture Manual for the State of Michigan*. Michigan Department of Natural Resources, Lansing, MI (unpublished), 1981.
23. J.C. Harris, *Pollutional Characteristics of Channel Catfish Culture*. Master's Thesis, Envir. Eng., University of Texas, Austin, TX, 1971.
24. C.E. Boyd, R.P. Romaire, and E. Johnston, *Trans. Am. Fish. Soc.* **107**, 484–492 (1978).
25. C.P. Madenjian, G.L. Rogers, *Aquacult. Eng.* **6**, 209–225 (1987).
26. D.I. Meyer and D.E. Brune, *Aquacult. Eng.* **1**, 245–261 (1982).

See also BIOCHEMICAL OXYGEN DEMAND; BLOWERS AND COMPRESSORS; CHEMICAL OXYGEN DEMAND; DISSOLVED OXYGEN; PURE OXYGEN SYSTEMS.

ALGAE: TOXIC ALGAE AND ALGAL TOXINS

C.O. PATTERSON
Texas A&M University
College Station, Texas

OUTLINE

Introduction
Increasing Frequency of Harmful Algal Blooms

INTRODUCTION

Although algae form the base of most aquatic food chains and are vital to both freshwater and marine ecosystems, certain species frequently become nuisances. Their presence, especially in high cell concentrations, may discolor water or produce unpleasant odors or flavors. Extremely high cell concentrations may produce episodes of hypoxia or anoxia in water bodies, either because of high respiratory oxygen demand during hours of darkness or because of chemical oxygen demand when the cells die and begin to decay. Algal cells may clog water filtration and purification equipment. Though all these effects cause problems, such problems rarely become life-threatening to humans or domestic animals. More dangerous effects occur when algal species produce chemical compounds which are actively toxic. The terms "red tide" and "brown tide" are increasingly associated in the public mind with outbreaks of toxin-producing algae, but these are misnomers. In many cases, toxin concentrations may reach dangerous levels with no apparent color change in the water, while in other cases discolored water is caused by species that produce no toxins. In this entry, attention focuses on true toxin-producers, ignoring milder nuisances of filter clogging, taste and odor production, and water discoloration. Such toxin-producers are coming to be known as Harmful Algal Blooms (HAB).

INCREASING FREQUENCY OF HARMFUL ALGAL BLOOMS

In recent years, more frequent and more serious outbreaks of toxin-producing algae have generated increasing concern among public health officials and increasing publicity in news media. Funding for the study of toxic algae and algal toxins has increased markedly within the past two decades. As investigators have sought to understand HABs, much literature has been generated, both as journal articles and as monographs (1–10). At least one semipopular book has focused on a specific toxic alga, *Pfiesteria* in the Albemarle–Pamlico–Neuse estuarine system on the Atlantic coast of North Carolina (11). The need for readily available sources of information and rapid dissemination of information and public health warnings has led to establishment of several Web sites for HABs. These sites should be consulted for the most recent information. Woods Hole Oceanographic Institute maintains the Toxic Marine Algae Web site at *http://www.redtide.whoi.edu/hab/*. The Cyanotox Web site [specializing in blue-green algae (cyanobacteria)] originates from La Trobe University (Victoria, Australia) and can be found at *http://luff.latrobe.edu.au/~botbml/cyanotox.html*. The Canadian Department of Fisheries and Oceans maintains the phycotoxins site at *http://www.maritimes.dfo.ca/science/mesd/he/lists/phycotoxins/index.html*.

It is often asked whether outbreaks of toxic algae are becoming more frequent or whether the increased numbers of reports merely reflect increased scientific and public awareness of these events, improved monitoring and characterization techniques, more detailed observation, and better reporting of episodes. No conclusive answers to this question have been reached, but there is a growing consensus among investigators that the frequency and severity of such outbreaks are increasing and blooms of toxic algae are occurring in locations from which they have been absent in the past (12,13). Kao (14) points out that symptoms of algal toxin poisoning are so distinctive and so startling to observers that it seems unlikely that episodes would have gone unnoticed or unreported if they had occurred in the past. Love and Stephens (15) mention that cases of ciguatera poisoning first began to appear in the islands of Midway, Johnson, Palmyra, Fanning (now Tabuaeran), and Christmas (now Kiritimati) in the early 1940s and were caused by eating fishes that had previously been known to be edible. The World Health Organization recorded approximately 900 cases of human paralytic shellfish poisoning between 1970 and 1983, with many of these cases occurring in regions where paralytic shellfish poisoning had been unknown (16). For example, the readily identified dinoflagellate *Gymnodinium catenatum*, which produces paralytic shellfish poisoning, was reported only twice (in the Gulf of California and off Argentina) between 1940 and 1970. Then, between 1976 and 1994, it was reported 12 times from widely scattered locations around the world, usually in connection with toxic outbreaks (17,18).

ECOLOGICAL FACTORS THAT CONTRIBUTE TO HABS

The causes of increased frequency and severity of toxic outbreaks are still under investigation. A number of possible factors have been identified. Improved transportation methods make it possible for materials, including algal cells, to be moved inadvertently (such as in ballast water) and with unprecedented speed from their ancestral habitats into new locales. Thus seed stocks or inocula transported into new habitats may be released from control by grazers or competing species that had previously held numbers low. In addition, many investigators suspect that ecological factors, such as nutrient availability, water temperatures, water turbidity, and growth-regulator analogs, have been altered in many habitats in ways that stimulate growth of the nuisance species. However, in no case do we currently have sufficient data or robust enough models to identify specific causes of HAB outbreaks.

Detailed information is now available about the molecular structure and mode of action of most of the algal toxins. We know the cellular-level effects in some detail. Much less is known about the genetic and environmental basis for toxin production. No complete biosynthetic pathway is known for any algal toxin. Similarly, the genetic basis for every algal toxin remains to be elucidated. We cannot yet identify with certainty the strains that produce toxins nor can we define the particular conditions under which toxins are produced. Factors that produce, sustain, and terminate blooms, and that control toxin production during blooms, are not understood. Outbreaks tend to be irregular and unpredictable. Consequently, we have very little predictive capability. Public health officials and agencies are limited almost entirely to reacting after a toxic incident is underway (sometimes even after the incident is essentially finished), rather than having the tools to anticipate incidents. We are even further from being able to manipulate situations to prevent such outbreaks. Disappearance of blooms is often as sudden and mysterious as their appearance. Sexual reproduction and formation of resting cysts seem frequently to be associated with the ending of blooms, especially among dinoflagellates. Sexual mating in some species has been induced under lab conditions by nutrient starvation.

Although every known algal toxin can be produced by more than one species, it is almost always found that any specific harmful bloom is composed of a single species. Factors responsible for bloom formation, and detailed mechanisms of bloom development, remain poorly understood for virtually all species. Investigators have focused on nutrient availability (19–21), effects of vitamins and chelators (22–25), fluctuations of nutrient concentrations produced by upwelling (26,27) or by runoff from land after heavy rains, on the ability of different species to reach nutrient supplies by vertical migration (especially diel migration) (28), on salinity, temperature, light intensity (29–31), competition with other species (32), effects of grazers (33), excystment or resuspension of resting stages or epiphytic forms (34,35), advection or concentration of cells into restricted geographic areas (36,37), and on availability of trace elements (38). Turner et al. (33) noted that "interactions between toxic phytoplankton and their grazers are complex, variable, and situation-specific. An overall synthesis of these interactions is elusive and premature because present results are still too disparate. Accordingly, information from one experimental study or natural bloom should be extrapolated to another with caution." Similar statements might be made with regard to each of the other biotic and abiotic factors influencing blooms of toxic species. Although detailed understanding of the causes and interactions is still in the future, sufficient information has been accumulated to say that each toxic species displays its own pattern of response to physical, chemical, and biological factors in its environment. That is, each toxic species exploits certain environmental parameters more efficiently than any other species. When specific combinations of features come together, a bloom results. Hallegraeff has provided a very useful summary of the "niche-defining factors" for the major marine species of toxin-producers (39). He categorizes such factors as either responses to the physicochemical features of the environment (temperature, salinity, inorganic nutrients, micronutrients, etc.) or as properties of the organisms themselves (mixotrophy/ability to utilize organic nutrients, allelopathy, or grazer avoidance, parameters of life history, ability for vertical migration, or response to turbulence in the water column, etc.). For example, the dinoflagellate *Alexandrium* (a producer of paralytic shellfish poisoning toxins) is sensitive to temperature, responds strikingly to micronutrient availability, and shows vertical migratory behavior. In contrast, most cyanobacteria are strongly sensitive to temperature and to availability of inorganic macronutrients (especially nitrogen and phosphorus). Further investigation will undoubtedly allow us to refine these categories and develop more detailed descriptions of the parameters that define the "toxic bloom niche" of each HAB species.

Prediction and monitoring of HABs assume great importance for public health officials, fisheries managers, academic investigators, and the general population who might encounter these occasionally dangerous organisms. Therefore, a number of studies are underway, seeking to overcome our present frustrating inability to anticipate toxic outbreaks. Some investigators focus on developing predictive models, based on numerical and statistical descriptions of blooms (40). Other groups (41,42) are developing methods to detect HABs by optical methods (using floating buoys, airborne or satellite surveillance, etc.). These techniques emphasize recognition of HAB-forming species from distinctive patterns of light absorbance by chlorophyll and other pigments or by fluorescence emission spectra. Results have been mixed, with no groups yet reporting complete success. Efforts continue in this area, and some optimism seems justified.

TAXONOMY AND IDENTIFICATION OF TOXIC ALGAE

Controversy abounds over the correct identification of virtually all toxin producers. Characteristics that have been used for identification include morphology, isozyme patterns, toxin profiles, life history patterns, sexual mating ability, and nucleic acid sequences. No one of these has proven entirely satisfactory. Additional difficulty results from taxonomic changes; most toxin-producing species are known by several names. Steidinger and Vargo (43) provide a useful list of synonyms for toxic dinoflagellates. The dinoflagellate genus *Alexandrium* is one of the most important producers of potent toxins responsible for paralytic shellfish poisoning. This genus includes about 30 species. Of these, some species produce toxins, while other species have failed to show toxin production. Of the toxin producers, some strains within a single species show toxin production, while others do not. It is clear that danger of toxic outbreaks cannot yet be assessed from taxonomic information alone. In late 1987, a major poisoning event occurred in eastern Canada. Intense efforts by public health authorities established that the toxin-producer was a diatom, initially identified as *Nitzschia pungens* forma *multiseries*. It required almost 10 years of work before the nomenclature of the organism was resolved; it is now known as *Pseudo-nitzschia multiseries* (44). The situation is further complicated by the growing recognition that different geographic strains (subspecies?) of what seems to be the same species may show sharply different toxin-producing characteristics. For example, the New England strain of the cyanobacterium *Aphanizomenon flos-aquae* is known to produce the paralytic shellfish poisons saxitoxin and neosaxitoxin, while the Oregon strain of the same organism has never shown any sign of toxin production.

Most (perhaps all) toxins are secondary metabolites, with very complex chemical structures. They are the end products of elaborate, multistep biochemical pathways. Very little is known about most of these pathways. Only the biosynthetic pathway for saxitoxins (etiologic agents of Paralytic Shellfish Poisoning) is relatively well understood. However, enough is known about virtually all the pathways to say that it is certain that multiple, probably unique, enzymes are required. The only exception to this general rule appears to be the synthesis of domoic acid by diatoms, where only a few (one to three) unique enzymes may be required (45). Although almost no data are yet available for the enzymes themselves, it can be predicted that the genes encoding these enzymes may occupy large segments of the organisms' DNA, and that collectively, possession of such genetic arrays may serve as a distinctive "signature," which might be used to identify toxin-producers (46).

The taxonomy of all cyanobacteria, including toxin producers, is chaotic and currently undergoing extensive revision. Morphological characters of cyanobacteria are notoriously variable. Many of the features that were thought to distinguish species and even genera are now known to depend on previous growth conditions and are unreliable for accurate identification of field-collected specimens. Consider one extreme example: the cyanobacterium *Spirulina*, which is widely sold commercially as a human nutritional supplement. It has long been thought that *Spirulina* was one of the most easily identified cyanobacterial genera, recognizable by its tightly coiled helical filaments. Patterson and Hearn (unpublished) observed that when *Spirulina* is maintained at high growth rates, it loses its helically coiled growth habit; the filaments straighten and become linear strands, morphologically indistinguishable from strains of *Oscillatoria*. Certain strains of *Oscillatoria* are known to be producers of hepatotoxins and neurotoxins (47,48). Skulberg et al. (49) offered a provisional key for identification of potentially toxigenic cyanobacteria, but it should be used with caution. The presence, position, and form of heterocysts is widely used for identification of filamentous cyanobacteria. Heterocysts, modified cells wherein nitrogen fixation occurs, differentiate only when the organisms are starved of nitrogen. In eutrophic (nitrogen-replete) waters, heterocysts may be completely absent, even from genera capable of forming them. Furthermore, toxic strains of cyanobacteria do not differ morphologically from nontoxic strains, so identification, even to species level, does not reliably predict whether a particular bloom will become toxic.

Analysis of nucleic acid features holds promise for precise identification of toxic species and strains. Scholin (50) provided a useful review of the techniques and difficulties involved. Manhart et al. (51) were able to distinguish Atlantic, Pacific, and Gulf of Mexico isolates of *P. multiseries* from isolates of *P. pungens*, based on distinctive differences in restriction fragment patterns (RFPs) of nuclear DNA. *P. multiseries* is a toxin producer (domoic acid), while *P. pungens* is nontoxic. This technique is not presently suited for field application nor for rapid identification of organisms, but it may provide the basis for development of dependable methods for distinguishing species that are morphologically very similar. Rouhiainen et al. (52) examined 37 strains of toxic and nontoxic cyanobacterial strains from northern Europe. Restriction fragment patterns and Southern blot analyses were shown to distinguish hepatotoxic *Anabaena* isolates from neurotoxic forms and from *Nostoc* strains. Further work is needed to determine whether the distinctive patterns recognized among these isolates from a fairly small geographic area are reliable on a worldwide basis. Successful application of such nucleic acid-based identification techniques will require cultivation of axenic cultures of many species and strains that have not yet been successfully grown in the laboratory. Many toxic strains, especially among marine forms, are notoriously difficult to maintain under laboratory conditions. Culture techniques must improve along with our abilities to carry out analyses of genetic structures.

It is now established that certain nonphotosynthetic bacteria can synthesize saxitoxins and perhaps other toxins as well. It is also known that the presence of certain bacteria enhances production of domoic acid by the diatom *P. multiseries*. Evidence is accumulating that intracellular symbiotic bacteria are present in many strains of toxigenic algae, but it is not yet clear what role these symbionts play in the formation or release of toxins (53,54). Involvement of symbiotic complexes may further complicate the problems of identifying toxic species or strains.

ALGAL TOXINS: THEIR CHEMISTRY AND PHYSIOLOGICAL EFFECTS

Toxins are known to be produced by at least four groups of algae: the Haptophyta (sometimes regarded as Class Prymnesiophyceae within Division Chromophyta), the dinoflagellates (Division Dinophyta or Class Dinophyceae within Division Chromophyta), the diatoms (Class Bacillariophyceae within Division Chromophyta), and the Cyanobacteria ("blue-green algae"). The first three groups show eukaryotic cell structure and hence are true algae. The cyanobacteria possess prokaryotic (bacterial) cell structure, though their old name of "blue-green algae" is still widely used. Dinoflagellates are probably the best known and longest studied of the algal toxin producers. These "red tide" organisms are the most familiar to the general public. Diatoms are the most recent additions to the list of toxin producers, with the first known outbreak of diatom-related poisoning occurring in eastern Canada in 1987 (55,56). Since 1987, toxic diatom blooms have become familiar to the public as "brown tides." Cyanobacteria are typically the source of toxins in freshwater environments (57,58), while the other groups are almost exclusively found in marine or brackish waters.

Toxicity testing is usually by mouse bioassay. Typically, a sample of the suspected toxin is injected intraperitoneally into the mouse, followed by 24 hours of observation. After 24 hours, any surviving mice are sacrificed for postmortem examination for tissue and cytological injury. This bioassay has formed the basis for the discovery, identification, study, and regulation of all the algal toxins. Like all bioassays, this technique suffers from several drawbacks, including inherent variability in the mice, difficulty of distinguishing effects when several toxins are present in a sample, expense, time delays, and increasing public opposition to testing of this sort. Investigators have developed several alternative assay methods in recent years. Gas chromatography, thin-layer chromatography, high-performance liquid chromatography, radioimmunoassay techniques, and enzyme-linked immunosorbent assay techniques all show promising results. Among the greatest challenges to the use of any of these techniques is the supply of purified toxins to use as standards (59).

Cyanobacterial Toxins

Among the cyanobacteria (blue-green algae), several genera have been shown to be toxin-producers. Cyanobacterial toxins are contained within the living cells and are not released into water until senescence or death of the cells (60). It is still not clear what factors influence or control production of cyanobacterial toxins. Many investigators have suspected that nutrient supplies are critically important, but no agreement has been reached as to which nutrients exert the dominant effect. Most attention has focused on nitrogen, phosphorus, and N:P ratios, since it has long been known that cyanobacteria bloom in waters enriched in N and P. There is still no clear evidence that nutrients influence toxin production other than via their general effect on growth rate. In other words, in nutrient-rich waters, growth rates are high, leading to rapid accumulation of dense populations of toxin-producing cells. But the concentration of toxin/cell appears to change very little.

Among the prokaryotic cyanobacteria (blue-green algae) production of both hepatotoxic and neurotoxic compounds has been extensively studied (61). Additional toxins, producing dermatitis, respiratory distress, and other symptoms have been reported, but are less well characterized. The cyanobacterial toxins include both neurotoxins and hepatotoxins. The most common hepatotoxins seem to be cyclic peptides containing a unique hydrophobic amino acid whose chemical name is usually abbreviated ADDA (3-amino-9-methoxy-2,6,8-trimethyl-10-phenyl-4,6-decadienoic acid). These toxins are called microcystin (seven amino acids in the ring) or nodularin (five amino acids in the ring). These were named for the genera from which they were first isolated (*Microcystis and Nodularia*, respectively), but it is now known that other genera also produce the toxins. The peptides are synthesized nonribosomally, but their biosynthetic pathways and normal function in the cell are not well understood. Both microcystin and nodularin are rapidly taken into vertebrate liver cells via the bile transport system and are also taken into epithelial cells of the small intestine, using the same bile transport mechanism. Once inside the target cells, the toxins act as potent inhibitors of protein phosphatases (classes 1 and 2A). This produces hyperphosphorylation of cell proteins, with a wide range of effects. One effect, almost immediately observable microscopically, is deformation of liver cells resulting from collapse of the cytoskeleton. Extensive hemorrhage and hepatocyte necrosis follow; the acute cause of death is shock due to blood loss (62). The effect of microcystin and nodularin in inhibiting protein phosphatases is strongly reminiscent of the mode of action of the dinoflagellate toxin, okadaic acid. However, the molecular structure and sites of action of okadaic acid are distinctly different and will be discussed next. Microcystin is degraded slowly (10–30 days) after release, probably by microbial action. Chlorination, flocculation, and filtration do not remove the toxin, although there are indications that ozonation or absorption on activated charcoal effectively remove the toxin.

A wide variety of cellular events and metabolic processes are regulated by the level of phosphorylation of cellular proteins. Proteins are phosphorylated by the action of protein kinases and dephosphorylated by action of phosphatases. Among other processes regulated by the level of phosphorylation is control of cell division and proliferation. Inhibition of phosphatases, leading to hyperphosphorylation, may increase cell proliferation, leading to tumor formation.

In mice, the lethal dose of microcystins via intraperitoneal injection is about 2–3 μg for a 30 g mouse, i.e., about 60–70 μg/kg body weight. Because of widespread consumption of cyanobacterial material as health supplements, concerns have been raised about acceptable levels of toxins in dried cyanobacterial biomass. The Oregon State Health Division has determined that 1 μg/g (1 ppm) is a safe level for microcystins in cyanobacterial material sold for human consumption.

A second type of cyanobacterial hepatotoxin, known as cylindrospermopsin, was originally isolated from *Cylindrospermopsis* in an Australian water supply. The toxin is a novel alkaloid with a cyclic guanidine unit (63). It appears to act by inhibiting protein synthesis. In addition to microcystin, nodularin, and cylindrospermopsin, another group of compounds that act as tumor promoters have also been isolated from cyanobacteria, especially from *Lyngbya*. These toxins, known as aplysiatoxin and lyngbyatoxin, activate protein kinase C, leading to cell proliferation (64).

Dinoflagellate Toxins: Saxitoxins

Paralytic shellfish poisoning (PSP) results from ingestion of any of a family of compounds produced by dinoflagellates. At this time, members of the genera *Alexandrium (Gonyaulax)*, *Pyrodinium*, and *Gymnodinium* have been shown to produce these toxic compounds. In *Alexandrium*, toxin production is enhanced when cells are phosphate limited, but decreases when cells are nitrogen limited (65,66).

The compounds are derivatives of saxitoxin, a tetrahydropurine with a unique 3-carbon ring between C4 and N3 (67,68). The biosynthetic pathway has been extensively studied, and is reasonably well understood (69,70). At least 18 naturally occurring derivatives of saxitoxin have been isolated. All produce their paralytic effect by binding to and blocking the voltage-gated sodium channel of neurons, skeletal muscle cells, and cardiac muscle cells. In normal, unpoisoned cells, the inward flow of sodium ions produces the action potential that is necessary in the transmission of nerve impulses and the contraction of muscle cells. In the presence of saxitoxin and its derivatives, the sodium channel is blocked, no action potential can be generated, and paralysis results.

PSP is potentially life-threatening. Onset of symptoms is rapid, within a few minutes to a few hours after consumption of the toxin. Symptoms include tingling, numbness, or burning of the mouth region, followed by giddiness, drowsiness, and staggering. Severe cases result in respiratory arrest. No antidote is known.

Dinoflagellate Toxins: Brevetoxins

Neurotoxic shellfish poisoning is produced by brevetoxins, products of the dinoflagellate *Ptychodiscus brevis (Gymnodinium breve)*. Brevetoxins are polyethers, with at least nine derivatives now known to be toxic (71). In contrast to the saxitoxins, which act by blocking sodium influx into excitable cells, brevetoxins specifically induce an irreversible channel-mediated sodium ion influx (72). Thus, brevetoxins depolarize both nerve and muscle cells; nerve cells are much more sensitive to these toxins. Depolarization induces neurotransmitter release in neuromuscular preparations, but the essential effect results from the opening of the sodium channels.

Dinoflagellate Toxins: Ciguatoxin and Gambiertoxins

Ciguatera poisoning is probably the most widespread and the least understood of the algal toxin-related syndromes. It is known that several closely related compounds are involved, and that these compounds are produced by several genera and species of dinoflagellates, notably *Gambierdiscus toxicus*, *Coolia monotis*, *Amphidinium carterae*, *Prorocentrum* spp., *Ostreopsis* spp., *Thecadinium* spp., and perhaps others. *G. toxicus* is the causative organism in most cases of ciguatera poisoning. The molecular structures of only a few of the toxins have been determined. Ciguatoxin and gambiertoxin have been shown to be complex cyclic polyethers, structurally reminiscent of brevetoxin and okadaic acid (73). Although some uncertainty remains as to the exact mode of action of ciguatoxins, it appears very likely that they act in much the same way as brevetoxin, that is, by binding to the sodium channel of neuronal membranes and triggering sodium influx (73,74). This irreversibly depolarizes the nerve cell. Effects of ciguatoxin can be blocked by tetrodotoxin, by excess extracellular Ca^{2+}, or by cholinesterase inhibitors.

Maitotoxin coexists with ciguatoxin in ciguateric fish and is one of the most potent marine toxins. It is produced by *G. toxicus* in more abundant quantity than ciguatoxin (75). In smooth muscle and skeletal muscle preparations, maitotoxin causes calcium ion-dependent contraction (76). Maitotoxin may act by changing configuration of a membrane protein, transforming it into a pore that allows Ca^{2+} to flow through (75,77).

Dinoflagellate Toxins: Okadaic Acid

The dinoflagellate genera *Dinophysis* and *Prorocentrum* include species that produce the toxin okadaic acid. This toxin causes diarrhetic shellfish poisoning (DSP), a nonfatal, but temporarily incapacitating illness, involving severe abdominal cramps, vomiting, diarrhea, and chills. Like brevetoxins and ciguatoxins, okadaic acid is a cyclic polyether. However, the mode of action of okadaic acid appears to differ from that of the other polyethers. Okadaic acid is thought to act as an inhibitor of protein phosphatases (78); in this respect, it is similar to microcystin and nodularin, although the molecular structure and the producing organisms are quite different. Okadaic acid apparently produces its effects by inhibiting dephosphorylation of myosin light chains in smooth muscle and thus producing tonic contractions (79). There is uncertainty as to whether okadaic acid is typically released from healthy cells into the water. Carlsson et al. (80), studying a *Dinophysis* bloom, found that it was not.

Diatom Toxins: Domoic Acid

Certain marine diatoms produce the toxin known as domoic acid. Domoic acid poisoning, also becoming known as amnesic shellfish poisoning, produces symptoms of vomiting, diarrhea, abdominal cramps, disorientation, and memory loss. Although the toxin is produced by diatoms (and by certain species of red algae), most cases of poisoning result from consumption of mussels (*Mytilus edulis*) that have accumulated the toxin from the diatoms on which they have fed. The toxin was originally isolated from the diatom genus now known as *Pseudo-nitzschia* and has been subsequently identified in at least seven additional diatom species (81). It has also been found in the red algal genera *Chondria*, *Alsidium*, *Amansia*, Digenea, and *Vidalia*. All reported poisoning incidents appear to have been associated with diatoms, none with reds. Production of domoic acid by *Pseudo-nitzschia* appears to increase when cells are silicate or nitrogen limited (82,83) or with higher levels of temperature and light (84,85).

The molecular structure of domoic acid is a water-soluble tricarboxylic amino acid (82). It appears to act by binding to glutamate receptors of neurons in the central nervous system, especially the hippocampus (86–88). Glutamate (or glutamic acid) is a well-known excitatory neurotransmitter. Domoic acid excites the neuronal membrane, leading to irreversible depolarization, accumulation of intracellular Ca^{2+}, neuronal swelling, and death (89).

Dinoflagellate Toxins: New and Unidentified Forms

The toxic dinoflagellate *Pfiesteria piscicida* and at least two other *Pfiesteria*-like species produce toxins that remain poorly characterized. *Pfiesteria* and the similar species were first discovered in the 1980s and thus far have only been found in the western Atlantic and Gulf regions. These organisms apparently behave as "ambush predators," releasing toxins specifically when prey organisms are near (90,91). The toxins narcotize finfish, and cause sloughing of epidermis and formation of open ulcerative lesions in finfish and shellfish. The toxins can be aerosolized and produce muscular pain, abdominal distress, dizziness, disorientation, and memory loss in humans when inhaled (92,93). The chemical structures of these toxins are not yet known. It is known that the toxins include both lipophilic and water-soluble compounds (94,95). Likewise, the mode(s) of action of these toxins have not yet been characterized.

Chrysophyte (Haptophyte) Toxins

The genera *Prymnesium*, *Chrysochromulina*, and *Phaeocystis* include species and strains well known for killing finfish and shellfish, both farmed and free-ranging. Toxin production appears to be promoted by phosphorus deficiency, but expression of the toxin is extremely variable. Dramatic effects seen in nature (massive, wide-ranging fish kills) have been difficult to reproduce under laboratory conditions. Much remains to be learned about the factors that influence or control toxin production and/or release. The toxin is believed to be a glycoside or a family of related glycosides (96). The toxins display a generalized effect on membrane permeability and disturb ionic balances in cells of a wide range of marine organisms by eliciting marked increases in membrane permeability, with resulting leakage of cell contents (97,98). Blood cells are disrupted, so the toxins are often described as hemolytic. These toxins appear to be especially dangerous to gill-breathing animals such as fish, tadpoles, and molluscs.

BIBLIOGRAPHY

1. D.L. Taylor and H.H. Seliger, eds., *Toxic Dinoflagellate Blooms*, Elsevier, New York, 1979.
2. D.M. Anderson, A.W. White, and D.G. Baden, eds., *Toxic Dinoflagellates*, Elsevier, New York, 1985.
3. E.M. Cosper, V.M. Bricelj, and E.J. Carpenter, eds., *Novel Phytoplankton Blooms. Causes and Impacts of Recurrent Brown Tides and Other Unusual Blooms*, Springer-Verlag, New York, 1989.
4. T. Okaichi, D.M. Anderson, and T. Nemoto, eds., *Red Tides: Biology, Environmental Science, and Toxicology*, Elsevier, New York, 1989.
5. E. Graneli, B. Sundstrom, L. Edler, and D.M. Anderson, eds., *Toxic Marine Phytoplankton*, Elsevier, New York, 1990.
6. I.R. Falconer, ed., *Algal Toxins in Seafood and Drinking Water*, Academic Press, New York, 1993.
7. T.J. Smayda and Y. Shimizu, eds., *Toxic Phytoplankton Blooms in the Sea*, Elsevier, New York, 1993.
8. P. Lassus, G. Arzul, E. Erard-LeDenn, P. Gentien, and C. Marcaillou-Le Baut, eds., *Harmful Marine Algal Blooms*, Lavoisier, Paris, 1995.
9. T. Yasumoto, Y. Oshima, and Y. Fukuyo, eds., *Harmful and Toxic Algal Blooms*, Intergovernmental Oceanographic Commission of UNESCO, Paris, 1996.
10. D.M. Anderson, A.D. Cembella, and G.M. Hallegraeff, eds., *Physiological Ecology of Harmful Algal Blooms*, Springer-Verlag, New York, 1998.
11. R. Barker, *And the Waters Turned to Blood*, Simon and Schuster, New York, 1997.
12. T.J. Smayda, in E. Graneli, B. Sundstrom, L. Edler, and D.M. Anderson, eds., *Toxic Marine Phytoplankton*, Elsevier, New York, 1990, pp. 29–41.
13. G.M. Hallegraeff, *Phycologia* **32**, 79–99 (1993).
14. C.Y. Kao, in I.R. Falconer, ed., *Algal Toxins in Seafood and Drinking Water*, Academic Press, New York, 1993, pp. 75–86.
15. H.G. Love and L.L. Stephens, in M.D. Ellis, ed., *Dangerous Plants, Snakes, Arthropods & Marine Life of Texas*, U.S. Department of Health, Education, and Welfare, Washington, 1975, pp. 185–238.
16. World Health Organization (WHO), *Aquatic (Marine and Freshwater) Biotoxins*, Environmental Health Criteria 37, International Programme on Chemical Safety, World Health Organization, Geneva, 1984.
17. S. Nehring, *J. Plankton Res.* **17**, 85–102 (1995).
18. G.M. Hallegraeff and S. Fraga, in D.M. Anderson, A.D. Cembella, and G.M. Hallegraeff, eds., *Physiological Ecology of Harmful Algal Blooms*, Springer-Verlag, New York, 1998, pp. 59–80.
19. R. Riegman, A.A.M. Noordeloos, and G.C. Cadee, *Mar. Biol.* **112**, 479–484 (1992).
20. R. Riegman, *Water Sci. Technol.* **32**, 63–75 (1995).
21. R. Riegman, M. de Boer, I. de Senerpont Domis, *J. Plankton Res.* **18**, 1851–1866 (1996).
22. J.W. Bomber and K.E. Aikman, *Biol. Oceanogr.* **6**, 291–311 (1989).
23. M. Durand-Clement, *Biol. Bull.* **172**(1, Papers from the First International Workshop on Ciguatera), 108–121 (1987).
24. T.R. Tosteson, D.L. Ballantine, C.G. Tosteson, V. Hensley, and A.T. Bardales, *Appl. Env. Microbiol.* **55**, 137–141 (1989).
25. S.L. Morton, D.R. Norris, and J.W. Bomber, *J. Exp. Mar. Biol. Ecol.* **157**, 79–90 (1992).
26. J.L. MacLean, *Limnol. Oceanogr.* **22**, 234–254, (1977).
27. G.H. Tilstone, F.G. Figueiras, and S. Fraga, *Mar. Ecol. Prog. Ser.* **112**, 241–253 (1994).
28. L. MacKenzie, *J. Appl. Phycol.* **3**, 19–34 (1991).
29. M.V. Nielsen, *J. Plankton Res.* **14**, 261–269 (1992).
30. P.A. Matrai, M. Vernet, R. Hood, A. Jennings, E. Brody, and S. Saemundsdottir, *Mar. Biol.* **124**, 157–167 (1995).
31. C. Langdon, *J. Plankton Res.* **10**, 1291–1312 (1988).
32. C. Lancelot, M.D. Keller, V. Rousseau, W.O. Smith, and S. Mathot, in D.M. Anderson, A.D. Cembella, and G.M. Hallegraeff, eds., *Physiological Ecology of Harmful Algal Blooms*, Springer-Verlag, New York, 1998, pp. 209–224.
33. J.T. Turner, P.A. Tester, and P.J. Hansen, in D.M. Anderson, A.D. Cembella, and G.M. Hallegraeff, eds., *Physiological Ecology of Harmful Algal Blooms*, Springer-Verlag, New York, 1998, pp. 453–474.
34. P.A. Tester and K. Steidinger, *Limnol. Oceanogr.* **42**(5, Part 2: Special Issue on the Ecology and Oceanography of Harmful Algal Blooms), 1039–1051 (1997).
35. D.R. Tindall, and S.L. Morton, in D.M. Anderson, A.D. Cembella, and G.M. Hallegraeff, eds., *Physiological Ecology of Harmful Algal Blooms*, Springer-Verlag, New York, 1998, pp. 293–313.
36. O. Lindahl, in T. Smayda and Y. Shimizu, eds., *Toxic Phytoplankton in the Sea*, Elsevier, Amsterdam, 1993, pp. 775–782.
37. R. Raine, B. Joyce, J. Richard, Y. Pazos, M. Moloney, K.J. Jones, and J.W. Patching, *ICES J. Mar. Sci.* **50**, 461–469 (1993).

38. G.L. Boyer and L.E. Brand, in D.M. Anderson, A.D. Cembella, and G.M. Hallegraeff, eds., *Physiological Ecology of Harmful Algal Blooms*, Springer-Verlag, New York, 1998, pp. 489–508.
39. G.M. Hallegraeff, in D.M. Anderson, A.D. Cembella, and G.M. Hallegraeff, eds., *Physiological Ecology of Harmful Algal Blooms*, Springer-Verlag, New York, 1998, pp. 371–378.
40. P.L. Donaghay and T.R. Osborn, *Limnol. Oceanogr.* **42**(5, Part 2: Special Issue on the Ecology and Oceanography of Harmful Algal Blooms), 1283–1296 (1997).
41. J.J. Cullen, A.M. Ciotti, R.F. Davis, and M.R. Lewis, *Limnol. Oceanogr.* **42**(5, Part 2: Special Issue on the Ecology and Oceanography of Harmful Algal Blooms), 1223–1239 (1997).
42. D.F. Millie, O.M. Schofield, G.J. Kirkpatrick, G. Johnsen, P.A. Tester, and B.T. Vinyard, *Limnol. Oceanogr.* **42**(5, Part 2: Special Issue on the Ecology and Oceanography of Harmful Algal Blooms), 1240–1251 (1997).
43. K.A. Steidinger and G.A. Vargo, in C.A. Lembi and J.R. Waaland, eds., *Algae and Human Affairs*, Cambridge University Press, Cambridge, 1988, pp. 373–402.
44. G.R. Hasle, C.B. Lange, and E.E. Syvertsen, *Helgol. Meeresuntersuchengen* **50**, 131–175 (1996).
45. D.J. Douglas, U.P. Ramsey, J.A. Walter, and J.L.C. Wright, *J. Chem. Soc. Chem. Commun.*, N9: 714–716 (1992).
46. F.G. Plumley, *Limnol. Oceanogr.* **42**(5, Part 2: Special Issue on the Ecology and Oceanography of Harmful Algal Blooms), 1252–1264 (1997).
47. O. Ostensvik, O.M. Skulberg, and N. Soli, in W.W. Carmichael, ed., *The Water Environment: Algal Toxins and Health*, Plenum Press, New York, 1981, pp. 315–324.
48. O.M. Skulberg, W.W. Carmichael, R.A. Andersen, S. Matsunaga, R.E. Moore, and R. Skulberg, *Environ. Toxicol. Chem.* **11**, 321–329 (1992).
49. O.M. Skulberg, W.W. Carmichael, G.A. Codd, and R. Skulberg, in I.R. Falconer, ed., *Algal Toxins in Seafood and Drinking Water*, Academic Press, New York, 1993, pp. 145–164.
50. C.A. Scholin, in D.M. Anderson, A.D. Cembella, and G.M. Hallegraeff, eds., *Physiological Ecology of Harmful Algal Blooms*, Springer-Verlag, New York, 1998, pp. 337–349.
51. J.R. Manhart, G.A. Fryxell, M.C. Villac, and L.Y. Segura, *J. Phycol.* **31**, 421–427 (1995).
52. L. Rouhiainen, K. Sivonen, W.J. Buikema, and R. Haselkorn, *J. Bacteriol.* **177**, 6021–6026 (1995).
53. G.J. Doucette, *Nat. Toxins* **3**, 65–74 (1995).
54. G.J. Doucette, M. Kodama, S. Franca, and S. Gallacher, in D.M. Anderson, A.D. Cembella, and G.M. Hallegraeff, eds., *Physiological Ecology of Harmful Algal Blooms*, Springer-Verlag, New York, 1998, pp. 619–647.
55. T.M. Perl, L. Bedard, T. Kosatsky, J.C. Hockin, E. Todd, and R.S. Remis, *New England J. Med.* **322**, 1775–1780 (1990).
56. E.C.D. Todd, *J. Food Protection* **56**, 69–83 (1993).
57. W.W. Carmichael, in A.T. Tu, ed., *Handbook of Natural Toxins, Vol. 3 Marine Toxins and Venoms*, Marcel Dekker, New York, 1988, pp. 121–147.
58. W.W. Carmichael, in C.I. Ownby and G.V. Odell, eds., *Natural Toxins: Characterization, Pharmacology and Therapeutics*, Pergamon Press, London, 1989, pp. 3–16.
59. J.J. Sullivan, in I.R. Falconer, ed., *Algal Toxins in Seafood and Drinking Water*, Academic Press, New York, 1993, pp. 29–48.
60. I.R. Falconer, in I.R. Falconer, ed., *Algal Toxins in Seafood and Drinking Water*, Academic Press, New York, 1993, pp. 165–175.
61. G.A. Codd, C. Edwards, K.A. Beattie, L.A. Lawton, D.L. Campbell, and S.G. Bell, in W. Wiessner, E. Schnepf, and R.C. Starr, eds., *Algae, Environment and Human Affairs*, Biopress Ltd., Bristol, UK, 1995, pp. 1–17.
62. I.R. Falconer, A.R.B. Jackson, J. Langley, and M.T. Runnegar, *Aust. J. Biol. Sci.* **34**, 179–187 (1981).
63. I. Ohtani, R.E. Moore, and M.T.C. Runnegar, *J. Amer. Chem. Soc.* **114**, 7942–7944 (1992).
64. H. Fujiki, M. Suganuma, H. Suguri, S. Yoahizawa, K. Takagi, M. Nakayasu, M. Ojika, K. Yamada, T. Yasumoto, R.E. Moore, and T. Sugimura, in *Marine Toxins: Origins, Structure, and Molecular Pharmacology, ACS Symposium Series* **418**, 232–240 (1990).
65. G.L. Boyer, J.J. Sullivan, R.J. Andersen, P.J. Harrison, and F.J.R. Taylor, *Mar. Biol.* **96**, 123–128 (1987).
66. J.G. MacIntyre, J.J. Cullen, and A.D. Cembella, *Mar. Ecol. Prog. Ser.* **148**, 201–216 (1997).
67. Y. Shimizu, C. Hsu, W.E. Fallon, Y. Oshima, L. Miura, and K. Nakanishi, *J. Amer. Chem. Soc.* **100**, 6791–6793 (1978).
68. Y. Shimizu, in F.J.R. Taylor, ed., *The Biology of Dinoflagellates*, Blackwell Scientific, Oxford, 1987, pp. 282–315.
69. Y. Shimizu, S. Gupta, M. Norte, A. Hori, A. Genenah, and M. Kobayashi, in D.M. Anderson, A.W. White, and D.G. Baden, eds., *Toxic Dinoflagellates*, Elsevier, New York, 1985, pp. 271–274.
70. Y. Shimizu, *Ann. Rev. Microbiol.* **50**, 431–465 (1996).
71. Y. Shimizu, H.N. Chou, H. Bando, G. VanDuyne, and J. Clardy, *J. Amer. Chem. Soc.* **108**, 514–515 (1986).
72. D.G. Baden and V.L. Trainer, in I.R. Falconer, ed., *Algal Toxins in Seafood and Drinking Water*, Academic Press, New York, 1993, pp. 49–74.
73. D.M. Miller, R.W. Dickey, and D.R. Tindall, in E.P. Ragelis, ed., *Seafood Toxins*, American Chemical Society, Washington, DC, 1984, pp. 241–255.
74. E. Benoit, A.M. Legrand, and J.M. Dubois, *Toxicon* **24**, 357–364 (1986).
75. C.H. Wu and T. Narahashi, *Ann. Rev. Pharmacol. Toxicol.* **28**, 141–161 (1988).
76. S.B. Freedman, R.J. Miller, D.M. Miller, and D.R. Tindall, *PNAS* **81**, 4582–4585 (1984).
77. M. Murata, F. Gusovsky, M. Sasaki, A. Yokoyama, T. Yasumoto, and J.W. Daly, *Toxicon* **29**, 1085–1096 (1991).
78. T.A. Haystead, A.T. Sim, D. Carling, R.C. Honnor, Y. Tsukitani, P. Cohen, and D.G. Hardie, *Nature* **337**, 78–81 (1989).
79. C. Bialojan, J.C. Ruegg, and A. Takai, *J. Physiol. (London)* **398**, 81–95, (1988).
80. P. Carlsson, E. Graneli, G. Finenko, and S.Y. Maestrini, *J. Plankton Res.* **17**, 1925–1938 (1995).
81. S.S. Bates, D.L. Garrison, and R.A. Horner, in D.M. Anderson, A.D. Cembella, and G.M. Hallegraeff, eds., *Physiological Ecology of Harmful Algal Blooms*, Springer-Verlag, New York, 1998, pp. 267–292.
82. S.S. Bates, A.S.W. de Freitas, J.E. Milley, R. Pocklington, M.A. Quilliam, J.C. Smith, and J. Worms, *Can. J. Fish. Aquat. Sci.* **48**, 1136–1144 (1991).
83. S.S. Bates, J. Worms, and J.C. Smith, *Can. J. Fish. Aquat. Sci.* **50**, 1248–1254 (1993).

84. N.I. Lewis, S.S. Bates, J.L. McLachlan, and J.C. Smith, in T.J. Smayda and Y. Shimizu, eds., *Toxic Phytoplankton Blooms in the Sea*, Elsevier, New York, 1993, pp. 601–606.
85. Y. Pan, D.V. Subba Rao, K.H. Mann, W.K.W. Li, and R.E. Warnock, in T.J. Smayda and Y. Shimizu, eds., *Toxic Phytoplankton Blooms in the Sea*, Elsevier, New York, 1993, pp. 619–624.
86. D. Hampson and R. Wenthold, *J. Biol. Chem.* **263**, 2500–2505 (1988).
87. C. Angst and M. Williams, *Transmissions* **3**, 1–4 (1987).
88. G. Debonnel, L. Beaushesne, C. Demonigny, *Can. J. Physiol. Pharmacol.* **67**, 29–33 (1989).
89. T. Nakajima, K. Nomoto, Y. Ohfune, Y. Shiratory, T. Takemoto, H. Takeuchi, and K. Watanabe, *Br. J. Pharmacol.* **86**, 645–654 (1985).
90. J.M. Burkholder, E.J. Noga, C.H. Hobbs, and H.B. Glasgow, *Nature* **358**, 407–410 (1992).
91. T.J. Smayda, *Nature* **358**, 374–375 (1992).
92. J.M. Burkholder, H.B. Glasgow, and W.E. Hobbs, *Mar. Ecol. Prog. Ser.* **124**, 43–61 (1995).
93. E.J. Noga, L. Khoo, J.B. Stevens, Z. Fan, and J.M. Burkholder, *Mar. Pollut. Bull.* **32**, 219–224 (1996).
94. H.B. Glasgow, J.M. Burkholder, D.E. Schmechel, P.A. Tester, and P.A. Rublee, *J. Toxicol. Environ. Health* **46**, 101–122 (1995).
95. J.M. Burkholder and H.B. Glasgow, *Limnol. Oceanogr.* **42**(5, Part 2: Special Issue on the Ecology and Oceanography of Harmful Algal Blooms), 1052–1075 (1997).
96. T. Igarashi, M. Satake, and T. Yasumoto, *J. Amer. Chem. Soc.* **118**, 479–480 (1996).
97. M. Shilo, in S. Kadis, A. Ciegler, and S.J. Ajl, eds., *Microbial Toxins Vol. VII: Algal and Fungal Toxins*, Academic Press, New York, 1971, pp. 67–103.
98. Y. Hashimoto, *Marine Toxins and Other Bioactive Marine Metabolites*, Japan Scientific Societies Press, Tokyo, 1979.

ALKALINITY

GARY A. WEDEMEYER
Western Fisheries Research Center
Seattle, Washington

OUTLINE

The alkalinity of water is its capacity to chemically neutralize acids. It is commonly expressed as mg/L (or parts per million, ppm) of equivalent calcium carbonate, $CaCO_3$ (1). In water suitable for aquaculture alkalinity is usually due to naturally occurring dissolved-mineral bicarbonates (HCO_3^-), carbonates (CO_3^{-2}), and hydroxides (OH^-), often from limestone deposits; and to a lesser extent, borates, phosphates, and silicates. However, pollution from industrial and municipal effluents or irrigation drain water can also contribute to alkalinity. Alkalinity is sometimes confused with the related concept, water hardness, which is also expressed in mg/L as $CaCO_3$. Hardness, however, is primarily a measure of the calcium and magnesium concentration, and is independent of alkalinity. Water originating from areas with limestone rock formations may contain both calcium and magnesium carbonates and can therefore reach high levels of both alkalinity and hardness. Waters of high alkalinity usually also have an alkaline pH (pH > 7) and a high concentration of total dissolved solids (TDS). The alkalinity of water supplies used for aquaculture can range from less than 10 mg/L (soft, freshwater ponds, streams) to as high as several hundred mg/L (sea water, hard alkaline fresh water).

ANALYSIS

Alkalinity is most conveniently determined by titrating a water sample with standardized acid (usually 0.1 N HCl) to the methyl orange end point (pH 4.3), using a portable water quality test kit. For highly accurate work, specialized equipment under controlled laboratory conditions should be used (1). In either case, the result is the total alkalinity—commonly expressed in mg/L (or ppm) as $CaCO_3$. In limnology and oceanography research, and in Europe, alkalinity is often expressed in units of milliequivalents per liter (meq/L), where

$$1\ \text{meq/L} = 50\ \text{mg/L as } CaCO_3$$

A total alkalinity determination effectively measures all the bicarbonates, carbonates, and hydroxides present. Many water chemistry test kits used in aquaculture also offer the opportunity to measure the phenolphthalein alkalinity (end point pH 8.3), in addition to the total alkalinity. The phenolphthalein alkalinity measures the total hydroxides (OH^-) and carbonates (CO_3^{-2}) present. If this number is zero or near zero, as it usually will be in ponds with aquatic plants, then the total alkalinity is nearly all due to bicarbonates. Thus, an alkalinity determination can provide useful information on both the total concentration and the identity of the alkaline (basic) substances dissolved in a particular water supply. In addition, the alkalinity, together with the water temperature and pH, can be used to calculate the dissolved CO_2 concentration. Tables or nomographs for this purpose can be found in standard reference works (1).

IMPORTANCE TO AQUACULTURE

Although fish do not have a direct physiological requirement for dissolved carbonates or bicarbonates, the alkalinity of a hatchery water supply can nonetheless strongly influence the health and physiological quality of production fish. Most importantly, the alkalinity provides a buffer against wide fluctuations in water pH that would otherwise occur due to the daily cycle of CO_2 addition and removal by animal and plant respiration and plant photosynthesis. The pH of natural waters is determined by the interactions between the dissolved CO_2 and carbonic acid produced by plant and animal respiration and the bicarbonate and carbonate minerals from which the alkalinity is derived:

$$CO_2(gas) \rightleftharpoons CO_2\,(aq) + H_2O \rightleftharpoons H_2CO_3$$
$$\rightleftharpoons H^+ + HCO_3^- \rightleftharpoons 2H^+ + CO_3^{-2}$$

Because this system can neutralize added acids or bases, water supplies with a sufficient degree of alkalinity are therefore buffered against the pH increases or decreases that would otherwise occur.

An alkalinity of 10–20 mg/L is usually considered the minimum level needed to stabilize water pH and protect the health and physiological quality of production fish in flow-through raceway culture systems (2). However, pond fish that depend on natural food may grow slowly in waters of such low alkalinity because the production of phytoplankton and zooplankton will be low. Wide fluctuations in pH will also occur. For example, the water available for warmwater pond-fish culture in the southeastern United States is often low in alkalinity and therefore poorly buffered. At night, respiring phytoplankton and fish add CO_2 to the pond water decreasing the pH to as low as 5.5. During daylight hours, rapid algal growth in the intense sunlight can consume dissolved carbon dioxide faster than it can be replaced by fish respiration and diffusion from the atmosphere. The pond pH can increase to 9.5–10 in a matter of hours. Such pH increases, by themselves, can usually be tolerated by warmwater fish if the increases are temporary. However, ammonia is usually also present, and the high pH may increase the proportion of toxic NH_3 to greater than the 0.02 mg/L level generally considered safe (2). Water with an alkalinity of 40 mg/L or more (and a total hardness of 20–200 mg/L) is considered more desirable for both extensive (static ponds) and intensive (flowing water) aquaculture systems (2,3). In addition, pond fertilization is more likely to be successful under such conditions.

Alkalinities in the 100–200 mg/L range offer several additional advantages, including (a) reducing the toxicity of heavy metals, (b) allowing the safe use of copper sulfate as an algicide and fish disease therapeutant, (c) providing adequate buffering capacity against the fluctuations in water pH that would otherwise occur as a result of the natural daily cycles of photosynthesis and respiration, and (d) affording the stable pH and carbon source needed by nitrifying bacteria in the biofilters used in recirculating aquaculture systems (2,3). The latter is especially important in high-density recirculating systems where large amounts of ammonia are produced that must be oxidized to nitrate. The accompanying acid production may be sufficient to consume the carbonates present, and the pH will progressively decrease unless the alkalinity is replenished (4).

In saltwater aquaculture, alkalinity is not normally a consideration. Because of the high concentrations of carbonates, ocean water is strongly buffered at about pH 8.2.

MANAGEMENT RECOMMENDATIONS

Water supplies with a minimum alkalinity of 20–40 mg/L as $CaCO_3$ are considered highly desirable for both cold- and warmwater fish in either intensive or extensive culture systems (2,3). Alkalinities in the 100–200 mg/L range will provide the additional buffering capacity needed to prevent wide pH fluctuations in pond culture systems, prevent leaching of toxic metals bound to soils and sediments, allow the use of copper compounds for fish disease control, and provide the carbon needed to assure biological productivity. In recirculating aquaculture systems, alkalinities in this range will also assure an adequate supply of carbon for the nitrifying bacteria in the biofilters, as well as provide a stable pH.

Low alkalinity is not usually a problem in earthen ponds constructed with soils rich in limestone or in ponds made of concrete. However, many such ponds are lined with plastic or rubber, which effectively cuts off the source of carbonate. If desired, the alkalinity of pond waters can be increased to adequate levels by adding either hydroxides (such as sodium or calcium hydroxide) or carbonate compounds (such as agricultural limestone or sodium bicarbonate). However, only the latter are safe for aquaculture. Sodium bicarbonate is readily soluble in water and will not cause areas of locally high pH. Applied at 10–20 lb per acre-foot of water, it will temporarily correct low alkalinity and mitigate CO_2 and NH_3 problems arising from low or high pH. For longer term alkalinity management, agricultural limestone can be used. However, the amount of lime needed to raise the alkalinity by a particular amount is difficult to calculate directly. As a guideline, Terlizzi (5) has reported that added limestone will increase the alkalinity by about 1 mg/L per 2 kg (4.5 lb) of lime per 1233 m^3 (one acre-foot) of water. In practice, limestone is usually simply applied at 1–2 tons per 1233 m^3 (one acre-foot) at monthly intervals, and the alkalinity monitored until it rises above 20 mg/L. This process is very slow, however, and liming may not succeed at all if pond sediments have been allowed to become strongly acidic (6).

BIBLIOGRAPHY

1. *Standard Methods for the Examination of Water and Wastewater*, 17th ed., American Public Health Association, Washington, DC, 1989.
2. G. Wedemeyer, *Physiology of Fish in Intensive Culture Systems*, Chapman Hall, New York, 1996.
3. R.R. Stickney, *Principles of Aquaculture*, John Wiley and Sons, New York, 1994.
4. J.C. Loyless and R.F. Malone, *Prog. Fish-Cult.* **59**, 198–205 (1997).
5. D. Terlizzi, *Maryland Aquafarmer*, Cooperative Extension Service, Maryland Sea Grant Program, University of Maryland, College Park, MD, 1997.
6. C.E. Boyd, *Water Quality in Warmwater Fish Ponds*, Auburn University Agricultural Experiment Station, Auburn, AL, 1989.

See also WATER HARDNESS.

ALLIGATOR AQUACULTURE

Michael P. Masser
Texas A&M University
College Station, Texas

OUTLINE

Alligators and other members of the order Crocodilia (crocodiles and caimans) have long been valued for their hides and meat. The leather from crocodilian hides is used to make attractive luxury apparel items like belts, wallets, purses, briefcases, and shoes. The high value of these leather products led to extensive hunting of these creatures in the wild. By the 1960s, this exploitation, combined with habitat destruction, had depleted many wild populations of crocodilians. Research into the life history, reproduction, nutrition, and environmental requirements of the American alligator (*Alligator mississippiensis*), coupled with rapid recovery of wild populations led to the establishment of commercial farms in the United States in the 1980s. Worldwide, several other species of crocodilians are also cultured using similar methods (1,2).

HISTORICAL PERSPECTIVE

The American alligator was once native to coastal plain and lowland river bottoms from North Carolina to Mexico. Historical records show that the American alligator can grow to 16 feet or more. The only other species of alligator, (*A. sinensis*) is found in China and is endangered.

American alligators were hunted for their hides beginning in the 19th century (3). At the turn of the century, the annual alligator harvest in the US was around 150,000 per year. Overharvesting from the wild, combined with habitat destruction, slowly depleted the wild population. Most states stopped alligator hunting by the 1960s. Under the 1973 Endangered Species Act, the U.S. Fish and Wildlife Service designated alligators as "endangered" or "threatened" species throughout most of their range (with the exception of Louisiana) to protect them from further exploitation (4).

Once protected, alligator populations recovered. Recovery was dramatic in some areas, particularly in Louisiana, which had stopped legal harvesting in 1962. Louisiana reopened limited harvesting of wild alligators based on sustainable yield in 1972. The Louisiana alligator population continued to increase even with sustained harvesting, and by 1984 the Louisiana population was estimated to be near turn-of-the-century numbers (4). Most other southern states also experienced population increases after federal protection.

In 1983, under the Convention on International Trade in Endangered Species of Wild Fauna and Flora (CITES), the U.S. Fish and Wildlife Service changed the classification of the American alligator to what is called "threatened for reasons of similarity in appearance." This classification means that the American alligator is *not* threatened or endangered in its native US range. However, the sale of its products must be strictly regulated so that the products of other crocodilian species are not sold illegally as those of American alligators. Today nuisance control is allowed in several southern states, and limited harvesting from the wild is permitted in Louisiana, Texas, and Florida. In 1996 wild harvest and farm-raised alligators supplied over 240,000 hides to world markets. Approximately 83% of these were from alligator farms (5).

ECOLOGY AND LIFE HISTORY

Alligators inhabit all types of fresh to slightly brackish aquatic habitats. Males grow larger than females, although the growth rate of the sexes is similar up to approximately 1.1 m (3.5 ft) in length (4). Growth and sexual maturity are dependent on climate and the availability of food. Along the Gulf coast, females usually reach sexual maturity at a length of 2 m (6.5 ft) and an age of 9 to 10 years. Sexual maturity is not reached until 18 to 19 years in North Carolina [still 2 m (6.5 ft) in length]. Like other cold-blooded animals, this difference in maturation age is related to temperature. Optimum growth occurs at temperatures between 29 and 33 °C (85–91 °F). No apparent growth takes place below 21 °C (70 °F), while temperatures above 34 °C (93 °F) cause severe metabolic stress and, sometimes, death.

Research has shown that young alligators primarily consume invertebrates like crayfish and insects (6), and as they grow, fish become part of their diet. Mammals such as muskrats and nutria become a substantial portion of the adult diet. Large adult alligators even consume birds and other reptiles, including smaller alligators. Carrion is consumed whenever available (7).

Females do not move or migrate over long distances once they have reached breeding age. They prefer heavily vegetated marsh-type habitat (8). Males move about extensively, but prefer to establish territories in areas of open water (9). Males longer than nine feet are the most successful breeders.

Alligator courtship and breeding are correlated to air temperature and occur between April and July, depending on weather conditions. Courtship and breeding take place in deep (at least 1.8 m or 6 ft), open water. Courtship behavior includes vigorous swimming and bellowing. Both males and females bellow, but the male bellow is much more bass and vocal than that of the female. Most courtship occurs just after sunrise and takes about

45 minutes from precopulatory behavior through the first copulation (10). Repeated copulation is commonly observed.

After courtship and mating, females move to isolated ponds, surrounded by dense vegetation, for nesting, which occurs about two or three weeks after mating. Nest building and egg laying occur at night. Females build nests by raking up surrounding vegetation and soil into a mound. From 20 to 60 eggs are laid from above, into the center of the mound. All the eggs are deposited at one time. When egg laying is completed, the female covers the nest with about 25 cm (1 ft) of vegetation. Nesting occurs only once a year, and not all females nest every year. Females guard their nests against predators.

Warm summertime temperatures, combined with heat generated from the decaying mound of vegetation, maintains temperatures between 24 and 33 °C (75–91 °F) and relative humidities of 94 to 99% in the nest. The eggs hatch in 65 days if the temperature in the nest is consistently above 28 °C (82 °F). The young make grunting or peeping sounds after hatching, and the female often claws open the nest to help release them. Hatching success is generally less than 60% (11). Research done in Louisiana suggests that the survival of young alligators to 1.2 m (4 ft) long averages 17% or less (12). After an alligator reaches this length, it has few enemies other than larger alligators and human beings. A good review of general ecological considerations and information on the natural history of the American alligator is (13).

CONTROLLED BREEDING AND EGG INCUBATION

In Louisiana, Florida, and Texas, eggs and/or hatchlings may be taken from the wild under special permitting regulations. In all other states it is illegal to take eggs or hatchlings from the wild. Therefore, prospective alligator farmers must purchase eggs or hatchlings from existing farms in Louisiana, Florida, or Texas or must produce their own through captive breeding.

Management of Breeding Alligators

Maintaining adult alligators and achieving successful and consistent reproduction has proved difficult and expensive. The exact environmental, social, and dietary needs of adult alligators are poorly understood. Adult alligators that have been reared entirely in captivity behave differently from wild stock (14,15). Farm-raised alligators seem to accept confinement and crowding as adults better than alligators captured from the wild. Also, adult alligators that have been raised together develop a social structure and probably adapt more quickly and breed more consistently than animals from the wild, which lack an established social structure.

Breeding pen design, particularly with respect to the land-to-water ratio and configuration, is very important. The ratio of land area to water area within the pen should be approximately 3 : 1. The shape of ponds needs to maximize the shoreline, utilizing an *M*, *S*, *W*, *Z*, or similar shape. The reason these shapes work best is that male alligators fight less during the breeding season if they cannot see each other.

A water depth of at least 1.8 m (6 ft) must be maintained during the breeding season. Whenever possible, ponds should also be constructed with drains so that water can be removed if the animals need to be captured. The pond shoreline should be no closer than 23 to 30 m (75–100 ft) from fences. Alligators are good climbers and diggers. Most states that license alligator farming have specific requirements pertaining to fencing in the construction of pens.

Dense vegetation around the pond is needed to provide cover, shade, and nesting material. The natural invasion of wetland plants may be sufficient for cover. Tall, deep grasses can be planted to increase vegetation that can be used for incubation material. Many producers add bales of hay to the breeding pens in June to supplement natural vegetation for nest building.

Shade is important to prevent overheating during the summer. Alligators will burrow into the banks of a pond if adequate shade is not provided. Awnings that provide shade will reduce burrowing activity.

The stocking density of adult alligators is usually between 25 and 50 of the animals per ha (10 to 20 per acre), in pens that are at least 5 ha (2 acres) in area. Adults between 6 and 20 years old are reliable breeders, and females 8 to 10 years old are the most consistent breeders (4,16). The female-to-male ratio should be near 3 to 1, but less than 4 : 1.

Each pen should have several feeding stations to keep the adult alligators spread out. Feeding stations should be established near basking areas or along the shoreline of the pond. Feeding should begin each spring when the temperature rises above 21 °C (70 °F). Alligators should be fed four to six percent of their body weight per week (definitely 6% throughout the summer) (4). Adults should usually be fed only once per week. Early fall feeding appears to be particularly important to enable the females to be in good condition for egg development. Adults do not need to be fed during the late fall and winter when temperatures are below 21 °C (70 °F). It is important that adult alligators not be overfed; they should be trim, not fat, for enhanced reproductive capabilities (16).

Adult breeders should be disturbed as little as possible from February through August, during egg maturation, courting, and nesting. Activities such as moving animals or maintaining ponds should be performed between September and January.

Nesting success in captive alligators has been highly variable. Wild versus farm-raised origin, pen design, density, the development of a social structure within the group, and diet all affect nesting success. Nesting rates for adult females in the wild averages around 60 to 70% where habitat and environmental conditions are excellent (17,18). Nesting rates in captivity are usually much lower, depending on the management skill of the producer.

Clutch size varies with age and condition of the female. Larger and older females generally lay more eggs. Clutch size should average 35 to 40 eggs. Egg fertility can vary from 70 to 95%. Survival of the embryo also varies from 70 to 95%, hatching rate from 50 to 90%. Egg fertility, the survival of the embryo, and the hatching rate of eggs taken from the wild and incubated artificially are 95, 95, and 90%, respectively (19).

Land costs, long-term care and maintenance of adults, and low egg production contribute significantly to the cost of maintaining breeding stock.

Egg Collection

The method and timing of egg collection are very important. Alligator embryos are extremely sensitive to handling from 7 to 28 days after the eggs are laid (20). Many embryos will die if handled during that period. Eggs should be collected within the first week or after the fourth week of natural incubation.

Unlike bird eggs, alligator eggs cannot be turned or repositioned when taken from the nest, except during the first 24 hours after being laid. The top of the eggs should be marked before removing them from the nest, so that they can be maintained in the same position during transport and incubation. Eggs that are laid upright in the nest (with the long axis perpendicular to the ground) will expire unless they are repositioned correctly (with the long axis parallel to the ground) within the first day after nesting.

During collection, the eggs should be supported by 20 to 30 cm (8–12 in.) of moistened nesting material or grass hay, placed in the bottom of the collection container. The marked eggs should be placed in a single layer in the container and in the same position that they were in the nest and should be covered with 5–7.5 cm (2–3 in.) of nesting material (20).

The age of the eggs and their development can be observed by means of changes in the opaque banding that occur during incubation. Figure 1 shows the sequence of banding associated with proper egg development (21).

Incubation and Hatching

Compared with wild nesting, artificial incubation improves hatching rates because of the elimination of predation and weather-related mortality. The best hatching rates for eggs left in the wild are less than 70% (22). Hatching rates for eggs taken from the wild and incubated artificially average 90% or higher.

Eggs should be transferred into incubation baskets and placed in an incubator within three or four hours after collection. Air circulation around the eggs is critical during incubation (19). Egg baskets can be made from plastic-coated 2.5 × 1.3-cm (1 × 1/2-in.) steel wire mesh or 1.3 cm (1/2 in.) heavy-duty plastic mesh. Dimensions for egg baskets can vary [30 × 60 cm (1 ft × 2 ft) and 60 × 90 cm (2 ft × 3 ft) are common], but should be 15 cm (6 in.) deep to accommodate both eggs and nesting material. Eggs must be completely surrounded by nesting material, the decomposition of which aids in the breakdown of the eggshell (21). Without this natural decomposition, hatching alligators will have a difficult time breaking

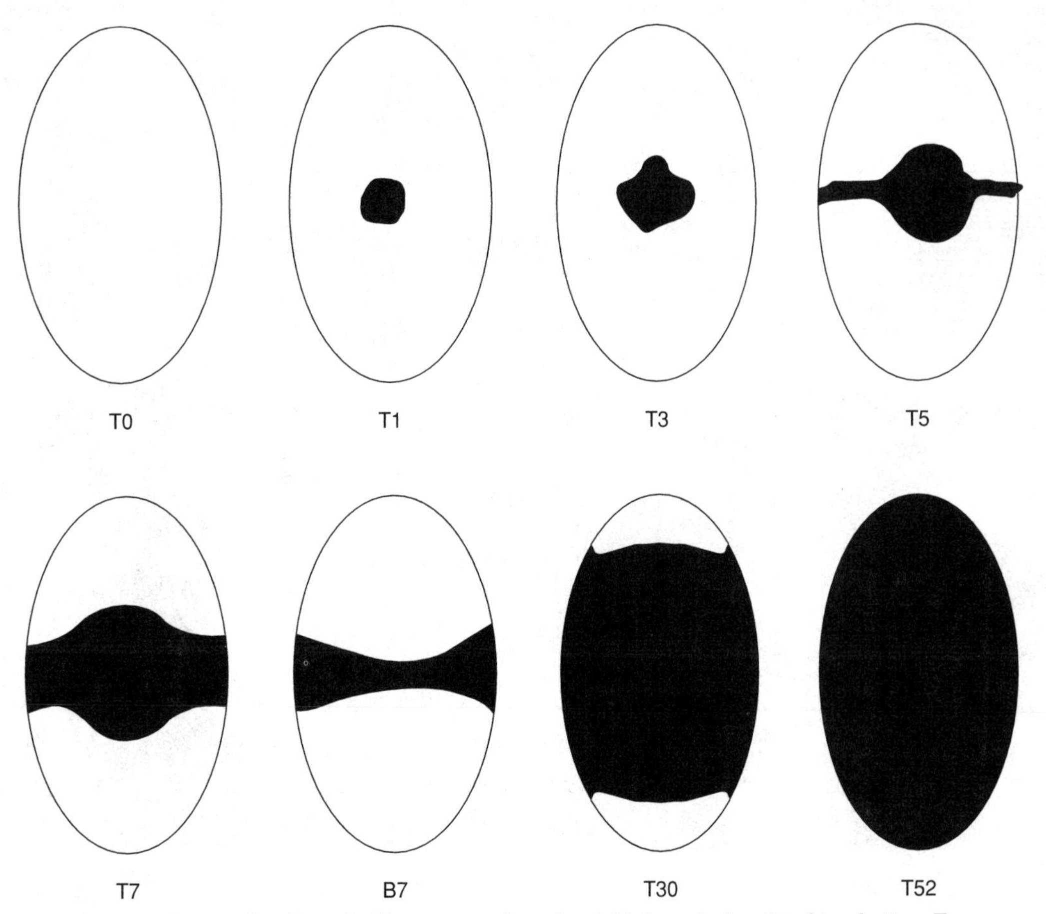

Figure 1. Opaque banding of alligator eggs from day laid through day 52 of incubation. From Ferguson, 1981. T = top view; B = bottom view. Numbers represent days of incubation.

out of the shell and may die. Fresh natural nesting material composed mostly of grasses is best. If natural nest material is not available, grasses which have been soaked in water for about a week prior to incubation can be used.

Hatching baskets should be set about 7.5 cm (3 in.) above heated water in an incubator, in which the temperature, humidity, and water level must be controlled. The relative humidity should be kept above 90% within the chamber, and incubation media should be moistened with warm water as necessary to maintain dampness.

The incubation temperature is critical to the survival and proper development of the hatchlings, even determining their sex (23). Temperatures of 30 °C (86 °F) or below produce all females, while temperatures of 33 °C (91 °F) or above produce all males. Temperatures much above or below these limits cause abnormal development that usually results in high mortality. Both sexes are produced at temperatures between 30 and 33 °C (86–91 °F). The critical period for sex determination is around 20 to 35 days after the eggs are laid.

Hatchling alligators (Fig. 2) make peeping or chirping sounds after hatching. Unhatched eggs can be carefully opened to release hatchlings. Eggs can be opened at one end, to free the baby alligators without detaching or damaging the umbilical cord. If the umbilical cord is broken the hatchling is likely to bleed to death or develop an infection. Hatchlings should be retained in their hatching baskets for 24 hours to allow the umbilical cord to separate naturally (19). After 24 hours the hatchlings should be removed from the egg baskets, sorted into uniform size groups, and moved into environmentally controlled growout facilities. Size grouping the baby alligators is important. Smaller, weaker individuals will not compete well with their larger siblings.

Hatchlings should be moved into small tanks, 60 × 60 cm (2 × 2 ft) or larger, heated to 30–32 °C (86–89 °F). Maintaining hatchlings at 32 °C (89 °F) for the first week

Figure 2. Newly hatched alligators in hatching tray with natural nesting material, note umbilical cords still attached to egg cases.

aids in increasing their ability to absorb the yolk (4). Usually, hatchlings will start to feed within three days at that temperature. Young that do not start feeding on their own can be force-fed using a large syringe (24). Hatchling tanks should be cleaned daily to prevent outbreaks of disease. Once hatchlings are actively feeding, they are ready to be moved into growout facilities.

GROWOUT

Many different designs of growout facilities have been employed. Growout buildings are basically heavily insulated concrete block, wood, or metal buildings with heated foundations. The concrete slab foundation is laced with hot-water piping or, less commonly, electric heating coils (25). A constant internal temperature is maintained by pumping hot water through the pipes. The slab is insulated to reduce heat loss. Pools, drains, and feeding areas are built into the foundation. Covering about two thirds of each pen is a pool of water, about 30 cm (1 ft) deep at the drain, toward which the bottom of the pool is sloped to facilitate cleaning. The remaining third of each pen, above the water, is used for a feeding and basking deck. Separate pens are constructed within a single building by using concrete block walls at least 90 cm (3 ft) tall.

Pens can be almost any size. In general, smaller pens are used for rearing small alligators, and progressively larger pens are used as the alligators grow. Many producers employ small fiberglass or metal tanks (for small alligators) stacked above the larger floor pens, an approach that maximizes the use of space and heat within the growout houses. Pens and tanks must be "climbproofed" to prevent the nimble young from escaping. Table 1 gives examples of pen size to alligator size and corresponding densities.

Most producers construct only a few sizes of growout pens and simply reduce the density by moving the animals as they grow. Commonly used stocking regimes are as follows:

- 9.2 cm^2 (1 ft^2) per animal until the animal reaches 50 cm (2 ft) in length.
- 27.9 cm^2 (3 ft^2) per animal until the animal reaches 1.2 m (4 ft) in length.
- 55.7 cm^2 (6 ft^2) per animal until the animal reaches 1.8 m (6 ft) in length.

A common construction plan uses an approximately 465-m^2 (5000-ft^2) building (e.g., 10 × 45m or 33 × 150 ft) with an aisle down the middle and pens on either side (25). A 1.2-m (4-ft) aisle leaves the pens roughly 4.3 m (14 ft) wide. Pens are usually about 4 m (13 ft) long. A 0.9-m (3-ft) high concrete block wall separates individual pens from the aisle.

Within the 4.3 × 4-m (14 × 13-ft) pen is a 1.5-m (5-ft)-wide deck next to the service aisle and a 2.7-m (9-ft)-wide pool. Food is placed on the deck, and the pen is hosed clean from the aisle without entering it. The pool edge slopes rapidly to a depth of 10 in next to the deck, and the pool bottom slopes from there to the drain.

Pens are easily divided by the construction of additional walls down the center. The large pen (4.3 × 4 m) can hold around 160 alligators 60 cm (2 ft) long or 50 alligators 1.2 m (4 ft) long. Some state laws require that alligators less than two feet long be separated from those over 60 cm (2 ft) in length.

Another popular building design is a single "roundhouse" (25), a structure 4.5 to 7.5 m (15–25 ft) in diameter constructed as a single pen. Round houses have also been built from concrete blocks or from a single section, and the roof from a prefabricated metal silo (used for storing grain). The round concrete slab on which the house sits is sloped (at an inclination of about 10 : 1) from the outer edge to a central drain. The roundhouse is filled with water to a depth such that about one-third of the outer floor is left above the water level. Some producers prefer this design because it is a single pen and, therefore, does not disturb alligators in other pens during feeding, cleaning, and handling operations.

Part of any alligator facility is the heating system, which usually consists of water heaters and pumps that circulate warm water through the concrete slab. The warm water is needed to heat the building, fill the pools, and clean the pens. Some heating systems consist of several industrial-sized water heaters. Other systems consist of a flash-type heater to heat water for cleaning and standard water heaters to circulate warm water through the slab. Both systems use thermostats to turn on the heaters and circulation pumps. The temperature in growout buildings must be maintained between 30 and 31 °C (86–88 °F) for optimal growth.

Growout buildings almost never contain any windows, and many producers prefer no skylights. In fact, most animals are kept in near or total darkness, except at feeding and cleaning times.

Feeding and Nutrition

Research reveals that the diet of a wild alligator changes as the animal grows, but, in general, alligators

Table 1. Recommended Pen Sizes for Growout Operations[a]

Gator Length [cm (in.)]	Pen Size [m^2 (ft^2)]	Gators per Pen	cm^2 (ft^2) per Gator	m^2 (ft^2) Needed per 350 Gators
18–38 (7–15)	0.8 (9)	20	4.0 (0.45)	14.7 (158)
38–76 (15–30)	11.1 (120)	80	14.0 (1.50)	48.8 (525)
76–122 (30–48)	15.6 (168)	50	32.0 (3.36)	109.3 (1176)
122–152 (48–60)	17.8 (192)	50	36.0 (3.84)	124.9 (1344)
152–183 (60–72)	20.1 (216)	40	50.0 (5.40)	175.6 (1890)

[a]From Ref. 26.

consume a diet high in protein and low in fat. Early producers fed their animals diets high in fish. Later research showed that wild populations of medium to large alligators eat mostly higher protein prey (i.e., birds and mammals).

Early producers manufactured their own feeds using inexpensive sources of meat, including nutria, beefcattle, horse, chicken, muskrat, fish, beaver, and deer (25). Today, however, artificial diets are available that provide adequate nutrition. These diets have eliminated the need to keep fresh-frozen meat products on hand.

Several feed mills are currently manufacturing pelleted alligator feeds. Commercial feeds, approximately 45% crude protein and 8% fat, are blends of fish meal, meat and bone meal, blood meal, and some vegetable protein, fortified with vitamins and minerals.

At present, most producers feed their animals only commercially available diets, although some continue to feed them a combination of meats and commercial diets.

Feed should be spread out on the deck in small piles to reduce competition and territoriality. Feeding should be done at least 5 days per week; some producers feed their animals 6 or 7 days per week. Alligators are normally fed at rates of 25% of body weight per week the first year; then the rate is gradually reduced to 18% by three years of age or a length of about 1.8 m (6 ft) (25). Feed conversion efficiency decreases as alligators grow larger, but averages about 40% (or between 2 : 1 and 3 : 1 when presented as a food conversion ratio), up to a length of 1.8 m (6 ft) (4). Overfeeding wastes money and can lead to gout, which is fairly common in pen-raised alligators, but can be cured by taking the animals off their feed for 7 to 10 days (27). No antibiotics are approved for use on alligators; therefore, any antibiotics that are needed can be obtained only through a prescription from a veterinarian.

Pen cleaning should be coordinated so that the animals are not disturbed just before, during, or soon after feeding. Many producers clean in the morning and feed their animals in the afternoon.

Growth rates of young alligators can be as great as 7.5 cm (3 in.) or more per month when the temperature is held at a constant 30 to 31.5 °C (86 to 89 °F) and the animals are fed a quality diet and protected from stress. Many producers rear alligators from hatchlings to 1.2 m (4 ft) in 14 months, and a few producers have grown alligators to 1.8 m (6 ft) in 24 months. Farm-raised alligators are generally 10% heavier than wild alligators of the same length. Table 2 gives average lengths and weights of wild and farm-raised alligators.

In an effort to reduce costs and still produce a larger (and more valuable) animal, some producers are utilizing outside, or ambient, growout facilities. In this system, alligators are moved into outdoor fenced ponds after the first year of growth in indoor facilities. The alligators are fed a commercial diet during warm weather and are allowed to hibernate during cool seasons. After approximately two years the ponds are drained, usually during the winter to facilitate handling, and the alligators are harvested.

Table 2. Length–Weight Relationships for Wild and Farm-Raised Alligators[a]

Length, (in.)	Weight, Wild, lb (oz)	Weight, Farm Raised, lb (oz)
12	0.15/(2.4)	0.16/(2.6)
18	0.42/(6.7)	0.47/(7.5)
24	0.68/(10.8)	0.75/(12.1)
30	3.5	3.9
36	8.6	9.5
42	13.0	14.7
48	17.7	19.8
54	28.0	31.1
60	39.6	44.0
66	45.4	50.4
72	49.6	55.1

[a]From Ref. 25.

Stress

Alligators are wild creatures that have been thrust into captivity. In the wild, they are relatively shy and reclusive creatures that do not normally aggregate together, except during the breeding season. The artificial conditions that are imposed upon them are unnatural and, therefore, stressful. Stress can lead to slow growth, disease, and aggressive behavior.

Alligators that are crowded into pens appear to be very sensitive to light and sound. Many producers like to keep the animals in the dark or at least in very reduced light. Toward that end, they try to locate and insulate facilities to minimize external noise.

Signs of stress include piling up of the animals, reduced feeding, "stargazing," and fighting (25). Piling up usually occurs in the corners of the pens and can lead to suffocation of those animals on the bottom of the pile. Reduced feeding is a sign of stress. "Stargazing" is a position wherein the alligator rises up on its front feet, arches its back and neck, and points its snout into the air. Fighting among animals that have been penned together, but are not overcrowded, is a definite sign of stress.

Stress often impels the larger animals to fight, causing scarring. A skin condition known as "brown spot" results in cosmetic blemishes to the hide and subsequent scarring. In either case, the quality of the hide is diminished and its value reduced.

Each producer must keep good records on environmental conditions, the animals' consumption of feed, and their general health. When signs of stress appear, the cause must be identified and remedied as soon as possible. Overcrowding, excessive disturbance, and poor feeding practices are common causes of stress.

Harvesting

Written approval and hide tags must be obtained from the appropriate state regulatory agency (e.g., the Department of Conservation and Natural Resources) before any alligators may be harvested. Some states also have a minimum length requirement at harvest (e.g., at least 1.8 m (6 ft), unless the animal has died from natural causes). All alligators must be labeled with tags from

Table 3. Percent Yield of Deboned Alligator Meat on a Live-Weight Basis[a]

Tail	Leg	Torso	Ribs[b]	Jaw
16–17	4–5	6–12	7–10	1

[a]From Ref. 25.
[b]With bones.

the state regulatory agency immediately after slaughter. Alligators may be skinned only at approved sites, using specific skinning instructions issued by the state agency.

Skinning, scraping, and curing must be done carefully to assure quality. Hides that are cut, scratched, or stretched—particularly on the belly—have reduced value.

Hides are scraped carefully to remove all meat and fat and then are washed to eliminate all blood, etc. Fine-grained mixing salt, not rock salt, is used to preserve the hide. Salt is rubbed thoroughly into the skin, with particular attention paid to all creases and flaps, to start the curing process.

Most hides are sold to brokers, who purchase and hold large numbers of hides and then sell them to tanneries for processing. A few farms are large enough to sell directly to the tanneries, the best of which are in Asia and Europe.

Producers who process alligator meat must comply with all sanitation requirements of federal, state, and local authorities. Local health departments can supply guidelines and assistance in complying with sanitation standards. Specific state laws regulate the size of meat cartons, labeling of the cartons (with the names of the seller and buyer), the date of sale, and the tag number that corresponds to the hide. Average deboned dress-out percentages for alligators in the 1.2 to 1.8-m (4 to 6-ft) range are given in Table 3.

It is interesting to note that while the prices of hides fluctuate, meat prices have stayed consistent, and it appears that the supply of alligator meat is well below market demand.

BIBLIOGRAPHY

1. A.C. Pooley, *J. South Africa Wildl. Mgmt. Assoc.* **3**, 101–103 (1973).
2. D.K. Blake and J.P. Loverage, *Biol. Conserv.* **8**, 261–272 (1975).
3. O.H. Stevenson, *US Comm. Fish and Fisheries Report* **1902**, 281–352 (1904).
4. T. Joanen and L. McNease, in J.W. Webb, S.C. Manolis, and P.J. Whitehead, eds., *Wildlife Management: Crocodiles and Alligators*, Surrey Beatty and Sons Pty. Limited, 1987, pp. 329–340.
5. M. Shirley, Louisiana Cooperative Extension Service, Louisiana State University Agricultural Center, Personal Communication, 1999.
6. R.M. Elsey, L. McNease, T. Joanen, and N. Kinler, *Proc. Southeast. Assoc. Fish and Wildl. Ags.* **46**, 57–66 (1992).
7. J.L. Wolfe, D.K. Bradshaw, and R.H. Chabreck, *Northeast Gulf Sci.* **9**, 1–8 (1987).
8. T. Joanen and L. McNease, *Proc. Southeast. Assoc. Game and Fish Comm. Conf.* **24**, 175–193 (1970).
9. T. Joanen and L. McNease, *Proc. Southeast. Assoc. Game and Fish Comm. Conf.* **26**, 252–275 (1972).
10. T. Joanen and L. McNease, *Proc. Southeast. Assoc. Game and Fish Comm. Conf.* **29**, 407–415 (1975).
11. T. Joanen, *Proc. Southeast. Assoc. Game and Fish Comm. Conf.* **23**, 141–151 (1969).
12. D. Taylor and W. Neal, *Wild. Soc. Bull.* **12**, 312–319 (1984).
13. G. Webb, S. Manolis, and P. Whitehead, eds., *Wildlife Management: Crocodiles and Alligators*, Surrey Beatty and Sons Pty. Limited, 1987.
14. R.M. Elsey, T. Joanen, and L. McNease, *Proc. 2nd Regional Conf. of the Crocodile Specialist Group*, Darwin, Australia, 1993.
15. R.M. Elsey, T. Joanen, and L. McNease, *Proc. 12th Working Meeting of the Crocodile Specialist Group*, Pattaya, Thailand, 1994.
16. P. Cardeilhac, *Aquaculture Report Series*, Florida Department of Agriculture and Consumer Services, 1988.
17. T. Joanen and L. McNease, in J.B. Murphy and J.T. Collins, eds., *Reproductive Biology and Disease of Captive Reptiles*, Society for the Study of Amphibians and Reptiles, Lawrence, KS, 1980, pp. 153–159.
18. R.H. Chabreck, *Proc. Southeast. Assoc. Game and Fish Comm. Conf.* **20**, 105–112 (1966).
19. T. Joanen and L. McNease, *Proc. World Marcult. Soc.* **8**, 483–490 (1977).
20. T. Joanen and L. McNease, *Proc. Intensive Tropical Animal Production Seminar*, Townsville, Australia, 1981, pp. 193–205.
21. M.W.J. Ferguson, *Proc. Alligator Production Conf.* Gainesville, FL **1**, 129–145 (1981).
22. T. Joanen, *Proc. Southeast. Assoc. Game and Fish Comm. Conf.* **23**, 141–151 (1969).
23. M.W.J. Ferguson and T. Joanen, *Nature* **296**, 850–853 (1982).
24. M.P. Masser, *Southern Regional Aquaculture Center* **231**, 1–7 (1993).
25. M.P. Masser, *Southern Regional Aquaculture Center* **232**, 1–4 (1993).
26. J. Smith and P.T. Cardeilhac, *Proc. First Ann. Alligator Production Conf.* Gainesville, FL, 10–15 (1983).
27. P.T. Cardeilhac, *Proc. First Ann. Alligator Production Conf.* Gainesville, FL, 58–64 (1983).

ANESTHETICS

PATRICIA W. VARNER
Texas Veterinary Medical Diagnostic Lab
College Station, Texas

OUTLINE

Common aquacultural practices conducted by the private, commercial, and research sectors can be stressful to fish and oftentimes result in immunosuppression or physical injury to handled fish. Both chemical and nonchemical methods of anesthesia are frequently used to minimize stress and facilitate animal restraint during routine husbandry practices and transport, as well as for spawning, surgical, and diagnostic purposes (1–4). The levels of restraint, efficacy, cost, ease of handling, and safety to the animal, handler, and environment must be considered in the selection of the most appropriate anesthetic.

An ideal anesthetic agent should have a wide margin of safety, with rapid induction and recovery periods, while providing consistently effective immobilization or analgesia. Anesthesia involves a combination of narcosis, analgesia, and skeletal muscle relaxation that results from sensory nerve block, motor nerve block, or reduction in reflex activity (4). Since different procedures require different levels of anesthesia, this point should be considered in the selection of the most appropriate anesthetic method. Sedation may be required only for transport or short, simple procedures, such as tagging and injections, while more invasive procedures require full surgical anesthesia. The degree of anesthesia attained is often dictated by either the molecular structure of the anesthetic agent itself, the concentration of anesthetic used, or the duration of exposure to the anesthetic. Other important considerations in selecting an anesthetic include species variation, body mass, health status, age, water chemistry factors, and the withdrawal time of the drug (1,5–7).

The basic stages of anesthesia exhibited by fish are similar to the stages observed in mammalian species (4,8). A hyperexcitable stage occurs during induction and is characterized by erratic swimming, disorientation, increased respiration, and loss of equilibrium. The sedative stage is characterized by loss of reactivity, slow swimming, and decreased respiration. The anesthetic stage is characterized by complete loss of equilibrium and slowing of respiration that progresses to a surgical plane denoted by an inability to swim, shallow respiration, and no response to stimuli. The deepest plane of anesthesia is marked by cessation of opercular movements, which can lead to cardiac failure and death. Since the anesthetic dosage for different preparations will vary with the species, preliminary anesthetic trials with an unfamiliar fish species are recommended prior to enacting the intended use of the anesthetic. For example, metomidate, a commonly used aquatic anesthetic, has been demonstrated to be efficacious in the Atlantic salmon and cod (9,10), but is undesirable for use in red drum and goldfish larvae (6).

PHARMACEUTICAL METHODS OF ANESTHESIA

Route of Administration

Chemically induced methods of immobilization and restraint can be administered by bath immersion, gill perfusion, parenteral injection, or oral administration (1,4,11,12). It is important to remember that both induction and recovery times for some drugs will vary according to the dosage used, the duration of exposure time to the anesthetic agent, and the total body-fat content of the fish (8).

Bath Immersion. For immersion methods, simultaneous preparation of both induction and recovery tanks of water is recommended. Water quality parameters (e.g., pH, temperature, salinity, hardness) in the tanks should closely match those of the natural habitat waters of the fish to be anesthetized. Aeration of these waters is advisable, due to the common development of hypoxia during anesthesia, which occurs secondarily to respiratory depression. Opercular movement is a good indicator of the plane of anesthesia attained and should be monitored throughout any procedure. A fish that enters too deep a plane of anesthesia can be resuscitated by immediate transfer to the recovery water. Oxygen exchange and anesthetic elimination can be enhanced at the gill level by either using open-mouth propulsion of the fish through the water or positioning the fish near an airstone (1,4).

Parenteral Administration. Injection of anesthetics is a possible alternative for larger fish, but this method of anesthesia has been reported to be inconsistent for both maintenance and recovery. Disparities associated with intramuscular administration of anesthetics have often been attributed to the slower uptake or possible leakage of anesthetic agents from injection sites due to anatomical or mechanical factors. The sterile granulomas that may develop at intramuscular injection sites or intraabdominal adhesions from intraperitoneal administration are additional possible sequelae that can occur secondarily to drug-induced tissue irritation or damage. A clinical report regarding intramuscular ketamine administration in various fish species, however, cites no deleterious side effects in the more than 50 trials conducted (1,4,13).

Oral Administration. There have been reports of oral administration of anesthetics in the food fed to fish, by capsule or by using gavage (4). Observations of delayed induction times associated with this method of drug administration have been attributed to probable slow drug absorption by the gastrointestinal tract and the possible difficulties related to incorporation of the drug into the diet. The inability to assess accurately the quantity of drug-treated feed consumed on an individual basis has been another concern.

Anesthetic Agents

The selection of some drugs currently used for anesthetic purposes in aquatic species can be traced to human and veterinary medical literature where the routine use of the drugs developed over the years within the research setting. The approved use of the drugs in commercially important fish species, however, is restricted and varies between countries. For example, Aqui-S is legally approved in New Zealand for use in foodfish with no withdrawal period, but is not approved for this use in the United States. On the other hand, MS-222 is legally approved in Canada and the United States with a 21-day withdrawal period in foodfish species. Table 1 lists chemical preparations commonly used for fish anesthesia. A short review of the recent literature for these drugs is presented in the paragraphs that follow.

Benzocaine. Benzocaine (ethyl aminobenzene), although not legally approved for use in foodfish in the United States, is a relatively safe, routinely used fish anesthetic. It is supplied as a water-insoluble white crystalline powder that requires reconstitution in ethanol or ether prior to adding it to water. A more water-soluble salt form, benzocaine hydrochloride, is also available, but is more expensive. Stock solutions of this preparation, if made in advance, should be buffered and stored in dark containers to prevent inactivation (4,8). Its solubility and efficacy in freshwater appear not to be affected by variations in water hardness and pH, but increases in temperature do seem to enhance its solubility (14). Strong aeration of anesthetic waters is important, due to the hypoxic effect of this drug, resulting from reduced gill ventilation subsequent to depression of the medullary respiratory centers. Since this agent is lipophilic, recovery times and residue levels in body tissues will vary according to the amount of stored body fat. The drug withdrawal time in young, nongravid trout (Salmonidae) and largemouth bass (*Micropterus salmoides*), however, has been demonstrated experimentally to be only 24 hr (8).

Tricaine Methane Sulfonate. Tricaine methane sulfonate (3-aminobenzoic acid ethyl ester, or MS-222) is a commonly

Table 1. Commonly Used Fish Anesthetics

Anesthetic	Species	Dosage (ppm)	Water Temperature (°C)	Water Soluble?	Induction Time <5 min?	Recovery Time <10 min?
Benzocaine (as 100 g/l ethanol stock solution)	Species variation (4,8,17,28)	50–500	(More toxic at warm temperatures)	Not soluble; salt form more soluble	Yes	Prolonged, due to fat solubility
	Large fish (as gill spray) (17)	1000				
	Trout and salmon (8)	25–45				
	Northern pike (8)	100–200				
Tricaine methane sulfonate (MS-222)	Cod (9)	75	8.4	Yes	Yes	Yes
	Koi (12)	150–200	24			
	Salmonids and tropical fish (11,34)	50–100	—			
	Halibut (35)	250	9.5–10.5			
	Red drum (36)	80	26			
	Porgy (5)	100	20			
Quinaldine sulfate	Red drum (6,36)	20–35	26	—	Yes	Yes
	Goldfish (6)	60	24	—	—	—
Metomidate	Salmon (8,10)	5	5	Yes	Yes	Recovery time correlates with exposure time
	Cod (9)	5	9.6			
	Halibut (35)	20–30	9.5–10.5			
	Red drum (36)	7	26			
	Catfish (8)	1–2.5	—			
	Tropical fish (8)	2.5–5	—			
Phenoxyethanol	Goldfish (18,37)	0.1–0.2 0.3–0.4	20	—	Yes (concentration dependent)	Yes (concentration dependent)
	Acanthopagrus schlegeli (38)	400	—		Yes	
	Lateolabrax japonicus (38)	400	—		Yes	
	Oreochromis mossambicus (38)	≥600	—		Yes	
	Poecilia velifera (38)	≥600	—		Yes	
Clove oil (as a 1% solution) (39)	Juvenile rabbitfish (39)	100	27–29	—	2 min	<3 min
	Rabbitfish (24)	50–100	—		2 min	<5 min
	Milkfish (24)	50–100	—		<2 min	<5 min
	Striped mullet (24)	50–100	—		<2 min	<5 min
	Freshwater and saltwater species (12)	40–120	—		3–5 min	Prolonged
	Pomacentrus amboinensis (26)	—	—		—	Prolonged
Aqui-S	Most species (7)	20	—	—	Yes	Yes

used fish anesthetic that is FDA approved in the United States for use in most freshwater and marine species. It is more water soluble than its parent compound, benzocaine, but is also more acidic in nature, requiring its solutions to be buffered with sodium hydroxide or sodium bicarbonate. This agent is eliminated at the level of the gill, kidney, and gall bladder, with no detectable metabolites in the mucus, feces, or gametes (15). It is considered to be a relatively safe drug, but its range of toxicity varies with fish species, fish size, water temperature, and water hardness. Aeration is recommended during induction, as hypoxia is a common sequela, as well as other physiologic effects (see Table 2). Individual variation with regard to dosage of this agent can also be quite broad. One reported human case of retinopathy has been associated with chronic cutaneous exposure to MS-222 (16). Due to this potential occupational hazard, wearing gloves is recommended as a measure to prevent systemic absorption of this compound.

Metomidate. Metomidate (1-(1-phenylethyp)-1H-imidazole-5-carboxylic acid methyl ester) is a rapidly acting, water-soluble, nonbarbiturate, imidazole-based hypnotic that has been found to be effective in various species (1,4,8,9). Due to its lack of analgesic properties in humans, it is recommended only for sedation in fish, and not for use in surgical procedures. Muscle fasciculations can commonly occur with its use, so it is not recommended for detailed procedures as well. It is a potent anesthetic in Atlantic salmon (*Salmo salar*) at low temperatures and induces rapid anesthesia in freshwater salmon parr, but it has been demonstrated to be most efficacious in large Atlantic salmon that are acclimated to sea water (10). Inhibition of cortisol release has been noted in metomidate-anesthetized fish (4,8), with increases seen in both the hematocrit and blood lactate levels (10). The "chemically induced interrenalectomy" caused by this agent at the interrenal cell level has been suggested to have research potential as a tool to separate catecholamine from the effects of cortisol in stress-related studies (10).

2-Phenoxyethanol. 2-Phenoxyethanol is an inexpensive anesthetic that has been reported to have a narrow margin of safety and to cause hyperactivity during induction or recovery (17). Considerable variation in its activity has been described according to fish species, body size, and density, as well as relative to water quality parameters. It occurs as an oily liquid with both bactericidal and fungicidal properties. In goldfish (*Carassius auratus*), the concentration used has a marked effect on the induction and recovery times. Concentrations of 0.3–0.4 g/L are useful for short procedures, and lower concentrations, of 0.1 and 0.2 g/L, are considered safe for prolonged sedation (18). Its potential use for transport purposes is supported by a study demonstrating its increased effectiveness over other anesthetics in suppressing oxygen consumption in platyfish (*Xiphophorus maculatus*) (19). However, in another transport study, with gilthead sea bream (*Sparus aurata*), a marked stress response was demonstrated subsequent to exposure to the drug, where the return of cortisol and hematologic parameters to normal levels followed metabolic elimination of the drug (20). No detrimental effects on sperm motility

Table 2. Miscellaneous Physiological Side Effects Associated with the Use of Anesthetics

Anesthetic	Effect on Respiration	Effect on Blood Chemistry	Effect on Cortisol	Elimination of Metabolic Wastes	Effect on Heart	Species
Metomidate	↓	↑ Hct[a]; ↑ lactate	No effect	—	—	Salmon (10)
	—	—	No effect	—	—	Red drum (36)
	—	—	—	No effect	—	Platyfish (40)
	—	No effect	—	—	—	Halibut (35)
MS-222	↓	↑ Hct	↑	—	—	Salmon (10)
	—	↑ glucose	↑	—	—	Red drum (36)
	—	↑ lactate	—	↓ NH_3	—	Platyfish (40)
	—	—	—	—	↓ CO[b]	Rainbow trout (41)
	—	No effect	—	—	—	Halibut (35)
2-Phenoxyethanol	↓O_2 consumption	—	—	—	—	Platyfish (18)
		—	—	↓ NH_3; ↓ CO_2	—	Platyfish (40)
	—	↑ Hct[a]	—	—	—	Red snapper (42)
	—	↑ Hb[c]	—	—	—	Grey mullet (42)
	—	↑ Osm[d]	—	—	—	Black porgy (42)
	—	—	—	—	↓ HR[e]; ↓ BP[f]	Rainbow trout (43)
	↓ O_2 consumption	—	—	—	—	*Poecilia reticulata* (44)
Quinaldine sulfate	↓	—	↑	↓ NH_3; ↓ CO_2	—	Platyfish (40)
	—	↑ Glucose	↑	—	—	Red drum (36)

[a]Hct = hematocrit.
[b]CO = cardiac output.
[c]Hb = hemoglobin.
[d]Osm = osmolarity.
[e]HR = heart rate.
[f]BP = blood pressure.

were reported with the use of 2-Phenoxyethanol during spawning of grass and silver carp (*Ctenopharyngodon idella* and *Hypopthalmichthys molitrix*, respectively) at a concentration of 0.2 mg/L (21).

Quinaldine Sulfate. Quinaldine sulfate is a water-soluble yellow powder that requires buffering with sodium bicarbonate upon its addition to water. Stock solutions, of 10 g/L, should be protected from exposure to light and air. Its potency is affected by water pH, temperature, and hardness, where it exhibits a decreased toxicity at lower pH and temperature values. It provides rapid induction and recovery rates, but reflex activity is not lost at the time of loss of equilibrium. Disappearance of this touch response generally occurs after approximately 20 seconds of contact, so most procedures can be performed under this anesthetic (1,8). Dosages of 15–70 mg/L have been used for warmwater species (4). The required dosage varies between species, where largemouth bass are most sensitive and carp (Cyprinidae) are relatively resistant. More rapid induction times have been achieved for combinations of quinaldine sulfate and MS-222 than for either drug alone (8,22). Quinaldine sulfate is not metabolized by fish and is excreted mainly at the gill level, but also by way of the kidney and bile, similar to the excretion of MS-222 (15). The parent compound, quinaldine, is not water soluble and requires dissolution in an organic solvent prior to mixing with water. Since it is considered a suspect carcinogen, precautions should be taken during its use (17).

Clove Oil. Although not approved for general use as a fish anesthetic in the United States, clove oil (eugenol derivative) has been demonstrated, both experimentally and clinically, to be as effective as MS-222 in most freshwater and marine species at concentrations of 40–120 ppm (12,23). At 25 ppm, sedation was demonstrated to be adequate in rabbitfish (Siganidae), milkfish (*Chanos chanos*), and striped mullet (*Mugil cephalus*) for conduction of short, simple procedures with induction and recovery times of less than or equal to 2 and 5 minutes, respectively (24); however, there have been reports of prolonged recovery times as well (25,26). The primary constituent of clove oil, eugenol, is similar in structure to MS-222 and 2-phenoxyethanol.

Aqui-S. This particular anesthetic/sedative is approved for use in aquatic species in New Zealand with no withdrawal time and is presently undergoing the New Animal Drug Act (NADA) approval process for use in the United States. The formulation of Aqui-S contains a series of synthetic flavoring ingredients that are individually approved for use in food; one of these ingredients is similar to a major component present in clove oil. Experimentally, induction and recovery times of less than 3 and 10 minutes, respectively, have been demonstrated in most species of adult fish at a dosage of 20 ppm Aqui-S (7).

NONPHARMACEUTICAL METHODS OF ANESTHESIA

Nonpharmaceutical methods of anesthesia are employed in situations where drug withdrawal times may interfere with the impending marketability of some foodfish species, in field studies, and in locations where no approved drugs are commercially available. In the past, carbon dioxide was used in the field setting. Disadvantages of this method are the prolonged induction and recovery periods, shallow anesthesia, and metabolic imbalances associated with the lowered pH (1,17,23). Combined use of sodium bicarbonate and acetic acid in salmon has been suggested as an alternative anesthetic method for laboratory, farm, and field use (27). Electrically induced anesthesia, using alternating, direct, and pulsating currents, is another possibility; however, there is an increased risk to the safety of the operator (28,29). The subsequent rapid immobilization reduces netting stress, but the physiological effects on the hematological parameters are similar to those associated with anesthetic agents (28). Use of AC voltages has resulted in variable and unpredictable levels of narcosis and physical damage (30). Electroanesthesia has been used as a viable anesthetic alternative for tagging and spawning purposes (30,31).

PHYSIOLOGICAL EFFECTS OF FISH RESTRAINT

Reduction of handling stress and physical trauma is a major consideration for fish sedation and anesthesia; however, the physiological and hematological side effects of some anesthetics can closely mimic the effects associated with the stress response. Using different species and varying the duration of exposure to the anesthetic may also cause variation in certain blood parameters. Within the research setting, therefore, any hematological or biochemical measurements obtained from anesthetized fish must be evaluated for potential confounding associated with anesthetic use. Benzocaine, quinaldine, and MS-222 have been reported to affect hepatic enzyme functions; therefore, their use is not recommended prior to hepatic enzyme analytic studies (32,33). Table 2 gives a partial listing of the potential side effects associated with the use of certain anesthetics.

BIBLIOGRAPHY

1. L.J. Detolla, S. Srinivas, B.R. Whitaker, C. Andrews, B. Hecker, A.S. Kane, and R. Reimschuessel, *ILAR* **37**, 159–173 (1995).
2. L. Swann, "Transport of Fish," *Aquaculture Extension*, Illinois—Indiana Sea Grant Program, IL-IN-SG-FS-91-3, (1998).
3. American Society of Ichthyologists and Herpetologists, American Fisheries Society, and American Institute of Fisheries Research Biologists, *Guidelines for Use of Fishes in Field Research*, American Society of Ichthyologists and Herpetologists, Lawrence, KS, 1987.
4. L.A. Brown, in ed., M.K. Stoskopf, *Fish Medicine* W.B. Saunders, Philadelphia, 1993.
5. S. Oikawa, T. Takeda, and Y. Itazawa, *Aquaculture* **121**, 369–379 (1994).
6. K.C. Massee, M.B. Rust, R.W. Hardy, and R.R. Stickney, *Aquaculture* **134**, 351–359 (1995).
7. G.R. Stehly and W.H. Gingerich, in *Book of Abstracts: Aquaculture '98*, World Aquaculture Society, February 15–19, 1998, pp. 518.

8. M.K. Stoskopf, in ed., L. Brown, *Aquaculture for Veterinarians.*, Pergamon Press, Taneytown, NY, 1993.
9. N.S. Mattson and T.H. Riple, *Aquaculture* **83**, 89–94 (1989).
10. Y.A. Olsen, I.E. Einarsdottir, and K.J. Nilssen, *Aquaculture* **134**, 155–168 (1995).
11. L.S. Brown, *Veterinary Clinics of North America: Small Animal Practice* **18**, 317–330 (1988).
12. G.A. Lewbart, *Suppl. Compend. Contin. Educ. Pract. Vet.* **20**(3A), 5–12 (1998).
13. T.D. Williams, J. Christiansen, and S. Nygren, *IAAAM Proceedings* **24**, 6 (1993).
14. J.L. Allen, G. Vang, S. Steege, and S. Xiong, *Prog. Fish-Cult.* **56**, 145–146 (1994).
15. J.L. Allen and J.B. Hunn, *Vet. Hum. Toxicol.* **28**, 21–24 (1986).
16. P.S. Bernstein, K.B. Digre, and D.J. Creel, *Am J. Ophthalmol.* **124**, 843–844 (1997).
17. E.J. Noga, *Fish Disease: Diagnosis and Treatment*, Mosby-Year Book, Inc., St. Louis, MO, USA, 1996.
18. A. Josa, E. Espinosa, J.I. Cruz, L. Gil, M.V. Falceto, and R. Lozano, *Vet. Rec.* **131**, 468 (1992).
19. F.C. Guo, L.H. Teo, and T.W. Chen, *Aquaculture Research* **26**, 887–894 (1995b).
20. A. Molinero and J. Gonzalez, *Comp. Biochem. Physiol.* **111**, 405–414 (1995).
21. N. McCarter, *Prog. Fish-Cult.* **54**, 263–265 (1992).
22. R.R. Stickney, *Principles of Warmwater Aquaculture*, John Wiley & Sons, New York, 1979.
23. W.G. Anderson, R.S. McKinley, and M. Colavecchia, *N. Am. J. Fish. Managem.* **17**, 301–307 (1997).
24. C.S. Tamaru, C. Carlstram-Trick, and W.J. Fitzgerald, in *Book of Abstracts: Aquaculture '98*, World Aquaculture Society, February 15–19, 1998, p. 532.
25. J.L. Keene, D.L.G. Noakes, R.D. Moccia, and C.G. Soto, *Aquaculture Research* **29**, 89–101 (1998).
26. P.L. Munday and S.K. Wilson, *J. Fish Biol.* **51**, 931–938 (1997).
27. A.M.J. Prince, S.E. Low, T.J. Lissimore, R.E. Diewert, and S.G. Hinch, *N. Am. J. Fish. Managem.* **15**, 170–172 (1995).
28. L.G. Ross and B.R. Ross, *Anaesthetic and Sedative Techniques for Fish*, Institute of Aquaculture, University of Stirling, Stirling, Scotland, 1984.
29. D.A. Sterritt, S.T. Elliott, and A.E. Schmidt, *N. Am. J. Fish. Managem.* **14**, 453–456 (1994).
30. M.K. Walker, E.A. Yanke, and W.H. Gingerich, *Prog. Fish-Cult.* **56**, 237–243 (1994).
31. J.M. Tipping and G.J. Gilhuly, *N. Am. J. Fish. Managem.* **16**, 469–472 (1996).
32. D.L. Fabacher, *Comp. Biochem. Physiol: Comp. Pharmacol.* **73**, 285–288 (1982).
33. E. Arinc and A. Sen, *Comp. Biochem. Physiol: Comp. Pharmacol. Toxicol.* **107**, 399–404 (1994).
34. B.R. Whitaker, *Compend. Contin. Educ. Pract. Vet. (Small Animal)* **13**, 960–966 (1991).
35. T. Malmstrom, R. Salte, H.M. Gjoen, and A. Linseth, *Aquaculture* **113**, 331–338 (1993).
36. P. Thomas and L. Robertson, *Aquaculture* **96**, 69–86 (1991).
37. O. Weyl, H. Kaiser, and T. Hecht, *Aquaculture Research* **27**, 757–764 (1996).
38. J.R. Hseu, S.L. Yeh, Y.T. Chu, and Y.Y. Ting, *J. Fish. Soc. Taiwan* **24**, 185–191 (1997).
39. C.G. Soto and Burhanuddin, *Aquaculture* **136**, 145–152 (1995).
40. F.C. Guo, L.H. Teo, and T.W. Chen, *Aquaculture Research* **26**, 265–271 (1995).
41. S.N. Ryan, P.S. Davie, H. Gesser, and R.M.G. Wells, *Comp. Biochem. Physiol. Comp. Pharmacol. Toxicol.* **106**, 549–553 (1993).
42. J.R. Hseu, S.L. Yeh, Y.T. Chu, and Y.Y. Ting, *J. Fish. Soc. Taiwan* **23**, 43–48 (1996).
43. K.T. Fredericks, W.H. Gingerich, and D.C. Fater, *Comp. Biochem. Physiol: Comp. Pharmacol. Toxicol.* **104**, 477–483 (1993).
44. L.H. Teo and T.W. Chen, *Aquaculture and Fisheries Management* **24**, 109–117 (1993).

See also LIVE TRANSPORT.

ANTIBIOTICS

BEVERLY A. DIXON
California State University
Hayward, California

OUTLINE

The increasing intensity of fish and shellfish farming, combined with the introduction of new species into culture, was inevitably paralleled by increase in disease. Bacteria represent a large variety and proportion of the pathogenic agents acting as both primary and secondary invaders in intensive culture. Antibiotic therapy for systemic bacterial infections began in the late 1930's, and was shortly followed by the introduction of sulfamerazine into the United States in 1948. As new drugs were introduced in human and veterinary medicine, suitable compounds were investigated for application in fish farming (1). The use of chemotherapy in aquaculture has always been limited by both legal and practical constraints. Considerations such as cost, route of application, labor, drug absorption, toxicity, and environmental impact have forced many potentially useful drugs to remain unavailable for aquaculture application (2).

Ideally, the selection of a chemotherapeutics against bacterial disease should include the process of isolating and identifying the specific pathogen responsible for the infection, followed by determining of the most efficacious drug. However, this practice is often time consuming, always costly, and mostly overlooked. Selection should also take into account the bioavailability of the drug, the concentrations at which it accumulates in the host tissue, the elimination rate (especially for food fish), and the route of administration. The current method of administering therapy for bacterial infections of food fish is incorporating an approved drug as a premix into feed (3). The treatment of nonfood fish is affected either through the use of medicated feed or, more commonly, by dispensing antibacterial drugs directly into the water as bath treatments. Both of these approaches can be problematic. For example, formulated feeds may not be readily accepted by young stages of fish, particularly larval fish raised on cultured live foods. Drugs may leach out of feed and affect the environment, or they may decompose. However, antibacterial therapy, particularly prophylactic treatment, has lead to unanswered questions concerning the development of antibiotic resistant bacteria, an emerging problem in aquaculture.

DEFINITIONS

Traditionally, the term *antimicrobial drug* was applied only to compounds such as dyes, and synthetic or organic substances used in the treatment of microorganisms. More recently, the term is used synonymously, and has come to overlap in meaning, with the term antibiotic. An *antibiotic* is a natural substance produced by the metabolic processes of microorganisms that can inhibit or kill other microorganisms. Put simply, antibiotics are a kind of chemical warfare used against microorganisms in the same habitat to compete for nutrients and space. Most antibiotics in use today are derived from spore-forming bacteria in the genus *Bacillus* and two actinomycetes *Streptomyces* and *Micromonospora*. Fungi of the genera *Penicillium* and *Cephalosporium* also produce antibiotics. The ability of antibiotics to affect microorganisms differs and is usually related to the mechanism of action and concentration. *Broad-spectrum* antibiotics, such as tetracyclines, inhibit bacterial protein synthesis, a mechanism that affects a variety of bacteria. In contrast, a *narrow-spectrum* antibiotic, such as penicillin acts to inhibit formation of the bacterial cell wall, a process more efficacious against Gram positive bacteria.

An important property of an antimicrobial drug is termed specificity, or *selective toxicity*. Ideally, the antimicrobial agent should act at a target site that is present in the microbe, but absent in the host, resulting in adverse effects on bacterial cells without simultaneously damaging host tissue. Selective toxicity is achieved by exploiting the differences in structure and metabolism of bacteria and host cells. As procaryotes, bacteria are structurally more distinct from eukaryotic host cells than are fungal or protozoan pathogens. Structural differences between procaryotic and eukaryotic ribosomes and DNA, and the presence and specific chemical composition of bacterial cell walls are likely targets for selective toxicity of many antibacterial drugs (4,5). Depending on concentration a drug may be bacteriostatic or bactericidal. *Bacteriostatic drugs* interfere with the machinery required for cell division, inhibiting reproduction. Their importance is to inhibit the growth of bacteria, thereby preventing the bacterial population from increasing and allowing host defense mechanisms to destroy the static population. *Bactericidal drugs* lyse and kill bacteria by inflicting direct damage on specific cellular targets. The distinction between the two mechanisms has become blurred because some bacteriostatic drugs may be bactericidal at higher concentrations. Conversely, drugs considered bactericidal such as chloramphenicol may be bacteriostatic for *Escherichia coli*, but bactericidal for *Haemophilus influenzae* (4).

MODES OF ACTION

There are four main targets for antibacterial action:

- cell wall synthesis
- protein synthesis
- nucleic acid synthesis
- cell membrane function

Inhibitors of Cell Wall Synthesis

Four classes of antibiotics act as inhibitors of bacterial cell wall synthesis: the beta-lactam family, which includes penicillins, cephalosporins, monobactams, and carbapenems; glycopeptides like vancomycin, cycloserine, and bacitracin. For the most part, these groups are not widely used in aquaculture, probably because their mode of action is more efficacious against Gram positive bacteria. Drugs that inhibit cell wall synthesis act at one of three stages in the process of wall formation. In the typical Gram positive bacterial cell, the cell wall encases the membrane as a continuous, highly cross-linked molecular network that provides structure and keeps the cell from rupturing. This structure, also known as the peptidoglycan, is unique to bacteria and provides an optimum target for selective toxicity. Synthesis of the peptidoglycan invokes three stages: synthesis of the subunits in the bacterial cytoplasm, transport of the subunits outside the cell, and the final cross-linking. Penicillin and cephalosporin act at the final stage by preventing the formation of cross-links between the units of the cell wall, resulting in a weakened network. Strands of peptidoglycan that cannot cross-link have been observed by electron microscopy to accumulate on the cell wall (6). Eventually, the cell membrane extrudes thought the weak points in the cell wall, and the cell ruptures. The bactericidal drugs penicillin and cephalosporin utilize the same mechanisms to cause bacterial cell death.

Glycopeptides, such as vancomycin, act at an earlier stage in cell wall formation than do the penicillin and cephalosporin. During the second stage of peptidoglycan synthesis, the basic repeating units of the cell wall are put together to form a long polymer. Vancomycin interferes with this mechanism by binding to the end of the pentaglycine polymers that are used to cross-link the growing peptidoglycan structure. This binding prevents new subunits from being incorporated into the growing cell wall.

Bacitracin and cycloserine act at the second and first stages of cell wall synthesis, respectively. Both are of limited clinical use and are not used at all aquaculture.

Penicillin. Penicillin was first isolated by Alexander Fleming in the late 1930s from the fungus *Penicillium notatum*. Now the penicillin family contains almost a dozen drugs all of which contain a beta-lactam ring attached to another five-membered ring containing a sulfur molecule and the side chains attached to those rings. Natural penicillin, chemically known as benzyl penicillin, is administered by injection and cannot be given by mouth. Biochemists have improved upon the original drug by adding chemical side chains to the penicillin core. The addition of side chains affects such chemical properties as solubility and absorption and allows these newer generation or semi-synthetic penicillins to be given orally. The newer also resist acid hydrolysis and counter some mechanisms of bacterial resistance. Some of the newer semi-synthetic penicillin are broad spectrum antibiotics and often are used against Gram negative bacteria, in contrast to the earlier generation of narrow spectrum drugs.

Cephalosporins. The first cephalosporins were isolated in the late 1940s from the fungus *Cephalosporium acremonium*. The family is characterized by a beta-lactam ring attached to another six membered ring. Like penicillin cephalosporin acts at the third stage of bacterial cell wall synthesis. The cephalosporins have a broader spectrum of action and are active against a variety of Gram negative bacteria. Newer generations of cephalosporin have been developed, and these account for the majority of antibiotics administered; however, they are relatively expensive and not readily used in aquaculture.

Vancomycin. Vancomycin is a large molecule that has difficulty penetrating the Gram-negative cell wall. It is a narrow-spectrum drug used mainly for the treatment of Gram-positive cocci and Gram-negative rods that are resistant to beta-lactam drugs.

Inhibitors of Protein Synthesis

Although protein synthesis proceeds in a similar manner in prokaryotic and eukaryotic cells, effective antibiotics can exploit the differences that do exist. Most inhibitors of protein synthesis act at the level of translation by reacting with the ribosome-messenger (mRNA) complex. The size and structure of prokaryotic ribosomes differ from those of eukaryotes affording selective toxicity. Two possible targets of ribosomal inhibition are the 30S and the 50S subunits of prokaryotic ribosomes. Attachment to the 30S subunit by aminoglycoside antibiotics interferes with the binding of transfer RNA (tRNA) to the ribosome, preventing the initiation of protein synthesis. Other animoglycosides like streptomycin also cause misreading of the mRNA. Both mechanisms lead to the formation of abnormal proteins.

Tetracyclines inhibit protein synthesis by binding to the 30S subunit further blocking the binding of tRNA to the mRNA complex. Tetracycline binding inhibits virtually 100% of protein synthesis. Another inhibitor of protein synthesis, chloramphenicol, acts by binding to the 50S ribosomal subunit preventing the formation of peptide bonds. Erythromycin, the best known and most widely used of the macrolide antibiotics, binds to the 50S subunit blocking the translocation step and preventing the release of tRNA after peptide bond formation. Tetracyclines, chloramphenicol, and erythromycin are all bacteriostatic inhibitors of protein synthesis, and all can bind to the ribosomes in eukaryotic mitochondria, causing a variety of toxicities.

Aminoglycosides. Aminoglycosides are derived from different species of the actinomycetes *Streptomyces* and *Micromonospora*. The most frequently used aminoglycosides are neomycin, kanamycin, gentamicin, tobramycin, streptomycin, and amikacin. The original compounds have been chemically altered by the addition of side chains to produce a newer generation of compounds, such as amikacin. The newer compounds are active against organisms that have developed resistance to earlier aminoglycosides. The aminoglycosides have broad spectrum activity and are highly efficacious against most Gram-negative rods. All are bactericidal and therefore rapidly stop protein synthesis. However, they are not readily absorbed from the gut and do not penetrate tissues well. Aminoglycosides are routinely administered by intramuscular injection. Neomycin and kanamycin are more water soluble and are often used as bath treatments in marine aquaria for infections caused by *Vibrio* spp. (7,8). Reports in the literature indicate that neomycin can disrupt nitrifying bacteria leading to an increase in both ammonia and nitrite levels in waters (9). Aminoglycosides generally should not be used sequentially or in combination with tetracycline or each other.

Tetracyclines. Tetracyclines are a large family of cyclic structures that inhibit protein synthesis by preventing animoacyl transfer RNA from entering the acceptor sites on the ribosome. Unfortunately, this action is not selectively toxic for prokaryotes and tetracyclines also will inhibit protein synthesis in eukaryotes. Some selective toxicity is afforded by the greater uptake of tetracyclines by bacterial cells. Tetracyclines are usually administered orally, are well distributed in the tissue, and penetrate host cells to inhibit intracellular bacteria, mycoplasma, rickettsiae, and chlamydiae. Doxycycline and minocycline are absorbed better than tetracycline, oxytetracycline or chlortetracycline. For this reason both doxycycline and minocycline cause less gastrointestinal upset, because there is less inhibition of normal gut flora. Oxytetracycline treatment has been reported to cause immunosuppression in some fish whether the drug is administered by injection, in feed or as a bath. In Rainbow trout (*Oncorhynchus mykiss*), the number of antibody-producing cells was reduced, while both the number of antibody-producing cells and the mitogenic response were reduced in carp (*Cyprinus*) treated with oxytetracycline (10–12). Tetracycline, minocycline, and doxycycline are suggested treatments against Gram positive infections of mycobacteriosis and streptococcosis (13).

Tetracycline readily binds to calcium and magnesium forming an insoluble chelate, and therefore may not be as effective as a bath treatment in marine waters or fresh waters of higher pH (harder water).

Erythromycin. Erythromycin is a member of a family of large cyclic molecules called macrolides. Members of this family all contain a macro cyclic lactone ring structure to which sugars are attached. Erythromycin specifically binds to the 50S subunit of the ribosome and blocks the translocation step in protein synthesis, thereby preventing the release of transfer RNA after peptide bond formation. Erythromycin is most active against Gram-positive bacterial infections, chlamydiae, rickettsiae, and mycoplasma. For yet unknown reasons, the concentration of erythromycin achieved in Gram-positive bacteria is greater than that in some Gram-negative bacteria (6). It is thought that erythromycin cannot readily penetrate the cell envelope of Gram-negative bacteria.

Erythromycin is readily absorbed at a neutral or a slightly alkaline pH, which makes it particularly useful for the treatment of African Rift Valley lake fish and live bearers. However, bacterial nitrifies will be severely hampered by the use of erythromycin (14).

Chloramphenicol. Chloramphenicol is a broad-spectrum antibiotic with a unique nitrobenzene structure that is responsible for some of the toxic problems associated with its use. Chloramphenicol blocks the action of peptidyl transferase, preventing peptide bond synthesis on the ribosome. It inhibits bacterial protein synthesis selectively, since it has a higher affinity for the transferase of the 50S bacterial subunit than that of the 60S subunit of the eukaryotic ribosome. Because of its unusual stability, chloramphenicol is poorly absorbed through water, and is commonly administered by injection (15). Chloramphenicol is highly illegal to use on foodfish in the United States.

Inhibitors of Nucleic Acid Synthesis

For the most part, inhibitors of nucleic acid synthesis are antimicrobial agents rather than true antibiotics. The exception is rifampin, a chemically altered compound produced by the genus *Streptomyces*. The chemical agents act to inhibit nucleic acid synthesis in one of three main ways: by inhibiting precursor synthesis, by inhibiting DNA replication, or by inhibiting RNA polymerase. The inhibitory effects may be so basic to cell function that protein synthesis and other metabolic pathways appear as targets (4).

Inhibitors of Precursor Synthesis

Sulfonamides. Sulfonamides, also known as sulfa drugs, and trimethoprim are often referred to as antimetabolites, because they interfere with the metabolic synthesis of precursor compounds. The sulfonamides are a group of molecules produced exclusively by biochemical synthesis; and they act as bacteriostatic inhibitors of bacterial growth. Many sulfonamides have been formulated, but only a few have proved useful in the treatment of bacterial and protozoal infections. The sulfonamides are structural analogs to, and act in competition with, para-aminobenzoic acid (PABA), an essential component in cell synthesis of folic acid (folate). A reduced form of folic acid functions as a coenzyme that transports one-carbon units from molecule to molecule. These one-carbon transfer reactions are required for the synthesis of thymidine, all purines (adenine, guanine), pyrimidines (cytosine, thymine), and several amino acids. Thymidine is necessary for DNA synthesis, and the purines and pyrimidines are necessary for all nucleic acid synthesis in the cell. When folate synthesis is inhibited, cell growth is arrested due to the inability to synthesize these macromolecules. Although the sulfonamides rapidly block folate synthesis, bacteria may continue to grow until their stored pools of folate are depleted.

Folic acid is required for growth by both prokaryotic and eukaryotic cells. Animal cells are unable to synthesize folate, but it enters the cells by active transport from dietary supplementation. Since folate does not readily enter bacterial cells, bacteria must manufacture their own folate intracellularly. The difference in membrane permeability accounts for the selective toxicity of sulfonamides to bacteria. Sulfonamides vary in solubility and are usually administered in feed. Although sulfonamides possess broad-spectrum activity, especially against Gram-negative bacteria, resistance is widespread. Sulfamerazine is approved for use in the United States against furunculosis in salmonids, but the drug was removed from the market by the manufacturer.

Trimethoprim has a pyrimidine-like structure analogous to a moiety of the folic acid molecule. In contrast to the sulfonamides, trimethoprim rapidly inhibits bacterial growth. When trimethoprim is added to a culture of growing bacteria, DNA, RNA, and protein synthesis are all affected. Trimethoprim was initially synthesized to maximize selective toxicity by exploiting the differences between eukaryotic and prokaryotic enzymes (6). The combination of trimethoprim or ormethoprim, a similar compound, with a sulfonamide affords a synergistic effect on cell growth. This effect, known as potentiation, is substantial and is attributed to the fact that the drugs inhibit different enzymes in the same biosynthetic pathway. Romet 30^R (Hoffman-LaRoche), a combination of ormethoprim and sulfadimethoxine is a potentiated sulfonamide drug approved for use in foodfish in the United States to treat *Aeromonas* and *Edwardsiella*. A similar combination of trimethoprim and sulfamethoxazole, while not approved for foodfish in the United states is widely used overseas (16,17).

Inhibitors of DNA and RNA Synthesis

Quinolones. Quinolones form a large family of synthetic agents originating from nalidixic acid, which was discovered as a distillate product of chloroquine synthesis, an antimalarial compound. Nalidixic acid and all newer quinolone agents are completely synthetic, and structurally related compounds have not been identified as products of living organisms (18). Analogs of nalidixic acid include oxolinic acid and the newer fluoroquinolones, such as sarafloxacin, enrofloxacin, and flumequine. Nalidixic acid and oxolinic acid are active principally against Gram-negative bacteria, while the fluoroquinolones have a broader spectrum that includes some Gram-positive bacteria and intracellular organisms like rickettsiae and chlamydiae. The newer compounds are better absorbed across the intestine. Most quinolones are rapidly bactericidal. All quinolones chelate divalent cations and are inhibited by hard water, and possibly, divalent cations in the diet.

Quinolones inhibit the activity of DNA gyrase, an enzyme involved in the initiation, elongation, and termination phases of DNA replication. Quinolones also inhibit the transcription of certain operons, and aspects of DNA repair, recombination, and transposition (18). Quinolone inhibition is specific to bacterial gyrase and does not affect similar enzymes in eukaryotic cells.

Metronidazole. Metronidazole belongs to a family of compounds called imidazoles, some of which have antifungal and anthelminthic activity. Metronidazole acts against microaerophilic and anaerobic bacteria and some intestinal protozoans by inhibiting hydrogen production. Metronidazole is reduced on entering the susceptible cell, to generate cytotoxic intermediates that cause a loss of the helical structure of DNA, a process that eventually leads to breakage of the cell's DNA (4,19). The selective toxicity of metronidazole is due to the lower redox systems that only anaerobic organisms possess. Metronidazole is bactericidal for anaerobic bacteria and is considered the universal drug of choice for these organisms. The compound is quite insoluble in water and must be thoroughly dissolved before being added to water or feed (20,21).

Rifampin. Rifampin (rifampicin) binds to DNA-dependent RNA polymerase, blocking the synthesis of mRNA and inhibiting cell growth. The selective toxicity of the drug is based on its greater affinity for bacterial polymerases than for equivalent eukaryotic enzymes. Rifampin is a chemical derivative of the fermentation products of the bacterium *Streptomyces mediterranei* and is an effective anti-mycobacterial antibiotic. Although used almost exclusively against mycobacteria, rifampin has activity for other Gram-positive and some Gram-negative bacteria. The efficacy of the drug in treating mycobacteriosis in fish is unproven, but rifampin was shown to be effective in experimental infection (22).

Interference with Enzyme Systems

Nitrofurans. Nitrofurans are synthetic antimicrobial compounds, several of which are commonly used as chemotherapeutants for bacterial fish pathogens. All the derivatives have broad-spectrum activity against Gram-positive and Gram-negative bacteria. Nitrofurans are known to inhibit a variety of enzyme systems in the cell; however, the exact mechanisms are not well understood. Nitrofurantoin generally is more effective at lower pH levels, and other nitrofurans are stable in both fresh and marine waters. Nitrofurans are photosensitive and are inactivated by bright light (8). Nifurpirinol (Furanace) and nitrofurazone (Furacyn^{R}) are used as single bath treatments or by oral administration. Palatability problems, however, can occur with oral usage of the drug (23). Nitrofurans are strictly illegal for use on foodfish in the United States because they were shown to be carcinogenic and mutagenic. Nitrofurans antagonize the antibacterial action of nalidixic acid and oxolinic acid and should not be used together with these compounds (6). Some scaleless fishes may be sensitive to the nitrofurans (24).

Interference with Mycolic Acid Synthesis

Isoniazid. Isoniazid is an isonicotinic acid hydrazide, a compound that inhibits mycobacteria, but does not affect other types of bacteria or animals to any great degree (4). Isoniazid is bactericidal against many actively growing mycobacteria and is thought to inhibit the synthesis of mycolic acid in the mycobacterial cell wall. Isoniazid prevents the elongation of a 26-carbon fatty acid, inhibiting the synthesis of long-chain fatty acids that are precursors of mycolic acids (6). This primary effect of the drug would explain both its limited spectrum of action and its selective toxicity. The use of isoniazid for mycobacteria in aquaculture remains controversial. Some authors have suggested that there is no suitable treatment for fish infected with mycobacteria, and thus, diseased stocks should be destroyed (25,26).

Inhibitors of Cytoplasmic Membrane Activities

Polymyxins. Polymyxins are true antibiotics synthesized by the bacterium *Bacillus polymyxa*. They are bactericidal, narrow-spectrum polypeptide antibiotics that act primarily against Gram-negative bacteria. The polymyxins exert a cationic detergent-like effect on the lipid bilayers of the cell membranes. Associating with phospholipids, they disrupt the structure of the cell membrane, resulting in leakage of small molecules, such as nucleosides, from the cell. The greater sensitivity of Gram-negative bacteria to polymyxins is most likely explained by the higher phospholipid content present in the cell envelope of these bacteria compared with Gram-positive bacteria.

ANTIMICROBIAL SUSCEPTIBILITY TESTING

Definitions

Following the isolation and identification of a bacterial pathogen, the next standard procedure is the determination of the organism's susceptibility to various antimicrobial drugs. The sensitivity pattern, also referred to as an *antibiogram*, determines which drugs are effective against the bacterium and will likely afford successful treatment.

Antimicrobial susceptibility tests for bacteria can be either quantitative or qualitative. In quantitative tests the minimum amount of drug that inhibits the visible growth of a bacterial isolate is determined. This lowest concentration termed the *minimum inhibitory concentration* (MIC), is determined by preparing serial dilutions of the test drug in either broth or agar and inoculating with the bacterial isolate. Following incubation, the MIC is recorded as the highest dilution in which there is no macroscopic growth. The MIC is also used as a comparative index for other antimicrobial agents.

Qualitative tests categorize a bacterial isolate as susceptible, intermediate, or resistant to a concentration of a particular antimicrobial agent. The most common qualitative method is the disk diffusion, or Kirby-Bauer, technique. Utilizing this method, fresh (18–24 hours old) bacterial colonies are placed into sterile physiological saline to match a turbidity standard (0.5 McFarland) representing 1.5×10^8 colony forming units/ml (CFU/ml). The suspension is then swabbed onto the surface of a Mueller-Hinton agar plate to create a lawn of bacteria. Antimicrobial-containing disks are then placed on the plate surface. During incubation the drug diffuses thought the agar, establishing a concentration gradient around each disk. Upon observation of the plates, the diameters of *zones of no growth* also known as *zones of inhibition* are measured. The sizes of the zones are compared to standard sizes and interpreted as susceptible, intermediate, or resistant (19). When working with marine bacteria, 1–3% sodium chloride is usually added to the broths or agars, to allow for the growth of halophilic organisms. The addition of salt, may interfere with zone sizes and should be interpreted only on a relative basis. The disk diffusion method should not be performed on selective or differential agars. *In vitro* susceptibility test results do not always correlate with successful treatment of the pathogen. Factors such as host response, drug dynamics, and microbial activity affect the eventual outcome of chemotherapy. Economic factors including cost, ease of handling and delivery, as well as government regulation, impact the availability and subsequent usefulness of drugs in aquaculture.

BACTERIAL RESISTANCE TO ANTIMICROBIAL DRUGS

Bacterial resistance to antimicrobial drugs has become widespread in aquaculture. Cultured fish have been reported to be infected with resistant strains of *Aeromonas hydrophila*, *Aeromonas salmonicida*, *Edwardsiella ictaluri*, *Pasteurella piscicida*, *Vibrio anguillarum*, *Yersinia ruckeri*, and streptococci resistant to many drugs used in aquaculture, including nalidixic acid, oxolinic acid, tetracycline, sulfa drugs, and chloramphenicol (27).

Antimicrobial resistance emerges as a result of a genetic change that is favored in the process of natural selection and may be *intrinsic* or *acquired*. Intrinsic resistance is an inherent physiological, biochemical or morphological feature of the cell that prevents antibiotic action. Intrinsic resistance is present in the cell and not dependent on drug exposure. It results from spontaneous chromosomal mutations that may alter physical or biochemical properties of the bacterial cell. Properties conferring intrinsic resistance include absence or insensitivity of a target site, antibiotic inactivation, and impaired antibiotic uptake.

Acquired resistance implies that an organism has developed resistance to a drug to which it was previously susceptible, following exposure to the drug. Exposure exerts the selective pressure, often resulting in the rapid overgrowth of resistant cells, from which emerges a new population of resistant cells. Acquired resistance confers properties similar to those of intrinsic resistance, as well as, enhancing antibiotic efflux, and causing overproduction of target sites requiring higher drug concentrations to inhibit bacterial growth (28). Acquired resistance can occur as a result of spontaneous chromosomal mutation or by the acquisition of extrachromosomal elements, such as plasmids and transposons. Spontaneous chromosomal mutations occur at a frequency of one gene mutation per 10^5 to 10^7 cells per cell division. These mutations can occur in the presence or absence of the drug, resulting in both intrinsic and acquired resistance.

Antibiotic resistance can be determined by genes that reside in the cell chromosome, on plasmids, or on transposons. Chromosomal mutations include both single and multiple DNA base pair alterations. However, in the vast majority of cases, the precise molecular basis of the mutational events leading to antibiotic resistance is unknown (28). As extrachromosomal genetic elements, plasmids replicate independently of the bacterial chromosome. Plasmid mediated antibiotic resistance is more common than chromosomal resistance in pathogenic bacteria. Genes encoded by plasmids are more mobile than chromosomal genes, because plasmids can be transferred both within and between certain bacteria. Transposons are mobile DNA sequences capable of transferring themselves from one DNA molecule (donor) to another (recipient). Transposons are not able to replicate independently and must be maintained either in the host chromosome or in plasmids. It is believed that transposons account for the emergence of multiple drug resistant bacteria.

Numerous reports in the literature document the emergence of antimicrobial drug resistance in bacterial pathogens of fish and shrimp (29). In the case of quinolone antibacterial drugs resistance is conferred by chromosomal mutation (18). Resistance to older quinolones such as nalidixic and oxolinic acids has been evidenced in such fish pathogens as *A. salmonicida* (30–32), *V. anguillarum* (27,33), and *Y. ruckeri* (34). The newer quinolones such as sarafloxacin and enrofloxacin have shown to be effective against multiple resistant strains of *Aeromonas* spp. (35–37) and *A. salmonicida* (38).

Antibiotic resistance to all other antimicrobial drugs is known to be conferred by chromosomal mutation, but primarily through plasmid acquisition. Plasmids transferring resistance to as many as five antimicrobial drugs have been identified from fish pathogens including *Vibrio salmonicida*, *V. anguillarum*, *A. salmonicida*, *A. hydrophila*, *Edwardsiella tarda*, *Citrobacter freundii*, and *Y. ruckeri* (31,33,39–45). Promising compounds such as florfenicol, a chloramphenicol analog, and a new generation of cephalosporins currently are being evaluated for use against bacterial fish pathogens (35,46,47).

IMMUNOMODULATION BY ANTIMICROBIAL DRUGS

Immunomodulation is a consequence of a change in the number or function of the cells involved in the immune response. Reports have described positive and negative effects, as well as no effect at all, of antimicrobial drugs on the immune response (48). Most studied are the immunomodulating effects of oxytetracycline (OTC), which appears to impair cellular immunity by decreasing mitogen response. Humoral immunity was suppressed in carp evidenced by a significant decrease in antibody producing cells during a primary immune response; however, no effect was observed during the secondary response (49). Some researchers speculate that in cases where the specific response is blocked, the phagocytic defense system becomes more active (49). Pharmacological studies have demonstrated that oxytetracycline (OTC) accumulates in lymphoid tissue which may explain the negative influence of the antibiotic on the immune response of carp. OTC also is known to suppress the antibody response on rainbow trout, while in certain experiments, potentiated sulfonamides did not. Grondel et al. (50) showed that at low concentrations of some chemotherapeutants can stimulate, while higher doses can inhibit, the immune response.

CHEMOTHERAPEUTIC REGULATION

Most of the drugs discussed are NOT approved for use in foodfish by the U.S. Food and Drug Administration and represent experimental or extralabel use. However, oxytetracycline is currently approved for use in certain species for the treatment of some Gram negative bacteria and *Streptococcus iniae*, and Romet[30] is approved for the treatment of *Aeromonas* and *Edwardsiella* infections in foodfish (13). Investigational New Animal Drug protocols (INAD) are being developed, such as erythromycin for use in salmonids to treat infections caused by *Renibacterium salmoninarum*.

Individual fish sensitivity may vary with the drug, treatment time, and dosage. Fish should always be treated in clean water, and water should be changed following the termination of treatment. Aeration and filtration should be maintained during treatment, but activated may have to be removed.

BIBLIOGRAPHY

1. J.S. Gutsell, *Trans. Am. Fish. Soc.* **75**, 186–199 (1948).
2. D.J. Alderman, J.F. Muir, and R.J. Roberts, eds., *Recent Advances in Aquaculture*, Vol. 3, Croom Helm, London, and Timber Press, Portland, OR, 1988.
3. D.J. Alderman and C. Michel, eds., *Chemotherapy in Aquaculture: From Theory to Reality*, O.I.E., Paris, 1992.
4. C.A. Mims, J.H.L. Playfair, I.M. Roitt, D. Wakelin, and R. Williams, *Medical Microbiology*, Mosby International Limited, London, 1998.
5. K. Talaro and A. Talaro, *Foundations in Microbiology*, W.C. Brown Publishers, Dubuque, IA, 1996.
6. W.B. Pratt, *Chemotherapy of Infection*, Oxford University Press, New York, 1977.
7. M.K. Stoskopf, *Fish Medicine*, W.B. Saunders Company, Philadelphia, 1993.
8. E.J. Noga, *Fish Disease, Diagnosis and Treatment*, Mosby-Yearbook, Inc, St. Louis, MO, 1996.
9. C.E. Bower and D.T. Turner, *Aquacultures* **29**, 331–345 (1982).
10. J.L. Grondel and H.J.A.M. Boesten, *Dev. Comp. Immunol.* **2**, 211–216 (1982).
11. W.B. van Muiswinkel, D.P. Anderson, C.H.J. Lamers, E. Egberts, J.J.A. van Loon, and J.P. Ijssel, in M.J. Manning and M.F. Tatner, eds., *Fish Immunology*, Academic Press, London, 1985.
12. A.K. Siwicki, D.P. Anderson, and O.W. Dixon, *Vet Immunol. Immunopath.* **23**, 195–200 (1989).
13. Tech Note/Veterinary Services, Streptococcus iniae: *Infection in Fish and Humans*, Animal and Plant Health Inspection Service (APHIS), United States Department of Agriculture (USDA), 1997.
14. M.T. Collins, J.B. Gratzek, D.L. Dawe, and T.G. Nemetz, *J. Fish. Res. Bd. Can.* **33**, 215–218 (1976).
15. K.E. Nusbaum and E.B. Shotts Jr., *Can. J. Fish. Aquatic. Sci.* **38**, 993–996 (1981).
16. J.A. Plumb, *Health Maintenance of Cultured Fishes: Principal Microbial Diseases*, CRC Press, Inc., Boca Raton, FL, 1994.
17. G.L. Lewbart, *Compend. Contin. Ed. Prac. Vet.* **13**, 109–116 (1991).
18. J.S. Wolfson and D.C. Hooper, eds., *Quinolone Antimicrobial Agents*, American Society for Microbiology, Washington, DC, 1989.
19. B.J. Howard, J.F. Keiser, T.F. Smith, A.S. Weissfeld, and R.C. Tilton, *Clinical and Pathogenic Microbiology*, Mosby-Yearbook, Inc., St. Louis, MO, 1994.
20. J.B. Gratzek and G. Blasiola, in J.B. Gratzek and J.R. Matthews, eds., *Aquariology: The Science of Fish Health Management*, Tetra Press, Morris Plains, NJ, 1992, pp. 301–315.
21. J.B. Gratzek, *Vet. Clin. North Am. (Small Animal Practice)* **18**, 375–399 (1988).
22. K. Kawakami and R. Kusada, *Nippon Suisan Gakkaishi* **56**, 51–53 (1990).
23. D.F. Amend, *Technical Paper No. 62*, United States Department of Internal Bureau of Sport Fish and Wildlife, 1972, p. 13.
24. Anonymous, Furanace: a new chemotherapeutic agent for fish disease. Dainippon Pharmaceutical Company Ltd., Osaka, Japan, Undated, p. 57.
25. G.N. Frerichs and R.J. Roberts, in R.J. Roberts, ed., *Fish Pathology*, 2nd ed., Bailliere Tindall, London, 1989, pp. 289–319.
26. V. Inglis, R.J. Roberts, and N.R. Bromage, eds., *Bacterial Diseases of Fish*, Halsted Press, New York, 1993.
27. C.S. Lewin, in C. Michel and D.J. Alderman, eds., *Chemotherapy in Aquaculture: From Theory to Reality*, O.I.E., Paris, 1992, pp. 288–301.
28. A.D. Russel and I. Chopra, *Understanding Antibacterial Action and Resistance*, Ellis Horwood, New York, 1990.
29. J.H. Brown, *World Aquacult.* **20**, 34–43 (1989).
30. S.C. Wood, R.N. McCashion, and W.H. Lynch, *Antimicrobial Agents and Chemotherapy* **29**, 992–996 (1986).
31. T.S. Hastings and A. McKay, *Aquaculture* **61**, 165–171 (1987).
32. A. Tsoumas, D.J. Alderman, and C.D. Rodges, *J. Fish Dis.* **12**, 493–507 (1989).

33. T. Aoki, T. Kanazawa, and T. Kitano, *Fish Path* **20**, 199–208 (1985).
34. W. Meier, T. Schmitt, and T. Wahli, in C. Michel and D.J. Alderman, *Chemotherapy in Aquaculture: From Theory to Reality*, eds., O.I.E., Paris, 1992, pp. 263–275.
35. B.A. Dixon, *J. World Aquacult. Soc.* **25**, 60–63 (1994).
36. B.A. Dixon and G. Issvoran, *J. World Aquaculture Soc.* **24**, 102–104 (1993).
37. B.A. Dixon, J. Yamashita, and F. Evelyn, *J. Aquatic Animal Health* **2**, 295–297 (1990).
38. P.R. Bowser, J.H. Schachte Jr., G.A. Wooster, and J.G. Babish, *J. Aquatic Animal Health* **2**, 198–203 (1990).
39. S.A. De Grandis and R.M. Stevenson, *Antimicrobial Agents and Chemotherapy* **27**, 938–942 (1985).
40. R.W. Hedges, P. Smith, and G. Brazil, *J. Gen. Micro.* **131**, 2091–2095 (1985).
41. A. DePaola, P.A. Flynn, R.M. McPhearson, and S.B. Levy, *Appl. Envir. Micro.* **54**, 1861–1863 (1988).
42. H. Sorum, T.T. Poppe, and O. Olsvik, *J. Clin. Micro.* **26**, 1679–1683 (1988).
43. S.G. Griffiths and W.H. Lynch, *Antimicrobial Agents and Chemotherapy* **33**, 19–26 (1989).
44. D.V. Straub, B. Contreras, and B.A. Dixon, *J. Aquariculture and Aquatic Sciences* **VI**, 76–77 (1993).
45. B.A. Dixon and B. Contreras, *J. Aquariculture and Aquatic Sciences* **VI**, 31 (1992).
46. O.B. Samuelson, B. Hjeltnes, and J. Glette, *J. Aquatic Animal Health* **10**, 56–61 (1998).
47. B.A. Dixon and G.S. Issvoran, *J. Wildlife Dis.* **28**, 453–456 (1992).
48. J.L. Grondel and W.B. van Muiswinkel, in Van Miert, ed., *Comparative Veterinary Pharmacology, Toxicology and Therapy*, 3rd EAVPT Congress, Ghent, Belgium, MPT Press, Ltd., Lancaster, England, 1986, pp. 263–282.
49. M.H.T. van der Heijden, W.B. van Muiswinkel, J.L. Grondel, and J.H. Boon, in C. Michel and D.J. Alderman, eds., *Chemotherapy in Aquaculture: From Theory to Reality*, O.I.E., Paris, 1992, pp. 219–230.
50. J.L. Grondel, A.M.G. Gloudemans, and W.B. van Muiswinkel, *Vet. Immunol. Immunopathol.* **9**, 251–260 (1985).

See also BACTERIAL DISEASE AGENTS; DISEASE TREATMENTS; DRUGS.

ANTINUTRITIONAL FACTORS

FAYE M. DONG
University of Washington
Seattle, Washington

RONALD W. HARDY
Hagerman Fish Culture Experiment Station
Hagerman, Idaho

DAVE A. HIGGS
West Vancouver Laboratory
West Vancouver, Canada

OUTLINE

Plants have developed a number of survival mechanisms to reduce the chances of their seeds being eaten or digested by insects, birds, or other animals. One defense mechanism is the production of compounds called antinutrients or antinutritional factors (ANFs), which are toxic to the animals or inhibit digestion of the seeds. Other compounds, for which the primary purpose in the seeds is not protection, can lower the bioavailability of nutrients to animals or otherwise affect their health. Research with mammalian models in the past 10–15 years has revealed some beneficial effects of certain ANFs, such as the reduction of blood lipids and the reduction in the incidence of cancer (1). Health benefits to fish have not been reported. Therefore, antinutritional factors are an important consideration in animal and fish nutrition because many feed ingredients used in animal and fish feeds are produced from grains, legumes, and oilseeds.

Two main approaches to eliminating ANFs in feed ingredients are the destruction/removal of ANFs during seed processing and changing specific genetic characteristics of the seeds. Some antinutritional factors can be destroyed or removed through certain processing conditions (e.g., high-temperature extrusion processing of soybean or canola meal and solvent extraction of whole oilseeds or their protein-rich by-products following oil extraction). The ANF concentrations in other products have higher lowered to acceptable levels by selective breeding of some oilseeds. Selective breeding has led to the development of new plant cultivars that have higher nutritive value for monogastric finfish species and other animals than the meals derived from their progenitor seeds. More recently, molecular-biology procedures have been employed to develop transgenic seeds with decreased levels of deleterious compounds. Nevertheless, antinutritional factors in feed ingredients of plant origin remain a source of concern, especially in fish feeds, where much less is known about the sensitivity of the fish to these compounds.

GRAINS

Grains are the foundation of most animal and some fish feeds. Ground whole-cereal grains (e.g., corn, wheat, rice, barley, and sorghum) constitute at least 50% of feeds for channel catfish (*Ictalurus punctatus*), tilapia, carp, and many other omnivorous fishes. Feeds for carnivorous fish, such as salmon, trout, and sea bass, contain ground whole grains at 10–20% of feed formulations and often include products made from grains, such as gluten meals (typically wheat and corn) or milling by-products. The antinutritional factors present in grains are protease inhibitor, hemagglutinins, cyanogen, phytic acid, tannins, estrogenic factors, antivitamin B-1, amylase inhibitor, invertase inhibitor, and dihydroxyphenylalanine (Tables 1 and 2) (2). Of these, phytic acid and hemagglutinin are of principle concern to fish feed producers.

Phytic acid, or phytate, is the hexaphosphate of myoinositol and is the storage form of phosphorus in seeds. In this form, phosphorus is unavailable to monogastric animals, including fish, because these animals lack intestinal phytase. Research with fish has demonstrated that phytic acid can lower the bioavailability of zinc and some other divalent ions, making it necessary to fortify fish diets with supplemental zinc when feeds contain phytic acid. Gatlin and Wilson (3) found that in order to ensure good growth and feed efficiency, the zinc concentration in catfish feeds must be increased five times when phytate-containing ingredients are used. Further, Richardson et al. (4) showed that the addition of phytate to chinook salmon feeds, which contained high levels of calcium and phosphorus as well as an adequate dietary zinc level (in the absence of phytate) induced zinc deficiency. The zinc deficiency led to bilateral-lens cataracts, anomalies in the structure of the pyloric-caecal region of the intestine, reduced growth, lowered feed efficiency, and reduced thyroid function. Spinelli et al. (5) reported that adding phytates to rainbow-trout feeds reduced protein bioavailability, presumably through the formation of protein-phytic-acid complexes. However, depressed protein bioavailability has not been observed in trout and salmon which were fed diets containing ingredients inherently rich in phytic acid due to processing (6). While phytate can also chelate other divalent cations (e.g., manganese, copper, magnesium, and iron), there is no evidence that nutritional deficiencies of these elements can occur in fish. Certainly, reduction of the phytate levels in fish feed ingredients is desirable for the reasons mentioned previously and to decrease excretion of unavailable phytate phosphorus into the environment. Phytate phosphorus excretion is of particular concern in freshwater systems where the excess phosphorus can stimulate the growth of phytoplankton and algae.

Table 1. Key to Antinutritional Factors in Feed Ingredients[a]

Number	Antinutrient	Heat Sensitive
1	Trypsin inhibitors	Yes
2	Hemagglutinin	Yes
3	Glucosinolates	No
4	Cyanogen	No
5	Phytic acid	Yes/No
6	Saponins	No
7	Tannins	No
8	Phytoestrogens	No
9	Gossypol	No
10	Antivitamins	Yes
11	Amylase inhibitor	Yes
12	Invertase inhibitor	Yes
13	Arginase inhibitor	Yes
14	Cholinesterase inhibitor	?
15	Dihydroxyphenylalanine	No
16	Mimosine	No
17	Cyclopropenoic acids	No

[a]Adapted from Jauncey (2).

Table 2. Antinutritional Factors in Grains[a]

Grain	Compounds[b]
Corn	1, 5, 8, 12
Wheat	1, 2, 5, 8, 11, 15
Rice	1, 2, 4, 5, 7, 11
Barley	1, 2, 5, 8,
Sorghum	1, 2, 4, 5, 7, 11

[a]Adapted from Jauncey (2).
[b]See Table 1 for the key to antinutritional compounds.

Recently, plant geneticists have developed varieties of corn and barley that are lower in phytate content than common varieties. These newer varieties contain the typical amount of phosphorus, but much less of the phosphorus is bound to phytic acid. The bioavailability of phosphorus to rainbow trout from low-phytate corn and low-phytate barley was significantly higher than in common corn and barley (7). Similar results will likely be found in other fish species. High-temperature extrusion also appears to be a practical way to reduce the levels of phytate in some plant-protein products before their dietary incorporation, while having beneficial effects on fish performance (8).

Hemagglutinins, also known as lectins, are proteins which cause agglutination of erythrocytes (red blood cells) *in vitro*. Hemagglutinins are destroyed by the acidic conditions in the stomach, suggesting that species of fish possessing acid stomachs will not be affected by them. For carp and other agastric fish species, hemagglutinins could be a problem; however, heat treatment inactivates them. The heat involved in pelletizing fish feeds likely eliminates these compounds as practical problems in fish farming.

Estrogenic factors in plants, also called phytoestrogens, are compounds having weak estrogenic activity which are found in several grains (e.g., barley, oats, rice, and wheat). The compounds having estrogenic activity in grains are isoflavones, coumestans, and resorcyclic acid lactones (9). The amounts of these compounds present in grains are low, and it is unlikely that they have any significant effect in fish at the levels present in fish feeds. However, studies to confirm this have not been done.

Amylase is an intestinal enzyme used to digest carbohydrates. Amylase inhibitors are found in wheat and beans. They have been reported to be heat sensitive [e.g., 95–100 °C (203–212 °F) for 15 min] and digested by pepsin (10). Carnivorous fish (e.g., salmon and trout) have low levels of amylase in their intestine, but herbivorous fish (e.g., carp) have much higher levels, presumably

Table 3. Antinutritional Factors in Oilseeds[a]

Oilseed	Compounds[b]
Soybean	1, 2, 3, 5, 6, 8, 10
Cottonseed	5, 8, 9, 11, 17
Rapeseed/canola	3, 5, 7
Peanut	1, 2, 5, 6, 8
Sunflower	1, 7, 13,
Sesame	5

[a]Adapted from Jauncey (2).
[b]See Table 1 for the key to antinutritional compounds.

associated with their natural diet (9). In both trout and carp, wheat flour inhibits about 80% of the amylase activity of intestinal fluid. Ground wheat is less inhibitory than wheat flour. The use of wheat flour in fish feeds likely results in reduced amylase activity and thus reduced rates of carbohydrate digestion, especially in fish with little endogenous amylase and in agastric fish lacking pepsin.

OILSEEDS

A number of feed ingredients produced from oilseeds are important in fish feeds, including soybean meal, cottonseed meal, and canola (rapeseed) meal. To a lesser extent, other oilseed meals, including peanut meal, sunflower meal, and sesame seed meal, are included in fish feeds where availability and economics support their use. The antinutritional compounds found in oilseeds are trypsin inhibitors, glucosinolates, phytic acid, saponins, tannins, phytoestrogens, gossypol, antivitamin E, A, D, B-12, arginase inhibitor, and cyclopropenoic acids (Table 3).

SOYBEAN PRODUCTS

Soybean meal contains antinutritional compounds, which must be removed or inactivated by processing before the meal can be used successfully in animal or fish feeds. The principal compounds of concern in soybean meal are trypsin inhibitors, which reduce protein digestibility by binding with the digestive enzyme trypsin in the intestine of the animal. Trypsin inhibitors are sensitive to heat, and once oil is extracted from soybeans, ordinary processing lowers the level of trypsin inhibitors in the dried meal to levels that do not affect the growth of most domestic animals and some species of fish (e.g., catfish). Salmon and trout, however, are more sensitive to the trypsin inhibitor level. Thus, more extensive heat treatment is necessary to reduce residual trypsin inhibitor levels below 5 mg/g, the level above which protein digestibility and growth performance are affected (11). However, overheating the soybean meal may reduce protein quality by fostering reactions between amino acid residues and portions of the carbohydrate fraction in soybeans. Trypsin inhibitor levels were rapidly lowered in unheated soy flakes from 181 trypsin units inhibited (TUI)/mg sample to 1.8 TUI after 20 minutes of heat treatment (120 °C, 25 psi) (12). Protein solubility was reduced from 98 to 70% by this treatment, but further heating to 40 minutes or more reduced protein solubility to less than 33%, an indication of overheating (13). Protein digestibility, measured in vivo using rainbow trout (*Oncorhynchus mykiss*), was increased from 74 to 91% by 20 minutes of heat treatment. This difference was presumably the result of heat inactivation of trypsin inhibitors. Channel catfish and carp are reported to be less sensitive than salmonids to trypsin inhibitors in soybean meal (14–16).

Regular solvent-extracted soybean meal, the most commonly used soybean product in feeds, is heat-treated to some extent during its manufacture, resulting in values of about 3.0 to 3.5 mg trypsin inhibited/g sample (17). Further heating occurs during feed pelleting, especially during cooking-extrusion pelleting (see the entry "Feed manufacturing technology"). This process presumably lowers trypsin inhibitor activity further. Full-fat soybeans (toasted whole soybeans) containing 46.5 mg TUI had TUI values of 7.6 and 8.5 after being extruded (18), illustrating the effects of cooking-extrusion on trypsin inhibitor levels.

Soy protein concentrates have low levels of trypsin inhibitor (12,19), but contain levels of phytic acid that are at least as high as those in soybean meal (12). The enzyme phytase releases phosphorus from phytic acid, and the addition of phytase to catfish feeds significantly improves the phosphorus availability in soybean-meal-based feeds.

Chemical Tests for Detecting Underheated-Soybean Meal

The various chemical tests used to determine the adequacy of heat treatment of soybean meal can be divided into two groups: those that detect underheated soybean meal and those that detect overheated meal (20). Chemical tests used to detect underheated soybean meal are tests of urease activity, trypsin activity, and protein solubility. Urease is an enzyme naturally-present in soybeans that does not have any substantial nutritional relevance, except that it is heat-sensitive and its activity correlates with residual trypsin activity in dried soybean meal. It is also relatively easy to measure (21). Urease activity in commercial soybean meal ranges from a 0.02 to 0.1 increase in pH (20). Values over a 0.5 increase in pH indicate insufficient heat treatment of the soybean meal. If no increase in pH is detected with the urease test, the soybean meal may have been overheated. Thus, some residual urease activity in the meal is preferred, especially for soybean meal intended for use in poultry feeds. Unheated soybean meal has a >2.25 pH rise in urease activity (22).

Another method for measuring the extent of heat treatment of soybean meal is the water solubility test. This test involves measuring Kjeldahl nitrogen levels in the soybean meal and in a water extract of the soybean meal (20). The method has been slightly modified by extracting the sample in 0.2% KOH (13). Heating decreases the percentage of 0.2% KOH-extractable proteins from about 99% in raw soybean meal to about 72% after 20 minutes of autoclaving. The corresponding decrease in trypsin inhibitor units is from 21.1 to 1.0 (13).

Soy products contain compounds that influence feed intake, gut histology, and immunological function (11). Complete replacement of fish meal with soybean meal in trout feeds lowered growth, primarily by lowering feed intake, but partial replacement of fish meal with soybean meal (e.g., 29% soybean meal and 42% fish meal in the diet) had no effect on trout feed intake or growth (11). Tolerance of trout to dietary soybean meal appeared to be higher in larger fish and at higher water temperatures. Intestinal mucosa of trout, which were fed soybean-meal-containing diets, were blunted or flattened, thus decreasing the absorptive surface of the proximal and distal intestine. However, it is not known if these intestinal changes are responsible for differences in growth associated with feeding diets containing high levels of soybean meal (11). Antigens present in soybean products stimulate the nonspecific defense mechanisms of trout, but it is unknown if such stimulation of the immune system results in higher resistance to infectious disease (11).

Other Oilseed Products

The other major antinutritional factors associated with other oilseed meals are as follows: glucosinolates and erucic acid in canola/rapeseed meals and gossypol and cyclopropenoic acid in cottonseed meal (Table 3). Glucosinolates interfere with the function of the thyroid gland in fish, posing problems during metamorphosis, smoltification, and maturation (23). Glucosinolates alone are not harmful compounds, but when they are hydrolyzed by the enzyme myrosinase in poorly-processed meals or by intestinal microorganisms, an array of products can result such as thiocyanate ions, isothiocyanates, nitriles and goitrin, depending upon the hydrolysis conditions. Isothiocyanates and nitriles are precursors to thiocyanate, which inhibits uptake of iodide, by the thyroid gland. Extra supplementation of the diet with iodine can overcome this. Goitrin actually inhibits the ability of the thyroid to bind iodide, and providing supplemental iodine cannot overcome this problem. Typically, the major consequences of thyroid function impairment include thyroid hypertrophy and hyperplasia and depressed thyroid hormone synthesis and plasma thyroid hormone titres. These effects lead to reductions in growth, feed intake, and feed utilization. Fortunately, varieties of rapeseed (called canola), which contain very low levels of glucosinolates, have been developed through genetical selection, and residual levels of the glucosinolates can be further decreased by solvent extraction and other means. Hence, the glucosinolates present in some canola protein products (e.g., concentrates and isolates) are very low in concentration and do not pose a problem.

Erucic acid is a 22-carbon monounsaturated fatty acid, and may constitute 20–55% of rapeseed oil (23). This compound causes lipid accumulation and necrosis of the heart tissue (24) and is toxic to coho salmon when fed at 3–6% of the diet (9). However, no erucic acid problems have been reported when rapeseed or canola meals are included in fish feeds, presumably because nearly all of the oil has been removed from these meals. The varieties of canola that have been developed contain lower levels of glucosinolates and erucic acid than in earlier rapeseed varieties. The use of rapeseed/canola-protein products in fish feeds has recently been thoroughly reviewed (6).

The presence of fiber, both independently and perhaps in combination with phytate, in oilseed meals such as commercial canola meal, appears to have the greatest adverse effects on feed digestibility in salmonids like the rainbow trout (25). Indeed, the insoluble fibers in the meal may depress the gut transit time and the absorption of amino acids and peptides, whereas the soluble fibers could inhibit digestive enzymes and restrict the diffusion of hydrolysis products (26).

Gossypol causes a number of problems in fish, including anorexia and increased lipid deposition in the liver. Gossypol is contained in the pigment glands of cotton, therefore, glandless cotton is free of gossypol. Fish species differ in their sensitivity to gossypol, with trout being sensitive, channel catfish more sensitive, and blue tilapia the most sensitive of the species for which information is available (23). Growth depression occurred in trout, which were fed more than 290 mg gossypol/kg diet, with more than 900 mg/kg for channel catfish and 1800 mg/kg for tilapia needed to lower growth. Solvent-extracted cottonseed meal samples from the United States contained between 400 and 800 mg gossypol/kg (27). Thus, catfish feeds containing no more than 20% cottonseed meal would not deliver enough gossypol to affect the fish. The use of cottonseed meal in fish feeds has recently been thoroughly reviewed (27). Cottonseed meal also contains cyclopropenoic fatty acids (sterculic and malvalic acids). These fatty acids are quite toxic in their own right and powerful carcinogens when fed to rainbow trout or salmon in combination with aflatoxins, fairly common toxins produced by *Aspergillis flavis*, a common mold found on grains (9). Cottonseed meal contains residual levels of cyclopropenoic fatty acids, even after oil extraction. Thus, it should not be used in feeds for trout or salmon.

LEGUMES

Legumes include peas, beans, alfalfa and ipil-ipil. Many contain trypsin inhibitors, hemagglutinin, cyanogens, phytic acid, saponin, antivitamin factors, and mimosins (Table 4). Most of these antinutritional factors have been discussed previously, with the exception of saponin and mimosin. Saponin is a nonprotein constituent of soybeans, which makes up about 0.5% of the weight of the soybean and is extractable from soybean globulins with alcohol. Isolated saponins do not harm chicks, rats, or mice when fed up to three times the amount found in soybean meal, but a crude saponin extract of soybean meal is reported to lower the feed intake of chinook salmon fingerlings and to

Table 4. Antinutritional Factors in Legumes[a]

Legume	Compounds[b]
Faba bean	1, 2, 5, 7
Chick pea	1, 4, 5, 8,
Lentil	1, 2, 6
Lupin	1
Ipil-ipil	16
Alfalfa	1, 6, 8, 12

[a] Adapted from Jauncey (2).
[b] See Table 1 for the key to antinutritional compounds.

Table 5. Antinutritional Factors in Tubers[a]

Tuber	Compounds[b]
Potato	1, 2, 4, 8, 11, 12, 14
Sweet potato	1, 12
Cassava	1, 4

[a]Adapted from Jauncey (2).
[b]See Table 1 for the key to antinutritional compounds.

reduce the growth of rainbow trout (28). Mimosin is found in ipil-ipil, and its effects on fish are unknown.

TUBERS

Tubers are potatoes, sweet potatoes, and cassava. The most important antinutritional factors in tubers are trypsin inhibitors, although potatoes also contain hemagglutinin, cyanogen, phytoestrogens, amylase inhibitor, invertase inhibitor, and cholinesterase inhibitor (Table 5). Cooking tubers before including them in feeds reduces the levels of these antinutritional factors to insignificant levels, especially considering the small amount of tuber products used in fish feeds.

OTHER ANTINUTRITIONAL FACTORS FOUND IN FEED INGREDIENTS

Thiaminase

Many species of fish, mainly freshwater species, contain thiaminase, an enzyme that hydrolyzes thiamin in feed preparations (23). Thiaminase is destroyed by heat treatment and also by the acidification used to produce fish silage. Thiaminase is a relatively slow-acting enzyme, therefore, feeding raw fish combined with dry mash within hours of preparation is not a high risk. An alternative strategy is to feed raw fish and dry mash containing thiamin separately.

Histamine and Gizzerosine

Histamine, 4-(2-aminoethyl)imidazol, is a primary amine arising from the decarboxylation of the amino acid, L-histidine. The toxic effects of histamine in the food supply of humans has been positively associated with scombroid-fish poisoning (29), an allergic reaction resulting from the ingestion of spoiled fish, usually of the families Scombridae and Scomberesocidae. The U.S. Food and Drug Administration has established a hazard-action level of 50 mg histamine/100 g canned tuna (30), and for fresh and frozen fish, the level is 20 mg histamine/100 g fish (31).

Histamine production in fish muscle can be controlled by low storage temperatures, which often help to reduce enzymatic and most microbial activities. However, if onboard refrigeration is inadequate, many of the fish captured could undergo significant deterioration prior to processing into fish meal. During the processing of fish meal (capture, transport, and storage), microbial and enzymatic degradation can take place rapidly at warm temperatures to produce high concentrations of histamine. If a significant portion of the catch is delivered to processing plants in a partially-decomposed state, then further degradation can occur after the fish have been unloaded, especially if the raw material is stored in nonrefrigerated concrete pits prior to processing (32).

Gizzard erosion and the resultant black vomit disease of chickens has long been associated with feeding thermally-abused fish meal, but it was not until 1983 that an active substance from fish meal was isolated (33). The active compound named gizzerosine was presumably formed through the condensation of histamine with the epsilon amino group of lysine during high-temperature processing of raw fish into fish meal. Mori et al. (34,35) synthesized several forms of the compound and established the L-form as a potent inducer of gizzard erosion in chickens. In one study, Mori et al. (34) observed severe gizzard erosion in chickens fed less than 50 μg gizzerosine/day (about 2.2 mg/kg diet) for one week. Considerable experimental evidence indicates that gizzerosine induces gizzard erosion in chickens by hyperstimulating the gastric-acid-secreting cells of the proventriculus.

Putrescine and cadaverine, as well as tyramine, β-phenylethylamine, and tryptamine, have been shown to potentiate histamine toxicity in vivo by inhibition of the histamine-metabolizing enzymes, diamine oxidase and histamine-N-methyltransferase (36). These data strongly suggest that fish meal toxicity to chickens is most likely related to the complex interactions among many chemical compounds found in degraded fish products.

In a study by Fairgrieve et al. (37), growth, feed intake, and the development of gastric abnormalities were assessed in juvenile rainbow trout. The fish were fed diets containing fish meal, which was acutely toxic to chickens, or they were fed casein or fish meal diets supplemented with histamine and two suspected potentiators of histamine toxicity (putrescine and cadaverine) and abusively heated. Rainbow trout were less sensitive than chickens to gastric erosion (GE)-positive fish meal, and there was no correlation between the GE score and the nutritional value of the fish meal for rainbow trout. Fish, which were fed diets containing GE-positive fish meal, had distended stomachs but no gastric lesions or cellular abnormalities. Similar effects were obtained by feeding diets containing casein or GE-negative fish meal supplemented with histamine (2,000 mg/kg dry diet). The addition of putrescine and cadaverine (500 mg/kg dry diet each) to the histamine-supplemented diets had no further effect. Feed consumption, feed efficiency, and growth were similar among dietary treatments, indicating that stomach distension did not reduce feed intake or impair gastric function. This study also showed that stomach distension resulting from feeding diets containing GE-positive fish meal could be duplicated by feeding diets supplemented with 2,000 mg histamine/kg diet.

Phytotoxins (Toxins of Algal Origin)

Many species of marine algae are capable of producing toxins, including paralytic shellfish toxin (PSP), diarrheic shellfish toxin (DSP) and domoic acid, which causes amnesic shellfish poisoning (ASP) (38). Shellfish, specifically bivalve mollusks, concentrate these toxins and are the principal vector through which the toxins are transferred to humans and possibly to farmed fish. Very little is known about the direct effects of these toxins on farmed fish or on the accumulation and/or biotransformations that may occur if farmed fish are fed diets containing contaminated shellfish or other marine product containing toxins. However, the potential human health risk is serious (38).

Sardines, and presumably other fish utilizing algae as food, consume algae containing domoic acid without apparent signs of toxicity (39). If these fish are harvested and used to produce fish meal, domoic acid, which is heat stable, can be concentrated in the fish meal by a factor of at least three, reaching 130 μg domoic acid per g fish meal (40). Rainbow trout, which were fed diets containing contaminated fish meal, did not exhibit any signs of toxicity or growth retardation, even though they consumed 50 μg domoic acid per kg body weight, much more than is required to cause illness in humans. Domoic acid was present in the GI tract of the trout, but not in the edible tissues (40).

MICROBIAL TOXINS

Molds that grow on feed ingredients and on prepared feeds are an important group of toxins affecting fish (9). In particular, aflatoxins, which are produced by the mold *Aspergillis flavis*, cause serious health problems in fish at much lower intake levels than in terrestrial animals. Cottonseed meal, peanut meal, and corn products are the most problematic feed ingredients with respect to aflatoxins, and grains including wheat, rice, barley, and oats are the next most problematic feed ingredients (9). Prolonged intake of very low levels of aflatoxin (<1 ppb) causes liver cancer after one year in rainbow trout, the most sensitive vertebrate to aflatoxin intake. Acute toxicity in rainbow trout is observed when fish are fed diets containing between 0.8 μg and 1.9 μg aflatoxin per g feed, depending upon the type of aflatoxin. Differences in sensitivity exist among strains of rainbow trout, among trout species, among other salmonids, and among other species of fish. Coho salmon and channel catfish, for example, are much more resistant to aflatoxin exposure than are rainbow trout (9). Other mold toxins of concern in fish feeding include ochratoxin, sometimes found as a contaminant of corn and wheat, vomitoxin, found in cereal grains, and T-2 toxin, also found in cereal grains. Of these, only vomitoxin has been evaluated in fish, with dietary levels of 20 μg toxin/g feed or higher causing feed refusal in trout (41). Obviously, fish should never be fed moldy feed.

SUMMARY

The biological significance of antinutritional factors in feed ingredients varies among factors and within factors among fish species. Research to document the relative importance of antinutritional factors in fish is extensive for some (e.g., glucosinolates), less extensive for others (e.g., phytic acid, trypsin inhibitors, gossypol), and nearly absent for others. If compounds that lower feed palatability are included in the antinutritional factor category, the practical significance of antinutritional factors increases. As use of feed ingredients from grains and oilseeds in fish feeds increases, the importance of understanding the biological effects of antinutritional factors on farmed fish and of developing methods for inactivating the antinutritional factors or overcoming their effects will become a critical element in the expansion of aquaculture production.

BIBLIOGRAPHY

1. F. Shahidi, *Antinutrients and Phytochemicals in Food*, American Chemical Society, Washington, DC, 1997, pp. 1–9.
2. K. Jauncey, *Tilapia Feeds and Feeding*, Pices Press, Ltd., Stirling, Scotland, 1998.
3. D.M. Gatlin III and R.P. Wilson, *Aquaculture* **41**, 31–36 (1984).
4. N.L. Richardson, D.A. Higgs, R.M. Beames, and J.R. McBride, *J. Nutr.* **115**, 553–567 (1985).
5. J. Spinelli, C.R. Houle, and J.C. Wekell, *Aquaculture* **30**, 71–83 (1983).
6. D.A. Higgs, B.S. Dosanjh, A.F. Prendergast, R.M. Beames, R.W. Hardy, W. Riley, and G. Deacon, in C.E. Lim and D.J. Sessa, eds., *Nutrition and Utilization Technology in Aquaculture*, AOCS Press, Champaign, 1995, pp. 130–156.
7. S. Sugiura, V. Raboy, K.A. Young, F.M. Dong, and R.W. Hardy, *Aquaculture* **170**, 285–296 (1998).
8. S. Satoh, D.A. Higgs, B.S. Dosanjh, R.W. Hardy, J.G. Eales, and G. Deacon, *Aquaculture Nutrition* **4**, 115–122 (1998).
9. J. Hendricks and J.S. Bailey, in J.E. Halver, ed., *Fish Nutrition*, 2nd ed., Academic Press, New York, 1989, pp. 605–651.
10. J.R. Whittaker and R.E. Feeney, *Toxicants Occurring Naturally in Foods*, 2nd ed., National Academy of Sciences, Washington, DC, 1973, pp. 276–298.
11. G.L. Rumsey, in C.E. Lim and D.J. Sessa, eds., *Nutrition and Utilization Technology in Aquaculture*, AOCS Press, Champaign, 1995, pp. 166–188.
12. R.E. Arndt, R.W. Hardy, S.H. Sugiura, and F.M. Dong, *Aquaculture* **180**, 129–145 (1999).
13. M. Araba and N.M. Dale, *Poultry Sci.* **69**, 76–83 (1990).
14. E.M. Robinson, J.K. Miller, U.M. Vergara, and G.A. Ducharme, *Prog. Fish-Cult.* **47**, 107–109 (1985).
15. R.P. Wilson and W.E. Poe, *Aquaculture* **46**, 19–25 (1985).
16. K. Dabrowski and B. Kozak, *Aquaculture* **18**, 107–114 (1979).
17. A.G.J. Tacon, J.V. Haaster, P.B. Featherstone, K. Kerr, and A.J. Jackson, *Bull. Jpn. Soc. Sci. Fisheries* **49**, 1437–1443 (1983).
18. T.R. Wilson, Ph.D., Dissertation, University of Washington, Seattle, WA, 1992.
19. J. Olli, Å. Krogdahl, and T. Berg-Lea, *Proc. Third Int. Symp. on Feeding and Nutr. in Fish, Toba, Japan, Aug. 28–Sept. 1, 1989*, 1989, pp. 263–271.
20. P. Vohra and F.H. Kratzer, *Feedstuffs* **63**, 22–28 (1991).
21. Association of Official Analytical Chemists (AOAC), *Official Methods of Analysis of the Association of Official Analytical Chemists*, 15th ed., Association of Official Analytical Chemists, Inc., Arlington, VA, 1990.
22. P. Waldroup, B.E. Ramsey, H.M. Helwig, and N.K. Smith, *Poultry Sci.* **64**, 2314–2320 (1985).
23. National Research Council (NRC), *Nutrient Requirements of Fish*, National Academy Press, Washington, DC, 1993.

24. S.J. Slinger, *J. Am. Oil Chem. Soc.* **54**, 94A–99A (1977).
25. S.A. Mwachireya, R.M. Beames, D.A. Higgs, and B.S. Dosanjh, *Aquaculture Nutrition* **5**, 73–82 (1999).
26. S.A.K. Mwachireya, Master's of Science thesis, University of British Columbia, Vancouver, BC, 1995.
27. E.H. Robinson and M.H. Li, in C.E. Lim and D.J. Sessa, eds., *Nutrition and Utilization Technology in Aquaculture*, AOCS Press, Champaign, 1995, pp. 157–165.
28. D.P. Bureau, A.M. Harris, and C.Y. Cho, *Abstracts of the Proc. VI. Int. Symp. on Feeding and Nutrition in Fish*, College Station, TX, 1996.
29. J.D. Morrow, G.R. Margolies, J. Rowland, and L.J. Roberts, *New England J. Med.* **324**, 716–720 (1991).
30. U.S.F.D.A., *Defect action levels for histamine in tuna; availability of guide. Federal Register* **47**, 40478 (1982).
31. U.C.S.G.E.P., *University of California Sea Grant Extension Program Publication 90-12, Seafood Safety*, 1990, p. 20.
32. W.T. Fairgrieve, Ph.D., Dissertation, University of Washington, Seattle, WA, 1992.
33. T. Okazaki, T. Noguchi, K. Igarashi, Y. Sakagami, H. Seto, K. Mori, H. Naito, T. Masumura, and M. Sugahara, *Agric. Biol. Chem.* **47**, 2949–2952 (1983).
34. K. Mori, T. Okazaki, T. Noguchi, and H. Naito, *Agric. Biol. Chem.* **47**, 2131–2132 (1983).
35. K. Mori, T. Sugai, Y. Maeda, T. Okazaki, T. Noguchi, and H. Naito, *Tetrahedron* **41**, 5307–5311 (1985).
36. J.Y. Hui and S.L. Taylor, *Toxicol. Applied Pharmacol.* **81**, 241–249 (1985).
37. W.T. Fairgrieve, M.S. Myers, R.W. Hardy, and F.M. Dong, *Aquaculture* **127**, 219–232 (1994).
38. V.M. Bricelj and S.E. Shumway, *Rev. Fish. Sci.* **6**(4), 315–383 (1998).
39. T.M. Work, B. Barr, A.M. Beale, L. Fritz, M.A. Quilliam, and J.L.C. Wright, *J. Zool. Wildl. Med.* **24**, 54–62 (1993).
40. R.W. Hardy, T.M. Scott, C.L. Hatfield, H.J. Barnett, E.J. Gauglitz, Jr., J.C. Wekell, and M.W. Eklund, *Aquaculture* **131**, 253–260 (1995).
41. B. Woodward, L.G. Young, and A.K. Lun, *Aquaculture* **35**, 93–101 (1983).

See also ALGAE: TOXIC ALGAE AND ALGAL TOXINS; OFF-FLAVOR.

AQUATIC VEGETATION CONTROL

MICHAEL P. MASSER
Texas A&M University
College Station, Texas

OUTLINE

Aquatic vegetation control can be one of the most perplexing problems facing any pond manager. Aquatic vegetation, particularly rooted and floating vegetation, may limit production and complicate feeding, harvesting, and predator control (1). Many water quality problems such as low dissolved oxygen, high carbon dioxide, and toxic nitrogenous compounds are driven by aquatic vegetation (2). Water-level manipulation (through clogging drains and intakes) in ponds and raceways can also be complicated by aquatic vegetation. Most water quality concerns in heavily fed and fertilized aquaculture ponds are driven by planktonic algae. Certain algae produce off-flavor compounds that result in unmarketable products and disrupt cash flow, production sequencing, and marketing strategies.

CLASSIFICATION OF AQUATIC VEGETATION

Aquatic vegetation is generally classified as either (*1*) algae, (*2*) floating, (*3*) submerged, or (*4*) emergent. These classification groups are somewhat arbitrary, and a few species appear to overlap more than a single group.

Some aquatic plants now common to many countries are not native (nonindigenous) to that continent. These include water hyacinth, hydrilla, Eurasian watermilfoil, giant water fern, and water lettuce. Aquarium and water garden enthusiasts have spread many of these species because of their attractiveness or because of accidental inclusion in shipments of other ornamental plants. However, in aquaculture even native species can become nuisances, because their rapid growth is stimulated by the high nutrient content common to culture ponds (2). For purposes of this discussion, forms or species will be limited to those that are most commonly problematic in aquaculture. Identification is the first key step in developing a management strategy for aquatic vegetation. Many publications, web sites, and computer software packages are available to assist in identifying aquatic vegetation (3–6).

Algae

Algae are the most common type of aquatic vegetation in aquaculture ponds (7). There are thousands of species of algae. Various algal species inhabit all known ranges of temperature and salinity in aquatic environments. Algae vary greatly in size and body form, ranging from microscopic single-celled species to groups of cells called colonies, or chains of cells called filaments (Fig. 1). Based on size and shape, algae are usually subdivided

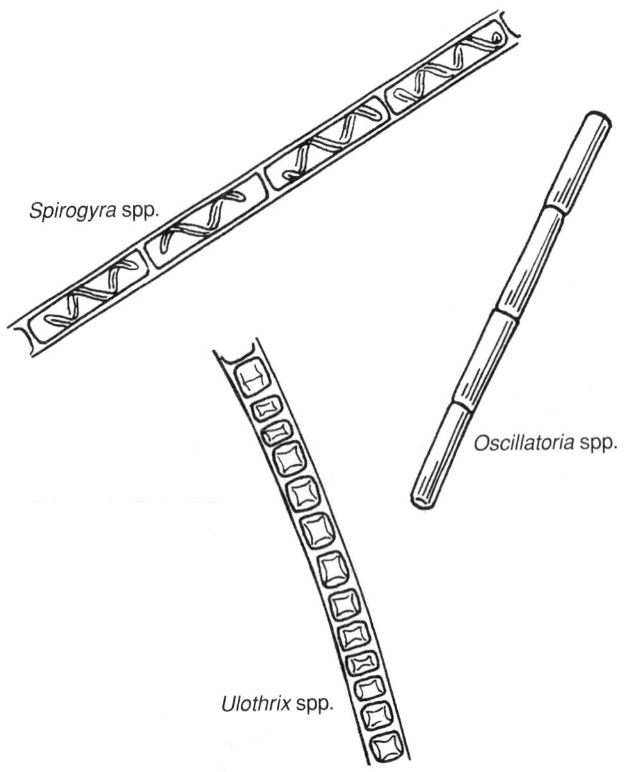

Figure 1. Structures of three types of algae common to aquaculture ponds. (Used by permission from IFAS, Center for Aquatic Plants, University of Florida, Gainesville, 1990.)

into three groups: (*1*) planktonic, (*2*) filamentous, and (*3*) macroalgae. The largest multicellular or macroalgae species are the marine kelps.

Algae do not flower or bear seeds, unlike most other aquatic plants. Algal reproduction, depending on the species, can be by asexual division, by resting cell or "cyst" formation, or by a simple form of sexual reproduction that results in spore formation. In some species spores and cysts can resist desiccation and can be borne by winds or birds to colonize other bodies of water (8). The production of new plants or colonies by fragmentation of multicellular forms is also common.

In pond aquaculture, some of the most problematic algae are single-celled and small multicellular species that proliferate in vast numbers [*Anabaena*, *Anacystis* (formerly *Microcystis*), *Cosmarium*, *Coelastrum*, etc.]. These dense algal or phytoplankton populations are referred to as "blooms." While the blooms produce most of the dissolved oxygen in ponds, they also can cause wide fluctuations in dissolved oxygen, carbon dioxide, nitrogen, and phosphorus levels in ponds, as their populations rapidly expand and die (2). Some species of algae can be toxic to other aquatic organisms and even man; these include *Anacystis*, *Anabaena*, *Gonyaulax*, *Gymnodinium*, and *Pfiesteria* (9).

Filamentous algae (*Spirogyra*, *Lyngbya*, *Pithophora*, *Oedogonium*, etc.) are often problems in pond, cage (or net pen), and raceway culture. Filamentous algae usually start growing along the pond bottom or become attached to structures in the water (e.g., cage mesh, raceway walls, etc.) wherever sunlight penetrates. Filamentous algae often form thick mats along the pond bottom that, as they grow, trap gasses among the filaments. This trapped gas eventually causes the mats to peel off the bottom and float to the surface. The floating mats of filamentous algae are often referred to as "pond scum or pond moss." Filamentous mats can entrap and kill larval fish, trap feed and make it unavailable to culture organisms, reduce culture chamber volume, complicate seining (or make it impossible), and if they cover most of the surface they can cause oxygen depletions (10).

In freshwater, there a few macroalgaes that resemble rooted, submerged aquatic macrophytes. These include members of the stonewort family. Common among these are *Chara* (Fig. 2) and *Nitella* species. These macroalgae do not have true roots and are attached to the pond bottom with rhizoids. Kelp, planktonic algae, and floating filamentous algae get their nutrients directly from the water column.

Marine macroalgaes (i.e., kelps) are an integral part of certain marine ecosystems. Some varieties of kelp are cultured, or harvested from the wild, as food or food supplements for humans and animals. Marine kelps can foul net pens, causing water circulation problems and reducing functional culture volume (11).

In aquaculture, algae growth is sometimes promoted through fertilization to stimulate the natural food chain for the species being cultured. Elaborate steps are taken, and selected fertilizers applied, to produce algal blooms or mats

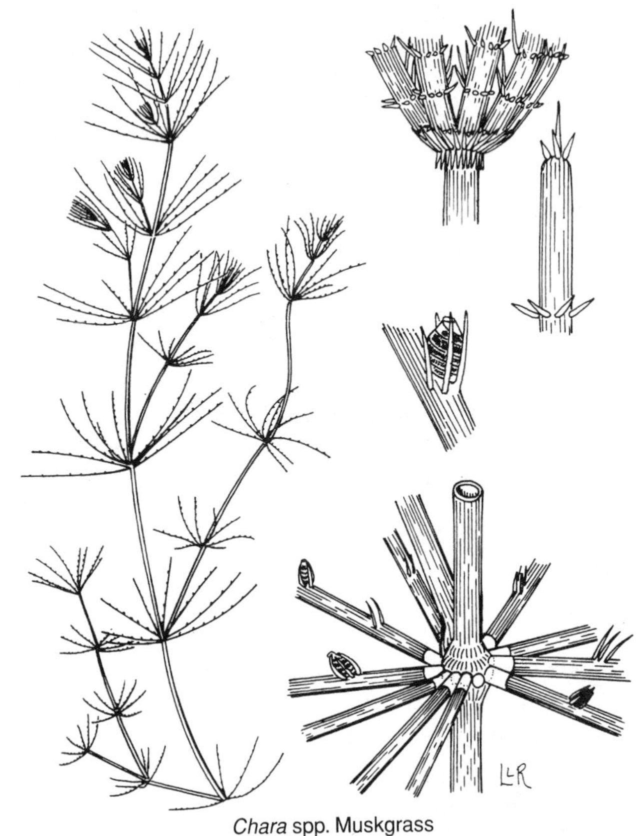

Chara spp. Muskgrass

Figure 2. *Chara* spp., muskgrass, is a member of the stonewort family of macroalgaes. (Used by permission from IFAS, Center for Aquatic Plants, University of Florida, Gainesville, 1990.)

for particular culture situations (12). Pond production of many species of fish and shellfish larvae and fry depend on fertilization to stimulate phytoplankton and zooplankton blooms for the cultured species to feed on (see the entry "Fertilization of fish ponds").

Floating Plants

Free-floating plants include such groups as the duckweeds (*Lemna* and *Spirodela*; Fig. 3), water hyacinth (*Eichhornia crassipes*; Fig. 4), water lettuce (*Pistia stratiotes*), some forms of bladderwort (*Utricularia* spp.), and water

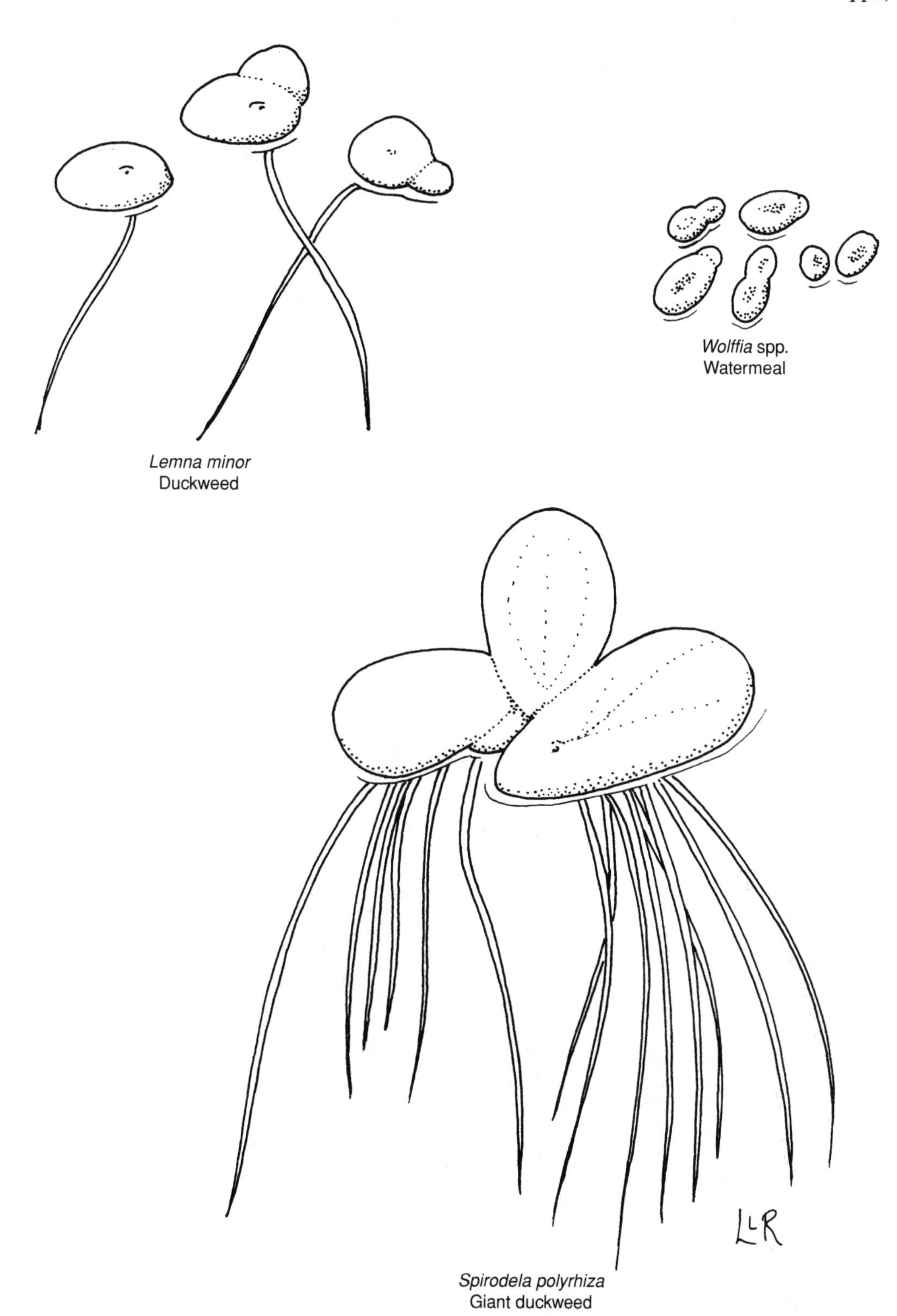

Figure 3. As free-floating plants, duckweeds and watermeal can become problematic when they cover the water surface. (Used by permission from IFAS, Center for Aquatic Plants, University of Florida, Gainesville, 1990.)

Eichhornia crassipes
Water hyacinth

Figure 4. The water hyacinth is considered one of the most noxious aquatic weeds in the world. (Used by permission from IFAS, Center for Aquatic Plants, University of Florida, Gainesville, 1990.)

ferns (*Azolla* and *Salvinia* spp.) (13,14). Water hyacinth, water lettuce, and giant salvinia are examples of aquatic plants that have been carried by man into nonindigenous environments. The water hyacinth is considered one of the most noxious aquatic weeds in the world. Floating plants like water hyacinth, water lettuce, and bladderwort are not usually a problem in aquaculture situations and can be removed by physical means (i.e., seining). The duckweeds (Family Lemnaceae), and particularly watermeal (*Wolffia columbiana*), are the smallest flowering plants (angiosperms) in the world. Duckweeds, watermeal, and water ferns can be problematic in aquaculture where they can rapidly cover the water surface, thus reducing oxygen production and transfer (15). Duckweeds are occasionally cultured to feed the young of species like tilapia and grass carp (16).

Submerged Plants

Submerged, or submersed, aquatic vegetation refers to those plants that grow underwater and, in some species, up to the water surface. Some submerged plants have small vegetative structures and/or flowers, as well as seed heads that extend above the surface; these plants can be confused with emergent plants. Submerged plants are usually flaccid and dependent on the water for support. They generally do not have rigid stems like emergent aquatic plants.

Hydrilla (*Hydrilla verticillata*), egeria (*Egeria densa*), and Eurasian watermilfoil (*Myriophyllum spicatum*; see Fig. 5) are examples of submerged aquatic plants that have been spread to nonindigenous environments, where they have caused severe problems in many parts of the world (Fig. 6) (17). These species typically spread rapidly by fragmentation, outcompete native species, and form dense stands that limit access and impact many water use activities.

In aquaculture, submerged aquatic vegetation can cause problems by reducing the effective culture volume of the production system, by entanglement of larvae, by trapping feed, by restricting the dispersal of feed, and

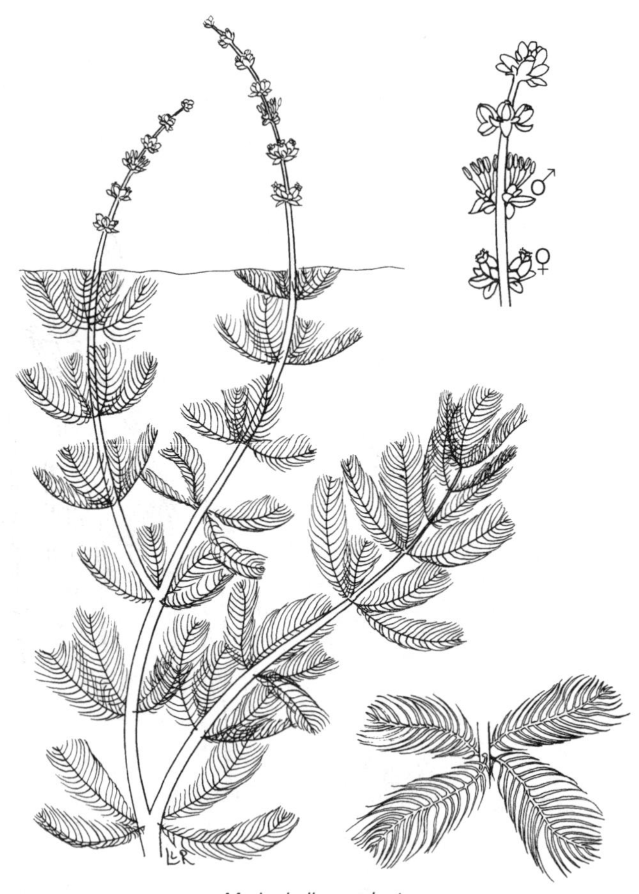

Myriophyllum spicatum
Eurasian watermilfoil

Figure 5. The Eurasian watermilfoil has spread to many nonindigenous environments. (Used by permission from IFAS, Center for Aquatic Plants, University of Florida, Gainesville, 1990.)

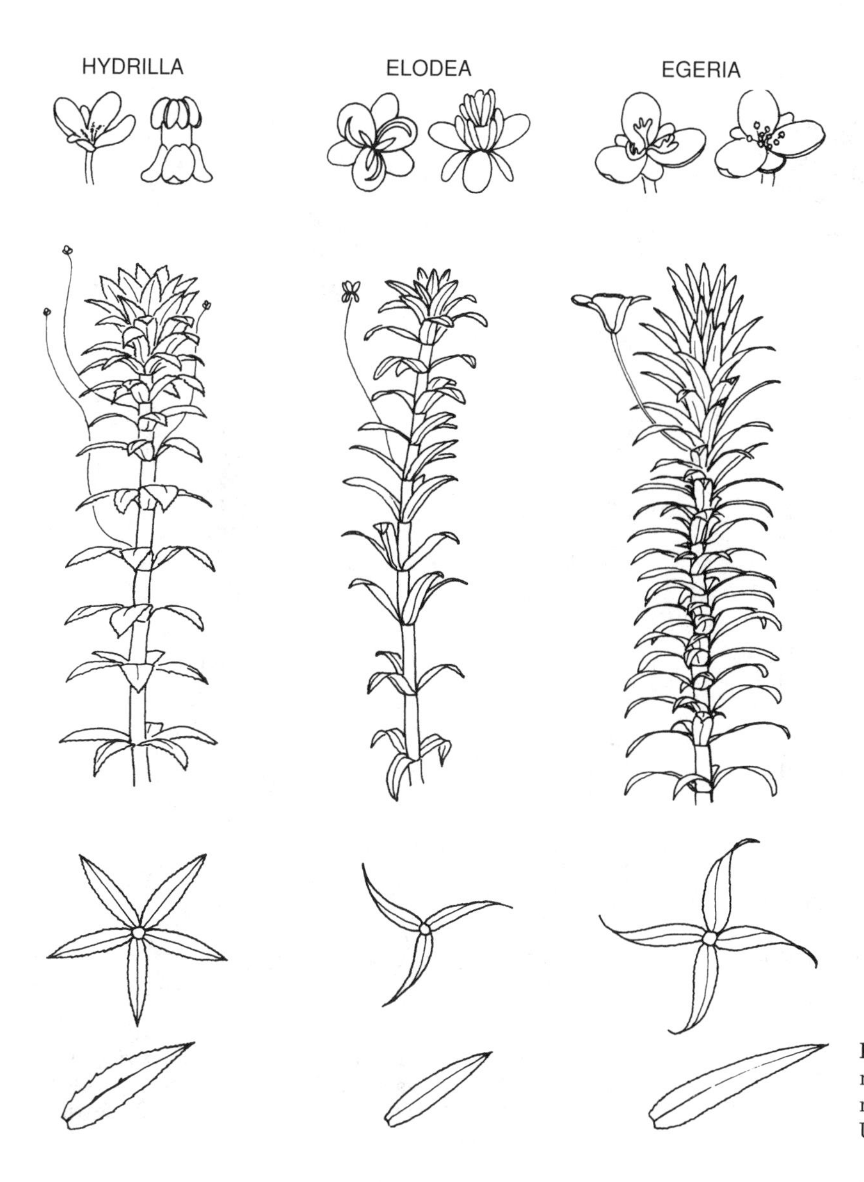

Figure 6. These species spread rapidly by fragmentation and form dense stands. (Used by permission from IFAS, Center for Aquatic Plants, University of Florida, Gainesville, 1990.)

by restricting or halting seining activities. Submerged aquatic vegetation can cause oxygen depletions due to nighttime respiration and die-offs. Submerged macrophytes that often cause problems in aquaculture include naiads (*Najas* spp.; Fig. 7), pond weeds (*Potamogeton* spp.; Fig. 8), coontail (*Ceratophyllum demersum*), and widgeongrass (*Ruppia maritima*). These macrophytes compete with planktonic algae for nutrients and light (10). Usually the most critical factor regulating submerged aquatic vegetation growth is turbidity.

Emergent Plants

Emergent, or emersed, plants are rooted aquatic plants that have stems or leaves that are rigid enough to rise above the surface of the water (13). Emergent aquatic plants are usually shallow water (<3 m) or shoreline inhabitants. Water lilies and lotus belong in this group, but are seldom a problem in aquaculture. Species that often cause problems in ponds include cattails (*Typha* spp.), rushes (*Juncus* spp.; Fig. 9), sedges (*Cyperus* spp.), spikerushes (*Eleocharis* spp.), water primrose (*Ludwigia* spp.; Fig. 10), and water pennywort (*Hydrocotyle ranunculoides*; Fig. 11).

Like most other aquatic vegetation, emergent plants can restrict access, reduce the effective culture volume, interfere with feeding and seining, and the quality the quality of the water.

MANAGEMENT METHODS

There is no single way to manage or control the wide range of aquatic plants. Effective management efforts begin with prevention. Once aquatic vegetation is established then management measures can include mechanical or physical, biological, and/or chemical control. Generally, an integrated pest management (IPM) approach is the best way to develop a sustainable, long-term aquatic strategy for management of vegetation. IPM involves combining or integrating several management methods to achieve effective, long-lasting control. Obviously, all methods of management must be compatible with the culture system and species in production.

Figure 7. Naiads and other macrophytes compete with planktonic algae for nutrients and light. (Used by permission from IFAS, Center for Aquatic Plants, University of Florida, Gainesville, 1990.)

Figure 8. Submerged aquatic plants such as pondweed can cause oxygen depletion due to nighttime respiration and die-offs. (Used by permission from IFAS, Center for Aquatic Plants, University of Florida, Gainesville, 1990.)

Figure 9. Emergent species such as soft rush can interfere with water access, feeding and seining. (Used by permission from IFAS, Center for Aquatic Plants, University of Florida, Gainesville, 1990.)

Prevention

In pond culture, prevention includes proper pond location, design, construction, and possibly, fertilization, stocking of aquatic herbivores, and drawdowns.

Ponds should not be located where they will get heavy runoff of organic matter. The enrichment of organic matter from excessive livestock, fertilized fields, septic tanks, or other sources tends to promote the growth of aquatic vegetation. Filling ponds from surface-water sources that have existing aquatic vegetation or receive heavy organic enrichment can also cause problems.

Aquatic vegetation can be discouraged if ponds are constructed with banks that slope quickly to a depth of 75 cm or more. This depth, coupled with a good fertilization program, can effectively shade most of the pond bottom and minimize the area where rooted aquatic vegetation can establish (10).

Shading by, and competition from, planktonic algae for nutrients is possibly the best overall management strategy for preventing rooted aquatic vegetation from becoming established. While planktonic algal blooms may not discourage all forms of emergent shoreline

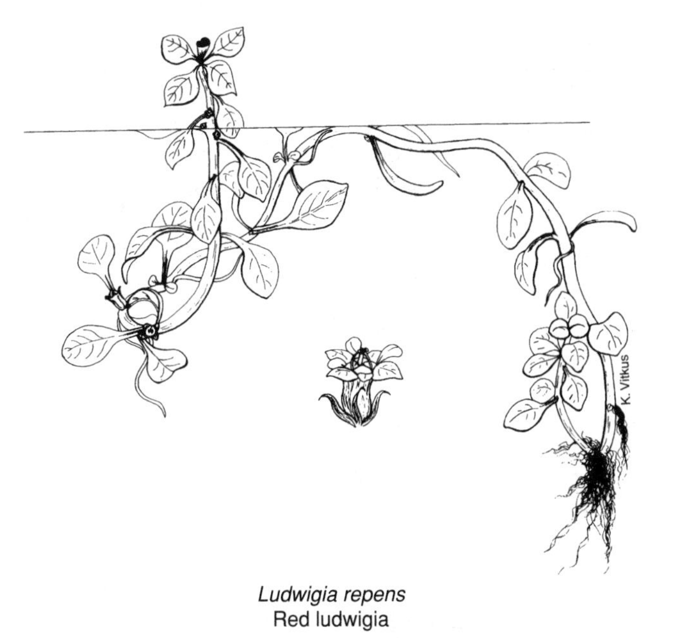

Figure 10. Water primrose and other emergent aquatic plants usually inhabit shallow water or shoreline. (Used by permission from IFAS, Center for Aquatic Plants, University of Florida, Gainesville, 1990.)

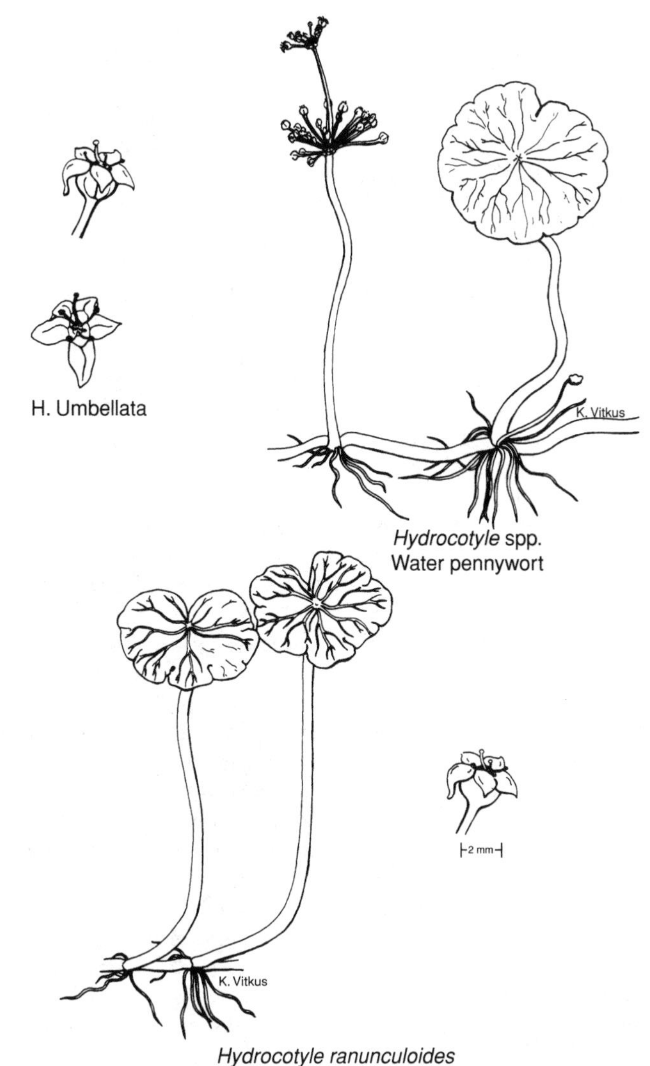

Figure 11. Water pennywort is another emergent species. (Used by permission from IFAS, Center for Aquatic Plants, University of Florida, Gainesville, 1990.)

vegetation, they will prevent submerged types and those emergent weeds that establish themselves in deeper water (19). Organic and inorganic fertilizers can be used in pond environments to produce plankton blooms (12). Fertilization techniques will vary depending on the species cultured, chemical composition of the fertilizer, and characteristics of the pond to be fertilized. However, the initiation of a fertilization program after noxious vegetation is established may increase the rapid growth and spread of the problem species. Nontoxic chemical dyes have also been developed that can be added to pond water to reduce the penetration of light and suppress the growth of aquatic plants.

Another common method to prevent some forms of aquatic vegetation from establishing themselves is the stocking of herbivorous fish. Grass carp (*Ctenopharyngodon idella*) are probably the most commonly stocked herbivorous fish, but tilapia (e.g., *Tilapia zillii*), common carp (*Cyprinus carpio*), and other less known species have been utilized in the control of aquatic vegetation. (See the section "Biological control.")

Drawdowns are another method to control or prevent certain aquatic vegetation problems. Winter drawdowns, which expose shallow pond bottom areas to freezing and desiccation can successfully prevent the establishment and spread of many species of submerged vegetation (20). Drawdowns are most effective if repeated for three to five consecutive years.

Mechanical or Physical Control

Mechanical or physical control methods include the simple removal of aquatic vegetation by hand (pulling, raking, and seining), the use of mechanical cutters or harvesters, and the use of physical barriers along the pond bottom (21). Pulling, raking, and seining usually work in aquaculture ponds when aquatic vegetation is confined to a small area. Mechanical cutters and harvesters are utilized in some public waters, but are seldom used in aquaculture because of the costs of these systems. Pond liners and bottom barriers made from plastic sheets or woven fiberglass cloth can be effective in eliminating rooted vegetation (22). Liners and barriers are seldom used in aquaculture because of the expense and incompatibility with some species or culture techniques.

Biological Control

Biological control is usually desirable in aquaculture because of the relatively low costs and the opportunity to control vegetation without the use of chemicals. Chemicals can in some cases stress the cultured organism, contaminate its flesh, cause a deterioration in the quality of water, or be directly toxic. While host-specific insects (e.g., hydrilla fly and alligatorweed flea beetle) have been used to control some types of aquatic vegetation in public waters, these methods are not generally utilized in aquaculture. Biological control utilizing a desirable food species is the most commonly practiced method of control of aquatic vegetation in aquaculture. Using a desirable food species also has the added benefit of producing another saleable crop. However, consumption of feed, competition, predation, or some other adverse effect of biocontrol agents must be considered.

Crayfish (*Procambarus*, *Orconectes*, and *Cherax* spp.) will control some types of submerged and emergent aquatic vegetation if stocked singularly in ponds or with species with which they are compatible (23). Stocking rates for crayfish are usually between 28 to 84 kg/ha (25 to 75 lb/ac).

Fish species generally chosen for biological control are the grass carp, tilapia, common carp, and a few other native species used in specific regions like India (e.g., rohu, *Labeo rohita*) or South America (e.g., pacu, *Piaractus* sp.) that are not generally available in other regions.

The grass carp is a native of eastern China and Siberia (24). It is a highly adaptable species that has been introduced into many countries for the control of aquatic vegetation. Grass carp tolerate a wide range of environmental extremes in temperature, pH, dissolved oxygen, and salinity (up to 7.5–8 ppt). They prefer soft submerged vegetation, including filamentous algae, but also control duckweeds (25). Grass carp will not reproduce in ponds (i.e., stagnant water), so they must be stocked at a density that will be sufficient to control the problem vegetation. Stocking densities of grass carp for the control of aquatic vegetation vary widely with the type of vegetation, water temperature, weight of fish stocked, and legal restrictions. Grass carp are generally stocked between 12 to 100/ha (5 to 40/ac) to control aquatic vegetation. In ponds containing predatory fish species, grass carp should be large enough at stocking to preclude predation.

Tilapia (*Sarotherodon*, *Oreochromis*, and *Tilapia* spp.) have been used successfully in the control of some types of aquatic vegetation. Tilapia are known to control filamentous algae, duckweeds, watermeal, and most submersed macrophytes if they are stocked or reproduce in sufficient numbers. Stocking rates are usually around 200 mixed-sex adults/ha (80/ac) (26). One drawback, or attribute of tilapias (depending on your long-term goals), is their inability to survive water temperatures below 55 °F (13 °C), necessitating annually restocking them in temperate climates. Tilapia culture ponds seldom have aquatic vegetation problems. Mixed-sex tilapia are not usually stocked in aquaculture situations with other species because of their tendency to overpopulate and compete with more desirable species. In a few cases tilapia have been stocked for vegetation control along with predaceous fish species that control the tilapia population.

Common carp, including Koi (*C. carpio*), will consume filamentous algae and some types of soft, submerged aquatic vegetation and duckweeds. Their benthic foraging habit also uproots vegetation and increases turbidity in ponds, effectively shading out submerged vegetation. Common carp are usually stocked at 125 to 250/ha (50 to 100/ac). Common carp, like tilapias, can overpopulate a pond quickly and are not generally used in aquaculture for control of aquatic vegetation. Carp production ponds seldom have submerged weed problems.

Ducks, geese, and swans will eat many types of aquatic vegetation and will help reduce or control some aquatic weeds (23). However, they are rarely desirable around aquaculture facilities at the population levels that would be required to control vegetation.

Many mammals, including cattle, water buffalo, goats, swine, moose, nutria, muskrats, capybara, and manatee eat aquatic plants. Cattle, water buffalo, swine, and goats can be grazed on emergent and marginal plants around ponds. Unlike cattle, water buffalo, and swine, goats are particularly effective in controlling many emergent and brush species without damaging dams and levees. The burrowing habits of nutria and muskrats are problematic around aquaculture facilities, and therefore, those mammals are usually discouraged from inhabiting areas near such facilities.

In many countries, the introduction of species like the grass carp, tilapia, and other nonindigenous fishes is regulated. Development of techniques to produce sterile individuals of these species (e.g., triploid grass carp) have led to the legalization of these sterile fishes in many countries (or states) where normal representatives of the species are illegal. Before stocking these species the aquaculturist should check local regulations.

Chemical Control

Most countries have an approval process for all pesticides, including aquatic herbicides. In some cases, registered herbicides differ from country to country. For purposes of this discussion, aquatic herbicides registered by the U.S. Environmental Protection Agency are presented. In general, herbicides should be considered a temporary solution. Many weed species will return, unless an integrated management approach is used. In most cases in aquaculture, herbicides are used as a temporary control measure while preventive, mechanical, or biological methods for long-term solutions are sought.

Herbicides are usually classified on the basis of their properties, including absorption, selectivity, and plant processes affected (27). Absorption properties are characterized as either contact or systemic herbicides. Contact herbicides (copper, diquat, and endothall) act quickly and generally kill all plants that they come in contact with. Contact herbicides are most effective on annual plants, but often multiple treatments are necessary to control perennial vegetation. Systemic herbicides (2,4-D, fluridone, and glyphosate) are absorbed and move within the plant to a site at which they go into action. Systemic herbicides tend to act more slowly, but are generally more effective against perennial and woody vegetation.

Selectivity refers to whether a herbicide affects certain plant species more than others. Physical factors, application methods, formulations, and morphological characteristics of the target plant can influence selectivity. Herbicides that are toxic to any plant are nonselective and often referred to as "broad spectrum." Broad spectrum herbicides include copper, diquat, endothall, and glyphosate. The truly selective herbicides are fluridone and 2,4-D.

Plant processes affected by herbicides include tissue development, respiration, photosynthesis, and enzyme activity. Herbicides that disrupt tissue development (2,4-D) are also called plant growth regulators and cause abnormal tissue development. Plant respiration is disrupted by endothall herbicides. Photosynthesis is disrupted by copper compounds, diquat, and fluridone. Glyphosate disrupts the enzymes involved in nitrogen metabolism.

Herbicide formulations can be liquid, powder, or granular. All herbicides carry warnings regarding drinking water, livestock watering, swimming, fish consumption, and irrigation (14). Local, state, or national laws may also require special certification or notification requirements when utilizing certain herbicides. Label directions and precautions must be followed completely.

Factors that can affect herbicidal control include temperature, light intensity, water quality variables (e.g., pH, alkalinity, hardness, and turbidity), and seasonal timing. Herbicide labels contain important information regarding interactions with water quality and treatment timing. Generally herbicides are most effective if applied while plants are actively growing, under sunny conditions, and when water temperatures are above 16 °C (60 °F). Some species are best controlled before they seed (e.g., alligatorweed) while others are best controlled late in the growing season when they are actively storing food reserves in their roots (e.g., cattails and willows).

Probably the most common problem in chemical control is failure to accurately calculate the area or volume to be treated (28,29). This results in either limited or no control due to inadequate or excessive concentration. Overdosing can cause rapid vegetation die-off, leading to oxygen depletion and stress or mortality of the cultured organisms. An overdose may also be directly toxic to the cultured species.

Most of the herbicides discussed here (except for possibly copper and some endothalls) are not themselves toxic to fish if used according to label directions. However, the danger of treating with any herbicide is that the decomposing vegetation will cause a reduction in dissolved oxygen. Bacterial and fungal respiration during decomposition can reduce dissolved oxygen to critical or lethal levels for cultured species (30). The risk of depletion of oxygen can be reduced by treating only a portion of the pond at a time (usually less than 25%) and then allowing the treated vegetation to decompose before continuing the treatment. Usually 10 to 14 days are sufficient to allow for decomposition. In many cases the lowest dissolved oxygen levels will occur 7 to 10 days after treatment. Providing mechanical aeration during and after herbicide treatments can be critical.

The first step in chemical control is the accurate identification of the aquatic weed or weeds to be controlled. Herbicides that work on one species often will not be effective on other similar species. The second most important step in chemical control of vegetation is to carefully follow all directions on the herbicide label. Table 1 gives specific recommendations on the effectiveness of selected aquatic herbicide formulations on specific types of aquatic vegetation.

Copper Compounds

Copper sulfate, or "blue stone," is a commonly available inorganic copper compound. Complexed copper compounds bind copper sulfate to carrier molecules that keep the active copper in solution longer. Chelated copper compounds contain copper, but are organic compounds. Copper compounds are contact herbicides that interfere with photosynthesis and are generally effective on algae and a few submerged weeds but have few water use restrictions.

Copper sulfate is very corrosive to metals (e.g., steel) and can irritate eyes and nasal passages. The effectiveness of copper sulfate depends on the alkalinity of the water. In waters of less than 50 ppm total alkalinity as $CaCO_3$, the amount that will kill algae may also kill fish. Some species, like rainbow trout (*Oncorhynchus mykiss*), are very sensitive to copper and are easily killed. At high alkalinity (>200 ppm) copper sulfate is often not effective because it precipitates too quickly.

Chelated copper compounds are somewhat less toxic to fish and less corrosive. Chelated copper compounds are available in liquid and granular formulations but are more expensive than copper sulfate. However, chelated copper compounds are more stable in water and generally are active longer in water than inorganic copper. Check the product label for water use restrictions and other precautions.

2,4-D Compounds

Many different herbicides contain 2,4-D, but only a few are approved and labeled for aquatic use. 2,4-D is a selective, systemic herbicide that is effective on some aquatic broadleaf vegetation. It is available in granular and liquid formulations. Granular formulations are relatively nontoxic to fish and will control many submerged and some emergent (e.g., watershield and waterlilies) types of vegetation. Liquid formulations are generally used to control emergent vegetation. The liquid ester formulations are often more toxic to fish than the other forms. Check the product label for water use restrictions and other precautions.

Diquat Compounds

Diquat compounds are contact herbicides that interfere with photosynthesis and are only available in liquid formulations. Diquat is effective in controlling most filamentous algae and many submerged plants if injected uniformly into the water column. It is also effective on duckweed and other floating plants if sprayed directly on the plant with the addition of a registered aquatic surfactant.

Diquat is rapidly tied up or neutralized by clay particles or organic matter in the water. Therefore, diquat is only used in clear and nonmuddy waters. Diquat is often mixed with chelated copper compounds or endothall to broaden the variety of aquatic vegetation controlled. Diquat can cause severe eye damage and must be used with caution.

Table 1. Treatment Response of Common Aquatic Plants to Registered Herbicides and Grass Carp[a]

	Aquatic Herbicide[b]						
Aquatic Group and Vegetation	Copper and Copper Complexes	2,4-D	Reward (diquat)	Aquathol Hydrothol (endothall)	Rodeo (glyphosate)	Sonar (floridone)	Grass Carp
Algae							
Planktonic	E	P	P	G[c]	P	P	
Filamentous	E	P	G	G[c]–P[d]	P	P	F
Chara/Nitella	E	P	P	G[c]–P[d]	P	P	G
Floating plants							
Duckweeds	P	F[f]	G	P	P	E	F
Salvinia	P	G	G		P	E	P
Water hyacinth	P	E	E		G	P	
Watermeal	P	F	F			G	P
Submerged plants							
Coontail	P	G	E	E	P	E	F–G
Elodea	P		E	F	P	E	E
Fanwort	P	F	G	E	P	E	F
Naiads	P	F	E	E	P	E	E
Parrotfeather	P	E	E	E	F	E	G
Pondweeds	P	P	G	E	P	E	E
Emergent plants							
Alders	P	E	F	P	E	P	
Arrowhead	P	E	G	G		E	
Buttonbush	P	F	F	P	G	P	
Cattails	P	F	G	P	E	F	
Common reed	P	F	F		E	F	
Waterlilies	P	E[e]	P		G	E	
Frogbit	P	E	E				
Pickerelweed	P	G	G		F	P	
Sedges and rushes	P	F	F		G	P	
Slender spikerush	P		G		P	G	
Smartweed	P	E	F		E	F	
Southern watergrass	P	P			E	G	
Water pennywort	P	G	G		G	P	
Water primrose	P	E	F	P	E	F	
Willows	P	E	F	P	E	P	

[a]Registered as of 4/99 by the U.S. Environmental Protection Agency (EPA).
[b]E = excellent control, G = good control, F = fair control, P = poor control, a blank space indicates unknown weed response.
[c]Hydrothol formulations.
[d]Aquathol formulations.
[e]Granular 2,4-D formulations.
[f]Liquid 2,4-D formulations.

Check the product label for water use restrictions and other precautions.

Endothall Compounds

Endothall compounds are contact herbicides that interfere with plant respiration. They are available in liquid and granular formulations. These herbicides are effective in controlling many types of submerged aquatic vegetation. Endothalls are applied over the surface or injected uniformly below the water surface. Endothalls generally work best in the spring when vegetation is actively growing and water temperatures are around 18 °C (65 °F). The Aquathol formulations work best on submerged aquatic vegetation, while the Hydrothol formulations are most effective for algae control. Hydrothol can be toxic to fish and is seldom used in culture situations. Check the product label for water use restrictions.

Glyphosate Compounds

Glyphosate is a liquid systemic herbicide that disrupts enzymes important in nitrogen metabolism. Glyphosate must be applied directly to the dry foliage of vegetation. It will control many floating and emergent types of aquatic vegetation. A registered aquatic nonionic surfactant must be added to glyphosate for it to be effective in aquatic situations. Rate and time of application of glyphosate depends on the targeted species. Check the product label for water use restrictions.

Fluridone Compounds

Fluridone is a systemic herbicide that controls most submerged and many emergent species of aquatic vegetation. It comes in granular and liquid formulations that can be sprayed, broadcast, or injected into the water.

Fluridone is slow acting and should not be applied in flowing water. It can injure desired shoreline vegetation if their roots extend into the water. Check the product label for water use restrictions.

CONCLUSIONS

Noxious aquatic vegetation is usually a reccurring problem in aquaculture. It is difficult to find any control that has no ecological, sociological, or economic drawbacks. For this reason an integrated approach is best, but most integrated approaches are site specific and must be developed individually by each aquaculturist.

The main reasons aquatic vegetation control fails are (*1*) misidentification of the problem species, (*2*) misapplication of the control technique (e.g., chemical), and (*3*) miscalculation of the area or volume of the treatment site. Pond managers that correctly address these three tenets should substantially improve their aquatic vegetation control success.

BIBLIOGRAPHY

1. H.K. Dupree and J.V. Huner, eds., *The Third Report of the Fish Farmer*, U.S. Fish and Wildlife Service, Washington, DC, 1984.
2. C.E. Boyd, *Water Quality in Warmwater Fish Ponds*, Agricultural Experiment Station, Auburn University, AL, 1979.
3. Center for Aquatic and Invasive Plants, Web site, Institute of Food and Agricultural Sciences, University of Florida, *http://aquat1.ifas.ufl.edu/*, 1999.
4. M.V. Hoyer, D.E. Canfield, Jr., C.A. Horsburgh, and K. Brown, *Florida Freshwater Plants*, Institute of Food and Agricultural Sciences, University of FL, Gainesville, FL, 1996.
5. J.T. Davis, *Aquatic Plants Field Identification Guide* (CD), Texas Ag. Ext. Service, Texas A&M University, 1995.
6. Aquatic Plant Information System—APIS (CD), U.S. Army, Corps of Engineers Waterways Experiment Station, Vicksburg, MS, 1998.
7. R.R. Stickney, *Principles of Warmwater Aquaculture*, John Wiley & Sons, 1979.
8. G.M. Smith, *The Fresh-Water Algae of the United States*, McGraw-Hill, New York, 1950.
9. P.R. Gorham, in D.F. Jackson, eds., *Algae and Man*, Plenum Press, New York, 1964.
10. C.E. Boyd, *Water Quality in Ponds for Aquaculture*, AL Agriculture Experiment Station, Auburn, AL, 1990.
11. M. Beveridge, *Cage Aquaculture*, Fishing News (Books) Ltd., Surrey, England, 1987.
12. C.F. Knud-Hansen, *Pond Fertilization: Ecological Approach and Practical Application*, The Pond Dynamics/Aquaculture Collaborative Research Support Program, 1998.
13. H.E. Westerdahl and K.D. Getsinger, *Aquatic Plant Identification and Herbicide Use Guide*, Volume II: Aquatic Plants and Susceptibility to Herbicides, U.S. Army Corps of Eng., Washington, DC, 1988.
14. H.E. Westerdahl and K.D. Getsinger, *Aquatic Plant Identification and Herbicide Use Guide*, Volume I: Aquatic Herbicides and Application Equipment, U.S. Army Corps of Eng., 1988.
15. W.M. Lewis and M. Bender, *Ecology* **42**, 602–603 (1961).
16. P. Skillicorn, W. Spira, and W. Journey, *Duckweed Aquaculture*, A World Bank Publication, Washington, DC, 1993.
17. Non-Indigenous Aquatic Species—Web site, U.S. Geological Service, Department of the Interior, *http://nas.er.usgs.gov/-nas.htm*, 1999.
18. C.E. Boyd, in G.E. Hall, eds., *Reservoir Fisheries Limnology*, Spec. Publ. No. 8, American Fisheries Society.
19. H.S. Swingle, *Management of Farm Fish Ponds*, Alabama Agricultural Experiment Station Bulletin 254, 1947.
20. A.J. Leslie, Jr., *Aquatics* **10**(1), 12–18 (1988).
21. S. McComas, *Lake Smarts*, Terrene Institute, Washington, DC, 1993.
22. G.D. Cooke, in G.D. Cooke, E.B. Welch, S.A. Peterson, and P.R. Newroth, eds., *Lake and Reservoir Restoration*, Butterworth Publ., Stoneham, MA, 1986.
23. J.W. Avault, Jr., *Fundamentals of Aquaculture*, AVA, Baton Rouge, LA, 1996.
24. J.V. Shireman and C.R. Smith, *Synopsis of Biological Data on the Grass Carp Ctenopharyngodon idella* (Cuvier and Valenciennes, 1844), FAO Fisheries Synopsis No. 135, Rome, 1983.
25. N. Zonneveld and H. Van Zon, in J.F. Muir and R.J. Roberts, eds., *Recent Advances in Aquaculture*, vol. 2, Westview Press, Boulder, CO, 1982, pp. 119–190.
26. H.S. Swingle, *Proc. Tenth Ann Meeting of the Southern Weed Conf.*, 11–17 (1957).
27. K.A. Langeland and D.D. Thayer, *Aquatic Pest Control Applicator Training Manual*, University of FL, Gainsville, 1998.
28. M.P. Masser and J.W. Jensen, *Southern Regional Aquaculture Center*, No. 103, 1991.
29. M.P. Masser and J.W. Jensen, *Southern Regional Aquaculture Center*, No. 410, 1991.
30. G. Almazan and C.E. Boyd, *Aquatic Botany* **5**, 119–126 (1978).

See also FERTILIZATION OF FISH PONDS.

AUSTRALIAN RED CLAW CRAYFISH

DAVID B. ROUSE
Auburn University
Auburn, Alabama

OUTLINE

With the exception of North America, Australia has the largest and most diverse freshwater crayfish fauna in the world. Over 100 species of crayfish live in Australian rivers, lakes, and swamps. All Australian crayfish belong to the family Parastacidae. Aquaculture interest has been directed toward three species, *Cherax tenuimanus*, or marron, *C. albidus-destructor*, or yabbie, and *C. quadricarinatus*, or red claw (1). Of these three species, the red claw is considered to have the best traits for commercial aquaculture and has been the primary species cultured outside of Australia on a commercial scale (2). Scientists did not become aware of the red claw's potential for culture until the mid-1980s. Since then, the red claw has been introduced to many countries in the Americas and Southeast Asia.

Red claw crayfish are native to northern Australia. They are reported to reach a maximum size of 400 to 600 g (0.9 to 1.3 lb) and attain sizes of 50 to 100 g (1.8 to 3.5 oz) in six months (2–5). Red claws reach sexual maturity at about six to eight months of age and are considered to be multiple spawners, spawning three to five times a year (4,6). Reproduction is easily obtained in tanks and ponds under culture conditions. Red claw crayfish are tolerant of a wide range of water quality conditions commonly occurring in culture operations. They survive water temperatures between 12 and 34 °C (54 and 93 °F), but grow best within a range of 22 to 30 °C (71.6 to 86 °F) (2). Red claws are reported to be nonburrowers; however, some have been reported to make shallow burrows (3,7).

INDOOR HATCHERIES FOR RED CLAW CRAYFISH

Spawning

Red claw crayfish will not survive winter conditions in cool, temperate climates, so indoor spawning techniques are needed in such areas. The photoperiod and temperature have a significant influence on spawning rates (8). Spawning activity begins at about 20 °C (68 °F) and increases until about 30 °C (86 °F). Average monthly spawning rates at 28 °C (83 °F) may range from 15% with less than 12 hours of light to over 35% with more than 12 hours of light. Peak spawning occurs with day lengths of 14 hours, but high spawning rates can be maintained only for about three months (9).

Evaluation of the effects of stocking density and sex ratios on red claw reproduction in indoor tanks indicates no suppression of spawning at densities from 20 to 32 red claw brooders/m^2 (10.8/brooders/ft^2) (8,10). Male-to-female sex ratios from 1 : 1 to 1 : 5 have been used successfully (8). The size of the males does not appear to affect spawning success when the males are within 28 g (1 oz) of each other, but mixed sizes (small, medium, and large, together) results in as much as a 49% reduction in spawning success. The size of the females does not appear to affect spawning activity, as long as mature females are used (10).

Mature red claw crayfish accept a wide variety of feeds, from fresh vegetable matter to chopped meats and manufactured diets. They are slow feeders, usually foraging for food throughout the afternoon and early evening. Varying dietary protein levels, from 30 to 45% and the addition of supplements such as beef liver or soybeans were found to have no effect on the number of eggs per spawn, but did decrease mortality rates, because of cannibalism among broodstock (10).

After spawning, females carry their eggs attached to their pleopods until the eggs hatch. At 28 °C (83 °F), incubation requires as few as 30 days (11), but has lasted 40 to 70 days under different environmental conditions (4,10,12). During incubation, changes in egg color appear to be consistent and predictable (4,11): Eggs are cream colored when first spawned, khaki or olive green after one weeks, dark brown after two weeks, reddish after three weeks, and hatch after four weeks. After hatching, juveniles cling to the female's pleopods for 7 to 10 days before dispersing.

Fecundity is an important part of the measurement of the reproductive potential of a culture animal. Fecundity in crayfish is usually defined as the number of eggs produced per spawn. Female red claw crayfish with newly spawned eggs bear an average of 10 eggs/g (284 eggs/oz) of the female's body weight (11). About 30% of the eggs are lost during incubation, resulting in an average of 7 eggs/g (200 eggs/oz) of the female's body weight at hatching (11). Broodstock held under stressful conditions and first-time spawners usually have fewer eggs.

Juvenile Rearing

Spawning in indoor tanks has been relatively successful, but rearing juveniles at high densities has been more difficult. Stocking densities from 50 to 1,250 juveniles/m^2 (10.8 juveniles/ft^2) have been reported (13). With good rearing conditions, survival rates range from 84 to 95% at densities up to 250 juveniles/m^2 (10.8 juveniles/ft^2), but decreased rapidly at higher densities. Final juvenile weights were significantly higher at densities of 100 juveniles/m^2 (10.8 juveniles/ft^2) and lower. No difference in final juvenile weights at densities of 250 juveniles/m^2 (10.8 juveniles/ft^2) and greater have been found (13). The use of additional substrate such as window screen or netting, and size grading after four weeks has been used to increase the survival rate (13). Commercially available diets for juveniles have been evaluated, and the best results (80% survival after 28 days) have been obtained by feeding the juveniles commercial shrimp pellets, supplemented during the first week with *Artemia* nauplii (14). Dietary levels of 33% protein appear to be adequate for juveniles in indoor systems (15,16).

Red claw crayfish juveniles have the ability to regulate their oxygen consumption rate over a wide range of environmental conditions, yet do not utilize anaerobic pathways to withstand hypoxic or anoxic conditions (17). Weight gain and survival have occurred over a temperature range from 16 to 32 °C (61–90 °F), but maximum weight gain and survival have been observed at 28 °C (82 °F) (17). Weight gain and survival have also been observed over a salinity range from 0 to 20 ppt, but are reduced at a salinity of 5 ppt and above (17,18).

Juvenile red claw's tolerances of ammonia, nitrite, and nitrate are similar to those reported for other crayfish species. The 96-hour LC_{50} value for nitrite toxicity for hatchlings is 1.03 ppm and for older juveniles is

4.7 ppm (19,20). Growth in juveniles is reduced at 0.4 ppm nitrite and 1.7 ppm total ammonia (19,20).

OUTDOOR CULTURE IN TEMPERATE CLIMATES

A water temperature of about 20 °C (68 °F) is considered safe for moving red claw crayfish to outdoor ponds. Culture ponds generally range from small ponds of 0.1 ha (0.25 acre) up to larger ponds of 1 ha (2.5 acre). Pond depths generally range from 0.7 to 1.0 m (2.3 to 3.3 ft). The red claws, like most crayfish, are detritivores and opportunistic carnivores and feed on most types of organic matter (4). Feeding strategies have combined forage-based systems typical of those used for North-American red swamp crayfish with pelleted rations commonly used for other crustaceans. Dry grasses added at amounts up to 500 kg/ha (445 lb/acre) per month, divided into several applications, have been used effectively. Pelleted shrimp or crayfish rations fed at 2–3% of the estimated red claw biomass per day are usually added to culture ponds, in addition to the forage material (5,21).

Different stocking densities for use in a five-month growing season have been evaluated. When stocked at 10,000 juveniles/ha (4,050 juveniles/acre), red claws averaged 67 g (2.4 oz), with a total yield of 475 kg/ha (423 lb/acre) at harvest. As densities increase, yields increase and average size decreases. At 50,000 juveniles/ha (20,250 juveniles/acre), the average weight of a juvenile was 38 g (1.3 oz), with a yield of 1,422 kg/ha (1,266 lb/acre) at harvest (5). An economic evaluation of these results indicated that the commercial potential for red claw culture is most sensitive to the cost of juveniles, the percentage of harvestable biomass in the large-size classes, the price that the large-size classes receive, and the length of the growing season (22). Typical stocking densities have been about 30,000 juveniles/ha (12,150 juveniles/acre) (23).

To increase yields and profits, red claw polyculture has been evaluated (24–26). Red claw polyculture experiments have been conducted with tilapia, *Oreochromis* species. Early experiences have suggested that a red claw/tilapia polyculture may not be beneficial. Competition between the two species for food and space, and the negative impact of reduced water quality, appear to be the primary factors affecting red claw growth in polyculture (26).

RED CLAW CULTURE IN TROPICAL COUNTRIES

Red claw crayfish farming began in tropical regions around 1993. Farms now exist in Central and South America and Southeast Asia, ranging in size from about 2 ha (5 acres) to 50 ha (120 acres). Most farms use pond spawning techniques similar to those developed in Australia, combined with large-scale growout (23).

Spawning and nursery operations are performed in small ponds [0.1 to 0.25 ha (0.25 to 0.6 acre)]. Spawning ponds are stocked with mature red claws at densities of about 2 crayfish/m^2 (10.8 crayfish/ft^2). Spawning usually begins within a few months after stocking. Bag material like that used for holding vegetables and citrus fruit is gathered into bundles of about four bags each. Each bundle is then weighted on one corner with a rock and placed in 0.5 m (1.5 ft) of water, with 3 m (10 ft) separating each bundle, around the edge of the pond. Juveniles are harvested by gently lifting the folded bags until a fine-mesh dip net can be slipped under the bundle, at which point the juveniles are shaken off. Estimates vary widely, but juvenile collections may run between 500,000 and 1,000,000 juveniles/ha (200,000 to 420,000 juveniles/acre) every four to five months (23).

Some farm managers stock 2.5- to 5-cm (1 to 2-in.) juveniles directly into 0.5- to 1.0-ha (1.2 to 2.5 acre) growout ponds at densities of 30,000 to 70,000 juveniles/ha (12,150 to 28,350 juveniles/acre), while others stock slightly higher densities into medium-sized ponds for about three months before harvesting and then restock at about 30,000 juveniles/ha (12,150 juveniles/acre) for another three to four months (23). Feeding strategies are similar to those used in temperate climates. Locally available forage materials, such as dry grasses, are provided at rates of 100 to 200 kg/ha (90 to 180 lb/acre) monthly. Daily applications of pelleted crustacean rations are added at 2–3% of the estimated red claw biomass. Water exchange is usually held to a minimum, with only enough water added to replace that lost from evaporation and seepage. Some farms have begun to use mechanical aeration to maintain satisfactory levels of dissolved oxygen and to increase water circulation (23).

Harvesting is accomplished with flow traps. Ponds are drained until only 10–20% of the water remains. Fresh water is pumped from a nearby supply canal or pond into a box, which overflows through an enclosed ramp positioned between the top of the box and the pond bottom. Red claw crayfish following the flow move up the ramp and fall into the box, from which they cannot escape. Some producers have reported that 90% of the crop can be harvested from a production pond during a single night (23).

Most producers expect to harvest about 1,500 to 2,000 kg/ha (1,300 to 1,800 lb/acre) per crop with two crops a year when 70- to 100-g (2.5 to 3.5-oz) red claws are produced. Yields can be increased with higher densities, but harvest size usually decreases as density increases (23).

ECOLOGICAL IMPACTS AND DISEASES

Potential ecological impacts from the introduction of red claw crayfish have been considered, and experiments have been conducted in year-round culture conditions to determine possible interactions between red claw crayfish and North American red swamp crayfish (7). The general health of red claw and red swamp crayfish was monitored throughout the year as the two species were cultured together and separately, respectively. No significant disease outbreaks occurred. Australian red claw crayfish grew well from May through October, but did not survive the winter. Red swamp crayfish, on the other hand, survived and grew well only during the cooler months. No negative effects on red swamp crawfish were observed (7).

No disease outbreaks have been reported from red claw farms. A number of bacteria and ectocommencals have been identified on farmed red claws, but have thus far been similar to those reported from other crayfish (7,27–29). Of major concern have been the potential effects of the fungus *Aphanomyces astaci*, known to cause significant mortalities in crayfish from most regions of the world, except North America. Challenge studies with *A. astaci* revealed the fungus to be pathogenic to red claws at 14 °C (57 °F) (30). No mortalities occurred at 20 °C (68 °F). Furthermore, no detrimental effects from *A. astaci* were observed when red claw crayfish were challenged at 20 °C (68 °F), nor could the fungus be isolated from the host red claw (30).

MARKETING

Red claw crayfish markets have developed in two directions. In temperate climates for which outdoor growing seasons are limited, producers have begun to develop markets in the ornamental pet trade. Indoor recirculating systems are used for spawning and juvenile rearing. Juveniles raised in indoor tanks and fed commercial shrimp diets usually have an attractive blue appearance. Small blue red claw crayfish are sold after about three months, when they are between 38 mm (1.5 in.) and 76 mm (3 in.) long.

Most producers culture red claw crayfish as a food animal. Farms are usually located in subtropical and tropical zones. Target market sizes are 70 g (2.5 oz) and above, so that they are larger than North American crawfish and fall into what might be considered a small-lobster-size category. A highly desirable aspect of the red claw, from a marketing point of view, is the large dressout percentage (30 to 35%), which more closely approximates that of lobster than it does North American crawfish (20 to 25%). Red claw crayfish can be marketed either as whole (live or frozen) animals, frozen processed (shell-on) tails, or processed tail meat. Tails from larger animals may be considered as a substitute for 54- to 84-g (2- to 3-oz) spiny lobster tails (e.g., *Panulirus* spp.) and small clawed lobsters, such as the Norway lobster, *Nephrops norvegicus*. Whether the red claw can command similar market prices as lobsters remains to be seen, since market research is limited (31).

BIBLIOGRAPHY

1. J.R. Merrick and C.N. Lambert, *The Yabby, Marron, and Red Claw: Production and Marketing*, Macarthur Press Pty., Limited, Parramatta, NSW, Australia, 1991.
2. P.G. Semple, D.B. Rouse, and K.R. McLain, *Freshwater Crayfish* **8**, 495–503 (1995).
3. R. Hutchings, *Austasia Aquaculture Mag.* **1**, 12–13 (1987).
4. C.M. Jones, *The Biology and Aquaculture Potential of* Cherax quadricarinatus, Queensland Department of Primary Industries, Fisheries Branch, Research Station, Walkamin, Queensland, Australia, 1990.
5. G.F. Pinto and D.B. Rouse, *J. World Aquacult. Soc.* **27**, 187–193 (1996).
6. N. Sammy, in L.H. Evans and D. O'Sullivan, eds., *Proc. First Australian Shellfish Aquacult. Conf.*, Curtin University of Technology, Perth, Australia, 1988, pp. 79–88.
7. P.B. Medley, D.B. Rouse, and Y.J. Brady, *Freshwater Crayfish* **9**, 50–56 (1993).
8. H.S. Yeh and D.B. Rouse, *J. World Aquacult. Soc.* **26**, 160–164 (1995).
9. D.B. Rouse and H.S. Yeh, *Freshwater Crayfish* **10**, 605–610 (1995).
10. P.B. Medley, *Production Capabilities and Economic Potential of an Australian Red Claw Crayfish* (Cherax quadricarinatus) *Hatchery in the United States*, Ph.D. Dissertation, Louisiana State University, Baton Rouge, LA, 1994.
11. H.S. Yeh and D.B. Rouse, *J. World Aquacult. Soc.* **25**, 297–302 (1994).
12. C.R. King, *Aquaculture* **109**, 275–280 (1993).
13. C.C. Raisbeck, *An Examination of Stocking Density and Substrate Type for Survival and Growth of Juvenile Australian Red Claw Crayfish*, Cherax quadricarinatus, *Reared in an Indoor Recirculating System*, M.S. Thesis, Auburn University, Auburn, AL, 1994.
14. K. Anson and D. Rouse, *J. Applied Aquacult.* **6**, 65–76 (1995).
15. C. Webster, L. Goodgame-Tue, J. Tidwell, and D. Rouse, *Trans. Kentucky Acad. Sci.* **55**, 108–112 (1994).
16. A.M. Keefe, *Protein Requirements for Juvenile Australian Red Claw Crayfish* Cherax quadricarinatus, M.S. Thesis, Auburn University, Auburn, AL, 1998.
17. M.E. Meade, *Effects of Diet and Environmental Factors on Growth, Survival and Physiology of the Juvenile Crayfish*, Cherax quadricarinatus, Ph.D. Dissertation, University of Alabama, Birmingham, AL, 1995.
18. K.J. Anson and D.B. Rouse, *J. World Aquacult. Soc.* **25**, 277–280 (1994).
19. D. Rouse, R. Kastner, and K. Reddy, *Freshwater Crayfish* **10**, 298–303 (1995).
20. H. Liu, *Lethal and Sublethal Effects of Ammonia and Nitrite to Juvenile Red Claw Crawfish*, Cherax quadricarinatus, M.S. Thesis, Louisiana State University, Baton Rouge, LA, 1994.
21. B.M. Kahn, *Effects of Tilapia*, Oreochromis niloticus, *Stocking Density on Production of Australian Red Claw Crayfish*, Cherax quadricarinatus, M.S. Thesis, Auburn University, Auburn, AL, 1995.
22. P.B. Medley, R.G. Nelson, D.B. Rouse, L.U. Hatch, and G.F. Pinto, *J. World Aquacult. Soc.* **25**, 135–146 (1994).
23. D.B. Rouse, *J. of Shellfish Res.* **14**, 569–572 (1995).
24. I.A. Karplus, A. Barki, S. Cohen, and G. Hulata, *Bamidgeh* **47**, 6–16 (1995).
25. R.E. Brummett and N.C. Alon, *Aquaculture* **122**, 47–54 (1994).
26. D.B. Rouse and B.M. Kahn, *J. World Aquacult. Soc.* **29**, 340–344 (1998).
27. X. Romero and R. Jimenize, *J. World Aquacult. Soc.* **28**, 432–435 (1997).
28. H. Villareal and R. Hutchings, *Aquaculture* **58**, 309–312 (1986).
29. B.W. Herbert, *Aquaculture* **64**, 165–173 (1987).
30. J.S. Roy, *Effects of* Aphanomyces astaci *and* Aeromonas hydrophila *on the Australian Red Claw Crayfish* Cherax quadricarinatus, M.S. Thesis, Auburn University, Auburn, AL, 1993.
31. D.B. Rouse, C.M. Austin, and P.B. Medley, *Aquacult. Mag.* **17**, 46–56 (1991).

See also CRAWFISH CULTURE; HARVESTING.

AYU CULTURE

HEISUKE NAKAGAWA
Hiroshima University
Higashi-hiroshima
Japan

OUTLINE

The ayu, *Plecoglossus altivelis* well known in Japan, is also called the "pond smelt" or "sweet fish." It is an anadromous member of the Plecoglossidae (Fig. 1). The ayu is native to east Asia, family is taxonomically close to salmonids, and has a life span of only one year. Ayu is one of the most popular and economically important freshwater fish in Japan, because of its favored taste and odor. Japan widely produces ayu for traditional culinary use. Taiwan produces ayu mainly for restaurant consumption. The cultural production of ayu in east Asian countries such as Korea, China, and Taiwan, is capable of further growth.

D.S. Jordan, an ichthyologist, said, "the one fish of all its fishes is the ayu its flesh is white and tender, and so very delicate in taste and odor that one who tastes it crisply fried or boiled feels that he has never tasted real fish before." A subspecies of this fish, Ryukyu ayu (*P. altivelis ryukyuensis*), distributed in the Okinawa islands

Figure 1. Wild ayu in Japan.

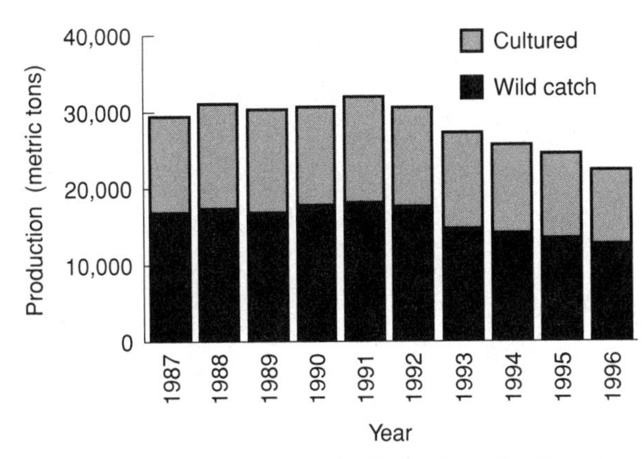

Figure 2. Annual production of wild-caught and cultured ayu.

in southwest Japan, is now on the brink of extinction in that area, because of environmental deterioration.

The merits of culturing ayu come from its high market value and growth rate. In the past, the ayu was a luxury fish in Japan, but it is now inexpensive, due to increasing production in aquaculture [Fig. 2(1)]. The amount of aquacultural production of ayu has been close to that of capture fisheries in rivers.

Natural History of Ayu

The downstream movement of mature ayu for spawning begins in early autumn. Following the completion of spawning, spent males and females die. The eggs are adhesive and are laid on sand in the lower reaches of rivers, from September to November. The eggs can hatch after about two weeks. The newly hatched larvae, which are 7 mm (0.28 in.) in length, are carried downstream to the sea, where they remain during winter. Larval-stage ayu are planktonivorous. In the following spring, the young individuals, which are 6–8 cm (2.4–3.1 in.) in total length and 3–5 g (0.10–0.17 oz) in weight, start the anadromous migration back to the rivers. After upstream movement, the fish change their feeding habitats to an herbivorous diet. The main sources of food in rivers are adherent blue-green algae, and diatoms, which contain 45–48% crude protein. Detailed embryological and physiological descriptions of ayu have been provided by Iwai (2).

Construction of artificial dams to meet industrial and agricultural demands has led to serious problems in the upstream movement of anadromous fishes. Therefore, young individuals [5–6 cm (2.01–2.4 in.) in total length] are released into the upper reaches of rivers, to enhance the natural supply of ayu, by Fisherman's Cooperative Associations in Japan. For example, in the Ohta-gawa river in Hiroshima prefecture, more than 70% of released ayu can be recovered annually by fishing. Almost all of the ayu caught in rivers by commercial and sport fishermen are used for food in restaurants and homes.

SEEDLING ACQUISITION AND PRODUCTION

About 80–90% of the fry for culture and stock enhancement are landlocked ayu, which are captured mainly in Lake Biwa, in Shiga prefecture. The remainder (10–20%)

are artificially produced. In addition, a small amount of wild fry caught along the seashore are used.

Seashore Fry

Wild fry 3.5–8.0 cm (1.38–3.15 in.) in total length and weighing 0.1–5.0 g (0.04–0.18 oz) caught along the seashore by seine nets during December and April are acclimated to freshwater. The fry can be reared with formulated feed.

Landlocked Fry

Landlocked fry, which weigh 0.3–5 g (0.01–0.18 oz) can be caught in Lake Biwa from February to June by using pound nets or scoop nets. These fry do not vary much in size. They are transported in tanks filled with sea water and can be fed with formulated feeds.

Artificially Produced Fry

In hatcheries, artificial fertilization is carried out by the same method used for trout. Laid eggs, which usually number from 20,000 to 50,000 (the maximum is 100,000) per female, adhere to hatchery nets. The suitable water temperature range for hatching is 12–20 °C (54–68 °F). The eggs hatch after 16 to 17 days at 15 °C (59 °F). It is necessary to acclimate the artificially produced fry larvae to sea water before feeding them live organisms. First foods include rotifers (*Brachionus plicatilis*) for fry of 7–35 mm (0.28–1.38 in.) in total length and brine shrimp (*Artemia salina*) for fry of 13–35 mm (0.51–1.38 in.). Fry larger than 10 mm (0.39 in.) are gradually adjusted to formulated feed. Rearing density is about 10 kg (22 lb) of fish per metric ton. The mortality rate is less than 2% from hatching to a size of 4–6 g (0.14–0.21 oz).

INTENSIVE CULTURE

Production techniques for ayu in Japan are well established. Production is about 20–70 kg/m^2 (4.09–14.31 lb/ft^2), with a depth of 1 m (3 ft). Mortality over 90 days may be only 10% during the period that ayu grow from 4–6 g (0.14–0.21 oz) to a market weight of 50–80 g (1.34–2.82 oz), in the absence of disease.

Culture System

A water supply system and a sufficient volume of water are essential for ayu culture. Ayu are reared in specially built concrete ponds, usually measuring 100–400 m^2 (1076–4304 ft^2) and averaging 150 m^2 (1615 ft^2), that are supplied with running water. Both river and well water are used for finishing production and cultivation. Octangular and round ponds are popular for cultivation (Fig. 3). The optimum water flow rate is at least 10 l/sec in 100-m^2 ponds with a depth of 1 m (3.28 ft). Round tanks are beneficial for producing water currents in rearing ponds and for discharging solid waste from a center outlet. Most ayu culture is carried out in outdoor tanks, but is also done in indoor tanks, because of the facility of water temperature control.

Figure 3. Typical ayu culture farm with outdoor ponds.

Environmental Factors

The empirically determined optimum rearing density is 150–200 fish/m^2 (13.94–18.59 fish/ft^2), but the culture density in farms is actually 500–600 fish/m^2 (46.47–55.76 fish/ft^2). Young individuals weighing 2–4 g (0.07–0.14 oz) reach 10 g (0.35 oz) in one month, 25–40 g (0.88–1.41 oz) in two months, 50–60 g (1.76–2.12 oz) in three months, and 70–100 g (2.47–3.53 oz) in four months.

The dissolved-oxygen requirement of ayu is higher than that of warmwater fish. Dissolved-oxygen levels should be higher than 5 mg/L (ppm). Ammonium and nitrate levels should be maintained at less than 2–3 mg/L and 0.2 mg/L (ppm), respectively. Ayu farmers use air pumps to aerate the water, although this process is expensive.

Sudden temperature changes in rearing ponds render ayu more susceptible to disease. The thermal tolerance range of ayu is from 13 to 30 °C (55 to 86 °F) and the optimum range is 15 to 25 °C (59 to 77 °F). The growth rate at 18 to 20 °C (64 to 68 °F), is twice as high as that at 13 to 16 °C (55 to 61 °F), indicating a high dependence on rearing temperature for optimum growth.

The fact that wild ayu live in waters with strong currents implies a need for water current in rearing tanks. A strong water current elevates body-weight gain and feed efficiency, but lightly suppresses lipid reserves. Body-weight loss during starvation before marketing is suppressed by rearing ayu in stronger water currents (35–45 cm/s) (1.15–1.48 ft/s). Thus, a strong water currents improve lipid metabolism, by mobilizing lipid reserves for various energy demands. Therefore, rearing tanks should have controlled water currents of about 45 cm/s (1.48 ft/s) (3). Under such conditions, the fish face the current and disperse throughout the rearing ponds. However, some expenses, such as electricity for controlling water current, are restricting factors.

Photoperiod affects hormonal regulation and influences reproductive cycles in ayu. Farmers use their knowledge of these relationships to control maturation in male fish. Roe-containing females are sold commercially. However, as gonadal maturation of males toward autumn is accompanied by darkening of the skin, male fish have a depressed market value. Maturation of male fish can be suppressed by maintaining light at greater than 50 lux for more than 16 hours daily.

Nutritional Requirements

After the fry stage, granular and crumble-type formulated feeds [ϕ 0.07–2.4 mm (0.003–0.094 in.)] are used throughout the rearing period. Extruded-type feeds have been recently employed as well. Formulated feeds for fry typically contain the following ingredients as protein and lipid sources: salmon egg and milt powder, krill meal, clam powder, casein and its hydrolyzed products, yeast, squid soluble, zein, fish meal, scallop powder, shrimp meal, gluten, amino acids, *Streptomyces* fermentation product, hen's egg powder, soybean lecithin, starch, and squid oil.

The proximate composition of commercial feeds is shown in Table 1. Ingredients in commercial feeds for growout are brown fish meal, krill meal, corn, wheat, soybean meal, peas, cotton seed meal, and kelp meal, as well as minerals and vitamins. Values of the proximate composition of commercial feeds for juveniles larger than 10 g (0.35 oz) are officially standardized as follows: crude protein $\geq$45.0%, crude fat $\geq$3.0%, crude fiber $\leq$4.0, and crude ash $\leq$15.0%. Ayu of more than 5–6 cm (1.97–2.36 in.) are herbivorous, feeding on algae, diatoms, and blue-green algae in the freshwater habitat. Nevertheless, commercial feeds that contain plants as the only protein source have not been fully used by ayu farmers.

The protein requirement of ayu is relatively high. Crude protein levels of 45–48% in the feed appear to satisfy the requirement. To improve growth and feed efficiency, oil, which is usually obtained from Alaskan pollack liver, has occasionally been supplemented, at about 2% of the diet. However, supplementation with oil leads to increased lipid reserves in muscle and viscera (4) and may deteriorate carcass quality. Essential fatty acids are n-3 fatty acids, such as C18:3n3 and C20:5n3, and should be provided at 1% of dietary lipids (5). In addition, phospholipids (e.g., phosphatidyl choline) are indispensable for growth and survival in the larval stage (6). Medium-chain triglycerides or plant oils are occasionally employed in formulated feeds. Medium-chain triglycerides can be consumed as an energy source without deposition in the body (7).

The vitamin and mineral requirements of ayu are not well established. Supplementation of vitamin C to commercial feed increases the aggression of ayu, a trait that is characteristic of wild ayu (8).

Various feed supplements have recently been used as trace nutrients in order to improve carcass quality.

Table 1. Composition of Commercial Ayu Feeds

	Starter	Juvenile	Adult
Fish size (cm)	<3.5	4.3–12	>9
Type of feed	Granule	Crumble	Crumble
Feed size (mm)	0.07–0.45	0.3–1.5	0.9–2.4
Crude protein (%)	50.0–60.0	47.0–55.0	45.0–50.0
Crude fat (%)	3.0–10.0	3.0–10.0	5.0–8.0
Crude fiber (%)	1.0–1.4	1.0–3.0	3.0–4.0
Crude ash (%)	14.0–17.0	15.0–17.5	15.0–19.0

Supplementation of *Chlorella* extract and *Spirulina* to formulated feed improves physiological condition, stress response, and resistance to disease (9–11). Wild ayu are tinged with a yellowish-orange color and a characteristic light yellow spot near the pectoral fin. They exhibit an orange band below the lateral line when mature. Surface color can be improved by dietary *Spirulina*, which contains β-carotene and zeaxanthin. Products such as marigolds and krill meal have been used as other sources of carotenoids.

Feeding Regimes

The most serious problem in maintaining quality ayu is excessive lipid accumulation in muscle and viscera. This leads to a loss of appetite and to the deterioration of carcass quality. High feeding frequency seems to result in greater overall food intake. However, feeding to apparent satiation does not always improve growth. A proper feeding regime depresses lipogenesis and activates lipolysis (12), allowing dietary protein to be spared, resulting in an improvement of carcass quality, and better growth performance. Feeding frequency is controlled with the use of automatic feeders: Fish less than 1 g (0.04 oz) are fed four or five times per day, fish 1–10 g (0.04–0.35 oz) three or four times, fish 10–80 g (0.35–2.82 oz) three times; and fish 180 g (6.35 oz) or larger two times.

Diseases

Overfeeding and a high rearing density, which are generally unavoidable in the culture of commercial fish, can lead to a variety of diseases. *Cytophaga psychrophila* infection, known in salmonid culture, has recently been widespread and caused considerable damage to ayu production. The extent of damage caused by *C. psychrophila* occasionally accounts for more than 50% of total disease outbreaks.

Ayu infected with *Vibrio anguillarum* show the following symptoms: tumescence of the body surface, dermal ulcers, and hemorrhaging of the base of pectoral fins. Vaccination is effective for preventing the disease.

Aeromonas hydrophila characteristically infects young ayu weighing less than 50 g (1.76 oz) at water temperatures greater than 20 °C (68 °F). Bacterial gill disease, associated with *Flavobacterium* sp., can be prevented by avoiding high-density rearing and overfeeding and can be cured by bathing fish in a 0.7 to 1% NaCl solution for one to two hours.

Streptococcus sp. infection is characterized by hemorrhages on the body surface, ocular disease, abdominal dropsy, etc.

Other problems include stomach mycosis, mycotic granulomatosis from *Aphanomyces piscicida*, ichthyophoniasis from *Ichthyophonus hoferi*, and phoma infection from *Phoma* sp. Phoma infection occurs in young ayu, and no method for curing the disease has been found yet.

Parasitic diseases include *Glugea plecoglossi* infection, which occurs at water temperatures greater than 18 °C (64 °F). Gyrodactylosis is associated with high-density rearing. The symptoms of fish attacked by the flukes *Gyrodactylus japonicus*, *G. tominagai*, and *G. plecoglossi* are poor appetite and hemorrhages on the skin. The flukes are found on the skin, fins, and gills of infected fish. A gill disease caused by *Pseudergasilus zucconis* is not lethal.

A typical disease of ayu is the so-called "chochin disease," which morphologically resembles ulcerative dermal necrosis in salmonids. The disease is caused by overcrowded rearing conditions and inadequate feeding regimes. It is not lethal, however, and reduction of stocking density can allow recovery to take place.

MARKETING AND QUALITY CONTROL

The market value of cultured ayu is distinctly different from that of ayu captured in rivers. The general market size of ayu is 50–150 g (1.76–5.29 oz); fish of this size can be obtained between June and October. Smaller fish [20–60 g (0.71–2.12 oz)] can be sold by May, which is before the opening of the capture fisheries in rivers. The market prices of cultured ayu ranged from ¥1200 to ¥2800/kg in 1995–1996. The price fluctuates with the season, and peak prices are to be found between April and September. Most cultured ayu are sold live, chilled, or frozen. Fry and feed costs make up the largest production costs at 17–22% and 22–24%, respectively. Blowers and water pumping can account for as much as 25% of the total production cost.

Cultured ayu are considerably different from wild ayu in appearance, especially skin color and shape. The odor, muscle, taste muscle hardness, and lipid content also clearly distinguish between wild and cultured ayu. The wild ayu have proven to be most highly favored, for their nonoily, plain taste.

"Semiwild ayu" have been produced under special rearing techniques. Although the quality of semiwild ayu is not yet officially standardized, the higher quality products are differentiated by skin color and odor from regularly cultured ayu. Special techniques to produce semiwild ayu are supplementation of carotenoids, plant oil, medium-chain triglycerides, or any combination thereof, in formulated feeds; starvation for several days before marketing; rearing in strong water current; reduction of rearing density, etc. Trace nutrients, such as *Chlorella* extract, garlic extract, *Spirulina*, and Chinese drugs, are sometimes used as well.

Wild ayu have a sweet smell, like the aroma of watermelon and cucumber. The characteristic odor is caused by the sessile algae growing on stones in the rivers where ayu dwell. The algae are the main food source for ayu. The main chemical components associated with the odor are 2-trans-6-cis-nonadienal and cis-3-hexenol (13). Cultured ayu are not as palatable as wild ayu, because they lack this specific odor. Supplementation of the odor-causing chemicals in formulated feed has not been successful in giving this special odor to cultured ayu. A bitter taste in wild ayu is attributable to anserin (14).

Ayu accumulate energy as adipose tissues in the peritoneal cavity and along the dorsal neural spine. A comparison of lipid content of cultured ayu (from 12 hatcheries) and wild ayu (from nine rivers) showed that the lipid level in the muscle of the cultured fish ($8.2 \pm 2.5\%$) was much higher than that of wild fish ($3.4 \pm 1.7\%$) (15). Research is needed to find a way to suppress lipid accumulation without sacrifice of growth.

BIBLIOGRAPHY

1. Anonymous, in *Annual Reports: Year Book of Japanese Fisheries and Culture Production*, Ministry of Agriculture, Forestry and Fisheries, Tokyo, 1997.
2. T. Iwai, *Bull. Misaki Mar. Biol. Inst.* **2**, 1–101 (1962).
3. H. Nakagawa, H. Nishino, Gh.R. Nematipour, S. Ohya, T. Shimizu, Y. Horikawa, and S. Yamamoto, *Nippon Suisan Gakkaishi* **57**, 1737–1741 (1991).
4. M. Takeuchi, *Bull. Tokai Reg. Fish. Res. Lab.* **93**, 103–109 (1978).
5. A. Oka, N. Suzuki, and T. Watanabe, *Bull. Jpn. Soc. Sci. Fish.* **46**, 1413–418 (1980).
6. A. Kanazawa, S. Teshima, S. Inamori, T. Iwashita, and A. Nagoya, *Mem. Fac. Fish. Kagoshima Univ.* **30**, 301–309 (1981).
7. Gh.R. Nematipour, H. Nishino, and H. Nakagawa, *Proc. Third Int. Symp. on Feeding and Nutr. in Fish.*, 233–244 (1989).
8. S. Koshio, Y. Sakakura, Y. Iida, K. Tsukamoto, T. Kida, and K. Dabrowski, *Fisheries Sci.* **63**, 619–624 (1997).
9. T. Hirano and M. Suyama, *J. Tokyo Univ. Fish.* **72**, 21–41 (1985).
10. Md.G. Mustafa and H. Nakagawa, *Israeli J. Aquacult.* **47**, 155–162 (1995).
11. H. Nakagawa, *Téthys* **11**, 328–334 (1985).
12. S.-J. Yao, T. Umino, and H. Nakagawa, *Fisheries Sci.* **60**, 667–671 (1994).
13. M. Suyama, T. Hirano, and S. Yamazaki, *Bull. Jpn. Soc. Sci. Fish.* **51**, 287–294 (1985).
14. T. Hirano and M. Suyama, *Bull. Jpn. Soc. Sci. Fish.* **46**, 215–219 (1980).
15. H. Nakagawa, Y. Takahara, and G.R. Nematipour, *Nippon Suisan Gakkaishi* **57**, 1965–1971 (1991).

B

BACTERIAL DISEASE AGENTS

JOHN P. HAWKE
Louisiana State University
Baton Rouge, Louisiana

OUTLINE

Gram-Negative Agents
Gram-Positive Agents
Bibliography

Bacterial diseases have historically been among the most significant problems facing aquaculturists worldwide. The oldest record of bacterial disease in aquaculture names "redpest" in eels in the Mediterranean as early as 1718 (1), and the causative agent *Vibrio anguillarum* was first cultured in 1883 (2). It has been estimated that 10% of fish loss in aquaculture is due to disease and that, of this portion, more than half is due to bacterial disease (3,4). Economic losses resulting from bacterial disease are difficult to assess, because many cases go unreported; however, a single bacterial disease, Enteric Septicemia of Catfish (ESC), is believed to cost the U.S. catfish industry US$19 million in direct fish losses annually (5,6). Bacterial pathogens are, in some cases, highly host-specific, infecting and causing disease in only one genus or even one species of fish. Others have a broad host range. Some are obligate pathogens, not normal free-living components of the aquatic ecosystem but ones requiring a fish host to survive. Others are facultative pathogens that are normal inhabitants of the aquatic environment and require stress to assist in initiating disease in a fish population. Despite our current knowledge of approximately 70 species of bacterial agents having the capability of causing disease in aquaculture (7), new diseases caused by previously undocumented bacterial fish pathogens continue to surface almost yearly. The aquaculture industry has experienced tremendous growth in recent years, and, as culture practices have become more intensive in order to increase profitability and as new species of fish are cultured, new, highly virulent bacterial pathogens have emerged. Currently, bacterial taxonomy is in transition, due to the utilization of more precise molecular methods of classification, and the names of many previously recognized pathogens have changed. This entry addresses only those bacteria that are established as pathogens in aquaculture. The most currently approved names, as well as the older names with which readers may be more familiar, are included. Older names or taxonomically invalid names are highlighted by quotation marks, (i.e., "*Haemophilus piscium*"). Despite the importance of bacterial disease in aquaculture, only one book is available that covers management techniques (7), and only two books are available that deal specifically with description of bacterial pathogens and of the diseases that they cause (8,9). Review articles that deal with various mechanisms of bacterial pathogenesis, have been written and several books are available that contain chapters devoted to methods of diagnosis of bacterial diseases and of identification of bacterial pathogens (10–13).

GRAM-NEGATIVE AGENTS

A. *Enterobacteriaceae*. Facultatively anaerobic, oxidase negative, gram-negative rods.

***I. Edwardsiella tarda*.** Causative agent of *Edwardsiella* Septicemia. Synonyms: Emphysematous putrefactive disease of catfish; red disease of eels; edwardsiellosis of salmon, of tilapia, and of striped bass.

a. History — The bacterium was originally described in 1962, from 256 isolates obtained from various sources in Japan; however most of the isolates were from snakes. Five isolates from human gastrointestinal infections were referred to as the "Asakusa group" (14). In the same year, a new disease of eels, "red disease," was described, and the causative agent was named "*Paracolobactrum anguillimortiferum*" by Hoshina (15). A new genus, *Edwardsiella*, including a new species, *E. tarda*, was described from 37 human isolates in the U.S. by Ewing, in 1965 (16). In 1973, *E. tarda* was almost simultaneously recognized as the cause of "red disease" of eels in Japan and Taiwan (17) and of "emphysematous putrefactive disease of catfish" in the U.S. (18). Cultures of "*P. anguillimortiferum*" were no longer available for comparison, so *E. tarda* became the valid scientific name.

b. Culture — Primary isolation of the bacterium from diseased fish is achieved on standard media, such as brain heart infusion agar (BHIA) or tryptic soy agar (TSA) with 5% sheep blood and incubation at 25–37 °C. Growth is rapid, with 1–2 mm colonies present in 24 hours. On EIM-selective medium (19), *E. tarda* forms a small green colony with a black center.

c. Description — Short-gram negative rod, $0.6 \times 2.0\ \mu m$, motile by peritrichous flagella at 25 and 37 °C, oxidase-negative, indole-positive, fermentative in semisolid glucose motility deeps (GMD), K/A with hydrogen sulfide production and gas in triple sugar iron (TSI) agar slants. The bacterium is differentiated from closely related organisms in Table 1.

d. Epizootiology — *Edwardsiella* septicemia appears to be favored by high water temperatures [28 °C (82 °F) and above] and by the presence of high levels of organic matter in catfish ponds. Infections in eels in Taiwan occur when water temperatures are fluctuating between 10 and 18 °C (50–64 °F). The incidence of *E. tarda* infections is relatively rare in channel catfish (*Ictalurus punctatus*) ponds and, mortality rates are usually low (~5%), but, when infected fish are moved into confined areas such as holding tanks, mortalities can reach levels as high as 50%.

Table 1. Biochemical and Biophysical Characteristics Important in Differentiating Three Fish Pathogens Belonging to the Family *Enterobacteriaceae*

Characteristic	*E. ictaluri*	*E. tarda*	*Y. ruckeri*
Motility at 25 °C	+	+	+
Motility at 35 °C	−	+	−
Indole	−	+	−
Citrate (simmons)	−	−	+
Trehalose	−	−	+
Gelatin (22 °C)	−	−	+
Gas from glucose (25 °C)	+	+	−
H_2S	−	+	−

The reservoir of infection is unclear, but the bacterium has been associated with a variety of aquatic invertebrates and of aquatic and terrestrial vertebrates. Fecal contamination of water from human or animal sources, may be a source of the bacterium. There is also some speculation that *E. tarda* may comprose a part of the normal microflora of the surfaces of certain fishes or snakes. In the U.S., *E. tarda* was isolated from as many as 88% of domestic dressed channel catfish (20) and was found in 30% of imported dressed fish. The bacterium was also found in 75% of catfish pond water samples, in 64% of pond mud samples, and in 100% of frogs, turtles, and crayfish from catfish ponds.

e. Geographic range and host susceptibility—*E. tarda* is found in both freshwater and brackish-water environments. It has been reported from 25 countries in North and Central America, Europe, Asia, Australia, Africa, and the Middle East. The bacterium has been isolated from over 20 species of freshwater and marine fish; it occurs most commonly in the following: the channel catfish, *I. punctatus*; the common carp, *Cyprinus carpio*; the largemouth bass, *Micropterus salmoides*; the striped bass, *Morone saxatilis*; the red sea bream, *Chrysophrys major*; the Japanese flounder, *Paralichthys olivaceus*; tilapia, *Oreochromis* sp.; yellowtail, *Seriola quinqueradiata*; and rarely in salmonids. The bacterium also causes disease in snakes, alligators, and sea lions and in various birds, cattle, and swine (8).

f. Clinical signs of disease and treatment—*Edwardsiella* septicemia in eels is characterized by hemorrhagic fins, by petechiae on the belly, by a swollen protruding anus, and by necrotic foci in the internal organs. Gas filled pockets may form between the skin and muscle. In channel catfish with emphesematous putrefactive disease, 3–5 mm (0.12–0.20 in.) cutaneous ulcers form in the skin. These lesions progress into larger abcesses deep in the tissues and emit a foul odor when ruptured. Eye disease is common in tilapia and striped bass, and infected fish exhibit exophthalmia and corneal opacity. Treatment in catfish is best achieved with medicated feeds: Romet 50 mg/kg/day (23.0 mg/lb/day) for 5 days and Terramycin 23.0–34.0 mg/lb/day for 10–14 days, both of which are approved for use in catfish by the United States Food and Drug Administration (USFDA). Japanese strains of *E. tarda* are resistant to many antibiotics; however, nalidixic acid has been used with success (7). *E. tarda* is serologically heterogeneous, with 49 O antigens and 37 H antigens; therefore, production of effective vaccines has met with difficulty (55).

II. Edwardsiella ictaluri. Causative agent of Enteric Septicemia of Catfish

a. History—The disease was first documented from cases involving diseased channel catfish submitted to the Southeastern Cooperative Fish Disease Laboratory at Auburn University in 1976. The laboratory, under the direction of Drs. W.A. Rogers and J.A. Plumb, recorded 26 cases of this new disease syndrome, from ponds primarily in Alabama and Georgia, between January 1976 and October 1979. The disease was described by Hawke in a published account in 1979 (21), and the causative organism was described as a new species, *E. ictaluri*, in 1981 (22). Records from fish disease diagnostic laboratories indicate that the disease was not prevalent in the industry immediately following its discovery. Enteric septicemia of catfish (ESC) occurred in only 8% of the total cases reported by the Mississippi Cooperative Extension Service in 1980 and 1981. Between 1982 and 1986, however, the increase in ESC incidence was explosive, and the economic impact on the catfish industry was significant (3). Enteric septicemia of catfish is believed to have been present in cultured catfish prior to its description, because archived tissues from the Fish Farming Experiment Station at Stuttgart, Arkansas reacted with a monoclonal antibody specific for *E. ictaluri* (23). Currently, ESC is the most important disease of farm raised catfish, accounting for approximately 30% of all disease cases submitted to fish diagnostic laboratories in the Southeastern United States. In Mississippi, where catfish make up the majority of case submissions, it has been reported at frequencies as high as 47% of the yearly total (3). It is estimated that ESC costs the catfish industry $19 million yearly in direct fish losses (6) and ESC is considered a disease of current or potential international significance by the OIE (65).

b. Culture—Primary isolation from the kidney, liver, spleen, or brain of diseased fish is achieved on standard media such as BHIA or TSA with 5% sheep blood. The bacterium has a narrow temperature range for optimum growth (25–28 °C), and primary isolation plates should be incubated within this range. Growth is slow even at these temperatures, with 1–2 mm colonies present in 48 hours. *E. ictaluri* forms a small pale green colony on EIM-selective medium (19).

c. Description—The bacterium is a short Gram-negative rod, 0.75 × 1.25 μm, motile by peritrichous flagella at 25–30 °C but not at 35 °C, oxidase-negative, fermentative in GMD, indole-negative, K/A with H_2S negative in TSI slant, and citrate-negative. The bacterium is differentiated from closely related fish pathogens in Table 1. The species is antigenically and physiologically homogeneous, regardless of the geographic source of the isolate. Confirmatory identification can be made with serological tests, among them the slide agglutination test, the indirect florescent antibody test (IFAT), the enzyme immunoassay (EIA), and the enzyme linked immunosorbent assay (ELISA). *E. ictaluri* may be identified with the API 20E system (BioMerieux Vitek Inc.) by generation of the code number 4004000.

d. Epizootiology — Outbreaks of ESC are strongly correlated with pond water temperature, with high stocking densities, and with stress resulting from poor water quality. The so-called "ESC window" occurs in the spring and in the fall, when water temperatures are in the range 22–28 °C (72–82 °F). When initially described, *E. ictaluri* was thought to be an obligate pathogen, because its survival was limited to approximately 8 days in sterile pond water (21); however, it was later found to survive for 95 days in pond mud at 25 °C (77 °F) (24). *E. ictaluri* infections are very common in commercial catfish ponds in the southeastern U.S. but are rarely found as the cause of natural fish kills. Mortality rates in cultured fish may vary from less than 10% to 50% or more of the population. The disease is most common, and causes the highest mortalities, in fingerling channel catfish, but production-size catfish with no previous exposure are also highly susceptible. Survivors of outbreaks are believed to become asymptomatic carriers that can transmit the disease if stocked into a pond containing a population of naive fish.

e. Geographic range and host susceptibility — *E. ictaluri* is known to occur primarily in the southeastern U.S. Isolated instances of disease have been reported from Thailand and Australia, in atypical species. The bacterium is specific for channel catfish, with other species of catfish apparently being less susceptible. Rare outbreaks have been documented in the blue catfish (*Ictalurus furcatus*), in the white catfish (*Ictalurus melas*), and in the brown bullhead (*Ictalurus nebulosus*). Natural infections have been found in the walking catfish (*Clarias batrachus*) in Thailand, and in two aquarium species, the Bengal danio (*Danio devario*) and the glass knife fish (*Eigemannia virescens*). Experimental infection and disease have been induced in chinook salmon (*Oncorhynchus tshawytscha*) and in rainbow trout (*Oncorhynchus mykiss*), but many species of warm and coldwater fish are refractory (7,25).

f. Clinical signs of disease and treatment — ESC occurs in acute, subacute, and chronic forms in channel catfish. The acute form is characterized by only a few clinical signs; clear or straw colored fluid in the body cavity, exophthalmia, petechial hemorrhage around the head and operculum, and enlargement of the kidney and spleen. The subacute form is characterized by more obvious external signs; small 2–3 mm (0.2–0.25 in.) ulcerative lesions (red or white spots) in the skin, hemorrhage and necrotic foci in the liver, hemorrhage in the intestine, and bloody ascites. The chronic form is characterized by an ulcer in the top of the head in the area of the sutura fontanelle between the two frontal bones of the skull (a phenomenon resulting in the name "hole in the head disease"). This lesion develops from inflammatory accumulations in the brain cavity, as a result of meningoencephalitis. The meningoencephalitis results in such behavioral signs as spinning or spiraling, prior to death. Treatment of channel catfish with ESC is best achieved orally with medicated feeds (Romet for 5 days with 3 days of withdrawal or Terramycin for 10–14 days with 21 days of withdrawal). The emergence of resistant strains in the late 1980s prompted investigations into alternative antimicrobials (sarafloxacin, amoxicillin, florfenicol); however, none has achieved USFDA clearance. Currently, experimental live attenuated vaccines show promise, but an effective commercial vaccine is not available.

III. Yersinia ruckeri. Causative agent of Enteric Redmouth Disease. Synonyms: "Hagerman redmouth," "salmonid blood spot."

a. History — The disease was first observed in rainbow trout aquaculture in the Hagerman Valley, Idaho, in the 1950s and was described by Rucker in 1966 (26). A description of the causative bacterium was first published by Ross et al. in 1966 (27), and the pathogen was named *Y. ruckeri* by Ewing et al., in 1978 (28). The common name, enteric redmouth disease (ERM), was adopted by the Fish Health Section of the American Fisheries Society in 1975 (29). Enteric redmouth disease has historically been one of the most significant diseases in salmonid aquaculture, with the potential for cumulative losses of 70% of some populations. The economic impact on the trout industry in the late 1970s was significant, with over US$2 million in losses annually; however, the current practice of vaccinating trout fingerlings with commercially available vaccines has reduced the impact on the industry.

b. Culture — Primary isolation of *Y. ruckeri* is achieved on general purpose bacteriological media (BHI or TSA w/5% sheep blood) and incubation in a normal atmosphere at 22–25 °C. Colonies (1–2 mm) are visible in 24–48 hours. The bacterium is capable of growth over a wide temperature range (9–37 °C); however, the strains that grow at 37 °C are avirulent for trout. *Y. ruckeri* forms a green colony surrounded by a zone of hydrolysis on Shotts and Waltman (SW) selective medium (30), and it produces a yellow colony on the selective medium of Rodgers (31). These media, although useful in selective isolation of the organism from exposed surfaces or intestinal cultures, are not specific enough for confirmatory identification.

c. Description — The bacterium is gram-negative and oxidase-negative. Cells of *Y. ruckeri* are rod shaped (1.0 × 1.0–3.0 μm), becoming filamentous in older cultures. The bacterium is motile by peritrichous flagella in the optimum temperature range, but non-motile strains occasionally occur (27). With the exception of sorbitol fermentation (approximately 32% of strains are positive), the strains are fairly homogeneous in biochemical phenotype. Most strains are positive in the following reactions: glucose fermentation in GMD; ONPG; lysine and ornithine decarboxylase; citrate utilization; Jordans tartrate; and gelatin liquefaction. The bacterium gives a negative reaction for H_2S, indole, and Voges–Proskauer. Features useful in differentiating *Y. ruckeri* from related fish pathogens are given in Table 1. Genetically, *Y. ruckeri* is distantly related to other members of the genus, being only 38% homologous by DNA–DNA hybridization (28). Five different serovars of *Y. ruckeri* are recognized, on the basis of formalin-killed whole-cell serology, and their occurrence seems to be correlated with the geographic region from which they are isolated. The serovars are described as follows: Serovar I (Idaho strains), Serovar II (Oregon strains), Serovar III (Australian strains), Serovar IV (excluded), Serovar V (Colorado strains), and Serovar VI (Ontario strains).

d. Epizootiology—Although *Y. ruckeri* is thought to be an obligate pathogen of salmonids, the role of stress is very important in the initiation of outbreaks (27). The bacterium has also been shown to survive up to two months in mud. Experimentally, the efficiency of transmission of ERM from carriers to susceptible fish and subsequent disease is greatly enhanced by the addition of stress. Reservoirs, other than survivors of outbreaks, have not been identified. Survivors carry the bacterium in their intestine and shed organisms into the water on a cyclical basis every 35–40 days; shedding precedes recurrent outbreaks of the disease by several days. Vertical transmission of *Y. ruckeri* has not been demonstrated. Fingerlings are the most susceptible to infection, and mortality in acute outbreaks may range from 30–70%. Enteric redmouth commonly occurs as a chronic disease, particularly in older fish >12 cm (4.75 in.), with cumulative mortality being approximately 30%. Most outbreaks occur when water temperatures are in the range 11–18 °C (52–64 °F), with the greatest severity occurring between 15 and 18 °C (59–64 °F).

e. Geographic range and host susceptibility—Since its initial isolation in the Hagerman Valley of Idaho, the disease has been found throughout U.S. trout-growing areas. Fish disease caused by *Y. ruckeri* has been confirmed in many regions throughout the world, including North America, Europe, Australia, Finland, Norway, and South Africa. The rainbow trout is the species most susceptible; however, all salmonids can be infected. Some nonsalmonid species, such as the fathead minnow, *Pimephales promelas*, the cisco, *Coregonus artedii*, the whitefish, *Coregonus clupesformis*, the sturgeon, *Acipenser* spp., and the turbot, *Scophthalmus maximus*, are susceptible. The bacterium has also been isolated from invertebrates, sea gulls, and muskrats, from sewage, and from river water. One human clinical isolate was reported (7,8).

f. Clinical signs of disease and treatment—As the common name of the disease suggests, reddening around the mouth and operculum (due to subcutaneous hemorrhage) is the most obvious of the gross external clinical signs. Other external signs that are commonly reported are exophthalmia and a general darkening of body pigmentation. Internal clinical signs are indistinguishable from other gram-negative bacterial septicemias. Treatment of ERM is best accomplished by feeding Romet or Terramycin medicated feeds at 50 mg/kg (23 mg/lb/day) body weight/day for 5 or 14 days respectively. ERM is successfully managed by vaccination of fingerlings with commercial bath vaccines.

IV. Other Enterobacteriaceae. Other members of the *Enterobacteriaceae* have, rarely, been implicated as the cause of disease in fish. Bacteria of the genera *Proteus*, *Serratia*, *Citrobacter*, *Enterobacter*, and *Hafnia* have been implicated as causative agents of fish disease; however, these diseases have not appeared with any consistency in aquaculture settings (9).

B. *Aeromonadaceae.* Facultatively anaerobic, oxidase-positive, 0/129-resistant, Gram-negative rods.

I. Aeromonas hydrophila ("A. punctata, A. liquefaciens"), A. sobria, A. caviae. Causative agents of Motile Aeromonad Septicemia. Synonyms: hemorrhagic septicemia, infectious dropsy, infectious abdominal dropsy, red sore. The motile aeromonads are treated here as a group, because the number of species involved is in question and because the taxonomy is in a state of transition.

a. History—Although reference is made as early as 1891 to fish with hemorrhagic septicemia (32), the description of a bacterial disease in cultured carp with infectious dropsy caused by "*Pseudomonas punctata*" (*A. punctata*), by Schaperclaus in 1930, is believed to be the first documented case of motile aeromonad septicemia (MAS) (33). Ewing et al., in 1961 (34), proposed that *A. punctata*, *A. hydrophila*, and *A. liquefaciens* were all variants of the same species and included them all in a single species, *A. hydrophila*. The name of the disease syndrome was changed to "motile aeromonad septicemia" in 1975, to reflect the multispecies etiology of the disease. Seven species of motile aeromonads were recognized by Carnahan et al., in 1991; however, only *A. hydrophila*, *A. sobria*, and *A. caviae* are currently recognized as representative of the motile aeromonad fish pathogens.

b. Culture—Primary isolation is achieved by streaking on standard media such as BHIA or TSA and incubation at 30–37 °C. Growth is rapid at these temperatures, with 2–3 mm, cream-colored colonies visible in 18–24 hrs. Most strains are beta–hemolytic on blood agar. Motile *Aeromonas* spp., form yellow–orange colonies on Rimler–Shotts selective medium when incubated at 35 °C (36).

c. Description—Short, gram-negative rods, 0.8 × 1.0 μm, motile by a single polar flagellum, oxidase-positive, fermentative in glucose motility deeps. Aeromonads also uniformly reduce nitrate to nitrite. Motile aeromonads are easily differentiated from the motile vibrios and *Photobacterium* by resistance to vibriostatic agent, 0/129 (2,4-diaminiop-6,7-diisopropyl pteridine phosphate). The motile aeromonad fish pathogens are differentiated from related fish pathogens in Table 2.

d. Epizootiology—Motile aeromonad septicemia is considered one of the most common diseases of cultured warm-water fish in freshwater environments. The bacterium is also known to cause disease in brackish-water fish culture at salinities up to 15 ppt. Aeromonads are common members of the microflora of natural waters, are ubiquitous in a variety of environments, and are commonly found as part of the normal microflora of the intestine and skin of fish. Motile aeromonad septicemia occurs most frequently in the spring of the year but can occur year round (7). Natural fish kills on lakes and reservoirs in the spring are often a result of MAS epizootics acting either alone or in concert with parasitic infestation (37). The motile aeromonads are considered opportunistic pathogens, and MAS is usually considered a stress related disease. Such stress factors as rising water temperatures, handling, transport, poor water quality, and parasitic load may contribute to outbreaks. Those outbreaks are chronic

Table 2. Biochemical and Biophysical Characteristics Important in Differentiating Some Fish Pathogens in the Family *Aeromonadaceae*

Characteristic	*A. hydrophila*	*A. sobria*	*A. caviae*	*A. salmonicida* subsp. *salmonicida*	*A. salmonicida* subsp. acromogenes
Motility at 25 °C	+	+	+	−	−
Gas in GMD[a]	+	+	−	+	−
Esculin hydrolysis	+	−	+	+	−
Brown pigment	−	−	−	+	−
Voges–Proskauer	+	+	−	−	−
Growth at 37 °C	+	+	+	−	−
Acid from arabinose	+	−	+	+	−

[a]GMD = glucose motility deep, incubated at optimum growth temperature of bacterium tested.

in nature, with low mortality rates. "Poststocking syndrome" is a disease that is characterized by a peak in mortality approximately three days following the transport and stocking of fish in a new pond or unit. Infection, disease, and mortality may continue for a few days; then the losses diminish and stop. Motile aeromonads are often a major component in "winter mortality syndrome" or "winterkill," a disease in catfish ponds related to rapid temperature drops in the winter months, to fungal disease, and to secondary bacterial infection. The exception to this may be certain strains of *A. sobria* and *A. hydrophila* that possess a surface protein layer (S-layer) that serves as a virulence factor (38). Highly virulent strains may cause a typical hemorrhagic septicemia, with rapid progress and high mortality rates.

e. Geographic range and species susceptibility—Motile aeromonads are found in warm-water freshwater and brackish-water environments throughout the world. This group of bacteria shows little or no host specificity; it infects a variety of freshwater and brackish-water fish hosts. Aquatic animals other than fish are susceptible to motile aeromonad infections; Compare redleg disease of frogs and fatal disease of reptiles (39). Although not considered a serious health concern, *A. hydrophila* has occasionally been responsible for wound infections and even fatal septicemia in humans (7).

f. Clinical signs of disease and treatment—MAS occurs in acute and chronic forms. The acute form is a typical bacterial septicemia characterized by exophthalmia, abdominal distension due to ascites, petechial hemorrhage in the skin, and diffuse necrosis in the internal organs. The progress of infection may be so rapid that few clinical signs are evident (38). The chronic form is characterized by the formation of deep muscular ulcerations, with associated hemorrhage and inflammation. Treatment for acute disease is with medicated feeds (Romet® or Terramycin®) at 50 mg/kg/day. The chronic form of the disease usually responds to improvement of water quality and to removal of stress. Vaccination is not considered feasible, because of the multiplicity of strains and serotypes encountered.

II. Aeromonas salmonicida. Nonmotile Aeromonads.

Three subspecies or groups of *A. salmonicida* have been proposed by McCarthy and Roberts (40), and supported by various taxonomists (41) on the basis of epizootiology and phenotype, even though DNA homology studies do not support the necessity for subspecies.

Group 1. *A. salmonicida* subsp. *salmonicida*
Causative agent of furunculosis in salmonids.

Group 2. *A. salmonicida* subsp. acromogenes ("*Haemophilus piscium*")
Causative agent of ulcer disease in salmonids and bacterial septicemia of salmonids (phenotypically atypical strains).

Group 3. *A. salmonicida* subsp. *nova*
Causative agent of carp erythrodermatitis and goldfish ulcer disease (phenotypically atypical strains).

a. History—The first report of furunculosis is from 1894, by Emmerich and Weibel, from a trout hatchery in Germany (42). The bacterium was referred to as *Bacillus der Forellenseuche* in Germany and *Bacillus salmonicida* in England in the early 1900s and was reclassified as *A. salmonicida* by Griffin in 1953 (43). The name furunculosis is somewhat of a misnomer and was given to describe the "boil-like" lesions of the original form of the disease in trout; however, from a pathologic standpoint, there is little similarity to the pus-filled boils that occur in humans (44). Judging from clinical signs, one can speculate that ulcer disease of trout was probably first seen in 1899 by Calkins (45). Ulcer disease of trout (UD) was described by Snieszko (46), and the causative bacterium was named *H. piscium* (from diseased brook trout) in 1980 (47). The relationship between *A. salmonicida* and *H. piscium* was investigated by Paterson (48) and by McCarthy and Roberts (40), and it was concluded that *H. piscium* was simply an atypical, achromogenic strain of *A. salmonicida*. Carp erythrodermatitis (CE) was described in 1972 by Fijan (49) from common carp, and goldfish ulcer disease (GUD) was described in 1980 by Elliott and Shotts (50).

b. Culture—Primary isolation of typical strains is achieved on BHI agar, TSA agar, or TSA agar with 5% sheep blood and incubation at 20–25 °C. Growth is fairly rapid even at these temperatures, with 2-mm cream-colored colonies visible in 48 hours. Some atypical strains from ulcer disease may be somewhat fastidious, and growth may be enhanced by the addition of fish peptone to the primary isolation medium (45).

c. Description—*A. salmonicida* is a gram-negative, facultatively anaerobic, nonmotile rod, 0.8–1.3 ×

1.3–2.0 μm. Typical strains, when grown on media containing tyrosine, produce a brown, water-soluble pigment. Atypical strains do not produce this pigment; they also vary on a few tests involving enzyme production and carbohydrate fermentation. Fresh isolates from fish are virulent, produce "rough" colonies (because of the presence of a surface protein "A-layer"), and autoagglutinate in broth culture. Colonies that have been subcultured, particularly at higher temperatures 25 °C (77 °F), are nonvirulent, become "smooth," are A-layer negative, and do not autoagglutinate (51). Once the bacterium has lost the A-layer via high-temperature incubation, it cannot be induced to revert to being A-layer positive. The nonmotile subspecies of *A. salmonicida* are differentiated from related fish pathogens in Table 2.

d. Epizootiology—*A. salmonicida* is considered an obligate pathogen, even though recent studies have shown it can survive for months in fresh water and in muds. The current belief is that furunculosis and other related diseases caused by the subspecies of *A. salmonicida* are transmitted horizontally from fish to fish by close contact through the skin or by ingestion of water containing high numbers of the organism (40). Skin damage due to abrasion or to external parasites (such as salmon lice) may open portals of entry for the bacterium. Asymptomatic infected carriers are vehicles for persistence within a population. Vertical transmission is not believed to be important in the spread of furunculosis. The disease has always been an important economic problem in salmonid aquaculture, and it has recently been cited as a problem in such new aquaculture venues as the sea-ranching of salmon in Scotland and Norway, where losses as high as 20% have been reported (8).

e. Geographic range and host susceptibility—Furunculosis of salmonids is found in almost every region of the world in which trout are grown, most notably in North America, Great Britain, Europe, Asia, and South Africa (7). All salmonids are susceptible to the disease. Atypical strains produce carp erythrodermatitis in common carp (*C. carpio*) in Europe and Great Britain but not in North America or Asia. Goldfish ulcer disease has been diagnosed from the goldfish, (*Carassius auratus*), in the U.S., Italy, Great Britain, Japan, and Australia (7).

f. Clinical signs of disease and treatment—Gross external clinical signs in salmonids with furunculosis vary depending on the time course of the disease process. In acute disease, fish show very few clinical signs, and mortality is high. In chronic disease, the following signs become apparent: darkening, lethargy, anorexia, petechial hemorrhage, swellings in the skin/muscle containing necrotic debris (furuncles), and hemorrhage in the gill filaments. In nonsalmonids infected with atypical strains, the disease begins with small localized skin infections that progress into larger ulcerations, which may be secondarily infected by motile aeromonads or pseudomonads. The disease ultimately becomes a septicemia, if the fish does not die first from a secondary infection. Medicated feeds with Terramycin® or Romet® administered at 50 mg/kg/day (23 mg/lb/day) have traditionally been the treatment of choice in the U.S. During the 1980s, drug-resistant strains of *A. salmonicida* began to appear in Europe and Norway, and additional antimicrobials had to be explored. In Scotland, there are currently four antibacterial agents licenced for control of furunculosis: oxytetracycline, oxolinic acid, trimethoprim-sulfadiazine, and amoxicillin. Multiple resistance to these drugs is common now, with the exception of amoxicillin (52); however, some isolates of *A. salmonicida* subsp. acromogenes have recently been reported to be resistant to amoxicillin (53). Florphenicol has been reported to be a very effective chemotherapeutic in experimental trials (54). *A. salmonicida* is an antigenically homogeneous species, with variability only in the presence or absence of A-layer; however, there have over the last 50 years been many failed attempts at producing successful commercial vaccines (55). The use of oil-adjuvanted vaccines has given the first encouraging results (56).

C. Vibrionaceae. Facultatively anaerobic, oxidase positive, 0/129 sensitive, gram-negative curved rods. All but two species are halophilic and thus are known primarily as pathogens of marine fish.

I. Vibrio anguillarum, Listonella anguillarum and Vibrio ordalii. Causative agents of vibriosis. Synonyms: salt water furunculosis, boil disease, ulcer disease, red pest of eels.

a. History—As stated in the introduction to this chapter, vibriosis may well be the longest known bacterial disease in aquaculture. The clinical signs of a disease in eels cultured in the Mediterreanean as early as 1718 are consistent with vibriosis (1). The causative agent was first cultured from diseased eels with "red pest" in 1883 (2) and named *V. anguillarum* in 1909 by Bergeman (57). Two distinct biochemical phenotypes of *V. anguillarum* were recognized by Nybelin in 1935; that work led to the designation of *V. anguillarum biovar I* and biovar II (9). The more fastidious and nonreactive strains composing biotype II received their own species status in 1981, after the work of Schiewe (58). Today, the taxonomy of this group of organisms is in a confused state, after the suggestion of MacDonell and Colwell in 1985 that *V. anguillarum* and *V. ordalii* be reclassified in the genus *Listonella* (59). Currently references to both names are found in the literature.

b. Culture—*V. anguillarum* may be cultured from diseased fish tissues on standard culture media such as TSA with or without 5% sheep blood or BHIA and incubation at 25–30 °C. The organism grows rapidly and 2-mm colonies are visible within 24 hours. *V. anguillarum* and *V. ordalii* are halophilic and grow on culture media with a final salt concentration of 1–3%. Primary isolation of *V. ordalii* is improved by using seawater agar with 3% salt and incubation at 15–25 °C. Growth is slow, and up to seven days may be required for typical colonies to develop (7). *V. anguillarum*, but not *V. ordalii*, can be isolated on thiosulfate citrate bile sucrose (TCBS) selective medium, on which it produces a yellow colony. This medium is not differential, so additional testing is required to speciate the organisms forming yellow colonies. A new selective and differential medium, *V. anguillarum* medium (VAM), used alone or in combination with dot-blot hybridization,

Table 3. Biochemical and Biophysical Characteristics Important in Differentiating Some Fish Pathogens of the Family Vibrionaceae

Character	*V. anguillarum*	*V. ordalii*	*V. salmonicida*	*V. vulnificus* 2	*P. damselae* subsp. *damselae*	*P. damselae* subsp. *piscicida.*	*P. shigelloides*
Motility	+	+	−	+	+	−	+
Sensitive to 0/129	+	+	+	+	+	+	+
Production of:							
Arginine dihydrolase	+	−	−	−	+	+	+
Lysine decarboxylase	−	−	nr	+	−	−	+
Ornithine decarboxylase	−	−	nr	−	−	−	+
β-galactosidase	+	−	−	+	−	−	+
Indole	+	−	−	−	−	−	+
Degredation of:							
Gelatin	+	+	−	+	−	V	−
Starch	+	−	−	+	+	+	nr
Urea	−	−	−	−	+	−	−
Lipids	+	−	−	+	−	−	nr
Nitrate reduction	+	V	−	+	+	−	+
Voges–Proskauer	+	−	−	−	+	+	−
Growth at 37 °C	+	−	−	+	nr	−	+

Symbols: V = variable results, nr = no record.

has been employed for environmental enumeration of *V. anguillarum* (60).

c. Description — Short, gram-negative, slightly curved rods, 0.5 × 1.4–2.6 μm, motile by monotrichous or multitrichous sheathed polar flagella at 25–30 °C, oxidase-positive, fermentative in glucose deeps, and sensitive to vibriostat 0/129. *V. anguillarum* is more reactive in biochemical tests than *V. ordalii*, being positive for arginine dihydrolase, Voges–Proskauer, beta-galactosidase, indole, citrate, arabinose, and sorbitol. The vibrionaceae pathogenic to fish are differentiated in Table 3.

d. Epizootiology — *V. anguillarum* is regarded as the marine counterpart to *A. hydrophila*, being a part of the normal microflora of the marine aquatic environment and marine fish (61). The bacterium exhibits long-term survival in aquatic environments, and virulent and nonvirulent strains may occur. Transmission is horizontal, via the water column, and infection is a result either of direct invasion of the skin or intestine by virulent strains, or of colonization of skin abrasions by less virulent strains, or of both. Perhaps the best-characterized virulence mechanism in a fish pathogen is the siderophore-mediated, plasmid-encoded, high-affinity iron-transport system in *V. anguillarum* strain 775 (62). The expression of this system requires a stretch of about 25 kilobases (kb) of a 65-kb plasmid, pJM1, and curing *V. anguillarum* strain 775 of pJM1 results in loss of virulence (63). *V. anguillarum* is a serious problem in marine aquaculture, particularly in stressed populations, and it is therefore considered an opportunistic pathogen. Temperature also plays a role, with mortalities in salmonids being highest (60%) at 18–20 °C (64–68 °F) and lowest (4%) at 6 °C (43 °F) (64).

e. Geographic range and host susceptibility — *V. anguillarum* is found worldwide in a variety of marine fish species, and on rare occasions in freshwater species. It is a significant fish pathogen in all coastal areas of North America, the North Sea, the Atlantic and Mediterreanean coasts of Europe and North Africa, and Asia. *V. ordalii* outbreaks are confined to the Northwest Pacific Coast of North America and Japan (7). Approximately 50 species of marine fish have been listed as being susceptible to vibriosis, but salmonids, eels (*Anguilla* spp.), hybrid striped bass (*Morone* sp.), milkfish (*Chanos chanos*), and ayu (*Plecoglossus altivelis*) are the most important aquaculture species. *V. ordalii* occurs most commonly in salmon and trout.

f. Clinical signs of disease and treatment — Clinical signs of vibriosis are similar to motile aeromonad septicemia. Acute disease is accompanied by few clinical signs other than erratic swimming behavior and lethargy. The spleen may be enlarged, and the liver, kidney, and spleen may contain necrotic foci. In chronically infected fish, skin/muscle ulcerations become prominent, as does anemia. In *V. ordalii*-infected fish, the bacteria are less dispersed in the tissue, and microcolonies can be observed in sections of heart and skeletal muscle. Control of vibriosis is best achieved by maintaining good husbandry practices and by using commercially available multivalent vaccines. Treatment of the disease in salmonids is by medicated feed (Romet® 50 mg/kg/day (23 mg/lb/day) for 5 days or Terramycin® 50–75 mg/kg/day (23–34 mg/lb/day) for 10–14 days). Clearance for Romet® in salmonids is 42 days that for Terramycin® is 21 days.

II. Vibrio salmonicida. Causative agent of coldwater vibriosis. Synonyms: Hitra disease, hemorrhagic syndrome.

a. History — Coldwater vibriosis (CV) is a disease that affects primarily the Atlantic-salmon industry in Norway. The disease was first described as a multifactorial malady by Poppe in 1977 (66), from Atlantic salmon cultured on the island of Hitra off the coast of Norway. Egidius (67) suggested a bacterial etiology for the condition and, along with work by Holm (68), provided a description of the etiologic agent of CV. The bacterium was fully described, and named *V. salmonicida*, by Egidius et al., in 1986 (69). Monoclonal antibodies to a major surface antigen, VS-P1, were used to confirm the identity of *V. salmonicida* (by immunohistochemistry) in archived tissues from original outbreaks of "Hitra disease" in 1977 (70).

b. Culture — Primary isolation of *V. salmonicida* is difficult, but it can be accomplished on TSA with 5% sheep blood or BHIA, provided 1.5–2% salt is added to the medium. Optimum temperature for incubation of primary isolation plates is 15–17 °C, and growth is slow, requiring 72 hrs to form 1–2 mm, smooth, grey colonies.

c. Description — The bacterium is a gram-negative, slightly curved or pleomorphic, motile rod 0.5–2.0 μm long. *V. salmonicida* is psychrophilic, and fails to grow above 25 °C. It is oxidase-positive, motile, and sensitive to vibriostat 0/129. The bacterium is generally non-reactive on many biochemical tests; it is differentiated in Table 3 from related organisms. Strains of *V. salmonicida* are serologically and biochemically homogeneous (68) and are distinct from *V. anguillarum*, *V. ordalii*, and other vibrios. Rare isolates from Atlantic cod belong to a different serotype.

d. Epizootiology — Coldwater vibriosis is transmitted via the water from carrier fish (71), and infection is primarily through the gills. The organism is capable of survival in seawater and sediments for over one year. Mortalities in natural CV outbreaks are high, losses of 5% per day being common. Mortality in experimentally infected Atlantic salmon was 90% at 45 days postinfection. Most natural infections on fish farms occur in the autumn and winter, when water temperatures are between 4 and 9 °C. All ages and size classes of salmonids are susceptible.

e. Geographic range and host susceptibility — CV has been reported from the coast of Norway, the Shetland Islands of northern Scotland, the Faroe Islands, eastern Canada, and the northeastern U.S. (7). Atlantic salmon grown in salt water or brackish-water net-pens are most susceptible. An outbreak of CV has been reported from a highly stressed, net-pen cultured population of juvenile Atlantic cod (*Gadus morhua*) (72).

f. Clinical signs of disease and treatment — Acutely infected fish come to the surface of net-pens, swim erratically, and exhibit few (if any) external clinical signs. In subacute to chronic stages of the disease, pale gills, hemorrhage at the base of the fins and in the muscle, reddish and prolapsed anus, bloody ascetic fluid in the peritoneal cavity, and hemorrhage and watery contents in the posterior intestine are all notable clinical signs. Histopathologically, CV is similar to vibriosis caused by *V. anguillarum*, except with more pronounced heart and muscle damage (13). CV is believed to be a multifactorial disease, so reducing stress on the population is a primary management strategy. Effective commercial vaccines have had a positive impact on the Norwegian Atlantic salmon aquaculture industry. In one field trial, mortalities were reduced from 24.9% to 1.87% (73). Feeds medicated with Tribrissen, Terramycin®, or furazolidone were used, initially at 75–100 mg/kg (34–45 mg/lb/day) body weight/day for 10 days, with positive results. Recent development of resistant strains, however, has led to the evaluation of new drugs, such as Florfenicol (54).

III. Vibrio vulnificus Biogroup 2. Causative agent of *V. vulnificus* infection of eel.

a. History — *V. vulnificus* biogroup 2 was originally documented as the cause of serious outbreaks of bacterial disease in eel populations in Japan between 1975 and 1977 (74). The disease is now known from Europe, where it has caused similar outbreaks in Spain (75), the Netherlands, and England (9). The bacterium was properly classified as *V. vulnificus*, and a new biotype (biogroup 2) was established by Tison et al. in 1982 (76).

b. Culture — Primary isolation is accomplished on seawater agar or TSA with an additional 2% NaCl. Growth is slow, requiring 4–7 days at 20–25 °C.

c. Description — Short gram-negative rods 0.5–2.0 μm long, oxidase-positive, sensitive to vibriostat 0/129, motile by a single polar flagellum, with growth between 20 and 37 °C. *V. vulnificus* biogroup 2 differs from typical *V. vulnificus* by negative results in tests for indole, for ornithine decarboxylase, and for acid from mannitol or sorbitol and by growth at 42 °C (9). Characteristics that differentiate *V. vulnificus* are found in Table 3.

d. Epizootiology — Little has been reported on the epizootiology of this disease, but it is believed to be an opportunistic infection, because the organism has low virulence in experimental infections and is ubiquitous in the marine environment.

e. Geographic range and host susceptibility — The disease is currently known to be a problem in eel mariculture in Japan, England, the Netherlands, and Spain. Thus far, only Japanese eels (*Anguilla japonica*) and European eels (*Anguilla anguilla*) are known to be susceptible. All strains pathogenic to fish form a homogeneous group in terms of serology and of biochemical phenotype.

f. Clinical signs of disease and treatment — The disease is characterized by hemorrhage in the gills and along the flank or tail. Internally necrotic lesions may be found in the liver, the spleen, the heart, and the intestine; they bear close resemblance to lesions seen with classical vibriosis in eel.

IV. Other Vibrio spp. Several additional species of vibrios have been found on various occasions to be pathogenic for fish and shellfish (7,9,13). Uncertainty about their overall impact on aquaculture has led to listing these additional species here as potential pathogens, without going into detail. With time, some of these bacteria may emerge as significant pathogens in aquaculture:

V. alginolyticus — gilthead seabream, mullet, penaed shrimp.

V. vulnificus biogroup 1 or typical strains — hybrid striped bass, red drum, tilapia, prawns.

V. carchariae — sharks.

V. cholerae, *V. mimicus* — ayu, channel catfish, red drum, crayfish.

V. fischeri—turbot, penaed shrimp.

V. harveyi—snook.

***V. Photobacterium damselae* subsp. *damselae* ("*Vibrio damsela*," "*Listonella damsela*").** Causative agent of ulcerative disease of damselfish and of hemorrhagic septicemia of various other species. Synonym: Vibriosis

a. History—*P. damselae* subsp. *damselae* ("*Vibrio damsela*") was initially discovered as the causative agent of an ulcerative disease in damselfish (*Chromis punctipinnis*) inhabiting the coastal waters of southern California (77). Additional work by Grimes in 1984, on isolates from captive sharks, confirmed the new species (78). The organism has also been isolated from human wound infections (77). The taxonomy of the organism has changed since its initial description, having been moved to *Listonella* in 1985 (59) and later to *Photobacterium* in 1991 (79).

b. Culture—Primary isolation is accomplished on BHIA or TSA with 5% sheep blood and incubation at 25 °C for 48 hours. Addition of 2% NaCl to primary isolation medium is suggested. Growth is rapid, with typical 2-mm colonies visible after 48 hours incubation at 25 °C. TCBS agar may be used for selective isolation; however, additional testing is required for speciation.

c. Description—Short (0.5–2.0 μm), gram-negative, slightly curved or pleomorphic rods weakly motile by one or more sheathed, polar flagella. *P. damselae* subsp. *damselae* is positive for oxidase, urease, and arginine dihydrolase and negative for acid from sucrose. Additional characteristics that differentiate it from related organisms are found in Table 4.

d. Epizootiology—*P. damselae* subsp. *damselae* is a normal inhabitant of marine waters; it may invade highly susceptible fish directly, and less susceptible fish secondarily to injury. Host specificity was investigated in laboratory experiments by scarifying the dermis of a variety of fish species and swabbing the area with a solution containing 10^8 viable cells of *P. damselae*. Only damselfish succumbed to the disease by this method.

e. Geographic range and host susceptibility—*P. damselae* subsp. *damselae* has been documented as a pathogen of natural populations of damselfish in coastal southern California (77), in sharks and dolphins held in captivity (78), in cultured turbot and sea bream in Europe (80,81), in yellowtail (*S. quinqueradiata*) in Japan (82), in rainbow trout in Denmark (83), and in barramundi (*Lates calcarifer*) in Australia (84). Stress associated with temperatures higher than normal is a common theme in aquaculture outbreaks.

f. Clinical signs of disease and treatment—*P. damselae* subsp. *damselae* causes ulcerative lesions in the skin of affected damselfish, usually near the pectoral fin and caudal peduncle. Ulcers sometimes increase in size to ~20 mm (1.75 in.), prior to death of the fish. In turbot and seabream, ulcers are not noticed, but hemorrhages in the eyes, at the base of the fins, and around the anus, as well as abdominal distension are typical clinical signs. Treatment of *P. damselae* infections in Europe and Japan is done with oxytetracycline, oxolinic acid, and trimethoprim–sulfamethoxazole.

***VI. Photobacterium damselae* subsp. *piscicida* ("*Pasteurella piscicida*").** Causative agent of photobacteriosis. Synonyms: Pasteurellosis, pseudotuberculosis.

a. History—Photobacteriosis was first described by Snieszko et al., in 1964 (85), from a massive natural fish kill involving white perch (*Morone americana*) and striped bass (*M. saxatilis*) in the upper Chesapeake Bay, U.S. The authors placed the causative organism in the genus *Pasteurella*, on the basis of physiological, morphological, and staining characteristics. The bacterium was

Table 4. Biochemical and Biophysical Characteristics Important in Differentiating Among Gram-Positive Cocci and Cocco Bacilli which have been Isolated from Fish

	S. iniae	*S. difficilis*	*L. garvieae*	*L. piscium*	*V. salmoninarum*	*E. faecium*	*C. piscicola*
Shape	cocci	cocci	ovoid	cocci	ovoid	ovoid	cocco-bacilli
Hemolysis	β	γ	α	γ	α	α, β, or γ	γ
Esculin	+	−	+	+	+	+	+
Hippurate	−	+	−	−	d	−	−
ADH	+	+	+	−	−	+	+
PYR	+	−	+	−	+	+	+
β-Gur	+	−	ND	ND	−	−	ND
PAL	+	+	−	ND	+	−	ND
H_2S	−	−	−	−	+	−	−
Growth at/in:							
45 °C	−	−	+	−	−	+	−
10 °C	v	−	+	+	+	+	+
pH 9.6	v	+	+	−	+	+	+
40% bile	−	+	+	ND	+	+	−
L-Arabinose	−	−	−	+	−	−	−
Inulin	−	−	−	−	−	−	+
Lactose	−	−	v	+	−	−	v
Mannitol	+	−	+	+	−	+	+
Raffinose	−	−	−	+	−	−	v

studied morphologically, physiologically, and serologically in 1968 by Janssen and Surgalla (86), who concluded that it represented a new species and proposed the name "*P. piscicida*." Outbreaks of "pseudotuberculosis" in cultured yellowtail in the late 1960s in Japan were attributed to this bacterium (87). Photobacteriosis remains one of the most serious diseases affecting Japanese mariculture; losses of yellowtail in excess of 2000 tons were reported in 1989 (88). Prior to 1990, there were no reports of photobacteriosis from Europe. After an account by Toranzo et al., in 1991 (89) of the disease in gilthead seabream (*Sparus aurata*) cultured in Spain, the disease seemed to spread throughout the region. Currently, photobacteriosis represents a significant economic problem in European and Meditereanean mariculture. The first report of photobacteriosis from cultured fish in the U.S. was from striped bass cultured in brackish water ponds on the Alabama Gulf Coast by Hawke et al., in 1987 (90). In 1990, the disease recurred on the U.S. Gulf Coast, in Louisiana, and yearly outbreaks have severely damaged an emerging hybrid striped bass mariculture industry (91). "*P. piscicida*" was never accepted as a valid name by bacterial taxonomists, and ultimately the bacterium was renamed *P. damselae* subsp. *piscicida* in 1995, on the basis of 16 sRNA sequencing (92); the name was later corrected to *P. damselae* subsp. *piscicida*.

b. Culture — Primary isolation is accomplished on TSA with 5% sheep blood or BHIA with 1–2.5% NaCl and incubation at 25–28 °C. Typical 1–2 mm, grey, nonhemolytic colonies appear after 48 hours of incubation. A selective medium is not available.

c. Description — *P. damselae* subsp. *piscicida* is a gram-negative, slightly curved or pleomorphic, nonflagellated, nonmotile rod (0.7 × 0.7–2.6 μm), often exhibiting bipolar staining. Coccoid forms predominate in older cultures. It is positive in tests for oxidase, catalase, Voges–Proskauer, and arginine dihydrolase, and it ferments glucose, mannose, galactose, and fructose. It is halophilic, requiring 0.5% salt to grow, and it is sensitive to vibriostatic agent 0/129. Strains of *P. damselae* subsp. *damselae* are serologically and phenotypically homogeneous regardless of the geographic source (93) and are easily differentiated from *P. damselae* subsp. *damselae* and other Vibrionaceae (Table 3).

d. Epizootiology — *P. damselae* subsp. *piscicida* was initially considered an obligate pathogen, because it fails to survive in brackish water for more than five days (94); however, Magarinos et al., demonstrated long term survival in seawater and in sediment, in a viable but nonculturable form that retains virulence (95). A carrier or latent state has not been demonstrated in susceptible hosts, but it has been theorized that other species of fish in the vicinity of fish farms, or perhaps an invertebrate, may harbor the pathogen (96). Most susceptible hosts can be infected via the water, so horizontal transmission from fish to fish within a culture unit is the most likely method of spread during epizootics. Outbreaks on fish farms are explosive and are characterized by sudden reduction in feeding response and rapid onset of mortality. Cumulative mortalities in striped bass (80%), gilthead seabream (40%), and striped jack (34%) have all been documented as occurring over a four week period. Outbreaks of photobacteriosis are correlated with water temperatures of 18–25 °C (64–77 °F) and salinities of 5–25 ppt.

e. Geographic range and host susceptibility — Since its initial description from white perch in Chesapeake Bay, reports of new hosts and of an extented range have continued to increase. At present, the disease is known from Japan, the Atlantic and Gulf Coasts of the U.S., Europe, the Mediterranean, and Israel. The list of susceptible hosts of aquaculture importance includes the yellowtail, the striped jack (*Pseudocaranx dentex*), the ayu, the black seabream (*Mylio macrocephalus*), the red seabream (*Pagrus major*), striped bass (and hybrids), the gilthead seabream (*S. aurata*), and the sea bass (*Dicentrarchus labrax*) (7).

f. Clinical signs of disease and treatment — Photobacteriosis is an acute bacterial septicemia with a striking lack of gross clinical signs in most susceptible hosts. In hybrid striped bass and striped bass, redness in the operculum, enlargement of the spleen, and pallor of the gills are the only consistent signs. Necrotic foci are seen microscopically in the spleen and kidney, but an inflammatory response is lacking. In yellowtail, the disease progresses slightly more slowly, and white miliary lesions of about 1–2 mm (0.1–0.2 in.) become visible in the spleen and kidney. Necrotic foci containing bacterial colonies in the splenic parenchyma, and an associated chronic inflammatory response, has led to the use of the misleading pathological term "pseudotuberculosis" (13). Because of the rapid onset of disease, medicated feeds are usually not offered early enough in the infection to be effective. If timely application is achieved, the pathogen responds well to treatment with oxytetracycline, Romet, oxolinic acid, ampicillin, amoxicillin, and florfenicol medicated feeds. Currently, there are no drugs approved for treatment of photobacteriosis in hybrid striped bass in the U.S. With the widespread use of antibiotics on Japanese fish farms, resistant strains of the pathogen, carrying R-plasmids marking multiple drug resistance, have been isolated. Because of the general ineffectiveness of medicated feeds in combating the disease, vaccination is a logical approach for future management. Commercial vaccines are currently in the developmental stage.

VII. Plesiomonas shigelloides and

VIII. Shewanella putrefaciens (Pseudomonas putrefaciens, Alteromonas putrefaciens). *P. shigelloides* and *S. putrefaciens* are bacteria of only questionable importance as pathogens in aquaculture. Only one published report lists *P. shigelloides* as a fish pathogen. The bacterium was isolated from diseased, farmed rainbow trout in Portugal (98). *S. putrefaciens* has also been reported on only one occasion, from rabbitfish grown in sea cages in the Red Sea (99). There are other anecdotal accounts of *P. shigelloides* and *S. putrefaciens* as suspected pathogens in a variety of fresh and marine fish species (9). Both species are commonly isolated as post-mortem contaminants from decomposing fish tissue and from the intestines of healthy fish. With this in mind, further description of the organisms is not given, other than to include data on identification of *P. shigelloides* to allow comparison with closely related organisms with which it might be confused (Table 3). The bacteria are presently classified in the Vibrionaceae but may be subject to a reclassification, one pending acceptance of the recommendations of MacDonell and Colwell (59).

D. *Pseudomonadaceae*

I. Pseudomonas anguilliseptica. Causative agent of red spot disease of eels, hemorrhagic septicemia of marine fish. Synonym for Japanese eel disease, Sekiten-byo.

a. History—*P. anguilliseptica* was first described in 1972, as the causative agent of red spot disease of cultured eels (*A. japonica*) in Japan, by Wakabayashi and Egusa (100). The disease has become one of the most significant problems in eel culture in Japan and recently has been reported from European eels (*A. anguilla*) cultured in Scotland (101). The disease seems to be increasing in importance in eel growing areas throughout many parts of the world (9). Recent outbreaks of disease on mariculture farms in Finland (102), in Malaysia (103), in Japan (104,105), and along the Mediterranean and Atlantic Coasts of France in a variety of fish species (106) are referred to as "hemorrhagic septicemia," and *P. anguilliseptica* is the causative agent. The lack of host specificity represents a potential hazard for many farmed species in the future.

b. Culture—Primary isolation can be made on nutrient agar with 10% horse blood or nutrient agar containing 0.5% (w/v) NaCl adjusted to pH 7.4. Incubation should be between 20–25 °C, and the resulting growth is slow, requiring 72 hours for small (~1 mm), shiny, pale-grey colonies to form.

c. Description—*P. anguilliseptica* is a Gram-negative rod (5.0–10.0 × 0.8 μm), motile by a single polar flagellum at 15 °C but not at 25 °C. The bacterium is oxidase and catalase-positive, does not produce acid from glucose or any of a number of other carbohydrates oxidatively or fermentatively, and is resistant to vibriostatic agent (0/129). Growth occurs between 5 and 30 °C at 0–4% NaCl. The occurrence of pleomorphic, filamentous rods 5–10 μm in length makes this bacterium unique in appearance.

d. Epizootiology—In Japan, the disease occurs in eels during April and May, when temperatures are between 15 and 20 °C (59–68 °F). Mortalities decline in the summer months, when water temperatures are between 20 and 25 °C (68–77 °F); however, outbreaks may recur in the fall, when temperatures decrease. Because of the salt tolerance of the organism, disease outbreaks are prevalent in brackish-water ponds (107). In experimental infections, Japanese eels are most susceptible to infection and disease at around 19–20 °C (66–68 °F), and very little mortality occurs above 25 °C (77 °F). Juvenile European eels (elvers) are much more susceptible to infection than adults (96% mortality vs 3.9% mortality) (101), and applications of copper sulfate at 25 to 100 μg/L increase the susceptibility of eels to infection and disease.

e. Geographic range and host susceptibility—Red spot disease primarily affects the Japanese eel in Japan and the European eel in Scotland (100,101). The Japanese eel has been shown to be the more susceptible of the two species. *P. anguilliseptica* has also been reported from eels in Taiwan (7). Recently, outbreaks of the disease have occurred in farmed black sea bream and ayu in Japan (104,105), in salmonids in Finland (102), and in sea bass, sea bream, and turbot cultured on the French Mediterreanean and Atlantic coasts (106). Heavy mortalities were reported in 1987 during infections of the giant sea perch (*L. calcarifer*) and in grouper cultured in offshore sea cages in Malaysia (103). *P. anguilliseptica* has also been isolated from wild Baltic herring on the southwest coast of Finland (108). Wild fish were suspected to serving as a vector to transmit the disease to farmed salmonids. Experimental infections have been achieved in ayu, bluegill, carp, goldfish, and loach (109).

f. Clinical signs of disease and treatment—Japanese eels exhibit extensive petechial hemorrhage in the subepidermal layer of the jaws, on the underside of the head, and along the ventral body surface. Internally, there are petechiae in the peritoneum, enlargement of the liver, and atrophy of the kidney and spleen. Pericarditis is also commonly observed. Clinical signs in other fish species are typical of hemorrhagic septicemia. Treatment of the disease in eels is achieved in a variety of ways, including the following: administering feed medicated with Romet® [50 mg/kg/day (23 mg/lb/day) for five days] or oxolinic acid [5–20 mg/kg/day (0.9–9.0 mg/lb) for three days]; antibiotic baths with oxolinic acid [2–10 mg/L (ppm)]; or raising the water temperature to >27 °C (80.6 °F). Formalin-killed bacterins are effective vaccines against red spot when administered by injection, but not when delivered by immersion (110).

II. Pseudomonas fluorescens. Causative agent of Pseudomoniasis. Synonyms: bacterial tail rot, fin rot, hemorrhagic septicemia.

a. History—*P. fluorescens* was described originally (as *Bacillus fluorescens*) by Trevisan in 1889 (111). The bacterium is a normal inhabitant of soil and water and has been associated with the spoilage of such foods as eggs, fish, and milk (111). Early accounts of hemorrhagic septicemia in cultured carp by Otte in 1963 included *P. fluorescens* along with the motile aeromonads as potential causative agents (9). The importance of *P. fluorescens* in aquaculture is not clear, and its occurrence over the years has been sporadic.

b. Culture—Primary isolation is achieved on nutrient agar, TSA with 5% sheep blood, or BHIA and incubation at 22–25 °C. Cetrimide agar (Difco) or Pseudosel (BBL) are selective media used to isolate pseudomonads from mixed populations of gram negative rods (112). Growth on standard isolation media is rapid, with 1–2 mm, cream-colored colonies visible in 24 hours at 25 °C.

c. Description—*P. fluorescens* is a gram-negative rod (0.5 × 1.5–4.0 μm) that is motile by 1–3 polar flagella and is positive in tests for oxidase, catalase and arginine dihydrolase. *P. fluorescens* produces acid from glucose aerobically, but is not fermentative (giving a +/no change reaction in GMD). Colonies on TSA, Cetrimide agar, or Mueller Hinton agar plates produce a diffusible yellow-green pigment (pyoverdin) that is fluorescent under short wavelength (ca. 254 nm) ultraviolet light (112). Growth occurs in the range of 4–37 °C but not at 42 °C. Five biovars are reported in Bergey's Manual of Systematic Bacteriology (111), but strains pathogenic to fish have not been classified by this method.

d. Epizootiology—*P. fluorescens* is an opportunistic pathogen that usually invades the host secondarily to stress or injury. Ghittino, in 1966, stated that degraded environmental conditions were vital to initiation of outbreaks of hemorrhagic septicemia caused by *P. fluorescens* and that often the infection was secondary to viral infection (45). *P. fluorescens* is most often seen causing fin rot and skin lesions in a variety of fish species reared under stressful conditions (45). Pseudomonad infections display many similarities to motile aeromonad septicemia. One exception is that *P. fluorescens* infections can also occur at low temperatures. For instance, 100% mortality was observed in tench fry over a 10-day period at 10 °C (50 °F) (113). Winter mortality in silver and bighead carp at temperatures near freezing has also been reported by Csaba (114).

e. Geographic range and host susceptibility—*P. fluorescens* is ubiquitous in aquatic environments world wide. Published accounts of pseudomoniasis have listed a wide range of susceptible fish species, including silver and bighead carp, goldfish, koi, tench, grass carp, black carp, white catfish, rainbow trout, and freshwater tropical aquarium species (12).

f. Clinical signs of disease and treatment—The most common clinical sign in pseudomoniasis is fin or tail rot, in which large portions of the fins are lost due to necrosis. In carp, the disease is usually indistinguishable from motile aeromonad septicemia, displaying hemorrhages in the skin and ascetic fluid accumulations in the body cavity. In rainbow trout and koi, the presence of skin lesions similar to those seen in ulcer disease are common. Treatment of pseudomonas septicemia is best achieved by improvement of water quality conditions and by elimination of environmental stress. On nonfood fish, bath treatments with benzalkonium chloride [1–2 mg/L (ppm) for 1 hour] and furnace [0.1 mg/L (ppm) for 24 hours] may be effective (9). Pseudomonads are notorious for multiple drug resistance; therefore, isolation of the pathogen and determination of antimicrobial susceptibility is essential before a recommendation can be made. Vaccines are currently not in use, presumably because of the relationship of stress to disease outbreaks.

***III. Other Pseudomonas* spp.** A variety of pseudomonads have been implicated as fish pathogens (9), but, because of a lack of consistency in their occurrence, they are simply listed here, along with their susceptible host(s) and geographic location:

P. chlororaphis—amago trout (*Oncorhynchus rhodurus*); Japan (9).

P. pseudoalcaligenes—rainbow trout; England (9).

P. sp. ("hemorrhagic ascites")—ayu; Japan (115).

GRAM-POSITIVE AGENTS

A. *Streptococcaceae.* Small cocci, usually in chains, facultatively anaerobic, catalase negative.

Recent changes in the classification of the streptococci, stemming from molecular taxonomic studies, from the taxonomic reassignment of previously known fish pathogens, and from the recent discovery of several new fish pathogens, have led to a great many changes in this group, and, hopefully, to elimination of some of the confusion. Streptococci have been recognized as fish pathogens since 1958 (116); they have traditionally been grouped on the basis of a few phenotypic characteristics (biochemical test reactions), of hemolytic reactions, and of Lancefield serology. This system breaks down because some streptococci do not possess a known Lancefield antigen, because members of different species may belong to the same Lancefield group, and because strains within a species may have heterogeneous Lancefield antigens. Also, hemolysis patterns may be interpreted differently in different laboratories. In spite of these problems, these methods are still useful as initial steps in the identification process (117). Because the older references use this system, many of the early descriptions of streptococci pathogenic to fish are incomplete. This defect has resulted in a multitude of published reports of streptococcal disease in a variety of fish species, giving the impression that many different species are the etiologic agents of disease. Some reviewers, out of frustration, simply lump them all together as causative agents of a single disease: "streptococcosis." In cases where strains were archived, these agents have been renamed and (often) reclassified; however, information on other unavailable strains is too sketchy for them to be classified properly. Discussed under this heading are diseases caused by members of the genera *Streptococcus*, *Enterococcus* (previously group D streptococci), *Lactococcus* (formerly group N streptococci), and *Vagococcus* (formerly motile group N streptococci).

I. Streptococcus iniae ("Streptococcus shiloi"). Causative agent of streptococcosis. Synonyms: -hemolytic streptococcal disease, bacterial meningoencephalitis, mad fish disease, golf ball disease of freshwater dolphin.

a. History—*S. iniae* was first described from diseased Amazon freshwater dolphin (*Inia geoffrensis*) housed at a public aquarium in San Francisco, California (118). The dolphin (an aquatic mammal) had numerous subcutaneous abcesses (a condition called "golf ball disease") from which β-hemolytic streptococci were isolated. The disease was reported from the same species at another aquarium in New York two years later (119). There were no further reports of this organism causing disease in the U.S. until 1994, when Perera et al. (120) described *S. iniae* as the causative agent of an epizootic in hybrid tilapia (*Oreochromis niloticus* × *O. aureus*) cultured in Texas. Eldar et al. (121), described a new species in 1994, *S. shiloi*, as the causative agent of "bacterial meningoencephalitis" in cultured tilapia and trout in

Israel; however, *S. shiloi* was later shown to be a junior synonym of *S. iniae* (122). Epizootics caused by β-hemolytic streptococci occurred on Japanese fish farms between 1979 and 1986 in a variety of species. Although the descriptions are difficult to compare, because of the different methods used, all appear consistent with *S. iniae*, and one archived strain has been identified by serology as *S. iniae* (8). The first report of *S. iniae* infection in commercially reared hybrid striped bass in the U.S. was by Stoffregen et al., in 1996, from fish cultured in closed recirculating systems (123). With the development of closed-recirculating-system technology, *S. iniae* has also emerged as the first significant disease of cultured tilapia in the U.S., having been identified in 14 states (124). In Israel and Japan, it has become an even more significant problem, causing losses in the millions of dollars yearly (125).

b. Culture—Primary isolation is on TSA with 5% sheep blood. Incubation should be between 30–35 °C in a normal atmosphere. Growth is slow, requiring 48 hours for typical opaque-white colonies to develop on blood agar. Selective isolation is best achieved on Columbia CNA agar with 5% sheep blood. The bacterium fails to grow on enterococcal-selective media containing sodium aside.

c. Description—*S. iniae* is a gram-positive coccus (0.6–0.8 μm) in pairs or chains. Typical colonies on TSA blood agar are surrounded by a very narrow zone of β-hemolysis and a broader zone of α-hemolysis, although more strongly β-hemolytic strains can occur. Hemolysis is best demonstrated by stabbing the blood agar with the inoculating loop. Growth is between 10, 40 and 50 °C, and optimum growth is at 37 °C. The bacterium is nonmotile, fermentative in GMD, negative for catalase and bile esculin, and positive for esculin, leucine arylamidase (LAP), pyrrolidonyl-arylamidase (PYR), and the CAMP test. *S. iniae* does not react with any of the available Lancefield typing antisera. Biochemical tests useful in differentiating *S. iniae* from other gram-positive cocci are found in Table 4.

d. Epizootiology—Opinions vary as to the source of *S. iniae* in the aquatic environment and its degree of pathogenicity. One school of thought is that the organism is widespread in the environment, possibly disseminated by homoiothermic animals, and that it causes disease in fish only under conditions of stress. Others feel that the bacterium is well adapted as a fish pathogen with limited host specificity and that it must be transmitted horizontally from sick or carrier fish to susceptible hosts. Currently this question is unresolved. What is known is that disease outbreaks are more common in high-density aquaculture environments and that epizootics are often preceded by degraded water quality conditions or by injury resulting from handling. Tilapia appear to become more susceptible when they are subjected to temperature extremes or rapid temperature shifts. Experimental infections were achieved in tilapia by injection, by oral intubation, and by immersion at 15 °C (59 °F) to 35 °C (95 °F), with the highest rate of mortality at 20 °C (68 °F) (126). The bacterium survives for only two days in sterile saline (0.85% NaCl), although it is able to survive in muds around sea farms in Japan from year to year (127). The zoonotic potential of *S. iniae* was realized in the winter of 1995–96 in Toronto, Canada, when eight cases of invasive disease were reported in people who sustained minor injuries (cuts or scratches) while preparing fresh whole tilapia purchased in local markets but originating from fish farms in the U.S. A single clone (determined by genetic analysis) is believed to be responsible for the human disease (128).

e. Geographic range and host susceptibility—*S. iniae* is currently known in the U.S. and Canada as a pathogen of tilapia and of hybrid striped bass, in Japan as a pathogen of yellowtail, ayu, tilapia, and flounder (8), and in Israel as a pathogen of tilapia and trout. The host specificity has not been well defined, but red drum (*Sciaenops ocellatus*), channel catfish, carp, and black seabream are refractory to experimental infection (126,129). The *Streptococcus* sp., described by Al-Harbi from hybrid tilapia in Saudi Arabia (130) and the *Streptococcus* sp., described from spinefoot (*Siganus canaliculatus*) cultured in Singapore (131) are believed to be biotypes of *S. iniae* (Al-Harbi, personal communication) (8).

f. Clinical signs of disease and treatment—Tilapia and hybrid striped bass infected with *S. iniae* circle listlessly at the surface, spiral, spin, or swim erratically. The disease is a typical bacterial septicemia in the acute phase. Infected fish often have petechiae around the anus and mouth, have congestion in the fins, and may exhibit bilateral or unilateral exophthalmia accompanied by corneal opacity. The liver, kidney, and spleen are pale and enlarged, and ascites can cause abdominal swelling. The chronic stages of the disease are characterized by infection of the brain, the optic nerve, and the eye, by subcutaneous abcesses, by fibrin deposits in the peritoneal cavity, and by pericarditis. Clinical signs in other species are very similar to those described above. Although the bacterium is sensitive to ampicillin, amoxicillin, oxytetracycline, and erythromycin, feeds medicated with these antimicrobic agents have met with only limited success, with recurrence of disease several weeks following treatment. Enrofloxacin in feed at 5 or 10 mg/kg/day (2.3–4.5 mg/lb) was an effective chemotherapeutic (123) when used to treat cultured hybrid striped bass. There are currently no USFDA-approved antimicrobic agents for treatment of streptoccocal disease in hybrid striped bass or tilapia in U.S. aquaculture. In Japan, erythromycin at 50 mg/kg/day (23 mg per lb) for four to seven days has been effective in cultured yellowtail. Formalin-killed injectable vaccines have been investigated by Eldar and have induced protection lasting four months in rainbow trout (132).

***II. Streptococcus difficilis ("Streptococcus difficile")—Group B, Type Ib Streptococcus* sp.** Causative agent of group B Streptococcosis. Synonyms: nonhemolytic streptococcal disease, bacterial meningoencephalitis.

a. History—Nonhemolytic group B, type Ib streptococci were first described as fish pathogens by Robinson and Meyer in 1966 (133) from diseased golden shiners (*Notemigonus crysoleucas*) cultured in farm ponds in

Arkansas, U.S. Koch's postulates were fulfilled in shiners, and several other species of fish were found to be susceptible to experimental infection with the streptococci. Unspeciated group B, type Ib nonhemolytic *Streptococcus* sp., were documented as causative agents of large fish kills in estuarine bays along the Florida and Alabama U.S. Gulf Coast in the fall of 1972 by Plumb et al., (134), in Lake Pontchartrain, Louisiana, U.S. in the fall of 1978 (135), and in the Chesapeake Bay, Maryland, U.S. in the fall of 1988 by Baya et al. (136). A variety of marine and estuarine species were affected during these outbreaks. A group B, type Ib *Streptococcus* has been responsible for mortality in bull minnows (*Fundulus grandis*) and in hybrid striped bass at mariculture facilities on the Alabama and Louisiana Gulf Coast from 1984 to the present (124,137). In 1994 Eldar et al., described a new species, *S. difficile*, as the causative agent of bacterial meningoencephalitis in cultured tilapia in Israel (121). *S. difficile* was later shown to be a group B, type Ib nonhemolytic *Streptococcus* that by whole cell protein electrophoretic analysis was identical to *S. agalactiae* (138). Biochemically, the bacterium was more similar to other group B, type Ib streptococci isolated from fish. In a study conducted in 1990, Elliott demonstrated the similarity in whole cell protein electrophoretic profiles of group B streptococci from humans, mice, cattle, frogs and fish (139). The group B streptococci isolated from shiners, tilapia, and estuarine fish are probably biotypes of the same species, although this has yet to be demonstrated by genetic analysis. The name was corrected to *S. difficilis* in 1998 (140).

b. Culture—Primary isolation is on TSA with 5% sheep blood or BHIA. Incubation is between 25–30 °C in a normal atmosphere; however, growth may be enhanced in an atmosphere of 5% O_2, 10% CO_2, and 85% N_2. Growth is slow, requiring 48 hours for typical opaque–white colonies to develop on blood agar. Selective isolation is best achieved on Columbia CNA agar with 5% sheep blood, and the bacterium fails to grow on enterococcal selective media containing sodium aside.

c. Description—*S. difficilis* is a gram-positive coccus (0.6–0.8 μm) in pairs or chains. The bacterium is nonmotile, and colonies on blood agar are (non) γ-hemolytic. Growth occurs between 20–30 °C, and the organism fails to grow above 35 °C or below 15 °C in a normal atmosphere. Growth occurs in BHI broth with 0.5–4.0% salt, but not in 6.5% salt. The bacterium is fermentative in GMD, negative for catalase and esculin hydrolysis, and positive for Voges–Proskauer, hippurate, alkaline phosphatase (PAL), leucine arylamidase (LAP), arginine dihydrolase (ADH), and acid from ribose. The CAMP reaction is variable. Biochemical tests useful in differentiating *S. difficilis* from other gram-positive cocci are found in Table 4.

d. Epizootiology—In natural fish kills caused by group B type Ib streptococci, high water temperatures, combined with poor tidal flushing in tributaries leading into estuaries, were speculated to be the stressors that initiated infections. In Alabama and Florida, the infections seemed to begin in menhaden (*Brevoortia* sp.), a schooling planktivorous fish, and spread to carnivores and scavengers that fed on the moribund and dead menhaden. Outbreaks on mariculture farms in Louisiana and Alabama were strongly correlated with high water temperatures in late summer and early fall and with bouts of low dissolved oxygen. Experimental infections have been conducted with tilapia (125) and bullminnows (137). The LD_{50} of *S. difficilis* in tilapia by IP injection, recorded over a 6-week period, was 10^7–10^8 colony-forming units (CFU) for cultures that had been passed repeatedly on culture plates and 10^2 CFU for cultures that had been passed 3× in fish. Rasheed and Plumb found the 96-hour LD_{50} for group B type Ib streptococci to be 1.4×10^4 CFU and the 7-day LD_{50} to be 7.5×10 CFU by IP injection, in bullminnows, but could not establish experimental infections by oral intubation with 10^6 cells or bath immersion in 10^{10} CFU/mL (137). When fish were injured by scratching the skin with a scalpel blade prior to immersion, mortality rates of 75–100% resulted after 7 days.

e. Geographic range and host susceptibility—*S. difficilis* and other group B type Ib streptococci are known in Israel as pathogens of tilapia and mullet, in Japan as pathogens of yellowtail, and in the U.S. as pathogens of the following wild marine and estuarine species: menhaden (*Brevoortia patronus*), sea catfish (*Arius felis*), striped mullet (*Mugil cephalus*), pinfish (*Lagodon rhomboides*), Atlantic croaker (*Micropogonias undulatus*), spot (*Leiostomus xanthurus*), stingray (*Dasyatis* sp.), silver sea trout (*Cynoscion nothus*), spotted sea trout (*Cynoscion nebulosus*), bluefish (*Pomatomus saltatrix*), red snapper (*Lutjanus campechanus*), striped bass (*M. saxatilis*), and weakfish (*Cynoscion regalis*) (7). Susceptible species in U.S. aquaculture/mariculture are tilapia, hybrid striped bass, bullminnows, and golden shiners. Freshwater species that have been experimentally infected are bluegill (*Lepomis macrochirus*), green sunfish (*Lepomis cyanellus*), buffalo (*Ictiobus cyprinellus*), goldfish (*C. auratus*), black crappie (*Pomoxis nigromaculatus*), and largemouth bass (*M. salmoides*). Channel catfish are refractory to experimental infection (133). American toads (*Bufo americanus*) were susceptible to experimental infection, and 80% mortality was reported in bullfrogs (*Rana catesbeiana*) cultured in Brazil (141).

f. Clinical signs of disease and treatment—The clinical signs of streptococcal disease in tilapia and hybrid striped bass are similar to those of *S. iniae* infection. Bacterial septicemia in the acute phase and meningoencephalitis in the chronic phase are the most common pathological conditions. Eye disease characterized by exophthalmia, corneal opacity, hemorrhage, and rupture of the globe are commonly observed in chronic infections. Death does not always result following infection, and a certain percentage of fish become blind as a result of infections of the eye or of the optic nerve. Enlargement of the spleen, exophthalmia, and focal necrosis of the liver are characteristic of the disease in bullminnows. Group B streptococci have been shown to be sensitive to terramycin, erythromycin, ampicillin, and amoxicillin, but difficulties with recurrent infections, similar to those encountered with *S. iniae* following antibiotic treatment, are common. Erythromycin-medicated feed, 25–50 mg/kg (11–23 mg/lb/day) of body weight per day for four to seven days, was effective in controlling *Streptococcus* in yellowtail in Japan. Vaccines have been used experimentally to control the disease in tilapia, yellowtail, and hybrid striped bass.

***III. Other Streptococcus* sp.** Other species of *Streptococcus* have been implicated as fish pathogens, but information about these infections is limited; therefore, they will only be listed, along with their hosts and geographic locations:

Streptococcus dysgalactiae—tilapia; U.S.
S. sp. *similar to dysgalactiae*—ayu (*P. altivelis*); Japan.
S. parauberis—turbot (*S. maximus*); Spain.

B. *Streptococcus*—Like Bacteria

I. Lactococcus garvieae ("Enterococcus seriolicida"). Causative agent of *Lactococcus* septicemia. Synonyms: streptococcosis, enterococcosis, enterococcal infection.

a. History—Gram-positive cocci, producing -hemolysis on blood agar plates, have been implicated as causative agents of fish disease since the first description from rainbow trout in Japan by Hoshina in 1958 (116). The streptococcus of Hoshina has never been classified properly, on account of the lack of complete phenotypic data; however, it shares many characteristics with the streptococcus strains isolated from yellowtail mariculture farms near Shikoku Island, Japan by Kusuda in 1974 and from eels in Japan in 1977 (142,143). Since 1974, the disease has spread throughout Japan and has become the most economically important disease in Japanese mariculture (88). In 1989, the reported losses were 8,240 tons of yellowtail and 180.8 tons of other maricultured species such as sea bream, saurel (*Trachurus japonicus*), and flounder. The classification of the bacterium has gone through several changes in recent years. Initially it was referred to as *Streptococcus* sp., in the Japanese literature, because it did not conform to any of the described species as determined by biochemical and cultural characteristics or by Lancefield serotyping (144). In 1991, the organism was described by Kusuda as a new species, *E. seriolicida*, on the basis of DNA–DNA hybridization, despite its lack of a Lancefield group D antigen (145). In 1996, during studies of the DNA relatedness between strains of enterococci from various sources and lactococci isolated from water buffalos with mastitis, *L. garvieae* and *E. seriolicida* were found to be related at the species level. Because *L. garvieae* was the senior synonym of *E. seriolicida*, the name was retained (146).

b. Culture—Primary isolation is on TSA with 5% sheep blood, or BHIA, and incubation is at 30–35 °C in a normal atmosphere. Growth is rapid, but 48 hours may be required for full visualization of the small (1.0 mm diameter) white colonies. Selective isolation is achieved on phenyl ethyl alcohol (PEA) blood agar or on blood agar containing .01% sodium aside.

c. Description—*L. garvieae* is a Gram-positive ovoid coccus (1.4 × 0.7 μm) forming short chains. The bacterium is nonmotile, and colonies on blood agar are α-hemolytic. Growth occurs between 10 and 45 °C, between 0 and 6.5% NaCl, and between pH of 4.5 and 9.6. Optimum growth is at 37 °C, 0% NaCl, and pH 7.5. The bacterium is fermentative in GMD, negative for catalase and hippurate hydrolysis, and positive for Voges–Proskauer, arginine dihydrolase, tetrazolium reduction, and bile-esculin hydrolysis. Biochemical tests useful in differentiating *L. garvieae* from other gram positive cocci are found in Table 4.

d. Epizootiology—*L. garvieae* may be detected in the aquatic environment year-round in Japan, but the highest counts are detected in sea water in the summer months in the vicinity of sea cages. Transmission of *L. garvieae* is horizontal, from one infected fish to another, in high-density aquaculture; however, the feeding of contaminated raw fish was determined to be a common source of infection (147). The bacterium was shown to survive for up to six months in frozen sandlance, a food commonly used in yellowtail culture. The bacterium has also been isolated from muds in the vicinity of sea cages during the cooler months of the year; these muds serve as a reservoir for infection (148). Wild fish such as sardine, anchovy and round herring are known to harbor the bacterium, and asymptomatic yellowtail can serve as carriers (147).

e. Geographic range and host susceptibility—*L. garvieae* is primarily a disease that affects cultured yellowtail and eel in Japan. Although the history is difficult to trace, because of problems with the taxonomy of streptococci pathogenic to fish, the initial outbreaks are believed to have been in the populations of yellowtail cultured near Shikoku Island from July to September of 1974 (88). Since 1974, the disease has occurred with increasing frequency and has become a disease of great economic importance in Japanese aquaculture. In cultured eels, the disease was first reported (by Kusuda) in 1978 (143).

f. Clinical signs of disease and treatment—Clinical signs of *L. garvieae* infections are similar to those due to other streptococcal infections in fish. The typical gross external clinical signs are exophthalmia, petechiae on the inside walls of the operculum, and congestion and hemorrhage in the intestine. Necrotic areas may be noted in the enlarged spleen and kidney. The first effective chemotherapeutant for the treatment of *L. garvieae* infections was erythromycin [25 mg/kg (11 mg/lb) fish/day for four to seven days in medicated feed], and, since 1984, erythromycin and spiramycin have been used to control the disease in Japan. Resistant strains have emerged in recent years; to combat infections due to them, josamycin has been used effectively at 30 mg/kg/day (13.6 mg/lb) for three days (149). Experimental vaccines have been effective at inducing protection, by immersion or by injection, but an effective vaccine has not been produced on a commercial scale. Nonspecific immunostimulants, such as the β-1,3 glucans, have produced increased resistance to infection by lactococci when injected at 2–10 mg/kg fish (150).

II. Lactococcus piscium, Carnobacterium piscicola ("Lactobacillus piscicola") and Vagococcus salmoninarum. Causative agents of pseudokidney disease in salmonids and of bacteremia in striped bass and channel catfish.

a. History—These three species are placed together on account of the similarity of the disease condition that they produce in salmonid fishes. Pseudokidney disease was named for the propensity to misdiagnose it as kidney disease, a condition in salmonids caused by the gram-positive bacterium *R. salmoninarum*, when only gram-stained tissue smears are examined. The latter is a common method of diagnosis, because *Renibacterium* is fastidious, and so special media and prolonged incubation times are required for isolation. Pseudokidney disease was described by Ross and Toth in 1974 (151) as a disease of adult salmonid broodfish that had undergone stress associated with handling or/or spawning. The causative organism was originally isolated and identified as *L. piscicola* by Hui, 1984 (152), but other unidentified coccobacilli were also isolated from certain fish. *L. piscicola* was renamed *C. piscicola* by Collins et al., in 1990 (153), who used molecular taxonomic methods. A new species, *V. salmoninarum*, was described in 1990 by Wallbanks et al. (154), from two cultures originally considered atypical lactobacilli isolated from diseased trout in Oregon and Idaho 1968 and 1981. *V. salmoninarum* infections were not again reported until yearly outbreaks of disease were recorded on a trout farm in the southwest of France from 1993–1997 (155). Another bacterium associated with pseudokidney disease of salmonids, *L. piscium*, was described as a new species by Williams et al., in 1990 (156), but very little information concerning pathogenesis or range of occurrence is available. Isolates of *Carnobacterium* spp., were reported by Baya et al., in 1991 (157), from cultured populations of striped bass and channel catfish and from wild populations of brown bullheads.

b. Culture—*Carnobacterium*, *Vagococcus*, and *Lactococcus* are not fastidious and are easily isolated on standard media such as TSA, TSA with 5% sheep blood, or BHIA. Incubation is between 20–25 °C, and small, white, round, entire, 1-mm colonies appear within 48 hours.

c. Description—1. *V. salmoninarum* is a short, ovoid, nonmotile, gram-positive coccobacillus (0.5 × 2 μm) occurring singly, in pairs, or in short chains. The bacterium is facultatively anaerobic; it grows between 5 and 37 °C but fails to grow at 40 °C. Tests for esculin hydrolysis, hippurate, and H_2S are positive; tests for arginine dihydrolase (ADH), catalase, oxidase, and urease are negative. Positive results are obtained in tests for the enzymes alkaline phosphatase (PAL), leucine arylamidase (LAP), pyrrolidonyl–arylamidase (PYR), and para–nitrophenyl–galactosidase (PNPG). Most strains are weakly α-hemolytic after two to three days of incubation. Growth occurs at pH 9.6 and in the presence of 40% bile, but not in 6.5% NaCl.

2. *C. piscicola* is a short, nonmotile, gram-positive rod (0.5 × 1.5 μm) occurring singly, in pairs or in short chains. The bacterium is facultatively anaerobic; it grows between 10 and 37 °C but fails to grow at 42 °C. Tests for esculin hydrolysis, ADH, and PYR are positive; tests for hippurate, urease, and H_2S are negative. Most strains are α-hemolytic and grow at pH 9.6 and in 6.5% salt but not in the presence of 40% bile. Results of PAL and LAP enzyme tests are not available.

3. *L. piscium* is a nonmotile, gram-positive coccus (0.5 × 1.0 μm) occurring singly, in pairs or short chains. The bacterium is nonhemolytic and facultatively anaerobic; it grows between 5 and 30 °C but fails to grow at 40 °C or at pH 9.6. Tests for esculin hydrolysis and starch hydrolysis are positive; tests for ADH, urease, H_2S, hippurate, and PYR are negative. Test results for growth in 40% bile are not available. Biochemical tests useful in differentiating this group of gram-positive organisms are included in Table 4.

d. Epizootiology—Pseudokidney disease is normally a chronic condition in salmonids, one resulting in low mortality rates and generally regarded as an opportunistic infection caused by one or more of the gram-positive organisms listed in this section. The bacteria have traditionally been considered to be normal components of the aquatic microflora and to gain entrance into the fish following trauma associated with spawning. Losses can be substantial, however, as is evidenced by mortality rates as high as 50% on trout farms in France caused by *V. salmoninarum* infections. The affected fish were adult trout weighing 600–4000 g (1.3 lb to 8.8 lb), and peak mortality followed the handling associated with the sorting and stripping of fish for spawning (155). Water temperatures during the outbreaks were lower than in other cases of pseudokidney disease, with peak mortality at around 9–10 °C (48–50 °F). A report by Michel in 1986 (158) indicated that young salmonids and carp reared in western Europe were susceptible to infection with *C. piscicola* and *L. piscicola*, although they noted that adult fish were the ones most commonly affected.

e. Geographic range and host susceptibility—*C. piscicola* is most commonly associated with adult salmonids with pseudokidney disease from the Pacific Northwest region of North America (152,159) and from Newfoundland, Canada (160); it has also been reported from brown trout (*Salmo trutta*), rainbow trout (*O. mykiss*), and common carp (*C. carpio*) cultured in France and Belgium (158). Isolates of *V. salmoninarum* have been obtained (*1*) from salmonids from the Pacific Northwest of the U.S. (154) and (*2*) from the southwest of France from diseased rainbow trout (155). The *L. piscium* type culture, HR1A-68, was isolated by R.A. Holt from rainbow trout at the Hood River Hatchery, Oregon (156). *Carnobacterium* spp., similar to *C. piscicola* were isolated from cultured striped bass and channel catfish and from wild populations of brown bullheads from Chesapeake Bay, Maryland (157). Attempts to confirm Koch's postulates with Maryland strains were successful in trout, but striped bass and catfish seemed to enter into a carrier state after being injected with the bacterium. Difficulty in recreating pseudokidney disease by experimental infection has been noted by several authors;

it is probably due to the low virulence of the bacterial strains (158,159).

f. Clinical signs of disease and treatment—*Carnobacterium* infections in salmonids are systemic and usually chronic in nature, with one or more of the following pathological signs: abdominal distension due to ascites, splenomegaly, muscle granulomata, internal hemorrhages, subcutaneous sanguineous vesciculation, and renal granulomata (152,158). In trout infected with *V. salmoninarum*, the prominent pathological manifestations were septicemia (with epicarditis) and meningitis (155). Treatment of pseudokidney disease is not well documented; however, strains of *Carnobacterium* have been shown to be susceptible to tetracycline, ampicillin, and erythromycin. Treatment of *V. salmoninarum* infections in France with ampicillin, amoxicillin, or erythromycin failed to prevent or decrease trout mortality. Vaccines for pseudokidney disease have not been considered, because of the opportunistic nature of the disease and the variety of causative bacteria.

***III. Enterococcus faecium, E. faecalis and Enterococcus* spp.** Causative agents of enterococcosis of rainbow trout and other species.

a. History—Enterococcus-like bacteria were recorded as fish pathogens as early as 1974 by Boomker et al., in 1979 (161) from rainbow trout cultured in South Africa; additional reports from this same geographic area were published in 1986 by Bragg and Broere (162). Carson and Munday described a disease in farmed rainbow trout in Australia caused by a similar organism, from outbreaks between 1982 and 1990 (163). Today, the disease is considered one of the major diseases of farmed fish in Australia and South Africa. Other published accounts of organisms presumptively identified as *Enterococcus* sp., or *E. faecalis* have been described from farmed rainbow trout in Italy by Ceschia (164) and Ghittino (cited in 164). In an unpublished case report, *E. faecium* was determined to be the causative agent of enterococcosis in hybrid striped bass cultured in semiclosed intensive culture systems in the U.S. (7), and unidentified enterococci have been isolated from diseased channel catfish (165). An organism presumptively identified as *Enterococcus* sp., has been described from diseased turbot cultured in Galicia, northwest Spain (166).

b. Culture—Primary isolation is on TSA with 5% sheep blood or BHIA and incubation at 30–35 °C in a normal atmosphere. Growth is relatively fast, but 48 hours may be required for formation of typical 1 mm, smooth, white colonies. Selective isolation is achieved on aside blood agar (0.2 g/L sodium aside) (Difco Laboratories, Detroit, Michigan) or by using a selective isolation procedure outlined by Bragg (167): In this method, tissue samples are placed in nutrient broth containing 100 μg/mL nalidixic acid and 160 μg/mL oxolinic acid or 200 μg/mL of sodium aside, and they are incubated for three days at room temperature; samples from the broth are then plated on tetrazolium agar plates, and small red colonies appearing on this medium are presumptively enterococci. Cephalexin–Aztreonam–Arabinose Agar (CAA) has also been used for the selective isolation and differentiation of *E. faecium* (168).

c. Description—The genus *Enterococcus* was created in 1984 to include the enteric streptococci *Streptococcus faecalis* and *S. faecium*, and, since that time, 17 additional species have been proposed for inclusion in the genus. The enterococci are gram-positive cocci occurring singly, in pairs, or short chains. The organisms are facultatively anaerobic and catalase-negative, and optimum growth occurs at 35 °C, however, most strains grow at 10 and 45 °C. The bacteria also grow in 6.5% NaCl and at pH 9.6 and hydrolyze esculin in the presence of 40% bile salts. Most hydrolyze pyrrolidonyl–naphthylamide (PYR), produce leucine arylamidase (LAP), and possess the Lancefield group D antigen. Biochemical tests useful in differentiating *Enterococcus* spp., from other gram-positive cocci are included in Table 4. Recent isolates from Italy (164), Spain (166), Australia, and South Africa (169) may ultimately be identified as biotypes or serotypes of *L. garvieae*, because they fail to grow at 45 °C, do not possess the group D antigen, and are otherwise biochemically identical to *L. garvieae*. This problem cannot be solved until genetic analysis is done on the strains in question. Strains from Spanish turbot differ serologically from the Italian trout strains as visualized by immunoblots and ELISA, and trout strains are not immunogenic in turbot (170).

d. Epizootiology—Temperatures were in the range of 19 °C (66 °F) on South African farms where outbreaks first occurred, and temperatures did not fluctuate due to the groundwater source; however, in subsequent cases reported in 1986 by Bragg and Broere, temperatures ranged from 18 to 25 °C (64–77 °F) and were believed to be a stress factor. Other factors such as water quality, feed, and feeding practices were found to be acceptable, but temporary overcrowding during periods of new pond construction did seem to be a predisposing factor to disease. On Italian farms, poor water quality was implicated as a factor leading to increased susceptibility to infection. The source or reservoir of the pathogen has not been determined, but, once it is established in a fish population, horizontal transmission can occur through damaged skin or by the fecal-oral route. Infectivity studies with turbot (mean weight 10 g) produced an LD_{50} in the range of 10^4 cells by intraperitoneal injection. Chronic disease in hybrid striped bass occurred at temperatures between 26 and 28 °C (79–82 °F) in fish stocked at high densities; this case was complicated by a parasitic infection with *Epistylis* sp.

e. Geographic range and host susceptibility—Thus far, enterococci have been implicated as causative agents of disease for rainbow trout in Australia, South Africa, and Italy, for turbot in Spain, and for hybrid striped bass and channel catfish in the U.S. Outbreaks in South Africa have occurred in cultured populations of rainbow trout in the Eastern Transvaal area; however, an extensive survey of fish populations in eight river systems in Natal yielded no isolations of enterococci or streptococci (171).

f. Clinical signs of disease and treatment—In trout, turbot, and hybrid striped bass, the clinical signs are reminiscent of other streptococcal diseases in fish.

Bilateral exophthalmia, corneal opacity, subcutaneous abcesses in periorbital spaces and at the base of fins, and caudal peduncle are common external signs. Internal clinical signs include pale liver, dark red enlarged spleen, hemorrhage in the intestine, and ascites in the peritoneal cavity. Treatment depends on the sensitivity of the individual strain isolated, because enterococci are often resistant to many of the common antibiotics used in aquaculture. Antibiotics that have been used in various cases include oxytetracycline, erythromycin, ampicillin, and enrofloxacin; however, the disease is chronic in nature, and short term medicated feed treatments are rarely effective. Vaccines are not yet available for enterococcal infections in fish.

C. Other Aerobic and Facultatively Anaerobic Gram-Positive Cocci. Organisms in this group have uncertain status as fish pathogens and will therefore just be listed with their hosts:

Micrococcus luteus — rainbow trout; UK.
Aerococcus viridans — lobsters.
Planococcus sp. — Atlantic salmon rainbow trout; UK.
Staphylococcus aureus — silver carp; India.
S. epidermidis — yellowtail, red sea bream; Japan.

D. Aerobic Gram-Positive Rods

I. Renibacterium salmoninarum. Causative agent of bacterial kidney disease. Synonyms: Dee disease, corynebacterial kidney disease, salmonid kidney disease.

a. History — Bacterial kidney disease (BKD) was first described in the early 1930s in wild populations of Atlantic salmon from the Dee river in Scotland and was thus named "Dee disease" (172). It was subsequently found in rainbow trout, brook trout (*Salvelinus fontinalis*), and brown trout (*Salmo trutta*) in the U.S.: in Massachusetts by Belding and Merrill, and in California by Wales (as cited by Earp (173)). The first isolations from Canada were in 1937, in cutthroat trout (*Salmo clarki*), by Duff (as cited by Evelyn et al., in 1986 (174)). Now BKD is known from hatchery and farmed salmonid populations in many parts of the world, from both fresh and marine waters, but only rarely has it been observed causing mortality in wild fish stocks. The causative agent of BKD was originally classified as a *Corynebacterium* (175), on the basis of morphology and staining. As additional information was gathered following the first successful culturing of the organism, investigators realized that it belonged in a new, previously undescribed genus. The new genus and species was described, and it was named *R. salmoninarum* by Sanders and Fryer in 1980 (176).

b. Culture — Culture of the fastidious causative agent of bacterial kidney disease was not achieved until Earp used a specially formulated nutrient-rich medium. Growth on this medium was poor, requiring more than 14 days to obtain small colonies. Continued improvements were made on the isolation medium, but most significant was the discovery by Ordal and Earp in 1956 (175) that L-cysteine added to nutrient-rich blood agar provided a significant boost in growth. This medium (cysteine blood agar) was used until the development of KDM2 medium by Evelyn in 1977 (177), which utilized the addition of fetal calf serum and replaced blood and many of the other nutrients with peptone, cysteine, and yeast extract. KDM2 and a selective medium, SKDM, derived from it by Austin et al., in 1983 (178) are currently in use for the primary isolation of *R. salmoninarum*. At 15 °C incubation, the organism requires 20 days for the appearance of smooth, white, creamy, 2-mm colonies. Because the organism grows slowly, plates must be kept moist during the incubation process. Research on improved culture techniques is ongoing, and rich, serum-free, semidefined media have been developed by Embley (179) and Sheih (9).

c. Description — *R. salmoninarum* is a gram-positive rod (0.5 × 1.0 µm) occurring singly or in pairs. The bacterium is a nonmotile, not spore-forming, not acid-fast, fastidious bacterium that grows best at 15–18 °C. The bacterium fails to grow at temperatures above 25 or below 5 °C. *R. salmoninarum* strains from different geographic locations are homogeneous, typically being positive in tests for catalase, litmus milk, alkaline phosphatase, caprylate esterase, glucosidase, leucine arylamidase, α-mannosidase, and trypsinase. The organism fails to liquefy gelatin and does not grow in 1% sodium chloride. Serologically, isolates appear to be homogeneous when tested using polyclonal antisera, however, differences among strains can be detected with monoclonal antibodies (180). All strains apparently possess a heat-stable 57-kd surface protein. Antibodies to this antigen have been useful for detecting *Renibacterium* by ELISA, and it has been determined that the observed strain differences result from the recognition by monoclonal antibodies of different epitopes on the 57 kd surface protein (181).

d. Epizootiology — It is generally felt that *R. salmoninarum* is an obligate pathogen of salmonids and that its reservoir of infection is other infected salmonids. Outbreaks of BKD have not been reported from nonsalmonids. Survival of the pathogen can occur outside the host, but it is short-lived in water and mud. Bacterial kidney disease can be present as an overt infection, or it can exist in a carrier-only state. Water quality seems to influence cumulative mortality and severity of disease; BKD is more common at soft-water hatcheries than at those with high total hardness. A serious problem is the effect of BKD on the survival of salmon smolts as they are moved from freshwater to seawater. Mortality rates as high as 17% have occurred in smolts moved to saltwater, as against 4% in those kept in freshwater (182). BKD can occur over a wide range of temperatures, but most epizootics occur in the fall between 12 and 18 °C (54–64 °F) and in the winter between 8 and 11 °C (46–52 °F). Water temperature influences the time between experimental exposure and death. Mortality occurs at 30–35 days postexposure at 11 °C (52 °F) at 60 to 90 days postexposure at 7–10 °C (45–50 °F). Bullock, in 1978 (183), was able to demonstrate vertical transmission of BKD and theorized that the bacterium was carried within the egg, because surface disinfection of the egg had no effect on transmission efficiency. This hypothesis was later proven to be true by Evelyn et al. (174), who demonstrated that *Renibacterium*-infected coelomic fluid is the source of infection for the egg. Several studies have shown that the dietary composition of feed influences the susceptibility of fish to *Renibacterium* infections (8).

e. Geographic range and host susceptibility — BKD has been reported from most areas of the world where salmonids occur, with the exception of Australia, New Zealand, and the Soviet Union. The disease has been reported from the United Kingdom, Europe, Japan, North America, and South America (Chile). All salmonids are considered susceptible, but brook trout and chinook salmon are the most susceptible. Natural outbreaks have been documented only in salmonids; however, a few fish species, such as the sable fish, the Pacific herring, the shiner perch, the common shiner, and the fathead minnow, have been infected experimentally.

f. Clinical signs of disease and treatment — External clinical signs of BKD are uncommon, except in the terminal stages of disease. The most common early signs of infection are dark pigmentation, exophthalmia, hemorrhages at the base of fins, and abdominal distension associated with a widely disseminated bacteremia. As the disease progresses, pale gills, exophthalmia, cutaneous blisters, and ulcerative abcesses may be seen externally; internally, cavitations in the musculature, bloody/turbid fluid, and creamy white granulomatous lesions are common in the kidney. Treatment of BKD by chemotherapy is not particularly effective, presumably because of the intracellular nature of the pathogen. Erythromycin is the antimicrobial of choice for treatment and prophylaxis against BKD. Treatment recommendations vary from oral administration in medicated feed at 100 mg/kg (45 mg/lb/day) of fish/day for 21 days for fingerlings [Wolf and Dunbar 1959 (184)] to injection of prespawn female salmon with 10–20 mg/kg (4.5–9.0 mg/lb) body weight and 1–2 mg/L (ppm) for 30 min as an additive during water-hardening of eggs. Disinfection of egg surfaces using iodophors at 25–100 mg/L (ppm) for 5 minutes has proven beneficial in reducing transmission of the disease. Although research on vaccination of salmonids for BKD has been ongoing since 1971 (185), there has been only limited success, and a commercial vaccine is currently not available.

II. Other Aerobic Gram-Positive Rods. Several other species of aerobic gram-positive, rod-shaped bacteria have been implicated as pathogens of fish, but because of uncertainty as to their significance, they are only mentioned here along with their hosts:

Corynebacterium sp. — rainbow trout (186).
Corynebacterium aquaticum — striped bass (187).
Rhodococcus sp. — chinook salmon (188).

E. Gram-Positive, Acid-Fast Rods

I. Mycobacterium marinum ("Mycobacterium piscium, M. platypoecilus, M. anabanti"), M. fortuitum, M. chelonei ("M. chelonei subsp. piscarium, M. salmoniphilum"). Causative agent of mycobacteriosis of fish. Synonym: fish tuberculosis.

Three species of mycobacteria are recognized as fish pathogens; they are discussed together here because of the similarities in the pathologic condition they produce in fish.

a. History — Acid-fast staining bacteria were first described as fish pathogens of common carp in 1987 in Europe by Bataillon (189); this is one of the oldest known fish diseases. The name "fish tuberculosis" was coined to reflect the similarity to human tuberculosis: the acid-fast staining reaction, and the granulomatous lesions commonly seen in infections. This old name has given way over the years to the less threatening name "mycobacteriosis." The first isolation of an organism in this group was reported by Bataillon in 1902 (190), but the cultures of that organism, named *Mycobacterium piscium*, have been lost, and the name lacks validity. *Mycobacterium marinum* was described by Aronson in 1926 (191) from a tropical coral fish kept at the Philadelphia Aquarium. *M. marinum* was originally thought to cause disease only in marine fish; however, it has subsequently been reported from many freshwater species as well, and the original isolates of *M. piscium* are believed to have been a strain of *M. marinum*. *Mycobacterium fortuitum* was isolated from diseased neon fish in 1953 (192) and continues to occur commonly in tropical aquarium species. Mycobacteriosis of Pacific salmon was originally described by Earp et al., in 1953 (173), and the causative organism, *Mycobacterium salmoniphilum*, was cultured, described, and named by Ross in 1960 (193). *M. salmoniphilum* was later shown to be a strain of *M. chelonei* (194), with the species name being corrected to *M. chelonei* (195).

b. Culture — Isolation and culture of the mycobacteria is difficult due to their fastidious nature and slow growth rate. Primary isolation is best achieved on Dorsett egg, Lowenstein–Jensen, Petragnani, or Middlebrook 7H10, with subcultures maintained in capped tubes. If the cell number per gram of tissue is high and contamination is low, general-purpose media such as TSA with 5% sheep blood may be used for primary isolation. Incubation should be at 20–30 °C for 2–30 days. *Mycobacterium fortuitum* is classified as a rapid grower, and colonies should appear in less than seven days on culture media. *M. chelonei* is also classified as a fast grower on subculture, but primary isolation may take several weeks. *M. marinum* is a slow grower, requiring 7–10 days for visible growth and several weeks for typical colonies to develop at 25 °C. Of the three species, only *M. fortuitum* is capable of growth at 37 °C. Selective isolation from contaminated sites on or in the fish may be enhanced by homogenizing the tissue in a grinder and treating it with 0.3% solution of the disinfectant Zepheran® (benzalkonium chloride 17%) for 20 minutes, prior to the adding of several drops of the homogenate to culture media (196). Because of the extended incubation times, care must be taken to keep plated media moist, by incubating in a humid atmosphere.

c. Description — Representatives of all the species of mycobacteria pathogenic to fish are Gram-positive, acid-alcohol-fast, non-motile, non-sporing, pleomorphic rods (1.0–4.0 × 0.2–0.6 μm), with occasional filamentous or coccoid forms. Colonies of *M. marinum* may vary in their appearance, depending on the culture medium and incubation time. *M. marinum* is said to be photochromogenic, because cultures incubated in the dark will produce white colonies, whereas cultures exposed to light will form yellow to yellow/orange colonies. Other mycobacteria can be photochromogenic, so this characteristic alone cannot be used for identification; however, *M. marinum* does not reduce nitrate and cannot grow at 37 °C, to separate it from similar organisms. Both *M. fortuitum* and *M. chelonei* are nonchromogenic and grow more rapidly than *M. marinum*, producing smoother, white- to buff-colored colonies, and are differentiated from other mycobacteria by their ability to grow on MacConkey agar. *M. fortuitum* is positive on iron-uptake, sucrose-utilization, and nitrate-reduction tests; they separate it from *M. chelonei*, which is negative on these tests. In addition to pigment and growth at 37 °C, *M. marinum* may be differentiated from *M. fortuitum* and *M. chelonei* by positive results for production of nicotinamidase and pyrazinamidase.

d. Epizootiology—Little is known about the epizootiology of mycobacterial infections in fish. The organisms mentioned in this section are all common in the soil and in freshwater and marine environments, and the factors that lead to the development and spread of disease are unknown. Mycobacterial infections are more common in aquarium fish and in food fish cultured in closed or semi-closed recirculating tank systems (197); they are very rare in pond fish. Stress due to overcrowding, to poor water quality, and to ingestion of contaminated food or aquatic detritus have all been mentioned as factors leading to mycobacterial infections. Feeding infected trash fish to cultured Pacific salmon (192) and snakehead (*Channa* sp.) (198) resulted in mycobacteriosis. Vertical transmission has been confirmed in the Mexican platyfish (*Xiphophorus maculatus*), but it has not been confirmed in salmonids (199). Historically, mycobacteriosis has not been the cause of disease in natural fish populations, and it has only rarely been seen as subclinical infections in feral fish (200). In 1997 and 1998, outbreaks of mycobacteriosis in wild populations of striped bass in Chesapeake Bay were documented by Vogelbein (201), with as many as 30 to 50% of the striped bass in certain tributaries reported as having skin lesions. As many as five different species of mycobacteria, including *M. marinum* were isolated from fish in this study, and the underlying cause of the epizootic is under investigation.

e. Geographic range and host susceptibility—Nigrelli and Vogel (202) published a list of 151 species of freshwater and marine fish that were susceptible to mycobacteriosis. That list has continued to grow, and the opinion is generally held that most, if not all fish species are susceptible to the disease. The occurrence of mycobacteriosis is worldwide in both marine and freshwater locations. *Mycobacterium marinum* and *M. fortuitum* are pathogenic to humans, and *M. marinum* causes the condition known as "fish handler's disease" or "swimming pool granuloma." The names result from the tendency for cutaneous granulomas to develop in the skin at the extremities of the body, such as the fingers, the toes, and the outsides of the elbows and the knees. Human infections are confined to the extremities by the optimum temperature range of the bacterium, which is normally 30–33 °C (86–91 °F).

f. Clinical signs of disease and treatment—Mycobacteriosis is a chronic, systemic, progressive, wasting disease that often goes unnoticed in a fish population until a high percentage of fish have become infected. Gross clinical signs include anorexia, emaciation, shallow hemorrhagic skin lesions, ulcers, and fin erosion. Cold-water salmonids may not exhibit all of these signs. Internally, greyish-white miliary granulomas are seen primarily in the spleen and the head kidney, but they may be found in most of the tissues of the fish. Organs such as the spleen and the kidney may be greatly enlarged and have a granular appearance. Treatment of mycobacterial infections, even when an early diagnosis is made, is not economically feasible for food fish. Infected stocks are best destroyed, and the systems disinfected, before new fish are brought in. Cleaning and disinfection of fish systems with HTH, chlorox solution, or chlorine dioxide-based sterilants is best. Care should be taken by fish culturists when handling infected fish, to avoid spine wounds, skin abrasions, and prolonged contact with water containing mycobacteria. More valuable ornamental fish may be treated in accordance with the antimicrobic susceptibility of the strain isolated.

II. Nocardia asteroides, N. seriolae ("N. kampachi"). Causative agents of nocardiosis.

The species are discussed together because of the similarity in the disease condition they cause in fish.

a. History—Nocardiosis was first described (from the neon tetra, a tropical freshwater fish), in 1963, by Valdez and Conroy (203). Snieszko et al., described the disease in rainbow trout in 1964 (204) and Campbell and MacKelvie found the disease in brook trout in 1968 (205). The species responsible for these early cases was identified as *Nocardia asteroides*. A second species, *N. kampachi*, was described by Kariya (206) in 1968 as the causative agent of disease in cultured yellowtail in Japan. The name *N. kampachi* was never fully accepted by taxonomists, and in 1988 Kudo et al. (207), proposed the name *N. seriolae* for the causative agent of nocardiosis of Japanese yellowtail. Nocardiosis has become one of the more important diseases in Japanese mariculture, with 262 tons of yellowtail reported lost to the disease in 1989 (88).

b. Culture—Nocardiae are typically less fastidious than the mycobacteria, and they can be cultured on standard media such as TSA with 5% blood and BHIA, although they also grow on Lowenstein–Jensen and Ogawa egg media. Incubation should be between 20 and 30 °C. Colonies of *Nocardia asteroides* appear in 4–5 days; 10 days may be required for *N. seriolae* colonies to appear. Colonies of *N. asteroides* are ridged and folded, are pigmented pinkish-white to yellow-orange, and produce aerial hyphae along the colony margin. *N. seriolae* produces flat and wrinkled colonies.

c. Description—*Nocardia* sp., are gram-positive, weakly acid-fast, nonmotile, long-branching bacilli that, in tissue imprints, resemble mycobacteria. Identification of isolates as *N. asteroides* is based on the physiological and biochemical description of Gordon and Mihm in 1962 (208). Acid is produced from glucose and glycerol but not from adonitol, arabinose, erythritol, inositol, lactose, maltose, mannitol, raffinose, sorbitol, or xylose. A characteristic of *N. asteroides* is the ability to grow at 37 °C; however, not all isolates of this species from fish are capable of growing at this temperature, and *N. seriolae* does not grow at 37 °C. *N. seriolae* is biochemically very similar to *N. asteroides*, but differs in the sugar fermentation pattern and organic acid utilization patterns reported by Kusuda (209).

d. Epizootiology — Norcardiosis is a slowly developing chronic infection, with accompanying low mortality. It is known to be a normal inhabitant of soil and water, and carrier fish may serve as a reservoir of infection. On rare occasions, mortality rates can be as high as 20% over a two week period, as was reported from cultured Formosan snakehead in ponds on Taiwan (210). Snieszko et al. (204), were unable to transmit *N. asteroides* to rainbow trout by feeding, but injection reproduced the disease in one to three months. Kusuda (209) also succeeded in transmitting the disease in yellowtail by injection or by smearing surface wounds with *N. seriolae*, results indicating a preference for this route of infection.

e. Geographic range and host susceptibility — Nocardiosis has been reported from the United States, Argentina, Germany, Japan, and Taiwan, and the potential exists for its occcurrence on a world wide basis. Thus far, infections with *N. asteroides* have been reported in rainbow trout, brook trout, neon tetra, snakehead, and giant gourami, and *N. seriolae* ("*N. kampachi*") has been reported in the yellowtail.

f. Clinical signs of disease and treatment — Nocardiosis shares many characteristics with mycobacteriosis, and the two diseases are often confused, particularly when histopathology is the only tool used for diagnosis. Young fish are the most vulnerable, but all ages may be affected, particularly in the late summer and early fall. In early stages of the disease in yellowtail, the clinical signs are emaciation, inactivity, and discoloration of the skin. Subcutaneous abcesses appear, later in the infection, in the skin and gill; they are white and about 5 mm (0.44 in.) in diameter. These lesions become histologically tuberculoid, with a fibrous capsule and bacterial filaments located in the center of the lesion. Abdominal distension resulting from internal granulomata may occur. Diffuse granulomatous lesions may be found in the skeletal muscle, in all of the visceral organs (particularly the spleen), and in the mesentery. Little is known concerning the treatment of norcardiosis. Food fish such as the yellowtail have been treated with 4–100 mg (1.8–45.4 mg/lb/day) of sulfamonomethoxine/kg/day for five to seven days, and mortality has been slowed; but a cure has not been reported. In most cases, removal and disposal of infected fish is the proper course of action followed by disinfection of facilities.

F. Gram-Positive Anaerobic Rods

I. Eubacterium tarantellus. Causative agent of eubacterial meningitis of marine and estuarine fish.

a. History — An anaerobic gram-positive rod was determined to be the causative agent of a large fish kill on Biscayne Bay, Florida in 1976 (211). The striped mullet (*M. cephalus*) was the species primarily affected; it exhibited a neurological condition resulting from a bacterial meningitis. In 1977, Udey et al. (212), described the bacterium as a new species, *Eubacterium tarantellus*. The *Catenabacterium* described from epizootics in red drum and grey mullet along the Texas coast is most likely the same organism (9,213).

b. Culture — The organism is readily cultured from the brain and liver on standard media, such as TSA with 5% sheep blood or BHIA plates with anaerobic incubation in a gas pack or other suitable anaerobic chamber at 20 °C for three to five days. Culture plates should be prereduced prior to primary isolation. Tissues from fish, including brain tissue, may be inoculated directly into fluid thioglycollate medium with 1% NaCl and 100 mcg/mL gentamycin incubated at 30 °C.

c. Description — *Eubacterium tarantellus* is a long, unbranched, filamentous, gram-positive, asporogenous rod that can fragment into shorter bacilli of 1.3–1.6 × 1.0–17.0 μm. Good growth occurs between 20 and 37 °C under anaerobic conditions, and colonies on agar media are flat, translucent, approximately 2–5 mm in diameter, colorless, rhizoid, and slightly mucoid. Colonies on blood agar are surrounded by a zone of β-hemolysis.

d. Epizootiology — Under laboratory conditions, the disease cannot be transmitted by direct fish-to-fish contact but can be reproduced by intraperitoneal injection of susceptible hosts, such as the mullet, with the bacterium. This fact had led some investigators to suspect that transmission occurs via external parasites. The organism has an affinity for the brain, the spinal cord, and the liver, regardless of the mode of infection. The salinity tolerance of the organism is 2 ppt, so it would be restricted to low-salinity environments. The bacterium has been found in several species of marine and estuarine fish showing no signs of disease, so some species may serve as carriers of the organism.

e. Geographic range and host susceptibility — The bacterium has been found in a variety of estuarine fishes, including the striped mullet, the snook (*Centropomus undecimalis*), the red drum, and the Gulf flounder (*Paralichthys albigutta*) (213). The range is not well known, but outbreaks of the disease have been restricted to estuaries of the Gulf of Mexico along the coasts of Florida and Texas.

f. Clinical signs of disease and treatment — Eubacterial meningitis is characterized by a neurological impairment that causes erratic swimming behavior of fish near death. Following experimental infection, the progress of the disease is slow, with clinical signs not appearing until 14 to 30 days postinoculation. Affected fish show darkened pigmentation, uncoordinated swimming movements, and inability to maintain proper orientation in the water column. Moribund fish may hang vertically, lie still on the bottom of the tank, swim slowly while rotating about their long axis, or whirl. Little is available in the literature concerning treatment of this disease; however, suggestions include the controlling of ectoparasites and the use of erythromycin- and oxytetracycline-medicated feed at 100 mg/kg/day (45.4 mg/lb/day) for 5 days.

***II. Clostridium botulinum, Clostridium* spp.** *Clostridium botulinum* and other *Clostridium* species have been isolated from diseased cultured fish on several occasions, but their significance in aquaculture is uncertain and therefore is not covered in detail. Isolation of strictly anaerobic, gram-positive rods with subterminal or terminal ovoid endospores would indicate the possibility of *Clostridium* spp.

G. Miscellaneous Bacterial Agents: Gliding, Flexing, and Yellow-Pigmented Gram-Negative Bacteria

I. Flavobacterium columnare ("Bacillus columnaris, Chondrococcus columnaris, Cytophaga columnaris, Flexibacter columnaris"). Causative agent of columnaris disease.

a. History — Columnaris disease was first described by Davis in 1922 (214) in warm-water fish from the Mississippi River, but it was not until the work of Ordal and Rucker in 1944 (215) that the organism was isolated, characterized, and named *Chondrococcus columnaris*. Davis originally named the organism *Bacillus columnaris*, inspired by the formation of columnar masses of cells that were visible microscopically on wet mounts of infected fish tissue. A year following the work of Ordal and Rucker, Garnjobst published an account of isolation of the bacterium and assigned it to the genus *Cytophaga* (216). The bacterium has been renamed and reclassified several times over the years on the basis of morphological and biochemical features, having been referred to at various times as *Chondrococcus columnaris*, *Cytophaga columnaris*, and *Flexibacter columnaris* (9). The currently accepted name for the organism, *Flavobacterium columnare*, was adopted in 1996, as suggested by Bernardet (217); however, total agreement has not been achieved, as evidenced by the opinion of Bader and Shotts, following their work on sequence analysis of the 16s ribosomal-RNA genes of *F. columnare* and other closely related organisms in this group (218), that the genus *Flexibacter* should be retained.

b. Culture — Culture of *F. columnare* may be accomplished on a variety of low-nutrient, low-agar-content media. The commonly used medium is the cytophaga agar (CA) of Anacker and Ordals (219); however, shorter incubation times and higher yields may be obtained in media containing salts, such as Shieh medium (220). Selective isolation of columnaris can be achieved from contaminated external sites such as the skin and gills on selective cytophaga agar (SCA) (221) or on Hsu-Shotts (HS) medium (222). *Flavobacterium columnare* may be incubated at 10–33 °C, but primary isolation plates are typically incubated at 28–30 °C. Pale yellow rhizoid colonies are present after 48 hours of incubation. A dilute form of Mueller–Hinton medium must be used for antimicrobic susceptibility testing (with disc diffusion methods) or for determination of minimal inhibitory concentration (MIC) in broth (221).

c. Description — Long, slender, gram-negative rods (0.3–0.5 × 3–10 μm), motile by gliding on surfaces, nonmotile in suspension except for flexing movement, with formation of columnar aggregates of cells on infected tissue (often referred to as "haystacks"). The physiological characteristics of *F. columnare*, as described by Bernardet and Grimont (223), are the following: strict aerobic growth; no acid produced from carbohydrates; positive for cytochrome oxidase and catalase; nitrate reduced to nitrite; positive for hydrogen sulfide; cellulose, chitin, starch, esculin, and agar not hydrolyzed; gelatin, casein, and tyrosine hydrolyzed; arginine, lysine, and ornithine not decarboxylated; flexirubin pigments produced. The NaCl tolerance is reported as 0.5%, but this may be variable. Griffin (224) devised a simple method of identifying *F. columnare*, using five characteristics that separate it from other yellow-pigment-producing aquatic bacteria:

1. Ability to grow in the presence of neomycin sulfate and polymyxin B
2. Colonies on CA plates typically rhizoid and pigmented pale yellow
3. Production of gelatin-degrading enzymes
4. Binding of congo red dye to the colony
5. Production of a chondroitin sulfate-degrading enzyme.

d. Epizootiology — Columnaris disease is one of the most common diseases in freshwater aquaculture worldwide, affecting a wide variety of species. It may occur as a secondary infection following stress or injury or as a primary infection. Columnaris disease most often occurs as an external infection of the skin, the fins, or the gills, but it has been isolated systemically from fish showing no external clinical signs. These internal isolates may indicate a systemic form of the disease; however, histopathological lesions are often lacking. Mixed infections involving other bacterial pathogens are also common. Hawke and Thune (221) found from 99 different cases of bacterial disease in catfish ponds in Louisiana,U.S.A. that columnaris disease was diagnosed in 53.5% of the submissions. In this study, *F. columnare* was the sole etiological agent in 7.0% of the cases; it was present in mixed infections with other pathogens (i.e., *Aeromonas* spp., *E. ictaluri*, and *E. tarda*) in 46.5% of the cases. Transmission of columnaris disease is typically from fish to fish via the water, but stress resulting from handling and poor water quality exacerbates the disease. Survival of *F. columnare* is poor in water with pH less than 7.0, with hardness less than 50 mg/L (ppm), or with low organic matter. Only 35% of inoculated cells survive for one week in sterile pond water at 20 °C (68 °F); this result indicates a need for carrier fish to maintain the organism in the aquatic environment (225).

e. Geographic range and host susceptibility — Columnaris exists worldwide in a variety of freshwater habitats. The channel catfish is the most severely affected aquaculture species in the U.S.; however, golden shiners, fathead minnows, hybrid striped bass, and goldfish are susceptible. In Europe and Asia, columnaris is a problem in cultured eels and in common carp. Salmonids and cultured centrarchids are also susceptible under permissive water temperatures. Anderson and Conroy (226) listed 36 species of fish from which columnaris disease had been described, and most species of fish are considered susceptible to infection.

f. Clinical signs of disease and treatment—Columnaris disease is characterized by shallow, white-to-yellowish focal patches of necrosis in the skin, often extending around the dorsal fin (hence, the term "saddleback"). Necrotic foci in the gills may also be observed, and lesions often appear brown or muddy, a coloration due to clay particles or detritus trapped in the slime secreted by the bacteria. Secondary infection of skin lesions by *Aeromonas* spp., is common and results in deeper, liquefactive lesions in the muscle. On catfish, the infected skin loses its natural sheen, and a grey-to-white margin surrounds the lesion. The mouth and inner walls of the oral cavity may be covered with a yellowish mucoid material. Scraping those areas and observing wet mounts of the tissue microscopically at 400× allows the investigator to observe the typical thin, filamentous, flexing rods and "haystack" formation described previously. This method of diagnosis is commonly used, although it is advisable to isolate the bacterium in pure culture and to run confirmatory biochemical tests or serology, for reliable identification. Improved husbandry and water quality conditions can help prevent columnaris disease, but, once an outbreak has occurred, only two basic methods of treatment are available. External columnaris may be treated by adding potassium permanganate to the water at 2 ppm over the permanganate demand. Potassium permanganate imparts a red color to the water and this color should persist for four to five hours for an effective treatment. Potassium permanganate is on deferred status by the USFDA: though currently permitted for use on food fish, it is not yet fully approved, and it may be made illegal in the future if new evidence finds it to be unsafe. Feeds medicated with Romet® or Terramycin® are effective for controlling columnaris disease, although they are not labeled specifically for treatment of this disease in the United States (227). Maintaining a salinity of 5 ppt is inhibitory to the development of columnaris disease.

II. Flavobacterium branchiophilum ("F. branchiophila"). Causative agent of bacterial gill disease.

a. History—Bacterial gill disease (BGD) of salmonids was originally described by Davis in 1926, from a salmonid fish hatchery (228). The primary etiological agent of BGD, *Flavobacterium branchiophila*, was described in 1989 by Wakabayashi (229), and the name of the bacterium was revised in 1990 to *F. branchiophilum* (230). *Cytophaga aquatilis* was also described as a causative agent of BGD by Strohl and Tait (231); this bacterium has been renamed *Flavobacterium hydatis* (217). It is suspected that other related *Flavobacterium* spp., *Cytophaga* spp., and *Flexibacter* spp., play an opportunistic role in BGD (7). The disease is now known to be multifactorial; degraded environmental conditions work in concert with bacteria to produce the classic lesions of BGD.

b. Culture—Primary isolation of *Flavobacterium branchiophilum* and related organisms associated with BGD is on cytophaga agar (CA). *Flavobacterium branchiophilum* is somewhat fastidious and grows slowly, producing 0.1–1.0 mm yellow, round, transparent, colonies on CA after two to five days of incubation at 25 °C. The organism grows between 10 and 30 °C, but not at 37 °C.

c. Description—Long, thin, gram-negative rods (0.5 × 5 to 8 μm); nonmotile by any mechanism; oxidase-positive; gelatin, casein, and starch hydrolyzed, but not chitin or esculin; acid produced from glucose, fructose, sucrose, maltose, and trehalose.

d. Epizootiology—The epizootiology of BGD is confusing, because many different etiologic agents have been implicated, and attempts to fulfill Koch's postulates with various isolates have failed. *Flavobacterium branchiophilum* has been successfully transmitted, and BGD was experimentally induced in the laboratory, but extreme environmental stress was required to initiate infection. It is now accepted that BGD is initiated following injury to the gills by chemical or physical irritants (7). If excess feed or detritus is present in the water column, these particles may become impinged in the hyperplastic and mucus-slime-covered gills and cause asphyxiation. Salmonid fry and fingerlings less than 5 cm (2 in.) in length are most susceptible to BGD, and mortality rates in affected populations can be as high as 80% over a one-week period. The survivability of *F. branchiophilum* in water is unknown, but infection is believed to be transmitted from fish to fish. Disease outbreaks commonly occur between 12 and 19 °C.

e. Geographic range and host susceptibility—Bacterial gill disease has been found in most parts of the world where fish are cultured in intensive freshwater systems. *Flavobacterium branchiophilum* has been reported from Japan, the United States, Hungary, the Netherlands, and Canada. Many species of fish are susceptible to BGD; however, the cultured species most affected are all of the salmonids, the common carp, goldfish, channel catfish, eels, and fathead minnows (232).

f. Clinical signs of disease and treatment—Bacterial gill disease is diagnosed by microscopic examination of diseased gill tissue. The presence of tufts of filamentous bacteria between hyperplastic secondary lamellae are indicative of the disease. Culture on CA is recommended for isolation and identification of *F. branchiophilum* and other related organisms. Bacterial colonization and fusion of the gill lamellae begins on the distal end of filaments and results in "clubbing" of the gills. Maintenance of good water quality and reduction of stress is the best method for controlling BGD. Disinfectants such as benzalkonium chloride have been used successfully at 1–2 mg/L (ppm) for one hour for treating infected salmonids in raceways, but the margin of safety is small. Chloramine-T at 8–10 mg/L (ppm) for one hour is the most effective treatment for BGD; however, neither of these compounds is currently approved by the USFDA for treatment of food fish (7,8).

III. Flavobacterium psychrophilum ("Cytophaga psychrophila, Flexibacter psychrophilus"). Causative agent of bacterial cold-water disease. Synonym: peduncle disease.

a. History—A disease of salmonids originally referred to as "peduncle disease" was described by Davis in 1946 from the eastern United States (233). This observation of a characteristic open lesion on or near the caudal peduncle led to the common name first used for the disease. The causative bacterium was isolated, was described, and was named *Cytophaga psychrophila* by Borg in 1960 (234); however, Bernardet and Grimont, using DNA relatedness techniques, renamed the organism *Flexibacter psychrophilus* (223). The bacterium has recently received additional reclassification; it is currently named *Flavobacterium psychrophilum* (217).

b. Culture—Primary isolation is on cytophaga agar (CA) incubated at 15–20 °C. After 48–96 hours incubation, 1–5 mm, bright yellow, raised, convex colonies with a thin spreading irregular edge are visible. Improved growth is reported for this organism in Shieh broth and Shieh agar media (220) and in tryptone-yeast extract (TYE) (8). Generally, growth is limited on standard media such as TSA. The organism grows at temperatures between 4 and 25 °C. No growth is detectable at 30 °C.

c. Description—Long, slender, Gram-negative rods, (0.3–0.75 × 2–7 µm), but filamentous forms 10–40 µm in length are occasionally observed. The bacterium is motile by gliding on solid surfaces, nonmotile in suspension. The physiological characteristics of *F. psychrophilum*, as described by Bernardet and Grimont, are: strict aerobic growth; no acid produced from carbohydrates; oxidase-negative; catalase-positive; negative for hydrogen sulfide; hydrolysis of casein, gelatin, albumin, elastin, tyrosine, and collagen; no hydrolysis of agar, cellulose, starch, or chitin; flexirubin pigments are produced. The NaCl tolerance is 0.5 to 1.0%.

d. Epizootiology—Bacterial cold-water disease (BCWD) appears in the early spring at hatcheries, when water temperatures are between 4 and 10 °C (39–50 °F), and the severity of the disease depends on the age and development of the affected fry. In sac fry, the mortality rate may be as high as 50%, but, when the disease occurs in fingerlings in rearing units, losses are usually less than 20%. Experimental infection was induced by Holt (235) in coho and chinook salmon and in rainbow trout at temperatures of 3 to 15 °C (37–59 °F), but, above 15 °C, the severity of the disease was greatly reduced. The natural reservoir of the bacterium is not understood, but it is generally believed that transmission is from fish that serve as carriers of the organism.

e. Geographic range and host susceptibility—Bacterial cold-water disease occurs throughout the trout- and salmon-growing regions of North America, Europe, and Japan. It is a serious problem in the northwestern United States and western Canada. All salmonid species are believed to be affected, but coho salmon are particularly susceptible (8). The nonsalmonids affected include European eels, carp, tench (*Tinca tinca*), and crucian carp (*Carassius carassius*).

f. Clinical signs of disease and treatment—The clinical signs of BCWD vary with the size and the age of affected fish. When the disease occurs in sac fry, the primary sign is erosion of the skin covering the yolk sac. Older fish exhibit lethargy, spiral swimming behavior, and the classic shallow peduncle lesion. The caudal peduncle of affected fingerlings first turns white; the skin then becomes necrotic and is sloughed off, leaving exposed underlying muscle tissue. Lesions can also occur laterally, dorsally, or on the isthmus. Internally, petechiae may be observed in the liver, pyloric cecae, adipose tissue, heart, swim bladder, and peritoneal lining. Outbreaks of BCWD are difficult to treat, because the infection is systemic and because affected fry do not feed. Bath treatments with water-soluble Terramycin® at 10–50 mg/L (ppm) and with quaternary ammonium compounds at 2 mg/L (ppm) are effective when infections are confined to the skin. For systemic infections, oxytetracycline in the diet at 50–75 mg/kg of fish/day (23–34 mg/lb) for 10 days has been found to be effective (7,8).

IV. Flexibacter maritimus ("Cytophaga marina"). Causative agent of salt water columnaris. Synonym: black patch necrosis of Dover sole (*Solea solea*).

a. History—A marine form of columnaris disease was described as early as 1960 by Borg (234), and a detailed account of a columnaris-like disease in juvenile red sea bream in Japan was given by Masumura and Wakabayashi (236). *Flexibacter maritimus* was first described in 1986 from a variety of marine fish species in Japanese aquaculture by Wakabayashi (237). An organism described as the causative agent of "black patch necrosis" of cultured Dover sole in Scotland (238) was later determined to be *F. maritimus*. Recently, *F. maritimus* has been found to be the cause of mortality in farm-reared sea bass in France (239) and in turbot, Atlantic salmon, and coho salmon cultured in Spain (240). Chinook salmon reared in sea cages in southern California, as well as white sea bass (*Atractoscion nobilis*), northern anchovy (*Engraulis mordax*), and Pacific sardine (*Sardinops sagax*), have been diagnosed with *F. maritimus* infections (241).

b. Culture—This bacterium must be cultured on cytophaga agar (CA) prepared with at least 30% sea water. Addition of sodium chloride to CA is not sufficient to culture the organism. Incubation is at 25–30 °C; however, growth can occur between 15 and 34 °C (8). A selective medium for the isolation of *F. maritimus* from external lesions has been developed in Spain (240).

c. Description—Long, slender, gram-negative rods, 0.5 × 2.0–10.0 µm, with occasional filamentous forms 30 µm in length; motile by gliding on solid surfaces, nonmotile in suspension except for flexing; and formation of columnar aggregates of cells on infected tissue. *Flexibacter maritimus* produces catalase and cytochrome oxidase and hydrolyses casein, gelatin, tributyrin, and tyrosin. Hydrogen sulfide is not produced; acid is not produced from carbohydrates and agar; cellulose, chitin, starch, and esculin are not degraded. Nitrate is reduced to nitrite. Flexirubin pigments are not produced.

d. Epizootiology—Saltwater columnaris of red sea bream occurs in the spring after transfer of juveniles from the hatchery to inshore net cages. The disease rarely affects sea bream greater than 60 mm (2.3 in.) in length. Stress occurring during the transfer period is believed to be responsible for initiation of infection. Black patch necrosis occurs in juvenile Dover sole between 60 and 100 days of age and occurs more frequently in the summer.

e. Geographic range and host susceptibility—*Flexibacter maritimus* is confirmed from red sea bream, black sea bream, and flounder in Japan and from the Dover sole in Scotland; however, it is speculated that many species of marine fish are susceptible to the disease. Sea-cage reared salmonids are also susceptible on the Pacific coast of the United States (241) and off the coast of Spain (239).

f. Clinical signs of disease and treatment—The clinical signs of saltwater columnaris and of black patch necrosis are similar to those of columnaris disease in fresh water species. (See sections G, I, and F of this entry). Avoiding overcrowding, overfeeding, and stress is the best means of managing the disease; antibiotic feeds have met with little success. Addition of a sand substrate to tanks containing juvenile Dover sole greatly reduced the incidence of black patch necrosis (238).

H. Obligate Intracellular Bacteria

I. Piscirickettsia salmonis. Causative agent of salmonid rickettsial septicemia. Synonyms: coho salmon syndrome, Huito disease.

a. History—Salmonid rickettsial septicemia (SRS) of coho salmon was first noticed in Chile as early as 1981, but it was not until 1991, in the published account of Cvitanich (242), that a formal description of the disease was given and the disease was formally named. The incidence of the disease has increased over the years; currently, it is the major problem affecting salmonid aquaculture in Chile. The organism was formally recognized as belonging to a new taxon, and the name *Piscirickettsia salmonis* was proposed by Fryer (243). This name currently has valid standing. Since 1988, the disease has been recognized as a problem on 51 farms on the west coast of Norway in cage-reared Atlantic salmon (244).

b. Culture—Attempts to culture *P. salmonis* on artificial media have been unsuccessful. The organism must be grown in cell culture, on cell lines of salmonid and non-salmonid origin; chinook salmon embryo CHSE-214 is the cell line of choice for isolation from infected kidney tissue. Incubation is at 15–18 °C, and cytopathic effect is witnessed within five to six days.

c. Description—*Piscirickettsia salmonis* is an obligate intracellular pathogen. The cells are pleomorphic but are predominantly gram-negative cocci ($0.5 \times 1.5–2.0$ μm) that are nonmotile.

d. Epizootiology—Salmonid rickettsial septicemia was first reported from the southern coast of Chile in salmonids reared in sea cages. There are no indigenous salmonids on the Pacific coast of South America; it is theorized that *P. salmonis* originated from local marine fish species and infected the introduced salmon. Samples of various ectoparasites of Chilean salmon revealed positive identification of *P. salmonis* from *Cerathothoa gaudichaudii*, a hematophagous external parasite now believed to be a vector for the disease (245).

e. Geographic range and host susceptibility—Once thought to be restricted to salmonids in Chile, the disease has now been reported from Atlantic salmon in Norway (244). Experimental infections have been achieved in coho salmon and rainbow trout (246).

f. Clinical signs of disease and treatment—Infected fish gather at the surface of cages and exhibit lethargy and inappetence. External signs include darker-than-normal coloration, pale gills, and hemorrhagic skin lesions. Internally, the fish exhibit ascites, peritonitis, and pale discoloration of swollen organs. The gastrointestinal tract, swim bladder, and fat deposits may have petechial hemorrhages. The disease is not apparent in the freshwater stage of culture; it does not appear until 6–12 weeks after transfer to seawater. The organism is sensitive to some antibiotics, among them clarithromycin, chloramphenicol, erythromycin, gentamycin, oxytetracycline, sarafloxacin, and streptomycin, but it is not sensitive to penicillin (8,9).

BIBLIOGRAPHY

1. G.F. Bonaveri (1907), as referenced by Austin and Austin [9].
2. G. Canestrini, *Atti Istitute Venito Service* **7**, 809–814 (1883).
3. J.D. Freund, R.M. Durborow, J.R. MacMillan, M.D. Crosby, T.L. Wellborn, P.W. Taylor, and T.L. Schwedler, *J. Aquat. Anim. Health* **2**, 207–211 (1990).
4. R.L. Thune, *Vet. Human Toxicol.* **33**, 14–18.
5. U.S. Dept. of Agriculture, Animal and Plant Inspection Service, (1997).
6. National Animal Health Monitoring System, Catfish '97, Centers for Epidemiology and Community Health, Animal and Plant Health Inspection Service, U.S. Department of Agriculture, 1997.
7. J.A. Plumb, *Health Maintenance and Principal Microbial Diseases of Cultured Fishes*, Iowa State University Press, Ames, Iowa, 1999.
8. V. Inglis, R.J. Roberts, and N.R. Bromage, *Bacterial Diseases of Fish*, John Wiley and Sons, New York, 1993.
9. B. Austin and D.A. Austin, *Bacterial Fish Pathogens: Disease in Farmed and Wild Fish*, 2nd ed., Ellis Horwood, Ltd., New York, 1993.
10. R.L. Thune, L.A. Stanley, and R.K. Cooper, *Ann. Rev. Fish Dis.*, 1993, pp. 37–68.
11. J.C. Thoesen, *Suggested Procedures for the Detection and Identification of Certain Finfish and Shellfish Pathogens*, 4th ed., Fish Health Section, American Fisheries Society, U.S. Fish and Wildlife Service, Bethesda, Maryland, 1994.
12. M.K. Stoskopf, *Fish Medicine*, W.B. Saunders Co., Philadelphia, 1993.
13. E. Noga, *Fish Disease: Diagnosis and Treatment*, Mosby, St. Louis, 1996.
14. R. Sakazaki, *Jpn. J. Bact.* **17**, 616–617 (1962).
15. T. Hoshina, *Bull. Jpn. Soc. Sci. Fish.* **28**, 162–164 (1962).
16. W.H. Ewing, A.C. McWhorter, M.R. Escobar, and A.H. Lubin, *Int. Bull. Bacteriol. Nomencl. Taxon.* **15**, 33–38 (1965).
17. H. Wakabayashi and S. Egusa, *Bull. Jpn. Soc. Sci. Fish.* **39**, 931–936 (1973).
18. F.P. Meyer and G.L. Bullock, *Appl. Micro.* **25**, 135–156 (1973).
19. E.B. Shotts and W.D. Waltman, *J. Wild. Dis.* **26**, 214–218 (1990).

20. L.E. Wyatt, R. Nicholson, II, and C. Vanderzant, *Appl. Env. Micro.* **38**, 710–714 (1979).
21. J.P. Hawke, *J. Fish. Res. Bd. of Can.* **36**, 1508–1512 (1979).
22. J.P. Hawke, A.C. McWhorter, A.G. Steigerwalt, and D.J. Brenner, *Int. J. Syst. Bact.* **31**, 396–400 (1981).
23. A.J. Mitchell, National Aquaculture Research Center, Personal Communication, 1998.
24. J.A. Plumb and E.E. Quinlan, *Progr. Fish Cult.* **48**, 212–214 (1986).
25. D.V. Baxa and R.P. Hedrick, Fish Health Section, *American Fisheries Society Newsletter* **17**, 4 (1989).
26. R.R. Rucker, *Bulletin de L'Office International des Epizooties* **65**, 825–830 (1966).
27. A.J. Ross, R.R. Rucker, and W.H. Ewing, *Can. J. Micro.* **12**, 763–770 (1966).
28. E.W. Ewing, A.J. Ross, D.J. Brenner, and G.R. Fanning, *Int. J. Syst. Bact.* **28**, 37–44 (1978).
29. American Fisheries Society, *Bluebook*, 3rd ed., Fish Health Section, U.S. Fish and Wildlife Service, Washington, D.C., 1975.
30. W.D. Waltman and E.B. Shotts, *Can. J. Fish. and Aquatic Sci.* **41**, 804–806 (1984).
31. C.J. Rodgers, *J. Fish. Dis.* **15**, 243 (1992).
32. G. Sanarelli, *Zentralblatt für Bakteriologie, Parasitenkunde, Infectionskrankheiten, und Hygeine* **9**, 193–228 (1891).
33. W. Schaperclaus, *Zeitung fur Fisherei* **28**, 289–370 (1930).
34. W.H. Ewing, R. Hugh, and J.G. Johnson, *Studies on the Aeromonas Group*, Monograph, Atlanta, Georgia, 1961.
35. A.M. Carnahan, S. Behram, and S.W. Joseph, *J. Clin. Microbiol.* **29**, 2843–2849 (1991).
36. E.B. Shotts and R. Rimler, *Appl. Microbiol.* **26**, 550–553 (1973).
37. J.P. Hawke, *Factors Contributing to Bacterial Fish Kills in Large Impoundments*, Master's thesis, Auburn University, Dept. of Fisheries and Allied Aquacultures, 1974.
38. R.L. Thune, L.A. Stanley, and R.K. Cooper, *Ann. Rev. Fish Dis.* 37–68 (1993).
39. E.B. Shotts, J.L. Gaines, L. Martin, and A.K. Prestwood, *J. Am. Vet. Med. Asso.* **162**, 603–607 (1972).
40. D.H. McCarthy and R.J. Roberts, in M.R. Droop and H.W. Jannasch, eds., *Advances in Aquatic Microbiology*, Academic Press, London, 1980, pp. 293–341.
41. R. Belland and T.J. Trust, *J. Gen. Micro.* **134**, 307–315 (1988).
42. R. Emmerich and E. Weibel, *Archives für Hygiene und Bacteriologie* **21**, 1–21 (1894).
43. P.J. Griffin, S.F. Snieszko, and S.B. Friddle, *Trans. Amer. Fish. Soc.* **82**, 241–253 (1953).
44. H.W. Ferguson and D.H. McCarthy, *J. Fish. Dis.* **1**, 165–174 (1978).
45. G.L. Bullock, D.A. Conroy, and S.F. Snieszko, *Diseases of Fishes*, Book 2A: Bacterial Diseases of Fishes, TFH publications, Neptune City, New Jersey, 1971.
46. S.F. Snieszko, *Trans. Amer. Fish. Soc.* **78**, 56–63 (1950).
47. P.J. Griffin and S.B. Friddle, *J. Bact.* **59**, 699–710 (1950).
48. W.D. Paterson, D. Douey, and D. Desautels, *Can. J. Micro.* **26**, 588–598 (1980).
49. N.N. Fijan, *Proc. Symp. Zool. Soc. London* **30**, 39–57 (1972).
50. D.G. Elliott and E.B. Shotts, *J. Fish Dis.* **3**, 133–143 (1980).
51. E.E. Ishiguro, W.W. Kay, T. Ainsworth, J. Chamberlain, J.T. Buckley, and T.J. Trust, *J. Bact.* **148**, 333–340 (1981).
52. R.H. Richards, V. Inglis, G.N. Frerichs, and S.D. Millar, in C. Michel and D.J. Alderman, eds., *Chemotherapy in Aquaculture: From Theory to Reality*, Office International des Epizooties (OIE), Paris, 1992, pp. 276–284.
53. A.C. Barnes, C.S. Lewin, S.G.B. Amyes, and T.S. Hastings, *ICES CM 1991/F* **28**, (1991).
54. R. Nordmo, J.M. Holth-Riseth, K.J. Varma, I.H. Sutherland, and E.S. Brokken, *J. Fish Dis.* **21**, 289–297 (1998).
55. A.E. Ellis, ed., *Fish Vaccination*, Academic Press, London, 1988.
56. C.M. Press and A. Lillehaug, *British Veterinary Journal* **151**, 45–69 (1995).
57. P. Baumann and R.H.W. Schubert, in N.R. Krieg and J.G. Holt, eds., *Bergey's Manual of Systematic Bacteriology*, Williams and Wilkins, Baltimore, 1984, pp. 516–550.
58. M.H. Schiewe, T.J. Trust, and J.H. Crosa, *Curr. Micro.* **6**, 343–348 (1981).
59. M.T. MacDonell and R.R. Colwell, *Syst. and Appl. Micro.* **6**, 171–182 (1985).
60. J. Martinez-Picado, M. Alsina, A.R. Blanch, M. Cerda, and J. Jofre, *Appl. Environ. Microbiol.* **62**, 443–449 (1996).
61. P.A. West and J.V. Lee, *J. Appl. Bacteriol.* **52**, 435–448 (1982).
62. J.H. Crosa, *Microbiol. Rev.* **53**(4), 517–530 (1989).
63. J.H. Crosa, L.L. Hodges, and M.H. Schiewe, *Infection and Immunity* **27**(3), 897–902 (1980).
64. W.J. Groberg, J.S. Rohovec, and J.L. Fryer, *J. World Maricult. Soc.* **14**, 240–248 (1983).
65. Anon, *OIE Diagnostic Manual for Aquatic Animal Disease*, Office International Des Epizooties, Paris, 1995.
66. T.T. Poppe, T. Hastein, and R. Salte, in A.E. Ellis, ed., *Fish and Shellfish Pathology*, Academic Press, New York, 1985, pp. 223–229.
67. E. Egidius, K. Andersen, E. Clausen, and J. Raa, *J. Fish Dis.* **4**, 353–354 (1981).
68. K.O. Holm, E. Strom, K. Stemsvag, J. Raa, and T. Jorgensen, *Fish Pathol.* **20**, 125–129 (1985).
69. E. Egidius, R. Wiik, K. Andersen, K.A. Hoff, and B. Hjeltnes, *Int. J. Syst. Bact.* **36**, 518–520.
70. O. Evensen, S. Espelid, and T. Hastein, *Dis. of Aquat. Org.* **10**, 185–189 (1991).
71. B. Hjeltnes, K. Andersen, H.M. Ellingsen, and E. Egidius, *J. Fish Dis.* **10**, 21–27 (1987).
72. T. Jorgensen, K. Midling, S. Espelid, R. Nilsen, and K. Stensvag, *Bull. of Eur. Assoc. of Fish Pathol.* **9**, 42–44 (1989).
73. A. Lillehaug, *Aquaculture* **84**, 1–12 (1990).
74. M. Nishibuchi and K. Muroga, *Fish Pathol.* **12**, 87–92 (1977).
75. E.G. Bioscia, C. Amaro, C. Esteve, E. Alcaise, and E. Garay, *J. Fish Dis.* **14**, 103–109 (1991).
76. D.L. Tison, M. Nishibuchi, J.D. Greenwood, and R.J. Seidler, *Appl. Env. Micro.* **44**, 640–646 (1982).
77. M. Love, D. Teebkin-Fisher, J.E. Hose, J.J. Farmer, III, F.W. Hickman, and G.R. Fanning, *Science* **214**, 1139–1140 (1981).
78. D.J. Grimes, R.R. Colwell, J. Stemmler, H. Hada, D. Maneval, F.M. Hetrick, E.B. May, R.T. Jones, and M. Stoskopf, *Helgolander Meeresunters* **37**, 309–315 (1984).

79. S.K. Smith, D.C. Sutton, J.A. Fuerst, and J.L. Reichelt, *Int. J. Syst. Bact.* **41**, 529–534 (1991).
80. B. Fouz, J.L. Larsen, and A.E. Toranzo, *Bull. Eur. Assoc. Fish Pathol.* **11**, 80–812 (1991).
81. P. Vera, J.L. Navas, and B. Fouz, *Bull. Eur. Assoc. Fish Pathol.* **11**, 112–113 (1991).
82. T. Sakata, M. Matsuura, and Y. Shimikawa, *Nippon Suisan Gakkaishi* **55**, 135–141 (1989).
83. K. Pedersen, I. Dalsgaard, and J.L. Larsen, *Appl. Env. Micro.* **63**, 3711–3715 (1997).
84. T. Renault, P. Haffner, C. Malfondet, and M. Weppe, *Bull. Eur. Assoc. Fish Pathol.* **14**, 117–119 (1994).
85. S.F. Snieszko, G.L. Bullock, E. Hollis, and J.G. Boone, *J. Bact.* **88**, 1814–1815 (1964).
86. W.A. Janssen and M.J. Surgalla, *J. Bact.* **96**, 1606–1610 (1968).
87. S.S. Kubota, M. Kimura, and S. Egusa, *Fish Pathology* **4**, 111–118 (1970).
88. R. Kusuda and F. Salati, *Ann. Rev. Fish Dis.*, 1993, pp. 69–85.
89. A.E. Toranzo, S. Barreiro, J.F. Casal, A. Figueras, B. Magarinos, and J.L. Barja, *Aquaculture* **99**, 1–15 (1991).
90. J.P. Hawke, S.M. Plakas, R.V. Minton, R.M. McPhearson, T.G. Snider, and A.M. Guarino, *Aquaculture* **65**, 193–204 (1987).
91. J.P. Hawke, Ph.D. dissertation, 154 pp., Department of Veterinary Microbiology and Parasitology, Louisana State University, Baton Rouge, LA, 1996.
92. G. Gauthier, B. LaFay, R. Ruimy, V. Breittmayer, J.L. Nicolas, M. Gauthier, and R. Christen, *Int. J. Syst. Bact.* **45**, 139–144 (1995).
93. B. Magarinos, J.L. Romalde, I. Bandin, B. Fouz, and A.E. Toranzo, *Appl. Env. Micro.* **58**, 3316–3322.
94. A.E. Toranzo, J.L. Barja, and F.H. Hetrick, *Bull. Eur. Assoc. Fish Pathol.* **3**, 43–45 (1982).
95. B. Magarinos, J.L. Romalde, J.L. Barja, and A.E. Toranzo, *Appl. Environ. Microbiol.* **60**, 180–186 (1994).
96. R.A. Robohm, in D.P. Anderson, M. Dorson, and P. Daborget, ed., *Antigens of Fish Pathogens*, Collection Foundation Marcel Merieux, Lyon, 1983, pp. 161–175.
97. T. Aoki and T. Kitao, *J. Fish Dis.* **8**, 345–350 (1985).
98. J.M. Cruz, A. Saraiva, J.C. Eiras, R. Branco, and J.C. Sousa, *Bull. Eur. Assoc. Fish Pathol.* **6**, 20–22 (1986).
99. M.O. Saeed, M.M. Alamoudi, and A.H. Al-Harbi, *Dis. Aquat. Org.* **3**, 177–180 (1987).
100. H. Wakabayashi and S. Egusa, *Bull. Jpn. Soc. Sci. Fish.* **38**, 577–587 (1972).
101. D.J. Stewart, K. Woldermarian, G. Dear, and F.M. Mochaba, *J. Fish Dis.* **6**, 75–76 (1983).
102. T. Wicklund and G. Bylund, *Dis. Aquat. Org.* **8**, 13–19 (1990).
103. G. Nash, I.G. Anderson, M. Schariff, and M.N. Shamsudin, *Aquaculture* **67**, 105–111 (1987).
104. K. Nakajima, K. Muroga, and R. Hancock, *Int. J. Syst. Bact.* **33**, 1–8 (1983).
105. T. Nakai, H. Hanada, and K. Muroga, *Fish Pathol.* **20**, 481–484 (1985).
106. F.C.J. Berthe, C. Michel, and J.-F. Bernadet, *Dis. Aquat. Org.* **21**, 151–155 (1995).
107. K. Muroga, *Fish Pathol.* **13**, 35–36 (1978).
108. L. Loennstroem, T. Wiklund, and G. Byland, *Dis. Aquat. Org.* **18**, 143–147 (1994).
109. K. Muroga, Y. Jo, and T. Sawada, *Fish Pathol.* **9**, 107–114 (1975).
110. T. Nakai, K. Muroga, K. Ohnishi, Y. Jo, and H. Tanimoto, *Aquaculture* **30**, 131–135 (1982).
111. N.J. Palleroni, in N.R. Kreig and J.G. Holt, ed., *Bergey's Manual of Systematic Bacteriology* Williams and Wilkins, Baltimore, 1984, pp. 141–219.
112. G.L. Gilardi, in E.H. Lenette, A. Balows, W.J. Hausler, and H.J. Shadomy, eds., *Manual of Clinical Microbiology*, 4th ed., ASM, Washington, DC, 1985, pp. 350–372.
113. W. Ahne, W. Popp, and R. Hoffmann, *Bull. Eur. Assoc. Fish. Pathol.* **4**, 56–57 (1982).
114. G. Csaba, M. Prigli, L. Bekesi, E. Kovacs-Gayer, E. Bajmocy, and B. Fazekas, in J. Olah, K. Molnar, and S. Jeney, eds., *Fish Pathogens and Environment in European Polyculture*, Szarvas, F. Muller Fisheries Research Institute, 1981, pp. 111–123.
115. H. Wakabayashi, K. Sawada, K. Nomiya, and E. Nishimori, *Fish Pathol.* **31**, 239–240 (1996).
116. T. Hoshina, T. Sano, and Y. Morimoto, *J. Tokyo Univ. Fish.* **44**, 57–68 (1958).
117. K.L. Rouff, in P.R. Murray, ed., *Manual of Clinical Microbiology*, 6th ed., ASM Press, Washington, DC, 1995, pp. 299–307.
118. G.B. Pier and S.H. Madin, *Int. J. Syst. Bact.* **26**, 545–553 (1976).
119. G.B. Pier, S.H. Madin, and S. Al-Nakeeb, *Int. J. Syst. Bact.* **28**, 311–314 (1978).
120. R.P. Perera, S.K. Johnson, M.D. Collins, and D.H. Lewis, *J. Aquat. Anim. Health* **6**, 335–340 (1994).
121. A. Eldar, Y. Bejerano, and H. Bercovier, *Curr. Microbiol.* **28**, 139–143 (1994).
122. A. Eldar, P.F. Frelier, L. Asanta, P.W. Varner, S. Lawhon, and H. Bercovier, *Int. J. Syst. Bact.* **45**, 840–842 (1995).
123. D.A. Stoffregen, S.C. Backman, R.E. Perham, P.R. Bowser, and J.G. Babish, *J. World Aquacult. Soc.* **27**, 420–434 (1996).
124. J.P. Hawke, Annual Records of the Louisiana Aquatic Animal Disease Diagnostic Laboratory, School of Veterinary Medicine, Louisiana State University, Baton Rouge, LA, 1990–1998 (unpublished).
125. A. Eldar, Y. Bejerano, A. Livoff, A. Horovitcz, and H. Bercovier, *Vet. Micro.* **43**, 33–40 (1995).
126. R.P. Perera, S.K. Johnson, and D.H. Lewis, *Aquaculture* **152**, 25–33 (1997).
127. T. Kitao, T. Aoki, and K. Iwata, *Bull. Jpn. Soc. Sci. Fish.* **45**, 567–572 (1979).
128. M.R. Weinstein, M. Litt, D.A. Kertesz, P. Wyper, D. Rose, M. Coulter, A. McGeer, R. Facklam, C. Ostach, B.M. Willey, A. Borczyk, and D.E. Low, *New Eng. J. Med.* **337**, 589–594 (1997).
129. K. Ohnishi and Y. Jo, *Fish Pathol.* **16**, 63–67 (1981).
130. A.H. Al-Harbi, *Aquaculture* **128**, 195–201 (1994).
131. J.T.W. Foo, B. Ho, and T.J. Lam, *Aquaculture* **49**, 185–195 (1985).
132. A. Eldar, A. Horovitcz, and H. Bercovier, *Vet. Immunol. and Immunopath.* **56**, 175–183 (1997).
133. J.A. Robinson and F.P. Meyer, *J. Bact.* **92**, 512 (1966).
134. J.A. Plumb, J.H. Schachte, J.L. Gaines, W. Peltier, and B. Carroll, *Trans. Amer. Fish. Soc.* **103**, 358–361 (1974).
135. P.H. Chang and J.A. Plumb, *J. Fish Dis.* **19**, 235–241 (1996).

136. A.M. Baya, B. Lupiani, F.M. Hetrick, B.S. Roberson, R. Lukacovic, E. May, and C. Poukish, *J. Fish Dis.* **13**, 251–253 (1990).
137. V. Rasheed and J.A. Plumb, *Aquaculture* **37**, 97–105 (1984).
138. P. Vandamme, L.A. Devriese, B. Pot, K. Kersters, and P. Melin, *Int. J. Syst. Bact.* **47**, 81–85 (1997).
139. J.A. Elliot, R.R. Facklam, and C.B. Richter, *J. Clin. Micro.* **28**, 628–630 (1990).
140. J.P. Euzeby, *Int. J. Syst. Bact.* **48**, 1073–1075 (1998).
141. R.L. Amborski, T.G. Snider, III, R.L. Thune, and D.D. Culley, Jr., *J. Wild. Dis.* **19**, 180–184 (1983).
142. R. Kusuda, K. Kawai, T. Toyoshima, and I. Komatsu, *Bull. Jpn. Soc. Sci. Fish.* **42**, 1345–1352 (1979).
143. R. Kusuda, I. Komatsu, and K. Kawai, *Bull. Jpn. Soc. Sci. Fish.* **44**, 295 (1978).
144. R. Kusuda, K. Kawai, and T. Shirakawa, *Bull. Jpn. Soc. Sci. Fish.* **48**, 1731–1738 (1982).
145. R. Kusuda, K. Kawai, F. Salati, C.R. Banner, and J.L. Fryer, *Int. J. Syst. Bact.* **41**, 406–409 (1991).
146. L.M. Teixeira, V.L.C. Merquior, M.E. Vianni, M.S. Carvalho, S.E.L. Francalanzza, A.G. Steigerwalt, D.J. Brenner, and R.R. Facklam, *Int. J. Syst. Bact.* **46**, 664–668 (1996).
147. T. Minami, *Fish Pathol.* **14**, 15–19 (1979).
148. T. Kitao, T. Aoki, and K. Iwata, *Bull. Jpn. Soc. Sci. Fish.* **45**, 567–572 (1979).
149. R. Kusuda and I. Takemaru, *Nippon Suisan Gakkaishi* **53**, 1519–1523 (1987).
150. H. Matsuyama, R.E.P. Mangindaan, and Y. Yano, *Aquaculture* **101**, 197–203 (1992).
151. A. Ross and R.J. Toth, *Prog. Fish Cult.* **36**, 191 (1974).
152. S.F. Hui, R.A. Holt, N. Sriranganathan, R.J. Seidler, and J.L. Fryer, *Int. J. Syst. Bact.* **34**, 393–400 (1984).
153. M.D. Collins, J.A.E. Farrow, B.A. Phillips, S. Ferusu, and D. Jones, *Int. J. Syst. Bact.* **37**, 310–316 (1987).
154. S. Wallbanks, A.J. Martinez-Murcia, J.L. Fryer, B.A. Phillips, and M.D. Collins, *Int. J. Syst. Bact.* **40**, 224–230 (1990).
155. C. Michel, P. Nougayrede, A. Eldar, E. Sochon, and P. DeKinkelin, *Dis. Aquat. Org.* **30**, 199–208 (1997).
156. A.H. Williams, J.L. Fryer, and M.D. Collins, *FEMS Micro. Lett.* **68**, 109–114 (1990).
157. A.M. Baya, A.E. Toranzo, B. Lupiani, T. Li, B.S. Roberson, and F.M. Hetrick, *App. Env. Micro.* **57**, 3114–3120 (1991).
158. C. Michel, B. Faivre, and B. Kerouault, *Dis. Aquat. Org.* **2**, 27–30 (1986).
159. C.E. Starliper, E.B. Shotts, and J. Brown, *Dis. Aquat. Org.* **13**, 181–187 (1992).
160. D.K. Cone, *J. Fish Dis.* **5**, 479–485 (1982).
161. J. Boomker, G.D. Imes, Jr., C.M. Cameron, T.W. Naude, and H.J. Schoonbee, *Onderstepoort J. Vet. Res.* **46**, 71–77 (1979).
162. R.R. Bragg and J.S.E. Broere, *Bull. Eur. Assoc. Fish Pathol.* **6**, 89–91 (1986).
163. J. Carson and B. Munday, *Austasia Aquaculture* **5**, 32–33 (1990).
164. G. Ceschia, G. Giorgetti, R. Giavenni, and M. Sarti, *Bull. Eur. Assoc. Fish Pathol.* **12**, 71–72.
165. M. Johnson, Mississippi Cooperative Extension Service, Stoneville, MS, personal communication.
166. A.E. Toranzo, S. Devesa, P. Heinen, A. Riaza, S. Nunez, and J.L. Barja, *Bull. Eur. Assoc. Fish Pathol.* **14**, 19–23 (1994).
167. R.R. Bragg, J.M. Todd, S.M. Lordan, and M.E. Combrink, *Onderstepoort J. Vet. Res.* **56**, 179–184 (1989).
168. M. Ford, J.D. Perry, and F.K. Gould, *J. Clin. Micro.* **32**, 2999–3001 (1994).
169. J. Carson, N. Gudkovs, and B. Austin, *J. Fish Dis.* **16**, 381–388 (1993).
170. J. Leiro, A.E. Toranzo, J. Estevez, L.J. Lamas, J.L. Barja, and F.M. Ubeira, *Vet. Micro.* **48**, 29–39 (1996).
171. R.R. Bragg, *Onderstepoort J. Vet. Res.* **58**, 67–70 (1991).
172. I.W. Smith, Dept. of Agr. and Fish. for Scotland, *Freshwater Fish. Salmon Res.* **34**, 1–12 (1964).
173. B.J. Earp, C.H. Ellis, and E.J. Ordal, Washington Dept. of Fisheries, *Special Report* **1**, 1–74 (1953).
174. T.P.T. Evelyn, Fish Health Section, *Amer. Fish. Soc. Newsletter* **14**, 6 (1986).
175. E.J. Ordal and B.J. Earp, *Proc. Soc. Exp. Biol. Med.* **92**, 85–88 (1956).
176. J.E. Sanders and J.L. Fryer, *Int. J. Syst. Bact.* **30**, 496–502 (1980).
177. T.P.T. Evelyn, *Bull. Off. Int. des Epizoot.* **87**, 511–513 (1977).
178. B. Austin, T.M. Embley, and M. Goodfellow, *FEMS Micro. Lett.* **17**, 111–114 (1983).
179. T.M. Embley, M. Goodfellow, and B. Austin, *FEMS Micro. Lett.* **14**, 299–301 (1982).
180. C.K. Arakawa, J.E. Sanders, and J.L. Fryer, *J. Fish. Dis.* **10**, 249–253 (1987).
181. G.D. Weins and S.L. Kaattari, *Fish Path.* **24**, 1–7 (1987).
182. J.L. Fryer and J.E. Sanders, *Ann. Rev. Micro.* **35**, 273–298 (1981).
183. G.L. Bullock, H.M. Stuckey, and D. Mulcahy, *Fish Health News* **7**, 51–52 (1978).
184. K. Wolf and C.E. Dunbar, *Trans. Amer. Fish. Soc.* **88**, 117–134 (1959).
185. T.P.T. Evelyn, *J. Wild. Dis.* **7**, 328–335 (1971).
186. B. Austin, D. Bucke, S. Feist, and J. Rayment, *Bull. Eur. Assoc. Fish Pathol.* **5**, 8–9 (1985).
187. A.M. Baya, B. Lupiani, I. Bandin, F.M. Hetrick, A. Figueras, A. Carnahan, E.M. May, and A.E. Toranzo, *Dis. Aquat. Org.* **14**, 115–126 (1992).
188. S. Backman, H.W. Ferguson, J.F. Prescott, and B.P. Wilcock, *J. Fish Dis.* **13**, 345–353 (1990).
189. E. Bataillon, Dubard, and L. Terre, *Comptes rendus des Seances de la Societe Biologie* **49**, 446–449 (1897).
190. E. Bataillon, A. Moeller, and L. Terre, *Zentrallblatt fur Tuberkulose* **3**, 467–468 (1902).
191. J.D. Aronson, *J. Inf. Dis.* **39**, 315–320 (1926).
192. A.J. Ross and F.P. Brancato, *J. Bact.* **78**, 392–395 (1959).
193. A.J. Ross, *Amer. Rev. Resp. Dis.* **81**, 241–250 (1960).
194. C.K. Arakawa, J.L. Fryer, and J.E. Sanders, *J. Fish Dis.* **9**, 269–271 (1986).
195. L.G. Wayne and G.P. Kubica, *Bergey's Manual of Determinative Bacteriology*, Vol. 2, 1986, pp. 1436–1457.
196. H.M. Sommers and R.C. Good, in E.H. Lennette, ed., *Manual of Clinical Microbiology*, 4th ed., 1985, pp. 216–248.
197. R.P. Hedrick, T. McDowell, and J. Groff, *J. Wild. Dis.* **23**, 391–395 (1987).
198. S. Chinabut, C. Limsuwan, and P. Chanratchakool, *J. Fish Dis.* **13**, 531–535 (1990).
199. A.J. Ross and H.E. Johnson, *Prog. Fish Cult.* **24**, 147–149 (1962).
200. J.A. Sakanari, C.A. Reilly, and M. Moser, *Trans. Amer. Fish. Soc.* **112**, 565–566 (1983).

201. W.K. Vogelbein, D.E. Zwerner, H. Kator, M. Rhodes, and J. Cardinal, *Abstracts of the 23rd Annual Eastern Fish Health Workshop*, 1999.
202. R.F. Nigrelli and H. Vogel, *Zoologica* **48**, 131–144 (1963).
203. I.E. Valdez and D.A. Conroy, *Microbiología española* **16**, 245–253 (1963).
204. S.F. Snieszko, G.L. Bullock, C.E. Dunbar, and L.L. Pettijohn, *J. Bact.* **88**, 1809–1810 (1964).
205. G. Campbell and R.M. MacKelvie, *J. Fish. Res. Bd. Can.* **25**, 423–425 (1968).
206. T. Kariya, S. Kubota, Y. Nakamura, and K. Kira, *Fish Pathol.* **3**, 16–23 (1968).
207. T. Kudo, K. Hatai, and A. Seino, *Int. J. Syst. Bact.* **38**, 173–178 (1988).
208. R.E. Gordon and J.M. Mihm, *J. Gen. Micro.* **27**, 1–10 (1962).
209. R. Kusuda, H. Taki, and T. Takeuchi, *Bull. Jpn. Soc. Sci. Fish.* **40**, 369–373 (1974).
210. S.C. Chen, M.C. Tung, and W.C. Tsai, *COA Fish. Ser. No. 15, Fish Dis. Res. (IX)* **6**, 42 (1989).
211. L. Udey, E. Young, and B. Sallman, *Fish Health News* **5**, 3–4 (1976).
212. L.R. Udey, E. Young, and B. Sallman, *J. Fish. Res. Bd. Can.* **34**, 402–409 (1977).
213. D.H. Lewis and L.R. Udey, *Fish Disease Leaflet #56*, U.S. Department of Interior, Fish and Wildlife Service, 1978.
214. H.S. Davis, *Bull. U.S. Bureau of Fish.* **38**, 261–280 (1922).
215. E.J. Ordal and R.R. Rucker, *Proc. Soc. Exp. Biol. Med.* **56**, 15–18 (1944).
216. L. Garnjobst, *J. Bact.* **49**, 113–128 (1945).
217. J. Bernardet, P. Segers, M. Vancanneyt, F. Berthe, K. Kersters, and V. Vandammee, *Int. J. Syst. Bact.* **46**, 128–148 (1996).
218. J.A. Bader and E.B. Shotts, *J. Aquat. Anim. Health* **10**, 320–327 (1998).
219. R.L. Anacker and E.J. Ordal, *J. Bact.* **78**, 33–40 (1959).
220. H.S. Shieh, *Microbios Lett.* **13**, 129–133 (1980).
221. J.P. Hawke and R.L. Thune, *J. Aquat. Anim. Health* **4**, 109–113 (1992).
222. T. Hsu, E.B. Shotts, and W.D. Waltman, *Newsletter for the Flavobacterium-Cytophaga Group* **3**, 29–30 (1983).
223. J.F. Bernardet and P.A.D. Grimont, *Int. J. Syst. Bact.* **39**, 346–354 (1989).
224. B.R. Griffin, *J. Aquat. Anim. Health* **4**, 63–66 (1992).
225. C.D. Becker and M.P. Fujihara, *Amer. Fish. Soc. Monograph* **2**, 92 (1978).
226. J.I.W. Anderson and D.A. Conroy, *J. Appl. Bact.* **32**, 30–39 (1969).
227. R.M. Durborow, R.L. Thune, J.P. Hawke, and A.C. Camus, *Southern Regional Aquaculture Center*, Publication No. 479, 1998.
228. H.S. Davis, *Trans. Amer. Fish. Soc.* **56**, 156–160 (1926).
229. H. Wakabayashi, G.J. Huh, and N. Kimura, *Int. J. Syst. Bact.* **39**, 213–216 (1989).
230. A. von Graevenitz, *Int. J. Syst. Bact.* **40**, 211 (1990).
231. W.R. Strohl and L.R. Tait, *Int. J. Syst. Bact.* **28**, 293–303 (1978).
232. V.E. Ostland, H.W. Ferguson, and R.M.V. Stevenson, *Dis. Aquat. Org.* **6**, 179–184 (1989).
233. H.S. Davis, *U.S. Dept. of the Interior Research Report No. 12*, U.S. Govt. Printing Office, Washington, DC, 1946.
234. A.F. Borg, *J. Wild. Dis.* **8**, 1–85 (1960).
235. R.A. Holt, A. Amandi, J.S. Rohovec, and J.L. Fryer, *J. Aquat. Anim. Health* **1**, 94–101 (1989).
236. K. Masumura and H. Wakabayashi, *Fish Pathol.* **12**, 171–177 (1977).
237. H. Wakabayashi, M. Hikilda, and K. Masumura, *Int. J. Syst. Bact.* **36**, 396–398 (1986).
238. A.C. Campbell and J.A. Buswell, *J. Fish Dis.* **5**, 495–508 (1982).
239. J.F. Bernardet, B. Kerouault, and C. Michel, *Fish Pathol.* **29**, 105–111 (1994).
240. F. Pazos, Y. Santos, A.R. Macias, S. Nunez, and A.E. Toranzo, *J. Fish Dis.* **19**, 193–197 (1996).
241. M.F. Chen and D. Henry-Ford, *J. Aquat. Anim. Health* **7**, 318–326 (1995).
242. J. Cvitanich, O. Garate, and C.E. Smith, *J. Fish Dis.* **14**, 121–145 (1991).
243. J.L. Fryer, C.N. Lannan, S.J. Giovannoni, and N.D. Wood, *Int. J. Syst. Bact.* **42**, 120–126 (1992).
244. A.B. Olsen, H.P. Melby, L. Speilberg, O. Evensen, and T. Haastein, *Dis. Aquat. Org.* **31**, 35–48 (1997).
245. L.H. Garces, P. Correal, J. Larenas, J. Contreras, S. Oyanedel, J.L. Fryer, and P.A. Smith-Schuster, *Abstracts of the International Symposium on Aquatic Animal Health*, Seattle, WA, 1994.
246. P.A. Smith, J.R. Contreras, L.H. Garces, J.J. Larenas, S. Oyanedel, P. Caswell-Reno, and J.L. Fryer, *J. Aquat. Anim. Health* **8**, 130–134 (1996).

See also ANTIBIOTICS; CHLORINATION/DECHLORINATION; DISEASE TREATMENTS; DRUGS.

BAITFISH CULTURE

NATHAN STONE
University of Arkansas at Pine Bluff
Pine Bluff, Arkansas

OUTLINE

Baitfish is a general term used to describe live fish sold for fishing bait or as "feeders," fish fed to ornamental fish and to invertebrates with piscivorous food habits. Over 20 species of fish are caught from the wild and used for bait, while relatively few, ubiquitous species are raised on farms. Major farm-raised species are the golden shiner (*Notemigonus crysoleucas*), the fathead minnow (*Pimephales promelas*) and the goldfish (*Carassius auratus*). Retail sales of baitfish for recreational fishing in North America have been estimated at $1 billion annually (1). Unique aspects of baitfish farming, as compared to food fish culture, include the vast numbers of individual fish produced, the variety of sizes required by the market, and the relatively greater importance of market forces. Arkansas farmers alone sell over six billion baitfish annually, so it is easy to see why fish propagation and early rearing of fry are critical aspects of baitfish culture. To produce a variety of fish sizes, farmers manipulate stocking and feeding rates to control growth while maintaining fish health. Fish are judged on appearance and must remain vigorous through the often-complex marketing process. Demand for bait is highly volatile, seasonal, and sensitive to weather conditions. Customer preferences vary widely in different regions. There is fierce competition in marketing baitfish.

DEVELOPMENT OF BAITFISH FARMING

Use of natural fish foods as bait is an ancient and common practice and over 75% of freshwater game fish caught are captured using some form of live bait (2). As an aquacultural product, baitfish rank third or fourth in value in the United States, and estimates based on industry surveys are thought to be of low accuracy, (3) due to under-reporting by farmers and wholesalers (4). In addition to recreational fishing and feeder fish markets, baitfish are also used in certain commercial saltwater fishing operations (5,6).

Sources of baitfish include wild capture, extensive culture, and intensive culture. Statistics are limited, but perhaps only half of all baitfish are farm raised. In the past, virtually all baitfish were captured from the wild. The Great Lakes supported a large fishery for emerald (*Notropis atherinoides*) and spottail (*Notropis hudsonius*) shiners (7) and schools of minnows were said to be so dense that upwards of 100,000 could be captured in one scoop of a large dipnet (8). In many areas, collecting small fish for bait from the wild is still legal and commercial fishermen use seines or traps to remove fish (3,9,10). Research by Brandt and Schreck (11) evaluated the effects of baitfish harvesting and found no significant impact on the densities of bait or game fishes in the study stream. However, wild-caught bait is often a mix of fish species, and its use can potentially affect recipient water bodies (1). Introduction of nonindigenous species by bait bucket transfer was judged to be almost a certainty by Ludwig and Leitch (12). Concerns regarding undesirable fish introductions from wild-caught bait were expressed as early as 1952 by Miller (13), who developed a key to 32 species of fish sold as bait along the lower Colorado River.

Extensive culture refers to the practice, in northern states, of raising seasonal crops of fathead minnows or white suckers in shallow (pothole) lakes. Adult fathead minnows or white sucker fry are stocked in the spring, allowed to grow (and spawn, in the case of fatheads), and then harvested before the next winter. These "winter-kill" lakes are so-named because they usually freeze solid during the winter, and remaining fish are killed. Yields are relatively low, as fish are raised on natural foods alone, but costs are also minimal.

Intensive baitfish farming was initially promoted as a solution to the shortages of minnows experienced in the 1930s and 1940s in Michigan and other north central states. As early as 1934, the Michigan Department of Conservation conducted research on the propagation of minnows (14). The first baitfish farms in Arkansas began in the late 1940s. Baitfish are raised in many states, but production is concentrated in Arkansas. Arkansas had an estimated 11,251 hectares (27,800 acres) of ponds that produced fish in 1997, with a farm-gate value of $51.6 million (15).

Relatively little research has been conducted on baitfish culture, perhaps due to the highly insular and competitive nature of the baitfish industry. Davis (16) and Brown and Gratzek (17) conducted comprehensive reviews of baitfish culture and marketing. Extension manuals on baitfish farming published over the past 60 years have served as the primary sources of industry information (18).

CULTURE METHODS

A wide variety of culture techniques are used for producing baitfish. Methods vary with species and from farm to farm. Marketing and distribution networks and market volatility exert such a powerful influence on the feasibility of the baitfish business that efficiency in production is of secondary importance. Individual farms may specialize in one species or raise a variety of baitfish species. Fish species raised for bait are easily propagated, hardy, fast growing, widely distributed, and attractive to game fish. Most baitfish are minnows, members of the family Cyprinidae. In a food selectivity study, northern pike were found to prefer soft-rayed fish (e.g., minnows) to spiny-rayed fish (e.g., perch or bluegill) of similar size (19). Golden shiners are the primary bait species, and goldfish the primary feeder fish species. Fathead minnows are sold mainly as bait, although limited quantities of small "rosy-red" fathead minnows are also marketed as feeders. A color variant of the normal fathead minnow, rosy red fatheads sell for more than twice the price of normal-colored fish, but commercial production in outdoor ponds is problematic due to poor survival (20,21). A limited number of larger goldfish are sold as trotline bait (also "trot-line" or "trout-line," a method of fishing where unattended lines are attached to limbs or floats) or as ornamental fish for garden pools.

Facilities

Most baitfish are raised in earthen ponds similar to those used for channel catfish culture. Typical pond size varies with the species being raised. Ponds used for golden shiners are often 2 to 8 hectares (5 to 20 acres) in size, while ponds up to 4 hectares (10 acres) are used for fathead minnows. Feeder goldfish are best raised in smaller ponds (0.8 hectares or 2 acres), as fish density must be carefully controlled to keep fish small. The trend is toward smaller ponds, as farmers increase stocking and feeding rates. Water depth is kept shallow (0.8 to 1.8 m; $2\frac{1}{2}$ to 6 ft) so that farm crews can harvest fish without draining. Pond levees are normally a minimum of 3.7 m (12 ft) wide to permit passage of farm trucks, and at least one levee of each pond should have a good gravel surface for all-weather access to fish, for the purposes of feeding and harvest.

Groundwater (well water) is used almost exclusively in baitfish culture. Surface water often contains wild fish, which, if introduced into production ponds, would compete with, or prey upon, the small baitfish. Wild fish would also require manual removal from bait species before shipment, a difficult and costly procedure. Fine mesh, self-cleaning mechanical filters are used to screen surface water, when it is used.

Estimates of water requirements for baitfish farming vary but at least 190 to 470 L/min/ha (20 to 50 gal/min/acre) is desirable. To prevent aquatic weeds from becoming established, enough water should be available to fill a pond within 7 to 10 days. Baitfish farmers routinely capture and reuse pond water in order to reduce groundwater consumption. As ponds are drained, water is relifted (using pumps) into other ponds. Farmers also fill ponds to less than the maximum capacity, leaving room for the capture of rainwater.

Harvested baitfish are held in vats, in a minnow shed, before shipment to market. Vats are shallow tanks, usually constructed of cement-covered blocks or concrete. Minnow sheds vary in size from small, open-sided shelters covering a dozen or so vats to fully enclosed buildings with dozens of vats. Larger sheds may have offices and break rooms as well as drive-through doors that allow workers to load hauling trucks indoors.

Propagation

Spawning techniques are similar for golden shiners and goldfish. Both species reproduce by scattering adhesive eggs over vegetation, and no parental care is given to the eggs. Both species will spawn multiple times during the spawning season. Goldfish begin spawning when the water temperature exceeds 16 to 18 °C (60 to 64 °F), while golden shiners start spawning at 20 to 21 °C (68 to 70 °F).

Traditionally, golden shiners and goldfish have been propagated using either the wild-spawn or egg-transfer method. In the wild-spawn method, broodfish are stocked into newly flooded ponds, which have margins or strips of standing vegetation. Fish spawn freely on the vegetation, and the resulting young may be raised with the parent fish or juveniles may be transferred to separate ponds. Reproductive success using this method is unpredictable. After the initial spawning, subsequent generations of eggs and fry are subject to intense predation by their older siblings. Young fish are often attacked by parasites spread from the broodfish (22). To increase fry production, most baitfish farmers use the egg-transfer method.

In the egg-transfer method, 3.8-cm (1.5 in.) thick spawning mats are used to collect eggs. Spanish moss was once commonly used, but producers have switched to a mat material made of latex-coated coconut fibers on a polyester mesh backing. This material was originally designed for (and is still used) as a heating and air-conditioning filter. Typically, 46 × 76 centimeter (18 × 30 in.) sections of mat material are sandwiched within a supporting wire mesh. Mats are placed level with the pond surface along the edges of brood ponds, at a depth of 2 to 5 centimeters (1 to 2 in.) (Fig. 1). Wooden stakes are used to hold the mats in place, or mats may be placed on floating racks. Alternatively, eggs are collected on weighted 3- to 4-m (10 to 12 ft) lengths of mat material placed in shallow water. After 24 hours, mats with eggs are transferred into a rearing pond containing a shallow layer of fresh well water, for incubation and hatching. Eggs hatch in 3 to 7 days, with warmer temperatures promoting faster development.

Stocking rates for eggs vary widely. Ponds are stocked lightly if rapid growth is desired, while other ponds may be stocked heavily with the aim of spreading out fish later. A single mat may contain from 5,000 to 250,000 eggs, and farmers transfer from 125 to 500 mats per hectare (50 to 200 mats per acre). As a result, ponds may be stocked with

Figure 1. Spawning mats staked along the edge of a golden-shiner brood pond.

upwards of 5 to 10 million eggs per hectare (2 to 4 million eggs per acre) for golden shiners, resulting in 1 to 4 million fry per hectare (0.5 to 1.5 million per acre). Feeder goldfish eggs are stocked at up to 25 million per hectare (10 million eggs per acre).

Incubating eggs are attacked by fungus, and eggs within the mats may be subjected to low dissolved-oxygen levels, even when pond levels are adequate (23). Fungus is thought to be the major cause of egg loss, and hatching rates were found to be less than 32% (average of 22.5%) even when mats were held in aerated tanks (24). Hatching eggs in indoor tanks was proposed by Morrison and Burtle (25) as a way to increase hatching rates and to allow stocking known numbers of fry into rearing ponds. Additional research has been conducted over the past decade, and within the last two years, a number of major Arkansas baitfish producers have adopted tank hatching methods. Egg survival is greatly improved, as eggs can be treated for fungus, and thus fewer brooders are needed. In commercial hatcheries, eggs are generally left attached to spawning material for incubation. Alternatively, sodium sulfite has been shown to be effective in removing golden shiner eggs from spawning mats (26), and detached eggs can be incubated in hatching jars (27).

In the tank hatching method, eggs are treated daily for fungus until eggs are eyed (the eye of the developing larvae becomes dark with pigment). Water is sprayed into tanks for aeration and to flush ammonia. Once fry hatch, they are held in tanks a day or two until the majority are able to swim freely. A 100- to 150-micron particle size feed (28) is offered to sustain older fry, but the efficacy of this practice is questionable, as newly hatched cyprinid fry are apparently unable to utilize prepared feeds fully (29). Fry are then harvested from hatching tanks, numbers are estimated, and fry are placed in plastic bags for transport to rearing ponds.

Newly hatched goldfish fry are approximately 4.5 to 5 mm (0.2 in.) in length (30,31), larval golden shiners range in length from 3 to 6 mm (0.1 to 0.2 in.) (32,33), and fathead minnows, 4 to 6 mm (0.2 in.) (32). These tiny fish are subject to predation by aquatic insects, especially backswimmers (34) and predaceous copepods (35). If necessary, baitfish farmers prepare fry ponds using the insecticide Baytex, for controlling predaceous insects, or Dylox, for copepod control (36).

A recent trend is to spawn broodfish in shallow tanks instead of ponds. This practice helps farmers extend the natural spawning season and keep mats free of debris that would interfere with tank hatching. Fish can be spawned indoors in the spring before they normally can be spawned in ponds, or fish can be spawned throughout the summer months, when reproduction does not occur naturally (37,38). Feeder goldfish producers developed the first commercial systems, which have since been used for golden shiners as well.

Fathead minnows reproduce by laying adhesive eggs on the undersides of floating and submerged objects. Eggs are small, slightly over 1 millimeter (0.04 in.) in diameter (39,40). Fathead minnows start spawning at 15 to 18 °C (59 to 64 °F), and will continue to spawn until the water temperature exceeds 29 °C (85 °F) (41). Gale and Buynak (42) determined fecundity and spawning frequency for five pairs of fathead minnows. On average, females spawned every 4 days (range, 2 to 16 days) during the spawning season. Total number of clutches per female over the 3-month study was 16 to 26, and total egg production per female for that period was 6,803 to 10,164 (mean = 8,604). Fathead males have been shown to release chemical compounds that are attractive to females (43).

Typically, fathead minnow producers use the wild-spawn method of reproduction (41,18). Broodfish are stocked into ponds and allowed to spawn. The optimum sex ratio for broodfish in ponds is 5 females:1 male (44,45). Male fatheads are generally larger than the females, and a #15 or #16 grader (slot width of 0.60 to 0.64 cm, 15/64 to 16/64 in.) is used to separate broodstock by sex. Farmers use plastic tarps, plastic irrigation tubing, wooden pallets, plywood, or even cardboard as spawning substrate. Young fish are left in the same pond with the adults, and fish for sale are periodically harvested. Yields are low (on average, 392 kilograms per hectare; 350 pounds per acre) (15), probably because older siblings or other predators eat many new fry. Producers may also move the spawning substrate with the eggs or transfer juvenile fish to new ponds at rates of 124,000 to 740,000 per hectare (50,000 to 300,000 per acre). Feeder fish are stocked at higher rates.

Fry Transfer Method

Although commonly listed as a method of propagation, "fry transfer" is simply stocking of ponds through transfer of juvenile fish. Once a new crop of fish reaches about 2 to 3 cm (0.8 to 1.2 in.) in length, the juvenile fish are hardy enough to withstand being harvested and transferred to other ponds. Golden shiners are stocked at 124,000 to 500,000 per hectare (50,000 to 200,000 per acre), while goldfish are stocked at 1 to 4 million per hectare (0.5 to 1.5 million per acre).

Fertilization

Rearing ponds are fertilized after the transferred eggs hatch or in preparation for stocking fry to promote the development of natural foods. A combination of organic and inorganic fertilizer is applied to fry ponds, while inorganic fertilizer alone is applied to maintain algae blooms in ponds with older fish. In the past, baitfish farmers used fertilizers extensively throughout the growing season, but usage has decreased, as feeding rates have increased.

Natural Foods, Nutrition, Feeds, and Feeding

Golden shiners are sight feeders adapted to feeding on zooplanktons, but they also eat a wide variety of other animal and plant materials (46–48). Young goldfish feed heavily on zooplankton, then become more omnivorous with age, feeding on algae and detritus as well as small invertebrates (49,50). The goldfish has a long digestive tract, twice that of its body length (51). Fathead minnows are primarily algae eaters, but animal foods such as zooplankton and insect larvae are also consumed (52,53). All three species readily accept prepared feeds.

Nutritional requirements for baitfish species are poorly understood and what knowledge exists usually is not reflected in commercial diets. Lochmann and Phillips (54) summarized current baitfish nutrition and feeding practices. Commercially-available baitfish feeds are similar, if not identical, to those used for channel catfish, yet high-fat diets may be beneficial for baitfish, to improve fish vigor and stress resistance (55). Amino acid requirements for baitfish are similar to those of common carp and channel catfish (56). In the absence of natural foods, optimal protein level for golden shiners and goldfish was found to be 29% (57). Formulation of practical diets for baitfish is complicated by the presence of natural foods, which form an important part of the diet of cultured fish even when complete feeds are provided (58,59). In a pond study, Lochmann and Phillips (60) found that growth, yield, and survival of golden shiners was unaffected by the presence or absence of a vitamin and mineral supplement in the prepared feed.

Feeding is now a standard practice in baitfish culture, as it can double or triple yields over those produced from natural foods alone. In raising baitfish, as compared to catfish production, feed is a relatively small proportion of total costs (61). Farmers feed newly hatched fish finely ground powdered feed called meal several times a day at rates of 2 to 11 kg/ha (2–10 lb/acre) per day. As fish size increases, farmers switch to crumbles (crumbled extruded pellets), and then to pelleted feeds. Feeding rates increase with fish size up to 22 to 28 kg/ha (20 to 25 lb/acre) per day for golden shiners and 45 to 56 kg/ha (40 to 50 lb/acre) per day for goldfish. Over the winter, fish are fed on warmer days, and feeding at 2% body weight per day is recommended for high fish production (21). Feeding rates for fish vary widely depending on the intended market.

In an aquarium study that measured weight gain of golden shiners, pelleted feeds were found to be superior to meal, but that difference was not seen in a corresponding pond trial (62). Floating feed offers farmers the advantage of being able to see their fish, and feeding response is used to estimate pond inventories. Increasing the daily feeding rate generally increases the size of baitfish, but not always the yield (21,63,64). In ponds that are fed and managed identically, differences in water quality, natural foods, and fish survival result in wide variations in fish sizes and yields.

Water Quality

During the production season, farmers monitor water quality in production ponds. Baitfish farmers have increased their use of aeration as stocking and feeding rates have increased. Many farms have permanent electric aerators in at least some ponds, and most baitfish farmers utilize portable tractor-driven aerators or pumps to provide emergency aeration.

In production ponds golden shiners can withstand dissolved-oxygen levels as low as 1.0 to 1.5 mg/liter (ppm) for short periods. They are apparently very tolerant of the low oxygen levels found under the ice during northern winters and have high survival rates even when dissolved-oxygen levels reach 0.2 or 0.3 ppm (65). Fish farmers report that adult-fathead minnows have similar tolerance levels to golden shiners; however, the fry are quite sensitive to low oxygen, and survival is reduced at levels below 4 ppm (66). Goldfish are more resistant to low-oxygen levels than the other two species, and can withstand levels of 0.5 ppm for several hours.

In addition to low oxygen, farmers occasionally encounter high-ammonia levels, especially after algal blooms crash. In these cases, feeding rates are reduced and new water may be added to create a zone of improved quality water. Circulating water within baitfish ponds was proposed as a method of improving water quality and fish production, but has been shown ineffective in increasing yields (67,68).

Murai and Andrews (69) examined the salinity tolerance of golden shiner and goldfish eggs and fry. Eggs were only slightly affected by salinity levels as high as 8 parts per thousand (ppt). However, levels as low as 2 ppt reduced survival of fry.

Diseases

Baitfish are susceptible to a wide variety of parasites and other diseases. Crowded conditions and restricted feed promote disease. Farmers carefully check ponds for signs of sick fish and routinely submit fish samples to diagnostic laboratories for fish health checks. Common parasites are single-celled protozoans and monogenic trematodes (flukes) (18,70,71). Fathead minnows are particularly susceptible to infestation by digenic trematodes (grubs). The crustacean parasites *Lernaea*, *Argulus*, and *Ergasilus* are also found on occasion, and two species of tapeworms can also occur in baitfish. One baitfish virus (Golden shiner virus) (72) has been described. Incidence of virus infection was found to increase with fish density (73). Golden shiners are also afflicted by a microsporidian parasite, *Pleistophora ovariae*, that grows in the ovaries and progressively destroys them (74,75). The progressive impact of this disease on older fish has prompted farmers to use one-year-old broodstock. Bacterial infections by *Aeromonas* sp. or *Flavobacterium columnare* (columnaris) can also cause fish losses, especially after fish have been subjected to suboptimal environmental conditions (77). Rarer diseases include myxosporidians (78).

Other Pond Management Problems

Fish-eating birds are a major problem for baitfish farmers. Despite expensive and time-consuming scare programs, birds consume hundreds of thousands of dollars worth of baitfish each year. Wading birds can also infect ponds with parasitic grubs. Snakes also feed on minnows, especially when fish are spawning and insensitive to danger.

Annual Production Cycle in Arkansas

The annual production cycle on baitfish (golden shiner and goldfish) farms starts with the spring spawning season (April and May), a period of intense activity. Ponds are drained to make room for the new crop. Spawning mats are placed in brood ponds daily for egg deposition and then moved the next morning to rearing ponds or a hatchery. As eggs hatch, the new fry must be fed several times a day. Meanwhile, ponds with fish from the previous year are harvested to meet spring markets, especially for crappie (*Pomoxis* spp.) bait. In June, warm weather brings a close to spawning, but farmers remain busy harvesting, as the market shifts toward a larger (bass) minnow, demand for which peaks around the July 4th holiday. Farmers spread out the new crop of juvenile fish, harvesting small, stunted fish from overcrowded ponds and moving them to other ponds with light loads of fish. Fish are fed daily, and ponds are monitored to detect disease or water quality problems. The demand for bait slows in August with the heat of summer, then picks back up again with cooler fall temperatures. Larger minnows for bass bait are removed from ponds with the new crop of fish by seining and pond grading (capturing fish with a net and running them through a floating grader to retain the large fish). Ponds may be fertilized in the fall, if necessary, to ensure a phytoplankton bloom that will last the winter. Winter markets require large minnows used in ice fishing. Although prolonged ice cover is rare in Arkansas, farmers will harvest by seining under the ice, if necessary, to meet market demands. Fish are fed on warmer winter days, and preparations are made for the next year's production cycle.

Harvesting

Golden-shiner and fathead-minnow producers usually harvest only a portion of the fish in a pond at a time (partial harvest) to obtain needed fish. Feeder-goldfish producers will usually completely harvest a pond within a week after harvesting commences, since the remaining fish will quickly grow too large for the feeder market if left at low densities. Fish are baited into corners using a sinking feed, then captured using knotless nylon seines (Fig. 2). The fish are transferred from the harvest seine to hauling tanks in buckets, each containing about 11 kilograms (25 lb) of minnows. Oxygen-equipped transport trucks carry the fish to the minnow shed, where they are unloaded into vats through a plastic pipe.

Figure 2. Harvesting golden shiners with a seine.

Holding, Grading, and Transport

Baitfish are held in shallow vats (0.5 m or 20 in. deep) before shipment to market. Typical vats contain 5,700 to 7,600 L of water (1,500 to 2,000 gal), and can hold 160 to 180 kg (350 to 400 lb) of fish. Fish are left untouched for the first 24 hours to allow them time to adjust to vat conditions and to empty their digestive tracts (purge), a process referred to as "hardening." Salt is often added at a rate of 0.5%. Vats are flushed initially to remove residual pond water, then at daily or twice daily intervals, and a light flow of water is left to prevent the buildup of ammonia. Groundwater is used in vats because it is clean and cool, but it may contain excess dissolved gases and high concentrations of iron. Spray towers or packed columns are used to aerate the water and, if necessary, pressurized sand filters are used to remove flocculated iron.

Most farmers use low-pressure blowers and 5 to 7 airstones per vat to maintain adequate dissolved-oxygen levels. A few operations still use 110-volt hanging agitators for vat aeration. A new procedure found on several farms equips each vat with an oxygen saturator (where water is pumped downwards through pipes of increasing diameter, as oxygen bubbles upwards, providing for a long contact time for gas transfer). Use of oxygen saturators at least doubles the weight of fish that can be held in a vat. Costs of operating saturator units are reduced because farms keep storage tanks of liquid oxygen for refilling cylinders on fish hauling trucks, and the units use oxygen gas that would otherwise be lost due to venting from storage tanks.

Fish are graded (separated into size categories) the second day in the vats. Farmers typically use drag graders, panels of evenly spaced aluminum bars that are pulled through the tanks (Fig. 3). Large fish are removed first, and then fish are separated by using progressively smaller graders. Graders are numbered by the width of the grader bar, spacing in units of 0.4 centimeter (1/64 in.), the industry standard.

Fish are graded into several size categories, depending on species. Golden shiners are usually separated into jumbos, mediums, and crappie bait. Tables of average

Figure 3. Holding vats for baitfish. The men are crowding golden shiners with a grader, in preparation for loading a hauling truck.

golden shiner lengths and weights for different graders have been developed (79). Feeder fish are classified by the weight of 1,000 fish; the standard category that makes up 90% of sales is 1.4 kilograms (3 lb) per 1,000 (18). Fish that are unsold several days after grading are returned to ponds. However, returned fish, or "turn backs," generally do not remain healthy and fish losses may be high.

Most baitfish are transported to market by hauling trucks equipped with insulated tanks and liquid-oxygen aeration systems. Feeder fish may be sent by truckload or in insulated shipping boxes. Fish sent by box are placed in a small amount of chilled water inside a double layer of plastic bags. The inner bag is then filled with oxygen and the bags are sealed. Ice is added to keep fish cool. Customers may pick up boxed fish at the closest airport or an overnight deliver company will transport the boxes directly to the retail shop.

MARKETING

Marketing is the most difficult aspect of baitfish farming. Live fish are a highly perishable product. Demand for bait varies widely with the weather, and farmers must monitor weekend weather forecasts for regions where their fish are sold in order to decide how many fish to harvest, grade, and harden in vats in anticipation of sales orders. In addition, live haulers (people who drive tractor-trailer loads of fish) face complex regulations governing transport and sales that vary widely from state to state (10). Baitfish farmers must also adjust quickly to varying demands for different sizes of baitfish, and small, isolated farms may have difficulty meeting these changing market requirements. Despite the highly competitive nature of the business, fish are routinely bought and sold among farms so that for individual farms can meet marketing obligations without losing customers.

Fish farms may sell to distributors, wholesalers, or directly to the public. Distributors (jobbers) haul fish to major wholesale outlet centers around the country. These distribution centers have their own holding facilities and retail networks. Fish are then delivered to retail outlets on a weekly or twice weekly schedule or are sold to other wholesalers. Fish losses are absorbed by the farmer and are usually made up in the next shipment.

ECONOMICS

Few economic analyses of baitfish farming have been conducted. Pounds et al. (61) examined economic effects of intensifying golden shiner production. Fixed costs in baitfish farming (costs incurred whether fish are grown or not, such as payments on land, facilities, ponds and equipment) are relatively high when contrasted to catfish production, so there are strong economic incentives to increase yield. Returns to investment in baitfish farming were found to be relatively low. Established farmers with high levels of equity capital (debt-free land and ponds) were able to produce fish at a profit (80).

Confounding factors affecting economic analyses of baitfish farming include diverse culture methods, multiple species and market sizes, variable demand, and market restrictions and barriers. Yields from individual ponds can exceed 1,100 kilograms per hectare (1,000 pounds per acre) for golden shiners and 1,700 kilograms per hectare (1,500 pounds per acre) for goldfish. A common mistake of novices is to develop budgets that assume high yields and that all the fish will be sold. In fact, average yields are considerably lower and market constraints can leave farmers with unsold fish. Baitfish farming is a challenging business.

BIBLIOGRAPHY

1. M.K. Litvak and N.E. Mandrak, *Fisheries* **18**(12), 6–13 (1993).
2. D. Sternberg, *Fishing with Live Bait*, Cowles Creative Publishing, Inc., Minnetonka, MN, 1982.
3. L.A. Nielsen, *North Amer. J. Fish. Mgt.* **2**, 232–238 (1982).
4. T.G. Meronek, F.A. Copes, and D.W. Coble, *The Bait Industry in Illinois, Michigan, Minnesota, Ohio, South Dakota, and Wisconsin*, North Central Regional Aquaculture Center Technical Bulletin Series #105, East Lansing, MI, 1997.
5. R.S. Shomura, ed., *Collection of Tuna Baitfish Papers*, NOAA Technical Report NMFS Circular 408, U.S. Department of Commerce, Washington, DC, 1977.
6. R.N. Uchida and J.E. King, *U.S. Fishery Bulletin* **199**(62), 21–47 (1962).
7. W.G. Gordon, *Fishery Leaflet 608*, U.S. Fish and Wildlife Service, Washington, DC, 1986.
8. C.L. Hubbs and G.P. Cooper, *Minnows of Michigan*, Bulletin No. 8, Cranbrook Institute of Science, Bloomfield Hills, MI, 1936.
9. L.E. Noel and W.A. Hubert, *North Amer. J. Fish. Mgt.* **8**, 511–515 (1985).
10. T.G. Meronek, F.A. Copes, and D.W. Coble, *Fisheries* **20**(11), 16–23 (1995).
11. T.M. Brandt and C.B. Schreck, *Trans. Amer. Fish. Soc.* **104**, 446–453 (1975).
12. H.R. Ludwig, Jr. and J.A. Leitch, *Fisheries* **21**(7), 14–18 (1996).
13. R.R. Miller, *Calif. Fish and Game* **38**(1), 7–42 (1952).
14. G.P. Cooper, *Trans. Amer. Fish. Soc.* **65**, 132–142 (1935).
15. C. Collins and N. Stone, *Aquaculture Magazine* **24**(2), 75–77 (1998).
16. J.T. Davis, in R.R. Stickney, ed., *Culture of Nonsalmonid Freshwater Fishes*, 2nd ed., CRC Press, Inc., Boca Raton, FL, 1993, pp. 307–321.
17. E.E. Brown and J.B. Gratzek, *Fish Farming Handbook: Food, Bait, Tropicals and Goldfish*, AVI Publishing Co., Westport, CT, 1980.
18. N. Stone, E. Park, L. Dorman, and H. Thomforde, *MP 386*, Cooperative Extension Program, University of Arkansas at Pine Bluff, Pine Bluff, AR, 1997.
19. G.B. Beyerle and J.E. Williams, *Trans. Amer. Fish. Soc.* **97**(1), 28–31 (1968).
20. G.E. Ludwig, *Prog. Fish-Cult.* **57**, 213–218 (1995).
21. G.E. Ludwig, *Prog. Fish-Cult.* **58**, 160–166 (1996).
22. F.P. Meyer, in S.F. Snieszko, ed., *A Symposium on Diseases of Fishes and Shellfishes*, Special Publication No. 5, American Fisheries Society, Washington, DC, 1970, pp. 21–29.
23. N. Stone, E. Park, and H. Thomforde, *North Amer. J. Aqua.* **61**, 107–114 (1999).
24. N. Stone and G.M. Ludwig, *Prog. Fish-Cult.* **55**, 55–56 (1993).

25. J.R. Morrison and G.J. Burtle, *Prog. Fish-Cult.* **51**, 229–231 (1989).
26. N. Stone and G.M. Ludwig, *Prog. Fish-Cult.* **55**, 53–54 (1993).
27. N. Stone, E. McNulty, and E. Park, *FSA9081*, Cooperative Extension Program, University of Arkansas at Pine Bluff, Pine Bluff, AR, 1998.
28. M. Rowan and N. Stone, *Prog. Fish-Cult.* **57**, 242–244 (1995).
29. K. Dabrowski and D. Culver, *Aquaculture Magazine* **17**(2), 49–61 (1991).
30. H. Battle, *Ohio J. Sci.* **40**(2), 82–93 (1940).
31. J.M. Gerlach, *Copeia* **1983**(1), 116–121 (1983).
32. D.E. Snyder, M.B. Mulhall Snyder, and S.C. Douglas, *J. Fish. Res. Board Can.* **34**, 1397–1409 (1977).
33. G.L. Buynak and H.W. Mohr, Jr., *Prog. Fish-Cult.* **42**, 206–211 (1980).
34. J.G. Burleigh, R.W. Katayama, and N. Elkassabany, *J. Applied Aqua.* **2**(3/4), 243–256 (1993).
35. A.A. Labay and T.M. Brandt, *Prog. Fish-Cult.* **56**, 37–39 (1994).
36. N. Stone, S. Lochmann, and E. Park, *FSA9080*, Cooperative Extension Program, University of Arkansas at Pine Bluff, Pine Bluff, AR, 1998.
37. V.L. De Vlaming, *Biol. Bull.* **148**, 402–415 (1975).
38. M. Rowan and N. Stone, *Prog. Fish-Cult.* **58**, 62–64 (1996).
39. V.C. Wynne-Edwards, *Trans. Amer. Fish. Soc.* **62**, 382–383 (1933).
40. H. Markus, *Copeia* **3**, 116–122 (1934).
41. S.A. Flickinger, *Bulletin 478A*, Cooperative Extension Service, Colorado State University, Fort Collins, CO, 1971.
42. W. Gale and G. Buynak, *Trans. Amer. Fish. Soc.* **111**, 35–40 (1982).
43. K.S. Cole and R.J.F. Smith, *J. Chem. Ecol.* **18**(7), 1269–1284 (1992).
44. S. Flickinger, *Proc. S. E. Assoc. Game Fish Comm.* **26**, 376–391 (1973).
45. A. Andrews and S. Flickinger, *Proc. S. E. Assoc. Game Fish Comm.* **27**, 759–765 (1974).
46. H. Cassidy, A. Dobkin, and R. Wetzel, *Ohio J. Sci.* **30**(3), 194–198 (1930).
47. A. Keast and D. Webb, *J. Fish. Res. Board Can.* **23**, 1845–1874 (1966).
48. D.J. Hall, E.E. Werner, J.F. Gilliam, G.G. Mittelbach, D. Howard, C.G. Doner, J.A. Dickerman, and A.J. Stewart, *J. Fish. Res. Board Can.* **36**, 1029–1039 (1979).
49. W.B. Scott and E.J. Crossman, *Freshwater Fishes of Canada*, Bulletin 184, Fisheries Research Board of Canada, Ottawa, 1973.
50. Y. Abe, in T. Kafuku and H. Ikenoue, eds., *Modern Methods of Aquaculture in Japan*, Elsevier Scientific Publishing Co., Amsterdam, 1983, pp. 79–90.
51. J.A. McVay and H.W. Kaan, *Biol. Bull.* **78**(1), 53–67 (1940).
52. E.E. Coyle, *Ohio J. Sci.* **30**(1), 23–35 (1930).
53. W.C. Starrett, *Ecology* **31**(2), 216–233 (1950).
54. R. Lochmann and H. Phillips, *Aquaculture Magazine* **22**(4), 87–89 (1996).
55. R. Lochmann, *Lab Animal* **27**(1), 36–39 (1998).
56. D.M. Gatlin, III, *Aquaculture* **60**, 223–229 (1987).
57. R.T. Lochmann and H. Phillips, *Aquaculture* **128**, 277–285 (1994).
58. G.M. Ludwig, *J. World Aquacult. Soc.* **20**(2), 46–52 (1989).
59. R. Lochmann and H. Phillips, *J. World Aquacult. Soc.* **27**, 168–177 (1996).
60. R. Lochmann and H. Phillips, *Arkansas Farm Research* **43**(3), 8–9 (1994).
61. G.L. Pounds, C.R. Engle, and L.W. Dorman, *J. World Aquacult. Soc.* **23**(1), 64–76 (1992).
62. D.W. Gatlin, III and H. Phillips, *J. World Aquacult. Soc.* **19**(2), 47–50 (1988).
63. H.T. Smith, C.B. Schrech, and O.E. Maughan, *J. Fish Biol.* **12**, 449–455 (1978).
64. M. Rowan and N. Stone, *J. World Aquacult. Soc.* **26**, 460–464 (1995).
65. G.P. Cooper and G.N. Washburn, *Trans. Amer. Fish. Soc.* **76**(1946), 23–33 (1949).
66. W.A. Brungs, *J. Fish. Res. Board Can.* **28**, 1119–1123 (1971).
67. G.M. Ludwig, *Aquaculture* **144**, 177–187 (1996).
68. N. Stone and M. Rowan, *J. World Aquacult. Soc.* **29**, 510–517 (1998).
69. T. Murai and J.W. Andrews, *Prog. Fish-Cult.* **39**(3), 121–122 (1977).
70. W.M. Lewis and M.G. Ulrich, *Prog. Fish-Cult.* **29**(4), 229–231 (1967).
71. W.M. Lewis and S.D. Lewis, in S.F. Snieszko, ed., *A Symposium on Diseases of Fishes and Shellfishes*, Special Publication No. 5, American Fisheries Society, Washington, DC, 1970, pp. 174–176.
72. J.A. Plumb, P.R. Bowser, J.M. Grizzle, and A.J. Mitchell, *J. Fish. Res. Board Can.* **36**, 1390–1394 (1979).
73. T.E. Schwedler and J.A. Plumb, *Prog. Fish-Cult.* **44**(3), 151–152 (1982).
74. R.C. Summerfelt, *Trans. Amer. Fish. Soc.* **93**(1), 6–10 (1964).
75. R.C. Summerfelt and M.C. Warner, in S.F. Snieszko, ed., *A Symposium on Diseases of Fishes and Shellfishes*, Special Publication No. 5, American Fisheries Society, Washington, DC, 1970, pp. 142–160.
76. W.M. Lewis and M. Bender, *Prog. Fish-Cult.* **22**(1), 11–14 (1960).
77. F.P. Meyer, *Prog. Fish-Cult.* **26**(1), 33–35 (1964).
78. S.D. Lewis, *J. Parasitology* **54**, 1034–1037 (1968).
79. G.M. Ludwig and N. Stone, *Prog. Fish-Cult.* **59**, 312–316 (1997).
80. C.R. Engle, *The Southern Business & Economic Journal* **16**, 213–231 (1993).

See also GULF KILLIFISH CULTURE.

BARRAMUNDI CULTURE

GEOFF ALLAN
Port Stephens Research Centre
Taylors Beach, Australia

ROBERT R. STICKNEY
Texas Sea Grant College Program
Bryan, Texas

OUTLINE

A relative newcomer on the aquaculture scene, the barramundi, or Asian sea bass (*Lates calcarifer*), is being farmed in southeast Asia and Australia. Production was estimated at nearly 20,000 tons in 1997 and included about 500 tons from Australia (1). Barramundi is a member of the family Latidae and is renowned as an excellent sportfish, especially in the rivers of northern Australia. It has fine eating qualities and is prized on the live-fish market in several southeast Asian cities and in Asian communities within Australia.

CULTURE OF BARRAMUNDI

Barramundi are protandrous hermaphrodites; that is, they first develop as males, but later change into females with functional ovaries. In the wild, males mature at about four years of age and convert to females at between six and eight years old. In culture, precocious development occurs and males and females mature as early as one or two years, respectively. Year-round hatchery production in possible through the manipulation of temperature (2), but hatchery production is limited by the inability of culturists to maintain functional males in the breeding population (3).

When barramundi culture was initiated in the mid-1980s, it was common practice to collect fry from nature for stocking into culture systems. This was followed by collecting gametes from wild adults, although that practice was unreliable, so the current emphasis is on establishing and maintaining captive broodstock.

Barramundi broodstock can be kept in reproductive condition year-round through environmental manipulation. Spawning usually requires administration of reproductive hormones. Hatching takes about 14 to 17 hours and larvae commence feeding one to two days after hatching (4). Larvae can be reared intensively in hatchery tanks, or extensively, in fertilized marine ponds. Larval barramundi feed on zooplankton. Copepods and rotifers are among their prey, although brine shrimp (*Artemia* sp.) can be successfully used as well, particularly if they are enhanced with n-3 fatty acids.

Fingerlings are typically maintained in nursery tanks or cages until they are about 80 mm (3.1 in.) total length, at which time they are introduced to formulated diets (4). Regular grading during the fingerling and early juvenile stages is essential to reduce cannibalism.

Growout facilities include tanks, cages, and ponds where water temperatures exceed 20 °C (68 °F), and preferably, are about 25 °C (77 °F). There are several intensive, indoor, environment-controlled recirculating systems producing barramundi in Australia. In addition, some fish are produced in earthen ponds. However, the majority of production occurs in cages located in estuaries or rivers, or within ponds (4). The largest farmer of barramundi in Australia produces approximately 140 tons/year, mainly in 5 m × 5 m (14 ft × 14 ft) cages within large freshwater ponds. In southeast Asia, most fish are produced in cages located in rivers, estuaries, or protected marine areas.

Barramundi are subject to various bacterial, fungal, and viral diseases, usually following stress such as sub-optimal water quality, poor nutrition, or harvesting (4). Columnaris is a particularly common bacterial disease, frequently associated with reduced temperature. Similarly, fungal outbreaks occur most often in cold water and appear as white blotches. Viral diseases have been reported, and one, the picornalike virus, can cause devastating hatchery losses. Concerns that this virus can be transferred to other native fish has led to the imposition of stringent controls on barramundi farming outside its natural range in some parts of Australia (e.g., New South Wales).

Nutritional research on barramundi is not extensive. The fish are carnivorous, and recent research has indicated that nutrient-dense formulated diets with about 20 mJ/kg OE and greater than 45% protein outperformed lower protein energy and diets that formed the basis of early commercial diets (5,6). Field-based research clearly indicated that barramundi grew equally well on diets where either meat meal or fish meal supplied the majority of the protein.

ACKNOWLEDGMENTS

The authors thank C.G. Barlow for providing some of the information used in this entry.

BIBLIOGRAPHY

1. G. Allan, *World Aquaculture* **30**(1), 39.
2. R.N. Garrett, in *Abstracts from Australian Barramundi Farming Workshop*, Department of Primary Industries, Brisbane Queensland, Australia, 1994, p. 16.
3. S.J. Matthews, P. Appleford, and T.A. Anderson, in *Book of Abstracts, World Aquaculture '99*, World Aquaculture Society, Baton Rouge, LA, 1999, p. 24.
4. C.G. Barlow, in K.W. Hyde (editor), *The New Rural Industries—A Handbook for Farmers and Investors*, Rural Industries Research and Development Corporation, Canberra, 1998, pp. 93–100.
5. G.L. Allan, K.C. Williams, D.M. Smith, and C.G. Barlow, *Proceedings of the Fifth International Symposium of Australian Renderers' Association*, Australian Renderers' Association, Sydney, 1999.
6. K.C. Williams, G.L. Allan, D.M. Smith, and C.G. Barlow, *Proceedings of the Fourth International Symposium of Australian Renderers' Association*, Australian Renderers' Association, Sydney, 1999, pp. 13–26.

BIOCHEMICAL OXYGEN DEMAND

David E. Owsley
Dworshak Fisheries Complex
Ahsahka, Idaho

OUTLINE

DEFINITION

Biochemical oxygen demand (BOD) is defined as the amount of oxygen required by bacteria to stabilize organic matter under aerobic conditions. BOD is the major criterion used in monitoring and controlling the pollution of natural waters. It is a limiting factor when organic matter must be restricted to maintain desired levels of dissolved oxygen. It is used by engineers to design treatment systems for sewage and industrial wastes, by regulatory agencies to control pollution of receiving waters, and by aquaculturists to maintain fish rearing environments and effluent control from hatcheries.

DISCUSSION

The term "biochemical oxygen demand" means the same thing to everyone who is involved in monitoring or controlling waste discharges. It is a regulatory parameter, no matter whether you work at a hatchery or at a sewage treatment facility. It is often referred to as "biological oxygen demand's" as well, and it is closely related to chemical oxygen demand (COD), which is the amount of oxygen required to chemically oxidize all of the organic matter in a water sample.

The BOD test is a wet oxidation procedure wherein living organisms serve as the medium for oxidation of the organic matter to carbon dioxide and water. The organic matter serves as "food" for the organisms, and energy is derived from its oxidation. The primary organisms are soil bacteria. The BOD test is essentially a bioassay procedure in which the oxygen consumed by bacteria is measured while the bacteria oxidize the organic matter under a controlled environment during a 5-day incubation period at 20 °C (68 °F). This temperature is considered to be a median value for natural bodies of water. Most of the BOD is everted after 5 days, although it takes 20 days for complete oxidation. The BOD everted throughout the first 5 days is approximately 70 to 80% of the total BOD. The BOD test procedures can be found in (1).

The levels of pollutant in a flow-through hatchery effluent can be determined with the following general equation (2):

$$\text{Average ppm pollutant} = \frac{\text{Pollutant factor} \times \text{amt of food fed (lb)}}{\text{Water flow (gpm)}}.$$

The following pollutant factors should be used in the equation:

Total ammonia	2.67
Nitrate	7.25
Phosphate	0.417
Settleable solids	25.0
Biochemical oxygen demand	28.3

Liao (3) has reported on the extent of pollution problems occurring at fish hatcheries in the United States. Principal parameters measured included BOD, ammonia, nitrate, suspended solids, and phosphates. Attempts were made to correlate the values encountered with fish loadings, water supply rates, and fish feeding rates. Liao's observations are summarized in Table 1.

Later, studies were carried out by Brisbin (4), who measured the wastewater discharge from the Summerland Hatchery and gave the following ranges for polluting substances:

Table 1. Polluting Characteristics of Hatchery Wastewater

Flow rates	0.52–26.0 lb of fish per gal per min
Fish loadings	0.069–1.14 lb of fish per ft^3 of holding capacity
Feeding rates[a]	1.14–7.15 lb per 100 fish per day
BOD	0.645–2.496 lb per 100 lb of fish per day
Suspended solids	0.5–3.0 lb per 100 lb of fish per day
Ammonia	0.031–0.404 lb of nitrogen per 100 lb of fish per day
Nitrate	0.006–1.501 lb of nitrogen per 100 lb of fish per day
Phosphate	0.004–0.094 lb of phosphate per 100 lb of fish per day

[a]Nine of the 15 hatcheries studied used feeding rates exceeding 3.7 lb of food per 100 lb of fish per day.

Table 2. Wastewater Discharge Parameters

Pollutant	Concentration (mg/L)	Concentration (lb/100 lb of fish)
BOD	3.5	0.47–0.99
Suspended solids	4.3–8.1	0.29–2.61
Volatile suspended solids	4.1–6.5	0.27–2.56
Ammonia	0.26–0.7	0.05–0.18
Nitrates	0.05–0.26	0.01–0.11
Nitrates	0.04–1.1	0.18–0.24
Phosphates	0.11–0.14	0.02–0.027

POLLUTION ABATEMENT

Liao et al. (5) reported that for pollution abatement, the systems were evaluated by comparing the amounts of pollutants discharged (10% bypass) to waste by the reuse systems to the amounts of pollutants discharged from a control single-pass rearing unit.

The percentages of pollutants reduced are calculated based on the number of pounds of pollutants reduced per 100 lbs of fish per day, as indicated in Table 3. The control single-pass rearing tank, installed at the Bozeman Station, was loaded and operated in the same manner as the fish tanks connected to the treatment systems. Therefore, the percentage calculated for pollutant discharge, based on the discharge from the control single-pass unit, should be reasonable. However, for comparison, the pollutants discharged from various single-pass and reuse systems are listed in Table 4. Also, it must be pointed out that the

Table 3. Average Reduction of Pollutant Discharge by the Bozeman Pilot Plant Reuse Systems[a]

	Activated Sludge	Extended Aeration	Upflow Filter	New Upflow Filter	Trickling Filter
BOD	97	93	89	91	86
Suspended solids	88	95	79	—	91
NH_4–N	23	10	49	49	69
PO_4–P	24	25	+25[b]	+33[b]	+33[b]
Orthophosphate					

[a]Reduction is expressed as a percent and is based on pollutant production rates measured in a single-pass system, in lb/100 lb of fish/day. (See Table 4.)
[b]Plus (+) sign represents an increase.

Table 4. Metabolites (pollutants) Discharged from Single-Pass and Reuse Systems[a,b]

Source	Orthophosphate PO_4–P	NH_4–N	NO_2–N	NO_3–N	BOD	Suspended Solids
Single Pass						
Bozeman circular pond	0.023–0.057	0.034–0.037	0.0000–0.0002		1.36 *	1.04
Bodien	0.015	0.058	—	—	1.3*	
Liao	0.011	0.113	—	0.02	1.34 2	0.5–3
BSFW Study						
Bozeman	0.012	0.077	0.004	0.009	1.10	
Coleman	0.006	0.280	0.029	—	1.31	5.8
Ennis	0.020	0.141	0.015	0.013	1.61	
Kooskia	0.025	0.171	0.039	0.057	1.64	
Quilcene	0.004	0.076	0.015	0.006	0.47	
Winthrop	0.001	0.078	0.002	0.030	1.06	
Reconditioning System						
Bozeman activated sludge	0.0297	0.0274	0.00066	0.00347	0.030[c]	0.121
Extended aeration	0.0292	0.032	0.00084	0.00522	0.093[c]	0.056
NUF Bozeman	0.034–0.064	0.01–0.026	0.00108–0.00158	0.0088–0.0114 0.0114	0.144[c]	—
OUF Bozeman	0.035–0.070	0.0075–0.028	0.0009–0.00156	0.0105–0.016 0.016	0.119[c]	0.220
TF Bozeman	0.04–0.064	0.0088–0.014	0.0015–0.0023	0.011–0.025 0.025	0.187[c]	0.092
Abernathy	—	0.0067	0.0016	0.021	—	—
Dworshak	—	0.0975	0.00062	0.0357	—	—

[a]Metabolite (pollutants) discharge rates are expressed in lb/100 fish/day.
[b]The following formula can be used to convert pollutant discharge in lb of pollutant/100 lb of fish/day to mg/L: mg/L = 0.834 × (fish carrying capacity, in lb of fish/gpm) × (value in the table, in lb/100 lb of fish/day).
[c]Calculated from measured COD. (BOD = 0.3 × COD.)

Table 5. Quality of Effluents from Catfish Ponds During the Two Phases of Fish Harvest

Analysis	Harvest Phase	
	Draining	Seining
Settleable matter (mg/L)	0.08	28.5
Biochemical oxygen demand (mg/L)	4.31	28.9
Chemical oxygen demand (mg/L)	30.2	342
Soluble orthophosphate (mg/L as P)	16	59
Total phosphorus (mg/L as P)	0.11	0.49
Total ammonia (mg/L as N)	0.98	2.34
Nitrate (mg/L as N)	0.16	0.14

Table 6. Percentage Losses of Several Pollutants from a Hypothetical Catfish Pond During the Seining Phase of Fish Harvest[a]

Analysis	% of Total Released During Seining Phase
Settleable matter	94.9
Biochemical oxygen demand	26.0
Chemical oxygen demand	37.3
Soluble orthophosphate	66.7
Total phosphorus	18.8
Total ammonia	11.2
Nitrate	4.3

[a] After Boyd (7).

phosphate indicated in Tables 3 and 4 is orthophosphate, rather than total phosphate.

Boyd (6 and 7) reported the following BOD levels for warmwater rearing ponds in Tables 5 and 6.

CONCLUSION

The biochemical oxygen demand is an important parameter for aquaculturists. It is an indicator of levels of pollution for receiving streams and rearing ponds. BOD levels are easy to determine, understand, and apply to studies of the environment.

BIBLIOGRAPHY

1. *Standard Methods for the Examination of Water and Wastewater*, 17th ed., American Public Health Association, Washington, DC, 1989.
2. R.G. Piper, *Fish Hatchery Management*, US Department of the Interior, US Fish and Wildlife Service, Washington, DC, 1982.
3. P. Liao, *Pollution Potential of Salmonid Fish Hatcheries*, Technical Reprint number 1-A, Kramer, Chin and Mayo, Consulting Engineers, Seattle, 1970.
4. K.J. Brisbin, *Pollutional Aspects of Trout Hatcheries in British Columbia*, Fish and Wildlife Branch, Victoria, BC, Canada, 1971.
5. P. Liao and R. Mayo, *Intensified Fish Culture Combining Water Reconditioning with Pollution Abatement*, Technical Reprint Number 24, Kramer, Chin and Mayo, Consulting Engineers, Seattle, 1974.
6. C.E. Boyd, *Water Quality in Warmwater Fish Ponds*, Agricultural Experimental Station, Auburn, AL, 1979.
7. C.E. Boyd, *J. Environ. Qual.* **7**, 59–62 (1978).

See also CHEMICAL OXYGEN DEMAND; DISSOLVED OXYGEN.

BLACK BASS/LARGEMOUTH BASS CULTURE

ROY HEIDINGER
Southern Illinois University
Carbondale, Illinois

OUTLINE

Largemouth bass (*Micropterus salmoides*) are the most sought-after freshwater sportfish in the United States. This species was originally found only in North America and has been cultured in the United States since the late 1800s (1). Currently, it is cultured both for stocking as a sportfish and for use as a foodfish. A survey of state and federal agencies on production of largemouth bass was conducted in 1997. The data were obtained from either 1995 or 1996. State and federal hatcheries produced approximately 21 million largemouth bass (LMB) for sportfish stocking. Most (89%) of the bass were stocked as fingerlings. States do not keep records on the exact number of bass produced each year in private commercial hatcheries. However, I estimate that at least as many bass are raised by commercial hatcheries as by state and federal hatcheries. The 1998 *Aquaculture Magazine Buyer's Guide* lists 103 hatcheries that produce LMB. Twenty of the hatcheries indicate that they sell "market-size" fish. Most commercial producers of LMB also produce other species, such as sunfish and channel catfish; only a few units produce LMB exclusively.

Some states require an aquaculture license to rear and sell fish, but others do not. In a few states, it is not legal to rear largemouth bass for stocking. Also, there are a few states in which it is not legal to sell LMB as foodfish. Since regulations are continuously changing, one should check with the appropriate state regulatory agency before culturing LMB or for further information.

The current wholesale value of LMB depends on the geographic location of the hatchery, the size of the fish, and the number of fish in the sale. As a rough guide, 3.8- to 5.1-cm (1.5- to 2.0-in.) bass are sold for $0.12 each, 10.2- to 15.2-cm (4- to 6-in.) fish for $0.55 each, 15.2- to 20.3-cm (6- to 8-in.) fish for $1.05 each, and larger (live) fish for $7.70/kg ($3.50/lb).

BIOLOGY

The basic biology of the largemouth bass has been reviewed by Heidinger (2). Additional information can be found in a symposium on black bass edited by Clepper (3). There are two recognized subspecies of largemouth bass: the northern largemouth bass, *M. salmoides salmoides*, and the southern Florida largemouth bass, *M. salmoides floridanus*. Contrary to popular belief, these two subspecies cannot be visually distinguished from one another; genetic tests are required to tell the two subspecies apart. Knowing which subspecies one is working with is important, because the southern subspecies cannot tolerate as cold a temperature as can the northern subspecies. In 2-m-deep (6-ft-deep) culture ponds located at the latitude of southern Illinois, young-of-the-year largemouth bass of the Florida subspecies die during normal winters. During winter, the ponds reach 4 °C (39 °F) for several months. Some researchers believe that even the northern subspecies is stressed at temperatures of 2 to 3 °C (36 to 37 °F). Such low temperatures occur at northern latitudes when winds keep shallow ponds from freezing.

LMB, like most fish, have to swallow their prey whole. In nature, the diet of LMB changes as they grow, but all sizes feed on live organisms. As bass increase in size, they eat larger organisms, which they forage for. Initially, fry eat primarily zooplankton. Then they eat insects and zooplankton until they reach 25 to 51 mm (1 to 2 in.) in total length. Bass over 51 mm tend to eat small fish. Adults feed heavily on fish and crayfish. Their large mouth, unwillingness to accept a prepared diet unless trained, and highly cannibalistic nature make them more difficult to culture than rainbow trout or channel catfish. Approximately 5 kg (11 lb) of prey fish, such as golden shiners, are required to produce 1 kg (2.2 lb) of LMB. Additionally, LMB do not eat plants, as they do not have the enzymes necessary to break down the cellulose in the walls of plants. Plant material is seldom found in the stomachs of bass. However, their elastic stomachs may contain prey organisms that weigh up to 10% of the weight of the bass. In nature, bass do not feed continuously, which is why it is not uncommon to find that 50% of a population at any given time have empty stomachs.

LMB spawn at 24 to 30 °C (75 to 86 °F). Male LMB usually build their nests at depths of 0.33 to 1.33 m (1 to 4 ft) on any firm substrate. The male bass places his head in the center of the nest and sweeps debris out in front of him. He then returns and, with his head in the middle of the nest, pivots around in a circle (4). Thus, the diameter of the nest approximates twice the length of the bass. Males are solitary nesters, and they chase other males out of their nesting territory. Unless some line-of-sight obstruction exists, nests are usually at least 2 m (6.5 ft) apart (4). Male LMB guard their nest and over a period of several days may entice more than one female to lay eggs in it. After the enticement period is over, the male defends the nest area against almost all other fish, including female LMB.

Not all of the eggs that a female LMB lays in a season mature in her ovary at the same time. Thus, she may lay her eggs in more than one nest (5). This nesting behavior has led to the recommendation that to best use the eggs present in females, two to three males should be stocked into a brood pond for each female stocked. This technique probably also shortens the duration of spawning. Successful bass nests have been reported to contain 5,000 to 43,000 eggs (6,7). Fertilized eggs are yellow to orange (they may be creamy white in pellet-fed females), spherical [1.4–1.8 mm (0.05–0.07 in.) in diameter], semiopaque, contain one large oil globule [0.5–0.7 mm (0.02–0.03 in.) in diameter], and are adhesive and demersal (4,8,9). The diameter of the egg increases with the size of the female (10). The eggs water harden within 15 minutes after fertilization, and at 10, 18, and 28 °C (50, 64, and 82 °F), the eggs hatch in 317, 55, and 49 hours, respectively (11,12). Upon hatching, the prelarvae are 3 to 5.5 mm (0.12 to 0.22 in.) in total length. They do not have mouths, gills, paired fins, or inflated gas bladders (4,8,10). Prior to inflating their gas bladder, larvae are found in debris at the bottom of the nest. At 22 °C (71 °F), it takes approximately five days for the gas bladder to inflate (9). Even though a connection between the gas bladder and the foregut, called the pneumatic duct, exists in 3- to 8-mm (0.12- to 0.31-in.) larvae, the larvae do not have to gulp air to fill their gas bladders (12). Fry must eat within six days after becoming free swimming, or else they will die (13). Fry do not feed at night, and they pass food through their stomachs much faster (3 hours) than adults (18 hours).

Male LMB guard their brood of fingerlings for 14 to 28 days, after which the young disperse. Female bass grow larger and live longer than male bass. Bass tend to live longer in the northern portion of their range than they do in the southern portion. A 24-year-old female bass from New York is the oldest bass on record (14).

CULTURE

LMB culture has been reviewed by Clepper (3), Heidinger (2), and Stickney (15). There is no single best method to produce LMB; almost all producers have developed some procedures that are unique to their operation. The procedures described later are composites from many agencies and private growers, and only a few of the variations are presented. Most adult LMB are placed in ponds to spawn, but to gain more control, there is a movement toward raceway spawning. The fry are either left in the spawning ponds or, more commonly, moved to rearing ponds. In either case, they initially feed on zooplankton and then zooplankton and aquatic insects until they are 25 to 51 mm (1 to 2 in.) long. Fingerlings 7.6 to 10.2 cm (3 to 4 in.) long can be raised at a low density on insects. If larger fish are desired, 25- to 51-mm (1 to 2 in.) fingerlings are trained in tanks to take a prepared diet, and then the trained bass are reared in ponds, net pens, or raceways to the appropriate size.

BROODSTOCK

Broodstock LMB are normally maintained in the hatchery. The genetic makeup of broodstock should be considered. Most fish geneticists recommend that the initial broodstock come from local sources, especially if the offspring are going to be used for stocking in the region. If the bass are going to be used only as foodfish, then the development of genetic stocks (strains) for this use would make sense. However, at this time, such stocks have not been developed. In order to prevent inbreeding, it is necessary to maintain a fairly large number of broodfish that are not closely related (16). Most mathematical calculations indicate that it is desirable to have 1,000 or more broodfish if one is going to introduce the offspring into natural populations. In a small hatchery, this is not economically feasible; however, one should try to maintain 300 to 400 broodfish of an equal sex ratio.

The sexual maturity of LMB depends more on size than age (17). Female bass reach maturity when approximately 25 cm (10 in.) in total length, while males may be mature at 22 cm (9 in.) (17–19). Thus, in the southern portion of their range, LMB mature in one year, whereas slower growing bass in the northern portion of their range may take three to four years to reach sexual maturity. Normally, larger 0.7- to 1.8-kg (1.5- to 4-lb) bass are kept as broodfish in a hatchery. Externally, bass greater than 35 cm (14 in.) in total length have been sexed correctly 92% of the time by looking at the scaleless area surrounding and immediately adjacent to the urogenital opening (20). However, Manns and Whiteside (21) were able to sex bass only 67% correctly using Parker's characteristics. In the male, the aforementioned area is nearly circular in shape, while in the female it is elliptical or pear shaped. One must hold the fish carefully in order not to change the shape of the scaleless area. Prior to spring spawning, the females that are heavy with eggs, which may weigh up to 10% of the female's body weight, are easy to separate from the thinner males. Just prior to spawning, ripe females have a swollen, red, and protruding vent. Semen may be expressed from the males by placing pressure on the abdomen.

Most producers replace their broodfish when they reach four to five years of age, when they weight 1.8 to 2.3 kg (4 to 5 lb). Larger fish are more difficult to handle. Also, some producers seem to have significant mortality in their older broodfish. The cause of this mortality is not well understood. However, at the San Marcos, Texas, National Fish Hatchery and Technology Center, eight-year-old LMB broodfish have been successfully used to produce offspring (15). It is desirable to keep genetically selected broodfish for as long as possible.

Broodfish are held throughout much of the year in ponds at rates of 336 to 448 kg/ha (300 to 400 lb/acre) (22). To maintain the fish in good physical condition, it is extremely important to feed them adequately. Live prey fish are frequently used; however, pelleted rations or combinations of prey fish and pellets are also used. Three kg of prey fish/kg (3 lb/lb) of LMB are required for maintenance, and 5 kg of preyfish/kg (11 lb/lb) of LMB are required for growth (22).

Rosenblum et al. (23) and Snow and Maxwell (24) found no difference in gonad development and fecundity between pellet- and prey-fed LMB. Pellet feed usually contains less than 10% water, whereas fish contain 80% water. Thus, per unit weight of food, there is much more caloric energy in pelleted food than in fish. For this reason, in terms of weight, less pelleted food needs to be fed than prey fish. Pelleted food is also desirable because it is relatively easy to store, and the amount fed daily can be adjusted to the feeding response of the bass. As the water cools, metabolism decreases, and food consumption also decreases. One good rule of thumb is to feed bass what they will consume once a day in 10 or 15 minutes, or 3% of their body weight daily. Feeding should be continued throughout the ice-free portion of the winter, especially during warm periods, but at a reduced rate of 1% of their body weight daily. Many aquaculturists prefer to have prey fish in the ponds during the winter, especially if long periods of ice cover persist. However, one potential problem of introducing prey fish is that they may carry parasites to the brood bass.

Commercially available fish feed comes in a variety of sizes. As the size of the bass increases from fingerling to adult, the size of the pellet should also increase (Table 1). Adult bass should be fed pellets that are 6 to 19 mm (0.25 to 0.75 in.) in diameter. Minimum protein requirements (≈40%) for 1-year-old and younger largemouth bass (25) are probably closer to those of salmon than those of channel catfish, in terms of their nutrient energy and protein requirement. Thus, a fairly high-quality fish food needs to be used for bass.

Large numbers of eggs may be lost in the spring, due to unstable temperatures. LMB spawn during the first warm spell. If water temperatures subsequently drop due to a cold spell, the LMB will leave the nest, and the eggs will be lost. Renesting will usually occur, but fewer progeny are produced. Spawning can be delayed until the weather stabilizes by holding broodstock in cages, holding sexes in separate ponds, or flushing ponds or raceways with enough cool water to hold the water temperatures below spawning temperatures.

Table 1. Pellet Sizes and Approximate Percentage of Body Weight for Feeding Largemouth Bass Fingerlings Under Summer Condition[a]

Fish Length		Pellet Diameter		Body Weight
(cm)	(in.)	(mm)	(in.)	(%)
<2.5	<1	0.8–1.0	0.03–0.04	15.0
2.5–3.2	1–1.3	1.0–1.3	0.04–0.05	15.0
3.2–4.4	1.3–1.7	1.5	0.06	15.0
4.4–8.9	1.7–3.5	2.5	0.10	10.0
8.9–11.4	3.5–4.5	3.0	0.12	10.0
11.4–12.7	4.5–5.0	4.0	0.16	7.5
12.7–15.2	5.0–6.0	5.0	0.20	5.0
15.2–20.3	6.0–8.0	6.0	0.24	3.0
20.3–25.4	8.0–10.0	9.0	0.35	2.0
25.4–30.5	10.0–12.0	12.0	0.47	2.0
>30.5	>12.0	19.0	0.75	2.0

[a] This practical feeding guideline was developed from (15) and data collected at the San Marcos National Fish Hatchery, using Biodiet®.

FINGERLING PRODUCTION

Spawning–Rearing Pond Method

The simplest method of producing fingerlings is to stock broodfish at a relatively low density, 25 to 100 per hectare (10 to 40 per acre). The pond should be fertilized with an organic fertilizer, such as manure, cottonseed meal, or alfalfa meal, to produce zooplankton (26). The adult fish spawn, and either both the young and adults are left in the ponds or as many adults as is practical are removed by angling after the bass stop spawning. When these fingerlings reach 25 to 51 mm (1 to 2 in.) in length, which takes 40 to 70 days, they are usually harvested by seining.

Production will vary considerably, from a few hundred fish to 123,500 per ha (50,000 per acre). Not all of the bass will spawn at the same time, and the older, slightly larger fingerlings may prey on the younger, slightly smaller ones. Adult bass left in the pond will also prey on the fingerlings. Once the zooplankton and insects are eaten, the fingerling LMB will turn cannibalistic. Although this technique is the simplest way to produce small fingerlings, the number of fry produced is unpredictable, since there are very few controls. Usually, less than a thousand larger sized fingerling can be produced per ha (2.5 acres), unless some type of prey fish is stocked into the juvenile-rearing pond. Unless prey fish are available at a very cheap price, this procedure is not economical.

Fry-Transfer Method

Before the mid-1900s culturists began to remove fry from the spawning pond at some stage and place them into a fertilized rearing pond. Spawning ponds that are free of other fish and predacious insects are usually stocked with 99 to 247 brood bass per ha (40 to 100 per acre) when water temperatures reach 17 to 20 °C (63 to 68 °F). Much higher stocking densities have also been used (3). Since a female may spawn from one to five times over the two- to eight-week spawning season, a male-to-female sex ratio of 2 : 1 or 3 : 1 is recommended (27). On average, with broodfish in good condition, 247,000 to 494,000 fry per ha (100,000 to 200,000 per acre) can be removed from a spawning pond (3). If the pond has a mud bottom, it may be necessary to place 2- to 5-cm (0.75- to 2-in.) gravel at 2- to 3-m (7- to 10-ft) intervals. Light-colored gravel allows the culturist to observe the spawning bass more easily than dark-colored gravel. The patches of gravel should be about 0.9 m (3 ft) in diameter and placed at a water depth of 0.6 to 1.2 m (2 to 4 ft). To keep the gravel from scattering, it is sometimes confined in boxes or old tires.

Depending on the water temperature, eggs take 48 to 96 hours to hatch. By day five to seven, fry inflate their gas bladder and rise from the nest. LMB fry aggregate in a rather compact school for two to three weeks, until they are 12 to 25 mm (0.5 to 1.0 in.) in total length. Schooling fry are captured either with fry traps, a lift net, or by short seine hauls along the shoreline with a small, fine-mesh seine. Fewer fry can be recovered from a spawning pond that is turbid or contains abundant vegetation. It may be necessary to treat the pond with organic material and lime to reduce turbidity or with an herbicide to reduce algae. To produce 38- to 51-mm (1.5- to 2.0-in.) fingerlings, prepared rearing ponds are stocked with 99,000 to 198,000 fry/ha (40,000 to 80,000 fry/acre). Several methods can be used to estimate the number of fry being moved from the spawning pond to the rearing pond. For example, a sample of fry can be counted into a container. Fry are then added to a similar container until the numbers in the containers visually appear to be equal. This procedure is then repeated until the desired number of fish have been stocked. Also, a known number of fish can be weighed or volumetrically measured. For example, a 500-mL (17-oz) beaker is filled with 250 mL (8.5 oz) of water, followed by 250 mL (8.5 oz) of fry. Five ml of fry are counted (28). The number–weight or number–volume relationship can then be used to calculate the number of fish for stocking. These techniques should estimate within $\pm 10\%$ the actual number stocked. The visual technique is probably the least stressful for small fish (3). If the culturist has an excess of fry and is, therefore, more interested in maximizing the yield of fingerlings rather than the survival rate, it is not necessary to stock as accurate numbers of fry.

The rearing ponds are filled with water 10 to 20 days before the 10- to 18-mm (0.5- to 0.75-in.) fry are stocked. This technique allows time for the zooplankton population to build up, but does not usually allow time for predacious insects, such as notonectids (back swimmers), which feed on small bass, to reach large numbers in the pond. Ponds should also be fertilized to produce a heavy crop of zooplankton. Snow (3) has recommended planting ryegrass in drained rearing ponds in the fall, to produce 2,240 to 2,688 kg per ha (2,000 to 2,500 lb per acre) of green organic material before the pond is flooded in the spring. Too much organic material that decays rapidly can cause an oxygen depletion. Snow also recommended one to three applications of inorganic 8-8-0 fertilizer at a rate of 112 kg per ha (100 lb per acre) per application, to supplement the ryegrass. In actuality, culturists have to experiment with what works best at their hatcheries. For additional information on fertilizing and reducing turbidity in ponds, see Boyd (26).

Harvesting should start when bass reach 38 mm (1.5 in.) in length, which occurs 2 to 4 weeks after stocking. With good management, it should be possible to recover 75 to 80% of the small bass stocked. Fish are usually harvested by trapping, seining, or draining the pond. Failure to harvest when the fish reach 38 to 50 mm (1.5 to 2 in.) in length will lead to cannibalism and low survival rates; however, cannibalism can happen at any time. Cannibalism is aggrevated by stocking fry at different ages, and hence different sizes, and by a lack of food. Ponds should be checked frequently for appropriate invertebrates. If the invertebrates are gone, then the bass need to be harvested, regardless of their size. Heavy filamentous algae or attached vegetation can make harvesting very difficult and should be controlled with an approved herbicide. Grass carp cannot be used for vegetation control during this phase, because they will damage the fingerling bass when they are confined with them in the seine. Many types of aquatic vegetation start to grow from the bottom of ponds. Properly constructed ponds that have a water depth primarily greater than

Table 2. Requirements for the Production of Various Numbers of Florida Bass Fingerlings, Using the Fry Transfer Method[a]

Size of LMB Needed		Spawning Area Needed		Broodfish Needed[b]		Fry Stocked	Fry Produced	Fingerling Pond Area		Fingerling Produced	Return
(cm)	(in.)	(ha)	(acre)	(kg)	(lb)	(no.)	(no.)	(ha)	(acre)	(no.)	(%)
2.5	1.00	0.42	1.05	60.5	133	105,600	80,000	0.53	1.32	76,000	95
3.2	1.25	0.45	1.12	63.6	140	111,300	70,000	0.64	1.59	63,000	90
3.8	1.50	0.47	1.18	67.3	148	117,600	60,000	0.78	1.96	51,000	85
4.4	1.75	0.50	1.25	70.9	156	125,000	50,000	1.00	2.50	40,000	80
5.1	2.00	0.53	1.33	75.5	166	133,200	40,000	1.33	3.33	30,000	75
6.4	2.50	0.62	1.54	87.7	193	153,750	25,000	2.46	6.15	16,250	65
7.6	3.00	0.73	1.82	103.6	228	181,800	15,000	4.84	12.12	8,250	55
10.2	4.00	0.80	2.00	113.6	250	200,000	10,000	8.00	20.00	5,000	50

[a]Adapted from White (22), cited in (15).
[b]Based on fry production of 250,000 fry per hectare (100,000 per ac).

0.9 m (3 ft) also help to reduce vegetation, as light on the bottom is required for vegetation to start to grow.

White (22) has estimated the size of the fish production unit and the weight of broodfish needed to produce 2.5- to 10.2-cm (1- to 4-in.) fingerlings on natural food (Table 2). As the size of the bass increases, the number of fingerlings produced decreases and the amount of resources greatly increases. The number of bass per kg (0.45 lb) also decreases rapidly as the size of the bass increases (Table 3).

Table 3. Number of Largemouth Bass per Unit of Weight at Various Lengths

Total Length[a]		Fish per Unit of Weight	
(cm)	(in.)	(kg)	(lb)
0.6	0.25	310,510	140,845
0.8	0.31	160,922	72,993
0.9	0.38	92,631	42,017
1.1	0.44	58,016	26,316
1.3	0.50	39,089	17,730
1.4	0.56	27,421	12,438
1.6	0.62	20,042	9,091
1.7	0.69	15,100	6,849
1.9	0.75	11,603	5,263
2.1	0.81	9,110	4,132
2.2	0.88	7,300	3,311
2.4	0.94	5,926	2,688
2.5	1.00	4,899	2,222
2.9	1.12	3,434	1,558
3.2	1.25	2,511	1,139
3.5	1.38	1,868	847
3.8	1.50	1,450	658
4.1	1.63	1,136	515
4.5	1.75	911	413
4.8	1.88	740	336
5.1	2.00	612	278
7.6	3.00	181.0	82.0
10.2	4.00	76.0	34.0
12.8	5.00	38.7	17.5
15.2	6.00	22.5	10.2
17.9	7.00	13.9	6.3
20.3	8.00	9.6	4.3
22.8	9.00	6.8	3.1
25.3	10.00	4.9	2.2

[a]For bass less than 5.1 cm (2 in.) the data are adapted from White (22), cited in (15). For bass greater than 5.1 cm (2 in.), the data are adapted from (43).

Intensive Method

A significant commercial market exists for 15- to 20-cm (6- to 8-in.) LMB for stocking into private lakes that have established fish communities. Advanced fingerling or yearling LMB are usually raised on pelleted food. The intensive method normally involves seining 25- to 51-mm (1- to 2-in.) fingerlings from their rearing pond, concentrating them in tanks, training them to accept a prepared diet, transferring the trained fish into a rearing pond or raceway, and feeding them a prepared diet until they reach the desired size. LMB have been trained to eat ground fish, fish eggs, krill, beef heart, etc. (29–31). Snow (32) has trained LMB to accept Oregon Moist Pellet. Other, similar diets are now on the market. Such diets are relatively expensive, contain approximately 10% water, and are best stored under refrigeration. To feed the trained fingerlings in rearing ponds, most growers switch to a less expensive diet that contains less water.

Fingerlings measuring 25 to 55 mm (1 to 2 in.) in length are moved from rearing ponds into training tanks. Circular and rectangular tanks of 845 to 1,691 L (200 to 400 gal) are often used. Although training facilities and procedures vary, successful ones all tend to have common characteristics. The fingerlings need to be concentrated in 21 to 30 °C (75 to 86 °F) water devoid of natural food. Frequently, initial training densities range from 3 to 12 fingerlings per liter (10 to 40 fingerlings per gal). To account for variation in the size of the bass trained, some culturists base their training densities on weight. Ranges of 5 to 7 g/L (0.04 to 0.06 lb/gal) are often used. The key to avoiding cannibalism is to stock and maintain the fingerlings at a uniform size and offer them food frequently. Grading with a bar grader may be necessary two or three times during the 7- to 10-day training period.

Culturists feed at different frequencies. Belt feeders are used to continuously offer the commercial diet (Fig. 1), but some training diets tend to clog the belt. Many culturists hand-feed fingerlings every 5 to 10 minutes throughout

Figure 1. Belt-type automatic feeder on a circular training tank.

Figure 2. Battery-operated pelleted food feeder anchored in the middle of an advanced-fingerling production pond.

at least part of the day, while others feed the fingerlings five times a day if they are using a training sequence involving ground fish or krill as a starter diet. Some culturists continuously illuminate the training facilities, while others do not. One can expect in excess of 60% of the fingerlings to learn to accept the prepared diet. Fingerlings that have been produced from first- or second-generation hatchery-reared broodfish tend to have training rates in excess of 90%. Oxygen levels should be maintained above 5 ppm in the training facilities. Complete flush rates (turnover rates) of one to four times per hour help maintain suitable oxygen levels without excessive current and unionized ammonia below 0.025 ppm (15). Aeration using gaseous oxygen or liquid oxygen can be used to supplement the oxygen added from the flush water. U-tube technology can be used to enhance diffusion efficiency of the oxygen into the water (33). Packed columns are also used by some culturalists.

After learning to accept the pelleted food, the trained fingerlings are graded to remove nonbreeders and cannibals and then moved into growout ponds at a stocking rate of 37,000 to 49,000 fingerlings per ha (15,000 to 20,000 fingerlings per acre). One can expect 40 to 50% of the bass to continue to accept the pelleted feed in the pond. Many of the smaller fish that decline pellets will survive. This percentage can be increased to 90 to 95% if the trained fingerlings are confined by a net or screen in a portion of the pond or in cages for 7 to 10 days. Some culturists use small ponds of 0.2 ha (0.5 acre). If placed in the middle of the pond, battery-operated mechanical feeders can be used to offer pelleted feed essentially over the entire pond at hourly intervals (Fig. 2). Ponds are frequently or continuously aerated with a paddlewheel aerator or blower, to maintain oxygen levels at higher feeding rates that are needed to produce higher standing stocks of bass (Fig. 3).

Other variations of intensive fry production involve removing the fry before they leave the nest site (3) or removing the fertilized eggs from the nest and incubating them in jars (Fig. 4) or in Heath Vertical Incubators (Fig. 5). In either case, the bass are encouraged to spawn in nest boxes provided by the culturist. Nest boxes have been used in ponds lined with plastic (15), in earthen ponds, and in raceways. In general, spawning boxes placed 1.8 to 3.7 m (6 to 12 ft) apart are used to collect either fry or eggs. A spawning box for the collection of pre-swim-up fry is actually a box within a box, where each box has a wire-mesh bottom. The smaller, interior box contains spawning gravel [pebbles from 2 to 5 cm (0.75 to 2 in.) in diameter] and has a wire-mesh bottom large enough [10 to 20 mm (0.39 to 0.79 in.)] for sac fry to wiggle through

Figure 3. Paddlewheel aerator powered by the PTO of a tractor.

Figure 4. Jar-type egg-incubation system.

Figure 5. Heath Vertical egg-incubation system.

easily following hatching. The outside box has a wire-mesh bottom that is too small for fry to move through (15). Fry are collected from the outer box by removing the inner box. For more details on spawning boxes see Huston (34). To collect bass eggs, some culturists use a spawning box that has a 46 × 46-cm (18 × 18-in.) bottom and is 9 cm (3.5 in.) deep. The box is filled with 1.3- to 2.5-cm (0.5- to 1-in.) gravel (see Fig. 6) and has a coarse screen on the bottom and 7.6-cm (3-in.) legs. A 0.1-mm (24-mesh-per-in.) fine mesh saran screen attached to a 1.3 m × 1.3-m (0.5 × 0.5-in.) plastic pipe frame is placed on top of the gravel inside the box. Bass lay their adhesive eggs on the saran screen (35). The eggs are removed from the screen by a slight rolling action into a pan containing water and then are poured into a Heath Vertical Incubator tray or into jars.

Spawning mats of similar size, weighted indoor–outdoor carpet, or nylon felt have also been used to collect eggs (36,37). Mats can be dipped in a 1.5% buffered sodium sulfite solution for two to three minutes to remove the eggs (3).

LMB tend to spawn within a week if male and female broodfish have been separated in the spring before spawning if they are placed together after spawning

Figure 6. Spawning nest box used to transfer eggs, half-filled with gravel.

temperatures have been reached, and if there are more mats or nest boxes than males. An advantage of using nest boxes and mats is that fry of the same age, and thus the same size, can be stocked, a practice that reduces cannibalism. Separating the sexes also has the advantage of delaying spawning until the water temperature has stabilized. Male bass will desert their nest, and the eggs will die, if a severe cold front sufficiently reduces the water temperature. This problem is particularly prevalent at northern latitudes. Taking the eggs from spawning structures is another way to alleviates the problem.

FOOD-SIZE BASS

Certain ethnic groups have created a primary demand for adult bass as foodfish. The customers for this market are centered in large cities, such as San Francisco, Chicago, and New York, and demand live fish. It takes two to three years, depending upon latitude, to raise fish of an adequate size for these markets. Basically, trained advanced fingerling or yearling largemouth bass are stocked into rearing ponds at rates of 3,700 per hectare (1,500 per acre) and raised to 0.7 kg (1.5 lb) or larger.

The food supply initially limits the amount of LMB one can raise per unit area. This limit is overcome by feeding the fish. The second limit is the pond's ability to process the waste materials of the LMB without depleting all of the dissolved oxygen. This limit is increased by mechanically aerating the pond. Without feeding, one might expect a pond to produce 112 kg of LMB per hectare (100 lb per acre). With feeding, 1,684 to 2,245 kg of LMB per hectare (1,500 to 2,000 lb per acre) or more might be raised before the amount of oxygen becomes a limiting factor. Without aeration or flushing, oxygen depletion frequently occurs when feeding rates reach 20 kg per ha (17.8 lb per acre) per day. By aerating or flushing, a production of 3,368 kg of LMB per hectare (3,000 lb per acre) can be reached.

HANDLING AND TRANSPORT

Healthy LMB are not particularly difficult to handle, but any severe stress, including a heavy parasite load, should be avoided. In general, LMB's tolerance of handling and physical and chemical conditions is intermediate between that of salmon/shad and minnows/catfish (2). Within 72 hours, 50% of the bass die when the unionized ammonia-N level is 0.82 mg/L, the alkalinity is 232 mg/L, and the hardness is 272 ppm (15). With in 96 hours, 50% of the bass die when the nitrite-N level is 140 ppm (38). Ideally, LMB should be seined and transported when water temperatures are cool. LMB can withstand a wide range of temperatures if acclimated, but handling below 4.5 °C (40 °F) makes it more difficult to avoid stressing the fish (39).

Williamson et al. (15) gave a more detailed review of stress in LMB. Carmichael et al. (40) presented a complex procedure for safely hauling 200-g (0.44-lb) LMB at a density of 180 g/L (1.5 lb/gal) for up to 30 hours, that virtually eliminated mortality due to hauling stress. Their method involves prophylactic disease treatments, withholding food from the fish before hauling, hauling at cool temperatures, anesthetizing the fish, hauling with salts in the water, etc. In general, most culturists stop feeding the LMB one to two days before seining fish seine them. After the fish have been seined, they are in raceways (Fig. 7) with well-aerated, 10- to −19 °C (50- to −66 °F) water for 24 to 48 hours and are then moved to the desired location.

LMB fry can be shipped in plastic bags that contain one fourth to one third of the bag's volume in water and are inflated with oxygen and sealed with rubber or, preferably, elastic bands or a heat sealer. The sealed bags are placed in an insulated container to keep them cool (41). Densities of 2,000 to 12,000 fry per L (7,600 to 46,600 fry per gal) of water are held for one to four days. When individual pond owners pick up small quantities of fingerlings at the hatchery or from a truck at a delivery point, the fingerlings are frequently placed in plastic bags. Densities of 100 to 200 fish/3.8 L (1 gal) of water can safely be hauled for at least four to six hours if the water temperature is maintained below 20 °C (68 °F). When LMB are moved from one place to another, they should be tempered if the difference in water temperature is greater than 10 °C (18 °F). If the fish are in plastic bags, they can be floated

Figure 7. Typical concrete raceway used for holding fish.

Figure 8. Oxygen-inflated plastic bag used to transport small LMB.

in the receiving water (Fig. 8). Plastic bags exposed to the sun warm up very quickly; this practice should be avoided. Note that when fish die in a plastic bag filled with oxygen, they do not die from a lack of oxygen, but from the buildup of carbon dioxide.

If the fish are in a hauling tank, water can be slowly pumped into the hauling tank from the receiving water. Frequently, a 10 °C (18 °F) difference in water temperature is brought to an equilibrium over a one-hour period. Larger bass and higher densities than can be transported in plastic bags are moved in truck-mounted hauling tanks equipped with aeration devices. The maximum density that can be hauled without mortality depends on the size of the fish, the condition of the fish, the water temperature, the overall water quality (including hardness), the aeration system, and the time of haul. In general, the larger the fish, the cooler the water, the higher the oxygen levels, and the shorter the amount of that time more kg (pounds) of fish per liter (gallon) of water can be hauled (Table 4). Some hauling tanks are equipped with mechanical agitators or an airblower to aerate the water, but most units now use a system that delivers oxygen. Oxygen can be purchased as a compressed gas or as a liquid. If a large amount of oxygen is used in the hatchery, the liquid form is the most economical. For a discussion of diffusion efficiency, see Williamson et al. (15).

Table 4. Normal Loading Capacity for LMB Hauling Tanks Equipped with an Agitator or Blower System in Hard Water at 18 °C (65 °F)[a,b]

Size of Fish		Duration of Transport (hours)							
		1		6		12		24	
(cm)	(in.)	kg/L	no./g	kg/L	no./g	kg/L	no./g	kg/L	no./g
5	2	0.24	2	0.18	1.5	0.12	1	0.12	1
20	8	0.36	3	0.36	3	0.24	2	0.28	1.5
112	14	0.48	4	0.48	4	0.36	3	0.24	2

[a]Modified from White (22), citing Johnson (45).
[b]For each 5.5 °C (10 °F) increase in water temperature, decrease the load by 25%.

PESTS AND DISEASES

Certain insects, such as notonectids (back swimmers), can kill large numbers of LMB smaller than 3.8 cm (1.5 in.), by attaching to their backs and sucking out their body fluids. Such insects have been controlled by chemicals that are not approved for aquatic use on foodfish. Since the insects are air breathers, during calm weather they can be controlled by adding a small amount of fuel oil to the water surface. Pond management is probably the best way to control insects. If the pond is filled with well or filtered water approximately 10 to 20 days before the fry are stocked, the insects will not have time to build up, and by the time that they do, the bass will be large enough to eat them. Crayfish and bullfrog tadpoles can also be a problem when harvesting small bass. Bullfrog tadpoles have to be separated from the bass in order to obtain an accurate count of the fish harvested when the weight–number or volume-to-number method is used. Crayfish also have to be separated from the bass, but, in addition, they can damage or kill a large proportion of the small bass when they are crowded too closely in the net during harvesting. Neither crayfish nor bullfrog tadpoles are a problem when raising larger bass or in bass brood ponds, since larger bass eat them.

In some areas of the United States, fish-eating birds, such as cormorants, water turkey, and blue herons, are major predators of bass. In general, killing those pests without a permit is illegal, and scare tactics, such as carbide cannons, are of limited use. The blackbird–grackle group can become quite proficient in obtaining small bass from an outdoor raceway system, but bird screens will eliminate the problem.

Bass are susceptible to a number of diseases. When stressed, bass can be killed by bacteria such as *Aeromonas* and *Cytophaga* (*Flexibacter*). The bacterial infections are frequently followed by a secondary invasion of fungus. The best solution to the problem is to avoid stressing the fish. However, in the spring, just before and during spawning, broodfish may exhibit bacterial and fungal infections. Oxytetracycline-treated food has been used to treat bacteria, and some culturalists train bass with food that contains oxytetracycline. Anchorworms, *Lernaea* spp., are often treated with Dylox® at 0.25 to 0.50 ppm at weekly intervals for four to six weeks. However, as of this writing, Dylox® is not approved for the treatment of *Lernaea* on LMB unless the treatment is done under an investigation of a new animal-drug permit. Gill flukes in the genus *Dactylogyrus* (trematodes) are treated with 250 ppm formalin for 0.5 to 1 hour. A 3% salt dip for an hour can also be used (15). Likewise, ciliated protozoans, such as *Trichodina* and *Costia*, can be treated with formalin and salt. There is no effective legal treatment for *Ichthyophthirius* or for bass tapeworm, *Proteocephalus*, at this time. Overall, very few chemicals are approved for use on foodfish (42,43).

CONCLUSION

It is possible to raise LMB economically, but only by people with experience in, or knowledge of, aquaculture. Some other books on aquaculture are provided by (42–44). The culturist must not only raise the fish, but also develop markets where the fish will be sold. Those who are serious about entering the business should consult experts before beginning. A good place to start is the state aquaculture extension agent. Remember, a private producer is not in the business of telling you how to compete with him or her.

BIBLIOGRAPHY

1. J. Arnold and M. Isacc, *Bull. U.S. Fish Comm.* **2**, 113–115 (1882).
2. R.C. Heidinger, *Synopsis of Biological Data on the Largemouth Bass*, Micropterus Salmoides (Lacepede) *1802*, FAO Fisheries Synopsia No. 115, 1976.
3. J.R. Snow, in H. Clepper, eds., *Black Bass Biology and Management*, Sport Fishing Institute, Washington, DC, 1975, pp. 344–356.
4. M.H. Carr, *Proc. New England Zool. Club* **20**, 43–77 (1942).
5. J.B. Lamkin, *Trans. Amer. Fish. Soc.* **29**(1900), 129–155 (1901).
6. J.R. Snow, *Proc. S.E. Assoc. Game Fish Comm.* **24**(1970), 550–559 (1971).
7. R.H. Kramer and L.L. Smith, *Trans. Amer. Fish. Soc.* **91**, 29–41 (1962).
8. F.A. Meyer, *Prog. Fish-Cult.* **32**(3), 130–136 (1970).
9. R.L. Chew, *Fish. Bull. Fla. Game Fresh Water Comm.* **7**, 1–76 (1974).
10. J.V. Merriner, *Trans. Amer. Fish. Soc.* **100**, 611–618 (1971).
11. T.R. Badenhuizen, *M.S. Thesis*, Cornell University, Ithaca, NY, 1969.
12. P.M. Johnston, *J. Morphol.* **93**(1), 45–67 (1953).
13. G.C. Lawrence, *N.Y. Fish Game J.* **18**(1), 52–56 (1971).
14. D.M. Green and R.C. Heidinger, *N.A.J.F.M.* **14**(2), 464–465 (1994).
15. J.H. Williams, G.J. Carmichael, K.G. Graves, B.A. Simio, and J.R. Tomasso, in R.R. Stickney, eds., *Culture of Nonsalmonid Freshwater Fishes*, 2nd eds., CRC Press, New York, 1993, pp. 145–197.
16. D. Tave, *Genetics for Fish Hatchery Managers*, AVI, Westport, CT, 1986.
17. R.B. Moorman, *Iowa State College J. Sci.* **32**(1), 71–88 (1957).
18. M.F. James, *Ph.D. Thesis*, University of Illinois, Urbana, IL, 1942.
19. G.W. Bennett, *Bull. Il. Nat. Hist. Surv.* **24**, 377–412 (1948).
20. W.D. Parker, *Prog. Fish-Cult.* **33**(1), 55–56 (1971).
21. R.E. Manns and B.G. Whiteside, *Prog. Fish-Cult.* **42**(2), 116–117 (1980).
22. B.L. White, *Inland Aquaculture Handbook*, Texas Aquaculture Association, College Station, TX, 1988, p. A0303.
23. P.M. Rosenblum, H. Horne, N. Chatterjee, and T.M. Brandt, in A.P. Scott, J.P. Sumpter, D.E. Klime, and M.S. Rolfe, eds., *Proceedings of the 4th International Symposium on the Reproductive Physiology of Fish*, University of East Anglia, Norwich, United Kingdom, 1991, pp. 265–267.
24. J.R. Snow and J.L. Maxwell, *Prog. Fish-Cult.* **32**(2), 101–102 (1970).
25. R.J. Anderson, E.W. Kienholz, and S.A. Flickinger, *J. Nutr.* **111**(6), 1085–1097 (1981).
26. C. Boyd, *Water Quality in Ponds for Aquaculture*, Birmingham Publishing Company, Birmingham, AL, 1990.

27. H. Bishop, *Proc. North Central Warmwater Fish Culture Workshops*, Iowa State University, Ames, IA, 1968, pp. 24–27.
28. R. Swanson, *Midwest Black Bass Culture*, Texas Parks and Wildlife Department (Austin, TX) and Kansas Fish and Game Commission (Pratt, KS), 1982, pp. 25–63.
29. A.M. Brandeburg, M.S. Ray, and W.M. Lewis, *Prog. Fish-Cult.* **41**(2), 97–98 (1979).
30. T.H. Langlois, *Trans. Amer. Fish. Soc.* **61**, 106–115 (1931).
31. J.R. Snow, *Proc. S.E. Assoc. Game Fish Comm.* **14**, 253–257 (1960).
32. J.R. Snow, *Prog. Fish-Cult.* **30**(4), 235 (1968).
33. L. Allen and E. Kinney, eds., *Proceedings of the Bio-Engineering Symposium for Fish Culture*, Fish Culture Section Pub. 1, American Fisheries Society, Bethesda, MD, 1981, pp. 53–62.
34. P.L. Hutson, *Prog. Fish-Cult.* **45**(3), 169–171 (1983).
35. W.F. Lingle, A.M. Brandenburg, and R. Smith, *Midwest Black Bass Culture*, Indiana Department of Natural Resources, 1986, pp. 37–40.
36. T. Trudeau, M. Sarti, and D. Woolard, *Midwest Black Bass Culture*, Illinois Department of Conservation, 1985, pp. 72–75.
37. G.A. Chastain and J.R. Snow, *Proc. S.E. Assoc. Game Fish Comm.* **19**(1965), 405–408 (1966).
38. R.M. Palacheck and J.R. Tomasso, *Can. J. Fish. Aquat. Sci.* **41**(12), 1739–1744 (1984).
39. A.J. Wilson, *Prog. Fish-Cult.* **12**(4), 211–213, (1950).
40. G.J. Carmichael, J.R. Tomasso, B.A. Simco, and K.B. Davis, *Trans. Amer. Fish. Soc.* **113**(6), 778–785 (1984).
41. J.R. Snow, *Proc. S.E. Assoc. Fish Wildlife Agen.* **22**(1968), 380–387 (1969).
42. J.W. Avault, *Fundamentals of Aquaculture: A Step-by-Step Guide to Commercial Aquaculture*, AVA Publishing Company, Baton Rouge, LA, 1996.
43. R.G. Piper, I.B. McElwain, L.E. Orme, J.P. McCraren, L.G. Fowler, and J.R. Leonard, *Fish Hatchery Management*, U.S. Department of the Interior, Fish and Wildlife Services, Washington, DC, 1982.
44. R.R. Stickney, *Principles of Aquaculture*, John Wiley & Sons, New York, 1994.
45. S.K. Johnson, Unpublished manuscript Texas A&M University, College Station, TX, 1979.

BLOWERS AND COMPRESSORS

John Colt
Northwest Fisheries Science Center
Seattle, Washington

OUTLINE

INTRODUCTION

Compressed air is commonly used for aeration, mixing, and pumping in aquatic systems. The design of compressed air systems depends strongly on the flow and pressure required. It is possible to obtain high-pressure compressed air in cylinders, but most compressed air is produced on-site.

CLASSIFICATION OF BLOWERS AND COMPRESSORS

The following four general types of units have been used in aquaculture to provide compressed air:

Type of Unit	Head Range, m (ft)	Aquaculture Application
Fan	0.006–0.06 (0.02–0.2)	Ventilation in building; forced air packed columns
Blower	0.30–2.4 (1–8)	Aeration in ponds and tanks; shallow airlift pumps
Multistage blower	1.2–5.2 (4–17)	Aeration in deeper ponds and rearing units
Compressor	12–100 (40–350)	Building compressed air; lake aeration; deep airlift pumps

Blowers are the most common type of unit used in aquaculture because of the limited depth of most rearing systems. The pressure provided by compressors is higher than that required for most aquaculture uses, but compressors are employed for many applications because they are commonly used to provide building air and therefore, they are available.

Because pressures and flows are specified in a variety of units, conversion information is needed. Pressure may be reported in millimeters of mercury (mm Hg), pounds per square inch (psi), Pascals (Pa), and as water head (feet or meters). Water head is the pressure exerted by a column of water and depends slightly on temperature. The following list may be used to convert between the bewildering pressure units:

1 standard atmosphere (atm) = 760 mm Hg
= 14.73 psi
= 10.34 m (15 °C)
= 33.93 ft (15 °C)
= 101.325 kPa
= 101,325 Pa
= 1.013 bar
= 1,013 millibar
= 1.033 kg_f/cm^2

The conversions between head and pressure units are equal to 73.49 mm Hg/m, 0.434 psi/ft, and 9.799 kPa/m at 15 °C. Barometric pressures are typically reported in mm Hg or millibar.

Air flow is typically expressed in cubic meters per minute (m^3/min) or cubic feet per minute (ft^3/min). The most common method of specifying airflow is in terms of standard volumetric flow. In the United States, the standard conditions are 20 °C (68 °F), 101.325 kPa (14.73 psi), and 36% relative humidity, resulting in a standard density of 1.20 kg/m^3 (0.0750 lb/ft^3). The standard conditions for reporting airflow are different in other countries, and the differences can introduce significant errors if not considered. Ignoring the minor impact of differences in moisture content, the actual (free air) volumetric flow rate is related to standard volumetric flow rate by (1)

$$q_{act} = q_{std}\left[\left(\frac{P_{std}}{P_{act}}\right)\left(\frac{T_{act}}{T_{std}}\right)\right] \tag{1}$$

where q_{act} is the actual volumetric flow rate (m^3/min, ft^3/min), q_{std} is the standard volumetric flow rate (standard m^3/min, standard ft^3/min), P_{std} is the standard pressure (101.325 kPa, 14.73 psi), P_{act} is the actual pressure (kPa, psi), T_{act} is the actual absolute temperature (°K, °R); °K = °C + 273.15; °R = °F + 459.69, and T_{std} is the standard absolute temperature (273.15 °K, 527.90 °R).

BLOWER AND COMPRESSOR DESIGN

The theoretical power for adiabatic compression (1) is given by

$$P_{ad} = \frac{q_{act}\gamma R T_1}{An}\left[\left(\frac{p_2}{p_1}\right)^n - 1\right] \tag{2}$$

where P_{ad} is the adiabatic power (kW, hp), q_{act} is the actual of air flow (m^3/min, ft^3/min), γ is the specific weight of air (kgf/m^3, lb/ft^3), R is the gas constant (287.1, 53.3), T_1 is the absolute inlet temperature (°K, °R), p_1 is the absolute inlet pressure (kPa absolute, psi absolute), p_2 is the absolute outlet pressure (kPa absolute, psi absolute), A is a constant (60,000, 33,000), n is $(k-1)/k = 0.283$ for air, and k is 1.395 for air.

The theoretical power for adiabatic compression (Eq. 2) is the power needed for a 100% efficient unit. The actual power need by the compression is called the brake power and is given by

$$\mathrm{BP} = \frac{P_{ad}}{e_c} \tag{3}$$

where BP = brake power (kW, hp) and e_c = efficiency of compressor (decimal). The actual power used by the motor, drive and gearbox, and compressor is called the wire power and is given by

$$\mathrm{WP} = \frac{P_{ad}}{e_c e_d e_m} \tag{4}$$

where WP = wire power (kW, hp), e_d = efficiency of drive and gearbox (decimal), and e_m = efficiency of motor (decimal). The wire horsepower is the power billed for if this unit were connected to an electrical utility. Commonly, compressor manufacturers only report the brake horsepower because they may not supply the unit with the motor or drive. Due to the significant difference between these parameters, it is important to understand clearly how the power is being reported.

The adiabatic temperature rise during compression is given by the approximation

$$\Delta T_{ad} \approx \frac{T_1}{e_c}\left[\left(\frac{p_2}{p_1}\right)^n - 1\right] \tag{5}$$

where ΔT_{ad} = increase in temperature (°C, °F).

The horsepower needed to supply a given airflow will depend on the inlet air temperature (T_1), the inlet air pressure (p_1), and the outlet air pressure (p_2). It is important to note that Eqs. 1–5 are written in terms of absolute pressures (barometric pressure + gauge pressure). Most pressure gauges measure the pressure relative to the local barometric pressure.

The density of air (2) can be computed from ideal gas relationships, which are presented in Table 1 for different air temperatures and elevations. The theoretical horsepower required to supply 28.3 m^3/min (1000 ft^3/min) air at 2 m (6.56 ft) total head and the resulting airflow rate at standard conditions are presented in Table 2. For the supply of a given flow rate expressed in terms of standard volumetric flow, the design capacity must be based on the maximum summer temperature, but the motor must be designed for the minimum winter temperature. The capacity of a compressor decreases at higher elevations.

SYSTEM DESIGN

The preceding equations can be used for preliminary sizing of equipment, but a different procedure is used for the actual design. Many of the specific system components will depend on the actual type of blower or compressor selected.

Head-Capacity Curve

The design of compressor and blower systems is based on the manufacturer's head-capacity curves (Fig. 1). These curves show the standard volumetric flow rate as a function of water head at a single elevation, air temperature, and power frequency. While most blowers

Table 1. Specific Weight of Air in kg_f/m^3 (lb/ft^3) as a Function of Elevation and Temperature[a]

Elevation, m (ft)	Pressure mbar (mm Hg)	Air Temperature		
		10 °C (50 °F)	20 °C (68 °F)	30 °C (86 °F)
0 (0)	1013 (760)	1.24 (0.0777)	1.20 (0.0750)	1.16 (0.0723)
305 (1000)	977 (733)	1.20 (0.0748)	1.15 (0.0720)	1.11 (0.0692)
610 (2000)	942 (707)	1.16 (0.0723)	1.11 (0.0695)	1.07 (0.0668)
914 (3000)	910 (683)	1.12 (0.0697)	1.07 (0.0671)	1.03 (0.0644)
1219 (4000)	878 (659)	1.08 (0.0673)	1.04 (0.0647)	1.00 (0.0621)
1524 (5000)	848 (636)	1.04 (0.0649)	1.00 (0.0624)	0.96 (0.0599)
1829 (6000)	818 (614)	1.00 (0.0626)	0.97 (0.0602)	0.93 (0.0578)
2134 (7000)	789 (592)	0.97 (0.0605)	0.93 (0.0581)	0.89 (0.0558)

[a] Assumes relative humidity = 32% at each temperature.

Table 2. Theoretical Power Needed to Supply 28.3 m^3/min (1000 ft^3/min) of Air at a Total Head of 2 m (6.56 ft) at the Stated Temperature and Pressure[a]

Elevation, m (ft)	Pressure mbar (mm Hg)	Theoretical Power (kw, hp)			Standard Flow Rate of Air Provided (standard m^3/min, standard ft^3/min)		
		10 °C (50 °F)	20 °C (68 °F)	30 °C (86 °F)	10 °C (50 °F)	20 °C (68 °F)	30 °C (86 °F)
0 (0)	1013 (760)	8.78 (11.77)	8.77 (11.75)	8.74 (11.73)	29 (1037)	28 (1000)	27 (965)
305 (1000)	977 (733)	8.74 (11.71)	8.70 (11.67)	8.65 (11.60)	28 (998)	27 (960)	26 (923)
610 (2000)	942 (707)	8.72 (11.70)	8.69 (11.65)	8.63 (11.58)	27 (964)	26 (928)	25 (891)
914 (3000)	910 (683)	8.70 (11.67)	8.67 (11.62)	8.61 (11.54)	26 (930)	25 (895)	24 (860)
1219 (4000)	878 (659)	8.68 (11.64)	8.64 (11.59)	8.58 (11.51)	25 (898)	24 (863)	23 (829)
1524 (5000)	848 (636)	8.66 (11.61)	8.62 (11.56)	8.56 (11.47)	25 (866)	24 (833)	23 (800)
1829 (6000)	818 (614)	8.63 (11.58)	8.60 (11.53)	8.53 (11.44)	24 (836)	23 (804)	22 (771)
2134 (7000)	789 (592)	8.61 (11.54)	8.57 (11.49)	8.50 (11.40)	23 (806)	22 (775)	21 (744)

[a] Standard flow rate of air is also presented.

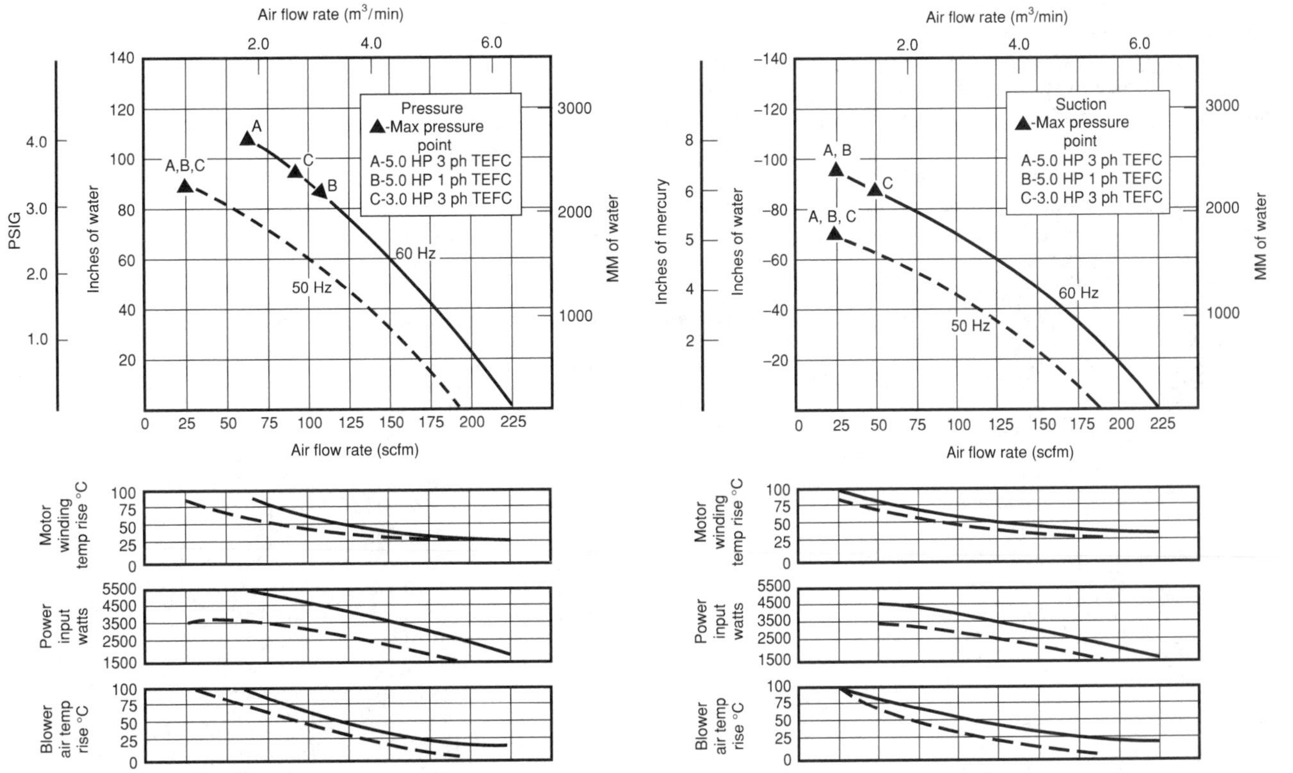

Figure 1. Head-capacity curve for a typical regenerative blower (Model DR/CP 6; Courtesy of AMETEK Rotron-Industrial Products, Saugerties, NY).

and compressors are powered by electrical motors, it is possible to use gasoline or diesel engines as backup or when electrical power is not available. The output depends on the discharge pressure and varies from high flow at low pressures to low flows at high pressures.

System Head Curve

To determine the actual amount of air that will be supplied by a unit, it is necessary to compute the system head curve over the potential range of air flows. The system head curve defines head losses to airflow in the overall system. The total pressure that a blower has to produce depends on the depth of submergence of the diffuser, losses in the piping system, and losses in the fittings and diffusers. This relationship can be expressed as

$$\text{Total head (ft)} = \text{Submergence} + HL_{\text{pipe}} + HL_{\text{fitting}} + HL_{\text{diffuser}} \quad (6)$$

The submergence term is simply the submergence depth of the diffuser. The HL_{pipe} value depends on airflow, pipe size, pipe roughness, air temperature, and pipe length. The HL_{pipe} term can be computed from the Darcy-Weisbach equation and the Moody diagram (3) or from a tabular listing such as that found in Table 3. The computation of HL_{pipe} is complicated by the fact that air temperature, pressure, and flows will change within the piping system. Therefore, it is prudent to design an air distribution system conservatively by oversizing the primary distribution headers. If there is significant head loss along the primary distribution header, the rearing units farther away from the blower or compressor will receive less airflow. The HL_{fitting} term accounts for head losses in elbows, tees, and valves. This term is generally very small and can be ignored for many systems. Commonly, the submergence and HL_{diffuser} terms are the largest losses in indoor systems.

Head losses in diffusers depend on both airflow and pore size. Diffusers, which produce small bubbles, have larger head losses. The head losses of many diffusers are provided by the manufacturer (see Fig. 2). Many diffusers used in aquaculture have head losses in the range of 15–25 cm (6–10 in.); fine-bubble diffusers used for pure oxygen applications may have head losses as high as 14–18 m (46–58 ft).

While it is desirable to select a diffuser with low head losses, unstable operation may result if the diffuser losses are less than 10 cm (4 in.) of head (3). Diffuser head losses can increase with time due to clogging (two major sources are particulates in the air supply and corrosion products from metal pipes) and the growth of microorganisms on the external surfaces. It is very difficult to quantify the long-term increase in diffuser head losses because of the dependence on operational procedures, but doubling the published values gives a conservative estimate.

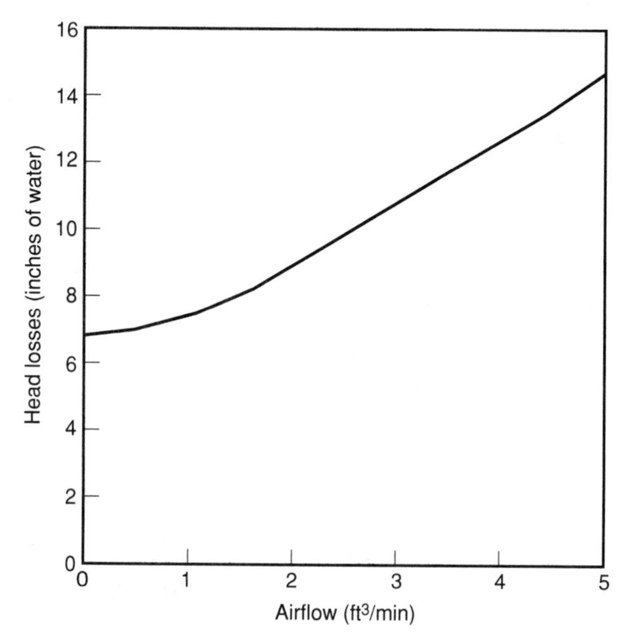

Figure 2. Head losses for a disc diffuser (cm = in. × 2.54, $m^3/min = ft^3/min \times 0.02831$).

Table 3. Frictional Head Losses for Air as a Function of Flow and Pipe Size[a]

Flow Rate in scmm (scfm)	Nominal Pipe Size 12.7 mm (1/2 in.)	19.1 mm (3/4 in.)	25.4 mm (1 in.)	38.1 mm (2 in.)	50.8 mm (2 in.)
0.14 (5)	15 (0.68)	4 (0.18)	1 (0.06)	0 (0.01)	0 (0)
0.28 (10)	55 (2.43)	14 (0.61)	4 (0.19)	0 (0.02)	0 (0)
0.42 (15)	117 (5.17)	27 (1.18)	9 (0.40)	1 (0.05)	0 (0.02)
0.57 (20)	195 (8.66)	47 (2.07)	15 (0.67)	2 (0.08)	1 (0.03)
0.85 (30)		102 (4.51)	32 (1.40)	4 (0.17)	1 (0.05)
1.42 (50)		262 (11.6)	80 (3.56)	10 (0.43)	3 (0.13)
2.12 (75)			167 (7.42)	20 (0.90)	6 (0.27)
2.83 (100)			291 (12.9)	35 (1.54)	10 (0.46)
3.54 (125)				45 (1.99)	15 (0.68)
4.25 (150)				74 (3.26)	21 (0.93)
5.66 (200)				127 (5.63)	37 (1.62)
7.08 (250)					55 (2.44)

[a]Head losses are expressed in terms of kilopascal/100 m of pipe (psi/100 ft of pipe).

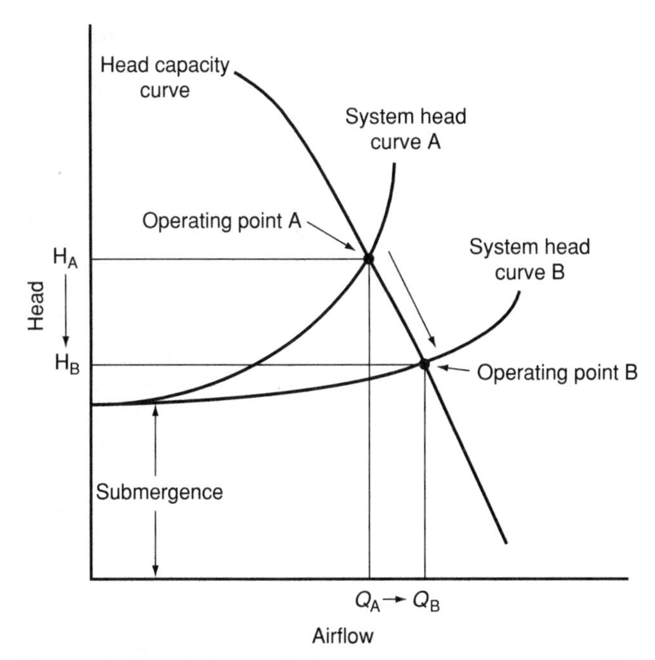

Figure 3. System design of a blower or compressor system. The operating point is where the system head curve intersects the manufacturer's head-capacity curve. If a larger pipe is used, the operating point will move from point A to point B and the flow will increase from Q_A to Q_B.

Operating Point

To determine the actual pressure and flow produced, the system head loss curve is computed from Eq. 6 over the expected operating range. This curve is plotted over the manufacturer's head-capacity curve (Fig. 3). The intersection of the two curves is the operating point (Operating Point A in Fig. 3). Changes in the system head curve will change the operating point. For example, increasing the pipe size will decrease the system head curve and move the operating point from point A to point B, resulting in an increase in airflow from Q_A to Q_B.

GENERAL DESIGN CONSIDERATIONS

A number of design considerations apply to all blowers and compressors and are discussed in this section. Design considerations that apply to specific types of units are presented in the next section.

Air Temperature

The density of air decreases with increasing air temperature. At higher air temperatures, less air is supplied in terms of q_{std} (standard volumetric flow rate). Therefore, the capacity of a system must be based on providing the required airflow rate during the maximum temperature period. For a given physical system, the maximum power consumption will occur at the lowest temperature. The motor must be sized large enough to supply the required power, or the discharge can be valved back to reduce power consumption.

Elevation

Increasing elevation reduces the output of compressors and blowers by about 5% per 300 m (1000 ft). This effect must be considered for high-elevation applications.

Diffuser Depth

If rearing units or holding tanks are of different depths, the system must be designed to provide air to the deepest unit. Valves must be installed on the distribution line to the shallower units to increase head losses. If this is not done, all of the air will flow to the shallower units.

Noise

Many compressors and blowers generate a significant amount of noise. These units should be placed in a separate room, if possible.

System Requirements

System requirements are very difficult to determine accurately, and they may change with time. To allow for future flexibility, the compressor building and main distribution system should be oversized. Unless the system output is constant over the year, two or more different-sized units are more economical than a single large unit. For critical applications, 100% standby capacity should be provided.

Pipe Selection

PVC pipe is probably the most commonly used pipe for primary and secondary air lines in blower systems. It is common to use plastic tubing between the secondary air line and the point of use. Air compressors and blowers result in a significant increase in air temperature (Eq. 5). The discharge temperature may be high enough to substantially reduce the strength of the discharge piping and to release potentially toxic compounds. Such problems are most likely to occur when two parallel compressors are turned on at the same time. Considerable care should be taken with the piping in the immediate area of the compressor, and a section of steel pipe may be needed.

In addition, the strength of PVC pipe is inadequate for most compressor systems, and PVC may not be allowed by some building codes. Black iron, galvanized iron, or copper pipe is commonly used in building systems. While condensation of water and oil is not a problem close to the compressor because of elevated temperature, condensation may occur in other parts of the distribution system, especially in unused feeder lines. The discharge of this condensed liquid mixture into the rearing units can be toxic to aquatic animals, and the rust and scale products will clog the diffusers.

SPECIFIC DESIGN CONSIDERATIONS

The most common types of blowers and compressors used in aquaculture are the positive displacement compressor, the rotary vane compressor, the liquid ring compressor, and the regenerative blower. Information on their specific design and operation is presented below.

Positive Displacement Compressors

Many compressors designed to provide building air at 400–700 kPa (60 to 100 psi) are not rated for continuous duty, and high-volume demand from an aeration system may result in early compressor failure as well as high power costs. A typical layout for a positive-displacement compressor system is presented in Figure 4a. While very desirable for aquaculture systems, most general building compressed air systems do not have a standby compressor, a refrigerated dehumidifier, or a particulate/oil filter. In these cases, it is necessary to install water/oil removal filters on individual air lines. One particularly useful system uses toilet paper as the filter element.

Regenerative Blower

The regenerative blower system is very commonly used to provide low-pressure air for aeration and mixing applications (Fig. 4b). A regenerative blower system produces oil-free air and is very reliable because of the small number of moving parts. Because of the low pressures produced by this type of unit, head losses in the distribution system and through diffusers must be carefully considered.

Rotary Vane Compressors

Rotary vane compressors have been used in the United States for aquaculture applications. When the rotary vanes are constructed from carbon, a small amount of fine carbon dust may be present in the air. This material can be filtered out, but is inert and should have no effect on most aquatic animals.

Liquid Ring Compressors

The liquid ring compressor is the only system that can produce oil-free and particulate-free air in the range of 400–700 kPa (60–100 psi). The unit uses a fixed-blade rotor in an elliptical casing. The casing is partially filled with water. This unit is widely used for hospital and laboratory applications but is more expensive than

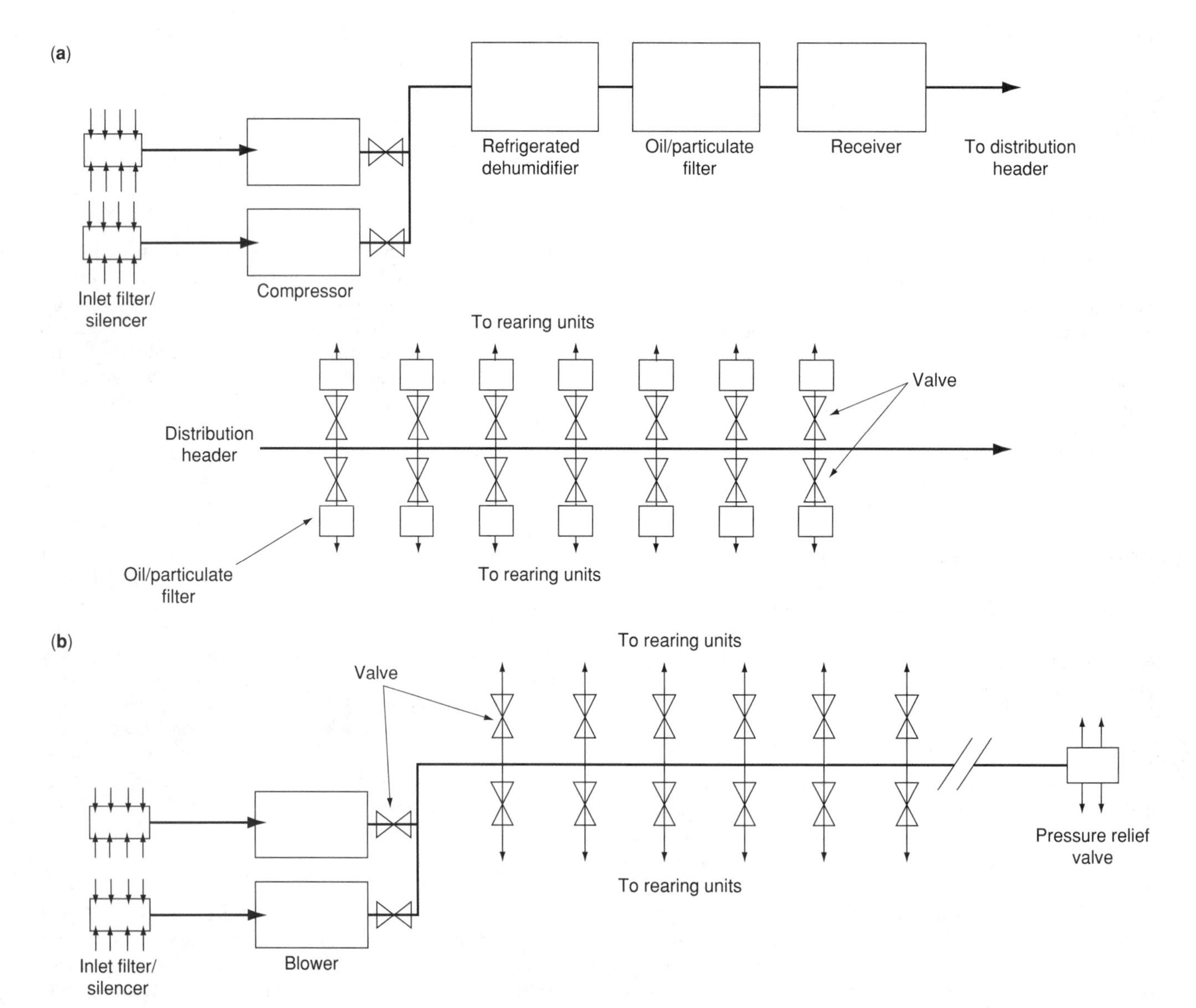

Figure 4. Process flowsheet for (**a**) compressor system and (**b**) blower system.

conventional compressors. This type of compressor has been used in critical shrimp hatchery applications.

General Operation Consideration

As with all rotating equipment, proper routine maintenance will greatly increase reliability and reduce overall costs. The clogging of diffusers is a serious problem in many applications. Clogging may result from the formation of scale in the piping system or from the growth of microorganisms on the external surfaces of diffusers. Most diffusers need to be removed from the water when the air system is turned off. Clogged diffusers can be cleared with varying degrees of success; this may involve the use of strong acids and dangerous chemicals. Many times clogged units are simply replaced with new ones.

The operation of compressors is generally controlled by a pressure sensor that turns on the unit at the low-pressure setpoint and turns it off when the pressure reaches the high-pressure setpoint. Commonly, a large storage tank (receiver) is provided to reduce the cycling of the compressor.

Many blowers have a maximum operating pressure (Fig. 3). For these units, an air-relief valve is provided to protect the blower from excessive pressures (Fig. 4b). Blowers are generally operated continuously because it is not practical to provide significant receiver capacity at lower operating pressures.

BIBLIOGRAPHY

1. R.P. Lapina, *Estimating Centrifugal Compressor Performance, Process Compressor Technology*, Volume 1, Gulf Publishing, Houston, TX, 1982.
2. J.E. Huguenin and J. Colt, *Design and Operating Guide for Aquaculture Seawater Systems*, Development in Aquaculture and Fisheries Science, 20, Elsevier, Amsterdam, 1989.
3. Metcalf and Eddy, Inc., *Wastewater Engineering — Collection, Treatment, and Disposal*, McGraw-Hill, New York, 1972.

See also Aeration systems; Dissolved oxygen.

BRAZIL FISH CULTURE

Newton Castagnolli
Independent Consultant
San Paulo, Brazil

OUTLINE

INTRODUCTION

Rodolpho von Ihering, the "father of Brazilian aquaculture," predicted early in the twentieth century that we should raise fish as chickens. His first attempts to induce breeding of native fish species were in the 1930s (1,2). Yet, fish farming in Brazil only reached commercial status during the last decade of the century. The industry developed soon after induced breeding and fingerling production of tambaquí (*Colossoma macropornum*, Fig. 1) (3) at the Pentecoste Fishery Station (Ceará state) and pacu (*Piaractus mesopotamicus*; Fig. 2) (4) in São Paulo state. Those Characoidei species are well known and appreciated for their good flavor. Popular recreational fishes, tambaquí and pacu were responsible, in part, for the success of a large number of fee fishing establishments that first appeared near the city of São Paulo and then spread to all

Figure 1. Tambaquí (*Colossoma macropornum*).

Figure 2. Pacu (*Piaractus mesopotamicus*).

the southeastern provinces and are now being developed in the central and eastern regions of Brazil.

Ihering's attempts to induce breeding (which he called hypophysation) in native fish species continued in the northeastern region of Brazil, where he conducted experiments and attempted to enhance fish production by releasing the fingerlings obtained through hypophysation in reservoirs built for irrigation and as urban water supplies (5). Reservoir stocking still continues under the auspices of technicians of the Serviço de Piscicultura, a part of the agency responsible for the dam construction in the region, which is subject to severe droughts.

The first fish culture experimental station in Brazil was established in São Paulo state, at Pirassununga, near the Mogi Guaçu river. The station was put into operation in 1938, shortly before the death of Rodolpho von Ihering, its founder. Even with such facilities, the work and heritage left by Ihering was not expanded upon as might have been expected. A lack of governmental vision and, as a consequence, the absence of motivated students and researchers in the aquaculture field meant that there was little progress for many years.

Interest in fish culture rose again the 1970s with the foundation of São Paulo state Fishery Institute (Instituto de Pesca) and later when courses in fish culture began to be offered in Agronomy and Animal Sciences Colleges. In Pernambuco and Ceará states, courses in fishery engineering were established. The Pirassununga station was revitalized with support from the Food and Agriculture Organization of the United Nations. The result was the production of a number of students and researchers who became involved in efforts to develop tropical fish farming technology. As a result, farmers came to believe that fish culture could be a profitable activity. While still in its infancy, aquaculture has become the fastest growing agroindustry in Brazil. Much of the information that follows was obtained from a review by St. Paul (6).

FIRST ATTEMPTS AT AQUACULTURE IN BRAZIL

In the early 1960s, what might be called the amateur stage of Brazilian aquaculture started with the first pond rearing trials with tilapia (*Tilapia rendalli*) and common carp (*Cyprinus carpio*). Those activities can be classified as preliminary observation-level studies due to the lack of scientific method employed. In fact, the first trials consisted of testing the effect of different forages and various levels of penicillin as supplements to rations for the common carp (7,8).

At that time carp and tilapia were extensively reared in small reservoirs. Commercial production did not exist. For a period of about 20 years, fish culture in Brazil existed without any special involvement from the government or private investors, probably due to the lack of technology for rearing the two species and because of their low economic value. The SUDEPE (Superintendencia do Desenvolvimento da Pesca), at that time the Fisheries Agency of the Agriculture Ministry, only focused on capture fisheries and was not interested in the development of fish culture technology.

During the 1970s, experiments continued with carp and tilapia. Studies consisted of examining the effects of pond fertilization (organic, inorganic, or combinations) on plankton and fish production (9). Some research was conducted at the Animal Husbandry Fish Culture Section of the Agronomy and Veterinary Medicine College in Jaboticabal. There researchers were able to replace corn meal in rations for carp and tilapia (10). Additional studies at Jaboticabal were conducted to evaluate the influence of seasons on fish performance. It was found that harvested fish biomass was 60% higher during the summer (January and February) as compared with that obtained during the colder winter months (July and August) (11). An interesting trial related to the management system, and fish production was performed in a polyculture experiment with common carp, Nile tilapia (*Oreochromis niloticus*) and Brazilian catfish (*Rhamdia hilarii*) in ponds that received organic fertilization, pelleted food, or both. It was found that the latter treatment resulted in the production of almost 9 tons/ha/year in comparison with only 2.5 tons/ha/yr produced in ponds that received only chicken manure (12).

THE ROUND FISHES PACU AND TAMBAQUÍ

The so-called round fishes, pacu (from the Paraná river basin) and tambaquí (from the Amazon region), were the first native fish species to be reared commercially in Brazil. Those species received special attention from researchers in aquaculture, biology, and applied physiology. There has been some focus by biochemical geneticists on the taxonomy of those species, which were once both classified as belonging to the genus *Colossoma*. Today only tambaquí remains in the genus, though both are in the family Characidae (subfamily Mileinae). Pacu (Paraná river) and pirapitinga (Amazon river) are now reclassified as belonging to the genus *Piaractus*. These fish, at fingerling stage, are quite similar in many respects to the infamous piranha (*Serrasalmus* spp.), which belongs to the subfamily Serrasalminae (13).

The goal of producing these round fishes is to meet the growing demand by the fee-fishing establishments. In São Paulo state there currently exist more than 2,000 such operations. Two workshops relating to these species have been held in São Paulo state to provide a forum for the exchange of experiences and to transfer technology that could be used for commercial production. The first workshop was held in February, 1985 on the campus of UNESP (Universidade Estadual de São Paulo) in Jaboticabal City. The information available at the time was summarized during the workshop (14). Researchers from Mato Grosso and Mato Grosso do Sul Federal Universities presented studies related to the food habits of pacu and the characteristics of the natural environment of the species. Other papers related to the breeding season (from November to February) and compared age and growth of pacu in nature where females usually mature at 48 months and 2 kg (4.4 lb), whereas in culture where the fish are fed prepared rations in ponds, that size can be reached in 18 to 20 months.

The second workshop was held in CEPTA (Centro de Pesquisas e Treinamento em Aquicultura), formerly CERLA (Regional Latin American Training Center). Management systems for commercial production, larviculture, experimental rearing in earthen ponds, polyculture in pens and cages, and feeding and nutrition of pacu were among the topics covered (15). Papers were also presented on the qualitative and quantitative aspects of plankton in experimental ponds and on chemicals that can be used to control *Dactylogyrus* sp., a monogenetic trematode that is commonly seen in pacu-rearing ponds. Sodium chloride in a 2.5 ppt solution for 20 minutes was found to be the most effective treatment.

CESP, the energy company of the São Paulo state government, presented several papers on studies that were conducted through an agreement (convenium) with the Fishery Institute of the São Paulo state government. The papers involved induced breeding and larviculture in relation to management techniques for raising pacu in ponds. Also presented was a paper that showed a high level of acceptability of small-size fish [around 1.0 kg (2.2 lb)] as compared with those caught in the wild that were usually more than 2.0 kg (4.4 lb).

In the Symposium on *Colossoma* culture, several review papers concerning reproductive physiology and induced breeding (16), genetics (17), nutrition (18), and infectious and parasitic diseases (19–20) in the genus *Colossoma* were presented.

Other meetings in which research results on these species have been presented have been held at the São Paulo state Fishery Institute and the UNESP Campus of Jaboticabal city, as well as by the DNOCS (National Department of Actions Against Draughts) and CODEVASF (The Development Company of São Francisco River Valley) in the northeastern region of Brazil.

THE BRYCONINAE SUBFAMILY

Species in the subfamily Bryconinae are, according to Mendonça (20), widespread in most South American river basins. More than 40 species of these characins exist, which are highly valued by anglers. They can be called tropical salmonids due to their similarity in body shape and fighting behavior to fishes in the family Salmonidae.

The first attempts to induce breeding of the matrinxã (*Brycon* spp.) from the São Francisco river were performed at a Fish Culture Experimental Station near the dam of Três-Marias, in Minas Gerais state. Even before that first step toward Bryconinae domestication, the first trial related to nutrition and management with a member of this group of fishes, *B. cephalus*, had been conducted in Manaus by using wild fingerlings stocked into a recirculating water system (21).

At the UNESP Campus of Jaboticabal, a nutrition experiment was conducted in which isocaloric and isonitrogenous rations with different levels of protein from animal or vegetable sources were evaluated. Fish performance and protein digestibility were found to be independent of protein source (22).

In June, 1994 at the CEPTA meeting in Pirassununga city, São Paulo state, a seminar was held to review existing information on the subfamily Bryconinae (23). A review paper summarized the breeding work conducted at CEPTA's Fishery Experimental Station. Nutrition management systems, pathology, and commercialization of piracanjuba (*B. orbignyanus*) and matrinxã (*B. cephalus*) were described. Interestingly, no papers were presented relating to the breeding, nutrition, and physiology of piraputanga, *B. lundii*, the natural habitat of which is the Paraguai river basin in the central region of Brazil. In that basin there is an annual production of hundreds of thousands of fingerlings of piraputanga, a species that is quite similar to piracanjuba (*B. orbignyanus*), endemic to the Paraná river and a few other streams. Several papers related to the use of hormones to induce breeding of piracanjuba (*B. orbignyanus*) and piabanha (*B. insignis*), a species endemic to the Paraíba do Sul river, were presented. The seminar demonstrated that the level of knowledge about the subfamily was sufficient for moving forward with culturing for the fee-fishing industry.

Figure 3. Pirarucu (*Arapaima gigas*) or Brazilian cod.

In addition to the species mentioned, there are other carnivorous fishes with great potential for aquaculture in Brazil, including the pirarucu (*Arapaima gigas*; Fig. 3) Brazilian cod, black-spotted, and striped catfish. Both produce good-quality boneless fillets and have high dressout percentages. However, there is still a lack of technology for raising these fish, especially with regard to diet development.

LARVICULTURE

Larviculture has played an important role in allowing Brazilian native fish culture to reach even its current modest production level, but there is still much more that can be accomplished in this area. At the end of 1989, a workshop on Larval Rearing of Finfish (22) was held in CEPTA, at Pirassununga city. The work done up to that time in Brazil, Colombia, and Central America was reviewed. Invited speakers presented talks on larval nutrition, health management, and zooplankton production in ponds. Later, several dissertations and theses on the larviculture of pacu, tambaquí, and Bryconinae (Fig. 4) species were given.

Yamanaka (24) has presented a detailed description of the early developmental stages of pacu, *Piaractus mesopotamicus*, from fry to fingerling. It was possible to obtain up to an 80% survival rate of pacu fingerlings reared within cages of fine mesh screen. Sipauba-Tavares (25) produced Chlorophycean algae to feed zooplankters

Figure 4. A school of fish in the subfamily Bryconinae.

(Cladocerans, Daphnidae, and calanoid copepod) as live feed for raising lambari, *Astyanax scabripinnis paranae*, and tambaquí, *Colosssoma macropomum*. High survival rates were obtained.

Zaniboni Filho (26) compared techniques for incubating tambaquí eggs and also identified the factors that affect qualitative and quantitative fingerling production. Funnel-shaped incubators with aeration enhanced fry hatchability. There was high variability in survival rates, but it was concluded that fingerling productivity increased with increasing doses of organic fertilization up to 620 kg/ha (528 lb/acre). Tambaquí fingerlings showed a clear preference for cladocerans. In fertilized ponds, fry only started consuming supplemental feed after 20 days of life.

Senhorini (27) experimented with chicken and bovine manure as organic fertilizers with and without supplemental feeding, compared with a control pond (without feed or fertilization). Survival rates in the fertilized ponds were around 50%, compared with 17.7% in the pond that only received supplemental feed, and only 3.6% in the pond without fertilization or supplemental feeding. Fregadolli (28) studied the effects of the availability of food and the presence of a predator, dragonfly nymphs (Odonata), on the growth and survival of tambaquí fry, in an experiment conducted in a 0.1-ha (0.04 ac) pond of CEPTA in the city of Pirassununga, in which fine-screen cages were set up for rearing fry up to the fingerling stage during a 40-day experiment. Treatments consisted of the introduction of cladocerans or chironomids as live feed and dragonfly nymphs as the predator. Tambaquí fry survival was not affected by the presence of the predator and only the availability of live feed (Cladocerans) had a significant influence on fry survival.

Cecarelli (29) set out to learn which species of fry from among pacu, tambaquí, and curimbata (*Prochilodus* sp.) would be the best to use as live feed for the production of Bryconinae. Results showed that pacu fry enhanced survival rates of Bryconinae fry due to their (pacu's) small size and slower swimming speed.

FISH NUTRITION

This area of research, considering its contribution to aquaculture development in Brazil, and a reasonable number of events sponsored by the Brazilian Committee on Animal Nutrition, deserves special comment. Experiments in fish nutrition started on the Jaboticabal campus of UNESP with a study in which the crude protein requirement for growth of tambaquí, an omnivorous Amazon species that can exceed 30 kg (66 lb), was found to be around 22% (30). On the same campus it has been determined that the crude protein required for growth of pacu is around 26% (31). In a subsequent experiment, it was shown that protein digestibility for pacu was 87% (32).

In the first symposium sponsored in Campinas city in 1986 by the Brazilian Committee of Animal Nutrition, R.T. Lovell from Auburn University, Alabama presented a paper on research in fish nutrition in which all aspects of the topic were covered (33). In 1988, the Third Mini Symposium of the Brazilian Animal Nutrition Committee was held on the Botucatu Campus of UNESP. At that event, Pezzato (34), speaking on technology of food processing for aquatic organisms, explained about the care needed during the grinding and mixing of formulated-feed ingredients and the effects of processing on the nutritive value and water stability of the ration. It is interesting to note that at that meeting nothing was mentioned about extruded rations.

In 1995 and 1997, two more symposia were sponsored by the Brazilian Committee of Animal Nutrition in Campos do Jordão and Piracicaba cities; both events were dedicated to the subject of nutrition. At the first of the two meetings (35), comprehensive overviews on fish and shrimp were presented, followed by presentations on prospects for the world's aquaculture feed supply, fish feed formulation and processing, regulation of growth, ration preparation, feeding strategies for raising carnivorous fish species, and nutrition management of nonsalmonid species.

At the second symposium (36), several aspects of fish production were covered. Topics included fish culture in Brazilian agribusiness, feeding strategies and water and feed quality in fish production, perspectives on industrial production of fish rations in Brazil, intensive and sustainable Brazilian native fish production, recent advances in the industrialization of freshwater fishes, and the raising of tilapia as an example of agroindustrial production.

CONCLUDING COMMENTS

Recently, the Brazilian National Research Council (Conselho Nacional de Pesquisas) organized a special workshop aimed at analyzing Brazilian aquaculture as a first step toward organizing this rapidly growing activity (37). The workshop pointed out that, in 1995, Brazilian aquaculture production was estimated at around 40,000 tons of freshwater fish, marine molluscs, and marine shrimp. Frogs (200 tons) and freshwater shrimp, *Macrobrachium rosenbergii* (250 tons), were also produced.

It was concluded that a more organized approach to technology development was needed (which is also under consideration by EMBRAPA, the national Agriculture Ministry Research Agency). EMBRAPA will be responsible for increasing aquaculture production, but there is a serious lack of any organized extension (technology transfer) program, a fact which is a major constraint to increasing overall aquaculture production in Brazil. Another serious problem in Brazilian aquaculture began to be resolved in 1998 with the transfer of the administration of aquaculture and fisheries from the Ministry of Environment to the Agriculture Ministry, where the Department of Fisheries and Aquaculture was created.

In mid-December of 1998 a law was passed that allows the use of public water bodies (lakes, reservoirs, and also the littoral area of the marine environment) for aquaculture projects. This opportunity and credit stimuli, especially in the central, north, and northeastern regions, should bring Brazilian aquaculture production to over 100,000 tons by the first one or two years of the twenty-first century. Several projects involving intensive production of tilapia and *Colossoma* sp. are under consideration for establishment in lakes, reservoirs, and irrigation canals of the northeastern drought region, and also a few in the southeastern reservoirs built for hydroelectric power generation.

The last Brazilian Symposium on Aquaculture held in Pernambuco state in December, 1998 featured some 1,200 presentations and around 300 posters and demonstrated that aquaculture is becoming the fastest growing agribusiness industry in Brazil (38). Both the Brazilian Aquaculture and Fisheries Engineering Associations have sponsored symposia every two years at which the progress of aquaculture in Brazil was documented, despite the absence of a governmental policy.

Through governmental efforts, including training through the organization of formal postmedium (soon after high school) courses in aquaculture, the real start of the "blue revolution" in Brazil is anticipated. The first plants are beginning to produce tilapia fillets in Paraná and São Paulo states. That activity represents the real onset of the third stage of Aquaculture in Brazil.

BIBLIOGRAPHY

1. R. von Ihering and P. Azevedo, *Arq. Inst Biol. São Paulo* **7**, 107–118 (1936).
2. R. von Ihering, *The Progressive Fish-Culturist* **34**, 15–16 (1937).
3. A.B. da Silva, A. Carneiro, Sobrinho, A. Mello, and L.L. Lovshin, in 1er Simposio de la asociación Latinoamericana de acuacultura, *Induced Spawning of Tambaquí*, Maracay, Venezuela, 1977.
4. N. Castagnolli and E.M. Donaldson, *Aquaculture* **25**, 275–279 (1981).
5. O. Fontenelle and A.J. Não Silva, *B. Tec. DNOCS* Fortaleza **33**(1), (1975).
6. U. St. Paul, *Aquaculture in Latin America*, European Aquaculture Society Special Publication No. 13, 1990, pp. 484.
7. P. de Azevedo, E. Millen, G.J. Alckrnim, and H.L. Stempniewski, *Bolm. Ind.* An. 22 n. s. (único): 69–79 (1964).
8. P. de Azevedo, E. Millen, and H.L. Stempniewski, *Rev. Med. Vet.* **1**(2), 88–99 (1965).
9. S. Sobue, N. Castagnolli, and R.A. Pitelli, *Rev. Bras. Biol.* **37**, 761–769 (1975).
10. N. Castagnolli and P.E. Fellício, *Rev. Ci e Cult.* **27**, 532–537 (1975).
11. N. Castagnolli, A.F. Camargo, G.T. Oliveira, and S. Ostini, *Bolm. Inst. Pesca* **10**, 101–112 (1983).
12. C.L. Justo and N. Castagnolli, *Bolm. Inst. Pesca* **12**, 21–30 (1985).
13. J. Gery, *Revue Fr. Aquariol.* **12**(4), 97–102 (1986).
14. N. Castagnolli and S.M.F. Zuim, Consolidação do conhecimento adquirido sobre o pacu, (*Colossoma mitre*); Berg, 1985, *Bol. Tec. n. 0 05 Fac. Ci. Agr. e Vet.*, Jaboticabal, SP, 1985, p. 30.
15. R.A. Hernandez, ed., in *Sudepe, Colciências, CIID*, Guadalupe Ltda., Bogotá, Andes, 1989, pp. 475.
16. J. Carolsfeld, in R.A. Hermandez, ed., *Sudepe, Colciências, CIID*, Guadalupe Ltda., Bogotá, Andes, 1989, pp. 475.
17. S.A. Toledo and Filho, in R.A. Hermandez, ed., *Sudepe, Colciências, CIID*, Guadalupe Ltda., Bogotá, Colombia, 1989, pp. 475.
18. D.J. Menton, in R.A. Hermandez, ed., *Sudepe, Colciências, CIID*, Guadalupe Ltda., Bogotá, Andes, 1989, pp. 475.
19. D.A. Conroy, in R.A. Hermandez, ed., *Sudepe, Colciências, CIID*, Guadalupe Ltda., Bogotá, Andes, 1989, pp. 475.
20. J.O.J. Mendonça, *Brycon*, Anais, 1994.
21. U. Werder and U. St. Paul, *Crescimento e Produção de Matrinxã, Brycon sp. em Viveiros e Pequenas Represas*, in Simpósio Brasileiro de Aquicultura, II, Jaboticabal, October 1980, SUDEPE, Brasilia, 1981, pp. 71–72.
22. J.E.P. Cyrino, Masters Thesis, *INPAIFUA. Programa em Biologia de Agua Doce e Pesca Interior*, Manaus, AM, 1985.
23. B. Harvey and J. Carosfeld, eds., *Workshop on Larval Rearing of Finfish*, Pirassununga, SP, November, 1989. Cida, Canada, 1990.
24. N. Yamanaka, Ph.D. thesis, São Paulo SP, 1988, pp. 125.
25. M.L. Sipauba-Tavares, Ph.D. thesis, UFSCar, Programa em Ecologia e Recursos Naturais, São Carlos, SP, 1988, pp. 191.
26. E. Zaniboni Filho, Ph.D. thesis, UFSCar Programa em Ecologia e Recursos Naturais, 1992, pp. 188.
27. J.A. Senhorine, Masters thesis, UNESP, Campus de Botucatu, SP, Programa em Zoologia, 1995, pp. 112.
28. C.H. Fregadolli, Ph.D. thesis, UFSCar, Programa em Ecologia e Recursos Naturais, 1996, pp. 152.
29. P.S. Ceccarelli, Ph.D. thesis, UNESP, Campus de Botucatu, SP, Programa em Zoologia, 1997, p. 92.
30. E.M. Macedo-Viegas, N. Castagnolli, and D.J. Carneiro, *Unimar* **18**, 321–333 (1996).
31. D.J. Carneiro, N. Castagnolli, and C.R. Machado, in *Simp. Bras. Aquicultura III*, São Carlos, SP, 1984, pp. 105–124.
32. D.J. Carneiro and N. Castagnolli, in *Simp. Bras, Aquicultura III*, São Carlos, SP, 1984, pp. 125–132.
33. R.T. Lovell, in *Anais do I Simpósio do Colégio Brasileiro de Nutrição Animal*, Campinas, SP, 1986, pp. 31–66.
34. L.E. Pezzato, in *Anais do II Mini-Simpósio do Colégio Brasileiro de Nutrição Animal*, Botucatu, SP, 1987, pp. 9–21.
35. Colégio Brasileiro de Nutrição Animal, in *Simpósio Internacional sobre Nutrição de Peixes e Crustáceos*, Campos do Jordão, SP, 1995, pp. 126.
36. Colégio Brasileiro de Nutrição Animal, *Simpósio sobre Manejo e Nutrição de Peixes*, Piracicaba, SP, 1997, pp. 164.

37. CNPq (National Research Council), *Aquicultura para o Ano 2000*, Proceedings of workshop organized for diagnostic of aquaculture in Brazil. São Carlos, SP, 1996, p. 95.
38. Abraq, Latin American Chapter of WAS Congresso Sulamericano de Aquicultura, X Simpósio Brasileiro de Aquicultura. Recife, Pe. Abstracts: 353 pp.; Proc. (Vol. 1), Conferences, 447 pp. (Vol. 2), Scientific Papers: 804 pp. 1998.

See also FEE FISHING.

BRINE SHRIMP CULTURE

GRANVIL D. TREECE
Texas Sea Grant College Program
Bryan, Texas

OUTLINE

"Brine shrimp" is the common name of the genus *Artemia*. *Artemia* is a genus of brachiopod crustacean belonging to the order Anostracan. The genus and species name *Artemia salina* was shortened to the genus name *Artemia*, since it was proven that *salina* was the only species in the genus. Additional common names given to *Artemia* from worldwide sources include brineworm, Salztierchen, verme de sale, and Fezzanwurm. During the second half of the 19th century, several studies were published dealing with *Artemia*, and soon after, it was found to be an ideal animal for all types of scientific testing.

Today, *Artemia* is probably the single most important live food source for aquaculture hatcheries, especially for freshwater shrimp, marine shrimp, and most marine finfish hatcheries. *Artemia* eggs, or cysts, are mostly collected in the wild and sold to aquaculturists, but they have been produced under controlled laboratory conditions as well. By using particulate or emulsified products, rich in highly unsaturated fatty acids, the nutritional quality of *Artemia* can be further tailored to suit the predators' requirements by bioencapsulating specific amounts of these products in the *Artemia* metanauplii. Application of this method of bioencapsulation, also called *Artemia* enrichment or boosting, has had a major role in improving larviculture outputs in the 1990s. This contribution discusses the different aspects and importance of the brine shrimp, *Artemia*.

HISTORICAL NOTES ON BRINE SHRIMP

Schlosser first described adult *Artemia* (see Fig. 1), or brine shrimp, in 1755 (1). The average adult *Artemia*, usually after 10 days growth, is 8–10 mm (0.3–0.4 in.) in length, but it may take as long as three weeks for it to reach that length, depending upon culture conditions. Linnaeus (2) followed Schlosser in describing *Artemia*. Schlosser's drawings turned out to be more accurate than Linnaeus's description, but with the primitive eyepieces used at the time of Linnaeus's writing, this determination was not made until 1836. Long before being scientifically described, brine shrimp had been associated with better salt production in brine pools, because brine shrimp are filter feeders that remove algae and other organic particles, resulting in cleaner salt crystals (3). Seal (4) and Rollefsen (5) reported the value of freshly hatched *Artemia* nauplii as food for fish fry, and ever since, the exploitation of *Artemia* cysts has gradually increased.

Until the late 1970s, commercial supplies of *Artemia* cysts were available only from the United States and Canada. Salt pools, ponds, lakes, and salterns (also called *salinas*) with *Artemia* populations are found worldwide; however the distribution of these *salinas* is not continuous. Certain bodies of salt water lack brine shrimp, either because of a failure of natural dispersion or because of periodically unfavorable climatic conditions. However, their distribution is rapidly expanding. Since 1977, successful inoculations have been achieved in areas that previously did not have *Artemia* populations (e.g., Brazil, India, the Philippines, Thailand, and other countries). To give one example, nauplii from 250 g (0.55 lb) of San Francisco Bay cysts were introduced into a limited number of evaporation ponds on a brine production farm in the Rio Grande de Norte area, Macau, Brazil. The ecological conditions in the Macau *salinas*, with intake waters coming from a rich mangrove area, turned out to be very favorable for *Artemia* production. The population spread over several thousand hectares, and within one

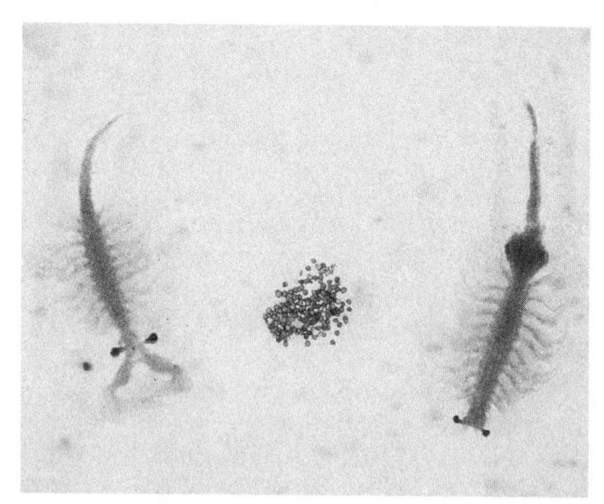

Figure 1. Adult *Artemia* with cysts in the middle, male (left) and female (right).

year, over 15 MT of cysts were harvested. For several years, the yields exceeded 30 MT of cysts per year (6).

Sorgeloos (7) demonstrated the technical feasibility of *Artemia* production in temporary salt ponds in Southeast Asia. Inoculation tests with various geographical strains provided new information on genotypes and phenotypes (8). The application of these inoculation principles helped alleviate a cyst production problem, which occurred in the late 1970s. Although cyst production was the primary goal, exploitation of the adult *Artemia* biomass as a protein source for many aquaculture organisms was also vigorously pursued. By 1979, previous shortages were over, and the price of high quality cysts dropped from US$70/kg (US$32/lb) to US$35/kg (US$16/lb).

Today, there are many geographical strains of *Artemia*. More than 60 strains have been registered from areas including Algeria, Argentina, Australia, Brazil, Bulgaria, Canada, China, France, India, Israel, Iran, Iraq, Italy, Japan, Kenya, Mexico, Peru, Puerto Rico, Spain, Tunisia, the United States, the former USSR, and Venezuela. Numerous commercial harvesters and distributors exist that sell brands of different qualities. The present cost of good-quality cysts can range from US$66/kg (US$30/lb) to US$88/kg (US$40/lb), and the buyer can expect to have 200,000 to 300,000 nauplii hatch from each gram (6–9 million/oz) of cysts.

ENVIRONMENTAL REQUIREMENTS OF *ARTEMIA*

In nature, *Artemia* are found only in natural or man-made brine lakes and salterns. They have been observed swimming among precipitated crystals of sodium chloride in saturated brine (9). In the laboratory, *Artemia* have been shown to be extremely euryhaline, withstanding salinities from 3 ppt to 300 ppt (10). It has been suggested that the reason brine shrimp survive only at high salinities in nature is that competitors and predators are absent at such high salinities (11).

Artemia show good survival in temperatures ranging from 15° to 55 °C (59° to 131 °F) (12). The animal is also quite tolerant of high ammonia levels, remaining active at levels near 4.4 to 5.0 μg atoms NH_4–N/L (80 to 90 ppm) (13,14). Brine shrimp show good growth on a variety of dried, frozen, and live microscopic algae, yeast and bacteria and have even been grown on organic aggregates formed by the bubbling of water.

Larvae and adults can be reared at extremely high densities—e.g., 3,000/L (11,355/gal)—with only a moderate amount of special treatment, such as being fed plenty of algae (15,16). The environmental and nutritional requirements of brine shrimp remain approximately the same throughout their life. *Artemia* can achieve sexual maturity two weeks after hatching and are able to produce 40 larvae per day throughout their six-month to one-year life span. Their food conversion efficiencies are high in comparison with those of other animals, ranging between 20 and 80%.

RESEARCH ON THE USE OF *ARTEMIA* IN AQUACULTURE

Most of the research on brine shrimp for use in aquaculture has been done at the University of Ghent, Belgium, since 1970. In 1970, research was conducted in the Laboratory of Ecology under Director and Dr. J. Hublé and was further expanded in 1972. It was continued in the Laboratory of Mariculture under the direction of Dr. G. Persoone in 1978. Under the direction of Dr. Patrick Sorgeloos, the facility became an independent research center, and, in view of an expansion of research and training activities, the name "Laboratory of Aquaculture & *Artemia* Reference Center (ARC)" was adopted in 1989 (Web site address: http://www.rug.ac.be/aquaculture). The University of Ghent has contributed much to our knowledge of brine shrimp. Research directly related to aquaculture has included the artificial inoculation of ponds with brine shrimp, optimization of the use of brine shrimp cysts in aquaculture facilities, controlled mass production of *Artemia* adults and cysts, comparative studies of the various geographical strains, and many other topics. Extensive literature exists on the hatching of brine shrimp (6,17).

THE TRADE OF *ARTEMIA*

As a marketing scheme, pet shops have sold *Artemia* cysts and referred to the hatching brine shrimp as "sea monkeys." With a stretch of the imagination, an *Artemia* adult could resemble a monkey to some people. Adult *Artemia* are often frozen or freeze dried and sold in the aquarium trade as fish food.

More advanced techniques for harvesting *Artemia* have evolved in recent years. As opposed to collecting the cysts on shore (Fig. 2) and separating them from the sand grains, spotter planes are now used to locate harvestable quantities of cysts, before the cysts wash on shore. Word of the cysts' location from the spotter plane is sent to a harvesting crew, and the crew generally approaches the area to be harvested by boat. The cysts windrow at the surface and can be concentrated and harvested using the

Figure 2. *Artemia* cysts on the bank or bottom of a *salina*.

Figure 3. Crew harvesting *Artemia* cysts by boat on the Great Salt Lake, Utah, USA.

same techniques used to clean up oil spills. Rubber, air-inflated booms or floating lines encircle and concentrate the cysts, which are then skimmed from the surface with vacuum heads (see Fig. 3). The cysts are placed in burlap sacks, sometimes weighing up to 2,000 pounds each. The sacks can be stored in warehouses until the cysts are processed. Processing involves separating the cysts from sand and other debris, drying the cysts, and vacuum sealing them in cans.

THE VALUE OF *ARTEMIA* IN AQUACULTURE; PHYSIOLOGICAL AND NUTRITIONAL MAKEUP

The value of *Artemia* in aquaculture is immeasurable. Due to the unique characteristics of its reproduction, development, physiology, and nutritional value, it is currently indispensable in aquaculture. Brine shrimp have two modes of reproduction: (*1*) ovoviviparous, when nauplii hatch in the ovisac of the mother and are born live; (*2*) oviparous, when embryos at the gastrula stage of development are encased in a hard capsule, or cyst. Reproduction is also parthenogenetic in some strains of *Artemia*, but sexual in most. Certain environmental conditions may trigger the adult female *Artemia* to produce cysts instead of live young. For example, the live animals may be threatened with cold weather, or the salinities of the salina may rise rapidly, thereby sending a message to the female that the water may soon evaporate. The dehydrated cysts can be stored for months or years without loss of hatchability (18). Each cyst is 200 to 300 µm (0.0078 to 0.01 in.) in diameter, depending upon the strain. Its external layer is composed of a hard, dark-brown, lipoproteinaceous chorion (19). Osmotic withdrawal of water, dehydration by air, or anoxia causes the encysted embryo to enter a resting stage with little or no sign of life. The cryptobiotic state of the cyst allows it to withstand complete desiccation, temperatures over 100 °C (212 °F) and near absolute zero, high-energy radiation, and a variety of organic solvents (20). Yet only water and oxygen are required to initiate the normal development of the embryo. This durable, easily hatched diapause stage makes *Artemia* cysts a convenient, constantly accessible source of live animals for the aquaculture hatchery operator.

Within 24 hours after being placed in seawater at 28 °C (82 °F), the chorion, or hard coating of the cyst, breaks (Fig. 4a), and the embryo, still surrounded by a transparent hatching membrane, is released (Fig. 4b). The prenauplius in E-1 stage can be seen in Fig. 4a, and the prenauplius in the E-2 stage can be seen in Fig. 4b. At this point, the embryo can be seen moving within the membrane, but still has not hatched. However, within a few hours, the nauplius breaks free of the hatching membrane and becomes free swimming (21) (Fig. 4c). It is then referred to as a freshly hatched Instar I nauplius. Within hours, the nauplius passes through the instar stages. The fifth instar stage can be seen in Fig. 4d.

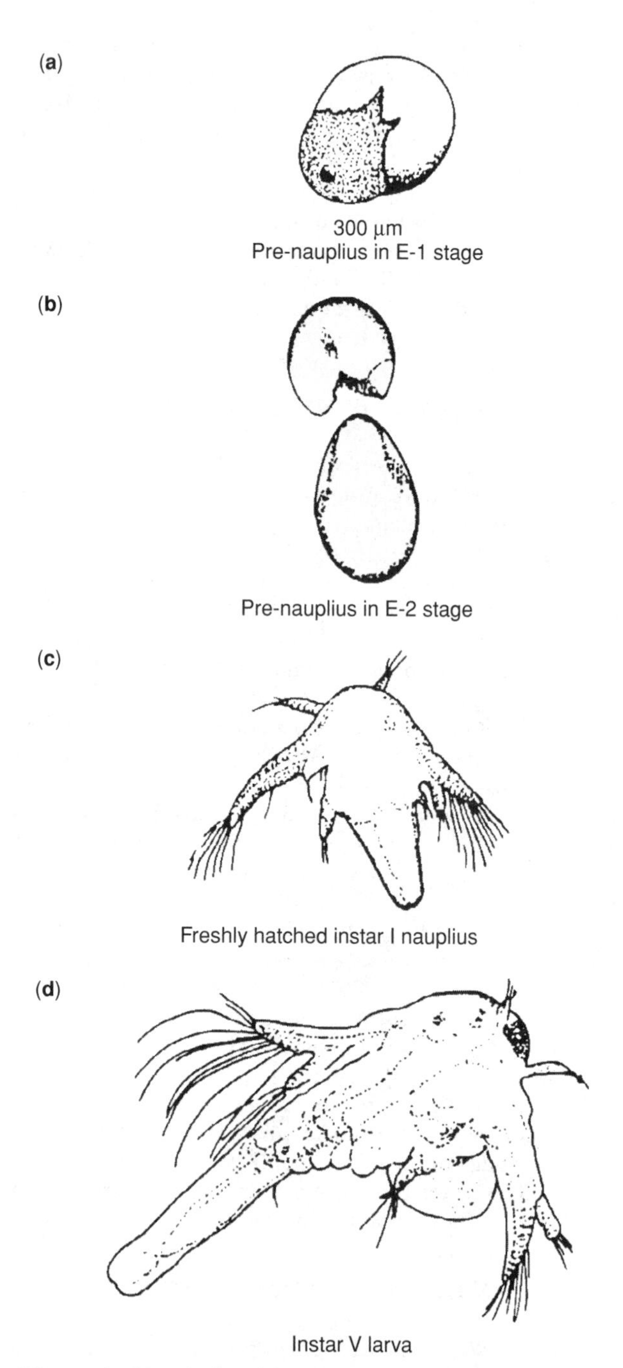

Figure 4. *Artemia*: hatching cyst, embryo, nauplius and larva.

Artemia nauplii can live on yolk and stored reserves for up to five days (22), but the yolk's caloric and protein content constantly diminish during this time (23). This yolk sac is one of the main reasons that *Artemia* nauplii provide such a good food source to other animals.

The lipid level and fatty acid composition of newly hatched *Artemia* nauplii can be highly variable, depending upon the strain (24–29). The presence of highly unsaturated fatty acids (HUFAs) in *Artemia* has been the source of many studies. Most of these studies have revealed one thing: The level of HUFAs in *Artemia* is directly related to the culture performance in larval fish and crustaceans (29–31). This factor confirmed the theory by Watanabe (32), which states that the presence of essential fatty acids is the principal factor for the food value of brine shrimp. Low levels of HUFAs result in low survival and vary from strain to strain, but also vary within a strain from one harvest to another (32–34). It has also been shown, in other studies, that the type of food consumed by the parent *Artemia* greatly influences the fatty acid profiles in the cysts. Manipulation of the food conditions has thus far been limited to small-scale operations such as intensive tank systems (34–36).

The chemical breakdown of one well-known brand of *Artemia* cyst is as follows: On a dry-weight basis, the cysts contain 28.8% crude protein, 10.0% crude fiber, and 10.0% crude fat. The fatty acid profile of the cyst is as follows:

Percent Unsaturated Fatty Acids	
Arachidonic	1.18
Clupanodonic	0.47
Linolenic	26.42
Oleic	27.09
Palmitoleic	6.23

Percent Saturated Fatty Acids	
Arachidic	5.13
Myristic	2.20
Palmitic	10.96
Stearic	7.28

It is important that the *Artemia* nauplii are harvested and fed to fish and crustacean larvae in their most energetic form—i.e., as soon as possible. Holding nauplii in seawater at room temperature (mostly in outdoor conditions) results in a continual decrease in the energy content of the nauplii. The nutritional value is significantly lower for older, starved nauplii, such as those in the Instar V stage (see Fig. 4d), versus freshly hatched Instar I nauplii (see Fig. 4c). The nutritional problems are associated not only with the decrease in nutritional value (a drop in dry weight and caloric content), but also with the increasing size of the nauplii (they become too large for the larvae to consume). Other problems associated with feeding older nauplii to larvae are swimming rates (the nauplii become too fast for the larvae to catch) and color perception (freshly hatched nauplii are dark orange and are much easier to see than the starved nauplii, which are transparent) (6).

Nutritionally, *Artemia* nauplii seem to meet the necessary requirements of most fish and crustacean larvae cultured today, but, as previously mentioned, the nutritional value among different *Artemia* strains is highly variable. To document these variations in strains, an international interdisciplinary study on brine shrimp was initiated in 1978. Several techniques were used to enhance the nutritional value of the "poorer" strains. Even though these techniques are not recommended as part of best management practices for hatcheries, they may help overcome problems with inferior cysts.

The technique for improving the nutritional value of *Artemia* nauplii consists of hatching and separating the nauplii from the debris and then holding the nauplii for up to three days in an enriched medium containing marine algae, encapsulated diets, yeast, and/or oil emulsions (37). This technique applies only if the *Artemia* are being fed to older fish and shrimp.

Without *Artemia*, finfish and shrimp hatcheries would have difficulty finding a dependable, nutritionally complete, economical food source for their larval culture. Saltwater finfish and crustacean hatcheries and freshwater crustacean hatcheries generally give live feeds to the larvae in order to obtain better survival. Some producers of hybrid striped bass and freshwater perch have used live feeds such as rotifers and brine shrimp to successfully raise freshwater fish. But this method is generally considered too expensive and is not practiced as often in freshwater finfish hatcheries as it is in saltwater finfish and crustacean hatcheries and in freshwater crustacean (*Macrobrachium* spp.) hatcheries. *Artemia* are fed to numerous commercially important cultured species in all stages of development (larvae, juveniles, and adults). Even though zooplankton contain their own enzymes that aid fish and crustaceans in digesting them, *Artemia* usually are deficient in several essential nutrients, especially the n-3 HUFAs, required for good growth and development of marine fish and shrimp larvae (38). These findings have led to further research, and methods were developed that enabled the nutritional condition of live feeds to be improved. The proximate analysis and n-3 HUFA levels of some of the common feeds used in larval fish and crustacean production are listed in Table 1.

Finfish producers have become concerned with improving the quality, quantity and cost effectiveness of their live-feed production facilities. Many producers worldwide now supplement *Artemia* cultures with omega yeast, vitamins (E, D, C, and B_{12}), marine oils, and vitamin B_{12}–producing bacteria to improve the quality. Today, the *Artemia* that are fed to fish larvae are being improved, and their biochemistry is routinely adjusted by controlling their diet and supplementing the cultures with microencapsulated feeds or emulsified oils. Such adjustments have resulted in better growth and development of fish larvae, but, in some cases, malpigmentation still occurs in the juvenile stage (40). While *Artemia* is one of most widely used live food items in aquaculture, its use is not without problems and limitations. For example, when the brine shrimp harvest is cut short on the Great Salt Lake in Utah, the price of *Artemia* cysts increases.

CRITICAL PARAMETERS FOR OPTIMUM HATCHING OF *ARTEMIA*

While extensive literature exists on the hatching of brine shrimp (17), the actual procedure of hatching *Artemia* cysts is simple. When working with large numbers and

Table 1. Proximate Analysis And n-3 HUFA Levels of Some of the Common Feeds Used in Larval Fish and Crustacean Production[a]

Species/Feed	Protein[b]	Carbohydrate[b]	Lipid[b]	n-3 HUFA (% total fatty acid)
Algae				
(Selected species)	6.9–49	15.2–24.7	4.3–8.1	3.4–25.8
Rotifers				
(*Brachionus* sp.)	52			3.1
Brine Shrimp				
(*Artemia* sp.)	51–55	14.4–14.8	16–18.9	3.7–15
Commonly Used				
Microencapulated Diet	52	13–14	12	2

[a]From D'Abramo et al. (39).
[b]Units are % dry weight.

high densities of cysts, the following parameters should be considered to assure maximum hatching efficiency:

- Keep the cysts dry. (Place them in a desiccator after the can is open.)
- Hold the temperature constant during hatching at 25° to 30 °C (77 to 86 °F). Below 25 °C (77 °F), the cysts will hatch slowly, and above 33 °C (91 °F), cyst metabolism is greatly affected.
- Maintain the salinity at 5 ppt. [Dilute natural seawater to 5 ppt with dechlorinated tapwater, or use 2 g of technical grade salt per liter (2.2 oz/10 gal) of tapwater.] Determine the salinity with a refractometer. Research has shown that nauplii have a higher energy content when hatched at low salinity (6).
- Oxygen levels should be above 2 mg/L (2 ppm). Constant aeration is necessary during hydration and hatching; it helps disperse the cysts.
- The cyst density should be 5 g of cysts per liter (2.2 oz/10 gal) of water.
- A bright, continuous light above the cysts is necessary during hatching. [Place 2,000 lux, 186-foot candles 20 cm (7.87 in.) from the hatching containers.]
- Cyst disinfecting is highly recommended to assist with improving hatching yields and killing bacteria that are often found on the cysts and could be harmful to the larvae being fed the *Artemia* nauplii.

Note the following precautions and limitations:

- Do not use airstones to aerate cysts, because the airstones will create foam.
- Do not put more than the recommended 5 g (0.16 oz) of cysts per liter (33 fl. oz) of water, (as foaming may also be caused by the addition of too many cysts to the hatching container.)
- If a sufficient oxygen level cannot be maintained without foam formation or mechanical injury of hatching nauplii, add a few drops of a nontoxic, food-grade antifoaming agent (e.g., silicone).
- A buffer may be necessary during hatching to keep the pH above 8.0.

According to Sorgeloos et al. (6), some strains of *Artemia* nauplii (e.g., Chaplin Lake, Canada) are very difficult to separate from debris. This problem may be overcome by decapsulation. The technique is described by Sorgeloos et al. (6) and later herein. The use of decapsulated cysts eliminates naupliar separation problems, reduces surface bacterial populations and makes it possible for fish larvae to ingest and digest the *Artemia* before they are hatched. One disadvantage, however, is that decapsulated cysts are not buoyant and will settle out if extra circulation or aeration is not provided. Most hatchery managers prefer to aerate gently and not circulate or exchange water at all during the first week, since the larvae are quite fragile. The decapsulation of cysts improves the hatchability of *Artemia* (41).

RECOMMENDED PROCEDURES FOR HATCHING *ARTEMIA*

- Obtain good-quality *Artemia* cysts from a reputable company that offers a hatch guarantee.
- Set up hatching containers as follows:

(a) Determine the amount of *Artemia* required, according to the size and demand of the hatchery.

(b) Size the hatching containers accordingly. Sizing is site specific, so only the important parameters of the hatching procedure are presented here, and determining the number and sizes of containers is left to each individual hatchery. Any funnel-shaped, transparent container will do for hatching.

(c) A typical *Artemia* hatching stand for a small-scale laboratory consists of four or five Imhoff Cones (sediment settling cones), either clear plastic or glass. (See Fig. 5.) The cones can be purchased open ended, with a plastic valve inserted at the small end for draining, or the cones can be purchased without the opening, and a siphon can be used to remove hatched *Artemia* after they settle to the bottom. Aeration to the cones is provided by connecting a glass pipette to an air line and allowing the weight of the glass pipette to hold the air supply in place, at the bottom of

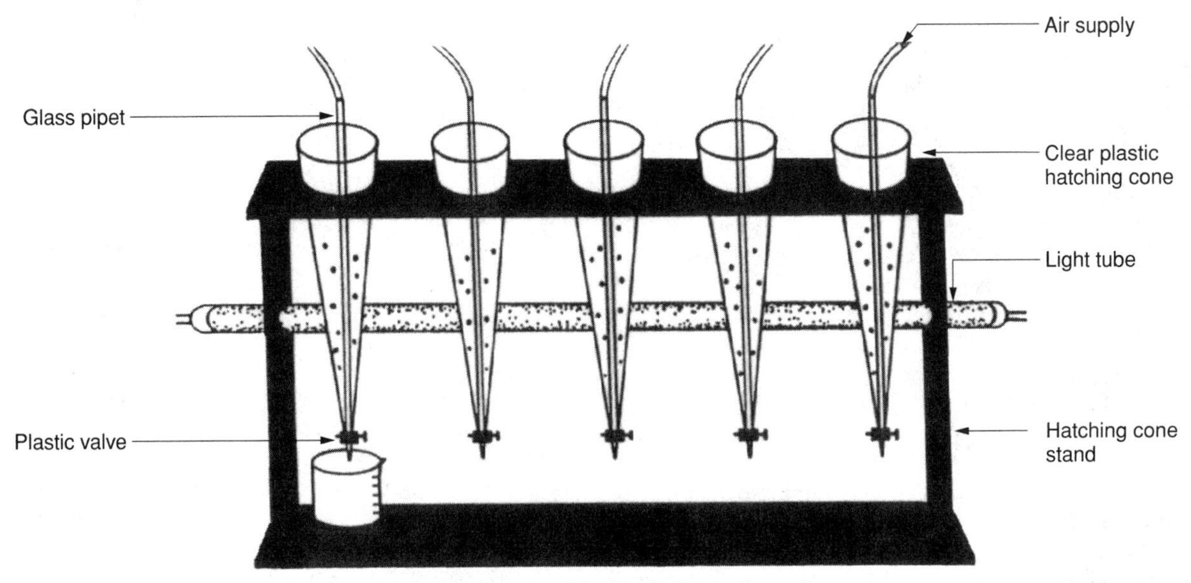

Figure 5. *Artemia* hatching cones and stand.

the cone. A fluorescent light is placed near the hatching containers.

Conical-shaped, clear plastic bags also work well as hatching containers, but seawater has a tendency to corrode the clips and metal ring stands used to support these bags. A wooden Imhoff cone support stand is often used for small-scale hatching of cysts. These stands can be homemade very easily. For additional ideas on the use of larger hatching containers, see Sorgeloos et al. (6). Some hatcheries utilize clear plastic, 19-L (5 gal) drinking water bottles with the bottoms removed. The bottles are inverted, and a rubber stopper is firmly placed in the mouth. A hole is drilled in the stopper to allow the placement of a tight-fitting air line, which is later used as a drain tube when harvesting nauplii. Crow (42) described the hatching of *Artemia* nauplii in 57-L (15 gal) cylindrical, conical-bottom, polyethylene tanks. That method used 100 g (3.5 oz) of cysts in each tank, which were hatched and fed to *Macrobrachium*. Large-scale efforts may involve 1,000-L (264 gal) or larger fiberglass conical tanks.

To disinfect cysts, soak the cysts for one hour in 20-ppm hypochlorite (bleach) in tapwater, or in a 200-ppm mixture for 20 minutes, if you are in a hurry. For example, adding 4 mL (0.03 fl. oz) of household bleach solution to 10 L (2.64 gal) of tapwater is effective and provides enough solution to disinfect one kg (2.2 lb) of cysts. Be sure to aerate the disinfecting solution so that each cyst is exposed to the disinfectant.

- Pour the cysts on to a 120-μm (0.004 in.) sieve, and wash them with tapwater.
- Then the cysts are ready to be placed into the hatching container.
- Start aeration, and leave the cysts in hatching medium (5 ppt seawater) for 24 hours before taking the first harvest. Remember to keep the temperature from 25 to 30 °C (77 to 86 °F) and turn the light on, placing a 60-watt fluorescent bulb within 20 cm (9 in.) of small hatching containers. Use two bulbs for a 19-L (5 gal) container, use four bulbs for a 76-L (20 gal) container, and so forth.

RECOMMENDED PROCEDURES FOR HARVESTING AND COUNTING *ARTEMIA*

After 24 hours, remove the aeration, and let the cysts settle for 5 to 10 minutes. A distinct separation should occur. Empty cyst shells will float to the surface, and the nauplii will tend to concentrate near the bottom. There may also be some debris and unhatched cysts at the very bottom. Open the valve on the hatching tank to drain the debris. Close the valve when the newly hatched nauplii begin coming out. A graduated beaker can be used to collect the now-concentrated nauplii for a small-scale hatching or a 19-L (5 gal) bucket can be used on a larger scale. This procedure should be repeated later to ensure that all of the newly hatched nauplii are harvested. Floatation of the cyst shells can also be improved by raising the salinity. The sudden salinity change will not harm Instar I nauplii.

Cysts from some strains do not all hatch at the same time, and sometimes a number of cysts are still unhatched after 24 hours. If this is the case, place more seawater into the hatching container, aerate, and try the harvesting routine again at 36 hours and finally at 48 hours. Nauplii can also be concentrated with light, since they are negatively phototactic at this stage. One may want to take advantage of this factor during harvesting, or one may prefer just to turn off the overhead light and let gravity do the concentrating.

Some strains of *Artemia* present more difficulty than others in the separation of nauplii from their old egg cases. With these particular cysts, or whenever contamination with empty shells and debris becomes a problem, use decapsulated cysts. This is an added step that most hatchery managers try to avoid but the procedure is well documented in the literature (6).

The decapsulation procedure involves the following steps: (*1*) hydration of the cysts, (*2*) treatment in a decapsulation solution, (*3*) washing and deactivation of chlorine used in the decapsulation, and (*4*) feeding the eggs directly to larvae, or waiting until the nauplii hatch and then feeding them to larvae.

In order to prevent contamination of the larval culture tank with glycerol (which is produced by *Artemia*), hatching metabolites, and excessive bacteria, harvested nauplii are placed on a 125-μm (0.004 in.) sieve and washed with tap water prior to being fed to larvae.

Ideally, *Artemia* should be fed to larvae immediately after hatching but if this is not possible, then freshly hatched nauplii should be stored in the refrigerator at 0 to 4 °C (32 to 39 °F) in aerated containers. According to Sorgeloos et al. (6), nauplii can be maintained in this state at densities up to 15,000/mL (0.03 fl. oz) for up to 48 hours, with nauplii viability remaining at more than 90%.

After nauplii have been washed and concentrated into a container of known volume, they can be counted by mixing the thick nauplii solution and subsampling it with a graduated pipette. A count of nauplii can be made by holding the pipette horizontally, counting the number of nauplii seen swimming between two of the graduated hash marks [with a known volume, of 0.1 mL (0.0034 fl. oz), for example], and extrapolating to give the total number of nauplii harvested. A small-volume, automatic pipette can also be used to trap a known volume of concentrated nauplii. A method of determining the amount of *Artemia* nauplii to feed finfish and crustacean larvae is illustrated in Table 2.

For most larvae, a food organism has to meet certain physical and nutritional requirements. Physically, *Artemia* are relatively free of extraneous material and disease-producing bacteria (after the separation and disinfecting techniques described previously). The acceptability of *Artemia* by fish and crustacean larvae is facilitated by their good perceptibility, catchability, and palatability (if fed very soon after they have hatched). However, in some cases, *Artemia* (even freshly hatched nauplii) can be difficult to ingest, due to their size. The size of a food organism indeed determines whether a larval fish or crustacean can successfully catch and ingest it. For this reason, it is extremely important to continue feeding larvae smaller food organisms such as algae and/or rotifers, while introducing to them the larger food organisms (i.e., *Artemia* nauplii) as well. The rotifer is generally between 99 to 281 μm (0.003 to 0.011 in.) in length, whereas the brine shrimp nauplius is 428 to 517 μm (0.0168 to 0.02 in.) in length, depending upon the strain. Figure 6 depicts the size of a freshly hatched *Artemia* nauplius relative to a 12- to 13-day posthatch larval fish. The larval sketch was redrawn from Johnson (43).

Table 2. Method of Determining the Amount of *Artemia* Nauplii to Feed Finfish and Crustacean Larvae

One way to determine the amount of *Artemia* to be added to a rearing tank is by following six easy steps:

1. Number of *Artemia*/mL (0.03 fl. oz) required = A.
2. Present density in larval rearing tank [number of *Artemia*/mL (0.03 fl. oz)] = B.
3. $A - B = C$.
4. $C \times$ [Volume of larval rearing tank in mL (0.03 fl. oz)] = D.
5. Number of *Artemia*/liter (33 fl. oz) in feed container = E.
6. $\frac{D}{E} \times 1{,}000$ = no. of mL (no. of 0.03 fl. oz) of *Artemia* to add.

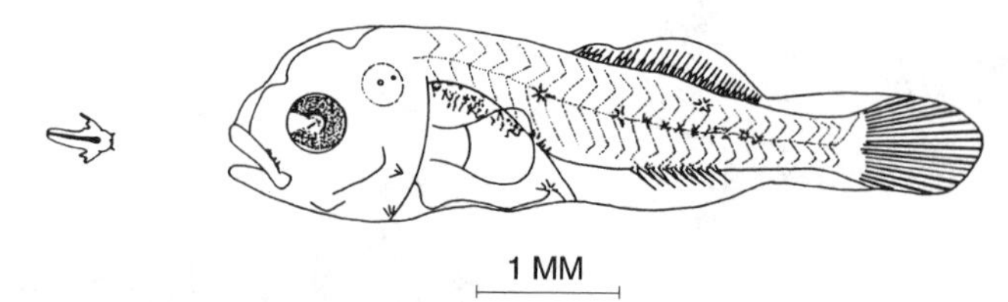

Figure 6. Schematic presentation of respective sizes of 12 to 13 day old predator larval red drum and freshly hatched *Artemia* nauplius.

Considerable differences in the size of *Artemia* have been found from strain to strain. Feeding an oversized *Artemia* strain may, therefore, explain poor growth and even mortality due to starvation of the predator larvae. Before selecting a strain of brine shrimp for use in a hatchery, it would be wise to ask other hatchery managers for their suggestions. Also read (31), which describes results of international studies on *Artemia*; the sections dealing with the culture success obtained with various marine animals fed *Artemia* nauplii from different geographical origins provide results on survival and growth.

WHEN *ARTEMIA* ARE FED: THE IMPORTANCE OF TIMING

Generally speaking, *Artemia* are fed to larval fish and some crustaceans until they are approximately 15 to 20 days old. However, some species, such as *Macrobrachium*, are fed *Artemia* during the entire hatchery phase, which could be as long as 45 days. Other finfish, such as the striped mullet and gray mullet (*Mugil* spp.), are often fed *Artemia* until they are 60 days old, but eventually hatchery managers all try to wean their species to other, less expensive foods.

Red drum (*Sciaenops ocellatus*) larvae are generally fed rotifers from day 3 posthatch to day 9 or 10, depending on the temperature and the size of the red drum larvae. It is important not to feed them the *Artemia* nauplii too soon; a gradual transition should be made to larger foods (i.e., from rotifers to *Artemia* nauplii). *Artemia* nauplii are then fed to the larvae from day 11 to day 15, and the *Artemia* nauplii are maintained in the culture tank at densities between 0.5 to 2.0 per mL (0.03 fl. oz). A shrimp puree and dry fry food diet is fed from day 15 to day 21. Weaning larvae from one feed to the next is commonly practiced in the industry for better larval survivals. Striped mullet larvae, to give another example, are fed rotifers at 5 to 20/mL (0.03 fl. oz), starting on day 2 posthatch and continuing until day 40, while *Artemia* nauplii are started on day 12 to 15 and continued through day 60. Artificial food is started as early as day 20. Grey mullet are routinely raised in fresh and brackishwater ponds in China, Hawaii, Hong Kong, India, and Israel and are fed *Artemia* in the hatchery phase. However, heavy mortality can occur, because the mullet ingest empty cysts, or shells, until they are four weeks old. Therefore, care must be taken to remove the empty cysts before feeding. In Hawaii, the culture of the threadfin (*Polydactylus sexfilis*) involves feeding them *Artemia* starting at day 10. Juveniles are weaned to chopped squid by day 44, after a short period of being fed frozen adult *Artemia*.

Saltwater shrimp are generally fed *Artemia* nauplii starting at the late zoea stage to the mysis stage and continuing until the shrimp are stocked in ponds. Saltwater shrimp do not have the capability to seek out *Artemia*, nor do they develop the mouth parts necessary to eat *Artemia* before the late zoea stage; therefore, feeding them *Artemia* too soon would be a waste. The Kuruma shrimp (*Marsupenaeus japonicus)* can consume 50 *Artemia* nauplii per day when it is in the mysis stage and 80–100 per day in the postlarval stage. Unlike saltwater shrimp, freshwater shrimp do have the capability to catch and eat larger prey earlier; therefore, they are started on *Artemia* right away, at concentrations between 5–10/mL (0.03 fl. oz) of water, and are generally fed nauplii twice daily for 20 days or throughout the entire hatchery phase. The King Crab (*Paralithodes camtschatica*) is fed *Artemia* nauplii starting with its zoea stage. However, other creatures eat *Artemia* nauplii as well and can cause problems in the hatchery. For example, occasional blooms of inadvertently introduced jellyfish, *Moerisia lyonsi*, have caused high mortalities among *Macrobrachium* spp. larvae, because the hydrozoan feeds on *Artemia* nauplii and establishes high population densities in a matter of weeks. In the process, the medusae of the jellyfish also ingest the shrimp larvae.

Juvenile lobsters have been fed frozen *Artemia* with good results, but the greatest survival and growth are obtained when lobsters are fed live adult *Artemia*. In 1979, the author of this contribution was in charge of a shrimp feeding trial conducted in St. Croix, USVI, wherein shrimp *Litopenaeus stylirostris*, were fed 15 different diets. The best results came from feeding the shrimp live adult *Artemia*. Unfortunately, this is not practical or economical in most locations.

For most finfish species being reared, weaning from live *Artemia* to dry food should begin a few days before transformation and should be finished by the time the fish are juveniles. This process might be done in three days or take as long as two weeks. Young fish have small stomachs and must be fed often, but once they are 100 mm (4 in.) in length, fish need only one feeding per day.

SOURCES AND COST OF *ARTEMIA*

Approximately 90% of the world's commercial harvest of brine shrimp eggs come from the Great Salt Lake, Utah, USA. With wholesale egg prices at US$22/kg (US$10/lb), the Salt Lake industry sells approximately US$30 million worth of the product per year. In an average year, the lake produces 4.5 million kg (10 million lb) of raw, wet eggs (cysts), which processes into 1.1 to 1.3 million kg (2.5 to 3 million lb) of Grade A product, with a hatch rate of greater than 90%. Growth in the brine shrimp harvest industry in Utah has been steady since 1950. Brine shrimp production from the Great Salt Lake reached a peak in 1995–1997, when the lake produced 6.75 million kg (15 million lb). In 1997, at least 32 firms had at least 80 crews on the lake harvesting cysts. Excessive rainfall in 1998 and 1999 caused a collapse in this industry and prices to rise. The cost of good-quality cysts fluctuates with supply and demand, and the buyer can expect to pay US$26.40–$88/kg (US$12–40/lb). The most common packaging form of *Artemia* cysts on the world market is the 0.45-kg (1 lb) can (see Fig. 7), similar to the size of a coffee can. Additionally, the cysts come packed in 2.2-, 4.5-, 6.7 and 11-kg (5, 10, 15 and 25 lb) pails and 103- and 126-kg (230 and 280 lb) barrels. Usually, orders of 454 kg (1,000 lb) or more qualify the buyer for substantial discounts from the wholesale suppliers. The buyer should expect 200,000–300,000 nauplii to hatch from each gram (0.03 oz) of cysts.

Once *Artemia* cysts are hatched, they can easily be separated from the shells and other debris and fed to larvae. (See Fig. 8.) The ease of feeding *Artemia* to larvae and the superior nutritional value of *Artemia* ensure that brine shrimp will be used in hatcheries for many years to come.

Figure 7. *Artemia* cysts in a one pound can.

Figure 8. Separating freshly hatched *Artemia* nauplii from eggs, egg shells and debris.

BIBLIOGRAPHY

1. D.J. Kuenen and L.G.M. Baas-Becking, *Zool. Med.* **20**, 222–230 (1938).
2. C. Linnaeus, *Systema Naturae.*, Xth ed., Hafniae, 1758.
3. C.S. Clark and S.T. Bowen, *J. Hered.* **67**(6), 385–388 (1976).
4. A. Seal, *Trans. Am. Fish. Soc.* **63**, 129–130 (1933).
5. G. Rollefsen, *Artificial Rearing of Fry of Seawater Fish*, Preliminary Communication. Rapp. P.-V. Reun. Cons. Perm. Int. Explor. Mer., 1939, pp. 109–133.
6. P. Sorgeloos, P. Lavens, P. Leger, W. Tackaert, and D. Versichele, *Manual for the Culture and Use of Brine Shrimp* Artemia *in Aquaculture*, *Artemia* Reference Center, Ghent, Belgium, 1986.
7. P. Sorgeloos, *The Culture and Use of Brine Shrimp* Artemia salina *as Food for Hatchery Raised Larval Prawns, Shrimp and Fish in South East Asia*, FAO Report THA/75/008/78/WP/3, 1978.
8. P. Vanhaecke and P. Sorgeloos, *International Study on* Artemia *XIII: The Biometrics of* Artemia *Strains of Different Geographical Origin*, International Symposium on the Brine Shrimp *Artemia*, Universa Press, Wetteren, Belgium, 1979.
9. R.P. Conte, *J. Comp. Physiol.* **80**, 239–246 (1972).
10. I.A.E. Bayly, *Ann. Rev. Ecol. Syst.* **3**, 233–268 (1972).
11. R.W. Morris, *Bull. Inst. Oceanogr.* (Monaco), No. 1082 (1956).
12. M. McShan, *J. Water Pollut. Control Fed.* **46**, 1742–1750 (1974).
13. E. Bossuyt and P. Sorgeloos, *Culture of* Artemia *Larvae in Air–Water–Life Raceways with Ricebran Suspension*, unpublished manuscript, 1978.
14. E. Bossuyt and P. Sorgeloos, in *The Brine Shrimp* Artemia, Vol. 3, Universa Press, Wetteren, Belgium, 1980, pp. 133–152.
15. P. Sorgeloos, *Proceedings of FAO Technical Conf. On Aquaculture*, Kyoto, Japan, June 1976, 1976, pp. 321–324.
16. D.E. Soloman, *Growth Efficiency of the Brine Shrimp* Artemia *Fed a Unialgal Diet*, M.S. Thesis, The University of Texas at Austin, Austin, TX, 1980.
17. P. Sorgeloos, in *The Brine Shrimp* Artemia, Vol. 3, Universa Press, Wetteren, Belgium, 1980, pp. 25–46.
18. J. Dutrieu, *Arch. Zool. Exp. Gen* **99**, 1–34 (1960).
19. E. Anderson, *J. Ultrastruct. Res.* **32**, 497–525 (1970).
20. J.S. Clegg, *Trans. Amer. Microsc. Soc.* **93**, 481–490 (1974).
21. P. Sorgeloos, in *International Symposium on the Brine Shrimp* Artemia, Universa Press, Wetteren, Belgium, 1979, pp. 1–6.
22. C.S. Olson, *Crustacean* **36**, 302–308 (1979).
23. F. Benijts, *Proc. 10th Eur. Symp. Mar. Biol.* **1**, 1–9, (1976).
24. P.S. Schauer, in *The Brine Shrimp* Artemia, Vol. 3, Universa Press, Wetteren, Belgium, 1980, pp. 365–373.
25. C.E. Olney, in *The Brine Shrimp* Artemia, Vol. 3, Universa Press, Wetteren, Belgium, 1986.
26. C.R. Seidel, *Bull. Jpn. Soc. Scient. Fish.* **46**(2), 237–245 (1980).
27. C.R. Seidel, *Mar. Ecol. Prog. Ser.* **8**(3), 309–312 (1982).
28. T. Soejima, in *The Brine Shrimp* Artemia, Vol. 2, Universa Press, Wetteren, Belgium, 1980, pp. 613–622.
29. Ph. Leger, *J. Exp. Mar. Biol. Ecol.* **93**, 71–82 (1985).
30. Ph. Leger, *Oceanogr. Mar. Biol. Ann. Rev.* **24**, 521–623 (1986).
31. Ph. Leger, in Artemia *Research and its Applications*, Vol. 3, Universa Press, Wetteren, Belgium, 1986.
32. T. Watanabe, *Bull. Jpn. Soc. Scient. Fish.* **44**(10), 1115–1121 (1978).
33. T. Watanabe, *Bull. Jpn. Soc. Scient. Fish.* **46**, 35–41 (1980).
34. J. Vos, *Hydrobiologia* **108**(1), 17–23 (1984).
35. P. Lavens, in Artemia *Research and its Applications*, Vol. 3, Universa Press Wetteren, Belgium, 1986.
36. M.E. Yates and G. Chamberlain, *Effect of Long Chain Highly Unsaturated Fatty Acids on Penaeid Larval Nutrition*, presented at the 18th Meeting of the World Aquaculture Society, 1987.
37. Ph. Leger, International Study on *Artemia* XXXV, in Artemia *Research and Its Applications*, Vol. 3, Universa Press, Wetteren, Belgium, 1986.
38. T. Watanabe, C. Kitajima, and S. Fujita, *Aquaculture* **34**, 115–143 (1983).
39. L.R. D'Abramo, D.E. Conklin, and D.M. Akiyama, eds., *Crustacean Nutrition*, World Aquaculture Society, Baton Rouge, LA, 1997.
40. T. Naess, M. Germain-Henry, and K.E. Naas, *Aquaculture* **130**, 235–250 (1995).
41. R.A. Browne, P. Sorgeloos, and C.N.A. Trotman, Artemia *Biology*, CRC Press, Boca Raton, FL, 1991.
42. C.W. Crow, *Effects of Brine Shrimp (*Artemia salina*) Density on Development and Survival of Two Densities of Larval Malaysian Prawns (*Macrobrachium rosenbergii*)*, M.S. Thesis, Louisiana State University, Baton Rouge, LA, 1987.
43. A.G. Johnson, in *Final Report, Federal Aid in Fisheries Restoration Act*, Federal Aid Project F-31-R, Statewide Fisheries Research Objective 21, Marine Fish Propagation Study, 1977.

See also GILTHEAD SEA BREAM CULTURE; HALIBUT CULTURE; LARVAL FEEDING—FISH; PLAICE CULTURE; SHRIMP CULTURE; SOLE CULTURE; SUMMER FLOUNDER CULTURE; WINTER FLOUNDER CULTURE.

BUFFER SYSTEMS

GARY A. WEDEMEYER
Western Fisheries Research Center
Seattle, Washington

OUTLINE

Chemically, a buffer is a mixture of a weak acid and its conjugate base (salt), the function of which is to prevent the changes in pH that would otherwise occur when external acids or bases are added to a solution. In aquaculture, the buffering system of interest is provided by dissolved carbon dioxide and the bicarbonate/carbonate mineral salts naturally present in water. In freshwater with low buffering capacity, (alkalinity <20 mg/L), the pH can fluctuate widely, due to the CO_2 addition and removal caused by the natural daily cycles of respiration of fish and respiration and photosynthesis of aquatic plants, algae, and phytoplankton. In pond aquaculture, for example, nighttime production of CO_2 can be quite high, because both aquatic plants and animals are respiring. During daylight hours, intense sunlight causes rapid algal growth that may consume dissolved carbon dioxide faster than it can be replaced by fish respiration and diffusion from the atmosphere. The pH of a pond can increase from a nighttime low of 5.0 up to 9.5–10 in a matter of hours. In contrast, the pH may change by only one or two units in ponds with good buffering capacity (alkalinity >100 mg/L). Similarly, CO_2 production in fish transport tanks can be quite high, and water with good buffering capacity can help prevent the pH fluctuations that would otherwise occur if CO_2 stripping by the aeration system is inadequate. Finally, buffering capacity is important in ponds and recirculating aquaculture systems because the nitrate and H^+ (nitric acid) produced as nitrifying bacteria oxidize ammonia would otherwise cause a progressive decline in the pH of the water.

The carbonate buffer system also exists in seawater, where it is even more effective, because of the higher concentrations involved. Surface ocean water is strongly buffered near pH 8.2, and pH fluctuations are not normally an issue in mariculture operations.

CHEMISTRY OF BUFFER SYSTEMS

The buffering capacity of water in aquaculture systems is provided by the natural alkalinity of the water, the major components of which are carbon dioxide (CO_2), carbonic acid (H_2CO_3), and bicarbonate (HCO_3^-) and carbonate (CO_3^{-2}) ions. The equilibrium reaction is:

$$CO_2(gas) \rightleftharpoons CO_2(aq) + H_2O \rightleftharpoons H_2CO_3 \rightleftharpoons H^+ + HCO_3^- \rightleftharpoons 2H^+ + CO_3^{-2}.$$

In soft-water areas, CO_2 dissolved from the atmosphere provides these components by reacting with water to form the weak acid H_2CO_3, which, in turn, dissociates to form bicarbonate and carbonate ions. Water high in carbonate hardness and alkalinity (from limestone in soils and bedrock, or the use of concrete ponds) provides additional bicarbonate and carbonate, which supplement the natural buffering capacity provided by dissolved atmospheric carbon dioxide. Over the pH range important to aquaculture (6.5–9), bicarbonate is the major species present in the water. Little carbonate ion is present unless the pH is greater than 10. Carbon dioxide and H_2CO_3 predominate at pH values less than 5.

Because the carbonate system can react with both acids and bases, it provides a relatively strong buffer against increases or decreases in pH. If alkaline (OH^-) substances are added, the equilibrium shifts to the right, OH^- ions are consumed, and the pH does not increase. The increases in pH that would otherwise occur, due to the consumption of dissolved CO_2 by algae and phytoplankton, are prevented in a similar manner.

A major source of acid in aquaculture systems is the metabolic CO_2 produced by fish and plant respiration. The decrease in pH that would occur due to CO_2 production is prevented, because the equilibrium of the carbonate buffer system shifts to the left. The added acid (H^+) is tied up in H_2O, H_2CO_3, and HCO_3^-, and the pH remains constant. The oxidation of ammonia by nitrifying bacteria in biofilters or pond sediments is another significant source of acid:

$$NH_4^+ + 2O_2 \longrightarrow NO_3^- + H_2O + 2H^+.$$

Again, the bicarbonate component of the buffer system reacts with the H^+, the added acid is removed from solution, and the pH remains relatively stable.

The bicarbonate buffer is classified as an open system, because one of its components (CO_2) can enter or leave the water relatively freely. Consequently, shifting the equilibrium to the left increases the H_2CO_3 concentration, and CO_2 is released and then stripped into the atmosphere by the aeration system. Although a slow process, carbonate and bicarbonate are steadily lost to the system and must be replaced, or else the buffer capacity will be reduced. The water exchange rate may be sufficient to replenish the alkalinity, or the CO_2 produced by fish respiration may be adequate. If not, sodium bicarbonate may be added (1).

The pH of a buffered system can be calculated from the Henderson–Hasselbach equation:

$$pH = pK + \log [base]/[acid].$$

The pK, the dissociation constant of the weak acid or base in question, is temperature and salinity dependent, and its values are available from standard handbooks. Substituting appropriate numbers for the carbonate and bicarbonate buffer in freshwater yields

$$pH = 6.37 + \log[HCO_3^-]/[H_2CO_3]; \quad (1)$$

$$pH = 10.25 + \log[CO_3^{-2}]/[HCO_3^-]. \quad (2)$$

Inspecting equations 1 and 2 reveals that the pH is determined by the ratios of the buffer components, rather than by their concentrations. Conversely, the ratio of $[H_2CO_3]$ to $[HCO_3^-]$, and $[HCO_3^-]$ to $[CO_2]$, is fixed at any given pH. When the concentrations of the two components are equal, the pH will equal the pK value. This is also the pH region of maximum buffering capacity (i.e., resistance to pH change). In freshwater of moderate alkalinity, the resulting pH is usually stabilized in the mid-7 range (i.e., 7.2–7.8). Because of its high carbonate concentration, seawater is strongly buffered at about pH 8.2.

As previously mentioned, the buffering capacity of freshwater over the pH range important to aquaculture is largely due to bicarbonate. As a first approximation, the $[CO_3^{-2}]$ can be neglected unless the pH is greater than 9, and the $[H_2CO_3]$ can be taken as the dissolved CO_2 concentration (P_{CO_2}), giving us

$$pH = pK + \log[HCO_3^-]/P_{CO_2}.$$

Table 1.

pH	Bicarbonate Alkalinity (mg/L as $CaCO_3$)				
	30	50	70	90	110
6.5	18.9	31.6	44.3	56.9	69.6
7.0	6.0	10.0	14.0	18.0	22.0
7.5	1.9	3.2	4.4	5.8	6.9
8.0	0.6	1.0	1.4	1.8	2.2

Thus, the dissolved CO_2 concentration is fixed by the buffer and can be calculated from the bicarbonate alkalinity and the pH (2). Alternatively, the alkalinity can be used to determine the concentration of CO_2 needed to adjust the pH of the water to a particular value. Automatic CO_2 injection equipment is commercially available for this purpose and is widely used to stabilize the pH in heavily planted aquaria, because the added CO_2 stimulates plant growth. Approximate values of free CO_2 (mg/L) for a range of alkalinity and pH values are shown in Table 1 (at 25 °C; no correction for total dissolved solids).

Note that allowing the pH to drop below about 7.0 in waters with even moderate degrees of alkalinity will generate free-CO_2 levels greater than the 20 mg/L generally regarded as safe (3), and adequate air stripping will be required. Detailed information for the complete range of temperature and pH important in both cold- and warm-water aquaculture can be found in Boyd (4).

ASSESSING BUFFERING CAPACITY

Most of the buffering capacity of freshwater and saltwater comes from the carbon dioxide, carbonates, and bicarbonates naturally present. Thus, either alkalinity or carbonate hardness (KH) determinations can be used to assess buffering capacity. Portable water-quality test kits give adequate results for routine assessments, but the results may be expressed in unfamiliar units. Milligrams of $CaCO_3$ per liter (mg/L) is most commonly used in aquaculture; degrees of hardness (dKH) is widely used in the aquarium industry; and milliequivalents per liter (meq/L) is used exclusively in the modern scientific literature. The conversion factors are

$$1 \text{ meq/L} = 2.8 \text{ dKH} = 50 \text{ mg/L } CaCO_3.$$

In practice, the terms alkalinity, KH, and buffering capacity are often used interchangeably, although they are technically distinct.

MANAGEMENT RECOMMENDATIONS

The buffering capacity provided by a minimum alkalinity of about 40 mg/L is widely considered necessary to provide the concentrations of bicarbonate and carbonate needed to stabilize water against changes in pH caused by (a) the addition and removal of carbon dioxide due to the daily cycle of photosynthesis and respiration and (b) the H^+ produced by nitrifying bacteria in biofilters or pond sediments. Carbonates also provide the carbon source for nitrifying bacteria in biofilters, and the alkalinity should be maintained at greater than 80 mg/L for proper ammonia oxidation in recirculating systems. Because carbonates are steadily consumed by this process, they must be replenished by either the incoming water, CO_2 injection, or addition of sodium bicarbonate. In intensive aquaculture systems, CO_2 injection is rarely considered, because economic constraints dictate high fish loadings, and thus high, rather than low, CO_2 values are the normal concern. Tables for the required sodium bicarbonate dosing levels can be found in Loyless and Malone (1).

BIBLIOGRAPHY

1. J.C. Loyless and R.F. Malone, *Prog. Fish-Cult.* **59**, 198–205 (1997).
2. *Standard Methods for the Examination of Water and Wastewater*, 17th ed., American Public Health Association, Washington, DC, 1989.
3. G. Wedemeyer, *Physiology of Fish in Intensive Culture Systems*, Chapman Hall, New York, 1996.
4. C.E. Boyd, *Water Quality in Warmwater Fish Ponds*, Auburn University Agricultural Experiment Station, Auburn, AL, 1989.

See also pH.

C

CAGE CULTURE

Robert R. Stickney
Texas Sea Grant College Program
Bryan, Texas

OUTLINE

Confining aquatic animals in small floating cages is an attractive option in some aquaculture situations, such as ponds that cannot be drained, reservoirs, lakes, and streams. Cage culture is used primarily in freshwater situations, but also has applications in coastal regions. Cages differ from net pens in that they are much smaller and, in most instances, have rigid frames. (See the entry "Net pen culture.") Confinement of fish in cages facilitates feeding and harvesting, though crowding can have negative impacts on water quality within the cages and may increase problems associated with diseases. Cage culture has been the subject of at least one comprehensive book (1).

CAGE DESIGN AND USE

Small floating structures covered with materials that allow water to freely flow through while retaining confined animals have been used for the culture of fish and other aquatic organisms in a variety of situations. Such structures, called cages, are generally relatively small (Fig. 1), commonly no more than a few meters (1 m = approximately 3 ft) on a side and perhaps 1 to 2 m (3 to 6 ft) deep. Larger structures, such as those used for the commercial culture of salmon, may be 20 to 40 m (66 to 131 ft) on a side and are called net pens. (See the entry "Net pen culture.") In most instances, cages have rigid frames, so that they hold their shape if lifted from the water.[1]

Figure 1. Small cages of the type used for research.

Cages have been used to some extent by aquaculturists for many decades, not only for rearing animals, but also for holding fish in advance of spawning, for rearing early life stages of various species, and for other purposes. Most cage culture involves finfish, but there have been instances wherein shrimp and other invertebrates have been reared in cages as well.

Hapas, which are small cages covered with very fine mesh netting, are used in the Philippines and other countries for spawning tilapia and rearing fry. (See the entry "Tilapia culture.") Cages woven from grasses or constructed from wood have also been used to hold fish for various periods of time and, perhaps, for at least limited growout.

In most cases, cages have ridged frames wrapped with material that allows free passage of water while retaining the aquatic animals in confinement. Frames may be constructed of bamboo, wood, or various types of metal and plastic. Hardware cloth, plastic or wire mesh, and braided nylon netting are common wrapping materials. The cost of cages can be very low (a few dollars each, in some instance) or quite expensive. Costs vary considerably, depending on the size of the cage and on the materials used to make it. One variable of cost that is associated with cage materials is whether the materials have to be purchased or can be obtained at little or no cost. Another consideration is the cost of labor associated with construction of the cage.

The standard approach to cage culture is to float cages at the surface of the water, with the cage bottoms kept above the substrate. Each individual cage can be fitted with floats, or floating platforms, from which the cages are hung, can be constructed. Popular float materials are Styrofoam, cork, and plastic. Cages can also be elevated above the bottom of the substrate from poles driven into the substrate, although this approach is suitable only in relatively shallow water.

Each cage needs to be provided with a top, to keep the animals from jumping out. Tops may be solid or constructed of the same type of mesh that wraps the sides and bottom of the cages.

Cages make sense in a number of aquaculture situations. They provide convenient experimental units and have been widely used by researchers as replicates in

[1] Cages can be lifted from the water intact when empty, but may rupture if any attempt is made to lift them out of the water when they are stocked with fish.

Figure 2. Commercial catfish cages in a reservoir.

various types of studies. Diet testing has been conducted in cage experiments, as have many studies on the feasibility of rearing fish in the heated-water effluents of power plants.

Fossil fuel and nuclear power plants heat water to produce the steam that turns the turbines which generate electricity. The steam is then condensed by being piped through water from a river, lake, or reservoir.[2] Water that is used to condense the steam is heated several degrees and can be used to extend the growing season for fish in temperate regions if the caged fish are moved from a large cooling reservoir into a warm discharge canal below a power plant as the temperature cools in the fall. However, problems associated with gas bubble disease (see the entry "Gas bubble disease") have occurred due to gas supersaturation when fish are confined in discharge canals. Since gas supersaturation occurs when water is rapidly heated, the problem is most severe during precisely the time that the culturist can best take advantage of the warm water in a discharge canal.

In commercial situations (Fig. 2), as well as in conjunction with research, cages can be used in any situation wherein the fish are difficult to capture or cannot be contained. Some research, and even commercial, culture has been conducted in relatively small ponds, including ponds that can be drained and seined. Once again, in the case of research, the purpose is often to provide replication for experiments. In commercial situations (and also in research), two or more species of fish that might not be compatible if released in an open pond have been separated by allowing one species to roam freely while the others were confined in cages. This technique is a form of polyculture. (See the entry "Polyculture.")

Cage culture has been conducted in streams in which confinement is necessary to keep the fish from escaping the region being used for aquaculture. Cages have also been used in irrigation canals and in impoundments. Many impoundments, including some small ponds that would otherwise be suitable water bodies for aquaculture, cannot be drained, have debris (including trees) in them, are too deep to be seined, or have some other factor associated with them that makes capture of freely ranging fish difficult or virtually impossible. Cages provide an option in such situations.

In most regions of the United States, public waters cannot be used for private aquaculture. Leases have been granted in some coastal waters (primarily in association with net-pen salmon culture in Washington and Maine) and in portions of some public reservoirs in Arkansas. Cage culture in public waters is far more common in some other nations.

ADVANTAGES AND DISADVANTAGES

Besides providing the ability to contain fish and other mobile aquatic animals within a known volume of water, cage culture greatly facilitates harvesting fish, as compared with open ponds. For example, fish can be dipnetted from cages relatively easily (Fig. 3). Also, cages may be taken to a shoreside location, where they are tilted to further concentrate the fish, making harvest even easier. In most cases, cages are not sufficiently strong to be lifted entirely from the water without becoming ruptured, particularly as the fish approach harvest size.

There are both advantages and disadvantages associated with feeding fish in cages as compared with feeding them in raceways or ponds. The fish culturist can easily deposit feed directly into a cage, either from a boat or by walking along a platform or other structure from which cages are suspended, depending upon how the cages are moored. However, feed placed in a cage will not necessarily stay there. Sinking feed can fall through the bottom of the cage by gravity or be carried out the sides of a cage by currents before being consumed by the fish inside the cage. Floating feeds may be thrown out of cages by the actions of actively feeding fish and will also be carried out the downstream side of cages unless a fringe of fine-mesh netting is placed around the circumference of the cage; the netting must extend sufficiently above and below the waterline to retain floating feed. Feeding rings — solid metal or plastic tubes or rectangles that are typically built into the top of a cage and extend into the water column within the

Figure 3. Channel catfish cages in the discharge canal of a power plant.

[2] In the case of nuclear power plants, a more complex system is used wherein a water jacket in the containment vessel holds the condenser tubes. Surface water is then used to cool the water in the containment vessel, thereby reducing the chance of radiation leaks.

cage — are also effective at keeping floating pellets in the desired location until they are consumed. Because aquatic animals confined in cages have little natural food available to them, it is necessary to provide nutritionally complete rations in order to avoid nutritional deficiency diseases and obtain a good rate of growth.

The water quality in cages can be very high if the cage site is properly selected. Cages placed in currents constantly receive water exchanges that help maintain good water quality, unless the incoming water is not of high quality. If several cages are placed in a string within a current, the most upstream cage will have the best water quality. As the water passes through each cage, the amount of dissolved oxygen will be reduced, the concentration of ammonia will increase, and fecal wastes will build up in a downstream direction. Ultimately, the water quality may be degraded to the point that growth in downstream cages is reduced. Stress associated with degraded water quality can be a precursor to diseases and even directly cause mortality. Proper separation of cages to allow maintenance of water quality is important. The appropriate number of cages in pods or strings needs to be determined for each water body. Those numbers will vary as a function of prevailing currents, tidal or stream flows, stocking densities, temperature, the level of dissolved oxygen, and the species being reared.

Animals are much more densely crowded in cages than in open ponds. As a result, there have been instances of fighting; scraping of the integument on cage sides, leading to scale loss and skin lesions; and increased levels of cannibalism and disease associated with close confinement. In theory, it is easy to treat diseases in cages. One method involves fitting a plastic bag around a cage containing diseased fish and placing therapeutants in the water. Placement of such bags around cages is, however, much more easily described than actually put into practice. Even if such a bag can be put in place, it cannot remain there long before water-quality deterioration will exacerbate the situation. Removing fish from cages to apply dip or bath treatments may not be difficult, but is an additional source of stress. If efficacious medications can be provided in feed, they can be administered quite easily to caged fish.

Biofouling of cages can be a problem in freshwater, though it tends to be a more significant source of trouble in the marine environment. Algae, bryozoans, and a few types of freshwater clams and mussels (e.g., zebra mussels) can cause significant problems in freshwater, while a wide variety of organisms, notably barnacles and bryozoans, foul cages in the marine environment. The situation can become so severe that the flow of water through cages becomes severely restricted or even stopped. As the fouling problem increases, the water quality will be negatively impacted. Frequent cleaning may be required to keep the mesh of the cage walls open.

SPECIES CULTURED IN CAGES

Nearly any aquatic animal can, at least theoretically, be reared in cages. Because cages are generally fairly small, species that are marketed at a large size, such as tuna, are not good candidates for cages, but can be reared, as are salmon, in net pens. (See the entry "Net pen culture.") Cages have been used to rear channel catfish in reservoirs, including cooling reservoirs and the discharge canals associated with power plants (2). Tilapia are reared in very large net pens in the Philippines and are also reared in cages in various countries. Cage culture in lakes and reservoirs is often permitted in developing countries.

Figure 4. Small commercial marine cages suspended from floating walkways in Malaysia.

In the marine environment, cages have been used for rearing fish of various species and penaeid shrimp in nearshore waters. Commonly, cages are attached to walkways in shallow water, allowing ease of access by the fish farmers (Fig. 4).

Cages have also been used to hold blue crabs caught in the wild and retained until molting. Newly molted crabs are removed and sold as soft-shell crabs, which bring a premium price.

BIBLIOGRAPHY

1. M.C.M. Beveridge, *Cage Aquaculture*, Fishing News Books, Ltd., Farnham, Surrey, United Kingdom, 1987.
2. R.R. Stickney, *Principles of Aquaculture*, John Wiley & Sons, New York, 1994.

See also Net pen culture.

CARBON DIOXIDE

Steven T. Summerfelt
The Conservation Fund's Freshwater Institute
Shepherdstown, West Virginia

OUTLINE

Carbon dioxide (CO_2) is a gas that is relatively soluble in water; it is much more soluble than oxygen and nitrogen. When dissolved, carbon dioxide is in acid–base equilibrium with the total carbonate system, which means that its dissolved concentration is affected by the pH. Therefore, all natural waters in contact with the atmosphere or with an inorganic carbon substrate, such as limestone, will contain carbon dioxide. Carbon dioxide is also excreted by fish through their gills as a by-product of metabolism (along with ammonia). Because carbon dioxide is produced by the fish, it can accumulate in the water, depending upon the rate at which it is produced, the water exchange rate, the gas exchange rate, and shifts in pH. Shifts in the concentration of dissolved carbon dioxide can be quite dynamic; high concentrations for short periods can be toxic, and moderate concentrations over extended periods can limit fish growth and feed conversion. Elevated levels of carbon dioxide can occur in aquaculture systems using alkaline ground waters (if adequate aeration has not yet been achieved), in fish hauling tanks with no water replacement and insufficient venting, and in intensive fish culture systems that have high fish densities, inadequate water exchange, and inadequate aeration. Low-intensity aquaculture systems generally have sufficient water exchange rates and/or aeration rates to keep carbon dioxide from accumulating above safe levels. Aeration processes and alkaline chemical addition are both viable options that can be used to increase the pH and control carbon dioxide accumulation in intensive aquaculture systems. To avoid or recognize carbon dioxide problems requires an understanding of carbon dioxide production, toxicity, chemical equilibrium, and treatment options.

CARBON DIOXIDE PRODUCTION AND TOXICITY

Carbon dioxide is the end product of most catabolic pathways within fish tissue. The volume of carbon dioxide produced in respiration is about the same as the volume of oxygen consumed (1). Based on the molecular weight of carbon dioxide (44 g/mole) and oxygen (32 g/mole), the mass of carbon dioxide produced is about 38% greater than the mass of oxygen consumed; therefore, fish such as salmon and trout produce 0.3 to 0.4 g (0.0106 to 0.0141 oz) of carbon dioxide per g (0.0353 oz) of feed (2). Fish also excrete about 10 times more carbon dioxide than ammonia on a molar basis (3).

Respiration and carbon dioxide production also occur in the biological treatment processes that are used to control ammonia and organic matter when culture water is treated for reuse. The organic matter trapped in the biological filter can be metabolized by heterotrophic microorganisms, producing additional carbon dioxide and releasing nutrients. Carbon dioxide is also produced during nitrification, the two-step process in which bacteria convert ammonia to nitrate. Autotrophic organisms are mainly responsible for nitrification, and carbon dioxide is their primary source of carbon. However, the free acid produced during nitrification reacts with bicarbonate alkalinity in the water to release more carbon dioxide than the autotrophs consume (4). The net result is a loss of 6.0 to 7.4 mg (0.00021 to 0.00026 oz) of calcium carbonate ($CaCO_3$) alkalinity for every 1 mg (3.5×10^{-5} oz) of ammonia nitrogen removed. In recirculating aquaculture systems, an alkalinity of at least 50 mg/L (50 ppm) of calcium carbonate ($CaCO_3$) should be maintained to support nitrification (5) and to prevent pH instability (6–8).

Fish excrete the carbon dioxide that they produce through their gills. The mechanism for carbon dioxide elimination across the gill surface (Fig. 1) has been summarized by Perry (9), Perry and Wood (10), Walsh and Henry (11), and Wedemeyer (12). Within fish tissue, carbon dioxide diffuses into circulating red blood cells, where the enzyme carbonic anhydrase converts the carbon dioxide into bicarbonate. Bicarbonate is the initial substrate of many biosynthetic pathways. As blood is pumped through capillaries within the gill epithelium, the enzyme-mediated process is reversed, and bicarbonate is rapidly converted back into carbon dioxide. A large portion of the carbon dioxide diffuses out of the plasma and across the cell membrane into the mucus layer coating the outsides of the epithelium. (See Fig. 1.) Extracellular carbonic anhydrase, contained within the mucus layer of the gill, rapidly catalyzes the conversion of carbon dioxide into bicarbonate to maintain a low boundary-layer concentration of carbon dioxide and an outward-directed concentration gradient. Carbon dioxide elimination through the gill has to be rapid, because a volume of blood is exchanged through the gill epithelium every few seconds (12).

Fish move relatively large volumes of water over their gills to obtain sufficient oxygen for aerobic metabolism and, at the same time, eliminate ammonia and carbon

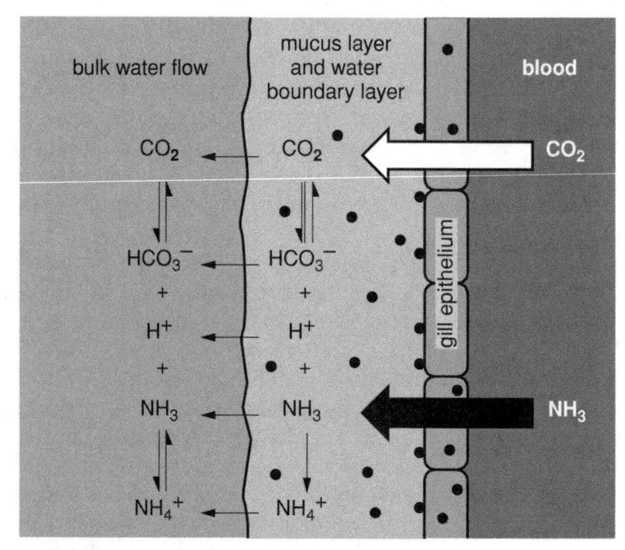

Figure 1. Illustration of the excretion of carbon dioxide (CO_2) and ammonia (NH_3) through the gill epithelium. Carbon dioxide and ammonia freely diffuse through the cell membrane into the mucus layer coating the gill. Carbonic anhydrase (•) within the mucus layer catalyzes the hydration of carbon dioxide into bicarbonate (HCO_3^-) and hydrogen (H^+) ions, promoting the removal of carbon dioxide. Thus, the carbon dioxide and ammonia diffusion gradients are maintained in a direction away from the gill epithelium. [From Wright et al. (13).]

dioxide (9,13,14). Carbon dioxide unloading at the gill is a function of both the ventilation rate and the difference between the carbon dioxide concentration in the water and in the fish's blood (11). However, fish such as rainbow trout (*Oncorhynchus mykiss*) regulate the rate at which they ventilate and pump blood to different portions of their gills based upon the blood's oxygen content (14), not on the blood's carbon dioxide or ammonia concentrations. If levels of carbon dioxide in the water are elevated, less carbon dioxide can be transferred from the gills into the water, even with carbonic-anhydrase-catalyzing transfer. Elevated carbon dioxide levels in the fish (i.e., hypercapnia) increase blood acidity, decrease the ability of hemoglobin to transport oxygen (the Bohr effect), and, in some species, also decrease the maximum oxygen binding capacity of blood (the Root effect). In salmonids, the Bohr effect begins to impair oxygen transport when the dissolved carbon dioxide concentration in the water rises to 20 mg/L (20 ppm) (12). Tilapia (Cichlidae) and catfish (Ictaluridae) are generally less sensitive to elevated concentrations of dissolved carbon dioxide in their culture environment.

Elevated concentrations of dissolved carbon dioxide can be encountered during fish transport or during intensive fish culture, especially when pure oxygen is added to increase the concentration of dissolved oxygen. The combination of supersaturation with dissolved oxygen and a high concentration of dissolved carbon dioxide in the environment can produce hypercapnia, sedate the fish, and may be lethal. Blood oxygen transport may be unaffected, since high concentrations of external dissolved oxygen maintain a sufficient driving force for oxygen transfer into the fish, even as the hyperoxic conditions reduce the fish's ventilation rate (12). However, at concentrations of dissolved carbon dioxide approaching 30–40 mg/L (30–40 ppm), the oxygen carrying capacity of blood will be depressed to the point at which even high concentrations of environmental dissolved oxygen may be insufficient to prevent decreased blood oxygen levels (12). In rainbow trout, the clinical signs of a carbon dioxide problem are moribund fish, gaping mouths, flared operculums, and extra-bright, maraschino-red gill lamellae (15).

Carbon dioxide may also be toxic to fish by contributing to nephrocalcinosis, the formation of calcareous deposits in their kidneys (16). These deposits have been reported in salmonids, catfish, and some marine fishes and are composed of precipitates containing calcium, phosphate, fluoride, and oxides of magnesium (12). Smart et al. (17) found that concentrations of dissolved carbon dioxide that increased from 12 to 55 mg/L (12 to 55 ppm) could increase the number of calcareous deposits in rainbow trout; additionally, growth was seriously impaired at the elevated concentrations of carbon dioxide. Nephrocalcinosis may also appear at relatively lower levels of carbon dioxide if hyperoxic conditions cause hypercapnia. However, other factors, such as the mineral composition and protein content of the diet and (to a lesser degree) the bicarbonate hardness of the water, may also contribute to nephrocalcinosis in trout (12).

Although carbon dioxide is a gas that is dissolved in water, it is extremely soluble and does not contribute significantly to gas supersaturation in water; consequently, carbon dioxide is not a major contributor to gas bubble disease (18).

To culture trout or salmon safely, the concentration of dissolved carbon dioxide should be limited to 20 mg/L (20 ppm) (18). In rainbow trout, the safe upper limit for chronic exposure to carbon dioxide has been reported to range from less than 9 to as much as 30 mg/L (9 to 30 ppm) (19). However, if the concentration of dissolved oxygen in the water is near or greater than saturation levels, the 20-mg/L (20 ppm) recommended safe level may be conservative (12,18). Safe levels of dissolved carbon dioxide can also depend upon the fish species, the developmental stage of the fish, and other water-quality variables. Fish cultured in waters of low alkalinity have a lower blood buffering capacity than fish cultured in more alkaline waters (20). Due to the buffering capacity of blood and acid–base chemistry, water with a high alkalinity, pH, or both may enable fish to tolerate higher concentrations of free carbon dioxide (21).

ADDRESSING PROBLEMS CAUSED BY CARBON DIOXIDE

Problems caused by carbon dioxide in aquaculture systems are occurring more frequently as more intensive production technologies are used to boost carrying capacities and increase transport and production efficiencies (22,23). Techniques used to increase carrying capacity include increased water exchange, improved feed management, and the implementation of oxygen supplementation, pH control, and/or waste removal unit processes (24). Carbon dioxide can be especially problematic when pure oxygen is added to hauling tanks or to the flow entering intensive culture systems and carbon dioxide removal or pH control processes are not considered. The carbon dioxide accumulation in systems that supplement oxygen is exacerbated, because these systems support higher rates of fish loading, the oxygen dissolution processes that are used provide insufficient gas exchange to strip the quantities of carbon dioxide produced, and these systems often do not use standard aeration processes (7,25,26). When standard aeration units alone are used to supply oxygen, however, far less carbon dioxide can accumulate, because levels of available oxygen are limited by the saturation concentration of oxygen, and the aeration provides enough air–water contact to strip carbon dioxide before it can accumulate to toxic levels (27).

Avoiding toxic accumulations of carbon dioxide and the associated pH shift in hauling tanks and intensive water-use systems requires that the rate of carbon dioxide removal through dilution, pH control, or an aeration process be at least as great as the rate of carbon dioxide generation at steady-state operating conditions. Combining mass balances and acid–base equilibrium relations, Colt et al. (23) reported that, under intensive conditions, there are no carbon dioxide limitations (with no aeration or pH control) when the cumulative consumption of dissolved oxygen is less than about 10 to 22 mg/L (10 to 22 ppm), depending upon the pH, alkalinity, temperature, species, and life stage. After this cumulative oxygen consumption level has been reached, the water flow cannot be used again unless it is passed through an air-stripping

Table 1. Equilibrium Types, Relationships, and Constants that Control Carbon Dioxide Concentrations in Water

Equilibrium Type	Equilibrium Relationships	Equilibrium Constants at 25 °C (77 °F) (35)	
Gas–liquid	CO_2 (g) $\longleftrightarrow$ CO_2 (dissolved)	$K_H = P_{CO_2}/X_{CO_2}$	$\approx 6.11 \times 10^{-4}$ atm^{-1}
Hydration–dehydration	CO_2 (dissolved) + H_2O $\longleftrightarrow$ H_2CO_3	$K_0 = [H_2CO_3]/[CO_2]$	$\approx 1.58 \times 10^{-3}$
Acid–base	$H_2CO_3 \longleftrightarrow HCO_3^- + H^+$	$K_1 = [H^+][HCO_3^-]/[H_2CO_3]$	$\approx 2.83 \times 10^{-4}$ mol/L
Acid–base	$HCO_3^- \longleftrightarrow CO_3^{2-} + H^+$	$K_2 = [CO_3^{2-}][H^+]/[HCO_3^{2-}]$	$\approx 4.68 \times 10^{-11}$ mol/L
Acid–base	$H_2O \longleftrightarrow OH^{2-} + H^+$	$K_W = [OH^-][H^+]$	$\approx 1.00 \times 10^{-14}$ mol^2/L^2
Dissolution–precipitation	$CaCO_3 \longleftrightarrow CO_3^{2-} + Ca^{2+}$	$K_{sp} = [CO_3^{2-}][Ca^{2+}]$	$\approx 4.57 \times 10^{-9}$ mol^2/L^2

unit or some form of alkaline chemical is added to reduce carbon dioxide accumulations.

Aeration processes and/or pH control can be implemented to reduce carbon dioxide levels and avoid potential problems caused by carbon dioxide, even under conditions of intensive water use (6–8,26).

Aeration

During aeration, air is mixed with water, so that oxygen is transferred from the air into the water. Aeration also strips carbon dioxide and nitrogen if the water is supersaturated with either of these gases.

When water is in contact with the atmosphere, the equilibrium concentration of carbon dioxide in the water is proportional to the amount of carbon dioxide in the air, according to Henry's law. (See Table 1.) Air contains a mole fraction of about 0.00032 mole of carbon dioxide per mole of air, which is equivalent to a partial pressure of 0.00032 atm. According to Henry's law, water in contact with the atmosphere at 25 °C (77 °F) has an equilibrium concentration of about 0.5 mg/L (0.5 ppm). That is,

$$\begin{aligned} x_{CO_2} &= K_H \cdot P_{CO_2} \\ &\approx \frac{0.000611 \text{ mol } CO_2}{\text{atm} \cdot \text{mol water}} \cdot 0.00032 \text{ atm} \\ &\times \frac{55.6 \text{ mol water}}{\text{L}} \cdot \frac{44 \text{ g}}{\text{mol } CO_2} \cdot \frac{10^3 \text{ mg}}{\text{g}} \\ &\approx 0.5 \text{ mg/L } CO_2. \end{aligned}$$

When water is not at equilibrium with the atmosphere, the concentration of dissolved carbon dioxide in water is controlled by the closed-system carbonate equilibrium. (See Table 1.) Therefore, the concentration of carbon dioxide will depend on the pH and temperature of the water and the concentration of carbonate and bicarbonate present (28), as described in the section in this entry titled "Alkaline Addition and Chemical Equilibrium."

Oxygen and carbon dioxide can be transferred into and out of water, respectively, with any open aeration system (22,27,29,30). However, because carbon dioxide is so much more soluble in water than is oxygen (18), it takes more air–water contact to strip carbon dioxide than to dissolve oxygen. Additionally, the rate at which carbon dioxide is stripped decreases as the partial pressure of carbon dioxide in the air passing through the water increases. Because the concentration of carbon dioxide in water can be 20–40 times the ambient saturation concentration, stripping carbon dioxide from water can rapidly and significantly increase the partial pressure of carbon dioxide in the passing air flow (7). Therefore, enormous volumes of air contact the water, as compared with the air–flow rates required for oxygen transfer alone. Effective carbon dioxide stripping requires contacting from 3 to 10 volumes of air flow for every 1 volume of water flow treated. This air–water contact is often accomplished by forced ventilation of the air through 1.0- to 1.5-m-tall (3 to 4.5 ft tall) cascade columns (Fig. 2) that are sized to treat 60–84 m^3/hr (2100–3000 ft^3/hr) of water flow per square meter (11 ft^2) of column cross-sectional area (26); however, hydraulic loading rates as high as 100–250 m^3/hr/m^2 (330–820 ft^3/hr/ft^2) are also suggested (7). Carbon dioxide stripping and aeration can be improved by packing the cascade columns with high-voidage plastic media or stacked screens that break up the water droplets, to increase the air–water contact area. If high-solids loadings are expected, a stripping tower with screens (Fig. 2) may be easier to maintain than a tower packed with media.

Air discharged from stripping columns should be vented from buildings, to prevent carbon dioxide from accumulating inside the building's airspace. Concentrations of airspace carbon dioxide of 50,000 ppm are immediately dangerous to life and health, and the Occupational Safety

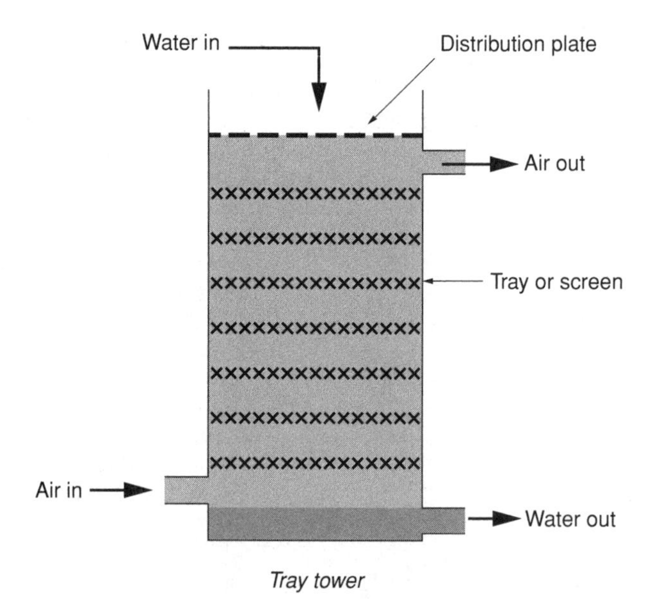

Figure 2. Illustration of an air-stripping tower that uses screens instead of plastic media to break up the fall of water.

and Health Administration (OSHA) limits the allowable time-weighted average exposure to carbon dioxide over an eight-hour workday to concentrations less than 5,000 ppm (31). In contrast, concentrations of ambient atmospheric carbon dioxide are about 350 ppm. However, venting this air from buildings in cold climates can result in considerable heat loss and higher operating costs, especially at the temperatures used to raise warmwater species in semiclosed aquaculture systems (31). Vinci et al. (31) have published computer software to estimate carbon dioxide stripping efficiencies, building ventilation rates, and the corresponding heating and ventilating costs. Under some conditions, a portion of the heat in air can be recovered by venting the air through an air–air heat exchanger. Alternatively, the carbon dioxide can be scrubbed from the air and the air conserved, which reduces the need for building air exchange (B. Watten, USGS Leetown Science Center, Kearneysville, WV, personal communication).

Fish transport tanks should also have a mechanism to rapidly vent air from their head space above the water level; this type of system has been shown to help maintain safe levels of dissolved carbon dioxide during long transport trips (32).

In practice, carbon dioxide stripping equipment is operated at a fixed level near its maximum capacity (8). Therefore, unless more stripping equipment is installed, any further carbon dioxide adjustments must be met through the addition of alkaline chemicals (8).

Alkaline Addition and Chemical Equilibrium

The concentration of carbon dioxide dissolved in water is governed by four types of equilibrium relationships (28): gas–liquid, hydration–dehydration, acid–base, and dissolution–precipitation. (See Table 1.) As described in the section titled "Aeration," gas–liquid equilibrium is the principal mechanism behind the transfer of carbon dioxide between air and water. On the other hand, dissolution of inorganic carbon compounds, such as sodium bicarbonate and limestone, increases a water body's pH, alkalinity, and total inorganic carbon content, thereby increasing the capacity of the water to neutralize an acid. Conversely, precipitation of inorganic carbon compounds of calcium and magnesium occurs after carbon dioxide is stripped from hard water and the water's pH has risen. Deposition of a limestone scale on tanks and equipment and of marl sediments in quiescent zones can both present operational problems to fish farms.

Several authors have reviewed the acid–base equilibrium relationships between pH and carbon dioxide within freshwater and seawater aquaculture systems (6–8,26,33,34). Dissolved carbon dioxide combines with water in a reversible hydration reaction to form carbonic acid (H_2CO_3). (See Table 1.) However, there is approximately 630 times more dissolved carbon dioxide than carbonic acid in water. Carbonic acid dissociates, releasing hydrogen ions (H^+) and bicarbonate ions (HCO_3^-). The bicarbonate ions then dissociate, releasing additional hydrogen ions and carbonate ions (CO_3^{2-}). Therefore, carbon dioxide is only one component within the dissolved inorganic carbon system $CtCO_3$, defined as

$$[CtCO_3] = [CO_2] + [H_2CO_3] + [HCO_3^-] + [CO_3^{2-}],$$

where the [] species represent molar concentrations.

Acid–base equilibrium and thus pH (i.e., $-\log_{10}[H^+]$) control the relative concentrations of each species in the inorganic-carbon system. (See Table 1.) Dissolved carbon dioxide can be calculated from $[CtCO_3]$ and the pH-dependent ionization fraction:

$$[CO_2] = [CtCO_3] \cdot \frac{1}{(1 + K_0K_1/[H^+] + K_0K_1K_2/[H^+]^2)}.$$

It is more common to classify a water based on its alkalinity than on its $CtCO_3$, because alkalinity is a measure of the capacity of a solution to neutralize an acid. Alkalinity is a function of the concentrations of the bicarbonate, carbonate, hydroxide (OH^-), and hydrogen ions:

$$\frac{\text{Alk}}{50{,}000} = [HCO_3^-] + 2[CO_3^{2-}] + [OH^-] - [H^+]$$

In this equation, the alkalinity is expressed in mg of $CaCO_3$/L.

Note that alkalinity is not affected by adding or removing carbon dioxide. Accumulation of carbon dioxide due to fish respiration will increase $[CtCO_3]$, build up the concentration of carbonic acid, and decrease the concentration of carbonate, causing the pH to drop to a lower value. Conversely, stripping carbon dioxide will decrease $[CtCO_3]$, shift the equilibrium as bicarbonate releases carbonate ions, and shift the pH to a higher value. Within minutes of the water exiting the stripping column, carbonic acid will have dehydrated to a new equilibrium level and thus will have replenished some of the carbon dioxide that was removed (7). This chemical equilibrium makes it difficult to strip a large fraction of carbon dioxide from well-buffered waters. The molar concentration of dissolved carbon dioxide at a given temperature can be estimated from the alkalinity and pH as follows:

$$[CO_2] = \left\{\frac{\text{Alk}}{50{,}000} - \frac{K_W}{[H^+]} - [H^+]\right\} \times \left\{\frac{1}{K_0K_1/[H^+] + 2K_0K_1K_2/[H^+]^2}\right\}.$$

Methods that increase the pH will reduce the proportion of $[CtCO_3]$ that exists as carbon dioxide. Addition of a source of alkalinity, such as lime, caustic soda, soda ash, and sodium bicarbonate, to the water will increase the water's alkalinity and raise its pH (6,8,26). Lime, caustic soda, and soda ash react directly with carbon dioxide to produce bicarbonate alkalinity and increase the pH. Sodium bicarbonate is simply a source of bicarbonate alkalinity and a means to increase the pH and level of $[CtCO_3]$, resulting in a net decrease in levels of dissolved carbon dioxide.

The pH also controls the acid–base equilibrium between ammonia (NH_3) and ammonium (NH_4^+) in the total ammonia nitrogen (TAN) system. Methods used to

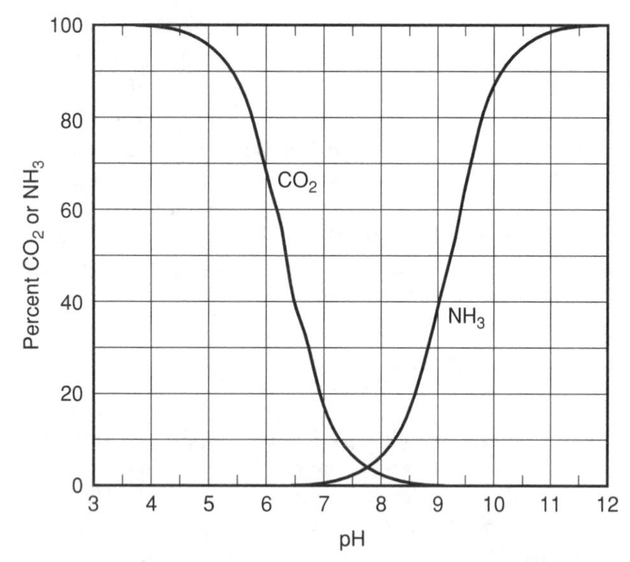

Figure 3. The pH dependence of the percent of total inorganic carbon as aqueous carbon dioxide (CO_2) and the percent of total ammonia nitrogen as ammonia (NH_3), assuming equilibrium at 25 °C (77 °F). [From Summerfelt (26).]

increase the pH and reduce levels of carbon dioxide also increase the portion of total ammonia nitrogen existing as ammonia, which is the more toxic form to aquatic life. Therefore, controlling levels of dissolved carbon dioxide by methods that increase the pH is limited to a pH range wherein unionized-ammonia concentrations are considered safe. Comparing the fractions of ammonia and carbon dioxide that exist as a function of pH (Fig. 3) indicates that levels of both can be minimized by operating in a pH range of 7.2–8.2 (26). However, the optimum pH range depends upon temperature, TAN concentration, $CtCO_3$ levels, and the relative toxicity of ammonia and carbon dioxide to the fish.

BIBLIOGRAPHY

1. M.N. Kutty, *J. Fish. Res. Board Can.* **25**, 1689–1728 (1968).
2. P.B. Liao and R.D. Mayo, *Aquaculture* **3**, 61–85 (1974).
3. D.J. Randall and P.A. Wright, *Can. J. Zool.* **67**, 2936–2942 (1989).
4. Environmental Protection Agency, *Process Design Manual for Nitrogen Control*, Office of Technology Transfer, Washington, DC, 1975.
5. W. Gujer and M. Boller, *Water Sci. Tech.* **16**, 201 (1984).
6. J.J. Bisogni and M.B. Timmons, in M.B. Timmons and T.M. Losordo, eds., *Aquaculture Water Systems: Engineering Design and Management*, Elsevier Science, New York, 1994, pp. 235–246.
7. G.R. Grace and R.H. Piedrahita, in M.B. Timmons and T.M. Losordo, eds., *Aquaculture Water Systems: Engineering Design and Management*, Elsevier Science, New York, 1994, pp. 209–234.
8. J.C. Loyless and R.F. Malone, *Prog. Fish-Cult.* **59**, 198–205 (1997).
9. S.F. Perry, *Can. J. Zool.* **64**, 565–572 (1986).
10. S.F. Perry and C.M. Wood, *Can. J. Zool.* **67**, 2961–2970 (1989).
11. P.J. Walsh and R.P. Henry, in P.W. Hochachka and T.P. Mommsen, eds., *Biochemistry and Molecular Biology of Fishes, Phylogenetic and Biochemical Perspectives*, Elsevier Science, New York, 1991, pp. 181–208.
12. G.A. Wedemeyer, *Physiology of Fish in Intensive Culture*, Chapman and Hall, New York, 1996.
13. P.A. Wright, D.J. Randall, and S.F. Perry II, *J. Comp. Physiol.* **158**, 627–635 (1989).
14. D.J. Randall and C. Daxboeck, in W.S. Hoar and D.J. Randall, eds., *Fish Physiology*, Vol. XA, Academic Press, New York, 1984, pp. 263–314.
15. A.C. Noble and S.T. Summerfelt, *An. Rev. Fish Dis.* Elsevier Science **6**, 65–92 (1996).
16. M.L. Landolt, in W.E. Ribelin and G. Migaki, eds., *The Pathology of Fishes*, The University of Wisconsin Press, Madison, WI, 1975, pp. 793–799.
17. G.R. Smart, D. Knox, J.G. Harrison, J.A. Ralph, R.H. Richards, and C.B. Cowey, *J. Fish Dis.* **2**, 279–289 (1979).
18. J.E. Colt and K. Orwicz, in D.E. Brune and J.R. Tomasso, eds., *Aquaculture and Water Quality*, World Aquaculture Society, Louisiana State University, Baton Rouge, LA, 1991, pp. 198–271.
19. J.M. Heinen, J.A. Hankins, A.L. Weber, and B.J. Watten, *Prog. Fish-Cult.* **58**, 11–22 (1996).
20. D.J. Randall, in D.E. Brune and J.R. Tomasso, eds., *Aquaculture and Water Quality*, World Aquaculture Society, Louisiana State University, Baton Rouge, LA, 1991, pp. 90–104.
21. W. Schaperclaus, *Fish Diseases*, Vol. 2, Amerind Publishing Co. Pvt., Ltd., New Delhi, India, 1991.
22. J.E. Colt and V. Tchobanoglous, in L.J. Allen and E.C. Kinney, eds., *Proceedings of the Bio-engineering Symposium for Fish Culture*, American Fisheries Society, Bethesda, MD, 1981, pp. 138–148.
23. J.E. Colt, K. Orwicz, and G. Bouck, in J. Colt and R.J. White, eds., *Fisheries Bioengineering Symposium 10*, American Fisheries Society, Bethesda, MD, 1991, pp. 372–385.
24. T.M. Losordo and H. Westers, in M.B. Timmons and T.M. Losordo, eds., *Aquaculture Water Systems: Engineering Design and Management*, Elsevier, New York, 1994, pp. 9–60.
25. B.J. Watten, J.E. Colt, and C.E. Boyd, in J. Colt and R.J. White, eds., *Fisheries Bioengineering Symposium 10*, Bethesda, MD, 1991, pp. 474–481.
26. S.T. Summerfelt, in R.C. Summerfelt, ed., *Walleye Culture Manual*, North Central Regional Aquaculture Center Publication Center, Iowa State University, Ames, IA, 1996, pp. 277–309.
27. R.E. Speece, in L.J. Allen and E.C. Kinney, eds., *Proceedings of the Bio-engineering Symposium for Fish Culture*, American Fisheries Society, Bethesda, MD, 1981, pp. 53–62.
28. J.N. Butler, *Carbon Dioxide Equilibria and their Applications*, Lewis Publishers, Chelsea, MI, 1991.
29. C.E. Boyd and B.J. Watten, *Reviews in Aquatic Sciences* **1**, 425–473 (1989).
30. B.J. Watten, in M.B. Timmons and T.M. Losordo, eds., *Aquaculture Water Systems: Engineering Design and Management*, Elsevier Science, New York, 1994, pp. 173–208.
31. B.J. Vinci, M.B. Timmons, S.T. Summerfelt, and B.J. Watten, *Carbon Dioxide Control in Intensive Aquaculture*, Northeast Regional Agricultural Engineer Service, Ithaca, NY, 1998.
32. J.A. Forsberg, R.C. Summefelt, and B.A. Barton, *Nor. Amer. J. Aquacult.* **61**, 220–229 (1999).

33. R.H. Piedrahita and A. Seland, *Aquacult. Engng.* **14**, 331–346 (1995).
34. S. Sanni and O.I. Forsberg, *Aquacult. Engng.* **15**, 91–110 (1996).
35. G. Tchobanoglous and E.D. Schroeder, *Water Quality: Characteristics, Modeling, Modifications*, Addison-Wesley Publishing Co., Reading, MA, 1987.

See also ALKALINITY; BUFFER SYSTEMS; pH.

CARP CULTURE

ROBERT R. STICKNEY
Texas Sea Grant College Program
Bryan, Texas

OUTLINE

Several species of carp (family Cyprinidae, the minnow family) are being cultured around the world. Carp may have been the first fishes cultured in the world, with a history of captive production going back millennia (see the entry "History of aquaculture") in China (Fig. 1). In addition to several species of carp popular in China, there are carp species native to the Indian subcontinent and cultured almost exclusively in that part of the world. The common carp (*Cyprinus carpio*) has been extensively cultured throughout most of the world. Eastern Europe became involved in carp culture several centuries ago. Carp culture also developed in Latin America within the last several decades. However, even though carp were introduced to the United States in the late 19th century, there is very little culture interest or activity in North America.

Figure 1. Carp ponds in China.

Carp culture, particularly with respect to common carp and grass carp (*Ctenopharyngodon idella*), was reviewed by McGeachin (1). Additional information can also be found in Avault's recent volume on general aquaculture, which includes information on Indian carp species (2). Those sources provide the foundation for this contribution and should be consulted for additional details and references. Carp production represents more than 50% of all the animal biomass being produced by aquaculture worldwide.

CHINESE AND COMMON CARP CULTURE

History

The first known document associated with fish culture was produced during the 5th century B.C. in China by Fan Li. It was only a few pages long and reported on the methods involved in rearing carp. During the period when the Holy Roman Empire ruled much of the known world, wild carp were captured from the Danube River and shipped to Rome. The practice continued into the 6th century A.D., when monks began producing carp in ponds at their monasteries. Culturing fish in ponds, rather than having them transported to the monasteries, was a considerable advancement as the fish were readily available for the many meatless fasting days dictated by the church. Carp culture spread across eastern and central Europe and was well established by the late middle ages. Carp culture in Japan did not appear until early in the 19th century (3).

Common carp were successfully introduced into the United States by the government in the latter half of the 19th century (4), though apparently unsuccessful introductions had been made earlier by private citizens. Carp culture and distribution were actively promoted by the U.S. Fish and Fisheries Commission under its first Commissioner, Spencer F. Baird, who seemed to be convinced that spreading carp around the nation was a beneficial activity. Some arguments for the introduction were based on the fact that many European immigrants to the U.S. recognized the common carp as a foodfish. Various problems associated with carp (their propensity for digging in pond banks, their often poor flavor, and their excessive number of small bones, among others) led to public rejection of carp, resulting in the cessation of government stocking in the U.S. by the end of the 19th century. However, by that time, carp had been widely distributed and many reproducing populations had become established. Today, carp appear to be here to stay.

Other carp species were later introduced to the U.S. for weed control and foodfish culture. Those introductions have been more controversial than the introduction of common carp. Various species of carp are currently banned in a number of states.

The common carp has been the subject of intense selective breeding in Europe, where several varieties have been developed. Selective breeding of European carp has been largely aimed at improving growth rate and improving other responses of the fish to culture conditions, as well as changing scale patterns. The standard common carp is covered with scales, but three other strains exist:

Figure 2. A mirror carp (*C. carpio*) produced in Israel.

(*1*) mirror carp (Fig. 2), which have a few scales located below the dorsal fin, (*2*) line carp with scales only along the lateral line, and (*3*) leather carp, which are scaleless. In China, on the other hand, there has been little activity aimed at improving the species.

Application of scientific principles to common carp breeding appears to have been started almost simultaneously in Russia (5) and Israel (6) in the mid-20th century. The studies were initiated when the observation was made that mass selection of carp was not a highly effective means of obtaining the improvements sought. In Israel, both mass and family (progeny) selection have been used.

The first attempts to induce gynogenesis in carp were in Hungary (7). Hybridization involving gynogenetic and other groups of carp led to development of a "Hungarian" race of common carp that has been widely distributed in various countries.

There have been crosses between the European and Asian races of common carp. The resulting fish have been shown to grow more rapidly than the Asian race, are easier to capture by seining, have increased disease resistance, and still perform well when subjected to polyculture.

There appears to have been less attention paid to selective breeding of the various other carp species. A great deal of information has been developed on producing triploid grass carp to produce sterile fish as discussed later.

Much more information is available on the common carp, and to a somewhat lesser extent grass carp, than on the other species in culture. In this entry, the information on culture techniques concentrates primarily on common and grass carp, and the final section briefly describes Indian carp culture.

Production

Not only does fish culture, with a primary focus on carp, appear to have originated in China, but that country continues to dominate world production of fish. Polyculture, the technique of rearing two or more compatible species in the same water body (see the entry "Polyculture"), was developed in China and continues to be widely employed there in ponds. A typical polyculture pond might be stocked with some combination of common carp, mud carp (*Cirrhinus mulitorella*), black carp (*Mylopharyngodon piceus*), bighead carp (*Aristichthys nobilis*), silver carp (*Hypopthalmichthys molitrix*), and grass carp, as well as crucian carp (*Carassius carassius*), which are also reared in China. Table 1 presents the total production of each species in China during 1995 according to FAO (8). In terms of global production rankings, silver carp ranks #1, followed in order by grass carp, common carp, bighead carp, and crucian carp. Noncarp species do not show up until sixth place (which is held by Nile tilapia, *Oreochromis niloticus*).

Various other countries produce significant quantities of carps, though none approaches China in terms of total production (Table 1). Common carp are produced in various countries, although mud carp are produced only in China and Taiwan. Silver, bighead, crucian, and grass carp have been produced in many nations, but production tends to be limited outside of China. Bighead carp are produced, for example, in Cambodia, Taiwan, Hong Kong, Iran, Laos, Malaysia, Myanmar, and Nepal (all in Asia), as well as Croatia, the Czech Republic, Germany, Hungary, Romania, and Ukraine (all in Europe) still, total production in these nations is only 22,673 tons.

The figure listed in Table 1 for the former USSR represents the combined production in the various now-independent carp producing nations that once made up the USSR. Included in that total are carp produced in Armenia, Azerbaijan, Belarus, Estonia, Georgia, Kazakstan, Kyrgyzstan, Latvia, Lithuania, Moldova, the Russian Federation, Tajikistan, Turkmenistan, Ukraine, and Uzbekistan. Each of those nations produces common carp, while fewer nations produce the other species listed in the table.

Total production of common carp from the former USSR has been declining in recent years. Between 1986 and 1995, a high of 319,766 tons was reached in 1990. A year later, production had fallen precipitously to 194,427 tons. On the other hand, in Indonesia, the production of common carp has increased annually from 1986 through 1995.

Table 1. Carp Production in Metric Tons (approximately 2,000 lb/ton) for Selected Countries During 1995, According the Food and Agricultural Organization of the United Nations (8)

Country	Species	Production (tons)
China	Common carp	1,398,618
	Mud carp	110,000
	Bighead carp	1,236,667
	Silver carp	2,473,333
	Grass carp	2,070,988
	Crucian carp	533,740
Japan	Common carp	18,272
	Crucian carp	945
Former USSR	Common carp	100,264
	Silver carp	47,660
	Grass carp	647
Indonesia	Common carp	145,500
Mexico	Common carp	27,506
Egypt	Common carp	30,895
Israel	Common carp	7,089
Poland	Common carp	19,720
	Silver carp	198
	Grass carp	3

Mexico is the leading carp-producing nation in Latin America; however, only common carp are being cultured in sufficient quantities to make the FAO statistical report (Table 1). Egypt is the largest common-carp-producing nation in Africa. Israel was a major carp producer in years past, but production was fairly constant over the period from 1986 through 1995, leaving other countries surpass the Jewish state. Tilapia (*Oreochromis* spp.) have largely replaced carp in Israeli aquaculture, although significant culture of several other freshwater and marine species also occurs. In Europe, the production of common carp exceeds 10,000 tons each in the Czech Republic, Germany, Hungary, and Poland (only Poland is included in Table 1).

In the United States, there is very little carp culture activity. Most of the small amount of common carp consumed are taken in the capture fisheries. There is a small ethnic demand associated with historical or religious consumption practices. Grass carp have been used primarily for aquatic weed control by aquaculturists, as well as small pond owners and state and federal management agencies. A small amount of commercial bighead and silver carp production exists in the United States, but it is so insignificant that it does not appear in the FAO statistics (8).

With changes that have occurred in the social structure and economy in the Peoples Republic of China in recent years, the focus on carp production has also begun to change (Li Yingren, Chinese Academy of Fisheries Science, personal communication). City dwellers are apparently demanding higher quality seafood products (shrimp and other seafoods of excellent quality can be found in the large cities of eastern China), so the primary demand for carp is associated with the rural population. Because that population is so large, demand remains high, though annual production increases can be expected to become static or even decline in the future. The government of the Peoples Republic of China has moved from a policy of stressing increased production to a policy that stresses product quality.

Common carp will root around on pond banks and cause damage, but that has not been a major problem in carp culture. On the other hand, the silver crucian carp (*Carassius auratus gibelio*), also known as the chiton, is a subspecies of goldfish (*C. auratus*) that has a reputation for causing extensive damage on fish farms. It also competes with the more desirable species. Apparently, entry of silver crucian carp into culture ponds is the result of the fish entering with incoming water, so proper filtration is recommended to reduce or eliminate the problem.

Culture Methods

Culture Systems. As the chinese begin to adopt modern aquaculture technologies, the situation is changing, but historically, prepared feeds have not been employed in Chinese polyculture ponds. Natural productivity is increased markedly by frequent, or even continuous, additions of manure and nightsoil. Organic fertilizers have been depended upon as sources of nutrition for the fishes stocked. Organic matter, in the form of agricultural wastes, has also been used, primarily for feeding grass carp, which will feed on various types of plants of aquatic or terrestrial origin. Fertilization promotes phytoplankton, zooplankton, and benthos production in ponds. Common carp are bottom feeders that ingest benthic organisms such as worms, insects, and molluscs. Mud carp are omnivorous and will consume detritus, including decaying vegetation. Black carp feed on snails. Silver carp are able to filter phytoplankton from the water, though they also ingest zooplankton. Bighead carp selectively feed on zooplankton, while crucian carp consume plant fragments and zooplankton.

Using the traditional polyculture approach, the Chinese have found ways of recycling livestock and human wastes (some ponds actually have privies suspended over them) and of utilizing all the food resources that are available. Production rates of about 8,000 kg/ha (7,140 lb/acre) are possible, using the Chinese approach to polyculture.

Polyculture is not limited to various carp species. To some extent in China, but more common in various other nations, is culture of one or two species of carp with fishes from other families. For example, in Israel, common carp have been polycultured with grass carp, tilapia, and mullet (*Mugil* spp.). In the United States there has been some culture of bighead, silver, and grass carp (at low stocking rates) with channel catfish (*Ictalurus punctatus*).

Outside of China, monoculture is much more common than polyculture. Regardless, the vast majority of carp produced, using either approach, are reared in earthen ponds. Production levels in ponds vary considerably from country to country, and even within countries, because of the levels of intensity that exist. With no pond fertilization or supplemental feeding, a yield of only a few hundred kg/ha (lb/acre) can be expected. Fertilization with manure increases natural productivity, and consequently, fish production levels (Fig. 3). Supplemental feeding with grains has led to production levels of over 1,000 kg/ha (approximately 1,000 kg/ha) in countries such as Poland. Polyculture in Chinese ponds fertilized with manure and supplemented with agricultural wastes can produce 1,500 kg/ha (about 1,500 lb/acre). Provision of high quality prepared feeds and the use of supplemental

Figure 3. Swine, cattle, and poultry are among the terrestrial animals that have been produced in association with fish. In this case, ducks are housed adjacent to a carp pond in Nepal. The duck droppings fertilize the water.

aeration to maintain dissolved oxygen levels can increase production dramatically (Fig. 4). In Israel, for example, monoculture common carp ponds have produced as much as 30,000 kg/ha/yr (about 30,000 lb/acre/yr).

Carp production in raceways is not common, though a moderate percentage of the carp produced in Japan are reared in such systems. Cages are sometimes used to rear carp, since large natural or manmade water bodies (rivers, lakes, and reservoirs) can be utilized effectively as aquaculture systems when cages are employed (Fig. 5). For cage culture to be successful, some type of prepared feed should be provided, as natural productivity will usually not be adequate, except perhaps in the case of silver or bighead carp if plankton densities are sufficient.

Similar to cage culture, in that it employs large water bodies for fish culture, is the blocking of bays and inlets with nets that confine stocked fish. In those situations, either natural food can be relied upon, if low stocking densities are employed and natural productivity is high, or culturists can provide supplemental feed to increase fish production.

Figure 4. At the highest level of intensity, the fish are offered prepared feeds and provided with supplemental aeration, as shown here in Israel, where paddlewheel aerators are commonly employed.

Figure 5. Simple cages, such as those shown here in a reservoir in Nepal, provide an alternative type of culture system for carp and other species.

Carp have also been produced in rice–fish culture, another form of polyculture. Rice–fish culture has a long history in Asia and elsewhere in the world, though strong research programs on rice–fish farming systems dates only from the 1980s. In China alone, some 500,000 ha (1,250,000 acres) of rice paddies have been used in rice–fish culture. The approach involves stocking rice paddies with fish that will forage on insects and other organisms that become established in the rice fields. It is important, of course, that the fish do not negatively impact rice production, which is not a problem if common carp are employed. Because fish, such as carp, cannot be reared to market size at the same rate that a rice crop can be produced, it may be necessary to retain fish in the paddy while two or more rice crops are grown. That can be accomplished by digging a trench in the paddy (usually down the middle) that provides sufficient water to support the fish while the paddy is drained for harvesting. This approach works well in theory, but if it becomes necessary to apply pesticides to treat for rice pest invasions, the fish may also be killed, even if the paddy is drained and the fish are isolated in the trench during spraying. Spray drift can be a major problem.

Rice–fish polyculture is often practiced with combinations of grass carp, common carp, and crucian carp, associated with high rice production. Rice–azolla–fish polyculture has also been employed. Yet another approach employed involves rotating crops between rice and fish.

Carp ponds are typically harvested with seines (Fig. 6). Seines may also be employed to harvest fish from portions of large water bodies, such as bays, that are blocked off with nets to provide captive rearing areas. Dip nets are usually the method of choice for harvesting fish from cages. (See the entry "Harvesting.") If a trench is provided in a rice–fish farming operation, the rice paddy can be drained, thereby forcing the majority of the fish into the trench, from which they can be removed with hand nets.

Reproduction. In their native habitat, common carp spawn during the spring, when the rivers are rising due to annual floods and the water temperature reaches 18 to 24 °C (64 to 75 °F). The fish can be allowed to spawn naturally in ponds (broodfish are stocked in newly filled ponds, which mimic rising water level and help induce

Figure 6. A seine full of carp in Israel.

spawning) or they can be induced to spawn with hormones, one of the most popular and effective being carp pituitary.

Typically, spawning ponds are stocked with broodfish in the ratio of 1:2 or 2:3 (females:males). The eggs of common carp are adhesive. When spawning ponds are utilized, some type of spawning mat is provided upon which the eggs can be deposited. Various materials can be used for spawning mats, with plastics of various kinds representing the most modern materials. Once spawning has occurred, the mats may be transferred to nursery ponds; or, when large spawning ponds are employed, once spawning has been completed, the mats may be left in place and the broodfish seined from the ponds. If the eggs are obtained through hormone induction, adhesion can be broken up through the use of various chemicals.

While common carp will spawn in ponds if the appropriate techniques are employed, grass carp can only be spawned through the use of hormone injections. Induced spawning involves injecting broodfish with carp pituitary extract, human chorionic gonadotropin, or leuteinizing hormone-releasing hormone. Injections of both males and females may be required.

Once eggs from the hormone-injected females begin to flow freely, they can be expressed into a dry round basin, after which the milt from one or more running-ripe males is added. The mixture is then stirred, and water is added after about one minute of stirring.

Chemicals, which include salt and urea, salt and milk solution, and others, can be used to break down the adhesion of carp eggs during incubation. Upwelling hatching chambers are employed during incubation, which requires 28 hours or less for grass carp and 55 hours or less for common carp, depending on temperature.

Yolk sac absorption requires a few days, after which the fish will swim to the water surface to inflate their swim bladders and begin feeding. Brine shrimp nauplii (see the entry "Brine shrimp culture"), rotifers, and yeast have been used as first feeds for carp produced in hatcheries. In ponds, natural foods—the production of which is encouraged through fertilization—are depended upon during the initial phases of culture, after which supplemental or complete feeds may be introduced, depending upon the level of technology being employed.

Fertilization, Feeding, and Nutrition. Carp can, as previously mentioned, be produced at modest levels in ponds that are fertilized, but receive no supplemental feed. Fertilization promotes algae blooms that support the production of zooplankton and benthic invertebrates upon which the carp feed. Submerged aquatic vegetation will be consumed by grass carp, though fresh agricultural waste (for example, leafy waste from vegetable production) is often provided, as ponds properly fertilized to induce algae blooms often have limited, if any, rooted plants growing in them. Supplementation with live foods is not common, though silkworm pupae have been used in Japan. The pupae are readily available at low cost in the silk producing regions of that nation.

Without fertilization, polyculture carp ponds may produce from 20 to 40 kg/ha/yr (about 20 to 40 lb/acre/yr). Production can easily be increased through the addition of organic or inorganic fertilizers, though the Chinese, as previously indicated, have pushed production in intensive polyculture ponds to the thousands of kg/ha/yr (lb/acre/yr).

The next step is to provide supplemental or complete formulated feeds. Such feeds have only been developed for common carp, since in most cases, the other species are expected to consume natural foods. Most species will, in fact, accept prepared feeds, but feeds specifically designed to meet their nutritional requirements have not been produced. In fact, it is generally assumed that all the carp species have similar nutritional requirements and will perform similarly on rations developed for common carp.

A great deal of nutritional research has been conducted, particularly in Israel and Japan, on common carp and satisfactory feeds have been developed (9). Ingredients that are commonly found in carp feeds include wheat, barley, corn, rye, rice, soybeans, sorghum, cottonseeds, and fish meal. Vitamin and mineral premixes are also employed in complete feeds.

Carp appear to require between 30 and 38% protein. A dozen amino acids have been found essential for proper carp nutrition and the dietary levels of those amino acids required by the fish have been quantified.

A good deal of research has also been conducted with respect to fatty acid requirements. Both 18:2n-6 and 18:3n-3 appear to be required by common carp (9). Carp diets do not typically contain high levels of lipid, but it is known that carp employ both lipids and carbohydrates as energy sources. Starch is more effectively utilized by carp than are dextrin or glucose.

No deficiency signs in carp fingerlings have been associated with diets lacking in vitamin B_{12}, C, D, or K, though some deformities have been observed at the larval stage in fish that do not receive vitamin C (9). The requirements for various other vitamins have been determined. It has been shown that the requirement for vitamin E increases as the level of dietary polyunsaturated fatty acids increases.

Required dietary minerals include phosphorus, magnesium, zinc, manganese, copper, cobalt, and iron (9). Deficiencies are generally associated with poor growth, as is the case with vitamin deficiencies.

Diseases. Carp, like other aquatic animals, are subject to various diseases. Proper management of culture systems to reduce or eliminate environmental stress on the fish will go a long way in reducing the incidence of diseases.

Carp are susceptible to an array of disease types. Viruses include *Rhabdovirus carpio*, or spring viremia of carp (10). The signs of that disease are similar to those associated with *Aeromonas salmonicida* bacterial infections (discussed in the next paragraph), though the virus may be a secondary infection. There is also a virus, *Herpesvirus cyprini*, that is the causative agent of the so-called fish pox, a disease which was first reported in the 16th century. A disease known as grass carp hemorrhagic virus disease has been reported from China. Signs are severe hemorrhaging of the intestinal tract and various internal organs.

A common problem is hemorrhagic septicemia or infectious abdominal dropsy that is associated with bacterial infections with *A. salmonicida* (10). Signs of infection include the distension of the abdomen and the accumulation of clear fluid in the abdominal cavity. There may also be ulcers on the liver, reduced hemoglobin levels, increased levels of leucocytes, and lesions in the intestinal tract. Columnaris disease resulting from bacterial infections of *Flexibacter columnaris* have also been reported from carp. *Pseudomonas* infections have also occurred. Terramycin can be used to treat these bacterial problems.

External carp parasites include protozoans, monogenetic trematodes, parasitic copepods, and anchor worms. Secondary fungal infections from *Saprolegnia* sp., have also been reported. Various chemicals have been found effective in treating parasites, but there are severe restrictions on chemicals approved for use in the United States. The fish culturist should attempt to remain knowlegable with regard to the chemicals that are currently approved for foodfish use, as the list is subject to change.

Controversy Surrounding Grass Carp

Grass carp are native to the Amur River in Siberia and the Yangtze River in China. They were first brought to the United States in 1963 by U.S. Fish and Wildlife Service biologists interested in evaluating the species as a biological control for aquatic vegetation. The fish were maintained at the Fish Farming Experimental Station at Stuttgart, Arkansas (now operated as a U.S. Department of Agriculture facility). Research demonstrated that stocking rates of 49 to 99 fish/ha (20 to 40 fish/acre) could effectively control aquatic vegetation in fish ponds.

Commercial production began in Arkansas during 1972, and there were virtually no restrictions on stocking the fish in that state. Other states, expressing concern that the exotic grass carp might consume desirable vegetation and, worse, could reproduce and become established in areas where they were not desired, banned the introduction of grass carp.

Studies of the life history of grass carp concluded that while the fish might spawn in large rivers in North America and elsewhere, the proper conditions for survival did not exist outside of the native range of the species. The lack of sufficiently long river reaches with the proper current speed and the presence of predators in North America rivers—which apparently are virtually nonexistent in the rivers that grass carp are native—were among the conditions cited as mitigating against survival of grass carp in the United States. Yet, within a few years, grass carp fry and juveniles began to appear in the Mississippi River. Based on the size of those fish, it was clear that grass carp were spawning successfully. Reports also came of successful spawning from the Rio Grande. Texas was among over 30 states that had banned grass carp, yet retailers were advertising and marketing the fish nationwide in trade magazines. Individual landowners usually had little difficulty having a few grass carp shipped to them for use in ponds. Escapement of those fish, and the reported presence of at least a few grass carp fingerlings in many minnow shipments from Arkansas into other states led to the diffusion of grass carp into many states where they were prohibited.

There were instances where exceptions were made to bans on grass carp in states that had once outlawed the species, or even retained a general ban on its introduction. An example is associated with Lake Conroe, Texas. By the late 1970s, that large reservoir north of Houston was becoming choked with an exotic, rooted-aquatic weed (*Hydrilla* sp.) from the shoreline out to a depth of about 10 m (30 feet) of water. Lakeside residents found it difficult or impossible to maneuver their boats through the weeds and at least one drowning was attributed to a child becoming entangled in the weeds. After mechanical harvesting and chemical control failed to ameliorate the problem, the Texas Parks and Wildlife Department commissioners voted to stock grass carp. That decision was challenged by bass fishing groups that were concerned that elimination of the weeds would negatively impact angling. The issue wound up in court, and the state lost the suit, based on a technicality surrounding a violation of its own administrative procedures (failure to provide sufficient time for comment between announcing the proposed course of action and making the final determination). The case was appealed, ultimately, to the Supreme Court of Texas. Before the Supreme Court acted, the Texas Legislature passed legislation requiring that the fish be stocked in Lake Conroe. All of this took many months, during which time the fish were being held in Arkansas, where they were outgrowing the farmer's ability to retain them.

The fish were ultimately stocked, and, as predicted, the vegetation was controlled. Fishing was going to be impacted negatively whether grass carp were present or not. With the dense weeds present, the anglers, who were concentrating on catching largemouth bass (*Micropterus salmoides*) had been enjoying unusually good fishing conditions. The bass had been impeded from getting into shallow water by the dense weeds, while their prey (smaller fish) could hide effectively in the weeds. The hungry bass were highly susceptible to angling, so for a while, at least, fishing had been excellent. However, the bass could not reach their spawning beds under these conditions, and the population was due to collapse if the weeds were not removed.

With the removal of the weeds, water clarity changed (was reduced), as plankton blooms replaced the rooted vegetation, creating the need for additional water treatment before the water could be used as a domestic supply in Houston. As it turned out, the water district had actually benefited from the presence of the weeds.

Surprisingly, there was documentation of successful spawning and survival of offspring downstream of the lake. The occurrence of young grass carp in the marshes along the coast fueled fears that valuable habitat might be destroyed. At present, the final outcome remains to be determined.

With the banning of grass carp in so many states and the intense controversy that surrounded the fish, commercial culturists soon began producing triploid grass carp (fish with three pairs of chromosomes), by hybridizing grass carp and bighead carp. The benefit of the triploid is that it will consume aquatic vegetation, but cannot reproduce. Triploids can be stocked at the appropriate levels and will control vegetation, and after several years they will die, and the decision may be made as to whether or not restocking is desirable. The same applies to diploid grass carp, but there is little or no concern when triploids escape, as they cannot establish reproducing populations.

With the availability of sterile triploids, many states began making exceptions to their bans and now allow triploid stocking, though the use of triploids tends to be strictly controlled. Permits may be required from the appropriate state agency. The agency may require that provision is made to prevent loss of fish over pond spillways during heavy runoff periods, even if the fish stocked are triploids. Also, the agency may dictate the maximum number of fish that can be stocked in a particular water body. In most cases, the producer must certify that the fish being sold are 100% triploids. This involves examining blood (the blood cells are larger in triploid than diploid fish), which significantly adds to the cost of stocking triploids.

Indian Carp Culture

A group of carp species known as the major Indian carps are produced predominantly on the Indian subcontinent. Included are the roho or rohu (*Labeo rohita*), mrigal (*Cirrhinus mrigala*), and catla (*Catla catla*). In addition to India (Table 2), roho and mrigal are being reared in Laos, Myanmar, and Thailand; while catla is produced in India, Laos, and Myanmar. With the exception of roho production in Myanmar, where 74,000 tons were produced in 1995, production levels of these carp species are low outside of India. There has been some research on the use of Indian carps in Egypt, but production there, if it is still underway, is not sufficient to make the FAO statistical reports (8).

All three species are native to India and spawn in the rivers of that country. Roho and mrigal have a preference for vegetation and decaying plant matter as food, while catla feed primarily on zooplankton.

Historically, pond or bund spawning provides for environmental manipulations that induce natural spawning (11). Grass is grown in large shallow ponds during the dry season, in which time, the broodfish are held in a pool within the pond. When the monsoon rains arrive, the levee between the holding pool and bulk of the pond is opened, and the fish enter the grassy area where they spawn. The eggs are collected with nets for incubation in small water bodies. Alternatively, hormone injections can be used to induce spawning. Growout involves rearing in fertilized ponds, which may also receive supplemental feed.

Table 2. Production of Indian Carps (metric tons) in 1995 (8)

Species	Production (tons)
Rohu	382,050
Mrigal	370,960
Catla	379,338

BIBLIOGRAPHY

1. R.B. McGeachin, in R.R. Stickney, ed., *Culture of Non-salmonid Freshwater Fishes*, CRC Press, Boca Raton, FL, 1993, pp. 117–143.
2. J.W. Avault, Jr., *Fundamentals of Aquaculture*, AVA, Baton Rouge, LA, 889 p.
3. T. Kakuku and H. Ikenoue, *Modern Methods of Aquaculture in Japan*, Elsevier Science Publishing, New York, 1983.
4. R.R. Stickney, *Aquaculture in the United States: A Historical Survey*, John Wiley & Sons, New York, 1996.
5. V.S. Kirpistchinikov, *FAO Fisheries Reports* **44**, 179–194 (1968).
6. A. Wohlfarth, R. Moav, and M. Lahman, *Bamidgeh* **13**(2), 40–54 (1961).
7. K.A. Golovinskaia, *FAO Fisheries Reports* **44**, 215–222 (1968).
8. FAO, *Fisheries Statistics: Catches and Landings*, Food and Agriculture Organization of the United Nations, Rome, 1995.
9. S. Satoh, in R.P. Wilson, ed., *Handbook of Nutrient Requirements of Finfish*, CRC Press, Boca Raton, FL, 1991, pp. 55–67.
10. J.A. Plumb, *Health Maintenance and Principal Microbial Diseases of Cultured Fishes*, Iowa State University Press, Ames, 1999.
11. J.E. Bardach, J.H. Ryther, and W.O. McLarney, *Aquaculture*, Wiley-Interscience, New York, 1972.

CHANNEL CATFISH CULTURE

Craig S. Tucker
Mississippi State University
Stoneville, Mississippi

OUTLINE

The channel catfish (*Ictalurus punctatus*) is a native North-American freshwater fish in the family Ictaluridae, or bullhead catfishes. The original range of the channel catfish extended from northern Mexico and the states bordering the Gulf of Mexico up the Mississippi River and its tributaries. Long esteemed as a foodfish in the southern part of its original range, the species has been introduced throughout the world as a sport fish and for aquaculture. In fact, channel catfish aquaculture has expanded rapidly since 1975 to become the largest aquaculture industry in the United States. In 1997, 250,000 metric tons (560 million pounds) of channel catfish were produced commercially, which accounted for about half of the total United States aquaculture production for all species.

Channel catfish are typical ictalurids, with an elongated, cylindrical body, a depressed head, and scaleless skin. All ictalurid catfishes possess an adipose fin and soft-rayed fins, although the pectoral and dorsal fins have sharp, hardened spines. The barbels of ictalurid catfishes are arranged such that four are under the jaws, two are above the jaws, and one is on each tip of the maxilla. Characteristics that distinguish channel catfish from other ictalurid catfishes include a deeply forked tail and a moderately rounded anal fin with 24–30 rays. Normally pigmented channel catfish are white to silvery on the undersides, grading to grayish blue or olivaceous to nearly black dorsally. (See Fig. 1.) Albinism is rare in wild fish, but is not uncommon in certain domesticated strains. Irregular dark spots are present on the sides of young fish, but are absent in albino fish and are often lost in normally pigmented fish over about 0.45 kg (1 lb).

Channel catfish are bottom dwellers that prefer a substrate of sand and gravel. Their natural habitat is sluggish to moderately swift rivers and streams, although they also thrive in lakes and ponds. Fish less than 10 cm (4 in.) long feed on detritus, aquatic insects, and zooplankton. Larger fish feed primarily on aquatic insects, crawfish, and small fish. The optimum temperature for growth is 25 to 30 °C (77 to 86 °F), and the fish feed poorly at temperatures below 10 °C (50 °F) and above 35 °C (95 °F). In nature, from two to five years may be required for a fish to reach a weight of 0.45 kg (1 lb), although much faster growth is achieved in aquaculture. Channel catfish may live for more than 40 years and attain weights in excess of 18 kg (40 lb).

Channel catfish possess a combination of desirable qualities for commercial aquaculture. They usually do not reproduce in culture ponds, a trait that gives the culturist control over pond populations. Sexually mature fish are, however, easily spawned under proper conditions, and large numbers of fry can be obtained using simple methods. Fry accept manufactured feeds at first feeding after absorbing their yolk sac, and growth and feed conversion efficiency on relatively simple manufactured feeds are satisfactory at all phases of production. Channel catfish are hardy fish that tolerate crowding and a wide range of environmental conditions. They also adapt well to all commonly used aquaculture production systems: ponds, cages, and raceways. Channel catfish have firm, white flesh with a mild flavor that retains high sensory quality after a variety of processing methods.

Although channel catfish account for virtually all of the catfish production in the United States, there is also some interest in commercial production of blue catfish, *Ictalurus furcatus*, a close relative of channel catfish. Blue catfish (see Fig. 2) resemble channel catfish in their general appearance, although blue catfish have a smaller head and the anal fin, which contains 30 to 36 rays, is longer and less rounded than that of the channel catfish. Blue catfish grow slower than channel catfish during the first two years of life, although strain effects are important and some strains of blue catfish grow faster than many strains of channel catfish (1). Blue catfish also mature sexually at an older age and larger size than channel catfish, which is considered an undesirable trait, because large broodfish are difficult to manage. On the positive side, blue catfish have a better dressout percentage than channel catfish (2),

Figure 1. Normally pigmented and albino channel catfish fingerlings of about 0.04 kg (0.1 lb).

Figure 2. Channel catfish (top), blue catfish (middle), and white catfish (bottom), all of about 0.25 kg (0.5 lb).

are easier to harvest by seining (3), and are more uniform in size at harvest (4). Blue catfish are less tolerant of poor water quality than are channel catfish (1), but are more resistant than channel catfish to the infectious diseases enteric septicemia of catfish (5) and channel catfish virus disease (6).

Attempts have been made to take advantage of the best traits of the blue and channel catfish by making interspecific hybrids. The female channel catfish × male blue catfish hybrid possesses many of the best features of the parents and is a highly desirable fish for commercial culture (7). However, the major obstacle to the commercial use of this hybrid is the low hybridization rate, which makes it difficult to obtain adequate numbers of fry for commercial use.

At one time, there was some interest in aquaculture of the white catfish, *Ameiurus catus* (8–10). The white catfish (see Fig. 2) has a moderately forked tail, is unspotted, and the anal fin contains 19 to 23 rays. Its color is bluish to gray on the dorsal surface and silvery below. White catfish are among the hardier of the ictalurid catfishes. They tolerate low concentrations of dissolved oxygen and high water temperatures better than channel catfish do. However, relative to channel and blue catfish, the white catfish has a poor dressout percentage and grows slowly. There is little current interest in growing that species commercially.

The fish species collectively known as bullheads are the smallest of the ictalurid catfishes commonly used as human food. The brown bullhead (*Ameiurus nebulosus*) and yellow bullhead (*Ameiurus natalis*) offer some potential for culture, and the yellow bullhead has been advocated as a good candidate for small-scale, "backyard" aquaculture (11). Bullheads tolerate poor water quality, and, under the right culture conditions, they have a desirable, mild flavor. On the other hand, bullheads are small fish that have a poor dressout percentage. They also readily reproduce in ponds and may overpopulate production ponds, resulting in large numbers of small fish. Consumer demand and prices paid for bullheads are low, and they are seldom cultured for food.

DEVELOPMENT OF THE CATFISH INDUSTRY

Many of the production practices used in channel catfish aquaculture were developed prior to 1960—well before there was significant commercial culture of the fish for food. Reliable spawning and hatchery techniques were developed between 1910 and 1960 by personnel at state and federal hatcheries, so that large numbers of fry and small fingerlings could be produced for stocking into reservoirs or sport-fishing ponds (12–14). The fundamentals of pond culture of adult fish were developed largely through the efforts of Dr. H.S. Swingle and his colleagues and students at Auburn University. In a series of studies conducted in the 1950s and early 1960s, the Auburn research group explored the potential for small-scale aquaculture of channel catfish in farm ponds in the southeastern United States. Initial efforts at growing catfish in ponds centered around increasing fish yields by enhancing natural pond productivity with inorganic fertilizers (15). Annual yields of catfish in fertilized ponds were around 100 to 200 kg/ha (90 to 180 lb/acre). Use of a crude supplemental feed, such as soybean cake, increased annual production to more than 250 kg/ha (225 lb/acre). Somewhat later, a dry, powdered diet—originally formulated for minnows—was fed to channel catfish, and an annual yield of 1,400 kg/ha (1,250 lb/acre) was achieved (16). Further developments in the nutrition and feeding of catfish included pelleting the feed and formulating the diet according to the nutritional needs of the fish. Early on, it was discovered that daily feeding rates greater than about 34 to 45 kg/ha (30 to 40 lb/acre) were not possible without causing oxygen depletions and fish kills (17). By providing mechanical aeration, oxygen depletion can be avoided, and summertime feeding rates as high as 140 kg/ha (125 lb/acre) per day are now possible. Annual fish yields on commercial fish farms now range from 3,300 to over 7,800 kg/ha (3,000 to over 7,000 lb/acre).

From 1955 to 1965, most of the growth in commercial catfish culture occurred in southeast Arkansas, where farmers found that raising fish could be a profitable alternative to growing traditional crops such as rice and cotton. The initial interest in foodfish aquaculture in Arkansas centered around growing buffalofish (*Ictiobus* spp.), a regionally popular table fish. In 1960, there were 1,500 ha (3,700 acres) of ponds used to raise buffalofish and 100 ha (250 acres) of catfish ponds in Arkansas, but by 1963, the production of channel catfish far exceeded that of buffalofish. In 1965, there were 4,000 ha (10,000 acres) of catfish ponds in Arkansas and another 1,000 ha (2,500 acres) of catfish ponds in other states in the southeastern United States. Annual production of catfish in the United States that year was about 7,500 metric tons (16.5 million lb).

Around 1975, the industry began to expand at a rapid rate, especially in Mississippi. That expansion was stimulated, in part, by declining profits from traditional agriculture (mostly cotton and soybeans) and a desire to diversify agricultural production and make use of land only marginally suited for row crops. Strong cooperation among farmers, particularly in the development of large feed mills and fish processing plants, was central to the early development of the industry. Growth was also facilitated by the formation of a national grower's association (The Catfish Farmers of America) in 1968 and by an extraordinarily effective national marketing effort directed by The Catfish Institute that began in 1986. In 1980, there were 15,000 ha (37,000 acres) of ponds devoted to catfish farming, and about 35,000 metric tons (77 million lb) of catfish were produced. By 1997, about 75,000 ha (185,000 acres) of ponds were in production, and over 250,000 metric tons (560 million lb) of fish were processed, which represented about half the total U.S. aquaculture production for all species. Over 95% of present-day catfish aquaculture occurs in the four southeastern states of Mississippi, Alabama, Arkansas, and Louisiana. Mississippi is the leading catfish-producing state and accounts for over 70% of the total U.S. production.

The success and rapid growth of channel catfish aquaculture in the southeastern United States has stimulated interest in using the farm-raised catfish industry as a model for the development of large-scale aquaculture in other regions and with other species. However, catfish farming may not be a good general model for aquaculture development, because the establishment and growth of the industry depended on conditions that may be difficult to duplicate with other species and in other locations. The success of catfish farming can be attributed to six general factors:

(1) Channel catfish are hardy and simple to grow.
(2) Techniques for reproduction and growout of channel catfish were known prior to large-scale industry growth. That is, lack of culture technology did not constrain early growth of the industry.
(3) Channel catfish were widely accepted as a food item in the region, which provided a ready market for initial production.
(4) Appropriate physical resources were available in the region that allowed the development of large farm operations that captured economies of size. These resources included a suitable climate, large tracts of flat land, and abundant underground water.
(5) Many farmers in the region had the capital needed to invest in large-scale aquaculture, and they were willing to take risks on a new enterprise. Farmers were also willing to cooperate in the development of the infrastructure, such as the feed mills and processing plants, needed to support commercial production.
(6) Rapid growth in production occurred when the industry moved from regional to national marketing of the fish. That transition was successful because of a well-conceived and extraordinarily effective generic marketing campaign.

CULTURE METHODS

Nearly all channel catfish produced commercially are grown in earthen ponds, because production costs are generally lower for catfish grown in ponds than in any other culture system. Production of channel catfish in systems other than ponds is economically viable only when some special circumstance exists, such as the opportunity to sell fish to a local market at an exceptional price or the availability of an unusual resource that makes production profitable. For instance, a unique channel catfish aquaculture industry is present in the arid, intermountain region of the western United States. In the Snake River Canyon of Idaho, large artesian springs supply geothermal water that allow production of catfish in flow-through raceways under nearly optimum temperature conditions throughout the year.

Catfish aquaculture in ponds is a straightforward process that typically involves four phases:

- Broodfish are held in ponds at relatively low-standing crops and allowed to mate randomly each spring when water temperatures rise above 20 °C (68 °F).
- The fertilized eggs are taken to a hatchery, where they hatch under controlled conditions. The fry are held in the hatchery for 5 to 10 days.
- Fry are transferred from the hatchery to a nursery pond, where they are fed a manufactured feed daily through the summer and autumn.
- Fingerlings weighing 20 to 40 g/fish (0.7 to 1.4 oz/fish) are seined from the nursery pond in winter or spring and transferred to foodfish growout ponds, where they are fed a manufactured feed until they reach a size desirable for processing [0.45–0.90 kg/fish (1–2 lb/fish)].

In the southeastern United States, 18 to 30 months are required to produce a food-size channel catfish from an egg.

In practice, the foregoing simple production scheme is complicated by a number of management decisions that must be made to optimize the production strategy for each farm. A few farmers specialize in producing fingerlings, which are then sold to farmers specializing in the production of food-size fish. Many farmers combine all aspects of production. Those farms have broodfish ponds, a hatchery, fry nursery ponds, and foodfish growout ponds. Specific management practices also differ from farm to farm. This factor is particularly evident in the variety of management practices used in foodfish growout ponds. Culture practices for pond-raised channel catfish were reviewed by Busch (18), Stickney (19), and Tucker and Robinson (20).

Facilities

Most ponds used for commercial channel catfish culture are built on flat land by removing soil from the area that will be the bottom of the pond and using that soil to form levees around the perimeter of the pond. (See Fig. 3.) The average pond size is between 4 to 6 water ha (10 to 15 acres), built on 4.5 to 7 ha (11 to 17 acres) of land. The depth of the water is 0.9 to 1.5 m (3 to 5 feet).

Figure 3. Channel catfish ponds in Mississippi.

Levee ponds built on flat land have little watershed to supply water by runoff, so a source of pumped water is required (21). Ideally, the water source should supply at least 140 L/min for each hectare of water (15 gal/min/acre). This figure is roughly twice the maximum daily pond evaporation rate in the southeastern United States and allows for maintenance of pond levels during periods of drought, with some excess water to meet moderate seepage losses. Groundwater is preferred over surface-water supplies, because groundwater supplies are of dependable availability and consistent quality over time and are free of wild fish and less prone to pollution than are surface waters. The most important water-quality criteria for catfish pond water supplies are salinity (it should be less than 4–6 ppt) and the absence of pesticides or other potentially harmful pollutants. Other desirable characteristics are moderately high concentrations of total alkalinity and calcium hardness (both at least 20 to 30 mg/L as $CaCO_3$). Ample total alkalinity provides pH-buffering capacity to the pond water, and calcium benefits fish osmoregulation and stress resistance.

Reproduction and Breeding

An important consideration in choosing a fish species for aquaculture is the ease with which the reproductive phase can be controlled or manipulated. In that respect, channel catfish are an ideal species for aquaculture. Channel catfish broodfish are easy to maintain in pond culture, and the spawning efficiency (the percentage of female broodfish that spawn in a given year) is reasonably good without any special manipulation of environmental conditions or the need for hormone treatments to induce spawning. Sexually mature channel catfish will not reproduce in ponds unless they are provided with an enclosed nesting site in which to mate. This is a highly desirable trait for a fish used in aquaculture, because uncontrolled reproduction in growout ponds leads to overpopulation and reduced yield of marketable fish. Eggs and fry of channel catfish are hardy and tolerate handling relatively well, and excellent egg hatchability and fry survival can be obtained in simple, inexpensive hatcheries. Eggs and fry develop rapidly at the proper temperature, so there is no need for a lengthy stay in the hatchery. At a water temperature of 27 °C (80 °F), yolk sac absorption by fry occurs 10 to 14 days after egg fertilization. After absorbing their yolk sac, channel catfish fry accept simple manufactured feeds at first feeding.

Channel catfish may mature sexually at two years of age and at weights as low as 0.3 kg (0.7 lb), but for reliable spawning, fish should be at least three years old and weigh at least 1.4 kg (3 lb). The most desirable broodfish are four to six years old and weigh between 1.8 and 3.6 kg (4 to 8 lb). Older fish produce fewer eggs per body weight, and larger fish may have difficulty entering the containers commonly used as nesting sites. At present, genetically improved stocks of fish are not widely available for commercial use, although some producers offer select strains of fish known to have good production characteristics. Commercial broodfish are sometimes obtained from foodfish growout ponds that contain large fish or from existing broodstock that appear to perform well on other farms. Broodfish are selected based on general health, size, and the development of robust secondary sexual characteristics. Fish sex is determined during broodfish selection, so that females and males can be stocked into brood ponds in the desired ratio. Determining the sex of fish and adjusting sex ratios is important, because males grow faster than females, so if broodfish selection is based only on size, then relatively few females will be selected.

In large-scale, commercial reproduction of channel catfish, broodfish are held in ponds provided with enclosed nesting sites in which fish are allowed to mate randomly. This method, while giving the culturist little control over the mating process, requires minimal facilities and less technical skill than other methods and can provide large numbers of eggs at low cost. Spawning success, which is measured as the percentage of females that spawn each year, varies from 25 to 75%. Success depends on the condition and age of the broodfish, water temperatures during the spawning season, and other factors yet to be identified. The culturist has little or no control over water temperatures, but spawning success can be maximized by choosing good broodfish and by ensuring good environmental conditions by using new or recently renovated ponds as brood ponds. Other keys to good spawning include maintaining proper standing crops and sex ratios in broodfish ponds and providing broodfish with an adequate food supply.

Maintaining a relatively low broodfish standing crop is necessary to provide good environmental conditions and to minimize suppression of spawning by overcrowding. Broodfish standing crops in ponds should not exceed about 2,250 kg of fish/ha (2,000 lb/acre). Broodfish are seined from ponds and inspected every two or three years. Large fish, which may be poor spawners, are culled at that time and replaced with smaller, younger broodfish. Brood stock replacement ensures a vigorous brood population and reestablishes proper fish standing crops. Periodic inspection of broodfish also provides an opportunity for adjusting the sex ratios within brood populations. Female channel catfish spawn once a year, but males can spawn two or more times a year. Consequently, stocking more females than males makes more efficient use of pond space. Male-to-female ratios in brood ponds are typically maintained at between 2 : 3 and 1 : 2.

Poor broodfish nutrition may result in poor egg quality or reduced spawning success, so broodfish must be provided with adequate food at all times. Also, an inadequate supply of food can result in poor-quality female broodfish, because males, which tend to be more aggressive, may consume most of the limited ration. When water temperatures are consistently above 20 °C (68 °F), broodfish are fed a nutritionally complete manufactured feed with 28 to 32% crude protein (22). Most farmers use the same feed for growout of food-sized fish. Feed is offered daily at about 0.5 to 1% of the total body weight of the fish in the pond. At water temperatures of about 13 to 20 °C (55 to 68 °F), feed is offered at 0.5 to 1% of the total body weight of the fish in the pond every other day. Channel catfish do not actively feed at water temperatures below about 10–13 °C (50–55 °F). Some producers stock forage fish (such as fathead minnows, *Pimephales promelas*)

Figure 4. Three types of containers used as artificial nesting sites for channel catfish.

into brood ponds, to provide a food source in addition to manufactured feeds.

Seasonal changes in water temperature control the reproductive cycle in channel catfish (23). Exposure to water temperatures below about 15 °C (60 °F) for a month or more over winter stimulates gametogenesis. A subsequent slow rise in the average water temperature to 20–25 °C (68–77 °F) usually initiates spawning in the spring. Water temperatures of around 25–27 °C (77–80 °F) are considered optimum for spawning.

Channel catfish must be provided with an enclosed nesting site for spawning. A variety of containers have been used successfully as artificial nesting sites. (See Fig. 4.) Most containers have an internal volume of about 75 L (20 gal) and an opening 15–25 cm (6–9 in.) across. Containers are placed in the brood pond shortly before water temperatures are expected to rise into the range for spawning. When water temperatures enter the range for spawning, the male chooses a container and cleans the inside of debris and sediment. The female is attracted to the container by olfactory signals, and mating begins. Spawning occurs over a period of several hours as several layers of adhesive eggs are deposited. Females weighing between 2 to 4.5 kg (4 to 10 lb) typically lay between 6,500 to 9,000 eggs/kg body weight (3,000 to 4,000 eggs/lb). Once spawning is complete, the male chases the female from the nest and guards the eggs. The eggs are about 0.45 cm (0.2 in.) in diameter, initially are light yellow, and become brownish yellow with age.

Culturists check the spawning containers every two or three days for the presence of eggs. A container holding eggs is gently brought to the surface and drained of water. If the male remains in the container, he is allowed to swim out before the egg mass is retrieved. (See Fig. 5.) The egg mass often sticks to the floor of the container and is gently removed by scraping it from the container. The eggs collected from the brood pond are placed in insulated, aerated containers and transported to the hatchery.

Hatchery Practices

Sophisticated hatchery facilities are not needed for producing channel catfish fry. The most critical factor for a successful hatchery is an adequate supply of high-quality water of the appropriate temperature (24,25). The optimum water temperature for egg incubation and fry rearing is 25 to 28 °C (77 to 82 °F). If water temperatures are below 25 °C (77 °F), egg hatching and fry development are prolonged, and fungi may invade the egg masses. If the temperature is above 28 °C (82 °F), the embryos may develop too rapidly, and there will be higher incidences of malformed and nonviable fry. Also, bacterial diseases of eggs and fry and channel catfish virus disease of fry are more common if the water temperature is too warm. Heating or cooling water is expensive, so most hatcheries in the major channel-catfish-producing areas of the southeastern United States obtain water from wells 300 to 400 m (1,000 to 1,300 ft) deep that naturally provide water at a constant temperature of 25 to 28 °C (77 to 82 °F).

Figure 5. A channel catfish egg mass being retrieved from a spawning container.

Shallow tanks holding about 350–400 L (90–110 gal) of water are used to incubate eggs and rear fry. The water flow through tanks is about 10–15 L/min (3–4 gal/min). Egg-hatching tanks have a series of paddles attached to a shaft running the length of the tank; the paddles are spaced along the length of the tank to allow wire-mesh baskets to fit between them. The paddlewheel shaft is then attached to a low-speed electric motor. One or two egg masses are placed in each basket, and the paddles gently rotate through the water to provide water circulation and aeration. The motion of the paddles mimics the action of the male parent, which, under natural conditions, circulates water over the eggs by fanning the water with his tail. Some culturists have discarded the traditional paddlewheel egg-hatching tank in favor of tanks with a set of airstones or a perforated pipe that delivers low-pressure, forced air from a blower. The vigorous bubbling produced by such systems aerates and circulates water around the egg mass.

The incubation time for channel catfish eggs varies from five to eight days, depending upon the temperature of the water. At hatching, the fry, called sac fry at this point, fall or swim through the wire-mesh basket and form schools in the corners of the tank. Fry are then siphoned into a bucket and transferred to a fry-rearing tank. Fry-rearing tanks are usually of the same dimensions as egg-hatching tanks, but are not equipped with paddles. Aeration in fry-rearing tanks is provided by one or two small surface agitators or by air bubbled through airstones.

Sac fry initially are golden in color and are not fed, because they derive nourishment from the attached yolk sac. Over a 3- to 5-day period after hatching, they absorb the yolk sac and turn black. At that time, the fry, now called swim-up fry, swim to the surface of the water, seeking food. Swim-up fry are fed at a daily rate an amount of feed equal to about 25% of the total body weight of the fry in the tank, divided into 6 to 12 equal feedings a day. Fry feeds are dry, finely ground meal or flour-type feeds, containing 45 to 50% crude protein (22). Most of the protein in fry feeds is supplied by fish meal. Fry are fed in the hatchery for 2 to 10 days before they are transferred to a nursery pond.

Nursery Pond Management

After their brief stay in a hatchery, fry are moved to a nursery pond for further growth. Nursery ponds are stocked with 250,000 to 750,000 fry/ha (100,000 to 300,000 fry/acre). It is difficult to feed small catfish fry recently transferred from the hatchery to a nursery pond, because fry are weak swimmers and are not able to move large distances to areas where feed is offered. It is therefore important to prepare nursery ponds so that they contain abundant natural foods, to promote growth until the fish are large enough to feed effectively on manufactured feeds. The best fertilization program for catfish nursery ponds uses a combination of high-phosphorus inorganic fertilizer and an organic fertilizer, such as cottonseed meal or alfalfa pellets. Fertilization is initiated two or three weeks before the pond is stocked with fry and continues until the fry are vigorously accepting manufactured feed. Even though fry can meet all of their nutrient requirement from the natural foods present in a well-fertilized pond, they should be offered a finely ground feed one or two times a day at 25 to 50% of the fry biomass. Fry may not make good use of the feed for several weeks, but this regimen will help train the fish to seek feed, and the unused feed is not wasted, because it acts as additional fertilizer. A month or so after stocking, the fry, now called fingerlings, will have grown to 3 to 4 cm (1 to 1.5 in.) in length and can be fed once or twice daily to satiation, using a crumbled feed or small pellet, containing 32 to 35% crude protein. The small feed can be fed to the fry throughout the nursery phase, or, as the fish grow, the size of the feed can be increased to the same size used for foodfish growout (22).

Fingerlings five to nine months of age, weighing 20–40 g each (0.7–1.4 oz), are harvested from nursery ponds in the autumn, winter, and early spring and transferred to foodfish growout ponds. Survival of fry to the fingerling stage in excess of 75% is considered good. Nursery ponds are harvested by seining each pond several times over a period of one to three months. Fingerlings from nursery ponds vary considerably in size and are usually graded to obtain a more uniform population for stocking. Grading is accomplished by using a seine with a mesh size that selectively retains fish of a certain minimum size, while allowing smaller fish to escape. Partial harvest of nursery ponds also serves to reduce the fingerling standing crop, which allows the smaller fish remaining in the pond to grow more rapidly to a desired harvest size.

After as many fingerlings as possible have been removed by seining, the pond is drained and allowed to dry, so that all remaining fish are eliminated. Removal of all fish from the nursery pond is important to prevent cannibalism of fry stocked in the subsequent cycle of fingerling production. Allowing the pond bottom to dry also helps reduce populations of predacious aquatic insects, such as dragonfly nymphs (order Odonata) and back swimmers (order Hemiptera, family Notonectidae), that may prey on catfish fry.

Foodfish Production

Cultural practices used to produce foodfish differ among farms to a much greater extent than do practices used in broodfish ponds or nursery ponds. This situation is due, in part, to differences in production goals between farms, as well as a lack of information on the economics associated with the various foodfish production strategies. Individual farmers have developed and used various production schemes, based on experience, personal preference, and perceived productivity and profitability.

The two fundamental production variables in foodfish growout are fish stocking density and the cropping system (20,26,27). Stocking densities in foodfish growout ponds range from about 10,000 to over 30,000 fish/ha (4,000 to over 12,000 fish/acre) and average about 15,000 fish/ha (6,000 fish/acre). The term "cropping system" refers to the stocking–harvest–restocking schedule. The two cropping systems used in commercial production are "clean harvesting" and "understocking." The two systems differ in the number of year-classes, or cohorts, of fish present in a pond at any one time.

In the clean-harvest cropping system, the goal is to have only one year-class of fish in the pond at one time. In a typical production cycle, fingerlings are stocked, grown to the desired harvest size [usually 0.45–0.90 kg/fish (1–2 lb/fish)], and an attempt is then made to harvest all fish from the pond before adding new fish. Fish are removed in two to four separate harvests, spaced over several months. At each harvest, faster growing fish are selectively removed with a large-mesh seine, and the smaller fish remain in the pond for further growth. After harvesting as much of the crop as possible, the pond is either drained and refilled or, more commonly, restocked without draining, to conserve water and reduce the time lost between crops.

In the understocking system, more than one year-class of fish is present after the first year of production. Initially, the pond is stocked with a single year-class of fingerlings. The faster growing individuals are selectively harvested using a large-mesh seine, and fingerlings are added ("understocked") to replace the harvested fish plus any losses suffered during growout. The process of selective harvesting and understocking continues for years without draining the pond. After a few cycles of harvesting and understocking, the pond contains several year-classes of fish and a continuum of fish sizes ranging from recently stocked fingerlings to fish that may be several years old, weighing over 1.4 kg (3 lb).

Both cropping systems have advantages and disadvantages. For example, fish harvested from clean-harvested ponds tend to be more uniform in size than those harvested from understocked ponds (27), and uniform fish size is highly desired by fish processors. It is also easier to maintain accurate inventory records when using the clean-harvest cropping system, because fish populations are "zeroed out" after each round of cropping. Feed conversion efficiencies also tend to be better for fish in clean-harvested ponds (27), probably because there is little or no year-to-year carryover of big fish, which convert feed to flesh less efficiently than do small fish. On the other hand, when all of the ponds on a farm are managed with understocking, more ponds will contain fish of marketable size at any one time than if the clean-harvest system were used. This aspect is important, because pond-raised catfish often are temporarily unacceptable for processing, because of algae-related off-flavors. So, if the timely harvest of fish from a particular pond is constrained by the presence of off-flavors (or other factors, such as ongoing losses to an infectious disease), there is a greater probability of having acceptable fish to sell from another pond when ponds are managed with the understocking strategy. Economic analyses indicate that clean-harvest cropping systems generate greater net revenues than understocked ponds when fish can be harvested and sold without constraint. But in the real world, where the presence of off-flavors and other market constraints often prevent timely fish harvest, the understocking system is the more desirable of the two cropping systems (26,28). Not surprisingly, the understocking system is by far the most common cropping system in use and is likely to remain so until solutions to off-flavor problems are developed.

Although long-term management strategies differ between the two cropping systems, daily pond management practices for foodfish growout are similar in both systems. When water temperatures are above about 15 °C (60 °F), fish are offered an extruded, floating feed of 28 to 32% crude protein (22). The feed is blown over a wide area of the pond from mechanical feeders that are mounted on or pulled by vehicles. The usual practice is to feed the fish once or twice daily to near satiation, based on visual assessment of fish feeding activity. Relatively large amounts of feed are offered daily because fish densities are high in commercial ponds. Feed allowances in commercial foodfish growout ponds average between 85 and 140 kg of feed/ha (75 and 125 lb/acre) per day during the late-spring-to-early-summer period of maximum feeding activity. Feeding activity declines as water temperatures drop in late fall, and feeding rates average less than 30 kg of feed/ha (25 lb/acre) per day during midwinter, although feed allowances may be considerably higher during abnormally mild winters. Feed allowances in individual ponds vary, depending on fish biomass (they are lower after harvest, for example) and fish health. In fact, day-to-day changes in appetite are an important indicator of the general health of fish, and reduced feeding activity is often the first sign of poor water quality or an infectious disease outbreak.

Foodfish growout ponds are usually drained only when pond levees need to be renovated or when there is a need to adjust fish inventory by completely harvesting the crop. Most commercial ponds remain in production for 5 to 15 years between pond renovations. Renovation usually involves drying the pond bottom, disking to break up the dried clay, and scraping to smooth the bottom profile and restore drainage slope. Accumulated sediment on pond bottoms is derived from erosion of levee slopes by waves, so the dried material scraped from the bottom is used to rebuild the levee and restore the proper slope. Other than measures to restore proper pond morphology, no treatment of the bottom soils is undertaken during renovation.

NUTRITION AND FEED FORMULATION

Research by H.S. Swingle in the 1950s (15) showed that production of natural foods in fertilized ponds was sufficient to produce about 200 kg/ha (180 lb/acre) of channel catfish annually. Crude feeds that provided protein and energy were later used to supplement natural productivity and increase fish yields. Early supplemental feeds were deficient in some essential nutrients, but, as long as fish were stocked at relatively low densities, the requirements for those nutrients could be met by natural food organisms eaten by the fish. As commercial catfish culture intensified, the contribution of natural foods to the nutrition of the fish became less significant, and feed quality and feeding practices became increasingly important to profitable production. Feed cost is the major cost of producing channel catfish, accounting for slightly over half of the annual operating expenses for a typical catfish farm. Efficient and economical feeds, together with effective feeding practices, are therefore essential to profitable catfish farming.

Fish nutrition is the most advanced area of channel catfish culture. The general nutritional requirements of channel catfish have been established, and practical diets can be formulated from relatively few ingredients to meet those requirements. Feeding practices, which have been described in previous sections, are less well standardized than nutritional requirements or feed manufacture. Reviews of nutrition, feed manufacture, and feeding of channel catfish are provided by Tucker and Robinson (20),

Robinson and Li (22), Lovell (29), Robinson (30), and the National Research Council (31).

Nutritional Requirements

The qualitative nutritional requirements of channel catfish are similar to those of other animals. However, the absolute and relative amounts of nutrients needed by catfish differ from those needed by other animals. Also, quantitative nutritional requirements for channel catfish may vary with fish size, growth rate, stage of sexual maturity, water temperature, diet formulation, feeding rate, and other factors. Many of these interrelationships are poorly understood.

Protein. Regular intake of dietary protein is required for animal growth and repair of tissues. Inadequate intake of protein results in poor growth or weight loss. If too much protein is consumed, part of the protein will be used as a source of energy, which is wasteful under commercial conditions, because there are less expensive sources of energy, such as lipids and carbohydrates. Protein must therefore be provided in the proper overall quantity and it must provide adequate amounts of the 10 indispensible amino acids required by catfish. (See Table 1.)

Commercial feeds for growout of foodfish contain 26–32% crude protein. Smaller fish have a higher protein requirement: Feeds for fry in hatcheries contain 45–50% crude protein, and feeds for small fingerlings in nursery ponds usually contain 32–36% crude protein. The proper balance of dispensible and indispensible amino acids in practical feeds is achieved by using a mixture of high-quality protein supplements in the feed. In practice, if catfish feeds are formulated from a mixture of plant and animal protein supplements to meet the minimum requirement for the indispensible amino acid lysine, all other indispensible amino acids will be present in adequate amounts as well.

The nonprotein energy content of feed has a significant influence on the protein requirement of catfish. Catfish apparently eat to satisfy an energy requirement, so excess dietary energy may cause the fish to eat less, which will reduce their intake of protein (and other nutrients). On the other hand, insufficient nonprotein energy results in the use of expensive dietary protein for energy rather than for tissue synthesis and growth. The best ratio of energy to protein in catfish feeds is about 8 to 9 kcal of digestible energy per gram of protein (32–34). Higher levels of dietary energy may produce greater weight gain, but may also result in increased deposition of body fat.

Table 1. Channel Catfish Requirements for Indispensible Amino Acids[a]

Amino Acid	Requirement (Percent of Dietary Protein)
Arginine	4.3
Histidine	1.5
Isoleucine	2.6
Leucine	3.5
Lysine	5.1
Methionine plus cystine[b]	2.4
Phenylalanine plus tyrosine[b]	5.0
Threonine	2.0
Tryptophan	0.5
Valine	3.0

[a]See [31].
[b]The requirement for the indispensible amino acids methionine and phenylalanine can be partially replaced by the dispensible amino acids cystine and tyrosine, respectively.

Lipids. Lipids are important as sources of essential fatty acids and as relatively inexpensive and highly digestible sources of energy. Dietary lipids also improve the absorption of fat-soluble vitamins and may enhance the flavor of feeds.

The essential fatty acid requirements of channel catfish are not precisely known, although it appears that small amounts of n-3 fatty acids are required in the diet (35). The n-3 notation refers to the family of fatty acids in which the first double bond is three carbon atoms from the methyl (CH_3) end of the molecule. The n-3 family is sometimes referred to as the $\omega 3$ (omega three) family. Because the essential fatty acid requirements of catfish can be met with only a small amount of lipids in the diet, the amount of lipids in catfish feeds is not based on a specific nutritional requirement for lipids, but rather on achieving the proper energy-to-protein balance in the feed so that protein is spared. The source of lipids used in feeds is important, because certain lipids can have a negative effect on product quality and fish health.

Lipid levels in feeds for growout of foodfish seldom exceed 5 to 6% of the diet, although there is no conclusive evidence regarding an optimum level for growth. High-lipid feeds are, however, difficult to manufacture into pellets, and their use may cause excessive fattiness in fish, which reduces dressout and affects the storage quality of processed products. About 75% of the total lipids in practical catfish feeds is inherent in the feed ingredients themselves. The remaining 25% is derived from fat that is sprayed (top dressed) on the finished feed to increase feed energy and help stabilize the pellet, so that less dust is produced when the feed is transported. Various animal and plant lipids have been used to top dress catfish feeds. Marine fish oils are not recommended for this use, because they may impart a "fishy" flavor to the otherwise mild-tasting catfish flesh. Also, there is evidence that some component of menhaden oil (probably highly unsaturated n-3 fatty acids) suppresses the immune system of channel catfish, making them more susceptible to the bacterial pathogen *Edwardsiella ictaluri* (36).

Carbohydrates. Catfish do not have a dietary requirement for carbohydrates, but carbohydrates are the least expensive source of energy in catfish feeds and are therefore important in catfish feeds to spare protein. Channel catfish use starch or dextrin more efficiently than simple sugars such as glucose and sucrose (37), and fiber is assumed to have essentially no food value to catfish (31). Commercial catfish feeds have 30 to 40% carbohydrates, primarily in the form of starch. Carbohydrates in the feed also serve a nonnutritional role, in that starch and other complex carbohydrates aid in manufacturing a stable floating pellet.

Vitamins. Fourteen vitamins are considered to be metabolically essential for channel catfish. Although the characteristic signs of vitamin deficiency listed in Table 2 can be induced under laboratory conditions, overt signs of vitamin deficiency are rare in nature or commercial culture. Vitamin deficiencies in pond-raised catfish have been documented only for vitamin C (38–40) and pantothenic acid (41). Apparently, vitamins naturally present in feed ingredients and in natural food organisms in the pond provide adequate levels of most vitamins during culture. Nevertheless, commercial feeds are supplemented with a vitamin premix that provides sufficient quantities to meet the requirement and to compensate for any vitamin losses that occur during feed manufacture.

Table 2. Vitamin Deficiency Signs and Minimum Dietary Requirements for Channel Catfish[a]

Vitamin	Deficiency Signs	Requirement
Fat soluble		
A	Exophthalmia, edema, ascites	1000 IU/Kg
D	Reduced bone mineralization	500 IU/Kg
E	Skin depigmentation, muscular dystrophy, anemia	50 IU/Kg
K	Hemorrhage, prolonged blood-clotting time	Required
Water soluble		
Thiamin	Dark skin pigmentation, neurological disorders	1 ppm
Riboflavin	Short-body dwarfism	9 ppm
Pyridoxine	Greenish-blue coloration, neurological disorders	3 ppm
Pantothenic acid	Clubbed gills, anemia	15 ppm
Niacin	Anemia, exophthalmia, skin lesions	14 ppm
Biotin	Anemia, skin depigmentation	Required
Folic acid	Anemia	1.5 ppm
B_{12}	Anemia	Required
Choline	Hemorrhagic kidney and intestine, fatty liver	400 ppm
Ascorbic acid	Scoliosis, lordosis, anemia, internal and external hemorrhage	60 ppm

[a]Generalized deficiency signs, such as death, reduced weight gain, and loss of appetite, are common to most vitamins and are not listed. [See also (22,31).]

Minerals. Fourteen minerals are considered essential for channel catfish. These include seven major minerals (calcium, phosphorus, magnesium, sodium, potassium, chlorine, and sulfur) and seven trace minerals (iron, zinc, copper, manganese, cobalt, selenium, and iodine). Most of the required minerals can be obtained from the water or from feed ingredients used to formulate commercial diets. Natural pond foods may also be significant sources of certain trace minerals. Phosphorus and calcium are required by catfish in relatively large quantities, and feeds are usually supplemented with dicalcium phosphate to ensure that adequate amounts of both are present in the diet. Commercial feeds are also supplemented with a trace-mineral premix, although there is good evidence that supplemental trace minerals are not needed in feeds for pond-cultured fish, especially if the diet contains some animal protein (22).

Feed Formulation and Manufacture

Feeds used in intensive pond culture of channel catfish are formulated to provide all of the required nutrients in the proper proportions. However, no single feedstuff can supply the optimum mix of nutrients, so a combination of feed ingredients is used to meet nutritional needs. Fortunately, the number of different feedstuffs needed to formulate effective channel catfish feeds is small. Examples of formulations for practical channel catfish feeds are presented in Table 3.

Feeds for fry are prepared as meals or flours of small particle size. (See Fig. 6.) They are formulated to contain high levels of dietary protein and high levels of fish meal. Most feeds for fingerlings and foodfish growout are prepared by extrusion cooking, followed by drying. That process produces a hard, expanded pellet that floats in water. (See Fig. 6.) Extruded feeds must contain about 25% corn or other high-carbohydrate grains

Table 3. Examples of Practical Feeds for Channel Catfish Culture

	Percent of Feed			
	Fry	Fingerling	Foodfish	Foodfish
Ingredient	(50% protein)	(35% protein)	(32% protein)	(28% protein)
Soybean meal		40	36	26
Cottonseed meal		10	10	10
Corn grain		15	23	31
Wheat middlings	20	20	20	22
Menhaden meal	60	6	4	4
Meat/bone/blood meal	15	6	4	4
Dicalcium phosphate		1	1	1
Top-dressed fat	5	2	2	2
Vitamin premix	added	added	added	added
Mineral premix	added	added	added	added

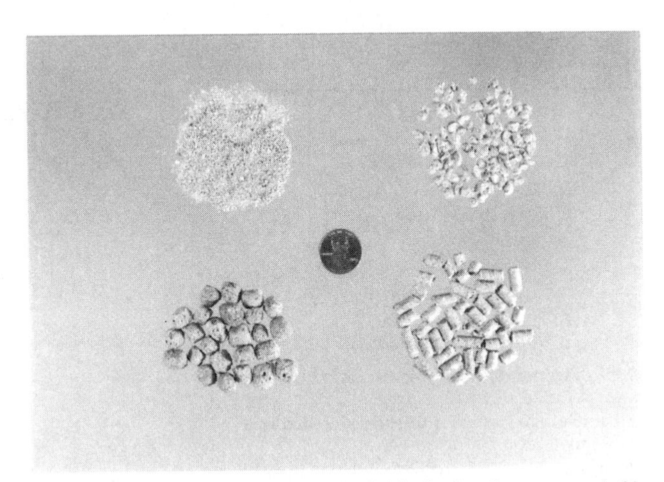

Figure 6. Four types of catfish feed (clockwise from upper left), a meal-type feed; a crumbled feed; a steam-pelleted sinking feed; and an extruded, floating pellet.

for proper gelatinization of pellets during extrusion. Soybean meal provides most of the protein in fingerling and foodfish feeds. Small amounts of fish meal or other animal protein are added to improve amino acid balance and enhance palatability. Commercial trace-mineral and vitamin premixes are added to all channel catfish feeds to ensure nutritional adequacy.

WATER QUALITY MANAGEMENT

The best environmental conditions for fish growth exist immediately after a pond is filled with water. At that time, the level of dissolved oxygen is near saturation, and the water contains negligible levels of ammonia, nitrite, and other potentially toxic substances that accumulate during culture. But as soon as fish are stocked and fed, the environment begins to deteriorate, and conditions become less suitable for fish growth.

Water quality deteriorates over time in catfish ponds for two reasons. First, the metabolic activities of fish directly affect water quality. For example, fish consume oxygen and produce ammonia as a waste product of protein catabolism. Second, and more important, water quality deteriorates because nutrients contained in fish wastes stimulate the growth of phytoplankton blooms (42). The presence of phytoplankton is not necessarily bad; in fact, low to moderate standing crops of phytoplankton are considered to be beneficial in aquaculture ponds, because they produce oxygen in photosynthesis and prevent the growth of noxious aquatic weeds by competing for light and nutrients. Phytoplankton growth is, however, difficult to control in large ponds, and phytoplankton blooms often become excessively dense during the summer months, when fish feeding rates are high. Dense phytoplankton blooms can cause serious water quality problems, including depletion of dissolved oxygen and the production of odorous algal metabolites that give fish unpleasant tastes (43).

The extent to which environmental conditions in catfish ponds deteriorate is proportional to the amount of fish waste entering the water, which is, in turn, related to the feeding rate (44). Thus, the key to profitable catfish aquaculture is to find the optimum balance between two opposing relationships that affect fish production: To grow more fish, you must provide more feed; but as more feed is offered, the water quality deteriorates and fish grow slower. The overall goal of pond management, then, is to feed fish at the highest possible rate without causing the environment to degrade to the point at which net economic returns decrease as a result of poor fish growth, loss to disease, or excessive costs incurred in attempts to manage water quality at an acceptable level (such as aeration to supply oxygen). The point of diminishing economic returns is not known with certainty, although the foodfish culture practices outlined previously appear to be profitable. Those practices involve fish densities of 10,000 to over 30,000 fish/ha (4,000 to over 12,000 fish/acre) and summertime feeding rates of 85 to 140 kg of feed/ha (75 to 125 lb/acre) per day. Production at that level of intensification can be sustained in ponds with little or no water exchange and no technological intervention other than use of supplemental aeration. Water quality management in catfish ponds has been reviewed by Tucker and Robinson (20), Boyd (45), Boyd and Tucker (46,47), Tucker and Boyd (48), and Tucker (49).

Water Quality Requirements

Channel catfish are relatively tolerant of poor environmental conditions, which makes them a good species for pond aquaculture. Optimum and tolerated ranges for individual water quality variables have been identified in laboratory studies (see Table 4), but that information is of limited value under commercial conditions, because environmental stressors interact, and a combination of stressors may kill fish even though individual values are within a tolerable range. For example, carbon dioxide reduces the oxygen transport capacity of blood, so its presence aggravates the stress imposed by exposure to low oxygen. While catfish may survive for hours at dissolved oxygen concentrations of 2 mg/L if carbon dioxide levels are low, they may quickly die if carbon dioxide levels are high.

Although hundreds of water quality variables may potentially affect the health of channel catfish, only a few are important under routine commercial conditions. Initially, the water supply for ponds should be checked to assure that it is free of pollutants and that the salinity, alkalinity, and hardness of the water are within the desired range. Once the water is impounded and used for culture, the important variables are those affected by biological processes, because those variables can change rapidly. The key variables in pond culture of channel catfish are dissolved oxygen, ammonia, and nitrite. Algae-related off-flavors, another important water quality problem encountered in catfish aquaculture, is unique, because it affects product quality rather than fish health.

Dissolved Oxygen

Maintaining adequate levels of dissolved oxygen is the most common water quality problem in pond culture of channel catfish, because the respiration of fish, plankton,

Table 4. Optimum and Tolerated Ranges of Some Water Quality Variables for Growth of Fingerling and Adult Channel Catfish[a]

Variable	Optimal Range	Tolerated Range
Water temperature	25–30 °C (77–86 °F)	0–40 °C (32–104 °F)
Salinity	0.5–4 ppt	<0.1–11 ppt
Dissolved oxygen	5–15 mg/L	1–20 mg/L
pH	6–9	5–10
Total alkalinity	20–400 mg/L as $CaCO_3$	<1 to >400 mg/L as $CaCO_3$
Total hardness	20–400 mg/L as $CaCO_3$	<1 to >400 mg/L as $CaCO_3$
Carbon dioxide	0 mg/L	0–20 mg/L[b]
Unionized ammonia	0 mg/L	0–0.2 mg/L as nitrogen
Nitrite	0 mg/L	Depends on chloride concentration[c]

[a]Prolonged exposure to nonoptimal conditions may not lead to death, but could result in reduced growth, impaired reproductive performance, or increased susceptibility to disease.
[b]Tolerance to carbon dioxide depends strongly on the concentration of dissolved oxygen, but the relationship is poorly quantified.
[c]Maintaining a ratio of at least 20 mg of chloride/L of water for every 1 mg of nitrite–nitrogen/L of water protects catfish from nitrite toxicosis.

and benthic organisms in the bottom sediment exerts a tremendous demand for oxygen (50,51). After the nutritional requirements of the fish, the availability of dissolved oxygen is the next factor that limits the intensification of catfish pond aquaculture.

Oxygen Requirements of Catfish. Channel catfish are considered to be relatively tolerant of low concentration of dissolved oxygen (hypoxia), but they are healthiest and grow best when concentrations of dissolved oxygen are near saturation. Channel catfish eggs are particularly susceptible to hypoxia, because eggs are not motile and therefore cannot move to water with a higher concentration of dissolved oxygen. Also, channel catfish eggs are laid in a large, adhesive mass, and the oxygen supply to the interior of the mass depends on diffusion of the gas through the outside layers of eggs. The rate of oxygen diffusion into the egg mass depends not only on the ambient concentration of dissolved oxygen, but also on water currents to continuously replenish the layer of oxygen-rich water at the surface of the egg mass. In nature, a steady supply of oxygen is ensured by the male parent, who constantly fans the eggs with his fins. Eggs will begin to die after several hours of exposure to concentrations of dissolved oxygen below about 2 to 3 mg/L, and the oxygen requirement of eggs increases as the eggs mature and the embryos become more active. Healthy adult fish can survive indefinitely when the concentration of dissolved oxygen is above 2 mg/L but they may feed poorly, grow slower, and be more susceptible to infectious diseases when the concentration is below about 3 to 4 mg/L. Adult channel catfish can survive for several hours at concentrations as low as 0.5 mg/L. Fingerlings are somewhat more tolerant of low concentrations of dissolved oxygen than are larger fish and may survive short exposures to even lower levels of dissolved oxygen.

Assessing the effect of dissolved oxygen on fish health under commercial conditions is complicated by interactions with other environmental and physiological factors. Hypoxia at the cellular level is not always related simply to low environmental concentrations of dissolved oxygen. Any set of conditions that increase the demand for cellular oxygen decrease the rate of diffusion of oxygen from water to blood, or decrease the amount of oxygen carried by blood can result in tissue hypoxia, even when the environmental concentration of dissolved oxygen is near saturation. Some common factors that interact with environmental concentrations of dissolved oxygen include exposure to high concentrations of carbon dioxide, unionized ammonia, or nitrite. Structural alterations of the gill, caused by bacterial infection or infestations of protozoan parasites, also reduce respiratory efficiency.

Management of Dissolved Oxygen. The concentration of dissolved oxygen is affected by photosynthesis, respiration, and gas transfer between water and atmosphere. Oxygen is produced in photosynthesis by plants and is consumed in respiration by fish, plankton, and benthic organisms. Gas transfer with the atmosphere can be either a source or loss of oxygen. In catfish ponds, phytoplankton metabolism dominates the oxygen budget, because phytoplankton photosynthesis is the major source of oxygen and phytoplankton respiration (together with bacterial decomposition of phytoplankton-based detritus) is the major cause of oxygen loss (50,51). The metabolic activities of phytoplankton cause the concentration of dissolved oxygen to cycle over a 24-hour period: The concentration rises during daylight periods, when plants photosynthesize, and falls at night, when photosynthesis stops, but respiration continues. Water temperature profoundly affects all processes involved in pond oxygen budgets, and problems with low concentrations of dissolved oxygen are most common when temperatures are high. Oxygen budgets also vary with the standing crop of fish, plankton, and benthic organisms. Each pond therefore has its own unique oxygen budget, because standing crops of fish and microorganisms differ from pond to pond.

The critical period for managing dissolved oxygen is at night during the summer, when oxygen concentrations are falling rapidly. The key to successful management of dissolved oxygen in channel catfish ponds is early identification of those ponds that may require supplemental mechanical aeration to keep fish alive. Culturists use portable electronic oxygen meters to measure the concentration of dissolved oxygen in all ponds at frequent intervals throughout the night. Aeration is initiated in a pond when the concentration of dissolved oxygen falls to a level considered critical by the individual farmer (usually around 3–4 mg/L). Aeration is continued until past dawn, when measurements indicate that the concentration of dissolved oxygen is increasing, as a result of photosynthetic activity. Problems with low concentrations of dissolved oxygen are rare in channel catfish ponds when water temperatures fall below 15 °C (60 °F), and most producers discontinue monitoring the concentration of dissolved oxygen in the winter, when water temperatures are expected to remain cool.

There is a trend in the catfish industry towards automation of monitoring dissolved oxygen and of aeration. Oxygen sensors can be permanently installed in each pond to continuously measure concentration of the dissolved oxygen. The data are then fed to a computer or microprocessor that is programmed to turn aerators on when the concentration of dissolved oxygen falls below a preset level.

Paddlewheel aerators (see Fig. 7) are the most common type of aerator used in channel catfish ponds (52). These devices consist of a hub with paddles attached in a staggered arrangement. Most paddlewheel aerators used in catfish farming are powered by a 7.5-kW (10-hp) electric motor and are mounted on floats and anchored to the pond bank. Current practice is to provide aeration at about 2 to 4 kW/ha (1 to 2 hp/acre); for instance, two 7.5-kW (10-hp) aerators may be placed in a 6-ha (15-acre) pond.

Ammonia and Nitrite

The use of mechanical aeration allows farmers to use high fish stocking densities and feeding rates without losing fish to depletion of dissolved oxygen. Nevertheless, aeration does not allow unlimited fish production, because accumulation of toxic nitrogenous wastes affects fish health and decreases fish growth rates as the intensity of the culture increases (44,53).

Virtually all of the combined inorganic nitrogen in unfertilized channel catfish ponds originates from nitrogen in feed protein and is excreted by fish as ammonia.

Figure 7. An electric paddlewheel aerator.

The ammonia excreted by fish is proportional to the amount of feed consumed and comes out to about 0.03 kg of ammonia–nitrogen/kg of feed consumed, so at a feeding rate of 100 kg/ha (90 lb/acre) per day, the input of ammonia–nitrogen to a typical catfish culture pond amounts to about 0.3 mg/L per day. If ammonia were to accumulate unabated, the concentration of unionized ammonia (the form of ammonia that is toxic to fish) would increase in a few days to lethal levels. The fact that culture is possible at that feeding rate indicates that transformations and losses of nitrogen act to reduce ammonia concentrations and allow the continual input of relatively large amounts of feed to ponds without dangerous accumulation of ammonia (54).

The maximum nitrogen loading rate possible without unreasonable accumulation of ammonia in the water is governed over the long term by the rate at which natural physical and microbiological processes remove nitrogen from the pond (54). Although the rates and mechanisms of nitrogen fluxes and transformations in channel catfish ponds are poorly understood, practical experience indicates that feeding rates up to about 85 to 140 kg of feed/ha (75 to 125 lb/acre) per day are possible during the summer growing season, without unreasonable accumulation of ammonia (55). This range of rates is generally accepted as the upper range for long-term feeding rates in channel catfish culture ponds managed with no water exchange. Feeding rates (and therefore fish production) could safely be increased above the ranges that are currently used if the rate of nitrogen removal from the pond could be increased. Although it is technologically possible to enhance rates of inorganic nitrogen removal from pond water, cost considerations limit the practicality of using formal filtration or water exchange to remove nitrogen from large commercial culture ponds.

Nitrite (NO_2^-) is the intermediate product in the two-step, bacteria-mediated oxidation of ammonia to nitrate. That process, termed "nitrification," occurs in all aerobic aquatic and terrestrial environments. Nitrite occasionally accumulates in channel catfish pond waters and poses a threat to fish health. Waterborne nitrite enters the bloodstream of channel catfish and oxidizes hemoglobin to methemoglobin (56). Methemoglobin is a brown-colored pigment that is incapable of binding oxygen. Fish with high levels of methemoglobin in their blood may suffocate even when the concentration of dissolved oxygen is high.

In the southeastern United States, episodes of elevated nitrite levels are most common in the spring and fall, when water temperatures are rapidly changing (55). Nitrite toxicosis is easily prevented by assuring that adequate chloride is present in the water because chloride competes with nitrite for uptake at the gills (56,57). Maintaining a ratio of 20 mg of chloride of water for every 1 mg of nitrite–nitrogen of water prevents nitrite toxicosis in channel catfish (58). Chloride is inexpensively added to channel catfish ponds as common salt, sodium chloride.

Off-Flavors

Certain algae and bacteria that grow in catfish ponds produce odorous organic compounds that can give fish undesirable "off-flavors" (59). The most common off-flavors in pond-cultured catfish are caused by geosmin, an earthy-smelling compound, and 2-methylisoborneol, which has a musty odor. The two compounds are synthesized by species of blue-green algae (60–62) and are taken up from the water across the fish's gills. Geosmin and 2-methylisoborneol are not toxic to fish or to humans eating tainted fish, but the earthy and musty flavors that they impart to fish are highly objectionable to consumers.

Algae-related off-flavors are relatively common in pond-raised catfish, and samples of fish are always tested for taste prior to harvest, to ensure that fish with off-flavors will not reach the marketplace. The testing is conducted by trained personnel at the fish processing plants. Typically, fish samples are submitted a week or two before the desired harvest date, a day or two before harvest, and immediately prior to unloading live fish from transport trucks at the processing plant. The entire pond population of fish is rejected for processing if off-flavors are detected in any sample from that pond.

The occurrence of earthy and musty off-flavors in pond-raised channel catfish is sporadic and coincides with the appearance and eventual disappearance of the blue-green algal species responsible for synthesis of the odorous compounds (63). It is not consistently possible to prevent off-flavors, because the specific environmental conditions leading to the occurrence of odor-producing blue-green algal species are not known. Furthermore, the use of algicides is not always successful, because algicides that are legal to use, such as copper sulfate, lack selective algicidal or algistatic activity against nuisance species.

Management of off-flavors in fish relies upon natural elimination of the odorous compound from the flesh once the fish is no longer in the presence of the organism producing the compound. In other words, fish with off-flavors often are simply left in the culture pond until the odor-producing algae disappears from the plankton community. Elimination of geosmin or 2-methylisoborneol from fish flesh is rapid (64); however, an unpredictable period (weeks to months) may be required for the odor-producing algae to disappear from the community. A somewhat more dependable approach to managing off-flavors is to transfer fish to a "clean" environment, such as a different pond. The flavor quality of the fish often improves within four to seven days after moving the fish. However, moving fish from pond to pond to eliminate flavor problems is labor intensive and stresses the fish. Also, the costs associated with holding market-sized fish in ponds past their proper harvest date are a significant economic burden to catfish farmers (28).

INFECTIOUS DISEASES

High fish densities and stressful environmental conditions in channel catfish culture ponds are conducive to the outbreak and rapid spread of infectious diseases. Infectious diseases may be caused by viral, bacterial, fungal, or protozoan pathogens. Bacterial diseases account for most of the losses of fingerlings in nursery ponds, while the major diseases of fish in foodfish growout ponds are proliferative gill disease (caused by a myxosporean parasite) and "winter-kill syndrome" (a disease associated with external fungal infections). Infectious diseases of channel catfish are reviewed by Tucker and Robinson (20), MacMillan (65), Johnson (66), and Thune (67,68).

Viral Diseases

Two viruses are known to cause disease in channel catfish. Channel catfish virus (CCV) causes an infectious disease of young catfish that can lead to large losses in hatcheries or fry nursery ponds (69). The other virus—channel catfish reovirus (CRV)—appears to be of low pathogenicity and is not considered economically important (69,70).

Channel catfish virus is a highly virulent herpesvirus (71). The virus is quite host specific, and channel catfish is the only species known to sustain natural epizootics. Fish with channel catfish virus disease (CCVD) usually have a distended abdomen and exophthalmia, due to accumulation of fluid in the body cavity (72). Infected fish feed poorly and swim erratically, often in a spiraling fashion. The course and outcome of the disease are strongly affected by fish size and water temperature (73). Fish less than 1 month old are considered very susceptible to the disease. Older fish are more resistant, and fish over about 20 cm (8 in.) long (typically 9 to 12 months old) are considered resistant to CCVD. Fish do not develop CCVD at water temperatures below 15 °C (60 °F); disease occurrence is irregular and mortalities are usually low at temperatures between 20 and 25 °C (68 and 77 °F). Losses to CCVD can be devastating at temperatures above 30 °C (86 °F). There is no cure for CCVD, but losses can be reduced by controlling the water temperature in hatcheries (when possible) and by minimizing stress in susceptible fry or fingerling populations. The incidence of this disease can also be reduced by using hygienic hatchery practices to reduce the spread of the disease.

Bacterial Diseases

Three bacterial diseases are significant in channel catfish aquaculture: enteric septicemia of catfish, columnaris disease, and motile aeromonad septicemia. Enteric septicemia of catfish and columnaris disease are especially important in the nursery phase of catfish production, when they can cause large losses.

Enteric septicemia of catfish (ESC) is caused by the gram-negative bacterium *E. ictaluri* (74). The bacterium is fairly host specific for channel catfish, although experimental infections can be established in several other fish species as well (75). Fish with ESC are listless and often swim in slow, erratic spirals at the surface of the water. Infected fish may have a rash of pinpoint hemorrhages around their fins and on their ventral surface. In acute infections, fish may develop an ulcerative lesion on the top of the head, giving the disease one of its trivial names, "hole-in-head disease" (76,77). The occurrence of the disease is highly temperature dependant (78). In the southeastern United States, epizootics occur only in the late spring and early autumn when the water temperature is between 22 and 28 °C (72 and 82 °F). Channel catfish of all sizes and ages are susceptible to ESC, but fingerlings account for most of the fish lost to the disease. Prompt diagnosis and treatment are critical to successful treatment of the disease. Therapy consists of oral delivery of the appropriate antibiotic, incorporated into a manufactured feed.

Columnaris disease is caused by the gram-negative bacterium *Flavobacterium columnare* (*Cytophaga columnaris*). The bacterium is ubiquitous in aquatic environments, and outbreaks of the disease are usually associated with some predisposing stressor, such as handling or poor environmental conditions (75). Epizootics occur throughout the warmer months of the year, particularly in spring and autumn when the water temperature is between 20 and 25 °C (68 and 77 °F). Columnaris usually begins as an external infection of the gills or body that progresses to an internal, systemic bacteremia (75). Gross lesions on the gills are characterized by yellow-brown areas of necrosis at the distal end of the gill filaments. Skin lesions appear as areas of depigmentation that progress to large necrotic ulcers. Infections of *F. columnare* can be controlled by antibiotic therapy, using the appropriate medicated feed.

Motile aeromonad septicemia (MAS) is caused by the gram-negative bacteria *Aeromonas hydrophila* and *A. sobria*. Epizootics of MAS may occur throughout the year, but are most common in spring and autumn when the water temperature is between 15 and 25 °C (60 and 77 °F). As with columnaris disease, outbreaks of MAS are usually associated with predisposing stress (79). Symptoms of the disease vary greatly, depending on the virulence of the strain of *Aeromonas* involved in the infection. Clinical signs range from external manifestations such as areas of external hemorrhage and ulcerative skin lesions to internal lesions characteristic of systemic bacteremias (80). Medicated feed therapy is used to treat MAS, but control may only be temporary unless the predisposing stressor is corrected.

Fungal Diseases

Water molds of the genus *Saprolegnia* are important pathogens of channel catfish eggs in hatcheries (20,65) and are also associated with a serious disease problem that affects adult fish in ponds during the colder seasons of the year (81). Water molds are common saprophytic microorganisms that usually cause problems only under suboptimum environmental conditions.

Problems with fungal infections of eggs are most common in hatcheries that use water at a temperature less than 25 °C (77 °F). Fungal growth begins on infertile or dead eggs and may then spread to and kill healthy eggs. Infected eggs may be treated with a 15-minute bath of 100 ppm formalin to kill the fungus. Use of water at the proper temperature for egg incubation is, however, the best solution to the problem.

External fungal infections of channel catfish are common in the winter months when water temperatures are below 15 °C (60 °F). Infections are easily diagnosed by the presence of brownish, cottony patches on the external surfaces. Areas of depigmentation and loss of the mucus layer may be present before masses of fungal mycelia are obvious. The conditions responsible for such infections are not clear (82), and mortality rates may be quite high. When losses are associated with the fungal infection, the disease is colloquially referred to as "winter-kill syndrome." There is no cost-effective treatment for fungal infections of fish in large commercial ponds (83).

Protozoan Parasites

Species of the protozoans *Trichodina*, *Trichophrya*, *Ambiphrya*, and *Ichthyobodo* are commonly found in low numbers on the external surfaces (particularly the gill filaments) of channel catfish and usually cause insignificant problems for the fish. Under certain conditions, however, the abundance of the parasite increases dramatically, to the detriment of the fish (67,68). Factors regulating fish parasitism by these organisms are poorly understood, but epizootics are commonly associated with overcrowding and poor environmental conditions. Most of these infestations can be treated with chemical therapeutants, such as copper sulfate or potassium permanganate, applied to the water. Treatments must be carefully considered, however, because the chemicals are toxic to fish at concentrations only slightly higher than the therapeutic dosage. Prevention of disease outbreaks by maintaining good environmental conditions is the best management practice for these protozoans.

Ichthyophthirius multifiliis ("Ich") can be a devastating protozoan parasite of channel catfish. Epizootics involving Ich are relatively rare, but losses may approach 100% of the affected population when conditions are optimal for spread of the disease (65). Ich is a ciliated protozoan with a biphasic life cycle. The adults, called "trophozoites," are relatively large [up to 1 mm (0.04 in.) in diameter], with a C-shaped nucleus. Trophozoites cause great tissue damage as they migrate through the epidermis, feeding on tissue fluids and cell debris. The trophozoite leaves the fish after maturing and settles to the bottom of the pond, where it encysts and divides to produce up to 2,000 infective cells, called "tomites." Ich is an obligate fish parasite, and the tomites will die if they do not locate a fish host within a few days. Epizootics of Ich are most common when the water temperature is between 20 and 25 °C (68 and 77 °F). Fingerlings are particularly susceptible, because they are held at high densities, which enhances the spread of the disease. The most obvious sign of Ich is the presence of many raised, white, pinhead-sized spots on the skin. Ich is difficult to control, because the parasite resides beneath the skin of the fish. Treatment consists of breaking the life cycle by killing trophozoites after they leave the fish or by killing the tomites before they infect the fish. Five to seven treatments of copper sulfate applied to the water daily may control the disease if the treatment is initiated early in the epizootic.

A species of the myxozoan *Aurantiactinomyxon* causes proliferative gill disease (PGD), a severe disease of pond-raised channel catfish (84). This protozoan is also found in the gut of a common mud-dwelling oligochaete, *Dero digitata*, which apparently serves as a reservoir for the parasite in ponds. The gills appear to be the initial route of infection of channel catfish. The disease is characterized by severe inflammation of the gills and varying degrees of tissue necrosis. The gill lesions reduce respiratory efficiency, and infected fish may suffocate. Outbreaks of PGD occur most commonly in spring and autumn when the water temperature is between 15 and 20 °C (60 and 68 °F), although parasites may be found in fish throughout the year. There is no treatment for the disease; however, losses can be reduced by assuring that high levels of dissolved oxygen are maintained during an epizootic.

HARVESTING

Harvesting fish from levee ponds is relatively simple, because most ponds have regular shapes (usually rectangular) and are shallow, with smooth bottoms that are devoid of snags (20,85). Seines used for harvesting are about 2.75 m (9 ft) deep and about 1.5 times as long as the maximum width of the pond. They are usually made of knotted polyethylene, have a float line at the top and a weighted lead line on the bottom, and are built with a tapered tunnel about 45 m (150 ft) from one end. The tunnel has a drawstring closure that can be connected to the metal frame of a live car (also called a sock), which is an open-topped bag made of netting material. The netting used to construct seines and live cars is sized to capture the smallest fish size desired. Seines used to harvest foodfish use a square or diamond mesh size of about 4 cm (1.5 in.), which retains fish greater than 0.35 kg (0.75 lb) and allows smaller fish to escape and remain in the pond for further growth.

Two tractors are used to pull the seine through the pond. Once the two tractors have moved the length of the pond, the ends of the seine are brought together, and the seine is wound onto a seine reel. As the seine is pulled in, the fish become increasingly crowded in the remaining area. At that point, a live car with the proper mesh size for grading is attached to the seine. The fish crowded inside the seine then swim into the live car through the tunnel connecting the live car to the seine. The live car is detached from the seine and closed after it is filled to the recommended capacity. Another live car may then be attached to receive additional fish if the harvest is large. The detached live cars are placed near an aerator, and the fish are allowed to grade for several hours. After size grading is complete, a net attached to a hydraulic boom is used to load fish onto transport trucks. (See Fig. 8.)

Figure 8. Channel catfish crowded into one end of a live car are loaded onto a transport truck, using a net attached to a boom.

PROCESSING

Fish are transported alive to processing plants. Transport trucks carry four to nine aerated tanks, each of about 3,000 L (800 gal). The tanks are filled with about 1,500 L (400 gal) of water, and each tank can hold about 1,000 kg (2,200 lb) of fish. At the processing plant, a sample of fish from the transport truck is tested for flavor quality. If the fish are of acceptable quality, they are unloaded into large concrete tanks containing flowing, aerated water. When needed in the processing plant, fish are removed from the tank, weighed, and stunned with alternating current electricity. Stunning renders the fish immobile and makes them easier to handle in the processing line.

The individual steps involved in processing channel catfish have been reviewed by Ammerman (86). Much of the processing is done by hand, but there is a trend towards increased automation to reduce processing costs. A variety of products are marketed, with common products being fillets, steaks, and whole dressed fish. Processed catfish are available either as a fresh, ice-packed product or frozen. Frozen fillets account for the largest portion (about a third by weight) of sales.

THE FUTURE

Channel catfish production has increased over eighty-fold since 1970, and per-capita consumption, which stood at 0.45 kg (1 lb) in 1997, ranks fifth among all seafood products in the United States, behind tuna, shrimp, pollock, and salmon. Although per-capita consumption is highly skewed towards the southeastern states, where catfish have long been a traditional food, more than 0.23 kg (0.5 lb) of catfish are consumed annually per capita in states throughout the midwest and southwest and in California. Because farm-raised catfish have become a widely accepted food item throughout much of the United States, the demand for catfish should continue to increase as American consumers increasingly turn towards fish and shellfish as part of their overall diet.

The continued market demand for farm-raised catfish appears to offer a bright future for the industry. However, the market demand for farm-raised channel catfish has consistently exceeded the supply over the last few years. This situation could pose a problem in the future, because the market share could be lost to other fish species that are obtained either from harvest fisheries or aquaculture unless the supply of catfish increases to meet the steadily increasing demand. Greater production of catfish must come about through expansion of facilities and improved production efficiency in existing facilities.

Large tracts of land suitable for pond culture of catfish remain available in the major catfish-producing states in the southeast, and it is unlikely that limitations of land or water will constrain industry growth in the near future. However, significant limitation on increased facility expansion is the large capital investment required to enter the industry, especially considering that economies of size greatly favor large farms. Much of the current expansion of catfish farming therefore derives from existing producers adding new ponds, rather than from new producers entering the industry. Another possible constraint on industry growth is increasingly strict governmental regulation of aquaculture, especially with respect to environmental effects. Pond culture of catfish can be a relatively benign form of agriculture, because current management practices are such that the volume of effluent discharged from ponds is low (87), and pesticides and drugs are used infrequently, if at all (88). Nevertheless, regulations promulgated without adequate consideration of the true effects of pond aquaculture on the environment could severely hamper growth of the industry.

Industrywide pond yields of catfish probably average about 4,000 kg/ha (3,500 lb/acre). That value is about half the yield that can be achieved from ponds using common culture practices and commercial feeds, but under controlled, experimental conditions (24). Frequent episodes of off-flavor, which reduce long-term production by delaying harvest, and difficulties in controlling infectious diseases in large commercial ponds account for most of the differences between commercial yields and yields attained under experimental conditions. Clearly, improvements in disease management and off-flavor abatement offer great potential for improving production efficiency. Also, the potential for increased yields from genetic improvement has long been recognized, but has yet to be widely implemented within the industry. This situation may change soon, however, as the products of the catfish breeding program supported by the United States Department of Agriculture become commercially available (89). Fish nutrition is the most advanced, and thus least limiting, area of knowledge in catfish aquaculture. As such, future developments in feed formulation and feeding practices may not result in great increases in production. Advances in feeding technology do, however, offer potential for substantial improvements in profitability, because feed costs represent most of the overall cost of production.

BIBLIOGRAPHY

In addition to the following references, publications on various aspects of catfish farming and warmwater aquaculture in general are available at the website maintained by the United States Department of Agriculture Southern Regional Aquaculture Center http://www.msstate.edu/dept/srac.

1. R.A. Dunham, C. Hyde, M. Masser, J.A. Plumb, R.O. Smitherman, R. Perez, and A.C. Ramboux, in D. Tave and C.S. Tucker, eds., *Recent Developments in Catfish Aquaculture*, Haworth Press, New York, 1993, pp. 257–268.
2. J.C. Grant and H.R. Robinette, *Aquaculture* **105**, 37–45 (1992).
3. J.H. Tidwell and S.D. Mims, *Prog. Fish-Cult.* **52**, 203–204 (1990).
4. M.J. Brooks, R.O. Smitherman, J.A. Chappell, J.C. Williams, and R.A. Dunham, *Proc. S.E. Assoc. Game Fish Comm.* **36**, 190–195 (1982).
5. W.R. Wolters and M.R. Johnson, *J. Aquat. Animal Health* **6**, 329–334 (1994).
6. J.A. Plumb and J. Chappell, *Proc. S.E. Assoc. Game Fish Comm.* **32**, 680–685 (1978).
7. R.A. Dunham and R.O. Smitherman, in C.S. Tucker, ed., *Channel Catfish Culture*, Elsevier Science Publishers B.V., Amsterdam, 1985, pp. 283–321.
8. E.E. Prather and H.S. Swingle, *Proc. S.E. Assoc. Game Fish Comm.* **14**, 143–145 (1960).
9. W.G. Perry, Jr. and J.W. Avault, *Proc. S.E. Assoc. Game Fish Comm.* **25**, 466–479 (1971).
10. H.K. Dupree and J.V. Huner, eds., *Third Report to Fish Farmers*, U.S. Fish and Wildlife Service, Washington, DC, 1984.
11. W. McLarney, *The Freshwater Aquaculture Book*, Hartley and Marks, Point Roberts, WA, 1984.
12. A. Clapp, *Trans. Am. Fish. Soc.* **59**, 114–117 (1929).
13. G. Lenz, *Prog. Fish-Cult.* **9**, 231–233 (1947).
14. B. Crawford, *Proc. S.E. Assoc. Game Fish Comm.* **11**, 132–141 (1957).
15. H.S. Swingle, *Proc. S.E. Assoc. Game Fish Comm.* **8**, 69–74 (1954).
16. H.S. Swingle, *Proc. S.E. Assoc. Game Fish Comm.* **10**, 160–162 (1956).
17. H.S. Swingle, *Proc. S.E. Assoc. Game Fish Comm.* **12**, 63–72 (1958).
18. R.L. Busch, in C.S. Tucker, ed., *Channel Catfish Culture*, Elsevier Science Publishers B.V., Amsterdam, 1985, pp. 13–84.
19. R.R. Stickney, in R.R. Stickney, ed., *Culture of Nonsalmonid Freshwater Fishes.*, CRC Press, Boca Raton, FL, 1986, pp. 19–42.
20. C.S. Tucker and E.H. Robinson, *Channel Catfish Farming Handbook*, Van Nostrand Reinhold, New York, 1990.
21. C.E. Boyd, in C.S. Tucker, ed., *Channel Catfish Culture*, Elsevier Science Publishers B.V., Amsterdam, 1985, pp. 107–133.
22. E.H. Robinson and M. Li, *A Practical Guide to Nutrition, Feeds, and Feeding of Catfish*, Mississippi Agricultural and Forestry Experiment Station, Mississippi State University, MS, 1996.
23. K.B. Davis, C.A. Goudie, B.A. Simco, R. MacGregor, and N.C. Parker, *Physiol. Zool.* **59**, 717–724 (1986).
24. C.S. Tucker, *Water Quantity and Quality Requirements for Channel Catfish Hatcheries*, Southern Regional Aquaculture Center, Stoneville, MS, 1991.
25. C.S. Tucker and J.A. Steeby, *J. World Aquacult. Soc.* **24**, 396–401 (1993).
26. C.R. Engle and G.L. Pounds, *J. App. Aquacult.* **3**, 311–332 (1994).
27. C.S. Tucker, J.A. Steeby, J.E. Waldrop, and A.B. Garrard, *J. Appl. Aquacult.* **3**, 333–352 (1994).
28. C.R. Engle, G.L. Pounds, and M. van der Ploeg, *J. World Aquacult. Soc.* **26**, 297–306 (1995).
29. R.T. Lovell, *Nutrition and Feeding of Fish*, Van Nostrand Reinhold, New York, 1989.
30. E.H. Robinson, *Rev. Aquat. Sci.* **1**, 365–391 (1989).
31. National Research Council, *Nutrient Requirements of Fish*, National Academy Press, Washington, DC, 1993.
32. R.T. Lovell and E.E. Prather, *Proc. S.E. Assoc. Game Fish Comm.* **27**, 455–459 (1973).
33. D.L. Garling and R.P. Wilson, *J. Nutr.* **106**, 1368–1375 (1976).
34. D.M. Gatlin, III, W.E. Poe, and R.P. Wilson, *J. Nutr.* **116**, 2121–2128 (1986).
35. S. Satoh, W.E. Poe, and R.P. Wilson, *Aquaculture* **79**, 121–128 (1989).

36. D.M. Fracalossi and R.T. Lovell, *Aquaculture* **119**, 287–298 (1994).
37. R.P. Wilson and W.E. Poe, *J. Nutr.* **117**, 280–285 (1987).
38. R.T. Lovell and C. Lim, *Trans. Am. Fish. Soc.* **107**, 321–325 (1978).
39. R.T. Lovell, *J. Nutr.* **103**, 134–139 (1973).
40. R.P. Wilson and W.E. Poe, *J. Nutr.* **103**, 1359–1364 (1973).
41. R.P. Wilson, P.R. Bowser, and W.E. Poe, *J. Nutr.* **113**, 2124–2128 (1983).
42. H.W. Paerl and C.S. Tucker, *J. World Aquacult. Soc.* **26**, 109–131 (1995)
43. D.W. Smith, *Aquaculture* **74**, 167–189 (1988).
44. B.A. Cole and C.E. Boyd, *Prog. Fish-Cult.* **48**, 25–29 (1986).
45. C.E. Boyd, *Water Quality in Ponds for Aquaculture*, Alabama Agricultural Experiment Station, Auburn, AL, 1990.
46. C.E. Boyd and C.S. Tucker, *World Aquacult.* **26**, 45–53 (1995).
47. C.E. Boyd and C.S. Tucker, *Water Quality Management in Warmwater Aquaculture Ponds*, Kluwer Academic Publishers, Amsterdam, 1998.
48. C.S. Tucker and C.E. Boyd, in C.S. Tucker, ed., *Channel Catfish Culture*, Elsevier Science Publishers B.V., Amsterdam, 1985, pp. 135–227.
49. C.S. Tucker, *Rev. Aquat. Sci.* **4**, 1–55 (1996).
50. C.E. Boyd, R.P. Romaire, and E. Johnston, *Trans. Am. Fish. Soc.* **107**, 484–492 (1978).
51. D.W. Smith and R. Piedrahita, *Aquaculture* **68**, 249–265 (1988).
52. C.E. Boyd and B.J. Watten, *Rev. Aquat. Sci.* **1**, 425–472 (1989).
53. J. Colt and G. Tchobanoglous, *Aquaculture* **15**, 353–372 (1978).
54. J.A. Hargreaves, *Aquacult. Eng.* **16**, 27–43 (1997).
55. C.S. Tucker and M. van der Ploeg, *J. World Aquacult. Soc.* **24**, 473–481 (1993).
56. J. Tomasso, B.A. Simco, and K.B. Davis, *J. Fish. Res. Bd. Can.* **36**, 1141–1144 (1979).
57. T.E. Schwedler and C.S. Tucker, *Trans. Am. Fish. Soc.* **112**, 117–119 (1983).
58. C.S. Tucker, R. Francis-Floyd, and M.H. Beleau, *Bull. Environ. Cont. Toxicol.* **43**, 295–301 (1989).
59. C.S. Tucker and J.F. Martin, in D. Brune and J. Tomasso, eds., *Aquaculture and Water Quality*, World Aquaculture Society, Baton Rouge, LA, 1991, pp. 133–179.
60. J.F. Martin, G. Izzaguire, and P. Waterstrat, *Water Res.* **25**, 1447–1451 (1991).
61. M. van der Ploeg and C.E. Boyd, *J. World Aquacult. Soc.* **22**, 207–216 (1991).
62. M. van der Ploeg, C.S. Tucker, and C.E. Boyd, *Water Sci. Technol.* **25**, 283–290 (1992).
63. M. van der Ploeg and C.S. Tucker, *J. Appl. Aquacult.* **3**, 121–140 (1994).
64. J.F. Martin, S.M. Plakas, J.H. Holley, and J.V. Kitzman, *Can. J. Fish. Aquat. Sci.* **47**, 544–547 (1990).
65. J.R. MacMillan, in C.S. Tucker, ed., *Channel Catfish Culture*, Elsevier Science Publishers B.V., Amsterdam, 1985, pp. 405–496.
66. M.R. Johnson, in L. Brown, ed., *Aquaculture for Veterinarians: Fish Husbandry and Medicine*, Pergammon Press, Oxford, United Kingdom, 1993, pp. 249–270.
67. R.L. Thune, *Vet. Human Toxicol.* **33**, 14–18 (1991).
68. R.L. Thune, in M.K. Stoshopf, ed., *Fish Medicine*, W.B. Saunders, Philadelphia, 1993, pp. 511–520.
69. K. Wolf, *Fish Viruses and Fish Viral Diseases*, Cornell University Press, Ithaca, NY, 1988.
70. D.F. Amend, T. McDowell, and R.P. Hedrick, *Can. J. Fish. Aquat. Sci.* **41**, 807–811 (1984).
71. K. Wolf and R.W. Darlington, *J. Virol.* **8**, 525–533 (1971).
72. J.A. Plumb and J.L. Gaines, in W.E. Ribelin and G. Migaki, eds., *The Pathology of Fishes*, University of Wisconsin Press, Madison, WI, 1975, pp. 287–302.
73. J.A. Plumb, *J. Fish Res. Board. Can.* **30**, 568–570 (1973).
74. J.P. Hawke, *J. Fish. Res. Board Can.* **36**, 1508–1512 (1979).
75. R.L. Thune, L.A. Stanley, and R.K. Cooper, *Ann. Rev. Fish Dis.* **3**, 37–68 (1993).
76. E.M. Shotts and V.S. Blazer, *Can. J. Fish. Aquat. Sci.* **43**, 36–42 (1986).
77. H.H. Jarboe, P.R. Bowser, and H.R. Robinette, *J. Wildl. Dis.* **20**, 353–354 (1984).
78. R. Francis-Floyd, M.H. Beleau, P.R. Waterstraat, and P.R. Bowser, *J. Amer. Vet. Med. Assoc.* **191**, 1413–1416 (1987).
79. G.R. Walters and J.A. Plumb, *J. Fish Biol.* **17**, 177–186 (1980).
80. M.T. Ventura and J.M. Grizzle, *J. Fish Dis.* **11**, 397–408 (1988).
81. J.E. Bly, L.A. Dawson, D.J. Dale, A.J. Szalai, R.M. Durburow, and L.W. Clem, *Dis. Aquat. Org.* **13**, 155–164 (1992).
82. J.E. Bly, L.A. Dawson, A.J. Szalai, and L.W. Clem, *J. Fish Dis.* **16**, 541–549 (1993).
83. M. Li, D.J. Wise, and E.H. Robinson, *J. World Aquacult. Soc.* **27**, 1–6 (1996).
84. E.L. Styer, L.R. Harrison, and G.J. Burtle, *J. Aquat. Anim. Health* **4**, 288–291 (1991).
85. R.L. Busch, in C.S. Tucker, ed., *Channel Catfish Culture*, Elsevier Science Publishers B.V., Amsterdam, 1985, pp. 543–567.
86. G.R. Ammerman, in C.S. Tucker, ed., *Channel Catfish Culture*, Elsevier Science Publishers B.V., Amsterdam, 1985, pp. 569–620.
87. C.S. Tucker, S.K. Kingsbury, J.W. Pote, and C.L. Wax, *Aquaculture* **147**, 57–69 (1996).
88. National Animal Health Monitoring System, *Part II: Reference of 1996 U.S. Catfish Management Practices*, USDA/APHIS, Fort Collins, CO, 1997.
89. W.R. Wolters, in K.L. Main and B. Reynolds, eds., *Selective Breeding of Fishes in Asia and the United States*, The Oceanic Institute, Honolulu, 1993.

CHEMICAL OXYGEN DEMAND

David E. Owsley
Dworshak Fisheries Complex
Ahsahka, Idaho

OUTLINE

DEFINITION

By definition, the chemical oxygen demand (COD) is a measure of the oxygen equivalent of the portion of organic matter that can be oxidized by a strong chemical oxidizing agent. COD analysis is an important parameter for measuring the total oxygen demand in wastewater from hatcheries and other aquaculture facilities. It is an especially important tool for measuring and controlling industrial wastes. The COD also gives a better estimate of the total oxygen demand, whereas the biochemical oxygen demand (BOD) measures only 50 to 70% of the total oxygen demand present. In addition, the COD test requires only a few hours, as compared to five days for the BOD test.

DISCUSSION

COD is used to measure the pollutional strength of domestic and industrial wastes. It is based upon the fact that all organic matter can be chemically oxidized using a strong-enough oxidizing agent under acidic conditions. As with BOD, organic matter is converted to carbon dioxide and water. However, COD values are usually greater than BOD values and may be significantly greater when high amounts of biologically resistant organic matter are present. For this reason, the COD method is widely used to characterize industrial wastes.

The analytical procedure for the COD test can be found in (1). The most widely used test is the potassium dichromate method. Potassium dichromate is capable of oxidizing a wide variety of organic substances, and the excess potassium dichromate remaining is easy to measure.

Liao and Mayo (2) developed an equation for the COD of trout culture systems that use a high percentage (up to 90%) of recycled water:

$$\mathrm{COD} = 1.89F.$$

Here, COD is the COD production rate at 10 to 15 °C (50 to 59 °F) in kg of COD/100 kg of fish per day, and F is the feeding rate, in kg of food/100 kg of fish per day.

For warmwater aquaculture, Boyd (3) reported that, even though a large amount of the feed applied to a pond is not converted to fish flesh and reaches the water as waste, most of the COD in fed ponds does not directly result from feeding waste. In data reported by Boyd (4), the total COD of phytoplankton produced during a 180-day growing season in channel catfish ponds in Auburn, AL, was 12,400 kg/hectare. The COD of all metabolic waste excreted by fish and uneaten feed during this period was 2,400 kg/hectare. Thus, the COD attributable to metabolic waste and uneaten feed was only about 20% of the COD attributable to phytoplankton: $(2{,}400/12{,}400) \times 100 \approx 20\%$. However, nutrients from excretory products and unconsumed feed were responsible for most of the phytoplankton growth. Boyd (5) also lists in Tables 1 and 2, the COD during the harvest phase of catfish in ponds.

COD can be removed from water by chemical filtration. Parkhurst et al. (6) reported on contaminant removal using granular activated carbon, as indicated in Table 3.

Table 1. Quality of Effluents from Catfish Ponds During the Two Phases of Fish Harvest[a]

Analysis	Harvest Phase	
	Draining	Seining
Settleable matter (mg/L)	0.08	28.5
Biochemical oxygen demand (mg/L)	4.31	28.9
Chemical oxygen demand (mg/L)	30.2	342
Soluble orthophosphate (mg/L as P)	16	59
Total phosphorus (mg/L as P)	0.11	0.49
Total ammonia (mg/L as N)	0.98	2.34
Nitrate (mg/L as N)	0.16	0.14

[a]From Boyd (5).

Table 2. Percentage Losses of Several Pollutants from A Hypothetical Catfish Pond During the Seining Phase of Fish Harvest[a]

Analysis	% of Total Released During Seining Phase
Settleable matter	94.9
Biochemical oxygen demand	26.0
Chemical oxygen demand	37.3
Soluble orthophosphate	66.7
Total phosphorus	18.8
Total ammonia	11.2
Nitrate	4.3

[a]From Boyd (5).

Table 3. Removal of Contaminants from Waste Water Using Activated Carbon

Contaminant	Influent		Effluent
Suspended solids	10.0	mg/L	<1.0
COD	47.0	mg/L	9.5
Dissolved COD	31.0	mg/L	7.0
Total organic carbon (TOC)	13.0	mg/L	2.5
Nitrate (as N)	6.7	mg/L	3.7
Turbidity	10.3	JTU	1.6
Color	30.0		3.0
Odor	12.0		1.0

Rubin et al. (7) reported that airstripping removes dissolved organic compounds by two mechanisms: (a) Dissolved, surface-active organic compounds may be absorbed at the gas–liquid interface and be concentrated in the foam; (b) dissolved, non-surface-active organic compounds may combine with surface-active solutes and be concentrated in the foam. These investigators found that airstripping removed up to 40% of the COD from sewage effluent. Also, Nebel et al. (8) reported that the COD could greatly be reduced by using ozone. The same study also showed a high reduction in the BOD using this method.

CONCLUSION

Although the COD test gives a better analysis of the actual total oxygen demand, it will not replace the BOD test, as most aquaculture facilities are equipped to run tests on dissolved oxygen and can also do the BOD test. The COD test requires a more specialized chemical procedure that most aquaculture facilities are not equipped to handle.

BIBLIOGRAPHY

1. *Standard Methods for the Examination of Water and Wastewater*, 17th ed., American Public Health Association, Washington, DC, 1989.
2. P.B. Liao and R.D. Mayo, *Aquaculture* **3**, 61–85 (1974).
3. C.E. Boyd, *Water Quality in Warmwater Fish Ponds*, Agricultural Experimental Station, Auburn University, Auburn, AL, 1979.
4. C.E. Boyd, *Amer. Fish Soc.* **102**, 606–611 (1973).
5. C.E. Boyd, *J. Environ. Qual.* **7**, 59–62 (1978).
6. J.D. Parkhurst, F.D. Dryden, G.N. McDermott, and J. English, *Pomona 0.3 MGD F. Water Pollution Control Federation* **30**, R70–R81 (1967).
7. E. Rubin, R. Everett Jr., J.J. Weinstock, and H.M. Schoen, *Contaminant Removal from Sewage Plant Effluents by Foaming* PHS Publ. No. 999-WF-5, 1963.
8. C. Nebel, P.C. Unangst, and R.D. Gottschling, "Ozone Disinfection of Secondary Effluents: Laboratory Studies," in *First International Symposium on Ozone for Water and Wastewater Treatment*, International Ozone Institute, Waterbury, CT, 1975, pp. 383–404.

See also Biochemical oxygen demand; Dissolved oxygen.

CHLORINATION/DECHLORINATION

Gary A. Wedemeyer
Western Fisheries Research Center
Seattle, Washington

OUTLINE

Chlorine is not a natural constituent of surface or groundwater and therefore should never be present in fish-rearing systems, unless they have been contaminated in some way. However, chlorine does play an important role in aquaculture. It is widely used to disinfect equipment, tanks, ponds, and entire facilities, and water discharged from hatcheries may be chlorinated to destroy endemic or exotic pathogens. Methods for neutralizing or otherwise removing chlorine from water are also important in fisheries work. Applications of such methods include dechlorinating municipal water so that it can be used for fish rearing, neutralizing chlorine residues on equipment that has been previously disinfected, or dechlorinating hatchery wastewater to prevent toxicity to organisms in the receiving water.

EQUIPMENT DISINFECTION

Chlorine for disinfecting nets, troughs, tanks, and other equipment is usually applied as a solution of sodium or calcium hypochlorite ($NaOCl$ and $Ca(OCl)_2$, respectively) diluted to 100–200 mg of active ingredient/L in water. Granular calcium hypochlorite (HTH), which is more stable, can also be used at the same concentration. If large volumes of water, such as hatchery effluents, must be treated, chlorine gas (Cl_2) injected at 1–3 mg/L is normally employed, because of its lower cost.

Whether chlorine is added to water as Cl_2 gas or as a hypochlorite salt, it reacts with the water to form an equilibrium mixture of molecular chlorine (Cl_2), hypochlorous acid (HOCl), and hypochlorite ion (OCl^-). The relative proportions of these species in the chlorine disinfectant solution is primarily dependent on the pH of the water. At pH values greater than 3, very little molecular Cl_2 is present, and the solution will be primarily a mixture of HOCl and OCl^-. Both of these chemicals are strong oxidants, but HOCl is the stronger germicide, because the electrical charge of the OCl^- ion impedes its penetration into microbial cells. Consequently, the pH of the water has an important effect on the efficacy of disinfection, because it controls the amount of HOCl that is present. At pH 6, about 95% of the chlorine will be in the more effective HOCl form, whereas at pH 9, about 95% of the HOCl has dissociated into the weakly germicidal OCl^-. In areas with alkaline water, dilute acetic acid is sometimes added to chlorine disinfectant solutions to reduce the pH to about 6, to maximize the formation of HOCl. Suspended particulates such as clay and organic matter tend to interfere with disinfection by protecting microorganisms from direct exposure to the chlorine. However, dissolved minerals such as calcium, magnesium, and iron generally have little effect.

For equipment and tank disinfection, a chlorine concentration of 100–200 mg/L with a contact time of at least 30 minutes is recommended (1). The amount of NaOCl or $Ca(OCl)_2$ to add is calculated from the percentage of active ingredient listed on the label of the product used. For NaOCl (SuperChlor®), this percentage is typically 11. HTH may contain up to 70% available chlorine. As previously mentioned, a sufficient amount of acetic acid to decrease the pH to about 6 may be added to the water in order to activate the chlorine. Never add acid to dry HTH. For fish transport tanks, a sufficient volume of water to cover the intakes of the recirculating pumps or spray agitator propellers is poured in and the calculated amount of liquid or dry chlorine added. For convenience, the chlorine is usually applied as an NaOCl solution or as granular $Ca(OCl)_2$ [e.g., 15 g of HTH, with 70% available chlorine, per 100 L (0.5 oz/38 gal)]. The solution is pumped through the system for at least 30 minutes with air or oxygen flowing through the diffusers to prevent backflow. After disinfection, the equipment must be thoroughly rinsed with chlorine-free water, and any residual chlorine should be allowed to dissipate naturally (24–48 hours) or should be neutralized with a sodium thiosulfate rinse [7.4 mg/L per mg/L (ppm) of chlorine to be neutralized (2)] before it is used to handle fish. As a precaution, commercially available chlorine test paper or a water chemistry test kit should always be used to confirm that chlorine residuals have dissipated to safe levels or have been adequately neutralized.

Occasionally, entire facilities must be decontaminated using chlorine. A detailed method for this procedure can be found in Piper et al. (1).

When chlorination is used to destroy pathogens in wastewater from aquaculture facilities, a free residual chlorine concentration of 1–3 mg/L [ppm] with a contact time of 10–15 minutes is typically recommended. If exotic pathogens are involved, residuals of 3–5 mg/L [ppm] are often suggested, with a contact time of 15–30 minutes to allow a margin of safety (3). Parasite spores, such as *Myxobolus cerebralis*, may require up to 1,500 ppm chlorine, with a contact time of several hours.

CHLORINE ANALYSIS

For most purposes, chlorine concentrations can be conveniently determined using any of several commercially available water chemistry test kits. For more precise work, specialized laboratory methods are available (4).

The individual concentrations of OCl^- and HOCl are somewhat difficult to measure, and methods used to analyze chlorine, such as DPD (N,N-diethyl-p-phenylenediamine) determination, measure the sum of their respective concentrations, usually termed the "free residual chlorine concentration" (4). In addition to reacting with water to form HOCl and OCl^-, chlorine also reacts with any ammonia or other nitrogenous materials present in the water to form chloramine compounds such as monochloroamine and nitrogen trichloride.

Chloramines are weaker germicides than chlorine, but their toxicity to fish is usually greater. In determinations of the amount of chlorine, the concentration of chloramines themselves is termed the "combined chlorine residual." Again, separating individual concentrations is somewhat difficult, and commonly used analytical methods measure the "total residual chlorine concentration," that is, the total concentration of the free chlorine and the combined chlorine residuals (4). If significant amounts of chloramines are present, this measurement will overstate the germicidal activity and understate the toxicity to fish. Chlorinated hatchery effluents may contain mainly combined chlorine residuals (chloramines), because of the ammonia present.

DECHLORINATION

The applications of dechlorination in aquaculture include removing chlorine from water so the water can be used for rearing purposes, neutralizing chlorine residues on equipment that has been previously disinfected, and dechlorination of hatchery wastewater streams to prevent toxicity to organisms receiving the water. The chosen dechlorination methods for a particular application should remove both chlorine and chloramines in order to be useful in aquaculture.

Small concentrations of chlorine [e.g., 0.05 mg/L (ppm)] can be removed from water by simple aeration, but ample ventilation is required, or else the chlorine will simply redissolve. Chlorine can also be allowed to dissipate naturally. As a guideline, about 20 hours per mg/L of chlorine [ppm] should be allowed. Unfortunately, both of these methods are too variable to be completely safe, and neither removes chloramines. Photochemical decomposition by ultraviolet irradiation, and oxidation using ozone, have considerable promise for removing chlorine, but high costs have limited their use. In most cases, neutralization with chemical reducing agents such as sulfur dioxide gas and sodium thiosulfate, and filtration through adsorption media such as activated carbon, are the most practical alternatives.

For large volumes of water, neutralization with sulfur dioxide (SO_2) gas is the most practical method. Dechlorination with SO_2 is inexpensive, easy to control, and the required equipment is commercially available. If necessary, commercial gas-chlorination equipment can be modified to inject SO_2 instead of Cl_2. Chemically, SO_2 reacts with water to form sulfite ions (SO_3^{-2}), which then reduce the chlorine and chloramines to nontoxic chloride ions and inconsequential amounts of acids in a 1 : 1 stoichiometric ratio. These reactions are so rapid that contact time is not an important consideration.

For facility wastewater discharges in the 190–380 L/min (50–100 gpm) range, the cost of sodium thiosulfate ($Na_2S_2O_3$) treatment may be about the same as that of the SO_2 needed for gas dechlorination, and with the former, the possibility of toxic gas leaks is eliminated. Equipment needs for sodium thiosulfate treatment are also simpler: a solution reservoir tank and a metering pump. Sodium thiosulfate rinses are also widely used for neutralizing chlorine residuals on disinfected equipment. This compound hydrolyzes to produce sulfite ions, which then react with chlorine in the same way as does SO_2. As mentioned earlier, a concentration of about 7.4 mg/L (ppm) of sodium thiosulfate solution will neutralize 1 mg/L (ppm) of chlorine (1,2). However, thiosulfate solutions are not completely stable and should be used within a day or two of preparation. Using a concentration of 8 mg of thiosulfate/L (ppm) per mg of chlorine to be neutralized/L is recommended to provide a small margin of safety (3).

Filtration through activated charcoal (carbon) is perhaps the most common method for dechlorinating water to be used for small-scale fish rearing. Activated carbon (C^*) chemically reacts with both chlorine and chloramines, converting them into innocuous amounts of carbon dioxide and ammonium salts. Fresh activated charcoal reliably reduces the concentration of free and bound chlorine down to the range of 10–20 μg/L (ppb) but complete removal almost always requires supplemental thiosulfite injection, especially as the filters age (3). A flow-through filter containing 12 ft^3 (0.339 m^3) of granular activated carbon will reliably dechlorinate about 950 L/min 25 gpm of water containing a chlorine residual of 0.05 mg/L (ppm) for a period of about one year before needing replacement (3).

TOXICITY

The toxicity of chlorine is due to both its HOCl and its OCl^- forms. Both forms are strong oxidants that destroy gill and other tissues by penetrating cell membranes and damaging cell structures, enzymes, DNA, and RNA. The OCl^- ion has about the same molecular weight as HOCl, but, as previously mentioned, its electrical charge impedes penetration through cell membranes, making it slightly less toxic to fish and invertebrates. The practical importance of this difference is slight, however, partly because comparatively little OCl^- is present over pH 6–8, the range that is typical of fish-rearing conditions. Environmental conditions such as temperature and the concentration of dissolved oxygen also influence chlorine toxicity, but again, the practical importance of any protection gained is slight. The end result is that the toxicity of chlorine is high. At the concentrations usually found in drinking-water supplies [0.1–0.3 mg/L (ppm)], chlorine will kill most commercially important aquatic species within minutes at any pH. Chronic sublethal chlorine exposures can induce gill damage and other pathological changes that will seriously compromise ultimate health, quality, and survival.

To adequately protect fish and invertebrates from gill and tissue damage, chronic exposure to chlorine should not exceed 3–5 μg/L (ppb). For short periods of time (up to 30 minutes), exposures as high as 0.05 mg/L (ppm) can usually be tolerated by most species important to aquaculture (3). Aquatic plants are relatively resistant to chlorine.

BIBLIOGRAPHY

1. R.G. Piper, I.B. McElwain, L.E. Orme, J.P. McCraren, L.G. Fowler, and J.R. Leonard, *Fish Hatchery Management*, Fish and Wildlife Service, US Department of Interior, Washington, DC, 1982.
2. G. Jensen, *Handbook of Common Calculations in Finfish Aquaculture*, Louisiana State University, Agricultural Experiment Station, New Orleans, LA, 1989.
3. G. Wedemeyer, *Physiology of Fish in Intensive Culture Systems*, Chapman Hall, New York, 1996.
4. *Standard Methods for the Examination of Water and Wastewater*, 17th ed., American Public Health Association, Washington, DC, 1989.

See also Disease treatments; Disinfection and sterilization.

CONSERVATION HATCHERIES

Thomas A. Flagg
Desmond J. Maynard
Conrad V.W. Mahnken
National Marine Fisheries Service
Manchester, Washington

OUTLINE

INTRODUCTION

Artificial propagation has been suggested as a potential mechanism to aid in the recovery of U.S. Endangered Species Act (ESA)-listed stocks of Pacific salmon on the West Coast of the United States (1–4). Theoretically, one of the fastest ways to amplify population numbers for depleted stocks of Pacific salmon is through culture and release of hatchery-propagated fish (2). However, most past attempts to use supplementation (i.e., the use of artificial propagation in an attempt to maintain or increase natural production) to rebuild naturally spawning populations of Pacific salmon have yielded poor results (5). The challenge is in developing protocols that increase fitness of hatchery-reared salmonids, thereby improving survival. This article describes the potential impacts of artificial propagation on the biology and behavior of fish and a conceptual framework of conservation hatchery strategies to help mitigate the unnatural conditioning provided by hatchery rearing.

POTENTIAL IMPACTS OF HATCHERY REARING

Hatcheries figure prominently in the management of Pacific Northwest salmon. For the most part, hatcheries have been successful in producing fish for the fishery (6). However, hatcheries and hatchery management practices have often worked to the detriment of wild fish (7–10). Present hatchery practices are geared toward mass production under unnatural conditions. In Pacific Northwest salmon hatcheries, fish are most often reared in the open, over a uniform concrete substrate; conditioned to minimal raceway flow regimes; provided no structure in which to seek refuge from water current, predators, or dominant cohorts; held at high, stress-producing densities; surface fed; and conditioned to approach large, moving objects at the surface (11). Although the protective nature of hatchery rearing increases egg-to-smolt survival (12–14), the postrelease survival and reproductive success of cultured salmonids is often considerably lower than that of wild-reared fish (15–19). Hatchery practices that induced genetic changes (domestication, etc.) are often considered prime factors in reducing fitness of hatchery fish in natural ecosystems (7,10,19–22). Rearing practices that disrupt innate behavioral repertoires may also play a major role in reduced performance of hatchery fish after release.

Behavioral deficiencies in released animals have been blamed for the failure to re-establish wild populations of higher vertebrates (23–26). Current fish culture techniques (12–14) may impart similar behavioral deficiencies in hatchery-reared salmon. Studies indicate that the hatchery rearing environment can profoundly influence social behavior of Pacific salmon, (11) and the social divergence of cultured fish may begin as early as the incubation stage. Lack of substrate and high light levels, which may occur in the hatchery incubation environment, induce excess alevin movement, lowered energetic efficiency, reduced size, and, in some wild stocks, death (27–31). Food availability and rearing densities in hatcheries far exceed those found in natural streams and may contribute to differences in agnostic behavior between hatchery- and wild-reared fish (32–38). Hatchery-rearing environments may also deprive salmon of the psychosensory stimuli necessary to fully develop antipredation behaviors (36). Evidence indicates that hatchery strains of salmonids have increased risk-taking behavior and lowered fright responses compared to wild fish (11,36). Surface feeding may condition hatchery fish to approach the surface of the water column (11,36,37), and this behavior may increase susceptibility to avian predation. Studies have also attributed increased avian and piscivorous predator vulnerability of hatchery fish in stream environments to decreased crypsis (camouflage coloration) caused by rearing against uniform (e.g., concrete) hatchery backgrounds (11,39,40). In addition, cultured and naturally reared salmonids may respond differently to habitat. In most cases, wild fish utilize both riffles and pools in streams, while newly released hatchery fish primarily use pool environments that are similar to their raceway rearing experience (33,36–38,41,42).

Seemingly, the only similarities in hatchery and wild environments for salmonids are water and photoperiod (43). Most other components of the hatchery rearing environment (food, substrate, density, temperature, flow regime, competitors, and predators) differ from what wild fish experience. It has been suggested that when a hatchery purpose includes protection of wild stocks, the operational strategy must switch from a production to a conservation mandate (4,9–11,44). A conceptual framework of conservation hatchery protocols for Pacific salmon follows.

THE CONSERVATION HATCHERY CONCEPT

The strategic role of a conservation hatchery will be to promote restoration of wild stocks of fish. This requires fish rearing be conducted in a manner that mimics the natural life history patterns, improves the quality and survival of hatchery-reared juveniles, and lessens the genetic and ecological impacts of hatchery releases on wild stocks. Conservation hatchery operational guidelines presented in the following include protocol recommendations designed to improve the survival of hatchery-reared fish in the wild.

Genetic Considerations

In order to maintain long-term adaptive traits, conservation hatcheries should provide fish with minimal genetic divergence from their natural counterparts (44). Protocols for fish culture and enhancement that reduce genetic risks (such as domestication, hybridization, inbreeding depression, and outbreeding depression) are well described (7,10,45,46). Mating strategies include random pairing, pairing in as many different combinations as possible, avoidance of pairing between siblings, crossings between different year-classes, and fertilization with cryopreserved sperm from other generations or from outside stock sources (1,14,45,46). Genetic risks of a particular strategy must be calculated on a case-by-case basis.

Broodstock Sourcing

Conservation hatcheries should use local wild broodstock whenever possible (44). For extirpated populations, the donor stock should be chosen following careful analysis of the environmental relationships and the life history parameters of the original stocks. Broodstocks should be representative of the genetics of the entire population to avoid potential reductions in effective population size by dramatically increasing only a fraction of the available genotypes in the parent population (1,14,45,46). Broodstocks can be sourced from all available life stages, including eyed eggs mined from redds, fry, smolts captured from the wild, and pre-spawning adults captured and artificially spawned (1,2). In cases where critical populations are dangerously close to extinction, a captive broodstock approach should be considered to maximize population size in the shortest time frame (2,3). Broodstock should be kept in a secure manner at one or more facilities to ensure against catastrophic loss (3). Rearing water supplies should be treated (sterilized) to remove pathogens (3,4,44), and natural water temperature profiles should be maintained to provide optimum maturation and gamete development (44).

Enriched Rearing Environments

Conservation hatcheries should use low rearing densities and base their goals for growth and size at emigration on natural population parameters (44). It is recommended they (*1*) determine spawning, hatching, and emergence times of the local population and duplicate these conditions in the hatchery by controlling water temperature to natural profiles; (*2*) simulate growth rate, body size, and body (proximate) composition by controlling water temperature, diet composition, and feeding rates. Conservation hatcheries should have incubation and rearing vessels with options for habitat complexity to produce fish more wild-like in appearance and with natural behaviors and higher survival (44). Darkened incubation systems containing substrates will produce larger fry that are more energetically efficient and alert (27–31). Fry-to-smolt rearing in enriched habitats containing cover, structure, and substrate can dramatically increase (up to 50%) postrelease survival in the stream environment (39). Natural stream-side cover can be created by suspending camouflage netting over 75% of each vessel approximately 1 m above the water surface along the margins of the raceways (44). Internal structures can be created by suspending small defoliated fir trees in rearing vessels occupying 30–60% of the surface area of the water (44). Substrates have been configured in several ways using sand, gravel, artificial rugose inserts, or painted patterns (39,44). Every effort should be made to match the color of the substrate (which produces cryptic coloration patterns in fish) to that of the receiving-stream environment to produce the body camouflage patterns most likely to reduce vulnerability to predators.

Other potential components of an enriched rearing environment for salmonids, such as foraging training, feed delivery systems modeled after natural feeding situations, and changing flow velocities to exercise the fish, can also offer advantages for increased survival and behavioral fitness (11). In addition, studies have shown that antipredator conditioning may increase postrelease survival (36,47–50).

Reintroduction Strategies

Conservation hatcheries should release smolts with a population size distribution equivalent to the size distribution of smolts in the wild population (44). The greatest risk of releasing oversized hatchery fish is that they will outcompete the smaller wild fish. In intraspecific contests over food and space, all else being equal, the largest fish usually wins (51–54). Fish from conservation hatcheries should be released of their own volition and outmigrate during windows for natural downstream migration of the stock (44). The key assumption of volitional release is that fish will not leave the hatchery until certain physiological processes, such as smoltification, trigger their downstream migratory behavior (9,55,56). The technique is simply to provide windows of opportunity for outmigration which mimic time and age patterns found in the wild populations. Within these windows, fish may leave if they wish or remain behind to fend for themselves and smolt, residualize, or perish as natural selection takes it course.

Conservation hatcheries should adopt practices to reduce straying to no more than 5% (44,57). To maximize imprinting opportunity, juvenile salmon must experience the odors of their natal system at various times and physiological states when the odors can be learned (58–62). Conservation hatcheries should, therefore, rear fish for their entire juvenile freshwater lives in water from the intended return location (44). When this is not possible, a period of acclimation on intended return water should improve imprinting and homing and reduce straying (44). Conservation hatcheries should program their releases to accommodate the natural spatial and temporal patterns of abundance in wild fish populations and release numbers should not exceed carrying capacities of (freshwater and oceanic) receiving waters (44).

CONCLUSIONS

With the emphasis on wild fish recovery required by the ESA, there is opportunity to transfer the role of certain hatcheries from production to wild stock enhancement. Simple and practical changes to hatchery-rearing protocols will allow conservation hatcheries to produce fish with the wild-like attributes required for aiding in restoration of depleted stocks. Conservation hatcheries should employ the latest scientific information and conservation practices to maintain genetic diversity and natural behavior and thereby reduce the short-term risk of extinction. Exact application of the conservation hatchery strategies outlined in this entry will be dependent on physical and management limitations of individual hatcheries.

BIBLIOGRAPHY

1. J.J. Hard, R.P. Jones, Jr., M.R. Delarm, and R.S. Waples, *NOAA Tech. Memo.*, NMFS-NWFSC-2, 1992.
2. T.A. Flagg, C.V.W. Mahnken, and K.A. Johnson, *Am. Fish. Soc. Symp.* **15**, 81–90 (1995).
3. M.H. Schiewe, T.A. Flagg, and B.A. Berejikian, *Bull. Natl. Res. Inst. Aquacult.* Suppl. **3**, 29–34 (1997).
4. P.J. Anders, *Fisheries* **23**(11), 28–31 (1998).
5. M.L. Cuenco, T.W.H. Backman, and P.R. Mundy, in J.C. Cloud and G.H. Thorgaard, eds., *Genetic Conservation of Salmonid Fishes*, Plenum Press, New York, 1993, pp. 269–293.
6. C. Mahnken, G. Ruggerone, W. Waknitz, and T. Flagg, *N. Pac. Anadr. Fish Comm. Bull.* **1**, 38–53 (1998).
7. R.S. Waples, *Can. J. Fish. Aquat. Sci.* **48**, 124–133 (1991).
8. T.A. Flagg, F.W. Waknitz, D.J. Maynard, G.B. Milner, and C.V.W. Mahnken, *Am. Fish. Soc. Symp.* **15**, 366–375 (1995).
9. National Research Council Committee on Protection and Management of Pacific Northwest Anadromous Salmonids, *Upstream: Salmon and Society in the Pacific Northwest*, National Academy Press, Washington, DC, 1996.
10. R.S. Waples, *Fisheries* **24**(2), 12–21 (1999).
11. D.J. Maynard, T.A. Flagg, and C.V.W. Mahnken, *Am. Fish. Soc. Symp.* **15**, 307–314 (1995).
12. E. Leitritz and R.C. Lewis, *Calif. Fish and Game Bull.* 164 (1976).
13. R.G. Piper et al., *Fish Hatchery Management*, U.S. Department of Interior, U.S. Printing Office, Washington, DC, 1982.
14. W. Pennell and B.A. Barton, eds., *Principles of Salmonid Culture*, Elsevier, Amsterdam, 1996.
15. C.W. Greene, *Trans. Am. Fish. Soc.* **81**, 43–52 (1952).
16. R.B. Miller, *Trans. Am. Fish. Soc.* **81**, 35–42 (1952).
17. N. Reimers, *Trans. Am. Fish. Soc.* **92**, 39–46 (1963).
18. M.W. Chilcote, M.W. Leider, S.A., and J.J. Loch, *Trans. Am. Fish. Soc.* **115**, 726–735 (1986).
19. T.E. Nickelson, M.F. Solazzi, and S.L. Johnson, *Can. J. Fish. Aquat. Sci.* **43**, 2443–2449 (1986).
20. R.R. Reisenbichler and J.D. McIntyre, *J. Fish. Res. Board Can.* **34**, 123–128 (1977).
21. M.L. Goodman, *Environ. Law* **20**, 111–166 (1990).
22. R. Hilborn, *Fisheries* **17**, 5–8 (1992).
23. J.H.W. Gipps, ed., *Beyond Captive Breeding: Reintroducing Endangered Species through Captive Breeding*, Zool. Soc. London Symp. 62, London, 1991.
24. J.E. Johnson and B.L. Jensen, in W.L. Minckley and J.E. Deacon, eds., *Battle Against Extinction*, Univ. Arizona Press, Tucson, 1991, pp. 199–217.
25. J. DeBlieu, *Meant to Be Wild: The Struggle to Save Endangered Species through Captive Breeding*, Fulcrum Publishing, Golden, CO, 1993.
26. P.J.S. Olney, G.M. Mace, and A.T.C. Feistner, eds., *Creative Conservation: Interactive Management of Wild and Captive Animals*, Chapman and Hall, London, 1994.
27. D.C. Poon, Ph.D. thesis, Oregon State University, Corvallis, 1977.

28. K.A. Leon and W.A. Bonney, *Prog. Fish-Cult.* **41**, 20–25 (1979).
29. J.L. Mighell, *Mar. Fish. Rev.* **43**(2), 1–8 (1981).
30. C.B. Murray and T.D. Beacham, *Prog. Fish-Cult.* **48**, 242–249 (1986).
31. H.J. Fuss and C. Johnson, *Prog. Fish-Cult.* **50**, 232–237 (1988).
32. P.E. Symons, *J. Fish. Res. Board Can.* **25**, 2387–2401 (1968).
33. R.A. Bachman, *Trans. Am. Fish. Soc.* **113**, 1–32 (1984).
34. J.W.A. Grant and D.L. Kramer, *Can. J. Fish. Aquat. Sci.* **47**, 1724–1737 (1990).
35. B.L. Olla, M.W. Davis, and C.H. Ryer, in J.E. Thorpe and F.A. Huntingford, eds., *The Importance of Feeding Behavior for the Efficient Culture of Salmonid Fishes*, World Aquaculture Society, Baton Rouge, LA, 1990, pp. 5–12.
36. B.L. Olla, M.W. Davis, and C.H. Ryer, *Bull. Mar. Sci.* **62**(2), 531–550 (1998).
37. K. Uchida, K. Tsukamotot, S. Ishii, R. Ishida, and T. Kajihara, *J. Fish. Biol.* **34**, 399–407 (1989).
38. B.A. Berejikian, Ph.D. thesis, Univ. Washington, Seattle, 1995.
39. D.J. Maynard, T.A. Flagg, C.V.W. Mahnken, and S.L. Schroder, *Bull. Natl. Res. Inst. Aquacult.* Suppl. **2**, 71–77 (1996).
40. W.A. Donnelly and F.G. Whoriskey, Jr., *N. Am. J. Fish. Manage.* **11**, 206–211 (1991).
41. B.J. Allee, Ph.D. thesis, Univ. Washington, Seattle, 1974.
42. T.A. Dickson and H.R. MacCrimmon, *Can. J. Fish. Aquat. Sci.* **39**, 1453–1458 (1982).
43. R.R. Reisenbichler and S.P. Rubin, *ICES J. Mar. Sci.* (in press).
44. T.A. Flagg and C.E. Nash, eds., *NOAA Tech. Memo.*, NMFS-NWFSC-38, 1999.
45. N. Ryman and F.M. Utter, eds., *Population Genetics and Fishery Management*, Univ. Washington Press, Seattle, 1986.
46. D. Tave, *Genetics for Fish Hatchery Managers*, 2nd ed., AVI (Van Nostrand Reinhold), New York, 1993.
47. B.G. Patten, *Fish. Bull.* **75**, 451–459 (1977).
48. B.L. Olla and M.W. Davis, *Aquaculture* **76**, 209–214 (1989).
49. B.A. Berejikian, *Can. J. Fish. Aquat. Sci.* **52**, 2076–2082 (1995).
50. M.C. Healey and U. Reinhardt, *Can. J. Fish. Aquat. Sci.* **52**, 614–622 (1995).
51. W.S. Hoar, *J. Fish. Res. Board Can.* **12**, 178–185 (1951).
52. D.W. Chapman, *J. Fish. Res. Board Can.* **19**, 1047–1080 (1962).
53. J.C. Mason and D.W. Chapman, *J. Fish. Res. Board Can.* **22**, 172–190 (1965).
54. J.C. Abbot, R.L. Dunbrack, and C.D. Orr, *Behavior* **92**, 241–253 (1985).
55. E.L. Brannon, C. Feldman, and L. Donaldson, *Aquaculture* **28**, 195–200 (1982).
56. A.R. Kapuscinski, in D.J. Stouder, P.A. Bisson, and R.J. Naiman, eds., *Pacific Salmon and their Ecosystems*, Chapman Hall, NY, 1997, pp. 493–512.
57. W.S. Grant, ed., *NOAA Tech. Memo.*, NMFS-NWFSC-30, 1995.
58. W.H. Sholes and R.J. Hallock, *Calif. Dep. Fish Game Fish Bull.* **64**, 239–255 (1979).
59. M. Labelle, *Can. J. Fish. Aquat. Sci.* **49**, 1843–1855 (1992).
60. M.J. Unwin and T.P. Quinn, *Can. J. Fish. Aquat. Sci.* **50**, 1168–1175 (1993).
61. M.A. Pascual, T.P. Quinn, and H. Fuss, *Trans. Am. Fish. Soc.* **124**, 308–320 (1995).
62. A.H. Dittman, T.P. Quinn, and G.A. Nevitt, *Can. J. Fish. Aquat. Sci.* **53**, 434–442 (1995).

CRAB CULTURE

David B. Rouse
Richard K. Wallace
Auburn University
Auburn, Alabama

OUTLINE

Crabs are the focus of fishery in many regions of the United States. Most crab fisheries exhibit dramatic fluctuations in stock abundance or harvest. This phenomenon, more than any other, has prompted investigations into the possibilities for culturing regionally available species. Species that have received the most attention include the blue crab (*Callinectes sapidus*), Dungeness crab (*Cancer magister*), Jonah crab (*Cancer borealis*), and rock crab (*Cancer irroratus*).

BLUE CRABS

This economically valuable crustacean ranges from Nova Scotia to northern Argentina, including Bermuda and the Antiles (1). The blue crab is an important commercially captured species along the eastern seaboard of the United States and the Gulf of Mexico. Blue crabs are considered coastal inhabitants ranging from the shoreline to approximately 90 m (295 ft) in depth (2). Blue crabs are scavengers, and their normal diet consists of a variety of materials, including fish, benthic invertebrates, and plant material (2–4).

Blue crabs usually reach the adult stage after 12 to 18 months (1,4,5) and may live 2 to 3 years (2,6,7). Female blue crabs mate only once. Mating occurs after a terminal molt, while the crab is still in the soft shell stage. Enough sperm is received and stored in the female's spermathecae, or "sperm pouch," for two and possibly three spawns (8). Van Engel (3) described the coloration of the abdomen of immature female crabs as grayish white and that of adult females as blue green. During the last few days before the female's final molt, the dark green of the soft, inner, mature abdomen shows through the translucent whiteness of the hard, outer, immature abdomen. This change, in addition to the red line stage on the border of the swimming paddles of the premolt female, will indicate that terminal molt is approaching. The abdomen of the immature male is tightly sealed on the ventral surface of the shell, while on a sexually mature male the abdomen hangs free or is held in place by a pair of tubercles (3). Induced mating and spawning under laboratory conditions should include a holding tank with a mud or gravel bottom that simulates the natural environment (8). A minimum of 2 to 3 months is required after mating, before ovulation and spawning occurs (8).

EGG DEVELOPMENT AND LARVAL REARING

Blue crabs produce 0.7 to 2 million eggs, which attach to swimmerets under the female's abdomen during incubation (1). Egg color can be used to determine the approximate age of the eggs (9). A yellow to orange color is characteristic of eggs that have been on the swimmerets 1 to 7 days. Eight to 15 days after spawning, the eggs appear brown to black in color. Hatching usually occurs after about two weeks.

Blue crab larvae normally go through seven zoeal stages to reach a prejuvenile stage, but they can survive with only six zoeal stages (10). Zoeae are free-swimming, but are classified as planktonic, since they lack much control over their position in tidal and strong current areas. In natural environments, eggs appear to hatch successfully in salinities of 23 to 30 ppt. Optimal salinity for zoeal development ranges from 26 to 30 ppt (10,11). Optimal temperature for zoeal development is considered to be 25 °C (77 °F) (11,12). Total time for zoeal development ranges from 31 to 49 days (11). The last zoeal stage metamorphoses into a single megalopal stage, which has both planktonic and benthic affinities (13,14). The megalopal stage lasts 6 to 20 days and precedes the first crab stage. External anatomy of the seven zoeal stages and the megalopal stage is described in detail by Costlow and Bookhout (11).

A number of unicellular algal species have been evaluated for use as larval foods (11,15). While some ingestion and development have been observed, no algae source has been sufficient as a sole food. Whitney (16) and Millikin (8) suggest that the reason algal species have failed to promote growth of zoea is that algal diets are devoid of animal sterols, which appear to be essential in crab diets. Millikin (8) recommends feeding strategies that include rotifers during Zoea I and II stages and *Artemia* nauplii for Zoea III to VII. Sulkin and Epifanio (12) concluded that food organisms no larger than 110 microns (0.004 in.) are optimal for the first and second zoeal stages.

GROWOUT

Few data are available on the culture of juvenile and adult crabs. Springborn (17) investigated the effects of salinity and temperature on juveniles and found little difference in survival or growth at salinities between 2 and 21 ppt. Mortality rates increased significantly at 1 ppt and below. Growth rates increased in the temperature range 20° to 30 °C (68° to 86 °F), with optimal growth between 28 and 30 °C (82° and 86 °F). Mortality rates increased rapidly above 30 °C (86 °F). Holland et al. (18) demonstrated that crabs readily accept pelleted commercial diets, but also concluded that cannibalism remains one of the major impediments to crab culture.

Although considerable information about the blue crab is available, culture from egg to adult has not been economical. A prolonged larval period, high mortality rates, and relatively low market value for hard crabs are the main reasons that hard-shell crab aquaculture has not been successful (1).

SOFT-SHELLED CRAB PRODUCTION

Blue crab biology and consumer tastes provide opportunities for a specialized form of crab aquaculture that has been practiced successfully. Blue crabs, as with other crustaceans, molt or shed their exoskeleton in order to grow (Fig. 1). Immediately after molting they are soft for several hours and can be consumed whole. Crabs in this state are referred to as soft-shells or soft crabs and command a higher price than hard crabs in the market.

Blue crabs are held for shedding for a relatively short period of time. During holding, they are not fed and, therefore, are not cultured in the conventional sense. Wild-caught crabs are examined for indications that they are close to molting. Premolt crabs (peelers) have a thin white, pink, or red line on the inside edge of the swimming paddles. Additionally, immature females have a mauve colored triangular apron on the abdomen. Mature females that have a semicircular apron will not molt. White-line crabs may shed within 7 to 14 days, pink-line crabs within 3 to 6 days, and red-line crabs within 1 to 3 days (19).

Premolt crabs are held in a variety of ways that range from floating containers secured to a dock to sophisticated water-recirculating systems located some distance from saltwater. Crabs are sorted daily in order to keep red-line crabs segregated from the other crabs and reduce cannibalism during shedding. Molting tends to take place at night, and operators must periodically check peeler crabs so that recently molted crabs can be removed while still soft. Once removed from the water, soft-shells remain soft and alive for 4 to 5 days if kept moist and cool. Storage temperatures of 9° to 10 °C (48° to 50 °F) are recommended (20). Soft-shell crabs left in the water more than a few hours begin to harden and lose value.

Soft-shell crab production is constrained by several factors. The supply of wild-caught premolt crabs is often irregular, seasonal, and inadequate for meeting the apparent market demand. Not all premolt crabs go on to molt, and not all that start to molt survive molting. Production is a labor-intensive effort that includes constant sorting and monitoring of crabs both day and night. Detailed manuals on blue crab shedding techniques are available from several state sea grant programs in the United States (19–21).

Most research on producing soft-shell crabs has focused on water-recirculating systems to hold crabs. Historically, crabs were held in floating trays in natural waters. Later developments included land-based systems of shallow trays with flow-through water provided by pumping.

Variations in water quality, high cost of waterfront property, and a desire to have more control led to the development of water-recirculating systems specifically for soft-shell crab production.

Recirculating systems for soft-shell crabs are similar in principle to other recirculating systems. A typical unit consists of three or four 2.4 × 0.9 × 0.3 m (8 × 3 × 1 ft) holding trays, a reservoir, and a biological filter and pump. Each tray supports 150 to 200 crabs in a system with 2271 to 3407 L (600 to 900 gal) of water and a flow of 34 to 45 l (9 to 12 gal) per minute, depending on the type of biological filter (22).

Simple biological filters can be made by using small clam shells or limestone pieces approximately 1.2 cm to 3.8 cm (0.5 to 1.5 in.) in size and 0.25 m^3 (9 ft^3) of material for every 300 crabs in the system (22). More sophisticated systems may employ combinations of upflow sand filters and fluidized bed filters (22). Systems employing shell or limestone benefit from the buffering capacity of calcium carbonate, while systems using sand or other noncarbonate materials need to be monitored closely for low pH. Protein skimmers are employed in some systems to remove dissolved organic material, stabilize pH, and provide additional aeration (20).

Carrying capacity and successful shedding are dependent on maintaining good water quality in the system (23). Water quality guidelines have been developed by experimentation and practical experience (Table 1).

Salinity is usually maintained above 5 ppt and within 5 ppt of the harvesting waters. Crabs from salinities as high as 15 ppt have successfully shed in freshwater systems (24).

Hardening in soft-shell crabs can be delayed by using low-calcium water (25). Molting crabs remain soft 3 to 4 hours longer than controls when held in shedding systems maintained at 18 to 40% of normal calcium levels (26). This refinement can reduce human observation, particularly at night, but pH should be monitored and, if below 7, adjusted periodically with the use of sodium bicarbonate (22).

Table 1. Water-Quality Guidelines for a Recirculating System when Used for a Crab Shedding Operation (22)

Parameter	Recommended Range
Dissolved oxygen	Above 5 ppm O_2 in holding trays; water leaving filters must contain above 2 ppm O_2.
Total ammonia	Below 1 ppm ($NH_3 + NH_4$)–N in holding trays
Nitrite	Below 0.5 ppm NO_2–N in holding trays
Nitrate	Below 500 ppm NO_3–N in sump
Temperature	24–27 °C (75–80 °F) in holding trays
pH	Hold between 7 and 8 for normal operation
	Hold between 7.5 and 8.0 during peak loading
Alkalinity	Above 100 ppm $CaCO_3$ at all times
Salinity	Match salinity of harvesting waters within 5 ppt

Dependence on wild-caught blue crabs and the seasonal nature of supply have hindered large-scale development of shedding operations. Most soft-shell crabs are produced in small, family run operations. An economic analysis indicated that a three-tray system costing about $2,700 to build would return about $5,100 for an eight-week season after operating costs were deducted, but it did not take labor costs into account (27).

CANCER CRABS

Crabs of the genus *Cancer* occur worldwide in temperate regions with several species exploited commercially (1). In the United States, east and west coast species are harvested directly in crab fisheries or as incidental catches in other fisheries. On the Pacific Coast, the Dungeness crab (*C. magister*) is important, and on the Atlantic Coast, the rock crab (*C. irroratus*) and the Jonah crab (*C. borealis*) are of interest. Because of widely fluctuating harvests, consideration has been given to the culture potential of all three species.

Life histories of *Cancer* crabs have been described (28–31). Mating usually occurs from April through September, followed by spawning from October to June. Fecundity of the Dungeness crab is as high as 1.5 million eggs, while rock crabs produce up to about 300,000 eggs per spawn (28,30). Five larval stages and one megalopal stage have been identified (30–32). Optimum temperature and salinity for larval development have been reported to be 10° to 13.9 °C (50° to 57 °F) and 25 to 30 ppt, respectively (33). Larval development may last 80 to 90 days and megalopal development another 30 days before metamorphosis into the first crab instar occurs (28,34). Oesterling and Provenzano (1) discuss various efforts in the laboratory to develop culture techniques. Larvae of all three species have been raised on a variety of diets, including *Artemia* nauplii, *Balanus glandula nauplii*, *Skeletonema coastatum*, individually and in various combinations (31,33–37). Despite all the information available, complete culture from egg to market appears to have a low potential for any of the three *Cancer* crabs studied. Reasons for low culture potential included problems with cannibalism during larval stages and later crab stages, slow growth rates requiring two to three years for crabs to reach market size, the need to rely on wild-caught Dungeness broodstock, and the relatively low fecundity of rock crabs (1).

Even though culture of hard-shelled *Cancer* crabs is impractical at this time, there is potential for the production of soft-shelled crabs similar to what has been accomplished with soft-shell blue crabs (1). Similar procedures used for shedding blue crabs would apply to *Cancer* crabs.

BIBLIOGRAPHY

1. M. Oesterling and A. Provenzano, in J. Huner and E. Brown, eds., *Crustacean and Mollusk Aquaculture in the United States*, AVI Publishing Company, Inc., Westport, CT, 1985, pp. 203–219.

2. A.B. Williams, *U.S. Fish. and Wildl. Serv. Fish. Bull.* **73**(3), 685–798 (1974).
3. W.A. Van Engle, *Comm. Fish. Rev.* **20**(6), 6–17 (1958).
4. R.M. Darnell, *Trans. Am. Fish. Soc.* **88**(4), 294–304 (1959).
5. F.J. Fischler, *Trans. Am. Fish. Soc.* **94**(4), 287–310 (1965).
6. E. Jaworski, Center for Wetland Res., L.S.U., Baton Rouge, LA, (1972).
7. A.B. Williams, *U.S. Fish. and Wild. Serv. Fish. Bull.* **65**(1), 1–298 (1965).
8. M.R. Millikin, *Mar. Fish. Rev.* **1353**, 10–17 (1978).
9. C.E. Bland and H.V. Amerson, *Chesapeake Sci.* **15**, 232–235 (1974).
10. S.D. Sulkin, E.S. Branscomb, and R.E. Miller, *Aquaculture* **8**, 103–113 (1976).
11. J.D. Costlow and C.G. Bookhout, *Biol. Bull.* **116**, 373–396 (1959).
12. S.D. Sulkin and C.E. Epifanio, *Esturine Coastal Mar. Sci.* **3**, 109–113 (1975).
13. A.B. Williams, *Chesapeake Sci.* **12**(2), 53–61 (1971).
14. S.D. Sulkin, *J. Exp. Mar. Biol. Ecol.* **20**, 119–135 (1975).
15. J.D. Rush and F. Carlson, *Chesapeake Sci.* **1**, 196–197 (1960).
16. J. Whitney, *J. Exp. Mar. Biol. Ecol.* **4**, 229–237 (1970).
17. R.R. Springborn, Masters thesis, Texas A&M, College Station, TX, 1984.
18. J.S. Holland, D.V. Aldrich, and K. Strawn, *Sea Grant Publications-SG-71-222*, Texas A&M Univ., 1971.
19. H.M. Perry, J.T. Ogle, and L.C. Nicholson, in H.M. Perry and W.A. Van Engel, eds., *Proceedings of the Blue Crab Colloquium, Gulf States Mar. Fish. Comm. No. 7*, Ocean Springs, MS, 1979, pp. 137–150.
20. M.J. Oesterling, *Special Report in Applied Marine Science and Ocean Engineering No. 271*, Virginia Institute of Marine Science, College of William and Mary, Gloucester Point, VA, 1984.
21. W. Wescott, *Sea Grant Publication UNC-56-84-01*, North Carolina State University, Raleigh, NC, 1984.
22. R.F. Malone and D.G. Burden, Louisiana Sea Grant College Program, Center for Wetl. Res., L.S.U., Baton Rouge, LA, (1988).
23. D.P. Manthe, R.F. Malone, and H. Perry, *Shellfish Res.* **3**(2), 175–182 (1983).
24. W. Wescott, *Univ. North Carolina Sea Grant Blueprint. UNC-SG-BP-88-1*, North Carolina State University, Raleigh, NC, 1988.
25. A.J. Freeman, D.L. Laurendeau, G. Kilgus, and H.M. Perry, *Northeast Gulf Sci.* **8**(2), 177–179 (1986).
26. H.M. Perry, C.B. Trigg, R.F. Malone, and J. Freeman, *Ann. Prog. Summary, MS-AL Sea Grant Consor.*, Ocean Springs, MS, 1994.
27. C. Adams and M. Oesterlings, *Florida Sea Grant. SGEB-38*, Univ. of FL, Gainesville, FL, 1997.
28. D.C. Mackay, *Fish. Res. Board of Can. Bull.* **62**, (1942).
29. K.W. Waldron, *Fish. Comm. of Oregon* **24**, (1958).
30. T.E. Bigford, *NOAA Tech. Rep. Circ. 426*, (1979).
31. A.N. Sastry, *Crustaceana* **32**(2), 155–168 (1977).
32. A.N. Sastry, *Proc. Joint Oceanography*, Assembly, Tokyo, Japan, 1970.
33. P.H. Reed, *J. Fish. Res. Board Can.* **26**(2), 289–397 (1969).
34. R.L. Poole, *Crustaceana* **11**(1), 83–97 (1966).
35. A.N. Sastry, *Proc. Joint Oceanography*, Assembly, Tokyo, Japan, 1971.
36. R.A. Shleser, *European Symp. Mar. Biol.* (1975).
37. M.C. Hatman, *Proc. World Mar. Soc.* **8**, 147–156 (1977).

See also CRAB CULTURE: WEST INDIAN RED SPIDER CRAB; MUD CRAB CULTURE.

CRAB CULTURE: WEST INDIAN RED SPIDER CRAB

R. LEROY CRESWELL
Harbor Branch Oceanographic Institution, Inc.
Fort Pierce, Florida

BJORN TUNBERG
Kristineberg Marine Biological Station
Fiskebäckskil, Sweden

OUTLINE

The West Indian red spider crab, *Mithrax spinosissimus*, is a large majid crab that inhabits coral reefs and rocky outcrops of the tropical western Atlantic Ocean. Its known distribution is from the Carolinas on the east coast of the United States, through the Bahamas and the islands of the eastern Caribbean Sea, and along the continental shelf of Latin America as far south as Venezuela (1,2). It occurs from shallow waters to depths of 180 m (600 ft), and it commonly inhabits manmade canals cut in oolitic limestone, such as the Florida Keys in the United States (3,4). As with most members of the family Majidae, *M. spinosissimus* remain in hiding during the day and venture several meters (yards) from their refuge at night to browse on benthic algae and associated epifauna. The sexes are dimorphic; the males reach a greater size overall, and their chelae attain massive proportions compared to those of the female. Size frequency distributions for male and female *M. spinosissimus* captured in fish traps indicated a mean size of 133.4 mm (5.3 in.) carapace length (CL) for males and 122.8 mm (4.8 in.) for females (5). Females usually average 50% less in weight than males.

Despite its large size, this crab is taken only occasionally by fisherman, for either home consumption or local marketing. Because of their paucity and sporadic distribution, a commercial fishery for the spider crab has never developed (an exception being in Panama, where they are locally abundant along the walls of the Panama Canal) (6).

Interest in the mariculture potential for the West Indian red spider crab, *M. spinosissimus*, began in the late

1970s at a small marine laboratory on the remote island of Dos Mosquises in the Los Roques Archipelago off the coast of Venezuela. The scientists at the Fundación Scientifica de Los Roques were pioneering the mariculture of several tropical marine species that compose the most valuable fisheries resources of the Caribbean region, among them the queen conch and the spiny lobster. Although the West Indian red spider crab was not fished commercially, its large size, delicious taste, and popularity in local fishing markets attracted attention as a potential mariculture candidate. The researchers at Los Roques reported their preliminary results, including larval development and growth of early juveniles, in the Proceedings of the World Aquaculture Society in 1977 (7).

In 1983, the Marine Systems Laboratory of the Smithsonian Institution began research to develop a full life cycle mariculture system for *M. spinosissimus*, one that emphasized low-technology methods appropriate for developing countries. Their technology was based on a cage culture system, in which crabs were fed a diet of algal turfs that were grown on screens placed in the open sea. Turf algae, characterized by blue-green, filamentous red algae, and benthic diatoms colonized floating screens and produced as much as 30 g (1 oz) dry weight of algal food per day (8). After being allowed a few weeks for the algae to grow, the screens were moved to floating growout cages that housed the crabs. Algal turf mariculture of *Mithrax* crabs was initiated on Grand Turk (in the Turks and Caicos Islands), in the Dominican Republic, and on the island of Antigua, with the intention of transferring the algal turf/cage culture system as appropriate technology to island fishers.

Despite the substantial resources dedicated to the project, positive results were not forthcoming. In retrospect, the concept of rearing *M. spinosissimus* exclusively on algal turf was fundamentally flawed. The flaws can be summarized as follows:

(i) The production of algal turf on floating screens in tropical, oligotrophic (nutrient-poor) waters was greatly overestimated, particularly because inorganic sediments, which comprised more than 25% of the deposition on the screens, were not excluded from the organic weight measurements. The amount of vegetable matter provided from the algal turf screens was insufficient to sustain the biomass of crabs housed in the floating cages.

(ii) *Mithrax* crabs are not docile, obligatory herbivores, as the investigators purported, but rather they are omnivorous, cannibalistic, and highly aggressive crustaceans.

(iii) *Mithrax* crabs undergo a terminal molt at puberty, after which no additional growth occurs. The marketable size of crabs, 1 kg (2.2 lb), based on an economic analysis from the Smithsonian study (9), would be achieved by only a small fraction of crabs that were the progeny of wild-caught broodstock. Only through a selective breeding program, over several generations of crabs, would a significant proportion of the population reach one kilogram prior to terminal molt.

In 1984, the Harbor Branch Oceanographic Institution initiated a research program to evaluate the potential for *Mithrax* crab mariculture, which was to include the algal turf/cage culture system as well as alternative methods (10). The information presented below is a synthesis of the results from various studies.

REPRODUCTIVE BIOLOGY AND LARVICULTURE

Sex and maturity among female *M. spinosissimus* can be distinguished by the shape of the abdominal apron; in males it is thin and narrow, while in mature females it is broad and full enough to cover the egg mass. The apron of immature females is intermediate in width. Spider crabs undergo a terminal molt at puberty, after which no further growth occurs (11). The molt to mature female form occurs by about 75 mm (3 in.) CL. Copulation can take place any time after the final molt—a soft-shelled condition is not required. Females are able to produce successive egg masses for extended periods, to be fertilized by stored sperm in the spermatheca. However, successive spawns from females kept in the laboratory produce smaller broods, and larval mortality is higher than from spawns collected from the field.

Fecundity estimates vary widely from "tens of thousands" (7) to much higher (up to 100,000) (12,13). Creswell et al. (6) reported $18{,}826 \pm 3{,}304$ ova/female, or 59.5 ± 8.8 ova/g ($\sim$1700/oz) body weight. The ova are approximately 1 mm (0.04 in.) in diameter and weigh 1 mg (0.00004 oz). Newly fertilized eggs are deposited on the pleopods and are bright orange in color, but the gross appearance of the egg mass changes as the embryos develop. As the yolk is absorbed, yolk pigments accumulate in the embryo's integument and eyes, turning the egg mass bright red after the first week, then burgundy, and finally tan to grey when hatching is imminent, 21 to 25 days after fertilization. Orange eggs on four crabs hatched after a period of 18 days (mean value). The average hatching period of red eggs on eight crabs was 9.5 days. (14).

M. spinosissimus has an abbreviated and essentially lecithotrophic larval cycle, the egg's yolky reserves providing all the necessary nutrition required for full development of the larva to first crab stage (15,16). Hatching usually occurs at night and continues for several hours. The larvae are released by a fanning motion of the pleopods of the crab, accompanied by vigorous jerking of the abdomen. Larvae usually hatch as swimming *first zoeae* and display positive phototaxis immediately after hatching. First zoeae molt to *second zoeae* within 10 to 12 hours after hatching. Provenzano and Brownell (15) and others have also described a non-swimming prezoela stage. This stage is, however, probably an aberrant occurrence associated with stress on the females or on the developing larvae. The zoeae undergo two molts within 36–48 hours, before metamorphosing to the megalopa (post-larval stage), when they first begin to feed. After three to four days, the megalopa molts to the first crab stage, six to eight days after hatch (Fig. 1) (6,7).

Larvae have been reared intensively in shore-based hatcheries and extensively in cages floating at sea. Brownell and Provenzano (7) reared larvae in 400–600 L

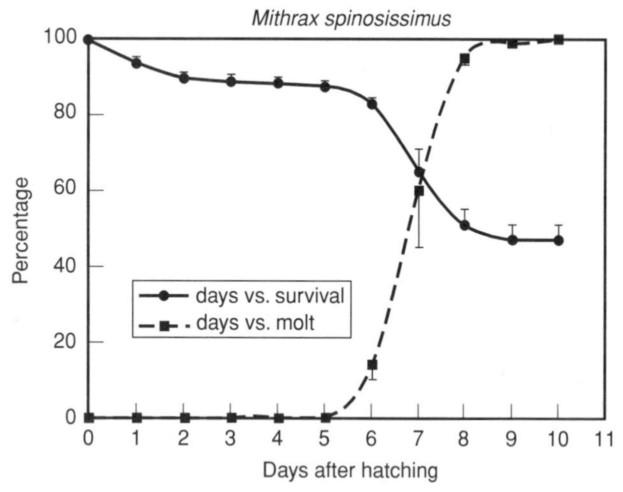

Figure 1. Survival rates of *M. spinosissimus* using the shallow tray system.

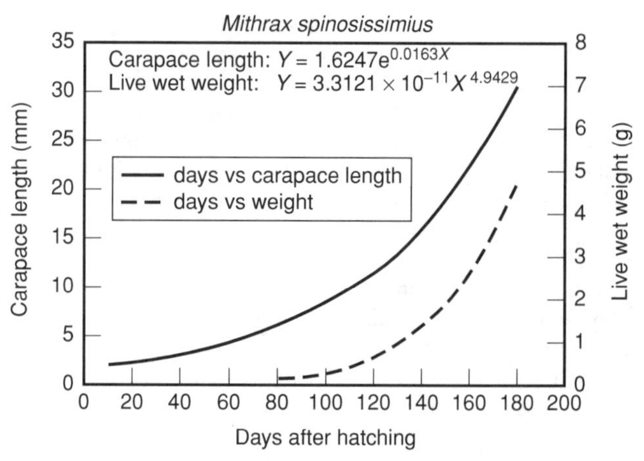

Figure 2. Growth rate of *M. spinosissimus* during first 180 days after hatching.

(105–158 gal) tanks supplied with mixed cultures of phytoplankton. They reported significant mortalities during the molt from second zoeae to megalopae, with higher survival during the molt to first crab (total survival to first crab was 2–5%). The Smithsonian Institution, in Turks and Caicos and in Antigua, B.W.I., attempting to raise crab larvae in small "kreisel" tanks, reported similar discouraging results (<4% survival). The authors experienced similar high mortalities at molt from zoea to megalopa when using fiberglass tanks with upwelling currents (i.e., cone-bottomed kreisel tanks). An alternative larviculture system, utilizing screen-bottomed [500 μm (0.04×10^{-3} in.)] trays floating in shallow water tables, resulted in survival rates exceeding 85% during molt from zoea to megalopa and approximately 50% to first crab stage (Fig. 1), indicating that the presence of substrate may be critical for successful completion of the molt to postlarvae. By utilizing the shallow tray system, zoea can be stocked at densities of 25,000/m^2 (10.8 ft^2), with survival exceeding 70% to first crab stage (10).

EARLY GROWTH

The 2nd and 3rd crab stages occur at around 15–20 days and 25–30 days after hatching, respectively. The water temperature during the studies described above was ca. 27–28 °C (81–83 °F). Molting frequency is dependent upon temperature (17–20), so it is likely that the intermolt period would decrease at higher temperatures [30 °C (86 °F)] and that more rapid growth would result.

Studies by Tunberg and Creswell (21) on early development (up to stage 12 crabs) showed that the intermolt period increases with stage (except between stages five and six). Stage two lasted for about 10 days; stage eleven lasted for about 20 days. Figure 2 shows the growth rate (carapace length and live wet weight) during the first 180 days after hatching. The growth rate (carapace length) at molt varies between approximately 22 and 40%, and the corresponding mean weight (live wet weight) increase varies between 90 and 135%.

Abdominal length and width measurements indicate that it should be possible, by visual observations, to distinguish between the sexes at a carapace length of approximately 12–15 mm (0.48–0.6 in.), that is, at an age of about 120–140 days.

During a long-term study (180 days) (21), only about 20% of the crabs survived throughout the period.

NUTRITION

Although the larvae of *M. spinosissimus* are facultative lecithotrophs, growth and survival were enhanced when feeding was initiated five to eight days after hatching (14). Benthic diatoms, as well as various types of macroalgae (e.g., *Enteromorpha*, *Gracilaria*, *Ulva*) and seagrasses (e.g., *Thalassia*), are provided as food for megalopae and early crab stages.

Winfree and Weinstein (22) reported that juvenile and adult *M. spinosissimus* are omnivorous and opportunistic feeders upon turf algae and seaweeds (24 species tested), marine fish and invertebrates (23 species tested), fresh beef (heart), pork (liver), chicken (muscle) and dry prepared feeds (12 varieties tested). Small juveniles [≤17 g (0.6 oz)] consume primarily macroalgae (78% of diet) when offered both algae and meat on a free-choice basis. Larger crabs are truly omnivorous and exhibit a strong tendency to supplement their macroalgal diet with meat, the relative proportions of algae and meat chosen being reversed. Crabs will consume a range of commercially available dry feeds, including homarid lobster and penaeid shrimp feeds, tropical fish flakes, and guinea pig pellets. Whether feeding on fresh or dried diets, the daily ration for *Mithrax* crabs is approximately 2.7% (dry feed to live body weight) (10).

AGRESSIVE AND CANNIBALISTIC BEHAVIOR

From postlarvae to large juveniles [approximately 50 mm (2 in.) CL; 50 g (1.7 oz)], *Mithrax* are aggressive and highly cannibalistic, even if well fed. Larger crabs appear to be less vulnerable to cannibalism, probably because of their thicker, more protective shells, except immediately following the molt. Evidence suggests that the severity of "cannibalistic" confrontations can be mitigated by providing the crabs adequate space and protective cover, although it is unlikely that manipulating these factors alone will totally eliminate this behavior.

Most crustacea become aggressive and/or cannibalistic when crowded beyond a certain density. This crowding factor, or density index (DI), may be expressed numerically as the ratio of the area of available habitat to the area of crustacea housed there (measured as the sum of the squares of the carapace lengths of those animals). The smaller the DI, the more crowded the animal, until a critical density index (CDI) is reached and aggression begins. Ryther (10) reported that cannibalistic encounters were observed far more frequently in females and that whereas the CDI was 50 to 60 for males [8 crabs/m^2 (0.8 ft^2) for 5 mm (0.2 in.) CL], it was 90 to 100 for females. These data suggest that elimination of females from the population at the earliest time possible [approximately 15 mm (0.6 in.) CL] would both increase survival and provide maximum system capacity for growout. Providing protective cover and/or complex, three-dimensional habitats, particularly during the early juvenile stage, may improve survival by reducing cannibalistic encounters. Unprotected open area systems, such as cages, raceways, or ponds and simple, two-dimensional structures afford little protection against aggressive behavior and are inappropriate for large-scale production of *Mithrax* crabs.

ENVIRONMENTAL CONDITIONS

Mithrax crabs tolerate a rather narrow range of environmental parameters, a fact that has practical implications for commercial production, particularly for site and stock selection (10). Survival of juvenile and adult crabs in captivity is directly proportional to salinity and temperature. Although crabs easily adapt to hypersaline conditions (40 ppt), stress becomes evident at 25 ppt and complete mortality occurs at 20 ppt. The optimal temperature range for culture occurs at 28–29 °C (83–84 °F), with poor growth at temperatures below this range and poor survival at temperatures above. Below 25 °C (77 °F), crabs are inactive, consume little feed, and are seldom agressive towards each other. At temperatures exceeding 30 °C (86 °F), lethargy, anorexia (despite active feeding), and proneness to disease and abnormal shell development occur. *Mithrax* reared in floating cages or coastal ponds may be subject to environmental fluctuations beyond ranges that are minimally required for healthy growth and survival. Terrestrial runoff could depress salinities over short time periods and cause mortality (or added stress leading to disease). Increased production time to marketable size should be expected in areas where water temperatures fall below 25 °C (77 °F) for extended periods, and unacceptably high mortality may occur in locations that experience temperatures above 30 °C (86 °F). This situation may eliminate the potential of production in northern areas, shallow bays, and tropical saltwater ponds, unless specific strains of *Mithrax* are identified that can tolerate these conditions.

INFECTION AND DISEASE

Ryther et al. (10) reported chronic bacterimia, similar to that described for other crustacea by Tubiash et al. (23), in *Mithrax* crabs cultured in recirculating seawater systems. Early signs include a loss of vigor and reduced feeding activity, which is followed by decreased joint mobility, particularly of the claws (chelipeds). Death occurs without warning, or occasionally preceded by limb loss. Postmortem examination of crabs succumbing to the infection reveals complete atrophy of the musculature and viscera, which are replaced by a spongy white bacterial mass. Application of nitrofurazone (furacin bath) and oxytetracycline (terramycin injections or bath) may be used to treat infections; however, improved water quality and nutrition will likely decrease the incidence of bacterial pathogens.

GROWTH TO HARVEST

All known species of majid crabs lose the ability to molt, and hence to grow, after they attain a specified point in the life cycle. The maricultural significance of a terminal molt in *Mithrax* depends upon whether the animal is large enough for sale at the time it occurs. Mean size of male crabs in terminal molt collected from Antigua was 125 mm (4.9 in.) CL (10); from other islands (Grenadines, Dominican Republic, Turks and Caicos), it was 131 mm (5.2 in.) CL (24).

Growth studies conducted at Harbor Branch Oceanographic Institution (10) indicate that postmolt increase in length of *Mithrax* crabs is relatively constant with age (27%), while weight increase fluctuated from 94 to 136%, with an overall average of 125%. Percentage postmolt weight change decreases as the crab increases in size, while the intermolt period tends to increase with age in a curvilinear fashion, with a greater percentage change during the later molt cycles.

Growth models suggest that male crabs grown to 650 grams (1.4 lb) (the predominant size class reached before terminal molt) would require approximately 20 months in culture (slower-growing female crabs would be discarded as small juveniles). Meat yield averages 20% of total weight (10).

CURRENT STATUS

Interest in the culture of the West Indian red spider crab, *M. spinosissimus*, has waned since the late 1980s, in large part because of some of the constraints on culture outlined herein. One commercial venture, West Indies Mariculture, Inc. (WIM), operated 1988–1994 on North Caicos, in the Turks and Caicos Islands. The project was managed by alumni from an earlier program on Grand Turk Island conducted by the Smithsonian Institution, and the culture methods employed a hybrid of algal turf screens and those later developed by Harbor Branch Oceanographic Institution. West Indies Mariculture produced softshell *Mithrax* crabs and marketed them directly to restaurants throughout the Turks and Caicos Islands.

Adult crabs were held in floating cages attached to docks in a tidal creek with strong currents. Juvenile crabs were fed algal turf until they reached 25 mm (1 in.) CL; thereafter they received a pelleted ration specifically formulated by WIM. Crabs reached 70–80 mm (2.8–3.1 in.) CL in 7–10 months and were transferred to shedding trays where, after molting, they were iced for transport to buyers.

Research into the mariculture of the West Indian red spider crab continues at the University de Oriente, Venezuela (25), but to the authors' knowledge no commercial operations exist today.

BIBLIOGRAPHY

1. M.J. Rathbun, *Bull. U.S. Nat. Mus.* **129**, 1–613 (1925).
2. A.B. Williams, *Shrimps, Lobsters and Crabs of the Atlantic Coast of the Eastern United States*, Smithsonian Institution Press, Washington, DC, 1984.
3. J.A. Bohnsack, *Florida Scientist* **39**, 259–266.
4. B. Hazlett and D. Rittschof, *Mar. Behav. Physiol.* **3**, 101–118 (1975).
5. J. Munro, *Caribbean Coral Reef Fishery Resources*, ICLARM, Manila, Philippines, 1983, pp. 218–222.
6. L. Creswell et al., *Proc. Gulf. Carib. Fish. Inst.* **39**, 469–476 (1986).
7. W. Brownell and A. Provenzano, *Proc. World. Mari. Soc.* **8**, 157–168 (1977).
8. W.H. Adey and R.S. Steneck, Smithsonian Institution, Washington, DC (unpublished), 1984.
9. M.C. Rubino et al., Smithsonian Institution, Washington, DC, (unpbl. ms.), 1985.
10. J. Ryther et al., *Final Report, USAID grant #538-014003*, 1988.
11. H.G. Hartnoll, *Crustaceana* **9**, 1–16 (1965).
12. W.H. Adey, *Mar. Syst. Lab. Smithsonian Inst.* (unpubl.), 1985.
13. W.L. Bernard and K.B. Bernard, *Smithsonian Inst.* (unpubl.) 1985.
14. B.G. Tunberg and R.L. Creswell, *Mar. Biol.* **98**, 337–343 (1988).
15. A.J. Provenzano and W.N. Brownell, *Proc. Biol. Soc. Wash.* **90**(3), 735–752 (1977).
16. C.A. Boulton, Master's thesis, Florida Institute of Technology, 1987.
17. J.A. Allen, *Oceanogr. Mar. Biol. A. Rev.* **10**, 415–436 (1972).
18. M.E. Christiansen, *Norw. J. Zool.* **21**(2), 63–89 (1973).
19. K. Anger, *Mar. Ecol. Prog. Ser.* **19**, 115–123 (1984).
20. R.R. Dawirs, *Mar. Ecol. Prog. Ser.* **31**, 301–308 (1986).
21. B.G. Tunberg and R.L. Creswell, *J. Crust Biol.* **11**(1), 138–149 (1991).
22. R. Winfree and S. Weinstein, *Proc. Gulf. Carib. Fish. Inst.* **39**, 458–468 (1986).
23. H.S. Tubiash et al., *Applied Microbiology* **29**(3), 388–392 (1975).
24. J.M. Inglehart et al., *USAID Report #LAC-0695-G-SS-3070*, 1986.
25. J. Rosas et al., *Proc. Gulf Carib. Fish. Inst.* (1998).

See also CRAB CULTURE; MUD CRAB CULTURE.

CRAPPIE CULTURE

ROBERT R. STICKNEY
Texas Sea Grant College Program
Bryan, Texas

OUTLINE

Crappie, members of the family Centrarchidae and therefore relations of largemouth bass and various sunfish species, are popular recreational fish in the regions of the United States and Canada in which they are found. No commercial foodfish production of crappies exists, but crappies are cultured for stocking ponds, lakes, and reservoirs. Crappies tend to be so prolific in small ponds that overpopulation and stunting often occur, though the problem seems less severe with white crappies than with black crappies (1). The two species, black crappie (*Pomoxis nigromaculatus*) and white crappie (*P. annularis*), have similar culture requirements and respond similarly to the culture environment.

DESCRIPTION AND LIFE HISTORY

The black crappie is common in the Canadian provinces of Quebec and Manitoba and is found in the northern and eastern United States and as far south as Florida and Texas. Black crappies are more abundant than white crappies in the northern part of their range. Typically weighing up to about 900 g (2 lb), black crappies as large as 2.3 kg (5 lb) have been collected (Fig. 1).

White crappies can be found from Minnesota eastward, including into southern Ontario, Canada. Their distribution is then southward to the Gulf of Mexico and includes South Carolina along the east coast. White crappies tend to weigh from 450 to 900 g (1 to 2 lb), with some specimens reaching as much as 1800 g (4 lb).

The two crappie species have about the same overall shape. Black crappies tend to have more black pigment on their scales than do white crappies, but pigmentation is not a suitable way to distinguish between the two species.

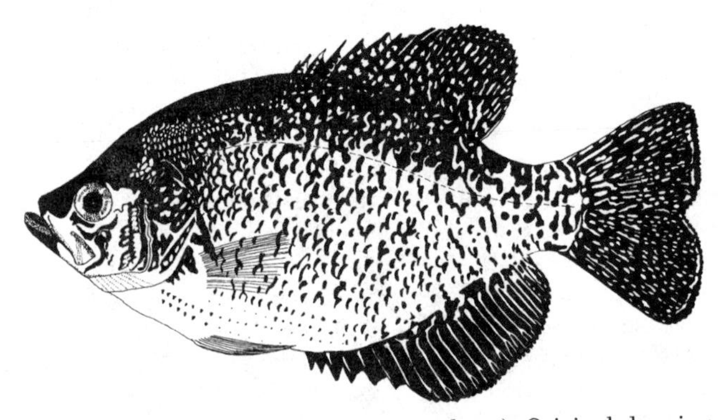

Figure 1. Black crappie (*P. nigromaculatus*). Original drawing by Michele McGrady.

The best way to differentiate the two is by examination of the dorsal fin. In the black crappie, the length of the dorsal fin is about the same as the distance from the anterior end of that fin to the eye. In the white crappie, the dorsal fin is shorter than the distance from the anterior end of the dorsal fin to the eye (2).

Crappies generally do not live longer than four years and become reproductively active at one year of age (3). They are carnivorous and consume a variety of invertebrates along with their primary food which is minnows. Large crappies also consume young bass, so stocking crappies and bass in the same pond is not recommended (1). Male crappies, like other fishes in the family Centrarchidae, construct nests into which the female lays eggs. A large female may produce as many as 140,000 eggs (2). Spawning occurs in the spring.

CULTURE PRACTICES

Culture practices are virtually identical for black and white crappies. Two-year-old fish are usually used as spawners. The two species may, in fact, be stocked in the same pond. Fathead minnows are often provided as forage, though threadfin and gizzard shad have also been stocked as forage in areas where those species are available.

Spawning and egg incubation, which occur when the water temperature rises to above 18 °C (64 °F), are allowed to proceed naturally in open ponds. There are conflicting reports as to the most desirable age for broodfish. Some culturists support the use two year olds, while others indicate that three-year-old broodfish are preferred (1,3). Adult stocking rates have varied, with about 250 adults per hectare (625 adults per acre) being the maximum number, above which fingerling production seems to be inhibited. Once the female has deposited her eggs in the nest and they have been fertilized, they will hatch in about 48 hours at 19 °C (66 °F) (3). Black crappies are preferred by fish farmers in that the stunting problem seems to be less than with white crappies, and both black crappies and hybrid crappies withstand handling stress better than do white crappies.

Hybrid crappies can be produced by crossing female black crappies with male white crappies. This crossbreeding can be done simply by stocking the fish in ponds, since hybridization has been observed in nature (1). Alternatively, adults can be manually stripped in the laboratory. Hybrids have been observed to grow more rapidly than either parental species and, during the first generation (F_1), to strongly resemble black crappies.

In at least two instances, triploid white crappies have been produced (4,5). Triploids have one extra set of chromosomes as compared with diploid, or "normal," fish. Heat shock was not very successful in inducing triploidy (4), but exposure to cold was effective. Subjecting eggs from gravid females to 5 °C (41 °F) water for 60 minutes produced 72–92% triploids (4), while in three out of seven attempts that involved exposure to 5 °C (41 °F) for 90 minutes, over 90% of the resulting fish were triploids (5). Since triploids cannot successfully reproduce, they could hold promise in small ponds for preventing overcrowding.

Handling stress is a significant problem with crappies (1,3). The incidence of columnaris disease (see the entry "Bacterial disease agents") is high following harvesting. The stress seems to be reduced if harvesting of fingerlings takes place during winter.

Training some species of fish to accept pelleted feeds is often difficult, though crappies can easily be trained to accept prepared rations. A seven-day training period, beginning with carp eggs as a starter diet and then gradually replacing the carp eggs with a commercial feed, has resulted in a high success rate (6). Crappies have also been trained to accept a pelleted diet, after an initial period during which they were fed krill (7). When a semimoist diet is used, small fingerlings are more easily trained than large fish, while the opposite is true when dry diets are provided (7). Small raceways provide a good culture environment for training crappies to accept prepared feeds. There is no specific information available on the nutritional requirements of crappies, so feeds that meet the needs of species such as trout and salmon are employed, since experience has demonstrated that a good trout or salmon diet will meet the nutritional needs of most other types of fishes.

BIBLIOGRAPHY

1. M. Martin, *Aquaculture Magazine*, May/June, 35–41 (1988).
2. S. Eddy and J.C. Underhill, *Northern Fishes*, University of Minnesota Press, Minneapolis, 1974.
3. J.H. Williamson, G.J. Carmichael, K.G. Graves, B.A. Simco, and J.R. Tomasso, in R.R. Stickney, ed., *Culture of Non-salmonid Fishes*, CRC Press, Boca Raton, FL, 1993, pp. 145–197.
4. N.W. Baldwin, C.A. Busack, and K.O. Meals, *Trans. Am. Fish. Soc.* **119**, 438–444 (1990).
5. G.R. Parsons, *Trans. Am. Fish. Soc.* **122**, 237–243 (1993).
6. J.F. Smeltzer and S.A. Flickinger, *N. Am. J. Fish. Man.* **11**, 485–491 (1991).
7. J.R. Triplett, T. Amspacker, and G. Thomas, in G. Libey, ed., *Successes and Failures in Commercial Recirculating Aquaculture*, 1996, pp. 507–520.

CRAWFISH CULTURE

ARNOLD G. EVERSOLE
Clemson University
Clemson, South Carolina

W. RAY MCCLAIN
Rice Research Station
Crowley, Louisiana

OUTLINE

Crawfish, crawdads, and crayfish are all common names for a large group of invertebrates in the phylum Arthropoda, class Crustacea, and order Decapoda. The preferred common name in the southern United States, where the aquaculture industry is centered, is crawfish (1). The term "crawfish" in this discussion is used exclusively for the red swamp crawfish, *Procambarus clarkii*, and the white (river) crawfish, *P. zonangulus*. References to other taxa are identified by genus and species.

Crawfish are characterized as having a hard, but flexible, exoskeleton; pairs of jointed appendages, including five pairs of pereopods; and gills for acquiring oxygen in their aquatic environment. Most people that have an opportunity to handle crawfish are familiar with the first pair of pereopods with enlarged pincers, called chelipeds. (See Fig. 1.) The front half of the crawfish's body is covered with an unsegmented shell, or carapace. The carapace covers the head and thorax, collectively referred to as the cephalothorax. In contrast, the posterior portion of the body, or abdomen, is clearly segmented. Appendages on the abdomen called swimmerets — or, more correctly, pleopods — are diagnostic characteristics for identifying the sex, level of maturity, and species of the crawfish. The first two pairs of pleopods on a male crawfish are modified to transfer sperm. These male reproductive structures, called gonopods, are strikingly different in reproductively active (Form I) individuals than in reproductively inactive (Form II) individuals. Gonopods in Form I males are cornified, structurally longer, and more pointed than in Form II males. Female pleopods are uniform, regardless of reproductive activity, and are adapted for carrying eggs and hatchlings. The last segment of the abdomen is expanded to form the tail, which is technically referred to as the telson. The eloquent treatise *The Crayfish: An Introduction to the Study of Zoology*, by T.H. Huxley, originally published in 1880 and reprinted several times (2), is an excellent source of detailed information about the structure and function of crawfish.

When startled, a crawfish contracts the large muscles in its abdomen rapidly, pulling the telson under the body. Repetition of this action will cause the animal to dart quickly backwards or to "jump" up from the surface on which it is standing. In addition to providing the crawfish with an escape mechanism, this large muscle provides humans with a prized food item. Typically, consumers and producers in the crawfish industry call this muscle "tail meat" when it is removed from the abdomen, rather than using the correct term, "abdominal meat."

CRAWFISH AS A FOOD ITEM

Although crawfish are harvested for fish bait and study specimens, most of the crawfish produced in the southern United States are sold for food. In Louisiana, where crawfish culture is centered, live crawfish constitutes 45–50% of the annual retail sales (3,4). Live crawfish are purchased at retail seafood stores, from wholesalers, or directly from farmers at the pond bank. Traditionally, live crawfish are cooked with ample seasoning, onions, potatoes, and corn in a common pot. (See Fig. 2.) Peeling crawfish takes some practice, but an experienced attendee at a crawfish boil can deftly remove the meat from several abdomens in less than a minute. Often adhering to the tail meat is the hepatopancreas, which is routinely consumed and is referred to as "fat" in the industry.

The ability to process crawfish allowed for the development of markets outside the traditional crawfish-farming areas. One processed product is cooked and frozen whole crawfish. Those crawfish destined for the domestic

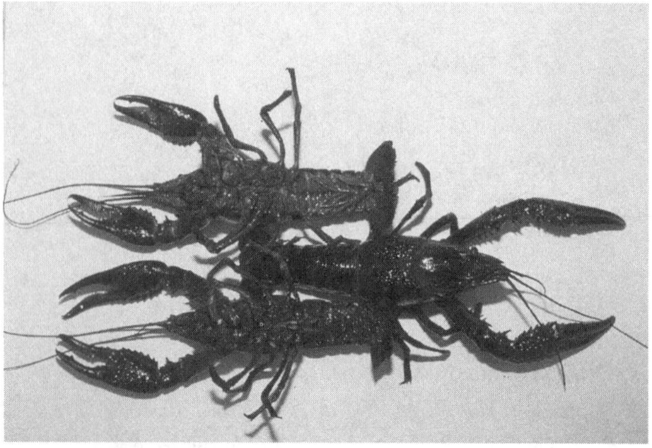

Figure 1. Dorsal and ventral views of mature *P. clarkii*. The ventral views show the sexual structures of the female (top) and male (bottom).

Figure 2. Traditional Cajun style of cooking crawfish: boiling with vegetables and spicy seasoning.

market are cooked with Cajun seasoning, whereas dill seasoning is used with crawfish for the European market (3). However, the most common processed product is cooked, hand-peeled, deveined tail meat. The meat may be packed with or without hepatopancreatic tissue, which is an important ingredient in Louisiana (Cajun) cuisine. Abdominal meat is a much more diversified and easier product to use than live crawfish. Recipes for crawfish, such as étouffée and jambalaya, are available in most comprehensive cookbooks and special publications by the industry, such as *Louisiana Crawfish: Heads & Tails Above the Rest*! (5).

Meat yield varies greatly with factors such as sexual maturity and size. For example, Huner (6) observed that immature male and female crawfish have 4–5% higher meat yield than mature males. Immature crawfish have smaller chelipeds than mature individuals; consequently, less weight is lost when the head portion of the processed animal is discarded. Smaller individuals, regardless of sex, usually yield a higher percentage of meat than larger crawfish. Equally important to achieving maximum meat yield is cooking time and peeling technique (3). In general, abdominal meat yield is assumed to be about 15% (3).

Softshell crawfish production technology and markets were developed in the mid-1980s in Louisiana, but that industry has since waned. Processing of softshell crawfish requires only that the gastroliths, or "stones," be removed, because these hard calcium carbonate structures are retained when the crawfish molts. Consequently, the yield of edible product is much higher than for hardshell crawfish. The yield from softshell crawfish varies from 92%, if only the stones are removed, to 72%, if the carapace and hepatopancreas are also removed in processing (7). Softshell crawfish and blue crabs are similar table fare, which is one of the reasons that softshell crawfish were readily accepted.

HISTORICAL DEVELOPMENT OF CRAWFISH CULTURE

Early immigrants to the southern United States took advantage of the extensive riparian wetlands to harvest game and fish. Crawfish grow particularly well in areas that experienced periodic flooding and drying. Characteristic of this type of riparian habitat is the Atchafalaya Basin, a region extending 130 km by 30 km (80 by 20 mi) in the heart of the Mississippi River flood plain in Louisiana. Harvests from the basin vary greatly with discharge; 27,000-metric-ton (60-million-lb) harvests may occur in high-water years, whereas, during years of low discharge, the harvest might reach only 4,000–8,000 metric tons (8.8–17.6 million lb) (6). These unpredictable harvests, combined with increased consumer demand for crawfish, provided much of the impetus for crawfish aquaculture development.

The beginning of crawfish aquaculture was somewhat humble, but has grown from an incidental crop to a cultured crop from permanent ponds. Crawfish are also cultured as an integral component of rice rotational systems in Louisiana and east Texas. Viosca (8), LaCaze (9), and Thomas (10) outlined the basic management procedure for culturing crawfish in rice fields, which is still the model from which current culture practices are adapted. Thomas (10) also published the first scientific evaluation of the growth of crawfish in rice fields in 1965.

Since rice farmers had suitable flat land and much of the necessary equipment, it was an easy jump to include crawfish as a second, off-season crop. At that time, the future of the crawfish industry was promising, and the acreage in crawfish culture grew rapidly until the mid-1970s, when it slowed (11). A second expansion in the industry took place in the early 1980s, when low profits from row crops made crawfish farming more attractive. The area devoted to crawfish culture has been relatively constant in Louisiana, at about 45,000 ha (112,000 acres), since this latest expansion and through the 1990s (4). Although outside Louisiana there are only about 2,000 ha (5,000 acres) used for crawfish production (4), considerable opportunity exists for expansion in other rice-growing regions, such as Texas, Mississippi, California, and Arkansas, as well as in areas with extensive wetlands. In states such as South Carolina, the wetlands are managed with the primary purpose of attracting waterfowl, with crawfish being a secondary product of the management system (12). The economics of supply and demand will govern the future expansion of the crawfish industry in those states and other areas.

Other taxa of crayfish are cultured to a limited extent in the United States. Members of the genus *Orconectes* have been cultured for fish bait in the midwest and northeastern United States for a long time, but, recently, there has been increased interest in culturing them for food (13). A limiting factor in culturing the larger *Orconectes* species as a food item may be the time required for them to reach market size (14). *Pacifastacus* spp. are harvested from wild populations in the northwestern United States, but are not grown commercially for food. Several *Cherax* species from Australia have been promoted as freshwater lobsters for culture in the United States. Experiences with culturing those crayfish in the United States are mixed, but research on the subject is continuing (see the entry "Australian red claw crayfish").

CRAWFISH LIFE CYCLE AND RELATED ASPECTS OF CULTURE

Crawfish culture in Louisiana is dominated by two species: the red swamp crawfish, *P. clarkii*, and the white river crawfish, *P. zonangulus*. The white river crawfish was formally known as *P. acutus acutus* until Hobbs and Hobbs (15) revised the taxonomy, separating the previous complex into two different species. *P. acutus acutus* still occurs in the eastern United States, whereas *P. zonangulus* is found in southern states along the Gulf of Mexico. The native range of the red swamp crawfish overlaps that of *P. zonangulus*, but extends farther north in the Mississippi River drainage. These two species are frequently found together in culture ponds and are often indistinguishable by persons not experienced with crawfish. Although the abundance of one species may vary among ponds and within ponds over years, red swamp crawfish usually dominate culture ponds in Louisiana (16). In a study of two commercial ponds,

Romaire and Lutz (17) observed that red swamp crawfish made up 93 and 76% of the population, respectively. The investigators hypothesized that the difference in abundance of the two species was linked to the red swamp crawfish's greater reproductive potential, because growth and survival were not different between the two species. In some cases, *P. zonangulus* increase in abundance as the culture pond ages; Huner (16) observed a decline in the composition of the experimental harvests from about 90% red swamp crawfish to only 5 to 35% over a five-year period. He suggested that later flooding favors *P. zonangulus*, and repeated late floodings may be one reason for the increased proportion of *P. zonangulus* in his experimental culture ponds. *P. zonangulus* were found to regulate oxygen consumption at lower levels of dissolved oxygen than *P. clarkii*, which may also increase survival and the subsequent abundance of *P. zonangulus* in some ponds (18).

In general, the two species have environmental requirements that are conducive to low-input aquaculture systems, but there are differences between the two species. For example, *P. zonangulus* produce fewer, but larger, eggs (19,20); lay eggs (oviposite) once a year (17); may grow faster at cooler temperatures (17); reach a greater maximum size (17,19); and, overall, appear to do better in less eutrophic waters than red swamp crawfish (20,21). Some care must be exercised when reviewing the literature before the white river crawfish was reclassified, because references may represent *P. acutus acutus*, *P. zonangulus*, or a complex of more than one species.

These procambarid crawfish evolved in seasonally flooded wetlands and have a life cycle that is well adapted to fluctuating periods of flooding and drying. Wet periods permit crawfish to grow and thrive, while temporary dry periods *(1)* promote aeration of the bottom sediments, *(2)* reduce the abundance of aquatic predators, and *(3)* allow for better establishment of vegetation, which serves as cover for crawfish and fuel for the food web during periods of flooding. Crawfish survive dry periods by digging or retreating to burrows (see Fig. 3), where they can avoid predators and maintain the moist environment necessary for survival. Crawfish reproduce within the protection of the burrow. Current methodology for pond production is based on the annual hydrological cycle to which crawfish have become adapted. Pond culture, however, allows for greater control of important environmental variables.

Figure 3. Typical chimney resulting from the construction of a burrow, which crawfish use to facilitate reproduction and survival during the dry season.

Dry Phase

In the southern United States, seasonally flooded wetlands are normally dry during late summer and autumn. Burrowing activity is initiated when females that have mated seek refuge to begin ovipositing or when the habitat begins to dry. These conditions are often synchronous, because decreasing water levels and increasing temperatures promote maturity and mating in crawfish. Crawfish mate in open water, and the female stores sperm in the seminal receptacle until she lays the eggs while in the burrow. Huner and Barr (22) give a good account of burrow ecology; in brief, crawfish dig simple, nearly vertical burrows that extend to the water table [approximately 40–90 cm (16–36 in.)]. Burrow entrances are often associated with items of cover, such as vegetation and woody debris, and may be covered with a simple chimney and/or mud plug. Burrows usually contain a single female or a male and a female, but occasionally contain additional crawfish. Crawfish are confined to burrows during dry periods until surface or rain waters reach them.

Water is usually held in crawfish culture ponds until late spring to early summer. Prior to dropping of the water level, some crawfish burrow near the waterline. As the water level drops, crawfish burrows appear lower on the levee and less frequently on the pond bottom when dry. Ovarian development requires three to five months, begins prior to burrowing, and is completed within the burrow. During ovarian development, oocytes become spherical in shape and change color from light to dark when mature. To assess the state of ovarian maturation prior to pond draining, sample crawfish can be dissected and staged by these criteria (23).

Although adults and hatchlings can survive in high-humidity environments within the burrow, free water appears necessary for successful reproduction (24). Eggs are fertilized externally with sperm from the seminal receptacle and are then attached to the female's pleopods with a cement called glair. (See Fig. 4.) The number of eggs oviposited varies with the size and condition of the female. It is not unusual for a large, healthy female to attach more than 500 eggs. The incubation period is temperature dependent and takes about two to three weeks at 23 °C (73 °F) (25). Hatched crawfish cling to the female's pleopods through two instar molts. The maintenance energy for this period is provided by the egg. Hatchlings instinctively remain with the female for several weeks after the second molt. It is critical that the female and her young leave the burrow within a reasonable time, because little food (other than siblings) is available in burrows.

With pond-reared crawfish, summer reproduction is somewhat synchronized; therefore, ponds are routinely flooded in autumn to coincide with emergence. Survival of broodstock, spawning success, and survival of progeny during the burrow period can have a significant impact

Figure 4. Female crawfish with eggs attached to the pleopodal setae on the abdomen. This condition is known as ovigerous, or "in berry."

on subsequent crawfish production, especially during harvests early in the season, when prices are usually the highest.

Since timing and duration of the dry period are controlled in pond culture, aeration of bottom sediments, elimination of predator fish, and propagation of a vegetative crop can be maximized. Two strategies exist for propagation of the forage crop. One practice uses volunteer plants, with producers exerting little control or additional input. A different strategy uses cultivated crops (such as rice, *Oryza sativa*). The latter strategy, when managed properly, ensures a greater predictability of crop establishment and vegetative biomass with desirable characteristics.

Wet Phase

Flooding or heavy rainfall is usually necessary to encourage emergence of crawfish from the burrow. Brood females emerge with young (and sometimes eggs) attached or clinging to their abdomen. These crawfish are highly susceptible to predators, because the attached brood prevent the typical tail-flipping escape response. As the female moves about, unattached hatchlings are left behind and become independent. The hatchlings, or young-of-the-year (YOY) recruits, and "holdover" crawfish from the previous wet cycle feed and grow in open water. Mature crawfish either enter the reproductive phase or the nonsexual, growing stage after molting. *P. clarkii* may reproduce more than once and in any season of the year in Louisiana, but *P. zonangulus* is classified chiefly as a seasonal spawner and reproduces mostly in the fall (17).

In crawfish aquaculture, autumn flooding is timed to coincide with the major recruitment period. The evidence of free-swimming young in burrows can serve as a gauge for timing of the flood. Recruitment is normally composed of several waves of YOY, with the primary wave occurring within the first four to six weeks of flooding. Multiple recruitment and differential growth result in several classes of different size within the population. Growth rate is affected by a number of variables, including water temperature, population density, level of dissolved oxygen, food quality and quantity, and (probably) genetic influences; however, environmental factors seem to be the most important variable (26).

Inundation begins the chain of events that establishes the food web from which crawfish obtain most of their nutrients. Crawfish have been classified as herbivores, detritivores, omnivores, and, more recently, obligate carnivores (27). Crawfish have been known to ingest living and decomposing plants, seeds, algae, microbes, and a myriad assortment of animal matter, from the smallest invertebrates to vertebrates such as small fish. However, those sources of food vary considerably in quality and quantity in the habitat. Vascular plants, often the most abundant food resource, probably contribute little to the direct nourishment of crawfish. Intact plant matter is consumed mainly when other food sources are in short supply and provides a limited amount of nutrients to growing crawfish. Decomposing plant material, with its associated microorganisms (collectively referred to as detritus), is consumed to a much greater degree and has a higher nutritional value. However, the ability of crawfish to use detritus as a mainstay food item in controlled studies was very limited (28–30). In the aquatic environment, there are numerous other animals that depend on detritus as their main food source. Molluscs, insects, worms, small crustaceans, and tadpoles consume detritus and also furnish crawfish with sources of high-quality food. Only recently have scientists realized that in order for crawfish to grow at their maximum rate, they must feed to a greater extent on these and other high-protein sources (27,31). Momot (27) established sufficient evidence for the classification of crawfish as an obligate carnivore and facultative detritivore–herbivore. Although crawfish must consume high-protein foods to achieve maximum growth, it appears that crawfish can sustain themselves by eating intact and detrital plant sources and bottom sediments containing organic debris.

Commercial aquaculture relies mostly on a self-sustaining food system, such as occurs in natural habitats, for growing crawfish. This system requires the establishment of a forage crop to serve as the basis of a detrital food chain, with crawfish as the top carnivore. Because commercial production dictates high crawfish densities and because flood duration is long (7–10 months), the food chain becomes highly taxed and requires careful management. High commercial production without supplemental feed requires adequate quantities of aquatic invertebrates in the pond, fueled by a constant supply of detritus. A continual supply of vegetative matter for decomposition is necessary throughout the production season, thereby requiring a forage crop that yields small portions of its material on a consistent basis over the duration of the season. Too much detrital material at any one time is wasted, because it cannot be stockpiled; thus, a large portion deteriorates without being consumed. The decomposition of excess detritus can also cause poor water quality conditions. On the other hand, too little detritus results in food shortages for crawfish and for organisms that crawfish rely on as high-quality food items. In severe times of shortages, the crawfish may decimate the invertebrate populations, thus

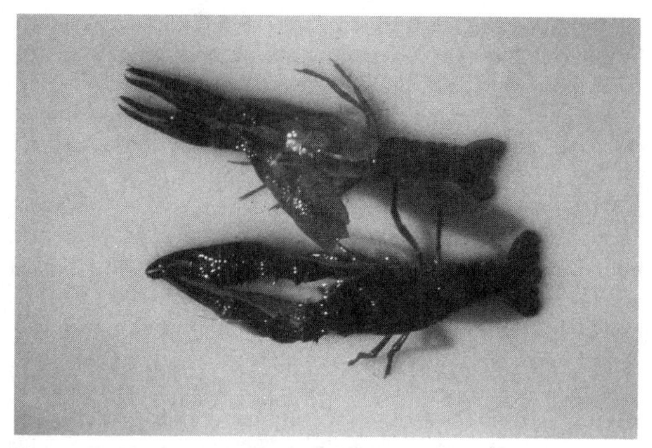

Figure 5. Newly molted (soft) crawfish (top) and the cast exoskeleton (bottom).

eliminating high-quality food sources for the remainder of the season.

As with all arthropods, crawfish must molt, or shed their hard exoskeletons, to increase in size. (See Fig. 5.) The growth process involves periodic molting (ecdysis) with intermolt periods between shedding episodes. There are five recognized phases of the molt cycle. The intermolt period is characterized by a fully formed and hardened exoskeleton. During that phase, crawfish eat and increase tissue and energy reserves. The premolt phase involves formation of the new underlying (uncalcified) exoskeleton and the dissolution of the old shell. During the latter part of the premolt stage, crawfish cease feeding and seek shelter. The molt phase involves actual shedding of the old exoskeleton, which is usually accomplished in minutes. The brittle exoskeleton splits between the carapace and abdomen on the dorsal side, and the crawfish usually withdraws by tail flipping. It is during the soft phase that the new exoskeleton expands to its new dimensions. Crawfish are most vulnerable to loss from cannibalism and predators during the soft phase, especially when the population density is high and cover is sparse. Calcification of the exoskeleton occurs during the postmolt phase. Initial hardening occurs using calcium from stored sources within the body (e.g., gastroliths) and absorbed from the water. As crawfish resume feeding, further mineralization occurs. Lowery (32) and Huner and Barr (22) provide further details of the complex processes of ecdysis.

Molting is hormonally controlled and occurs more frequently in younger animals than older ones. Increases in length and weight during molting and the interval of the intermolt period vary greatly and are affected by environmental factors, such as water temperature, water quality, amount of available food, and crawfish density. After a period of growth, both males and females molt to a sexually active phase and cease growth. Huner and Barr (22) have advised that a minimum of 11 molts is necessary for crawfish to reach maturity. Mature individuals exhibit distinct secondary sexual characteristics, including darker coloration, enlarged chelae, and cornified seminal receptacles for females and cornified gonopodia and prominent hooks at the bases of the third and fourth pair of pereopods for males (33).

In culture ponds, frequent molting and rapid growth occur during spring, due to warmer waters and adequate food sources. Crawfish can increase up to 15% in length and 40% in weight with each molt under optimum conditions (34). Therefore, it is important to pay careful attention to environmental conditions, including overcrowding, because certain factors can negatively impact growth. The abundance of mature crawfish increases as the season progresses. Rapid increases in temperature [>21 °C (>70 °F)] under conditions of overcrowding and food shortages stimulate onset of maturity at smaller sizes. "Stunting," the condition whereby crawfish mature at an undesirable marketing size, is a problem for some producers.

Factors that can affect the growth and well-being of crawfish, other than those mentioned previously, include diseases and toxicants. Individual crawfish are susceptible to various pathogens, such as bacteria, viruses, fungi, protozoans, and parasites (21). However, serious disease problems associated with extensive culture of crawfish have not been documented and are thought to be rare. Procambarid crawfish are also known carriers of the *Aphanomyces* fungi that caused high mortalities in European crayfish, but are not affected by the fungus themselves. Crawfish, however, are highly susceptible to many pesticides (35), which is of great concern to producers, because much of the crawfish culture is associated with other agriculture enterprises that commonly employ pesticides.

CULTURE SYSTEMS

Crawfish are amenable to culture because they are hardy, prolific, adaptable, and do not require highly technical cultivation practices. Unlike the culture of aquatic species that require hatcheries and formulated feeds, crawfish culture is based on self-sustaining populations that use a forage-based food system (Fig. 6). Since crawfish are grown in shallow earthen ponds [25- to 76-cm (10- to 30-in.) water depth], relatively flat but drainable land with suitable levees is required. Soil with sufficient clay to hold water and to accommodate burrows is needed. Water requirements for crawfish production are similar to those

Figure 6. Crawfish pond with rice as the forage base. Note the trapping lanes used to harvest crawfish.

for other aquaculture enterprises, with the exception of the quantity of water. With the entire pond area covered with substantial amounts of vegetation, the biological oxygen demand (BOD) is sometimes great, and its effects must be overcome with timely water exchanges. Equipment requirements for culturing crawfish include irrigation systems, harvesting equipment (boats and traps), and agricultural implements to establish the forage crop and maintain levees. Sufficient labor and marketing opportunities are also essential for successful commercial operations.

Although crawfish culture systems have been categorized by pond type and dominant vegetation (22,36), categorization by major production strategy is perhaps a better alternative. The two major strategies for crawfish production are monoculture and crop-rotation systems. Whereas there are many similarities with regard to management between the two strategies, different production goals dictate different management concerns. Crawfish is the sole crop harvested in monoculture systems, and production typically occurs in permanent ponds. Crop-rotation systems involving crawfish include one or more agronomic crops (e.g., rice and soybeans). Crawfish either are rotated with another seasonal crop in the same physical location year after year or are cultured in different locations each year, to conform to typical field rotations of crops.

Monoculture Systems

Permanent ponds devoted entirely to crawfish production vary in size and intensity of production. Pond strategies range from large [>120 ha (>300 acres)], impounded wetlands with little management to small [<6 ha (<15 acres)], intensively managed systems (36). Monoculture systems are the method of choice for many small farms or where marginal lands are available and unsuited for other crops. The main advantage of this type of system is that it allows producers to design and manage for optimal crawfish production without concern for other crops or about pesticide exposure. A single-crop production system is also easier to manage. Disadvantages often include *(1)* the need to construct ponds, *(2)* the fact that the cost must be amortized over one crop only, and *(3)* crawfish overcrowding, which frequently occurs after several annual cycles.

Crawfish yields typically range from <225 kg/ha (200 lb/acre) in large, low-input systems to >1,120 kg/ha (1,000 lb/acre) in intensively managed ponds. Some ponds have even yielded in excess of 3,030 kg/ha (2,700 lb/acre) (36). Pond yields tend to increase with consecutive production, because resident populations usually expand annually. Ponds with higher ratios of linear levee area to pond surface area (i.e., ponds with increased burrowing space) usually have more recruitment and increased yields. Earlier and more intense harvesting is often justified in older, permanent ponds, because of the dense populations and increased numbers of "holdover" crawfish. Also, dense populations put more pressure on the forage crop; thus, forage is often depleted prematurely, which can have an adverse effect on the overall yield and size of harvested crawfish.

Production schedules vary within and between geographical regions, but permanent monocropping ponds in the Southern United States generally follow the schedule shown in Table 1. Since crawfish populations are self-sustaining, stocking usually is needed only in new ponds, when a pond has been idle for a year or more, or after extensive levee renovation. Ponds should be thoroughly drained to aid in predator control, soil aeration, and proper establishment of forage. Harvesting should be initiated and maintained when catch per unit effort (CPUE) and marketing conditions justify the effort and expense.

Rotational Systems

Crawfish are often rotated with rice and, sometimes, other crops as well in two basic farming rotational systems. In each system, crawfish culture follows the rice harvest, and the forage crop used for growing crawfish is derived from regrowth of rice stubble after grain harvest. Advantages of rice–crawfish rotational systems include the efficient use of land, labor, and farm equipment. Moreover, some fixed costs and the cost of establishing rice can be amortized over two crops instead of just one.

One rotational system, referred to as rice–crawfish double-cropping, takes advantage of the seasonality of each crop to obtain two crops in one year. Rice is grown and harvested during summer, while crawfish are grown during autumn, winter, and early spring in the same field. (See Table 2.) As with permanent monocropping systems, crawfish are stocked only initially; however, stocking occurs after the rice crop has been established (45 to 60 days after planting). Subsequent crawfish crops

Table 1. Typical Permanent Monocropping Schedule

Time	Procedures
April–May	Flood new ponds
Late April–early June	Stock mature crawfish (new ponds only)
Late May–June	Drain ponds over a two- to four-week period
July–mid-August	Plant (or encourage) vegetative crop
July–September	Fertilize and irrigate forage crop as needed
October	Flood ponds
November–May	Harvest crawfish
Late May–June	Drain ponds and repeat the cycle

Table 2. Typical Rice–Crawfish Double-Cropping Schedule

Time	Procedures
April	Plant rice and maintain the crop under a shallow flood [<15 cm (<6 in.)]
June	Stock mature crawfish (new ponds only)
August	Drain fields and harvest rice
August	Fertilize rice stubble; irrigate as needed
October	Flood ponds
November–March	Harvest crawfish
March	Drain ponds and repeat the cycle

rely on holdover broodstock from a previous cycle. An almost certain disadvantage with this production strategy is that neither crop can be managed to yield maximum production. The best rice yields in the southern United States are achieved when rice is planted in March or April. However, draining crawfish ponds prematurely to accommodate ideal rice-growing conditions decreases the total crawfish yield. Pesticide use is also a major management consideration with this system. Overall, crawfish and rice yields are variable and depend on management. Systems managed mainly for crawfish can expect crawfish yields similar to those of well-managed monocropping systems, but at the expense of rice yield, and vice versa.

The other major rotational strategy employs crawfish in field rotations of crops (e.g., rice and soybeans). Rice is often not cultivated in the same field during consecutive years, to aid in control of diseases and weeds. As with rice–crawfish double-cropping, crawfish culture follows rice cultivation; therefore, crawfish production does not occur in the same location in consecutive years. (See Table 3.) In lieu of draining crawfish ponds in March to plant rice, crawfish harvest can proceed until early May, when the pond is drained to plant soybeans, or longer if plans are to leave the field fallow. Under this system, three crops per field can be realized in two years, and a field with crawfish will be fallow for at least one year before crawfish culture resumes. Some producers elect to leave the field fallow instead of planting soybeans, but another crop could follow as well.

This type of farming system is common to south Louisiana and east Texas, where farms are large and producers are accustomed to field rotations. The advantage of field rotation systems is that all crops can be better managed. For example, crawfish can be harvested over the entire season in lieu of draining the ponds early to plant rice. Furthermore, by rotating physical locations each year, overpopulation is rarely a problem, and often, crawfish size is larger, due to lower population densities. Some disadvantages, however, are the need to restock every year and that, frequently, the bulk of the harvest occurs late in the season, when seasonal declines in prices and marketing difficulties are common.

Crawfish yields under this management approach are not commonly as high as with monocropping systems, but with proper management, yields can routinely exceed 900 kg/ha (800 lb/acre). Note that not all management strategies fit exactly under these categories as described here; there are many variations. For example, a producer may alternate rice and crawfish production in the same field for two or three years and then rotate them into a different field.

Table 3. Typical Rice–Crawfish–Soybean (or fallow) Rotational Schedule

Time	Procedures
March–April	Plant rice and maintain the crop under a shallow flood [<15 cm (<6 in.)]
May–June	Stock mature crawfish
July–August	Drain fields and harvest rice
August	Fertilize rice stubble; irrigate as needed
October	Flood ponds
January–April/May	Harvest crawfish
May	Drain ponds
May–June	Plant soybeans (or leave the field fallow)
October–November	Harvest soybeans
March–April	Plant rice and repeat the cycle

MANAGEMENT CONSIDERATIONS

Stocking and Population Management

Unlike many aquaculture ventures, which stock juveniles, crawfish aquaculture relies on the stocking of adults. It is preferable to stock new ponds between April and June, when broodstock are sexually mature and ovarian development has commenced. *P. clarkii* is the recommended species to stock in the south, because of its preference in the marketplace and well-known culture methodology. Stocked crawfish should be in good health, sexually mature, and composed of 50–60% females. The size of the broodstock should be of little concern. Large crawfish produce high numbers of young, but there are fewer large crawfish purchased per unit weight of broodstock, whereas small crawfish produce fewer young, but more adults are purchased per unit weight. Broodstock should not be stored in a cooler and should be handled carefully and stocked within a few hours after their capture. Recommended stocking rates vary, depending on the amount of native crawfish and cover in and around the pond edge. Stocking rates of 45–56 kg/ha (40–50 lb/acre) are recommended for areas that lack native crawfish and have sufficient protection from predators.

Population management is the most elusive aspect of crawfish production. Aside from intentional stocking of crawfish, recruitment depends on survival of broodstock, successful reproduction, and survival of offspring. These factors are affected by management practices, although the practices provide little control overall, and are also largely influenced by environmental conditions. Because of continual recruitment and difficulties in sampling the population, crawfish producers have difficulty assessing the population density and structure. Sampling is crude and is currently accomplished by dip-net sweeps and baited traps. Population management entails little more than controlling the timing of the flood to coincide with normal reproduction peaks, ensuring adequate water quality and food resources, and carrying out adequate harvesting to remove market-size animals. Reducing the density to control overpopulation in ponds shows promise (37) as well, but has not yet become a common practice. In addition, supplemental stocking in underpopulated ponds sometimes occurs, but may not always be feasible.

Forage Crops and Management

Forage crops in crawfish aquaculture serve to provide cover for crawfish, a substrate for attachment of food organisms, and vertical structure. Vertical structure is important to crawfish, a bottom-dwelling animal, because it provides access to the water column and air interface and may ease crowding and increase growth (38). Notwithstanding, the most important role of forage is to furnish small amounts of vegetative material continuously and consistently to the detrital pool as fuel for the complex food web.

Establishing rooted plants that will adequately provide a consistent supply of material to the detrital pool over the entire season is difficult. Native voluntary vegetation is the least expensive type of plant to establish and can sometimes perform satisfactorily, but is often unreliable and insufficient for maximum crawfish production. Terrestrial plants usually die when flooded, resulting in poor water quality and food shortages. Semiaquatic plants (such as alligator weed, *Alternanthera philoxeroides*, and smartweed, *Polygonum* spp.) normally thrive in crawfish ponds, but often, growth is unpredictable, and adequate biomass is not always achieved. Also, there frequently are long periods when semiaquatic plants remain intact, and thus little material enters the food chain. During winter, however, the emergent portion of the plant dies and large "slugs" of detritus result, usually at a time when need for it is reduced, because of low water temperatures. Native vegetation can sometimes be effectively used when there is an appropriate mixture of aquatic, semiaquatic, and terrestrial species; however, little control is achievable with voluntary stands.

The most dependable means of obtaining sufficient forage is to rely on planted agronomic crops and follow recommended management practices. Rice has become the standard forage crop for the industry (39). Because of its semiaquatic nature, it tends to persist well in flooded crawfish ponds. Yet, it furnishes plant fragments to the detrital pool in a consistent manner. When immature rice is flooded, older leaves gradually die and drop off, furnishing detrital material as the plant continues to grow and mature. When the portion of the plant above water dies during winter, it gradually fragments, providing a steady supply of detrital fuel until it is depleted. In contrast, mature rice ages and begins its fragmentation shortly after flooding, resulting in faster depletion and possible food shortage (40).

Sorghum–sudangrass hybrid (*Sorghum bicolor* × *S. sudanense*), first developed for use as a hay crop, also appears to be a well-suited forage for crawfish production (41). This plant displays extremely fast growth rates, produces a remarkable amount of forage dry matter (see Fig. 7), is very hardy and drought resistant, and may prove to be more reliable than rice for late-summer stand establishment. It also exhibits good persistence in crawfish ponds, with consistent fragmentation of material well into the season. Few other crops have been thoroughly screened for use in crawfish production, but preliminary evaluations of millets and other sorghums have not been especially encouraging (42).

Recommendations for establishing forage crops vary from area to area. Considerations should be given to choosing species that produce high vegetative biomass and do not rapidly deteriorate, but instead provide a steady rate of disappearance, under prolonged flooded conditions. Tall, late-maturing rice varieties that have low grain-to-forage ratios and are adapted to local conditions generally prove best for crawfish production (43). The time of planting is also an important consideration. Planting should occur early enough to achieve maximum vegetative biomass, but not so early that the plant matures and begins to die prior to frost. For rice and sorghum–sudangrass planted in the southern United States, the most appropriate planting time ranges from mid-July to early August for rice and mid- to late August for sorghum–sudangrass.

Figure 7. Sorghum–sudangrass hybrid, a suitable forage crop for crawfish production, produces abundant vegetative biomass.

Multicropping of rice and crawfish impose different forage management strategies from monocropping systems. Varietal choices of rice are limited for this purpose, since the varieties should have good yielding and milling characteristics as well as the ability for adequate regrowth from the stubble (ratooning). Management is virtually limited to ensuring proper growth from the ratoon crop (via fertilization and irrigation) and mitigating the effects on water quality from straw and debris generated from grain harvest. The "combine tailings" can be baled and removed, burned, or chopped (to speed up degradation) prior to flooding, or timing of the flood can be delayed until cooler weather.

Water Quality and Management

Both surface and well water are commonly used in crawfish aquaculture, and either is satisfactory, provided that the quality and quantity are acceptable. Wells provide predator- and pesticide-free water, but associated costs are usually higher. Subsurface water is generally low in dissolved oxygen, may be high in dissolved iron and hydrogen sulfide, and should be aerated (see Fig. 8) for rectification. Surface water may be less dependable in quantity and quality and should be screened for predators. Water-quality variables of importance and their desirable ranges are listed in Table 4.

As with most aquaculture endeavors, dissolved oxygen is usually the most critical water-quality concern. Problems with dissolved oxygen in crawfish aquaculture are compounded by the presence of huge amounts of decomposing vegetation, which impede ready remedies (e.g., emergency aerators). Management of dissolved

Table 4. Water-Quality Variables Important to Crawfish Culture

Variable	Lethal Low	Desirable	Lethal High	Reference
Temperature (°C)	0	21–27	34	44,45
Dissolved oxygen (mg/L)	0.5–1.0	≥3	—	43,46
pH	3.0	6.5–8.5	>10.0	45,47
Total hardness (mg/L as $CaCO_3$)	*	>50	*	48,49
Total alkalinity (mg/L as $CaCO_3$)	*	>20	*	45,49
Unionized ammonia (mg/L)	—	<0.06	2.65	47
Nitrite (mg/L)	—	<0.6	5.95	47
Ferrous iron (mg/L)	—	<0.3	*	9,45
Hydrogen sulfide (mg/L)	—	<0.002	>5.0	44,49
Salinity (g/L)	—	<6	15	50,51

*Indicates that data are unavailable.

Figure 8. Often, water entering a crawfish pond is first passed through aeration screens.

oxygen in crawfish culture must entail preventative measures rather than corrective measures. Management options include choice of forage type and flooding dates, close monitoring and water exchange or circulation, and proper pond design. Ponds constructed with interior levees to divert water flow to all areas of the pond are effective in improving levels of dissolved oxygen throughout the pond (43). Paddlewheel aerators designed to move water, in concert with diversion levees, are especially effective (45).

Turbid water is often encountered in crawfish culture in spring, when forage biomass has declined. This event is normal and indicates a healthy crawfish population, and turbid water is not harmful to crawfish. However, pesticides in the culture environment can be detrimental to crawfish and other aquatic fauna that serve as food sources for crawfish. Refer to Eversole et al. (52), Toth et al. (53), and Huner and Barr (22) for details on the relative toxicity of various pesticides to crawfish.

Harvesting

Romaire (54,55) has provided excellent overviews of harvesting methods and strategies used in commercial crawfish aquaculture and has helped to develop many of those strategies. This subsection briefly summarizes current practices in the industry. There are several factors unique to crawfish culture that limit options and dictate harvest methodologies. Since crawfish recruitment to the harvestable population is somewhat continual, regular, frequent harvests must be carried out, as opposed to batch harvests. Seine harvesting, the most common method used for many aquatic species, is ineffective in traditional crawfish aquaculture, because of the dense vegetation normally present in culture ponds. Additionally, the presence of soft crawfish lends difficulty to seining and other active methods. Thus, crawfish removal relies on the passive technique of attracting the crawfish to baited traps. Flow trapping, a passive technique used to capture *C. quadricarinatus* (see the entry "Australian red claw crayfish"), probably also is not effective, due to the large size of crawfish ponds.

Although different types of traps have been used in the past, one trap design has emerged as the standard, used by the majority of culturists. The three-sided, three-funnel, wire-mesh trap (Fig. 9), commonly referred to as a "pyramid trap," is efficient to operate and effective in crawfish capture. Mesh size and shape govern the size of crawfish retained by the trap. Most traps are made of 1.9-cm (0.75-in.) plastic-coated hexagonal mesh, although some producers use 1.9-cm square mesh to retain smaller crawfish. The pyramid trap is designed to be positioned upright in the pond, with the top extending above water.

Figure 9. The efficient "pyramid" trap has become an industry standard for harvesting crawfish in culture ponds.

The top is open, to facilitate rapid removal of crawfish and rebating; contains a retainer band to minimize crawfish escape; and serves as a handle. The length of the neck varies to accommodate use in different water depths, and, occasionally, a metal rod is attached to the trap to increase stability in high winds.

Crawfish are attracted to the trap with bait. Two categories of bait are used: *(1)* natural baits of fish and *(2)* formulated bait manufactured from dried fish and grain by-products, attractants, and binders. Fish baits include gizzard shad, menhaden, herring, carp, suckers, mullet, and fresh by-products from fish processing facilities. Formulated baits are manufactured by several companies and are proprietary in their formulations. These baits are cylindrical pellets that weigh approximately 50–75 g (1.8–2.6 oz). Beecher (56) determined that fish is the most effective bait at water temperatures below about 16 °C (61 °F) and that manufactured baits are more effective at temperatures above 20 °C (68 °F). A combination of the two bait types seem to work best at 16 to 20 °C (61 to 68 °F). Approximately 150 g (5.3 oz) of bait per trap set, regardless of bait type, has been recommended as the most cost-effective amount (56). Most baiting regimes are based on a 24-hour trap set, but 12- and 48-hour (or longer) sets are sometimes used.

To be effective, traps must be distributed throughout the pond. The pond size often dictates the method used to empty the traps. Small ponds are often harvested by people pulling, poling, or paddling a small boat. However, most commercial ponds are harvested by one or two persons in a motorized, flat-bottom, aluminum boat designed for shallow-water propulsion. An ordinary "jon boat" on which an inexpensive, outboard motor that uses a weedless propeller attached to a long, horizontal shaft is sometimes used. Others use a larger boat designed for and equipped with a hydraulic-driven metal wheel that either pushes or pulls the boat through the pond (Fig. 10). A gasoline engine provides hydraulic pressure for power and steering. Traps are set in rows to accommodate harvest by boat. Commonly, the boat travels down the lane, and fishermen "run" the traps from the side of the boat, often without stopping at traps.

Figure 10. Hydraulic propulsion boats are commonly used to harvest crawfish in ponds. The broadcast distributor mounted on the rear was used for experimental feeding studies.

The average number of traps used is about 30 traps/ha (12 traps/acre), but can range from 25 to 100 traps/ha (10 to 40 traps/acre) for individual farms. Trapping frequency is highly variable and depends on factors such as trap density, crawfish size and density, water temperatures, and marketing conditions. Basically, because harvesting accounts for 50 to 70% of the total direct expenses of crawfish farming (55), trapping frequency is influenced largely by CPUE and marketing price. Traps are emptied two to seven days per week and for three to eight months, beginning as soon as the CPUE is justifiable after flooding. CPUE is influenced by pond population density and structure, trapping effort, indigenous food resources, bait quantity and quality, and environmental conditions (57), but rarely exceeds 1.5 kg (3.3 lb) per trap per day.

Research (55) has shown that in well-managed ponds, harvesting efficiency is most cost efficient with 50 to 60 pyramid traps per ha (20 to 24 traps/acre) fished three to four consecutive days per week. Use of a rotational or intermittent harvest strategy may also increase the CPUE. With this approach, trapping occurs in one part of a pond for one to two weeks, allowing other parts to "rest." The trapping sequence is then rotated to another part of the pond, and so on. Under this strategy, crawfish are given additional time to grow and reach harvestable size between trapping episodes. Refinement and recommendations for this approach are not yet available. Preliminary investigations also show that by alternating the bait type or manufacturer, the CPUE may increase. Finally, using nets or catch basins at the drain site when draining the pond allows recovery of some crawfish that have not been harvested.

Managing for Large Crawfish

Marketing developments during the late 1980s established a demand for large crawfish that swayed the industry to begin grading crawfish according to size. Production priorities shifted from maximizing the total yield, with little regard to harvest size above an acceptable minimum, to an emphasis on large, high-value crawfish (58). Factors that affect crawfish size at harvest have not been thoroughly characterized, but are thought to include harvesting strategy, water conditions, food quality and quantity, population density, and genetic influences. Most data indicate that environmental factors are more influential than genetic factors (26). McClain and Romaire (58) have discussed several management considerations that may increase crawfish size at harvest. These factors include supplemental feeding, intermittent harvesting strategies, reductions in population density prior to harvesting, and relay (transfer) of stunted crawfish to ponds where food and space are more abundant.

Feeding

Feeding is not regularly practiced in the crawfish industry. In the past, crawfish producers provided supplemental feeds to counteract forage depletion, which is characteristic of some ponds in the spring. Hay was placed in the pond to provide substrate and energy for the microbial-based food chain; however, this technique often proved to be ineffective, and the logistics of frequency and distribution of hay prevented its effective use. It was not cost effective to meet current recommendations of about 7,800 kg/ha (7,000 lb/acre) of hay in more or less equal proportions across the bottom of the pond (43).

Although formulated feeds have been shown to increase crawfish growth and production under controlled conditions (59–61), feeding practices in commercial ponds frequently have proved to be uneconomical, and positive results were inconsistent and difficult to repeat. Feeding expensive pelleted feeds, formulated to meet the nutritional needs of crawfish, may not be the most cost-effective means of supplementing the natural crawfish diet with a relatively low-valued product. Since production currently relies on a forage-based system, low-cost feedstuffs used as true supplements to the indigenous food web may provide nutritional augmentation on a cost-effective basis. Recent studies found that whole rice seeds and raw soybeans provided improved growth when fed to crawfish grown in experimental units (62). Pond studies, however, indicated that feeding while harvesting reduced trap efficiency; therefore, feeding schedules would need to be altered to accommodate harvesting.

For year-round crawfish production, as some have proposed, formulated rations must meet all nutritional requirements in lieu of just supplementing forage-based systems. Recent reviews by D'Abramo and colleagues (63,64) underscored the lack of information on nutritional requirements and feeding rates of crawfish. Currently, there is not enough evidence to make a sound recommendation for the use of formulated feeds in traditional crawfish-growing areas where prices are lowest.

Handling and Marketing

Crawfish are harvested and leave the farm as a live product in 18-kg (40-lb) open-mesh plastic sacks (see Fig. 11); consequently, some stress and mortality may occur before the product reaches the consumer or processor. Some simple procedures during and following harvest can go a long way in reducing stress and subsequent mortality. For example, minimizing trap soak time; avoiding exposure of crawfish to direct sunlight; maintaining crawfish in a cool, moist environment; and tightly bagging sacks, without overfilling them, all help prevent stress and mortality. Crawfish in tightly packed sacks, are restricted in their movement and thus have less opportunity to inflict damage to each other with their chelae. Immature and recently molted crawfish do not fare well during harvesting and handling and require special care. Crawfish in good health can be stored at cool temperatures [1.6–4 °C (35–39 °F)] for four to five days without excessive mortality if the sacked crawfish are kept moist and are not overstacked (3). In one study, losses during cold storage [4 °C (39 °F)] increased from 7% to almost 22% after storing from two to six days (65). McClain (65) also observed that purging did not improve survival during cold storage; rather, it increased mortality rates. Shipping live crawfish to distant markets requires the same careful storage and handling techniques.

Figure 11. Open-mesh sacks that confine crawfish and restrict their movements, thereby preventing chelae-inflicted damage, are the accepted method for transporting and storing live crawfish.

Crawfish markets have changed considerably since the mid-1980s, when there was no economic incentive to grade crawfish by size. The Swedish market demand for large crawfish [>30 g (>1 oz)] provided some of the impetus for grading, as well as did higher prices for large crawfish in the domestic market; therefore, grading became a common industry practice. Grading is one step in the initiation of a crawfish quality assurance program in the industry. The most commonly accepted size-grade categories are listed in Table 5. It should be noted that larger crawfish are worth up to five times more per unit weight than the smallest crawfish. Market channels are often based on size grades, with the small crawfish going to peeling plants, medium crawfish being directed to restaurants and live markets, and the largest animals heading for export. Nearly all grading occurs at processing plants and wholesaling outlets, using vegetable or self-built graders.

Another step in establishing quality assurance is purging. Purging, or depuration, is achieved by maintaining animals after harvest in a holding system without food for 24 to 48 hours. Basically, during the purging process, the digestive tract is voided and the gill chamber and exoskeleton are cleansed of mud, rendering a cleaner, more attractive product (Fig. 12). Purging leaves the characteristically dark intestine, found within the abdominal muscle, opaque and less conspicuous. This method should not be confused with the practice of bathing crawfish in saltwater before boiling, because the latter is not effective in evacuating the gut and is little more than a wash.

Table 5. Common Crawfish Size-Grade Categories[a] and Price Multiples in Relation to Small Crawfish[b]

Category	Count/kg (no./lb)	Weight (g)	Price Multiplier (%)
Large	≤31 (≤14)	≥32	200–500
Medium	30–48 (15–22)	20–31	100–300
Small	>48 (>22)	<20	—

[a] From (43).
[b] From (66).

Figure 12. Purged crawfish (left) are more appealing than nonpurged crawfish (right) when consumed as a whole, boiled product, because the intestine of the former is mostly devoid of contents and is less conspicuous than the distended intestine of the nonpurged crawfish.

Figure 13. Nonearthen raceways with suspended baskets are effective and efficient purging systems when adequate water exchange and supplemental aeration are used.

Purged crawfish have higher consumer acceptance, particularly outside of Louisiana, and purging contributes to repeat sales.

Crawfish are purged in both spray (67) and submerged (66,67) systems. In submerged systems, crawfish are held in screened baskets, usually in raceways (see Fig. 13); however, static systems have also been used. In maintaining high crawfish densities in submerged systems, concerns, regarding the level of dissolved oxygen are mitigated by high flow rates, supplemental aeration, or a combination of the two. Spray systems involve confining the crawfish under a constant shower or mist wherein no more than 2.5 cm (1 in.) of water accumulates.

Louisiana farmers sell the vast majority of harvested crawfish to either wholesalers or processing facilities. Crawfish are often graded at these locations. Wholesalers, in turn, market the crawfish to the retail public, brokers, and processors. Approximately 60% of crawfish are sold as a shell-on product and the other 40% are processed, although these percentage may vary significantly among years. In Louisiana, it is estimated that 70 to 80% of crawfish are consumed locally and the remaining 20 to 30% are sold out of state.

BIBLIOGRAPHY

1. G.H. Penn, *Am. Midl. Nat.* **56**, 406–422 (1956).
2. T.H. Huxley, *The Crayfish: An Introduction to the Study of Zoology*, D. Appleton and Co., New York, 1880, 1973, 1974, 1977.
3. M.W. Moody, *J. Shellfish Res.* **6**, 293–302 (1989).
4. J.V. Huner, *Fisheries* **22**, 28–100 (1997).
5. Anonymous, *Louisiana Crawfish: Heads & Tails Above the Rest!*, Louisiana Crawfish Promotion & Research Board, Baton Rouge, LA, 1997.
6. J.V. Huner, in D.M. Holdich and R.S. Lowery, eds., *Freshwater Crayfish: Biology, Management and Exploitation*, Timber Press, Portland, OR, 1988, pp. 239–241.
7. D.D. Culley and L. Duobinis-Gray, *J. Shellfish Res.* **6**, 293–302 (1989).
8. P. Viosca, Jr., *Educ. Bull., No. 2*, Louisiana Department of Wildlife and Fisheries, Baton Rouge, LA, 1966.
9. C.G. LaCaze, *Fish. Bull., No. 7*, Louisiana Department of Wildlife and Fisheries, Baton Rouge, LA, 1981 revised.
10. C.H. Thomas, *Proceed. Annu. Conf. Southeast Assoc. Game and Fish Comm.* **17**, 180–185 (1965).
11. J.W. Avault, Jr. and J.V. Huner, in J.V. Huner and E.E. Brown, eds., *Crustacean and Mollusks Aquaculture in the United States*, Avi Publishing Co., Westport, CT, 1985, pp. 1–62.
12. A.G. Eversole and R.S. Pomeroy, *J. Shellfish Res.* **6**, 309–314 (1989).
13. P.B. Brown, M.L. Hove, and W.G. Blythe, *J. World Aquacult. Soc.* **21**, 53–58 (1990).
14. W.T. Momot, in D.M. Holdich and R.S. Lowery, eds., *Freshwater Crayfish: Biology, Management and Exploitation*, Timber Press, Portland, OR, 1988, pp. 262–282.
15. H.H. Hobbs, Jr. and H.H. Hobbs III, *Proceed. Biol. Soc. Wash.* **103**, 608–613 (1990).
16. J.V. Huner, *Freshwater Crayfish* **10**, 456–468 (1994).
17. R.P. Romaire and C.G. Lutz, *Aquaculture* **81**, 253–274 (1990).
18. M.L. Powel, D.W. Kraus, and S.A. Watts, *Freshwater Crayfish* **11**, 243–248 (1997).
19. X. Deng, D.L. Bechler, and K.R. Lee, *J. Shellfish Res.* **12**, 343–350 (1995).
20. S.B. Noblitt, J.F. Payne, and M. Delong, *Crustaceana* **68**, 575–582 (1995).
21. J.V. Huner, M. Moody, and R. Thune, in J.V. Huner, ed., *Freshwater Crayfish Aquaculture in North America, Europe and Australia*, Food Products Press, New York, 1995, pp. 1–156.
22. J.V. Huner and J.E. Barr, *Red Swamp Crayfish: Biology and Exploitation*, Louisiana. Sea Grant Program, Baton Rouge, LA, 1991.
23. L.W. de la Bretonne, Jr. and J.W. Avault, Jr., *Freshwater Crayfish* **3**, 133–140 (1977).
24. J.V. Huner, T.B. Shields, Jr., J.P. Bohannon, M. Konikoff, and D. Guilmet, *J. Shellfish Res.* **16**, 318 (1997).
25. T. Suko, *Sci. Rep. Saitama Univ. (Japan)* **2B**, 213–219 (1956).
26. C.G. Lutz, Ph.D. Dissertation, Louisiana State University, Baton Rouge, LA, 1987.

27. W.T. Momot, *Rev. Fish. Sci.* **3**, 33–65 (1995).
28. P.D. Jones and W.T. Momot, *Can. J. Fish. Aquat. Sci.* **38**, 175–183 (1981).
29. M. Sanguanruang, Ph.D. Dissertation, Louisiana State University, Baton Rouge, LA, 1988.
30. W.R. McClain, W.H. Neill, and D.M. Gatlin III, *Aquaculture* **101**, 251–265 (1992).
31. W.R. McClain, W.H. Neill, and D.M. Gatlin III, *Aquaculture* **101**, 267–281 (1992).
32. R.S. Lowery, in D.M. Holdich and R.S. Lowery, eds., *Freshwater Crayfish: Biology, Management and Exploitation*, Timber Press, Portland, OR, 1988, pp. 83–113.
33. D.M. Holdich and R.S. Lowery, in D.M. Holdich and R.S. Lowery, eds., *Freshwater Crayfish: Biology, Management and Exploitation*, Timber Press, Portland, OR, 1988, pp. 1–10.
34. R.P. Romaire, *Crawfish Tales* **8**(28), 31–32 (1989).
35. A.G. Eversole and B.C. Sellers, *Freshwater Crayfish* **11**, 274–285 (1997).
36. L.W. de la Bretonne and R.P. Romaire, *J. Shellfish Res.* **8**, 267–276 (1989).
37. W.R. McClain and R.P. Romaire, *J. App. Aquacult.* **5**, 1–15 (1995).
38. W.R. McClain, *Book of Abstracts-World Aquacult.* **97**, 316–317 (1997).
39. J.W. Avault, Jr. and M.W. Brunson, *Rev. Aquatic Sci.* **3**, 1–10 (1990).
40. W.R. McClain and R.T. Dunand, *Proceed. 25th Rice Tech. Work. Group* **25**, 140–141 (1994).
41. W.R. McClain, *Progress. Fish-Cult.* **59**, 206–212 (1997).
42. M.W. Brunson, *J. World Aquacult. Soc.* **18**, 186–189 (1987).
43. J. Avery and W. Lorio, *Crawfish Production Manual*, Louisiana Cooperative Extension Services, Baton Rouge, LA, 1996.
44. S.K. Johnson, in J.T. Davis, ed., *Proceedings of the Crawfish Production and Marketing Workshop*, Texas Agricultural External Services, Orange, TX, 1982, pp. 16–30.
45. J.V. Huner, *Rev. Aquatic Sci.* **2**, 229–254 (1990).
46. E. Melancon, Jr. and J.W. Avault, Jr., *Freshwater Crayfish* **2**, 371–380 (1977).
47. T.H. Hymel, M.S. Thesis, Louisiana State University, Baton Rouge, LA, 1985.
48. L.W. de la Bretonne, Jr., J.W. Avault, Jr., and R.O. Smitherman, *Proceed. Annu. Conf. Southeast Assoc. Game and Fish Comm.* **23**, 626–633 (1969).
49. C.E. Boyd, *Water Quality in Warmwater Fish Ponds*, Auburn University Agricultural Experiment Station, Auburn, AL, 1990.
50. H. Loyacano, *Proceed. Annu. Conf. Southeast. Assoc. Game and Fish Comm.* **21**, 423–424 (1967).
51. W.G. Perry, Jr. and C.G. LaCaze, *Proceed. Annu. Conf. Southeast. Assoc. Game and Fish Comm.* **23**, 293–302 (1969).
52. A.G. Eversole, J.M. Whetstone, and B.C. Sellers, *Handbook of Relative Acute Toxicity Values for Crayfish*, South Carolina Sea Grant Consortium Publishers, Charleston, SC, 1995.
53. S.J. Toth, Jr., G.L. Jensen, and M.L. Grodner, *Acute Toxicity of Agricultural Chemicals to Commercially Important Aquatic Organisms*, Louisiana Cooperative External Services, Baton Rouge, LA, 1988.
54. R.P. Romaire, *J. Shellfish Res.* **8**, 281–286 (1989).
55. R.P. Romaire, *J. Shellfish Res.* **14**, 545–552 (1995).
56. L. Beecher, M.S. Thesis, Louisiana State University, Baton Rouge, LA, 1997.
57. M. Araujo and R.P. Romaire, *J. World Aquacult. Soc.* **20**, 199–207 (1989).
58. W.R. McClain and R.P. Romaire, *J. Shellfish Res.* **14**, 553–559 (1995).
59. S. Cange, M. Miltner, and J.W. Avault, Jr., *Progress. Fish-Cult.* **44**, 23–24 (1982).
60. H.H. Jarboe and R.P. Romaire, *J. World Aquacult. Soc.* **26**, 29–37 (1995).
61. W.R. McClain, *J. World Aquacult. Soc.* **26**, 14–23 (1995).
62. W.R. McClain, *Book of Abstracts-World Aquacult.* **98**, 355 (1998).
63. L.R. D'Abramo and E.H. Robinson, *Rev. Aquatic Sci.* **1**, 711–728 (1989).
64. P.B. Brown, *J. Shellfish Res.* **14**, 561–568 (1995).
65. W.R. McClain, *J. Shellfish Res.* **13**, 217–220 (1994).
66. D. Landreneau, *Louisiana Crawfish Farmers Assoc. Newsletter*, May 1995.
67. T.B. Lawson and C.M. Drapcho, *Aquacult. Engr.* **8**, 339–347 (1989).

See also AUSTRALIAN RED CLAW CRAYFISH; HARVESTING.

D

DEGASSING SYSTEMS

John Colt
Northwest Fisheries Science Center
Seattle, Washington

OUTLINE

Supersaturated levels of dissolved gases are common in water from wells, springs, streams, and lakes and may vary considerably between seasons and aquaculture operations. Chronic exposures of hatchery fish to supersaturated levels of gases may result in developmental problems, increased incidence of disease, and mortality. Effective degassing of hatchery water is needed for many applications.

TYPES OF AERATORS USED FOR DEGASSING

A number of different types of aerators have been used to degas water, including submerged, gravity, and surface aerators. The most commonly used aerator for degassing is the packed-column aerator (PCA). The PCA consists of a column filled with high surface-area packing (Fig. 1). Water flows down over the medium in a thin film. Because of the large gas–liquid surface area, this system is highly efficient for transfer of oxygen and nitrogen, but will not produce dissolved-gas supersaturation.

Other types of surface aerators (cascade, inclined plane, tray, screen, and lattice) may be useful in some applications, but are not as effective as the PCA and require substantially more space and head. Submerged aerators can actually produce gas supersaturation and should be avoided for most applications.

SINGLE-COMPONENT DEGASSING BASED ON CONCENTRATION

Integration of the basic two-film mass-transfer model (1) for slightly soluble gases such as nitrogen, argon, and oxygen yields

$$\ln\left(\frac{C^* - C_{\text{in}}}{C^* - C_{\text{out}}}\right) = (K_L a)t \tag{1}$$

where

C^* = equilibrium saturation concentration at local temperature and pressure (mg/L),
C_{in} = influent concentration (mg/L),
C_{out} = effluent concentration (mg/L),
K_L = overall liquid-phase mass-transfer coefficient (m/hr),
a = interfacial surface area (m^2/m^3),
t = aeration time (hr).

Equation 1 applies to both degassing and aeration applications. If $C_{\text{in}} > C^*$, oxygen will be transferred from the water into the air. If $C_{\text{in}} < C^*$, oxygen will be transferred into the water. In both cases, C_{out} will approach the value of C^*, but an infinitely long column is needed for $C_{\text{out}} = C^*$. Because of the logarithmic transfer relationship

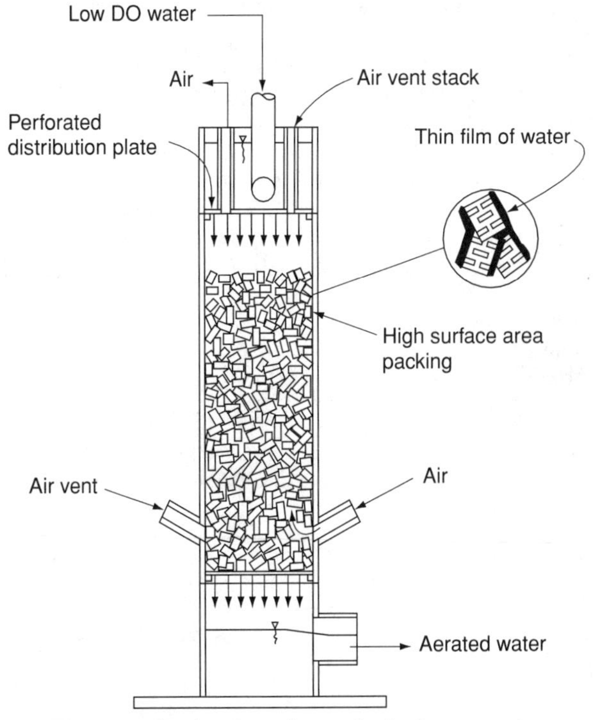

Figure 1. Section through a packed-column aerator.

(Eq. 1), it is generally not economically possible to reduce $|C_{out} - C^*|$ to within 0.05 C^* units. Most degassing columns are designed to operate at atmospheric pressure, so C^* is equal to the local atmospheric saturation concentration. In this case, the value of C_{out} will approach the value of the local atmospheric saturation concentration.

Based on experimental work with oxygen (2), it has been found that $(K_L a)t$ depends directly on height of the medium and, over a wide range of loadings, is independent of hydraulic loading:

$$(K_L a)t = b + KZ \tag{2}$$

where

- b = a constant that depends on incidental aeration that occurs due to the distribution plate (dimensionless),
- K = a constant that depends on the type and size of medium used (1/m),
- Z = depth of medium (m).

K and b values are reported at 20 °C for clean water conditions. The values of b and K at other temperatures can be computed by using

$$(K_L at)_T = 1.024^{(T-20\,^\circ\mathrm{C})}\{b_{20\,^\circ\mathrm{C}} + (K_{20\,^\circ\mathrm{C}})Z\} \tag{3}$$

where

- $(K_L at)_T$ = value of parameter at general temperature,
- T = water temperature (°C),
- $b_{20\,^\circ\mathrm{C}}$ = value of b at 20 °C and clean water conditions,
- $K_{20\,^\circ\mathrm{C}}$ = value of K at 20 °C and clean water conditions.

Values of $K_{20\,^\circ\mathrm{C}}$, $b_{20\,^\circ\mathrm{C}}$, maximum loading rates, and minimum column diameters are presented in Table 1 for various medium sizes and types (2,3). The performance of the column depends on the even distribution of flow over the medium, and the distributor should have at least 45 distribution points/m^2. Columns higher than 2 m should provide a redistribution plate to reduce water flow down the wall. Most degassing columns are designed to add oxygen and remove nitrogen, argon, and carbon dioxide. To prevent depletion of oxygen in the gas phase (and low-effluent dissolved-oxygen concentration), the gas-to-liquid ratio should be maintained above approximately 2 to 3 volume of gas/volume of water. For most degassing applications, forced-air ventilation is not required as long as the columns are designed to allow air flow through the top and bottom of the columns.

Table 1. Recommended Design Parameters for Packed Column Degassing Columns (1,3)

Media Type/Size (cm)	$b_{20\infty C}$	$K_{20\infty C}$ (l/m)[b]	Maximum Loading[a] (m^3/m$^2\cdot$h)	Minimum Column Diameter (m)
Pall ring/2.54	0.40	2.50	150	0.2
Pall ring/3.81	0.40	1.71	300	0.3
Pall ring/5.03	0.40	1.58	>340	0.4
Pall ring/8.89	0.40	1.05	>340	0.7
Tri-Pack/2.54	0.30	1.62	>220	0.2
Nor-Pack/3.81	0.36	1.49	>220	0.3
Nor-Pack/5.08	0.20	1.57	>220	0.4

[a] The maximum loading rates were computed from head loss considerations or field experiments; these values assume even distribution of the flow over the media.

[b] Feet = m × 0.3048; gpm/ft^2 = m^3/m$^2\cdot$h × 0.4090.

MULTICOMPONENT DEGASSING BASED ON PRESSURE

Equation 1 is written for a single gas, however, the risk from gas supersaturation depends on the total gas pressure, not the concentration of a single gas. A multicomponent model for degassing of oxygen, nitrogen, and argon has been developed in terms of total gas pressure (4):

$$\Delta P_{out} = \Delta P_{in(O_2)} e^{-G(t)} + \Delta P_{in(N_2+Ar)} e^{-0.85G(t)} \tag{4}$$

where

- ΔP_{out} = effluent ΔP from column (mm Hg) (see the entry "Gas bubble disease" for a discussion of the measurement of total gas pressure and ΔP),
- $\Delta P_{in(O_2)}$ = influent ΔP for oxygen gas (mm Hg),
- $\Delta P_{in(N_2+Ar)}$ = influent ΔP for nitrogen + argon gas (mm Hg),
- e = base (2.718...) of natural system of logarithms,
- $G_{(t)}$ = overall packed-column mass-transfer coefficient computed from Equation 2 and adjusted to the actual temperature by using Equation 3 (dimensionless),
- 0.85 = a constant that accounts for the slower mass transfer of nitrogen + argon gases.

Ignoring the pressure contribution of carbon dioxide, the overall ΔP is equal to the sum of the ΔP's for the major component gases (5):

$$\Delta P = \Delta P_{(O_2)} + \Delta P_{(N_2+Ar)} \tag{5}$$

and

$$\Delta P_{(O_2)} = \left(\frac{C}{\beta_{o_2}}\right) 0.5318 - 0.20946(\mathrm{BP} - P_{H_2O}) \tag{6}$$

where

- C = concentration of dissolved oxygen (mg/L),
- β_{o_2} = Bunsen coefficient of oxygen (L/L · atm; see reference 6),
- BP = local barometric pressure (mm Hg),
- P_{H_2O} = vapor pressure of water (mm Hg; see reference 6).

Once $\Delta P_{(O_2)}$ has been computed from Equation 6, $\Delta P_{(N_2+Ar)}$ can be computed from Equation 5. For gas-supersaturation work, nitrogen and argon gases are considered a single gas and referred to as dissolved nitrogen (DN).

DESIGN OF PACKED COLUMNS FOR DEGASSING

The performance of packed columns can be computed from Equation 4 as a function of temperature, barometric pressure, dissolved oxygen, and ΔP. For most degassing applications, the effluent must meet both a dissolved oxygen and ΔP criteria.

Column Height

Column height has a significant impact on the effluent ΔP. For an influent $\Delta P = 250$ mm Hg, the effluent ΔP was reduced to 58–68 mm Hg at 1 m and to 18–23 mm Hg for the two cases presented in (Fig. 2). For reasonable column heights, packed-column aerators can reduce the ΔP to 10–20 mm Hg. To achieve ΔP's below this range, a very high column or a pure oxygen column is needed (3). A pure oxygen column operated at a vacuum is the only type of system that can reduce the DN concentration below saturation, while maintaining high DO's. A vacuum column without the addition of pure oxygen can also reduce DN below saturation, but will, at the same time, reduce DO. Neither type of columns will remove a significant amount of carbon dioxide because of the low gas-exchange rate.

Hydraulic Loading

For a given flow, the cross-section area of the column can be computed from the water flow and maximum loading rates listed in Table 1. For example, a water flow of 500 m^3/h and a loading of 150 $m^3/m^2 \cdot h$ would require a crosssectional area of 3.33 m^2 (column diameter = 2.06 m).

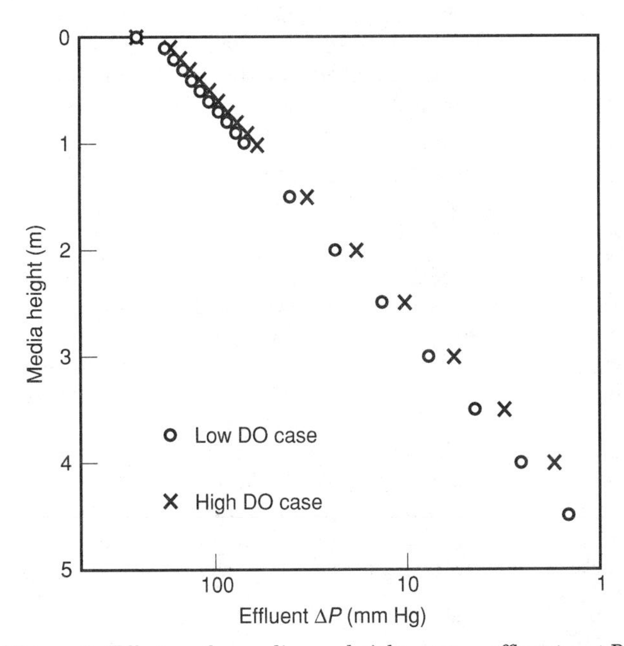

Figure 2. Effect of medium height on effluent ΔP (Influent $\Delta P = 250$ mm Hg, temperature = 10 °C, barometric pressure = 760 mm Hg, $K_{20} = 1.71$/m, $b_{20} = 0.40$). Low DO case is based an influent DO = 2.00 mg/L and dissolved nitrogen + argon (DN) = 26.94 mg/L; the high DO case is based on an influent DO = 15.00 mg/L and DN = 19.81.

Media Type

The required column height depends on the medium selected (Fig. 3). While in general, the small-sized media are more efficient at transferring gas, the column diameter for those media must be increased due to their reduced hydraulic capacity (Table 1).

Impact of DO Concentration

At low ΔP values, the column height is controlled by the dissolved-oxygen criteria (Fig. 3). Above an influent ΔP of approximately 60 mm Hg, the column height is control by the ΔP criterion.

DEGASSING OF OTHER GASES

A number of other gases may need to be stripped out of aquaculture waters. These include methane and hydrogen, carbon dioxide, hydrogen sulfide, and radon.

Methane and Hydrogen

Methane and hydrogen are two insoluble gases that may be found in groundwaters and bottom water from lakes and reservoirs. Conventional packed-column aerators are effective in removing these gases. In applications where these gases may be present, the columns should be vented to the outside.

Carbon Dioxide

The concentration of carbon dioxide gas in groundwaters is highly variable and can typically range from below saturation (less than 0.5 mg/L) to 30–40 mg/L. The concentration depends on the depth and characteristics of the soil, the subsurface geology, and the chemical reactions of dissolved carbon dioxide gas. Carbon dioxide gas is

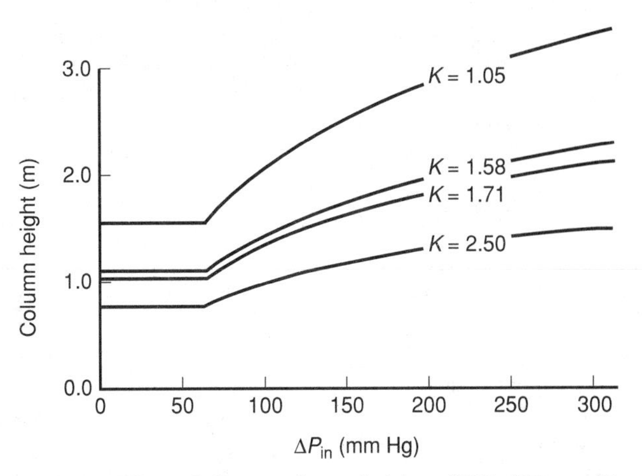

Figure 3. Effect of K on column height (15 °C, 760 mm Hg, $DO_{in} = 5.0$ mg/L, $\Delta P_{max} = 20$ mm Hg, $DO_{min} = 0.90C^* = 9.08$ mg/L).

difficult to remove because of its high solubility. Gas-to-liquid ratios in the range of 5–10 are needed to remove a significant amount of carbon dioxide (7). When carbon dioxide gas is added to water, a portion is converted to bicarbonate and carbonate ions. Only carbon dioxide gas can be removed in a gas transfer system. The kinetics of the carbonate system are slow compared with hydraulic transit time through a packed column. Therefore, while 80–90% of the carbon dioxide gas can be removed by packed column (7), this may be only 40–50% of the added carbon dioxide gas.

Hydrogen Sulfide

Hydrogen sulfide is similar to carbon dioxide, with high solubility and liquid-phase reactions with water. A forced-draft packed column is needed to strip significant amounts of hydrogen sulfide (8). Because of the liquid-phase reactions, the amount of hydrogen sulfide removed depends very strongly on pH.

Radon

High levels of radon gas have been found in hatchery buildings, resulting from degassing of groundwaters prior to use (9). The off-gas from these applications should be discharged to the outside.

BIBLIOGRAPHY

1. W.K. Lewis and W.C. Whitman, *J. Ind. Eng.* **16**, 1215–1220 (1924).
2. G.E. Hackney and J.E. Colt, *Aquacult. Eng.* **1**, 275–295 (1982).
3. B.J Watten, *Aquacult. Eng.* **9**, 305–328 (1990).
4. J. Colt and G. Bouck, *Aquacult. Eng.* **3**, 251–273 (1984).
5. J. Colt, *Wat. Res.* **17**, 841–849 (1983).
6. A.E. Greenberg, L.S. Clesceri, and A.D. Eaton, eds., *Standard Methods for the Examination of Water and Wastewater*, American Public Health Association, Washington, DC, 1992.
7. R.H. Piedrahita and G.R. Grace, in M.B. Timmons and T.M. Losordo, eds., *Aquacultural Reuse Systems: Engineering Design and Management*, Elsevier Science, New York, 1994.
8. K.J. Howe and D.F. Lawler, *J. Am. Wat. Works Assoc.* **81**, 61–66 (1989).
9. W.P. Dwyer and W.H. Orr, *Prog. Fish-Cult.* **54**, 57–58 (1992).

DIETARY PROTEIN REQUIREMENTS

Meng H. Li
Edwin H. Robinson
Mississippi State University
Stoneville, Mississippi

Ronald W. Hardy
Hagerman Fish Culture Experiment Station
Hagerman, Idaho

OUTLINE

Proteins include a large group of chemically similar, but physiologically distinct, molecules that are important in the structure and function of all living organisms. Proteins account for about 65% to 80% of the dry weight of the soft tissues that make up the organs and muscles of fish. All proteins are composed of subunits of amino acids. Animals actually do not require the protein molecule, but rather have a dietary requirement for amino acids and nonspecific nitrogen. Generally, the most economical source of these chemicals is a mixture of proteins. Since protein metabolism is a dynamic process in which tissue proteins are continually being catabolized and resynthesized, a dietary source of amino acids and nonspecific nitrogen is required throughout the life of the organism. Ingested proteins are hydrolyzed to release amino acids that may be used for synthesis of tissue proteins. Fish rations should be balanced to assure that adequate levels of nonspecific nitrogen, amino acids, and nonprotein energy are supplied in proper proportions necessary to maximize protein deposition.

DIETARY AMINO ACID REQUIREMENTS OF FISH

Amino Acids, Defined

Amino acids, which are the structural units of protein, are generally classified as indispensable (essential) or dispensable (nonessential). An indispensable amino acid is one that the animal cannot synthesize or cannot synthesize in quantities sufficient for body needs; thus, it must be provided in the diet. A dispensable amino acid is one that can be synthesized by the animal in quantities sufficient for maximum growth. Most simple-stomached animals, including fish, require the same 10 indispensable amino acids: arginine, histidine, isoleucine, leucine, lysine, methionine, phenylalanine, threonine, tryptophan, and valine.

Determining Amino Acid Requirements

Qualitative amino acid requirements have been determined by measuring weight gain and feed efficiency of fish fed diets containing crystalline amino acids in which a single amino acid was omitted. When fish fed a diet without a specific amino acid did not gain weight or had a markedly reduced weight gain compared with fish fed a control diet, this indicated the essentiality of that single amino acid in their diet (1). Quantitative amino acid requirements were determined by feeding graded levels of crystalline amino acids or a combination of chemically defined proteins, such as casein and gelatin, and crystalline amino acids. The amino acid composition of the test diets was usually similar to that found in whole chicken eggs, fish eggs, or fish muscle. Growth curves generated from such feeding trials have typically been used to estimate amino acid requirements.

There has been disagreement concerning the reliability of quantitative amino acid requirements determined for various species of fish in which highly purified diets containing predominately crystalline amino acids were used. This disagreement arose because fish fed such diets grew slowly compared with those raised on diets with similar amino acid profiles prepared from intact proteins (2–5). For example, early research with juvenile salmon fed highly purified diets generally resulted in the salmon gaining 0.8% to 1.6% of their body weight per day. Fish raised in comparable conditions, but fed practical diets, had weight gains double this rate (6). The argument was that amino acid requirements determined using crystalline amino acid diets may not be representative of dietary requirements of fish fed practical diets. Thus, other methods of determining amino acid requirements have been used. For example, an inferior protein feedstuff, such as corn gluten meal (which is deficient in certain amino acids) has been used as the basic protein to which crystalline amino acids were supplemented to quantify the amino acid requirements (7,8). Similarly, diet formulations using wheat gluten (deficient in arginine) have been used to study the arginine requirement and lysine-arginine interactions (9–11). However, this method is problematic because amino acid requirements may be affected by protein digestibility, amino acid imbalance, or transit and absorption rates of crystalline amino acids as compared with those from intact proteins. Another method suggested by Ogino (12) to estimate amino acid requirements of fish was to use the increase in retention of indispensable amino acids in whole body protein. This method is questionable because it does not include all the amino acids needed for maintenance.

Perhaps there is no perfect method to determine amino acid requirements. However, the requirements obtained from feeding trials with crystalline amino acids and a combination of crystalline amino acids and purified proteins have been widely used to formulate commercial fish diets. This method appears to be reliable. Robinson et al. (13) reevaluated the lysine requirement of channel catfish (*Ictalurus punctatus*) using both purified and practical diets containing 30% protein, and the requirement was 5.0% of protein, similar to 5.1% of protein previously determined using crystalline amino acids at 24% dietary protein (14). Similarly, the dietary requirement for arginine for coho salmon (*Oncorhynchus kisutch*), estimated using practical ingredients, was similar to that determined using semipurified diets (15). Quantitative indispensable amino acid requirements of several fish species are presented in Table 1. The values are fairly similar among species, primarily because the amino acid composition of lean fish tissue does not vary much among species. However, some variations in amino acid requirements exist among species due to differences in physiological needs. The amino acid requirements are

Table 1. Essential Amino Acid Requirements (expressed as percentage of dietary protein) of Juvenile Fish Determined in a Feeding Trial Using Chemically Defined Diets

Amino Acid	Channel Catfish[a]	Common Carp[b]	European Sea Bass[c]	Japanese Eel[d]	Milkfish[e]	Nile Tilapia[f]	Rainbow Trout[g]	Red Drum[h]	Sunshine Bass[i]
Arginine	4.3	4.3	3.9	4.5	5.3	4.2	4.4	—	—
Histidine	1.5	2.1	—	2.1	—	1.7	2.1	—	—
Isoleucine	2.6	2.5	—	4.0	—	3.1	2.6	—	—
Leucine	3.5	3.3	—	5.3	—	3.4	4.1	—	—
Lysine	5.0–5.1	5.7	4.8	5.3	4.0	5.1	5.3	5.7	4.0
Methionine/cystine	2.3	2.1–3.1	2.0	3.2	—	2.7	2.9	—	2.0–2.9
Phenylalanine/tyrosine	5.0	6.5	—	5.8	6.9	3.8	5.3	—	—
Threonine	2.0	3.9	—	4.0	4.5	3.8	2.4	2.3	9.0
Tryptophan	0.5	0.8	—	1.1	—	1.0	0.6	—	—
Valine	3.0	3.6	—	4.0	—	2.8	3.5	—	—
% dietary protein used	24–30	38.5–42	46–50	38	40–45	28	34	35	35
Reference	(2,3,13,14, 18,37,66)	(20,67)	(68–70)	(20)	(71,72)	(73)	(17)	(74,75)	(76–79)

[a] *I. punctatus.*
[b] *Cyprinus carpio.*
[c] *Scophthalmus maximus.*
[d] *Anguilla japonica.*
[e] *Chanos chanos.*
[f] *Tilapia nilotica.*
[g] *Oncorhynchus mykiss.*
[h] *Sciaenops ocellatus.*
[i] *Morone chrysops* × *M. saxatilis.*

also affected by the type of test diet offered and the method used to calculate the requirements (1,16,17).

Expressing Amino Acid Requirements

Amino acid requirements may be expressed as the amount of the amino acid needed per animal per day, as a percentage of diet, or as a percentage of dietary protein. The most precise method might be to express amino acid requirements as the amount needed per animal per day, but this is very difficult to estimate in growing animals, and there is a lack of information to accurately express amino acid requirements for fish in this manner. Perhaps the most practical way to express amino acid requirements is as a percentage of dietary protein, because a constant relationship exists between amino acids and dietary protein.

Amino Acid Interactions

The presence of dietary cystine reduces the dietary methionine requirement because methionine is the precursor of cystine. Cystine can replace 60% of the methionine requirement of channel catfish (18) and 42% of the methionine requirement of rainbow trout (*Oncorhynchus mykiss*) (19). A similar relationship exists between aromatic amino acids: Fish can convert phenylalanine to tyrosine. Tyrosine can replace about 60% of the phenylalanine requirement in common carp (*Cyprinus carpio*) (20) and 50% in channel catfish (13).

Adverse interactions may occur among amino acids that have similar chemical structures if their concentrations in the diet are unbalanced. The lysine-arginine antagonism common in certain animals does not appear to be a problem in fish. This antagonism was not apparent in channel catfish (3) or rainbow trout in some studies (11,19). However, Kaushik and Fauconneau (21) demonstrated that a metabolic antagonism between lysine and arginine may exist in rainbow trout. They found that increasing lysine intake increased plasma arginine levels and attributed that observation to a decrease in arginine degradation rate as dietary lysine increased.

Another common amino acid interaction occurs among the branched-chain amino acids leucine, isoleucine, and valine. Excesses of leucine or isoleucine in diets deficient in one of the other branched-chain amino acids cause a reduction in weight gain and feed efficiency. This interaction has been observed in common carp (20), chinook salmon (*Oncorhynchus tshawytscha*) (22), and channel catfish (23). Hughes et al. (24) found that plasma leucine and isoleucine were increased in valine-deficient rainbow trout. However, Choo (25) showed that free isoleucine and valine in plasma, liver, and muscle were not affected by an increase in dietary leucine.

Meeting Amino Acid Requirements

Amino acid requirements may be met by feeding an excess of protein, so that the level of the most limiting amino acid meets the dietary requirement, by supplementing deficient proteins with crystalline amino acids, or by feeding a mixture of complementary sources. Feeding a mixture of complementary proteins is almost always the most economical choice. To mix protein sources effectively to meet the amino acid requirements of various species of fish, one needs to know the amino acid composition of the feed ingredients and the biological availability of amino acids from each ingredient. Data on the composition of amino acid are available in feed tables for commonly used feedstuffs (17). Data on the availability of amino acid are not readily obtainable. However, data on availability for several commonly used dietary ingredients are known for a number of farmed species. For further information, (see the entry "Protein sources for feeds.") Protein digestibility coefficients may be used to estimate the availability of amino acids when data on the availability of individual amino acids are lacking. However, protein digestibility coefficients are simply the average of individual amino acid digestibility coefficients, with some being above and others below the protein digestibility coefficients. For example, average protein digestibility of cottonseed meal to channel catfish is 83% (26); whereas, lysine from cottonseed meal is only about 66% available to channel catfish (27). This is a fairly extreme example, but if protein digestibility were used to formulate a feed containing cottonseed meal, a lysine deficiency could result. Feed formulations should be designed to contain a slight excess of limiting amino acids, unless digestibility coefficients for the limiting amino acids in the feed ingredients used in the formulation are known.

Utilization of Crystalline Amino Acids

The practice of using supplemental crystalline amino acids to improve the quality of inferior protein feedstuffs in fish diets has been a subject of debate by fish nutrition researchers. As previously discussed, crystalline amino acid test diets are often poorly utilized, compared with diets containing intact proteins. In early studies with common carp, Aoe et al. (28) reported that the fish did not gain weight when fed crystalline amino acid test diets at 3% body weight divided into four equal feedings. However, Nose et al. (29) showed improved utilization of crystalline amino acid test diets by feeding the fish to satiation six times daily. Also, Yamada et al. (30) obtained a similar weight gain with crystalline amino acid test diets to that with a control diet when common carp fry were fed 10% body weight divided into 18 equal feedings per day. These researchers suggested that feeding frequency affects utilization of crystalline amino acids. It has been demonstrated that the amino acids from the crystalline amino acid test diets were absorbed much more rapidly from the intestine of common carp than the amino acids from a casein diet (31). In addition, crystalline amino acids appeared to be removed from plasma at a more rapid rate after feeding a crystalline amino acid diet than after feeding a casein diet (32). The same may be true where diets contain a high percentage of enzymatically-hydrolyzed protein, e.g., fish silage or hydrolysates (33). Increasing feeding frequency may provide a continual supply of the amino acids for protein synthesis and partially overcome this problem (1).

Murai et al. (34) reported that 36% of the dietary amino acids were recovered in the rearing water within 24 hours after feeding common carp an all-crystalline amino acid diet. About 13% of the amino acids were recovered after feeding the fish a diet composed of both casein and crystalline amino acids, while only 1% of the amino acids were recovered in the water of fish receiving a casein or gelatin diet. The authors suggested that the poor utilization of crystalline amino acids was mainly caused by excretion of unutilized amino acids. However, Wilson (1) proposed that part of the amino acids recovered in the water may have been caused by amino acid leaching from the diet. Recently, Zarate and Lovell (35) showed that after adjusting the data for leaching, the bioavailability of crystalline lysine by channel catfish was 57% to 68% of that of intact lysine in soybean meal. Leaching accounted for 13% and 2% of the dietary lysine concentration for crystalline lysine and bound lysine, respectively. The lower bioavailability of crystalline lysine as compared with that of protein-bound lysine was supported by Zarate (36) who demonstrated that crystalline lysine passed out of the stomach of channel catfish earlier than protein-bound lysine. Thus, the crystalline lysine was absorbed from the intestine before the protein-based lysine was available for absorption. It appears that crystalline amino acids should be bound or coated to delay their absorption in order to match the absorption of intact protein.

Unlike common carp (29,30), channel catfish appear to be able to effectively utilize crystalline lysine (35,37–40) when the fish are fed once or twice daily. Zarate (36) found no differences in utilization of supplemental lysine by channel catfish fed two and five times daily. Feeding frequency apparently has little effect on lysine utilization in channel catfish.

There is evidence in land animals that supplemental amino acids improve weight gain and feed efficiency in low-protein, amino acid-replete diets. In channel catfish, Bai and Gatlin (41) showed that supplementation of low protein diets (assumed to be replete in amino acids) with crystalline lysine improved weight gain and feed efficiency. However, Li and Robinson (42) were unable to demonstrate any benefit of lysine, methionine, or both when supplemented in low protein, amino acid-sufficient diets for channel catfish. The difference between those two studies may have been that the diets used by Bai and Gatlin (41) were lysine deficient. The authors pointed out that the positive effect of supplemental lysine may have been due to the lower lysine availability of the soy protein isolate used in their study. There is ample evidence that supplemental lysine is efficacious when added to lysine-deficient channel catfish diets.

DIETARY PROTEIN REQUIREMENT

Quantitative Requirements

Because protein is the most expensive nutrient in fish diets, many studies have been conducted to establish protein requirements for various cultured fish. The dietary protein requirements of several species of fish are listed in Table 2. Most of the requirements were estimated from growth responses of fish fed various levels of protein that were presumed to be balanced in respect to amino acids. Dietary protein requirements, ranging from 23% to 55% (Table 2), vary with species and size of fish. In juvenile rainbow trout, weight gain increases linearly with dietary protein intake over a wide range of feed intake, approximately 6 g to 20 g protein/kg fish/day (43), so protein intake ought not to influence the protein requirement studies provided that feed intake is maintained within the range listed earlier.

Factors Influencing Protein Requirement

Dietary protein requirements determined in various studies vary considerably even within a species. For example, dietary protein requirements for channel catfish have been reported to be 24% to 55%. The wide variation in protein requirements for channel catfish is not surprising because of the different conditions under which the studies were conducted. Factors that may affect the dietary protein requirement include fish size and age, diet composition, feeding rate, presence of natural foods, methods used to estimate the requirement, water temperature, and stocking density.

The dietary digestible-energy-to-protein ratio (DE/P) has a profound influence on protein requirements in fish. Since the concentration of dietary energy appears to affect feed intake, diets that contain excess energy may reduce feed intake, while diets deficient in energy may result in protein being used to meet energy needs, rather than for protein synthesis. Thus, it is important that energy and protein be supplied in the proper proportions and that adequate levels of nonprotein energy be included in the diet. Nonprotein energy sources, such as lipids and digestible carbohydrates, have been shown to spare dietary protein in various species (44–49). When feed is restricted, higher protein diets usually result in better growth. Minton (50) reported that weight gain of pond-raised channel catfish was not different when the fish were fed to satiation with 30% and 36% protein diets, but feeding at approximately 75% of satiation, the fish fed the 36% protein diet gained more weight than fish fed the 30% protein diet. Similarly, Li and Lovell (51,52) also found that a dietary protein concentration of 24% to 26% was adequate for optimum weight gain when fish were fed as much as they would consume, while fish fed at a predetermined maximum level (60 kg/ha/day), a minimum dietary protein level of 32% was necessary for optimum growth. A study with milkfish (*Chanos chanos*) appears to support that contention.

Sumagaysay and Borlongan (53) demonstrated that a 24% protein diet was adequate for optimum growth of milkfish fed at 4% body weight, while a 31% protein diet provided for better growth than the 24% protein diet when the fish were fed at 2% of body weight. The reason for the interaction between dietary protein level and feeding rate may be that the protein requirement for maintenance accounts for a higher proportion of the total protein requirement in fish fed a low-protein diet compared with a higher protein diet when fish are underfed. In a recent study with pond-raised channel catfish, Robinson and Li (54) showed that a 28% protein diet supported weight gain equivalent to a 32% protein when the fish were fed at a predetermined feeding rate equal to or greater than 90 kg/ha/day.

Table 2. Estimated Dietary Protein Requirement for Optimum Weight Gain of Fish (as-fed basis)

Species	Requirement (%)	Size (g)	Environment	Reference
Atlantic salmon (*Salmo salar*)	45			80
Blue tilapia (*Tilapia aurea*)	36	0.4–9.7	Wood tank	81
Channel catfish (*I. punctatus*)	55	0.02–0.2	Aquarium	82
	32–36	7–29	Aquarium	45
	29	9.1–26.6	Aquarium	83
	25	114–500	Fiberglass tank	44
	35	14–100	Fiberglass tank	44
	24	20–517	Pond	51
	24	594–1859	Pond	51
	24	27–354	Pond	62
Common carp (*C. carpio*)	38	1.0–2.0	Aquarium	84
	31	4.3–9	Aquarium	46
	34	2.6–13	Aquarium	34
Gilthead seabream (*Sparus aurata*)	40	2.6–18	Plastic tank	85
Gold fish (*Carassius auratus*)	29	0.2–1.3	Aquarium	86
Golden shiner (*Notemigonus crysoleucas*)	29	0.2–1.4	Aquarium	86
Grass carp (*Ctenopharyngodon idella*)	41–43	0.2–0.6	Aquarium	87
	23–28	2.4–8.0	Aquarium	88
Grouper (*Epinephelus malabaricus*)	48	3.8–9.2	Aquarium	89
	44	9.2–53	Aquarium	49
Hybrid tilapia (*Tilapia nilotica* × *T. aurea*)	24	2.9–8.4	Aquarium	90
	28	21–53	Aquarium	91
Milkfish (*C. chanos*)	40	0.04–0.18	Aquarium	92
Mozambique tilapia (*Tilapia mossambica*)	40	1.8–10.3	Aquarium	48
Nile tilapia (*T. nilotica*)	30	4–10	Aquarium	93
Rainbow trout (*O. mykiss*)	40	2–10	Aquarium	94
Red drum (*Sciaenops ocellatus*)	35–44	4.1–28	Aquarium	64
	40	2.0–27	Aquarium	95
Sunshine bass (*Morone chrysops* × *M. saxatilis*)	41	125–550	Cage	65
White sturgeon (*Acipenser transmontanus*)	40	145–300	Fiberglass tank	96

The presence of natural food organisms may reduce the need to supply high levels of protein in prepared diets for certain fish. For example, tilapia and shrimp grow as well on low-protein diets (25% and below) as on higher protein diets when natural food is abundant, but they require higher protein diets when natural food is limited (16). Natural food organisms are also abundant in channel catfish ponds, but their contribution to growth of food-size fish stocked intensively may be minimal. It has been estimated that only 2.5% of the protein requirement and 0.8% of the energy needed for channel catfish grown in relatively high density in ponds was obtained from natural food (55). Fish raised in intensive production systems, e.g., salmon and trout, do not obtain any measurable contribution to their dietary needs from natural food organisms.

There are indications that statistical methods used to estimate protein requirements affect the results (56,57). Several methods have been used to estimate nutrient requirements, including analysis of variance and multiple range tests, broken-line analysis (58), and regression analysis (58–60). Multiple range tests may not provide a precise level in some cases and may not be statistically justified (56). Broken-line analysis is based on the intersection of two lines, but not all responses are linear. Certain data may be curvilinear and a quadratic or

Table 3. Optimum Dietary Digestible Energy to Protein (DE/P) Ratio for Optimum Growth and Body Composition for Selected Species (expressed as kcal DE/g crude protein)

Species	DE/P Ratio	Size (g)	Environment	Reference
Channel catfish	11.0 (28)	20–517	Pond	51
(*I. punctatus*)	9.7 (28)	27–354	Pond	62
Common carp (*C. carpio*)	10.0 (34)	2.6–13	Aquarium	34
Gold fish (*C. auratus*)	9.7 (29)	0.2–1.3	Aquarium	86
Golden shiner (*N. crysoleucas*)	9.7 (29)	0.2–1.4	Aquarium	86
Rainbow trout (*O. mykiss*)	9.5 (38)			17
Red drum (*S. ocellatus*)	8.8 (40)	2.0–27	Aquarium	95
Sunshine bass (*M. chrysops* × *M. saxatilis*)	10.1 (41)	125–550	Cage	65

an asystotic curve may be more appropriate. Quadratic regression tends to result in a higher requirement than broken-line analysis (60). In an evaluation of nine data sets, Robbins et al. (58) found that the asymptotic curve method provided adequate fits for all data sets, while the broken-line analysis gave adequate fits only for six data sets. When both methods provided adequate fits, the estimated requirements were nearly the same. While arguments can be made in support of the various methods used to estimate nutrient requirements, broken-line analysis appears to be used most often.

Digestible Energy to Protein Ratio

The concentration of dietary protein not only affects the weight gain of fish, but also affects the composition of the gain. Research with channel catfish and trout has shown that, as the dietary protein level is decreased, body fat generally increases. The increased body fat in channel catfish fed low-protein diets (below 24%) appears to be caused by an imbalance that occurs between dietary energy and protein when protein is reduced and energy is not adjusted accordingly. Some body fat is inevitable in growing fish, and in rainbow trout the proportion of fat in their body increases with fish size (61). Past a certain dietary fat threshold, however, body fat increases as dietary fat increases. Fattiness is generally increased as the DE/P ratio of the diet is increased from the optimum range in channel catfish (44,45,51,52,62), blue tilapia (*T. aurea*) (47), common carp (63), red drum (*S. ocellatus*) (64), and sunshine bass (*M. chrysops* × *M. saxatilis*) (65). Optimum DE/P ratios for various fishes are given in Table 3.

PROTEIN RETENTION AND EXCRETION

Protein retention in farm-raised salmon and trout has increased significantly over the past decade. Protein retention refers to the percentage of ingested protein that is retained by fish as protein gain. Retention values have increased from 25%–30% to 40%–50% through the use of high-fat feeds, better formulation to match dietary amino acid levels with requirements, and the use of high quality protein feedstuffs. The theoretical maximum protein retention is approximately 65% to 75%, indicating that further improvement may be possible. Protein retention is an important response variable in research associated with protein and amino acid requirements, optimum DE/P ratios, and ingredient and diet formulation studies. It is also an important consideration in the development of "Environmentally friendly feeds" diets (see separate entry in this volume) and diets to control pond water quality. In channel catfish, increasing the dietary protein concentration within a certain range generally results in increased nitrogen retention, but also increases nitrogen excretion. However, it is not economical to use dietary protein concentrations that exceed 28% to 32% to improve nitrogen retention and reduce body fattiness in commercial channel catfish culture.

BIBLIOGRAPHY

1. R.P. Wilson, in J.E. Halver, ed., *Fish Nutrition*, Academic Press, Inc., New York, 1989, pp. 111–149.
2. R.P. Wilson, O.W. Allen, Jr., E.H. Robinson, and W.E. Poe, *J. Nutr.* **108**, 1595–1599 (1978).
3. E.H. Robinson, R.P. Wilson, and W.E. Poe, *J. Nutr.* **111**, 46–52 (1981).
4. M.J. Walton, C.B. Cowey, and J.W. Adron, *J. Nutr.* **112**, 1525–1535 (1982).
5. M.J. Walton, C.B. Cowey, R.M. Coloso, and J.W. Adron, *Fish Physiol. Biochem.* **2**, 161–169 (1986).
6. R.W. Hardy, in R.P. Wilson, ed., *Nutrient Requirements of Fish*, CRC Press, Inc., Boca Raton, FL, 1991, pp. 105–121.
7. J.E. Halver, D.C. DeLong, and E.T. Mertz, *FASEB* **17**, 1873 (abstract) (1958).
8. H.G. Ketola, *J. Anim. Sci.* **56**, 101–107 (1983).

9. S.J. Kaushik, B. Fauconneau, L. Terrier, and J. Gras, *Aquaculture* **70**, 75–95 (1988).
10. Y.N. Chiu, R.E. Austic, and G.L. Rumsey, *Fish Physiol. Biochem.* **4**, 45–55 (1987).
11. I.P. Forster, Ph.D. dissertation, University of Washington, Seattle, WA, 1993.
12. C. Ogino, *Bull. Jpn. Soc. Sci. Fish.* **46**, 171–175 (1980).
13. E.H. Robinson, R.P. Wilson, and W.E. Poe, *J. Nutr.* **110**, 1805–1812 (1980).
14. R.P. Wilson, D.E. Harding, and D.L. Garling, Jr., *J. Nutr.* **107**, 166–170 (1977).
15. U. Luzzana, R.W. Hardy, and J.E. Halver, *Aquaculture* **163**, 137–150 (1998).
16. T. Lovell, *Nutrition and Feeding of Fish*, Van Nostrand Reinhold, New York, NY, 1989.
17. National Research Council, *Nutrient Requirement of Fish*, National Academy Press, Washington, DC, 1993.
18. D.E. Harding, O.W. Allen, Jr., and R.P. Wilson, *J. Nutr.* **107**, 2031–2035 (1977).
19. K.I. Kim, T.B. Kayes, and C.H. Amundson, *FASEB* **41**, 716 (abstract) (1983).
20. T. Nose, in K. Tiews and J.E. Halver, eds., *Finfish Nutrition and Fish Feed Technology*, Heenemann, Berlin, Germany, 1979, pp. 145–156.
21. S.J. Kaushik and B. Fauconneau, *Comp. Biochem. Physiol.* **79A**, 459–467 (1984).
22. R.E. Chance, E.T. Mertz, and J.E. Halver, *J. Nutr.* **83**, 177–185 (1964).
23. E.H. Robinson, W.E. Poe, and R.P. Wilson, *Aquaculture* **37**, 51–62 (1983).
24. S.G. Hughes, G.L. Rumsey, and M.C. Nesheim, *Trans. Am. Fish. Soc.* **112**, 812–817 (1983).
25. P.S. Choo, Master's thesis, University of Guelph, Guelph, Ontario, Canada, (1990).
26. R.P. Wilson and W.E. Poe, *Prog. Fish-Cult.* **47**, 154–158 (1985).
27. R.P. Wilson, E.H. Robinson, and W.E. Poe, *Trans. Am. Fish. Soc.* **111**, 923–927 (1982).
28. H. Aoe, I. Masuda, I. Abe, T. Saito, T. Toyota, and S. Kitamura, *Bull. Jpn. Soc. Sci. Fish.* **36**, 407–412 (1970).
29. T. Nose, S. Arai, D. Lee, and Y. Hashimoto, *Bull. Jpn. Soc. Sci. Fish.* **40**, 903–908 (1974).
30. S. Yamada, Y. Tanaka, and T. Katayama, *Bull. Jpn. Soc. Sci. Fish.* **47**, 1247–1247 (1981).
31. S.M. Plakas and T. Katayama, *Aquaculture* **24**, 309–314 (1981).
32. S.M. Plakas, T. Katayama, Y. Tanaka, and O. Deshimaru, *Aquaculture* **21**, 307–313 (1980).
33. F.E. Stone and R.W. Hardy, in *Proceedings of Aquaculture International Congress*, Vancouver, BC, Canada, 1989, pp. 419–426.
34. T. Murai, H. Ogata, T. Takeuchi, T. Watanabe, and T. Nose, *Bull. Jpn. Soc. Sci. Fish.* **50**, 1957–1962 (1985).
35. D.D. Zarate and R.T. Lovell, *Aquaculture* **159**, 87–100 (1997).
36. D.D. Zarate, Ph.D. dissertation, Auburn University, AL, (1997).
37. E.H. Robinson, R.P. Wilson, and W.E. Poe, *J. Nutr.* **110**, 2313–2316 (1980).
38. P. Munsiri and R.T. Lovell, *J. World Aquacult. Soc.* **24**, 459–465 (1993).
39. E.H. Robinson, *J. Appl. Aquacult.* **1**, 1–14 (1991).
40. E.H. Robinson and M.H. Li, *J. World Aquacult. Soc.* **25**, 217–276 (1994).
41. S.C. Bai and D.M. Gatlin, III, *Aquacult. Fish. Manage.* **25**, 465–474 (1994).
42. M.H. Li and E.H. Robinson, *Aquaculture* **163**, 297–307 (1998).
43. W.T. Fairgrieve, Ph.D. dissertation, University of Washington, Seattle, WA, 1992.
44. J.W. Page and J.W. Andrews, *J. Nutr.* **103**, 1339–1346 (1973).
45. D.L. Garling, Jr. and R.P. Wilson, *J. Nutr.* **106**, 1368–1375 (1976).
46. T. Takeuchi, T. Watanabe, and C. Ogino, *Bull. Jpn. Soc. Sci. Fish.* **45**, 983–987 (1979).
47. R.A. Winfree and R.R. Stickney, *J. Nutr.* **111**, 1001–1012 (1981).
48. K. Jauncey, *Aquaculture* **27**, 43–54 (1982).
49. S.-Y. Shiau and C.-W. Lan, *Aquaculture* **145**, 259–266 (1996).
50. R.V. Minton, Master's thesis, Auburn University, AL, (1978).
51. M. Li and R.T. Lovell, *Aquaculture* **103**, 153–163 (1992).
52. M. Li and R.T. Lovell, *Aquaculture* **103**, 165–175 (1992).
53. N.S. Sumagaysay and I.G. Borlongan, *Aquaculture* **132**, 273–283 (1995).
54. E.H. Robinson and M.H. Li, *J. World Aquacult. Soc.* **30**, 311–318 (1999).
55. C. Wiang, Ph.D. dissertation, Auburn University, AL, (1977).
56. D.H. Baker, *J. Nutr.* **116**, 2339–2349 (1986).
57. C.B. Cowey, *Aquaculture* **100**, 177–189 (1992).
58. K.R. Robbins, H.W. Norton, and D.H. Baker, *J. Nutr.* **109**, 1710–1714 (1979).
59. I.H. Zeitoun, D.E. Ullery, W.T. Magee, J.L. Gill, and W.G. Bergen, *J. Fish. Res. Board Can.* **33**, 167–172 (1976).
60. K.E. Were, Master's thesis, University of Guelph, Guelph, Canada, (1989).
61. K.D. Shearer, *Can. J. Fish. Aquat. Sci.* **41**, 1592–1600 (1984).
62. E.H. Robinson and M.H. Li, *J. World Aquacult. Soc.* **28**, 224–229 (1997).
63. M.H. Zeitler, M. Kirchgessner, and F.J. Schwarz, *Aquaculture* **36**, 37–48 (1984).
64. W.H. Daniels and E.H. Robinson, *Aquaculture* **53**, 243–252 (1986).
65. C.D. Webster, L.G. Tiu, J.H. Tidwell, P.V. Wyk, and R.D. Howerton, *Aquaculture* **131**, 291–301 (1995).
66. R.P. Wilson, W.E. Poe, and E.H. Robinson, *J. Nutr.* **110**, 627–633 (1980).
67. F.J. Schwarz, M. Kirchgessner, and U. Deuringer, *Aquaculture* **161**, 121–129 (1998).
68. H. Thebault, E. Alliot, and A. Pastoured, *Aquaculture* **50**, 75–87 (1985).
69. E. Tibaldi and D. Lanari, *Aquaculture* **95**, 297–304 (1991).
70. E. Tibaldi, F. Tulli, and D. Lanari, *Aquaculture* **127**, 207–218 (1994).
71. I.G. Borlongan and L.V. Benitez, *Aquaculture* **87**, 341–347 (1990).
72. I.G. Borlongan, *Aquaculture* **93**, 313–322 (1991).
73. C.B. Santiago and R.T. Lovell, *J. Nutr.* **118**, 1540–1546 (1988).
74. P.B. Brown, D.A. Davis, and E.H. Robinson, *J. World Aquacult. Soc.* **19**, 109–112 (1988).
75. R.S. Boren and D.M. Gatlin, III, *J. World Aquacult. Soc.* **26**, 279–183 (1995).

76. M.E. Griffin, P.B. Brown, and A.L. Grant, *J. Nutr.* **122**, 1332–1337 (1992).
77. M.E. Griffin, M.R. White, and P.B. Brown, *Comp. Biochem. Physiol.* **108A**, 423–429 (1994).
78. C.N. Keembiyehetty and D.M. Gatlin, III, *Aquaculture* **104**, 217–277 (1992).
79. C.N. Keembiyehetty and D.M. Gatlin, III, *Aquaculture* **110**, 331–339 (1993).
80. S.P. Lall and F.J. Bishop, Technical Report No. 688. Fisheries and Marine Service, Environment Canada, Ottawa, ON, 1977.
81. A.T. Davis and R.R. Stickney, *Trans. Am. Fish. Soc.* **107**, 479–483 (1978).
82. R.A. Winfree and R.R. Stickney, *Prog. Fish-Cult.* **46**, 79–86 (1984).
83. D.M. Gatlin, III, W.E. Poe, and R.P. Wilson, *J. Nutr.* **116**, 2121–2131 (1986).
84. C. Ogino and K. Saito, *Bull. Jpn. Soc. Sci. Fish.* **36**, 250–254 (1970).
85. J.J. Sabaut and P. Luquet, *Marine Biol.* **18**, 50–54 (1973).
86. R.T. Lochmann and H. Phillips, *Aquaculture* **128**, 277–285 (1994).
87. K.R. Dabrowski, *Aquaculture* **12**, 63–73 (1977).
88. D. Lin, in R.P. Wilson, ed., *Handbook of Nutrient Requirements of Finfish*, CRC Press, Boca Raton, FL, 1991, pp. 89–96.
89. H.Y. Chen and J.C. Tsai, *Aquaculture* **119**, 265–271 (1994).
90. S.Y. Shiau and S.L. Huang, *Aquaculture* **81**, 119–127 (1989).
91. R.G. Twibell and P.B. Brown, *J. World Aquacult. Soc.* **29**, 9–16 (1998).
92. C. Lim, S. Sukhawongs, and F.P. Pascual, *Aquaculture* **17**, 195–201 (1979).
93. K.W. Wang, T. Takeuchi, and T. Watanabe, *Bull. Jpn. Soc. Sci. Fish.* **51**, 133–140 (1985).
94. B.P. Satia, *Prog. Fish-Cult.* **36**, 80–85 (1974).
95. J.A. Serrano, G.R. Nematipour, and D.M. Gatlin, III, *Aquaculture* **101**, 283–291 (1992).
96. B.J. Moore, S.S.O. Hung, and J.F. Medrano, *Aquaculture* **71**, 235–245 (1988).

See also DIGESTIBILITY; FEED EVALUATION, CHEMICAL; NUTRIENT REQUIREMENTS; PROTEIN SOURCES FOR FEEDS.

DIGESTIBILITY

SHOZO H. SUGIURA
Hagerman Fish Culture Experiment Station
Hagerman, Idaho

OUTLINE

For fish to use dietary nutrients for various physiological purposes, the nutrients must be absorbed by the fish. Therefore, knowing the digestibility of nutrients in feeds is critical to estimating the nutritive value of feeds for fish. Feeds are often the single largest variable cost in intensive aquaculture operations. For fish farmers to reduce the cost of feed without reducing the amount of fish produced, a reliable index is necessary to evaluate or compare varieties of feeds and feed ingredients. Knowing the digestibility of nutrients in feeds, as well as the nutrient content and the price, is therefore critical for selecting best-buy feeds and feed ingredients. Knowing the digestibility of nutrients in feeds is also essential to reduce excretion of undigested nutrients that pollute the aquatic environment. The digestibility values of dietary nutrients provide basic information for evaluating fish feeds and ingredients that meet biological, economical, and environmental requirements.

DEFINITION

Digestibility, Availability, and Bioavailability

The term "digestibility" is used for dietary nutrients that need to be digested before they can be absorbed from the gastrointestinal tract of animals (e.g., proteins, carbohydrates, and fats), whereas "availability" is used for nutrients that are absorbed without being digested or decomposed in the gastrointestinal tract (e.g., vitamins, minerals, and amino acids). Digestibility and availability are often expressed as the percentage of dietary nutrients absorbed by fish, which is determined by analyzing the respective nutrient contents of feeds (input) and feces (output). The difference between these values is the amount of nutrients assumed to have been absorbed by the fish. "Fractional absorption" and "percentage absorption" (of dietary nutrients) are other terms used to describe this process.

"Bioavailability" is defined as the proportion of a nutrient in a diet that is absorbed and used for one or more biological functions (1,2). It is usually determined using biological responses, such as growth, nutrient retention, tissue levels, bone strength, enzyme activities, and other criteria that measure the degree of use of a test nutrient. However, it is difficult to measure the actual utilization of some nutrients, and, therefore, absorption is sometimes used to approximate the bioavailability of dietary nutrients, assuming that all or almost all of the absorbed nutrients can be used. This assumption is, however, sometimes questionable. For example, free-amino-acid diets have a higher absorption of total nitrogen (91%) than does the corresponding casein–gelatin diet (82%) in carp (*Cyprinus carpio*); however, growth has been found to be markedly lower in fish fed a free-amino-acid diet than in those fed a casein diet (3). This finding suggests that, in some cases, absorption is not a good criterion for expressing bioavailability of dietary nutrients. The relative, or comparative, (bio)availability is the value of the bioavailability of a nutrient when expressed in relation to a response obtained with a standard reference material (normally of a highly available source).

Apparent and True Digestibility

The apparent digestibility (or apparent availability) of nutrients in diets, also called the (fractional) net absorption or the apparent absorption of nutrients in diets, is defined as follows:

$$\text{Apparent digestibility (\%)} = \frac{\left(\begin{array}{l}\text{nutrient intake from diet}\\ -\text{ nutrient excreted into feces}\end{array}\right) \times 100}{\text{nutrient intake from diet}}$$

Even if a dietary nutrient, such as protein, is completely digestible (and absorbed), or even if the diet is protein free, the feces will still contain a small amount of protein. This portion of fecal protein is not of dietary origin, but of endogenous origin, and includes sloughed intestinal epithelial cells, digestive enzymes, bile salts, intestinal fluid, and bacterial flora.

The true digestibility (or true availability) of nutrients in diets, also called the (fractional) true absorption of nutrients in diets, is the value for which nutrients of endogenous origin have been accounted for and is therefore defined as follows:

$$\text{True digestibility (\%)} = \frac{\left\{\begin{array}{l}\text{nutrient intake from diet}\\ \quad -\text{ (nutrient excreted into feces}\\ \quad -\text{ endogenous nutrient excreted into feces)}\end{array}\right\} \times 100}{\text{nutrient intake from diet}}$$

The endogenous excretion of nutrients is measured by either feeding the organism diets free of the test nutrient or by feeding the organism diets of various nutrient levels and then extrapolating the zero level by regressing the measured points (4,5). For major nutrients, such as protein, carbohydrates, and lipids, the apparent digestibility and the true digestibility are normally close to one another, unless the dietary level is unreasonably low. The distinction between apparent and true digestibility, therefore, has little practical meaning in most feeding practices (6). Studies with higher animals show that this is not the case for some minerals, such as copper, manganese, and zinc, since considerable amounts of absorbed (endogenous) portions are known to be excreted back into the gastrointestinal tract. In addition, this endogenous amount differs at different levels of intake (7). For these minerals, therefore, there could be large differences between the apparent and the true availability (absorption). For other minerals, such as iron and calcium, intestinal absorption per se is regulated by the body's need. Thus, estimating the availability of minerals by means of their intestinal absorption requires carefully controlled experimental procedures, as discussed later. Throughout the rest of this contribution, the apparent digestibility and apparent availability are generally called digestibility and availability, respectively, unless otherwise stated.

METHODS OF DIGESTIBILITY MEASUREMENTS

Direct and Indirect Method

The direct method of measuring digestibility and availability is based on the total amount of nutrients ingested by test animals and excreted as feces. It is therefore essential when using the direct method to collect all feces excreted by the animal, as well as to record the exact amount of feed ingested by the animal. In omnivorous and carnivorous animals, some indigestible, easily distinguishable inert substance, called a marker, is sometimes used. The marker, usually a colored substance, is fed to the animal just before the beginning of ration ingestion and again at its close. Feces collection is begun when the first marker appears and is ended with the appearance of the second marker. Carmine is a frequently used marker. Ferric oxide, chromic oxide, and soot have also been employed. In herbivorous animals, with their much longer and more complicated digestive tracts, the use of a marker is not a suitable method. Experimenters have, therefore, resorted to "time collection," which entails following an extended (preliminary) period during which the ration to be tested is fed daily in constant amounts (8,9). In experiments on fish feed, however, it is technically difficult to collect feces quantitatively without losing fecal components (both water-soluble and disintegrated solid matter) in the water. Also, the direct method is a laborious and time-consuming procedure as compared with the indirect method (described next), and therefore it is no longer in common use for measuring the digestibility of nutrients in fish feeds.

The indirect method is overwhelmingly popular in nutrient digestibility studies of fish feeds. This method requires the inclusion of an indigestible indicator substance (described next) in the diet. By measuring the ratio of the amount of nutrients to the amount of indicator substance in the diet and in the feces, the percentage of dietary nutrients that were absorbed by the fish can be determined. Thus, this technique requires only a small fecal sample instead of a complete collection of feces. Also, there is no need to measure the amount of feed ingested by the fish.

Researchers have compared the apparent digestibility values determined by the direct method (complete collection of feces) with those determined by the indirect method (using chromic oxide as the indicator substance). They obtained similar values of digestibility for dry matter, crude protein, and energy using rainbow trout and chinook salmon (10,11).

Indicators

Indicator substances, also called reference substances, must be nonabsorbable, nontoxic (for both test animals and experimenters), and tasteless; must not affect normal digestion, absorption, and physiological functions and bacterial flora in the intestine; and must have a rate of passage through the gastrointestinal tract equal to that of the nutrient being tested. In addition, they have to be precisely quantified; must be easily mixed with feeds; and must have no interaction, reaction, or catalytic effect with other dietary components (i.e., they must exhibit inertness) during storage of the diet or in the digestive tract. Indicators are either exogenous (added to the diet; e.g., chromic oxide) or indigenous (contained in feed ingredients; e.g., acid-insoluble ash and fiber). The use of indigenous indicators is convenient for commercial feeds; however, since the amount of indigenous indicator in feeds is generally very small, larger amounts of feed and fecal samples are required for chemical analyses in order to reduce measurement errors.

Indicator substances used in digestibility trials with mammals and birds have been adapted for experiments with fish species. By far, the most commonly used indicator substance in the measurement of nutrient digestibility in fish and animal feeds is chromic oxide (Cr_2O_3). It is typically included in diets at levels of 0.5 or 1%. Diet and fecal samples containing chromic oxide can be analyzed colorimetrically with relatively simple and inexpensive procedures (12,13). Although chromic oxide has been questioned for its validity as an appropriate indicator, for various reasons (14–17), it is still the indicator of choice.

Various other indicators have also been used or recommended as the inert indicator for digestibility trials in fish or higher animals, including acid-insoluble ash, silica or SiO_2 (18,19), crude fiber or cellulose (20–22), lignin (23), titanium oxide (24), polyethylene (21,25), chromium-51-EDTA (26,27), magnesium ferrite ($MgO{\cdot}Fe_2O_3$) (28,29), cerium-144 (30,31), barium sulfate or carbonate (32,33), ytterbium acetate or oxide (34,35); yttrium-91 (27,36); yttrium oxide (35); ammonium ^{32}P-molybdate (37), cholestane ($C_{27}H_{40}$) (38), and n-alkanes (39). Each of these substances has one or more advantages over chromic oxide. Nevertheless, none of them has yet successfully replaced chromic oxide, due to various disadvantages or uncertainties (e.g., insufficient information).

Collection of Feces

Since fish live in water, feeding them diets and collecting their feces without losing soluble components requires different approaches from those used in digestibility studies in terrestrial animal species. Isotopes, which are commonly used in studies involving higher animals, are not suited for aquatic species, since the loss of such substances in water is difficult to prevent. Also, the use of a closed or recirculating water system permits fish to absorb various substances from water. Much effort has thus been directed toward collecting fecal samples with minimal losses of soluble components. The following methods of feces collection are currently used in digestibility experiments with aquatic species, all with varied efficiency and success, depending on the type of equipment and techniques used: settling, netting, stripping, dissection, anal suction, and metabolic chamber methods.

It is important to select an appropriate method, depending on the nutrient of interest in the feces. Some investigators have developed unique apparati for collecting fish feces. Smith (40) and Smith et al. (41,42) collected feces from fish confined in metabolic chambers. Choubert et al. (43) collected feces by netting, using a mechanically rotating screen that captures material in effluent water in tanks. Cho et al. (44) used a settling column which was modified by Hajen et al. (11), to collect voided feces. Since the composition of the diet largely affects the consistency of voided feces, various indigestible binders and certain feed ingredients with binding properties (for feces) are sometimes used when formulating diets to increase the stability of voided feces in water. This technique reduces the measurement error associated with leaching of soluble components in feces before collection.

The feces of goldfish (*Carassius auratus*) collected by stripping were not different from those collected by dissection from the rectal part of the intestine, in terms of measured protein digestibility of *Chlorella* (45). Other researchers, reported that fecal samples from rainbow trout (*Oncorhynchus mykiss*), chinook salmon (*Oncorhynchus tshawytscha*), and sea bass (*Dicentrarchus labrax*) collected by stripping gave digestibility values lower than those for samples collected by intestinal dissection, anal suction, settling, immediate pipetting, continuous filtration, decantation, and complete collection (11,46,47).

Fish should be fed test diets containing an indicator for five to seven days (under normal feeding regimes and temperatures) before fecal samples are collected. The collection of feces may be extended over a few days or longer, to collect sufficient material for analyses and to reduce day-to-day variations of nutrient digestibility. Diurnal variation in the excretion of indicator substances, especially chromic oxide, has long been recognized in higher animals, suggesting that continuous collection of feces may have an advantage over a single collection (e.g., stripping and dissection). Collecting feces by stripping should not be repeated frequently, since it potentially alters the passage rate of intestinal content, and the handling stress of the previous day could affect nutrient digestibility and feed intake of the fish in the following days. Feces (plus water, if feces are collected in a settling column) are dried using an oven or freeze-dryer, depending on the nutrient being tested, and the dried material is finely ground for chemical analysis.

Measurement of Nutrient Digestibility for Each Feed Ingredient

When formulating feeds by mixing various ingredients, it is essential to know the digestibility or availability of nutrients in each ingredient. This information is critical for minimizing the risk of nutrient deficiency (a biological problem) or excess (economical and environmental problems). Three methods are currently used to measure ingredient nutrient digestibility. The first method is to feed a single ingredient to fish, often by force feeding (40–42). Thus, the diet is generally nutritionally imbalanced, sometimes toxic, and not similar to actual feeds, and therefore it is rarely used to measure the digestibility of feed ingredients. Advantages of this method, however, are its simplicity of concept, sensitivity, and that there is no leaching loss of dietary nutrients. The second method is to feed a formulated diet in which the nutrient of interest is supplied solely from the test ingredient, but not from the other sources in the diet. The strength of this method is its sensitivity and reliability, since the diet can be formulated in a nutritionally more balanced manner than in the first method. When the concentration of test nutrient in an ingredient is low, this is the method of choice.

The third method uses a basal (reference) diet of known nutrient digestibility (9,44). The basal diet (as the complete ingredient mixture) is combined with a test ingredient at a certain ratio, generally 70% of the basal diet and 30% of the test ingredient. The combined diet is called the "test diet." By measuring the nutrient digestibility of the test diet and that of the basal diet, the digestibility of nutrients in the test ingredient can be estimated by subtraction, using the formula given in the next section. When the basal diet is free of the test nutrient, this method is analogous to the second method.

An implicit assumption for the third method is that the digestibility of the basal diet is unchanged when it is combined with test ingredients. This assumption cannot be true if there are severe nutrient interactions between the basal diet and the test ingredient, which is in fact a distinct advantage of this method over the first two methods. It thereby allows the measurement of an interactive property of a feed ingredient for each nutrient within a compound feed. For micronutrients, this information is equally, or sometimes more, important than the nutrient availability of the ingredient per se. Also, the use of a nutritionally adequate basal diet can buffer large differences in the nutrient composition among test ingredients, and therefore the digestibility can be measured without disturbing the normal digestive and physiological systems of fish. This method is more reliable and advantageous when studying nutrient availabilities in various feed ingredients of different nutrient concentrations, especially with respect to trace minerals (and carbohydrates) for which fish regulate intestinal absorption or intestinal endogenous excretion, depending on the level of intake. A serious disadvantage of this method, however, is its high vulnerability to experimental errors, especially when the percentage of the ingredient of interest in the test diet is low or the concentration of test nutrient in the ingredient is low (relative to that in the basal diet).

Calculation of Nutrient Digestibility for Diets

The first step in calculating the digestibility of nutrients in diets is to determine the concentration factor (CF), which can be obtained as follows:

$$\mathrm{CF} = \frac{\text{indicator concentration in feces}}{\text{indicator concentration in diet}}$$

This factor indicates the portion of the feces that corresponds to a unit amount of the diet. Then the nutrient content in the feces needs to be divided by the CF, to determine the amount of nutrient excreted in the feces per diet. The overall calculation procedure of nutrient digestibility in diets is expressed as follows:

$$\text{Digestibility (\%)} = \frac{\left(\begin{array}{c}\text{nutrient concentration in diet} \\ -\ \text{nutrient concentration in feces/CF}\end{array}\right) \times 100}{\text{nutrient concentration in diet}}$$

Example 1. Given that the protein concentration in a diet is 50%, the protein concentration in the feces is 25%, the indicator concentration in the diet is 0.5%, and the indicator concentration in the feces is 2.5%, the CF can be calculated as $2.5/0.5 = 5$, and the digestibility of protein in the diet is $(50 - 25/5) \times 100/50 = 90\%$. For the calculation of dry-matter digestibility, simply replace the nutrient concentration in feed and feces with 100%. The calculation is therefore $(100 - 100/5) \times 100/100 = 80\%$.

Calculation of Nutrient Digestibility for Ingredients

Calculation of nutrient digestibility for ingredients is the same as that of the diet just shown if the test ingredient is fed alone by force feeding (first method) or when the ingredient is the sole source of the test nutrient in the formulated feed (second method). When the basal diet is used to measure nutrient digestibility of ingredients (third method), however, the nutrient concentration and the digestibility of the basal diet need to be subtracted from those of the test diet (i.e., basal diet plus test ingredient) (48). The nutrient digestibility in test ingredients is therefore calculated as follows:

$$\text{Digestibility (\%)} = \frac{\left[\left(\begin{array}{c}\text{nutrient concentration in the test diet} \\ \times\ \text{digestibility of the nutrient in} \\ \text{the test diet}\end{array}\right) - \left(\begin{array}{c}0.7 \times \text{nutrient concentration in the} \\ \text{basal diet} \times \text{digestibility of the} \\ \text{nutrient in the basal diet}\end{array}\right)\right] \times 100}{\left(\begin{array}{c}\text{nutrient concentration in the test diet} \\ -\ 0.7 \times \text{nutrient concentration in the basal diet}\end{array}\right)}$$

Example 2 (in concept). Basal diet, casein–gelatin semipurified diet (60% protein, dry basis); Test ingredient, wheat middlings (20% protein, dry basis); Ratio of wheat middlings and the basal diet in the test diet: 3 : 7 (dry basis); Protein content (dry basis) in the test diet (should be) $20 \times 0.3 + 60 \times 0.7 = 48\%$; Digestibility of protein in the basal diet: 95%; Digestibility of protein in the test diet: 85%. Then, the digestibility of protein in the test ingredient can be calculated as follows:

$$\begin{aligned}\text{Digestibility} &= \frac{(48 \times 0.85 - 0.7 \times 60 \times 0.95) \times 100}{(48 - 0.7 \times 60)} \\ &= 15\%\end{aligned}$$

The first part of the numerator in this formula indicates the amount of protein absorbed from the test diet. The second half of the numerator is the amount of protein absorbed from the basal diet portion of the test diet. The numerator of the formula thus represents the total amount of protein absorbed from the test-ingredient portion of the test diet. The denominator of the formula indicates the amount of protein contributed from the test ingredient in the test diet and can also be written as (0.3 × protein concentration in the test ingredient). The coefficient 0.7 in the formula for digestibility is, of course, the dry-matter ratio (proportion of the basal diet in the test diet). When a test ingredient is mixed in the basal diet at ratios other than 30% (on a dry basis), the coefficient needs to be changed accordingly. If the protein (contribution) ratio is used instead of the dry matter ratio, the calculation in Example 2 will proceed as follows:

$$\text{Digestibility} = \frac{(0.85 - 0.875 \times 0.95) \times 100}{0.125} = 15\%$$

The coefficient $\frac{(60 \times 0.7)}{(60 \times 0.7 + 20 \times 0.3)}$ or 0.875, is the protein contribution ratio (i.e., the amount of protein contributed from the basal-diet portion in the test diet/the total protein content in the test diet); the denominator (1 – 0.875), or 0.125, is the ratio of protein contributed from the test-ingredient portion in the test diet.

The two different methods for doing the preceding calculation should yield the same value of digestibility. If the indicator substance is included in the basal-diet mixture before being combined with a test ingredient, the dry-matter ratio (coefficient) can be determined precisely and conveniently by analyzing the concentration of the indicator in dried diets (i.e., the concentration of indicator in the test diet/the concentration of indicator in the basal diet).

Example 3 (in practice)

Procedure. The basal diet mixture, containing an indicator at 1%, was mixed with a test ingredient at a 7 : 3 ratio (not on a dry basis).

Analytical Data (minimum required data). The indicator concentration (dry basis) In the basal diet is 1.05%, in the test diet is 0.76%, in the feces of fish fed the basal diet is 8%, and in the feces of fish fed the test diet is 4%.

The protein content (dry basis) in the basal diet is 50%, in the test diet is 40%, in the feces of fish fed the basal diet is 10%, and in the feces of fish fed the test diet is 20%.

Then, the digestibility of protein in the test ingredient can be calculated as follows:

$$\text{Digestibility} = \frac{\{40 \times (40 - 20/(4/0.76))/40 - (0.76/1.05) \times 50 \times (50 - 10/(8/1.05))/50\} \times 100}{\{40 - (0.76/1.05) \times 50\}} = 25.19\%$$

where (40 – 20/(4/0.76))/40 (=90.5%) represents the digestibility of protein in the test diet; 0.76/1.05 (= 72.38%) is the precise proportion (dry basis) of the basal diet in the test diet; and (50 – 10/(8/1.05))/50 (=97.38%) represents the digestibility of protein in the basal diet.

The foregoing information also provides us with the following:

- Digestibility of dry matter = 65.60% (by replacing the protein contents of feeds and feces in the foregoing formula with 100%);
- Protein content in the test ingredient (dry basis) = {40 – (0.76/1.05) × 50}/(1 – 0.76/1.05) = 13.79%;
- Moisture content of the basal diet mixture (x) is found from $1/(1 - x) = 1.05$, so $x = 4.76\%$;
- Moisture content of the test ingredient (y); is found from $70 \times (1 - x)/\{30 \times (1 - y) + 70 \times (1 - x)\} = 0.76/1.05$, so $y = 15.20\%$.

It is not worthwhile to simplify the formulas just presented, because it is essential to understand the calculation process and thereby to understand the meaning of the obtained values. Also, it should be noted that these estimated values are less accurate than those measured directly. The degree of error associated with indirect determination (calculation) is dependent upon measurement (analytical) errors and the nutrient contribution ratio in the test diet.

Although a precise ratio of a test ingredient to the basal diet needs to be known for calculation purposes, the ratio per se is not critical for measuring nutrient digestibility of test ingredients. When the inclusion level of the test ingredient is too low, however, it will cause a larger experimental error than when the inclusion level of the ingredient is high. The same problem will arise when the concentration of test nutrient in the test ingredient is very low relative to that in the basal diet. This is because a small difference between the measured nutrient digestibility values of the test diet will be responsible for the small amount of nutrient contributed from the test-ingredient portion. Using Example 2, we find that the protein digestibility of the test diet cannot be below 83.1% or above 95.6% when the protein digestibility of the test ingredient is calculated to be 0% and 100%, respectively. This result indicates that a small experimental error of the measured digestibility value of the test diet will be magnified into the calculated digestibility value of the test ingredient. When the inclusion level of the test ingredient into the test diet is too high, however, it will increase differences in the nutrient composition between test diets, depending on the test ingredient incorporated into the diet. Such diets may well be nutritionally inadequate or imbalanced, thereby lessening the advantage of this method. The appropriate ratio of test ingredients to the basal diet therefore depends on their nutrient-contribution ratio within the test diet.

Interpretation of Digestibility Data

The level of one nutrient (or antinutrient) in a diet always affects the absorption (availability) of other nutrients in the diet by antagonistic or synergistic interactions and dilution or concentration of dietary nutrients. These processes may happen during feed processing and storage, as well as after the ingestion of feeds within the diet or between the diet and endogenous excreta in the gut. Nutrient interactions, however, become obvious only when the degree of interactions is relatively high. For example, when the basal diet contains 10 mg of available nutrient A per 100 g of diet, and this diet is to be mixed with a test ingredient at a 7 : 3 ratio (the basal diet to the test ingredient), then the mixed diet should contain at least 7 mg of the available nutrient A per 100 g of diet, even if the same nutrient A supplied from the test ingredient is totally unavailable. If the net absorption of nutrient A from the diet is below this level (7 mg/100 g), then this is clear evidence of (antagonistic) interaction. In other words, the test ingredient reduced the absorption of nutrient A supplied from the basal-diet portion in the test diet. In this case, the original assumption of this method—i.e., that the digestibility of nutrients of the basal diet is unchanged—can no longer be supported. Alternatively, assuming that the digestibility of nutrients in the test ingredient is zero, the nutrient digestibility of the basal-diet portion in the test diet can be recalculated to express the degree of inhibition (antagonistic interaction) by the test ingredient (48). The substances in the test ingredient that might be responsible for the lowered absorption of the nutrient A could be other minerals (e.g., calcium, phosphorus, and iron); phytic acid; or fiber or other dietary components that have binding, precipitating, or oxidation/reduction properties. The absorption of some trace elements, especially that of iron and manganese, is substantially reduced by many feed ingredients, compared with the absorption from the casein basal diet alone (48).

In formulating feeds, the interacting property of each feed ingredient needs to be considered, especially for the antagonistic interactions, as seen in many ingredients having negative availability values. The calcium, manganese, and iron contained in many feed ingredients are not only unavailable, but also reduce the absorption of calcium, manganese, and iron supplied from the other part of the diet, including inorganic mineral supplements. Thus, feed ingredients have two distinct and opposing effects: (*1*) supplying minerals and (*2*) supplying substances that reduce or increase the absorption (availability) of minerals. Available mineral contents in feed ingredients could, therefore, be below zero (i.e., negative digestibility or availability) or exceed the total amount (i.e., higher than 100% digestibility or availability), due to this second effect of the ingredients in formulated feeds.

When fish are fed diets that contain nutrients in an amount higher than that required the fish may regulate their absorption of the nutrients. For protein, fats, and some minerals whose regulatory sites are not at the gastrointestinal level, fish metabolize any excess as an energy source or simply excrete the excess portion via urine or through their gills. This process does not interfere with the measurement of digestibility or availability values. For carbohydrates and some other minerals, such as iron, (water insoluble) phosphorus, and calcium, however, absorption appears to be regulated at the gastrointestinal level, due to limited digestive capacity, gastric acid output, or homeostatic regulations. Other minerals, such as manganese, copper, and zinc, have been shown to be absorbed in higher animals (49), and the excess amount is subsequently excreted into the intestinal lumen (endogenous loss). Long-term feeding will eventually allow fish to adjust the amount of net absorption of the elements, bringing it toward a more balanced level, which may be close to the required levels of the elements.

Measuring nutrient availability at this stage, however, cannot be justified, since fish regulate either the absorption or the excretion of the minerals. Conversely, taking a long acclimation period with diets that are marginally deficient in test nutrients may also cause problems, such as the depression of growth, feed intake, or normal physiology in metabolic processes. The acclimation (adaptation) period to the test diet therefore should be minimized when measuring the availability of minerals, and the collection of fecal samples should be completed before the fish start regulating the absorption of dietary minerals. However, there should be a sufficiently long conditioning period preceding the feeding of test diets (50). This period is critical to neutralize any carry-over effect of previous dietary regimes, to replenish body pools for each of the minerals and vitamins, and to normalize the growth rate and feed intake. The use of the 30/70 ratio to determine mineral availability coefficients for feed ingredients has an advantage of buffering the direct effect of the fish's regulatory system by standardizing (calibrating) the availability coefficients using the basal diet which is normally fed at the same time to a separate group of fish from the same initial that which have the same nutritional and physiological background. For the same reason, using relative availability is more appropriate than using absolute availability for studies involving micronutrients, because the conditioning of fish cannot be complete or confirmed in reality, and there always are physiological differences between fish from one study to another.

The basal diet should be formulated carefully, to minimize any unwanted interactions between the basal diet and test ingredients. In a sense, the basal diet is a dietary indicator unlike other indicators; it is digestible, but, because its digestibility is precisely known, we can calculate the digestibility of test ingredients based on the basal diet. The basal diet, therefore, needs to meet the requirements of the dietary indicator (given in the preceding section). The use of acidic salts of minerals (e.g., monobasic potassium and sodium phosphates) in the basal diet may not affect the availability of minerals in the basal diet per se, but when combined with a test ingredient, these substances may increase the availability of minerals in the test-ingredient portion in the test diet, due to the acidity of the basal diet and increased solubilization of minerals in the test ingredient. Thus, adding these substances causes an overestimation of the availability of minerals in test ingredients, especially for ingredients of high ash content. The use of fish meal in the basal diet, on the other hand, may precipitate or coprecipitate minerals in the intestine, due to the high fraction of calcium from bone in the fish meal, causing an underestimation of mineral availabilities in test ingredients. Due to various interactions, plant ingredients containing phytic acid may also have a similar effect. For minerals and carbohydrates, for which fish appear to have some regulatory functions, the level of nutrients in the test diet affects the absorption of nutrients. The low availability of phosphorus in high-ash ingredients, for example, is primarily due to the high (greater than dietary requirements) calcium and phosphorus content in the ingredients rather than to their inherent availability, since reducing the fraction of bone increases the absorption (in the percentage of intake, not in the absolute amount) of calcium and phosphorus in the ingredients (51).

FACTORS THAT AFFECT MEASUREMENT OF DIGESTIBILITY

Various biological, environmental, and dietary factors have been shown to affect the absorption and use of dietary nutrients. It is necessary to understand and control these variables in order to measure the digestibility of dietary nutrients and to interpret and use digestibility data in a meaningful way.

Biological Factors

Different species have different abilities to digest dietary nutrients, especially carbohydrates and minerals. These variations are associated with differences in feeding habits, as well as in digestive and metabolic systems (e.g., carnivores, herbivores, agastric fish). In addition, genetic differences within species appear to influence protein and carbohydrate digestibilities in salmonid fishes (52,53).

In the early stages of life, typically the postlarval stage, the digestive functions of many fish are not fully developed. This is especially the case for many seawater fish, which lay large numbers of small pelagic eggs. The digestion of food during this stage is largely aided by the autolysis of prey plankton. The poor performance of microparticulate diets in marine fish larvae is due to an incompletely developed digestive tract and a lower attractability of microparticulate diets than that of live foods (54). In herbivorous fish, an increase in the relative gut length in the early stages of life is associated with an increase in the capability to digest macroalgae (55).

Homeostatic mechanisms of organisms have profound effects on the absorption of some dietary nutrients, especially trace minerals. Animal species, including fish, regulate the uptake of some nutrients (e.g., iron) from the intestine when the amount of intake is in excess of the dietary requirement. For some nutrients (e.g., trace minerals), excess portions of absorbed nutrients are excreted into feces. Thus, true availability $\gg$ apparent availability. Diet history (i.e., previous diet regime or habit) also has important effects on the absorption of nutrients, especially trace minerals, which may cause a negative balance (excretion $>$ intake) for an extended period.

Environmental Factors

In rainbow trout, high salinities have been shown to decrease the absorption efficiency of dietary protein (56). In Arctic charr (*Salvelinus alpinus*), digestibilities of protein and lipids were found to be lower in seawater than in freshwater (57). In Atlantic salmon (*Salmo salar*), digestibility of nitrogen [sic] was found to be 9–10% lower in seawater-adapted smolts than in smolts adapted to freshwater (58). Digestibilities of energy and protein were lower in rainbow trout reared in saltwater as compared with those reared in freshwater (59). The rate of food movement was faster and the protein digestibility was lower when milkfish (*Chanos chanos*) were raised in seawater, rather than in freshwater (60). No difference was observed in protein digestibility between cod (*Gadus morhua*) acclimated to low salinity (14 ppt) and to seawater, even though low salinity provided higher growth rates and higher feed conversion ratios (61). Low salinity (10 ppt) did not affect the digestibility of protein by tilapia (*Oreochromis niloticus*) fry (62). The digestibility of casein and fish meal by prawn (*Penaeus monodon*) did not change with salinity (16 ppt vs. 32 ppt), but the digestibility of soybean meal significantly decreased at 32 ppt (63).

Water temperature did not alter protein digestibility in pacu (*Piaractus mesopotamicus*) (64). The digestibilities of protein, lipid, and energy in rainbow trout were lower at 7 °C (45 °F) than at 11 or 15 °C (52 or 59 °F) in small fish [18 g (0.040 lb)], but not in larger fish [207–586 g (0.456–1.292 lb)] (65). Digestibility of protein was higher in rainbow trout at 18 °C (64 °F) than at 9–15 °C (48–59 °F) (66).

Hypoxia did not change the apparent digestibility of protein, energy, and dry matter in rainbow trout (67,68). In prawns, a significant reduction in growth was obtained at a dissolved-oxygen (DO) level of 1 ppm; however, no significant differences were observed for apparent feed digestibility (69).

Ration size (feeding rate) had no significant effect on the digestibility of protein or lipid in rainbow trout (65). However, other observers noted that a larger daily meal size reduced the apparent digestibility of most amino acids and crude protein in rainbow trout (70) and that increasing the feeding level reduced the apparent digestibilities of dry matter and protein in commercial feeds fed to African catfish (*Clarias gariepinus*) (71). Digestibility of protein in Atlantic salmon smolts was about 10% higher when the fish were fed to excess in four hours daily than when the fish were fed to excess continuously (58).

Feeding frequency (two, four, or six times a day) did not affect digestibilities of dry matter, protein, and energy in rainbow trout with a body weight of about 100 g (0.22 lb) (72). Reducing the frequency of feeding increased the availability of phytate-phosphorus in a soybean-meal-based diet supplemented with microbial phytase (51).

Infection of rainbow trout by the bacteria *Aeromonas salmonicida* reduced the feed intake and the apparent digestibility of all nutrients and several amino acids (68). In largemouth bass (*Micropterus salmoides*), the presence of the intestinal worms (acanthocephalans) lowered both the protein digestibility and the amino acid availability of herring meal (73).

Dietary Factors

Heat (cooking extrusion) processing increases the digestibility of raw starch by many fish and animal species. For many fish, however, the digestibility of cooked starch and dextrin decreases as their content in the diet increases. Cooked starch also reduces protein digestibility when included at high levels in the diet, especially for carnivorous fish. The availability of niacin from corn was increased by about 57% when the corn was extrusion cooked (74). Crude (raw) starch in the diet reduces amylase activity in rainbow trout. Amylase is adsorbed to crude starch, so that starch hydrolysis is inhibited. Crude starch also accelerates the passage of the feed through the intestine, thus reducing the time available for absorption (75). Retrogradation of cooked starch (known as resistant starch) may reduce digestibility, as shown in higher animals. This change may be important in moist pellets, since moisture is necessary for retrogradation of gelatinized starch. Heating is necessary to inactivate trypsin inhibitors in soybeans, which otherwise reduce the digestibility of protein and amino acids (76,77). Excessive heating, especially with reducing sugars or aldehydes of lipid oxidation, causes serious losses of lysine, arginine, and cystine, due to the Maillard browning reaction, which reduces protein digestibility and amino acid availability.

Phytate (phytic acid and phytin), contained in all plant seeds, reduces protein digestibility and mineral availabilities in many monogastric animals, including fish. Phytase is an enzyme specific to phytate hydrolysis. This enzyme occurs naturally in some feed ingredients, such as wheat bran, and is also produced in a small amount by microbial flora in the intestine, but by far the most common source is fungal phytase, which is now commercially available as a feed additive. High processing temperatures of feeds generally inactivate endogenous or supplemental phytases, while low temperatures may preserve phytase activity. Phytate in soybean meal is very resistant to dry heating, steam heating, pressure cooking, and microwaving, regardless of the heating intensity (51). Certain nonlethal mutant grains containing lower levels of phytate have a higher availability for phosphorus and some other minerals than do ordinary grains (78).

The digestibility of fats generally decreases as the melting point increases. The digestibility of hydrogenated (hardened) oil is lower than that of liquid oils. Oxidized fish oil has a lower digestibility than that of the fresh oil in red seabream (*Chrysophrys major*) (79). Lipid hydroperoxide of methyl linoleate was not well absorbed by carp; however, the addition of vitamin E increased the absorption of the hydroperoxide to a level equivalent to that of methyl linoleate. But, ethoxyquin was not effective for increasing the absorption of hydroperoxide (80). Cholesterol was absorbed at a rate of only 28% in a lipid-free diet; however, the inclusion of palmitic acid, tripalmitin, or chicken-egg lecithin markedly increases the absorption of cholesterol to 85–98% in prawns (81).

The apparent absorptions of chelated minerals (i.e., copper proteinate, iron proteinate, manganese proteinate, selenium proteinate, and zinc proteinate) are all higher than those of the corresponding inorganic sources (i.e., copper sulfate, ferrous sulfate, manganese sulfate, sodium selenite, and zinc sulfate, respectively) in both semipurified diets and soybean meal diets fed to channel catfish (*Ictalurus punctatus*) (82). The bioavailability of heme-bound iron (e.g., blood meal) is much higher than that of inorganic sources of iron in fish and higher animals. Various chelating agents can either increase or decrease the absorption of dietary minerals, depending on the stability constant of the chelate formed.

Dietary calcium appears to reduce the availability of many minerals in diets, as the dietary concentration of calcium (as well as that of phosphorus, ash, and bone) is inversely correlated to the apparent absorption of other minerals (48). Tricalcium phosphate decreases zinc and manganese availability in carp (83) and zinc availability in rainbow trout (84). Citric acid (17) and inorganic acids (51) markedly increase the availability of phosphorus and some other minerals in fish meal. Ascorbic acid increases the absorption of inorganic iron in diets. Further, numerous interactions among minerals or other dietary components have been shown to affect the intestinal absorption of dietary minerals in various animal species (85).

Crude fiber or cellulose (up to 15%) appears to have no major effect on protein digestibility in goldfish (86) and carp (87). Alginate and guar gum reduced the apparent digestibilities of protein and fats in the diet fed to rainbow trout (88). Various alginates differing in gelling property all reduced the apparent digestibilities of nitrogen, fat, ash, and calcium when fed at a concentration of 5.0% in moist diets for rainbow trout (89). Other researchers, however, found that sodium alginate does not have a negative effect on protein and lipid digestibilities in sea bass at concentrations up to 8% in the diet, while at a concentration of 15% in the diet, the apparent digestibilities of both protein and lipids decreased (90). Agar included in the diet at a concentration of 10% reduced carbohydrate and dry-matter digestibilities in fingerling tilapias (91).

Particle sizes of feed ingredients may affect the digestibility of nutrients. With channel catfish, higher digestibilities of carbohydrates and protein were reported for feeds formulated with fine ingredients than for feeds formulated with coarse ingredients (92).

For protein, fats, and some minerals, the digestibility coefficients in individual ingredients are additive, so that the content of digestible nutrients in a diet and in fecal waste can be predicted from the digestibility coefficients of each ingredient and the nutrient concentrations in the diet. For carbohydrates and many minerals, however, the same prediction is more difficult to prove, due to the regulation of digestion or absorption of dietary nutrients, which depend on the dietary levels of the nutrients and the various interactions of the nutrients with other compounds in the diet and in the digestive tract.

CONCLUSION

Digestible (available) nutrients in diets are either used or, when the amount of nutrients is greater than the dietary requirement, are excreted or not absorbed by the fish, to avoid abnormal accumulation or overdose toxicity. The route of excretion can be via either feces (thus reducing the measured digestibility), urine, or gills. Due to increasing concern about environmental pollution, formulated feeds should not contain digestible or available nutrients in an amount greater than the dietary requirement. This standard indicates that data on both digestibility and dietary requirements will be critical sources of information for formulating biologically, economically, and environmentally optimal feeds for successful aquaculture operations.

BIBLIOGRAPHY

1. B.L. O'Dell, *Nutr. Rev.* **42**, 301–308 (1984).
2. S.J. Fairweather-Tait, *Biochem. Soc. Trans.* **24**, 775–780 (1996).
3. H. Aoe, I. Masuda, I. Abe, T. Saito, T. Toyoda, and S. Kitamura, *Bull. Jpn. Soc. Sci. Fish.* **36**, 407–413 (1970).
4. T. Nose, *Bull. Freshwater Fisheries Res. Lab.* **17**, 97–105 (1967).
5. C. Ogino and M. Chen, *Bull. Jpn. Soc. Sci. Fish.* **39**, 649–651 (1973).
6. National Research Council, *Nutrient Requirements of Fish*, National Academy Press, Washington, DC, 1993.
7. C.B. Ammerman, in C.B. Ammerman, D.H. Baker, and A.J. Lewis, eds., *Bioavailability of Nutrients for Animals*, Academic Press, San Diego, CA, 1995, pp. 83–94.
8. L.A. Maynard and J.K. Loosli, *Animal Nutrition*, 5th ed., McGraw-Hill Book Company, New York, 1962.
9. E.W. Crampton and L.E. Harris, *Applied Animal Nutrition*, 2nd ed., W.H. Freeman and Company, San Francisco, 1969.
10. J. De la Noue and G. Choubert, *Prog. Fish-Cult.* **48**, 190–195 (1986).
11. W.E. Hajen, R.M. Beames, D.A. Higgs, and B.S. Dosanjh, *Aquaculture* **112**, 321–332 (1993).
12. D.W. Bolin, R.P. King, and E.W. Klosterman, *Science* **116**, 634–635 (1952).
13. A. Furukawa and H. Tsukahara, *Bull. Jpn. Soc. Sci. Fish.* **32**, 502–506 (1966).
14. J.J. Knapka, K.M. Barth, D.G. Brown, and R.G. Cragle, *J. Nutr.* **92**, 79–85 (1967).
15. S.H. Bowen, *Trans. Am. Fish. Soc.* **107**, 755–756 (1978).
16. S.Y. Shiau and M.J. Chen, *J. Nutr.* **123**, 1747–1753 (1993).
17. S.H. Sugiura, F.M. Dong, and R.W. Hardy, *Aquaculture* **160**, 283–303 (1998).
18. C.F. Hickling, *J. Zool.* **148**, 408–419 (1966).
19. J.L. Atkinson, J.W. Hilton, and S.J. Slinger, *Can. J. Fish. Aquat. Sci.* **41**, 1384–1386 (1984).
20. R.K. Buddington, *Trans. Am. Fish. Soc.* **106**, 653–656 (1980).
21. A.G.J. Tacon and A.M.P. Rodrigues, *Aquaculture* **43**, 391–399 (1984).
22. S.S. De Silva, K.F. Shim, and A. Khim-Ong, *Reprod. Nutr. Dev.* **30**, 215–226 (1990).
23. E.A. Kane, W.C. Jacobson, and L.A. Moore, *J. Nutr.* **47**, 263–273 (1952).
24. E. Lied, K. Julshamn, and O.R. Braekkan, *Can. J. Fish. Aquat. Sci.* **39**, 854–861 (1982).
25. P.T. Chandler, E.M. Kesler, and J.J. McCarthy, *J. Dairy Sci.* **47**, 1426 (1964).
26. A.M. Downes and I.W. McDonald, *Brit. J. Nutr.* **18**, 153–162 (1964).
27. D. Sklan, D. Dubrov, U. Eisner, and S. Hurwitz, *J. Nutr.* **105**, 1549–1552 (1975).
28. H. Neumark, A. Halevi, S. Amir, and S. Yerushalmi, *J. Dairy Sci.* **58**, 1476–1481 (1975).
29. H. Barash, H. Neumark, and E. Heffer, *Nutr. Rep. Int.* **29**, 527–532 (1984).
30. R.J. Garner, H.G. Jones, and L. Ekman, *J. Agr. Sci.* **55**, 107 (1960).
31. W.C. Ellis and J.E. Huston, *J. Nutr.* **95**, 67–78 (1968).
32. F. Hoelzel, *Am. J. Physiol.* **92**, 466 (1930).
33. M. Riche, M.R. White, and P.B. Brown, *Nutr. Res.* **15**, 1323–1331 (1995).
34. M.J. Deering, D.R. Hewitt, and H.Z. Sarac, *J. World Aquacult. Soc.* **27**, 103–106 (1996).
35. S. Refstie, S.J. Helland, and T. Storebakken, *Aquaculture* **153**, 263–272 (1997).
36. C.S. Marcus and F.W. Lengemann, *J. Nutr.* **76**, 179–182 (1962).
37. J. Yamada, R. Kikuchi, M. Matsushima, and H. Ogami, *Bull. Jpn. Soc. Sci. Fish.* **28**, 905–908 (1962).
38. S. Sigurgisladottir, S.P. Lall, C.C. Parrish, and R.G. Ackman, *Aquaculture Association of Canada Conference*, No. 90–94 pp. 41–44, 1990.
39. O. Gudmundsson and K. Halldorsdottir, *J. Appl. Ichthyol.* **11**, 354–358 (1995).
40. R.R. Smith, *Prog. Fish-Cult.* **33**, 132–134 (1971).
41. R.R. Smith, M.C. Peterson, and A.C. Allred, *Prog. Fish-Cult.* **42**, 195–199 (1980).
42. R.R. Smith, R.A. Winfree, G.W. Rumsey, A. Allred, and M. Peterson, *J. World Aquacult. Soc.* **26**, 432–437 (1995).
43. G. Choubert, J. De La Noue, and P. Luquet, *Aquaculture* **29**, 185–189 (1982).
44. C.Y. Cho, S.J. Slinger, and H.S. Bayley, *Comp. Biochem. Physiol.* **73B**, 25–41 (1982).
45. T. Nose, *Bull. Freshwater Fish. Res. Lab.* **10**, 1–10 (1960).
46. J.T. Windell, J.W. Foltz, and J.A. Sarokon, *Prog. Fish-Cult.* **402**, 51–55 (1978).
47. P. Spyridakis, R. Metailler, J. Gabaudan, and A. Riaza, *Aquaculture* **77**, 61–70 (1989).
48. S.H. Sugiura, F.M. Dong, C.K. Rathbone, and R.W. Hardy, *Aquaculture* **159**, 177–202 (1998).
49. B. Sandström, *Proc. Nutr. Soc.* **47**, 161–167 (1988).
50. M.J. Jackson, *Eur. J. Clin. Nutr.* **51**(S1), S1–S2 (1997).
51. S.H. Sugiura, unpublished data.
52. E. Austreng and T. Refstie, *Aquaculture* **18**, 145–156 (1979).
53. K.R. Torrissen, E. Lied, and M. Espe, *J. Fish Biol.* **45**, 1087–1104 (1994).
54. A. Tandler and S. Kolkovski, *Isr. J. Aquacult. Bamidgeh* **44**, 128–129 (1992).
55. A.G. Benavides, J.M. Cancino, and F.P. Ojeda, *Funct. Ecol.* **8**, 46–51 (1994).
56. M.G. MacLeod, *Mar. Biol.* **43**, 93–102 (1977).
57. E. Ringø, *Aquaculture* **93**, 135–142 (1991).
58. M.L. Usher, C. Talbot, and F.B. Eddy, *Aquaculture* **90**, 85–96 (1990).
59. A. Aksnes and J. Opstvedt, *Aquaculture* **161**, 45–53 (1998).
60. R.P. Ferraris, M.R. Catacutan, R.L. Mabelin, and A.P. Jazul, *Aquaculture* **59**, 93–105 (1986).
61. J.D. Dutil, Y. Lambert, and E. Boucher, *Can. J. Fish. Aqua. Sci.* **54**(suppl. 1), 99–103 (1997).
62. S.S. De Silva and M.K. Perera, *Aquaculture* **38**, 293–306 (1984).

63. S.Y. Shiau, K.P. Lin, and C.L. Chiou, *J. Appl. Aquacult.* **1**, 47–54 (1992).
64. D.J. Carneiro, R.T. Rantin, T.C.R. Dias, and E.B. Malheiros, *Aquaculture* **124**, 131 (1994).
65. J.T. Windell, J.W. Foltz, and J.A. Sarokon, *Trans. Am. Fish. Soc.* **107**, 613–616 (1978).
66. C.Y. Cho and S.J. Slinger, *Proc. World Symp. on Finfish Nutrition and Fishfeed Technology*, Vol. II, Hamburg, Germany, 1979, pp. 239–247.
67. T. Pouliot and J. De la Noue, *Aquaculture* **79**, 317–327 (1989).
68. H. Neji, N. Naimi, R. Lallier, and J. De la Noue, in S.J. Kanshik and P. Luquet, eds., *Fish Nutrition in Practice*, colloq. no. 61, INRA, France, 1993, pp. 187–197.
69. E.R. Seidman and A.L. Lawrence, *J. World Maricult. Soc.* **16**, 333–346 (1986).
70. B. Hudon and J. De la Noue, *J. World Maricult. Soc.* **16**, 101–103 (1986).
71. A.M. Henken, D.W. Kleingeld, and P.A.T. Tijssen, *Aquaculture* **51**, 1–11 (1985).
72. B. Hudon and J. De la Noue, *Bull. Fr. Piscic.* **293–294**, 49–51 (1984).
73. C.E. Ayala, C.C. Kohler, and R.R. Stickney, *Prog. Fish-Cult.* **55**, 275–279 (1993).
74. W.K. Ng, C.N. Keembiyehetty, and R.P. Wilson, *Aquaculture* **161**, 393–404 (1998).
75. L. Spannhof and H. Plantikow, *Aquaculture* **30**, 95–108 (1983).
76. J.J. Olli, K. Hjelmeland, and A. Krogdahl, *Comp. Biochem. Physiol.* **109**, 923–928 (1994).
77. A. Krogdahl, T.B. Lea, and J.J. Olli, *Comp. Biochem. Physiol.* **107A**, 215–219 (1994).
78. S.H. Sugiura, V. Raboy, K.A. Young, F.M. Dong, and R.W. Hardy, *Aquaculture* **170**, 285–296 (1999).
79. H. Sakaguchi and A. Hamaguchi, *Bull. Jpn. Soc. Sci. Fish.* **45**, 545–548 (1979).
80. M. Takeuchi, *Bull. Jpn. Soc. Sci. Fish.* **38**, 155–159 (1972).
81. S. Teshima and A. Kanazawa, *Bull. Jpn. Soc. Sci. Fish.* **49**, 963–966 (1983).
82. T. Paripatananont and R.T. Lovell, *J. World Aquacult. Soc.* **28**, 62–67 (1997).
83. S. Satoh, K. Izume, T. Takeuchi, and T. Watanabe, *Bull. Jpn. Soc. Sci. Fish.* **58**, 539–545 (1992).
84. S. Satoh, K. Tabata, K. Izume, T. Takeuchi, and T. Watanabe, *Bull. Jpn. Soc. Sci. Fish.* **53**, 1199–1205 (1987).
85. C.B. Ammerman, D.H. Baker, and A.J. Lewis, *Bioavailability of Nutrients for Animals*, Academic Press, San Diego, CA, 1995.
86. A. Furukawa and Y. Ogasawara, *Bull Jpn. Soc. Sci. Fish.* **17**, 255–258 (1952).
87. M. Pereira-Filho, N. Castagnolli, and S.N. Kronka, *Aquaculture* **124**, 61–62 (1994).
88. T. Storebakken, *Aquaculture* **47**, 11–26 (1985).
89. T. Storebakken and E. Austreng, *Aquaculture* **60**, 121–131 (1987).
90. P. Spyridakis, R. Metailler, and J. Gabaudan, *Aquaculture* **77**, 71–73 (1989).
91. S.Y. Shiau and H.S. Liang, *Aquaculture* **127**, 41–48 (1994).
92. R.T. Lovell, *Feedstuffs* **51**, 28–32 (1984).

See also DIETARY PROTEIN REQUIREMENTS; ENVIRONMENTALLY FRIENDLY FEEDS.

DISEASE TREATMENTS

JOHN H. SCHACHTE
New York State Department of Environmental Conservation
Rome, New York

OUTLINE

INTRODUCTION

Treatments for diseases of fish generally refer to the administration of chemical and/or antimicrobial agents to combat infectious disease. These may involve the administration of chemicals directly to the water to eliminate ectoparasites and external bacterial infections or the oral administration of antibiotics in or on the feed. The latter are usually a treatment for one or more systemic bacterial infections. They are usually employed as a result of failure of management practices that would otherwise prevent or avoid the introduction of clinical disease. Once the onset of clinical signs of disease is evident, the fish culturist must respond with the appropriate therapy. Treatment of infectious diseases of fish requires knowledge of a number of areas of aquatic health management as well as the nature of the cultural system and the species of fish being treated. It is not sufficient to only identify the pathogen, look up the appropriate therapeutic agent, and then conduct the treatment. This article attempts to identify those critical factors that must be considered to conduct successful fish disease treatments. Additionally, those compounds frequently used for specific disease agents, their method of administration, sample calculations, and specific precautionary details are discussed.

PRETREATMENT CONSIDERATIONS

Choice and Availability of Therapeutic Agents

Historically fish culturists have used a wide variety of chemical and antimicrobial compounds to treat infectious diseases of fish. Salt was probably the earliest therapeutic agent used as a dip to treat external diseases. Formalin was first reported for use as a parasiticide in 1909 (1). Until the past ten years or so, many compounds were tried experimentally and, following successful field trials, were employed by fish culturists as acceptable therapy. More recently however, the U.S. Food and Drug Administration (FDA) Center for Veterinary Medicine (CVM) has become actively involved in the enforcement of laws regulating the use of therapeutic agents on fish. As the federal agency responsible for regulation of such activity with food animals, the FDA has established a framework to interact with aquaculture. Specific guidelines identifying compounds available for use on food fish have been generated. A mechanism to facilitate the evaluation and approval of additional compounds has been put into place in the last several years. The latter, the investigational new animal drug (INAD), exemption process provides a vehicle for legal field and laboratory testing of therapeutants to allow accumulation of safety and efficacy data sufficient for the FDA to make a determination of approval or rejection of use on food fish. Under this process, government agency hatcheries and/or academic institutions are named as coordinators with cooperators authorized to use and collect data on certain fish disease treatment agents. Presently a number of agencies cooperate under INADs issued for such compounds as chloramine-t and diquat, used for external bacterial infections of fish. At present there are very few therapeutic chemicals and antimicrobials available with FDA approval for use on food fish. Schnick et al. (2) have provided a list of the current status of therapeutic agents. Additionally the International Association of Fish and Wildlife Agencies is currently sponsoring a multistate-funded initiative to support research to facilitate the FDA approval of a list of key fish disease treatment compounds. It is incumbent upon the fish culturist to determine the appropriate chemical for a particular use and to adhere to label instructions regarding use and withdrawal times prior to harvest.

A final note with respect to regulation of fish disease treatment chemicals deals with discharge of those chemicals into adjacent waters. In the mid-1970s the Environmental Protection Agency (EPA) established the National Pollution Elimination Discharge System (NPDES) that sets standards for the quality and quantity of chemical discharges into adjacent waters. These standards, which may be more stringent depending upon in which state you reside, may influence which disease treatment you use. In flowing water aquaculture, depending upon the size of your annual production, you may be restricted in the concentration and or duration of a particular treatment. Amounts may be dictated by your state regulatory agency based upon receiving stream flows and resultant dilution. It would be prudent to check with the appropriate agency in your state to see if your operation must comply.

The Host, Pathogen, and Environment

Any discussion of fish diseases, whether it be diagnosis and identification or treatment and control, requires some basic understanding of the host pathogen relationship. Here I refer to necessary information about the characteristics of the host fish, the disease agent involved, and the environmental factors that influence the outbreak and severity of disease or lack thereof. It is well established that in the absence of clinical disease in fish, equilibrium exists between the host and a pathogen that may be present in the water or in the fish. In 1973, Snieszko (3) described the host, pathogen, environment relationship, and equilibrium. Any management failure on the part of the culturist or other event that upsets this balance can precipitate clinical disease. Specifically with regard to the host, some infectious diseases may be host specific. Some species of fish may be more or less resistant to a particular disease. Certain pathogens, protozoan parasites for example, require specific temperature ranges for maximum activity. The same applies for certain bacterial pathogens. Various environmental factors such as dissolved oxygen levels, temperature, and free ammonia levels can influence the outbreak and/or severity of a particular disease. Many of these same factors must be evaluated by the fish culturist prior to treatment. He or she must also answer some basic questions with regard to the impending treatment. What is the species of fish I'm treating and for which particular disease? What is this fish's reaction to the treatment of choice? Do some combinations of treatments offer advantages or disadvantages? What about multiple infections? Is there any order of priority in the treatment regime? Are there any complicating environmental factors that could influence the toxicity of the treatment chemical to the fish? Could any of these factors render the treatment ineffective?

METHODS OF ADMINISTRATION

Depending upon the type of fish culture facility, it may be necessary to employ one or more of the different techniques of delivering therapeutic compounds to fish. Earthen pond fish culture, for example, may require the use of a method quite different from that used in flowing water aquaculture. The following represent the most common methods for administration of therapeutic agents to fish.

Prolonged Bath Treatment

Prolonged bath treatment is a static treatment method where water flow is interrupted. The volume of water to be treated is determined and the appropriate amount of chemical is added for a predetermined duration, usually up to 60 minutes. Following the time expiration, the chemical bath is subjected to a rapid flush returning the fish to fresh water. This technique will usually result in a very accurate administration of the treatment chemical. Some precautions must be observed when employing this method:

- Oxygen levels should be monitored and provisions made for aeration if needed.
- Treatment should be conducted at a time when minimum stress due to environmental factors is present and personnel are available for continuous observation.
- Uniform distribution of chemical should be ensured.

This method, most frequently used in flowing water aquaculture, is usually administered in concrete raceways or other tanks of similar configuration. The size and shape of these units may influence uniform mixing. It is usually advisable to administer the chemical in several locations and mix thoroughly where possible.

Indefinite Bath Treatment

This technique is most often used to treat ponds where essentially static water fish culture is conducted. Large warmwater fish ponds are most frequently treated in this fashion. This method employs a lower level concentration than the previous method relying on natural decomposition and/or detoxification to terminate the treatment. The chemical may be added to the pond water by a variety of methods. Once the volume of water to be treated is determined, a standard solution of known concentration is delivered via pumps, gravity flow, or venturi apparatus from a power boat. Certain precautions with regard to the pond environment must be adhered to when using this method. Time of day and concomitant water quality should be considered. Low dissolved oxygen and/or heavy algal blooms may contribute to adverse effects or overt toxicity to fish. Chemicals such as formalin can further depress oxygen levels through its toxicity to aquatic plants. Depending upon the severity of the infection and the volume of the pond, such treatments may become prohibitively expensive. As a result, the culturist may want to consider an available alternative.

Constant Flow Treatment

This method is most frequently used in flowing water aquaculture. Because of hatchery design or weight of fish present, interruption of flow may not be feasible in this situation. Such treatments consist of administration of a standard solution of the therapeutic chemical by a constant flow device calibrated to deliver for a predetermined duration. In trout and salmon culture this is the most frequently used method. The device used may be as simple as a chicken waterer consisting of an inverted plastic or glass jar with a screw-on base that has a small hole drilled near the edge of the horizontal portion of the base. Since the base maintains a constant level of solution until the last few seconds of delivery, a constant flow is maintained. Where even smaller volumes of chemical are required, such as egg incubators, flexible plastic intravenous solution bags fitted with a flow metering valve can be effective. The alternate delivery device is some type of constant flow siphon vessel. The latter is usually a larger volume container which allows treatment of larger volume flow situations. We have found that a 5-gal plastic pail fitted with a simple wooden circular float and a calibrated plastic siphon tube provides a very satisfactory treatment unit. Alternatively, metering pumps of various design have been employed to accurately deliver the desired chemical directly into the water supply for a series of fish holding tanks. This method seems to have particular application on indoor or covered early life-stage rearing operations.

The use of the constant flow siphon method requires the accurate determination of water flow and accurate calibration of the delivery container. It should be noted that large flows may require prohibitively large amounts of chemical and concurrent expense. Where chemical volume delivery becomes a problem, multiple delivery units may be employed simultaneously. One should also be aware of the potential influence of the shape of the fish rearing unit. Such units of irregular configuration may result in so-called cold spots where fish may escape the desired concentration of the treatment chemical.

Feeding

In most situations where systemic bacterial infections occur, the administration of antibiotics can be accomplished orally by incorporation either in or on the feed. The antibiotics terramycin and Romet may already be incorporated into the diet in feed obtained from feed manufacturers. This is obviously the most efficient method of purchase and administration. Terramycin is incorporated into the feed at such a rate that if feed is administered at the recommended rate (usually 0.1% body weight per day) based upon 2 to 4 mg of active terramycin per 100 lb of fish fed, the desired dosage rate of active antibiotic is achieved. There are some precautions that should be considered with premedicated feed. Only the amount of feed expected to be used in a 30- to 90-day period should be ordered and stored. Depending upon time of year and ambient storage temperatures, such feed has a finite shelf life with regard to the potency of the antibiotic. At high summer temperatures where cool storage is unavailable, the antibiotic may be degraded below desired dosage level after only 30 days. Premedicated feed from the manufacturer usually is available in one or very limited sizes. Consequently, fish feeding on a small pellet experiencing a bacterial disease epizootic may not be able to utilize the premedicated feed size available.

Where premedicated feed is unavailable or undesirable, the fish culturist may buy the FDA-approved antibiotic of choice and add it to the feed. Such substances as gelatin and vegetable oils have been used. The latter, as described by Petrie and Ehlinger (4), requires the preparation of an oil–antibiotic premix slurry in a pail, wheelbarrow, or small cement mixer. Upon the addition of the appropriate amount and size feed, the fish culturist has a fresh preparation of medicated diet of the size needed.

In all cases where antibiotics are used, accurate diagnosis is paramount. Several of the common systemic infections may be precipitated by strains resistant to the usual antibiotic treatment. Failure to recognize this fact can result in excessive loss of fish and waste of valuable antibiotic. It is also important to remember that it is vital to complete the duration of treatment. In a 5- or 10-day treatment regime, it is not unusual to see marked improvement in 3 to 6 days. In the interest of economy some may be tempted to cease treatment. Such action may well result in reappearance of clinical signs in a few days. The purpose of the 5- or 10-day regime is to ensure that the level of infective organisms is reduced below that which might allow the disease to regain epizootic proportions. Conversely, one should be aware of the fallacy that if a large amount of antibiotic will cure the disease in 5- or 10-days, then a smaller amount periodically should keep it away permanently! Such a medicated feed regime encourages the development of resistant forms of bacterial pathogens by eliminating the dominant and more abundant sensitive individuals in a population. As a result, the less dominant antibiotic resistant individuals become the dominant strain present. Subsequently, when the culturist needs the compound to treat an active epizootic, the antibiotic may not be effective.

Injection

Where manageable numbers of large valuable fish are at risk for infection with certain bacterial diseases, injection with antimicrobials may be appropriate. Examples of this method can be found in adult spawning Atlantic salmon where such diseases as furunculosis caused by the bacterium *Aeromonas salmonicida* threaten restoration efforts. Intramuscular (IM) or intraperitoneal (IP) injection of an antibiotic can result in ability to hold and spawn such fish. Other examples may be found where subclinically infected adults held for spawning may be candidates for injection. Chinook salmon in the Great Lakes infected subclinically with *Renibacterium salmoninarum*, the causative agent of bacterial kidney disease (BKD), have been shown to experience a marked reduction of vertical transmission of the disease following experimental injections with erythromycin sulfate. Care must be taken not to cause internal injury to vital organs during administration. In the case of IP injections, fish should be held with ventral surface up. Internal organs settle to the backbone and allow injection in the region of the ventral fins with minimal risk.

Vaccination

This method of disease control relies upon taking a proactive approach to prevent the outbreak of clinical disease by the administration of a biological agent designed to stimulate the fish's immune system. Vaccines generally contain inactivated forms of the causal organism with the addition of other proprietary agents to enhance the stimulation of the fish's response. Most vaccines available today for use in both warmwater and coldwater fish are referred to as bacterins. These contain inactivated bacteria of the causal organism for the disease in question. Inactivated viral protein has also been employed for vaccination against viral disease. Plumb (6) provides an up-to-date list of the vaccines currently available for finfish aquaculture. The vaccine is usually administered to the fish at a suitable time in the early stage of development by one or more routes. These may include injection, oral or immersion techniques, or some combination thereof.

TREATMENT GUIDELINES

The conduct of any disease treatment must be preceded by attention to certain factors that can influence success or failure. Lassee (5) and Poupard (7) have provided lists of precautions that should be considered both prior to and during disease treatments administered to fish.

1. Fish should be examined prior to treatment to determine the disease agent to be treated. If in doubt, consult a trained fish health professional. The "shotgun" approach to treatment is poor policy which may hasten the development of drug resistant microorganisms.
2. Determine the general condition of the fish and existing water quality. Environmental conditions such as low dissolved oxygen (DO) and hardness may cause further losses if certain chemicals are used. Formalin, for example, may cause a further reduction of DO where existing levels are less than 5 mg/L (ppm).
3. Any parasitism or infection of the gills should be treated first. Failure to do so may affect the respiratory capability of the fish resulting in increased treatment toxicity.
4. Eliminate feeding 24 to 48 hours prior to treatment. Oxygen consumption is reduced and resultant stress reduction enables fish to better withstand any negative treatment effects.
5. In the case of concrete raceways, tanks, etc., rearing units should be as clean as is practical prior to treatment. Excess organic waste may contribute to lack of treatment effect by tying up treatment chemicals.
6. Where fish densities are excessive they should be reduced, if possible, prior to static treatment. Supplemental aeration and/or ability to provide rapid inflow of fresh water should be available.
7. During warm weather, treatments should be administered early in the day during the period of coolest temperatures. Such action also facilitates ability to observe and respond to any adverse treatment effects.
8. Always check calculations before commencing any treatment. Take care to consider the stock chemical concentration and adjust calculations accordingly to achieve desired final concentration. In a chemical bath treatment, an error factor of ten may result in no treatment effect or acute chemical toxicity.
9. Prior to treatment, check the condition of the chemicals. Improper storage or length of storage may alter chemical composition and result in errors in final treatment concentration. Ensure that any container used to deliver the chemical is clean and properly rinsed prior to use. Inadvertent mixture

of certain therapeutic agents such as quaternary ammonium compounds and formalin can result in acute toxicity to fish.

10. Always comply with local chemical discharge regulations to prevent any inadvertent toxicity to nontarget aquatic organisms.
11. Ensure proper mixing and distribution of chemicals in the water column of the rearing unit. This is particularly important in static water treatments of larger ponds.
12. Continually monitor fish throughout the treatment period for signs of distress.
13. Ensure that predisposing factors for stress mediated infection are identified and remedied. Failure to do so will usually result in reappearance of clinical signs of disease in a short period following initial treatment.
14. Adhere to all label required clearance times following treatment.
15. Keep records of all treatments, their purpose, and results for future reference.

APPROVED TREATMENT THERAPEUTIC AGENTS

Previous discussion has emphasized the importance of using FDA label-approved compounds when treating fish intended for human consumption. The federal Joint Subcommittee on Aquaculture (8) has generated a list of approved therapeutic agents current as of 1994. Table 1 is modified from that list with categories established by the USFDA.

Table 2 presents a list of unapproved drugs of low regulatory priority for the FDA. The FDA is unlikely to object at present to the use of these substances if the following conditions are met:

1. The drugs are used for the prescribed indications, including species and life stages where specified.
2. The drugs are used at the prescribed dosages.
3. The drugs are used according to good management practices.
4. The product is of an appropriate grade for use in food animals.
5. An adverse effect on the environment is unlikely.

The FDA's enforcement position on the use of these substances should be considered neither an approval nor or an affirmation of their safety and effectiveness. Based upon information available in the future, the FDA may take a different position on their use.

Classification of substances as new animal drugs of low regulatory priority does not exempt facilities from complying with other federal, state, and local

Table 1. FDA-Approved New Animal Drugs

Trade Name	Sponsor	Active Drug	Species	Uses
Finquel (MS-222)	Argent Chemical Laboratories	Tricaine methanesulfonate	Ictaluridae, Salmonidae, Esocidae, and Percidae	Temporary immobilization (anesthetic)
Formalin-F	Natchez Animal Supply	Formalin	Trout, salmon, catfish, largemouth bass, and bluegill; salmon, trout, and esocid eggs	Control of external protozoa and monogenetic trematodes; control of fungi of the family Saprolegniacea
Paracide-F	Argent Chemical Laboratories	Formalin	Trout, salmon, catfish, largemouth bass, and bluegill; salmon, trout, and esocid eggs	Control of external protozoa and monogenetic trematodes; control of fungi of the family Saprolegniacea
Parasite-S	Western Chemical	Formalin	Trout, salmon, largemouth bass, and bluegill; trout, salmon, and esocid eggs; cultured pained shrimp	Control of external protozoa and monogenetic trematodes; control of fungi of the family Saprolegniacea; control of external protozoan parasites
Romet 30	Hoffmann-LaRoche	Sulfamethoxine and ormetoprim	Catfish, salmonids	Control of enteric septicemia and furunculosis
Sulfamerizine in Fish Grade[a]	American Cyanimid	Sulfamerizine	Rainbow, brook, and brown trout	Control of furunculosis
Terramycin for fish	Pfizer	Oxytetracycline	Catfish	Control of bacterial septicemia and pseudomonas disease
			Lobster	Control of gafkemia
			Salmonids	Control of ulcer disease, furunculosis, bacterial hemorrhagic septicemia, and pseudomonas disease
			Pacific salmon	Marking of skeletal tissue

[a] According to sponsor, this drug is not presently being distributed.

Table 2. Unapproved Drugs of Low Regulatory Priority for the FDA

Common Name	Permitted Use
Acetic acid	Used as a dip at a concentration of 1,000–2,000 mg/L (ppm) for 1–10 min as a parasiticide for fish
Calcium chloride	Used to increase water calcium concentration to ensure proper egg water hardening; dosages used to raise hardness to 10–20 mg/L (ppm) calcium carbonate
Calcium oxide	Used as external protozoacide for fingerling to adult fish at 200 mg/L (ppm) for 5 sec
Carbon dioxide gas	Used as an anesthetic for cold, cool, and warm water fish
Fuller's earth	To reduce the adhesiveness of fish eggs to improve hatchability
Garlic (whole)	For the control of helminth and sea lice infestations in marine salmonids at all life stages
Hydrogen peroxide	Used at 250–500 mg/L (ppm) to control fungi on all species and all life stages of fish, including eggs
Ice	To reduce metabolic rate of fish during transport
Magnesium sulfate (Epsom salts)	Used to treat external monogenetic trematode and crustacean infestations in freshwater fish at all life stages; immersion at 30,000 mg/L (ppm) magnesium sulfate and 7,000 mg/L (ppm) sodium chloride for 5–10 sec
Onion (whole)	To treat external crustacean parasites and deter sea lice from infesting external surface of fish at all life stages
Papain	To remove the gelatinous matrix of fish egg masses to improve hatchability and decrease incidence of disease; 2% solution
Potassium chloride	Used as an aid in osmoregulation to relieve stress and prevent shock; dosages used to increase chloride concentration to 10–2,000 mg/L (ppm)
Povidone iodine compounds	Used as a fish egg disinfectant at rates of 50 mg/L (ppm) for 30 min, during water hardening and 100 mg/L (ppm) for 10 min, after water hardening
Sodium bicarbonate (baking soda)	Used at 142–642 mg/L (ppm) for 5 min as a means of introducing carbon dioxide into the water to anesthetize fish
Sodium chloride (salt)	Used as a 0.5–1% solution for an indefinite period as an osmoregulatory aid for the relief of stress and prevention of shock; used as a parasiticide at 3% solution for 10–30 min
Sodium sulfite	Used as a 15% solution for 5–8 min to treat eggs in order to improve hatchability
Urea and tannic acid	Used to denature the adhesive component of fish eggs at a concentration of 5 g urea and 20 g NaCl/5 L of water for approximately 6 min, followed by a separate solution of 0.75 g tannic acid/5 L of water for an additional 6 min; volumes sufficient for 400,000 eggs

environmental requirements. For example, facilities using these substances would still be required to comply with National Pollution Discharge Elimination System requirements.

DISEASE TREATMENT CALCULATIONS

The preceding discussion of treatment applications revealed that therapeutic agents are added directly to the water or, in certain situations, may be added to or incorporated into the feed. In the former case, a final concentration of the chemical agent is achieved by the direct addition of the chemical or a standard solution of that agent to the water. In general, calculations for such treatments are expressed as weight-to-weight or volume-to-volume relationships. They require, for example, that one determine the weight of the water in relation to the weight of the chemical added to the water to achieve the final desired concentration of the chemical in the solution to which the fish are exposed. In the case of flowing water aquaculture where it is inadvisable to interrupt water flow, the volume of water per unit of time must be determined with respect to the volume of therapeutic agent administered per unit of time. Both methods imply the equivalent weight of the solvent (chemical/drug) in the weight of the solution (water). Concentrations of chemicals administered to fish in water are usually expressed as parts per million (ppm). They may also be referred to in the metric as milligrams per liter (mg/L). In a weight-to-weight relationship, 1 ppm (mg/L) is that quantity of a substance that makes a concentration of one pound per million pounds of water. Lasee (5) has provided a table (Table 3) which presents the amount of chemical that should be added to various volumes of water to result in a final concentration of 1 ppm (mg/L).

Prolonged or Indefinite Bath Method Calculations

In both cases, prolonged or indefinite bath method calculations require the treatment of a static water situation. The volume of water to be treated must be determined. The prolonged bath treatment in a tank or raceway where water flow is interrupted requires the calculation of the volume and corresponding weight of the water column to be treated. Volume calculations are as

Table 3. Amount of Chemical Added to Various Volumes to Achieve a Final Concentration of 1 ppm (mg/L)

1 ppm (mg/L) =	1 ppm (mg/L) =
2.72 lb/ac-ft	0.0584 grains/gal
1233 g/ac-ft	1.0 mg/L
1.0 oz/100 ft^3 (cubic feet)	0.001 g/L
0.13 oz/1,000 gal	8.34 lb/1,000,000 gal
3.78 g/1,000 gal	1 g/m^3 (cubic meter)
0.0283 g/ft^3	1 mg/kg
0.0000624 lb/ft^3	10 kg/ha-m
0.00378 g/gal	

follows for a rectangular raceway, tank, or pond:

$$V = L \times W \times D$$

Here L = length (ft), W = width (ft), and D = depth of water column (ft).

Example. A rectangular raceway is 20 ft long by 10 ft wide with a water depth of 18 in.

$$V = 20 \times 10 \times 1.5$$

$$V = 300 \text{ ft}^3$$

In large earthen ponds, the calculation of the volume of the water requires determining the number of acre-feet present. An acre-foot is one surface acre, one-foot deep. Since large ponds of this type may not have uniformly deep bottoms or margins, one may determine the volume by multiplying the number of surface acres by the average depth of the pond. The volume is expressed as ac-ft.

Example. A two-acre pond has an average depth of three feet.

$$V = 2 \text{ ac} \times 3 \text{ ft average depth}$$

$$V = 6 \text{ ac-ft}$$

Calculation of Concentration (Dosage Rates) Examples

In the calculation of dosage rates for the treatment of fish diseases, it is useful to determine the number of gallons of water to be treated. The following conversions facilitate that determination: 1 ft^3 = 7.48 gal; 1 ac-ft = 325,851 gal; and 1 L = 0.26 gal. It is also important to remember that many therapeutic agents are supplied in concentrations other than 100% active ingredient. Active ingredient, usually expressed as a percentage of the final packaged product, can be found on the product label. All recommendations for treatment imply a 100% active product for administration to the fish. Where such product concentrations are not 100% active, calculations must be adjusted to account for that difference. Where 2 lb of a chemical at 100% active ingredient is required to treat a tank at 1 ppm, 4 lb would be required to treat that tank if only a 50% active ingredient product is available. It is important to note at this point that there is a common exception to this rule for a frequently used FDA approved chemical. Formalin, which is supplied as a 37% solution, is considered to be 100% active when calculating concentrations for treatment.

Example of a Prolonged Bath Treatment

A rectangular holding tank that is 3 ft wide and 10 ft long with a 1.5 ft water depth requires a static hydrogen peroxide treatment at 75 ppm (hydrogen peroxide is supplied in a 35% active solution). Thus,

$$\text{Volume} = 3 \times 10 \times 1.5$$

$$= 45 \text{ ft}^3$$

Referring back to Table 3, we find that 1 ppm = 0.0283 g/ft^3. Therefore,

$$\text{Amount of } H_2O_2 = (45 \text{ ft}^3 \times 0.0283 \text{ g/ft}^3 \times 75 \text{ ppm})$$

$$\div \text{ AI (active ingredient)}$$

$$= 95.5 \text{ g} \div 0.35 \text{ active}$$

$$= 272 \text{ mL hydrogen peroxide}$$

For liquids, divide grams by specific gravity of liquid to obtain ml required (i.e., for H_2O sp. Gr. = 1).

Peroxide should be diluted in 1 to 2 gal of water and mixed into the tank. At the end of the treatment period the fresh water should be turned on to flush the tank. The same general calculation may be used to treat a circular tank. Only the determination of volume differs in that the formula for the volume of a cylinder is used. Thus,

$$\text{Volume} = \pi \times R^2 \times D$$

where π = 3.14 (pi constant), R = radius of tank (one-half the diameter), and D = diameter of the tank.

The volume of a round tank 10 ft in diameter with a depth of water of 4 ft would be

$$V = 3.14 \times 5^2 \times 4$$

$$V = 3.14 \times 25 \text{ ft}^2 \times 4 \text{ ft}$$

$$V = 314 \text{ ft}^3$$

Example of an Indefinite Bath Treatment

An earthen pond of two surface acres with an average depth of three feet requires a formalin treatment at a concentration of 15 ppm:

$$\text{Volume} = 2 \text{ ac} \times 3 \text{ ft average depth}$$

$$= 6 \text{ ac-ft}$$

From Table 3, we determine that 1 ppm = 2.72 lb/ac-ft. Therefore

$$\text{Amount of formalin} = 2.27 \text{ lb/ac-ft} \times 15 \text{ ppm}$$

$$= 40.8 \text{ lb}$$

One gallon of formalin weighs 10.56 lb. Therefore 40.8 lb ÷ 10.56 lb = 3.86 gal of formalin required per ac-ft to achieve a 15 ppm concentration, or 6 × 3.86 gal for the entire pond.

Constant Flow Treatment

This method of treatment administration requires the calculation of the weight of the chemical agent based upon the flow rate of the water to be treated and the time. The basic formula for such a treatment is as follows:

$$\text{Wt} = [\text{Flow (gpm)} \times \text{(ppm)} \times \text{T} \times \text{CF}] \div \text{AI}$$

where flow = rate of water inflow (gal/min), ppm = concentration in parts per million, T = treatment time (min), CF = correction factor (no. g/gal), and AI = active ingredient.

Example of a Constant Flow Treatment

A rectangular trough 4 ft long, 1 ft wide, and 6 in. deep with a flow of 5 gpm containing eyed-trout eggs requires a 1667 ppm treatment of formalin for 15 min.

$$\text{Flow} = 5 \text{ gpm} \times 15 \text{ min} = 75 \text{ gal}$$

$$\text{Trough volume} = 4 \text{ ft} \times 1 \text{ ft} \times 0.5 \text{ ft} = 2 \text{ ft}^3$$

$$\text{Volume} = 2 \text{ ft}^3 \times 7.48 \text{ gal/ft}^3 = 14.96 \text{ gal}$$

These values illustrate the rate of exchange of water during the treatment.

The amount of formalin required to treat a flow of 5 gpm for 15 min is as follows:

$$\text{Wt (vol) of formalin} = 5 \text{ gpm} \times 1667 \times 15 \text{ min} \times 0.00378 \text{ g/gal} = 472.5 \text{ g (mL)}$$

Since formalin is considered to be 100% active, the results just obtained are divided by one. The formalin used (472.5 mL) should be diluted in a volume of water sufficient to flow from the delivery vessel in 15 min. As described earlier, a plastic chicken waterer with a small hole drilled in the base of the delivery cup provides a satisfactory vessel to administer this type of treatment.

Drugs Added to the Feed

The previous discussion of treating systemic bacterial infections with drugs in the feed requires the culturist to be familiar with adding medication to existing rations. Time or unavailability of the appropriate size premedicated diet may dictate the need to prepare a medicated ration on site. Oxytetracycline–HCl, generally referred to by the trade name Terramycin (TM 50), and Romet B are the approved antibiotics usually added to fish feed. The treatments are administered on the basis of weight of drug (in grams) per unit weight of fish per day. This will usually be expressed in grams of active drug per 100 lb of fish fed per day. The calculation for determining the amount of antibiotic required is

$$\text{Amount of drug/day} = \text{Wt} \div (100 \times D) \qquad (1)$$

where Wt is the weight of fish in pounds and D is the daily dosage rate in grams.

$$\text{Adjustment for active ingredient (AI)} = [\text{Amount of drug (g/day)}] \div (\text{g AI/lb premix}) \qquad (2)$$

where AI is the active ingredient in grams active per pound of premix.

$$\text{Amount of drug/lb feed/day} = [\text{Amount of drug/day (g)}] \div \text{lb feed consumed/day} \qquad (3)$$

Fish may be fed between 1 and 3% body wt/day. It is usually advisable to use the lowest rate possible to ensure that fish consume all of the medicated ration. A 1% rate is usually acceptable.

Example 1. What is the amount of terramycin required to treat 4000 lb of fish at a rate of 4 g active terramycin per day? Assume TM 50 available at 50 g active ingredient per lb of premix. Standard treatment is 10 days duration.

Using Eq. 1,

$$\text{Amount of drug/day} = 4{,}000 \text{ lb}/100 \times 4 \text{ g} = 160 \text{ g/day}$$

Using Eq. 2,

$$\text{Adjustment for AI} = (160 \text{ g/day}) \div (50 \text{ g/lb}) = 3.2 \text{ lb (1453 g)}$$

Using Eq. 3,

$$\text{Amount of drug/lb food} = 3.2 \text{ lb} \div 40 \text{ lb} = 0.08 \text{ lb (36.3 g)/lb of food}$$

The drug should be added to feed with corn oil. A slurry of oil and drug is prepared using approximately one quart of oil per 50 lb of feed to be medicated. This mixture can then be applied manually in a wheelbarrow or similar sized container for 50 lb or less of feed. Larger amounts of feed may be mixed with the drug slurry in a small portable cement mixer.

As a final note in the determination of treatment rates, one should always double check calculations. If another party is available, he or she should also check the results. A slight math error can either result in no treatment effect or one of toxicity to the fish. If the fish culture operation has access to a computer, it may be desirable to acquire a program such as SAMCALC (9) to assist in treatment calculations. The latter DOS-based program provides calculations for standing and flowing water treatments as well as other useful fish culture calculations.

SUMMARY

Some infectious diseases of fish will not respond to treatment. There are no effective therapeutic agents to treat or control viral diseases. Certain parasites such as the causal organism for whirling disease, *Myxobolus cerebralis*, will not respond to any approved therapy. In some cases, as with the aforementioned pathogens, quarantine, destruction of fish, and facility disinfection are the only choices for control.

There are many chemical agents and drugs that have been employed through the years to treat various diseases. Many have been effective in controlling serious infectious diseases and parasites. However, with the aforementioned entry of the FDA's Center for Veterinary Medicine in the regulation of aquaculture therapeutic agents, many of those compounds used in the past are now prohibited for use on food fish. Chemical discharge regulations of federal and state agencies may also restrict uses that have been acceptable in the past. One should also be aware that ongoing research on chemicals used in fish health may reveal new information on efficacy and/or safety. Such information may change the status of current FDA approval. Consequently, the fish culturist should always read and follow label instructions on any therapeutic agent used to treat food fish.

BIBLIOGRAPHY

1. R.A. Schnick, U.S. Fish Wild. Serv., *LaCrosse, WI. NTIS No. 237 198*. p. 145 (1973).
2. R.A. Schnick, F.P. Meyer, and D.L. Gray, *U.S. Fish Wild. Serv. and Univ. of Arkansas Cooperative Extension Service, Little Rock, Arkansas. MP 2421-11M-1-86*. p. 24 (1986).
3. S.F. Snieszko, *Adv. Vet. Sci. Comp. Med.* **17**, 291 (1973).
4. C.J. Petrie and N.F. Ehlinger, *Prog. Fish Cult.* **37**, 236 (1975).
5. B.A. Lassee, ed., *Introduction to Fish Health Management*. U.S. Fish Wild. Serv., Onalaska, WI, 1995, p. 139.
6. J.A. Plumb, *Health Maintenance and Principal Microbial Diseases of Cultured Fish*, Iowa State University Press, Ames IA, 1999, p. 328.
7. C.J. Poupard, in R.J. Roberts, ed., *Fish Pathology*, Bailliere Tindall, London, 1978, pp. 269–275.
8. Federal Joint Subcommittee on Aquaculture, Working Group on Quality Assurance in Aquaculture Production, *Guide to Drug, Vaccine and Pesticide Use in Aquaculture*, Texas Agricultural Extension Service, College Station, TX, Aug. 1994, p. 68.
9. J.N. Fries, *Prog. Fish Cult.* **55**, 62–64 (1994).

See also Drugs; Vaccines.

DISINFECTION AND STERILIZATION

S.K. Johnson
Texas Veterinary Medical Diagnostic Laboratory
College Station, Texas

OUTLINE

Sterilization involves the destruction of all forms of life on and in an object. Viable microorganisms, including their spores, are the usual targets. To kill the highly resistant spores of bacteria is challenging and is usually accomplished by steam under pressure or, in the case of removal from fluids, by microfiltration. Some chemicals also sterilize objects under suitable conditions, given that the chemicals are strong enough and in contact with the objects for a sufficient length of time. Effective sterilization prevents spoilage of products and the occurrence of disease.

Disinfection destroys pathogenic microorganisms associated with inanimate objects. Physical or chemical means are used to treat solid surfaces, solutions, or air. Chemical agents called disinfectants kill the vegetative stage of pathogenic microorganisms, but not necessarily their spore stage. Antiseptics are chemicals applied to living tissue for the purpose of preventing or retarding the growth of microorganisms without necessarily killing them.

In aquaculture, disinfection has widespread application as a method of sanitation. With regard to epidemiology, sanitation attempts to eliminate or suppress the presence of the initial infective unit, the inoculum. The presence of an inoculum in sufficient quantities is a requisite for disease. Other methods of sanitation include quarantine, the removal of alternative hosts, the rotation of species, and letting systems dry out or lie fallow. As a sanitary application, a narrow use of disinfection could be a foot bath, and a broader one might be the periodic distribution of chlorinated lime in growing units.

Eradication may be the goal of disinfection when intolerable disease agents are present within stocks. The decision to use eradication may come from farm management or by regulatory directive.

The term "disinfection" as commonly used in aquaculture is not as restrictive as the definition that limits the application to inanimate objects. Disinfection in the first instance could mean, for example, chemical applications to living eggs, fish, and food, as well as to equipment, supply water, and facilities. Perhaps aquaculturists use the term "disinfection" instead of "antisepsis" or some other term because the objective is usually to destroy the inoculum rather than to restore animal health. Thus, we usually find that in aquaculture, an effort to destroy all vegetative stages of pathogenic microorganisms, particularly in one short interval of time, is called disinfection. Strong chemical application to the fishless water medium of a confinement system, for example, is disinfection. On the other hand, chemical applications intended to destroy surface-dwelling parasites on swimming hosts are said to be treatments.

The effectiveness of disinfection is scalar in many situations in aquaculture. An effective method must not be too expensive and cannot involve unacceptable human or environmental risk. Further, the use of an effective, safe, and inexpensive application is not practical in cases where reinfection is highly probable. Contaminated surface water supplies, replacement stocks, and vectors are common sources of reinfection.

Survivability and longevity of pathogens in host-free environments are an important part of disinfection strategy. *Edwardsiella ictaluri* and *Edwardsiella tarda* infections, for example, may become serious over time, due to the ability of these pathogens to survive for long periods in pond mud. Following release from a fish, myxosporan spores and nematode eggs are able to survive for relatively long periods prior to consumption by a second host. Disinfection may not be an appropriate option if environmental persistence of the pathogen is very short. On the other hand, if the pathogen can survive for a long time, one must determine the economic feasibility of disinfection.

Following the removal of stock from culture systems, environmental conditions, especially temperature, affect the persistence of agents. Generally speaking, obligate microbial parasites (i.e., viruses) remain viable for several days in water of 25 °C (77 °F), but at 4 °C (39 °F), they may be viable for weeks or months. Particularly for viruses, and also for many other obligate parasites, knowledge is greatly lacking about reservoir organisms and their possible influence on environmental persistence of the viruses or parasites.

Except when a facility is out of production, disinfection of growing systems must take into account the infected stock. Market-size animals that have been infected may be sold commercially under certain circumstances. Otherwise, an acceptable practice is removal of the animals and burial or incineration. If the target of disinfection is a parasite that uses a secondary or a reservoir host in its developmental cycle, the application must be able to destroy any remaining alternative hosts.

A variety of system components are subject to disinfection: equipment, such as buildings, pipes, tanks, biofilters, nets, floors, boots, and a variety of fomites; culture water and pond soil; in put of water food, stock, and air; and output of effluent, pests, debris, and dead fish.

CLEANING

A preliminary cleaning accompanies many kinds of equipment disinfection. Cleaning removes the garden of sessile organisms and the buildup of organic grime that coat solid surfaces and make up part of sediments. Organic coatings, both living and dead, cover, occlude, and shield potential pathogens, often sustaining their viability and making them less vulnerable to actions that aim to disinfect. Tanks and other equipment are normally scrubbed with brushes or scrub pads of various designs. Extensions such as poles on brushes or pads make for easier scrubbing. Following the initial scrubdown, further cleaning with detergents may be beneficial if detergent properties include compatibility with ionic constituents of the water and ease of removal.

FREEZING AND HEATING

Freezing temperatures in drained pond systems have a disinfecting effect on certain cold-sensitive parasites, such as coccidians. Otherwise, viruses, bacteria, and many parasites tolerate freezing conditions for rather lengthy periods.

Heat exchangers of some hatchery systems heat water to temperatures that are intolerable to some microbial pathogens. Heating of water is suitable only for very small systems or system components. The cost of heating is higher than that of applying of ultraviolet radiation.

Applications that employ steam sterilization are rarely used. Application of moist heat (steam) at room atmospheres to tanks, floors, and object surfaces typical of aquaculture operations is ineffective in destroying many kinds of microbes, because the high temperature is not held for a sufficient duration.

FILTRATION

The primary use of filtration in aquaculture is to remove solids from supply water; it rarely is used for water sterilization. However, the effect of filtration may greatly determine the success of subsequent applications of disinfection. Removing particulate matter also removes a large complement of microbial associates, adding efficiency and economy to methods that follow.

RADIATION

Ultraviolet light at wavelengths of approximately 2500 to 2650 Å effectively destroys vegetative stages of many pathogens. However, practicality, output, and cost limit application to small systems, such as hatcheries, where the use of ultraviolet radiation is especially common. Surface water supplies must receive some pretreatment by sedimentation or filtration to make the water suitable for radiation, as ultraviolet light does not penetrate very far into water, and solids can block its path to a target.

CHLORINATION AND OZONATION

Chlorination and ozonation are useful for disinfecting water supplies. Introduction of chlorine to a water supply is accomplished by feeding it directly or, more commonly, by introducing it as a solution. In either case, an effort is made to attain a concentration of 0.2 to 0.5 mg of chlorine per liter of water (0.2 to 0.5 ppm) for 20 to 30 min of contact time or 3 to 5 mg/L (5 ppm) for 1 to 5 min. The water may then be passed over charcoal for chlorine removal. Calcium hypochlorite is used as a chlorine source in the disinfection of some small systems. Application is achieved by addition to head tanks or by use of dispensers.

When chlorine dissolves in water, it reacts to form hypochlorous acid, hydrogen, and chloride. Hypochlorous acid then forms an equilibrium with hypochlorite ions. Water acidity affects the presence of the stronger disinfectant, hypochlorous acid, producing a more favorable level in acidic pH ranges. The alkaline nature of seawater and, to some degree, the ionic constituents therein diminish the effectiveness of chlorine.

Electric discharge and ultraviolet irradiation of a suitable wavelength produce a three-atom form of oxygen (ozone) from the ubiquitous two-atom form. Ozone has strong oxidative properties. As air flows past the irradiation tubes, some conversion of oxygen to ozone occurs. Ozone output from electric discharge (corona discharge) equipment is much greater. The electric-discharge method uses either gaseous oxygen or atmospheric air. If air is used, it goes through pretreatment to remove impurities and moisture. Contacting devices are of many designs, but bubbling of the ozonator output is a common feature of most. As with chlorine, the concentrations of ozone needed to kill pathogens are near those that are harmful to culture animals. A special head tank is commonly placed between a supply source and the animal unit to control contact and avoid over-exposure of the animals. Contact concentrations are around 3 mg/L (3 ppm) of water, with residuals of below 0.15 mg/L (0.15 ppm). Ozone may also be removed with charcoal prior to use.

Ozone produces clear water and improves biofilter function. In good management design, it is safe for fish. However, the cost of equipment and use is relatively high. The diligence required to monitor residuals can also be costly.

The general organic content of water quickly reduces the oxidative power of both ozone and chlorine. In practice, an effective attack on viable pathogens requires certain adjustments to compensate for the influence of dissolved and suspended organic matter.

Chlorine, in the form of calcium and sodium hypochlorites, and chloramines are commonly used for disinfection of equipment, standing water, and, occasionally, soils. However, organic substances greatly compromise the effectiveness of these agents in ponds. Nevertheless, typical applications attempt to maintain available chlorine at 200 mg/L (200 ppm) of water for 30 minutes, 100 mg/L (100 ppm) for 1 hour, and 10 mg/L (50 ppm) for 24 hours. Calcium hypochlorite, also known as chlorinated lime, has wide use and contains 50 to 70% available chlorine. All chlorine compounds pose a risk to human safety.

FORMALIN AND FORMALDEHYDE

Formalin, water containing dissolved formaldehyde, is a readily available industrial chemical. It has promise for virus destruction in many kinds of systems, including ponds. Concentrations for virus disinfection are poorly established in aquaculture, but some disinfecting activity may take place at less than 50 mg of formalin/L (50 ppm) of water. Formalin is not very effective for disinfection of gram-positive bacteria. A 1:100 mixture of formalin and water is sometime used to disinfect standing water in tanks. Formaldehyde gas has been used for the disinfection of sealed buildings. Usually, the gas is generated by heating paraformaldehyde, the polymerized solid of formaldehyde.

QUARTERNARY AMMONIUM COMPOUNDS

Quarternary ammonium compounds are widely available in numerous brands and strengths. They have widespread use in aquaculture for equipment disinfection at active-ingredient concentrations that approximate 1200 mg/L (1200 ppm) of water. They are effective to a degree in standing water at 2 mg/L (2 ppm) for one hour of contact. Quaternary ammonium compounds are effective on gram-positive bacteria and some viruses, but are less effective on fungi and gram-negative bacteria.

IODOPHORES

Iodophores are generally available at 1% solution. They are organic compounds that release free iodine. Application of iodophores gives an amber tint to the water and causes the pH of soft water to drop. Acidity of the water promotes disinfectant activity, the iodine being more effective in more acidic conditions. A customary concentration for disinfection of standing water is 50 mg of active ingredient per liter (50 ppm) of water and, for equipment, 250 mg/L (250 ppm). Iodophores effectively disinfect bacteria and some viruses.

MISCELLANEOUS DISINFECTANTS

Application of calcium hydroxide and calcium oxide to ponds provides a degree of disinfection to the bottom mud of empty ponds. Their effectiveness is uncertain, but surely the abrupt shift in pH destroys many life forms. Similarly, 2% sodium hydroxide exposes microbes to an extreme pH when applied to nonmetallic equipment. Phenols and cresols sometimes find use in foot baths, but troublesome residuals deter other uses in aquaculture. Ethyl and isopropyl alcohol at strengths of 50–70% are effective disinfectants for equipment, but their volatility and flammability may restrict some applications. A dilute solution of hydrochloric acid also has potential as a disinfectant.

DEHYDRATION AND FALLOWING

Dehydration of equipment or entire culture systems provides a degree of disinfection, and in some cases, it may be the most practical action. Drying easily destroys fragile stages of many parasites, and dry surfaces have greater exposure to the effects of ultraviolet light and oxidation. Excessive sun exposure, however, greatly shortens the life of equipment such as seines. The combination of dryouts and certain lengths of time is often successful against obligate disease agents, which have a limited amount of time for which they can survive without a host. Reservoir and secondary hosts also may perish using this method.

See also CHLORINATION/DECHLORINATION; DISEASE TREATMENTS.

DISSOLVED OXYGEN*

ROBERT R. STICKNEY
Texas Sea Grant College Program
Bryan, Texas

OUTLINE

Sources, Sinks, and Acceptable Levels
Measuring Dissolved Oxygen
Causes of Oxygen Depletions
Overcoming Oxygen Depletions
Bibliography

The level of dissolved oxygen (DO) available to the animals in an aquaculture system is perhaps the most critical among the water quality variables routinely monitored. If a sufficient level of DO is not maintained, animals become stressed and may not eat well. In addition, their susceptibility to disease can increase dramatically. In the worst instance, the animals may die. Even slight reductions in DO below the minimum desirable levels can lead to reduced growth rates and suboptimal food conversion efficiencies (FCE). DO depletions can be exacerbated when the culturist provides food to animals that are not eating properly since waste feed decomposes and increases the oxygen demand on the system.

A few aquaculture organisms such as walking catfish (*Clarias* sp.) and tilapia (*Oreochromis* spp.) are able to tolerate low DO without apparent consequence, but most are not. Aquaculturists must measure the level of DO in their water system frequently, be familiar with the minimum DO level tolerated by the species with which they are working, and be prepared to take remedial action when a DO deficiency is detected.

SOURCES, SINKS, AND ACCEPTABLE LEVELS

Oxygen is dissolved in water by diffusion from the atmosphere and through release into the water as a byproduct of photosynthesis in aquatic plants. Diffusion from the atmosphere is aided when turbulence occurs.

For example, during windy weather, more water surface area is placed in contact with the atmosphere than when the water is calm. The greater the surface area of water in contact with the atmosphere, the more diffusion will occur.

When air or pure oxygen is bubbled into a water body, the action increases the amount of water surface that is in contact with the gas. The small bubbles produced from an air stone in an aquarium provide a large amount of surface area. Oxygen transfers from the bubbles to the water by diffusion.

All aquaculturists recognize that respiration is a normal physiological function of animals and that the removal of oxygen from water through respiratory activity can greatly impact the DO level. Many do not realize, however, that plants respire continuously. During daylight and above the compensation depth (1% light level), photosynthesis should produce more oxygen than is being consumed by both plant and animal respiration. In some instances, oxygen production is so great that the water becomes supersaturated. Oxygen may then be lost to the atmosphere. At night, no oxygen is produced, so only diffusion is operating to replace the oxygen removed through plant and animal respiration. Usually, the respiratory demand during the night is not sufficient to reduce the DO to critical levels, but under some circumstances a deficiency can occur.

Oxygen can be removed from water as a result of certain inorganic chemical reactions (chemical oxygen demand or COD) and through the decomposition of organic matter by microorganisms. The oxygen required by the microbial decomposition plus the oxygen demand associated with plant and animal respiration is called the biochemical oxygen demand (BOD). The BOD test is empirical and has been standardized for use in many types of water (2). The BOD in aquaculture systems becomes important when large amounts of aquatic vegetation are decaying (e.g., after a pond has been treated with a herbicide) or when dead animals are allowed to decompose in the water system. A high BOD can trigger an oxygen depletion.

As a general rule, if the DO concentration is greater than or equal to 5 mg/L (ppm), conditions relative to this parameter should be acceptable for aquatic organisms (3). None of the species currently being reared by commercial aquaculturists seem to require a level above 5 mg/L (ppm), and some species can tolerate DO levels well below 5 mg/L (ppm) with little or no resultant stress or impact on growth. Even so, many culturists prefer to see DO levels at or near saturation at all times. Several species of tilapia survive at DO concentrations as low as 1 mg/L (ppm) (4,5) and continue to grow rapidly, even if exposed to periods of such low levels on a daily basis (6). Other species suffer severe stress or die if exposed to such low concentrations, though most perform well at 4 mg/L (ppm) and may survive for extended periods at 3 mg/L (ppm). While most fish species can tolerate 1 to 2 mg/L (ppm) for brief periods, mortality occurs if exposure to those levels exceeds a few hours (7). Salmonid culturists worry when DO levels fall below 5 mg/L (ppm), while catfish farmers usually find levels of 3 mg/L (ppm) or higher acceptable.

The amount of oxygen that can be dissolved in water under ambient equilibrium conditions is called the saturation concentration. DO solubility is dependent on various factors, with temperature, salinity, and altitude being the most important. The DO saturation concentration decreases as temperature, salinity and altitude increase. At high temperatures and salinities, the DO saturation value exceeds 5 mg/L (ppm) at sea level. Thus, unless some other factor leads to an oxygen depletion, the water utilized in most aquaculture facilities should have the capacity to hold sufficient oxygen to support the species being reared. A mariculture facility located at high altitude (e.g., a recirculating system located in the mountains), could face a chronic oxygen problem

* This entry was adapted from Stickney (1) with permission.

because of the inability to maintain the desired level at saturation without supplemental oxygen.

Supersaturated oxygen conditions can occur, as previously indicated, when the rate of photosynthesis by aquatic plants is high. Levels of DO as high as 30 mg/L (ppm) are not uncommon under some conditions.

Supersaturated oxygen is not usually a problem in aquaculture.

MEASURING DISSOLVED OXYGEN

The aquaculturist should routinely monitor DO to ensure that depletions in the water system are not occurring. Daily measurements should be made at or before dawn, particularly in culture chambers where oxygen depletions are anticipated. Details regarding the causes and steps that can be taken to alleviate problems are presented in the next two sections of this entry.

DO concentration can be measured by Winkler titration (2) or with an oxygen meter. Both methods can determine DO to within 0.1 mg/L (ppm). Oxygen meters provide readings within seconds, while the titration procedure requires a few minutes, the availability of several chemicals, and some items of glassware including a burette or calibrated eyedropper for titration. Oxygen meters are quite reliable. Maintenance of models used by scientists usually involves infrequent battery replacement and periodic replacement of the membrane covering the end of the probe and the fluid within the probe.

Selection of a DO meter should be made with consideration of the environment in which it is to be used. For freshwater culture systems, temperature and altitude compensation are required, while mariculturists should employ a DO meter that is salinity and temperature compensated. For DO meters that do not have built-in compensation features for each of the desired parameters, conversion tables are generally provided by the manufacturer.

A good-quality DO meter costs a few hundred dollars, while titration kits are less expensive. DO meters save time, however, and the need to replace chemicals and broken glassware used for titration may make the meter the more economical choice in the long run.

With the advent of computerized monitoring systems for aquaculture operations, continuous monitoring of various water quality parameters has become routine in some situations. DO meters are among the types of equipment that are used for constant monitoring of culture systems. Computers can be programmed to process data collected through the meters, and alert the culturist by telephone in the event that the DO dips below the critical level (as determined by the aquaculturist).

CAUSES OF OXYGEN DEPLETIONS

Changes in the behavior of many aquatic species can indicate to the aquaculturist that an oxygen depletion is occurring. Feeding activity often decreases or ceases completely (decreased feeding activity can also be a response to other problems or an aspect of normal behavior). When DO becomes critically low, many fish will rise to the water surface and appear to gulp air. Crawfish climb on emergent vegetation to place their gills in contact with the atmosphere. Shrimp often move to shallow water when the DO level is low. Shrimp farmers frequently check their ponds at night by shining a flashlight into the water. If shrimp are concentrated in the shallow water, the oxygen may be low. When the shrimp move quickly away from the light, the situation is typically not critical. However, when the animals ignore the light, the problem may be severe.

In ponds, sufficient oxygen is usually provided by photosynthesis during daylight and atmospheric absorption at all times to maintain the DO above the critical level. Flowing water systems may require continuous aeration in order to maintain the desired DO level. As biomass and the amount of feed put into the culture chambers increase during the growing season, maintenance of the required DO level may become increasingly difficult. In ponds, the problem is exacerbated by high primary production rates because of the added respiratory demand at night. Temperature plays an important role due to the direct effect it has on the solubility of oxygen in water and also because a rise in temperature increases the metabolic rate of the organisms in the system.

In temperate climates, DO problems in ponds are fairly predictable. During spring, primary production may be high, but animal biomass tends to be relatively low and the water temperature is cool, so DO depletions tend not to be a problem. In the fall, the water temperature cools once again, and primary production may show a somewhat lower peak than observed in the spring. Biomass may be reduced from the higher summer levels through harvesting and a reduction in feed rate, thus, DO problems are not too common. During winter, the temperature is low, the feeding rate may be reduced (or feed may not be offered), the biomass may be further reduced due to harvesting, and primary productivity is slowed because of temperature. Unless a pond is ice covered and subjected to winterkill from oxygen depletions, DO deficiencies should not occur.

Summer is the season when most factors work against the maintenance of high DO levels in ponds. The temperature is at a maximum, so the solubility of oxygen is lowest. Biomass and feeding rates are typically approaching maximum levels for the year, and while primary production may have declined considerably from the spring high, crashes in phytoplankton blooms can occur leading to greatly increased BOD as the cells decompose. Respiration by all the organisms in the pond is at its highest level, and the decomposer community places a considerable demand on the available oxygen.

Declines in phytoplankton are also associated with periods of cloudy weather that reduce the amount of available sunlight and restrict the photic zone. During cloudy periods, the water temperature might cool somewhat, thereby increasing the solubility of oxygen. However, the aquaculturist should be aware that cloudy weather can trigger the crash of a phytoplankton bloom.

Typically, the lowest level of DO that occurs in a pond over any 24-hour period will coincide with dawn because respiration takes place in the absence of photosynthesis from about dusk the day before. Throughout the day, oxygen should increase as photosynthesis occurs. In a system at equilibrium, the diel pattern in DO may lead to a minimum level that is acceptable [5.0 mg/L (ppm) or more] at dawn. At about dusk, the level may be a few milligrams per liter higher, and the pattern will continue with little change from day to day. Ponds do not tend to be at equilibrium. As discussed previously, the photosynthetic rate can change significantly with temperature, the level of nutrients present, and the amount of light as influenced by cloud cover. At the same time, the demand on oxygen from respiration is constantly being altered as biomass changes. These and other factors tend to lead to a net change in DO concentration from one day to the next. The change may be either upward or downward and varies among ponds.

By monitoring pond DO early in the morning, preferably, close to dawn, the culturist will be able to predict when a problem may be imminent. To the delight of aquaculturists, particularly those with a large number of ponds, few ponds respond in exactly the same way, even when stocked at the same rate and subjected to the same management practices. It has been said that no two ponds are alike, and that tends to be true with respect to DO curves. Thus, a culturist may see critically low DO levels in one or two ponds on a given day, but it is rare when a large percentage of ponds at a given facility experience depletion problems.

In aquaculture facilities other than ponds, primary productivity is not usually a major factor, although cages and net pens are sometimes an exception. For tanks and raceways in flow-through or recirculating configurations oxygen depletions often occur due to the increasing respiratory demands of biomass increases. Failure of a biofilter can quickly lead to DO problems in a recirculating system, as can a reduction or loss of water input in any flowing system. Routine determination of DO can be accomplished at any time of the day in systems that are not subjected to diel fluctuations related to the photosynthetic cycle.

OVERCOMING OXYGEN DEPLETIONS

When a DO depletion is detected, immediate action should be initiated to restore the DO to a safe level. Several means can be used to accomplish that restoration. The addition of large amounts of well-oxygenated new water to a culture system will quickly raise the DO level. New water often comes at a premium, and if it comes from a well, it may contain little or no oxygen. Thus, other means of aeration are commonly used.

Any technique that increases the surface area of water in contact with the atmosphere increases the amount of oxygen in the exposed water. Aeration with compressors or air blowers is typically used in small systems. While such systems can sometimes be used in ponds, they are often not practical. Other ways of getting oxygen into water include aeration with compressed air, bottled oxygen, and liquid oxygen. Again, while commonly employed in tank and raceway systems, those sources of oxygen are not usually used on ponds or with cages and net pens. Aquaculture aeration systems and their efficiencies have been reviewed by Boyd and Watten (8).

An excellent means of aerating ponds is to circulate the water. By continuously bringing deep water to the surface, the entire water volume can be put into motion. Splashing water increases the amount of surface area in contact with the atmosphere, but it may not completely mix the pond. Some commercial aeration devices draw water from below and throw it up in the air, where it becomes oxygenated. The water then falls back into the pond and is recycled through the aerator. The net result is an improved DO level in the immediate vicinity of the aeration device but not in the entire pond.

Total pond mixing can be accomplished with paddlewheel aerators which are typically operated by electric motors or from the power takeoff of a tractor. Tractor-driven paddlewheel aerators are commonly used in the United States where stocking levels are not so high that all ponds in a complex require daily aeration. Most farmers only have one or two such aerators available (no more than the available number of tractors) that can be quickly moved from pond to pond as the need arises. If very high densities are stocked, and the farmer knows that DO problems will be chronic in a number of ponds, it may be necessary to provide an aerator for each, in which case, electric devices are preferred. Engle (9) reported that when a pond requires less than 250 hours of aeration per year, a tractor-driven paddlewheel is more efficient from an economic standpoint than an electric one, but that the electric motors are more efficient when more than 250 hours of aeration are required. Pond size influences efficiency as well, with floating electric paddlewheels being the more efficient choice in large ponds.

Paddlewheel aerators do not need to splash a great deal of water to be effective since their primary purpose is to turn over the pond. The splashing action may provide some psychological benefit because the farmer can clearly see that something is happening. Boyd and Watten (8) discussed a number of paddlewheel designs along with other methods for aerating ponds.

When an oxygen depletion is anticipated, preventative measures can be taken. Paddlewheel or other types of aerators can be operated through the night. Oxygen-rich new water can also be added to flush the pond. Both the addition of new water and aeration are often used by the catfish farmers in Mississippi during the period of the year when biomass and feeding rates are at their highest levels and the water quality is difficult to maintain. The addition of new water reduces the amount of suspended organic material by flushing it from the pond. As a result, the BOD is significantly reduced.

Culture animals should not be fed when an oxygen depletion is anticipated or has recently occurred. The stress associated with oxygen depletion often causes the animals to reject feed, so the ration will only serve to increase the BOD when it decomposes. Stress that does not result in death of the animals may make them susceptible to diseases. Epizootics may occur as soon as 24 hours after an oxygen depletion or as much as two weeks later. The culturist should be aware of that time frame and watch the animals closely. Treatment is more effective if a disease is detected early rather than after a full-blown epizootic develops.

Potassium permanganate ($KMnO_4$), a strong oxidizing agent, has long been used by fish culturists at the rate of about 2 to 3 mg/L (ppm) in instances where oxygen levels are low. The theory supporting the use of potassium permanganate is that the oxidation of organic matter will lower the BOD and chemical oxygen demand (COD). Tucker and Boyd (10) found that 2 mg/L (ppm) of $KMnO_4$ killed over 99% of the gram-negative bacteria present and a considerable percentage of the gram-positive bacteria. In addition to the oxidizing activity, potassium permanganate will release free oxygen directly into the water.

Boyd (11) reported on studies of the effectiveness of potassium permanganate and concluded that ponds treated with the chemical early in the morning actually recover more slowly than untreated ponds. He also examined the chemistry of the permanganate ion in water and found that free oxygen is released; however, the number of milligrams per liter of potassium permanganate that would be required to release 1 mg/L (ppm) of O_2 is at least 6.58 mg/L (ppm). In order to increase the DO in a pond, the amount of $KMnO_4$ needed to oxidize the organic matter present would have to be supplemented by over 6 mg/L (ppm) for each milligram per liter (ppm) of O_2 increase required. Thus, before any meaningful increase in DO could be obtained, the level of potassium permanganate required would be toxic to the fish.

Unlike ponds, the oxygen dynamics in raceways are highly predictable. Because of the rapid turnover rate that exists in most raceway systems, there is little opportunity for phytoplankton and other photosynthetic organisms to become established. As indicated by Boyd and Watten (8), the sources of oxygen in raceways are the incoming water and supplemental aeration. The amount of oxygen used by fish in respiration can be related by a proportionality constant (K_1) that correlates with the weight of oxygen consumed per weight of feed offered. Oxygen demand increases dramatically during feeding and for a period thereafter because of the increased metabolic rate associated with feeding activity and digestion.

Hydrogen peroxide (H_2O_2) can be used to generate oxygen gas and has been evaluated as a source of oxygen in conjunction with live hauling. Innes Taylor and Ross (12) described a technique by which hydrogen peroxide was used to generate oxygen in a container that was separate from the fish hauling tank.

Pure oxygen in the form of compressed gas or liquid oxygen (LOX) is increasingly being used in aquaculture systems, particularly high-density tank and linear raceway systems. LOX is often available within easy trucking distance of even remotely located fish farms in the United States. It is generally more economical to have LOX delivered than to manufacture it on site. By continuously injecting oxygen into a raceway system, much higher standing crops can be supported with a given water flow rate than with the same system operating without such supplementation.

BIBLIOGRAPHY

1. R.R. Stickney, *Principles of Aquaculture*, John Wiley & Sons, New York, 1994.
2. APHA, *Standard Methods*, American Public Health Association, Washington, DC, 1989.
3. F.W. Wheaton, *Aquaculture Engineering*, Wiley-Interscience, New York, 1977.
4. R.M. Uchida and J.E. King, *Fishery Bulletin* **62**, 21–25 (1962).
5. H.W. Denzer, *FAO Fisheries Report* **44**, 357–366 (1968).
6. R.R. Stickney, H.B. Simmons, and L.O. Rowland, *Texas Academy of Science* **29**, 93–99 (1977).
7. R.G. Piper, I.B. McElwain, L.E. Orme, J.P. McCraren, L.G. Fowler, and J.R. Leonard, *Fish Hatchery Management*, U.S. Fish and Wildlife Service, Washington, DC, 1982.
8. C.E. Boyd and B.J. Watten, *Reviews in Aquatic Science* **1**, 425–472 (1989).
9. C.R. Engle, *Aquaculture Engineering* **8**, 193–207 (1989).
10. C.S. Tucker and C.E. Boyd, *Transactions of the American Fisheries Society* **106**, 481–488 (1977).
11. C.E. Boyd, *Water Quality in Ponds for Aquaculture*, Agricultural Experiment Station, Auburn University, Auburn, AL, 1990.
12. N. Innes Taylor and L.G. Ross, *Aquaculture* **70**, 183–192 (1988).

See also AERATION SYSTEMS; BIOCHEMICAL OXYGEN DEMAND; PURE OXYGEN SYSTEMS.

DOLPHIN (MAHIMAHI) CULTURE

ANTHONY C. OSTROWSKI
The Oceanic Institute
Waimanalo, Hawaii

OUTLINE

Aquaculture of the dolphinfish (*Coryphaena hippurus*), or mahimahi is probably one of the most challenging endeavors to be undertaken by the fish farmer. These highly aggressive, yet relatively sensitive, fish are one of the most difficult species to raise in controlled environments, because of stringent water quality, nutritional, and facility design needs, as well as the constant care required throughout all stages of their life. The mahimahi has generated interest as a potential aquaculture candidate because of its reputation as a high-value fish, large worldwide consumer demand, good feed conversion, spontaneous spawning, and rapid growth rate. Although difficult to culture, mahimahi can be raised using relatively streamlined methods. Successful culture requires placing emphasis on appropriate facility design and paying close attention to details throughout all stages of production.

WATER QUALITY

Mahimahi are a fast-swimming, primarily oceanic fish. They are distributed worldwide in tropical and subtropical oceans within 20 °C (68 °F) isotherms and usually within the 30°-latitude range (1). Culture is generally restricted to temperatures between 18 and 30 °C (64 and 86 °F), but most research has been conducted at temperatures between 24 and 28 °C (75 and 82 °F). The mahimahi's growth rate varies greatly with temperature, particularly during early developmental stages. Its growth rate to 40 days of age doubles with a 2 °C change in rearing temperature, from 26 to 28 °C (79 to 82 °F) (2). Differences in the growth rate during growout between 24 and 28 °C (75 and 82 °F) are less apparent (3), but temperatures much below 20 to 24 °C (68 to 75 °F) will limit growth (4,5).

Mahimahi are physiologically adapted to open-ocean environments that are generally highly oxygenated and virtually free of bacteria and sediment. This species can be successfully cultured only in pristine environments that have very low levels of dissolved organic and particulate matter and are devoid of parasites. Mahimahi appear to be poorly adapted to hypoxic (i.e., low oxygen) conditions, and it is recommended that they be maintained in waters as oxygen saturated as is practical, and not below 82% oxygen saturation [i.e., 5.5 parts per million (ppm)] (6). Signs of stress begin to occur at oxygen saturation levels below 90% (5.97 ppm) (7,8). Hypoxia-related mortality occurs at an oxygen concentration between 4.3 and 4.4 ppm (between 65 and 66% saturation) (6–8). Also particulate matter causes gill aneurysms in mahimahi, which lead to death (9).

BROODSTOCK

Mahimahi reach sexual maturity at approximately 5–6 months of age [with a standard length of 45–55 cm (17.7–21.6 in.)] in captivity and produce 100,000–250,000 eggs every other night (2,10–12), with spawns of greater than 500,000 eggs possible. Fecundity (the number of eggs released) is related to fish size (Fig. 1), but egg quality is highly variable (9,13). Mahimahi spawn in captivity without artificial inducement.

Both wild-caught and hatchery-reared mahimahi can be used as broodstock. Wild-caught animals adapt readily to life in captivity and, if sexually mature, may begin releasing eggs within a few weeks after being caught in the open ocean. Mahimahi are easily excitable and prone to rapid swimming bursts. Care must be taken to avoid mortality caused by fish ramming themselves head first into the side walls of tanks. Doughnut-shaped tank designs (10) for holding broodstock have proven especially

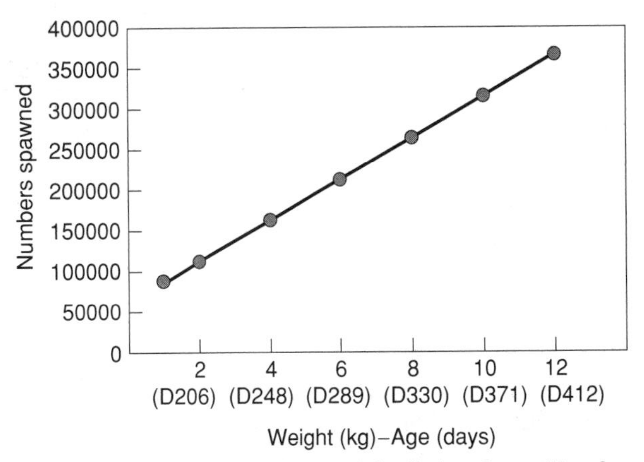

Figure 1. Relationship between weight (kg) and age (D = days after hatching) of female mahimahi and the number of eggs spawned. Data from The Oceanic Institute (9).

useful for preventing this problem, as they restrict perpendicular movement (Fig. 2). Reproductively active males are highly aggressive, and, typically, only one male is paired with one to three females. Broodstock are usually fed a raw diet of chopped squid and fish, either with or without supplemental vitamins (2,10,11,14), or are fed raw mashes of raw ingredients and some fishmeal. Successful spawns have been obtained from fish fed pelleted diets (2), but pelleted diets are not preferred by either wild-caught or hatchery-raised broodstock. Broodstock usually eat from 5 to 14% of their body weight daily in raw feed (2,9,11).

Mahimahi broodstock are easily stressed and intolerant of poor water quality, rapid changes in water quality, parasitic infection, and disease. Gaping of the jaws, labored swimming, and darkening of the body are indicators of potential problems in fish. Mahimahi are particularly susceptible to skin and gill parasites, including the dinoflagellate *Amyloodinium ocellatus* and the ciliate *Cryptocaryon irritans* (9). Mild infestations in which fish have not lost their appetite can be treated with brackish water [10 parts per thousand (ppt)] over a 12-hour period, or with freshwater dips for five minutes.

EGG INCUBATION AND HATCHING

Although mahimahi spawn frequently, poor egg quality (e.g., small yolk, multiple oil droplets, and asymmetry of embryos) is not uncommon among broodstock and is largely unpredictable between spawned batches (9,13,15). Survival prior to first feeding is typically monitored. Larval rearing will either be maintained or abandoned, depending on the survival rate, and a new batch of eggs or larvae stocked. Mahimahi eggs are 1.4–1.6 mm (1/18–1/16 in.) in diameter and contain a single oil droplet about 0.3 mm (1/83 in.) in diameter. Viable eggs are buoyant at 30–35 ppt salinity and hatch within approximately 40 hours after fertilization at 25–27 °C (77–82 °F). Kraul et al. (16) have reported that the hatch rate and yolk sac survival are higher in tanks with turbulence. High concentrations of calcium and magnesium are important for hatching and larval survival (17). Survival rates during

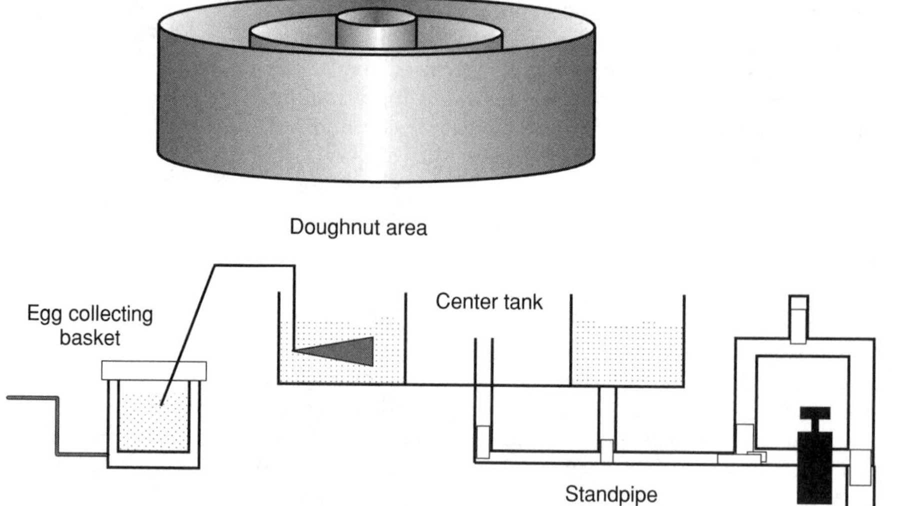

Figure 2. Mahimahi broodstock tank design with egg collector. The doughnut tank is 6.1 m (20 ft) in diameter, with an inner ring that is 3.0 m (10 ft) in diameter. Adapted from Kim et al. (2) and the Oceanic Institute (9).

incubation can range from 40 to 70% (2). Broodstock nutrition and stress are generally thought to play key roles in egg quality, while the physical and chemical factors of water quality compound the survival of prefeeding larvae.

LARVAL REARING

Widely diverse methods have been successfully employed in mahimahi larviculture. For example, publications by Kraul et al. (16), Kraul (10,18), and Szyper (10) describe feeding methods that employ rotifers (with algae), copepods, and yolk sac larvae of mahimahi as live foods. Kim et al. (2) have described a more streamlined method that utilizes *Artemia* (brine shrimp) as the sole live food source, and no algal inputs. Feeding trials have shown a 14% greater survival rate through first feeding when rotifers are offered instead of *Artemia*, although the additional cost of culturing rotifers and their algae feed is not offset by the potential increase in first-feeding success (19). Similarly, mass culture of copepods may not be practical for commercial hatcheries, due to low yields (20). Pilot farms have focused on use of *Artemia* as the first, and generally only, live food item, with no algal inputs during larviculture (21,22). Rotifers (with algae) are preferred when rearing temperatures are below 26 °C (79 °F) (9).

High mortality and fast growth rates are characteristic of larval mahimahi culture. Survival rates of up to 40% have been reported from stocked larvae (16), but rates of 10–20% are more common (9,10,22). Larvae are typically raised in 5,000-L (1,323-gal) circular tanks and stocked directly as eggs or prefeeding larvae. Separate, 1,500-L (397-gal) tanks have also been used for egg incubation and first feeding to concentrate prey, with larvae being transferred to clean 5,000-L (1,323-gal) tanks during the second week of development (9). Stocking densities in 5,000-L (1,323-gal) tanks range between 1 (10,18) and 5 larvae/L (4 and 19 larvae/gal) (9).

Mahimahi larvae undergo continuous, rapid organogenesis and tissue differentiation during the early larval stage and are extremely sensitive to environmental change. (See Table 1.) Mahimahi do not possess a swim bladder, and emphasis is placed on digestive system development to support rapid larval growth. During the third week, weight gain increases dramatically, corresponding with the completion of metamorphosis and an increase in feeding rates. Overall, larval mahimahi grow from 0.5 to 60.0 mg (1.8×10^{-5} to 2.2×10^{-3} oz) wet weight from day 0 to day 20, yielding an instantaneous gain in weight of approximately 25% wet body weight daily (9).

Dietary lipids (fats) play a key role in growth and survival of mahimahi larvae, and an understanding of lipid nutrition during early developmental stages has helped advance the use of more streamlined methods of production. Lipids are the primary endogenous energy source of eggs and prefeeding mahimahi larvae (23), and it appears that mahimahi eggs and larvae are well adapted to exploiting diets high in lipids upon transition to exogenous food sources. During the first week of development, the total amount of dietary lipids is more important in supporting good survival and growth of larvae than is the fatty acid profile (i.e., the exact makeup of the fat). The rate of survival to seven days of age was higher in mahimahi larvae fed rotifers high in lipid content than those with a low lipid content, regardless of the fatty acid profile (24). A higher algal-lipid

Table 1. Timetable for Mahimahi Larval Development at 26–28 °C (79–82 °F)

Days of Age	Morphology	Standard Length (mm)
0	Hatching	2.8
3	Onset of feeding	4.2
7	Gut coiling	6.0
8	Notochord flexion	6.3
10	Red blood pigmentation	6.7
12	Onset of metamorphosis	8.0
17–20	Completion of metamorphosis	14.8
20	Yellow pigmentation	18.5

content may have explained better first-week survival rates of mahimahi that were fed rotifers cultured on the algae *Tetraselmis chuii* than mahimahi fed rotifers cultured on *Nannochloropsis oculata* (25), despite the suggestion of a poor fatty acid profile in the former. During the second week of development, however, the fatty acid profile of the lipid becomes important, and larvae must obtain a dietary source of the essential highly unsaturated fatty acid (HUFA) docosahexaenoic acid (DHA), to support good growth, survival (26), and stress resistance (27) through metamorphosis. Like other marine species (28), mahimahi larvae are unable to manufacture DHA from endogenous (24) or dietary precursors (24,26), but instead contain an endogenous store that supports survival over the first week of life (24). Typically, the DHA content of rotifers and *Artemia* is fortified with commercially available enrichment products. The presence of replete endogenous stores allows unenriched rotifers or newly hatched *Artemia* (which cannot be enriched) to be used successfully as first-feed items. Ostrowski and Divakaran (24) have recommended that diets for first-feeding larvae be high in both lipids and DHA, for maximum benefit.

The timing of live-food additions and enrichments is of utmost importance in mahimahi culture. A composite feeding schedule from the various methods employed is presented in Figure 3. As indicated in the figure, mahimahi can be fed diets deficient in DHA through the first week of development, but must receive either copepods (which are high in DHA) or enriched *Artemia* toward the beginning of the second week (5–7 days of age). Preferential retention of DHA in body phospholipid (biomembrane) fractions appears to be related to survival over the first week of development (24), but the DHA becomes diluted to low body levels as fish grow (24,26). Mass mortality occurs between 8 and 12 days of age if larvae are continued on DHA-deficient diets (26,27,29). A pulse-feeding technique in which discrete, large doses of 2- to 3-day-old (36 to 48 hours) enriched *Artemia* are fed to larvae throughout the day, beginning around 5 days of age, has been described by Kim et al. (2). Since *Artemia* rapidly lose the benefits of enrichments, pulse feeding ensures consumption of nutritionally replete *Artemia* and delays gut evacuation time, to ensure that digestion is maximized. Larvae will typically gorge and ingest the majority of *Artemia* offered within a half hour of their presentation.

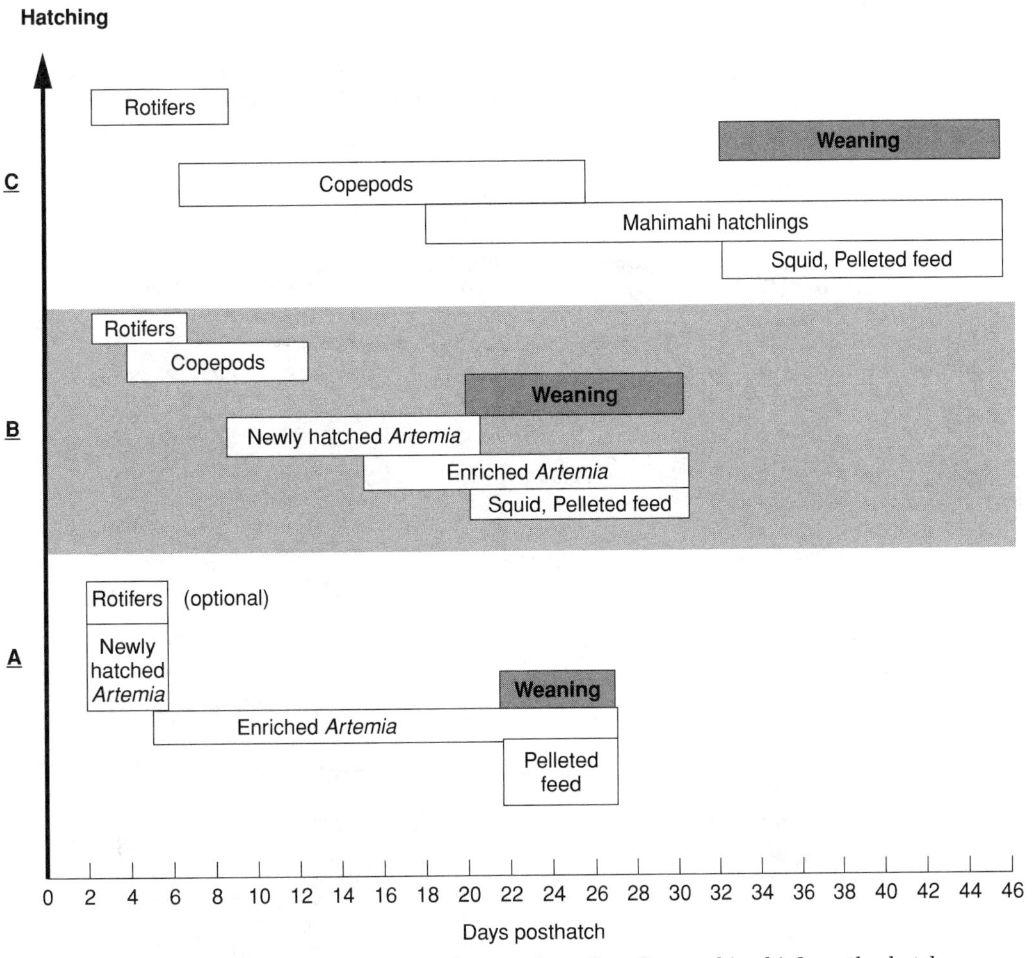

Figure 3. Various feeding schemes used to produce fingerling mahimahi from the hatchery. Adapted from Kim et al. (2) and the Oceanic Institute (9); schemes B and C are adapted from Kraul (10) and Szyper (11). Weaning denotes the period of transition from live or raw food to pelleted feeds.

A break from the normal grouping pattern of sated larvae indicates when a pulse of *Artemia* should be applied (9). It is generally believed that mahimahi have very high DHA requirements, and enrichments containing at least 30% DHA are recommended for use (2,9). By the beginning of the third week (ca. day 15), larvae are nearing completion of metamorphosis (postflexion stage), and emphasis should be placed on ensuring that larvae consume sufficient amounts of live feed, to support rapid growth. *Artemia* alone (2), or combinations of *Artemia*, newly hatched mahimahi or other fish larvae, and copepods are used (5,10,11,16,18) as feed.

NURSERY

The period between 20 and 45 days of age is often marked by unacceptably high mortality rates, often nearing 80% of all fish stocked (2,5,9,16,21,22,30). During that period, fish are trained to accept dry, pelleted feeds, and natural agonistic behaviors peak. Juvenile mahimahi frequently chase and nip siblings during daylight hours. Cannibalism typically occurs when there is a sufficient size disparity among siblings, although individuals who are attacked and killed are not always eaten. Fish that are too small or weak are especially vulnerable, although aggressive behavior can be exhibited by any individual, regardless of size. It is not uncommon for fish that cannabalize to die from attempting to consume too large a sibling that then becomes lodged in the fish's mouth.

Research has concentrated on ways to improve survival in the nursery stage through manipulating system design, feed type, feeding rate, and stocking density, but the issue has yet to be fully resolved. Brownell (31) has indicated that, although tail nipping was immediately halted by the introduction of food to fish that had been deprived for more than one hour, nipping occurred to prefeeding levels within 15 minutes, whether food remained or not. No effect on nipping frequency was observed when refuges or floating objects were introduced into tanks, or when the fish density increased. Kim et al. (32) obtained best growth and survival rates of fish at an initial stocking density of 0.5 fish/L (1.9 fish/gal); however, growth and survival rates were better at 3.0 fish/L (11.3 fish/gal) than at other intermediate densities [i.e., 1.0, 1.5, 2.0 fish/L (3.8, 5.7, 7.6 fish/gal, respectively)] tested. Crowding appears to be beneficial in a commercial setting, because aggressive individuals have difficulty targeting and chasing victims.

Shallow-water tank designs coupled with continuous feeding regimens have proven successful in improving nursery survival rates to 60% for mahimahi (2) and almost eliminating cannibalism in other marine carnivores (33) (Fig. 4). Shallow water with a strong, directional current facilitates the transition of fish from live to pelleted feeds by making pelleted feeds more attractive and available to the fish: Shallow water increases the contact between

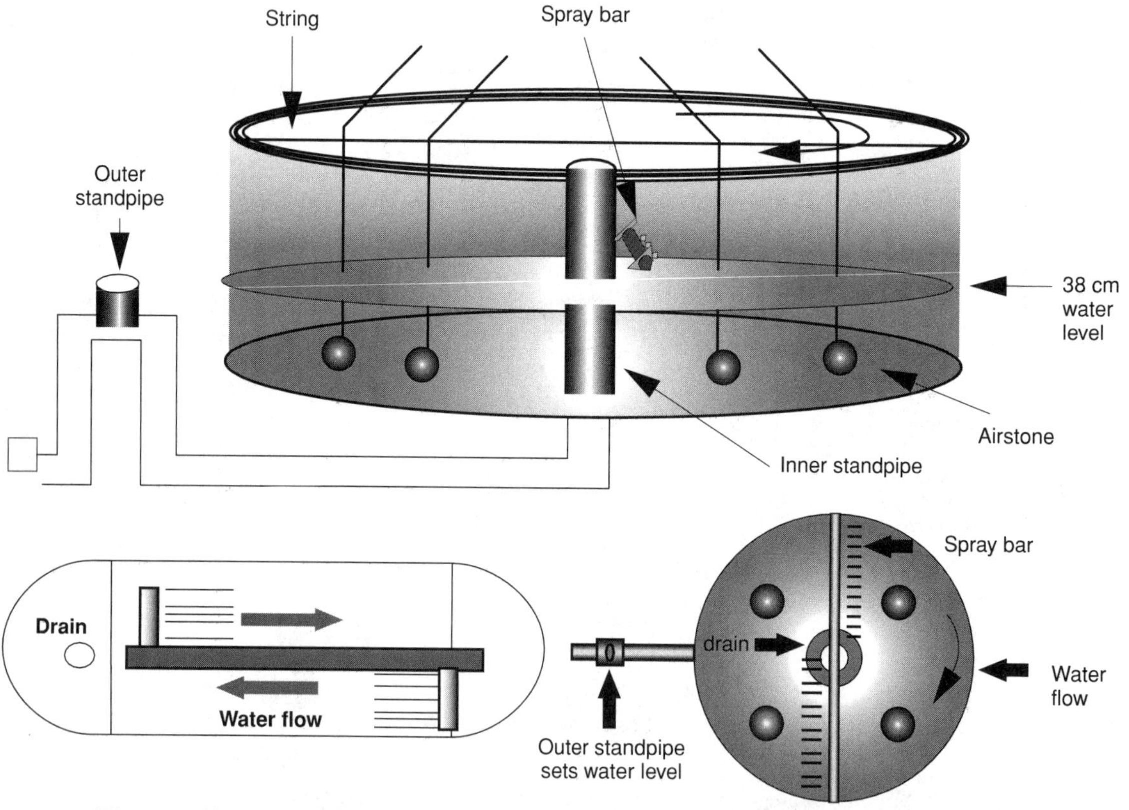

Figure 4. Shallow-water tank designs used to wean juvenile mahimahi to pelleted feeds in the nursery (20–40 days of age). Oval and circular designs are shown in the overhead view. Details of the circular design is shown in the side view. From the Oceanic Institute (9).

the feed particles and the fish, and strong water current animates the feed particles. The current also forces fish to occupy their time by swimming against the directional flow, which makes them less prone to attack their siblings, and provides a means by which an attacked individual can quickly escape an aggressor. Other methods, such as using protracted weaning periods and combinations of several live feed items, including yolk sac larvae, yield survival rates of up to 50% (5,10,11,18). Overall survival from stocked larvae through the nursery phase typically ranges from 2.5 to 10%.

GROWOUT

Successful growout is the most critical component of profitability in a mahimahi farm. During this phase, the cost and value of an individual fish increases ten- to twentyfold, depending upon the size of the fish at harvest. Thus, each fish represents a substantial investment in feed, labor, capital, equipment, and utilities.

Mahimahi are the most rapidly growing species yet that has been considered for intensive aquaculture development (34,35). They reach a harvest weight of about 1.7–2.0 kg (3.7–4.4 lb) at 180 days of age (12,21,22,30,35), yielding an instantaneous increase in the rate of weight gain of 4.0–4.2% body weight daily from an initial 5 grams (0.2 oz) at stocking at 40–45 days of age. Successful growouts to 4.5 kg (9.9 lb) have been achieved at eight months (240 days) of age (21), and growouts to 5.4 kg (11.9 lb) have been found at 8.7 months (261 days) of age (13). Ostrowski (30) has reported good growth of males to 4.0 kg (8.8 lb) to 240 days of age, but the average growth of females and the mixed population was much slower. On average, males are typically 15–35% larger than females by 180 days of age (9,30). The slower growth of females may be due to the energy required in gonadal development prior to and during maturation (120 to 180 days of age): At 150 days of age, ovaries makeup 6.0% of the total body weight of females, while testes make up only 0.5% of the total weight of males (30).

Mahimahi have been raised at commercial densities in sea cages and onshore in both flow-through and recirculating tank systems. Nel (21) raised 2-kg (4.4-lb) fish in onshore flow-through tanks 4.0–8.0 m (13.1–26.2 ft) in diameter to a harvest density of 25–30 kg/m^3 (1.6–1.9 lb/ft^3). Kraul and Ako (13) used doughnut-shaped tanks to raise fish to 10 kg/m^3 (0.6 lb/ft^3). Ostrowski (30) achieved densities as high as 20 kg/m^3 (1.2 lb/ft^3) at loading rates of 1.0 kg/L/min (8.3 lb/gal/min) in trials conducted in circular tanks 7.3 m (24.0 ft) in diameter. Hagood (22) raised fish to 12–15 kg/m^3 (0.7–0.9 lb/ft^3) to final harvest in circular tanks (both flow through and recirculating) 5.4 m (17.7 ft) in diameter. The Oceanic Institute (9) achieved densities of 19 kg/m^3 (1.2 lb/ft^3) and raw-water loading rates of 4.2 kg/L/min (34.9 lb/gal/min) with near-85% water recirculation in a semirecirculation system.

Several growout attempts in circular, salmon net pens [ca. 8 m (26.2 ft) in diameter, 4.0 m (13.1 ft) deep] have been conducted in Australia (21), the Bahamas (4,22), and Tahiti (37), and the use of these pens for growout of mahimahi appears promising. However, problems that have been encountered in these areas include parasitic infections (22,37), and predators (22), and low seasonal water temperatures [less than 21 °C (70 °F)] that limit growth (4,21).

Practical, pelleted diets have been successfully used in mahimahi growout, yielding very low feed conversion ratios (FCRs), or the amount of dry-matter feed consumed to the wet body weight gain of the fish. FCRs increase with fish age (Table 2), but overall, are less than 1.7 for growout to all sizes. Nel (21) has obtained an FCR of 1.2 from two to eight months of age, yielding a final mean weight at harvest of 4.5 kg (9.9 lb). Kraul and Ako (13) have reported an FCR of 1.6 for growout of 8.7-month-old fish to 5.4 kg (11.9 lb).

Feed type is of utmost importance in mahimahi culture, and because of stringent requirements, there are few commercial formulations presently available that promote good feeding and growth rates for this species. Pilot farms have used commercial diets, in-house formulations, or a combination of both (21,22). The most important factor in a mahimahi diet is the quality of the fishmeal (38). Only those diets made with high-quality, low-temperature-processed fishmeal are sufficiently palatable to mahimahi to promote high feeding rates, good feed utilization, and rapid growth. Chemical quality indices for choice of fishmeal for mahimahi culture are defined in (38).

Research has shown that mahimahi grow most rapidly and efficiently on diets that are high in protein and moderate in lipids and carbohydrates. Practical diets containing between 55 and 60% protein (dry-matter basis), 10 and 14% lipids, and less than 12% carbohydrate balanced at 32–35 mg of protein/MJ (132–147 mg of protein/kcal) metabolizable energy (values for rainbow trout) are recommended for rapidly growing juveniles (39). In practice, mahimahi have been successfully raised with diets containing between 53 and 60% protein and 12 and 22% lipids (5,22,40,41). Primary protein sources are either fishmeal or other marine animals products, although mahimahi can use some level of plant-based proteins as well. Soybean meal can be used effectively at up to 20% in diets (40–43). Soybean or corn oil can also be used,

Table 2. Composite of Estimated Feed Conversion Ratios[a] for Periods of Mahimahi Growth at ca. 26–28 °C (79–82 °F)[b]

Age (days)	Change in Weight (grams–grams)	Specific Growth Rate (% body weight/day)	Feed Conversion Ratio (dry feed fed/gain in wet body weight)
40–60	5–40	10.4	0.7
60–90	40–250	6.1	0.7–0.8
90–120	250–700	3.4	0.8–1.0
120–150	700–1250	1.9	1.0–1.4
150–180	1250–1800	1.2	1.5–2.1

[a] FCR = dry feed fed/wet body weight gain.

[b] The changes in weights have been rounded to the nearest whole number and are based on typical numbers reported. Specific growth rates {SGR = [(final weight/initial weight)/(days in period)] × 100} are calculated as a percentage of the body weight gain per day from the beginning of each period. The overall FCR from 40 to 180 days of age ranges from 1.0 to 1.2.

but the diet should contain between 1.0 and 1.3% HUFAs (30,44). The amino acid requirements of mahimahi have been estimated from body tissue analyses (45). Mahimahi are unable to synthesize taurine (46).

Mortality in growout is a key issue that must be overcome for commercial success of mahimahi culture. Reported growout survival rates to 180 days of age range from 62 to 80% (a mode of 70%), depending upon what age the fish were stocked at, the density of the stocked population, and the system used (13,21,22,30). Major causes of mortality have been identified (3,30) and are usually associated with a particular period or stage of development. Mortality is generally higher during the early stages of growout. Chronologically, the first potential problem encountered in mahimahi growout is caudal peduncle (47), or red-tail disease, which is most prevalent in nursery fish and 40- to 90-day-old fish in growout (30). Red-tail disease occurs after opportunistic *Vibrio* sp. (48) and *Myxobacteria* sp. (47) bacterial infections of wounds caused by tail-nipping. Bloating, or a gross enlargement of the stomach, is caused by an accumulation of fluid in the stomach one to two hours after feeding. It occurs in fish fed pelleted diets and most often in fish between 40 and 90 days of age (9,49). The exact cause of bloating is not known, and the percentage of affected fish varies, but the extent of the condition is accentuated under conditions of stress—e.g., low dissolved-oxygen levels, excessive handling (9), and poor diet quality (5,9).

Mortality due to physical trauma and shock associated with the impact of fish ramming head first into the side walls of tanks is especially prevalent during the latter stages of growout (after 120 days of age), when fish reach over 500 g (1.1 lb). Hemorrhages within the brain cavity have been observed in fish examined soon after death by such an incident. Foam padding installed in tanks 6.1 m (20 ft) in diameter reduced collision mortality from 34% to 13% of the total mortality observed in growout (9,30). Doughnut-shaped tanks proved very useful in nearly eliminating all collision mortality (13,37).

MARKET AND ECONOMICS OF PRODUCTION

Commercial-scale production of mahimahi is technically feasible, but profitability relies largely on mastering commercial-scale production methods and optimizing the market price received for the product. Wild-caught mahimahi support major commercial and sport fisheries within their range and are found on the menus of many upscale restaurants worldwide. Cultured mahimahi is considered a quality product, but there is reluctance to pay higher prices for the cultured product when wild fish are normally obtained cheaply through commodity channels (50). Additional problems encountered include the fact that a product smaller than 4.5–6.8 kg (10–15 lb) is generally considered inferior, largely due to processing concerns. Production of larger [4.5-kg (10-lb)] fish, establishment of niche markets ("pan-sized" mahimahi), and increased awareness of the advantages of the cultured product [e.g., consistent supply; quality control; and a 50–60% fillet yield (4,12,21), as compared with 40–45% in wild fish] need to be pursued (50). Market studies in Australia in 1995 (51) indicate that it is unlikely that the price for cultured fish will fall below A$8.00/kg (A$3.64/lb), yielding an internal rate of return of 26%, and that a price of about A$10.00/kg (A$4.55/lb) is more likely. Several attempts to establish farms in Hawaii (22,37), Australia (21), Florida, the Bahamas (4,22), Tahiti (37), and Greece (5) did not advance to fruition, because various technical, financial, and market assumptions were not met. Currently, one mahimahi pilot farm is in operation in Tahiti (37), and the potential for production is still being pursued in Australia (21) and Florida (22).

ACKNOWLEDGMENTS

The author's mahimahi culture technology development research was supported by National Oceanic and Atmospheric Administration grants NA-90-OAA-D-FM-238, NA16FV0505-01, NA36FV0068, and NAFV0082. This contribution was funded in part by a grant and cooperative agreement from the National Oceanic and Atmospheric Administration. The views expressed within this contribution are those of the author and do not necessarily reflect the views of NOAA or any of its subagencies. Unpublished reports by and personal communications with Randy Hagood of American Aqua Resources, Syd Kraul of Pacific Planktonics, Steve Nel of Mariculture Development PTY, Ltd., Dennis Peters of Science Applications International, and John and Elizabeth Sweetman of Ecomarine, Ltd. were valuable contributions to this work. These people shared their commercial experiences and data on mahimahi culture that would otherwise be unavailable.

BIBLIOGRAPHY

1. B.J. Palko, G.L. Beardsley, and W.J. Richards, *FAO Fisheries Synopsis No. 130*, NOAA Technical Report, NMFS Circular 443, 1982.
2. B.G. Kim, A.C. Ostrowski, and C. Brownell, in C.-S. Lee, M.S. Su, and I.C. Liao, eds., *Finfish Hatchery in Asia '91: TML Conference Proceedings 3*, Tungkang Marine Laboratory, Taiwan Fisheries Research Institute, Tungkang, Pingtung, Taiwan, 1993, pp. 179–190.
3. E. Schaleger and A.C. Ostrowski, in *Book of Abstracts, 20th Annual Meeting of the World Aquaculture Society (San Diego, CA)*, World Aquaculture Society, Baton Rouge, LA, 1995, p. 208.
4. D. Peters, Science Applications International Corporation, Ft. Walton Beach, FL, personal communication.
5. J. Sweetman and E. Sweetman, EcoMarine, Ltd., Cephalonia, Greece, personal communication.
6. M.M. Lutnesky and J.P. Szyper, *Prog. Fish-Cult.* **52**, 178–185 (1990).
7. D.D. Benetti, *Bioenergetics and Growth of Dolphin*, Coryphaena hippurus, Ph.D. Dissertation, University of Miami, Coral Gables, FL, 1992.
8. D.D. Benetti, R.W. Brill, and S.A. Kraul Jr., *J. Fish Biol.* **46**(6), 987–996 (1995).
9. The Oceanic Institute, in "Final Report: Development of Mahimahi (Coryphaena hippurus)," *Aquaculture Technology*

1990–1994, Section II National Oceanic and Atmospheric Administration, La Jolla, CA, 1996, pp. 1–106.

10. S. Kraul, in J.P. McVey, ed., *CRC Handbook of Mariculture*, Vol. 2, Finfish Aquaculture, Boca Raton, FL, 1991, pp. 241–250.
11. J. Szyper, in J.P. McVey, ed., *CRC Handbook of Mariculture*, Vol. 2, Finfish Aquaculture, Boca Raton, FL, 1991, pp. 228–240.
12. S. Nel, *InfoFish Internat.* **3**/90, 32–34 (1990).
13. S. Kraul and H. Ako, in *Book of Abstracts, 24th Annual Meeting of the World Aquaculture Society (Torremolinos, Spain): From Discovery to Commercialization, European Aquaculture Society Special Publication No. 19*, European Aquaculture Society, Oostende, Belgium, 1993, p. 403.
14. T. Iwai Jr., H. Ako, and L.E. Yasukochi, *World Aquacult.* **23**(3), 49–50 (1992).
15. S. Kraul, A. Nelson, K. Brittan, H. Ako, and A. Ogasawara, *J. World Aquacult. Soc.* **23**(4), 299–305 (1992).
16. S. Kraul, A. Nelson, K. Brittan, and D. Wenzel, *Sea Grant Quart.* **10**(3), 1–6 (1988).
17. C.-S. Lee and L. Krishnan, *J. World Maricult. Soc.* **16**, 95–100 (1985).
18. S. Kraul, *J. World Aquacult. Soc.* **24**(3), 410–421 (1993).
19. A.C. Ostrowski, L. Rao, and D. Benetti, in *Book of Abstracts, 21st Annual Meeting of the World Aquaculture Society (Halifax, Nova Scotia, Canada)*, World Aquaculture Society, Baton Rouge, LA, 1990.
20. S. Kraul, K. Brittan, R. Cantrell, T. Nagao, H. Ako, A. Ogasawara, and H. Kitagawa, *J. World Aquacult. Soc.* **24**(2), 186–193 (1993).
21. S. Nel, Mariculture Development Pty., Ltd., Yanchep, Western Australia, personal communication.
22. R.W. Hagood, American Aqua Resources, Ft. Pierce, FL, personal communication.
23. A.C. Ostrowski and S. Divakaran, *Mar. Biol.* **109**, 149–155 (1991).
24. A.C. Ostrowski and S. Divakaran, *Aquaculture* **89**, 112–115 (1990).
25. A.C. Ostrowski, *Prog. Fish-Cult.* **51**, 161–163 (1989).
26. A.C. Ostrowski, in *Book of Abstracts, 26th Annual Meeting of the World Aquaculture Society (San Diego, CA)*, World Aquaculture Society, Baton Rougen, LA, 1995, p. 184.
27. S. Kraul, H. Ako, K. Brittan, A. Ogasawara, R. Cantrell, T. Nagao, P. Lavens, P. Sorgeloos, E. Jaspers, and F. Olleveir, *Larvi '91: Special Pub. Euro. Aquacult. Soc.* **15**, 45–47 (1991).
28. M.V. Bell, R.J. Henderson, and J.R. Sargent, *Comp. Biochem. Physiol. B* **83**(4), 711–719 (1986).
29. C. Brownell and A.C. Ostrowski, in *Book of Abstracts, 20th Annual Meeting of the World Aquaculture Society (Los Angeles, CA)*, World Aquaculture Society, Baton Rouge, LA, 1989, p. 64.
30. A.C. Ostrowski, in K.L. Main and C. Rosenfeld, eds., *Culture of High-Value Marine Fishes in Asia and the United States: Proceedings of a Workshop (Honolulu, HI)*, The Oceanic Institute, Honolulu, HI, 1995, pp. 153–166.
31. C.L. Brownell, in J.H.S. Blaxter, J.C. Gamble, and H. von Westernhagen, eds., *The Early Life History of Fish*, Vol. 191, The Third ICES Symposium (Bergen, Norway), 1989, p. 485.
32. B.G. Kim, A.C. Ostrowski, and C. Brownell, in *Book of Abstracts, 22nd Annual Meeting of the World Aquaculture Society (San Juan, Puerto Rico)*, World Aquaculture Society, Baton Rouge, LA, 1991, p. 36.
33. A.C. Ostrowski, T. Iwai, S. Monahan, S. Unger, D. Dagdagan, P. Murakawa, A. Schivell, and C. Pigao, *Aquaculture* **139**, 19–29 (1996).
34. S. Kraul, in *Proceedings of the Second International Conference on Warmwater Aquaculture: Finfish, Laie, HI*, 1985, pp. 99–108.
35. D.D. Benetti, E.S. Iversen, and A.C. Ostrowski, *Fish. Bull.* **93**, 152–157 (1995).
36. H. Ako, S. Kraul, L. Fujikawa, K. Brittan, and M.C. Holland, in *Book of Abstracts, 24th Annual Meeting of the World Aquaculture Society (Torremolinos, Spain): From Discovery to Commercialization, European Aquaculture Society Special Publication No. 19*, European Aquaculture Society, Oostende, Belgium, 1993, p. 306.
37. S. Kraul, Pacific Planktonics, personal communication.
38. A.C. Ostrowski, S. Divakaran, B. Kim, and E.O. Duerr, *J. Appl. Aquacult.* **6**(2), 39–56 (1996).
39. A.C. Ostrowski, B. Kim, and E.O. Duerr, in *Book of Abstracts, 23rd Annual Meeting of the World Aquaculture Society (Orlando, FL)*, World Aquaculture Society, Boca Raton, FL, 1992, p. 176.
40. A.C. Ostrowski, E. Schaleger, and E.O. Duerr, in *Book of Abstracts, 23rd Annual Meeting of the World Aquaculture Society (Orlando, FL)*, World Aquaculture Society, Boca Raton, FL, 1992, p. 177.
41. A.C. Ostrowski, E. Schaleger, and E.O. Duerr, in *Book of Abstracts, 24th Annual Meeting of the World Aquaculture Society (Torremolinos, Spain): From Discovery to Commercialization, European Aquaculture Society Special Publication No. 19*, European Aquaculture Society, Oostende, Belgium, 1993, p. 426.
42. E. Suzuki, A.C. Ostrowski, and S. Divakaran, in *Book of Abstracts, 21st Annual Meeting of the World Aquaculture Society (Halifax, Nova Scotia, Canada)*, World Aquaculture Society, Baton Rouge, LA, 1990.
43. E. Suzuki, *Protein Accretion and Bioenergetics in Dolphin Fish* (Coryphaena hippurus), M.S. Thesis, University of Hawaii at Manoa, Honolulu, HI, 1992.
44. A.C. Ostrowski and B.G. Kim, in *Book of Abstracts, 24th Annual Meeting of the World Aquaculture Society (Torremolinos, Spain), From Discovery to Commercialization, European Aquaculture Society Special Publication No. 19*, European Aquaculture Society, Oostende, Belgium, 1993, p. 424.
45. A.C. Ostrowski and S. Divakaran, *Aquaculture* **80**, 285–299 (1989).
46. S. Divakaran, S. Ramanathan, and A.C. Ostrowski, *Compar. Biochem. Physiol.* **101B**(3), 321–322 (1992).
47. B. LeaMaster, B.G. Kim, and A.C. Ostrowski, in *Book of Abstracts, 24th Annual Meeting of the World Aquaculture Society (Torremolinos, Spain), From Discovery to Commercialization, European Aquaculture Society Special Publication No. 19*, European Aquaculture Society, Oostende, Belgium, 1993, p. 404.
48. B. LeaMaster and A.C. Ostrowski, *Prog. Fish-Cult.* **50**, 251–254 (1988).
49. A.C. Ostrowski, C. Brownell, and E.O. Duerr, *World Aquacult.* **20**(4), 104–105 (1989).
50. The Oceanic Institute, *Hawaiian Mahimahi: Current Market Analysis and Marketing Strategies for Increased Production*, The Oceanic Institute, Honolulu, HI, 1988.
51. S. Nel, *Austasia Aquacult.* **9**(6), 51–53 (1996).

See also NET PEN CULTURE.

DRUGS

Robert D. Armstrong
Schering-Plough Animal Health
Forestville, California

OUTLINE

This entry provides a general introduction to the use of drugs for achieving optimum aquatic animal health, including a discussion of drug approval regulations related to fish treatment. The intent is to provide essential background information that will help the managers of modern intensive aquaculture production facilities to understand the complex drug approval process and to participate effectively in the decision to treat their fish. As the author is most familiar with the North American situation, readers in other areas may wish to review requirements for their particular jurisdiction.

Various chemicals have been used as drugs to treat fish since the first person reared fish in a captive environment, and trial and error over many years has produced a cupboard full of these traditional chemical treatment options. Examples of such traditional treatments are malachite green to control fungal infections and potassium permanganate for external parasites. For a very good detailed discussion of these chemicals and their use, the reader is referred to Herwig (1). That comprehensive reference summarizes an exhaustive amount of material on treatments, dosages, and indications for a broad selection of chemicals.

Surprisingly, few of the traditional chemical treatments are approved for use in the treatment of aquatic animals today. The reasons for this situation are frequently more a consequence of the regulatory approach for drugs than the nature of the chemicals themselves. However, some traditional chemical treatments can be toxic and have untoward side effects. Caution is required when considering any chemical treatment of foodfish.

Selection of a drug for treatment requires first that the producer obtain a correct and complete diagnosis of the problem. Accurate diagnosis is usually the result of close cooperation between the producer and a professional consultant with fish health experience. Specific drugs and treatment dosages in use in contemporary aquaculture are not listed here, because the most straightforward and safest method for determining the dosage of drugs is to use only approved drugs and to follow precisely directions on the drug's label. The list of approved drugs also differs in each country, and a single list provided here would not be correct for all readers. The consequences of unapproved drug use can be devastating for an intensive food-producing aquaculture business. For specific information on drugs approved for use in aquaculture in the United States, the reader is referred to the *Guide to Drug, Vaccine and Pesticide Use in Aquaculture* (3). An excellent summary of drugs used internationally is also available (2).

TYPES OF DRUGS

A drug can be broadly defined as any chemical that modifies a physiological process in an animal. Regulatory agencies adopt this type of broad definition in deciding whether a compound is a drug and therefore whether it requires regulatory approval before it can be manufactured and sold for use. Food and Drug Administration (FDA) officials in the United States determined, after their opinion was sought, that some surprisingly common substances, such as ice and salt, are drugs. These common substances have since been classified by the FDA as of "low regulatory priority," as their use is not going to generate concern; a list of these compounds is readily available (3,5).

For the purposes of this article, a drug is considered to be a chemical that has, in addition to the aforementioned physiological effects, enough potential to cause harm such that its manufacture, sale, and use are controlled through regulatory restrictions. Pesticides are a very similar class of chemical, with the subtle distinction that they are generally applied externally to animals or the animals' environment with the intent to kill a specific targeted pest. However, this distinction between drugs and pesticides is not sufficient to merit independent consideration in this article, and the term "therapeutant" is occasionally used to include both drugs and pesticides that are used for animal treatment. A drug may be applied prophylactically, to prevent a disease outbreak, or therapeutically, to control an outbreak already underway or to produce a physiological change (e.g., as anesthesia or to induce spawning). Vaccines, also called biologics, consist of immunologically active material derived from modified remnants of disease-causing organisms suspended in a liquid carrier. Vaccines are almost always applied prophylactically, to prevent outbreak of a disease by "jumpstarting" the fish's immune system.

Therapeutants can be classified according to their different biological properties, including antibacterial, antiparasitic, antiviral, antifungal, anesthetic, and hormonal activities. Therapeutants from each of these groups are used in terrestrial agriculture and are finding similar opportunities for use in aquaculture (except for antiviral drugs, which have great potential, but are not used, because the few products available for animal use are cost prohibitive). Also, there are many other varieties of drugs used in human and companion-animal medicine that have not found a place in the treatment of food animals, for health and economic reasons.

Antibacterial drugs, often called antibiotics, are perhaps the best known group of drugs in use in agriculture and aquaculture and are discussed by another author in this text. Readers may also consult a recent scientific review of the use of antiinfective drugs in aquaculture (4).

Antifungal drugs are also very important in aquaculture, because of the rapid invasion of *Saprolegnia* fungi into the skin and eggs of freshwater fish. A traditional, and very effective, antifungal treatment has been topical exposure to malachite green (sometimes mixed with formalin) in a bath (1). However, malachite green has some drawbacks that make it unlikely ever to receive regulatory approval. First, it is not a pure substance, as it represents a mix of related dyes, and it has teratogenic potential (that is, it may cause defects in developing embryos). Additionally, malachite green residues are very persistent in the tissues of treated fish. Therefore, use of malachite green is now discouraged, even in some countries where it had previously been accepted. However, there are a few alternative candidates under development for fungal control as bath treatments, including hydrogen peroxide and formalin. In aquaculture, antifungal drugs are not presently used in feed, although systemic fungal infections occur sporadically.

Antiparasitic drugs are an increasingly important treatment category for fish producers. Parasitic infestations that occasionally need to be managed through chemotherapy include ectoparasitic (on the outside of the fish) crustaceans (e.g., sea lice), ectoparasitic protozoa (e.g., *Trichodina*), intestinal helminths (e.g., *Eubothrium*), and intestinal or systemic protozoa (e.g., *Hexamita*). Antiparasitic treatments may be administered topically (e.g., formalin baths for control of *Trichodina*) or as in-feed preparations (e.g., fumagillin for treatment of sporozoan infections). Organophosphates and pyrethroids are used as baths to control sea lice infections, while insect growth regulators (chitin inhibitors) and avermectins have been used systemically. Systemic anthelmintic and antiprotozoal drugs are likely to become increasingly important for future fish culture industries, just as helminth and coccidial controls are critical components of terrestrial agriculture health management.

Anesthetics and sedatives have important roles in aquaculture production. Fish anesthesia and sedation are essential for reducing the stress of management procedures, such as handling, grading, weighing, vaccinating, and transporting fish. Formulations of tricaine methane sulfonate are approved in many countries and are frequently used as fish anesthesia. A new development is the use of sedatives that do not require a withdrawal period because they do not result in unacceptable tissue residues. One example of this kind of drug is Aqui-S, licensed in New Zealand for sedating salmon. This type of drug can be used for sedating aquatic animals before they are transported to the processor, resulting in better flesh quality and shelf life.

Hormonal treatments alter the physiological state of the fish by causing an increase or decrease in levels of the chemicals that regulate normal bodily functions. These treatments can increase growth rates, induce a fish to initiate spawning, or force a young genetically female fish to develop into a male adult. Generally, hormonal treatments must be used at the right time to be effective, so that they work in conjunction with natural developments in the fish. For example, administration of additional growth hormone will not have the desirable effect on a fish that has reached its full body size and is not growing. Hormonal treatments have found uses in aquaculture in controlling the timing of spawning in intensive production facilities and assisting in the production of single-sex populations, which are essential for optimum fish production in some situations. Use of hormones for growth promotion is not a standard practice in aquaculture.

Drugs can also be classified according to their route of administration. Some are administered to aquatic animals topically, by application to the skin surface, while others are given systemically (that is, for distribution throughout the body). However, these distinctions are not as clear-cut for aquatic animals as for terrestrial animals. This is because a topical treatment for fish is applied as a bath, with the result that some of the drug will be absorbed by the gill tissue, and some of the drug could also be swallowed. The most practical method for systemic treatment of intensively cultured aquatic species is through addition of medication to the feed. This technique is known as oral administration and in most systems will also result in some drug exposure to the gills and skin. Systemic treatment by injection is rarely used in aquaculture, because of the labor cost, and is a procedure that is generally reserved for an all-out attempt to save particularly valuable animals, such as ornamental koi carp. However, injectable vaccines have become important in the last few years, because of the effectiveness of these products in controlling gram-negative bacterial infections of salmon — particularly furunculosis and vibriosis.

Pharmacokinetics is the study of what happens to a drug after it is applied to an animal. This discipline is still at an early stage for aquaculture species, and most of the available information was developed in the last 10 years. However, the general stages of a drug's journey through an aquatic animal are analogous to the stages described for all other animals and include absorption, distribution, metabolism, and excretion phases.

Drug entry into a fish begins with absorption through the skin, gills, intestine, or other surface tissue. In some cases, the drug's chemical may be changed before absorption — for example, through the action of intestinal enzymes and stomach acids — and these changes may have significant impacts on its biological activity. Following absorption, the drug is distributed throughout the body by blood flow, and most drugs will then selectively concentrate in particular tissues. Knowledge of this selective concentration is very important in choosing an appropriate drug treatment. An ideal treatment (sometimes termed the "silver bullet") will target the tissue affected by a particular disease problem, while building up minimal residue in the edible parts of the fish, particularly the muscle. Once the drug is absorbed into the body, the fish begins to metabolize it by breaking it down into components or binding it to other molecules. These changes convert the drug into an inactive, or perhaps more toxic, "metabolite." Finally, the fish eliminates the drug by excreting it and its metabolites. Elimination may occur through a variety of routes, including excretion in the feces or urine or across the skin or gills.

The pharmacokinetic behavior of drugs is usually extremely complex, with an unlimited possible range of different behaviors. The limited current knowledge of aquatic-animal pharmacokinetics is a factor that contributes to the uncertainty of regulatory agencies and increases the cost of developing new therapeutants for aquaculture. One example of such limitations is the lack of information on how well-detailed pharmacokinetic data for one aquatic species can be used to predict drug behavior in another species. For example, will the drug-residue depletion time measured in a salmon be the same in a halibut a catfish, or an oyster? The answer to this question has important implications for determining the amount of data required to develop a list of approved species for inclusion on the drug's label. Answers to such questions are being sought through significant "crop grouping" research projects currently underway and should help to reduce the cost of drug approvals for aquatic animals.

Another pharmacokinetic complexity is that aquatic animals are poikilotherms (cold blooded), and as such, their metabolic rate and activities are affected by changing water temperature. This complexity is not found in the much better understood mammalian and avian (homeotherms) physiological systems, which remain within a nearly constant temperature range. Decreasing the water temperature generally slows the metabolic actions of aquatic animals, causing slower drug absorption, metabolism, and excretion. One important consequence of this effect is the possibility of longer residue retention in aquatic animal tissues following treatment. Also, fish eat less when the temperature cools below their optimum level, leading to lower dietary intake of a drug in feed. Lower intake, combined with slower drug absorption because of a reduced metabolic rate, may lead to both reduced drug effectiveness and lower tissue residue levels. These factors require that the metabolic state of the fish and the predicted changes in the environment of the fish over the weeks following treatment be considered carefully when planning the treatment. Scientists have begun to develop physiologically based pharmacokinetic models that map out drug behavior under changing environmental conditions, and these models offer the promise of providing much more specific treatment plans and withdrawal periods for medicating aquatic animals, based on the specific circumstances anticipated for each treatment.

Increasing the dose of a drug does not lead to a proportionally greater change in the effect on the aquatic animal being treated. Biological systems have a very limited number of response options, and there tends to be a limited range of possible changes within these options. Negative responses, such as lost growth, poor reproductive performance, and increased mortality, will result if biological systems are pushed beyond their capabilities. A graph of the changing effects on a fish exposed to different drug doses is called a dose response curve. The one consistent feature of dose response curves is the indication that "if some is good, more is not necessarily better." The optimal treatment effect is achieved by carefully following the recommended drug dosages, which have been calculated based on scientific evaluation of the dose response curve.

The drug that a producer buys is rarely a pure substance. Most drugs are part of a formulation that includes the active ingredient together with other compounds that act as carriers, dispersants, stabilizers, etc. These other compounds are called excipients, and they may have an impact on how the drug behaves in treated fish. Regulatory agencies approve the complete drug formulation, rather than just the active ingredient, thereby taking into account the total effect that the treatment will have. The formulation will have a trade name that is different from the chemical name of the active ingredient, which is the pure drug substance. For example, the fish anesthetic drug tricaine methane sulfonate may be bought in two different formulations that have government approval in the United States: TMS and Finquel.

The drug dosage is often expressed on the label in terms of the amount of active ingredient to be applied per unit of body weight (e.g., milligrams active ingredient per kilogram fish weight). This dosage would be very different if it were expressed in terms of the amount of formulation per unit of body weight. Therefore, producers need to know the concentration of active ingredient in a drug formulation when calculating how much substance to add to the water or feed (the inclusion rate). A second key number when calculating in-feed medication dosages is the feeding rate, which is usually expressed as a percentage of body weight (e.g., 2% of body weight per day). This number indicates how much feed the animals will consume and therefore the weight of drug that must be included in the feed such that each fish will receive the correct dosage. These calculations always need to be rechecked before mixing feeds or administering treatments, and, ideally, a second knowledgeable, independent person should review the figures.

REGULATORY CONCERNS

Consumer concerns regarding drug safety and the potential for chemical contamination of foods are the driving force for substantial regulatory oversight of therapeutants worldwide (although there is no single regulatory system in use globally). These consumer concerns are perhaps greater than warranted, considering how remarkably little evidence there is that treatment of food animals has caused harm; instead, treatments help to provide abundant high-quality produce. Perhaps the success of treatments is a result of the stringent regulatory controls, which allow producers to ensure that their food products are safe by scrupulously complying with regulatory requirements regarding therapeutant use. Unfortunately, these regulatory approval requirements significantly increase the cost of approved therapeutants; however, confidence in the safety of approved therapeutants is a benefit to the aquaculture producer. Approved drugs have been shown through rigorous scientific examination to be safe and effective. The label on the container of a therapeutant lists appropriate cautions to follow in order to protect the workers administering the drug and to ensure the quality of aquatic-animal products subsequently shipped to consumers. The prudent producer will reread the label of every therapeutant before every use and then be sure to follow all of the precautions and heed all of the warnings listed on the label.

One problem for aquaculture producers around the world is the lack of approved safe and effective drugs for treatment of aquatic animals. This problem is the result of two conflicting forces: the high regulatory approval costs and the low potential sales values within different countries. The company that sponsors a therapeutant for approval is required to pay for development of the complete scientific data package needed to obtain approval, as well as to cover the costs of subsequent regulatory requirements associated with keeping the therapeutant on the market. As a result, it simply is not cost effective for some companies to market their therapeutants. For example, one formulation of the active ingredient sulfamerazine is approved for use in aquaculture in the United States; however, the manufacturer reportedly no longer makes this product available, because the market is insufficient to cover the costs of meeting ongoing regulatory requirements. If the estimated potential sales are insufficient to justify the research investment required for approval, then drug sponsors will not develop new treatments for aquaculture.

Regulatory agencies recognize the economic pressures that prevent drug approvals for aquatic animals. Similar problems are seen in other animal-production operations, and these disadvantaged groups are often grouped together and termed “minor species.” Drug manufacturers do not develop new treatments for minor species, with the result that drugs approved for use in aquatic animals are almost always formulations of different chemicals that have already received approval for use in other food animals, often called “major species.”

Aquatic animals are a minor species in most countries, including all of North America and Europe, and approved drug availability for aquatic animals is a problem that will not be easily resolved without a significant change in the present regulatory approach. A new public understanding is required that recognizes the potential benefits to society of ensuring that approved safe and effective drugs are available to aquatic-animal producers. Additional public funding, carefully directed, is also essential to complete the necessary scientific data for approval. Gaining this public understanding will be an important step forward in facilitating the development of new food-producing activities, such as aquaculture.

Research and data requirements for therapeutant approval have very similar general outlines in developed countries. However, there can be considerable variation in the specific details that regulatory agencies want to see. These differences are another factor that increases the costs of drug approval, and further advancements toward internationally harmonizing the specific approval requirements are needed. Streamlining the scientific research requirements among several countries will help to solve the economic dilemma that discourages sponsors for aquatic-animal therapeutants and will lead to a better climate for developing the necessary scientific data for approval.

Drug regulatory approval requirements are divided into the general categories of manufacturing, human safety, environmental impacts and target-species efficacy and safety. The manufacturing data prove that the drug's active ingredient is a pure and stable compound that is prepared under standardized conditions and that the active ingredient is mixed only with other standardized compounds in the final formulation. These data ensure that the producer who purchases a drug receives the exact product identified on the drug label and that the product has a known shelf life, as indicated by the expiration date.

Human safety data are frequently the most costly part of the data package. Specific short- and long-term toxicity tests must be performed on the drug and its major metabolites, to determine the level of risk that the drug poses to humans. If the results of the toxicity tests are acceptable, then the sponsor must provide data that show exactly how rapidly residue levels decrease in a treated aquatic animal and that identify the tissues that contain residues for the longest period of time.

Key numbers that are calculated from the drug residue depletion and safety data are the allowable tissue tolerance level [or maximum residue level (MRL)] and the withdrawal period for the drug. The withdrawal period is the amount of time required following the administration of the last treatment for measurable residues of the drug or a metabolite to decrease to below an established safe level. This safe level is the MRL, or tolerance. The withdrawal period begins on the day after the last treatment has been administered. In some countries, withdrawal periods are provided in degree–days—to account for the impact of water temperature on fish metabolism—and in other countries, they are provided as calendar days, sometimes with recommended changes in the withdrawal period for different water temperature ranges. Careful attention to the withdrawal period is a very important aspect of treatment administration, and producers need to maintain complete records that show the treatment completion dates on aquatic animal stocks. These stocks must be identifiable from the time of treatment until harvest. If a group with one withdrawal completion date is mixed with a second group that has a later date, then the whole group must now be assumed to have the later withdrawal date. If a particular group is unidentified or mislabeled, then it should be assumed to have the latest withdrawal date of any treated group on the production site.

Occupational safety data are another part of the regulatory human safety consideration. Producers reduce drug administration hazards through appropriate employee training on safe drug handling. Failure to ensure that necessary precautions are in place and to abide by all warnings and cautions on the drug label can have harmful consequences. For example, repeated exposure to some treatments without appropriate protection—for instance, the organophosphate parasite treatments developed for sea lice control—could result in injury to aquaculture production—site employees.

Approval data relating to the potential environmental impact of drugs for aquatic animals is the basis for increasing regulatory action—and confusion. The challenge is to develop data that will accurately measure the environmental effects of drug use and then to assess the acceptability of these effects to society. There is no doubt that drugs which are used because of their effects on the physiology of aquatic organisms will also affect other animals (nontarget organisms) in the environment. However, determining the level of impact on nontarget organisms that is acceptable is a challenging task and is more a question of subjective opinion than a scientifically measurable number. Laboratory tests can show that a drug has a toxic impact on selected aquatic animals at specific concentrations, but how is this information to be interpreted? At present, regulators look primarily for evidence that the chemical is broken down readily in the environment and does not accumulate in the food chain.

Further regulatory confusion arises in assessing the environmental impact of a drug when more than one government agency is involved in giving a decision on chemical use. Separate agencies frequently have very different perspectives on acceptable impacts, with one group of scientists typically having responsibility for drug approval and another having responsibility for pollution control. As a result of their different perspectives and responsibilities, the agencies may arrive at different answers to the same request for approval, based on the same data. For example, a drug may be approved for use in aquatic animals, but discharge of water will not be permitted if the drug is actually used at a production site. This situation has occurred, for example, in the United Kingdom, where drug approval authorities accepted a sea lice drug, while the Environmental Protection Agency did not approve discharge of the same drug. This contradiction will not be resolved until government authorities determine when an impact is acceptable, recognizing that it is impossible for any drug to produce no impact at all.

Drugs for aquaculture use have come under greater scrutiny for environmental impact than have drugs that have been submitted for approval for terrestrial food-animal agriculture. This is partly because of the increasing public concern about environmental degradation and partly because the addition of drugs to water is perceived to have a greater potential impact, due to greater distribution in the fluid medium. The validity of these concerns remains to be evaluated; however, the situation has substantially increased the costs for aquaculture drug development packages.

Finally, the regulatory data on the target species are usually the least expensive part of the approval package. These data are also of most interest to the producer and include results of research on the safety and effectiveness of the drug for its intended use in aquatic animal species. Treatment safety is assessed by exposing fish to multiples of the proposed dosage and through prolonged treatment beyond the proposed duration. Additionally, an adverse-reaction reporting system is in place in many countries, to collect relevant information on occasions when a treatment has apparently resulted in an unwanted side effect. Drug efficacy is shown through rigorous laboratory tests against the intended target, followed by dosage determination studies to find the most appropriate treatment level. These studies are followed by field trials that assess the performance of the drug for its intended use under production conditions.

The lack of approved aquaculture drugs in many countries has resulted in an increased need for interim regulatory mechanisms that allow drug use on a temporary or experimental basis. In the United States, the Investigational New Animal Drug (INAD) exemption has permitted producers to assist in the development of drug approval data, while providing controlled interim access to an unapproved drug. The Investigational New Drug (IND) exemption in Canada and Animal Test Certificate (ATC) in the United Kingdom provide similar opportunities in these countries. In Norway, the drug regulatory authority provides an exemption that allows a drug manufacturer to sell the drug for commercial use after several key portions of the data package have been reviewed, but before full approval is granted.

New changes in regulatory requirements for drug use in fish production facilities are occurring as a result of the introduction of mandatory hazard analysis critical control point (HACCP) regulations for seafood safety inspection. The European Union, Canada, and the United States have all recently introduced seafood safety regulations based on variants of the HACCP principles. These mandatory food safety regulations are directed at fish processors, but they have an effect on aquaculturist drug-use practices, because of the concern that chemical residues in treated fish could present a hazard.

The fish processor, under the HACCP regulation, is required to have in place a plan to evaluate aquaculture animals (including lobsters from a pound) received at the plant loading dock and ensure that these products are free of residues beyond allowable tolerances. This change puts more responsibility on the processor to ensure that producers comply with safe and effective drug-use procedures. Good record keeping, good communications between the producer and processor, and the use of approved drugs according to the directions on the label are important components for managing under the new regulatory requirements. For details on this topic, the reader is referred to Chapter 11 of the *Fish & Fisheries Products Hazards & Controls Guide* (5).

IMPLEMENTATION AND COST

Safe and effective chemical treatments need to be available as options for the producer to manage fish health when problems arise. However, the objective of the producer is to avoid drug treatment unless it is necessary, because even if the treatment is cost effective, avoidance of a health problem in the first place is even more cost effective. The producer can work with fish health professionals to develop a sound health maintenance program for the operation that reduces the probability of later intervention and treatment.

When treatment becomes necessary, there are many considerations to keep in mind while administering the medication, in order to achieve a positive result. The first step is to make sure that any preexisting conditions that predispose aquatic animals to the occurrence of disease are corrected. For example, a treatment for infectious bacteria or parasites will knock back the agents of a disease for a while, but it is important to ensure that preexisting problems or situations are corrected to allow the fish to continue to fight off the pathogens once the treatment is complete. Drug treatments cannot be a substitute for good management, and they cannot make a poorly cared-for animal prosper.

If the proposed treatment has not previously been tried on the production site or is not a standard industry treatment procedure, it is wise to pretest the intended dosage on a small group of representative animals. Treated animals should be closely observed during the hours after administration has been started, and farm workers need to be prepared to discontinue the treatment immediately if there is any evidence of an adverse effect. Fish mortality data can be closely monitored during a longer treatment application.

The producer also has a responsibility to minimize the development of resistance to treatment among local populations of pathogenic parasites and bacteria. Sensitivity testing is an essential part of effective management of infectious agents, although emergency treatment may sometimes need to be initiated before laboratory results are obtained. The limited range of therapeutants approved for use in aquaculture is an unfortunate circumstance that predisposes aquatic animals toward developing resistance to the available therapeutants. However, management approaches can help to avoid this problem — for example, salmon farmers using single year-class sites (all in or all out), farm-site fallowing, and area management agreements among farms to reduce the occurrence of resistance and to permit continuing effective control of diseases such as furunculosis and sea lice.

A treated population should be monitored closely during the drug administration period, and treatment should be stopped if an adverse effect is suspected. Monitoring may include extra dives to check for problems that are not visible from the surface. Regulatory agencies generally have a reporting requirement for registering and publicizing observed adverse effects. Knowledge of adverse effects helps to ensure that other producers are aware of the potential for a problem and helps the whole industry to effectively manage animal treatments.

A sufficient number of trained personnel must be available to conduct a treatment, while enough staff must remain to manage routine farm duties. A crew specifically trained for the procedure may best handle certain types of treatment applications, such as tarpaulin baths for floating net pens. (Administering a tarpaulin bath to a large net pen is a very difficult and dangerous task if there are currents or winds.) Applicator training needs to address the risks of drug treatment, the appropriate precautions to take, the correct use of safety equipment, typical problems that might come up, and essential first aid if toxicity is encountered. At a minimum, the staff needs ready access to, and familiarity with, the safety information sheet, such as the MSDS, and the label on the drug.

Availability of a range of safe and effective drug treatment options will improve the producer's ability to provide a correct treatment for any problem that occurs. Additional treatment options will also help to prevent the development of resistance in local populations of parasites and bacteria. Although the number of available approved drugs may be limited, use of approved drugs helps aquaculture producers in several ways. First, the drugs meet their specific responsibilities under the country's food safety regulations, and using them encourages drug sponsors to shepherd new treatment options through the regulatory process. Producers' associations can help their members by developing a list of their top-priority fish health treatment needs and ensuring that these priorities are brought to the attention of governments, research funding agencies, researchers, and pharmaceutical companies. This action will encourage these organizations to cooperate in providing the resources necessary to develop and approve additional treatments.

One important producer concern that is not comprehensively addressed by the rigorous regulatory process is drug efficacy. An approved drug is used under a great variety of field conditions, and experience is the only way to develop a thorough understanding of the best way of using the drug under the different conditions encountered. Therefore, the producer needs access to more information than can be made available through the regulatory process. The best way to obtain this information is to work with a professional who has experience with drug treatments on similar aquaculture production operations. The essential calculation is that the cost of treatment must be less than the benefits from the improved fish survival that results from the treatment. Some costs of treatment are obvious, including the cost of the drug and the labor to apply it. Other costs are less obvious, including the impact of reduced harvesting options because of withdrawal periods following treatment, and the setback weight gain when fish are taken off feed, or are on reduced rations, during treatment. Also, some costs of treatment may be hidden, including potential long-term impact on the production-site environment from multiple treatments, and reduced drug effectiveness through resistance development in the local population of target organisms.

The best strategic approach for producers is to use approved drugs and to work scrupulously within the restrictions imposed by the label of the drug. These requirements may at times seem onerous, but they are based entirely on sound scientific evaluation. Cheaper, unapproved drug formulations may be tempting to use and may appear, superficially, to be similar to approved drugs. However, using these products is a very shortsighted strategy, as negative public opinion, which is a very real potential consequence of consistent use of unapproved drugs, poses a very high cost to the whole aquaculture production industry.

BIBLIOGRAPHY

1. N. Herwig, *Handbook of Drugs and Chemicals Used in the Treatment of Fish Diseases*, Charles C. Thomas, Springfield, MA, 1979.
2. R.A. Schnick, D.J. Alderman, R. Armstrong, R. Le Gouvello, S. Ishihara, S. Percival, and M. Roth, *Bull. Eur. Ass. Fish. Pathol.* **17**(6), 251–260 (1994).
3. Texas Agricultural Extension Service, *Guide to Drug, Vaccine and Pesticide Use in Aquaculture*, B-5085, 1994.
4. J.F. Burka, K.L. Hammell, T.E. Horsberg, G.R. Johnson, D.J. Rainnie, and D.J. Speare, *J. Vet. Pharmacol. Therap.* **20**, 333–349 (1997).
5. US Department of Health and Human Services, Public Health Service, Food and Drug Administration, Center for Food Safety and Applied Nutrition, and Office of Seafood, Fish & Fisheries, *Products Hazards & Controls Guide*, 2nd ed., January 1998.

See also Disease treatments; Vaccines.

E

ECHINODERM CULTURE

NILS T. HAGEN*
Bodø College
Bodø, Norway

OUTLINE

The average wholesale price of fresh Japanese sea urchin roe is approximately 14,000 ¥/kg, making it one of the most valuable seafoods in the world. Increasing demand for sea urchin roe in Japan has spurred the development of extensive domestic fishery enhancement techniques (1,2). It has also provided the incentive for a worldwide expansion of sea urchin fisheries (3–9).

With a total production of 60,000 live-weight tons of whole sea urchin per year, the world's supply of wild sea urchins has reached a plateau. However, this production level is probably not sustainable at the current population of sea urchins, since the declining productivity of exploited sea urchin stocks no longer can be offset by further geographical expansion of fisheries. To maintain or expand the world's supply of high-quality sea urchin roe is a major aquacultural opportunity awaiting commercial-scale trials.

Echinoderm culture refers to the cultivation of both sea urchins (Echinoidea) and, to a lesser extent, sea cucumbers (*Holothuroidea*). As sea urchins are more valuable than sea cucumbers, and their cultivation is more advanced, this contribution concentrates on sea urchins, although the cultivation of sea cucumbers is considered as well.

*This contribution is adapted, with permission, from a paper entitled *Echinoculture: From Fishery Enhancement to Closed Cycle Cultivation*, which appeared in World Aquaculture, 27(4):6–19, 1996. The original article was edited to conform with the style and format of other contributions to this encyclopedia. — Editor.

CATCH AND CONSUMPTION

The total Japanese catch of sea urchins (of which there are six species; see Table 1) and sea cucumbers (*Stichopus japonicus*) peaked in the late 1960s at approximately 27,500 tons per annum for sea urchins and 13,000 tons per annum for sea cucumber. (See Fig. 1.) Over the two next decades sea urchin landings fluctuated between 20,000 and 27,000 tons, until the catch dropped to 14,000 tons in 1991. The cause of this recent decline remains unknown, although it coincides with observations of disease-related sea urchin mortality. Even so, sea urchin landings remain two to three times greater than the steadily declining sea cucumber landings. (See Fig. 1.)

The gonads are the primary soft tissue and the only edible part of the sea urchin. The harvested gonads of both female and male sea urchins are called "roe," regardless of sex. As the most important market for sea urchins, Japan imports approximately 5,000 tons of sea urchin gonads per annum (see Fig. 2), the equivalent of 40,000–50,000 tons of whole, live sea urchins. In addition Japan imports a moderate amount of whole, live sea

Table 1. Exploited Japanese Sea Urchins

Japanese Name	Scientific Name
Aka uni	*Pseudocentrotus depressus*
Bafun uni	*Heterocentrotus pulcherrimus*
Ezo bafun uni	*Strongylocentrotus intermedius*
Kita murasaki uni	*Strongylocentrotus nudus*
Murasaki uni	*Anthocidaris crassispina*
Shirahige uni	*Tripneustes gratilla*

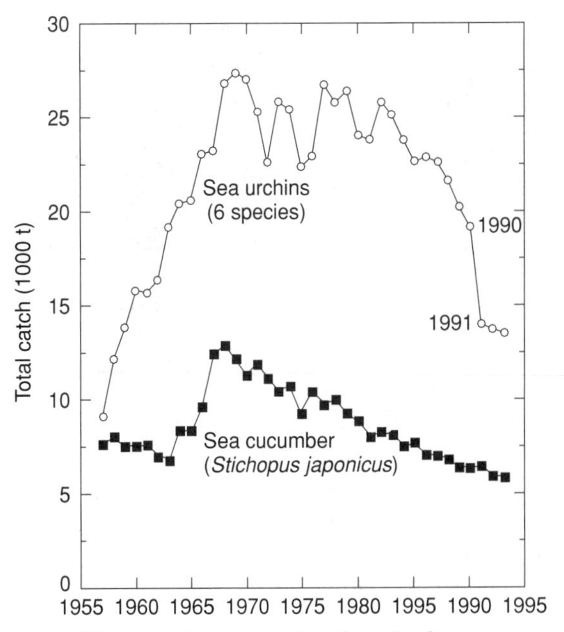

Figure 1. Japanese echinoderm landings.

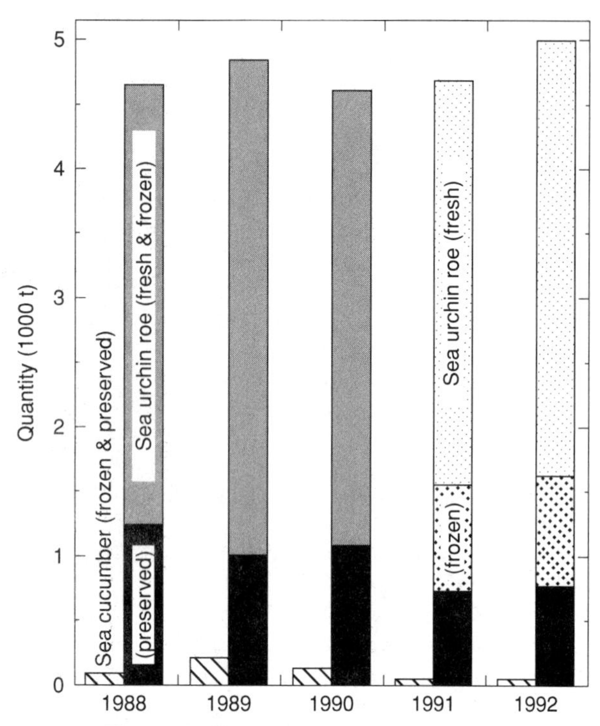

Figure 2. Japanese echinoderm imports.

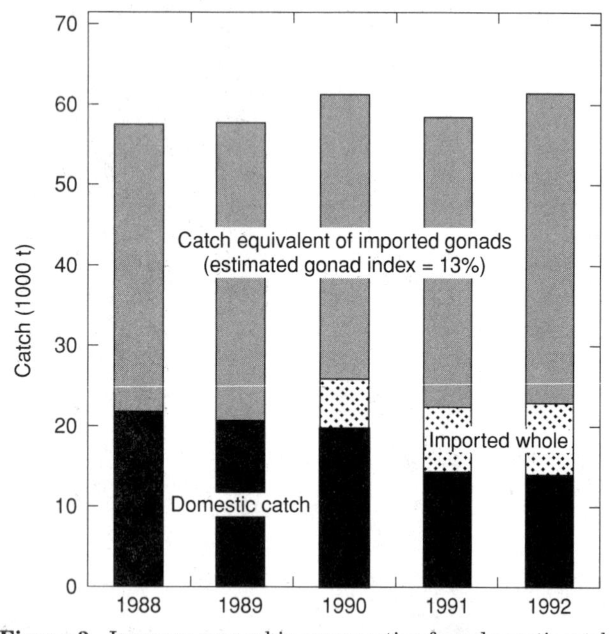

Figure 3. Japanese sea urchin consumption from domestic catch and imports. The gonad index value is a conservative estimate based on the average gonad yield in commercial sea urchin fisheries in northern Japan.

urchins. (See Fig. 3.) Thus, total Japanese consumption, including the domestic catch, is approximately 60,000 tons of whole sea urchins per annum. (See Fig. 3.) The second largest consumer nation is France, with an annual consumption of approximately 1,000 tons of whole sea urchins.

PRICE

The wholesale price of whole Japanese sea urchins (all six species) was similar to the price of the Japanese sea cucumber in the 1960s, but although the price of both sea urchins and sea cucumbers continued to increase, by 1990 sea urchins were twice the price of sea cucumbers, reflecting increasing demand for quality urchin gonads. (See Fig. 4.) Fresh Japanese sea urchin gonads fetch approximately 10 times the price of whole, live sea urchins, where the average wholesale price of the gonads was almost 14,000 ¥/kg in 1993. Fresh imported sea urchin gonads fetched an average wholesale price of only 6,000 ¥/kg the same year, due to their inferior quality. (See Fig. 5.) These quality problems are related to the nutritional and reproductive status of the source population, as well as the processing and shipping routines of the suppliers. Furthermore, some harvested species of sea urchins do not produce superior quality gonads under any circumstances.

SEA URCHIN CULTIVATION

Sea urchin aquaculture in Japan is part of a multispecies fisheries enhancement effort organized by local fishery cooperatives. The three limiting factors of the sea urchin fishing industry have been identified as (*1*) insufficient food supply, (*2*) lack of suitable habitat, and (*3*) insufficient recruitment.

Sea urchin transplantation and seaweed reforestation are aimed at improving the gonad yield of undernourished adult sea urchins by providing increased access to food. Undernourished sea urchins, transplanted from barren grounds off the northwestern coast of Japan to seaweed-dominated feeding grounds off the northeastern coast, are usually ready for recapture after three months (10). The objective of seaweed reforestation is to improve the

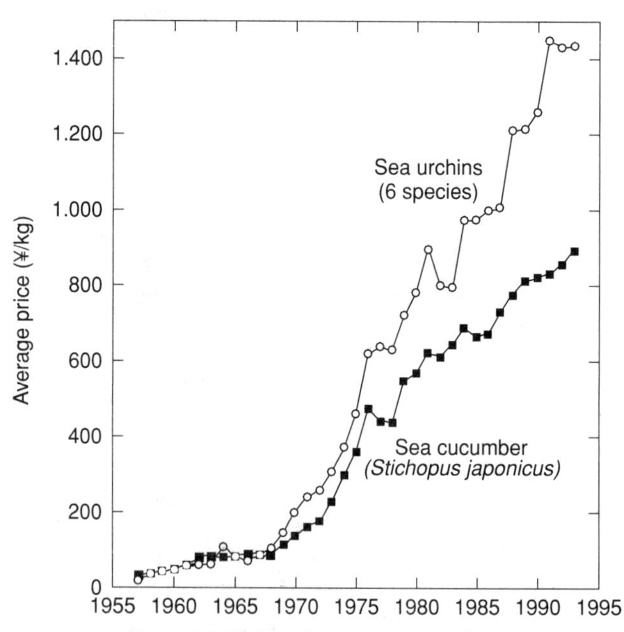

Figure 4. Echinoderm prices in Japan.

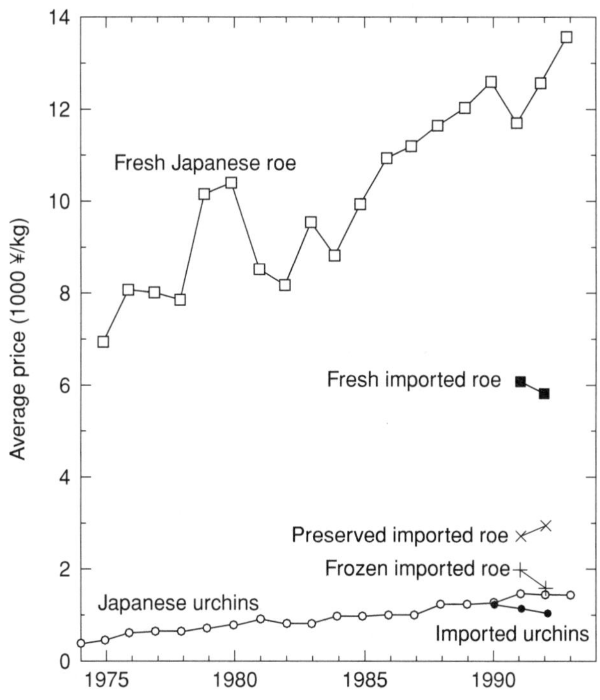

Figure 5. Sea urchin prices in Japan.

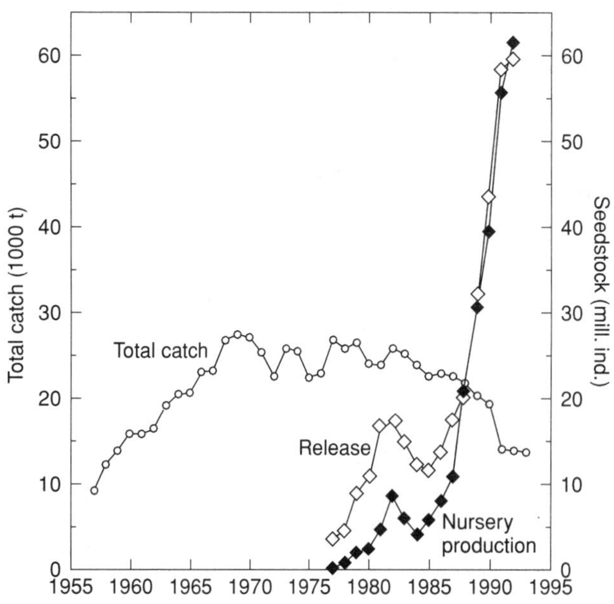

Figure 6. Catch and cultivation of Japanese sea urchins.

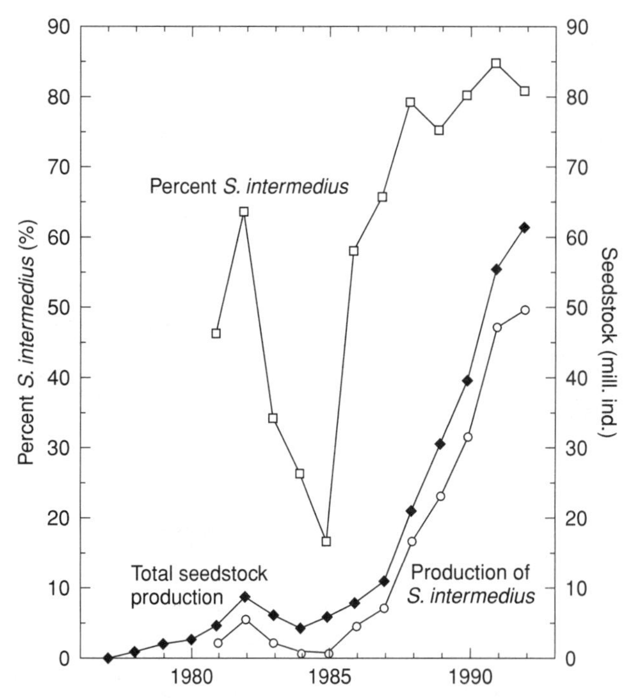

Figure 7. Seedstock production of *S. intermedius* in Japan.

local food supply of undernourished sea urchin stocks by establishing algae in barren areas, through a combination of algal cultivation and overgrazing control (11). Seaweed reforestation is still an experimental technique, and transplantation of adult sea urchins is gradually being replaced by release of juvenile seedstock. Feeding of captured sea urchins is also being attempted on a small scale.

Habitat improvement and habitat creation are part of a larger ongoing fisheries enhancement program that is aimed primarily at the construction of artificial reefs (4,12). The main purpose of artificial reefs is to improve the productivity of soft substrates by creating new fishing grounds for the kind of flora and fauna that are normally associated with rocky shore habitats (e.g., edible seaweed, abalone, top shells, sea urchins, and rockfish) In a related effort, the effective surface area and habitat complexity of existing natural reefs have been increased by blasting, adding rocks, and constructing shallow channels with wave-powered water circulation (2,13).

The third limiting factor, insufficient recruitment, is still a major problem, even though the Japanese sea urchin fishery is strictly regulated to ensure that spawning stocks are not depleted. To improve recruitment, a large-scale seedstock release program has been implemented. Juvenile seedstock is produced in land-based nurseries from hatchery-reared larvae and from wild larvae collected on suspended settlement plates (10). Although seedstock production has increased rapidly in the past decade, to the current level of more than 60 million individuals per annum, the total catch of sea urchins has remained at a relatively low level since 1991. (See Fig. 6.)

The single most important species of sea urchin in Japan, *S. intermedius* (*ezo bafun uni*), accounts for approximately 80% of total seedstock production. (See Fig. 7.) The remaining 20% is divided among the five other species, of which *P. depressus* (*aka uni*) is the most important. (See Table 1.)

BROODSTOCK MANAGEMENT

Japanese sea urchin cultivation is based on the spawning of wild broodstock (see Fig. 8), with the availability of mature sea urchins, in most places, restricted to the annual breeding season. However, a local population of

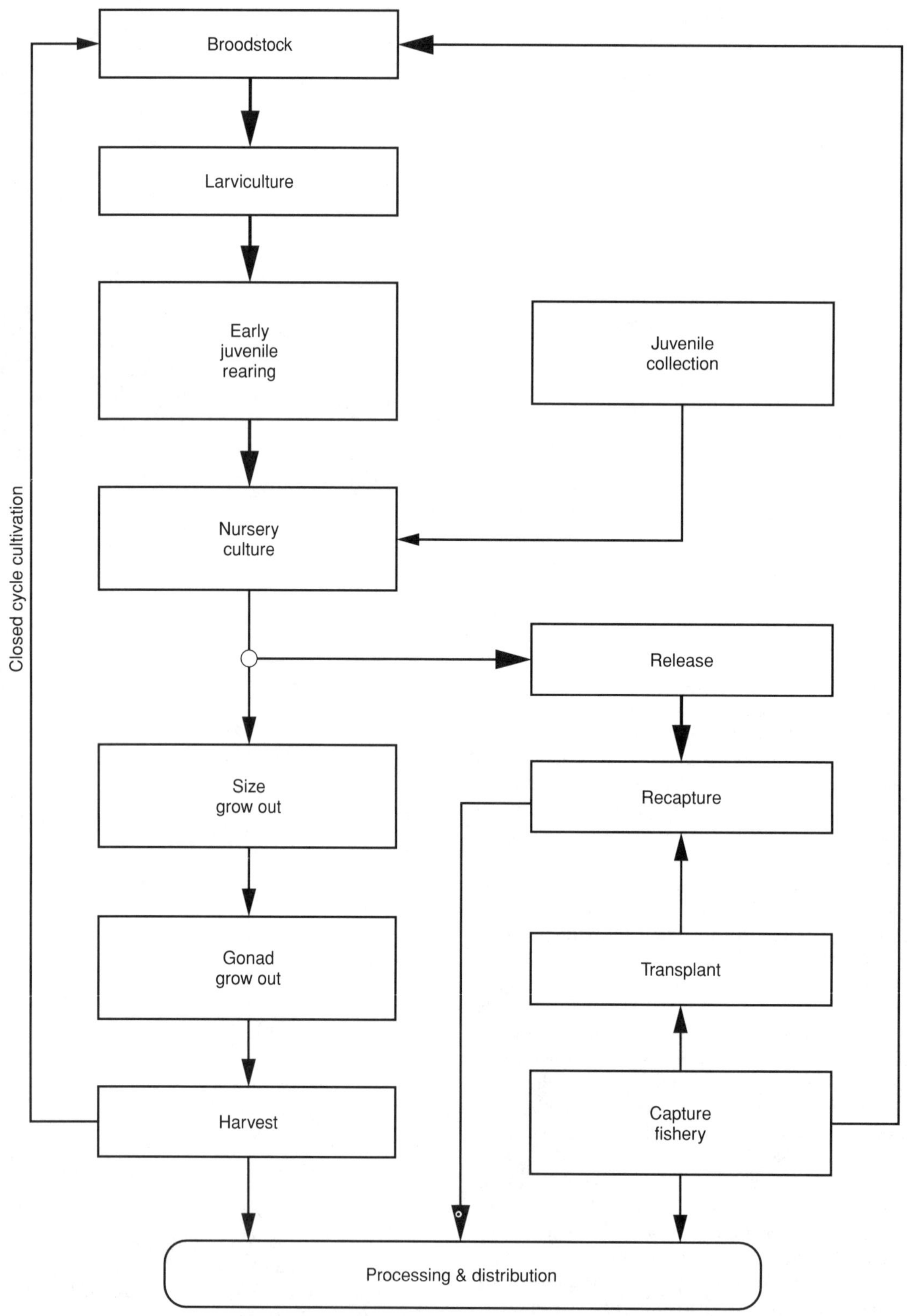

Figure 8. Principal sea urchin cultivation techniques.

S. intermedius, in the southern part of its range, has biannual reproduction, which allows some hatcheries to produce two annual batches of sea urchin larvae. For example, the largest sea urchin nursery in Japan, located in Shikabe municipality, southeastern Hokkaido, obtains broodstock from this source population and produces one batch of juveniles in the spring and another in the fall, for a total of 11 million juvenile *S. intermedius* per annum (10).

The factors the control reproductive maturation are not entirely known for most sea urchin species, but the photoperiod and water temperature are considered to be important (14–16). Experimental broodstock cultivation has shown that multiple spawning is possible when well-fed sea urchins are cultivated in darkness in relatively warm water (17).

Mature sea urchins are easily induced to spawn by injecting 1–2 ml (0.03–0.07 oz) of a 0.53-mol KCl solution (18,19). A new technique for individual identification of sea urchins, using electronic passive induced transponder (PIT) tags, has the potential to facilitate cultivated broodstock management (20).

LARVICULTURE

Larviculture in Japanese hatcheries commences with the mixing of gametes from several animals. Excess sperm is rinsed off, and the fertilized eggs hatch after approximately 20 hours. Three to four days later, the hatchlings have developed to the early pluteus stage and require planktonic microalgae as food. The diatom *Chaetoceros gracilis* is commonly used in commercial sea urchin hatcheries, whereas the green flagellate *Dunaliella tertiolecta* is popular in research laboratories (19). *C. gracilis* is cultivated in 3-L (0.52 gal) batches in a separate microalgae cultivation room and is fed to the larvae at an initial rate of 1.5 L (0.26 gal), or, 5000 cells/ml, per tank per day (142,857 cells/oz/tank/day). The amount is gradually increased to 10 L/tank/day (2.6 gal/tank/day) in the final stages of cultivation. The larvae are cultivated in 1,000 L (260-gal) tanks, with continuous flow of 1 μm (0.00004 in.) of filtered seawater. The water flow is increased from 15% water exchange/day to 100% at the time of settlement, which occurs 16–30 days after fertilization, depending on the water temperature. Circulation in the larvae tanks is provided by two large air stones with a gentle flow, one on the bottom and another near the surface. The central water outlet is covered by 100 μm (0.004 in.) of plankton net [later, 150 and then 200 μm (0.006 and then 0.008 in.)]. The density is initially 1.5 larvae/ml (42.8 larvae/oz), but decreases to 0.8 larvae/ml (22.8 larvae/oz) at the time of settlement. The metamorphosed juveniles are approximately 0.3 mm (0.01 in.) in diameter (2,21).

EARLY JUVENILE REARING

Settlement is induced by introducing wavy settlement plates covered with the minute green alga *Ulvella lens* (21) or the benthic diatom *Navicula ramosissima* (22), which serve as the initial food source for the juvenile sea urchins. The plates are made of transparent polycarbonate, which facilitates the growth of benthic microalgae on both sides of the plates. The settlement plates are prepared in tanks inoculated with the desired algae. Nutrient salts are added to stimulate algal growth, and the tanks are occasionally drained and rinsed to eliminate unwanted benthic diatoms. Feeding with soft seaweed, such as *Ulva lactuca*, commences when the juveniles reaches 3–4 mm (0.12–0.16 in.) in diameter (2,21).

NURSERY CULTURE

The larvae are transferred from the settlement plates to nursery culture when they reach 4–5 mm (0.16–0.2 in.) in diameter. Most juveniles are transferred to nursery tanks with open-mesh cages or to habitat modules made from nontransparent wavy plates, but some are transferred to hanging cages suspended 1–2 m (3–6 ft) below the surface of the water. The juveniles are fed with kelp (*Laminaria* spp., *Undaria pinnatifida*, *Eisenia bicyclis*) and other locally available seaweed, supplemented by food pellets or knotweed leaves (*Oita dori*, *Polygonum* sp.). Food pellets come in different sizes, but their food value is still inferior to that of fresh kelp. Some juveniles are released six months after fertilization, when they are 7–10 mm (0.3–0.4 in.) in diameter. Six months of additional nursery cultivation produces larger [15–20 mm (0.6–0.8 in.)] seedstock with higher survival rates. The large juveniles are ready for recapture two years after release, when they have reached a diameter of more than 40 mm (1.6 in.). Survival rates are variable, but have been estimated at 20–50% for large seedstock. Approximately 80% of the survivors are captured, yielding a total recovery rate of 16–40% (10).

CLOSED-CYCLE CULTIVATION

Closed-cycle cultivation requires growout facilities. These facilities can be constructed by expanding existing nursery techniques, adapting technology developed for the intensive cultivation of abalone (23), or developing of new technology. The French adopted the last alternative and developed a prototype of a multilayered growout tank that consists of four stacked, sloping shelves. Water is pumped to the top shelf from a reservoir tank under the shelves and then runs down through the stack of shelves in a zigzag pattern. The accumulation of sea urchin feces in the reservoir tank is siphoned off at regular intervals. The recirculated water is gradually replenished by marine groundwater (24,25).

An important factor when considering closed-cycle cultivation is the choice of a good target species. The gonads of the green sea urchin, *Strongylocentrotus droebachiensis*, for example, are popular in Japan, despite quality problems commonly caused by food limitation of wild stocks. Highly similar to the *S. intermedius*, which is found only in the northwestern Pacific, *S. droebachiensis* is a coldwater species with a wide distribution throughout the north Atlantic and the northeastern Pacific (26). The larvae of *S. droebachiensis* must be cultured at temperatures below 10–11 °C (50–52 °F) (27,28), but juveniles and adults can tolerate somewhat higher temperatures (20). Other potential candidates for closed-cycle cultivation are the Chilean sea urchin *Loxechinus albus* (29,30) and the European sea urchin *Paracentrotus lividus* (24,25).

A commercial-scale growout facility requires a stable food supply, with kelp as the major ingredient. Kelp is necessary for good flavor and coloring of the sea urchins' gonads, but a protein supplement can enhance growth and improve overall food conversion as well (30,31). The most important determinant of gonad quality, besides food, is the reproductive state of the sea urchin. Sexually mature gonads have an undesirable soft consistency and bitter taste, due to the reduced number of glycogen-containing nutrient cells. Closed-cycle production offers the potential to inhibit sexual maturation through manipulation of the photoperiod and water temperature, thereby extending the harvesting season and improving gonad quality, gonad yield, and food conversion rates. Closed-cycle cultivation also offers the opportunity for growth acceleration through systematic broodstock selection and breeding.

Closed-cycle cultivation is capital intensive and has high operational costs, but requires only a modest investment in research and development. Full-scale hatchery and nursery technology is well established in Japan, and prototype growout facilities already exist in France. A recent profitability analysis of a hypothetical Norwegian growout facility, using pessimistic, realistic, and optimistic parameter estimates, was unable to demonstrate nonprofitability. In fact, there was a large potential for profit when realistic and optimistic parameter estimates were used.

SEA CUCUMBER CULTIVATION

Sea cucumbers are consumed in Japan, as well as in Chinese markets, including Singapore, Hong Kong, and Taiwan. The Japanese sea cucumber market is largely self-sufficient, whereas the Chinese markets are more import oriented. Sea cucumber landings peaked at 13,000 tons in the late 1960s and subsequently declined to the present level of approximately 6,000 tons. To supplement natural recruitment, seedstock production commenced in the early 1980s, but is still at a modest level of 2–3 million individuals per annum. (See Fig. 9.) Most of the seedstock is being released, but some is being used for experimental closed-cycle aquaculture.

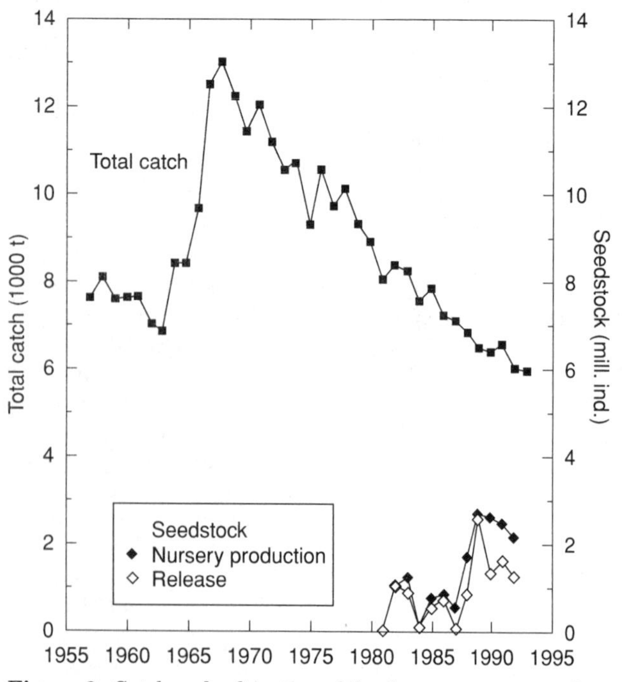

Figure 9. Catch and cultivation of the Japanese sea cucumber.

Sea cucumber broodstock is captured in spring when it is sexually mature. Spawning is induced by temperature shock—for example, by raising the water temperature from 16 to 21 °C (61–70 °F). The planktonic larvae are then cultivated in 1000-L (260-gal) tanks and fed a planktonic microalgae, most commonly the diatom *C. gracilis*. Settlement occurs after approximately two weeks. At this stage, the larvae measure approximately 800 μm (0.003 in.) in length, but after metamorphosis they measure only about 200 μm (0.0008 in.). These tiny juveniles are then fed an initial diet of mixed benthic diatoms (*Navicula* spp.) before being transferred to outdoor nursery tanks filled with unfiltered seawater. There, they feed on the natural growth of benthic diatoms on the tank walls, supplemented with dried seaweed powder (*U. pinnatifida*, or in Japanese, *Wakame*). Mortality is high during the hatchery and nursery stages. The juveniles are released after six months, when they measure 2–8 cm (0.8–3.1 in.) in length, and are recaptured one year later, when they measure approximately 20 cm (8 in.) in length.

Sea cucumbers feed on detritus and suspended particulate matter, including the fecal pellets of sea urchins. Intensive polyculture of sea urchins and sea cucumbers appears feasible, but has yet to be investigated.

In conclusion, it appears that closed-cycle cultivation of sea urchins for the Japanese market is an emerging coldwater aquaculture opportunity awaiting commercial trials, whereas sea cucumber cultivation is still in the experimental stage. However, joint development of urchin and cucumber polyculture is an interesting possibility, since sea urchin waste can be used as sea cucumber food. In addition, echinoculture has the potential to become a long-term sustainable industry, with ecologically and environmentally sound production, since both sea urchins and sea cucumbers are primarily herbivorous organisms.

ACKNOWLEDGMENTS

Thanks to my Japanese friends, teachers, and colleagues for their support during my visits as a MONBUSHO scholarship recipient, conference delegate, and STA fellow. Thanks also to H.K. Marshall for improving the logical flow and linguistic content of the manuscript. The National Research Institute of Aquaculture, Mie, Japan provided access to Japanese fishery statistics and time for manuscript preparation.

BIBLIOGRAPHY

1. I. Matsui, *Fish. Res. Bd. Can. Trans. Ser.* **1063**, 1–172 (1968).
2. M.J. Tegner, *Mar. Fish. Rev.* **51**, 1–22 (1989).
3. C. Conand and N.A. Sloan, in J.F. Caddy, ed., *Marine Invertebrate Fisheries: Their Assessment and Management*, John Wiley & Sons, New York, 1989, pp. 647–663.
4. Y. Hasegawa, *J. Fish. Res. Bd. Canada* **33**, 1002–1006 (1976).

5. J.H. Himmelman, Y. Lavergne, F. Axelsen, A. Cardinal, and E. Bourget, *Can. J. Fish. Aquat. Sci.* **40**, 474–486 (1983).
6. S. Kato, *Comm. Fish. Rev.* **34**, 36 (1972).
7. S. Kato, *Mar. Fish. Rev.* **35**, 23–30 (1973).
8. S. Kato and S.C. Schroeter, *Mar. Fish. Rev.* **47**, 1–20 (1985).
9. D.E. Kramer, *Can. Ind. Rep. Fish. Aquat. Sci.* **114**, 116 (1980).
10. K. Sato, in C.M. Dewees, ed., *The Management and Enhancement of Sea Urchins and Other Kelp Bed Resources: A Pacific Rim Perspective*, California Sea Grant Publication No. T-CSGCP-028, University of California, La Jolla, CA, 1992, pp. 1–38.
11. H. Kito, S. Kikuchi, and N. Uki, in *Proceedings: International Symposium on Coastal Pacific Marine Life*, Western Washington University, Bellingham, WA, 1980, pp. 55–66.
12. M.G. Mottet, *Washington Dept. Fish. Tech. Rep.* **20**, 1–66 (1976).
13. T. Funano, *Sci. Rep. Hokkaido Fish. Exp. Sta.* **23**, 9–52 (1981).
14. J.S. Pearse, V.B. Pearse, and K.K. Davis, *J. Exp. Zool.* **237**, 107–118 (1986).
15. K. Sakairi, M. Yamamoto, K. Ohtsu, and M. Yoshida, *Zool. Sci.* **6**, 721–730 (1989).
16. M. Yamamoto, M. Ishine, and M. Yoshida, *Zool. Sci.* **5**, 979–988 (1988).
17. P.S. Leahy, *J. Exp. Zool.* **204**, 369–380 (1978).
18. R.T. Hinegardner and M.M. Rocha Tuzzi, in *Laboratory Animal Management, Marine Invertebrates*, National Academy Press, Washington, DC, 1981, pp. 291–302.
19. N.W. Strathmann, in M.F. Strathmann, ed., *Reproduction and Development of Marine Invertebrates of the Northern Pacific Coast: Data and Methods for the Study of Eggs, Embryos, and Larvae*, University of Washington Press, Seattle, 1987, pp. 511–534.
20. N.T. Hagen, *Aquaculture* **239**, 271–284 (1995).
21. K. Saito, K. Yamashita, K. Tajima, A. Obara, Y. Nishihama, M. Kawarnata, and K. Kawamura, *Manual of Artificial Seed Production of Sea Urchin, Strongylocentrotus intermedius*, Northern Hokkaido Institute of Mariculture, Shikabe, Hokkaido, Japan, 1985.
22. Y. Ito, H. Kanamani, and K. Masaki, *Bull. Jpn. Soc. Sci. Fish.* **53**(10), 1735–1740 (1987).
23. E.E. Ebert, in S.A. Shepher, M.J. Tegner, and S.A. Guzman del Proo, eds., *Abalone of the World: Biology, Fisheries, and Culture*, Fishing News Books, Oxford, United Kingdom, 1992, pp. 570–582.
24. P. LeGall, in G. Barnbad, ed., *Aquaculture*, Ellis Horwood, New York, 1990, pp. 443–462.
25. P. LeGall and D. Bucaille, in N. DePauw, E. Jaspers, H. Ackefors, and N. Wilkins, eds., *Aquaculture: A Biotechnology in Progress*, European Aquaculture Society, Bredene, Belgium, 1989, pp. 53–59.
26. M. Jensen, *Sarsia* **57**, 113–148 (1974).
27. N.M. Hart and R.E. Scheibling, *Mar. Biol.* **99**, 167–176 (1988).
28. R.E. Stephens, *Biol. Bull.* **142**, 145–159 (1972).
29. L.P. Gonzales, J.C. Castilla, and C. Guisado, *J. Shellfish Res.* **6**(2), 109–115 (1987).
30. M.L. Gonzalez, M.C. Perez, D. López, and C.A. Pino, *Aquaculture* **115**, 87–95 (1993).
31. T.S. Klinger, J.M. Lawrence, and A.L. Lawrence, *J. World Aquacult. Soc.* **25**, 489–496 (1994).

ECONOMICS, BUSINESS PLANS

WADE L. GRIFFIN
Texas A&M University
College Station, Texas

OUTLINE

The development of many aquaculture businesses has been based on "gut feelings" rather than on sound business planning. In most cases, such businesses are predestined to fail. The attitude of the developers has often been "I want to raise fish," instead of "I want to run an aquaculture business." Potential new entrants typically ask, "How do I start a farm?" and "What does it cost?", and if they can afford to start the business, then off they charge, without asking any further questions. Later, as struggling fish farmers; they ask, "How can I pay my bills and keep the farm going?" The result of their endeavor is often frustration and financial failure. The intent of this entry is to help prospective aquaculturists evaluate the real potential of their "gut feelings" and communicate that potential to others, using a business plan. The purpose here is to provide a framework for preparing a business plan in the aquaculture industry (1–8).

BUSINESS PLAN

A business plan describes what we expect to happen in the future of our business and the things that have to be done to bring those expectations to reality. During preparation of a business plan, things that make the aquaculture venture a poor investment may be discovered. The result may be that considerable heartache, embarrassment, and financial distress are avoided.

A business plan explains how a particular idea can be accomplished. Everything from site selection for production to selling the product must be put to the "Will this really work?" test. The test simply examines the tasks in the aquaculture venture and asks what factors need to be considered and whether the tasks can be performed economically. Investors and lenders are more apt to provide capital to aquaculture ventures that have withstood the rigors of thorough analysis of a well-prepared plan than to ventures that have not.

A business plan can be used for many purposes. Among the more important uses are to:

(1) critically evaluate a business idea,
(2) obtain money from lenders or investors,
(3) establish a track record to demonstrate business expertise, and
(4) communicate plans internally within a firm.

It is important to understand that developing a business plan can be tedious and cumbersome. Many areas of business expertise and a considerable amount of time spent on gathering information are required for a particular idea to be properly evaluated and acted on. Unfortunately, many potential aquaculturists omit critical steps in developing an aquaculture business plan. Many have no clear direction or objectives. For others, the overwhelming desire to start the business has led to a distortion of facts. Yet others, intentionally misrepresent facts or are overly optimistic, in order to "sell the project" to investors or lenders. For the best chance of business success, a logical, honest effort from concept to business plan should be employed. The steps for developing an aquaculture business plan are as follows:

(1) Make sure that the business idea is worth pursuing.
(2) Convert the idea into what needs to be accomplished.
(3) Identify specific product(s) and market(s).
(4) Design the facility, and plan operations.
(5) Produce various alternatives for accomplishing the goals established in the initial facility and operation design.
(6) Analyze the alternatives, based on unfavorable events that could occur.
(7) Decide which alternative is the best, based on the particular circumstances.
(8) Finally, write down the business plan, with an organized purpose and logical plan of action.

IS THE IDEA ANY GOOD?

Many potential aquaculturalists begin by immediately calculating revenue and costs, to determine if the venture will be a success or a failure. Depending on how badly they desire to go into business, they might falsely raise the revenues or lower the costs in their calculations, in order to force a successful outcome. The excitement of a potential aquaculture project has too frequently overwhelmed the normally cautious, sensible individual, and many failures have resulted from using overly optimistic assumptions and expectations in developing a business plan. Therefore, "running the numbers" at the outset of preparing a business plan should not be done. Start instead by examining the source of the idea to go into aquaculture, and then confirm the idea as legitimate by checking it with a trustworthy expert. The following is a checklist for determining the reliability of the source that generated the idea to start an aquaculture business:

(1) Is the source affected by whether aquaculture in general develops or not?
(2) Is the source affected by whether or not you in particular are involved in aquaculture?
(3) Is the source involved directly in aquaculture? If yes,
   (a) Is the source a producer?
   (b) Does the source market aquaculture products?
   (c) Has the source been involved in aquaculture more than five years?
(4) Is the information provided by the source factual? (That is, does the information come from actual observation or documentation, or just from "a feeling?")
(5) Is the source's information less than three events away from actual observation? (That is, did the idea come from at least a reputable secondhand information source?)
(6) Can the source's information be substantiated through a knowledgeable, independent third party?
(7) Is the source's information less than one year old? If not, is the source's information still valid?
(8) Is this idea original? If yes, was the idea developed because of a market need? If your source does not check out, seek out other sources until a reliable one is found.

The following list of common providers of information on aquaculture may help in confirming or researching aquaculture business ideas:

(1) Actual acquaculturists, or people involved in businesses closely associated with acquaculture.
(2) Industry publications: *Aquaculture* magazine, *The Progressive Fish-Culturist (now North American Journal of Aquaculture), Journal of the World Aquaculture Society, Seafood Leader, and The Seafood Business Report.*
(3) Government personnel: state agricultural extension services, state agricultural experiment stations, state fish and game agencies, National Marine Fisheries Service, and Natural Resources Conservation Service.
(4) Private consulting services, which are available through many aquaculture consulting firms.
(5) Academic sources: textbooks and universities with aquaculture programs.

DESIGNING THE BUSINESS

After the idea has been determined to be worth pursuing, the aquaculture business should be designed. Before drafting the physical facility, however, complete the following three tasks, to provide direction: (*1*) decide what is wanted from the business, (*2*) define exactly the nature of the business, and (*3*) determine how much to spend. The results of these activities will help guide the physical design of the operation.

If the culturist identifies exactly what is wanted from his or her effort, a better operation to capture the desired outcome can be planned. All of the tasks and activities that have to be accomplished are more clearly defined, and therefore, unnecessary costs and changes can be avoided.

The prospective aquaculturist should ask the following questions of himself or herself:

(1) Financial
- (a) Can I stand to lose everything I put into the business? If not, how much can I afford to lose?
- (b) How long can the initial invested sum of money be tied up in the business?
- (c) How much of a return do I expect on my investment?
- (d) Do I require a salary?
- (e) Do I plan on selling the business? If so, at what stage?
- (f) Will I be seeking any outside financial support? If yes, am I planning on using investors or lenders? If yes, what are the terms, conditions, and expectations of the investors or lenders?
- (g) Do I plan on substantial growth of the business? If yes, must it grow on internally generated funds?

(2) Other
- (a) How much time do I have to succeed?
- (b) How much control (ownership) must I keep?
- (c) What position would I like to have within the industry (e.g., leader, follower, etc.)?
- (d) Do I plan on doing something that has never been done before in aquaculture?
- (e) Why is aquaculture a good idea for me?
- (f) Does aquaculture complement any other operations I might have, in terms of making the existing operations more profitable?
- (g) Am I looking at aquaculture because I have idle resources that I think could be used, such as land? If so, what do I expect from using the resources?
- (h) What are my strengths and weaknesses as a manager?
- (i) How have I compensated for or covered and perceived weaknesses?

It is best to confirm business goals through introspection and by contacting the expert or information provider that helped confirm the aquaculture business idea.

The goals of others either involved in or affected by your decision should also be considered, to ensure that future barriers are minimized. Some features of others' goals in an aquaculture business setting are given in the upcoming list. Clearly identifying these goals is essential to evaluating the feasibility of a venture. It is extremely important to accomplish this task before designing the physical attributes of the aquaculture operation.

(1) *Governmental agencies*. Depending on the type of aquatic species being cultured and the location of the aquaculture facility, various governmental agencies can have conflicting goals. Protected fish and wildlife may exist at the site. If exotic species are to be cultured, special procedures may have to be in place in order to prevent them from entering the environment. Special soil and water conservation practices may have to be used. In many cases, special licenses and permits will be required.

(2) *Competitors*. Competition may be friendly or hostile. If it is friendly, common objectives may lead to shared equipment, labor, or marketing efforts, providing a chance to reduce costs. If it is hostile, independent strength in production and marketing should be planned.

(3) *Customers*. The customer is always concerned about the product, and it is important to understand why the customer wants certain things and thereby determine how these desires can be met. For instance, a restaurant owner may need the product to be consistent in size, because of the price of a plate. Determining acceptable quality and consistency is another major consideration.

(4) *Suppliers*. Typically, suppliers of goods and services are after your business, which means they are more concerned about meeting your needs than you are about meeting theirs. However, they may have other goals that you can exploit, such as when a supplier needs your advice for technological improvements or assistance in entering the marketplace for the first time.

(5) *Investors*. Investors are interested in the amount and timing of returns. Therefore, their objectives may not be the same as those of management. For instance, selling the business may provide the best returns, but doing so means that managers lose their jobs or control over the operation.

(6) *Lenders*. Lenders are interested in the amount of risk involved and in getting their principal and interest in a timely fashion. Like suppliers, they may also have an interest in learning the nature of the aquaculture business.

(7) *Employees*. Often overlooked, but vitally important to aquaculture, are the skilled employees. Their goals usually include a higher salary, more benefits, increased responsibilities, and a sense of worth.

Clearly identified goals are essential to evaluating the feasibility of your venture. It is extremely important to accomplish this task before designing the physical attributes of the aquaculture operation.

What Business are You in?

A common mistake made in aquaculture business plans is the lack of sufficient attention placed on marketing.

Many plans are developed from the viewpoint of the aquaculturist who is interested in production, and sales frequently are an assumed item. However, selling the product in large quantities and at a price that will provide adequate returns is equally important.

The aquaculture business person must identify where and to whom the product will be sold. Local and regional markets are the usual places of sale in aquaculture. For instance, catfish are most popular in the southeast United States, whereas trout dominate in the northwest. Determining customer preference for the product and then identifying where those customers are located, where they shop, how much they buy, etc., are all parts of the process of identifying the market. The following is a list of several considerations for identifying an aquaculture-related market:

(1) Physical characteristics
 (a) Who are the buyers?
 (b) When do the buyers buy?
 (c) How many potential buyers are there?
 (d) Where are the buyers located?
 (e) How much can the buyers reasonably buy, in total?
 (f) How often do the buyers buy?
 (g) What are the buyers doing with the product?
(2) Buyers
 (a) Why are the buyers interested in the product?
 (b) What is important to the buyer?
 (c) Why should buyers buy from you in particular?
 (d) How long have the buyers been in business?
 (e) What is the future of the buyers interest in the product?
 (f) Do the buyers buy more at some times than at others?
 (g) Are there particular ethnic or religious groups that buy the product?
 (h) What proof is there that someone will buy the product at your price?
 (i) Is a dependable supply a major factor?
 (j) Do the buyers require purchase contracts? If so, what are the terms? Can you get a contract?

Many seemingly good ideas have not met with consumer acceptance. Therefore, once the product and market have been carefully identified, they should be confirmed. For relatively large endeavors (for instance, statewide), professional marketing agencies should be employed to conduct surveys and market tests and to substantiate the product and market. Other market confirmations could involve negotiating with a local processor or restaurant owner who might purchase the product.

How Much Money do You have to Spend?

Knowing how much you intend to spend on an aquaculture project helps direct planning, by keeping the business plan within reason. Your investment will be determined by the amount that you are willing to risk, either directly or through borrowing, and by the portion of the business that you are willing to share. The amount that you are willing to risk is the amount that you can afford to lose.... Not an easy decision! Also, the more of the business you are willing to share, the less funds you will need to invest. This decision requires serious thought, however, because outside investors will want something in return. Some of the considerations associated with sharing the business are as follows:

(1) Can the aquaculture business be created without bringing in outside investors?
(2) Could the size of the venture be reduced and objectives accomplished without outside investors? If so, what problems are there in doing so?
(3) How much control over operations will be lost if outside investors buy in?
(4) Can you obtain additional management expertise in exchange for some business ownership?
(5) Why do the outside investors want to be involved?
(6) What will be the reporting requirements for the outside investors?
(7) Are there any legal risks in bringing in outside investors?
(8) What form of legal organization is required?
(9) Are the investors risk and return requirements reasonable?
(10) Are the investors reputable?

If you are considering borrowing part of your investment, you need to give careful thought to how much you will be able to borrow. In particular, you must consider what the lender requires. The following questions, often asked by lenders, should be answered when converting your business plan into a financing proposal:

(1) How much money is to be borrowed?
(2) When will the money be needed?
(3) What is the money going to be used for?
(4) How will taking out a loan affect the borrower's financial position?
(5) How will the loan be secured?
(6) When will the loan be repaid?
(7) How will the loan be repaid?
(8) How will alternative possible outcomes for the business affect repayment ability?
(9) How will the loan be repaid if the first repayment plan fails?

Business Operations: Facilities and Processes

A business plan needs a clear description of the facilities (including equipment) and processes to be relied upon. There should also be a good fit between these items, such as matching equipment (e.g., pumps) of an appropriate size with the time and quantity of production.

Facilities. Many facility-design errors can be eliminated by knowing your business goals and limitations, detailing the facility, and confirming the design of the facility. Knowing your management and investing limitations will keep you from planning a facility that is too big to handle. It is usually preferable for the size of the operation to grow in line with management's physical and financial capabilities.

Design the facility with as much detail as possible, given the situation. A common mistake is that too many assumptions are made regarding the facility: "I assumed that anyone could use river water!" "I assumed these ponds would hold water." Don't get caught in these types of situations. Plan the facility in detail, and confirm the details with an expert. A checklist of areas to detail for many aquaculture facilities should include the following:

(1) The source of water
  (a) History
  (b) Physical characteristics
  (c) What it takes to obtain it (e.g., pumps, power, etc.)
(2) The site
  (a) History
  (b) Physical characteristics
  (c) Location
(3) The environment
  (a) Historical patterns
  (b) Labor conditions
  (c) Support businesses (e.g., processors, suppliers, etc.)
(4) The production facility
  (a) Size and type of water impoundments
  (b) Transport of water
  (c) Hatchery and/or tempering facilities
  (d) Special harvest provisions
  (e) Troubleshooting provisions
  (f) Feeding provisions
  (g) Storage
  (h) Miscellaneous equipment
(5) Permitting
(6) Administrative facilities
(7) Special marketing facilities
(8) The fit between facilities and provisions
(9) Provision for emergencies

Processes. After designing the facility, you must decide how you intend to operate it. Specifically, you will need to estimate the inputs, outputs, management, and marketing. It is imperative to know the limiting factors. For inputs and outputs, this means understanding what inputs (such as fingerlings, water, and feed) are available, how much the facility can produce, and the conversion process.

The conversion process (how you end up with more money than you started with each year, after you bought, produced, and sold the product) deserves special attention in the business plan. It is this process that the business depends upon for day-to-day survival. The technical aspects of this process should be well defined with respect to feed conversion ratios, survival rates, growth rates, etc. In addition, how you intend to accomplish this conversion should be described—that is, management and marketing.

Management will make or break the business. Management must see that the various tasks in the conversion process are coordinated and carried out. One way of presenting this aspect in the business plan is to write job descriptions for each manager, to organize your thoughts about operations. To do so, define major tasks, and assign them to the appropriate manager. Be careful not to give any manager too few tasks, that practice can result in "too many chiefs." The process of dividing up tasks and recording job descriptions will give you the basis for developing an organizational structure that shows who reports to whom and where ultimate responsibility and accountability lie. A clear set job descriptions and a chain of command should be documented.

Last but not least, it is important to state how you intend to accomplish your marketing ideas. Describe each step of the process, from where the product goes as soon as it is harvested to the point at which you receive payment. Include the amount of time that you are responsible for the product and what arrangements must be made at each step.

Monitoring and Control. A business entity needs a feedback system to monitor its activities, so that problems or potential problems are detected and appropriate remedies are taken. Feedback should take the form of a well-organized accounting and record-keeping system. For many aquaculture operations, a simple bookkeeper and a production manager, dedicated to documenting the financial and physical activities of the business, will be appropriate. More complicated operations may require the use of a computer and a formal accounting and records department. At minimum, records should be kept concerning the financial and production activities in your aquaculture business. These records include such items as payroll, purchases, sales, open accounts, debt and equity capitalization, quantities produced, and production inputs. Periodically, these records should be summarized in the form of financial statements and production reports and should be reviewed by management.

ANALYSIS

Once you have designed your aquaculture business, you must put it to the test. First, develop several alternative ways in which you might be able to obtain your goals. You can determine these alternatives by making changes in the initial design.

Next, decide what tests you want to make on each alternative and what results are acceptable. "What if" tests, such as "What if the growth rate is less than expected." should be considered. A useful set of tools for testing aquaculture business designs is *pro forma* (projected) financial statements. These statements typically include a *pro forma* cash flow, income statement, balance sheet, and statement of changes in financial position.[1]

The cash flow statement shows the movement of cash through the business. It is valuable for determining when cash is needed and when it can be withdrawn. By using a cash flow statement, timing and amounts to borrow may be better planned. The cash flow statement can point to weaknesses, such as when revenue from the sale of the product in a particular month is insufficient to cover a loan payment. It can also be used to determine the return on investment and the amount of time necessary to get your investment back (payback period).

The income statement is used for demonstrating the profitability of the business. It includes all revenues and expenses (including income tax) of the firm, whether or not (i.e., depreciation) they involve cash. The income statement shows which revenues and expenses are most important and tells how efficiently the operation is run, from a financial standpoint.

The balance sheet shows the financial position of the business at a particular point in time. It is useful in identifying the liquidity structure of various assets and liabilities (i.e., how quickly they could be turned into cash) and determining relative financial strengths. The balance sheet shows how much equity the business carries (i.e., what you own minus what you owe).

The statement of changes in financial position is used to show where the business got and used its cash in going from one balance sheet to the next. It can tell whether the sources of funds are properly matched with the uses of funds. For example, was the cash that you got from the short-term operating note used for purchasing operating supplies, or was it used to purchase a tractor instead?

All of the aforementioned financial statements are a good source of information for measuring and evaluating your aquaculture business designs. The types of information provided by the statements include such measures as the internal rate of return, (IRR), payback period, debt-to-equity ratios, times-interest-earned ratio, current ratio, and net-profit margins. Other measures that you may want to consider are the rate of growth in equity, risk measures, and production efficiency measures.

The final step in the analysis of the project is to select the best designs. Compare the designs by using the measurements you've taken from your *pro forma* financial statements. The best design will be the one that most closely matches your business goals. Testing may point to needed modifications in the designs, or it may be that none of the designs are good enough for you to commit time or money to the project. Keep an open mind to design failure! Better that you find out that your design is not feasible after spending only a small fraction of time and money; instead of actually building the aquaculture facility and then discovering that it doesn't work.

[1] Two sources of information that provide an in-depth understanding of financial statements are (*1*) *Coordinated Financial Statements for Agriculture*, by Thomas L. Frey and Danny A. Klinefelter, available through Agri Finance, Suite G, 5520 West Touhy Avenue, Skokie, IL 60077, and (*2*) *How to Analyze Financial Statements in Agriculture*, by John B. Penson and Clair J. Nixon, available through Agri-Information Corporation, P.O. Box M-21, College Station, TX 77844.

SOME OUTSIDE FACTORS TO CONSIDER

Most of the factors discussed thus far have been related to the internal aspects of an aquaculture business. This section describes factors that are outside of the business, but important to consider and include in the business plan. There are two areas outside of the aquaculture firm that need to be considered: the general environment and the industry. The general environment can be viewed through sociocultural, economic, governmental or legal, technological, and international issues. The industry can be examined by assessing competitive forces, including new entrants, supplier power, buyer power, substitute products, and competitive rivalry. Some of the questions that you should consider in analyzing outside influences to your business are as follows:

(1) How many suppliers of seafood similar to yours are there in the area?
(2) Can you produce what the market demands? Is there a minimum quantity to produce?
(3) Are you aware of other new ventures being planned or expansions of existing producers?
(4) Is there existing or potential foreign competition?
(5) What are your comparative advantages?
(6) Can you distinguish your product in the minds of the consumers?
(7) Are there trade or technological constraints restricting potential competitors? If so, are they likely to change? How soon?
(8) Are necessary processors, suppliers, and distributors available? How much market power do they have? What is their likelihood of staying in business? How dependent are you on those processors, suppliers, and distributors?

Sociocultural issues include general attitudes about farm-raised fish and seafood products, religious beliefs, dominant family forms, education, etc. Economic issues center around the end user's ability to purchase your seafood product. Disposable income, leisure time, priority spending, changes in interest rates, the rate of inflation, and the rate of unemployment are some specific economic considerations. Governmental or legal issues sometimes can play a major role in an aquaculture project and should be examined with great care. Rights to water, exotic-species permits, environmental control agencies, and taxing authorities must all be considered in the project. Aquaculture is an emerging industry, with constant innovation and technological improvements. You should weigh the costs of using the latest technology with its potential benefit. International issues are important for some products, such as shrimp and salmon, but are less important for products that are more local in nature, such as catfish and tilapia. Specific international considerations, if applicable, should include topics such as import/export potential; foreign exchange rates; foreign demand and supply; and practical barriers, such as language and customs.

The competitive forces in the industry deserve close attention. The ease with which new competitors can come into the industry should be considered; that is, if the industry is attractive, can outsiders come into the industry easily and compete away profits? If the aquaculture project relies on outside sources for different inputs, such as feed, stock, special management, etc., special attention should be given to the market power (e.g., number and quality) of those suppliers. For example, if there is only one feed company that can supply the type of feed required, the price of the feed will be relatively high, and it could be necessary to hold more feed in inventory. Likewise, buyer market power should be assessed. Buyers may have more or less power, depending on how much seafood is needed and how much is available for purchase. Also, the number of and relative power among buyers should be considered.

Competitive advantage is reduced when there are more substitute products in the market. The number of available substitutes should be examined, including surimi and other imitation seafood products. Finally, the degree of competitive rivalry in the industry should be closely examined. The more destructive the competition, in terms of price cutting and increased service, the more difficult it will be to attain higher profits. If you find an aquaculture business design that satisfies your goals, then you are now ready to begin writing the business plan.

WRITING THE BUSINESS PLAN

If you have followed the previous sections of this contribution closely and have collected the required information along the way, then you are ready to prepare a business plan that you can take to a lender or investor or use as a guide in developing your business. Above all, be sure that *you* are convinced that the aquaculture business is going to be a success *before* writing the business plan.

Business plans can have different formats, depending on the type and source of funding being sought and the general purpose. (If funding is already in place, a plan is still worth writing; as it can easily be used as a guide for strategy.) A suggested format for an aquaculture business plan is as follows:

- Title page
- Table of contents
- Statement of purpose
- Executive summary
- The business
    - History
    - Description
    - Market
    - Marketing
    - Competition
    - Operations
    - Management
    - Research and development (optional)
    - Personnel
    - Loan or investment application and effects
    - Development schedule
    - Summary
- Financial plan
    - Sources and applications
    - Capital equipment list
    - Break-even analysis
    - *Pro forma* balance sheet
    - *Pro forma* income statement
    - *Pro forma* cash budget
    - Historical financial statements
    - Equity capitalization
    - Debt capitalization
- Supporting documents

The rest of this contribution discusses the contents of each item of the suggested format.

Title Page

The title page should include at least four pieces of information: the name of the proposed project, the name of the business, the principals involved, and the address and phone number of the primary contact.

Table of Contents

The table of contents should include the major topics of the body of the business plan and list critical tables and figures of which the investor should take particular notice. Do not include every detail of the plan in the table of contents. The main function of the table of contents is to guide the reader to the critical areas of the business plan.

Statement of Purpose

This section is a brief statement of the aquaculture venture: what is to be accomplished, why this project was chosen, and how the project is to be done. The statement should include requirements for outside funding and describe the repayment plan and source of repayment.

Executive Summary

This section is dedicated to presenting the key elements of the business plan to prospective lenders or investors. Its length should be kept under five pages, in order to increase its likelihood of being read. The summary should begin with a brief restatement of the purpose of the project. Next, the venture's specific products, markets, and business objectives should be described, along with proof that the particular idea possesses a competitive advantage. The specific details of the management team should then be presented, along with the various strengths and weaknesses (and solutions for the weaknesses). Lastly, the financial aspects of the venture should be summarized, showing projected returns, outside financing requirements, and the basic timing of cash flows.

The Business

This section of the business plan provides details of the venture. The following points should be addressed:

History. Describe your present organization, including when it was founded, progress made to date, present form (e.g., partnership, corporation, etc.), past financing, and prior aquaculture successes and experiences.

Description. Concisely lay out the proposed venture. Describe the location, size, products, facility design, and installation procedures. This section discusses in detail what the venture intends to do, such as the type of production technique that will be used, the water conditions of the chosen site, etc. Certain justifications for the facility and products should be given when these aspects are a critical assumption of the business. For example, the production technique (intensive, extensive, or semiintensive) should carry a cost/benefit justification.

Market. This section provides a description of the market in which the aquaculture product will be sold.

Marketing. This section discusses how the venture will specifically cater its product to the market. It should include a product description and details on how the product will be taken to the market described in the "Market" section.

Competition. This section presents an analysis of significant competitors. It includes a listing of direct competitors (such as other aquaculturists within the market segment) and indirect competitors (such as fisheries products, imitation fish products, and fish imports). Possible reactions from competitors and the effects that the product will have on the market should be addressed, and strategic moves by the venture in relation to competitors should be discussed and explained.

Operations. This section provides a detailed explanation of production processes and limitations. Assumptions about the growth of the particular production, the population characteristics, and biomass levels should be identified. Managerial techniques in the different phases of the growth cycle should be discussed. The life cycle of the product should be described in detail, with the appropriate risks and strengths highlighted. A suggested technique for describing operations is to divide the project into its separate functions. Each function can be explained in detail along with its importance. The functions can then be brought together to form the whole system of operations.

Management. This section discusses how the venture will run. The organizational structure is described to show lines of authority and responsibility. An organizational chart should be presented that depicts clear functional duties and communication lines. Each critical functional area (e.g., pond, hatchery, maintenance) requires descriptions of managerial control and feedback. The type and style of management can also be discussed. For example, a fish farm may be run by a sole proprietor who subcontracts most of the labor or by a highly integrated team that performs all functions internally. Depending on the purpose of the business plan, you may want to discuss key personnel and include their credentials. Describing this information is standard when obtaining funds from lenders or investors.

Research and Development (Optional). If the aquaculture venture is involved with new techniques or new animals that have not been raised before, this section should be included. It should discuss whether expertise is to be hired or developed from within and should describe details such as costs, expected accomplishments, and timing.

Personnel. The complete personnel plan shows the expected hiring needs and personnel policies over the planning horizon of the venture. The expected number of full-and part-time employees, the amount of overtime, and any seasonal trends should be included. Also, fringe benefits and control policies need to be listed, as well as whether or not it is critical to keep employees. A demonstration of knowledge of any laws that may affect employees should be included as well.

Loan Application and Effects. This section presents any source of external financing and describes why it is necessary and what benefits are expected. Amounts to be borrow for equipment, land, feed, etc. should be listed, with an explanation of the terms, critical aspects (such as options), required collateral, and why borrowing makes sense. Also, any assumptions concerning financing should be spelled out.

Development Schedule. The timing of the venture development can be presented as a chart or in some other form that shows critical dates. This section should include the decision milestones and guide the reader through each stage of the venture. For example, this section could provide a calender chart that shows various stages of the venture and any completion dates. Any information such as dependence on governmental agencies, weather, or equipment manufacturers should be included.

Summary. This section should present, in an abbreviated form, the information provided in all of the previous sections and general information about the business. It is important to highlight the most important facts and assumptions and to exclude the details.

The Financial Plan

Once the basic business has been presented, it is necessary to demonstrate its economic feasibility over the planning horizon of the venture. Certain information is desired by banks and investors for determining the financial success of the venture. This information is as follows.

Sources and Applications. This section presents a summary of where funds are to be obtained and where they are to be applied. It is a good way to introduce the business to financial intermediaries, who are concerned with the use of the funds that they may be providing and what other funds you plan to obtain. There are four categories of investment: capital assets, initial inventories, working capital, and contingency reserve. The farm firm may reduce assets (a source of funds) or increase assets (a use of funds). Other sources of funds include increasing a liability (borrowing) or obtaining more paid-in capital. Other uses of funds include decreasing liabilities (paying back loans) or paying back of equity capital.

Capital Equipment List. This section provides the lender or investor with a detailed listing of the major equipment items. Some general categories can be used, such as hand tools, office furniture, etc., but all major items should be listed separately. A listing of tractors, hatchery tanks, gate valves, computers, feed silos, seines and nets, and so forth should be provided in order to inform the reader about the depreciability, maintenance, extent of involvement, and collateralization of assets.

Break-Even Analysis. This section gives a good idea about what minimum levels the venture must perform at to meet all of its obligations. The break-even point is the level of production that achieves zero profit (i.e., the revenue minus the variable and fixed costs equals zero). If there is more than one product, the analysis becomes somewhat more difficult, because it is necessary to find out each product's break-even point, which requires the allocation of certain fixed costs. Break-even analysis involves showing the break-even point for varying venture scenarios.

Pro Forma Balance Sheet. A balance sheet should be provided as of the end of each year in the venture's planning horizon. The purpose of this statement is to show the financial position of the venture at various points of time. The assets and liabilities should be as detailed as possible, with proper notes and assumptions provided.

Pro Forma Income Statement. An annualized accrual-based income statement should also be provided over the length of the planning horizon, in order to show profitability of the venture. Supportive data, such as production estimates, learning curves, and feeding efficiency ratios, should be listed. It may be necessary to give month-by-month or quarterly statements for the first few years of the operation, to demonstrate seasonality of the business.

Pro Forma Cash Budget. This financial statement indicates the cash flows of the operation and its ability to meet cash obligations. Each year of the planning horizon should be included in the business plan. The first few years should be presented on a monthly basis, to show seasonality and critical points of cash flows. The cash budget should also indicate the amount and timing of cash flows to and from investors and financial intermediaries.

Historical Financial Statements (Optional). If you have been in business on other projects, or if this venture is an existing business (e.g., for buyouts, expansions, etc.), it may be necessary to provide historical financial statements. This is particularly the case when personal or corporate guaranties are being relied upon for collateral.

Equity Capitalization. This section summarizes the total amount and timing requirements of invested capital. Its purpose is to show the lender or investor the degree of financial risk being taken by the principal(s). The method of payback of capital should also be reported.

Debt Capitalization. This section is similar to the previous one, and the two can even be combined. The purpose of this information is to highlight the extent of involvement by lenders. The types of borrowing, terms, amounts and timing, and percentage of the total capital that the borrowed money amounts to should be indicated. Also an annual ratio of debt to equity should be provided in this section for the planning horizon.

Supporting Documents. This is the final section of the business plan and should include any pertinent information left out of the main body. Documents such as personal resumés of the principal(s), personal financial statements (if not already provided), credit reports (if the venture is already in business), letters of reference, relevant articles, and copies of contracts and legal documents should be placed in this section.

BIBLIOGRAPHY

1. Arthur Andersen & Co., *How to Develop a High-Technology Business Plan*, AA & Co., New york, 1983.
2. P.J. Barry, J.A. Hopkin, and C.B. Baker, *Financial Management in Agriculture*, The Interstate Printers & Publishers, Danville, IL, 1979.
3. Inter-American Development Bank, Project Analysis Department, *Guide for the Preparation of Loan Applications: Fishery*, February, 1984.
4. J.R. Mancuso, *How to Start, Finance, and Manage Your Own Small Business*, Prentice-Hall, Inc., Englewood Cliffs, NJ, 1978.
5. W.R. Osgood, *Planning and Financing Your Business: A Complete Working Guide*, CBI Publications, Boston, 1983.
6. W.R. Osgood and D.P. Curtin, *Preparing Your Business Plan with Symphony*, Prentice-Hall, Englewood Cliffs, NJ, 1985.
7. M.E. Porter, *Competitive Advantage: Creating and Sustaining Superior Performance*, Free Press, New York, 1985.
8. S.R. Rich and D.E. Gumpert, *Business Plans That Win*, Harper and Row, New york, 1985.

ECONOMICS: CONTRAST WITH WILD CATCH FISHERIES

Walter R. Keithly
Kenneth J. Roberts
Louisiana State University
Baton Rouge, Louisiana

OUTLINE

How Do Culture Fisheries Compete with Capture Fisheries?

U.S. Consumption of Seafood and Markets

The introduction to a 1988 study conducted by the U.S. Department of Commerce (1) to comply with a Congressional request begins as follows:

> Recent strides in aquaculture have raised troublesome questions for commercial fishermen. At issue is whether gains in aquaculture output will disrupt markets now supplied principally by the traditional "capture"-type commercial fisheries. Cultured salmon and shrimp already have won important market positions, particularly in the United States and Western Europe. Advances in aquaculture technology promise success for other species.

The introduction concludes by stating the following:

> The impacts of aquaculture will be reflected in prices at all market levels for species that are supplied from both commercial fishing and aquaculture sources, such as salmon and shrimp. There is speculation that a surge in supplies of cultured products will devastate the market and jeopardize traditional suppliers. The salmon/shrimp experience is still too new to use as an empirical basis for determining whether cultured products are a significant threat to the interests of commercial fishing.

More than a decade has elapsed since that Congressionally requested study was conducted and published. During the intervening years, the salmon and shrimp culture industries, as well as some others, have moved from a stage of infancy toward one of maturity. Associated with the increased maturity of many of the culture industries is an increase of available information (data) and subsequent analyses. This entry examines the competition between capture- and culture-based fisheries in the market place and, in particular, the impact of increased cultured supply on captured prices, related markets, and support sectors. Constraints on and degradation of the natural environment, resulting from certain culture practices, can also result in competition with respect to the production of species. This aspect of competition is not treated in this paper, but the reader is referred to Naylor et al. (2) for information. The primary emphasis is placed on the U.S. markets but, where relevant, other markets are examined. All money amounts in this entry are in U.S. dollars.

HOW DO CULTURE FISHERIES COMPETE WITH CAPTURE FISHERIES?

To see how changes in the supply of cultured product affect the price of captured product, consider Figure 1. For a more detailed economic analysis of the effect of cultured seafood products on wild-based products, the reader is referred to (1). Economists generally portray consumer demand for a good (commodity) through the use of a "demand curve," graphically illustrated by the line labeled D–D. This curve, which is generally downward sloping, shows the quantity of a given good that consumers are willing and able to purchase at alternative market prices. The downward slope of the curve implies, as one would expect, that, as price declines (increases), the quantity of the good demanded by consumers increases (decreases). The overall position of the demand curve, reflecting the level of demand at any given price, is generally considered to be influenced by such factors as: income; tastes and preferences; and the price of substitute products. For most goods, increases in income will result in an upward shift of the demand curve. Increases (decreases) in the price of substitutes, by comparison, will generally result in a upward (downward) shift in the demand curve because the *relative* cost of purchasing the original good has now become lower (higher).

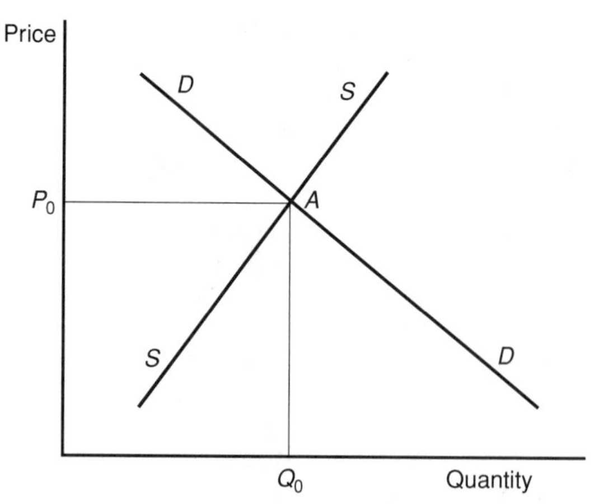

Figure 1. Hypothetical demand and supply curves for seafood from captured fisheries.

In counterpart to the demand curve, economists generally portray industry supply for a good (commodity) through the use of a "supply curve," graphically illustrated by the line labeled S–S in Figure 1. This curve, which is generally upward sloping, depicts the quantity of a given good that industry producers are willing to place on the market at alternative market prices. The upward sloping nature of the curve implies, as one would expect, that, as the market price increases (decreases), producers would be willing to place additional (less) quantity of the good on the market. The overall position of the supply curve, reflecting the level of supply at any given output price, is generally considered to be influenced by such factors as input prices and technology. Improvements in technology, for example, will allow for a higher level of production for any given level of inputs. This effect is represented by the supply curve's shifting downward (equivalently, to the right), a movement implying that more product will be placed on the market at any given output price. Advances in technology in both shrimp and salmon culture since the early 1980s have, in fact, resulted in significant increases in the quantity of the product being offered at a given output price [see, for example, Asche (3), with respect to salmon.]

The market *equilibrium* is determined by the intersection of the demand and supply curves. At the intersection, labeled as A in Figure 1, the quantity demanded by buyers at the stated market price (P_0) is equal to what the producers are willing to place on the market at that price. At any price above P_0, the quantity demanded by consumers is less than what producers are willing to provide. Conversely, at any price below P_0, the quantity demanded by consumers exceeds the amount that producers are willing to supply. Associated with the equilibrium price, P_0, is the equilibrium quantity, denoted Q_0.

Culture-based products can compete with capture-based products in the market in two ways: as perfect substitutes, or as imperfect substitutes. The effect of perfect substitutability is illustrated in Figure 2. The supply curve for the captured product is represented by the curve S–S. The supply curve for the captured plus cultured product is represented by the curve S'–S'. The natural market price of the captured product, in the absence of cultured product, is equal to P_0, and the quantity offered on the market is equal to Q_0. With the addition of the cultured product on the market, the equilibrium price is reduced to P_1, while the equilibrium quantity is increased to Q_1. The greater the advances in culturing technology (other factors being held constant), the more one would expect the equilibrium price to fall below that which would be observed in the absence of competition from a perfectly substitutable cultured product.

Cultured product can also be an imperfect substitute for the captured product. If so, changes in the quantity of cultured product produced, and the related change in what would be the market price of the cultured product alone, will result in a change in the position of the demand curve for the captured product. Assume, for example, that the quantity of a cultured seafood product experiences a significant increase that, in turn, causes a reduction in the product's market price. Under this scenario, consumer demand for the cultured product would be expected to increase and to, in turn, cause a reduction in the demand for the imperfectly substitutable captured product. While not illustrated, the reduction in demand for the captured product can be expressed by a downward shift in the demand curve for the captured product. As a result of the downward shift in the demand curve for the captured product, the equilibrium price received for the captured product will decline, as will the equilibrium quantity.

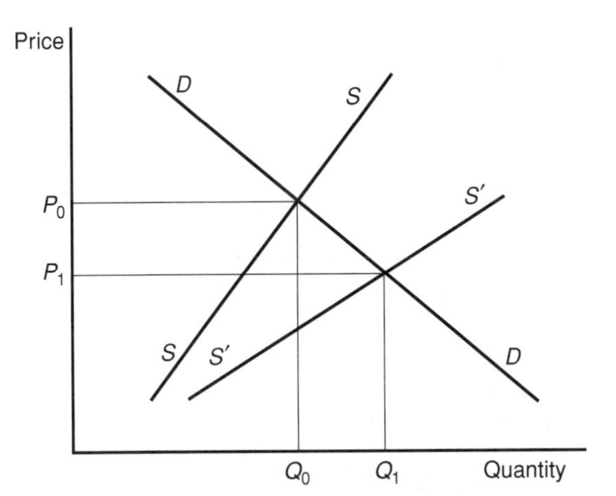

Figure 2. Hypothetical demand and supply curves for seafood, assuming perfect substitutability between captured and cultured products.

When dealing with the seafood market, it is important to recognize that competition between cultured and captured products is often global, rather than local, in nature. This situation reflects the large amount of international trade that occurs in fishery commodities. For example, as examined in greater detail in the next section, a sizeable portion of the seafood consumed in the United States tends to be of imported origin, and the percentage is considerably higher for some of the more "desired" species such as shrimp and spiny lobsters. Hence, when one is examining the impact on captured-product markets resulting from competition between cultured and captured species, it is necessary to take an international perspective. When evaluating competition between cultured and captured shrimp, for instance, one must first recognize that the majority of the shrimp consumed in the United States is of imported origin and that much of it is cultured. These imports can affect the domestic capture fishery through their impact on domestic prices. At the other extreme, much of the United States captured-salmon product is destined for overseas markets. Hence, cultured-salmon product that competes with the U.S. wild product in overseas markets is likely to reduce the dockside price of the U.S. product.

It is also important to recognize that competition between cultured and captured products can extend well beyond the species (or group-of-species) level. It is intuitively obvious, for example, that cultured shrimp is likely to compete with captured shrimp in the market place. Can it be said, however, that cultured shrimp competes with, say, cod or flounder? The answer is yes. Consumers have a limited food budget, a certain percentage of which tends to be allocated to seafood products. An increase in cultured production of shrimp that results in a reduction in the price of shrimp to consumers, therefore, is likely to result in a greater percentage of the seafood budget being directed toward the purchase of shrimp and a smaller percentage being directed toward other species, such as cod and flounder. This discussion tends to neglect potential income effects. Specifically, it is possible that the reduction in shrimp price is large enough that real income is enhanced to the extent that it allows consumers to purchase more of all seafood products. The extent to which this kind of very imperfect substitution occurs depends upon the *cross-price elasticity* of demand.

In the light of the foregoing discussion, the next section briefly examines some trends in U.S. seafood consumption and markets, trends from which some inferences are drawn regarding the competition between cultured and captured products. At the end of the next section, some attention is given to competition with respect to individual species.

U.S. CONSUMPTION OF SEAFOOD AND MARKETS

Per-capita consumption of fish and shellfish in the United States, which equaled 5.67 kilograms (12.5 pounds) of edible weight in 1980 (Fig. 3), peaked in 1987 at 7.35 kilograms (16.2 pounds). Since 1987, per-capita consumption has been gradually trending downward; the 1997 per-capita consumption, 6.62 kilograms (14.6 pounds), was only 90% of that observed in 1987 (4). Wessels and Anderson (5) suggest that the downward trend may reflect increases in relative prices of (or decreased demand for) luxury goods during the recessionary period of the early 1990s, as well as perceived increase in the health risk associated with the consumption of seafood. It is noteworthy that, despite large increases in the U.S. real disposable income since the early 1990s, no increase in U.S. per-capita consumption is evident. It has also been suggested that a lack of supply, a lack of marketing activities that would permit the seafood industry to compete better with other, more established industries (such as beef), and a lack of product development (to enhance the convenience of preparing the product for at-home consumption) also play a role in the observed decline in per-capita seafood consumption in the United States (6, 1998).

Despite the downward trend in U.S. per-capita consumption in recent years, total consumption, expressed on an edible-weight basis, has remained relatively stable in the 1.7 to 1.8 billion kilogram (3.75 to 4.0 billion pound) range, as a result of an expanding population base. The product required to meet these consumption needs consists of both domestic and imported product and of both captured and culture product. In 1980, for example, the U.S. supply of edible fishery products (round weight) consisted of 1.66 billion kilograms (3.65 billion pounds) of domestically harvested product and 1.97 billion kilograms (4.35 billion pounds) of imported product; these figures indicate that approximately 55% of the U.S. supply was import-based (Fig. 4). The percentage of imports reported herein likely under represents the proportion used in domestic consumption, because a sizeable percentage of the domestic supply is exported. The ratio of imported to domestic product in overall edible supply peaked in 1985 and 1986, when imports represented approximately 65% of the total supply. Since 1986, the ratio of imported to domestic product has fallen, with imports representing only 40% of the total edible seafood supply by 1993. In 1997, domestic product equaled 3.29 billion kilograms (7.25 billion pounds), or almost 60% of the U.S. edible seafood supply; imports equaled 2.94 billion kilograms

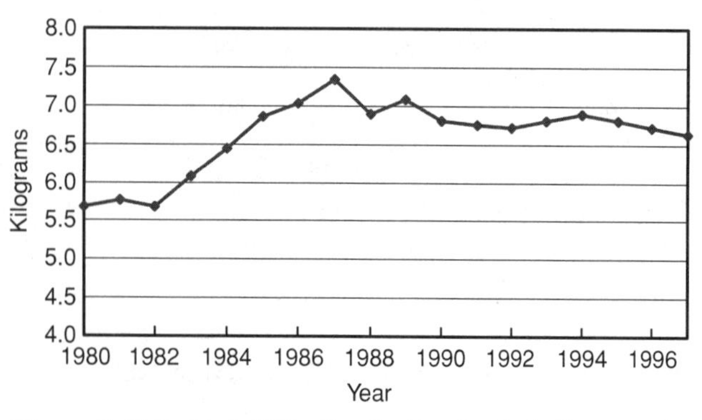

Figure 3. Estimated U.S. Per-Capita Seafood Consumption (edible weight), 1980–1997. *Source*: (4).

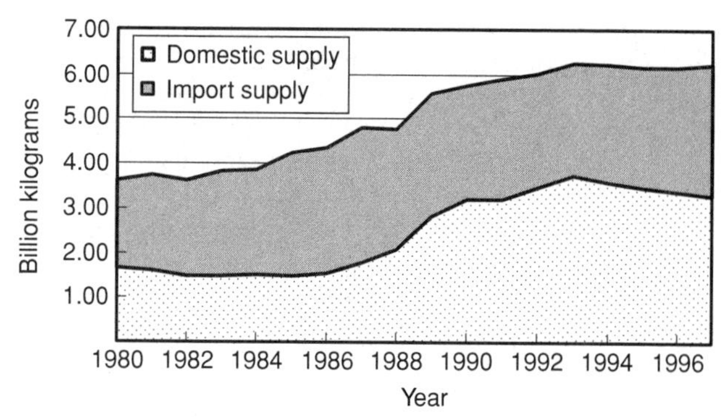

Figure 4. U.S. Supply of Edible Fishery Products (round weight), by domestic and imported sources, 1980–1997. *Source*: (4).

(6.5 billion pounds). It is noteworthy that much of the increase in the domestic supply has been the result of increased pollock landings off the coast of Alaska. Specifically, pollock landings increased from less than 2.3 million kilograms (five million pounds) annually in the early 1980s to in excess of 1.36 billion kilograms (three billion pounds) annually between 1990 and 1993, before falling to 1.13 billion kilograms (2.5 billion pounds) in 1997 (4).

U.S. per-capita consumption of selected seafood items is provided in Table 1 for the years 1987 and 1997. Several features are highlighted by the information. First, domestic consumption of canned tuna, though falling on a per-capita basis, dominates total consumption, accounting for in excess of 20% of the total of the 10 most-consumed species in both 1987 and 1997. Second, per-capita consumption of many of the "traditional" capture-based fishery products — such as cod, flatfish, and clams — has fallen sharply during the period of analysis, at least in part on account of overfishing and the subsequent reduction in the overall supply of these species. The one capture-based fishery wherein per-capita consumption has advanced by a substantial amount is that of Alaska pollock. The harvest of this species is used primarily in surimi-based products, such as crab analog. While only

Table 1. U.S. Per-Capita Consumption (in kilograms) of Selected Seafood Items

	1997	1987	% Change[b]
1. Canned Tuna	1.41	1.59	−11.4
2. Shrimp	1.23	1.04	17.4
3. AK Pollock	0.74	0.40	84.3
4. Salmon	0.59	0.20	203.5
5. Cod	0.48	0.76	−37.1
6. Catfish	0.46	0.27	69.3
7. Clams	0.21	0.30	−30.4
8. Crabs	0.19	0.15	31.3
9. Flatfish	0.15	0.33	−54.6
10. Halibut	0.13	0.11[a]	15.3

[a]Per-capita consumption of halibut was unavailable for 1987. The number presented here is from 1988 data.
[b]Percentage change may not quite agree with the numbers given in the table, because of rounding.
Source: (6, various issues).

recently developed, the fishery is already showing sign of overfishing: landings declined, from 1.5 billion kilograms in 1993 to 1.13 billion kilograms in 1997 (4). Finally, the information contained in Table 1 suggests that three of the top ten species — shrimp, salmon, and catfish — are all associated, to a greater or lesser extent, with culture activities. Per-capita consumption of these three species accounted for only 21% of the total per-capita consumption in 1987, but almost 35% of the total in 1997. Overall, U.S. per-capita consumption of salmon tripled during the period, while per-capita consumption of catfish and shrimp increased by 170% and almost 20%, respectively.

At least two inferences, with respect to market impacts resulting from competition between cultured and captured product, can be drawn from the information presented above. One is that cultured products appear to be increasingly being substituted for the traditional captured products in the American seafood diet. The implications of this process are two-fold. First, the increased consumption of cultured products, because they are in general imperfect substitutes, has likely resulted in a reduction in demand for some of the traditional captured species, all other things being equal. This reduction implies that prices for some of the more traditional captured species are likely lower than they would have been in the absence of the increased availability (and consumption) of the cultured species. The second implication is that the increased availability of cultured species, to the extent that their increased acceptance has resulted in a reduction in demand for the captured species, has helped to alleviate at least some of the fishing pressure (and resultant, overfishing) generally associated with many of the captured fisheries. As one example, Keithly et al. (7) estimated that the number of vessels in the U.S. Gulf of Mexico shrimp fishery would have been approximately 7,100 in 1988 rather than the reported 5,930, if there had been no shrimp aquaculture (and concurrent increase in U.S. imports) that subsequently led to a decline in the (constant dollar) dockside shrimp price.

The second inference that can be drawn from the information presented earlier is that U.S. consumption of seafood products is, to a large extent, driven by the import market — and that, in turn, is driven largely by global activities. Consequently, changes in activities in other countries that result in increased culture production and subsequent export of this product to the United States can significantly reduce captured production in the United States. As discussed in greater detail in later sections, this is the situation facing several of the larger capture-based U.S. fisheries, particularly shrimp and salmon. Specifically, increased cultured production of these species in other regions of the world has significantly reduced the U.S. prices of these captured species (as measured at the dockside).

ANALYSIS OF INDIVIDUAL SPECIES

Six species are examined in this section of the paper. Two species, shrimp and salmon, are examined in considerable detail, because of their significance in world trade as well as in the U.S. seafood budget. Other species considered include crawfish, alligator, catfish, and tilapia. These six species, while not all-inclusive, represent many of the more relevant species, for which cultured product competes with captured product, and their story provides adequate "flavor" regarding the different ways in which the cultured product can compete with the captured product.

Shrimp

World exports of shrimp, valued at $8.1 billion, constituted about 15% of the $52 billion international trade market in fisheries commodities in 1996 (8) (1996) and were the largest single item in seafood trade. Trade in shrimp has expanded considerably since the 1980s, in response both to increased world production of the commodity, primarily from culture activities, and to favorable economic conditions. To examine the extent to which cultured shrimp competes with the captured product in the market place, some basic production trends are first considered for the period 1980–1996. Then, attention is given to the trade market in shrimp. Finally, the competition between the two products is considered.

Production Trends. Shrimp production, as with the other seafood commodities discussed in this section, is a combination of wild harvest and farming activities. Estimated total annual shrimp production (i.e., capture and culture production) throughout the world, as indicated in Figure 5, expanded from 1.57 billion live-weight kilograms (3.48 billion pounds) in 1980 to 2.9 billion kilograms (6.41 billion pounds) in 1996. Production of *Acetes japanicus*, a species harvested primarily in China and used for paste, has been excluded from total production figures presented herein. This increase translates into a growth rate of approximately 78 million kilograms (170 million pounds) per year.

In general, shrimp production can be segmented between warm-water species and cold-water species. The warm-water species tend to grow considerably larger than the cold-water species and also tend to command a much higher price per unit weight on the world market (7). Production from Asia, Central America, and South America tends overwhelmingly to be of the warm-water species. The United States produces significant quantities of both warm-water and cold-water species. Shrimp production in most other producing areas of the world is primarily of cold-water species.

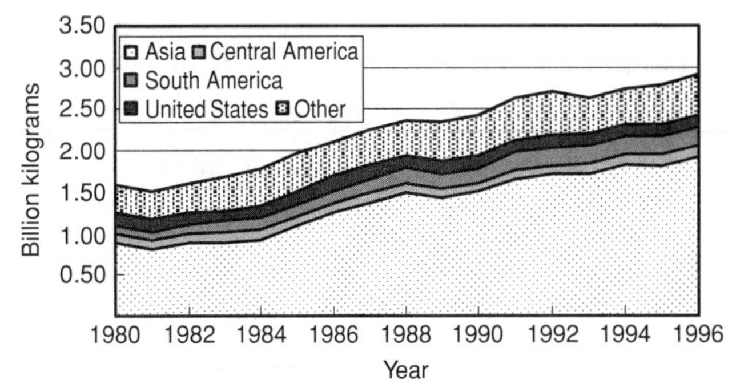

Figure 5. Estimated World Shrimp Production (live weight), by primary producing regions, 1980–1996. *Source*: (9).

The majority of world production, as is indicated in Figure 5, is Asian based. In 1980, an estimated 55% of the world shrimp production was of Asian origin. With a doubling in production since 1980, Asia's production of 1.91 billion kilograms (4.23 billion pounds) in 1996 represented 65% of the total world output of captured and cultured shrimp. Shrimp production in Central America generally fell into the relatively narrow range from 100 million to 125 million kilograms (220 million to 275 million pounds) annually during the period 1980–1992. After 1992, however, production advanced by a substantial amount, to about 133 million kilograms (294 million pounds) in 1996. South American production, more than doubled during the period of analysis; 85 million kilograms (187 million pounds) in 1980, compared to 232 million kilograms (512 million pounds) in 1996.

Combined production from Central and South America (194 million live-weight kilograms, or 428 million pounds) represented 13% of estimated world output in 1980. In general, little change in the share of world production from this area has been evident, despite substantially higher production in recent years (366 million kilograms in 1996). While production from Central and South America represents a relatively small percentage of the world total, Latin America plays a significant role in the international shrimp market because much of its output is destined for the U.S. market (10). This relationship is examined in greater detail next.

The United States, although one of the world's two largest importers of shrimp (along with Japan), nonetheless produces significant quantities of shrimp (Fig. 5). With exceptions, annual live-weight production of shrimp in the United States generally ranged from about 135 to 165 million kilograms (300 million to 365 million pounds), and no discernible trend in U.S. production was evident, during the period 1980–1996. Shrimp production from "other" areas tended to average 15% to 20% of the world production during 1980–1996. Most of the production is of cold-water species, with Europe generally accounting for approximately one-half of the total.

Much of the growth in world shrimp production since 1980 has been the result of successful culture activities throughout the world, particularly in Asia and, to a lesser extent, in South and Central America. World production of cultured shrimp equaled about 90 million live-weight kilograms (200 million pounds) in 1980, about 6% of total world production then. By 1996, culture production had advanced to about 0.91 billion live-weight kilograms (2.0 billion pounds), approximately one-third of the total world production of shrimp at that time. The penaeid shrimps account for more than 95% of the cultured production (11). Wild harvest during the same period advanced from about 1.6 billion kilograms (3.5 billion pounds) to 2.0 billion kilograms (4.4 billion pounds), or by about one-quarter. It is thus evident that the overwhelming majority of shrimp production growth during the period 1980–1996 has been the result of expanded farming activities.

When examined on a regional basis, world production and growth in cultured shrimp is dominated by Asia. Estimated Asian production of cultured shrimp in 1985 equaled about 180 million live-weight kilograms, which accounted for about 85% of the world supply of cultured shrimp for that year and about 15% of the region's total shrimp production (i.e., wild plus cultured). By 1996, Asia's production of cultured shrimp had increased to 0.78 billion kilograms, which represented almost 80% of the world's supply of cultured shrimp and about 40% of the region's total shrimp output (12). Major Asian producers of cultured shrimp in 1996 included Thailand (223 million kilograms; 491 million pounds), Indonesia (156 million kilograms; 343 million pounds), India (87 million kilograms; 192 million pounds), Philippines (78 million kilograms; 172 million pounds), and China (89 million kilograms; 195 million pounds).

South American shrimp culture advanced from an estimated 32 million live-weight kilograms (71 million pounds) in 1985 to 125 million kilograms (275 million live-weight pounds) in 1996 (12). The cultured production's share in relation to the total production from the region for the period ending in 1996 advanced from less than one quarter to more than one half. Ecuador, the primary producer of cultured shrimp in the region, accounts for about 85% of the South American total. Production of cultured shrimp in Central America is relatively minor compared to that in either Asia or South America. It did, however, increase from a negligible amount in the mid-1980s to about 35 million kilograms (80 million pounds) in 1996. Mexico accounted for more than a third of the 1996 farm-based output; Panama and Honduras also contributed sizeable amounts (12).

Trends in World Exports and Imports. World exports of fresh and frozen shrimp (the two categories constituting the overwhelming majority of trade) equaled 387 million product weight kilograms or 854 million pounds in 1980 (Fig. 6). By 1996, exports had increased by about 170%, to 1.09 billion kilograms (2.4 billion pounds). The fact that the percentage increase in the world shrimp trade from 1980 to 1996 greatly exceeded the 75% increase in world production during the corresponding period suggests that an increasing proportion of world production is being traded on the world market. Keithly and Diagne (10) estimated that about 40% of the world production entered the world trade market in 1980, compared to 60% currently.

The value of world shrimp exports in 1980 equaled $2.3 billion. The Food and Agriculture Organization of the United Nation's estimates that this value represented 15% of the total world trade in seafood products. By 1996, the current value of world shrimp trade had more than tripled, to $8.1 billion. Much of the apparent increase in the value of world shrimp trade during the period of study was due to currency inflation. After adjusting for inflation, and in terms of the 1982–1984 Consumer Price Index, the value of world shrimp trade advanced by 85%, from $2.8 billion to $5.2 billion. This 85% increase in constant-dollar value was significantly lower than the 175% increase in export quantity, indicating a sharp decline in the real price of the exported product during the period of analysis. Overall, the 1996 constant-dollar price, $4.78 per product-weight kilogram ($2.17 per pound), reflects a one-third decline

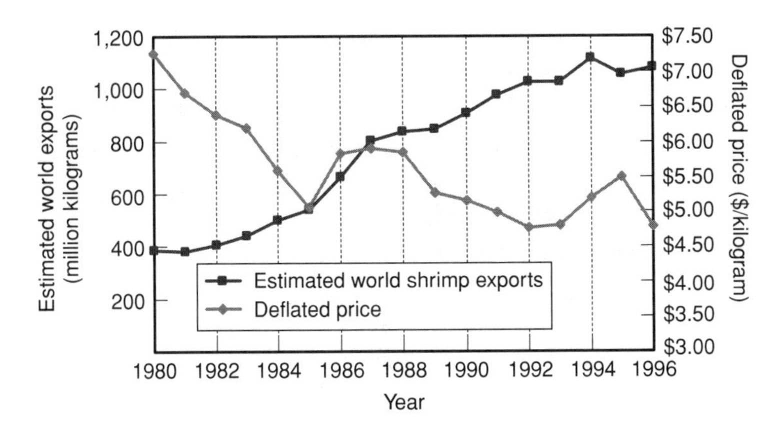

Figure 6. World Exports of Shrimp (product weight) and associated export price (in constant dollars), 1980–1996. *Source*: (8).

in export price when compared to the $7.25 per kilogram ($3.29 per pound) price observed in 1980 (Fig. 6). This decrease would tend, all other things being equal, to suggest that growth in supply from 1980 to 1996 exceeded the growth in demand and caused a downward trend in the real price of the product.

The five largest shrimp exporters by value in 1980, as is indicated in Table 2, were Mexico ($495 million), India ($233 million), China ($180 million), Indonesia ($178 million), and Australia ($131 million). Exports from these five countries represented 53% of 1980 global exports by value and 40% by weight. Of these five countries, only Indonesia and India had any sizable cultured-shrimp activities at the time. In conjunction with growth in cultured production in many countries throughout the world, export sources changed. By 1996, Mexico had fallen from being the largest exporting country to being only the fifth-largest, while Thailand, which was not even among the top five exporters in 1980, replaced Mexico as the largest exporter of shrimp. All of the five largest exporting countries in 1996, with the exception of Mexico, reported cultured production in excess of 68 million live-weight kilograms (150 million pounds), and two of the countries, Thailand and Indonesia, had cultured production surpassing the 136 million kilogram

Table 2. World Exports of Fresh and Frozen Shrimp by Principal Countries, 1980 and 1996[a]

Country	Million Kilograms	$ Million
	1980	
1. Mexico	43.6	495.0
2. India	47.8	233.3
3. China	21.7	180.2
4. Indonesia	30.5	177.9
5. Australia	12.1	130.7
	1996	
1. Thailand	162.2	1,721.9
2. Indonesia	84.3	848.3
3. India	95.7	659.8
4. Ecuador	85.7	627.4
5. Mexico	38.1	392.8

[a]Ranked by value of export sales.
Source: (8).

(300 million pound) mark. As noted by Csavas (13), cultured shrimp is of greater importance than wild shrimp in export markets. The reasons cited by the author include the following: (*1*) the cultured product has greater "freshness" than the wild product; (*2*) cultured shrimp production is less seasonal in nature, and more reliable, than its wild counterpart. As noted by Csavas, the seasonal nature of wild-based product results in idle capacity among processing establishments over extended periods of time and has increased storage costs for importers and exporters, who must keep higher supplies to satisfy consumer needs. (*3*) species and sizes can be controlled better in a cultured-based system than in a wild-based system; and (*4*) the current trend towards vertical integration in the cultured system lends itself to better adaption to consumer needs.

While the primary exporters of shrimp are many and have changed substantially over time, two countries, the United States and Japan, have long dominated the import market. These two countries accounted for almost three-quarters of world shrimp imports, by value, in 1980, and about 60% in 1996 (8). In general, Japan and the U.S. prefer the warm-water species that account for the vast majority of cultured shrimp production (10). Shrimp imports by the European countries, which account for much of the remaining trade in shrimp, have historically been dominated by the cold-water species, though this situation is gradually changing. In particular, the European countries are reportedly purchasing an increasing share of the Ecuadorian product. Jacobson (14), for example, reported that 31% of the 1992 Ecuadorian shrimp exports went to the European market, 67% to the U.S. market. Annual U.S. shrimp imports, expressed on a *headless shell-on equivalent* basis, approximately tripled during the 17-year period ending in 1996, from 118 million kilograms (260 million pounds) to 327 million kilograms (721 million pounds) (Fig. 7). Imports from Asia advanced from about 32 million kilograms (70 million pounds) in 1980 to 175 million kilograms (385 million pounds) in 1996, while imports from South America advanced from 14.5 million kilograms (32 million pounds) to 69 million kilograms (152 million pounds). Increased U.S. imports of shrimp from both of these regions have largely been in response to increased output, particularly cultured production (10). Annual U.S. shrimp imports from Central America, unlike those observed from Asia

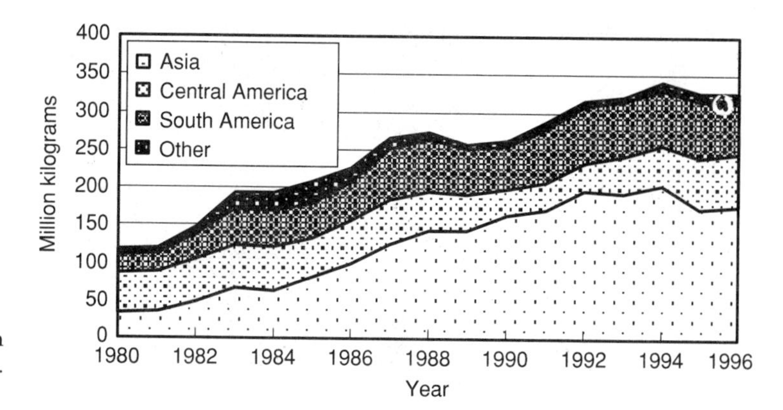

Figure 7. U.S. Imports of Shrimp (headless shell-on equivalent weight), by primary regions, 1980–1996. *Source*: (15).

and South America, exhibited little or no upward trend over the period 1980–1996. This period coincides with the relatively long-run stable production in the Region (see Fig. 5). Overall, the United States imported an estimated 13% of the total world shrimp production (excluding U.S. production) in the early 1980's, about 20% by the mid-1990s. This increase primarily reflects the greater U.S. utilization of cultured Asian and, to a lesser extent, South American production.

Japan's imports of fresh and frozen shrimp advanced from 144 million product-weight kilograms (318 million pounds) in 1980 to 289 million kilograms (636 million pounds) in 1996 (16). Japan's imports cannot be converted to a headless shell-on weight, as was done for the United States, because of insufficient information. Niemier and Walsh (17) report that about 70% of Japan's total shrimp imports are in the form of a headless shell-on product. The vast majority of Japan's shrimp imports are of Asian origin — from 75% to more than 80%. As importers, Japan and the U.S. compete primarily for the Asian product. As was the case for the United States, increased culturing activities in Asia have provided additional shrimp supply for the Japanese market.

Competition Between Wild and Cultured Shrimp. The United States, as noted, harvests sizeable quantities of shrimp. These harvests are composed of both warm-water (panaeid) and cold-water (pandalid) species. The warm-water species are harvested primarily in the Southeast (the coastal states extending from North Carolina through Texas); the cold-water species are harvested in the New England and Pacific Regions. Farm-raised shrimp are primarily of warm-water species and compete primarily with the U.S. captured warm-water shrimp production.

The annual capture of warm-water shrimp in the United States (i.e., Gulf of Mexico and South Atlantic landings) is illustrated in Figure 8 for the 1980–1997 period. It varies considerably on an annual basis, and no discernable trend in annual production is evident. It has been suggested that the large variation is primarily induced environmentally, largely by changes in salinity and water temperature during the shrimp's growth cycle — many factors contribute to the number and average size of shrimp caught (18).

While the U.S. annual capture of warm-water shrimp has exhibited no discernable trend between 1980 and 1997, the constant-dollar price of the landed product has clearly fallen; as with landings, however, considerable variation in the dockside price is also evident. Overall, the 1995–1997 constant-dollar dockside price, $2.82 per live weight kilogram ($1.28 per pound), was approximately 30% below the constant-dollar 1980–1982 average annual dockside price, $4.01 per live-weight kilogram ($1.82 per live-weight pound) (approximately $2.95 per headless pound). This decline is particularly striking in the light of the fact that U.S. per-capita income has risen sharply since the early 1990s. There is general agreement (*1*) that most, if not all, of the decline in real price since 1980 is the result of increased imports, primarily-farm based in nature, which compete directly with the domestic wild catch in the market place and (*2*) that, in the absence of the increased cultured production, prices in 2000 would be significantly higher than those observed in the early 1980s.

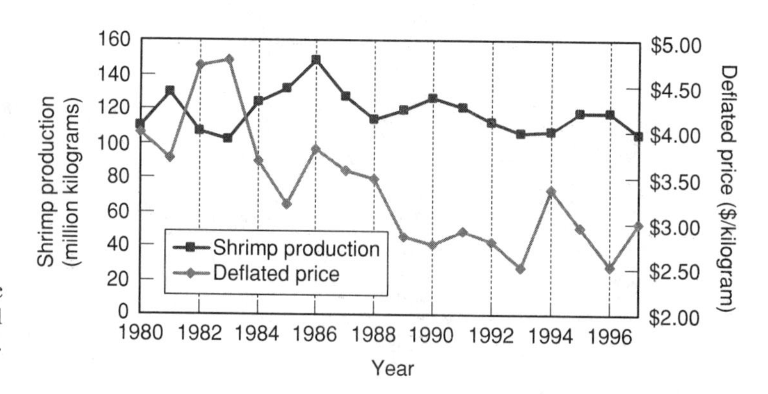

Figure 8. Annual U.S. Gulf of Mexico and South Atlantic (i.e., warm-water) shrimp landings (live weight), and associated dockside price (in constant dollars), 1980–1997. *Source*: (4).

In one of the few studies that attempted to evaluate the impact of cultured shrimp on U.S. dockside prices, Keithly et al. (7) estimated that, in the absence of farm product, the 1988–1989 Southeast U.S. dockside shrimp price would have been approximately 70% above that actually observed during the period, and import supply would have been approximately 30% below observed levels. The authors also suggested that, in the absence of significant income expansion in the United States, further increases in cultured shrimp production would result in further erosion of the domestic U.S. real dockside price.

Using monthly time-series data covering the period 1981–1995, Gillig et al. (19) analyzed the U.S. Gulf of Mexico shrimp *ex-vessel* prices by size classes. Overall, three size classes of shrimp were examined: under 30 count to the pound (0.45 kg), 30–67 count to the pound, and over 67 count to the pound. The authors found that South American exports to the United States (lagged one month) significantly influenced the dockside prices of all three classes, a fact implying that South American shrimp substituted for all size classes of domestic landings. The authors found that a 10% increase in South American exports to the United States would be expected to result in a decrease of less than 1% in the dockside price of the large shrimp, but one of almost 2% in the price of the small shrimp. The fact, that (a) South American exports to the United States approximately tripled during the period 1980–1996 (from 23 million kilograms to 69 million kilograms) and (b) the vast majority of this increase was in increased cultured product from Ecuador suggest that increases in cultured product in South America have significantly reduced the Gulf of Mexico dockside prices in all classes. Exports to the United States from Asia (lagged one month) were found to reduce the dockside price of only the midsize [i.e., 30–67 count to the pound (0.45 kg)] shrimp; exports from Central America (lagged one month) were found to reduce the Gulf of Mexico dockside price of only the small size [i.e., over 67 count to the pound (0.45 kg)].

Keithly and Diagne (10) used a slightly different technique to examine the impact of increasing imports on the Gulf of Mexico dockside price, using quarterly data covering the period 1980–1995. Assuming that the imported shrimp from the different regions were imperfect substitutes for domestic production, the authors specified the Gulf of Mexico dockside price (deflated to constant dollars) as a function of the deflated prices of the imported product from the three primary exporting regions (Central America, South America, and Asia), of the quantity landed in the Gulf of Mexico, of the average size of shrimp landed in the Gulf of Mexico (i.e., estimated number of shrimp to the pound), and of other exogenous variables. The authors found that the Central American import price had the greatest effect on the Gulf of Mexico dockside price: a 10% increase (decrease) in Central American price resulted in a 7.2% increase (decrease) in the deflated Gulf of Mexico dockside price, all other variables being held constant. By comparison, a 10% increase (decrease) in the South American import price was found to result in only a 4.7% increase (decrease) in the Gulf of Mexico dockside price; a 10% increase (decrease) in the Asian import price was estimated to result in only a 2.5% increase (decrease) in the Gulf of Mexico dockside price. Both the Central American and the South American real export prices to the United States fell sharply during the period 1980–1995; these declines help to explain the decline in the Gulf of Mexico dockside price.

The Southeast U.S. harvesting sector is one component of the shrimp industry that generally has been unable to make the adjustments needed in response to the growing import base and the resulting suppression of prices. Not unexpectedly, therefore, requests for relief from perceived problems associated with increasing imports originate primarily from the harvesting sector. In the southeastern United States, harvesters have frequently sought regulatory relief from burdensome imports.

The first of three such attempts occurred in 1975, when the U.S. International Trade Commission (USITC), through the public hearing process, reacted to a petition filed by the National Shrimp Congress. The subsequent investigation in 1976 sought to determine whether shrimp products identified in item 114.45 of the Tariff Schedules of the United States were being imported in quantities that caused serious injury to the domestic shrimp industry. The initial investigation, it should be noted, occurred before cultured shrimp production was a significant factor. The analyses and public testimony resulted in a finding of serious injury to the native capture fishery. Adjustment assistance permitted under Title II of the Trade Act was approved, to allow shrimp-boat operators to obtain loans or loan guarantees. This action, it was reasoned, would help domestic shrimp producers become competitive with foreign producers. Approximately five years later, it was pointed out that this action had actually failed to provide a remedy. In response, John Breaux of Louisiana (then a U.S. Representative and later a U.S. Senator), authored a bill to formulate policy, including a temporary quota combined with a 30% ad-valorem tariff, to provide for domestic shrimp industry protection (H.R. 4041). Although the bill failed to attain the support necessary for passage, it was significant, because it was introduced at the time that cultured shrimp was becoming an increasingly significant factor, and because attention was focused on the harvesting sector.

The focus remained on the shrimp harvesting sector when, in 1985, the International Trade Commission again evaluated the shrimp import situation. Renewed supply increases, primarily from cultured activities, were being experienced. The frequent forecasts of overseas successes of shrimp-farming companies were becoming a reality. The prospect of additional shrimp farming successes in Central America, South America, and Asia loomed on the business horizon. In explaining the situation to the U.S. International Trade Commission (20), the U.S. Gulf of Mexico and South Atlantic shrimp harvesters claimed (a) that harvesting businesses were being injured as a result of imports, and (b) that shrimp industries in foreign countries benefit from government assistance, which artificially allows their products to be more competitive in U.S. markets.

Following a staff review of the information and a public hearing, the USITC chose to issue a report rather than to recommend actions. Through these activities, it became apparent that United States policy and regulations would not be used to protect captune-fishery participants; the implication was that the harvesting sector would have to compete with a growing import base and further price suppression.

The impetus for trade investigations for shrimp, as noted, emanated from the southeastern U.S. shrimp harvesting sector. Other components in the industry, most notably the processing sector, did not actively pursue import restraints; they were benefitting from the increased imports at the time of the investigations. Some anecdotal evidence of this fact was presented by Roberts et al. (21), through the examination of secondary data available on the shrimp processing sector. The authors reported that during the period 1970–74, southeastern U.S. landings of shrimp averaged 66.7 million kilograms (147 million headless pounds) annually, while headless shell-on equivalent weight processing activities equaled 94.4 million kilograms (208 million pounds); the discrepancy suggested an annual deficit, in domestic supplies relative to processing requirements, of about 27.2 million kilograms (60 million pounds). This deficit had expanded to almost 32 million kilograms (70 million pounds) by 1980–1984, and it exceeded 40 million kilograms (90 million pounds) by 1985–1988. The increasing deficit between domestic landings and processing activities led the authors of that study to conclude that imports were playing an enhanced role in the southeastern U.S. processing activities during the period of analysis. In addition, the authors found, through collection of primary data, that (a) while Florida and Georgia shrimp processing establishments exhibited a long history of imported shrimp usage, firms in Alabama, Mississippi, and Louisiana had begun using imported shrimp in their processing establishments more recently, in conjunction with the increased cultured shrimp exports to the United States; (b) processing establishments that used imported shrimp more than doubled their output during the period 1974–1987, while firms not using imported shrimp, by comparison, showed no growth; and (c) among plants not using imports, growth occurred only in the raw peeled shrimp category, but among processors of imported shrimp, the shell-on, raw-peeled, and breaded activities all exhibited large quantity increases. In general, the authors concluded that imports of cultured shrimp permitted the processing sector to increase overall processing activities significantly, by providing an additional source of raw-material supply.

In a subsequent analysis, in 1991, Keithly and Roberts (22) found that, while the southeastern U.S. shrimp processing sector depended primarily on domestic landings for the production of raw headless and peeled raw product, production of the peeled cooked and breaded products relied primarily on imported raw material to satisfy processing capacities. Specifically, imported raw material constituted more than 90% of the input used in the production of both breaded and peeled cooked product, but only about 20% of the raw headless and peeled raw product. Overall, three countries (China, Ecuador, and Thailand) were found to account for approximately two-thirds of all the imported shrimp used in the southeastern U.S. shrimp processing activities in 1991. All of these countries were major producers of cultured shrimp at the time of the study. Year-round availability of the imported (primarily, cultured) product was one of the reasons most frequently cited for the use of imports.

Recent analysis by Diop (23) suggests that the southeastern shrimp processors may no longer be benefitting from the increased import supply, because of a continued narrowing of the marketing margin between the price of the processed product and the price of the raw input used to produce the processed product. Overall, this study found a pronounced decrease in the number of processing establishments between 1988 and 1996 (a 37% decline), along with little or no increase in processed quantity (product weight). To adjust to the declining marketing margins, the remaining processors have increased the output per establishment. This analysis suggests that, if increases in cultured shrimp production and subsequent export of the product to the U.S. market continue at the rate observed during the past decade, the marketing margin between raw and processed product, as well as the number of firms in the industry, will further erode.

Salmon

Growth in the per-capita consumption of salmon in the United States during the 1987–1997 period, as earlier noted, has been substantial. That growth reflects large increases in the production of cultured salmon throughout the world, much of which is destined for the U.S. market. There exists, however, a sizeable capture fishery for salmon in the United States, and much of its product is exported to European nations and Japan. The increased cultured production, on top of the increases in the captured product, has resulted in a significant decline in the constant-dollar U.S. dockside price of the different salmon species. To examine this phenomenon, trends in world production of cultured and captured salmon will first be reviewed; then, the export market for salmon will be analyzed; finally, competition between the captured and the cultured product will be investigated.

Production Trends. World production of salmon has expanded, from an estimated 568 million kilograms live weight (1.25 billion pounds) in 1980 to 1.56 billion kilograms (3.4 billion pounds) in 1996 (Fig. 9). As with shrimp, much of the increase has been the result of successful culture activities. In 1980, culture product equaled approximately 7.2 million kilograms (16 million pounds), approximately 1% of the total world production. By 1996, cultured production, equal to 643 million kilograms (1.4 billion pounds), accounted for 40% of the total world salmon supply. As noted by Anderson and Fong (25), a sizeable, but unknown, percentage of the wild salmon production is attributable to ranching. Inclusion of this ranched production would increase the culture share.

Three species of salmon — Atlantic, chinook, and coho — account for virtually all of the farming activities in recent years (12). Farm production of Atlantic salmon in 1996 equaled 556 million kilograms (round weight) and represented 85% of the global farm production of

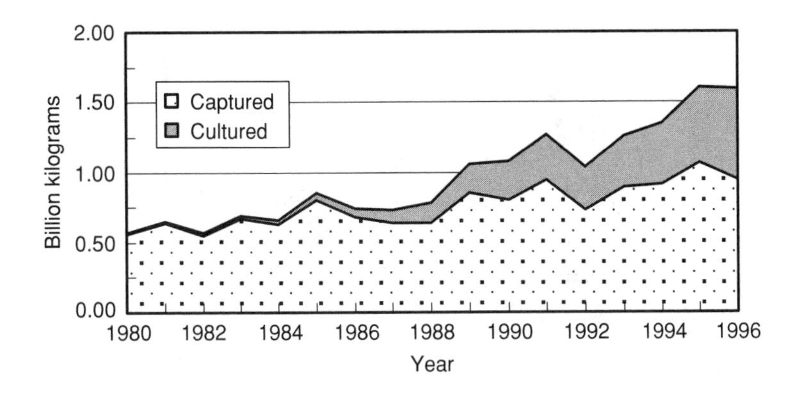

Figure 9. Estimated World Production of Salmon (live weight), by wild and cultured product, 1980–1996. *Source*: (24,9, and 12).

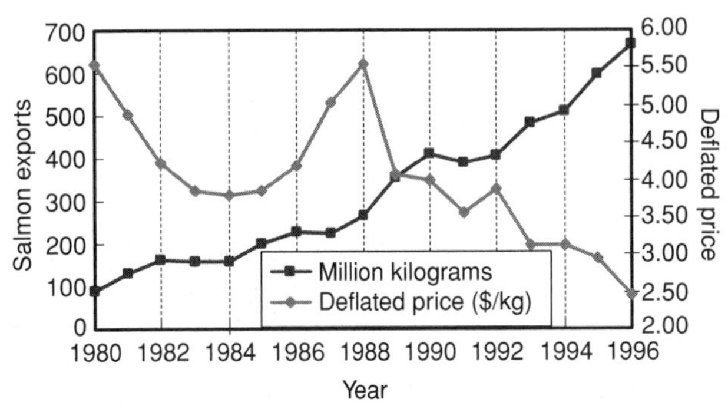

Figure 10. World exports of salmon (product weight), and associated export price (in constant dollars), 1980–1996. *Source*: (8).

salmon for the year. Production of farm-raised coho salmon equaled 76 million kilograms (167 million pounds) and represented approximately 11% of the farm-based output. Production of chinook salmon was relatively small (12 million kilograms) and accounted for less than three percent of the world farm-raised production of salmon in 1996.

Production of farm-raised salmon is dominated by Norway (301 million kilograms), which accounted for almost 50% of the world's total in 1996 (12). All of the product raised in Norway is Atlantic salmon. Chile, the world's second-largest producer of farm-raised salmon, produced 145 million kg (319 million lb) of product in 1996, approximately one-fifth of the world's farm-raised product. Almost one-half of Chile's farm-based output in 1996 was coho salmon; Chile's production of this species represented the vast majority of the world's production of this species. Other major world producers of farm-raised salmon include the United Kingdom (83 million kg 1,83 million lb), Canada (45 million kg, 99 million lb), the United States (14 million kg, 31 million lb), and the Faroe Islands.

Six species of salmon—pink, chum, sockeye, coho, chinook, and cherry—compose the majority of wild harvest (9). Pink salmon (295 million kg, 649 million lb) accounted for almost a third of the 1996 wild harvest; chum salmon production (411 million kg, 904 million lb) accounted for an additional 40% of the total. Sockeye and coho salmon, both of which are harvested primarily in the United States, accounted for 20% and 3%, respectively, of the 1996 global supply of wild product. Overall, approximately 40% of the 1996 world wild harvest of salmon was produced in the United States (398 million kg, 876 million lb); an additional 35% was produced in Japan (342 million kg, 752 million lb). Most of the remaining production was from the Russian Federation (163 million kg, 359 million lb).

World Exports and Imports. World exports of fresh and frozen salmon, which equaled 89.2 million product-weight kilograms (197 million lb) in 1980, advanced to 665.3 million kg (1.47 billion lb) in 1996 (Fig. 10); that change translates into a growth rate of about 35 million kg (77 million lb) per year. On a percentage basis, the increase exceeded 600%. With few exceptions, growth in exports occurred throughout the entire 17-year period ending in 1996; that growth parallels the growth in cultured salmon production.

The value of fresh and frozen salmon exports advanced from $408 million in 1980 to $2.6 billion in 1996 (8). After adjusting for inflation (based upon the 1982–1984 U.S. Consumer Price Index), the increase was from $495 million to $1.6 billion (230%). The fact that the percentage change in the quantity exported during the period 1980–1996 greatly exceeded the percentage change in constant-dollar value over the same period suggests a sharp decline in the constant-dollar price of the exported product. As indicated in Figure 10, the constant-dollar export product weight price fell from $5.54 per kilogram ($2.51 per pound) in 1980 to $3.86 per kilogram ($1.75 per pound) in 1985, before advancing sharply to the original price ($5.54 per kilogram) in 1988. Since 1988, however, the price has again fallen sharply, to just $2.46 per kilogram ($1.12 per pound) in 1996.

The five largest exporters of fresh and frozen salmon in 1980, as indicated in Table 3, were the United States [55 million kg (121 million lb)], Canada [20 million kg (44 million lb)], Norway [four million kg (8.8 million lb)], Denmark [two million kg (4.4 million lb)], and Greenland [one million kg (2.2 million lb)]. Combined, these five countries accounted for more than 90% of the world total of fresh and frozen salmon exports in 1980, on the basis of either poundage or value. By 1996, Norway's exports had advanced to 214 million kg (471 million lb), an amount constituting almost one-third of the world's exports of fresh and frozen salmon in that year. Essentially all of the growth in Norway exports was the result of increased

Table 3. World Exports of Fresh/Chilled and Frozen Salmon by Principal Countries, 1980 and 1996[a]

Country	Million Kilograms	$ Million
	1980	
1. United States	55.4	198.4
2. Canada	19.6	99.7
3. Norway	4.2	48.0
4. Denmark	2.0	17.9
5. Greenland	1.3	8.7
	1996	
1. Norway	214.1	882.6
2. United States	124.5	464.8
3. Denmark	57.2	243.2
4. Canada	47.8	240.4
5. Chile	70.2	240.0

[a]Ranked by value of export sales.
Source: (8).

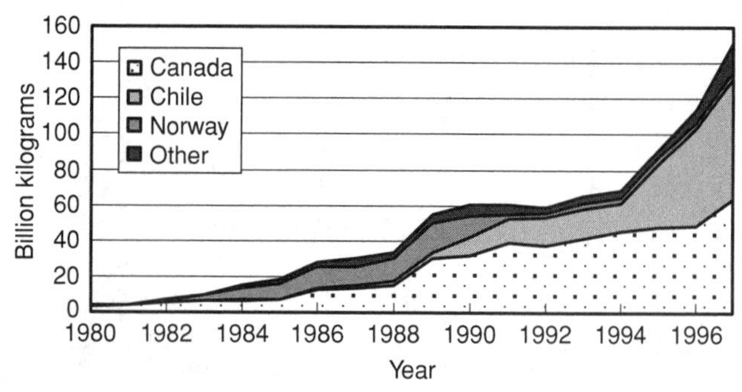

Figure 11. U.S. Imports of Salmon (round weight equivalent), by primary countries, 1980–1997. *Source*: (15).

culturing activities in the country. Similarly, 1996 exports from Chile, which ranked third in terms of volume and fifth by value, were essentially all of a farmed nature. While the available statistics do not allow the separation of the exported wild product from the cultured product, much of the growth in Canadian exports between 1980 and 1996 likely reflects increased cultured product entering into the international market. One can surmise that most of the increase in U.S. exports of salmon between 1980 and 1996 reflects increased exports of wild product, from the fact that U.S. production of cultured salmon equaled only 14 million kg (31 million lb) in 1996. In general, the information in Table 3 clearly demonstrates the increasing role of cultured salmon production in international trade, when it is evaluated on a country-by-country basis.

Japan and the United States have traditionally been the world's primary salmon markets. As a result of rapid expansion in culture activities, however the market for salmon in the European Community has advanced rapidly during the past decade (26). On the basis of imports, Japan, which imported almost 192 million kg (422 million lb) of fresh and frozen salmon product in 1996, accounted for almost 40% of the trade in these products by volume. The second and third largest importers of fresh and frozen product in 1996 were France and Denmark, respectively. The majority of the product imported by Denmark is subsequently re-exported. While being a major exporter of (mostly wild) salmon, the United States imports of (mostly farmed) salmon (round weight converted basis) expanded from 3.6 million kilograms (eight million pounds) in 1980 to 150 million kilograms (332 million pounds) in 1997 (Fig. 11). In 1980, Canada accounted for virtually all of the salmon imported by the United States. By 1997, Canada's share of salmon exports to the United States, 64 million kilograms (274 million pounds), had to approximately 40%.

In lockstep with advances in salmon culture activities in Norway, salmon exports from Norway to the United States expanded from virtually nothing in 1980 to almost 17 million kilograms (37 million pounds) in 1989. In 1990, a dispute between the United States and Norway arose, regarding allegations that the Norwegian salmon industry was dumping product onto the U.S. market and thereby potentially undermining the viability of the U.S. cultured salmon industry. The details of this dispute can be found in (25). On 25 February 1991, the U.S. International Trade Commission (USITC) ruled that the Norwegian industry was selling its farm-based product at below market value and that Norwegian imports were benefitting from subsidies provided to the industry. A countervailing duty of 2.27% was placed on fresh Norwegian product imported by the United States as a result of these findings by the U.S. International Trade Commission; in addition, an antidumping duty ranging from 15.65% to 31.81% was levied on selected companies. The effect of these actions was a significant decline in Norwegian exports of salmon to the United States. By 1991, specifically, Norwegian exports to the United States had fallen to less than 2.25 million kg (five million lb) and by 1997 had recovered only to the 3.6 million kg (eight million lb) mark.

Chilean exports to the United States were nonexistent in 1980. In conjunction with culturing activities in the country, however, exports of Chilean salmon to the United States advanced rapidly, exceeding the Norwegian peak of 17 million kilograms (38 million pounds live weight equivalent) by 1993. Chilean exports to the United States of 68 million kilograms (149 million pounds) in 1997 accounted for almost one-half of the total U.S. imports of 151 million live-weight kilograms (332 million pounds) for the year. The United States is by far the largest market for Chilean salmon. According to the United States International Trade Commission (27), the United States accounted for more than 90% of the dressed and cut Chilean salmon exports (i.e., steaks, fillets, etc.) in 1996, and one-third of the whole product that was exported, though the U.S. share of this latter product has been declining. Home-market consumption of the Chilean production is only about 2%.

As was the case with the Norwegian product, the rapid rise of Chilean salmon exports to the United States resulted in an investigation by the USITC, pursuant to section 735(b) of the Tariff Act of 1930. The purpose of that investigation, the petition for which was filed by the Coalition for Fair Atlantic Salmon Trade, was to determine whether the salmon industry in the United States was being materially injured, or was being threatened with material injury, by imports from Chile of fresh Atlantic salmon. On July 22, 1998, the Commission reported to the Secretary of Commerce that it had made a determination (pursuant to the anti-dumping investigation) that the industry in the United States was being materially injured or was being threatened with material injury by imports from Chile of fresh Atlantic salmon that were found by the U.S. Department of Commerce to have been sold at below fair value (27). In response, countervailing duties ranging from 0.2% to 11% were imposed on specific exporting companies; exporting companies not identified in the original petition were assessed a countervailing duty equal to 5.9%.

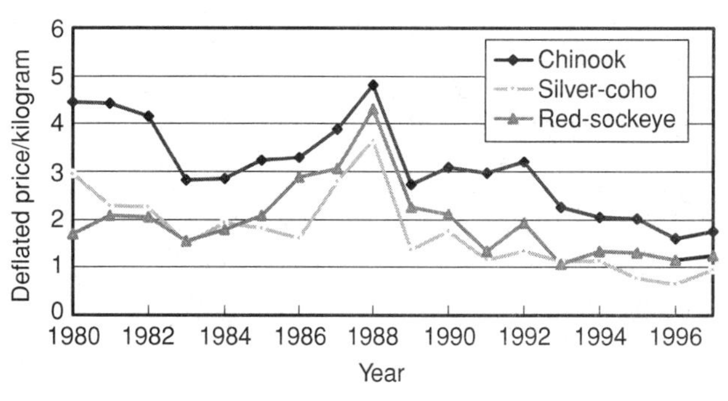

Figure 12. U.S. Dockside Price (in constant dollars) of high-value salmon species, 1980–1997. *Source*: (4).

Competition Between Wild and Cultured Salmon. The salmon fishery is one of the most valuable U.S. capture fisheries at dockside, valued at $270 million in 1997 (4). There are five commercially harvested species of Pacific salmon in the United States (28). U.S. landings of high-valued salmon (mainly chinook, coho, and sockeye) in 1997 equaled 108 million kilograms (237 million pounds) worth $221 million at dockside (4). U.S. production of low-valued salmon (such as chum and pink salmon) in 1997 equaled 150 million kilograms (330 million pounds), but the dockside value equaled only $49 million (4). Overall, the average price of the high-valued salmon, $2.05 per kilogram ($0.93 per pound), exceeded the average price of the low-valued salmon, $0.33 per kilogram ($0.15 per pound), by a factor exceeding five.

The differences in prices for the different salmon species reflect distinct characteristics of the different species that fill different niches in the market place (28). The chinook salmon, because of its high fat content, is valued for smoking; it is the preferred salmon species among high-end restaurants in Europe and the United States. Coho salmon, by comparison, are frequently purchased by the European market for smoking, but they are more affordable to U.S. consumers than chinook salmon. U.S. production of sockeye salmon is almost exclusively marketed in the Japanese market, whose consumers favor its flavor and its deep red color. Chum salmon, because of its low ex-vessel value, is often used as a fast-food product; pink salmon is often canned and is consumed both domestically and abroad.

The constant-dollar ex-vessel prices of the three high-valued salmon species harvested in the United States are presented in Figure 12 for the period 1980–1997. As indicated, the constant-dollar prices of all three species fell sharply after 1988. The marked fall in 1989 reflected the sharp increase in farmed production in 1989, which exceeded the 1988 production level by 45% (25) and paralleled the sharp decline in the export price for salmon products (see Fig. 10). The decline in constant-dollar prices since 1989 reflects the continued expansion of farm-raised product as well as increased harvests of wild product. Analysis by Herrmann (29, as cited in 28) indicates that 90% of the decline in the Alaska dockside price of salmon after 1988 and on into the early 1990s was attributable to increased world supplies of salmon rather than to such other factors as changes in exchange rates and worldwide recessions.

One of the earliest econometric studies that examined the role of cultured salmon was conducted by Herrmann and Lin (30). Using a set of simultaneous equations, the authors estimated the market conditions for Norwegian Atlantic salmon in the United States and in the European Community. Using monthly data covering the period from 1983 through March 1987, the authors estimated that chinook salmon were a weak substitute for Norwegian-produced salmon in the U.S. market. Specifically, the cross-price elasticity between chinook salmon and Norwegian Atlantic salmon in the United States (0.56) suggested that a 10% decrease (increase) in the price of the Norwegian product would result in a 5.6% decrease (increase) in the U.S. demand for chinook salmon. With respect to the European Community market, the authors estimated that European imports of chinook, sockeye, and coho salmon from North America were weak substitutes for Norwegian-produced Atlantic salmon. The authors also concluded that a 1% increase in the U.S. price of the Norwegian product resulted in a long run increase of about 6% in the supply of Norwegian salmon to the U.S. market, and the same percentage increase in the European price resulted in a roughly equivalent percentage decline in product directed to the U.S. market. Finally, the Norwegian product was found to be a luxury good in both the U.S. and European Community markets.

As noted by Herrmann (28), "[s]ince (the 1988 study), several others have found that farmed Atlantic salmon and high-valued Pacific salmon (chinook, coho, and sockeye) are substitutes in the markets of the United States, Europe, and Japan. The substitutional relationship has been expanding in both strength and significance (p. 12)." These studies include ones by Herrmann et al. (31,32) and by DeVoretz and Salavannes (33). More recently, Asche et al. (26) have also provided evidence of substitution between farmed Atlantic and wild Pacific salmon. While substitutability between Norwegian and high-valued Pacific salmon has been found in many of the econometric studies, substitutability of these products in the Japanese market is less well substantiated (32). Given the facts, however, that the U.S. production of coho salmon is destined primarily for the Japanese market, and that Chile is now producing large quantities of cultured coho salmon and targeting the Japanese market, it is likely that the Chilean product will compete with the Pacific coho product in the Japanese market.

These substitutional relationships that have been established for the different species and sources of salmon raise some doubt about whether countervailing duties placed on salmon products entering the U.S. market via trade will provide significant long-term benefits to the capture industry in the United States. Specifically, the countervailing duties cause a decrease in the amount of the cultured product being shipped to the U.S. in the short run, but an increase in shipments of this product to those countries that also tend to import the U.S.-produced wild Pacific product. To the extent that the wild Pacific product tends to compete with the cultured product in those countries, the increase in non-U.S. cultured product will result in a reduced demand for the U.S. Pacific product, if all other things remain unchanged. This reduced demand, in turn, implies a declining dockside price for the wild product.

In addition to the econometric studies that have analyzed substitutability between wild and cultured salmon, several studies, as noted, have been conducted to examine preferences between the two products. Rogness and Lin (34), in a survey of U.S. wholesalers, found that approximately 80% of the surveyed firms considered fresh Atlantic salmon to be a substitute for fresh chinook. In an expanded study of U.S. wholesalers, Herrmann et al. (35) found that the majority of firms considered cultured salmon to be a substitute for fresh chinook, coho, and sockeye in the marketplace. In general, wholesalers considered farmed salmon superior to wild salmon, because it offers more consistent supply, quality, shelf life, and appearance. The primary advantage of the wild salmon, by comparison, was in price.

Finally, conjoint analysis has been used to determine preferences for cultured versus captured salmon (36,37). The results, in general, tend to be mixed. Anderson and Bettencourt found that "expensive" New England restaurants tended to prefer cultured Atlantic product to chum or sockeye salmon, but not necessarily to coho or chinook. The buyers for the "expensive" restaurants, in accordance with their desire to vary their menus according to season, preferred seasonal to year-round salmon products. Buyers for the fish market segment, by comparison, preferred year-round to seasonal salmon, because of their need to maintain a stable supply. Overall, buyers for fish markets tended to be somewhat more indifferent than buyers for "expensive" restaurants with respect to preferred species.

In a recent study of northeastern and mid-Atlantic households, Wessels and Holland found that the average household consumed salmon at home approximately once every six months. More than 60% of respondents were unaware of cultured salmon. Based on conjoint analysis of consumer preferences, the authors concluded that "[t]he salmon industry's strategy of labelling product as to what method of production is used, farm or wild, seems to favor farmed salmon more than wild salmon in the region covered.... This may be the result of the perception that farmed is higher quality and safer than wild, possibly because a farmed product is connected with a product over which the harvester has some degree of control, unlike wild salmon" (p. 57). The authors caution, however, that the results may be specific only for the region considered in the study, and that the preference for wild versus farmed product may be reversed on the west coast, where the wild product dominates.

Crawfish

The production of crawfish or crayfish in ponds represents the largest crustacean aquaculture industry by weight or product, in the United States. Although several dozen species occur in the country, the red swamp crawfish (*Procambarus clarkii*) is the only species farmed. Small amounts of the white river crawfish (*Procambarus zonangulus*) occur in ponds even though it has not been stocked. United States production of crawfish is centered in Louisiana. Texas is the only other domestic producer over the million-pound level annually. Louisiana farmers account for 80% of the country's pond-produced crawfish (38). United States crawfish production from cultured and wild harvests is synonymous with Louisiana. The farmers and fishermen in the state account for more than 90% of the country's total pond and wild crawfish production and for over half of the world production (39).

Farmed production of crawfish occurs in shallow, owner-operated ponds. The pond harvest extends from November through May (as opposed to the batch harvest common in many other aquaculture systems). The wild harvest begins with small amounts in February, followed by a peak in April/May. Because production comes from flooded backwaters of the Atchafalaya River, year-to-year variation in production is significant. The range in wild production between 1990 and 1997 was from 3.2 to 23 million kg (7–51 million lb) (40, 1998). The corresponding farm-raised crawfish production for the period ranged from about 20 to 27.67 million kg (44 to 61 million lb). The advantage of a more reliable and stable supply was in part responsible for the higher producer price for farmed crawfish. A more important factor is the advantage aquaculturists have in supplying the markets earlier in the season. This early part of the season coincides with the Roman Catholic religion's Lenten period, marked by increased consumption of seafood. Farmers generally receive higher prices for their ability to supply crawfish during this period of peak demand. Farm-raised crawfish brought producers from \$1.10 to \$1.65 per kg (\$0.50 to \$0.75 per lb) during the period 1990–1997. The weighted average price was \$1.28 per kilogram (\$0.58 per pound), 11.5% higher than the weighted average price for wild-system crawfish.

The higher average weighted price received by farmers has its foundation in early-season and Lenten-period demand. Farmed and wild crawfish are not differentiated in markets. Each is subject to seasonal availability; farmed crawfish do not have a characteristic of most aquacultured species, year-round availability. This basis for market differentiation to justify higher prices is not available to farmers. When crawfish supply from ponds overlaps the wild-harvest supply, no distinction on the basis of price occurs. The Louisiana Crawfish Promotion and Research Board does not attempt to differentiate the two sources on the basis of price. The only price differentiation is rooted in market grades established on the basis of size.

Mechanical grading of live crawfish was initiated in 1990. The practice involves four grades: jumbo are 15 or fewer live crawfish per pound (0.45 kg); large are 16 to 20 per pound; medium are 21 to 25 per pound; peelers are 26 or more per pound. Ungraded crawfish are marketed as field run. Grading puts larger sizes into the export market and into live retail/restaurant usage. The smaller crawfish are commonly utilized in the processing industry to produce cooked crawfish meat.

Although crawfish from farmed and wild sources are not differentiated on the basis of price, Bell (41) demonstrated a unique linkage. Farmed supply reaches the market earlier in the year than does wild supply. Bell's research found that aquacultured crawfish increases supply and so causes lower prices for wild supply than would occur if no farm industry existed. In terms of natural-resource economics, farm-produced crawfish lowers the price for the wild supply, and that reduction reduces, in turn, the quantity harvested from public waters. It is unique in aquaculture for a farmed supply to be linked to lower price for production from natural areas. The reverse is usually claimed as an interaction between the wild and the aquaculture supplies of various species.

A noteworthy linkage to lower crawfish prices experienced by both farmers and fishermen was low-priced imported crawfish tail meat. An alliance of crawfish processors petitioned the U.S. International Trade Commission for an antidumping investigation of processed crawfish tail meat from China. The 1996 investigation found the domestic industry to have been materially injured by imports because the imported product was being sold at less than fair value (42). Tariffs on processed crawfish meat from China were established. The tariff level varied by exporting company but averaged approximately 115%. This tariff will temporarily insulate U.S. crawfish producers from the deleterious effects of wild supply from China.

Alligator

Alligator (*Alligator mississippiensis*) production comes both from wild harvests and from farm systems. There is often a linkage between failed public management of a species and the development of culture systems. This is undeniably the case with the alligator. The legal hunting of wild alligators gave way to excesses and to poaching when regulations were stiffened. With the exception of the harvesting of nuisance alligators, hunting was banned in the 1960s. The populations were rapidly rebuilt, as a result of effective research and enforcement programs. Controlled harvest was initiated in Louisiana in the fall of 1972. Other states in the south, led by Florida, followed with programs allowing a tightly controlled harvest. The supply (less than 1,500) from harvests was far below the annual alligator harvest of 150,000 in the United States early in the twentieth century (43).

The harvest of wild alligators increased throughout the 1980s and 1990s. However, the approximately 35,000 harvested were an indication wild stocks would not support historical levels. Alligator farming in Louisiana alone grew to such an extent after 1985 that 160,000 animals were produced in 1996. Florida farms produce at about one-third of the Louisiana level; Texas, Georgia, Mississippi, and Alabama combine for less than 10,000 animals. Collectively, wild harvests and farmed production were projected to be approximately 250,000 in 1997 (44).

The farmed production of alligators may have arisen out of a need to replace reduced harvests of wild animals. Farmed supply is, however, not a perfect substitute for wild alligators. The basic difference stems from the average length and width of wild alligators. Louisiana data reflecting 28,433 wild animals indicated that 76% were from 2 to 2.4 m (6 to 8 ft) in length. The best data on farmed alligators also relates to Louisiana animals. Ninety-seven percent of the production was in the 1–1.3 m (3 to 4 ft) range (45). Farmed alligators seldom bring the same price as wild ones. After 1987, wild animals brought from 33% to 138% more per foot (0.3 m) than did farmed animals (45). The longer and wider wild animals mean more square measure (surface) units for final-product manufacturers. Alligator meat is also a factor in determining the average price. More meat is a feature of the wild alligators, because of their larger size.

The interaction of the wild and the farmed industries occurs domestically and internationally. Markets for alligator skins are overseas, meat is utilized domestically. The dependence on export markets for most United States production puts farmers in competition with foreign producers. Classic species such as American alligators and crocodiles dominate world supply. Data are troublesome to interpret, because trade statistics are often outdated. Also, skins can be stored and sold in another production year. This flexibility can affect producer prices in ways not familiar to aquaculturists producing food species less conducive to long term storage.

Alligator farmers are also subject to interaction with the wild harvest industry via imperfect substitution. Caiman skins from wild sources, once heavily regulated, are gaining market share via improved compliance of exporters and importers. Increasing supplies of these smaller skins will have an impact on farmed alligator production in the United States. Prices to U.S. farmers peaked at 112/m ($36/ft) in 1988. Since then prices have trended into the low teens, and they will be restrained from recovery as supplies continue to increase. The demand side of the market is highly linked to economic conditions in Europe and Asia. France, Italy, Japan and Singapore are among the leading importers. Economic conditions in these countries must be favorable for there to be supply-clearing prices for skins and the goods manufactured from them. Unlike United States aquaculturists of catfish, crawfish, and tilapia, alligator farmers are dependent on export markets. These markets are subject to domestic and international wild-supply competition.

Catfish

Aquaculture of channel catfish (*Ictalurus punctatus*) is concentrated in Mississippi, Arkansas, Alabama, and Louisiana. Pond-culture technologies dominate over raceway and cage-culture approaches. The farm-raised industry, in its formative years of the 1960s, was thought to be constrained by competition by domestic and imported wild catfish (46). In 1965, catfish from the wild accounted for 71% of supply (47). Imports from the Amazon River system in Brazil were identified as problematical; however, the farmed-catfish industry established a reputation for reliability and quality of production, and so lower-priced wild fish, whether imported or domestic, actually never posed a threat to growth. The aquaculture industry, essentially, produced a product that the market quickly differentiated from wild-source catfish. Year-round supply and relative size uniformity were traits that the U.S. marketing system evidently valued as seafood consumption increased. To date, catfish aquaculture technology has not been successfully utilized overseas to produce competing products. The prediction that foreign-aquaculture catfish supply would compete effectively in U.S. markets was in error (46). Essentially, all catfish product imports originate in Brazil, from river-system sources. The peak of imports from this source during the 1990s occurred in 1991. The equivalent of 5.9 million kg (13 million lb) live weight was imported at a frozen fillet price of $2.42/kg ($1.10/lb). This type of product is clearly serving a lower-valued market in the United States.

The growth in farm-raised supply was also not impeded by domestic wild catfish catches. Pond supply more than doubled from 1985 to 1997, yet producer prices increased. Wild supply of all catfish species has never exceeded 6.8 million kg (15 million lb). There is no market advantage to the 1.10/kg ($.50/lb) received by fishermen versus the farm supply price of approximately $1.50/kg (0.68/lb) (4, 1998). Farm-raised catfish supply exceeded 180 million kg (400 million lb) for the first time in 1992 and has continued to move to the >225 million kg (>500 million lb) level. The industry has correctly shed concern over production competition.

Realization that market expansion was linked to industry organization (to address domestic consumer demand) has yielded results. A farm-raised catfish promotional program began in 1987. Funded by an assessment on catfish feed purchases, the Catfish Institute began efforts to increase consumer awareness and attitudes. Purchase frequencies for at-home and restaurant consumption increased 11–12% (48). Another evaluation of the promotional program estimated a 7% average increase in wholesale sales, attributed to industry-sponsored advertising (49). Catfish farmers are intent on retaining these gains and on further differentiating their products from wild supplies. A Catfish Quality Assurance Program was developed by the Catfish Farmers of America in 1993 (50). This program assures the safety and quality of farm-raised catfish. Gains in favorable consumer perceptions about farm-raised catfish are a significant achievement and an asset to be protected.

Tilapia

Supplies of domestically produced tilapia come from aquaculture businesses and from wild fisheries. The situation is in no way comparable to the wild and cultured supplies of shrimp, salmon, crawfish, or alligators. Aquaculture businesses have produced uninterrupted increases in tilapia supplies ever since government data were first reported in 1991. The beginning report was 2.25 million kilograms (5 million pounds); supplies had risen to 7.7 million kilograms (17 million pounds) by 1997. Wild supply reached 0.14 million kilograms (0.31 million lb) only once, 1991. Thereafter, wild tilapia levels were between 0.045 and 0.12 million kg (0.1 to 0.26 million lb). This production, primarily from Florida, is a low-value fish: approximately $0.88/kg ($0.40 per pound), compared to the 1997-average farmed price of $3.31/kg (1.49/lb) (4, 1998). Supply from wild sources will not be a determinant of any price, supply, or quality elements associated with domestic tilapia aquaculture.

Domestic tilapia supplies are primarily coming from intensive recirculating systems, not ponds. This situation is likely a result of regulations and economics. Some states allow production only from recirculating systems, to minimize the escapement potential of exotics. Florida and Arkansas do allow tilapia culture from pond systems, but these fish are generally lower-priced than the supply from recirculating systems. Thus, some aquacultured fish have the same effect on others as do wild-caught fish. Intensive recirculating systems accounted for about 70% of domestic supply (51). Expansion of the industry via this technology will reduce, the potential of pond raised tilapia for depressing prices. Recirculating systems are conducive to the efficient marketing of live tilapia. Live fish account for 85% of sales. Projections for 1998 domestic tilapia production 9.45 million kg of (21 million lb) (52) indicate that the live markets historically used will be more price-sensitive in the future. The outlook is for the growth in domestic supply to be increasingly used as processed product.

Imports reached 36.8 million kilograms (81 million pounds) live weight equivalent in 1997. This level was attained after uninterrupted increases since 1993. Approximately half comes from Taiwan, as frozen whole fish valued at less than $1.54 per kilogram ($0.70 per pound). Costa Rica and Ecuador are leading exporters of fresh fillets. A notable aspect of the import situation is that virtually all supply originates from aquaculture. Thus, domestic aquaculturists are minimally influenced by wild supplies from all possible sources. The future growth of the domestic tilapia business depends on competitiveness with imported fresh fillets. These will be the closest product substitute as domestic producers turn to the fillet market form when live-market sales slow.

BIBLIOGRAPHY

1. U.S. Department of Commerce, *Aquaculture and Capture Fisheries: Impacts in the U.S. Seafood Markets*, 1988.
2. R.M. Naylor, R.J. Goldburg, H. Mooney, M. Beveridge, J. Clay, C. Folke, N. Kautsky, J. Lubchenco, J. Primavera, and M. Williams, *Science* **282**(October), 883–884 (1998).
3. F. Asche, *Marine Resource Economics* **12**(1), 67–73 (1997).
4. U.S. Department of Commerce, National Marine Fisheries Service (1980–1997). Fisheries of the United States.
5. C.R. Wessels and J.G. Anderson, *J. Cons. Aff.* **29**(1), 85–107 (1995).
6. H.M. Johnson & Associates (various issues). *Annual Report on the United States Seafood Industry*. H.M. Johnson & Associates, Bellevue, WA.
7. W.R. Keithly, K.J. Roberts, and J.M. Ward, in U. Hatch and H. Kinnucan, eds., *Aquaculture: Models and Economics*, Westview Press, Inc., Boulder, CO, 1993, pp. 123–156.

8. Food and Agriculture Organization of the United Nations (1980–1996). *Yearbook of Fishery Statistics*: Fishery Commodities, Rome, Italy.
9. Food and Agriculture Organization of the United Nations (1980–1996). *Yearbook of Fishery Statistics: Catches and Landings*, Rome, Italy.
10. W.R. Keithly and A. Diagne, *An Economic Analysis of the U.S. Shrimp Market and Impacts of Management Measures*. National Marine Fisheries Service Saltonstall-Kennedy Program, Silver Spring, Maryland, 1998.
11. K. Rana and A. Immink, *Trends in Global Aquaculture Production: 1984–96*. FAO Fisheries Department, Fishery Information, Data and Statistics Service, 1998, Rome, Italy.
12. Food and Agriculture Organization of the United Nations (1985–1996). *Aquaculture Statistics*, Rome, Italy.
13. I. Csavas, *World Aquaculture* **25**(1), 34–56 (1995).
14. P. Jacobson, in K. Chauvin, ed., *Proceedings of the Fifth Shrimp World Conference*, Shrimp World Inc., New Orleans, LA, 1993, pp. 159–171.
15. U.S. Department of Commerce. Bureau of Census, *Foreign Trade Statistics*, Washington, DC, 1980–1997.
16. Japan Tariff Association, *Japan Exports & Imports: Commodity by Country*, Tokyo, Japan, 1980–1996.
17. P.E. Niemeier and R. Walsh, in P. Menesses, W. Chauvin, and A. Cuccia, eds., *Proceedings of the Third Shrimp World Market Conference*, Shrimp World Inc., New Orleans, LA, pp. 157–175.
18. B.J. Rothschild and S.L. Brunenmeister, in J.A. Gulland and B.J. Rothschild, eds., *Penaeid Shrimps-their Biology and Management*, Fishing News Books, Ltd., Farnham, England, 1984, pp. 145–172.
19. D. Gillig, O. Capps, and W.L. Griffin, *Mar. Res. Econ.* **13**, 89–102 (1998).
20. U.S. International Trade Commission. *Conditions of Competition Affecting the U.S. Gulf and South Atlantic Shrimp Industry*. Washington, DC, 1985.
21. K.J. Roberts, W.R. Keithly, and C.M. Adams, *The Impact of Imports, Including Farm-Raised Shrimp, on the Southeast Shrimp Processing Sector*, Tampa, Florida, 1990.
22. W.R. Keithly and K.J. Roberts, Shrimp Closures and their Impact on the Gulf Region Processing and Wholesaling Sector (Expanded to Include South Atlantic), St. Petersburg, Florida, 1994.
23. H. Diop, *Impact of Shrimp Imports on the United States Southeastern Shrimp Processing Industry and Processed Shrimp Market*. Ph.D. dissertation, Louisiana State University, 1999.
24. Alaska Seafood Marketing Institute, *Salmon Wild Catch and Aquaculture* (unpublished).
25. J.L. Anderson and Q.S. Fong, *Aqua. Eco. and Mang.* **1**(1), 29–44 (1997).
26. F. Asche, T. Bjorndal, and K. Salvanes, *Can. J. Agric. Eco.* **46**, 69–81 (1998).
27. U.S. International Trade Commission, *Fresh Atlantic Salmon From Chile*, Washington, DC, 1998.
28. M. Herrmann, *J. Aquatic Food Product Tech.* **3**(3), 5–21 (1994).
29. M. Herrmann, *Arctic Res. of the U.S.* **6**, 34–36 (1992).
30. M. Herrmann and B. Lin, *Can. J. Agric. Eco.* **36**, 459–471 (1988).
31. M. Herrmann, R.C. Mittelhammer, and B. Lin, in U. Hatch and H. Kinnucan, eds., *Aquaculture: Models and Economics*, Westview Press, Inc., Boulder, CO, 1993, pp. 187–214.
32. M.L. Herrmann, R.C. Mittelhammer, and B. Lin, *Can. J. Agric. Eco.* **41**, 111–125 (1993).
33. D.J. DeVoretz and K.G. Salvanes, *Amer. J. Agric. Eco.* **75**, 227–233 (1993).
34. R.V. Rogness and B. Lin, *The Marketing Relationship Between Pacific and Pen-Reared Salmon: A Survey of U.S. Seafood Wholesalers*. Alaska Sea Grant Report 86–103, 1986.
35. M. Herrmann, B. Lin, and R. Mittelhammer, *U.S. Salmon Markets: A Survey of Seafood Wholesalers*. Alaska Sea Grant Report 90–01, 1990.
36. J.L. Anderson and S.U. Bettencourt, *Marine Res. Eco.* **8**, 31–49 (1993).
37. C.R. Wessels and D. Holland, *Aqua. Eco. and Man.* **2**(2), 49–59 (1998).
38. J.L. Avery and W. Lorio, *Crawfish production manual*. Louisiana State University Agricultural Center, Baton Rouge, LA, 1996.
39. J.V. Huner and J.M. Huner, *Crawfish—Louisiana's Crustacean Delight*. Louisiana Ecrevisse, Lafayette, LA, 1997.
40. LSUAC, *Louisiana Summary: Agriculture and Natural Resources*. Louisiana State University Agricultural Center, Baton Rouge, LA, 1998.
41. F.B. Bell, *Amer. J. Agric. Econo.* **68**(1), 95–101 (1986).
42. U.S. Department of Commerce, *Crawfish Tail Meat from China: Investigation no. 731–TA–752*. U.S. International Trade Commission, Publication 3057, Washington, DC, 1997.
43. M. Masser, *Alligator Commercial Production*. Alabama Cooperative Extension Service, Auburn, AL, 1990.
44. D. Ashley, *International Alligator Crocodilian Trade Study Report to Louisiana Department of Wildlife and Fisheries*, Ashley Associates, Inc., Tallahassee, FL, 1996.
45. L. McNease, *Louisiana's Alligator Management Program*. Louisiana Department of Wildlife and Fisheries, Baton Rouge, LA, 1998.
46. J.E. Greenfield, *Economic and Business Dimensions of the Catfish Farming Industry*, U.S. Bureau of Commercial Fisheries, Ann Arbor, MI, 1970.
47. M.A. Walters, *Wholesale Market Demand for Catfish*. College of Business Administration, University of Arkansas, Little Rock, AK, 1967.
48. H.W. Kinnucan and M. Venkateswaran, *J. App. Aqua.* **1**(1), 3–31 (1991).
49. W. Zidack and U. Hatch, *J. World Aquacult. Soc.* **22**(1), 10–23 (1991).
50. CFA, *Catfish Quality Assurance*. Cooperative Extension Service, Mississippi State University, Mississippi State, MS, 1993.
51. Anonymous, *American Tilapia Association Situation and Outlook report*, Arlington, VA, 1998.
52. Anonymous, *Aqua. Mag.* July/August, (1998).

EEL CULTURE

Margie Lee Gallagher Ph.D.
East Carolina University
Greenville, North Carolina

Gad Degani
Galilee Technological Center
Qiryat Shemona, Israel

OUTLINE

Eel culture is probably the first and most long-lived example of intensive aquaculture. The Japanese have successfully practiced intensive eel culture since the 1880s, well over a hundred years. The mainland China fishery (where eels are caught and cultured) is the largest eel fishery at present, with a production of 147,316 tons in 1996. However, significant production of cultured eels also comes from Japan and Taiwan. Throughout the 1990s, cultured eel production in Japan (24,171 tons in 1997, for example) dramatically exceeded the wild catch (860 tons). In Asia, the net worth of cultured eel production is second only to that of yellowfin and makes up 13% of the total value of aquaculture products produced in the region. In addition to Asia, Europe is also an important eel-culturing region of the world, with the highest production being in Italy, where 3,000 tons were grown in 1996 (1). In spite of the importance of cultured eels, however, the industry overall still depends upon wild-caught elvers (young eels) to supply young for culture.

Of the sixteen species of eel in the world, four are currently important in aquaculture production: the Japanese eel (*Anguilla japonica*), the European eel (*A. anguilla*), the American eel (*A. rostrata*), and the Australian eel (*A. australis*). According to Usui (2), these four species are so similar that they may be considered to be the same for marketing purposes. Some investigators believe that *A. anguilla* and *A. rostrata* are, in fact, the same species, because of their similarities in terms of anatomy and life history (3).

Although there has been interest in eel culture in the United States since the early 1980s and some individuals are experimenting with commercial production, there is currently no significant commercial production in the United States. However, a wild fishery for eel culture does exist. Preliminary data from the National Marine Fisheries Service for 1996 indicate that total landings of American eels from the Atlantic and Gulf states were near 646 tons worth approximately US$5 million (4). These figures represent the catch of both market-sized eels, which are generally exported to European markets, and young eels (elvers), which are sold to foreign commercial culture operations. The price of elvers in the United States varies with the demand from out-of-country operations, which has been high for the last several years. For example, in Japan, the so-called "seedling eel" (elver) catch declined from 40.3 tons in 1993 to 22.5 tons in 1997. This decline has increased the market demand for elvers from the United States.

LIFE HISTORY

One of the most striking features of the biology of eels is their catadromous migration, which, in the case of the European eel, covers as much as 7,000 km, from inland European waters to the spawning grounds of the Sargasso Sea. The biology of eels has been thoroughly reviewed by Tesch (5). Tsukamoto (6) and Umezawa and Tsukamoto (7) have reviewed the life history of *A. japonica*. Usui (2), in his practical book on eel culture, records the story of the classic work of the Danish biologist Johannes Schmidt in discovering the elusive life cycle of the European eel.

Adult silver eels migrate from freshwater to the area of the Atlantic Ocean known as the Sargasso Sea for spawning. After spawning, floating eggs hatch in about 24 hours. As the tiny larvae grow into flat, leaf-shaped leptocephali, they are carried along by the currents (of the Gulf Stream in the case of European and American eels) toward land. This journey can take from 10 months to as long as three years, depending on the origin of the larvae and the eventual destination. During this time, the leptocephali metamorphose into unpigmented, cylindrical, "glass" eels. These eels, apparently guided by their olfactory sense to the natural odors of freshwater and/or very specific salinity changes, migrate into inland rivers and lakes. When elvers enter freshwater, they become pigmented. As growth continues and pigmentation is completed, the eels are referred to as brown- or yellow-stage eels (8). This coloration is in contrast to that of the more commercially valuable silver eels, a name that refers to the silver color of large eels that are about to migrate back to sea for spawning. Migrating eels have increased body fat (25–30%) and do not feed during migration. It is assumed that eels that have spawned die, since silvering of the skin, maturation of the gonads, and increases in body fat are also accompanied by degeneration of the gastrointestinal tract (9,10). Although Japanese researchers have had some success in artificially inducing egg production and spawning in eels (11–14), as yet the life cycle of eels has not been closed under captive conditions. Therefore, eel culture depends upon a steady supply of young eels (elvers) from the wild. These populations are highly exploited at present.

TRAINING AND SORTING OF ELVERS

Wild-caught elvers do not feed readily and must be trained to feed. Training is usually accomplished by feeding them minced fish, bivalve flesh, earthworm flesh (in Japan), or *Tubifex* worms and then gradually mixing in increasing amounts of artificial feed. In most commercial operations, feeds are fed as a paste, made by mixing a fishmeal-based powdered diet with water, 5–10% oil, and vitamins. (See Fig. 1.) Elvers may be trained to eat moist feeds, dry crumbles, and pellets just as easily (15,16). Arai (17) noted, however, that eels that are to be fed a dry diet should be trained on a dry diet from the elver stage and that sorting by size must take place more frequently for such eels. In addition, chicken and chicken meal are viable substitutes for fishmeal in artificial diets for elvers (18,19).

Figure 1. Elvers eating an artificial diet. The diet is fed as a paste in a feeding tray that is suspended into the elver tank. (Courtesy of W.L. Rickards.)

High mortality (up to 30%) and widely varying growth rates characterize the initial period of training and growth. Degani and Gallagher (9) have reviewed the phenomenon of widely varying growth rates in American eels. Some elvers do not ever learn to eat the provided diet; they lose weight and eventually die or are cannibalized. Some elvers that do learn to eat grow so slowly and have such poor food conversion ratios that they are of no economic value and must be sorted from those that are growing at appropriate rates (20). Although attractants, such as amino acids, have been shown to be effective in getting eels to eat (21), knowledge of attractants is still too scant to incorporate the technology into culture techniques.

Although cannibalism is not a particularly bad problem in elvers, since only those of little or no economic value tend to be cannibalized, continual sorting of eels must take place during their growth, so that eels in the same pond are of approximately the same size and so that undersized and stunted eels are removed (2). Sorting increases feed efficiency and reduces cannibalism and stress. The eel is a solitary animal in its natural habitat, but exhibits aggressive behavior in culture when there is a size differential, thus causing stress that can affect the health and feeding of all eels (9).

Usui (2) estimated that, given normal mortality rates, careful handling, and continuous sorting, one can expect to harvest 400 kg (880 lb) of eels weighing 150–200 g (5–6.7 oz) each from 1 kg (2.2 lb) of elvers over a 2-year period. If 1 kg (2.2 lb) of elvers contains approximately 3,500 individuals then this harvest rate indicates an overall survival rate of about 60%. It should also be noted that proper training, management, and sorting allow the culturist to bring eels to a marketable size in 2 years, which is much faster than in the wild, where such growth may take from 5 to 10 years.

CULTURE METHODS

Because eel culture has been practiced for so long, much is known about the parameters for successful culture. However, different temperature ranges have been reported as appropriate for eel culture in different *Anguilla* species. Arai (22) reported the optimum range for Japanese eels is 23–30 °C (73–86 °F), while that for European eels is 20–23 °C (68–73 °F). Degani and Gallagher (9) reported the optimum culture temperatures for American and European eels to range from 23 to 25 °C (73 to 77 °F). When the temperature drops below these ranges, the eels consume less food, and they stop eating altogether between 8 and 15 °C (46 and 59 °F).

Eels appear to be less sensitive to low oxygen levels than some other cultured species. Degani et al. (23) found no significant differences in growth rates of European eels within a dissolved-oxygen range of 4–8 mg/L (ppm). Both Arai (22) and Usui (2) reported that the level of dissolved oxygen for Japanese eels must remain above only 1 mg/L (ppm). However, excess oxygen or nitrogen can lead to serious problems with gas embolism in eels under 6 cm (2.4 in.) in size (24).

Current eel culture systems vary from large, extensive static-water ponds to much smaller, highly intensive recirculating systems. Elvers are almost always grown in indoor tanks, where the temperature can be controlled, and as mentioned previously, elvers can be trained to eat artificial diets. When elvers reach 8 to 12 cm (3.2 to 4.8 in.), they are stocked into adult-rearing ponds. The stocking density depends upon the culture method used.

The advantage of recirculating systems, either as tanks or greenhouse-enclosed ponds, is that the water requirement is relatively small, stocking densities can be increased [up to 60 kg/m^3 (0.0037 lb/ft^3)], and the water quality (temperature, oxygen content, pH, and solids content) is controlled. (See Fig. 2.) The ability to heat fingerling ponds in the spring using greenhouses with or without auxiliary heating increases the effective growing season (9). Disadvantages include the need for constant monitoring since oxygen levels can become critical and blockage of filters and build-up of sediment occur with poor feeding practices. These systems must have a biological filter and a source of oxygen. A sedimentation tank or pond is also advisable. Although recirculating systems are becoming more economically viable, more traditional pond culture methods are still the norm.

Figure 2. Poured concrete recirculating tanks used for intensive rearing of eels. (Courtesy of G. Degani.)

The simplest eel ponds in use have no temperature control or water replacement and are essentially static. Stocking densities must be kept much lower in these ponds [1–5 kg/m^3 (0.00006–0.0003 lb/ft^3)]. However, most commercial pond operations use some variation of a flow-through (running water) or partial-recirculating flow-through system. The stocking densities of these operations are around 10 kg/m^3 (0.0006 lb/ft^3) (2,9,24).

The basic pond design (earthen or concrete) has one main water inlet and one main outlet, with the bottom sloping toward the outlet sluice. Sluice gates usually have three doors: The board door permits water to be discharged from the bottom or top of the pond; a mesh door prevents the escape of the eels; and a final sluice controls the amount of water leaving the pond. Ponds may have mud bottoms or solid bottoms. Solid bottoms are preferable, since they do not provide eels with hiding places during harvest. Earthen ponds may have solid walls made of concrete slats fitted into grooved concrete posts or boulders cemented together to form a solid wall. These walls support the pond and help prevent elvers and fingerling eels from escaping. In Japan, ponds also have a specific feeding area, which is located on the side of the pond opposite to the source of the prevailing wind. The area is often shaded to facilitate feeding. Larger ponds [>600 m^2 (71,700 yd^2)] in Japan also have specifically constructed "resting areas." "Resting areas" have their own water inlet and outlet, allowing for a limited area of running water. There is also usually a source of aeration, such as a splasher or vertical pump. The concept is that during short periods of poor water quality in the larger pond, eels may enter the resting area, where the water quality is better (2,9).

SEX AND BODY COMPOSITION

Generally, larger eels with higher fat content have a greater market value. Female eels grow to much bigger sizes [60 cm, 300 g (24 in., 10 oz)] and have a higher fat content as compared with males [35 cm, 70 g (14 in., 2.3 oz)]. Thus, females are of greater commercial importance (2), and therefore it is important in aquaculture operations to maximize the number of females that emerge when sexual development begins at around 30 cm (12 in.). Some investigators have suggested that a high population density may lead to a higher proportion of males in wild populations (5,25,26). Degani and Kushnirov (27) found that 77% of communally reared eels became male when injected with human chorionic gonadotropin (hCG), while 60% of eels raised in isolation became female after hCG injection. Hormones can also be used to influence sexual differentiation in eels. Degani and Kushnirov (27) also showed that 70% of eels fed 60 mg/kg (ppm) 17-beta-estradiol (E) for one year became female while of those fed 30 mg/kg (ppm) E or no E, only 32% and 26%, respectively, became female.

In addition, the lipid content of females can significantly influence the marketability of the eels produced. The silver stage of the sexually mature, ready-to-migrate female eel is usually much more desirable than somewhat smaller brown eels, not only because of the difference in size, but also because the lipid content of the silver eel (25–30%) is much higher than that of the brown or yellow eel (5–15%). Size and culture temperature have a profound effect on the total lipid composition of eels. Gallagher et al. (28) found that lipid content was linearly and significantly correlated to weight ($y = 0.09x + 25.99$, $r = 0.62$, where x is weight and y is lipid content). Degani (23) found that eels accumulated higher amounts of body fat at 25 °C (77 °F) and 27 °C (81 °F) than at 23 °C (73 °F).

The amount of effort put into raising eels of a particular sex and the market in which the culturist wishes to sell, since the market size for eels is variable, will determine the lipid content. In Europe, consumers have traditionally preferred eels larger than 0.5 kg (1.1 lb), while in Asia there are markets for both 0.5-kg (1.1-lb) eels and much larger eels, of 1 kg (2.2 lb) or more.

NUTRITION AND FEEDING

Many studies have investigated the nutritional requirements and feeding habits of eels. Arai (17,22) and Lovell (29) reviewed the nutritional requirements and feeding practices of Japanese and European eels, as elucidated in a series of studies by Arai and colleagues. In addition, Cowey and Walton (30) have reviewed the intermediary metabolism of fish, including that of eels. Arai et al. (31) and Nose and Arai (32) investigated the qualitative and quantitative amino acid requirements of Japanese and European eels. The qualitative requirement is similar to that of other species (33,34). However, quantitatively, Japanese eels appear to require higher amounts of many of the 10 essential amino acids — including threonine, tryptophan, and the branched-chain amino acids, leucine, isoleucine and valine — as compared with channel catfish and chinook salmon. Since there is often interaction between branched-chained amino acids, it is likely that an increased requirement for one influences the requirements for the other two (21). In addition, L-cysteine, usually a dispensable amino acid, promotes growth in eels when it is used to replace DL-methionine (35). Therefore, interactions of indispensable amino acids with dispensable amino acids may be of significance in diets for eels.

More recently, Degani and Gallagher (9) reviewed growth and nutrition of *Anguilla* sp. Optimum protein values for eel diets range from 35 to 45% protein, depending upon the energy content of the diet, the source of protein, the age of the fish, and the degree of digestibility of the protein sources used. Elvers may have higher protein requirements. Tibbets (36) reported that the optimum protein requirement for 8-g (0.3-oz) elvers is 48% protein. The protein-to-energy ratio can affect protein use. Gallagher and Matthews (37) and Degani and Gallagher (38) showed that when this ratio is below optimum (140 mg/kcal), there are significant increases in protein catabolism (i.e., the protein is used for energy rather than for protein synthesis). However, as in all fish, this increased use of protein has no significant energy cost. Eels grow best when animal protein sources are used. Gallagher and Degani (39) found that replacing fishmeal with poultry meal did not significantly influence weight gain in elvers, as long as both diets had a source of fish oil. However, Degani (40) found that replacing fish meal or poultry meal with 10 or 20% soybean meal reduced growth in elvers. On the other hand, Schmitz et al. (41) reported digestibility coefficients for soy protein (0.96) and soybean meal (0.94) that were similar to those for fish meal (0.94) and casein (0.99). Degani and Gallagher (9) reported that protein digestibility varies with size and dietary protein content. When eels (six months from the glass-eel stage) were compared with slow-growing eels that were a year older but of the same size [mean weight of 5 g (0.2 oz)], the protein digestibilities of the two stages of eels were found to be similar (92–94%). But when the former eels were compared to eels of the same age (six months) that were smaller [2.5 g (0.1 oz)], the protein digestibility in the smaller eels was found to be lower (86–94%). Protein digestibility decreased as the protein level of eel diets increased from 30–35% to 40–45%.

Arai et al. (42) reported that eels require both n-6 and n-3 fatty acids and showed that eels grew best on a 2:1 lipid mixture of corn oil to cod liver oil. Takeuchi et al. (43) concluded that the fatty acid requirement for eels can be met with either 0.5% of alpha-linolenic acid and 0.5% of linolenic acid or 1% of alpha-linolenic acid alone. Kanazawa (44) reported that eels can elongate and desaturate alpha-linolenic acid to 20:5 n-3 and 22:6 n-3. In addition, eels are able to use large amounts of fat in their diets, apparently due to their ability to deposit large amounts of fat for their long migrations in the wild. Dosoretz and Degani (45) showed that large eels fed diets with 30% fat grew faster than those fed diets with lower amounts of fat (20%). In addition, the difference in weight gain was due to fat gain. In many cultured species, increased fat gain is not desirable, but in eels, an increased fat content makes the product more marketable. The source of fat in the diet is also important. Gallagher and Degani (9) and Degani et al. (18) showed that (*1*) some fish oil is necessary in the diet of eels (probably as a source of alpha-linolenic acid) and (*2*) lipids of plant origin (e.g., soybean oil) are not as effective in achieving weight or lipid gain as compared with lipids of animal origin (e.g., poultry oil).

Eels can also use carbohydrates better than can coldwater species such as trout. With careful formulation, it is possible to replace part of the protein in eel diets with carbohydrates. In a series of studies, Degani and coworkers showed that *A. anguilla* grew better on diets lower in protein (40% vs. 50%) but higher in carbohydrates (38% vs. 20%) (46). Wheat meal was superior to corn starch, sorghum meal, and potato starch as a source of carbohydrates for eels (47). Simple sugars, such as glucose and sucrose, also improved growth when added to diets in levels up to 30% (48). The metabolic role of dietary carbohydrates is unclear. Carbohydrates are stored as glycogen in eels, and they possess glycogenolytic enzymes, but those enzymes do not respond to glucagon. Rather, glucagon increases gluconeogenic enzyme activity in eels, indicating that this is the most important pathway for maintaining blood glucose, at least under normal conditions (21,49). However, glycogen metabolism does respond to acute stress. For example, the amount of liver glycogen decreased when eels were exposed to sublethal doses of pesticides (50).

Lovell (51) summarized the known vitamin requirements of eels. The Japanese eel has been shown to produce deficiency symptoms when fed diets lacking the vitamins required for energy production: niacin, thiamin, riboflavin, and pantothenic acid. A lack of pyridoxine, which is required for metabolism of amino acids and production of neurotransmitters, causes convulsions and nervous disorders in eels. Also required are biotin, folic acid, cyanocobalamine, and ascorbic acid, vitamins that participate in synthesis reactions in the body. Both choline and inositol are also required; these compounds are structural components of cell membranes. Inositol may also play an important regulatory role in the fertilization of eggs during spawning (52). Of the fat-soluble vitamins—A, E, D, and K—only E (alpha-tocopherol) has been studied and found to be essential.

Arai et al. (53) found that eels grew best on artificial diets containing a mineral mix at 2% of the diet. Increasing the level of minerals (to 3–5%) decreased the eels' weight gain. Trace mineral requirements for eels have not yet been determined, but it is likely that eels require all or most of the minerals required by other finfish. Park and Shimizu (54) found that eels grew best when their diets were supplemented with 50–100 μg/g Zn. However, trace mineral supplements should not be added indiscriminately to diet formulations for eels, as excessive levels can suppress growth, and some mineral supplements may be toxic to temperate species of fish.

The feeding rate necessary for good growth is dependent upon the size of the eel. Arai (22) recommended feeding dry commercial feeds at a rate of 6–8% of body weight per day for elvers and small eels, but reducing the feeding rate to 2–3% for larger eels. Matsui (24) estimated that glass elvers should be feed wet feeds at 10–15% of their body weight per day, and larger eels should receive the feed at 7–10% of their body weight per day. The conversion ratios for dry feed to weight gain is about 1.4:1, while for a feed of raw fish, it is around 7–9:1 (2,9).

Knights (55) reviewed the feeding behavior of eels (*A. anguilla*). It is essential to feed eels food of an appropriate particle size, where a 1:1 ratio of particle size to mouth width is appropriate for soft foods. For harder, pelleted foods the ratio must be reduced to 0.4–0.6:1. In elvers, chance encounters with food particles is the way that food is recognized, while in larger eels, vision becomes more important, although eels can also feed in dark or turbid waters. Knights also noted the effects of stressors on appetite and feeding. Most notable among such stressors are temperature shock and intraspecific behaviors. Eels stop or decrease their feeding in response to acute low-temperature stress. The longer the duration of the stress, the longer the recovery period. Also, smaller eels become subordinate to larger eels if size differences become too large. Subordinate eels and elvers may stop eating altogether. As mentioned previously, size grading is crucial to maintaining a homogenous population, and thus reducing aggressive encounters.

DISEASES

Several authors have reviewed diseases of eels (2,5,23,56,57). Because of its intensive nature, disease can be a major problem in eel culture. One of the common bacterial diseases is red spot disease, which affects the intestine and liver and causes red spots to appear on the eel, due to ulceration of the fins and trunk. (See Fig. 3.) It is commonly caused by *Pseudomonas anguilliseptica*. A similar infection, known as red fin disease, is caused by *Aeromonas liquefaciens* and *Paracolobactrum anguillimortiferum*. Other species of *Vibrio*, *Pseudomonas*, and *Aeromonas* have also been implicated in that disease. Columnaris disease is a bacterial disease caused by the myxobacterium *Chondrococcus columnaris*. When the bacteria attack the gills and cause their eventual disintegration, it is called gill disease. But when the bacteria attacks the tail, causing lesions and sloughing of diseased tissue, it is known as tail rot. These common bacterial infections can now be treated effectively by adding antibacterial drugs to feeds.

Cotton cap, or fungus disease, is caused by the aquatic fungus *Saprolegnia parasitica* and often occurs secondarily to a primary bacterial infection. The occurrence of this disease is linked to high or low pH and can be carried over in ponds from year to year. Antibacterial drugs used in conjunction with malachite green are an effective treatment, but the use of malachite green is prohibited in most culture situations. Infected ponds should be drained and disinfected before they are used again.

The parasitic protozoa *Ichthyophthirius multifiliis* also infects eels. The appearance of grey or white spots on the fins and skin of the eel characterize the infection. Since the parasite cannot tolerate saline conditions, the addition of salt water to culture ponds is usually effective in eliminating it. *Myxidium* sp. also infects eels. This protozoal infection is often confused with *Ichthyophthirius* or columnaris infections, because it produces white cysts on the gills, which destroy the gills. The only method for dealing with an outbreak is to isolate or remove infected and dead eels quickly.

Viruses also infect eels. "Cauliflower disease" is characterized by tumors, usually on the jaw, that resemble heads of cauliflower. Mortality is not high from this viral infection, but the eels cannot eat and become emaciated.

Figure 3. Eels with red spot disease, caused by *P. anguilliseptica*. (From Rickards (56), used with permission.)

A parasitic copepod, *Lernaea cyprinacea*, causes anchor worm disease in Japan. This worm attaches itself inside the mouth of the eel, preventing feeding.

Other organisms have also been shown to infect eels. Those listed previously are presently the most economically important in the culture of eels. In addition, diseases that commonly occur in one species of eel can often readily be introduced into other species. For example, the swimbladder nematode *Anguillicola orassus* was introduced into European eels from Japanese eels (58).

Not all significant diseases of eels are caused by organisms. Gas bubble disease, or gas embolism, (usually occurring in elvers) is caused by supersaturated oxygen or nitrogen levels in the culture water. Bubbles appear on the head of the elver, and it will not eat. Adding new water that is not supersaturated alleviates the problem.

SUMMARY

Eel culture is a strong and viable industry in many parts of the world. However, the life cycle of eels is not closed in captive stocks, and the industry remains dependent upon capture of wild elvers to sustain it. This fact limits exploration into broodstock development and genetic manipulation of wild stocks, which would improve growth rates and disease resistance. Nevertheless, as long as the wild sources of elvers remain plentiful, the industry will prosper.

BIBLIOGRAPHY

1. Food and Agricultural Organization of the United Nations, FAO Statistics, personal communication from the World Wide Web, 1998.
2. A. Usui, *Eel Culture*, 2nd ed., Fishing News Books, Oxford United Kingdom, 1991.
3. R.K. Koehn, *Marine Biol.* **14**, 179–181 (1972).
4. The National Marine Fisheries Service, Fisheries Statistics and Economics Division, personal communication from the World Wide Web, 1998.
5. F.W. Tesch, *The Eel*, Chapman and Hall, Ltd., Edinburgh, United Kingdom, 1979.
6. K.J. Tsukamoto, *Fish Biol.* **36**, 659–672 (1990).
7. A. Umezawa and K. Tsukamoto, *J. Fish Biol.* **39**, 211–223 (1991).
8. K. Beulleas, E.H. Eding, F. Oilevier, J. Komen, and C.J.J. Richter, *Aquaculture* **153**, 151–162 (1997).
9. G. Degani and M.L. Gallagher, *Growth and Nutrition of Eels*, MIGAL-Galilee Technological Center, Rosh Pina, Israel, 1995.
10. N. Pankhurst, *Can. J. Zoo.* **62**, 1143–1149 (1984).
11. H. Ohta, H. Kaquawa, H. Tanaka, K. Okuzawa, and K. Hirose, *Aquaculture* **139**, 291–301 (1996).
12. H. Satoh, K. Yajamori, and T. Hibiya, *Nippon Suisan Gakkashi* **58**, 825–832 (1992).
13. P.W. Sorensen, *J. Fish Biol.* **25**, 261–268 (1984).
14. P.P. Todd, *New Zealand J. of Marine and Freshwater Research* **15**, 237–246 (1981).

15. B. Knights, *Aquaculture* **30**, 173–190 (1983).
16. S. Appelbaum, *Arch. Fisch Wiss.* **31**, 15–20 (1979).
17. S. Arai, in R.P. Wilson, ed., *Handbook of Nutrient Requirements of Finfish*, CRC Press, Boca Raton, FL, 1991, pp. 69–75.
18. G. Degani, D. Levanon, and G. Triger, *Aquacult. Bamidgeh* **36**, 47–52 (1984).
19. G. Degani, H. Hahamu, and D. Levanon, *Comp. Biochem. Physiol.* **84A**, 739–745 (1986).
20. M.L. Gallagher, *Prog. Fish Cult.* **46**, 157–158 (1984).
21. S.S. DeSilva and T.A. Anderson, *Fish Nutrition in Aquaculture*, Chapman and Hall, Ltd., Edinburgh, United Kingdom, 1995.
22. S. Arai, in T. Lovell, ed., *Nutrition and Feeding of Fish*, Reinhold, New York, 1991, pp. 223–230.
23. G. Degani, A. Horowitz, and D. Levanon, *Aquaculture* **46**, 193–200 (1985).
24. I. Matsui, *Theory and Practice of Eel Culture*, A.A. Balkema, Rotterdam, The Netherlands, 1986.
25. H.E. Winn, W.A. Richkus, and L.K. Winn, *Helgel. Wiss. Meeresunters* **27**, 156–166 (1975).
26. J. Parson, K.V. Vickers, and Y. Warden, *J. Fish Biol.* **10**, 211–229 (1977).
27. G. Degani and D. Kushnirov, *Prog. Fish Cult.* **54**, 88–91 (1992).
28. M.L. Gallagher, E. Kane, and R. Beringer, *Comp. Biochem. Physiol.* **78A**, 533–536 (1984).
29. R.T. Lovell, *J. Animal Science* **69**, 4193–4200 (1991).
30. C.B. Cowey and M.J. Walton, in J. Halver, ed., *Fish Nutrition*, Academic Press, New York, 1989, pp. 259–329.
31. S. Arai, T. Nose, and Y. Hashimoto, *Bull. Jap. Soc. Sci. Fish.* **38**, 753–759 (1972).
32. T. Nose and S. Aria, *Bull. Freshwater Fisheries Res. Lab.* **22**, 145–147 (1972).
33. National Research Council, *Nutrient Requirements of Warmwater Fish*, National Academy of Sciences, Washington, DC, 1983.
34. National Research Council, *Nutrient Requirements of Coldwater Fish*, National Academy of Sciences, Washington, DC, 1981.
35. S. Arai, T. Nose, and Y. Hashimoto, *Bull. Freshwater Fish. Res. Lab.* **21**, 161 (1971).
36. S. Tibbets, S.P. Lall, D.M. Anderson, and R.H. Peterson, *Aquaculture '98 Book of Abstracts*, 1998, p. 540.
37. M.L. Gallagher and A.M. Matthews, *J. World Aquacult. Soc.* **18**, 107–112 (1987).
38. G. Degani and M.L. Gallagher, *Bol. Fisiol. Anim.* **12**, 71–79 (1988).
39. M.L. Gallagher and G. Degani, *Aquaculture* **73**, 177–187 (1988).
40. G. Degani, *Ind. J. Fish.* **34**, 213–217 (1987).
41. O. Schmitz, E. Grevel, and E. Pfeffer, *Aquaculture* **41**, 21–30 (1984).
42. S. Arai, T. Nose, and Y. Hashimoto, *Bull. Freshwater Fish Res. Lab.* **21**, 161–178 (1971).
43. T. Takeuchi, S. Arai, T. Watanabe, and Y. Shimma, *Bull. Japan. Soc. Sci. Fish.* **46**, 345–353 (1980).
44. A. Kanazawa, in C.B. Cowery, A.M. Mackie, and J.G. Bell, eds., *Nutrition and Feeding in Fish*, Academic Press, Orlando, FL, 1985, pp. 281–298.
45. C. Dosoretz and G. Degani, *Comp. Biochem. Physiol* **87A**, 733–736 (1987).
46. G. Degani and S. Viola, *Aquaculture* **64**, 283–291 (1987).
47. G. Degani, S. Viola, and D. Levanon, *Aquaculture* **52**, 97–104 (1986).
48. G. Degani and D. Levanon, *Envir. Biol. Fish.* **18**, 149–154 (1987).
49. A.J. Matty and K.P. Lone, in C.B. Cowery, A.M. Mackie, and J.G. Bell, eds., *Nutrition and Feeding in Fish*, Academic Press, Orlando, FL, 1985, pp. 147–167.
50. M.D. Ferrando, *Comp. Biochem. and Physiol C: Comp. Pharm. and Tox.* **101**, 437–441 (1992).
51. T. Lovell, *Nutrition and Feeding of Fish*, Reinhold, New York, 1989.
52. M.L. Gallagher, L. Paramore, D. Alves, and R. Rulifson, *J. Fish Biol.* **52**, 1218–1228 (1998).
53. S. Arai, T. Nose, and H. Karatsu, *Gen. Comp. Endoctrinol.* **43**, 211–217 (1974).
54. C.W. Park and C. Shimizu, *Nippon Suisan Gakkaishi.* **55**, 2134–2141 (1989).
55. B. Knights, in C.B. Cowery, A.M. Mackie, and J.G. Bell, eds., *Nutrition and Feeding in Fish*, Academic Press, Orlando, FL, 1985, pp. 223–242.
56. W.L. Rickards, *A Diagnostic Manual of Eel Diseases Occurring Under Culture Conditions in Japan*, UNC Sea Grant Publication UNC-SG-78-06, 1978.
57. D.M. Forrest, *Eel Capture Culture Processing and Marketing*, Fishing News, Ltd., Surrey, United Kingdom, 1976.
58. O.L.M. Haenen, T.A.M. van Wijngaarden, H.H.T. van der Heijden, J. Hoglund, J.B.J.W. Cornelissen, L.A.M.G. van Leengoed, F.H.M. Borgsteede, and W.B. van Muiswinkel, *Aquacultue* **141**, 41–57 (1996).

EFFLUENTS: DISSOLVED COMPOUNDS

Shulin Chen
Washington State University
Pullman, Washington

Gary C.G. Fornshell
University of Idaho Cooperative Extension System
Twin Falls, Idaho

OUTLINE

INTRODUCTION

Aquaculture effluent refers to either a continuous or intermittent discharge of wastewater from an aquaculture facility. All types of aquaculture production systems generate effluents, but the amount of discharge differs significantly between systems. For a unit of fish production, the frequency of discharge, the amount of flow, and the concentration of pollutants in the effluent determine the effluent load generated by the aquaculture system. In general, systems located in public waters, such as net pens, and systems such as raceways, which discharge continuously, generate more effluent than recirculating or pond systems.

Aquaculture effluents may have adverse environmental impacts because they contain dissolved and suspended pollutants. This entry deals primarily with the dissolved components. Suspended components are covered in the entry entitled "Effluents: sludge."

MAJOR EFFLUENT COMPONENTS AND THEIR ENVIRONMENTAL IMPACTS

The major substances contained in aquaculture effluent, which have potential environmental significance, include phosphorus (P), nitrogen (N), and organic matter. Phosphorus is an essential element for aquatic life and is often regarded as a limiting production factor in freshwater aquatic environments. Biologically available P in aquaculture discharge can stimulate the primary productivity of receiving waters and accelerate the natural eutrophication process. Phosphorus in fish farm discharge water is present as either soluble P, e.g., orthophosphate, or insoluble P. Orthophosphate originates from P excreted in fish urine and from the conversion of insoluble P contained in feces and uneaten feed to soluble P by microbial action.

Nitrogen is another important element supporting aquatic life. Like P, an excess concentration of N increases aquatic productivity and possible eutrophication. The nitrogen compounds of concern to fish health include ammonia and nitrite, which are a concern with respect to receiving waters. Ammonia is a direct product of fish excretion and has high solubility in water. Nitrifying bacteria can biologically oxidize ammonia to nitrite and then to nitrate under aerobic conditions. Both ammonia and nitrate can be used directly by algae and other plants as nutrients. Leaching of nitrate to the groundwater can cause contamination. Ammonia exists in two forms in water: the un-ionized fraction (NH_3) and the ionized fraction (NH_4^+). The ratio of the two fractions is controlled by pH and temperature. The higher the pH and temperature, the higher the un-ionized fraction. The un-ionized fractions of ammonia and nitrite are highly toxic to aquatic life. For example, when the un-ionized ammonia concentration exceeds 0.0125 to 0.025 mg/L (ppm), growth rates of rainbow trout (*Oncorhynchus mykiss*) are reduced and damage to gill, kidney, and liver tissue may occur (1).

Organic substances are also important in evaluating the environmental impacts of aquaculture effluent. Organic decay consumes oxygen and may reduce dissolved oxygen concentration in receiving waters. The impact of organic decay on oxygen concentration can be more directly represented by biochemical oxygen demand (BOD). BOD is the amount of dissolved oxygen necessary to oxidize the readily decomposable organic matter. In the measurement of the standard BOD, a water sample is sealed in a filled bottle, the bottle is incubated in the dark at 20 °C (68 °F) for 5 days, and the measured loss of dissolved oxygen in the bottle can be converted to BOD contained in the water sample in mg/L (ppm).

The primary source of P, N, and organic matter in aquaculture effluent is fish feed. Feed ingredients of animal origin, fish meal, and bone meal are rich sources of P. Phosphorus is also available from cereal grains and other plant protein sources. However, most of the P is present as phytate, which is not well utilized by fish. The concentration and availability of P in the feed have the greatest influence on P retention in the fish and, consequently, on the amount excreted. Nitrogenous compounds, primarily ammonia, are the result of protein catabolism. Quality and quantity of dietary protein and the protein-to-energy ratio of the diet influence nitrogen retention and excretion.

WASTE PRODUCTION

Waste production is typically related to the feeding rate since virtually all of the wastes generated from intensive aquaculture systems originate from fish feed. The majority of the feed will eventually be wasted as fish excretion products, wasted feed, and feces. For instance, in typical salmonid diets, approximately 7.2 to 7.7% of the feed is nitrogen. Of the N in feed, 67 to 75% will be lost to the environment, either as excretion products or as uneaten feed. Nitrogen excretion occurs in two forms, the dissolved form, such as ammonia, and the particulate form, such as organic N in feces. Between 70 and 90% of the nitrogenous catabolites are ammonia with a typical production rate of 3% of the feeding rate.

Phosphorus excretion is also significant. Ketola and Harland (2) reported that the retention of P in several salmonid diets ranged from 14 to 22%, meaning that approximately 80% of dietary P was discharged into the water. Phosphorus is excreted in soluble and particulate forms, and the form of P consumed by the fish will affect the excretion amount of each form. Persson (3) reported 30% of total P from feed was excreted as soluble P.

The amount of feces excreted by fish is typically reported as total suspended solids (TSS), which is defined according to the standard analysis method (4) as particles greater than 2 μm in size. Based on reports of a variety of systems, feces generation values in the range of 0.2 to 0.4 kg of TSS per kg feed appear to be typical (5). The N and P excreted in particulate form can be converted to dissolved form through biological decomposition processes, during which organic nitrogen will be transformed to ammonia and organic P to orthophosphate. In other words, decay of the TSS also contributes to the dissolved components.

Mass balance calculations are a relatively easy method for determining the amount of waste produced through intensive aquaculture. The amount of organic matter, N, and P produced is equal to the difference between the amount added through the feed and the amount is retained by the fish. This approach is applicable to those aquaculture systems based on complete feeding. Data required for a mass balance calculation include feed and fish composition, feed consumed, and fish production. An example of a calculation based on mass balance indicates that 1 kg of trout growing at a rate of 0.007 kg/day (fed dry feed at about 1.1% of body weight/day) will consume 6.9 g O_2, and produce 4.0 g BOD, 0.63 g N, and 0.07 g P/day (6).

EFFLUENT GENERATION

The amount of effluent generated by an aquaculture operation is largely determined by the type of culture system. Among the principal aquaculture systems, cage culture produces the largest amount of effluent because of the direct water exchange between the culture system and the surrounding environment. Flow through aquaculture systems produce the second highest amount, where typical water flow rates are 8–25 L/s per ton of production (6). Recirculating aquaculture systems greatly reduce effluent volume. Typical systems operate with a water exchange rate less than 10% of the system volume per day. Pond systems generally produce low amounts of effluent. Water discharged from ponds occurs primarily during harvesting, water exchange, or levee repairs.

ENVIRONMENTAL REGULATIONS

In response to concerns about the environmental impact of aquaculture effluents, aquaculture has drawn the attention of regulatory agencies and is facing increased regulatory control. The U.S. Environmental Protection Agency (EPA) issues the National Pollution Discharge Elimination System (NPDES) permit to regulate various pollutants from point sources, including aquaculture. NPDES permits are required for fish hatcheries, fish farms, or other facilities that raise aquatic animals under the following conditions:

1. Coldwater fish species or other coldwater aquatic animals in ponds, raceways, or similar structures that discharge at least 30 days per year, produce more than 9,090 kg (20,000 lb) of aquatic animals per year, or receive more than 2,273 kg (5,000 lb) of food during the month of maximum feeding.
2. Warmwater fish species or other warmwater aquatic animals in ponds, raceways, or similar structures that discharge at least 30 days per year. This does not include closed ponds which discharge only during periods of excess run-off or warmwater facilities which produce less than 45,454.5 kg (100,000 lb) of aquatic animals per year.
3. Facilities determined on a case-by-case basis by the permitting authority to be significant contributors of pollution to waters of the United States. Discharge of pollutants to receiving waters from aquaculture production facilities, except as provided in the permit, is a violation of the Clean Water Act and may be subject to enforcement action by EPA.

NPDES permits for aquaculture operations can set discharge limits on solids, nutrients, and chemical compounds used for water treatments. For example, the current NPDES permit for aquaculture operations in Idaho limits the maximum average net total suspended solids to 5 mg/L (ppm). However, future permits will include nutrient restrictions. The EPA is conducting a preliminary study on water discharged from aquaculture facilities nationwide. Two possible outcomes from that study are the development of national effluent limitation guidelines and standards or the development of technical guidelines for one or more production systems.

Various state agencies also regulate the discharge of aquaculture effluent. Individual states designate beneficial uses for water bodies within their jurisdictions and establish water quality criteria to protect those uses. The state agencies may establish stricter standards than the EPA. Some argue that the current NPDES permit program is adequate, and national effluent guidelines are not necessary. Others feel that several state permit programs are inadequate or inconsistent with the Clean Water Act.

EFFLUENT POLLUTION CONTROL

Controlling the discharge of dissolved compounds from aquaculture effluent presents a challenge, especially in flow through systems. Although typical wastewater treatment technologies for nutrient control and dissolved compound removal are available from other industries, the capital and operational costs of the technologies are often prohibitive. Thus, the technologies are often impractical for aquaculture systems, primarily because of the large effluent volume that requires treatment. The two basic approaches for effluent pollution control are source reduction and interception and removal of the pollutants prior to their discharge into receiving waters.

Source reduction includes the development of more efficient feeds and improved feed management to minimize waste generation. Scientists and the aquaculture industry have made significant progress in fish nutrition, feed formulation, and feed manufacture to reduce the excretion of nutrients. Examples include the selection of ingredients that increase the bioavailability and digestibility of macronutrients such as phosphorus (7), lowering the P content in feeds, and the reduction of urinary excretion of P by fish. Research results have demonstrated that formulating feeds to minimize the excess levels of available nutrients in the fish diets can significantly reduce the excretion of nutrients. However, realizing the benefits of high-performance feeds requires minimizing feed wastage through proper feed management.

Effluent treatment research concerned with pollutant interception and removal has primarily focused on the treatment of suspended solids. Solids need to be removed from systems with continuous discharge as soon as possible to avoid nutrient solubilization. Several filtration processes are available for particle removal that may be feasible in certain aquaculture systems. Other methods include settling basins and constructed wetlands. In pond systems, natural, biological, and chemical processes remove nutrients and organic matter.

One effective method of effluent reduction is recycling the water. In the past two decades, significant research efforts have been devoted to developing recirculating aquaculture systems. The advancements in aeration, biofiltration, and solids removal technology have made it possible for commercial scale recirculating systems to be profitable under the right conditions. Typical recirculating systems release less than 10% of the total system volume as effluent daily. The effluent can be treated before being discharged to water bodies, or disposed of on land through irrigation.

BIBLIOGRAPHY

1. Idaho DEQ, Idaho Waste Management Guidelines for Aquaculture Operations. Idaho Department of Environmental Quality, Boise, Idaho, 1995.
2. H.G. Ketola and B.F. Harland, *Trans. Am. Fish. Soc.* **122**, 1120–1126 (1993).
3. G. Persson, *Relationships between Feed, Productivity and Pollution in the Farming of Large Rainbow Trout* (Salmo gairdneri), Swedish Environmental Protection Agency, Stockholm, PM3534, 1988.
4. APHA. *Standard Methods for the Examination of Water and Wastewater*, 19th ed., American Public Health Association, New York, 1995.
5. S. Chen, D.E. Coffin, and R.F. Malone, *J. World Aquacult. Soc.* **28**(4), 303–315 (1997).
6. T.V.R. Pillay, *Aquaculture and the Environment*. Fishing News Books, Oxford, England, 1992.
7. S.H. Sugiura, F.M. Dong, C.K. Rathbone, and R.W. Hardy, *Aquaculture* **159**, 177–202 (1998).

EFFLUENTS: SLUDGE

SHULIN CHEN
Washington State University
Pullman, Washington

OUTLINE

INTRODUCTION

Aquacultural effluent refers to either a continuous or intermittent discharge of a liquid stream from an aquaculture facility. Aquaculture effluent can potentially cause adverse environmental impact because of both the dissolved and suspended pollutants it contains. This entry deals primarily with the particulate components of the pollutants. The dissolved compounds are covered under the entry entitled "Effluents: dissolved compounds."

For the purpose of either water conservation or more efficient wastewater handling and treatment, it is desirable in an aquaculture operation to concentrate the particulate waste as much as possible through a solid/liquid separation process. The concentrated waste stream, typically with a solids content greater than 5%, is often called sludge.

Although waste management strategies are likely different for different culture systems, solids removal and sludge management are often the most important components of the strategy because the majority of the pollutants are associated with the particles in the sludge. This entry covers source, production rate, characteristics, regulations, and treatment and disposal of aquacultural sludge.

SOURCES OF PRODUCTION

Waste Excretion

The first step in waste management for an aquacultural facility is to estimate the quantity of waste excreted by fish. Waste excretion rate is typically related to feeding rate since virtually all the wastes generated from an intensive aquacultural system originate from fish feed. Assuming a typical feed conversion ratio (feed input: fish gain) of 1 to 2 and neglecting the impact of uneaten food, 80% of feed (on a dry weight basis) input to an aquacultural system will eventually become waste as fish excretion products (1). Typical waste forms include CO_2, ammonia, feces, and other dissolved substances. The amount of waste produced is often closely related to feeding rate.

The amount of feces excreted by fish is typically reported as total suspended solids (TSS), which is defined according to the standard analysis method (2) as particles greater than 2 μm in size. TSS excretion rates for trout and catfish have been a topic of many studies. The direct TSS excretion rate varied from one report to another, ranging from 0.30 to 0.52 kg per kg feed for trout and from 0.18 to 0.69 kg per kg feed for catfish. Clearly, TSS excretion rate may vary with species, temperature, feeding rate, and management. Based on reports on a variety of systems, however, feces generation values in the range of 0.2 to 0.4 kg TSS per kg feed appear to be typical (3).

One of the major factors that contribute to the differences in reported excretion values is the amount of uneaten feed that ends up in the water. The amount of uneaten feed is usually minimized under laboratory conditions but is less controllable under commercial conditions. Consequently, the actual solids generation rate from commercial aquacultural facilities may be higher than the excretion rates obtained under laboratory conditions. Naturally, effective feeding management can minimize the amount of solids that result from uneaten feed.

Waste Production as Sludge

The amount of solids and the associated sludge volume are two major factors in aquacultural waste management.

The solids mass production rate is determined by the waste excretion rate of the fish within the system, as well as the system's internal waste treatment capability. The volume of sludge generated, on the other hand, is controlled not only by the amount of solids produced but also by the degree to which the TSS is concentrated in the sludge stream. When solids are removed from a system either through backwashing a filter or cleaning a sedimentation tank, the resultant sludge is usually still relatively dilute. Clearly, the disadvantage in dealing with dilute sludge is the large waste volume involved. Thus, it is desirable to design the solids separation and removal system to obtain a sludge that is as highly concentrated as possible. For example, 5% of solids concentration is achievable for sludge thickening using a sedimentation process.

Fish culture produces less waste than large land animals for a unit weight gain because aquatic animals are very efficient at feed conversion. However, if daily waste generation is evaluated on a live weight basis, the amount of waste produced by fish is comparable to that of other animals, but with a much higher sludge volume. The higher sludge volume production is primarily due to the dilute nature of aquaculture waste.

CHARACTERIZATION

Pertinent Environmental Parameters

The potential environmental impact of aquaculture sludge can be described using several key parameters, e.g., concentrations of TSS, biochemical oxygen demand (BOD), and nutrients in the sludge. TSS concentration represents the amount of particulate matter present. In an aquatic environment, elevated TSS concentration increases water turbidity. Additionally, TSS can deposit in the receiving water body causing alteration of habitat conditions. Moreover, because of the organic nature of the TSS from aquacultural operations, the breakdown of TSS creates an oxygen demand that will reduce the concentration of dissolved oxygen in the receiving water.

The second parameter is BOD_5, representing the amount of oxygen consumed by the organics in the waste that biodegrade within five days under standard conditions [20 °C (68 °F)]. BOD_5 excretion rate is also generally expressed as a ratio to the feeding rate. BOD_5 excretion rate varies more widely and is typically lower than that of TSS. Of the BOD_5 excreted, over half is typically in particulate form and the rest is dissolved in water.

Nutrient contents are also important parameters in aquacultural sludge management. The major concerns are nitrogen and phosphorus. The majority of the nutrients are in organic forms when sludge is first discharged from a solids separation and removal process of an aquaculture operation. Mineralization occurs later under a variety of conditions and transforms the organic nitrogen and phosphorous to inorganic forms that are available to plants and are mobile in the environment. Discharge of the nutrients into surface waters can stimulate algae growth, causing eutrophication, while leaching of nitrate nitrogen into groundwater can cause groundwater contamination.

Chemical Characteristics

Aquacultural sludge can be characterized both by the concentration of waste constituents or by the ratios of given constituents to total solids (TS) in the sludge. TS concentration in aquacultural sludge is mainly from TSS. For example, the BOD_5/TS ratio is a measure of the degree of sludge stabilization. A high BOD_5/TS ratio implies a sludge that will rapidly decay and potentially cause oxygen depletion and odor problems if not properly managed. Similarly, the nutrient content of the sludge can be described by total Kjedahl nitrogen (TKN) and total phosphorus (TP) to TS ratios. Higher ratios represent better fertilizer values or strong pollution potential. The characteristics of the aquacultural sludge from a recirculating system using a plastic beads filter are illustrated in Table 1, with reference to domestic sludge.

Compared with typical municipal sludge, most aquacultural sludge has lower solid and BOD_5 concentrations. Of the solids contained in the aquacultural sludge, more than 80% are volatile, 20% higher than those in municipal sludge. Aquacultural sludge also has higher nitrogen and phosphorus contents. Olson (6) found that fish wastes contain a higher percentage of nitrogen than cattle, pig, and sheep wastes. The percentages may change with respect to the stabilization time it will take for sludge as organic nitrogen and phosphorous to convert to ammonia and soluble phosphate, respectively, during the decay processes. The high TP and TKN contents in the fish sludge originate from the feed; most fish feeds contain 7.2 to 7.7% of nitrogen by weight. Of the nitrogen in feeds, 67 to 75% will be lost to the aquatic environment. The phosphorus content of the commercial fish diet ranges from 1.2 to 2.5%, with as much as 80% being lost to the aquatic environment (7).

Table 1. The Chemical Compositions of Aquacultural Sludge

	Aquacultural Sludge (4)			Domestic Sludge (5)	
Parameter	Range	Mean	Standard Deviation	Range	Typical
TS (%)	1.4–2.6	1.8	0.35	2.0–8.0	5.0
BOD_5 [mg/L (ppm)]	1,588–3,867	2756	212	2,000–30,000	6,000
TKN (N, % of TS)	3.7–4.7	4.0	0.5	1.5–4	2.5
TP (P, % of TS)	0.6–2.6	1.3	0.7	0.4–1.2	0.7
pH	6.0–7.2	6.7	0.4	5.0–8.0	6.0

Physical Characteristics

Major physical sludge characteristics that affect the design of a treatment system include particle settling velocity, density, and size distribution. Ning (4) observed that the solids particles in aquacultural sludge from a bead filter settled out fairly quickly with an average zone settling velocity of 1.37×10^{-3} m/s (8 cm/min) (3.1 m/min). The wet density of the sludge was measured as 1.004 g/mL, which is close to the typical value of a municipal sludge. The particle size distribution of the aquacultural sludge showed that particles less than 60 μm and greater than 1,000 μm accounted for 15% and 17% of the total dry weight, while particles ranging from 60 to 105, 105 to 500, and 500 to 1,000 μm represented 8%, 29%, and 31%, respectively.

Biochemical Characteristics

One of the major biochemical characteristics of sludge is the decomposition (or decay) rate constant that represents how fast waste material decays under given conditions. The rate constant referred to here is defined as a first-order reaction ($dC/dt = -kC$, where C is the concentration, t is time, and k is the rate constant). A laboratory study (4) found that oxygen availability and temperature had significant impact upon decay rate constants. From 10 to 30 °C, the anaerobic (without oxygen) decay rate constants of BOD_5 varied from 0.004/d to 0.037/d. For the same temperature range, the aerobic (with oxygen) decay rate constants varied from 0.188/d to 0.329/d. The impact of temperature on digestion rates was more significant for anaerobic digestions than for aerobic digestions. The study also found that aquacultural sludge had digestion rate constants comparable to domestic sludge. For example, the maximum aerobic digestion rate constant of BOD_5 for aquacultural sludge at 20 °C was 0.14 to 0.32/d, whereas the reported value for municipal sludge was 0.05 to 0.3/d (5).

TREATMENT AND DISPOSAL

Environmental Regulations

Federal and state pollution control regulations apply to aquacultural operations. These regulations prohibit direct discharge of aquacultural sludge to receiving water bodies. In most cases, aquaculture facilities are required to obtain permits from the appropriate state agencies for the handling and disposal of aquacultural sludge. Typically, a permit is not going to be issued unless the aquaculture operation has a waste management plan that meets the environmental regulations.

Minimizing the environmental impact of nutrients and solids produced from aquacultural facilities is the primary objective of most aquacultural waste management plans. A typical waste management plan will include, at a minimum, (*1*) a description of the solids handling and removal systems components, (*2*) schedules for cleaning frequency of the various waste collection components, (*3*) a plan including rate and schedule for land application or other approved method of utilization for the waste material, and (*4*) a monitoring plan that evaluates the effectiveness of the overall system.

It is important during the development of a waste management plan to incorporate best management practices that have been proven effective by the industry and that are acceptable to the regulatory agencies. In fact, the aquaculture industry, with help from the research community, has been making constant efforts to address both pollution reduction and waste management and has developed and demonstrated some best management practices for pollution minimization. Examples include reducing phosphorus through using low phosphorous feed and improving phosphorus digestion efficiency, effluent minimization through the adoption of recirculating systems, effluent treatment through solids separation, land application of fish waste to utilize the fertilizer values, and sludge storage for winter months. A waste management plan should incorporate these practices as much as possible.

Treatment and Disposal

As environmental regulations become more stringent, sludge management becomes one of the major tasks in an aquaculture operation. A proper waste management strategy is now considered critical for maintaining the legality, profitability, and sustainability of an aquaculture facility. Treatment and ultimate disposal are the two major steps in aquacultural waste management. Before the final disposal, two treatment processes (thickening and stabilization) may be necessary for some applications.

Gravity Thickening. The purpose of thickening is to increase solids content. A thickening process is typically employed between the point of effluent discharge from the culture system and the sludge storage and stabilization units. In virtually all applications, it is always desirable to increase solids concentrations to a higher level to improve the economics of treatment and disposal. Clarification, often in settling tanks or ponds, is a common thickening process during which particles settle to the bottom by gravity. When the sediments are removed from the settling unit, solids concentrations are usually at 2 to 5%.

Sludge Stabilization and Storage. A stabilization process is necessary in environmentally sensitive areas where offensive odors need to be minimized. There are typically two major benefits that are related to sludge stabilization: organics decay and volume reduction. Stabilization processes can provide for complete oxidation of readily degraded organics, resulting in a sludge that is unoffensive in nature. Stabilized sludge poses few problems when disposed through land application or landfilling. In addition, stabilization can reduce sludge volumes by 50–75% (8).

Sludge storage is required in cold climates. During wintertime, frozen ground, along with wet winter weather, decreases waste utilization on land and increases surface runoff potential. Therefore, land application of waste in these regions is limited only to certain times of the year when the vegetation and crops on land are active. As a result, waste has to be stored during the winter months.

Lagoons are the most feasible technology for stabilizing aquacultural sludge. Anaerobic lagoons have been used to treat waste discharges from all phases of the vast agricultural industry and have also been considered suitable treatment processes for aquacultural wastes. In an anaerobic lagoon, organic loading is so high that no appreciable oxygen concentration exists. Sludge introduced into the lagoon ranges from that containing relatively light solids concentrations (approximately 0.1% solids) to slurries containing just enough water to transport the solids into the lagoon. Anaerobic lagoons function successfully over a wide solids-loading range with little maintenance. The major parameters used for anaerobic lagoon design are volatile suspended solids (VSS).

Whenever possible, a two-stage lagoon system should be used for sludge stabilization. A two-stage lagoon system is typically designed in such a way that the first stage is anaerobic, and the second stage is facultative. The main objective of an anaerobic lagoon is BOD reduction through organic decay, while the main objective of a facultative lagoon is nitrogen reduction as well as additional BOD removal. Typical effluent quality from a facultative lagoon is not adequate for direct discharge but is more appropriate for irrigation.

Clearly, a lagoon system can also be used for sludge storage in wintertime. However, the design for the two functions is usually different. Storage capacity is the major design criteria for storage lagoons (or ponds) where sludge and precipitation volume is more important. Organic (VSS) loading rate, on the other hand, is the major design criteria for anaerobic lagoons where the amount of organics and the decomposition rate of the organics are more important.

Sludge Disposal. Currently, the most often used aquacultural sludge disposal process is direct land application. Application methods include using sprinklers and tank trucks. Because high-rate land application of animal manure as a waste has been proven to cause adverse environmental impacts (9), a better approach for aquacultural sludge management is to use the waste only according to its fertilizer value for crops. The high nitrogen content (4 to 6%) of aquacultural waste makes it valuable to crops as a fertilizer, but limitations for such application have also been identified (6). The first is odor, which prohibits this option in populated areas. The second is the propensity for the applied sludge to form a crust. If the sludge is not thoroughly plowed into the soil, some plant seedlings may be unable to push through the crust. The third limitation is the expense of hauling and spreading. The fourth is the slow nitrogen release rate. Since about 90% of the total nitrogen is in organic form, only one-third of the nutrients can be used in the first year (6). This makes application in high rainfall areas questionable, since runoff of the unused nitrogen may cause water quality problems in local surface waters.

The guidelines for application rates of aquacultural sludge on cropland have not yet been established. However, it would be reasonable to manage aquacultural waste following guidelines similar to those for managing other types of waste. Most animal waste management plans are based on nitrogen. The fundamental premise is that the rate of animal waste applied on land should not provide more plant available N than crops need, in order to avoid contaminating groundwater with nitrate. Studies on animal manure indicate that crops typically remove between 100 to 200 kg of nitrogen/ha (9). Therefore, a similar rate for application may eliminate nitrogen accumulation in the soil and avoid adverse impacts on the environment. The high nitrogen content that makes aquacultural waste valuable to crops as a fertilizer could, of course, also make overapplication of nitrogen more likely. Olson (6) tested three application rates of trout manure in a greenhouse (111, 222, 336 kg N/ha) for growing spring wheat. Satisfactory results were obtained from application rates of both 222 and 336 kg N/ha. Subject to further experimental verifications, Olson (6) recommends 222 kg N/ha as a design criterion for aquacultural sludge application on land. The information presented here would be useful for estimating the amount of land needed for sludge land application for a given operation. The characteristics of aquaculture sludge, however, may vary substantially depending upon waste management systems and many other factors (10,11). Thus, the nutrient values and corresponding rate for land application may be different for aquaculture sludge from different operations.

BIBLIOGRAPHY

1. T.A. Hopkins and W.E. Manci, *Aquaculture Mag.* **15**(2), 30, 32–36 (1989).
2. APHA, *Standard Methods for the Examination of Water and Wastewater*, 19th ed., American Public Health Association, New York, 1989.
3. S. Chen, D.E. Coffin, and R.F. Malone, *J. World Aquacult. Soc.* **28**(4), 303–315 (1997).
4. Z. Ning, *Characteristics and Digestibility of Aquacultural Sludge*, Master's thesis. Department of Civil and Environmental Engineering, Louisiana State University, Baton Rouge, LA, 1996.
5. Metcalf and Eddy, Inc., *Wastewater Engineering, Treatment / Disposal / Reuse*, 3rd ed., McGraw Hill, 1991.
6. G.L. Olson, *The Use of Trout Manure as a Fertilizer for Idaho Crops*, Paper presented at National Livestock, Poultry and Aquaculture Waste Management Workshop, July 29–31, 1991, Kansas City, MS, 1991.
7. G.K. Iwama, *Crit. Rev. Environ. Control* **21**(2), 177–216 (1991).
8. T.D. Reynolds, *Unit Operations and Processes in Environmental Engineering*, Brooks/Cole Engineering Division, Monterey, CA, 1982.
9. M.R. Overcash, F.J. Humenik, and J.R. Miner, *Livestock Waste Management, Volume II*, CRC Press, Inc., 1983.
10. P.W. Westerman, J.M. Hinshaw, and J.C. Barker, in J.K. Wang, ed., *Techniques for Modern Aquaculture*, Proceedings of an Aquacultural Engineering Conference, June 1–23, 1993, Spokane, WA, ASAE, St. Joseph, MI, 1993.
11. J.G. Twarowska, P.W. Westerman, and T.M. Losordo, *Aquacultural Eng.* **16**, 133–147 (1997).

ENDANGERED SPECIES RECOVERY: CAPTIVE BROODSTOCKS TO AID RECOVERY OF ENDANGERED SALMON STOCKS

Thomas A. Flagg
Conrad V.W. Mahnken
National Marine Fisheries Service
Manchester, Washington

OUTLINE

INTRODUCTION

A major obstacle often facing endangered species recovery programs is overcoming the demographic extinction risks posed by a small population size and limited breeding pairs. Captive rearing of animals to produce adults or offspring to supplement wild populations is a gene maintenance and population amplification technique that has gained worldwide popularity as a component of species enhancement (1–4). Currently, over 105 species of mammals, 40 species of birds, 29 species of fish, 14 species of invertebrates, and 12 species of reptiles are being maintained or enhanced through forms of captive breeding (3,5). Captive broodstocks are an especially attractive alternative for highly fecund animals such as fish. In this entry, we examine the potential use of captive broodstock technology to aid the recovery of depleted stocks of Pacific salmon.

THE CAPTIVE BROODSTOCK CONCEPT

A number of stocks of anadromous salmonids in the Pacific Northwest are currently listed as threatened or endangered under the U.S. Endangered Species Act (ESA) (6–8). In addition, over 200 stocks have been identified as being "of special concern" (9). The ESA recognizes that the conservation of listed species may be facilitated by artificial means, such as captive broodstocks, while factors impeding population recovery are identified and corrected (10).

Captive broodstock programs differ from conventional salmon culture in that fish of wild origin are maintained in captivity throughout their life to produce offspring for the purpose of supplementing wild populations. The relatively short generation time (2–7 years) (11–13) and the potential to produce large numbers of offspring (an average of 1,500–5,000 eggs per female, depending on the species) (11–14) make Pacific salmon ideal for captive broodstock rearing. Survival advantages offered through the protective culture of such large numbers of eggs can be profound. The potential benefits of captive culture over natural production can best be viewed in terms of the two near-independent stages of anadromous Pacific salmon life history: the egg-to-smolt stage and the smolt-to-adult stage.

Wild Pacific salmon generally have high natural mortality through the early life-history stage. For instance, naturally spawned ESA-listed endangered Snake River sockeye salmon generally experienced less than 6% egg-to-smolt survival (15). In contrast, the protective environment of hatcheries can produce many more juveniles than are expected in the wild: egg-to-smolt survival for hatchery-reared sockeye salmon is generally at least 75% and frequently greater (14,16,17). Survival ratios are similar for other Pacific salmon species. Thus, successful hatchery rearing through the juvenile stages alone may provide up to twelvefold survival advantage compared to natural production.

The survival advantages of protective culture offer the greatest benefits during the smolt-to-adult phase. For instance, under current environmental conditions in the Columbia River Basin, wild smolt-to-adult survival of anadromous Snake River sockeye salmon has been estimated at less than 0.2% (16). However, smolt-to-adult survival of Pacific salmon in protective captive culture may easily exceed 50%, more than a 250-fold survival advantage over natural production during these life stages (17). Theoretically, the captive culture of Pacific salmon through both the egg-to-smolt and smolt-to-adult life stages could provide a survival rate more than 3,000 times higher than the rate of natural production of a depleted stock.

Artificial propagation and captive broodstock technologies are not without potential complications and risk. Captive rearing programs have been criticized as "halfway technologies" that address the effects of endangerment but not its underlying causes (18,19). The potential genetic and environmental hazards of using captive culture (inbreeding, genetic drift, domestication, selection, behavioral conditioning, and exposure to disease) and the possible negative interactions of hatchery and wild fish have been well documented (20–24). Some authors argue that the primary course of recovery for depleted populations should be through habitat improvements, after which populations should be left to rebound naturally. Others point out the potential for catastrophic loss of a potentially major portion of the gene pool in captivity through failure of the culture facility or disease outbreak (1–16).

Captive breeding is also widely regarded as less cost-effective in the long-term than in situ preservation (25). However, most fisheries researchers and managers recognize that even aggressive habitat improvements will take several fish generations to complete. Pragmatically, captive broodstocks may offer the best chance for continued existence of endangered populations through the enhanced survival during protective culture (16). As Johnson and Jensen (2) pointed out, "If the gene pool is lost, no amount of habitat protection will help the species." Thus, in many cases, it appears that the risk of extinction in waiting for natural recovery through habitat improvements is greater than the risk to the population from husbandry intervention.

Fish for captive broodstocks can be sourced from all available life stages: eyed eggs, fry, smolts captured from the wild, and prespawning adults, captured and artificially spawned. As a practical matter, the capture of adult Pacific salmon at weirs during their upstream migration and the capture of smolts at weirs during their downstream migration are the easiest to accomplish. However, redd sampling and fry trapping may also be employed. Fish for captive broodstocks should be representative of all portions of the gene pool to avoid artificial amplification of only a portion of a population and subsequent inadvertent reduction in the overall effective population size (10).

Two captive broodstock approaches are being applied to salmon recovery in the Pacific Northwest (17). One strategy involves rearing the populations to maturity in hatcheries. The first or second generation offspring are then stocked into ancestral lakes or streams at one or more juvenile life stages (i.e., fry, parr, smolt). Another strategy involves rearing the broodstock in captivity to adulthood, then releasing the adults back into their natal habitats to spawn naturally. In either case, fish can be reared using simple modifications of standard fish culture practices (16). It is advisable to rear duplicate groups of captive broodstocks at low density in disease free water at two or more secure facilities to avoid potential catastrophic loss of valuable gene pools (16). All fish should be passive integrated transponder (PIT) tagged (26,27) or otherwise marked to allow identification of individuals from each family.

Maintenance of multiple lineages and year classes in culture allows the development of mating strategies that can maintain and potentially increase genetic diversity. Such mating strategies can include random pairing, pairing in as many different combinations as possible, avoidance of pairing between siblings, crossings between different year-classes, and fertilization with cryopreserved sperm from other generations (10). In any event, each discrete year-class of captive broodstock should be maintained for only a limited number of generations to guard against domestication and to help assure that genetic integrity and adaptability to native habitats are preserved.

PERFORMANCE OF CAPTIVE BROODSTOCKS

The dramatic difference between the natural environments experienced by wild Pacific salmon and the artificial environments experienced by captively reared fish appears to create a number of differences in their relative reproductive potential. Schiewe et al. (17) indicated that egg-to-adult survival during captive broodstock culture has been observed to range from 0 to 88% for sockeye salmon, 2 to 78% for chinook salmon, and 80% or more for coho salmon. Survival may generally be expected to be in the upper range in the absence of disease outbreaks. The size and age of maturity of captively reared adults is generally less than their wild cohorts. For instance, Joyce et al. (28) indicated that captively reared chinook salmon females from the Unuk River in Alaska, matured at 4 years of age and 6.8 kg, while wild cohorts matured at 5 years and 12.8 kg. Similarly, Schiewe et al. (17) reported that captively reared Redfish Lake sockeye salmon matured at 3 years of age and 1.2 kg compared to 4 years and 2.0–3.0 kg for wild fish. Average viability of eggs from captively reared spawners (30–70%) has also been found to be commonly lower than the 75–95% viability of similar strains of hatchery-spawned wild fish (16,17,29). Captively reared salmon have been shown to demonstrate a full range of reproductive behaviors and the ability to naturally reproduce (29–31). Nevertheless, recent behavioral studies indicate that captively reared Pacific salmon released to spawn in streams may have lower breeding success than comingling wild salmon (32).

The reasons for the poorer reproductive performance of artificially propagated captively reared fish versus ocean-ranched and wild cohorts are not well understood. Most captive broodstock programs we reviewed used spawners collected from a wild population. Therefore, it seems intuitive that much of the poor performance, at least in first-generation offspring, can be attributed to the effects of artificial culture environments. We speculate that the development of nutritionally complete species-specific brood diets would improve the reproductive performance of captive broodstocks by improving gamete quality. However, the effects of genetic change in the captively reared populations as a basis for reduced spawner size, egg viability, and reproductive behavior in fish remain a possibility.

Although in many cases, the average survival and eyed-egg viability of captive broodstocks have not met optimum expectations yet, they still are fulfilling population-amplification goals, at least from the juvenile-production perspective (16,17). For instance, since Snake River sockeye salmon from Redfish Lake (Idaho) were ESA-listed as endangered in 1991, only 16 adults had returned to spawn through 1998. All were captured and spawned, and the eggs were reared as captive broodstock. Those broodstocks have subsequently produced over 800,000 animals that have been released (as either eyed eggs, fry, presmolts, smolts, or prespawning adults) into historic habitats (Flagg, unpublished data). For those Redfish Lake sockeye salmon captive broodstocks, most individual group population amplifications have represented one-to-several thousand times the current estimated natural egg-to-adult survivals of this endangered stock (Flagg, unpublished data). Using captive broodstock techniques, the program has been able to maintain all gene pool segments of Redfish Lake sockeye salmon that existed at the time of listing (6,16,17). A few wild breeding pairs should return in the next few years to augment the population. However, it should be emphasized that for the Redfish Lake sockeye salmon, the major barriers to survival are downstream of the lacustrine rearing habitat. Captive broodstocks alone cannot recover the Redfish Lake sockeye salmon population; habitat improvements must occur for recovery to be a possibility. In addition, it should be noted that captive broodstocking of Pacific salmon is still in the initial stages of development. Years of monitoring and evaluation of adult returns will be necessary to fully evaluate the technology.

CONCLUSIONS

Since a multitude of factors affect both the decline and potential for recovery of a stock, exacting rules cannot presently be defined for the implementation of captive broodstocks. However, knowledge of survival, reproductive success, and offspring fitness is critical to determining levels of risk in implementing a salmonid captive broodstock program. In general, the use of captive broodstocks should be restricted to situations in which the natural population is dangerously close to extinction. Proper precautions should be taken to minimize genetic impacts during the collection, mating, and rearing of captive broodstocks, as any alteration to the original genetic composition of the population in captivity may reduce the efficacy of supplementation in rebuilding the natural population. Furthermore, the liberation of fish from captive broodstocks should be consistent with the known behavior of existing wild fish and with the available knowledge of the life-history characteristics of the wild fish.

In some cases captive broodstocks may provide the only mechanism to prevent extinction of a stock and may be undertaken regardless of prospects for immediate habitat improvement. However, captive broodstocks should be viewed as a short-term measure to aid in population recovery—never as a substitute for reestablishing naturally spawning fish in the ecosystem. Because the benefits and risks have not been established through long-term monitoring and evaluation, captive broodstock development should be considered an experimental approach and used with caution. Salmonid captive broodstocks can provide an egg base to help "jump start" a population, but these efforts must go hand in hand with scientifically sound resource management (e.g., habitat restoration, harvest reform) to fully aid in recovery. The primary consideration should be restoring the fish species to its native habitat.

BIBLIOGRAPHY

1. J.H.W. Gipps, ed., *Beyond Captive Breeding: Reintroducing Endangered Species through Captive Breeding*. Zool. Soc. London Symp. 62, London, 1991.
2. J.E. Johnson and B.L. Jensen, in W.L. Minckley and J.E. Deacon, eds., *Battle against Extinction*, Univ. Arizona Press, Tucson, 1991, pp. 199–217.
3. P.J.S. Olney, G.M. Mace, and A.T.C. Feistner, eds., *Creative Conservation: Interactive Management of Wild and Captive Animals*, Chapman and Hall, London, 1994.
4. J. DeBlieu, *Meant to be Wild: The Struggle to Save Endangered Species through Captive Breeding*, Fulcrum Publishing, Golden, CO, 1993.
5. Captive Breeding Specialist Group (CBSG), *CBSG News* **2**(4), 12 (1991).
6. R.S. Waples, O.W. Johnson, and R.P. Jones Jr., U.S. Dept. Commerce, *NOAA Tech. Memo.* NMFS F/NWC-195 (1991).
7. R.S. Waples, R.P. Jones Jr., B.R. Beckman, and G.A. Swan, U.S. Dept. Commerce, *NOAA Tech. Memo.* NMFS F/NWC-201 (1991).
8. G.M. Matthews and R.S. Waples, U.S. Dept. Commerce, *NOAA Tech. Memo.* NMFS F/NWC-200 (1991).
9. W. Nehlson, J.E. Williams, and J.A. Lichatowich, *Fisheries* **16**(2), 4–21 (1991).
10. J.J. Hard, R.P. Jones, Jr., M.R. Delarm, and R.S. Waples, U.S. Dept. Commerce, *NOAA Tech. Memo.* NMFS-NWFSC-2 (1992).
11. R.E. Foerster, *The Sockage Salmon*, Bull. Fish. Res. Board Can., 162, 1968.
12. R.L. Burgner, in C. Groot and L. Margolis, eds., *Pacific Salmon Life Histories*, Univ. British Columbia Press, Vancouver, BC, Canada, 1991, pp. 1–118.
13. M.C. Healey, in C. Groot and L. Margolis, eds., *Pacific Salmon Life Histories*, Univ. British Columbia Press, Vancouver, BC, 1991, pp. 311–394.
14. J.W. Mullan, U.S. Fish Wildl. Serv., *Biol. Rep.* **86**(12), (1986).
15. T.C. Bjornn, D.R. Craddock, and D.R. Corley, *Trans. Am. Fish. Soc.* **97**, 360–375 (1968).
16. T.A. Flagg, C.V.W. Mahnken, and K.A. Johnson, *Am. Fish. Soc. Symp.* **15**, 81–90 (1995).
17. M.H. Schiewe, T.A. Flagg, and B.A. Berejikian, *Bull. Natl. Res. Inst. Aquacult. Suppl.* **3**, 29–34 (1997).
18. N.B. Frazer, *Conserv. Biol.* **6**(2), 179–184 (1992).
19. G.K. Meffe, *Conserv. Biol.* **6**(3), 350–354 (1992).
20. H.L. Kincaid, *Aquaculture* **3**, 215–227 (1983).
21. F.W. Allendorf and N. Ryman, in N. Ryman and F. Utter, eds., *Population Genetics and Fisheries Management*, Univ. Washington Press, Seattle, 1987, pp. 141–159.
22. D. Tave, *Genetics for Fish Hatchery Managers*, Van Nostrand Reinhold, New York, 1993.
23. R.S. Waples, *Can. J. Fish. Aquat. Sci.* **48**(Supplement 1), 124–133 (1991).
24. R.S. Waples, *Fisheries* **24**(2), 12–21 (1999).
25. C.D. Magin, T.H. Johnson, B. Groombridge, M. Jenkins, and H. Smith, in P.J.S. Olney, G.M. Mace, and A.T.C. Feistner, eds., *Creative Conservation: Interactive Management of Wild and Captive Animals*, Chapman and Hall, London, 1994, pp. 1–31.
26. E.F. Prentice, T.A. Flagg, and C.S. McCutcheon, *Am. Fish. Soc. Symp.* **7**, 317–322 (1990).
27. E.F. Prentice, T.A. Flagg, C.S. McCutcheon, D.F. Brastow, and D.C. Cross, *Am. Fish. Soc. Symp.* **7**, 335–340 (1990).
28. J.E. Joyce, R.M. Martin, and F.P. Thrower, *Prog. Fish-Cult.* **55**, 191–194 (1993).
29. C. Groot and L. Margolis, eds., *Pacific Salmon Life Histories*, Univ. British Columbia Press, Vancouver, BC, 1991.
30. I.A. Fleming and M.R. Gross, *Evolution* **48**, 637–657 (1994).
31. I.A. Fleming, B. Jonsson, and M.R. Gross, *Can. J. Fish. Aquat. Sci.* **51**, 2808–2824 (1994).
32. B.A. Berejikian, E.P. Tezak, S.L. Schroder, C.M. Knudsen, and J.J. Hard, *ICES Journal of Marine Science* **54**, 1040–1050 (1997).

ENERGY

Ian Forster
The Oceanic Institute
Waimanalo, Hawaii

Ronald W. Hardy
Hagerman Fish Culture Experiment Station
Hagerman, Idaho

OUTLINE

Energy is not a nutrient; rather, it is the capacity to do work. While most people think of this as mechanical work, in biological systems, energy is the currency driving the chemical reactions of metabolism, including those associated with growth, locomotion, and health.

Energy for biological systems ultimately comes from the sun in the form of radiant energy, which is captured by green plants and used to synthesize glucose from carbon dioxide and water (photosynthesis). Glucose is then used as a source of energy and substrate to manufacture more complex molecules needed for life, including fatty acids, amino acids (protein), and carbohydrate. Animals, of course, cannot obtain energy in this way. They must obtain it through their diet. Thus, animals, including fish, are net consumers of energy to support growth, activity, reproduction, and general metabolism.

The study of the balance between accumulation, consumption, and loss of energy is known as nutritional energetics, or bioenergetics. This entry considers the dietary requirements of energy by aquatic animals and the various dietary sources used to meet those requirements. Several reviews of this topic, as it relates to aquatic animals, are available (1–4) and these should be consulted for more in-depth information.

UNITS OF MEASUREMENT

All forms of energy can be expressed in terms of heat. Until recently, the calorie has been the standard unit for energy. The British thermal unit (BTU) has not been used in nutritional sciences to any great extent. One calorie (cal), also called gram-calorie (gcal), is defined as the amount of energy required to raise the temperature of 1 gram of water from 14.5 to 15.5 °C (58 to 60 °F). In 1960, the Conférence Général de Poids et Mesures formally adopted a Système Internationale d'Unités (SI), which is a refinement of the metric system. In the SI system, the calorie is replaced by the joule (J), which is used for all forms of energy, including chemical, electrical, mechanical, and thermal. The joule is defined as the "work done when the point of application of a force of one newton is displaced a distance of one meter (3 ft) in the direction of the force." One calorie is equivalent to 4.1855 joules. Although the calorie is still widely used by nutritionists, it is increasingly being replaced by the joule as a unit of measure of energy among nutritionists. For convenience, it is common practice to refer to the energy of a substance in terms of kilocalories (kcal or Cal) or kilojoules (kJ) per gram of feed. A typical trout feed contains 4000 kcal (16,740 kJ) of total (gross) energy per kg (2.2 lb).

ENERGY FLOW

It is useful to consider the energy content of a feed in terms of its relation to the nutrients in the feed. The four proximate principals — crude protein, crude lipid, fiber, and nitrogen-free extract — account for the energy content of the feeds (the minerals, as measured by the ash content, provide negligible energy, although some of them are necessary for its utilization by the animal). Once ingested, the energy of a feed is either absorbed into the body, or it is lost via the feces. Once in the body, a proportion of the energy is lost to the environment via the urine or gills, and the rest is used for metabolism, activity, and tissue growth.

The energy contained in a feed or feedstuff is usually expressed either as the gross energy (GE), the digestible energy (DE), or the metabolizable energy (ME). The GE is defined as the energy released by total combustion of a food into water, carbon dioxide, and other gases. This is known as the heat of combustion, and it is measured as the heat captured by water from the controlled combustion of a known quantity of the substance in pure oxygen. The device commonly used for this measurement is the bomb calorimeter, which consists of a metal chamber that is charged with pure oxygen under elevated pressure and suspended in a known quantity of water. The temperature of the water is then carefully monitored after ignition of the feed.

The GE of substances are additive. That is, if the GE of the individual components of a feed is known, then the GE of the overall mixture can be calculated as the sum of its GEs of constituent ingredients, adjusted for their percentage in a feed. The DE is the proportion of the energy content of the feed that is absorbed by the animal after ingestion. It is generally measured by subtracting the calorie content of food lost via feces from GE. The DE of a feed can also be calculated as the sum of the DE of the constituents, although, under some conditions, the digestibility of constituents can be influenced by the level of the constituent itself, or by the presence of other components of the diet. For example, high levels of carbohydrate may reduce the availability of protein and fats.

The proportion of ingested energy that is available for useful action by the animal (growth, movement, metabolic activity, etc.) is called the metabolic energy (ME). This is measured as the difference between GE and the energy loss via feces, urine, and gill excretions. As with DE, the ME of a mixture can be very much influenced by the relative level of the components. This is so because the efficient utilization of feedstuffs requires that all essential nutrients be present. For example, endogenous protein synthesis requires the presence of suitable levels of each essential amino acid. If there is a deficiency of even one of these, protein synthesis is impaired, resulting in conversion of the amino acids into energy-yielding compounds and waste products with resulting reduction in ME.

ME includes the energy required for assimilation of food, called the heat increment, which is energy not available for metabolism and growth. Net energy (NE) is defined as the ME minus heat increment. Heat increment manifests itself as heat lost through the body and can be measured directly by using a respiratory calorimeter in terrestrial animals. Only limited data exist for the NE of feeds aquatic animals. The NE of ingested food either is used for growth (NEg), maintenance (NEm), or voluntary activity.

The GE of the energy-yielding proximate principals is not uniform; fat contains more GE per gram than protein or carbohydrates. Research conducted with different animal species and a variety of feedstuffs has provided enough information so that nutritionists can calculate whether or not a particular feed contains sufficient energy without having to resort continually to biological testing. This is less true with aquatic species, however, especially with the recent interest in novel feedstuffs.

The term *physiological fuel value* is generally reserved for use in human nutrition, but is sometimes used in aquatic animal nutrition. The term refers to the portion of the GE that is ultimately available for use in the body. The physiological fuel value is equal to the digestible energy adjusted in the case of protein for the loss of energy in the urine, and is an estimate of the metabolizable energy. The physiological fuel values for humans for carbohydrate, lipid, and protein are, approximately, 16.7, 37.7, and 16.7 J/g (4, 9, and 4 kcal/g), respectively.

The value for protein was based on the energy content of urea, the principal nitrogen containing product of protein degradation in mammals (birds excrete uric acid). In aquatic animals, the principal nitrogenous end product of protein catabolism is ammonia, which requires less energy to excrete than urea requires. The ME value of protein in aquatic species is 21.3 kJ/g of digested protein. Of course, to correctly calculate the ME of feed or feedstuff, the digestibility of the components must be known, and this can be highly variable.

The GE content of the components of crude lipid is not uniform. For example, the energy content of phospholipids is considerably less than that of triglycerides (39.7 kJ/g for triglycerides and 33.5 kJ/g for phospholipids). These values should be used in cases where there is a high level of phospholipids in relation to triglycerides. Also, in general, long-chain fatty acids contain more energy than do short chain ones. The GE of butyric acid (8 carbon chain), for example, is 25.1 kJ/g, compared to 39.3 kJ/g for palmitic acid (16 carbon chain).

The digestibility of different types of carbohydrate is highly variable. Some aquatic animal species are better at digesting complex carbohydrates than others. Not surprisingly, in general, herbivores are better at digesting complex carbohydrates than are carnivores. Many aquatic species are unable to tolerate large influxes of simple carbohydrates, such as glucose. The ME of monosaccharide is 15.9 kJ/g of digestible energy; for disaccharides, it is 16.7 kJ/g and for digestible polysaccharides, it is 17.6 kJ/g.

By making the aforementioned adjustments and by applying the digestibility coefficients for various feed ingredients, it is possible to calculate a fairly accurate ME value for a diet. Direct measurement of ME for aquatic animals requires the use of special chambers to house the animals and quantitatively collect gill excretions. This has been done for some species, notably rainbow trout, but the difficulty of the work and the necessity of specialized equipment has precluded its widespread application.

REQUIREMENTS

The energy requirements of aquatic animals are affected by a variety of factors, including level of metabolic activity, growth rate, and reproductive state. The metabolic rate of fish increases with increasing water temperature. Fish thus require more dietary energy, all other things being equal, when their water temperature is high. Fish adjust their feed intake relative to water temperature; thereby taking in more energy at higher water temperatures. Growth rate is also affected by water temperature, as is feed intake. Maturation in fish results in diversion of dietary energy from somatic (muscle) tissue synthesis to gonadal synthesis. Spawning migrations in wild fish often occur during periods of voluntary starvation. Fish store energy in anticipation of periods of voluntary starvation and food scarcity. Fish raised in hatcheries do not undertake spawning migrations, but do have periods of low feed intake or voluntary starvation as spawning approaches. The energy requirement, expressed in terms of kJ/day, is higher in fish prior to those spawning periods than during regular growth.

Crustacea and finfish differ radically in their tolerance of dietary lipid. Studies with a variety of shrimp species have shown that crustacea fed diets containing lipid at levels in excess of about 10% have reduced growth. For this reason, crustacean feeds derive a greater proportion of their energy from carbohydrate relative to lipid than do finfish feeds. Many species of finfish, e.g., yellowtail (*Seriola quinqueradiata*) and salmonids, grow best under culture conditions when fed diets containing 20–35% lipid. Warmwater species, such as channel catfish (*Ictalurus punctatus*), are intermediate in tolerance of dietary lipid, but tend to acquire lipid deposits in the musculature if fed diets containing more than 10% lipid (5). Farmed marine species, e.g., sea bass (*Dicentrarchus labrax*), sea bream (*Sparus aurata*), and striped bass (*Morone saxatilis*) do not appear to tolerate high dietary lipid levels. Thus, their feeds are limited to 14–16% lipid, with energy being supplied by carbohydrates, similar to channel catfish feeds.

ENERGY BALANCE

Nutrient balance is a widely applied method for assessing nutritional status. To determine nutrient balance, intake is compared with the sum of the losses from the body from all channels. A positive balance occurs when the amount ingested exceeds the amount lost. A negative balance refers to greater loss than intake. A healthy growing animal will have a positive balance. In some cases, however, where intake is very low or nonexistent, such as occurs for salmon prior to spawning, a negative balance is a normal condition.

Clearly the aquaculturist is interested in maintaining as high an energy balance as economically possible. In animal tissues, the principal repositories of energy are protein and lipid. Carbohydrate, which occurs mostly in the form of glucose or glycogen, accounts for a small proportion of energy stored in the body. Of these, protein accretion is most tightly linked to growth. For a given species and size, the protein content of an animal tends to be less variable than the lipid content, under normal conditions. Protein accretion in animals depends on a variety of factors, including the presence at the sites of protein synthesis of the types, or species, of amino acids required. Twenty-two amino acids are used to make most proteins, and these are obtained either from endogenous sources (catabolism of existing protein, or from de novo production) or from the diet. Amino acids arriving in the body following ingestion are not necessarily available for protein synthesis, but may embark on a variety of competing pathways, including energy production via lipids or carbohydrate. The demands for energy by the body supersede those for protein synthesis, and amino acids are preferentially catabolized to yield energy, when ingested energy from other sources is low. As a consequence, in order to maximize dietary protein retention by aquatic animals, it is important that the nonprotein available energy levels are high.

The improvement in protein retention resulting from increasing the nonprotein energy content of feeds, is referred to as the protein-sparing effect. Modern feeds for several species (including Atlantic and Pacific salmon and yellowtail) are made with very high levels of fat in order to minimize the utilization of dietary protein as an energy source. On the other hand, if the protein to energy ratio is too much tilted toward energy, feeding rate and feed efficiency are impaired. A goal of feed formulators for intensive aquaculture (i.e., where manufactured feed is the principal or exclusive source of nutrients and energy) is to achieve the optimum protein to energy ratio, or more meaningfully, digestible protein to ME ratio, for growth and dietary protein efficiency for various cultured species under specific culture conditions.

The dietary protein to energy ratio that yields the optimum growth and feed efficiency is species dependent and can be affected by a number of factors, including water temperature, swimming level, and source of energy. The current level of knowledge concerning the effect that these factors have on protein and energy utilization is quite rudimentary, as summarized in Table 1. In general, the optimum protein to energy ratio is about two to three times higher than that for swine and poultry. This does not indicate a high protein requirement, but rather a low energy requirement relative to these terrestrial animals. An estimated 3,560 kcals are needed to produce 1 kg of trout, containing 1,910 kcals of energy (6). Thus, trout retain approximately 54% of dietary energy, comparable to 56% reported for channel catfish (7).

Table 1. Optimum Ratio of Dietary Digestible Protein (DP) to Digestible Energy (DE) in Various Species of Fish, Poultry, and Swine (11)

Species	DP/DE (g/MJ)
Channel catfish	19.4–23.2
Red drum	23.4
Hybrid striped bass	26.8
Nile tilapia	24.6
Common carp	25.8
Rainbow trout	22.0–25.1
Swine and poultry	9.6–14.3

SOURCES

As mentioned previously, animals obtain the energy they need from the protein, lipid, and carbohydrate fractions of their food. Each of these components contributes differently to the available dietary energy and has different implications for the overall cost of feed manufacture. Aquatic animal nutritionists must balance the value of the nutrients and energy against the cost of the ingredients when formulating feeds.

Protein

In general, protein is the most costly component of aquatic animal feeds. Strictly speaking, animals do not require protein per se, but instead require a well-balanced supply of amino acids. Ten of the 22 amino acids that are used in the endogenous production of protein cannot be produced by the animal and need to be obtained through the diet (the indispensable amino acids, IAA). The other 12 amino acids can be endogenously manufactured by the animal (the dispensable amino acids, DAA). Most of the DAA are synthesized from compounds that are intermediates in the TCA cycle within each cell. Others are produced from amino acid precursors.

Theoretically, animals could be raised on feeds that supply sufficient levels of all the IAA, but there is an energy cost involved with this. In practice, aquatic animal feeds supply all amino acids, both IAAs and DAAs, thereby reducing the energy required for amino acid manufacture. A feed that supplies all the amino acids (IAA and DAA) at appropriate levels will minimize the energy required to meet the needs of the animal and will optimize protein utilization (assuming that all other nutrients are also present at required levels). Aquatic animals have a higher tendency to use protein for energy than do many terrestrial animals. Amino acids can be used to generate a variety of products, including other amino acids, carbohydrates, and lipids. Some amino acid species can only be used to generate one or the other of these, and some may be used to make all three. In each case, however, all can be used to generate energy.

Lipid

Lipids consist of a variety of substances that are distinguished by their nonpolarity, and thereby their high

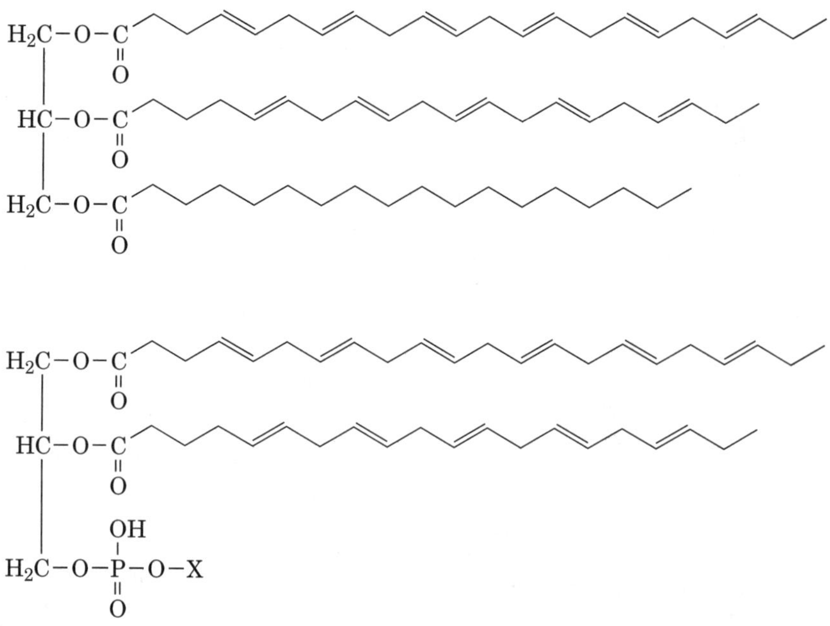

Figure 1. Two-dimensional representation of a triacylglyceride (top) and a phospholipid (bottom). The three carbons arranged vertically on the left of both molecules are derived from a glycerol molecule, while the series of serrated line segments on the right are fatty acids. The intersections of the line segments in the fatty acids portions represent the locations of carbons in these molecules. The double lines indicate double bonds between adjacent carbons and are regions of unsaturation. In phospholipids (bottom figure), one of the fatty acids is replaced by a phosphate group, which is further bonded to some other molecule (e.g., choline, inositol, ethanolamine), which is represented in this figure by an "X." None of the bonds in this figure are to scale.

solubility in nonpolar solvents (notably in ether). In the lipid class, the most important energy sources are the fatty acids, which are largely present as triglycerides. Triglycerides are composed of three fatty acids joined to a molecule of glycerol (Fig. 1) and are a prevalent structural component of cell membranes, as well as the principal source of fatty acids in the diet and the principal storage form of energy in the body. On a weight basis, catabolism of fatty acids yields more useful energy than either amino acids or carbohydrate. The digestibility of the lipid sources used in animal feeds is generally high, greater than 90%. Most lipid in commercial feeds is supplied as oil from plants (corn oil, soybean oil), terrestrial animal (tallow, poultry fat), or marine animal sources (fish oil). Marine oils increase feed palatability and are excellent sources of essential fatty acids, e.g., n-3 fatty acids.

Some aquatic animals, such as salmonids and yellowtail, have a high tolerance for dietary lipid, whereas others, such as shrimp do not. This fact has obvious consequences for the feed formulator. Modern commercial salmon diets have very high fat levels (greater than 30%), which spares dietary protein by reducing the use of dietary protein for metabolism energy. Shrimp, on the other hand, cannot tolerate diets containing more than about 10–12% lipid, forcing shrimp feed formulators to look for other sources of dietary energy.

Carbohydrate

Carbohydrates consist of a variety of constituents and are distinguished by their composition of carbon, hydrogen, and oxygen (in a ratio of roughly 1 : 2 : 1) and their lack of solubility in nonpolar solvents. The nitrogen-free extract and fiber proximate fractions are basically measures of different kinds of carbohydrates. The basic structure of carbohydrates are six or five chain carbon sugars, with the most relevant as a source of energy being the six carbon ring of glucose (Fig. 2). Carbohydrates are also present in feedstuffs as double or polymeric sugar (saccharide) units. In feeds, polysaccharides are the most important carbohydrate component, while mono- and disaccharides are also important sources of energy in animals. The majority of polysaccharides in plant ingredients are made of either starch or cellulose (Fig. 3), which are composed of long chains of glucose connected by α 1–4 linkages (starch) or β 1–4 linkages (cellulose). Both starch and cellulose polysaccharide chains may also contain small proportions of α or β 1–6 linkages. Two forms of starch are amylose, which contains essentially only straight chain 1–4 linkages, and amylopectin, which also contains 1–6 linkages.

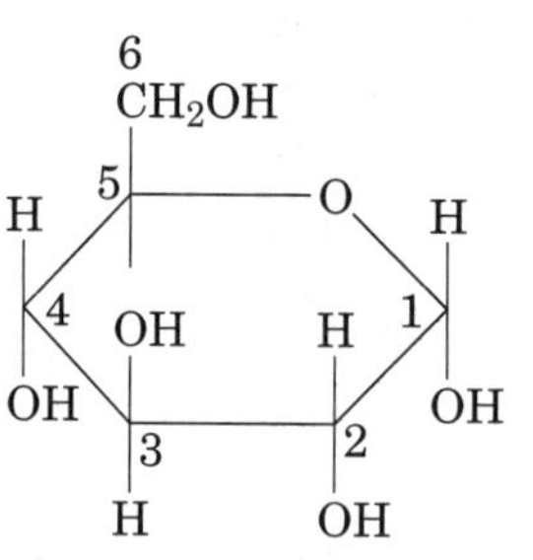

Figure 2. Two-dimensional representation of glucose. The intersections of the line segments are the locations of the carbons in this molecule. The numbers adjacent to the carbon atoms are the conventional numbering of these atoms. None of the bonds in this figure are to scale.

Plants use polysaccharides as energy stores (starch) and as structural components (cellulose). Storing glucose in polymeric chains reduces the osmotic pressure of the stored sugar. A polysaccharide consisting of 1,000 glucose units exerts an osmotic pressure that is only 1/1,000 of the pressure that would result if the glucose units were all present as separate molecules. In polymeric form, glucose can be stored compactly until needed. Animals also store glucose in polymeric form as glycogen, a compound related to amylopectin, but with shorter 1–6-linked side chains. Glycogen is primarily stored in the liver and is used to

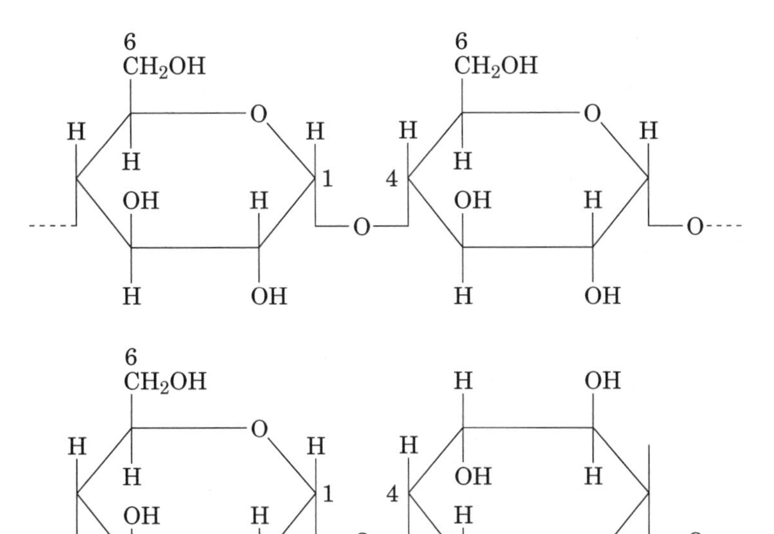

Figure 3. Two-dimensional representations of starch and cellulose. Starch is a linear series of glucose molecules in identical spatial orientation. Cellulose is similar to starch, except that alternate glucose molecules are inverted. A low percentage of both starch and cellulose of linkages are between carbons 1 and 6 of adjacent glucose molecules (the numbers next to the carbon atoms are the conventional numbering of these atoms). The proportion of these 1–6 linkages in starch or cellulose influences characteristics of the material. None of the bonds in this figure are to scale.

supply short-term demand for energy. Animals (including fish) mainly store energy as lipid and do not use cellulose at all. Some animals utilize carbohydrate as part of their structural components (e.g., chitin, a component of the exoskeleton of insects and crustaceans).

All animals readily absorb simple sugars, like glucose, but in order to utilize the energy contained in the various types of dietary carbohydrates, animals must have the ability to digest the polysaccharides into single glucose units. The glucose contained in cellulose is essentially unavailable to aquatic animals. Carnivorous fish, salmonids for example, produce only small quantities of the enzymes necessary to digest starch (8), and it is recommended that only limited levels of carbohydrates be included in feeds for these animals. Omnivorous or herbivorous animals [e.g., tilapia (*Oreochromis* spp.) and carp (*Cyprinus carpio*)], however, can digest starch and can utilize feeds containing higher levels.

Carbohydrate utilization as an energy source is affected by more than just digestibility. Even highly digestible carbohydrates (i.e., those that are readily absorbed from the digestive tract), may not be effectively utilized. For example, Wilson and Poe (9) compared the ability of channel catfish to utilize six types of carbohydrates, including two monosaccharides (glucose and fructose), two disaccharides (sucrose and maltose), corn starch, and dextrin (a short-chain polysaccharide) as dietary energy sources. Cellulose, which is assumed to be completely unavailable, was used in the control diet. These researchers found that growth, feed efficiency, and proportion of ingested energy retained in the tissues were highest for fish fed the diet containing dextrin, followed by those fed starch. The fish fed mono- or disaccharides as energy sources grew less well than those fed the longer chain polysaccharides (other than cellulose), and, in fact, grew no better than the fish fed the low DE control diet. The fish fed the fructose diet performed significantly worse than the others. The inability of several species of fish to tolerate rapid influxes of simple sugars is well known, and may, in part, explain the inability of catfish to utilize the simple sugars in this trial. The hormone insulin is released in response to increases in blood glucose levels and is responsible for stimulating the transportation of circulating glucose into cells, where its energy is utilized. The slower rate of intestinal digestion of dextrin and starch into glucose results in an attenuated influx of glucose into the blood stream, with elevated levels coinciding with maximum insulin secretion, facilitating maximum utilization of the dietary energy from glucose.

In plants, starch is contained in granules, which are relatively undigestible. Under conditions of heat and pressure, such as occur during feed manufacture, these granules rupture in a process called gelatinization. Gelatinized starch is more readily digestible to aquatic animals (resulting in higher glucose availability) than is nongelatinized starch and is also important for its binding characteristics. By ensuring that adequate levels of starch are present in a mash, feed manufactures take advantage of the binding characteristics of this gelatinized starch to ensure adequate strength of the pellets.

CHEMICAL FACTORS AFFECTING ENERGY UTILIZATION

Useful energy is derived from carbohydrates, lipids, and amino acids through the action of very specific series of enzyme-controlled reactions. Several of the enzymes involved in these reaction series require the presence of other molecules for their activity; these are known as cofactors. Cofactors include minerals and water-soluble vitamins. Unlike terrestrial animals, aquatic animals can obtain some or all of the minerals they require for the metabolic reactions involved in energy storage or usage from their environment. In common with monogastric animals, however, they need a dietary supply of the vitamins for energy metabolism. In fact, since many of the

B vitamins are closely related to the transfer of energy, animal requirements for them vary in proportion to the intake of energy. Published requirements for vitamins are often best expressed in terms of DE content of the diet. There is some limitation to this principle, however. For example, if a very high proportion of the DE content of a diet is from the lipid fraction, the thiamin (a B vitamin) requirement is actually reduced, because the energy from free fatty acids is obtained via a pathway that does not have a thiamin-dependant step.

As mentioned earlier, insulin is important in utilization of carbohydrates for energy, but the physiology of hormonal action and energy is very complex. For example, insulin secretion in fish species, in contrast to many terrestrial animals, may be even more highly affected by dietary levels of certain amino acids than by glucose (10). Utilization of dietary energy requires the coordinated operation of a great many enzyme systems under hormonal control. The current state of knowledge of the regulation of energy utilization and metabolism is very spare, but recent work in this area has shown promise. A more complete understanding of this field may have profound implications for formulation of future commercial feeds.

BIBLIOGRAPHY

1. C.Y. Cho, S.J. Slinger, and H.S. Baley, *Comp. Biochem. Physiol.* **73B**, 25–41 (1982).
2. R.T. Lovell, *Nutrition and Feeding of Fish*, 2nd Edition, Kluwer Academic Press, Dordrecht, the Netherlands, 1998.
3. S. Kaushik and F. Médale, *Aquaculture* **124**, 81–97 (1994).
4. R.R. Smith, in J.E. Halver, ed., *Fish Nutrition* 2nd ed., Academic Press, New York, 1989, pp. 2–28.
5. R.R. Stickney and R.W. Hardy, *Aquaculture* **79**, 145–156 (1988).
6. C.Y. Cho and S.J. Kaushik, *World Rev. Nutr. Diet.* **61**, 132–173 (1990).
7. D.L. Gatlin, III, W.E. Poe, and R.P. Wilson, *J. Nutr.* **116**, 2121–2131 (1989).
8. Å. Krogdahl, S. Nordrum, M. Sørensen, L. Brudeseth, and C. Røsjø, *Aquacult. Nutr.* **5**, 121–133 (1999).
9. R.P. Wilson and W.E. Poe, *J. Nutr.* **117**, 280–285 (1987).
10. E.M. Plisetskaya, L.I. Buchelli-Narvaez, R.W. Hardy, and W.W. Dickhoff, *Comp. Biochem. Physiol.* **98A**, 165–170 (1991).
11. National Research Council (NRC), *Nutrient Requirements of Fish*, National Academy Press, Washington, DC, 1993.

ENHANCEMENT

Robert R. Stickney
Texas Sea Grant College Program
Bryan, Texas

OUTLINE

INTRODUCTION

Enhancement is the stocking of hatchery-reared fish and shellfish into public waters to increase the natural population of the same species inhabiting those waters. One purpose of enhancement is to restore overfished stocks of commercially or recreationally important species. Another is to help recover threatened or endangered species.

The U.S. Fish and Fisheries Commission was created in 1870 and has since evolved into the National Marine Fisheries Service. It was created, in part, to establish hatcheries that produced fish for restocking the nation's inland and marine waters and for introducing new species. Billions and perhaps trillions of fry, fingerling, and larger fishes, along with molluscs and crustaceans, were stocked beginning in 1872. The stocking of marine fishes was largely discontinued after several decades, due to lack of evidence that the stocking programs — which almost exclusively involved stocking newly hatched animals with little chance of survival — were having the desired effect.

State and federal hatcheries continued stocking species that did demonstrate measurable survival, along with producing fish for stocking new bodies of water (e.g., ponds and reservoirs). Most of those programs involved inland water bodies, though hatchery programs, to produce and release Atlantic salmon (*Salmo salar*) and Pacific salmon (*Oncorhynchus* spp.), have continued for over a century and still produce fish that live most of their lives in the ocean.

MARINE ENHANCEMENT RECONSIDERED

Advances in aquaculture over the past few decades, have involved the development of the required technology to bring a variety of marine fishes through the difficult early life stages and to rear them in captivity to sizes where they stand a reasonable chance of survival if released into the natural environment. An important purpose of the research and technology development was to produce animals for captive rearing to market, but a logical alternate use is in conjunction with well-planned enhancement programs.

A well-planned enhancement program should involve determining sufficient information about the environment into which the animals are to be released, and includes a high level of assurance that no significant disruption of the ecosystem will occur. Prior to releasing additional animals into the marine environment, knowledge should be gained about the carrying capacity of that environment and about how the hatchery fish will interact with wild conspecifics. Every effort should be made to ensure that the stocked fish or shellfish have the same genetic profile as the wild ones. Most of the enhancement programs conducted to date were initiated without considering any consequences except meeting the need of increasing the species population size. That is now changing.

The Japanese have been involved with marine fish and invertebrate enhancement for over 30 years. Several dozen species have been involved, though, until recently, few data were available to show the program's effectiveness. Over 20 years ago in the United States, red drum (*Sciaenops ocellatus*) enhancement was initiated in Texas after the commercial fishery for that species was closed. A private organization has provided funding to the Texas Parks and Wildlife Department to construct and operate hatcheries that release several million fingerlings annually. Due, in part, to the enhancement program, and helped by the commercial fisheries ban, the species has recovered and now provides excellent recreational fishing along the Texas coast.

More recently, along the eastern seaboard of the United States a crash in the population of striped bass (*Morone saxatilis*) led to a total fishing ban and the establishment of an enhancement program. Within a few years, a recreational fishery was once again in place. In Maine and elsewhere, clam-stocking programs have either created new industries or reinvigorated existing ones.

The National Marine Fisheries Service in the United States has, as one of its goals, to build sustainable fisheries. Recovery of overfished stocks to historical levels can be accomplished through regulation, but it may be aided by enhancement stocking programs. Discussions on how such programs might be put into place, and about the types of research that will be required, in conjunction with those programs, are currently underway.

A NEW MODEL

Historically, in the United States, enhancement programs have been operated by government. In Japan, enhancement is conducted, in many cases, through collaboration of Prefectural governments and the private sector. A private-sector role for enhancement of marine species could be an attractive option in the United States.

In the future, an enhancement program might work like this. The organism selected (fish or invertebrate) would be reproduced and reared to release size in a private hatchery. A state or the federal government would determine how many animals should be released and where the releases should occur. Commercial or recreational fishermen would pay access or license fees to participate in the fishery, thereby providing revenue to the commercial aquaculturist who produced the animals for release. Government assistance (perhaps through an up-front fee program) would be required during the period between the establishment of a hatchery and the time the enhanced species recruited into the fishery. Also, the private aquaculturist would have to recognize that, if and when the fishery recovered to the desired sustainable level, further enhancement might be curtailed. That level of recovery may, of course, never occur. However, since many species have been overfished, the private producer could switch from one species to another. Alternatively, the private producer could expand into growout activities or supply stock to other growout aquaculturists.

See also RECIRCULATING WATER SYSTEMS.

ENVIRONMENTALLY FRIENDLY FEEDS

SHOZO H. SUGIURA
RONALD W. HARDY
Hagerman Fish Culture Experiment Station
Hagerman, Idaho

OUTLINE

INTRODUCTION

In order to cope with an ever-increasing world human population, it will be necessary to increase world food production. Aquaculture is one of the fastest growing food-producing activities in the world. Aquaculture utilizes water resources but is a nonconsumptive use of these resources. Nevertheless, aquaculture adds nutrients to aquatic resources that can cause environmental pollution. The nutrients added to the aquatic environment originate in the feeds used to raise fish, shrimp, and other farmed aquatic species. Feeds that minimize the nutrient output from aquaculture are called environmentally friendly feeds. Environmentally friendly feed is, therefore, of critical importance for the aquaculture industry, for the aquatic environment, and for the well-being of humankind in the twenty-first century.

METHODS AND TECHNOLOGY

Principal Approaches

Sources of environmental pollution associated with aquaculture are diverse: phosphorus, nitrogen, solid matter, trace metals, chemotherapeutants, and uneaten feeds including fines. Genetic and ecological pollution, such as with net pen escapees and transplantation of nonindigenous species including invertebrates, parasites, and aquatic plants to other locations, are also of serious concern for their unpredictable long-term impacts to local ecosystems (genetic and ecological pollution is not in the scope of this entry). Aquaculture production, particularly intensive production in raceway ponds and net pens, is increasingly subjected to environmental regulations and waste discharge guidelines. Various regulations, including limitations of production itself, have already been set in many countries. Extensive or semi-intensive aquaculture production in closed systems (e.g., large stagnant-water ponds, indoor water-recycling tanks) may be exempt from environmental regulations. They are, however, more vulnerable to self-pollution and destruction of the internal environment. Waste products that accumulate in a pond deteriorate soil quality by anaerobic fermentation and proliferation of bacteria or pathogens; while in the water column, dissolved wastes not only stress fish directly but also cause an excessive growth of phytoplankton that deplete oxygen at night. Minimizing the excretion of wastes from fish is therefore critical in both intensive and extensive aquaculture systems.

Phosphorus is generally the first limiting nutrient for plants and algae in the freshwater environment, while it is usually nitrogen in seawater and brackish environments. Each can lead to eutrophication of the ecosystem. Solid waste, uneaten feeds, trace metals, and chemical wastes are important in all waters. The amount of uneaten feed in aquaculture production can be reduced by good management practices. Voided fish feces may be collected by settling or screening and can be decomposed in an annexed facility in a manner similar to the treatment of municipal sewage. This, of course, is an effective approach to reducing waste discharge and thus reducing environmental pollution. Feeds are, however, the ultimate source of pollution in most aquaculture operations. Generally, fish excrete all indigestible portions of feeds as feces and the excess portion of absorbed nutrients via urine, gills, or feces. Environmentally friendly feeds must, therefore, be formulated in a way that fish can digest, absorb, and retain as much of the feed nutrients as possible. This implies that increasing the retention of dietary phosphorus (and to a limited extent nitrogen) by improving feed quality is a logical and efficient strategy to reduce environmental impacts of aquaculture.

Data from several studies consistently indicate that approximately 80% of dietary phosphorus in typical commercial salmonid feeds is excreted into water as soluble and fecal forms. The amount of unretained phosphorus is largely influenced by the amount of phosphorus in feeds and its bioavailability. The ultimate source of nitrogen excreted by fish is protein in the diet. Any excess portion of dietary protein not used to synthesize body protein is utilized as an energy source. The nitrogen moiety is excreted as ammonia primarily via the gills, while the carbon moiety enters the Krebs cycle and is metabolized as an energy source. This excess portion of dietary protein is not only wasted as far as fish growth is concerned but it is also environmentally destructive. Reducing the level of dietary (digestible) protein to the minimum required level for fish is, therefore, critical in commercial aquaculture. The digestibility of dry matter and the availability (retention) of trace minerals in the diet involves the same principles as in phosphorus and nitrogen. In summary, the following factors will be critical in formulating environmentally friendly feeds:

1. Reduce nutrient levels in the diet to the minimum requirement levels for fish.
2. Select highly digestible feed ingredients (avoid ingredients of low digestibility).
3. Process feed ingredients to improve the digestibility or availability of nutrients.

Minimum Dietary Requirements of Nutrients for Commercial-Size Fish

In commercial aquaculture, large fish consume predominant portions of feeds and contribute much more waste in the effluent than do small fish. Phosphorus allowances used for commercial fish feeds are, however, based upon data obtained with very small fish reared in laboratories. Nutrient requirements in most animals are known to decrease as they become older or larger because, as they grow larger, the growth rate (retention of dietary nutrients) decreases and increasing portions of dietary nutrients, including phosphorus, are used for maintenance (substantial portions of which can be recycled). For example, the dietary phosphorus requirement of broiler chickens is 50% higher for 0 to 3-week-old birds than for market-size (6–8-week-old) birds (1). Since, in commercial aquaculture, fish are often raised up to 3 kg (6.6 lb) to 5 kg (11.0 lb) in body weight before being harvested, estimates of nutrient requirements made using small fish are unsatisfactory for market-size fish. Obtaining an accurate estimation of the dietary requirement of phosphorus for large fish is, therefore, necessary to minimize phosphorus excretion from fish into the aquatic environment.

Ogino and Takeda (2) reported the dietary phosphorus requirement for optimum growth of rainbow trout (*Oncorhynchus mykiss*) to be 0.7–0.8% of dry diet using juvenile fish of 1.2 g (0.04 oz) in body weight. Numerous researchers have investigated dietary phosphorus requirements for other fish species including channel catfish (*Ictalurus punctatus*), tilapia (*Oreochromis* sp.), Pacific salmon (*Oncorhynchus* sp.), Atlantic salmon (*Salmo salar*), Japanese eel (*Anguilla japonica*), and carp (*Cyprinus carpio*) using young fish of less than 10 g (0.35 oz) in body weight (3). Because young fish are sensitive to nutrient deficiencies, their response to dietary concentrations of nutrients is normally rapid and clear in virtually any response variables, and even the mortality rate may suffice to establish dietary requirement for young fish. The requirement values determined with young fish, however, are not predictive for the requirement for larger fish as mentioned previously. Rodehutscord (4), working on larger trout (initial body weight 53 g or 0.117 lb), found large differences in estimated requirement for phosphorus when weight gain was used and when phosphorus retention was used as the response criteria. For large fish, using an insensitive response variable such as growth to establish the minimum dietary requirement of phosphorus is questionable, especially when the duration of feeding is not sufficient. Large fish can subsist for an extended period without reducing the growth by using body stores unless the level of intake is exceedingly low. The minimum dietary requirement measured using large fish, therefore, tends to be an underestimation and misleading. Eya and Lovell (5), studying post-juvenile catfish (initial body weight 61 g or 0.135 lb) for their dietary phosphorus requirements, failed to see any effect of dietary phosphorus concentrations on growth of the fish in 140 days of feeding. Numerous other response criteria for estimating the dietary requirement of phosphorus, including plasma, tissue, or body saturation levels, have no rational basis or practical meaning. Changes of enzyme activities could well be a normal adaptive response of the fish to a lowered dietary intake, which do not necessarily imply clinical deficiency for the species.

For estimating dietary requirement of nutrients, the feeding duration must be long enough to obtain legitimate deficiency signs (6). However, because large fish already have substantial body stores (pool) and because their growth rate is low, studying nutrient requirements by means of those conventional approaches requires months of feeding before any response of fish to the dietary treatment can be detected. Also, it is impractical to feed large fish with expensive semipurified research diets for an extended period. With these constraints, there are few data currently available regarding the minimum dietary requirements of nutrients including phosphorus for commercial-size fish. Sugiura (7) used a different approach to estimating the minimum dietary requirement of phosphorus for large fish. The method determines the phosphorus requirement in a few days based upon urinary excretion of phosphorus of fish in a metabolic tank. At low dietary phosphorus concentrations, fish absorb and retain essentially all the available phosphorus contained in the diet (Fig. 1). Once dietary phosphorus concentrations exceed a certain level, fish excrete excess phosphorus via urine. Excretion increases linearly and proportionally to the dietary phosphorus concentration. The point where fish start to excrete phosphorus via urine is assumed to be the minimum requirement level of phosphorus in the diet. The rapidity and sensitivity of this method make it possible to estimate the requirement of phosphorus in fish of various nutritional, physiological and environmental conditions. The minimum dietary requirement of available phosphorus determined with large trout (body weight 200 to 400 g or 0.44 to 0.88 lb) using this method has been shown to be variable, ranging between 0.41 and 0.66% of dry diet, depending on the nutritional and physiological status of the fish (unpublished data).

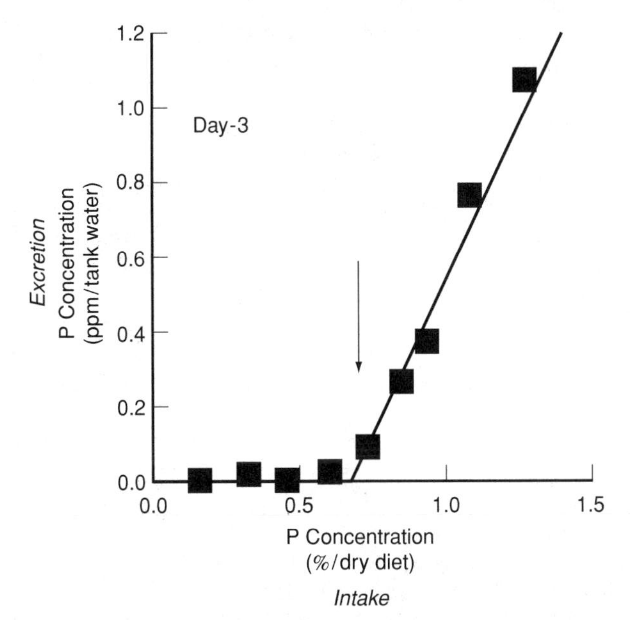

Figure 1. Estimating the minimum dietary requirement of phosphorus for large rainbow trout (body weight 203 g or 0.45 lb) based on the urinary excretion of phosphorus of fish fed incremental concentrations of P in diets for 3 days. The value remained constant for days 6, 9, and 12. Phosphorus excreted into the urine was indirectly measured by analyzing recirculating tank water. *Source*: Sugiura (7).

The dietary requirement of a nutrient expressed as a percentage of the diet, however, is not always a reliable value. Growth is the major factor that governs the dietary requirement of many nutrients, especially those constituting muscle and skeletal tissues such as protein (nitrogen), phosphorus, calcium, and several other minerals. Various factors have been shown to influence fish growth (e.g., environmental stresses, cultural conditions, fish strain, life stage, physiological and endocrine factors, energy density, nutrient balance, interactions, digestibility, feed processing and storage conditions, deficiency of trace nutrients, presence of antinutritional factors, feeding frequency, and feeding rate) (8,9). Expression of nutrient requirement per nitrogen retention as representing the growth of fish has been proposed as an accurate way of expressing nutrient requirement (7). Similarly, standardizing (dividing) the measured requirement value (g/g dry diet) by feed efficiency (g fish weight gain/g dry diet) of the diet to obtain Standardized Requirement or Requirement Coefficient (i.e., requirement value when feed efficiency) is a valid procedure to eliminate various confounding factors among different experiments represented by different feed efficiency. Also, it is important to express nutrient requirement in diet on an available nutrient basis rather than as total amount to eliminate differences in nutrient availability (digestibility) among diversified test (basal) diets. Currently, dietary requirements of nutrients are expressed as total amount on a diet basis (without any standardization) or on a digestible energy basis (3). Digestible energy levels are difficult to estimate from the data available in the literature. This is because nutrient content (composition) and the digestibility of an ingredient are variable among manufacturers, and because digestibility of energy sources, especially carbohydrates, varies at different dietary levels and pelleting methods. Also, digestible energy intakes are not always correlated to growth since digested energy is also used for various other ways such as basal metabolism, activity, and heat increment with varied proportions at different feed intakes.

Digestibility and Availability of Nutrients in Feed Ingredients

In order for dietary nutrients to be utilized for various biological functions, including growth, they first need to be absorbed (with or without digestion) from the gastrointestinal tract. Knowing the content of digestible or available nutrients in ingredients (and in the diet) is one of the critical pieces of information for formulating environmentally friendly feeds.

Phosphorus. Many factors are known to influence dietary phosphorus utilization by fish, including the level and the form of phosphorus in the diet and its interaction with other dietary components. Major sources of phosphorus in fish feeds are organic and inorganic forms of phosphorus found in muscle and bone tissues of fish, poultry, or animal by-product meals and inorganic and phytate phosphorus in various plant protein sources. Some fish feeds contain added phosphorus supplements such as di-calcium phosphate. Several investigators have researched the availability of phosphates from various sources to different species of fish (3). In spite of the increasing importance of this subject, however, currently available data are quite limited. Knowing precisely the available phosphorus content in common feed ingredients is essential for selecting feed ingredients based on available phosphorus content, reducing available phosphorus in formulated diets to the minimum required level for fish and eliminating the risk of potential deficiency problems.

The apparent availability of phosphorus in feed ingredients commonly used for fish feeds is highly variable (Table 1). The apparent availability of phosphorus in high-ash ingredients such as meat and bone meal and high-ash fish meal is very low, whereas that of low-ash ingredients such as blood meal and feather meal is high. The low availability of phosphorus in meat meal is presumably due to the bone particle size and the source (animal vs. fish).

The apparent availabilities of calcium, phosphorus, magnesium, and iron in fish bone decrease as the bone content in the diet increases (7). The availability of phosphorus in fish bone was predicted to be above 90% when its concentration is very low, whereas it decreases as the bone content in the diet increases. Although the absorption of phosphorus and many other minerals from diets was significantly and negatively correlated to the concentrations of ash, calcium, phosphorus, and bone in the diet (as the fractional net absorption), the actual net absorption of bone minerals remained fairly stable. This suggests that reducing bone minerals in diets is an essential approach in formulating low-pollution feeds. Once this requirement is met, then the interfering effect of other dietary components having binding properties with minerals, notably phytic acid (discussed later) in plant ingredients, will more likely reveal their nutritionally significant potentials. The use of fish meal replacers that contain less phosphorus is therefore essential to reduce overall phosphorus content in formulated feeds. To replace major portions of fish meal in aquaculture feeds with other protein sources, however, involves several practical problems, including reduction of total protein (and some essential amino acid) content, creation of amino acid imbalance, reduction of palatability, and increases of carbohydrate and fiber contents (i.e., reduction of dry matter digestibility). This is particularly the case when plant protein sources are used in place of fish meal in the feed. Animal byproduct meals, if their ash content is low, offer many advantages over plant sources to offset the previously mentioned shortcomings.

Nitrogen (Protein). Generally, protein digestibility of animal by-products is highly variable depending on the quality of raw materials and processing conditions (Table 1). Apparent digestibility of protein in ring-dried or spray-dried blood meal and deboned whitefish meal are high, while those of feather meal and menhaden meal are relatively low. Blood meal dried in a continuous dryer appears to be very low in protein and energy digestibilities (10). Chilled or ensiled blood are highly digestible (97%), indicating that the poor digestibility of blood meal is due to the processing temperature during the drying process (11). Poultry by-product meals differed markedly in percentages of digestible protein and energy (10,12). Other dietary components, notably gelatinized starch, have been shown to reduce protein digestibility, especially in carnivorous fishes, when the dietary level is too high.

Solid Matter. Apparent digestibility of solid (dry) matter is also highly variable among feed ingredients (Table 1). Animal ingredients of low ash content such as blood meal and deboned whitefish meal are highly digestible. High ash ingredients such as low quality fish meals and high ash animal byproducts have low digestibility of dry matter due to the limited digestibility of the ash portion of the ingredient. Feather meal, despite its low ash content, may not be well digested if the processing condition of the ingredient is not appropriate. Plant ingredients generally have lower digestibility of dry matter than animal ingredients due to their high fiber content and the incomplete digestion of the ingredient, especially when the starch portion is uncooked.

Other Minerals. Apparent digestibility of ash largely differs among ingredients. Animal byproduct meals containing a high percentage of ash have low ash digestibility. Since the ash portion of most ingredients is mainly composed of calcium and phosphorus, reducing the ash content in feed ingredients is essential in reducing phosphorus excretion into feces. Also, the levels of calcium, phosphorus, and ash (and the bone) in diets containing various animal protein sources showed inverse correlations with net absorption (percentage) of calcium, iron, magnesium, manganese, phosphorus, strontium, and zinc (7). Reducing the bone content of fish meals as studied by Babbitt et al. (13) appears to be a rational procedure for increasing availability of dietary minerals in fish meal-based feeds.

Apparent availability of magnesium in fish feed ingredients ranged between 0 and 100%, but most values were between 50 and 70%. Apparent availability of potassium in feed ingredients was high in all ingredients, while that of sodium was lower and more variable than that observed with potassium. Apparent availability (absorption) of iron in fish meals and plant ingredients was very low. In wheat gluten meal, iron absorption was high but the amount of iron in wheat gluten was very low. Blood meal and feather meal contained high levels of iron, and the apparent availability (absorption) was also high, resulting in much higher levels of net absorption than with the other ingredients (7). Net

Table 1. Total and Digestible (available) Crude Protein (CP) and Phosphorus (P) Contents (%) in Feed Ingredients Determined with Rainbow Trout[a]

		Protein			Phosphorus		
Ingredient	Dry Matter Digestibility	Total CP	Digestibility	Digestible CP	Total P	Availability	Available P
Herring meal-A	89.2	73.6	94.6	69.7	2.05	44.4	0.91
Herring meal-B[b]	—	—	—	—	2.41	52.1	1.26
Anchovy meal	87.7	73.7	93.7	69.0	2.90	50.4	1.46
Menhaden meal-A	78.6	67.7	89.8	60.8	3.43	36.5	1.25
Menhaden meal-B	70.3	66.6	84.8	56.5	3.61	35.0	1.27
Peruvian fish meal	79.0	61.7	85.6	52.9	2.92	43.9	1.28
Whitefish meal, deboned meal-A	92.6	78.2	96.7	75.7	1.69	46.8	0.79
Whitefish meal, deboned meal-B	86.2	71.5	93.7	67.0	1.57	36.0	0.57
Whitefish meal, whole meal	74.0	71.7	88.4	63.4	3.50	17.2	0.60
Whitefish meal, skin and bone meal	50.5	46.9	76.1	35.7	7.41	11.8	0.87
Poultry by-product meal-A	91.6	81.0	95.9	77.6	2.17	63.5	1.38
Poultry by-product meal-B	81.4	68.2	85.8	58.5	2.50	47.7	1.19
Poultry by-product meal, low ash	84.6	72.7	89.5	65.1	1.65	50.8	0.84
Poultry by-product meal, deboned	73.7	70.2	82.7	58.1	2.09	47.4	0.99
Poultry by-product meal, low ash (averages of 4 products)	68.6	70.7	73.4	51.8	1.68	63.0	1.06
Poultry by-product meal, regular (averages of 9 products)	60.4	64.3	69.4	44.6	2.36	38.3	0.90
Poultry by-product meal, high ash (averages of 4 products)	43.8	56.7	55.0	31.2	3.45	14.7	0.51
Meat meal	59.5	63.0	80.9	50.9	2.76	2.5	0.07
Meat meal, low ash	66.0	67.3	80.9	54.5	2.28	44.7	1.02
Meat and bone meal-A	55.9	58.5	79.8	46.6	5.59	26.9	1.51
Meat and bone meal-B	61.6	58.9	79.0	46.5	2.68	21.8	0.58
Meat and bone meal, low ash	56.5	59.8	78.3	46.8	2.49	35.0	0.87
Feather meal-A	83.8	77.3	85.9	66.4	0.75	61.7	0.46
Feather meal-B	77.7	75.6	83.3	63.0	1.26	79.4	1.00
Blood meal, ring-dried-A	93.5	93.0	94.1	87.5	0.12	107.4	0.13
Blood meal, ring-dried-B	95.8	103.4	94.8	98.0	0.08	118.4	0.10
Blood meal, spray-dried	88.3	94.9	91.2	86.6	0.72	103.5	0.74
Soybean meal, dehulled, solv-ext.-A	71.2	53.2	90.1	47.9	0.76	22.0	0.17
Soybean meal, dehulled, solv-ext.-B[b]	58.8	50.9	85.1	43.3	0.85	26.6	0.23
Soybean meal, dephytinized[b]	61.8	50.9	85.1	43.3	0.85	92.5	0.79
Wheat gluten meal	94.7	85.0	100.5	85.5	0.18	74.7	0.13
Corn gluten meal	87.7	72.3	97.3	70.4	0.54	8.5	0.05
Wheat middling	45.0	20.5	90.7	18.6	1.17	55.3	0.65
Wheat flour	43.0	15.5	100.4	15.6	0.32	47.0	0.15
Barley, grain[b,c]	67.3	—	—	—	0.33	47.1	0.16
Corn, dent yellow, grain[b,c]	62.1	—	—	—	0.26	36.7	0.10
Corn, flint yellow, grain[b,c]	64.9	—	—	—	0.34	36.3	0.12

[a] *Source*: Sugiura (7), Sugiura and Hardy (unpublished data). The digestibility or availability values were expressed as fractional net absorption of nutrients (% per intake). The dietary concentrations of protein and phosphorus (total, digestible or available) were expressed on a dry basis. The fecal samples were collected either by stripping (ingredients indicated with an asterisk) or settling (without asterisk). Dash indicates that the value was not determined. Ingredients of the same trade name but obtained from different suppliers were identified with the letter "A" or "B." The letter does not indicate the grade or quality of the product.

[b] Determined in purified diets (diets low in phosphorus and calcium).

[c] Ingredients were heat extruded.

absorption of copper was significantly correlated to the amount of intake, and the amount of copper absorbed from whole diets was relatively unaffected regardless of source (ingredient). Net absorption of manganese was found to be quite low in all ingredients except for wheat gluten meal, blood meal, and the casein basal diet. This indicates these ingredients or diets contain only minor levels of substances which interfere with the absorption of manganese. Unlike many other elements, net absorption of manganese was not increased by the amount of intake, yet the overall balance of manganese always remained positive in all dietary treatments (7,18). Dabrowski and Schwartz (14) also noted that there was no significant apparent absorption of manganese in the carp digestive tract. In human studies, the percentage absorption of manganese is often below 10% (15). Dietary tri-calcium phosphate reduced the availability of manganese and zinc in carp (16). Availability of manganese in whitefish meal appears to be low in rainbow trout (17).

Apparent availability of zinc was low, especially that from fish meals (18). Rainbow trout fed a whitefish meal-based diet showed growth depression, bone malformation, and cataract incidence, which was suspectedly due to zinc deficiency (19). High bone content in diets has been shown to reduce zinc availability in rainbow trout (20) and Atlantic salmon (21). Also, tri-calcium phosphate reduced bioavailability of zinc in the diet fed to rainbow trout (22–24). Dietary phosphorus has been shown to decrease absorption of zinc in rats (25) and in rainbow trout (26). Menhaden meal, which had higher calcium and phosphorus levels than other fish meals, had lower zinc availability than the others. The availability of zinc in plant sources is generally lower than that in animal sources despite their lower calcium and phosphorus contents (18). Phytate or phytic acid contained in many plant ingredients have been shown to reduce the absorption of zinc in fish (27,28) and higher animals (29–31).

Antibiotics. Apparent digestibility of chloramphenicol was close to 99%, whereas it was in the 7 to 9% range for oxytetracycline (OTC) at two different concentrations (0.1 and 0.5%) in a dry diet fed to rainbow trout. For oxolinic acid, these percentages were 38.1 and 14.3% for the 0.1 and 0.5% doses, respectively (32). They suggested the low digestibility of oxytetracycline (7–9%), compared with the 60% absorption level usually described for OTC in man, may be explained by the great affinity of tetracyclines for calcium and the much higher concentration of calcium in fish feeds than in human diets. Others, however, reported 25% and 30% oral bioavailability of oxytetracycline in seawater chinook salmon (*Oncorhynchus tshawytscha*) and freshwater rainbow trout, respectively (33). The apparent digestibility of virginiamycin fed to rainbow trout at 40 ppm/diet was 98% for factor M and 79% for factor S (34).

Approaches to Increasing Digestibility of Dietary Nutrients

Many feed ingredients having fairly low digestibilities (availabilities) of phosphorus, protein, dry matter, or other nutrients may have other advantages, such as total protein content, amino acid balance, palatability, and price that make them desirable ingredients in aquatic feeds. If this is the case, one might want to use the ingredient but somehow enhance the inherent low digestibility of the feed ingredients. Various methods currently available for this purpose follow.

Dietary Level. Levels of nutrients in diets have an effect on the digestibility or absorption of the nutrient. Notable examples are the digestibility of carbohydrates and the availabilities of many minerals. Dietary phosphorus can be almost completely absorbed regardless of dietary concentration when it is supplied as a water-soluble form except for phytate phosphorus. The excess portion of absorbed phosphorus, however, is subsequently excreted via the urine (7). When dietary phosphorus is supplied in a water-insoluble form such as bone or calcium phosphates, an excess portion of dietary phosphorus that is above the minimum dietary requirement may not be absorbed. Thus, the higher the bone phosphorus in diets, the lower the resulting digestibility of phosphorus. Fish appear to absorb bone (water-insoluble) phosphorus up to their requirement but not in excess, which may be related to gastric acidity and acid output since fish absorb an excess amount of phosphorus when the diet is acidified with organic or inorganic acids (7). Reducing the concentration of phosphorus in diets, therefore, increases the digestibility of phosphorus by fish. Levels of protein have no significant effect on the digestibility of protein in diets; however, any excess portion of protein in diets will be deaminated and the nitrogen moiety excreted as ammonia, while carbon moieties will be utilized as an energy source. This suggests that excess protein in diet, while having no effect on protein digestibility, cannot be justified from either economical and environmental standpoints. Fish, or any other animal species, cannot retain excess amounts of minerals over time. Available minerals in diets, when they are in excess of the dietary requirement, have to be excreted either via urine, gills, or feces, or intestinal absorption has to be regulated. In all cases, reducing the amounts of available minerals in diets is critical to reduce their excretion by the fish.

Phytase. About two-thirds of total phosphorus in soybean meal or most other plant ingredients made from seeds is present as phytate or phytic acid, which is not efficiently utilized by nonruminant animals, including fish (3,35). Besides its low availability, phytate has been shown to interact directly and indirectly with various dietary components to reduce their availability to animals. Calcium compounds, including bone and the dietary supplements, have a strong affinity to phytate to form acid-insoluble precipitates and reduce both calcium and phytate availability. Calcium-bound phytate increases chelation with trace minerals such as zinc to form coprecipitates (36) or with protein to form either phytate–protein or phytate–mineral–protein complexes that are resistant to proteolytic digestion (37). Phytate may decrease endogenous zinc reabsorption as well as affect availability of dietary zinc (38). Increasing the level of phytate from 1.1 to 2.2% in channel catfish diets containing 50 mg zinc/kg (50 ppm) showed decreased weight gain, feed efficiency, and zinc content in the vertebrae (39). With 1.1% phytate in diets, channel catfish require about 200 mg zinc/kg feed (200 ppm), which is 10 times higher than the dietary requirement of available zinc (40). High phytate levels in semipurified diets (2.58%) depressed the growth and feed efficiency of chinook salmon (28).

Phytase is an enzyme specific to phytate hydrolysis. This enzyme is present in the digestive tract of many animals; however, the amount is normally too small to digest dietary phytate to a significant extent. The enzyme present in some plant ingredients, such as wheat bran, is generally inactivated by the processing temperature of feed manufacturing. Development of technology to produce phytase at a low cost offered an opportunity to use this enzyme in commercial animal feeds. Recent studies have shown that supplementing diets with commercially available fungal phytase or pretreating plant feed ingredients with phytase effectively increase availability of dietary phosphorus in rainbow trout (7,41–43), channel catfish (44–46), carp (47), as well as in higher animals such as pigs and chickens.

Various factors affect the efficiency of supplemental phytase. The rate of phytate hydrolysis by phytase enzyme varies not only with pH, but also with temperature. The optimum temperature varies among different phytases in the range of 45 to 57 °C (48). For homeotherms, body temperature appears to be favorable for the enzyme reaction. For poikilotherms, especially coldwater fishes, however, body temperature could be a limiting factor for the enzyme activity (7), suggesting an economical disadvantage for the use of this enzyme as a dietary supplement in fish feeds. Increasing water temperature increases enzyme activity but also simultaneously increases the rate of food passage through the gastrointestinal tract, offsetting the net effect. Reducing water temperature increases the retention time of foods in the digestive tract; however, the reduction of enzyme activity, feed intake, and fish growth are inevitable at lower temperatures. Alternative methods are those which employ preliminary digestion of phytate in plant ingredients by phytase before mixing with other feed ingredients (41,49,50). Adding water to dry ingredients, which is essential for the enzyme reaction, may be prohibitive as a practical application because of the extra energy and cost required in the succeeding drying process.

An important consideration for the use of supplemental phytase in commercial fish feeds is that the effect may be negligible when the diet contains phosphorus in an amount more than the dietary requirement (typical in most fish meal-based feeds). Supplemental phytase effectively reduced fecal phytate content in fish meal–soybean meal combined diet, while it increased the fecal non-phytate phosphorus. Total phosphorus levels in feces were, therefore, not significantly reduced. In contrast to this observation, supplemental phytase significantly increased the net absorption (availability) of phosphorus, calcium, magnesium, sodium, iron, copper, zinc, manganese, strontium, and the digestibility of protein, ash, and dry matter when the diet was formulated with soybean meal and low-ash ingredients containing phosphorus near the dietary requirement (7). Apparent availability of phosphorus increased proportionately with the supplemental phytase from 26.6% (no phytase added) up to 90.1% (4,000 units of phytase added per kg or 2.2 lb of dry diet) or 92.5% when soybean meal was pretreated or dephytinized (Fig. 2).

Because of its pH optimum, phytase is active only in the stomach. It should, therefore, be important to increase the stomach retention time to keep added phytase active. It was well recognized before the advent of scientific experimentation that a large single meal stays in the stomach longer than small meals. Researchers demonstrated that small meals could be more rapidly evacuated from the stomach than large meals (51,52). Feeding a single large meal should, therefore, be recommended for phytase-supplemented feeds. This simple practice increased the effect of supplemental phytase about two times (7). Multiple feeding or the use of automatic feeders or demand feeders

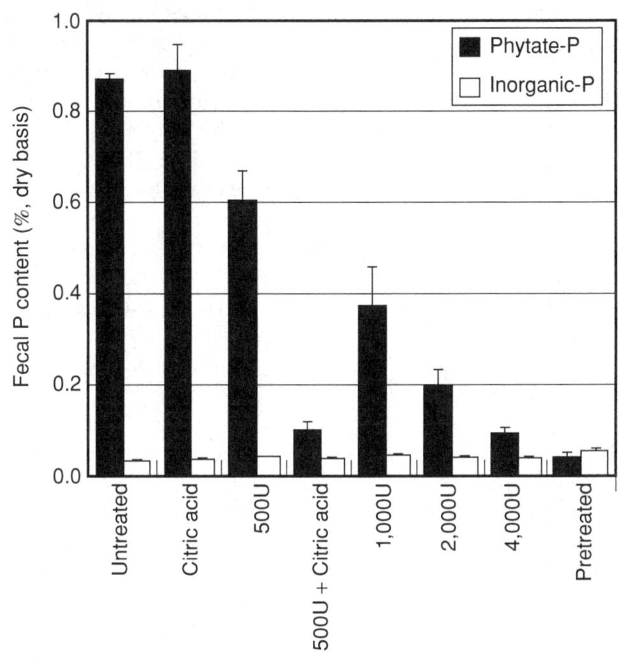

Figure 2. Phosphorus content in the feces of rainbow trout fed soybean meal-based diet supplemented with varied levels of phytase with/without citric acid. Error bars indicate standard errors of three replicate tanks. Pretreated (right end columns in the figure) indicates that soybean meal was treated with phytase before mixing with other ingredients (200U phytase/kg soybean meal; equivalent to 100U/kg dry diet, for 24 hours at 50 °C or 122 °F, pH 5.3, soybean meal/water ratio 2 : 1). The amount of citric acid added was 5% in the diets (dry basis). *Source*: Sugiura (7).

may result in a significant loss of phytase activity. For plant protein feeds, however, not only phytase but also free amino acids may be supplemented to correct amino acid imbalance. In this case, introducing a single feeding practice may not be the best way since frequent feeding is favorable to counteract different absorption rates between supplemented free amino acids and the amino acids in intact protein of feed ingredients (53,54). Another way to increase gastric retention time of feeds, and thus potentially increase hydrolysis of phytate, is to increase the energy density of the feeds (55,56). Also, gastric evacuation time is affected by the size of fish; that is, slower in large fish and faster in small fish (thus small fish require more frequent feeding than large fish). The overall effect of supplemental phytase is, therefore, predicted to be most efficient in large fish fed high-energy feeds in a single feeding per day.

While fungal phytase is now commercially available as a feed supplement, plant phytases inherent in wheat, barley, and rye should be considered as a low-cost alternative when the cost of feeds is one of the limiting factors in feed manufacturing. Stone et al. (50) reduced phytic acid content in canola meal by blending it with acidified fish silage and wheat bran (source of phytase) and keeping the mixture at room temperature for five weeks. During that period, low pH protected the wet material from bacterial spoilage and also facilitated phytate hydrolysis. Also, if endogenous phytase that is present in some feed ingredients is not heat-inactivated during feed processing, the active enzyme could hydrolyze phytate in the stomach after it is ingested by the animal. The hydrolysis of phytate in this way, however, appears to be less efficient, due to the amount of phytase, limited gastric retention time of food, and the low ambient temperature, particularly in coldwater fish. While microbial phytase produced by intestinal bacterial flora may have a significant effect on the decomposition of phytate in ruminant and other homeotherms, it may have negligible effect in fish species due to low bacterial population in the intestine and their low "body" temperatures.

Acidification. Dietary acidification has long been a common practice in weaning pigs to support their insufficient digestive capacity and to improve their weaning performance (57). Dietary acidification in aquatic animal feed, however, has not drawn deserved attention until recently with regard to the digestibility of dietary nutrients, especially for minerals. Fish meal is one of the main ingredients in commercial feeds for many aquaculture species. The form of phosphorus in fish meal is mostly hydroxyapatite, which is not efficiently utilized by fish (3,35). Several dietary supplements appear to influence the availability of phosphorus and other minerals in fish meal. Supplementing fish meal-based diets with citric acid at a 5% level (dietary pH 4.0) increased the availability of phosphorus, calcium, magnesium, iron, manganese, and strontium in rainbow trout (7,58). Supplementation of fish meal-based diets with citric acid at a 10% level (dietary pH 3.5) further increased the absorption (availability) of phosphorus (Fig. 3), but did not affect feed intake, feed utilization,

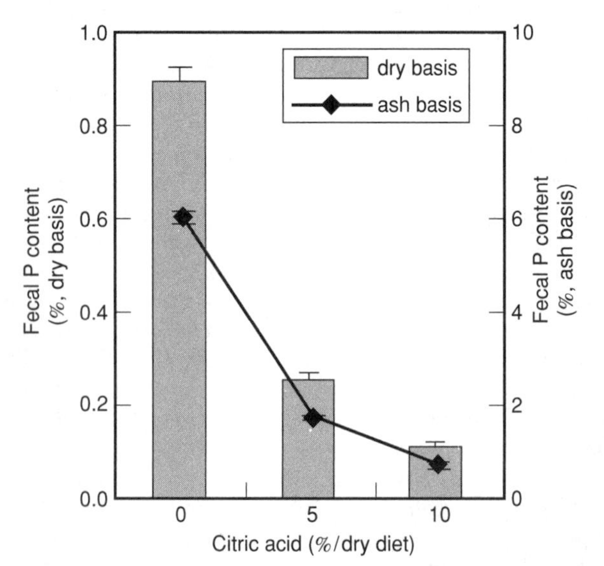

Figure 3. Phosphorus content in the feces of rainbow trout fed fish meal-based diet supplemented with citric acid at 0, 5 or 10% in the diet (dry basis). Error bars indicate standard errors of three replicate tanks. The apparent availability of phosphorus in the diet was 70.6% (0% citric acid), 90.9% (5% citric acid), and 96.6% (10% citric acid). *Source*: Sugiura (7).

protein digestibility, or weight gain of rainbow trout during 35 days of satiation feeding (7). Sulfuric acid and, to a lesser extent, hydrochloric acid were also very effective in increasing the availability of phosphorus in fish meal-based diets when they were added at 38 g and 76 g per kg dry diet (ppt), respectively (the apparent availability of phosphorus was 94.8% and 87.8% with sulfuric and hydrochloric acids, respectively, compared with 71.4% in the control or non-acidified group). The diets supplemented with these inorganic acids, however, had very low dietary pH (2.0–2.4) and had lower feed intakes of the fish than non-acidified diet during 23 days of satiation feeding (7).

Dietary acidification is also effective to increase efficacy of supplemental phytase in diets composed primarily of plant ingredients. Fungal phytases have their pH optima in an acidic range (48). Because of this, supplemental phytase is active only in the stomach of fish. A previous study showed that the pH of the feces was slightly lower when rainbow trout were fed citric acid at 5% in diets than when they were fed nonacidified control diet, suggesting that dietary acidification may preserve phytase activity throughout the digestive tract. In addition, phytate-mineral complexes, which are less soluble at a neutral pH, have increased solubilities at lower pH (59–62). When the diet contained fish meal, however, citric acid strongly disabled the effect of supplemental phytase, and the phytate supplied from soybean meal in the diet was little hydrolyzed. In contrast to this observation, citric acid markedly increased the efficacy of supplemented phytase when the diet was formulated with ingredients of low calcium content and without fish meal. Supplementing a soybean meal-based diet with 500 units of phytase and citric acid amplified the effect of supplemental phytase up to a level equivalent to 4,000 units of phytase even though citric acid per se had no effect on the hydrolysis of phytate (Fig. 2). Dietary acidification might increase the solubility of calcium phosphates in fish meal, which might precipitate phytate as calcium–phytate complex that is known to be resistant to enzymatic hydrolysis (63).

Heating. The digestibility of carbohydrates (and thus that of energy) varies substantially depending on processing temperature and the inclusion level in the feed (64,65). If feeds are processed at relatively low temperatures to conserve endogenous phytase, the carbohydrate portion of feeds may not be well-utilized by fish. This increases gastric passage rate of the chyme and reduces the time available for the digestion of phytate phosphorus as well as other dietary nutrients (66). The low digestibility of carbohydrates also indicates the low digestibility of dry matter (organic matter) and energy. Undigested carbohydrates and other organic matters in feces, instead of stimulating the growth of phytoplankton, will be subjected to bacterial fermentation and mold infestation in a pond sediment and create a noxious, unwholesome environment.

Phytate is resistant to heat treatment. Little decomposition of phytate appears to occur under the processing temperature effective for reducing other antinutritional factors (e.g., trypsin inhibitors) in soybean meal. Microwave heating, which was demonstrated effective to reduce phytate in full-fat soybean (67), appears to be ineffective for solvent-extracted soybean meal. Autoclaving at high temperatures has also been reported to reduce phytate but at the expense of some heat labile amino acids (68,69). A high loss of phytate in milled rice by cooking was also reported (70). Little or no loss was recorded for phytate in soybean meal, regardless of the method of heat treatment and the degree of intensity (7).

Heating is also critical to increase digestibility of protein in soybean meals by inactivating trypsin inhibitors and other antinutritional factors in the ingredient or in feather meal by hydrolyzing proteins at appropriate levels of heat treatments. In all cases, excessive heating leads to a reduction of protein digestibility and amino acid availability due to the reaction of amino acids with reducing sugars (known as Maillard browning) or with aldehydes in oxidized lipids. Maillard browning caused an approximately 80% loss in bioavailable lysine in fish protein isolate [incubated at 37 °C (99 °F) for 40 days] in rainbow trout (71). True digestibilities of amino acids in various meat and bone meals varied substantially, depending on the processing systems and temperatures, ranging from 68 to 92% for lysine and from 20 to 71% for cystine (72).

Chelation. An additional benefit of using citric acid and other organic ligands, such as EDTA (ethylenediaminetetraacetate), is that they may prevent precipitation or coprecipitation of minerals at a neutral pH in the presence of calcium and other minerals (73–75). Supplementing diets with EDTA or sodium citrate increased the apparent availability (absorption) of manganese but not other minerals examined (7,58). Hardy and Shearer (22) also reported that supplementing diets with EDTA did not increase availability (retention) of inorganic zinc in the diets for rainbow trout. Chelated zinc (zinc proteinate), however, has been found to be more available than inorganic sources for rainbow trout (22) and channel catfish (76). The apparent availabilities of chelated minerals (copper proteinate, iron proteinate, manganese proteinate, selenium proteinate, zinc proteinate) are all higher than those of the corresponding inorganic sources (copper sulfate, ferrous sulfate, manganese sulfate, sodium selenite, zinc sulfate) in both semipurified diets and soybean meal diets fed to channel catfish (76).

FORMULATION OF THE ENVIRONMENTALLY FRIENDLY FEEDS

Concept of Sustainable Feeds

It is not difficult to prepare environmentally *very* friendly feeds with purified feed ingredients used for laboratory experiments. In order to provide environmentally friendly feeds for commercial aquaculture, however, it will be imperative to meet other standards such as the cost of the feed (economical requirement), fish growth, feed efficiency, disease resistance of the fish (biological requirement), and the final product quality and safety (marketing requirement) (Fig. 4).

Aquaculture of carnivorous fishes, including salmonids and many seawater fishes, relies heavily on the supply of fish meal, which comprises the major portion of the

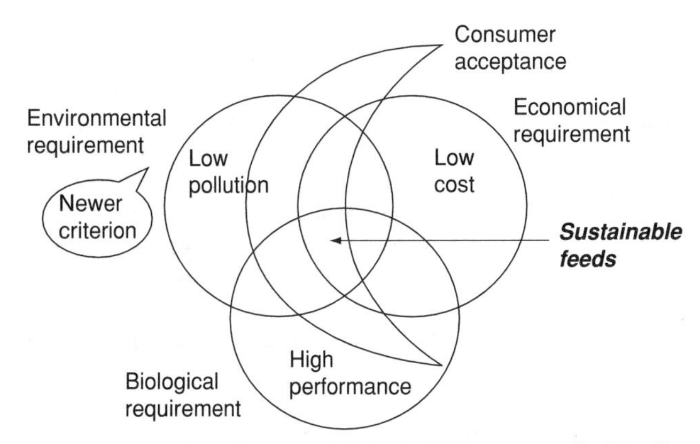

Figure 4. Concept of sustainable feeds. Environmentally friendly feeds have to meet various other standards.

feed for those species. As a result of the rapid growth of world aquaculture production in the past few decades, there has been a concomitant increase of demand for fish meal (77). The increasing demand, however, cannot be met due to the limited production of fish meal from capture fisheries which are approaching the plateau or maximum sustainable limit (78). Most fish meals contain phosphorus in an amount far excess of the minimum dietary requirement for fish. In order to reduce phosphorus in fish diets, it will be necessary to reduce the fish meal content in the diet by replacing it with other ingredients containing less phosphorus. Also, there has been an increasing contention and skepticism among the general public in regard to the aquaculture of carnivorous fishes since there is a substantial loss of animal protein in this process. Considering the ever-increasing world human population, feeding animal protein sources (fish meal) to fish on any significant scale will not be an affordable or a sustainable practice. Consequently, replacing fish meal with other ingredients that are not directly usable for human consumption should receive increasing priority in formulating aquaculture feeds in the future. This issue is particularly important in developing countries where the use of fish meal in aquaculture feeds is often economically prohibitive (79).

Ingredient Selection

There are numbers of ingredients that may replace a significant portion of fish meal in aquaculture feeds. The availability of alternate protein sources is largely dependent upon the region or country. Since the supply of plant protein sources is more stable and feasible compared with animal by-product materials, the use of plant ingredients should receive high priority, especially when considering the sustainability of aquaculture in terms of the future supply of ingredients for fish feeds. In this regard, Hardy (77) suggested that soybean meal is likely to be the most promising alternate protein source for fish feeds. Also, most plant protein sources contain less phosphorus than fish meal. Replacing major portions of fish meal with plant protein sources, however, encounters several practical problems (e.g., reduction in the growth of fish due to low palatability (low feed intake), low protein content or amino acid imbalance, and low dry matter digestibility due to high carbohydrate and fiber contents).

Since many animal by-product meals contain high levels of phosphorus, the maximum inclusion level in low-pollution feeds needs to be substantially reduced. The absorption of phosphorus in high-ash (high-phosphorus) ingredients increases as the dietary concentration decreases in terms of the percentage of absorption per intake. The phosphorus content in blood meal is low and the digestibility of protein and the availability of many minerals in it are high. The use of blood meal in practical feeds, however, should be limited because of its amino acid imbalance, which could likely reduce retention of dietary protein and increase nitrogen (ammonia) excretion. Feather meal, deboned fishmeal, and low-ash animal byproducts meals may be feasible alternate protein sources due to their relatively low phosphorus content, high protein content, high digestibility for both phosphorus and protein, and high palatability.

Nonlethal, low phytic acid mutations in corn and barley cause the seed to store most of the phosphorus as inorganic phosphorus instead of as phytate phosphorus (80). Using mutant grains containing lower levels of phytate in fish feeds can reduce phosphorus excretion by the fish (7,81).

Balanced Formula

Protein Levels. Optimum digestible protein (DP)/digestible energy (DE) ratio in rainbow trout feeds has been reported to be 92 mg/kcal (82), 105 mg/kcal (83), or 22.6 g/MJ DE (94.56 mg/kcal) (84). Similar DP/DE values have been found in the literature with other fishes (3,84). Nitrogen excretion in rainbow trout was reported to be 39 to 40% of digestible nitrogen intake for the diet having a DP/DE ratio of 18 mg/kJ (75.31 mg/kcal) and 44% of digestible nitrogen intake for the diet having a DP/DE ratio of 23 mg/kJ (96.23 mg/kcal) (85). Others reported that growth and feed utilization in rainbow trout improved markedly as dietary DP/DE ratio increased from 16.35 to 18.23 g/MJ (68.41 to 76.27 mg/kcal) with no significant effect on nitrogen discharge per kg of weight gain (mean values 29.1 to 29.9 g nitrogen/kg weight gain) (86). These results suggest that rainbow trout require DP in an amount at around 95 mg/kcal DE to support maximum growth, while the requirement of DP to provide maximum nitrogen retention (minimum nitrogen excretion) may be around 75 mg/kcal DE.

Fat Levels. A high fat content is generally preferable in low-polluting feeds. Fish oil and plant seed oils are commonly used for fish feeds. Those sources are highly digestible, essentially free of phosphorus and nitrogen, and supply high energy to the diet. Increasing dietary fat level is, therefore, a simple yet very efficient approach to reduce phosphorus and nitrogen excretions by the fish. Extrusion-processed feeds are generally preferable to compressed pellets for low-polluting feeds. Feeds processed by cooking extrusion can retain higher percentages of fats (25 to 35%) in the pellets due to their high porosity than feeds made by compressed (steam) pelleting (15 to 20%). In addition, carbohydrate digestibility (and also energy and dry matter digestibilities) of extruded feeds is much higher than that of the compressed pellets due to high moisture and temperature applied in the manufacturing process (although this raises the processing cost). When fish are deficient in phosphorus, however, they appear to increase the retention of dietary fats (87) due to the inhibition of the beta-oxidation of fatty acids. Since high levels of dietary fats and low levels of dietary phosphorus synergistically elevate the fat content of the fish, it appears to be essential to adjust the levels of dietary phosphorus or fats depending on the desired fat content of the final product.

Carbohydrate Levels. Increasing DE value by increasing digestible carbohydrate may reduce feed intake and reduce growth rate of the fish, especially in carnivorous species (65,88). Low-protein (38%) diets containing high levels (30 to 40%) of digestible carbohydrate, however, did not adversely affect overall growth or nutrient retention efficiencies in rainbow trout (89–92). These findings indicate that where fats are unavailable as an energy source for aquaculture feeds, high levels of digestible carbohydrates (often less expensive than fats) can be tolerated even for carnivorous species like rainbow trout. If fish are deficient in phosphorus, however, it may cause glucose intolerance and the high level of digestible carbohydrates in the diet may result in pathological consequences as reported in higher animals (93).

Finishing Feeds

Deficiencies of micronutrients in diets do not cause immediate clinical deficiency of fish. Although it is largely dependent on the nutrient, the degree of deficiency, physiological demand, interactions, and the body store (diet history), it takes at least two weeks with small fish or much longer when fish are large before any signs of clinical deficiency arises. If the nutrient is only marginally deficient, fish can subsist considerable periods using the body store without showing any deficiency signs. Hardy et al. (94) proposed a periodic feeding of low- and high-phosphorus feeds to increase retention of dietary phosphorus and to reduce excretion of phosphorus into water. An extension of this method is the use of low-phosphorus feeds as a finishing diet. It is, of course, imperative to harvest fish before signs of phosphorus deficiency arise. Eya and Lovell (5) reported that year-two channel catfish fed commercial-type feeds containing only 0.2% available phosphorus did not reduce weight gain and feed efficiency in a 140-day feeding period compared with those fed diets containing higher concentrations of available phosphorus.

Commercial trout production feeds generally contain phosphorus at levels much higher than the dietary requirement. The finishing feeds can also be used to dilute phosphorus concentration in the commercial feeds (7). Thus, the finishing feed and the commercial feed need to be mixed in an appropriate ratio to minimize excretion of phosphorus. The ratio needs to be adjusted according to the size of fish in addition to the phosphorus content of both feeds. The ratio may be best determined by actually monitoring the level of phosphorus excreted by the fish and the performance of the fish.

Quality Assurance

Environmentally friendly feeds generally contain high levels of unsaturated fatty acids from fish oil or similar sources. In formulated feeds, varieties of compounds can catalyze the oxidation of fish oil, while others counteract it. Inorganic copper and iron are known to be strong pro-oxidants for unsaturated fatty acids and for certain labile vitamins, to increase rancidity of dietary fats, and to cause a loss of vitamins in feeds during storage. Feeding rancid fats causes undesirable effects; for example, impairment of fish health, increase of vitamin E requirements, decrease in vitamin E stored in tissues and possibly reduction in frozen storage stability of fillets. Since the availability of copper is relatively high in many ingredients and the availability of iron is high in blood meal and feather meal, the redundant supplementation of these metals (as inorganic supplements) should be avoided. Use of chelated minerals may protect labile fatty acids and vitamins in the diet. Phytic acid in plant ingredients may offer similar effects (chelating minerals) and thereby protect labile compounds in the diet, while it also reduces the availability of the mineral to the fish.

Low acceptability of diets containing high levels of soybean meal needs to be further investigated if it is intended to be used as the major protein source in production feeds. Raw (unheated) soybean flour containing high levels of trypsin inhibitor depressed feed intake of rainbow trout at 30% level and completely deterred the feed intake at 60% level in diets. Conversely, heated soy flour and soy protein concentrate containing negligible levels of trypsin inhibitor did not affect the feed intake at either 30 or 60% inclusion levels in diets (unpublished data). This suggests that trypsin inhibitor levels in soybean might have a major effect on the palatability of the ingredient. The diet history of fish, particularly at earlier stages (normally fish meal-based commercial feeds), may be another factor for the preference of feeds (95).

CONCLUSION

Earlier in the twentieth century, when Clive M. McCay pioneered fish nutrition research in the United States, the goals of nutrition in aquaculture were (*1*) to develop feeds that support optimum growth of fish (biological requirement) and (*2*) to use inexpensive feed ingredients whose supply are stable (economical requirement) (96). In the second half of the twentieth century, however, additional needs have emerged: (*1*) to control final product (fish) quality (marketing requirement) and (*2*) to reduce excretion of wastes (environmental requirement). Fish feeds nowadays have to meet all of these requirements: maximizing fish growth and fish quality while minimizing feed cost and waste excretion. These requirements are all endless, yet have critical and direct effects for the sustainability and the development of aquaculture for the future.

BIBLIOGRAPHY

1. National Research Council (NRC), *Nutrient Requirements of Poultry* (8th rev. ed.), National Academy Press, Washington, DC, 1984.
2. C. Ogino and H. Takeda, *Bull. Jpn. Soc. Sci. Fish.* **44**, 1019–1022 (1978).
3. NRC, *Nutrient Requirements of Fish*, National Academy Press, Washington, DC, 1993.
4. M. Rodehutscord, *J. Nutr.* **126**, 324–331 (1996).
5. J.C. Eya and R.T. Lovell, *Aquaculture* **154**, 283–291 (1997).
6. D.H. Baker, *J. Nutr.* **116**, 2339–2349 (1986).
7. S.H. Sugiura, *Development of Low-Pollution Feeds for Sustainable Aquaculture*, Ph.D. dissertation, University of Washington, Seattle, 1998.
8. J.C. Abbott and L.M. Dill, *Behaviour* **108**, 104–111 (1989).
9. G.K. Iwama, in W. Pennell and B.A. Barton, eds., *Principles of Salmonid Culture*, Elsevier, New York, 1996.
10. W.E. Hajen, R.M. Beames, D.A. Higgs, and B.S. Dosanjh, *Aquaculture* **112**, 321–332 (1993).
11. T. Asgard and E. Austreng, *Aquaculture* **55**, 263–284 (1986).
12. F.M. Dong, R.W. Hardy, N.F. Haard, F.T. Barrows, B.A. Rasco, W.T. Fairgrieve, and I.P. Forster, *Aquaculture* **116**, 149–158 (1993).
13. J.K. Babbitt, R.W. Hardy, K.D. Reppond, and T.M. Scott, *J. Aquat. Food Prod. Technol.* **3**, 59–68 (1994).
14. K.R. Dabrowski and F.J. Schwartz, *Zool. Jb. Physiol.* **90**, 193–200 (1986).
15. B. Sandstrom, *Proc. Nutr. Soc.* **51**, 211–218 (1992).
16. S. Satoh, K. Izume, T. Takeuchi, and T. Watanabe, *Nippon Suisan Gakkaishi* **58**, 539–545 (1992).
17. S. Satoh, T. Takeuchi, and T. Watanabe, *Nippon Suisan Gakkaishi* **57**, 99–104 (1991).
18. S.H. Sugiura, F.M. Dong, C.K. Rathbone, and R.W. Hardy, *Aquaculture* **159**, 177–202 (1998).
19. S. Satoh, T. Takeuchi, and T. Watanabe, *Nippon Suisan Gakkaishi* **53**, 595–599 (1987).
20. H.G. Ketola, *J. Nutr.* **109**, 965–969 (1979).
21. K.D. Shearer, A. Maage, J. Opstvedt, and H. Mundheim, *Aquaculture* **106**, 345–355 (1992).
22. R.W. Hardy and K.D. Shearer, *Can. J. Fish. Aquat. Sci.* **42**, 181–184 (1985).
23. S. Satoh, K. Tabata, K. Izume, T. Takeuchi, and T. Watanabe, *Bull. Jpn. Soc. Sci. Fish.* **53**, 1199–1205 (1987).
24. S. Satoh, N. Porn-Ngam, T. Takeuchi, and T. Watanabe, *Nippon Suisan Gakkaishi* **59**, 1395–1400 (1993).
25. D.A. Heth, W.M. Becker, and W.G. Hoekstra, *J. Nutr.* **88**, 331–337 (1966).
26. N. Porn-Ngam, S. Satoh, T. Takeuchi, and T. Watanabe, *Nippon Suisan Gakkaishi* **59**, 2065–2070 (1993).
27. J. Spinelli, C.R. Houle, and J.C. Wekell, *Aquaculture* **30**, 71–83 (1983).
28. N.L. Richardson, D.A. Higgs, R.M. Beames, and J.R. McBride, *J. Nutr.* **115**, 553–567 (1985).
29. P. Saltman, J. Hegenauer, and L. Strause, in O.M. Rennert and W. Chan, eds., *Metabolism of Trace Metals in Man*, (Vol. I), CRC Press, Inc., Boca Raton, FL, 1984, pp. 1–16.
30. K.T. Smith and J.T. Rotruck, in L.S. Hurley, ed., *Trace Elements in Man and Animals* Vol. 6, Plenum Press, New York, 1988, pp. 221–228.
31. R.S. Gibson, *Am. J. Clin. Nutr.* **59**, 1223s–1232s (1994).
32. J.P. Cravedi, G. Choubert, and G. Delous, *Aquaculture* **60**, 133–141 (1987).
33. S. Abedini, R. Namdari, and F.C.P. Law, *Aquaculture* **162**, 23–32 (1998).

34. J.P. Cravedi, M. Baradat, and G. Choubert, *Aquaculture* **97**, 73–83 (1991).
35. C. Ogino, L. Takeuchi, H. Takeda, and T. Watanabe, *Bull. Jpn. Soc. Sci. Fish.* **45**, 1527–1532 (1979).
36. Anonymous, *Nutr. Rev.* **25**, 215–218 (1967).
37. M. Cheryan, *Crit. Rev. Food Sci. Nutr.* **13**, 297–335 (1980).
38. E.R. Morris, in E. Graf, ed., *Phytic Acid: Chemistry and Applications*, Pilatus Press, Minneapolis, MN, 1986, pp. 57–76.
39. S. Satoh, W.E. Poe, and R.P. Wilson, *Aquaculture* **80**, 155–161 (1989).
40. D.M. Gatlin, III and R.P. Wilson, *Aquaculture* **41**, 31–36 (1984).
41. K.D. Cain and D.L. Garling, *Prog. Fish-Cult.* **57**, 114–119 (1995).
42. M. Rodehutscord and E. Pfeffer, *Water Sci. Technol.* **31**, 143–147 (1995).
43. M. Riche and P.B. Brown, *Aquaculture* **142**, 269–282 (1996).
44. L.S. Jackson, M.H. Li, and E.H. Robinson, *J. World Aquacult. Soc.* **27**, 309–313 (1996).
45. J.C. Eya and R.T. Lovell, *J. World Aquacult. Soc.* **28**, 386–391 (1997).
46. M.H. Li and E.H. Robinson, *J. World Aquacult. Soc.* **28**, 402–406 (1997).
47. A. Schaefer, W.M. Koppe, K.H. Meyer-Burgdorff, and K.D. Guenther, *Water Sci. Technol.* **31**, 149–155 (1995).
48. N.R. Nayini and P. Markakis, in E. Graf, ed., *Phytic Acid: Chemistry and Applications*, Pilatus Press, Minneapolis, MN, 1986, pp. 101–118.
49. T.S. Nelson, T.R. Shieh, R.J. Wodzinski, and J.H. Ware, *Poultry Sci.* **47**, 1842–1848 (1968).
50. F.E. Stone, R.W. Hardy, and J. Spinelli, *J. Sci. Food Agric.* **35**, 513–519 (1984).
51. M. Jobling, D. Gwyther, and D.J. Grove, *J. Fish Biol.* **10**, 291–298 (1977).
52. E. He and W.A. Wurtsbaugh, *Trans. Am. Fish. Soc.* **122**, 717–730 (1993).
53. S. Yamada, Y. Tanaka, and T. Katayama, *Bull. Jpn. Soc. Sci. Fish.* **47**, 1247 (1981).
54. S. Yamada, K.L. Simpson, Y. Tanaka, and T. Katayama, *Bull. Jpn. Soc. Sci. Fish.* **47**, 1035–1040 (1981).
55. D.J. Lee and G.B. Putnam, *J. Nutr.* **103**, 916–922 (1973).
56. D.J. Grove, L.G. Loizides, and J. Nott, *J. Fish. Biol.* **12**, 507–516 (1978).
57. V. Ravindran and E.T. Kornegay, *J. Sci. Food Agric.* **62**, 313–322 (1993).
58. S.H. Sugiura, F.M. Dong, and R.W. Hardy, *Aquaculture* **160**, 283–303 (1998).
59. H. Møllgaard, *Biochem. J.* **40**, 589–603 (1946).
60. U. Tangkongchitr, P.A. Seib, and R.C. Hoseney, *Cereal Chem.* **59**, 216–221 (1982).
61. F. Grynspan and M. Cheryan, *JAOCS* **60**, 1761–1764 (1983).
62. K.B. Nolan, P.A. Duffin, and D.J. McWeeny, *J. Sci. Food Agric.* **40**, 79–85 (1987).
63. R. Lasztity and L. Lasztity, in Y. Pomerantz, ed., *Advances in Cereal Science and Technology*, (Vol. X), American Association of Cereal Chemists, Incorporated, St. Paul, MN, 1990, pp. 309–371.
64. R.P. Singh and T. Nose, *Bull. Freshwater Fish. Res. Lab.* **17**, 21–25 (1967).
65. E. Pfeffer, J. Beckmann-Toussaint, B. Henrichfreise, and H.D. Jansen, *Aquaculture* **96**, 293–303 (1991).
66. L. Spannhof and H. Plantikow, *Aquaculture* **30**, 95–108 (1983).
67. Y.S. Hafez, A.I. Mohammed, P.A. Perera, G. Singh, and A.S. Hussein, *J. Food Sci.* **54**, 958–962 (1989).
68. J.J. Rackis, *J. AOCS* **51**, 161A–174A (1974).
69. A.R. De Boland, G.B. Garner, and B.L. O'Dell, *J. Agric. Food Chem.* **23**, 1186–1189 (1975).
70. R.B. Toma and M.M. Tabekhia, *J. Food Sci.* **44**, 629–632 (1979).
71. S.M. Plakas, T.C. Lee, and R.E. Wolke, *J. Nutr.* **118**, 19–22 (1988).
72. X. Wang and C.M. Parsons, *Poultry Sci.* **77**, 834–841 (1998).
73. P. Vohra and F.H. Kratzer, *J. Nutr.* **82**, 249–256 (1964).
74. F.H. Nielsen, M.L. Sunde, and W.G. Hoekstra, *J. Nutr.* **89**, 35–42 (1966).
75. D.B. Lyon, *Am. J. Clin. Nutr.* **39**, 190–195 (1984).
76. T. Paripatananont and R.T. Lovell, *J. World Aquacult. Soc.* **28**, 62–67 (1997).
77. R.W. Hardy, in C. Lim and D.J. Sessa, eds., *Nutrition and Utilization Technology in Aquaculture*, AOCS Press, Campaign, IL, 1995, pp. 26–35.
78. FAO, *The State of World Fisheries and Aquaculture*, 1996, FAO fisheries department, FAO, Rome, Italy, 1997.
79. A.G.J. Tacon, *World Aquacult.* **27**(3), 20–32 (1996).
80. V. Raboy and P. Gerbasi, in B.B. Biswas and S. Biswas, eds., *Subcellular Biochemistry* (Vol. 26), Plenum Press, New York, 1996, pp. 257–285.
81. S.H. Sugiura, V. Raboy, K.A. Young, F.M. Dong, and R.W. Hardy, *Aquaculture* **170**, 285–296 (1999).
82. C.Y. Cho and S.J. Kaushik, in C.B. Cowey, A.M. Mackie, and J.G. Bell, eds., *Nutrition and Feeding in Fish*, Academic Press, London, 1985, pp. 95–117.
83. C.Y. Cho and B. Woodward, in *Proceedings of the Eleventh Symposium on Energy Metabolism*, European Association for Animal Production Publication 43, Wageningen, Netherlands, 1989, pp. 37–40 (cited in NRC, 1993).
84. C.B. Cowey, *Water Sci. Technol.* **31**, 21–28 (1995).
85. F. Medale, C. Brauge, F. Vallee, and S.J. Kaushik, *Water Sci. Technol.* **31**, 185–194 (1995).
86. D. Lanari, E.D. D'Agaro, and R. Ballestrazzi, *Aquacult. Nutr.* **1**, 105–110 (1995).
87. S. Sakamoto and Y. Yone, *Bull. Jpn. Soc. Sci. Fish.* **46**, 1227–1230 (1980).
88. J.W. Hilton, C.Y. Cho, and S.J. Slinger, *Aquaculture* **25**, 185–194 (1981).
89. A. Pieper and E. Pfeffer, *Aquaculture* **20**, 323–332 (1980).
90. A. Pieper and E. Pfeffer, *Aquaculture* **20**, 333–342 (1980).
91. S.J. Kaushik, F. Medale, B. Fauconneau, and D. Blanc, *Aquaculture* **79**, 63–74 (1989).
92. J.D. Kim and S.J. Kaushik, *Aquaculture* **106**, 161–169 (1992).
93. R.A. DeFronzo and R. Lang, *N. Engl. J. Med.* **303**, 1259–1263 (1980).
94. R.W. Hardy, W.T. Fairgrieve, and T.M. Scott, in S.J. Kaushik and P. Luquet, eds., *Fish Nutrition in Practice*. colloq. no. 61, INRA, France, 1993, pp. 403–412.
95. S. Refstie, S.J. Helland, and T. Storebakken, *Aquaculture* **153**, 263–272 (1997).
96. C.M. McCay, *Trans. Am. Fish. Soc.* **57**, 261–265 (1927).

See also NITROGEN.

EXOTIC INTRODUCTIONS

William L. Shelton
University of Oklahoma
Norman, Oklahoma

OUTLINE

Aquaculture is the farming of aquatic organisms, including fish, invertebrates, and plants. Its objective may be to produce organisms for sport, conservation, ornamental purposes, food, or some specific product. Aquaculture practices range from low-input extensive systems, including ranching, to intensive, highly managed operations. Contemporary aquaculture usually includes artificial propagation, which permits domestication, stock improvement, reproductive manipulation, and, most significantly, assures a supply of animals for stocking. However, the capacity to manage reproduction has greatly increased the opportunity for translocation of species, which can be a double-edged sword: On the one hand, it offers multiple beneficial applications, but on the other hand, it may have associated potential adversities as well. Human-instigated movement of organisms outside their native range is not new. Plants and animals have been moved intentionally or passively with humans during their emigrations throughout the world; in fact, humans are the most widespread exotic species (1). However, our basis for defining an exotic organism is human involvement in the translocation. Food crops have been dispersed throughout suitable climates for their cultivation, and similarly, domesticated animals have had enhanced vagility associated with human resettlement. These companion plants and animals have been greatly modified from their native ancestral types through intentional breeding and passive selection. Consequently, nonnative domesticated organisms, even though they are not indigenous to their contemporary locale, are no longer generally perceived as exotic.

Fish have been moved outside their native ranges, but only a few are greatly changed from ancestral types. The common carp, *Cyprinus carpio*, and goldfish, *Carassius auratus*, probably have had the longest controlled association with humans. The common carp may have been the earliest fish that was transplanted (2,3); both the common carp and goldfish have distinct varieties that differ significantly from the wild type (4), and rainbow trout, *Oncorhynchus mykiss*, has over 200 different stocks. Nonnative fish are a part of aquaculture to different degrees in various countries, although not to the extent that other cultivars and breeds are used in agriculture and animal husbandry. Actually, limiting fish farming to indigenous species or culturing native fishes only within their historic range can be a serious constraint on aquatic food production. Agriculture practices have evolved over centuries and have been modified to adapt to various climates and economies, and nonnative organisms have been fully accepted.

Aquaculture is a practice dating back at least 30 centuries (see the entry "History of aquaculture"), but the scientific bases for modern activities are relatively recent developments (5,6). Most of the adaptive practices in the United States for economically important species have been incorporated within the past four to five decades, but concomitantly, there also has been a growing environmental sensitivity. Thus, aquaculture has been compelled to develop under the scrutiny of many ethical and environmental considerations that were not applied during the formative period of agriculture.

While the goals of an unmodified environment are laudable, it is not realistic to aspire to return the faunistic distribution in North America to that of pre-European arrival. We can be custodians of natural systems without halting appropriate development. If aquaculture is to meet the expanding aquatic food deficit associated with the current capture-fisheries shortfall, if it is to provide biocontrol options in natural-resource management, and if it is to reduce overexploitation of ornamental fish in their native range, then introduced species must be recognized as a necessary option of managed production in aquatic systems. Yet, we cannot ignore the potential adverse impact of exotics on aquatic habitats and the native fauna; rather, we should assess and evaluate introductions using controls and safeguards and, if indicated, proceed responsibly. This contribution addresses exotics largely from the perspective of the United States, but since aquaculture is a global issue and the definition of "exotic" is from a relative perspective, a thorough understanding of exotics in aquaculture must include information for other countries as well.

BACKGROUND

A general understanding of aquaculture is somewhat intuitive as a parallel to agriculture, since aquatic culture is usually associated with foodfish production, which is essentially aquatic-animal husbandry. Historically, eggs or larvae were collected from natural spawning areas and transferred to culture systems for growout, but as propagation techniques were developed, all phases of the life history became controlled (7–10). Today, aquaculture includes management of broodstock, controlled propagation, nursing, growout, processing, and marketing, as well as genetic manipulation (11). A nonfood commercial application of the culture of fish is the production of ornamentals, which is also a positive development for the conservation of species within their native ranges. In a slightly different application of the culture of fish, juveniles are produced for stocking into sport fisheries, although the growout phase is usually in public waters and harvest is through angling. The cultivation of endangered species for restocking into open systems is similar to the culture of sportfishes in that it encompasses only spawning and the early phases of life history.

The facilities where aquaculture activities are conducted generally characterize the focus of the operations. Commercial aquaculture is associated with fish farms, while the facilities dedicated to the culture of sportfish or for conservation programs are usually called hatcheries. The latter are most often operated by government organizations, while the former are managed by owners and operators in the private sector. Huet (7) has distinguished between fish culture for restocking and fish culture for food, while Bardach et al. (8) have defined aquaculture to be the rearing of aquatic organisms for human food. Stickney (9) and Pillay (10) have incorporated a broader interpretation of aquaculture, stating that it is not limited to vertebrates, finfishes, food fishes, or even animals, but also includes plants and "byproducts," such as pearls. All of these activities include introductions, either exotic or transplanted.

The concept of exotic organisms must be defined, especially within the context of aquaculture. The terminology used in this chapter is based on the standardized system proposed by the introduced fish section (the exotic fish section, before 1985) of the American Fisheries Society (12). A fish or organism that has been intentionally or accidentally moved outside its natural range by humans is defined as introduced. A species is considered native or indigenous in its historically natural range (13). Introductions can be made either within a country (the organism is considered a transplant) or between countries (the organism is considered an exotic). Thus, a transplanted aquatic organism has been moved by man between watersheds within the country of origin (14). Further, an introduction is considered to include the release, escape (15), or establishment of an exotic species into a natural ecosystem—i.e., one in which the species does not naturally occur, either presently or historically (16). Since the introduction of an exotic species from another country and the movement of a species into a new drainage or range can have comparable ecological effects, even the formerly widely accepted practice of managers transplanting fish in stocking programs is becoming increasingly scrutinized (17). Definition of an introduction as established or naturalized is based on successful reproduction and recruitment and provides some concept of long-term considerations.

It is important to understand that even though the terms under which an introduction is defined as established or naturalized may seem biologically illogical, they carry ecological implications and determine legal issues for the regulation of translocations. Despite the best of intentions and concerted efforts, escape of transplanted or exotic organisms from research or commercial facilities must be anticipated (18,19); consequently, confinement under such conditions is still considered an introduction. The potential impact of fish introductions into a new environment are speculative at best, and while an assessment requires detailed studies, even the results of such studies are usually inconclusive (20,21). Nevertheless, appropriate dialogue should precede an introduction, and if an assessment is considered, it should be done in an ecologically responsible way, using fish that are reproductively limited (22,23). A period of observation within a particular aquatic habitat is the only realistic means of accurately evaluating an introduction. Management of reproduction through artificial propagation has been the foundation for commercial culture (24), but now the capability to manipulate phenotypic sex, chromosome composition, and the genome itself have provided a quantum leap forward (25,26).

MANAGEMENT OF REPRODUCTION

Management of reproduction has been critical to the development of contemporary aquaculture, and it has also been a major factor in the increasing number of introductions. On the other hand, manipulation of organisms' reproductive systems can offer new tools for using the organisms in an ecologically responsible manner. Thus, management of fish reproduction can be considered from two perspectives: One involves the production of seedstock under controlled conditions, while the other limits reproductive success. Both approaches are valuable for aquaculture. The control of fish reproduction through artificial propagation has provided tremendous opportunities. The capability to spawn fish under controlled conditions assures an adequate supply of young for growout, whether for food or stocking, and removes the constraints of limiting culture to the geographic proximity of the native range of the fish (27). Artificial propagation also has been one of the most important milestones in catholic aquaculture, in that it has facilitated the capability to move fish to new areas and to establish culture for these species far outside of their natural range, such as is the case for Chinese carps (28). Traditional culture of Chinese and Indian carps, as well as of milkfish, was dependent on capturing wild-spawned fry or fingerlings. Artificial propagation also opens new vistas for the culture of species formerly available only from capture fisheries, such as penaeid shrimps, various characins, and many marine species (29–31). Further, even for species that reproduce under culture conditions, such as common carp, induced ovulation has permitted more efficient management and has provided a greatly enhanced capacity to conduct breeding programs (4,32–38).

Development of reproductive controls has direct applications to aquaculture in cases for which management of recruitment is desirable or for quality considerations (39), as well as for introduced fish for which no reproduction is wanted (40,41). Reproductively limited populations have been instrumental in efficient culture of several species, and while limitation of reproduction is not commonly used as a component of introduction protocol, the capability for it is already developed (42).

Artificial Propagation

Reproduction of fish represents one of the most diverse assemblages of modalities and strategies among vertebrates (43–45). Artificial propagation can be considered across the spectrum, from simple environmental manipulation to more sophisticated physiological control (24,25). Photoperiod and temperature programming have been effective for inducing spawning in a variety of marine species. Hormonal therapy was started in the 1930s and was initially restricted to the use of homologous or heterologous injection of pituitary glands. Today, gonadotropins, such as salmon and carp gonadotropic hormone (GtH), have been extracted, purified, and packaged for more convenient use. They are calibrated and bioassayed for efficacy (27,35). Pituitary extracts and simple dried glands are currently still in wide use for artificial propagation. Time-release carrier systems, which greatly enhance the treatment of fish that are sensitive to handling or that need multiple injections (32,46), are also being perfected for long-term hormone delivery. Various mammalian hormones such as luteinizing hormone (LH), follicle-stimulating hormone (FSH), and human chorionic gonadotropin (HCG), were tried in the 1950s and 1960s, but only the latter is still in common use, and it is not as broadly effective as GtH. Gonadotropin-releasing hormone (GnRH), the hormone produced by the hypothalamus, regulates the production and release of GtH from the pituitary gland; it was purified in the 1970s and reported as luteinizing hormone-releasing hormone (LH-RH), which is a mammalian term. For fish, GnRH is the more appropriate descriptor. Superactive analogues (GnRHa and LH-RHa) now have been synthesized and are increasingly being used (27).

Artificial propagation through controlled final maturation, ovulation, and spermiation permit genetic selection to improve stocks for various desirable traits in animal breeding programs (5). Genetic improvement of fish growth in aquaculture is the primary focus, but factors such as disease resistance, color selection, and others may also be the object of selection. Fish have a few biological characteristics that are conducive to selective breeding: external fertilization, high fecundity, and potential for hybridization.

Domestication is the transfer of an organism from its natural area to a captive environment, and its genetic adaptation to the new environment. Common carp have been the most intensively domesticated fish, but many strains of channel catfish, *Ictalurus punctatus*, and rainbow trout have been developed by multiple hatcheries, farms, and research institutes as well. Carp have been cultured for centuries, while most domesticated strains of channel catfish originated from the wild only in 1949. Culture of tilapia may date back to ancient Egypt, but little stock selection has occurred. Artificial reproduction of naturally ripened broodstock of rainbow trout originated in the 18th and 19th centuries, and there are over 200 different stocks of the trout species.

Artificial propagation of foodfish has been instrumental in enlarging the geographic range for the culture of exotic species, such as Chinese carps, and also of native species. This technology has greatly expanded the culture of species used in sport fisheries and for conservation-oriented reintroductions. The effectiveness of induced spawning in culture for sport fisheries is exemplified by the widespread introduction of the striped bass, *Morone saxatilis* (47). The primary impetus of using artificial propagation on striped bass was to develop inland fisheries, but in the interim, this species and its hybrid with the white bass have become increasingly important in foodfish culture (see the entry "Striped bass and hybrid striped bass culture"). Native to the Atlantic and Gulf coasts of the United States, the striped bass has been widely disseminated into inland waters. The current sport fisheries and foodfish producers are totally dependent on artificial propagation, with the exception of about 10 naturally reproducing reservoir populations. This example also illustrates the contrasting attitudes of transplanting sport fish and using an exotic species. Widespread stocking of striped bass was possible only after the techniques for induced spawning were developed (48). Beginning in the 1960s, reservoirs throughout the United States were stocked with striped bass, which have provided countless hours of recreational fishing and millions of dollars of increased revenues for sport fisheries (49). However, even though reservoirs are artificial systems, little consideration was given to the potential impact that the introductions might have on resident piscivores. Although there was little opposition to these earlier transplants, this type of action in the future probably will be as thoroughly scrutinized as exotic introductions.

Introduction of an exotic fish during the same period of time did not benefit from a similar "honeymoon." The grass carp, *Ctenopharyngodon idella*, was imported in 1963 as a potential biocontrol for nuisance aquatic plants, but opposition to its introduction developed immediately. The importance of artificial propagation again was the key factor that made the introduction possible, as grass carp culture had been restricted to China until induced spawning techniques were developed in 1961 (28,50). The dichotomy in opinion concerning this introduction presented little negotiable middle ground. The controversy that ensued epitomizes the adversarial atmosphere that can develop between proponents and opponents in such a conflict and that interferes with a rational evaluation. However, the controversy did add impetus to investigations of techniques for the control of unwanted reproduction. The results from some of these studies have contributed to the potential for an orderly system of assessment. Thus, the quandary that can result from the advances in artificial propagation may also be the source of solutions to rectify conflicts and provide a basis for effective assessment.

Manipulation of Reproduction

Early efforts to manage unwanted reproduction in fish were initiated as a means of controlling recruitment of tilapia for aquaculture (5,24,25,36,51). These techniques include monosexing, sterility, hybridization, and various combinations (see the entry "Tilapia culture"). The proposition of using reproductively limited fish in aquaculture was the stimulus to apply these techniques to exotic introductions (18,22,23,25,40,41); some of these control measures were developed and practiced only after the introduction, as for grass carp (42). The effectiveness of these techniques was obviously compromised in these cases, but the debate was instrumental in stimulating research on the protocols. While most of the techniques have been most widely applied to grass carp, they have also been tested with several other species (52,53).

Reproductively limited fish can be developed by a number of techniques, with differing complexities, and thus the techniques involve various time constraints and/or degrees of security against possible reproduction. Direct induction of monosex or sterile fish can be applied after the specific details of the technique are developed for the species of concern. Although these direct methods are rapid, their disadvantages include the fact that the treatment must be applied to each individual and that innate biological variability prevents uniform attainment of 100% efficacy. On the other hand, if induction is used as a means to develop broodstock, then absolute efficacy is not necessary. The fish developed with the successful treatment will be the agents for production of the reproductively limited progeny. Only the offspring will be used in environmentally sensitive situations. In fact, reproductively limited progeny can be used in stocking, or reproductive control can be limited to the actual assessment.

The sex-determining mechanism is not important in direct steroid monosexing, unless the breeding approach is to be used, in which case it becomes pivotal to success. Sex determination is not a defined issue, particularly in tilapia, although it is generally considered to be monofactoral (5,54–59). Trombka and Avtalion (60) indicated a high degree of crossing over where sex genes may be transposed, and Muller-Belecke and Horstgen-Schwark (61) reported that two or more sex-determining factors may override the XX–XY mechanism. Sex determination was considered to be more predictable among cyprinids (42), but Komen et al. (62) identified a masculinizing gene in common carp that modifies sex determination in XX females. The following option chart summarizes categories of reproductively limited fish that can be developed and is based on a model in which sex determination is considered to be controlled by homogametic (XX) females and heterogametic (XY) males.

I. MONOSEX
- (1) Hybridization (tilapia)
- (2) Direct
  - (a) Sex reversal (hormone)
  - (b) Gynogenesis (all females)
- (3) Breeding
  - (a) Sex-reversed males (all females)
  - (b) Androgenesis (YY males)

II. STERILE
- (1) Most fish hybrids are fertile
- (2) Direct
  - (a) Triploidization (3N)[1]
  - (b) Tetraploidization (4N is fertile)
- (3) Breeding: Tetraploids (4N) × (2N) = 3N

Hybridization. Hybridization has been used in aquaculture and stocking programs for a number of reasons, such as to modify an organism for particular culture situations or to meet alternative sport fishery needs. Interspecific hybridization among fish within the context of fish culture has been reviewed broadly (63–66) and for coldwater fish (67,68), warmwater species (69,70), molluscs (71), and crustacea (72). Specific families have been the focus of more intensive investigations, including salmonids (73), ictalurids (74), cyprinids (75), cichlids (55,76), esocids (77), centrarchids (78), and percichtheids (79,80). The application for the latter three groups has been primarily in sport fisheries.

Monosexing for reproductive control was not the primary objective among the majority of interspecific hybrid crosses, but has been extremely valuable with reference to aquaculture, monosex, and sterile progeny. The triploid hybrid grass carp was a particular application in the United States (81,82), and monosexing of cichlids by hybridization was a major advancement in tilapia aquaculture. Hickling (83) discovered that hybridization between particular cichlid species yields all-male progeny. Without management of excessive recruitment, the culture of tilapia would be reduced to a low-quality, subsistence activity. However, when unwanted reproduction is controlled, tilapia are among the most desirable culture species. Interspecific hybridization of tilapia has been studied further, and numerous other crosses of interest have been identified (55,76,84–86). Eight interspecific crosses yield all-male, or nearly all-male, progeny (5). Hybridization was the primary means of monosexing for tilapia until hormone-induced sex reversal was developed, which today has become the primary means of reproductive management in most tilapia culture systems. However, hybridization is still a primary tool in Israeli aquaculture, although sex reversal is used in combination with the interspecific crosses (87,88).

Hormone-Induced Sex Reversal. Sex reversal through administration of exogenous steroids has been developed most extensively for tilapia culture. The assumptions of this technique are that (*1*) steroids (androgens and estrogens) mimic natural induction by genetic sex-determining factors to alter development of the phenotypic or gonadal sex; (*2*) the exogenous steroid must be efficacious, adequately concentrated, and efficiently delivered, so as to provide the physiological or pharmacological effect; (*3*) treatment must proceed during a critical period of gonadal differentiation; and (*4*) steroid-induced development of the gonadal sex does not spontaneously revert. While the details deviate for particular species or under certain circumstances, these generalized guidelines have resulted in the development of effective programs for sex control (25,36).

[1] The normal, or diploid, chromosome number is designated by 2N.

Studies of sex reversal in tilapia were initiated in the mid-1960s, underwent rapid experimental development in the 1970s, and attained commercialized application during the 1980s (41,87,89). Major control factors have included the selection of the steroid, the concentration of steroid, the mode of the steroid's delivery, treatment initiation, treatment duration, and treatment conditions (90–92). Given the selection of an efficacious steroid and the appropriate concentration for the delivery mode, one of the major considerations is the temporal relationship of the treatment. Under the assumption that a physiological or pharmacological level of the steroid must be administered throughout the period of gonadal differentiation, the length of that period must be determined either histologically or by way of a range of empirical treatments. Furthermore, factors that affect growth are considered pertinent to sex reversal, since physiological processes such as gonadal differentiation are altered by a balance between chronological age (time) and growth (size) (93). Therefore, the period of treatment must consider age and growth; this window of opportunity for treatment optimization has been conceptualized for tilapia (51) and expanded for common carp (93–95) and grass carp (96,97). The genetic basis for the development of phenotypic sex is determined at fertilization, which usually translates gonadal sex with fidelity, but because of the sexual bipotentiality of premeiotic germ cells, the gonadal sex can be exogenously influenced (25).

Functional sex reversal can be considered as a programmatic component, rather than a direct means of producing monosex fish. Postulated breeding programs for tilapia (51,98,99) have been demonstrated (59,100,101). All-female rainbow trout culture has been similarly developed through sex reversal and breeding and is being practiced on a commercial scale in Europe (102,103). Treatments for tilapia through oral delivery can be 100% effective, but frequently are somewhat lower, and treatments for grass carp via hormone implants have been over 90% effective. Thus, these protocols can be used in developing broodstock for breeding monosex progeny (42,51).

Chromosome Manipulations

Gynogenesis. Chromosome manipulation, including gynogenesis, androgenesis, and polyploid induction, provides numerous options for genetic selection, sterility induction, and sex control (25,37,40,41,104–107). Gynogenesis is the development of ova without paternal genetic contribution (see the entry "Gynogenesis"). Sex control for monosexing is dependent on the mechanism of sex determination and permits all-female production for homogametic female species through gynogenesis and, indirectly, all-male populations through androgenesis by producing YY males. In contrast, induced sterility by ploidy manipulation is based on odd chromosome sets (3N), and its effectiveness is not dependent on genetic sex. While our understanding of sex determination is far from complete, we can consider certain aspects as somewhat basic. Heteromorphic sex chromosomes in fish are rare, but functional heterogametic sex appears relatively well established, despite evidence for autosomal influence in tilapia (5,54,57–61,108) and common carp (62,95,109).

Induced gynogenesis and polyploidy are based on optimized manipulation of at least four variables: shock type (cold, hot, or pressure), intensity of shock, time of application postinsemination, and duration of shock (104,110,111). Shock must be applied either early (polar body), to retain the chromosome set usually lost with the second meiotic polar body, or late (endomitotic), to interfere with first mitosis. Further, improved standardization of treatment can result by referencing time of shock to biological age (tau), thereby incorporating temperature-affected changes into the rate of development (111,112). Artificial gynogenesis and polyploidization have been accomplished in tilapia (113,114), common carp (62,115–117), and Chinese carp, with no shock (42,118), heat shock (52,53,119), cold shock, and pressure shock (120,121).

Gynogenesis can be an end or a means. All-female progeny can be directly produced by gynogenesis, which was the basis of one of the first biological assessments using reproductively limited fish. Stanley (118,122,123) used gynogenesis to supply monosex grass carp for a large-scale study in Lake Conway, Florida. No shocking was used in the induction, and only about 45,000 diploids were produced from over 58 million eggs. While this effort was inefficient, the reproductive limitation was sufficient to overcome opposition for testing. Shock techniques would have greatly enhanced the efficiency, but were not well studied at the time. With the monosex breeding system for grass carp (42), the total number of monosex fish needed for the Lake Conway study were produced 10 times over by using a single XX male to fertilize eggs from only one female; unfortunately, this program of induction was developed a couple of years too late.

Gynogenesis also can be a valuable means of estimating an optimum treatment protocol for induction of polyploidy (22,107,124). Early-shock gynogenesis optimization provides a protocol for optimum triploid production, while a late-shock protocol will estimate the best treatment for tetraploidization. The evaluation of optimal treatment relationships is facilitated, since assessment is based only on the hatch of viable diploid larvae, whereas polyploidy (3N or 4N) must be verified by more involved methods. Haploids may survive to hatch, but they die before swim-up; therefore, a direct count of viable larvae indicates the best treatment. Several features are conveniently incorporated into most gynogenetic studies, to permit verification that diploids are gynogens and not progeny from normal fertilization. First, heterologous sperm are usually used. Second, these gametes are treated to inactivate the paternal DNA; this feature assumes thorough elimination of the male genome. Ultraviolet irradiation is the most convenient and safest means of sperm treatment. As a third level of assurance, a genetic marker is used, if available, during protocol development. For example, using a female with a homozygous recessive trait, such as the scattered-scale pattern in common carp, or one of the color mutations provides a visual confirmation of progeny with only maternal inheritance. Gynogenetic progeny of a light-color koi carp female lack pigmentation at hatching, while any diploids that might result from normal fertilization by a nonkoi male have dark pigmentation at hatching. Albinism is an excellent genetic marker and is available for several species, including the grass carp (125). The recessive marker is used in the parent for which the genome will be represented in the progeny—i.e., female albino for gynogenetic studies and male albino for androgenetic studies.

Triploidization. The induction of polyploidy is analogous to gynogenesis, except that untreated sperm from conspecifics is used for fertilization, and a genomic contribution, rather than a heterologous DNA-deactivated spermatozoa, is employed. Triploids are produced by early shock for polar-body retention, while tetraploids can be induced by interference with first mitosis by late shock. Evaluation of polyploid induction requires determining the ploidy by karyotyping, red-blood cell nuclear analysis, or quantitative DNA determination (126,127). Triploidization has tremendous potential in the production of sterile fishes for water resource management or for assessment studies. The level of gonadal development of triploid fish differs from that of diploids. In general, triploid fish have smaller gonads than diploids; males may have near-normal size testes, but with limited spermatogenesis, while the ovaries of 3N fish are poorly developed and typically have few ova (128).

Tetraploidization. Another means of producing triploids is through the initial induction of tetraploidy and the subsequent use of the 4N fish in a breeding program with diploids. Progeny from a 4N-by-2N cross all will be triploids. The induction of tetraploidy follows the protocol characterized by endomitotic gynogenesis. Normal fertilization is accomplished, and development proceeds, but first mitosis is interrupted by shocking. Tetraploid males and females are produced and develop as fertile adults; therefore, theoretically, a 4N line can be produced (104,107,110). Further, and perhaps more significantly, all-triploid progeny can be faithfully produced by breeding a tetraploid (male or female) with a diploid (female or male, respectively). Thus, production of tetraploids has its greatest potential in the development of broodstock for production of all-triploid progeny. Tetraploidy has been induced in relatively few fish—for example, various salmonids (129), channel catfish (130), common carp (62), grass carp (121), and tilapia (131,132). Chourrout et al. (129) and Thorgaard et al. (110) produced triploid rainbow trout by breeding tetraploids with diploids.

Androgenesis. Androgenesis generally can be considered as the reciprocal of gynogenesis: It is the production of progeny with only the paternal genome; the female genome must be eliminated in order to induce androgenesis. In this process, non-genome-bearing ova are activated with normal spermatozoa, and diploidization is accomplished by shocking prior to first mitosis. A late-shock gynogenesis protocol can be used to estimate the treatment to produce diploid androgenotes. However, elimination of the female genome is more difficult than in sperm treatment, since the egg is larger, and penetration of UV radiation is limited. Thus, while gynogenesis required the appropriate treatment of spermatozoa, androgenesis depends first on effective ova treatment. As the viability of eggs is reduced during storage, treatment must proceed soon after ovulation. Despite these complications, there are several advantages for ploidy manipulations that result in progeny with paternal inheritance. The expected production of progeny with a 1 : 1 sex ratio is based on segregation of X- and Y-bearing sperm. Diploid induction results in XX females and YY males. The latter will be functional as broodstock and produce only male (XY) offspring. Also, androgenesis has potential for restoration of endangered species, through the induction of diploid progeny from cryopreserved sperm.

Androgenesis and the resulting production and viability of YY male progeny have been verified in several coldwater fish (133–135). Androgenesis has been demonstrated in tilapia (136) and common carp (137–140) and all-male progeny have been produced by breeding YY males of common carp, loach, and sturgeon.

HISTORY AND STATUS OF AQUACULTURE WITH REGARD TO EXOTIC SPECIES

Fish culture is an ancient practice. It originated in China about 3,000 years ago and then developed in Europe somewhat later. In North America, fish culture can be traced to the mid- to late 19th century (141,142). Aquatic resources are an important source of animal protein for human nutrition; the worldwide commercial harvest of aquatic animals was about 97 million metric tons (MMT; 1 metric ton = 2,200 lb = 1 long ton) in 1991, with about 71% being directly used for human consumption (143). The potential world wide production from all sources is estimated to be 100–150 MMT, of which about 27 MMT is discarded bycatch. At present, the growth of capture fisheries is insufficient to maintain the needed food supply sustainably, in fact, the harvest has already entered into a period of decline (144,145).

Fish culture must meet the future growing needs for aquatic products, since the capture fishery supply is overexploited and the harvest is currently static at best, despite increased effort to expand it. Worldwide aquaculture production has steadily increased in recent years, at about 15% per annum; in 1992, 84% of production was in developing countries. Aquaculture production in 1994 was 18.5 MMT, 42% of which was from the production of carp (146). Much of the increasing production has been due to worldwide transplants of carp, tilapia, catfish, and salmonids into areas where they are now cultured (144). After centuries of development, animal husbandry has generally focused on four herbivorous mammals and four omnivorous birds, while in aquaculture, about 300 species are produced in large quantities. Yet, limited scientific data on indigenous species generally has discouraged the use of these species, in favor of accepting the more tested nonnative species that have well-developed culture systems. Two groups of fish—carp and tilapia—have been most important globally in fish culture (147). These fish are widely cultivated outside of their historic native ranges. The rate of exotic transfers has increased since about 1945, somewhat in conjunction with expanded developments in techniques for artificial propagation (148).

Exotic transfers have occurred worldwide in four general waves. Prior to 1900, movements were made primarily by salmonids; then common carp was disseminated in the early part of the 20th century. Tilapia were commonly transferred during the early part of the second half of the 20th century. Finally, the most recent trend has been the movement of Chinese carp, during the 1960s and 1970s (149). A total of 1,354 introductions of 237 species into 140 countries have been recorded; only 98 species have been introduced for aquaculture, however. Also, relatively few of the transferred species have been widely distributed; only 10 species have been introduced into more than 10 countries. Seventy-eight exotic species have been introduced for sport fishing, but this figure does not include intranational transplants. Over 298 (22%) of the introductions did not result in establishment of the species, and 246 (18%) of the introductions did not breed under natural conditions. A total of 321 (24%) of the introductions resulted in established populations, but only 89 had sufficient impact to cause serious concern, while the other 232 have been judged to be neutral or beneficial to the area (149).

The first half of the 1800s was an active period for fish transfer in the United States. During that time, fish were being cultured largely for restocking, due to depletion of natural populations from overfishing. Native fish were also being transplanted outside of their natural range by government agencies; for example, striped bass were moved to California and rainbow trout and Pacific salmon to the east. The common carp was the first recorded introduction into the United States, followed by the brown trout, *Salmo trutta*. During the same period of time, rainbow trout were being exported from the United States to other countries. Today, rainbow trout have been introduced into 44 countries, which is nearly as widespread a range as that of carp and tilapia. Also, many centrarchids have been exported to numerous countries, and, more recently, paddlefish, buffalofish, and channel catfish have been introduced into Russia and China, as well as 10 other countries. Thus, exotic fish transfers have not been a one-way street. Tilapia were imported into the United States in the 1950s and 1960s, and Chinese carp were brought in the early 1960s. The brown trout of Europe has been introduced into 33 states as a sportfish and is generally considered a positive addition; there is virtually no contemporary foodfish culture for this species in the United States (150). Forty-six species have been transferred into North America, 39 of which were considered established in the early 1980s (151); the United States now has at least 70 established exotic species (21), 42 of which were considered to have reproducing populations in the mid-1960s (150). Well over half of the these 42 species (23) are established in Florida, and all but 3 were related to the aquarium trade (152).

Foodfish

Exotic. World aquaculture production was about 13.9 MMT in 1992, 51% of which was finfish, 20% molluscs, and 4% crustaceans; the rest was plants and various other products (143,153). In 1994, the total production was estimated to be 18.5 MMT (146). Culture of the Pacific oyster, *Crassostrea gigas*, in the United States takes place primarily in the northwest, where the Pacific oyster was imported early in the 1900s, to supplement dwindling stocks of native species (154). The Pacific oyster makes up about 40% of the 20,000 MT of oysters marketed in the United States (155) and about 56% of the world production of oysters (156). The Pacific oyster was introduced to Britain from the United States in 1965, again because overexploitation had depleted the native fishery; today, a commercial industry has developed around the species. The most widely cultured freshwater shrimp in the world is the giant freshwater prawn, *Macrobrachium rosenbergii*. Most of its culture occurs in the Indopacific, but it was introduced into Hawaii in the mid-1960s and by 1988 was a $18.2 million crop (157).

Tilapia production in 1995 was 700,000 MT worldwide, and over half (473,000 MT) of the tilapia were farm raised. The natural distribution in Africa of aquaculturally important species has been described by Philippart and Ruwet (158). Despite introduction worldwide, tilapia production is only about 5% of the total production of farmed fishes, but because of the management of reproduction through hand sexing, hybridization, and sex reversal, they are becoming increasingly used (159). The Nile tilapia, *Oreochromis niloticus*, is the most popular species, making up 64% of the world production of tilapia, and *O. mossambicus* makes up about 10% of the worldwide production (160–162). Tilapia are cultured in about 70 countries; China has the greatest production (157,000 MT) and the Philippines the next greatest (63,000 MT). The United States ranks third in world production of tilapia (6,800 MT), with California culturing more than any other state.

Tilapia culture is one of the most rapidly growing foodfish components in the United States, but imports currently far exceed production on domestic farms. Although domestic production doubled between 1986 and 1992, only 20% of the demand for tilapia was satisfied; 24,000 MT were imported in 1995, making tilapia the third largest imported aquatic product, behind salmon and shrimp (163,164). About US$1 million worth of tilapia were imported from Honduras in 1996 alone. In 1989, about 3,000 MT of tilapia were produced in Jamaica, all of which were sex-reversed monosex males (165). Farmed tilapia are almost exclusively monosex, either through hybridization, hormone-induced sex reversal, or both. Israel produced about 15 million monosex fry in 1996 (166). While some concern has been raised over human consumption of steroid-treated tilapia and although the Food, and Drug Administration (FDA) is presently regulating the use of hormone treatment in the United States, there is no evidence of any human health hazard. Studies have demonstrated rapid posttreatment tissue clearance in fry, which are only a few grams (28 g = 1 oz) in weight, and that no residual hormones can be detected within one month of the termination of monosexing treatment (167,168).

Common carp was the first fish known to be cultured for food, the first fish to be transported outside its native range, and the first exotic species to be introduced into the United States. It is farmed in 60–70 countries, but in only 11 of them is production in excess of 10,000 MT/year (35). The total worldwide carp production in 1991 was in excess of 987,000 MT (6,169). Carp as a group is the most widely cultured fish, and tilapia culture is nearly as widespread. Exotic finfish production in the United States during the 1980s has been summarized by Shelton and Smitherman (18). During that period, chinese carp were cultured in eight states by more than 30 producers, but grass carp were used primarily to stock culture areas for aquatic vegetation control; tilapia were grown on 24 farms in 13 states for the food market. Current commercial production of exotic aquatic organisms in the United States includes shrimp, Pacific oyster, tilapia, common carp, and Chinese carp. Foodfish production for the latter two groups is mainly for ethnic markets, but a primary focus in the culture of common carp is the koi variety, for the ornamental market.

Transplants. Some critics of exotic species in aquaculture have suggested that native species should be used preferentially (170). In fact, the primary basis for U.S. aquaculture is endemic channel catfish and rainbow trout; however, much production is outside their native ranges. Production from the culture of exotic species in the United States is relatively little compared with that of the culture of native species. In 1985, over 200,000 MT of catfish were produced in 50,000 ha (123,500 acres) of water. During that year, about 2% of the channel catfish grown in the United States were cultured in California (171), where it is a transplanted species (21). About 30% of the rainbow trout farmed in the United States are outside their natural range, but the majority are cultured in the Snake River Valley (141), which is within their native range. The development of artificial propagation for salmonids in the mid-1800s initiated a period of movement, both as transplants in the United States and as introductions into other countries. Denmark received rainbow trout in 1870 and soon established a commercial facility for them. Rainbow trout are currently produced in 40–55 countries and are the major freshwater salmonid produced for food. In 1994, production of rainbow trout was 300,000 MT; Denmark led the production, with 41,000 MT, while production in the United States and Chile was about 15,000 MT each (172).

Another endemic species that has had major advances in its culture is the striped bass. Striped bass were first transplanted to the west coast of the United States between 1871 and 1881, and a naturalized sport fishery was established during that time. The species has been exploited in both a sport and a commercial capture fishery within its natural range on the east coast. In the 1940s, a landlocked population was discovered that was reproducing in the newly impounded Santee-Cooper reservoir system. Investigation of artificial propagation techniques followed, and in the early 1960s, a successful protocol was reported. During the period that the propagation techniques were being developed, the natural populations of striped bass on the Atlantic and Gulf coasts were declining. Curtailment of harvest created a deficit in the ongoing demand, and commercial farming developed rapidly. The hybrid striped bass × white bass, *Morone chrysops*, has emerged as the primary culture organism, favored over either parental species. U.S. production of the hybrid in 1989 was estimated at 450 MT and was projected to be about 1,350 MT by 1995; 152 MT were produced in California in 1987, an area not within the natural range of either species (173,174). Concern for maintaining genetic purity in natural waters has stimulated the production of sterile triploid hybrids for culture, so that no escapees will contaminate the gene pool of the wild population (175).

Exotic Foodfish in Other Countries. The culture of exotic fish outside the United States is an even larger proportion of the total production. Rainbow trout, as previously mentioned, were exported from the United States beginning in the latter half of the 19th century; today, rainbow trout is cultured in over 40 countries, and, with the exception of rainbow trout cultured in the United States, all represent the farming of an exotic species. Chile is second to Denmark in the production of farmed salmonids, including coho, *Oncorhynchus kisutch*, and Atlantic salmon, *Salmo salar* (176).

Tilapia are cultured worldwide and are exotic in all countries outside the African continent. Transplanting tilapia around the African continent has also been a common practice (76). The numerous introductions worldwide and subsequent redistributions have resulted in several genetic bottlenecks, with the resultant loss of genotypic variability (177,178). The total world production of tilapia is second only to that of carp.

In China, 109 species of fish have been introduced, and 16–18 of the 40 freshwater species cultivated in China are exotic, including rainbow trout, channel catfish, and paddlefish — all from the United States — and common carp and tilapia (179–181). In addition, 88 tropical exotic fish are cultured for the ornamental trade. Pond culture in China has been developed to a virtual art form, but it is also common for fish to be stocked into more extensive reservoir ranching systems (182,183). Ocean ranching of transplanted sturgeon in the Black Sea and the Caspian Sea has maintained the caviar industry for decades (184), although the fisheries are now in a rapid state of collapse, due to the breakup of the Soviet Union (185).

In the 1960s, India imported grass carp and silver carp, *Hypopthalmichthys molitrix*, from China; these plus common carp, in the polyculture of its native carp, is now called "composite culture" (186–188). This combination is used in traditional pond culture, but also to supplement reservoir capture fisheries (189). Introduction into Sri Lanka of *O. mossambicus* in the 1950s and *O. niloticus* in the 1970s has also been the basis of a major capture fishery in reservoirs (190). Balayut (191) has discussed some of the failures to establish fisheries by stocking exotics in reservoirs of southeast Asia.

Israel introduced common carp from Europe in the 1930s. Common carp was the basis for fish culture until tilapia was introduced in the 1950s, which developed into polyculture by the 1960s (192). The production of common carp in Israel has been stable over the past couple of decades, but the production of tilapia has been increasing; in the mid-1990s, the amounts produced of the two species were nearly equivalent (193,194). *Oreochromis aureus* is native to the Jordan Valley, but *O. niloticus* is exotic, and about 60% of the tilapia cultured are hybrids between those species (166). In 1996, 15,000 MT of five groups of freshwater finfish were produced in Israel [common and Chinese carp (38%), tilapia (45%), rainbow trout, and hybrid striped bass]; all of the species are exotic (195). The total production of farmed fish in Israel occurs almost exclusively on *kibbutzim*; the 55 farms in Israel, with about 3,000 ha of water area, had an average annual yield of 4.5 MT/ha (1 ha = 2.47a; 1 MT/ha = 1.12 T/a). Foodfish production has intensified through polyculture, improved diets, and aeration; the yield has increased from about 2 MT/ha in the 1950s to an average of about 5 MT/ha in the 1990s (195). Annual fish consumption in Israel is about 10 kg/person (22 lb/person), for an annual consumption of about 60,000 MT (196). Approximately 5,000 MT come from capture fisheries in marine and freshwaters, and about 15,000 MT (25%) is from culture; however, 40,000 MT (67%) must still be imported.

Sportfish

Sport fishing in the United States is important recreationally and economically; 50 million Americans fish each year and generate US$69 billion in economic output (197). Stocking has been an important component of sport fishery management, and fish culture is the foundation that supports the entire infrastructure: Stocking programs would cease without artificial propagation and culture. In the early days of fish culture in the United States, hundreds of millions of fry were produced, and the public accepted stocking as the answer to management.

Exotic fish have been a significant, although comparatively small, segment of the aforementioned stocking programs. Most of the exotic species that have been used for sport fishing in the United States have been experimental, and virtually none have developed into lasting fisheries (198). On the other hand, some of the fisheries based on transplants have been spectacularly successful. The vast majority of stocked sportfish are native, but not necessarily to the area to which they have been released. Thus, introductions through transplanting endemic species have been by far the primary focus of culture for sport fishing. Supplemental stocking of native species has proven largely ineffectual, but transplanting native fish into new areas has produced good results, as well as disastrous ones. The Colorado River reservoirs provide a particularly poignant example of the latter. Rainbow trout, red shiner (*Cyprinella lutrensis*), threadfin shad, (*Dorosoma petenense*), channel catfish, striped bass, bluegill (*Lepomis macrochirus*), largemouth bass, smallmouth bass, (*Micropterus dolomieu*), black crappie, (*Pomoxis nigromaculatus*), walleye (*Stizostedion vitreum*), and yellow perch, (*Perca flavescens*) are among the fish that have been transplanted outside their native ranges into this system. The positive outcomes are generally well known, while the disastrous consequences of these actions may be generally less familiar, even though they have been effectively articulated (199). The closure of the Hoover Dam in 1935 initiated drastic changes in habitat that affected the fish of the Colorado River; however, the introduction of the aforementioned array of predatory and prey species into the newly created reservoirs further limited the survival of native fish and subsequently complicated recovery efforts for the unique species that were recognized as near extinction.

More than 206 species in the United States occupy waters beyond their native range (21), 75% of which were introduced as sportfish. Thirty-six percent of the states have fewer native than nonnative sportfish species; on the average, nonnative fish make up about 38% of the fish that go into state recreational fisheries (200). Forty-nine of the 50 states use nonnative sportfish in their management programs. California epitomizes the significance of transplanted fish in sport fishery management: Most of the important North-American sport species are not native to California, including centrarchids, ictalurids, and percichtheids. Only 10 inland species native to California are considered sportfish; during the past 125 years, 30 nonnative fishes have been introduced for recreational purposes (201). Fishing for nonnative species has supplied between 40 and 75% of the angling effort over the past 50 years.

Coho and chinook salmon, *Oncorhynchus tshawytscha*, were introduced into Lake Michigan after the decline of the lake trout, *Salvelinus namaycush*, fishery, and sport fisheries for the two types of salmon were established. In some warmwater reservoirs, rainbow trout have supported two-story fisheries far outside of the trout's natural range, both geographically and climatologically, and unique tailwater fisheries also have been maintained. The success of the striped bass in this regard is legendary. Dispersal of striped bass into 456 inland reservoirs in 36 states has established viable sport fisheries. Hybrid culture was developed simultaneously with these fisheries programs, and hybrids were introduced into 264 reservoirs (47,141); about 2.3 million ha (5.7 million acres) have been stocked with hybrids (79). These stocking programs have been an unprecedented success. It is significant that in all but 10 reservoirs, the fisheries must be maintained by annual stocking from hatchery production. Each state is responsible for perpetuating its own stocking program and must cultivate the expertise and maintain the facilities to propagate striped bass and/or hybrids. One of the primary reasons for the success of the program has been the strong support of governmental agencies and the dissemination of developed technology through production manuals (48). The programs represent significant transplanting activity and total commitment to the culture for continuance.

A component that is vital to sport fisheries, but which is often overlooked in the context of fish culture, is the production of baitfish. About 11,000 MT of baitfish were cultured in the mid-1980s, ranking third in value, behind catfish and trout (141). The annual farm-level sales for baitfish were about US$71 million in the mid-1980s (202). While this culture activity is supportive of sport fisheries, the use of live baitfish itself can be a source of transplanting. On the downside, bait bucket introductions have been indicated for 58 species transplants (21).

Resource Management

Introductions of fish as agents for biological control have not been great in terms of the number of species that have been introduced, but perhaps no other activity has elicited such conflict among various stakeholders. The National Academy of Science (NAS) has proposed that biological control become the primary pest-control method in the United States (203). Biological control is considered to be a more environmentally friendly alternative than chemical controls. However, most of the introductions have involved exotics, and proposals for their use in varying degrees to open waters have resulted in extreme opposition. Biological control of nuisance aquatic plants has been the primary target, although other aquatic management aspects have also been considered.

Several species of tilapias were introduced for plant control, most notably *O. mossambicus* and *Tilapia zillii* (50,204). Those two species have been effective in some situations, but they are infrequently used today for this purpose. The grass carp is the primary species used for control of nuisance aquatic plants. Its introduction in 1963 for investigative purposes was soon followed by its employment for pest control. A special session at the 1977 American Fisheries Society (AFS) meeting discussed the status of the grass carp in the United States (205). The AFS sponsored another, more general symposium in 1985, called "Strategies for Reducing Risks from Introduction of Aquatic Organisms," the contributions of which were published in a 1996 issue of *Fisheries*. The cover of the issue depicted three exotic fish and the movie title "The Good, the Bad, and the Ugly." This expression and the perspectives in the papers illustrate the polarity that had developed over the use of exotic fish for pest control. The proponents and opponents of this method are probably no closer to meeting at some midground today than in the 1960s and if anything, the chasm is widening. Several symposia have been convened in the intervening years, and the result to date is the creation of a set of literature on grass carp that is probably unrivaled by the literature on any other species. This information is best accessed by starting with the biological synopsis (206). For more details on the biology, culture, and efficacy of grass carp as a biocontrol, consult other major contributions (8,207–209). The single most significant outcome of this exotic introduction was the stimulus to examine reproductive controls.

Original stockings of diploid grass carp involved about 40 states in the United States by 1972. During the 1970s, controversy over potential natural reproduction stimulated the development of techniques to produce monosex grass carp (42,118) and a triploid hybrid (female grass carp × male bighead carp, *Aristichthys nobilis*) (210), and then, in 1982, an Arkansas fish farmer produced triploid grass carp. Direct induction of triploidy has been accomplished for various cold- and warmwater species, including salmonids, catfish, cyprinids, and cichlids, by thermal and/or pressure shock (25,32,113).

In 1984, the U.S. Fish and Wildlife Service issued an opinion that female triploid grass carp are functionally sterile and that sperm from triploids are probably nonfunctional (211). The triploid reproductive potential is characterized by extremely low development of viable spermatozoa, low fertilization success, and extremely low survival of embryos (212,213). The production of triploid grass carp in the United States has greatly expanded in the last five years to meet the demands for biological control of nuisance aquatic plants; triploids are sanctioned by most state governments and are considered ecologically safe (28). Stocking has encompassed farm or watershed ponds as well as large water bodies [i.e., >50,000 ha (123,500 acres)], with one-time stockings of over 400,000 triploids. Verification of triploidy is necessary, since some diploids are also produced via direct induction, and some states permit only triploids to be stocked. Consequently, each individual fish is tested by the producer, then independently verified through subsampling before shipment (214), and a subsample is usually rechecked at the destination. The system of verification, certification, and use of the triploids has been described (215). Despite the problem of diploid contamination and the need to cull the progeny, direct 3N production of grass carp developed commercially in the United States following the shortfall of expectations for the spontaneous triploid–hybrid grass carp (214,215). Most triploid grass carp used in the United States are produced in Arkansas, where the direct value to fish farmers is about US$1 million annually (216). Optimization of triploid induction has been studied by scientists (119,120), although the specific protocols used by commercial producers are proprietary information.

The collection of eggs and larvae of grass carp in open systems since the 1980s (28,217,218) has documented that natural reproduction is occurring in North America as predicted (219); however, it is important to emphasize that this spawning is related to the use of diploid fish in open systems for nearly two decades before reproductively limited stocks were produced. This fact stresses the need to use reproductive controls before testing or stocking exotics; had this process been undertaken with grass carp prior to their release, naturalization would not have occurred in U.S. waters.

Chemical and mechanical controls are more than twice, and up to 20 times, more expensive than biological controls (209). Further, the environmental impacts attributable to grass carp would have occurred even if the vegetation were removed mechanically, since the primary biological effect is due to the removal of cover. Stocking for nuisance aquatic plant control is related to weed biomass and type, and to the growth of fish. In a large-scale stocking test in Lake Conroe, Texas, triploids were stocked at 74/ha (30/acre) of vegetated area. Submersed vegetation was eliminated in two years, and the sport fishery changed, as would be expected, to less cover-dependent species and a more open-water fishery. A thorough analysis of weed control effectiveness and reported impacts for this case has been performed (28,209).

In addition to grass carp, other Chinese carp have been considered for water quality management. For example, silver carp and bighead carp have been investigated for use in systems to manage plankton, as reported by various workers (14,28,220–224). Black (snail) carp, (*Mylopharyngodon piceus*) introduction has been erroneously linked to importation for control of the zebra mussel (*Dreissena polymorpha*) (225). In fact, the black carp was imported in the early 1980s as a biocontrol for snails (M. Freeze, Keo Fish Farm, AR, personal communication, 30 September, 1994), as certain snails are intermediate hosts for some parasites of wading birds; the parasites subsequently infest fish muscles as yellow or white grubs. Some processors have rejected catfish because of heavy grub infestations. Triploid black carp stocked at 5–10/ha (2–4/acre) successfully eliminated yellow grubs from a North Carolina hybrid striped bass culture (226). Commercial producers have already developed induction techniques for producing sterile triploid black carp, and gynogenesis has also been reported for this species (53).

The black carp is the first fish species to be considered under the Generic Nonindigenous Aquatic Organism Risk Analysis Review Process, which was authorized by the U.S. Congress in the Nonindigenous Aquatic Nuisance Control and Prevention Act of 1990 (15). The recommendation of this process was that the black carp should not be used for controlling zebra mussels in open waters unless research demonstrated the effectiveness of the method, but its use as a biocontrol of yellow grubs in fish farm facilities should be allowed (227). The primary concern with the use of the black carp in open systems in the United States is the vulnerability of endangered molluscs. Black carp also have been suggested as a potential biocontrol of snails in areas of bilharzia endemicity, but only preliminary studies on this use have been completed (228). Again, consideration in this case was also with the proviso that reproductively limited fish be developed prior to use.

Ornamental Fish

About 6,000 freshwater species are in the world trade for ornamental fish, and 59 of the most popular species are currently commercially reared in the United States (229). Ornamental fish culture is the most profitable aquacultural enterprise. Worldwide, it is farm valued at about US$400 million per year. A very positive aspect involving conservation is that fish from breeding farms are replacing fish from wild-caught sources (144). Ornamental fish production is among the leading cash crops in U.S. aquaculture, with a retail value of approximately US$1 billion (230). There were 193 tropical fish growers in Florida in 1987, with sales of about US$21 million (50).

Twenty-eight species of ornamental fish have escaped and established breeding populations in Florida (229,231) 10 of which have since been extirpated (152). While ornamental fish are often considered tropical species, considerable culture is directed toward two temperate species: goldfish and the fancy carp, or koi ("the living rainbows"). The estimated farm-level value of goldfish alone is US$10–20 million annually, a figure that includes aquarium fish as well as feeder and bait fish (232). More recently, backyard water gardens have increased in popularity with the growing interest in koi and fancy goldfish. Local garden clubs have further enhanced participation in this leisure activity. In Israel, production of ornamental fish does not appear in statistics for foodfish, but this component of fish culture is important economically. MagNoy, a marketing cooperative among four kibbutzim, exports goldfish and koi and accounts for about 30% of the European market, or about US$20 million per year.

ENVIRONMENTAL IMPACT

Many forms of human activity, including food production, whether terrestrial or aquatic, affect the environment (19). Human perturbation has had the effect that few freshwater habitats are undisturbed (21). This scenario is forcefully summarized by Warren and Burr as follows:

> ...nearly all major streams in the East are dammed and regulated. Groundwaters across much of the country are being pumped at rates exceeding those at which aquifers can be recharged. Problems of point source and general pollution continue in the Great Lakes, Midwest, South and Northeast. Acid rain is destroying northern lakes, and poor land-use practices in the agricultural states of the Midwest and South have doubled sedimentation rates in the past 20 years. Drainage of wetlands, primarily for agriculture and urban expansion, throughout the eastern United States continues at unabated rates in many regions. Predation and competition for resources with nonnative species are increasingly recognized problems in the West and subtropical states such as Florida (233)

While this summary is appalling, it is not possible to completely avoid environmental damage or resource exploitation (19).

If we consider aquatic extinctions as a measure of environmental impact, then habitat alteration has been a contributing factor to the extinction of 73% of the 40 extirpated North American fish (234). It should be noted that one of the factors of aquatic ecosystem degeneration mentioned by Warren and Burr (233) was nonnative species, which has also been identified as contributing factor in 68% of the species extirpations (234). An important point to remember, however, is that nonnative fish have been considered as a *contributing factor to*, not the *cause of*, extinctions. As has been described for the case of aquarium fish in Florida, the impacts of exotic fish are more symptomatic of environmental problems than they are the primary cause of the problems (235). There are numerous examples in which invading species have become established and integrated into local biota without an associated extinction of native species (236). Warren and Burr further state that few successful invaders have become pests or have caused major changes in the receiving system. Most invasions do not cause extirpation of organisms directly. However, predicting whether a species will be benign or become a pest is not possible; it is in a large part conjecture (50,170) or speculation (20). However, paucity of ecological damage cannot be considered proof of general safety (203).

Courtenay (21) has stated that any introduction is a risk and will result in impacts on native biota; however, an impact may be negligible or it may be major, and predicting its magnitude is nearly impossible. Seventy percent of 31 studies conducted by Ross (237) have documented a decline in the population of native fish following the introduction of an exotic or transplanted species, but most introductions lack adequate prestocking data. Introduced forms impact native ecosystems through (*1*) habitat alteration, (*2*) introduction of disease or parasites, (*3*) hybridization with native species, (*4*) trophic alteration, and (*5*) spatial alteration (238). Many examples of fish-introduction "cause and effect" may be, in reality, correlative events related to habitat perturbation or overexploitation of a fishery, rather than causation. For example, rainbow trout introduction to Lake Titicaca, Peru, has been considered negative (239), but concurrent changes were not adequately documented and may have been contributing causal factors as well (143). Also, regarding the environmental effects attributed to fish introductions into Colorado River impoundments (199,240), while the numerous introductions may have been misguided, the effects on the native fauna were certainly first precipitated by the ecological shift from a riverine habitat to the semilacustrine system of reservoirs.

Despite the potential of farmed fish to escape, hatchery-reared individuals generally have lower survival rates and a lesser ability to compete with native species in open systems. Hybridization potential is a greater risk with transplanted native species than with introduced exotic species (143). Statistics for worldwide introductions suggest a very low success rate for naturalization and an even lower level of pest status for introduced species (149,241).

Why introduce a species and thus run the risk of adversity? There are several reasons. First, humans have created numerous altered habitats (e.g., reservoirs), and it is reasonable to use exotic or nonnative species in such systems (143). Second, in aquaculture, introducing or transplanting nonnative species is justifiable, because relatively few of the vast number of species with potential for aquaculture have a developed biological technology adequate to assure success in culture. Also, aquaculture is not a recreation, nor an avocation; it is a business with financial success at risk. Without the introduction of transplanted channel catfish, striped bass, and exotic tilapias, what would be the aquaculture profile in California? Also in California, without the introduction of centrarchids, what would be the sport fishery?

Much of the published criticism of exotic introductions has not been repeated here; such sources are harbingers of the drastic mistakes made by introductions and repeatedly warn of the likelihood of a "Frankenstein effect." Yet, most of those sources indicate that examples of damage or pest status are not common, and are even rare, considering worldwide statistics. While the total number of species that have become established is relatively small, the number that has become problematic is much smaller. Opposition to stocking exotics is usually on the basis of their potential to cause environmental degradation or to compete with native fauna (238). These concerns can be addressed for nonnative fishes using reproductively limited populations, so as to retain the option of terminating continued population support if further stocking is contraindicated.

CONTROL MEASURES AND RISK ASSESSMENT

Many species transfers are related to utilitarian motivation. For fish, such motivations may include sport fishing interests; biological control applications; aquaculture considerations; or simply aesthetics, as with ornamentals (242). The potential environmental effects of introductions are not altered by whether access to natural aquatic habitats is by way of purposeful stocking or through escape from controlled environments, since the impact can be similar. "Assessment" implies an *a posteriori* situation, whereas "risk assessment" attempts to predict or evaluate the likelihood of damage (243). Therefore the evaluation process becomes basically a modeling exercise for risk assessments. If we subscribe to the concept of "environmental roulette" proposed by opponents of exotic fish, and if the risk assessment process takes into account previous statistics on introductions, the odds are greatly in favor of predicting that an introduction will cause no adverse impact. Obviously, however, the process of risk assessment cannot depend simply on the odds. Assessment further implies that a commitment for an introduction may have been made, or that an unintentional introduction has occurred. In either case, the evaluation process may be affected by the potential for serious environmental or professional repercussions. However, this need not be the case if a system of reproductive limitation is developed before the introduction takes place.

Species transplants can have adverse ecological effects, and it may be impossible to eliminate the species from the area. However, if the introduction does not result in naturalization—that is, reproduction and recruitment are prevented—then any adverse impact will be temporary. Assessment using reproductively limited fish is a means of evaluating the impact of an introduction without making an irrevocable commitment. Fish with reproductive limitations cannot spawn, even if a suitable habitat is present: They are sterile or monosex. Both of the restrictions can be induced directly, or may be developed through a breeding program that produces only offspring with the reproductive limitation, as described in previous sections. The direct means of inducing monosex or sterile fish can be developed quickly and used in a preliminary assessment, but with some constraints, while a breeding system requires more time to develop, but can be perpetuated and accomplished on a commercial scale.

The outcome of introductions is not precisely predictable. If none of the introduced individuals survive, there should be no impact on the environment, while on the other hand, naturalization can result if reproduction and recruitment occur. Colonization may involve competition with natives or various other interactions within the local community, some of which can be catastrophic. Thus, purposeful animal transfers are fraught with uncertainties, and consequently, anticipation of the myriad of potential biotic interactions is nearly impossible.

Introduction of genetically modified organisms (GMOs) has been considered in the same environmental context as introduction of an exotic or a transplant. Depending on the degree of change and the phenotypic effect in a GMO, the same range of potential environmental impacts might be anticipated as would be hypothesized for exotic species

(244,245). However, in the US guidelines for engineering GMOs, ploidy manipulated fish are considered to be GMOs, which contradicts the use of the proposed system (246). On the other hand, Thorgaard and Allen (247) indicate that the use of sterile organisms has little negative impact on native populations; therefore, using monosex or triploid populations would seem to be logical means of evaluating the impact of GMOs.

Radonski and Loftus (17) and Horak (248) have indicated that the U.S. Nonindigenous Aquatic Nuisance Prevention and Control Act of 1990 generally ignores or minimizes the benefits of introductions, while emphasizing or exagerating the risks, which is typical of much of the literature on introductions. A program that can functionally encompass the assessment of the introduction of either a GMO or an exotic fish would be useful. Presently, exotic-fish introductions are being considered under the umbrella of various guidelines that are recommended for use, most of which incorporate a series of decision boxes (249) or codes of practice (250,251). These processess are literature-review driven and also may include a survey of experts opinions, called "opinionnaires." The ICES (International Council for the Exploration of the Sea) guidelines are perceived as idealistic and restrictive, and they do not bind the signatories; therefore, they might even be ignored (15,252). The U.S. Nonindigenous Aquatic Nuisance Prevention and Control Act of 1990 (15) provides for a slightly more proactive risk assessment process and has been used in a first-case consideration for a fish, the black carp, by an aquatic nuisance species task force (227). The process is modeled after experience in agriculture pest assessment, which still consists primarily of basically literature review and panel hearings.

Selecting a course of action without an option to reverse the decision is not logical, but essentially has been the modus operandi for introducing species. Accidental escapes will continue to happen, but this eventuality, as well as purposeful introductions, can be approached pragmatically, scientifically, and realistically. In the final analysis, an introduction can be assessed only within the actual habitat or community under consideration. The binary decision either to forgo making the introduction or to try it out does not need to be such a decisive dichotomy; assessment can be done in an ecologically safe manner. Since colonization depends on reproduction and recruitment, evaluation of a fish introduction (transplant, exotic, or GMO) can be made in a real context without making a permanent, irretrievable commitment. If tests are conducted using reproductively limited fish, then effects can be documented without concern for recruitment. Further, if the reproductively limited fish are developed outside of the United States—for example, in the country of their origin—then the threat of accidental escape of nonlimited fish is also eliminated.

The level of security associated with various approaches to sex control can be considered in the context of acceptable risk. Stocking into a habitat that lacks some essential environmental element for spawning may be considered to provide adequate security; however, this safeguard is minimal, considering the capability of the fish to emigrate and gain access to conditions that meet reproductive requirements. This fallacious premise was posed in the original grass carp introduction into the United States. Additionally, recruitment may be limited by predation or other factors (219), but such factors should not be perceived as primary security measures on which to base an introduction. On the other hand, we can consider monosexing or sterilization as primary means of reproductive control (38,39,123).

Stocking only one sex can provide considerable security against reproduction, but the monosexing effort must be totally effective. Monosexing can be accomplished through a combination of gynogenesis, sex reversal and breeding for the production of single-sex progeny, or through androgenesis and subsequent breeding for the production of (all-male) progeny. Direct steroid induction by sex reversal is not recommended as a reproductive limitation for an assessment process, only as a means of developing a breeding program. In stocking monosex fish, we must consider that no prior introduction of mixed-sex populations has occurred. Monosexing may be appropriate for exotic fish, but is of limited value for transplanting native species into a new watershed, depending on the resident fauna, as closely related species with similar reproductive biology present the opportunity for hybridization. However, considering the selective disadvantage for successful reproduction of hybrids in a natural community (64), this method may still have some value.

Genetic-based sterility involves polyploidization, specifically triploidy, which has been achieved through some intergeneric hybridizations, resulting in triploid hybrids, or by induced chromosome manipulation (123,253). Hybrid reproductive potential may vary from complete sterility in triploids to apparently normal fertility (64). However, there may still be cause for concern regarding the basis of differential effects between sexes: Triploid females usually have poorly developed ovaries, while males have more extensive morphological development, but still with general gametic inviability (212,254,255). The specific mechanism for producing triploidy might occur by direct induction, as has been and still is being applied for grass carp, or through a breeding system (tetraploid × diploids) developed with polyploid induction.

We must consider introduction protocol when determining the appropriate time for applying a selected technique. The codes of practice and guidelines for the transfer of exotic fish (249–251) all have one basic shortcoming: No provision is made for actually assessing the effects of the transfer within the habitat of consideration. However, this deficiency can be rectified by incorporating the use of reproductive limitations in the process. To be most effective, the sex control mechanism should be developed prior to the introduction. Then only reproductively limited fish need be imported for assessment, or alternatively, broodstock that produce the reproductively limited progeny can be developed overseas and imported for breeding. At slightly lower levels of security, reproductive limitations can be induced directly in fish to be tested. Techniques for induction of triploidy were not developed for use with grass carp until after their introduction to the United States, but the methods have been successfully applied to black carp so as to reduce the risk associated with any proposed introduction. This may avert some of the controversy that occurred with the grass carp. Methods to produce reproductively limited fish can be used to test introductions, as they have been developed sufficiently to be considered operational and available for incorporation into a functional assessment protocol. Additional techniques still await refinement or development, after which they can be added to our repertoire of options. Tetraploid induction is potentially one of the most logical systems to be refined in the future. However, this process will require several years to optimize, as will efforts to develop processes for androgenesis.

BIBLIOGRAPHY

1. B. Morton, in F.M. D'Itri, ed., *Zebra Mussels and Aquatic Nuisance Species*, Ann Arbor Press, Chelsea, MI, 1997, pp. 1–54.
2. G.W. Wohlfarth, in I.L. Mason, ed., *Evolution of Domesticated Animals*, Longman, London, 1984, pp. 375–380.
3. E.K. Balon, *Aquaculture* **129**, 3–48 (1995).
4. G. Hulata, *Aquaculture* **129**, 143–155 (1995).
5. G.W. Wohlfarth and G. Hulata, in M. Shilo and S. Sarig, eds., *Fish Culture in Warm Water Systems: Problems and Trends*, CRC Press, Boca Raton, FL, 1989, pp. 21–63.
6. R. Billard, in C.E. Nash and A.J. Novotny, eds., *Production of Aquatic Animals: Fishes*, Volume C8, Elsevier, Amsterdam, 1995, pp. 21–55.
7. M. Huet, *Textbook of Fish Culture: Breeding and Cultivation of Fish*, 2nd ed., Fishing News Books, Ltd., Surrey, England, 1986.
8. J.E. Bardach, J.H. Ryther, and W.O. McLarney, *Aquaculture: The Farming and Husbandry of Freshwater and Marine Organisms*, Wiley Interscience, New York, 1972.
9. R.R. Stickney, *Principles of Aquaculture*, John Wiley & Sons, New York, 1994.
10. T.V.R. Pillay, *Aquaculture: Principles and Practices*, Fishing News Books, London, 1993.
11. N.R. Bromage and R.J. Roberts, eds., *Broodstock Management and Egg and Larval Quality*, Blackwell Science, London, 1995.
12. P.L. Shafland and W.M. Lewis, *Fisheries* **9**(4), 17–18 (1979).
13. Aquatic Nuisance Species Task Force (ANSTF), *Findings, Conclusions and Recommendations of the Intentional Introductions Policy Review: Report to Congress*, 1994.
14. E.A. Lachner, C.R. Robins, and W.R. Courtenay, Jr., *Smithson. Contrib. Zool.* **59**, 1–29 (1970).
15. Nonindigenous Aquatic Nuisance Prevention and Control Act (NANP & CA), U.S. Public Law, 1990, pp. 101–646.
16. Executive Order 11987, *Exotic Organisms*, Fed. Reg. **42**(101), (1977).
17. G.C. Radonski and A.J. Loftus, in H.L. Schramm and R.G. Piper, eds., *Uses and Effects of Cultured Fishes in Aquatic Ecosystems*, American Fisheries Society Symposium 15, Bethesda, MD, 1995, pp. 1–4.
18. W.L. Shelton and R.O. Smitherman, in W.R. Courtenay and J.R. Stauffer, eds., *Distribution, Biology, and Management of Exotic Fishes*, John Hopkins University Press, Baltimore, MD, 1984, pp. 262–301.
19. T.V.R. Pillay, *Aquaculture and the Environment*, John Wiley & Sons, New York, 1992.
20. C.C. Kohler, in A. Rosenfield and R. Mann, eds., *Dispersal of Living Organisms into Aquatic Ecosystems*, Maryland Sea Grant College Program, College Park, MD, 1992, pp. 393–404.
21. W.R. Courtenay, Jr., in H.L. Schramm and R.G. Piper, eds., *Uses and Effects of Cultured Fishes in Aquatic Ecosystems*, American Fisheries Society Symposium 15, Bethesda, MD, 1995, pp. 413–424.
22. W.L. Shelton, in R.H. Stroud, ed., *Fish Culture in Fisheries Managment*, American Fisheries Society, Bethesda, MD, 1986, pp. 427–434.
23. W.L. Shelton, *Fisheries* **11**(2), 16–19 (1986).
24. Y. Zohar, in M. Shilo and S. Sarig, eds., *Fish Culture in Warm Water Systems: Problems and Trends*, CRC Press, Boca Raton, FL. 1989, pp. 65–119.
25. W.L. Shelton, *CRC Rev. Aquat. Sci.* **1**, 497–535 (1989).
26. R. Patino, *Prog. Fish-Cult.* **57**, 118–128 (1997).
27. Z. Yaron and Y. Zohar, in J.F. Muir and R.J. Roberts, eds., *Recent Advances in Aquaculture*, Volume IV, Blackwell Science, London, 1993, pp. 3–10.
28. K. Opuszynski and J.V. Shireman, *Herbivorous Fishes: Culture and Use for Weed Management*, CRC Press, Boca Raton, FL, 1995.
29. J.P. McVey, *Handbook of Mariculture: Crustacean Aquaculture*, Volume 1, CRC Press, Boca Raton, FL, 1983.
30. J.P. McVey, *Handbook of Mariculture: Finfish Aquaculture*, Volume 2, CRC Press, Boca Raton, FL, 1991.
31. G. Barnabe and R. Billard, eds., *L'Aquaculture du Bar et des Sparids*, INRA, Paris, 1984.
32. E.M. Donaldson, *Anim. Reprod. Sci.* **42**, 381–392 (1996).
33. G.W. Wohlfarth, in R. Billard and J. Marcel, eds., *Aquaculture of Cyprinids*, INRA, Paris, 1986, pp. 195–208.
34. Z. Yaron, *Aquaculture* **129**, 49–73 (1995).
35. S. Rothbard and Z. Yaron, in N.R. Bromage and R.J. Roberts, eds., *Broodstock Management and Eggs and Larval Quality*, Blackwell Scientific, London, 1995, pp. 321–352.
36. R.A. Dunham, *CRC Rev. Aquat. Sci.* **2**, 1–17 (1990).
37. G.H. Thorgaard, in N.R. Bromage and R.J. Roberts, eds., *Broodstock Management and Eggs and Larval Quality*, Blackwell Scientific, London, 1995, pp. 76–117.
38. E.M. Donaldson and G.A. Hunter, *Can. J. Fish. Aquat. Sci.* **39**, 99–110 (1982).
39. F. Yamazaki, *Aquaculture* **33**, 329–354 (1983).
40. W.L. Shelton, in R. Billard and J. Marcel, eds., *Aquaculture of Cyprinids*, INRA, Paris, 1986, pp. 179–194.
41. W.L. Shelton, in K. Tiews, ed., *Selection, Hybridization and Genetic Engineering in Aquaculture*, Volume 2, Heenemann, Berlin, 1987, pp. 175–194.
42. W.L. Shelton, *Aquaculture* **57**, 311–319 (1986).
43. C.M. Breder and D.E. Rosen, *Modes of Reproduction in Fishes*, Natural History Press, Garden City, NJ, 1966.
44. G.W. Potts and R.J. Wooton, eds., *Fish Reproduction: Strategies and Tactics*, Academic Press, New York, 1984.
45. A.D. Munro, A.P. Scott, and T.J. Lam, eds., *Reproductive Seasonality in Teleosts: Environmental Influences*, CRC Press, Boca Raton, FL, 1990.
46. Y. Zohar, *Bull. Natl. Res. Instit. Aquacult.*, Suppl. **2**, 43–48 (1996).
47. R.E. Stevens, in J.P. McCraren, ed., *The Aquaculture of Striped Bass*, University Maryland Sea Grant, College Park, MD, 1984, pp. 1–15.

48. R.M. Harrell, J.H. Kerby, and R.V. Minton, eds., *Culture and Propagation of Striped Bass and Its Hybrids*, American Fisheries Society, Bethesda, MD, 1990.
49. D.K. Whitehurst and R.E. Stevens, in R.M. Harrell, J.H. Kerby, and R.V. Minton, eds., *Culture of Striped Bass and Its Hybrids*, American Fisheries Society, Bethesda, MD, 1990, pp. 1–5.
50. J.P. Clugston, *CRC Rev. Aquat. Sci.* **2**, 481–489 (1990).
51. W.L. Shelton, K.D. Hopkins, and G.L. Jensen, in R.O. Smitherman, W.L. Shelton, and J.H. Grover, eds., *Culture of Exotic Fishes Symposium Proceedings*, American Fisheries Society, Bethesda, MD, 1978, pp. 10–33.
52. J.A. Mirza and W.L. Shelton, *Aquaculture* **68**, 1–14 (1988).
53. S. Rothbard and W.L. Shelton, *Aquacult. Internat.* **5**, 51–64 (1997).
54. W.L. Shelton, F.H. Meriwether, K.J. Semmens, and W.E. Calhoun, in L. Fishelson and Z. Yaron, comp. *International Symposium on Tilapia in Aquaculture*, Tel Aviv University Press, Israel, 1983, pp. 270–280.
55. G.W. Wohlfarth and G.I. Hulata, *Applied Genetics of Tilapias*, International Center for Living Aquatic Resources Management Studies and Reviews 6, Manila, 1983.
56. L.J. Lester, K.S. Lawson, T.A. Abella, and M.S. Palada, *Aquacult. Fish. Manage.* **20**, 369–380 (1989).
57. G.C. Mair, A.G. Scott, D.J. Penman, J.A. Beardmore, and D.O.F. Skibinsksi, *Theor. Appl. Genet.* **82**, 144–152 (1991).
58. G.C. Mair, A.G. Scott, D.J. Penman, D.O.F. Skibinski, and J.A. Beardmore, *Theoret. Appl. Genet.* **82**, 153–160 (1991).
59. G.C. Mair, J.S. Abucay, D.O.F. Skibinski, T.A. Abella, and J.A. Beardmore, *Can. J. Fish. Aquat. Sci.* **54**, 396–404 (1997).
60. D. Trombka and R. Avtalion, *Israeli J. Aquacult.—Bamidgeh* **45**, 26–37 (1993).
61. A. Muller-Belecke and G. Horstgen-Schwark, *Aquaculture* **137**, 57–65 (1995).
62. J. Komen, A.B.J. Bongers, C.J.J. Richter, W.B. von Muiswinkel, and E.A. Huisman, *Aquaculture* **92**, 127–142 (1991).
63. K.E. Sneed, in *Genetic Selection and Hybridization of Cultivated Fishes*, FAO/UNDP **2926**, 143–150 (1971).
64. B. Chevassus, *Aquaculture* **33**, 245–262 (1983).
65. D. Chourrout, in K. Tiews, ed., *Selection, Hybridization and Genetic Engineering in Aquaculture*, Volume 2, Heenemann, Berlin, 1987, pp. 111–126.
66. C.A. Longwell, in K. Tiews, ed., *Selection, Hybridization and Genetic Engineering in Aquaculture*, Volume 2, Heenemann, Berlin, 1987, pp. 3–21.
67. R. Suzuki and Y. Fukuda, *Bull. Freshwater Fish. Res. Lab. (Tokyo)* **21**, 117–138 (1972).
68. T. Refstie, in K. Tiews, ed., *Selection, Hybridization and Genetic Engineering in Aquaculture*, Volume 1, Heenemann, Berlin, 1972, pp. 293–302.
69. J. Bakos, in K. Tiews, ed., *Selection, Hybridization and Genetic Engineering in Aquaculture*, Volume 1, Heenemann, Berlin, 1987, pp. 303–311.
70. Z.L. Krasznai, in K. Tiews, ed., *Selection, Hybridization and Genetic Engineering in Aquaculture*, Volume 2, Heenemann, Berlin, 1987, pp. 35–45.
71. K.T. Wada, in K. Tiews, ed., *Selection, Hybridization and Genetic Engineering in Aquaculture*, Volume 1, Heenemann, Berlin, 1987, pp. 313–322.
72. S.R. Melecha, in K. Tiews, ed., *Selection, Hybridization and Genetic Engineering in Aquaculture*, Volume 1, Heenemann, Berlin, 1987, pp. 323–336.
73. J.R. Dangel, P.T. Macy, and F.C. Whitler, *Annotated Bibliography of Interspecific Hybridization of Fishes in the Subfamily Salmonidae*, U.S. Department of Commerce, NOAA Tech. Memo WFC-1, Washington, DC, 1987.
74. R.A. Dunham, in K. Tiews, ed., *Selection, Hybridization and Genetic Engineering in Aquaculture*, Volume 2, Heenemann, Berlin, 1987, pp. 393–416.
75. J. Bakos, Z. Krasznai, and T. Marian, *Aquacult. Hung.* **1**, 51–57 (1978).
76. J.D. Balarin and J.P. Hatton, *Tilapia: A Guide to their Biology and Culture in Africa*, University Stirling, Scotland, 1979.
77. K. Buss, J. Meade, and D.R. Graff, in R.L. Kendall, ed., *Selected Coolwater Fishes of North America*, American Fisheries Society Special Publication 11, Washington, DC, 1978, pp. 210–216.
78. W.F. Childers, *Ill. Nat. Hist. Surv. Bull. 29*, Article 3, Urbana, IL, 1967.
79. J.H. Kerby, in R.R. Stickney, ed., *Culture of Nonsalmonid Freshwater Fishes*, CRC Press, Boca Raton, FL, 1986, pp. 128–147.
80. J.H. Kerby and C.E. Nash, in C.E. Nash and A.J. Novotny, eds., *Production of Aquatic Animals: Fishes*, Volume C8, Elsevier, New York, 1995, pp. 161–174.
81. Z. Krasznai, T. Marian, L. Buris, and F. Ditroi, *Aquacult. Hung.* **4**, 33–38 (1984).
82. D.L. Sutton, J.G. Stanley, and W.W. Miley, *J. Plant Manage.* **19**, 37–39 (1981).
83. C.F. Hickling, *J. Genet.* **57**, 1–10 (1960).
84. B. Jalabert, P. Kammacher, and P. Lessent, *Ann. Biol. Anim. Bioch. Biophys.* **11**, 155–165 (1971).
85. Y. Pruginin, S. Rothbard, G. Wohlfarth, A. Halevy, R. Moav, and G. Hulota, *Aquaculture* **6**, 11–21 (1975).
86. L.L. Lovshin, in R.S.V. Pullin and R.H. Lowe-McConnell, eds., *The Biology and Culture of Tilapias*, ICLARM, Manila, 1982, pp. 279–308.
87. S. Rothbard, E. Solnik, S. Shabbath, R. Amado, and I. Grabie, in L. Fishelson and Z. Yaron, comp. *International Symposium on Tilapia in Aquaculture*, Tel Aviv University Press, Israel, 1983, pp. 425–434.
88. G. Hulata, *Israeli J. Aquacult.—Bamidgeh* **49**, 174–179 (1997).
89. T.J. Pandian and S.G. Sheela, *Aquaculture* **138**, 1–22 (1995).
90. W.L. Shelton, D. Rodrigez-Guerrero, and J. Lopez-Macias, *Aquaculture* **25**, 59–65 (1981).
91. W.L. Shelton, R.O. Smitherman, and G.L. Jensen, *J. Fish Biol.* **18**, 45–51 (1981).
92. R.P. Phelps, G. Conteras Salazar, V. Abe, and B.J. Argue, *Aquacult. Res.* **26**, 293–295 (1995).
93. W.L. Shelton, V. Wanniasingham, and A.E. Hiott, *Aquaculture* **137**, 203–211 (1995).
94. J. Komen, P.A.J. Lodder, F. Huskens, C.J.J. Richter, and E.A. Huisman, *Aquaculture* **78**, 349–363 (1989).
95. J. Komen and C.J.J. Richter, in J.F. Muir and R.J. Roberts, eds., *Recent Advances in Aquaculture*, Volume IV, Blackwell Scientific, Oxford, 1993, pp. 78–86.
96. G.L. Jensen, W.L. Shelton, S.L.T. Yang, and L.O. Wilken, *Trans. Amer. Fish. Soc.* **112**, 79–85 (1983).
97. S.E. Boney, W.L. Shelton, S.L. Yang, and L.O. Wilken, *Trans. Amer. Fish. Soc.* **113**, 348–353 (1984).
98. G.L. Jensen and W.L. Shelton, *Aquaculture* **16**, 233–242 (1979).

99. K.D. Hopkins, W.L. Shelton, and C.R. Engle, *Aquaculture* **18**, 263–268 (1979).
100. E. Lahav, *Israeli J. Aquacult.—Bamidgeh* **45**, 131–136 (1993).
101. C. Melard, *Aquaculture* **130**, 25–34 (1995).
102. M. Cousin-Gerber, G. Burger, C. Boisseau, and B. Chevassus, *Aquat. Living Resour.* **2**, 225–230 (1989).
103. V.J. Bye and R.F. Lincoln, *Aquaculture* **57**, 299–309 (1986).
104. G.H. Thorgaard, in W.S. Hoar, D.J. Randall, and E.M. Donaldson, eds., *Fish Physiology*, Volume 9B, Academic Press, New York, 1983, pp. 405–434.
105. G.H. Thorgaard, *Aquaculture* **57**, 57–64 (1986).
106. R. Billard, in M. Bilio, H. Rosenthal, and C.J. Sinderman, eds., *Aquaculture: Achievements, Constraints, and Perspectives*, Europ. Maricult. Soc., Bredene, Belgium, 1986, pp. 309–349.
107. D. Chourrout, in K. Tiews, ed., *Selection, Hybridization and Genetic Engineering in Aquaculture*, Volume 2, Heenemann, Berlin, 1987, pp. 111–126.
108. M.G. Hussain, *Asian Fish. Soc.* **8**, 133–142 (1995).
109. J. Komen, G.F. Wiegertijes, V.J.T. van Ginneken, E.H. Eding, and C.J.J. Richter, *Aquaculture* **104**, 51–66 (1992).
110. G.H. Thorgaard, P.D. Scheerer, W.K. Hershberger, and J.M. Myers, *Aquaculture* **85**, 215–221 (1990).
111. N.B. Cherfas, O. Kozinsky, S. Rothbard, and G. Hulata, *Israeli Journal of Aquaculture—Badmigeh* **42**, 3–9 (1990).
112. W.L. Shelton and S. Rothbard, *Israeli Journal of Aquaculture—Bamidgeh* **45**, 73–81 (1993).
113. J. Don and R.R. Avtalion, in Y. Zohar and B. Breton, eds., *Reproduction in Fish: Basic and Applied Aspects in Endocrinology and Genetics*, Les Colloques de l'INRA 44, Paris, 1988, pp. 199–205.
114. G.C. Mair, *Aquaculture* **111**, 227–244 (1993).
115. S. Rothbard, *The Israeli Journal of Aquaculture—Bamidgeh* **43**, 145–155 (1991).
116. L. Horvath, G. Tamas, and I. Tolg, in J.E. Halver, ed., *Special Methods in Pond Fish Husbandry*, Akademiai Kiado, Budapest, 1994, pp. 1–147.
117. L. Horvath and L. Organ, *Aquaculture* **129**, 157–181 (1995).
118. J.G. Stanley, *Trans. Amer. Fish. Soc.* **105**, 10–16 (1976).
119. J.R. Cassani and W.E. Caton, *Aquaculture* **46**, 37–44 (1985).
120. J.R. Cassani and W.E. Caton, *Aquaculture* **55**, 43–50 (1986).
121. J.R. Cassani, D.R. Malony, H.P. Allaire, and J.H. Kerby, *Aquaculture* **88**, 273–284 (1990).
122. J.G. Stanley, *J. Aquat. Plant Manage.* **14**, 68–70 (1976).
123. J.G. Stanley, in J.V. Shireman, ed., *Proceedings of the Grass Carp Conference*, University of Florida, Gainesville, FL, 1979, pp. 201–242.
124. O. Linhart, P. Kvasnicka, V. Slechtova, and D. Pokorny, *Aquaculture* **54**, 63–67 (1986).
125. S. Rothbard and G.W. Wohlfarth, *Aquaculture* **115**, 13–17 (1993).
126. T.J. Benfey, A.M. Sutterlin, and R.J. Thompson, *Can. J. Fish. Aquat. Sci.* **41**, 980–984 (1984).
127. O.W. Johnson, P.R. Rabinovich, and F.M. Utter, *Aquaculture* **43**, 99–103 (1984).
128. O.W. Johnson, W.W. Dickhoff, and F.M. Utter, *Aquaculture* **57**, 329–336 (1986).
129. D. Chourrout, B. Chevassus, F. Krieg, G. Burger, and P. Renard, *Theor. Appl. Genet.* **72**, 193–206 (1986).
130. C.A. Bidwell, C.L. Chrismana, and G.S. Libey, *Aquaculture* **51**, 25–32 (1985).
131. J. Don and R.R. Avtalion, *J. Fish Biol.* **32**, 665–672 (1988).
132. J. Don and R.R. Avtalion, *Israeli Journal of Aquaculture—Bamidgeh* **40**, 17–21 (1988).
133. J.E. Parsons and G.H. Thorgaard, *J. Heredity* **76**, 177–181 (1985).
134. B. May, K.J. Henley, C.C. Krueger, and S.P. Gloss, *Aquaculture* **75**, 57–70 (1988).
135. K. Arai, K. Matsubara, and R. Suzuki, *Nippon Suisan Gakkaishi* **57**, 2173–2178 (1992).
136. J.M. Myers, D.J. Penman, Y. Basavaraju, S.F. Powell, P. Baoprasertkul, K.J. Rana, N. Browage, and B.J. McAndrew, *Theor. Appl. Genet.* **90**, 205–210 (1995).
137. S. Kondoh, S. Satoh, and M. Tomita, *Rep. Niigata Pref. Inland Water Fish. Exp. Stn.* **15**, 19–23 (1989).
138. A.B.J. Bongers, E.P.C. in't Veld, K. Abo-Hashema, I.M. Bremmer, E.H. Eding, J. Komen, and C.J.J. Richter, *Aquaculture* **122**, 119–132 (1994).
139. A.S. Grunina, B.I. Gomelskii, and A.A. Neyfakh, *Soviet Genetics* **26**, 1336–1341 (1991) (Genetika **26**, 2037–2043 (1990).
140. A.S. Grunina, A.A. Neyfakh, and B. I. Gomelsky, *Aquaculture* **129**, 218–219 (1995).
141. N.C. Parker, in C.J. Shephard and N.R. Bromage, eds., *Intensive Fish Farming*, BSP Professional Books, Oxford, 1988, pp. 268–301.
142. N.C. Parker, *CRC Rev. Aquat. Sci.* **1**, 97–109 (1989).
143. J.E. Thorpe, G.A.E. Gall, J.E. Lannan, and C.E. Nash, eds., *Conservation of Fish and Shellfish Resources: Managing Diversity*, Academic Press, New York, 1995.
144. W.H.L. Allsopp, in E.K. Pikitch, D.D. Huppert, and M.P. Sissenwine, eds., *Global Trends: Fisheries Management*, American Fisheries Society Symposium 20, Bethesda, MD, 1997, pp. 153–165.
145. S.M. Garcia and C. Newton, in E.K. Pikitch, D.D. Huppert, and M.P. Sissenwine, eds., *Global Trends: Fisheries Management*, American Fisheries Society Symposium 20, Bethesda, MD, 1997, pp. 3–27.
146. M.B. New, *World Aquacult.* **28**(2), 11–30 (1997).
147. G. Borgstrom, in S.D. Gerking, ed., *Ecology of Freshwater Fish Production*, Blackwell, Oxford, 1978, pp. 469–491.
148. R.L. Welcomme, in W.R. Courtenay, Jr., and J.R. Stauffer, eds., *Distribution, Biology, and Management of Exotic Fishes*, John Hopkins University Press, Baltimore, MD, 1984, pp. 22–40.
149. R.L. Welcomme, *FAO Fish. Tech.*, Paper 294, Rome, 1988.
150. W.R. Courtenay, Jr., and C.C. Kohler, in R.H. Stroud, ed., *Fish Culture in Fisheries Management*, American Fisheries Society Bethesda, MD, 1986, pp. 401–413.
151. W.R. Courtenay, Jr., D.A. Hensley, J.N. Taylor, and J.A. McCann, in W.R. Courtenay and J.R. Stauffer, eds., *Distribution, Biology, and Management of Exotic Fishes*, John Hopkins University Press, Baltimore, MD, 1984, pp. 41–77.
152. P.L. Shafland, *CRC Rev. Fish. Sci.* **4**, 101–122 (1996).
153. C.E. Nash, ed., *Production of Aquatic Animals: Crustaceans, Molluscs, Amphibians and Reptiles*, Volume C4, Elsevier, New York, 1991.
154. K.K. Chew, in R. Mann, ed., *Exotic Species in Mariculture*, MIT Press, Cambridge, MA, 1979, pp. 54–82.
155. V.G. Burrell, Jr., in J.V. Huner and E.E. Brown, eds., *Crustacean and Mollusk Aquaculture in the United States*, AVI Publishing, Westport, CT, 1985, pp. 235–273.

156. G.C. Matthiessen, in C.E. Nash, ed., *Production of Aquatic Animals: Crustacea, Molluscs, Amphibians and Reptiles*, Volume C4, Elsevier, New York, 1991, pp. 89–119.
157. J.R. Davidson, J.A. Brock, and L.G.L. Young, in A. Rosenfield and R. Mann, eds., *Dispersal of Living Organisms into Aquatic Ecosystems*, Maryland Sea Grant College Program, College Park, MD, 1992, pp. 83–101.
158. J.Cl. Philippart and J.Cl. Ruwet, in R.S.V. Pullin and R.H. Lowe-McConnell, eds., *The Biology and Culture of Tilapias*, ICLARM, Manila, 1982, pp. 15–59.
159. D.J. MacIntosh and D.C. Little, in N.R. Bromage and R.J. Roberts, eds., *Broodstock Management and Egg and Larval Quality*, Blackwell Science, London, 1995, pp. 277–320.
160. C.R. Engle, in B.A. Costa-Pierce and J.E. Rakocy, eds., *Tilapia Aquaculture in the Americas*, Volume 1, World Aquaculture Society, Baton Rouge, LA, 1997, pp. 229–243.
161. D. Mires, in C.E. Nash and A.J. Novotny, eds., *Production of Aquatic Animals: Fishes*, Volume C8, Elsevier, New York, 1995, pp. 133–159.
162. A.E. Eknath, in J.E. Thorpe, G.A.E. Gall, J.E. Lannan, and C.E. Nash, eds., *Conservation of Fish and Shellfish Resources: Managing Diversity*, Academic Press, New York, 1995, pp. 177–194.
163. R.R. Stickney, *World Aquacult.* **28**(3), 20–25 (1997).
164. B.W. Green, K.L. Veverica, and M.S. Fitzpatrick, in H.S. Egna and C.E. Boyd, eds., *Dynamics of Pond Aquaculture*, CRC Press, Boca Raton, FL, 1997, pp. 215–243.
165. F. Hanley, *World Aquacult.* **22**(1), 42–4 (1991).
166. G. Hulata, *Israeli J. Aquacult.—Bamidgeh* **49**, 174–179 (1997).
167. R. Johnstone, D.J. MacIntosh, and R.S. Wright, *Aquaculture* **35**, 249–257 (1983).
168. C.A. Goudie, W.L. Shelton, and N.C. Parker, *Aquaculture* **58**, 215–226 (1986).
169. G.W. Wohlfarth, in J.E. Thorpe, G.A.E. Gall, J.E. Lannan, and C.E. Nash, eds., *Conservation of Fish and Shellfish Resources: Managing Diversity*, Academic Press, New York, 1995, pp. 138–176.
170. H. Li and P.B. Moyle, in C.C. Kohler and W.A. Hubert, eds., *Inland Fisheries Management*, American Fisheries Society, Bethesda, MD, 1993, pp. 287–307.
171. *USDA Aquaculture: Situation and Outlook Report*, Economic Research Service, Aqua-4, Rockville, MD, 1990.
172. A.J. Novotny and C.E. Nash, in C.E. Nash and A.J. Novotny, eds., *Production of Aquatic Animals: Fishes*, Volume C8, Elsevier, New York, 1995, pp. 175–238.
173. L.C. Woods, B. Ely, G. Leclerc, and R.M. Harrell, *Aquaculture* **137**, 41–44 (1995).
174. J.H. Kerby and C.E. Nash, in C.E. Nash and A.J. Novotny, eds., *Production of Aquatic Animals: Fishes*, Volume C8, Elsevier, New York, 1995, pp. 161–174.
175. J.H. Kerby, J.M. Everson, R.M. Harrell, C.C. Starling, H. Revels and J.G. Geiger, *Aquaculture* **137**, 355–358 (1995).
176. C. Lever, *Naturalized Fishes of the World*, Academic Press, New York, 1996.
177. B.J. McAndrew and K.C. Majundar, *Aquaculture* **30**, 249–261 (1983).
178. F.W. Allendorf and N. Ryman, in N. Ryman and F. Utter, eds., *Population Genetics and Fishery Management*, Washington Sea Grant Program, Seattle, 1987, pp. 141–159.
179. X. Lu, in E.K. Pikitch, D.D. Huppert, and M.P. Sissenwine, eds., *Global Trends: Fisheries Management*, American Fisheries Society Symposium 20, Bethesda, MD, 1997, pp. 185–194.
180. S. Li and J. Mathias, *Freshwater Fish Culture in China: Principles and Practices*, Developments in Aquaculture and Fisheries Science 28, Elsevier, New York, 1994.
181. S.S. de Silva, ed., *Exotic Aquatic Organisms in Asia*, Asian Fisheries Society Special Publication 3, Manila, 1989.
182. S. Li, in R. Billard and J. Marcel, eds., *Aquaculture of Cyprinids*, INRA, Paris, 1986, pp. 347–355.
183. S. Li and Z. Biyu, *Asian Fish. Sci.* **3**, 185–196 (1990).
184. W.J. McNeil, in T.V.R. Pillay and W.A. Dill, eds., *Advances in Aquaculture*, Fishing News, Ltd., Surrey, England, 1979, pp. 547–554.
185. R.P. Khodorevskaya, G.F. Dovgopol, O.L. Zhuravleva, and A.D. Vlasenko, *Environ. Biol. Fish.* **48**, 209–219 (1997).
186. B.S. Bhimachar and S.D. Tripathi, *A Review of Fisheries Activities in India*, FAO Fish Report 44 **2**, 1–33 Rome, 1967.
187. S.B. Singh, K.K. Sukumaran, P.C. Chakrabarti, and M.M. Baagchi, *J. Inland Fish. Soc. India* **4**, 38–50 (1972).
188. M.C. Nandeesha, in C.E. Nash and A.J. Novotny, eds., *Production of Aquatic Animals: Fishes*, Volume C8, Elsevier, New York, 1995, pp. 57–74.
189. V.G. Jhingran, in R. Billard and J. Marcel, eds., *Aquaculture of Cyprinids*, INRA, Paris, 1986, pp. 335–346.
190. C.D. de Silva, *Aquacult. Internat.* **5**, 330–349 (1997).
191. E.A. Balayut, *Stocking and Introduction of Fish in Lakes and Reservoirs in the ASEAN (Association of South Asian Nations) Countries*, FAO Technical Paper 236, Rome, 1983.
192. S. Tal and I. Ziv, in R.O. Smitherman, W.L. Shelton, and J.H. Grover, eds., *Culture of Exotic Fishes Symposium Proceedings*, American Fisheries Society, Bethesda, MD, 1978, pp. 1–9.
193. S. Rothbard, *Advances en Biotechnologia 3*, Havana, Cuba, 1995.
194. S. Sarig, *Israeli J. Aquacult.—Bamidgeh* **48**, 158–164 (1996).
195. S. Sarig, *Israeli J. Aquacult.—Bamidgeh* **49**, 84–89 (1997).
196. D. Mires, *Israeli J. Aquacult.—Bamidgeh* **47**, 78–83 (1995).
197. G.C. Radonski and R.G. Martin, in R.H. Stroud, ed., *Fish Culture in Fisheries Management*, American Fisheries Society, Bethesda, MD, 1986, pp. 7–13.
198. G.C. Radonski, N.S. Prosser, R.G. Martin, and R.H. Stroud, in W.R. Courtenay and J.R. Stauffer, eds., *Distribution, Biology and Management of Exotic Fishes*, John Hopkins University Press, Baltimore, MD, 1984, pp. 313–321.
199. W.R. Courtenay, Jr., and C.R. Robins, *CRC Reviews in Aquatic Sci.* **1**, 159–172 (1989).
200. D. Horak, in H.L. Schramm and R.G. Piper, eds., *Uses and Effects of Cultured Fishes in Aquatic Ecosystems*, American Fisheries Society, Bethesda, MD, 1995, pp. 61–67.
201. D.P. Lee, in H.L. Schramm and R.G. Piper, eds., *Uses and Effects of Cultured Fishes in Aquatic Ecosystems*, American Fisheries Society, Bethesda, MD, 1995, pp. 16–20.
202. M. Rowan and N. Stone, *Prog. Fish-Cult.* **58**, 62–64 (1996).
203. D. Simberloff and P. Stiling, *Biol. Cons.* **78**, 185–192 (1996).
204. J.V. Shireman, in W.R. Courtenay and J.R. Stauffer, eds., *Distribution, Biology, and Management of Exotic Fishes*, The Johns Hopkins University Press, Baltimore, MD, 1984, pp. 302–312.
205. J.G. Stanley, *Trans. Amer. Fish. Soc.* **107**, 104 (1978).

206. J.V. Shireman and C.R. Smith, *Synopsis of Biological Data on Grass Carp Ctenopharyngodon idella (Cuvier & Valenciennes, 1844)*, FAO Fisheries Synopsis 135, Rome, 1983.
207. N. Zonneveld and H. Van Zon, in J.F. Muir and R.J. Roberts, eds., *Recent Advances in Aquaculture*, Volume 2, Croom-Helm, London, 1985, pp. 119–292.
208. L. Horvath, G. Tamas, and C. Seagrave, *Carp and Pond Fish Culture*, Fishing News Books, Oxford, 1992.
209. J.R. Cassani, ed., *Managing Aquatic Vegetation with Grass Carp: A Guide for Water Resource Managers*, American Fisheries Society, Bethesda, MD, 1996.
210. T. Marian and Z. Krasznai, *Aquacult. Hungarica* **1**, 44–50 (1978).
211. J.P. Clugston and J.V. Shireman, *Triploid Grass Carp for Aquatic Plant Control*, U.S. Fish and Wildlife Service Leaflet 8, Washington, DC, 1987.
212. S.K. Allen, R.G. Thiery, and N.T. Hagstrom, *Trans. Amer. Fish. Soc.* **115**, 841–848 (1986).
213. J.P. Van Eenennaam, R.K. Stocker, R.G. Tiery, N.T. Hagstrom, and S.I. Doroshov, *Aquaculture* **86**, 111–125 (1990).
214. B.R. Griffin, *Aquacult. Magazine* January/February, 1–2 (1991).
215. S.K. Allen and R.J. Wattendorf, *Fisheries* **12**(4), 20–24 (1987).
216. C. Collins, *Aquacult. Magazine* May/June, 51–52 (1986).
217. J.V. Conner, R.P. Gallagher, and M.F. Chatry, in L.A. Fuiman, ed., *Fourth Annual Larval Fish Conference*, FWS/OBS 80/43, Ann Arbor, MI, 1980, pp. 1–19.
218. D.J. Brown and T.G. Coon, *N. Amer. J. Fish. Manage.* **11**, 62–66 (1991).
219. J.G. Stanley, W.W. Miley, and D.L. Sutton, *Trans. Amer. Fish. Soc.* **107**, 119–128 (1978).
220. H. Leventer, *Biological Control of Reservoirs by Fish*, Mekroth Water Co., Nazareth, Israel, 1984.
221. S. Henderson, in R.O. Smitherman, W.L. Shelton, and J.H. Grover, eds., *Culture of Exotic Fishes Symposium Proceedings*, American Fisheries Society, Bethesda, MD, 1978, pp. 121–136.
222. H. Buck, R.J. Baur, and C.R. Rose, in R.O. Smitherman, W.L. Shelton, and J.H. Grover, eds., *Culture of Exotic Fishes Symposium Proceedings*, American Fisheries Society, Bethesda, MD, 1978, pp. 144–155.
223. A. Milstein, B. Hepher, and B. Telch, *Aquacult. Fish. Manage.* **19**, 127–137 (1988).
224. A. Milstein, *Hydrobiologia* **231**, 177–186 (1992).
225. J.R.P. French, *Fisheries* **18**(6), 13–19 (1993).
226. A.J. Mitchell, *Aquacult. Magazine* July/August, 93–97 (1995).
227. L.G. Nico and J.D. Williams, *Risk Assessment on Black Carp (Pisces: Cyprinidae)*, Florida/Caribbean Science Center, National Biological Service, Gainesville, FL, 1996.
228. W.L. Shelton, A. Soloman, and S. Rothbard, *Israeli J. Aquacult.—Bamidgeh* **47**, 59–67 (1995).
229. D.A. Conroy, *An Evaluation of the Present Status of World Trade in Ornamental Fish*, FAO Fisheries Technical Paper 146, Rome, 1975.
230. F.A. Chapman, S.A. Fitz-Coy, E.M. Thunberg, and C.M. Adams, *J. World Aquacult. Soc.* **28**, 1–10 (1997).
231. W.R. Courtenay, Jr., and J.R. Stauffer, *J. World Aquacult. Soc.* **21**, 145–159 (1990).
232. M. Martin, *Aquacult. Magazine* May/June, 38–39 (1983).
233. M.L. Warren and B.M. Burr, *Fisheries* **19**(1), 6–21 (1994).
234. R.R. Miller, J.D. Williams, and J.E. Williams, *Fisheries* **14**(6), 22–38 (1989).
235. P.L. Shafland, *CRC Rev. Aquat. Sci.* **4**, 123–132 (1996).
236. P.B. Moyle and T. Light, *Biol. Cons.* **78**, 149–161 (1996).
237. S.T. Ross, *Environ. Biol. Fish.* **30**, 359–368 (1991).
238. J.N. Taylor, W.R. Courtenay, and J.A. McCann, in W.R. Courtenay and J.R. Stauffer, eds., *Distribution, Biology, and Management of Exotic Fishes*, The Johns Hopkins University Press, Baltimore, MD, 1984, pp. 322–373.
239. G.V. Everett, *J. Fish Biol.* **5**, 429–440 (1973).
240. P.B. Moyle, H. Li, and B.A. Barton, in R.H. Stroud, ed., *Fish Culture in Fisheries Management*, American Fisheries Society, Bethesda, MD, 1986, pp. 415–426.
241. P.A. Larkin, in H. Clepper, ed., *Predator–Prey Systems in Fishery Management*, Sport Fishing Institute, Washington, DC, 1979, pp. 13–22.
242. W.R. Courtenay, Jr. and J.D. Williams, in A. Rosenfield and R. Mann, ed., *Dispersal of Living Organisms into Aquatic Ecosystems*, Maryland Sea Grant College Program, College Park, MD, 1992, pp. 49–81.
243. U.S. Environmental Protection Agency (USEPA), *Framework for Ecological Risk Assessment*, EPA/630/R-92/001, Washington, DC, 1992.
244. R.S.V. Pullin, in R.S.V. Pullin, H. Rosenthal, and J.L. McClean, eds., *Environment and Aquaculture in Developing Countries*, ICLARM Conference Proceedings 31, Manila, 1993, pp. 1–19.
245. R.S.V. Pullin, *NAGA: The ICLARM Quarterly* **17**(4), 19–24 (1994).
246. E.M. Hallerman and A.R. Kapuscinski, *Aquaculture* **137**, 9–17 (1993).
247. G.H. Thorgaard and S.K. Allen, in A. Rosenfield and R. Mann, eds., *Dispersal of Living Organisms in Aquatic Ecosystems*, Maryland Seagrant College Program, College Park, MD, 1992, pp. 281–288.
248. D. Horak, *Fisheries* **19**(4), 18–21 (1994).
249. International Council for the Exploration of the Sea (ICES), *Proposed Guidelines for Implementing the ICES Code of Practice Concerning Introductions and Transfers of Marine Species*, ICES, Copenhagen, Denmark, 1982.
250. European Inland Fisheries Advisory Commission (EIFAC), *Codes of Practice and Manual of Procedures for Consideration of Introductions and Transfers of Marine and Freshwater Organisms*, FAO, EIFAC Occasional Paper 23, Rome, 1988.
251. C.C. Kohler and J.G. Stanley, in W.R. Courtenay and J.R. Stauffer, eds., *Distribution, Biology, and Management of Exotic Fishes*, The Johns Hopkins University Press, Baltimore, MD, 1984, pp. 387–406.
252. R. Mann, in R. Mann, ed., *Exotic Species in Mariculture*, MIT Press, Cambridge, MA, 1979, pp. 331–354.
253. B. Chevassus, R. Guyomard, D. Chourrout, and E. Quillet, *Genet. Sel. Evol.* **15**, 519–532 (1983).
254. J. Gervai, S. Peter, A. Nagy, L. Horvath, and V. Csanyi, *J. Fish Biol.* **17**, 667–671 (1978).
255. W.R. Wolters, G.S. Libey, and C.L. Chrisman, *Trans. Amer. Fish. Soc.* **110**, 310–312 (1982).

See also SHRIMP CULTURE; TILAPIA CULTURE.

EYESTALK ABLATION

Joe Fox
Texas A&M University at Corpus Christi
Corpus Christi, Texas

Granvil D. Treece
Texas Sea Grant College Program
Bryan, Texas

OUTLINE

The use of eyestalk ablation for induction of maturation in crustacean-rearing facilities is considered by most aquaculturists to be essential for predictable hatchery operation. Recent efforts at achieving maturation via naturally induced manipulation of physiochemical and nutritional parameters have also shown signs of success in some species. In female penaeid shrimp and female crabs, the production and storage sites of the gonad-inhibiting hormone (GIH) are located in the eyestalks. This hormone inhibits the maturation of the ovaries. In nature, some environmental factor or factors cause the decrease of this substance as the shrimp migrate from the estuaries to offshore areas, where they normally spawn. Eyestalk ablation, removal, or extirpation eliminates, or at least reduces, this inhibitory hormone to a level where full and accelerated maturation of the ovaries can take place. Ablation is the preferred methodology for captive maturation of penaeid shrimp due to (*1*) inhibitions inherent to most culture situations, (*2*) increased fecundity of ablated females, and (*3*) enhanced spawning frequency. The ablation process is discussed in this entry.

BACKGROUND

The stimulating effect of eyestalk ablation on reproduction of decapod crustacea was first evaluated for penaeid aquaculture in the early 1970s, when bilateral eyestalk ablation (both eyes) was attempted by French researchers (1). This methodology was found to stimulate rapid ovarian maturation; however, ablated females suffered high mortality (probably due to hormone desynchronization), and ova were typically reabsorbed without subsequent spawning (2–4). Those problems were alleviated by the ablation of only one eyestalk (unilateral eyestalk ablation), which provided moderate hormonal stimulus without reabsorption of ova or excessive mortality (3,5,6). Consequently, unilateral eyestalk ablation rapidly emerged worldwide as a simple procedure for inducing reproduction of numerous species of penaeid shrimp reared in captivity. Some researchers have even used ablation to improve growth rate of shrimp (7).

THE ABLATION PROCESS

The ablation process requires that only hard-shelled (intermolt) shrimp be used. Postmolt (Stage V) females are not recommended for ablation, due to increased potential for handling mortality associated with the softened exoskeleton. Premolt (Stage IV) individuals are also not used because of potential for molting during recovery from the ablation process. One study of note (8) has shown that ablation undertaken between 8 and 10 days postmolt resulted in significantly greater egg production versus ablation at 13–15 days postmolt.

Eyestalk ablation has been performed using a variety of methods, including simple severing with scissors, enucleation (9), cautery (4), and ligation (10). Each of these methods has been effective in removing or destroying the X organ/sinus gland complex. Stress can be reduced and losses minimized if shrimp are held in chilled water before and after ablation (2). To minimize stress, the ablation should be performed as quickly as possible and can be done under chilled water, but this is not always necessary or practical. However, in commercial hatcheries ablation is usually done in the early mornings, when temperatures are lowest. Female mortality due to ablation should be very low, but some mortality should be expected. The effects of eyestalk ablation vary with season of the year and stage in the molt cycle. Shrimp that are ablated as they prepare to enter their reproductive peak are more conditioned to yield a reproductive (as opposed to molting) response than those entering a reproductively dormant period (11,12). Within a molt cycle, ablation performed during premolt leads to molting, ablation immediately after molting causes death, and ablation during intermolt leads to maturation (13).

Ablation is performed on either the left or right compound eyestalk. The eyestalk chosen for ablation can, if possible, be one that is obviously infected or otherwise damaged. The damaged eye should be ablated in order to leave the shrimp with one unablated functional eye.

There are various methodologies used to accomplish the ablation procedure. All procedures require holding the female shrimp gently, yet firmly, and partially submerged in a separate ablation tank. The following methods are typically employed:

(a) **Enucleation** — Grasp the eyestalk just behind the eyeball, using the thumb and index finger. Squeeze hard and roll the thumb and finger outwards away from the body, thus crushing the eyestalk and squeezing out the contents of both it and the eye. The objective is to squeeze the contents outwards and not let them follow the eyestalk back into the head region. To aid in the removal of eyestalk contents, an incision on the front of the eye with a sharp blade may be made. Many penaeid aquaculturists prefer this method, however, it is simply a matter of preference. The authors have found incision of the compound eye to be unnecessary. Enucleation has the advantages of simplicity and rapid clotting of hemolymph within the empty eyestalk. The main objective of any ablation methodology should be speed and reduced stress. The only detrimental aspect of enucleation is that it results in an open wound, which increases the potential for subsequent infection.

(b) **Ligation**—By this method, a string is tied around the base of the eyestalk as close to the carapace as possible. The string should be drawn fairly tight, causing the eyestalk to fall off in a few days (14). The procedure does not leave the shrimp with an open wound; however, successful ablation is often unpredictable.

(c) **Cautery**—This method calls for severing the eyestalk, followed by sealing of the wound via electrocauterization, heated forceps, or the application of a silver nitrate bar. Some practitioners simply cauterize the compound eye itself. In either case, the wound is rapidly sealed by scar tissue.

Pinching is considered the most readily applicable method of ablation. It can be undertaken by one person, and the wound should heal rapidly, without application of antibiotics. Ligation requires two persons, one to hold the shrimp and the other to tie the eyestalk. Cautery requires either a cauterizer or silver nitrate bar, either of which is often unavailable. Ultimately, the latter methods involve increased handling of subject animals.

FECUNDITY AND VIABILITY OF SPAWNS

Ovarian development in sexually mature females can commence within three days postablation, followed by a first spawn within approximately one week. The ablated population should be in full production three weeks after ablation. If ablated during the intermolt stage, the females will mature and spawn immediately. However, if ablated during early premolt, they will molt before maturing.

The fecundity and viability of spawns from ablated females have sometimes been inferior to spawns from females matured in the wild (13,15–17). Furthermore, commercial producers prefer postlarvae produced from wild, rather than captive spawns, which suggests that embryonic characteristics may influence juvenile survival and growth. For example, researchers found that eyestalk ablation of the crab *Paratelphusa hydrodromous* during the prebreeding season resulted in precocious ovarian growth (18). However, in comparison to normal mature ovaries, the ovaries from ablated females were smaller, higher in lipid composition, and more variable in distribution of yolk among oocytes. These differences presumably are consequences of hormonal insensitivity of ablated shrimp to physiological or environmental limitations, such as improper oocyte differentiation, nutrient storage, food supply, or temperature.

After spawning, unilaterally ablated females immediately reinitiate ovarian maturation and consequently spawn more frequently under nonnatural conditions than unablated females. This increased frequency can deplete female nutrient stores and is probably partially responsible for the lower survival rates of ablated to unablated females: Emmerson (16) reported that ablated *Fenneropenaeus indicus*, having a spawning frequency of five days, eventually spawned fewer eggs and died; however, it was suggested that this problem could be alleviated by proper nutrition.

ABLATION OF MALE PENAEID SHRIMP

Male ablation causes precocious maturation of *Marsupenaeus monodon* and *Fenneropenaeus merguiensis* (19); however, it has also been shown to increase gonad size and to double the mating frequency of smaller (25–30 g; 2–10.5 oz) *Litopenaeus vannamei* in comparison to similar-sized, unablated control shrimp (20). Eyestalk ablation of male shrimp has rarely been considered useful, and the authors do not recommend its use under practical culture conditions.

Further information on ablation and shrimp reproductive physiology can be found in References 21 and 22.

MATURATION WITHOUT ABLATION

A few researchers and one commercial hatchery in Venezuela have been able to achieve egg development, mating, and spawning of captive penaeids without ablation, utilizing temperature and photoperiod manipulation. However, as a rule, hatcheries have not been able to base a long-lasting, profitable, highly productive commercial operation on this approach to maturation.

MATURATION RESEARCH INTO HORMONAL CONTROL

Researchers in the 1980s were able to isolate and characterize hormonal systems involved in maturation and reproduction of the spiny lobster (*Panulirus* sp.) (23). Those same researchers later undertook the same study with penaeid shrimp (24–29). The only penaeid shrimp investigated was the pink shrimp, *Farfantepenaeus duorarum*. Progress with respect to elimination of some of the husbandry problems associated with maturation (egg fertility, decreased spawning rate over time) appears slow in coming. Unilateral eyestalk ablation is still the method of choice for maturation of penaeid shrimp. Until ablation can be eliminated successfully and replaced with a superior method, hatcheries must depend on it to sustain production.

Other important works dealing with crustacean hormonal control, ablation, and crustacean reproduction can be found in Refs. 30–67.

BIBLIOGRAPHY

1. J.B. Panouse, *C. R. Acad. Sci.* **217**, 553–555 (1943).
2. C.W. Caillouet, *Proc. World Maricul. Soc.* **3**, 205–225 (1972).
3. Aquacop, *Proc. World Maricul. Soc.* **6**, 123–132 (1975).
4. M. Duronslet, A.I. Yudin, R.S. Wheller, and W.H. Clark, Jr., *Proc. World Maricul. Soc.* **6**, 105–122 (1975).
5. D.R. Arnstein and T.W. Beard, *Aquaculture* **5**, 411–412 (1975).
6. R.G. Wear and A. Santiago, Jr., *Crustaceana* **31**(2), 218–220 (1976).

7. A.K. Hameed and S.N. Dwivedi, *J. India Fish. Assoc.* **3**(1), 136–138 (1977).
8. C. Browdy and T.M. Samocha, *J. World Aquacult. Soc.* **16**, 236–249 (1985).
9. J.H. Primavera, in Proc. of 1st Int. Conf. on the culture of penaeid shrimp, *A review of maturation and reproduction in closed-thelycum penaeids* SEAFDEC, the Philippines, 1985.
10. J.H. Primavera, *Aquaculture* **13**, 355–359 (1978).
11. M.L. Schade and R.R. Shivers, *J. Morphol.* **163**, 13–26 (1980).
12. D.E. Bliss, *Am. Zool.* **6**, 231–233 (1966).
13. K.G. Adiyodi and R.G. Adiyodi, *Biol. Rev.* **45**, 121–165 (1970).
14. Aquacop, *Proc. World Maricul. Soc.* **8**, 927–945 (1977).
15. T.W. Beard and J.F. Wickins, *Aquaculture* **20**, 79–89 (1980).
16. W.D. Emmerson, *Mar. Ecol. Prog. Ser.* **2**, 121–131 (1980).
17. F. Lumare, *J. World Maricul. Soc.* **12**(2), 335–344 (1981).
18. G. Anilkumar and K.G. Adiyodi, *Int. J. Invertebr. Reprod.* **2**, 95–105 (1980).
19. K.H. Alikunhi, A. Poernomo, S. Adisufresno, M. Budiono, and S. Busman, *Bull. Shrimp Cult. Res. Cent.*, Jepara **1**(1), 1–11 (1975).
20. G.W. Chamberlain and A.L. Lawrence, *J. World Maricul. Soc.* **12**(1), 209–224 (1981).
21. G.D. Treece and M.E. Yates, *Laboratory Manual for the Culture of Penaeid Shrimp Larvae* Publication 88–202(R), Texas A&M University, Sea Grant College Program, Bryan, Texas, 1993.
22. G.D. Treece and J.M. Fox, *Design, Operation and Training Manual for an Intensive Culture Shrimp Hatchery* Publication 93–505, Texas A&M University, Sea Grant College Program, Bryan, Texas, 1993.
23. L.S. Quackenbush and W.F. Herrnkind, *J. Crust. Biol.* **3**, 34–44 (1983).
24. S. Chan, S.M. Rankin, and L.L. Keeley, *Biol. Bull.* **175**, 185–192 (1988).
25. J.Y. Bradfield, R.L. Berlin, S.M. Rankin, and L.L. Keeley, *Biol. Bull.* **177**, 344–349 (1989).
26. S.M. Rankin, J.Y. Bradfield, and L.L. Keeley, *Int. Invert.* Reproduction and Development, (1989).
27. S. Chan, S.M. Rankin, and L.L. Keeley, *Comp. Biochem. Physiol.* (1991).
28. S.M. Rankin and R.W. Davis, *Tissue and Cell Journal* (1990).
29. S.M. Rankin, J.Y. Bradfield, and L.L. Keeley, *UNJR* (1990).
30. R.G. Adiyodi and T. Subramoniam, in K.G. and R.G. Adiyodi, eds., *Reproductive Biology of Invertebrates*, John Wiley & Sons, New York, 1983, pp. 443–495.
31. R.D. Andrew and A.S.M. Saleuddin, *J. Comp. Physiol. B* **134**(4), 303–314 (1979).
32. Aquacop, *IFREMER*, p. 132–152 (1984).
33. Aquacop, *Proc. World Maricul. Soc.* **10**, 445–452 (1979).
34. C. Bellon-Humbert, F. Van Herp, G.E.C.M. Strolenberg, and J.M. Denuce, *Natantia. Biol. Bull.* **160**, 11–30 (1981).
35. A. Bomirski, M. Arendarczyk, E. Kawinska, and L.H. Kleinholz, *Int. J. Invert. Reprod.* **3**, 213–219 (1981).
36. W.A. Bray Jr. and A.L. Lawrence, in A. Fast and L.J. Lester, eds., *Culture of Marine Shrimp: Principles and Practices*, Elsevier Scientific Publishing Company, Amsterdam, The Netherlands, 1991, pp. 93–170.
37. P.J. Dunham, *Biol. Rev.* **53**, 555–583 (1978).
38. Y. Faure, C. Bellon-Humbert, and H. Charniaux-Cotton, *C.R. Acad. Sc. Paris* **293**(3), 461–466 (1981).
39. M. Fingerman, *Scientia* **105**, 1–23 (1970).
40. K.C. Highnam, in P.J. Gaillard and H.H. Boer, eds., *Comparative Endocrinology. Proceedings of the 8th Intl. Symp. on Comparative Endocrinology*, Elsevier/North-Holland Biomedical Press, Amsterdam, 1978, pp. 3–12.
41. L.H. Kleinholz, in P.J. Gaillard and H.H. Boer, eds., *Comparative Endocrinology. Proceedings of the 8th Intl. Symp. on Comparative Endocrinology*, Elsevier/North-Holland Biomedical Press, Amsterdam, 1978, pp. 397–400.
42. E.E. Kulakovskii and Y.A. Baturin, *Tsitologiya* **21**, 1200–1203 (1979).
43. A. Laubier-Bonichon and L. Laubier, *Reproduction Control with the Shrimp*, Marsupenaeus japonicus, F.A.O. Tech. Conf. Aqua. Kyoto, Japan FIR: AQ/Conf./76/E.38, 1976.
44. T. Otsu, *Embryologia* **8**, 1–20 (1963).
45. J.H. Primavera, T. Young and C. de los Reyes, *Survival, Maturation, Fecundity and Hatching Rate of Unablated and Ablated Penaeus indicus. Contrib. No. 59*, SEAFDEC Aquaculture Dept., Philippines, 1980.
46. A.C. Santiago, Jr., *Aquaculture* **11**(3), 185–196 (1977).
47. S.U. Silverthorn, *Comp. Biochem. Physiol.* **50A**, 281–283 (1975).
48. F. Van Herp, C. Bellon-Humbert, J.T.M. Luub, and A. Van Wormhoudt, *Arch. Biol.* **88**, 257–278 (1977).
49. B.S. Beltz, in H. Laufer and R.G.H. Downer, ed. *Endocrinology of Selected Invertebrate Types., Vol. 2*, Alan R. Liss, New York, 1988, 235–258.
50. D.P. De Kleijn and F. Van Herp, *Comp. Biochem. Physiol. B Biochem. Mol. Biol.* **112**, 573–579 (1995).
51. H. Dircksen, *Frontiers in Crustacean Neurobiology*, 485–491 (1990).
52. M. Fingerman, *Scientia* **105**, 1–23 (1970).
53. M. Fingerman, R. Nagabhushanam, and R. Sarojini, *Zoological Science* **10**, 13–29 (1993).
54. M.L. Grieneisen, *Insect Biochem. Mol. Biol.* **24**, 115–132 (1994).
55. Y. Hasegawa, E. Hirose, and Y. Katakura, *Amer. Zoo.* **33**, 403–411 (1993).
56. E. Homola and E.S. Chang, *Comp. Biochem. & Physiol. B: Comp. Biochem.* **117**, 347–356 (1997).
57. F.I. Kamemoto and S.N. Oyama, *Neuroendocrine Influence on Effector Tissues of Hydromineral Balance in Crustaceans*, Hong Kong University Press, Hong Kong, 1985.
58. R. Keller, *Experientia* **48**, 439–448 (1992).
59. A. Krishnakumaran, **25**, 79–92 (1962).
60. H. Laufer, J.S.B. Ahl, and A. Sagi, *American Zoologist* **33**, 365–374 (1993).
61. L. Liu and H. Laufer, *Archives Insect Biochem. Physiol.* **32**, 375–385 (1996).
62. D.R. Nassel, *Cell and Tissue Research* **273**, 1–29 (1993).
63. G.D. Prestwich, K. Touhara, L.M. Riddiford, and B.D. Hammock, *Insect Biochem. Mole. Biol.* **24**, 747–761 (1994).
64. K.R. Rao, *Pigment Cell Res. Suppl.* **2**, 266–270 (1992).
65. E.A. Santos and R. Keller, *Comp. Biochem. Physiol. A—Comp. Physiol.* **106**, 405–411 (1993).
66. D. Soyez, *Acad. Sci.* **814**, 319–23 (1997).
67. F. Vanherp, *Invert. Reproduction Dev.* **22**, 21–30 (1992).

See also SHRIMP CULTURE.

F

FEE FISHING

ROBERT R. STICKNEY
Texas Sea Grant College Program
Bryan, Texas

OUTLINE

In many parts of the United States and some other countries, it is possible to find privately owned ponds and lakes that have been stocked with fish and to which access can be gained by anglers. Charges for recreational fishing are usually assessed on the basis of the weight of fish caught; this type of fishing is called "Fee Fishing." Many fee fishing operators are also fish culturists. They may produce fish only for stocking their facilities, or they may both stock their own water bodies and sell excess fish to other culturists or fee fishing operations. Fee fishing has been present in the United States for at least a few decades (1). In Brazil, fee fishing is a more recent development—one that absorbs a large percentage of the fish produced by commercial aquaculturists in that nation.

PROVIDING A GOOD FISHING EXPERIENCE

While the majority of anglers may consider taking a boat or hiking to some place on their favorite stream, lake, or reservoir to be one of the most satisfying parts of the fishing experience, merely getting away from a hectic life for a few hours also has appeal for an increasing number of people. That may be why many of the fee fishing operations established in the United States are within short drives of large urban areas. Fee fishing can provide an angling opportunity without requiring a major expenditure in time for preparation or money for travel and gear. As important, perhaps, is that persons who frequent fee fishing operations can almost be guaranteed that they will catch fish.

The quality of fee fishing operations varies widely. (See Fig. 1.) The more successful ones have well-maintained facilities and offer a variety of amenities, in addition to well-stocked bodies of water filled with hungry fish. Fee fishing operations do not have to be located on expansive sites or even outside of city limits. Small earthen or concrete ponds, circular tanks, and raceways constructed on large lots within a city have been used successfully for fee fishing in some instances. More serious anglers may not be as much interested in such facilities as in larger ones located in more pastoral settings, but local, virtually backyard operations have some appeal, particularly to children's groups.

It is often not possible to obtain permits to establish a fee fishing operation on a stream, because such waters are considered to be public in many countries. Thus, fee fishing is typically restricted to small ponds and reservoirs on private land. A fee fishing operation may have from one to several bodies of water associated with it. Successful operations may rotate ponds, allowing fishing in some parts of the facility while closing others. The parts of the facility open for fishing may be altered periodically. Ponds or reservoirs might undergo renovation and restocking or just be closed, in order to move fishing pressure around and provide anglers with new surroundings.

Ponds or reservoirs may be square or rectangular in shape (see the entry "Pond culture"), or, to make them more aesthetically appealing, they may have irregular shapes. Pond banks are often landscaped to provide at least the impression of being far removed from the urban environment, even when they are located in or very close to a major city. Grassy, well maintained sites, with trees to provide shade and additional beauty, are common features.

SPECIES OF FISH

Fishes commonly found at fee fishing operations in the United States are channel catfish (*Ictalurus punctatus*), largemouth bass (*Micropterus salmoides*), and trout (rainbow trout, *Oncorhynchus mykiss*, being perhaps the most popular). Historically, some crawfish (*Procambarus* spp.) producers in Louisiana have opened their ponds to those willing to pay for the privilege of trapping the desirable crustaceans (2).

Catfish and bass fee fishing operations exist in most states, while trout operations are seen only at altitudes and latitudes where the water is sufficiently cold to allow the fish to survive throughout the year. Some state fish

Figure 1. A rural fee fishing operation in Brazil provides a restful setting.

and game agencies stock public waters with trout in the spring and sell special licenses that allow anglers to fish for them, with the idea that by the time the water temperature becomes too warm to support the trout, most of the fish will have been caught. It would be possible for fee fishing operations to do the same thing, but most depend on having year-round fishing for one or two primary species, so one would rarely find trout in a facility that is better suited for the maintenance of bass or catfish.

In Brazil, various native species, including some of the Amazon River region catfish, have been stocked in fee fishing ponds. Tilapias (*Oreochromis* spp.), which are exotic to Brazil, can also be found. Fishing for tilapias involves attaching a feed pellet to a hook, since the fish consume plankton in nature. Cheese and some other types of bait have also been successfully employed, but tilapias, which readily accept prepared feeds, find feed pellets to be a familiar dietary item. An individual Brazilian fee fishing pond may contain a mixture of species, or each species might be stocked exclusively in its own ponds, thereby allowing the angler to select the particular fish he or she wishes to catch. (See Figs. 2 and 3.)

HOW IT WORKS

For small operations, it makes sense to purchase fish for stocking. Small facilities may not have the physical space to maintain broodstock, hatch eggs, and rear fish to catchable size. If the operator is sufficiently trained and has the space and personnel, maintaining spawning and rearing facilities may be the most desirable approach. Having the complete life cycle under control of the culturist helps ensure a good supply of fish of known quality and, in large facilities, may provide significant cost savings relative to purchasing fish of catchable size. A third option is to purchase fingerlings and rear them for whatever period is required for them to reach a size at which they can be stocked into fee fishing ponds.

Figure 2. This Brazilian fee fishing operation features an array of native and exotic species.

Figure 3. Ponds may contain multiple species or, as in this case, be stocked with only one species (in this case, red tilapia).

While the fish within a fee fishing pond vary in size, any that are caught should all be considered "keepers" by the anglers who fish for them. This is because, in most instances, fees are assessed based on the weight of the fish caught. At most fee fishing locations, anglers are required to keep all fish landed, a practice that could become contentious if the fish were considered too small to take home. Another alternative is to charge a set rate and allow anglers to take home a specified maximum number of fish.

By stocking heavily and limiting the amount of supplemental food provided in fee fishing ponds, the fee fishing operator can ensure that most anglers will catch fish relatively quickly and easily. As an incentive to keep anglers fishing once they have put a number of fish in the creel, as well as to keep the anglers returning for more, fee fishing operators may tag one or more fish that, if caught, will carry a reward rather than a cost to the angler.

As time passes, fish that have been caught and surreptitiously released or hooked only to escape, will learn to avoid a lure or baited hook. As a result, over a period of time, angler success may decline, even though there are plenty of fish in the pond. Removing fishing pressure by rotating ponds open to anglers may help reduce the problem by allowing time for the fish to "forget" what they have learned about lures and bait, but in some cases, it is best to drain the pond and remove all of the fish prior to restocking. The fish that carry reward tags, however, would probably be retained and restocked, since the operator is not particularly interested in having them landed; the more wary those fish are, the better.

AMENITIES

The wise fee fishing operator should cater to all of the needs of the anglers who frequent the facility. (See Fig. 4.) A snack bar can provide food and drink, while a bait shop

Figure 4. Amenities may be fancy or modest. Stools placed around this pond provide an extra touch of comfort for anglers.

should supply not only live bait, if used, but also fishing tackle for purchase or rent. Ice and coolers, sunscreen, first-aid supplies, and a number of other products that anglers might need are often available. Fishing licenses should also be available for sale to customers. Tours of hatchery facilities may or may not be made available.

Picnic tables and fire pits or barbecue grills can provide an added attraction for family outings. Toilet facilities, typically in the form of portable outhouses, should be readily available to customers. Coin-operated fish-feed dispensers may also be provided near rearing ponds, so that, for a cost, anglers also have the opportunity to feed fish that are being grown for future stocking.

Once the anglers decide to depart, they have their catch weighed, and the fee is assessed. Many operators will also clean the catch for an additional charge.

The success of fee fishing operations depends not only on providing anglers with a successful fishing experience, but also on location and expectations of anglers. Careful study of angler preferences, in terms of the species available for capture and the amenities considered desirable or indispensable, along with the level of interest, the willingness of potential customers to use the facility, and the frequency of use, help ensure success.

BIBLIOGRAPHY

1. D. Saults, M. Walker, B. Hines, and R.G. Schmidt, eds., *Sport Fishing USA*, United States Department of the Interior, Washington, DC, 1971.
2. J.E. Bardach, J.H. Ryther, and W.O. McLarney, *Aquaculture*, Wiley-Interscience, New York, 1972.

See also Black bass/largemouth bass culture; Brazil fish culture; Channel catfish culture.

FEED ADDITIVES

Frederic T. Barrows
USFWS, Fish Technology Center
Bozeman, Montana

Ronald W. Hardy
Hagerman Fish Culture Experiment Station
Hagerman, Idaho

OUTLINE

INTRODUCTION

Feed additives are nonnutritive ingredients or nonnutritive components of ingredients that are included in feed formulations. Since the goal in formulating and manufacturing feeds is to supply nutrition to the animal, it appears to be contradictory to include nonnutritive ingredients in feeds. However, feed additives are necessary to stabilize feeds to prevent deterioration during storage, or to improve fish health, nutrition, or product quality. Feed additives augment the nutritional value of feeds or otherwise increase feed efficiency or fish production.

COMPONENTS OF INGREDIENTS

Water and fiber are constituents of many feed ingredients and their levels in feeds need to be considered during feed formulation. Water is an essential substance for terrestrial animals, since intake of water is necessary for life and for nutrient metabolism to occur. Fish live in water, so it is not necessary to supply water in the diet. Natural food for fish contains 65–85% water. However, fish must drink water to hydrate and dissolve manufactured feed particles in the gut. In addition, some fish species consume moist feeds more readily than dry feeds. Newly hatched salmon are a noteworthy example (1,2). Whether this difference is associated with feed texture or a physiological response in the digestive tract is unknown. Growth rates of brown trout (*Salmo trutta*) and turbot fingerlings were not affected by dietary moisture content (3,4). The need for dietary water may be related to species and the salinity of the environment. Fish must swallow water to hydrate dry feeds, and fish living in saltwater must excrete the salt. More research is needed to determine the relationship between water in the feed and salinity of the rearing environment. Regardless of the nutritional or feed consumption effects of feed moisture level, high levels of water in feeds necessitates special storage methods (e.g., freezing) or the addition of antimicrobial or fungal compounds to prevent mold growth.

Fiber is a component of feed ingredients that is nonnutritive to most fish, has some beneficial effects at moderate inclusion levels, and can be detrimental to both feed quality and fish performance at high dietary levels. Fiber is indigestible to salmonids and other carnivorous fish and is highest in ingredients from plant origin because plants contain cellulose, hemicellulose, and lignins. Digestibility of fiber occurs as a result of enzymes produced by intestinal microflora in the digestive tract of species such as channel catfish (5). Digestive enzymes such as cellulase, which are needed to utilize the energy from fiber, are not produced by fish.

The beneficial effects of fiber are associated with its ability to stabilize or blend pellets. Low levels of fiber increase the stability of pellets produced by some processing methods. Fiber is added to semipurified diets used in research. Some types of fiber are added to practical diets, such as those manufactured by compression pelleting, which enhance pellet hardness and durability. However, high levels of fiber in a feed lower pellet stability, causing the feed to disintegrate quickly in water or fracture during handling.

ADDITIVES THAT AFFECT FEED QUALITY

Binders

A variety of substances are added to feeds to facilitate the binding of ingredients into a feed particle. Increasing the strength of the pellet increases feed efficiency by reducing the production of fines during shipping, handling, and after the feed is put in the water. Also, for species that require high feeding rates, or when feed is supplied improperly, a water-stable pellet will not disintegrate and foul the water. Some binders act as a lubricant during processing, thereby increasing production rates and decreasing horsepower requirements and wear on pellet dies.

Binders should be selected to match the requirements of the species fed, type of feed, and manufacturing method used to make pellets. Some binders are activated by heat and pressure while others require only moisture. Some binders provide additional nutritional value to the feed. Inclusion rates of binders vary; nonnutritive binders are limited to only 2–5% of the feed, while nutritive binders may be included up to 20–30%. Higher inclusion rates are appropriate for nutritive ingredients that have binding properties, such as fish hydrolysates.

The primary ingredients used in aquatic feeds vary in their ability to form a pellet, or pelletability. Inclusion of cereal grains (e.g., wheat flour and soybean meal) facilitates pellet formation. Other ingredients (e.g., wheat gluten, krill meal, and liver meal) also have high pelletability, but they are relatively expensive and are generally used only in larval, starter, or specialty feeds. Fish meal is the main protein source used in many aquatic feeds and has neutral to negative effects on feed pelletability. Use of low-ash fish meals (e.g., products with the bones removed either before or after drying) seem to make feed mixtures more difficult to pellet. Higher moisture levels can overcome this problem, but require more energy to dry the pellet, thus increasing processing costs.

Many types of materials have been used for binding pellets and more are being developed. Aquatic feeds are just one of dozens of uses where pelleting is necessary. Other industries requiring pelleted products include terrestrial animal feeds, pharmaceuticals, human nutritional supplements, plastics, agricultural chemicals, fertilizers, and others. Binders used in aquatic feeds can be classified as plant extracts, animal extracts, minerals, polymers, and wood processing by-products. Carrageenan, alginates, and agar are extracts from marine plants and are used as binders in aquatic feeds. They are particularly useful in larval feeds and moist feeds. Products from plants include starches, pectins, molasses, and a wide variety of gums. Again the type of manufacturing used in pellet production must be considered when choosing a binder. Starches are the primary binder used for pellets produced by cooking extrusion, and molasses is most effective binder in compression pellets. Purified and semipurified diets used in nutritional research often use gelatin (extracted from terrestrial animal by-products) as a binder. Gelatin is also used in many specialty feeds. Collagen, also a by-product of terrestrial animal farming, is an effective binder in some applications. Urea formaldehyde condensation polymer is a pellet binder used in animal feeds at low levels (0.5–2%). It is not currently approved for use in feeds for aquatic species in the United States. However, the polymer is used in shrimp feeds in many parts of the world and is very effective.

Wood-processing by-products used in aquatic feeds include lignin sulfonate (Permapel), hemicellulose, and carboxymethylcellulose. These binders are used only with compression pelleting and are indigestible to fish. Besides contributing the binding of the feed pellet, lignin sulfonate allows more steam to be added during processing. The additional steam increases gelatinization of the starch, increasing the energy value to most species. Lignin sulfonate is added at levels of 2–4% of the feed. Spray-drying of the soluble by-product from pressed wood manufacturing produces a dry hemicellulose product. It is not used as often in fish feeds as lignin sulfonate and is limited to 2% of the feed. Carboxymethylcellulose is used more widely in applications other than in aquatic feeds, and its upper inclusion level is 2–3%.

Bentonite is a mineral (clay) that is mined for many purposes, and has been used in commercial aquaculture feeds for some time. It consists of trilayered aluminum silicate and is supplied in either a sodium or calcium salt form. Inclusion levels for both products are typically 1–2%. Sodium bentonite swells when water is added but calcium bentonite does not. These characteristics can affect final pellet quality and performance. Bentonite is also reported to bind with aflatoxins in the gut, thus carrying them through the digestion tract and preventing their absorption and subsequent toxic effects (6).

Moist and semimoist feeds require different binders than do dry feeds (<10% moisture). Moist feeds contain 25–70% water. For example, the H440 semipurified research diet and other such diets use gelatin as a binder, usually at 4–10% of the diet, depending on the specific feed formulation (7). Moist diets containing a mixture of dry meals and a wet-fish component sometimes use alginates to hold pellets together. A comparison of binders using a 41% moisture feed showed that calcium and alginate were more effective than corn starch, gum, agar, carboxymethylcellulose, chitosan, carragenaan, and collagen (8). Alginates require calcium ions to be activated. Calcium can be provided by fish bones or inorganic calcium sources added to the feed.

Antimicrobial Compounds

Mold, yeast, and bacteria can grow in feeds containing more than 12% moisture, unless the feed is frozen or the water activity of the feed is lowered by the addition of sugar, salt, glycerol, or propylene glycol. Water (i.e., steam) is added to feed mixtures during pelleting, and the resulting pellets are dried before being sold. Pellets generally are dried to 8–9% moisture, but exposure to humid conditions or rain can increase pellet moisture levels sufficiently to support mold growth. If mold is visible in a sack of feed, it should be discarded. Before mold is visible, it is well established. Many molds produce compounds that are toxic to fish (9) and, at the very least, will decrease feed consumption.

Mold inhibitors are routinely added to fish and animal feeds. Many proprietary products containing a combination of mold inhibitors are available. A list of antimicrobial compounds used in feeds is presented in Table 1.

Some fish feeds contain moisture levels in excess of 12%. The Oregon moist pellet (OMP) contains 28–32% moisture and must be stored frozen to prevent spoilage. Semimoist feeds containing 17–25% moisture are used as starter feeds for a variety of fish fry. In these feeds, microbial spoilage is controlled by a combination of approaches. The first element of microbial control is to begin with a feed mixture that has a low microbial load. Pasteurizing wet fish ingredients accomplishes this. A second element of microbial control is to add antimicrobial compounds to the feed. Naturally, the antimicrobial compounds must be ones that do not lower feed palatability or otherwise affect fish health or performance. The third element of control is to lower water activity, as mentioned earlier. This is the same principle used to preserve jam and jelly; sugar lowers water activity in these products, reducing the available or unbound water below levels sufficient to support microbial growth. The final element of microbial control is to package feed in a modified atmosphere of inert gases, such as nitrogen or carbon dioxide. Aerobic microorganisms cannot grow in an atmosphere lacking oxygen. Once the feed package is opened, it must be used rapidly or frozen.

Table 1. Antimicrobial Compounds Used in Feeds[a]

Compound	Limit
Benzoic acid	0.1%
Calcium propionate	None
Calcium sorbate	None
Distearyl thiodipropionate	0.005%
Formic acid	2.5%
Methylparaben	0.1%
Potassium bisulfate	Not for use in B_1 sources
Potassium metabisulfite	Not for use in B_1 sources
Potassium sorbate	None
Propionic acid	None
Propylparaben	0.1%
Sodium benzoate	0.1%
Sodium bisulfite	Not for use in B_1 sources
Sodium metabisulfite	Not for use in B_1 sources
Sodium nitrite	0.002%
Sodium propionate	None
Sodium sorbate	None
Sodium sulfite	Not for use in B_1 sources
Sorbic acid	None

[a]From Reference 10.

Antioxidants

Feeds for many fish species contain very high levels of polyunsaturated fatty acids relative to terrestrial animal feeds. Unsaturated oils are susceptible to oxidation, and oxidizing lipids decrease feed quality by forming harmful free radicals and other compounds and by destroying other nutrients. Once polyunsaturated fatty acids begin to oxidize, the oxidation process is self-sustaining (autoxidation). Antioxidants are chemical compounds that prevent or interrupt autoxidation.

The occurrence of autoxidation requires a substrate (e.g., polyunsaturated fatty acids), but is accelerated by a variety of environmental factors, including increased temperature and exposure to pro-oxidants e.g., copper and iron, light, UV radiation, and oxygen (10). Antioxidants protect the feeds from autoxidation, but an abusive feed storage condition will eventually result in autoxidation, regardless of antioxidant content of feeds.

Autoxidation consists of three steps that result in an autocatalytic reaction that will continue as long as substrate is available and no antioxidants are present. The initiation step requires oxygen, a substrate, and a catalyst and results in free radical production (10). The propagation step itself forms more free radicals, which is why the process is described as autocatalytic. The quantity of free radicals produced by the oxidation of one fatty acid increases with the number of double bonds possessed by the fatty acid. Free radicals and peroxyradicals combine to form stable products in the third step of the reaction, known as termination.

There are two types of antioxidants, hydrogen donors and chelates, that tie up metallic pro-oxidants, (e.g., copper and iron). Thus, by preventing the formation of free radicals (chelation) or by quickly making the free radicals unreactive (hydrogen donors), the antioxidants prevent oxidation from continuing. After antioxidants donate hydrogen atoms, they no longer have antioxidant properties, which is why they are called sacrificial antioxidants. When all of the sacrificial antioxidants in a feed are used up, oxidation proceeds rapidly.

Sacrificial antioxidants commonly used in aquatic feeds include ethoxyquin (santoquin), butylated hydroxytoluene (BHT), and butylatede hydroxyanisole (BHA) (11). Other antioxidants of this type include thiodipropionate, propyl gallate, and thiodipropionate, but those are not commonly used in commercial aquatic feeds. Ethoxyquin is an amine antioxidant and is usually added directly to oils, whereas BHT and BHA are added to mixed feed ingredients. Legal limits have been established for these additives in oils and feeds, but since these antioxidants are sacrificial, their detectable levels diminish as they react. Inclusion levels in finished feeds are typically 0.1% for BHA and BHT, and slightly lower for ethoxyquin. Tocopherols are natural antioxidants found mainly in plant oils, and alpha-tocopherol is a primary source of vitamin E. Beta, delta, and gamma tocopherols are effective antioxidants, but have less vitamin E activity than alpha-tocopherol.

Chelating antioxidants include both natural and synthetic compounds. Ethlyenediaminetetraacetic acid (EDTA) is a synthetic compound that chelates pro-oxidant metals and is commonly used to prevent oxidation. Ascorbic acid also functions as a chelate. A synergistic effect has been observed when phenolic or amine (sacrificial) antioxidants are combined with antioxidants that chelate pro-oxidants. This combination helps prevent oxidative damage by inhibiting participation of the pro-oxidant in the reaction and by reacting with any free radicals that are formed.

Some nutrients commonly founds in feeds also have antioxidant activity including, ascorbic acid, phytic acid, lecithin, and selenium (11). These sources can be expensive to add as antioxidants, and depletion in the feed over time may indicate that oxidation is in progress.

Feeding Stimulants

Circumstances are encountered in aquaculture whereby fish will not readily consume formulated feeds. These situations include, but are not limited to, first-feeding larvae, fry, or fingerlings being weaned from live food to formulated feed, low water temperatures, disease, culture of wild-strain fish, and addition of ingredients that reduce the palatability of feeds. Feed consumption is influenced by several factors, with flavor, taste, or smell of the feed being very important. Flavorings to enhance feed palatability have been used extensively in the pet food industry for many years and their use in aquaculture is increasing. Feeds for farmed fish species do not often use flavoring except in specific situations.

Single compounds and synthetic ingredients that increase fish feed consumption have been identified. Not all species of fish respond to all ingredients. Many natural ingredients are highly palatable and can increase feeding response and feed intake. Several feeding stimulants work synergistically. Fish meal and fish hydrolysates, krill meal and krill hydrolysates, shrimp meal, liver meal, fish solubles, and fish oils all have a positive response on the feeding behavior of many species of fish. Very high inclusion levels of the hydrolysates can decrease palatability and feed pH. High levels of liver meal can interfere with pellet formation depending on the processing method used. Thus, the use of these ingredients must be matched with feed pelleting constraints and fish feeding response.

ADDITIVES THAT AFFECT FISH PERFORMANCE AND QUALITY

Fish performance can be affected by feed additives in a number of ways. Metabolic reactions can be stimulated and directed by feeding enzymes and hormones. Pathogenic organisms in the gut can be controlled by feeding chemotherapeutants. Probiotics can enhance the immune response of fish. The addition of pigments to the feed results in flesh with appropriate color to meet consumer expectations, as is the case with salmon. In addition, pigment supplementation to feeds can alter external coloration of fish.

Enzymes

Increased growth, improved nutrient utilization, and pollution reduction are all reasons for adding enzymes to formulated feeds. If feeds only contained the ingredients that fish consume in the natural environment, enzyme supplementation would not be necessary. However, natural foods are expensive and not practical for most aquaculture operations.

Young animals typically feed on a variety of zooplankton, and it is thought that the endogenous enzymes of the zooplankton aid digestion. The growth rate of *Marsupenaeus japonicus* (shrimp) larvae was improved with the addition of microencapsulated bovine trypsin to a formulated feed (12). This observation alone was significant, but an increase in total endogenous protease activity was also observed. The supplemented enzymes seemed to initiate the metabolic process in the digestive system. Microencapsulated amylase, a starch digesting enzyme, also increased growth of shrimp larvae (12).

Enzymes can also be beneficial in adult animals with fully functional digestive systems. Phytate is the storage form of phosphorus in seeds, such as corn and wheat, can bind minerals making them unavailable to fish. Phytase is an enzyme that releases phosphorus from phytate, thus increasing its availability to monogastric animals (e.g., poultry, swine, and fish). Phytase is heat sensitive, and its activity is lost if it is added to feeds before extrusion pelleting. When it is added to fish feeds after pelleting by topdressing, it increases phosphorus availability, especially in catfish (13–17).

Hormones

Some feed ingredients contain anabolic steroids and others contain phytoestrogens, compounds that mimic the activity of estrogen. Fish meal produced from mature fish contains biologically significant quantities of testosterone compounds that stimulate muscle growth (18). Adding testes to fish carcasses increases the growth rate of salmon fed diets containing fish meal produced from the fish carcasses (19). The effects of phytoestrogens on fish growth or maturation is not well known. Phytoestrogens are present in several grains (9).

Hormones have been studied as feed additives and their use in fish production has been reviewed (20,21). Hormones are used to influence fish growth rates, sexual development, and osmoregulation. Sex reversal is the main use of hormones added to fish feeds at the present time, particularly for species such as tilapia where one sex grows faster than the other (11).

Recently, researchers have been investigating the effects of growth hormone on fish growth. Injections of recombinant bovine somatotrophin (bST) dramatically increases growth in rainbow trout, catfish, and sturgeon (*Acipenser* sp.) (22,23). An oral form of bST has not yet been developed. In the future, it may be feasible to add growth hormone to feed to influence growth and reduce the time required to raise long-lived fish, such as sturgeon, to maturation. Growth hormone is unlikely to be used in food fish production but might have application in stock restoration and reintroduction efforts.

Pigments

The color of flesh, skin, or eggs of farm-reared fish has often been used as a measurement of quality. Pigments found in salmon and other fishes are not synthesized but must be provided in the diet. A variety of natural pigments is found in both plant and animal feedstuffs. Some of these pigments improve product quality, but other pigments have a negative influence on product quality. Yellow pigment from xanthophyll in corn is undesirable in channel catfish (24) and trout (25), but is desirable for yellowtail (*Seriola quinqueradiata*), red sea bream, (7) and muskellunge (*Esox masquinongy*) (26). Differences among species have been observed in utilization of pigments (27–29).

Because of the relationship of color to perceived quality, pigmentation in farm-reared salmonids has been extensively studied (7,30,31). Carotenoid pigments are responsible for coloration in salmonids, and they must be added to feeds. Besides being important for consumer acceptance, carophylls (i.e., B-carotene) may serve as a precursor to vitamin A. Both canthaxanthin and astaxanthin are carotenoid pigments that provide a reddish-orange color to salmonids. Astaxanthin is the primary carotenoid consumed by wild salmon, and is available either in synthetic form or from natural sources.

Chemotherapeutants

Antibiotics can be effectively administered through the feed to combat pathogens. Prophylactic use of antibiotics, and their use to increase growth and feed efficiency is very controversial. Concerns for stimulating resistant strains of pathogens to antibiotics and introducing the drugs or their metabolites into human foods are the primary concerns. Because of these concerns and the economic cost, antibiotics are not frequently used in aquaculture to increase animal performance. Feed-borne therapeutants are often effective in treating disease. Currently, only oxytetracycline is approved for use in the United States, but efforts are underway to gain approval for florfenicol and there is some interest in pursuing approval for amoxycillin. In the United States, medicated feeds for both aquatic and terrestrial animal production have specific labeling requirements, withdrawal periods, and other restrictions required by the Food and Drug Administration of the U.S. Department of Agriculture. The aquaculture industry has a very limited number of antibiotics available for use, relative to terrestrial animals, due to regulatory restrictions and the cost of obtaining regulatory approval for minor use species.

BIBLIOGRAPHY

1. D.L. Crawford, D.K. Law, T.B. McKee, and J.W. Westgate, *Prog. Fish Cult.* **35**, 33–38 (1973).
2. D.A. Higgs, J.R. Markert, M.D. Plotkinoff, J.R. McBride, and B.S. Dosanjh, *Aquaculture* **47**, 113–130 (1985).
3. H.A. Poston, *J. Fish Res. Boad. Can.* **31**, 1824–1826 (1974).
4. P.J. Bromley, *Aquaculture* **20**, 91–99 (1980).
5. R.R. Stickney and S.E. Shumway, *J. Fish Biol.* **6**, 779–790 (1974).
6. R.W. Ellis and R.R. Smith, *Prog. Fish Cult.* **46**, 116–119 (1984).
7. National Research Council (NRC), *Nutrient Requirements of Fish*, National Academy Press, Washington, DC, 1993.
8. J.M. Heinen, *Prog. Fish Cult.* **43**, 142–145 (1981).
9. J.D. Hendricks and G.S. Bailey, *Fish Nutrition*, 2nd ed., Academic Press, New York, 1989, pp. 605–651.
10. R.W. Hardy, *Fish Nutrition*, 2nd ed., Academic Press, New York, 1989, pp. 475–548.
11. K. Jauncy, *Tilapia Feeds and Feeding*, Pisces Press Ltd., Stirling, Scotland, 1989, 146–155 (1998).
12. P.D. Maugle, O. Deshimaru, T. Katayama, T. Nagatani, and K.L. Simpson, *Bull. Jpn. Soc. Sci. Fish.* **49**, 1421–1427 (1983).
13. K.D. Cain and D.L. Garling, *Prog. Fish Cult.* **57**, 114–119 (1995).
14. L.S. Jackson, M.H.L., and E.H. Robinson, *J. World Aquacult.* **27**, 309–313 (1996).
15. M. Rodehutscord and E. Pfeffer, *Water Sci. Tech.* **31**(10), 141–147 (1995).
16. J.C. Eya and R.T. Lovell, *J. World Aquacult.* **28**(4), 386–391 (1997).
17. M.H. Li and E.H. Robinson, *J. World Aquacult.* **28**(4), 402–406 (1997).
18. S. Sower and R.N. Iwamoto, *Aquaculture* **48**, 11–18 (1985).
19. J.R. Borghetti, R.N. Iwamoto, R.W. Hardy, and S. Sower, *Aquaculture* **77**, 51–60 (1988).
20. E.M. Donaldson, U.H.M. Fagerlund, D.A. Higgs, and J.R. McBride, *Fish Physiol.* 455–497 (1979).
21. D.A. Higgs, J.R. Markert, M.D. Plotkinoff, J.R. McBride, and B.S. Dosanjh, *Comp. Biochem. Physiol.* **73b**, 143–176 (1982).
22. G.T. Schelling, R.A. Roeder, E.L. Brannon, J.C. Byatt, and R.E. Rompala, *World Aquacult. Soc.* Book of Absts., p. 476.
23. G.T. Schelling, R.M. Silflow, M.T. Casten, R.W. Hardy, and R.A. Roeder, *World Aquacult. Soc.* Book of Absts., p. 680 (1999).
24. P. Lee, *Carotenoids in Channel Catfish*, Ph.D. dissertation, Auburn University, Auburn, AL, 1986.
25. D.I. Skonberg, R.W. Hardy, F.T. Barrows, and F.M. Dong, *Aquaculture* **166**, 269–277 (1998).
26. A. Moore, *Federal Aid to Sport Fisheries*, Annual Report, Des Moines, Iowa, 1997.
27. M. Hata and M. Hata, *Bull. Jpn. Soc. Sci. Fish.* **38**, 339–343 (1972).
28. M. Hata and M. Hata, *Bull. Jpn. Soc. Sci. Fish.* **39**, 189–192 (1973).

29. M. Hata and M. Hata, *Bull. Jpn. Soc. Sci. Fish.* **38**, 203–205 (1976).
30. K.L. Simpson, C.O. Chichester, and T. Katayama, in *Carotenoids as Colorants and Vitamin A Precursers*, Academic Press New York, 1981, pp. 463–538.
31. O.J. Torrissen, R.W. Hardy, and K.D. Shearer, *Rev. Aquat. Sci.* **1**(2), 209–225 (1989).

See also Antibiotics; Drugs; Lipid oxidation and antioxidants.

FEED EVALUATION, CHEMICAL

Faye M. Dong
University of Washington
Seattle, Washington

Ronald W. Hardy
Hagerman Fish Culture Experiment Station
Hagerman, Idaho

OUTLINE

It is evident that the quality of fish feeds is determined in large part by the quality of the feed ingredients used to make the feed. High-quality fish feed cannot be produced from low-quality ingredients. The challenge for fish-feed manufacturers is to determine which measurable characteristics contribute to high-quality feed ingredients. Feed ingredients are complex mixtures of organic and inorganic compounds that are subjected to a variety of cooking, extracting, and drying processes, each of which can vary between manufacturers and between individual batches. Variation in plant products (from crop to crop), in animal or fish products (from species to species), and in the components of the raw primary material (plant or animal) that make up the utilizable by-product further complicate the task of making consistent the properties, composition, and quality of feed ingredients. Superimposed upon these variables are the potential changes that can occur during storage, first after a feed ingredient is produced but before it is used in feed manufacture, and then after a feed has been produced. Enumerating all of the possible factors that affect the quality of feed ingredients is an imposing task, but limiting our concerns to the major factors affecting the quality of principal feed ingredients is generally sufficient for the needs of the aquaculture feed industry.

It is also evident that the aquaculture feed industry needs to have rapid, simple, inexpensive chemical tests that accurately predict the nutritional quality of feed ingredients for fish and shrimp. For simple feed ingredients (e.g., fats and oils) that are subject to just a few major quality problems, such tests exist. At the other end of the spectrum is fish meal, which is subject to nearly all the conceivable quality problems. For decades, scientists have devised chemical tests to predict the nutritional quality of fish meal, and, while these tests have proven useful, no single test has been developed that measures all of the quality parameters of fish meal. Between these two extremes lie most other feed ingredients; they often have relatively complicated quality problems, ones that nonetheless are well-defined and are detectable with appropriate testing.

This entry covers the main factors that govern the quality of the major feed ingredients used in aquatic feeds, and the various ways in which specific problems can be identified and, if possible, quantified. The quantification is important, because no feed ingredient is perfect with respect to quality, and price should correlate with quality. As the sensitivity of analytical methods improves, and as new tests for measuring quality are developed, the aquatic feed industry must learn what limits are acceptable in feed ingredient quality testing, and which test results most accurately predict nutritional quality and enable the buyer to avoid poor-quality or potentially lethal batches of feed ingredients.

GENERAL CONSIDERATIONS

Feed Ingredient Definitions

Feed ingredients all have standard definitions that follow the International Feed Vocabulary, which is designed to give a comprehensive and concise name for each ingredient (1). Each name is based on six components: (a) the origin, including scientific name, common name, and chemical formula (if appropriate); (b) the part fed to animals, as affected by process(es); (c) the process(es) and treatment(s) to which the part has been subjected; (d) the stage of maturity or development; (e) cutting (applicable to forages), and (f) grade. An international feed name for each ingredient is based upon these components. For example, anchovy meal is named as follows:

Fish, Anchovy, *Engraulis ringen*, meal, mechanically extruded.

Soybean meal is named as follows:

Soybean, *Glycine max*, seeds without hulls, meal, solvent extracted.

Each feed ingredient also has been assigned an international feed number, which is a six-digit number used as identification, especially in computer databases. The feed identification numbers for the anchovy meal and soybean meal are 5–01–985 and 5–04–612. These feed ingredient names and numbers and the nutritional data can be found in a number of publications (2,3). In the United States, the Association of Feed Control Officials (AAFCO) provides definitions of feed ingredients that are more descriptive, having been developed over many years. The AAFCO utilizes international feed names and numbers in its ingredient identification systems. This information is available through the Feed Composition Data Bank, National Agricultural Library, 5th Floor, 10301 Baltimore Blvd., Beltsville, MD, 20705, USA.

Chemical Tests: Limits and Value

The results of chemical tests are used to evaluate and rank feed ingredient quality. The aims of the tests are to determine whether batches of feed ingredients meet the specifications associated with high quality. For the most part, the results of these tests enable ingredient buyers to avoid poor-quality feed ingredients: those that do not meet specifications because of adulteration, poor-quality raw material, inadequate or excessive processing, deterioration associated with storage, or microbial contamination. The value of these tests may be compromised by a number of factors, including improper or non-representative sampling of the feed ingredient, mistakes made in the laboratory while conducting the tests, inadequate replication of measurements, imprecise lab technique, errors in calculations, and lack of understanding of the value and limits of the results. What follows are descriptions of the major chemical tests conducted on aquatic feed ingredients, with emphasis on why each test is done, what it is intended to measure, and what the results mean with respect to ingredient quality, all as they pertain to feeds for aquatic species.

FISH MEAL

Steps in Manufacturing that Affect Quality

World production of fish meal has remained relatively constant in recent years, at an annual production level of 6–7 million metric tons, and it is not expected to increase significantly in the future (4). Currently, aquaculture uses about 15% of the world supply of fish meal each year, while 50% is used for poultry, 25% for swine, and 10% for all other uses. Fish meal remains the principal protein source in feeds for the salmonids, for shrimp, and for most marine species, constituting between 35 and 50% of most feed formulations (5). Because of its high proportion in aquatic feeds, its relatively high cost, and its importance in supplying amino acids for tissue synthesis, fish meal is the feed ingredient most closely scrutinized. Despite this scrutiny, defining high quality in fish meal remains a complex undertaking.

"Fish meal" is actually a generic term for a range of protein meals prepared from numerous species of fish and fish offal. Most fish meal is produced from whole fish of species not generally used directly for human consumption. The major species of fish used to produce fish meal and the countries in which various fish meals are produced are listed in Table 1. Because the species of fish used to produce fish meal determines many of its quality characteristics, fish meals are generally defined by species. Proximate composition of fish meal varies with species, primarily in the proportion of protein and ash in the final meal (Table 2). Fish meals produced from bony fish, such as mackerel or menhaden, contain a higher percentage of ash (minerals) and lower percentage of protein than do meals produced from less bony fish, such as herring, capelin, pilchard, and anchovy. The percentage of crude lipid is usually in the range 7–10% for herring, anchovy, and menhaden meals, and around 5% for white fish meal (Table 2). Another defining quality characteristic that depends upon the species of fish used to produce the meals is the color of the meal. Fish meals are sometimes categorized as either "white" fish meals or "brown" fish meals. White fish meals are produced from white-fleshed fish, such as pollock and whiting, while brown fish meals are produced from mackerel, menhaden, and anchovy. White fish meals are usually produced on fishing vessels from very fresh, cold-water fish; brown fish meals are produced in tropical or subtropical areas, where fish spoilage is rapid. In the past, white fish meals were considered superior to brown fish meals for use in aquatic feeds, most likely because of differences in quality associated with raw material freshness and/or conditions of manufacture. Because brown fish meals were typically transported by ship in bulk, the meals were allowed to age (oxidize) outdoors, to prevent combustion caused by

Table 1. The Types of Fish Meal Produced by Countries that Export It[a]

Country	Type(s) of Meals
USA	Menhaden (FD and SD)
	Pollock (white fish meal) (SD)
Canada	Herring (SD)
Peru	Anchovy (FD and SD)
Chile	Anchovy and horse mackerel (FD and SD)
South Africa	Pilchard
Norway, Iceland	Herring and capelin (LT)
Denmark	Sand eel, pout, and sprat (LT)
Japan	Sardine

[a]FD = flame-dried, SD = steam-dried, LT = low-temperature dried.

Table 2. Average Percentage of Protein and Ash (on an as-fed basis) in Various Fish Meals[a]

Type of Fish	Crude Protein (%)	Crude Lipid (%)	Ash (%)
Herring, capelin, sand eel	70–72	8–9	10–11
Anchovy, horse mackerel	65	7–8	14–15
Menhaden	64–65	9–10	17–19
White fish meal	62–63	5	21–22
Fish processing waste	55–60	9–10	18–24

[a]It is reasonable to expect that values for crude protein and ash should be within 1% of the published values.

heat production during shipping. Thus, the quality of the lipids in brown fish meals produced in South America was lower than that in white fish meals. The addition of antioxidants has reduced that problem. Recent studies indicate that there is little or no difference in nutritional quality between white fish meal and brown fish meal, provided that the freshness of the raw material and the conditions of manufacture are similar (6).

To understand the factors that determine the quality of fish meals and the rationale behind the chemical tests used to measure fish meal quality, one needs to be familiar with the manufacturing processes used in fish meal production. Fish are caught, then held onboard the catcher boats for various periods, before being delivered to fish meal factories. The choice of the net used to catch the fish and the way in which the fish are brought on board the fishing boat are the first steps in which quality can be affected. Large nets that allow fish to be crushed and net unloading practices that further physically damage the fish can reduce quality by rupturing cells and releasing proteolytic enzymes that begin to hydrolyze cellular and structural proteins in the fish. Enzymatic protein hydrolysis is accelerated by warm temperatures, meaning that the longer the fish remain in the hold of the fishing boat, especially without refrigeration or ice, the more hydrolysis will occur. The fish are then delivered to the fish meal factory, where they are unloaded into large pits prior to being processed. Again, thermal and physical abuse of fish at the bottom of pits increase the likelihood and extent of protein hydrolysis.

Once the fish are taken into the factory, they are rendered and dried to produce fish meal (Fig. 1). First, the fish are cooked, to release fish oil and to denature protein. Cooking is a continuous process, beginning with heaters and ending in a cooking vessel when the fish reach about 90 °C (194 °F). This thermal pressing is sufficient to denature and inactivate proteolytic enzymes in the fish, so there is no further enzymatic hydrolysis of the protein. No solid material, oil, or water is lost during this stage, because the cooking takes place in a closed system. After cooking, the fish is pressed on a screen to remove water and oil. The residue, or presscake, then moves to a dryer, where more water is removed by heat in the form of a flame, hot air, or steam-jacketed surfaces, the latter with or without vacuum. The liquid fraction removed by pressing is separated by centrifugation into water and oil fractions. The water fraction (stickwater), which contains soluble protein, is concentrated by heating. After concentration to about 50% moisture, the stickwater, now called "solubles," is generally added to the dryer where presscake material is being dried. Fish meal containing both presscake and solubles is called "whole fish meal"; fish meal made from dried presscake only is called "presscake meal." On some fish processing vessels containing fish meal factories, there is no room for the equipment for recovering and concentrating the stickwater. On these vessels, the stickwater is discarded. Most white fish meals made in the past were actually presscake meals.

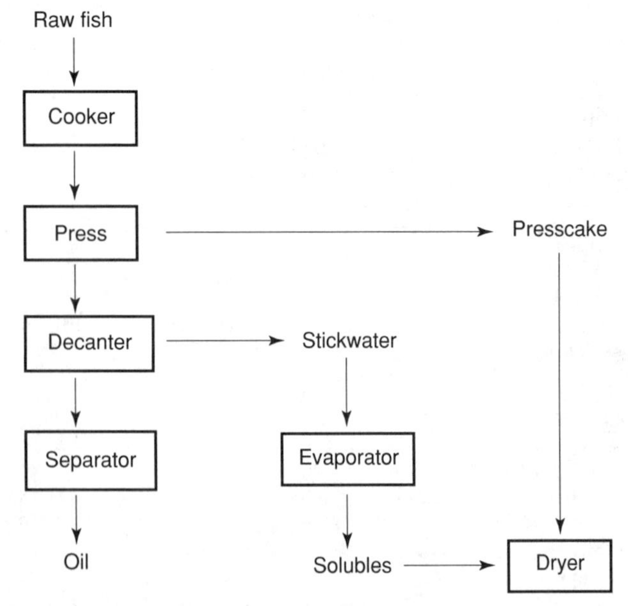

Figure 1. Simplified schematic presentation of the steps in the manufacturing of fish meal.

From this brief description of fish meal production, one can identify critical steps or stages where quality can be influenced by variability in the freshness of the raw material or by differences in processing. Chemical tests for fish meal quality yield results that relate to one or more of the critical steps in the process. The first stage affecting quality is raw material freshness, which is judged by the levels of volatile and biogenic amines both in the raw material and in the finished fish meal. The second stage affecting quality is presscake drying, which involves drying temperature and time. The third stage affecting quality involves the solubles: specifically, the conditions (time and temperature) of heating during concentration. The final stage of fish meal manufacturing affecting quality is whether or not solubles are added back to presscake (either from the same or from a different batch of raw material). This last point is often overlooked, and it can explain contradictory results with respect to fish meal quality that are sometimes obtained from different chemical tests on a single batch of fish meal, because some tests measure changes associated with the soluble fraction, others those with the presscake fraction.

Freshness Indices for Fishery Products

Degree of freshness of the starting material is an important factor contributing to the quality of the finished fish meal. The scale of degree of freshness runs from "fresh" (meaning a product having a chemical composition nearly identical to that of the live fish) to "deteriorated" or "spoiled" (meaning that chemical and microbial changes postmortem have proceeded significantly). While there are many factors that contribute to the rate of decomposition and to the chemical changes characteristic of spoilage (e.g., species of fish, handling, sorting of the product), the most important factor affecting quality is the postcapture, preprocessing holding temperature (7,8).

"Fresh vs. spoiled" should be distinguished from "frozen vs. not frozen," because it is a common practice to equate "frozen" with "not fresh." Proper freezing of fish postcapture includes having the fish muscle temperature pass through the critical freezing zone (0 to −5 °C, 32 to 23 °F) as rapidly as possible, to minimize (i) the formation of large ice crystals within tissue cells, (ii) the residence time of concentrated salt solutions with the proteins, and (iii) effects on intracellular pH changes (9). Proper freezing and storage of fish at −20 °C (−4 °F) or lower can help keep a product fresh. Therefore, a fresh product that has been properly frozen will still have many of the characteristics (chemical especially, and usually — depending on species — textural) of the freshly caught product; conversely, a spoiled product that has never been frozen should not be called "fresh."

The ideal test for fish freshness should be accurate, sensitive to the early stages of product decomposition, repeatable, rapid, relatively inexpensive, and applicable to a large number of species. The tests that have been developed and applied historically are the following: for microbial decomposition — total volatile nitrogen (TVN), total volatile bases (TVB), biogenic amines (e.g., histamine), and microbial profiles; for reduction of raw food fish freshness — nucleotides formed from the breakdown of adenosine triphosphate (ATP), principally inosine monophosphate and hypoxanthine (K_1 value); for overall microbial proliferation — microbial identification and numeration; and for general reduction of product freshness — degree of lipid oxidation. Although the K_1 value is the only truly sensitive index of early product freshness, it is typically used to assess fish for the sashimi market rather than fish destined to be processed for fish meal.

Total Volatile Nitrogen, and K_1 Value. There have been several tests used in the seafood industry to estimate the freshness of fish and shellfish. Several of these rely on the measurement of nitrogenous compounds; others attempt to estimate the level of lipid oxidation (discussed elsewhere in this volume) or degradation products of adenosine triphosphate (ATP). Total volatile nitrogen (TVN) can be measured; it includes both total volatile bases (TVB) and total volatile acids (TVA). TVB analysis measures low-molecular-weight volatile bases and amine compounds produced by microbial decarboxylation of amino acids; this test is commonly used as a microbial spoilage indicator. TVB also includes trimethylamineoxide (TMAO), which is a molecule in marine fish and shellfish associated with osmoregulation. In frozen fish, TMAO is reduced by endogenous enzymes to dimethylamine (DMA) and formaldehyde, whereas in fresh or iced fish, it is reduced by bacterial enzymes to trimethylamine (TMA) (10,11).

Although the various fractions of TVN can be measured by methods such as distillation/titration (12), flow injection analysis with gas diffusion (13), and gas chromatography (14), the volatility of these compounds affects their levels in products (e.g., fish meal) that have undergone thermal processing, or in pelleted feeds. In fish meal production, for example, the TVN compounds can evaporate during the drying stage. Pike (15) held herring at about 7 °C for up to seven days before processing into fish meal. TVN values in the raw material increased from 22 to 143 mg N per 100 g fish during this period. Norwegian LT-94 fish meal specifies TVN values no higher than 40 mg N per 100 g fish at the time of processing, compared to allowable levels for lower grade fish meal no higher than 90 mg N per 100 g fish (16). If there were no loss of TVN during the fish meal drying process, TVN values should be concentrated by four to five times in dried meal. However, TVN values in herring, menhaden, anchovy and LT-94 fish meals typically range from about 28 to 155 mg per 100 g sample (17), because of the loss of TVN compounds during drying. In frozen herring, TVN values measured in the fish before drying increased, from 83.5 mg N per 100 g after thawing to between 337 and 412 mg N per 100 g, after 12 days of storage at 2–5 °C (35.6–41 °F) (15). Values in dried presscake meal ranged from 82 to 100 mg N per 100 g, far below the theoretical range of 1350 to 2060 mg N per 100 g another demonstration that TVN compounds are lost during drying.

Biogenic Amines. Conditions that favor the formation of biogenic amines and the analytical methods to measure them are discussed in the section "Tests for fish meal quality performed on dried meal."

Nucleotide Degradation in Fresh Fish. The K_1 value is an index that is widely used in Japan for estimating the freshness of fish (18). The K_1 value is a ratio of the concentration of ATP to that of its degradation products (Fig. 2); it provides an assessment of freshness prior to bacterial spoilage. The assay of nucleotide degradation products required for the calculation of the K_1 value can be performed with high pressure liquid chromatography or by polarographic methods using immobilized enzymes (19). For fresh fish, K_1 values range from <0.2–0.4. The reactions that degrade ATP occur rapidly after harvest, especially if the fish are stressed during capture. The little ATP that is left is usually degraded during the first few days of refrigerated storage. Therefore, the K_1 value can be simplified by eliminating the adenosine nucleotide compounds from the calculation:

$$K_1 = 100 * ([HxR] + [Hx])/([HxR] + [Hx] + [IMP])$$

where HxR = inosine, Hx = hypoxanthine, and IMP = inosine 5′ − monophosphate (20).

Degree of Lipid Oxidation. Methods for measuring the degree of lipid oxidation are discussed in the entry "Lipid oxidation and antioxidants."

ATP → ADP → AMP → IMP → Inosine (HxR) →

Hypoxanthine (Hx) → Xanthine (Xa) → Uric acid

where:

ATP = adenosine triphosphate

ADP = adenosine diphosphate

AMP = adenosine monophosphate

IMP = inosine-5-monophosphate

Figure 2. Degradation pathway of ATP.

Tests for Fish Meal Quality Performed on Dried Meal

Proximate and Amino Acid Composition. The proximate composition of fish meal is the first characteristic upon which fish meal quality is judged. The composition depends upon the species of fish used as raw material, upon the amount of solubles added during drying, and upon whether fish processing waste (very high in ash) was used to produce the meal. The proximate composition of fish meals should be compared to published values (3), and meals having either protein levels lower than expected (by ≥1%) or ash levels higher than expected (by ≥1%) should be avoided (or else purchased at a discount).

The amino acid composition of fish meal can vary with the degree of thermal abuse during drying and with the proportion of processing by-product included in the raw material. Thermal abuse can reduce the protein digestibility of fish meal, and such a reduction is usually associated with decreased bioavailability of essential amino acids such as lysine and methionine (20a); thus, the nutritional value of the meal will be reduced by thermal abuse. The amino acid composition of fish meals produced from fish processing by-products, such as the by-products of filleting or surimi, can vary substantially, depending upon the proportions of muscle, viscera, skin, and bones in the raw material. In both cases, examination of the amino acid profiles can be useful. The concentrations of individual amino acids can be determined by acid or alkaline hydrolysis of the protein, followed by separation and quantification of the amino acids by ion-exchange, gas-liquid, or high-performance-liquid chromatography (20b). It is important to remember, however, that although the total amino acid compositions may be similar between two different fish meals, the *in vitro* digestibilities of their specific amino acids may differ (Table 3).

Water-Soluble Protein Test. The water-soluble protein test is sometimes used to provide insight into the extent of enzymatic hydrolysis (decomposition) in the raw material before processing, the assumption being that the products of hydrolysis (e.g., peptides and free amino acids) are soluble compounds. In practice, however, this test is not an accurate measure of the freshness of the raw material. In the fish meal manufacturing process, water-soluble protein is removed from the cooked fish during the pressing process, along with fish oil. After oil removal and concentration of the water fraction, the solubles may or may not be added back to the presscake in the dryer. Thus, applying the water-soluble protein test to fish meal will indicate whether the fish meal contains solubles, rather than indicate the freshness of the raw material. If solubles have been added back in order to produce whole fish meal, then the value for water-soluble protein should be around 30–35% of total protein. If the value is significantly higher, either more solubles were added than normal, or the raw material had undergone significant enzymatic hydrolysis before processing. If the value is very low (<5%), the fish meal is not whole meal, but rather presscake meal (without solubles). If the value is between 5 and 30%, the most likely cause is that a low proportion of solubles was added to the presscake during the drying process. This test is affected by the amount of fish solubles contained by the fish meal, so it does not necessarily measure the degree of enzymatic hydrolysis of the raw material or predict the nutritional quality of the fish meal.

Table 3. The Digestibility, by Catfish (C) and Atlantic Salmon (AS), of Several Amino Acids in Various Protein Ingredients (in %)[a]

Ingredient	ARG	LYS	MET	VAL
Canola meal (AS)	91.4	92.0	99.9	83.8
Herring meal (AS)	95.3	92.3	87.6	76.1
Menhaden meal (AS)	88.5	87.6	83.1	86.3
Menhaden meal (C)	—	86.4	83.1	87.1
Meat and bone meal (C)	87.9	86.7	80.4	80.8
Peanut meal (C)	97.7	94.1	91.2	93.3
Soybean meal (AS)	88.3	83.6	94.0	77.3
Soybean meal (C)	—	94.1	84.6	78.5

[a] Reference 3.

Biogenic Amines. Biogenic amines are compounds produced from amino acids by spoilage bacteria (Table 4). Their levels in a fish meal reflect both the extent of the enzymatic hydrolysis of the raw material and the conditions of processing. Although all fish meals likely contain biogenic amines at low levels that vary according to the species of fish used to produce the fish meal, elevated levels suggest that the raw material used to make the fish meal had undergone proteolytic and/or microbial degradation. Unlike TVN compounds, biogenic amines are heat-stable and do not volatilize or evaporate during the typical drying process of fish meal. Biogenic amines are water-soluble; they separate from presscake along with the soluble fraction and remain with the solubles. Typical levels of biogenic amines in fish meal produced from fresh, from moderately fresh, and from spoiled fish show a dramatic upward trend (Table 5). High levels of added histamine in fish feeds have been reported to cause abnormalities in the digestive tract of rainbow trout (21,22). These changes are not necessarily associated with reduced growth rates or mortality in fish, in contrast to findings with poultry. Of particular interest is the biogenic amine gizzerosine, so named because it causes gizzard erosion and death in chicks when present in their feed. Gizzerosine is a concern in fish meals made from scombroid fish such as mackerel, albacore, and tuna. Fish meal produced from stale scombroid fish contains high levels of histamine, which can combine with lysine during processing to form gizzerosine. Gizzerosine exerts its toxicity on chicks by stimulating excessive gastric

Table 4. Biogenic Amines and their Amino Acid Precursors

Biogenic Amine	Amino Acid Precursor
Cadaverine	Lysine
Gizzerosine	Histamine and lysine
Histamine	Histidine
Putrescine	Arginine
Tyramine	Tyrosine

Table 5. Levels of Amines in Fish Meal Produced from Herring Held for 12 Hours in Ice (Fresh), for 48 Hours at 7 °C (44.6 °F; Moderately Fresh), or for Seven Days at 7 °C (44.6 °F Stale)[a,b]

Category	Fresh	Moderately Fresh	Stale
Histamine (μg/g)	<30	440	830
Cadaverine (μg/g)	330	1,000	1,600
Putrescine (μg/g)	30	230	630
Tyramine (μg/g)	<30	400	800
TVN (mg/100 g)[2]	22	62	143
NH_3-N (g/16 g N)	0.12	0.16	0.25

[a]Reference 15.
[b]As measured in the raw material at the time of processing.

acid secretion. Because gizzerosine is difficult to measure chemically, bioassays using young chicks are conducted to identify fish meals containing gizzerosine (23). While the effects of gizzerosine on salmonids and other farmed fishes are unknown, the morphological changes in the stomachs of fish resulting from histamine addition to fish feeds are identical to those observed when fish meal containing high levels of biogenic amines, as determined by chick bioassay, are included in fish feed (22).

Methods available for the assay of biogenic amines include high-pressure-liquid chromatography (HPLC), gas-liquid chromatography, flow-injection analysis (24), and enzyme biosensors (24a). The HPLC method is becoming the method of choice for the quantification of histamine and the other biogenic amines. Conventional amino acid analysis has also been used for histamine analysis. For fish meals produced from scombroid fishes, histamine levels clearly reflect the freshness of the raw material. For other species, such as herring, reports in the literature are not in agreement. Some researchers report that histamine levels increase dramatically as the fish spoils (15); other researchers report that levels of cadaverine increase more dramatically than those of histamine (8).

***In Vitro* Digestibility (Pepsin and pH-stat. Digestibility).** Exposure of fish meal to high temperatures for long periods during the drying process can reduce the nutritional quality, by causing chemical linkages to form between amino acids that make them indigestible by fish and other animals (25,26). For years, the animal and fish feed industries have relied upon the pepsin digestibility test, performed in the laboratory to detect fish meals that have been subjected to thermal abuse during manufacture. This test relies upon the enzyme pepsin, usually obtained from pig stomachs, to digest the protein in the fish meal. If the protein has been damaged by thermal abuse during drying, then the pepsin digestibility value will be lower than the values typically obtained from high-quality fish meal (not thermally abused). The original pepsin digestibility test (27) was modified, by diluting the concentration of pepsin to increase the accuracy of the test, so that fish meals of average and high quality could be distinguished from each other (28). This modification is known as the Torry method. The original method could detect differences only between high- and low-quality fish meals; comparisons between average and high-quality fish meals gave pepsin-digestibility values of 97.7, 96.8 and 98.5% for menhaden, anchovy, and Norse LT-94 fish meals, respectively (17). By the Torry method, however, pepsin-digestibility values for the same fish meals were 84.0, 87.4, and 96.8%. No correlation was found between the AOAC pepsin-digestibility test and biological tests of fish meal quality, but a significant positive correlation was found between the results of the Torry pepsin-digestibility method and the protein quality of the fish meals as determined by biological tests (17).

A similar method for measuring the *in vitro* digestibility of fish meals and other protein sources is the multienzyme pH-stat method. This method uses a combination of proteolytic enzymes rather than a single enzyme (pepsin) to digest the sample, and it maintains a constant pH in the test solution during digestion. Recently, this method has been applied to aquaculture feed ingredients for salmonids, by substituting proteolytic enzymes extracted from the pyloric cacae of rainbow trout in place of enzymes from land animals (29). The use of digestive enzymes from shrimp hepatopancreas to perform an *in vitro* evaluation of protein in feeds for white shrimp also has been reported (29a). Several studies suggest that multienzyme tests accurately predict the biological value of fish meals (17,30).

Other Chemical Tests. Beside the recommended values of chemical tests discussed previously (Table 6), several other chemical tests designed to measure the effects of overheating of fish meal are usually listed as predictive of nutritional quality; however, recent research has not found them to be particularly accurate for fish. These are the available lysine test and the measurement of sulfhydryl groups and disulfide bonds in fish meals. Lysine is a reactive amino acid; under conditions of overheating and in the presence of suitable reactive compounds, such as glucose, lysine can form undigestible chemical linkages, which can be indirectly measured by a chemical assay (31). Similarly, overheating of fish meals causes the number of SH groups to decrease and the number of S–S bonds to increase (32). Anderson et al. (17) compared 10 fish meals and found differences in available lysine level, sulfhydryl groups, and disulfide bonds among the fish meals. Percentage available lysine in the fish meals was correlated with fish performance in feeding trials, but the test did not rank the fish meals in the exact order of nutritional value — that is, levels of sulfhydryl groups and

Table 6. Summary of Chemical Tests to Measure Fish Meal Quality for Aquatic Feeds

Chemical Test	Recommended Values
Total volatile nitrogen (TVN)	<60 mg N/100 g sample (raw material)
Total volatile nitrogen (TVN)	<150 mg N/100 g sample (meal)
Pepsin digestibility (Torry)	>87.5%
Histamine	<800 μg/g
In vivo "apparent digestibility" coefficient (protein)	>90% (>94%, with "feces settling" method)

disulfide bonds in the fish meals did not appear to be strongly correlated with fish performance. In summary, these tests are useful in characterizing fish meals for aquatic feeds, but they do not appear to provide any information on the nutritional quality of fish meals beyond that provided by the Torry pepsin digestibility test and by multienzyme digestibility tests.

MEALS MADE FROM ANIMAL BY-PRODUCT

General Considerations

Blood meal, poultry by-product meal, feather meal, and meat-and-bone meal are protein ingredients that have been used in fish feed formulations for many species. Each of these protein sources has at least one undesirable characteristic that has limited its use (Table 7). For example, poultry by-product meals from different manufacturers vary widely in ash content and in apparent protein digestibility (33). Meat-and-bone meal also varies in ash content between producers, depending upon the proportion of bone in the raw material used to make the meal (34). Variability in ash content makes it difficult to formulate a low-ash, low-phosphorus (environmentally friendly) fish feed from these ingredients. Feather meal, produced by hydrolysis of chicken feathers in a strong base under pressure, is a high-protein feed ingredient containing a relatively low ash content. Its use in fish feeds is, nevertheless, limited by variability in its amino acid profile and in its apparent protein digestibility among product from different manufacturers (35,36). Blood meal is another high-protein feed ingredient that has historically been used at levels up to 10% in many salmonid diets. Again, variability in apparent protein digestibility among products from different manufacturers (and occasional contamination with salmonella) limit its use in aquatic feeds (35).

Tests for Quality

For animal by-product meals, the proximate composition and the appearance are the first characteristics upon which quality assessment can be based. Meals having crude protein and ash values different from tabled values (3) should be avoided. The color and odor of animal by-product meals both suggest the degree of heating used during processing. Dark meals have likely been overheated; they will generally have lower apparent protein digestibility values than lighter meals. A burned odor can sometimes be detected in dark meals — another suggestion that the product has been overheated. Pepsin digestibility should be measured on animal by-product meals, and batches with substandard values should be avoided. Rendered products are available in various grades (e.g., regular, pet-food, and low-ash). The pet-food and low-ash grades are value-added products, produced by a combination of ash removal with blending to a specified protein and ash percentage. These specifications are important for pet-food manufacturers, but they are not predictive of ingredient quality for fish.

Table 7. Disadvantages of Some Alternate Protein Sources Potentially Suitable for Aquatic Feeds

Ingredient	Negative Qualities
Blood meal	Variable digestibility coefficients
Canola meal	High in fiber and in phytic acid
Corn gluten meal	Adds fiber; colors fish flesh yellow
Feather meal	Variable digestibility coefficients
Meat and bone meal	High ash level
Poultry by-product meal	Variable in quality; high ash
Rapeseed/soy protein conc.	High level of phytic acid
Soybean meal	Poor palatability, antinutritional factors
Wheat gluten meal	Too expensive

Other Animal/Fish Protein Sources

Molluscan and crustacean meals are often used in feeds for shrimp and marine fish. These specialty meals are produced, in relatively small amounts, by a variety of producers. One potential nutritional problem is the very high ash content of crustacean meals — especially shrimp and crab meals (~40% ash; Refs. 3,36a), if used at high substitution levels. Another problem is that variability in quality, especially in squid meal, can be very high. Most variation in quality results from the type of processing, mainly drying, used in meal production. Drying methods range from those used in food production (e.g., spray drying, freeze drying), through those used in fish meal production, to primitive methods such as sun-drying. There are no well-defined standards for these molluscan and crustacean meals; therefore, buyers must judge the quality of products by appearance, odor, proximate composition, and, possibly, pepsin digestibility. These ingredients are, however, usually used in feeds only at low substitution levels: as a source of trace minerals and carotenoid pigments (36a), to improve palatability and feed intake, and to supply specific essential nutrients (e.g., sterols for shrimp). They are less important in feeds as protein sources supplying amino acids for growth.

OILSEED MEALS AND CONCENTRATES

Manufacturing Considerations

Oilseed meals are produced from the residue or presscake of oilseeds grown and processed for their oil. The most common oilseed meals are soybean meal, canola (rapeseed) meal, sunflower meal, peanut meal, and cottonseed meal; soybeans constitute 50% of the world oilseed production (37). After the seeds are crushed, oil is extracted from the seed and the residue is dried. In some products, the hull of the seed is removed. The hull contains fiber; removing the hulls increases the percentage of protein in dried oilseed meal. Oilseed protein concentrates are products made from the de-oiled presscake by removing the soluble carbohydrate fraction and concentrating the protein fraction by a series of extractions (38). The proximate composition, nutritional value, and functional properties of oilseed protein concentrates vary with the method of production.

Generally speaking, the apparent protein digestibilities of oilseed protein concentrates are higher than those of oilseed meals. Typical apparent protein digestibility values for oilseed meals and concentrates, as measured in salmonids, are shown in Table 8.

Soybean Products: Quality Problems and Detection

Soybean meal, like many oilseed meals, contains antinutritional compounds that must be removed or inactivated by processing before the meal can be used in animal or fish feeds. The principal antinutritional components in soybean meal are trypsin inhibitors, which reduce protein digestibility by binding with the digestive enzyme trypsin in the intestine of the animal. Trypsin inhibitors are sensitive to heat, and ordinary presscake drying lowers the level of trypsin inhibitors in the dried meal to levels that do not affect the growth of most domestic animals and of some species of fish (e.g., catfish). Salmonids, particularly juveniles, are more sensitive than catfish to trypsin inhibitor level, and thus more extensive heat treatment is necessary to reduce residual trypsin inhibitor levels in soybean meal for salmon and trout (39). Overheating soybean meal, however, may reduce protein quality by causing reactions between amino acid residues, such as lysine, and portions of the carbohydrate fraction in soybeans, a process often leading to the Maillard reaction (40). The products of these reactions are generally indigestible. In soy protein concentrates, where the levels of trypsin inhibitor are usually low, the major concern is the level of phytic acid, which can form strong bonds with divalent cations.

Vohra and Kratzer (37) summarized the various chemical tests used to determine the adequacy of heat treatment of soybean meal. They divided the chemical tests into two groups: those that detect underheated soybean meal, and those that detect overheated meal.

Chemical Tests for Detecting Underheated Soybean Meal. Chemical tests to detect underheated soybean meal mentioned by Vohra and Kratzer (37) were the measurement of urease activity, of trypsin activity, and of protein solubility. Urease is an enzyme naturally present in soybeans; it does not have any known substantial nutritional relevance, but it is heat-sensitive, and its activity correlates positively with the activity of residual trypsin in dried soybean meal. It is also relatively easy to measure (27). Urease activity in commercial soybean meal ranges from 0.02 to 0.1 increase in pH (37); values over 0.5 increase in pH indicate insufficient heat treatment of the soybean meal. If no increase in pH is detected with the urease test, then this result might mean that the soybean meal has instead been over-heated, so some residual urease activity in the meal is preferred, at least for soybean meal intended for use in poultry feeds. Unheated soybean meal has a urease activity of >2.25 pH rise (41).

As mentioned, trypsin inhibitor activity in soybean meal decreases with heat treatment, in proportion to urease activity. Unheated, solvent-extracted soybean meal can contain more than 21 trypsin inhibitor units/mg sample (42), but commercial soybean meal subjected to normal heating during the presscake drying process generally contains about half the trypsin inhibitor activity of unheated meal. Additional heating further reduces trypsin inhibitor activity; the amount of reduction depends upon the temperature and the duration of heat treatment (Table 9). For salmonids, additional heat treatment increases fish growth rates when the treated meal is added to feeds, as compared to no heat treatment (45). The heat generated by steam extrusion is sufficient to lower trypsin inhibitor levels, at least in full-fat soybeans (44). Wilson (45) extruded full-fat soybeans having an initial trypsin inhibitor activity of 46.5 (in trypsin units inhibited per mg sample) and succeeded in reducing it to 8.1 after steam extrusion.

A third method for measuring the extent of heat treatment of soybean meal is the water solubility test; this test involves measuring Kjeldahl nitrogen levels in the soybean meal and in a water extract of the soybean meal (37). The method has been slightly modified by performing the extraction in 0.2% KOH (42). Heating decreases the percentage of 0.2% KOH-extractable protein, from about 99% in raw soybean meal to about 72% after 20 minutes of autoclaving; this decrease corresponds to a decrease in trypsin inhibitor units from 21.1 to 1.0 (42).

Chemical Tests for Detecting Overheated Soybean Meal. Excessive heat treatment of soybean meal is thought to reduce protein digestibility by causing the creation of protein-carbohydrate linkages that are indigestible by most animals. Tests for detecting overheated soybean meal are based upon the number of free functional groups in the protein fraction of the soybean meal; these groups can be detected by several dye-binding tests, by formaldehyde titration, or by a fluorescent derivative (37). Cresol Red dye binding is a relatively simple, rapid test requiring

Table 8. Digestibility by Salmonids of Oilseed Meals and Concentrates (in %)[a]

Ingredient	Dry Matter	Protein
Canola meal	53.5	83.5
Rapeseed protein concentrate	69.8	95.6
Soybean meal	61.7	77.0
Soybean protein isolate	68.4	86.3

[a]Reference 35.

Table 9. Results (± SD of Two Replications) of Two Chemical Tests Used to Detect Underheated and Overheated Soybean Meal[a]

Autoclaving Time (min)[b]	Trypsin Inhibitor Units/mg Sample	Protein Solubility (%)
0	181.1 ± 18.2	98.3 ± 0.5
5	123.7 ± 16.5	no data
10	16.2 ± 4.1	70.5 ± 1.1
20	1.8 ± 0.6	70.0 ± 0.5
40	0.9 ± 0.6	32.8 ± 1.1
60	0.9 ± 0.1	28.0 ± 0.6
90	1.1 ± 0.1	20.6 ± 0.0
120	1.0 ± 0.7	17.8 ± 0.5

[a]Reference 43.
[b]Heating at 120 °C (248 °F), 25 psi, and 17% moisture.

Table 10. Results of Various Chemical Tests Used to Detect Underheated and Overheated Soybean Meal[a]

Autoclaving Time (min)	Urease (pH Inhib. Change)	Trypsin Binding (activity/g)	Formalin (mL 0.05 N NaOH/2g)	Commassie Blue (mg RPE/ 100 mg prot)	Orange G Binding (mg/g meal)
0	1.95	39,000	8.8	67.3	47.8
15	0.10	480	7.1	44.7	40.2
30	0.10	0	6.6	32.9	38.9
45	0.07	0	6.0	30.0	35.3
60	0.05	0	5.5	28.9	34.4
120	0	0	5.3	15.7	33.5
180	0	0	5.2	12.6	29.1

[a]Reference 37.

only a spectrophotometer. Olomucki and Bornstein (44) reported that values in the range 3.8–4.3 mg Cresol Red absorbed per g meal indicate properly heated soybean meal, and that values over 4.3 indicate that overheating has occurred. Orange G dye binding and Commassie Blue are two other rapid, simple tests to detect overheated soybean meal; of the two, the Commassie Blue test is preferred, because it is more rapid and sensitive (Table 10) (46).

Other Oilseed Products: Quality Problems and Detection

The two major quality problems associated with other oilseed meals are glucosinolates in canola/rapeseed meals and gossypol in cottonseed meal. Glucosinolates interfere with the function of the thyroid gland in fish, and so pose problems during metamorphosis and maturation. The use of rapeseed/canola protein products in fish feeds has been thoroughly reviewed (47). Gossypol causes a number of problems in fish, including anorexia and increased lipid deposition in the liver. The use of cottonseed meal in fish feeds has also been thoroughly reviewed (48).

FATS AND OILS

Fish oils contain higher levels of polyunsaturated fatty acids than do oils extracted from oilseeds and other plants; therefore, they are more susceptible to oxidation. The use of high-energy feeds in salmon farming requires the addition of over 20% fish oil; that supplementation level increases the potential for oxidative problems, both during lipid storage in the feed manufacturing plant and after pelleting, before the feed is used. Oxidation of polyunsaturated oils produces free radicals, peroxides, and other potentially toxic or reactive compounds, and feeding fish a diet containing oil that is in the process of oxidizing may cause signs of vitamin E deficiency in fish of previously marginal vitamin E status (3). Most fish and plant oils contain naturally occurring antioxidants that, up to a point, prevent oxidation of fatty acids; beyond that point, oxidation occurs very rapidly. The antioxidants ethoxyquin, BHA, BHT, and propyl gallate are commonly added to fish oils to prevent lipid oxidation. Other quality concerns with fish oils are the percentages of free fatty acids, moisture, and nitrogen. Free fatty acids are always present in fish oils, but high levels of them suggest that the oil has been subjected to abusive processing and has undergone the hydrolysis of triglycerides to free fatty acids. High levels of moisture and nitrogen suggest that the separation of the oil fraction from the stickwater fraction during manufacturing was not done properly, and so there was some carryover of water and water-soluble protein into the oil. For a more thorough discussion of fish oil quality and oxidation, see the separate entry in this volume.

Table 11. Specifications for Fish Oils Used in Aquatic Feeds[a]

Category	Recommended Value
Free fatty acids	<3%
Moisture	<1%
Nitrogen	<1%
TBARS[b]	<25 nm malonaldehyde equivalents/g
Peroxide value (PV)	<5 meq/kg
Anisidine value (AV)	<15 meq/kg
Totox[c]	<20

[a]References 3, 16.
[b]TBARS = thiobarbituric acid reactive substances.
[c]Totox = 2 × PV + AV.

BIBLIOGRAPHY

1. L.E. Harris, H. Haendler, R. Riviere, and L. Rechaussat, *International feed databank system; an introduction into the system with instructions for describing feeds and recording data.* Publ. 2. Prepared on behalf of INFIC by the International Feedstuffs Institute, Utah State University, Logan, UT, 1980.
2. National Research Council (NRC), *United States–Canadian Tables of Feed Composition*, National Academy Press, Washington, DC, 1982, p. 148.
3. National Research Council (NRC), *Nutrient Requirements of Fish*, National Academy Press, Washington, DC, 1993, p. 114.
4. H.M. Johnson, *Annual Reports on the United States Seafood Industry*, 6th ed., H.M. Johnson and Associates, Bellevue, WA, 1998, p. 88.
5. G.L. Rumsey, *Fisheries* **18**, 14–19 (1993).
6. T. Watanabe, H. Nanri, S. Satoh, M. Takeuchi, and T. Nose, *Bull. Jpn. Soc. Sci. Fish.* **49**(7), 1083–1087 (1983).

7. N.C. Jensen, in S. Keller, ed., *International Conference on Fish By-Products, Alaska Sea Grant Program, 1990 Report No. 90-07* Fairbanks, AK, 1990, pp. 127–130.
8. S. Clancy, R. Beames, D.A. Higgs, B. Dosanjh, N.F. Haard, and B. Toy, *Aquaculture Nutrition*, 1995.
9. J.J. Licciardello, Freezing, in R.E. Martin and G.J. Flick, eds., *The Seafood Industry*, Van Nostrand Reinhold, New York, 1990, pp. 205–218.
10. J.M. Regenstein, M.S. Schlosser, A. Samson, and M. Fey, in R.E. Martin, G.J. Flick, C.E. Hebard, and D.R. Ward, eds., *Chemistry and Biochemistry of Marine Food Products*, AVI Publishing Co., Westport, CT, 1982, p. 137.
11. C.E. Hebard, G.J. Flick, and R.E. Martin, in: G.F. Flick, C.E. Hebard, and D.R. Ward, eds., *Chemistry and Biochemistry of Marine Food Products*, AVI Publishing Co., Westport, CT, 1982, p. 149.
12. A.D. Woyewoda, S.J. Shaw, P.J. Ke, and B.G. Burns, *Can. Tech. Rep. Fish. Aquacult. Sci.*, No. 1448 (1986).
13. M.M. Wekell, T.A. Hollingsworth, and J.J. Sullivan, in: D.E. Kramer and J. Liston, eds., *Seafood Quality Determination*, Elsevier Science Publishers B.V., The Netherlands, 1987, pp. 17–26.
14. M.E. Krzymien and L. Elias, *J. Food Sci.* **55**, 1228–1232 (1990).
15. I.H. Pike, in S.J. Kaushik and P. Luquet, eds., *Fish Nutrition in Practice*, INRA, Paris, 1993, pp. 843–846.
16. R.W. Hardy and T. Masumoto, in S. Keller, eds., *Proc. International Conference on Fish By-Products, Alaska Sea Grant Program, 1990 Report No. 90-07*, Fairbanks, 1990, pp. 109–120.
17. J.S. Anderson, S.P. Lall, D.M. Anderson, and M.A. McNiven, *Aquaculture* **115**, 305–325 (1993).
18. S. Ehira and H. Uchiyama, in D.E. Kramer and J. Liston, eds., *Seafood Quality Determination*, Elsevier Science Publishers B.V., The Netherlands, 1987, pp. 185–220.
19. J.F. Morin, in M.N. Voight and J.R. Botta, eds., *Advances in Fisheries Technology and Biotechnology for Increased Profitability*, Papers from the 34th Atlantic Fisheries Technology Conference and Seafood Biotechnology Workshop, Technomic Publishing Co., Inc. Lancaster, PA, 1990, pp. 481–485.
20. I. Karube, H. Matsuoka, S. Suzuki, E. Watanabe, and K. Toyama, *J. Agricult. Food Chem.* **32**, 314–319 (1984).
20a. Z. Sikorski, N. Haard, T. Motohiro, and B. Sun Pan, in P.E. Doe, ed., *Fish Drying and Smoking*, Technomic Publishing Co., Inc. Lancaster, PA, 1998, pp. 89–115.
20b. FAO/WHO. *Protein Quality Evaluation.* Report of a Joint FAO/WHO Expert Consultation on Protein Quality Evaluation. Bethesda, MD, Dec. 4–8, 1989. World Health Organization—Geneva and Food and Agriculture Organization of the United Nations—Rome, 1990, pp. 66.
21. T. Watanabe, T. Takeuchi, S. Satoh, K. Toyama, and M. Okuzumi, *Nippon Suisan Gakkaishi* **53**(7), 1207–1214 (1987).
22. W.T. Fairgrieve, M.S. Myers, R.W. Hardy, and F.M. Dong, *Aquaculture* **127**, 219–232 (1994).
23. J.J. Romero, E. Castro, A.M. Diaz, M. Roveco, and J. Zaldivar, *Aquaculture* **124**, 351–358 (1994).
24. J.M. Hungerford, K.D. Walker, M.M. Wekell, J.E. LaRose, and H.R. Throm, *Analytical Chem.* **62**, 1971–1976 (1990).
24a. K.B. Male, P. Bouvrette, J.H.T. Luong, and B.F. Gibbs, *J. Food Sci.* **61**, 1012–1016 (1996).
25. A.E. Bender, *J. Food Technol.* **7**, 239–250 (1972).
26. I.M. McCallum and D.A. Higgs, *Aquaculture* **77**, 181–200 (1989).
27. Association of Official Analytical Chemists (AOAC), *Official Methods of Analysis of the Association of Official Analytical Chemists*, 15th ed., Association of Official Analytical Chemists, Inc., Arlington, VA, 1990, p. 1298.
28. J. Olley and R. Pirie, *Int. Fish. News* **5**, 27–29 (1966).
29. L.E. Dimes and N.F. Haard, *Comp. Biochem. Physiol.* **108A**, 349–362 (1994).
29a. J.M. Ezquerra, F.L. García-Carreño, R. Civera, and N.F. Haard, *Aquaculture* **157**, 251–262 (1997).
30. L.E. Dimes, N.F. Haard, F.M. Dong, B.A. Rasco, I.P. Forster, W.T. Fairgrieve, R. Arndt, R.W. Hardy, F.T. Barrows, and D.A. Higgs, *Comp. Biochem. Physiol.* **108A**, 363–370 (1995).
31. K.J. Carpenter, *Biochem. J.* **77**, 604–610 (1960).
32. J. Opstvedt, R. Miller, R.W. Hardy, and J. Spinelli, *J. Agricult. Food Chem.* **32**, 929–935 (1984).
33. F.M. Dong, R.W. Hardy, N.F. Haard, F.T. Barrows, B.A. Rasco, W.T. Fairgrieve, and I.P. Forster, *Aquaculture* **116**, 149–158 (1993).
34. P. Brooks, *Feedstuffs* **63**(28), 13–16,22 (1991).
35. W.E. Hajen and D.A. Higgs, *Aquaculture* **112**, 333–348 (1993).
36. D.P. Bureau, A.M. Harris, and C.Y. Cho, *Aquaculture* **180**, 345–358 (1999).
36a. R.W. Hardy, in J.E. Halver, ed., *Fish Nutrition, Academic Press*, New York, 1989, pp. 475–548.
37. P. Vohra and F.H. Kratzer, *Feedstuffs* **63**(8), 22–28 (1991).
38. R.R. Stickney, R.W. Hardy, K. Koch, R. Harrold, D.A. Seawright, and K.A. Massee, *J. World Aquacult. Soc.* **27**, 57–63 (1996).
39. G.L. Rumsey, J.G. Endres, P.R. Bowser, K.A. Earnest-Koons, D.P. Anderson, and A.K. Siwicki, in C.E. Lim and D.J. Sessa, eds., *Nutrition and Utilization Technology in Aquaculture*, AOCS Press, Champaign, IL, 1995, pp. 166–188.
40. A.K. Smith and S.J. Circle, *Soybeans: Chemistry and Technology.* Volume 1., The Avi Publishing Co., Inc., Westport, CT, 1972.
41. P.W. Waldroup, B.E. Ramsey, H.M. Helwig, and N.K. Smith, *Poultry Sci.* **64**, 2314–2320 (1985).
42. M. Araba and N.M. Dale, *Poultry Sci.* **69**, 76–83 (1990).
43. R.E. Arndt, R.W. Hardy, S.H. Sugiura, and F.M. Dong, *Aquaculture* **180**, 129–145 (1999).
44. E. Olomucki and S. Bornstein, *J.A.O.A.C.* **43**, 440–441 (1960).
45. T.R. Wilson, *Full-fat soybean meal—an acceptable, economical ingredient in chinook salmon grower feeds.* Ph.D. Dissertation, University of Washington, 1992, p. 181.
46. F.H. Krazer, S. Bersch, and P. Vohra, *J. Food Sci.* **55**, 805–807 (1990).
47. D.A. Higgs, B.S. Dosanjh, A.F. Prendergast, R.M. Beames, R.W. Hardy, W. Riley, and G. Deacon, in C.E. Lim and D.J. Sessa, eds., *Nutrition and Utilization Technology in Aquaculture*, AOCS Press, Champaign, IL, 1995, pp. 130–156.
48. E.H. Robinson and M.H. Li, in C.E. Lim and D.J. Sessa, ed., *Nutrition and Utilization Technology in Aquaculture*, AOCS Press, Champaign, IL, 1995, pp. 157–165.

See also DIETARY PROTEIN REQUIREMENTS; ENERGY; LIPIDS AND FATTY ACIDS; MINERALS.

FEED HANDLING AND STORAGE

Timothy O'Keefe
Aqua-Food Technologies, Inc.
Buhl, Idaho

OUTLINE

INTRODUCTION

Prepared feeds for fish and shrimp are perishable products. Depending on the type of feed, they are also more or less fragile. Feed processors attempt to formulate and manufacture aquaculture feeds to extend their shelf life and improve durability. However, the degree to which aquaculturists can reduce wasted feed and realize its full purchase value is ultimately dependent on how well the basic principles of feed storage and handling are understood and applied.

Little practical information has been published specifically on proper storage and handling of the most common types of feed currently used in aquaculture. Even though feed most often represents the greatest percentage of the total cost of producing aquatic species, and substantial amounts can potentially be wasted through spoilage and breakage, proper storage, and handling of dry, semimoist, and moist feeds are usually only addressed in the literature in a general sense. The specifics are left to assumption.

This article is intended to provide some detailed discussion, and information references where possible, on the most common causes of degradation and waste of aquaculture feed on the farm. It is impractical to address every conceivable storage and handling situation that may occur with each type of feed. However, the guidelines presented here, along with some practical recommendations, should help in those instances where judgments or compromises are required.

STORAGE

For reasons of cost and convenience, dry and semimoist diets are currently the most widely used feeds in aquaculture. The general rule for preservation of both types of feeds is to store them in a dry, well-ventilated area that affords some protection from rapid changes in temperature. Cooler temperatures are best, although actual ambient temperature is less important than minimizing extreme changes. Any storage facility should also provide adequate containment for control of pests.

Moist feeds, such as the Oregon Moist Pellet (OMP), require more demanding and costly storage conditions. With moisture levels averaging between 30 and 35%, these feeds must be received in freezer vans or containers and maintained in a frozen state until ready for use. The length of time they can be held in good condition is highly dependent on the storage temperature. The recommended temperature for maximum, long-term storage of OMP feed is −18 °C (1). Under practical conditions, however, temperatures below −11 °C are sufficient to hold OMP for up to 3 months (2).

No matter what type of feed is used, there is little or nothing that can be done to enhance its potential storage stability once it has been delivered to the farm. Much of what is subsequently done during storage, however, can substantially affect whether a feed can remain acceptable over the intended shelf life. A practical knowledge of some factors contributing to feed degradation and a little attention to maintaining proper storage conditions for dry, semimoist, and moist feeds can significantly minimize the loss of vitamin potency, mold growth, fat rancidity, and infestation by insects and rodents.

Vitamin Potency

The potency of most vitamins contained in formulated feeds declines during storage, because many of the organic compounds are highly reactive and unstable. Under certain conditions, they can be easily denatured by heat, oxygen, moisture, and even ultraviolet light (3). The rate of vitamin activity loss in a given feed formulation is dependent on the particular vitamin, its source, and the conditions under which feed is stored. The average storage stability values of different vitamins and vitamin sources in dry feeds are summarized in Table 1. These data can be used to estimate normal vitamin activity losses under proper storage conditions (4,5). Most manufacturers of aquaculture feeds recognize these potential losses. They attempt to fortify their diets with sufficient overages of each vitamin to provide the intended levels of activity within the declared product shelf life.

Typical changes in vitamin activity levels that occur in both dry and moist feeds have been studied during prolonged storage (2). Results showed that after three months, if stored under proper conditions, vitamin levels in well-formulated diets of both types of feeds could meet or exceed the National Research Council (NRC) recommendations for Pacific salmon. Further monitoring of vitamin activity losses in these feeds revealed that, after doubling the recommended storage time, only vitamin C activity declined below minimum acceptable levels.

It is important to recognize that even significant losses of vitamin activity during storage need not render feed unusable. Vitamin requirements are actually a function of feed consumption and desired biological response of the fish, rather than a specific concentration in the feed (6). As long as storage deterioration of feed is restricted to vitamin loss, meaning that there are no other quality problems such as molding or fat rancidity, feed stored over

Table 1. Average Vitamin Stability in Stored Feeds

		Percentage of Vitamin Retention at Month:		
Vitamin	Ingredient Source	1	3	6
A	Beadlet	83	69	43
D_3	Beadlet	88	78	55
E	Acetate	96	92	88
	Alcohol	59	20	0
K	MSBC[a]	75	52	32
	MPB[b]	76	54	37
Thiamin	Hydrochloride	86	65	47
	Mononitrate	97	83	65
Riboflavin	Riboflavin	93	88	82
Pyridoxine	Hydrochloride	91	84	76
B_{12}	Cyanocobalamin	97	95	92
Pantothenic acid	Calcium d-pantothenate	94	90	86
Folic acid	Folic acid	97	83	65
Biotin	Biotin	90	82	74
Niacin	Nicotinic acid	88	80	72
Vitamin C	Ascorbic acid	64	31	7
	Fat-coated ascorbic acid	95	82	50
	Ascorbyl phosphate	98	90	80
Choline	Chloride	99	98	97

[a]MSBC, Menadione sodium bisulfite complex.
[b]MPB, Menadione dimethyl pyrimidinol bisulfate.

a long period of time can still be put to beneficial use. Some appropriate applications of feed with low vitamin activity levels would be short-term feeding of harvest-size fish, or prolonged feeding of fish or shrimp raised under extensive culture conditions. Conversely, those feeds should not be used where increased vitamin activity is required to promote an adaptive response such as disease resistance, or to achieve maximum tissue storage as required in broodstock feed.

Mold Growth

All too often, feed stored in fish hatcheries and on farms is destroyed by common molds. The potential for this to occur is always present because of the fact that mold producing fungi and other microorganisms exist naturally throughout the environment. They are present in grains after harvest and in animal carcasses prior to rendering. Food processing operations involved in stabilizing feedstuffs and in manufacturing feeds typically use heat and dehydration steps, that are sufficiently destructive to eliminate the original contaminating microflora. However, some fungal spores can survive harsh processing conditions. Other airborne spores may also recontaminate the feed during handling and storage. All of these spores then remain dormant in and on the feed until conditions exist that are favorable for growth.

Contaminating fungi grow best when the moisture content of the feed is 14.5 to 20% and in equilibrium with a relative humidity of 70 to 90% (7). Most dry feeds are manufactured at considerably lower moisture levels, allowing a safety margin for variability among individual feed particles. The maximum recommended moisture content for extruded pet feeds is 12% (8). Most aquaculture feed manufacturers take this a step further, keeping moisture levels at or below 10%. This is generally done because of the superior handling characteristics of low-moisture pellets in bulk bins, and the tendency for fish feeds to be stored over prolonged periods.

Special additives, which reduce water activity and inhibit germination of fungal spores within the feed, can be used to further diminish the possibility of mold growth. However, many of these additives are relatively expensive. Cost-effective application is mostly in specialty diets such as semimoist feeds that have moisture levels of 14 to 20%, but do not require storage conditions different from dry feed.

On a large production scale, there is no economical way of eliminating fungi spores in feed. The most effective mold-prevention strategy, therefore, is to maintain moisture levels in stored feed below requirements for fungal growth. To do this, it is obviously necessary to provide a dry area where feed can be protected from rain. Less obvious is the need to control moisture migration within the feed. Sufficient temperature differentials, even in feed with only 10% average moisture, can cause that moisture to concentrate at much higher levels in the cooler areas of a sack or bulk bin.

It is usually not practical to provide climate-controlled storage for large quantities of feed. However, every effort should be made to avoid conditions that allow extreme temperature changes to occur over a short period of time. In situations where bagged feed is stored outdoors under tarpaulins or bulk feed is held in dark-colored and poorly ventilated bins, moisture in the feed can volatilize during the heat of the day and condense near the top and surrounding container surfaces when the temperature rapidly decreases at nightfall. Similar moisture migration can even occur when bags of feed at ambient temperature are stacked on or against cool concrete floors and walls. Once feed has been subjected to these kinds of storage conditions, it is only a matter of time before mold growth begins in localized areas of high moisture.

The first species of fungi to develop in feed is usually *Aspergillus glaucus*, which has a minimum environmental moisture requirement of only 14.5% (7). If identified at an early stage, feed containing trace amounts of pellets with this type of mold can usually be fed to fish with little risk of adverse consequence. With more time, however, the number of mold colonies multiply quickly, creating higher temperature and moisture conditions. As the environmental conditions within the feed undergo a succession of changes caused by the growth of these spoilage microorganisms, other species quickly emerge and proliferate.

At moisture levels near 18%, there is a possibility that molding feed will become infested with *Aspergillis flavis* (7). This is an especially dangerous species of mold because it is capable of producing aflatoxins. Rainbow trout are particularly sensitive to these carcinogenic metabolites (9,10). Consumption of only 0.5 mg of aflatoxin B_1 per kg of body weight causes mortality within 3 to 10 days. Feeding aflatoxin-contaminated feeds with as little as 0.1 to 0.5 ppb aflatoxin B_1 results in hepatomas after 4 to 6 months. Other aquatic species, such as coho salmon (9), catfish, (11) and shrimp (12–14), are believed to be more tolerant, though similarly affected.

The probability of aflatoxin production in complete feed is actually quite low. It is much more likely to occur in high-moisture crops such as peanuts, cottonseed, and corn. Studies have shown that the presence of other microorganisms in a complex substrate like fish feed tends to interfere with aflatoxin production (7). However, among these interfering microorganisms, there are also species of *Fusarium* and *Penicillium* fungi that can produce their own mycotoxins. For this reason, the practice of using feed that is obviously molded should be avoided.

Lipid Rancidity

Lipids used in aquaculture feeds are usually the type that contain significant levels of unsaturated fatty acids, which are required for good health and growth of most species of fish and shrimp. The high degree of unsaturation of these fatty acids causes them to be particularly prone to oxidative rancidity. Feed manufacturers attempt to prevent oxidation in lipid sources such as fish oil by stabilizing them with antioxidants. However, the commonly used antioxidants such as ethoxyquin, butylated hydroxyanasole, and butylated hydroxytoluene, are sacrificial in the way that they protect the oil. Once they are used up, free radicals that are already present in the oil begin to react with unsaturated fatty acid components and the process of oxidation begins.

It is often thought that freezing is the best method of long-term preservation. However, cold temperature in the range achievable with most freezers is not effective in reducing the rate of free radical formation or the resulting lipid oxidation. In actuality, the experience with low-moisture feeds has been that freezing accelerates lipid oxidation (15). It is believed that the reason for this is that only free water is frozen at ordinary freezer temperatures. This results in the concentration of metal salts and other pro-oxidants in an unfrozen phase, making interaction with lipids more probable. It is also thought that the further reduction of water activity caused by freezing dry feed allows oxygen to penetrate the pellets more freely.

What all of this means is that there is very little that can be done on the farm to improve lipid stability in stored feed. Rotating the feed inventory as quickly as possible is the only effective strategy to avoid having feed go rancid before it is used. This can usually be accomplished easily with feeds that are fed in high volume. However, inventories of starter feeds, crumbles, and broodstock pellets are usually more difficult to manage. Animals that eat these feeds are also most likely to be at a point in their life stage where they are extremely vulnerable to the negative effects caused by consuming rancid lipids.

Pest Infestation

The presence of insects and rodents in feed storage areas can often be an overlooked but serious problem in aquaculture. These pests not only consume feed, but also cause additional and sometimes greater feed losses through packaging damage and the creation of environmental storage conditions that promote mold growth. They also have the potential to serve as vectors for transmission of disease to humans.

Insects. Insect infestation can be a very serious problem in feeds stored over a prolonged period of time. Jokes about how insects "just add a little protein to the feed" tend to divert attention from the magnitude of feed loss that they can cause. An actively reproducing population of insects can quickly consume significant amounts of food and deteriorate the physical quality of the remaining feed (16). Internal infesting species such as grain weevils and warehouse beetles can bore through feed sacks, providing a port of entry for other insects. If present in sufficient numbers in bulk feed, they also have the potential to create localized heating, moisture migration, and molding. External infesting species, however, are more frequently the cause of problems in complete feeds. These include Indian meal moths and flour beetles, which prefer to obtain nourishment from processed grain products, along with carpet beetles that feed on meat meal, feather meal, and other ingredients of animal origin.

Most of these insects thrive on food containing 12 to 14% moisture. They are capable of completely developing from an egg to a reproductively active adult within 30 days when temperatures are between 20 and 30 °C. At 16 °C, most of these species cease to lay eggs. They usually become dormant at about 4.5 °C. Under optimal environmental conditions, propagation of tremendous numbers of insects can occur in a very short period of time because of their short maturation time and relatively high fecundity.

With the knowledge and ability to recognize conditions that promote insect infestation and rapid population growth, it is easy to see that effective control requires a sustained and concerted effort with several prevention and housekeeping tasks. Rapid inventory rotation is perhaps the most important control process. However, regular inspection of feed and early detection of bugs, along with good sanitation in storage areas, are proactive practices that can greatly reduce the incidence of feed contamination with bugs. As a final resort, insecticides can be used to eliminate a persistent infestation.

In the United States, all chemicals used as insecticides must be registered for this purpose and be properly labeled according to Environmental Protection Agency (EPA) regulations. Fumigants such as hydrogen phosphide (phosphine gas), methyl bromide, and chlorpyrifos-methyl require application by individuals that are certified by controlling state agencies. These insecticides are highly effective and leave no residue in the feed. However, the vapors are very toxic. Fumigant insecticides should always be used with extreme caution and only according to the manufacturer's recommendations for treatment of feed.

Rodents. Populations of rats and mice that become established in storage areas obviously consume some amount of feed. However, the losses they cause through packaging damage and the resultant feed spillage, exposure to insects, and molding conditions are probably far greater. They also pose a substantial health hazard to workers handling the feed.

As with insect pests, several methods of control must be employed in a concerted manner to be effective. The basis for a rodent control program should always be good housekeeping, both inside the warehouse as well as around the exterior perimeter. Combining this with maintenance of physical barriers that limit entry and an aggressive trapping effort will noticeably minimize feed loses caused by rodents.

Use of poisons should only be considered as a last resort to control rodent populations in feed storage areas. Baits containing strychnine or other acute rodenticides, in close proximity to stored feed, impose an increased risk of feed contamination and dangerous contact with humans or pets. These same risks exist with the use of anticoagulant rodenticides, such as warfarin, even though they are much less dangerous.

HANDLING

Movement of feed on the farm can only be considered as a necessary evil. Some amount of feed or nutrient loss occurs each time it is handled in the processes of receiving, storing, and feeding. These chronic losses are usually small, but they accumulate over time. A good general control strategy is to identify the causes of greatest loss, and make any practical modifications necessary to handle feed as gently and as little as possible. More specifically, there are distinct differences between the physical and nutritional characteristics of moist and semimoist feeds and those of dry feeds. These different characteristics impose some handling requirements that are unique to each type of feed.

Moist and Semimoist Feeds

Moist feeds such as OMP probably present the greatest number of handling challenges. They are reasonably durable and shelf stable at temperatures below 0 °C. When thawed, however, their soft texture and highly perishable nature cause a number of handling problems. Relatively small compression forces cause moist feeds to clump and compact. Pellets are easily broken or smashed by conventional feed-handling equipment. Perhaps more importantly, the high-moisture content of the feed promotes vitamin activity loss and growth of bacterial contaminants. When combined with warm temperatures, this process of nutrient loss and food spoilage becomes greatly accelerated.

Good feed-handling procedures for moist feeds therefore include some means of keeping it frozen until the last few minutes before feeding. At many farms and hatcheries, feed for each pond is weighed into individual containers on the day before it is required and then kept in a freezer until feeding begins on the next day. The amount of time between removal from the freezer and feeding is kept to a minimum. To accommodate multiple feedings per day, containers holding each daily ration of feed are either returned to the freezer after every feeding or equipped with insulation to maintain feed in a frozen state. A further requirement to limit growth of contaminating bacteria and mold is the routine and thorough cleaning of these containers and other handling equipment used to distribute the feed.

Over the past several years, use of moist feed has been steadily declining in favor of feed products that are not as perishable and much easier to handle. Semimoist diets have subsequently attained widespread application where soft textured feed particles are required. The very low water activity in these feeds provides much better nutrient stability at ambient temperatures. However, clumping and compacting of pellets is still a problem because of their soft texture. These problems can be minimized by using low-profile containers, which help avoid having feed pile up and being compressed by its own weight. Movement and dispersion of feed can also be done by gentle means such as belt conveyors or vibratory feeders that allow feed to move freely without damage.

Dry Feeds

Both pelleted and extruded dry feeds have excellent handling characteristics compared to any other type of feed. Unlike moist feeds, handling procedures on the farm have no effect on nutrient quality of dry feeds. Pellet durability of both types of dry feed is also usually quite good, although variability in the consistency of feed ingredients may cause some batches of feed to be softer and more fragile. In addition to these attributes, the cylindrical or spherical particle shapes allow dry feed to flow easily from trucks, bins, and feeders.

The physical characteristics of dry feeds are so well suited to the handling and distribution requirements of aquaculture that inherent limitations are often challenged. It is easy to overlook the fact that even the most durable crumbles and pellets can break down into dust and fines when subjected to sufficient amounts of compression and abrasion. In handling any feed in bag or bulk form, it is important to give ample consideration to moving the feed as little as possible and as gently as possible.

With bagged feed, the challenge is to reduce the amount of particle size attrition that occurs when pellets or crumbles are forced to rub against each other. Use of forklifts and pallets, or hand trucks and minipallets, allow bags to be handled in multiple units. This minimizes the amount of feed movement within each bag and reduces the creation of dust and fines. When it is necessary to handle single bags, the process should be done as gently as possible. Obviously, rough treatment such as throwing or walking on sacks of feed should be avoided.

Pellet-against-pellet abrasion in bulk feed is more difficult to control. The very nature of this method of storing and handling feed requires that pellets flow from the delivery vehicle to a bin, and from the bin throughout the farm. It also necessitates the use of conveying equipment. These mechanical devices are often the source of, or solution to, most problems with excessive levels of dust and fines in bulk feed.

Among the types of conventionally used feed conveying equipment, bucket elevators, belt conveyors, and drag conveyors are the least destructive (17). These work well because they control movement of feed against feed and minimize the potential of shearing or pinching pellets in conveying mechanisms. Pneumatic, oscillating, and vibratory conveyors cause only slightly more abrasion. However, they almost eliminate losses from pellet breakage when properly maintained and operated.

The most potentially destructive conveyance mechanism for feed is the auger. Tube-type screw conveyors, as well as flexible augers, are widely used in feed-handling systems on farms because of their low cost and simplicity of operation. Their most frequent use is in unloading bulk bins. In this application, the equipment design is usually more appropriate for handling mash feeds or whole grain, where the auger turns at a high rate of speed and has an inclined discharge. Most are also "choke loaded," meaning that the feed completely covers the inlet to the conveyor, causing compression and breakage as pellets enter the tube. While proper equipment design can minimize many of these problems, the added expense usually ends up favoring the selection of conveyors that are more appropriate for use with feed.

SUMMARY

Aquaculture feeds, like most food products, have a finite shelf life and special handling requirements. In order to realize the full economic and nutritional value of these feeds, it is necessary to store and handle them properly. Deterioration of feed quality during storage can be minimized by frequent rotation of the inventory, and a concerted effort to maintain good housekeeping and environmental conditions that discourage mold growth and insect and rodent infestation. Proper handling techniques can also reduce nutrient loss and pellet breakage just prior to feeding.

The importance of careful attention to the specific requirements for proper storage and handling of aquaculture feeds cannot be overstated. At most farms that raise fish or shrimp, feed cost is the largest single expense item. Therefore, even a small reduction in wasted feed can significantly affect production cost and directly impact bottom-line profitability.

BIBLIOGRAPHY

1. E. Leitritz and R.C. Lewis, *Trout and Salmon Culture—Fish Bulletin No. 164*, University of California, 1980.
2. L.G. Fowler, W.M. Thorson, W.L. Wallien, G.R. White, and P.E. Martin, *USFWS Technology Transfer Series* **89**(1), (1990).
3. M. Gadient, in R.A. Erdman, ed., *Proceedings Maryland Nutrition Conference for Feed Manufacturers*, University of Maryland, 1986, pp. 73–79.
4. M.B. Coelho, *Feed Management* **42**(10), 24–35 (1991).
5. BASF, *Keeping Current KC 9138*, 5th ed., BASF Corporation, Mount Olive, NJ, 1994.
6. National Research Council, *Nutrient Requirements of Trout, Salmon, and Catfish*, National Academy of Science, Washington, DC, 1973.
7. K.W. Chow, in K.W. Chow, ed., *Fish Feed Technology*, UNFAO, Rome, Italy, 1980, pp. 216–224.
8. G.J. Rokey, J.R. Krehbiel, K.E. Matson, and G.R. Huber, in R.R. McEllhiney, ed., *Feed Manufacturing Technology III*, American Feed Industry Association, Arlington, VA, 1985, pp. 222–237.
9. L.M. Ashley, in J.E. Halver, ed., *Fish Nutrition*, Academic Press, New York, 1972, pp. 439–530.
10. L. Friedman and S.I. Shibko, in J.E. Halver, ed., *Fish Nutrition*, Academic Press, New York, 1972, pp. 182–239.
11. W. Jantrarotai and R.T. Lovell, *J. Aquat. Anim. Health* **2**, 248–254 (1991).
12. M.D. Wiseman, R.L. Price, D.V. Lightner, and R.R. Williams, *Applied Environmental Microbiology* **44**, 1479–1481 (1982).
13. D.V. Lightner, in C.J. Sindermann and D.V. Lightner, eds., *Disease Diagnosis and Control in North American Marine Aquaculture*, Elsevier, Amsterdam, 1988, pp. 96–99.
14. H.T. Ostrowski-Meissner, B.R. LeaMaster, E.O. Duerr, and W.A. Walsh, *Aquaculture* **131**, 155–164 (1995).
15. R.W. Hardy, *Aquaculture Magazine*, Sep./Oct. (1998).
16. J.R. Pedersen, in R.R. McEllhiney, ed., *Feed Manufacturing Technology III*, American Feed Industry Association, Arlington, VA, 1985, pp. 380–389.
17. D. Falk, in R.R. McEllhiney, ed., *Feed Manufacturing Technology III*, American Feed Industry Association, Arlington, VA, 1985, pp. 167–190.

See also ANTINUTRITIONAL FACTORS; LIPID OXIDATION AND ANTIOXIDANTS; TEMPERATURE; VITAMIN REQUIREMENTS.

FEED MANUFACTURING TECHNOLOGY

FREDERIC T. BARROWS
USFWS, Fish Technology Center
Bozeman, Montana

RONALD W. HARDY
Hagerman Fish Culture Experiment Station
Hagerman, Idaho

OUTLINE

INTRODUCTION

Supplying complete nutrition in a formulated feed for fish is more of a challenge than doing so for terrestrial species. A major component of this challenge is obvious. Feed for aquatic animals is exposed to water before it is eaten, often for long periods of time. This can result in loss of water-soluble nutrients due to leaching, and breakdown of the feed pellets, which degrades water quality. This is a greater problem for larval fish, crustaceans, and slow-feeding species like sturgeon than for species like trout, which consume feed rapidly. Another factor that makes aquatic feed manufacturing a challenge is the necessity of processing fish feeds in some fashion. For many terrestrial species (e.g., poultry and cattle), ingredients can be just mixed together and fed without any further processing. Feeds for fish must always be processed into pellets, flakes, or some other particle.

The aquaculture feed industry uses several feed-processing technologies. Each processing method imparts different characteristics to the feed (e.g., floating, sinking, high nutrient stability, or high palatability). Various fish-rearing situations often require unique feed particle characteristics. No single feed-processing method can produce the best feed for all situations. The effect of the processing method on the physical characteristics of fish feed must be understood to develop an effective fish-rearing program.

TYPES OF FEED

Aquaculture feeds can be categorized in many ways, but the most common categorization is based upon the life-history stage of the fish for which the feed is designed. In this manner, feeds are designated as larval, starter, grower, and broodstock feeds. Feeds for those life-history stages differ in particle size and in formulation (choice of ingredients and the percentage in the feed mixture). For example, larval feeds are usually less than 500 μm in size and are manufactured with specialized processing equipment. The processing methods used to produce starter feeds are not as exacting as those used for larval feeds, and the feeds are correspondingly less expensive. Trout, salmon, catfish, and tilapia fry are all species that thrive when given starter feeds. Species that produce very small larvae (e.g., striped bass and walleye) do not flourish on starter feeds and require the more expensive larval feeds.

Grower feeds are fed after fish have passed the juvenile stage. Grower feeds constitute the largest proportion of feed used during a production cycle, over 90% in some species. Since feed costs account for a large portion of production costs and grower feeds represent the major portion of feeds used in a production cycle, the cost of grower feeds greatly affects the overall cost of producing fish. However, feed cost should be compared on the basis of unit (kg, lb) of product produced per unit feed rather than just on the unit cost of feed. Broodstock feeds are formulated to support high levels of egg production. Fecundity and gamete viability are more important considerations with these feeds than weight gain of the brood fish.

There are other types of special-application diets that are generally grouped into the category of specialty diets. These include transition feeds and low-pollution (i.e., environmentally friendly) feeds. Transition feeds were developed to ease the transition of young fish being switched from live feeds to formulated (artificial) feeds. Low-pollution feeds are designed to decrease the level of nutrients, leaving a hatchery in the effluent water. Even though the overall nutrient contribution of hatcheries and fish farms is relatively low, they are a point source of nutrients and subject to increasing regulation by state and federal authorities.

Some overlap exists in the processing technology used to produce the different types of feed. The methods that are appropriate for producing one type of feed may not be right for another type of feed. For example, a common processing method for grower diets (cooking extrusion) increases carbohydrate availability to fish. This method is not desirable for larval fish that have a limited capacity to utilize carbohydrates. The additional expense and effort of processing the feed to increase carbohydrate digestibility may not be justified for larval feed production.

MANUFACTURING METHODS

Manufacturing methods are categorized based on the equipment used to produce the feed particle or pellet, but in all feed manufacturing processes, ingredients are ground and mixed before particle formation or pelleting takes place (1). Complete mixing of the ingredients is crucial for the production of a homogeneous product, and both the design and operation of the mixer are critical (2). After mixing, the feed blend is conditioned by adding steam and then is pelleted and cooled, or dried, depending on the pelleting method. Also, it is often top-dressed with additional oil. Next, pellets are placed into bags, stacked on pallets, stored, and eventually shipped to the fish farm. Some feeds, such as those for catfish, are delivered in bulk without being bagged. This simplifies handling large quantities of feed.

Particle-Size Reduction (Grinding)

Most feed ingredients arrive at feed mills as coarse particles. Ingredients are commonly ground before mixing. This step of feed preparation is time-consuming and expensive (3). Some feed ingredients are difficult to grind alone and must be combined with other ingredients. Thus, in some feed mills, particle-size reduction is performed on feed mixtures rather than on single ingredients. Many types of equipment are used to reduce particle size by impact, crushing, and cutting. Feed ingredients and mixtures used in aquaculture feeds are most commonly ground using impact mills, also known as hammer mills. Hammer mills operate by impacting the feed mixture until it shatters into particles small enough to pass through a screen. The screen size determines the degree of particle-size reduction. Hammer mills are effective and efficient grinders, unless the ingredients contain high levels of oil (fat), as do some types of fish meal. The heat generated by grinding can separate residual fish oil in fish meal, thereby clogging the screens. High-oil feed ingredients must be mixed with other low-oil ingredients, to allow grinding to proceed efficiently without clogging, or ground using a pulverizer. Pulverizers can be identified by the lack of retaining screens and a very large volume of air passing through the mill (2). The high volume of air passing through the pulverizer keeps ingredient temperatures low and thus inhibits separation of residual oil in fish meals.

The degree of grinding or particle-size reduction depends on the type of pelleting or particle formation that will be produced. For example, larval feeds require very fine particles while grower feeds do not need to be ground as finely. Grinding requires energy and generates heat, which can damage some nutrients. Therefore, feed ingredients are ground to pass through the largest screen opening allowance for the feed being produced. A common rule of thumb is to reduce the ingredient particle size to 20% of the diameter of the holes in pelleting dies. This degree of grinding is sufficient to prevent holes from becoming plugged during pelleting. Ingredient grinding also facilitates the production of more durable and water-stable pellets (4). Another benefit of grinding is that it allows the feed ingredients to be mixed into a homogeneous mixture. This presumably results in nutrients being evenly distributed within feed pellets or particles.

After the feed ingredients are ground, they are mixed together in proportions dictated by the feed formulation. Mixing is carefully timed to ensure even particle distribution and to avoid overmixing, whereby ingredients can become separated based upon particle density. Oils are generally not added to feed mixtures, at least when the mixtures contain more than 4 to 5% residual oil. Rather, oil is added after pelleting. Too much oil in feed mixtures reduces compression during pelleting, resulting in soft, breakable pellets.

Feed mixtures are then subject to conditioning, pelleting, drying, and, often, top-dressing (adding fish or plant oils to cooled pellets). Conditioning involves the addition of moisture, generally in the form of steam. The amount of moisture and degree of heat added varies with the pelleting method. The purpose of conditioning is to add sufficient moisture to enable the feed mixture to pellet and stick together, and to gelatinize (cook) a portion of the starch in the feed mixture, both to facilitate pellet binding and to increase the digestibility of starch. Common pelleting methods used to produce fish feeds include cold extrusion, steam pelleting, and cooking extrusion.

Cold Extrusion

Pelleting is accomplished by forcing the feed mixture through a plate or die containing holes. Cold extrusion, a process that does not use steam, but simply adds water or water-containing ingredients, is accomplished by forcing the mixture through a plate with holes in it, similar to the production of pasta. The feed mixture must contain at least 25% moisture for this process to work properly. The feed mixture is carried down a barrel by a rotating auger, and as the feed squeezes out of the holes in the die, it is cut by a rotating knife into the desired pellet length, which is usually similar to the diameter of the pellets. After the pellets are cut, they are quickly frozen or dried. Cold extrusion is used to make the semipurified feeds used in research studies, larval feeds, and semimoist feeds, such as the Oregon moist pellet or its derivatives.

Several designs of cold extruders are available, differing in configuration to allow for the production of very small particles. The most common design is the front-discharge extruder, also called an axial-discharge extruder, which can produce pellets as small as 1.5 mm diameter. With this design, the pellets exit the machine at the end of the extruder barrel through a flat plate. Radial-discharge and twin-dome extruders can produce pellets down to 0.5 and 0.3 mm diameter, respectively. (See the entry "Microbound feeds.") Even though these extruders have fixed-screw configurations, they are usually equipped with a variable-speed drive on the main screw which enhances the versatility of the machine.

Because of the range of characteristics of products that can be produced using cold extrusion, it is difficult to define the particle characteristics for all cold extruded products. In general, however, these feeds have a soft texture, are highly palatable, and have not been exposed to heat. The low-temperature processing reduces or eliminates destruction of heat-labile nutrients. High palatability and soft texture are particularly advantageous for larval and starter feeds, and are also beneficial in starter, grower, and broodstock feeds for undomesticated strains or species of fish.

Pellet binders used in cold extrusion include both nutritive and nonnutritive materials. Nutritive binders include cooked oats, wheat gluten, pregelatinized starches, and gelatin. Nonnutritive binders include tapioca, carboxymethlycellulose, alginates, agar, and various gums (5).

Compression Pelleting

Compression pelleting, also known as steam pelleting, is the most common type of feed pelleting used to produce animal and fish feeds. In this process, the feed mixture is conditioned by steam to achieve a moisture content of 16–18% and a temperature of 65–80 °C (6). Conditioning partially cooks or gelatinizes starch and activates binders, such as lignin sulfonate, a wood product. Conditioning is a continuous process that takes approximately 30 seconds, after which the feed mixture is conveyed into the middle of a rotating die, similar in shape to a doughnut. On the inside of the die is a stationary, rotating roller, which forces the feed mixture into holes that radiate from the inside of the rotating die to the outside. The feed mixture emerges from the outside edge of the rotating die and is cut off by a stationary knife. The length of the holes in the die dictate the amount of compression and frictional heat to which the feed mixture is exposed, and thus, the density and hardness of the resulting pellets. As pellets emerge from the die, they typically have a glazed, smooth surface. This is the result of gelatinization of starch on the pellet surface caused by frictional heat, generated as the mixture goes through the die. The pellets are hot when they emerge from the die, typically about 90 °C. They are conveyed to a cooler, where the moisture level is reduced to less than 12% by the residual heat of the pellets in combination with flowing air. After cooling, the pellets can be top-dressed to a maximum of 18% total fat. The upper limit of oil addition by top-dressing is a consequence of pellet density and the glazing that occurs on the outside of the pellets. Compressed pellets are hard, but they are subject to fracture, thus making conveying, bagging, and subsequent handling procedures important to avoid excessive pellet disintegration into dust, also known as fines.

Expanders

Expansion is a feed mixture conditioning step that occurs after mixing, but before compression pelleting. The process involves steam injection and mixing in a preconditioning chamber, followed by applying pressure (shear) along a barrel. The mixture is then forced through a narrow gap, created by the presence of a cone in a tapered outlet of a chamber. Heat from steam, pressure (shear), and frictional energy, generated as the feed mixture squeezes through the gap, causes starch gelatinization. As the pressure is lost when the mixture exits the gap, moisture is lost. The mixture is then conveyed through a normal compression pelleting system. The resulting pellets are identical to steam pellets, except for the degree of gelatinization, which also affects pellet density and the amount of oil that can be top-dressed. Pellets made by this process can be top-dressed to achieve up to 22% total fat.

Cooking Extrusion

This conditioning and pelleting process is more sophisticated, versatile, and more expensive than compression pelleting. Cooking extrusion technology is used to produce puffed breakfast cereals, snack foods, and dog foods. Feed or foods are actually cooked using this method and the starch in the mix is gelatinized, binding the pellet together. No extra binders are added to fish feeds made using cooking extrusion pelleting. Gelatinized starch is sufficient.

Cooking extruders are effective at heating and thus, cooking the feed. Heat is applied to the ingredients in three main ways. First, steam is added to the feed mixture in a condition chamber prior to entering the extruder. Preconditioning the mix is an important to insure starch gelatization in any processing method (4). After being vigorously mixed for several minutes, the feed mixture enters the extruder barrel, usually consisting of sections which can be heated or cooled. If additional heat is required, steam is injected into barrel sections, or, less commonly, the feed mixture is heated with electric barrel heaters. The third and greatest burst of energy is through shear, and the frictional energy contributed by the screw in the extruder barrel.

Extruders are available both in single- and twin-screw configurations. With both types of extruders, the screw consists of multiple segments that can be assembled in different ways to add more or less energy to the product. The screw segments are available in different lengths, pitch, and design (7). The single-screw configuration is less expensive and capable of producing a high-quality product. The twin-screw design is more expensive, but is capable of adding more energy to the feed mix with shear produced by self-wiping screws.

The mixture moves down the barrel, transported by the screw, which compresses the mixture as it gets close to the die. This pressure can be sufficient to change the steam in the feed into a liquid state. The mixture then exits the barrel through a die that shapes and further compresses the mixture, forming a noodle that is cut into pellets upon emerging from the die. The sudden release of pressure as the feed exits the die causes the feed mixture to expand as super-heated water turns suddenly to steam. This, in turn, creates small air pockets within the pellet, mainly associated with gelatinized starch, which then bind the mixture into a water-stable, hard pellet. The moist pellets are then conveyed to a hot-air dryer to lower the moisture level to less than 10%. After drying and cooling, pellets are top-dressed to achieve high fat levels. Extruded pellets can absorb high quantities of fish or plant oils due to the relatively low density associated with their tiny air pockets.

The density of the pellets can be controlled by adjusting the formulation and the amount of moisture added during conditioning. The feed can be cooked, but also can be made to sink by reducing the pressure and/or the temperature of the feed mix before it leaves the extruder (7). Many times a neutrally buoyant feed is desired for particular feeding situations. This is very difficult to produce consistently due to the many variables that are difficult to control. The variables include atmospheric conditions, ingredient composition and characteristics (which vary from batch to batch of the same ingredient from the same source), and the addition of oil, which affects buoyancy. It is easier to produce either floating or slowly sinking pellets with cooking extrusion. Extruded pellets can be very hard and resist fracture or compression during conveying, bagging, and shipping to fish farms. Because of these qualities [e.g., high fat content, floating or slowly sinking, water stability, and a low percentage of fines (pellet hardness) (7)], extruded pellets are increasingly being used in aquaculture. The disadvantage of extruded pelleting is that it increases the cost of feed when compared to steam pelleting, due to lower throughput per horsepower of energy expended per unit feed produced. Extruded pellets contain more moisture than compressed pellets, increasing the cost of drying compared to compressed pellets. As previously mentioned, the cost of feed is less important to fish farmers than the feed cost per unit of product grown and sold.

Universal Pellet Cooker

Recently, a new system of feed pelleting was introduced which combines some aspects of compressed pelleting and cooking extrusion. This system is known as the universal pellet cooker, or UPC, and the pelleting equipment resembles cooking-extrusion equipment. Basically, the UPC process involves enhanced preconditioning for 2–3 minutes, during which the addition of steam results in 40–50% starch gelatinization. Less steam and water is added during preconditioning than in cooking extrusion, resulting in a feed mixture with 16–18% moisture. The other major difference in the UPC system is the modified action in the barrel of the pelleter, where the auger or rotor turns 2–3 times faster than a conventional cooking extruder. This increases production levels to 18–20 Mt/hr, nearly the same as steam pelleting. The fast-turning rotor adds much more energy to the feed mixture, and the frictional energy associated with the faster-turning rotor and lower moisture in the feed mixture further gelatinizes the starch, to 60–80%. In addition, the combination of frictional energy and steam pasteurizes the feed mixture, unlike compressed pelleting, even when expansion is used. Because the moisture content of the feed mixture is lower than in cooking extrusion, there is less water entrapped in the feed mixture and thus, less expansion when pellets exit the die. This results in a higher bulk density for UPC pellets than for cooking extrusion pellets, but the UPC system can produce pellets from 400 g/L to 600 g/L, depending on how it is operated (Table 1). Pellets can be dried in a cooler rather than in a dryer. This reduces the amount of equipment and lowers the cost of production. The UPC produces higher-density pellets than cooking extrusion, but without the glazed surface associated with compressed pelleting. By top-dressing, up to 30% higher levels fat can be added to the dried pellets. An additional benefit of the UPC pelleting system is that the dense particles for starter feeds can be produced by turning the external knife faster. This eliminates conventional crumbling by rollers and the inevitable production of fines. As is the case with cooking extrusion, gelatinized starch acts as a pellet binder.

Comparing the Types of Pellets

The main issues important to fish farmers concerning feed pellet type are starch digestibility (determined by the degree of gelatinization), pellet density (buoyancy), water stability, durability, nutrient destruction during pelleting, and cost. How the types of pelleting processes affect these issues is summarized in Tables 1 and 2. Each pelleting process has advantages and disadvantages, but for each major species of farmed fish, there is generally a clear advantage of one type of pelleting process (Table 3). The UPC process is not likely to replace either compressed pelleting or cooking extrusion, but for some sectors of aquaculture, it may produce a product that fills a niche between compressed and extruded pellets.

Crumbles (Starter Feeds)

Steam pelleting is used mostly to produce grower feeds, but starter feeds have also been produced for many years using the same method. Since the minimum size of a

Table 1. Typical Physical Qualities of Various Types of Fish–Feed Pellets

Physical Quality	Compressed	Annular Gap	Extruded	UPC
Density (g/L)[a]	590	680	400–550	400–600
Maximum temperature (C)	95	135	150	150
Time exposed to steam	<1 min	<1 min	2–5 min	2–3 min
Starch gelatinization (%)	<40	65–70	>80	60–80
Maximum fines (%)	2–3	<1	<1	<1
Maximum fat level (%)	18	25	38	30

[a]480 g/L is the breakpoint for pellets to float; higher-density pellets sink, lower-density pellets float in freshwater.

Table 2. Attributes of Pellet Types Important to Fish Farmers

Attribute	Compressed	Annular Gap	Extruded	UPC
Starch digestibility[a]	Low	High	High	Medium to high
Pellet buoyancy	Sinking	Sinking	Floating/sinking	Floating/sinking
Water stability	Low	Low	High	High
Durability	Low	Medium	High	High
Nutrient destruction[b]	Low	Low	Medium	Medium/low
Cost of pelleting	Lowest	Low	Highest	Medium

[a]Starch digestibility is a function of the degree of gelatinization.
[b]Nutrient destruction during pelleting is caused by high temperature, high pressure, length of time that feed mixture is exposed to high temperature, and most greatly affects certain vitamins and the carotenoid pigment, astaxanthin.

Table 3. Current Applications of Various Pellets for Selected Types of Aquaculture[a]

Aquaculture Sector	Pellet Type	Important Qualities
Catfish, pond culture	Cooker extruded	Starch, floating
Salmon, marine net-pen culture level	Cooker extruded	Slowly sinking, high fat
Rainbow trout, raceway culture durability, medium fat	Compressed, extruded	Cost, density, level
Shrimp, semiintensive pond culture	Compressed, extruded	Water stability, cost
Sea bass, sea bream	Cooker extruded	Slowly sinking, starch

[a] UPC pelleting has just been introduced and is not yet widely used.

steam pellet is usually limited to 3 mm, crumbling is necessary to produce smaller feeds. Crumbles are made by crushing steam pellets between rollers moving at different speeds, and the resulting pellet fragments are screened to produce several size ranges of particles (4). Crumbling creates particles that have jagged irregular shapes, with a high surface area-to-volume ratio. These feeds disintegrate in water relatively quickly and thus care must be taken to avoid water-quality deterioration when crumbles are being fed to fry.

LARVAL FEEDS

Many technologies have been used and are being developed for the production of aquatic larval feeds. The technology for larval feed production (8–10) is necessarily different from that used for the production of starter feeds (1) due to the extremely small size of larval feeds (less than 400 μm), and thus the high surface area-to-volume ratio of such small particles. Leaching of water-soluble nutrients is a major problem in larval feeds (11). These feeds also need to be highly palatable, and the nutrients must be digestible to larvae that may not have a completely functional digestive system at first feeding. Some methods of larval-feed binding lower nutrient availability and palatability of the feed.

The methods for producing larval feeds can be classified into three major categories. These are microbound, microencapsulated, and complex particles. The first two feed types are differentiated by the type of binding that holds the particles together. Microbound particles are held together from the inside of the particle by a variety of different binders (12,13). Microbound feeds can be produced in the appropriate particle size with some processing methods, or a cake or flake is formed and then crumbled to the appropriate size. Ornamental fish feeds have been traditionally produced using flaking technology, a microbound processing method.

Microencapsulated feeds are surrounded by a layer of material, a capsule that retains the feed-ingredient mixture inside the particle (11,14). These particles can be designed to have a slow release of the material inside the capsule, or to totally prevent leaching of the water-soluble nutrients. Complex particles consist of embedded and carrier particles. A smaller embedded particle is placed inside a larger carrier particle. The embedded particle can be produced using the same manufacturing method that is used to produced the carrier particle, or by a different method.

Combining particle types takes advantage of the benefits from different processing methods and hopefully, eliminates some of the disadvantages of each. For a detailed discussion of the production of microbound and microencapsulated feeds and complex particles, please refer to the articles on microbound feeds, microparticulate feeds and complex particles, and microencapsulated feeds.

BIBLIOGRAPHY

1. R.W. Hardy, *Fish Nutrition*, 2d ed., Academic Press, New York, 1989.
2. K.C. Behnke, C. Fahrenholz, and E.J. Bartnoe, *Feed Manufacturing Technology IV*, American Feed Industry Association, Arlington, VA, 1994.
3. J. Sorenson and D. Phillips, *Feed International* **12**, 28–32 (1992).
4. R.K.H. Tan and W.G. Dominy, *Crustacean Nutrition*, World Aquaculture Society, 1997.
5. J.M. Heinen, *Prog. Fish-Cult.* **43**, 142–145 (1981).
6. R.G. Piper, I.B. McElwain, L.E. Orme, J.P. McCraren, L.G. Fowler, and J.R. Leonard, *Fish Hatchery Management*, U.S. Department of Interior, Washington DC, 1982.
7. R. Hauck, G. Rokey, O. Smith, J. Herbster, and R. Sunderland, *Feed Manufacturing Technology IV*, American Feed Industry Association, Arlington, VA, 1994.
8. D.A. Jones, D.L. Holland, and S. Jabborie, *Appl. Biochem. Biotech.* **9**(10), 275–288 (1984).
9. A. Kanazawa, S.I. Teshima, and H. Sasada, *Bull. Jpn. Soc. Sci. Fish.* **48**, 195–199 (1982).
10. S. Teshima and A. Kanazawa, *Bull. Jpn. Soc. Sci. Fish.* **49**, 1893–1896 (1982).
11. D.F. Villamar and C.J. Langdon, *Marine Biol.* **115**, 635–642 (1993).
12. A. Kanazawa, *World Mariculture Society*, Special Publication No. 2 (1981).
13. F.T. Barrows, R.E. Zitzow, and G.K. Kindschi, *Prog. Fish Cult.* **55**, 224–228 (1993).
14. C.J. Langdon, D.M. Levine, and D.A. Jones, *J. Micro.* **2**(1), 1–11 (1985).

See also ANTINUTRITIONAL FACTORS, ENVIRONMENTALLY FRIENDLY FEEDS, FEED HANDLING AND STORAGE, INGREDIENT AND FEED EVALUATION, LIPID OXIDATION AND ANTIOXIDANTS.

FERTILIZATION OF FISH PONDS

MARTIN W. BRUNSON
JOHN HARGREAVES
Mississippi State University
Mississippi State, Mississippi

NATHAN STONE
University of Arkansas at Pine Bluff
Pine Bluff, Arkansas

OUTLINE

Fish farmers and recreational farm pond owners fertilize ponds to increase fish production and to prevent rooted aquatic weeds from becoming established. Aquaculture ponds are fertilized to increase the available natural food organisms (phytoplankton and zooplankton) for fry or larval fish, or for use by species that are efficient filter feeders. Recreational ponds are also fertilized to increase the available natural food organisms. A well-managed fertilized recreational pond can produce 227 to 455 kg of fish/ha (200 to 400 lb/acre) annually (1,2). This is three to four times the fish production that can be obtained without fertilization. Fertilizer nutrients stimulate the growth of microscopic plants (phytoplankton) in the water. Phytoplankton, in turn, serve as the plant food base for other organisms (zooplankton and larger animals) that are then consumed by fish. Abundant growth of phytoplankton gives water a greenish color called a "bloom" that can prevent light from reaching the pond bottom and thereby reduce the potential for growth of rooted aquatic weeds.

Not every pond should be fertilized. In many cases, increased production of fish is not desirable. If a pond's primary purpose is for watering cattle or wildlife habitat, fertilization is unnecessary (3). Similarly, pond owners desiring clear water should not fertilize. In clear water, however, rooted aquatic vegetation is more likely to become abundant than in waters where transparency is reduced either by plankton blooms or clay turbidity. Fertilization of recreational ponds to increase fish production is of little value if the angling pressure is not high enough to utilize the increased fish biomass.

Fertilization is a common practice to increase yields of bass and sunfish from ponds. An overcrowded sunfish population or a pond that is otherwise out of balance in terms of the relative proportions of bass and sunfish should be corrected before initiating a fertilization program. Also, ponds dominated by undesirable fish species should not be fertilized until the contaminating fish are eradicated. It is usually not necessary to fertilize catfish ponds in which the fish are fed regularly, since the provided feed represents a supplemental nutrient source, especially from uneaten or wasted feed. Raising catfish with feed is an excellent choice for producing fish in ponds that have low alkalinity water and where liming is not an option.

BEFORE FERTILIZING

Before beginning a fertilization program, test the alkalinity, total hardness, and calcium hardness of the pond water. Waters that are low in alkalinity or total hardness (below 20 ppm) will need liming for fertilizers to be effective (4,5). Most ponds that receive runoff from watersheds with acid soils will have low-alkalinity and/or low-hardness water. Typical application rates for agricultural or dolomitic limestone are 1 to 3 tons/acre (6). The recommended liming rate is based on the lime requirement of the pond bottom soils, as determined by soil testing. Ponds should not be fertilized at the same time that lime is applied, since the calcium in lime will remove phosphorus from the water.

Ponds that are muddy, infested with weeds, or subject to excessive water flow should not be fertilized until the problem is corrected. Mud prevents light from entering the water, thereby inhibiting phytoplankton growth. Weedy ponds should never be fertilized, as the nutrients will simply stimulate the growth of more weeds, rather than phytoplankton. Excessive water flow (where the pond water volume is exchanged in less than two weeks) dilutes the fertilizer nutrients to the point where they are ineffective. In addition, nutrients flushed from the pond can pollute downstream waters.

Although the plankton blooms that result from fertilization can be highly desirable, excessive fertilization can cause problems. Dense plankton blooms [Secchi disk visibility <30 cm (12 in.)] can lead to plankton die-offs (and, in turn, dissolved oxygen depletions), critically lower dissolved oxygen readings in the morning, and elevated afternoon pH levels, which increase the concentration of un-ionized (toxic) ammonia in the water and can stress fish. In some cases, dense algae blooms result in direct production of toxins in the water. These factors should be considered when weighing the benefits of implementing a fertilization program for a pond.

TYPES OF FERTILIZER

The formulation of a fertilizer, or grade, indicates the percentage by weight of nitrogen (N), phosphorus (as P_2O_5), and potassium (as K_2O). For example, an 11-37-0 fertilizer contains 11% nitrogen, 37% phosphorus (as P_2O_5), and 0% potassium (as K_2O). Phosphorus is the most important nutrient in pond fertilization (4,7,8), but occasionally nitrogen or potassium may limit plankton production (9,10). In new ponds, some nitrogen may be beneficial, while potassium is rarely, if ever, needed. In selecting a fertilizer, choose a formulation that is high in phosphorus.

Inorganic fertilizer comes in liquid, powdered, or granular forms. Liquid fertilizer dissolves the most readily (11), followed by powdered, and then granular forms. Powders are generally more expensive than liquid or granular forms, but are relatively easy to apply.

Although more expensive than other fertilizer forms, controlled-release fertilizer is now available for pond owners (12). The resin-coated granules slowly release nutrients into the pond water, with the rate of release corresponding to water temperature and water movement. Ideally, one application of controlled-release fertilizer in the spring will be sufficient for the entire growing season. However, this is subject to other environmental factors that may result in reduced blooms or bloom die-offs, and supplementation with a more readily available source of nutrients may be necessary following these events.

Organic fertilizers, such as cottonseed meal, are used in combination with inorganic fertilizers for preparing larval fish ponds (13,14). Organic materials are generally not recommended as recreational farm pond fertilizers, since excessive amounts of organic fertilizer may result in critically low dissolved oxygen levels, possibly killing fish. In addition, they can promote the growth of undesirable filamentous algae (commonly known as pond moss or pond scum).

All of these pond fertilizers should be readily available through any farm supply dealer. Numerous name brands exist, and some are formulated specifically for pond fertilization. However, any fertilizer formulation providing the appropriate nutrient levels can be used, unless the product contains other ingredients that may be harmful to fish or other aquatic organisms. For example, some fertilizer products marketed primarily to homeowners and intended for lawn or turf application may contain either herbicides or insecticides. Although the nutrients supplied may be suitable for pond applications, these should be avoided.

APPLYING FERTILIZER

Table 1 gives suggested fertilization rates for ponds based on water calcium hardness and type of fertilizer (7,15). Provided that the alkalinity of the water exceeds 20 ppm, the rates listed in the table are based upon calcium hardness, since some phosphorus applied in the fertilizers can be removed by calcium before it is taken up by plankton. This becomes a greater problem as hardness increases. Rates should be adjusted based on the response of each individual pond. For example, ponds that receive runoff from active pastures are likely to require less fertilizer, due to nutrient influx from the surrounding watershed.

The risk of low dissolved oxygen conditions in a pond is increased somewhat with the application of fertilizer, although the benefits likely outweigh the added risks. Even unfertilized ponds will experience turnovers or bloom die-offs that can lead to low dissolved oxygen (16). Excessive fertilization should be avoided as it can produce such a dense bloom that the risk of oxygen depletion increases greatly. Allow at least one week, preferably two, between fertilizer applications to evaluate the result of each application. As pond water becomes warmer, the response to a fertilizer application will be stronger and more rapid.

Begin fertilizing in the spring when the water temperature stabilizes above 15 °C (60 °F). Make three applications of fertilizer two weeks apart, then make additional applications whenever water transparency exceeds 45 cm (18 in.). A Secchi disk can provide a consistent means of evaluating the density of the pond bloom. Recreational ponds will require additional applications of fertilizer at intervals during the summer months and into the fall. In some cases, depending upon the weather and rainfall amounts, as well as water hardness, up to 10 to 12 applications are required (15). Aquaculture ponds, especially if fish are fed a commercial ration, will not require many applications. In ponds where fish are fed, a few fertilizer applications in the spring may be all that is needed to get a bloom established.

If the decision is made to fertilize a pond, it is important to follow a fertilization schedule and to continue to monitor the pond and add fertilizer as needed. Especially in recreational ponds, the increased weight of fish produced as a result of the initial fertilizer applications cannot be sustained without maintaining a good bloom. Fish will

Table 1. Suggested Fertilization Rates (per application)[a]

Fertilizer		Water Calcium Hardness[b]		
Type	Grade	Low Hardness	Moderate Hardness	High Hardness
Liquid:	11-37-0	0.5–1 gal/ac	1–2 gal/ac	2–4 gal/ac
	13-37-0			
	10-34-0			
Powder:	12-52-4	4–8 lb/ac	8–16 lb/ac	16–32 lb/ac
	12-49-6			
	10-52-0			
Granular:	0-46-0	4–8 lb/ac	8–16 lb/ac	16–32 lb/ac
	0-20-0	8–16 lb/ac	16–32 lb/ac	32–64 lb/ac
Time release:	10-52-0	25 lb/ac	30–40 lb/ac	50 lb/ac
	14-14-14	75 lb/ac	100–125 lb/ac	150 lb/ac

[a]Use this table as a starting point and modify for pond conditions by adding more or less fertilizer per application.
[b]For pond waters with calcium hardness below 50 mg/L, use the low rates. For water with calcium hardness between 50 and 100 mg/L, use the moderate rates. For waters with calcium hardness above 100 mg/L, use the high rates. Most recreational farm ponds will be low in hardness. After the initial application, apply one-half of the recommended application rate. It is likely that high hardness waters will require more frequent fertilizer applications to maintain pond blooms.

lose weight and be in poor condition if the bloom is not maintained. Discontinue fertilization for the year when the water temperature drops below 15 °F (60 °F) in the fall. Fertilization of ponds during the winter is ineffective (17) and can lead to excessive growth of undesirable filamentous algae the following spring.

Fertilizers are generally caustic materials. In applying fertilizer, take precautions to avoid unnecessary exposure to the fertilizer and clean the equipment thoroughly after each application. Always read and follow label directions for the product that you are applying. Wearing protective eyewear and clothing is advisable when handling any fertilizer formulation.

Methods for applying fertilizers vary with the form of the product. Mix one part liquid fertilizer with 5 to 10 parts water and splash or spray over as much of the pond surface as is practical. This dilution is essential, since liquid formulations are more dense than water and will sink to the bottom and become lost in the soils if not prediluted. For ease of application in larger ponds, diluted fertilizer can be poured into the prop wash of a boat as it is driven around the pond. Broadcast powdered fertilizers over as much of the pond surface as is practical. Powders are highly water soluble and most of the fertilizer will dissolve before reaching the pond bottom.

Granular fertilizers, such as triple superphosphate (0-46-0), are the least desirable choice for pond fertilization when a rapid bloom response is needed because they dissolve relatively slowly and will sink rapidly to the pond bottom if they are broadcast (18). However, triple superphosphate is also one of the least expensive pond fertilizers, and can be used with great success in recreational fish ponds. If a granular fertilizer is used, it must be applied in a manner that avoids soil contact. Granular fertilizers should not be broadcast onto a pond. Granules can be poured onto an adjustable platform that can be maintained at a depth of 4–12 in. below the water surface. One properly placed platform will serve for a pond up to 2.0 to 2.4 ha (5 to 6 ac) (19). Although the design of a platform is not critical, and many shapes and configurations can be utilized, platform construction and placement can be difficult in existing ponds. Alternatively, fertilizer bags may be slit on the larger, flat side in an "x" fashion, corner to corner, so that one side of the bag can be removed. The bags can be slit prior to being placed in the water, or they can be placed in shallow water and then slit to reduce spillage of granules. Controlled release granules must also be kept from contact with the mud and should be applied in the same manner as other granular fertilizers.

TROUBLESHOOTING

Every pond is different and will respond differently to an identical fertilizer application schedule. The recommendations given in Table 1 are suggested rates only. The number and frequency of fertilizer applications necessary to obtain a satisfactory bloom will vary from pond to pond. Fertilization to stimulate the development of an algae bloom in ponds will not be effective in many situations, and corrective action will be required. Table 2 lists many of the common problematic situations and corrective actions.

Table 2. Trouble-Shooting Common Problems

Situation	Corrective Action
Water flow to pond is such that water is exchanged in less than two weeks.	Divert runoff or stream around pond.
Pond is excessively muddy (turbid).	Treat the pond with alum or gypsum (20–22).
Pond is heavily infested with aquatic weeds.	Control weeds through mechanical, biological, or chemical means. (23,24)
Alkalinity of the water is low (<20 ppm as $CaCO_3$).	Lime the pond. Contact your county extension office for assistance.
Water temperature is <15 °C (<60 °F.)	Delay applying fertilizer until the water warms.
Pond is heavily shaded by surrounding trees.	Clear overhanging vegetation from the shoreline.

BIBLIOGRAPHY

1. E.V. Smith and H.S. Swingle, *Trans. Am. Fish. Soc.* **68**, 309–315 (1938).
2. H.S. Swingle and E.V. Smith, *Trans. Am. Fish. Soc.* **68**, 126–135 (1938).
3. C.E. Boyd, *Trans. Am. Fish. Soc.* **105**, 634–636 (1976).
4. C.F. Hickling, *Fish Culture*, Faber and Faber, London, UK, 1962.
5. C.E. Boyd and E. Scarsbrook, *Archiv für Hydrobiologie* **74**, 336–349 (1974).
6. C.E. Boyd, *Bulletin No. 459*, Auburn University/Alabama Agricultural Experiment Station, Auburn, AL, 1974, pp. 20.
7. B. Hepher, *Bamidgeh* **14**, 29–38 (1962).
8. H.S. Swingle, B.C. Gooch, and H.R. Rabanal, *Proceedings of the Annual Conference of Southeastern Game and Fish Commissioners* **17**, 213–218 (1963).
9. C.E. Boyd, *Aquaculture* **7**, 385–390 (1976).
10. C.E. Boyd, *Trans. Am. Fish. Soc.* **110**, 541–545 (1981).
11. R.J. Metzger and C.E. Boyd, *Trans. Am. Fish. Soc.* **109**, 563–570 (1980).
12. R.J. Kastner and C.E. Boyd, *J. World Aquacult. Soc.* **27**, 228–234 (1996).
13. J.C. Geiger, *Aquaculture* **35**, 353–369 (1983).
14. G.M. Ludwig, N.M. Stone, and C. Collins, *Southern Regional Aquaculture Center*, Publication 469 (1998).
15. H.S. Swingle and E.V. Smith, *Management of Farm Fish Ponds*, Bulletin No. 254, Alabama Polytechnical Institute/Alabama Agricultural Experiment Station, Auburn, AL, 1947.
16. C.E. Boyd, E.E. Prather, and R.W. Parks, *Weed Science* **23**, 61–67 (1975).
17. L. Yuehua and C.E. Boyd, *Prog. Fish-Cult.* **51**, 270–272 (1990).
18. C.E. Boyd, *Trans. Am. Fish. Soc.* **110**, 451–454 (1981).
19. J.M. Lawrence, *Prog. Fish-Cult.* **16**, 176–178 (1954).
20. C.E. Boyd, *Trans. Am. Fish. Soc.* **108**, 307–313 (1979).
21. R. Wu and C.E. Boyd, *Prog. Fish-Cult.* **52**, 26–31 (1990).

22. J.A. Hargreaves, *Southern Regional Aquaculture Center*, Publication 460 (1999).
23. J.L. Shelton and T.R. Murphy, *Southern Regional Aquaculture Center*, Publication 360 (1992).
24. T.R. Murphy and J.L. Shelton, *Southern Regional Aquaculture Center*, Publication 361 (1992).

See also CARP CULTURE; MILKFISH CULTURE; NITROGEN; TILAPIA CULTURE.

FILTRATION: MECHANICAL

SHULIN CHEN
Washington State University
Pullman, Washington

OUTLINE

Particulate matter removal is a major design consideration and management task for intensive aquaculture systems. Externally, particulates discharged from an aquaculture system can cause environmental pollution due to the organic nature and nutrient content of the material. Internally, accumulation of suspended particles in a recirculating system can cause fish gill damage (1), mechanical clogging of biofilters, additional ammonia production due to mineralization, and increased oxygen demand as the particles decay. Consequently, particulate matter has to be removed from intensive aquaculture systems such as raceways and water recirculating systems.

The removal of suspended particles from an aquaculture system is typically accomplished in a filtration process, where the particles are mechanically separated from the liquid stream, and removed later. The term "mechanical filtration," as used in this entry, is broadly defined as a physical process that separates solid particles from the culture water by gravity or physical restrictions. Thus, the filtration processes discussed here include mechanical filtration and sedimentation. This entry discusses the characteristics of particles that are related to filtration, filtration mechanisms, typical filtration processes, and process selection.

PARTICLE SOURCES AND CHARACTERISTICS

The amount of particulate matter in an aquaculture system is typically represented by the total suspended solids (TSS) concentration, which is defined as the mass of particles [in mg/L (ppm)] that are larger than 2 μm in diameter (2) contained in a known volume of water [1 L (0.26 gal)]. Virtually all the wastes generated in an intensive aquaculture system originate from feed. Of the different forms of fish excretory products, TSS primarily in the form of feces, is a major component. The mass production rate of feces is determined by the feeding rate. Quantitative investigations on several finfish species indicate that an estimation of feces production ranging from 20 to 30% of the feed consumed is well accepted. Other components of suspended solids in an aquaculture system include uneaten food and bacteria biomass.

The two most important physical characteristics of suspended solids, with regard to filtration, are particle density and size distribution. Density can be represented by specific gravity, which is the ratio of the density of a wet particle to that of water (2). The specific gravity is determined by the source of the particles, and depends largely on the characteristics of the feed and the feces. Size distribution is determined by the source of particles, the fish size, the temperature, and the turbulence in the system. Reported specific gravities of the particulate matter in aquaculture systems ranged from 1.004 to 1.19 (3,4).

Information on size distribution of fish feces is difficult to obtain because size distribution may vary according to certain conditions. When fecal material is first excreted from the fish, the particle size of the material is relatively large. The large particles may soon break down into smaller ones depending on the turbulent conditions and the stability of the fecal particles. Table 1 shows the particle size distribution of fecal particles 24 hours after being excreted from catfish fed with four types of feeds using different binders. Other reports on particle size distribution indicate that large particles (>30 microns) were dominant in fish feed, and fine particles (<30 microns) dominated the culture water of recirculating systems (5).

FILTRATION MECHANISMS

Suspended solids removal by filtration is a solid/liquid separation process, which is accomplished when the

Table 1. Particle Size Distribution (%) of Fish Excretion in Response to Different Feed Binders

Particle Size (μm)	Feed 1	Feed 2	Feed 3	Feed 4
1–30	18.8	18.5	18.6	18.3
30–105	76.3	77.8	76.5	76.5
105–1000	4.8	3.7	5.0	5.2

particles attach to an interface that provides a stronger attractive force to the particles than water. The dominant mechanisms in filtration processes include sedimentation, straining, interception, and diffusion.

Sedimentation

In the sedimentation process, solid particles are separated from water as a result of the density difference between the particles and water. For the typical TSS concentration range in an intensive aquaculture system, sedimentation is typically discrete. Each particle has a constant settling velocity that is independent of the settling velocities of other particles. Thus, the sedimentation efficiency depends on settling velocity. For the majority of the particles in aquaculture systems, the theoretical settling velocity, V_s (m/s), can be calculated using the equation

$$V_s = \frac{g(\rho_p - \rho)D_p^2}{18\mu} \quad (1)$$

where

g = gravitational acceleration (m/s^2),
ρ_p = density of particles (kg/m^3),
ρ = density of water (kg/m^3),
D_p = diameter of particles (m), and (3 ft),
μ = dynamic viscosity (Pa sec).

Equation (1) shows a direct relation between acceleration, particle density and particle size and the velocity with which the particles will settle out of water.

When sedimentation occurs in filtration processes that use granular material as a filter medium, the particles tend to deviate from the flow streamline due to the density difference. A particle's deviation from the streamline results in contact between the solid particle and the filtration medium. The efficiency of transport of the particles due to sedimentation in a granular filtration process can be calculated according to the equation (6)

$$\eta_s = \frac{(\rho - \rho)gD_p^2}{18\mu \cdot V_{sp}} \quad (2)$$

where

η_s = transport efficiency by sedimentation and
V_{sp} = superficial velocity (m/s).

Straining

Straining occurs when particles larger than the pore size of the filter medium (or opening of the screen) are strained out mechanically as the water passes through. In some cases, smaller particles can also be separated from the flow when several particles bridge together to form a cluster that is larger than the pore size.

Interception

If a particle has no significant settling velocity, it will follow the streamline of the flow while the liquid passes by an interface (e.g., a filter medium surface). When the distance between the streamline and the surface is less than the radius of the particle, the particle will collide with the surface, and the collision may result in subsequent attachment. The particle is then said to be intercepted by the interface. The efficiency of interception can be estimated by (7)

$$\eta_I = \frac{3}{2}\left(\frac{D_p}{D_s}\right)^2 \quad (3)$$

where

η_I = interception efficiency,
D_p = diameter of the particle (m),
D_s = diameter of the media or the interface (m).

Equation (3) indicates that the bigger the particle and the smaller the size of the interface, the higher the efficiency due to interception.

Diffusion

Diffusion due to Brownian motion is another important transport mechanism in filtration. This transport mechanism is most significant, for particles less than several microns. The efficiency of particle transport onto an individual interface can be estimated by the equation (6)

$$\eta_d = 0.9\left(\frac{kT}{\mu D_p D_s V_{sp}}\right) \quad (4)$$

where

η_d = diffusion efficiency,
k = the Boltzmann constant (1.38×10^{-23} J °K), and
T = absolute temperature, °K.

Equation (4) shows that at a given temperature, the efficiency of transport due to diffusion is inversely related to both particle and interface sizes.

In a granular filtration or floatation process, the ultimate separation efficiency also depends on the attachment efficiency after the solid particles have been transferred onto the interfaces (or the media). The attachment process is complicated since it is affected by many factors, including particle size and density, Reynolds number, and particle surface properties. Because the transport of a particle onto an interface is the first step in separation, Equations 1–4 should provide insight for evaluating the effectiveness of different processes for solids separation.

PROCESS DESCRIPTION

A variety of filtration unit processes have been applied to aquacultural operations. These processes, based on the filtration mechanisms, can be divided into the following main categories: gravity separation (clarifiers, tube settlers, and hydrocyclones), granular filtration [granular media (GM) filters, porous media (PM) filters], screening (rotating drum filters, triangular filters), and floatation units (foam fractionation).

Settling Basins

Sedimentation is among the simplest of technologies available to separate solid particles in wastewater treatment. A sedimentation process is usually accomplished in a settling basin or tank (also called a clarifier). The key design parameter for settling basins is the overflow rate (V_o), which is defined as the volumetric flow rate per unit surface area of the basin. The overflow rate can be directly related to the settling velocity of the particles to determine if a particle will settle out. Any particle with a settling velocity (V_s) greater than the overflow rate (V_o) will settle out of the suspension. Finer particles, for which $V_s < V_o$, will be removed in the ratio V_s/V_o, depending upon their vertical position at the inlet of the tank.

Field investigations revealed that an appropriate overflow rate (V_o) for the design of settling basins in intensive salmonid aquaculture systems typically ranges between 40–80 $m^3/m^2/d$ (982–1964 gpd/ft^2). These overflow rates translate to particle settling rates (V_s) of 0.046–0.092 cm/s. At these settling rates, approximately 65 to 85% of the total suspended solids will be removed in the settling basin. A number of factors, such as turbulance, scour and short-circuiting, lower the performance of settling basins.

The main advantages of sedimentation processes include minimal energy input, low installation and maintenance costs, and low operational skills requirement. The main disadvantages are low hydraulic loading rate and poor removal efficiency of small particles (<100 μm). Settling basins (or tanks) have been widely used in effluent treatment for flow-through aquaculture systems. For recirculating systems, however, sedimentation alone may not be sufficient for solid removal as fine solids tend to accumulate.

The performance of settling tanks can be improved by installing tube settlers consisting of inclined tubes or plates (8). In a tube settler, sedimentation is enhanced in three ways: the multiple tubes stacked above one another provide a large effective settling area; the small hydraulic radius of the tubes maintains laminar flow and promotes uniform flow distribution; and the movement of particles against the direction of flow favors particle contact in steeply inclined tubes.

Hydrocyclones and swirl separators are another type of gravitational solids separator. The hydrocyclone is a cone-shaped structure. Water is introduced tangentially toward the top of the unit. The tangential flow causes a swirling motion within the unit. At sufficient velocities 15.2–18.3 m/s (50–60 ft/s) a vortex forms within the center of the hydrocyclone. Hydrocyclones employ the principle of centrifugal sedimentation whereby the weight difference between the particles and water is amplified, thus, enhancing the gravity separation by increasing the value of the acceleration. The density difference between the solids and water is the main factor determining the separation efficiency (9). The fact that solid particles found in aquaculture systems have low densities detracts from the benefits of the centrifugal separation approach. Although the continuous operation of the hydrocyclone is advantageous, drawbacks to this system include poor removal of fine particles (<35 μm) and the need for large pumps to maintain flows with high head loss.

A swirl separator is a version of the hydrocyclone in which flow is introduced in a much slower velocity, and a vortex does not develop in the center. Water flows out of the unit by means of a cylindrical weir. A swirl separator can best be described as an accelerated settling basin. The swirl separator augments the natural gravitational force through utilizing the centrifugal force generated by the gentle swirling within the unit. As with the hydrocylclone, specific gravity becomes the limiting factor in a swirl separator.

Microscreen Filters

Screen filters are widely-used filtration devices that operate by straining particles from the water. There are two types of screens, stationary and rotary, which are used according to the mode of operation. Stationary screen filters are usually placed perpendicular to the direction of the flow. As water flows across the screen, the particles suspended in the water are strained onto the screen surface and become separated from the water. Stationary screens are one of the simplest filtration methods. When the screens become clogged, however manual backwashing is usually needed, thus the operation is intermittent. Triangle filters address this problem by passing the influent down an inclined screen surface. The water flows through the screen, while the particulates resting on the screen surface are pushed by the influent and by high-pressure cleaning water to the end of screen. In a triangle filter, the backwashing is automatic, and the process is continuous.

Rotary microscreen filters are designed to mitigate the clogging potential. A typical configuration of a rotary microscreen filter consists of a motor-driven rotating drum mounted horizontally in a rectangular chamber. A fine-screen medium covers the periphery of the drum. Influent water enters the drum at one end and passes radially through the screen with the accompanying deposition of suspended solids on the inner surface of the screen. When the pressure drop across the screen reaches a given level, pressure jets of clean (or effluent) water, at the top of the drum, are directed onto the screen to remove the mat of deposited solids. The dislodged solids, together with that portion of the backwash stream penetrating the screen, are captured in a waste hopper that discharges to a settling basin.

The design of a microscreen filter includes the selection of screen opening size, hydraulic loading rate, drum rotating speed, and backwash pressure. The desired screen opening size is determined by the specified operation. Typically, screen opening sizes of greater than 60 μm are used for aquaculture applications. Although smaller pore sizes increase suspended particulate removal, the associated increase in backwash frequency and high pressure requirements partially offset the benefits. Hydraulic loading rate to a microscreen filter is very sensitive to screen pore size and solids concentration in the influent. The finer the opening size and the higher the solids concentration, the lower the hydraulic loading.

Screen filters have been used in aquaculture systems for both effluent polishing and solids removal in recirculating systems. Additionally, screening can be especially useful as a pretreatment before sand filters (or other biological filters) in recirculating systems. Compared to other solids removal processes, microscreen filters are more suitable for effluent polishing because they can handle large hydraulic loading, and the effluent quality can be controlled to meet specific requirements by selecting the right screen size.

Granular Media Filters

A granular (GM) media filter typically consists of a cylindrical vessel that contains a solid granular medium such as sand or plastic beads. A screen capable of retaining the medium, while still allowing water to pass, is placed at the effluent side of the medium. When water flows through the bed of the granular material, suspended particles deposit onto the medium surface through either sedimentation, straining, interception, or diffusion. The filters are operated in the filtration mode until the hydraulic conductivity of the bed drops due to solid accumulations. The filter is then bypassed, and the bed is washed (or backwashed) by increasing or reversing the flow through the bed until the bed expands and fluidizes, releasing the captured solids.

Three types of granular filters are commonly used in aquaculture applications: downflow pressurized sand filter, upflow sand filter, and floating plastic beads filters. The main advantage of upflow sand filters over downflow filters is that the upflow sand filters can avoid caking problems that plague traditionally designed downflow filters. The main advantage of bead filters is their ability to reduce water losses associated with backwashing the filter bed.

Solid removal efficiency of a GM filter depends mainly upon the size of the particles, the size of the filter medium, and the flow rate. For particle sizes above approximately 40 μm, straining is the controlling removal mechanism. For particles smaller than 1 μm, diffusion will control particle transport. Between the two limits, interception and gravity sedimentation dominate particle capture.

A major advantage to using granular filters in fish culture systems is that such filters can also be used for biofiltration as the medium provides surface areas for bacteria to grow on. It combines the two processes into one unit. The main disadvantage of using granular filters for solid removal is the high energy consumption. Granular media filters require between 3.3–10 m (10–30 ft) of head pressure to operate. Another disadvantage of some granular filters is the water loss during backwashing processes.

Porous Media Filters

Porous media (PM) filters usually have thicker media than screen filters and finer pore sizes than granular filters. Porous media filters are similar to screen filters since straining is the main mechanism for solid removal. Porous filters usually have lower hydraulic loading rates and higher head losses than screen filters. Two examples of porous filters are diatomaceous earth filters (DE) and cartridge filters. A DE filter consists of a vessel in which a septum is supported. Deposited on the septum are thin layers of a filter aid known as a precoat, which is the effective filtering medium. Cartridge filters use disposable cartridges of different sizes. Due to their fine pore size, both DE and cartridge filters can remove very small solid particles (<1 μm). Consequently, they are very susceptible to clogging even at low influent TSS concentrations, and they experience high head losses. Recharging or replacement demands effectively prohibit the use of porous filters as the principal solid capture device in intensive production systems.

Foam Fractionation

A common name for suspended solids removal using dispersed air bubbles is foam floatation or, in a broader sense, bubble separation. A typical foam fractionation unit is a bubble column where bubbles generated at the bottom rise through a liquid in which solids are suspended. Solid particles attach to the bubble surface with the help of surfactants (e.g., a protein) in the water, forming foam that is removed at the top of the column. Like granular filtration, solid mass transfer from the liquid to the bubble surface is mainly through diffusion, interception, and sedimentation (10). Although foam fractionation is primarily designed for dissolved, surface active organics removal, it can also remove fine suspended solids with diameters smaller than 30 μm. Foam fractionation can be a continuous operation with no backwashing necessary. However, since foam fractionation is affected by the uncontrollable chemical properties of the water and the solids in a system, solids removal may be erratic.

PROCESS SELECTION

Understanding filtration processes is critical to the development of an integrated treatment scheme for either effluent polishing or wastewater treatment. One of the major differences among the processes is fine solids removal capability. Figure 1 illustrates particle size ranges over which each process is effective. The major goal in selecting a process for TSS control is to achieve the required water quality while minimizing capital and operating costs. The main factors to keep in mind are

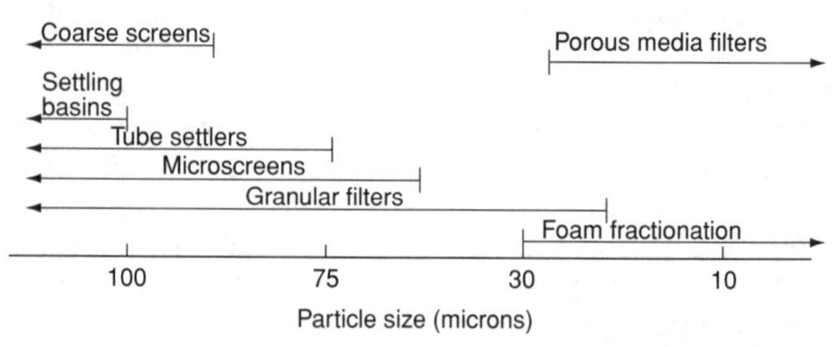

Figure 1. Suitable particle size range for different filtration processes.

the objective of the filtration process, the integration with other system components and the simplicity of operation. Main criteria for evaluation of a TSS control process used in recirculating systems should include the following: (*1*) hydraulic loading rate, (*2*) fine solids removal capability, (*3*) head loss, (*4*) water loss during filter backwash, and (*5*) resistance to clogging.

Clearly, one filtration process may be more appropriate than others for a given application. Adequate consideration of alternatives, including an evaluation of the capabilities and limitations of each process, will assure a successful application. Additional information can be obtained in other literature (11,12).

BIBLIOGRAPHY

1. P.E. Chapman, J.D. Popham, J. Griffin, and J. Michaelson, *Water, Air, and Soil Pollution* **33**, 295–308 (1987).
2. APHA, *Standard Methods for the Examination of Water and Wastewater*, 19th ed., American Public Health Association, New York, 1995.
3. S. Chen, M.B. Timmons, D.J. Aneshansley, and J.J. Bisogni, *Aquaculture* **112**, 143–155 (1993).
4. Z. Ning, *Characteristics and Digestibility of Aquacultural Sludge*, Master's thesis, Department of Civil and Environmental Engineering, Louisiana State University, Baton Rouge, LA, 1996.
5. C. Easter, *System Component Performance and Water Quality in a Recirculating System Produce Hybrid Striped Bass*, Paper Presented at Aquacultural Expo V, New Orleans, January 11–16, 1992.
6. J.M. Montgomery, Consulting Engineers, Inc., *Water Treatment Principles and Design*, John Wiley & Sons, New York, 1985.
7. G.E. Jackson, *CRC Critical Reviews in Environmental Control* **10**, 339–373 (1980).
8. Environmental Protection Agency (EPA), *Process Design Manual for Suspended Solids Removal*, U.S. EPA Technology Transfer Publication, 1975.
9. L. Svarovsky, in L. Svarovsky, ed., *Solids–Liquid Separation*, Butterworths, 1977.
10. S. Chen, *Theoretical and Experimental Investigation of Foam Separation Applied to Aquaculture*, Ph.D., Dissertation, Cornell University, Ithaca, NY, 1991.
11. S. Chen and R.F. Malone, in *Engineering Aspect of Intensive Aquaculture*, Northwest Regional Agricultural Engineering Services, NRAES–49, 1991.
12. S. Chen, D. Stechey, and R.F. Malone, in M.B. Timmons and T.M. Losordo, eds., *Aquaculture Water Reuse Systems: Engineering Design and Management*, Elsevier, 1995.

See also RECIRCULATING WATER SYSTEMS.

FINANCING

DANNY KLINEFELTER
Texas A&M University
College Station, Texas

OUTLINE

As in any developing industry, financing for the aquaculture industry has been limited by the lack of experience of both lenders and producers, as well as by the rate at which the aquaculture industry infrastructure has developed. Despite the obstacles that still must be overcome, the prospects for aquaculture financing are brighter today than ever before.

This entry addresses the following topics: (*1*) current changes occurring in financing requirements, (*2*) financing issues specific to aquaculture, (*3*) sources of financing, and (*4*) recommendations for enhancing the availability of financing.

THE CHANGING FINANCING ENVIRONMENT

Although there are many specific concerns related to the availability of financing for aquaculture, the most significant changes in lending practices and policies are not unique to aquaculture financing and are applicable to all agricultural borrowers. The changes are primarily due to significant loan losses, the number of financial institution failures, and tighter regulatory requirements, all of which resulted from the financial experiences of the 1980s.

Most of the changes borrowers are experiencing fall into the following five categories:

1. Lender's more rigorous requirements for additional and accurate information.
2. More thorough analysis and verification of the information provided.
3. Greater emphasis on both repayment ability and risk management.
4. Increased requirements for monitoring business performance after loans are made.
5. Stricter adherence to the lending institution's policy guidelines, i.e., fewer exceptions to the rules.

Agricultural producers, including aquaculture producers, are beginning to be treated like any other commercial borrowers. They will be required to develop detailed business plans that rely on trends and past performance and incorporate general economic and specific enterprise outlook analysis. Borrowing will become increasingly complex whenever operations are vertically integrated or involve multiple ownership. In cases where ownership interests involve a variety of businesses, loan analysis also will require more thorough evaluation of contractual arrangements between entities and financial statement consolidations.

Many borrowers tend to view much of the information lenders request as just more red tape but the marketing, production, and financial information a lender needs are important to the borrower if he is going to successfully manage his business.

When preparing a loan request and a business plan, prospective borrowers need to recognize and address the following questions:

1. How much is to be borrowed over the planning period?

   The loan request should not be an initial figure, but should account for the entire period, and the final purpose of, the loan request. Lenders don't want to loan the full amount they feel comfortable extending and then find they need to lend significantly more in order to see the situation to completion. Borrowers' estimates of repayment ability need to be realistic and conservative, and cost estimates need to address typical contingencies.

2. What is the loan going to be used for?

   Borrowers must be specific. It's not enough to say "operating expenses." In the past, too many operating loans have been used to subsidize lifestyles, refinance carryover debt, and finance capital purchases. Plans need to be supported not just by budgets, but by historical documentation showing that they represent past experience. Many projections appear to be based on realistic estimates, but further investigation often reveals that estimates represent performance levels out of line with what the business has actually been able to achieve in the past.

3. How will the loan affect the borrower's financial position?

   Borrowers should know their net worth, financial structure, historical cash flows, profitability, and risk exposure at the time of the loan request, and estimate what things will look like after the loan is made.

4. How will the loan be secured?

   Borrowers need to recognize the fact that collateral is adequate only if, under the worst conditions, enough collateral could be collected to generate sufficient cash to repay the loan and cover all the costs involved. Except for control purposes, collateral's primary purpose is to provide insurance in the event of default. Therefore, the important lending consideration is not the collateral's worth at the time of the loan request, but the expected value at the due date of the note or at the date of the next scheduled payment. The lender needs to account for the period of time involved, potential changes in collateral value and condition, legal and selling costs, and the fact that a distress sale will bring less than an arm's length transaction under normal market conditions. The changing nature of security has been one of the most significant factors affecting agricultural lending. More loans are now based on soft, rather than hard assets, i.e., contracts and leases versus land and chattel. There are also more joint ownership arrangements and market risks related to specific attribute raw materials than to straight commodities. All of these factors make it more difficult for the lender to assess a net realizable value.

5. How will the loan be repaid?

   Borrowers must decide if repayment will come from operating profits, from nonfarm income, from the sale of the asset being financed, from refinancing, or from the liquidation of other assets.

6. When will the money be needed and when will it be repaid?

   This two-part question should be answered by the projected cash-flow budget. Answering the question establishes that both the borrower and the lender know how the business operates. Almost as many credit problems have resulted from a lack of understanding and communication as from unrealistic expectations. Marketing plans and trigger points, contract terms and conditions, and various pooling arrangements are often not adequately communicated or documented.

7. Are projections reasonable and supported by documented historical information?

   Too many producers still do not have the production, marketing, and financial records necessary to demonstrate their financial track record and support their numbers. Many loans have not been made that probably could have been repaid, simply because of a borrower's inability or unwillingness to provide the lender with complete and well documented historical information on his or her financial position and performance.

8. How will alternative possible outcomes, in terms of both prices and quantities, affect repayment ability?

   Because cash-flow projections are based on expected values, the actual outcome is frequently very uncertain. Even under marketing and production contracts with established price bases, quality discounts, and premiums can still result in a great deal of uncertainty. However, the most common error occurs when borrowers and lenders actually believe they are addressing the issue and evaluate the impact of standard scenarios such as a 10 or 25 percent decrease in revenues. For some businesses this practice overstates the risk involved, while for others it may seriously understate the potential risks. Alternatives considered should reflect the business's actual historical performance variability as well a the range of current forecasts.

9. How will the loan be repaid if the first repayment plan fails?

   No commercial lender wants to enter into a situation in which foreclosure is the only alternative if things do not go as planned. Contingency planning is critical. Every plan should have a backup plan, and every entry strategy should have an exit strategy. This latter point is particularly true where niche markets are involved.

10. How much can the borrower afford to lose and still maintain a viable business?

    The borrower must recognize that a viable net worth is not any amount above zero. Most commercial lenders require a minimum equity position, e.g., 30 percent, below which they will not continue financing without an external guarantee. With this in mind, the answer to the preceding question must be based on the effect of various combinations of both potential operating losses and declines in asset values. Then, both the borrower and the lender need to determine how likely it is that such a situation will occur.

11. What risk management measures have been, or are to be, implemented?

    Risk management measures can cover anything from formal risk management tools to management strategies. The most important issue is that both the borrower and the lender understand how these measures work. It is also critical that the lender is supportive and committed. For example, incorrect use of commodity futures and options can increase, rather than reduce, risk. A lender's unwillingness to finance margin calls can also destroy a successful hedge.

12. What have been the trends in the business's key financial position and performance indicators? If these trends are adverse, what are the specific plans for turning things around?

    Timely action and the ability to manage problems are hard to measure, but as risk increases, they become critical in the credit decision process.

In addition to the changes discussed previously, a more subtle but significant shift in lending practices is occurring in response to legislation. This shift provides for additional borrowers' rights, more liberalized bankruptcy laws, and the threat of lender liability lawsuits. Lenders are becoming more selective in terms of whom they finance. Litigation usually arises from situations where the borrower is in financial trouble, therefore, it will become increasingly difficult for marginal and higher risk borrowers to qualify for credit. Just as malpractice lawsuits have changed the practice of medicine, the fear of legal action has changed the lending environment and caused lenders to be more cautious and conservative.

ISSUES IN FINANCING AQUACULTURE

The following are some of the broader issues specific to financing aquaculture.

Lack of Aquacultural Experience

Because risk is a function of uncertainty, the less a lender understands about a business or an industry, the greater is the risk he perceives. This risk is compounded if both the lender and the management of the aquacultural operation lack prior experience. While part of the problem is perceptual, the risk of something going wrong is actually greater, until an adequate amount of experience is gained, because mistakes are naturally made during the learning process.

Early Stage of Development

The rate of development of the industry infrastructure and the size of the market create a second problem in aquaculture financing. This problem manifests itself in the collateral value of the specialized equipment and improvements required for aquacultural production. If a market is expanding or is well established, specialized items tend to have a more ready market. However, the current situation in aquaculture often requires a large discount from the construction or purchase price in order to protect the lender from a limited or illiquid market.

The marketability of an aquacultural operation can be judged roughly by the number of processors or marketing channels bidding for the farm's product. If there is only one processor or marketing channel, there is less competition and a lower assurance of a continuing market. Therefore, in a market in which there is only one processor of the farm's production, improvements may be valued at as little as 10 to 20 percent of cost or book value, while the existence of three or more processors may increase this value to 40 to 60 percent. Obviously, these valuation factors are also influenced by the size, financial strength, and reputation of the processors involved. The same factors affect the collateral value of contractual arrangements between producers and processors.

Inventory Questions

A third problem that affects the availability of financing for aquaculture is the difficulty in establishing a value for growing products. Despite many jokes about lenders using glass-bottom boats and scuba gear, there are significant limitations on inventorying the growing products, both in terms of quantity and quality.

These problems differ significantly with the species produced. There are both new and established aquacultural products. For example, channel catfish *Ictalurus punctatus*, have been produced successfully on a large commercial scale for year, and the market is well established. While individual lenders or producers may lack experience with catfish, there is at least information and experience available. This experience can be accessed through the study of published materials, the employment of consultants, and the hiring of experienced management or loan officers. The same situation does not exist, however, for many aquacultural enterprises.

Factors Outside the Business

The importance of a well-developed business plan in obtaining credit has already been mentioned. However, many of the plans developed by aquacultural producers have focused almost entirely on the internal or technical aspects of the business. The concern of many lenders is that the greatest risks may be related to factors outside the business. Thus, prospective borrowers have to address these outside factors in their plans.

Two particular areas outside the aquacultural firm that need to be considered are the general environment and the specific industry. The general environment needs to be evaluated in terms of social, cultural, economic, government/legal, technological, and international issues. The specific industry needs to be evaluated in terms of market forces represented by potential new entrants, supplier market power, buyer market power, substitute products, and the existing degree of competition.

Social and cultural issues include general attitudes toward the health and safety of farm-raised aquacultural products, consumers' religious beliefs, consumers' education, etc. Economic issues are based upon the end-users' ability to purchase the product. Disposable income, leisure time, spending priorities, changes in interest rates, inflation rate, and unemployment rate are some specific economic considerations. Government/legal issues can play a major role in an aquacultural project and should be examined carefully. Rights to water, exotic species permits, environmental regulations, and taxing authorities all need to be addressed in the planning process.

Competition

The competitive forces in the industry also require close examination. The ease with which new competitors can enter the industry should be considered. How rapidly will new entrants come, or existing capacity expand, in response to favorable prices levels? How much will prices fall if production increases significantly? While price response is difficult to estimate, producers should at least conduct a breakeven analysis to determine how far prices could fall and still allow the project to be viable. If the aquaculture project relies on outside sources for inputs such as feed, seed stock, etc., particular attention needs to be given to the potential market power of those suppliers. The same issues arise relative to the market power of buyers or processors. How able and how likely are suppliers or buyers to squeeze margins if they have or obtain significant market power because of either their size or limited numbers? Competitive advantage is also a function of the number of available substitute products. How sensitive is the market to price differences between competitive products?

Market Contracts

In addition to the factors listed in the previous section and the biological risks involved in production, one of the factors that most affects a lender's willingness to finance an aquaculture project is the ability of the borrower to obtain market contracts for his or her production. Contracts will be examined in terms of their length, pricing terms, and quantity and quality restrictions, as well as the reputation and financial strength of the contracting firm. Currently, only limited contracting opportunities exist for aquacultural producers. However, processing capacity is increasing, and many of the new aquaculturists are interested in contracting.

SOURCES OF FINANCING

The third area to be addressed relates to sources of financing. Unlike the other areas addressed here, sources of financing tend to be specific to each country. In the developing countries of the world, development banks, the World Bank, and government lenders are often involved in financing aquaculture projects. In the developed countries, however, most of the financing comes from private investors and commercial lenders. This section focuses on describing how in the United States different types of institutions and different types of programs are used to provide financial support to aquaculturists.

Aquaculturists with established operations or sufficient financial strength are usually able to qualify for credit from the various types of commercial lending institutions, such as commercial banks, agricultural credit associations, production credit associations, federal land bank associations, and life insurance companies. Others, however, wishing to enter aquacultural ventures involving products that have a successful track record, e.g., catfish, may be unable to secure loans from such sources because the borrower lacks management experience or financial strength or because the lenders are unwilling to lend because the borrowers have no previous experience. Hence, these producers may be able to obtain assistance through the Small Business Administration (SBA) or the Farm Service Agency (FSA, formerly FmHA). There are also programs in several states that can provide limited assistance.

A third group of aquaculturists are those interested in high-risk, but potentially high-profit, operations such as shrimp fanning. It is possible to obtain support for such ventures from SBA or FSA, but many of these operations will be forced to seek venture capital or obtain outside guarantors to provide additional financial strength. Finally, there are aquaculturists who are interested in obtaining funding for the development of commercial operations based on technology that has not been demonstrated outside of the research laboratory. Projects of this nature include artificial upwelling and closed culture of various species. There is little or no credit available for these types of ventures. Funding must be obtained almost entirely through venture capital or by placing the developer's own equity capital at risk.

Nongovernment Funding Sources

Commercial Banks. Commercial banks lend for operating expenses, capital improvements, and real estate purchases. To receive such financing, a loan guarantee is sometimes required, depending upon the financial strength and previous experience of the borrower and the riskiness of the project as perceived by the bank. Guarantees, which may be personal or through a state or federal program, assure repayment of a certain percentage of the loan. FSA and SBA, for example, can guarantee loans for up to 90 percent of their value for qualified borrowers.

There are two factors will make commercial banks more interested in diversifying their loan portfolios, but at the same time may make them more reluctant to take risk: the reform of the federal deposit insurance system designed to vary FDIC premium rates according to perceived risks and the raising of capital requirements for "higher risk" banks. These changes will encourage greater reliance on loan guarantees and reinforce the need for more education and a better understanding of aquaculture by both lenders and regulators (bank examiners).

Farm Credit System. The banks and associations that comprise the borrower-owned cooperative Farm Credit System provide credit and related services to farmers, ranchers, producers and harvesters of aquatic products, agricultural and aquacultural cooperatives, rural homeowners, and certain businesses involved in the processing of agricultural and aquacultural products.

The United States is currently divided into seven farm credit districts. The seven Farm Credit Banks provide a source of funds, as well as supervision and support services to Federal Land Bank Associations (FLBAs), Production Credit Associations (PCAs), and Agricultural Credit Associations (ACAs).

FLBAs make 5- to 40-year first-mortgage loans for land and capital improvements. Loans may not exceed 85 percent of the market value of the property taken as security unless guaranteed by a federal agency. PCAs make short and intermediate term loans for operating expenses, capital purchases, and capital improvements. Producers and harvesters of aquatic products may receive terms of up to 15 years. ACAs are associations created by the merger of one or more FLBAs and PCAs and therefore, have the ability to make short-, intermediate-, and long-term loans.

The other lending arm of the Farm Credit System is composed of the Banks for Cooperatives (BCs). The BCs offer a complete line of credit and leasing services to agricultural cooperatives, rural utility systems, and other eligible customers. BCs require at least 80 percent of the voting control of the cooperative be in the hands of farmers, ranchers, or producers and harvesters of aquatic products. A cooperative must also do at least 50 percent of its business with or for its members. The BCs may also finance joint ventures between eligible cooperatives and private firms, as long as the cooperative has a controlling interest. Two banks, each with a national charter, comprise the BC system. CoBank, the National Bank for Cooperatives, which is headquartered in Denver, Colorado, also finances agricultural exports and provides international banking services for the benefit of US farmer-owned cooperatives. The other BC is the St. Paul Bank for Cooperatives, which is headquartered in St. Paul, Minnesota. The St. Paul Bank is in the process of being merged into the National Bank for Cooperatives.

Life Insurance Companies. In the past, life insurance companies were primarily real estate mortgage lenders. Recently, however, several companies have broadened their lending activities to cover all phases of agricultural and aquacultural lending activities. The primary limitation for many borrowers is that these companies tend to limit their lending to larger loans and concentrate on only the most creditworthy borrowers.

Government Funding Sources

The Small Business Administration. The SBA provides both guarantees and direct loans to aquaculture operators. SBA loans may be used for the purchase and improvement of land or buildings, construction, machines and equipment, operating expenses, and refinancing of debts. SBA also provides disaster loans in authorized areas.

Farm Service Agency. The FSA provides both guarantees and direct loans to aquaculture operators. The various types of FSA loans that can be obtained for aquacultural purposes are as follows:

a. **Farm Ownership and Loan Guarantees** are made to help eligible applicants become owner-operators of family farms; to make efficient use of land, labor and, other resources; and to enable farm families to have a reasonable standard of living. These loans can be made for the purchase and development of real estate, including water resources.
b. **Operating Loans and Loan Guarantees** are made to operators of family farms and to applicants wanting to become operators of such farms. These loans can be used for financing and refinancing equipment, for livestock or fish purchases, for family living and farm operating expenses, and for minor land or water improvements. Objectives of the program are to improve living and economic conditions and to help operators become established in a sound system of aquaculture or agriculture. The combined farm ownership and operating loan limit is $400,000 for direct loans and $700,000 for guaranteed loans.
c. **Emergency Loans** are made in counties where property damage or severe production losses occur as a result of a natural disaster or because of other emergency situations. The funds can be used for major adjustments, operating expenses, and other essentials, to enable borrowers to continue their operations. This program involves only direct loans and has a loan limit of $500,000 or the amount of loss sustained, whichever is less.

Borrowers under the direct farm ownership and operating loan programs may qualify for the special limited resource loan program. Eligible borrowers qualify for initial interest rates, which are approximately half the normal loan rate, but this rate will adjust upward as the borrower's ability to pay improves.

Rural Development. Rural development provides only loan guarantees under the Business and Industrial

(B and I) Loan Program. The B and I guarantees promote development of business and industry, including aquaculture, in cities and towns with less than 50,000 population. However, applications for projects in open country, rural communities, and towns of 25,000 people or fewer receive preference. These loans can be made for conservation, recreation, tourism, and the development and utilization of water for aquacultural purposes, as well as for aquacultural related businesses, such as processing plants. Loans of less than $2 million are eligible for a 90 percent guarantee; those between $2 and $5 million, an 80 percent guarantee; those between $5 and $10 million, a 70 percent guarantee; and those between $10 million and $25 million, a 60 percent guarantee.

State Loan Programs. Many states offer financial assistance programs that target beginning farmers, nontraditional enterprises, and rural economic development. Some states offer direct loans, others, loan guarantees, and still others, linked deposit programs. In linked-deposit programs, the state makes deposits in participating financial institutions at less than the going market rate of interest. The financial institution, in turn, passes these reduced rates on to borrowers who qualify for the loans under the program.

RECOMMENDATIONS

At this time, several things must be done to improve the ability of aquacultural producers to obtain necessary financing. The first and most important is a coordinated effort to educate lenders, producers, potential investors, and financial regulators about the aquaculture industry. Second, there is a need for qualified appraisers with the experience and training to assess the collateral value of equipment and improvements employed in aquaculture. Finally, the aquaculture industry needs a readily available insurance program to insure producers of established aquacultural products against potential disasters.

Three other areas that merit further study and education include alternative uses of assets if a venture fails, alternatives to the ownership of land and capital improvements, and market or production contracts. One obvious example of an alternative use of assets is the use of ponds for water storage for agricultural irrigation or municipal use. Alternatives to land and capital purchases that need exploration include long-term renewable leases for land and leasehold improvements. This would include an analysis of the risks to the leassor and the leassee and studies of alternative lease terms and arrangements. Contractual arrangements need to be studied and evaluated in terms of the risks involved, their impact on market performance, and the economic costs and benefits to the parties involved. Additional information is available in the references listed (1–3).

BIBLIOGRAPHY

1. Danny A. Klinefelter, *Being Prepared to Borrow*, Texas Agricultural Extension Service, Texas A&M University System, College Station, TX, L-5071, May, 1993.
2. Danny A. Klinefelter, *Farm and Ranch Credit*, Texas Agricultural Extension Service, Texas A&M University System, College Station, TX, B-1464, February, 1992.
3. Danny A. Klinefelter and Greg Clary, *Financing Aquaculture in Texas*, Texas Agricultural Extension Service, Texas A&M University System, College Station, Texas, L-2426, August, 1997.

FISHERIES MANAGEMENT AND AQUACULTURE

Nick C. Parker
U.S. Geological Survey
Lubbock, Texas

OUTLINE

FISHERIES MANAGEMENT AND AQUACULTURE

National and international discussions and debates have been held to compare the value of wild native fish, especially trout and salmon with those produced in hatcheries (1,2). Strong debate has raged for years on the U.S. west coast concerning the management of Pacific salmon stocks (3). Concerns have been expanded to question nearly all aspects of fisheries management and especially the role of hatcheries and fish culturists (4).

Fisheries scientists and anglers alike are biased by their personal experiences, and the biases start with the most fundamental of terms. What do we mean when we refer to "wild" fish and particularly to "wild trout" or "wild salmon?"

Webster's New World Dictionary defines wild as "living or growing in nature; not tamed or cultivated by man." Does a fish captured from the wild and moved into a hatchery as broodstock immediately cease to be wild? Would progeny from such a fish be wild if they were immediately returned to the waters from which the parent stock were taken? Would they be wild if maintained in captivity for one day, one month, one year, or one generation?

WILD TROUT

A survey of fish culturists on national fish hatcheries found that 82% believed that wild trout could be produced in hatcheries (5). The majority, 65% of respondents, would expect the progeny of wild fish to become domesticated in five generations. The definition of "wild trout" as viewed by 51% of those fish culturists would not preclude fish from being produced in public hatcheries or in commercial fish farms.

OUR CHANGING WORLD

Continual adaptation and change are necessary for biological organisms, political systems, businesses, and managers of natural resources to maintain their positions in the world. Just as Darwin's finches adapted to minimize competition and to survive side by side, aquaculturists, anglers, natural resource managers, including fisheries managers, and others must adapt to rapidly changing conditions. Aquaculture, which has been described as "the emerging giant," has drastically altered management of some aquatic resources and has the potential to alter others. Change in U.S. fisheries (and global fisheries) has been ongoing for hundreds of years. However, the rapid rise in the human population is forcing change at an unprecedented rate.

THE FIRST FISHERIES MANAGERS

The first professional managers of natural aquatic resources in the United States were probably the fish culturists, now known as aquaculturists, who lobbied Congress in 1871 to establish the U.S. Fish and Fisheries Commission, with Spencer F. Baird, Assistant Secretary of the Smithsonian Institute, as the first commissioner (6). These early managers also persuaded Congress to appropriate funds for the propagation and introduction of shad (*Alosa* spp.), trout (Salmonidae), and other valuable fishes throughout the country. The goals of both public officials and private producers were to provide fish for the food market and to provide species acceptable to Europeans in the New World. To help meet this goal, they first introduced the common carp (*Cyprinus carpio*) into this country from France in 1831 or 1832 and, by 1875, the fish were well established in California and New York (7). With some fanfare and at the request of Spencer Baird, Congress appropriated $5,000 in 1877 to construct culture ponds to rear common carp on the grounds of the Washington Monument. Over the next several years, farm-raised fish, including brown trout (*Salmo trutta*) and common carp, were promoted by private aquaculturists and government officials to increase angling opportunities. These culturists voiced their concerns about dams, siltation and pollution from manufacturing plants, sawmills, and other industrial and municipal sources. Fish culturists sought to preserve the aquatic environment to support fish and fishing.

After the death of Spencer Baird in 1887, the value of cultured fish in the management of natural resources began to be questioned and funding for research became more limited. By 1897, when *A Manual of Fish Culture* (8) was published by the Commission, fish culture was a well-developed practice in federal fish hatcheries as well as in the private business sector. The manual contained descriptions for the culture of more than 40 species or groups of finfish, plus lobsters, oysters, clams, and frogs. Fry produced in hatcheries were widely stocked along the Atlantic Coast from about 1870 to 1900; the programs then were abandoned, because they did not appear to influence commercial landings of American shad (*Alosa sapidissima*), striped bass (*Morone saxatilis*), or other estuarine fishes.

Conflicts of Aquaculturists and Natural Resource Managers

Now, about 100 years later, resource managers view aquatic resources as part of the public domain and consider hatchery-produced fish an important tool to manage these aquatic resources (9). Private producers of fish view aquaculture as agriculture in the aquatic environment and believe the products produced should be treated no differently from poultry, beef, or soybeans. The opposing views of natural resource managers and private fish producers are a major source of conflict. For example, aquaculturists rear striped bass as a food fish for sale in restaurants and grocery stores, whereas fisheries managers rear striped bass in hatcheries for release into public waters as game fish for anglers.

Role of Hatcheries

The popularity of public hatchery programs has cycled from high to low to high in about 30-year periods as support for stocking fish in public waters and the perceived benefits of managing aquatic resources have been alternately promoted and questioned. Habitat alteration—including dams, diversions, channelization, and pollution—and the growing demand for fish and fishery products have increased the need for hatchery-produced and farm-raised fish. State and federal hatcheries are expected to become increasingly important as tools to preserve biodiversity by maintaining rare, threatened, and endangered genotypes. Hatcheries are important educational tools and will most likely continue to be used to produce fish in support of innovative programs such as fishing in urban parks, shopping malls, and other nontraditional environments as resource managers seek to broaden support for fishery programs, attract new anglers, and educate the public about aquatic resources and the increasing demand for those resources.

INTERNATIONAL DEFICITS IN FISH

In recent years, the import of fish and fishery products into the United States (US$9.4 billion in 1997) has been exceeded in value only by the import of petroleum products (US$49 billion in 1996). The value per unit weight of fish and fishery products has increased steadily since the 1950s, reflecting the growing demand in world markets. Recent records from the Food and Agricultural Organization (FAO) of the United Nations show that the volume of fish and fish products traded among 167 nations was about 14% greater than the volume produced, indicating multiple exchanges at the wholesale level before final sale to consumers.

The rapid expansion of the aquaculture industry to produce farm-raised catfish, salmon, shrimp, and other species has been driven by increased demands on finite stocks. The concentration of much of the world's population along seacoasts and inland waterways has resulted in loss of aquatic habitat, environmental degradation, and intense fishing pressure; together, these perturbations have decimated many wild stocks. Resource managers and aquaculturists are evaluating both the risks and benefits of using introduced, transgenic, polyploid, hybrid, and reproductively sterile aquatic organisms to provide the fish and fisheries products demanded by world markets for food and recreation. The greatest challenge in fisheries management may be to maintain genetic stocks of native fishes and still provide the recreational fishing opportunities and food fish production demanded by growing world markets.

Channel catfish (*Ictalurus punctatus*) and rainbow trout (*Oncorhynchus mykiss*), the most commonly cultured fishes, are defined as agriculture crops in some states—channel catfish in Mississippi, trout in Idaho, and all farm-raised aquatic organisms in Missouri. These farm-raised fish are not regulated by the state conservation and natural resources agencies. Other species, such as striped bass, red drum (*Sciaenops ocellatus*), and largemouth bass (*Micropterus salmoides*), are now attracting the interest of commercial producers (10) and may ultimately force changes in existing regulations. Will these changes harm or benefit anglers, commercial operators, consumers, and the nation's aquatic resources?

Demand for Game Fish Increases

According to a survey conducted by the U.S. Fish and Wildlife Service in 1996, 35 million anglers spent $72 billion in 620 million-angler days fishing (11). Sport fishing is projected to double by the year 2030. The fishing pressure on public waters is expected to increase much more rapidly than the ability of the resource to produce. Even today, some anglers have abandoned public waters to fish in more productive private waters. The public waters of Texas yield slightly less than one legal-size bass in 10 hours of fishing, whereas 30–40 bass can be taken from privately owned and managed lakes in only 3 or 4 hours. Land owners are willing to buy catchable-sized fish to be stocked into private ponds for recreational purposes. For example, a company in Bryan, Texas has sold 1 kg (2.2 lb) rainbow trout to landowners in Texas to stock for recreational fishing during the winter months, and warmwater species such as channel catfish demand a premium price when available in large size.

Other aquaculturists are producing hybrids of striped bass and white bass, Florida strain largemouth bass, and other warmwater species for food and recreation. In Danbury, Texas, some anglers have paid $900 per day for the opportunity to catch 3 kg (6.6 lb) trophy-sized bass in private waters, and other fishermen routinely pay $90 in the off-season and $165 per day in the peak season to catch 1 to 2 kg (2.2 to 4.4 lb) fish (12). Largemouth bass are reared in protected nursery ponds and then transferred into larger ponds for recreational use.

At a fish farm in Danbury, Texas, anglers may catch and release multiple fish, but keep only their largest one as a trophy. Various personal services, such as catering of meals and use of lodges and equipment, are included in the $300–900 daily charges, but fewer services are provided for the $90 per day charge. Are these practices the same as sport fishing in public waters? Are they indicative of future practices? Or are they just an innovative way to market an agricultural crop as a recreational event?

States require most anglers fishing in state waters to have a valid recreational fishing license. However, in some states, such as Alabama, Arkansas, and Missouri, anglers fishing in privately owned water for farm-raised fish are exempt from licensing requirements. Although private fishing establishments provide recreational opportunities, the catch of fish is frequently so high that it resembles a supermarket activity. In a single day, anglers have harvested more than 500 kg (1,100 lb) of channel catfish from a 0.1 ha (0.4 ac) pond in Longview, Texas. Do activities such as these relieve the fishing pressure on public waters? Aquaculturists believe that they do—at least, they provide recreational opportunities in excess of those available on public waters. To the extent that fishing in private waters removes fishermen from public waters, these programs work in concert with, not in conflict with, public programs.

Game Fish or Food Fish?

Natural resource managers have often been reluctant to allow the sale of farm-raised fish. The two major obstacles preventing the sale of the hybrids of striped bass (*M. saxatilis*) × white bass (*Morone chrysops*) were identified (13) as (*1*) the inability of enforcement agencies to distinguish farm-raised hybrids from wild-caught striped bass that are prohibited from sale in many states, and (*2*) laws prohibiting the sale of striped bass because it is a game fish. Producers of channel catfish and trout faced these same obstacles as they developed commercial markets for farm-raised fish. How were conflicts resolved? Today, both farm-raised catfish and trout can be legally sold in all states. Before 1960, when there were only 160 ha (400 ac) of commercial catfish ponds, there were significant commercial fisheries for catfish in inland waters, such as the Tennessee and Mississippi Rivers. Today, after numerous incidences of environmental degradation, including release or spills of mercury, PCBs, chlorinated hydrocarbons, and other toxicants and contaminants, the public no longer associates quality and safety with wild-caught species, but more frequently with farm-raised products. In 1987, about 182,000 MT of farm-raised catfish and 23,000 MT of farm-raised trout were processed as food fish. Existing catfish processing plants had the capacity to process more than 227,000 MT/yr. Many other farm-raised fish were sold directly to consumers as food, while still others were sold for restocking as recreational fish. The development of the catfish and trout industries appears to have almost eliminated the importance of the commercial catch.

Fishing pressure on wild stocks of these inland species may continue until the last fish is caught, but the relative value of the wild fish as food fish will decline as farm production increases. The commercial production of Atlantic salmon has similarly expanded at a phenomenal rate in the 1980s and 1990s. About 50,000 MT of farm-raised Atlantic salmon were imported into the United States in 1987, whereas commercial landings from the Atlantic Ocean were only about 10,000 MT. Based principally on aquaculture production, the import of Atlantic salmon exceeded 55,000 MT in 1997. Foreign investors from Norway, Sweden, Iceland, and Japan have recently established several netpen fish farms to culture both Atlantic and Pacific salmonids in North America. Most of these farms have been placed in Canadian waters, primarily British Columbia, because regulations there are less restrictive than in the United States. Alaska fishermen recognize the threat of this foreign competition, and even though they would prefer to continue to fish for wild stocks, they recognize the potential profit of netpen culture and are ready to "jump on the band wagon and farm salmon" when state laws are modified to permit that activity.

Aquaculture Production

In 1996, aquaculturists produced about $41.5 billion worth (26.4 million MT) of farm-raised fish globally, and United States aquaculture accounted for $736 million (393,000 MT) of the world total (14). This production is based on the techniques pioneered by early culturists, as recorded in the 1897 *Manual of Fish Culture*. Today, of the 24,618 species of finfishes (15), only 179 species are cultured for food. Another 134 species are used for bait, and 19,800 species are used in the ornamental trade (15). However, we are learning to culture additional species, and one of the most valuable exports of United States science today is aquaculture technology. In 1989, a survey of 275 computer bases revealed that the terms "fish culture" or "aquaculture" occurred fewer than 2,000 times cumulative in the top seven databases (16). In 1998, an electronic search of one database, AGRICOLA, for the terms "aquaculture," "fish culture," and "fish hatchery" yielded 2,442 entries, while a search of the Excite® database for these same terms on the World Wide Web located 23,706 documents. Although today's fish culturists no longer hang dead carcasses above the raceways to feed fish with insect larvae, many of the techniques now available to aquaculturists and the training received by fish culturists in the United States can be traced back to the pioneering works of fish culturists and their practices as described in the *Manual of Fish Culture*. Today's fish culturists must be proficient in other subject areas not found in the 1897 manual, such as the use of fisheries chemicals, knowledge of genetics, use of computers, awareness of regulations, and, most importantly, the ability to work not in isolation but in view of an increasingly demanding public — customers, regulators, and anglers.

Demand for Food Increases

The demands for fish and fishery products in the United States are expected to expand faster than the supply of fish will expand. Imports of fish and fishery products into the United States were valued at $365 million in 1960, $7.6 billion in 1986, and $14.5 billion in 1997 when the imports consisted of $6.7 billion worth of nonedible products (animal feeds, industrial products, etc.) and $7.8 billion for edible fishery products. The annual per-capita consumption of fish increased over 20% from 1975 to 1986, when the per-capita rate reached 6.7 kg (14.7 lb); it is expected to be 13.6 kg (29.9 lb) by the year 2020. The world's catch of fish (millions of metric tons) was 27 in 1954, 57 in 1966, 74 in 1976, 83 in 1984, 90 in 1986, 101 in 1989, and has remained level or declined slightly since then. The catch has increased with the demand only because previously unused resources — those formerly classified as trash fish — are now being captured and processed into consumer-acceptable forms, such as imitation crab, lobster, shrimp, and scallops. The ocean's resources are recognized as finite, having an estimated maximum sustainable yield of about 100–120 million MT. The expansion of demand in a market with limited supply is expected to continue to drive prices up and make fish farming even more lucrative than it is today, when more than 25% of the global fish supply (fish and shellfish) is produced by aquaculture. Global aquaculture has increased from 7% of the total fish in 1950 to 29% by 1996. Aquaculture production of nonfood items has increased from 3 million MT in 1950 to 31 million MT in 1997. Since 1960, worldwide aquaculture has grown at an annual percentage rate of 10.9%.

Distinguishing Between Wild and Farm-Reared Fish

Aquaculturists, resource managers, and law enforcement personnel have tools and techniques today that were not available when many of the rules and regulations were written to protect aquatic resources. For example, few of the states' laws address hybrid fish. In some states (e.g., Maryland), hybrid striped bass were considered to be striped bass for purposes of regulation. In other states (e.g., New Jersey and Massachusetts), hybrids were not even mentioned in the regulations. Some states (e.g., Florida, Virginia, North Carolina, Georgia, and Mississippi) began to reexamine and modify their laws in 1987 to allow for possession and sale of farm-raised striped bass and hybrids (17).

The availability of reproductively sterile hybrids and triploid fish has promoted some states, in response to anglers, boaters, and owners of lakefront property, to reevaluate and modify their regulations for some species. For example, in 1978, grass carp (*Ctenopharyngodon idella*) had been in at least 35 states at one time or another, regardless of the species legal status. By 1987, however, 12 states had no restrictions on the species, 15 had specific policies for them, 4 permitted research with triploid forms, and the other 19 states technically prohibited all grass carp, even though some exceptions were made (18). Presumably sterile hybrids of grass carp × bighead carp (*Aristichthys nobilis*) were first legalized in some states, but preference shifted to the use of triploid grass carp. For grass carp, the impetus to modify state regulations came not only from the producers, but also from state fishery biologists seeking to use these herbivorous fishes for biological control of aquatic vegetation in public waters.

Producers of triploid grass carp use specialized medical equipment [a coulter counter device that is used in clinics and hospitals to count blood cells (about $20,000 each)] to verify that each fish certified as a triploid does indeed have three and not two (diploid) sets of chromosomes. The equipment and procedures for this test are expensive; the test cannot be performed in the field by law enforcement personnel. Nevertheless, the procedures and paper trails established seem to provide a workable solution acceptable to law enforcement personnel, resource managers, and aquaculturists.

The production of monosex (either all-male or all-female) populations is another tool available to aquaculturists and resource managers to limit reproduction of fish. A considerable amount of research has been conducted to develop monosex populations of tilapia to limit reproduction in culture ponds (19). Techniques used include the production of hybrids with highly skewed ratios of males and females and the production of an all-male population by feeding androgenic steroids to immature genetic females to induce sex reversal (20). Other techniques used to alter sex ratios include gynogenetic production of female fish by fertilizing eggs with sperm that has been irradiated with ultraviolet light to denature the genetic material, the DNA. Once development of the egg has been activated by the irradiated sperm, eggs are shocked by exposure to heat, cold or pressure to disrupt normal cell development and produce a diploid zygote with no genetic contribution from the male (21). Some of these resulting all-female fish can then be fed androgenic hormones to produce functional males (genetically female) and mated back with their siblings to produce a second generation of all-female fish (22).

When these techniques are further perfected, they will allow aquaculturists to produce not only monosex fish, but also fish with selected traits, such as rapid growth, disease resistance, and tolerance to high or low temperature. It is expected that desirable traits can be propagated in cultured species much more rapidly with these techniques than through the normal process of selective mating.

Similar techniques will almost assuredly be used to produce sterile exotic fish for recreational fishermen. If anglers will pay $300 to $900 per day for the opportunity to catch a trophy-sized largemouth bass weighing 3 to 4 kg (6.6–8.8 lb), what would they pay to catch a 50 kg (110 lb) freshwater fish? Several exotic species, including Nile perch (*Lates niloticus*), reach or surpass this size. How long will it be before reproductively sterile exotic fish are available in private waters to anglers? Once trophy-sized sterile exotics are available, will there be a demand for put-grow-and-take fisheries in public waters? Will sterile classification reduce the threat to native stocks enough to make these fish acceptable? If grass carp can be used as an example, it is reasonable to expect to see other sterile exotics produced by fish farmers and in-state hatcheries stocked by resource managers for sport fishermen.

Law Enforcement

In several states, laws have been modified or regulations developed to allow the possession, culture, and sale of farm-raised fish as food fish, bait, or for restocking. In other states, regulations prohibit the sale of all game fish as food fish, but do not limit the possession, culture, and sale of fish for restocking. Laws designed to protect game fish appear to be one of the major restraints limiting expansion of aquaculture in many states and the expansion of culture of species other than catfish and trout. Law enforcement personnel are probably not unwilling to distinguish between wild-caught and farm-raised fish, but cannot do so within their existing budgets and with field techniques now available to them. Prohibition on sale of game fish as food fish creates an illegal market somewhat similar to that for alcohol during the era when all sales of alcoholic beverages were illegal in the United States. The increasing demand for fish and fishery products is expected to stimulate the illegal market. However, farm-raised fish could fill the market and reduce poaching of wild stocks. Poaching pressure would decline only when economic incentives were lowered to make it unattractive; some enforcement personnel do not expect the aquaculture production of highly desirable fish, such as hybrid striped bass, ever to be abundant enough to meet the market demand. Some enforcement personnel further believe that protecting wild stocks of fish from poaching in the presence of legal sales of farm-raised fish creates enforcement problems that could be even greater than the enforcement problems resulting from alcohol sales during prohibition.

Tools that law enforcement personnel could use to distinguish between wild and cultured fish are now being used in other areas of law enforcement. Forensic scientists are developing techniques (i.e., capillary zone electrophoresis-mass spectrometry, CZE-MS) that will sample volumes one-billionth of a liter in size to detect and analyze compounds at concentrations one-thousandth of that possible with existing techniques. The analyses of fragmented DNA molecules yields genetic fingerprints that have been used to provide positive identification of individuals when the only evidence at the scene of a crime was a small sample of dried blood, semen, or other tissue. The evolution of modern molecular genetic techniques and associated technical terms such as DNA probes, plasmids, recombinant DNA, transgenics, ploidy manipulation, and sex reversal, may seem foreign and confusing to many biologists, and even more so to the layperson. However, special applications of molecular genetic techniques in the field of fishery science include production of monosex populations, polyploid fish, and use of DNA probes to identify or classify unknown tissue samples and biopsies by species, sex, geographic origin, or specific genotypic traits. The technique of electrophoresis with isoletric focusing has been used to determine species of raw muscle tissue to distinguish between trout and salmon and to identify the four species of *Morone* and their congeneric hybrids.

The equipment for electrophoresis and other analytical assay techniques is commonly available in most universities and even in some fish hatcheries. Other techniques to distinguish between wild and cultured fish include analysis of daily growth rings on scales or otholiths, scale-shape analysis, elemental composition of scales and bone, lipid profile analysis, detection of such tracer compounds as tetracycline and calacine in farm-raised fish, and morphometric differences. Some of these techniques have been used to distinguish striped bass taken from freshwater from those taken from saltwater and between Chesapeake Bay and Hudson River stocks. These and other sophisticated techniques are routinely used in forensic laboratories (23), but may require more laboratory support than current fish and wildlife enforcement budgets provide unless priorities are altered. These techniques, in conjunction with paper trails to document source and movement of farm-raised fish, are tools that are now available to enforcement personnel. It is expected that, if there is a real market for field test kits that would allow enforcement personnel to distinguish between farm-reared and wild fish, such products will be developed. Law enforcement personnel now use a field kit that detects lead, to distinguish between game taken with a bow and arrow and that taken by gunshot. Similar field tests may be developed to detect trace elements in farm-raised fish. For several years, kits have been available that enable diabetics to monitor their blood sugar and for women to test for pregnancy. Pharmaceutical firms are attempting to develop a simple and rapid test to detect the virus responsible for AIDS—the fatal autoimmune deficiency syndrome. Similar on-the-spot tests used in the medical field can very likely be adapted to detect chemical and biochemical differences in cultured and wild fish. It is even conceivable that, through bioengineering, future cultured fish may have hidden marks that become visible or undergo a color change when exposed to ultraviolet light or some other activator. A blue microbe has been developed (and patented) that will degrade PCBs; the blue color allows the organism to be easily traced in the environment and distinguished from other wild-type bacteria.

Techniques available to law enforcement personnel today include the use of various markers and tags to identify fish and trace them through the market system. For example, fish have been marked with tetracycline (which fluoresces under ultraviolet light), fluorescent pigments, and color-coded plastic chips (Microtaggants) (24). Some law enforcement personnel have used these materials to mark fish in illegal nets and then to trace them through the market system. In at least one case, passive inductive transponders (PIT) tags were used to identify fish stolen from a U.S. Fish and Wildlife Service (USFWS) laboratory in Marion, Alabama. The PIT tag is a small glass-encased electronic device that can be implanted, with the aid of a special syringe, in fish or other animals. The PIT tag derives power from an external transmitter to drive an internal transmitter and broadcast a unique 12-digit hexadecimal number detected and displayed on a monitor. Another type of small internal tag, the coded-wire tag, was used to mark about 1.2 million hatchery-reared striped bass that were released into Chesapeake Bay from 1985 through 1987, and it has also been used to mark hundreds of millions of salmonids along the Pacific Northwest coast of the United States and throughout the world. Similar techniques might be used by management biologists to access stocks in inland reservoirs. Law enforcement personnel could then use these tags to identify fish taken from state waters.

In August 1987, at a conference on aquaculture in South Carolina, law enforcement personnel proposed a resolution that all states establish a 12-digit numbering system to individually identify all hybrid striped bass produced on farms for the food fish market. Under the proposal, each fish would bear a unique tag. This proposal failed to gain support of the aquaculture industry and strengthened the resolve of aquaculturists that aquaculture is agriculture. Pressure by aquaculturists increased to have hybrid striped bass classified as farm products not controlled by natural resource managers.

How will aquaculturists and resource managers deal with other cultured species—red drum, orangemouth corvina (*Cynoscion xanthulus*), Florida pompano (*Tranchinotus carolinas*)—and the numerous hybrid crosses that will surely be made? It seems reasonable that cultured fish will become much more important in the American economy. The culture of Florida pompano was of interest a few years ago, but never developed into an economically viable commercial industry. The Florida pompano fishery is now almost nonexistent due to overharvest. Florida pompano have reportedly retailed for $26/kg ($11.82/lb), when particularly scarce, and in 1987, commonly sold for $11/kg ($5/lb) [up from $4.40/kg ($2/lb) in 1974]. Even though earlier attempts to produce Florida pompano on farms failed financially and they are not farm products today, it seems very likely that production will become economically feasible as the demand increases.

Common Goals

Natural resource managers and aquaculturists share important common goals: They continue to evaluate the condition of our aquatic resources and the supply and demand for those resources. For example, trout, catfish, salmon, beef steaks, pork loins, and chicken are not individually numbered as a condition of marketing. Food fish and shellfish, unless specifically identified as an injurious species, may be imported into the United States and are not required to be numbered individually or even federally inspected. Boxes must be clearly labeled to identify contents, shipper, and consignee. So why, the aquaculturists ask, should trade in American-produced farm products be restricted by the cumbersome process of individually tagging selected farm-reared aquatic species? Although some aquaculturists have attempted to identify all farm-raised crops as agriculture, they recognize the potential impact of cultured fish on wild stocks and work with resource managers to reduce the chances for damage. Until domesticated brood stocks are developed, aquaculturists must depend on wild stocks and work closely with natural resource managers to protect those stocks from overexploitation and genetic alteration. Maintaining high quality in the aquatic environment is of prime importance to both aquaculturists and fisheries managers.

As demand for aquatic resources increases with the continual growth of the world's population, strict preservationists and potential exploiters must both alter their positions and agree to some changes. The combined demands of recreational and commercial fishers exceed the supply of fish on a worldwide basis. Either catches must become severely limited as demand increases, or aquaculture must expand to fill the void. Projections have been made that, within the next 50 years, aquaculture products will equal or surpass the wild production of fish (25). Some sport fishermen in Texas, and other states, have already recognized the value of aquaculture to their interests and have helped to fund and establish hatcheries to culture red drum and other species to be stocked in coastal and inland waters. It appears that aquaculture holds tremendous potential for recreational anglers, consumers, and producers, and it could become an even stronger tool for natural resource managers.

STRIPED BASS RESTORATION ALONG THE ATLANTIC COAST

An example of the role of aquaculture in fisheries management is found in the story of striped bass along the Atlantic Coast. Stocks of adult striped bass (*M. saxatilis*) were extremely low from 1980 to 1987 along the Atlantic seaboard, especially in Chesapeake Bay. In an effort to rebuild those stocks, the Atlantic States Marine Fisheries Commission (ASMFC) developed a coastwide management plan for anadromous striped bass. The plan included a stocking and evaluation program developed by a technical advisory committee composed of representatives from all coastal states from Maine to North Carolina, the U.S. Fish and Wildlife Service (USFWS), and the National Marine Fisheries Service. The committee prepared a report that provided guidance for restoration programs for striped bass along the Atlantic coast. In 1985, the USFWS, Maryland Department of Natural Resources (MDNR), and the state of Virginia entered into a cooperative agreement to develop a striped bass stocking program in Chesapeake Bay. The USFWS assigned a coordinator to assist the coastal states in implementing this program. By January 1988, 1.35 million striped bass, reared in hatcheries and tagged with binary-coded wire tags, of which 23,250 were also marked with internal anchor tags, were stocked back into Chesapeake Bay. Tags from that program and others along the Atlantic coast were collected by personnel of the coastal states and returned to evaluate the effectiveness of the program.

Several factors that may have contributed to the decline in abundance of striped bass in Chesapeake Bay were reviewed in recent studies, summarized in the Emergency Striped Bass Research Study Reports of USDI and USDC (26,27). Because of the potentially synergistic and masking effects of interacting causes, no single reason for the decline was identified.

Suggested causes of the decline included contaminants, predation and competition, availability of acceptable and nutritionally adequate prey for younger fish, water-use practices, disease, natural climatic or random environmental events, and overexploitation. Of these factors, contaminants, prey availability, nutrient overenrichment, and water-use practices were important in localized situations, on either a temporary or sustained basis. However, two primary factors appeared to exert significant control over striped bass populations: (*1*) A large component of random environmental or abiotic events that influenced, either positively or negatively, the survival of eggs to the juvenile stage, and (*2*) overexploitation or excessive fishing mortality, which reduced survival from the juvenile to the spawning adult stage.

Because of the limited stock of adult striped bass (28) and extensive reproductive failure within Chesapeake Bay (29), USFWS and MDNR signed a cooperative agreement in 1985. That agreement was to implement an experimental program to evaluate hatchery-reared striped bass in Chesapeake Bay. In 1986, the state of Virginia and USFWS also signed a cooperative agreement, the goal of which was to investigate the feasibility of using artificial propagation to supplement the spawning stocks of striped bass in Maryland and Virginia. The program was considered a pilot program and not a full restoration program based on stocking hatchery-reared fish. Under cooperative agreements, USFWS committed six federal hatcheries to the production of striped bass 15- to 20-cm (6–8 in.) long, commonly known as phase II fish, to be reared from fry provided by MDNR and Virginia.

The intent of these efforts was to maintain the viability of the resource by artificial means, (i.e., by stocking hatchery-reared fish) until the quality of the habitat improved, the fishery was brought under coordinated control, and natural reproduction and recruitment were restored.

Biologists and managers voiced concerns about the potential effects of Atlantic coast striped bass restoration actions on native stocks. A committee, working under direction of the ASMFC, was formed to represent the states bordering the migratory range of striped bass along the Atlantic coast, from Maine to North Carolina. Seven charges were assigned the committee (30).

The first charge was to develop an inspection system to ensure that no pathogens were present on eggs or larvae shipped into other states and then returned to Maryland to be stocked in Chesapeake Bay. The committee's second charge was to review tagging programs for striped bass and recommend a coordinated tagging system for all stocked fish. The third charge was to develop procedures to evaluate the stocking restoration program to determine its effectiveness and when it should be terminated. In charges four and five, the committee was to assess the threat posed by stocking programs to the genetic integrity of native striped bass along the Atlantic coast, and to make recommendations regarding time, size, and strain of fish to be stocked. The sixth charge was to review stocking programs in each Atlantic state to ensure they did not conflict. The final charge was to determine if hatchery reared fish stocked along the Atlantic coast would mature, return to the areas where stocked, and spawn.

Actions to Control Disease

Because infectious diseases may cause mortality under certain conditions found in a hatchery or a large-scale tagging center, it was extremely important to have all stocks of striped bass sampled for pathogens before a tagging program was started. Obviously, neither aquaculturists nor fisheries managers wanted to release infected juvenile striped bass into the natural environment where they could pass diseases to uninfected wild fish. Subsequent undetected releases of large numbers of infected fish would also bias tag returns for that particular cohort. An undetected kill of a large percentage of pathogen-positive fish could result in a critical bias if known numbers of tagged striped bass were used in extensive mark-recapture experiments.

Reduced Effort Effect. Because of the Maryland moratorium of taking striped bass in Chesapeake Bay, and because of greatly reduced commercial and sport harvests along the Atlantic coast, more help was needed from state and federal agencies to sample the wild stock of coastal migratory striped bass. The temporary loss of samples, formerly supplied by fishermen, required public agencies to develop a large-scale assessment program to evaluate the status of the striped bass population. Therefore, more time, energy, and money were spent by state and federal agencies to obtain fishery-independent data on the coastal migratory stock. Excellent interagency cooperation and use of one state's field-sampling program to obtain data to meet another agency's needs became the normal operational procedure. Through 1987, all adult striped bass used in the restoration program were screened for the infectious pancreatic necrosis (IPN) virus. Gamete samples were obtained by biologists during manual spawning, and fry samples were collected and analyzed where natural spawning had occurred and gamete samples were not readily available. No positive IPN samples were identified, and only fish free of the IPN pathogen were stocked.

A Tagging Program

The traditional method of marking, by clipping adipose fins, in salmonids that carry binary-coded wire tags does not work with striped bass, which lack an adipose fin and quickly regenerate other fin clips. Instead, biologists must sample or subsample large numbers of fish to obtain tag returns. Portable detection units must be used by field crews to document the presence or absence of tags in fish.

Because portable tag detectors are capable of detecting minute disturbances in a magnetic field, they are difficult to use in rolling seas, in the presence of large metal booms and winches, and on board vessels where the vibrating diesel engines produce positive readings. Thus, with certain vessels used in fishery research along the Atlantic seaboard, biologists were not able to use existing portable tag detectors. Because the problem was unique to striped bass being tagged with binary-coded wire tags, biologists worked closely with the manufacturer of the equipment to document problems encountered in field sampling. The subsequent generation of portable detectors for coded wire tags contained a shielding mechanism to prevent interference from vibration and movement. Also, the new design enabled field crews to quickly sample large numbers of striped bass, whether they were captured commercially or during the biological sampling program.

All striped bass released in 1986–1988 were marked with binary-coded wire tags, a method (31) that allows identification of thousands of different groups. A tagging center was developed to tag fish returning to Maryland from various hatcheries along the Atlantic coast. Large circular holding tanks (2.1 m × 0.9 m) supplied with 10 ppt saltwater and an extensive recirculating, liquid-oxygen injection system [10–15 mg/L (ppm)] were used to hold fish for as long as 12 hours before tagging. Coded wire tags were placed in the adductor mandibularis muscle (a muscle below the eye) of phase II striped bass (32).

For the tagging operation, a specially modified trailer (12 m × 2.5 m) was used that held six coded wire tagging machines and quality control units, as well as large temporary holding tanks for anesthetizing fish. Large tanks outside of the trailer were used to hold fish for post-tagging recovery from the anesthetic. Tagging equipment was used to inject a binary-coded wire tag 1.07 mm long and 0.254 mm in diameter into each fish through a 24-gauge hypodermic needle. Fish were aligned visually to the hypodermic needle, without the aid of a guiding head mold, and impaled manually on the injector needle; the tag was then injected through the hypodermic needle into the adductor muscle. The tagging machine was adjusted so that the needle was extended and stationary at the start of each cycle. Tags were magnetized in the needle before the machine was cycled. Up to 25,000 fish (over 4,000 per machine) were tagged daily with binary-coded wire tags, of which 500 also were marked with internal anchor tags. The typical tagging crew consisted of 12 members: six to operate the six machines, three to handle fish for the machine operators, two to insert the internal anchor tags, and one to maintain the records. Tagging mortality was less than 1% when water temperature was less than 15 °C. Fish were held in highly oxygenated salt water, and 1 mg/L (ppm) of tricaine methanesulfonate was used as an anesthetic during the tagging. Tags were placed only in the left adductor mandibularis muscle to ensure proper tag placement and to localize the area to be searched for coded wire tags during coastal sampling programs. All binary-coded wire tags showed agency code, year stocked, hatchery producing the fish, and stocking location. Short-term tag retention of coded wire tags averaged about 96%. X-rays showed that the coded wire tag was almost immediately encapsulated within the adductor muscle mass, thus the potential for tag loss was low. Apparently the few tags lost during this procedure worked loose within 24 hours of injection.

Subsamples were marked with an internal anchor tag (a 5-mm × 20-mm toggle attached to a 75-mm streamer) inserted just posterior to the left pectoral fin, while it was compressed to the body. A scale was removed at the point of tag insertion, and a vertical incision about 5 mm long was made with a curved scalpel blade, through the peritoneum, but not deep enough to damage internal organs. The anchor of the tag was inserted through the incision and set in place with a gentle pull on the streamer. All streamers were treated with an algicide. Anchor tags were used to obtain additional information on coded wire tag retention in the wild, exploitation rates, movement, and growth. They also served as indicators of movement of fish marked with binary-coded wire tags outside the sampling areas.

Numerous surveys were conducted coastwide to obtain tag returns for program evaluation. Survey techniques included coordinated sampling by state, private, and federal agencies. Sampling in Chesapeake Bay was conducted with beach seines, gill nets, trawls, and by electrofishing. All fish within the possible size-range of stocked tagged fish were checked for the presence of tags. At this point in the cooperative restoration program, fish that tested positive for binary-coded wire tags were not sacrificed to obtain additional data.

Along the Atlantic seaboard, additional surveys and adult striped bass tagging programs were established. All striped bass four years old or less when captured were surveyed for binary-coded wire tags as part of a coordinated coastal effort to obtain information on movements, migration, and exploitation of hatchery-reared and tagged fish.

Maintenance of Genetic Integrity

All hatchery striped bass involved in this large-scale tagging program were identified through the hatchery phase by lot numbers and parental rivers of origin. All fish were tagged and released only into their natal rivers. Techniques used to ensure this arrangement included careful record keeping, coordination of assigned binary-coded wire tag codes with the tag manufacturer to ensure correct stocking location designation, use of a central tagging location, and coordination to help ensure that the correct tagging codes were used each day.

Size, Source, and Time of Stocking

In the Chesapeake Bay striped bass restoration program, all stocking was with tagged phase II (15–20 cm TL) striped bass, except for an experimental stocking in June–July 1987, of phase I (35–50 mm long) fish performed to test marking methods and tag retention. In phase I striped bass, tag placement in the adductor muscle was perpendicular to the body axis. Tagging machines were used with the tag injection needle in the stationary mode. Standard length (1 mm) binary-coded wire tags were used. About 15,800 phase I fish were tagged and released into the Patuxent River, Maryland. Posttagging, overnight mortality rates averaged 1% during two test periods. In a control group of 3,000 phase I fish, tag retention was 97% after 6 months.

State Programs Are Nonconflicting

Each year, coastal states engaged in stocking and tagging of Atlantic coast striped bass held a review meeting. This enabled the subcommittee to verify compliance with stocking and tagging guide lines.

Recommendations for an Evaluation Program

All coded wire and internal anchor tag returns have been coordinated for the cooperative tagging program by the USFWS. Most internal anchor tag returns were obtained by collect phone calls to the USFWS, Annapolis, Maryland. Each caller was asked by trained personnel to answer a standard questionnaire over the phone. A central depository and database organizational protocol was developed.

Application to Other Programs

The pilot restoration program for striped bass operated smoothly. Its operational success was attributed largely to the decision-making process and cooperative efforts of all Atlantic coast states, the federal agencies involved, and the private sector. Both fisheries managers (state and federal) and aquaculturists (hatchery managers, fish culturists) jointly developed and modified the program to accommodate special needs or address specific problems. The success of the program was greatly increased by the willingness of the states and federal agencies to collect and share data, to collect and spawn brood fish, and to provide fry to be reared in federal and private hatcheries.

Public support and involvement in the program was encouraged and maintained through a series of educational activities, including news releases, press conferences, videotapes, and conspicuous involvement of high-level public officials at ceremonial releases of striped bass in the Chesapeake Bay or other coastal waters. The reward system established for return of the external portion of the internal anchor tag was increased awareness and public support for the program. Establishment of a central processing point for all tags increased chances for success of the program by reducing the confusion and duplication associated with multiple tag-return sites.

CONCLUSION

It should be obvious that not all fish culturists hold the same views regarding the value of hatchery or farm-raised fish for stocking in public waters or the role of hatcheries in striped bass or wild trout programs. However, given the opportunity, almost all culturists would like to produce fish of a quality equal to that of wild fish (some apparently now believe that they do so). Major obstacles recognized by most respondents in the survey of hatchery managers producing trout for stocking were inadequate budgets, excessively high production quotas, and compliance with established procedures for control of selected fish diseases. Hatcheries were seen as an aid to manage and maintain populations of wild trout as they have been for striped bass and other species. Culturists recognized the importance of proper habitat to maintain healthy populations of wild trout and other aquatic species. They also recognized that the growth of the human population and the increasing demands of anglers cannot be met by wild fish alone. There is a recognized need for put-and-take fisheries in urban environments and the need to use hatchery-reared fish to restore stocks in areas where habitat has been degraded.

As beauty is in the eye of the beholder, so may wild fish be in the eye of the angler, culturist, or manager. No fish culturist wants to produce fish without fins or those described as "swimming sausages with scales." However, they do want to use hatcheries as appropriate to support programs for restoration of wild trout, striped bass, and other species, and to provide angling opportunities for those unable to fish in pristine waters for wild trout.

BIBLIOGRAPHY

1. H.L. Schramm, Jr. and V. Mudrak, *Fisheries* **19**(8), 6–7 (1994).
2. R.R. Stickney, *Fisheries* **19**(5), 6–13 (1994).
3. S. Wright, *Fisheries* **18**(5), 3–4 (1993).
4. D.P. Philipp, J.M. Epifanio, and M.J. Jennings, *Fisheries* **18**(12), 14–16 (1993).
5. N.C. Parker, in R. Bamhart, B. Shake, and R.H. Hamre, eds., *Wild Trout V: Wild Trout in the 21st Century*, Symposium sponsored by American Fisheries Society, USEPA, Trout Unlimited, USDA, Federation of Fly Fishers, and USFWS. Yellowstone National Park, September 26–27, 1994, pp. 93–96.
6. P.E. Thompson, in N.G. Benson, ed., *A Century of Fisheries in North America. Special Publication No. 7*, American Fisheries Society Washington, DC, 1970, pp. 1–11.
7. J.T. Bowen, in N.G. Benson, ed., *A Century of Fisheries in North America. Special Publication No. 7*, American Fisheries Society, Washington, DC, 1970, pp. 71–93.
8. J.J. Brice, *A Manual of Fish Culture*. United States Commission of Fish and Fisheries, Government Printing Office, Washington, DC, 1897.
9. H.L. Schramm, Jr. and R.G. Piper, *Uses and Effects of Cultured Fishes in Aquatic Ecosystems*, American Fisheries Society Symposium 15, American Fisheries Society. Bethesda, MO, 1995.
10. H.K. Dupree and J.V. Huner, eds., *Third Report to the Fish Farmer: The Status of Warmwater Fish Farming and Progress in Fish Farming Research*. U.S. Fish and Wildlife Service, Washington, DC, 1984.
11. U.S. Fish and Wildlife Service, *1996 National Survey of Fishing, Hunting, and Wildlife-Associated Recreation*, U.S. Government Printing Office. Washington, DC, 1997.
12. N.C. Parker, *Trans. No. Amer. Wild. Natl. Res. Conf.* **53**, 584–593 (1988).
13. J.M. Carlburg and J.C. Van Olst, in R. Hodson, T. Smith, J. McVey, R. Harrell, and N. Davis, eds., *Hybrid Striped Bass Culture: Status and Prospective*, University of North Carolina Sea Grant College Publication UNC-SG-87-03, North Carolina State University. Raleigh, NC, 1987, pp. 73–82.
14. Food and Agricultural Organization of the United Nations. *Review of the State of the World Aquaculture*, FAO Fisheries Department, 1998.
15. M.J. Williams, in J.E. Bardach, ed., *Sustainable Aquaculture*, John Wiley & Sons, New York, 1997, pp. 14–52.
16. N.C. Parker, *CRC Critical Rev. Aquatic Sci.* **1**(1), 97–109.
17. T.I.J. Smith, *Aquacult. Mag.* **14**(1), 40–49 (1988).
18. S.K. Allen, Jr. and R.J. Wattendorf, *Fisheries* **12**(4), 20–24 (1987).
19. J.F. Baroiller, in R.S.V. Pullin, J. Lazard, M. Legendre, J.B. Kothias, and D. Pauly, eds., *The Third International Symposium on Tilapia in Aquaculture*, ICLARM conf. Proc. 41, 1996, pp. 229–208.
20. J.Y. Jo, R.O. Smitherman, and L.L. Behrend, in R.S.V. Pullin, T. Bhukaswan, K. Tonguthai, and J.L. MacLean, eds., *The Second International Symposium on Tilapia in Aquaculture*, ICLARM Conf. Proc. 15, 1988, pp. 203–208.
21. T.J. Benfey and A.M. Sutterlin, *Aquaculture* **36**, 359–367 (1984).
22. C.J. Gilling, D.O.F. Skibinski, and J.A. Beardmore, in R.S.V. Pullin, J. Lazard, M. Legendre, J.B. Kothias, and D. Pauly, eds., *The Third International Symposium on Tilapia in Aquaculture*, ICLARM conf. Proc. 41, 1996, pp. 314–319.
23. M.D. Weiss, *Induct. Chem.* **9**(2), 28–34 (1988).
24. G.T. Klar and N.C. Parker, *No. Amer. J. Fish. Manage.* **6**, 439–444 (1986).
25. P.A. Larkin, *Fisheries* **13**(1), 39 (1988).
26. U.S. Department of the Interior (USDI) and U.S. Department of Commerce (USDC). *Emergency Striped Bass Research Study*, USDI and USDC Report for 1985, Washington, DC, 1986.
27. U.S. Department of the Interior (USDI) and U.S. Department of Commerce (USDC), *Emergency Striped Bass Research Study*, USDI and USDC, Report for 1986, Washington, DC, 1987.
28. C.P. Goodyear, J.E. Cohen, and S.W. Christensen, *Trans. Amer. Fish. Soc.* **114**, 146–151 (1985).
29. J. Boreman and H.M. Austin, *Trans. Amer. Fish. Soc.* **114**, 3–7 (1985).
30. N.C. Parker and R.W. Miller, *Recommendations Concerning the Striped Bass Restoration Program for the Atlantic Coast with Emphasis on Chesapeake Bay*, Special Report 10 to the Atlantic States Marine Fisheries Commission. Striped Bass Stocking Subcommittee, Washington, DC, 1986.
31. K.B. Jefferts, P.K. Bergman, and H.F. Fiscus, *Nature (London)* **198**, 460–462 (1963).
32. G.T. Klar and N.C. Parker, *No. Amer. J. Fish. Manage.* **6**, 439–444 (1986).

ADDITIONAL READING

S.E. Belanger, D.S. Cherry, I.I. Ney, and D.K. Whitehurst, *Trans. Amer. Fish. Soc.* **116**, 594–600 (1987).

A.L. D'Andrea and V.M.T. Leedham, *Electrophoresis* **6**, 468–469 (1985).

M.C. Fabrizio, *Trans. Amer. Fish. Soc.* **116**, 588–593 (1987).

V. Guillory and R.D. Gasaway, *Trans. Amer. Fish. Soc.* **107**, 105–112 (1978).

W.D. Harvey and L.T. Fries, *Identification of Four Morone Species and Congeneric Hybrids Using Isoelectric Focusing Techniques*. Proceedings of the Annual Conference of the Southeastern Association of Fish and Wildlife Agencies, in press.

S. Hattangadi, *Industrial Chemist* **9**(2), 11–12 (1988).

Maryland Department of Natural Resources (MDNR), *Second Annual Report on Striped Bass*, 1986. MDNR, Tidewater Administration, Fisheries Division, Annapolis, Maryland, 1986.

J.P. McCraren, in R.H. Stroud, ed., *Fish Culture in Fisheries Management*. Fish Culture Section and Fisheries Management Section, American Fisheries Society, Bethesda, MD, 1986, pp. 161–172.

K. Menke, *Fisheries* **18**(11), 18–20 (1993).

B. Shupp, *Fisheries* **19**(4), 24–25 (1994).

J.G. Sutton, *J. Forensic Science Soc.* **23**(3), 241–243 (1982).

C. Trosclair, ed., *Water Farming J.* **2**(8), 1 (1987a).

C. Trosclair, ed., *Water Farming J.* **2**(8), 5 (1987b).

F.M. Utter, *Fisheries* **19**(8), 8–9 (1994).

J. Van Tassel, *Culture of Maryland Striped Bass*, Maryland Department of Natural Resources, Tidewater Administration, Annapolis, MD, 1986.

W. Van Winkle, K.D. Kumar, and D.S. Vaughan, *Amer. Fish. Soc. Monograph* **4**, 255–266 (1988).

R.J. Wattendorf, *Prog. Fish-Cult.* **48**, 125–132 (1986).

J.E. Weaver, R.B. Fairbanks, and C.M. Wooley, *Mar. Recreational Fish.* **11**, 71–95 (1986).

See also ENHANCEMENT; HISTORY OF AQUACULTURE.

FLOUNDER CULTURE, JAPANESE

TADAHISA SEIKAI
Fukui Prefectural University
Fukui, Japan

OUTLINE

Japanese flounder, *Paralichthys olivaceus*, belonging to the family Paralichthyidae, can be found throughout Japanese waters, with the exception of the Pacific coast of Hokkaido (1) (Fig. 1). This fish is one of the most familiar species to the Japanese and is traded at high prices. Therefore, Japanese flounder is one of the most important target species for stock enhancement and aquaculture in Japan. Flounder culture is a relative newcomer to aquaculture, but has already established a third position ranking in marine finfish culture, with production exceeding landings since 1990 (2) (Fig. 2).

BIOLOGICAL CHARACTERISTICS OF JAPANESE FLOUNDER

Japanese flounder can be found from the coasts of Sakhalin and Kuril in Russia to the Yellow Sea and the East China Sea of China (1), and Japan's natural resources may consist of several regional populations (3). Especially dense populations exist along the coast of the Sea of Japan affected by Tsushima current, the north Pacific coast of Honshu, and the Seto Inland Sea. Fishermen catch flounders by hook and line, gill-net, trawl, and fixed nets. Annual landings fluctuate between 5,500 and 8,300 tons (Fig. 2).

Because of the wide distribution, there is a large variation in growth and maturation (3). Wild fish grow to 15–30 cm (5.9–11.8 in.) and 50–76 cm (19.7–27.6 in.) total length (TL) within one year and five years, respectively (2), and females show significantly better growth performance than males (3). Flounder spawn in coastal waters shallower than 50 m (160 ft) depth at 12–17 °C (54–63 °F) (3). The spawning season starts in late winter in the southern waters, and finishes in early summer in the northern waters, as temperature increases (3). During a spawning season, males greater than 30 cm (12 in.) and females greater than 40 cm (16 in.) spawn repeatedly (3).

Fertilized eggs [0.9 mm (0.035 in.) in diameter] are buoyant and nonadhesive. The larvae are planktonic and have bilaterally symmetrical body forms for about one month after hatching. During the latter part of the larval stage, larvae start to form asymmetric bodies and shift to a benthic life through a remarkable metamorphosis, in which the right eye migrates to the opposite side of the head (4). The juveniles inhabit sandy shallow waters and feed mainly on mysid shrimp (3,5). When juveniles grow to 5–10 cm (2–4 in.) TL, their feeding preference shifts to piscivorous (3,5). They grow to adulthood in the coastal waters, and then some migrate to the southwest to spawn.

HISTORY AND OUTLINE OF FLOUNDER CULTURE

First in production among maricultured fishes in Japan is yellowtail, *Seriola quinqueradiata*, and second is red sea bream, *Pagrus major*. Thereafter come Japanese flounder, tiger puffer, *Takifugu rubripes*, and striped jack, *Pseudocaranx dentex* (Fig. 3). Japanese flounder culture began on the assumption that juveniles could be produced in hatcheries. The first successful flounder fry production was by Kinki University in 1965 (6). That success opened a new window for aquaculture, and the technology for mass fingerling production was established during the 1980s. From 1980, cultured flounder production in Japan expanded very quickly (Fig. 2) and appeared in the fisheries statistics in 1983 for the first time (7). Since 1990, cultured production numbers have exceeded those of wild fishlandings, and the production by aquaculture was 7,292 tons in 1994. The marketable size of cultured flounder is 500 g–1 kg (1.1–2.2 lb). Flounder culture started both in land-based tanks and net cages, but net cage culture nearly disappeared within 2 to 3 years because of the inability to avoid high temperatures during the summer season and a relatively low selling price. Therefore, most flounder are now produced in land-based tanks (Fig. 4). Flounder grow at an optimum temperature range of 10 to 25 °C (50–77 °F), and shows highest growth at around 21 °C (70 °F). High mortality is observed at temperatures higher

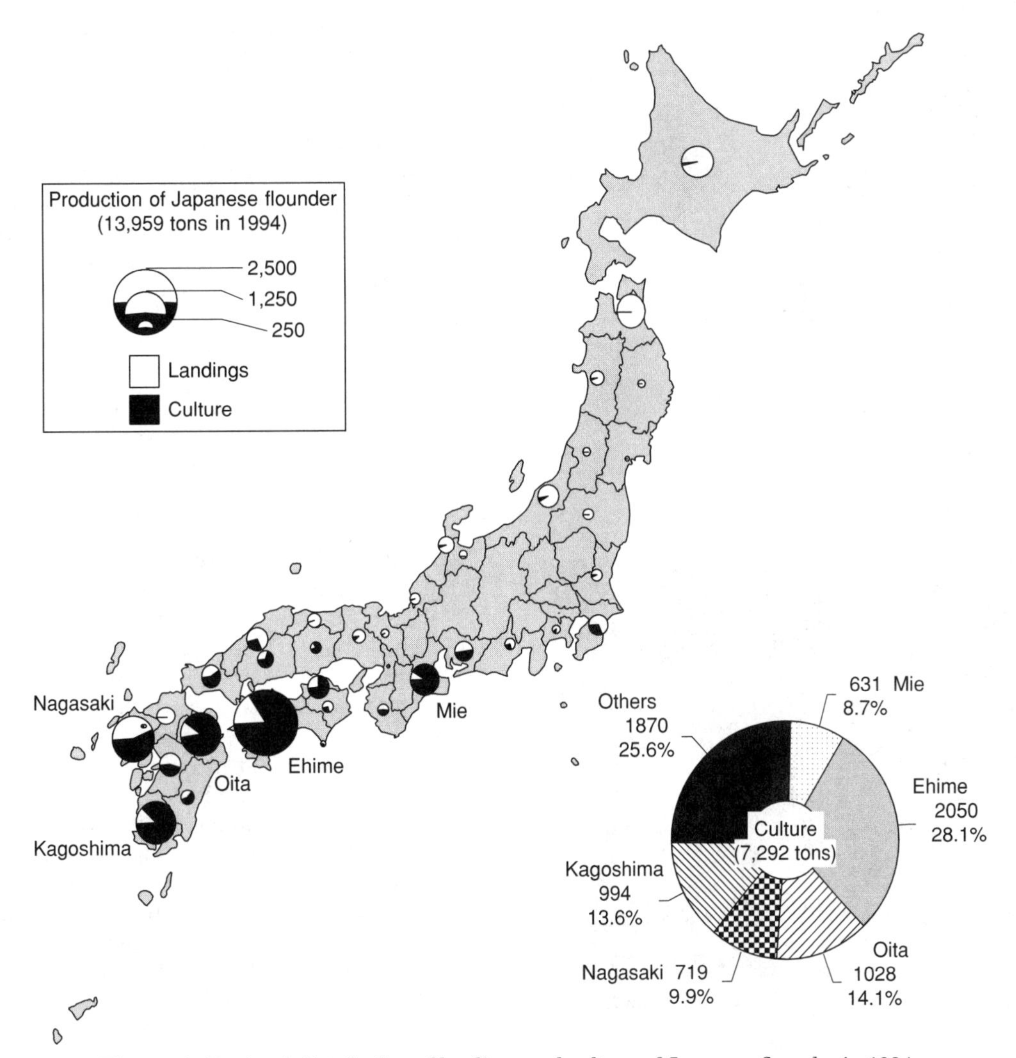

Figure 1. Regional distribution of landings and culture of Japanese flounder in 1994.

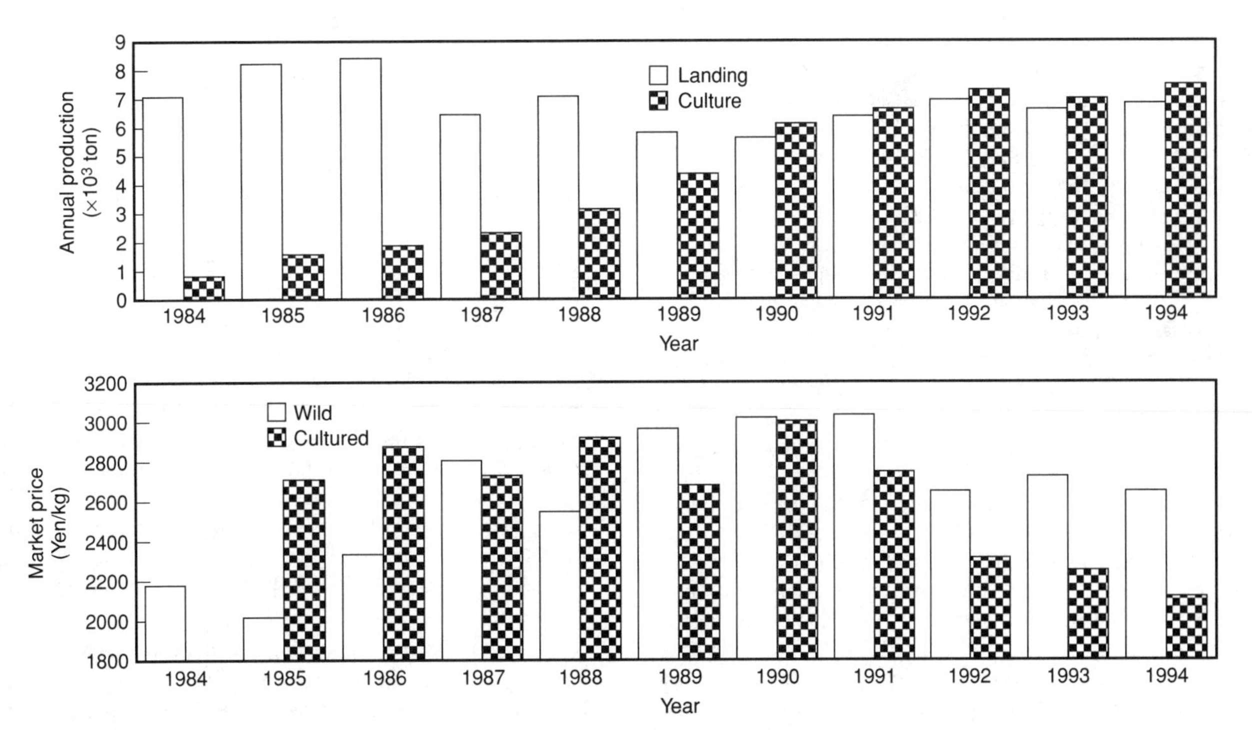

Figure 2. Annual landings, culture production, and market prices of Japanese flounder from 1984 to 1994.

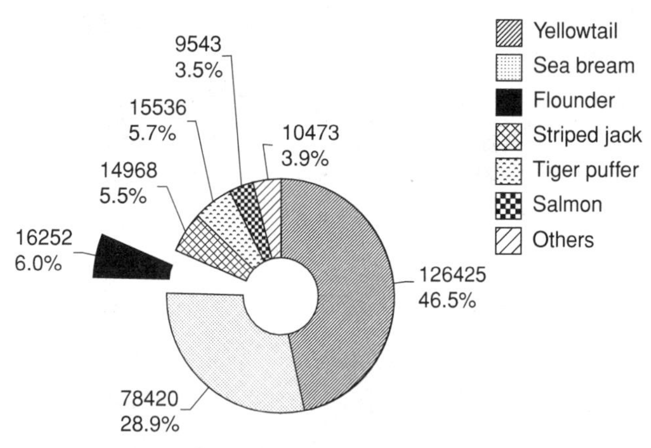

Figure 3. Cultured marine finfish in 1994. [Total amount is 271,617 million Yen (US$2,263 million).]

Figure 4. Land-based tank culture of Japanese flounder in Mie prefecture.

than 25 °C (77 °F) during the summer season, but the fish can withstand low temperatures during the winter.

Japanese flounder culture is based mainly in the western part of Japan. Five prefectures, Ehime (28.2%), Kagoshima (13.7%), Nagasaki (9.9%), Oita (14.1%) and Mie (8.7%) produced about 75% of the total production in 1994 (Fig. 1) (2). Japanese flounder has the following advantages for aquaculture: (*1*) ease of obtaining fingerlings supported by an established hatchery system, (*2*) high efficiency of feed utilization and good growth, (*3*) high market price, and (*4*) no requirement for fishery rights because of the land-based tank culture (8). When cultured flounder first appeared in the statistics, their selling price was higher than that of the wild fish for two reasons: (*1*) cultured fish were sold as live fish with uniform sizes, and (*2*) wild fish of various sizes were not always sold alive (Fig. 2). Later, as flounder culture increased, the selling price became stable and similar to that of the wild fish. In 1989, the price of wild fish surpassed that of cultured fish and then both prices dropped. The price of cultured fish dropped more quickly, to about half that of wild fish. These price changes were attributed to the following reasons: (*1*) shift of wild flounder from fresh to live fish, (*2*) overproduction of cultured fish, and (*3*) deterioration of economic conditions in Japan.

PRESENT STATE OF FLOUNDER CULTURE

Fry Production

Egg Taking and Hatching. Fertilized eggs are collected from natural spawning by brood stock raised in captivity (9,10). Tank-spawned eggs are usually of higher quality than stripped and artificially fertilized eggs. Maturation is typically controlled by photoperiod and temperature so that fertilized eggs can be obtained nearly year round (11). After the elimination of dead eggs, fertilized eggs are incubated in nets or polycarbonate tanks. Clean seawater and gentle aeration are supplied to the incubated eggs. Water temperature is maintained at 15 °C (59 °F) to achieve better hatching rates and lower rates of deformity (12). Usually, rearing tanks are stocked with eggs just before hatching, or recently hatched larvae.

Larval Rearing. Stocking densities of newly hatched larvae are adjusted to $15-20 \times 10^3$ ind./m^3 and $30-50 \times 10^3$ ind./m^3 in large (20–100 m^3) and small rearing tanks (1–10 m^3), respectively (14). When high-density culture is begun in small tanks, restocking to several tanks or transfer to the larger tanks is necessary to reduce stocking density as larvae grow. Rearing temperature should be maintained at 18–20 °C (64–68 °F) during this period. Newly hatched larvae develop by themselves, with their own yolks and oil globules, for 2 to 3 days after hatching. At 7 to 9 days after hatching [7.0 mm (0.28 in.) TL], the anterior three dorsal fin rays start to elongate (15). At about 20 days after hatching [larger than 10.0 mm (0.4 in.) TL], larvae begin to metamorphose with right eye migration. At 25 to 30 days after hatching, larvae start to settle as the climax of metamorphosis is reached. At 30 to 35 days after hatching, larvae transform to juveniles.

An extremely simple feeding regime is established, beginning with rotifers (3 to 20 days after hatching), brine shrimp (10 to 35 days after hatching), and then artificial diets (20 to 35 days after hatching) as fish grow (13). The survival rate from hatching to juvenile in mass production is sometimes higher than 90% in large tanks (13). Albinism had been frequently observed, with an extremely high percentage of occurrence (50–80%), but now has been nearly overcome with the improvement of larval diets (14).

Juvenile Rearing. Juveniles are raised on artificial diets in land-based tanks, or net cages installed in land-based tanks, until reaching 3–5 cm (1.2–2.0 in.) TL. In case of excess stocking density or a deficiency of feed, inferior juveniles with darker pigmentation are forced to swim near the surface, and mortality increases as a result of biting or cannibalism by larger fish. During this phase, it is important to maintain a reasonable stocking density, feed sufficiently, and grade fish by size, to avoid cannibalism. Most of the juveniles used in culture are provided by private hatcheries. The selling prize per individual is 100–150 yen (US$0.83–1.25).

Diseases During Fingerling Production. Diseases contracted during fingerling production can sometimes become very serious, because of the difficulties in detecting the reasons for mortality and in providing proper treatments. Recently, disease problems during this phase have become more prevalent, due to the frequent outbreaks of viral diseases. The most common diseases are vibriosis (caused by *Vibrio anguillarum*), bacterial enteritis (*Vibrio ichthyoenteri*), epidermal hyperplasia (flounder herpes virus), and nervous necrosis (striped jack nervous necrosis virus) (15).

Culture to Marketable Size

Fingerlings are raised to marketable size in land-based tanks and shipped as live fish. Advantages of land-based tank culture are as follows: (*1*) easy observation and better care, (*2*) agreement with the flounder's settling behavior, (*3*) easy environmental control, and (*4*) easy entry into the industry by investors and culturists who do not have fishery rights. Disadvantages of land-based tank culture include (*1*) the high costs of facilities, (e.g., tanks and pumps), (*2*) the high operating costs for electricity, (*3*) the possibility of extermination of a crop by accident, and (*4*) the necessity to renew facilities every several years because of fouling organisms and corrosion by sea water (16).

Selection of Sea Water and Site. The seawater and site for flounder culture should be considered from both biological and economic aspects. Seawater should satisfy a number of conditions: (*1*) the temperature should be within the optimum range as long as possible, (*2*) dissolved oxygen should be at a high enough level, (*3*) there should be no risk of extreme decreases in salinity by inflowing rivers, and (*4*) there should be no effects from red tide or pollutants. Offshore bottom water or saline well water fit these criteria. However, saline well water usually needs additional aeration because of low dissolved-oxygen levels. A suitable site for flounder culture should satisfy the following demands: (*1*) low land cost, (*2*) good access to the shore line, (*3*) small change in elevation from sea level, to save electricity for pumping, (*4*) easy access to market and supply outlets, and (*5*) protection from natural disasters (16).

Facilities. The minimum facilities requirements for land-based tank culture include a pump system, plumbing for water supply and drainage, aeration, and tanks. Tank size ranges from 4–10 m^2 (40–100 ft^2) for juveniles to 30–100 m^2 (320–1080 ft^2) for growout populations. Because Japanese flounders sometime jump when they are frightened, the rim of the tank should be 30–50 cm (12–20 in.) above the water surface [50–100 cm (20–40 in.)]. The bottom of the tank slopes downward to the center to drain uneaten feed and feces. Tanks need roofs to shade them from direct sunlight to prevent algae growth and over heating and to keep fish calm.

Stocking Density. A rough standard for stocking density is 5–15 kg/m^2 (120–340 lb/ft^2), within optimum temperature ranges and at a flow rate of 10–12 exchanges/day (16). Overstocking decreases growth and feed efficiency and increases the possibility of disease outbreaks. A commercial farmer in the Mie prefecture established the stocking standard in relation to fish size shown in Table 1.

Table 1. Standard Stocking Density in Land-Based Tanks

Total Length (mm)	Body Weight (g)	Stocking Density	
		ind./m^2	kg/m^2
5	1.5	800	1.2
10	10	200	2
15	60	95	5.7
20	85	50	4.3
25	140	35	4.9
30	320	22	7
35	460	17	7.8
40	800	13	10.4

Feeding and Growth. Prepared diets are popularly used for both juveniles and larger fish. Dry pellets typically contain protein (48–56%), lipids (6–14%), and ash (<17%). The size of the feed pellets should be as large as the fish can swallow. Frequency of feeding is 3 to 4 times per day during the juvenile stage and decreases to 1 to 2 times per day as the fish grow. Feeding rates should be adjusted to 6–10% of body weight at juvenile stage, 0.4–2.2% at 20 cm (8 in.), 0.3–1.8% at larger than 30 cm (12 in.), and less than 0.2% in the winter at 12–13 °C (54–55 °F). Fish stop feeding at temperatures higher than 27–28 °C (81–82 °F). Flounder culture in Japan is widely distributed across western Japan, so growth is different from place to place because of the timing of fingerling introduction, seasonal changes in water temperature, culturing procedures, and environmental conditions. Several examples of growth in land-based tanks are shown in Figure 5 (17).

Disease. Common bacterial diseases contracted during culture are vibriosis, *V. anguillarum*, gliding bacterial disease, *Flexibacter maritimus*, and edwardsiellosis, *Edwardsiella tarda* (Fig. 6). White spot disease, *Cryptocaryon irritans*, and ichthyobodosis, *Ichthyobodo* sp., are typical examples of protozoan diseases. These diseases frequently occur during the summer season because disease resistance is lower and the culture conditions deteriorate at high temperatures (16).

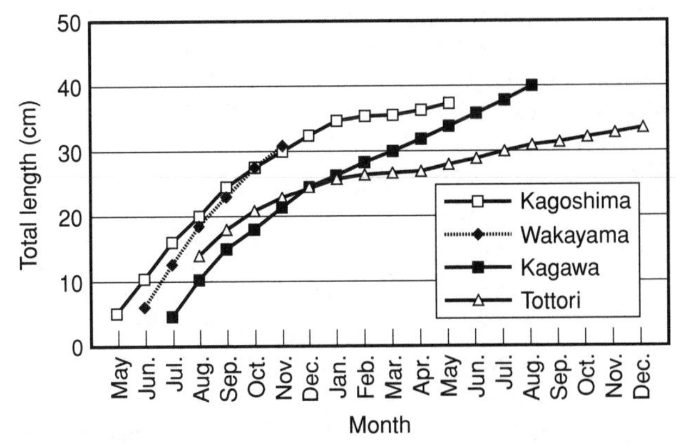

Figure 5. Growth of Japanese flounder cultured in land-based tanks.

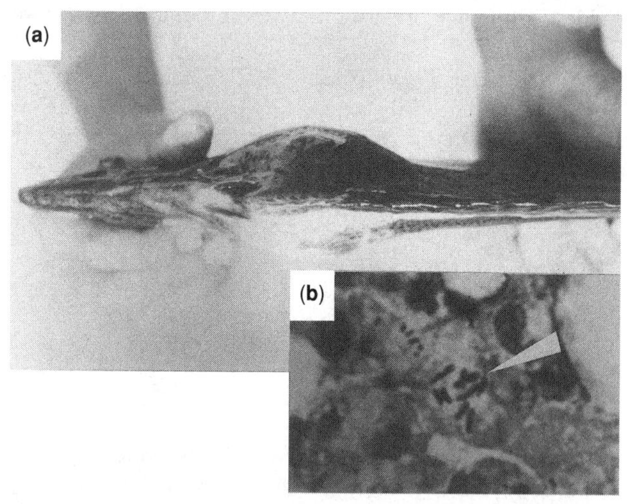

Figure 6. Cultured flounder infected with edwardsiellosis (**a**), the most serious disease in Japanese flounder culture. Accumulated mortality by this disease may reach 30%. Stamp smear of liver (**b**). Arrow shows the colony of *E. tarda* [major axis is about 2 μm (8×10^{-5} in.)]. (Photographs were provided by Ehime Prefectural Fish Disease Control Center.)

Shipping and Economic Analyses

The scale of culture ranges from several thousands to tens of thousands of fish per farm. The largest farm, in Mie prefecture, cultures 550,000 fish and employs 50 laborers (Fig. 4). Most farmers sell one-year-old fish until September for the following reasons: (*1*) during the summer season, water temperature will begin to exceed the upper part of the optimum range [25 °C (77 °F)], (*2*) cultured fish can attain marketable size within one year, (*3*) the amount of fish caught in landings becomes extremely small during the summer season, and (*4*) quick rotation of tanks increases economical effectiveness for the fish farmer.

All cultured fish are distributed to consumers as live fish. Fish are shipped to markets in big cities or are bought by brokers from fish farmers and sold directly to restaurants.

In the case of land-based tank culture, the cost for facilities and the cost of electricity make up 15–20% of annual expenses. The selling price and survival rate greatly affect the economic balance, so the early introduction of fingerlings for better growth and better survival until shipping should be considered more carefully. The potential for increasing the selling price of cultured flounder, by controlling the timing and destination of the product, should also be given more attention.

NEW TECHNOLOGY

All Female Production

As females of this species grow much faster than males (3), production of all-female fingerlings for culture has been developed (17). In this species, the phenotypic expression of sex is determined by environmental factors, namely, temperature and feed (fresh fish or artificial diets) during the juvenile stage, together with genetic information (18,19). Recently, the successful production of all-female fingerlings on a mass scale was realized (the first step was the creation of gynogenetic diploids by chromosome manipulation and the retaining of the second polar body). The second step was the creation of sex-reversed functional males from the gynogenetic diploids by steroid hormone treatment. Hybridization between the sex-reversed females (functional males) and normal females was the third step. Finally, adjustment of rearing temperature during the juvenile stage is used to produce all-female fingerlings. These techniques have already been applied in commercial aquaculture (18).

Cloning

Completely homozygous flounders can be produced from the gynogenetic diploids by the blocking of the first cleavage. Each homozygous fish can produce a first generation of the same homozygous offspring from the gynogenetic diploids by retaining the second polar body. A portion of these fish can be transformed to functional males by exposing the fish to hormones or high temperature during the sex determination stage. Hybridization between females and sex-reversed females (functional males) within a cloned population results in the production of homozygous cloned progeny (18). Hybridization between two different cloned populations results in the production of heterozygous cloned progeny (18). Recently, it was revealed that each strain has a different growth potential and disease resistance. Therefore, this method may be useful in the future in selective breeding by commercial aquaculturists (19).

BIBLIOGRAPHY

1. N. Maki, H. Terashima, and H. Nakamura, in M. Abe and Y. Honma, eds., *Modern Encyclopedia of Fishes*, NTS, Inc., Tokyo, 1997, pp. 557–561.
2. Ministry of Agriculture, Forestry and Fisheries, *Showa 58th Statistics on Fisheries and Water Culture Production*, Association of Agriculture Statistics, 1983, Tokyo, pp. 1–312.
3. T. Minami, in T. Minami and M. Tanaka, eds., *Biology and Stock Enhancement of Japanese Flounder*, Koseisha Koseikaku, Tokyo, 1997, pp. 9–24.
4. M. Okiyama, *Bull. Jpn. Sea Reg. Fish. Res. Lab.* **17**, 1–12 (1967).
5. T. Noichi, in T. Minami and M. Tanaka, eds., *Biology and Stock Enhancement of Japanese Flounder*, Koseisha Koseikaku, Tokyo, 1997, pp. 25–40.
6. T. Harada, S. Umeda, O. Murata, H. Kumai, and K. Mizuno, *Rep. Fish. Res. Sta. Kinki Univ.* **1**, 9–303 (1966).
7. Ministry of Agriculture, Forestry and Fisheries, *Heisei 6th Statistics on Fisheries and Water Culture Production*, Association of Agriculture Statistics, 1997, Tokyo, pp. 1–292.
8. T. Harada, *Yoshoku* **17**(4), 48–53 (1980).
9. K. Takahashi, Y. Hayakawa, D. Ogura, and H. Nakanishi, *Saibai Giken* **9**, 41–46 (1980).
10. Y. Hiramoto, N. Miki, and K. Kobayashi, *Rep. Tottori Pref. Fish. Sta.* **23**, 7–12 (1981).

11. T. Ijima, T. Abe, R. Hirakawa, and Y. Torishima, *Saibai Giken* **15**(1), 57–62 (1986).
12. N. Yasunaga, *Bull. Tokai Reg. Fish. Res. Lab.* **81**, 151–169 (1975).
13. S. Torii, Y. Shiokawa, Y. Hayakawa, S. Siota, K. Igarashi, Y. Kuji, K. Sato, A. Ito, H. Wakui, M. Kawanobe, A. Iwamoto, and T. Fukunaga, *Corpus of Hirame Seed Production Technologies in North Pacific Area*, Jpn Sea-Farm. Assoc., Tokyo, 1994, pp. 5–16.
14. T. Seikai, in *Proceed. 26th U.S.-Japan Aquacult. Panel Sympo.*, 1998.
15. K. Muroga, in I.M. Shariff, R.P. Subasinghe, and J.R. Arthur, eds., *Diseases in Asian 12 Aquaculture*, Fish Health Section, Asian Fisheries Society, Manila, Philippines, 1992, pp. 215–222.
16. T. Seikai, in Y. Oshima et al., eds., *Shallow Water Aquaculture*, Taisei Shuppan, Tokyo, 1986, pp. 246–265.
17. K. Tabata, *Bull. Hyogo Pref. Fish. Expt. Stn.* **28**, 1–134 (1991).
18. E. Yamamoto, *Bull. Tottori Pref. Fish. Exp. Stn.* **34**, 1–145 (1995).
19. E. Yamamoto, in T. Minami and M. Tanaka, eds., *Biology and Stock Enhancement of Japanese Flounder*, Koseisha Koseikaku, Tokyo, 1997, pp. 83–95.

See also Brine shrimp culture; Halibut culture; Larval feeding—fish; Plaice culture; Sole culture; Summer flounder culture; Winter flounder culture.

FROG CULTURE

Gianluigi Negroni
Alveo Co-operative Society
Bologna, Italy

OUTLINE

INTRODUCTION

A majority of the frogs that reach the markets of the world are either frozen or live wild frogs. Wild frogs are in decline due to such factors as pollution, destruction of natural wetlands, and overfishing. Major frog-importing countries are the European Union (EU), Japan, and the United States. Frogs come from many countries where there are ample natural or artificial wetlands, such as rice fields and inundated areas (1).

Capturing frogs is a labor-intensive activity that requires inexpensive labor in nations that have the appropriate natural conditions to produce large numbers of them. For many frog hunters, 2 kg (4.4 lb) a day is considered a typical catch. Capture techniques vary, but all require skill.

One system is to tie a piece of cotton to a hook that is shaken over the frog. Attracted by the motion of the cotton, the frog strikes and is hooked. Another method involves catching frogs at night by blinding them with a light and netting them.

As world consumption has increased and resources have declined, frog farming has come to be of considerable interest to culturists. While there has been a considerable amount of research in this area and some significant results have been produced, raising frogs in large numbers requires a considerable amount of technical expertise (1) as well as the appropriate environmental conditions. Techniques for industrial production of frogs have been developed; these are most widely used in nations with warm climates, since it is possible to produce more crops each year and increase the profitability of frog farming in a warm climate.

FROG BIOLOGY

The information presented here relates primarily to frogs in the family Ranidae, which is the primary family of frogs currently being cultured. Frogs in the family Ranidae, also known as true frogs, number about 250 species (2) and are distributed throughout the world. They generally live close to water and take refuge there when alarmed. They can survive for long periods out of water if the environment is sufficiently humid. Most cultured frogs are of the genus *Rana*. Frogs in the genera *Xenopus* and *Leptodactylus* have been raised, but never in commercial numbers.

The Goliath frog (*Conraua goliath*) can exceed 34 cm (13.3 in.) in body length. It lives in African rain forests, and is very difficult to find. The author worked for one year in Goliath frog habitat and did not see one. Another unusual frog is the burrowing frog (*Rana adspersa*), which lives much of its life cycle in the ground awaiting the rainy season. The burrowing frog is common in the southern desert area of Africa and is considered to be a local delicacy.

Table 1 provides a list of the most commonly consumed and cultured frogs on each continent where frogs occur (3). The table was developed with contributions from many frog farmers, scientists, colleagues, and the author's experience.

Table 1. Scientific Name, Length from Nose to Cloaca, Breeding Approaches, and Supplementary Information by Continent

Species	Length, cm (in.)	Breeding	Notes
South American Frogs			
Batrachophrynus microphtalmus	20 (7.8)	Under study	Totally aquatic, cold resistant
Caudiverbera caudiverbera	20 (7.8)	None	Locally consumed
Ceratophrys cornuta	15 (5.9)	None	Locally consumed
Leptodactylus fallax	15 (5.9)	None	Locally consumed
Leptodactylus labyrinthicus	20 (7.8)	Under study	Locally consumed
Leptodactylus ocellatus	25 (9.8)	Intensive	Locally consumed
Leptodactylus pentadacylus	25 (9.8)	Under study	Locally consumed
Leptodactylus macrosternum	20 (7.8)	None	Locally consumed
Telmatobius celeus	20 (7.8)	None	Completely aquatic, cold resistant
North American Frogs			
Rana aurora	12(4.7)	None	Not often eaten
Rana catesbeiana	20 (7.8)	Intensive	Most widely used for culture
Rana grylio	16 (6.2)	None	Locally consumed
Rana pipens	15 (5.9)	Intensive	Used in research; locally consumed
Rana hecksheri	15 (5.9)	None	Locally consumed
European Frogs			
Rana ridibunda	10–15 (3.9–5.9)	Extensive	Locally consumed
Rana esculenta	10–15 (3.9–5.9)	Extensive	Locally consumed
Rana dalmatina	10–15 (3.9–5.9)	Extensive	Locally consumed
Rana lessonae	10–15 (3.9–5.9)	Extensive	Locally consumed
Rana temporaria	10–15 (3.9–5.9)	Extensive	Locally consumed
Rana graeca	10–15 (3.9–5.9)	Extensive	Locally consumed
Rana latastei	10–15 (3.9–5.9)	Extensive	Locally consumed
Asian Frogs			
Glyphoglossus molossus	10 (3.9)	None	Locally consumed
Rana acanthi	10 (3.9)	Extensive	Locally consumed
Rana blythi	10 (3.9)	None	Locally consumed
Rana hexadactyla	13 (5.1)	None	Locally consumed
Rana magna	13 (5.1)	None	Locally consumed
Rana moodei	13 (5.1)	Extensive	Locally consumed
Rana tigrina	25 (9.8)	Intensive and extensive	Most widely grown in Asia
Rana crassa	20 (9.8)	Intensive	Exported
Rana limnocharis	15 (5.9)	Intensive	Exported
African Frogs			
Conraua goliath	>30 (>11.8)	None	World's largest
Conraua robusta	14 (5.5)	None	Locally consumed
Pyxicephalus adspersus	22 (8.6)	None	Aggressive with painful bites; locally consumed
Rana fuscigola	>8 (>3.1)	None	Locally consumed
Rana vertebralis	15 (5.9)	None	Locally consumed
Xenopus mulleri	>20 (>7.8)	Intensive in laboratory	Locally consumed
Xenopus laevis	20 (7.8)	Intensive in laboratory	Locally consumed

The most commonly raised frogs are the American bullfrog, *Rana catesbeiana*, and the Asian bullfrog, *R. tigrina*, though many others are reared as pets and for scientific purposes. *R. catesbeiana* (hereafter referred to as the bullfrog) is the most widely raised of the top two species; most of the information contained in this entry is based on work conducted with that species. It has adapted to conditions from temperate to tropical, and can survive in cold water or at extremely low temperatures.

It is possible to determine the sex of a bullfrog by examining its tympanum, which in males is larger than in females. Male bullfrogs are known for their loud calls during the breeding season. Males have a callus thumb to assist in the making of mating noises. Breeding is prompted by proper temperature, atmospheric pressure, photoperiod, and humidity. Proper conditions can stimulate breeding in culture so long as the frogs are well fed and not diseased.

Female bullfrogs generally produce about 5,000 eggs, though it is reported that older females can produce more than 20,000 eggs (2). Normally, the male is ready to breed when it starts calling; the female is ready to breed when it exhibits an enlarged belly. The eggs are laid in still water and will hatch in 48–72 hours depending on water temperature. Tadpole bullfrogs may remain in the tadpole stage for as long as two years in temperate climates and can reach a length of 18 cm (7.2 in.) (2). Tadpoles have very long intestines and can digest plant material, while metamorphosed bullfrogs are carnivorous and require high–protein foods.

The stages in the life cycle of the bullfrog are as follows:

- Egg
- Larva
- Phase G1 tadpole
- Phase G2 tadpole (starting metamorphosis with hind legs showing)
- Phase G3 tadpole (front leg developing)
- Phase G4 tadpole (tail being absorbed, front legs out, gills being absorbed, increase in length, change in intestinal system)
- Small frog (froglet) similar to adult, but immature
- Adult

Tadpoles live in the water; metamorphosed frogs can live in the water or on land.

FARMING TECHNIQUES

A considerable amount of information on frog culture is currently available (e.g., Refs. 3–7). Technology is developing rapidly, so it is necessary for those interested in the topic to stay well informed on the subject.

Production methods can be divided into the categories of extensive, semi-intensive, and intensive (4). Each method depends on both environmental conditions and such factors as the cost of feed and labor. An integrated intensive frog farm will have the following components:

- Broodstock area
- Reproduction area
- Hatchery
- Tadpole tank
- Metamorphosis facility
- Growout area
- Processing plant
- Offices and laboratory
- Warehouse

Not all frog farms have all of the components listed. Many purchase only tadpoles or froglets and grow them out to market size. There are other farmers who specialize in producing tadpoles and froglets to supply the growout operators. Frog farmers have been known to form cooperatives or collectives which allow production with reduced risk, more efficiency, and improved sanitary control (6). Collectives usually involve many small farmers.

Extensive Frog Farming

This approach to frog culture can involve rearing frogs in open areas, such as rice fields, or the capture of frog and tadpole specimens from ponds. In many fish farm ponds, frogs and tadpoles are unpopular because they feed on fish food or on the fish that are being cultured. Many fish farmers do not realize that they can obtain additional income by making use of the frogs and tadpoles that live in their ponds.

Extensive rearing is characterized by a shortage of enclosure structures that will protect the frogs from predators; consequently, survival of introduced tadpoles is typically very low (often under 10%). Unless confined, tadpoles and frogs will often leave the area in which they are stocked to search for food. Eggs, tadpoles, and frogs are preyed upon by aquatic birds, fish, snakes, insects, and other creatures that can decimate a frog population.

In developing countries, the capture and restocking of frogs in rice fields can produce a remarkable quantity of marketable frogs. Extensive culture depends on the proper climate, food availability, and limited predation to be successful. It also must be remembered that rice fields and marshes that are subjected to pollution or to chemical treatment are not suitable for frog culture. In China extensive frog-cum-rice farming is common in large rice farms and swampy areas. Production varies from less than 100 kg (220 lb) to a few hundred kilograms per hectare (2.4 ac), depending on the season of the year and food availability. Feed is rarely provided, though the availability of natural feed can be increased by providing lights to attract insects.

Frogs are important to the ecology of rice fields because they control insect pests. Removal of frogs from rice fields can lead to a reduction in rice harvest. India has instituted a ban on frog exports to help ameliorate its problem with low frog populations.

Semi-Intensive Frog Rearing

The semi-intensive approach to frog rearing began in the 1980s and its techniques have remained largely unchanged ever since. To defend frogs from external predators and to prevent their escape, barriers are installed in the rearing areas. This approach can accommodate a considerable number of tadpoles in a central pond, where they will remain after metamorphosis to be captured or reared to market size (5). Because frogs of various sizes may be present in the same system, cannibalism can be a significant problem.

Heavy fertilization is implemented during the tadpole stage to produce phytoplankton and zooplankton, which provide food. The frogs will also feed on macrophytes. The feeding of metamorphosed frogs is based principally on insects that occur naturally in association with the water and those that can be attracted by night lights properly placed inside the enclosures. The frogs will also eat insect larvae and small fish and will take pieces of liver or kidney that are placed on the water surface. Survival of tadpoles reared to the marketable adult frog stage (around 150 g; 5.2 oz) is usually not more than 15%.

Intensive Rearing

As a result of a decline in the availability of wild frogs, intensive rearing methods have been developed since the 1980s. The previously mentioned systems have always resulted in very low survival rates. Intensive, closed cycle rearing systems involve all phases of culture (production of eggs, tadpoles, and adult frogs) so that there will be no problems with availability from the wild. Some farms also process frogs and produce value added frog products.

With the development of new and better breeding systems, it has become possible to increase survival and produce frogs of standard sizes to meet the demands of both domestic and export markets (1). The largest frog farms are usually found in developing countries because of their favorable climatic conditions and production cost factors. An exception can be found in Taiwan, which has high production levels as a result of heated indoor facilities.

Each stage in the frog's life cycle requires adequate containers designed to minimize stress. Frog farmers often demonstrate a great deal of originality in the technical solutions they adopt for the different frog life stages.

It is possible to split the intensive system into different phases, each of which requires appropriate facilities. Those phases are reproduction and hatchery, tadpoles, and metamorphosis to growout, each of which is described with respect to bullfrog culture.

Reproduction and Hatchery Phase. Selection of broodstock is important because those animals will form the gene pool from which the marketed frogs will be produced. Adult frogs are normally kept separated by sex and age. Frogs are generally ready to mate at one year of age in tropical climates and at two years in temperate areas. Production records of broodstock should be kept for selective breeding purposes.

During the reproductive period, males have a yellow colored throat, a callous nuptial thumb developed to assist in mating, and a nuptial embrace reflex that, together with the loud calls, signal that the male is ready to mate. Each male will actively defend his territory against possible intrusion by other males.

The reproductive condition of females can be determined by calculating an index that relates abdomen perimeter (a) and the distance between the eyes (b). When (a):(b) is more than 3.0, the female is ready to mate. A round-shaped abdomen is a clear indicator of egg development.

Fertilization in bullfrogs is external; the male clasps the female and releases sperm over the eggs as they are extruded. Areas where frogs mate must be protected, isolated, shaded, and quiet. In the Brazilian system, mating cages have 150-L (39.6-gal) water pits where the frogs can spawn. Mating can also be successfully achieved in tanks. Selected pairs of adults are stocked in such systems.

Commercial reproduction systems sometimes employ long open raceways filled with standing water. These systems require less human work than some others, but there is also less genetic control. Short-cut grass and shade are provided between the raceways. Bullfrog males signal their readiness to mate using the powerful bellowing sounds they make. The sound generated by thousands of mating frogs can be quite annoying to people.

Mating generally takes less than 24 hours, after which a new pair can be placed in the reproduction area. This system provides a constant supply of eggs. The spawning pit should be disinfected before each pair of frogs is stocked. Each adult female will produce an average of 5,000 eggs (the maximum is about 20,000). Egg to larvae survival rate often reaches about 80% under good hatchery conditions (7).

Eggs should be transferred on a screen and incubated in stagnant water in the hatchery. The eggs and larvae from each spawn are retained for 10 to 15 days in the hatchery in individual basins or in collective tanks. Individual basins permit the farmer to evaluate hatching success from a particular pair of frogs and to compare growth and other factors.

Environmental factors, such as temperature, photoperiod, solar radiation, and barometric pressure, influence egg production and spawning in bullfrogs. Broodstock maintained in environmentally controlled greenhouses may spawn year-round. Simulated rain can be used to prompt breeding. Air temperature should be maintained around 30–32 °C (86–90 °F). A constant photoperiod of 10 hours light and 14 hours dark should be maintained during the reproduction period.

When female frogs are mature it is possible to inject hormones (such as frog pituitary) to stimulate spawning. As a rule of thumb, the pituitary from one adult female is enough to stimulate a female of similar weight. Trained technicians are necessary to extract, preserve, and inject pituitaries.

Tadpole and Metamorphosis Phase. Tadpoles are kept in tanks or raceways. In less intensive systems, earthen ponds can be used, though these allow for high mortality due to predation. The tadpole phase of the rearing cycle is exclusively aquatic. Optimum density for tadpole growing is 1/L (1/0.25 gal). Under the proper conditions, tadpoles begin to metamorphose after one to three months. During metamorphosis they stop feeding and utilize tail fat reserves for energy. After the tadpole phase, the froglets are collected and transferred to growout facilities. Survival rate from tadpole to metamorphosed froglet is often 80% in tanks and 50% in open-pond systems.

Time of metamorphosis and size during metamorphosis may be influenced by crowding, water pH (8), water temperature, feed availability, or photoperiod. Large tadpoles result in larger froglets. By maintaining temperature in the range of 16 to 20 °C (61–68 °F) metamorphosis will be prevented and large tadpoles can be produced.

Growout Phase. Growout involves rearing frogs to a market size, which can range from 70 to 200 g (2.4 to 7 oz), depending on demand. Growth rates as high as 1 to 2 g/day (0.03 to 0.06 oz/day) (7) have been reached on some frog farms. A balanced pelleted diet, sometimes manufactured through extrusion, is provided. Good hygiene is maintained throughout the growing period. Survival from froglet to market size can reach 85% under the proper conditions. During the growout period frogs are particularly sensitive to loud noise and direct sunlight, so many facilities are under shade or in greenhouses. Air and water temperature should range from 20 to 32°C (66 to 88°F) for good performance. Intensive systems can be stocked at between 10 and 60/m^2 (1.2 yd^2). Densities even higher [as high as 120/m^2 (1.2 yd^2)] have been reported. Table 2 summarizes survival and performance estimates for a well-run intensive frog farming operation.

General Description of Growout Technologies. The growout stage is considered to be the most costly, risky, and difficult phase of frog farming. Frog farming intensities can reach 10 to 120 frogs/m^2 (1.2 yd^2).

The growout phase may be conducted using one of two general techniques. In the wet technique the frogs sit on the floor of a shallow water region of the rearing area, which is covered with a few centimeters (1 cm = 0.4 in.) of water, or they can swim freely in a deeper section of the rearing area. In the semi-wet technique the frogs can choose to stay in or out of water. System selection is based on local climate. In warm climates the semiwet method is used, while in colder climates the wet method with temperature control is mandatory for year-round production. Frogs cannot perform well at low temperatures when out of water. The following are examples of some of the systems currently in use.

Brazilian Systems. Brazil has developed an intensive frog farming industry over the past three decades. There have been as many as 2,000 frog farms in Brazil, though the number may be lower today because of large farms' buyouts of some of the smaller ones. In temperate areas, frogs are raised in greenhouses.

One Brazilian system employs water channels and dry areas, with or without refuge. Shade is provided to prevent exposure to direct sunlight. Groups of channels and refuges are separated by 1 m (3 ft) high walls. Frogs are often selected by size to avoid competition and cannibalism. The channels are flushed daily to eliminate feces and feed debris. Wood or concrete refuge structures are frequently cleaned to maintain sanitary conditions. Normally, there are rows of water channels between rows of refugia. The production areas are usually fabricated from concrete. The Brazilian systems are classified as semiwet. Various institutes and universities are involved in the development of frog farming technology in Brazil. During recent years, a pilot hyperintensive vertical system where froglets are raised to market size of 150 to 170 g (5.2 to 5.9 oz) has been developed. Froglets are placed in 20 cm (7.8 in.) high and 1 m^2 (1.2 yd^2) plastic boxes at densities that are reportedly as high as 1,000/m^2 (1.2 yr^2) (10).

The frogs sit in trays of water with their heads protruding into the atmosphere. The water is frequently changed so it stays clean. The frogs' diet is composed of high-protein pellets. Although the system is still under development, the equipment producer claims a very high productivity rate and high economic return. Pathology is still a problem in this system, however.

Taiwanese System. In Taiwan, frogs are maintained during the winter in heated sheds in 5- to 8-m^3 (6.5- to 10.5-yd^3) tanks at high densities. The water is from 0.3 to 1 m (1 to 3 ft) deep in the tanks and high-protein extruded floating pellets are provided. This is considered a wet system. The system uses water recirculation to save water and energy. A platform may be placed a few centimeters (1 cm = 0.4 in.) under the water surface in growout tanks where the frogs can sit with just their heads out of the water. All operations are highly mechanized, which makes it possible for two to three people to produce more than 100 tons/yr.

Pathology laboratories are incorporated into the facilities to monitor frog health status, which is often a limiting factor if not properly managed. Hyperintensive Taiwanese systems can generally rear frogs at densities of 50 to 90/m^2 (3.2 ft^2), though densities of up to 120/m^2 (3.2 ft^2) have been realized.

Chinese Systems. China has a long tradition in aquaculture, including frog farming. Frog culture that can be considered intensive is generally conducted in small units, in rural areas that are concentrated in southern and central China where the climate is suitable. Earthen tanks that are 100 to 200 m^2 (120 to 240 yd^2) and are 1 m (3.2 ft) deep are enclosed by 1 m (3.2 ft) high walls. Normally, an inlet channel provides each facility with several daily pondwater changes. Dry fish and other protein-rich feed materials are provided on a little wooden island in the pond where the frogs rest and feed.

Large cooperative frog farms of more than 1,000 ha (2,500 ac) employ the extensive-system approach in swampy areas and rice fields. Production rates range from 100 to 300 kg/ha (2.5 ac). In these cooperatives, hatcheries provide the tadpoles that are used to stock the growout areas. Fertilization, and the reduction of the pests that the frogs prey upon, are believed to increase rice production.

Mexican System. The Mexican system is considered to be of the intensive wet type. The system involves rearing frogs on the bottom of tanks that are filled with water

Table 2. Expected Survival Rate for Various Stages in Frog Culture and Anticipated Performance Parameters in a Well-Managed Intensive Frog Culture Facility

Parameter	Value or Result
Survival	
Egg to larvae	80%
Tadpole to froglet	75%
Froglet to 170 g (5.9 oz)	85%
Age at Maturity	1–2 years
Eggs in average spawn	5,000
Tadpole stocking density	1–20/L (0.26 gal)
Froglet stocking density	3–50/m^2 (1.2 yd^2)
Food conversion ratio for tadpoles[a]	1.5
Food conversion ratio for froglets	1.2
Time for tadpole stage	30–90 days
Time from froglet to 170 g (5.2 oz) frog	100–180 days

[a]Food conversion ratio is calculated as dry weight of feed offered per unit time divided by wet weight of gain during the same period of time.

to the shoulder level of the animals. Frogs are reared in sheds that provide quiet, shaded areas to reduce stress. Tarpaulin covers are used to provide the shade. Extruded pellets produced in the United States and that cost US$1/kg (2.2 Ib) are fed to the frogs.

Patented reproduction units permit accurate selection and can be operated with low manpower costs. The system has been sold abroad and employed with good results.

PATHOLOGY

Frogs, like other aquatic animals, are subject to various types of pathology, including problems caused by viruses, bacteria, fungi, and parasites (9,10). Presently, there are no vaccines available for use by frog producers (10).

Maintaining clean conditions in the culturing environment is critical to health maintenance. Frogs are also particularly sensitive to stress. Cannibalism can become a major problem when frogs are not well fed or are stocked as mixed sizes. Ulcerations or skin lesions (particularly on the head and extremities) are signs of improper nutrition, bacterial infections, parasitism, or mechanical trauma. "Red-leg disease," the best known and most common bacterial problem, can cause mass mortality on commercial farms. The disease is related to overcrowding, diet deficiency, poor quality water, inadequate lighting, or improper temperature. Rectal prolapse, where the large intestine protrudes from the cloaca, is also a condition related to several factors, including temperature, diet, and parasitism.

Bacterial diseases are common in frog farming. Bacteria isolated from farmed bullfrog blood have included *Streptococcus* spp., *Flavobacterium ranicida*, *Pseudomonas* sp., *Aeromonas* sp., *Mima polymorpha*, *Citrobacter freundii*, and *Staphylococcus* sp. (11–15). Studies support the theory that red leg is caused by a complex interaction of different bacteria that are common inhabitants of the aquatic environment, and the frogs that become pathogenic when they are stressed (14,15). Bacteria isolated from frogs with red leg have included *Aeromonas hydrophila*, *Pseudomonas aeruginosa* and others. Those bacteria may also occur in tadpoles and can cause metamorphosis problems. Skin ulcerations help the proliferation of septicemic disease. In red-leg infected frogs, fat tissue analyses show higher concentrations of bacteria than in blood (15).

Other less common bacterial diseases associated with frog farming include encephalitis, conjunctivitis, mycobacteriosis, bacterial mouth disease, and general edema. Taiwanese specialists have come to the conclusion that proper culture system management significantly influences the development of bacterial epizootics. The three most important factors are stocking density, feed type, and prophylaxis.

Because of their delicate skin, frogs are very sensitive to bacterial diseases; consequently, appropriate cleaning and disinfecting operations must take place frequently, particularly in intensive systems. It is very important to maintain high water quality in the tanks. For dry systems, cleaning and disinfecting all cage surfaces fairly often is a good practice (11–13).

Frogs suspected of harboring pathogens or showing signs of disease should be isolated and treated. It may be necessary to contact a specialized laboratory to obtain definite identification of a disease. An isolated area should be constructed for quarantine of frogs brought to the facility. This will reduce the chance of the disease being introduced from outside. Dead frogs should be incinerated to avoid the possible spread of infectious diseases.

Saprolegniasis is the most common fungal disease in frog. It is common among eggs and tadpoles in low-quality water. It can be treated by dipping the eggs in a 5% salt (sodium chloride) solution for two minutes (9). Parasitic nematodes, cestodes, and trematodes are present in wild frogs but cause few problems in farmed frogs (16).

Though frogs naturally live in stagnant water, the water quality in culture tanks must be controlled. Frogs are very sensitive to coliform bacteria in water. Growth of bullfrogs slows at temperatures below 15 °C (60 °F); the optimum temperature is between 28 and 32 °C (84 to 88 °F) for normal reproduction and physiological development. A pH range of 6.5 to 7.5 is best during tadpole metamorphosis. Natural citrus extract is successfully used to disinfect frog farms.

FEEDS AND FEEDING

Important factors in feeds and feeding include feed costs, feeding techniques, and feed conversion efficiency (17–19). All of these factors influence the economics of a frog farm. It is very important to know about changes in the digestive system and food habits at various stages in the life cycle and to provide the most appropriate feed (20–23). In some systems, feeds may include insect larvae, worms, or other live foods. Intensive systems employ pellets.

Feeding Tadpoles

After absorbing the yolk sac, tadpole larvae are omnivorous and begin feeding on phytoplankton, bacteria, and protozoa. Some frog farmers encourage algae development on the walls of rearing tanks to enhance the amount of available food. At a later stage, tadpoles begin to feed on rotifers and crustaceans.

Physiologically, tadpoles have a very long digestive system compared to body length. The tadpole's digestive system is adaptable to several types of diet (17). Studies have shown that high-protein diets can considerably accelerate growth and metamorphosis (22). A raw-protein level of 30 (24,25) to 44% (20) is considered appropriate for optimal tadpole growth. Optimal particle size is 0.21 mm (0.008 in.) in diameter (24).

A ratio of 1 : 1 for animal protein versus plant protein (at 40% protein) has led to good results with tadpoles (25). A feed formula that has been successfully used by the author is presented in Table 3. The diet results in optimum growth of healthy tadpoles, but is costly.

The São Paulo Fishery Institute (Brazil) recommends the following feeding protocol for tadpoles:

- 10% of live weight during the first month
- 5% of live weight during the second month
- 2% of live weight during the third month

Table 3. Feed Formulation for Tadpoles in Intensive-Rearing Systems

Ingredient	Percentage of Diet
Fish meal	38.5
Soybean protein	5.5
Brewery waste	16.5
Rice bran	21.0
Powdered milk	5.0
Fish oil and solubles	7.0
Fatty acids	0.5
Vitamin and mineral premix	3.25
Binder	2.15

Water temperature affects tadpole feed utilization, growth, and metamorphosis. Optimum water temperature should range from 20 to 32 °C (68 to 88 °F). During metamorphosis, particularly at the G4 phase, tadpoles stop feeding and rely only on their tail-fat reserves.

Feeding Metamorphosed Frogs

Frogs that have metamorphosed have large, muscular stomachs and short intestinal tracts. They require a high-protein diet. Frogs prefer to feed on moving prey and therefore need to be trained to feed on prepared diets. Frogs have large mouths and can swallow prey that is large relative to their body size. These characteristics have been considered as methods for feeding frogs were developed (26).

One technology consists of feeding live food, such as fish, adult insects, and tadpoles. Problems associated with the use of live feeds include lack of availability and high costs due to manpower demands the method requires. South American frog raisers like to mix 2 to 20% insect larvae (e.g., *Musca domestica*) with feed pellets. The movement of larvae in the pellets induces the frogs to feed on the mixture. Putting pieces of meat, liver, or kidney in the water or on feeding trays is another option, although the author considers it risky because it facilitates the spread of diseases, especially in tropical climates. New technology allows the use of vibrators, which provide motion to pellets without necessitating the use of insects.

The first period after metamorphosis is the most critical. High mortality can occur, particularly in conjunction with newly metamorphosed frogs. Providing insect larvae and brine shrimp (*Artemia* sp.) can help get the frogs through the critical period. Another method is to put extruded floating pellets on the water surface, which move with currents and entice the frogs to feed. This feed is very well accepted by the frogs and can lead to an excellent food conversion ratio. The method is used in intensive systems and requires high-quality pellets. The cost of high-protein extruded and floating pellets can exceed US$1/kg (2.2 lb) but can produce food conversion ratios (FCR = dry weight of feed offered per unit time divided by wet weight gain) of 1.1 to 2.0, depending on pellet quality.

Some studies on adult frog nutritional requirements have been conducted, but more research is needed. The author has successfully used extruded pelleted trout feed that contained 42% raw protein. Pellet size should vary in relation to frog mouth size, but will normally range from 3 to 7 mm (1.4 to 3.3 in.) in diameter.

The frequency and quantity of feedings are related to temperature and frog size. Detailed studies have not yet been conducted, but judging from practical experience, adult frogs of 150 g (5.2 oz) consume from 3 to 3.5% of their body weight daily when fed *ad libitum*. Younger frogs may eat as much as 6% of their body weight daily.

A specimen formula developed by the author for feeding frogs in growout tanks contains the following ingredients: fish meal, blood meal, gelatinized wheat starch, meat meal, fish oil, dry milk, a vitamin and mineral mixture, binder, and dL-methionine. The cost of the feed was US$1/kg (2.2 lb) in Italy. The results were very good; frogs had more than 1 g (0.04 oz) weight gain daily. The proximate composition of the formula is presented in Table 4. It is possible to formulate a less expensive pelleted feed, but frog performance is reduced.

FROG ECOLOGY

With the advent of water pollution and the clearing of swamps, habitat for frogs is disappearing and their cells are heard less and less around the world. Frogs are vitally important in the ecosystem of swamps and humid areas. They are often used as indicators of environmental quality. Where there are frogs, the environment is usually still in good condition. Many countries, both developing and developed, are now protecting frogs. Frog farming can play a key role in the conservation of frogs because cultured frogs can reduce the pressure on their wild counterparts and even supplement wild stocks through stocking programs (7).

It is well-known that frogs are moved around the world and sold for farming and as pets, research animals, and biological pest control agents. The American bullfrog has been introduced to many countries in Latin America (Brazil, Argentina, Venezuela, Uruguay, Paraguay, and others), Europe (Spain, Italy, France, United Kingdom, and Greece), and the Far East (Thailand, Malaysia, Indonesia, and Taiwan) as pets and for farming. In some of these countries there are well-established populations of American bullfrogs that do not appear to have caused problems in local ecosystems (27). Some EU nations have bans against bullfrogs and other species. Certain biologists claim that environmental damage could result from the introduction of frogs in these areas. On the other side of the issue is the fact that imported frogs provide food for various predators.

Table 4. Proximate Analysis of a Frog Growout Pelleted Diet Used in Intensive Frog Farming

Parameter	Percentage in Feed
Moisture	10
Crude protein	44
Crude fat	11
Cellulose	1.5
Ash	14

Excessive wild-frog overfishing can create problems in rice areas, such as those that have occurred in India and Bangladesh where an increase in rice pests was correlated with frog overfishing. The countries reacted with a ban on frog exports. Extensive and semi-intensive frog rearing systems can help contain the rice-pest attacks.

GENETICS

Frogs that are captively bred today have not undergone a great deal of selection, since they were wild a few generations ago. Mass selection is the most common form employed by frog farmers. For certain types of biomedical research, cloned and gynogenetic frogs (28) are selected, but the cost for these is very high; only a large frog farm can provide a budget for genetic research.

A wide variety of sizes of same-aged frogs has been seen on commercial farms and indicates a lack of genetic selection. To increase the gene pool on a frog farm, the farmer can breed bullfrogs from existing farms that are located in widely separated regions or on separate continents. Some frog farms in southeast Asia have developed frogs with improved performance that are also well adapted to frog farming. Progeny and performance testing can produce improved frog strains at low cost. Improvements from selective breeding programs occur more rapidly with frogs than with many other terrestrial animals because of the large numbers of eggs produced per year and the short generation time. It is now possible to control the frog reproductive cycle through the use of environmental control and hormones. When these methods are coupled with genetic selection, we should see dramatic improvement in frog performance in the future (1,29–31).

PROCESSING

Frog slaughtering and processing must respect the internationally accepted HACCP (Hazard Analysis Critical Control Point) system, which helps to ensure that frogs are processed under good sanitary conditions (32–34). The use of a HACCP system is mandatory for frogs shipped to the EU and the United States (35). All seven principles of the HACCP system apply to frog processing:

- Conduct a hazard analysis
- Determine the critical control points
- Establish critical limit
- Establish a system to monitor control of the critical control points
- Establish the corrective action to be taken when monitoring indicates that a particular critical control point is not under control
- Establish procedures for verification to confirm that the HACCP system is working effectively
- Establish documentation concerning all procedures and keep records appropriate to these principles and their application

These principles help frog-processing companies develop programs specific to their plants. Frog processing must be accurately controlled in both abattoir construction design and in the management of the processing activities (7,31).

A good frog-processing plant will have the following facilities:

- Reception room
- Slaughter room
- By-products collection area
- Mechanical room
- Packing room
- Cold storage
- Shipping area
- Employee dressing room
- Water filter (for water used in the processing line)

The last point is often neglected and has been the cause of many problems in frog-processing plants.

The processing plant must be isolated to the greatest extent possible from the exterior environment. In the most advanced frog-processing plants, air is conditioned before entering the exterior environment. The processed frogs must be handled as little as possible, to avoid contamination. *Salmonella*, *Enterococcus*, and *Clostridium* bacteria have been isolated in imported frogs. *Salmonella* in the digestive tract of frogs can contaminate the meat during processing if care is not taken. Only a well-applied HACCP system can prevent dangerous microbial contamination during frog processing.

Ready-to-market frogs are left on the farm for one to two days without eating and are then transferred to the frog abattoir and placed in a tank with ice, water, and a mild disinfectant. The cold stops the activity of the frogs so that they may be easily processed without suffering. Generally only the legs are sold as IQF (individually quick frozen) products. They are packed in polyethylene bags and stored in cold room (32). Frog dressout is usually around 54% in a modern processing plant.

TRANSPORT

Live frogs and tadpoles can be transported by air, truck, or ship. Before packing, the live tadpoles and frogs must be conditioned. They are kept off food for 24 to 48 hours. The author recommends disinfection just prior to packing. Tadpoles of certain sizes can be shipped dry for as long as 12 hours. Small tadpoles are transported in plastic bags that contain water and are charged with oxygen. The time of the travel and the size of the tadpoles influence the number of tadpoles per unit volume of water. Tadpoles of 1 cm (0.4 in.) can be transported at 20/L (0.26 gal) for one day with very low mortality. Temperature is an important factor in tadpole transportation. It is important to keep the temperature low during the transport period.

Live adult frogs are transported in various ways. One way is to use a mesh plastic sac with 2 kg (4.4 lb) of live frogs inside. Several bags can be placed inside a large cardboard box that has holes in it to allow the frogs to breathe. Mortality tends to be very low for this method.

MARKETING

There is very little advertising for frogs, because demand exceeds supply. At the same time, many people have had very few occasions to taste frog meat; many have never tasted frog at all. Frog meat is a traditional food in some areas; in others, it is rather exotic. It is well known around the world as a gourmet food item that can generally be found in upscale restaurants. It is also possible to find fresh frog legs in fish shops and frozen ones in supermarkets. There is still prejudice against frog legs by some consumers.

The majority of farmed and captured frogs in the world are utilized for food consumption. The meat is white, rich in protein, easily digestible, and very low in fat. In European countries, only small frog legs are utilized since Europeans have traditionally had those sizes on their tables (5). In fact, in other frog-consuming countries, you can find larger legs imported from South American and Asian countries, where the customers prefer larger legs. Frog thighs are sold and imported fresh or frozen, depending on the availability, labor cost, and market demand.

The highest frog demand is in traditional frog-fishing areas. People living in humid and wetland regions are traditionally frog consumers: from the Danube to Mississippi and the Yang-Tze Delta (5,36), areas that have been influenced by the French culture (37) are those where frogs are considered a delicacy. Other major consumers and importers are the EU member countries of Belgium, the Netherlands, Italy, and Spain. Frog thighs are considered a gourmet food and always bring a high price in restaurants. In some countries, frogs are eaten whole, after being slaughtered and skinned.

In Europe, frogs are slaughtered at a weight of 60 to 90 g (2.1–3.1 oz), but American and Asiatic consumers often prefer frogs of 170 g (5.9 oz) or larger. European consumers believe that the small frogs have a better taste than the larger ones.

The selling price in the United States was about US$7/kg (2.2 lb) in 1997 for imported live frogs of 6–7 individuals/kg weight. In Italy, in 1998, the price was 7,000 lira (US$4)/kg for imported live frogs of 11–12 individuals/kg but 34,000 lira (around US$19) for a clean fresh leg in the supermarket. In France, in February 1997, clean legs in the Rungis market were 28 francs/kg (US$5.20/kg).

Indonesia and China are the biggest exporters to the EU, some countries of which are reexporters of processed frogs. Indonesia and China are major suppliers of frogs to France, which is the largest EU consumer.

Frog legs are popular and are widely marketed, but other products can be made from frogs. Brazil has been a leader in developing such value-added products as frog vinegar, frog pizza, and many other products that permit the use of all the edible frog parts.

Frog farmers have discovered some niche markets that can provide high revenues. Frogs raised in captivity and sold for research and student education are sold not by weight, but by number. There are many tens of thousands of frogs sold for scientific and educational purposes at an average retail price of approximately $15/frog in the United States each year. There is a lack of producers or distributors of frogs for scientific research in Europe. The typical price for one adult frog for use in research in Italy is US$13. Frog skeletons may also be sold for courses in the natural sciences.

Although it is not the first use the author can think of, the utilization of frog skin is a very promising business. Obviously, the infrastructure will need to be developed for skinning and tanning frog skins if a product with consistent quality is to be produced. Frog skin is used by the fashion industry and could also have applications in the treating of burns.

The author has worked in frog bracelet production and faced the problems of inconsistent supply, high skin cost, low quality of raw skin, and poor processing. Only one quality bracelet can be made with one large skin. Often the skins have holes in them and must be discarded. The Italian market accepted frog skin bracelets tanned in several colors; the gross price of one bracelet sold to the retail shop was US$10. Wallets, purses, belts, bikinis, jackets, and other products can be manufactured from processed skins (5). The cosmetics industry has extracted a special oil from frogs that is used to produce creams and other cosmetics (7).

There are good markets for tadpoles, broodstock, frog feed, and equipment associated with rearing the animals (38). A good pair of adult frogs selected for high quality as breeders will cost US$20 or more. Tadpole prices vary depending on size and number purchased.

The pet animal market is very promising. A frog in a small aquarium can be enjoyed by both adults and children. *Xenopus* sp. albinos now appear in pet shops. Frogs with strange shapes, colors, and sizes are valued by hobbyists. Pet animal market demands are expected to rise in the next few years.

Frogs are also utilized to organize games; one of the oldest is the Sacramento International Jumping Frog Competition held yearly in Sacramento, California. Local television broadcasts the event and competitors come from all over the world to participate.

FINANCIAL CONSIDERATIONS

There are many types of frog farming, and one example of the costs involved will not begin to cover all the possibilities. Costs vary among countries, so it is important to do appropriate economic planning before actually initiating development in a particular country; economics are very important in the development of a frog farm (39). Frog farming is more complex than many types of aquaculture. All the investment and management costs must be identified during the development of the business plan.

Table 5 presents an investment cost estimate for a partially integrated frog farm, without processing, with a projected production of 100 tons/yr of live frogs. Total surface area of the facility is 5.5 ha (13.75 ac), using the Taiwanese indoor system of tanks. Table 6 provides an annual operating cost estimate for the facility. The estimates shown in the operation outlined in Tables 5 and 6 are an example of only one type of system. Actual costs

Table 5. Cost Estimates for Construction of a Hypothetical 100-Ton Taiwanese-System Frog Farm

Item	Area	Cost (US$)
Construction costs		
Breeding and hatchery facility	300 m^2 (360 yd^2)	40,000
Tadpole tanks	20 at 10 m^3 (13 yd^3)	40,000
Growout tanks with feeders	200 tanks at 50 m^3 (66 yd^3)	250,000
Laboratory and offices	60 m^2 (72 yd^2)	20,000
Workshop and warehouse	60 m^2 (72 yd^2)	20,000
Packing and shipping room	30 m^2 (36 yd^2)	15,000
Dressing room for personnel	30 m^2 (36 yd^2)	20,000
Equipment		
Electrical and mechanical		20,000
Water supply and filtration		15,000
Pumps		10,000
Other equipment		20,000
Total capital and equipment costs		470,000
Design, legal fees, surveying, etc.		50,000
Total investment		520,000

Table 6. Annual Operating Expense Estimates for Frog Culture Facility in Table 5

Item	Quantity	Cost (US$)
Variable costs		
Feed	150 tons	200,000
Medications		30,000
Repairs (including labor)		10,000
Utilities		15,000
Miscellaneous supplies		50,000
Fixed costs		
Personnel	2 individuals	80,000
Consulting and administrative services		20,000
Overhead		15,000
Maintenance (3% of investment cost)		15,600
Interest payment		10,000
Insurance (1% of investment cost)		5,200
Land lease		5,000
Depreciation (5% of investment cost)		26,000
Total operating costs		481,800
Taxes (10%)		48,180
Grand total		529,980
Gross revenue		700,000
Annual net profit		170,020

will vary depending on specific locale and changes that might occur in the economy, both locally and globally.

The payback period for the hypothetical facility, based on the figures presented, is calculated as the initial investment divided by the average net income before depreciation (40), or

$$\text{Payback period} = \text{US\$520,000/US\$455,800} = 1.14 \text{ years}$$

The payback period indicator of 1.14 years is considered good for an aquaculture facility.

The rate of return on the investment is calculated as the average annual profit divided by the initial investment × 100 (40), or

$$\text{Return on investment} = \text{US\$170,020/US\$520,000} = 32\%$$

A rate of return of 32% can be considered excellent performance for frog farming.

The good economic projections shown in the example are results of a farm that employs high technology and good managers. It requires a high capital investment. Another alternative would be to form a cooperative among a few farmers, each of whom specializes in one aspect of frog culture. In the future, well-managed intensive frog farms may prove to be the only ones that are technically and economically viable.

CONCLUSIONS

This entry has presented a general, brief, and incomplete description of the different possibilities for rearing frogs. New technology is constantly emerging; in fact, the frog industry works hard to develop new approaches. Because of the decline in wild frog populations and the strong increase in demand, frog rearing has become a good investment opportunity. Due to the complexity of frog farming, the reader is reminded that it is good practice to obtain the advice of experts before investing in frog aquaculture.

ACKNOWLEDGMENT

Thanks to Dr. Claudia Maris and Prof. Dos Lantos for general advice on this entry and detailed information about Brazilian frog farming. Dr. A. Flores-Nava provided information on Mexican frog systems.

BIBLIOGRAPHY

1. G. Negroni, *World Aquaculture* **28**(1), (1997).
2. J.F. Breen, *Encyclopedia of Reptiles and Amphibians*, T.F.H. Publications, Neptune City, NJ, 1974.
3. G. Negroni, *Informatore Zootecnico* **24**, 51–52 (1991).
4. M. Rodriguez-Serna, A. Flores-Nava, M.A. Olvera-Novoa, and C. Carmona-Osalde, *Aquacult. Eng.* **15**, 233–242 (1996).
5. G. Negroni, *Infofish Int.* **4**, 96 (1996).
6. A. Hilken, G. Dimigen, and J. Iglauer, *Lab. Animals* **29**, 152–162 (1995).
7. R.D. Teixeria, *Infofish Int.* **3**, 59–62 (1992).
8. I.P. Martinez, R. Alvarez, and M.P. Herraez, *Aquaculture* **142**, 163–170 (1996).
9. G. Post, *Textbook of Fish Health*, T.F.H. Publications, Neptune City, NJ, 1987.
10. M. Pearson, *Aquacult. News* **6**, 23–24 (1996).
11. M. Crumlish and V. Inglis, *Aquacult. News* **11**, 25–28 (1995).
12. R. Bennati, M. Bonetti, A. Lavazza, and D. Gelmetti, *Vet.-Rec.* **135**, 625–626 (1994).
13. A. Barney, *Anphibia Disease*, T.F.H. Publications, Neptune City, NJ, 1989.
14. C. Glorioso, R.L. Amborski, J.M. Larkin, G.F. Amborski, and D.C. Culley, *Am. J. Vet. Res.* **35**, 447–457 (1974).
15. C. Glorioso, R.L. Amborski, G.F. Amborski, and D.C. Culley, *Am. J. Vet. Res.* **35**, 1241–1245 (1974).
16. R.J. Roberts, *Fish Pathology*, Cassel, Ltd., London, 1990.
17. A. Flores-Nava and E. Leyva-Gasca, *Aquaculture* **152**, 91–101 (1997).
18. C.C. Osalde, M.A. Olvera-Novoa, M.R. Serna, and A. Flores-Nava, *Aquaculture* **141**, 223–231 (1996).
19. L.E. Brown and R.R. Rosati, *Prog. Fish-Cult.* **59**, 54–58 (1997).
20. C. Carmona, M. Osalde, A. Olvera-Novoa, M. Rodriguez-Serna, and A. Flores-Nava, *Aquaculture* **141**, 223–231 (1996).
21. I.P. Martinez, M.P. Herraez, and R. Alvarez, *Aquaculture* **128**, 235–244 (1994).
22. I.P. Martinez, M.P. Herraez, and M.C. Dominguez, *Aquacult. Fish. Manage.* **24**, 507–516 (1993).
23. D. Culley and S.P. Meyers, *Feedstuffs* **44**(31), 26–27 (1972).
24. C.L. Mandelli, Jr., L.A. Justo, L.A. Penteado, D. Fontanello, H. Arruda Shoares, and B.E.S. de Campos, *Bol. Inst. Pesca* **12**, 61 (1985).
25. D. Fontanello, H. Arruda Soares, C.L. Mandelli, Jr., J. Justo, L.A. Penteado, and B.E.S. de Campos, *Bol. Inst. Pesca* **12**, 43–47 (1985).
26. D. Lester, *Aquaculture Magazine*, March-April (1988).
27. J. Baker, *World Aquacult.* **29**(1), 14–17 (1998).
28. B.G. Brackett, G.E. Seidel, Jr., and S.M. Seidel, *New Technologies in Animal Breeding*, Academic Press, New York, 1981.
29. D.D. Culley, A.M. Othman, and R.A. Easley, *J. Herpetology* **16**, 311–314 (1982).
30. S. Minucci, L. Di Matteo, G. Chieffi Baccari, S. Fasano, M. D'Antonio, and R. Pierantoni, *Gen. Comp. Endocrin.* **79**, 147–153 (1990).
31. J. Elmberg, *Can. J. Zool.* **68**, 121–127 (1988).
32. R.D. Teixera, G.F. Borges, and G.A.C. Junior, *Infofish Int.* **2**, 29–30 (1987).
33. J. Baker and A. Mackenzie, *Infofish Int.* **6**, 44–48 (1996).
34. H.J. Iben and B. Knurr, *Eastfish Mag.* **1**, 38–42 (1997).
35. D. Ward, ed., *Microbiology of Marine Food Products*, Van Nostrand Reinhold, Amsterdam, 1989.
36. G. Negroni, *Resource* April (1997).
37. A.D. Cabero, *Bullfrog* **10**, 25–30 (1980).
38. R. Donizete-Teixera, *Aquaculture Mag.* **19**(2), 42–48 (1993).
39. K. Jagadees, *Seafood Export J.* **11**(39), 22–33 (1979).
40. Y.C. Shang, *Advances in World Aquaculture, Vol. 2*, The World Aquaculture Society, Louisiana State University, Baton Rouge, LA.

FUNGAL DISEASES

Edward J. Noga
North Carolina State University
Raleigh, North Carolina

OUTLINE

Fungi (mushrooms, yeasts, molds, etc.) are primitive organisms that are characterized by the absence of chlorophyll and the presence of a rigid cell wall. They range in form from a single cell to a mass of branched, filamentous structures. Fungi usually feed saprophytically on dead organic matter (soil, dead plants, and animals), and thus virtually none require a living host to survive. However, some species will infect a living host if its normal defenses are impaired. Therefore, fungi are classical opportunists. Wounds, especially on the skin or cuticle, but also other body surfaces (gills, gut), due to trauma or other pathogens provide a portal of entry for fungi (1,2). Handling, crowding, heavy feeding rates, and high organic loads also appear to increase the risk of some fungal infections. In other cases, there is no discernible reason for an outbreak.

A fungal infection in a sick animal is usually easily detected using simple microscopic examination of the diseased tissue, as most pathogenic fungi produce long, filamentous structures, known as "hyphae," that are easily identifiable in tissue squashes (see Fig. 1) or histological sections. However, identification of the specific fungus (genus and species) responsible usually requires culture of the agent, in order to stimulate production of reproductive structures, which are needed for taxonomic identification. Gene and antibody tests have been experimentally used in diagnosis, but are not commercially available.

Fungal infections can be extremely difficult to treat, because, unlike bacteria, their metabolism is very similar to that of higher animals (both invertebrates and vertebrates). Thus, drugs that inhibit fungi are often toxic to the host at the levels required for the drugs to be effective. There are no highly useful treatments that are approved for aquaculture species. Thus, prevention of stress, and quarantine, when possible, are the best means for managing fungal problems. For more details on fungal infections, see Hatai (3), Chako (4), and Noga (5–7).

IMPORTANT FUNGAL DISEASES OF FINFISH

Fungal infections of finfish can be artificially divided into two types, based upon whether they cause mainly superficial infections of the skin or gills or penetrate deeply into the body.

Superficial Infections

Typical Water Mold Infections (*Saprolegniosis*). *Saprolegniosis* is the most common fungal infection of freshwater fish. Water molds are distributed worldwide; virtually every freshwater fish is susceptible to at least one species of water mold. Water molds are members of the Class Oomycetes; over 30 species have been isolated from diseased fish. While water molds are classified as animals in the Kingdom Protoctista (8), they look and behave like true fungi and are managed in the same way as the true fungi. The great majority of fish pathogens are in the Family Saprolegniaceae (Order Saprolegniales). Most infections are caused by *Saprolegnia* (which is why the disease is called saprolegniosis), but other Oomycetes cause an identical disease.

Water molds are ubiquitous saprophytes in soil and freshwater. They are appropriately named, reproducing in water or moist soil via formation of swimming spores (sporulation); this mode of reproduction distinguishes them from terrestrial fungi, which produce aerial (airborne) spores. Water molds are transmitted by their swimming spores. Most infections of fish are probably acquired from inanimate sources (i.e., from fungi that produce spores on dead organic matter) rather than from live-fish-to-live-fish transmission. Outbreaks often occur after a drop in temperature (3). This trend may be due to lower immunity of the host (9) and because many Oomycetes are more active in the cooler months of the year (5).

Typical water mold infections appear as a relatively superficial, cottony growth on the skin or gills that can rapidly spread over the body surface. Newly formed infections are white (the color of the hyphae); with time, they often become red, brown, or green as they trap sediment, algae, or debris in the hyphae. When observed on a fish removed from the water, the fungus appears as a slimy, matted mass on the body. (see Fig. 2.) "Winter kill" is a water mold infection of channel catfish (*Ictalurus punctatus*) during winter. The disease typically appears soon after the passage of a cold front, which rapidly drops water temperatures and reduces the resistance of the fish (9).

Although superficial, skin or gill damage from water molds is often fatal, due to loss of body fluids and salts from the open wounds. Mortality increases as the area of affected skin or gill tissue increases. With acute outbreaks, fish usually die within several days or recover within several weeks. Oomycetes are important pathogens of fish eggs. Infections usually begin on dead eggs and can rapidly spread to healthy eggs, eventually resulting in complete loss of the brood.

Figure 1. Wet mount of water mold. (Photo courtesy of A. Colorni.)

Figure 2. Photograph of a water mold infection. Note the slimelike appearance of the infection when viewed out of the water.

Diagnosis of a water mold infection requires that affected fish be alive when examined, because water molds are ubiquitous saprophytes in soil, freshwater, and, to some extent, estuarine environments; dead fish are quickly colonized, resulting in a misdiagnosis. Oomycetes will also invade wounds caused by other pathogens (e.g., bacteria and parasites). Always look for other initiating causes when water molds are identified in a wound.

Water molds are among the most difficult diseases to treat. Except for salt, agents legally approved for foodfish (10) are of limited effectiveness. Malachite green is the most effective agent for treating water mold infections in fish, but it is not approved for use with foodfish in many countries, because it is a teratogen and mutagen. Most water molds that are pathogenic to fish are inhibited even by low salt concentrations (less than 3 ppt), which is probably why most species do not affect marine fish. (However, see the upcoming section on atypical water mold infections.) Salt also helps to counteract osmotic stress due to skin damage and subsequent ion loss, though long-term salt baths are impractical in most commercial production situations. Short-term hydrogen peroxide or formalin baths may be useful for treating some infections (11,12).

Because of the often rapid development of oomycete infections, as well as their resistance to drugs, prophylaxis is the best strategy for treating the infections. Avoid skin damage and predisposing stresses. Salt baths are an effective prophylactic when transporting fish or acclimating them to a new environment. It is probably not possible to eliminate water molds from a culture system.

Deep Infections

Atypical Water Mold Infections. Atypical water mold infections (e.g., epizootic ulcerative syndrome [EUS], red-spot disease [RSD], mycotic granulomatosis [MG], and ulcerative mycosis [UM]) differ from typical water mold infections in that the former cause an extremely deep, penetrating wound. (see Fig. 3.) While atypical water mold infections are less common than typical water mold infections, they are a serious disease in many parts of the world (13,14).

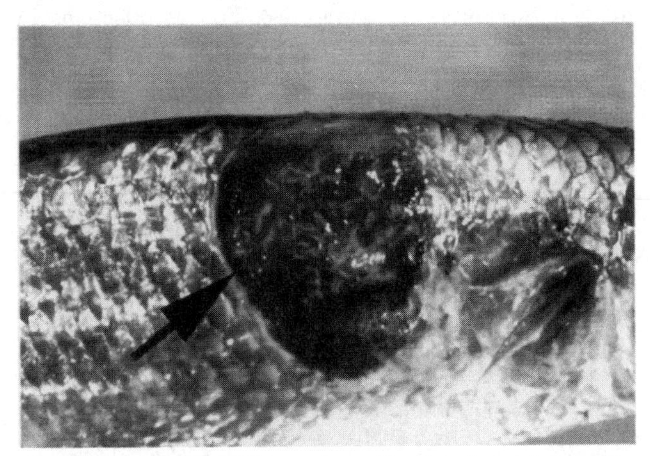

Figure 3. Photograph a large, deep skin ulcer (arrow) caused by EUS, also known as RSD. (Photo courtesy of R.B. Callinan.)

Atypical water mold infections occur in numerous estuarine and freshwater fish populations, including snakehead, walking catfish (*Clarias* spp.), mullet (*Mugil* spp.), and gouramies. In the Australo–Pacific and southern Asia, they make up one of the most important diseases of cultured fish (13,15). Morbidity and mortality can be very high, and epidemics can develop rapidly. However, once an epidemic has occurred in an area, the prevalence and severity of future outbreaks often subside. The infections are also a problem in wild estuarine fish of the western Atlantic Ocean (7).

Unlike typical water mold infections, atypical water mold infections usually produce reddened, deep wounds (see Fig. 3) that may even penetrate into the body cavity. *Aphanomyces* is most commonly isolated from wounds (7,16). A presumptive diagnosis of atypical water mold infections is based on a microscopic identification of typical hyphae surrounded by many host immune cells in the wound.

While atypical water mold infections can be diagnosed as a disease by confirming that a water mold is present, the primary cause of the ulcers is largely unknown. While the fungi and bacteria present in lesions probably play an important role in killing fish, they are probably not responsible for initiating any of the lesions. There is evidence that skin damage caused by a toxic dinoflagellate, *Pfiesteria*, can lead to atypical water mold infections in Atlantic coast estuarine fish, including hybrid striped bass (17). The cause of the clinically similar RSD and EUS is unknown, although a rapid drop in pH has been suspected to play a role in some cases (18). There is also evidence for the spread of some agent, since EUS and RSD have spread from the Australo–Pacific region to encompass most of southern Asia (13). There is no proven treatment for atypical water mold infections.

Branchiomycosis. Branchiomycosis has sporadically caused acute infection and, often, high mortality rates in several freshwater fish, including cultured American eel (*Anguilla rostrata*), European eel (*A. anguilla*), and centrarchids (19). It has been reported mainly from Europe and Taiwan, but isolated cases have also occurred in the United States. The gills of infected fish are "mottled" in appearance, due to the formation of clots in blood vessels. Wet mounts or histopathology of characteristic branched hyphae causing deep gill infection are diagnostic. There is no known treatment. Reducing organic loading and reducing the temperature below 20 °C (68 °F) might be helpful.

Miscellaneous Deep Fungal Infections. All other deep fungal infections in fish are rare. Most have been encountered as sporadic cases, although some have caused localized epidemics. Virtually all such diseases are chronic infections. Among the most prevalent are *Exophiala* (found in trout, salmon, cod, and flounder) and *Ochroconis* (found in trout and salmon).

IMPORTANT FUNGAL DISEASES OF SHELLFISH

Fungal diseases occur in most major groups of cultured invertebrates, including shrimp, lobster, crabs, clams, oysters, and cephalopods (3,4,6). While they occasionally are localized problems, they are usually much less important than other infectious agents, such as viruses, bacteria, and parasites.

Water Molds

Oomycetes are the most important fungal pathogens of aquatic invertebrates. They usually infect shellfish in the early stages of life, affecting eggs, larvae, and postlarvae. Particular species are restricted to either marine or freshwater environments. This is due mainly to the inability of freshwater species to withstand the osmotic conditions of the marine environment, and vice versa (20).

Crayfish Plague. The most important fungal infection of aquatic invertebrates is *Aphanomyces astaci*, or krebspest, the cause of crayfish plague. All native species of European crayfish are highly susceptible (with usually 100% mortality) to this fungus, and its spread throughout Europe, to which it was introduced from North America with shipments of imported crayfish, has lead to the extinction of many native crayfish populations. Species of North American crayfish, such as *Pacifastacus lenisculus*, are much more resistant and can carry *A. astaci* and transfer it to other crayfish species (21).

A. astaci is transmitted by motile spores that settle on and infect the cuticle, eventually penetrating into deeper tissues. Crayfish may die within two weeks of becoming infected. Infected crayfish are hyperactive at first, but then become lethargic. The tail is extended instead of tucked under the abdomen, and the crayfish may lie on its back and move its legs continuously. The claws hang down when the crayfish is taken out of the water. Normally noctural, the crayfish will begin to move around during the day. Its abdominal muscles may turn white. Infections may be difficult to detect grossly, but hyphae are visible in infected tissue through microscopic examination.

Resistant species of crayfish can also become infected and may become sick, but the infection is usually successfully limited by their host defense, mainly the production of toxic chemicals that melanize the hyphae, turning them black and killing them. It is advisable to avoid moving crayfish from infected areas. Traps and other gear used in areas that have *A. astaci* should be disinfected to kill the fungus. Unlike other Oomycetes, *A. astaci* supposedly requires live crayfish to survive and thus should eventually die out in waters that are free of crayfish (21).

Other Oomycete Infections. Other important water mold pathogens of aquaculture species include *Lagenidium*, which is responsible for larval mycosis of shrimp and crabs, and *Sirolpidium*, which causes larval mycosis of oysters, clams, and shrimp. Many Oomycetes have a broad range of hosts. Examples include *Lagenidium callinectes*, which infects many species of crabs and shrimp, and

Figure 4. *Fusarium* infection causing "black gill" (arrow) in penaeid shrimp. (Photo courtesy of C. Bland.)

Haliphthoros milfordensis, which infects crabs, shrimp, and lobsters (6,22).

Hyphomycetes

With the hyphomycete fungi, there is no true distinction between marine and freshwater pathogens. Unlike the Oomycetes, the Hyphomycetes are not well adapted to survive in water. However, they can still cause serious disease in aquatic animals. Unlike Oomycetes, which infect mostly larval invertebrates, Hyphomycetes usually affect subadult and adult crustaceans. They are weakly pathogenic. Thus, they often cause disease in concert with some stressful event. The most clinically important hyphomycete is *Fusarium*, which is responsible for black gill disease of shrimp (see Fig. 4) and black (burn) spot disease of shrimp, lobsters, and crayfish (22,23). *Fusarium* produces intensely dark (melanized) areas, especially on the appendages and gills. Diagnosis is via wet mount or histology showing characteristic fungal hyphae and spores.

FUNGAL DISEASES OF FROGS AND ALLIGATORS

Water molds have occasionally caused disease outbreaks in immature frogs (i.e., tadpoles). Fungal infections in reptiles such as alligators or turtles are rare, although *Aphanomyces* water mold has recently caused epidemics in softshelled turtles (*Pelodiscus sinensis*) (24).

BIBLIOGRAPHY

1. W.W. Scott and A.H. O'Bier Jr., *Prog. Fish-Cult.* **24**, 3–15 (1962).
2. W.N. Tiffney, *Mycologia* **31**, 310–321 (1939).
3. K. Hatai, in B. Austin and D.A. Austin, eds., *Methods for the Microbiological Examination of Fish and Shellfish*, John Wiley and Sons, New York, 1989, pp. 240–272.
4. A.J. Chako, in D.K. Arora, L. Ajello, and K.G. Mukerji, eds., *Handbook of Applied Mycology*, Marcel Dekker, New York, Vol. 2, Marcel Dekker, New York, 1991, pp. 735–755.
5. E.J. Noga, *Annual Review of Fish Diseases* **3**, 291–304 (1993).

6. E.J. Noga, in F.O. Perkins and T.C. Cheng, eds. *Pathology in Marine Science*, Academic Press, New York, 1990, pp. 143–160.
7. E.J. Noga, in J.A. Couch and J.W. Fournie, eds., *Pathobiology of Marine and Estuarine Organisms*, CRC Press, Boca Raton, FL, 1993, pp. 85–109.
8. M.W. Dick, in L. Margulis, J.O. Corliss, M. Melkonias, D.J. Chapman, and H.I. McKhann, eds., *Handbook of Protoctista*, Jones and Bartlett Publishers, Boston, 1990, pp. 661–685.
9. J.E. Bly, S.M.-A. Quiniou, L.A. Lawson, and L.W. Clem, *Dis. Aquatic Org.* **24**, 25–33 (1996).
10. R.A. Schnick, D.J. Alderman, R. Armstrong, R. Le Gouvello, S. Ishihara, E.C. Lacierda, S. Percival, and M. Roth, *Bull. Eur. Assn. Fish Pathol.* **17**, 251–260 (1997).
11. L.L. Marking, J.J. Rach, and T.M. Schreire, *Prog. Fish-Cult.* **56**, 225–231 (1994).
12. E.J. Noga, *Fish Disease: Diagnosis and Treatment*, Mosby-Year Book, St. Louis, MO, 1996.
13. R.J. Roberts, B. Campbell, and I. MacRae, eds., *ODA Regional Seminar on Epizootic Ulcerative Syndrome*, Overseas Development Administration and Aquatic Animal Health Research Institute, Bangkok, 1994.
14. R.J. Roberts, *Fish Farming International* August: 6 (1989).
15. R.J. Roberts, *Fish Pathology*, Balliere-Tindall, London, 1989.
16. J.H. Lilley and R.J. Roberts, *J. Fish Dis.* **20**, 135–144 (1997).
17. E.J. Noga, L. Khoo, J. Stevens, Z. Fan, and J. Burkholder, *Marine Pollution Bulletin* **32**, 219–224 (1996).
18. R.B. Callinan, J. Sammut, and G.C. Fraser, in R.J. Smith and H.J. Smith, eds., *Proceedings of the Second National Conference on Acid Sufate Soils*, Robert J. Smith and Associates Publishers, Australia, 1996, pp. 146–151.
19. G.A. Neish and G.C. Hughes, *Fungal Diseases of Fishes*, TFH Publications, Neptune, NJ, 1980.
20. J.L. Harrison and E.B.G. Jones, *Trans. Brit. Mycol. Soc.* **65**, 389–394 (1975).
21. H. Ackefors and O.V. Lindqvist, in J.V. Huner, ed., *Freshwater Crayfish Aquaculture in North America, Europe and Australia*, Food Products Press, New York, 1994, pp. 157–216.
22. D.V. Lightner, ed., *A Handbook of Shrimp Pathology and Diagnostic Procedures for Disease of Cultured Penaeid Shrimp*, World Aquaculture Society, Baton Rouge, LA, 1996.
23. C.J. Sindermann and D.V. Lightner, eds., *Disease Diagnosis and Control in North American Marine Aquaculture*, 2d ed., Elsevier, New York, 1988.
24. S. Sinmuk, H. Suda, and K. Hatai, *Mycoscience* **37**, 249–254 (1996).

See also DISEASE TREATMENTS.

G

GAS BUBBLE DISEASE

John Colt
Northwest Fisheries Science Center
Seattle, Washington

OUTLINE

Supersaturation of atmospheric gases in water is common due to a number of natural and man-induced processes. The exposure of aquatic animals to gas supersaturation can result in the formation of gas bubbles on body surfaces or within the vascular system and tissues. This condition is called gas-bubble disease or trauma. Gas bubble-disease (GBD) can reduce growth, increase mortality, and result in buoyancy problems in small larval fish and amphibians. The formation of gas bubbles and resultant inflammatory reaction provides an ideal environment for opportunistic secondary bacterial invaders. The variability in sensitivity to GBD within a population is significant and is not fully understood.

WHAT IS GAS SUPERSATURATION?

The potential formation of gas bubbles depends on the difference between the measured barometric pressure and the total gas pressure (TGP) in the water (1), given as:

$$\text{Total gas pressure} = \text{PP}^1_{O_2} + \text{PP}^1_{N_2+Ar} + \text{PP}^1_{CO_2} + \text{PP}_{H_2O} \quad (1)$$

where

$\text{PP}^1_{O_2}$ = partial pressure of oxygen gas in the water (mm Hg),
$\text{PP}^1_{N_2+Ar}$ = partial pressure of nitrogen gas in the water (mm Hg),
$\text{PP}^1_{CO_2}$ = partial pressure of carbon dioxide in the water (mm Hg), and
PP_{H_2O} = vapor pressure of water (mm Hg).

In gas supersaturation work, nitrogen gas and argon gas are generally treated as a single gas ($N_2 + Ar$ or simply N_2) because the two gases are not individually determined (2).

Three conditions may exist:

Total gas pressure = barometric pressure (equilibrium)
Total gas pressure > barometric pressure (supersaturated), and
Total gas pressure < barometric pressure (undersaturated).

The difference between total gas pressure and barometric pressure (TPG-BP) is called the differential pressure and is indicated by the symbol ΔP, which can be measured directly. ΔP is the best predictor of the danger from gas supersaturation. If $\Delta P < 0$, then gas bubbles cannot form, regardless of the degree of supersaturation of a single gas.

Gas supersaturation may also be reported as a percentage of the local barometric pressure BP:

$$\text{TGP\%} = \left[\frac{\text{BP} + \Delta P}{\text{BP}}\right] 100 \quad (2)$$

The actual risk to an individual animal depends on both ΔP and the animal's position in the water column (3). The ΔP an animal experiences is equal to the difference between the total dissolved gas pressure and the local pressure (barometric + hydrostatic pressures). The uncompensated ΔP is equal to

$$\Delta P_{\text{uncomp}} = \Delta P - \rho g Z \quad (3)$$

where

ΔP = measured ΔP (mm Hg),
ρg = hydrostatic pressure of water (mm Hg/meter of submergence),
Z = depth in water column (m).

The value of ρg depends both on temperature and on salinity. At 20 °C and 0 g/kg salinity, the value of ρg is equal to 74.3 mm Hg/m, and other values can be found in the literature (2).

SOURCES OF GAS SUPERSATURATION

Gas supersaturation can be produced by a variety of physical and biological processes. Eight mechanisms can produce gas supersaturation (4): heating of waters, ice formation, mixing of waters of different temperatures, air entrainment, photosynthesis, pressure changes, physiological process, and bacterial action.

In a specific situation several mechanisms may be involved. Many surface waters, groundwaters, and springs are naturally supersaturated during some part of the year. Gas supersaturation can also be produced within the hatchery or rearing unit. Not all supersaturation problems can be prevented by simple degassing of influent water.

MEASUREMENT OF GAS SUPERSATURATION

The preferred method of gas analysis is the direct-sensing membrane-diffusion method. The instruments for this analysis use a variable length of gas-permeable tubing (e.g., dimethyl silicone rubber) connected to a pressure measuring device (5). Silicone rubber tubing is highly permeable to dissolved gases, including water vapor. At steady state, the gauge pressure inside the tubing is equal to the difference in gas pressure (ΔP) between the total dissolved gas pressure and local barometric pressure. Several types of membrane-diffusion instruments are commercially available. These instruments are field portable, and all data collection is completed in the field.

CLINICAL SIGNS OF GAS-BUBBLE DISEASE IN AQUATIC ANIMALS

General information on the sensitivity of aquatic animals to gas supersaturation is available in published sources (6). The clinical signs of GBD are best described in fish (7–9); less information is available for other aquatic culture animals.

The clinical signs of GBD are very dynamic. The suite of clinical signs observed may depend more on the rapid disappearance of some clinical signs than the exposure levels or exposure time. A hydrostatic pressurization of 5 min to 30.5 m of head resulted in a substantial reduction in clinical signs of GBD in the fins, lateral line, and gills of yearling spring chinook salmon (*Oncorhynchus tshawytscha*). Clinical signs of GBD were lost most rapidly in gills, followed by the lateral line. The rate of bubble loss was much less for the primarily extravascular bubbles found in the fins (10). In work conducted on salmon smolts (*Oncorhynchus* sp.), it was found that gill filaments must be examined within 2 min after the gill arch is excised (11), and care must be exercised to differentiate bubbles from similar-appearing lipid bodies. Examination of moribund or freshly dead fishes for subcutaneous bubbles on fins and body surfaces is easily done in the field, providing representative specimens can be collected (8). Histopathologic examination of fish from suspected supersaturation kills has often been found to be inconclusive and has offered no specific diagnosis. The major clinical signs of GBD that can cause death or lead to high levels of stress in aquatic animals are discussed next.

Subcutaneous Emphysema on Body Surfaces, Including the Lining of the Mouth

Subcutaneous emphysema (gas bubbles within skin tissues) is commonly found on fins and tail, inside the mouth and operculum, and on the body surface. The formation of gas bubbles and resultant inflammatory reaction provides an ideal environment for opportunistic secondary bacterial invaders. Emphysema of tissue in the mouth may also contribute to the blockage of respiratory water flow and death by asphyxiation (12). Subcutaneous emphysema on the dorsal fin of a chinook salmon is presented in Figure 1a.

Bubble Formation in the Vascular System

Formation of bubbles in the vascular system may result in petechial (pinpoint) hemorrhaging, restricted blood flow, necrosis, and death. Long, tubular bubbles may be observed in the gill vessels of fish and are shown in Figure 1b. Bubbles may be found in the blood vessels, heart, kidney, spleen, and liver. Gill bubbles can be observed with a binocular or compound microscope, but bubbles in internal organs are sometimes difficult to see. In animals exposed to high levels of gas supersaturation, incision of internal organs may produce a significant amount of bubbles (8). *Xenopus laevis* exposed to gas supersaturation accumulate a massive amount of gas under the skin (Fig. 1c); when the skin on legs is cut, the incision can bubble (13). Increased mortality due to secondary bacterial infections such as *Aeromonas hydrophila* and *Vibrio alginolyticus* has been observed (14,15).

Bubble Formation in the Eyes of Fish

Exophthalmia or “popeye” results from the accumulation of gas in the eye. While this clinical sign is quite distinctive, it is not as common as subcutaneous emphysema and may be caused by a variety of other diseases. Exophthalmia in itself does not cause blindness and may be reversible if the animal is transferred to nonsupersaturation conditions (8). Exophthalmia is common in captively held cod (*Gadus morhua*) (16) and rock fish (*Sebastes* spp.) (17). It is thought that this condition is caused by malfunctioning of the choroid gland-pseudobranch complex (16) and may be due to a lack of adequate hydrostatic pressure.

Overinflation and Possible Rupture of the Swim Bladder in Fish

The impact of swim bladder overinflation depends strongly on species and size. Fish can be classified as either physostomes (swim bladder is connected to the gut by the pneumatic duct) or physoclists (closed swim bladder).

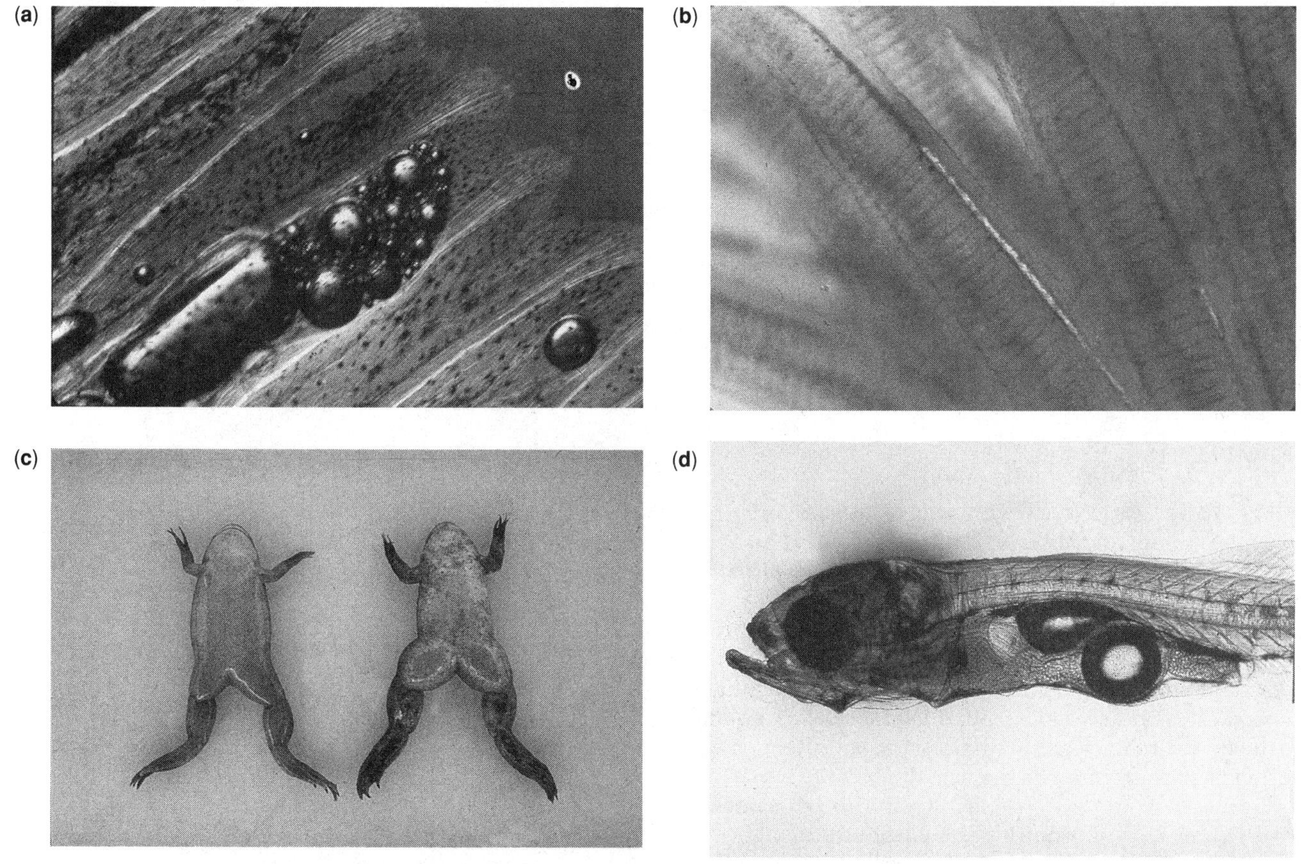

Figure 1. Common clinical signs of gas-bubble disease in aquatic animals. (**a**) Subcutaneous emphysema on the dorsal fin of chinook salmon *O. tshawytscha* (11); (**b**) bubbles in gills of chinook salmon, *O. tshawytscha*, (11); (**c**) accumulation of gas in *X. laevis* (29); and (**d**) overinflation of the swim bladder and intestinal bubble in larval striped bass, *Morone saxatilis* (21).

Some fish are physostomes as larvae and lose their pneumatic duct as they grow larger. The ability to actively secrete gas into the swim bladder is better developed in the physoclists (18).

The overinflation of swim bladders of small marine fish such as cod (19), sea bass (*Lates calcarifer*) (20), striped bass (*M. saxatilis*) (21), *Siganus lineatus* (22), and mullet (*Mugil cephalus*) (23) is common in culture. Once the swim bladder is inflated, very little gas supersaturation is needed to overinflate the swim bladder. The ability of small larval fish to regulate the volume of gas in the swim bladder may be limited, and formation of a single bubble may float them to the surface. Because of the significant forces exerted against the swim bladder and the resulting physical damage, survival of floating larval fish may be limited even if the swim bladder can be deflated. An example of an overinflated swim bladder and intestinal bubbles in larval striped bass is presented in Figure 1d.

Swim bladder overinflation is also a problem in salmonids. In rainbow trout (*Oncorhynchus mykiss*), the swim bladder can become overinflated as a result of dissolved gases diffusing from the water to the bladder by way of the gills and vascular system. When this happens, fish may become severely overbuoyant. In small rainbow trout ($<$10 g), the pressure required to vent air out the swim bladder exceeds the swim bladder rupture pressure (24). Larger fish are able to expel gas out the pneumatic duct and regulate swim bladder size.

Formation of Bubbles in the Gastrointestinal Tract or in Other Organs

The formation of bubbles in the gut of fish has been observed in herring (*Clupea harengus*) (25), plaice (*Pleuronectes platessa*) (19), *Acanthopagrus cuvieri* (26), channel catfish (*Ictalurus punctatus*) (27), white sturgeon (28), and striped bass (21). The accumulation of gas in the gut of bullfrog tadpoles causes them to float with their left sides elevated or on their backs (14). The accumulation of gas in the body and legs of adult bullfrogs and African clawed frogs (*X. laevis*) has resulted in floating animals (13,29). Flotation problems have been observed in small surf clams (*Spisula solidissima*) (30) and *Daphnia magna* (31).

Formation of Bubbles in the Lateral Line of Fish

In salmonids, one of the more common clinical signs of exposure to gas supersaturation is the formation of bubbles in the scale pockets of the lateral line. The formation of bubbles in the lateral line results in reduced or eliminated ability of the fish to detect near-field water displacements and may decrease the fish's ability to avoid predation (32).

Formation of Bubbles Attached to Body Surfaces or in the Mouth

In aquatic insects, exposure to high levels of gas supersaturation may result in the growth of gas bubbles on external as well as internal surfaces. External bubbles affect the buoyancy of the insect, but can be dislodged if the animal rises to the water surface (31).

In steelhead fry (*O. mykiss*), large bubbles may form in the buccal cavity. While these bubbles are not physically attached to the fish, their presence may result in reduced water flow through the gill, development of opercular deformities, and may force the fish to swim in a heads-up position (33).

LETHAL LEVELS

Acute GBD is associated with bubble formation in the vascular system and tissues. It produces a large array of clinical signs, leads to damage to tissue and vascular occlusions, and results in high rates of mortality (up to 100%) with even short exposures (2 h to 10 days). Lethal concentration information is useful for assessing the potential impact of short-term exposures to high gas supersaturation levels. The lethal tolerance of an animal will decrease as the exposure time increases, although the variability among animals is surprisingly large.

Eggs and newly hatched fry appear to be resistant to high ΔP's. In a study of steelhead trout, ΔP's up to 200 mm Hg had no effect on hatching (34). The lack of sensitivity of eggs is probably due to the fact that pressure within the eggs is higher than 1 atmosphere. After 1,000 h of incubation, pressure inside the eggs of different salmonids species ranged from 51 to 76 mm Hg above barometric pressure (35). This pressure reduces the ΔP_{uncomp} within the egg (Eq. 3). In the incubation of semibuoyant eggs, such as those of striped bass, attachment of bubbles to the external surface of the egg may float the egg out of incubation systems and down the drain.

Newly hatched steelhead trout are resistant to ΔP until about day 16 posthatch (34). According to the study, at day 16 bubbles formed in the mouth, gill cavity, and yolk sac. Accumulation of bubbles in the larvae prevented normal swimming and feeding and eventually trapped the fish at the surface. Over a 90-day exposure, starting with fertilization, a ΔP of approximately 130 mm Hg resulted in 50% mortality.

The 4-day LC_{50} value of juvenile and adult fish ranges from 53 to 230 mm Hg. The 30- to 35-day LC_{50} of fish ranges from 106 to 117 mm Hg. The lethal tolerance of salmonids depends strongly on size or age (Fig. 2). The tolerance of Atlantic salmon (*Salmo salar*) and lake trout (*Salvelinus namaycush*) to gas supersaturation is high for small fish (<50 mm) and decreases for larger fish. For lake trout, the tolerance to gas supersaturation also decreases for larger fish (>150 mm Hg). The tolerance of marine species is highly variable; no quantitative data are available for larval stages of the small marine and estuarine species known to be very sensitive to gas supersaturation. Crustaceans and insects have a wide variation in sensitivity to gas supersaturation, at least over a standard 4-day exposure. Molluscs have a high tolerance to gas supersaturation, some of which may be due to the ability to "clam up" for an extended period of time. Amphibians appear to have a sensitivity to gas supersaturation comparable to that of freshwater fish.

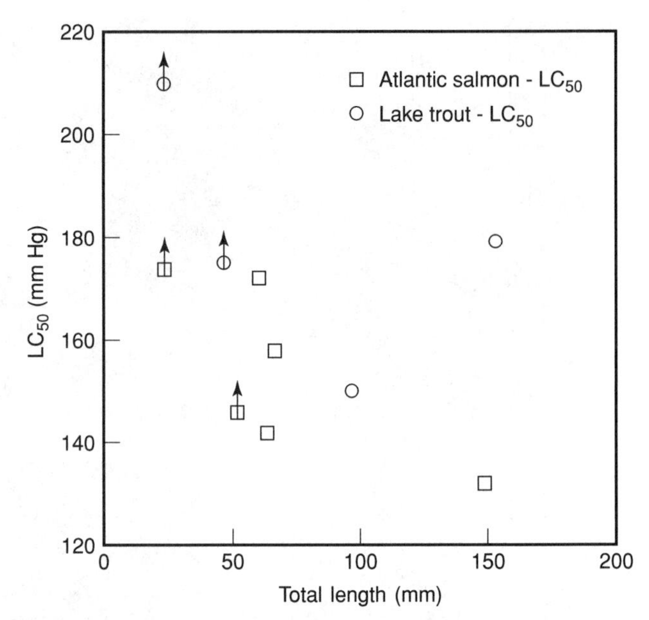

Figure 2. Tolerance of Atlantic salmon and lake trout to gas supersaturation as a function of total length (45). The vertical arrow symbol represents exposures levels where mortality was insufficient to allow computation of the LC_{50} value. The 4-d LC_{50} levels is the level of gas supersaturation that will kill 50% of the animals in 4 days.

CHRONIC GBD

When aquatic animals are continuously exposed to ΔP's in the range of 20 to 100 mm Hg, a chronic type of GBD develops and is associated with extravascular symptoms such as bubble formation in the gut and buccal cavity, hyperinflation or rupture of the swim bladder, and low-level mortality in juvenile animals over extended periods of time.

In a study of striped bass larvae, gas supersaturation resulted in hyperinflation of the swim bladder and formation of gas in the gut (21). Clinical signs of GBD were observed at ΔP's as low as 22 mm Hg, and mortality was increased at 42 mm Hg. However, mortality may have been due to rupture of the swim bladder, because none of the typical clinical signs of acute GBD were observed. The period of maximum sensitivity appeared to occur near the first filling of the swim bladder and the beginning of feeding. The accumulation of gas in these small larvae commonly floated them to the surface. Flotation problems, overinflation of the swim bladder, and the accumulation of gas in the gut are the most common clinical signs of GBD observed in small marine fish larvae.

Growth is not a good indicator of sublethal impacts of gas supersaturation in some fish. In studies of channel catfish and rainbow trout, gas supersaturation levels resulting in significant mortality from gas supersaturation had no impact on the growth of survival animals (33,36). In lake trout, corneal swelling was observed at ΔP's greater than 17 mm Hg, although the total incidence of ocular abnormalities did not increase for ΔP's up to 43 mm Hg (37).

ANCILLARY OR MODIFYING FACTORS

Physical and biological factors can significantly modify the effects of a given ΔP. Some of the most important factors in aquatic systems are discussed next.

Depth

The ΔP_{uncomp} decreases approximately 74.3 mm Hg/m (Eq. 3), and a few meters of submergence can protect aquatic animals from high levels of gas supersaturation as long as they remain at this depth. The impact of hydrostatic pressure on an acutely lethal gas supersaturation level of 250 mm Hg is presented in the following table:

Depth (m)	ΔP_{uncomp} (mm Hg)
0	250
1	176
2	101
3	27
4	−47

As long as an aquatic animal remains below the compensation depth of 3.36 m, there is no tendency for gas bubbles to form. However, if this animal is forced to the surface, GBD can develop rapidly.

It does not appear that fish can directly detect gas supersaturation, but they will respond to overinflation of the swim bladder. Work with coho salmon, (*Oncorhynchus kisutch*), shows that fish increased their mean depth to alleviate the symptoms of GBD up to a ΔP of 84 mm Hg (38). Above this ΔP, the fish no longer remained below a depth that compensated for the ΔP, and increased signs of stress were evident. In the laboratory environment, Shrimpton (39) also found that given the opportunity to use water depth to compensate for overbuoyancy, small rainbow trout would spend a significant amount of time at a water depth where they were neutrally buoyant. Furthermore, as ΔP increased, fish would move deeper in the water column to overcome the effects of swim bladder overinflation. Increasing the depth of culture systems can offer passive depth compensation by allowing the fish access to greater hydrostatic pressure (40).

(Nitrogen + Argon): Oxygen Pressure Ratio

Increasing the partial pressure ratio of (nitrogen + argon): oxygen, increases the lethal effects of a given ΔP (41). Regardless of the value of this ratio, the ΔP_{uncomp} inside the animal must be greater than 0 for the formation of gas bubbles.

Feeding and Diet

Fasting appears to increase the impact of a given ΔP (14,21).

Intermittent Exposure

Juvenile salmon and trout can tolerate an acutely lethal ΔP (170 mm Hg) for 16 hours/day if they are returned to saturated water for the other 8 hours (42).

WATER QUALITY CRITERIA FOR GAS SUPERSATURATION

The water criterion for gas supersaturation established by the U.S. Environmental Protection Agency (43) is 110% of barometric pressure or a $\Delta P = 76$ mm Hg. This criterion is inadequate to protect the more sensitive species of nonsalmonid fish or salmonids exposed to chronic gas supersaturation, particularly when water depth is limited.

The following chronic water-quality criteria are proposed for gas supersaturation:

Characteristics	ΔP (mm Hg)
Very sensitive animals and experimental trials	<10
Sensitive animals	<20
Average animals under production conditions	<40

An example of a very sensitive animal is the lake trout if ocular lesions are considered. A sensitive animal would include small marine fish larvae, especially when reared in shallow containers (<0.5 m). Average animals would include most other fish, crustaceans, and mollusks reared under production conditions. Criteria in the range of 10–20 mm Hg are consistent with GBD threshold levels based on bubble formation and growth (44). These criteria should be very protective on a chronic basis, and higher values may be tolerated for shorter periods of time. A criterion of 40 mm Hg may not offer absolute protection for all species and life stages, but is a reasonable goal when economics are a consideration.

PREVENTION OF GAS BUBBLE DISEASE

The prevention of GBD within a hatchery will depend on degassing of influent water and some process waters, the design and operation of aquatic systems to prevent production of gas supersaturation within the hatchery, and possibly changes in facility design and management practices for sensitive animals. The monitoring of ΔP's of influent waters and at key points in the hatchery may help to identify problems and allow correction before major mortality occurs.

BIBLIOGRAPHY

1. J.E. Colt, *Wat. Res.* **17**, 841–849 (1983).
2. J.E. Colt, *Computation of Dissolved Gas Concentrations in Water as Functions of Temperature, Salinity, and Pressure*, Spec. Pub. No. 14, Am. Fish. Soc.

3. M.D. Knittel, G.A. Chapman, and R.R. Garton, *Trans. Am. Fish. Soc.* **109**, 755–759 (1980).
4. J. Colt, *Aquacult. Eng.* **5**, 49–85 (1986).
5. A.E. Greenberg, L.S. Clesceri, and A.D. Eaton, eds., *Standard Methods for the Examination of Water and Wastewater*, 18th ed., American Public Health Association, Washington DC, 1992, pp. 2–75 to 2–80.
6. D.E. Weitkamp and M. Katz, *Trans. Am. Fish. Soc.* **109**, 659–702 (1980).
7. J.P. Machado, D.L. Garling, Jr., N.R. Kevern, A.L. Trapp, and T.G. Bell, *Can. J. Fish. Aquat. Sci.* **44**, 1985–1994 (1987).
8. R.E. Wolke, G.R. Bouck, and R.K. Stroud, in S.B. Saila, ed., *Fisheries and Energy Production: A Symposium*. Lexington Books, Lexington, MA, 1975, pp. 239–265.
9. C.E. Smith, *Prog. Fish-Cult.* **50**, 98–103 (1998).
10. R. Elston, J. Colt, S. Abernethy, and W. Maslen, *J. Aquat. An. Health* **9**, 317–321 (1997).
11. R. Elston, J. Colt, P. Frelier, M. Mayberry, and W. Maslen, *J. Aquat. An. Health* **9**, 258–264 (1997).
12. R.G. White, G. Phillips, G. Liknes, J. Brammer, W. Conner, L. Fidler, T. Williams, and W. Dwyer, *Effects of Supersaturation of Dissolved Gases on the Fishery of the Bighorn River Downstream of the Yellowtail Afterbay Dam*, Final Report to the U.S. Bureau of Reclamation, Montana Cooperative Fishery Research Unit, Montana State University, Bozeman, MT, 1991.
13. J. Colt, K. Orwicz, and D.L. Brooks, *J. World. Aquacult. Soc.* **18**, 229–236 (1987).
14. J. Colt, K. Orwicz, and D.L. Brooks, *Aquaculture* **38**, 127–136 (1984).
15. R. Elston, *J. Fish Dis.* **6**, 101–110 (1983).
16. P.V. Dehadrai, *J. Fish. Res. Bd. Can.* **23**, 909–914 (1966).
17. R.W. Engelman, L.L. Collier, and J.B. Marliave, *J. Fish Dis.* **7**, 467–476 (1984).
18. R. Fänge, *Rev. Physiol. Biochem. Pharmacol.* **97**, 111–158 (1983).
19. E. Henly, *Rapports et Procès-Verbaux des Réunions* **131**, 24–27 (1952).
20. T. Bagarinao and P. Kungvankij, *Aquaculture* **51**, 181–188 (1986).
21. J. Cornacchia and J.E. Colt, *J. Fish Dis.* **7**, 15–27 (1984).
22. P.G. Bryan and B.B. Madraisau, *Aquaculture* **10**, 243–252 (1977).
23. S. Kraul, *Aquaculture* **30**, 273–284 (1983).
24. J.M. Shrimption, D.J. Randall, and L.E. Fidler, *Can. J. Zool.* **68**, 962–968 (1990).
25. H.M. Bishai, Zeitschrift für Wissenschaftliche Zoologie **163**, 37–64 (1960).
26. N. Hussain, S. Akatsu, and C. El-Zahr, *Aquaculture* **22**, 125–136 (1981).
27. P.R. Bowser, R. Toal, R. Robinette, and M.W. Brunson, *Prog. Fish-Cult.*, 208–209 (1983).
28. J.L. Callman and J.M. Macy, *Arch. Microbiol.* **140**, 57–65 (1984).
29. J. Colt, K. Orwicz, and D. Brooks, *J. Herpetology* **18**, 131–137 (1984).
30. R. Goldberg, *Aquaculture* **14**, 281–287 (1978).
31. A.V. Nebeker, D.G. Stevens, and J.R. Brett, in D.H. Fickeisen and M.J. Schneider, eds., *Gas Bubble Disease*, USERDA CONF–741033, 1975, pp. 51–65.
32. D.D. Weber and M.H. Schiewe, *J. Fish Biol.* **9**, 217–233 (1976).
33. J.O.T. Jensen, *Aquaculture* **68**, 131–139 (1988).
34. A.V. Nebeker, J.D. Andros, J.K. McCrady, and D.G. Stevens, *J. Fish. Res. Bd. Can.* **35**, 261–264 (1978).
35. D.F. Alderdice and J.O.T. Jensen, *Aquaculture* **49**, 85–88 (1985).
36. J. Colt, K. Orwicz, and D. Brooks, *Aquaculture* **50**, 153–160 (1985).
37. W.F. Krise and R.A. Smith, *Prog. Fish-Cult.* **55**, 177–179 (1993).
38. J.M. Shrimpton, Bachelor of Science Thesis, University of Victoria, BC, 1985.
39. J.M. Shrimpton, D.J. Randall, and L.E. Fidler, *Can. J. Zool.* **68**, 969–973 (1990).
40. M. Lund and T.G. Heggeberget, *J. Fish. Biol.* **26**, 193–200 (1985).
41. A.V. Nebeker, A.K. Hauck, and F.D. Baker, *Wat. Res.* **13**, 299–303 (1979).
42. T.K. Meekin and B.K. Turner, *Wash. Dept. of Fish. Tech. Rep.* **12**, 78–126 (1974).
43. U.S. Environmental Protection Agency, *Quality Criteria for Water*, Washinton, DC, 1976.
44. L.E. Fidler, *Gas Bubble in Fish*, Ph.D. dissertation, Department of Zoology, University of British Columbia, 1988.
45. W. Krise and R.L. Herman, *J. Aquat. An. Health* **3**, 2480–253 (1991).

See also AERATION SYSTEMS; DISEASE TREATMENTS.

GILTHEAD SEA BREAM CULTURE

GEORGE WM. KISSIL
AMOS TANDLER
ABIGAIL ELIZUR
ANGELO COLORNI
National Center for Mariculture
Elat, Israel

YONATHAN ZOHAR
University of Maryland Biotechnology Institute
Baltimore, Maryland

OUTLINE

Life History
Reproduction and Larval Culture
Commercial Culture
Nutritional Requirements
Disease Problems
  Viral Diseases
  Bacterial Diseases
  Protistan Parasities
  Metazoan Parasites
Bibliography

Gilthead sea bream, a member of the family Sparida, is found in the Mediterranean and Black Seas and extends into the Atlantic Ocean from the British Isles south to Senegal. Sparidae is represented by approximately

26 species in this region, including two migrants from the Red Sea that have become established in the eastern Mediterranean Sea since the opening of the Suez Canal (1,2). The reported fisheries catch for this family in the Mediterranean and Black Seas during 1996 reached close to 70,000 metric tons (77,000 short tons); the gilthead sea bream contributed almost 7% (3). Farming of sea bream is carried out in seawater ponds and lagoons; the bulk of production occurs in sea cages of various types. Most countries around the Mediterranean culture sea bream; Greece, Turkey, and Spain are the major producers in the region, accounting for over 70% of production. The culturing of sea bream has made impressive strides in much less than two decades, going from an estimated 110 metric tons (121 short tons) of fish in 1985 (4) to 41,900 metric tons (46,200 short tons) in 1998 (5). This success is the result of strong research and development programs in many of the region's countries, fueled by a persistently strong market demand for sea bream.

LIFE HISTORY

The gilthead frequents coastal and lagoonal waters during most of its life, moving into deeper waters during the spawning season in late fall and winter (6). The adults return to the shallower areas after spawning and are later joined by the young fry, which remain in the area until they reach sexual maturity. Sea bream generally feed on shrimp and shellfish, which they find on or in the sandy bottom sediments. Fish can be observed digging their heads into the sand and removing shellfish and crustaceans, which they easily crush with their strong molar-like teeth.

The gilthead sea bream has a typically oval shaped body, which is deep and compressed laterally. Its body color is mainly silver grey, with a large black blotch starting behind the gill cover and extending forward over the upper part of the gill cover. A yellowish area appears below this black blotch, and bright yellow lines both between the eyes and on the stomach, starting behind the pelvic fins. These bright yellow colors are the origin of its Latin species name *aurata*.

REPRODUCTION AND LARVAL CULTURE

Early attempts at sea bream culture started in Italy and France (4), in sea water lagoons and tidal ponds under extensive conditions, with wild-caught fry, juveniles, and growout stage fish of various sizes. The successful reproduction of sea bream in captivity resulted in barely enough production of stockable fry for the initiation of intensification of sea bream culture. The following decades saw the production of stockable fry increase significantly, to 161 million by 1997 (7). This increase was paralleled by the development of feeds specifically formulated for sea bream and by the development of the intensive culture and management techniques that have made sea bream culture a major form of mariculture in the Mediterranean region today. In addition, the gilthead sea bream has established itself as an important model species in the study of fish reproductive physiology and molecular biology (8–16).

The gilthead sea bream, a protandrous hermaphrodite (17), in captivity develops as a functional male by the end of its first year. After the spawning season, the males begin to develop ovaries; this process continues until early fall (September in the eastern Mediterranean region); at this point, a fish either develops as a functional female or absorbs the ovarian tissue and redevelops male gonadal tissue (17). The proportion of males that undergo sex reversal is dependent primarily on the male-to-female ratio in the population (18). Males do not lose their potential for sex-reversal, and can do so at any age if placed in the right social environment in terms of male-to-female ratio.

Sea bream reproduction is controlled by light (or photoperiod), thermal, and social cues. Fish respond primarily to a shortening of the day's length and secondarily to a reduction in water temperature. Spawning in captivity in the eastern Mediterranean region starts at around the shortest day of the year, provided that fish are in a balanced sex ratio—optimally 1 : 2 (male : female) in groups over 9 fish, 1 : 1 in smaller groups. Single-pair matings are difficult to obtain on a reliable basis (19) and should be avoided. Sea bream will spawn in a wide range of tank sizes, from 1 m^3 (264 gal) upward, but their density should not exceed 5 kg/m^3 (11 lb/264 gal). A minimum of 2–3 females, with a similar number of males, provides successful spawnings and a high rate of egg fertilization.

The act of spawning usually takes place in the late afternoon, although some groups can adopt a morning pattern. Prior to spawning, the fish undergo a change in both body coloration and behavior; then, a typical group courting swim occurs, and finally spawning. Sea bream are batch spawners; females undergo daily cycles of final oocyte maturation, ovulation, and spawning during the reproductive season (15,17). The net result is that each female can produce some 1–3 million eggs during the spawning season (20,21).

The strong influence of photoperiod (day : night ratio) on sea bream reproduction makes the fish susceptible to a manipulation favoring the goal of extending its reproductive season. In fact, year-round spawning has been accomplished under artificial lighting, that mimics changes in day length, coupled with the shifting of the shortest day of the year to March, June, or September. This photoperiod shift, coupled with corresponding control of water temperature, can provide three groups of broodstock that will spawn outside of the natural reproductive season and yield year-round egg production.

Parent stock derived from fish bred in captivity for a number of generations usually spawn naturally and do not require spawning induction. This situation has been true for the past four to five years; previously, however, hormonal therapy for spawning induction was essential in order to obtain reliable spawning. The most effective treatment for the sea bream is the use of controlled release gonadotropin releasing hormone (GnRH) devices, administered either as implants or as microspheres (22–24). If it is important that all the females in the broodstock spawn (such as in cases of genetic selection), then it is recommended that the females be treated using one of the methods mentioned.

Extensive research into the influence of environmental and biotic factors on the rearing of gilthead sea bream larvae have resulted in the successful mass rearing that is carried out in many commercial hatcheries throughout the Mediterranean region. The techniques used vary among the different protocols, but most are based on R&D carried out over the past two decades. The following research findings helped in the development of one of these protocols, which will be described later.

Intermediate photoperiods of 15L : 8D (15 hours light and 8 dark) provide better survival than either more or less daylight; however, best larval growth seems to occur under continuous light (25). An intermediate light intensity, 1000–1500 lux at the water's surface, favors better growth and survival than lower (100–250 lux) or higher intensities (10,000 lux) (26,27).

A reduced salinity (25 ppt) also promotes the growth and survival of sea bream larvae, causing up to a 13% improvement in growth and a three-fold increase in survival, to 32 days, as compared with that of larvae reared in 40-ppt seawater (28,29). This inverse relationship between survival and salinity suggests that a portion of the larval population may be predisposed to having insufficient reserves of energy, which would result in reduced survival, and, further, that any environmental factor associated with energy sparing may therefore affect larval survival. In addition, reduced salinity (25 ppt) is associated with an almost 50% improvement in the development of a functional swimbladder (28).

Sea bream larvae demonstrate a size preference in rotifer consumption that is associated with their size. Newly pigmented larvae prefer 180 μm (7.1×10^{-3} in.) rotifers, even avoiding larger, 340 μm (1.34×10^{-2} in.) ones, while six-day-old larvae demonstrate a reverse preference (30). The enrichment of rotifers and Artemia is designed to modify the feed organisms' essential fatty acid composition to better suit the larval sea bream requirements for maximum growth and survival. Sea bream up to 22 days old require rotifers enriched to 4–8 mg *n*-3 highly unsaturated fatty acids (HUFA)/g dry weight ($1.4–2.8 \times 10^{-4}$ oz/3.5×10^{-2} oz dry weight); older fish require 29 mg *n*-3(HUFA)/g dry weight (1×10^{-3} oz/3.5×10^{-2} oz dry weight) of Artemia (31–33).

Management factors in larval rearing—for example, the type of tanks used, larval stocking densities, the water temperature regime, and the species and concentration of microalgae used—all play important roles in the success of a rearing regime. Under conditions of high salinity (40–41 ppt), cylindroconical tanks 400–1700 L (104–442 gal) in volume have provided consistent larval survivals of 25% up to day 32 after hatching. This survival has also been related to rate of water exchange; two to four exchanges per day are commonly used for these tanks (34).

The following rearing protocol, based on the research findings discussed above, was devised and is currently in commercial use. Fertilized eggs are stocked at 100/L (90/qt) in the cylindro-conical tanks, and hatching success is estimated with aliquot measurements. The initial water temperature is set at $19 \pm 0.5\,^{\circ}$C ($66.2 \pm 0.9\,^{\circ}$F); the temperature is raised gradually, to $24.5 \pm 0.5\,^{\circ}$C ($76.1 \pm 0.9\,^{\circ}$F) by the last nine days of the 32 day rearing period. The tanks are supplied continuously with filtered (to 10 μm = 3.9×10^{-4} in.) water, and temperature and salinity are controlled. Seawater is exchanged at a rate of two to four times a day, depending upon larval age. Freshly enriched live food (rotifers and Artemia nauplii) as well as single cell algae (*Nannochloropsis* sp.) are supplied continuously to the rearing tanks through a controlled delivery system designed to maintain rotifers at 10/mL (296/oz), algae at 5×10^5 cells/mL (148×10^5/oz), and nauplii at 1/mL (29.6/oz) during the 15 hours of light. Rearing tanks require only one cleaning during the 32 days to produce 15–35 larvae/L (13.5–31.5 larvae/qt) at the end of the cycle, a 20–40% survival rate.

Although current protocols for the nutrition of larval sea bream are based on live food organisms, there has been emphasis on developing formulated microdiets that can replace live food. The initial work has indicated that microdiets are eaten and assimilated much less than live food (35); however, the addition of exogeneous digestive enzymes promotes a 30% improvement in assimilation (36). The combination of a microdiet with live food promotes the best growth, which may result from the attraction of the live food (35). The addition of attractants (glycine, alanine, arginine or betaine) to the microdiet raises ingestion to the same levels as live feed (37).

Recent work has shown that the phospholipid content of soybean lecithin, in particular phosphatidyl choline (PC), acts as a feeding attractant (38) in sea bream larvae and increases the assimilation rate of dietary lipids into body tissues by almost 50% (39). PC is thought to increase lipoprotein production, the function of which is to carry absorbed lipids to the rest of the body (39). Arachidonic acid has been suggested as playing an important role in promoting growth and survival in young sea bream larvae, through its role as a precursor of different eicosanoids in the body.

Once larvae metamorphose and reach 32 days of age, the problem of cannibalism becomes important and must be dealt with to minimize mortality (40). Homogeneity in the size of the larvae being cultured significantly reduces mortality by reducing the extreme size differences among the individual larvae that facilitate predation. In the above protocol, larvae are routinely graded mechanically into three size groups: 5, 10, and 32 mg wet weight (2, 4, and 12×10^{-4} oz); this segregation by size helps increase the overall survival of the population.

COMMERCIAL CULTURE

Juvenile sea bream weighing 1–5 g ($3.5–17.6 \times 10^{-2}$ oz), when stocked in culture systems, reached commercial sizes of 250–300 g (9–10.5 oz) over a period of 18–24 months during the early years of intensification of sea bream culture. Domestication, improved feeds and feeding techniques, and a limited amount of selective breeding have reduced the growout time for sea bream to 12–14 months, yet yield a larger fish: 400–500 g (14–17.6 oz) average weight. Fish are cultured to sizes of up to 1.5 kg (3.3 lb), but the majority of sea bream are marketed in the 250–800 g (8.8–28 oz) weight range.

The intensive culture of sea bream today is carried out mainly in sea cages, although land-based cement ponds or raceways are also in use where this type of culture is appropriate. Sea-cage culture is usually less costly to initiate and operate: there is no need to pump seawater, to provide aeration, or to undertake costly construction of ponds or raceways. Sea cages do, however, require protection from heavy seas and storms that can wreak havoc upon, even destroy, an established cage farm. The use of sinkable cages, or of cages specifically designed to withstand adverse sea conditions, has afforded some protection to sea bream cage farms. In many cases, however, the lack of sheltered lagoons or of a protected coastline has made establishment of the more expensive land-based systems necessary. Pollution caused by cage farms and the resulting environmental deterioration are another reason for farming sea bream on land. Effluents from ponds and raceways can readily be treated to reduce their effect on the surrounding environment before the water returns to the sea; sea-cage wastes are more difficult to treat. A comparison of the characteristics of cage- and land-based culture systems is presented in Table 1.

NUTRITIONAL REQUIREMENTS

The nutritional requirements of sea bream (Table 2) have been studied over the past 25 years, and the findings have been used to develop optimal feeds for their culture. Commercial feeds have been available to farmers since the 1970s and those feeds have undergone gradual improvement using research findings to refine feed formulations. Feed improvement has been characterized by better feed use by fish (lower feed conversion ratio), faster growth, healthier fish, and lower feed costs to the farmer. The present trend is the development of high-energy extruded feeds that are used more efficiently by the fish, for better growth and less environmental impact. Such feeds encourage the use of lipids as an alternative source of energy for the fish and, in so doing, spare some of the dietary protein for the preferred use: building and maintaining body tissues. Lipid levels approaching 20% are currently being used in commercial diets, and even higher levels are being investigated. The major limitation on the maximum level of lipid that can be incorporated in such diets is its effect on body composition: excessive levels of fat in the fish reduce their consumer appeal and so, ultimately, their price in the marketplace.

DISEASE PROBLEMS

The rapid development of sea bream farming has not been paralleled by adequate progress in the veterinary aspects of its culture. As a result, health management remains one of the major concerns of the aquaculturist, because diseases can cause major losses to commercial crops. The following are the diseases most commonly identified in gilthead sea bream, to date.

Viral Diseases

Lymphocystis. This highly contagious infection is caused by a cytoplasmic-DNA iridovirus. The disease

Table 1. Characteristics of Sea Bream Culture Systems[a]

Type of System	Ponds		Sea Cages
Approach	Semi-intensive	Intensive	(Standard)
Size (m^3)	100–1000 (130–1300 yd^3)	100–500 (130–650 yd^3)	1000–5000$^+$ (1300–6500$^+$ yd^3)
Construction	Earthern — can be plastic/rubber lined	Cement or block — can have liner	Solid or flexible mesh — fish net common
Relative cost[b] ($/$m^3$ are $/1.3 yd^3)	50–100	100–200	20–80
Max. fish culture density (kg/m^3)	10–25 (22–55 lb/264 gal)	>50 (>110 lb/264 gal)	15–20 (33–44 lb/264 gal)
Water exchange (vol/day)	0.5–5 (open system)	10^{+c} (open system)	Function of area's current flow
Supplementary oxygen supply	Aeration by mechanical means	Pure oxygen enrichment	None
Feeding systems	Hand or automatic	Same	Same
Effects on environment	Effluent can be controlled or treated	Same	Little control — no significant treatment of waste
Sensitivity to environmental factors	Water quality and temperature	Reduced sensitivity — water temperature and quality	Vulnerable to sea conditions, water temperature and quality
Routine maintenance	Periodic emptying and treatment of sediments	Periodic emptying and cleaning; removal of organic accumulation	Frequent cage inspection and repair; removal of fouling growth; periodic cage removal for treatment

[a]N. Mozes, NCM — personal communication.
[b]Total investment including all system components.
[c]< 1/day in recirculating systems.

Table 2. Summary of the Known Nutritional Needs[a] of the Gilthead Sea Bream

	Percentage of Diet Unless Otherwise Indicated
Protein	
Total dietary level	
Larval/juvenile	50–60[41]
Growout	45–50[41]
Amino acids	
Arginine	<2.6[42] (% of dietary protein)
Lysine	5.0[42] (% of dietary protein)
Methionine + cysteine	4.0[42] (% of dietary protein)
Tryptophan	0.6[42] (% of dietary protein)
Estimates of remaining (IAA)[b,43]	
Histidine	1.7[43] (% of dietary protein)
Isoleucine	2.6[43] (% of dietary protein)
Leucine	4.5[43] (% of dietary protein)
Valine	3.0[43] (% of dietary protein)
Phenylalanine + tyrosine	2.9[43] (% of dietary protein)
Threonine	2.8[43] (% of dietary protein)
Lipid	
Total dietary level	12–24[cf]
(n-3) HUFA (EPA + DHA)	
Larvae: 7–23 mg ($2.5–8.1 \times 10^{-4}$ oz)	2.98[44] (% dry Artemia)
Juveniles: 1–11 g ($3.52–38.7 \times 10^{-2}$ oz)	≥0.9[41]
12–30 g ($4.22–10.6 \times 10^{-1}$ oz)	1[45]
Growout to commercial size	1.5–2.7[cf]
Broodstock	0.42[c,46]
Energy	
Daily maintenance requirement	55.8 kJ · BW (kg)$^{-0.83}$(47)
Growth requirement	23 MJ/kg live weight gain[d,47]
Carbohydrate[e]	20[49]
Vitamins mg/kg diet—pyridoxine (B_6)	3–5[51] ($5–8 \times 10^{-5}$ oz/lb)
Biotin	0.37[52] (5.9×10^{-6} oz/lb)
Nicotinic acid	63–83[53] ($1–1.33 \times 10^{-3}$ oz/lb)
Thiamin (B_1)	>5.0[49] (8×10^{-5} oz/lb)
Riboflavin, and pantothenic and ascorbic acids	Essential: levels unknown[54,55]
Minerals[f]	Commercial premixes[rv]
P:E ratios	28–19 g digestible protein/MJ dig. Energy[g]

[a] Superscripts on table values indicate: "numbers" = research data from published literature; "rv" = recommended values; "cf" = values found in commercial feeds for sea bream.
[b] Estimates of IAA based on ratios of whole-body IAA to total IAA (43).
[c] Minimum level required to provide 50% viability of spawned eggs.
[d] Calculated from function.
[e] Low digestibility.[48–50]
[f] No information.
[g] Calculated range for sea bream from 10–500 g (0.35–17.6 oz) (I. Lupatsch, NCM—unpublished data).

is characterized by tumor-like tissue masses on the body surface. These external growths are clusters of extremely hypertrophied fibroblastic dermal cells. Occasionally, internal organs will become infected (56). The disease follows a chronic course, and, although it is usually nonlethal, it causes an unsightly appearance of the skin that makes the fish unmarketable. Diagnosis of lymphocystis can be confirmed by histological sections and the appropriate staining of the tissue lesions. Extensive infections occur mainly in juveniles, with limited mortalities. Infected fish recover within a few weeks, retaining little or no scar tissue. Although the disease is widespread, affecting at least 30 families of marine fish (57), lymphocystis is a species-specific disease. The virus from one fish species will not infect another fish species, so the disease is most likely caused by a group of different viral strains. No effective therapy is known; reduction in stocking density and the removal of heavily infected individuals are the only measures that can be adopted to reduce the impact of this disease.

Bacterial Diseases

The clinical symptoms of many bacterial diseases are similar; therefore, clear diagnosis requires the isolation and culture of the organism involved in the disease. Subsequent identification is not always simple, especially of aquatic bacteria, because their taxonomy is not well known. Many aquatic bacteria are opportunistic; they become virulent only when the delicate balance between the fish and its environment is disturbed. This imbalance can result from poor culture conditions due to high stocking densities of fish, to improperly balanced feed, or to deteriorating water quality, from a high load of pathogens, from rough handling, and from a variety of other factors that cause stress. The following diseases are caused by "true" bacterial pathogens.

Epitheliocystis. Epitheliocystis is caused by a *Chlamydia*-like, obligate, intracellular prokaryote that produces a chronic gill infection. The epithelial cells of the gills become packed with minute coccoid organisms, and hyperplasia and fusion of adjacent lamellae soon follow. Affected fish have flared opercula and display fast, shallow breathing. Epitheliocystis infections in juvenile fish tend to be both extensive and lethal. In histological sections, the infected cells, are of dimensions up to 220 × 100 μm ($8.66 \times 10^{-3} \times 3.9 \times 10^{-3}$ in.), are basophilic, and are uniformly granular in appearance. The disease is highly infectious but species-specific, so different strains of closely related bacteria probably cause these infections. The bacteria cannot be cultured on ordinary laboratory media, and no effective treatment is known. Infections in farmed sea bream have been reported from the Red Sea (58–60).

Vibriosis. Vibriosis is the name given to a disease caused by a large group of bacteria belonging to the family Vibrionaceae. These organisms are gram-negative rods with motile polar flagella; they are noncapsulated and nonspore-producing. They are positive when assayed for oxidase and catalase enzymes. This bacterial family is widespread in marine environments; many of them are facultative pathogens. Their taxonomy is still controversial, in particular as a result of recent phylogenetic studies based on DNA homology among strains. Mortalities of sea bream cultured in the Red Sea have been associated with *V. alginolyticus*, *V. parahaemolyticus*, and *V. anguillarum* (or *V. anguillarum*-like) bacteria.

The disease is characterized by a systemic hemorrhagic septicemia; the clinical symptoms are the appearance of hemorrhages on the surface of the body. Vibrios produce a wide variety of proteases and extracellular enzymes, which are responsible for the extensive tissue damage. As the disease progresses, intestinal hemorrhage, destruction of the tunica mucosa, congestion and hemorrhage of the liver, enlargement and liquefaction of the spleen, the liver, and the kidney are often observed.

The same vibrios that cause vibriosis exist in the aquatic habitat and are often part of the normal intestinal flora of the fish. Good animal husbandry and adequate nutrition are essential to prevent the development of the disease. Treatment of vibriosis with medicated feed can be effective if done at the initial stage of the disease, while the fish are still eating.

Pasteurellosis. Pasteurellosis, a disease widespread in the USA, Japan, and the Mediterranean basin, is caused by *Pasteurella piscicida*, recently renamed *Photobacterium damselae* subsp. *piscicida* (61), another member of the Vibrionaceae. The infection caused by this bacterium develops rapidly into an acute septicemic condition that is characterized by an enlarged spleen containing typical foci of bacterial microcolonies. In advanced infections, these lesions appear as whitish spots and patches on the spleen surface. Sea bream suffer large mortalities from this bacterium, especially during their post-larval and juvenile stages.

Vaccines have been developed against *P. damselae*, but their effectiveness has yet to be proven. Early detection and administration of medicated feeds, while fish are still feeding, seem to offer the best chance of saving infected individuals. *P. damselae* rapidly develops resistance to antibiotics, which gradually become less effective; therefore, it is advisable to perform an antibiogram before treating fish for this disease. This bacterium is highly infectious, so strict measures should be taken to limit its spread.

Protistan Parasities

The protistans associated with fish form a large, heterogeneous group of single-cell organisms; some are ectoparasites, others follow an endoparasitic life cycle. Both types can cause severe damage to intensively cultured sea bream.

Amyloodiniosis. Amyloodiniosis is one of the most devastating parasitic diseases in temperate (62,63) and tropical mariculture (64–67). *Amyloodinium ocellatus*, which causes the disease, is a dinoflagellate that is highly adapted to parasitism. In the wild, damage to the host is limited by the shortness of its parasitic stage. In the confined volume of a tank or pond, however, this organism finds ideal conditions to reproduce and infect the entire population of fish in a matter of days. Its life cycle has three main phases: a parasitic feeding stage (trophont), an encysted reproductive stage (tomont), and a free-swimming infective stage (dinospore) (67).

Copper compounds are effective against *A. ocellatum*, but care must be taken when using them, because of the toxic effects of copper on fish. The concentrations used to kill the pathogen are close to the levels at which copper acts as a membrane poison and harms the gills, the liver, the kidney, and the nervous system. In addition, its immunosuppressive nature is a major drawback to its use in treating fish (68). Despite these drawbacks, treatment of infected fish is carried out over a 12–14 day period by maintaining a concentration of 0.75 mg/L (2.39×10^{-3} oz/qt) of $CuSO_4$ in the water (66), using a slow drip of a concentrated solution and closely monitoring its concentration in the water.

Cryptocaryonosis. Cryptocaryonosis is caused by the ciliate *Cryptocaryon irritans*, class Colpodea (69), which is known to have intraspecific variants (70–72). Although typical of tropical seas, this parasite has a worldwide distribution and extends well into temperate environments. In the Mediterranean sea, it was diagnosed in sea bream from Israel, Italy, and Spain (70).

The ciliate invades the skin, the eyes, and the gills of its host and impairs the functioning of these organs.

External signs of the disease are the appearance of pinhead-size whitish "blisters" on the skin, an increase in mucus production, and, in the case of heavily infested fish, frequent surfacing and gasping for air. Diagnosis of cryptocaryonosis is made by microscopic examination of gill, fin, or skin tissues to determine the presence of the large, revolving ciliate. The ciliate's life cycle consists of four phases. The first is parasitic (trophont); it feeds on the fish's epithelia. After three to seven days of growth, it leaves its host, loses its cilia (protomont), encysts, and starts dividing (tomont), eventually producing up to 200 free-swimming infective organisms (theront). Theronts have a life span of 24 hours, but their ability to infect a host decreases rapidly after 6–8 hours (73,74).

Brooklynellosis. *Brooklynella hostilis* is a ciliate easily recognizable: by its oval, dorsoventrally flattened shape; by its notched oral area; and by its size 36–86 × 32–50 μm (1.4–3.4 × 10^{-3} × 1.26–1.97 × 10^{-3} in.) (75). As a gill pathogen, *B. hostilis* can cause serious skin lesions (76), destroying the host's surface tissue with its cytopharyngeal armature, feeding on tissue debris, ingesting blood cells, and causing hemorrhages in the gills (75). *B. hostilis* was recently diagnosed in cage-cultured sea bream, *Sparus aurata*, in the Red Sea (77).

Myxosporean Infections. Myxosporeans are endoparasites that either can reside in visceral cavities such as the gall bladder, the swim bladder, and the urinary tract (celozoic species) or can settle as inter- or intracellular parasites in the blood, in muscle, or in connective tissue (histozoic species). Spores typical of the genus *Kudoa* [6.4 to 13.6 mm (0.25–0.54 in.) long] were found in the viscera of gilthead sea bream cultured in the Red Sea (78) and, on occasion, reappear in the same species. This parasite causes relatively benign infections, one usually limited to a few individuals.

A debilitating myxosporean disease caused by *Myxidium leei*, a histozoic species, has been described in sea bream (79,80). The parasite settles in the intestinal mucosa; in cases of heavy infections, affected fish have an enlarged abdomen, and the intestinal tract is filled with a foul-smelling liquid. Histological examination of the intestine shows the presence of spores between the epithelial cells of the mucosa lining the entire tract. The same parasite was later discovered in Mediterranean fish (sparids and grey mullet) (81), suggesting that it was imported into the Red Sea with its host. Recent literature has shown that it can be transmitted directly (82).

Metazoan Parasites

Monogenean Infections. The monogeneans form a very diverse group of (mostly ectoparasitic) worms that feed on epithelial cells and mucus. Monogeneans are hermaphroditic, and they do not require an intermediate host to complete their life cycle. A free-swimming ciliated larva emerges from an egg, the shape of which is species-dependent. Most of the marine monogeneans have a long polar filament, for attachment to the substratum (83). This direct life cycle, coupled with the availability of stressed fish in high-density culture systems, facilitates infestation (84–86).

The body of these parasites is relatively large and flat; it has a conspicuous muscular disc haptor at the posterior end and a pair of large disc-like adhesive organs on the anterior end. The intestinal caeca are diverticulate and end blindly. *Neobenedenia melleni* has been detected on the body of sea bream cultured in the Red Sea (87). The active feeding of the monogeneans on mucus and on epithelial cells leads to hemorrhage, inflammation, and the over-production of mucus (88,89). Monogeneans often settle on or around the eyes, damaging the cornea and causing blindness (89,90). Despite their size [up to a few mm (1 m = 0.04 in.)], monogeneans may go unnoticed. A freshwater dip of about five minutes is sufficient to dislodge the parasite from its host and kill it, causing it to turn opaque and become more visible.

Diplectanid monogeneans, which infest only the gills of their host and feed on mucus, are elongated and are characterized by a large flat opisthaptor with squamodiscs at the posterior end. The diplectanid species commonly infect farmed fish of the sea bass and sea bream families. They are host-specific—each species will infect only one particular species of fish, even if other fish are being cultured in the same water.

BIBLIOGRAPHY

1. M.L. Bauchot and J.C. Hureau, in P.J.P. Whitehead, M.L. Bauchot, J.C. Hureau, J. Nielsen, and E. Tortonese, eds., *Fishes of the North-Eastern Atlantic and the Mediterranean*, Unesco, Paris, 1984, pp. 883–907.
2. D. Golani, *Isr. J. Zool.* **42**, 15–55 (1996).
3. FAO, *Yearbook of Fishery Statistics*, capture production 1996, Rome, vol. 82.
4. D. Popper and Y. Zohar, *Proc. Aquacult. Int. Cong. Expo.*, 1988, pp. 319.
5. FEAP, http://www.feap.org/seabreams.html.
6. A. Ben-Tuvia, *Investigación Pesquera* **43**(1), 43–67 (1979).
7. FEAP, http://www.feap.org/juvenil.html.
8. Y. Zohar, A. Goren, M. Fridkin, E. Elhanati, and Y. Koch, *Gen. Comp. Endocrinol.* **79**, 306–319 (1990).
9. Y. Zohar, A. Elizur, N.M. Sherwood, J.F.F. Powell, J.E. Rivier, and N. Zemora, *Gen. Comp. Endocrinol.* **97**, 289–299 (1995).
10. A. Elizur, I. Meiri, H. Rosenfeld, N. Zemora, W.R. Knibb, and Y. Zohar, *Proc. Intern. Symp. Reproductive Physiology of Fish*, 1995, pp. 13–15.
11. Y. Gothilf, M. Chow, A. Elizur, T.T. Chen, and Y. Zohar, *Mol. Marine Biol. Biotech.* **4**, 27–35 (1995).
12. I. Meiri, Y. Gothilf, W.R. Knibb, Y. Zohar, and A. Elizur, *Proc. Intern. Symp. Reproductive Physiology of Fish*, 1995, p. 37.
13. A. Elizur, N. Zemora, H. Rosenfeld, I. Meiri, S. Hassin, H. Gordin, and Y. Zohar, *Gen. Comp. Endocrinol.* **102**, 39–46 (1996).
14. Y. Gothilf, J.A. Muñoz-Cueto, C.A. Sagrillo, M. Selmanoff, A. Elizur, and Y. Zohar, *Biol. Reprod.* **55**, 636–645 (1996).
15. Y. Gothilf, I. Meiri, A. Elizur, and Y. Zohar, *Biol. Reprod.* **57**, 1145–1154 (1997).
16. J.F.F. Powell, Y. Zohar, A. Elizur, M. Park, W.H. Fischer, A.G. Craig, J.E. Rivier, D.A. Lovejoy, and N.M. Sherwood, *Proc. Natl. Acad. Sci.* **91**, 12081–12085 (1994).
17. Y. Zohar, M. Abraham, and H. Gordin, *Annales Biologie Animale Biochimie Biophysique* **18**, 877–882 (1978).

18. A. Happe and Y. Zohar, in Y. Zohar and B. Breton, eds., *Reproduction in Fish-Basic and Applied Aspects in Endocrinology and Genetics*, INRA, Paris, 1988, pp. 177–180.
19. S. Gorshkov, H. Gordin, G. Gorshkova, and W. Knibb, *Israeli J. Aquacult.-Bamidgeh* **49**(3), 124–134 (1997).
20. H. Gordin and Y. Zohar, *Annales Biologie Animale Biochimie Biophysique* **18**, 985–990 (1978).
21. Y. Zohar and H. Gordin, *J. Fish Biol.* **15**, 665–670 (1979).
22. Y. Zohar, G. Pagelson, Y. Gothilf, W.W. Dickhoff, P. Swanson, S. Dugary, W. Gombotz, J. Kost, and R. Langer, *Proc. Intern. Symp. Control. Rel. Bioact. Mater.* **17**, 51–52 (1990).
23. A. Barbero, A. Francescon, G. Bozzato, A. Merlin, P. Belvedere, and L. Colombo, *Aquaculutre* **154**(33–4), 349–359 (1997).
24. Y. Zohar, M. Harel, S. Hassin, and A. Tandler, in N. Bromage and R. Roberts, eds., *Broodstock Management and Egg and Larval Quality*, Blackwell Science, Oxford, UK, 1995, pp. 94–117.
25. C.L. Peguin, Master's thesis, *Hebrew University of Jerusalem*, the effect of photoperiod and prey density on the growth and survival of larval gilthead seabream *Sparus aurata* L. (Perciformes, Teleostei), 1984, pp. 93.
26. A. Tandler and C. Mason, in R.R. Stickney and S.P. Meyers, eds., *World Mariculture Society Special Publications* **3**, 103–116 (1983).
27. A. Tandler and C. Mason, *European Mariculture Society, Special Publication* **8**, 241–259 (1984).
28. A.F. Anav, Master's thesis. *Tel Aviv University*, the effect of salinity on growth, respiration and Na^+/K^+-ATPase activity in gilthead seabream (*Sparus aurata*) larvae, 1991, pp. 54.
29. A. Tandler, F.A. Anav, and I. Choshniak, *Aquaculture* **135**, 343–353 (1995).
30. S. Helps, Master's thesis, *Plymouth Polytechnic*, An examination of prey size selection and its subsequent effect on survival and growth of larval gilthead seabream (*Sparus aurata*), 1982, pp. 50.
31. W.M. Koven, G.Wm. Kissil, and A. Tandler, *Aquaculture* **79**, 185–191 (1989).
32. W.M. Koven, A. Tandler, G.Wm. Kissil, D. Sklan, O. Friezlander, and M. Harel, *Aquaculture* **91**, 131–141 (1990).
33. W.M. Koven, A. Tandler, G.Wm. Kissil, and D. Sklan, *Aquaculture* **104**(1–2), 91–104 (1992).
34. A. Tandler and S. Helps, *Aquaculture* **48**, 71–82 (1985).
35. A. Tandler and S. Kolkovski, *European Aquaculture Society, Special Publication* **15**, 169–171 (1991).
36. S. Kolkovski, A. Tandler, and G.Wm. Kissil, in S.J. Kaushik and P. Luquet, eds., *Fish Nutrition in Practice*, INRA, Paris, 1993, pp. 569–578.
37. S. Kolkovski, A. Arieli, and A. Tandler, 1997. *Aquaculture International* **5**(6), 527–536 (1997).
38. W.M. Koven, G. Parra, S. Kolkovski, and A. Tandler, *Aquaculture Nutrition* **4**, 39–45.
39. E. Hadas, Master's thesis, *Hebrew University of Jerusalem*, influence of dietary phosphatidyl choline on feeding rate and lipid metabolism in gilthead seabream (*Sparus aurata*), 1998, pp. 62.
40. Y. Shezifi, Master's thesis, *Tel Aviv University*, variability measures of body weight in young *Sparus aurata* populations, 1990, pp. 53.
41. P. Morris, *Fish Farmer-International File* December: 6–9 (1997).
42. P. Luquet and J.J. Sabaut, *Actes de Colloques, Colloques sur L'Aquaculture* **1**, 243–253 (1974).
43. S.J. Kaushik, *Aquat. Living Resour.* **11**(5), 355–358 (1998).
44. W.M. Koven, A. Tandler, G.Wm. Kissil, and D. Sklan, *Aquaculture* **104**, 91–104 (1992).
45. C. Ibeas, J. Cejas, T. Gomez, S. Jerez, and A. Lorenzo, *Aquaculture* **142**, 221–235 (1996).
46. M. Harel, A. Tandler, G.Wm. Kissil, and S.W. Applebaum, *Brit. J. Nutr.* **72**, 45–58 (1984).
47. I. Lupatsch, G.Wm. Kissil, D. Sklan, and E. Pfeffer, *Aquaculture Nutrition* **4**(3), 165–174 (1998).
48. J.M. Vergara and K. Jauncey, in S.J. Kaushik and P. Luquet, eds., *Fish Nutrition in Practice*, INRA, Paris, 1993, pp. 453–458.
49. P. Morris and S.J. Davies, *Animal Sci.* **61**, 597–603 (1995).
50. I. Lupatsch, G.Wm. Kissil, D. Sklan, and E. Pfeffer, *Aquaculture Nutrition* **3**(2), 81–89 (1997).
51. G.Wm. Kissil, C.B. Cowey, J.W. Adron, and R.H. Richards, *Aquaculture* **23**, 243–255 (1981).
52. G.Wm. Kissil, *EMS Spec. Publ.* **6**, 49–55 (1981).
53. P.C. Morris and S.J. Davies, *Animal Sci.* **61**, 437–444 (1995).
54. P.C. Morris, S.J. Davies, and D.M. Lowe, *Animal Science* **61**, 419–426 (1995).
55. M.N. Alexis, K.K. Karanikolas, and R.H. Richards, *Aquaculture* **151**, 209–218 (1997).
56. A. Colorni and A. Diamant, *J. Fish Dis.* **18**, 467–471 (1995).
57. K. Wolf, *Fish Viruses and Fish Viral Diseases*, Cornell University Press, Ithaca, NY, 1988.
58. I. Paperna, *Aquaculture* **10**, 169–176 (1977).
59. I. Paperna, I. Sabnai, and M. Castel, *J. Fish Dis.* **1**, 181–189 (1978).
60. I. Paperna, I. Sabnai, and A. Zachary, *J. Fish Dis.* **4**, 459–472 (1981).
61. G. Gauthier, B. Lafay, R. Ruimy, V. Breittmayer, J.L. Nicolas, M. Gauthier, and R. Christen, *J. Sys. Bac.* **45**, 139–144 (1995).
62. A.R. Lawer, in C.J. Sindermann, ed., *Disease Diagnosis and Control in North American Marine Aquaculture*, Elsevier, Amsterdam, 1977, pp. 257–264.
63. A. Barbaro and A. Francescon, *Oebalia* XI–2, N.S., 745–752 (1985).
64. M.C.L. Baticados and G.F. Quinitio, *Helgoländer Meeresuntersuchungen, Helgoländer Meeresunters* **37**, 595–601 (1984).
65. I. Paperna, *J. Fish Dis.* **3**, 363–372 (1980)
66. I. Paperna, *Aquaculture* **38**, 1–18 (1984).
67. I. Paperna, *Annales de Parasitologie Humaine et Comparée* **59**, 7–30 (1984).
68. P. Cheng, in M.K. Stoskopf, ed., *Fish Medicine*, W.B. Saunders, Philadelphia, PA, 1993, pp. 646–658.
69. B.K. Diggles and R.D. Adlard, *Dis. Aquat. Organisms* **22**, 39–43 (1995).
70. A. Diamant, G. Issar, A. Colorni, and I. Paperna, *Bulletin of the European Association of Fish Pathologists* **11**, 122–124 (1991).
71. A. Colorni and A. Diamant, *Eur. J. Protis.* **29**, 425–434 (1993).
72. B.K. Diggles and J.G. Lester, *J. Parasitol.* **82**, 384–388 (1996).
73. T. Yoshinaga and H.W. Dickerson, *J. Aquat. An. Health* **6**, 197–201 (1994).
74. B.K. Diggles and J.G. Lester, *J. Parasitol.* **82**, 45–51 (1996a).
75. J. Lom and I. Dyková, *Protozoan Parasites of Fishes*, Elsevier, Amsterdam, 1992.

76. E.J. Noga, *Fish Disease, Diagnosis and Treatment*, Mosby, St. Louis, MO, 1996.
77. A. Diamant, *Bull. Eur. Assoc. Fish Pathol.* **18**, 33–36 (1998).
78. I. Paperna, *J. Fish Dis.* **5**, 539–543 (1982).
79. A. Diamant, *Bull. Eur. Assoc. Fish Pathol.* **12**, 64–66 (1992).
80. A. Diamant, J. Lom, and I. Dyková, *Dis. Aquat. Organisms* **20**, 137–141 (1994).
81. J. Lom and G. Bouix, in G. Brugerolle and J.P. Mignot, eds., *Protistological Actualities*, Proc. 2nd European Congress of Protistology, Clermont-Ferrand, France, 21–26 July 1995.
82. A. Diamant, *Dis. Aquat. Organisms* **30**, 99–105 (1997).
83. G.C. Kearn, *Adv. Parasitol.* **25**, 175–273 (1986).
84. I. Paperna, A. Diamant, and R.M. Overstreet, *Helgoländer Meeresuntersuchungen, Helgoländer Meeresunters* **37**, 445–462 (1984).
85. T.S. Leong and S.Y. Wong, *Aquaculture* **68**, 203–207 (1988).
86. D.K. Cone, in P.T.K. Woo, ed., *Fish Diseases and Disorders*, CAB International, Oxon, England, 1995, pp. 289–327.
87. A. Colorni, *Dis. Aquat. Organisms* **19**, 157–159 (1994).
88. I. Paperna, *Ann. Rev. Fish Dis.* **1**, 155–194 (1991).
89. S. Egusa, *Infectious Diseases of Fish*, Amerind Publ. Co., New Delhi, 1992.
90. A. Colorni, in S.S. De Silva, ed., *Tropical Mariculture*, Academic Press, San Diego, CA, 1998, pp. 209–255.

See also Brine shrimp culture.

GOLDFISH CULTURE

Robert R. Stickney
Texas Sea Grant College Program
Bryan, Texas

OUTLINE

The goldfish (*Carassius auratus*) is a member of the minnow family (Cyprinidae) and is one of the most widely recognized of all fishes, at least in countries where ornamental species are maintained in homes, restaurants, and other types of business establishments. Goldfish are extremely hardy, so they make excellent aquarium species as well as good laboratory species. They are easy to culture and come in several varieties, some with unusual names, such as the blue shubunkin and the telescopic-eye black goldfish. Their hardiness and ready availability give them scientific value for genetic and physiological research. Fish, such as goldfish, can sometimes be used in place of mammals for research, including biomedical research that may ultimately have implications for human health.

ATTRIBUTES OF GOLDFISH

Selective breeding has produced a wide variety of goldfish, some of which carry unusual names and attributes. Goldfish come in a number of colors, from red and orange to black, olive-green, and the popular gold. There are goldfish with greatly protruding eyes, odd body shapes, and unusual tail fin configurations. All are extremely hardy, which is one of the reasons that they are very popular as ornamentals.

Goldfish can frequently be found in backyard ponds, pools built in conjunction with restaurants, aquaria located in medical and professional offices and many other types of businesses, and, of course, in the home fish bowl or aquarium. Most children, at least in the United States, have had at least some experience in owning and caring for a goldfish. Being tolerant of degraded water quality and able to survive long periods of abuse, goldfish can survive infrequent feeding, as well as overfeeding, and will often live in unaerated water that has not been changed in months. They can also tolerate temperature extremes.

Goldfish were introduced to the United States in 1878 and have become widely dispersed around the country (1). While many people think primarily of goldfish as ornamentals, they have had some use as forage fish and as bait (1,2), though use of goldfish for bait is not legal in some states. They are also produced as feeder fish for carnivorous aquarium species. Arkansas is now the leading goldfish-producing state in the United States (3). Large numbers of goldfish are also produced in Missouri and a few other states. Various other nations also produce goldfish, including countries in both Europe and Asia.

When confined in a small fish bowl or aquarium, densely crowded, stressed by poor water quality, or not provided with sufficient amounts of food, goldfish tend to grow very slowly and often may not increase much in size, even after one or more years following purchase at a retail outlet. Given sufficient space, good water quality, and a good diet, goldfish will reach nearly a foot (30 cm) in length. Large goldfish are most commonly seen in very large aquaria or in outdoor ponds.

CULTURE TECHNIQUES

The basic techniques for rearing goldfish generally do not differ from those for culturing other freshwater ornamentals and warmwater fish. (See the entries "Carp culture," "Ornamental fish culture, freshwater," and "Tilapia culture," for example.) Goldfish are usually produced in ponds, captured as fingerlings, and shipped throughout the world, usually in oxygen-charged plastic bags packed in insulated boxes. Spawning techniques used in conjunction with goldfish are dissimilar from those used with many other species, though the techniques for goldfish and another member of the minnow family, the golden shiner (*Notemigonus crysoleucas*), are virtually identical (4).

Goldfish can be induced to spawn at any time of the year, through environmental manipulation, but the normal spawning season is in the spring when the water temperature reaches 64 °F (18 °C). Natural spawning is the dominant method used by commercial culturists who rear goldfish in ponds and depends on nature to provide the proper temperature for spawning. A large female may produce from 2,000 to 4,000 eggs (1).

There are three spawning methods employed by the goldfish industry (3,4). One technique, which is an extensive method known as free spawning, involves lowering the water level in spawning ponds during the late winter and early spring, to allow grass to grow along the shoreline. Rye grass may actually be seeded to ensure that plants will become established. As spawning season approaches later in the spring, the ponds are refilled and the grass is used as a spawning substrate.

Induction of spawning and extension of the spawning season can be achieved by flowing cool water into the ponds and rapidly raising the water level. Once the fry reach a size at which they begin to compete for food with the broodfish, the latter may be removed from the spawning ponds. The adult fish may be sold as large baitfish or retained as broodfish for subsequent years.

A second, more intensive form of spawning goldfish, called the egg transfer method, is commonly practiced by large farms and is virtually the same as the method used for spawning golden shiners (4). Rather than encouraging plant growth in spawning ponds, those who employ the egg transfer method diligently manage their ponds to avoid the establishment of vegetation. Instead of plants, spawning mats are placed in the brood ponds. The mats are typically about 1 × 2 ft (30 × 60 cm) in area, though other sizes are also used. Mats can be constructed by sandwiching Spanish moss between pieces of welded wire with a mesh size of 4 × 4 in (10 × 10 cm). Hog rings are typically used to tie the two pieces of welded wire together. Alternatively, mat material is available in rolls of latex-coated coconut fibers on a polyester net backing. The latter resembles air-conditioning filter material.

The spawning mats are placed on the bottom of the pond when the broodfish approach spawning condition. The mats should be placed on level areas, so that there is 1 in. (2.5 cm) of water above them. This technique requires modification from the standard construction of pond levees. (See the entry "Pond culture.") The mats should be placed end to end, parallel to the shoreline. Once they are uniformly covered with eggs, the mats should be removed and placed in nursery ponds. New mats may be installed as replacements if spawning continues to occur. Allowing too many eggs to accumulate on mats can give rise to fungus problems. Mats can be checked each morning to determine the density of eggs. Goldfish usually spawn just after dawn and stop spawning when sunlight directly strikes the water.

Up to 167 mats/acre (400 mats/ha) can be placed in the nursery ponds. The mats are allowed to remain in the nursery ponds for 10 days, after which they are washed thoroughly and reused for egg collection, if necessary.

A third spawning method used by goldfish producers is known as "fry transfer." This method has some advantages over the other two, in that the number of fish stocked in nursery ponds can be determined with a higher degree of accuracy (it is possible to estimate the number of eggs on a spawning mat, but such estimates are not very accurate in most cases). Fry transfer involves removing fry, instead of broodfish, from spawning ponds; removing both fry and broodfish and separating them; or capturing fry from ponds in which they have been incubated on mats transferred from spawning ponds. The fry are restocked at densities that typically range from about 20,000 to 1,000,000 fry/acre (50,000 to 2,500,000 fry/ha). The density used depends on whether the fish are to be reared quickly for sale (low density) or overwintered for growout to market size the following spring (high density).

Nursery ponds are fertilized to induce algae blooms, which retard, through shading, the development of unwanted aquatic vegetation and to produce natural food for the fry. Liquid inorganic fertilizer, such as 10-34-0, is commonly used and may be used in combination with an organic fertilizer, such as cottonseed meal or manure. Sufficient fertilizer is applied to provide a Secchi disc[1] reading of about 8 in. (20 cm).

When fry are observed swimming freely in the nursery ponds, prepared feed should be provided in the form of finely ground meal. Initially, a 48% protein feed is appropriate, and ponds should receive 1 kg/ha/day (1 lb/acre/day). After a few weeks, the protein level can be reduced to 33%, and the feeding rate can be increased to 2–6 kg/ha/day (2–6 lb/acre/day). When the fish reach about 1 in. (2.5 cm) in length, they can be fed small pellets or crumbles. Goldfish fingerlings are usually fed pellets or crumbles at the rate of 20–60 kg/ha/day (20–60 lb/acre/day). (2). The feeding rate depends on the stocking density, water temperature, and other factors. It is adjusted as needed, based on the basic condition that all feed is consumed in a reasonable period of time (i.e., no more than two hours). Goldfish producers may provide hard-boiled egg yolk filtered through cotton cloth as a first feed, to promote rapid growth. The egg yolk is soon replaced by more conventional prepared feed.

Predators can be a significant problem. Bird predation is a constant threat, and there are a number of kinds of birds that may be involved. In addition, carnivorous fishes, mink, raccoons, frogs, snakes, turtles, insects, and alligators can be problems. Human poachers may also cut into the inventory if proper security is not provided.

Diseases of goldfish include various of the more common bacteria; the protozoans *Trichodina*, *Ichthyobodo*, *Chilodonella*, *Cryptobia*, and *Ichthyophthirius*; the sporozoan parasite *Mitraspora cyprini*; and the crustacean parasite *Argulus*, which is also known as the fish louse (1,4). Viruses have also been reported from goldfish. (For more information on each disease, see the entry "Bacterial disease agents.")

To avoid damage, goldfish should be harvested with soft-mesh seines. Lift nets are also used. Lift nets are pieces of fine-mesh netting suspended from long poles by ropes. They are baited with fish feed and then lowered into the water and allowed to remain for a some time, after which they are lifted and the fish that have been attracted to the feed are removed. The fish may be held unfed in raceways for 24 hours prior to shipment, to allow them to void feces and acclimate to a more confined environment.

[1] A Secchi disc is a flat, circular plate, typically painted in pie-shaped wedges of alternating black and white, that is lowered into the water column by a rope. The Secchi disc reading is the depth at which the disc just disappears from sight.

BIBLIOGRAPHY

1. N. Stone, E. Park, L. Dorman, and H. Thomforde, *Baitfish Culture in Arkansas: Golden Shiners, Goldfish and Fathead Minnows*, University of Arkansas at Pine Bluff, Pine Bluff, AK; 1997.
2. R.G. Piper, I.B. McElwain, L.E. Orme, J.P. McCraren, L.G. Fowler, and J.R. Leonard, *Fish Hatchery Management*, US Department of the Interior, Washington, DC, 1982.
3. N. Stone, E. Park, L. Dorman, and H. Thomforde, *World Aquaculture*, December 5–13 (1997).
4. J.T. Davis, in R.R. Stickney, ed., *Culture of Nonsalmonid Freshwater Fishes*, CRC Press, Boca Raton, FL, 1993, pp. 307–321.

See also Baitfish culture.

GROUPER CULTURE

John W. Tucker, Jr.
Harbor Branch Oceanographic Institution
Fort Pierce, Florida

Arietta Venizelos
National Marine Fisheries Service, NOAA
Virginia Key, Florida

Daniel D. Benetti
University of Miami
Miami, Florida

OUTLINE

Groupers are classified in 14 genera of the subfamily Epinephelinae, which comprises at least half the approximately 449 species in the family Serranidae. Throughout most warm and temperate marine regions, serranids are highly valued for food, and both small and large species are kept in aquariums. Maximum size ranges from 12 cm (4.7 in.) total length (TL) for the Pacific creole-fish (*Paranthias colonus*) to more than 4 m (13 ft) TL (440 kg, 968 lb) for the groper, or brindlebass, (*Epinephelus lanceolatus*). Several grouper species have been raised commercially (mainly in Hong Kong, Taiwan, and the Southeast Asian region), usually by growing out captured wild juveniles. Some species can grow from 15–20 g (0.5–0.7 oz) to 1 kg (2.2 lb) in about a year. Research has been conducted on spawning and rearing of dozens of serranid species (Table 1). The main accomplishments are reviewed here. Additional information can be found in various publications (1–7).

REPRODUCTION

Most groupers studied mature within 2 to 6 yrs (7). Many serranids are protogynous hermaphrodites (8,9). Some species, as a rule, change from female to male with age, while others might change only if there is a shortage of males. In nature, Nassau groupers (*Epinephelus striatus*) spawn in large aggregations (100s to 1,000s of fish) with a sex ratio near 1 : 1. Gag groupers (*Mycteroperca microlepis*) spawn in harems, with a sex ratio often near 1 male : 10 females. For both species, individual spawning events usually involve small numbers of fish (e.g., 2 to 5). Small serranids often spawn in pairs without aggregating. A few species are simultaneous hermaphrodites, but self-fertilization seems to be rare.

Voluntary spawning of captive groupers has occurred mostly with well-fed, uncrowded fish during the natural spawning season under conditions of ambient temperature and partial or total natural light (7,10). Day length seems to be a less important stimulus than temperature. At least 27 serranid species have spawned voluntarily in captivity, with groupers spawning in 1- to 21,200-m^3 tanks or ponds and 26- to 75-m^3 cages. In Kuwait, 40 female and 9 male orangespotted groupers (*Epinephelus coioides*) held in a concrete tank spawned almost continuously for 50 days during April to June (11). In the Philippines, 1 female orangespotted grouper with 2 males held in a 48-m^3 cage spawned 5 to 10 times a month for 4 months (12). In Singapore, during December 1989 to October 1990, 10 female and 10 male brownmarbled groupers held in a 75-m^3 cage spawned 2 to 5 times during each of nine periods of 2 to 6 days, usually starting between the last quarter moon and new moon (13). In Taiwan, 8 female leopard coraltrout held in a pond produced eggs 110 times during May to October (Chen et al., 1991a, cited in 7). In Florida, 3 or 4 female Nassau groupers and 2 males held in a 37-m^3 raceway spawned near the full moon in March and April, with each female spawning as many as 9 times a day for 1 to 4 days (14).

Hormone-induced ovulation of ripe, wild, or captive groupers also is reliable (7,10). At least 31 serranid species have been induced to ovulate. Typically, a female with fully yolked oocytes will ovulate within 24 to 72 hours (usually 36 to 50 hours) after the first of 1 to 3 injections of 500 to 1,000 IU human chorionic gonadotropin/kg body weight. Similar results have been obtained for several species given 1 to 3 injections of 10 to 50 μg gonadotropin releasing hormone analog/kg body weight. GnRH-analogue implants were effective for spawning white groupers (*Epinephelus aeneus*). For six grouper species with egg diameters of 800 to 1,000 μm, the minimum effective oocyte diameter before injection was in the range 41 to 61%. For Nassau groupers, the time from ovulation to overripeness is only 1 to 2 hours at 26 °C.

Nassau groupers at 6 kg can produce about 900,000 eggs per day by natural or hormone-induced ovulation. At the same size, hormone-treated brownmarbled groupers can produce 1.7 million eggs. Hormone-treated 1.5-kg squaretail coraltrout can produce 400,000 eggs. A 1-kg (2.2 lb) redspotted grouper can produce more than

Table 1. Some Characteristics of Representative Groupers Raised Commercially (C) and Experimentally (E)

Species	Type of Culture[a]	Locations	Egg Diameter (μm)	Larval Duration (d)	Source of Juveniles[b]	Market Size (kg)	Maximum Size[c]	Maximum Age (yr)
Polka dot grouper *Cromileptes altivelis*	C	SE Asia	890		W	0.5	90 cm	
Redspotted grouper *Epinephelus akaara*	C	Japan Hong Kong	825	45–50	H, W	0.5	5 kg, 60 cm	>6
Squaretail grouper *Epinephelus areolatus*	C	Hong Kong			W		60 cm	
Orangespotted grouper *Epinephelus coioides*	C	SE Asia Middle East	807	35–40	H, W		≥95 cm	
Brownmarbled grouper *Epinephelus fuscoguttatus*	C	SE Asia	840	35–40	H, W	0.6	120 cm	
Malabar grouper *Epinephelus malabaricus*	C	SE Asia	852	36–60	H, W	0.6	>25 kg, 115 cm	
Sevenband grouper *Epinephelus septemfasciatus*	C	Japan	820	~60			120 cm	
Nassau grouper *Epinephelus striatus*	E	Caribbean	920	46–70	H	2	>25 kg, 120 cm	16
Greasy grouper *Epinephelus tauvina*	C	SE Asia	900	36–50+	H, W	0.6	75 cm	~25
Leopard coraltrout *Plectropomus leopardus*	C	SE Asia	875	~55	H, W		≥20 kg, 80 cm	
Chinese perch *Siniperca chuatsi*	C	PR China	~2000		H	0.45	>5 kg	

[a]C = commercial, E = experimental.
[b]H = from a hatchery, W = from wild stocks.
[c]Whole weight or total length.

5 million eggs in a season, and a 6-kg Nassau grouper can produce 3.3 million eggs in a 4-day period.

With good timing and luck, groupers have been caught just before spawning and held in tanks or cages until they ovulate naturally. The eggs are stripped, or rarely, the fish are left in the tank for voluntary or accidental fertilization to occur.

LARVAL FOODS

With the notable exception of *Siniperca* spp., which have large eggs and hatchlings (~5 mm) and are easy to feed and rear, grouper larvae usually are small and fragile and have relatively small mouths at first feeding. Yolk and oil tend to be exhausted quickly (7,15). Typically, the larval period is long, and groupers tend to require live food longer than most marine fishes that have been reared.

Grouper larvae usually are raised in green water (*Nannochloropsis, Tetraselmis, Chlorella* spp.). At first feeding, most species can eat small rotifers, but oyster or clam eggs and trochophore larvae sometimes are used as a supplement. Growth and survival rates tend to increase if copepods or mixed zooplankton are included in the diet, but care must be used to avoid introduction of pathogens or predators. Enriched *Artemia* can be a staple food beginning at 10 to 30 days, but their density should be controlled to minimize gorging. Microfeeds (artificial diets) have been tried as a supplement during the first week, but probably are not digested well until at least 2 to 4 weeks. Weaning can be completed just before or during transformation into the juvenile stage, which occurs at 35 to 70 days after hatching, depending on species.

COMPOUND FEEDS

In nature, juvenile and adult groupers eat mainly fish, crabs, shrimp, mantis shrimp, lobsters, and molluscs (16). Red groupers (*Epinephelus morio*) seem to prefer crabs first, then shrimp. In Thailand and other areas, groupers have been fed mainly trash fish (with vitamins and minerals) secondarily moist or semimoist pellets, and rarely high-protein dry pellets (17). A suitable starter feed for groupers would contain 50 to 60% high-quality protein, 12 to 16% fat, no more than 15% carbohydrate, less than 3% fiber, and less than 16% ash (7). Groupers larger than 500 g (1.1 lb) can be given a feed with approximately 45% protein, about 9% fat, and no more than 20% carbohydrate, 4% fiber, and 22% ash. Lower quality feeds likely would result in a higher feed conversion ratio and possibly slower growth.

RAISING GROUPERS TO MARKET SIZE

In Indo-Pacific and Middle Eastern regions, several species of grouper are farmed in cages, ponds, and tanks, but usually they are raised from wild juveniles and are fed trash fish. They sometimes are fed small tilapia (*Oreochromis* spp.) and occasionally are polycultured with them. Typical market size is 500 to 1,000 g (1.1–2.2 lb), which can be reached in 6 to 8 months of grow-out. The minimum size to begin grow-out, 75 to 100 mm, can be obtained in nursery tanks, cages, or ponds. They are stocked up to 60 fish/m^3 (<1 kg/m^3) in cages. In Taiwan, a pond farm typically stocks 60,000–80,000 groupers/ha and harvests 80% of them for a production of 30,000 to 40,000 kg/ha; the groupers are fed mostly trash fish and grow from 46 mm to 600 g in 12 months and 2 kg (4.4 lb) in 19 months (18). When fed pellets only, Nassau groupers can reach at least 450 g (1 lb) at 12 months and 2 kg (4.4 lb) at 24 months of age.

HEALTH

For groupers, snappers, and similar warmwater fish, gram-negative bacteria (*Vibrio, Aeromonas, Pseudomonas, Pasteurella* spp.), *Streptococcus, Mycobacterium*, ectoparasitic protozoans (*Amyloodinium ocellatus*, sporozoans, *Cryptocaryon irritans, Brooklynella* spp.; *Ichthyophthirius* sp.), and monogeneans (*Neobenedenia melleni, Diplectanum* spp.) are among the most important pathogens (19–23). In Singapore, sleepy grouper disease (lethal) probably was caused by a virus introduced with wild juvenile groupers imported for cage farming (24). Other viral pathogens and diseases include golden eye disease, red grouper reovirus, spinning grouper disease, and viral nervous necrosis. Rancid dietary lipids are thought to cause nervous suffering disease of groupers, which could result in gill, blood, gas bladder, liver, heart, brain, and nerve damage (25). In Japan, pasteurellosis has been a major disease of young redspotted groupers (26).

ANNUAL PRODUCTION

In 1996, the Peoples' Republic of China produced 58,437 metric tons (mt) of Chinese perch (mandarin fish), which has been considered a serranid, but has affinities with the centropomids (snooks). Malaysia produced 837 mt of greasy groupers and Hong Kong 360 mt. Hong Kong produced 750 mt of squaretail groupers. Taiwan produced 1,883 mt of miscellaneous groupers, Thailand 600 mt, Philippines 595 mt, Singapore 93 mt, and Republic of Korea 9 mt. In 1995, Hong Kong produced 30 mt of redspotted groupers (27).

GENERAL COMMENTS

Commercial-scale hatcheries for redspotted grouper and kelp grouper (*E. bruneus*) in Japan and Malabar grouper in Taiwan and Thailand have raised large batches of juveniles, with survival as high as 34% from hatchlings (7). The larval period is longer than for most cultured fishes. Some groupers need small rotifers, trochophores, or copepods at first feeding. Proper aeration is critical. Early grouper larvae, especially when stressed, sometimes exude an excess of mucus, which can cause them to stick to each other, to the surface film, or to solid objects. Too little turbulence in larval tanks can allow the water to stratify and zooplankton and fish to aggregate dangerously. With too much turbulence, the fish are battered. Gorging on *Artemia* is another source of mortality, and cannibalism among early juveniles can be a problem. Larvae are fragile, and survival from eggs to juveniles often has been only 0 to 1%, but juveniles and adults are among the hardiest of fish.

Grouper farming fluctuates because of variability in the (mostly decreasing) supply of wild juveniles and lack of sustained hatchery production for most species. Variability in quantity and quality of trash fish and the lack of economical compound feeds also has been a constraint in some areas. Nevertheless, the commercial feasibility of grouper culture has been proven in several countries, including Malaysia, Hong Kong, Taiwan, China, Philippines, Singapore, Japan, and Korea.

BIBLIOGRAPHY

1. Anonymous, *Manual on Floating Netcage Fish Farming in Singapore's Coastal Waters*, Singapore Prim. Prod. Dept., 1986.
2. J.J. Polovina and S. Ralston, eds., *Tropical Snappers and Groupers: Biology and Fisheries Management*, Westview Press, Boulder, 1987.
3. S. Tookwinas, *Actes Colloq.* **9**, 429–435 (1990).
4. M. Doi, M.B.H.M. Nawi, N.R.B.N. Lah, and Z.B. Talib, *Artificial Propagation of the Grouper,* Epinephelus suillus *at the Marine Finfish Hatchery in Tanjong Demong, Terengganu, Malaysia*, Dept. Fish., Kuala Lumpur, Malaysia, 1991.
5. K. Maruyama, K. Nogami, Y. Yoshida, and K. Fukunaga, *Third Asian Fisheries Forum*, Asian Fish. Soc., Manila, 1994, pp. 446–449.
6. F. Arreguín-Sánchez, J.L. Munro, M.C. Balgos, and D. Pauly, eds., *Biology, Fisheries and Culture of Tropical Groupers and Snappers*, ICLARM Conf. Proc. 48, Manila, 1994.
7. J.W. Tucker, Jr., *Marine Fish Culture*, Kluwer Academic Publishers, Boston, 1998.
8. D.Y. Shapiro, in J.J. Polovina and S. Ralston, eds., *Tropical Snappers and Groupers: Biology and Fisheries Management*, Westview Press, Boulder, 1987, pp. 295–327.
9. D.Y. Shapiro, *J. Exp. Zool.* **261**, 194–203 (1992).
10. J.W. Tucker, Jr., *J. World Aquacult. Soc.* **25**, 345–359 (1994).
11. N.A. Hussain and M. Higuchi, *Aquaculture* **19**, 339–350 (1980).
12. J.D. Toledo, A. Nagai, and D. Javellana, *Aquaculture* **115**, 361–367 (1993).
13. L.C. Lim, T.M. Chao, and L.T. Khoo, *Singapore J. Prim. Ind.* **18**, 66–84 (1990).
14. J.W. Tucker, Jr., P.N. Woodward, and D.S. Sennett, *J. World Aquacult. Soc.* **27**, 373–383 (1996).
15. H. Kohno, A. Ohno, and Y. Taki, *Third Asian Fisheries Forum*, Asian Fish. Soc., Manila, 1994, pp. 450–453.
16. T. Brulé, D.O. Avila, M.S. Crespo, and C. Déniel, *Bull. Mar. Sci.* **55**, 255–262 (1994).
17. N. Ruangpanit and R. Yashiro, in K.L. Main and C. Rosenfeld, eds., *Culture of High-value Marine Fishes in Asia and the United States*, Oceanic Inst., Honolulu, 1995, pp. 167–183.
18. Anonymous, *Aqua Farm News* **10**(3), 9 (1992).
19. Y.C. Chong and T.M. Chao, *Common Diseases of Marine Foodfish*, Singapore Prim. Prod. Dept., 1986.

20. W.-Y. Tseng and S.K. Ho, *Grouper Culture—A Practical Manual*, Chien Chieng Publisher, Kaohsiung, Taiwan, 1988.
21. D. Gallet de Saint Aurin, J.C. Raymond, and V. Vianas, *Actes Colloq.* **9**, 143–160 (1990).
22. W.G. Dyer, E.H. Williams, Jr., and L. Bunkley-Williams, *J. Parasitol.* **73**, 399–401 (1992).
23. K.-K. Lee, *Microbial Pathogenesis* **19**, 39–48 (1995).
24. F.H.C. Chua, M.L. Ng, K.L. Ng, J.J. Loo, and J.Y. Wee, *J. Fish Dis.* **17**, 417–427 (1994).
25. D.-K. Hua, M.-L. Cai, and Z.-Y. Zhang, *Third Asian Fisheries Forum*, Asian Fish. Soc., Manila, 1994, pp. 357–360.
26. H. Sako, in K.L. Main and C. Rosenfeld, eds., *Aquaculture Health Management Strategies for Marine Fishes*, Oceanic Inst., Honolulu, 1996, pp. 81–90.
27. FAO, Aquaculture Production Statistics 1987–1996. FAO Fish. Circ. No. 815, Rev. 10, 1998.

GULF KILLIFISH CULTURE

Robert R. Stickney
Texas Sea Grant College Program
Bryan, Texas

OUTLINE

INTRODUCTION

The Gulf killifish (*Fundulus grandis*), also commonly known in Texas as the mudfish or mudminnow and in Alabama as the bull minnow, is a widely used baitfish within the marine recreational fishing community along the Florida and northern Gulf of Mexico coasts of the United States. Historically, the species has been captured from nature and sold through retail bait outlets; however, supplies from that source have waned in recent years, and both interest and activity associated with the culture of Gulf killifish have developed in areas to which that fish is native.

Gulf killifish are found in coastal marshland areas. They are common in grassy bays and canals, and even in adjacent freshwater areas, because they have a very wide tolerance for salinity. While their maximum length (1) is reportedly 180 mm (7.1 in.), fish in excess of 15 cm (6 in.) are not commonly seen. Bait dealers prefer fish of 7.5 cm (3 in.) or slightly less. Bait-sized fish can be produced from eggs in a few months. A brief guide to their culture, by Strawn et al., in 1986, continues to be a highly useful publication, which, unless otherwise noted, forms the basis of the following discussion (2).

LIFE HISTORY

Gulf killifish can be found along both the Atlantic and Gulf coasts of Florida, in the Florida Keys, and along the Gulf of Mexico coast to eastern Mexico. The fish is also found in Cuba (1).

Male Gulf killifish are a uniform greenish-silver in color, while the females are darker and have prominent spots on their bodies and fins. Eggs are laid on suitable substrates, such as submerged or emergent vegetation in nature. Spawning can occur over a period of several months, from early spring to early fall. Reproduction has been known to occur in the salinity range of 3–20 parts per thousand (ppt).

Eggs hatch in 10–21 days; the time depends on temperature and salinity. Higher temperatures reduce incubation time, as do lower salinities. Egg hatching has been reported over a salinity range from 0 to 40 ppt with subsequent good fish growth; survival has been reported over a salinity range from 5 to 40 ppt. Fry survival and growth are reduced in freshwater or at salinities in excess of 60 ppt, yet juveniles and adults can tolerate very low salinities (<1 ppt) and near-zero dissolved oxygen concentrations for at least brief periods of time. Anecdotal reports of high mortalities associated with high pH (9.0–9.5) have occurred (Jack Booth, personal communication), but the mortalities may have been related to high percentages of un-ionized ammonia (NH_3). Un-ionized ammonia is the more toxic form. In solution, total ammonia is a combination of ionized (NH_4^+) and un-ionized ammonia. The percentage of un-ionized ammonia in the mixture increases with increased pH. In any case, one desirable feature of Gulf killifish is their ability to survive under the stressful conditions that exist in bait buckets.

CULTURE TECHNIQUES

The recommended stocking rate for Gulf killifish of 5–7.5 cm (2–3 in.) is 30,000/ha (12,000/ac) in a ratio of two females for each male in advance of the spawning season. Commercial minnow feeds have been used successfully, as have floating catfish feeds and agricultural by-products.

In ponds, natural vegetation can provide suitable spawning substrate, but the standard practice is to employ spawning mats that consist of about 1 kg (2.2 lb) of Spanish moss sandwiched at a thickness of 5–7.5 cm (2–3 in.) between pieces of plastic-coated wire mesh. The pieces of wire mesh are typically 0.7 × 1 m (2 × 3 ft) in size, though any convenient size is suitable. Materials other than Spanish moss have been used, but Spanish moss continues to be commonly employed and is locally available along much of the Gulf coast.

Spawning mats should be evenly distributed along the edges of each brood pond at a rate of no less than 125/ha (50/a). In windy areas, the mats should be placed along the upwind side of the ponds. Placement in that less turbid pond region helps avoid silting in of the mats, which would result in smothering of the eggs.

The mats should be fully submerged and should be suspended a few cm (1 cm = 2.5 in.) above the bottom, as the fish will deposit their adhesive eggs on all mat surfaces. If the mats lie on the bottom, access to their undersides is denied and the mud can work up into the mats and cause eggs to die.

It is possible to allow the eggs to hatch in spawning ponds and then to harvest the young fish after a few months for marketing, but that approach requires sorting (which is not difficult) and can result in high mortality due to cannibalism. The preferred method is to remove the mats after about one week and put them in a fertilized, empty pond. Clean spawning mats can be reintroduced to the brood pond after the egg-laden mats are removed so that more eggs can be collected.

Ideally, the culturist should transfer spawning mats on which 1,000 to 2,000 eggs have been deposited. An estimate of the number of eggs present can be obtained by counting the eggs found in several random meshes of the wire and multiplying the average egg count by the total number of meshes on the upper and lower surfaces of the mat. Ponds should be stocked with a goal of producing fry at the rate of 480,000/ha (160,000/ac). To make the calculation, you can assume that the hatching rate will be between 50 and 80%, and that the percentage will increase as the total number of eggs on a particular spawning mat decreases (i.e., 50% hatch when 2,000 eggs are present on a mat, but as much as 80% when the mat contains about 1,000 eggs).

An alternative to the method described earlier is to stock much larger numbers of eggs [up to 3.75 million/ha (1.5 million/ac)]. When the fry reach approximately 1.25 cm (1/2 in.), they should be transferred to growout ponds in which the stocking rate should be no more than 400,000/ha (160,000/ac). A major advantage of this approach is that the farmer can better control the number of fish stocked in growout ponds. In addition, the fish can be graded so they can be stocked at uniform sizes in the receiving growout ponds.

Commercial feeds are available, as previously mentioned. However, it may be possible to improve fish growth rate and shade out unwanted submerged vegetation by fertilizing ponds to stimulate phytoplankton blooms. One study indicated that fertilization of ponds with 45.5 kg/ha (40 lb/ac) of 12-12-12 (N-P-K) fertilizer was effective to meet this goal (3). Fertilizer was added at weekly intervals in amounts that resulted in maintenance of a Secchi disk reading of 30 cm (12 in.).

Harvesting is typically accomplished by placing baited minnow traps in growout ponds, with feed pellets being used as bait. Seines can also be used to harvest marketable fish.

BIBLIOGRAPHY

1. J.D. McEachran and J.D. Fechhelm, *Fishes of the Gulf of Mexico, Volume 1*, University of Texas Press, Austin, 1998.
2. K. Strawn, P.W. Perschbacher, R. Nailon, and G. Chamberlain, *Raising mudminnows*, Texas Sea Grant College Program publication TAMU-SG-86-506, Texas A&M University, College Station, 1986.
3. P.W. Perschbacher and K. Strawn, *Proc. Annual Con. Southeastern Assoc. Fish Wild. Agencies* **37**, 355–342 (1983).

See also BAITFISH CULTURE.

GYNOGENESIS

ROBERT R. STICKNEY
Texas Sea Grant College Program
Bryan, Texas

OUTLINE

Gynogenesis is the production of viable offspring by a mature female from ova that develop without any paternal contribution to the genome (1,2). Gynogenesis sometimes occurs naturally, and it can be induced artificially in various species through proper manipulation of unfertilized eggs. The offspring, which are all female, may be desirable when a culturist wants to avoid reproduction or when females grow more rapidly than males.

REPRODUCTIVE PLASTICITY

Fish and many invertebrates are particularly diverse with respect to reproductive system function. In fish, one can find nearly every imaginable variation. Included are species that are hermaphroditic (i.e., contain both testes and ovaries), that spend part of their lives as males before becoming females, that spend part of their lives as females before becoming males, and that have distinct sexes throughout their adult lives. In addition, the sex of some fish may not be determined at egg fertilization, or even during embryogenesis. In fact, for many species, it is possible to force the majority of the fish in any population to become either males or females by feeding them hormones for a few weeks at the time of first feeding. (See "Reproduction, fertilization, and selection.")

Aquaculturists often want to produce fish of only one sex, to enable the fish grow more rapidly or to prevent reproduction. For example, female channel catfish tend to grow more rapidly than males, while male tilapia grow more rapidly than females and, in the latter case, often become reproductively active prior to reaching market size. (See the entries "Channel catfish culture" and "Tilapia culture.") Also, all of the fish stocked are of a given sex, they obviously cannot reproduce, which is often a consideration when exotic species are to be stocked.

In addition to using sex reversal to produce animals that cannot reproduce, techniques have been developed that produce sterile polyploid fishes (i.e., fish that have more than two pair of chromosomes in their cells) and gynogenetic fishes.

THE PROCESS OF GYNOGENESIS

In the typical scheme of reproduction for vertebrates, the sexes are separate and diploid; that is, they have two sets of chromosomes in the cells of their bodies. One set of those chromosomes is obtained from the male parent and the other from the female. Various, but usually constant, numbers of chromosomes are present in every species, with one pair being sex chromosomes. Diploid females, in general, have two X sex chromosomes, while males have an X sex chromosome and a Y sex chromosome. As body (somatic) cells divide through a process known as mitosis, the pairs of chromosomes in the cells are replicated. Eggs and sperm cells are different, however. Those cells divide through a process called meiosis and, at the time of reproduction, have only one set of chromosomes and are known as haploid cells. Each female haploid cell has one X chromosome in its set, while haploid sperm may have an X or a Y chromosome. Offspring produced by the joining of the chromosomes in the two gametes are normally diploid, with the sex of the progeny being determined by the male contribution to the pair of sex chromosomes.

During meiosis, also known as reduction division, one set of the chromosomes produces what are known as polar bodies. The process, simplified for this discussion, involves the production of two polar bodies at different stages, with the second polar body being the one of primary interest. Neither of the polar bodies becomes involved in normal reproduction, but both are a means by which the second set of chromosomes is ultimately eliminated from the gamete. In gynogenetic animals, haploid eggs are produced and development occurs, but without any contribution of genetic material from the male sperm. Gynogenetic fish are, however, diploid, because the second polar body is not ejected; its genetic material is recombined with that in the egg to produce the diploid embryo.

In most cases wherein aquaculturists induce artificial gynogenetic development, they do so by eliminating the ability of sperm from the same species to fertilize the eggs (often through irradiation with radioisotopes or, more commonly, through exposure to ultraviolet light), or they use sperm from a distantly related species (3,4). The sperm penetrates the eggs, but no combination of the genetic material occurs. Instead, the second polar body is induced to recombine within the egg to form a diploid cell that can divide to produce all female offspring. Induction of recombination of the ova and polar body can be achieved by heat shock, cold shock, increased pressure, or exposure of the eggs to certain chemicals.

An optional approach, and one that will create clones of the adult female, involves allowing sufficient time to pass after mock fertilization for the polar body to be ejected. The culturist then applies the proper type of physiological shock to the haploid egg to induce the chromosomes to divide within the egg, producing a diploid cell that will, thereafter, divide normally through mitosis and produce viable embryos. While the technique is relatively simple, details of the process vary from species to species (5,6), and it is not widely used by commercial aquaculturists, because often only small numbers of viable gynogenetic animals are produced from any batch of eggs. Other methods to produce a single-sex set of animals have proven to be more effective and often less difficult to undertake in a commercial culture setting.

Gynogenesis continues to be the subject of researchers interested in understanding sex determination and other aspects of genetics and reproduction (6). Fishes and molluscs have been the primary aquaculture species studied by researchers to date.

BIBLIOGRAPHY

1. R.R. Stickney, *Principles of Aquaculture*, John Wiley & Sons, New York, 1994.
2. C.G. Lutz, *Aquaculture Magazine*, July/August, 67–71 (1997).
3. W. Pennell and B.A. Barton, eds., *Principles of Salmonid Culture*, Elsevier, New York, 1996.
4. C.E. Bond, *Biology of Fishes*, Saunders, New York, 1996.
5. C.G. Lutz, *Aquaculture Magazine*, September/October, 73–78 (1997).
6. C.G. Lutz, *Aquaculture Magazine*, January/February, 75–78 (1998).

See also REPRODUCTION, FERTILIZATION, AND SELECTION.

H

HALIBUT CULTURE

Niall Bromage
Carlos Mazorra
Mike Bruce
Nick Brown
University of Stirling
Stirling, Scotland

Robin Shields
Sea Fish Industry Authority
Argyll, Scotland

OUTLINE

INTRODUCTION

The halibut (*Hippoglossus* spp.) is the largest of the pleuronectid flatfish and, in life spans reported to be as long as 50 years, may reach lengths in excess of 2 m and weights of 300 kg. Two major species have interested farmers: the Atlantic halibut (*H. hippoglossus*) and the Pacific halibut (*H. stenolepis*), although at present only the Atlantic species is the subject of significant commercial production. Consequently, much of this review will concern the Atlantic halibut. However, the two species appear very similar in shape, size, and biology, so much so that over the years at times they have been classified as a single species (see Ref. 1 and the section on "Taxonomy" that follows). Consequently, it is likely that the research carried out on the Atlantic halibut will be directly applicable to any future culture of the Pacific species.

Halibut have been identified as ideal fish for farming at higher latitudes as they maintain good growth rates in relatively cold northern waters (0–14 °C) and can reach a predicted market size for farmed stocks of 3–5 kg, approximately 2–3 years postweaning. The Atlantic halibut is one of the most highly priced fish at retail first-sale, fetching two to three times that of salmon, which is the principal farmed fish in European markets. Reductions in wild fish to less than 3,000 tons per year over the past 15–20 years have only served to strengthen their market value.

High fecundity accompanied by the docile nature of the broodfish, a batch-spawning habit and a large pelagic egg, compared to other marine species have facilitated controlled spawning under farm conditions. Although there is now a fast-developing hatchery production of juveniles in Norway, Scotland, Canada, and more recently in Chile, survival through the long egg and yolk-sac incubation periods, plus the time to first-feeding and metamorphosis up to weaning, is relatively poor. As a consequence, there has been only limited growout of the fish, and up to the present, only small quantities of farmed product have reached the market. A major constraint to the farming of halibut is the length of the production cycle from broodfish to marketable fish. Figure 1 summarizes the different stages of the life cycle and their approximate duration. However, much work still remains to be carried out on (*1*) identifying the optimum conditions and requirements of broodstock, (*2*) establishing a reliable supply of good-quality eggs, (*3*) increasing the number and quality of metamorphosed juveniles, and (*4*) appropriate husbandry for the growout phase of the production cycle.

TAXONOMY

The Atlantic halibut was first described by Linnaeus in 1758 as *Pleuronectes hippoglossus*, at which time all flatfish were placed in the same genus. Following more detailed descriptions of their fin structure in the nineteenth century, Cuvier varied the taxonomic position of flatfish and classified the halibut in a new genus: *Hippoglossus*. Initially, the Atlantic and Pacific halibut were both considered to be the same species, first as *H. vulgaris* and then as *H. hippoglossus*. Subsequently, Schmidt in 1930, using studies of meristic characters and scales, placed the Atlantic and Pacific forms as separate species: *H. hippoglossus* and *H. stenolepis*, respectively, which, despite a series of other studies and genetic analyses, remains the taxonomic position today (1).

BROODSTOCK MAINTENANCE

In the wild, mature halibut gather every year in winter in defined spawning grounds at depths of up to 700 m (2,300 ft). Halibut are batch spawners that produce up to 15 but more usually 5–10 batches of 100–200,000 eggs at 3- to 4-day intervals over a 1- to 2-month spawning period (2); different geographical stocks spawn at different times over the period February through May, usually at water temperatures of 5–7 °C (41–45 °F) or less. Males reach maturity at a younger age or size than females; e.g., Faeroese males mature at 4–5 years of age, 55 cm (22 in.) in length, and 1–3 kg (2.2–6.6 lb) in weight, whereas corresponding females mature at 7–9 years of age, 110+ cm (43 in.) in length, and at weights of >15 kg (33 lb). Most large fish captured in the wild are females, with males rarely exceeding 50 kg (110 lb).

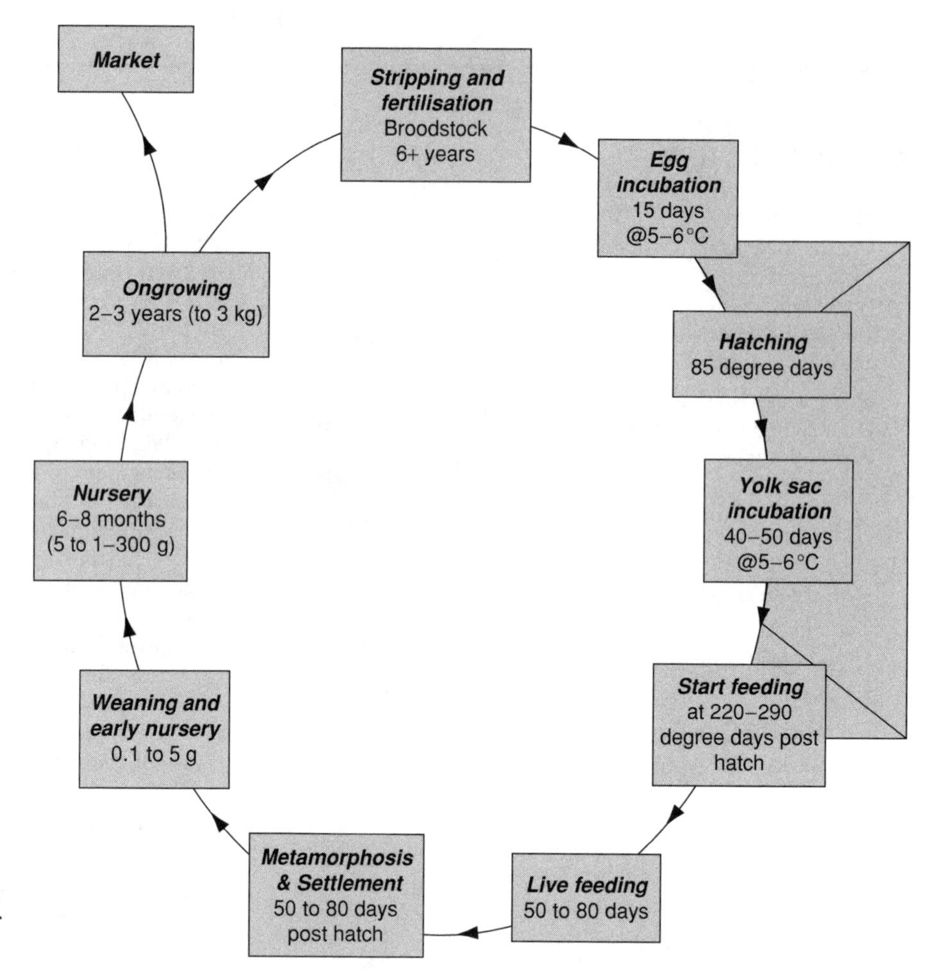

Figure 1. Life cycle of farmed halibut showing approximate times and sizes of the different stages of development.

The physical and environmental requirements of broodstock halibut impose considerable problems on their culture. In order to produce good-quality eggs, temperatures are usually maintained at <6 °C (43 °F) for at least a month before spawning and are continued until all fish in the tank have completed spawning (3). At present, it is not entirely clear whether the necessity for low-temperature water needs to be extended beyond this period. Certainly there is little feeding above 12 °C (54 °F), and higher temperatures should be avoided. In certain locations, either water chilling or heating is necessary in order to achieve a constant temperature of 6 °C (43 °F) or less, which because of the relatively high costs, inevitably has led to investigations of possible water reuse or recirculation (4).

There is also a parallel requirement for full-salinity seawater supplies (32–34 ppt). This has meant that at sites where dilution of seawater has occurred due to freshwater inputs or high rainfall, salinity control is necessary. Most broodstock facilities for halibut rely on pumped seawater supplies, preferably with deep-water inlets because these provide more stable temperature, high-salinity waters, and are less susceptible to the problems of surface water contamination and storm detritus.

Fish are usually held in 5–10 m (16–33 ft) diameter circular tanks at densities of 5–10 kg/m^3, with a tangential inflow, central drain, and water levels of 1–2 m (3–6 ft) depth. Oxygen levels should be at least 6 ppm in the outflow. Flow rates and water exchanges should be as high as is economically possible, with replacement rates of 10–15 water exchanges each day sometimes quoted. Pumping and chilling costs may, however, prove prohibitive and fewer exchanges (2–3/day) or water reuse with biological treatment may be both necessary and feasible (4). Tanks should be covered with netting, for despite their large size, halibut broodstock may jump out. If the tanks are not in a covered building, the netting also serves to protect, in part, against sunburn, which is a particular problem for halibut. Sections of the floors of the tanks are also now commonly covered with a plastic-coated netting which reduces the development of sores that somewhat surprisingly are found on the undersurfaces of fish maintained in smooth-bottomed tanks; the netting is also easier to clean than the gravel previously used.

Generally, halibut are exposed to subdued artificial lighting (approximately 100 lux appears generally suitable), with different broodstock groups exposed to a series of out-of-phase seasonal photoperiod regimes to both advance and delay spawnings (5,6). In this way egg and fry availability are extended beyond that derived from broodstock maintained under simulated natural daylengths.

Broodstock are commonly fed 3–4 times a week to satiation (0.1–0.4% body weight/day) on moist diets containing 60% protein, 17% lipid, and 6–7% moisture. Fresh or frozen wet fish has historically been a major component of broodstock diets. However, because of the variable nutritional composition of this component and the possible inadvertent introduction of pathogens, many operators have switched to using dry ingredients and fully fabricated diets (7,8).

Experience in the United Kingdom has shown that captive halibut are very susceptible to eye damage. Williams et al. (9) described the occurrence of gas-filled cysts and cataracts in captive halibut. There is no clear explanation as to why the eye damage appears, but handling is the most likely reason. The condition frequently progresses until there is a total loss of vision from the affected eye. Some fish become blind in both eyes, with consequent difficulties in feeding and detrimental effects on spawning condition.

A range of parasitic infections is found in broodstock, including the "halibut louse" (*Lepeophtheirus hippoglossi*), *Entobdella hippoglossi*, *Trichodina* spp., and costia (*Ichthyobodo necator*). In captivity, infestations are usually confined to *Entobdella*, which can be treated successfully with either formaldehyde or freshwater baths. Halibut also suffer from *Vibrio* spp., infectious pancreas necrosis (IPN) (10), nodavirus (VNN) (11), and *Cytophaga* spp. (12), all of which can cause mortalities. They may also act as carriers to furunculosis (13,14).

Readers interested in the studies that have been carried out on broodstock and early larval rearing of Pacific halibut should refer to the review of Stickney and Liu (15).

SPAWNING

Historically, broodstock tanks comprise 3–4 pairs of fish. However, much larger groups are now maintained successfully in Norway, Iceland, and Canada. Where possible these groups should be established at least one year before spawning, as it has been suggested that spawning success is influenced by behavioral and social group interactions. Some farms, however, routinely move females from tank to tank, particularly around spawning time, without any apparent ill effects on egg quality.

Impending spawning time can be estimated from the visible abdominal (gonadal) swelling of the females. A more recent development has been the use of ultrasound scanning, which can be used to differentiate the gender of immature fish as well as to monitor maturational state (16).

A major difficulty with spawning marine flatfish is that of predicting the optimal time for stripping of the broodfish (i.e., removal of eggs by gentle hand manipulation) and, in turn, fertilization of the eggs (17,18). Halibut are batch spawners producing a series of egg batches at 70- to 90-hour (often more at lower temperatures, up to 110 hours) intervals (17,18). However, each fish has its own specific ovulatory rhythm, and eggs must be stripped and fertilized within six hours of ovulation at 6–7 °C (43–45 °F); otherwise there is a rapid decline in fertility (17,18). If eggs are kept at 1–3 °C (34–37 °F), they can be kept for up to 12 hours poststripping without noticeable reductions in fertilization rates. Before stripping is attempted one needs to establish the timing of this rhythm. This is achieved by placing some form of egg collector in the outflow water. Eggs are collected from the first 2–3 spawnings of each fish, and the average intervening time between spawnings of that individual broodfish is then used to predict the time interval at which that brood female is monitored and examined for the occurrence of ovulation, over the remainder of her spawnings. Ovulatory periodicity can be disturbed by changes in temperature and handling stress (17). Later egg batches from individual females also tend to occur earlier than might be expected from the recorded periodicity of the first few batches.

Once ovulatory rhythms are established, each broodfish should be examined at the appropriate interval and an attempt made to strip eggs. Eggs have to be stripped from the female fish without coming into contact with seawater, i.e., the fish have to be removed from the water at least until the gonadal pore is visible and out of the water. However, the large size of the brood females and their shape preclude manual lifting, and the following procedure, which requires a minimum of two persons, is commonly adopted. First the tank water level is lowered to approximately 45 cm (18 in.). A table constructed from aluminum, plastic, or other suitable material, free from any sharp edges, is often used to position the broodfish for stripping (Fig. 2). The table, which should be approximately 2 m (6 ft) long, has two fixed and two adjustable legs whose lengths are such that the fixed end of the table is above the water level in the tank while the other adjustable end is submerged. The selected broodfish is gently guided onto the inclined surface of the table, whereupon the submerged end is raised and the adjustable legs lengthened, lifting the fish clear of the water. This procedure is possible with smaller broodstock. With larger fish some farms use a pulley- or winch-operated system to raise the table. After carefully drying the area around the gonopore, eggs are stripped

Figure 2. Stripping of 30 kg (66 lb) halibut female.

from the brood female into 2 to 3-L (~1–3 qt) graduated jugs by gentle posterior–anterior hand pressure.

In some instances females fail to release eggs, which have been ovulated into the ovarian lumen, even when assisted by stripping. The retention of eggs can cause blockages when this material starts to degenerate inside the fish. Blocked females can be injected intramuscularly with antibiotics (amoxycillin or oxytetracycline), which generally induce the release of the egg debris within three days.

A similar procedure is used to strip male fish. Males generally continue to produce milt for the duration of the spawning seasons of the different females in the same tank. Again the fish must be carefully dried and the milt then stripped into 250 mL (8 oz) graduated containers. The gametes should then be transferred to the hatchery in a light-proof insulated box as temperatures must be maintained below 6 °C (43 °F) and light affects osmolarity and, in turn, the buoyancy and handling of the eggs. Ideally the egg incubation facility should be a purpose-built room whose temperature is maintained at 6 °C (43 °F). Before the eggs are fertilized the milt should be checked for motility under a microscope (400×) after adding a few drops of seawater to a drop of milt. Although generally males continue to produce milt throughout the spawning period of the females, toward the end of the season some males may become spent or their milt becomes viscous and very difficult to handle. Providing the fish are not spent, high spermatocrits can be diluted and hence made more fluid by treating brood males with 25 μg/kg of GnRHa (19). Milt can also be stored in the refrigerator for seven days or an adjustable freezer at −4 °C (25 °F) for 29 days without any loss of viability (20). Once a ready supply of milt is assured, then one can proceed with fertilization.

The eggs are "wet fertilized" in a ratio of 250 : 1 : 250 egg to milt to UV sterilized seawater (we use 1000 : 1 : 1000) by first diluting the milt with the sterilized seawater and then quickly adding the mixture to the eggs. The eggs and diluted milt should be gently stirred by hand and then left undisturbed for 20 minutes. Sperm motility ceases after 2–3 minutes although fertilization is complete in less than a minute. After 20 minutes the eggs should have absorbed water and become fully water-hardened. They are then washed in two further changes of sterile seawater to remove excess milt and any egg debris or ovarian fluid. The fertilized eggs are then stocked in egg incubation systems in a dark room at 6 °C (43 °F). A small sample (about 100 eggs) of each egg batch is retained in full-strength seawater in a screw top jar for assessments of fertilization rate or egg quality, as described in the following section.

EGG QUALITY ASSESSMENT

After 16 hours at 6 °C (43 °F), i.e., 96-degree hours, fertilized eggs will have undergone cleavage and reached the eight-cell stage. At this point an assessment can be made of fertilization rate and the predicted future "quality" of the egg batch (i.e., the potential of the eggs to produce viable fry), thus enabling decisions to be made as to whether to discard the egg batch or to stock the incubation systems. Egg and yolk-sac incubation and early rearing are relatively long processes in halibut (50+ days) compared to other marine species. Informed predictive decisions to discard poor batches of eggs, which are likely to exhibit poor survival later in development, can be of great benefit in the saving of staff time and the optimization of usage of limited rearing facilities. For some marine species, egg buoyancy provides an accurate prediction of subsequent egg quality, whereas for other species fertilization rate alone suffices and shows good correlation with survival rates at hatch and first-feeding. However, neither of these is particularly helpful for halibut, and alternative methods have been sought. In the United Kingdom, a useful assessment has been developed which relies on the morphology of the early cell divisions or blastomeres at the eight-cell stage (21). High viability has been demonstrated in egg batches in which the divided cells are closely adherent to one another, have clear, well-defined margins, and show symmetrical cleavages with few or no vacuolar inclusions or droplets (Fig. 3). In practice fertilization rates of 70% or more of stripped eggs can be readily achieved and hatcheries generally incubate all egg batches with rates above 40%.

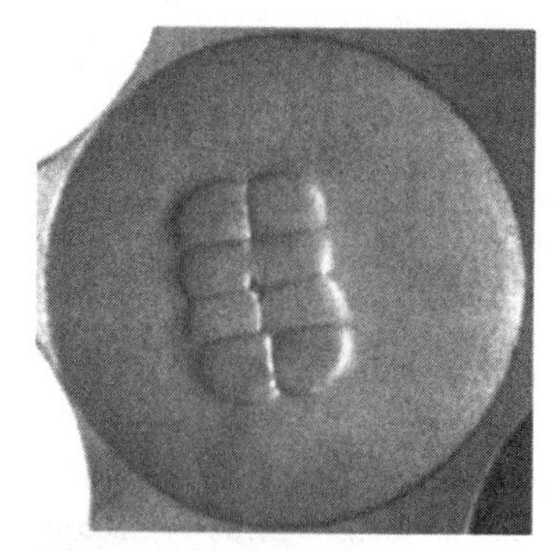
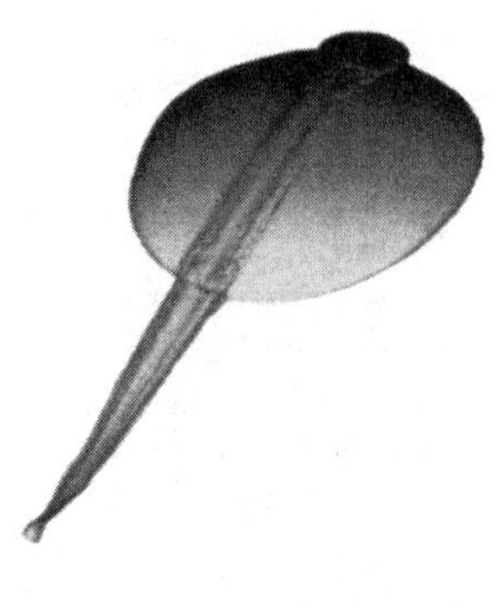

Figure 3. Halibut blastomere at the 8-cell stage of development (left) and yolk-sac larva (right).

EGG AND YOLK-SAC INCUBATION

Procedures for halibut egg incubation and yolk-sac rearing differ considerably from those of many other marine fish and most cultured flatfishes because of the large size of the eggs [3.5 mm (0.14 in.) diameter] and the extended time involved in these periods of development. It takes 85 degree days [15 days at 5–6 °C (41–43 °F)] for the egg to hatch and a further 220–290 degree days to reach first-feeding, with temperatures maintained throughout at 5–6 °C (41–43 °F) (22). Generally, eggs up to hatch are incubated in 100 to 500-L (26–132 gal) cylindroconical tanks in an upwelling current of 1 to 5-μm filtered, UV-sterilized and temperature-controlled seawater. Because of the cost consideration of ensuring a supply of 5–6 °C (41–43 °F) seawater, some hatcheries operate their egg incubation facilities with recirculated water supplies. The eggs are generally neutrally buoyant in full-strength seawater and the aim of the upwelling current is to distribute the eggs evenly throughout the tank; flow rates of 2–5 L (0.5–1.1 gal) per minute are commonly used with stocking densities of 200–1000 eggs L (760–3800 eggs/gal). Usually, each incubator is stocked with a batch of eggs from a single female. Dead eggs are removed from the incubation system on a regular basis using a technique involving a hypersaline (40 ppm) "salt plug" (23); this procedure relies on the fact that dead eggs are nonbuoyant and will sink rapidly through a salt plug whereas live eggs do not pass into this layer. The salt plug is applied by turning off the upwelling seawater flow and introducing approximately 2–10 L (0.5–2.6 gal) of the hypersaline solution into the conical base of the tank. After a few minutes most of the dead eggs enter the "salt plug" and this layer can be removed and the upwelling seawater inflow restarted. Generally, 40–50% of the stocked eggs are lost from each batch during incubation.

After 65 degree days of incubation at 6 °C (43 °F), the eggs are disinfected by immersion in either gluteraldehyde (200–400 ppm) or peracetic acid (200 ppm), for 10 minutes. Disinfection helps avoid the carryover of bacterial contamination, which has been implicated as a major cause of losses in yolk-sac fry (24). Some hatcheries move eggs directly to a yolk-sac production system after disinfection; others initially move them to a clean egg incubation system and only move to the yolk-sac system after hatching. This has the advantage of not transferring any of the egg debris from hatching to the yolk-sac system.

A wide range of tank sizes has been used for yolk-sac rearing from 400 to 10,000 L (105 to 2630 gal) capacity (23,25). All are cylindroconical and use 5–6 °C, (41–43 °F) filtered, sterile seawater in an upwelling current (Fig. 4). Water inflow rates range from 1 to 5 L (0.25 to 1.3 gal) per minute depending on tank capacity (23). Inlet pipe diameters are critical to the flow characteristics of the upwelling current; generally most hatcheries use 15–25 mm (0.6–1.0 in.) diameter in the smaller-volume tank inlets, but diameters may reach 100 mm in the larger-volume system. All systems should be in the dark.

After disinfection the eggs or larvae are stocked into the yolk-sac tanks at densities 15–40 eggs or larvae/L (57–152/gal). Hatching occurs after 85 degree days (see Fig. 3); it can be synchronized by applying light (20 lux) to the eggs (26) for 24 hours and then switching off the lights; hatching then occurs after a few hours.

After hatching the yolk-sac larvae remain in the system for a further 200 degree days (up to 270 degree days), with any dead larvae being removed daily using the salt plug technique. With use of this procedure, survivals through the yolk-sac stage are usually 40–50%. However, substantial losses have occurred in both Norwegian and Scottish hatcheries approximately 150 degree days after hatching, at which time some of the larvae are seen to collect at the surface and then gradually sink down to the conical base of the tank and die. In an attempt to avoid these losses, many hatcheries are transferring larvae into first-feeding tanks up to 100 degree days before they are expected to take first feed (27). The larvae start to develop a positive phototaxis around 160 degree days posthatch and this behavioral response is used to advantage in helping to draw the larvae to the surface of the tank for collection and subsequent transfer to the first-feeding tanks.

LARVAL REARING

Larval rearing can be usefully divided into two phases. During the first of these the larvae are fed on live feed organisms of which the main ones are the brine shrimp (*Artemia* sp.) and copepods e.g., *Eurytemora* (28). In addition to the live feed, a range of microalgae, including *Nanochloris, Nannochloropsis*, and *Isochrysis* spp., are added to the tanks before stocking the fish. This "green water" approach significantly improves the initiation of start feeding in the larvae probably by its influence on light levels and/or on the behavior of the live feed organisms or the halibut larvae, rather than any effect on nutrient intake. Figure 5 shows larvae at the onset and at a later stage of live feeding.

The second phase of larval rearing involves the weaning of the larvae onto inert dry diets (29). This is carried out after the larvae have completed metamorphosis. Metamorphosis comprises a major developmental reorganization of

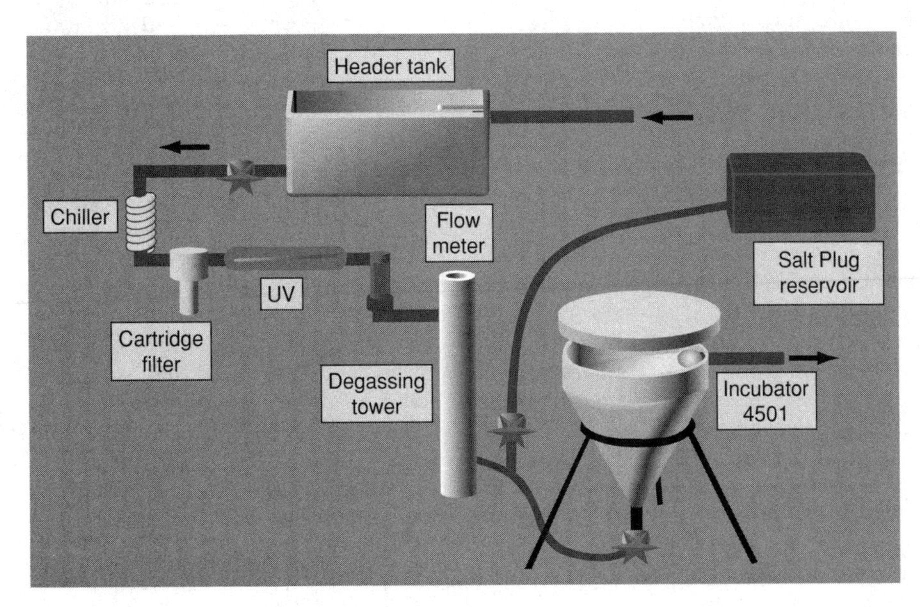

Figure 4. Tank system for yolk-sac larvae.

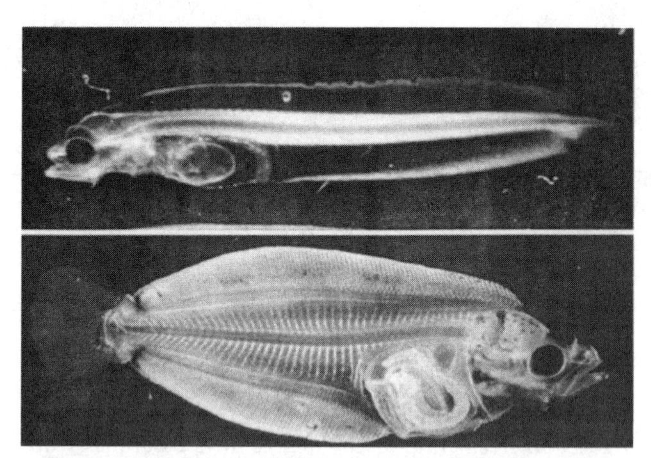

Figure 5. Early (top) and later (bottom) live-feeding larvae.

tissues and physiology in flatfish, during which the larva changes from a bilaterally symmetrical, free-swimming, pelagic form to an asymmetric, flattened, and demersal or bottom-living form (30). This involves the left side of the fish effectively becoming the "ventral" or bottom-facing surface with the left eye having migrated from the left to the right sides so that both eyes are on the upper surface (31). The upper and lower surfaces of the fish also change their pigmentation, becoming darker and lighter, respectively. There are also profound changes in the structure and physiology of the digestive system (32). Metamorphosis is usually completed at about 650 degree days and the larvae are then transferred to purpose-built weaning tanks over the next 15–20 days.

The tanks used for initial feeding with live feeds are usually circular, 1.5–6.0 m (4.5–20 ft) in diameter, 1–2 m (3–6 ft) deep, and flat bottomed with a tangential inflow and central gridded outflow drain. Unlike the earlier egg and yolk-sac stages, the first-feeding larvae require light for feeding (100 lux is commonly used but some hatcheries use brighter light). Tanks are often provided with a peripheral collar, which serves to shade the walls of the tanks from the above-tank lighting (33). The photoperiod is 24 hr continuous light (LL) and the water temperature ideally 6–8 °C (43–46 °F). Light intensity levels are gradually increased during the period of larval rearing up to 1000 lux by day seven post-first feeding. Temperature can also be gradually increased up to 12 °C (54 °F).

Before stocking with larvae, the tanks are "greened" with appropriate microalgae at concentrations up to 10^7 algal cells/L; these concentrations are maintained by further daily additions of algae up to approximately 500 degree days posthatching. Larvae are stocked at 200 degree days posthatch by some hatcheries, although recently there has been a move to delay first-feeding as late as 270 degree days. Stocking densities range from 2 to 7 larvae/L (8–27 gal). Water flow rates are increased from 1 to 5 L/min (0.3–1.3 gal/min) during first-feeding, although some hatcheries also use aeration to maintain an upwelling current and a good dispersal of larvae, live feed organisms, and microalgae within the tank.

By 220–270 degree days posthatch larvae are ready to take first feed. Different hatcheries use a range of different protocols principally because procedures are still being optimized (4,27). Generally, hatcheries that have access to supplies of copepods prefer to use them because it is widely recognized that they provide the ideal nutritional source for halibut (34). However, because of difficulties in culturing or obtaining supplies from the wild, copepods are invariably in short supply. Hence, their use has to be rationed. Most hatcheries start-feed with enriched *Artemia* as halibut larvae at 13–14 mm (0.51–0.55 in.) in length are already larger in size than most marine larvae at start-feeding (4,35). A few units initially use rotifers, although these are generally considered to be too small for halibut to prey upon.

Artemia can only be used successfully as a first-feed for halibut larvae if its nutritional composition is first supplemented with enrichment of additional nutrients, in particular, sources of essential polyunsaturated fatty acids and amino acids (36,37). Enrichment aims, in part, to simulate the nutrition provided by copepods or by the natural zooplankton diet of wild halibut. Enrichment procedures for *Artemia* show considerable variation among hatcheries because the detailed nutrient requirements of first-feeding halibut, and in turn, the appropriate enrichment levels needed to meet those requirements, are not fully known. In a number of hatcheries the *Artemia* are enriched with mixtures of Super Selco (INVE Aquaculture, Belgium) and a marine heterotroph Algamac 2000 (Aquafauna Biomarine Inc., USA). These are then added to the tanks at the rate of 1,000 *Artemia* nauplii/L (3,800/gal) twice a day; this corresponds to a daily ingestion rate for each halibut larva of approximately 2000–3000 prey organisms/day. As the nutrient composition of the enriched *Artemia*, and in particular, the polyunsaturated lipid levels, deteriorate over time, it is essential that the fish eat all the added *Artemia* within a few hours. Accordingly, *Artemia* additions are matched closely to consumption levels. Where possible the *Artemia* are supplemented with 20% copepods. Using such proportions, hatcheries have achieved 30% survival through first-feeding and 95% fully pigmented fish. Most of the mortalities occur during the initial phases of first-feeding, and after 350 degree days, losses are minimal. Tank hygiene can become a problem, so every day dead larvae and debris are removed by siphoning.

By 600–700 degree days, the fish are fully pigmented, metamorphosed, and starting to settle on the bottom of the tanks. They are then ready for transfer to weaning systems. If the fish are in good condition and around 100–130 mg in size, the process of weaning can generally be accomplished in 2–3 weeks (29,38).

Weaning tanks are usually circular 2–4 m (6–12 ft) tanks preferably with a conical sump covered by a perforated grille and a central double-sleeved standpipe, both tank adaptations which aim to improve the self-cleaning properties of the tank, as the inert diets can cause hygiene problems. Water temperature should be 10–15 °C (50–59 °F), with 5 μm filtered, UV sterilized water and continuous 1000 lux white light. Fish are stocked at 2–5 larvae/L (8–19/gal). As fish at this stage rest on the bottom, it is more appropriate to define stock levels in terms of bottom surface area. Fish stocked at 2–5/L (8–19/gal) equates to 1–2,000 fish/m^2.

A wide range of dry feeds is available and these can be fed continuously to satiation via automatic feeders. Over the first week of weaning some live food should be given with the dry feed (1000–3000 *Artemia*/fish/day). As soon as the fish are weaned, they grow rapidly and should reach 500 mg (0.02 oz) in size by 80 days post-first feeding. At 20 g (0.67 oz) in size the fish can be moved into the larger circular tanks [up to 5 m (16.5 ft) in diameter]. Aggression can be a significant problem; consequently, the fish should be regularly graded to avoid disparities in size within single tanks.

GROWOUT

It is considered that much of the initial growout of halibut will occur in sea cages or pens, mainly because many of the operators who have become involved with halibut are salmon farmers wishing to diversify their production. However, there are difficulties in the cage design and construction of cages for halibut because of their bottom-living habit. Hence, tank-based holding systems with solid floors may be more appropriate. This development becomes increasingly likely with the continuing reductions in costs of pump ashore systems and advances in recirculation technology. At present three of the biggest producers of market-size halibut in Norway and Scotland are using land-based pumped seawater facilities. One of those farms has constructed a series of shelves within a deep tank to maximize the stocking capacities of the tanks.

Problems associated with cage culture of halibut have included loss of appetite or mortality following exposure to excessive current speeds and swells; mortalities resulting from high surface water temperatures; sunburn in shallow cages without adequate shading; predation from seals and otters attacking fish lying on cage bottoms; and difficulties with removal of mortalities from cage bottoms covered with other live fish.

Most of the modified salmonid cages used for halibut utilize some form of rigid frame and anchor ropes around the four sides of the cage bottom in order to provide a flat and square floor on which the halibut can settle. There have been trials with cages with solid floors but these are highly subject to tidal damage and movement. Although we might have expected that the halibut would prefer a solid floor, it appears that netted floors are quite acceptable to the fish. It is clear, however, that the cage sites must be in sheltered positions preferably with waves no more than 0.5 m (1.5 ft) in height and currents less than 5 cm/sec (2 in./sec). For this reason some farms have experimented with submersible cages as the surface effects of wind are significantly reduced at only a few meters (yards) of depth.

Fish can be safely transferred to sheltered sea cage sites when they have reached 100 g (3.3 oz) in weight. Initially, stocking densities are 10–15 kg/m^2 with further reductions generally when stock levels reach 40–50 kg/m^2, although some farms have stocked up to 100 kg/m^2 without any problems. Fish continue to feed and grow down to 3–4 °C (37–39 °F), although the optimum temperatures for growth are 13–16 °C up to 25 g; 11–13 °C, 25–100 g; 10–12 °C, 100–500 g; 9–11 °C, 500–1000 g; and 7–11 °C, 1000+ g (38,39). Artificially extended daylength can, however, increase growth at the lower temperatures. Fish up to 500 g and 1–2 kg in weight grow at 0.5–1.0% body weight/day, respectively, and will reach 2 kg (4.4 lb) in size 1.5 years after the transfer to sea cage and market size (3–5 kg) (6.6–11 lb) after 2–2.5 years. Mortalities range from 3 to 9% during growout. FCRs of 1 : 1 are achievable with a 45–50% protein and 25% oil-extruded pellet offering 22–25 mj kg^{-1} of energy. Work on different dietary formulations and their effects on body composition and sensory assessments for halibut is only just beginning (40).

A major problem with the growout of flatfish, and halibut in particular, is the poorer growth of male fish. Most male halibut fail to reach a market size of 3 kg (6.6 lb) by the time they reach three years of age. This constraint has led to preliminary work on producing all-female or sterile triploid lines of fish (38,41). However, a complication for halibut at present is that it is not clear how gender is controlled and whether the male or the female is the homogametic sex. Clearly, commercial production of gender-manipulated stocks must await clarification of these issues.

OVERALL COMMENT

It is clear that halibut will be the next major cultured species in colder, higher-latitude waters. At present a number of problems still remain. However, significant numbers of weaned larvae are now beginning to appear on the market, and it is likely that world production will approach 1000 tons early in the twenty-first century. The continuing high prices achieved for halibut together with the diminishing wild fishery and the improvements in culture will no doubt ensure the future growth of halibut farming.

ACKNOWLEDGMENTS

The authors are grateful to the BBSRC Foresight Challenge, the Fishmongers' Company, the NERC, MAFF, the BHA, and Trouw Aquaculture who provided support for some of the studies published here, and to a range of commercial halibut hatcheries in Norway, Canada, and the United Kingdom who provided additional data and information.

BIBLIOGRAPHY

1. R.J. Trumble, J.D. Neilson, W.R. Bowering, and D.A. McCaughran, *Can. Bull. Fish Aquat. Sci.* **227**, 84 (1993).
2. E. Kjorsvik and I. Holmefjord, in N.R. Bromage and R. Roberts, eds., *Broodstock Management and Egg and Larval Quality*, Blackwell Science, Oxford, 1995, pp. 169–196.
3. N.P. Brown, N.R. Bromage, and R.J. Shields, in F.W. Goetz and P. Thomas, eds., *Proceedings of the Fifth International Symposium on Reproductive Physiology of Fish*, University of Texas, Austin, 1995, p. 181.
4. R.J. Shields, B. Gara, and M. Gillespie, *Aquaculture* **176**, 15–25 (1999).
5. P. Smith, N. Bromage, R. Shields, L. Ford, J. Gamble, M. Gillespie, J. Dye, C. Young, and M. Bruce, in A.P. Scott,

J.P. Sumpter, D.E. Kime, and H.S. Rolfe, eds., *Proceedings of the Fourth International Symposium on Reproductive Physiology of Fish*, University of East Anglia, Norwich, England, 1991, p. 172.
6. B.T. Björnsson, O. Halldorsson, C. Haux, B. Norberg, and C. Brown, *Aquaculture* **166**, 117–140 (1998).
7. M. Bruce, F. Oyen, G. Bell, J. Asturiano, B. Farndale, M. Carrillo, S. Zanuy, J. Ramos, and N. Bromage, *Aquaculture* **177**, 85–98 (1999).
8. M. Bruce, C. Mazorra, N. Jordan, W. Roy, G. Bell, and N. Bromage, in *Proceedings of the Sixth International Symposium on Reproductive Physiology of Fish*, Bergen, 1999 (in press).
9. D.L. Williams, A.E. Wall, E. Branson, T. Hopcroft, A. Poole, and W.M. Brancker, *Vet. Rec.* **136**, 610–612 (1995).
10. E. Biering and Ø. Bergh, *J. Fish Dis.* **19**, 405–413 (1996).
11. S. Grotmol, G.K. Totland, K. Thorud, and B.K. Hjeltnes, *Dis. Aquat. Org.* **29**, 85–97 (1997).
12. G. Hansen, Ø. Bergh, J. Michaelsen, and D. Knapskog, *Int. J. Syst. Bacteriol.* **42**, 451–458 (1992).
13. Ø. Bergh, B. Hjeltnes, and A.B. Skiftesvik, *Dis. Aquat. Org.* **29**, 13–20 (1997).
14. I. Bricknell, T. Bawden, D. Bruno, P. MacLachlan, R. Johnstone, and A. Ellis, *Aquaculture* **175**, 1–14 (1999).
15. R.R. Stickney and H.W. Liu, *Aquaculture* **176**, 75–86 (1999).
16. C. Mazorra, R. Shields, P. Smith, and N. Bromage, in *Proceedings of the Sixth International Symposium on Reproductive Physiology of Fish*, Bergen, 1999 (in press).
17. B. Norberg, V. Valkner, J. Huse, I. Karlsen, and G.L. Grung, *Aquaculture* **97**, 365–371 (1991).
18. N. Bromage, M. Bruce, N. Basavaraja, K. Rana, R. Shields, C. Young, J. Dye, P. Smith, M. Gillespie, and J. Gamble, *J. World Aquacult. Soc.* **25**, 13–21 (1994).
19. E. Vermeirssen, R. Shields, C. Mazorra, and A. Scott, *Fish Physiol. Biochem.* (in press).
20. K.J. Rana, S. Edwardes, and R. Shields, in P. Lavens, E. Jaspers, and I. Roelants, eds., *Larvi '95, Fish and Crustacean Larviculture Symposium*, Eur. Aquacult. Soc. Spec. Pub. No. 24, Gent, Belgium, 1995, pp. 53–56.
21. R.J. Shields, N.P. Brown, and N.R. Bromage, *Aquaculture* **155**, 1–12 (1997).
22. I. Lein, I. Holmefjord, and M. Rye, *Aquaculture* **157**, 123–135 (1997).
23. A. Mangor-Jensen, T. Harboe, J.S. Hennø, and R. Troland, *Aquacult. Res.* **29**, 887–892 (1998).
24. T. Harboe, I. Huse, and G. Øie, *Aquaculture* **119**, 157–165 (1994).
25. T. Harboe, S. Tuene, A. Mangor-Jensen, H. Rabben, and I. Huse, *Prog. Fish-Cult.* **56**, 188–193 (1994).
26. J.V. Helvik and B.T. Walther, *J. Exp. Zool.* **263**, 204–209 (1992).
27. T. Harboe and A. Mangor-Jensen, *Aquacult. Res.* **29**, 913–918 (1998).
28. B. Gara, R. Shields, and L. McEvoy, *Aquacult. Res.* **29**, 935–948 (1998).
29. G. Rosenlund, J. Stoss, and C. Talbot, *Aquaculture* **155**, 183–191 (1997).
30. K. Pittman, A.B. Skiftesvik, and L. Berg, *J. Fish Biol.* **37**, 455–472 (1990).
31. A.M. Kvenseth, K. Pittman, and J.V. Helvik, *Can. J. Fish. Aquat. Sci.* **53**, 2524–2532 (1996).
32. F.S. Luizi, B. Gara, R.J. Shields, and N. Bromage, *Aquaculture* **176**, 101–116 (1999).
33. T. Harboe, A. Mangor-Jensen, K.E. Naas, and T. Naess, *Aquacult. Res.* **29**, 919–923 (1998).
34. R.J. Shields, J.G. Bell, F. Luizi, B. Gara, N.R. Bromage, and J.R. Sargent, *J. Nutr.* (in press).
35. T. Naess, M. Germain-Henry, and K.E. Naas, *Aquaculture* **130**, 235–250 (1995).
36. L.A. McEvoy, T. Naess, J.B. Bell, and Ø. Lie, *Aquaculture* **163**, 237–250 (1998).
37. I. Rønnestad, *Aquaculture* **177**, 201–216 (1999).
38. N.P. Brown, Ph.D. Thesis, University of Stirling, Scotland, 1999.
39. B. Björnsson and S.V. Tryggvadottir, *Aquaculture* **142**, 33–42 (1996).
40. R. Nortvedt and S. Tuene, *Aquaculture* **161**, 295–313 (1997).
41. I. Holmefjord and T. Refstie, *Aquacult. Int.* **5**, 169–173 (1997).

See also Brine shrimp culture; Flounder culture, Japanese; Larval feeding—fish; Plaice culture; Sole culture; Summer flounder culture; Winter flounder culture.

HARVESTING

Robert R. Stickney
Texas Sea Grant College Program
Bryan, Texas

OUTLINE

INTRODUCTION

Once an aquatic species has been reared to sufficient size for marketing, the animals must be harvested and, in many instances, hauled some distance for processing or for live sale. At a minimum, harvesting usually involves capturing a significant proportion of the animals within a given culture chamber (pond, raceway, tank, netpen) within a short time period. This can be accomplished by reducing the water volume and netting the animals, seining, trapping, and by other means. For animals that are to be kept alive after harvest, it is important to minimize stress to the extent possible.

Some foodfish aquaculturists incorporate processing into their operations, but many send the live animals to processing plants in water or on ice. Ornamental fish are usually packaged for shipping at the site of production and taken to airports for dispersal to retail dealers. The information presented here is adapted from Stickney (1).

HARVESTING INTENSIVE CULTURE SYSTEMS

One of the primary advantages of intensive culture is the ease with which harvesting can be accomplished, in systems of virtually all types. In closed recirculating water systems, as well as in open systems, utilizing tanks, or raceways, harvesting is often merely a matter of draining the culture chambers and collecting the animals in dip nets. In large raceways and circular tanks, the water may be partially drained and the animals herded into a relatively small volume with the aid of movable screens (Fig. 1). As the number of animals is reduced through dipnetting, the volume confining those remaining can be decreased further by repositioning the screens and lowering the water level.

Some aquaculturists, including those who rear fish in ponds, employ a technique known as continuous harvesting. As the culture animals reach market size, they are removed from the system. Replacement juveniles are added either after each such harvest or less frequently. While no given pond, raceway, or tank is harvested every day (days or weeks may separate harvesting activity in any particular culture chamber), harvesting is conducted almost every day in large facilities.

If continuous harvesting is practiced in intensive culture systems, grader screens can be used in tanks or raceways. This allows submarketable fish to escape while crowding the marketable fish into a small space from which they can be dipnetted, often without requiring raceway draining. Another option is to rapidly drain the culture chamber through a net bag or some type of structure that will retain the animals while allowing the water to flow away.

In most cases, animals harvested from intensive culture systems can be loaded directly into hauling tanks for transport to the processing plant. All such systems, including those located in buildings, should be designed to provide easy access to all culture chambers by hauling vehicles; or some other suitable technique, such as fish pumps, (Fig. 2), should be used to move the fish from the tanks to the hauling truck. A fish pump can remove fish from culture chambers after they have been crowded. Fish pumps can move fish relatively long distances, can sort the fish in some cases, and when properly designed and operated, will not damage the animals.

Figure 1. Workers crowding fish in a raceway (Idaho, USA). The fish are removed from the crowding screen with dip nets.

Figure 2. A large fish pump (Israel). The fish are sorted by workers who separate them into marketable and submarketable groups. Submarketable fish are returned to a pond for further growout.

Harvesting aquaculture animals from cages can be a relatively simple matter. Cages can be towed to shallow water or to a dock where the fish can be removed with dip nets or fish pumps (Fig. 3). If dip nets are used, the fish are typically placed in baskets and carried by hand or lifted by means of a gantry fitted with a block and tackle to the hauling truck or to a frame attached to the dock. A scale hung between the block and tackle and the basket provides a means of weighing each basket of fish as it is being loaded.

Lifting cages from the water is not generally feasible, since most are not sufficiently strong enough to be removed from the water when filled with harvestable fish. If a cage ruptures during the process, the fish will be lost. An exception is small cages [typically no larger than 1 m^3 (9 ft^3)] like those used for research (Fig. 4). The small cages are not usually heavily stocked and are commonly constructed of sturdy materials that will accommodate removal from the water while the fish are present.

Netpen harvesting is more involved than harvesting associated with cages, because netpens are much larger in both surface area and depth. Typically, harvesting involves pulling up the netting to reduce the volume of the netpen, dipnetting out some of the fish, pulling up more net to concentrate the remaining fish, dipping more

Figure 3. For harvest, cages can be towed to a dock where the fish are removed with dip nets or by using a fish pump (Arkansas, USA).

Figure 4. Research cages tend to be small enough to allow removal from the water for harvest (Texas, USA).

fish, and so forth, until all the fish have been harvested. It may be necessary to grade the fish during the process. Submarketable individuals can be placed in a different netpen for additional growout. If the facility is attached to the land (e.g., attached to some access structure such as a rigid or floating dock), it may be possible to drive vehicles relatively close to pens that are being harvested. For offshore facilities, the fish are transported to the shore on boats.

HARVESTING EXTENSIVE CULTURE SYSTEM

The harvesting of ponds is easier when they are constructed in regular shapes, with properly sloped banks, proper depth, easy access by vehicles, large drain lines that allow rapid and complete emptying, and the incorporation of a harvest basin near the drain. Many ponds lack one or more of these features. Each is important, but the regularity of pond shape is the least concern. In some instances, it is more convenient to lay out a series of ponds that conform to natural variations in terrain than to greatly alter a site to accommodate square or rectangular ponds. It is important to have smooth and clean pond bottoms. If culture ponds are not fitted with drains, the water can be pumped out of them at the time of harvest.

It is common practice in the catfish industry to use the continuous harvesting approach previously described. Ponds can be kept in production for a period of three or more years by harvesting marketable fish every few weeks and by restocking with replacement fingerlings a few times a year. Many other species of fish are maintained in ponds through a single growing season and are then harvested *en masse*. Whatever technique is used, at some point it becomes necessary to drain each pond. As time passes, ponds that are used for continuous harvesting accumulate organic matter, impairing water quality and reducing productivity. For example, phosphorus levels increase from year to year in undrained ponds (2). In addition, stunted fish continue to eat but may never reach market size. As their numbers increase (a result of intermittent restocking), the impact on culture economics can become significant. In most cases, even ponds used for continuous harvesting are drained and completely harvested every few years, although some farmers have apparently kept ponds in continuous production for 15 or more years. When ponds are drained, the bottoms are allowed to dry and are disked to promote oxidation of organic matter. The ponds are then placed back in production.

Harvesting should be planned in advance, and precautions should be taken to avoid stress to the extent possible, particularly when the harvested animals are to be live-hauled to market. Feeding should be discontinued at least 24 hours before harvesting. Automatic feeders and other obstacles to seining should be removed, and water quality should be examined to ensure that optimum conditions are maintained. Temperature is of particular importance. To help avoid stress, harvesting should be undertaken during the coolest part of the day (very early in the morning) during warm months.

While seining of full ponds may be undertaken when subsamples of fish are collected or in conjunction with the continuous harvest technique, harvesting is usually conducted with partial draining prior to initial seining, even when harvest basins are present. Typically, the water level is reduced, perhaps by half, and a seine is then passed through the pond to capture a portion of the fish present. The volume of the pond is then further reduced, and depending on the size of the pond, subsequent seine hauls may be made before the final harvest occurs in the harvest basin. In ponds with no harvest basins, seine hauls are made until the pond volume is reduced to 10% or less of the original volume, then the rest of the water is removed, and the remaining fish are harvested by hand.

Small ponds can be seined by hand (Fig. 5). Tractors or trucks are required to pull seines in large ponds. When the fish have been concentrated in a harvest basin, they may be dipnetted into baskets or pumped, eventually ending up in live-hauling trucks.

Harvest seines should be approximately 1.5 times longer then the pond is wide to ensure that they will bow out during harvesting operations. For additional capacity, seines may be equipped with a bag located in the middle. Seine bags provide additional capacity while reducing escapement. The depth of the seine should be

Figure 5. A seining operation (Philippines).

at least twice the depth of the water in the pond being harvested. The mesh size should be small enough to retain harvestable animals but no smaller, since reduction in mesh size increases the resistance of the net in the water making it more difficult to pull. In addition, the cost of netting increases with decreasing mesh size.

The upper rope on a seine (float or head line) typically consists of a rope to which floats made from cork, plastic, Styrofoam, or some other buoyant material are strung at intervals. The bottom rope (lead line) is designed to keep the seine in contact with the sediments. Lead lines may be ropes to which lead weights are attached at intervals or ropes that have a lead core. In ponds with muddy bottoms, traditional lead lines are often not efficient since they tend to burrow into the sediments and dig up mud, which weighs down the seine and causes it to roll up. In soft-bottomed ponds, seines with mud lines tend to be more effective. A mud line is composed of a number of relatively small diameter ropes wound loosely together. The ropes are made from a material that readily absorbs water (e.g., cotton). Mud lines tend to maintain contact with pond bottoms without digging into the sediments. Thus, escapement under seines, so equipped, is reduced. Such lines tend to wear rapidly if used in ponds with firm sediments.

Harvesting is the most labor-intensive activity associated with an aquaculture operation. Several people are required on the typical seine crew, even if trucks or tractors are used to do the bulk of the work. The aquaculturist may be required to live extra help during harvesting, and this added expense should be taken into consideration during business planning.

SPECIALIZED HARVESTING TECHNIQUES

Harvesting sessile marine animals such as oysters and mussels requires techniques different from those used for motile species. Oysters reared on the bottom floor can be harvested manually by picking intertidal oysters up at low tide or by tonging or dredging for subtidal animals. A mechanical oyster harvester developed in South Carolina was described by Collier and McLaughlin (3,4). Oysters grown in trays and on long lines suspended from rafts or floats are manually harvested. Mussels are grown on the bottom, on poles, and on long lines. The harvesting technique is based on how the mussels are grown, but generally mirrors the methods used for oysters. Clams can be dredged or grown in trays and manually harvested. Scuba divers may also be used for harvesting benthic animals such as abalone.

Crawfish are harvested by trapping, although mechanical harvesting devices are being developed (5). During the harvest season, which extends from 60 to 180 days in Louisiana, traps are set out at intervals over the pond bottom. Various factors influence the efficiency with which crawfish are trapped. The catch can be affected by water quality, the amount of forage present, crawfish density, climate, trap design, the bait being used, and of course, trap density (6). Romaire and Pfister (7) compared catch rates with trap densities of 25, 50, 75, and 100 traps/ha (10, 20, 30, and 40 traps/ac) and found that 3.3 times more crawfish were captured at the highest density as compared to the lowest. The catch rates at 50 and 75 traps/ha (20 and 30 traps/ac) were 2.8 and 1.6 times more than the rate at 25 traps/ha (10 traps/ac). A variety of trap designs have been used in such studies.

Traps are baited with dead fish or commercially manufactured bait and are usually checked once or twice a day. One impetus for developing manufactured baits is the increasing cost of the fresh fish that have traditionally been used, such as gizzard shad (8). Rach and Bills (9) evaluated the effectiveness of three types of commercial crawfish baits, comparing them with dead fish bait. Results showed that commercial baits were easier to handle, did not have noxious odors, and did not require refrigeration—all advantages over dead fish. Meriwether (10) found that commercial crawfish bait containing about 20% protein resulted in the highest crawfish yields.

In small ponds, the culturist may wade through the pond to check traps, but in most instances boats are used. The boat may be operated by one or two people. If two people are involved, one drives the boat while the other handles the traps. As each capture location is approached, a newly baited trap is placed in the water and the trap that has been in the pond is removed, emptied, and rebaited. The newly baited trap is then placed in position at the location of the next trap in the line. With the technique described, it is possible to keep the boat in continuous motion during harvesting.

Following harvest, crawfish are often held in flowing water long enough to allow their digestive tracts to be purged. This process leads to improvement in overall appearance and increases the market value of the animals. Purging in spray systems works as well as flowthrough systems (11). Purging in spray systems requires no more than 40 hours (12). If the proper spray rates and crawfish densities are employed, mortality is less than 5% during the purging period.

Crawfish producers began producing soft-shell crawfish in 1985 (13). The technique involves trapping immature animals, placing them in culture trays at high density, and feeding them. Individuals that are about to molt are identified and removed to molting trays to prevent cannibalism. Newly molted crawfish are packaged and frozen. Soft-shell crawfish bring a premium price and can be eaten whole, whereas processing for tail meat results in only about a 15% dress-out.

Some ponds have drains that empty into a receiving ditch immediately adjacent to the pond levee. In such ponds, an alternative approach to collecting fish or shrimp in a harvest basin during pond draining is to place a bag of appropriate mesh size over the effluent end of the drain pipe. As water is released from the pond, many of the aquaculture animals are swept into the bag. The technique works fairly well with shrimp and other species that are not particularly strong swimmers in strong current. Workers must still walk the pond bottom to pick up animals left behind when the pond was drained.

BIBLIOGRAPHY

1. R.R. Stickney, *Principles of Aquaculture*, John Wiley & Sons, New York, 1994.
2. W.D. Hollerman and C.E. Boyd, *Aquaculture* **46**, 45–54 (1985).
3. J.A. Collier and D.M. McLaughlin, *J. World Maricult. Soc.* **14**, 297–301 (1983).
4. J.A. Collier and D.M. McLaughlin, *J. Shell. Res.* **4**, 85 (1984).
5. R.P. Romaire, *J. Shell. Res.* **7**, 210–211 (1988).
6. R.P. Romaire, *J. Shell. Res.* **8**, 281–286 (1989).
7. R.P. Romaire and V.A. Pfister, *N. Am. J. Fish. Manage.* **3**, 419–424 (1983).
8. C. Burns and J.W. Avault, Jr., *J. World Maricult. Soc.* **16**, 368–374 (1986).
9. J.J. Rach and T.D. Bills, *N. Am. J. Fish. Manage.* **7**, 601–603 (1987).
10. F.H. Meriwether, *J. World Aquacult. Soc.* **19**, 166 (1988).
11. T.B. Lawson and C.M. Drapcho, *Aquaculture Engineering* **8**, 339–347 (1989).
12. T.B. Lawson, H. Lalla, and R.P. Romaire, *J. Shell. Res.* **9**, 383–387 (1990).
13. D.D. Culley and L. Doubinis-Gray, *J. Shell. Res.* **8**, 287–291 (1989).

See also CAGE CULTURE; NET PEN CULTURE; RECIRCULATING WATER SYSTEMS; TANK AND RACEWAY CULTURE.

HISTORY OF AQUACULTURE

ROBERT R. STICKNEY
Texas Sea Grant College Program
Bryan, Texas

OUTLINE

Aquaculture has a history that likely goes back over 4,000 years, possibly beginning in Egypt with tilapias or China with carp. Until the past few decades, aquaculture was conducted almost exclusively in ponds, at relatively low densities. In the United States, aquaculture began in the 19th century, mostly in association with state and federal government hatcheries that produced fish to stock public waters. Little changed until the 1960s, when commercial aquaculture began to grow at an exponential rate, not only in the United States, but around the world. Today, channel catfish is the leading species of fish cultured in the United States, while marine shrimp culture dominates in many tropical nations. Various species of algae, shellfish, and finfish are produced worldwide, with salmonids (trout and salmon) dominating cold-water areas. The public is familiar with cultured finfish (many species), molluscs (clams, oysters, and mussels), and crustaceans (marine and freshwater shrimp), but seaweeds are also widely cultured, as are, to a lesser extent, echinoderms (sea urchins), cephalopods (cuttlefish, squid, and octopus), amphibians (frogs), and reptiles (sea turtles and alligators).

THE FIRST FEW THOUSAND YEARS

The first known written document on aquaculture appeared in China in 475 B.C. (1). That very short volume by Fan Li, called *Fish Breeding*, discussed carp culture, which may have already been going on for centuries when the book was written. History reports that oyster culture was being undertaken during the period the Roman Empire was flourishing. Even earlier, the ancient Egyptians may have been involved in fish culture: Hieroglyphs in the tombs of the pharaohs depict what appear to be tilapia. (See the entry "Tilapia culture.") Whether the fish were being cultured or captured is not known, but pond culture is a distinct possibility. If that was the case, then perhaps both the Egyptians and the Chinese were involved in aquaculture as much as 4,000 years ago.

Early aquaculture was not restricted to the great civilizations of the world: Coastal fishponds were constructed in Hawaii many centuries ago by people thought to have been settlers from Polynesia. The ponds were undoubtedly stocked by the tides; then the inflow channels were closed, and the animals within the ponds were allowed to grow prior to harvesting them.

The Chinese may have grown marine shrimp as early as the eighth century B.C., and the Japanese were referring to shrimp culture by 730 A.D. (1). The basis for modern shrimp culture was established by Japanese scientists in the 1930s, although the first commercial shrimp farms in Japan were not constructed until the early 1960s. Latin America and various Asian nations quickly adopted shrimp production and now produce the majority of the world's cultured shrimp. Total production exceeded 900,000 tons in 1995 (2).

Polyculture (see the entry "Polyculture") — the rearing of two or more compatible species in the same culture system — appears to have been developed by the Chinese. Various species of carp with different food habits have long been cultured together to take advantage of all the natural food in the pond, in addition to supplemental feeds provided by culturists. (See the entry "Carp culture.") This form of culture has survived for millennia and continues to be pursued in China today.

In Europe, at least primitive forms of aquaculture, including attempts at enhancing existing fish populations, were put in place by the eighteenth century (3). By the late nineteenth century, culture methods as sophisticated as or more sophisticated than, those employed in the United States had been adopted (4).

PUBLIC AQUACULTURE IN THE UNITED STATES

Aquaculture in the United States began as a commercial enterprise. In the 1850s, fish culturists in the eastern part of the nation developed the technology necessary to spawn and rear brook trout. Several became proficient in the technique and found that they could sell their fish for a reasonable profit.

During the latter half of the nineteenth century, it became apparent to at least some individuals that valuable commercially fish species were becoming depleted. One of those concerned about the problem was Spencer F. Baird, who was affiliated with the Smithsonian Institution (Fig. 1). Baird had a vision to create a federal fisheries agency and worked with Congress for its establishment. Ultimately, he was not only successful, but was named the first commissioner of the US Fish and Fisheries Commission in 1871 (3).

Figure 1. Spencer F. Baird as he appeared as a young man.

Baird was responsible for the establishment of the Woods Hole Oceanographic Institution, in addition to being the visionary behind the development of the US government's fish hatchery system. He enlisted the assistance of the practicing fish culturists of the day to spawn and distribute species of interest. Among those early culturists were Seth Green and Livingston Stone.

Species that initially received the most attention were the Atlantic salmon and the American shad. Others that held interest were the flounder, largemouth bass, striped bass, and walleye, as well as various species of oysters, and lobsters. Within a few years of the establishment of the commission, large numbers of fishes and invertebrates were being transported across the country on railroad cars.

Livingston Stone (Fig. 2) was dispatched by Baird to California to spawn chinook salmon (then known by their Indian name of Quinnat salmon, Fig. 3), as well as rainbow trout, and to ship eggs and fry to the east coast. Since Atlantic salmon were declining, it was thought that Pacific

Figure 2. Livingston Stone.

Figure 3. Drawing of Quinnat salmon that appeared in the *Manual of Fish Culture*, produced by the US Fish and Fisheries Commission in 1897.

salmon might be successfully introduced to create a new fishery. At the same time, efforts to reestablish Atlantic salmon on the east coast were being pursued. Over the years, Atlantic salmon were also shipped to the west coast in attempts to establish the species there. While those attempts failed, rainbow trout were successfully introduced well outside of their west-coast range. In addition, spawning populations of striped bass and a few other Atlantic coast species were successfully introduced to the Pacific coast of the United States.

Soon after the US Fish and Fisheries Commission was established, it became involved in shipping fish to, and receiving them from, various foreign nations. Brown trout were imported from Europe, while rainbow trout and Pacific salmon were shipped to Europe and New Zealand. Baird, a strong proponent of establishing the common carp in US waters, was not the first to bring carp into the United States, but he was the first to launch a campaign to establish the species on a massive scale. Initially, he arranged to have carp shipped in from Europe, and then he established hatcheries for the production of the species and saw to it that carp were widely distributed throughout the country. His praise of carp resulted in orders from numerous states and territories. Trains were kept busy for several years trying to meet the demand for a fish that ultimately fell into disrepute.

The commission expanded over the years and, early in 1904, became the US Bureau of Fisheries, which was housed within the Department of Commerce. Things remained the same until 1950, when the Bureau of Fisheries was divided into the Bureau of Commercial Fisheries (BCF) and the Bureau of Sport Fisheries and Wildlife (BSFW), which later became the US Fish and Wildlife Service (USFWS). The BSFW was placed in the Department of the Interior, while the BCF remained in the Department of Commerce. In 1971, the BCF was renamed the National Marine Fisheries Service (NMFS), a division within the National Oceanic and Atmospheric Administration, still within the Department of Commerce. Basically, the NMFS has responsibility (including regulatory responsibility) for marine commercial fisheries, while the USFWS is primarily involved with freshwater fishes, whose use is chiefly recreational. The two agencies share responsibility for anadromous fishes, some species of which (e.g., chinook and coho salmon) are of interest to both recreational anglers and commercial fishermen.

Throughout the years, the various federal fisheries agencies have established hatcheries across the nation. Fish have been produced in the hundreds of billions for stocking both coastal and inland waters. During that time, state fish and game commissions also actively constructed hatcheries and distributed fish. Indeed, some state hatchery programs preceded federal efforts.

Both state governments and the federal government also established research facilities that made many contributions to advancing the art of fish culture in the United States. Among them were the USFWS laboratories in Hagerman, Idaho, and Cortland, New York, which concentrated on fish nutrition and dietary considerations; the Western Fish Disease Laboratory in Seattle, Washington; and the Stuttgart, Arkansas, laboratory, established in 1958, with a satellite facility in Marion, Alabama, established in 1959. The latter two laboratories initially conducted research to assist in the development of the then fledgling commercial channel catfish industry. Later, their activities were expanded to other species. The Marion laboratory was closed in 1995, and the Stuttgart facility was transferred to the US Department of Agriculture, although its basic mission did not change. Aquaculture research has also been, and continues to be, conducted at various NMFS laboratories located along the east and west coasts of the United States, as well as the coast of the Gulf of Mexico.

Agency hatcheries produced fish of many species. Easily reared species like trout, salmon, and, eventually, catfish could be grown to fingerling size and released. Those species had a relatively good chance of survival after they were liberated into streams, lakes, or reservoirs. The technology was also developed to rear largemouth bass and sunfishes for stocking as fingerlings. Many other species were successfully spawned, but rearing beyond hatching was difficult or impossible given the technology available in the late nineteenth and early twenteeth centuries.

A major problem was associated with the fact that many important commercial and recreational species have very small eggs and, as a result, similarly small larvae. Feeding fish and invertebrates that are nearly microscopic was beyond the ability of the early fish culturists and, in fact, continues to be a major impediment to the commercial culture of many species today. The approach was to release newly hatched fry into inland and coastal waters. It is unlikely that many of the billions of walleye, shad, lobsters, and various other species of fishes and invertebrates released into these waters survived. The majority undoubtedly served as food for predators or entered the food chain as detritus. There is no evidence that stocking US waters with such larvae ever increased the numbers of animals that were entering the commercial or recreational fisheries.

State fish hatcheries, some of which were established prior to the formation of the US Fish and Fisheries Commission, were also actively producing fish and stocking their local waters with them. Among the many major achievements of the state public hatchery system were advances in the development of prepared feeds, the identification of diseases and the development of treatment protocols for them, engineering advances associated with water systems, and the identification of methods to measure and control water quality.

Fish production continues in both state and federal hatcheries, although the number of facilities, particularly in the federal hatchery system, has declined in recent years. An exception is the Columbia River basin, where federal hatcheries continue to produce hundreds of millions of salmon in Washington, Oregon, and Idaho in an attempt to mitigate the loss of natural spawning habitat caused by the construction of dams, agricultural practices, and industrial development. Each of the states mentioned operates hatcheries in the Columbia River basin.

State hatcheries throughout the nation produce fish for stocking waters within their jurisdiction. At one time, nearly any landowner in the country who had a farm pond could obtain free fish from the state, but that activity is increasingly being turned over to the private sector. Streams, lakes, and reservoirs within the various states continue to receive hatchery fish, however.

Pacific salmon (coho and chinook) were successfully introduced into the Great Lakes during the 1980s to replace top predators that had been decimated for a variety of reasons. A multibillion-dollar recreational fishing industry resulted, and a number of hatcheries were established to maintain the fisheries that developed.

DEVELOPMENT OF COMMERCIAL AQUACULTURE IN THE UNITED STATES

Academic Programs

Around 1960, commercial foodfish production, largely restricted to trout prior to that time, began to grow in the United States. The trout industry, centered in Idaho, expanded rapidly, and interest developed in foodfish production with warmwater species in the southern states. First among the warmwater fishes to be evaluated were buffalo (*Ictiobus* sp.), and a modest amount of production resulted. The warmwater fish farmers quickly turned to catfish, however, and buffalo culture waned.

Dr. H.S. Swingle was an entomologist with a passion for fishing that ultimately caused him to refocus his research activities on fish culture. In 1957 and 1958, he presented information which demonstrated that channel catfish could be reared profitably in ponds (5,6). Within a few years, the catfish industry, helped by research at Auburn and a few other universities, as well as the Stuttgart and Marion federal laboratories, began to develop and expand.

Commercial aquaculture caught the attention of researchers in several universities, particularly in the southern and western states. The University of Washington, which had been involved in salmon culture for decades, expanded its program significantly. Dr. Lauren Donaldson, who began his pioneering and universally recognized research on trout and salmon at that university in 1930, had established a salmon run that returned salmon to a pond adjacent to the campus hatchery. The fish, thus, literally returned to the classroom. Dozens of graduate students received advanced training and hundreds of thousands of children were introduced to the unique life cycle of Pacific salmon at that university's facility. Expertise in nutrition was provided by John Halver, who had been a leader in the development of prepared salmonid feeds during his tenure with the US Fish and Wildlife Service. William Hershberger was retained to run the hatchery after Donaldson's retirement in 1973 and to conduct research in fish genetics. Ken Chew, a world-renowned expert on molluscs, conducted research and trained students in the culture of oysters, mussels, and geoducks.

While the University of Washington had already developed a powerful program in aquaculture even prior the growth of the catfish industry, Auburn University, with a slower start, made up ground rapidly. Swingle got the Department of Fisheries and Allied Aquacultures established and became its head. Wayne Shell assumed that position when Swingle retired. Between them, they created a program with a highly respected faculty and pond facilities that were not paralleled anywhere in the country. An international program was established too, and it obtained numerous grants and contracts to assist in the creation or expansion of aquaculture programs and facilities in developing nations and to train personnel to manage those programs and facilities.

Auburn's faculty has been unique in that its core membership remained constant until the 1990s, when several faculty members retired. Among those who were active in the department for more than 20 years were Wilmer Rogers and John Plumb (fish diseases), R. Thomas Lovell (nutrition), Claude Boyd (water quality), and R. O'Neal Smitherman (reproduction). A number of other aquaculturists with backgrounds in complementary disciplines were added over the years, so that the core faculty not only failed to decline in numbers, but actually rose significantly.

During the 1970s, other universities expanded their faculties' expertise and research activities in the discipline of commercial aquaculture. In the west, Oregon State University became a leader, along with the University of Washington, in conducting research that led to the development of ocean ranching and salmon net-pen culture. The National Marine Fisheries Service in Washington, which collaborated with the University of Washington, was also heavily involved in net-pen culture development at the federal level. In California, the most visible program was developed at the University of California at Davis, where various warmwater species were subjects of research. In recent years, paddlefish and sturgeon culture have been studied in detail as well. The University of Idaho made numerous contributions, and further west, the University of Hawaii developed a program involving the culture of tropical fishes and invertebrates. The University of Arizona houses the world's leading program in shrimp viral diseases, despite the fact that the state had neither a seacoast nor, until recently, any commercial shrimp culture! Still, Donald Lightner's program is recognized throughout the world, and his expertise is in great demand.

William Lewis at Southern Illinois University (SIU) established the Cooperative Fisheries Research Laboratory at about the same time that Swingle was forging the Auburn program. Lewis retained Roy Heidinger as his assistant director, and together they conducted a large number of studies in fisheries and aquaculture and taught numerous students who went to other universities and became well known in their own right. SIU was among the leading universities involved in striped-bass and walleye culture research.

In the north central United States, the University of Wisconsin became a leader in yellow-perch culture. Purdue University in Indiana and Michigan State University developed active aquaculture programs. Schools in Minnesota became involved in research on what turned out to be an abortive attempt to develop aquaculture facilities in association with lakes created in abandoned iron mines in that state. The program was terminated due to complaints that nutrients added to the lakes from fish feed and waste products were entering the region's groundwater.

In the south, a strong program was developed at Mississippi State University, both on the main campus in Starkville and at the school's Stoneville research station. Robert Wilson developed a widely recognized catfish nutrition research program in the biochemistry department on the main campus of Mississippi State during the 1970s. H. Randall Robinette and Louis d'Abramo of the agriculture college were among the faculty who conducted research on fish and crustaceans. Craig Tucker, joined later by Edwin Robinson, led the program in Stoneville.

Active aquaculture programs were also established at the University of Tennessee, North Carolina State University, Clemson University in South Carolina, the University of Georgia, and the University of Florida. The University of Arkansas was engaged in aquaculture for a few years, but much of the activity in that state ultimately was centered at the branch campus at Pine Bluff rather than at the main Fayetteville, campus. Texas A&M University and the University of Texas, particularly at its Marine Science Institute in Port Aransas, developed aquaculture programs in the 1970s that remain active today.

With the development of technology for producing Atlantic salmon in net pens and the collapse of New England cod fishery in the 1990s, interest in aquaculture has been increasing in portions of the region. Universities in Maine and New Hampshire, in particular, have become increasingly active in aquaculture research and education.

In the mid-1980s, the US Department of Agriculture (USDA) created six regional aquaculture centers: a tropical center in Hawaii and US territories in the Pacific, and five others that represent the western, midwestern, central, southern, and north Atlantic regions, respectively. Several hundred thousands of dollars are allocated annually to each region to support university research. Partnerships with state and federal agencies provide additional input into the system.

The Department of Commerce operates the Sea Grant College Program, with 29 participants in the coastal and Great Lakes states. Over its more than 30-year history, Sea Grant has provided millions of dollars in aquaculture research at universities, and today it even operates (as does the USDA) a Small Business Innovative Research Program that provides funding on a competitive basis to promising private enterprises involved in aquaculture development.

Many other universities and colleges have programs in aquaculture research. In some states junior colleges and trade schools offer certificates in aquaculture.

Parallel with the development of academic programs at universities and colleges was that of professional aquaculture associations. On December 20, 1870, the American Fish Culturists' Association was formed. In 1878 it changed its name to the American Fish Cultural Association and in 1884 the American Fisheries Society (AFS).

The Fish Culture Section of the AFS was established in 1973. The World Mariculture Society was founded in 1969 and became the World Aquaculture Society in 1986. Both societies enable scientists to present their research findings at annual meetings, and both publish respected scientific journals in which scientists can communicate their findings.

Agricultural research in the United States developed in response to pleas from farmers for more information that could be provided only through a scientific approach. The result, the creation of the land-grant college system, revolutionized agriculture. When foodfish aquaculture was beginning to grow in the United States, the scientists tended to be somewhat ahead of the practitioners. Research results often demonstrated the feasibility of an approach before the industry even asked the appropriate questions about it. That state of affairs quickly turned around with the growth of the commercial aquaculture industry.

As the 1980s waned, opposition to aquaculture development began to be voiced. Discharge water was blamed for degrading the quality of receiving waters, while marine culture systems were seen to interfere with navigation, access to fishing grounds, and the views of property owners. Various other issues, including the use of exotics and excessive noise, have also been raised. It is clear that future aquaculture development will have to be environmentally sensitive. With heavy competition for coastal lands and increasing demands on the nation's surface and ground waters, many feel that the future of commercial aquaculture rests in the use of recirculating water systems on land and offshore systems at sea.

Trout Culture

Brook trout culture in the United States was commercial in nature as far back as the 1850s (4). However, while a modest amount of commercial fish culture took place during the intervening decades, including some development of rainbow trout farming in the Pacific Northwest, it was not until the 1960s that foodfish aquaculture became visible to the general public.

The so-called Magic Valley of Idaho on the Snake River is also famous as the Thousand Springs region. Pure, cold water flows freely into the Snake River in the vicinity of Twin Falls, Idaho, from a number of subterranean rivers that originate far to the north in Canada. By harnessing some of that water, fish farmers in Idaho monopolized the commercial rainbow trout farming industry for many years. The water was essentially funneled through raceways and then released into the river. While the trout culture industry has become more dispersed in recent years, at one time over 90% of all cultured trout produced in the United States came from the Magic Valley. A relatively small number of producers was responsible for developing the industry. Vertically integrated, those producers not only spawned and grew trout, but even processed the fish and delivered them to the marketplace.

Catfish Culture

Three species of catfish can be readily reared in aquaculture: the channel catfish (*Ictalurus punctatus*), white catfish (*I. catus*), and blue catfish (*I. furcatus*). Research was conducted on each of these species and on crosses among them by the BSFW and, later, the USFWS at its Stuttgart, Arkansas, laboratory, and in Marion, Alabama. John Guidice in Stuttgart and Kermit Sneed in

Marion were the main researchers on the project. While each of the species was adaptable to culture, it quickly became clear that the channel catfish was the species of choice.

The Stuttgart laboratory was established by Congress in 1958 to conduct research in support of the fledgling catfish industry. During its early development, the industry was centered in Arkansas in rice country where water was plentiful and it was relatively easy for rice farmers to convert all or part of their land to catfish farming. By the early 1970s, it became clear from falling water tables that the potential for expanding catfish farming in Arkansas was limited.

In their search for new locations, catfish farmers discovered the Mississippi Delta region, a flat area dominated by cotton production and featuring a relatively shallow water table that seemed to be inexhaustible. The industry soon became centered in the state. Production continued to expand in Arkansas, however, and other states also became involved. While Mississippi dominates production today, channel catfish are produced in all the southern states, including Texas. Limited commercial catfish production occurs in many other states as well, including California and Idaho. (The latter state has geothermal water, which allows year-round production.)

Research in federal, state, and university laboratories, together with studies conducted within commercial facilities, led to the technology upon which the industry depends today. Surprisingly, when in 1958, Swingle demonstrated the feasibility of producing catfish profitably (5,6), he placed the pond-bank market price at $0.50/lb ($1.10/kg). In 1997, the pond-bank value of catfish, was less than $0.80/lb ($1.76/kg) after 40 years and significant inflation. The fact that catfish farmers continue to make a profit relates to the fact that production was much more efficient in 1997 than in 1958. In 1997, sales of catfish were in excess of $350 million. The total number of operations in the United States (the only nation that produces channel catfish to any extent) was in excess of 1200. Catfish were being produced on over 170,000 acres (68,000 ha), of which some 100,000 acres (40,000 ha) were in Mississippi.

Catfish were once of interest as food only in the southern states. Through aggressive marketing and the production of a high-quality product, the catfish farming industry has developed a national market and is expanding on the international front as well.

Other Freshwater Species

In addition to trout and catfish, various other freshwater foodfishes and invertebrates have been reared commercially in the United States since the 1960s. Among the most widely cultured are various species of tilapia (in particular, *Oreochromis aureus* and *O. niloticus*) and crayfish (see the entries "Tilapia culture" and "Crawfish culture").

Tilapias, exotic tropical fishes from Africa and the Middle East, were introduced into the United States in the 1960s and are being reared in ponds in locations where they can survive during winter (Hawaii, south Florida, and, during most years, extreme south Texas), in outdoor facilities supplied by geothermal water and in indoor facilities where the water can be kept warm enough for their survival with electrical or fossil-fuel heaters. While not widely recognized in the United States even a few years ago, tilapias (sometimes called St. Peter's fish) are highly marketable and increasingly in demand for their excellent flavor and texture.

Crawfish have long been consumed in large numbers in Louisiana, but in recent years they have come to be accepted in markets throughout the United States. Most of the crawfish that enter the American market come from the Atchafalaya River basin in Louisiana and are captured from nature. That supply is augmented by a significant level of production in shallow ponds (often converted rice ponds) in the state. Additional crawfish production has come from adjacent states, but is minor in comparison with the activity in Louisiana.

A considerable amount of interest, but limited production, has been associated with such fishes as walleye, yellow perch, and sturgeon. All three are marketed, but because of difficulties associated with their culture or limited popularity, no major industries have developed around them. That situation could change, however, as appropriate technology is developed and the fish are more widely and aggressively marketed.

During the 1970s, a great deal of interest arose in the culture of freshwater shrimp (in particular, *Macrobrachium rosenbergii*, the so-called Malaysian giant prawn). Several commercial operations were established in Hawaii and a few other states. Ultimately, nearly all of those ventures, which were augmented by a significant amount of university research, failed. Freshwater shrimp did not keep well when frozen and had to be reared to very large sizes if premium prices were to be obtained. When marine fish culture came on the scene in the 1980s, interest in freshwater shrimp culture waned, not only in the United States, but worldwide. Currently, little or no commercial freshwater shrimp is cultured in the United States.

Other species that have received attention by researchers and commercial aquaculturists are recreationally important fishes, such as largemouth bass, sunfish, and crappies, and species that are of interest primarily as bait — for example, minnows and goldfish.

The ornamental fish industry in the United Stated is centered in Florida, where well over 100 species of freshwater ornamentals are routinely produced in a number of commercial operations. Cultured ornamental fishes entering the aquarium trade are indistinct from fishes imported from abroad.

Finally, at least some level of plant culture is taking place in the United States. Decorative plants for aquariums, water lilies, and plants used in restoring or creating wetlands are produced commercially, although the industry is not large compared with various other freshwater aquaculture enterprises.

Anadromous Fishes

Interest in the commercial culture of salmon arose in the Pacific Northwest during the 1970s. Net-pen technology had been developed, and a large facility was put in place in Puget Sound, Washington, in which pansized coho salmon were produced. Research conducted in government and university laboratories first focused on coho and chinook salmon. In the meantime, the NMFS was charged with the task of maintaining and spawning threatened stocks of Atlantic salmon that were shipped from Maine to the Puget Sound region of Washington where the work was conducted. The researchers learned that Atlantic salmon were better able to adapt to being reared in captivity than their Pacific counterparts were, and by the mid-1980s, several net-pen facilities were in place in Puget Sound.

Interest in Atlantic salmon culture also developed in Maine. In Washington, opposition to net-pen culture by property owners and environmentalists became very strong and led to long delays in permitting as prospective fish farmers attempted to address the claims that they were responsible for pollution, excessive noise, the transmission of disease to wild populations, and various other unacceptable consequences. As a result, the industry remained small. Opposition led by environmentalists also existed in Maine, but with the collapse of the cod fishery and the desire to retrain fishermen, salmon aquaculture quickly became an accepted practice, and those who were willing to become farmers of salmon or other aquatic species became eligible for federal assistance. Today, salmon culture continues to grow in Maine, while it is virtually stagnant or declining in Washington.

Spawning and rearing of Atlantic salmon for the first year are conducted in land-based facilities using freshwater. Smolts weighing about 1 1/3 ounces (40 g) are introduced into net pens, which they grow out of in approximately two years.

Net-pen salmon culture is illegal in Oregon and Alaska, but salmon ranching can be practiced in both states. Salmon ranching involves the spawning and rearing of the fish (Pacific salmon) up to the smolt stage, after which they are released to grow to adulthood at sea. When the fish return to spawn, they come back to the hatchery of their origin, where they can be captured and marketed (with a portion being retained as brood stock for the next generation).

In Oregon, for-profit salmon ranching has been attempted, but with only limited success, in part because commercial and recreational fishermen have access to the returning fish. Thus, there are often insufficient numbers coming back to the hatcheries to afford a profit once the required numbers of brood fish are collected. The quality of the flesh of fish that are allowed to reach full maturity is poor, so those fish have little value as human food.

In Alaska, not-for-profit ocean ranching is practiced. Hatcheries contract with commercial fishermen to capture the returning fish (less the number needed for brood stock). Ocean-ranching hatcheries utilize species not historically found in the rivers on which the hatcheries are located, so fish returning to the immediate vicinity of a hatchery represent those that were produced in that hatchery. In Oregon, ocean-ranched fish mix freely with wild and government hatchery fish, so there is no easy way to sort them out or charge a fee for the ocean-ranched fish.

Another anadromous fish that has been the subject of a considerable amount of research and commercialization is the striped bass. There is also culture of hybrids between striped bass and white bass. While it is possible to rear brood stock, most hatcheries depend on wild stock as brood animals.

Striped-bass and hybrid striped-bass farming in ponds is concentrated in the southeastern United States, although the fish can be reared virtually anywhere in recirculating water systems. Such systems have been established in many states, with modest production following. Rearing can be in either freshwater or saltwater.

Marine Species

Commercial marine fish culture in the United States is a fledgling industry. A few commercial operations produce red drum (primarily in Texas), and opportunities for expansion of that industry seem good, particularly in ponds. High-intensity systems may not be profitable, and the cost of rearing red drum in offshore facilities seems much too high, given the current value of the fish.

There is considerable interest in flounder culture, particularly in the northeastern and southeastern United States, but also along portions of the Gulf coast. The technology for spawning and rearing flounders, primarily the summer flounder and the southern flounder, has been developed, and a few commercial enterprises have been established.

American eels have been cultured, and at least one commercial producer grows Japanese eels. Since eels cannot currently be spawned and reared to the elver stage in captivity, elvers are collected from nature and introduced into culture facilities for growout. There is a limited, though significant, market for eels in the United States, and extensive markets for eels exist in Europe and Asia.

Tuna have been spawned in captivity, and early rearing has been successfully achieved, although the technology has yet to reach the commercial aquaculture sector. Tuna can bring very high prices in the marketplace, so there is interest in their culture. Current commercial culture depends on capturing juvenile tuna and rearing them in net pens, although that will change with time as hatchery technology improves.

Similarly, the basic information required for rearing dolphins (mahimahi) is now known, and the species sustains some commercial activity in Hawaii. Milkfish, Japanese flounder, and other species are also being reared or evaluated as candidates for culture in Hawaii. Other marine fishes that are being studied and that hold potential as commercial species are the Atlantic halibut, the Pacific halibut, various members of the snapper family, and the pompano.

Among the invertebrates, oysters have long been cultured in the United States. Much of the culture has been extremely extensive in nature, often involving little more than control of predators on leased beds. Oyster hatcheries do exist, however, and are responsible for supplying many of the oysters that are grown on the west coast of the United States. Triploid oysters — those carrying an extra pair of chromosomes — were developed several years ago and are now being stocked in many areas. Triploids do not become sexually mature, so they do not divert energy into gonad development, thereby maintaining excellent eating quality throughout the year.

Diseases have devastated oyster production along the east and Gulf coasts of the United States. Research has resulted in disease-resistant oysters that could eventually lead to explosive expansion of production of these molluscs. Offsetting the demand is public fears of transmission of disease from oysters that become contaminated in coastal regions where sewage pollution occurs. Enhanced inspection, successful efforts to reduce or eliminate sewage runoff into coastal waters, and the recent introduction of pasteurized (pathogen-free) oysters into the marketplace should help overcome the negative attitudes.

Other molluscs that are reared commercially in the United States include mussels, clams, and, to a lesser extent, scallops and abalone. Those molluscs, as well as oysters, can be reared on the sea bottom or suspended from ropes or rafts.

Marine shrimp culture is a major industry in portions of Latin America and Asia, and while much of the research that led to successful shrimp culture was developed at the Galveston, Texas, laboratory of the National Marine Fisheries Service in the 1960s and 1970s, the US commercial industry is small. Commercial growout efforts began during the 1960s, with much of the activity centered in Florida. Many operations came and went, but success with domestic species was elusive. When the Latin American industry achieved success by using species native to that part of the world, US shrimp farmers shifted from Gulf of Mexico species (white shrimp, *Litopenaeus setiferus*; brown shrimp, *Farfantepenaeus aztecus*; and pink shrimp, *Farfantepenaeus duorarum*) to exotic Latin American species (whiteleg shrimp, *Litopenaeus vannamei*; and blue shrimp, *L. stylirostris*) and, in a few cases, species from Asia (primarily the giant tiger prawn, *Penaeus monodon*).

Interest in shrimp culture has remained high in the United States, but strict permitting regulations, viral diseases, and problems associated with nutrients and levels of suspended solids in effluents from shrimp farms have been major issues in some areas where shrimp are reared. Another problem is the inability of pond shrimp farmers to produce two or more crops a year in most of North America because of low water temperatures over several months in the late fall through early spring. Some interest has developed in rearing shrimp in closed water systems, and a few commercial facilities have been established, but production remains at the pilot-scale level. Pond facilities for rearing shrimp can be found in South Carolina, Texas, and Hawaii; all of those farms use exotic species as the primary culture organism, but interest is once again turning to domestic species, although a considerable amount of research may be required before domestic shrimp can supplant exotics in the US shrimp-farming industry.

There is some seaweed culture in the United States, although it is not a large or very visible industry. Additional information on culture techniques for most of the species discussed in this article can be found elsewhere in the volume.

GLOBAL AQUACULTURE

In 1995, global aquaculture production exceeded 27,750,000 tons. (See Table 1.) Of that amount, 52.8% was finfish, 24.5% aquatic plants, 18.3% molluscs, and 4.1% crustaceans. Minor species made up the remaining 0.3% (2). The leading species produced are listed in Table 2.

Table 1. Top 20 Aquaculture Nations in Terms of Production in 1995[a]

Nation	Production (tons)
People's Republic of China	17,600,000
India	1,609,000
Japan	1,405,500
South Korea	1,017,250
Philippines	812,300
Indonesia	719,400
Thailand	464,200
United States	413,400
Bangladesh	321,500
Taiwan	286,200
Norway	282,500
France	280,800
Italy	224,900
Vietnam	219,400
North Korea	216,000
Chile	206,300
Spain	138,300
Malaysia	132,700
United Kingdom	93,800
Ecuador	91,200

[a]From (2).

Table 2. Top 15 Aquatic Animals Produced in Aquaculture During 1995[a]

Species	Production (tons)
Silver carp (*Hypopthalmichthys molitrix*)	2,555,000
Grass carp (*Ctenopharyngodon idella*)	2,102,700
Common carp (*Cyprinus carpio*)	1,783,400
Bighead carp (*Aristichthys nobilis*)	1,256,900
Crucian carp (*Carassius carassius*)	537,500
Giant tiger prawn (*P. monodon*)	502,700
Nile tilapia (*O. niloticus*)	473,600
Atlantic salmon (*Salmo salar*)	471,800
Roho labeo (*Labeo rohita*)	458,800
Catla (*Catla catla*)	381,400
Mrigal (*Cirrhinus mrigala*)	373,000
Rainbow trout (*Oncorhynchus mykiss*)	358,500
Milkfish (*Chanos chanos*)	358,100
White amur bream (*Parabramis pekinensis*)	335,900
Channel catfish (*I. punctatus*)	206,100

[a]From (2).

As with the United States, the beginnings of modern aquaculture everywhere can be placed in the 1960s. The Chinese, who were responsible for the creation of aquaculture, did not begin to adopt modern technology until several years after most other nations had done so.

Tracing the development of aquaculture throughout the world would require hundreds of pages, and much of that history can be found in the contributions on various species that are included in this encyclopedia. Only a brief summary is given here.

Europe and the Middle East

Among the freshwater species that are of commercial aquaculture interest in Europe are trout, walking catfish, and crayfish. Trout (including rainbow trout introduced to Europe from the United States) have long been popular in many European nations. Walking catfish were introduced from Africa and Asia. Most of the activity associated with walking catfish culture is in the Netherlands. Crayfish culture has been plagued by disease in recent years.

In the Middle East, the tilapia is the dominant freshwater species cultured in many countries, although a significant level of carp production takes place in Israel, a leading aquaculture nation in the Middle East. Northern Israel also has at least one trout farm. Tilapias are native to north Africa and the Middle East, so they are naturally of interest to aquaculturists in that part of the world. Interest has also been sparked in freshwater shrimp in Saudi Arabia, Israel, and other Middle Eastern nations.

Marine aquaculture in Europe is quite well developed. Norway and Scotland both produce Atlantic salmon, of whose culture Norway is the world's leading nation. The same two countries also produce Atlantic halibut. The United Kingdom pioneered the development of culture technology associated with plaice, sole, and turbot. Each of those fishes is being reared commercially, with considerable prospects for increasing production dramatically. France has also been involved in flatfish culture, though not with halibut, as that species cannot tolerate the warm waters off the coast of France. Sea bass and sea bream have been of interest in Europe, and some production of those fishes occurs.

Mussels and other shellfishes are grown commercially in Europe, with Spain being a leader in their production. Raft culture of mussels is particularly well developed in Spain, while mussels are grown on poles in France.

Sea bream and, to a lesser extent, sea bass have been of interest in Israel. Much of the early research that led to the commercial sea bream industry was conducted there. Sea bream and sea bass could also be commercially grown in other Middle Eastern nations, in which some commercial marine shrimp culture is already being conducted.

Africa

Tilapias and walking catfish are the primary fish species being cultured throughout much of Africa. A large part of the continent is involved primarily with aquaculture at a subsistence level. Notable exceptions are Egypt and South Africa, each of which has developed commercial culture. In Egypt, interest is strong in the culture of tilapias, various species of carp, walking catfish, freshwater shrimp, and frogs. There is also a potential for culturing various marine species, including sea bream, mullet, sea bass, rabbitfish, flatfish, groupers, snappers, molluscs, and shrimp (7).

Rainbow trout were introduced to South Africa in the late 19th century to support recreational fisheries. The first trout farm was established there in 1945 (8). Today, in addition to trout, freshwater culture species include catfish, ornamentals, tilapias, and carp. Marine culture involves mussels, oysters, clams, shrimp, and redbait (a species of tunicate sold as bait).

Latin America and the Caribbean

The most visible and lucrative aquaculture activity in Latin America is the production of marine shrimp. Shrimp farming in Latin America started on a banana plantation in Ecuador in the 1960s and spread from there. The explosion in Latin American shrimp farming began in the 1980s, led by Ecuador, which was later joined by several other nations, including Mexico, Costa Rica, and Panama. Other marine organisms of interest in Ecuador include molluscs, flatfishes, red drum, snook, pompano, and yellowtail (9). Freshwater species that are of interest in Latin America include tilapias, carp of various species, rainbow trout, walking catfish, and a number of native species (10).

Peru and Chile have a considerable involvement in conjunction with the rearing of molluscs. In southern Chile, the emphasis has been on salmonid culture, with both the commercial trout and the commercial salmon industries having grown to the point that their production has a major impact on world markets for those commodities. Indeed, Chile is now second only to Norway in the production of cultured salmon.

In the Caribbean, a limited amount of aquaculture production occurs on various islands. Puerto Rico produces shrimp and tilapias, and the Bahamas have experimented with rearing tilapias in seawater and at one time had some interest in shrimp culture. A considerable amount of the tilapia that enters US markets comes from Jamaica. A small amount of subsistence-level aquaculture takes place in Haiti, where there is also a professed interest in commercial culture. Haiti's neighbor, the Dominican Republic, on the island of Hispañola, raises tilapias and marine shrimp. Attempts to raise local crabs, queen conchs, and various species of fishes have not led to significant production to date.

Asia

The majority of the world's aquaculture production comes from Asia. Far and away in the lead in terms of total fish production is China (Fig. 4), although in terms of employing technology, nations like Japan and Taiwan have assumed leadership roles. China is associated primarily with an enormous production of freshwater fish,

Figure 4. Carp culture pond in China.

including some seven million tons of various species of carp annually. Tilapias and other freshwater species are produced, too, but pale in comparison with carp production. In recent years, China has become a major nation in the culture of marine shrimp, and the Chinese also produce other marine species.

In India, several native carp species are produced, as are other freshwater fishes, including tilapias and freshwater shrimp. By 1985, marine shrimp culture had become an important industry in India. Important marine finfish and shellfish that are cultured are mullet, milkfish, pearl and edible oysters, and clams. (11). Seaweed is cultured as well.

Throughout Southeast Asia (Figs. 5 and 6), aquaculture has become an important activity. Thailand is the world's leading marine shrimp-producing nation, and Indonesia, Malaysia, and the Philippines are also major producers of shrimp. Milkfish are important in most Southeast Asian nations, as are tilapias, exotics first introduced in the 1930s. Seaweed culture also is important in various southeast Asian countries.

Korean aquaculture is varied and includes species from seaweeds to finfish. Russia and other countries that made up the former USSR also produce significant amounts of cultured products, mainly finfish. Carp is a popular culture species in some regions, sturgeon in others. Eastern Russia is much involved in salmon culture.

Figure 5. Shrimp farm in the Philippines (photo by Victor Mancebo).

Figure 6. Commercial marine cages in Malaysia.

Figure 7. Marine net pen in southern Japan for rearing yellowtail.

Figure 8. Nets for attachment and growth of seaweed in Japan. The nets ride up and down the poles with the tide.

Japan is perhaps the world's most advanced nation with respect to aquaculture. Relying heavily on protein from the sea to feed its people, Japan has long been a nation of fishermen and aquaculturists. From the southern part of that nation, where yellowtail, sea bream, and warm-water molluscs are reared, to the north, where salmon and cold-water mussels predominate, aquaculture is an important aspect of food production. Mussels, clams, scallops, oysters (including pearl oysters), abalones, shrimp, echinoderms, and a wide variety of finfish are produced by Japanese aquaculturists (Fig. 7). Japan is also a world leader in seaweed culture, with large areas of certain bays employed exclusively for that purpose (Fig. 8).

Australia and New Zealand have active aquaculture programs. Native species are of primary interest in both nations, although in New Zealand the introduced chinook (Quinnat) salmon and rainbow trout are still produced, primarily for recreation. A number of different fishes and shellfishes are receiving attention in both nations.

BIBLIOGRAPHY

1. E.M. Borgese, *Seafarm*, Harry N. Abrams, Inc., New York, 1977.

2. FAO, *Aquaculture Production Statistics, 1986–1995*, Food and Agriculture Organization of the United Nations, Circular 815, Rev. 9, Rome, 1997.
3. R.R. Kirk, *A History of Marine Fish Culture in Europe and North America*, Fishing News Book, Ltd., Farnham, Surrey, England, 1987.
4. R.R. Stickney, *History of Aquaculture in the United States*, John Wiley & Sons, New York, 1996.
5. H.S. Swingle, *Proceedings of the Southeastern Association of Game and Fish Commission* **10**, 160–162 (1957).
6. H.S. Swingle, *Proceedings of the Southeastern Association of Game and Fish Commission* **12**, 63–72 (1958).
7. A.K. Hamza, *World Aquaculture* **27**(1), 14–19 (1996).
8. P. Cook, *World Aquaculture* **26**(4), 14–19 (1995).
9. D.D. Benetti, C.A. Acosta, and J.C. Ayala, *World Aquaculture* **26**(4), 7–13 (1995).
10. N. Castagnolli, *World Aquaculture* **26**(4), 35–39 (1995).
11. H.P.C. Shetty and G.P. Satyanarayana Rao, *World Aquaculture* **27**(1), 20–24 (1996).

HORMONES IN FINFISH AQUACULTURE

EDWARD M. DONALDSON
Aquaculture and Fisheries Consultant
West Vancouver, Canada

OUTLINE

INTRODUCTION

The word *hormone* is derived from a Greek word meaning "to rouse or set in motion." Hormones are organic compounds that are synthesized by the endocrine glands and are typically transported in the blood to other tissues or organs where they interact with hormone receptors to alter cell function. Key functions such as growth, reproduction, osmoregulation, metabolism, and the stress response are all regulated by hormones. Some hormones are involved in the maintenance of homeostasis while others are involved in developmental processes such as the coordination of gametogenesis. Many hormones are controlled by the central nervous system (CNS). The hypothalamus produces releasing hormones and release-inhibiting factors that interact at the pituitary level to regulate the synthesis and release of individual pituitary hormones. These pituitary hormones in turn target specific endocrine organs regulating the synthesis of the hormones that they produce. For information on early studies on hormones in fish, refer to the treatise by Pickford and Atz (1).

The structure of peptide and protein hormones such as those produced in the hypothalamus and pituitary gland has evolved over time. Many piscine hormones of this type are biologically active in fish but not in mammals. On the other hand, mammalian peptide and protein hormones are often active in fish, although not necessarily as active as the native piscine hormone. In the case of steroid hormones, the key steroids may be identical to those found in mammals. Thus, the stress hormone cortisol and the estrogen estradiol 17 are present in both fish and many mammals while the male-specific androgen in many fish is 11-ketotestosterone rather than testosterone. The steroids mainly responsible for final maturation in fish are either 17,20-dihydroxy-4-pregnen-3-one or 17,20,21-trihydroxy-4-pregnen-3-one (2).

Of all the hormones that are known to exist in finfish, relatively few are used in aquaculture at the present time. However, those that are used have a major impact on the overall success of aquaculture production systems and within appropriate regulatory frameworks, the use of hormones in a safe and sustainable manner can be expected to increase in the future. We focus attention here on those hormones that are currently used in aquaculture to regulate reproduction and sex differentiation and hormones that may be used in the future to regulate growth.

FINAL MATURATION, OVULATION, AND SPERMIATION

The first and, at present, perhaps still the major application of hormones in aquaculture is in the induction of sexual maturity in captive finfish. There are several objectives, including the induction of ovulation and spermiation in fish which do not undergo final maturation in captivity, the acceleration of spawning date in fish that do mature in captivity, the synchronization of spawning in fish that would otherwise spawn over an extended period, and the synchronization of spawning dates between related species that are being hybridized. In species where culture was previously based on the capture of wild juveniles, the development of induced spawning technologies has contributed to the sustainability of wild stocks.

Hormones were first used in aquaculture in Brazil in the 1930s when von Ihering (3) injected fish pituitary homogenates to induce spawning. The use of pituitary homogenates has continued to the present day. However, the process has some drawbacks, including the lack of standardization of most pituitary preparations, the presence of other hormones in addition to the gonadotropins, the high cost of fish pituitaries, and, in some cases, poor quality control in the production of dried pituitary preparations. Pituitaries are preferably collected from mature fish when gonadotropin content is maximal and either used immediately, frozen directly on dry ice, or processed through several changes of cold acetone to remove all water before drying. Pituitary donor species that have proved successful include common and other carps and several salmon species. Owing to the phenomenon of species specificity in piscine gonadotropins, the use of pituitaries from homologous or closely related species is recommended.

Partially purified and purified gonadotropins have been prepared from fish pituitaries, especially gonadotropin II, which is homologous with mammalian LH (luteinizing hormone). However, the use of these has been mainly restricted for economic reasons to research and development. Human chorionic gonadotropin HCG, which is extracted from pregnant human urine, is effective in several finfish species including carps, mullet, and sea bream. See Refs. 4–8 for reviews on induced spawning in fish.

GnRH AND LHRH ANALOGS IN INDUCED SPAWNING

The most important development in the advancement of induced spawning technologies in finfish has been the application of gonadotropin releasing hormone (GnRH) analogs over the past two decades. These hormones stimulate the synthesis and release of the endogenous piscine gonadotropins and are therefore believed to mimic more closely the natural maturation process. The natural gonadotropin releasing hormones are decapeptides, i.e., they contain ten amino acids. Several natural forms have been identified, which vary in one or more amino acids. In some fish species, two or three different forms have been found within the same species. Thus, the sea bream contains sea bream GnRH, salmon GnRH, and chicken GnRH II (9). The separate functions of these several forms have not been elucidated. The natural GnRHs have relatively weak activity when administered to fish however; potent synthetic analogs have been produced which have greater receptor affinity (10) and greater resistance to enzymatic degradation (11). Typically, the analogs are substituted in position 6 with an appropriate D-amino acid, such as D-Ala or D-Arg, and the terminal glycine in position 10 is deleted and replaced with an ethylamide group (5). GnRH analogs that have seen considerable use in aquaculture include [D-Ala6, des-Gly10] mammalian GnRH (otherwise known as [D-Ala6, des-Gly10] LHRH) and [D-Arg6, des-Gly10] salmon GnRH. Several of the potent analogs are effective in fish *in vivo*, providing that they are from a reliable source and are of high peptide purity. Typical dosages range between 5 and 100 g/kg, depending on species, the maturity of the broodstock, the nature and purity of the GnRH analog, and, in species where dopamine inhibition occurs, whether a dopamine antagonist has been administered.

Modes of Administration

A variety of modes of administration has been developed for GnRH analogs (12). These include intraperitoneal or intramuscular injection in aqueous solution, injection in a slow release form, e.g., microencapsulated, implantation of GnRH incorporated into a cholesterol or polymer pellet, oral administration in solution or in the diet, and immersion in a solution of GnRH with or without exposure to ultrasound. Injection is effective in many fish, especially warm water species. Implantation is useful for cold water species, such as salmonids, as it removes the requirement for two spaced injections and for repeat spawners such as sea bream. Oral administration and immersion offer possibilities to regulate spawning in species that are stressed by handling (13,14).

Dopamine Antagonists

A number of fish species especially carps respond poorly to GnRH/LHRH injection as a result of the effect of dopaminergic inhibition of gonadotropin release (15). This inhibition can be overcome by the co-administration of a dopamine antagonist with the GnRH analog (16). Dopamine antagonists that have been used successfully for this purpose in fish include domperidone (7), pimozide (15), metoclopramide (17) and sulpiride (18). Appropriate dosages are in the 5–20 mg/kg range depending on the antagonist, the species, the maturity of the broodstock and the dose of GnRH. Domperidone (Motilium), which is not water-soluble is injected in aqueous suspension or dissolved in propylene glycol or other suitable solvent (19). Oral administration in conjunction with GnRH is also possible (14).

Antiestrogens and Aromatase Inhibitors

It has been known for some time that ovulation can be induced by manipulation of the feedback mechanisms that control endogenous GnRH and gonadotropin release. Thus, it is possible to use antiestrogens such as tamoxifen to induce ovulation (19a). Recently, we have demonstrated induction of both ovulation (20) and spermiation in Pacific salmon using the aromatase inhibitor fadrozole that is capable of inhibiting estrogen biosynthesis (21,22). However, further research will be required to determine whether this can be developed into a practical technique.

CONTROLLED SEX DIFFERENTIATION

Hormones, specifically androgens and estrogens, have played an important role in the development of technologies for the production of monosex and sterile fish stocks. In several species, one sex is more valuable than the other, e.g., where the roe is valued. In other species monosex culture facilitates the implementation of a more efficient production system, e.g., where one sex grows to market size faster than the other or where one sex is prone to precocious maturation before reaching full market size. The production of monosex or sterile stocks also provides a means of reproductive containment for exotic species or genetically modified stocks (23–27).

Species where monosex production has been implemented on a production scale include chinook salmon, a species where all commercial culture in Canada has been monosex female for over a decade, and rainbow trout where much of the production in several countries is now monosex female or monosex female triploid. Monosex male tilapia are also widely grown in several countries. Species where monosex technology may be applied in the future include other salmonids, such as Atlantic salmon (27) and coho salmon, flatfish, such as the turbot and halibut where the female grows faster than the male, and the carps, mullets, sea bass, and catfishes. There are two distinct approaches to sex control, direct and indirect (28). In the direct method, production fish are treated with an appropriate androgen or estrogen, usually during early development, to induce gonadal differentiation into the desired sex. This method is widely used for the production of monosex male tilapia (29,30). However, it is being gradually replaced by the newly developed indirect methods for tilapia (31). In the indirect method, hormone treatment occurs in the previous generation and not on production fish. The method depends on the production of monosex sperm which, when used to fertilize normal eggs, results in the production of monosex fish. Thus in the salmonids, which are female homogametic, genetic females are treated by immersion in the alevin stage and in some cases by diet in the early feeding stage to produce phenotypic males that are genetically female. When these fish mature, they produce monosex female sperm which, when used to fertilize normal eggs, results in the production of monosex female offspring. The development of new monosex female salmonid stocks requires the separation of masculinized females from normal males. This has traditionally been accomplished by progeny testing; however, the development of Y-specific DNA probes has facilitated this process in a number of salmonids (32,33). The indirect production of monosex male stocks in female homogametic species requires more steps than the production of monosex female stocks as it depends on the production of YY supermales (31,34). In catfish, this can be accomplished by dietary treatment of genetic males with estrogen (or androgen) to produce XY phenotypic females. When mature, these XY females are mated with normal XY males. The offspring have a 3 : 1 male–female ratio. One-third of these males would be YY males (identified by progeny testing) which, when mature, produce monosex male sperm. When used to fertilize normal ova the result is a monosex male population. Reviews on sex control technology in fish include Hunter and Donaldson (35), Shelton (36), Pandian and Sheela (37), Donaldson et al. (38), Donaldson (6), and Piferrer (39).

CRITICAL VARIABLES IN CONTROLLED SEX DIFFERENTIATION

Choice of Androgen or Estrogen

A number of factors influence the success of hormonal sex control. The choice of a suitable androgen or estrogen is important. In the case of androgens testosterone is only weakly androgenic, while the natural nonaromatizable androgen 11-ketotestosterone is effective (40) but costly to use. The synthetic androgen 17-methyltestosterone is cost-effective in many species. However, it can lead to paradoxical feminization when administered at higher dosages. The term *paradoxical feminization* refers to the appearance of increasing proportions of females as the dose of androgen increases. This phenomenon has been suggested to occur when the administered androgen is aromatized to estrogen *in vivo*. The nonaromatizable synthetic androgen 17-methyldihydrotestosterone is an effective masculinizing agent in both salmonids (40) and tilapia (41). In some species, such as the channel catfish, masculinization is not possible with any androgen tested to date; however, feminization poses no problem (34,42). In the case of estrogens, the natural estrogen estradiol 17 is effective and the synthetic estrogen 17-ethynylestradiol is also effective and more potent (43).

Dosage and Timing

The dose varies according to species, steroid used, and mode of administration. Treatment must occur during the labile period, i.e., the period during which the fish is responsive to exogenous androgen or estrogen treatment (28). In new species the first step is to determine through histological studies the time of morphological sex differentiation. For species treated by immersion before first feeding, androgen dosages in the range of 400–2000 g/L in the immersion water are effective in, for example, salmonids. Typically, the androgen is dissolved in ethanol or other suitable solvent before mixing with the water. It is important that the immersion water be circulated over the eggs or larvae by aeration or recirculation pump and that the number of eggs or larvae per liter is not excessive. A typical immersion duration is 2 hours; however, isotope studies indicate that uptake is still underway at this time (44). Success has also been achieved with longer immersion periods at low temperatures. Some fish species respond to a single immersion during the labile period while others require repeated treatments during the labile period or a combination of immersion and dietary treatment (45).

Dietary dosages for masculinization vary from 1 to 3 mg/kg diet in salmonids (46) to up to 60 mg/kg diet in tilapia (30). In species such as the Mediterranean sea bass sex differentiation occurs late. Success has been achieved in this species with dietary administration of 17-methyltestosterone at 10 mg/kg initiated 126 days after fertilization and continued for 100 days (47). Androgen dosages higher than those used for masculinization can result in sterility (48–50).

Health and Environmental Issues

Androgens and estrogens utilized in aquaculture are potent steroids that must be treated with great respect. Persons working with these compounds should avoid skin contact or inhalation. The use of hormone mimics such as diethylstilbestrol, which is a known carcinogen of the human reproductive system, should be avoided. Effluent water containing androgens or estrogens should be disposed into a large volume or flow of water where it will be diluted to an insignificant concentration or disposed to ground well away from domestic water supplies. The use of androgens or estrogens in closed water systems where untreated fish are on the same water supply can lead to unexpected effects on "untreated" fish (51).

GROWTH

The average production cycle for cultured finfish greatly exceeds the production cycle for avian and mammalian species. This results in capital facilities being occupied for longer periods and exposes the fish in each production cycle to greater risk from loss due to a variety of factors, including storm damage, predation, disease, and harmful algal blooms. Progress in reducing production cycles has been achieved through selective breeding, improved nutrition, and improved husbandry. However, it remains of considerable interest to determine whether further reductions in the production cycle can be achieved on a practical and sustainable basis by using our knowledge of the endocrine system. The hormone that has received the most attention in these studies and to date the most effective in stimulating growth has been pituitary growth hormone or somatotropin (52,53) and related hormones such as bovine placental lactogen (54). Somatotropin is a protein with a molecular weight in the region of 20,000 Daltons. Mammalian growth hormones are effective when administered to fish, while fish growth hormones are ineffective in mammals. In addition to stimulating growth, somatotropin facilitates the smoltification process in salmonids (55,56) and may improve feed conversion (54). Its production and secretion from the pituitary gland is under both stimulatory (somatocrinin or growth hormone releasing hormone) and inhibitory control (somatostatin) (57) by the hypothalamus. The actions of growth hormone are believed to be mediated through insulin-like growth factor (IGF I) (somatomedin). However, the effects of growth hormone on growth in fish have not yet been fully replicated by treatment with IGF I (58), suggesting that growth hormone may also act though other mechanisms or that it has not yet been possible to administer IGF I in a fully effective manner for growth stimulation without causing physiological problems normally associated with high levels of insulin (59). Other hormones that have been investigated for their effects on growth include the androgens and thyroid hormones (59a). There are two methods by which growth hormone can be administered to fish. Exogenous hormone can be administered during appropriate periods of the production cycle or the fish itself can be engineered to produce a greater amount of growth hormone, which is not under hypothalamic control, throughout the production cycle (60).

Administration of Exogenous Growth Hormone

There are several means by which growth hormone is administered to fish. It can be given by weekly or biweekly intraperitoneal injection in aqueous form or at much longer intervals in a slow-release formulation (56,61). It can also be supplied as an intraperitoneal or intramuscular slow-release implant such as a cholesterol- (62) or polymer-coated pellet (63,64) or a mini-osmotic pump (62,65). Immersion in a solution of growth hormone has also been tested (66). Probably the most feasible means of providing growth hormone is either by dietary administration (67,68) or by injection of a slow-release formulation. Finfish are able to absorb intact proteins and peptides from the digestive tract (69,70) and means for protecting the hormones during gut transit and enhancing uptake have been investigated (70a).

Advantages of administration of exogenous growth hormone versus growth hormone transgenic fish include the following: application is possible at a specific life stage; a withdrawal period is possible; reproductive or physical containment is not required; and there is no need to sterilize production fish and no need to maintain a specific broodstock in quarantine. Disadvantages include the following: performance enhancement is not yet as impressive as the transgenics; cost of the growth hormone; treatment requires handling (except the oral route); repeat treatment may be required; regulatory issues, especially with non-homologous proteins; and potential concern over residues in the product and in the environment.

Development of Fish with Enhanced Endogenous Growth Hormone

The development of transgenic fish with enhanced ability to synthesize growth hormone provides another means of increasing growth rate. Success has been achieved with DNA constructs containing homologous or closely related growth hormone genes driven by piscine promoters such as the ocean pout antifreeze promoter (70b,71), and the sockeye salmon metallothionein-B promoter (60). In some fast-growing transgenic fish the increased level of growth hormone has been sufficient to induce deformities in the head region (71); however, fish with moderate acceleration are normal in appearance. Transgenic technology offers the possibility of enhancing or suppressing other parts of the endocrine system (72).

Advantages of growth-enhanced transgenic fish include the following: exceptional growth performance; peptide production is endogenous; the peptides used can be homologous; no fish handling is required; performance enhancement is continuous; performance traits are inherited; and there is the future possibility of controlling growth hormone expression at specific life stages.

Disadvantages of growth enhanced transgenic fish include the following: development of a true breeding stock takes several generations; control over level of expression is only possible through selection and timing of expression is not yet possible; public perception of genetic engineering in export markets; possible concerns over use of nonhomologous constructs; public concern over risk to ecosystem associated with possible escape of genetically modified aquatic organisms; need for broodstock quarantine; and reliable sterilization of production fish.

Many of the possible problems associated with the use of either exogenous growth hormones or transgenic fish do have solutions, and it is probable that both of these technologies will be implemented in the not too distant future.

CONCLUSIONS

This brief review of the use of hormones if fish has focused on those hormones that are actually used or are close to being used in aquaculture. It is important to note that the ability to simply measure hormone levels in wild, hatchery-supported, and aquacultured fish provides powerful tools to assess, for example, the stress and reproductive status of fish.

BIBLIOGRAPHY

1. G.E. Pickford and J.W. Atz, *The Physiology of the Pituitary Gland of Fishes*, N.Y. Zool. Soc., New York, 1957.
2. A.P. Scott and A.V.M. Canario, in D.I. Idler, L.W. Crim, and J.M. Walsh, eds., *Proceedings of the Third International Symposium on Reproductive Physiology of Fish*, M.S.R.L. St. John's, Newfoundland, 1987, pp. 224–234.
3. R. von Ihering, *Prog. Fish-Cult.* **34**, 15–16 (1937).
4. E.M. Donaldson and G.A. Hunter, *Fish Physiol.* **9**(Part B), 351–403 (1983).
5. E.M. Donaldson, *Anim. Reprod. Sci.* **42**, 381–392 (1996).
6. E.M. Donaldson, in J. Coimbra, ed., *Proceedings of the NATO Advanced Research Workshop on Modern Aquaculture in the Coastal Zone: Lessons and Opportunities*, Plenum, Porto, Portugal, 1998 (in press).
7. H.R. Lin and R.E. Peter, *Asian Fish. Sci.* **9**(1), 21–33 (1996).
8. Y. Zohar, in V. Shito and S. Sarig, eds., *Fish Culture in Warm Water Systems: Problems and Trends*, CRC Press, Boca Raton, FL, 1989, pp. 65–119.
9. J.F.F. Powell, Y. Zohar, A. Elizur, M. Park, W.H. Fischer, A.G. Craig, J.E. Rivier, D.A. Lovejoy, and N.M. Sherwood, *Proc. Natl. Acad. Sci. U.S.A.* **91**, 12081–12085 (1994).
10. H.R. Habibi, T.A. Marchant, C.S. Nahorniak, H. Van Der Loo, R.E. Peter, J.E. Rivier, and W.W. Vale, *Biol. Reprod.* **40**, 1152–1161 (1989).
11. Y. Zohar, A. Goren, M. Fridkin, E. Elhanati, and Y. Koch, *Gen. Comp. Endocrinol.* **79**, 306–319 (1990).
12. E.M. Donaldson, I.I. Solar, and B. Harvey, in T. Tiersch and P. Mazik, eds., *Cryopreservation in Aquatic Species*, World Aquacult. Soc., Baton Rouge, LA (2000).
13. P. Thomas and N.W. Boyd, *Aquaculture* **80**, 363–370 (1989).
14. N. Sukumasavin, W. Leelapatra, E. McLean, and E.M. Donaldson, *J. Fish Biol.* **40**, 477–479 (1992).
15. J.P. Chang and R.E. Peter, *Gen. Comp. Endocrinol.* **52**, 30–37 (1983).
16. R.E. Peter, H.R. Lin, and G. Van der Kraak, *Aquaculture* **74**, 1–10 (1988).
17. S. Drori, M. Ofir, B. Levavi-Sivan, and Z. Yaron, *Aquaculture* **119**, 393–407 (1994).
18. A.I. Glubokov, J. Kouril, E.V. Mikodina, and T. Barth, *Aquacult. Fish. Manage.* **25**(4), 419–425 (1994).
19. R.J. Omeljaniuk, S.H. Shih, and R.E. Peter, *J. Endocrinol.* **114**, 449–458 (1987).

19a. E.M. Donaldson, G.A. Hunter, and H.M. Dye, *Aquaculture* **26**, 143–154 (1981).

20. L.O.B. Afonso, G.K. Iwama, J. Smith, and E.M. Donaldson, *Gen. Comp. Endocrinol.* **113**, 221–229 (1999).
21. L.O.B. Afonso, P.M. Campbell, G.K. Iwama, R.H. Devlin, and E.M. Donaldson, *Gen. Comp. Endocrinol.* **106**, 169–174 (1997).
22. L.O.B. Afonso, G.K. Iwama, J. Smith, and E.M. Donaldson, *Fish Physiol. Endocrinol.* **20**, 231–241 (1999).
23. W.L. Shelton, *Fisheries* **11**(2), 18–19 (1986).
24. R.H. Devlin and E.M. Donaldson, in C.L. Hew and G. Fletcher, eds., *Transgenic Fish*, World Scientific Press, Singapore, 1992, pp. 229–266.
25. J.A. Beardmore, G.C. Mair, and R.I. Lewis, *Aquacult. Res.* **28**, 829–839 (1997).
26. E.M. Donaldson, R.H. Devlin, I.I. Solar, and F. Piferrer, in J.G. Cloud and G.H. Thorgaard, eds., *Genetic Conservation of Salmonid Fishes*, Plenum, New York, 1993, pp. 113–129.
27. E.M. Donaldson, I.I. Solar, W. Harrower, and D.L. Tillapaugh, *World Aquacult. Soc. (WAS) Aquacult. Am. '99*, Tampa, FL, 1999, Book Abstr., p. 36.
28. F. Piferrer and E.M. Donaldson, *Recent Adv. Aquacult.* **4**, 69–77 (1993).
29. A.E. Hiott and R.P. Phelps, *Aquaculture* **112**, 301–308 (1993).
30. B.J. McAndrew, *Recent Adv. Aquacult.* **4**, 87–98 (1993).
31. G.C. Mair, J.S. Abucay, D.O.F. Skibinski, T.A. Abella, and J.A. Beardmore, *Can. J. Fish. Aquat. Sci.* **54**, 396–404 (1997).
32. R.H. Devlin, B.K. McNeil, I.I. Solar, and E.M. Donaldson, *Aquaculture* **128**, 211–220 (1994).
33. S.J. Du, R.H. Devlin, and C.L. Hew, *DNA Cell Biol.* **12**, 739–751 (1993).
34. W.A. Simco, K.B. Davis, and C.A. Goudie, *World Aquacult. Soc. (WAS) Aquacult. Am. '99*, Tampa, FL, 1999, Book Abstr., p. 174.
35. G.A. Hunter and E.M. Donaldson, *Fish Physiol.* **9**(Part B), 223–303 (1983).
36. W.L. Shelton, *Rev. Aquat. Sci.* **1**, 497–535 (1989).
37. T.J. Pandian and S.G. Sheela, *Aquaculture* **138**, 1–22 (1995).
38. E.M. Donaldson, R.H. Devlin, F. Piferrer, and I.I. Solar, *Asian Fish. Sci.* **9**, 1–8 (1996).
39. F. Piferrer, in E.M. Donaldson and C.-S. Lee, eds., *Proceedings of Aquaculture Interchange Program (AIP) Workshop on Reproductive Biotechnology for Marine Finfish*, Honolulu, HI, 1999 (in preparation).
40. F. Piferrer, I.J. Baker, and E.M. Donaldson, *Gen. Comp. Endocrinol.* **91**(1), 59–65 (1993).
41. W.L. Gale, M.S. Fitzpatrick, and C.B. Schreck, in F.W. Goetz and P. Thomas, eds., *Proceedings of the Fifth International Symposium on Reproductive Physiology of Fish*, University of Texas, Austin, 1995, p. 117.
42. K.B. Davis, B.A. Simco, C.A. Goudie, N.C. Parker, W. Cauldwell, and R. Snellgrove, *Gen. Comp. Endocrinol.* **78**, 218–223 (1990).
43. F. Piferrer and E.M. Donaldson, *Aquaculture* **106**(2), 183–193 (1992).
44. F. Piferrer and E.M. Donaldson, *Fish Physiol. Biochem.* **13**, 219–232 (1994).
45. G. Feist, C.-G. Yeoh, M.S. Fitzpatrick, and C.B. Schreck, *Aquaculture* **131**, 145–152 (1995).
46. I.I. Solar, E.M. Donaldson, and G.A. Hunter, *Aquaculture* **42**, 129–139 (1984).
47. M. Blazquez, F. Piferrer, S. Zanuy, M. Carrillo, and E M. Donaldson, *Aquaculture* **135**, 329–342 (1995).
48. E.M. Donaldson, F. Piferrer, I.I. Solar, and R.H. Devlin, *Can. Tech. Rep. Fish. Aquat. Sci.* **1789**, 37–45 (1991).

49. J.E. Shelbourn, W.C. Clarke, J.R. McBride, U.H.M. Fagerlund, and E.M. Donaldson, *Aquaculture* **103**, 85–99 (1992).
50. F. Piferrer, M. Carrillo, S. Zanuy, I.I. Solar, and E.M. Donaldson, *Aquaculture* **119**, 409–423 (1994).
51. J.S. Abucay and G.C. Mair, *Aquacult. Res.* **28**, 841–845 (1997).
52. J.A. Gill, J.P. Sumpter, E.M. Donaldson, H.M. Dye, L. Souza, B.J. Wypych, and K. Langley, *Bio/Technology* **3**, 643–646 (1985).
53. E. McLean and E.M. Donaldson, in M.P. Schriebman, C.G. Scanes, and P.K.T. Pang, eds., *The Endocrinology of Growth, Development and Metabolism in Vertebrates*, Academic Press, San Diego, CA, 1993, pp. 43–71.
54. R.H. Devlin, J.C. Byatt, E. McLean, T.Y. Yesaki, G.G. Krivi, E.G. Jaworski, W.C. Clarke, and E.M. Donaldson, *Gen. Comp. Endocrinol.* **95**, 31–41 (1994).
55. J.M. Shrimpton, R.H. Devlin, E. McLean, J.C. Byatt, E.M. Donaldson, and D.J. Randall, *Gen. Comp. Endocrinol.* **98**(1), 1–15 (1995).
56. E. McLean, R.H. Devlin, J.C. Byatt, W.C. Clarke, and E.M. Donaldson, *Aquaculture* **156**, 113–128 (1997).
57. I. Mayer, E. Mclean, T.J. Kieffer, L.M. Souza, and E.M. Donaldson, *Fish Physiol. Biochem.* **13**(4), 295–300 (1994).
58. S.D. McCormick, K.M. Kelly, G. Young, R.S. Nishioka, and H.A. Bern, *Gen. Comp. Endocrinol.* **86**, 398–407 (1992).
59. T. Skyrud, O. Andersen, P. Alestrom, and K.M. Gautvik, *Gen. Comp. Endocrinol.* **75**, 247–255 (1989).
59a. D.A. Higgs, U.H.M. Fagerlund, J.G. Eales, and J.R. McBride, *Comp. Biochem. Physiol.* **73B**, 143–176 (1982).
60. R.H. Devlin, T.Y. Yesaki, C.A. Biagi, E.M. Donaldson, P. Swanson, and W.K. Chan, *Nature (London)* **371**, 209–210 (1994).
61. E. McLean, R.H. Devlin, E.M. Donaldson, and J.C. Byatt, *Aquacult. Nut. (Oxford)* **2**, 243–248 (1996).
62. N.E. Down, E.M. Donaldson, H.M. Dye, K. Langley, and L.M. Souza, *Aquaculture* **68**, 141–155 (1988).
63. E. McLean, E. Teskeredzic, E.M. Donaldson, Z. Teskeredzic, Y. Cha, R. Sittner, and C.G. Pitt, *Aquaculture* **103**, 377–387 (1992).
64. E. McLean, E.M. Donaldson, I. Mayer, E. Teskeredzic, Z. Teskeredzic, C. Pitt, and L.M. Souza, 1994 *Aquaculture* **122**, 359–368 (1994).
65. N.E. Down, P.M. Schulte, E.M. Donaldson, and H.M. Dye, *J. World Aquacult. Soc.* **20**, 181–187 (1989).
66. P.M. Schulte, N.E. Down, E.M. Donaldson, and L.M. Souza, *Aquaculture* **76**, 145–156 (1989).
67. E. McLean, E.M. Donaldson, H.M. Dye, and L. Souza, *Aquaculture* **91**, 197–203 (1990).
68. E. McLean, E.M. Donaldson, E. Teskeredzic, and L.M. Souza, *Fish Physiol. Biochem.* **11**, 363–369 (1993).
69. E. McLean and E.M. Donaldson, *J. Aquat. Anim. Health* **2**, 1–11 (1990).
70. E. McLean, A.C. Von Der Meden, and E.M. Donaldson, *J. Fish Biol.* **36**, 489–498 (1990).
70a. E. McLean, B. Roensholdt, C. Sten, and Najamuddin, *Aquaculture* **177**, 231–247 (1999).
70b. S.J. Du, Z. Gong, G.L. Fletcher, M.A. Shears, M.J. King, D.R. Idler, and C.L. Hew, *Bio/Technol.* **10**, 176–180 (1992).
71. R.H. Devlin, T.Y. Yesaki, E.M. Donaldson, S.J. Du, and C.L. Hew, *Can. J. Fish. Aquat. Sci.* **52**(7), 1376–1384 (1995).
72. E.M. Donaldson and R.H. Devlin, in W. Pennell and B.A. Barton, eds., *Principles of Salmonid Culture*, Elsevier, Amsterdam, 969–1020 (1996).

I

INGREDIENT AND FEED EVALUATION

J.P. Lazo
D.A. Davis
Marine Science Institute
Port Aransas, Texas

OUTLINE

INTRODUCTION

The development of satisfactory diets for the production of farmed fish and shrimp demands a comprehensive understanding of nutritional requirements, feed processing, and methods to assess the quality of the ingredients that comprise a feed. In the framework of this article, ingredient quality mainly refers to the ability of the ingredient to allow optimum growth and health of the animals in culture. No single feed ingredient can supply all the nutrients and energy required for optimal growth; therefore, a mixture of feed ingredients is combined to produce a diet with the desired nutrient profile. Unfortunately, feed quality can vary considerably even for species for which nutritional requirements and commercial processing techniques are well established. In most cases, variability of the feed can be traced back to ingredient quality and/or quality control at the mill. Consequently, it is essential that the quality of feed ingredients, as well as the final product, are well defined.

Feed and ingredient evaluations can be approached as a stepwise process in which ingredients are analyzed by a series of simple physical and chemical tests, blended to produce a product with the desired nutrient profile, and then evaluated with the target species. A variety of tests that aid in the evaluation of the nutritional quality of ingredients and feeds are currently available. No single test will provide the necessary data for adequate feed evaluation. Hence, a variety of physical, chemical, and biological methods of evaluation are utilized to provide the information required to assess quality and nutritional value of single ingredients and feeds.

Physical and chemical tests include a variety of practical methods that provide a fast and simple way of screening ingredients. Physical methods rely on the characteristics of the feedstuff in question, such as particle size, density, and stability in water. Physical examinations, particularly microscopic evaluations, serve to evaluate the purity or composition of a particular ingredient, but provide little information on the nutritional value of the ingredient. Chemical methods are utilized to further characterize or define the chemical composition of the test substance and can give a very precise chemical definition of nutrients that are present. However, it is important to realize that both types of evaluations do not measure the real nutritional value of the feed or feed ingredient to aquatic animals, but provide only an estimate of its quality and gross composition.

Biological tests do not provide any information about the chemical composition, but offer a more accurate estimate of nutritional value and efficiency to produce growth and maintain a healthy organism. Live organisms are utilized to conduct well-designed feeding trials to evaluate the specific effect of a particular ingredient, nutrient, or feed formulation. Therefore, biological methods provide information that ascertains the true value of the feedstuff to the organism. Although biological tests are the preferred method and the ultimate test of performance, they have the disadvantages of being time-consuming, expensive to conduct, and require specialized facilities for holding live animals.

Physical, chemical, and biological methods will be discussed in the context of evaluating the quality of ingredients. The majority of methods described are equally applicable for the determination of nutritional requirements and the evaluation of prepared feeds.

PHYSICAL EVALUATION

The first step in assessing the quality of ingredients and/or a processed feed is to examine or inspect its physical characteristics. With the aid of a microscope one can confirm the purity of the ingredients (identify foreign contaminants or adulterants) or the variety of ingredients in a feed (1). In addition, some of the common physical characteristics to evaluate are (*1*) particle size and distribution, (*2*) density, (*3*) water stability, (*4*) texture, (*5*) feed shape and pellet quality, (*6*) homogeneity of ingredients, (*7*) color and contrast, and (*8*) smell (e.g., rancidity of oil or ammonia).

The optimal (desired) characteristics to look for will depend on the species of interest, the type of diet, and the size of the organisms in culture. For aquatic feeds, evaluation of pellet stability, shape, and density is very important because these properties will greatly influence the extent of feed utilization. Detection of any possible contaminants present in the ingredient or feed is desirable. Examples are feathers in meat and bone meals, large quantities of sand in fish meals, or toxin-producing models in grains. Physical evaluations can be conducted relatively quickly and represent an easy method to help ensure the purity of ingredients and hence minimize adulteration of the final product.

CHEMICAL EVALUATION

Chemical evaluation assays are those completed *in vitro*. The most common method used to analyze the nutrient composition of feeds or feedstuffs is known as the Proximate Analysis System or Weende Method. Developed in Germany over 100 years ago, proximate analysis is applied routinely in laboratories throughout the world. Although this system provides a general overview of the chemical composition, other chemical methods of evaluation provide more specific nutrient analysis of the ingredients and feed such as amino acid profiles, fatty acid composition, and essential vitamins and minerals.

Proximate Composition

Proximate analysis is based on simple assays that separate the feed or feed ingredients in question into different classes of nutrients such as moisture, protein, lipid, ash, crude fiber, and N-free extract. Detailed descriptions of the following assays are presented in *Official Methods of Analysis of the American Association of Analytical Chemists* (1). These nutrient groups can then be further defined by specific analytical techniques.

A brief description of each method follows, and a flowchart of the sequential procedure can be found in Figure 1.

Dry Matter. Estimates of dry matter (expressed as a percentage of total wet weight) allow for comparison of ingredient composition without the potentially confounding effects caused by differences in moisture content. Dry matter content is determined by heating a sample to a constant weight at temperatures above the boiling point of water (100 to 105 °C). Other methods for determination of water in feeds include toluene distillation, drying under vacuum, and freeze drying. Although estimates of dry matter tend to be simple and accurate, this technique can be a potential source of error for the stepwise procedure of proximate analysis. For example, if heated at temperatures that are too high, some fatty acids or fermented products such as silage can be volatilized, thereby resulting in an underestimate of actual dry weight (2). Conversely, the dry weight of a sample can be overestimated because some liquids do not volatilize when they become oxidized upon heating, and therefore become part of the dry weight of the sample. Since this overestimate can be a source of error that will be amplified in the next steps of the proximate analysis, care must be taken to optimize drying procedures relative to the type of sample being analyzed.

Crude Protein. The standard method used to determine crude protein content is the Kjeldahl technique. This method destroys organic matter, and all forms of nitrogen are then reduced to ammonium sulfate in the presence of sulfuric acid and various catalysts. After addition of excess alkali, ammonia is liberated and subsequently distilled and trapped in boric acid. The boric acid is then titrated with standard acid solution to determine the total nitrogen content (3). Crude protein is calculated by multiplying the amount of nitrogen by the empirically derived conversion factor of 6.25, which is based on the assumption that protein contains 16% nitrogen. Although this value is a good estimate, the actual range of nitrogen content in different proteins is between 12 and 19% (2,4). The calculated protein content can then be normalized to the wet or dry weight (dry matter) of the sample to obtain the percentage of protein content.

The Kjeldahl method does not distinguish between nitrogen originating from protein (amino acids) in the sample and that originating from either single-cell protein sources (bacteria, algae, and yeast) or nonprotein sources (amines and urea). In general, fish do not efficiently utilize single-cell protein sources and cannot utilize nonprotein sources of nitrogen. Consequently, the actual protein available for growth may be overestimated.

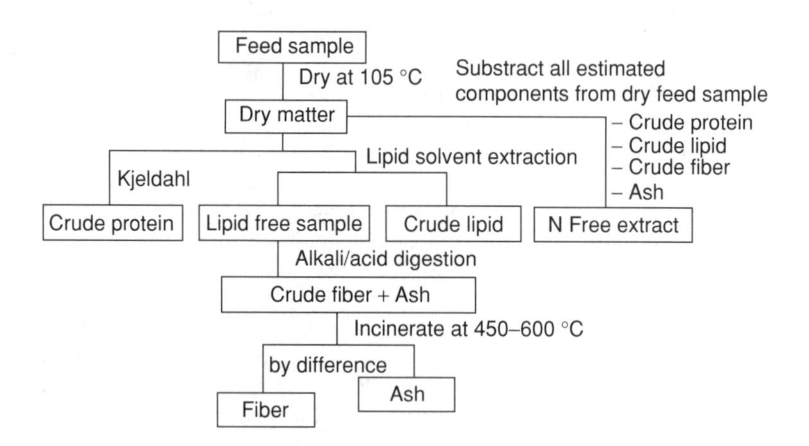

Figure 1. Diagram flow of proximate analysis.

Lipid Content. Lipid content is determined by extracting lipids from a sample with organic solvents. The solvent can be ether (1) or other mixtures of organic solvents, such as chloroform and methanol, in a ratio of 2 : 1, vol/vol, respectively (5), or hexane : methanol 4 : 1, vol/vol (6). Once lipids are extracted, the solvent is evaporated under nitrogen and the remaining amount of lipid is determined gravimetrically. The lipid weight can then be normalized to either the wet or dry weight of the ingredient to determine a percentage of lipid content.

Some potential sources of error exist for lipid extraction techniques. Any ether-soluble or chloroform : methanol-soluble compounds will be included in the estimate. Many of these compounds (e.g., chlorophyll, volatile oils, resins, pigments, and plant waxes) are of limited nutritional value to animals. In some cases, as much as 50% of the extract can be composed of these compounds (4). Alternately, lipids in some single-celled protein sources and processed feeds are bound to compounds (i.e., polysaccharide structures of bacterial cell walls) by covalent association and cannot be extracted with organic solvents. Such samples require a pretreatment with 4-N-HCl before extraction (acid hydrolysis) to cleave the lipid–polysaccharide complex (7).

Crude Fiber. Crude fiber is generally regarded as the indigestible part of the carbohydrates, having no nutritional value. To estimate crude fiber, a lipid-free sample is first digested by boiling in weak acid (0.255 N H_2SO_4) followed by boiling in a weak alkali (0.312 N NaOH) solution. This stepwise procedure removes the proteins, sugars, and starches from the sample. The sample is then dried and weighed, burned in a furnace at 600 °C, and reweighed. The material (weight) lost during the oxidation or burning corresponds to the content of crude fiber. This fraction consists primarily of hemicellulose, cellulose, and some insoluble lignin.

Some components of crude fiber, such as hemicellulose, can be partially dissolved by the acid/base treatment and lost from the sample, thereby resulting in an underestimate of crude fiber content. An overestimate of the fiber content can occur if protein is bound to lignin or other insoluble chemical forms, because the protein will remain in the sample and its weight will be included in the crude fiber estimate (4).

Ash. The material remaining after a sample is burned by combustion at 600 °C represents the inorganic or mineral content (8). The combustive process oxidizes all organic material, volatilizing and removing it from the sample. Because some minerals such as chlorine, zinc, selenium, and iodine can be lost through volatilization during combustion (9), caution must be exercised if the ash sample is to be analyzed subsequently for mineral determinations.

Nitrogen-Free Extract (NFE). Once the moisture, protein, lipid, fiber, and ash contents of an ingredient have been determined, the carbohydrate content or NFE can be estimated as the difference between the weight of the whole dry sample and the sum of the protein, lipid, fiber, and ash content. NFE primarily consists of available carbohydrates, such as sugars and starches, but can also contain some hemicellulose and lignin. The levels of specific carbohydrates can be determined by following the techniques described in the AOAC (1).

Energy Content

The gross energy content of a feed or an ingredient can be estimated indirectly from its chemical composition or directly by oxidizing all organic materials and measuring the heat produced (10,11). The direct method is performed with an instrument called an adiabatic bomb calorimeter, which measures heat released during combustion. Using this technique, the average energy values for the major nutrients have been estimated to be 5.6 kcal/g (23.4 kJ/g) for proteins, 9.5 kcal/g (39.8 kJ/g) for lipids, and 4.1 kcal/g (17.2 kJ/g) for carbohydrates (12).

In the indirect method, the estimated proximate composition of the feed or feed ingredient (i.e., protein, lipid, and carbohydrate) is used to determine the gross energy content by multiplying the appropriate average energy conversion factor for each nutrient by the amount of each nutrient and summing them. Both techniques assume that all the dietary energy is available to the organism.

Specific Nutrient Content

Proximate analysis separates general nutrient classes. It does not separately identify nutrients that cannot be utilized by the animal, compounds that have a low nutritive value, or ingredients with an inadequate balance of specific nutrients. Hence, more specific analyses are often warranted. Nutrients such as protein, lipids, carbohydrates, vitamins, and minerals can be analyzed by identifying and quantifying their respective components. Although chemical analyses of specific nutrients provide useful compositional information, such methods do not quantify actual nutrient availability to the animals themselves.

Amino Acid Analysis. Proteins are the principal components of the organs and soft structures of an animal's body. They are large molecules which differ widely in chemical composition, physical properties, solubility, and biological function. All proteins have one common property: their basic structure is composed of repeating units of amino acids. There are 22 amino acids that are commonly found in proteins. Some amino acids (nonessential or dispensable) can be synthesized and hence may not be required in the diet, whereas other amino acids (essential or indispensable) cannot be synthesized or synthesized in sufficient quantities to meet physiological demands.

Since amino acids are the building blocks of proteins, the amino acid profile (identification and quantification of amino acids) of a particular ingredient must be determined to formulate a feed that will meet essential amino acid requirements of the target species. Methods for determining the amino acid composition of ingredients and diets can be found in AOAC (1), and include microbial methods, column chromatography, and HPLC techniques.

However, these methods do not provide information about the chemical form of the amino acids, i.e., free, bound in proteins, or the degree of availability to the organisms.

Chemical Scores, Indispensable Amino Acids Index (*IAAI*), and Essential Amino Acid Index (*EAAI*). Several chemically based methods have been developed to assess the nutritional and potential biological value of feed ingredients to organisms in lieu of the use of live feeding trials. Chemical scores, *IAAI*, and the *EAAI* are based on a comparison of the amino acid content of the ingredient to that of a complete protein or reference protein such as whole egg (chicken) or tissue samples (muscle) of the target species (2).

The chemical score is the simplest and least accurate method and is calculated as follows:

$$\text{Chemical score} = \frac{\text{g amino acid in test ingredient}}{\text{g amino acid in reference protein}} \times 100$$

A score of 100 means that the quantity of the amino acid is equal to that present in the reference protein source. The lowest chemical score of those essential amino acids examined in an ingredient determines the final chemical score. This method measures the potential nutritive value of the protein based on the most limiting essential amino acid (4).

When quantitative data on the essential amino acid requirements of a fish or shrimp species of interest have not been determined, a helpful initial approach is to utilize feed ingredients that approximate the amino acid profile of either the carcass of the animal or a good protein source. For example, the *IAAI* is determined by adding the ratios of the level of each indispensable amino acid present in the ingredient to the level of the same indispensable amino acid in whole egg (2) or a reference protein (*RP*). The higher the *IAAI* value of the ingredient, the greater potential biological value it represents. *IAAI* is calculated as follows:

$$IAAI = \frac{\text{Arg(ing)}}{\text{Arg}(RP)} + \frac{\text{His(ing)}}{\text{His}(RP)} + \frac{\text{Iso(ing)}}{\text{Iso}(RP)} + \cdots \times 100$$

In addition, the *EAAI* (13) can be used to estimate the potential nutritive value of protein sources by calculating the ratio of essential amino acid (*aa*) content in the ingredient to the content in the animal carcass (*AA*) as follows:

$$EAAI = \sqrt[n]{\frac{aa_1}{AA_1} \times \frac{aa_2}{AA_2} \times \frac{aa_3}{AA_3} \cdots \frac{aa_n}{AA_n}}$$

The closer the *EAAI* is to 1.0, the higher the potential nutritive value of the ingredient. In general, ingredients scoring 0.9 are considered good quality, 0.8 are medium quality, and less than 0.7 are considered poor quality (14). This technique is especially useful for the development of diets for larval stages of marine fish and shrimp, for which essential *AA* requirements are very difficult to determine. These methods have in some cases led to a better understanding of the requirement of essential *AA* for some fish species. Significant correlation has been found between the essential *AA* requirements of fish and the *AA* profiles of fish carcass.

All of these methods assume that the amino acids determined by chemical analyses are totally available to the animal and that this ingredient is the exclusive source of amino acids. In the case of formulated feeds, multiple sources of proteins (amino acids) are combined to produce a final amino acid profile for the feed. Consequently, two protein sources that independently have poor *AA* profiles may, when combined, produce high-quality dietary protein. Additionally, the digestibility of proteins and amino acids can vary considerably; therefore, a portion of the amino acids determined by chemical analyses will not be available to the animal.

Fatty Acids. Triglycerides are the most common form of lipids in animals and plants. They are esters of fatty acids and glycerol and serve as a source of energy for many physiological functions in organisms. In addition, fatty acids are an essential component of the membrane structure of cells and have important metabolic functions (i.e., as precursors of hormones and prostaglandins). Fish and shrimp have limited or no capabilities to synthesize certain highly unsaturated fatty acids (HUFA) and thus have an essential dietary requirement for them. Knowledge of the fatty acid composition of an ingredient is important to the formulation of a diet that will provide the required level of HUFA for the species being cultured. As with the crude protein content, the total lipid content of a diet or ingredient does not provide information on fatty acid composition. Methods for the determination of the fatty acid composition of feedstuffs can be found in the *Official Methods of Analysis* (1). For example, once the lipid is extracted with an organic solvent, polar and neutral lipids are separated by column chromatography (i.e., using silicic acid columns; 15). Polar and neutral lipids are then esterified with boron trifluoride. The resulting fatty acid methyl esters are then determined qualitatively and quantitatively by gas chromatography against known standards. In general, digestibility values for lipids are very high and do not vary widely between sources; hence, the lipid content and fatty acid profile are generally true indicators of the nutritional values.

Vitamins. A variety of nutrients, which do no include proteins, lipids, and carbohydrates, are required by animals in small amounts for normal health and growth. These organic micronutrients are referred to as vitamins. Since animals are not capable or have limited capabilities for synthesizing these compounds, they must be obtained from the diet.

The vitamins may be classified according to their solubility in water as water-soluble (e.g., thiamine, riboflavin, pyridoxine, and ascorbic acid) or fat-soluble (e.g., retinol, cholecalciferol, and tocopherol; 11,16). Although the amount of vitamins included in dietary formulations usually comprises a relatively small fraction of the bulk diet, vitamins significantly add to the cost of complete feed formulations. The high cost and potential loss of vitamins during the manufacturing process renders determination of the final vitamin content essential. Due to the uncertainty in vitamin content and bioavailability in ingredients, as well as the potential interactions of vitamins with ingredients in a diet, vitamin requirements are usually met by supplementation with a fortified vitamin premix. A vitamin deficiency may cause several pathological signs (e.g., for vitamin C, reduced growth, impaired collagen formation, scoliosis, and lordosis; 17). The level of individual vitamins can be determined by several chemical and biological techniques. For a detailed explanation of the methods, see Methods AOAC (1).

Minerals. The inorganic or ash component is composed of minerals. There are about 22 mineral elements that have been found to be essential in at least one species. They are classified according to the level of their requirement; those required in large quantities are termed macro or major, and those required in trace amounts are termed minor or micro (4,11). Some examples of the macrominerals are calcium, phosphorus, magnesium, sodium, potassium, chlorine, and sulfur. Iron, iodine, manganese, copper, cobalt, and zinc are examples of trace minerals. There are marked differences in mineral requirements for fresh and saltwater fish. Differences are primarily attributed to those minerals that play a major role in osmoregulation, but differences are also due to the relative amounts of minerals in the water and the relative ability of the animals to absorb minerals directly from the water. Additionally, all mineral elements, whether essential or nonessential, can adversely affect an animal if included in the diet at excessively high levels (18). Consequently, determining the mineral profile of the feeds and feed ingredients is essential. Prior to analyses, the feed ingredient is either oxidized by a strong acid (acid digestion) or combusted in a muffle furnace, thus removing the organic component. The minerals are then dissolved in a weak acid for analyses. The two most common methods for mineral analyses utilize an atomic absorption spectrophotometer or an inductively coupled plasma spectrophotometer. Additionally, there are colorimetric methods (e.g., phosphorus) and fluorometric (e.g., selenium) methods for determination of specific minerals.

Antinutrients

When evaluating a feed or feedstuffs, in addition to knowledge about the nutritional value, one must be concerned about the possible presence of antinutrients. These substances can affect the health and normal performance of the test animal. Antinutrients include substances that are either toxic to the organism or inhibit metabolic reactions that are catalyzed by enzymes. Antinutrients may be classified as endogenous if they are a natural component of the ingredient or exogenous if they are the result of external contamination. Examples of endogenous antinutrients are urease, gossypol, trypsin inhibitors, and thiaminase. Exogenous contaminants may be aflatoxins, pesticides, heavy metals, and others. Most pesticides are lipid-soluble and bioaccumulate in the lipids of different organisms. Therefore, ingredients with high lipid content should be routinely monitored for pesticides. Antinutrients within the feedstuffs can be determined by several chemical methods. For example, trypsin inhibitors can be detected by adding commercial trypsin and measuring the decrease in enzyme activity (19). Aflatoxins can be measured by thin layer chromatography and pesticides can be detected by gas chromatography. For a detailed explanation of the analytical methods, the reader is referred to Refs. 8, 20, and 21.

Other Methods for Feedstuff Evaluation

Near-Infrared Analysis. Near-infrared spectroscopy (NIRS) is a new instrumental method that uses light of a near-infrared wavelength to determine the composition and quality of ingredients without causing any physical or chemical damage. This rapid and simple method screens a large number of samples without the need for time-consuming chemical assays. This technique is based on the principle that every organic compound has a unique absorbance spectrum in the near-infrared range. In the analysis, a light source consisting of different wavelengths is directed at the sample and the resulting spectra are collected, registered, and compared to known standards (22). This method of analysis is currently being utilized to determine moisture, protein, and lipid content as well as fatty acid composition of various feedstuffs (23,24).

BIOLOGICAL EVALUATION

For biological evaluation feeding trials are conducted to determine the performance of a complete diet or a particular ingredient of interest. Some measure of performance, such as growth, survival, or feed utilization, is used to evaluate the adequacy of an ingredient or diet. In most cases, other, more specific measures of performance such as the retention or loss of a specific nutrient in the body of the organism, shifts in enzyme activity, or the ability of the organism to survive a specific environmental challenge (i.e., shifts in temperature or salinity and exposure to pathogenic organisms) are needed. Feeding trials should be conducted under strict experimental conditions, which include environmental monitoring, adequate replication, and the manipulation of only one or a few variables at a time. To minimize any nutritional contribution from external sources such as primary production or bacteria, the design of an experimental system must include a means to remove external sources of nutrients from the water prior to entering the culture containers.

Measurements of Performance

Given adequate survival, the two most common measures of response to a particular ingredient or diet are growth and feed utilization. Growth can be measured as function of weight, length, or specific nutrient gain (e.g., protein). One of the goals of the aquaculturist is to maximize production in terms of biomass; hence, biomass increase is a common measure of performance. A distinction should be made between gains in weight or length and true growth. Increases in weight gain can be achieved through the growth of muscle, bone, organs, and the deposition of specific biochemical components such as proteins or lipids. However, some diets result in a gain in weight caused by the excessive deposition of lipid reserves in the adipose tissue. In fish, this type of lipid deposition is generally undesirable because it decreases dress-out percentage (fillet yield) and may adversely influence shelf life, resulting in human health concerns. Therefore, when a diet is evaluated by means of a growth trial, the performance parameter measured (e.g., growth as weight gain) should be complemented by an analysis of the proximate composition of the carcass of the organism prior to and following feed administration.

The second indicator of performance in feeding trials is feed utilization. Feed utilization describes to what extent food eaten by the organism is actually converted into growth. This information is particularly critical when comparing the economic cost of feeds and their potential for polluting the culture environment. Since the method of feeding will influence the degree of feed utilization, the type of feeding strategy should be well documented in terms of ration size (restricted vs. excess) as well as the number of feedings per day. Several indices are used to express feed utilization and will be described later.

Survival is often another indicator of nutritional status of the animal and should always be high (e.g., >90%) for a positive control or reference diet. In some experimental designs reductions in survival are unavoidable and are a clear indicator of poor nutrition. However, reductions in survival introduce variability into an experiment and thereby can create problems in interpreting the experimental results. This variability may be due to a number of factors which include reductions in density, unequal distribution of feed rations due to the dominance of healthy animals, and secondary sources of nutrients (i.e., cannibalism).

The following calculations are used to derive the indicated indices for weight gain or growth, feed utilization, and survival:

Weight Gain or Growth

Absolute growth or weight gain:

$$\text{Weight gain} = W_f - W_i$$

Percentage weight gain (relative growth):

$$\%\ \text{Weight gain} = \frac{W_f - W_i}{W_i} \times 100$$

Absolute growth rate (AGR):

$$AGR = \frac{(W_f - W_i)}{(t_f - t_i)}$$

Relative growth rate (RGR):

$$RGR = \frac{(W_f - W_i)}{[W_i \times (t_f - t_i)]}$$

Instantaneous growth rate (g):

$$g = \frac{(\ln W_f - \ln W_i)}{(t_f - t_i)}$$

Specific growth rate (G):

$$G = g \times 100$$

where W_i is initial weight; W_f is final weight; t_i is initial time (in days); and t_f is final time (in days).

Feed Utilization

Feed conversion ratio (FCR):

$$FCR = \frac{\text{Weight of feed fed}}{\text{Weight gain}}$$

Feed efficiency ratio (FE):

$$FE = \frac{\text{Weight gain}}{\text{Weight of feed fed}}$$

Survival

Percentage survival:

$$\%\ \text{Survival} = \frac{\begin{pmatrix}\text{Initial number fish}\\ -\ \text{Final number fish}\end{pmatrix}}{\text{Initial number fish}} \times 100$$

Digestibility

In a strict sense, digestion is the breakdown of feeds in the gut by mechanical, chemical, and enzymatic processes into its constituents parts, rendering them soluble and available for absorption by the gut. Absorption is the process by which molecules and ions are absorbed by the cells lining the gut. Digestibility is therefore a measure of biological availability of the nutrients or energy in the ingredient (i.e., how much is available for absorption), whereas absorption refers to actual uptake. The methods commonly used to measure digestibility, including apparent digestibility and true digestibility, are in fact measures of combined digestion and absorption.

Apparent Digestibility. The most direct method to estimate digestibility involves feeding a specific amount of an experimental diet and carefully recording the quantity of feed consumed and feces produced. The amount of a particular nutrient remaining in the feces is then subtracted from the initial quantity in the test feed; the difference represents the amount of nutrient absorbed by the animal (11).

$$AD = 100 \times \frac{(\text{Nutrient in diet} - \text{Nutrient in feces})}{\text{Nutrient in diet}}$$

This method is termed apparent digestibility (AD) because the feces also contain endogenous fecal excretion (i.e., about 3% for trout; 25) in addition to unabsorbed feed. Hence, the digestibility estimate can be an underestimate because some of the nutrients present in the feces could have originated from endogenously produced waste.

True Digestibility. To obtain a true estimate of digestibility (TD), an accounting of the amount of endogenous fecal excretion is needed. Estimates of endogenous fecal excretion can be obtained by feeding a diet that does not contain the nutrient being tested and then determining the amount of the nutrient in the feces.

$$TD = 100 \times \frac{\begin{pmatrix}\text{Nutrient in diet} - (\text{Nutrient in feces} \\ - \text{Nonfeed nutrient in feces})\end{pmatrix}}{\text{Nutrient in diet}}$$

Determination with Inert Markers. It is often difficult to determine feed intake accurately and/or to collect all fecal material produced by aquatic animals. A number of techniques, utilizing inert markers, have been developed to avoid tracking all feed consumed and all feces produced (26,27). This method is less time-consuming for obtaining estimates of digestibility. Animals are fed a diet that contains an inert indigestible marker such as chromic oxide (0.5 to 1%) for several days. The quantity of the nutrient of interest relative to the inert marker can be determined in the feed and feces. Therefore, percentage digestibility can be calculated as follows:

$$AD = 100 - 100 \times \frac{\text{Marker in feed}}{\text{Marker in feces}} \times \frac{\text{Nutrient in feces}}{\text{Nutrient in feed}}$$

Problems and Sources of Error with Digestibility Measurements. Many problems are associated with the determination of digestibility coefficients. Digestibility values can vary relative to many factors such as size and age of the animal, type of feed processing and processing conditions, environmental parameters, interactions with other ingredients and/or nutrients in the diets, method of collecting the feces, type of inert marker used, as well as leaching of nutrients from feed and feces (28,29). Although many technical problems are associated with the use of inert markers, this method remains one of the easiest and best to determine the apparent digestibility of an ingredient and its associated nutrients quickly.

Tracer Studies. Many researchers have utilized radio and stable isotopes as tracers of either ingestion, digestion, and assimilation of dietary nutrients, as well as their metabolism (30–34). Radio (i.e., ^{14}C) or stable (i.e., ^{15}N) isotope-labeled nutrient can be added to the diet and utilized as tracer by determining the amount deposited in the animal tissue. In addition, if an appropriate labeled substrate is utilized, metabolic pathways can also be determined by tracing the incorporation of the labeled substrate into those compounds that the organism can synthesize. Many difficulties such as the possible loss and recycling of the tracer through metabolism are associated with these techniques. Once the tracer is absorbed, it can be utilized for synthesis of new tissue or metabolized and excreted as waste. One possible solution to these problems is to account for labeled nutrient losses through metabolism by using a twin tracer ($^{51}Cr{-}^{14}C$) technique (35).

***In Vitro* Digestibility.** Recently developed assays to measure protein quality, such as *in vitro* methods, are simple, rapid, and inexpensive techniques to screen potentially useful ingredients. These techniques rely on the use of digestive enzymes extracted from the organism under study or those readily available commercial enzymes (i.e., extracted from pig, sheep, or bacteria). The enzymes are added to a sample of the ingredient being tested and digestion is measured *in vitro* by one of the following analytical methods:

1. pH-drop method: As proteolytic enzymes attack the peptide bonds of proteins, hydrogen is released and the pH of the protein solution reduced. The pH reduction is highly and positively correlated with the degree of protein digestion (36).
2. pH-stat method: To keep digestive enzymes close to their optimal pH, pH can be maintained by adding NaOH. The amount of NaOH consumed is proportional to the degree of protein hydrolysis and is highly correlated to *in vivo* apparent protein digestibility (37).

Other methods based on *in vitro* digestibility have been developed to measure the degree of digestion in several ways and include the release of free amino acids after digestion (38), the percentage of soluble nitrogen (39), or the absorbance at 280 nm of the soluble fraction after precipitation with trichloroacetic acid (40). Dimes and Haard have reviewed the various methods (41). The accuracy of *in vitro* digestibility methods can be affected by several factors. The source of enzymes (e.g., mammalian vs. fish) can give different digestibility values. In general, *in vitro* assays utilizing enzymes from the animal under study are better correlated to *in vivo* digestion (37,40,42). Due to differences in enzyme concentration, environment, and duration of digestion, *in vitro* assays can overestimate protein digestion of some ingredients that are actually poorly digested *in vivo* (36). In addition, the buffering capacity of the protein source itself may influence the change in pH and thus affect the measurement of pH decline or change. Biological methods to evaluate ingredient digestibility still provide the most accurate nutritional value, but are time consuming and expensive to perform.

Nutrient Retention or Deposition

The daily deposition of a nutrient or its components in the animal carcass can be another useful way of evaluating the availability of a specific nutrient such as the proper proportions of dietary amino acids and fatty acids (43,44). The following derives apparent nutrient retention:

$$\text{Apparent retention} = \frac{\begin{pmatrix}\text{Nutrient content in} \\ \text{fish at } T_f - \text{Nutrient} \\ \text{content in fish at } T_i\end{pmatrix}}{\text{Nutrient intake}} \times 100$$

where T_i is initial time, and T_f is final time.

Electrical Conductivity

When working with specimens for which physical samples cannot be removed (e.g., intermittent samples, brood stock evaluations, or evaluations of endangered species), traditional chemical methods for measuring lipids and water content of animals are not suitable. An alternative method based on the different electrical properties of lipids and water can be used. Fat tissue has approximately 20 times less electrical conductivity than lean tissue. By applying an electromagnetic field and measuring the different electrical conductivity the amount of lipid and water in a live organism can be estimated (45). This approach has been successfully used to measure body composition of several aquatic species such as red drum (46) and catfish (47).

Protein Quality

In addition to the previously mentioned chemical assays for evaluating protein and amino acid content, a frequently utilized procedure assesses the quality of proteins by comparing different protein sources in terms of fish weight gain per unit of protein fed. The following calculations are based on this concept.

Protein Efficiency Ratio (*PER*):

$$PER = \frac{\text{Weight gain}}{\text{Protein intake}}$$

Protein Conversion Efficiency (*PCE*):

$$PCE = \frac{PC_f - PC_i}{\text{Protein fed}} \times 100$$

where PC_f is final carcass protein, and PC_i is initial carcass protein.

Apparent Net Protein Utilization (*ANPU*):

Measurements of the protein content of the test animal at the start and end of the experiment, combined with an estimate of the digestibility value of the protein of interest (digestibility coefficient), can be used to estimate *ANPU* (2):

$$ANPU = \frac{(PC_f - PC_i)}{(\text{Protein fed} \times \text{Digestibility coefficient})}$$

To determine true net protein utilization (*TNPU*), endogenous protein changes must also be accounted for by feeding a protein-free diet for the same length of time and then determining the change in carcass protein. True net protein utilization is calculated by subtracting the change in protein carcass of fish fed the protein-free diet from the change in carcass protein of the fish fed the protein diet as follows (2):

$$TNPU = \frac{(PC_f - PC_i) - (PCF_f - PCF_i)}{(\text{Protein fed} \times \text{Digestibility coefficient})}$$

where PC_f is final carcass protein; PC_i is initial carcass protein; PCF_f is final carcass protein of protein-free dietary treatment; and PCF_i is initial carcass protein of protein-free dietary treatment.

Biological Value (*BV*):

Instead of measuring nutrient deposition as in the comparative carcass methods, nutrient excretion during a period of time can be measured. For example, all nitrogen excreted in the feces, urine, and gills is measured and compared to the total nitrogen fed (2):

$$\text{Apparent } BV = \frac{(NF - FN - UN - GN)}{NF} \times 100$$

As with the digestibility method, to obtain a "true" biological value (*TBV*), we have to estimate the endogenous loss of the nutrient in question by feeding a nitrogen-free diet:

$$TBV = \frac{\begin{pmatrix} NF - (FN - \text{endogenous } FN) \\ -(UN - \text{endogenous } UN) \\ -(GN - \text{endogenous } GN) \end{pmatrix}}{NF} \times 100$$

where *NF* is nitrogen fed; *FN* is fecal nitrogen; *UN* is urinary nitrogen; and *GN* is gill nitrogen.

Other Condition Measurements

For some stages of development, such as the larval period, many measurements of performance cannot be easily conducted due to the small size of the organism and the difficulties in measuring feed consumption or feces produced. Consequently, alternative measurements are critical for the determination of the nutritional status of the test animal. For a review of some measurements of larval condition, see Ferron and Leggett (48).

RNA/DNA Ratio. The quantity of deoxyribonucleic acid (DNA), the compound responsible for carrying genetic information, is relatively constant in somatic tissues and has been shown to reflect cell number (49). The quantity of ribonucleic acid (RNA), the transcriptor and translator of genetic information, is directly proportional to protein synthesis inside the cell. Consequently, the ratio of RNA to DNA has been found to correlate well with recent growth (i.e., protein synthesis) and, therefore, the nutritional status of several species of fish (50–52). High RNA/DNA ratios indicate adequate growth and nutritional status while low ratios indicate poor nutritional condition. For a review of the use of this index see Clemmensen (53) and Bergeron (54). For detailed information on analytical techniques, the reader is referred to Buckley (50) and Clemmensen (55).

Challenge Tests for Larvae. Several tests to evaluate the physiological condition of fish larvae in nutritional studies have been proposed (56,57). In these tests, fish larvae are exposed to stressful conditions, such as removing them from the water for a few seconds or exposing them to high or low salinity for a period of time. Thereafter, the cumulative mortality through time is determined. These tests assume that weak fish larvae (poor nutritional condition) will not be able to survive extreme conditions as well as healthy fish larvae (good nutritional status). This approach has been used in the study of the quantitative requirements of fatty acids in marine fish larvae (55,58). In the latter study, Brinkmeyer and Holt (58) used the stress index and detected significantly different responses among red drum larvae fed diets containing different DHA to EPA ratios. These differences were not detected by comparing larval growth and survival.

FEEDING EXPERIMENTS

Feeding experiments are conducted to evaluate objectively the response of the culture animal to different dietary formulations. Experiments must be conducted under conditions in which biotic (i.e., size, weight) and abiotic (i.e., temperature, light, and water quality) variables are controlled and suitable replication is employed. Such conditions minimize the confounding effect of variables other than those that are being studied while providing the test animal with optimal conditions for growth. Many possible sources of error can arise when conducting and interpreting data from nutrient requirement experiments. Interpretation of nutrient requirements must be conducted within the specific context of the experimental design, including stage of development, size of the animals, sex, and previous nutritional condition. For a detailed critical review of the problems that are often encountered, the reader is referred to Baker (59) and Jobling (60). Additionally, when considering the design of larval feeding experiments, one must recognize that the nutritional status and condition of the broodstock will influence the quality of eggs and larvae (61). In an attempt to standardize larval quality, several egg quality tests that can be performed prior to conducting a larval feeding experiment have been suggested. For example, the percentage of fertilized eggs, buoyancy, appearance of the chorion, distribution of oil globules, transparency, microbial colonization, size, consistency, and shape have been used to determine egg quality. For a review see Kjorsvik et al. (62) and Bromage (63).

The ultimate test of a practical diet is its successful use by commercial growers; thus, evaluation of diets under less stringent conditions may ultimately be necessary. For such purposes, experimental conditions should simulate those found in commercial aquaculture enterprises, such as ponds, as much as possible. In all cases, control should be exercised over as many variables as possible.

Controlled Environments

Unless one of the variables of interest is environmental (e.g., temperature), it is of paramount importance to maintain complete consistency among the environmental parameters within treatments and vary only the specific nutrient of interest. In general, the basic aspects to consider when conducting a feeding trial are the experimental animals, the rearing facilities, and the experimental diets. The appropriate methods can vary considerably between species; hence, the following descriptions should be viewed as basic guidelines for conducting feeding experiments that will require modification relative to the species of interest or specific life stage.

Animals. Same generation animals of similar nutritional history should be used in feeding trials and should be graded to a uniform size before stocking into rearing tanks. Each replicate should be stocked at equal densities with a sufficiently large number of animals to minimize any effects of biased sex ratios (sex-related growth rate) or dominance feeding patterns (hierarchical behavior) and to provide an adequate tissue sample if biochemical analyses are planned. The number of replicates required will vary depending on the expected coefficient of variation for the response(s) to be measured but should never fall below three. In most situations the test population should be allowed to acclimate to the culture system and the basal diet over a 1 to 2-week period. The need for an acclimation period will depend on the species and size of the test animal. This preconditioning practice will be best served by initially stocking an excess of animals and then feeding them a basal diet similar to that which will be used in the experiment. To avoid any confounding effects related to health and disease, the health of the animals should be monitored prior to and during all experiments. If possible, periodic sampling should be conducted to determine growth rates and inspect the animals for signs of disease or stress.

Rearing Facilities and Culture Period. Holding facilities should be of equal characteristics (size and material) and large enough to allow normal growth throughout the experimental period, thereby minimizing any density-dependent effects. The culture system must be suitably designed to allow maximum growth and provide adequate water quality throughout the experiment. Assurance of suitable water quality is particularly important near the end of an experiment when nutrient loading is highest. The system should be isolated or shielded from outside disturbances. Water entering the tanks should come from the same source and be of equal temperature, chemical composition, and flow rate. The system can be designed as a single-pass, flow-through system or as a semiclosed recirculating system. Regardless of the type of system used, uneaten food and fecal material should be removed from the culture chambers daily and no alternate food sources should be available. Additionally, water quality conditions must be maintained within the suitable ranges for the species being tested. Maintaining well-characterized and suitable water quality parameters is critical to the design of any experiment. Therefore, water quality parameters should be monitored either intermittently (e.g., biweekly measurements for ammonia, nitrite, and nitrate) or daily (e.g., dissolved oxygen, temperature, and salinity). Fluctuations in temperature and salinity should be minimized and photoperiod should be kept constant throughout the experiment.

The length of time required to conduct a given study is dependent on a variety of factors that include the objectives of the experiment, species utilized, and the age of the test animal. In general, the growth trial should be of sufficient duration to produce relatively large increases in growth and statistically significant differences between some of the dietary treatments. Examples of experimental periods for different life stages include 14–28 days for larvae, 6–8 weeks for juveniles, and 14–18 weeks for larger fish. Although purified diets often produce slower growth, in general growth rates in the laboratory should be comparable to those achieved under more natural conditions (e.g., pond) where both high-quality feeds and natural productivity are available as food sources.

Diets. Ingredients for making the experimental diets should be as pure and well characterized as possible so that composition is well defined. Diets prepared for the different treatments should be as close to the same physical (texture, size, water stability), chemical (attractants and taste), and nutritional characteristics as possible. The only differences among the diets should be the amount of nutrient or feedstuff of interest. It should be noted that if a response curve is to be determined, a minimum of four and preferably six dietary levels should be tested. If less than four levels are utilized, fitting the data to a response curve cannot be conducted accurately.

When preparing a diet, all dry ingredients (sources of proteins, carbohydrates, vitamins, and minerals) should be mixed thoroughly before adding and mixing in the lipid ingredients and water. Once the feed has been thoroughly homogenized, it is shaped (spray-dried, microencapsulated, flaked, or pelleted) and then, if desired, dried to a moisture content of 8–10%. The final product should be of a suitable size and texture for the animals to consume easily. Once the diet is made, it can be temporarily stored under refrigeration or frozen until used.

Animals should be fed according to a well-defined feeding rate (ration size) which must be adjusted as the animal grows. Types of feeding include (*1*) restrictive ration, i.e., feed is offered based on a fixed rate of the animal body weight that is below satiation; (*2*) in excess, i.e., feed is offered at a fixed rate that is in excess of what the animals will consume; and, (*3*) to apparent satiation, in which case the food is offered during a specified period of time until the test animals stop consuming feed. The number of feedings and time of day the animals are fed should also be appropriate for the species and size of the animal. Generally, food should be offered on a semicontinuous basis to larvae, 4 times a day for small juveniles, twice daily for large juveniles, and once daily to subadults.

Practical Environments

Nutritional requirements can only be determined when all nutrient sources are quantified. This is quite difficult to accomplish when an animal is exposed to natural food sources. However, under commercial production conditions prepared feeds are not the only source of nutrients. Quite often natural productivity contributes a significant portion of the daily intake of nutrients. Consequently, it is often desirable to evaluate the influence of changes in practical diet formulations when natural productivity is present. The relative contributions of natural productivity and prepared feed can be estimated with the use of stable isotopes of biologically important elements such as carbon, nitrogen, and sulfur. Anderson et al. (64) fed Pacific white shrimp (*Litopenaeus vannamei*) diets that were isotopically different from those of natural pond organisms and were able to determine the relative contribution of each carbon source. However, natural productivity varies considerably within and between sites, with time of year, and from year to year. Hence, results derived from such studies must be interpreted with caution. From a commercial point of view, the testing of practical diets under actual commercial conditions is desirable. However, replication of commercial production systems is very expensive besides being unrealistic. Therefore, several techniques are used to approximate these conditions. Examples of such systems include (*1*) tank studies where outdoor tanks, in which natural productivity is allowed to become established, are used as replicates; (*2*) cage studies where cages are either floated in a pond or affixed to the pond bottom and used as replicates; and (*3*) pond studies where small ponds are used as replicates (11). When conducting such an experiment, one must realize that a number of variables cannot be controlled in an outdoor situation; hence, appropriate statistical design becomes crucial. To account for high variation, the number of replicates per treatment is often increased. When conducting an outdoor experiment, the stocking and feeding procedures should be similar to those previously described. Additionally, methods to quantify the type and quantity of natural food sources should be considered as part of the experimental design.

BIBLIOGRAPHY

1. W. Horwitz, ed., *Official Methods of Analysis of the American Association of Analytical Chemists*, Am. Assoc. Anal. Chem. (AOAC), Washington, DC, 1984.
2. R.W. Hardy, in J.E. Halver, ed., *Fish Nutrition*, Academic Press, San Diego, CA, 1989, pp. 475–548.
3. L.W. Winckler, *Z. Angew. Chem.* **26**, 231 (1913).
4. A.G. Tacon, *Standard Methods for the Nutrition and Feeding of Farmed Fish and Shrimp*, Argent Laboratory Press, Washington, DC, 1990.
5. J. Folch, R. Lees, and G.H. Sloane-Stanley, *J. Biol. Chem.* **226**, 497–507 (1957).
6. G.R. Nematipour and D.M. Gatlin, *Aquaculture* **114**, 141–154 (1993).
7. M. Kates, *Techniques of Lipidology*, Elsevier, Amsterdam, 1986.
8. W. Horwitz, ed., *Official Methods of Analysis of the American Association of Analytical Chemists*, Am. Assoc. Anal. Chem. (AOAC), Washington, DC, 1980.
9. S.A. Katz, S.W. Jenniss, and T. Mount, *Int. J. Environ. Anal. Chem.* **9**, 209–220 (1981).
10. R. Smith, in J.E. Halver, ed., *Fish Nutrition*, Academic Press, San Diego, CA, 1989, pp. 1–29.
11. R.T. Lovell, *Nutrition and Feeding of Fish*, Van Nostrand-Reinhold, New York, 1989.
12. C.Y. Cho, S.J. Slinger, and H.S. Bayley, *Comp. Biochem. Physiol. B* **73B**, 25–41 (1982).
13. V. Penaflorida, *Aquaculture* **83**, 319–330 (1989).
14. A.G. Tacon, *FAO Fish. Circ.* **866**, 1–25 (1993).
15. S. Satoh, W.E. Poe, and R.P. Wilson, *Aquaculture* **79**, 121–128 (1989).
16. National Research Council, *Nutrient Requirements of Fish*, National Academic Press, Washington, DC, 1993.

17. A.G. Tacon, *FAO Fish. Tech. Pap.* **330**, 20–39 (1992).
18. National Research Council, *Mineral Tolerance of Domestic Animals*, National Academic Press, Washington, DC, 1980.
19. L.M. Millan, I.C. Guerrero, and P. Fernandez, in J. Espinosa de los Monteros and U. Labarta, eds., *Alimentaction en Acuicultura*, Plan de Formación de técnicos Superiores en Acuicultura, Madrid, Spain, 1987, pp. 1–22.
20. I.E. Liener, *Toxic Constituents of Plant Foodstuffs*, Academic Press, New York and London, 1980.
21. Ministry of Agriculture, Fisheries and Food, *Food Standards Committee Report on Novel Protein Foods*, FSC/REP/62, MAFF, London, 1981.
22. D.E. Root and J.W. Hall, *Med. Electron.* **28**, 64–66 (1994).
23. J.A. Panford, P.C. Williams, and J.M. deMan, *J. Am. Oil Chem. Soc.* **65**, 1627–1634 (1988).
24. T. Sato, H. Abe, and S. Kawano, *J. Am. Oil Chem. Soc.* **71**, 1049–1055 (1994).
25. T. Nose, *Bull. Freshwater Fish. Res. Lab.* **13**, 41–43 (1967).
26. A. Furukawa and H. Tsukahara, *Bull. Jpn. Soc. Sci. Fish.* **45**, 983–987 (1966).
27. J.L. Atkinson, J.W. Hilton, and S.J. Slinger, *Can. J. Fish. Aqua. Sci.* **41**, 1384–1386 (1984).
28. S.R. Coelho, Master's Thesis, Texas A&M, College Station, 1984.
29. J. De La Noue and G. Choubert, *Prog. Fish-Cult.* **48**, 190–195 (1986).
30. C.B. Cowey, J.W. Adron, and A. Blair, *J. Mar. Biol. Assoc.* **50**, 87 (1970).
31. S.I. Teshima and A. Kanzawa, *Bull. Jpn. Soc. Sci. Fish.* **46**, 51–55 (1980).
32. R.W. Ellis, J. Long, L. Leitner, and J. Parsons, *Prog. Fish-Cult.* **49**, 305–305 (1987).
33. T. Forseth, B. Jonsson, R. Naeumann, and O. Ugedal, *Can. J. Fish. Aqua. Sci.* **49**, 1328–1335 (1992).
34. H. Vincent, J.N. Thibault, and S. Bernard, *J. Nutr.* **128**, 1969–1977 (1998).
35. J.A. Wightman, *Oecologia* **19**, 273–284 (1975).
36. J.P. Lazo, R.P. Romaire, and R.C. Reigh, *J. World Aquacult. Soc.* **29**, 774–450 (1998).
37. J.M. Ezquerra, F.L. Gracia-Carreno, and O. Carrillo, *Aquaculture* **163**, 123–136 (1998).
38. G. Bitterlich, *Aquaculture* **50**, 123–131 (1985).
39. R.A. Buchanan, *Br. J. Nutr.* **23**, 533–544 (1969).
40. C.C. Lan and B.S. Pan, *Aquaculture* **79**, 59–70 (1993).
41. L.E. Dimes and N. Haard, *Comp. Biochem. Physiol.* **108**, 349–362 (1994).
42. F.M. Dong, R.W. Hardy, N. Haard, F. Barrows, B. Rasco, W. Fairgrieve, and I. Foster, *Aquaculture* **116**, 149–158 (1993).
43. C. Ogino, *Bull. Jpn. Soc. Sci. Fish.* **43**, 171–174 (1980).
44. R.W. Hardy and K.D. Shearer, *Can. J. Fish. Aquat. Sci.* **42**, 181–184 (1985).
45. W. Piasecki, P. Koteja, and J. Weiner, *Rev. Sci. Instrum.* **66**, 3037–3041 (1995).
46. S.C. Bai, G.R. Nematipour, R.P. Perera, and F. Jaramillo, *Prog. Fish-Cult.* **56**, 232 (1994).
47. F. Jamarillo, S.C. Bai, B.R. Murphy, and D.M. Gatlin, *Aquat. Living Resour.* **7**, 87 (1994).
48. A. Ferron and W.C. Leggett, *Adv. Mar. Biol.* **30**, 217–303 (1994).
49. Q. Dortch, T.L. Roberts, J.R. Clayton, and S.I. Ahmed, *Mar. Ecol.: Prog. Ser.* **13**, 61–71 (1983).
50. L.J. Buckley, *Mar. Biol.* **80**, 291–298 (1984).
51. G.J. Holt, *J. World Aquacult. Soc.* **42**, 225–240 (1993).
52. C. Clemmensen, *Mar. Biol.* **118**, 377–382 (1994).
53. C. Clemmensen, in Y. Watanabe, Y. Yamashita, and Y. Oozeki, eds., *Survival Strategies in Early Life Stages of Marine Resources*, A.A. Balkema Publishers, Netherlands, 1996, pp. 67–82.
54. J.P. Bergeron, *J. Fish Biol.* **51**, 284–302 (1997).
55. C. Clemmensen, *Mar. Ecol.: Prog. Ser.* **100**, 177–183 (1993).
56. T. Watanabe, F. Oowa, and S. Fujita, *Aquaculture* **34**, 115–143 (1983).
57. P. Dhert, P. Lavens, and P. Sorgeloos, *Meded. Fac. Landbouwwet., Rijksuniv. Gent* **57**, 2135–2142 (1992).
58. R.L. Brinkmeyer and G.J. Holt, *Aquaculture* **161**, 253–268 (1983).
59. D.H. Baker, *J. Nutr.* **116**, 2339–2349 (1986).
60. M. Jobling, *J. Fish Biol.* **23**, 685–703 (1983).
61. T. Watanabe and V. Kiron, in N.R. Bromage and R.J. Roberts, eds., *Broodstock Management and Egg and Larval Quality*, Blackwell Science, Oxford, 1995, pp. 398–412.
62. E. Kjorsivk, A. Mangor-Jensen, and I. Homefjord, *Adv. Mar. Biol.* **26**, 71–113 (1990).
63. N.R. Bromage, in N.R. Bromage and R.J. Roberts, eds., *Broodstock Management and Egg and Larval Quality*, Blackwell Science, Oxford, 1995, pp. 1–24.
64. R.K. Anderson, P.L. Parker, and A. Lawrence, *J. World Aquacult. Soc.* **18**, 148–155 (1987).

See also FEED MANUFACTURING TECHNOLOGY; NUTRIENT REQUIREMENTS.

LARVAL FEEDING — FISH

Frederic T. Barrows
USFWS, Fish Technology Center
Bozeman, Montana

Michael B. Rust
Northwest Fisheries Science Center
Seattle, Washington

OUTLINE

INTRODUCTION

Feeding microparticulate feeds to species with small larvae will aid in the development of reliable, economically sustainable fingerling production industries. However, the culture of larval fish with formulated feeds requires rearing methods different from those used for feeding live feeds. Factors under the control of the larval fish culturist include water flow, lighting, tank color, tank shape, feeding rate, and environmental factors. The effect of feed processing on feed quality, nutrient availability, and feed consumption must also be considered for a larval production program to be effective.

LIVE OR FORMULATED?

Because of the inherent differences between formulated and live diets, use of the same rearing techniques that are adequate for live foods with formulated feeds will likely result in a loss of livestock. Simple modifications in tank design, water flow rate, feeding method and rate, and water flow pattern can result in a much enhanced response of the fish to formulated feeds. Live feeds can be fed less often, at lower rates, and uneaten feed does not deteriorate as rapidly as formulated feeds. Even with these differences, there are several reasons to strive for a production program that utilizes formulated feeds instead of live foods.

Rearing organisms to feed production species has its own inherent problems. If the live feed culture fails for any reason (i.e., power outage, low hatch rate of cysts, etc.), the production of the target species will suffer. Collection of wild feed for larval fish is susceptible to availability due to weather or other factors that limit zooplankton production. Conversely, formulated feeds can be purchased in advance of the hatch of the larval animals and can be stored for relatively long periods of time, ensuring an adequate supply of feed. Also, wild food supplies may introduce predators, toxins, and/or diseases to the larvae. Feeding formulated feeds avoids that risk. The cost of live foods can also be much greater than with formulated feeds when evaluated on a dry matter basis. Although there are still numerous species of fishes where culture with formulated feeds has been unsuccessful, there have also been many improvements in larval feed manufacturing methods that appear to be increasing survival, growth, and quality of larvae fed formulated feeds (see the entries "Microbound feeds," "Microparticulate feeds, complex microparticles," and "Microparticulate feeds, micro encapsulated particles"). These technological improvements in feed manufacturing should alleviate some of the problems observed with feeding formulated feeds in the past.

FORMULATED FEED AND FEEDING

By evaluating the characteristics of formulated feeds and the larval fish's ability to ingest these feeds, the reasons for modifying rearing techniques becomes clear. Many factors affect feeding in larval fish. Development of culture systems that utilize microparticulate feeds for new species will require attention to both the feed and the response of the larval fish to feed before and after it is ingested. Modifying feed characteristics and culture system design so that fish response is optimized will improve the chances for success. Figure 1 presents factors to consider when feeding larval fish. Some factors are best addressed in the diet formulation and manufacture process, while others need to be addressed by management of the culture system in which the diets are fed. Factors to be addressed in diet manufacture include appearance, attractiveness, texture, formulation, and physical properties. Manufacturing and formulation changes can be used to address problems associated with ingestion, nutrient requirements, digestibility, and leaching. Factors affected by culture system design and management include: feed availability, larval distribution, sensory acuity, and environmental quality. Husbandry and system design changes can improve larval feed consumption, lower cannibalism, reduce deformities, and maintain high water quality. Optimal conditions established for one species or feeding strategy (live vs. formulated) for a given factor may not be optimal for

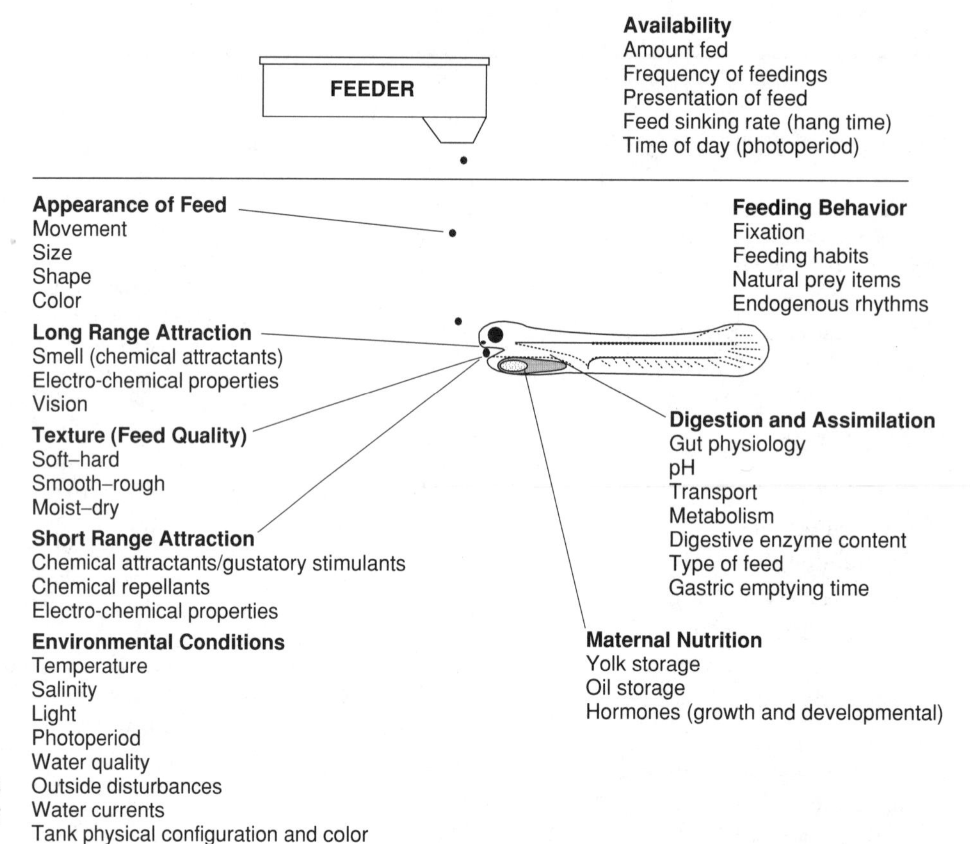

Figure 1. Some considerations that are important for larval fish feeding and nutrition.

other species. Likewise, a microparticulate diet that works well for one species may or may not work well for other species.

FACTORS AFFECTING NUTRIENT AVAILABILITY

Digestibility

Rust and co-workers determined how nutrient assimilation was influenced by the structural complexity of nutrients (1,2). The studies specifically investigated the assimilation of protein and lipid in various chemical forms by larvae of different species. The rationale behind the trials was to increase understanding of larval fish digestion so that future formulated larval feeds can be produced which deliver nutrients to the gut in a form that is easily assimilated by the developing digestive system.

Differences in nutrient assimilation efficiencies were found among species and among developmental stages within the same species. For species that start feeding prior to the development of gastric digestion, small molecular weight nutrients were initially assimilated more efficiently than complex forms of the same nutrients, with differences becoming less pronounced as the larvae reached metamorphosis. For example, amino acids were assimilated at higher efficiencies than polypeptides and polypeptides higher than proteins during the first-feeding stage. In striped bass (*Morone saxatilis*), assimilation of protein- and polypeptide-bound methionine increased (from 30 to 62% and 49 to 82%, respectively) over the period from first feeding to metamorphosis. At metamorphosis, all structural forms were assimilated at similar efficiencies. This pattern held for striped bass, walleye (*Stizostedion vitreum vitreum*), and zebrafish (*Brachydanio rerio*) but not for goldfish (*Carassius auratus*) (which do not develop gastric digestion) or salmon (*Oncorhynchus* spp.) (which start feeding after gastric digestion is developed). It is likely that assimilation efficiency of complex nutrients correlates with the presence, location, and amount of key digestive enzymes in the larval gut. Larval microparticulate diets that can deliver simple forms of nutrients to the gut during developmental stages (that lack some of the digestive processes of adults) may improve larval growth and survival because of greater assimilation efficiencies during the critical first-feeding stage.

Nutrient Requirements

No quantitative requirement for any nutrient has yet been determined for the larvae of any species of fish. This is partly due to the difficulty of working with such small animals that typically have high and variable mortality during this life stage. Feeds have been formulated based upon the composition of the fish larvae or the composition of zooplankton. Cowey and Tacon (3) observed that in many juvenile and adult fish, there is good correlation between dietary amino acid requirements and the pattern of amino acids in the muscle tissue of the consuming animal. This approach assumes that the bioavailability of dietary nutrients is equal, an assumption that does not hold for larvae (1). Other scientists have preferred to base diet formulations on the composition of zooplankton, the natural prey of most larval fish (4,5). That can also lead to misleading formulations because typically only nutrient composition is taken into consideration and not nutrient form.

FACTORS AFFECTING FEEDING

Our understanding of factors that control larval feeding behavior is limited. Before nutrient digestion can occur, the feed must first be ingested by the larval fish. A key research area with formulated feeds is to increase the palatability of the microparticle so the larval fish will identify it as food and ingest it. Physical properties of the feed, such as color and size, along with chemical properties, such as flavor and odor, may be key attributes that stimulate the larvae to approach and ingest the diet in sufficient quantities to sustain growth. In order for microparticulate feeds to be ingested, they must be attractive and visible to the larvae and must be presented under the proper environmental conditions. Many species of fish larvae are visual feeders, though taste buds and olfaction are often also functional at this time (6). This makes vision and chemoreception two of the most important sensory systems used by first-feeding larvae to locate and ingest feed (6,7).

Vision

Physical properties of the diet and rearing environment, such as color, intensity, and wavelength of light, and degree of polarization and contrast, may be key attributes that stimulate the larvae to approach and ingest feed. Three or four cones in the eye may be operational in fish larvae, each with a different light wavelength of maximum sensitivity (8–15). As the larvae develop, visual acuity increases (6,7). Specialized retinal structures such as a tapetum lucidum and macroreceptors (which are adaptations to low-light environments) often do not develop until much later (8).

A heart rate conditioning technique used by Hawryshyn and co-workers (9) has led to the identification of up to four cones in rainbow trout that have different regions of maximal photosensitivity. In addition to the normal blue (400–450 nm), green (500–550 nm), and red (600–650 nm) sensitive cones, a transient UV-sensitive (340–370 nm) cone was also identified. The UV cone was only found in larval and early juvenile fish and was completely absent in adults. This cone has been identified in a number of planktivorous fish (10–15). The ecological significance of this cone is unknown for most fish, but it appears to be involved in prey identification and feeding in fish which feed on zooplankton (15–17). The chitonous exoskeleton of zooplankton is highly reflective to UV light (15). Ultraviolet vision may greatly increase the larval fish's ability to see "Zooplankton culture". If this is the case, then microparticulate feeds that also reflect UV may be advantageous. Tank lighting that lacks UV may be problematic for feeding. For example, lingcod (*Ophiodon elongatus*) larvae reared outdoors under full-strength sunlight feed and behave much differently from larvae reared indoors under incandescent or fluorescent light. These artificial light sources are lacking in the near UV ranges and are strong in the yellow bands. Fish reared indoors under these lights feed poorly and display "nosing" behavior typical of stressed larvae. In several trials, all larvae reared indoors died within three weeks of first feeding (living only slightly longer than starved larvae), while larvae reared outdoors fed normally and survived beyond the juvenile stage (Ken Massee, personal communication).

Chemosensory Systems

Fish have two major chemosensory systems: olfaction and gustation (18). Olfaction is involved in diverse teleost behaviors, including migration, reproduction, fright reactions, and feeding. Gustation is involved primarily in feeding. The olfactory receptors develop early—usually before hatching—from an ectodermal anlagen, while gustatory receptors develop later and are endodermal in origin (18).

Both olfaction and gustation are likely involved in feeding, although the integrated functioning of these two systems remains ambiguous. A likely pattern is that olfactory receptors and external taste buds serve to locate feed at a distance and to trigger the feeding response (19). Internal taste buds serve to screen for palatability and control swallowing (19).

Hara and Zielinski (18), working with rainbow trout, showed that the olfactory mucosa could be stimulated by amino acids in embryos as young as 20 days postfertilization (well before first feeding). At that stage, ciliated receptor cells are sparse and only have 1–3 short cilia, yet are functional. The detection thresholds for amino acids decrease (from 10^{-3} to 10^{-9} M) as the embryos develop until just before first feeding where they remain through adulthood (18). The first immature taste buds (gustatory system) develop later than the olfactory mucosa (20), however, most species have functional taste buds by first feeding (20).

Chemical feeding stimulants have proven effective in feeds for young fish which are being trained to eat formulated feeds after being reared on live foods (21). Stimulants have also been shown to be advantageous for microparticulate feeds used for first feeding carp (*Cyprinus carpio*) larvae (22). Several chemicals have been identified as feeding stimulants in fish, including L-amino acids, betaine, nucleotides, and others (23–31). The specific chemical or combination of these chemicals which is most effective for a given species is often related to the type of prey item (and the content of those chemicals in the prey items) that the given species consumes in nature (26). Betaine plus amino acids tend to be more effective with species consuming invertebrates (worms, molluscs, and crustaceans), and nucleotides plus amino acids are more effective with species consuming vertebrates (26). In most cases, these substances exhibit synergistic effects when used in combination. Because most fish larvae eat similar diets (i.e., zooplankton), it is reasonable to expect fish larvae to exhibit similar preferences for gustatory stimulants.

REARING SYSTEMS

Culture methods such as tank design, effluent screen size and area, lighting, feed distribution, and other factors combine to produce a rearing system. Analyzing the effect on fish performance that each of these factors has with specific types of formulated feed is a key to successful larval culture. Even though specific variations in a rearing system can be important for a given species, a system developed by Barrows and co-workers for larval walleye has also been effective with other species (32).

Feeding System

One of the most important aspects of larval culture is providing large quantities of high-quality feed that is available at all times. Live feeds may swim in the water column for long periods of time, remaining available as a feed source to the larvae. Most formulated feeds will sink, though at different rates, and will not be eaten by most fish once on the bottom of the tank. Different feed manufacturing technologies can produce feeds with vastly different sinking rates. A feed with an appropriate sinking rate should be chosen for each species (see the entries "Microbound feeds" and "Microparticulate feeds, micro encapsulated particles"). The sinking nature of formulated feeds requires several culture technique modifications. The first modification is the use of automatic feeders to provide a steady introduction of feed into the tank. Since the feed is constantly settling out of the feeding area, more feed must be introduced to assure availability of the feed to the fish. The second modification is the use of extremely high feeding rates relative to juvenile fish culture or with live foods. Since the formulated feed does not move like live feeds, attraction of the fish to the formulated feed is presumed to be reduced compared to live feed. To make up for this, a higher feeding rate is used, thus increasing feeding opportunities. The high feeding rate will, over time, deposit significant quantities of uneaten feed on the bottom of the tank, which necessitates several other changes in rearing techniques. A rule of thumb for the early larval rearing stages is to feed as much feed as water quality conditions will allow. Different types of larval feeds break down and foul the water at different rates (33), so feeding rate will be somewhat dependent on feed type and exchange rate.

Water Flow and Effluent Screens

The water flow rate and the design of the effluent screen need to be optimized when feeding formulated feeds. High water flow rates will help reduce water quality degradation caused by uneaten feed on the bottom of the tank. High water flow into the tank may also sweep some of the uneaten feed out of the tank. To facilitate the removal of uneaten feed from the tank, the effluent screen can be modified in two ways. First, openings in the mesh should be as large as possible and still retain the larvae. Simple studies can be conducted to determine the optimal size for a particular species and stage (34). Second, to avoid impingement of fry upon the effluent screen, the surface area of the screen should be maximized. For example, a circular tank with an interior standpipe can be covered with a larger pipe that has mesh from the top to the bottom of the tank. In this case, flow rate through each mesh opening will be minimized while allowing for a high overall flow rate through the tank.

Gas Bladder Inflation

Feeding formulated feeds can create problems for species that inflate their gas bladders after initiation of feeding (35), even though lack of a gas bladder has also been observed in walleye reared in ponds feeding on natural foods (36). Gas bladder inflation problems have been reported for intensively reared walleye (37), red sea bream (*Pagrus major*) (38), northern anchovies (*Engraulis mordax*) (39), and striped bass (40,41). Several of these species were found to require unobstructed access to the surface or bubbles for initial inflation of the gas bladder. Colesante et al. (35) compared larval walleye fed live foods to fish fed formulated feeds and observed little inflation with the formulated feeds and high levels of gas bladder inflation in the tanks fed live food. To one degree or another all formulated feeds create a surface film when introduced to the rearing container. Feed formulation and manufacturing method will affect the film created when formulated feeds are placed in water. Several methods have been developed to reduce or eliminate the surface film caused by the formulated feed (32,37). The most practical and effective method involves a simple water spray on the surface to disperse the film (32) or the provision of a slight fine bubble stream (41). Selecting a feed that produces a minimal film is also beneficial. A combination of feed selection and modified rearing method can result in high levels of gas bladder inflation when feeding formulated feeds.

Tank Maintenance

Once excess feed accumulates on the bottom of the tank, it should be removed as frequently as possible. Siphoning excess feed from the bottom of the tank two or three times per day during the early stages of rearing is recommended. The type of larval feed used will also impact the frequency of siphoning required. As the particle size of the feed is increased, the need for frequent siphoning decreases. Higher water temperatures, however, will increase the need for siphoning. Siphoning can be facilitated by using a strong circular water flow pattern in the tank. This will cause the feed to accumulate in a circle next to the center. This feed can be quickly siphoned off compared with having feed scattered equally across the bottom of the tank. Siphoning feed and feces from a tank is difficult when larval fish are near the bottom. If the fish are photopositive, use of a deep tank (>60 cm or 24 in. deep) will separate the fish from the excess feed (32). The photopositive fry are attracted to the upper portions of the tank and are not removed during siphoning. Some larval tank designs have incorporated an upwelling water flow in order to suspend the feed in the water column. The desired effects with this tank design is to increase feed consumption and decrease feeding rate. A problem with this approach is that most larval feeds will lose significant portions of the water-soluble nutrients over a relatively short time period. It is counterproductive to have fish eating feed that has been in the tank for long periods of time. A system designed to separate uneaten feed from the fish, even though feed cost may increase, will in many cases result in higher quality larvae.

Not all of these suggestions will work or are appropriate for all species in every situation. The fish culturist should be aware, however, that the rearing conditions (tank configuration, water flow rate, feeding rate, etc.) can determine the success of a larval rearing program when using formulated feeds. No matter how much technology is employed in manufacturing a larval feed, poor culture techniques can make even the best formulated feed appear ineffective.

BIBLIOGRAPHY

1. M.B. Rust, R.W. Hardy, and R.R. Stickney, *Aquaculture* **116**, 341–352 (1993).
2. M.B. Rust, *Quantitative Aspects of Nutrient Assimilation in Six Species of Fish Larvae*, Ph.D. dissertation, University of Washington, Seattle, 1995.
3. C.B. Cowey and A.G. Tacon, in G.D. Pruden, C.J. Langdon, and D.E. Conklin, eds., *Proceedings of the Second International Conference on Aquaculture Nutrition: Biochemical and Physiological Approaches to Shellfish Nutrition*, Louisiana State University, Baton Rouge, LA, 1983, pp. 13–30.
4. K. Dabrowski and M. Rusiecki, *Aquaculture* **30**, 31–42 (1983).
5. T. Watanabe, T. Tamiya, A. Oka, M. Hirata, C. Kitajima, and S. Fujita, *Bull. Jpn. Soc. Sci. Fish* **49**, 471 (1983).
6. D.L. Noakes and J.-G.J. Godin, in W.S. Hoar and D.J. Randall, eds., *Fish Physiology, Vol. XI, B*, Academic Press, San Diego, CA, 1988, pp. 345–384.
7. J.H.S. Blaxter, in W.S. Hoar and D.J. Randall, eds., *Fish Physiology, Vol. XI, A*, Academic Press, San Diego, CA, 1988, pp. 1–84.
8. L. Vandenbyllaardt, F.J. Ward, C.R. Braekevelt, and D.B. McIntyre, *Trans. Am. Fish. Soc.* **120**, 382–390 (1991).
9. C.W. Hawryshyn, M.G. Arnold, D.J. Chaisson, and P.C. Martin, *Vis. Neurosci.* **2**(3), 247–254 (1989).
10. F.I. Harosi and Y. Hashimoto, *Science* **222**, 1021–1023 (1983).
11. K.V. Singarajah and F.I. Harosi, *Biological Bulletin* **182**, 135–144 (1992).
12. F.I. Harosi, in A. Fein and J. Levine, ed., *The Visual System*, 1985, pp. 41–55.
13. A.V. Whitmore and J.K. Bowmaker, *J. Comp. Phy. A* **166**, 103–115 (1989).
14. W.N. McFarland and E.R. Loew, *Vision Res.* **34**, 1393–1396 (1995).
15. G.S. Losey, T.W. Cronin, T.H. Goldsmith, D. Hyde, N.J. Marshall, and W.N. McFarland, *J. Fish Biol.* **54**, 921–943 (1999).
16. E.R. Loew, W.N. McFarland, E. Mills, and D. Hunter, *Can. J. Zoo.* **71**, 384–386 (1993).
17. E.R. Loew, F.A. McAlary, and W.N. McFarland, in P.H. Lenz, D.K. Hartline, J.E. Purcell, and D.L. Macmillan, eds., *Zooplankton: Sensory Ecology and Physiology*, Gordon and Breach, Australia, 1996, 95–209.
18. T.J. Hara and B. Zielinski, *Trans. Am. Fish Soc.* **118**, 183–194 (1989).
19. J. Atema, in J.E. Bardach, J.J. Magnuson, R.C. May, and J.M. Reinhart, eds., *Fish Behavior and Its Use in the Capture and Culture of Fishes*, ICLARM, Manila, 1980, pp. 57–101.
20. T. Iwai, in J.E. Bardach, J.J. Magnuson, R.C. May, and J.M. Reinhart, eds., *Fish Behavior and its use in the Capture and Culture of Fishes*, ICLARM, Manila, 1980, pp. 124–145.
21. L.L. Lovshin and J.H. Rushing, *Prog. Fish-Cult.* **51**, 73–78 (1989).
22. A.O. Kasumyan and V.Y. Ponomarev, *Rybn. Khoz* **12**, 25–27 (1985).
23. S. Fuke, S. Konosu, and K. Ina, *Bull. Jpn. Soc. Sci. Fish.* **47**(12), 1631–1635 (1981).
24. S. Murofushi and K. Ina, *Agric. Biol. Chem.* **45**, 1501–1504 (1981).
25. A.M. Mackie and A.I. Mitchell, *Comp. Biochem. Physiol.* **73A**(1), 89–93 (1982).
26. A.M. Mackie and A.I. Mitchell, in C.B. Cowey, A.M. Mackie, and J.G. Bell, *Nutrition and Feeding in Fish*, Academic Press, London, 1985, pp. 177–190.
27. A.I. Mitchell and A.M. Mackie, *Comp. Biochem. Physiol.* **75A**(3), 471–474 (1983).
28. I. Hidaka, T. Ohsugi, and Y. Yamamoto, *Bull. Jpn. Soc. Sci. Fish.* **51**(1), 21–24 (1985).
29. I. Ikeda, H. Hosokawa, S. Shimeno, and M. Takeda, *Bull. Jpn. Soc. Fish* **54**(2), 229–233 (1988).
30. I. Ikeda, H. Hosokawa, S. Shimeno, and M. Takeda, *Bull. Jpn. Soc. Fish* **54**(2), 235–238 (1988).
31. K.A. Jones, *J. Fish Biol.* **34**(1), 149–160 (1989).
32. F.T. Barrows, R.E. Zitzow, and G.K. Kindschi, *Progressive-Fish Culturist* **55**, 224–228 (1993).
33. J. Lopez-Alvarado, C.J. Langdon, S. Teshima, and A. Kanazawa, *Aquaculture* **122**, 335–346 (1994).
34. G.A. Kindschi and F.T. Barrows, *Progressive-Fish Culturist* **53**, 53–55 (1991).
35. R.T. Colesante, N.B. Youmans, and B. Zioloski, *Progressive-Fish Culturist* **48**, 33–37 (1986).
36. G.A. Kindschi and F.T. Barrows, *Progressive-Fish Culturist* **55**, 219–223 (1991).
37. F.T. Barrows, W.A. Lellis, and J.G. Nickum, *Progressive-Fish Culturist* **50**, 160–166 (1988).
38. C.Y. Kitajima, Y. Tsukashima, S. Fujita, T. Watanabe, and Y. Yone, *Bull. Jpn. Soc. Fish* **47**, 1289–1294 (1981).
39. J.R. Hunter and S. Sanchez, *U.S. National Marine Fisheries Service Bulletin* **74**, 847–855 (1976).
40. S.I. Doroshov and J.W. Cornacchia, *Aquaculture* **16**, 57–66 (1979).
41. M.B. Rust, *Factors Affecting Density and Initial Swimbladder Inflation in Striped Bass (*Morone saxatilis*)*, MS thesis, University of California, Davis, 1987.

See also Brine shrimp culture; Feed manufacturing technology; Halibut culture; Plaice culture; Sole culture; Summer flounder culture; Winter flounder culture.

LIPID OXIDATION AND ANTIOXIDANTS

RONALD W. HARDY
Hagerman Fish Culture Experiment Station
Hagerman, Idaho

D.D. ROLEY
Bio-Oregon, Inc.
Warrenton, Oregon

OUTLINE

INTRODUCTION

Lipid oxidation is a multistep chemical process by which double bonds on fatty acids become single bonds, thereby converting unsaturated fatty acids to saturated ones. This process is similar to hydrogenation which is used to manufacture margarine from corn oil, soybean oil, safflower oil, and other oils. Hydrogenation introduces hydrogen atoms at carbon double bonds, making them single bonds. The melting point is higher for saturated lipids than for unsaturated ones, and hydrogenation makes margarine produced from liquid plant or fish oils become solid at room temperature but melt when slightly heated. Oxidation involves the addition of oxygen to carbon double bonds but in the process converts them to other compounds. In addition, the oxidation process creates free radicals as a by-product, and these free radicals steal hydrogen atoms from other carbon double bonds on fatty acids, making the double bonds unstable and open for oxygen to attach. Oxidation can occur in any unsaturated lipid if free radicals are available to initiate the oxidation process. Once oxidation reaches a certain stage, it becomes a self-sustaining process, called autoxidation, and proceeds very rapidly, especially in fish oils that typically contain high proportions of polyunsaturated fatty acids having numerous carbon double bonds.

Oxidation of plant and fish oils used in animal and fish feeds is undesirable for a number of reasons. First, oxidation destroys vitamins C and E in the feed unless they are supplemented in chemically protected forms. Second, oxidation produces toxic compounds which cause specific pathological changes in animals and fish. Third, when animal and fish consume feed containing oxidizing lipids, it places a burden on detoxifying mechanisms in tissues and membranes, causing further pathological changes. Finally, oxidation generates heat, sometimes sufficient to cause feed ingredients or feeds to catch fire. Fish oils are highly unsaturated lipids and particularly susceptible to autoxidation. Fish oils are also important ingredients in fish feeds, supplying essential fatty acids and energy. Oxidation must be detected in fish oils before they are used in feeds to maintain feed quality and to ensure fish health. Preventing fish oil oxidation is the most sensible approach in fish feed production, and strategies to prevent oxidation begin with fish oil production and end when fish consume fish feeds.

LIPID OXIDATION

Lipid oxidation is so named because it involves the addition of oxygen along the carbon chains that form all fatty acids at the point where there are double bonds between carbon atoms. Fatty acids differ in the number of carbon atoms and the number of double bonds along their carbon chains. These differences give each fatty acid its unique chemical and physical properties. Fatty acids in plants, animals, and fish generally range from 14 to 22 carbons in length, with a carboxyl group (COOH) at one end and a methyl group (CH_3) at the other. Shorthand notation for fatty acids identifies the number of carbons and double bonds and also the number of the first carbon from the methyl end where a double bond occurs. For example, a saturated fatty acid containing 16 carbons would be noted as C : 16, while an unsaturated fatty acid containing 18 carbons and two double bonds, with the first double bond appearing at carbon number 6 from the methyl end, would be noted as C : 18 : 2,n-6. Typically, lipids from terrestrial animals are highly saturated (few double bonds), making them solid at room temperature and relatively resistant to oxidation. Plant oils are typically liquid at room temperature because they contain fairly high levels of fatty acids having one or two double bonds. Fish oils, particularly from temperate water marine species, tend to have high proportions of long-chain, polyunsaturated fatty acids (PUFAs), of which 20 to 25% have three to six double bonds (1). Thus, of the lipids used in feeds, fish oils are generally the most unsaturated, have the lowest melting point, and are the most susceptible to lipid oxidation (Table 1).

The process of lipid oxidation involves three general stages: initiation, propagation, and termination (Fig. 1). Initiation of oxidation involves the creation of free radicals; without free radical formation, oxidation cannot occur. Oxygen by itself cannot start the process of lipid oxidation

Table 1. Major Fatty Acids Groups (%) in Lipid Sources Used in Fish Feeds

Lipid Source	Sats[a]	Mono[a]	Dienes[a]	PUFAs[a]
Sardine	26	30	2	27
Menhaden	29	23	2	22
Herring	23	42	2	11
Anchovy	28	29	2	26
Mackerel	26	30	2	23
Soybean	15	23	51	7
Corn	15	36	48	1
Tallow	55	41	3	<1

[a]Sats, no double bonds (saturated); mono, one double bond; dienes, two double bonds; PUFAs, three or more double bonds.

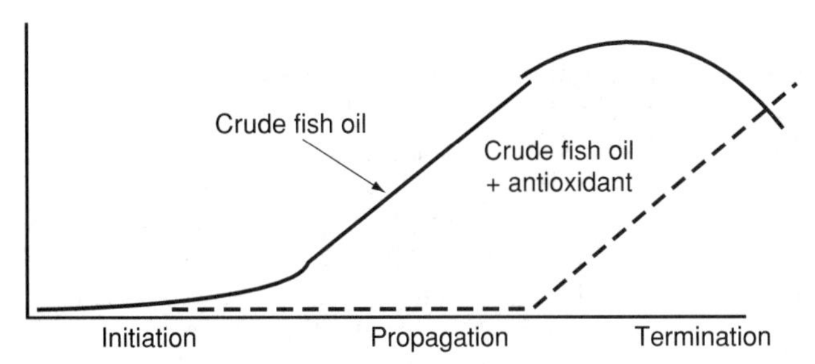

Figure 1. Generalized scheme for oxidation of fish oils.

because the activation energy required to create free radicals from oxygen is very high. Rather, free radical formation begins with hydroperoxide decomposition by metal catalysis or by exposure to light (2). Iron from heme is thought to react with oxygen, converting it from Fe^{2+} to Fe^{3+}, the active form of iron, forming superoxide in the process. Superoxide reacts with two hydrogen atoms to form hydrogen peroxide and oxygen. Fe^{2+} reacts with hydrogen peroxide to form Fe^{3+} and a hydroxy (free) radical (−OH). This free radical then reacts with double bonds between carbon atoms along fatty acids, changing the double bond to a single bond in a series of steps that result in formation of various chemical products and more free radicals. Until the oxidation process begins to generate more free radicals than it consumes, the process is still in the initiation stage. The initiation stage can be a short or very long period, depending upon temperature, the presence of metal catalysts, and the concentration of antioxidants in an oil (3). One category of antioxidants works by converting free radicals to stable compounds, thus limiting the level of free radicals in a lipid and keeping oxidation in check. However, once the antioxidants in the oil are used up, free radicals accumulate rapidly and lipid oxidation enters the propagation stage.

As mentioned, hydroxy (free) radicals remove hydrogen atoms from carbons adjacent to double bonds on fatty acids, creating instability at the double bond and an opportunity for oxygen to attach to the carbons (2). The resulting compound is also unstable, and, depending upon the particular fatty acid and which double bond is destabilized, further reactions occur which create peroxides, aldehydes, ketones, furans, numerous alcohols, and epoxides from the fatty acid. More importantly, the breakdown of fatty acids into other compounds creates more free radicals. At this point, lipid oxidation has reached the propagation stage, meaning that free radicals are being propagated by the oxidation process itself rather than by other mechanisms. Oxidation now resembles a nuclear reaction in that it is self-sustaining, proceeds rapidly, and releases heat. The number of free radicals formed during breakdown of peroxides increases with the number of double bonds in a given fatty acid. Linoleic acid (C18 : 2n-6), the most common unsaturated fatty acid in soybean oil, forms one free radical as it oxidizes. Thus, oxidation of linoleic acid is self-sustaining once it starts. Oxidation of linolenic acid (C18 : 3n-3), the most common fatty acid in linseed oil, forms two free radicals as it oxidizes, thus accelerating the rate of lipid oxidation. This explains why linseed oil has always been an important component of paints and varnishes; it oxidizes rapidly, especially after it has been boiled to destroy naturally present antioxidants, forming polymers that coat and protect wood. Eicosapentanoic (EPA, C20 : 5n-3) and docosahexanoic acids (DHA, C22 : 6n-3) in fish oils can form four or five free radicals for every fatty acid undergoing oxidative breakdown, making the rate of peroxide formation and oxidation extremely rapid in fish oils during the propagation stage. Termination, the last step of oxidation, occurs when breakdown products combine to form stable end products, and the rate of free radical formation slows as the number of double bonds not yet oxidized on fatty acids decreases.

When fatty acid oxidation is in the propagation stage, the levels of free radicals and intermediate breakdown products of fatty acid oxidation are high. When an oil becomes completely oxidized in the termination stage, the level of free radicals declines, as does the concentration of intermediate by-products. Saturated fatty acids, some unsaturated fatty acids, and final end products of oxidation are left. The end products of oxidation vary depending upon the fatty acids present in the oil and the temperature at which oxidation takes place. In pure lipids, the end products are somewhat predictable. For example, oleic acid (C18 : 1) breaks down via a relatively simple mechanism to four hydroperoxides, the proportion of which depends upon temperature (4). Linoleic acid (C18 : 2) breakdown is more complicated because it has two double bonds. Thus it can break down into a variety of final products ranging in size from two or three carbon compounds to longer molecules, including dimers, polymers, cyclic peroxides, hydroperoxides, aldehydes, ketones, alcohols, acids, and epoxides. Crude fish oils contain up to 50 different fatty acids, cholesterol and oxysterols, protein, and other reactive compounds, making the list of possible breakdown products almost infinite.

DETECTING OXIDATION IN FISH OILS

As described, oxidation of fish oils involves the addition of oxygen to double bonds on fatty acids and the production of breakdown products such as aldehydes and peroxides from fatty acids. The chemical tests for detecting oxidation and quantifying the extent of oxidation of fish oils are based upon these changes. One approach to detect oxidative status of lipids involves measuring oxygen uptake while a lipid sample undergoes accelerated oxidation at elevated temperature, using a Warburg apparatus (5) or by measuring small changes in weight of a lipid sample (6). When weight gain is measured, the number of hours needed to achieve a 0.6% gain in weight is the usual unit of measurement. Common laboratory tests for oxidation in lipid samples measure the concentration of intermediate products of lipid oxidation (e.g., aldehydes or peroxides). Aldehyde concentrations are measured by thiobarbaturic acid reactive compound concentration (TBARs) or anisidine value (7). TBARs measure malonaldehyde, a breakdown product of polyunsaturated fatty acids (PUFAs) found in relatively high concentrations in fish oils. Anisidine value is similar to TBARs in that it measures both malonaldehye concentration and other secondary oxidation products of unsaturated aldehydes. Both anisidine and TBARs are quantitative, highly reproducible, relatively inexpensive, and rapid. However, TBAR values go up during the initial phase of the propagation stage and go down in the termination stage as intermediate products are converted to final products of oxidation (8). Thus, it is difficult to determine exactly where a lipid sample is on the continuum between the propagation stage and the termination stage based upon a single TBAR value, in that a given value could mean that the oxidation is increasing or decreasing, depending on the stage. Another chemical test, peroxide value (POV), measures the concentration of peroxides in an oil sample. As is the case with the TBARs test, POVs go up during the propagation stage and down again during the termination stage. The POV is a highly reproducible, inexpensive, and rapid test and is particularly suited to lipid samples. Lipids are sometimes characterized with respect to their oxidation status by the Totox number, which is simply the anisidine value plus 2X the peroxide value.

Other chemical tests to gauge the oxidative status of lipid samples measure the change in numbers of double bonds in fatty acids. The traditional test for this is to measure iodine number. The method involves exposing oils to iodine vapor in a closed chamber and determining the amount taken up by double bonds (7). High iodine numbers are characteristic of polyunsaturated fatty acids, typical of fish oils. Thus, fish oils will have iodine numbers of 120 to 160, depending on the type of fish oil. As oils oxidize, iodine numbers decrease (fewer double bonds). Changes in fatty acid profiles also occur as oxygen is taken up by double bonds along fatty acid chains. Thus, unsaturated fatty acid concentrations decrease as a consequence of oxidation, and this decrease can be measured using a gas chromatograph. Values for some of these measurements in fish oil indicate a wide range of acceptable values (Table 2).

Based upon a single POV or TBAR value, it is impossible to know if an oil is in the early or later stages of initiation or propagation, which is critical information for a fish feed manufacturer. The amount of time required to reach the propagation stage is called the induction time. Without additional information, one does not know if an oil will be stable for a suitable period, if antioxidants should be added to extend the induction time, or if it is about to enter the propagation stage and rapidly oxidize. To overcome this problem and determine the induction time remaining before an oil oxidizes, the oil must be subjected to an accelerated oxidation test by exposing a sample to air and heat for a specified period of time and testing again. If the POV or TBAR value after stress testing is substantially higher than the initial value, then the oil is near the end of the initiation stage. The Schaal oven test is the best choice for such measurements (8). After an initial POV or TBAR measurement, the oil is placed in an oven at 40 to 45 °C and tested again at daily intervals (Fig. 2). This method is very useful to judge how much longer the oil will remain in the initiation stage during storage. It is not uncommon for an oil to yield a relatively low POV value upon arrival at a fish feed manufacturing plant and be used in production of fish feed, only to have the oil in the fish feed oxidize shortly after it is manufactured. Conducting a Schaal oven test on the oil at arrival will clearly show the quality of the oil with respect to resistance to oxidation provided by natural or added antioxidants. POV values should be less than 20 after five days of the Schall oven test for the oil to resist oxidation through the storage, pelleting, and shipping stages and until used at a fish farm. If the oil is close to entering the propagation stage, antioxidants must be added to extend the initiation stage and thus prevent oxidation.

The chemical tests used to measure oxidation status of fish oils are less accurate when they are used to measure the oxidation status of fish feeds. This is due in part to the difficulty of extracting lipids from fish feeds, particularly feeds that have been pelleted by cooking extrusion. The temperature and pressure used to extrude fish feeds causes some of the lipid in the feed mixture to become bound, making it necessary to subject the feed sample to acid hydrolysis before organic solvents are used to extract the lipid. Feeds for some species of fish (e.g., salmon, trout, and yellowtail) are sprayed (top-dressed) with fish oil after pelleting, but some of the lipid in pellets is present in the mixture before pelleting. It is the latter that is difficult to extract. Another potential problem associated with measuring the oxidation status of fish feeds is that the conditions of pelleting may destroy the intermediate products of oxidation, leading to low TBAR or POV values that underestimate the true oxidation status. A final problem is that the POV test is based upon a titration which causes a color change in the oil sample. This color change is relatively easy to observe in samples

Table 2. Typical Crude Fish Oil Specifications[a]

Measurement	Range of Values
Free fatty acids, %	2–5
Moisture and impurities, %	0.5–1.0
Peroxide value, Meq/kg	3–20
Anisidine number	4–60
Iron, mg/kg	0.5–7.0
Iodine value	
Capelin	95–160
Herring	115–160
Anchovy	180–220

[a]From Ref. 16.

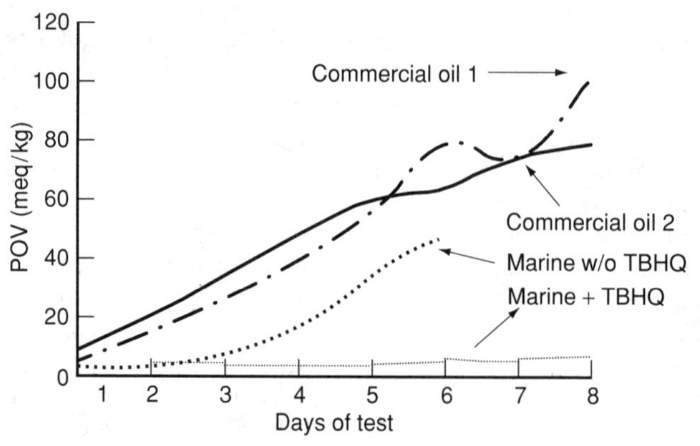

Figure 2. Schaal oven test with various fish oils.

taken from fish oils, but samples from feeds sometimes contain pigments that mask the distinctive end point (color change) during titration.

Based upon studies in which oil of a known POV value was added to a poultry feed, the POV value of the feed after pelleting was significantly lower than the theoretical calculated value (9). A single petroleum ether extraction removed only 60% of the oil in the feed; but after acid hydrolysis and three extractions, 90% of the oil was recovered. In spite of this more complete extraction, the POV value of the oil in the feed remained significantly below theoretical values, most likely because of peroxide destruction during pelleting (9). TBAR values are not affected by factors that make the color change in POV testing difficult to measure accurately, but they are affected by malonaldehyde destruction during pelleting and incomplete lipid extraction from pellets. The safest approach to ascertain the oxidation status of lipids in fish feeds is to conduct thorough extraction, test the oil, heat the pellets in an oven, and conduct repeated testing to determine if AV, POV, or TBAR values increase. This approach will yield information on induction time and status regarding propagation and termination stages of oxidation.

ANTIOXIDANTS

Antioxidants are compounds that interfere with or interrupt the process of oxidation in lipids. There are three categories of antioxidants: preventative antioxidants, sacrificial antioxidants, and peroxide destroyers, and these antioxidants exert their protective effects by affecting oxidation at different stages. Preventative antioxidants operate by chelating or tying up metals (e.g., iron and copper) that initiate free radical formation, thus removing them from circulation. Examples of preventative antioxidants are citric acid, phosphoric acid, ascorbic acid, phytic acid, and EDTA (10). Preventative antioxidants influence the initiation stage of lipid oxidation, acting to extend this stage. Once lipid oxidation moves into the propagation stage, preventative antioxidants do not influence the rate or extent of oxidation because free radical generation is independent of reactions involving copper or iron.

Sacrificial antioxidants interrupt the propagation stage of oxidation by donating a hydrogen to free radicals, converting them to stable and nonreactive forms. Thus, they operate by catching free radicals before they can react with double bonds on fatty acid chains or by donating a hydrogen atom to the double bond after it has lost one to a free radical, thus returning the double bond to its original state and preventing further oxidation reactions. Each molecule of antioxidant can donate two hydrogens, after which it is out of extra hydrogen atoms and no longer able to interrupt oxidation. Thus, it has sacrificed its hydrogens and been altered in the process; hence the name sacrificial antioxidants. In natural oils from plants and fish, tocopherols (various forms of vitamin E) are the sacrificial antioxidants that protect the oils from oxidation.

The third type of antioxidants (peroxide destroyers) operate by reducing peroxides formed in the propagation stage. Their antioxidant properties are associated with removing peroxides, which helps limit the generation of new free radicals and reduces the extent to which oxidation remains a self-sustaining process. Peroxide destroyers are not normally used to protect fish oils and fish feeds because they react with other feed compounds, altering the nutritional value of the feed.

If conditions in a fish oil are such that free radicals are being formed during the initiation stage, oxidation will be limited as long as there are sacrificial antioxidants available. When the antioxidants in fish oil have donated all of their hydrogen atoms to free radicals, there is nothing to stop propagation from proceeding to the propagation stage and the oil from oxidizing. If a batch of fish oil has entered the propagation stage of oxidation and antioxidants are added, propagation will stop until the added antioxidants are used up, at which time propagation will begin again.

Fish and plant oils contain tocopherols and other naturally occurring compounds that protect oils against oxidation (Table 3). In fish oils, alpha-tocopherol is the primary antioxidant, although carotenoid pigments, primarily astaxanthin, are hypothesized to offer some protection against oxidation. The tocopherol content of fish oils varies with fish species, conditions of manufacture, and length of storage of the oil. The alpha-tocopherol content of crude menhaden oil ranges from 20 to 70 mg/kg, while herring and tuna oil are reported to contain 140 to 160 mg alpha-tocopherol/kg (10). Plant oils contain alpha, beta, delta, and gamma tocopherols, plus tocotrienes (11). Soybean oil contains 1,078 mg total tocopherols/kg oil, of which alpha tocopherol accounts for only 93 mg/kg. Gamma-tocopherol accounts for 695 mg and delta-tocopherol for 277 mg/kg oil. These tocopherols are potent antioxidants, but their vitamin E activity is relatively low for animals compared to alpha-tocopherol (12).

There are a number of synthetic, sacrificial antioxidants that are used to protect fish oil and other foods from oxidation. These include ethoxyquin (1,2-dihydro-6-ethoxy-2,-2,4-trimethyl quinoline), BHA (butylated hydroxyanisole), BHT (butylated hydroxytoluene), propyl gallate, TBHQ (tertiary-butyl-hydroquinone), and ascorbyl palmitate (12). Ethoxyquin is not permitted in human food but is approved for use in animal feeds not to exceed 150 mg/kg. BHA and BHT are approved for use in human food and are occasionally used in animal feeds to a maximum level of 0.02% of the lipid content of the food

Table 3. Tocopherol Levels (mg/kg) in Fish Oil and Soybean Oil

Oil Source	Tocopherol Content
Sardine	40
Menhaden	20–70
Herring	140
Tuna	160
Soybean	1,078[a]

[a] Alpha, 93; beta, 12; delta, 277; gamma, 695.

Table 4. Oxidative Uptake to 0.6% Weight Gain of Capelin Oil with Added Antioxidants (6)

Treatment	TBAR (mg malonaldehye/kg oil)	Duration (hr)	Efficiency
None	10	30	1.0
Tocopherol acetate, 0.03%	19	34	1.1
Ascorbyl palmitate, 0.02%	20	42	1.4
BHT, 0.02%	11	55	1.8
Ethoxyquin, 0.02%	10	56	1.9
BHA, 0.02%	9	67	2.2
TBHQ, 0.01%	9	119	4.0

or feed. Propyl gallate and ascorbyl palmitate are used in human foods.

Antioxidants are not equally effective at preventing oxidation of fish oils. Several antioxidants were added to fish oil from capelin, and the oil samples were subjected to an accelerated oxidation test until the oil had gained 0.6% weight (Table 4). Unsupplemented oil reached this gain in weight after 30 hours, at which time it had a TBAR value of 10, a relatively low value. Ethoxyquin extended protection of the oil to 56 hours, longer than tocopherol acetate or ascorbyl palmitate (34 and 42 hours, respectively). BHT offered nearly identical protection to ethoxyquin, but BHA extended protection to 67 hours. TBHQ was the most effective antioxidant, extending protection to the oil for 119 hours. Similar but less dramatic findings are reported for fish feeds to which various antioxidants have been added (13).

Blends of natural antioxidants effectively extend the shelf life (induction time) of fish oils (14). Combinations of tocopherol, lecithin, ascorbic acid, and a small percentage of water to help disperse ascorbic acid increased storage time of fish oil by 22X compared to the storage time of fish oil supplemented with tocopherol alone. Commercial blends of natural antioxidants are available for protecting fish oil and add about US$0.11/kg to the price of the oil, increasing the cost of the oil by about 20%. Because plant oils are rich sources of various tocopherols, mixing plant and fish oils plus adding ascorbic acid (a synergist with tocopherol) and water may be a promising option for protecting fish oils against oxidation without using synthetic antioxidants.

PREVENTING OXIDATION IN FISH OIL

Fish oil is the second largest source of edible and feed-grade oil produced in the world, and currently 25% of world production is used in fish feeds. Of the amount used in fish feeds (380,000 MT), half is used in salmon feeds, with trout the second largest aquaculture user (15). A decade from now, estimates are that nearly 500,000 MT will be used in salmon feeds alone, with carp, shrimp, and yellowtail aquaculture feeds also predicted to increase dramatically (15). Given these predictions, it is clear that fish oil will continue to be a key constituent of fish feeds and that preventing oxidation of fish oils will require a better understanding of the critical points between fish oil production and its use in fish feeds where the potential for oxidation is enhanced.

Oxidation of fish oil requires the production of free radicals and the presence of oxygen. Free radicals are formed as a consequence of enzymatic breakdown of fish tissue, mainly through release of iron from hemoglobin and from contamination of oil with iron and copper from equipment used to manufacture, pump, store, and transport the oil. Therefore, minimizing tissue damage of fish used to produce oil and avoiding iron and copper contamination from storage tanks, processing equipment, and especially valves used to pump oil between tanks and trucks or railcars, are key factors to produce stable fish oil.

Most fish oil and fish meal is produced using the wet reduction process (16,17). The wet reduction process is a continuous process involving cooking, pressing, decanting, separating, and polishing. Cooking and pressing do not enhance oxidation, but the decanting, separating, and polishing steps, which separate and clean the fish oil, can influence oxidative stability of the oil. Crude fish oil is not a pure substance. Insoluble impurities and free fatty acids are commonly present, as are small amounts of water and water-soluble proteins, trace metals, oxidation products, pigments, and tocopherols. Fish oil is then stored in large tanks where antioxidants may or may not be added, depending on the final use of the oil. Approximately 75% of the fish oil produced in the world is used to make edible products, such as margarine, and addition of antioxidants to oil that will be used to make edible products would interfere with their manufacture. Consequently, bulk fish oil often does not have antioxidants added to it until an order is placed from a feed manufacturer, and it is shipped. Certain conditions in storage tanks can accelerate the oxidation process. These conditions include a large head space containing air, contamination with iron or copper, and direct sunlight on the storage tank, which causes thermal convection and turnover of oil in the tank. Circulation of oil within a tank brings more oil into contact with the surface, thereby exposing more of the oil to oxygen in the air in the head space. Further, entrapment of air into the oil can occur as a consequence of thermal convection. Lastly, tanks should contain a sump in which water and insoluble impurities settle. The sumps should contain drains through which these materials can be regularly removed as they collect. If this mixture of water and protein is not removed, it can act as a substrate for bacteria that generate acid, thus hydrolyzing triglycerides to free fatty acids, which are more susceptible to oxidation.

The consequence of poor oil storage conditions is that they reduce the concentration of natural antioxidants present in the oil, thereby shortening the induction time of the oil once it reaches a feed plant. The same conditions that are stressful to fish oil in storage can occur when the oil is transported to distribution facilities and to the feed plant. In addition, oil is sometimes transported in rail cars or trucks that have previously contained other materials, thereby potentially contaminating the oil. Cleaning rail cars and tanks between shipments is important, even if the tank is only used to ship fish oil; because after oil is drained from a tank, the inside of the tank remains coated. This coating is fully exposed to air and is very likely to oxidize, thus contaminating new shipments with oxidizing oil.

Once fish oil reaches a fish feed plant, it is stored in tanks where oxidation can be enhanced by the same conditions already described. In particular, it is critical to clean out storage tanks when they are emptied before a new shipment of oil is added. Oil tanks should have conical bottoms to eliminate dead spaces and ensure that all the oil in the tank can be drained out. The stearine fraction of fish oil, which is rich in saturated fatty acids and thus solid at temperatures where fish oil is liquid, tends to make a relatively solid coating on the insides of fish oil storage tanks. This material is less susceptible to oxidation than fish oil, but because it coats the inside of tanks and is thus exposed to air when oil tanks are emptied, it can oxidize. Stearines must be removed from the inside surfaces of storage tanks and pipes by pressure washing with water, detergents, or biodegradable solvents to reduce the chance of contaminating new shipments of oil with oxidizing oil. Other practices that can be employed to reduce the rate of lipid oxidation in storage tanks include coating the inside of iron or steel tanks with epoxy and injecting nitrogen into the oil. Injecting nitrogen strips air from the oil and displaces air in the headspace as the oil in the tank is used.

Fish feed pelleting involves heat, moisture, and pressure, all of which one would think would likely accelerate destruction of antioxidants that protect fish oils from oxidation. However, most damage to lipids in fish feeds is associated with overdrying the pellets after they are produced. Heat is less of a contributor to oxidation than is drying to a moisture content of $<8.5\%$ moisture. After pellet drying, fish oil is sometimes added by top-dressing. Oxidizing oil present in pipes and top-dressing equipment can contaminate clean oil as it is used to coat pellets. Thus, this equipment should be cleaned regularly.

After pellets are manufactured, oxidation can be accelerated by several factors. These include abusive storage conditions (primarily high-temperature storage), the presence of pro-oxidants (e.g., iron and copper) coming from pelleting equipment and mineral supplements, and, especially, freezing of dry pellets. Most freezers at fish farms are not cold enough to freeze all of the water in pellets. Thus, pure water is frozen, but water containing pro-oxidants remains a liquid and is concentrated in parts of pellets. Besides concentrating water-soluble pro-oxidants, freezing allows oxygen to penetrate more freely into the pellets. Thus, dry fish feed should never be frozen; rather it should be stored in a cool, dry place.

HEALTH EFFECTS OF FEEDING DIETS CONTAINING OXIDIZING FISH OIL

Oxidation of lipids is one of the more common problems in fish feeds. The trend is toward high-energy (high-lipid) feeds, especially in salmon farming, and periodic dips in global fish oil production, which causes feed manufacturers to seek fish oil of any quality. This combination of factors increases the likelihood that fish feeds will contain oxidizing oil. Feeds containing oxidizing lipids can cause nutritional disease by depleting tissue antioxidant levels, producing toxic compounds and destroying essential fatty acids (18). The levels of tissue and membrane antioxidants (e.g., tocopherol and vitamin C) are reduced when feeds containing oxidizing lipid are fed because they are used to control free radical levels in tissues faster than they can be regenerated. This results in biomembrane changes, mainly alterations in membrane permeability and fragility. In addition, the free radicals, peroxides, aldehydes, and ketones produced during lipid oxidation are all toxic to fish and capable of reacting with other dietary components. While essential fatty acids (PUFAs) are destroyed in feeds as a consequence of oxidation, the health problems associated with feeding oxidizing lipids are most often the result of depleted tissue antioxidants and the toxic products named previously.

The primary health effects associated with feeding oxidizing fish oils in feeds to fish are liver degeneration, anemia, and spleen abnormalities (19–22). The most serious problem associated with rancidity is lipoid liver disease. As mentioned, rancid lipids are toxic per se, but they also have a deleterious effect on tissue levels of vitamins E and C. Consequently, the features of lipoid liver disease often vary from outbreak to outbreak, depending on the contribution of each of the components to the degeneration. Fish suffering from lipoid liver disease have extreme anemia (manifested by pallor of the gills and erythrocyte fragility), a bronzed, rounded heart, and a swollen liver with rounded edges (19). Histologically, the main feature is the extreme infiltration of hepatocytes by lipid, which causes loss of cytoplasmic staining and distortion of hepatic muralia. There is degeneration of splenic and renal haemopoietic tissue with high levels of pale-staining pigment in melanomacrophage centers. There is also often auxiliary haemopoiesis in the subepicardial tissues and the periportal areas. Depending on the length of time the condition has been extant, the degree of oxidation, and the type of fat in the diet, there is a varying degree of infiltration of the liver by macrophages containing ceroid, a pigmented breakdown product of phospholipid metabolism (23).

All salmonids are susceptible to lipoid liver degeneration, but it is a particularly significant problem in rainbow trout culture (23). Slightly affected fish are usually capable of complete recovery, but once there is severe anemia and hepatic ceroidosis has developed, the fish is rarely capable of satisfactory recovery to its previous efficiency of feed conversion. Other conditions reported to be associated with the feeding of rancid lipids include exophthalmia, steatitis, darkening, splenic hemosiderosis, and skeletal myopathy (21,24–26).

Because one of the primary health effects of feeding oxidizing fish oil is depletion of tissue vitamins C and E, it is logical to conclude that elevating the dietary intake of these vitamins might alleviate the condition. In fact, in catfish and trout, many of the pathological effects associated with feeding oxidizing fish oil can be prevented by increasing the dietary intake of tocopherol (21,25). However, acute toxicity is not always prevented by increasing dietary tocopherol intake, especially in small fish which have not yet accumulated substantial tissue stores of ascorbic acid and tocopherol. Growth rates of Atlantic and coho salmon fry are reported to be inversely proportional to the level of oxidation in starter feeds, with reduced growth being observed at levels of oxidation that do not reduce growth in larger fish (20,27). There is also some suggestion in the literature that there may be species differences in sensitivity to oxidizing feeds (27). The link between tocopherol deficiency signs and those associated with consumption of oxidizing lipids is weak. If oxidizing lipids caused pathological problems in fish solely by reducing tissue tocopherol levels, one might expect that signs of tocopherol deficiency (e.g., muscular dystrophy), exudative diathesis, and depigmentation, would be prevalent in fish consuming feeds containing oxidizing lipids. However, this does not occur.

A very distinctive syndrome closely resembling diabetes has been described in carp fed diets containing high levels of silkworm pupae (28). This syndrome, called Sekoke disease, is characterized by destruction of the endocrine pancreatic islets of Langerhans, concomitant lipid infiltration of parenchymatous organs, bilateral cataracts, and degenerative alterations to the extrinsic eye muscles, the retina, and the choroid (23). Sekoke disease was found to be caused by oxidation of lipids in the silk worm pupae and can be prevented by increasing the dietary intake of vitamin E.

BIBLIOGRAPHY

1. E.H. Gruger, Jr., in M.E. Stansby, ed., *Fish Oils*, Avi, Westport, CT, 1967.
2. H.O. Hultin, in F. Shahidi and J.R. Botta, eds., *Seafoods: Chemistry, Processing Technology and Quality*, Chapman & Hall, London, 1994.
3. S.S.O. Hung, C.Y. Cho, and S.L. Slinger, *Can. J. Fish. Aquat. Sci.* **37**, 248–1253 (1980).
4. W.W. Nawar, in O.R. Fennema, ed., *Food Chemistry*, 3rd ed., Marcel Dekker, New York, 1996.
5. S. Thorisson, F. Gunstone, and R. Hardy, *J. Amer. Oil Chem. Soc.* **69**(8), 806–809 (1992).
6. J.K. Kaitaranta, *J. Amer. Oil Soc.* **69**(8), 810–813 (1992).
7. D. Firestone, ed., *Official Methods and Recommended Practices of the American Oil Chemists' Society*, 4th ed., American Oil Chemical Society, Champaign, IL, 1990.
8. E.N. Frankel, *Trends Food Sci. Tech.* **4**, 220–225 (1993).
9. W.D. Shermer and D.F. Calabotta, *Feedstuffs* **57**(45), 19–20 (1985).
10. M.E. Stansby, ed., *Fish Oils in Nutrition*, Van Nostrand Reinhold, New York, 1990.
11. S.W. Souci, W. Fachmann, and H. Kraut, in Deutsche Forschungsanstalt für Lebensmittelchemie, Garching b. Munchen, eds., compiled by H. Scherz and F. Senser, *Food Composition and Nutrition Tables*, 5th ed., Medpharm Scientific, Stuttgart, 1994.
12. H.S. Olcott and M.E. Stansby, ed., *Fish Oils*, Avi, Westport, CT, 1967.
13. D. Inaba, K. Saruya, and K. Toyama, *J. Tokyo Univ. Fish.* **52**(1), 71–76 (1972).
14. D. Han, O.-S Yi and H.-K. Shin, *J. Amer. Oil Chem. Soc.* **68**(10), 740–743 (1991).
15. A.P. Bimbo and I.H. Pike, *Fishmeal and Oil: Current and Future Supplies for Asian Aquafeeds*, Proc. Vietnam-Asia '96 Conference, Bangkok, Thailand, Nov. 12–15, 1996.
16. A.P. Bimbo, in M.E. Stansby, ed., *Fish Oils in Nutrition*, Van Nostrand Reinhold, New York, 1990.
17. A.P. Bimbo, *Inform* **9**(5), 473–483 (1998).
18. J.G. Bell and C.B. Cowey, C.B. Cowey, A.M. Mackie, and J.G. Bell, eds., *Nutrition and Feeding in Fish*, Academic Press, London, 1985.
19. C.E. Smith, *J. Fish Dis.* **2**, 429–437 (1979).
20. S.S.O. Hung, C.Y. Cho, and S.L. Slinger, *J. Nutr.* **111**, 648–557 (1981).
21. R.D. Moccia, S.S.O. Hung, S.L. Slinger, and H.W. Ferguson, *J. Fish Dis.* **7**, 269–282 (1984).
22. J. Rehulka, *Aqua. Fish. Mgmt.* **21**, 419–434 (1990).
23. R.J. Roberts and R.W. Hardy, in R.J. Roberts, ed., *Fish Pathology*, 3rd ed., Academic Press, San Diego, 1999.
24. A.K. Soliman, R.J. Roberts, and K. Jauncey, in Fisherson, ed., *Proceedings Int. Symp. Tilapia in Aquaculture*, Tel Aviv University Press, 1983.
25. T. Murai and J.W. Andrews, *J. Nutr.* **104**, 1416–1431 (1974).
26. S.I. Park, *Bull. Korean Fish. Sec.* **11**, 1–4 (1978).
27. H.G. Ketola, C.E. Smith, and G.A. Kindschi, *Aquaculture* **79**, 417–423 (1989).
28. M. Yokoke, *Bull. Freshwater Fish Res. Lab.* **20**, 39–72 (1970).

See also ANTINUTRITIONAL FACTORS; LIPIDS AND FATTY ACIDS.

LIPIDS AND FATTY ACIDS

DAVE A. HIGGS
West Vancouver Laboratory
West Vancouver, Canada

FAYE M. DONG
University of Washington
Seattle, Washington

OUTLINE

Types, Classification, General Functions
  Lipids
Digestion, Absorption, Transport, and Storage
Lipid and Fatty Acid Requirements
Dietary Lipid Composition and Fish Reproduction
Influence of Dietary Lipid Composition on Fish Health
Marine and Nonmarine Lipid Sources in Feeds for Finfish
  Lipid Sources

Lipids and their constituent fatty acids, as well as the metabolic derivatives of some of the latter that are termed eicosanoids and other associated compounds, play central and dynamic roles in the maintenance of optimum growth, the reproduction, the health, and the flesh quality (market size) of finfish species. In this review, our first intent is to introduce the reader to the various types of lipids and fatty acids that are present in the diet and the bodies of finfish species and to the general functions of those compounds. Thereafter, we describe how finfish species digest, absorb, transport, and store lipids in their bodies, emphasizing some interspecific differences in this regard. Then, we highlight the extent to which the various species differ in their requirements for dietary lipids and fatty acids, taking into consideration the lipid compositions of their natural diets in the freshwater and marine environments and at each life history stage—larvae or fry, juveniles, post-juveniles, and adults (nonmature or maturing). Next, we describe in general terms the effects that dietary fatty acids have on finfish reproduction and health (cardiovascular function and immunocompetence). Lastly, we consider their impacts on the chemical composition and sensory attributes of the flesh, especially when non-marine sources are used as partial replacements for marine lipids during times when the latter are of high price and low quality.

TYPES, CLASSIFICATION, GENERAL FUNCTIONS

Lipids

"Lipids" refers to compounds that are relatively insoluble in water but are soluble in organic solvents such as chloroform, ether, hexane, and benzene. There are many types of lipids, and their classification has been undertaken in several ways. For instance, they have been differentiated either by the presence or absence of fatty acids or of the alcohol glycerol in their basic structure or according to their polarity. With respect to the latter, some lipids such as triacylglycerols, wax esters, alkyl diacylglycerols, and sterol (e.g., cholesterol) esters are insoluble in water because of nonpolar hydrocarbon groups. Consequently, they are called nonpolar lipids. By contrast, other lipids such as phosphatidyl choline (PC), phosphatidyl ethanolamine (PE), phosphatidyl serine (PS), and phosphatidyl inositol (PI)—collectively called phosphoglycerides—along with plasmologens, shingomyelin, cerebrosides, and gangliosides, contain polar groups as part of their basic structure. Accordingly, these are referred to as polar lipids, and they are essential components of biological membranes. Many finfish species such as salmonids (salmon, trout, and charr) derive most of their dietary non-protein energy from triacylglycerols and, in the case of wild salmon in the ocean and marine species, also from wax esters that are present in several marine prey species.

The basic structures of the preceding compounds differ considerably from each other. Triacylglycerols, for instance, consist of three fatty acids esterified to the alcohol glycerol. The three different fatty acid hydrocarbon chains vary in the number of carbon atoms, the degree of unsaturation and, where applicable, the position of the first double bond in relation to the methyl end of the molecule. Numerous combinations of fatty acids are possible within the triacylglycerol structure, because more than 40 different fatty acids are known to occur in nature (1). This variety results in differences in such chemical and physical properties as melting point, dependent upon which fatty acids are affixed to the glycerol moiety. For example, the melting points of the saturated (no double bonds in the carbon chain) fatty acids 14 : 0 (myristic acid), 16 : 0 (palmitic acid), and 18 : 0 (stearic acid) are respectively 54, 63, and 70 °C. The corresponding value for a typical monounsaturated (possessing one double bond) fatty acid, 18 : 1*n*-9 (oleic acid), is lower: 16.3 °C. Moreover, those for polyunsaturated (two or more double bonds) fatty acids, such as 18 : 2*n*-6 (linoleic acid), 18 : 3*n*-3 (linolenic acid), 20 : 4*n*-6 (arachidonic acid), 20 : 5*n*-3 (eicosapentaenoic acid), and 22 : 6*n*-3 (docosahexaenoic acid) are even lower, being (respectively) −5, −10, −49.5, −54.4, and −44.5 °C (2).

At this point, it is necessary to provide some background information about fatty acids before proceeding further. In this regard, it should be mentioned that fatty acids differ from each other not only in their degree of unsaturation but also with respect to the number of carbon atoms in the chain and the family or series of fatty acids to which they belong. The products of fatty acid synthetase are saturated fatty acids. However, fish lipids contain high levels of unsaturated fatty acids, derived in part from food ingestion. The polyunsaturated fatty acids belong to one of three major families or series, namely, the oleic or *n*-9, the linoleic or *n*-6, and the linolenic or *n*-3 series. The majority of members of the two most important ones from a nutritional standpoint are depicted in Figure 1. The individual members of each of the families can therefore be differentiated by the number of carbon atoms and double bonds in the chain and by the position of the first double bond, counting from the terminal methyl (CH_3) group carbon to the carbon atom of the first double bond. Thus, for example, the complete structure of 18 : 3*n*-3 is as follows:

$$CH_3CH_2CH{=}CHCH_2CH{=}CHCH_2CH{=}CH(CH_2)_7COOH.$$

With reference to Figure 1, it should be noted that finfish, like other vertebrates, but unlike plants, do not possess the Δ12 and Δ15 desaturase enzymes required for the synthesis of 18 : 2*n*-6, the parent acid of the *n*-6 series, and 18 : 3*n*-3, the precursor of the *n*-3 series (3). Consequently, these fatty acids (or their metabolic derivatives) must be of dietary origin. Also, depending upon the finfish species, the parent acid of the

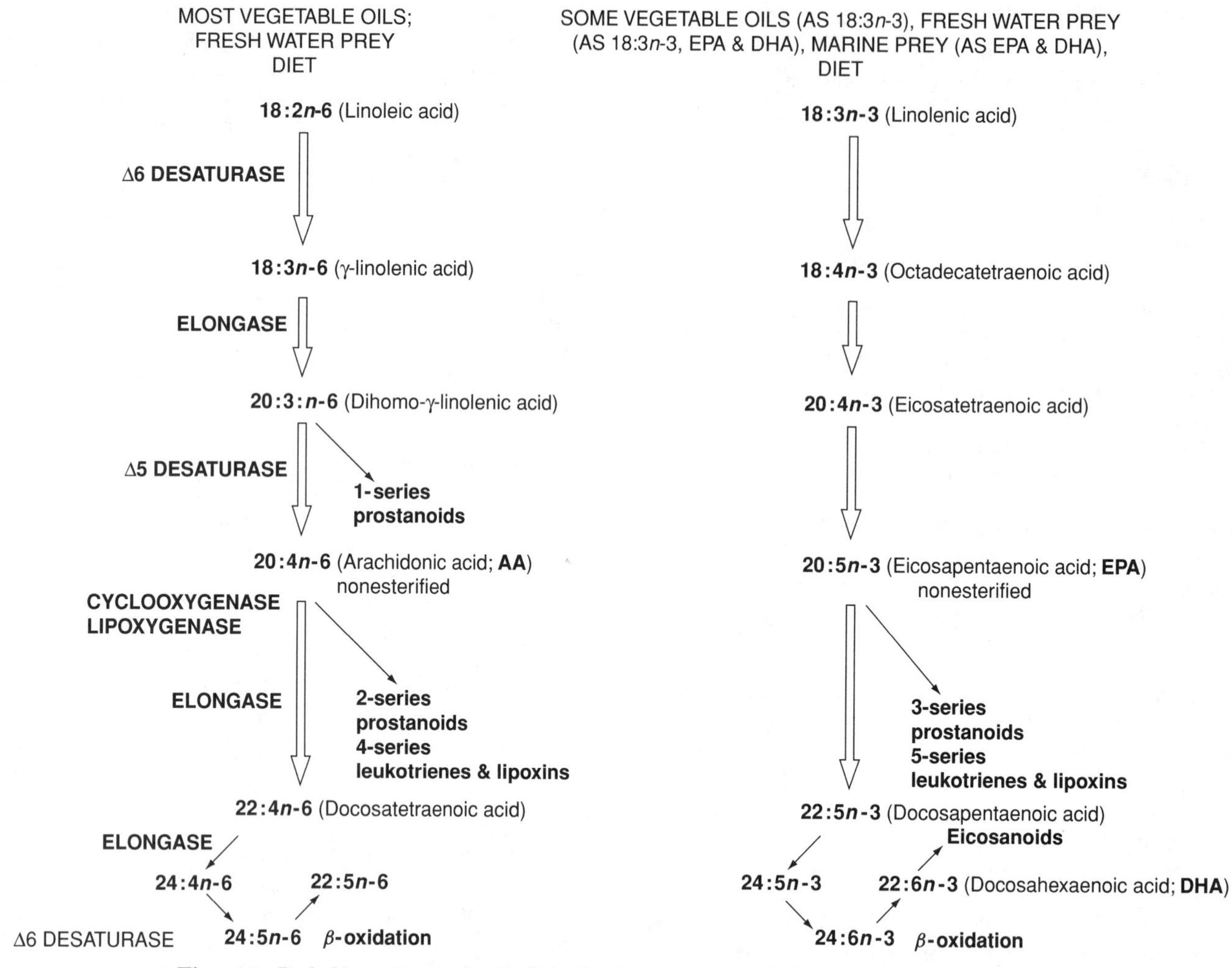

Figure 1. Probable pathways involved in the desaturation and elongation of the n-6 and n-3 series of fatty acids in freshwater fish. The parent acids of each family, as well as their respective nutritionally important highly unsaturated fatty acids and series of cyclooxygenase and lipoxygenase-derived prostanoids (prostaglandins and thromboxanes), leukotrienes, and lipoxins (collectively termed eicosanoid compounds) are indicated. The production of 22 : 5n-6 and 22 : 6n-3 is now believed to occur as shown rather than through Δ4 desaturation of 22 : 4n-6 to yield 22 : 5n-6 and 22 : 5n-3 to produce 22 : 6n-3 (148,149). There is some evidence that eicosanoids are formed from DHA (e.g., 14-hydroxydocosahexaenoic acid), through the action of 12-lipoxygenase (150).

n-3 series, the proper levels and proportions of some of the highly unsaturated members of the n-3 series, and the counterparts of the n-6 series of fatty acids are considered to be essential for normal growth, food utilization, health, and reproductive viability (refer to the later section on lipid and fatty acid requirements). The members of each of the families of fatty acids are created from their respective parent acids by a common enzyme system of alternating desaturases and elongases that yield series of fatty acids of increasing unsaturation and length. Further, the members of one family are not interconvertible with those of another. The highly unsaturated fatty acids (HUFAs) of the n-9, n-6 and n-3 families of nutritional significance are (respectively) eicosatrienoic acid (20 : 3n-9), dihomo-γ-linolenic acid (20 : 3n-6), arachidonic acid (20 : 4n-6; AA), eicosapentaenoic acid (20 : 5n-3; EPA), and docosahexaenoic acid (22 : 6n-3; DHA). The latter four fatty acids are progenitors of a series of compounds (collectively called eicosanoids) that are essential for the regulation of many physiological processes in the body. The last two of them are frequently referred to as n-3 HUFAs (EPA + DHA), although this term can also include other C20 members of the n-3 family.

With this background information, it is now possible to continue the discussion of triacylglycerols. In addition to differences in the melting points of the individual fatty acids affixed to the glycerol moiety, the fatty acids also show differences with respect to their positional distribution. This distinction has been reviewed thoroughly by Polvi 1989 (4), who observed that the n-3 fatty acids in the triacylglycerols of Atlantic salmon (*Salmo salar*) were divided almost equally between positions 2 and 3. Thus, depending upon their fatty acid composition, triacylglycerols may exist in either a liquid or a semisolid state at room temperature, in which cases they are called oils and fats, respectively.

The wax esters present in the prey—for example, calanoid copepods, bathypelagic mysids, some euphausiid species, deep-sea squid and myctophids—of wild salmon in sea water and of marine species are unlike triacylglycerols, because they are primary esters of long-chain fatty alcohols (not glycerol) and long-chain fatty acids. Furthermore, the alcohols are rich in 20 : 1*n*-9 and 22 : 1*n*-11, and the fatty acids can be represented well by *n*-3 polyunsaturated fatty acids and 14 : 0 (5).

Cholesterol and cholesteryl esters (cholesterol esterified to a fatty acid) are the third group of nonpolar lipids that should be considered before discussing the essential fatty acid needs of some of the finfish species and the influence of lipids on their cardiovascular health. Cholesterol is ubiquitous within the body of fish and other animals. It is an essential component, along with phospholipids (discussed later) and proteins, of all cellular and subcellular membranes (6). In addition, cholesterol is involved in lipid transport, because both free and esterified cholesterol are constituents of lipoproteins (as discussed in the lipid transport section). Moreover, cholesterol is a precursor of adrenal and reproductive hormones (androgens and estrogens), of vitamin D_3, and of bile acids that facilitate dietary lipid digestion and absorption.

The basic structure of polar lipids consists of a glycerol or sphingosine (an amino alcohol) moiety coupled with one or two fatty acids and a polar head group. All polar lipids except glycolipids (cerebrosides and gangliosides) contain a phosphate group and generally a nitrogenous base; they are therefore referred to as phospholipids (7). Phospholipids contain two fatty acids esterified to glycerol. Usually the fatty acids in position 1 are saturated or monounsaturated (e.g., C16 : 0 or C18 : 1, present in marine poikilotherms) (8). By contrast, polyunsaturated fatty acids, especially of the *n*-3 series, are generally found in position 2 (4). Because of the high occurrence of C_{20} and C_{22} polyunsaturated fatty acids, such as EPA and DHA at position 2, the fatty acids of phospholipids are more unsaturated than those of triacylglycerols. Phospholipids are classified according to the nitrogenous base moiety attached to phosphoric acid. Lecithins, for example, have choline (phosphatidylcholine; PC) as their nitrogenous base, whereas cephalins have ethanolamine (phosphatidylethanolamine; PE). Other cephalins have serine (phosphatidylserine; PS) or inositol (phosphatidylinositol; PI) (6).

Phospholipids have hydrophilic (because of the polar phosphoric acid and the nitrogenous base region) and hydrophobic (because of the nonpolar fatty acid chain) properties. Together with proteins and cholesterol, they form the basic structures of cellular and subcellular membranes. Within the membranes, the polar lipids are organized in the form of a bilayer, with the hydrocarbon chains providing a hydrophobic environment in the interior of the bilayer and the lipid polar head groups encountering the outer aqueous phase. The proteins within the membranes are bound to the bilayer surface, or they are integrated into the lipid bilayer, with hydrophobic amino acids in the interior and charged amino acids on the exterior. The lipid bilayer region is generally well ordered, and only a small region is liquidlike (9).

The physical and functional properties of the membranes are determined by the following: (*1*) the levels and types of the constituent phospholipids; (*2*) the fatty acid compositions of the phospholipids, especially at position 2 of the glycerol backbone; (*3*) the interactions of the phospholipids with cholesterol and either enzymatic or structural proteins; and (*4*) the specific pairing of long-chain polyunsaturated fatty acids with $\Delta9$ (18 : 1) monounsaturated fatty acids in the *sn*-1 position, especially in PE (8,10–12). Biomembrane fluidity, for example, largely reflects the balance between polyunsaturated fatty acids on position 2 and saturated fatty acids or monounsaturated fatty acids on position 1 of the glycerol backbones of the phospholipids (8). In this regard, there is in the livers of winter (cold)-acclimated carp (*Cyprinus carpio*) a decrease in the level of 18 : 0/22 : 6 species and an attendant increase in the levels of 1-monounsaturated, 2-polyunsaturated (18 : 1/22 : 6, 18 : 1/20 : 4) species in PC and PE. The accumulation of 18 : 1/22 : 6 and to some degree 18 : 1/20 : 5 PE has also been demonstrated in marine fish inhabiting cold waters and in the brains of fresh water fish adapted to reduced temperatures (12). The elevation of PE and the concomitant decrease of PC in membranes may facilitate membrane fluidity at low temperatures, because PE does not readily form compact lamellae (3). Water temperature is a very important factor influencing the growth and physiology of poikilothermic finfish species, and it is important that the biomembranes exist in a liquid-crystalline state at body temperature. Indeed, the cells of the body must adapt to establish a new equilibrium between the environment and the physicochemical properties of their membranous structures in order to survive the new conditions. This process is known as homeoviscous adaptation (12).

Cellular function and metabolism in mammals can be influenced by alterations in the composition of the membrane constituents in several ways. First, the activities of membrane-bound enzymes such as Na^+,K^+ATPase, adenylate cyclase, and Ca^{2+}-ATPase can be changed (9,11). Second, the extent of hormone binding to membrane receptor sites, such as insulin to adipocyte membranes and triiodothyronine to hepatic nuclear membrane sites, can be varied (11). Third, there may be modification of the control and expression of cell nucleus activity (11). Lastly, there may be modulation of the types and levels of cyclooxygenase and lipoxygenase eicosanoid products (Fig. 1) produced in response to the levels of non-esterified AA, EPA, and DHA. All of the foregoing mechanisms appear to operate in fish, and disruption of their normal synchrony in cellular metabolism can lead to adverse physiological consequences.

For instance, overproduction of the highly bioactive *n*-6 polyunsaturated (PUFA)-derived eicosanoids may be involved in the induction of cardiac myopathy in Atlantic salmon postsmolts that ingest a diet rich in *n*-6 fatty acids (13,14). Also, suboptimal levels of di-22 : 6*n*-3 species in the retinal membranes of herring (*Clupea harengus*) are known to reduce their visual acuity at low light intensities, where retinal rods normally function (15). Further, reduced levels of di-22 : 6*n*-3 species in brain phospholipids of fish probably adversely influence their brain function, a phenomenon that has been demonstrated in mammals (including man) (16). In addition, dietary deficiencies of DHA and AA impair pigmentation in the Japanese flounder (*Paralichthys olivaceus*) (17). Kidney and gill function (osmoregulatory ability) in Atlantic salmon post-smolts can likely be influenced positively or negatively by the dietary balance between n-6 and n-3 fatty acids and by the ratios between their respective prostaglandins derived from AA (2-series) and EPA (3-series) (18). Likewise, pineal organ function in salmon probably is influenced by dietary levels of n-6 and n-3 fatty acids and by suboptimal ratios of 2- and 3-series prostaglandins that occur in response to varying nonesterified levels of AA in relation to EPA (19). As Clandinin et al. (1991) (11) pointed out, biological membranes should be viewed as dynamic and highly responsive structures whose constituents (and consequently functions) vary according to both intrinsic factors (e.g., fatty acid desaturation and elongation; phospholipid biosynthesis; types and levels of eicosanoid compounds elaborated) and extrinsic factors (e.g., dietary lipid composition). The interrelationship between these factors will be considered further in the next section.

DIGESTION, ABSORPTION, TRANSPORT, AND STORAGE

The hydrolysis of lipids in the intestinal lumen appears to be accomplished mainly by nonspecific and bile-salt dependent lipase, with perhaps some involvement of 1,3 specific pancreatic lipase (20). In many teleosts, lipid digestion and absorption take place in the anterior intestine, where the pyloric caecae are located and the pancreatic enzymes are secreted into the intestinal lumen. In some species, however, like the turbot (*Scophthalmus maximus*), there is evidence that these processes occur mainly in the hindgut and rectum (20). The process of lipid digestion yields free fatty acids, fatty alcohols from wax esters (marine species), glycerol, 2-mono-acylglycerol, sterols, and lysophospholipids. As a general trend, the absorption of the various fatty acid groups in the intestinal region of finfish shows the following sequence: saturates < monoenenes < polyunsaturates. Moreover, fatty acid digestibility (bioavailability) is known to decrease with increasing chain length. The bioavailability of marine lipids (rich in n-3 HUFAs) in most fish species appears to be $\geq$90% (21–23) using reliable digestibility procedures. (Refer to (24,25).)

The aforementioned products of lipid digestion are metabolized further in the intestinal mucosa enterocytes. Most of the lysophospholipids are re-esterified with fatty acids to phospholipids, and glycerol and 2-monoacylglycerol are re-esterified with free fatty acids into triacylglycerols (3).

Some of the exogenous fatty acids may also undergo desaturation and elongation in the intestinal enterocytes (26). The lipids are then transported to the liver as chylomicrons (lipoprotein complexes comprised of triacylglycerols and of minor amounts of cholesterol, cholesteryl esters, phospholipids, and protein components, termed apoproteins) and as very low density lipoproteins (VLDL), via the blood or the lymphatic system. Also, a significant quantity of free fatty acid may be transported as albumin complexes, via the portal blood (5). Within the liver, fatty acids synthesized endogenously—saturated and/or monounsaturated fatty acids—are combined with those of exogenous (dietary) origin in the form of VLDL. The liver is the main site for endogenous synthesis in fish. This organ is also highly active in modifying fatty acids of both endogenous and exogenous origin, through desaturation and elongation enzymes located in the microsomes (27). Dietary lipid composition, as well as such other factors as water temperature, influences the types of fatty acid derivatives that are elaborated.

In regard to dietary lipid composition, the substrate preference for $\Delta 6$ desaturase (Fig. 1) in fresh water salmonids is $18:3n$-3 > $18:2n$-6 > $18:1n$-9. Also, a surfeit of one series of dietary fatty acids (for example, high levels of n-6 versus n-3 fatty acids) may, depending upon the species, competitively inhibit the formation of the long chain highly unsaturated members of the series in lower concentration. Further, high dietary levels of n-3 HUFAs in rainbow trout (*Oncorhynchus mykiss*) are known to inhibit the hepatic bioconversion of $18:3n$-3 and $18:2n$-6 (27). Lastly, a deficit of n-3 and n-6 fatty acids in the diet of salmonids leads to the bioconversion of $18:1n$-9 to $20:3n$-9, and this situation indicates essential fatty acid deficiency (3). Hence, the character of the lipids composing hepatic VLDL can vary considerably. VLDL transports lipids of exogenous and endogenous origin from the liver to the extrahepatic tissues. Here, the VLDL triacylglycerols are hydrolyzed by lipoprotein lipase, with the consequent uptake of fatty acids into the cells and with the formation of cholesterol-rich low-density lipoprotein, which is specialized for transport of the esterified cholesterol to the extrahepatic tissues (28).

Within the cells, the new fatty acids supplied by VLDL can be esterified into triacylglycerols or incorporated into membrane lipids by *de novo* phospholipid synthesis and by acyl group turnover in the membrane phospholipids (11). Some metabolically active extrahepatic tissues such as the heart, gonad cells, and leukocytes may also be able to desaturate the fatty acids derived from VLDL. Other possible pathways for the nonesterified fatty acids originating from VLDL or from membrane phospholipids by the action of phospholipase A_2 include the formation of eicosanoid compounds, because of the action of cyclooxygenase and lipoxygenase enzymes (Fig. 1), and of acylcarnitine for mitochondrial β-oxidation (energy provision).

Two other lipoproteins that are also involved in lipid transport also require mention. These are high-density lipoprotein (HDL) and vitellogenin (VTG). HDL is derived from LDL and is rich in phospholipids and cholesterol (5). HDL is specialized for transport of cholesterol that has been taken up from the extrahepatic tissues to the liver (29). VTG is a specific female lipoprotein that is present in oviparous fish. VTG (as well as VLDL) is secreted by the liver in response to estrogen stimulation; these lipoproteins play a central role in gonadal development (30,31). HDL is generally the main lipoprotein in fish (5,32,33), although not always. For instance, in the Japanese eel (*Anguilla japonica*), a species that is known to accumulate considerably more lipid in its muscle than in the liver, VLDL was identified as the main plasma lipoprotein (31). As a general trend, plasma VLDL levels in fish are directly related to their ability to store lipid in the muscle as opposed to the liver. Thus, Ando and Mori (1993) (33) observed that the striped jack (*Caranx delicatissimus*), which has a high level of muscle lipid, had a high level of plasma VLDL, whereas pufferfish (*Takifugu rubripes*), which have high hepatic lipid stores, were noted to have low plasma VLDL levels. Nevertheless, as mentioned, HDL still remained the dominant plasma lipoprotein in these species.

Besides liver and muscle (red and white), lipid can be stored along the intestine and its mesentery. In *Oncorhynchus* species (Pacific salmon and rainbow trout), the stores of lipid in the preceding locations vary widely, especially seasonally. Most of this variation is due to the following factors: (*1*) the species and its size and sex; (*2*) the level of food (digestible energy) intake; (*3*) the proportions of total available dietary energy originating from protein and lipid; (*4*) the smoltification and reproductive status of the fish; (*5*) water temperature; (*6*) salinity; (*7*) the level of physical activity; (*8*) the extent of hepatic fatty acid synthesis; or (*9*) a combination of these factors (25).

LIPID AND FATTY ACID REQUIREMENTS

The known dietary lipid and fatty acid requirements for most species of fish of commercial importance are provided in Table 1. In regard to the needs of finfish for dietary lipid, it first should be stressed that these requirements were estimated in general, by using optimal dietary concentrations and sources of the other energy-yielding nutrients (protein and carbohydrate) and lipid sources of high digestibility ($\geq$90%).

The information presented in Table 1 reveals that there are wide differences in dietary lipid requirements within and between species. Salmonids, for example, often have high requirements for dietary lipid (usually $\geq$150 g/kg diet) relative to most fresh water non-salmonid species and many marine species, to spare dietary protein for growth, to enhance the efficiency of energy retention, and to meet the required dietary amounts of digestible energy. Lipid is the preferred dietary non-protein energy source in salmonids, because of their limited ability to utilize digestible carbohydrate as an energy source. This bias likely stems from the fact that salmonids in the wild derive most of their energy needs from the high levels of protein and lipid in their prey (25). Indeed, the needs of wild salmonids for glucose are met largely through the process of gluconeogenesis, which uses the glucogenic amino acids derived from the digestion of dietary protein or tissue proteolysis (e.g., alanine, serine, and glycine) plus lactate and glycerol as the substrates. Salmonids also have other metabolic deficiencies that restrict the utilization of high dietary levels of digestible carbohydrate [reviewed by (25)], and it is generally recommended that the dietary level of digestible carbohydrate should not exceed 150 g/kg and should, in some cases, be even lower. Other finfish species, such as channel catfish (*Ictalurus punctatus*), common carp (*C. carpio*), and tilapia (*Tilapia zillii* and *Oreochromis niloticus*), have greater ability than do salmonids to utilize digestible carbohydrate as a non-protein energy source (34–36); accordingly, they place less emphasis on digestible lipid to meet their non-protein energy demands.

The data in Table 1 also suggest that the dietary lipid needs of finfish may vary in relation to the stage of life history. This effect is clearly evident in the Atlantic salmon (*S. salar*), where it has been found that very high-energy ($\sim$330 g lipid/kg) diets support maximum performance (growth and feed efficiency) of post-juvenile salmon ($>$200 g) in sea water, whereas the dietary lipid needs of the juvenile salmon in fresh water are lower (240 g/kg).

The dietary essential fatty acid needs of finfish shown in Table 1 largely reflect the lipid compositions of their respective natural prey. *Oncorhynchus* species in fresh water, for example, ingest prey that contain substantial amounts of *n*-3 and *n*-6 fatty acids, mostly in the form of the parent acids and highly unsaturated members of each series (i.e., AA, EPA, and DHA, with EPA often greater than DHA). The levels of *n*-3 series fatty acids in the freshwater prey items always exceed the levels of the *n*-6 fatty acids. In the marine prey of these species, the levels of *n*-3 and *n*-6 fatty acids are respectively much higher and lower than those found in the fresh water prey. Also, the fatty acids of the *n*-3 series are largely represented by EPA and DHA, and frequently the level of DHA is equivalent to or greater than that of EPA (25). Accordingly, the essential fatty acid needs of salmonids in freshwater are mostly satisfied by 18:3*n*-3 alone, and in one instance (chum salmon, *Oncorhynchus keta*), by a combination of 18:3*n*-3 and 18:2*n*-6 or by *n*-3 HUFAs alone (present at $\geq$10% of the dietary lipid level). All of the species in freshwater appear to have good ability to convert 18:3*n*-3 to *n*-3 HUFAs, and, depending upon the speed of the bioconversion in each species, *n*-3 HUFAs may have greater essential fatty acid activity than 18:3*n*-3 (e.g., in rainbow trout) or equivalent essential fatty acid activity (e.g., in coho salmon, *Oncorhynchus kisutch*). The essential fatty acid needs of *Oncorhynchus* species in sea water have been studied only in chum salmon, and they did not differ from those established for this species in fresh water (i.e., 10% of the dietary lipid level as *n*-3 HUFAs). It is assumed that this similarity is also the case for the essential fatty acid requirement of the other *Oncorhynchus* species in the marine environment.

The natural diet of nonsalmonid freshwater fish (e.g., carps, tilapias, and eels) is comprised of terrestrial and aquatic plants and insects. Consequently, these species in the wild, like the anadromous salmon described above, consume considerable amounts of 18:3*n*-3 and 18:2*n*-6 and lower levels of 22:6*n*-3. In response, these species, like the salmonids, generally convert 18:3*n*-3 and 18:2*n*-6 readily to 22:6*n*-3 and 20:4*n*-6, respectively (37). Hence, the essential fatty acid needs of the freshwater species are generally satisfied by 18:3*n*-3 or 18:2*n*-6 alone or in combination. In some instances, however, the requirements are satisfied by a combination of 18:2*n*-6 and *n*-3 HUFAs, or by 20:4*n*-6 alone, or by *n*-3 HUFAs alone (Table 1).

Many marine species eat fish in the wild; others ingest zooplankton, and a small number consume unicellular

Table 1. Recommended Dietary Levels (g/kg dry weight basis and percentage of dietary lipid where established) of Lipid and Fatty Acids for Maximum Growth and Feed Efficiency, as well as for Reproductive Performance, in Finfish Species*

Species/ Life History Stage	Lipid (g/kg)	Fatty Acid: 18:3*n*-3 (g/kg)	18:3*n*-3 (%)	18:2*n*-6 (g/kg)	18:2*n*-6 (%)	*n*-3 HUFAs (g/kg)	*n*-3 HUFAs (%)	20:4*n*-6 (g/kg)	Source
A. Salmonids									
O. mykiss Freshwater (FW) (juvenile–adult)	150–230	8.3–16.6	≥20[a] ≤80	<10		20–30	≥10[a] ≤40	R?[a]	90–93
O. tshawytscha (juveniles in FW)	>63–200	R[b]	R	≤26		R	R		25
(postjuveniles in sea water (SW), <500 g)	150–200	R	R			R	R		25
O. kisutch (juveniles in FW)	160–180	10–25	10–25; <40	≤10		R	R		25
(maturing fish in FW)		R[b,c]					R[b,c]	R?	38
O. keta (juveniles in FW)[d]	55–109	10		10			10		95–96
(juveniles in SW)[d]		10		10			10		97
O. masou (juveniles in FW)[e]		10				5			98
S. salar (juveniles in FW; 80 g)	240	R				R			99
(postjuveniles in SW; >200 g-adults)	≥330	R				R			100,101
Salvelinus alpinus (juveniles in FW)	200	10–20	20–40	≤7					102–105
Salmo trutta (postjuveniles in SW; 1600 g)	290								106
B. Nonsalmonids									
I. punctatus (FW)	50–60	10–20				5–7.5[f]			34,107
C. carpio (FW; juvenile)	80–125	10		10					35
Ctenopharyngodon idella (FW; juveniles)	~40	10		10		5[g]			108,109
Clarias batrachus (FW; juveniles)	81								110
Clarias macrocephalus × *C. gariepinus* (FW; juveniles)	44–96								111
Catla catla (FW; juveniles)	40								112
Oreochromis and *Tilapia* spp.	50–60								36
O. niloticus				5					113
T. zillii				10				10[h]	114
Plecoglossus altivelis		10				10[i]			115

Table 1. *Continued*

Species/ Life History Stage	Lipid (g/kg)	Fatty Acid							Source
		18:3*n*-3		18:2*n*-6		*n*-3 HUFAs		20:4*n*-6 (g/kg)	
		(g/kg)	(%)	(g/kg)	(%)	(g/kg)	(%)		
Anguilla japonica		5		5					116
Coregonus lavaretus (juveniles)						10	20		117
Acipenser transmontanus (FW; underyearlings)	264–363								118
Morone chrysops × *M. saxatilis* (FW; juveniles)						10	20		119
M. chrysops × *M. saxatilis* (FW; juveniles)	65–96								120
Pseudocaranx dentex (juveniles)						17[j]			121
Pagrus major (juveniles in SW)	164						20		122
Sparus aurata									
(11.5 g juveniles in SW)						≥10[k]			123
(42.5 g juveniles in SW)	80–100					>8–≤19[k]			124
(17-day old larvae in SW)						15[l]			125
(450 g maturing males and females in SW)	115					≥4.2[m]			39
(740 g males and 1290 g females in SW)	142					16[n]	11.3		40
Seriola quinqueradiata (1.8 g juveniles in SW)	157–215						14–22		126
Scophthalmus maximus						8			127
(0.88 g juveniles in SW)	150					5.7–13[o]		3[o]	128
Chanos chanos									
(juveniles)	70–100								129
(8.6 g juveniles in SW)	70	10[p]				10[p]			130
Siganus guttatus (fry)	100					R			129
Lates calcarifer									
(juveniles in SW)	100–120								131
(0.9–1.3 g juveniles in SW)						10			132
Dicentrarchus labrax (2.8 g juveniles in diluted SW)	120–140								133
Sciaenops ocellatus									

(*continued*)

Table 1. *Continued*

Species/ Life History Stage	Lipid (g/kg)	Fatty Acid 18:3*n*-3 (g/kg)	18:3*n*-3 (%)	18:2*n*-6 (g/kg)	18:2*n*-6 (%)	*n*-3 HUFAs (g/kg)	*n*-3 HUFAs (%)	20:4*n*-6 (g/kg)	Source
Juveniles in brackish water	74–112								134
Larvae in SW						15–52[q]			135
Juveniles in brackish water	70					5	7		136
Sebastes schlegeli (5.9–0.8 g juveniles in SW)						9–10[r]			137,138
Sebastes thomposoni	100								139

[*]It is assumed that the dietary levels and sources of the other energy-yielding nutrients, viz., protein and carbohydrate, are optimal and that the digestibility of lipid is ≥90%. In many studies, "*n*-3 HUFAs" refers to 20:5*n*-3 (EPA) and 22:6*n*-3 (DHA); in others, this term also includes small amounts of 20:4*n*-3 and 22:5*n*-3, and sometimes 20:3*n*-3.

[a]The rainbow trout requires 20% of the dietary lipid content as C18:3*n*-3 or 10% as *n*-3 HUFAs. This provision appears to satisfy all needs for growth and reproduction, although there may be a small need for 20:4*n*-6 for optimal reproductive performance.

[b]Required.

[c]Maturing coho salmon in freshwater need ≥10 g of *n*-3 fatty acids/kg diet for optimal reproductive performance. It is unknown whether there is a small requirement for 20:4*n*-6.

[d]*O. keta* require either 1% 18:3*n*-3 and 1% 18:2*n*-6 or 10% of dietary lipid as *n*-3 HUFAs.

[e]*O. masou* need 1% 18:3*n*-3 or 0.5% *n*-3 HUFAs in their diet.

[f]*I. punctatus* require 1.0–2.0% 18:3*n*-3 or 0.5–0.75% *n*-3 HUFAs in their diet.

[g]*C. idella* require 1.0% 18:*n*-3 and 1.0% 18:2*n*-6 or 1.0% 18:2*n*-6 and 0.5% *n*-3 HUFAs in their diet.

[h]*T. zillii* need 1.0% of 18:2*n*-6 or 1.0% 20:4*n*-6 in the diet.

[i]*P. altivelis* require 1.0% 18:3*n*-3 or 1.0% 20:5*n*-3 in the diet.

[j]*P. dentex* require 1.7% *n*-3 HUFAs or 1.7% 22:6*n*-3 in their diet. (DHA has higher essential fatty acid activity than does EPA).

[k]The ratio of EPA to DHA in the dietary lipids was 2:1.

[l]The ratio of EPA to DHA in the dietary (rotifer) lipids was 0.77 or the ratio of DHA to EPA was 1.3.

[m]This dietary level is recommended for optimum egg quality. The ratio of DHA to EPA in the dietary lipids was 2.8.

[n]This dietary level is recommended for highest fecundity, hatching, and larval survival. The ratio of DHA to EPA in dietary lipids was 0.66.

[o]Deduced requirement for DHA and 20:4*n*-6, considering the results of this study in relation to those of previous ones on turbot.

[p]*C. chanos* required 1% 18:3*n*-3 and 0.5% EPA + 0.5% DHA in their diet.

[q]The ratio of DHA to EPA should exceed 2.5.

[r]*S. schlegeli* needs about 1% of EPA and/or DHA in the diet. DHA has higher essential fatty acid activity than does EPA, and the optimum dietary ratio of EPA to DHA is less than 1.

algae. In all cases, they consume large amounts of EPA and DHA and little 18:3*n*-3 in their natural diets (37). Consequently, these species have little or no requirement to biotransform 18:3*n*-3 to EPA, and they therefore have little or no Δ5 fatty acid desaturase activity. Hence, there is insufficient conversion of EPA to DHA to meet the requirements of these species for growth. In the diet, then, DHA must be supplied preformed (37). The dietary essential fatty acid needs of the marine species shown in Table 1 support the preceding scenario, and they clearly show that all species require *n*-3 HUFAs and generally optimal ratios of DHA and EPA in their diet for maximum growth and for optimal feed utilization. In some species, DHA has been shown to have higher essential fatty acid activity than EPA. Also, in the turbot (*Scophthalmus quinqueradiata*), it is noteworthy that a small requirement for AA has been found, in addition to that for DHA, for optimal performance.

The quantitative essential fatty acid requirements of finfish species undergoing gonadal maturation are not extensive. In salmonids, Hardy et al. (1989) (38) reported that maturing coho salmon in fresh water require ≥10 g of *n*-3 fatty acids/kg diet for optimal reproductive performance. In nonsalmonids, Harel et al. (1994) (39) reported that gilthead seabream (*S. aurata*) require ≥4.2 g of *n*-3 HUFAs/kg diet (ratio of DHA to EPA in dietary lipids, 2.8) for optimum egg quality. Further, Fernández-Palacios et al. (1995) (40) found that 16 g of *n*-3 HUFAs/kg of diet (ratio of DHA to EPA in the dietary lipids, 0.66) was necessary for highest fecundity, hatching, and larval survival of gilthead seabream.

As a general observation, larval marine fish species require high dietary levels of *n*-3 HUFAs [by most estimates, between 9 and 39 g/kg dry diet (41)], as well as optimal dietary ratios between DHA, EPA, and AA, for maximum growth and survival (42,43).

DIETARY LIPID COMPOSITION AND FISH REPRODUCTION

In fish, environmental signals mainly trigger oocyte growth; these signals are converted from electrical to chemical in the hypothalamus. Gonadotrophin releasing hormone is then released from the hypothalamus, and this stimulates the secretion of gonadotrophins (GtH 1 and GtH 11) from the anterior pituitary. GtH 1 stimulates the oocytes to produce estrogen (estradiol-17β), which subsequently promotes the production of yolk-protein precursors and egg-shell protein by the liver. GtH 1 also stimulates uptake of VTG into trout oocytes. GtH 11 functions later in oocyte development and acts in the follicle cells to promote the synthesis of progesterones, which, in turn, control termination of oocyte growth and ovulation of the egg (44).

Egg lipids are derived from nonpolar lipid stores in the body and from dietary (exogenous) sources. The latter are particularly important in species such as gilthead sea bream, which continue to eat during sexual maturation and throughout the spawning season. Triacylglycerols from body lipid stores (e.g., visceral and muscle lipid in rainbow trout and Pacific salmon), together with free amino acids originating from muscle-protein breakdown, are carried in the blood to the liver. Here, the free fatty acids are incorporated mainly into phospholipids that are rich in n-3 HUFAs, especially DHA, and to a lesser degree into triacylglycerols. The amino acids are used for hepatic synthesis of egg-specific apoproteins that are then combined with the newly synthesized lipids to form VTG. VTG is then transported to the developing oocytes via the blood. Thereafter, VTG is sequestered by the oocytes by a process of pinocytosis, and then it is cloven in the egg to generate the egg-yolk proteins, namely, lipovitellin and phosvitin. Most of the n-3 HUFA-rich phospholipids are located in the former protein (45). Eggs with short and long incubation times accumulate low and high levels of triacylglycerols, respectively, and both phospholipids and triacylglycerols are catabolized to provide metabolic energy during egg development and early larval rearing. The principle role of the n-3 HUFAs, particularly DHA, is in the elaboration of cellular membranes of neural tissues (e.g., brain and eyes). Low levels of AA are also present in (primarily) PI, and it is believed that this fatty acid has a specific role in eicosanoid formation (45).

Most studies that have assessed the influence of dietary lipid composition on the reproductive performance of finfish have found dramatic effects when marine species such as red (*Pagrus major*) and gilthead sea bream have been employed as the test fish (Table 2). These species rely extensively on exogenous lipid for oocyte development. Consequently, the lipid composition of the developing oocytes can be changed rapidly to reflect the dietary lipid composition (within 15 days in gilthead sea bream). Further, there can be an attendant decline in egg viability within 10 days in gilthead sea bream ingesting a diet deficient in n-3 HUFAs (39). In another study on this species, Fernández-Palacios et al. (1995) (40) reported improved spawning quality (fecundity, hatching, and larval quality) after only 3 weeks of feeding a diet containing an optimal concentration of n-3 HUFAs. Marine species generally require high dietary levels of n-3 HUFAs, especially DHA, relative to salmonids; consequently, suboptimal dietary levels and suboptimal ratios of these fatty acids to one another and to AA (46); (Table 2) appear to affect their reproductive success negatively to a greater extent than is noted in most studies on salmonids (Table 2). If all of the studies on marine species shown in Table 2 are viewed collectively, it is apparent that dietary deficiencies or excesses of n-3 HUFAs or suboptimal ratios of EPA and DHA themselves or with AA can markedly influence almost all aspects of reproductive performance. By contrast, the reproductive performance of salmonids appears to be affected adversely mainly in cases of extreme dietary deficiencies of n-3 fatty acids over a 3- to 12-month period. Moreover, there is little indication that wide variations in dietary levels of n-6 fatty acids (such as 18:2n-6) depress their reproductive performance, provided that their dietary needs for n-3 fatty acids are met (Table 2). Other consequences of feeding diets to broodstock fish that contain inappropriate lipid composition may include reduced sperm quality (47,48), depressed levels of serum VTG (49), decreased testicular steroidogenesis and development (50), and suboptimal levels of 2-series and 3-series prostaglandins (derived from AA and EPA, respectively) (40,51).

INFLUENCE OF DIETARY LIPID COMPOSITION ON FISH HEALTH

As mentioned previously, our intent here is to restrict our focus concerning the effects of dietary lipid composition on fish health to those effects concerned with cardiovascular function and disease resistance (immunocompetence). There is an extensive literature in the latter area, and only some of the highlights will be mentioned below. The purpose is to provide the reader with some appreciation of the possible consequences of using lipids other than of marine origin in fish diets.

In humans and other animals, high dietary intake of lipid (especially saturated fatty acids) and cholesterol, coupled with a diet rich in n-6 fatty acids (relative to the proportion of n-3 fatty acids), is known to increase the likelihood of atherosclerosis, heart attacks, strokes, and various inflammatory conditions. The etiology of atherosclerosis is believed to result from overproduction of eicosanoid compounds derived from AA via the cyclooxygenase and lipoxygenase enzymes. For instance, thromboxane A_2, which is produced from the blood platelet membranes, is a potent blood platelet aggregator and constrictor, and 4-series leukotrienes are proinflammatory. By contrast, depressed levels of unesterified AA in the tissues and elevation of EPA lead to the production of thromboxane A_3 and prostacyclin I_3 which reduce blood platelet aggregation and enhance vasodilation. Also, 5-series leukotrienes from EPA are antiinflammatory, and they attenuate the response of neutrophils and monocytes to inflammatory stimuli (25).

Studies on post-smolt Atlantic salmon suggest that metabolic and cardiovascular events similar to those described above for mammals occur in fish. Thus, Bell et al. (1991) (13) and Bell et al. (1993) (14) have shown that Atlantic salmon fed diets containing excessive quantities of n-6 fatty acids (but still adequate in dietary levels of n-3 fatty acids for growth) develop severe cardiomyopathy. This effect caused extensive thinning of the ventricular muscle and active necrosis in the atrium and ventricle. These pathological responses, in turn, were accompanied by elevation of AA in tissue phospholipids at the expense of EPA and by increases of the levels of 2-prostanoids (prostaglandins and thromboxane B_2, a stable metabolite of thromboxane A_2 derived from AA).

Table 2. Influence of Dietary Fatty Acid Composition on the Reproductive Performance of Finfish*

Species	Duration	Test Lipid or Fatty Acid Source	Dietary Level (g/kg) *n*-3 HUFA[a]	*n*-3	*n*-6	20 : 4*n*-6	Reproductive Performance Fecundity[b]	Egg Size[c]	Egg Viability[d]	Percentage Eggs Hatched[e]	Fry of Larval Quality[f]	Source
A. Salmonids												
O. mykiss	34 months	18 : 3*n*-3		10			0	0	0	0	0	(94)
		18 : 3*n*-3 & 18 : 2*n*-6		10	15		0	0	0	0	0	
O. mykiss	3 months	Fish oil	A[g]	A			0	0	0	0		(47)
		Methyl laurate	D[h]	D	D	D	–	–	–	–		
O. mykiss	12 months	Control (source unknown)		A					0	0	0	(140)
		Rape-seed oil	D	D					–	–	–	
O. mykiss	12 months	Cod liver oil	A				0	0	0	0	0	(141)
		Corn oil		A	E[i]		0	–[j]	0	0	0	
O. kisutch	5 months	Herring oil (HO)		>10				0	0			
		Soybean oil (SBO)		>10				0	0			
		Beef tallow (BT)		>10				0	0			(38)
		HO & SBO (1 : 1)		>10				0	0			
		HO & BT (1 : 1)		>10				0	0			
		SBO & BT (1 : 1)		>10				0	0			
B. Nonsalmonids												
Chrysophrys (Pagrus) major	2 months	Fish oil	17	A			0	0	0	0	0	
		Corn oil and beef tallow	0	D			0	0	–	–	–	(142)
C. (Pagrus) major	2–3 months	Fish oil	20					0	0	0	0	(143)
		Corn oil	8		E[i]			0	–	–	–	
S. aurata	60 days	Squid oil (SQO)	7.5	10	9.3		0		0			
		SQO and soybean oil (SBO) (1 : 1)	4.2	7.5	15.7		0		0			(39)
		SBO	0	4.7	29.3		0		–[k]			
S. aurata	3 months	Sardine oil and beef	10.6	13.7		0.46	–	0	–	–	–	
		Tallow varied to	14.6	18.0		0.64	+[l]	+	+	+[l]	+[l]	
		create different	19.6	26.9		0.86	–	0	+	–	–	(40)
		dietary levels of *n*-3 HUFAs	28.3	39.3		1.23	–	0	+	–	–	

Table 2. *Continued*

			Dietary Level (g/kg)				Reproductive Performance					
Species	Duration	Test Lipid or Fatty Acid Source	n-3 HUFA[a]	n-3	n-6	20 : 4n-6	Fecundity[b]	Egg Size[c]	Egg Viability[d]	Percentage Eggs Hatched[e]	Fry of Larval Quality[f]	Source
S. aurata	20 weeks	Fish oil from fish meal and cod liver oil	16.3	23.2		0.97	0		0	0		
		Mainly olive oil, linseed oil, and small amount of 20 : 4 n-6	0	17.5[m] mainly 18 : 3n-3		Trace (0.1 g/kg added)	–		–	–		(144)
S. aurata	7 months	Fish oil from fish meal and cod liver oil	16.3	23.2		0.97	0		0	0		
		Mainly olive oil, linseed oil, and small amount of 20 : 4 n-6	0	17.5[m] mainly 18 : 3n-3		Trace (0.1 g/kg added)	–		–[n]	–[n]		(145)
D. labrax	2 years (two spawning seasons)	Fish oil from Northern Hemisphere meal and oil	29.7			0.87			0	0	0	(46)[o]
		Fish oil from Northern Hemisphere meal and tuna orbital oil	52.7			2.94			+	+	+	

*The symbols 0, +, and – signify respectively, that the dietary treatment either had no effect (normal response), or enhanced or depressed the reproductive performance parameter.
[a]Refers to the sum of 20 : 5n-3 and 22 : 6n-3.
[b]Total numbers of eggs produced, relative to body weight.
[c]Egg volume, weight, or diameter.
[d]Percentages of fertilized eggs, eyed eggs, or buoyant eggs (sea bream) or egg development.
[e]Percentages of total eggs, eyed eggs, or buoyant (viable) eggs (sea bream and sea bass) that hatched into viable fry or larvae.
[f]Fry or larval size, growth, survival, and/or morphology.
[g]Essential-fatty-acid-sufficient diet.
[h]Essential-fatty-acid-deficient diet.
[i]Diet enriched with 18 : 2n-6.
[j]Noted when fish at 18 °C but not at 8 °C.
[k]Noted within 10 days of feeding a diet deficient in n-3 HUFAs and when the egg n-3 HUFA content was $<$17 mg/g dry weight.
[l]Results examined /kg female. Egg composition and spawning quality were affected by dietary essential fatty acid levels after 3 weeks of feeding.
[m]Diet also rich in 18 : 1n-9.
[n]Number of buoyant eggs significantly decreased from day 80 onwards. The hatching percentages decreased from day 70 onwards (response not significant). High negative correlations between 18 : 3n-3, 18 : 1n-9, and the 18 : 1n-9/(n-3 HUFA) ratio of polar lipids and the percentages of buoyant eggs were observed.
[o]Optimal levels and ratios of 20 : 5n-3, 22 : 6n-3, and 20 : 4n-6 are required to maximize reproductive performance.

In relation to immunocompetence, fish possess both non-specific and specific defence mechanisms. The specific defence system involves T and B (specific antibodies) lymphocytes; non-specific defence mechanisms include phagocytes or natural killing cells and several humoral components (e.g., lysozymes and complement) that facilitate the activity of phagocytes (52).

The results of three studies, two on catfish (*I. punctatus*) and one on Atlantic salmon (*S. salar*), serve to illustrate the importance of this area to the investigation of alternate lipid sources. In the studies on catfish, which are known to require either 1.0–2.0% 18 : 3*n*-3 or 0.5–0.75% *n*-3 HUFAs in their diet to meet their requirements for growth (Table 1), Fracalossi and Lovell (1994) (53) and Li et al. (1994) (54) found that excessive dietary levels of *n*-3 fatty acids, either from 18 : 3*n*-3 (from linseed oil) or from *n*-3 HUFAs (from menhaden oil), increased the mortality of the fish that were challenged with *Edwardsiella ictaluri*. The adverse response was found to be dependent upon the prevailing water temperature, and no differences in circulating antibody titers were observed in the former study. Interestingly, the fish ingesting the diets enriched in *n*-3 exhibited growth equivalent to or better than that of the fish fed the diets based on animal or plant lipid sources. In the study on Atlantic salmon parr, Thompson et al. (1996) (55) observed that salmon fed diets adequate in *n*-3 content, but containing a low ratio of *n*-3 to *n*-6 fatty acids, were less resistant to infection (when challenged with *Aeromonas salmonicida* and *Vibrio anguillarum*) than salmon fed diets with a high ratio of *n*-3 to *n*-6 fatty acids. Further, they found that vaccinated salmon fed the latter diet also exhibited higher numbers of B cells in the kidney and spleen following an *A. salmonicida* challenge.

Hence, there appears to be a need to carefully balance the dietary levels and ratios of *n*-3 and *n*-6 fatty acids in fish diets (specifically EPA, DHA, and AA) to ensure optimal growth and immunological performance, as well as freedom from pathology.

MARINE AND NONMARINE LIPID SOURCES IN FEEDS FOR FINFISH

Lipid Sources

Fish oil is the traditional source of lipid for fish feeds, because it is a rich source of the dietary essential fatty acids needed by fish and is a by-product of fish meal production. However, the challenge of finding environmentally and economically sustainable sources of fish-feed ingredients raises questions about the future suitability and availability of fish oil. As the demand for fish oil increases relative to supply, the price increases, making other lipid sources economically competitive. In addition, there has been a trend toward increasing the percentage of lipid in feeds for some species, such as salmon and trout (Fig. 2), and it is clear that lipid ingredients are a major part of and major expense item in fish feeds. The trend toward an increasing demand for fish oil in a market of static or dwindling supply (Fig. 3) further supports the need to investigate the suitability of

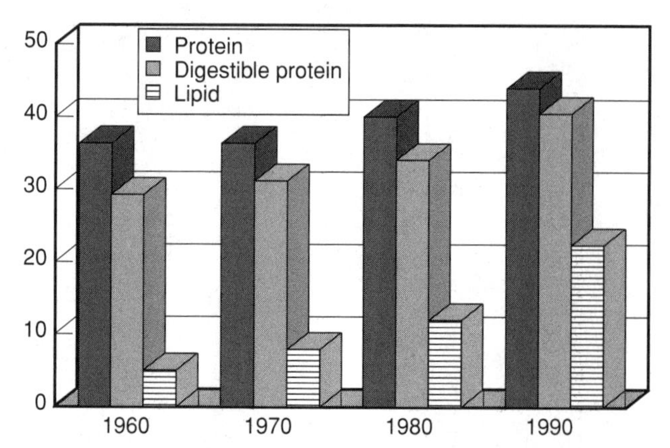

Figure 2. Changes in protein and lipid levels in trout feeds (% of feed).

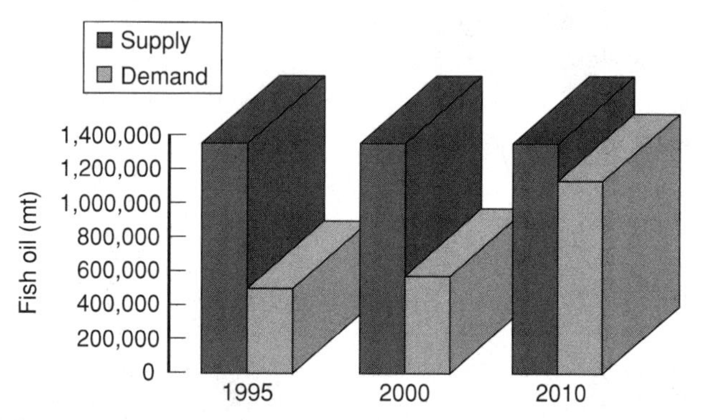

Figure 3. Predictions of future fish oil needs in aqua feeds versus global fish oil supply.

non-fish sources of lipid ingredients. In general, when the price of fish oil exceeds that of soybean oil, the use of plant oils offers an advantage. Animal fats are favorably priced relative to fish oil most of the time.

While both menhaden oil and herring oil currently are commonly used ingredients, other oils are already being included in feeds for some aquatic species (Fig. 4). Increased attention has been given to studying the nutritional value of other, more sustainable, animal lipid

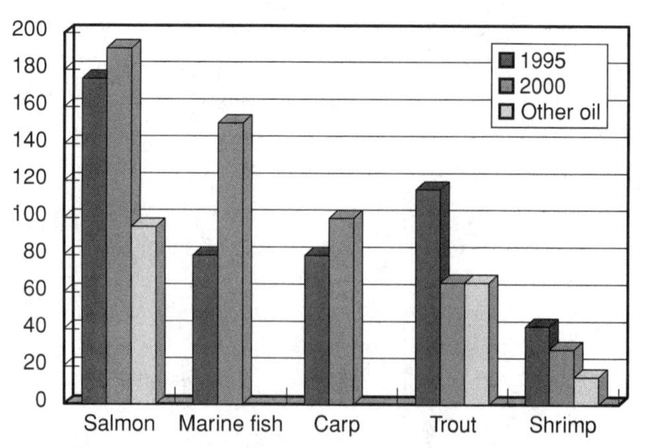

Figure 4. Use of fish oil in fish feeds (in 1000 mt).

Table 3. Studies on Lipid Ingredients in Fish Feeds

Species	Author(s)	Reference	Lipids Tested	Time Fed (wks.)
Arctic charr	Olsen et al., 1991	(78)	Coconut oil plus linoleic or linolenic	12
Atlantic salmon	Bell et al., 1991	(13)	Sunflower oil	16
	Hardy et al., 1987	(59)	Menhaden oil, soybean oil, tallow	23
	Heras et al., 1994	(60)	Dogfish silage, herring silage	9
	Koshio et al., 1994	(85)	Canola oil, herring oil	8
	Parrish et al., 1991	(62)	Fish silage, wet salmon feed, dry feeds	12, 24
	Polvi et al., 1991	(69)	Canola oil, herring oil	12
	Polvi and Ackman, 1992	(70)	Canola oil, herring oil	29
	Poston, 1990b	(81)	Choline, soy lecithin	12–16
	Thomassen and Røsjø, 1989	(73)	Soybean oil, rapeseed oil (low- and high-erucic acid), capelin oil	16, 20
	Waagbo et al., 1993	(83)	Soybean oil, capelin oil, sardine oil	16, 20
Ayu	Nematipour et al., 1989	(151)	Med. chain triglycerides	6
Brook charr	Guillou et al., 1995	(66)	Canola oil, soy oil	17
Brown trout	Arzel et al., 1994	(106)	Cod liver oil, corn oil	14
Catfish	Conrad et al., 1995	(146)	Chicken eggs, dried	12–16
	Mukhopadhyay and Mishra, 1998	(68)	Cod liver oil, sunflower oil, hydrogenated veg oil	6
	Sugiura and Lovell, 1996	(87)	High-oleic corn oil	10
Chinook salmon	Dosanjh et al., 1988	(77)	Canola oil, herring oil, pork lard	8
	Mugrditchian et al., 1981	(61)	Beef suet, linseed oil, salmon oil	16
Coho salmon	Dosanjh et al., 1984	(76)	Canola oil, herring oil, pork lard	12
	Hardy et al., 1989	(38)	Beef tallow, herring oil, soybean oil	8, 20
	Skonberg et al., 1993, 1994	(71,72)	Herring oil, high-oleic sunflower oil	6–8
	Yu and Sinnhuber, 1981	(65)	Beef tallow, salmon oil	14
Goldfish	Lochmann and Brown, 1997	(67)	Cod liver oil, soybean oil, soybean lecithin	6
Hybrid striped bass	Fowler et al., 1994	(82)	EPA and DHA supplemented	24
Rainbow trout	Boggio et al., 1985	(56)	Fish oil, pork lard	16
	Cowey et al., 1979	(84)	Hide fleshings with saturated fat	12
	Greene and Selivonchick, 1990	(58)	Beef tallow, chicken fat, linseed oil, pork lard, salmon oil, soybean oil	20
	Poston, 1990a	(80)	Choline, soy, lecithin	16, 20
	Reinitz and Yu, 1981	(63)	Beef fat, fish oil, pork lard, soy oil	
	Skonberg et al., 1993, 1994	(71,72)	Herring oil, high-oleic sunflower oil	6–8
Red drum	Craig and Gatlin, 1995	(57)	Beef tallow, coconut oil, corn oil, menhaden oil, tricaprylin	6
	Craig and Gatlin, 1997	(79)	Lecithin supplemental choline	6
Sturgeon	Xu et al., 1996	(64)	Canola oil, cod liver oil, corn oil, pork lard, linseed oil, safflower oil, soybean oil	9

resources and processing by-products, such as rendered fat from hogs (lard), from chickens (yellow grease), and from cattle and sheep (tallow) and the oil that comes with fish silages (38,56–65). Sustainable vegetable sources have also been studied, such as soybean oil, canola oil, soybean lecithin, corn oil, safflower oil, and linseed oil (13,38,59,61,63,64,66–73) (Table 3). Whether each of these oils, or a combination thereof, constitutes a suitable replacement for fish oils depends on whether the feed and lipid ingredients meet the requirements of the fish for essential fatty acids, on the oxidative stability of the lipids, on the extent of breakdown before and after incorporation into feed, on the cost of the ingredient, and on the effect, if any, of the lipid source on the fatty acid composition of the fillet and on the lipid deposition pattern in the whole fish. Fish health concerns, such as susceptibility to disease, also need to be considered.

Because marine oils are rich sources of the essential fatty acids (EFA) (Fig. 5) and almost all plant oils are not, some marine oil has to be blended with other lipid ingredients and added to the feeds. The fatty acid profiles and the cholesterol and phytosterol concentrations in lipids from plant sources are shown in Table 4, those from animal sources in Table 5. Both animal and plant lipid sources can vary in fatty acid profiles and in the extent of fatty acid hydrolysis and oxidation. Animal sources have effectors such as diet, species, and age; plant sources have variables such as variety, growing location, and harvesting time. Both animal and plant lipid sources can be affected by initial product quality,

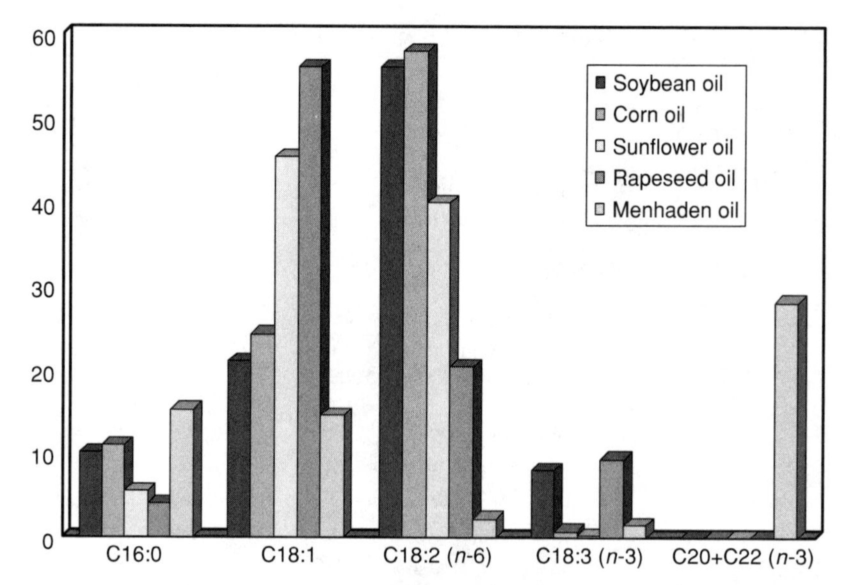

Figure 5. Differences in percentages of selected fatty acids between soybean, corn, sunflower, rapeseed, and menhaden oils.

Table 4. Concentrations of Selected Fatty Acids, Cholesterol, and Phytosterols in Lipid Ingredients from Plant Sources[a]

Nutrient	Canola	Coconut	Corn	Soybean	Soybean Lecithin	Sunflower[b]	Sunflower[c]
Saturated	7.100	86.500	12.700	14.400	15.005	10.100	9.748
12:0	0.000	44.600	0.000	0.100	0.000	0.000	NR
14:0	0.000	16.800	0.000	0.100	0.101	0.000	NR
16:0	4.000	8.200	10.900	10.300	11.984	5.400	3.682
18:0	1.800	2.800	1.800	3.800	2.920	3.500	4.320
20:0	0.700	NR[d]	NR	NR	NR	NR	NR
22:0	0.400	NR	NR	NR	NR	NR	1.000
24:0	0.200	NR	NR	NR	NR	NR	0.800
Monounsaturated	58.900	5.800	24.200	23.300	10.977	45.400	83.594
16:1	0.200	0.000	0.000	0.200	0.403	0.200	NR
18:1	56.100	5.800	24.200	22.800	10.574	45.300	82.630
20:1	1.700	0.000	0.000	0.200	0.000	0.000	0.964
22:1	0.600	0.000	0.000	0.000	0.000	0.000	NR
Polyunsaturated	29.600	1.800	58.700	57.900	45.318	40.100	3.798
18:2	20.300	1.800	58.000	51.000	40.182	39.800	3.606
18:3	9.300	0.000	0.700	6.800	5.136	0.200	0.192
18:4	0.000	0.000	0.000	0.000	0.000	0.000	NR
20:4	0.000	0.000	0.000	0.000	0.000	0.000	NR
20:5	0.000	0.000	0.000	0.000	0.000	0.000	NR
22:5	0.000	0.000	0.000	0.000	0.000	0.000	NR
22:6	0.000	0.000	0.000	0.000	0.000	0.000	NR
Cholesterol (mg)	0.000	0.000	0.000	0.000	0.000	0.000	0.000
Phytosterols (mg)	NR	86.000	968.000	250.000	NR	100.000	NR

[a]In g/100 g edible portion except where noted. Values were from (147).
[b]Linoleic <60%.
[c]Oleic 70% and over.
[d]Not reported.

by conditions of storage before lipid extraction, and by level of antioxidant present (naturally occurring or added). Also, during feed manufacture, the formulation of the feed, time/temperature treatments, and post-processing storage can significantly affect the fatty acid profile. For a discussion of the main contributors to lipid oxidation and of the effects on fish of feeding oxidized lipid, the reader is referred to the entry by Hardy and Roley on "Lipid Oxidation and Antioxidants" in this book.

The fatty acid profile of fish fillets largely reflects the dietary lipid composition, and that, in turn, is influenced by the fatty acid compositions of the dietary lipid sources. Thus, diet can potentially affect fillet storage quality, with fillets having high levels of highly unsaturated fatty acids being more susceptible to oxidation than those containing increased proportions of monounsaturated fatty acids. The fatty acid profile in the fillet can also affect the sensory properties of the fillet, in both the raw and the cooked

Table 5. Concentrations of Selected Fatty Acids and Cholesterol in Lipid Ingredients from Animal Sources[a]

Nutrient	Beef Tallow	Chicken Fat[b]	Pork Fat[b]	Cod Liver Oil	Herring Oil	Menhaden Oil	Salmon Oil	Sardine
Saturated	49.800	20.250	23.520	22.608	21.290	30.427	19.872	29.892
12:0	0.900	0.040	0.070	NR[c]	0.157	NR	NR	0.103
14:0	3.700	0.600	0.840	3.568	7.186	7.958	3.280	6.525
16:0	24.900	14.650	14.460	10.630	11.704	15.146	9.840	16.646
18:0	18.900	4.080	7.870	2.799	0.818	3.775	4.245	3.887
Monounsaturated	41.800	30.300	29.940	46.711	56.564	26.694	29.037	33.841
16:1	4.200	3.860	1.820	8.309	9.642	10.482	4.823	7.514
18:1	36.000	25.290	27.600	20.653	11.955	14.527	16.978	14.752
20:1	0.300	0.730	0.520	10.422	13.625	1.332	3.859	5.986
22:1	0.000	0.000	NR	7.328	20.613	0.352	3.376	5.589
Polyunsaturated	4.000	14.200	7.210	22.541	15.604	34.197	40.324	31.867
18:2	3.100	13.260	6.110	0.935	1.149	2.154	1.543	2.014
18:3	0.600	0.700	0.590	0.935	0.763	1.490	1.061	1.327
18:4	0.000	0.000	NR	0.935	2.305	2.739	2.798	3.025
20:4	0.000	0.040	0.170	0.935	0.289	1.169	0.675	1.756
20:5	0.000	0.000	NR	6.898	6.273	13.168	13.023	10.137
22:5	0.000	0.000	NR	0.935	0.619	4.915	2.991	1.973
22:6	0.000	0.000	NR	10.968	4.206	8.562	18.232	10.656
Cholesterol (mg)	109.000	58.000	93.000	570.000	766.000	521.000	485.000	710.000

[a]In g/100 g edible portion except where noted. Values were from (147).
[b]Raw, separable fat.
[c]Not reported.

form, and thus can affect consumer acceptability of the food fish product.

Numerous researchers have incorporated alternate lipid sources into fish feeds to determine a variety of outcomes most importantly the following: how well the feed sustained body weight gain; the effects on the fatty acid composition of the carcass, the muscle, the viscera, visceral fat, eggs at spawning, the liver, the heart, the brain, and the retina and on the fatty acid composition of triacylglycerols versus phospholipids; the effects on fillet quality in terms of sensory attributes; the degree of lipid oxidation during refrigerated and/or frozen storage; and the effects of dietary antioxidants. There has been particular interest in identifying lipid ingredients from plant-based sources that are cheaper than fish oils or that may help to extend the shelf life of the food fish product. Because, however, of the reports of lower levels of n-3 HUFAs in cultured fish compared to those in wild fish (74), there has also been interest in discovering the levels of marine oils in fish feeds that can help to enhance the n-3 fatty acid content of the fish fillets (75).

Effects on Growth and Body Composition

The most consistent findings in all of the reports in the literature are (a) that, in general, the various substitution levels of plant or animal-based lipid for fish oil were successful in supporting normal weight gain as long as the dietary levels of essential fatty acids were maintained and (b) that the fatty acid composition of the muscle and of the whole body often nearly reflected the composition of the diet, but usually within limits. Reports on salmonids and on other cold water fishes, such as sturgeon and ayu, have focused on such plant sources as sunflower oil (regular and high-oleic; Table 6), soybean oil, canola oil, soy lecithin, low- and high-erucic acid, rapeseed oil, linseed oil, and coconut oil (supplemented with specific fatty acids) and on such animal sources as chicken fat, beef tallow, fish silage, beef and swine fat, hide fleshings, and capelin oil (Table 3). The control diets typically have had menhaden oil, herring oil, or sardine oil for comparison. In warmwater fishes, such as catfish, red drum, hybrid striped bass, goldfish, and mackerel, plant lipid sources such as coconut oil, corn oil, canola oil, soybean lecithin, soybean oil, sunflower oil, high-oleic corn oil, and linseed oil and animal sources such as beef tallow, cod liver oil, and lard have been tested (Table 3). The majority of these studies determined the fatty acid composition of the whole body or of muscle after feeding of the different lipid sources for anywhere from 6 weeks to 10 months.

Table 6. Percentage of Oleic Acid (C18:1) in Regular Versus in High-Oleic Varieties of Selected Oils

Seed	Regular Seed	High-Oleic Seed
Corn	33	65–80
Peanut	59	75–80
Rapeseed	56	85–90
Safflower	13	75–80
Sunflower	24	80–90

In both coldwater and warmwater species, there have been reports that, although the composition of the flesh usually reflected the composition of the diet, the triacylglycerol fraction was more responsive than the phospholipid fraction to the lipid composition of the diet. Increases in the n-3 HUFAS tended to be in the phospholipid fraction, and the accumulations of n-3 HUFAS and saturated lipids were limited.

Mugrditchian et al. (1981) (61) fed juvenile chinook salmon feeds containing salmon oil, linseed oil, and beef suet in diets where EFA needs were met, and they reported no significant differences in weight after 16 weeks. The fish maintained a constant level of saturated fatty acids regardless of the amount in the diet. The fatty acid composition of nonpolar lipids generally reflected the composition of the diets, whereas the polar lipids selected for the more unsaturated fatty acids of the n-3 series. Although Mugrditchian et al. (1981) (61) reported that dietary 18:3n-3 was not deposited in body lipids as such, but instead was apparently converted to 22:6n-3, significant activity of this pathway could not be confirmed by Polvi et al. (1992) (70) in Atlantic salmon fed diets including canola oil.

In a study with sturgeon, Xu et al. (1996) (64) fed diets containing 15% of either canola oil, corn oil, cod liver oil, lard, linseed oil, soybean oil, or safflower oil and measured the phospholipids and triacylglycerols of muscle, liver, and brain. Like other researchers, they found that tissue triacylglycerol fatty acid composition ranged widely, in step with the dietary lipid composition, while phospholipid changes were more conservative. In particular, the brain phospholipid fatty acid composition was less responsive (more nearly conserved) than that in muscle and liver. Considerable amounts of n-6 and n-3 long chain PUFAS were found in the triacylglycerol and phospholipid fractions in fish fed all diets, showing that white sturgeon can desaturate and elongate linoleic and linolenic acids. The highest EPA and DHA levels in muscle triacylglycerol were found in fish fed the diet with fish oil rather than in those fed the diet with linseed oil. Therefore, it was concluded that the best dietary enhancement is with preformed EPA and DHA.

Hardy et al. (1989) (38) fed diets to coho salmon in which approximately 40% of the dietary lipid source was either herring oil, soybean oil, or beef tallow (or combinations thereof). Fatty acid profiles of the fish muscle and developing eggs reflected dietary fatty acid profiles after two months of feeding and at spawning for monoenoic, dienoic and n-3 fatty acids. Saturated fatty acid profiles of the muscle and the eggs did not reflect dietary levels and were similar among groups. Again, among dietary groups, the ranges in the fatty acid categories were larger in the nonpolar lipid fraction than in the polar lipid fraction. They observed no differences in fecundity, egg viability, or egg size among the dietary treatment groups.

When Greene and Selivonchick (1990) (58) fed diets to rainbow trout containing either salmon oil, soybean oil, linseed oil, chicken fat, pork lard, or beef tallow, they found that the dietary treatments supported similar growth rates and feed conversions during a 20-week trial. Although some of the diets provided 18:3n-3 that could be desaturated and elongated to EPA and DHA, there appeared to be a physiologically optimum level of long-chain fatty acids maintained in the muscle.

Dosanjh et al. (1984) (76) reported that coho salmon fed diets with either canola oil, pork lard, or herring oil had body lipids that generally reflected that of the diet, except that the levels of saturated and unsaturated fatty acids were less variable in the body lipids than in the dietary lipids. Dosanjh et al. (1988) (77) confirmed the findings of this study, using chinook salmon, and observed that, although body lipid generally reflected that of the diet, the percentage of DHA was higher in the body than in the diet. They also reported that the range for percentages of saturated fatty acids in body lipids was narrower (15.7–22.8%) than for the dietary lipids (12.5–29.5%).

Craig and Gatlin (1995) (57) reported that red drum (*Sciaenops ocellatus*) could not efficiently use tricaprylin (caprylic acid, 8:0), a medium-chain triacylglycerol, as an energy source and that consequently there were significantly lower weight gains, lower n-3, and greater n-6 fatty acid levels in the nonpolar lipid fraction of muscle tissue than in fish fed no tricaprylin. However, the dietary inclusion of coconut oil and beef tallow resulted in normal weight gain as long as the requirements for essential fatty acids were provided.

Alternate lipid ingredients have also been used to help determine the essential fatty acid requirements of particular species. Mukhopadhyay et al. (1998) (68) suggested the essential nature of both n-3 and n-6 in fingerling and fry catfish using cod liver oil, sunflower oil, a mixture of these, and hydrogenated vegetable oil. There were significant differences in weight gain, feed efficiency, and tissue fatty acid profiles, and the deposition of fatty acids in the carcass very closely followed the fatty acid composition of the diet. They concluded that a combination of n-3 and n-6 fatty acids resulted in best growth, best feed efficiency, and increased deposition of HUFAs.

After feeding Arctic charr diets with 1% coconut oil plus different PUFAs, Olsen et al. (1991) (78) suggested that charr require n-3 fatty acids, and they noted that these were used in preference to n-6 fatty acids for desaturation, elongation, and incorporation into phospholipid. They also reported that muscle phospholipids were less influenced by diet than those in the liver, where phopholipid PUFAS were significantly influenced by diet composition.

Effects of Soy Lecithin

The inclusion of soy lecithin in the feed has been reported to improve weight gain and feed efficiency of some species of warmwater and coldwater fishes. Lochmann and Brown (1997) (67) reported that, in goldfish, weight gain and feed-efficiency ratio were significantly higher with soybean lecithin than without. They hypothesized that lecithin provided myoinositol or phophatidylcholine to support rapid membrane proliferation, increased absorption of dietary lipid, and enhanced lipid transport.

Craig and Gatlin (1997) (79), using red drum, also reported that lecithin increased weight gain and feed efficiency and also increased liver lipid concentration. Choline alone increased muscle lipid as well as intraperitoneal fat content and decreased liver lipid concentrations.

Poston (1990a) (80) reported that rainbow trout needed at least 4% soy lecithin, either with or without choline, for maximum growth. With Atlantic salmon, Poston (1990b) (81) examined the effect of the initial size of the fish on response to food-grade soy lecithin and to feed-grade lecithin (corn–soy mix). He reported that dietary lecithin enhanced survival when the fish were small (0.18 and 1.0 g) but not large (7.5 g).

Effects on Food Fish Product Quality

In some studies, the inclusion of alternate lipid sources has been reported to result in significant differences in sensory analysis tests of the fillets. The usual difference was that feeding a non-fish-oil ingredient resulted in a less fishy aroma or flavor than was found in fillets from fish fed fish oil.

Fowler et al. (1994) (82) reported that hybrid striped bass fed diets supplemented with EPA and DHA and containing at least 0.5% EPA and 0.24% DHA in the fillets had a more fishy flavor in the cooked fillets compared to that in the fillets from unsupplemented controls. Skonberg et al. (1993) (71) reported that substituting high-oleic sunflower oil for herring oil in feeds for rainbow trout and coho salmon for 6–8 weeks resulted in significant differences in triangle tests, with the sunflower-oil fillets having less fishy aroma than the herring-oil fillets. Using Atlantic salmon, Thomassen and Røsjø (1989) (73) replaced up to 68% of the capelin oil in feeds with either soybean oil, low-erucic acid rapeseed oil (LERO), or high-erucic acid rapeseed oil (HERO). Although the heart and muscle lipids and the n-3 to n-6 ratio were affected, there was no difference in growth or mortality. Fillets from fish given the soybean treatment were reported to have less salmon taste than those from fish receiving the LERO treatment. Also, the fillets from fish given the rapeseed oil treatments had less salmon odor than those from the capelin-oil control fish, and, by sensory and instrumental analyses, the HERO fillets were less red than other groups.

Waagbø et al. (1993) (83) assessed the influence of varying dietary n-3 PUFA content (low, medium, or high PUFA content was achieved by blending soybean oil, capelin oil, and sardine oil), as well as that of two supplemental levels of vitamin E (0 and 300 mg alpha-tocopherol/kg), on the flesh quality of Atlantic salmon. They found that the total lipid fatty acid composition and the vitamin E content in the fillets reflected the diet composition; however, the vitamin E content of fillets did not influence the fatty acid composition. Sensory analysis was performed on cooked fillets from fresh, 4-day frozen, 5-week frozen (−18 °C), and traditionally smoked never-frozen fish. Rancid flavor, fattiness, juiciness, and taste intensity were significantly higher in fish raised on the diet with high n-3 PUFA content and on the low vitamin E diets, than in the controls.

There have also been studies indicating no significant differences in sensory qualities when alternate lipid ingredients have been given compared to fish oil controls: Boggio et al. (1985) (56), for example, when rainbow trout were fed diets with swine fat; Cowey et al. (1979) (84), when rainbow trout were fed diets with rendered hide fleshings that contained saturated fat; Guillou et al. (1995) (66), when brook trout were fed diets with soya or canola oils; Hardy et al. (1987) (59), when Atlantic salmon were fed diets containing soybean oil or tallow; Koshio et al. (1994) (85), when Atlantic salmon received diets either with canola oil or with herring oil or with these same oils submitted to oxidation treatments; Morris et al. (1995) (86), when catfish were fed diets supplemented with menhaden oil; and Sugiura and Lovell (1996) (87), when catfish were fed diets with high-oleic corn oil. The possible reasons for the discrepant sensory-analysis results are many, viz., differences in the lipid concentrations and/or protein/lipid ratios in the feeds, dissimilar diet formulations and lipid quality, and differences either in the types of sensory tests performed or the methods used to prepare the samples.

Effects of Dietary Antioxidants on Storage Quality of Cultured Fish

Gatlin et al. (1992) (88) examined the storage quality and body composition of catfish fed natural and synthetic antioxidants (AO) in the diet. They used a semipurified casein/gelatin diet, with either 60 or 240 mg dl-alpha-tocopherol and either with no synthetic AO, with ethoxyquin (150 mg/kg), with BHT (10 mg/kg), with BHA (10 mg/kg), or with Endox (125 mg/kg). None of the AO significantly affected growth, feed efficiency, proximate composition of whole body and fillets, or tissue levels of alpha-tocopherol. In forced-oxidation tests, however, fillets from fish ingesting the higher level of alpha-tocopherol had lower thiobarbituric acid reacting substances (one index of lipid oxidation), but the synthetic AO's did not reduce oxidation.

Ackman et al. (1997) (89) reported that gamma-tocopherol may potentially replace alpha-tocopherol to reduce autooxidation of fillets during storage of Atlantic salmon. Atlantic salmon have a requirement for dietary EPA and DHA for subcellular membrane lipids; the fish also deposits them in the muscle depot fats. Ackman et al. (1997) (89) showed that if both alpha- and gamma-tocopherol were fed, alpha-tocopherol was found in phospholipid-rich organ tissues and gamma-tocopherol was found in the lipid stores of the muscle. There may thus be a potential to feed gamma-tocopherol to reduce autooxidation of fillets during storage, so one approach would be to use natural mixtures of alpha- and gamma-tocopherol as a low-cost alternative to 100% alpha-tocopherol. They hypothesized that highly oxidized herring or canola oil was not toxic to salmonids as long as alpha-tocopherol was present, although feed intake may be reduced. Others have shown that flavors of farmed Atlantic salmon were not altered by different tocopherol types, by dietary fats and oil, or by the oxidation status of those fats/oils.

Concluding Remarks

Considerable research is still required to establish optimal dietary concentrations of n-3 and n-6 fatty acids in cultured finfish species and, in particular, the desirable levels and ratios of EPA, DHA, and AA for their growth, health, and reproductive viability. Research should concurrently involve assessments of how much the dietary treatments influence the types and levels of eicosanoid compounds elaborated from AA, EPA, DHA, and 20:3n-6 and whether these in turn adversely affect fish physiology, biochemistry, endocrine status, and histopathology. This information is required to facilitate establishment of acceptable dietary levels of various blends of alternate animal and plant lipid sources with marine lipids at different stages of the life history.

The global supply of marine lipids will be insufficient to meet the demands for aquafeeds at some point in the next 15 to 20 years, unless suitable alternate lipid sources are identified and/or developed. Additional studies are also required to assess whether diets containing various mixtures of alternate lipid sources with marine lipids negatively affect the sensory attributes and chemical composition of the flesh from market-size fish.

ACKNOWLEDGMENTS

The authors wish to thank Kenneth Liu and Russell Herwig for their assistance in editing the manuscript.

BIBLIOGRAPHY

1. J. Tinoco, *Prog. Lipid Res.* **21**, 1–45 (1982).
2. R. Ballestrazzi and A. Mion, *Rivista Italiana Acquacultura* **28**, 155–173 (1993).
3. R.J. Henderson and D.R. Tocher, *Prog. Lipid Res.* **26**, 281–347 (1987).
4. S. Polvi, M.Sc. thesis, Technical University of Nova Scotia, Halifax, NS, 1989.
5. J. Sargent, R.J. Henderson, and D.R. Tocher, *Fish Nutrition*, 2nd edition, Academic Press, London, 1989, pp. 153–218.
6. A.G.J. Tacon, *The Essential Nutrients. GCP/RLA/075/ITA Field document*, 2nd edition, Food and Agriculture Organization of the United Nations, Rome, 1987.
7. H. Ashton, M.Sc. thesis, University of British Columbia, Vancouver, BC, 1991.
8. J.R. Sargent and K.J. Whittle, *Analysis of Marine Ecosystems*, Academic Press, London, 1981, pp. 491–533.
9. P.J. Yeagle, *FASEB J.* **3**, 1833–1842 (1989).
10. M.V. Bell, R.J. Henderson, and J.R. Sargent, *Comp. Biochem. Physiol.* **83B**, 711–719 (1986).
11. M.T. Clandinin, S. Cheema, C.J. Field, M.L. Garg, J. Venkatraman, and T.R. Clandinin, *FASEB J.* **5**, 2761–2769 (1991).
12. E. Fodor, R.H. Jones, C. Buda, K. Kitajka, I. Dey, and T. Farkas, *Lipids* **30**, 1119–1126 (1995).
13. J.G. Bell, A.H. McVicar, M.T. Park, and J.R. Sargent, *J. Nutr.* **121**, 1163–1172 (1991).
14. J.G. Bell, J.R. Dick, A.H. McVicar, J.R. Sargent, and K.D. Thompson, *Prostaglandins Leukotrienes and Essential Fatty Acids*, Longman Group, UK Ltd., 1993, pp. 665–673.
15. M.V. Bell, R.S. Batty, J.R. Dick, K. Fretwell, J.C. Navarro, and J.R. Sargent, *Lipids* **30**, 443–449 (1995).
16. G. Mourente and D.R. Tocher, *Fish Physiol. and Biochem.* **18**, 149–165 (1998).
17. A. Estevez, M. Ishikawa, and A. Kanazawa, *Aquaculture Research* **28**, 279–289 (1997).
18. J.G. Bell, B.M. Farndale, J.R. Dick, and J.R. Sargent, *Lipids* **31**, 1163–1171 (1996a).
19. R.J. Henderson, J.G. Bell, and M.T. Park, *Biochimica et Biophysica Acta* **1299**, 289–298 (1996).
20. W.M. Koven, R.J. Henderson, and J.R. Sargent, *Aquaculture* **151**, 155–171 (1997).
21. P. Spyridakis, R. Metailler, J. Gabaudan, and A. Riaza, *Aquaculture* **77**, 61–70 (1989).
22. S.-M. Lee, *J. Korean Fish. Soc.* **30**, 62–71 (1997).
23. R.E. Olsen, R.J. Henderson, and E. Ringo, *Aquaculture Nutrition* **4**, 13–21 (1998).
24. C.Y. Cho, S.J. Slinger, and H.S. Bayley, *Comp. Biochem. Physiol.* **73B**, 25–41 (1982).
25. D.A. Higgs, J.S. Macdonald, C.D. Levings, and B.S. Dosanjh, *Physiological Ecology of Pacific Salmon*, UBC Press, Vancouver, BC, 1995, pp. 159–315.
26. X. Pelletier and C. Leray, *Lipids* **22**, 1053–1056 (1987).
27. C. Leger, L. Fremont, and M. Boudon, *Comp. Biochem. Physiol.* **69B**, 99–105 (1981).
28. M.A. Sheridan, *Comp. Biochem. Physiol.* **90B**, 679–690 (1988).
29. C. Leger, *Nutrition and Feeding in Fish*, Academic Press, London, 1985.
30. O. Lie, A. Sandvin, and R. Waagbo, *Fish Physiol. Biochem.* **12**, 249–260 (1993).
31. S. Ando and M. Matsuzaki, *Fish Physiol. Biochem.* **15**, 469–479 (1996).
32. M.C. McKay, R.F. Lee, and M.A.K. Smith, *Physiol. Zool.* **58**, 693–704 (1985).
33. S. Ando and Y. Mori, *Nippon Suisan Gakkaishi* **59**, 1565–1571 (1993).
34. R.P. Wilson, *Handbook of Nutrient Requirements of Finfish*, CRC Press Inc., Boca Raton, Florida, 1991, pp. 35–53.
35. S. Satoh, *Handbook of Nutrient Requirements of Finfish*, CRC Press Inc., Boca Raton, FL, 1991, pp. 55–67.
36. P. Luquet, *Handbook of Nutrient Requirements of Finfish*, CRC Press Inc., Boca Raton, FL, 1991, pp. 169–179.
37. J.R. Sargent, *Fish Oil: Technology, Nutrition and Marketing*, P.J. Barnes and Associates, 1995a, pp. 67–94.
38. R.W. Hardy, T. Masumoto, W.T. Fargrieve, and R.R. Stickney, *The Current Status of Fish Nutrition in Aquaculture. The Proc. Third Int. Symp. Feeding and Nutr. in Fish*, Laboratory of Fish Nutrition Department of Aquatic Biosciences Tokyo University of Fisheries, Tokyo, Japan, 1989, pp. 347–355.
39. M. Harel, A. Tandler, G.W. Kissil, and S.W. Applebaum, *British J. Nutr.* **72**, 45–58 (1994).
40. H. Fernandez-Palacios, M.S. Izquierdo, L. Robaina, A. Valencia, M. Salhi, and J.M. Vergara, *Aquaculture* **132**, 325–337 (1995).
41. M.S. Izquierdo, *Aquaculture Nutrition* **2**, 183–191 (1996).
42. T. Watanabe, *J. World Aquaculture Soc.* **24**, 152–161 (1993).
43. J.R. Sargent, L.A. McEvoy, and J.G. Bell, *Aquaculture* **155**, 117–127 (1997).
44. S. Brooks, C.R. Tyler, and J.P. Sumpter, *Reviews in Fish Biology and Fisheries* **7**, 387–416 (1997).
45. J.R. Sargent, *Broodstock Management and Egg and Larval Quality*, Oxford Blackwell Science Ltd., 1995b, pp. 353–372.
46. M. Bruce, F. Oyen, G. Bell, J.F. Asturiano, B. Farndale, M. Carillo, S. Zanuy, J. Ramos, and N. Bromage, *Aquaculture* **177**, 85–97 (1999).
47. T. Watanabe, T. Takeuchi, M. Saito, and K. Nishimura, *Bull. Jpn. Soc. Sci. Fish.* **50**, 1207–1215 (1984a).
48. M.V. Bell, J.R. Dick, M. Trush, and J.C. Navarro, *Aquaculture* **144**, 189–199 (1996b).
49. L. Fremont, C. Leger, B. Petridou, and M.T. Gozzelino, *Lipids* **19**, 522–528 (1984).

50. J. Cerda, S. Zanuy, and M. Carrillo, *Aquaculture International* **5**, 473–477 (1997).
51. J.G. Bell, B.M. Farndale, M.P. Bruce, J.M. Navas, and M. Carillo, *Aquaculture* **149**, 107–119 (1997).
52. R. Waagbo, *Aquaculture and Fisheries Management* **25**, 175–197 (1994).
53. D.M. Fracalossi and R.T. Lovell, *Aquaculture* **119**, 287–298 (1994).
54. M.H. Li, D.J. Wise, M.R. Johnson, and E.H. Robinson, *Aquaculture* **128**, 335–344 (1994).
55. K.D. Thompson, M.F. Tatner, and R.J. Henderson, *Aquaculture Nutrition* **2**, 21–31 (1996).
56. S.M. Boggio, R.W. Hardy, J.K. Babbitt, and E.L. Brannon, *Aquaculture* **51**, 13–24 (1985).
57. S.R. Craig and D.M. Gatlin, *J. Nutr.* **125**, 3041–3048 (1995).
58. D.H.S. Greene and D.P. Selivonchick, *Aquaculture* **89**, 165–182 (1990).
59. R.W. Hardy, T.M. Scott, and L.W. Harrell, *Aquaculture* **65**, 267–277 (1987).
60. H. Heras, C.A. McLeod, and R.G. Ackman, *Aquaculture* **125**, 93–106 (1994).
61. D.S. Mugrditchian, R.W. Hardy, and W.T. Iwaoka, *Aquaculture* **25**, 161–172 (1981).
62. C.C. Parrish, H. Li, W.M. Indrasena, and R.G. Ackman, *Bull. Aquacult. Assoc. Canada*, (March):75–84 (1991).
63. G.L. Reinitz and T.C. Yu, *Aquaculture* **22**, 359–366 (1981).
64. R. Xu, S.S.O. Hung, and J.B. German, *Aquacult. Nutr.* **2**, 101–109 (1996).
65. T.C. Yu and R.O. Sinnhuber, *Can. J. Fish Aquat. Sci.* **38**, 367–370 (1981).
66. A. Guillou, P. Soucy, M. Khalil, and L. Adambounou, *Aquaculture* **136**, 351–362 (1995).
67. R. Lochmann and R. Brown, *JAOCS* **74**, 149–152 (1997).
68. P.K. Mukhopadhyay and S. Mishra, *J. Appl. Ichthyol.* **14**, 105–107 (1998).
69. S.M. Polvi, R.G. Ackman, S.P. Lall, and R.L. Saunders, *J. Food Proc. Preserv.* **15**, 167–181 (1991).
70. S.M. Polvi and R.G. Ackman, *J. Agric. Food Chem.* **40**, 1001–1007 (1992).
71. D.I. Skonberg, B.A. Rasco, and F.M. Dong, *J. Aquatic Food Prod. Tech.* **2**, 117–133 (1993).
72. D.I. Skonberg, B.A. Rasco, and F.M. Dong, *J. Nutr.* **124**, 1628–1638 (1994).
73. M.S. Thomassen and C. Rosjo, *Aquaculture* **79**, 129–135 (1989).
74. T.V. Vliet and M.B. Katan, *Am. J. Clin. Nutr.* **51**, 1–2 (1990).
75. M.R. Turner, R.H. Lumb, J.L. West, and V. Brown, *Prog. Fish-Culturist* **52**, 130–133 (1990).
76. B.S. Dosanjh, D.A. Higgs, M.D. Plotnikoff, J.R. Markert, J.R. McBride, and J.T. Buckley, *Aquaculture* **36**, 333–345 (1984).
77. B.S. Dosanjh, D.A. Higgs, M.D. Plotnikoff, J.R. Markert, and J.T. Buckley, *Aquaculture* **68**, 325–343 (1988).
78. R.E. Olsen, R.J. Henderson, and E. Ringo, *Fish Physiol. and Biochem.* **9**, 151–164 (1991).
79. S.R. Craig and D.M. Gatlin, *Aquaculture* **151**, 259–267 (1997).
80. H.A. Poston, *Prog. Fish-Culturist* **52**, 218–225 (1990a).
81. H.A. Poston, *Prog. Fish-Culturist* **52**, 226–230 (1990b).
82. K.P. Fowler, C. Karahadian, N.J. Greenberg, and R.M. Harrell, *J. Food Science* **59**, 70–75, 90 (1994).
83. R. Waagbo, K. Sandnes, O.J. Torrissen, A. Sandvin, and O. Lie, *Food Chem.* **46**, 361–366 (1993).
84. C.B. Cowey, J.W. Adron, R. Hardy, J.G.M. Smith, and M.J. Walton, *Aquaculture* **16**, 199–209 (1979).
85. S. Koshio, R.G. Ackman, and S.P. Lall, *J. Agric. Food Chem.* **42**, 1164–1169 (1994).
86. C.A. Morris, K.C. Haynes, J.T. Keeton, and D.M. Gatlin, *J. Food Sci.* **60**, 1225–1227 (1995).
87. S.H. Sugiura and R.T. Lovell, *J. World Aquacult. Soc.* **27**, 74–81 (1996).
88. D.M. Gatlin, S.C. Bai, and M.C. Erickson, *Aquaculture* **106**, 323–332 (1992).
89. R.G. Ackman, M.P.M. Parazo, and S.P. Lall, *Flavor and Lipid Chemistry of Seafoods, ACS Symposium Series 674*, American Chemical Society, Washington, DC USA, 1997, pp. 148–165.
90. T. Watanabe, *Comp. Biochem. Physiol.* **73B**, 3–15 (1982).
91. C.Y. Cho, *Food Reviews International* **6**, 333–357 (1990).
92. C.Y. Cho, *Proceedings CFIA Eastern Nutrition Conference*, The Canadian Feed Industry Association, Dartmouth/Halifax, NS, 1996, pp. 171–178.
93. C.Y. Cho and C.B. Cowey, *Handbook of Nutrient Requirements of Finfish*, CRC Press Inc., Boca Raton, FL, 1991, pp. 131–143.
94. T.C. Yu, R.O. Sinnhuber, and J.D. Hendricks, *Lipids* **14**, 572–575 (1979).
95. T. Akiyama, I. Yagisawa, and T. Nose, *Bull. Natl. Res. Inst. Aquacult.* **2**, 35–42 (1981).
96. T. Takeuchi, T. Watanabe, and T. Nose, *Bull. Jpn. Soc. Sci. Fish.* **45**, 1319–1323 (1979).
97. T. Takeuchi and T. Watanabe, *Bull. Jpn. Soc. Sci. Fish.* **48**, 1745–1752 (1982).
98. T. Watanabe, *Intensive Fish Farming*, BSP Professional Books, London, 1988, pp. 154–197.
99. B. Grisdale-Helland and S.J. Helland, *Aquaculture* **152**, 167–180 (1997).
100. F. Johnsen and A. Wandsvik, *Proc. First Int. Symp. Nutritional Strategies in Management of Aquaculture Waste*, University of Guelph, Guelph, ON, 1991, pp. 51–63.
101. M. Hillestad, F. Johnsen, E. Austreng, and T. Asgard, *Aquaculture Nutrition* **4**, 89–97 (1998).
102. J.L. Tabachek, *J. Fish Biol.* **29**, 139–151 (1986).
103. X. Yang and T.A. Dick, *Aquaculture* **116**, 57–70 (1993).
104. X. Yang and T.A. Dick, *J. Nutr.* **124**, 1133–1145 (1994).
105. X. Yang, J.L. Tabachek, and T.A. Dick, *Fish Physiol. and Biochem.* **12**, 409–420 (1994).
106. J. Arzel, F.X.M. Lopez, R. Metailler, G. Stephan, M. Viau, G. Gandemer, and J. Guillaume, *Aquaculture* **123**, 361–375 (1994).
107. S. Satoh, W.E. Poe, and R.P. Wilson, *J. Nutr.* **119**, 23–28 (1989).
108. L. Ding, *Handbook of Nutrient Requirements of Finfish*, CRC Press, Boca Raton, FL, 1991, pp. 89–96.
109. T. Takeuchi, K. Watanabe, W.-Y. Yong, and T. Watanabe, *Nippon Suisan Gakkaishi* **57**, 467–473 (1991).
110. Erfanullah and A.K. Jafri, *Aquaculture* **161**, 159–168 (1998).
111. W. Jantrarotai, P. Sitasit, and S. Rajchapakdee, *Aquaculture* **127**, 61–68 (1994).
112. D. Seenappa and K.V. Devaraj, *Aquaculture* **129**, 243–249 (1995).

113. T. Takeuchi, S. Satoh, and T. Watanabe, *Bull. Jpn. Soc. Sci. Fish.* **49**, 1127–1134 (1983).
114. A. Kanazawa, T. S-I, M. Sakamoto, and M.A. Awal, *Bull. Jpn. Soc. Sci. Fish.* **46**, 1353–1356 (1980).
115. A. Kanazawa, *Handbook of Nutrient Requirements of Finfish*, CRC Press Inc., Boca Raton, FL, 1991, pp. 23–29.
116. S. Arai, *Handbook of Nutrient Requirements of Finfish*, CRC Press Inc., Boca Raton, FL, 1991, pp. 69–75.
117. S. Thongrod, T. Takeuchi, S. Satoh, and T. Watanabe, *Nippon Suisan Gakkaishi* **55**, 1983–1987 (1989).
118. S.S.O. Hung, T. Storebakken, Y. Cui, L. Tian, and O. Einen, *Aquaculture Nutrition* **3**, 282–286 (1997).
119. G.R. Nematipour and D.M. Gatlin, *J. Nutr.* **123**, 744–753 (1993).
120. C.D. Webster, L.G. Tiu, J.H. Tidwell, P.V. Wyk, and R.D. Howerton, *Aquaculture* **131**, 291–301 (1995).
121. NRC and N.R. Council, *Nutrient Requirements of Fish*, National Academy Press, Washington, DC, 1993, pp. 114.
122. T. Takeuchi, Y. Shiina, and T. Watanabe, *Nippon Suisan Gakkaishi* **58**, 509–514 (1992a).
123. C. Ibeas, J. Cejas, T. Gomez, S. Jerez, and A. Lorenzo, *Aquaculture* **142**, 221–235 (1996).
124. C. Ibeas, M.S. Izquierdo, and A. Lorenzo, *Aquaculture* **127**, 177–188 (1994).
125. C. Rodriguez, J.A. Perez, P. Badia, M.S. Izquierdo, H. Fernandez-Palacios, and A.L. Hernandez, *Aquaculture* **169**, 9–23 (1998).
126. T. Takeuchi, Y. Shiina, T. Watanabe, S. Sekiya, and K. Imaizumi, *Nippon Suisan Gakkaishi* **58**, 1341–1346 (1992b).
127. J. Guillaume, M.-F. Coustans, R. Metailler, J.P.-L. Ruyet, and J. Robin, *Handbook of Nutrient Requirements of Finfish*, CRC Press Inc., Boca Raton, FL, 1991, pp. 77–82.
128. J.D. Castell, J.G. Bell, D.R. Tocher, and J.R. Sargent, *Aquaculture* **128**, 315–333 (1994).
129. M. Boonyaratpalin, *Aquaculture* **151**, 283–313 (1997).
130. I.G. Borlongan, *Fish Physiology and Biochemistry* **9**, 401–407 (1992).
131. M.R. Catacutan and R.M. Coloso, *Aquaculture* **149**, 137–144 (1997).
132. J. Wanakowat, M. Boonyaratpalin, and T. Watanabe, *Fish Nutrition in Practice, IVth International Symposium on Fish Nutrition and Feeding*, INRA, Paris, 1993, pp. 807–817.
133. L. Perez, H. Gonzalez, M. Jover, and J. Fernandez-Carmona, *Aquaculture* **156**, 183–193 (1997).
134. E. Robinson, *Handbook of Nutrient Requirements of Finfish*, CRC Press Inc., Boca Raton, FL, 1991, pp. 145–152.
135. R.L. Brinkmeyer and G.J. Holt, *Aquaculture* **161**, 253–268 (1998).
136. R.T. Lochmann and D.M. Gatlin, *Fish Physiol. and Biochem.* **12**, 221–235 (1993).
137. S.-M. Lee, J.Y. Lee, Y.J. Kang, H.-D. Yoon, and S.B. Hur, *Bull. Korean Fish. Soc.* **26**, 477–492 (1993).
138. S.-M. Lee, J.Y. Lee, and S.B. Hur, *Bull. Korean Fish. Soc.* **27**, 712–726 (1994).
139. K. Ikehara and M. Nagahara, *Bull. Jpn. Sea Reg. Fish. Res. Lab.* **29**, 103–110 (1978).
140. C. Leray, G. Nonnotte, P. Roubaud, and C. Leger, *Reprod. Nutr. Develop.* **25**, 567–581 (1985).
141. G. Corraze, L. Larroquet, G. Maisse, D. Blanc, and S. Kaushik, *Fish Nutrition in Practice. IVth International Symposium on Fish Nutrition and Feeding*, INRA, Paris, 1993, pp. 61–66.
142. T. Watanabe, T. Arakawa, C. Kitajima, and S. Fujita, *Bull. Jpn. Soc. Sci. Fish.* **50**, 495–501 (1984b).
143. T. Watanabe, A. Itoh, A. Murakami, Y. Tsukashima, C. Kitajima, and S. Fujita, *Bull. Jpn. Soc. Sci. Fish.* **50**, 1023–1028 (1984c).
144. C. Rodriguez, J.R. Cejas, M.V. Martin, P. Badia, M. Samper, and A. Lorenzo, *Fish Physiology and Biochemistry* **18**, 177–187 (1998).
145. E. Almansa, M.J. Perez, J.R. Cejas, P. Badia, J.E. Villamandos, and A. Lorenzo, *Aquaculture* **170**, 323–336 (1999).
146. K.M. Conrad, M.G. Mast, and J.H. MacNeil, *J. Aquatic, Food Prod. Tech.* **4**, 23–33 (1995).
147. U.S.D.A., U.S. Department of Agriculture, Agricultural Research Service, *Nutrient Database for Standard Reference*, Release 12, http://www.nal.usda.gov.fnic/foodcomp (1998).
148. M. Buzzi, R.J. Henderson, and J.R. Sargent, *Comp. Biochem. Physiol.* **116B**, 261–267 (1997).
149. P.C. Calder, *Adv. Enzyme Regul.* **37**, 197–237 (1997).
150. J.B. German, G.G. Bruckner, and J.E. Kinsella, *Biochimica et Biophysica Acta* **875**, 12–20 (1986).
151. G.R. Nematipour, H. Nishino, and H. Nakagawa, *Proc. Third Int. Symp. Feeding and Nutr. in Fish*, Toba, Jpn., 1989, pp. 233–244.

See also DIGESTIBILITY; ENERGY; FEED EVALUATION, CHEMICAL; FEED MANUFACTURING TECHNOLOGY; LIPID OXIDATION AND ANTIOXIDANTS.

LIVE TRANSPORT

S.K. JOHNSON
Texas Veterinary Medical Diagnostic Laboratory
College Station, Texas

OUTLINE

Live transport is a method that is essential to all of the animal industries. The great diversity of aquatic livestock originates from aquacultural and natural systems. In the realm of international traffic, the total number of aquatic livestock probably exceeds that of terrestrial livestock. The size of aquatic animals makes their live transport practical. The volume of a chicken egg, for example, is roughly equivalent to that of 7,000 trout eggs or to the volume of a shipment of water containing 6,000 crustacean nauplii. Although fishes are carried alive in different stages, most fry and fingerling stages and many adults are smaller than the smallest stage at which terrestrial vertebrates are stocked.

Hatcheries ship eggs and larvae long distances to other hatcheries and nurseries, where they will grow to a stockable size. International commerce in juveniles for the purpose of stocking production units is feasible because a multitude of individuals may travel inside packages that weigh little have a small and volume. Local and regional commerce of many cultured fishes typically involves the sale of fingerlings.

Live transport serves other purposes as well. Ornamental, bait, and forage products have market value only as live animals. The sale of certain live food species brings more revenue than does the sale of a freshly killed or refrigerated product of the same species. Indeed, some processors require that the product arrive alive at their plant.

CONVENTIONAL SEALED CONTAINERS

Boxes of sizes suitable for hand loading are useful means of transporting the eggs and young of most species and adults of smaller species. The method is also sometimes convenient for distributing small lots of moderately sized adults intended for use as broodstock or for sale in specialty markets.

The typical unit consists of a corrugated shipping carton enclosing walls of Styrofoam™ insulation and sealable plastic bags. The Styrofoam may itself be a molded box for use within shipping cartons or, if it is of a more durable design, as the sole container. More sophisticated designs are commercially available, but the conventional box that contains sealed bags remains the norm for delivering small swimming animals by air.

Bags may be built to fit the dimensions of their container. Those with square bottoms prevent folds that could entrap smaller animals from forming. Bags are one to four mils in thickness and are often doubled; the thicker ones are the more widely used. Shippers of ornamentals pack several smaller bags within a box, each containing a preestablished count of a particular species. US wholesalers of minnows for bait commonly use a cylindrical bag of 46 × 81 cm (18 × 32 inches) in size.

Bags typically contain 25% water, but packers sometimes use up to 40%. A shallower level of water that barely covers the fish can be successful in some cases. The temperature of the bag water is adjusted prior to loading. A rubber band or other binding device seals the bag. Oxygen gas overlaying the water is preferable to air, but air is sufficient in certain applications. Frozen gel packets placed inside the box, but out of contact with the water, control the temperature while a package is en route. Heat packs prevent tropical species from over chilling in packages traveling to or through cold climates.

Lower water temperatures slow many biological processes and reduce the animals' release of waste in confinement. Lower temperatures also cause more oxygen to become saturated in the water and reduce the animals' need for oxygen. Consequently, many shippers pack boxes in a way that maintains the temperature near the lowest suitable temperature for a species.

The addition of pure oxygen to sealed bags supersaturates the transport water.

Other than oxygen, additives for bag transport are mostly meant to arrest microbial reproduction and give the animals that are being shipped an osmoregulatory advantage. At levels that are not toxic to the animals, antibacterials give only a modest benefit: Their effectiveness is soon overcome by resistant bacteria that flourish in the presence of the rising levels of organic compounds. The addition of calcium chloride or sodium chloride to the water medium usually provides an osmoregulatory benefit to freshwater species. The more equal internal and external osmotic states reduce the physiological adjustments that must be made to control waste and salt exchanges. In sufficient concentrations the sodium chloride has the additional advantage of blocking the reproduction of ectoparasitic protozoans. Marine- or brackish-water species benefit from salinity adjustments when the water at the destination differs sharply from that at the source. The adjustment is usually downward and is accomplished by the addition of freshwater.

Waste from animals accumulates in sealed containers. By not feeding the animals for an appropriate period prior to transport, shippers give them time to discharge solid waste that would otherwise foul the transport medium. Pond fishes of fingerling size are normally brought to holding facilities immediately after harvest. A further period without feeding enhances the survivability of the fishes. In the meantime, other operational chores, such as grading for size, culling, sorting for sale, and determining weights or volumes are performed. The outcome of the fasting process is better for species with shorter intestines than for species with longer ones.

During transport, the bag water retains considerable quantities of carbon dioxide, whose otherwise toxic effect is greatly offset by the ample quantity of dissolved oxygen. The accumulation of carbon dioxide causes an increased concentration of hydrogen ions and, hence, a lower pH. Accordingly, freshwater aquaculturists occasionally add buffering compounds to bag water to help stabilize the pH. Supplying the bags with relatively hard groundwater, however, is usually adequate. Still, species that are particularly intolerant of low pH may especially benefit from one of the widely available buffering compounds. Of course, the normal lowering of pH by dissolved carbon dioxide abates some of the harmful effect of increasing nitrogen waste by shifting the ionic state of ammonia (unionized) to a less harmful ammonium (ionized).

Ammonia is the most harmful waste. As the primary nitrogenous waste for marine and freshwater species, its accumulation proceeds almost unchecked in bag water. Controlling the travel temperature and selecting loading rates greatly determine the amount of ammonia released. In freshwater transport, packets of ion exchange media may be helpful for binding ammoniated nitrogen, but the presence of sodium chloride, so often a packaging additive, can interfere with the binding and diminish its benefit.

The conversion of ammonium nitrogen into nitrite by bacteria is usually not a problem, due to both efforts to keep ammonia production and bacterial development in check and the insufficient time for meaningful conversion to occur. Chloride competitively inhibits nitrite uptake by fishes. Enough chloride is present in transport water that is suitable for marine species and in freshwater that contains additional salt to negate nitrite effects. Some further benefit may result from the bacterial seeding of saline water, the intention of which is to convert accumulations of ammonia to nitrite, any harm from which is offset by the presence of chloride ions.

Bag water with an overlay of oxygen requires no special aeration prior to loading if the bag is closed immediately after fish are added. Depending on the circumstances, adjusting the temperature of the water or the fish prior to loading may be helpful. Within a known safe range, the sudden introduction of most animals into cooler water calms them and reduces injury due to handling.

Sharp points on certain animals' bodies, such as the rostra of large shrimps or the spines of some fishes, may puncture a bag. For protection, a hard, lightweight canister may be used to contain spiny fishes within their bags. Also, stuffing of suitable thickness may be utilized between the inner and outer bags. Crustaceans may be shipped within physical constraints that can prevent punctures, as well as injury from aggressive movements. In some cases, inserting the animals' sharp points into protective tubes or other coverings also helps to avoid damage.

Knowledge of the behavioral and physiological characteristics of transport animals is useful in predicting their response during loading and delivery. For example, the size of the animals during early growth may determine the rate of accumulation of nitrogenous waste in their containers (Table 1). If the shipper knows the characteristics and tolerances of a particular species and its developmental stages, he or she may anticipate the transport conditions and the length of time the animals will be in transport and, on the basis of these parameters, choose an appropriate loading rate. Predicting carbon dioxide and total ammonia concentrations may also help (Table 2). A conventional application of this knowledge according to the type of species shipped, provides a good starting point for selecting a loading rate (Table 3).

Boxes containing sealed packages usually move to their destinations in airplanes and a variety of land vehicles. If the shipment is by air, certain standards may be set by commercial carriers, particularly with respect to international shipments. Once packages (bags) are sealed and boxes closed, delivery may be directly to the airport. There the bags and boxes may sit in storerooms or outside on concrete until they are loaded into special airfreight containers.

Table 1. Production of Total Ammonia, in mg/L (ppm), by Fingerling Channel Catfish of Three Sizes at 21 °C (70 °F) in 24 Hours[a]

g/L of Fish	Group I (13–27 mm, mean 21 mm)	Group II (24–43 mm, mean 32.5 mm)	Group III (62–107 mm, mean 88.6 mm)
3.0	3.16 (2.63–3.50)	2.63 (2.61–2.66)	1.02 (0.91–1.11)

[a]From (1).

Table 2. Water Chemistry of Sealed Transport Bags (with water volume one-third of total volume) Maintained at 17 °C (63 °F) for 24 Hours and Containing Various Weights of 7.6-cm (3-in.) Channel Catfish[a]

Amounts of Fish (g/L)	Total Ammoniated Nitrogen (mg/L)	Carbon Dioxide (mg/L)
25	5	5
75	14	30
150	21	55

[a]From (1).

Table 3. Capacity (normal) Load, in Grams per Liter of Water or Number of Animals per Liter of Water, for Conventional Bag-Oxygen Method of Transporting Aquatic Animals in Good Condition at 18 °C (65 °F)[a]

	Duration of Transport (hr)		
Size or Stage of Animal	12	24	48
Fish fry with yolk sac (large)	4,000	3,000	2,500
Fish fry yolk with sac (small)	—	10,000	8,000
Fish fry, no yolk sac (large)	2,500	2,000	1,500
Fish fry, no yolk sac (small)	—	5,000	4,000
Fish fingerling, 1.25 cm (0.5 in.)	[50]	[40]	—
Fish fingerling, 2.5 cm (1.0 in.)	[75]	[50]	[25]
Fish fingerling, 5.0 cm (2.0 in.)	[75]	[50]	[25]
Fish fingerling, 10.0 cm (4.0 in.)	[100]	[75]	[50]
Fish fingerling, 20.0 cm (8.0 in.)	[200]	[150]	[75]
Adult fish	[200]	[150]	[75]
Crustacea nauplii	200,000	100,000	—
Crustacea, 1 week old, as postlarvae	2,500	2,000	—
Macrobrachium postlarvae	1,000	150	—

[a]From (1).

Some boxes may be opened for inspection by customs, either at the departure point or at the destination. In international or interjurisdictional transfers, inspectors who open boxes may seek to discover contraband, confirm the contents or determine of a shipment, whether the items that are shipped conform to packaging standards. Incomplete documentation — for example, with health certificates or official permits missing — may block deliveries.

Upon arrival of the shipment at the destination, unloading proceeds. First, attention is given to the state of the package and the condition of the animals in it. Then the animals must be transferred into water of good quality and of a type that is close enough to that of the package water to avoid shock. Important differences could exist in temperature, gas saturation, salinity, and pH.

Delays in transferring animals to receiving units can be highly detrimental. When sealed containers are opened, the benefit of a high level of saturated oxygen soon disappears. Unnecessary risk to animals may also be posed when bags are floated in receiving units for the purpose of equalizing temperatures. Once the temperature rises in a bag, a delay in removing the animals that are inside may result in serious damage to the animals due to the high content of waste toxins.

Whenever possible, new arrivals should not be placed directly into systems that already contain resident populations. Mixing the two populations runs the risk of inoculating the residents with disease agents.

HUMID CONTAINERS

Humid containers, airtight or not, allow sufficient oxygen to reach the animals inside, while preventing an undesirable absorption of heat. There is no "standard" humid container. Airtight boxes commonly contain a supply of pure oxygen and some form of water-soaked material to sustain a moist interior. The distribution of the animals is usually critical, and the use of trays makes greater loading possible. Reducing the temperature prior to transport and maintaining the lower temperature throughout the duration of shipment of the animals is usually beneficial.

Some fish species that are tolerant of cold do very well in this form of packaging. The most useful means of transporting salmonids over long distances is to send "eyed" eggs in humid boxes under ice. Salmonid eggs hatch after weeks of incubation, rather than the period of several days that is typical of many other fish species of interest to aquaculturists. This slow rate of development contributes to the successful transport of salmonids.

Though sometimes sent in conventional sealed bags, crabs, crawfish, and freshwater shrimp can be suitably transported in open, but humid, boxes, baskets, or bags. Sufficient ventilation must be present to allow moist air to reach all the animals inside. Transport is generally successful if the packages remain cool and the fragile animals withstand crushing from clumsy handling. Chilled humid boxes are used to ship cold-tolerant species of marine shrimp and lobsters to restaurants for display in live tanks. Molluscs from young to adult stages also transfer well in humid boxes.

TANKS

Tanks are a popular means of delivering fish that are to be transported within a reasonable driving range. The capacity of a typical transport tank is suitable for the delivery of large numbers of fingerling to adult-sized animals. Tanks are sometimes also useful for the delivery of very large numbers of small organisms, such as shrimp postlarvae. It is usually cheaper to haul aquatic animals for short distances than it is to package them for delivery within sealed containers.

Transport tanks are typically loaded and fastened to the back of a pickup or flatbed truck. Tanks that are made of wood, fiberglass, or aluminum are more common than those made of stainless steel or plastic. Some tanks contain urethane foam or some other insulating material for temperature control. Bottoms may be false, slanting toward the gate to make it easy to remove animals. An inside sliding gate controls the flow of water better and allows an outside gate to be removed without releasing the contents of the tank. Smooth surfaces on the interiors of tanks reduce abrasions that result from rubbing. Wide doors at the tops of tanks aid in netting transported animals. Vents at the top prevent the buildup of carbon dioxide. Agitators that run off the vehicle's electrical system enhance water circulation and gas exchange. Larger tanks may contain inside baffles to prevent sloshing, especially if the driver loses control of the vehicle when braking. Some larger tanks have compartments with individual gates. Tanks with double levels or floors are sometime used to carry bottom-hugging species.

For long transfers, bottles, cylinders, tanks, or dewars supply oxygen to replenish that consumed by the animal load. The oxygen originates from the container and feeds through lines bottom to the of the tank, where diffusers deliver the gas as fine bubbles. Flowmeters may be used to discover the gas delivery rate. A small tank or compartment typically receives a flow of two to five liters per minute.

Haulers that regularly transport large volumes of water and animals use liquid oxygen, one advantage of which is a lengthy duration of flow. With the gas flowing at more than 5 liters per minute, a standard-sized oxygen cylinder lasts less than one day. In contrast, a 160-liter liquid-oxygen dewar would flow for more than two weeks at a rate of 5 liters per minute. The cost of liquid oxygen is approximately one-third that of gaseous oxygen and requires less loading weight per unit of volume. The cost equation changes, however, according to frequency of use and leasing arrangements made for the container.

Lights mounted on the rear of the cab of the truck help at night. When salt is regularly used as a freshwater additive, or when marine species are being transported, truck or trailer beds might best be constructed of aluminum to reduce the corrosive effects of the salt. Local regulations may require that transport tanks or trucks be specially marked.

Smaller, shorter transfers in tanks require less sophistication. Common equipment consists of easily removable units in the backs of small trucks. Agitators provide aeration during delivery of the cargo. Cylinders containing compressed oxygen gas are kept on board for emergency use. In addition to custom made hauling tanks, aquaculturists use a variety of sturdy, light, and well-insulated containers for the shipment of produce and chilled animal products.

Water-displacement tests determine the total volume of the animals to be shipped, from which their numbers or weight may be derived by simple algebra. Tank windows, outside tubes, or interior markings are useful in determining the animals' volume. Scales are also useful; some trucks have a hinged boom that supports a scale and a basket for on-the-spot weighing when the animals are unloaded.

The air temperature can become a factor to be reckoned with if it differs considerably from that of the water in the tank. The latter may become too warm if its circulation by agitators promotes the uptake of heat. In contrast, the constant delivery of fine bubbles from a liquid-oxygen dewar promotes cooler-than-ambient water. Ice is widely used to chill tank water to a suitable temperature; refrigeration or heating units in connection with transport tanks or the airspace in which tanks travel are uncommon in aquaculture.

Water-quality constraints for transfers of aquatic animals in tanks are almost the same as those for transfer of the animals in bags. Tanks are not sealed, however, and carbon dioxide does not appreciably accumulate in tanks that receive normal aeration. Ammonia accumulates in tank water according to its output by the animals.

Sedatives have the potential for slowing the animals' activity, but in practice, they have relatively little application. In freshwater fishtanks equipped with agitators, mucus and organic matter will froth at the water surface and can interfere with gas exchange. The application of "defoaming" agents, and to some extent sodium chloride, resolves the problem. Sodium chloride and calcium chloride are widely used in conjunction with tank transport of freshwater species.

Tank water may be supersaturated with oxygen bubbles prior to loading fishes. Supersaturation satisfies the initial demand for oxygen by excited fish. Salts appropriate to the circumstances of transport may also be added prior to loading.

Nighttime and early morning are preferable delivery times during hot seasons. The driver of a delivery truck may monitor the dissolved-oxygen content of tanks on the truck bed if the cab is equipped with a meter that connects to a probe of suitable length. Accurate knowledge of the water quality upon arrival at the destination is helpful. When differences in important water parameters are great, adjustments must be made en route or at the destination to prevent harm to the animal load. Adjustments at the destination usually involve a slow blending of the receiving water with the delivery water. In some cases, adjustments to the receiving water are made prior to arrival of the load.

Many variables influence the selection of the animal load for tank transport. Table 4 gives some customary loading rates for several species of fish.

Careful attention must be paid to the selection of a transport temperature when atmospheric temperatures are extreme and deliveries are long. A water temperature suitable for cold-season transport is likely to be too low for summer transport. In either season, the temperature of the transport water should not differ by more than 10 °C (20 °F) from the ambient temperature. The lowest suitable transport temperature for tropical species is higher than that for temperate species. Tilapia, for example, should travel above 16 °C (60 °F).

Fish displace a certain amount of water when they are loaded. Interpreting the amount of displacement is easier when one uses metric units rather than English units. A pound of fish displaces roughly 0.12 gallon of water (one kilogram displaces one liter of water). If one knows the total weight requirement for a delivery of fish, a converting to gallons gives the volume of water displaced by the fish. This conversion helps to determine tank capacity requirements and load limitations according to an acceptable loading rate for a particular species.

Fish are sold according to their weight or number, and type or size. A common means of determining the number of weight of small animals is to estimate the volume of animals at the time of loading. Estimates of volumes for particular products are established with experience and, perhaps, the help of length–weight tables that are available for many species.

Table 4. Customary Loading Rates (proportion of fish in tank load as a percent fish) for Several Fish Stages in Transport Tank Water at 18 °C (65 °F) Aerated with Air

Fish and Stage	Duration (hr)	Percent Fish[a]
Catfish		
13 cm (5 in.)	8,16	26,18
23 cm (9 in.)	8,16	32,23
40 cm (16 in., 1.3 lb)	8,16	38,32
Sunfish		
2.5 cm (1 in.)	16	4
7.5 cm (3 in.)	16	7
13 cm (5 in.)	16	15
Carp		
<15 cm (<6 in.)	16	11
>15 cm (<6 in.)	16	15
Tilapias		
5 cm (2 in.)	12	23
>7.5 cm (>3 in.)	12	32
Morone		
5 cm (2 in.)	8	4
Drums		
5 cm (2 in.)	8	4

[a]Percent fish is given to avoid confusion that might otherwise arise from differing expressions of volume composition. Generally, the fish load (weight) increases 25% if a tank is supplied with liquid oxygen. The weight decreases or increases 25% if the water temperature is 5 °C (10 °F) warmer or cooler, respectively, than 18 °C (65 °F) for these species listed.

FARM DISTRIBUTION AND TRANSPORT

On farms, live animals are frequently transported and distributed to nearby sites. The conditions for this short transport of live animals differ considerably from those discussed in previous sections. The brief time involved does not allow to accumulate significantly metabolic wastes, in the transport units if reasonable loads are carried. Attention is paid primarily to providing adequate oxygen. Many variations, of course, exist, and a usual application is to replenish oxygen to a loaded container by bubbling oxygen from a sparger that is connected to a bottle of oxygen gas. Shallow water may also be useful in quick transfers. Water of a depth that barely covers fishes permits enough air to diffuse through the water to match that consumed by the fishes. The number of animals a sealed container is able to hold is greater for short deliveries.

Conditions in some areas cause the transport time to extend for several days. Adjusting the density of animals to account for the buildup of waste is not economical, so water must be exchanged at intervals during transport. Similarly, intercontinental transfers are often too long without the benefit of repacking. In either case, handlers may profit from planning and foresight regarding the conditions at key points of the delivery route.

LIVE TRANSPORT OF PARTICULAR ANIMAL GROUPS

Commercial traffic in live marine shrimp consists primarily of nauplii, postlarvae, and adults. Long-distance shipments use sealed containers. The destinations of nauplii and adults are other hatcheries. There, the nauplii are grown into postlarvae. The destination of postlarvae is a nursery or a growout system.

Sellers prefer to ship postlarvae of small size (5 to 12 days in the postlarvae stage), because more can be safely sent per container. Nauplii are even smaller and do not feed during the first day after hatching. Feeding peculiarities of larval stages other than nauplii discourage their transport.

Crawfish ponds are stocked with adults. The source is another farm where they are packed for delivery in sacks similar to those used to transport onions for marketing. When the animals are not allowed to dry out in the back of a vehicle or to overheat in sunlight, delivery is generally successful.

Like crawfish, adult freshwater shrimp may be carried to destinations in a moist container such as a covered basket. The young are shipped in a manner similar to the way young marine shrimp are shipped. Brine shrimp are important as food for many young aquaculture animals, particularly those of marine species. Following harvest of the buoyant eggs, a dehydration process makes the eggs suitable for distribution as a living dry product in a variety of types of packages. Crabs and lobsters are transported to the market primarily as adults.

Live molluscs are transported between aquaculture facilities mostly as larvae and juveniles. Edible market animals are shipped according to industry standards. In the United States, the Interstate Shellfish Sanitation Conference was instrumental in the development of the National Shellfish Sanitation Program's manual of operations. A manual also is available from the U.S. Food and Drug Administration.

Juvenile molluscs (postset) with shells 1 mm (0.04 in.) in length are often shipped in 500-mL (0.13 gal) volumes. Size is determined by screening. The handler spreads the animals on moist paper or food wrap and sprays them with a mist of cool saltwater. When the paper or wrap reaches 18 to 20 °C (65 to 68 °F), it is folded and placed into insulated containers with or without small chill packs. Delivery is by routine service of a courier.

Mature larvae (preset) are sometimes transported in a damp cloth in insulated boxes. Another means of transport, sometimes also used with oysters, is to allow the larvae to set individually on particles of ground shell prior to packaging. Smaller preset and postset animals have a nursery as their destination. Larger postsets (0.5 to 1.0 in. length) (1.25 to 2.5 cm length) are packaged in boxes as balls or layers. Transport is successful if temperature and moisture meet the requirements for the species.

All stages of fish are subject to live transport. Eggs are shipped in humid boxes and in water in sealed or open containers. Eggs should be able to be transported to their destinations without hatching. Some fish species are deliverable as fry during the period prior to absorption of the yolk sack. At that stage, the fish have not yet begun to feed. Large adults of some species are apt to become too aggressive in transport and may consequently be given individual space in tanks or sealed containers of a suitable size. Shippers normally pack only one fish per plastic bag if the fish are valuable and are to travel over a long distance or by air.

Tadpoles and aquatic stages of salamanders are transported in tanks or plastic bags. Adult frogs transfer well in vented boxes that retain reasonable humidity and temperature. Because of cannibalistic tendencies, they are shipped in uniform sizes. Turtles and alligators are packaged so as to reduce harm from aggressive behavior.

Shippers are mindful of local, national, and international regulations that may affect live-animal shipments. Differences in government strictures provide a complexity of often interrelated regulatory applications. Air shipment will normally include a shipper's certificate, an air waybill, a health certificate, and the necessary export–import permits or licenses. The package will likely have to conform to packaging standards. A couple of organizations produce helpful documentation (2,3) that explains the legalities of transport by air and road. Regulations in some areas may require that a bill of lading and special permits and licenses be on board a road vehicle.

BIBLIOGRAPHY

1. S.K. Johnson, in Texas Agricultural Extension Service, *Inland Aquaculture Handbook*, A1504, 1988.
2. Animal Transportation Association (AATA), *Manual for the Transportation of Live Animals by Road*, Surrey, England, 1996.
3. International Air Transport Association (IATA), *Live Animals Regulations*, 24th ed., Montreal, 1997.

See also ANESTHETICS; CARBON DIOXIDE; DISSOLVED OXYGEN; pH.

LOBSTER CULTURE

PAUL OLIN
University of California Sea Grant Extension
Santa Rose, California

OUTLINE

Classification and Distribution
Lobster Anatomy
Life History and Broodstock Management
Larval Husbandry
Lobster Growout
Future Prospects
Bibliography

The American lobster, *Homarus americanus*, is a highly sought-after seafood, and there has been considerable research on lobster aquaculture for seafood markets and fisheries stock enhancement. Lobsters were not always so highly appreciated, and, in colonial times in North America, indentured servants in Massachusetts insisted on a clause in their contracts that they be fed lobster no more than three times per week (1). Today, there is a thriving North American lobster fishery centered in New England and the Canadian Maritimes, with 90% of the landings occurring from the Gulf of Maine and coastal waters around Nova Scotia and New Brunswick. A similar fishery exists for the European lobster, *Homarus gammarus*, throughout its range in the Eastern Atlantic Ocean, the Mediterranean, and the North Sea. These fisheries respond to a strong market demand that piqued the interest of aquaculturists. As a result of this, in the 1970s, significant research in Canada, Europe, and the United States successfully refined techniques for the controlled reproduction and growth of both American and European lobsters for aquaculture.

Efforts to reproduce lobsters began early in this century, when government hatcheries were built in Canada, France, Norway, the UK, and the US. Hatchery managers held gravid broodstock until eggs hatched and then released these early larval stages, in an attempt to enhance the commercial fishery. This was a massive effort, and Atlantic Canada alone had 15 hatcheries that released around 900 million larvae between 1891 and 1917 (2). The Massachusetts State Lobster Hatchery also released millions of lobster larvae subsequent to opening in 1939 on Martha's Vineyard and is still a center of lobster culture and research.

In the 1970s, concerns about the lobster fishery, coupled with the lobster's high value, encouraged institutions in Europe and North America to develop research programs to close the life cycle of lobsters in captivity. This included research on lobster reproduction, nutrition, disease, larval rearing and grow-out. Researchers at the Ministry of Agriculture, Forestry and Fisheries Laboratory (MAFF) in Conway, North Wales succeeded in developing techniques to culture the European lobster to a market size of 350 g (~3/4 lb.) in 2 1/2 to 3 years (2).

During this same time, scientists at the University of California's Bodega Marine Laboratory, with funding from the California Sea Grant Program, embarked on a similar research program aimed at refining techniques for captive reproduction, larval rearing, and grow-out of the American lobster. This research was also successful and contributed significantly toward defining nutritional requirements and techniques for broodstock management, larval rearing, and commercial production of the American lobster (3–5).

While successful at closing the life cycle and demonstrating techniques for controlled production of lobsters in captivity, this research has not led to large-scale development of lobster aquaculture. Reasons given for this are a lack of comprehensive information on nutrition, disease control, and viable grow out systems (6,7). Computer modelling of a commercial lobster production facility suggested an economically viable production level to be just shy of one million animals per year or 450 metric tons of product annually (8,9). There have been no attempts at large-scale lobster culture that might validate this model. The greatest constraint to large-scale commercial production is the need to isolate animals in order to protect them from cannibalistic tank mates while molting. The labor and capital costs associated with the care, holding, and feeding of isolated animals will have to be overcome to allow economically viable commercialization of lobster aquaculture as a food product. There is, however, a strong interest in stock enhancement using cultured juveniles, and projects are currently underway, primarily in the United Kingdom.

CLASSIFICATION AND DISTRIBUTION

Clawed lobsters belong in the phylum Arthropoda and are members of the class Crustacea, which they share with the other crabs and shrimp. They belong in the order Decapoda and the suborder Reptantia. Two species have been the primary focus of aquaculturists, and these are the American lobster, *H. americanus*, and the European lobster, *H. gammarus*. Some interest has developed recently in culturing the clawless spiny lobsters of the genus *Panulirus*; however, due to the extremely long larval cycle, it is likely that this will involve capturing wild late-stage larvae or juveniles for further growout in order to be economically viable.

LOBSTER ANATOMY

Lobsters have a hard jointed exoskeleton covering their bodies that provide sites for muscle attachment and protection from predators. The pigmented exoskeleton is a mottled dark greenish-black on the dorsal surface and orange on the ventral surface. It is comprised primarily of chitin and minerals absorbed from the surrounding seawater. Three main body parts are the head, thorax, and abdomen, with the first two being fused into a cephalothorax that is covered by a hard-shelled carapace. The jointed abdomen houses the large abdominal muscle sometimes referred to as the tail. This is attached to the cephalothorax with a soft tissue suture. For growth to occur, lobsters must shed their rigid exoskeleton in a process called molting, and it is at this suture that the lobster withdraws from the old exoskeleton. Rapid absorption of water then increases the animal's size by as much as 20%, and the new exoskeleton hardens around this enlarged body. Lobsters are extremely vulnerable to predation after molting and must seek shelter from predators.

As decapod crustaceans, lobsters have ten appendages on their thorax: the conspicuous paired claws, and four pairs of walking legs or pereiopods (Fig. 1). The large crushing claw is used to crush shellfish and other prey, while the tearing claw rips soft tissue. Lobsters are primarily benthic organisms and use their eight walking legs to move around. When alarmed, however, they will propel themselves rapidly backwards through the water with powerful contractions of the abdominal muscle. Attached to the ventral surface of the abdominal segments are the paired pleopods or swimmerets (Fig. 2). These are used to help maintain position and swim when animals are in the water column but more importantly serve as a site for egg attachment. An egg bearing or "berried" female will constantly fan the egg laden pleopods back and forth to keep clean water circulating around the eggs (Fig. 1).

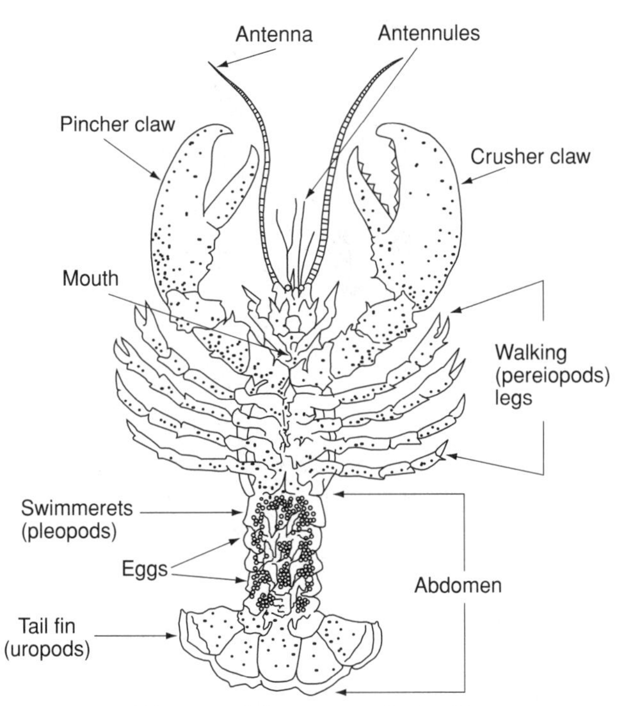

Figure 1. Lobster, ventral view (9).

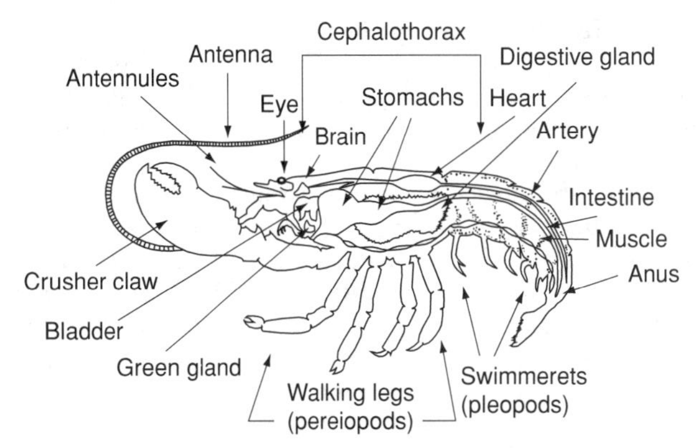

Figure 2. Lobster anatomy, lateral view (9).

It is important for lobsters to have a well developed sensory system that allows them to find food and seek out a mate when they become reproductively mature. The compound eyes located on anterior eyestalks allow lobsters to see motion and respond to different light intensities, but the chemosensory system is far more important as an aid in finding food and escaping predation. The long antennae are sensitive to touch and can detect slight changes in water pressure. The antennae and similarly sensitive hairs along most of the ventral surface of the body provide information on currents and potential predators and prey. The branched antennules are extremely sensitive chemoreceptors that aid in locating food, warn of predators, and help to locate mates (Fig. 1). Additional sensory hairs lining the feeding appendages allow animals to identify food items selectively.

Lobsters feed primarily at night and are opportunistic predators, consuming a wide variety of prey items as they are available. These include other crabs and shrimp, shellfish, polychaetes, urchins, and starfish. Juvenile lobsters generally seek refuge amidst rock crevasses and feed on drifting particles, animals growing on rock surfaces within the burrow, and other small prey found nearby. Adult lobsters will forage away from sites of refuge. When feeding, lobsters use their large claws to crush and tear apart prey but rely more on their first two pairs of clawed walking legs and accessory mouth parts to select and manipulate food to the mouth. These mouth parts include the paired mandibles, the first and second maxillae, and the first, second, and third maxillipeds. Ingested food passes from the mouth into the first (cardiac) stomach, where it is mechanically ground by three hardened chitinous teeth referred to as the gastric mill. Further tissue breakdown results from digestive enzymes. The ground particles then pass into the pyloric stomach and from there to the digestive gland or hepatopancreas, which is the principle site for nutrient absorption. The hepatopancreas is the large green organ known and loved by lobster aficionados as tomalley. Undigested material and waste passes into the intestine and leaves via the anus at the tip of the tail (Fig. 2).

LIFE HISTORY AND BROODSTOCK MANAGEMENT

Mature female lobsters can mate only following molting, and, in the wild, pre-molt females typically seek out a male's den the summer this occurs. The male cues in on sex pheromones and becomes less aggressive, allowing the female to enter the den with him for a few days until she molts. The male then transfers a sperm packet into a receptacle on the female's ventral surface, using a modified first pair of pleopods. The male provides protection until the female's shell hardens, and within about one week the female will depart, retaining the sperm packet for the next 9 to 12 months. The following summer, primarily in July and August, the female will roll on her back and cup her tail before extruding 10,000 to 20,000 eggs, which are fertilized as they pass over the sperm receptacle (9). These eggs attach to the pleopods, where they are kept clean and maintained by the female until they hatch into an early larval stage the following summer.

It is desirable in aquaculture to have healthy gravid lobsters and complete control over the reproductive cycle, so that larvae are consistently available. Fortunately, this can be accomplished through the control of temperature and photoperiod. Hedgecock and co-workers at the Bodega Marine Lab describe a technique for managing broodstock that consists of holding three separate broodstock populations under different photoperiod and temperature regimes and is summarized here (3). A reserve population is held in System I at 10 to 15 °C under a short-day photoperiod of 8hL : 16hD. This temperature promotes egg development and vitellogenesis, while the short day length inhibits extrusion of eggs. Females are placed in this tank after they have completed a nuptial molt and have been mated. Four months in this system allows eggs to develop fully, and the females are then transferred to System II, which is at the same temperature with a long-day photoperiod of 15hL : 9hD that promotes extrusion of eggs. The egg-bearing females are then transferred to

System III, which uses a long-day photoperiod and an elevated temperature of 20 °C. This elevated temperature accelerates embryo development, such that the eggs hatch in four months rather than the 9 to 10 months observed at 10 °C.

While this system can consistently produce lobster larvae, there are areas in which additional research could greatly improve the economics and efficiency of larval production. Only 64% of the females held in System I extruded eggs, and less than half were able to carry the eggs until hatch (3). In those females that carried eggs to hatch, clutch sizes were often greatly reduced. Areas for investigation to improve larval production include low mating success, poor sperm and egg quality, and inadequate nutrition. Other stresses in captivity could interfere with mating or female care and cleaning of eggs during brooding.

To alleviate stress, culturists must insure that broodstock holding tanks have excellent water quality and are in quiet locations, so normal brooding behavior is not disrupted. Dissolved oxygen levels should be at saturation, and the salinity and pH should be around 32 ppt and 8, respectively. Ammonia levels should be kept below 0.2 mg/L (13). Broodstock nutrition is critical to high health and egg production, and animals should be fed to satiation using a mixed diet of chopped squid, fish, mussels, and shrimp at 2% of body weight per day (3).

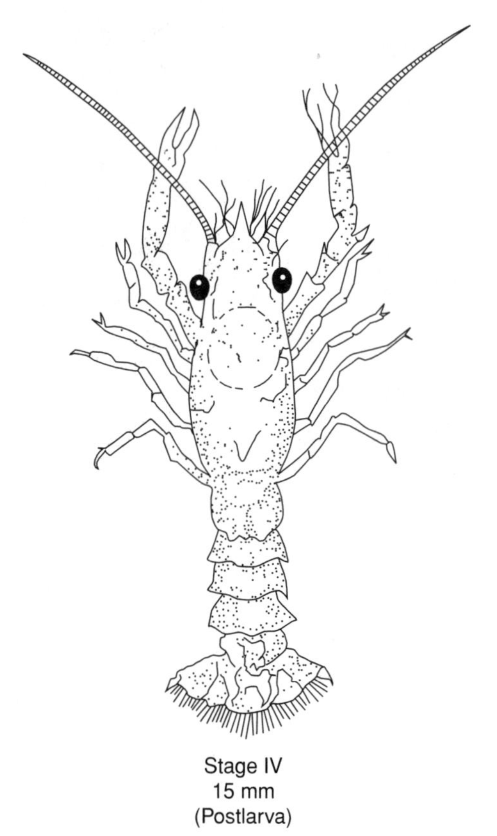

Figure 4. Lobster, postlarva (9).

LARVAL HUSBANDRY

The relatively short larval cycle of American and European lobsters is an attractive life history characteristic for aquaculturists. Lobster larvae have four planktonic stages, and at 20 °C they will complete their larval cycle in 10 days (5) (Fig. 3). The final larval molt involves a metamorphosis into a postlarva, which closely resembles the adult form, and, at this time, the small lobsters will settle to the bottom and seek shelter to avoid predation (Fig. 4). As with the culture of any larval crustacean, it is important to have ready availability of high-quality food and maintain excellent water quality during the larval cycle.

To obtain newly hatched larvae, "berried" females are moved two weeks before the eggs are ready to hatch and held in individual mesh containers in hatching tanks (4). These tanks receive UV-sterilized, 50 μm-filtered seawater at 20 °C provided with light aeration. Flow rates should be sufficient to maintain ammonia levels below 0.2 mg/L. Outflow from the hatching tanks enters a 1-mm-screen-bottomed container floating partially submerged in a larger tank fitted with an overflow pipe, so that the larvae are retained in the screened container. It is recommended that larvae for culture be collected over at most a 24-hour period, so that they are all approximately the same age and size, in order to reduce cannibalism. When sufficient larvae have been collected, they are transferred to the larval rearing tanks.

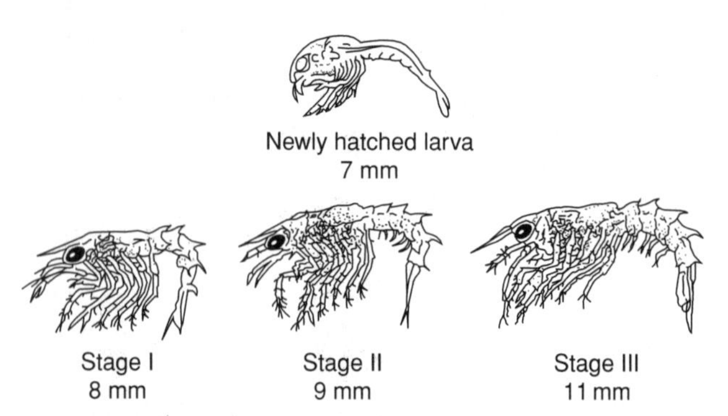

Figure 3. Lobster post-hatch and larval stages (9).

A variety of tank configurations can be successfully used for larval rearing, provided the water can be uniformly mixed. Some tanks use a "Kreisel" system, where incoming seawater enters from jets in the bottom that maintain a pattern of upwelling. This helps to keep the larvae uniformly distributed (10). Conical-bottomed tanks work well, and aeration can be used to generate an upwelling that serves the same purpose as the "Kreisel" jets. Rectangular flat-bottomed tanks have also been used, and strategic placement of aeration can accomplish uniform mixing in these tanks. This even mixing keeps larvae and feed organisms uniformly distributed and insures ready access to food for larvae, while minimizing cannibalism. Larval rearing tanks are stocked at densities ranging from 50 to 100/1, and the highest survivals, of 70%, are achieved at the lower end of the range (4,11).

Water quality in larval rearing containers is critical and can be maintained by using either flow-through systems or static systems with batch exchanges of water. Water quality parameters are similar to those required for the hatchery tanks, with flow rates or frequency of water exchanges set to maintain low ammonia levels. Mild aeration is used to insure oxygen saturation and provide mixing. Outlets are screened at 1 mm to retain larvae.

Nutrition is extremely important for larval organisms, to provide the energy necessary for successive molting and the protein and essential fatty acids for proper development. Newly hatched and adult Artemia provide an excellent source of feed. Citations in the literature recommend maintaining an Artemia ratio of four per lobster larva or providing eight Artemia/larva/day (4,12). The important thing to address in feeding larval crustaceans is availability of feed, and for this reason it is recommended that a particular prey density be maintained. Feeding levels from 0.5 to 1 Artemia/mL will provide enough food at a density that ensures it is readily available to the larvae. An additional benefit of live Artemia is that they will not decompose as rapidly as previously frozen Artemia or other prepared foods. Although survival is usually reduced, other feeds that have been successfully used are mixes of finely chopped fish, squid, other molluscs, and frozen Artemia. There are a number of larval diets that have been developed for use in penaeid shrimp hatcheries that should also work well for lobster larvae and should be evaluated. If not using live feeds, it is important to monitor water quality closely, and any build-up of uneaten food should be removed with a siphon.

Cultured algae such as *Isochrysis galbana* or *Chaetoceros gracilis* can also be added to the rearing tank, at densities from 10 to 100 cells/mL. While this algae will not be consumed directly by the carnivorous larval lobsters, it is eaten by the Artemia and will help to maintain their nutritional value. Algae will also contribute oxygen to the system through photosynthesis and will utilize nutrients that might otherwise be available to other bacteria and protozoans.

LOBSTER GROWOUT

Lobsters can be grown from postlarvae to six-month-old juveniles either communally or in separate containers. If they are grown communally, it is essential to provide complex habitat for individuals to seek shelter and avoid predation, especially after molting. A variety of materials can be used to provide this cover, including oyster shell, pieces of white fluorescent light baffling, or an assortment of biofilter media. While these materials provide adequate shelter, they complicate tank cleaning, and the decision for communal or individual rearing should be based on relative costs of materials and labor. Densities used in communal culture range from 100 to 200 per square meter. Other techniques to reduce cannibalism in communal rearing units include sorting to maintain a uniform size and immobilization of the claws or removal of the dactyls, all of which require considerable labor.

Tanks used for individual lobster grow-out are generally shallow and often situated in tiers to maximize utilization of floor space. Individual compartments are fashioned with plastic mesh, slotted PVC, or acrylic plastic (Fig. 5). A slotted false bottom is sometimes employed, to facilitate cleaning the tank and removing uneaten food. Container size is important, or growth will be reduced. A 25-mm-carapace-length (CL) lobster should be held in a space of at least 310 cm^2, while a 40-mm-CL animal requires a 620-cm^2 container (5).

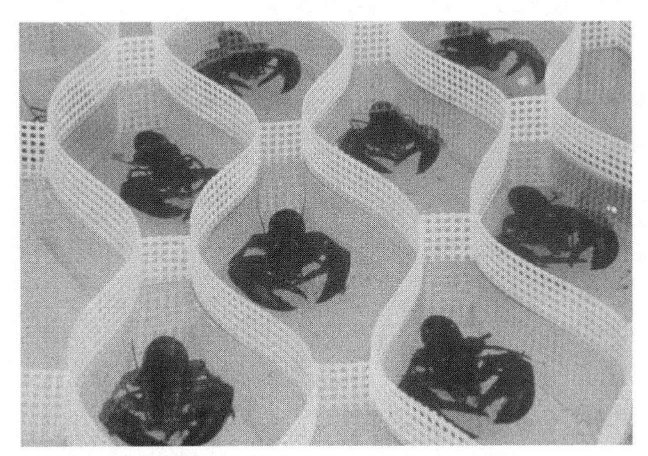

Figure 5. Individual lobster enclosures used in growout.

Water needs to be uniformly distributed to each rearing unit, and this can be accomplished by using overhead rotating spray bars (or submerged rotating slotted bars, in tanks with false bottoms). Sprinkler heads can also be used to distribute water over the surface. Water quality should be maintained as for broodstock, except that ammonia levels should be kept below 2 mg/L in growout systems. Water temperatures of 20 °C have been found to promote optimal growth, and a market-size lobster can be grown in 2 to 3 years under these conditions (2,5).

There are currently no commercial rations known to support growth in larval and juvenile lobsters comparable to that achieved using fresh and frozen Artemia. A similar situation exists for older animals, which exhibit the best growth rates on diets of fresh and frozen foods. This highlights the need for research and development to produce economically viable lobster diets. A feeding rate of 2% of body weight/day produces the best food conversion, but 4%/day maximizes growth (14). Given the difficulty of handling large volumes of fresh feeds and their impact on water quality, it is unlikely that a large venture will develop until a suitable prepared diet is commercially available.

FUTURE PROSPECTS

There is considerable recent interest in operating lobster hatcheries and nurseries to augment fishery stocks. In the U.K., there are government hatcheries, fisherman's associations, and supermarkets conducting outplantings of tens of thousands of juvenile lobsters for subsequent recapture in the fishery. In initial trials, five-year survival of outplanted juveniles was found to be 38%. Eight percent of these survivors were subsequently recaptured in the fishery (2). These results are encouraging, and, as the economics of juvenile production improve, this may become a viable technique to enhance the commercial fishery.

Lobster aquaculture from egg to market remains an elusive goal. The difficulties of raising a highly carnivorous crustacean in individual compartments and the resulting labor costs have, to date, precluded any successful business development. As long as the fishery remains stable and prices do not escalate significantly, it is unlikely a commercial industry will develop. Even so, the basic information to culture lobsters is available, and, if prices increase sufficiently to allow economically viable culture, a ready market is there to purchase the product.

BIBLIOGRAPHY

1. Smithsonian Magazine, October 1997.
2. Fish Farming International, July 1998.
3. D. Hedgecock, in J.P. McVey and J.R. Moore, eds., *CRC Handbook of Mariculture, Crustacean Aquaculture*, Vol. 1, 1983.
4. E.S. Chang and D.E. Conklin, in J.P. McVey and J.R. Moore, eds., *CRC Handbook of Mariculture, Crustacean Aquaculture*, Vol. 1, 1983.
5. D.E. Conklin and E.S. Chang, in J.P. McVey and J.R. Moore, eds., *CRC Handbook of Mariculture, Crustacean Aquaculture*, Vol. 1, 1983.
6. D.E. Aiken and S.L. Waddy, in J.R. Factor, ed., *Biology of the Lobster Homarus americanus*, Academic Press, London, 1995.
7. B.F. Phillips and L.H. Evans, *Mar. Freshwater Res.* **48**, 899–902 (1997).
8. P.G. Allen and W.E. Johnston, *Aquaculture* **9**, 144–180 (1976).
9. Gulf of Maine Aquarium Web site, http://octopus.gma.org/-lobsters
10. W. Greve, *Mar. Biol.* **1**, 201 (1968).
11. A. Schuur, W.S. Fisher, J.C. Van Olst, J.M. Carlberg, J.T. Hughes, R.A. Shleser, and R.F. Ford, *Hatchery Methods for the Production of Juvenile Lobsters Homarus Americanus*, Sea Grant Publ. No. 48, University of California, Institute of Marine Resources, La Jolla, 1976.
12. J.M. Carlberg and J.C. Van Olst, *Proc. World Maricult. Soc.* **7**, 379 (1976).
13. D.A. Delistraty, J.M. Carlberg, J.C. Van Olst, and R.F. Ford, *Proc. World Maricult. Soc.* **8**, 647 (1977).
14. D.M. Bartley, J.M. Carlberg, and J.C. Van Olst, *Proc. World Maricult. Soc.* **11**, 355 (1980).

MARKET ISSUES IN THE UNITED STATES AQUACULTURE INDUSTRY

Upton Hatch
Terry Hanson
Auburn University
Auburn, Alabama

OUTLINE

Aquaculture has grown dramatically worldwide in the past two decades. Although there have been culture systems developed for a wide array of aquatic species, the dominant species have been shrimp, salmon, catfish, carp, and tilapia. Southeast Asia has been the leading production area. The catfish industry in the south central United States has grown rapidly, while other production areas in the United States have experienced some success and some emerging species. New production areas may prove to be important in the near future.

Cultured fish and seafood products have established important market niches in the U.S. food supply. In only a few decades, production systems have been developed that provide a standardized product with availability extended across geography and seasons. Farm-raised shrimp, catfish, and trout, and pen-raised salmon are available virtually anywhere in the United States, any time of year. In addition, several emerging species (tilapia, molluscs, crawfish, and scallops) may reach this level of market development in the very near future.

The consumer image of fish and seafood supplies will have an important impact on the U.S. aquaculture industry. Although producers of cultured fish assert the relative quality and safety advantage their product has over wild-caught species, the general public may not perceive the difference. In addition, public perceptions of negative externalities through environmental or aesthetic degradation could play an important role in U.S. aquaculture production growth.

Nonetheless, the growth of aquaculture appears assured. Growing population, changing nutritional habits, increasing income, relatively constant catch from capture sources, and water pollution all suggest that the share of fish and seafood coming from culture sources will increase. In fact, if enhancement of wild populations is included as an aquaculture enterprise, the world fish and seafood market will be dominated by supplies dependent on culture sources in the near future. This entry focuses on the market issues that will largely determine the future growth of the U.S. aquaculture industry. The emphasis is placed on demand factors; however, related supply issues are also discussed. Comparative advantage underlies interregional and international competition for market share and profits. Market development and demand characteristics are intertwined with comparative advantage in complex ways that do not generally allow for a clean distinction between marketing and production influences on growth.

INDUSTRY TRENDS

Major Species

Wild-caught seafood dominates production from the ocean, however, in fresh and brackish water, aquaculture is a major source of commercial fish and seafood production. Even recreational catch is based on frequent restocking from aquaculture operations of private, state, or federal hatcheries.

For U.S. aquaculture to continue to grow, there must be an expanding demand for its products. Consumers have indicated that they consider seafood to be a healthy product, yet seafood consumption has remained steady over the past decade (Fig. 1a) (1); compared to other meat and poultry products, seafood is a distant fourth to poultry, beef, and pork (Fig. 1b). The 10 most important fish and seafood species consumed in the United States are presented in Figure 1c. Because most aquaculture products tend to be in the middle-to-high end of the price range for seafood products, the state of the economy plays a large role in their demand. The sales values A for the major culture species in the United States (catfish, trout, and salmon) are depicted in Figure 2, where catfish has dramatically increased and salmon has reached the level of trout production.

Catfish. The top 10 catfish-producing states are (in order of production) Mississippi, Alabama, Arkansas, Louisiana, California, South Carolina, Missouri, North Carolina, Kentucky, and Texas. Some production also occurs in Tennessee, Oklahoma, Kansas, and Florida; however, catfish production is dominated by the first four states in the first list. Dramatic growth in catfish production has occurred in the past two decades (Fig. 3a).

Catfish farm prices have ranged from \$1.32 to 1.74/kg (\$0.60 to 0.79/lb) in the 1990s, with an average price of \$1.58/kg (\$0.72/lb). On a real-price basis, the long-term trend for catfish farm prices has been downward (Fig. 3b). The number of catfish producers peaked in 1991 and

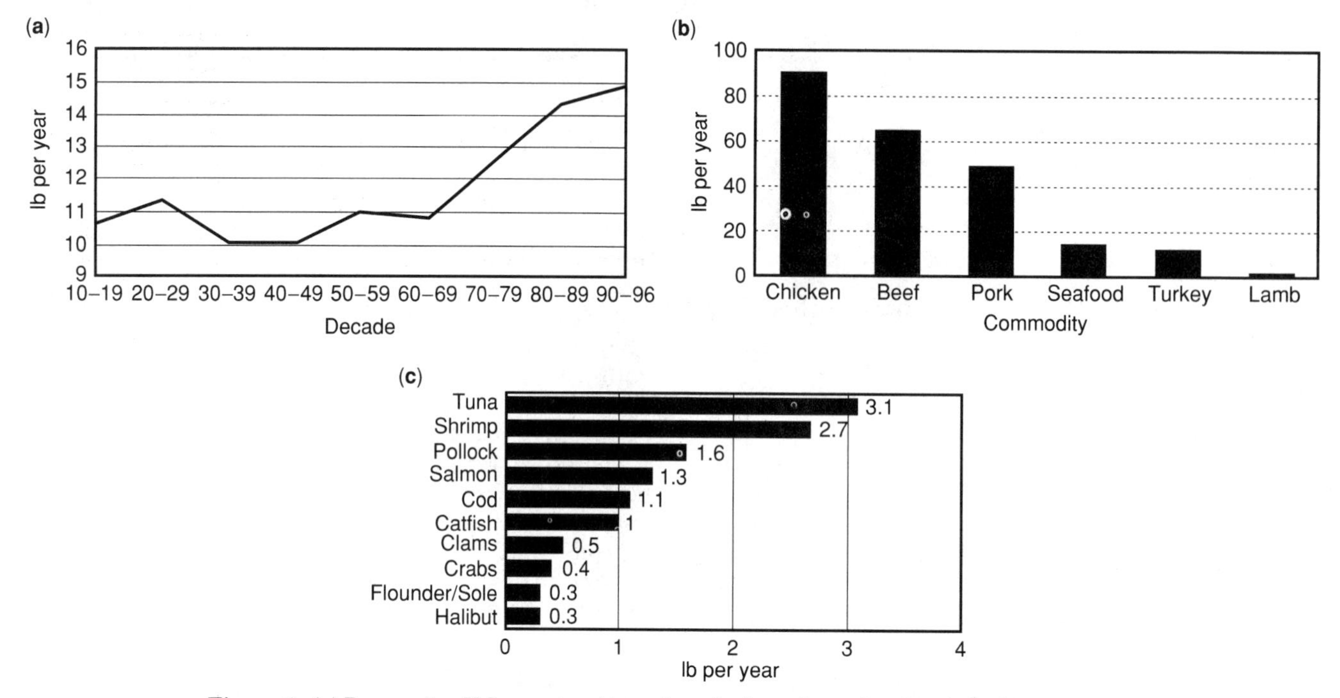

Figure 1. (**a**) Per capita U.S. consumption of seafood products by decade.[1] (**b**) Per capita consumption of poultry, meat and seafood in the United States (1998).[1] (**c**) Top 10 fish and seafood by per capita U.S. consumption, 1997.[2] *Source*: [1]Economic Research Service, U.S. Department of Agriculture, [2]National Fisheries Institute, 1998.

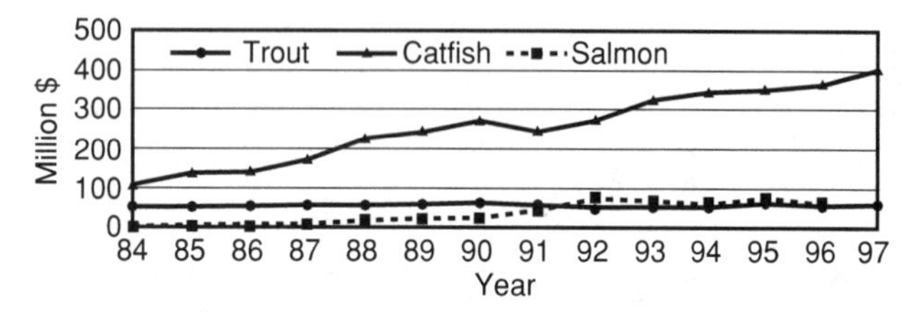

Figure 2. Value of farm-raised catfish, salmon and trout in U.S. *Source*: Economic Research Service, U.S. Department of Agriculture, 1994.

has since declined, with 955 operations in 1981, to 1,943 operations in 1991, and 1,250 operations in 1998 (Fig. 3c). Average farm size doubled, increasing from 165 ha (66 ac) in 1981 to 330 ha (132 ac) in 1998.

Exports of catfish products have been seen as a source of future market growth for the catfish industry. To develop foreign markets for catfish, The Catfish Institute has used grants from the U.S. Department of Agriculture's (USDA) Foreign Agricultural Service to determine how best to promote its product in places such as the European Community and Japan. In 1992, U.S. exports of catfish products totaled 111,360 kg (245,000 lb) and were valued at $561,000. Currently, the United Kingdom is the largest buyer of U.S. catfish products, with small amounts going to other European countries, Japan, and Singapore.

Imports peaked in 1980 at 6.75 million kg (15 million lb) and have since declined steadily as a percentage of domestic production. Catfish imports fell 44% from 1991 to 1992 to 1.3 million kg (2.9 million lb), the lowest in 13 years. The drop in 1992 imports can be attributed to low farm prices in the United States. Imports of catfish were up in 1993, aided by higher catfish prices at the farm and processor levels. Brazil was again the largest supplier of catfish imports, accounting for 90%. Mexico has been the second largest supplier of catfish to the United States. In 1998, catfish imports were 625,500 kg (1.39 million lb) with Vietnam and Guyana becoming important suppliers.

Trout. Idaho dominates trout production in the United States, comprising 75% of the market. The growth of trout culture has been somewhat variable during the 1990s but has averaged 25.2 million kg (56 million lb) and $71 million in sales annually, and the number of operations has remained steady during the 1990s. The primary states involved are Idaho, North Carolina, California, Pennsylvania, Washington, and Virginia. Trout farming requires colder water temperatures and flowing water with a high oxygen content (>5 ppm). The predominant trout species raised is the rainbow trout (*Oncorhynchus mykiss*), which is considered a hardier variety compared to other species such as brown, (*Salmo trutta*) and cutthroat trout (*Salmo clarki*). The United States has been a net importer of trout during the 1991 to 1997 period.

Salmon. Commercial landings of salmon in the United States are enormous (Fig. 4) and dwarf farm-raised salmon production in the U.S. (Fig. 2). Estimates of farm-raised Atlantic salmon (*Salmo salar*) production in the United States were 3.15 million kg (7 million lb) in 1988 and increased to 13.95 million kg (31 million lb) in 1996. Virtually all U.S. production is in Maine and Washington.

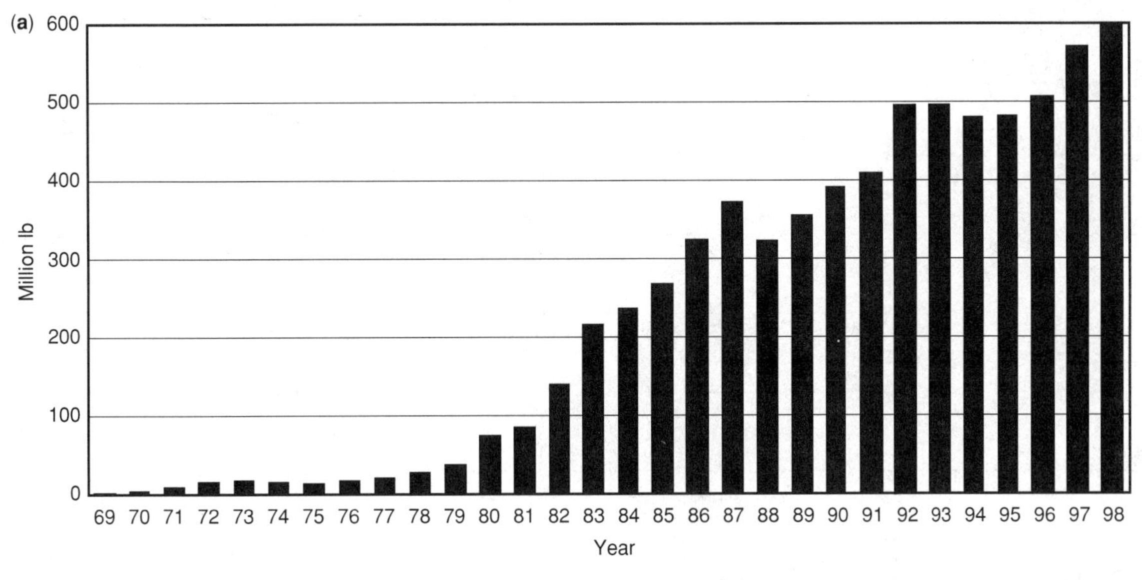

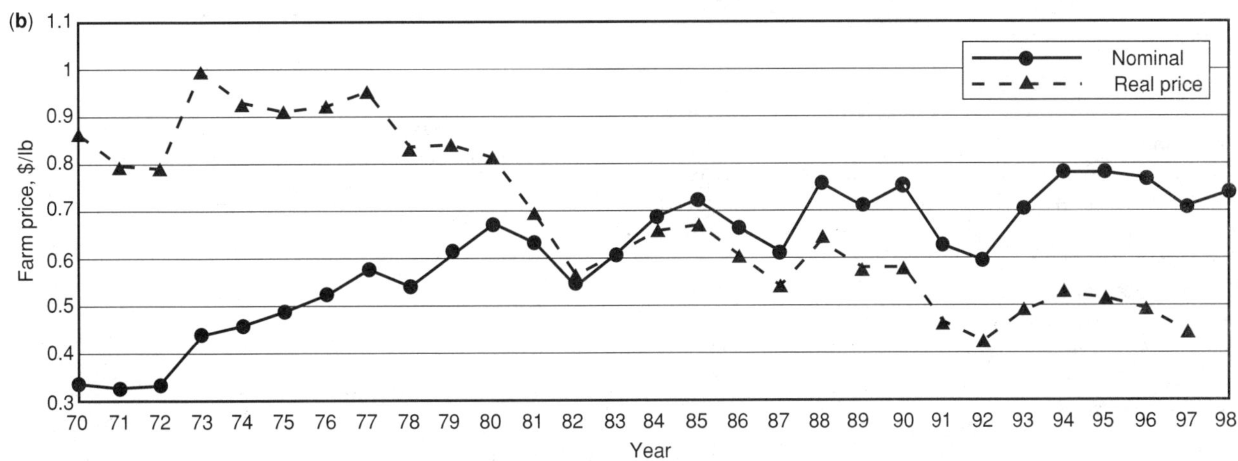

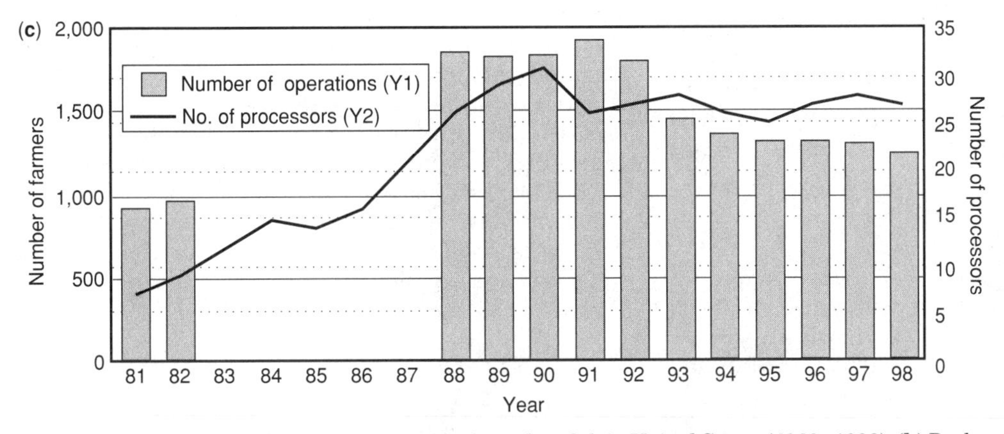

Figure 3. (**a**) Production of food-sized catfish in United States (1969–1998). (**b**) Real and nominal price of catfish 1970–1997, lb. (**c**) Number of catfish operations and processors, 1981–1998. *Source*: Economic Research Service, U.S. Department of Agriculture.

Maine's seafood industry ranks the contribution of 9 million kg (20 million lb) of farm-raised Atlantic salmon, valued at $55 million annually, second only to wild-caught lobster. This is remarkable considering that the industry only successfully experimented with Atlantic salmon in the early 1980s and expanded rapidly from 1987 to 1991.

While the domestic farm-raised salmon industry is expected to expand, its rate of growth in coming years will likely be much slower (in both Maine and Washington), due to the lack of high-quality sites and cost of obtaining new farming permits. Almost all of the increase in production in the last several years has been at existing leases, not

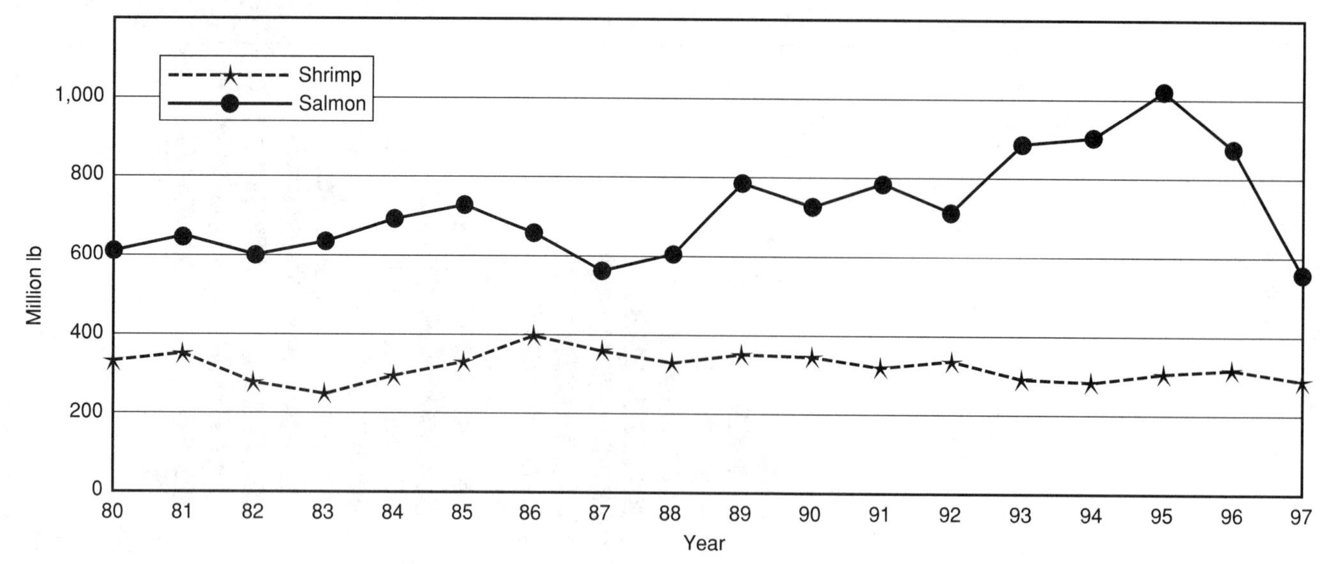

Figure 4. Commercial landings of shrimp and salmon in the United States (1980–1997). *Source*: Economic Research Service, U.S. Department of Agriculture.

through additional lease sites. Another major factor in the slowdown is increasing foreign competition in the salmon market. Canada is currently the largest supplier of farm-raised product, but imports from Chile have also expanded rapidly during the past few years. While Norway is not a large supplier in the fresh market, it remains the world's leading producer of farmed Atlantic salmon. Production in Norway has not been growing rapidly, but innovations in production practices promise to lower production costs. Production in Chile has expanded very rapidly over the last several years so that Chile is now probably the lowest cost salmon producer.

From 1993 to 1997, U.S. salmon exports (wild-caught Pacific salmon plus farm-raised Atlantic salmon) fell from $700 million to $436. The 1993 total included 1.8 million kg (4 million lb) of fresh cultured Atlantic salmon valued at $11 million; exports of fresh Pacific salmon, all wild caught, were 10.35 million kg (23 million lb), valued at $34.8 million. The majority of salmon exports were frozen Pacific salmon, 115.65 million kg (257 million lb), valued at $484 million.

Shrimp. In 1993, U.S. culture of shrimp was approximately 3 million kg (6.6 million lb), which is about 2% of the U.S. wild-caught shrimp quantity. Imports of shrimp were estimated to be $3.0 billion in 1997, and historically imports have always exceeded exports. Approximately 60% of the imported shrimp are thought to be farm raised. Expansion of U.S. shrimp aquaculture is limited because of the low cost of production in foreign countries and the large U.S. wild catch. Asian and Latin American countries that produce farm-raised shrimp have developed Western markets through the steady supply of consistently high-quality and uniformly sized shrimp at competitive prices. It has been suggested that the comparative U.S. advantage may lie in the technical aspects of producing disease-free postlarvae for the foreign shrimp industries or in providing hatchery technology for production of postlarvae.

Emerging Species

Molluscs. Culture of oysters, mussels, and clams is small compared to the wild catch, but aquaculture is growing because of reduction in available oyster and clam stocks and aquaculture's ability to supply a steady high-quality product that can command a premium price. U.S. production data for most molluscs are not available on any comprehensive basis that separates the species or distinguishes between wild and cultured. In 1993, total oyster sales in the United States were approximately 14.4 million kg (32 million lb), valued at $98 million. Mollusc imports ranged from 1.8 to 4.5 million kg (4 to 10 million lb) from 1989 to 1993, with value ranging from $52 to $64 million. Mollusc exports averaged approximately 2 million kg (4.4 million lb), valued at approximately $11 million during the 1990–1993 period; however, in recent years, this trade imbalance has grown dramatically (Fig. 5). In 1997, scallop imports exceeded $200 million, and the other three emerging species approached $50 million each. As these new species imports grow in combination with existing shrimp and other fish and seafood imports, it is clear that these sources are becoming an important element in the U.S. trade imbalance.

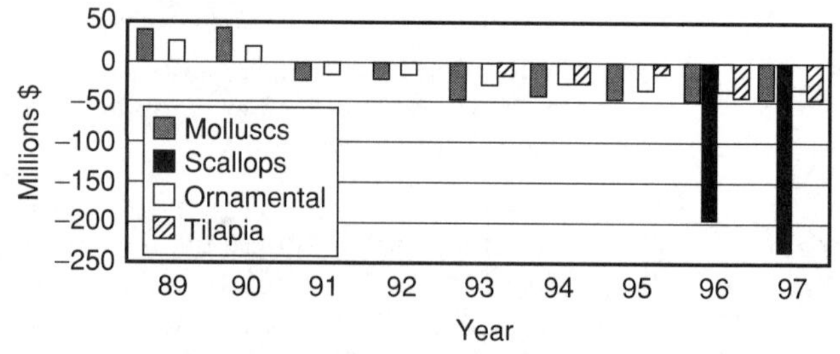

Figure 5. Net import–export of emerging species in the United States (1989–1997). *Source*: Economic Research Service, U.S. Department of Agriculture.

The outlook for mollusc culture is mixed. Advances continue in efficient culturing methods for some species such as abalone and oysters, while for other species such as soft clams, surf clams, and scallops, culture techniques are still in the experimental stages. The two greatest constraints to expansion of mollusc culture are the limited number of suitable sites and food safety related to consumption of raw molluscs, particularly oysters and clams.

Tilapia (*Oreochromis* sp.). The outlook for tilapia production in the United States is good for several reasons. First, tilapia can be polycultured (cultured in the same pond with other fish species) with no adverse effects. Second, it can eat many types of feed, and, of special importance, it can be grown profitably using diets of less expensive vegetable protein. Third, it can be bred easily and quickly. Last, the flesh is mild and can be substituted for a number of other traditional seafood species. However, there are limitations to U.S. culture of tilapia, foremost being its intolerance to water temperatures below 7 °C (45 °F) and reduced growth below 21 °C (70 °F). Geothermal water sources and recirculating systems using heated water are being used to circumvent the temperature constraint. Tilapia production is not concentrated in one region or area and is presently being undertaken in Arizona, California, Florida, Idaho, New Jersey, and Texas. Promotional efforts to make consumers more aware of the product and its quality are attempting to expand the market for tilapia. Some consumers prefer the lighter color of the golden hybrid to the darker color of the more common tilapia species. Consumer tests also indicate a preference for a larger size, such as the fillet from a 340 g (0.75 lb) fish.

Future growth of the industry will require expansion of the market, lowering of production costs to be competitive with other fish species, and development of cost-efficient water recycling production systems. Imports of tilapia have grown dramatically from $13 million in 1992 to $49 million in 1997 (Fig. 5). Most of the tilapia imported from Asian countries are cultured, while fresh fish imports from Mexico are primarily wild caught.

Ornamental Fish (Tropical Fish). The ornamental fish industry is centered in Florida, particularly in the Tampa and Miami areas. These products are considered luxury items and may decline during recessions, as in 1992, when imports were down 12%. Producers are concerned about the Food and Drug Administration applying the same rules for therapeutic chemical use in food fish production to the ornamental fish industry.

The forecast for U.S. exports of ornamental fish indicates continued expansion, but net trade will continue to be negative (Fig. 5). Export markets to the European Community (EC) are increasing but American producers must compete with Asian producers for this market. There are concerns about recent EC directives covering the importation of fish, which require U.S. exports to the EC to be certified by a U.S. agency that the product complies with EC regulations. Imports of ornamental fish have ranged from $12 to $53 million from 1988 to 1997, while exports have ranged from $6 to $41 million over this same period.

Crawfish. Crawfish culture is probably the U.S. aquaculture activity most heavily affected by wild production of the same species. Wild production occurs during the same season as aquaculture production. Wild harvests are dependent on water temperature and the volume of water moving through the swamp areas of Louisiana. Because of the variability of these factors, wild crawfish production has experienced wide year-to-year swings 8.1 million kg from (18 million lb) in 1991 to 31 million kg (69 million lb) in 1993. This supply fluctuation is a serious limitation to crawfish market development because the food industry wants a product that is in constant supply, has good quality, and has a relatively stable price. Crawfish are not available year round, which makes it difficult to build a steady market. Seasonal harvesting patterns, along with a considerable wild catch, means there are huge swings in prices throughout the season.

The advantages of crawfish culture are the relatively low fixed costs of production, natural reproduction, and the ability to be double-cropped with rice. United States consumer demand is still relatively restricted to the states where crawfish are found naturally and those areas where they are farmed. Large crawfish are exported to Sweden at a premium price, medium crawfish are sold to restaurants, and small crawfish are peeled for tail meat. Recent exports from China during the off-season for U.S. crawfish production may allow for a more continuous year-round supply. United States culture may actually expand as a result.

COMPARATIVE ADVANTAGE

Industry growth is largely determined by comparative advantage, with both production and marketing factors involved. Aquaculture competes for resources with other potential or existing production activities and competes in the marketplace with fish and seafood products from capture sources. On the production side, fish culture systems may often be able to use, and even thrive, on land resources that are not suitable for other agricultural enterprises. Examples include salt flats for shrimp, highly erosive watersheds for catfish, and pens for salmon.

The relative net returns to land, water, labor, management, and capital in one location versus other locations will ultimately determine aquacultural viability. It is not simply a matter of relative production costs. As long as the returns in a location provide an incentive, culture will continue. If the opportunity cost of resources is lower in a given area, the cost of production can be higher, allowing the industry to still maintain competitiveness. Returns to alternative uses of land, water, and other resources may be higher as a result of agricultural commodity programs or potential urban-industrial uses. Thus, the ability to compete with production from other areas will depend on relative input costs, including opportunity cost. For example, 0.4 ha (1 ac) of delta land in Arkansas may have a lower cost of production in catfish than the same area in the Black Belt of Alabama; but if the land in Arkansas can produce rice and receive a subsidized price, it will probably not be put in catfish while land in Alabama is available for catfish production because net returns to alternative enterprises is low.

The quantity that can be produced from a given area using a given technology will be limited by the availability of the resource. Often this will be land and water of a certain quality. Once the low-cost alternative reaches its capacity, the next more expensive technology and resource combination will come into production. New technology can increase the capacity by decreasing dependence on the most limiting resource. If water availability or quality were constraining the production acreage of catfish in the Mississippi delta region, then new techniques to conserve or re-use water could increase the area's capacity. Where the availability of postlarvae (PL) for shrimp production has become limiting, hatchery technology has been improved to decrease dependence on wild-caught PL. Waste assimilation in salmon culture has been an important constraint that has limited growth and has resulted in greater public and private scrutiny of location and intensity of pen culture.

Aquaculture is closely linked to capture fisheries. The development of aquaculture is greatly enhanced by selecting a species with a strong history of consumption. The shrimp and salmon aquaculture industries have grown in response to a deficit in the supply of these species from capture sources that allowed the cost of production to be competitive with the cost of capture. The consistent quality and quantity that can be made available to the market through culture is a further advantage. Catfish production has been hindered by the lack of a history of consumption in many areas of the United States. Catfish promotion and advertising programs have been needed to improve its image. Aquaculture is able to break the seasonal and geographical limits historically imposed by capture sources.

Culture industries that are favored by a ban or moratorium on a capture fishery are vulnerable to the possibility that the capture fishery could recover before the cost of culture has decreased sufficiently to compete with capture sources. Such bans on striped bass, salmon, and cod may provide opportunities for the culture of these species; but they may also prove to be either temporary opportunities or an opportunity better filled by other species that the consumer is willing to switch to. Consumer loyalty to a species is probably limited. At some point, higher prices, low quality, and relative lack of availability of the traditionally consumed but restricted species will lead the consumer to try alternatives. Thus, a restriction on the capture of a particular species does not necessarily imply a long-lasting opportunity for the culture of that species.

Externalities

Aquaculture generates and is affected by a number of externalities that could have profound impacts on the sustainability of the industry. Environmental pollution associated with pond effluent, discharge from processing plants, and sedimentation in coastal areas is a pressing issue that has to be resolved. There is some evidence that catfish aquaculture is lowering the water table in the Mississippi Delta. This raises pumping costs for current producers and threatens the long-term viability of the industry. Communities and other producers in the region that depend upon the water table are also affected.

Aesthetic objections represent another class of problems for U.S. aquaculture, especially in connection with salmon farming and the raft culture of shellfish in the Pacific Northwest. Coastal residents find rafts, pens, and other facilities unsightly. Zoning, licensing, and other restrictions have been placed on the industries. Although aesthetics represent a legitimate concern, there are tradeoffs involved in terms of reduced employment in the affected communities, reduced local tax revenues, and higher seafood prices. Economic research is needed to shed light on the magnitude of these tradeoffs so that all parties and costs are adequately represented in the debate.

Environmental quality problems experienced by overseas producers may improve the competitive position of U.S. aquaculture. For example, the shrimp culture industry in Taiwan was decimated by environmental and disease problems in the early 1990s. Shrimp production areas in Indonesia, the Philippines, Ecuador, and Thailand have experienced some of the same problems. Much of the problem is rooted in overexploitation of the resource base through intensive production practices located in highly exploited watersheds. Aquaculturists have espoused the advantages of culture relative to capture based on the notion that capture fisheries have reached their carrying capacity, but watersheds also have carrying capacities. Ill-advised intensity and density of production operations have led to conflicts with other users of the resources in the watershed and degradation in the quality of the pond environment. Not only has this overexploitation led to declining yields and profits, in some cases the intensive production practices, especially the use of some chemicals, have raised issues of consumer health concerns. If U.S. producers are better able to deal with environmental and food safety concerns, they may gain an important advantage relative to international competitors.

Common property resources such as wild-caught postlarval shrimp represent another class of externalities. Postlarval shrimp obtained from the oceans are used by shrimp farmers in many parts of the world to stock their ponds. Increased harvests of the postlarvae reduce the wild catch of adult shrimp. Thus, cost savings from using the wild postlarvae for aquaculture come at some expense to those involved in the ocean capture industry. There is growing concern that shrimp aquaculture is causing a rapid depletion of natural shrimp populations.

The importance of postlarval shrimp in the food chain of many marine fish species raises additional concerns. Reduced shrimp populations may adversely affect fish populations of important commercial, ecological, recreational, or aesthetic value. Increased development of shrimp hatcheries can reduce dependence on an uncertain seed supply and eliminate the threat to the regeneration of wild stocks.

Aquaculture may be adversely affected by externalities generated by other enterprises. Fish health, growth rate, and human safety may all be compromised by pesticides, fertilizers, and industrial pollutants in streams and groundwater. Catfish farms are often in close proximity to chemical-intensive row crop operations such as cotton, soybeans, and peanuts. In fact, much of the land converted to catfish ponds in Mississippi is former cotton and soybean land. Agricultural runoff, crop-dusting, chemical leaching into the groundwater, and chemical residues from former crops are all potential vectors of contamination.

Major problems for molluscs are generated by their dependence on water quality, because they are filter feeders that concentrate toxins. Dinoflagellate blooms produce toxins that are harmful to humans, for example, paralytic shellfish poisoning (PSP). Mollusc beds are continually monitored by state and federal agencies for dinoflagellates, and producers must tag all bags of molluscs for identification purposes. A solution commonly used in Europe is depuration facilities. They remove most harmful substances, including infectious bacteria, but not viruses or heavy metals. Such facilities will become more commonly used in the future as concerns increase about the safety of seafood products.

Growers must sometimes deal with restrictions concerning obstructions that interfere with navigable waters. In addition, in most states, growers will be faced with a long and elaborate process of leasing the rights to farm-specific areas of the bay or ocean bottom. Not owning the property may increase risks, limit planning horizons, and lower the amount of total capital invested. These limitations are also faced by salmon netpen culturists. Leasing ocean bottoms may cause potential conflict among aquaculture operators and fishermen, regulatory agencies, and recreational fishermen. Leasing arrangements may be even more of a problem for off-bottom operations, such as raft culture, as operators lease the water column as well as the bottom.

Infrastructure

Infrastructure facilitates the flow of aquacultural products between buyers and sellers. Specialized research facilities, industry trade associations, extension services, physical facilities such as feed mills, fish disease diagnostic laboratories, supply firms specializing in aquacultural inputs, live-haul and other transportation services, and processing plants are all important elements of the infrastructure supporting aquaculture. For example, growth in the shrimp aquaculture industry in South Carolina has been impeded occasionally due to limited stock availability from wild sources and hence, a lack of suitable infrastructure. Because aquaculture in the United States is relatively new, many of the components of infrastructure that are taken for granted in other industries (e.g., extension expertise and specialized research facilities) either do not exist or are still in developmental stages.

Market growth has permitted increased specialization and improved scale economies in production and processing (2). Processors are a pivotal link in the marketing chain that enables aquacultural production units to thrive in a particular locale.

Vertical integration, production and marketing contracts, and cooperatives can be implemented as a means of lowering costs or enhancing price. The chicken broiler industry may serve as a model for the catfish industry, suggesting its eventual evolution into a fully integrated production system. There are many situations in which contracts or integration are cost-saving alternatives to market exchange. For example, processing is an industrial activity that achieves lowest cost by operating at a steady rate throughout the year. Contracts or integration offer a mechanism for assuring a continuous supply of the size and type of fish desired by the processor. Farmers gain through year-round markets at known prices. Cooperatives and bargaining associations permit fish farmers to exert greater control over the pricing and marketing of their products.

Supply Response and "Boom and Bust" Cycles

Aquaculture has been described as simply another form of livestock production and, as such, should receive the same treatment as the more traditional forms of agriculture. Livestock development evolved from hunting to domestication of animals, and the fishing industry is following a similar path. The fishing industry is basically involved in the tracking and harvesting of wild populations. However, modern fishing fleets have the capacity to harvest fish faster than the animals can reproduce. In conjunction with the problems inherent in common property resources, the depletion of ocean fish populations is occurring. Farming trials to evaluate wild fish species to determine their suitability for aquaculture have been conducted over the past few millennia in China, and in the past few decades in many Western countries. Even though a number of fish species are now farmed, in many cases the genetic stock is still basically that of a wild population. Using the same type of genetic selection that has increased the efficiency of livestock and poultry production, aquaculture operations should be able to increase their efficiency.

Another comparison often made between aquaculture and livestock production involves cyclical or "boom and bust" production cycles. This type of cyclical production may be developing now in the catfish industry. That industry has gone through a series of slower and faster growth periods, but they have been overshadowed by the overall growth of the industry (Fig. 3a, b). Over the 1991 to 1993 period, the catfish industry saw production increase very rapidly, resulting in farm prices falling below production costs for many producers. In response to this, producers cut back on their stocking rates and some growers were forced out of the industry. The end result was that, after a period of time, stocks of available fish were reduced and farm prices rose. As prices improve, the growers remaining in the industry will gradually increase their production. If prices remain above the average cost of production for an extended period of time, new growers will enter the industry. With new growers entering the industry and established growers expanding production, prices will begin to fall and the cycle will start all over again.

Now that the aquaculture industry has gotten past the initial phase of working out basic production techniques, it seems to be following the type of development seen in the livestock and poultry industries. For the major livestock species, some of the trends have been for larger units of production, greater concentration of production, and vertical integration of production. Many of these trends are also evident in the growth of the shrimp industry in other countries and the catfish, trout, and salmon industries domestically. If the major aquaculture species follow the development path of traditional livestock and poultry industries, the outlook will be for larger production units, higher concentration of production, more vertical integration, rising production efficiencies, and declining real costs.

In the long term, real prices for livestock and poultry products have declined. This trend has also occurred in the catfish industry (Fig. 3b). As a result, growers will continue to be pressured to adopt or develop new methods of increasing their production efficiency. On the positive side, this long-term decline in real prices will likely mean that the price of farmed products will become more competitive with the wild harvest and that aquaculture production will become a major factor in those markets.

However, aquaculture may develop differently from the livestock and poultry industries because of its wider range of species. In this sense, aquaculture could be more closely compared with the fruit and vegetable industries. Some aquaculture species appear to be developing into major industries, such as catfish and trout, and they will likely follow the path taken by the livestock and poultry industries. For many other aquaculture species, production will remain much lower, and they will be marketed more as specialty products (3).

The response of aquacultural supplies to changes in price is a critical factor affecting the ability to understand and predict the economic consequences of technical change. Similarly, the effects of competing fish supplies, feed costs, industry advertising, and governmental regulation and policies are difficult to analyze without this information. While an increasing body of evidence is becoming available on the demand elasticities for fish and seafood products (See the section "Demand Characteristics"), including aquacultural products (4,5), relatively little is known about the size of supply elasticities for aquacultural products. Branch and Tilley (6) estimate harvest response elasticity for catfish of about 0.60. Zidack, Kinnucan, and Hatch (7) estimated the short-run supply elasticity for catfish over the 1980–1989 period to range from 0.15 to 0.72. Beyond those studies, little is known about supply response for major aquacultural products, such as shrimp and salmon.

Knowledge of supply is complicated greatly by the problem of monitoring on-farm inventories. At any given time, there is a large "floating inventory" of fish in ponds that often cannot be measured with much precision (8). The lags that are prevalent with the growout of land animals are also important in aquaculture supply. In addition, the rapid growth of an infant industry, such as aquaculture, can result in misallocation of resources and price instability.

MARKET DEVELOPMENT

The main tools of market development are price, product, place, and promotion. The demand for fish in general, and aquacultural products in particular, is price elastic, meaning that consumers are price sensitive. On the one hand, price increases for nonaquacultural fish products associated with the depletion of natural fisheries enhance market development efforts of aquaculture. At the same time, however, increased production costs are difficult to pass along to consumers without loss in market share for aquaculture. Product differentiation is one means for decreasing the price sensitivity of consumers.

"Product" refers to the physical attributes of aquacultural commodities in terms of the characteristics that consumers desire. Convenience, freshness, flavor, texture, product form, safety, and nutrient content are all characteristics of aquacultural products that have relevance to the consumer. Consumers place a value on each relevant product attribute. Knowledge of the implicit prices consumers attach to fish characteristics can be used to position aquacultural products against competing fish and non-fish food items. Consumer taste tests (9) delineate differences between captured and cultured fish and provide valuable information for developing marketing strategies.

Spatial dimension of marketing are referred to as "place." Areas or locations are identified where the product commands a premium price net of transportation costs. Although shrimp is a universal favorite, regional differences exist in consumer preferences for fish and seafood. For example, salmon, halibut, catfish, perch, haddock, and flounder are the most preferred finfish in the United States depending on the census region (10). The identification of regional preference patterns aid market development by pinpointing areas in which aquacultural products may be introduced with the least consumer resistance and, therefore, the highest probability of success.

In some cases, overseas markets are easier to penetrate than domestic markets. For example, in 1991, over 10% of the cash value of U.S. crawfish was exported, primarily to Sweden and Finland (11). Although export markets in general may be a less economical way to expand the demand for aquacultural products than domestic market development, subsidies available through federal export promotion programs may encourage foreign market development in some instances.

"Promotion" is an especially important market development tool in infant industries where little is known about the product. Classical product life-cycle curves consisting of introductory, growth, and maturity phases characterize the development of many aquaculture industries (8). Technical constraints, production costs, and competing products are fundamental factors governing how rapidly a particular industry moves through phases. Product promotion, however, plays a key role.

Consumers must be informed about the product's existence and unique characteristics if an industry is to gain a competitive edge in a complex market such as the United States. The mass media, both electronic and print, offer an efficient vehicle for communicating the merits of fish to a large potential market. In the case of catfish, print media advertising increased consumer awareness of farm-raised catfish, which in turn significantly increased the ability of the industry to expand markets both within and outside the traditional consuming areas (12).

Market development will be an important source of growth as the aquaculture industry matures. In general, there will be a continuous attempt to increase supply in order to take advantage of larger market opportunities. Individual growers, groups of growers, and processors will be in a constant struggle to expand to the next level of market size; however, there is "lumpiness" in the technologies and markets related to the additional size needed to exploit economies of size. Expansion must take place within the context of the minimum flow of product required in the larger market and of maintaining profitability.

In cases where the species is well established, infrastructure investment will be needed to augment the consistent quality and flow of product. When the cultured species is relatively new to the consumer, infrastructure development will have to be augmented with strategies to increase consumer awareness. Several studies have found that consumers are more likely to experiment with new food purchases in a restaurant than in a grocery store. Restaurant managers are often looking for new products to add to their menus, and new "exotic" entrees attract consumers.

Grocery store managers are more likely to select well-established products due to the competition for space in the store and consumer tendency to purchase products with which they are familiar in terms of taste, difficulty of home preparation, and safety. Managers of grocery stores that have specialized seafood sections may be more like restaurant managers in that they will desire diversity. Prepackaged, value-added cultured species may be located in the meat section along with beef, pork, and poultry as opposed to the fish and seafood section.

Market development can target two segments: domestic and foreign. Because the United States is a major market, it makes sense first to exploit the latent demand for aquacultural products within the United States before investing significant resources to expand foreign markets. The long-term trend of rising U.S. per capita fish consumption appears to have ceased in the early 1990s, based largely on safety and quality concerns. Since most, if not all the world's major fisheries are being exploited at or beyond sustainable yields, the opportunity for aquaculture to expand its market share is obvious, especially if renewed consumer confidence can result in a continuance of the long-term upward trend in per capita fish consumption.

Aquaculture's ability to provide a consistent, high-quality, standardized product for large commercial markets has made it attractive for institutional markets. Restaurants and retail supermarkets have had problems with the inconsistent quality and quantity of fish and seafood from capture sources (13,14). As a competitor with other sources of protein (principally beef, pork, and chicken), fish contains less fat (15).

Processors dominate the marketing of aquacultural products. Due to economies of size and the expense of transporting live fish, many aquaculturists are served by few processors (16,17) (Fig. 3c). Alternative marketing channels have been profitable for small numbers of producers; however, the limited market represented by direct sales to restaurants, live-haul, and fish-out creates a dependence on the processor for moving large quantities of product (18). The development of these smaller markets may be an appropriate transitional strategy to build up sufficient acreage in an area to attract a processor.

Because of the large number of fish species available in the market, the demand for any particular aquacultural product is likely to be price elastic (19). Price increases are resisted by consumers, who simply purchase a close substitute. This high substitutability also implies that increases in production or marketing efficiency that lower price offer an important avenue for accelerating the rate of aquaculture growth. If a capture species becomes more expensive, consumers are likely to switch to the cheaper cultured product.

Information dissemination and consumer education programs can play a vital role in expanding the demand for aquacultural products. Aquaculture's ability to control environmental factors that may affect the safety and nutritional quality of fish may not be perceived by the consumer unless the information is effectively communicated. The negative product image associated with some fish or mollusc species may be overcome with appropriately designed generic advertising campaigns (12). In other cases, product differentiation and market segmentation can be facilitated through the development of consumer information programs tailored to specific target groups, for example, Pacific coast versus Gulf coast oysters.

Aquacultural products are similar to other new products in that they are subject to the same S-shaped diffusion paths (8). Whether the diffusion proceeds slowly or rapidly is determined to a significant degree by the marketing effort to support the product. Although industry advertising offers an efficient mechanism for stepping up the rate of consumer acceptance of aquacultural products, the collective action required to obtain the funding and the attendant free-rider problems pose a challenge for the diverse and fragmented aquacultural industry (20).

Product Differentiation

Product differentiation is an especially important aspect of U.S. aquaculture that has received inadequate attention. The ability to convince consumers of the superior qualities of fish and seafood products in general, and aquacultural products in particular, is crucial to the industry's long-term growth prospects. Although being farm raised may not ensure a better-tasting product (9), consumers may find the farm-raised product appealing in other respects. For example, feeding farm-raised fish carefully formulated feeds may allay fears about the safety of eating fish. Kummer (21) stated that chefs find penned fish flesh flaccid because of lack of exercise and flavorless because of the fish's packaged feed diet. A bland taste, however, is what most U.S. consumers appear to want in fish. Kummer (21) attributed catfish's rapid growth in popularity to the "perfect blandness" of the farm-raised product.

Health-conscious consumers want a diet with less fat and cholesterol. The leanness of fish could be a targeted message of promotion campaigns. Better understanding of consumer perceptions of aquacultural versus nonaquacultural fish products will assist in identifying misconceptions and enhance the ability to educate consumers about aquaculture's contribution to a healthy and satisfying diet.

Food Safety

Food safety is one of the key consumer issues of the 1990s. Media attention given to contamination incidents (e.g., vegetables, fruits, milk, and meats) has sensitized consumers to food quality and safety. The perishability of fish, coupled with their tendency to absorb and concentrate some pollutants, make the issue of food safety especially germane to aquaculture. Brooks (22) found that 38% of consumers had seen or heard news stories on some negative aspect of seafood. Actual risks have been measured to be substantially less than consumer perceptions. Greater control over the culture process as compared with the total lack of control over the capture process should be emphasized to allay consumer safety concerns. Water quality in ponds is monitored daily and nutrient intake is controlled through diet. Fish is a highly nutritious food and has been associated with a lower risk of heart disease (23).

Consumers often weigh negative information more heavily in their decision-making than equal quantities of positive information (24). The knowledge that catfish is farm raised (and presumably safe) was found to be the single most important factor influencing consumer perceptions of product quality and, therefore, purchase behavior (12). However, advertising must try to project the positive attributes of cultured species without creating strong negatives concerning capture products. The concerns associated with the latter message could lead consumers to bypass fish and seafood altogether, even if they are convinced that cultured fish and seafood are better than supplies from capture sources.

Quality control and consumer apprehension concerning the supply of fish and seafood may become an important comparative advantage for U.S. aquaculture. For example, perceptions that shrimp produced under the intensive Taiwanese system may cause health problems are resulting in greater care on the part of processors and exporters in their purchasing decisions. Consumers may be willing to pay extra for fish and seafood that are certified to be of high quality and pathogen-free. Proximity to the final consumer may make this certification process easier to implement. This quality issue may result in a more distinct separation in the "bulk" market and high-quality market niches.

Demand Characteristics

An essential element of demand analysis is selection of appropriate substitute products, which is complicated by the large number of fish species and the potential competition with other meat and protein sources (5,10,25–30). Although the controlled environment in which cultured species are raised should be an advantage for cultured products relative to wild-caught species, especially during periods of heightened consumer awareness of specific or general problems with safety in wild-caught populations, a surprisingly close relationship between demand for farm-raised and wild fish has been empirically documented. Typically, the consumer does not seem to make the distinction between farmed and wild supply and tends to react to a scare associated with one particular species at one particular time as an indictment of fish and seafood generally. Thus, farm-raised species as yet have not benefited from safety concerns related to wild fish (9). Returns to promotion and advertising (31–33) may be particularly high for fish species, especially farm raised, in a market environment in which the consumer is deluged by reports of seafood safety problems (34,35).

Catfish. The Southern Regional Aquaculture Center of the U.S. Department of Agriculture commissioned a cooperative university research group to develop a comprehensive analysis of U.S. markets for catfish (10,14,36,37). The study, completed in 1988, included telephone surveys of 3,600 consumers, 1,800 restaurant managers, and 1,800 grocery managers. Some 87% of respondents reported they eat fish and seafood, and of those consumers, 60% had eaten catfish. The New England and the mid-Atlantic regions had the lowest percentage of catfish consumers and the south central states had the highest consumption. Catfish ranked third behind shrimp and lobster as the consumer's favorite fish or seafood. Catfish ranked

Table 1. Favorite Fish or Seafood for Consumers and Sales Leaders for Retail and Restaurant Outlets Selected by at Least 3% of Respondents

Consumer[a]		Retail[b]		Restaurant[c]	
Species	(%)	Species	(%)	Species	(%)
Shrimp	22	Shrimp	9	Shrimp	27
Lobster	10	Catfish	7	Cod	10
Catfish	7	Cod	7	Catfish	7
Crab	6	Perch	4	Scallops	5
Scallops	4	Orange roughy	4	Lobster	5
Flounder	4	Red Snapper	3	Flounder	5
Cod	4	Flounder	3	Crab	4
Salmon	3	Haddock	3	Salmon	3
Haddock	3	Sole	3	Haddock	3
Oysters	3	Salmon	3	Halibut	3
Halibut	3	Halibut	3		
Trout	3				
Perch	3				

[a]Percent of responses to the question, "What are your three favorite types of fish or seafood?"
[b]Percent of responses to the question, "What are the top five fish and seafood products in terms of sales?"
[c]Percent of responses to the question, "What are the three most popular fish or seafood items on your menu in terms of sales?"
Source: Hatch, L.U. "National Survey of U.S. Fish Consumption." Presented at Aquaculture International Congress and Exposition, Vancouver, Canada, September 1988.

(a)

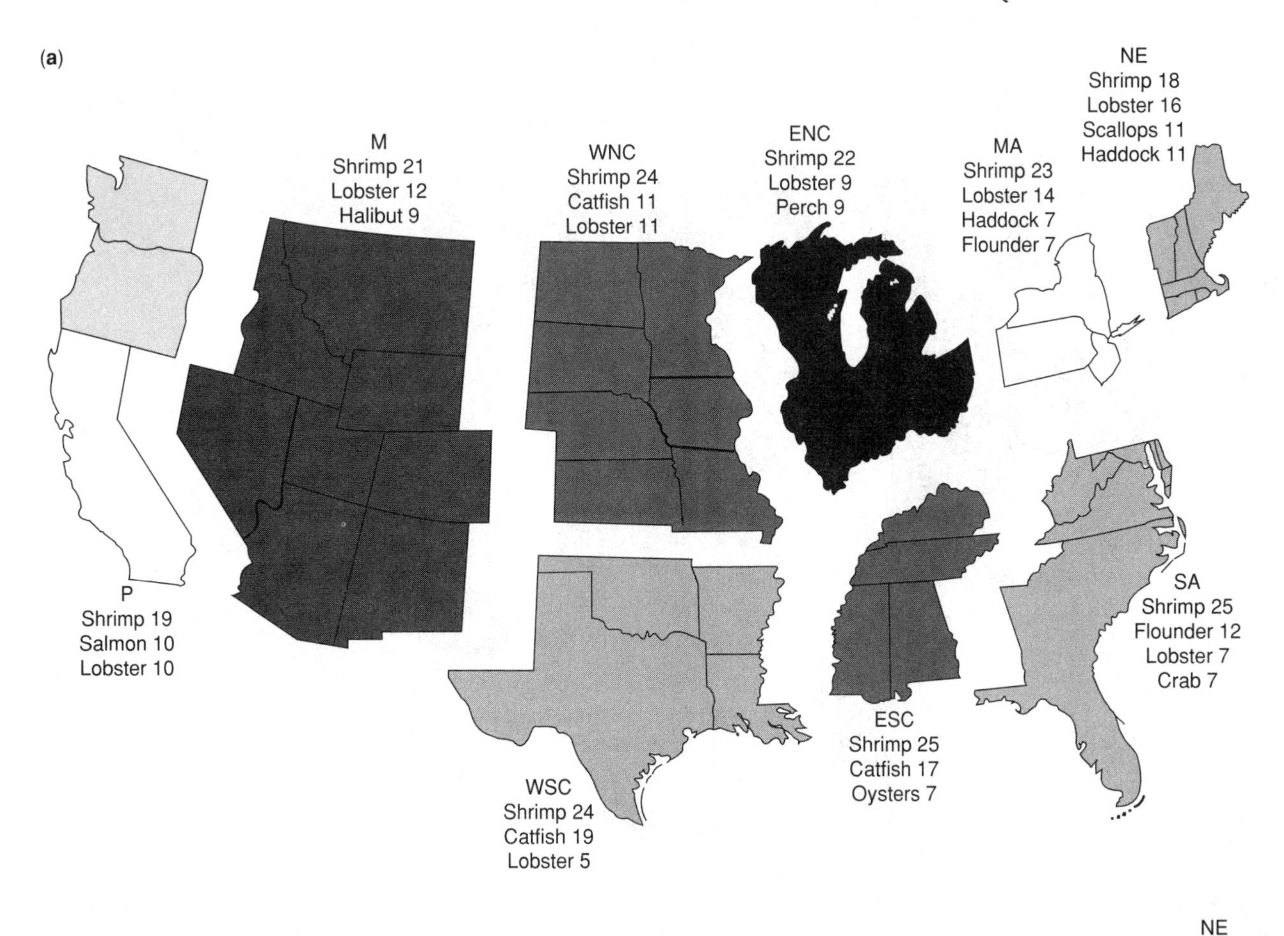

(b)

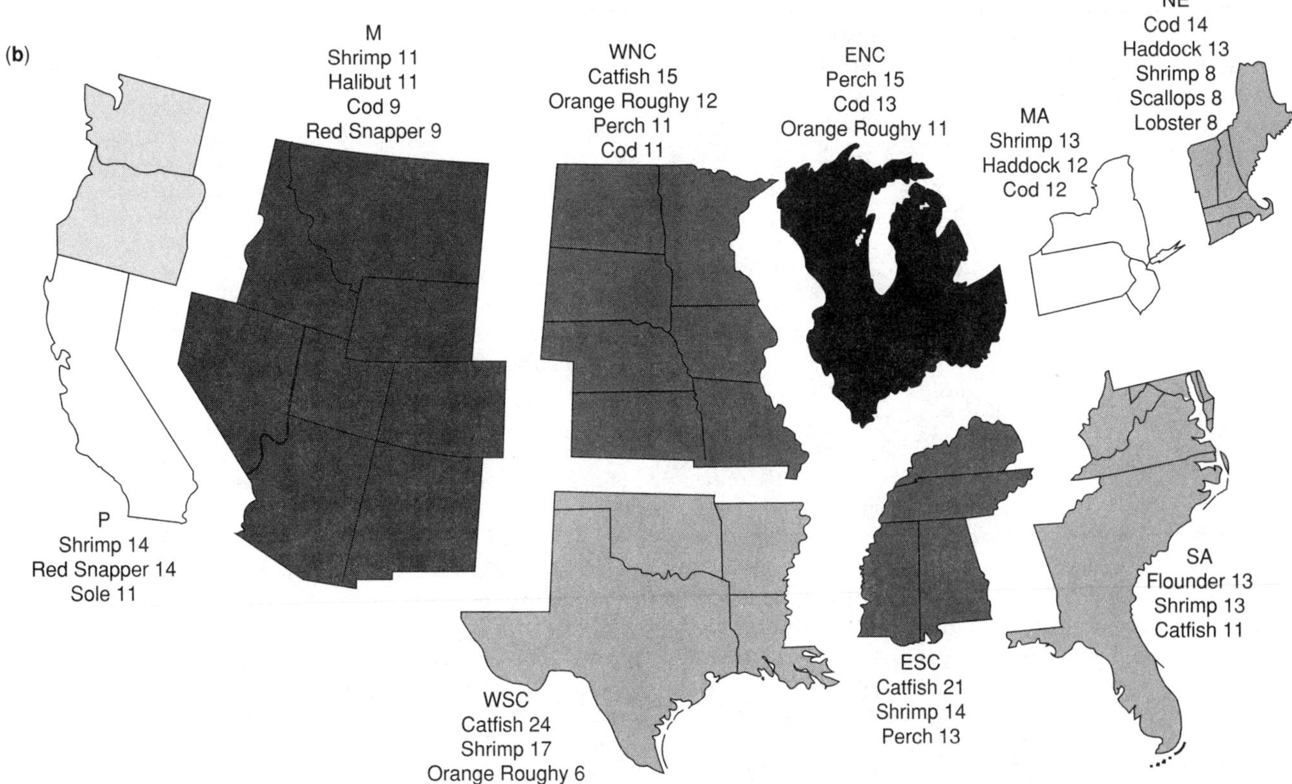

Figure 6. (**a**) Favorite fish and seafood of consumers by region, 1988. (**b**) Top-selling retail grocery fish and seafood products by region, 1988. (**c**) Top-selling fish and seafood products in restaurants by region, 1988. *Source*: Engle et al., "Retail Grocery Markets for Catfish" *Bulletin 611*, July 1991.

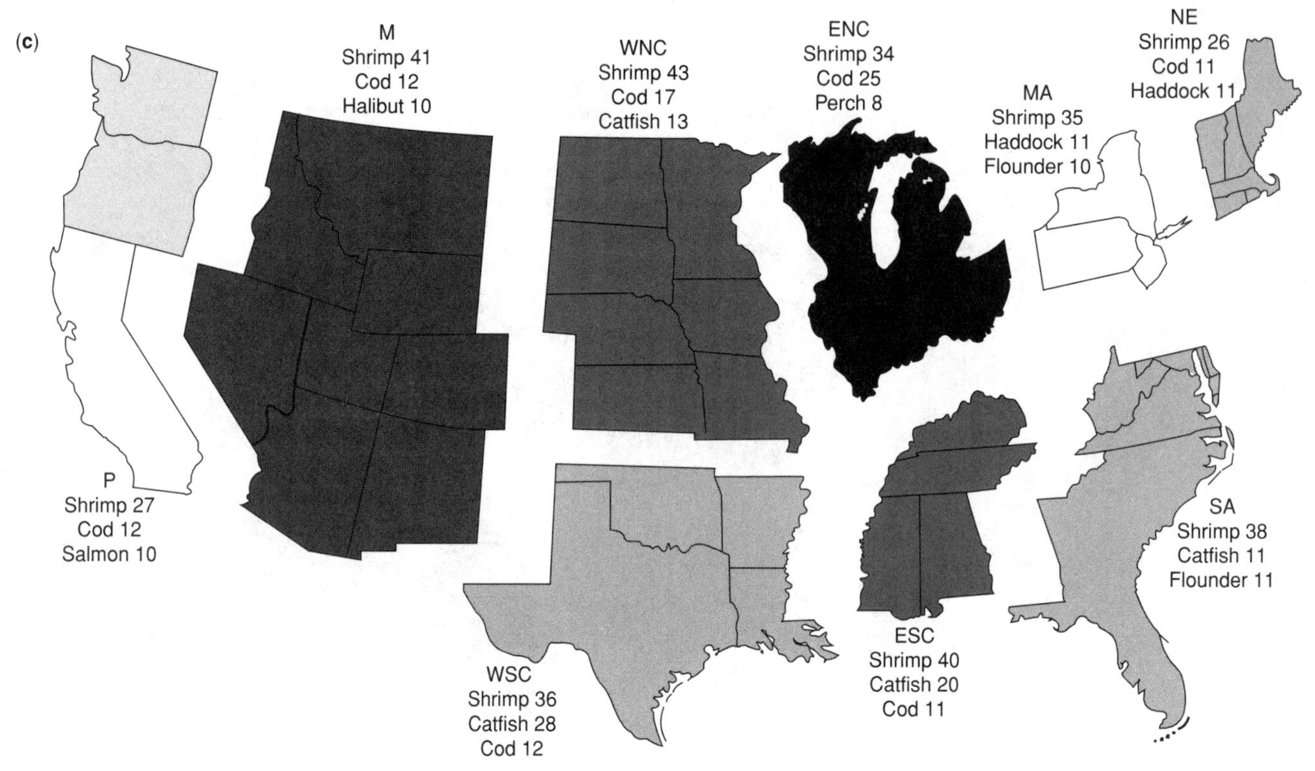

Figure 6. *Continued.*

second to shrimp in the retail grocery survey and third to shrimp and cod in the restaurant survey. The top 10 fish and seafood products ranked by consumers, excluding canned products, were shrimp, lobster, catfish, crab, scallops, flounder, cod, salmon, haddock, and oysters. For retail grocery managers, the ranking was shrimp, catfish, cod, perch, orange roughy, flounder, haddock, red snapper, sole, and salmon. For restaurant managers, the ranking was shrimp, cod, catfish, scallops, flounder, lobster, crab, haddock, salmon, and halibut (Table 1). Figure 6a–c shows the regional preferences for fish and seafood as judged by consumers and managers of restaurants and grocery stores. Those lists provide a good indication of competing products for cultured species.

Nutrition, ease of home preparation, flavor, and consistent quality were judged to be assets of catfish, while packaging, smell, and product availability were deemed its liabilities. Most consumers did not perceive a difference between farm-raised and wild-caught catfish. Relative cost to other meats was a distinct advantage for catfish over other fish and seafood.

Lambregts, Capps, and Griffin (38) estimated seasonal demand for catfish from 1987 to 1988 in a Houston, Texas retail market. Fillets appear to have less seasonality than whole dressed catfish. Beef is a gross substitute for catfish products. Shellfish appear to be a strong complement to whole dressed catfish. Kinnucan et al. (19) determined demand elasticities for catfish to be price elastic at the processor level (−1.28) and price inelastic at the farm level (−0.37).

Potential to expand the U.S. aquaculture industry through export to Europe was investigated by Lombardi and Anderson (39). The potential market for catfish in Germany was targeted using a conjoint analysis to provide estimates of relative value of fish attributes (e.g., size, form, and country of origin). Their analysis suggests that U.S. catfish producers may have an opportunity to export larger-sized fish to Germany.

Salmon. A survey of U.S. salmon markets for the Canadian Department of Fisheries and Oceans identified a 'salmon consumer' group of 43% and a nonsalmon consumer group of 47% (40). The majority of consumers did not perceive a difference between farmed and wild salmon. Price elasticities of −3.5 and −2.4 in the U.S. retail and restaurant sectors, respectively, were estimated. The majority of salmon consumption was concentrated in a small segment of the population. Salmon consumption would increase the most from year-round availability of fresh salmon with occasional, at-home consumers of salmon. All five areas surveyed (San Francisco, Los Angeles, Chicago, New York, and Dallas) showed good potential for increased consumption based on year-round availability of fresh salmon. In each area, gains in the retail sector are potentially higher than in the restaurant sector. Consumption patterns in Canada are similar to the United States except for a higher overall salmon consumption rate in Canada (66%). The Canadian sample indicates that at-home salmon consumption would increase if in-store recipes are made available.

Hermann et al. (41) found that demand for Norwegian Atlantic salmon was price and income elastic in both the United States and European Community and that demand was highly seasonal. Frozen chinook salmon was a weak substitute for Norwegian-raised salmon in European markets. Price elasticity was −1.97 in the U.S. and −1.83 in Europe; income elasticity was 4.51 in the U.S. and 2.73 in Europe.

Wessells and Holland (42) addressed consumer preferences for farm-raised or wild-harvested salmon using household data from the Northeast and mid-Atlantic regions. Consumer perceptions were greatly affected by assurances of quality, with farm-raised and federally inspected products to be preferred.

CONCLUDING COMMENTS

Marketing of aquaculture products has increasingly become an important part of the U.S. food supply. Several decades ago, farm-raised catfish, trout, shrimp and pen-raised salmon were restricted geographically and temporally; however those products, along with emerging cultured species, are becoming available nationwide thoughout the year. The ability to produce a standardized product, available year-round in environmentally controlled conditions, has allowed cultured fish and seafood products to establish important market niches. The concurrence of increasing population and associated environmental degradation with a rather stable supply of fish and seafood product from capture sources would suggest a rather bright future and rising importance of fish culture.

BIBLIOGRAPHY

1. U.S. Department of Agriculture, Economic Research Service, *Aquaculture: Situation and Outlook Report*, various issues.
2. M.J. Fuller and J.G. Dillard, *Mississippi State University Bulletin 930*, Mississippi State, MS, 1984.
3. D. Harvey, *Aquaculture: Situation and Outlook Report*, U.S. Department of Agriculture, Economic Research Service, Washington, DC, 1993.
4. K.F. Wellman, *Applied Econ.* **24**, 445–457 (1992).
5. H.T. Cheng and O. Capps, *Amer. J. Ag. Econ.* **70**, 533–541 (1988).
6. W. Branch and D.S. Tilley, *So. J. Ag. Econ.* **23**, 29–37 (1991).
7. W. Zidack, H. Kinnucan, and U. Hatch, *Applied Econ.* **24**, 959–968 (1992).
8. W. Zidack and U. Hatch, *J. World Aquacult. Soc.* **22**, 10–23 (1991).
9. G. Sylvia, in D. Liao, ed., *Proceedings of the Symposium on Seafood Advertising and Promotion: Research and Experience*, International Institute of Fisheries Economics and Trade, 1991.
10. C. Engle, O. Capps, Jr., L. Dellenbarger, J. Dillard, U. Hatch, H. Kinnucan, and R. Pomeroy, *Arkansas Agricultural Experiment Station Bulletin No. 925*, 1990.
11. D. Harvey, *Aquaculture: Situation and Outlook Report*, U.S. Department of Agriculture, Economic Research Service, Washington, DC, 1992.
12. H. Kinnucan and M. Venkateswaran, *So. J. Ag. Econ.* **22**, 137–151 (1990).
13. S.O. Olowolayemo, U. Hatch, and W. Zidack, *J. Appl. Aquacult.* **2**, 51–72 (1991).
14. R.S. Pomeroy, J.C.O. Nyankori, and D.C. Israel, Department of Agricultural Economics and Rural Sociology, AE 464, Clemson University, Clemson, SC, 1990.
15. T. Lovell, *Nutrition and Feeding of Fish*, Van Nostrand Reinhold, New York, 1989.
16. H. Kinnucan and G. Sullivan, *So. J. Ag. Econ.* **18**, 15–24 (1986).
17. J. Nyankori, *So. J. Ag. Econ.* **28**, 247–252 (1991).
18. C. Engle, U. Hatch, S. Swinton, and T. Thorpe, *Alabama Agricultural Experiment Station Bulletin No. 596*, Auburn University, AL, 1989.
19. H. Kinnucan, S. Sindelar, D. Wineholt, and U. Hatch, *So. J. Ag. Econ.* **20**, 81–92 (1988).
20. H. Kinnucan and M. Venkateswaran, *J. Appl. Aquacult.* **1**, 3–31 (1991).
21. C. Kummer, *The Atlantic* **270**(2), (1992).
22. P. Brooks, in U. Hatch and H. Kinnucan, eds., *Aquaculture: Models and Economics*, Westview Press, Boulder, CO, 1993.
23. University of California, Berkeley, CA School of Public Health, *Wellness Letter* **8**(10), (July 1992).
24. H.S. Chang and H.W. Kinnucan, *Amer. J. Ag. Econ.* **73**, 1195–1203 (1991).
25. U. Hatch, Paper presented at Aquaculture International Congress and Exposition, Vancouver, Canada, 1988.
26. P. Bird, *Marine Res. Econ.* **3**, 169–182 (1986).
27. D. De Voretz, *Can. J. Ag. Econ.* **30**, 49–59 (1982).
28. M. Kabir and N. Ridler, *Can. J. Ag. Econ.* **32**, 560–568 (1984).
29. E. Tsoa, W. Schrank, and N. Roy, *Amer. J. Ag. Econ.* **64**, 483–489 (1982).
30. F. Wirth, C. Halbrendt, and G. Vaughn, *So. J. Ag. Econ.* **23**, 155–164 (1991).
31. H. Kinnucan, M. Venkateswaran, and U. Hatch, *Alabama Agricultural Experiment Station Bulletin No. 607*, Auburn University, AL, 1990.
32. O. Capps and J. Lambregts, *So. J. Ag. Econ.* **23**, 181–194 (1991).
33. P. Brooks and J. Anderson, *J. Bus. Econ. Studies* **1**, 55–68 (1991).
34. J.G. Anderson and M.T. Morrissey, in V. Haldeman, ed., *The Proceedings of the American Council on Consumer Interests 37th Annual Conference*, Columbia, MO, University of Missouri, 1991.
35. U.S. Department of Commerce, *Analysis of Consumer Perspectives of Fish and Seafood*, Washington DC, National Fish and Seafood Promotion Council, 1991.
36. U. Hatch, R. Dunham, H. Hebicha, and J. Jensen, *Alabama Agricultural Experiment Station Circular 291*, Auburn University, AL, July, 1987.
37. W.M. McGee, L.E. Dellenbarger, and J.G. Dillard, *Mississippi Agricultural and Forestry Experiment Station Technical Bulletin No. 168*, Southern Regional Aquaculture Center, SRAC Publication No. 508, December 1989.
38. J. Lambregts, O. Capps, and W.L. Griffin, in U. Hatch and H. Kinnucan, eds., *Aquaculture: Models and Economics*, Westview Press, Boulder, CO, 1993.
39. W. Lombardi and J. Anderson, *Aquaculture Economics and Management* **2**, 43–48 (1998).
40. D. Egan, in U. Hatch and H. Kinnucan, eds., *Aquaculture: Models and Economics*, Westview Press, Boulder, CO, 1993.
41. M. Herrmann, R.C. Mittelhammer, and B.H. Lin, in U. Hatch and H. Kinnucan, eds., *Aquaculture: Models and Economics*, Westview Press, Boulder, CO, 1993.
42. C. Wesselles and D. Holland, *Aquacult. Econ. Management* **2**, 49–59 (1998).

MICROALGAL CULTURE

Gary H. Wikfors
Northeast Fisheries Science Center
Milford, Connecticut

OUTLINE

WHAT ARE MICROALGAE?

By direct translation from Latin, microalgae are "little seaweeds." However, defining microalgae further is not simple, because the microalgae represent a taxonomically diverse group of organisms, rather than a single, phylogenetic category. A functional definition of microalgae might be "photosynthetic single-celled or colonial microorganisms"; however, most of these microbes are able to grow without light if dissolved sugars are provided. Microalgal cells range in size from one micrometer — roughly the size of a bacterium — to several hundred micrometers (1 μm) — barely visible to the naked eye. Colonies and chains of some microalgal cells can attain a length of several centimeters (2.5 cm = 1 in.). This group of organisms contains remarkable morphological diversity, with shapes ranging from simple spheres to the ornate, silica shells of one group, the diatoms. Many microalgae are motile, propelling themselves with flagella, by amoeboid motion, or by gliding on extruded mucilage. Microalgae are found in an astonishing range of habitats, such as in fresh, saline and hypersaline waters, in polar ice, in soil, attached to plants and animals, and even in symbiotic relationships with fungi (e.g., lichens) and animals (e.g., corals). Historically, many of these organisms have been claimed and named by both zoologists and botanists; therefore, taxonomy has been, and remains, problematic. From the perspective of aquaculture, there are common characteristics that warrant their consideration as a functional group; the foregoing definition will suffice for this discussion.

As "microalgae" is a functional rather than phylogenetic group, a list of taxa that would reasonably fit within the category is necessary (see Table 1). Higher level systematics of these groups are under revision; hence, only the class level is specified. The diversity of the group is underscored by the presence of both prokaryotic (not containing a nucleus, Cyanophyceae, or cyanobacteria) and eukaryotic (nucleated) taxa. The eukaryotic groups are thought to have arisen from the incorporation of photosynthetic prokaryotes (or, subsequently, photosynthetic eukaryotes) within protozoan-like host organisms. This hypothesized process is referred to as the endosymbiosis theory. Evidence from both electron microscope studies of microalgal cells (usually focused upon numbers and types of membranes within the cell) and more recent molecular sequencing work indicates that endosymbiotic creation of "new" organisms may have occurred a number of times in evolutionary history, leading to the diversity in morphology and physiology seen today. This diversity, especially in terms of physiology, provides opportunities for the current and potential use of these organisms in the aquaculture industry.

Table 1. List of Currently Recognized "Microalgal" Classes and Representative Genera Used in Aquaculture

Class	Common Name	Representative Genera Used in Aquaculture
Cyanophyceae	Blue-green algae	*Spirulina*
Prochlorophyceae	Prochlorophytes	None
Rhodophyceae	Red algae	*Porphyridium*
Prasinophyceae	Scaled green algae	*Tetraselmis*
		Pyramimonas
Chlorophyceae	Green algae	*Chlorella*
		Dunaliella
		Haematococcus
Cryptophyceae	None	*Cryptomonas*
		Rhodomonas
Chlorarachniophyceae	None	None
Euglenophyceae	None	None
Dinophyceae (Pyrrophyceae)	Dinoflagellates	*Crypthecodinium*
Chrysophyceae	Golden-brown algae	None
Raphidophyceae	None	None
Eustigmatophyceae	None	*Nannochloropsis*
Xanthophyceae (Tribophyceae)	None	None
Bacillariophyceae	Diatoms	*Thalassiosira*
		Chaetoceros
		Nitzschia
Dictyophyceae	None	None
Pelagophyceae	None	None
Haptophyceae (Prymnesiophyceae)	None	*Isochrysis*
		Pavlova

WHY CULTURE MICROALGAE?

There are two main reasons for which microalgae are grown: (*1*) for extractable chemicals and (*2*) as feeds for aquacultured animals. Efforts to produce foods for direct human consumption have met with limited success, and crops such as the cyanobacterium *Spirulina* remain in the realm of "dietary supplement," rather than food crop; these cases are considered along with extractable chemicals.

Microalgae for Extractable Chemicals

Chemical analyses of microalgae have led to the discovery of many novel chemical compounds, some of which are useful as food additives, pharmaceuticals, cosmetics, or in other high-value applications. Examples of microalgae currently commercially cultured for extractable chemicals include (*1*) *Dunaliella* for β-carotene, a human nutritional supplement; (*2*) *Haematococcus* for the pigments astaxanthin and canthaxanthin, which are used as coloring agents in salmon feeds; (*3*) *Crypthecodinium* for docosahexaenoic acid (DHA), which is incorporated into infant formula; and (*4*) the aforementioned *Spirulina*, which is used as a dietary supplement or additive to cosmetic products. In most cases, the chemical compound of commercial interest is present as a small fraction of the total algal biomass; therefore, the scale of cultures for commercial production generally is in the range of many cubic meters ($1\ m^3 = 35.3\ ft^3$). When a selected portion of the algal biomass is extracted and refined, the presence of chemical and biological contaminants within production cultures is not relevant, unless overall production is constrained or an undesired chemical is coextracted with the product compound. Thus, in many cases microalgae for extractable compounds are cultured in large, open-pond facilities; *Dunaliella* ponds of several hectares (1 ha = 2.4 acres) in size are characteristic. Alternatively, industrial, "brewery-type" technology may be employed, as for *Crypthecodinium* cultured heterotrophically for DHA.

Although the list of successful commercial products that are extracted from cultured microalgae is short at present, the potential is enormous. Natural-products chemists have described numerous novel compounds from microalgae, including chemicals with antibiotic, antitumor, and neuroactive characteristics. As such products are developed for commercialization, considerable growth in microalgal culture for extractable compounds is predicted.

Microalgae as Aquaculture Feeds

The most common reason for culturing microalgae is as a feed in an aquaculture food chain (especially for marine or estuarine animals). Microalgae are consumed directly by bivalve molluscs (e.g., clams mussels, oysters, and scallops) throughout life, by young stages of crustaceans (shrimp), and even by first-feeding stages of some finfish. In addition, microalgae can be used to grow small invertebrates, such as rotifers (*Brachionus* sp.) and brine shrimp (*Artemia* sp.), that are fed to young stages of crustaceans and finfish and are used to enrich these small invertebrates with nutritional compounds, especially fatty acids and sterols, required by the larval crustaceans and finfish.

Criteria for selection of microalgae as aquaculture feeds include ease of culture, size, digestibility, and nutritional value. All but the first of these criteria are covered in another entry of this volume. Ease of culture under particular circumstances will be dependent upon the tolerance of the alga to physical, environmental conditions, such as temperature and salinity. An alga's ability to coexist with or exclude microbial contaminants, ranging from bacteria to fungi and protozoans, also is critically important in most commercial applications.

A number of specific, clonal strains of microalgae have been found, empirically, to possess desired characteristics and are in wide use; these can be obtained from aquaculture supply companies, academic institutions, and government-funded institutions that maintain microalgal culture collections. The strain level of identity is appropriate, because microalgal taxonomy remains in a state of continuing development and because different isolates of the same taxonomic species may differ widely in growth or nutritional characteristics relevant to their use as aquaculture feeds.

HOW DO MICROALGAE WORK?

Energy

Microalgae are referred to as autotrophs, a word that translates literally as self-feeding. This term is wholly appropriate because the process of photosynthesis, by which light energy is converted to chemical energy, provides sugars that are subsequently eaten, or burned, to support the heterotrophic processes of the rest of the cell. In cyanobacteria, photosynthesis occurs in cellular structures, called lamellae, that are not segregated by membranes from the rest of the cell. The photosynthetic apparatus of eukaryotic microalgae is contained within a membrane-bound structure, called a plastid or chloroplast, that is thought to have arisen, evolutionarily, from a cyanobacterial endosymbiont.

Regardless of its location within the cell, the photosynthetic process itself accomplishes nothing more than creating sugars and oxygen from carbon dioxide and water — a transforming process in the natural history of the earth, but insufficient in itself to sustain life. The reverse process, catabolism of sugars to release energy and carbon dioxide, is necessary for the cell to make use of the energy captured within the sugars. This concept is emphasized, because the energy and gas (oxygen and carbon dioxide) dynamics of microalgal cultures, particularly in natural diurnal light cycles, can vary considerably. Heterotrophic processes are active continuously to sustain life, while autotrophic processes occur only in the presence of light. In dim light, or in bright light with self-shading in dense cultures, light may be insufficient to counterbalance heterotrophic processes. The level of light energy input needed to just sustain the population without growth is called the compensation point. For the population of microalgae to increase, light energy above the compensation level must be provided.

Chlorophyll α, the chemical compound that catalyzes photosynthesis, absorbs light in the wavelength range of 400–700 nm, this range is referred to as photosynthetically active radiation, or PAR. Full, noon sunlight is in the range of 2,000 µmol photons per square meter per second of PAR; other units of PAR flux encountered may be micro-Einsteins per square meter per second, shortened by microalgal icon Ralph Lewin to Alberts. Thus, both the quantity and quality of light is important to energy acquisition by microalgae.

An often-overlooked aspect of the energy-input requirements of microalgae is the potential to supply sugars to the heterotrophic cellular machinery from a source other than photosynthesis. Indeed, most microalgae tested are physiologically able to grow on sugars added to the culture medium as the sole energy source (i.e., in the dark). This capability has only recently been exploited commercially, and only under highly controlled, bacteria-free conditions. Bacterial sugar uptake and growth rates generally are more rapid than those of microalgae. In a competition for dissolved sugars, bacteria have an advantage over microalgae; therefore, aseptic production is necessary. In addition, while it may be more cost-effective to provide artificial sugars than artificial light to algae, the biochemical composition of microalgae grown photo-autotrophically vs. heterotrophically may affect the ultimate cost–benefit decision.

Materials, or Nutrients

Life processes require two inputs: energy and materials. The energy needs of the microalgae just discussed interact with materials — either carbon dioxide and water, or sugars and oxygen — and the material needs, predictably, interact with energy status of microalgal cells as well. At least 24 chemical elements have been identified as being essential, (i.e., nutrients) in all living cells (see Table 2). Although all of these elements could pass through membranes surrounding cells by simple diffusion, active, energy-consuming uptake processes have been demonstrated for essentially all nutrients. Beyond the simple fact that diffusion would be highly rate limiting with dilute, natural nutrient concentrations, many required elements exist as chemical compounds or complexes in solution that must be modified before the element can be assimilated by the cell. Thus, nutrient uptake can be considered an active, energy-requiring process.

Table 2. Chemical Elements Considered to be "Essential" for Living Cells, Including Microalgae

Element	Chemical Symbol	Typical Source in Microalgal Culture Media
Carbon	C	Carbon dioxide in air
Hydrogen	H	Water
Oxygen	O	Water
Nitrogen	N	Nitrate, ammonia, urea
Phosphorus	P	Phosphate salts
Calcium	Ca	Calcium carbonate
Sodium	Na	Sodium chloride
Chlorine	Cl	Sodium chloride
Magnesium	Mg	Magnesium chloride or sulfate
Potassium	K	Potassium chloride
Sulfur	S	Sulfate salts
Boron	B	Boric acid
Iron	Fe	Ferric chloride
Selenium	Se	Selenous acid
Copper	Cu	Cupric chloride or sulfate
Manganese	Mn	Manganese chloride
Zinc	Zn	Zinc chloride or sulfate
Molybdenum	Mb	Molybdenum chloride
Cobalt	Co	Cobalt chloride or vitamin B_{12}
Iodine	I	Potassium iodide
Nickel	Ni	Nickel chloride
Silicon	Si	Sodium silicate
Fluorine	Fl	Sodium fluoride
Chromium	Cr	Dichromate salts

Microalgal nutrients can be segregated into two arbitrary categories: macronutrients and micronutrinents, or trace elements. These categories are arbitrary in that deficiency in any element, macro or micro, will constrain growth. For all microalgae, nitrogen (N) and phosphorus (P) are considered macronutrients and are major structural components of proteins, nucleic acids, and the energy-management chemicals (e.g., adenosine triphosphate, or ATP) of the cells. For one class of microalgae, Bacillariophyceae or the diatoms, silica (Si) also is a macronutrient, because it is required in relatively large amounts for the formation of cell walls.

Forms of nitrogen biologically available to microalgae include ammonium (all), nitrate (some cannot use), and organic compounds (urea, amino acids, etc., available to most microalgae). Ammonium is the nitrogen "currency" within the cell, while nitrate must be reduced and urea catabolized during uptake; therefore, thermodynamics favors ammonium uptake. Most commonly used microalgal nutrient enrichments provide nitrate, though, because it is more stable in solution (not volatile) and not available to many bacteria in contaminated cultures. Microalgae that are not able to produce nitrate reductase, the enzyme that catalyzes the reduction of nitrate to ammonium, generally must be grown on ammonium. As nitrogen is a major component of proteins, deficiency arrests protein synthesis and may lead to an accumulation of energy-storage products, such as starches and lipids, in microalgal cells; this response can be exploited if increased storage-product yield is desired in cultures.

Phosphorus exists in natural waters chiefly in the oxidized form phosphate, although some organic phosphorus compounds may exist as well. Phosphate is by far the most common form of phosphorus added to microalgal culture media, and amounts added generally are in the range of one phosphorus atom for every 16–25 nitrogen atoms. Phosphorus deficiency in algal cells may lead to physiological disruption of protein synthesis, similar to nitrogen deficiency, but also may disrupt energy management within the cell; therefore, phosphorus starvation is a less dependable culture-management strategy than nitrogen starvation.

As previously mentioned, silica is required for the formation of cell walls in diatoms. As for nitrogen and phosphorus, an oxidized form, silicate, is encountered in natural waters and added to culture media. Different diatoms require different amounts of silica, and most possess the ability to make thinner "shells" under silica deficiency. Extreme silica deficiency arrests cell division and may cause cells to accumulate storage products, providing a possible culture-management strategy.

Trace elements, or micronutrients, represent a group of elements for which cellular needs are several orders of magnitude lower than the macronutrients. Micronutrients tend to be minor but necessary components of enzymes involved in basic cell processes. Many of these elements are metals that are present in natural, oxygenated waters in highly oxidized forms, often complexed with dissolved organic molecules. In these forms, limited ability of microalgal uptake mechanisms to remove free ions from stable complexes may limit bioavailability and growth. In culture media, artificial complexing, or chelating, agents, such as EDTA, NTA, or citric acid, are added to moderate the bioavailability of trace metals, in particular. When chelating agents are used, metals must be added in excess of microalgal needs to saturate chelator-metal complex sites; thus, levels of trace metals added to media with artificial chelators may far exceed levels that would be toxic if metals were in the free-ion form. Indeed, the concentration range for many metals between deficiency and toxicity is relatively narrow; therefore, trace-element management of microalgal cultures can be quite challenging.

Some microalgae used widely in aquaculture (e.g., the prymnesiophytes) require vitamins, usually B_{12} and thiamin, and sometimes biotin. In open, bacterized cultures, bacteria generally supply these growth factors; however, pure cultures require that these vitamins be added in trace quantities for the algae that require them.

Microalgal Growth = Population Growth

If suitable and sufficient energy and materials are provided to an algal cell, it divides into two cells, which separate and become distinct individual organisms. Hence, microalgal growth is characterized in terms of increases in numbers of individual cells, rather than in terms of increases in an individual's size, which is the case for metazoan organisms. Production rate of microalgal biomass in a culture often is quantified in terms of average number of cell divisions per unit time (divisions per day), or the reciprocal, average time between cell divisions. A mathematical description of progressive cell divisions is a logarithmic function; calculation of this function from cell counts at two or more times is a straightforward procedure and a powerful tool for describing and managing a culture's performance.

There are two fundamental management strategies for microalgal cultures: batch culture and continuous culture. In a batch culture, a small population of cells (inoculum) is placed within a relatively large container supplied with sufficient energy and materials to produce a much larger number of cells. The production rate and the maximum number of cells that can be supported in the large container can be constrained by either energy or materials, but the main point is that algal growth will stop eventually, when there is no longer sufficient energy or materials to support further cell divisions. Yield and time of active growth are finite in a batch culture, and the container must be cleaned and reinoculated at the end of each growth cycle. In continuous culture, materials and energy are supplied continuously, and cells are removed at a constant rate equal to, or less than, the rate of production of new cells. Thus, a continuous culture has an indefinite lifespan, as long as the production of new cells is at least as fast as the removal of cells. There are a number of different categories of continuous cultures; the term chemostat, describing but one type of continuous culture, often is misapplied to cultures with constant fluid replacement rates (cyclostats), or constant standing biomass (turbidostats).

In theory, it would appear that continuous culture offers overwhelming advantages over batch culture, in terms of effort (labor), production-rate optimization, and management and control. In practice, however, it can be very difficult to maintain the steady-state conditions necessary for most continuous cultures. Furthermore, living contaminants, such as bacteria and protozoans, can divide more quickly than the cultivated alga, eventually replacing the intended crop. A compromise between batch and continuous culture management strategies, semicontinuous culture, offers some advantages of both. A single culture can be maintained for an extended period of time, replenishment rate (both amount and time between partial harvests) can be varied in response to culture performance, and partial harvests can be delayed to allow accumulation of lipids or carbohydrates if these are important in the application. Recent efforts to incorporate contemporary, computer process-control technology into microalgal culture have improved the effectiveness of continuous and semicontinuous microalgal culture methods.

HOW ARE MICROALGAE CULTURED?

Containers

Essentially, any vessel or structure that can contain water can be used to culture microalgae. Experimental and "seed" cultures routinely are grown in test tubes, flasks, jars, bottles, etc., made of glass and various kinds of plastic. Tanks, tubes, buckets, barrels, bags, pools, and ponds, constructed of nearly limitless materials, have been applied to production-scale microalgal cultures. For photosynthetic production, light can be provided from any direction if the container is transparent to light in the PAR range. Opaque containers with the artificial light sources immersed within the culture have been used with limited success. Alternatively, containers with an exposed surface lighted from above can have opaque sides and bottom. In addition to permitting light penetration, containers must provide for gas exchange with the atmosphere or incorporate an added gas stream. Tank and pond cultures generally are kept relatively shallow to provide for gas exchange with the atmosphere, by virtue of a high surface-to-volume ratio, and also to maintain light input above the compensation level. Gas exchange and culture mixing often are enhanced by introduction of a diffused gas stream (bubbles) through the culture. Although widely used, bubbling of open, bacterized microalgal cultures may encourage bacterial degradation of the culture; physical mixing with foils, paddles, etc., may be more successful and has been applied in many large systems. Tubular containers, consisting of many meters of narrow-diameter, transparent glass or plastic tubing have seen several periods of intense interest, as this design offers some advantages of both open containers (short light path) and closed systems (exclusion of contaminants). Gas-transfer limitations have constrained effectiveness of tubular systems. For heterotrophic microalgal production, standard industrial bioreactors used for other microorganisms have proven to be transferrable, with few modifications, to most microalgae.

Energy

Most commercial microalgal production is based upon photosynthetic growth. Light can be provided by natural sunlight or by artificial lights. Natural sunlight is inexpensive (a function of land cost), but varies seasonally and may be unreliable daily, depending upon location and climate. Artificial lights can be controlled very precisely, in terms of both quantity and quality, but can account for up to 95% of the cost of culturing microalgae. Many small-scale fish and shellfish hatcheries culture microalgal feeds using artificial lights because feed production must coincide reliably with seasonal life-history stages of the animals that are, in turn, often nonsyncronous with seasonal light and temperature cycles. Accordingly, the cost of producing microalgal feed cultures in these facilities is among the highest of all cultivated products, ranging from $100 up to $400 per dry kg ($45 to $180 lb) of microalgal biomass. In contrast, culture of *Dunaliella* for β-carotene in desert ponds can cost over 100 times less, because of the difference in energy cost alone. For heterotrophic culture, various waste and off-specification agricultural products can be used as energy sources for microalgal culture. The economics of using reduced-carbon energy sources for heterotrophic production appear to be intermediate between artificial light and solar energy; the success of each type of energy source is dependent upon the application of the microalgal product and the specific alga that is cultured.

Materials

Fertilization of microalgal cultures ranges, in commercial practice, from animal manure added directly to earthen ponds to complex formulations prepared with pharmaceutical care. In instances where extractable chemicals are the product, industrial-grade chemicals are generally sufficient, because chemical contamination of the product is not a practical concern. When cultured microalgae are part of a food chain leading to a human food product, more refined chemical fertilizers are recommended. Culture medium formulations can be assembled from component chemicals according to recipes found in standard texts (see Bibliography), but proven microalgal fertilizer mixtures, based upon the Guillard "f/2" formulation, now are being marketed by aquaculture-supply companies.

Culture Management

Batch culture still is used in most applications. Usually, small-scale flask cultures maintained under pure (bacteria-free) conditions are increased in volume by inoculating a series of progressively larger containers until the production-scale is reached; the term for this practice is "progressive batch culture." For many aquacultured animals fed microalgal cultures, the lipid content of the algal feed is the most critical nutritionally; therefore, many farmers allow cultures to enter a nutrient-deficient (usually nitrogen) phase, during which the nondividing microalgal cells often store photosynthetic energy as lipid. Continuous and semicontinuous culture methods are employed routinely in extractable-product operations and increasingly are being developed for marine-animal feeds applications.

WHAT INNOVATIONS ARE EXPECTED?

In reviewing the history of the development of modern microalgal culture, one is struck by the existence of two, nearly independent "heritages." One line of investigation, dominated by engineers and chemists, was motivated by the potential for industrial products, ranging from human foods to synthetic fuels and oxygen factories for extended space travel. Working almost exclusively with the freshwater chlorophyte, *Chlorella*, scientists and engineers developed and built pilot plants employing state-of-the-art electronics and fluidics innovations. While most of the perceived applications for *Chlorella* biomass have not proved successful, the legacy of this research effort provided the foundation for today's extractable-product microalgal technology. The second line of investigation, dominated by biologists and ecologists, arose from laboratory-scale culture and feeding apparatus designed to maintain experimental animals for investigation of their life cycles, feeding habits, etc. These types of methods were generalized and modified to culture a wide variety of microalgal types and were used in the elucidation of nutritional requirements of animals that were to subsequently enter into aquaculture production. Little thought was given, however, to economics or problems of scaling the design for commercial use. At the time of this writing, it appears that the two "lineages" of microalgal culturists finally are communicating and collaborating on engineered systems that satisfy the biological needs of the variety of microalgae used as aquaculture feeds, as well as for extractable compounds, and do so in a cost-effective way. This cooperation holds great promise for the future of microalgae in aquaculture.

BIBLIOGRAPHY

Author's Note: The following references were selected to provide background in general principles pertaining to microalgal culture, not as specific how-to manuals. What manuals have been produced, usually by academic or government research laboratories, tend to be narrow in application and availability, and all still require a firm grounding in basic principles to be useful.

I. Akatsuka, ed., *Introduction to Applied Phycology*, SPB Academic Publishing, The Hague, 1990.

E.W. Becker, *Microalgae Biotechnology and Microbiology*, Cambridge University Press, Cambridge, 1994.

J.S. Burlew, *Algal Culture from Laboratory to Pilot Plant*, Carnegie Institution of Washington Publ. 600, Washington, DC, 1961.

G.E. Fogg, *Algal Cultures and Phytoplankton Ecology*, The University of Wisconsin Press, Madison and Milwaukee, WI, 1965.

F.H. Hoff and T.W. Snell, *Plankton Culture Manual (Fourth Edition)*, Florida Aqua Farms, Inc., Dade City, FL, 1997.

J.T.O. Kirk, *Light and Photosynthesis in Aquatic Ecosystems*, Cambridge University Press, Cambridge, 1983.

A. Richmond, *Handbook of Microalgal Mass Culture*, CRC Press, Boca Raton, FL, 1986.

G. Shelef and C.J. Soeder, *Algae Biomass Production and Use*, Elsevier, Amsterdam, 1980.

A. Sournia, *Phytoplankton Manual*, UNESCO, Paris, 1978.

J.R. Stein, ed., *Handbook of Phycological Methods Culture Methods and Growth Measurements*, Cambridge University Press, Cambridge, 1973.

F.R. Trainor, *Introductory Phycology*, John Wiley & Sons, New York, 1978.

MICROBOUND FEEDS

FREDERIC T. BARROWS
USFWS, Fish Technology Center
Bozeman, Montana

WILLIAM A LELLIS
USGS, Research and Development Laboratory
Wellsboro, Pennsylvania

OUTLINE

INTRODUCTION

Microbound feed is a general category of small particulate feeds ranging in size from 50 to 700 μm that are typically fed to larval stages of fish and invertebrates. The defining characteristic of these feeds is that the particles are held together by an internal binder, which may be a complex carbohydrate or a protein having adsorptive and adhesive properties (1). This differs from microencapsulated feeds, which are characterized by a distinct wall or capsule surrounding a central core of material (2,3).

There are many types of larval feeds available on the market and more are being developed in laboratories around the world. The particular feeding situation (e.g., species, water temperature, or culture system) determines which type of larval feed to use. Different feeding situations may require different types of feed. To understand the processing methods used in manufacturing microbound feeds, an evaluation of the desired characteristics of an effective feed is necessary.

Palatability, nutrient stability, nutrient availability, and particle stability are all important traits of a microbound feed. A feed of outstanding nutritional quality is of little value if it has low palatability and the animal will not consume it. Palatability is affected by factors such as smell, flavor, and texture. The smell and flavor of the feed is caused by the leaching of nutrients into the water, and thus some leaching is necessary for adequate levels of feed consumption. Excessive leaching, however, can result in a poor quality feed due to reduced nutrient content. Nutrient stability is very important to ensure adequate nutrition, but if the feed is bound too tightly, consumption and/or nutrient absorption could decrease. Feeds for finfish must be formulated, manufactured, and selected with the goals of high palatability and high nutrient stability in mind, even though those objectives seem to work in opposite directions.

Another important characteristic of a microbound feed is particle stability. This refers to the loss of material from the feed particle in water. A feed with low particle stability will disintegrate and degrade water quality, which can decrease survival of the cultured animal. Gill damage caused by high levels of particulate matter in the tank, bacterial contamination, and low oxygen levels are all problems that can be caused by a feed with low particle stability (4).

Feed characteristics are strongly affected by the type of binders and processing methods used in manufacture, so each feeding situation may best be accommodated by a particular formulation or processing method. There can also be interactions between processing method and ingredient formula, including binder source. Sometimes a particular processing method is categorized as ineffective for use in a particular situation, when the formulation itself in combination with the processing method was actually inadequate. The culture technique used when feeding formulated feeds can also determine if the program is a success or a failure. Many factors need to be considered when selecting a larval feed, but by considering formulation, processing method, palatability, nutrient stability, particle stability, and culture methods, an effective larval production program using microbound feeds can be achieved.

CLASSES OF MICROBOUND FEEDS

Microbound feeds can be separated into three major classes according to production process: crumbled, on-size, and complex particles. Crumbled feeds are produced by manufacturing a pellet, flake, or cake that is fractured into smaller pieces and sifted to obtain the desired size (5,6). On-size feeds are manufactured directly to the correct size particle (2,7), which not only saves a production step, but produces physical characteristics that differ from crumbled feeds. Complex particles are produced by combining two or more techniques (including microencapsulation) to exploit the advantages and overcome the disadvantages of individual production methods (8,9).

CRUMBLED FEEDS

Pelleted Feeds

Pellets used in the manufacture of crumbled feeds can be produced through either steam pelleting, cooking extrusion, or cold extrusion. Steam pelleting refers to a process by which solid pellets are formed by forcing a feed mixture (mash) through holes in a rotating die after preconditioning with steam to a temperature of 70–85 °C and moisture content of 15–18% (5,6,10,11). As the pellets emerge from the die, they are cut to desired length by an adjustable knife. The steam partially gelatinizes dietary starch, which aids in binding the ingredients. Precooked corn, sorghum, potato, palm nut, wheat, and tapioca starches are sometimes added to the diet at 10–20% to increase pellet bond strength (6). Special organic hydrocolloids such as lignin sulfonate and carboxymethyl cellulose can also be added at 0.25–5.0% of the ingredient formula to aid in the pelleting process (6,12). The compressed pellets are dried, crumbled, and sifted to the desired size. The resulting particles have an irregular shape, are fairly dense, and generally have low particle and nutrient stability in water.

Cooking extrusion is a process by which a feed mash is moistened with 20–25% water, precooked at 100–150 °C, then forced through a die under high heat and pressure (5,6,13–16). As material passes through the die, pressure suddenly drops, causing water vaporization that traps air within the resulting pellets. After cooling and drying, pellet density may be such that the feed can float or sink slowly in water. The almost complete gelatinization of starch within extruded feeds produces pellets that are more tightly bound and have greater water stability than steam-pelleted feeds. The cooking of carbohydrates increases availability of dietary energy (17), but many larval fish may have limited ability to digest carbohydrates. Cooking extrusion is a very versatile processing method that can produce feeds with physical characteristics not possible with other pelleting methods.

Cold extrusion is the process of pressing a wet mash (>20% moisture) through holes in a plate without addition of heat, producing a noodle that is cut or broken to form pellets (5,15). This type of extrusion can generate pellets with diameters as small as 1.0–1.5 mm, depending upon the characteristics and particle size of the ingredients used (16). Pellet stability is accomplished through selection of ingredients with high-binding activity, such as wheat gluten or protein hydrolysates. This manufacturing method is often used to produce moist, semimoist, and soft-moist feeds, which typically do not have the long-term water stability of harder, more tightly bound feeds.

Steam-pelleted and extruded crumbles were developed for salmonid culture and have been instrumental in development of that industry. However, fish feed manufacturers are currently evaluating other processing methods for production of starter feeds. Even though steam pelleting and extrusion are different processes, the crumbled products produced with each method are fairly similar.

Flaked Feeds

For a long time, flaked feeds have been the most common type of feed fed to aquarium fish. Although a variety of methods can be used to produce flakes, the double-drum drier affords the greatest control of variables affecting flake quality (18). The equipment consists of two parallel drums rotating in opposite directions that are heated internally with steam. Feed ingredients are ground to approximately 0.1 mm and blended with water to form a slurry or dough that is coated onto the drum surface (6). The dough is flattened to a uniform thickness between the rotating drums and dried to a thin sheet. A blade scrapes the sheet from the drum at a point approximately two-thirds the drum circumference from the nip of the rolls (18). The thickness of the flake can be adjusted by changing the distance between the drums. The dried sheet is then crumbled to produce flakes or ground and screened to produce small particles. The resulting feed has a high surface area-to-volume ratio and will float for a long time before saturating with water and sinking. Long floating times are beneficial in many cases, since fish have more time to consume the feed. The high surface area-to-volume ratio, however, can also result in a rapid loss of water-soluble nutrients if they have not been stabilized. Carbohydrates are often used as binders in flake feeds, but other ingredients with good hydrocolloidal properties and tensile strength can be used (6).

Flaked feeds are often criticized because of the high temperature required for drying. Proteins can be burned, lipids oxidized, and vitamins lost by exposure to high temperatures during any manufacturing process (15,19–21). Flaked feed, however, is exposed to heat for only a short time, and studies evaluating the effect of processing methods that have not been confounded by ingredient formulation have been limited. High temperature may not be a problem if the time of heat exposure is short. The concept of flaking machines producing poor-quality products may be a result of comparisons of different feed formulations produced with different processing techniques.

Cake Feeds

Crumbled cake refers to a feed manufacturing process in which a mixture of feed ingredients and binding agents are gelled into a matrix that is dried, crushed, and sieved into appropriate size particles (1–3,22–25). Many different binders are used to produce cake feeds, including agars, alginate, carrageenan, cold-water gelling starches, egg albumin, gelatin, and zein. Each binding system is activated differently. For example, zein is an amino acid containing compound found in corn and is soluble in alcohol, but not water. Solubilizing this compound in alcohol, adding the base mix, and then evaporating the alcohol will result in a particle with high water stability. Egg albumin can also form a matrix, but it is activated by heat. Combinations of different binders are sometimes necessary to produce particles with the desired physical characteristics. The crumbled cake method is currently being used commercially to produce some very effective larval feeds.

ON-SIZE FEEDS

Microextruded Marumerized Feeds

Microextrusion marumerization (MEM) is a two-step process adapted from the pharmaceutical industry and is now being used to manufacture small, preshaped larval feeds (7,16,26). A wet mash of finely ground feed ingredients is first formed into thin noodles by using extruders designed to reduce the operating pressure. The lower operating pressures allow for smaller diameter noodles to be produced. Radial discharge and twin-dome extruders are capable of producing noodles as small as 500 and 300 μm, respectively. Noodles are then broken and shaped in a marumerizer, which consists of a cylindrical chamber with a high-speed rotating grooved plate at the bottom. Plates are available with different depths of grooves, which affects the amount of energy transferred to the feed during marumerization. A very strong noodle requires a deeply grooved plate, while a soft noodle is processed most effectively with a shallow-grooved plate. Feed discharged from the marumerizer can be fed wet or dried. Particle size range within a given production run is very narrow, and sifting is not as important as with other processing methods. The marumerizer imparts two effects to the feed. The first is to break the noodles into lengths roughly equivalent to noodle diameter and shape the resulting particles into spheres (16). The uniformity of particle length and the amount of shaping are affected by feed formulation and moisture level, among other factors. The second effect is to increase surface density of the feed particles. As the feed spins in the marumerizer, centrifugal forces cause migration of water and small ingredient particles to the pellet surface, thus increasing surface density proportional to the pellet interior. This effect can be visualized under a microscope and has been demonstrated to increase particle stability. It is also thought to increase nutrient stability — although this has not been proved. The speed of the plate also has a profound effect on the marumerization process. The higher the speed, the more energy is transferred to the feed, causing more shaping and more centrifugal force.

Many types of binders can be used with MEM particles, provided they are moisture and pressure activated. No heat is added during this process, but some is generated at the extrusion screen due to friction. The amount of heat generated is affected by formula and moisture level, but is normally quite low. Binding systems based on gums have been used, but protein hydrolysates are also effective. Hydrolysates have the advantage of added nutrition as well as binding. Care must be taken in formulation with hydrolysates, since high levels can result in agglomeration of feed in the marumerizer into very large particles.

Feed produced by MEM can be characterized as smooth and spheroid with high density. The smooth shape may decrease nutrient leaching by decreasing the surface area-to-volume ratio, relative to a rough particle of the same size (i.e., crumbles or flakes). The high density of the feed will result in a faster sinking particle, which is a negative effect for species that feed in the water column.

Particle-Assisted Rotationally Agglomerated Feeds

Particle-assisted rotational agglomeration (PARA) is a processing method that utilizes a marumerizer without extruded noodles (U.S. Patent 5,851,574) (7,26,27). Wet mash is placed directly into the marumerizer with a charge of inert particles. The rotation of the marumerizer imparts energy to the inert particle, which in turn transfers energy to the mash, producing spheroid particles in a wide range of sizes. Feed formulation and moisture content are very important with this method. Several other process variables can be controlled to affect particle size distribution and density. An advantage of this process as compared to MEM is lower capital expenditure and lower operating costs due to elimination of the extruder from the process. The same binders used with MEM particles are effective in PARA particles, with minor modifications. The PARA particles are not as uniform in shape as particles produced by MEM. Size distribution of PARA particles is also much greater than MEM particles, thus necessitating sifting, but allowing for the production of several sizes in a single run. The PARA process is a low-pressure agglomeration method that results in low-density feed particles with a slow sink rate, useful for species that feed in the water column.

Spray Beadlets

Spray beadlets are small microbound feeds that trap high-molecular-weight, water-soluble nutrients such as starch and protein within gels of calcium alginate and gelatin (2). Particles are produced by spraying a slurry of dietary components and a selected gelling agent (e.g., alginate) into a curing bath (e.g., calcium chloride solution) (28). Many types of binders can be used with this method and produce good results. This process creates a wide range of particle sizes, so sifting is often required. (See the entry "Microparticulate feeds, micro encapsulated particles.")

COMPLEX PARTICLES

Complex particles are produced by combining two or more manufacturing techniques to exploit the advantages and overcome the disadvantages of individual production methods. This is an exciting development in larval feed manufacturing that allows for production of feeds that more closely fit specific needs. For example, microcapsules or crumbled cake particles may be embedded within larger MEM, PARA, or spray beadlet particles. Villamar and Langdon (8) embedded lipid-wall microcapsules within alginate–gelatin particles to deliver both micro- and macronutrients to suspension-feeding larval shrimp within a single complex microcapsule. Ozkizilcik and Chu (9) prepared complex protein-walled microcapsules containing lipid-walled capsules for feeding striped bass larvae. The complex particle combines the advantages of several types of feeds and may be the larval feed type of the future.

BIBLIOGRAPHY

1. T. Watanabe and V. Kiron, *Aquaculture* **124**, 223–251 (1994).
2. C.J. Langdon, D.M. Levine, and D.A. Jones, *J. Microencapsulation* **2**, 1–11 (1985).
3. D.A. Jones, M.S. Kamarudin, and L.L. Vey, *J. World Aquacult. Soc.* **24**, 199–210 (1993).
4. C.Y. Cho, C.B. Cowey, and T. Watanabe, *Finfish Nutrition in Asia*, International Development Research Centre (IDRC-233e), Ottawa, Ontario, 1985.
5. R.W. Hardy, *Fish Nutrition*, 2nd ed., Academic Press, San Diego, 1989, pp. 475–548.
6. T. Lovell, *Nutrition and Feeding of Fish*, Van Nostrand Reinhold, New York, 1989.
7. C. Gill, *Feed International* **6**, 22–25 (1997).
8. D.F. Villamar and C.J. Langdon, *Mar. Biol.* **115**, 635–642 (1993).
9. S. Ozkizilcik and F.-L. Chu, *J. Microencapsulation* **13**, 331–343 (1996).
10. D. Falk, *Feed Manufacturing Technology III*, American Feed Industry Association, Inc., Arlington, 1985, pp. 167–190.
11. R.K.H. Tan, *Proc. Aquaculture Feed Processing and Nutrition Workshop*, Thailand and Indonesia, American Soybean Association, Singapore, 1992, pp. 138–148.
12. W.G. Dominy and C. Lim, *Proc. Aquaculture Feed Processing and Nutrition Workshop*, Thailand and Indonesia, American Soybean Association, Singapore, 1992, pp. 149–157.
13. O.B. Smith, *Feed Manufacturing Technology III*, American Feed Industry Association, Inc., Arlington, 1985, pp. 195–204.
14. C.C. Botting, *Proc. Aquaculture Feed Processing and Nutrition Workshop*, Thailand and Indonesia, American Soybean Association, Singapore, 1992, pp. 129–137.
15. G. Cuzon, J. Guillaume, and C. Cahu, *Aquaculture* **124**, 253–267 (1994).
16. F.T. Barrows and W.A. Lellis, *Walleye Culture Manual*, NCRAC Culture Series 101, Iowa State University, Ames, 1996, pp. 315–321.
17. J.W. Hilton, C.Y. Cho, and S.J. Slinger, *Aquaculture* **25**, 185–194 (1981).
18. G.M. Pigott and B.W. Tucker, *Fish Nutrition*, 2nd ed., Academic Press, San Diego, 1989, pp. 653–679.
19. S.J. Slinger, *Effect of Processing on the Nutritional Value of Feeds*, National Academy of Sciences, Washington, DC, 1993, pp. 48–66.
20. M. Boonyaratpalin and R.T. Lovell, *Aquaculture* **12**, 53–62 (1977).
21. J. Gabaudan, G.M. Pigott, and J.E. Halver, *Proc. World Maricult. Soc.* **11**, 424–432 (1980).
22. A. Kanazawa, S. Teshima, H. Sasada, and S.A. Rahman, *Bull. Jap. Soc. Sci. Fish.* **48**, 195–199 (1982).
23. S. Teshima, A. Kanazawa, and M. Sakamoto, *Mini Review and Data File of Fisheries Research, Kagoshima University* **2**, 67–86 (1982).
24. A. Kanazawa, *Proc. Aquaculture Nutrition Workshop*, Salamander Bay, 15–17, April 1991; 64–71 (1992).
25. D.A. Jones, *Rev. Fisher. Sci.* **6**, 41–54 (1998).
26. F.T. Barrows, R.E. Zitzow, and G.K. Kindschi, *Progressive Fish-Cult.* **55**, 224–228 (1993).
27. F.T. Barrows, *A Method for Agglomerating Fine Powders*. U.S. Patent 5,851,574, Washington, DC, 1998.
28. D.M. Levine, S.D. Sulkin, and L. Van Heukelem, *Culture of Marine Invertebrates: Selected Readings*, Hutchinson Ross Publishing Company, Stroudsburg, 1993, pp. 193–204.

See also FEED MANUFACTURING TECHNOLOGY.

MICROPARTICULATE FEEDS, COMPLEX MICROPARTICLES

CHRIS LANGDON
Oregon State University
Newport, Oregon

OUTLINE

Text
Bibliography

A complex microparticle for delivery of nutrients to aquatic or marine suspension feeders consists of two or more different microparticle or microcapsule types combined into a single complex microparticle (1). The advantage of using a complex microparticle is that a single food particle can potentially deliver all required nutrients to a targeted suspension feeder. This way, the suspension feeder cannot selectively feed on one particle type versus another, which would result in a modified composition of the ingested diet.

Villamar and Langdon (1) described a complex particle for use in feeding studies with mysis shrimp larvae (*Litopenaeus vannamei*). The complex particle consisted of lipid-walled microcapsules (2) incorporated within alginate–gelatin microbeads. Lipid-walled capsules are necessary for retention and delivery of low-molecular-weight, water-soluble nutrients, while alginate–gelatin beads supply other dietary ingredients, such as protein and carbohydrate. Villamar and Langdon (1) reported that complex particles retained 85% of encapsulated glucose after 18 hours suspension in seawater. The mysis shrimp larvae were able to physically break down the tripalmitin walls of the lipid-walled microcapsules by the action of their mouth parts and assimilate released glucose.

Ozkizilcik and Chu (3) described a second kind of complex particle developed for the delivery of nutrients to striped bass (*Morone saxatilis*) larvae. This complex particle consisted of lipid-walled capsules incorporated within cross-linked, protein-walled capsules. More than 70% of encapsulated lysine was retained by these complex particles after 2 hours of suspension in seawater compared with only about 1% lysine retention by cross-linked, protein-walled capsules.

The ability of complex particles to deliver complete artificial diets to suspension feeders is an important step in the development of economic microparticulate feeds. They may provide complete replacement of living diets in hatcheries, reduce food costs, and increase the reliability of supplies of high-quality food.

BIBLIOGRAPHY

1. D.F. Villamar and C.J. Langdon, *Mar. Biol.* **115**, 635–642 (1993).
2. C.J. Langdon and C.A. Siegfried, *Aquaculture* **39**, 135–153 (1984).
3. S. Ozkizilcik and F.-L.E. Chu, *J. Microencapsulation* **13**, 331–343 (1996).

See also FEED MANUFACTURING TECHNOLOGY.

MICROPARTICULATE FEEDS, MICRO ENCAPSULATED PARTICLES

CHRIS LANGDON
Oregon State University
Newport, Oregon

OUTLINE

Text
Bibliography

The structure of a microcapsule is characterized by a distinct wall surrounding core material. The chemical composition of the wall is usually different from that of the core material and is designed to meet specific functions of the microcapsule. In contrast, microbound or microgel particles lack a distinct wall and consist of a matrix of carrier material in which nutrients are bound. Microcapsules are used to deliver nutrients to suspension feeders, such as larval fish, shrimp, and bivalves [see reviews by Langdon et al. (1) and Robert and Trintignac (2)].

The objective of microencapsulating diets for freshwater and marine organisms is to reduce leakage of nutrients into the surrounding water. Leakage of nutrients results in elevated bacterial concentrations (3) and possible outbreaks of disease. The ideal microcapsule wall should be impermeable to nutrient leakage and remain intact until the microcapsule is consumed by the target organism. Once the microcapsule is eaten, then its wall should be readily broken down, liberating the capsule contents for digestion and assimilation.

The first microcapsule type to be used for the delivery of nutrients to marine suspension feeders was the nylon–protein-walled capsule described by Jones and colleagues (4). Subsequently, this capsule type has been used by many researchers and has undergone many modifications. The elimination of detergents in washing capsules free of organic solvents improved the capsule's acceptability by suspension feeders, while elimination of nylon from the wall resulted in a cross-linked, protein-walled capsule that was not potentially contaminated with toxic, nylon-forming monomers (5,6). Such modified capsules have been successfully used in feeding studies with fish larvae (7–9), crustacea (4,5,10–12), as well as bivalves (3,6,13–17). Incorporation of carbohydrate into the wall of cross-linked, protein-walled capsules (14) and coating capsules with triglycerides (16,18) allowed modification of capsule core retention efficiencies and digestibilities.

A disadvantage in the use of nylon–protein and protein-walled capsule types for diet delivery to suspension feeders is that both capsule types are typically formed by emulsifying dietary ingredients in an organic phase, such as cyclohexane. This emulsification process extracts lipids from the diet, reducing dietary lipid content. As an alternative for the delivery of dietary lipid to suspension feeders, Langdon and Waldock (19) used gelatin–acacia capsules to deliver dietary lipid supplements to oysters. Preparation of gelatin–acacia capsules does not involve organic solvents and results in capsules that are stable but digestible. Subsequently, this capsule type has been used by other researchers in feeding experiments with bivalves (20–23) and as a means of improving the fatty acid composition of *Artemia* (24,25).

A second disadvantage of nylon–protein and protein-walled capsule types is that the capsule walls are permeable to low-molecular-weight nutrients, such as amino acids and water-soluble vitamins (18). To deliver these water-soluble nutrients to suspension feeders, Langdon and colleagues (3,26) used a type of lipid-walled microcapsule to deliver water-soluble vitamins to oysters. Subsequently, Chu and colleagues successfully grew oyster larvae to metamorphosis on a diet that included lipid-walled capsules (20). A softer capsule wall that is partly made up of triglycerides with a low melting point, such as triolein or fish oil, is optimal for bivalve molluscs. (27). Lipid-walled capsules prepared with triglycerides with high melting points, such as tripalmitin, have been successfully used by Villamar and Langdon (28) to deliver glucose to shrimp larvae because the larvae are able to physically break down capsule walls with their mouth parts.

Liposomes are an alternative to lipid-walled microcapsules for delivery of low-molecular-weight, water-soluble substances to suspension feeders (29–32). Liposomes are prepared from phospholipids, while lipid-walled capsules are prepared from triglycerides. Chapman (33) described three types of liposomes: multimembrane liposomes made up of numerous layers of concentric spheres of phospholipid with the aqueous phase trapped between the layers; small, single-membrane liposomes that are 200 to 500 Å in diameter; and large, single-membrane liposomes that range from 600 Å to 10 μm in diameter. Liposomes are more fragile than lipid-walled capsules and tend to be more "leaky." Kulkarni et al. (34) described how the walls of liposomes can be strengthened and made less leaky with the addition of cholesterol and vitamin E to the phospholipid walls.

In summary, different microcapsule types have been successfully used to supply nutrients to suspension feeders. However, there are very few examples of microencapsulated diets that have been found to be complete replacements for living diets in rearing suspension feeders. Clearly, additional research is needed to improve microcapsule design and diet composition to achieve this goal.

BIBLIOGRAPHY

1. C.J. Langdon, D.A. Levine, and D.A. Jones, *J. Microencapsulation* **2**, 1–11 (1985).

2. R. Robert and P. Trintignac, *Aquat. Living Res.* **10**, 315–327 (1997).
3. C.J. Langdon and E.T. Bolton, *J. Exp. Mar. Biol. Ecol.* **82**, 239–258 (1984).
4. D.A. Jones, J.G. Munford, and P.G. Gabbott, *Nature* **247**, 233–235 (1974).
5. D.A. Jones, K. Kumaly, and A. Arshad, *Aquaculture* **64**, 133–146 (1987).
6. C.J. Langdon, *Mar. Biol.* **102**, 217–224 (1989).
7. S. Appelbaum, *Aquaculture* **49**, 209–221 (1985).
8. J. Walford, T.M. Lim, and T.J. Lam, *Aquaculture* **92**, 225–235 (1991).
9. C. Fernandez-Diaz and M. Yufera, *Aquaculture* **153**, 93–102 (1997).
10. A. Kanazawa, S.-I. Teshima, H. Sasada, and S.A. Rahman, *Bull. Jap. Soc. Sci. Fish.* **48**, 195–199 (1982).
11. K. Kurmaly, D.A. Jones, A.B. Yule, and J. East, *Aquaculture* **81**, 27–45 (1989).
12. M. Kumlu and D.A. Jones, *J. World Aquacult.* **26**, 406–415 (1995).
13. C.J. Langdon and A.E. DeBevoise, *Mar. Biol.* **105**, 437–443 (1990).
14. D.A. Kreeger and C.J. Langdon, *Biol. Bull., Woods Hole* **185**, 123–139 (1993).
15. D.A. Kreeger, A.J.S. Hawkins, B.L. Bayne, and D.M. Lowe, *Mar. Ecol. Prog. Ser.* **126**, 177–184 (1995).
16. D.A. Kreeger and C.J. Langdon, *Mar. Biol.* **118**, 479–488 (1994).
17. D.A. Kreeger, *Limnol. Oceanogr.* **41**, 208–215 (1996).
18. J. López-Alvarado, C.J. Langdon, S.-I. Teshima, and A. Kanazawa, *Aquaculture* **122**, 335–346 (1994).
19. C.J. Langdon and M.J. Waldock, *J. Mar. Biol. Assoc. U.K.* **61**, 431–448 (1981).
20. F.-L.E. Chu, K.L. Webb, D.A. Hepworth, and B.B. Casey, *Aquaculture* **64**, 185–197 (1987).
21. K. Numaguchi and J.A. Nell, *Aquaculture* **94**, 65–78 (1991).
22. J. Knauer and P.C. Southgate, *J. Shellfish Res.* **16**, 447–453 (1997).
23. J. Knauer and P.C. Southgate, *Aquaculture* **153**, 291–300 (1997).
24. S. Ozkizilcik and F.-L.E. Chu, *J. World Aquacult.* **25**, 147–154 (1994).
25. P.C. Southgate and C. Dong, *Aquaculture* **134**, 91–99 (1995).
26. C.J. Langdon and C.A. Siegfried, *Aquaculture* **39**, 135–153 (1984).
27. M.A. Buchal and C.J. Langdon, *J. Aquacult. Nutr.* **4**, 263–274 (1998).
28. D.F. Villamar and C.J. Langdon, *Mar. Biol.* **115**, 635–642 (1993).
29. R.S. Parker and D.P. Selivonchick, *Aquaculture* **53**, 215–228 (1986).
30. M.S. Madhyastha, I. Novaczek, R.F. Ablett, G. Johnson, M.S. Nijjar, and D.E. Sims, *Aquat. Toxicol.* **21**, 15–28 (1991).
31. I. Novaczek, M.S. Madhyastha, R.F. Ablett, G. Johnson, M.S. Nijjar, and D.E. Sims, *Aquat. Toxicol.* **21**, 103–118 (1991).
32. S. Ozkizilcik and F.-L.E. Chu, *Aquaculture* **128**, 131–141 (1994).
33. D. Chapman, *Liposome Technology*, CRC Press, Boca Raton, FL, 1984.
34. S.B. Kulkarni, G.V. Betageri, and M. Singh, *J. Microencapsulation* **12**, 229–246 (1995).

See also FEED MANUFACTURING TECHNOLOGY.

MILKFISH CULTURE

ROBERT R. STICKNEY
Texas Sea Grant College Program
Bryan, Texas

OUTLINE

Milkfish (*Chanos chanos*) are popular foodfish in the Indo-Pacific region. In some places, they are also used as bait by the tuna fishing industry. For centuries, milkfish have been widely cultured in the tropical Indo-Pacific, and they still are. For example, milkfish culture has been practiced in Indonesia for 700 years and in the Philippines for at least 300 years (1). Until the late 1980s, the culture of milkfish was entirely dependent upon the capture of fry or fingerlings from nature.

Milkfish are popular because they can be reared in a wide range of salinities, do not require high-quality prepared feeds, and grow rapidly. Milkfish can be grown from fry to harvest size in six to eight months. The major milkfish-producing countries are Indonesia, the Philippines, and Taiwan. A recent review of milkfish culture (2) forms the basis for much of the information presented here.

LIFE HISTORY

Adult milkfish can attain sizes of up to 4.5 ft (1.4 m) and 33 lb (15 kg). The fish will live up to 15 years, are found swimming in schools in the water column (that is, they are pelagic), and are highly migratory (3). The eggs, embryos, and larvae are also pelagic. Larvae reach sizes of 0.4 in. (10 mm) within two to three weeks after hatching. The larvae move with the aid of currents and by actively swimming to shallow-water areas such as mangrove swamps and lagoons associated with coral reefs and atolls. For several months, the fish remain in the coastal region, where they become juveniles. Eventually, the young milkfish migrate offshore, to inhabit the continental shelf for the remainder of their lives.

Adult female milkfish produce eggs averaging 0.05 inch (1.2 mm) in diameter and ranging in number from about 300,000 to 700,000. Natural spawning occurs from June through October in Hawaii, at the higher end of latitudes where milkfish occur, while in the southern part of the range the fish may spawn from August through May. Multiple spawning can occur at intervals of three to four weeks.

SPAWNING AND LARVAL REARING

Milkfish fry can be collected throughout the tropical Indo-Pacific in coastal waters. Historically, the milkfish culture industry has depended solely upon fry caught in the wild for stocking ponds and net pens. In Taiwan alone, the demand for milkfish fry exceeds 100 million annually. The fish farmers are not typically involved in collecting fry. Instead, fishermen employ nets to collect the fry, and then they transport them, often in small hand-carried waterproof containers of various kinds, to the fish farmers. Significant losses can occur during transit when the fish are carried long distances by human porters.

Since the late 1980s, Taiwan and the Philippines are among the milkfish-producing nations that have achieved some independence from the need for wild fry by developing hatchery systems. In Taiwan, hatchery activity is often conducted in two separate phases involving culturists who maintain and spawn broodfish and those that rear fry to the size at which they may be stocked.

Hatcheries often implant females with a combination of cholesterol, leuteinizing hormone-releasing hormone (LHRH), and methyltestosterone to help ensure proper egg development. Implants are effective only when the fish are simultaneously exposed to long photoperiods, so the technique does not eliminate the need for at least one environmental cue.

Once proper egg development has occurred, the females can be induced to spawn with an injection of LHRH. The timing of the injection is based on an examination of the developing eggs by sampling anesthetized females. When the eggs are of the proper size, the hormone is injected, and spawning will take place within 12 to 15 hours. Each injected female is placed in a spawning tank, often in the presence of two males (2).

In the Philippines, broodfish of five years or older held in floating cages will spontaneously spawn, and their eggs can be collected for hatching and rearing subsequent of the larvae. Milkfish will also spawn naturally in (canvas-lined, concrete, or fiberglass) tanks and in ponds. Eggs need to be removed from the spawning habitats soon after they are produced, to prevent them from being consumed by the adults.

Reports have surfaced of morphological abnormalities at rates of from 3 to 26% in hatchery fry (4). Abnormalities include clefts in the branchiostegal membrane and deformities in the opercula. High mortality rates are common in deformed fry.

Hatchery-produced fry are commonly reared in ponds and fed planktonic organisms, such as rotifers, reared specifically as food for the milkfish. Rotifers should be available from hatching until the fish are 25 days old. Phytoplankton should be present at least during the first 12 days after hatching. Brine shrimp nauplii and prepared feeds may be accepted beginning on the 12th day (2). Oyster eggs are sometimes used for the first feeding stage of milkfish fry. Naturally produced copepods are also a good food source for fry in extensive culture systems. Stocking rates vary, depending on the type of culture process used (extensive or intensive).

GROWOUT

Milkfish growout is typically conducted in ponds, although an interesting exception occurs in a large lake above Manila Bay in the Philippines known as the Laguna de Bay. There, where the salinity ranges from fresh to slightly brackish, large net pens have been constructed in which milkfish are reared. Many of the pens are located near houses on stilts. The fish culturists reside in the houses, primarily to maintain the nets and forest all poaching. No food is provided for the milkfish, as the Laguna de Bay is a highly productive body of water. The proliferation of net pens in the bay in the 1970s resulted in a loss of natural productivity due to overstocking with milkfish and tilapia. (See the entry "Tilapia culture.") Regulations on the number and size of operations were imposed to increase and maintain milkfish production, which reportedly reaches 1.6 tons per acre (4 tons per hectare) in the Laguna de Bay area (1).

Pond culture also often involves providing natural food. Prior to stocking, ponds are partially filled and fertilized to encourage the growth of a mixed benthic algal, microorganism, and animal community known in the Indo-Pacific as lab-lab. The lab-lab community often includes several species of bacteria, blue-green algae, diatoms, protozoa, copepods, amphipods, ostracods, nematodes, polychaetes, molluscs, cladocerans, isopods, and other organisms. As the community develops, additional water is added until the pond is full, after which the young milkfish are introduced.

Pond production varies greatly from one nation to another, depending upon the intensity of culture, the use of prepared feeds, and overall management strategies. In Taiwan production averages some 2,000 lb/acre (approximately 2,000 kg/ha), while in the Philippines average production is about 870 lb/acre and in Indonesia is even less, 450 lb/acre (450 kg/ha) (1).

Prepared feeds have also been developed for milkfish. Research by fish nutritionists have outlined the protein and lipid requirements of milkfish, and some information is available regarding substituting dietary fiber for other feed ingredients. Detailed information on nutritional requirements, however, is generally lacking. Because milkfish are largely herbivorous during the growout phase, prepared diets can be relatively low in protein compared with carnivore feeds.

Like other aquaculture species, milkfish are susceptible to a variety of diseases. In addition to bacterial diseases, milkfish may harbor such parasites as protozoa, copepods, trematodes, nematodes, and cestodes.

BIBLIOGRAPHY

1. C. Lim, in R.P. Wilson, ed. *Handbook of Nutrient Requirements of Finfish*, CRC Press, Boca Raton, FL, 1991, pp. 97–104.
2. C.S. Tamaru, F. Cholik, J.C.-M. Kuo, and W.J. FitzGerald, Jr., *Reviews in Fisheries Science* **3**, 249–273 (1995).
3. T. Bagarinao, *Environmental Biology of Fishes* **39**, 23–41 (1994).
4. G.V. Hilomen-Garcia, *Aquaculture* **152**, 155–156 (1997).

See also FERTILIZATION OF FISH PONDS; LARVAL FEEDING—FISH.

MINERALS

Delbert M. Gatlin, III
Texas A&M University
College Station, Texas

OUTLINE

INTRODUCTION

Minerals comprise a distinct group of compounds characterized by their inorganic nature. Many of these elements are required by animals to support various structural and metabolic functions and thus are considered essential. It has been established that fish and crustaceans generally require the same essential minerals as terrestrial animals to support various bodily processes (1). However, a considerable amount of dissolved minerals in the aquatic environment may contribute to satisfying some of the metabolic requirements of aquatic organisms and thus influence dietary requirements. The biochemical functions of minerals in aquatic species are generally similar to those in terrestrial animals with the exception of osmoregulation, which involves maintenance of osmotic balance between body fluids and the water in which aquatic animals live (2). In terms of osmoregulation, freshwater species lose ions to the hypotonic environment and therefore suffer from hydration; thus these organisms generally do not drink water but excrete large quantities of excess water as dilute urine. In contrast, marine species drink seawater to make up for a loss of fluid from their body to the environment. Fish and crustaceans in freshwater and marine environments may obtain dissolved minerals from the water by absorption across the gills or other body surfaces or from ingestion under certain conditions. Thus, the aquatic environment has tended to complicate and limit research on mineral nutrition of fish and crustaceans. However, research efforts over the past decade have continued to advance knowledge about various aspects related to mineral nutrition of aquatic species, much of which has been applied to the production of these organisms in aquaculture.

CLASSIFICATION OF MINERALS

There are over 100 elements listed in the periodic table, and many of these elements occur in nature in an inorganic form (unbound to carbon). Of all these elements, less than 40 are present in appreciable quantities in the animal body (3). Some of these mineral elements do not have established metabolic roles and a regular dietary supply is not required; thus, they are classified as nonessential. The presence of such minerals in the animal is generally attributed to their occurrence in the environment and transmission to the animal in its food. Examples of these nonessential minerals include aluminum, antimony, bismuth, boron, germanium, gold, lead, mercury, rubidium, silver, and titanium. Other mineral elements found in the body have been demonstrated to be nutritionally essential. Their essentiality is based on the development of reproducible structural and/or physiological abnormalities with the individual deletion of each specific mineral from an otherwise nutritionally adequate diet and reversal of these abnormalities by addition of the specific mineral back to the selectively deficient diet. At least 21 different mineral elements have been established as essential for certain animals (3). It is these nutritionally important minerals that will be the focus of this article.

In addition to minerals being classified as essential or nonessential, they also are generally categorized as macrominerals or microminerals based on the amounts found in the body and generally required in the diet. Microminerals also may be referred to as trace minerals. The macrominerals and microminerals will be discussed separately in the next two sections. The individual minerals within each group will be considered with regard to their specific functions, known aspects of metabolism, and dietary requirement levels in representative fish and crustacean species. Subsequent sections will consider other aspects of mineral nutrition such as potential toxicity of certain minerals, interactions between minerals and other compounds, sources and forms of minerals for dietary supplementation, and biological availability of minerals.

MACROMINERALS

Macrominerals are generally present in the body in appreciable amounts and required in the diet in relatively large quantities. Seven macrominerals are generally considered to be essential. These include the cations calcium (Ca^{2+}), magnesium (Mg^{2+}), potassium (K^{+}), sodium (Na^{+}), and the anions chloride (Cl^{-}), phosphorus (PO_4^{3-}) and sulfur (SO_4^{3-}) (3).

Calcium

Calcium is a primary structural component of hard tissues such as bones, exoskeleton, scales, and teeth of aquatic animals. Additionally, calcium is essential for a variety of physiological processes such as blood clotting, muscle function, nerve impulse transmission, osmoregulation, and as a cofactor for enzymatic processes (3).

Aquatic organisms differ from terrestrial organisms in that their metabolic requirements for calcium typically can be met by absorbing this mineral from the water in which they live. Both shrimp and fish can absorb some minerals from the water via drinking (primarily marine organisms), and by direct absorption via the gills, fins, and skin (1). The gills are the most important site of calcium regulation in freshwater and marine fish (2). Dietary calcium is primarily absorbed from the intestine by active transport. In vertebrates, the vitamin D metabolite 1,25-dihydroxycholecalciferol functions in the maintenance of serum calcium levels by altering the rate of intestinal absorption (via a Ca^{2+}-binding protein), renal resorption, and bone mobilization (3). There is some evidence that vitamin D and its metabolites affect calcium homeostasis in teleosts, although the importance of vitamin D in utilization of dietary calcium via intestinal uptake appears to be less critical in fish as compared to terrestrial animals (1,4,5).

Freshwater of moderate hardness (~50 mg/L as $CaCO_3$) as well as brackish water and seawater, which contain much higher levels of calcium, generally have been shown to provide fish and crustaceans with adequate calcium to sustain metabolic functions in the absence (or presence of very low levels) of dietary calcium (1). However, in the presence of low levels of waterborne calcium, the essentiality of dietary calcium has been established for various freshwater species such as channel catfish, *Ictalurus punctatus* (6), and tilapia, *Oreochromis aureus* (7). In such low-calcium water (<1 mg Ca/L), fish fed calcium-deficient diets manifested deficiency signs similar to terrestrial animals such as reduced growth and impaired ossification of bone and scale tissues. Under such conditions, dietary calcium levels ranging from 0.45 to 0.7% of diet were required to maintain normal growth and bone mineralization (6,7).

Some of the earliest studies with marine penaeid shrimp reported dietary calcium requirements of 1 to 2% of diet (8,9). However, other studies with *L. topenaeus* and *L. vannamei* raised in seawater were not able to demonstrate a dietary calcium requirement (10,11).

Adequate calcium hardness is generally required for freshwater to be suitable for aquaculture, and brackish water and seawater contain high levels of dissolved calcium; therefore, dietary supplementation of calcium for various fish and crustacean species is generally not necessary. In fact, excessive levels of dietary calcium should be avoided because it can negatively affect the utilization of other minerals, as will be discussed in subsequent sections.

Phosphorus

In contrast to calcium, phosphorus is typically limiting in water and must be provided in the diet to meet the metabolic requirement of aquatic organisms. Approximately 80% of phosphorus in the body is associated with hard tissues (3). Phosphorus also is a component of a variety of organic phosphates such as nucleotides, phospholipids, coenzymes, deoxyribonucleic acid, and ribonucleic acid. Inorganic phosphates also serve as important buffers to maintain normal pH of intracellular and extracellular fluids (3).

Information concerning phosphorus metabolism of fish and crustacea is rather limited. Uptake of waterborne phosphorus by fish has been demonstrated (2), although significant quantities are not obtained in this manner due to the low levels in freshwater and seawater. Levels generally remain low due to the rapid uptake of phosphorus by plants and microorganisms, as well as phosphorus binding to soils. Absorption of dietary phosphorus from the intestine is most critical in satisfying metabolic requirements. Storage and mobilization of phosphorus may be influenced by numerous factors including those that control calcium metabolism. Phosphorus is primarily excreted from the body via the urine although a considerable amount of unabsorbed phosphorus from the diet is lost in the feces (2).

Due to the importance of dietary phosphorus in meeting the metabolic requirements of fish, signs of phosphorus deficiency are generally manifested quite readily and have been well established in numerous fish species (1). Some of the most prominent deficiency signs include impaired growth and reduced mineralization of scales and skeletal tissues. Many studies have been conducted to determine dietary phosphorus requirements for various fish species. These requirement values range rather widely from 0.3 to 0.9% of diet (1). Even wider ranges in dietary phosphorus requirements have been reported for some crustacean species. For example, phosphorus requirements as low as 0.34% of diet (11) and as high as 1% (9), 1 to 2% (8) and 2% (10) of diet have been recommended for penaeid shrimp. Several factors may influence the phosphorus requirements determined in these various studies. Among these are the calcium content of the diet which at elevated levels may interfere with phosphorus absorption, the form of dietary phosphorus and its availability to the organism, and other experimental factors such as size and nutritional condition of the organism and criteria used in estimating requirements.

The eutrophying effects of phosphorus in aquaculture effluents have resulted in a considerable amount of research being focused on phosphorus nutrition of fish in recent years. These efforts have focused on increasing dietary phosphorus utilization and/or limiting phosphorus excretion while meeting metabolic requirements of the organism. Further consideration will be given to these aspects of phosphorus nutrition in subsequent sections of this entry.

Magnesium

Magnesium is another macromineral with diverse metabolic functions. In vertebrates, approximately 60% of total body magnesium is located in bone (3). In soft tissues, magnesium occurs both intracellularly and extracellularly. In addition, magnesium is essential for cellular respiration and neuromuscular transmission. It also activates many enzymatic systems involved in the metabolism of fats, carbohydrates and proteins (3). Excess magnesium is excreted in the urine of fish (2).

Freshwater and marine fish have been shown to utilize either waterborne or dietary magnesium to meet metabolic requirements (2). Dietary magnesium deficiencies have been documented for a variety of freshwater fishes (1). Deficiency signs include poor growth, loss of appetite, lethargy, muscle flaccidity, convulsions, vertebral curvature, high mortality, and depressed magnesium levels in the whole-body, blood serum, and bone (2). Fish in freshwater, which typically contains 1 to 3 mg of Mg/L, require magnesium at 0.04 to 0.06% of diet (1). Seawater typically contains high levels of magnesium (1,350 mg/L), and marine crustacea and fish generally have blood magnesium levels lower than that of the external medium. Thus, marine species may not require a dietary source of magnesium (12). Red sea bream, *Chrysops major*, reared in seawater showed no signs of deficiency when fed magnesium at 0.012% of diet (13). Supplementation of magnesium at 0.1 to 0.5% of diet was reported to be beneficial to *Marsupenaeus japonicus*; however, weight gain was very low in this experiment and dietary essentiality was not established (8). A depression in hepatopancreas magnesium levels was observed in *L. vannamei* in response to the deletion of magnesium from a semipurified diet; however, weight gain and magnesium levels of the carapace were unaffected (14). Based on the high levels of magnesium in seawater and the magnesium requirements of freshwater fish, a dietary magnesium requirement for marine species would not be expected. Because most feed ingredients, especially those of plant origin, are high in magnesium, supplementation of magnesium to practical diets is generally not necessary.

Sodium, Chloride, and Potassium

Sodium, chloride, and potassium occur principally in fluids and soft tissues of the body and are each involved in controlling osmotic pressure and acid–base equilibrium (3); thus, they are generally considered together as a group. Sodium is a primary electrolyte and the most abundant cation in extracellular fluids. Chloride is the most abundant anion in extracellular fluids while potassium is the major intracellular cation. Concentrations of these electrolytes in the body are principally controlled by the gills and kidney. Dietary deficiencies of sodium, chloride, and potassium have been difficult to demonstrate in fish due to the uptake of these elements from the water (2). The supplementation of high levels (4.5 to 11.6% of the diet) of sodium chloride (NaCl) to the diet has been reported to inhibit feed efficiency of rainbow trout raised in freshwater, presumably due to nutrient dilution (15). Additionally, there were no positive or negative effects of NaCl supplementation to the diet of channel catfish raised in freshwater (16) or Atlantic salmon, *Salmo salar*, raised in freshwater or seawater (17). In contrast, the supplementation of NaCl to the diet resulted in increased weight gain of the euryhaline red drum, *Sciaenops ocellatus*, when reared at low salinity (<6 ppt) (18). Similar benefits were not observed when red drum were fed the same diet in seawater (32 ppt). The addition of dietary NaCl at low salinities may increase the absorption of amino acids and/or satisfy other metabolic requirements, thus resulting in a physiological advantage to the red drum. The supplementation of NaCl to practical diet formulations at 7 to 10% also has been found to increase survival of fish being transferred from freshwater to seawater (19), presumably through the stimulation of osmoregulatory function and gill sodium- and potassium-ATPase activity (19).

A dietary potassium requirement has been identified for channel catfish (20) and chinook salmon, *Oncorhynchus tshawytscha*, in freshwater (21) but not for red sea bream in seawater (22), indicating that marine fish can obtain adequate levels of potassium from the water. Deficiency signs in chinook salmon included reduced feed intake, convulsions, tetany, and death (1). Dietary essentiality of potassium for marine shrimp is less clear, although a diet containing potassium at 0.9% was reported to improve growth of *M. japonicus* as compared to diets containing potassium at 1.8% (8). The individual deletion of potassium from a semipurified diet did not result in a significant depression in tissue potassium or growth of *L. vannamei*; however, tissue levels of magnesium were affected, indicating a potential interaction (14).

Most freshwater and all seawater probably contain sufficient amounts of sodium, potassium, and chloride ions to satisfy the physiological needs of fish (1). These ions also are found in substantial amounts in most feedstuffs, making the necessity of dietary supplementation unlikely.

Sulfur

This mineral occurs almost exclusively as a constituent of numerous organic molecules in the body, including the amino acids cystine and methionine. Ingestion of inorganic sulfur is not able to satisfy the organism's requirement for sulfur-containing compounds (3).

TRACE MINERALS

Trace minerals or microminerals are present in the body and required in the diet at much lower levels than macrominerals. Minerals included in this category include chromium (Cr^{3+}), cobalt (Co), copper (Cu^{2+}), iodine (I^-), iron (Fe^{2+}), manganese (Mn^{2+}), selenium (SeO_3^{2-}), and zinc (Zn^{2+}). Dietary deficiencies of these minerals have been produced in some fish species by feeding purified diets under controlled conditions for extended periods of time. The essentiality of other trace minerals such as fluorine, molybdenum, nickel, silicon, tin, and vanadium has not been established for fish or crustaceans (1).

Chromium

Chromium in the trivalent state has been established in terrestrial animals to be a cofactor with insulin and thus influences carbohydrate metabolism (3,23). It also has been implicated in affecting protein and lipid metabolism of terrestrial animals. The role of chromium in fish nutrition has been investigated to a much more limited extent than in terrestrial animals, although more attention has been focused on this mineral in recent years (23). One study (24) reported that chromium supplementation at 2 mg/kg diet caused significant increases in weight gain, energy deposition, and liver glycogen, as well as altered postprandial plasma glucose concentrations of hybrid tilapia (*Oreochromis niloticus* × *O. aureus*) when the diet contained glucose. In contrast, dietary chromium did not enhance the utilization of diets containing cornstarch by these fish. Also in that study, chromium in the form of Cr_2O_3 was much more effective in altering glucose utilization of hybrid tilapia as compared to $CrCl_3$ and Na_2CrO_4. In contrast, other studies such as a recent one with channel catfish reported no effect of Cr_2O_3 on dietary glucose utilization or chromium retention in the body (25). Specific mechanisms by which chromium influences dietary carbohydrate utilization of fish have not been elucidated. Chromium nutrition of crustaceans has not been investigated at this time. Supplementation of chromium in practical diets containing complex soluble carbohydrates (e.g., starch) does not appear warranted for various fish species.

Cobalt

Cobalt is a component of vitamin B_{12} and is required for microbial synthesis of that vitamin. Microbial synthesis of vitamin B_{12} in the intestine of channel catfish was reduced when cobalt was eliminated from the diet (26). In terrestrial monogastrics, a dietary supply of cobalt is generally dispensable, especially if vitamin B_{12} is provided. This would also apply to fish and crustaceans.

Copper

Copper functions in blood cell formation and in numerous copper-dependent enzymes including lysyl oxidase, cytochrome *c* oxidase (CCO), ferroxidase, tyrosinase, and superoxide dismutase (SOD) (3). Lysyl oxidase functions in the formation of cross-links during the synthesis of collagen and elastin. The failure of collagen maturation (cross-linking) in the organic matrix of bone accounts for increased fragility of bones and the associated abnormalities of copper deficiencies. Failure of collagen and elastin cross-linking and an undefined muscular defect result in enlargement of the heart and cardiac failure in copper-deficient animals (3). Copper is also involved in the absorption and metabolism of iron and functions in the formation of hemoglobin in vertebrates. In contrast to hemoglobin in vertebrates, crustaceans utilize copper-containing hemocyanin as the oxygen-carrying pigment (2). It has been estimated that, on a freshweight basis, 40% of the whole-body copper load in shrimp is found in hemocyanin (27). This suggests a considerable increase in the physiological demand for copper by crustaceans above that required by vertebrates.

Dietary deficiencies of copper have been documented for several freshwater fish (28–31), but the dietary essentiality of copper has not been evaluated in marine fish. Dietary requirements for copper range from 1.5 to 5 mg Cu/kg diet (1). In addition to growth and feed efficiency, copper-dependent enzymes such as ceruloplasmin, copper- and zinc-dependent SOD and CCO have been shown to be excellent indicators of copper nutriture (30).

The dual deletion of iron and copper had no significant effect on growth and survival of *M. japonicus* (8). However, in this series of experiments, weight gain was low and survival was poor; hence, the nutritional stress or the quality of the diet may not have been adequate to induce a dietary deficiency. A dietary copper deficiency was demonstrated in *L. vannamei* fed semipurified diets containing <34 mg Cu/kg diet (32). Deficiency signs included poor growth, reduced copper levels in the carapace, hepatopancreas, and hemolymph, and enlargement of the heart. Similarly, based on growth, survival, CCO activity and tissue mineralization, a dietary copper requirement of 53 mg Cu/kg diet was reported for *Penaeus orientalis* (33). These results indicate that shrimp cannot meet their physiological needs for copper from seawater and that a dietary source is required for maximum growth and tissue mineralization. This also indicates that species utilizing copper as a component of their respiratory pigment have an increased copper requirement over species utilizing iron-based respiratory pigments.

Iron

Iron is a trace element that is essential for the production and normal functioning of hemoglobin, myoglobin, cytochromes, and many other enzyme systems. In vertebrates, the principal role of iron is as a component of hemoglobin. Red blood cells are regenerated periodically, and most of the iron is recycled. That which is not recycled is excreted via the bile into the intestine. Like other elements of low solubility, iron is absorbed and transported in the body in a protein-bound form (3). In vertebrates, mucosal apoferritin binds Fe^{2+} in the intestinal lumen and transports it across the mucosal brush border. Within the cell, Fe^{3+} is bound to transferrin forming transferritin. Iron-bound transferritin is then transported in the blood, where the iron is again released at target tissues (liver and hematopoietic tissue) (3).

In crustaceans, the hepatopancreas has been found to be the organ richest in iron. Storage cells containing iron have been reported in crayfish, *Procambarus clarkii* (34), and the crab, *Cancer irroratus* (35). Iron-transporting proteins also have been found in the hemolymph of two species of crabs (36,37). These observations indicate the presence of a regulatory mechanism similar to that of vertebrates. In addition to the digestive system, gills appear to play an active role in iron metabolism. In *C. irroratus*, iron accumulates by forming a coating around the branchial lamellae during the intermolt cycle, which is then rejected at ecdysis along with the integument. Absorption of iron from the water through the gills may provide an additional source of iron.

Iron deficiencies have been documented for several species of fish; however, dietary deficiencies for shrimp have not been observed (8,12,38). Iron deficiency causing anemia has been reported for freshwater fish such as the brook trout *Salvelinus fontinalis* (39) and common carp, *Cyprinus carpio* (40), and in marine fish such as the red sea bream (41) and yellowtail, *Seriola quinqueradiata* (42). However, growth depression was not observed in these iron-deficient fish. Iron deficiency signs of channel catfish fed a semipurified basal diet (9.6 mg Fe/kg) included suppressed weight gain and feed efficiency as well as reduced hemoglobin, hematocrit, plasma iron, transferrin saturation, and erythrocyte-count values (43). In that study, a minimum of 20 mg supplemental Fe/kg diet (30 mg total Fe/kg) was required by channel catfish for best growth and hematological values. Although many practical diets may contain considerable levels of endogenous iron, little is known about its form and availability (2). Hence, a low level of supplementation (20–30 mg/kg) of an inorganic source is often recommended to ensure adequacy of the diet.

Iodine

Iodine is an essential element which is present in most cells of the body, although in vertebrates the thyroid gland is the main location of iodine reserves. Thyroid hormones, which contain iodine, are known to have roles in thermoregulation, intermediary metabolism, reproduction, growth and development, hematopoiesis and circulation, as well as neuromuscular functioning (3). The minimum dietary iodine requirement of fish has not been well defined; however, 1 to 5 mg I/kg diet have been found adequate (1). The physiological essentiality of iodine has not been evaluated in shrimp.

Manganese

Manganese functions as a cofactor in several enzyme systems, including those involved in urea synthesis from ammonia, amino acid metabolism, fatty acid metabolism, and glucose oxidation (3). Principal signs of manganese deficiency in terrestrial species include reduced growth rate, skeletal abnormalities, convulsions, reduced righting ability, abnormal reproductive function in males and females, and ataxia in the newborn (3).

Dietary manganese deficiencies in fish have included poor growth, skeletal abnormalities, high embryo mortalities, and poor hatch rates (1). A total dietary manganese content of 12 to 13 mg/kg has been recommended for the common carp and rainbow trout, *Oncorhynchus mykiss* (28); however, 2.4 mg Mn/kg diet was sufficient for normal growth and health of channel catfish (44). Supplementation of 10 and 100 mg Mn/kg diet did not improve the growth of *M. japonicus* (8). However, it should be noted that percent weight gain during that study was low and the nutritional stress placed on the shrimp may not have been severe enough to reduce body stores and induce a deficiency. Because the manganese content of seawater is very low (0.01 mg/L), significant absorption from the water is unlikely. Thus, a dietary source of manganese may be necessary for marine shrimp and fish.

Selenium

Selenium is a trace element which functions primarily as a component of the enzyme glutathione peroxidase which converts hydrogen peroxide and lipid hydroperoxides into water and lipid alcohols, respectively. Thus, this enzyme functions in protecting the cell from deleterious effects of peroxides (45). Glutathione peroxidase acts along with vitamin E to function as a biological antioxidant to protect polyunsaturated phospholipids in cellular and subcellular membranes from peroxidative damage (1). In addition to its enzymatic functions, selenium helps protect against mercury toxicosis by forming a mercuric–selenium complex. This protein-bound complex is diverted from the kidney (where inorganic mercury devoid of selenium is deposited) to the liver and spleen where its toxicity is considerably reduced (46).

Complementary functions of selenium and vitamin E may allow these nutrients to interact physiologically (47). Selenium and vitamin E interrelationships have been investigated in several animal species, and a variety of common and unique deficiency signs have been described (1). Differing responses, especially with respect to gross deficiency signs, were observed when Atlantic salmon (48), rainbow trout (49), and channel catfish (47) were fed diets without supplemental selenium, vitamin E, or both nutrients.

Levels of 0.15 to 0.38 mg Se/kg diet (50) and 0.25 mg Se/kg diet (51) were required to provide maximum growth and glutathione peroxidase activity in rainbow trout and channel catfish, respectively. In zooplanktonic daphnids, a selenium deficiency in the medium resulted in cuticle deformation and a depression in reproduction. In the presence of replete zinc, 1 ppb Se was adequate (52); however, in the absence of detectable zinc, 5 ppb Se in the medium was required to eliminate deficiencies characteristic of selenium deprivation (53). It was found that juvenile *L. vannamei* grew best when fed semipurified diets supplemented with 0.2 to 0.4 mg Se/kg diet (54). Although a specific level has not been quantified, it appears that shrimp have a dietary requirement for selenium.

Based on currently available dietary selenium requirements that have been quantified for aquatic animals, it appears that supplementation of selenium to practical diets is warranted. Selenium supplementation is regulated by the United States Food and Drug Administration and supplemental selenium at 0.1 mg/kg is approved for minor-use animals such as aquaculture species.

Zinc

Zinc is required for normal growth, development, and function in all animal species that have been studied (3). Zinc functions as a cofactor in several enzyme systems and is a component of a large number of metalloenzymes which include carbonic anhydrase, carboxypeptidases A and B, alcohol dehydrogenase, glutamic dehydrogenase, D-glyceraldehyde-3-phosphate dehydrogenase, lactate dehydrogenase, malic dehydrogenase, alkaline phosphatase, aldolase, superoxide dismutase, ribonuclease, and DNA polymerase (3).

A dietary requirement for zinc has been quantified for a variety of freshwater fishes fed semipurified diets: 20 mg Zn/kg diet for channel catfish (55) and blue tilapia (56); 15 to 30 mg Zn/kg diet for common carp (57); and 15 to 30 mg Zn/kg diet for rainbow trout (58). In daphnids reared under controlled trace-element exposure utilizing a controlled media system (59), the absence of detectable zinc resulted in decreased life span and increased demand on the organism's pool of available selenium (53). *L. vannamei* was found to require 33 mg Zn/kg diet to maintain normal tissue mineralization (60). The euryhaline red drum was determined to require approximately 20 mg Zn/kg diet (61).

POTENTIAL TOXICITY OF MINERALS

A number of nonessential minerals such as arsenic, cadmium, lead, and mercury are well known as being toxic to the body. Additionally, there are certain essential minerals such as copper, iron, and selenium which may be toxic or impart other detrimental effects if consumed in large enough quantities.

Toxicity of waterborne copper to aquatic organisms is well established (62). Dietary copper in excessive quantities also can be toxic to aquatic species. However, the metal-binding capacity of inducible metallothioneins in aquatic animals is considered to be protective of cellular function (62). Typically, concentrations of dietary copper in excess of 250 mg/kg are toxic to terrestrial animals, whereas dietary concentrations of almost three times that level has no adverse effects on fish (1).

Excessive levels of dietary iron have been shown to be toxic to some aquatic organisms. For example, excessive iron supplementation appeared to have potentially adverse effects on growth of *M. japonicus* (8,12). Additionally, iron-catalyzed lipid oxidation increases with iron supplementation, which in turn adversely affects feed stability (63). Iron is one of the primary metals involved in lipid oxidation, and ferrous iron is a more potent catalyst of lipid peroxidation than ferric iron (62). Ferrous iron catalyzes the formation of hydroperoxides and free radical peroxides by providing a free radical initiator in the presence of unsaturated fatty acids and oxygen. Marine fish and crustacean diets generally contain predominantly polyunsaturated lipids; therefore, supplementation of ferrous iron to the diet could be expected to affect the stability of the diet through increased lipid oxidation (rancidity) and reduced stability of ascorbic acid (64). Consequently, limited supplements and restriction of ingredients with high levels of iron are recommended in practical diet formulations.

Selenium is potentially the most toxic of all essential minerals as relatively low dietary concentrations have been shown to have detrimental effects on various animals (46). Selenium concentrations of approximately 15 mg/kg diet caused reduced growth and feed efficiency as well as elevated mortality in rainbow trout (50) and channel catfish (51).

NUTRITIONAL INTERACTIONS INVOLVING MINERALS

There are several minerals that interact with each other and/or with other dietary constituents such that their utilization by the animal is affected. Of the macrominerals, interactions between calcium and phosphorus are most prominent. It is well established in terrestrial animals that an excess of either calcium or phosphorus relative to the other will result in impaired absorption of both. This is due to the excessive mineral being present in the intestine in a free form such that it combines with the other mineral to form insoluble tricalcium phosphate. Typically, a calcium : phosphorus ratio (Ca : P) between 1 : 1 and 2 : 1 is desired (1). In some fish species, this ratio does not appear to be as critical as in terrestrial animals (65). However, juvenile lobsters, *Homarus americanus* were affected by varying the dietary Ca : P (66). Based on growth and histological studies of the endocuticle, a Ca : P of 0.51 (0.56 : 1.10) was found to be best for lobster juveniles, while Ca : P of 1.55 and greater resulted in abnormalities of the endocuticle. In *M. japonicus*, a dietary Ca : P of 1 : 1 has been recommended (8,9). Supplementation of calcium to a semipurified diet appeared to limit phosphorus bioavailability to *L. vannamei*, but the Ca : P did not totally explain the inhibitory effects of calcium (10). Based on these studies, it appears that calcium may affect phosphorus availability and that calcium levels in excess of 2.5% of diet should be avoided. Although there does not appear to be a Ca : P that will produce optimal results, a ratio of less than 2 : 1 provides good results in commercial formulations.

High levels of dietary calcium provided by fish meal also have been shown to reduce the availability of certain trace minerals such as manganese and zinc (67). The bioavailability of zinc in various fish meals has been found to be inversely related to the tricalcium phosphate content of the meal. Thus, zinc bioavailability is generally lowest in white fish meals, which contain the highest level of tricalcium phosphate, and slightly higher in brown fish meals (67). Reduced bioavailability of zinc in response to calcium phosphate supplementation also has been observed in rainbow trout (68–70), whereas high levels of calcium from calcium carbonate did not affect dietary zinc bioavailability to blue tilapia (56) or channel catfish (71).

Zinc bioavailability to fish also can be reduced by other dietary components such as phytate, a chelating compound which is commonly associated with plant feedstuffs. The adverse effect of phytate on zinc bioavailability has been documented in a variety of animals (3), including fish (56,71–74) and shrimp (60). Calcium also promotes the complexing of zinc to phytate (56,71–74). Practical diets, especially those containing plant feedstuffs or high-ash fish meal, should be supplemented with zinc at rather high levels (100 to 150 mg Zn/kg diet) to overcome the effects of inhibitory agents.

DIETARY SUPPLEMENTATION OF MINERALS

Of the macrominerals, phosphorus is generally the most critical to supplement in prepared diets for fish. Several trace minerals such as copper, iron, manganese, selenium, and zinc should also be supplemented to the diets of fish and crustaceans due to low levels in practical feedstuffs and/or interactions with other dietary components that reduce bioavailability as previously described (54). In order to meet an animal's physiological requirements for various minerals, dietary sources must be available to the animal.

Sources and Bioavailability of Minerals

Of the feed ingredients used in practical aquatic animal diets, fish meal is the richest source of minerals. Research on the bioavailability of minerals contained in fish meals has demonstrated that there is considerable variation among fish species (perhaps due to luminal pH) and that the bioavailability of minerals is affected by meal type and ash content. Diets based on fish meal generally require supplementation of available sources of copper, phosphorus, magnesium, manganese, and zinc to prevent dietary deficiencies and maximize fish growth (67). As the aquatic animal feed industry increases its use of less expensive plant feedstuffs, which are generally poor sources of minerals and may contain factors that reduce the bioavailability of minerals, the need for mineral supplements should increase.

Phytate, which constitutes approximately 67% of the phosphorus in grains, is one of the most undesirable components in many plant feedstuffs. The phosphorus from phytate is not readily available to fish (1) and shrimp (60). In addition, phytate may inhibit the availability of other minerals such as zinc as previously described. It should be noted that phytate phosphorus can account for a considerable portion of the phosphorus in practical diet formulations. The treatment of plant feedstuffs or diets with the microbial enzyme phytase, which can hydrolyze phytate, has been shown to increase the availability of this form of phosphorus to various fish species (5,75,76). However, the cost of phytase and the need to alter feed manufacturing procedures due to the instability of phytase to heat has limited its commercial use in aquaculture at this time. Nonetheless, the present concern to minimize phosphorus in effluents from aquaculture facilities in order to limit potentially adverse eutrophication has continued to stimulate investigations of various phosphorus supplements and strategies to meet the phosphorus needs of fish while lowering phosphorus output. The relative availability or apparent absorption of phosphorus from different supplements can vary considerably for fish (1) and crustaceans (77).

In recent years, there also has been a great deal of interest in chelated trace minerals, which in terrestrial animals as well as some fish species have been shown to have higher bioavailability than inorganic forms. The chelated forms are generally more expensive than inorganic forms. If an element is chelated by a compound that will release it in ionic form at the site of absorption or will be readily absorbed as the intact chelate, this form may greatly enhance the absorption of the element by preventing its conversion to insoluble compounds in the intestine or by preventing its strong adsorption on insoluble colloids. Research on chelated minerals has been rather limited with fish. However, compared with inorganic sources, chelated forms of copper, iron, manganese, selenium, and zinc (each as proteinates) were shown to have higher bioavailability to channel catfish in purified and practical diets (78). In that study, the average improvement in net absorption of chelated minerals over inorganic minerals was approximately 39% for the purified diets and approximately 80% for the practical diets. The greater percent improvement in availability of the chelated minerals relative to inorganic minerals in practical diets is attributed to the greater number of inhibitory compounds such as phytate and fiber in practical feedstuffs to which the chelated minerals are less susceptible. Organic forms of selenium including selenomethionine and selenoyeast also have been shown to have higher bioavailability than inorganic sodium selenite for channel catfish (79). In another study (80), zinc methionine was reported to have over four times the potency of zinc sulfate in practical diets for channel catfish; whereas, no apparent differences in bioavailability of these two compounds were noted in another study with channel catfish (81).

The potentially higher bioavailability of chelated minerals may allow for lower levels of dietary supplementation and a reduction in waste production from unassimilated minerals. As the potential benefits of chelated minerals become better defined, their use with aquatic organisms is likely to increase.

SUMMARY AND CONCLUSIONS

Information on the mineral requirements of aquatic organisms has expanded considerably but is still somewhat limiting for many species. Undoubtedly, as information about mineral requirements expands along with establishment of species and environmental differences, the ability to tailor mineral delivery systems will become more precise. This increased precision will assist in enhancing the economic viability and environmental sustainability of aquaculture.

BIBLIOGRAPHY

1. National Research Council, *Nutrient Requirements of Fish*, National Academy Press, Washington, DC, 1993.
2. S.P. Lall, in J.E. Halver, ed., *Fish Nutrition*, 2nd ed., Academic Press, San Diego, 1989, pp. 219–257.
3. L.E. Lloyd, B.E. McDonald, and E.W. Crampton, *Fundamental of Nutrition*, 2nd ed., W.H. Freeman, San Francisco, 1978.
4. J.P. O'Connell and D.M. Gatlin, III, *Aquaculture* **125**, 107–117 (1994).
5. J. Vielma, S.P. Lall, J. Koskela, F.-J. Schoner, and P. Mattila, *Aquaculture* **163**, 309–323 (1998).
6. E.H. Robinson, S.D. Rawles, P.B. Brown, H.E. Yette, and L.W. Greene, *Aquaculture* **53**, 263–270 (1986).
7. E.H. Robinson, D. LaBomascus, P.B. Brown, and T.L. Linton, *Aquaculture* **64**, 267–276 (1987).
8. A. Kanazawa, S. Teshima, and M. Sasaki, *Mem. Fac. Fish. Kagoshima Univ.* **33**, 63–71 (1984).
9. K. Kitabayashi, H. Kurata, K. Shudo, K. Nakamura, and S. Ishikawa, *Bull. Tokai Regl. Fish. Res. Lab., Tokyo* **65**, 91–108 (1971).
10. O. Deshimaru and Y. Yone, *Bull. Jpn. Soc. Sci. Fish.* **44**, 907–910 (1978).
11. D.A. Davis, A.L. Lawrence, and D.M. Gatlin, III. *J. World Aquacult. Soc.* **24**, 504–515 (1993).
12. W. Dall and J.W. Moriarty, in D.E. Bliss, ed., *The Biology of the Crustacea*, Vol. 5. Academic Press, New York, 1983, pp. 215–216.
13. S. Sakamoto and Y. Yone, *Bull. Jpn. Soc. Sci. Fish.* **45**, 57–60 (1979).

14. D.A. Davis, A.L. Lawrence, and D.M. Gatlin, III, *J. World Aquacult. Soc.* **23**, 8–14 (1992).
15. N.A. Salman and F.B. Eddy, *Aquaculture* **70**, 131–144 (1988).
16. M.W. Murray and J.W. Andrews, *Prog. Fish-Cult.* **41**, 151–152 (1979).
17. H.W. Shaw, R.L. Saunders, H.C. Hall, and E.B. Henderson, *J. Fish. Res. Bd. Canada* **32**, 1813–1819 (1975).
18. D.M. Gatlin, III, D.S. MacKenzie, S.R. Craig, and W.H. Neill, *Prog. Fish-Cult.* **54**, 220–227 (1992).
19. M.M. Al-Amoudi, *Aquaculture* **64**, 333–338 (1987).
20. R.P. Wilson and G. El Naggar, *Aquaculture* **108**, 169–175 (1992).
21. K.D. Shearer, *Aquaculture* **73**, 119–129 (1988).
22. S. Sakamoto and Y. Yone, *J. Fac. Agricult. Kyushu Univ.* **23**, 79–84 (1987).
23. Committe on Animal Nutrition, *The Role of Chromium in Animal Nutrition*, National Academy Press, Washington, DC, 1997.
24. S.Y. Shiau and M.J. Chen, *J. Nutr.* **123**, 1747–1753 (1993).
25. W.-K. Ng and R.P. Wilson, *J. Nutr.* **127**, 2357–2362 (1997).
26. T. Limsuwan and R.T. Lovell, *J. Nutr.* **111**, 2125–2132 (1981).
27. M.H. Depledge, *Mar. Environ. Res.* **27**, 115–126 (1989).
28. C. Ogino and G.-Y. Yang, *Bull. Jpn. Soc. Sci. Fish.* **46**, 455–458 (1980).
29. T. Murai, J.W. Andrews, and R.G. Smith, Jr., *Aquaculture* **22**, 353–357 (1981).
30. D.M. Gatlin, III and R.P. Wilson, *Aquaculture* **54**, 277–285 (1986).
31. K. Julshamn, K.-J. Andersen, O. Ringdal, and J. Brenna, *Aquaculture* **73**, 143–155 (1988).
32. D.A. Davis, A.L. Lawrence, and D.M. Gatlin, III. *Nippon Suisan Gakkaishi* **59**, 117–122 (1993).
33. F. Liu, D. Liang, F. Sun, H. Li, and X. Lan, 1990. *Oceanol. Limnol. Sinica.* **21**, 404–410 (1990).
34. M. Miyawaki, M. Matsuzaki, and N. Sasaki, *Kumamato J. Sci.* **5**, 170–172 (1961).
35. J.-L.M. Martin, *Comp. Biochem. Physiol.* **46A**, 123–129 (1973).
36. M.H. Depledge, R. Chan, and T.T. Loh, *Asian Mar. Biol.* **3**, 101–110 (1986).
37. W. Ghidalia, J.M. Fine, and M. Marneux, *Comp. Biochem. Physiol.* **41B**, 349–354 (1972).
38. D.A. Davis, A.L. Lawrence, and D.M. Gatlin, III, *J. World Aquacult. Soc.* **23**, 15–22 (1992).
39. H. Kawatsu, *Bull. Frshwtr. Fish. Res. Lab.* **22**, 59–67 (1972).
40. S. Sakamoto and Y. Yone, *Bull. Jpn. Soc. Sci. Fish.* **44**, 1157–1160 (1978).
41. S. Sakamoto and Y. Yone, *Rpt. Fish. Res. Lab., Kyushu Univ.* **3**, 53–58 (1976).
42. Y. Ikeda, H. Ozaki, and K. Uematasu, 1973. *J. Tokyo Univ. Fish.* **59**, 91–99 (1973).
43. D.M. Gatlin, III and R.P. Wilson, *Aquaculture* **52**, 191–198 (1986).
44. D.M. Gatlin, III and R.P. Wilson, *Aquaculture* **41**, 85–92 (1984).
45. C. Little, R. Olinescu, K.G. Reid, and P.J. O'Brien, *J. Biol. Chem.* **245**, 3632–3636 (1970).
46. National Research Council, *Mineral Tolerance of Domestic Animals*, National Academy Press, Washington, DC, 1980.
47. D.M. Gatlin, III, W.E. Poe, and R.P. Wilson, *J. Nutr.* **116**, 1061–1067 (1986).
48. H.A. Poston, G.F. Combs, Jr., and L. Leibovitz, *J. Nutr.* **106**, 892–904 (1976).
49. J.G. Bell, C.B. Cowey, J.W. Adron, and A.M. Shanks, *Br. J. Nutr.* **53**, 149–157 (1985).
50. J.W. Hilton, P.V. Hodson, and S.J. Slinger, *J. Nutr.* **110**, 2527–2535 (1980).
51. D.M. Gatlin, III and R.P. Wilson, *J. Nutr.* **114**, 627–633 (1984).
52. K.I. Keating and B.C. Dagbusan, *Proc. Natl. Acad. Sci. USA* **81**, 3433–3437 (1984).
53. K.I. Keating and P.B. Caffrey, *Proc. Natl. Acad. Sci. USA* **86**, 6436–6440 (1989).
54. D.A. Davis and D.M. Gatlin, III, *Rev. Fish Sci.* **4**, 75–99 (1996).
55. D.M. Gatlin, III and R.P. Wilson, *J. Nutr.* **113**, 630–635 (1983).
56. W.R. McClain and D.M. Gatlin, III, *J. World Aquacult. Soc.* **19**, 103–108 (1988).
57. C. Ogino and G.-Y. Yang, *Bull. Jpn. Soc. Sci. Fish.* **45**, 967–969 (1979).
58. C. Ogino and G.-Y. Yang, *Bull. Jpn. Soc. Sci. Fish.* **44**, 1015–1018 (1978).
59. K.I. Keating, *Water Resour.* **19**, 73–78 (1985).
60. D.A. Davis, A.L. Lawrence, and D.M. Gatlin, III, *J. World Aquacult. Soc.* **24**, 40–47 (1993).
61. D.M. Gatlin, III, J.P. O'Connell, and J. Scarpa, *Aquaculture* **92**, 259–265 (1991).
62. G. Roesijadi, *Aquatic Tox.* **22**, 81–114 (1992).
63. L.M. Desjardins, B.D. Hicks, and J.W. Hilton, *Fish Physiol. Biochem.* **3**, 173–182 (1987).
64. J.W. Hilton, *Aquaculture* **79**, 223–244 (1989).
65. R.P. Wilson, E.H. Robinson, D.M. Gatlin, III, and W.E. Poe, *J. Nutr.* **112**, 1197–1202 (1982).
66. M.L. Gallagher, W.D. Brown, D.E. Conklin, and M. Sifri, *Comp. Biochem. Physiol.* **60A**, 467–471 (1978).
67. T. Watanabe, S. Satoh, and T. Takeuchi, *Asian Fish. Soc.* **1**, 175–195 (1988).
68. H.G. Ketola, *J. Nutr.* **109**, 965–969 (1979).
69. R.W. Hardy and K.D. Shearer, *Can. J. Fish. Aquat. Sci.* **42**, 181–184 (1985).
70. S. Satoh, K. Tabata, K. Izume, T. Takeuchi, and T. Watanabe, *Nippon Suisan Gakkaishi* **53**, 1199–1205 (1987).
71. D.M. Gatlin, III and H.F. Phillips, *Aquaculture* **79**, 259–266 (1989).
72. D.M. Gatlin, III and R.P. Wilson, *Aquaculture* **41**, 31–36 (1984).
73. N.L. Richardson, D.A. Higgs, R.M. Beames, and J.R. McBride, *J. Nutr.* **115**, 553–567 (1985).
74. S. Satoh, W.E. Poe, and R.P. Wilson, *Aquaculture* **80**, 155–161 (1989).
75. J.C. Eya and R.T. Lovell, *J. World Aquacult. Soc.* **28**, 386–391 (1997).
76. M.H. Li and E.H. Robinson, *J. World Aquacult. Soc.* **28**, 402–406 (1997).
77. D.A. Davis and C.R. Arnold, *Aquaculture* **127**, 245–254 (1994).
78. T. Paripatananont and R.T. Lovell, *J. World Aquacult. Soc.* **28**, 62–67 (1997).
79. C. Wang and R.T. Lovell, *Aquaculture* **152**, 223–234 (1997).

80. T. Paripatananont and R.T. Lovell, *Aquaculture* **133**, 73–82 (1995).
81. M.H. Li and E.H. Robinson, *Aquaculture* **146**, 237–243 (1996).

MOLLUSCAN CULTURE

K.K. Chew
T.L. King
University of Washington
Seattle, Washington

OUTLINE

Mollusc culture and harvest have been practiced for centuries, but most techniques for rearing commercially exploited groups have been refined during the twentieth century. Descriptions of basic culture methods from 1900 to 1950 may be found in the literature, which is especially true for the most desired groups such as oysters, mussels, and clams. Between 1950 and 1970 mechanization and sophisticated farming practices were developed. Culture efforts over the past two decades have provided new techniques and concepts. Such efforts have been developed to restore fisheries that have declined due to pollution, disease, and socioeconomic factors. Moreover, new fisheries utilizing introduced species or unexploited indigenous stocks have been initiated. A wide range of topics on molluscan culture could be covered. Genetic manipulation, predation control, disease, collection of seed or juveniles, nutritional requirements, site selection, harvesting, storage, depuration, environmental concerns, and equipment and its maintenance are all important to the proper culture and handling of commercial shellfish. Furthermore, appropriate changes in culture practices must be viewed as dependent on the species and in the context of traditional culture patterns.

This article discusses examples of some culture concepts and innovations, with some inclusion of historically established techniques where appropriate. Seven groups of molluscs are addressed: oysters, mussels, clams, scallops, abalone, other gastropods, and pearl oysters. The countries with the greatest production for each group are also noted from United Nations Food and Agriculture Organization (FAO) statistical records.

OYSTERS

Production

One of the predominant molluscan groups in the world is the oyster, which is an important protein source in many regions and may become even more so in the future. Total world production of oysters has been increasing steadily since 1970, and increased dramatically between 1987 and 1994, with the current production of approximately 3 million mt in 1996. The top oyster producers, along with their 1996 production estimates in metric tons, are as follows (FAO): (*1*) People's Republic of China: 2,284,663; (*2*) Japan: 222,853; (*3*) Korea: 185,339; and (*4*) France: 152,129. The People's Republic of China has shown the largest rise in recent years, with production between 1987 and 1996 rising from 400,468 to 2,284,663 mt, almost a 2-million-t increase in nine years.

Culture practices and growout methods for oysters vary. However, it must be concluded that semi-intensive culture, from hatchery and nursery phases to growout and harvest, is a well-established practice in some locations such as the Pacific Coast of North America.

Extensive culture methods requiring handling and movement of the shell stock at various times during its life history remains the most widespread technique of producing oysters, probably accounting for more than 90% of all those produced in the world. For example, the U.S. West Coast growers must prepare shell (cultch) to obtain the Pacific oyster (*Crassostrea gigas*) spat from natural field locations or hatchery sources to then plant them for growout in another location. United States oyster growers in the East and Gulf of Mexico coastal areas take advantage of natural reefs for spat production. At times they collect the natural catches of spat on the reefs and deposit them in areas more favorable for growth and survival or protection from disease. Nearly all countries with significant oyster production require some handling of the shell stock prior to harvest time.

Culture Methods

Hatcheries have become important to oyster farmers in the United States over the past 40 years. Through government and private funding commercial oyster hatcheries have been initiated in Chile, Mexico, Tasmania, and other countries during more recent years. The basic concept behind hatcheries is to promote seed availability amid unpredictable or unobtainable natural catches. Several publications outline procedures for developing oyster hatcheries (1–4). Many oystermen on the Pacific Coast of the United States are building large tanks to catch their own spat from eyed larvae purchased from private hatcheries (5–7). One hatchery can produce several billion such larvae in any given year (8). This is resulting in significant worldwide interest in the ease of shipment and a fairly high success rate for seed settlement.

Proper food for hatchery-reared oyster larvae is essential. Many species of algae and diatoms are cultured to support the nutritional requirements of larval and adult oysters. Experiments have been conducted to provide artificial feed such as manufactured pellets (microcapsules) or simple carbohydrates. Such feed has yet to be proven practical in a commercial setting.

Innovations in oyster culture have also brought forth the four-season, all-season, or triploid Pacific oyster, which was developed and now is commercially produced in the northwest United States (9). Triploid means that instead of two sets of chromosomes, there are three sets. This altered oyster, which produces very little sex products during the summer, is sought after because of its high meat quality during the summer months. Details on how to produce the triploid Pacific oyster have been published (10).

Recent improvements in the cultivation of oysters have emphasized technique refinements, occasionally involving mechanization. Bottom culture in the United States and Europe can include the moving of seed oysters from nursery to growout areas where high growth and conditioning occur before harvest. Such gathering of oysters for transfer is still accomplished by hand picking during low tide, particularly among the smaller companies. Larger firms, mainly in the United States, are using drag dredges or specialized hydraulic or mechanical harvesters.

During the past 30 to 40 years, major emphasis has been placed on off-bottom culture as the beaches suitable for bottom culture are already under cultivation. Raft, rack, and longline methods for oyster cultivation are highly developed in Japan (11–14). Tray culture is established and used in parts of Australia, New Zealand, the United States, and Europe where new designs are continually being developed to suit the needs of growers. Tray construction has evolved from early designs using wood and wire screen to those with rubber or plastic-coated wire mesh, and, most recently, mass-produced polypropylene trays. Biofouling is still an important concern in the use of shellfish trays.

An older but extremely useful guide outlines off-bottom techniques in tropical areas, diagrams raft, rack stake, tray, and longline culture techniques, and discusses spatfall prediction systems, collectors for spat, and other subjects important to growers (15). Culture techniques from other parts of the world have been summarized (11–20).

In recent years, plastic mesh bags have become widely used in oyster culture. Growers along the Pacific Coast of the United States use these bags to hold shell (cultch) to catch seed. In Tasmania, oyster growers are using similar plastic mesh bags to hold mature animals placed on intertidal racks for three to four months prior to harvesting (21).

Lantern and pearl nets designed in Japan are used in several areas for growing cultchless seed or single oysters. The nets are constructed from durable UV-resistant polyethylene mesh and hung from rafts or longline systems. Although these nets are an accepted method, fouling of the mesh is a problem and other off-bottom hanging techniques have been investigated.

Off-bottom culture may incorporate the use of containers consisting of steel frames fitted with up to 100 trays (22,23). Each structure is supported on the sea floor with legs and covers nearly 3 m^2. Buoys delimit these containers, which are kept subtidally so as to allow the stock constant exposure to food. They are easily handled with boat-based winches and offer advantages over other off-bottom techniques that also allow for protection from predation. The system is effective against ice and turbulence and is particularly useful in muddy areas.

Some of the more sophisticated types of cultivation involve raceways and artificial ponds. Two former commercial systems in Hawaii, Aquatic Farms and Kahuku Seafood Plantation of System Culture Corporation, have grown oysters on land-based operations (24). Other experimental systems of this type have been tested with some success, but none has been adapted on a commercial scale (23,25).

Barge-based commercial mariculture has been described (26) in an attempt to reconcile conflicting environmental, legal, financial, and operating difficulties inherent in coastal facilities. This applies not only to potential oyster cultivation but also to shellfish in general. Large intensive growout systems on mobile barges, designed and operated like floating dry docks, appear well suited for attaining optimum environmental and growing conditions by selecting varied deployment sites (e.g., seasonally). Strong potential feasibility was discussed (26) along with the possibility of centralized modular mass construction with economies of scale.

Forced upwelling culture systems as part of a three-dimensional nursery and potential growout culture operation (27), a tide-powered upwelling system (28), and an offshore upwelling principle (29) were discussed in the early 1980s. An upwelling column culture technique for oysters was reported in the literature during the same time frame as the Ghent meeting (30). Since 1981, numerous performance studies on upwelling systems have been conducted for a variety of bivalve species in various countries to solve a myriad of problems (31–42).

MUSSELS

Production

The world production of marketable mussels has increased from 900,000 mt in 1987 to over 1,100,000 mt in 1996 with the People's Republic of China, Spain, Italy, and the Netherlands as the leading producers (FAO). During the same time frame France, Korean Republic, Korean Democratic People's Republic, New Zealand, and Thailand also had good production. The top mussel producers, along with their 1996 production estimates in metric tons, are as follows (FAO): (*1*) People's Republic of China: 366,251; (*2*) Spain: 188,462; (*3*) Italy: 100,000; and (*4*) Netherlands: 94,496.

Many species of the genus *Mytilus* are cultured throughout the world (43–52). *Mytilus edulis*, which has a wide cosmopolitan distribution, is the most often discussed in the literature. It is common in Europe, North America, Chile, Asia, and New Zealand. Among the other identified species are *M. coruscus* and *M. crassitestis* in Korea and *M. planulatus* in Australia (46,53,54). There are also the fast-growing green mussels under culture, such as *Perna canaliculus* in New Zealand (49,55,56), *P. perna* in Venezuela, and *P. viridus* in Thailand (57), the Philippines, Indonesia, and Singapore. The *Mytilus* sp. of the Indian coast has been redesignated *Perna (Fenneropenaeus indicus* and *P. viridus*) (58).

Culture Methods

The Netherlands has historically used bottom culture. Spain, which entered the fishery business in the 1960s, is using a raft culture method. France uses the bouchot method and rack culture. Italy has traditionally used a rack or stake method for mussel production. All of these countries can provide the basic requirements for successful production, including reasonable amounts of shelter and seawater of sufficient quality and phytoplankton content.

Growth rate is an important criterion in assessing the potential of mussel culture for a given area. Depending on geographic latitudes and environmental factors (light, salinity, temperature, primary productivity, currents, and tides), growth of mussels varies. In Galicia, Spain, mussels generally reach commercial size (80–90 mm) in 1–1.5 yr. In France they reach commercial size (approximately 40–50 mm) in 2 yr, and in the Netherlands commercial size (72 mm) is reached in 3 yr (55,63). A review of mussel cultivation in Spain and France has been published (59).

Off-bottom culture methods using rafts or longlines appear to be the most popular current approach to mussel culture. Basic improvements have been along technological lines, often introducing mechanization to facilitate the handling of seed mussel strings, and crop harvesting with transport to deputation and processing plants. Countries involved with this type of mussel farming include Korea, the People's Republic of China, the United States, Canada, Chile, the Philippines, India, Thailand, Indonesia, New Zealand, and Singapore. These and other countries have begun to appreciate the value of mussels as a food product for its nutritional value as well as its availability. Efforts have advanced toward using mussels as part of a polyculture system in the waters of Puget Sound, Washington. Natural mussels are harvested from the net pens that are used to cultivate salmon for the commercial market.

Modern techniques for depuration of mussels in Europe are well known and necessary. North America has only recently entered into the production of mussels and is already aware of the potential need for depuration in certain areas. This consideration relates to oysters and clams as well. The European system of depuration presents useful precedents and experience in such techniques for other countries that share this concern.

CLAMS

Production

Historically, clams have been cultured in Spain, Portugal, Japan, the People's Republic of China, and most Southeast Asian Countries (60). The first reference to clam culture is found in Chinese literature from 746 A.D. and mentions the transplantation of clams from one bay to another (a primitive form of culture). By the mid-1600s more direct forms of clam culture had been developed (61). The culture technique practiced during that period provided the basis for clam culture found throughout the modern world (62).

Many species of clams are harvested but few are actually cultivated or handled in some fashion by humans before harvest. Most species that are cultured fall in the families Arcidae and Veneridae. *Anadara* sp. of Arcidae are regarded favorably for cultivation in countries such as the People's Republic of China, Japan, Korea, Venezuela, Thailand, and the Malaysian peninsula. Of the Veneridae, *Venerupis (Tapes)* sp. are of importance to countries such as France, Italy, Japan, Korea, the Philippines, and the United States; *Meretrix* sp. to the People's Republic of China, Taiwan, and Japan; and *Mercenaria* and *Protothaca* to the United States. The culturing of other species is being investigated but none is as well established as those within the aforementioned genera.

The total world production of clams is steadily increasing, with 1,777,543 mt being harvested in 1996 (FAO). The People's Republic of China is clearly the largest clam producer in the world with 1,093,948 mt of *Venerupis japonica* harvested in 1996 alone (FAO). The United States, Korea, Japan, Thailand, and Chile have contributed to clam production as well. However, it is generally recognized that the majority of clams harvested come from wild stocks, and only a small fraction of transplanted juveniles are actually hatchery reared. Mariculture techniques have been considered with increasing interest as natural populations continue to decline in the face of pollution, natural catastrophes, and overfishing.

Culture Methods

Japan is the world leader in the intertidal culture of clams. Extensive culture has been undertaken with the Japanese littleneck or Manila clam, *V. japonica* (also known as *Tapes semidecussata*, *Tapes japonica*, and *Ruditapes philippinarium*). Hatcheries for this species are not needed in Japan due to the abundance of natural catches. The seed clams are collected and moved successfully to intertidal growout areas. This clam was introduced to the West Coast of the United States and Canada through the importation of Japanese oyster spat (*C. gigas*) and has adapted very well. It has since been introduced from these areas as spat or adults to Europe and is now well established in the United Kingdom, France, and Italy (8). It is also cultivated in Korea (63).

In some areas, such as the Indo-Pacific region, clam culture is labor intensive and lacks mechanization. The seed is provided by fishermen who gather juvenile arcid clams from highly productive naturally occurring clam beds and then sell them to clam farmers for planting and growout to market size (62,64–66). The ground is often improved by one of the following three methods: (*1*) leveling, (*2*) harrowing to remove macroalgae and loosen the substrate, or (*3*) the addition of sand or shell fragments, depending on the particular problems of growing at an optimum growth rate while predators and pests are later removed during the growout period. As clams reach marketable size, they are harvested by hand or with hand tools for the fresh market. Any mechanization usually involves the development of new procedures to harvest or handle the clams when they are transplanted. But for the most part, all labor and harvesting throughout the Indo-Pacific region is done by hand.

The primary culture species for the West Coast of the United States and Canada is the Japanese littleneck or Manila clam (*V. japonica*). Seed is produced at hatcheries in California and Washington and planted in the intertidal beds. The Manila clam grows best in the intertidal zone at or above the +0.5-in. to +2.0-in. tidal level. Several techniques have been tried to enhance its survival and growth. The use of gravel, screens, or other protective devices on the beds reduces predation and makes it possible to plant small seed for growout to commercial size (67–70). The use of plastic netting accords several advantages to the grower, including the exclusion of predators, stabilization of beach substrate, and the possible enhancement of natural settlement, thus allowing the use of both hatchery and wild seed (68–71).

The hard clam, *Mercenaria mercenaria*, is spawned in commercial hatcheries on the East Coast of the United States (72). Juveniles from the hatchery are transplanted to nursery growout systems until they are ca. 25 mm before being planted out into the natural shellfish beds for further growout and eventual harvest (73,74).

Venerupis decussata has been overexploited in Italy, and the gonadal cycle of this species has been studied as a preliminary step toward its culture (75). Although *V. decussata* is being investigated, Manila clam (*V. japonica*) production is growing rapidly because of ease in obtaining hatchery seed for planting and natural spawning populations being established in parts of France and Italy (8). The United Kingdom has also experimented with *M. mercenaria*, and researchers have successfully reared the clam to 10 mm in hatcheries (76). Clams planted in the field must then be protected from predators with some form of mesh covering, as already mentioned.

The use of upwelling systems for juvenile oysters as part of the nursery program has been mentioned in the section on oysters. This system is equally applicable to clams (8,27–30,42).

Investigations into the culturing of the giant clam, *Tridacna* sp., have occurred in the South Pacific since the late 1970s. There are currently four hatcheries that supply juveniles which are transplanted throughout the Indo-Pacific area. One attraction as a cultured animal is the fact that they do not need to be fed algae because they obtain energy through a symbiotic algae. After the hatchery phase, they are transferred to a land-based nursery system. After nine months, they are then put into an ocean nursery system where they must be protected from predators by being placed in enclosures. They are slow-growing animals and reach a harvestable size in four or five years.

SCALLOPS

Production

World production of scallops in 1996 reached approximately 1,275,958 mt. The FAO fisheries statistics records also indicate that the People's Republic of China, Japan, and Chile were the leading producers in 1996 with 999,573, 265,553, and 9,779 mt, respectively.

Japanese scallop harvests are dominated by *Patinopecten yessoensis*, and the recent rise in production can be attributed to well-developed methods for catching wild seed for culture. The Japanese have led world technology in scallop culture by a wide margin. However, researchers in other countries have been testing Japanese culture techniques for catching wild spat and have shown potential for local varieties. The People's Republic of China has introduced the bay scallop (*Argopecten irradians*) from the United States with great success. Scallop culture on the East Coast of the United States has centered on the bay scallop (78–82).

Culture Methods

As early as 1934, scallop larvae were collected on scallop shells hung from rafts in Japan for the purpose of later releasing the juveniles into favorable rearing grounds (83). By 1941, spat was being collected commercially. Bottom conditions during summer in Japan can be unfavorable, inducing mortality rates that can be above 99% (84). This problem soon stimulated a trend toward extending the culture operations for longer periods of time. By the late 1950s, collected larvae were routinely reared in hanging cages or holding ponds until the scallops exceeded 3 cm shell length (after about seven months) and were then released. Subsequently, increasing quantities of scallops were raised to market size in aquaculture systems. Although successful hatchery spawning techniques have been developed, the dependable and abundant set in certain bays has so far precluded the necessity for hatchery seed in Japan. Other species account for only a small fraction of effort relative to that devoted to *P. yessoensis*.

An excellent outline of commercial scallop culture in Japan describing larval development, different types of collector, float designs, and other useful information has been published (84,85). In the early 1980s, the scallop industry in Japan returned to the previous cultural practice of sowing juvenile seed scallops to growout on subtidal beds. One of the advantages of sowing scallop juveniles rather than culturing them in a lantern net is the reduction of deformed shells (only 10% as compared to 30% in lantern culture) (86).

Notwithstanding the general success of scallop culture in Japan, it should be noted that outbreaks of paralytic shellfish poison (PSP) in past years have reduced the scallop harvest for human consumption. The whole scallop is eaten when the shellfish does not contain PSP. However, when scallops do have PSP, some harvest still takes place for the adductor muscle only. The fishery is depressed with the persistence of this natural phenomenon. PSP, which recurred each year from 1977 to recent years, affected mostly hanging culture (86). This greatly impacted areas such as Mutsu Bay in Aomori Prefecture.

In regard to the subject of capturing larvae from natural spawnings, a study from the Isle of Man in the Irish Sea found two species of scallops, *Chlamys opercularis* and *Pecten maximus*, in collectors, which were essentially identical to those developed by the Japanese (87). Various factors affecting settlement and early survival of the larvae were investigated. In Newfoundland in the early 1970s, the natural setting of the deep-sea scallop *Placopecten magellanicus* was the subject of a pilot study by the Department of Fisheries aimed at developing a commercial-scale operation (92). It was observed that polyethylene film bags were more suitable than nylon netting, burlap, or fiberglass bags. The results of this study indicated the biological feasibility of collecting the deep sea scallop spat in sufficient numbers for cultivation.

The purple hinge or rock scallop *Hinnites multirugosus* is a prime candidate for mariculture along the Pacific Coast of the United States. Culturing this species from egg to harvest size (10–15 cm) within a period of 2–2.5 yr has been achieved by applying methods similar to those of oyster culture (89). A unique characteristic enables it to attach permanently to the substrate (like an oyster) after it has reached a size of approximately 3 cm. Although various experimental techniques have been used to grow this scallop to commercial size, the authors concluded that large-scale production by aquaculture must await the solution of two principal problems: (*1*) production of substantial numbers of seed stock for prospective scallop farmers, and (*2*) development of economical and effective procedures for containment and maintenance of stocks in natural waters until ready for harvest.

ABALONE

Production

The two major abalone-producing countries are the People's Republic of China and the United States. Mexico, South Africa, Korea, Chile, Australia, and the Channel Islands also show catches through the years, but considerably less than the two main countries (FAO). Total world production of abalone decreased from 14,300 t in 1987 to 2,185 mt in 1996.

Culture Methods

The Republic of China (Taiwan), Japan, and the United States are the principal countries culturing abalone. Japan has several government laboratories in which abalone research is being conducted. The culture of juveniles in the hatchery will continue to boost abalone production in the future.

Abalone culture in the United States is still generally in the experimental stage, even though some companies in California are producing small quantities for the commercial market. There are several companies in California producing abalone (90). In the past, firms sold juveniles to local volunteer groups and government agencies to plant in depleted abalone beds for enhancement purposes. Although land-based companies are producing small abalones (2–3 in. in diameter) for the specialty market with some success, a polycheate worm was inadvertently introduced impacting the California culture fisheries. This worm infests the shell, affecting growth and rendering the shell useless for sale as a value-added product.

Advances in the culture of abalone expanded greatly when successful spawning induction techniques were unlocked (91). Temperature was found to be important in conditioning abalone for spawning (92). Irradiated seawater (93) and hydrogen peroxide were found to induce spawning in adults (94,95). The later has also been tried on other species to determine the nature of its effect. It appears that both stimuli act to create oxides in the seawater. A combination of desiccation, elevation of temperature, hydrogen peroxide, and sperm from the male abalone has also proved successful in generating spawning (96).

A special chemical additive known as GABA (γ-aminobutyric acid), which acts as a potent neurotransmitter and has been commercially successful in the stimulation of larval setting, was derived from crustose coralline red algae (97). It was also reported that the development of an advanced biological engineering system for abalone seed production in Japan resulted in a technique to synchronize the spawning period (98). Suitable conditions for larvae to settle on the collector plates were produced by using the mucus of juvenile and adult animals together with certain types of diatoms.

There have also been studies demonstrating the importance of food type at all stages of development (99,100). Japan has made improvement in the use of artificial diets (pellets) in recent years (101). Furthermore, raising abalone at high temperatures was found to increase growth rates (102). Portable habitats have been designed to protect abalone seed when being planted at sea (103). Constructed of either concrete blocks or shelves, they serve as attachment sites during land transport and provide temporary refuge on the sea bed.

Areas of low production in Japan have been improved through a combination of longline kelp culture and predator removal. Sea urchins are grown first and then harvested, removing competition for the abalone seed while nearing the size required to eat the kelp. The urchins often pose strong competition in prime abalone sites, but fortunately their roe is a valuable commodity in Japan.

In the United States it is difficult to obtain private control of prime abalone growout areas. For this reason one company has grown abalone in large concrete conduits placed vertically in subtidal waters (104). Another has reared its stock in cages suspended from oil rigs (105). There is a trend toward growing abalone in the open ocean off California because of the high cost of land and seawater systems, but boat transportation for cleaning the cages and feeding the stocks is expensive and there is less control over the total operation.

OTHER GASTROPODS

Production

Gastropods other than abalone are consumed in various parts of the world, and in some areas provide an important protein source. These include *Helix pomatia* and *Helix aspersa*, the escargot of France; *Strombus gigas*, the queen conch of the Caribbean; *Littorina littorea* and *Bucciunum undatum* of France; *Turbo cornutus* of Japan and Korea; *Trochus niloticus* of the Caroline Islands; *Concholepas concholepas*, commonly called loco of Chile; and *Neptunea* sp., the marine snails from the Bering Sea.

Gastropods (excluding *Haliotis*, the abalones) make up a small percentage of the world total molluscan landings, amounting to only 1,164 mt annually (FAO). With the exception of *H. pomatia* and *H. aspersa*, production occurs almost entirely from the capture fisheries. Only recently has the culture of other gastropod species been attempted and then usually on an experimental basis.

Culture Methods

H. pomatia and *H. aspersa* have been cultured in France for nearly 2,000 yr and are still typically raised in fenced gardens called cochleria (106). They are also raised in caves and fed a variety of fungi (107). Although it appears that little innovation in culture methods has occurred in recent years, the area under cultivation has expanded. *H. aspersa*, considered a pest in the United States, has been raised and marketed in California. Snails were purchased from botanical gardens and farms that do not use pesticides. The snails were starved for 24 hours to purge them of pollutants and were then fed a combination of cabbage, wheat meal, and bran for two to three weeks before being processed.

Methods are also being developed to culture the queen conch *S. gigas*, but are thus far underway only on an experimental basis (108–110). A problem presently being solved involves the rearing of larvae through metamorphosis in sufficiently high densities to be economically feasible (111). This problem has been addressed by studying the feasibility of raising the queen conch in shallow-water fenced enclosures (112).

Although commercial culture of the topshell *T. cornutus* does occur along the Seowipo coast of Korea and the Japan coast (113,114), further research is needed. Similarly, another topshell, *T. niloticus*, has been cultured on an experimental basis in the Caroline Islands by the Micronesian Mariculture Demonstration Center (MMDC) and the Palau Marine Resources Division (115). Culture of this species using methods developed by the MMDC is also being considered in the Philippines.

The culture of gastropods has generally been limited to but a few species, essentially a condition of economics. With the exception of *H. pomatia*, *H. aspersa*, and *Haliotis*, the market price of gastropods is quite low compared to that of other shellfish. In addition, their mobility (ability to forage), together with a vulnerability to a variety of predators, does not make them a particularly attractive group to culture.

PEARL OYSTERS

Production

There has been little change in pearl culture techniques since the 1940s. The main change has been a decrease in the stocks of wild oysters for natural pearls. This is especially true in countries that have historically exploited this resource. If the industry is to survive, cultured pearls are the only hope. However, it has been pointed out that one cause of the hiatus in Japan's pearl farming was an unstable price situation induced by an increase in poor-quality pearls (116). A mere increase in pearl production in the future is not foreseen because the aim of the industry is to improve the quality, rather than quantity, of cultured pearls.

Four different genera are used in the cultivation of pearls: *Pinctada*, *Pteria*, *Cristaria*, and *Hyriopsis*. The latter two are freshwater genera. According to FAO statistics, the pearl and pearl shell production industry has depended on three major areas: Japan, Australia, and the Philippines to Fiji. Before 1969 both Ethiopia and Madagascar produced pearls, but no statistics were found for African countries since then. Japan has been a steady producer with about 100 t (live weight) through the 1960s, then dropped to 30 t by 1974, rose again to approximately 48 t in 1980, and dropped to 34 t in 1987. In 1980, Australia produced 310 t, then dropped to 42.5 t in 1987.

General pearl culture techniques have been discussed and generally presented (116–119). The majority of cultured pearls in Japan is produced by the pearl oyster *Pinctada fucata*. This is the species of choice for the cultivation of tiny round pearls, their size being generally less than 10 mm in diameter (118). A large pearl (to 15 mm) can be cultured in *Pinctada maxima*, which inhabits tropical and subtropical Pacific waters, such as off northern Australia and the Philippines. Major efforts are being made to culture the "black steel" pearl from the South Pacific species *P. margaritifera* (119).

Beyond the increased production of cultured pearls in the industry, there have been efforts to mechanize operations toward facilitating the handling of culture animals. There is also a need for more experimentation leading to higher quality and colored pearls. Steps must be taken to obtain good culture grounds and to exercise vigilance to preserve them from environmental deterioration (116).

CONCLUSIONS

This article has highlighted some of the techniques and concepts in the culture of molluscs. All of the techniques could not be covered here because the subject of molluscan culture is broad in scope. Any emphasis on a given group of animals depends on geographical location and the most desired local commercial species. Historic techniques developed in one country may be considered new innovations and concepts in another. In recent years the search for improved culture methods has predominated as the need for increased protein production and high-valued products are recognized in most countries.

From reviewing various groups of molluscs, it is clear that countries long involved in the culture of desired species continually seek to improve their technology and introduce mechanization to facilitate growing and harvesting. This includes an emphasis on improved hatchery, nursery, and growout techniques; off-bottom culture; diet and nutritional concerns; disease and predator control; genetic manipulation; and other phases of the total molluscan culture scenario. Depending on the country, modernization and changing techniques relate to sociopolitical concerns and, more specifically, to traditional cultural patterns.

Global molluscan production could be increased greatly by applying methods that are currently used in certain parts of the world. This is particularly true for oysters, mussels, clams, and select species of gastropods such as abalone and land snails (escargot). It is also evident that developing countries have begun to engage in molluscan culture because they have recently discovered unexploited indigenous stocks or commercial prospects for introduced species. This especially applies to the oyster species because of its worldwide distribution and acceptability in the commercial market. Other molluscan species available for potential culture are now being considered favorably. Given the growth potential for local and world markets, it is expected that the culture of molluscs will increase markedly in the years ahead.

BIBLIOGRAPHY

1. V. Loosanoff and H. Davies, *Rearing of Bivalve Molluscs, Adv. Mar. Bio.* **1**, 1–136 (1963).
2. J. Dupuy, N. Windsor, and C. Sutton, *Manual for Designing and Operation of an Oyster Seed Hatchery, Virginia Institute of Marine Science Special Report* 142, 1977.
3. W. Breese and R. Malouf, *Hatchery Manual for the Pacific Oyster*, Oregon State University Sea Grant Program Publication ORESU-H-75-002, 1975.
4. H. Hidu and M. Richmond, "Commercial Oyster Aquaculture in Maine," *University of Maine Sea Grant Bulletin* **2**, 1–60 (1974).
5. G. Jones and B. Jones, "Methods for Setting Hatchery Produced Larvae," *Inf. Rep. Mar. Res. Br., Min. Envir. Prov. Brit. Columbia* **4**, 1–61 (1982).
6. G. Jones and B. Jones, *Advances in the Remote Setting of Oyster Larvae*, Aquaculture Association of British Columbia/British Columbia Ministry of Agriculture and Fisheries Joint Special Publication, 1988.
7. T. Nosho and K. Chew, eds., *Remote Setting and Nursery Culture for Shellfish Growers*, Washington Sea Grant Program, WSG-WO 91-02, 1991.
8. K. Chew, "Global Bivalve Shellfish Introduction-Implications for Sustaining a Fishery or Strong Potential for Economic Gain," *World Aquacult.* **21**(3), 9–22 (1990).
9. K. Chew, "Oyster Aquaculture in the Pacific Northwest," *Fourth Alaska Aquaculture Conference Proceedings*, 1987, pp. 67–75.
10. S.K. Allen, Jr., S. Downing, and K. Chew, *Hatchery Manual for Producing Triploid Oysters*, Washington Sea Grant Technical Report WSG 89-3, 1989.
11. A. Cahn, *Oyster Culture in Japan, U.S. Fish. Leaflet Fish Wildl. Serv.* **383**, (1950).
12. T. Imai, *Aquaculture in Shallow Seas*, Amerind Publishing, New Delhi, 1977.
13. J. Glude, "Oyster Culture—A World Review," in T.V.R. Pillay and W.A. Dill, eds., *Advances in Aquaculture*, Fishing News Books, Farnham, 1979, pp. 325–331.
14. F.B. Davy and M. Graham, *Bivalve Culture in Asia and the Pacific*, Workshop Proceedings, Singapore, 16–19 Feb. 1982, International Development Research Centre IDRC-200e, Canada, 1982.
15. D. Quayle, *Tropical Oysters: Culture and Methods*, International Development Research Center IDRC-TS17e, Ottawa, Canada, 1980.
16. P. Korringa, *Farming the Cupped Oysters of the Genus Crassostrea, Dev. Aqua. Fish. Sci.* 2, Elsevier, Amsterdam, 1976.
17. P. Korringa, *Farming the Flat Oysters of the Genus Ostrea, Dev. Aqua. Fish. Sci.* 3, Elsevier, Amsterdam, 1976.
18. P. Walne, *Culture of Bivalve Molluscs*, Fishing News Books, London, 1974.
19. D. Quayle, *Pacific Oyster Culture in British Columbia, Can. Bull. Fish. Aquat. Sci.* **218**, (1988).
20. D. Quayle and G. Newkirk, *Farming Bivalve Molluscs: Methods for Study and Development*, The World Aquaculture Society, 1990.
21. "Tasmanian Oysterman Has Success in the Bag," *Aust. Fish.* **39**(9), 16–17 (1980).
22. R. Meixner, "Culture of Pacific Oysters *Crassostrea gigas* in Containers in German Coastal Water," in Ref. 8.
23. B. Wiseley, "Recent Developments in NSW Oyster Research," *Aust. Fish.* **39**(9), 5–10 (1980).
24. K. Lee, J. Corbin, and W. Brewer, *Oyster Culture in Hawaii and Various United States Pacific Island Territories*, North American Oyster Workshop, Seattle, Washington, March, 1981.
25. K. Lee and H. Yamauchi, *Economics of Land-Based Oyster Aquaculture in Hawaii*, World Mariculture Society Technology Session, Seattle, Washington, March, 1981.
26. J. Huguenin, S. Huguenin, and B. Tobiasson, "Barge-Based Commercial Marine Aquaculture," in Marine Technology Society, ed., *Marine Technology 16th Annual Conference, Oct.* 1980, MTS, Washington, D.C., 1980.
27. J. Bayes, "Forced Upwelling Nurseries for Oysters and Clams Using Impounded Water Systems," *EMS (Special Publication)* **7**, 73–83 (1981).
28. B. Spencer and B. Hepper, "Tide-Powered Upwelling: The Systems for Growing Nursery-Size Bivalves in the Sea," *EMS (Special Publication)* **7**, 283–309 (1981).
29. P. Williams, "Offshore Nursery Culture Using the Upwelling Principle," *EMS (Special Publication)* **7**, 311–315 (1981).
30. P. Radhouse and M. O'Kelly, "Flow Requirements of the Oyster *Ostrea edulis* L. and *Crassostrea gigas* Thunb. in an Upwelling Column System of Culture," *Aquaculture* **22**, 1–10 (1981).
31. O. Roels, "The Economics of Artificial Upwelling Mariculture," *J. Shellfish Res.* **1**(1), 122 (1981).
32. J. Glude, "The Applicability of Recent Innovations to Mollusc Culture in the Western Pacific Islands," *Aquaculture* **39**(1–4), 29–43 (1984).
33. "Department Develops Oyster Culture Techniques," *FINS.* **18**(2), 3–6 (1985).
34. E. Spencer, M. Akester, and I. Mayer, "Growth and Survival of Seed Oysters in Outdoor Pumped Upwelling Systems

Supplied with Fertilized Sea Water," *Aquaculture* **55**(3), 173–189 (1986).

35. B. Spencer, "Growth and Survival of Seed Oysters in Outdoor Pumped Upwelling Systems," *Aquaculture* **75**(1–2), 139–158 (1988).
36. A. Campos and M. Saveedra, "Nursery Culturing of Bivalve Molluscs in Two Upwelling Systems in San Pedro Channel (Bay of Cadiz, SW Spain)," *Spec. Publ. Eur. Aquacult. Soc.* **10**, (1989).
37. C. Bacher and J. Baud, "Intensive Rearing of Juvenile Oysters *Crassostrea gigas* in an Upwelling System: Optimization of Biological Production," *Aquat. Living Resour.* **5**(2), 89–98 (1992).
38. N. Hadley, R. Rhodes, R. Baldwin, and M. Devoe, "Performance of a Tidal-Powered Upwelling Nursery System for Juvenile Clams in South Carolina," *J. Shellfish Res.* **13**(1), 285 (1994).
39. R. Rhode, N. Hadley, R. Baldwin, and M. Devoe "Cost Analysis of a Tidal-Powered Upwelling Nursery System for Juvenile Clams,"*J. Shellfish Res.* **13**(1), 286 (1994).
40. T. Poomtong and K. Promchinda, "Growth and Survival of Oyster Spat (*Crassostrea belcheri* Sowerby) as a Function of Flow Rates in a Recirculated Upwelling System," *Spec. Publ. Phuket. Mar. Biol. Cent.* **15**, (1995).
41. M. Kilgen, E. Melancon, K. Rush, and R. Malone, "Evaluation of an Upwelling Injection Field Polishing System for Elimination of Fecal Coliforms and Enteric Viruses (F+ RNA phage)," *J. Shellfish Res.* **16**(1), 268 (1997).
42. J. Manzi and M. Castagna, eds., *Clam Mariculture in North America*, Dev. Aqua. Fish. Sci. **19**, Elsevier, Amsterdam, 1989.
43. P. Korringa, *Farming Marine Organisms Low in the Food Chain, Dev. Aqua. Fish. Sci.* **1**, Elsevier, Amsterdam, 1976.
44. P. Korringa, "Economic Aspects of Mussel Farming," in Ref. 8.
45. Central Marine Fisheries Research Institute, "Coast Aquaculture," *CMFRI Bull.* **29**, (1980).
46. A. Figueras, "Cultivo del mejillon (*Mytilus edulis* L.) y possibilidades para su expansion," in Ref. 8.
47. M. Aquirre, "Biologia del mejillon (*Mytilus edulis*) de cultivo de la Ria de Vigo," *Boln lnst. esp. Oceanagr.* **5**, (1979).
48. G. Jamieson, "Mussel Culture," *World Aquaculture* (Special Session) **20**(3), 8–100 (1989).
49. R. Jenkins, *Mussel Cultivation in the Marlborough Sounds (New Zealand)* Jones, Wellington, 1979.
50. R. Lutz, "Raft Cultivation of Mussels in Maine Waters — Its Practicality, Feasibility and Possible Advantages," *Univ. of Maine Sea Grant Bull.* **4**, (1974).
51. P. Dare, "Mussel Cultivation in England and Wales," *Lab. Leafl. Fish. Lab. Lowestoft* **50**, (1980).
52. D. Skidmore and K. Chew, *Mussel Aquaculture in Puget Sound*, Washington Sea Grant WSG 85-4, 1985.
53. J. Mason, "The Cultivation of the European Mussel *Mytilus edulis* L.," *Ocean Agr. Mar. Biol.* **10**, 15–18 (1972).
54. A. Paz-Andrade, "Raft Cultivation of Mussels Is Big Business in Spain," *World Fish.* **17**(3), 50–52 (1968).
55. R. Hickman, "Farming the Green Mussel in New Zealand," *World Aquacult.* **20**(4), 20–28 (1989).
56. R.J. Jenkins, 1988.
57. K. Chalermwat and R.A. Lutz, "Farming the Green Mussel in Thailand," *World Aquacult.* **20**(3), 41–55 (1989).
58. P. Kuriakose, "Mussels (Mytilidae): genus *Perna* of the India Coast," *CMRFI Bull.* **29**, (1980).
59. A. Figueras, "Mussel Culture in Spain and France," *World Aquacult.* **20**(3), 8–19 (1989).
60. E. Borgese, *Seafarm*, Abrams, New York, 1977.
61. A. Cahn, *Clam Culture in Japan, U.S. Fish. Leafl. Fish Wildl. Serv.* **399**, (1951).
62. J. Bardach, J. Ryther, and W. McLarney, *Aquaculture*, Wiley-Interscience, New York, 1972.
63. "Present Status and Problems of Coastal Aquaculture in the Republic of Korea," in T.V.R. Pillay, ed., *Coastal Aquaculture in the Indo-Pacific Region*, Fishing News Books, London, 1972, pp. 48–51.
64. S. Ling, "A Review of the Status and Problems of Coastal Aquaculture in the Indo-Pacific Region," in Ref. 73.
65. G. Blanco, "Status and Problems of Coastal Aquaculture in the Philippines," in Ref. 73.
66. A. Sribhibhadh, "Status and Problems of Coastal Aquaculture in Thailand," in Ref. 73.
67. M. Miller, K. Chew, C. Jones, L. Goodwin, and G. Magoon, *Manila Clam Seeding as an Approach to Clam Population Enhancement*, Washington Sea Grant WSG 78-1, 1978.
68. G. Anderson and K. Chew, *Intertidal Culture of the Manila Clam Tapes japonica Using Hatchery-Reared Seed Clams and Protective Net Enclosures*, paper presented at the 69th Statutory Meeting of ICES, Copenhagen, Oct. 1980.
69. G. Anderson, M. Miller, and K. Chew, *A Guide to Manila Clam Aquaculture in Puget Sound*, Washington Sea Grant WSG 82-4, 1982.
70. D. Toba, D. Thompson, K. Chew, G. Anderson, and M. Miller, *Guide to Manila Clam Culture in Washington*, Washington Sea Grant Program, WSG 92-01, 1992.
71. D. Magoon and R. Vining, *Introduction to Shellfish Aquaculture in the Puget Sound Region*, Washington Dept. Nat. Resources Handbook, 1980.
72. M. Castagna and J. Kraeuter, "Manual for Growing the Hard Clam Mercenaria," *Virginia Inst. Mar. Sci. Spec. Rep. Appl. Sci. Ocean Engng.* **249**, (1981).
73. J. Manzi, V. Burrell, F. Stevens, M. Maddox, W. Carson, and H. Clawson, *Hard Clam (*Mercenaria mercenaria*) Mariculture in South Carolina: A Commercial Scale Demonstration Project*, Nat. Shellfisheries Assoc., Shellfish Inst. North America Meeting, Williamsburg, Va., Aug. 1981.
74. J. Kassner, K. Strong, and G. Proios, *Raceway Culture and Field Growout for Mercenaria mercenaria in the Town of Brookhaven*, Nat. Shellfisheries Assoc., Shellfish Inst. North America Meeting, Poster Session, Williamsburg, Va., Aug. 1981.
75. P. Breber, "Annual Gonadal Cycle in the Carpet-Shell Clam *Venerupis decussata* in Venice Lagoon, Italy," *Proc. Nat. Shellfish. Assoc.* **70**(1), 31–35 (1980).
76. R. Kirk, "Marine Fish and Shellfish Culture in the Member States of the European Economic Community," *Aquaculture* **16**, 95–122 (1979).
77. O. Roels, J. Sunderlin, and S. Laurence, "Bivalve Molluscan Aquaculture in an Artificial Upwelling System," *Proc. World Maricult. Soc.* **10**, 122–138 (1979).
78. M. Castagna, "Culture of the Bay Scallop *Argopecten irradians* in Virginia," *Mar. Fish. Rev.* **37**, 19–24 (1975).
79. P. Heffernan, R. Walker, and D. Gillespie, "Biological Feasibility of Growing the Northern Bay Scallop, *Argopecten irradians irradians* (Lamarck, 1819) in the coastal waters of Georgia," *J. Shellfish Res.* **7**(1), 83–88 (1988).
80. E. Rhodes and J. Widman, "Some Aspects of the Controlled Production of the Bay Scallop (Argopecten irradians)," *Proc. World. Mar. Soc.* **11**, 235–246 (1980).

81. R. Walker, P. Heffernan, J.W. Crenshaw, Jr., and J. Hoats, "Mariculture of the Southern Bay Scallop, Argopecten irradians concentricus (Say, 1822), in the Southeastern U.S.," in S. Shumway and P. Sandifer, eds., *An International Compendium of Scallop Biology and Culture*, World Aquaculture Society, 1991, pp. 313–321.
82. R. Karney, "Ten Years of Scallop Culture on Martha's Vineyard," in S. Shumway and P. Sandifer, eds., *An International Compendium of Scallop Biology and Culture*, World Aquaculture Society, 1991, pp. 308–312.
83. T. Tamura, Marine Aquaculture, *Nat. Tech. Inf. Serv. Transl.* PB 195 OSIT, P 1,2, Springfield, Va., 1970.
84. V. Kasyanov, "Development of the Japanese Scallop Mizuhopecten yessoensis (Jay 1985)," in S. Shumway and P. Sandifer, eds., *An International Compendium of Scallop Biology and Culture*, World Aquaculture Society, 1991, pp. 1–9.
85. M. Mottet, *A Review of the Fishery Biology and Culture of Scallops, Wash. Dept. Fish. Tech. Rep.* **39**, (1979).
86. R. Ventilla, "The Scallop Industry in Japan," *Adv. Mar. Bio.* **20**, 309–382 (1982).
87. A. Brand, J. Paul, and J. Hoogesteger, "Spat Settlement of the Scallop *Chlamys opercularis* (L.) and *Pecten maximus* (L.) on Artificial Collectors," *J. Mar. Bio. Ass. U.K.* **60**, 379–390 (1980).
88. K. Naidu and R. Scaplen, "Settlement and Survival of Giant Scallop *Placopecten magellanicus* Larvae on Enclosed Polyethylene Film Collectors," in Ref. 8.
89. D. Leighton and C. Phleger, *The Suitability of the Purple Rock Scallop to Marine Aquaculture* San Diego State Univ. Sea Grant T-SCSGP (1981).
90. S. McBride, "Current Status of Abalone Aquaculture in the Californias," *J. Shellfish Res.* **17**(3), 593–600 (1998).
91. K. Hahn, ed., *Handbook of Culture of Abalone and Other Marine Gastropods*. CRC Press, Boca Raton, FL, 1989.
92. S. Kikuchi and N. Uki, "Technical Study on Artificial Spawning of Abalone Genus *Haliotis* 1. Relation Between Water Temperature and Advanced Sexual Maturity of *Haliotis discus hannai* Ino," *Bull. Tohoku Reg. Fish Res. Lab.* **33**, 79–96 (1974).
93. S. Kikuchi and N. Uki, "Technical Study on Artificial Spawning of Abalone Genus *Haliotis* 11. Effect of Irradiated Sea Water with Ultraviolet Rays on Inducing to Spawn," *Bull. Tohoku Reg. Fish. Res. Lab.* **33**, 79–86 (1974).
94. D. Morse, H. Duncan, N. Hooker, and A. Morse, "Hydrogen Peroxide Induces Spawning in Molluscs, with Activation of Prostaglandin Endoperoxide Synthetase," *Science* **196**, 298–300 (1977).
95. D. Morse and A. Morse, "Chemical Control of Reproduction in Bivalve and Gastropod Molluscs, III: An Inexpensive Technique for Mariculture of Many Species," *Proc. World Maricult. Soc.* **9**, 543–547 (1978).
96. S. Olsen, unpublished data.
97. D. Morse, H. Hooker, H. Duncan, and L. Jensen, "Gamma-Aminobutyric Acid, a Neurotransmitter, Induces Planktonic Abalone Larvae to Settle and Begin Metamorphosis," *Science* **204**, 407–410 (1979).
98. T. Seki, "An Advanced Biological Engineering System for Abalone Seed Production," in *Proceedings of the International Symposium on Coastal Pacific Marine Life*, Western Washington University, Bellingham, 1980, pp. 45–54.
99. S. Kikuchi, V. Sakurai, M. Sasaki, and I. Ito, "Food Value of Certain Marine Algae for the Growth of the Young Abalone *Haliotis discus hannai*" *Bull. Tohoku Reg. Fish. Res. Lab.* **27**, 93–100 (1967).
100. N. Uki and S. Kikuchi, "Food Value of Six Benthic Micro-Algae on Growth of Juvenile Abalone, *Haliotis discus hannai*," *Bull. Tohoku Reg. Fish Res. Lab.* **40**, 47–52 (1979).
101. N. Uki and T. Watanabe, "Review of the Nutritional Requirements of Abalone (*Haliotis* spp.) and Development of More Efficient Artificial Diets," in S. Shepherd, M. Tegner, and S. Guzman del Proo, eds., *Abalone of the World: Biology, Fisheris and Culture*, Fishing News Books, 1992.
102. D. Leighton, "Laboratory Observations on the Early Growth of the Abalone *Haliotis sorenseni* and the Effect of Temperature on Larval Development and Settling Success," *Fish Bull. U.S.* **70**, 373–381 (1972).
103. D. Sedwick, "Replanting the Ocean Garden: Abalone Farming off Santa Barbara, California," *Oceans* **11**(4), 61–62 (1978).
104. D. Klopfenstein and J. Klopfenstein, "U.S. Abalone Hatcheries Adapt a Japanese Technique," *Fish Farm. Int.* **3**(3), 46–48 (1976).
105. D. Rutherford, "California Farm Rears Prized Red Abalone," *Fish Farm. Int.* **3**(4), 9 (1976).
106. "Snails," *Encyclopedia Britannica*, Vol. 20, Encyclopedia Britannica, Chicago, 1971, pp. 708–715.
107. A. Kohn, unpublished data.
108. C. de Asaro, "Organogenesis, Development and Metamorphosis in the Queen Conch *Strombus gigas*, with Notes on Breeding Habits," *Bull. Mar. Sci.* **15**, 359–416 (1965).
109. C. Berg, "Growth of the Queen Conch *Strombus gigas* with a Discussion of the Practicality of Its Mariculture," *Mar. Biol.* **34**, 191–99 (1976).
110. S. Siddell, *Larviculture*, Proceedings of the Queen Conch Fisheries and Mariculture Meeting, Freeport, Bahamas, Jan. 1981.
111. S. Siddell, unpublished data.
112. "Mass Rearing of Queen Conch under Controlled Conditions Becomes Reality," *Aquacult. Mag.* **7**(3), 4 (1981).
113. J. Lee, K. Lee, and J. Lee, "On the Population, Growth and Environment of Habitats for Topshell *Turbo cornutus* Solander," *Bull. Mar. Biol. Stn. Jeju Nat. Univ.* **2**, 3–14 (1978).
114. L. Cresswell, unpublished data.
115. G. Helsinga and A. Hillmann, "Hatchery Culture of the Commercial Top Snail *Trochus niloticus* in Palau, Caroline Islands," *Aquaculture* **22**, 35–43 (1980).
116. S. Mizumoto, "Pearl Farming—A Review," in Ref. 8.
117. A. Cahn, *Pearl Culture in Japan, Fish. Leafl. Fish Wildl. Serv. U.S.* **357**, (1949).
118. K. Wada, "Modern and Traditional methods of Pearl Culture," *Underwater J.* **1973**(2), 28–33 (1973).
119. C. Fassler, "Farming Jewels, New Developments in Pearl Farming," *World Aquacult.* **26**(3), 5–10 (1995).

See also ABALONE CULTURE.

MUD CRAB CULTURE

DAN D. BALIAO
Southeast Asian Fisheries Development Center
Tigbauan, Philippines

OUTLINE

Mud crabs are one of the most widely sought crustacean species that inhabit the estuarine areas and tidal rivers and creeks of the Asian and Indo-Pacific regions. Hailed as "food for the gods," the mud crab is recognized as a candidate species for culture in brackishwater ponds and/or other suitable impounded brackishwater environments.

In the past, mud crabs were a secondary species to cultured finfishes or crustaceans. Larvae entered ponds with incoming water and became trapped. Although conceived as a fishpond crop, the mud crab has also been considered a nuisance in ponds it because it burrows into dikes and causes damage and leaks.

Farming of mud crab has been progressing rapidly due to a promising market and profitability.

With the availability of mud crab juveniles from the wild throughout the year and the recent development in hatchery technology, there is a strong indication that production of mud crabs on a commercial scale could be a lucrative industry.

The information presented here is based on the recently published extension manuals and literature on mud crab culture both in brackishwater ponds and pen enclosures in mangroves (1–3).

TAXONOMY, FOOD HABITS, AND DISTRIBUTION

Mud crabs are of the genus *Scylla* and reported to consist of three species: *Scylla serrata*, *S. oceanica*, and *S. tranquebarica*, and a variety of *S. serrata*, var *paramamosain*. It is generally called mud or mangrove crab in Australia, "Samoan" crab in Hawaii, *alimango* in the Philippines, *tsai jim* in Taiwan, *nokogiri gazami* in Japan, *kepiting* in Indonesia, *kalapu kakuluwa* in Sri Lanka, *haubba kankera* in Bangladesh, and *ketam nipah* or *ketam bakau* in Malaysia.

In its natural habitat mud crabs mainly feed on crustaceans, while adults and subadults consume molluscs (4,5). Fish remains are rarely found and it was concluded that *S. serrata* does not normally catch mobile forms such as fish and penaeid shrimp. Mud crabs usually remain buried during the day, emerging at sunset to spend the night feeding, which occurs intermittently even when unlimited food is available (2).

Mud crabs inhabit both marine and brackishwater environments and prefer muddy and sandy bottoms. They are found in many countries of the West Indo-Pacific region, from South Africa to Hawaii and from North Australia to Southern Japan.

SOURCE OF JUVENILES

In the Philippines, mud crab juveniles [10 to 40 g and 5 to 20 cm (0.3 to 1.3 oz or 2 to 8 in.) in carapace breadth] are available throughout the year, with a peak occurring from May to September. Most crab juveniles from the wild are sourced from coastal provinces of the country. Common collecting gear includes crab lift nets (*bintol*), bamboo cage traps (*panggal* or *bobo*), tube traps (*patibong*) and crab hooks (*panukot*). Mud crabs are also caught using fish corrals (*baklad*), baited lines with scoop nets, and even with bare hands. Depending on size, quantity, sex, and species, crab juveniles are ordered in advance from collectors. This gives ample time for collection, handling, storage, and transport. Technology in the hatchery of mud crabs has been developed in the Southeast Asian Fisheries Development Center, Aquaculture Department (SEAFDEC/AQD) based in Tigbauan, Iloilo, Philippines.

Young crabs (Fig. 1) obtained from the wild are normally transported inside bamboo wicker baskets, cardboard boxes, or palm (*pandan*) bags to the rearing ponds and pens. A pandan bag can carry 150 to 200 juvenile crabs, each weighing 20 to 50 g (0.7 to 1.7 oz) during a transport of up to 10 to 12 hr. No immobilization of chelipeds is required for crabs weighing less than 30 g (1 oz), but the chelipeds of larger crabs should be tied shut. Fronds of mangrove trees *Rhizophora* spp., (*pagatpat*) or *Avicennia* spp., (*bungalon*) are provided inside the basket to keep the temperature cool and to minimize fighting among crabs. It is recommended that active and healthy juveniles with complete body parts be procured. Juveniles falling short of these requirements should be discarded.

GROWOUT OF MUD CRAB IN BRACKISHWATER PONDS

Earthen or concrete-lined ponds, preferably rectangular in shape (Fig. 2) and of areas 250 m^2 to 1.0 ha (0.06 to 2.5 ac), could be utilized for culture of mud crabs. Enclosed areas of newly or partially developed fish ponds provided with water control structures could also serve the purpose. The soil type should be sandy clay or clay loam with a rich organic matter base and, preferably, alkaline. A good quality water coming directly from the sea or a tidal river should be available year-round. The most desirable

Figure 1. Mud crab juveniles ready for stocking.

Figure 2. Growout pond with a nylon fence.

ranges of water quality are as follows: salinity 10–34 ppt, temperature 23–30 °C (73–86 °F), and dissolved oxygen above 3 ppm and pH 8.0–8.5 (6). To prevent an increase in salinity, especially during the summer months, it is advantageous to have a freshwater source. This will enable the farmer to adjust the salinity to a level favorable to the growth of the crabs.

Other socioeconomic factors such as cheap and skilled labor, market accessibility, proximity to construction materials, and production inputs, as well as the peace and order situation in the locality should be considered.

The pond should be capable of holding water of depth 1 to 2 m (3 to 6 ft) and have a double gate system (one each for entrance and exit) made either of concrete or wood. The pond bottom must be level and clean to allow easy harvesting.

Each pond compartment is provided with about 12 earthen mounds [5 m^3 (6.7 yd^3)] installed in strategic areas (Fig. 3). These mounds serve as breathing spots where mud crabs can climb out during periods of low dissolved oxygen. The mounds should be installed in the middle of the pond and be high enough so that the peaks remain above water even when maximum water depth of 60 to 80 cm (24 to 31 in.) is reached. Used tires stacked and tied vertically in layers and separated horizontally by wooden or bamboo poles are also utilized as substitutes for earthen mounds. Shelters made of sawed bamboo or used PVC pipes [50 cm (20 in.) long with a 15 cm (6 in.) diameter opening at both ends] should be added to avoid mortality due to fighting and cannibalism.

Figure 3. Growout pond with bamboo fence and mud mounds.

To prevent the crabs from escaping, the area should be fenced with either bamboo slats or nylon net [1–2 cm (0.4–0.8 in.) mesh size] fence extending about 30 cm (1 ft) above the waterline. The fence is kept in place by supporting it vertically with bamboo or wooden posts driven 50 to 70 cm (20 to 27 in.) deep into the pond bottom sediment and supported horizontally with some bamboo splits. A plastic strip or sheet about 30 cm (1 ft) wide should be installed along the top edge to prevent crabs from climbing over the top of the net fence. For concrete-lined ponds with relatively steeper slopes, nylon net fence is no longer necessary. Catwalks and feeding trays may be provided for feed monitoring and sampling. Life-support systems, such as a water pump (axial or centrifugal), and paddlewheel aerators may be necessary for emergency water exchange and aeration, especially during neap tides and calm days at night, or when water conditions so require.

Stocking and Rearing

Preparation of ponds for the culture of mud crabs does not require such meticulous procedures as growing natural food. However, prior to stocking, proper installation of the net fence, earthen mounds, and other physical requirements should be considered. Ponds should be drained completely and the bottoms allowed to dry for a week or two. Eradication of pests and predators is done during the pond preparation stage by applying teaseed powder at the recommended rate of 15 to 30 ppm (depending on salinity) or a combination of hydrated lime, $Ca(OH)_2$, and ammonium sulfate fertilizer (21–0–0) at the ratio of 1:5. Other environmently friendly organic pesticides such as tobacco dust and derris root extracts are also recommended.

Stocking may be done early in the morning or late in the afternoon, but most preferably at night when the temperature is cool. During stocking, pincers are united and crabs are released directly into the ponds at densities of 5,000 to 10,000 juveniles 1 ha (2,000–4,000/ac).

It is advisable to stock uniform-sized mud crabs in order to obtain a relatively uniform size at the end of the rearing period mud crabs of may 50 g (1.7 oz) be grown in polyculture with milkfish, *Chanos chanos*, at the rate of 2,500 fingerlings 1 ha (1,000/ac).

It is essential to maintain good water quality during the culture period. When considerable numbers of crabs start to crawl on top of the earthen mounds or cling to the walls of bamboo or the net fence, this indicates that water conditions are not favorable. It is advisable, therefore, to change at least one-third of the pond water, especially during spring tides. An irrigation pump may be necessary in case water change is needed during neap tides. Dikes, gates, and net fences should be regularly inspected for possible leaks and dilapidation.

Trash fish, animal hides or entrails, snails, and other locally available and cheap protein sources may be provided as feed for mud crabs. Feeds are broadcast evenly throughout the pond each day. The recommended feeding rate is 10% of crab biomass initially. That rate is reduced by 1% each month until 5% is reached. One-half of the daily feed requirement is given in the morning and the other half in the afternoon.

Filamentous green algae, locally known as *lumut* or *gulaman* (*Gracillaria* spp.), when readily available in quantity may also be used as feed.

Partial harvesting of crabs should be done when some reach 200 to 250 g (6.7 to 8.3 oz) by using baited traps or hand lines with scoop nets. In the Philippines, this is commonly known as the *pasulang* method. The ponds are partially drained (50%) during low tide, and at high tide new seawater is admitted, thereby causing the crabs to swim against the current toward the water gate. The majority of the catch using this technique will be females with maturing eggs (*aligue*), because the marine phase of life or spawning stage of the animal is about to begin. While swimming against the current and concentrating along the gate, the crabs are caught by using scoop nets and the pincers are then securely tied by using strips of soaked coconut sheath (*suwak*) or plastic straws. Care coupled with skill in tying the pincers safeguards workers from being severely pinched. Baited traps (*bintol*) or baited hand lines can also be used if selective harvesting falls on an ebb or neap tide. This method minimizes competition for food and space by the remaining stock and reduces the incidence of cannibalism.

Harvesting by totally draining the pond is done during low tide, when the remaining crabs are collected by hand. Earthen mounds are examined to ensure complete retrieval of crabs. Normally, this requires a day or two using five workers.

Newly harvested mud crabs (Fig. 4), mixed or sorted by size, are always tied in bunches either by the kilogram (2.2 lb) or dozen. Sometimes, females with maturing eggs are sorted from the males with large pincers for delivery to discriminating customers. For long-distance travel, they are kept inside wooden or Styrofoam boxes or bamboo (*tiklis*) or palm (*buri* or *pandan*) baskets. Mud crabs are a hardy species that can stay alive for up to one week when sprinkled occasionally with water. Prolonged holding periods, however, will lessen the weight (*hagas*) or eventually cause death. Price varies from region to region in the Philippines. Crabs of average weight [150 to 200 g (5 to 6.7 oz)] bring from 120 to 180 pesos per kilogram (2.2 lb); larger crabs [200 g (6.7 oz)] bring 160 to 250 pesos per kilogram (2.2 lb). Generally, female crabs with developing gonads are more expensive than males. In the Philippines and abroad, the demand for mud crabs exceeds the supply.

PEN CULTURE OF MUD CRAB IN MANGROVES

Using net enclosures to grow mud crabs in mangroves or tidal zones offers the latest bright prospect for aquaculture. The technique is environmentally friendly, sustainable, and can provide additional income for coastal communities.

Net Enclosure Construction

The mangrove area selected for culturing mud crabs should be unpolluted and have a water depth of 0.8 to 1.0 m (2–4 to 3 ft) at high tide. Areas that are susceptible to large waves should be avoided. Salinity should range from 10 to 30 ppt, and temperature from 25° to 30 °C (77° to 85 °F).

Net enclosures (Fig. 5) can be constructed by driving 3 to 4 m (9 to 12 ft) long wooden or bamboo poles into the sediment. The tops of the poles should extend above the high water level by at least 30 cm (1 ft). Horizontal bamboo crossbars are secured between the vertical wooden poles, both at the high water level and 30 cm (1 ft) above the high water level, to provide structural support and attachment sites. Polyethylene netting with mesh size of 1 to 2 cm (0.4 to 0.8 in.) is hung from the top bamboo crossbars with the bottom of the netting buried in the sediment to a depth of 70 cm. Plastic sheeting is hung between the two bamboo crossbars to keep the crabs from climbing over the top of the enclosure. Ditches or depressions should be dug to a depth of at least 20 to 40 cm (7.9 to 15.8 in.) and placed in from 20 to 30% of the enclosed area to provide

Figure 4. Harvested mud crabs with drained pond in background.

Figure 5. A net-enclosed mangrove area. Repairs are being made on the plastic above the mesh.

refuge areas for crabs during extreme low tides. Caution should be taken to avoid cutting mangrove roots during the digging process. Catwalks should be constructed around the enclosure's perimeter or across the middle of the pen for feeding and observation of the crabs.

Enclosure size and shapes can vary. The Southeast Asian Fisheries Development Center, Aquaculture Department (SEAFDEC, AQD) test enclosures were of an area of 4,000 m^2 [(3,640 yd^2)]. Manageable sizes appear to range from about 0.2 to 1 ha (0.5 to 2.5 ac).

Stocking and Rearing

Prior to stocking, the net enclosure is cleared of debris and unwanted organisms, such as predators. The perimeter of the pen should be carefully checked prior to stocking to ensure that integrity of the structure is intact.

The recommended stocking rate for juvenile mud crabs weighing from 20 to 50 g (0.7 to 1.7 oz) is 5,000 to 10,000/ha (2,000 to 4,000/ac). Stocking should occur when the pen is filled with tidal water and during early morning, or late afternoon, to avoid the hottest part of the day. The crabs should be tempered (acclimated) to the temperature and salinity of the pen before release. Tempering can be achieved by placing the crabs in plastic basins and gradually exchanging the hauling water with water from the pen. The process should be completed in about one hour. Ties on chelipeds should be removed before the animals are released.

Crabs can be fed chopped trash fish, animal hides or entrails, mussels, snails, or prepared feed (commercially available in the Philippines and formulated for mud crabs). Feed should be broadcast as evenly as possible throughout the pen each day. If possible, this should be done on an incoming tide. The recommended initial feeding rate is 10% of crab biomass daily. The rate should be reduced by 1% each month until 5% is reached, which will be at the end of a culture cycle.

Baited lift nets should be used to collect samples each month. To provide an estimate of growth and biomass (for adjustment of feeding rate), 30 to 50 crabs should be collected, weighed, and measured.

SEAFDEC, AQD recommends that a determination of dissolved oxygen, temperature, salinity, and turbidity be made at least three times each week. If stressful conditions occur, good management strategy would involve cessation of feeding until the water quality improves. Enclosures should be inspected frequently. Debris should be removed and any tears in the netting should be repaired. Nightly inspections are recommended to deter poachers.

Harvest should begin after the crabs have been in the enclosure for three months (Fig. 6). Baited lift nets are used and only marketable animals are taken (submarketable crabs are returned to the enclosure). The average size of marketable mud crabs in the Philippines is 275 g (0.6 lb). Partial harvest should be conducted until the sixth month of culture, at which time, the remaining crabs are removed and marketed. The enclosure can then be restocked, giving two crops annually. Each crop should yield about 485 kg (1,067 lb) when an aquaculturist used a 0.4 = hectare (1 ac) enclosure.

Figure 6. Harvesting mud crabs from an enclosed mangrove area.

BIBLIOGRAPHY

1. D.D. Baliao, E.M. Rodriquez, and D.D. Gerochi, *Q. Res. Rep. SEAFDEC Aqua. Dept.* **5**(1), 10–14 (1981).
2. D.D. Baliao, M.A. De los Santos, and N.M. Franco, *Mudcrab Culture in Brackishwater Ponds*, Aquaculture Department, Southeast Asian Fisheries Development Center, Iloillo, Philippines, 1999.
3. D.D. Baliao, M.A. De los Santos, and N.M. Franco, *Pen Culture of Mudcrab in Mangroves*, Aquaculture Department, Southeast Asian Fisheries Development Center, Iloilo, Philippines, 1999, pp. 11.
4. S.C. Jayamanne and J. Jinadasa, *Vidyodaya J. Sci.* **3**(2), 61–70 (1991).
5. B.J. Hill, *Marine Biol.* **34**, 109–116 (1976).
6. C. Suseelan, *Bull. Cent. Mar. Fish. Res. Inst.* **48**, 99–102 (1996).

See also Crab culture.

MULLET CULTURE

Cheng-Sheng Lee
The Oceanic Institute
Waimanalo, Hawaii

OUTLINE

INTRODUCTION

Striped mullet (*Mugil cephalus*) is one of the most popular cultured species with both euryhaline and eurythermal characteristics. It is also known as grey, flathead, and jumping mullet. As a cosmopolitan species, mullet are found in all coastal waters between latitudes 42 °N and 42 °S. They are found as far north as Hokkaido, Japan and as far south as South Africa, Cuba, and India (1). From mitochondrial DNA genotype analysis, striped mullet from ocean basins around the world revealed pronounced population genetic structure (2). Taxonomically, striped mullet belong to the family Mugilidae (order Perciformes). A total of 14 genera and 64 species have been identified throughout the world (3). Presently, about 20 species are farmed in different regions of the world (4). Striped mullet is the most desired species among them.

Mullet are consumers of low trophic layers. They have a complex pharyngobranchia filter which, associated with the gill raker system, enables them to feed on a variety of microorganisms and decaying organic matter, as well as larger food such as algae, insect larvae, and small molluscs (4). Because of high tolerance to both temperature and salinity variations, mullet can be found in a wide range of waters, from freshwater to seawater. Thus, mullet farming has been practiced throughout history and has become a tradition in many parts of the world such as the Mediterranean, southeast Asia, east Asia, and some Pacific islands.

SOURCE OF FRY

Liao (1), in 1981, estimated that about 99% of fingerlings stocked in nursery ponds around the world are from the wild. Although no survey was conducted to identify the sources of fry for striped mullet farming, the majority of farms are still using wild-caught fry. The easy access to wild-caught fry and high cost of hatchery-produced fry may be the explanation. The labor cost for fry production, representing 51% of operational costs, must be reduced to make hatchery fry more affordable to farmers (5). At this time, hatchery-produced mullet fry are still uncommon among farmers.

Wild Supply

Collection Methods. Mullet fry appear seasonally in coastal areas at the age of approximately two months. They tend to concentrate in shallow estuarine environments, converge in the mouths of lagoons and rivers, and migrate in schools against currents with different water quality. Fry can be collected in gear with passive (fixed) or active methods. Active gears such as scoop nets, skimming nets, and beach seines are commonly used to collect wild fry. Lights with 600 lux, 400 W bulbs are used in addition to scoop nets to attract fry in Taiwan (1). Motor boats equipped with fry collection gear (Fig. 1) have also been used in recent years. Passive gear, such as fish traps with a short wing pointing toward shore and a longer wing protruding into deeper water to guide the fish into the trap, are placed in areas where fry naturally gather or migrate, usually along the banks of streams.

Different collection techniques are used due to the distinct behaviors among various sizes of fry (1). Fry smaller than 30 mm (1.2 in.) generally swim weakly in schools and can be collected by nets and seine. However, these smaller fry are fragile and survival after stocking is less than 30%. Fry larger than 50 mm (2 in.) do not swim in schools and are hardier and active; therefore, they are more difficult to collect. Because larger fry tend to swim against the freshwater current, they are commonly collected through sluice gates directly into fish farms.

Figure 1. Boat equipped with fry collection gear.

Handling, Transportation, and Stocking. Striped mullet fry are sensitive to handling. Improper handling can cause massive losses. Fry die from infections and shock due to careless handling, drying in air, abrasions from nets, and brain hemorrhages from knocking into hard walls of containers (6). Other factors such as extremely low or high dissolved oxygen, sudden temperature and/or salinity changes, disease, and starvation can also lead to mortalities. In Egypt, it was reported that only 1.6% of wild fry survived to market size. Most mortalities occurred during transportation and stocking.

Acclimation is a necessary step to reduce mortality caused by osmotic and thermal shock. Sarojini (7) reported that acclimation reduced mortality to less than 1% in *M. cephalus* and *M. seheli*. Tolerance of fry to salinity change varies according to fish size, water temperature, and initial salinity. Sudden transfer of fry from 100% seawater to 11 ppt resulted in negligible mortality; however, when fry were transferred from 4 ppt to freshwater, 100% mortality ensued (8). Normally, large fingerlings [35–42 mm SL (1.38–1.65 in.)] were more resistant to lower salinities than smaller fry [23–31 mm SL (0.91–1.22 in.)] (1). In spite of greater salinity changes at 17 °C (62.6 °F), *M. cephalus* fry had a better survival than at 10 °C (50 °F) (8). Sylvester (9) also found that juveniles [80–120 mm SL (3.15–4.72 in.)] require longer acclimation times at lower temperatures.

During the process of collection, large numbers of fry are forced into a small area where fry expend much energy. They jump, scrape against the net and each other, swirl in circles, and are exposed to air. The shock from capture can eventually lead to mortality. To increase the survival, some collectors use anesthetics to slow down fish activities or reduce the energy expenditure of the fry during collection (6). Anesthetics such as MS-222, chloral hydrate, tertiary amyl alcohol, chlorobutanol, quinaldine, paraldehyde, and tertiary butyl alcohol have been used during the transport of mullet seed (10). Using anesthetics is expensive and must be monitored closely to prevent death from overdosing. The drug must be used at sublethal dosages and for a specified amount of time, dependent on fish species and size.

Hatchery Supply

The interest in culturing this species as a source of food has resulted in numerous investigations regarding control of their reproductive cycle (11,12). Likewise, considerable attention has been given to developing techniques for mass rearing of early larvae to juvenile stage (13–16). Broodstock source has to be identified before the production of fry in the hatchery can be realized. Mature spawners can be obtained from the open water or culture facilities (17). Collection of broodstock during spawning migration by barrier net such as the one shown in Egypt (Fig. 2) results in less damage to fish than those collected from gillnet fishing in the open sea. However, the most reliable source is from culture facilities. Broodstock can be grown in either seawater or freshwater environments prior to the spawning season (18,19). Indoor facilities fail to provide mature fish that produce good-quality eggs. The shortage of arachidonic acid may be the cause (20). Induction of spawning should be carried out in full-strength seawater for the best fertilization and embryonic development (19). The life cycle of striped mullet has been closed under captive conditions (21). Natural spawning of this species in any holding facilities has yet to be documented.

The first step in propagating mullet begins with the successful induction of spawning using a variety of hormonal treatments (Table 1) (12,22). Control over the maturation process of males and females by the use of either environmental or hormonal methods has demonstrated that sexually mature broodstock can be produced throughout the year (23–25). In Hawaii, fish were held in outdoor tanks under photoperiod and water temperature control to reach final maturation at the desired time. Each outdoor tank was covered

Figure 2. Barrier net used to collect broodstock during spawning migration.

Table 1. Dosages and Cost of Hormones for Induced Spawning of Mullet Using Different Strategies (12)

Hormone Treatment	Dosage	Cost (US$)	Spawning Rate (%)	Fertilization Rate (%)	Time of Spawning After 2nd Injection (h : min)[c]
Saline			0 (6)[a]		
CPH/CPH	20 mg/40 mg	8.40	75 (8)	49.7 ± 17.8	12 : 18
				23.6–73.0	10 : 00–15 : 20
CPH/LHRH-a	20 mg/200 μg	3.81–6.95	90 (20)	86.9 ± 9.0	12 : 42
				66.0–100	10 : 36–17 : 30
LHRH-a/CPH	200 μg/20 mg	3.81–6.95	100 (5)	49.9 ± 29.9	11 : 12
				7.0–87.0	10 : 00–12 : 33
LHRH-a/LHRH-a	400 μg	2.02–8.30	87.5 (8)	46.8 ± 44.1	17 : 20
				0–97.9	11 : 48–21 : 00
HCG/LHRH-a	5,000 IU/200 μg	7.56–10.70	100 (5)	53.6 ± 34.8	14 : 24
				0–97.2	11 : 28–17 : 55
HCG/LHRH-a	10,000 IU/200 μg	14.11–17.25	83.3 (6)	65.8 ± 19.1	12 : 50
				40.1–97.0	11 : 00–14 : 40
LHRH-a/HCG	200 μg/5,000 IU	7.56–10.70	0 (4)	—	—
LHRH-a/HCG	200 μg/10,000 IU	14.11–17.25	0 (2)[b]	—	—

[a]Number of fish tested.
[b]Fish spawned after additional LHRH-a injection.
[c]Values in second row are range.

with a light-proof black fabric sheet to obtain a shorter photoperiod than under natural conditions. The black sheet was manually removed during the daytime cycle. To extend the photoperiod, fluorescent lamps were added. The water temperature was lowered by adding cold freshwater instead of using a chiller. This approach has avoided the requirement of high-cost environment-controlled facilities. Frequency of spawning has also been altered from once per year up to three times per year. When undergoing chronic hormonal treatment or placed under controlled environmental conditions, broodstock can be induced to spawn more than once without any appreciable loss in egg quality (26–28). It is obvious that major advances have occurred with regard to delivering fertilized eggs on demand (29).

Development of techniques that can produce a large number of fertilized eggs and larvae has been ongoing for over 20 years (13–16,30,31). It is now possible to consistently mass-produce mullet fry in hatcheries. However, production costs for mullet fry need to be reduced.

Fertilized Eggs. The fertilized eggs required in larval rearing can be easily obtained from induced spawning, utilizing the method described by Lee et al. (22). Females that possess oocytes with average diameters of 600 μm and above are selected and held in 170-L (180-qt) aquaria for hormonal induction of spawning. They are given a priming injection of carp pituitary homogenate [20 mg/kg (ppm) body weight] followed 24 hours later with a resolving injection of 200 μg/kg (ppb) body weight of LHRH-a. Two running ripe males are placed with the female when she receives the second hormonal injection and spawning usually takes place 12 hours after the second injection. Spawning and fertilization of the eggs occur naturally. After fertilization, eggs are removed from spawning tanks to incubation tanks or larval rearing tanks for hatching.

Live Feed Production. The green alga *Nannochloropsis oculata* has been identified as the most suitable phytoplankton to be used for feeding rotifers in mullet larval rearing. Stock cultures are initially grown on slightly modified Miquel culture media. Amplification of the phytoplankton volume is achieved with an outdoor batch system strategy using three 500-L (132-gal) intermediate tanks, four 5,000-L (1,321-gal) fiberglass raceways, and three 30,000-L (7,926-gal) fiberglass tanks, as described by Eda et al. (16).

Brachionus rountiform (S-type, 110–230 μm lorica length) is the zooplankton commonly used for first feeding of mullet. The culture method for rotifers varies among hatcheries and is selected based on a hatchery's specific needs. In Hawaii, rotifers are cultured indoors using a batch system strategy. Rotifers are initially stocked into 600-L (158.5 gal) of *N. oculata* (a minimum of 10 million cells/mL) at a density of 100 rotifers/mL. Twenty-four hours later, the volume is raised to 1,200 L (317 gal), and the culture is allowed to grow for another 24 hours. After 40 hours, the entire tank is harvested by draining the culture through a 60-μm nitex bag. The main food source for the rotifers is *N. oculata*, but is frequently supplemented by yeast. The nutritional quality of the rotifers should be manipulated to increase the highly unsaturated fatty acid (HUFA) content. When low content of HUFA in rotifers is detected, enrichment of rotifers prior to providing them to mullet larvae is necessary to obtain better larval survival.

Larval Rearing. Temperature was found to be the determinant in the rate of development of mullet embryos, as in other finfish species studied (32). Salinities between 15 and 45 ppt did not affect the time of hatching, nor was any consistent variability observed between the median time of hatching between females. Hatching times ranged between 65 and 73 hours when incubated at 20 °C (68 °F) to 25 to 27 hours when incubated at 32 °C (90 °F). Mortality of mullet embryos was observed to be affected by both temperature and salinity. The upper and lower thermal tolerances for normal development is above 30 °C (86 °F) and below 18–20 °C (64.4–68 °F), respectively. Salinities of 15 to 45 ppt examined did not appear to limit survival (33). The calculated optimum hatching (96.6%) was found to be at 25.5 °C (78 °F) and 37.4 ppt. When larval rearing was conducted in 5,000-L (1,321-gal) tanks from hatching to 50 days posthatching, no significant differences in survival and growth were found between the 22–23 ppt salinity group and the 32–35 ppt salinity group (34).

At 70 hours posthatching, rotifers were found in the gut of the fish larvae (16). Over the next 10 hours, the number of fish fed gradually increased, and by 104 hours posthatching 75% of the larvae were feeding (16). The number of rotifers eaten per larva ranged between 0 and 11. The incidence of feeding observed was influenced significantly by the number of rotifers introduced into the rearing tanks. By the second day posthatching, 40–50% of the larvae presented with 10 rotifers/mL began feeding. In contrast, less than 1% of the larvae presented with 1 rotifer/mL were found to be feeding. Food preference changes as fish larvae grow. Initially the gut contents mirrored the composition of food organisms found in the water column. However, the larger food organisms began to predominate as early as the fourth day posthatching and remained throughout the duration of the experiment (35). The forage ratio ranged from highest to lowest for *Artemia* nauplii, L-type and S-type rotifers, respectively.

Growth of larvae from the same spawning can vary when they are reared in a separate 5,000-L (1,321-gal) tank with a different feeding regime. Larvae fed according to their feed preference were found to be significantly ($p < 0.05$) larger by the 20th day posthatching than individuals from the tank where S-type rotifers were the only live food organism. The growth of mullet larva from hatching to 50th day posthatching is shown in Figure 3.

The quality of food organisms also affects larval survival. Larvae fed on rotifers grown on the combination of *N. oculata* and yeast or *N. oculata* alone exhibited significantly higher survival than those fed exclusively on yeast-fed rotifers (31). Survival was significantly higher at 42 days posthatching for larvae fed rotifers grown on *N. oculata* alone as compared to that of yeast. In contrast, at the end of the 42-day rearing trial no significant differences could be detected in total length among the

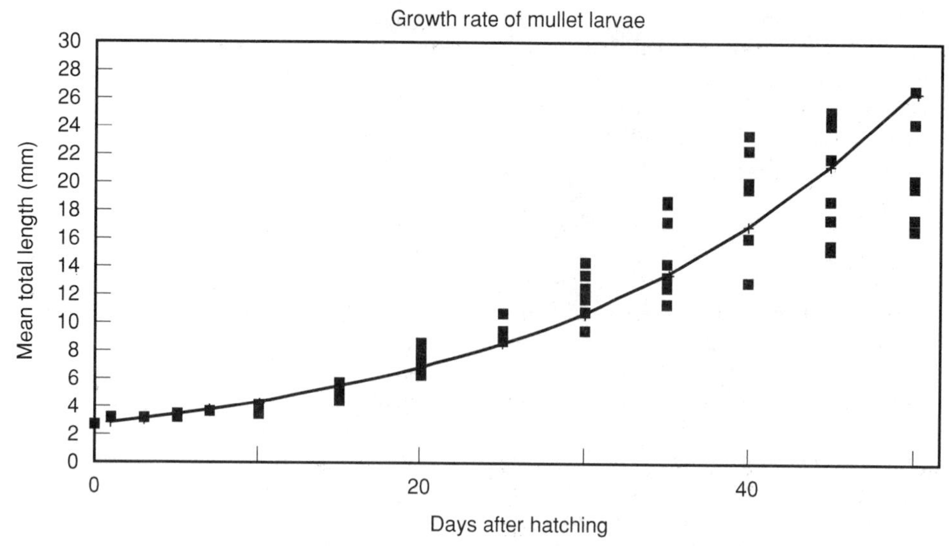

Figure 3. Growth rate of mullet larvae during intensive rearing trials.

treatment groups. There were significant differences in sizes of the larvae between days 15 and 30 posthatch. Enriching *Artemia* is not necessary for mullet larval rearing (16). While enrichment did not increase larval survival, it significantly ($p < 0.05$) enhanced larvae size (16). Similarly, larvae fed nonenriched *Artemia* were significantly ($p < 0.01$) larger than those fed exclusively on S-type rotifers.

Tamaru et al. (36) indicated the importance of using low levels of phytoplankton as background during the rearing trials. There is a significant ($p < 0.01$) improvement in the overall survival of mullet larvae (15 days posthatch) when the rearing process is conducted in the water with background algae. This is consistently observed even when the rotifers are of a high nutritional quality (i.e., have been reared on algae or yeast). There is also significant improvement in growth by using background algae.

Currently, the best feeding regimen (Fig. 4) developed for mullet larval rearing provides rotifers at a density of 20/mL for the first 25 days posthatch and green algae at 300,000 cell/mL in the larval rearing tank. From day 17 until day 35, brine shrimp are given at a density of 0.02 to 3 individuals/mL, while larvae are trained to feed on artificial feed from day 25 until fry are harvested on day 40. This feeding regime can provide consistent production of 5 mullet fry/L (19 mullet fry/2 gal).

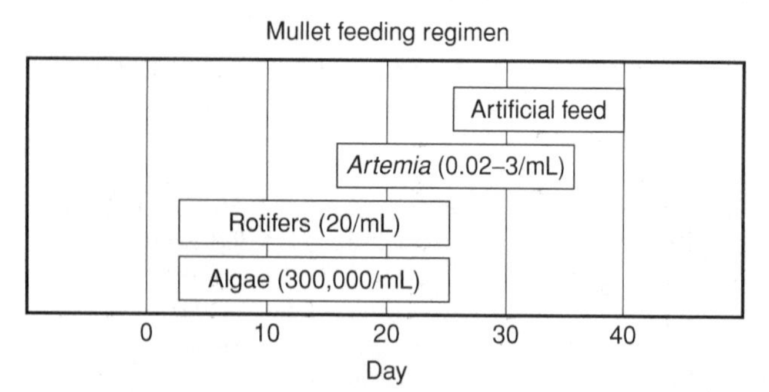

Figure 4. Recommended feeding regimen for mullet larvae.

GROWOUT

M. cephalus is successfully cultured in many countries such as China, India, Japan, and Israel, as mentioned by Hepher and Pruginin (37), on a smaller scale in Greece (a lagoon), India (experimental level) and Yugoslavia (the Vrana lagoon), reported by Uwate and Kunatuba (38); and in Vietnam (39), Japan (40), France, India, Indonesia, Hawaii, and Pakistan as reported by Hickling (41) and Ling (42). At different locations, different culture methods were employed.

The effect of water quality on survival and adaptability of mullet was studied by Azariah et al. (43). Fry survival was high in the first 24-hour period after release at high tide; however, survival was low at low tide. Low survival was due to lack of dissolved oxygen and the presence of sewage in the inland river water at low tide. Dissolved oxygen should be maintained above 50% of saturation (44). Liao and Chao (45) also reported the high sensitivity of striped mullet to oxygen depletion. The iron crush effluent (0.07 ± 0.25 mg Fe/L; pH 7.8 ± 0.60) from an iron crush factory in Egypt did not affect growth of *M. cephalus* (46). Growth rate of 0.70 g (0.02 oz)/day was similar to that in Egyptian fish farms. The appearance of a blue-green algal bloom (*Lyngbya limnetica*) at 775,000 organisms/L led to high mortality in one pond [death of 100 kg (220 lb) of fish].

Mullet can be cultured under conditions ranging from freshwater to seawater. They were successfully reared in salinities of 60–70 ppt (47). Bardach et al. (48) reported that mullet [50 g (1.61 oz)] and tilapia [50 g (1.61 oz)] were stocked at 214/ha (87/ac) and 139/ha (56/ac), respectively, into a pond close to the Dead Sea in Israel, which has a salinity of 36 to 145 ppt. After 109 days, harvest revealed a yield of 512 kg/ha (455 lb/ac) of mullet. Tungkang Marine Laboratory in Taiwan compared growth of mullet in seawater (16.4 to 32.7 ppt), brackishwater (11.8 to 20.6 ppt), and freshwater. Mullet with initial weight of 113.8 g (3.66 oz) attained the average weight of 278.6, 287.0, and 227.0 g (8.96, 9.22, and 7.30 oz), with daily weight gains of 0.53, 0.64, and 0.49 g (0.02, 0.02, 0.02 oz), respectively, after 190 days in seawater, brackishwater, and freshwater. There were no significant differences ($p > 0.05$) in seawater and brackishwater growth rates. The mullet cultured in freshwater were more prone to injuries during handling and were not tame and docile.

Cultivation Methods. Extensive culture, polyculture, and intensive culture are the three most common methods used throughout the world to culture mullet fry to marketable size. As a feeder of low trophic layers, mullet have been cultured extensively for centuries, mostly in polyculture with other species. Polyculture is the practice of stocking more than one species in the same pond to increase the total production.

Growth of mullet depends on many factors such as stocking density, and genetic strains. Generally, the growth of *M. cephalus* in the wild is much slower than in culture. In nature, 4-year-old mullet range from 229 g (7.4 oz) in the Black Sea to 1,217 g (39.1 oz) in the Mediterranean (49). *M. cephalus* on the Mediterranean coast of Egypt are 300 g (9.7 oz) at 2 years and 900 g (28.9 oz) the next year (50). In several lakes of India, Rangaswamy (50) found mullet that were 300 g (9.7 oz) at 2 years, and 600 g (19.3 oz) at 3 to 4 years. However, mullet fry of 20 to 35 mm (0.8 to 1.4 in.) stocked at the density of 25,000/ha grew to 100 to 140 mm (3.94 to 5.51 in.) in length in four months. An average size of 375.6 mm/505.9 g (14.8 in./16.3 oz) at the end of the first biological year was also reported in India (51). The monthly growth rate was 75.3 mm/70.6 g (3.0 in./2.3 oz). In Taiwan, striped mullet of 0.36 g (0.01 oz) grew to 471 g (15.1 oz) in a year (45). In Taiwan, mullet are harvested at two to three years of age (45). Mullet are also harvested at greater than three years to harvest the roe of females. Dried mullet roe (Fig. 5) are in high demand and can command a high market price. In Israel, market-size fish at one year is 400 g (12.9 oz), and at two years is 600 g (19.3 oz) (37). In general, cultured mullet attain marketable size [400–1,000 g (12.86–32.15 oz)] in a little over a year to four years.

Extensive Culture. Extensive culture, a common method for growout of mullet in the world, is usually practiced in ponds on shores, near estuaries, or in freshwater ponds located inland (45). Fenced lagoons, creeks, and swamps are also used (38). In Hawaii, mullet is traditionally cultured in fish ponds formed by lava rock near the shore. Fish are entirely dependent on food available in the pond. Yield is usually low and does not justify the investment (52). Most of the production is for subsistence purposes.

Figure 5. Dried mullet roe is processed in Taiwan.

Growth depends on stocking density and pond condition (1). In Taiwan, low stocking density and good weather conditions produce mullet with a weight range of 300–600 g (9.7–19.3 oz) after one year (45). A well-known method of extensive culture is "kawa culture" practiced in Aichi Prefecture, Japan. In kawa culture, fry are stocked in waterways at very high densities [15,800–19,750/ha (6,394–7,992/ac)]. Mullet is the main species and is sometimes polycultured with common carp (*Cyprinus carpio*), crucian carp, sea bass (*Dicentrarchus labrax*), perch, and eel (*Anguilidue*). Extensive culture ponds often include other species in addition to mullet. In Japan, various studies reported average mullet production of 480 kg/ha (428 lb/ac) (53), 392 kg/ha (349 lb/ac) (54), and 663 kg/ha (590 lb/ac). Water temperature was as high as 29–31 °C (84.2–87.8 °F) in summer and as low as 4–6 °C (39.2–42.8 °F) in winter. In Hawaii, mullet production varies from 100 to approximately 2,000 kg/ha (89 to 1,780 lb/ac), depending on the management style.

Polyculture. Polyculture is the predominant system used for growout of mullet. Striped mullet is most suitable for polyculture due to its feeding habits (herbivore and detritophore), peaceful nature (does not compete for food or attack other fish), and easy adaptation to various salinities. While mullet obtain food both from the benthos and plankton, they efficiently exploit fine detritus and small zooplankton not used by other cultural species (55). On the other hand, mullet have a detrimental effect on large zooplankton and chironomid midges, which are diets of common carp and tilapia (*Oreochromis* sp.). Thus, selection of the appropriate species for polyculture with mullet is an important factor for overall production. Aquatic species most commonly cultured with striped mullet are shown in Table 2 (45,56–60).

The species that have been polycultured with striped mullet in freshwater environment include grass carp (*Ctenopharyngodon idella*), silver carp (*Hypopthalmichthys molitrix*), bighead (*Aristichthys nobilis*), common carp (*C. carpio*), milkfish (*Chanos chanos*), and mullet (*Mugil capito*, *M. auratus*). Grass prawn (*Penaeus monodon*), milkfish (*C. chanos*), mullets (*Liza ramada*, *L. auratus*, *L. saliens* and *Chelon labrosus*), gilthead seabream (*Sparus aurata*), sea bass (*D. labrax*) and eel (*Anguilla anguilla*) were cultured with striped mullet in brackishwater. The role of striped mullet in polyculture is equally important as other species in total production or as a secondary crop to others. Striped mullet were produced in brackishwater ponds as a secondary crop to milkfish in Indonesia (4), and to milkfish and grass prawn in Taiwan. The following is an overview of the practice of polyculture in several countries.

In Egypt, production of *M. cephalus* in fertilized ponds gave yields of 192–350 kg/ha (170.9–311.5 lb/ac) or 36–77% higher as opposed to 131 kg/ha (116.6 lb/ac)

Table 2. Most Common Species Polycultured with *Mugil cephalus*

Species	Water Type	Country	Reference
Grass prawn, *Penaeus monodon*	Brackishwater	Taiwan	Liao and Chao (45)
Milkfish, *Chanos chanos*			
Grass carp, *Ctenopharyngodon idella*	Freshwater		
Silver carp, *Hypopthalmichthys molitrix*			
Bighead	Freshwater	Hong Kong and Taiwan	Sinha (56)
Grass carp			
Common carp			
Milkfish, *C. chanos*			
Mullets (*Liza ramada*, *L. auratus*, *L. saliens* and *Chelon labrosus*)	"Valli" (modified lagoons), brackishwater	Italy	Jhingran and Natarajan (57); Brown (58)
Gilthead seabream (*Sparus aurata*)			
Sea bass (*Dicentrarchus labrax*)			
Eel (*Anguilla anguilla*)			
Common carp	Freshwater	Israel	Pruginin and Kitai (59); Sinha (56); Yashouv (60)
M. capito			
M. auratus			

without fertilization. Additionally, 162–219 kg/ha (144.2–194.9 lb/ac) of other species were harvested. Similar results were achieved in Hong Kong and Taiwan (61). Yields of mullet were increased by 96% due to the simple application of fertilization.

In India, milkfish (*C. chanos*), grey mullet (*M. cephalus*), and prawn (*Fenneropenaeus indicus*) were cultured together in a salt pan reservoir (62). Mullet had the best survival and production among the three species. Marketable size was achieved in nine months.

In Israel, six species of mullet are abundant along the Mediterranean coast. Of the six, *M. cephalus* and *L. ramada* adapt easily to freshwater, have good growth rates, and thus are selected for polyculture (52). Immediately after collection, fry are held over the winter for 140 days and cultured with carp. From an initial weight of 0.2 g (0.01 oz), mullet fry attain a weight of 15–50 g (0.48–1.61 oz) at which time they are ready to be stocked at low densities into nursery ponds. Ponds are fertilized weekly for eight months and fish are fed daily, initially with sorghum, and followed later by pellets with 25% protein at 3,000 kcal. Feed is mainly aimed toward the carp and tilapia, with the mullet subsisting on food available in the pond. Water qualities are 18–32 °C (64.4–89.6 °F); pH 8.0–9.0; and dissolved oxygen level is variable due to phytoplankton blooms, 5–10% saturation in morning and 150% in afternoon. Aeration is supplied to offset variable oxygen levels. Mullet is the secondary crop to carp in freshwater ponds. However, if mullet are stocked at 800–1,200/ha (324–486/ac), carp growth is slightly lower when compared to those grown in monoculture (60).

In Italy, "valle" and "lagoon" culture had a total annual yield from 90 to 200 kg/ha (80–178 lb/ac) of fish (48). The "Vallicoltura" is an ancient practice of extensive culture of euryhaline species in confined areas of the Italian North Adriatic lagoons, which spread out to the Mediterranean. Mullet were stocked as juveniles at densities of 1,000–4,000/ha (405–1,619/ac) and reached marketable size in three to five years, with minimum survival of about 10%.

In Japan, polyculture of mullet with carp and eel or with carp and crucian carp did not show any relation between total production of other species and mullet production (54). It was concluded that mullet can be stocked at the maximum optimal density without considering stocking ratio of mullet to other fishes. Mullet had quite different feeding habits from other species in the system. Nakamura (63), however, showed that mullet cultured with eel and carp had better survival and generally higher weight gains when stocked at lower densities (0.6–1.8/m^2) rather than at high densities (1.8–2.5/m^2). Average survival and weight gain for low and high densities was 66.1% ± 17.4 and 104.9 g ± 93.9; and 42.9% ± 10.8 and 85.7 g ± 42.2, respectively. The only supplemental feed given was 57–227 g/m^2 of rice bran. These results indicate the stocking density for mullet was too high.

The marketable size of mullet is 400–1,000 g (12.9–32.2 oz), and the average time to grow mullet to marketable size from initial capture size [0.2 g (0.01 oz)] is approximately 450 days, when no supplemental feed is given. In one study, 30–70 g (0.1–2.3 oz) two-year-old mullet were stocked in carp ponds at 500–800/ha (202–324/ac) and harvested 120–150 days later at a final weight of 400–700 g (12.86–22.51 oz) (60). Stocking of mullet for two years to achieve 1 kg (2.2 lb) size fish is also practiced.

The typical practice of polyculture in freshwater ponds is as follows. After collection, fry may be directly stocked at 30,000 fry/ha (12,141/ac) into nursery ponds (37). If necessary, fry may be placed into adaptation tanks prior to stocking the nursery ponds. Here, farmers disinfect fry with formalin and clean them of parasites and protozoans (52). Grass carp [200–300/ha (81–121/ac)] are added to control filamentous algae. Fry should be transferred to larger ponds at 1–3 g (0.03–0.10 oz) after 60–100 days. As mullet fry are delicate in muddy waters and at temperatures higher than 30 °C (86 °F), handling of fry should occur in the early morning. Yashouv (60) recommends no more than 800 mullet/ha (324/ac) in a carp–mullet culture while Pruginin et al. (64) indicates density can be increased to 1,200/ha (486/ac) in a tilapia, carp, and silver carp culture due to advanced technology.

Intensive Culture. Intensive culture is the least common method used for growout. This technique requires advanced technical equipment to sustain high stocking density, but results in higher production. Several studies were conducted in the past to evaluate the performance of mullet under high stocking density and/or intensive feeding.

In Taiwan, Liao and Chao (45) reported the highest mean daily weight gain of 2.32 g (0.07 oz) was achieved when 50 juveniles were stocked in a round concrete pond [8.25 m (27.07 ft) diameter] and were provided aeration and artificial feed. That growth rate was achieved at salinity ranging from 11.8 to 20.6 ppt. Survival rates during the experiment ranged from 88 to 94%. Mean daily weight gain in other experiment groups ranged from 0.49 to 2.28 g (0.02 to 0.08 oz).

In Japan, fry [18.1–30.0 mm (7.13–11.81 in.)] stocked at a density of $2/m^2$ were fed silkworm pupae at $0.3–2.0\ kg/m^2$. In eight months, final body weight reached 75–360 g (2.65–12.70 oz) with total production of 1,290 kg/ha (1,148 lb/ac) (54). Survival rate was about 48%.

In another study, Linder et al. (65) cultured mullet using heated effluent from a power plant. Mullet were fed commercially prepared feed. The feed cost to yield 1 kg (2.2 lb) of mullet was US$0.61–0.87. The average daily weight and standard length gains ranged from 0.36 g (0.01 oz) to 0.67 g (0.02 oz) and from 0.17 mm (0.01 in.) to 1.03 mm (0.04 in.), respectively. Survival was 50–85% with a final production of 293–804 kg/ha (261–716 lb/ac). Feed conversion ratio was 2.24–3.31.

Intensive farming requires formulated feed to support the rapid growth. Nour et al. (66) concluded that a diet containing 30% crude protein supported normal growth of striped mullet in freshwater ponds. Commercial diet supplemented with vitamin E to 190 and 290 mg/kg (ppm) significantly increased body and ovary weights for mullet (67).

Table 3. 1996 World Mullet Production in Tons[a]

Country	*M. cephalus*	Other Mullet
Algeria		26
China, Taiwan	2,321	
Dominican Republic		20
Egypt	20,101	
Greece	502	
Guam		5
Hong Kong	1,561	
Indonesia		11,300
Israel	1,232	
Italy	3,100	
Korea	27	
Portugal		5
Russian Federation		70
Spain		125
Thailand		200
Tunisia	295	
Ukraine		157
Total:	29,139	11,908

[a] *Source*: Reference 68.

Production. World production of mullet in 1996 according to FAO (68) was about 41,047 tons (Table 3). The production data might not necessarily represent the actual annual yield due to the difficulty of species identification.

DISEASES/PARASITES AND TREATMENTS

Because of the small number of striped mullet cultured intensively, disease problems have been infrequent. Occurrences of disease usually result from polluted waters, a poor culture environment, and management (high stocking density, excessive handling). Lin et al. (69) investigated the diseases of cultured brackish and freshwater *M. cephalus*. From 39 diseased fish, 41.4% of the fish had a bacterial infection, 33% were infected by protozoa (*Trichodina* sp. and *Apiosoma* on the gills), 20.5% were infected by a combination of bacteria and protozoa, and 5.1% were initiated by oxygen deficiency and bad water quality. Disease outbreaks were more common during warmer temperatures (August to September) than cold temperatures (November to February).

For the successful prevention and treatment of the unicellular parasites *Chilondonella* sp., *Costia* sp., and *Trichodina* sp., *M. cephalus* fingerlings [5–30 g (0.16–0.96 oz)] may be treated with 30–40 ppm technical formalin for six hours (70). Aeration should be provided during treatment as oxygen levels drop at night due to phytoplankton mortality. As temperatures increase, formalin toxicity increases as well. Lahav and Sarig (71) reported that Bromex (0.2 ppm) can effectively treat mullet infested with the parasite *Ergasilus sieboldi*. Sparolegnia infection can be treated by maintaining fingerlings in brackishwater (seawater : freshwater ratio of 1 : 9 or 1 : 4) to reduce mortalities. Baticados and Quinitio (72) also reported the mortalities of cultured mullet caused by dinoflagellate-like parasites. *Amyloodinium* attached to the gill filaments led to the disruption of the lamellae and lamellar tissue degeneration. Infection by the copepod ectoparasite, *Caligus bombavensis*, was recorded in laboratory-reared striped mullet and was treated effectively by 0.02% formaldehyde solution (73).

Frequent inspection of broodstock for any skin and gill parasites is an important step in preventing the outbreak of disease. During the holding of broodstock, it is very common to encounter fish lice such as *Argulus* or gill parasites such as *Oodinium*. For treatment, a change of salinity is currently the only legal treatment accepted by the U.S. Food and Drug Administration (FDA). Other possible parasite treatments which are considered to be of low regulatory priority by the FDA but not tested for mullet are: 10 minute dip in 1,000–2,000 ppm acetic acid, 5 second dip in 2,000 ppm calcium oxide, 250–500 ppm hydrogen peroxide, 5–10 minute dip in 30,000 ppm magnesium sulfate followed by a similar dip in 7,000 ppm NaCl. Treatments approved for other species but not for mullet are oxytetracycline, sulfamethoxine, formalin, and sulfamerazine. The effective chemical treatments for mullet, but not approved by the FDA, include 0.5 ppm of Furacin (Nitrofurazone) for 1 hour and repeated for several days for bacterial infections, and 0.5 ppm of Trichlorfon (Masoten) once a week for four weeks for treatment of *Argulus*.

FUTURE PROSPECTS

Historically, striped mullet has been a traditional cultured species and an important subsistence species in many parts of the world (Fig. 6). Mullet will continue to be cultured and will play an important role in fish farming. However, the expansion of mullet culture requires efforts in the advancement of both culture technology and marketing. Mullet farming has depended primarily on wild-caught fry, which are inconsistent and not reliable. With the establishment of mullet fry production technology, farmers should be encouraged to use and upgrade the available technology. Further development of mullet hatchery technology can make the hatchery-produced fry more cost-effective.

Depending on the strain of striped mullet, the growing period from juvenile to market size will take one to three years. In order to be profitable, cultivated species should reach marketable size in one year, particularly in areas with cold winter seasons. Traditional genetic selection or advanced genetic engineering technology should be used to improve growth performance of mullet.

Figure 6. The harvesting of mullet is an important social and cultural event in countries such as Egypt.

Although mullet can be a desirable table species for more consumers, little effort has been made to market the species beyond its traditional consumers. The current market demand for mullet is very limited. Consequently, the price of mullet stays low and farmers are moving away from culturing subsistence species to high-valued species in response to profit-driven operations. A marketing effort for mullet is an essential step to encourage more farming. Developing the technology for producing special products such as dried ovaries could also expand the currently limited market.

BIBLIOGRAPHY

1. I.C. Liao, in O.H. Oren, ed., *Aquaculture of Grey Mullets*, Cambridge University Press, London, 1981, pp. 361–389.
2. D. Crosetti, W.S. Nelson, and J.C. Avise, *J. Fish Biol.* **44**, 47–58 (1994).
3. J.M. Thomson, in O.H. Oren, ed., *Aquaculture of Grey Mullets*, Cambridge University Press, London, 1981, pp. 1–16.
4. D. Crosetti and S. Cataudella, *World Animal Science Series*, Vol. 34B, Elsevier Science Publishers, Amsterdam, 1995, pp. 253–267.
5. P.-S. Leung, C.-S. Lee, and L.W. Rowland, *Proc. Finfish Hatchery Asia '91*, **3**, 239–244 (1993).
6. M. Ben-Yami, in O.H. Oren, ed., *Aquaculture of Grey Mullets*, Cambridge University Press, London, 1981, pp. 335–359.
7. K.K. Sarojini, *J. Zool. Soc. India* **3**(1), 159–179 (1951).
8. D. Mires, Y. Shak, and S. Shilo, *Bamidgeh* **26**(4), 104–109 (1974).
9. J.R. Sylvester, *J. Fish Biol.* **6**(6), 791–796 (1974).
10. V.S. Durve, *Aquaculture* **5**(1), 53–63 (1975).
11. C.E. Nash and Z.H. Shehadeh, *ICLARM Stud. Rev.* **3**, 87 (1980).
12. C.-S. Lee, C.S. Tamaru, and C.D. Kelley, *Aquaculture* **73**, 341–347 (1988).
13. I.C. Liao, Y.J. Yu, T.L. Huang, and M.C. Lin, in T.V.R. Pillay, ed., *Coastal Aquaculture in the Indo-Pacific Region*, Fishing News (Books) Ltd., London, 1972, pp. 213–243.
14. C.E. Nash, C.M. Kuo, and S.C. McConnel, *Aquaculture* **3**, 15–24 (1974).
15. S. Kraul, *Aquaculture* **30**, 273–284 (1983).
16. H. Eda, R. Murashige, Y. Oozeki, A. Hagiwara, B. Eastham, P. Bass, C.S. Tamaru, and C.-S. Lee, *Aquaculture* **91**, 281–294 (1990).
17. C.-S. Lee and C.D. Kelley, in J.P. McVey, ed., *CRC Handbook of Mariculture*, Vol. 2, CRC Press, Boca Raton, FL, 1991, pp. 193–209.
18. I.C. Liao, D.L. Lee, M.Y. Lim, and M.C. Lo, *Joint Commission on Rural Reconstruction Fish. Ser.* **11**, 30–35 (in Chinese, English abstr.) (1971).
19. C.S. Tamaru, C.-S. Lee, C.D. Kelley, G. Miyamoto, and A. Moriwake, *J. World Aquacult. Soc.* **25**(1), 109–115 (1994).
20. C.S. Tamaru, H. Ako, and C.-S. Lee, *Aquaculture* **104**(1), 83–94 (1992).
21. I.C. Liao, *J. Fish Soc. Taiwan* **5**, 1–10 (1977).
22. C.-S. Lee, C.S. Tamaru, G.T. Miyamoto, and C.D. Kelley, *Aquaculture* **62**, 327–336 (1987).
23. C.M. Kuo and C.E. Nash, *Aquaculture* **5**(2), 119–134 (1975).
24. G.M. Weber and C.-S. Lee, *J. Fish Biol.* **26**, 77–84 (1985).

25. C.D. Kelley, C.S. Tamaru, C.-S. Lee, and A.M. Moriwake, *Proc. 20th Annu. Meet. World Aquacult. Society (WAS)*, Los Angeles, 1989, p. 63.
26. C.D. Kelley, C.-S. Lee, and C.S. Tamaru, in D.I. Idler, L.W. Crim, and I.M. Walsh, eds., *Proceedings of the Third International Symposium on Reproductive Physiology of Fish*, M.S.R.L. St. John's, New Foundland, 1987, p. 203.
27. C.S. Tamaru, C.D. Kelley, C.-S. Lee, K. Aida, and I. Hanyu, *Gen. Comp. Endocrinol.* **76**, 114–127 (1989).
28. C.D. Kelley, C.S. Tamaru, C.-S. Lee, A. Moriwake, and G. Miyamoto, in A.P. Scott, J.P. Sumpter, D.E. Kime, and M.S. Rolfe, eds., *Proceedings of the Fourth International Symposium on Reproductive Physiology of Fish*, University of East Anglia, Norwich, England, 1991, pp. 142–144.
29. C.-S. Lee, C.D. Kelley, and C.S. Tamaru, *Asian Fish. Sci.* **9**(1), 9–20 (1996).
30. I.H. Tung, *Rep. Inst. Fish. Biol. Minist. Econ. Aff. Nat. Taiwan Univ.* **3**(1), 187–215 (1973).
31. C.S. Tamaru, R. Murashige, C.-S. Lee, H. Ako, and V. Sato, *Aquaculture* **110**, 361–372 (1993).
32. J.H.S. Blaxter, in W.S. Hoar and D.J. Randall, eds., *The Physiology of Developing Fish*, Part A; Vol. XI, Academic Press, San Diego, CA, 1988, pp. 1–58.
33. W.A. Walsh, C. Swanson, and C.-S. Lee, *Aquaculture* **97**, 281–289 (1991).
34. R. Murashige, P. Bass, L. Wallace, A. Molnar, B. Eastham, V. Sato, C. Tamaru, and C.-S. Lee, *Aquaculture* **96**(3–4), 249–254 (1991).
35. Y. Oozeki, A. Hagiwara, H. Eda, and C.S. Lee, *Nippon Suisan Gakkaishi/Bull. Jpn. Soc. Sci. Fish.* **58**(7), 1381 (1992).
36. C.S. Tamaru, R. Murashige, and C.-S. Lee, *Aquaculture* **119**, 167–174 (1994).
37. B. Hepher and Y. Pruginin, *Commercial Fish Farming*, Wiley, Toronto, 1981.
38. K.R. Uwate and P. Kunatuba, *A Short Review of World Mullet Culture*, East-West Center, Honolulu, HI, 1983, p. 17.
39. L.V. Dang, in T.V.R. Pillay, ed., *Coastal Aquaculture in the Indo-Pacific Region*, Fishing News (Books) Ltd., London, 1972, pp. 105–108.
40. A. Furukawa, in T.V.R. Pillay and W.A. Dill, eds., *Advances in Aquaculture*, FAO, Rome, 1979, pp. 45–50.
41. C.F. Hickling, *Fish Culture* Faber & Faber, London, 1971.
42. S.W. Ling, in *Coastal Aquaculture in the Indo-Pacific Region*, Fishing News (Books) Ltd., London, 1972, pp. 2–25.
43. J. Azariah, P. Nammalwar, and K. Narayanan, in *Proceedings of the Symposium on Coastal Aquaculture, Cochin, 1980*, Part 4, No. 6. Mar. Biol. Assoc. India, Cochin, 1986, pp. 1463–1464.
44. K. Chiba, *Proc. Spring Meet. Jpn. Soc. Sci. Fishs.*, 1974, p. 1979.
45. I.C. Liao and N.H. Chao, in J.P. McVey, ed., *CRC Handbook of Mariculture*, Vol. 2; CRC Press, Boca Raton, FL, 1991, pp. 117–132.
46. S. Sadek, *J. Aquacult. Trop.* **4**, 165–188 (1989). pp. 391–409.
47. J.H. Wallace, *Aquaculture* **5**, 111 (1975).
48. J.E. Bardach, J.H. Ryther, and W.O. McLarney, *Aquaculture: The Farming and Husbandry of Freshwater and Marine Organisms*, Wiley-Interscience, New York, 1972.
49. J.M. Thompson, *Fisheries*, Synopsis No. 1, CSIRO, 1963.
50. C.R. Rangaswamy, *J. Inland Fish. Soc. India* **5**, 9–22 (1973).
51. T. Rajyalakshmi and D.M. Chandra, *Indian J. Anim. Sci.* **57**(3), 229–240 (1987).
52. S. Sarig, in O.H. Oren, ed., *Aquaculture of Grey Mullets*, Cambridge University Press, London, 1981, pp. 391–409.
53. Y. Oshima, in Y. Oshima, ed., *Suisankouza Series 3: Inland Aquacuture*, Dai Nippon Suisan Kai, Tokyo, 1957, pp. 431–448.
54. N. Nakamura, in N.Y. Kawamoto, ed., *Particulars on Fish Culture*, Suisangaku Ser. 23, Kousesha Kousekaku, Tokyo, 1967, pp. 212–228.
55. L. Cardona, X. Torras, E. Gisbert, and F. Castello, *Isr. J. Aquacult. Bamidgeh* **48**, 179–185 (1996).
56. V.R.P. Sinha, in T.V.R. Pillay and W.A. Dill, eds., *Advances in Aquaculture*, FAO, Rome, 1979, pp. 123–126.
57. V.G. Jhingran and A.V. Natarajan, in T.V.R. Pillay and W.A. Dills, eds., *Advances in Aquaculture*, FAO, Rome, 1979, pp. 532–541.
58. E.E. Brown, *World Fish Farming: Cultivation and Economics*, AVI Publ. Co., Westport, CT, 1977.
59. Y. Pruginin and H. Kitai, *Bamidgeh* **9**(4), 70–75 (1957).
60. A. Yashouv, *FAO Fish. Rep.* **4**(4), 258–273 (1968).
61. T.P. Chen, T.V.R. Pillay, ed., in *Coastal Aquaculture in the Indo-Pacific Region*, Fishing News (Books) Ltd., London, 1972, pp. 410–416.
62. R. Marichamy and S. Rajapackiam, in *Proceedings of the Symposium on Coastal Aquaculture, Cochin, 1980*, Part 1; No. 6, Mar. Biol. Assoc. India, Cochin, 1986, pp. 256–265.
63. N. Nakamura, *Bull. Jpn. Soc. Sci. Fish.* **14**(4), 211–215 (1949).
64. Y. Pruginin, S. Shilo, and D. Mires, *Aquaculture* **5**, 291–298 (1975).
65. D.R. Linder, K. Strawn, and R.W. Luebke, *Aquaculture* **5**, 151–162 (1975).
66. A.E.A. Nour, H. Mabrouk, E. Omar, A.E.K.A. Akkada, and M.A. El Wafa, *Proc. 1st Int. Symp. Aquacult. Technol. Investment Oppor.*, 1993, pp. 560–569.
67. F.F. Shyu and B. Sun Pan, *J. Food Drug Anal.* **1**(2), 155–163 (1993).
68. Food and Agriculture Organization of the United Nations (FAO), *Aquaculture Production Statistics 1987–1996*, FAO Fish. Circ. No. 815, Rev. 10, FAO, Rome, 1998, p. 197.
69. C.L. Lin, S.N. Chen, K.H. Kou, C.L. Wu, and Y.Y. Ting, *Council of Agriculture Fish. Serv.* **4**, 46–60 (1993).
70. M. Lahav and S. Sarig, *Bamidgeh* **24**(1), 3–11 (1972).
71. M. Lahav and S. Sarig, *Bamidgeh* **19**(4), 69–80 (1967).
72. M.C.L. Baticados and G.F. Quinitio, *Dis. Mar. Org.* **37**(1–4), 595–601 (1984).
73. A. Chatterji, B.S. Ingole, and A.H. Parulekar, *Indian J. Mar. Sci.* **11**(4), 344–346 (1982).

See also LARVAL FEEDING—FISH.

MULTICELLULAR PARASITE (MACROPARASITE) PROBLEMS IN AQUACULTURE

ERNEST H. WILLIAMS, JR.
University of Puerto Rico
Lajas, Puerto Rico

LUCY BUNKLEY-WILLIAMS
University of Puerto Rico
Mayagüez, Puerto Rico

OUTLINE

We call the multicellular parasites in the animal kingdom macroparasites because they are, for the most part, large enough to be seen without the use of a microscope. Many species of macroparasites are important in aquaculture for causing destruction, disfigurement, and a reduction in the growth and vitality of cultured organisms.

Modern aquaculture encompasses an overwhelming multitude of cultured species. Even animals as exotic and diverse as marine mammals and reef-building corals are now raised, traded, and sold. The assortment of macroparasites that attack this huge variety of organisms is both numerous and complex. In this contribution, we discuss the macroparasites most important in aquaculture and provide a list of all the macroparasite groups known to bother aquaculture animals (see Table 1). We cite popular references for information on most of those groups we do not discuss for lack of room.

In wild animals, macroparasites are always present, and they constantly tax a portion of the hosts' growth and productivity. [For example, isopod parasites may, amazingly, infect 10% of a fish population, reduce its growth by 15%, and cause the loss of billions of kilograms (1 kg = 2.2 lb) of fish flesh annually in a single fishery (1). And that is just the loss caused by one species of the many that may infect each host!] When we take wild aquatic animals out of their natural environment and place them into aquaculture, most macroparasites are unable to make the transition and die or waste away. Unfortunately, though, this happy scenario does not always occur: Some natural parasites are able to survive and even flourish in the new, biologically peculiar set of circumstances.

Table 1. A Listing of All Known Macroparasites of Aquaculture Organisms[a]

Classification[b] — Common Names	Page
Kingdom Animalia — animals	
Phylum Mesozoa — urine microworms (33,35)	
Class Rhombozoa — cephalopod mesozoans	
Class Orthonectida — invertebrate mesozoans	
Phylum Porifera — sponges (35)	
Phylum Cnidaria — jellyfish etc.	
Class Myxozoa (?) — myxosporidians[c] (1)	
Class Hydrozoa — sturgeon hydroid, etc.	
Class Anthozoa — sea anemone, ctenophore parasites, etc.	
Phylum Ctenophora — tunicate parasite, etc. (35)	
Phylum Platyhelminthes — flatworms	564
Superclass Acoelomorpha — acoels	
Superclass Rhabditophora	
Order Rhabdocoela — mollusc/crustacea turbellarians, etc. (33)	
Order Alloeocoela — fish turbellarian, etc. (33)	
Order Tricladida — elasmobranch turbellarian, etc. (33)	
Order Polycladida — oyster turbellarian, etc. (33)	
Superclass Cercomeria	
Class Temnocephalida — *Macrobrachium* turbellarian, etc. (19)	
Class Udonellea — copepod worm (1)	
Class Cercomeridea	
Subclass Trematoda — flukes and soleworms (33)	
Infraclass Aspidobothrea — soleworms (33)	
Infraclass Digenea — flukes	564
Superfamily Didymozoidea — tissue flukes	565
Subclass Cercomeromorphae	
Infraclass Monogenea — gillworms	565
Order Gyrodactylidea — live-bearing gillworms	566
Order Dactylogyridea — simple gillworms	567
Order Montchadskyellidea (33)	
Order Capsalidea — capsalids	567
Order Monocotylidae (33)	
Order Polystomatidea (33)	
Order Mazocraeidea — complex gillworms	568
Order Diclybothriidea (33)	
Order Chimaericolidea (33)	
Infraclass Cestoidea — tapeworms	568
Cohort Gyrocotylidea (33)	
Cohort Cestoda — tapeworms	
Subcohort Amphilinidea — simple tapeworms (33)	
Subcohort Eucestoda — true tapeworms	
Phylum Nemertea — ribbonworms (35)	
Phylum Rotifera — rotifers (35)	
Phylum Nematoda — roundworms	569
Class Enoplea	
Class Rhabditea	
Phylum Nematomorpha — horsehair worms (35)	
Phylum Acanthocephala — spiny-headed worms	571
Class Archiacanthocephala — of higher vertebrates	
Class Palaeacanthocephala — of vertebrates	
Class Eoacanthocephala — of lower vertebrates	
Class Polyacanthocephala — of fish	
Phylum Annelida — segmented worms	
Class Polychaeta — polychaetes[d] (35)	

Table 1. *Continued*

Classification[b] — Common Names	Page
Class Myzostomida — symbiotic annelids (35)	
Class Hirudinida — leeches	571
Subclass Acanthobdellida — primative leeches	
Subclass Branchiobdellida — crayfish leeches (19)	
Subclass Hirudinea — true leeches	
Phylum Arthropoda — armored animals	
Subphylum Cheliceriformes — spiders, etc.	
Class Arachnida	
Subclass Acari — mites (2), bird and seal ticks (33), etc.	
Class Pycnogonida — sea spiders (35)	
Subphylum Uniramia	
Class Insecta — insects	
Order Anoplura — sucking lice (33,35)	
Order Mallophaga — bird lice (33,35)	
Order Diptera — goby mosquito, etc. (35)	
Order Trichoptera — seastar caddisfly, etc. (35)	
Subphylum Crustacea — aquatic armored animals	572
Class Maxillopoda	
Subclass Ostracoda — seed shrimp (1,19)	
Subclass Copepoda — copepods	572
Subclass Branchiura (?)	
Order Branchiura — fish lice	574
Order[c] Pentastomata — tongueworms	574
Subclass Cirripedia — barnacles	575
Order Thoracica — acorn and goose barnacles (1,19)	
Order Acrothoracica — burrowing barnacles	
Order Ascothoracica — coral, urchin, barnacles, etc.	
Order Rhizocephala — crustacean barnacles (33)	
Subclass Tantulocarida — crustacean ectoparasites (33)	
Class Malacostraca	
Order Decapoda — crabs, shrimp, etc. (33)	
Order Isopoda — isopods	576
Order Amphipoda — amphipods (20,33)	
Phylum Tardigrada — sea cucumber, water bear, etc. (35)	
Phylum Mollusca — seashells, etc.	577
Class Aplacophora — wormlike molluscs	
Class Bivalvia — clams, oysters, etc.	
Order Mytiloida — marine mussels[e] (17)	
Order Unionoida — freshwater mussels, glochida (2), etc.	
Order Veneroida — fingernail- and peaclams (2,14)	
Class Gastropoda — snails and slugs	
Order Mesogastropoda — echinoderm snails, etc. (35)	
Order Neogastropoda — elasmobranch nutmeg, etc.	
Order Pyramidelloida — giant-clam snail, etc.	
Order Nudibranchia — seaslugs	
Phylum Echinodermata — seastars, etc.	
Class Ophiuroidea — coral and crinoid brittlestars, etc. (35)	
Class Holothuroidea — fish, sea cucumber, etc. (35)	
Phylum Vertebrata — backboned animals	

Table 1. *Continued*

Classification[b] — Common Names	Page
Subphylum Pisces — fishes	
Class Agnatha — jawless fishes	
Order Petromyzontiformes — lampreys (1)	
Class Chondrichthyes — cartilaginous fish	
Order Squaliformes — cookiecutter sharks (1), etc.	
Class Osteichthyes — bony fishes	
Order Ophidiiformes — pearlfish (35), etc.	
Order Siluriformes — pencil catfishes, candiru, etc.	
Order Perciformes — remoras (1), pilotfish (1), etc.	

[a] Items with page numbers are discussed in this contribution, and most of the other items have citation numbers that refer to references where you can look up additional information.
[b] Classification systems are difficult, but they are necessary to understand the relationships among the parasites discussed in this chapter and those that were only listed, and to gauge some of the complexities from which this contribution was simplified. This classification system is particularly important, since so much of it has, rather recently, changed drastically and most available reference books are not up to date. [See discussion in (49).]
[c] This classification is disputed and is often placed in its own phylum.
[d] Very few parasitic forms exist, but one has caused mass mortalities of oysters in southern Australia, and another feeds on the blood of eels and other fish in the Mediterranean.
[e] Larval forms cause disease in pen-reared salmonids.

Often, our first reaction is to blindly throw powerful, generalized chemical treatments at the invaders. However, an overdependence on chemical treatments may, ironically, create stronger and more resistant parasites and may eventually render our chemical weapons obsolete and useless.

Then how can we avoid damage to our aquaculture stocks by macroparasites? First, if we keep our culture animals as "happy and healthy" as possible, we will probably see few macroparasite problems. Second, many macroparasites have vulnerable life cycle stages that can be easily eliminated by removing intermediate hosts in the system and by maintaining a flow of water that is sufficient to wash away the intermediate stages of the parasites. Chemical treatments are necessary in emergencies, but by knowing a particular parasite and its life cycle, you can control that parasite with rational and effective management.

Thus, the identification and biology of macroparasites are not mere curiosities and irrelevances to aquaculture. We need to know the identity of each macroparasite that is causing problems in order to chose the proper control measure, and we need to know the biology of the macroparasite in order to devise the simplest and most effective long-term management practices so as to avoid the problem in the future.

METHODS

The terms for the sizes of parasites and the levels and frequency of infection employed in this entry are defined and explained in (1). Treatments are mentioned only briefly in the sections that follow, because the topic is considered in detail elsewhere (see the entry "Disease treatments"). We cite only popular or semipopular books, papers, or electronic references with additional information, as scarce as these may be in the aquaculture-parasite literature. More technical information may be found in Refs. 1–47 or in the references cited in the bibliographies of these works. If all else fails, hire a parasitologist.

FLATWORMS (PLATYHELMINTHES)

Flatworms form a phylum of soft-bodied, bilaterally symmetrical, flattened wormlike animals. Usually, each worm has a set of both female and male reproductive organs (i.e., the worms are hermaphroditic). Flatworms either have a primitive blind gut and a mouth or absorb nutrients through their bodies. They respire through their "skin" (tegument) and possess specialized cells that maintain a water balance and secrete nitrogenous waste products. There are about 20,000 species of flatworms, including fish turbellarians, copepod worms, soleworms, flukes, tissue flukes, gill worms, and tapeworms.

FLUKES (DIGENEA)

Flukes or digeneans (formerly called digenetic trematodes) form an infraclass of flatworms. Flukes reproduce as adults and again as larvae — hence the name "digenetic" or "of two births." They cause serious and fatal diseases in many animals, including humans. Bilharzia diseases infect humans in fresh waters in many tropical regions and are a threat to aquaculturists. Flukes are important fish parasites, serving as an intermediate host for grubs and as a final stage. The adult host for fluke almost never causes problems in aquaculture animals, but grubs can encyst in vital organs and can often become abundant enough to be important. Figure 1 shows the yellow grub, *Clinostomum complanatum*, which causes problems in aquacultured fishes. More than 9000 species of digeneans have been described. Adults range in size from less than 0.2 mm to greater than 10 cm (0.008 to 3.9 in.). One of the largest, *Hirudinella ventricosa*, occurs in the stomachs of wahoos, *Acanthocybium solandri*, and other offshore big-game fishes.

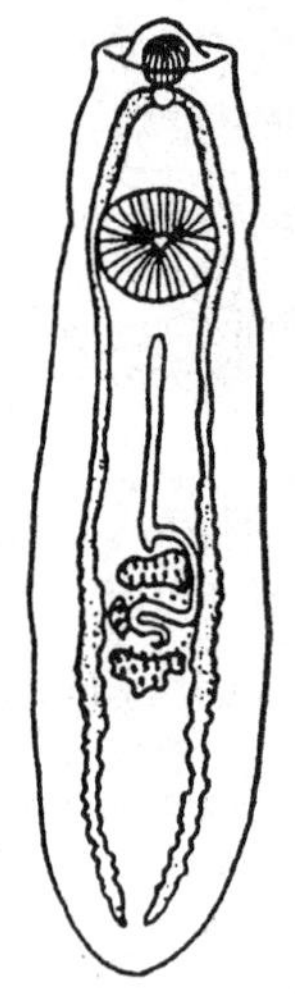

Figure 1. The yellow grub, *C. complanatum*.

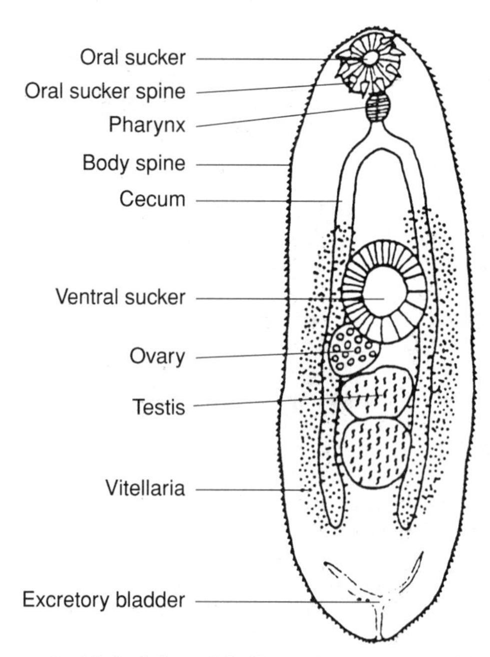

Figure 2. Adult fluke, with distinctive structures labeled.

Flukes usually look like typical flatworms, with a mouth in the anterior region and a blind gut and reproductive and other organs in the trunk region. Unlike generalized flatworms, many flukes have two suckerlike holdfast organs. One (the oral sucker) is located near the mouth, and the other (the ventral sucker, or acetabulum) is usually in the middle of the worm on the ventral side. Great differences in shape, size, and orientation of structures occur in different species. The internal organs illustrated in Figure 2 can be seen if worms are placed in saline wet mounts and viewed with a compound microscope. For study, flukes either must be relaxed and preserved in hot 5% formalin or must be arranged flat in a wet mount on a microscope slide with slight coverslip pressure; then 5% formalin can be slowly drawn into one edge of the coverslip by absorbing the water out of the opposite edge with a paper towel. To identify species of flukes, the specimens must be stained so that the internal organs are distinguished. The animals are usually placed on microscope slides in permanent mounting media under coverslips.

Flukes have a complex life cycle, usually with two or three intermediate hosts, possible transfer hosts, and a final host. Typically, eggs from the body of the fluke pass out of the intestine of the final (definitive) host. The eggs either are eaten by the first intermediate host or hatch into a swimming ciliated larva (miracidium) that infects the first intermediate host, almost always a snail. Once inside the snail, the miracidium transforms into a sporocyst. Each sporocyst asexually produces many larval parasites (rediae), which in turn produce many swimming infective larvae (cercariae) that leave the snail. The cercariae infect the second intermediate host, encyst, and become metacercariae. If the appropriate final host eats this infected host, the cyst is digested and the metacercariae emerge and become adult flukes. There are many variations in the life cycle of the fluke, especially in the asexual phases. Since each fluke has both female and male reproductive organs, self-fertilization is possible, but cross-fertilization is more common.

Flukes are permanent parasites in most marine fishes and in many freshwater fishes, amphibians, reptiles, mammals, and birds. Larval stages occur in a variety of invertebrates and vertebrates. Flukes usually inhabit the intestine, stomach, mouth, or, occasionally, the lungs and other organs. Larval forms occur in almost any tissue. Sometimes predators (false hosts) temporarily support flukes digested from prey, but these soon pass out of the predator.

Didymozoids, or tissue flukes, belong to the subclass Digenea, but their exact taxonomic positions and relationships are not clear. The classification issue will not be resolved until the life cycle *in toto* and especially its early stages are studied. We consider both of these topics next.

Popular references on this subject are (50–53).

TISSUE FLUKES (DIDYMOZOIDEA)

These vivid and often spectacular worms are found encapsulated in the tissues of marine fishes, including greater amberjack and yellowtail, big-game fishes (1), groupers (family Serranidae), and snappers (family Lutjanidae), but a few (probably of marine origin) infect freshwater fishes. The capsules (and the damage they cause) are frequently seen in fishery products, but have been of no consequence in aquaculture. Tissue flukes may cause considerable problems in the culture of dolphin, greater amberjack, and other big-game fishes. The scientific name is from "didymos," which means "double" or "twin," and "zoë," which signifies "life." The name refers to the characteristic two worms that usually occur in each capsule. The bright, usually yellow, color that makes the capsules so conspicuous is due to masses of eggs that occupy much of the hosts' bodies. Figure 3 illustrates *Koellikeria bipartita*, which occurs in greater amberjack and tunas.

Tissue flukes in fish muscle may cause it to deteriorate more rapidly, making the meat less desirable for

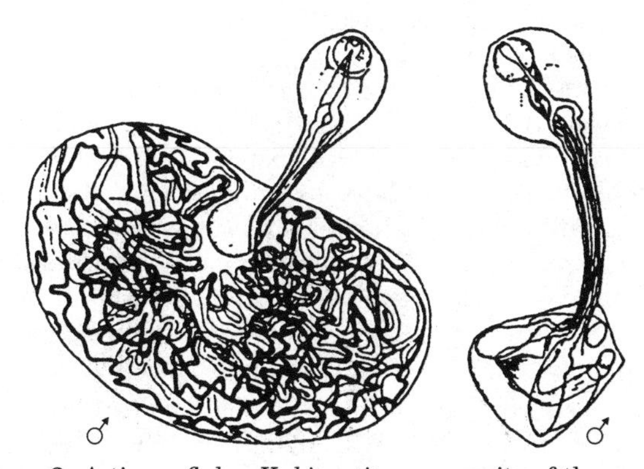

Figure 3. A tissue fluke, *K. bipartita*, a parasite of the greater amberjack and tunas.

consumers. Flesh from a heavily infected Atlantic mackerel, *Scomber scombrus*, has to be discarded in processing, and wahoo in eastern Australia is undervalued due to its reputation for having large tissue flukes (a third of the animals are infected). Heavy to very heavy infections are routinely found in big-game fishes. In the Indian Ocean, 100 or more capsules are found in every wahoo and in half the tuna. Superinfections of up to 1167 capsules have been reported on occasion in tuna from the southern Gulf of Mexico.

Approximately 200 species are known. Encapsulated pairs of various species range from a few millimeters (1 mm = 0.04 in.) to the size of an adult human fist. They are usually colored yellow or orange, but blue capsules occur in wahoos (*A. solandri*) from Australia. Worms vary in length from a few mm to over 12 m (0.08 in. to 13.2 yd). The body of the worm may be elongated, filamentous or ribbonlike, and intricately tangled, or it may be divided into a narrow anterior and a broadly swollen posterior.

The complete life cycle is not known. Eggs are released annually from encapsulated worms in the gills, ovaries, or other exposed locations. More deeply embedded or inaccessible worms may release eggs only after the death of the host and survive passage through the gut of the predator. In some cases, the capsule and surrounding tissue break down or ulcerate when the tissue flukes mature, releasing both eggs and adults. A nonciliated miracidium with two or more circles of spines around the oral sucker hatches from the egg. Tissue fluke cercariae occur in plankton, and metacercariae parasitize arrowworms, barnacles, copepods, krills, squids, small bony fishes, and sharks.

All species are hermaphroditic (have both sexes in each worm), but often associate in pairs, one individual of which has more developed female organs and a smaller partner that has more developed male organs. Other pairs are fused to each other in the genital region.

Popular references on this subject are (1,12).

GILLWORMS (MONOGENEA)

Simply stated, gillworms are the most important macroparasites in aquaculture. No other macroparasites come close in the amount of damage they do to fishes, the terror they pose to aquaculturalists, the diversity of their species, and the difficulty they present in controlling them. The name "Monogenea" means "born once" and refers to a simple life cycle. In very heavy infections, gillworms can kill captive fishes and, occasionally, wild ones. Heavy infections cause irritation to the fish, excess mucus production, cell proliferation, and smothering of the gills, resulting in death. Complex gillworms routinely cause tissue damage even in light infections. A treatment of 250 parts per million (ppm) of formalin for one hour will remove most gillworms (15). More than 1500 gillworm species have been described, but they are probably only a small percentage of those existing. Gillworms include more host-specific species than any other group of parasites. This property often makes them more easy to control in aquaculture, but we have found some breakdown in specificity after long culture with similar hosts (2).

Adults range from 30 μm to 20 mm (0.0012 to 0.79 in.) in length and are translucent, cream, or pink. In general, live-bearing gillworms are microscopic, with simple gillworms microscopic to tiny and capsalids and complex gillworms small to large (1). Both simple, live-bearing gillworms and capsalids have a distinct attachment organ on their posterior end called a haptor (or opisthaptor) with hardened anchors or specialized clamps to pierce the epithelium of the host and grab hold of it. Sclerotized marginal hooks often surround the haptor, and bars, disks, scales, or spines may occur on or near the haptor. The head may have eyespots or specialized holdfast organs (Fig. 4). Complex gillworms have large suckers or numerous clamps adapted for attachment. Most reproduce by laying eggs that hatch free-swimming, ciliated larvae (oncomiracidia) which quickly mature and then find and attach to a host. Live-bearing gillworms rapidly produce new worms from embryos that are born able to immediately attach to the parental host with no intervening free-swimming stage. Since none require intermediate hosts, they all multiply rapidly. When intensive culture crowds fish together, most gillworm offspring survive and can quickly kill fishes. Gillworms are permanent (33) parasites in the gills, in the mouths, or on the bodies of fishes. In general, the relatively smaller live-bearing and simple gillworms occur naturally on a host in greater numbers than the relatively larger capsalids and complex gillworms. This difference may have something to do with the carrying capacity of the host. Typically, live-bearing gillworms prefer the skin and fins of hosts; simple gillworms the gills; capsalids the skin and the gill, mouth, and nasal cavities; and complex gillworms the gills. Some occur in the nares, pockets in the lateral line, or, rarely, the gut of fishes. Some species occur in the urinary bladder of fishes, frogs, or turtles; the intestine of tilapias; or even the eye of hippopotamuses. Most gillworms feed on mucus or sloughed-off epithelial cells, though some complex gillworms feed on blood. Gillworms are common on fishes in all aquatic environments. Live-bearing gillworms are largely limited to freshwater fishes, but occur on some euryhaline and brackish-water hosts. Simple gillworms appear to have more species on freshwater fishes, but more genera on marine fishes, which could indicate a marine origin and a more recent spread to fresh waters. Simple gillworms are found in almost all aquatic habitats, with the exception of the offshore pelagic realm (1). Their absence from that region may be due to the inability of their free-swimming larval stages to find hosts in the open ocean. Capsalids and complex gillworms are marine animals, except for a few species that parasitize freshwater fishes of obvious marine origin that seem to have taken some of their parasites with them to fresh waters.

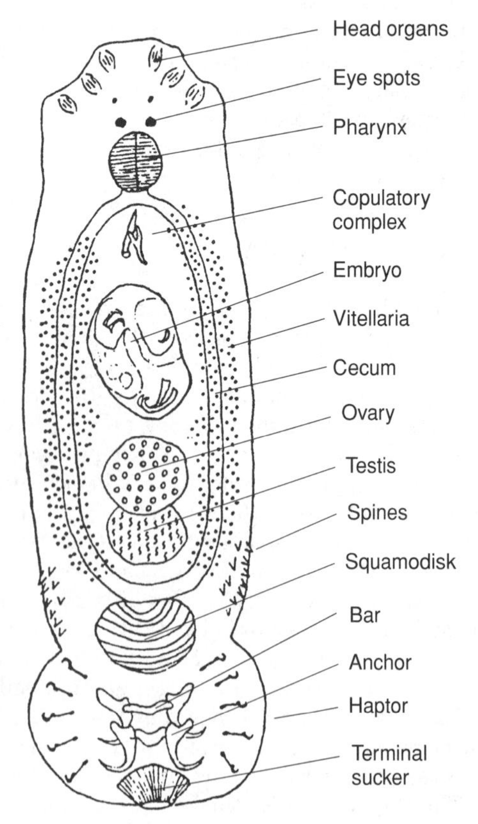

Figure 4. A generalized gillworm, with distinctive structures labeled.

A popular reference on this subject is (52).

LIVE-BEARING GILLWORMS (GYRODACTYLIDEA)

Gyrodactylids, largely in the genus *Gyrodactylus*, have simple attachment organs: a haptor with one pair of anchors, a connected shield, and, often, interconnecting bars and accessory bars or plates. These worms differ from other gillworms by having embryos in their bodies. They are generally smaller than simple gillworms and much smaller than capsalids and complex gillworms. Figure 5 illustrates *Gyrodactylus cichlidarum*, which causes problems in tilapia culture.

Live-bearing gillworms can normally be found in wet mounts of skin scrapings or clippings of gills observed with a compound microscope. Their absence in samples does not assure that the associated populations of fishes are free from these worms: Live-bearing gillworms are small, and when they occur in low numbers, they can often be difficult to detect. To identify species, the parasites can be relaxed in 1 part of formalin to 4000 parts of water until they are dead, after which they may be fixed in 5% formalin and mounted in glycerine jelly. Relaxed specimens can be stored in 5% formalin.

In heavy infections, the attachment of the hooks causes skin or gill irritation and a heavy production of mucus. The skin may be flecked with white patches, especially behind the fins, and the gill filaments may thicken. Fishes may also exhibit flashing behavior and scrape their bodies on the sides of containers. Swimming may alternate between wild activity and lethargy. Secondary bacterial infections can enter damaged areas, further weakening the fish. In confined tanks and raceways, live-bearing and simple gillworms sometimes rapidly increase their populations on fishes. Often, the simple management technique of increasing the flow of water will flush early life-cycle stages of these gillworms out of the tanks and slow their

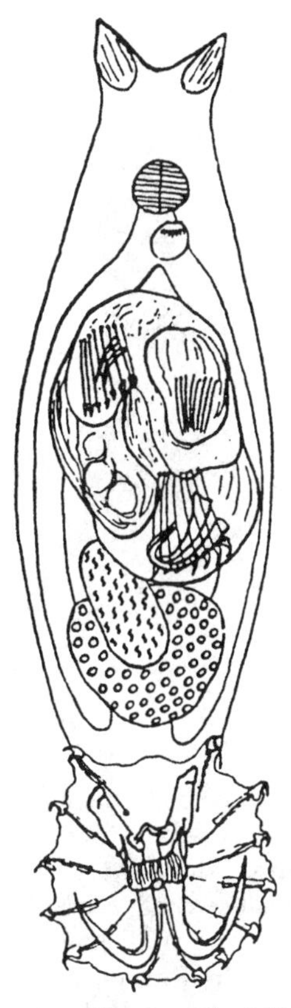

Figure 5. A live-bearing gillworm, *G. cichlidarum*, of tilapias.

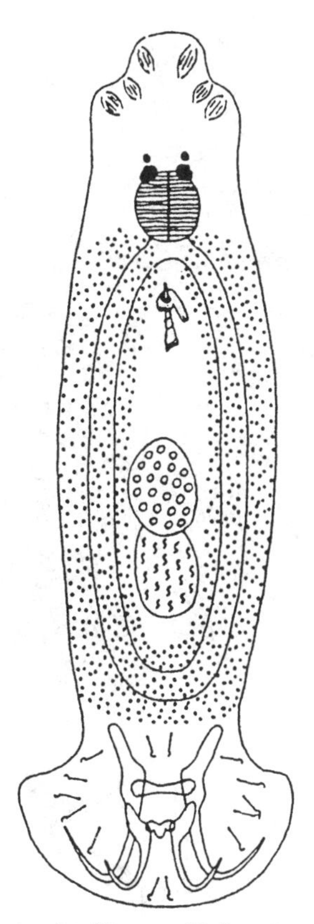

Figure 6. A simple gillworm, *H. furcatus*, of sunfishes.

accumulation on the bodies and gills of fishes. No simple treatment or U.S. aquaculture-legal series of treatments will eliminate all gillworms; treatments merely reduce the levels to tolerable loads for fishes. Gillworm-free stocks can be established in fishes with host-specific gillworms, albeit with difficulty, but this is pointless if everyone else is using "dirty" stocks. Once a gillworm becomes established in a widely cultured fish species, it is almost impossible to eliminate. This fact should be kept in mind when one is contemplating importing fishes to new geographic areas.

SIMPLE GILLWORMS (DACTYLOGYRIDEA)

Dactylogyrids occur in a great variety of shapes, sizes, and forms and in a myriad of different host genera and species. No single genus is representative, although *Dactylogyrus* may be most ubiquitous and important in fresh waters and *Ancyrocephalus* or *Haliotrema* in salty waters. The dactylogyrids have simple attachment organs, usually a haptor with one or two pairs of anchors with interconnecting bars. They are distinguished from live-bearing gillworms by the fact that they lack an embryo in their bodies and are generally much smaller than capsalids and complex gillworms. Figure 6 illustrates *Haplocleidus furcatus*, which causes problems in the culture of basses and bream (sunfishes, family Centrarchidae). Simple gillworms can be collected, examined, and treated in the same manner as the live-bearing gillworms.

CAPSALIDS (CAPSALIDEA)

These relatively large, broad, and flat gillworms attach to the host with a large cup-shaped haptor on the posterior of the worm and two smaller suckers on the anterior. A microscopic pair of anchors are found on the haptor. Figure 7 illustrates the notorious "killer capsalid," *Neobenedenia melleni*, which causes problems in the cage culture of fishes in Atlantic tropical and subtropical waters. We have seen thousands of these worms covering tilapias reared in seawater, upon which they look like the scales of the fish. This capsalid caused the failure of otherwise promising efforts to raise tilapias in seawater cages and ponds in the Bahamas, Jamaica, and Puerto Rico (2). It was unnecessarily introduced to Hawaii, Hong Kong, Okinawa, and the main islands of Japan along with greater amberjack fry in mariculture (1).

Larger adult worms are plainly visible on the host, and small or developing worms can be found in wet mounts of skin scrapings observed with a compound microscope. Capsalids will quickly leave a captured fish and can often be found in the sediment of plastic bags containing fish samples. Their absence in samples does not assure that the associated populations of fishes are free from capsalids: These worms can hide effectively in the nares, mouth, and gill chambers of fishes.

Capsalids and complex gillworms often do not relax well in 1/4000 formalin. Most can be relaxed by freezing. They can also be flattened in a wet mount on a microscope slide, as described previously for flukes. They may be mounted in glycerine jelly to distinguish their hard parts, but

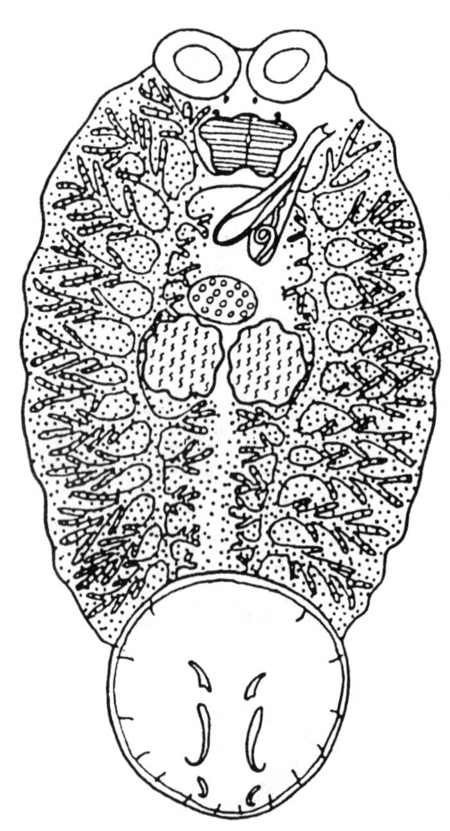

Figure 7. The killer capsalid, *N. melleni.*

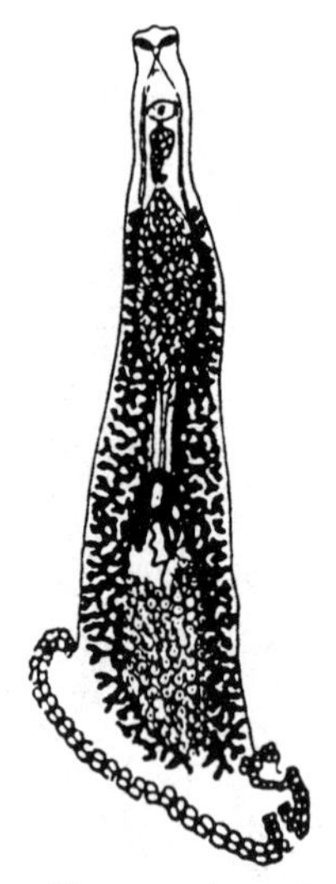

Figure 8. A complex gillworm, *A. mcintoshi*, of the greater amberjack.

additional specimens must be stained, again in a manner similar to the way flukes are stained, to discern internal structures. Like the live-bearing gillworms, hosts infected with capsalids can be treated with formalin. Freshwater dips also are effective.

Many capsalids are less host specific than other gillworms. Often, these worms infect fishes in a genus, a family, or many families. Those capsalids that lack host specificity are the most dangerous to aquaculture fishes.

COMPLEX GILLWORMS (MAZOCRAEIDEA)

These worms have more complex haptors than other gillworms. Most complex gillworms can easily be seen with the naked eye. They have intricate attachment organs composed of a series of complicated clamps or suckers, often on extensions of a complex haptor. Figure 8 illustrates *Allencotyla mcintoshi* from cultured greater amberjack. The complex gillworms produce few, relatively large eggs and thus increase in numbers more slowly than other gillworms. However, they feed on blood and cause considerable damage to the gills of their hosts. Only slight increases in abundance can injure or kill fishes. These worms can be prepared, studied, and controlled in the same manner as noted for capsalids.

TAPEWORMS (CESTOIDEA)

Tapeworms or cestodes form a large infraclass of the flatworms, or platyhelminths. The common name comes from the long series of body segments, which resemble a tape measure. Most adult tapeworms look like long, flat cooked noodles. Most larval forms look like the tiny pieces of noodle that stick in the colander. A giant, broad tapeworm that lives in the body cavity of freshwater fishes in Europe is apparently routinely eaten by humans, who mistake it for parts of the fish! Tapeworms can reduce the growth and affect the reproductive success of fishes. Most do not survive under culture conditions, but some destructive exceptions are the bass tapeworm, *Proteocephalus ambloplitis*, whose larvae infect a variety of freshwater fishes, and the Asian tapeworm, *Bothriocephalus acheilognathi* (Fig. 9), which parasitizes fishes cultured for bait in the United States and tilapias in Cuba.

Some tapeworms that infect humans occur as immature or larval forms in freshwater fishes. A small boat fishery for tunas on the northwest coast of Puerto Rico was closed by the government because of reports of "wormy" flesh in these food fishes. When we were contacted to examine these parasites, we found that they were larval tapeworms, and we recommended that the fishery be immediately reopened. We knew that previous experimental attempts to infect mammals with similar larvae failed because the parasites mature only in sharks and rays. The fishery was quickly reopened, averting major losses to local fishermen. Occasionally, we receive similar samples of wormy fillets from individuals. People do not like seeing larval tapeworms moving around in the flesh of their fish, but most of those that occur in marine fishes are relatively harmless (particularly when cooked). For example, *Pseudogrillotia zerbiae* larvae, which are found in the flesh of the greater amberjack (Fig. 10), offend humans, but these would not develop in cultured amberjack.

More than 5000 species of tapeworm are known. Adults range from less than a mm to more than 30 m (0.04 in. to 33.3 yd) in length. Tapeworms usually consist of a chain of segments (proglottids), each of which has a set of female and male reproductive organs. The segments

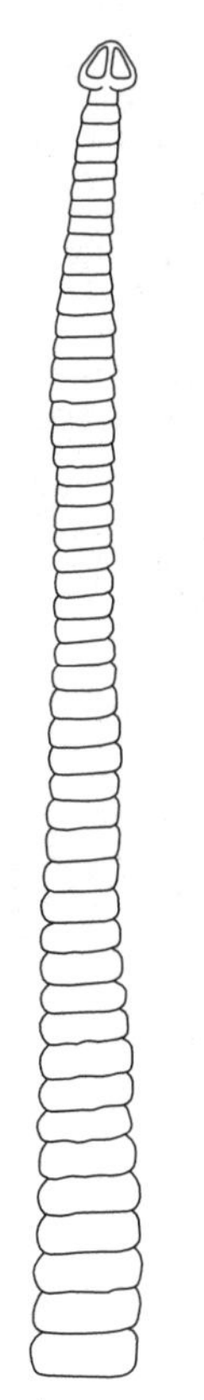

Figure 9. The Asian tapeworm, *B. acheilognathi*.

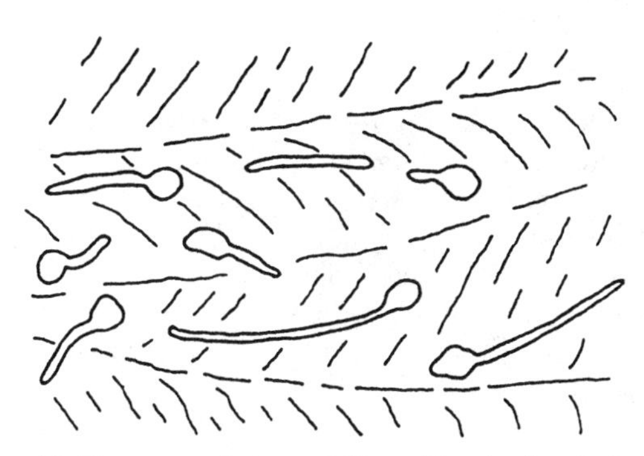

Figure 10. Tapeworm larvae of *P. zerbiae* in the flesh of the greater amberjack, *Seriola dumerili*.

are continuously budded in the anterior portion of the body or neck and enlarge and mature as they slowly move posteriorly. The scolex or "head" on the anterior end is usually armed with various combinations of suckers, hooks, bothridia (outgrowths), or bothria (sucking grooves) for attachment in the host intestine. Eggs escape through pores, or a whole, mature egg-filled segment may break off and pass out of the intestine. A successful tapeworm may produce millions of eggs over its lifetime.

In aquatic life cycles, larvae (ciliated coracidia) that hatch from the egg are eaten by the first intermediate host (insect, crustacean, or annelid) and become elongate procercoids. This host is subsequently eaten by a vertebrate (the second intermediate host), and the larvae develop into partially differentiated plerocercoids or plerocerci, which can be passed from one host to another when an infected fish is eaten by another fish. Feeding infected viscera to fish therefore can greatly concentrate or increase the intensity of infection by these worms. Indeed, we have seen this practice cause a superinfection in caged red hind, *Epinephelus guttatus*, in Puerto Rico and in caged fishes raised in the northern Gulf of Mexico. If viscera must be used as fish food, they should be cooked or frozen for several days to kill parasites. When the correct final or definitive vertebrate host eats the second intermediate host, the adult tapeworms develop in the intestine.

As mentioned, most tapeworms have both female and male sex organs in each proglottid. A few have separate sexes. Tapeworms occur in all kinds of vertebrates and in all habitats around the world. All are permanent (33) parasites. Because the intestine of these worms has been lost through evolution, food from the gut of the host is absorbed directly through the body wall. Few species of adult tapeworms are found in marine bony fishes, but many species of larval tapeworms are found in the intestinal tract, often in large numbers, and a few are encapsulated in the tissues. Most of these larval forms are found as adults in sharks or rays. A necropsy is necessary to find adults or larvae in the gut and encapsulated larvae in internal organs. Adult tapeworms can be relaxed in tap water until they no longer react to touch, after which they may be preserved in 5% formalin. Specimens must be stained as has been described for flukes in order to distinguish the internal organs for identification purposes.

Popular references on this subject are (38,54).

ROUNDWORMS (NEMATODA)

Roundworms, also called threadworms or nematodes, constitute a phylum. Along with flatworms and spiny-headed worms, they are sometimes called helminths. Few roundworms cause problems in aquaculture, because their complex life cycles cannot be completed in culture facilities, but some damaging exceptions occur. *Camallanus cotti* (Fig. 11) causes disease in cultured aquarium fishes and has been spread around the world. The Asian roundworm devastated the culture of European eels (46), *Anguilla anguilla*, and has spread to the American eel, *Anguilla rostrata* (48). Eel culture is increasing, and the potential for the introduction of this destructive parasite into new regions is great.

Roundworms cause serious diseases and even death in humans. The recent increase in popularity of Japanese raw-fish dishes has caused a concomitant increase in the number of fish-roundworm-related illnesses around the world. Modern refrigeration of fish catches has also allowed dangerous roundworms that would have been discarded by quick cleaning to migrate from the gut and mesenteries into the edible flesh. The unwise practice of swallowing live fish has produced severe gastric distress in humans when roundworms from fish burrowed through the intestinal wall into the person's body cavity. In one case, a roundworm from the flesh of a Hawaiian jack penetrated a wound in the hand of a man cleaning the fish.

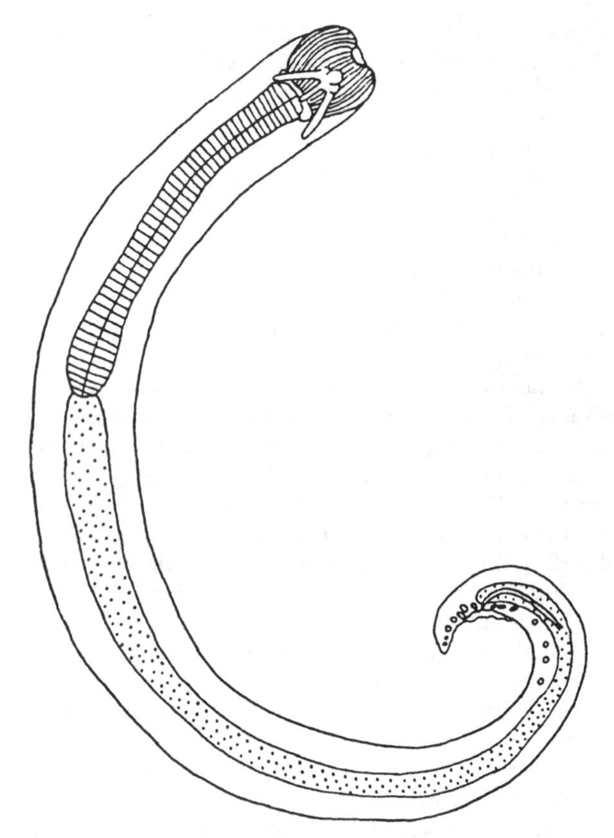

Figure 11. A roundworm, *C. cotti*, of cultured aquarium fish.

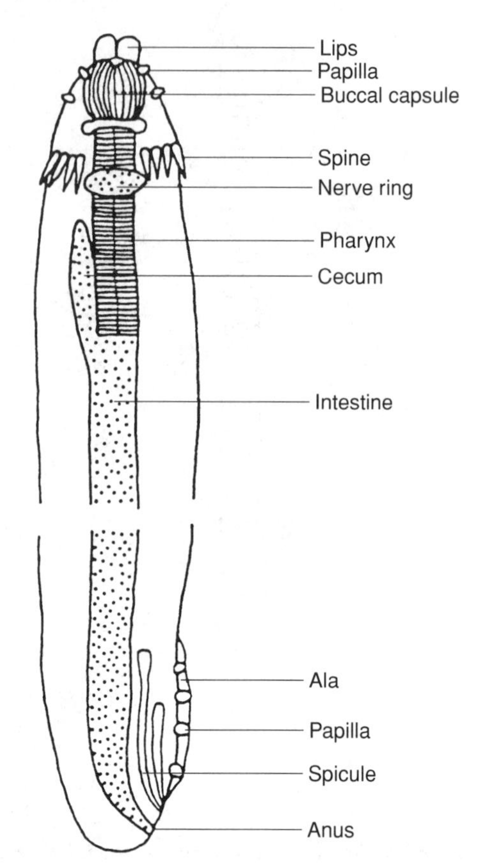

Figure 12. A generalized nematode, with the distinctive structures labeled.

The entry was painful, and the worm could be removed only with surgery.

More than 12,000 species have been described of the 500,000 to 2 million roundworms that probably exist. They are one of the most abundant multicellular organisms on earth, both in number of individuals and in number of species. Roundworms occur in such high numbers in almost all vertebrates and invertebrates, that it has been suggested that the shapes of all living animals could be seen from space merely by seeing the mass of worms in each animal!

Most adult free-living forms are small to microscopic, but parasitic forms are large, up to 8 m (8.6 yd) long. As the name implies, roundworms are circular in cross section. The body is nonsegmented, elongate and slender, often tapered near the ends, and covered with cuticle. Three to six lips of various shapes surround the mouth. The digestive tract is complete, the musculature has only longitudinal fibers, and a pseudocoel (false body cavity) is present. A nerve ring is usually visible in the anterior end of the body. The sexes are separate. The male has a cloaca, a pair of chitinized copulatory structures (usually spicules), often with a variety of papillae, alae (long, thin flaps of cuticle), and suckers in or on the posterior end. All of these male structures are important in identifying species (Fig. 12). Eggs are released into the intestine of fishes or, through holes in the host skin in tissue-dwelling roundworms, into the water. The eggs of some species contain developed roundworms, while those of others are expelled in a less developed stage. Some larvae are eaten by fishes and develop directly into an adult. Usually, larvae must go through two to five molts in one or more crustacean or fish intermediate hosts. Roundworms are found in all marine, freshwater, and terrestrial habitats. Flying insects, birds, and bats take them into the skies.

A necropsy is needed to find the larvae in the organs and mesenteries and the adults in the stomach and intestine. Roundworms can be relaxed in acetic acid and stored in a mixture of 70% ethanol with 5% glycerine. Adult roundworms from tissues of fishes are exceedingly delicate and tend to explode if placed in freshwater or preservatives. These small worms can be put into a steaming 0.8% saline, 5% formalin solution. Once fixed (15 minutes for small worms and up to 24 hours for large ones), worms can be rinsed in freshwater and slowly transferred into gradually increasing concentrations of ethanol, until they are stored in a mixture of 70% ethanol and 5% glycerine. Roundworms are usually examined in wet mounts. Semipermanent mounts may be prepared using glycerine jelly. Most larval species are difficult to identify, but most genera can be readily determined. Knowledge of the entire life cycle of a roundworm may be necessary to identify a given species of larvae.

No treatment is possible for roundworms in the body cavity or tissues of fishes, and treatment is seldom necessary for intestinal roundworms. Worms that perforate the intestine of humans must be surgically removed.

A popular reference on this subject is (41).

SPINY-HEADED WORMS (ACANTHOCEPHALA)

These worms form a small phylum in the animal kingdom. The name "acanthocephala" means "spiny headed." Despite their formidable and highly destructive armaments (Fig. 13), they seldom, if ever, cause problems in aquaculture animals, because their complex life cycles cannot be completed under culture conditions. Figure 13 illustrates *Rhadinorhynchus pristis*, which occurs in the dolphin (*Coryphaena hippurus*) and other big-game fishes. All spiny-headed worms are permanent (33) parasites in the intestine of most vertebrates, including humans. Over 1000 species are known.

Adult females vary from 1 mm (0.04 in.) to longer than 1 m (1 yd, 3.4 in.), but are usually about 2 cm (0.8 in.) long. Males of the same species are typically smaller than females. Spiny-headed worms may be white, yellow, orange, or red in color. (*Pomphorhynchus lucyae* seems to absorb orange pigments from crayfish in the intestines of fishes inhabiting southeastern U.S. coastal waters.) They are bilaterally symmetrical and unsegmented, and they attach to the gut of their host with a globular or cylindrical, protrusible spiny proboscis. The proboscis pops out like an everting plastic glove, and the spines fold out and lock like a compact umbrella. Muscles invert the proboscis, and a hydraulic system (lemnisci) pops it back out (Fig. 14). Some species have spines on the body as well (Fig. 13). The sexes are separate, fertilization is internal, and embryos develop in the body of the female. Shelled larvae (acanthors) are shed into the intestine of the host, pass out with the fecal material, are eaten by a crustacean, insect, mollusk, or fish intermediate host, and develop first into an acanthella and then into an encysted cystacanth larva in the second intermediate host. When the final host consumes an infected intermediate host, the cystacanth develops into an adult in the intestine. Adults absorb nutrients from the contents of their hosts' guts. Proboscis spines (or hooks) cause some mechanical damage, which is serious only in a heavy infection. Treatments seldom are necessary. Natural infections in most fishes usually consist of only a few worms per host. Some inshore New England fishes are routinely infected with hundreds to thousands of spiny-headed worms. The worms rarely harm humans, since they are usually discarded with the intestine and other internal organs when fish are cleaned, but contamination is possible. Thorough cooking kills these parasites. Tuna, salmon, and other fishes known to harbor dangerous worms are prized ingredients of Japanese raw-fish dishes, but most of these products are frozen long enough to kill the parasites before the dishes are served "fresh" in Japan.

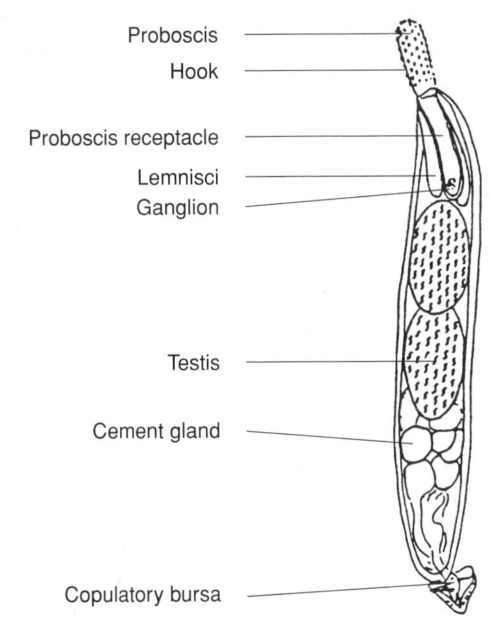

Figure 14. A spiny-headed worm, *Acanthocephalus alabamensis*, with distinctive structures labeled.

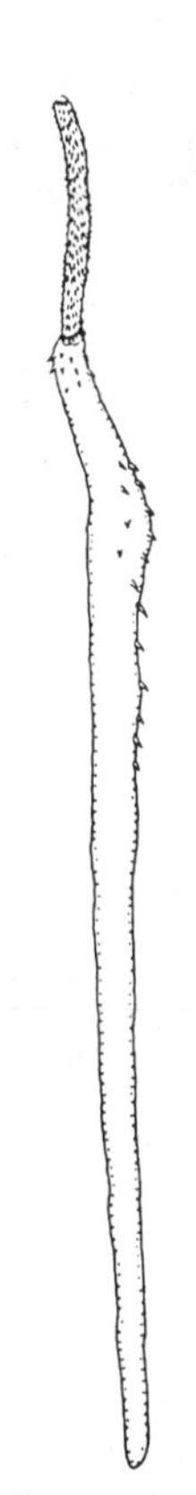

Figure 13. A spiny-headed worm, *R. pristis*, of the dolphinfish, *C. hippurus*.

A necropsy is necessary to find spiny-headed worms in the intestine. For routine examinations, the worms can be identified in wet mounts. For more detailed study, the proboscis must be fully everted before preserving the animals. Worms must be refrigerated in distilled water or freshwater for 12–24 hr before preserving them in 5% formalin. The thick cuticle of these worms does not allow alcohol solutions or stains to penetrate them readily. The cuticle must be pierced before dehydrating the worms in alcohol solutions and then staining them.

LEECHES (HIRUDINIDA)

Leeches form one of the classes of segmented worms, or annelids. The most familiar member of this phylum is the earthworm. Leeches are often called bloodsuckers for their feeding habits. They are sometimes used as hook-and-line bait for fishes. Predacious leeches have also been used to control snails which have parasites that cause swimmers itch (the cercariae of a bird fluke that attack and irritate the skin of humans). Leeches occasionally invade aquaculture systems, and their direct development allows them to build up their numbers and cause problems. Inexplicably, they rarely damage aquaculture organisms, but a few spectacular cases have occurred. Figure 15 illustrates *Myzobdella lugubris*, which has killed fishes and crabs in the continental United States and Puerto Rico (2).

More than 300 species of leeches have been described. Adults vary from 0.5 to 45 cm (0.2 in. to 18 in.) in length, although they can greatly change their size and shape through powerful muscle contractions and enlargement during feeding. A leech's body is pigmented (sometimes brightly), depressed (dorsoventrally flattened), and made up of 34 segments that may be further subdivided by shallow creases or lines into annuli. One sucker occurs on the posterior end of the body, and another usually is found on the anterior end surrounding the mouth. A poorly formed saddle-shaped midportion (clitellum) functions in copulation and cocoon formation. The body cavity (coelom) is reduced to a few channels. Eggs are usually deposited in cocoons formed of epithelial tissue and mucus. Development is direct (i.e., there are no larval stages). The complex reproductive system of the animal contains both female and male sexual organs (i.e., leeches are hermaphroditic). Leeches occur in fresh, brackish, and marine waters and also in moist conditions on land around the world, except in Antarctica. They can crawl and attach with their suckers, and many are able to swim. Leeches come in three varieties: permanent (33) to temporary parasites, predators, and scavengers. Parasitic forms feed on whole blood of crustaceans, fishes, amphibians, reptiles, birds, and mammals. They are often vectors of pathogenic protozoa, roundworms, and tapeworms. They have also been accused of spreading lymphocystis disease and bacterial diseases between hosts.

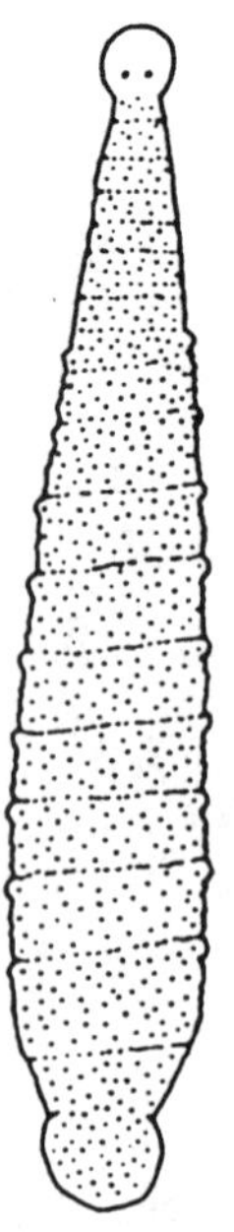

Figure 15. A leech, *M. lugubris*, that causes kills of fishes and crabs (2).

CRUSTACEANS (CRUSTACEA)

Crustaceans are one of the subphyla of animals with hard, segmented shells (exoskeletons), something like medieval knights in armor (arthropods). They are largely aquatic, whereas insects, spiders, and their allies (arachnids) are mostly terrestrial. Crustaceans generally have two pairs of antennae, respire through gills or the body surface, have paired, segmented, usually biramous appendages anteriorly or throughout their length. More than 40,000 living species and many fossil species have been described, including seed shrimp, copepods, fish lice, tongueworms, barnacles, isopods, and amphipods.

COPEPODS (COPEPODA)

Copepods form a subclass of the crustaceans. The common name "copepod" means "oar foot." Copepods occasionally cause significant losses of aquaculture fishes. They parasitize or otherwise associate with many kinds of invertebrates, but are not an important problem in the aquaculture of invertebrates. Two of the most destructive types of copepods in freshwater fish aquaculture are *Ergasilus* spp. (Fig. 16) and the anchorworm, *Lernaea cyprinacea* (Fig. 17). The anchorworm is an international superparasite that causes problems in fishes around the world, from culture ponds near the equator in Brazil to culture facilities near the Arctic Circle in Japan. *Ergasilus labracis* also damages pen-reared salmonids in brackish water and saltwater. Relatively large salmon copepods (caligids [sometimes erroneously called salmon lice or sea lice, common names reserved for branchiurans]) cause losses to salmonids (family Salmonidae) cultured in cages and pens, and these parasites transmit microbial diseases.

Copepods are found in marine water and freshwater; most are free living and are very important sources of food for a variety of aquatic life. Approximately 10,000 species have been described, and about 2,000 of these parasitize fishes. Many of the species that parasitize fishes remain to be named. Copepods range in length from 0.5 to 25 cm (0.2 in. to 10 in.), but most are less than 1 cm (0.4 in.) long. Egg strings of some species may exceed 60 cm (2 ft). Body shapes vary greatly, from a generally cylindrical shape to a flattened or saucer shape. Theoretically, the body is divided into 16 segments, or somites (five head or cephalic somites, seven thoracic somites, and four abdominal somites), but most of these are fused together, combined, or overlapped, so they cannot be seen. The first six to nine somites are fused into an expanded "head," cephalothorax, or cephalosome, and the remainder are variously fused or separated into thoracic and abdominal units. Appendages are modified into mouthparts and other structures in the cephalosome and legs and other appendages in the thorax. The abdomen has no appendages and usually terminates in a bifurcate tail with projections (caudal rami). We illustrate females,

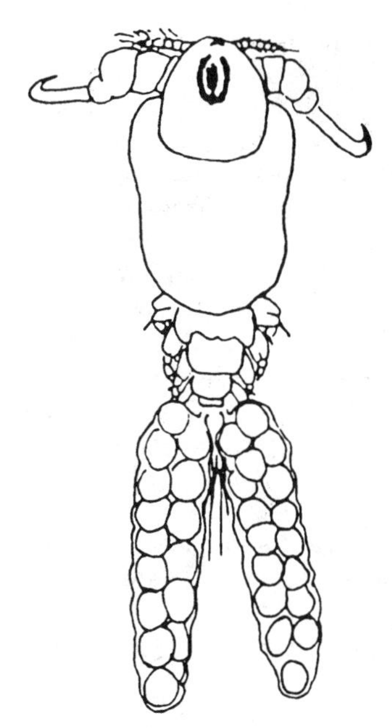

Figure 16. The copepod, *Ergasilus caeruleus*, of striped mullet, *Mugil cephalus*.

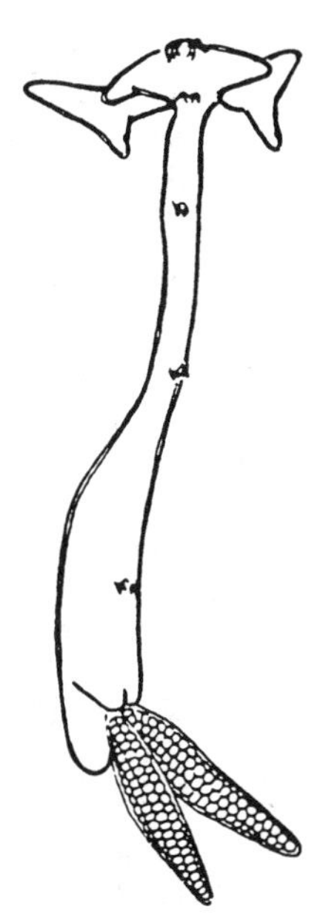

Figure 17. The anchorworm, *L. cyprinacea*, an international superparasite.

use their morphological characters, and seldom mention males, because females are usually larger, are more available, and have more diagnostic characteristics.

In most parasitic forms, the life cycle is direct, but typically involves a series of free-swimming planktonic stages. Some copepods have intermediary hosts, such as *Pennella* spp. on squids and shark copepods embedded in coral-reef fishes. The hatching nauplii can often be observed merely by holding egg strings in bowls of seawater until the eggs hatch. Nauplii of many species have never been described or drawn. Raising the other planktonic stages of parasitic copepods is more complicated, but can be accomplished. A newly hatched nauplius of *Caligus elongatus*; an early nauplius, later nauplius, and metanauplius of *C. bonito*; and a chalimus of *C. elongatus* attached to a host scale are illustrated in Figure 18. The sexes are separate, with males often much smaller than females. The life-cycle stage that first attaches to fishes (the copepodid—sometimes chalimus) can be very damaging to young fishes. Copepods frequently occur on the gills or skin of fishes, but highly specialized species burrow into the flesh or head sinuses or crawl into the nose (nares, nasal fossae, or lamellae) or eyes (orbits). They also associate with or parasitize a variety of invertebrates. Large copepods that are parasitic on fishes are capable of biting humans, but such injury has seldom been reported. Indeed, most aquaculturalists would not admit being attacked by a mere copepod! These parasites can be of value to humans: Eskimos eat the *Gadus morhua*, a giant parasitic copepod that attacks the gills of Atlantic cod. Those on fishes are usually permanent (33) parasites, feeding on mucus, sloughed epithelial cells, and

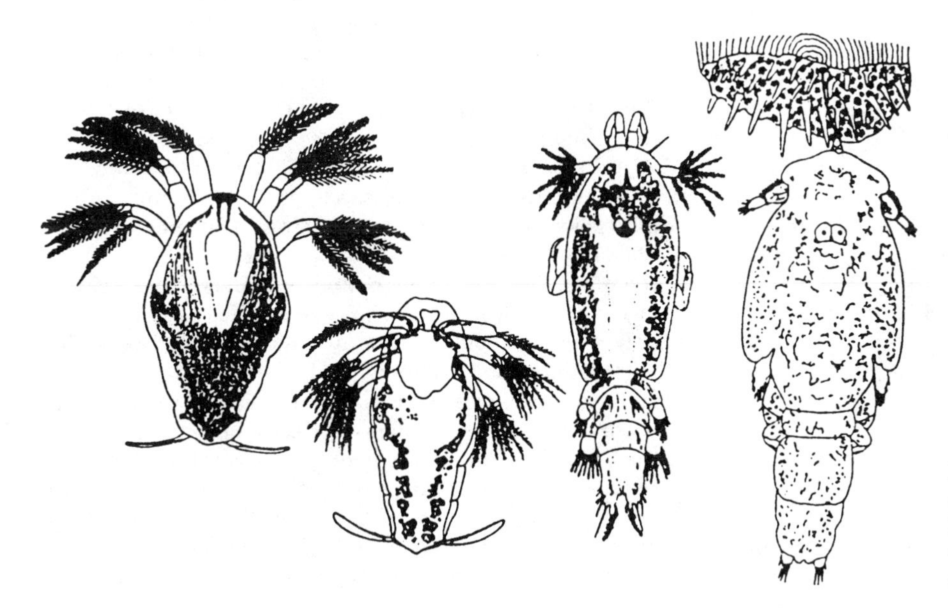

Figure 18. A newly hatched nauplius of *Caligus elongatus*; an early nauplius, later nauplius, and metanauplius of *C. bonito*; and chalimus of *C. elongatus* attached to a host scale.

tissue fluids. Copepods can directly transmit microbial diseases.

Copepod specimens can be preserved and stored in 70% ethanol. They can be examined in alcohol or in a mixture of alcohol and glycerine to clear some structures. Smaller specimens or parts can be mounted in glycerine jelly for convenience in handling.

Popular references on this subject are (20,55).

FISH LICE (BRANCHIURA)

Fish lice, argulids, or branchiurans form a small subclass of crustaceans. They can be very harmful to fishes, especially those in hatcheries or other rearing facilities. Fish lice can infect the eyes of humans who carelessly handle live fishes. Approximately 150 species have been described in four genera, more than 120 in the genus *Argulus*. *Argulus japonicus* (Fig. 19) is an international superparasite transported on goldfish and carp. Fish lice are relatively large parasites, varying from a few to 20 mm (0.1 to 0.8 in.) long.

The body is flattened (strongly depressed) and has a large, expanded head (carapace), a thorax, and an abdomen. The thorax has four ill-defined segments and the abdomen is completely fused. The two pairs of antennae are modified into attachment hooks. Some appendages of the carapace also are modified (maxillule into suckers and maxilla into leglike structures), and those of the thorax are four pairs of unmodified legs (Fig. 19). Most fish lice have large suckers on the underside (ventral surface) of the front of the carapace. Many have a long and vicious stinger (stylet) in front of the mouth and between the antennae. The abdomen terminates in a bifurcate tail with tiny rami.

Fish lice mate while free swimming, off the host. Eggs are held in the body of the female, which leaves the host to deposit eggs in clusters attached to the substrate. *Argulus* spp. hatch as nauplius larvae, but members of the other genera hatch as juveniles from eggs in 15–55 days and develop directly into adults. Swimming juveniles must find a host in 2–3 days and, once attached, develop into adults in 30–35 days. The sexes are separate.

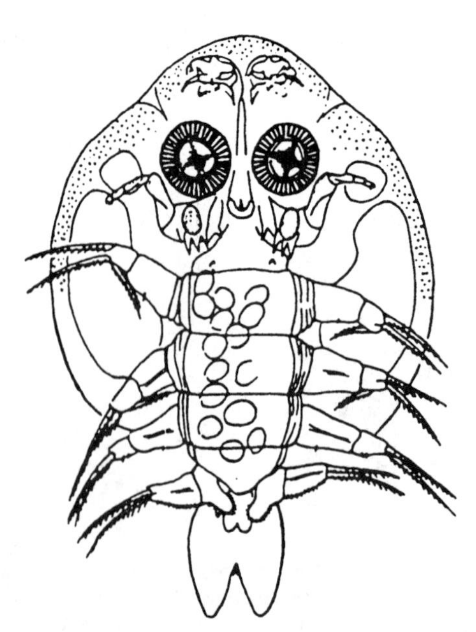

Figure 19. A fish louse, *A. japonicus*, an international superparasite [upside-down (dorsal) view].

These parasites attach to the body, fins, gills, and mouth of fishes and sometimes to frogs and tadpoles. Fish lice are more important in fresh, brackish, and inshore marine waters. They are obligate parasites, feeding on blood, but adults are capable of changing hosts and spending prolonged periods off any host. Fish lice may prefer some fishes, but are usually not host specific. Heavy infections can kill fishes. Combinations of moderate infections of *Argulus lepidostei*, *Anilocra acuta* (an isopod), a bacterial infection, and polluted conditions have caused mortalities in wild inshore fishes in the Gulf of Mexico. Fish lice directly transmit viral and bacterial diseases. They have introduced microbial diseases into culture facilities and caused epizootics.

Fish lice are the only crustacean fish parasites known to infect humans. Other parasites may bite or attack humans, but only *Argulus* spp. penetrate, survive in, and cause diseases in humans. The first reported case of human argulosis was in a child infected by *Argulus laticauda* while swimming in saltwater off the Atlantic coast of the United States (1). We recently interviewed an aquaculture specialist from South America who became infected with an *Argulus* sp. during her attempt to control losses of cultured tilapias caused by very heavy infections of this parasite in freshwater. One of the argulids she was working with was splashed into her face and lodged between her eye and the orbit, causing severe irritation and minor tissue damage for 24 hours before it was discovered and removed. She was able to sleep with the organism in her eye. All fish lice should be treated with caution.

Popular References on this subject are (20,33).

TONGUEWORMS (PENTASTOMATA)

Tongueworms, or pentastomes, form a small order of strange animals. They were dinosaur parasites, and most of them perished with these spectacular hosts. “Tongueworm” refers to their shape. “Pentastome” means “five mouths” and refers to the fingerlike projections supporting the four legs and a mouth in some species. Larval tongueworms (Fig. 20) could damage or kill fishes in aquaculture, but they are more likely to cause problems as adults in the culture of alligators and crocodiles. They can also infect humans as larvae or adults. About 95 species are known. Adults are a few mm to 15 cm (0.08 in. to 5.9 in.) long. They are flat and elongate, and their soft body is reduced and wormlike. Two pairs of legs occur under the anterior end and may be reduced to only single or double hooks or claws. The body covering is chitinous, highly porous, and marked with striations, rings (annuli), or segments. Molts occur between growth stages. The sexes are separate and males are smaller than females. The latter hold up to several million small thick-shelled eggs that pass up the trachea or down the nasal passages of the host, are swallowed, and emerge from the intestine. Eggs on vegetation or bottom mud are inadvertently eaten by a vertebrate, whereupon three

Figure 20. An unidentified larval tongueworm from largemouth bass, *Micropterus salmoides*, and peacock bass, *Cichla ocellaris*.

larval (nymphal) stages develop, and infective stages either break out or are digested out when an appropriate predator eats the intermediate host. Fishes may suffer massive infections. Adult parasites occur in the lungs, nostrils, and nasal sinuses of reptiles primarily, but may occur in some birds and mammals. Larval forms infect fishes, amphibians, reptiles, and a few mammals. In humans, the larvae become encysted in calcareous capsules, soon die, and usually do little damage. Many people who ingest them do not even realize that they have been parasitized. However, some species can develop in the nasal passages or the throat of humans, can damage the eyes and other vital organs, and can cause severe irritation or even death. Tongueworms are permanent (33) parasites that feed on mucus, tissue fluids, and blood of their host.

Popular references on this subject are (33,56).

BARNACLES (CIRRIPEDIA)

Barnacles form a subclass of the crustaceans. The common name "barnacle" is from the Middle English "bernak" (a goose) or the French "bernicle," apparently for the shape of goose barnacles. In Europe, the barnacle goose (a bird) was reputed by the ancient Greeks to spawn spontaneously from goose barnacles. Barnacles occur on or in fishes, sea turtles, and marine mammals. For example, the striped goose barnacle, *Conchoderma virgatum* (Fig. 21), occurs on all of these animals and on parasitic copepods and isopods. Barnacles do not cause problems in the culture of any vertebrates. They cause serious diseases in crustaceans that are important fishery products, but none of these crustaceans are involved in aquaculture. Barnacles may cause problems when crustaceans are more frequently cultured. They also associate with or parasitize a variety of other commercially important marine organisms. Famous for encrusting the bottoms of boats and other marine structures, barnacles are more important as pests to

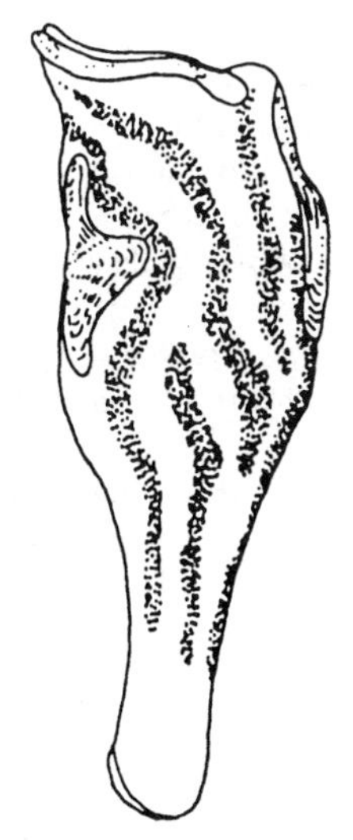

Figure 21. The striped goose barnacle, *C. virgatum*.

marine aquaculture facilities than as parasites. Cleaning barnacles from structures, antifouling methods, and the transport of exotic organisms involve serious economic and environmental problems.

A great variety of organisms eat these animals, including "barnacle eaters" (filefishes, *Alutera* spp.). Barnacles were eaten by native Americans and are a delicacy in Europe. The world's largest barnacle supports an important fishery in Chile.

More than 1000 living species have been described, and most are free living. Barnacles vary in size from the minute acrothoracicans, which burrow into the calcareous skeletons of corals and seashells, to parasitic rhizocephalans anastomosing throughout the body of large crabs. The crustacean body form is greatly modified in all barnacles, but is drastically altered in some parasitic species. Most barnacles have a heavy calcareous shell that is unique among crustaceans in that it is composed of several to many parts embedded in a soft mantle surrounding the animal. Free-living forms are attached either by long stalks (goose barnacle) or directly to the base (dorsum) of their shells (acorn or volcano barnacle). Free-living barnacles filter feed by sweeping slender, jointed appendages (cirri) through the water. Parasitic forms live on or in various marine crabs and other crustaceans, echinoderms (sea stars, etc.), and soft corals; most have lost the shell, appendages, and body segmentation of free-living barnacles. Some barnacles are specialized associates of particular crabs, seashells, other invertebrates, turtles, and whales; others attach to a variety of substrates or organisms.

The larvae are planktonic. Most barnacles have both female and male sexual organs (i.e., they are hermaphroditic), but some groups, particularly parasitic ones, have separate sexes. Parasitic barnacles attach externally and burrow inside the skeleton or endoparasitically in a variety of hosts. Barnacles vary from free living through various levels of association to permanent (33) parasites. Parasitic barnacles feed on the tissues of their host. The shells of barnacles fossilize well and have left a good fossil record. Fertilization occurs in the mantle cavity or at the base of the oviduct; eggs (ova) generally develop to the first-stage nauplius within the mantle cavity and are expelled by pumping movements of the body. The triangularly shaped nauplius larva has corners tipped with prominent spines. It molts through six stages in plankton and then seeks a host or substrate. The thin, transparent, bivalve, ostracodlike cyprid larva rapidly swims and crawls around a substrate, selects a site for attachment, cements itself in place, and metamorphoses into a subadult.

Popular references on this subject are (20,33).

ISOPODS (ISOPODA)

Isopods are an order of the crustaceans. The name means "all legs (pods) approximately similar in size and shape (iso)." Isopods kill, stunt, and damage commercially important fishes. Approximately 9.4% of chub mackerel (*Scomber japonicus*) along the Peruvian coast are parasitized by *Ceratothoa gaudichaudii*, causing a 15% loss in body weight and costing Peruvian fishermen approximately 1.3 billion kilograms (2.9 billion pounds) of fish annually. Other species of isopods also cause significant losses to commercial fishermen, including severe damage and losses in salmonid culture in southern South America. *Cymothoa oestrum* can cause superinfections in caged fishes in the Caribbean. Figure 22 shows a female in the mouth of a jack. Sea gnats (larval gnathid isopods) also kill caged cultured eels and salmonids. A larval *Gnathia* sp. is illustrated in Figure 23. Epicaridean isopods damage crustaceans that are important fishery products, including shrimp, freshwater prawns, crabs, and other hosts that are also involved in aquaculture or that may be cultured in the future.

A few isopods that are parasites of fish actively swim after and bite humans, sometimes alarmingly in mass attacks, but bites are more likely to occur in the handling infected fishes. Free-living isopods are reported to clean *Saprolegnia* spp. (water mold) from fishes. In Puerto Rico, *Anilocra* spp. are dried and used to make a tea to treat colds. New England fishermen use "salve bugs" (*Aega* spp.)

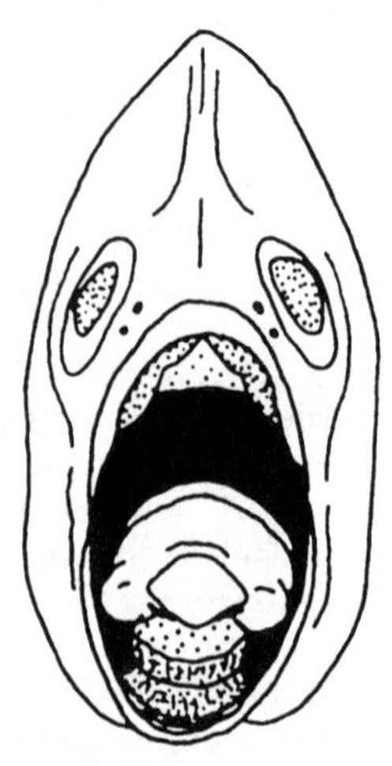

Figure 22. An isopod, *C. oestrum*, in the mouth of a jack.

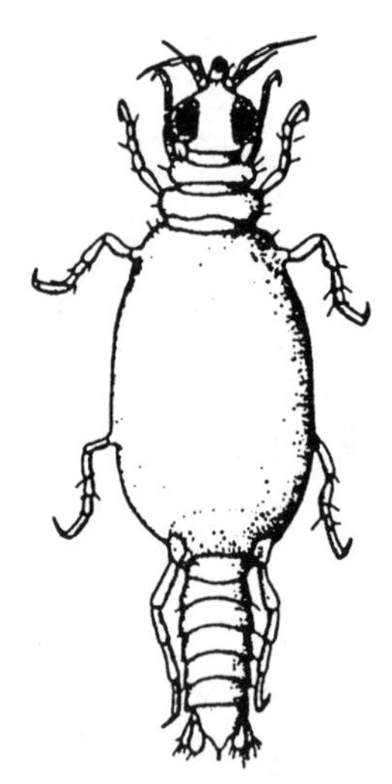

Figure 23. Sea gnats, larval gnathid isopods, kill cage-cultured fishes. A larval *Gnathia* sp. is illustrated.

for medicinal purposes. Isopods are eaten by a variety of animals. Giant isopods (*Bathynomus* spp.) are fished commercially for human food in Japan and Mexico, and Hawaiians eat a smaller species. The presence of parasitic isopods on marine tropical fishes supposedly indicates that these fishes do not contain high amounts of ciguatera, a poisonous toxin. This relationship is not proven, but is highly interesting, particularly because large barracuda and jacks are commonly implicated in ciguatera poisoning and often have attached isopods.

Approximately 4000 species of isopods have been described, of which more than 450 species are known to associate with fishes. They vary from 0.5 mm to 44 cm (0.02 in. to 17 in.) in length. The world's largest species, *Bathynomus giganteus*, is found in the Western Atlantic and the Caribbean Sea. The head of an isopod is fused with first thoracic segment (cephalothorax), and the animals have a seven-segmented thorax and six-segmented abdomen (often fused into two to five segments). One pair of thoracic appendages is modified into mouthparts, and seven pairs are unmodified. The abdomen has six pairs of appendages and ends in a terminal, often shield-shaped segment called the pleotelson (Fig. 24). Eggs, larval forms, and juveniles develop either in a brood pouch beneath the abdomen of the female or in pouches inside the abdomen. Most isopods have a stage of free-swimming juveniles that develop into adults, but gnathiid juveniles parasitize the gills and skin of fishes, and the adults are free living. The sexes are separate in most species, while others begin life as males and later become females (i.e., they are protandrous hermaphrodites). Isopods are common in most environments, including dry land. They parasitize fishes, crabs, shrimp, and other isopods. Isopods associated with fishes vary from accidental (cirolanids) and temporary or casual (corallanids and aegids) parasites to permanent (33) (cymothoids) parasites (57). They attach at a variety of locations, including the skin, the gills, inside the mouth, and on the fins, and some even burrow under the skin to form a cyst in the flanks of fish. They exhibit a broad range of food habits. The forms

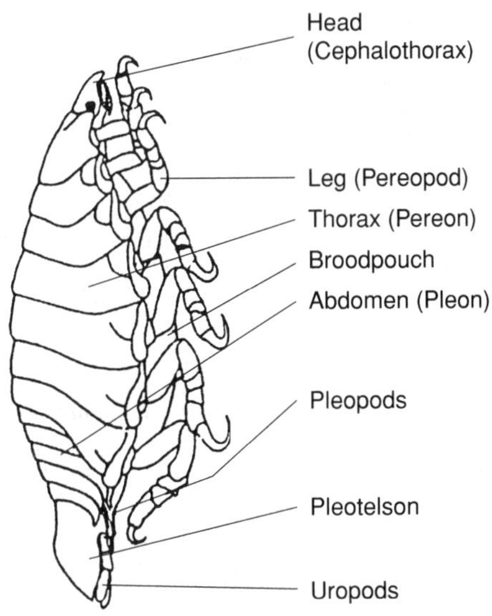

Figure 24. An isopod, *Vanamea symmetrica*, with distinctive structures labeled.

associated with fishes feed on blood or ooze from wounds. The wounds isopods cause may provide entry points for microbial diseases. Isopods can be preserved and stored in 70% ethanol (151-proof rum will do) or 40% isopropanol (rubbing alcohol).

Popular references on this subject are (20,33,58–60).

SEASHELLS AND ALLIES (MOLLUSCA)

Seashells, or molluscs, form a phylum in the animal kingdom. "Shell" refers to their hard calcium-carbonate coverings. "Mollusc" means soft and refers to their soft bodies. Seashells include the familiar black abalone, *Haliotis cracherodii*; eastern oyster, *Crassostrea virginica*; Japanese pearl oysters, *Pinctada martensii*; northern quahog, *Mercenaria mercenaria*; and queen conch, *Strombus gigas*, all aquaculture animals. Glochidia, the larvae of freshwater mussels (Fig. 25), infect the gills or skin of freshwater fishes and may cause problems in aquaculture. Similar larvae of marine mussels have severely damaged pen-reared salmonids (17). *Turbonilla* spp., ectoparasitic gastropods of giant clams, have caused mass mortalities under culture conditions. Some aquaculturists of giant clams in the West Indies have attempted to eliminate this pest and other dangerous parasites. Molluscs serve as intermediate hosts for many dangerous and harmful parasites of humans and aquaculture animals. Extraneous molluscs should be excluded from aquaculture facilities whenever possible.

More than 50,000 living species of molluscs have been described. They range from the size of sand grains to the 1.3-m (51-in.) giant clam, *Tridacna gigas* (also an aquaculture animal), and 20-m (65-ft) giant squid, *Architeuthis sanctipauli*. The shells of molluscs can be external or internal, or they may be absent; shells are produced by a fold in the body wall (mantle). Usually, the sexes are separate and fertilization occurs in the water, but some species have internal fertilization. Eggs are normally

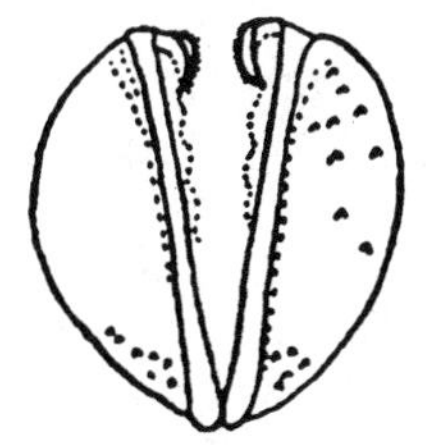
Figure 25. A "glochidia" of freshwater fish.

encased in various kinds of coverings. Typically, free-swimming larvae (trochophores) hatch and develop into veligers in plankton, but there are many other methods of development. The larvae (glochidia) of most freshwater clams and mussels develop as parasites in fishes. Many molluscs are predators, grazers, or filter feeders. Only a few adult molluscs associate with or parasitize other organisms. We described the only sea slug (nudibranch) known to associate with a fish. Molluscs inhabit marine waters, freshwaters, and moist areas on land.

GENERAL DISCUSSION

The vast majority of macroparasites have complex life cycles on multiple hosts. Placing the final host in aquaculture situations eliminates almost all of these parasites: The intermediate hosts and natural conditions that support multiple life stages are simply not available under culture conditions. Thus, most of the "gut worms," or helminths, are absent from aquaculture, and many diagnostic services that specialize in aquaculture diseases do not even look for intestinal helminths, because they assume that none will occur.

Those macroparasites with simple or direct life cycles (gillworms, fish lice, copepods, leeches, etc.) are the ones that are most damaging in aquaculture conditions. All parasites produce many times more offspring—often astronomically so—than are necessary to maintain their numbers in nature. This fecundity exists because being a parasite is a hazardous occupation. Most young parasites are washed away, lost, and never find a host, or else they get eaten. In aquaculture conditions, the hosts are crowded together in a small, safe environment, ensuring that most parasite offspring (that would ordinarily be lost in nature) thrive. Thus, the hosts quickly become overcrowded with, and damaged by, parasites. Cage culture in natural waters also allows parasites distributed in plankton (copepods and isopods) and migrating parasites (copepods, leeches, and isopods) to accumulate and cause problems.

Intermediate stages of quite damaging macroparasites can also be introduced to aquaculture animals through poor management practices, such as allowing the animals to come into contact with birds or other predators and allowing extraneous snails or other invertebrates into the system. Direct introductions can occur when macroparasites are fed to aquaculture animals along with raw and unfrozen living materials. For example, marine plankton often contains parasitic isopods, so feeding live plankton can result in direct infections of aquarium and pen-reared fishes.

Biological parasite control is far superior to chemical control, but has been little considered in aquaculture. A good example of such control is the simple removal of an intermediate snail host from the culture system; this measure can totally eliminate an otherwise difficult-to-control parasite. Recently, some attempts have been made to control "salmon copepods" and "killer capsalids" (Fig. 7) with cleaner wrasse (family Labridae). This tactic is something like adding the striped mullet, *M. cephalus*, to pompano (*Trachinotus carolinus*) culture cages to eat the encrusting materials off the mesh so that the cages do not have to be cleaned. Cleaners do not completely eliminate parasites, but can keep them at tolerable levels. If we can identify a parasite, we can accumulate biological information on the species that may be used to develop an effective, intelligent biological control.

Most parasites could, in fact, be eliminated from commercial aquaculture stocks; however, through economic considerations, time restraints, or carelessness, this is often not accomplished. Once a parasite is spread around in the world of aquaculture, it is almost impossible to eliminate. By placing animals into unnatural aquaculture conditions, we select for those parasites best able to thrive under those conditions, and chemical regimens may subsequently select for treatment-resistant parasites. Overcrowding and poor nutrition may weaken the resistance of cultured animals to parasites, and epizootics may ensue, further favoring the more destructive parasites. Aquaculture conditions in themselves sometimes select for and develop, albeit inadvertently, the "international superparasite."

Most cultured animals and almost all of the problematic macroparasites are exotic introductions. Conducting aquaculture with an exotic animal is the same thing as introducing it. Animals always escape. At this point, three things may occur, all of which are negative: Exotic hosts themselves can damage or endanger the environment, exotic parasites may infect native susceptible hosts, and native parasites may infect the exotic hosts. Native hosts infected with exotic parasites may have no evolutionary experience with, and thus no defenses against, an exotic parasite, resulting in drastic reductions in their numbers. Native parasites on exotic hosts may have the same effect, but escaped exotic hosts then serve as a reservoir for potential movement into the culture environment, again with drastic results. These scenarios should be avoided at almost all costs. Careful testing in "neutral" or controlled conditions should be carried out before introducing any exotic species.

Some macroparasites (roundworms, tapeworms, tongueworms) that are found in aquaculture animals can be injurious or fatal to humans, but such injuries or fatalities are much more likely to result from eating wild fishes. The danger can be avoided by *(1)* never eating a raw or improperly cooked crustacean (some shrimps, crabs, and crayfishes are particularly dangerous), *(2)* freezing fishes for several days before using them in "raw" fish dishes (even the Japanese freeze most of their sushi and sashimi fishes), and *(3)* avoiding native raw dishes and do-it-yourself raw dishes (professional preparations cost more, but are safer) (1). Isopods are the only macroparasites known to directly attack humans, but fish lice and tongueworms can infect aquaculturalists.

ACKNOWLEDGMENTS

We thank Drs. William G. Dyer, Department of Zoology, Southern Illinois University; S. Kenneth Johnson; and Robert R. Stickney for reviewing this contribution.

BIBLIOGRAPHY

1. E.H. Williams, Jr. and L. Bunkley-Williams, *Parasites of Offshore, Big Game Sport Fishes of Puerto Rico and the Western North Atlantic*, P.R. Dept. Nat. Environ. Res., San Juan, and Dept. Biol., Univ. Puerto Rico, Mayagüez, PR, 1996.
2. L. Bunkley-Williams and E.H. Williams, Jr., *Parasites of Puerto Rican Freshwater Sport Fishes*, P.R. Dept. Nat. Environ. Res., San Juan, and Dept. Mar. Sci., Univ. Puerto Rico, Mayagüez, PR, 1994 [Spanish ed., 1995].
3. E. Amlacher, *Textbook of Fish Diseases*, TFH Publ., Neptune City, NJ, 1976.
4. C.A.E. Andrew and N. Carrington, *The Manual of Fish Health*, Tetra Press, Morris Plains, NJ, 1988.
5. S.M. Bower and S.E. McGladdery, *Synopsis of Infectious Diseases and Parasites of Commercially Exploited Shellfish*, http://www.pac.dfo.ca/pac/sealane/aquac/pages/title.htm, 1998.
6. J.A. Brock and K.L. Main, *A Guide to the Common Problems and Diseases of Cultured* Litopenaeus vannamei, Ocean. Inst., Honolulu, HI, 1994.
7. L. Bunkley-Williams and E.H. Williams, Jr., *Evaluating Fishes for Parasite Problems: A Simplified Guide of Basic to Advanced Techniques*, P.R. Dept. Nat. Environ. Res., San Juan, and Dept. Biol., Univ. Puerto Rico, Mayagüez, PR, 2000, (in prep.).
8. M.D. Dailey, *Essentials of Parasitology*, W.C. Brown Publ., Dubuque, IA, 1996.
9. R.M. Durborow, *Diseases of Warmwater Fish*, Kentucky State Univ. Coop. Ext. Progr., 30 min., VHS, 1994.
10. R.M. Durborow, *Trout Diseases*, Kentucky State Univ. Coop. Ext. Progr., 30 min., VHS, 1995.
11. E. Elkan and H. Reichenback-Kinke, *Color Atlas of the Diseases of Fishes, Amphibians and Reptiles*, TFH Publ., Neptune City, NJ, 1974.
12. J. Grabda, *Marine Fish Parasitology*, VCH, Weinheim, Germany, 1991.
13. R. Heckmann, *Aquaculture Mag.* **21**(1), 43–49, 52, 54–57 (1995).
14. G.L. Hoffman, *Parasites of North American Freshwater Fishes*, Cornell Univ. Press, Ithaca, NY, 1999.
15. G.L. Hoffman and F.P. Meyer, *Parasites of Freshwater Fishes: A Review of their Control and Treatment*, TFH Publ., Neptune City, NJ, 1974.
16. D. Hoole, D. Bucke, and Burgess, *Diseases of Carps and Other Cyprinid Fishes*, Blackwell Sci., Ltd., 1999.
17. S.C. Johnson, M.L. Kent, and L. Margolis, *Aquaculture Mag.* **23**(2), 40, 42, 44–56, 58–64 (1997).
18. S.K. Johnson, *Handbook of Shrimp Diseases*, Sea Grant TAMU SG-75-603, 1978.
19. S.K. Johnson, *Crawfish and Freshwater Shrimp Diseases*, Sea Grant TAMU SG-75-605, 1977.

20. Z. Kabata, *Crustaceans as Enemies of Fishes*, TFH Publ., Neptune City, NJ, 1970.
21. Z. Kabata, *Parasites and Diseases of Fish Cultured in the Tropics*, Taylor and Francis, London, 1985.
22. Z. Lucký, *Methods for the Diagnosis of Fish Diseases*, Amerind Publ. Co., New Delhi, India, 1977.
23. R. MacMillan, *Aquaculture Mag.* **21**(2), 37–40 (1995).
24. L.E. Mawdesley-Thomas, *Diseases of Fish*, Sympos. Zool. Soc., No. 30, London, 1972.
25. P.M. Muzzall, *Aquaculture Mag.* **22**(10), 49–50, 51–61 (1996).
26. E.J. Noga, *Fish Disease, Diagnosis and Treatment*, Mosby, St. Louis, MO, 1996.
27. R.M. Overstreet, *Marine Maladies? Worms, Germs and Other Symbionts from the Northern Gulf of Mexico*, Miss.-Ala. Sea Grant Cons. MASGP-78-021, 1978.
28. J.A. Plumb, ed., *Principal Diseases of Farm-Raised Catfish*, Revised ed., Southern Cooperative Series Bull. 225, 1985.
29. J.A. Plumb, *Health Maintenance of Cultured Fishes: Principal Microbial Diseases*, CRC Press, Boca Raton, FL, 1994. [See also our review of this book, *Fish. Rev.* **40**, 535 (1995).]
30. G. Post, *Textbook of Fish Health*, 2d ed., TFH Publ., Neptune City, NJ, 1987.
31. W.E. Ribelin and G. Migaki, Eds., *The Pathology of Fishes*, Univ. Wis. Press, Madison, WI, 1975.
32. L.S. Roberts, *Fish Pathology*, Bailliere, Tindall Gassell, Ltd., London, 1978.
33. L.S. Roberts and J. Janovy, Jr., *Foundations of Parasitology*, 5th ed., W.C. Brown Publ., Dubuque, IA, 1996.
34. R.J. Roberts and C.J. Shepherd, *Handbook of Trout and Salmon Diseases*, 3d ed., Fishing News, Ltd., Surrey, England, 1977.
35. O.K. Rohde, *Ecology of Marine Parasites: An Introduction to Marine Parasitology*, CAB Intern., Oxon, United Kingdom, 1993.
36. S. Sarig, *The Prevention and Treatment of Diseases of Warmwater Fishes Under Subtropical Conditions, with Special Emphasis on Fish Farming*, TFH Publ., Neptune City, NJ, 1971.
37. W. Schäperclaus, *Fish Diseases*, Vols. 1 & 2, Oxonian Press, New Delhi, India, 1991.
38. G.D. Schmidt, *Handbook of Tapeworms Identification*, CRC Press, Boca Raton, FL, 1986.
39. C.J. Sindermann, *Principle Diseases of Marine Fish and Shellfish*, 2nd ed., Vols. 1 & 2, Academic Press, Inc., San Diego, CA, 1990.
40. C.J. Sindermann and D.V. Lightner, *Disease Diagnosis and Control in North American Marine Aquaculture*, Elsevier Science Publ., Amsterdam, 1988.
41. S.F. Snieszko, ed., *A Symposium on Diseases of Fishes and Shellfishes*, Amer. Fish. Soc. Spec. Publ. 5, 1970.
42. S. Spotte, *Captive Seawater Fishes: Science and Technology*, Wiley-Interscience Publ., New York, 1979.
43. M.K. Stoskopf, ed., *Fish Medicine*, W.B. Saunders, Philadelphia, PA, 1993. [See also our review of this book, *Fish. Rev.* **40**, 533 (1995).]
44. D. Untergasser, *Handbook of [Aquarium] Fish Diseases*, TFH Publ., Neptune City, NJ, 1989.
45. J.W. Warren, *Diseases of Hatchery Fish*, 6th ed., U.S. Fish Wildl. Ser., Pacific Region, 1991.
46. E.H. Williams, Jr., and C.J. Sindermann, in R. DeVore, ed., *Proceedings of Introductions and Transfers of Marine Species Conference and Workshop*, South Carolina Sea Grant, 1992, pp. 71–77.
47. P.T.K. Woo, *Fish Diseases and Disorders*, Vol. 1, CAB Intern., Wallingford, Oxon, United Kingdom, 1995.
48. A.M. Barse and D.H. Secor, *Fish. Mag.* **24**(2), 6–10 (1999).
49. E.H. Williams, Jr., in G.L. Hoffman, ed., *Parasites of North American Freshwater Fishes*, Cornell Univ. Press, Ithaca, NY, 1999, pp. ix–xi.
50. J.W. Avault, Jr., *Aquaculture Mag.* **22**(3), 87–92 (1996).
51. R. Heckmann, *Aquaculture Mag.* **23**(5), 43–54, 56–60 (1997).
52. S.C. Schell, *How to Know the Trematodes*, W.C. Brown Co., Dubuque, IA, 1970.
53. V.N. Stewart, *Sea-Stats*, No. 1, Fla. Mar. Res. Inst., http://www.epa.gov/gumpo/sea01.html, 1998.
54. G.D. Schmidt, *How to Know the Tapeworms*, W.C. Brown Co., Dubuque, IA, 1970.
55. J.W. Avault, Jr., *Aquaculture Mag.* **22**(2), 85–88 (1996).
56. E.H. Williams, Jr., *Nat. History Mag.* **104**, 4 (1995).
57. G.A. Boxshall and D. DeFaye, eds., *Pathogens of Wild and Farmed Fish: Sea Lice*, Ellis Horwood, Chichester, England, 1993.
58. L. Bunkley-Williams and E.H. Williams, Jr., *J. Parasit.* **84**, 893–896 (1998).
59. E.H. Williams, Jr., and L. Bunkley-Williams, in G.L. Hoffman, ed., *Parasites of North American Freshwater Fishes*, Cornell Univ. Press, Ithaca, NY, 1999, pp. 310–331.
60. L.B. Williams and E.H. Williams, Jr., *Nat. History Mag.* January: 40–41, 92 (1989).

MYCOTOXINS

R.T. Lovell
Auburn University
Auburn, Alabama

OUTLINE

Aflatoxins
Cyclopiazonic Acid
Ochratoxin
Fusarium Toxins
Conclusions
Bibliography

The mycotoxins of concern in the United States are produced by three genera of molds: *Aspergillus*, *Penicillium*, and *Fusarium*. These molds are ubiquitous (found widely in nature) and grow and produce toxins under the proper conditions, which include adequate substrate (carbohydrate), moisture greater than 14%, relative humidity greater than 70%, temperature, 15–35 °C (59–95 °F) and oxygen. The most familiar mycotoxins are the aflatoxins, ochratoxins, and some fusarium toxins. There are others, such as cyclopiazonic acid, which have received less notoriety, and some newly discovered toxins, such as moniliformin. Effects of mycotoxins on livestock and poultry are fairly well documented. However, much less is known about the effects of mycotoxins in fish. Mycotoxins are usually produced in feedstuffs prior to harvest, but can develop in finished feeds that are not properly dried or stored.

AFLATOXINS

In the late 1960s (1), Dr. John Halver at a U.S. Fish and Wildlife Laboratory in Washington state tested aflatoxin B_1 (AFB_1) in rainbow trout (*Oncorhynchus mykiss*) and found that young rainbow trout are one of the most sensitive animals tested. The oral LD_{50} for 50-g (1.7-oz) rainbow trout was 500 ppm (parts per million), and prolonged feeding (20 months) of a dose as low as 0.5 ppb (parts per billion) would cause liver cancer. Bailey et al. (2) demonstrated that rainbow trout were more sensitive than other salmonids to diets containing 20 ppb of AFB_1, and 62% of the fish developed liver tumors; however, they fed coho salmon 40 ppm of AFB_1 and found no tumors. The salmon tended to produce benign liver adenomas when exposed to AFB_1 in contrast to the trout, which produced malignant liver carcinomas. DNA binding by aflatoxin in the liver was 20 times faster in trout than salmon.

Channel catfish (*Ictalurus punctatus*), however, are much less sensitive to aflatoxin BI than are rainbow trout. Studies were conducted at Auburn University to demonstrate the effects of acute and subacute doses of purified AFB_1 on channel catfish. In these studies, both oral and intraperitoneal (IP) administration of 12 mg of AFB_1 per kg of body weight caused the stomachs of channel to catfish to regurgitate their contents (3). Ammonia detoxification of AFB_1 significantly reduced its effect on regurgitation. When AFB_1 was injected intraperitoneally, the median lethal dose (LD_{50}) for 35-g (1.2-oz) channel catfish was 11.5 mg/kg (2×10^5 oz/lb) of body weight. Fish examined before death showed extremely pale gills, liver, and other internal organs, and hemoglobin concentration was about 10% of that of the control fish. Histological lesions in these fish included sloughing of intestinal mucosa and necrosis of hematopoietic tissues, hepatocytes, pancreatic acinar cells, and gastric glands.

To determine the effects of prolonged feeding of subacute does of AFB_1 to channel catfish, Jantrarotai and Lovell (3) fed semipurified diets containing 0, 100, 500, 2,000, or 10,000 ppb to 7-g (0.2-oz) fish in flowing-water aquaria for 10 weeks. Only the highest concentration of AFB_1 had adverse effects on the fish. The growth rate, hematocrit, hemoglobin concentration, and red-blood-cell count were lower, and the white-blood-cell count was higher in fish fed 10,000 ppb of AFB_1 than in fish fed lower concentrations. At the highest concentration, AFB_1 caused necrosis and basophilia of hepatocytes, enlargement of blood sinusoids in the head kidney, accumulation of iron pigments in the intestinal mucosa epithelium, and necrosis of gastric glands.

The maximum concentration of aflatoxin allowed by FDA in feedstuffs (feed ingredients or finished feed) in interstate commerce is 20 ppb. However, concentrations of 400 ppb have been found in locally produced (noninspected) grain sold for use in catfish feed. This concentration is well below the oral toxicity level for channel catfish, which appears to be between 2,000 and 10,000 ppb.

CYCLOPIAZONIC ACID

Cyclopiazonic acid (CPA) is produced by several species of *Aspergillus* and *Penicillium* fungi. It is indigenous to warmer latitudes and appears to be fairly widespread. It is often found in combination with aflatoxins. Gallagher et al. (4) reported that CPA occurred more frequently (found in 52% of samples tested) than aflatoxins (found in 33% of samples tested) in peanuts contaminated with *Aspergillis flavis*. Leistner (5) identified 20 different mycotoxins from 1,481 *Penicillium* molds isolated from livestock feeds and found that CPA occurred with the highest frequency.

Because of the potential for CPA to occur in fish feeds made in the southern United States, laboratory studies were conducted at Auburn University with channel catfish to compare relative toxicity of CPA with AFB_1 at both acute and subacute doses (6). To determine acute toxicity (LD_{50}), channel catfish fingerlings weighting 19 g (0,602) were injected intraperitoneally with 0, 2.4, 2.8, 3.2, 3.6, 4.0 or 7.0 mg (1 mg $= 0.4 \times 10^{-5}$ oz) CPA per kg of body weight and observed for 96 hr. The intraperitoneal LD_{50} for CPA was 2.82 mg/kg (5.1×10^{-5} oz/lb). The effects of CPA were characteristic of a neurotoxin. Affected fish showed severe convulsions and were dead within 30 minutes after injection. There were no changes in organs of intoxicated fish examined. To determine subacute effects of prolonged feeding of CPA, channel catfish fingerlings were fed purified diets containing subacute concentrations (0, 100, 500, 2,000, or 10,000 ppb) of CPA for 10 weeks. The fish were evaluated for weight gain and signs of pathology. The lowest concentration of CPA, 100 ppb, significantly reduced the growth rate, but produced no other toxicity signs. The highest concentration, 10,000 ppb, caused necrosis of the gastric glands.

According to available data, CPA is more toxic to channel catfish than aflatoxin. The intraperitoneal LD_{50} for AFB_1 is 11.5 mg/kg body weight and for CPA is 2.82 mg/kg body weight. The subacute toxicity dietary doses are between 2,000 and 10,000 ppb for AFB_1 and approximately 100 ppb for CPA. The facts that CPA is more toxic than AFB_1 and is often found in combination with, and perhaps more frequently than, aflatoxins indicates that CPA may be a more serious contaminant than aflatoxins in fish feeds made in the southern United States and other regions of the world with a similar climate.

OCHRATOXIN

Ochratoxins are produced by *Aspergillus* and *Penicillium* mold species that are widely found in nature. Although not recognized as causing widespread problems in animal feeding in the United States, it is suspected that these toxins cause poor growth and feed conversion in livestock in undetected cases because of widespread occurrence. Ochratoxins are recognized as kidney toxins, causing pale, swollen kidneys and renal tubular failure in swine, rats, and mice. The intraperitoneal LD_{50} for ochratoxin A in six-month-old rainbow trout is 4.7 mg/kg (8.5×10^{-5} oz/lb) of bodyweight (7). Pathological signs in trout fed ochratoxin A are severe necrosis of liver and kidney tissues, pale kidneys, pale, swollen livers, and death.

FUSARIUM TOXINS

The *Fusarium* toxins that are most often associated with animal health are vomitoxin, T_2 toxins, and zearalenones. More recently, the effects of fumonsins and moniliformin have been reported. The zearalenones are a group of estrogenic metabolites, some of which cause reproductive problems in farm animals consuming 0.6 mg/kg to 5 mg/kg (1.1 to 9×10^{-5} oz/lb) in diets. The tricothecenes usually develop in corn in storage in the Midwest, with alternating cooling and warming trends in the fall. Toxicity signs in fish have not been described, but in livestock and poultry they are reduced growth, reduced red-blood-cell formation, widespread hemorrhage, slow blood clotting, and impaired immune responses. Rainbow trout are highly sensitive to the *Fusarium* toxin, vomitoxin (8). Trout fed diets with as low as 20 ppm of the toxin refused feed, and their growth rate decreased when they were fed diets containing 1 to 13 ppm. Poston (9) found that dietary doses of T_2 above 2.5 ppm reduced the weight gain of rainbow trout.

Very recently, fumonsins, produced by the mold *Fusarium moniliforme*, have received a great deal of attention. Apparently, fumonsins have been found to be fairly widespread, and they are highly toxic or carcinogenic to certain animals. In 1989, numerous episodes of equine leukoencephalomalacia in horses (ELEM) and porcine pulmonary edema in pigs (PPE) were reported in various areas of the United States, including the Midwest and the South. Case studies showed that horse feeds containing 8 ppm were associated with cases of ELEM and that swine feeds containing 20 ppm were linked to PPE (10,11). Both conditions are highly fatal to the affected animals. A fumonsin concentration of 8 ppm seems to be the minimum toxicity dose for horses and 20 ppm is the minimum toxicity dose for swine (10). These concentrations have been frequently found in commercial feedstuffs. In a recent report from the University of Illinois (11) in which swine feeds were sampled from 21 farms in an area where PPE was found, 12 of the farms had cases of PPE. Only one sample from the 21 farms was free of fumonsin. The lowest concentration of fumonsin was 6 ppm, and the highest concentration was 73 ppm.

Corn and corn screenings seem to be the most serious sources of fumonsins. These products are used in catfish feeds, so a study was conducted by Auburn University (12) to examine the sensitivity of channel catfish to fumonsins in feeds. A wide range of dietary levels of fumonsin B1 were used, 0, 20, 80, 320, and 740 ppm, because of uncertainty of the sensitivity of catfish to fumonsins. Some animals, such as rats and chickens, seem to be relatively insensitive, while others, such as horses, are highly sensitive.

Strain 826 of *F. moniliforme*, a high producer of fumonsins, was used to inoculate sterilized corn to produce cultures of toxic corn. Different combinations of clean and toxic corn were used to formulate diets containing the various concentrations of fumonsins. Year 1 [2-g (0.07 oz)] and year 2 [30-g (1.0 oz)] catfish fingerlings were fed. At six weeks, the small fish had shown significant repression in growth rate at the lowest dose of fumonsin, 20 ppm. The larger fish were not sensitive to 20 ppm, but did show reduced weight gain and lower resistance to bacterial infection when fed 80 ppm. Fish fed the two highest dose levels essentially stopped eating after the third week. Subclinical evaluations showed that the liver was the only organ damaged by fumonsin. Sphingolipid synthesis was suppressed, which seems to be a mechanism for the toxicity of fumonsins.

Moniliformin is produced by *F. moniliforme* and also by *Fusarium proliferatum*. This toxin has not been associated with practical problems in commercial animal production, but is found in corn and corn products from the Midwest. Moniliformin apparently interferes with enzymes in the tricarboxylic acid (TCA) cycle, such as pyruvate dehydroglucose. Laboratory studies have revealed lesions in the cardiac muscle of chickens fed the toxin. Studies with channel catfish indicate that the oral dose causing chronic toxicity is 20 to 80 ppm, similar to fumonsin B_1.

CONCLUSIONS

Comparative sensitivity of channel catfish and rainbow trout to various mycotoxins is presented in Table 1. Note that some values are related to fish body weight and other to dietary content.

Aflatoxin B_1 seems to be much more toxic to some fish, such as rainbow trout, than to others, such as channel catfish. This indicates that fish species vary in their sensitivity to aflatoxin. There are opinions that generally, warmwater fish are less sensitive than coldwater species; however, rainbow trout are markedly more sensitive than coho salmon, which also dwells in cold water. The study described in this report indicates that aflatoxin B_1 is not a major problem in channel catfish because (a) the fish apparently regurgitate acutely toxic concentrations of the toxin, and (b) subacute toxic concentrations (with prolonged feeding) would be between 2,000 and 10,000 ppb, which would be highly unusual in commercial feedstuffs and would only occur with severely molded batches of feed. It is recognized that crude aflatoxin (composed of a mixture of aflatoxins, synergists, and possibly other toxins) would likely be more toxic than the pure aflatoxin B_1. Also, the presence of aflatoxins may indicate the presence of other toxins in the feedstuff.

Table 1. Sensitivity of Fish to Acute and Subchronic Mycotoxins: MG of Toxin per kg of Diet or Body Weight[a]

	Acute		Subchronic	
Mycotoxin	Channel Catfish	Rainbow Trout	Channel Catfish	Rainbow Trout
Aflatoxin B_1	11.5(BW)	500	2–10	0.005–0.020
Cyclopiazonic acid (CPA)	2.8(BW)	—	0.10	—
Fumonsin				
Crude	>720	—	20–40	—
Pure FB_1	—	—	>250	—
Vomitoxin	—	—	—	13
Tricothecene (T_2)	—	—	—	2.5
Ochratoxin A	—	4.7 (BW)	—	—

Sources: Channel catfish: aflatoxin and cyclopiazonic acid (3,6); fumonsin (12); Rainbow trout: aflatoxin (1); vomitoxin (8); T_2 toxin (9); ochratoxin (7).
[a] 1 mg/kg = 1.8×10^{-5} oz/lb; BW, body weight.

Cyclopiazonic acid appears to be a more serious problem than aflatoxins for fish feeds and may be more prevalent in feedstuffs than aflatoxins. It is more toxic to catfish than aflatoxins; the subacute toxicity dose concentration is 1/20 to 1/100 of that of aflatoxin and is within a range that would be more likely to occur in commercial feeds.

Fumonsins could be the mycotoxin of greatest concern to catfish feed manufactures who use corn products. The subacute toxicity dose for catfish is 20 to 80 ppm, and based upon the limited surveys with animal feeds and feedstuffs, this concentration might occur often enough to cause concern. Episodes of fumonsin toxicity in livestock have been found in northern and southern areas of the United States, indicating that the toxins are ubiquitous.

Effective field screening tests are available for detecting aflatoxins, and fish-feed mills routinely use them. However, feed ingredients are not routinely screened for CPA, ochratoxin, or *Fusarium* toxins. ELISA screening tests are available for aflatoxin, vomitoxin, ochratoxin, fumonsin T_2 toxin, and zearalenone. Aflatoxin screening is used routinely by most segments of the feed industry, and other toxins are tested where problems are suspected.

Mycotoxins are generally heat tolerant, so even extrusion processing will not inactivate them. Because of this and the ubiquity of mycotoxins, fish feeds and ingredients should be carefully surveyed for the presence of mycotoxins. If certain mycotoxins are suspected of being associated with certain feedstuffs, geographic regions, or climatic conditions, screening tests for the suspected mycotoxin(s) should be implemented.

BIBLIOGRAPHY

1. J.E. Halver, in L.A. Goldblat, ed. *Aflatoxin; Scientific Background, Control, and Implications.* Academic Press, New York, 1969, pp. 265–306.
2. G.S. Bailey, D.E. Williams, J.S. Wilcox, P.M. Loveland, R.A. Coulombre, and J.D. Hendricks, *Carcinogenesis* **9**, 1919–1926 (1988).
3. W. Jantrarotai and R.T. Lovell, *J. Aquatic Animal Health* **2**, 248–254 (1990a).
4. R.T. Gallagher, J.L. Richard, H.M. Stahr, and R.J. Cole, *Mycopathologia* **66**, 31–36 (1978).
5. L. Leistner, in H. Kurata and Y. Ueno, eds., Toxigenic Fungi, Elsevier Science, Publications, Amsterdam, 1983, p. 162.
6. W. Jantrarotai and R.T. Lovell, *J. Aquatic Animal Health* **2**, 255 (1990b).
7. R.C. Doster, R.O. Sinnhuber, and J.H. Wales, *Fd. Cosmet. Toxicol.* **10**, 85–92 (1972).
8. B. Woodward, L.G. Young, and A.K. Lun, *Aquaculture* **35**, 93–101 (1983).
9. H.A. Poston, *Aquatic Toxicology* **2**, 79–88 (1983).
10. P.F. Ross and A.E. Ledet, *J. Vet. Diag. Invest.* **5**, 69–74 (1993).
11. D.P. Bane, E.J. Newman, and R.L.N. Slife. *Mycopathologia* **117**, 121–124 (1992).
12. S. Lumlertdacha and R.T. Lovell, *Aquaculture* **130**, 201–218 (1995).

See also ANTINUTRITIONAL FACTORS; FUNGAL DISEASES.

N

NET PEN CULTURE

William T. Fairgrieve
Northwest Fisheries Science Center
Seattle, Washington

OUTLINE

Traditionally, aquaculture of fish in seawater has been restricted to enclosed natural bays or small, fixed cages. However, few natural bays or protected near-shore locations have abundant, high-quality water with appropriate chemical and physical characteristics, while the demand for aquacultured species continues to grow. This constraint to the expansion of intensive marine aquaculture has been addressed by the development of movable buoyant enclosures, which are called net pens, specifically designed for use in large, open bodies of water. Because of design limitations, net pen farms have historically been located in areas protected from waves, wind, and high currents. However, recent technological advances in net pen design have permitted the expansion of net pen farming to open ocean, high-energy locations. This entry provides a basic overview of pen designs currently in use or under development and discusses their advantages and limitations. Netting selection, net bag construction, and pen-mooring systems are also discussed.

NET PEN SYSTEM DESIGNS

Modern net pen aquaculture probably originated in Japan in the early 1950s, where the first commercial culture of yellowtail (*Seriola quinqueradiata*) commenced in 1957 (1). During the 1960s, pen rearing of Atlantic salmon (*Salmo salar*) in Norway began (2), and experiments were initiated with coho salmon (*Oncorhynchus kisutch*) in net pens in seawater in southern Puget Sound, Washington, United States (3). At the same time, pen culture methods for plaice (*Pleuronectes platessa*), turbot (*Scophthalmus maximus*), and Dover sole (*Microstomus* sp.) were being developed by researchers in the United Kingdom (Colin E. Nash, personal communication). By 1970, the White Fish Authority, Scotland, had developed several successful designs for floating-net enclosures for marine fish culture (1). Today, net pens are the mainstay of industrial scale farming of Pacific and Atlantic salmon and rainbow trout (*O. mykiss*), as well as many other marine and freshwater species.

During the past 40 years, net pen aquaculture systems have evolved rapidly from their origins as simple net-enclosed cages surrounded by floating, wooden catwalks to a myriad of carefully engineered floating, submersible, and submerged structures, many of which are specifically designed to withstand extreme wind, wave, and water-current conditions encountered at open-water farming sites. Despite their intrinsic differences in design and construction, commonalities in the means used to maintain net volume and shape permit their convenient classification (4): Class I—Gravity (buoyancy and weight); Class II—Anchor–Tensioned (buoyancy and external rigging); Class III—Semirigid (flexible internal structure and special rigging); and Class IV—Rigid (rigid internal structure). The defining characteristics and attributes of pens in each classification are discussed next.

Class I: Gravity Pens

Gravity pens consist of a flexible net bag supported by a floating collar. Net bag shape and volume are maintained by weights attached at intervals along the lower perimeter of the net. Weight and buoyancy offset the effects of water currents, waves, and accumulation of fouling organisms on the net, and must be adjusted to accommodate local conditions. Gravity pens are the type most commonly used in commercial aquaculture today.

Floating collars used with gravity pens take two basic forms: circular or rectangular. Circular pens are most often constructed of two or three concentric rings of high-density polyethylene (HDPE) pipe interconnected at intervals by molded HDPE or galvanized steel collars. Rectangular pens generally are fabricated from articulated galvanized steel catwalks supported by polystyrene-filled steel or plastic floats. The net bag is attached by ropes or hooks to the collar, while stanchions and handrails located on the inner edge support light weight netting, which prevents fish loss due to jumping or predatory birds (2).

Circular pens commonly used today range from 15–30 m (49–98 ft) in diameter [ca. 175–700 m^2 (1,900–7,500 ft^2)] and are moored individually. Rectangular pens may be as large as 30 m (98 ft) on each side [900 m^2 (9,700 ft^2)] and are often interconnected to form large rafts. Net bag depth ranges from 5 to 20 m (16–66 ft), depending on the requirements of the species in culture and the depth of water under the pen complex.

Gravity pens are relatively low cost, can often be constructed from locally available materials and, in the case of rectangular pens, may be linked to form stable platforms. However, pens in this classification suffer from a number of disadvantages which limits their use to protected, low-current locations.

The principal forces acting on any net pen are those arising from the effects of wind, waves, and currents. In the case of Class I pens, the floating collar is exposed to wind and wave action that can be extremely destructive to the integrity of the structure. High-density polyethylene pipe used in modern circular-pen construction is flexible and resists breakage, but improperly moored pens are easily deformed and may collapse. Wave action is most destructive to rigid steel collars, because cyclic vertical and horizontal motion imposes severe bending and shear forces that must be absorbed by relatively few articulated linkages. In recent years, linkage designs incorporating rubber bushings, chain, or other flexible materials have been introduced, but fundamental problems with wave energy absorption continue to restrict the use of steel pen collars to protected locations.

The greatest problem with Class I pens lies in the difficulty of maintaining net-enclosed volume and shape. Because this class of pen relies on gravitational accelerations to offset horizontal forces, increasing current velocity requires that increasingly heavy weights and more buoyant collars be employed. In practice, even low or moderate currents cause net bag deformation that reduces rearing volume and imposes severe stresses on individual net twines. Consequently, net bags are usually reinforced by a grid of vertical and horizontal cables attached to both the collar and suspended weights. Detailed engineering studies have shown that gravity cages able to withstand stresses imposed by open ocean currents [>50 cm/sec (1.64 ft)] and adequately resist net collapse are not practical (4).

Class II: Anchor-Tensioned Pens

Anchor-tensioned pens are specifically designed for use in open ocean, high-energy locations. They differ from typical gravity cages in that they do not use a flotation collar, relying instead on fixed anchors, buoys, and special rigging to maintain shape and position in the water column. In general, four vertical semisubmerged spar buoys interconnected at the top and bottom form a boxlike grid which is tensioned by a system of anchors and buoys that compensate for tidal fluctuations and wave action (4). The net bag is closed with a sewn-on cover to prevent loss of fish should the pen become submerged by wind or wave action. The net is attached to the grid at the corners.

Anchor-tensioned pens have several major advantages over conventional gravity configurations. First, anchor line tension, grid stability, and thus resistance to net bag deformation increase proportionately with water current velocity. Second, hydrodynamic forces imposed by moving water on the net enclosure are distributed uniformly to the perimeter of the net, making tearing less likely. Third, because rigid or articulated floating structures are absent and tensioned pens tend to submerge as current increases, the risk of storm damage is greatly reduced.

Although anchor-tensioned systems are technologically superior to gravity pens, in many ways, they suffer from operational disadvantages that complicate some routine farming activities. For example, individual gravity pens may be detached from their moorings and moved to another location in the farm system to facilitate grading or harvesting, or to reduce exposure to waterborne disease organisms. In contrast, Class II pens are immobile, because they require constant tension to maintain their shape and volume. This factor also makes anchor positioning and holding characteristics more critical in tensioned systems, compared with self-supporting or gravity pens. Most reported failures of tensioned systems have been attributed to anchor movement during storms (7).

Class III: Semirigid Pens

In contrast with gravity and tensioned pens, which rely on external support systems to counteract static and dynamic loads, semirigid pens are self-supporting. In general, they are structurally similar to a typical bicycle wheel: a rigid steel ring (rim) is connected by equally tensioned ropes (spokes) to a central, buoyant column (axle). The entire structure is fully enclosed by taut netting, allowing the pen to be fully submerged without loss of fish. Buoyancy is controlled by changing the volume of air within the central column (6,7). Pen volume ranges from ca. 1000 m^3 to over 20,000 m^3 (35,000–706,000 ft^3) (5).

Design characteristics of Class III cages specifically address structural and operational problems with gravity and anchor-tensioned pens, making them ideally suited for use in high current, exposed locations. These include absence of joints or articulations that may fail under high static or dynamic loads, excellent resistance to lateral net deformation by currents, and control of operating depth to avoid storms or meet the environmental requirements of the species under cultivation. Finally, semirigid pens may use simplified single-point mooring systems that permit them to change position or submerge automatically when currents exceed a predetermined threshold, making them highly resistant to storm damage (6). Despite their many advantages, self-supporting cages of this type have not gained wide acceptance within the aquaculture industry, mainly due to their high initial cost and lack of a proven track record under commercial conditions.

Class IV: Rigid Pens

Rigid pens rely on a system of jointed columns and beams forming a boxlike or cylindrical structure that directly supports the net enclosure. Because the frame is not flexible, welded wire netting is often used in place of the synthetic materials used with other pen designs. Class IV pens are supported by buoys or floats or may be organized into large, floating "barges" (2). Some designs incorporate a collar that can be inflated to move the structure up and down within the water column (8).

Rigid pens have some advantages over conventional gravity pens, insofar as they are highly resistant to current-induced net deformation. However, light weight, resilient materials such as high density polyethylene pipe are not sufficiently rigid to resist compression, bending, and torsional forces encountered in high-energy environments. Metal cage frames are costly to build and maintain, and are prone to breakage during stormy conditions. Nevertheless, rigid metal cages, which utilize welded metal mesh in place of synthetic netting have proven useful in protected locations, where predation by marine mammals is problematic.

NET BAG MATERIALS

Net pens of all designs rely on mesh enclosures to contain the cultured species and exclude predatory fish, birds, and mammals.

Materials used in modern pen bag construction fall into two general categories: flexible netting, manufactured from various natural or synthetic fibers and semirigid or rigid mesh, from extruded plastic or welded-metal materials. Most pen bags in use today are manufactured from flexible, synthetic coal or oil-based materials, such as polyamide, polyester, polyethylene, and polypropylene. Of these, polyamide (nylon) net is perhaps the most commonly used, because of its wide availability, low cost, light weight, and high-breaking strength. Natural fibers are rare because they cannot be used to manufacture knotless netting, have a relatively high diameter to strength ratio, and are prone to sun damage and rotting.

Choosing appropriate netting materials for pen bag construction requires considerable knowledge of the characteristics of the various materials and experience in matching them to the pen type and environmental conditions that will be encountered. Factors to consider include weight of the net relative to the buoyancy or structural characteristics of the collar or frame; strength, extensibility, and resistance to abrasion; resistance to fouling; and the ability to accept antifouling paints or treatments. Knotless meshes are preferred over knotted types to minimize abrasion damage to eyes, fins, and scales of the fish. Additional information on net material characteristics, net weaving techniques, and net bag design and fabrication may be found in references 1 and 2.

Modern rigid frame pens may be covered with panels of flexible netting, but rigid or semirigid mesh fabricated from extruded plastic or welded metal are widely used. Extruded-plastic netting is available in a wide variety of mesh sizes, is light weight, and somewhat more resistant to fouling than flexible netting. Compared with flexible netting of the same mesh size, plastic meshes have a lower percentage of open area and thus permit less water exchange. They also have a higher drag coefficient.

Metal meshes have been used for many years in the construction of rigid-frame pens. They are strong and easily attached to the pen frame when new, but are subject to galvanic corrosion when placed in water. Metal meshes and their supporting structures are often galvanized, and sacrificial anodes are used to extend their service life. Chain-link wire netting used in the past has been largely replaced by wire meshes that are galvanized after welding. The service life of wire mesh varies according to the quality of galvanizing, composition of the wire, and the chemical characteristics of the water. In general, the service life of wire mesh in seawater is less than two years (2).

PEN MOORING

Mooring systems include dead weights or anchors; lines, cables, and chains connecting them to the pens; and buoys or floats that compensate for tidal fluctuations. The mooring system absorbs shock loads, and its design affects the behavior and stability of the pens in inclement weather and their ability to withstand the static and dynamic loading imposed by wind, waves, and currents. This factor is especially important in high energy, open ocean locations where extreme vertical and horizontal displacement must be accommodated. Consequently, modern pens and their moorings are engineered to function as systems and must be installed together to meet design specifications.

There are two main types of mooring systems: single and multipoint. Single-point mooring systems have been suggested for semirigid pens (Class III) used in open ocean environments (6,7). Single-point mooring has the advantage of allowing the pens to move to a position of least resistance to wind, waves, and currents. However, most gravity pen systems utilize multipoint moorings, especially those which are joined by articulated linkages into large rafts, or circular pens that rely on opposing forces to maintain the shape of the collar.

A variety of anchor types can be used, depending on bottom type and wind, wave, and water-current conditions. In protected sites, concrete dead weights, that can be fabricated on shore are often used. The holding power of concrete blocks is low, and in some situations may be attached to an embedded anchor to increase security. Where substrate permits, embedded anchors may be used alone. Many types of anchors exist and should be chosen based on a thorough evaluation of the substrate characteristics at the mooring site (2,9).

Mooring lines may be attached directly to the anchor, or to a ground chain, which increases holding power, absorbs shock loads, and compensates for the rise and fall of the tide. Steel mooring lines are sometimes used, but are heavy, prone to corrosion, lack elasticity, and will abrade plastic pen collars or floats. Nylon lines are often preferred because they effectively absorb cyclic and shock loads, which would otherwise be transmitted to the anchor and cage system. Other synthetic materials, such as polypropylene or polyester, are also commonly used (2,9). Mooring lines may be connected directly to the pen frame or attached first to one or more buoys sized to compensate for the weight of the line and increase mooring flexibility (10). Nylon lines or chains connect the pen to the mooring. The total length of the mooring line depends on the depth of the water at the site, substrate composition, and type of anchor used.

BIBLIOGRAPHY

1. P.H. Milne, *Fish and Shellfish Farming in Coastal Waters*, Fishing News (Books) Ltd., London, 1972.
2. M.C.M. Beveridge, *Cage Aquaculture*, Fishing News Books, Oxford, 1987.
3. C.V.M. Mahnken, *Proc. World Maricult. Soc.* **10**, 280–305 (1979).
4. G.F. Loverich and L. Gace, in C.E. Helsley, ed., *Open Ocean Aquaculture: Proceedings of an International Conference*, University of Hawaii Sea Grant Program #CP-98-08, 1997, pp. 131–144.
5. M.D. Chambers, in C.E. Helsley, ed., *Open Ocean Aquaculture: Proceedings of an International Conference*, University of Hawaii Sea Grant Program #CP-98-08, 1997, pp. 77–87.

6. G.F. Loverich and C. Goudey, in M. Polk, ed., *Open Ocean Aquaculture: Proceedings of an International Conference*, New Hampshire/Maine Sea Grant Program #UNHMP-CP-SG96-9, 1996, pp. 495–512.
7. L.Y. Bugrov, in M. Polk, ed., *Open Ocean Aquaculture: Proceedings of an International Conference*, New Hampshire/Maine Sea Grant Program #UNHMP-CP-SG96-9, 1996, pp. 209–295.
8. U. Ben-Efraim, in M. Polk, ed., *Open Ocean Aquaculture: Proceedings of an International Conference*, New Hampshire/Maine Sea Grant Program #UNHMP-CP-SG96-9, 1996, pp. 327–335.
9. S. Kery, in M. Polk, ed., *Open Ocean Aquaculture: Proceedings of an International Conference*, New Hampshire/Maine Sea Grant Program #UNHMP-CP-SG96-9, 1996, pp. 297–325.
10. E. Lein, H. Rudi, O. Slaattelid, and D. Kolberg, in M. Polk, ed., *Open Ocean Aquaculture: Proceedings of an International Conference*, New Hampshire/Maine Sea Grant Program #UNHMP-CP-SG96-9, 1996, pp. 93–105.

See also CAGE CULTURE; HARVESTING; REGULATION AND PERMITTING.

NITROGEN

ROBERT R. STICKNEY
Texas Sea Grant College Program
Bryan, Texas

OUTLINE

Nitrogen is required by all living organisms, because it is an important component of protein and of other essential chemical substances. Nitrogen is taken up by plants primarily in the form of nitrate (NO_3^-) ions. Animals satisfy their nitrogen requirements through the intake of food. Nitrogenous wastes are excreted by animals in several forms: ammonia, creatine, creatinine, free amino acids, urea, and uric acid. Nitrogenous compounds are also released during the bacteriological decomposition of plant and animal matter. The primary source of nitrogen from aquaculture animals is in the form of ammonia (NH_3). Bacteria in the genus *Nitrosomonas* are responsible for nitrifying ammonia to nitrite (NO_2^-); bacteria in the genus *Nitrobacter* are responsible for the step from nitrite to nitrate. The nitrification reactions are critical to efficient biofilter operation, but they also occur in ponds and in other types of culture systems.

Other bacteria are able to denitrify nitrate and convert it to elemental nitrogen (N_2) gas. Such denitrifying bacteria can be found in the genera *Pseudomonas*, *Achromobacter*, *Bacillus*, *Micrococcus*, and *Corynebacterium* (1). Energy for the reduction reactions involved may come from certain carbohydrates and alcohols. Recirculating water systems have been designed that employ special chambers to promote these reactions by supplying the proper substrate.

In pond environments, there is little concern over the accumulation of nitrate, because primary producers in the system generally remove it from the water nearly as rapidly as it is produced. In closed systems, including some hatchery systems where susceptible species such as shrimp are produced (2,3), nitrate levels may sometimes become sufficiently high to produce stress or even mortality. Various types of water systems that receive high levels of organic or inorganic fertilization can also exhibit high nitrate concentrations.

Although nitrogen is available to plants in the form of nitrate, it apparently must be reduced to ammonia once again before it can be absorbed into plant tissues (5). The reaction appears to be light-catalyzed and to proceed as follows:

$$NO_2^- + H_3O^+ \xrightarrow[\text{light}]{} NH_3 + 2O_2$$

This entry is adapted from an earlier publication (4). It is used with permission.

TOXICITY

Nitrite and ammonia are both toxic to aquatic animals at much lower concentrations than is nitrate. Nitrite is rare in natural waters, because it is an intermediate that is quickly transformed by bacteria to nitrate, but it sometimes occurs in high concentrations in aquaculture systems.

Historically, the problem has been found largely in flowing water systems and has been resolved through the incorporation of efficient biofiltration or through suitable exchange rates with new water. Nitrite toxicity has occurred in ponds when very high densities of animals are being maintained. Channel catfish (*Ictalurus punctatus*) farmers in Mississippi, for example, have experienced nitrite toxicity during the late summer or early fall, when fish biomass is at the highest level of the year, the water is warm, and the feeding rate is extremely high.

Because the growth of *Nitrobacter* is inhibited by the presence of ammonia, nitrite may become concentrated in biofilters until the ammonia concentration is greatly reduced (6). Once the biofilter begins to operate efficiently, nitrite usually ceases to be a problem unless some change occurs that results in destruction of the *Nitrobacter*. This problem can happen when chemicals are used to treat for diseases or if the system becomes anaerobic.

Nitrite toxicity in fish was reviewed by Lewis and Morris (7). Their review included species of aquaculture interest and other species. The authors indicated that salmonids were among the most sensitive fishes tested. Results from some of the studies that have been conducted with species of aquaculture importance are presented in Table 1.

Experiments to determine the LC_{50} of a chemical on an aquatic species over a discrete time period (usually not more than 96 hours) are called *acute* studies. Long-term or chronic exposure to much lower levels of a toxicant like nitrite might be lethal or could cause pathological changes. Thus, safe levels are often considered to be some fraction

Table 1. Nitrite Toxicity for Selected Species of Fish and Invertebrates of Aquaculture Importance

Species	Type of Trial	Lethal Level [mg/L (ppm)]	Citation
	Fishes		
Anguilla anguilla	96-hr LC_{50}	84–974[a]	(57)
Chanos chanos	48-hr LC_{50}	12,675[b]	(11)
Clarias batrachus	48-hr LC_{50}	35.6	(58)
	48-hr LC_{50}	15.8, 35.6[c]	(59)
Clarias lazera	96-hr LC_{50}	28,32[c]	(60)
Ctenopharyngodon idella	96-hr LC_{50}	4.62	(61)
Dicentrarchus labrax	96-hr LC_{50}	154–274[d]	(57)
I. punctatus	24-hr LC_{50}	33.8	(17)
	48-hr LC_{50}	28.8	(17)
	72-hr LC_{50}	27.3	(17)
	96-hr LC_{50}	24.8	(17)
	96-hr LC_{50}	7.1 1.9	(62)
Micropterus salmoides	96-hr LC_{50}	140.2 8.1	(62)
Morone saxatilis	24-hr LC_{50}	163	(63)
Tilapia aurea	96-hr LC_{50}	16.2 2.3	(62)
	Invertebrates		
Fenneropenaeus chinensis	24-hr LC_{50}	339	(2)
	96-hr LC_{50}	37.7	(2)
	120-hr LC_{50}	29.2	(2)
	144-hr LC_{50}	27.0	(2)
	192-hr LC_{50}	23.0	(2)
Penaeus monodon	96-hr LC_{50}	1.36[e]	(13)
	96-hr LC_{50}	0.11[e]	(13)
	24-hr LC_{50}	218[f]	(64)
	48-hr LC_{50}	193[f]	(64)
	96-hr LC_{50}	171[f]	(64)
	144-hr LC_{50}	140[f]	(64)
	192-hr LC_{50}	128[f]	(64)
	240-hr LC_{50}	106[f]	(64)

[a] Several values were obtained over a salinity range of from 0 to 36 parts per thousand. Tolerance to nitrite increased with increasing salinity. The values shown are for salinities of 0 and 36 parts per thousand.
[b] The low value was obtained in fresh water, the higher one in water of 16 parts per thousand salinity.
[c] Two sizes of fish were tested.
[d] Trials were run at temperatures ranging from 17 to 27 °C (From 63 to 81 °F). Toxicity decreased with increasing temperature.
[e] The value represents what the authors considered to be a safe level for postlarvae [1.36 mg/L (ppm)] and nauplii [0.11 mg/L (ppm)].
[f] Experiments were run on animals of 91–8 mm (3.5–0.3 in.).

(e.g., 1/10 or even 1/100) of the acute toxicity level. Most studies report the LC_{50} value, though some have developed from the LC_{50} data what they consider to be a safe level of exposure (8).

Effects of nitrite on the eggs, alevins, and fry of Atlantic salmon (*Salmo salar*) by Williams and Eddy (1989) have shown that the early development stages can tolerate very high levels: 24-hour LC_{50} values were 3,276 mg/L (ppm) for eggs and 2,940 mg/L for early alevins, decreasing to 121.8 mg/L as the alevins developed (9). However, exposure of eggs to as little as 14 mg/L (ppm) of nitrite in either fresh or brackish water delayed hatching and had measureable effects on the cardiovascular system.

The results of studies on the same species can vary considerably, as is shown in Table 1 for such species as *A. anguilla*, *I. punctatus*, and *P. monodon*. The duration of the studies is one factor, the size of the animals tested another. As indicated by Russo (10), pH, chloride concentration, and calcium concentration also affect nitrite toxicity. A study with eels and milkfish (*C. chanos*) demonstrated that salinity can have a significant influence on the tolerance of fish to nitrite (11). Different strains of channel catfish have been found to respond differently to nitrite, with some strains being susceptible to nitrite and others being resistant to nitrite (12).

Greater toxicity has been found in exposure of *P. monodon* to mixtures of ammonia and nitrate than in exposure to either chemical alone. The synergistic effects of the two forms of nitrogen became apparent after 96 hours of exposure (13).

In fish, nitrite combines with hemoglobin in the blood to produce methemoglobin and a condition known as methemoglobinemia. Hemoglobin that has been converted to methemoglobin is unable to carry oxygen, so affected animals are asphyxiated. Suspected incidents of nitrite toxicity can be confirmed quickly if the culturist sacrifices a fish and examines the blood. Nitrite is believed to enter the blood in conjunction with chloride/bicarbonate exchange. Eddy and Williams (14) concluded that fish such as salmonids that have high chloride uptake rates are much more susceptible to nitrite toxicity than fish such as carp that have low chloride uptake rates. In the common carp (*Cyprinus carpio*), there is a high correlation between chloride level and nitrite toxicity (15). On the other hand, while chloride ion increased the tolerance of channel catfish for nitrite, there was no such response in largemouth bass (*M. salmoides*) (16).

In fish with methemoglobinemia, the blood will be chocolate brown in color. In the case of channel catfish, affected fish will rest on the bottom of the culture chamber and will swim erratically for up to one minute immediately before dying. They die with their mouths open and their opercles closed (17).

Exposures to sublethal concentrations of nitrite can cause pathology in fish. Hemolytic anemia was reported in the sea bass (*D. labrax*) (18). Arillo et al. (19) concluded that liver hypoxia was the cause of mortalities in rainbow trout (*Oncorhynchus mykiss*) exposed to nitrite. Gill hypertophy has been reported in rainbow trout (20) and in tilapia of various species and their hybrids exposed to sublethal levels of nitrite (21).

Steelhead trout (*O. mykiss*) demonstrated increased tolerance to nitrite when the dietary level of vitamin C (ascorbic acid) was increased (22). It was speculated that ascorbic acid acted to reduce methemoglobin to hemoglobin, but the report also indicated that the vitamin has a protective effect against stress in fish; that effect may have played a role.

Ammonia is one of the variables in water that is of primary concern to aquaculturists. Other forms of nitrogenous waste are relatively unimportant in most cases, so it is the ammonia level in water that has generally been monitored. As we have seen, nitrite can also be a critical factor, so it is also sometimes closely watched. Colorimetric tests that can be conducted virtually anywhere have been developed and are in wide use.

Fishes excrete most of their nitrogenous waste through the gills, in the form of ammonium ion, NH_4^+ (23). Ammonium ion, or ionized ammonia, accounts for as much as 60 to 90% of the total nitrogen excreted (24–26). In addition to the ionized form, un-ionized ammonia (NH_3) occurs in water. The toxicity of ammonia to aquatic organisms is associated primarily with the level of un-ionized ammonia (27–32); the ionized form appears to be relatively harmless (33).

Both ionized and un-ionized ammonia can occur together, but the ratio between them is dependent on temperature, pH, DO, carbon dioxide concentration, bicarbonate alkalinity, and salinity (24,32,34–37). Un-ionized ammonia increases relative to ionized ammonia with increasing temperature and pH (Table 2), but it decreases as carbon dioxide increases and in hard and saline waters (36,37). Long-term exposure of aquatic animals to elevated ammonia levels can result in reduced growth, impaired stamina (38), gill abnormalities, and, ultimately, death.

Ammonia electrodes used in conjunction with a pH meter and colorimetric tests for ammonia can provide the aquaculturist with a total-ammonia value, which is satisfactory under most circumstances. Seawater interferes with the colorimetric technique for ammonia determination, so such samples should be distilled prior to testing. Ammonia concentration in the distillate is not changed, but the chemicals causing the interference are not passed in the condensed steam. Tables such as those produced by Emerson et al. (35) will provide the culturist with a means of determining that actual level of un-ionized ammonia.

Studies of ammonia toxicity have been conducted on a number of species of aquaculture interest and under a variety of conditions. Table 3 provides an indication of how such factors as pH, life-cycle stage, salinity, and the form in which ammonia is added for purposes of the bioassay influence the results. In general, coldwater fishes are less tolerant of ammonia than are warmwater species. Calamari et al. (39) proposed a water-quality standard of 0.02 mg/L (ppm) NH_3 for rainbow trout. Haywood (40), who reviewed the effects of ammonia on teleost fishes, recommended maximum total-ammonia exposure levels of

Table 2. Percentage of Total Ammonia in the Un-Ionized (NH_3) Form for a Few Temperatures and pH Values (36)

Temperature °C (F)	pH 6.5	7.0	7.5	8.0	8.5
16 (61)	0.1	0.3	0.9	2.9	8.5
18 (64)	0.1	0.3	1.1	3.3	9.8
20 (68)	0.1	0.4	1.2	3.8	11.2
22 (72)	0.1	0.5	1.4	4.4	12.7
24 (75)	0.2	0.5	1.7	5.0	14.4
26 (79)	0.2	0.6	1.9	5.8	16.2
28 (83)	0.2	0.7	2.2	6.6	18.2
30 (86)	0.3	0.8	2.5	7.5	20.3

Table 3. Ammonia Toxicity for Selected Species of Fishes and Invertebrates of Aquaculture Importance

Species and Conditions	Type of Trial	Lethal Level [mg/L (ppm)]	Citation
	Fishes		
Anguilla japonica			
pH = 5	24-hr LC_{50}	2844[a]	(44)
pH = 7	24-hr LC_{50}	820[a]	(44)
pH = 9	24-hr LC_{50}	16.8[a]	(44)
C. chanos	24-hr LC_{50}	1.89[b]	(65)
	48-hr LC_{50}	1.46[b]	(65)
	72-hr LC_{50}	1.25[b]	(65)
	96-hr LC_{50}	1.12[b]	(65)
C. batrachus	48-hr LC_{50}	15.78[b]	(59)
C. idella			
26 days old	96-hr LC_{50}	0.57[b]	(66)
47 days old	96-hr LC_{50}	1.61[b]	(66)
125 days old	96-hr LC_{50}	1.68[b]	(66)
47 days old	48-hr LC_{50}	1.73[b]	(66)
60 days old	48-hr LC_{50}	2.05[b]	(66)
125 days old	48-hr LC_{50}	2.14[b]	(66)
C. carpio			
small fry	48-hr LC_{50}	1.87[b]	(67)
small fry	96-hr LC_{50}	1.84[b]	(67)
small fry	168-hr LC_{50}	1.78[b]	(67)
larger fry	48-hr LC_{50}	1.76[b]	(67)
larger fry	96-hr LC_{50}	1.74[b]	(67)
larger fry	168-hr LC_{50}	1.68[b]	(67)
I. punctatus	24-hr LC_{50}	2.77[b]	(56)
pH 8.8[d]	24-hr LC_{50}	1.91[b]	(45)
pH 8.0[d]	24-hr LC_{50}	1.45[b]	(45)
pH 7.2[d]	24-hr LC_{50}	1.04[b]	(45)
pH 6.0[d]	24-hr LC_{50}	0.74[b]	(45)
pH 8.8[e]	24-hr LC_{50}	2.24[b]	(45)
pH 8.0[e]	24-hr LC_{50}	1.75[b]	(45)
pH 7.2[e]	24-hr LC_{50}	1.16[b]	(45)
pH 6.0[e]	24-hr LC_{50}	0.81[b]	(45)
	96-hr LC_{50}	1.5–3.1[b]	(68)
M. salmoides	96-hr LC_{50}	0.7–1.2[b]	(68)
O. mykiss			
Eggs to hatch	96-hr LC_{50}	0.49[b]	(39)
70-day-old fry	96-hr LC_{50}	0.16[b]	(39)
Fingerlings	96-hr LC_{50}	0.44[b]	(39)
	96-hr LC_{50}	0.3[b]	(68)
O. tshawytscha parr			
Freshwater	24-hr LC_{50}	0.36[b]	(69)
9.6 ppt salinity[c]	24-hr LC_{50}	2.2[b]	(69)
Sparus aurata	96-hr LC_{50}	23.7[a]	(50)
Oreochromis aureus	48-hr LC_{50}	2.40[b]	(42)
Oreochromis mossambicus × *O. niloticus* hybrid	48-hr LC_{50}	6.6[b]	(70)
	72-hr LC_{50}	4.07[b]	(70)
	96-hr LC_{50}	2.88[b]	(70)
	Invertebrates		
Mercenaria mercenaria			
4 mm (0.2 in.)	30-day LC_{50}	20.0[a]	(71)
6 mm (0.3 in.)	30-day LC_{50}	28.0[a]	(71)
10 mm (0.4 in.)	30-day LC_{50}	34.5[a]	(71)

Table 3. *Continued*

Species and Conditions	Type of Trial	Lethal Level [mg/L (ppm)]	Citation
Fenneropenaeus chinensis	24-hr LC_{50}	3.29[b]	(2)
	48-hr LC_{50}	2.10[b]	(2)
	90-hr LC_{50}	1.53[b]	(2)
	120-hr LC_{50}	1.44[b]	(2)
P. monodon	96-hr LC_{50}	1.69[b]	(48)
Penaeus semisulcatus	96-hr LC_{50}	23.7[a]	(46)

[a]Reported as mg/L (ppm) total ammonia.
[b]Reported as mg/L (ppm) un-ionized ammonia.
[c]ppt is parts per thousand.
[d]Ammonium chloride.
[e]Ammonium sulfate.

1.0 mg/L (ppm) for salmonids and 2.5 mg/L (ppm) for other freshwater and marine fishes.

Thurston et al. (41) found that rainbow trout and cutthroat trout (*Salmo clarki*) were more tolerant of constantly elevated ammonia levels than of fluctuating concentrations. Thurston and Russo (42) indicated that the median tolerance limit for un-ionized ammonia ranged from 0.16 to 1.1 mg/L (ppm) in rainbow trout, with susceptibility to ammonia decreasing as the fish developed from sac fry to the juvenile stage. They also found that toxicity decreased as temperature increased, over the range of 12 to 19 °C (54 to 66 °F). Redner and Stickney (43) found that *O. aureus* could develop an increased tolerance to ammonia when exposed to sublethal levels in advance of bioassays.

Both fish (44,45) and shrimp (46) become more tolerant of elevated ammonia with increasing pH. Shrimp appear to be more sensitive to ammonia in the periods just before, during, and after ecdysis (47,48). Further, the toxicity of ammonia is enhanced as DO concentration is reduced (49). The relationship between ammonia toxicity and DO has also been shown in gilthead seabream (50).

Sublethal concentrations of ammonia can cause histological changes in fish (39,43,51–53) and will also affect growth (54,55). Gill hyperplasia is a common sign of chronic ammonia toxicity (52,56).

BIBLIOGRAPHY

1. T.L. Meade, Marine Memorandum No. 40, Marine Advisory Service, University of Rhode Island, Kingston, 1966.
2. J.-C. Chen, Y.Y. Ting, J.-N. Lin, and M.-N. Lin, *Marine Biology* **107**, 427–431 (1990).
3. P.R. Muir, D.C. Sutton, and L. Owens, *Marine Biology* **108**, 67–71 (1991).
4. R.R. Stickney, *Principles of Aquaculture*, John Wiley & Sons, New York, 1994.
5. G.E. Fogg, *Photosynthesis*, Elsevier, New York, 1972.
6. H. Lees, *Biochemistry Journal* **52**, 134–139 (1952).
7. W.M. Lewis Jr. and D.P. Morris, *Transactions of the American Fisheries Society* **115**, 183–195 (1986).
8. J.-C. Chen and T.-S. Chin, *Aquaculture* **69**, 253–262 (1988).
9. E.M. Williams and F.B. Eddy, *Canadian Journal of Fisheries and Aquatic Science* **46**, 1726–1729 (1989).
10. R.C. Russo, in G.M. Rand and S.R. Petrocelli, eds., *The Fundamentals of Aquatic Toxicology: Methods and Applications*, Hemisphere, Washington, DC, 1984, pp. 455–474.
11. J.M.E. Almendras, *Aquaculture* **61**, 33–40 (1987).
12. J.R. Tomasso and G.J. Carmichael, *Journal of Aquatic Animal Health* **3**, 51–54 (1991).
13. J.-C. Chen and T.-S. Chin, *Journal of the World Aquaculture Society* **19**, 143–148 (1988).
14. F.B. Eddy and E.M. Williams, *Chemical Ecology* **3**, 1–38 (1987).
15. M.R. Hasan and D.J. Macintosh, *Aquaculture and Fishery Management* **17**, 19–30 (1986).
16. J.R. Tomasso, *Aquatic Toxicology* **8**, 129–137 (1986).
17. M. Konikoff, *The Progressive Fish-Culturist* **37**, 96–98 (1975).
18. G. Scarano, M.G. Saroglia, R.H. Gray, and E. Tibaldi, *Transactions of the American Fisheries Society* **113**, 360–364 (1984).
19. A. Arillo, E. Gaino, C. Margiocco, P. Mensi, and G. Schenone, *Environmental Research* **34**, 135–154 (1984).
20. E. Gaino, A. Arillo, and P. Mensi, *Comparative Biochemistry and Physiology* **77A**, 611–617 (1984).
21. D. Lightner, R. Redman, L. Mohney, G. Dickenson, and K. Fitzsimmons, in R.S.V. Pullin, T. Bhukaswan, K. Tonguthai, and J.L. Maclean, eds., *The Second International Symposium on Tilapia in Aquaculture, ICLARM Conf. Proc. No. 15*, International Center for Living Aquatic Resources Management, Manila, Philippines, 1988, pp. 111–116.
22. O. Blanco and T. Meade, *Reviews in Tropical Biology* **28**, 91–107 (1980).
23. P.W. Hochachka, in W.S. Hoar and D.J. Randall, eds., *Fish Physiology, Vol. 1*, Academic Press, New York, 1980, pp. 351–389.
24. G.M. Smith, *Fresh-Water Algae of the United States*, McGraw-Hill, New York, 1950.
25. J.D. Wood, *Journal of Biochemical Physiology* **36**, 1237–1242 (1958).
26. R.O. Fromm, *Comparative Biochemistry and Physiology* **10**, 121–128 (1963).
27. W.A. Chipman, Jr., *The Role of pH in Determining the Toxicity of Ammonium Compounds*. Ph.D., Dissertation, University of Missouri, Columbia, 1934.
28. K. Wuhrmann and H. Woker, *Schweiz. Zur Hydrol.* **11**, 210–214 (1948).
29. K. Wuhrmann, F. Zehender, and H. Woker, *Vierteljahrsschr. Naturforsch. Ges. Zür.* **92**, 198–204 (1947).
30. K.M. Downing and J.C. Merkens, *Annals of Applied Biology* **43**, 243–246 (1955).
31. J.C. Merkens and K.M. Downing, *Annals of Applied Biology* **45**, 521–527 (1957).
32. R. Lloyd, *Water & Waste Treatment Journal* **8**, 278–279 (1961).
33. K. Tabata, *Bulletin of the Tokai Regional Fisheries Research Laboratory* **34**, 67–74 (1962).
34. R. Lloyd and D.W.M. Herbert, *Annals of Applied Biology* **48**, 399–404 (1960).
35. V.M. Brown, *Water Research* **2**, 723–733 (1968).
36. K. Emerson, R.C. Russo, R.E. Lund, and R.V. Thurston, *J. Fish. Res. Board Can.* **32**, 2379–2383 (1975).
37. R.V. Thurston, R.C. Russo, and G.A. Vinogradov, *Environmental Science and Technology* **15**, 837–840 (1981).
38. R.E. Burrows, *Effects of Accumulated Excretory Products on Hatchery-Reared Salmonids, U.S. Bureau of Sport Fisheries and Wildlife Resources Report No. 66*, Washington, DC, 1964.

39. D. Calamari, R. Marchetti, and G. Vailati, *Rapp. P.-V. Reun. Ciem.* **178**, 81–85 (1981).
40. G.P. Haywood, *Canadian Technical Reports in Fisheries and Aquatic Sciences, No. 1177*, Canadian Department of Fisheries and Oceans, Ottawa, 1983.
41. R.V. Thurston, C. Chakoumakos, and R.C. Russo, *Water Research* **15**, 911–917 (1981).
42. R.V. Thurston and R.C. Russo, *Transactions of the American Fisheries Society* **112**, 696–704 (1983).
43. B.D. Redner and R.R. Stickney, *Transactions of the American Fisheries Society* **108**, 383–388 (1979).
44. Y. Yamagata and M. Niwa, *Bulletin of the Japanese Society of Scientific Fisheries* **48**, 171–176 (1982).
45. R.J. Sheehan and W.M. Lewis, *Transactions of the American Fisheries Society* **115**, 891–899 (1986).
46. J.-C. Chen and T.-S. Chin, *Asian Fisheries Science* **2**, 233–238 (1989).
47. N. Wajsbrot, A. Gasith, M.D. Krom, and T.M. Samocha, *Environ. Toxicol. Chem.* **9**, 497–504 (1990).
48. H.P. Lin, G. Charmantier, and J.-P. Trilles, *C.R. Acad. Sci., Ser. III Sci. Vie* **312**, 99–105 (1991).
49. G.L. Allan, G.B. Maguire, and S.J. Hopkins, *Aquaculture* **91**, 265–280 (1990).
50. N. Wajsbrot, A. Gasith, M.D. Krom, and D.M. Popper, *Aquaculture* **92**, 277–288 (1991).
51. J. Flis, *Acta Hydrobiol.* **10**, 205–238 (1968).
52. G. Smart, *Journal of Fish Biology* **8**, 471–475 (1976).
53. R.V. Thurston, R.C. Russo, R.J. Luedtke, C.E. Smith, E.L. Meyn, C. Chakoumakos, K.C. Wang, and C.J.D. Brown, *Transactions of the American Fisheries Society* **113**, 56–73 (1984).
54. H.R. Robinette, *The Progressive Fish-Culturist* **38**, 26–29 (1976).
55. K. Sadler, *Aquaculture* **28**, 173–181 (1981).
56. C.E. Smith and R.G. Piper, in W.E. Ribelin and G. Migaka, eds., *The Pathology of Fishes*, University of Wisconsin Press, Madison, 1975, pp. 497–514.
57. M.G. Saroglia, G. Scarano, and E. Tibaldi, *Journal of the World Mariculture Society* **12**, 121–126 (1981).
58. M. Duangsawasdi and C. Sripumun, *Acute Toxicities of Ammonia and Nitrite to Clarias Batrachus and their Interaction to Chlorides*, National Inland Fisheries Institute, Bangkok, Thailand, 1981.
59. C. Sripumun and C. Somsiri, *Thai Fisheries Gazette* **35**, 373–378 (1982).
60. A.M. Hilmy, N.A. El-Domiaty, and K. Wershana, *Comparative Biochemistry and Physiology.* **86C**, 247–253 (1987).
61. H. Wang and D. Hu, *Journal Fish. China* **13**, 207–214 (1989).
62. R.M. Palachek and J.R. Tomasso, *Canadian Journal of Fisheries and Aquatic Sciences.* **41**, 1739–1744 (1984).
63. P.M. Mazik, M.L. Hinman, D.A. Winkelmann, S.J. Klaine, B.A. Simco, and N.C. Parker, *Transactions of the American Fisheries Society* **120**, 247–254 (1991).
64. J.-C. Chen, P.-C. Liu, and S.-C. Lei, *Aquaculture* **89**, 127–137 (1990).
65. E.R. Cruz, *Fisheries Research Journal of the Philippines* **6**, 33–38 (1981).
66. Y.-X. Zhou, F.-Y. Zhang, and R.-Z. Zhou, *Acta Hydrobiol. Sin.* **10**, 32–38 (1986).
67. M.R. Hasan and D.J. Macintosh, *Aquaculture* **54**, 97–107 (1986).
68. P.J. Ruffier, W.C. Boyle, and J. Kleinschmidt, *Journal of the Water Pollution Control Federation* **53**, 367–377 (1981).
69. R.R. Harader, Jr. and G.H. Allen, *Transactions of the American Fisheries Society* **112**, 834–837 (1983).
70. S.K. Daud, D. Hasbollah, and A.T. Law, in R.S.V. Pullin, T. Bhukaswan, K. Tonguthai, and J.L. Maclean, eds., *The Second International Symposium on Tilapia in Aquaculture, ICLARM Conference Proceedings No. 15*, International Center for Living Aquatic Resources Management, Manila, Philippines, 1988, pp. 411–413.
71. F.S. Stevens, *Journal of Shellfisheries Research* **2**, 107 (1982).

See also ENVIRONMENTALLY FRIENDLY FEEDS; RECIRCULATING WATER SYSTEMS; RECIRCULATION SYSTEMS: PROCESS ENGINEERING.

NORTHERN PIKE AND MUSKELLUNGE CULTURE

ROBERT R. STICKNEY
Texas Sea Grant College Program
Bryan, Texas

OUTLINE

Northern pike (*Esox lucius*) and muskellunge (*E. masquinongy*) are members of the family Esocidae, which are popular freshwater sport fish that are reared in some state hatcheries for stocking. Northern pike can be found in temperate and arctic regions of the northern hemisphere, while the muskellunge has a limited distribution in the eastern United States and Canada. A hybrid between the two species, known as the tiger muskellunge, has been produced in hatcheries and distributed into selected water bodies.

Northern pike of 10 to 15 kg (22 to 33 lb) are common, and fish as large as 20 kg (44 lb) have been caught. Muskellunge that are caught are commonly in the same size range as northern pike, but can also reach 30 kg (66 lb). Large northern pike and muskellunge are sought after as trophy fish. Neither species is considered to be a foodfish.

In addition to the United States, esocids are produced in Canada, Germany, Austria, Switzerland, the Netherlands, France, Belgium, and Sweden. Their culture has been reviewed by Westers and Stickney (1). Low hatchability, cannibalism, and highly variable production rates in hatcheries have been cited as major barriers to large-scale propagation.

SPAWNING AND HATCHING

It is rare for captively reared broodstock to be used in the production of northern pike and muskellunge. More commonly, broodfish are collected from nature and taken to hatcheries for spawning. Adults, as well as fingerlings, can be readily sexed, through external examination of the urogenital region.

In nature, spawning of northern pike occurs in the spring at temperatures ranging from 5 to 10 °C (41 to 50 °F), while muskellunge spawn later, when the water temperature is between 10 and 14 °C (50 and 57 °F). Large females of both species produce about 100,000 eggs each year. Some hybridization has been observed in nature, though the two species do not commonly live in the same water bodies, because fingerling northern pike may already be present when muskellunge fry become available as food.

Low hatchability has been a problem for northern pike and muskellunge culturists. Broken eggs were once a significant problem, but the practice of anesthetizing broodfish to immobilize them during egg taking has alleviated the problem to a considerable extent. Anesthetizing both male and female broodfish also reduces the risk of injury to hatchery personnel by the thrashing about of the fish.

Carp pituitary hormone has been used to accelerate ovulation in females. Male northern pike and muskellunge tend to produce milt in small amounts, but milt production can be increased two-to threefold by injecting males with appropriate dosages of the hormone progesterone. Very small amounts of milt can successfully fertilize all of the eggs from a female, and the sperm can be stored for several days at temperatures from 3 to 5 °C (37 to 41 °F). Both eggs and milt can be obtained by stripping the adults, though milt can be more effectively obtained with the use of a catheter. Catheterization avoids contact between water and the milt; once exposed to water, sperm are viable only for a minute or two.

Incubation is usually conducted in some type of hatching jar. The eggs are very susceptible to mechanical stress during the first several days of incubation, so they are kept in static or nearly static conditions during that period. Water flow through the jars is then increased to remove metabolites and maintain a high level of dissolved oxygen. Under proper conditions, the percentage of eggs that reach the eyed stage can be increased dramatically. The optimum temperature range for incubating both species is from 9 to 13 °C (48 to 55 °F), though viable fry can be produced at somewhat lower and higher temperatures. Temperatures below about 3 °C (37 °F) and above 24 °C (75 °F) are lethal to embryos and sac fry.

Approximately 12 days at 10 °C (50 °F) are required to hatch northern pike eggs. Muskellunge eggs will hatch in 18 days at the same temperature. Under ambient conditions, it can take as long as six hours for a batch of eggs to hatch. This period can be greatly reduced by rapidly elevating the temperature several degrees [e.g., from an initial temperature of 10 °C to a final temperature of 16 °C (50 °F to 60 °F)] when the eggs first begin to hatch.

LARVAL REARING AND FINGERLING PRODUCTION

Once the fry hatch, they can be placed in incubators through the period of yolk sac absorption. The technique, which was developed by Michigan fish culturists, employs Heath trays, which are commonly used to incubate salmon eggs. The standard screens in the bottom of the trays need to be replaced with screens of a finer mesh size, to retain the fry. A typical Heath tray can hold from 30,000 to 35,000 fry. The fry are kept in the trays for 6 to 10 days and are provided with water at 16 to 17 °C (61 to 63 °F).

There is some production of fingerlings in raceways and cages, but the most common approach is to stock fry into ponds for rearing. Pond stocking rates have varied from 25,000 to 250,000 fry per ha (10,000 to 100,000 fry per acre), but it is recommended that stocking be within the range of 80,000 to 125,000 per ha (32,000 to 50,000 per acre).

Prior to stocking, ponds should be fertilized to induce zooplankton blooms. Fertilization rates vary, depending on the response of a given pond, but a typical scheme might involve the application of from 300 to 400 kg/ha (300 to 400 lb/acre) of alfalfa meal or pellets. Using 15 kg/ha (15 lb/acre) of dry-matter swine manure daily until the proper bloom is obtained has also worked well for some culturists.

Northern pike and muskellunge can become cannibalistic within a few days after first feeding. Cannibalism can be reduced by maintaining a good zooplankton bloom. In most cases, the fish are captured and stocked by the time they reach about 8 cm (3 in.) in length. At that size, they have a fairly good chance of survival after stocking, and losses in the ponds due to cannibalism may not have reached catastrophic levels.

Intensive culture of esocid fingerlings in raceways has been developed in recent years. The fish are initially fed live food, but can be trained to accept pelleted rations. The typical prepared feed for northern pike and muskellunge contains 50% or more protein, primarily from fish meal and other animal protein sources. Tiger muskellunge are often used in raceway culture, because they are much easier to convert from live to prepared feeds than are either northern pike or muskellunge.

Fry need to be fed to excess every 3 to 5 minutes for 15 hours daily to keep the rate of cannibalism low. Once the fish reach about 10 cm (4 in.), they can be fed every 15 minutes, though in most cases the fish will be stocked into lakes before reaching that size. Automatic feeders are usually used to supply the feed, since the time between feedings is so short. It is necessary to siphon waste feed and feces from the raceways at least daily to avoid deterioration of water quality.

CONCLUSIONS

A considerable amount of progress in the culture of northern pike and muskellunge has been made since these fishes were first spawned in captivity in the late 19th century. There is still much to be accomplished, however. The nutritional requirements of both species need to be better elucidated, so that nutritionally complete diets can be developed. Using the proper attractants and making pellets of the proper texture may help reduce the time required to train the fish during conversion from live to prepared feeds. It may even be possible, eventually, to produce prepared feeds that will be accepted at first feeding.

Selective breeding may be used in the future to improve the growth rates of esocids during the hatchery phase and sometime later could possibly lead to some reduction in the cannibalistic tendency of these fish. Before successful breeding programs can be developed, however, it is necessary to produce captive broodstock that contain sufficient genetic diversity, so that fish geneticists can select for desirable traits.

BIBLIOGRAPHY

1. H. Westers and R.R. Stickney, in R.R. Stickney, ed., *Culture of Nonsalmonid Freshwater Fishes*, CRC Press, Boca Raton, FL, 1993, pp. 199–213.

See also LARVAL FEEDING—FISH.

NUTRIENT REQUIREMENTS

IAN FORSTER
The Oceanic Institute
Waimanalo, Hawaii

OUTLINE

The dietary requirement for a nutrient can be defined as the minimum level of that nutrient in a complete diet that will meet the physiological needs of a healthy animal. In the past century, advances in chemistry and nutrition have led to the identification and quantification of the requirement for numerous nutrients. Some of the advances in our understanding of aquatic animal nutrition have come about through the application of methodology developed on mammals and birds. In recent years, a great deal of work has been done to quantify the requirement of those nutrients. Knowledge of the nutrient requirements for a particular species allows greater flexibility in feed design and permits prediction of the performance on a formulated feed. This is of particular importance to researchers investigating alternative nutrient sources for use in commercial feeds and to those responsible for ensuring that a commercial feed will meet the needs of the customers.

Although the nutrient requirements of aquatic animals have been investigated, there is considerable uncertainty in some of the values obtained. Indeed, in many instances there are several different values reported for the requirement of a particular nutrient by the same species. This variability is, in large part, caused by differences in the methodology used in the various studies. This entry describes some of the methodology used to determine nutrient requirements of aquatic animals and some of the techniques used to minimize variability in those estimates. Other reviews of this subject include those of Baker (1), Cowey (2) and D'Abramo and Castell (3).

EXPERIMENTAL PROCEDURES/METHODOLOGY

Most of the methods employed to measure dietary requirements for nutrients involve dose–response experiments. This type of trial involves three steps: (*1*) feeding a series of diets containing a range of concentrations of the nutrient for a period of time; (*2*) analyzing one or more measured responses with respect to a model or statistical method; and, (*3*) estimating a dietary requirement from the model. This entry is almost exclusively concerned with this dose–response-type methodology.

There is a great variety in the type of responses that are monitored. Some common ones include growth rate, feed efficiency, protein retention, absence of disease signs, survival, enzyme saturation, maximum tissue storage, blood-plasma levels, oxygen consumption, and excretion rate of the nutrient. Which of these are suitable in a given case depends on the nutrient under investigation and the species used. Generally, more than one response is measured to increase the confidence in the requirement value arrived at.

Aquatic animal nutrition work has developed more recently than that for other cultured animals, and the variability of reported values for requirement estimates is often considerable. The goal of research undertaken to investigate requirement values is to provide a true representation of the actual requirement. To accomplish this, nutritionists have attempted to standardize the conditions under which their research is performed. To make use of requirement values in practical situations, however, it is necessary to extend our knowledge of the relationship between dietary nutrient level and response to include conditions other than those obtained in research settings.

MODELS

Multiple Comparison of Means

One of the simplest ways of estimating a requirement based on dose response is to compare the mean response of animals fed diets containing various levels of the nutrient under investigation using analysis of variance procedures and a multiple-means comparison test, or a range test. Using this methodology, the requirement is estimated to be the minimum level of the nutrient producing a response that is not significantly different from the response obtained from the maximum level (4,5). This approach can be useful in preliminary investigations, but is unsuitable for quantitative requirement determination for several reasons. The most important of these reasons is that the estimate of the requirement is overly dependent on the variability of the responses within each treatment, as measured by the pooled standard error. Conducting a feeding trial under conditions of poor recording of feed intake or an inappropriate feeding regimen will result in high within-treatment variability in growth and feed efficiency data, leading to an underestimation of the dietary requirement. (The mean response of animals fed diets containing less than the requirement will not be statistically different than the maximum response.) A related difficulty with using range tests for requirement studies is that the value of the estimation is influenced by the value of the error probability used to access the statistical significance of differences between values of treatment means (i.e., alpha). For instance, an alpha of 95% will give a lower estimate of the requirement than a value of 99% will. Finally, using ANOVA and range tests to analyze data from a feeding trial consisting of a series of dietary treatments containing a graded level of a nutrient is statistically indefensible. Data from this type of study should be analyzed using some sort of regression analysis.

Quadratic Model

The quadratic model has been used to estimate the requirement of protein and amino acids in fish (6,7). In this method, a quadratic equation is used to fit the response data obtained from feeding a dietary series:

$$R = a + bI + cI^2 \quad (1)$$

where R is the measured response; I is the dietary nutrient concentration; and a, b, and c are constants that are calculated to provide the best fit of the data.

The value of I that produces the maximum response ($I_{\max}$) is calculated as follows:

$$I_{\max} = -0.5\,(b/c) \quad (2)$$

The requirement is then determined to be either the concentration of the nutrient that produces the maximum response (Imax) or the concentration of the nutrient that produces a response that is some arbitrary level below the maximum (e.g., 95% of maximum or 95% confidence limit of the maximum). This model assumes that feeding diets containing a nutrient at concentrations either above or below the requirement will produce responses that are less than the maximum and that the degree of the reduction will be symmetrical. Although this model has been successful in investigating optimum dietary levels for protein when a suitable response is measured, the assumptions of the model are unlikely to be met in many cases. Most nutrients do not become toxic at any level likely to be encountered by an animal, and even in cases where this may happen, there is no reason to suppose that the dose-response curve will similarly reflect this.

Enzyme Kinetic Models

A model based on enzyme kinetics has been developed to determine nutrient requirements (8,9). This model uses four parameters to fit a response equation for dietary nutrient concentrations.

$$R = [b(\mathrm{K}.5^n) + R_{\max}(I^n)]/[(\mathrm{K}.5^n) + (\mathrm{I}^n)] \quad (3)$$

where

R = measured response
I = dietary nutrient concentration
b = intercept on y-axis (y-intercept)
$R_{\max}$ = maximum theoretical response
n = apparent kinetic order
K.5 = nutrient concentration for $R = 1/2\ (R_{\max} + b)$.

The observed responses for the dietary concentrations in the experimental diets are then fitted to this equation, using standard nonlinear curve-fitting techniques, to obtain values for the four parameters (b, $R_{\max}$, n, and K.5). This model generates a sigmoid curve to characterize the dose–response of a dietary series (Fig. 1), where increasing levels of a nutrient produce a smooth, but diminishing, increase in response up to an asymptotic maximum ($R_{\max}$; Fig. 1). The requirement is then determined to be the nutrient concentration that produces a response within some arbitrary range of the asymptote (10), for example, the level (r; Fig. 1) that will reduce the maximum slope (line b; Fig. 1) of the relationship by 95% (line a; Fig. 1). For a variety of nutrient-specific reasons any nutrient is harmful at sufficiently high levels. For example, some amino acids are known to interfere with other amino acids for intestinal absorption, and if they are present at high levels, there can be an induced amino acid imbalance. Other examples include the toxicity of many minerals (and some vitamins) and the inhibitory effect of high levels of energy on feed intake. Each of these conditions will result in reduced growth, feed efficiency, and perhaps even health. To account for this, the enzyme kinetic model has been extended by the addition of a fifth parameter to include nutrient levels high enough to result in suboptimum response (11). One feature of this model claimed by its authors is that the level that produces the maximum

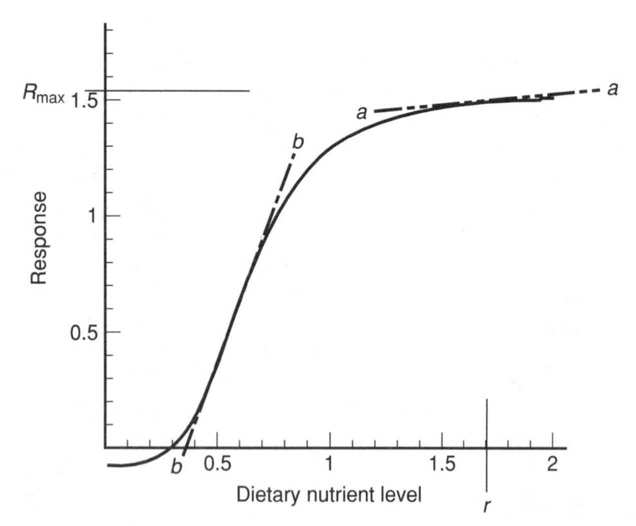

Figure 1. The relationship of physiological response to the concentration of a required nutrient, using the enzyme-kinetics model. The slope of line b has the maximum slope of the curve. The requirement (r) is arbitrarily defined as the dietary nutrient level corresponding to the point on the curve where the slope is 5% of maximum (i.e., the slope of line a).

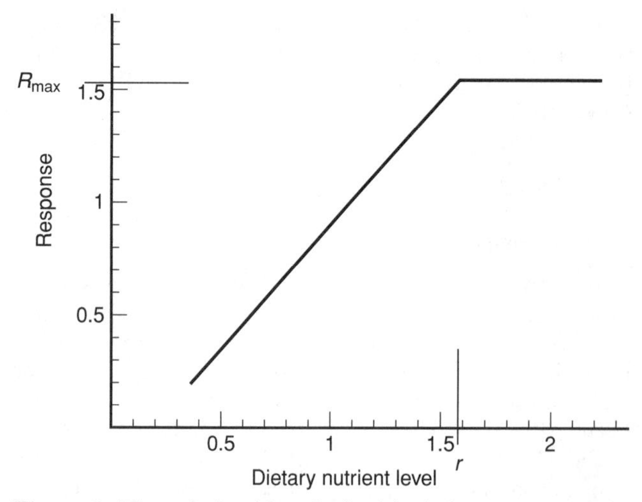

Figure 2. The relationship of physiological response to the concentration of a required nutrient, using the broken line model. The requirement (r) is determined to be the nutrient concentration corresponding to the intersection of the two parts of the line.

response (i.e., the requirement) can be calculated directly, instead of some arbitrary reduction in response or slope reduction, as previously mentioned. To this date, the five-parameter equation has been used in limited cases, and not at all in aquatic animals. Potentially, the information that can be obtained from this approach is more than merely determining nutrient requirements. If this extended model can be shown to be predictive of dose–responses over the full range of dietary nutrient concentration in aquatic animals, then this information can be used to ensure that acceptable diets are formulated. For example, this model can be used to determine if a 5% reduction in growth rate or feed efficiency is acceptable in order to utilize a cheaper, alternative nutrient source that contains suboptimum levels of some nutrients. The relative toxicity of the nutrients can also be calculated from this model (12,13).

Broken-Line Model

The broken-line model is the most common method used in estimating nutrient requirements of fish and other animals. It has the desirable features of fitting the data well in most cases, while providing a definite value for the requirement. This model assumes that the observed response is linearly related to levels of the nutrient below the requirement and that the response to dietary levels above the requirement is constant (i.e., a plateau with a slope of zero; Fig. 2). The physiological response (dependent variable) of animals fed diets containing the nutrient at concentrations below the putative requirement are fitted to a straight line against the dietary nutrient level (independent variable), while those above the putative level are averaged. The requirement is then determined to be the value of the independent variable that elicits the average value of the plateau region (i.e., the breakpoint). Researchers have sometimes used their own judgment to decide which data points are below the requirement and which are at or above this level. The use of continuous nonlinear regression analysis, available on many computer software packages, eliminates the subjective assignment of nutrient levels to the deficient or plateau region of the response (14). The model used is described by the next equation.

$$R = R_{\max} - a(r - I);\ \text{for } r > I; \qquad (4)$$
$$R = R_{\max};\ \text{for } r \leq I,$$

where

r = requirement
R = physiological response
I = dietary nutrient concentration
a = slope of the line below the requirement
R_{max} = the theoretical maximum response.

Despite its prevalence in requirement work, the use of the broken-line model has been criticized because of its discontinuity at the requirement level, where the slope abruptly becomes zero, instead of gradually leveling off. In some studies, a progressive change in the slope of the relationship of the response to the dietary nutrient level, especially at nutrient levels only slightly less than the requirement, seems to better describe of the physiological response pattern than does a broken line. Baker (1,3) proposed an explanation for this curvilinear response pattern based on the individual requirements of each member of the test population. At dietary nutrient levels slightly below the calculated requirement, the requirement will be met for some of the test animals, resulting in a change in the overall slope in response to higher levels. As the nutrient is increased to levels near the overall requirement, the slope of the response curve tends toward zero as the requirement is met for an increasing proportion of the test population. According to this explanation, reducing the variability within the test animals should reduce the deviation of the relationship from linearity. Reducing the variability in test animals is further discussed later. Where the deviation from linearity of the deficient region of the response curve is apparent, alternative models (e.g., the saturation kinetics model discussed here) may fit the data better and should be considered.

A/E Ratio

It has been observed that, in many species of aquatic animals, whole body essential amino acid (EAA) patterns are highly correlated to the reported requirement values. On the basis of this correlation, Arai (15) proposed the concept of A/E ratio to formulate diets for fish. The A/E ratio for a specific EAA relates the level of this amino acid in the whole body of fish to the total amount of all EAA, including cystine and tyrosine.

$$\text{A/E ratio} = 1000\,(\text{AA}_x/\text{EAA}_{\text{total}}) \tag{5}$$

where

AA_x = whole body content of a specific EAA
$\text{EAA}_{\text{total}}$ = total whole body content of all EAA, including cystine and tyrosine.

Animals fed diets containing amino acids in proportion to their A/E ratios have shown good growth and feed efficiency (16).

A/E ratios provide an index of relative balance among essential amino acids, but are not themselves requirements. Once the requirement for one EAA has been determined, the A/E ratios can be used to estimate the requirements for the others (17–19).

OTHER EXPERIMENTAL CONSIDERATIONS

Range of Nutrient Levels

To estimate a requirement value for a nutrient by the methods just described, it is necessary to measure responses over a range of dietary levels, including both those below and those above the requirement. Establishing the slope of the line in the deficient region (below the requirement) of the broken-line model requires at least two points, and an equal number of points in the plateau region (above the requirement) are required to ensure that the slope of the plateau is zero. Thus, a minimum of four dietary nutrient levels are required to use the broken line method, although to ensure that a sufficient numbers of points are above and below the requirement it is prudent to include five or more levels. A similar number of dietary levels are needed to use the enzyme kinetic model, as well. For ease of calculation, the levels should be equally spaced and cover a wide range. Nutrient levels that are considerably in excess of the requirement can be deleterious to the animal and result in depressed response. In this event, it is appropriate to select a model that can account for this (e.g., the five-parameter enzyme saturation kinetics model discussed earlier).

The nutrient content of each of the diets should be determined directly prior to commencement of the feeding trial. It is not sufficient to rely on calculated levels in most cases.

If even the approximate requirement is not known, a preliminary trial using a few widely spaced levels may be conducted to provide this information, prior to conducting a more detailed trial. In nutrition work, at least two replicate groups of animals are assigned to each treatment in most cases. This is not only insurance against the loss of a group during the course of the study, but it reduces the overall experimental error and improves the quality of the values obtained.

Duration of Trial

The duration of a trial is critically important to the reliability of the results. If a trial is carried out for an insufficient period, the differences in response to the dietary treatment may be too small to draw meaningful conclusions. The duration of the trial necessary for requirement studies depends on a variety of factors, including the response measured, the susceptibility of the animals to nutrient deficiency, and the nature of the nutrient under consideration.

Some physiological responses can be detected very rapidly and with sufficient accuracy to allow one to estimate the requirement. Enzyme activity and blood levels are examples of this type of response. Examples of responses that cannot be detected quickly include growth, feed efficiency, and survival.

The susceptibility of animals to a dietary nutrient deficiency is dependent on the initial level of the nutrient in the animal, the ability of the animal to sequester the nutrient, and the absolute need of the nutrient for growth and maintenance (i.e., the rate of nutrient turnover) by the animal. Sequestered fat-soluble vitamins and some minerals, for example, can be mobilized under conditions of dietary deficiency, with the effect that deficiency signs may be delayed. Water-soluble vitamins, essential amino acids, and some minerals, on the other hand, are essentially not stored, and deficiency signs may show more quickly.

Rearing the test animals under conditions conducive to good growth is the most effective way to reduce the time needed to detect response differences among treatments. Wherever possible, requirement studies should include a treatment where a feed that has proved to promote good growth is fed; often a high quality commercial feed. The growth rate of the animals fed that feed should be similar to what is known to be typical for the feed and the best growth rate of the animals fed the experimental diets should be comparable to that rate. In general, requirement studies should not be terminated until the average weight of the fastest growing treatment is at least three times the initial weight (i.e, a 200% increase in weight). Some researchers suggest even higher weight gains.

The time required to elicit differences in response can be minimized by maintaining the animals on the experimental diet containing the lowest level of the nutrient before the actual start of the trial, thereby partially depleting the nutrient. The duration of this preconditioning period will be longer for relatively slow-growing, adult organisms than for young, fast-growing ones. Obviously, knowledge of the dietary history of the animals used in a trial can be useful in deciding on the duration of preconditioning. Establishing the whole-body levels of the nutrient under investigation prior to the commencement and at the termination of feeding is important for some nutrients, for example, minerals and essential fatty acids, to determine if there has been accretion of the nutrient.

Variability of Test Animals

Reducing the initial variability among test animals improves the sensitivity of requirement trials (20). The simplest and most effective way of reducing the variability of the test animals is to obtain closely related individuals of the same age (ideally, siblings), that have been reared together and that are selected to be of the same size. As pointed out by D'Abramo and Castell (3), organisms of similar size may not be the same age, due to different histories of nutrition, disease or intraspecific interactions. Preconditioning the animals to be used in a requirement study is often effective in reducing the variability in response to dietary nutrient levels.

Interaction with Other Nutrients and Dietary Components

When designing a requirement study for a particular nutrient, interactions that the nutrient may have with other elements of the diet must be considered. Several such interactions are known to occur, although the effects can be species specific. A requirement for a particular nutrient obtained from a feeding trial can be over- or underestimated by the presence of other constituents. There are several ways that this can occur: The nutrient can be synthesized endogenously from precursors in the diet; the bioavailability of the nutrient can be affected by the presence of certain substances in the diet; or, the nutrient can work synergistically with other dietary components.

Some essential amino acids can interfere with the utilization of others. High dietary concentrations of lysine, for example, are known to interfere with arginine, while excess levels of the branched-chain amino acids (leucine, isoleucine, and valine) are known to interfere with each other in chickens, but not in channel catfish (*Ictalurus punctatus*) (21,22). Another example of an interaction between dietary constituents that can result in an over-estimation of requirement values is the reduction of the bioavailability of bivalent cations (e.g., zinc and calcium) in the presence of phytate (23), which is found in appreciable amounts in several plant meals (e.g., canola or rapeseed meal and soybean meal).

It is sometimes possible to account for the influence of a dietary constituent by expressing the requirement for a nutrient in terms of the level of the other constituent. For example, the minimum level of an essential amino acid that is necessary to maximize protein synthesis and growth is proportional to the level of the protein in the diet, up to the level of protein required for maximum growth. The requirement for essential amino acids is, therefore, most commonly expressed in terms of the percent of dietary protein. If the arginine requirement in a trial is found to be met by diets containing at least 2% of this amino acid, and if the test diets contained 40% protein, then the arginine requirement is reported as 5% of dietary protein. Similarly, some vitamins are expressed as amount (weight) per unit of available dietary energy. Diets containing high concentrations of crystalline amino acids, such as those commonly used in amino acid requirement trials, can have low pH. Some species of fish, especially those without an acidic stomach, cannot utilize those diets effectively, resulting in impaired growth and protein utilization. Amino acid test diets must, therefore, be neutralized before they can be used in requirement studies for these species (24,25).

Exogenous Nutrient Sources

To obtain meaningful results from requirement studies, it is important to ensure that the test animals cannot meet some or all of their requirement for the nutrient under investigation from exogenous sources, such as the water supply or from plant growth in the culture container. Many aquatic organisms can meet some or all of their requirements for some minerals directly from the surrounding water. This is particularly true of those animals that live in saltwater, of course, but it is also the case in fish reared in fresh water. Requirement estimates obtained from animals reared in conditions under which the requirements for some minerals are partially met from the water may result in deficient conditions when applied to the manufacture of feed for animals cultured in water containing only low mineral levels. It is therefore important to account for this possible source of error. One way to do this is to culture the animals in deionized water, although this is difficult to do in many cases. Generally, the level of the mineral under investigation in the culture water is reported in requirement studies. Shearer reviewed some of the methodology used to estimate mineral requirements (26).

Similarly, other required nutrients can be partially or totally met from organisms or particulate matter suspended in the water or from the consumption of algae or bacteria growing on the culture vessel. The degree to which this is a concern in requirement studies is related to the nature of the nutrient under investigation and the conditions of the trial. In many cases, passing the incoming water through a simple sand filter is sufficient to remove organisms that might interfere in this type of work. When studying animals that are capable of ingesting algae or bacteria growing on the culture vessel, it is essential to scrupulously clean the inside surfaces of the containers regularly. Algae growth can be reduced with suitable lighting, but the buildup of bacterial organisms can interfere with the trial.

Frequency of Feeding

Utilization of essential amino acids in the free form (i.e., not as part of protein) for protein synthesis is improved when the animals are fed frequently (27,28). The explanation for this is that amino acids are absorbed from the intestine much more rapidly when present in the free form than are those that are present as part of the dietary protein. As amino acids are absorbed, they become available for protein synthesis, but they are also susceptible to being used for other purposes, such as the production of high-energy compounds. If diets containing high levels of free amino acids are fed in large, infrequently spaced meals, the postprandial endogenous level of amino acids will exceed the ability of the animal to utilize them for protein synthesis, and these "excess" amino acids will be metabolized into other compounds. Feeding smaller levels of those same diets more frequently, evens out the endogenous levels of the amino acids, thereby permitting more efficient protein production. This is particularly important in amino acid requirement trials, where the majority of the dietary amino acids are often present as part of protein, except for the amino acid of interest, which is present, in large part, in the free form. If the diets are fed infrequently, then the free amino acid will arrive at the sites of endogenous protein synthesis ahead of the others and become susceptible to destruction, resulting in reduced utilization for protein synthesis and possibly inflating the estimate of the requirement. If, on the other hand, the diets are fed more frequently, then the rapidly absorbed amino acids from the diet will be met at the sites of protein synthesis by those amino acids derived from the dietary protein, which are absorbed more slowly and over a longer period of time, of previous meals. As a result, the endogenous free-amino-acid levels will be better balanced for efficient protein synthesis.

Another strategy to improve the utilization of dietary free amino acids is to blend them with a substance that will retard their absorption. Cho et al. (10) used agar in this manner to examine the dietary arginine requirement of rainbow trout.

Water Stability/Leaching

In trials conducted to ascertain the requirement for a nutrient, the form of the nutrient that is supplemented to the diet is often highly water soluble. This is the case with many amino acids, vitamins, and some minerals. Loss of appreciable amounts of the nutrient from the diet prior to ingestion will result in an overestimation of the requirement. This problem is of particular significance when studying species that ingest meals over a prolonged period of time, such as shrimp. The stability of feed pellets is of concern as well, since rapid disintegration of pellets in water increases the exposure of the nutrient to the aquatic environment, with a concomitant increase in the likelihood that the nutrient will be unavailable to the animal.

The loss of the nutrient due to leaching can be reduced to acceptable levels by ensuring that the diets are prepared using suitable binders. Embedding the nutrients directly in a binder, as previously mentioned, can further assist in decreasing the loss to leaching. Care must be taken in selecting the type and level of the binder for use in these diets, however, as some have been shown to diminish nutrient availability (29,30).

Substitution of Ingredients

In preparing a series of diets for requirement studies, nutrients or nutrient-containing ingredients are supplemented in replacement of some other ingredient. It is often important to consider this replaced ingredient. For example, when studying the requirement for an essential amino acid, the amino acid under investigation is usually added in substitution for a nonessential free amino acid, typically glutamic acid or a mixture of nonessential amino acids. In this way, the total level of free amino acids in the dietary feeds is equalized.

Form and Digestibility of Nutrient

The effective level of the nutrient in a diet may, in fact, be considerably lower than the amount provided by the ingredients used to make it. This can happen if the bioavailability of the nutrient is low or if the nutrient is partially destroyed during manufacture or storage of the diet.

The bioavailability of nutrients from different sources can vary considerably, and this can have major consequences on the results obtained from requirement trials. For example, if the nutrient from a source used in a trial is only 50% available, then the dietary level of the nutrient needed to satisfy the requirement will be twice the actual requirement. Conversely, to use data from requirement studies in the formulation of practical feeds, it is necessary to know the availability of the nutrients in the sources being contemplated for inclusion.

Essential amino acids are a somewhat controversial special case of this. As mentioned earlier, the minimum dietary amount of an essential amino acid that will maximize endogenous protein synthesis is dependent on the dietary protein content. In most amino acid requirement studies, graded amounts of an essential amino acid are added in crystalline form to a basal diet containing a mixture of protein and other amino acids to form a series of diets. The bioavailability of crystalline amino acids is generally higher than those derived from protein (approximately 100%), and the requirement, expressed as a proportion of dietary protein, based on this type of trial may underestimate the actual requirement. Instead, the requirement level should be based on the amount of available dietary protein. This value can be obtained by dividing the amount of supplemental free amino acid required to maximize growth and protein synthesis by the digestibility of the dietary protein.

Examples of nutrients that are susceptible to degradation during processing and storage are vitamin C (ascorbic acid) and essential fatty acids (highly unsaturated fatty acids). Ascorbic acid is particularly prone to destruction from the heat applied in the pelleting of diets. Using low temperature during the preparation of a diet or supplementing a diet with a chemical (L-ascorbyl-2-polyphosphate) that is not very heat labile, but which is readily transformed into active ascorbic acid once ingested, are examples of techniques that are used to counter this problem. Essential fatty acids, on the other hand, are commonly protected using a variety of commercial antioxidants (e.g., ethoxyquin, BHT, and BHA).

Precursors of Nutrients

When formulating dietary series for use in requirement studies, it is important to eliminate, or at least to account for, precursors of the nutrient. For example, beta-carotene is a precursor of vitamin A. If the feeds in a trial conducted to determine vitamin A requirement contain unaccounted levels of beta-carotene, the requirement level determined for this vitamin will be erroneously low. Two nonessential amino acids are capable of being transformed into essential ones; cysteine (into methionine) and tyrosine (into phenylalanine). Requirement studies with these amino acids usually report the levels of the sum of both amino acids.

PARAMETERS

When examining nutrient requirements using a series of diets, the responses that are appropriate depend on the experimental design and the species and size of the animals under consideration. In general, more than one response is measured. The following list describes some common ones:

Growth

Weight gain is considered by many nutritionists to be the most meaningful response in requirement studies. Weight gain is so important partly because of the unambiguity of measurement and partly because of it's comprehensiveness (all aspects of an animal's physiology must be optimized for maximum growth). There are, however, some limitations in the use of growth data. Crustaceans, for example, go through periods of molting (ecdysis), and their growth, as measured by weight gain, is discontinuous, adding to the complexity of using this criterion (3).

Growth is a function of, among other things, both the nutritional quality and the rate of consumption of the diet. If a nutrient contributes positively to the consumption of the diet at levels of inclusion in excess of its requirement, then the requirement level, as determined from growth data alone, will be overestimated. Taking this argument to it's logical conclusion, any substance that stimulates consumption rates and that is not toxic can be said to be required using growth as the sole criterion, even if there is no physiological necessity for this substance. As animals grow during the course of a requirement trial, their growth may become impaired by limitations of the culture environment. Within a species, larger animals consume higher amounts of oxygen than do smaller ones in similar conditions, and their need for living space is greater. The ability of a trial to distinguish differences in growth rates between dietary treatments can be seriously impaired if care is not taken to ensure that the culture conditions are conducive to good growth of all the groups of animals throughout the duration of a trial.

There are a number of descriptive measurements for growth used in requirement studies. The simplest is usually weight gain (WG).

$$\mathrm{WG} = (\mathrm{W_f} - \mathrm{W_i}) \tag{6}$$

where

$\mathrm{W_f}$ = the weight of the animals at the end of the trial
$\mathrm{W_i}$ = the initial weight of the animals.

The relative weight gain (RGW) is more commonly used:

$$\mathrm{RWG} = [(\mathrm{W_f} - \mathrm{W_i})/\mathrm{W_i}]100 \tag{7}$$

Another common descriptor of growth in fish is the specific growth rate (SGR). The weight gain of animals is different at different sizes. For example, the time required for a 1-g (0.03-oz) rainbow trout (*Oncorhynchus mykiss*) to double in size is less than for this same fish to grow from 100 to 200 g (4 to 8 oz), given the same conditions for optimum growth. The SGR index attempts to minimize the effect of differing growth rates of animals of the same species that are of different size.

$$\mathrm{SGR} = [(\ln \mathrm{W_f} - \ln \mathrm{W_i})/\Delta t]100 \tag{8}$$

where

$\ln \mathrm{W_f}$ = the natural logarithm of the final weight
$\ln \mathrm{W_i}$ = the natural logarithm of the initial weight
Δt = time period of the feeding in days.

Jobling points out that there is a reduction in the "growth potential" of fish with increasing body size, resulting in a gradual decrease in SGR as the fish increases in size (31). He therefore suggests the use of a linear equation that relates the logarithm of SGR with the logarithm of fish size as a more suitable approach to relating results of experiments with a specific nutrient conducted on animals of differing sizes or under different exeperimental conditions.

Another measurement of growth, the daily growth coefficient (DGC), was used by Cho et al. to estimate the arginine requirement of rainbow trout (10). The DGC is based on the cube root of the initial and final body weights, so it is little affected by differences in initial body weight:

$$\mathrm{DGC} = [(\sqrt[3]{\mathrm{wt}} - \sqrt[3]{\mathrm{wi}})/\Delta t]100 \tag{9}$$

Feed Efficiency

Feed efficiency (FE) is defined as gain in body weight divided by the dry weight of diet consumed. Using FE as a response eliminates the aformentioned difficulty for weight-gain data associated with different feed intakes occurring among animals in different treatments for reasons unrelated to the nutritional value of the diets. Calculating FE poses it's own difficulties, however. Because it is the ratio of weight gain and feed consumption, the experimental error of FE is higher than both of these. Feed consumption is difficult to measure accurately in many situations, principally because of problems in determining the amount of the diet that is presented to the animals, but not consumed. Such uncertainty seriously erodes the value of FE information. Under laboratory conditions, it is often possible to reduce the wasting of feed to acceptable levels by careful feeding. Indeed, it is possible to determine the actual consumption of the diets by accounting for feed wastage, by counting uneaten pellets, for example. When feed intake can be accurately measured, FE is generally preferable to weight gain in ascertaining the nutritional quality of diets (32).

A related technique is to equalize the feeding for all dietary treatments over the course of the trial. Under this type of feeding regime, the animals in the dietary treatments are fed each of the test diets amounts equal to the feed rate of the lowest consuming treatment. This technique provides information about the relative nutritional quality of diets, without the complication of differing feeding rates. Equalized feeding works best on animals that are housed singly (for example, some crustaceans), but is limited in its suitability in trials for which the animals are reared communally, which is the most common type of trial. Restricting the feed intake of groups of animals that are reared together may promote the establishment of dominance hierarchies within each culture vessel, resulting in some animals consuming the majority of the meal at the expense of others in the group and an increase the likelihood of cannibalism. Such situations reduce the accuracy of the results and should be avoided.

Feed conversion ratio (FCR), the inverse of FE, is sometimes reported in nutritional work. Baker discussed why the use of this parameter is inappropriate for this type of work and should be avoided (1).

Protein Retention

Protein retention is defined as the gain in animal protein divided by the protein consumed. This parameter is commonly used in amino-acid requirement trials, as it measures the efficiency of utilization of dietary protein for endogenous protein synthesis (ignoring differences in protein catabolism). For protein synthesis to occur, all amino acids need to be available at the appropriate locations and in suitable balance. A dietary deficiency of any one of the essential amino acids will be manifested by a reduced rate of protein synthesis. The other amino acids, which are not used for protein synthesis, will instead be directed into one of a myriad of biochemical pathways, where they will be catabolized. Thus, protein retention is a more direct measure the amino acid utilization than is growth or feed efficiency.

Survival

In some situations, survival of animals fed experimental diets is a suitable parameter to assist in determining nutrient requirements. In the case of the larval stage of some organisms, there is no practical way of measuring feed intake, and survival is a good backup indicator of the nutritional quality of the diets.

Animals may be more susceptible to environmental stressors when maintained on diets that are deficient in an essential nutrient. Forster provided an example of this (33). Juvenile rainbow trout were fed a series of diets containing graded levels of arginine. The water supply of the culture system was recirculated through a biofilter that was coupled with a chiller to maintain the water

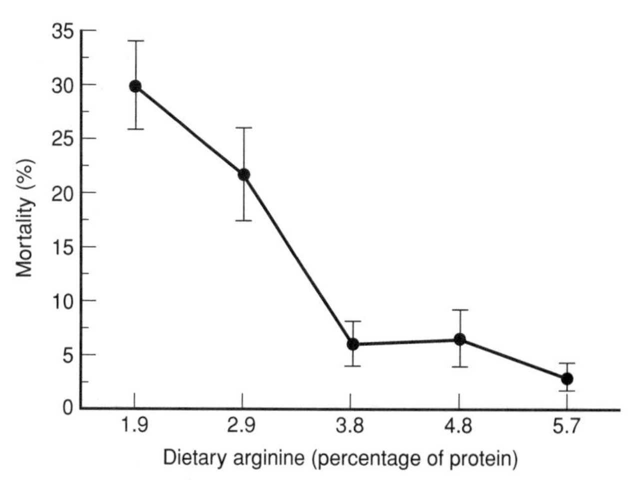

Figure 3. Cumulative mortality of rainbow trout fed diets containing various concentrations of arginine. The fish were held for four weeks in a partial recirculating fresh water system. The nitrite levels in the culture water during the fourth week frequently exceeded those known to cause stress in these fish. Each point represents the mean of 6 groups ± standard error bars.

at an acceptable temperature for this species. The water ammonia and nitrite levels were monitored and the nitrite levels were frequently close to or above the reported toxic level [0.5 mg/L (ppm) of water] for these fish in fresh water. After 28 days of feeding, the mortality was higher for the fish fed the arginine deficient diets and was linearly related to the degree of deficiency (see Fig. 3). The fish fed the diets containing arginine at or above requirement level (as determined by other studies), exhibited a much lower mortality rate.

Enzyme Activity

The activity of specific enzymes in response to dietary vitamin levels can sometimes be used to assess dietary nutrient requirements. This is especially true of some water-soluble vitamins, which, in modified form, function as coenzymes (2). For example, the hepatic enzyme FAD-dependent D-amino acid oxidase has been used to estimate the riboflavin requirement of channel catfish (*I. punctatus*) (34).

Liver and Blood Levels

The level of several water-soluble vitamins in the liver is positively related to dietary levels in excess of the amount required for maximum growth and absence of deficiency signs. The minimum dietary concentration that maximizes liver levels is used as an estimate of the requirement for these nutrients (35), even if this exceeds the requirement for maximum growth. Free essential amino acid concentration of the blood, especially of the serum or plasma, is sometimes used to estimate the requirement. At dietary levels below the requirement, the postprandial blood concentration of an essential amino acid is uniformly low, as the majority of this nutrient is utilized for synthesis of protein and other compounds (provided that all other nutrients are present in sufficient amounts). If the dietary level exceeds the requirement, then its concentration in the blood will be positively related to its dietary content. Several studies have used this model to provide information about the requirement for essential amino acids, but with variable success (10,36). Because of this uncertainty, such information can only be used in support of the values obtained using growth, FE, and PR data.

BIBLIOGRAPHY

1. D.H. Baker, *J. Nutr.* **116**, 2339–2349 (1986).
2. C.B. Cowey, *Aquaculture* **100**, 177–189 (1992).
3. L.R. D'Abramo and J.D. Castell, in L.R. D'Abramo, D.E. Conklin, and D.M. Akiyama, eds., *Crustacean Nutrition*, World Aquaculture Society, Baton Rouge, LA, 1997, pp. 3–25.
4. P.B. Brown, D.A. Davis, and E.H. Robinson, *J. World Aquacult. Soc.* **19**, 109–112 (1988).
5. M.E. Griffin, K.A. Wilson, M.R. White, and P.B. Brown, *J. Nutr.* **124**, 1685–1689 (1994).
6. Y.N. Chiu, R.E. Austic, and G.L. Rumsey, *Aquaculture* **69**, 79–91 (1988).
7. I.H. Zeitoun, D.E. Ullrey, W.T. Magee, J.L. Gill, and W.G. Bergen, *J. Fish. Res. Board. Can.* **33**, 167–172 (1976).
8. L.P. Mercer, *J. Nutr.* **112**, 560–566 (1982).
9. L.P. Mercer, S.J. Dodds, and J.M. Gustafson, *Nutrition Reports International* **34**, 337–350 (1986).
10. C.Y. Cho, S. Kaushik, and B. Woodward, *Comp. Biochem. Physiol.* **102A**, 211–216 (1992).
11. L.P. Mercer, H.E. May, and S.J. Dodds, *J. Nutr.* **119**, 1465–1471 (1989).
12. L.P. Mercer, *J. Nutr.* **122**, 706–708 (1992).
13. L.P Mercer, T. Yi, and S.J. Dodds, *J. Nutr.* **123**, 964–971 (1993).
14. K.R. Robbins, H.W. Norton, and D.H. Baker, *J. Nutr.* **109**, 1710–1714 (1979).
15. S. Arai, *Bull. Jpn. Soc. Sci. Fish.* **47**, 547–550 (1981).
16. H. Ogata, S. Arai, and T. Nose, *Bull. Jpn. Soc. Sci. Fish.* **49**, 1381–1385 (1983).
17. R.P. Wilson and W.E. Poe, *Comp. Biochem. Physiol.* **80B**, 385–388 (1985).
18. I. Forster and H.Y. Ogata, *Aquaculture* **161**, 131–142 (1998).
19. H.Y. Moon and D.M.G., III, *Aquaculture* **95**, 97–106 (1991).
20. B.E. March, C. MacMillan, and F.W. Ming, *Aquaculture* **47**, 275–292 (1985).
21. E.H. Robinson, R.P. Wilson, and W.E. Poe, *J. Nutr.* **111**, 46–52 (1981).
22. E.H. Robinson, W.E. Poe, and R.P. Wilson, *Aquaculture* **37**, 51–62 (1984).
23. N.L. Richardson, D.A. Higgs, R.M. Beams, and J.R. McBride, *J. Nutr.* **115**, 553–567 (1985).
24. R.P. Wilson, D.E. Harding, and D.L. Garling, *J. Nutr.* **107**, 166–170 (1977).
25. T. Nose, S. Arai, D. Lee, and Y. Hashimoto, *Bull. Jpn. Soc. Sci. Fish.* **40**, 903–908 (1974).
26. K.D. Shearer, *Aquaculture* **133**, 57–72 (1995).
27. E.S. Batterham and G.H. O'Neill, *Br. J. Nutr.* **39**, 265–270 (1978).
28. D.H. Baker and O.A. Izquierdo, *Nutr. Res.* **5**, 1103–1112 (1985).
29. T. Storebakken, *Aquaculture* **47**, 11–26 (1985).
30. T. Storebakken and E. Austreng, *Aquaculture* **60**, 121–131 (1987).
31. M. Jobling, *J. Fish. Biol.* **22**, 153–157 (1983).
32. D.H. Baker, *Nutr. Rev.* **42**, 269–273 (1984).
33. I.P. Forster, Ph.D., Dissertation, University of Washington, Seattle, 1993.
34. G. Serrini, Z. Zhang, and R.P. Wilson, *Aquaculture* **139**, 285–290 (1996).
35. J.E. Halver, in J.E. Halver, ed., *Fish Nutrition*, Academic Press, New York, 1989, pp. 31–109.
36. M.J. Walton, C.B. Cowey, R.M. Coloso, and J.W. Adron, *Fish Physiol. Biochem.* **2**, 161–169 (1986).

See also DIGESTIBILITY; ENVIRONMENTALLY FRIENDLY FEEDS; FEED EVALUATION, CHEMICAL; FEED HANDLING AND STORAGE; INGREDIENT AND FEED EVALUATION; LARVAL FEEDING—FISH; LIPIDS AND FATTY ACIDS; MICROALGAL CULTURE; MICROBOUND FEEDS; MICROPARTICULATE FEEDS, COMPLEX MICROPARTICLES; MICROPARTICULATE FEEDS, MICRO ENCAPSULATED PARTICLES.

OFF-FLAVOR

ROBERT R. STICKNEY
Texas Sea Grant College Program
Bryan, Texas

OUTLINE

Incidents and Causes
Amelioration

Undesirable flavors and odors sometimes occur in various species of fish and shellfish and can negatively affect the acceptability of affected seafoods to consumers. Many off-flavor problems, such as off-flavors associated with oxidative rancidity and bacterial spoilage, occur after the animals are processed. Since those problems occur, in the majority of cases, after the aquaculture species leave the control of the producer, they are not considered here. Rather, the emphasis here is on off-flavors that occur when the animals are under culture.

Off-flavors can occur in response to ingredients in prepared feeds. For example, if high levels of fish oil are added to a prepared feed, the animals that consume that feed may develop a "fishy" flavor, which is often considered to be undesirable by consumers who are looking for a more bland taste. That problem can easily be overcome by adjusting the feed formulation to reduce or remove the offending ingredient(s).

Another source of off-flavors is pollutants. Hydrocarbons, for example, can produce off-flavors in aquatic animals. If there is contamination of an aquaculture facility, for example, by an oil spill, it is unlikely that the affected animals would ever appear in the market. Aquaculturists may experience such problems, but in all cases take every precaution to avoid them.

The off-flavors discussed here are those associated with blooms of certain organisms, usually blue-green algae, which can occur naturally in aquaculture systems. Chemicals in the algae can give fish what has been referred to as an earthy or musty flavor. The presence of off-flavor from this source can have devastating economic ramifications to both producers and processors. Faced with off-flavored fish, consumers have tried to mask or neutralize the problem with mustard, lemon juice, and other substances before the meat is cooked, though usually to no avail. The best approach is to ensure that no off-flavor fish ever reach the marketplace.

INCIDENTS AND CAUSES

Off-flavors have been reported in a number of cultured fish and shellfish species. In the United States, the reports have been most common with respect to channel catfish (*Ictalurus punctatus*) reared in ponds. As high-density culture in recirculating water systems has developed in recent years, reports of off-flavors in fish reared in such systems have also appeared. Tilapia, which have rarely demonstrated off-flavors in ponds in the United States, have been known to develop the problem in recirculating systems. Catfish and other species also have developed off-flavors in recirculating systems when held at high density.

Flavor problems in fish can emanate from a number of sources, including, as previously mentioned, postharvest problems and pollution. While contamination of cultured fish with pollutant chemicals is a possibility, virtually all off-flavor problems of aquacultured fishes are natural in origin. While affected fish may have an objectionable taste, they do not appear to pose a health threat to consumers.

The dominant sources of the earthy or musty flavor that has been reported from cultured animals are 2-methylisoborneol and geosmin. Produced by many species of blue-green algae, possibly by actinomycetes, and perhaps by other types of organisms, 2-methylisoborneol and geosmin are absorbed into fish flesh primarily through diffusion across the gills. Secondarily, absorption may occur as a result of normal feeding activity and the incidental ingestion of organisms containing the chemicals.

The same chemicals that produce off-flavors in aquatic animals are, in fact, a common problem in drinking water. While there has been a significant economic price associated with off-flavors in aquaculture products, it is undoubtedly trivial compared with that associated with off-flavors in domestic water supplies.

With respect to the early years of the commercial catfish industry, off-flavor incidence often increased late in the growing season, when the weather was hot, pond biomass was high, and other conditions promoted the growth of blue-green algae. With the adoption of intermittent harvest, fish densities and biomasses continued to be high throughout much of the year, so the period during which the problem was expected to occur was extended.

Similarly, fish reared in recirculating systems are held at high density and biomass. In addition, if an off-flavor chemical is produced within a closed system, it will be retained within the system and can become concentrated, because of the low water-turnover rates within such systems. Frequent replacement of the water might reduce or eliminate the problem, but that approach runs counter to the very basis upon which closed-system aquaculture is established.

AMELIORATION

Geosmin, 2-methylisoborneol, and perhaps other chemicals responsible for off-flavors in fish and shellfish will be metabolized, excreted, or diffused back into the water within several hours or a few days if the source of the chemical is removed. Moving affected fish from a pond into flowing well water is an excellent way to deal with the problem. However, that approach may be impractical if appropriate facilities and supplies of water are not available, and it will add to the expense of rearing the fish.

A second approach is to wait. In most instances, algae blooms in ponds will eventually decline, and the source of the off-flavor chemicals will be removed. Once that happens, the chemicals in the fish will ultimately be eliminated. In closed systems, the dynamics of microorganisms may be quite different than those in ponds. The off-flavor chemicals may persist for long periods (a problem that can also occur in ponds that are harvested intermittently), so purging the fish by maintaining them in water free of the off-flavor chemicals prior to processing may be necessary. Alternatively, as previously mentioned, the system might be flushed with new water, to allow the fish to eliminate the off-flavor chemicals prior to harvest.

The channel catfish industry was the first to face the problem of off-flavors. Recognizing that the consumer who purchases or is served an off-flavor catfish may be reluctant to purchase or order the product in the future, a quality-control process was implemented by the fish processors. Farmers were instructed to collect a fish from each pond that was scheduled to be harvested and take it to the processor a week or two in advance of the harvest date. A portion of each fish is then cooked in a microwave oven and evaluated for odor and taste. If there is doubt on the part of the inspector as to whether a particular fish has an off-flavor, a second opinion is sought.

If the pond passes the first evaluation, the farmer brings another fish to the processor to undergo the evaluation process a few days (usually three) before harvest. If the pond passes the second test, the producer is given permission to harvest. On the day that the fish are delivered to the processor, but before they are unloaded, a third evaluation is made from a random sample of the fish on the truck. Only after the pond passes the third test is the load of fish accepted by the processor.

If an off-flavor is detected during any of the evaluations, the fish will not be accepted at the processing plant. If an off-flavor is detected during one of the first two evaluations, the farmer will merely have to wait until the problem is resolved through natural processes. Returning a truckload of fish to a partially harvested pond poses a more significant problem, however, since the stress of harvesting and hauling can lead to disease outbreaks and, subsequently high mortality rates.

The problem of off-flavors can be very severe. I visited a catfish processing plant in Alabama and was told that the percentage of fish that had to be rejected that day was quite low. It was about 50%, while on some days, 80% or more of the ponds scheduled for harvest demonstrated off-flavors.

ORNAMENTAL FISH CULTURE, FRESHWATER

Frank A. Chapman
University of Florida
Gainesville, Florida

OUTLINE

The culture of ornamental fish is primarily for the home aquarium. Considered a luxury item, ornamental, or tropical, aquarium fish are kept for hobby and are gaining popularity as pets and companion animals.

The retail value of the aquarium hobby worldwide is estimated between US$4,000 and US$7,200 million. The aquarium industry is worth some US$1,500 million, of which the United States has the largest share. The aquarium hobby is also a major segment of the Japanese, British, German, French, Italian, Belgian, South African, and Chinese pet industries.

Traditionally not recognized as a form of aquaculture, ornamental freshwater fish culture is one of the most economically profitable areas of fish farming activities. A list of the most valuable fishery commodities worldwide is summarized in Table 1. In Florida in 1997, ornamental freshwater fish sales totaled US$57.2 million and were cultured by 203 growers in approximately 622 ha (1,536 acres) of water surface area. Another US$57 million worth of ornamental fish were exported from Singapore. (See Table 2.)

Several thousand species and hundreds of varieties of ornamental fish are sold through the pet trade, from both freshwater and marine origin. Although the largest

Table 1. The Most Valuable Fishery Commodities Worldwide[a]

Selected Fishery Commodities	US$/ton
Fish for ornamental purposes	45,564
Abalone	28,553
Caviar and caviar substitutes	15,891
Lobster, frozen	14,531
Eel, live	14,485
Liver and roe, dried or salted	14,139
Salmon, smoked	13,429
Lobster, fresh or chilled	12,262
Crab, prepared or preserved	10,678
Shark fin, dried or salted	10,404

[a]The numbers represent the average unit value of imports for 1994–1996. The data are compiled from *FAO Yearbook 1996*, Fishery Statistics, Commodities Volume 83, 1998.

Table 2. Major World Suppliers of Fish for Ornamental Purposes and Their Destinations [a]

Suppliers		Destinations	
Country	Export Value (US$000)	Country	Import Value (US$000)
Singapore	56,872	United States	75,891
United States	18,223	Japan	73,367
China (Hong Kong)	15,784	United Kingdom	26,775
Czech Republic	9,402	Germany	25,929
Japan	8,782	France	25,859
Indonesia	8,664	Singapore	14,095
Philippines[b]	8,230	Italy	12,885
Israel	7,536	Belgium	11,692
Germany	7,227	The Netherlands	9,231
Malaysia	6,999	Spain	8,129
Colombia	4,767	China (Hong Kong)	7,988
World total	193,000	World total	328,095

[a]The data are compiled from the average value of exports and imports for 1994–1996, from *FAO Yearbook 1996*, Fishery Statistics, Commodities Volume 83, 1998. Value = US$1,000.
[b]Primarily fish of marine origin.

volume of fish (approximately 95%) in the aquarium trade is from farm-raised, freshwater species, the greatest diversity of species is collected from the wild. Major sources for wild-caught freshwater aquarium fishes are the basins of the Amazon and Congo Rivers and streams in the watersheds of major rivers in India and Southeast Asia. Ornamental fish are farm raised principally in Southeast Asia and the United States. (See Table 2.) In the United States, the major center for ornamental freshwater fish production is in Florida, near both Tampa and Miami. (See Fig. 1.) Many retail fish are also raised by advanced aquarists in home garage or "backyard" facilities.

Written information on the requirements and practices for ornamental fish production is limited. Although many techniques used for ornamental fish culture are similar to those used in the production of foodfish, the husbandry methods for specific ornamental fish species are closely guarded secrets. Farmers have operated almost entirely on their own, developing their own methods and relying on many years of experimentation.

Figure 1. A beautiful farm in Florida for the aquaculture of fish for ornamental purposes. Courtesy of 5-D Tropical, Inc., Plant City, Florida.

This contribution was written to provide an overview of ornamental freshwater fish culture. Because of the great diversity of species in production, only basic information is presented on husbandry technology and marketing. Emphasis is placed on providing a summary of essential biological characteristics and requirements that typify the wide variety of species. The intent is to provide a set of guidelines that blends theory with practical applications, to assist in the development or formulation of appropriate management practices for ornamental freshwater fish culture. Also, this contribution is designed to highlight industry and research needs.

The information in this contribution is based primarily on my own experiences with the ornamental fish industry in Florida, popular ornamental fish literature, and laboratory investigations dealing with the experimental culture of several dozen species representing the major fish groups in the trade. I have avoided the direct use of references in the text, because written information on the biology and husbandry of ornamental fish is primarily found in the popular or hobbyist's literature. Instead, the bibliography contains introductory sources for more detailed information on the culture of ornamental freshwater fish and their trade. Although koi and goldfish are important ornamental fish, their culture is not emphasized.

Historical Account

Little information exists that dates the origins of ornamental fish culture, but it can be assumed that it was developed in China, where the goldfish is the traditional ornamental fish and the culture of foodfish had its beginnings (believed to have been somewhere in the year 2000 BC). Accounts of several goldfish hatcheries in China date to the Sung Dynasty between 960 and 1279 AD. By the mid-nineteenth century, the aquaculture of goldfish and koi was well established in Japan.

Contemporary aquaculture of ornamental aquarium fish can be identified with the beginnings of the hobby in England, during the years 1841 to 1852. The Germans were also aquarists and without delay became prominent fish breeders. The Americans soon followed, with records for goldfish farms in Ohio and Maryland dating to 1878 and 1889, respectively. By the early 1900s, a significant number of fish were being bred and imported into the United States. It was not until the mid-1920s and 1930s that ornamental fish keeping became popular and spread throughout the world. To supply the growing market for ornamental fish, the first commercial farms and numerous hobbyists began to raise aquarium fish in indoor tanks and outdoor pools. In the United States, the largest commercial farm at the time (in the 1920s) was Schaumberg's Crescent Fish Farm, in New Orleans, Louisiana. In Florida, the first fish farms date to the late 1920s. In Singapore, the center for world trade in ornamental fish, commercial production of ornamental fish began in the late 1940s and early 1950s.

POPULAR ORNAMENTAL FRESHWATER FISH

Over 1,000 freshwater species in about 100 families are represented in the ornamental fish trade at any one time. Despite this diversity, only about 150 species in 30–35 families are in great demand and account for the largest volume in the trade. Those species in commercial production represent close to 15 taxonomic families (e.g., Cyprinidae, Characidae, Callichthyidae, Mochokidae, Pangasiidae, Loricariidae, Melanotaeniidae, Pseudomugilidae, Telmatherinidae, Poeciliidae, Cyprinodontidae, Cichlidae, Belontiidae, and Helostomatidae). New species are constantly entering production, and there is continuous development of new varieties from species such as goldfish (*Carassius* sp.), koi (*Cyprinus* sp.), danios (*Brachydanio* spp.), angelfish (*Pterophyllum scalare*), bettas (*Betta splendens*), guppies and mollies (*Poecilia* spp.), and swordtails and platies (*Xiphophorus* spp.). Several hundred strains or varieties of these species have already been developed. A list of the most popular species in the ornamental fish trade is presented in Table 3.

GENERAL CHARACTERISTICS OF AQUARIUM FISH

Ornamental or tropical aquarium fish are numerous, geographically widely distributed and reflect great taxonomic and ecological diversity. This diversity is displayed in an astonishing variety of colors, body shapes, locomotion, behavioral patterns, reproductive tactics, feeding strategies, and other unique environmental adaptations.

Except for a few popular species, like goldfish and koi, the majority of ornamental fish are native to tropical regions of the world and cannot tolerate water temperatures below 18 °C (64 °F); hence they are given the generic name "tropical fish."

A useful distinction between freshwater and marine fish is an ability of the species to tolerate or acclimate to salinities above or below 10 parts per thousand (ppt). Popular freshwater species do not do well above 2 ppt salinity. For example, characins, certain cyprinids, and catfish are especially intolerant of saline water. Freshwater species are more prevalent in the trade than saltwater species, because of ease in care and shipping. Saltwater, or marine, ornamental species will become more popular as their husbandry and the necessary technology for their care are simplified and further developed.

Aquarium fish are relatively small and usually attain their adult body proportions and coloration in the first two to eight months of their life. Most species appearing in the market weigh between 0.3 and 40 g (0.1 and 1.4 oz) and range from 2 to 15 cm (0.8 to 6 in.) in total length. Aquarium fish regularly live 6–10 years. Some koi have been recorded as living for 70 to 80 years.

Ornamental freshwater fish are conveniently divided into egg layers and livebearers. These two modes of reproduction reflect the way in which most fish are bred and raised commercially. Livebearing fish have internal fertilization and give birth to their young as larvae, commonly referred to as fry. Egg-laying fish deposit their eggs or broadcast them for external fertilization. Among egg layers, there are mouth brooders; bubble, cavity or material nesters; open-water spawners; and substrate spawners.

PRODUCTION TECHNOLOGY

Typically, an ornamental fish farm is small and is owned and operated by a family. An average-size farm ranges from 0.5 to 6.0 hectares (1.2 to 15 acres). Customarily, farms combine the use of indoor and outdoor facilities for fish production. (See Fig. 1.) More and more farmers are beginning to produce ornamental fish strictly in indoor facilities. The use of outdoor ponds and tanks is restricted to conditioning broodstock. Some of the most advanced, specialized, and innovative techniques in aquaculture and wastewater treatment are used in the production of these fish. Because of their small size and high individual value, ornamental fish can be raised intensively and profitably indoors.

Buildings for indoor production customarily consist of modified greenhouses and wooden or steel sheds that are insulated and protected from the elements. Fish are reared in tanks or aquaria no larger than 2 m^2 (21.5 ft^2). The air and water temperatures inside each facility are closely regulated, and the photoperiod is controlled. Operation, monitoring, and maintenance of each production unit is facilitated through alarm and computer systems.

The indoor area is used primarily for breeding, hatching eggs, and raising larvae. The remaining areas of the building are devoted to holding, sorting, packaging, and shipping fish. Office space, wash areas, and an occasional laboratory bench are available. Quarantined and sick fish are sometimes treated indoors. Specialized foods are prepared and stored indoors as well. Indoors, fish are maintained primarily in concrete vats (e.g., burial vaults) and glass aquaria of various sizes. Tanks made of plastic and fiberglass are also becoming common.

Table 3. Major Taxonomic Groups of Fish for Ornamental Purposes[a]

CYPRINODONTIFORMES

Poeciliidae: *Poecilia reticulata*(1) A, *P. velifera*(13) A, *P. sphenops*(14) A, *P. latipinna*(22) A, *Xiphophorus maculatus*(5) A, and *X. helleri*(11) A
Cyprinodontidae: *Aphyosemion australe*, *A. gardneri*, and *Nothobranchius* spp.

CHARACIFORMES

Characidae: *Paracheirodon innesi*(2) A,W, *P. axelrodi*(31) W, *Hyphessobrycon bentosi*, *Hyphessobrycon erythrostigma*, other *Hyphessobrycon* spp., *Metynnis* spp., *Nematobrycon palmeri*, *Pristella maxillaris*, *Thayeria boehlkei*, *Phenacogrammus interruptus*, and *Hemigrammus* spp.
Lebiasinidae: *Nannostomus* spp.
Anostomidae: *Anostomus anostomus* and *Chilodus punctatus*
Gasteropelecidae: *Thoracocharax* sp., *Gasteropelecus* sp., and *Carnegiella* sp.

CYPRINIFORMES

Cyprinidae:
Cyprinins: *Carassius auratus*(3) A and *Cyprinus carpio*
Other cyprinins or Barbins: *Barbus tetrazona*(15) A, *Barbus conchonius*, *Barbus titteya*, and other *Barbus* spp.
Systomins: *Balantiocheilus melanopterus*(16) A
Labeonins: *Labeo bicolor*(21) A and *L. erythrurus*(27) A
Banganas: *Epalzeorhynchos siamensis*
Rasborins: *Rasbora heteromorpha*(28) A, *Rasbora* spp., *Brachydanio rerio*, and *Tanichthys albonubes*
Gyrinocheilidae: *Gyrinocheilus aymonieri*(30) W
Cobitidae: *Botia macracantha*(6) W and *Acanthopthalmus kuhlii*(25) W

PERCIFORMES

Cichlidae: *Astronotus ocellatus*(7) A,W, *Pterophyllum scalare*(8) A,W, *Symphysodon discus*(9) A,W, *Apistogramma ramirezi*(10) A,W, *Pelmatochromis kribensis*(24) A, *Pseudotropheus* spp., *Labidochromis* spp., and *Melanochromis* spp.
Belontiidae: *Betta splendens*(4) A, *Colisa lalia*(12) A, and *Trichogaster* spp.
Helostomatidae: *Helostoma temmincki*
Chandidae: *Chanda lala*(23) W
Gobiidae: *Brachygobius* sp.(32) W,A

SILURIFORMES

Pimelodidae: *Pimelodus pictus*(19) W
Pangasiidae: *Pangasius sutchi*(26) A
Siluridae: *Kryptopterus bicirrhis*(18) W
Mochokidae: *Synodontis multipunctatus*, *S. angelicus*
Callichthyidae: *Corydoras paleatus*, *C. aeneus*, *C. panda*, and *C. trilineatus*
Loricariidae: *Hypostomus plecostomus*(17) A,W, *Ancistrus* sp., and *Peckoltia* sp.

ATHERINIFORMES

Melanotaeniidae: *Melanotaenia splendida*, *M. boesemani*, and *M. praecox*
Pseudomugilidae: *Pseudomugil furcatus*
Telmatherinidae: *Telmatherina ladigesi*(29) W,A

OSTEOGLOSSIFORMES

Osteoglossidae: *Osteoglossum bicirrhosum*(20) W, *O. ferreirai* and *Scleropages formosus*
Notopteridae: *Notopterus ornata* or *chitala*
Mormyridae: *Mormyrus petersii*

[a] Includes the most popular fish (top 32) in terms of value [data adapted from (5)], and the status of the fish [i.e., primarily from a fishery (wild caught, W, $n = 8$), primarily cultured (aquaculture, A, $n = 16$), or both from a fishery and cultured ($n = 8$)].

Outdoor production facilities typically consist of earthen ponds and concrete tanks. Some ponds have concrete sides, and net cages are placed inside each pond, for better management and handling of the fish. Pond sizes vary greatly; a Florida outdoor pond is 20–25 m (65–82 ft) in length, 6–9 m (20–30 ft) wide, and 1.5–1.8 m (5–6 ft) deep. Ponds are simply excavated, with no built levees, have steep slopes, and there are no drain lines. In south Florida, pools slightly larger than a bathtub are carved into the limestone bed for raising the fish. In Asia, a typical large cement pond may be 10–18 × 2–9 × 1.5 m (33–59 × 3.3–30 × 4.9 ft), a large tank 4–7 × 2–2.5 × 0.5 m (13–23 × 6.6–8.2 × 1.6 ft), and a square tank 1.5 × 1.5 × 0.5 m (4.9 × 4.9 × 1.6 ft). Net cages to place inside ponds measure 3 × 1 × 1 m (9.8 × 3.3 × 3.3 ft). Many farmers cover ponds and tanks with nets to protect the fish from predators. Depending on the location or time of year, farmers provide shade and enclose ponds with plastic to prevent the water temperatures from being too low during the winter.

Water Quality

Ornamental freshwater fish are highly adaptable to culture conditions and are capable of living under a wide range of environmental conditions. Despite the great number of species, the overall quality standards for culture water are similar to those for foodfish. Some of the most important factors are salinity, temperature, dissolved oxygen, hardness, conductivity, pH, carbon dioxide, and nitrogenous products (e.g., ammonia and nitrite). Another factor that may need to be considered are fish pheromones, such as crowding factors (e.g., in white cloud or zebra danios), which can inhibit growth or reproduction of other fish in the culture system.

Although freshwater fish cannot tolerate high salinities, the use of salt for transport and disease treatment is highly beneficial. The occurrence of disease also has been lowered when fish are raised in slightly saline waters (0.1–2 ppt). The preferred temperature for raising most ornamental freshwater fish is between 24 and 30 °C (75 and 86 °F). The incidence of diseases and water quality deterioration is increased drastically when the temperature is above 30 °C (86 °F). Although many species can tolerate low levels of dissolved oxygen (e.g., gouramies, several catfish), concentrations in production systems should be maintained above 5 mg/L or 5 parts per million [ppm] to support optimum growth and reproduction. High levels of free carbon dioxide (e.g., 100 mg/L or 100 ppm) are often encountered when water is pumped from shallow wells and when fish are raised or transported at high densities. Because of the influence of carbon dioxide on respiration, prolonged exposure to high levels of carbon dioxide may result in poor growth. Normal tolerance levels for carbon dioxide in foodfish have been established at 10–20 mg/L (10–20 ppm).

The hardness and pH of the waters from which many of the species originate can vary from a pH of 5 to 9 or 10 and a hardness of 5–20 mg/L (5–20 ppm) to over 300 mg/L (300 ppm) of calcium carbonate. However, most of these species can be acclimated to local conditions and raised successfully. Although not well understood, the stringent water requirements for temperature, pH, and hardness appear to be for breeding, hatching, and rearing the larvae. Black water directly from stream beds or small lakes is used for breeding and rearing larvae of certain species. This type of water is characteristically transparent, but dark brown in color (above 40 color units), because of dissolved organic matter, primarily from humic acids. The origin of these waters has been the same for many decades, and their specific locations have been kept secret.

Water Management

Well water is used to fill ponds and furnish the water to recycle systems. Water is supplied to each tank, often in spray form and flow-through fashion. Ground water, rainfall, and surface runoff typically maintain the water level in ponds. For species that require soft water, well water is usually treated with cation exchange resins and passed through reverse-osmosis units.

Intensive systems recycle 60–100% of their water. Each tank is supplied with water that often is injected below the surface and saturated with pure oxygen. Exchange of other gases is achieved with degassing chambers, air blowers, and oxygen. The water is treated with mechanical and biological filters. Removal of suspended and settleable solids (e.g., waste, feed, and feces) is accomplished with settling basins, baffles, screens, and upflow solids contact clarifiers. Microbes in trickle filters, modified upflow clarifiers, and fluidized media are used for nitrification and decomposition of organic wastes. Ozone is also used for further oxidation of the organic compounds. Fine suspended solids (scum) and other dissolved organics are stripped with dissolved air flotation or foam fractionation technology. For disinfection of treated water, ozone and ultraviolet light are often used. If necessary, the treated water is reconstituted with specific chemicals before it is used again. The effluent sludge is usually diverted and collected in small ponds to undergo further aerobic and anaerobic digestion.

Breeding and Propagation

Perhaps the greatest difficulties in the cultivation of ornamental fishes lie in obtaining and conditioning the broodstock for breeding. Also challenging are artificial incubation of embryos, initial feeding of larvae, and weaning from the larval stage. Reproductive conditioning and performance of ornamental freshwater species are greatly dependent on the species strain, age of the fish, body size, fish density, population structure, water quality characteristics (e.g., temperature, conductivity, and hardness), and other environmental conditions, such as availability, quantity, and quality of foods.

Most egg layers are artificially bred in indoor hatcheries. Broodfish are paired in tanks or spawned together in large groups. Fish are stimulated into breeding by using spawning mats and by manipulating environmental factors such as water flow, temperature, hardness, and pH. Hormone preparations from pituitaries or artificial sources, such as Ovaprim™, are also used to induce spawning. After spawning, the eggs are allowed to hatch where they are laid or are placed in various types of artificial incubators. The larvae that hatch are pooled and transferred to rearing tanks or outdoor ponds. This period is a critical time in the management of a hatchery. At this time and up to two weeks after stocking, incurred larval losses may reach or exceed 75%.

Standard temperatures for spawning and incubation fall between 23 and 29 °C (73 and 84 °F). Optimum spawning and incubation temperatures for individual species should not vary more than 2 to 3 °C (3 to 5 °F). Natural spawning takes between 30 to 90 minutes. The number of deposited eggs varies from hundreds to a few thousand or more. Eggs that are laid are mostly demersal or adhesive and are between 0.7 and 1.1 mm in diameter (slightly less than 1/16 in.). Some mouthbrooders (e.g., cichlids) and catfish may lay fewer than one hundred eggs, but these eggs are large (several millimeters in diameter), and the larvae are precocial. Depending on the species and water temperature, embryos may hatch in as little time as 12 hours (e.g., danios) or as much time as up to several months (e.g., killifishes). In general, embryos hatch within 1 to 4 days at 25–28 °C (77–82 °F). Curiously, this period is prolonged to seven days in rainbow fishes (Melanotaeniidae) and 14–24 days in blue-eyes (Pseudomugilidae). Upon hatching, larvae of egg-laying fish are generally 2–3 mm in length and 3–4 mm (approximately 1/8 in.) at the beginning of external feeding.

Larvae of many species are phototactic at hatch, but after a few hours avoid bright light. Swim-up, or swim bladder inflation, normally occurs close to the initiation of external feeding. Duration of the larval period is variable, but, in general, juvenile fish become apparent at three to five weeks.

Most species typically attain sexual maturity between five and seven months of age. Species that mature at older ages (e.g., labeos) enter puberty during their first and second year of life. Loaches and certain catfish may not mature until age three and four. Although the stage of sexual maturity is difficult to assess in many species, the sex of each fish is relatively easy to determine, particularly close to spawning: Males usually show brighter colors than females and display a variety of secondary sexual characteristics, such as longer fins, modified fin rays, and proportionally larger heads, sometimes with tubercles or bristles. Most characteristic is the alteration in sex behavior in males, who become more aggressive. The female's abdomen is distended when full of eggs, and the ovipore becomes distinct.

Broodfish are capable of spawning many times during their life. Completion of one reproductive cycle may take from two weeks in tetras and barbs and up to one year in labeos, goldfish, and koi; these annual species may spawn several times during a reproductive season. However, the frequency of spawning (i.e., in commercial production) greatly affects the reproductive output of the fish. For example, the reproductive performance (i.e., the number of eggs laid and the larval survival) of most species increases after the second or third spawning. Also, if the fish are spawned continuously (e.g., tetras that are spawned every two weeks), reproductive activity ceases after approximately one year.

The fertilization and spawning of livebearing fish is allowed to occur naturally in breeding ponds or tanks. Broodfish are typically stocked at 100–150 fish/m^3 (3–5 fish/ft^3), at a 1-to-4 : 7 ratio of males to females. Broodfish of four to six months in age are preferred. The number of young produced by a female ranges from 1 to less than 300. The average brood size is between 20 and 50 larvae every 26 to 63 days. In production, this fecundity estimate may be extrapolated to 0.7–1.1 larvae per female per day. The gestation period, or period of embryonic development, averages 30–35 days. At parturition, the larvae measure between 5 and 12 mm (about 3/16 to 1/2 in.) and consume a variety of foods, including brine shrimp (*Artemia*). In production, it is good practice to separate the parents from their offspring, as the parents will tend to eat the newly born.

Some species of livebearing fish (e.g., *Poecilia* sp.) can reach sexual maturity as early as four to five weeks from birth and 16–28 mm (about 5/8–1 in.) in body length. Swordtails (*Xiphophorus* sp.) may reach sexual maturity in 9–12 weeks of age or at 25–30 mm (about 1 in.). The age and body size at sexual maturity in livebearers have been demonstrated to be highly heritable traits.

Among the greatest potential factors for increasing the productivity of commercial farms are a better understanding of the nutritional and water quality requirements for conditioning egg-laying broodstocks and the development of breeding programs for livebearing fish. Because most species in the trade are imported or collected from the wild, the development of domestic broodstocks is perhaps the most important consideration for assuring the continual expansion of the industry.

Growout to Market Size

Most aquarium fish are grown for retail in outdoor ponds and tanks. Some fish are grown in indoor tanks, vats, and aquaria. In general, it takes three to six months to attain market-ready fish. Depending on the species, fish are sorted by age, sex, size, and color, beginning as early as three weeks from birth.

Stocking densities and survivorship estimates are difficult to assess and vary greatly between species and production systems. A typical-size pond [e.g., 200 m^2 (approximately 2,152 ft^2) of water surface] may be stocked with 10,000 to 80,000 fish from egg-laying parents or with a couple hundred livebearing broodfish. After two months, the livebearing population in the pond can reach approximately 30,000 fish, and harvest is initiated. A typical survival estimate for fish in a production pond is 40 to 70%. The survival rate of livebearing fish is slightly higher. The primary losses on an outdoor system are due to predation, deterioration of water quality, and diseases. High stocking densities in intensive culture systems may approach 4,000 fish/m^3 (15 fish/gal) without oxygen injection and 15,000 fish/m^3 (58 fish/gal) with oxygen injection. Losses under these conditions are minimized, and fish survival may increase to 85% and above.

Husbandry practices for the growout phase in ponds are designed primarily for obtaining fast growth, reducing predation, and maintaining proper health. Success is determined to a large extent by the amount and quality of natural food in the pond, the time of stocking, and proper handling at harvest time.

Ponds used for growout of juvenile fish are prepared for stocking by draining, washing, or removing the sludge; disinfecting; and fertilizing. Although species dependent, some ponds have remained in production unwashed for a year or two. When disinfected, ponds are treated with hydrated lime after cleaning. A variety of fertilizers is used in ponds. After a few days, the ponds are filled with water, and an algae bloom is allowed to develop. The ponds are then stocked with fish.

Management practices in intensive systems are aimed at maintaining a high exchange of good-quality water and at providing the proper nutrition, including pigments, to the fish. Only under these conditions can fish be raised at high densities, be in good health, and attain uniform size and coloration.

Many species exhibit schooling behavior, while others are solitary. Aggression toward members of the same species is common among solitary species and can be deterred by increasing the water flow in the culture tank or providing significant amounts of individual cover.

Fish are harvested with fine seine nets, dip nets, and traps. The process of harvesting ornamental fish differs from that for foodfish, because the fish are individually selected and must be kept alive and in good condition. The selection process includes sorting by color and size, which is labor intensive, as the sorting is done by hand and is subjective. The industry is in great need of a machine designed to count small fish and selectively grade and sort fish by color and size.

Feeding Fish

Very little information exists on the nutrition and feeding of ornamental fish. Knowledge of their dietary requirements has evolved primarily from trial-and-error tests by individual farmers and a few studies in research laboratories in universities and feed manufacturing companies. However, from analysis of these experiences and practices in the aquarium hobby, it is fair to conclude that the nutrient requirements of ornamental fish are, in general, similar to those of foodfish species.

Feeding habits among aquarium fish are highly diversified, but under culture conditions, most species become opportunistic feeders and take a variety of foods. Yolk-sac larvae in many species normally commence external feeding three to six days after hatching. Since many larvae have small mouths or esophagi, they require live foods between 50 and 150 μm in size (e.g., rotifers, trocophores, *Paramecium*). Larvae prefer to feed under dim light, and the initiation of external feeding occurs close to swim-up, or swim-bladder inflation. After one or two weeks, larvae will readily accept 250–400-μm particle feeds [e.g., brine shrimp (*Artemia* sp.) and formulated feeds]. Thereafter, the pellet size for feeding most fish varies between 1 and 3 mm (approx. 1/8 in.). The success in culturing many of the catfish, cichlid, and livebearing fishes is attributed to the ability of the larvae to eat large-particle foods, such as brine shrimp, at the initiation of external feeding.

The natural diet of fish raised in ponds is supplemented with formulated diets. However, many of the essential nutrients and pigments are obtained from the available natural foods. To stimulate the production of natural food in ponds, a combination of organic and inorganic fertilizers is used. A favored organic fertilizer is cottonseed meal.

Used for feeding fish cultured indoors, popular feed mixtures from commercial sources contain from 33 to 35% protein. Farm-made feeds usually are made up of 45% to 60% protein. Proteins from animal sources (e.g., fish meal, beef heart, and liver) are preferred ingredients in the formulation of diets. Ingredients from plant proteins (e.g., *Spirulina*, soybean, and alfalfa meals) and fiber are incorporated into diets for goldfish, koi, and herbivorous fishes (e.g., certain catfishes and cichlid species). To enhance coloration in fish, a combination of synthetic and natural carotenoid pigments is added, at a level of 0.04–2% of the diet.

Feed is delivered primarily by hand or automatic feeders. Because of the small particle size and the low volume of feed involved in growout, feed is allotted at a constant rate (3–10% of fish biomass per day). It is good practice to feed primarily floating or neutrally buoyant feeds, as sinking pellets are easily mixed with sediments or bottom debris and are not efficiently utilized. Fish are fed at least twice a day and sometimes almost continuously, depending on the species and culture system.

Fish Health Management

Poor health of the stock and loss of fish are due primarily to predation, diseases caused by substandard water quality, excessive handling, and inadequate nutrition. These losses can be minimized with the establishment of an integrated health management program that includes proper husbandry, the management of water quality, appropriate nutrition, proper sanitation, quarantine, diagnosis of illness, proper use of medications, and identification of the source of disease. A daily system for maintenance and monitoring is essential for early detection of diseased or malnourished fish. Symptoms of sick fish include physical changes, odd behavior, and a lack of feeding activity.

The industry has used a variety of chemicals, sedatives, and water additives to reduce the stress and possible injury of fish during handling, sorting, crowding, harvesting, and shipping. To prevent skin abrasions and scale loss, water additives containing polymer formulations (e.g., Polyaqua™) and salt have been very successful.

Ornamental freshwater fish are afflicted by diseases similar to those found in the foodfish industry, ranging from viruses to opportunistic aquatic bacteria, parasites, and fungal infections. However, an understanding of epidemiology, etiology, and pathology of infectious agents of ornamental fish is in its infancy. Also, knowledge of the immune system in these fish is limited.

Parasites and bacteria are two of the most common causes of infectious diseases in ornamental fish. The most prevalent of the external parasites are ciliated protozoans, primarily *Ichthyophthirius multifiliis*, or "ich," and *Trichodina*. Also, monogean trematodes, or flukes (e.g., *Dactylogyrus* and Gyrodactylus); nematodes; and the flagellated protozoans *Hexamita* and *Spironucleus* are frequently encountered internal parasites that debilitate and compromise the health of fish. Salt, formalin, copper sulfate, and potassium permanganate are used to treat external parasites. Fenbendazole and Metronidazole are recommended for internal nematodes and flagellates, respectively.

Among the most prevalent of the infectious bacteria are those in the aeromonad (*Aeromonas*) and columnaris groups (e.g., *Flavobacterium columnaris*). Common drugs used to treat bacterial infections are tetracycline, erythromycin, nitrofurazones, nalidixic acid, potassium permanganate, and copper sulfate. Correct use of antibiotics, based on bacterial isolation and sensitivity testing, is strongly recommended.

Fungal diseases (e.g., saprolegniasis) are widespread when fish are stressed, primarily during temperature shifts or following handling. Salt and potassium permanganate are two compounds commonly used in dip and bath treatments to minimize fungal invasions and other problems of an external nature. Once fungal infections become systemic, treatments using other compounds are necessary.

Only a few viral diseases have been identified in ornamental fish (e.g., iridovirus in gouramies). Although no specific medications are available for treating viruses, temperature manipulation and reduction of crowding may be methods to prevent the spread of the disease.

MARKETING

From the farm, fish are sold to large distributors or regional wholesalers, which, in turn, resell the fish to local retail stores. Fish collected from the wild, by fishers, are sold to a collector, who, in turn, resells the fish to the distributor or wholesaler. Although prices vary greatly, the wholesale price for many common egg-laying fish can average US$0.45 and the right livebearer US$0.20–0.30.

The centers for ornamental freshwater fish production and distribution are Singapore and Florida. From those two locations, fish are dispatched to all corners of the world. The fish are transported by air freight, and their survival and health are totally dependent on proper handling, packing, and timely arrival to their destinations. A shipment will arrive at its destination typically within 15 hours; however, the fish are packed for an extended trip duration of 48 to 72 hours. Fish are shipped inside insulated boxes, wherein they are placed in plastic bags, usually filled with one part water and three parts pure oxygen, on a volume-to-volume basis. The number of fish per box varies with species, the size or weight of the fish, and duration of the trip. A box may contain between 50 and 500 fish, or about 25 g/L (3 oz/gal). Cold or heat packs are placed inside containers to limit temperature fluctuations during shipment. Buffers, sedatives, and bacteriostatic chemicals may also be added to reduce water quality deterioration, the possibility of injury, and disease outbreaks. Fish are starved for at least 48 hours prior to shipping.

Competition in the ornamental fish market is keen, since many species are collected from the wild, and the bulk of the fish production is for a selected or preferred number of species. Another major source of competition, particularly in the United States, is from imported fish from southeast Asia, primarily in Singapore and Hong Kong. Markets for common fish are difficult to enter, since the structure and procedures for buying and selling fish are very complex and embedded in tradition. Perhaps the most important attribute of the market is the availability of a large number of the species in the trade. Access to an assortment of imported and cultured fish has allowed Singapore and Florida to become worldwide hubs for the purchase and distribution of ornamental fish.

Production of ornamental freshwater fish provides one of the best business opportunities in aquaculture, because most species in the trade are still collected from the wild, and the largest markets are in regions that must import fish. The fish are imported and exported either through interstate boundaries (e.g., from Florida to New York) or from international locations (e.g., from Singapore to the United States). Ornamental fish culture can be practiced on a small scale, with the business being both family owned and family operated. Although outdoor production can be practiced only in warm climates, many fish can be raised intensively indoors and close to the markets. On a per-weight basis, ornamental fish command among the highest prices of aquacultural products, often several hundred dollars per kilogram. The recognition of opportunities provided by ornamental fish aquaculture is best represented by the rapidly expanding industry in the Czech Republic and Israel. In only a few years, both countries have become major participants in the world market. (See Table 2.) Development of ornamental fish aquaculture has allowed for rapid diversification of traditional aquaculture practices and enhanced economic opportunities.

Essential for the establishment and development of an ornamental fish industry is a consistent supply of a wide variety of species. The expansion of and future opportunities in the ornamental fish industry are dependent on the continued development of new varieties and the establishment of domestic broodstocks for species that are imported or collected from the wild.

Additional information can be obtained by consulting (1–31).

BIBLIOGRAPHY

1. C. Andrews, *J. Fish Biol.* **37**, 53–59 (1990).
2. M. Boonyaratpalin and R.T. Lovell, *Aquaculture* **12**, 53–62 (1977).
3. D.E. Campton, *J. Heredity* **83**, 43–48 (1992).
4. F.A. Chapman, D.E. Colle, R.W. Rottmann, and J.V. Shireman, *Prog. Fish-Cult.* **60**, 32–37 (1998).
5. F.A. Chapman, S.A. Fitz-Coy, E.M. Thunberg, and C.M. Adams, *J. World Aquacult. Soc.* **28**, 1–10 (1997).
6. F.A. Chapman, *J. App. Aquacult.* **7**, 69–74 (1997).
7. D.A. Conroy, *An Evaluation of the Present State of World Trade in Ornamental Fish*, FAO Fisheries Technical Paper No. 146, Rome, 1975.
8. *Aquaculture*, Florida Agricultural Statistics Service, Orlando, FL, 1998.
9. G. Degani, *Bamidgeh* **41**, 67–73 (1989).
10. G. Degani, *Aquacult. Eng.* **9**, 367–375 (1990).
11. G. Degani, *Aquacult. Fish. Man.* **24**, 725–730 (1993).

12. A.A. Fernando and V.P.E. Phang, *Aquaculture* **51**, 49–63 (1985).
13. A.A. Fernando and V.P.E. Phang, *Freshwater Ornamental Fish Aquaculture in Singapore*, Singapore Polytechnic University, Singapore, 1994.
14. *Freshwater and Marine Aquarium*, R/C Modeler Corporation, Sierra Madre, CA.
15. G.F. Hervey and J. Hems, in G.F. Hervey and J. Hems, eds., *Freshwater Tropical Aquarium Fishes*, Spring Books, London, 1963, pp. 1–7.
16. A.J. Klee, *A History of the Aquarium Hobby in America*, American Cichlid Association, Raleigh, NC, 1987.
17. G.A. Lewbart, *The Compendium: Small Animal* **13**(6), 969–978 (1991).
18. National Research Council, *Nutrient Requirements of Fish*, National Academy Press, Washington, DC, 1993.
19. W.J. Ng, K. Kho, L.M. Ho, S.L. Ong, T.S. Sim, S.H. Tay, C.C. Goh, and L. Cheong, *Aquaculture* **103**, 123–134 (1992).
20. W.J. Ng, K. Kho, S.L. Ong, T.S. Sim, J.M. Ho, and S.H. Tay, *Aquaculture* **110**, 263–269 (1993).
21. M.C. Pannevis and K.E. Earle, *J. Nutr.* **124**, 2616S–2618S (1994).
22. B. Pénzes and I. Tölg, *Goldfish and Ornamental Carp*, Barron's Educational Series, Inc., Hauppauge, 1983.
23. J.S. Ramsey, *J. Alabama Acad. Sci.* **56**, 220–245 (1985).
24. S. Rothbard, *Koi Breeding*, T.F.H. Publications, Inc, Neptune, City, NJ 1997.
25. J.V. Shireman and J.A. Gildea, *Prog. Fish-Cult.* **51**, 104–108 (1989).
26. R. Socolof, in E.E. Brown and J.B. Gratzek, eds., *Fish Farming Handbook: Food, Bait, Tropicals and Goldfish*, AVI Publishing, Westport, CT, 1980, pp. 163–206.
27. R. Socolof, *Confessions of a Tropical Fish Addict*, Socolof Industries, Bradenton, 1996.
28. M.K. Stoskopf, *The Veterinary Clinics of North America: Small Animal Practice* **18**, 1–474 (1988).
29. C.S. Tamaru, B. Cole, R. Bailey, and Christopher Brown, *A Manual for Commercial Production of the Tiger Barb,* Capoeta tetrazona, *a Temporary Paired Tank Spawner*, Center for Tropical and Subtropical Aquaculture Publication Number 129, Honolulu. Also available from the USDA Center for Tropical and Subtropical Aquaculture are the following: *A Manual for Commercial Production of the Gourami,* Trichogaster trichopterus, *a Temporary Paired Tank Spawner*; *A Manual for Commercial Production of the Swordtail,* Xiphophorus helleri; *Report on the Economics of Ornamental Fish Culture in Hawaii*; *Shipping Practices in the Ornamental Fish Industry*.
30. *Tropical Fish Hobbyist*, T.F.H. Publications, Neptune City, NJ.
31. R.A. Winfree, *World Aquaculture* **20**, 24–30 (1989).

See also ORNAMENTAL FISH CULTURE, MARINE.

ORNAMENTAL FISH CULTURE, MARINE

G. JOAN HOLT
University of Texas
Port Aransas, Texas

OUTLINE

Keeping an aquarium is a hobby that engages an estimated 10–20 million enthusiasts, who own more than 90 million tropical fish (1). The retail value of the ornamental fish trade is approximately $1 billion (2). Although many of the freshwater ornamental fish sold to the public are farm raised, essentially all of the marine reef products (fish, invertebrates, live rock) are collected from the wild.

The Asian/Pacific region is the global center of marine diversity; it supports more species of coral and fish than does any other region on earth. This region is home to over 4,000 species of reef fish and more than one-third of the world's coral reefs. In this region, and throughout the tropics, natural populations of coral reef fish are increasingly threatened by development, dredging, coral collecting, and the live food-fish and aquarium-fish trade. Unfortunately, many of the common collection methods (dynamite, sodium cyanide) are destructive (3), and cause irreparable damage to coral reefs in addition to greatly reducing the area of natural habitat available for the settlement of new fish recruits.

Japan and the United States lead the market for aquarium fish; these two countries account for over half of the world's ornamental fish trade. The Philippines is a major exporter of marine aquarium products for the global aquarium trade, supplying an estimated 75–80% of the market (3). It is estimated that up to 90% of all aquarium fish imported by the United States from the Philippines have been collected by using cyanide (4). There are several cyanide fishing techniques, all of which damage corals. Cyanide is a broad-spectrum poison that acts on enzyme systems involved in respiratory metabolism (3). Exposing a fish to cyanide causes internal damage to its liver, intestine, and reproductive organs. Fish that survive exposure to cyanide, but are not taken by divers, can be expected to suffer impaired chances of survival, growth, and reproduction, and lowered disease resistance.

Less is known about the specific impacts of cyanide on reef invertebrates, but it seems likely that similar debilitating effects occur to them. In general, invertebrates seem to be more susceptible to cyanide than are fish. Loss of zooxanthellae (coral bleaching), inhibition of photosynthesis, and impaired respiration of coral in response to cyanide concentrations used in fishing have been documented in invertebrates (4).

BREEDING MARINE ORNAMENTAL FISH

The ability to meet demands for marine ornamentals by utilizing wild-caught fish is decreasing due to more stringent regulations on collections that deplete wild stocks and cause damage to fragile coral reef ecosystems. In the Philippines, one of the important suppliers of marine ornamental species to the United Kingdom, Europe, and the United States of America, destructive collecting techniques damage reef habitats. Furthermore, the marine environments in which these species exist under natural conditions are increasingly being threatened by various forms of human interference and natural disasters that result in extensive damage to coral reefs. Recently, some countries have taken steps to protect several species of marine fish and invertebrates.

Sri Lanka and other countries in the region are showing an increasing interest in the breeding of marine ornamental fish and invertebrates, both to reduce dependence on wild stocks and as a means of generating income for coastal communities. Increasing pressures on natural populations of coral reef organisms, and their expanding popularity in the aquarium trade, have spurred interest in the development of culture techniques for marine ornamentals. There is a strong call for tank-reared fish by both marine aquarium hobbyists and professional aquarists. Currently, few species of marine fishes are regularly reproduced in captivity and sold in the aquarium trade. The most commonly available are clownfish (*Amphiprion* spp. and *Premnas biaculeatus*), the neon goby (*Gobiosoma oceanops*) and relatives, and more recently the dottybacks (*Pseudochromis* spp. and *Ogilbyina novaehollandiae*) (5). Some other species have been bred and raised in captivity during the last 30 years (6), but not reliably or efficiently enough to sustain an industry. There are generally fewer problems in the spawning of tropical marine fish and invertebrates than in culturing the young. In fact, larval culture is the most difficult biological aspect of the culture of most marine species.

Why are some species raised successfully while most others are not? The easiest to raise, clownfish, have the advantage of being sequential hermaphrodites (they can change sex), which eliminates the problem of obtaining a spawning pair, and they produce demersal eggs for which the adults provide parental care (7). Although parental care ceases with the hatching of the eggs, the larvae are large enough at hatching to consume cultured rotifers (*Brachionus plicatilis*) (6).

Many of the more popular and costly marine aquarium fish, such as butterflyfish, angelfish, and wrasses, are more difficult to raise in captivity. They produce small free-floating eggs that are only about one-third the size of clownfish eggs, and the adults do not provide parental care. The newly hatched larvae are less than 2.5 mm (0.1 in.) long, and they have no eye or mouth development and very limited swimming ability. Traditional prey for the rearing of marine fish larvae (rotifers and *Artemia* spp. or brine shrimp) are not accepted, and may not be nutritionally adequate, for these marine ornamentals.

Early Life History

Survival through the early life stages has proved to be the major obstacle to large-scale production of a wide variety of marine ornamentals in captivity. Production of sufficient numbers of healthy juvenile fish for the market depends primarily on the refinement of larval rearing methods. Appropriate environmental conditions and proper prey for different stages during development are unknown and must be determined. Development of rearing systems for the tiny, delicate larvae is a challenge. Since they have rudimentary sensory and motor development, the larvae cannot swim and search large volumes of water to find food. This necessitates the feeding of dense blooms of live prey, and it often results in deteriorated water quality. Simple systems, such as the one described by Henny and coworkers (8), provide a confined volume for larvae and their prey, which facilitates high feeding density, and also allows for constant exchange of water. Water in the completely closed system is filtered, aerated, and heated, if necessary, in order to maintain adequate water quality (Fig. 1).

The next problem is that of finding appropriate prey. The first food for marine larvae is live plankton, generally zooplankton. Cultured rotifers are the most commonly-fed food in aquaculture, but many marine ornamentals do not grow and thrive on them. It may be because rotifers are too large, do not give appropriate behavioral signals, or are not nutritionally adequate. Natural plankton (including algae) would be the best choice, but they are not always accessible-and often when they are the species composition and nutrient quality vary over time. For reliable production, cultured prey must be available. Phytoplankton is relatively easy to culture but is not an adequate food by itself. Copepods are generally considered the best food overall for larval marine fish, however, culture success has been marginal at best. Gut analysis (9) showed tintinnids, dinoflagellates, protozoans, diatoms and copepods, in that order, as the most important prey for small (less than 3 mm or 0.12 in.) coral reef fish larvae. These findings suggest that the best first foods for rearing ornamentals are microzooplankton and large phytoplankton.

SPECIES COMMONLY REARED IN CAPTIVITY

Clownfish or Anemonefish (Pomacentridae)

Types of fish commonly raised in ornamental fish culture (10) are species that have three characteristics: they can change sex (sequential hermaphrodites), they have eggs that take several days to hatch, and they produce relatively large larvae (>2.5 mm or 1/10 in.) that have fast development. The clownfish is a good example. These fish can change sex from male to female (protandry). Generally two to four fish might occupy an anemone, including a large dominant female, a smaller male and one or two adolescents. If something happens to the female, the male, now dominant, will change into a female and one of the smaller individuals will become a male. Most commercial operations typically have only pair housed in the breeding tank. Clownfish spawn about twice a month by laying adhesive eggs onto a hard substrate. Eggs are tended

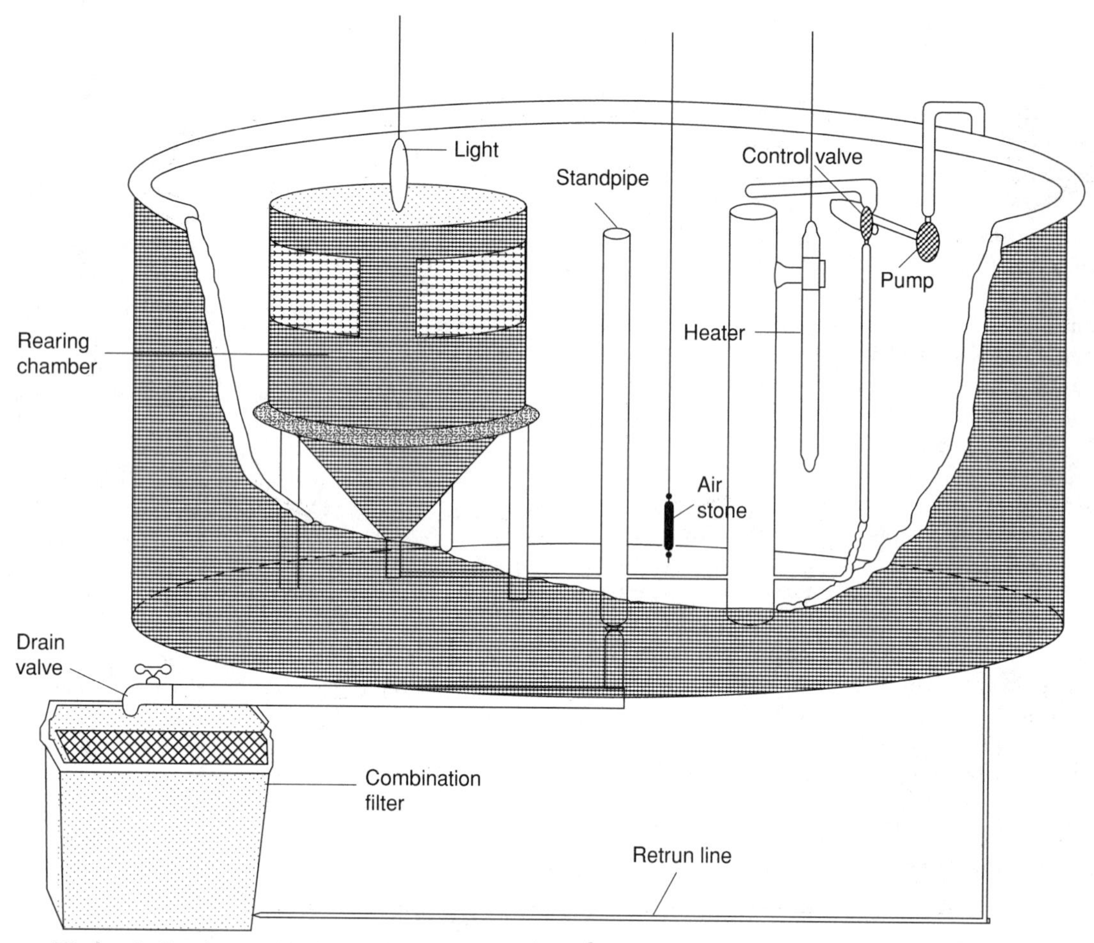

Figure 1. Rearing system for development of larvae of marine ornamentals. (From Hinney et al. 1995.)

by the male and hatch after a week. Newly hatched larvae feed on rotifers for several days, followed by brine shrimp. Within one to three weeks they metamorphose into juveniles. High survival of the young is possible with many of the 12-plus species that are regularly produced in captivity, and several commercial establishments produce and sell captive-bred clownfish. Joyce Wilkerson's new book *Clownfish: A Guide* (11) is an excellent source of information on the captive care and breeding of these fish.

Gobies (Gobiidae)

Gobies are typically small benthic fish that often act as cleaners, removing parasites from other fish. They are protogynous hermaphrodites, which means they can change sex from female to male but not the other direction. In captivity, gobies spawn regularly (every 2–3 weeks) and produce demersal adhesive eggs that are usually attached to the underside of a rock or to the roof of a small cavity. In most cases, spawning requires only the presence of a mated pair plus a suitable spawning site. The male guards and tends to the eggs for the 4–10 days between fertilization and hatching. Neon goby eggs are quite large; the larvae are 3–4 mm (0.1–0.2 in.) and are relatively advanced at hatch (12). Young can be raised on rotifers and wild zooplankton, and, later, brine shrimp nauplii. Metamorphosis to the juvenile stage takes about a month. Several species are regularly spawned in captivity. Commercially available gobies include the neon goby, the yellow-striped goby, the West Indian cleaner goby, the genie goby, and the masked goby (13).

Dottybacks (Pseudochromidae)

Dottybacks or fairy basslets are also protogynous hermaphrodites. They lay a gelatinous ball of eggs (about 2.5 cm or 1 in. in diameter), which is guarded by the male for a few days before the eggs hatch into 2–4 mm (0.1–0.2 in.) larvae. While the male is caring for the eggs, he chases away the female. When he wants to spawn, he aggressively invites the female to join him to spawn. The rotifer is readily accepted as first food by the larvae, and is replaced by brine shrimp nauplii after about two weeks. Metamorphosis occurs at about 30 days, followed closely by the development of full adult coloration. During the last three years, several species have been successfully reared (14) and are now offered for sale (13). M.A. Moe (15) has written a book that details the care and breeding of the orchid dottyback (*Pseudochromis fridmani*).

Seahorses (Syngnathidae)

Seahorses typically live in shallow areas and cling on to vegetation, rocks and other suitable substrates with their tails. Several species are monogamously pair-bonded and will not change spawning partners unless one of them

Table 1. Life History Characteristics of Popular Marine Aquarium Fishes that have been Spawned in Captivity

Mode of Reproduction	Egg Type	Number of Eggs	Egg Size mm (in.)	Incub Time in Days	Hatch Size mm (in.)	Larval Stage in Days	Reared in Captivity	Commercial	Species Reared
GONOCHORISIM (separate sexes)									
Syngnathidae—seahorses	Pouch brooder	50–1500	?	20–28	6–13 (.2–.5)	?	Yes	No	Six species (17)
Apogonidae—cardinalfish (nocturnal)	Oral brooder	20+	1–2.5 (.04–.1)	21–24	3–6 (.12–.24)	21	Yes	No	Banggai Cardinalfish (20)
Sciaenidae—drums	Pelagic	2000–5000	.8–1.2 (.03–.05)	1	2.5 (.1)	20–30	Yes	No	Jackknife fish, cubbyu (18)
Ephippidae—spadefish, batfish	Pelagic	1000–5000	1 (.04)	1	2.5 (.1)	15	Yes	No	Spadefish (21)
Chaetodontidae—butterfly fish (coral polyp feeders)	Pelagic	3000–4000	.7–.9 (.03–.035)	1	1.4–2 (.06–.08)	40–60	No		
Opistognathidae—jawfish (live in burrows)	Oral brooder	1000	.85 (.03)	7–9	4 (.16)	21–28	Yes	Yes	Yellowhead jawfish (22)
Blenniidae—blennies (benthic)	Demersal (attached)	1000	.8 × 1 (.03 × .04)	4–10	4 (.16)	18–20	Yes	No	Black-lined blenny (23) marbled blenny (24), + others (25,26)
Callionymidae—dragonets	Pelagic	?	.6–1 (.02–.04)	1	1–2 (.04–.08)	20–30	Yes	No	Mandarinfish (13)
SEQUENTIAL HEMAPHRODITISM (can change sex)									
PROTANDOUS (change from male to female)									
Pomacentridae (in part)—anemonefish or clownfish	Demersal (attached)	400–1500	1.5 (.06)	4–10	2.5–3 (.1–.12)	7–21	Yes	Yes	About 12 species of Amphiprion or clownfish (11 and references therein)
PROTOGYNOUS (change from female to male)									
Cirrhitidae—hawkfish	Pelagic	500	0.7 (.03)	1	1.5 (.06)	35–50	No		
Serranidae (in part)—swissguard basslet and reef basses	Pelagic	300	.6–.9 (.02–.035)	1	1.5–3 (.06–.12)	20	No		
Pseudochromidae—dottybacks or basslets	Demersal (egg mass)	500–2000	1–1.4 (.04–.055)	4–7	2–4 (.08–.15)	21–35	Yes	Yes	Dottybacks (13,14), royal gramma (27), blackcap basslet (28,29)
Plesiopsidae—comet or marine beta (secretive in caves/crevices)	Demersal (egg mass)	300–400	.95 (.04)	5–6	3–4 (.12–.16)	60	Yes	No	Marine beta (30–32)
Labridae—wrasses	Pelagic	500–7000	.5–1.2 (.02–.05)	1	1.5–2 (.06–.08)	30	Yes	No	Hogfish (33) spotfin hogfish to day 18 (32)
Pomacanthidae—angelfish	Pelagic	150–75,000	.6–.9 (.02–.035)	1	1.5–2.5 (.06–.1)	30	Yes	No	French, grey angelfish (34,35)
Pomacentridae (in part)—damselfish (intraspecific aggression)	Demersal (attached)	500–1000	.6–.9 (.02–.035)	3–5	2–4 (.08–.16)	30	Yes	No	Beaugregory (36), yellow-tailed damsel (37,38) threespot and white-tailed damsel (39)
Gobiidae (in part)—reef gobies	Demersal (adhesive)	100–1000	1–2 (.04–.08)	4–10	2–7 (.08–.28)	25–35	Yes	Yes	Green-banded, citron, neon goby + (40,41)
SIMULTANEOUS HEMAPHRODITISM									
Serranidae (in part)—hamlets, sea basses	Pelagic	1500	.7–1 (.03–.04)	1	2.2 (.09)	20–30	No		

dies or is lost (16). Seahorses exhibit an unusual type of parental care in which the male becomes 'pregnant.' After a courtship that extends over several days, the female deposits her eggs in the pouch of the male, where they are fertilized. The male carries the eggs in his pouch for several weeks. By the time young seahorses are released from the pouch, they are approximately 10 mm (0.4 in.) in length, are well developed, and are able to feed on small plankton like rotifers and crustacean nauplii. Young seahorses can triple in size in 3 to 4 weeks, and they reproduce within a year. Seahorses have been successfully bred and reared in captivity by researchers and hobbyists alike. Information on captive spawning and rearing of seahorses may be found in an article by Amanda Vincent (17).

Drums (Sciaenidae)

An exception to the general characteristics for easy-to-rear ornamentals are the sciaenids or drums. The drums are gonochorists, i.e. an individual is either male or female for its entire life. These fish produce floating or pelagic eggs that hatch in less than 1 day into 2.5 mm (0.1 in.) long larvae. The larvae are able to feed on small rotifers at first feeding, but may do better with added wild zooplankton. Growth is fairly rapid; metamorphosis to the juvenile stage occurs at about three weeks. Several species, including the jackknife fish and highhat, have been spawned and reared in captivity (6,18,19) but none are yet produced in sufficient quantity for the tropical fish industry.

SPECIES WITH POTENTIAL FOR CAPTIVE BREEDING

Many of the marine ornamentals that have been spawned in captivity are listed in Table 1. Some are already commercially available, and others have the potential to be reared successfully in captivity. Based on previous successes, the most obvious species for future work might include fish that change sex, care for their embryos, produce large larvae (>2 mm or 0.1 in.) that can feed on rotifers or brine shrimp nauplii, and quickly develop into juveniles.

BIBLIOGRAPHY

1. C. Andrews, *J. Fish Biology* **37**, 53–59 (1990).
2. F.A. Chapman, S.A. Fitz-Coy, E.M. Thunberg, and C.M. Adams, *J. World Aquaculture Soc.* **28**, 1–10 (1997).
3. P.J. Rubec, *Environmental Biology of Fishes* **23**, 141–154 (1988).
4. R.J. Jones and A.L. Steven, *Mar. Freshwater Res.* **48**, 517–522 (1997).
5. T. Gardner, *SeaScope* **14**, 1–2 (1997).
6. M.A. Moe, Jr., *Marine Aquarium Handbook*, 2nd ed., Green Turtle Publications, Plantation, FL, 1992.
7. R.E. Thresher, *Reproduction in Reef Fishes*, T. F. H. Publications, Inc., Ltd., Neptune City, NJ, 1984.
8. D.C. Henny, G.J. Holt, and C.M. Riley, *Progressive Fish-Culturist* **57**, 219–225 (1995).
9. C.M. Riley and G.J. Holt, *Rev. Biol. Trop.* **41**, 53–57 (1993).
10. Breeders' Registry Web site:www.breeders-registry.gen.ca.us.
11. J.D. Wilkerson, *Clownfishes: A Guide to their Captive Care, Breeding & Natural History*, Microcosm, Ltd., Shelburne, VT, 1998.
12. P. Colin, *The Neon Gobies*, T. F. H. Publications, Neptune City, NJ, 1975.
13. T.R. Gardner, *SeaScope* **14**, 1–2 (1997).
14. R. Brons, *Freshwater and Marine Aquarium* **19**(6), 1–8 (1996).
15. M.A. Moe, Jr., *Breeding the Orchid Dottyback Pseudochromis Fridmani*, Green Turtle Publications, Plantation, FL, 1997.
16. A.C.J. Vincent and L.M. Sadler, *Animal Behaviour* **50**, 1557–1569 (1995).
17. A.C.J. Vincent, *The Journal of MaquaCulture* **3**(1,2), 1–5 (1995).
18. G.J. Holt and C.M. Riley, *Fishery Bulletin* **65**, 825–838 (1999).
19. E.D. Houde and A.J. Ramsey, *Progressive Fish-Culturist* **33**, 156–157 (1971).
20. F.C. Marini, *Journal MaquaCulture* **4**(4), (1996).
21. S.D. Walker, *SeaScope* **8**, 1–2 (1991).
22. J. Walch, *SeaScope* **11**, 1–2 (1994).
23. L. Fishelson, *Copeia* **4**, 798–800 (1976).
24. C.M. Breder, *Zoologica* **26**, 233–235 (1941).
25. J.B. Jillett, *Australian Journal Marine and Freshwater Research* **19**, 9–18 (1968).
26. W. Watson, *Bulletin Marine Science* **41**, 856–888 (1987).
27. J. Walch, *The Breeders Registry Newsletter* **2**(1), 1–4 (1994).
28. P. Rosti, *Saltwater Aquarium* **2**, 106–108 (1967).
29. W.M. Addison, *The Breeders Registry Newsletter* **2**(3), 1–4 (1994).
30. H. Wassink and R. Brons, *SeaScope* **7**, 1–3 (1990).
31. P.L. Colin, *Fishery Bulletin* **80**, 853–862 (1982).
32. J. Baez, *SeaScope* **15**, 1–4 (1998).
33. G.J. Holt (unpublished).
34. R.P.L. Straughan, *The Aquarium* **28**, 211–212 (1959).
35. M.A. Moe, Jr., *Marine Aquarist* **7**, 17–26 (1976).
36. F.J. Brinley, *Copeia* **4**, 185–188 (1939).
37. M.A. Moe, Jr., *Freshwater and Marine Aquarium* **4**, 24–25 (1981).
38. T. Potthoff, S. Kelley, V. Saksena, M. Moe, and F. Young, *Bulletin Marine Science* **40**, 330–375 (1987).
39. B.S. Danilowicz and C.L. Brown, *Aquaculture* **106**, 141–149 (1992).
40. M.A. Moe, Jr., *Marine Aquarist* **6**, 4–10 (1975).
41. S.D. Brown, *The Journal of MaquaCulture* **4**, 1–3 (1996).

See also ORNAMENTAL FISH CULTURE, FRESHWATER.

OSMOREGULATION IN BONY FISHES

JOSEPH J. CECH, JR.
University of California, Davis
Davis, California

OUTLINE

The blood and body tissues of bony fishes (teleosts and chondrosteans) must have suitable proportions of water and the correct dissolved solutes to meet the requirements of their living cells (1–3). Solutes such as ionized substances (e.g., Na^+, Cl^-, Ca^{2+}, PO_4^{2+}, H^+, OH^-, NH_4^+) and dissolved organic compounds (e.g., amino acids, fatty acids, and sugars) must be regulated because of the vast solute concentration differences that typically exist between fish bodies and their environments. The regulation of body water and its total dissolved solutes is termed osmoregulation, for which two life-dependent strategies evolved among bony fishes that allow them to exist in marine and freshwater environments. Osmoregulation is inexorably linked to ionic concentration regulation (ionic regulation), the excretion of nitrogenous waste (principally ammonia), and acid-base balance. Although fish culture environments rarely undergo dramatic changes in salinity, routine fish-handling and transport procedures associated with culture operations usually stress fish. The consequences of these stresses often include increased internal acidity and increased movements of ions (via diffusion) and water (via osmosis) between fish tissues and the environment, due to increased gill permeability. Also, fish in high-density culture environments may be stressed by hypoxic (low dissolved O_2), hypercapnic (high dissolved CO_2), or hyperoxic (higher than atmospheric levels of dissolved O_2) conditions. Hypercapnia, especially, increases environmental acidity, affecting the movements (particularly across gill tissues) of several ions between the fish and its environment. Overall, adequate regulation of the body's water and dissolved solutes is critical for the well-being of fish. To maximize production efficiency, culturists seek to optimize fish well-being, components of which include the understanding of fish osmoregulatory structures and mechanisms, and the understanding of osmoregulatory consequences of culture-related activities.

OSMOREGULATION AND IONIC REGULATION OF MARINE FISH

Marine environments (like the oceans) have dissolved salt concentrations that are more concentrated (termed "hypertonic") that those in the bodies of marine bony fish, and the bodies of marine bony fish have salt concentrations approximating one-third that of their environment (Table 1). Consequently, they tend to continually lose water to the environment and gain monovalent ions from it by diffusion, especially over the thin membranes of the gills. Teleosts (2) and chondrosteans, such as sea water-acclimated sturgeon (4–6), continually replace lost water by drinking (ingesting) sea water, but this results in an even larger dissolved salt intake (Fig. 1). Excess salts must be excreted via special (chloride) cells in the gill and opercular skin tissues and via the kidney. When sea water is ingested to replace water that has diffused into the environment, many ions, as well as water, are absorbed into the bloodstream across the intestinal wall. Teleostean kidneys primarily excrete the excess divalent ions (e.g., Mg^{2+}, SO_4^{2-}) that are ingested with the sea water, but they are of little help in excreting the much larger quantity of ingested monovalent ions. In fact, many marine

Table 1. Major Ionic Constituents and Total Solutes (expressed as osmolality) in the Environment and Plasma of Several Bony Fish

Medium or Species (Salinity)	Ionic Constituents ($mEq \cdot L^{-1}$)				Osmolality	Reference
	[Na^+]	[Cl^-]	[K^+]	[Ca^{2+}]	($mOsm \cdot kg^{-1}$)	
Seawater	480	612	10	20	1,049	6,90
Gulf of Mexico sturgeon (35 ppt)	152	149	3	—	294	6
Chum salmon (SW)	167	144	2	4	370	91
Rainbow trout (30 ppt)	172	137	1	—	—	92
Atlantic cod (35 ppt)	176	164	6	4	—	93
Turbot (35 ppt)	158	136	—	2	336	94
Turbot (10 ppt)	156	129	—	2	330	94
Freshwater	1	1	<1	1	1–6.5	98
Gulf of Mexico sturgeon	136	107	3	—	261	6
Goldfish	133	109	3	2	—	96
Carp	137	122	2	2	270	97
Northern pike	122	102	—	—	—	95
Rainbow trout	137	92	3	—	—	92
Chum salmon	151	118	2	5	324	92

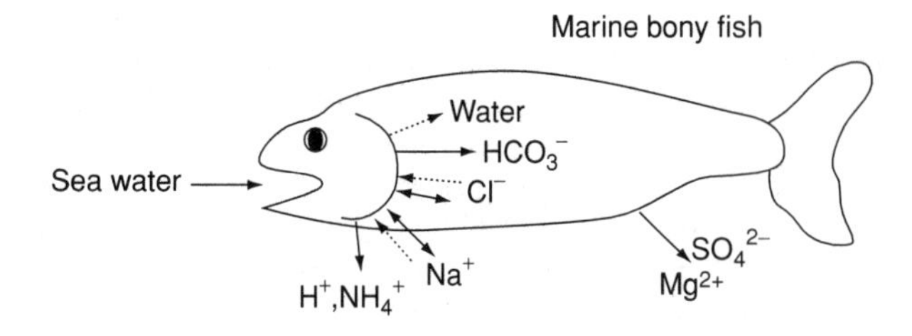

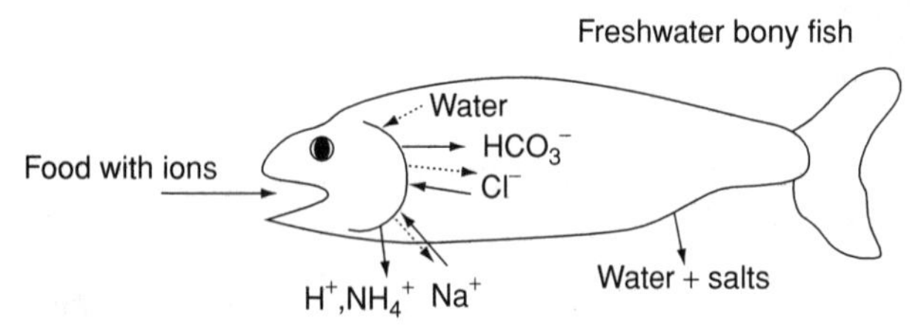

Figure 1. Diffusive (dashed lines) and carrier-mediated (solid lines) movements of major ions between bony fish and their hypertonic (seawater) and hypotonic (freshwater) environments. Arrows emanating from the posteriors of the fish models mostly reflect urinary excretions, although intestinal excretions may also play minor roles.

teleosts [e.g., oyster toadfish (*Opsanus tau*) and plainfin midshipman (*Porichthys notatus*)] have an aglomerular kidney to minimize water losses. Instead, the chloride cells, which are activated by cortisol and growth hormone secretions (see later), actively transport the excess monovalent anions against their overall concentration gradients back into the environment. The chloride cells are larger than the flat epithelial cells of the gills specialized for respiratory gas exchange (7). They are often termed alpha chloride cells to distinguish them from beta chloride cells, which are found in freshwater teleosts (8), and they are typically located on the gill filaments at the base of lamellae. Besides gill locations, chloride cells have also been found on the skin of several species, including the opercular epithelium of sea water-acclimated Mozambique tilapia (*Oreochromis mossambicus*) (9) and on larval turbot (*Scophthalmus maximus*) (10). Compared with most cells, chloride cells are packed with mitochondria, display extraordinary development of cytoplasmic microtubules, and contain an abundance of Na^+-K^+-ATPase (the enzyme system that facilitates salt pump function). The Na^+-K^+-ATPase is located along the basolateral areas and in the extensive microtubular system of the chloride cell; it actively transports Na^+ out of the cell in exchange for K^+ (11). In this way, this enzyme maintains a high trans-cell-membrane Na^+ gradient, with high [Na^+] in the tubules and the adjacent plasma and low [Na^+] in the chloride cell cytoplasm (12). This high transmembrane Na^+ gradient drives a linked Na^+-Cl^- carrier system, increasing the cytoplasmic [Cl^-]. The buildup of Cl^- inside the chloride cell also increases the cell's electronegativity, and Cl^- follows its electrochemical gradient (including a negative-to-positive transmembrane potential) by passively moving out of the apical pit area into sea water (Fig. 2). Na^+ probably exits passively to sea water via shallow tight junctions between chloride cells or between chloride cells and accessory cells, driven by the positive-to-negative transepithelial potential (11). Researchers Utida and Hirano (13) have shown that both gill Na^+-K^+-ATPase concentrations and numbers of gill alpha chloride cells increase with exposure to increasing environmental salinity in the Japanese eel (*Anguilla japonica*).

OSMOREGULATION AND IONIC REGULATION OF FRESHWATER FISH

Freshwater environments (including lakes and streams) have much lower (hypotonic) salt concentrations than those in the bodies of freshwater fish, which contain about one-fourth to one-third the salt concentration of seawater (Table 1). Freshwater fish continually gain water from their environment and lose monovalent ions to the environment through diffusion (Fig. 1). The excess water is continually excreted through the kidneys as a large volume of dilute urine (up to one-third of the body weight per day). Control of diuretic (urine-producing) processes is influenced by many factors, including blood pressure, glomerular filtration, and kidney tubular reabsorption changes, which are mediated by hormones such as the renin-angiotensin system (14–16), atrial natriuretic peptide (17–20), and arginine vasotocin (21). Diffusional losses of monovalent ions (e.g., across the gill membranes) are reduced in environments that have higher calcium concentrations (i.e., hard water) (22) and from secretions of the pituitary hormone, prolactin (23–25; See the section

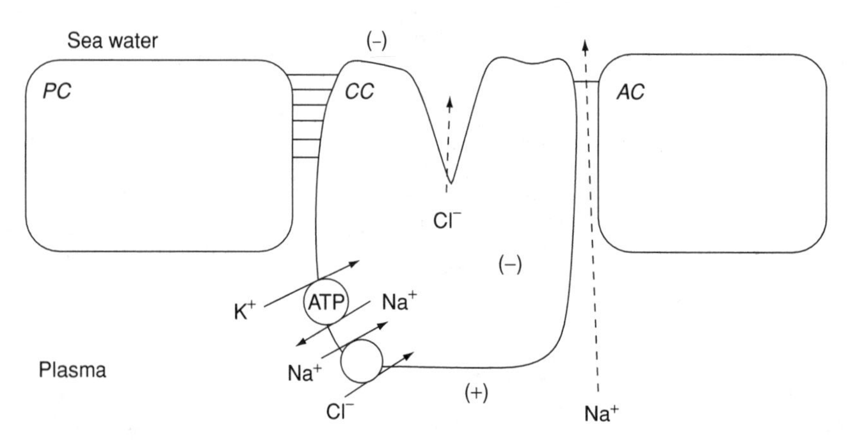

Figure 2. Current ideas on the mechanisms for Na^+ and Cl^- excretion across the gills of a seawater-acclimated fish via alpha chloride cells (CC) and adjacent pavement (PC) and accessory cells (AC). The many lines between the CC and the PC represent a deep tight junction, whereas the single line between the CC and the AC represents a shallow tight junction [based on (11)].

"Hormonal controls"). Lost internal salts are replaced with those taken in with food or those taken up at the gills, through the use of active transport mechanisms (Fig. 1). Therefore, an energy-requiring salt pump operates in special cells (including beta chloride cells) of the gills in freshwater fish as well, except that ions are pumped inward, rather than the reverse, as exhibited by the marine teleosts.

From the pioneering work of Krogh (26) and of Maetz and Romeu (27), a model has been formulated that describes ion-exchange mechanisms across the gills in freshwater teleosts. These ion-exchange mechanisms apparently reside in epithelial cells (including chloride cells). These beta chloride cells (8) occur on gill filaments between lamellae; and in fishes in very soft (low dissolved-ion concentrations) water, on gill lamellae (28). They resemble chloride cells in marine fish in that they contain mitochondria, the tubular system, and Na^+-K^+-ATPase; but they are different in that they are less numerous, usually occur singly, lack the apical crypt, and generally assist movement of Na^+ and Cl^- into the fish, rather than out (29). Figure 3 diagrammatically shows current ideas concerning these Na^+ and Cl^- movements. It is not known whether all of these movements take place in beta chloride cells or in other (e.g., flat epithelial) types. These ion-exchange mechanisms serve several functions besides maintenance of $[Na^+]$ and $[Cl^-]$ in the fish. The Na^+ exchange for NH_4^+ conveniently rids the fish of part of its ammonia production, the principal waste product of protein digestive breakdown. Experimental injections of NH_4^+ into freshwater goldfish (*Carassius auratus*) thereby stimulated Na^+ influx (28). Both the Na^+ exchange for H^+ and the Cl^- exchange for HCO_3^- tend to maintain internal acid-base homeostasis. Interestingly, these same exchanges also occur, presumably for NH_4^+ elimination and acid-base balance, in marine fish (30). These exchanges exacerbate the osmotic and ionic regulation problems of marine teleosts in hypertonic sea water by bringing in extra Na^+ and Cl^- for elimination. There is also evidence for Na^+ uptake via a Na^+ specific channel, which is driven by the electrical gradient created by the active (H^+-ATPase system) extrusion of H^+ (Fig. 3). Although the presence and relative magnitudes of these different routes of Na^+ uptake in various fish remain uncertain, they are functionally equivalent in terms of their effects on acid-base balance.

The general "exchanges" of H^+ (or NH_4^+) for Na^+ and of HCO_3^- for Cl^- generally balance the acid-base and ionic requirements in good fashion (Fig. 3). However, this link can force unwanted adjustments in one system, while compensating for a disturbance in another. The osmotic problems fish face in maintaining their acid-base balance are among the main reasons they have a hard time surviving in high-acid waters, such as streams that drain many mines, or lakes contaminated by acid rain or acid snow. Because high environmental $[H^+]$ can both

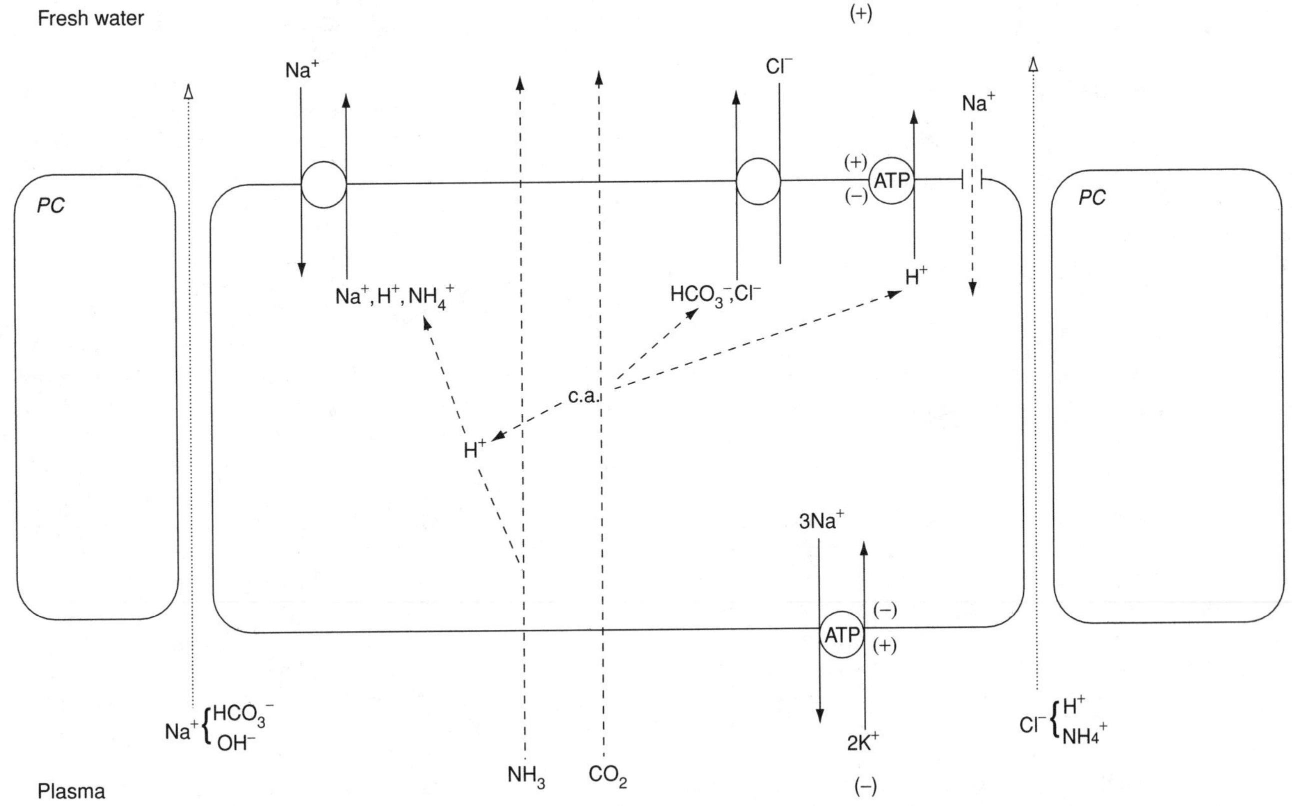

Figure 3. Current ideas on the mechanisms for Na^+ and Cl^- movements across the gills of a freshwater-acclimated fish via beta chloride (or other) cells, with allied movements of CO_2 and NH_3 in various forms, via diffusive (dashed and dotted lines) and carrier-mediated (solid lines) processes [based on (99)].

inhibit H^+ excretion associated with Na^+ uptake across gill epithelia and increase passive ionic losses via paracellular pathways, acid-exposed teleosts often die from insufficient plasma NaCl. For example, Leivestad and Muniz (31) attributed the death of brown trout that were exposed to low-pH conditions in the Tovdal River (southern Norway) to extreme reductions in plasma NaCl.

OSMOREGULATION AND IONIC REGULATION IN EURYHALINE AND DIADROMOUS FISH

Whereas most bony fish are stenohaline (meaning that they have narrow salt-tolerance range) and have evolved the exact osmotic structures (e.g., the appropriate sorts of chloride cells and intercellular junctions between cells) and mechanisms (e.g., hormonal controls) necessary to cope with the relatively constant salt concentration of either marine or freshwater habitats, euryhaline fish have broader salt tolerance ranges. Examples of euryhaline fish include the sheepshead minnow (*Cyprinodon variegatus*), Mozambique tilapia, and striped bass (*Morone saxatilis*). Diadromous (meaning "to run across") fish move between freshwater and marine environments at discreet life stages. Examples of diadromous species include the Pacific lamprey (*Lampetra tridentata*), Pacific salmon (*Oncorhynchus* spp.), and the American eel (*Anguilla rostrata*). Regardless of the range of salt tolerance and how it may change during the life cycle of a fish, osmoregulation requires energy. Indeed, growth rates and swimming performance of fish may be affected by how much energy they allocate toward osmoregulation or other energy costs, such as activity. For example, euryhaline sciaenid fish (drums) are able to maximize metabolic efficiency [e.g., for optimal growth (32) or swimming performance capacity (33), when held near the midpoints of their salinity tolerance ranges, compared with those near their tolerance extremes]. In contrast, cultured milkfish (*Chanos chanos*) in southeast Asia grow best at very high salinities (55 ppt), compared with lower salinities (15 and 35 ppt) for other fish, because the fish reduces its voluntary activity rates at high salinities (34). The cost of osmoregulation is generally considered to be a small fraction of the overall energy costs of fish (35), although it probably increases with increases in activity (36).

Hormonal Controls

Cortisol (a steroid hormone from the interrenal tissue associated with the kidney) is critical in ionic regulatory control in hypertonic environments. Anguillid eels are catadromous (meaning "to run down") fish that migrate down rivers as adults to spawn in the ocean. Forrest et al. (37) noted that plasma cortisol concentration showed a five- to seven-day increase in freshwater-adapted American eels (*A. rostrata*) when they were transferred to seawater. Cortisol injection experiments support the notion that this transient peak morphologically prepares diadromous or euryhaline fish for survival in hypertonic environments (38). For example, cortisol injections stimulated an increase in chloride cell density in freshwater-acclimated Mozambique tilapia (39). Conclusively, direct exposure of freshwater-acclimated tilapia opercular membranes to cortisol increased chloride cell density, size, and Na^+-K^+-ATPase activity (40). Adaptive mechanisms, such as increased Na^+ excretion and Na^+-K^+-ATPase activity at the gills, increased water permeability of the urinary bladder (to retain water), and the increased uptake of ions and water in the gut (related to seawater drinking) have been associated with such cortisol-stimulated changes (41).

Changes in the ion-regulatory abilities of diadromous fish are typically associated with ontogenetic changes, mediated by hormones. Adult salmon and some trout are anadromous (meaning "to run up") fish that migrate up rivers from the ocean (or a lake) to spawn. Juvenile salmon and trout migrate down the rivers, usually to saltwater, after changing from a freshwater (parr) form to a seawater-capable (smolt) form. Seawater readiness among young salmon and trout species ranges from a modest springtime rise in resistance shown by underyearling steelhead trout (*Oncorhynchus mykiss*), and other trouts and charrs, to complete tolerance and survival in recently hatched chum (*O. keta*) and pink (*O. gorbuscha*) salmon alevins. For most salmonids, tolerance of marine conditions develops in the spring, prior to the seaward migration of the silvery smolts. Hoar (42) pointed out the importance of lengthening springtime photoperiods for timing these changes. On a finer scale, Grau et al. (43) showed that plasma thyroxine surges, which occur in coho salmon (*O. kisutch*) during their smoltification period, correspond to new moon phases during spring. High and low tidal heights and resulting tidal currents are accentuated when the gravitational pulls of the moon and the sun are additive (i.e., during new moons or full moons). New-moon-related strong outflowing currents, and dark nights, may decrease the young coho's predation vulnerability (through faster transit and poorer visibility) during their movements down tidally influenced rivers to the estuary. A few weeks after the thyroxine peaks, increased plasma cortisol concentrations (44) and the cortisol-related changes in chloride cells, urinary bladder, and intestine are observed (45), and the coho smolts migrate to the sea. Besides cortisol, the pituitary hormone, growth hormone (GH), also plays an important role in the parr–smolt transformation among anadromous *Salmo* and *Oncorhynchus* species (46). For example, freshwater-acclimated sea-run brown trout (*S. trutta*) yearlings regulate plasma ions better and survive in greater numbers after seawater exposure with injections of both GH and cortisol, compared with noninjected controls (47). Further, GH's effectiveness regarding young salmonid's ionic regulation in hypertonic environments may be linked with insulinlike growth factor-I (48).

Prolactin and Ca^{2+} play important roles in regulating Na^+ gill permeability in freshwater-acclimated teleosts. Prolactin prevents Na^+ diffusive loss in freshwater-adapted bony fish and minimizes increases in passive Na^+ loss as euryhaline forms pass from seawater to freshwater. For example, hypophysectomy (removal of the anterior lobe of the pituitary, where prolactin is synthesized) of freshwater-acclimated mummichogs (*Fundulus heteroclitus*) promotes a marked drop in plasma ion concentrations compared with sham-operated controls (49). The many ion-regulatory roles of fish prolactin are essentially opposite those of cortisol regarding various physiological and morphological changes in the gills, gut, kidney, and urinary bladder of various species (50). Gill permeabilities in many species are also affected by the concentration of calcium (Ca^{2+}) in the water. It is known that addition of 10 mM Ca^{2+} to freshwater will reduce prolactin synthesis rates by 50% in tilapia (51). The sodium efflux from the plains killifish (*Fundulus kansae*) in freshwater is reduced by 50% when 1 mM Ca^{2+} is added to the water (52). Calcium can also reduce Na^+ permeability across the branchial epithelia of fish in seawater, such as *Anguilla* (53). Water permeability of rainbow trout gills is decreased as the number of calcium-binding sites in the gills increases (54). Gundersen and Curtis (54) showed that acclimation of rainbow trout to high-hardness (1.0 mM Ca^{2+}, as $CaCO_3$) water produced significantly fewer high affinity Ca binding sites in their gills, compared with those acclimated to water of low hardness (0.1 mM Ca^{2+}, as $CaCO_3$).

STRESS RESPONSES AND EFFECTS

Stressors such as extremely vigorous exercise, netting and handling, pronounced hypoxia (including air exposure), hypercapnia, hyperoxia, and dietary deficiencies may all have effects on the osmoregulation or ionoregulation of bony fish. Some of these factors may stimulate adaptive physiological changes (including enhanced aerobic performance capabilities). Mild and short-term stressors stimulate secretions of catecholamines, from chromaffin tissues, and cortisol, from the interrenal. The catecholamine (especially, epinephrine) and cortisol secretions lead to glucose mobilization for metabolic needs and cardiovascular adjustments to optimize aerobic efficiency (55,56). Epinephrine promotes shifting more of the gills' blood flow from filamental pathways to thin-walled lamellar pathways (a process termed lamellar recruitment) for enhanced gas exchange (57) at higher pressures due to increased cardiac output and systemic (body) vasoconstriction (58,59). In concert, cortisol mobilizes more epinephrine receptors on RBC membranes for maintenance of blood O_2 transport (60). However, prolonged or especially severe stressors overtax the stress response system and lead to ionic imbalances, decreased growth, increased metabolic exhaustion and disease incidence, and possible mortality (55). The increased lamellar blood flow increases gill permeability to water and small ions (61), as well as to oxygen, leading to the possible hydromineral imbalances in fish (62). Thus, stress will invoke water losses (and some monovalent ion gains) in marine fish and water gains (and some monovalent ion losses) in freshwater fish. Exercised *Oreochromis niloticus* that are acclimated to seawater tend to increase plasma osmotic pressure (from water loss) and tend to decrease osmotic pressure (from water uptake) after acclimation to freshwater (63). This trade-off between enhanced O_2 uptake and gill permeability is probably more costly for less-active species (e.g., sunfish) from lakes than for more-active ones (e.g., trout and shiner minnows) from streams (64). Gonzales and McDonald (64) showed that the former group has a much higher "Na^+ loss" to "O_2 gained" ratio after vigorous exercise (chasing in a tank) than the latter group.

Handling/Forced Swimming

If bony fish are stressed because of netting, tank transfer, or forced swimming, hormones are secreted and gill permeability to water and monovalent ions increases (62). When a stressed striped bass (*M. saxatilis*) is in freshwater, water rapidly diffuses into the fish (as revealed by a rapid gain in water weight) and overwhelms osmotic and ionic regulatory controls. The consequent mortality is high, compared with a group kept in brackish (near-isosmotic) salt concentrations (10 ppt salinity; 65,66). When intensively exercised and stressed from netting and handling, subadult striped bass incur an acidotic state. During stress recovery from exercise or handling in a hypertonic environment, both plasma Cl^- concentration and mortality rate increase, compared with a group recovered in 10 ppt salinity (near isosmotic) conditions (66). The addition of salt (NaCl, i.e., "rock salt") to the tanks or transport vessels holding stressed fish minimizes the osmotic gradient (and, consequently, water and ionic diffusion rates) between the environment and the fish's plasma, thus increasing survival. For striped bass, environmental salt (NaCl) concentration effects are more significant than those of water hardness ($CaCl_2$), regarding survival. Mazik et al. (65) found that juvenile striped bass that were transported and subsequently held in 1% NaCl solutions showed significantly more stable plasma Na^+ and Cl^- concentrations, and increased survival, compared with those in either soft freshwater or moderately hard (0.1% $CaCl_2$) freshwater. These authors attributed the observed striped bass mortalities to osmoregulatory dysfunction. Further, both cultured and wild young-of-the-year striped bass that had been exercise-conditioned (trained in flowing water) for 60 days at 1.2–2.4 body lengths/sec regulated plasma ions significantly better than unexercised striped bass, after capture and net-confinement stress (67). Truck-transport stress of largemouth bass (*Micropterus salmoides*) was reduced and related mortality was eliminated when fish were treated for diseases, starved for 72 hours before loading, lightly anesthetized, hauled at a cool temperature in physiological concentrations of salts, and allowed to recover in the same medium, minus the anesthetic (68). Transportation-related mortality of delicate delta smelt (*Hypomesus transpacificus*) was minimized after addition of salt and a commercial water conditioner that contained polymers (69).

Hypoxia

If significant decreases in environmental dissolved-oxygen levels (hypoxia) affect osmoregulatory function, the effects appear to be minor ones. Thomas and Hughes (70) found no change in plasma Cl^- concentration after 12-, 20-, and 120-minute exposures to mild hypoxia [60 mm Hg partial pressure of oxygen (PO_2); ca. 40% air-saturation] in freshwater rainbow trout. Claireaux and Dutil (71) show a slight, but significant, increase in Atlantic cod (*Gadus morhua*) plasma Cl^- (ca. 6 meq/L) after 6 hours exposure to mild hypoxia (ca. 60 mm Hg PO_2) in dilute seawater (21–28 ppt salinity), while Na^+ shows no significant change. In contrast, both plasma Cl^- and

Na^+ concentrations significantly increased (ca. 10 meq/L) when exposed to more-severe hypoxia (ca. 30 mm Hg PO_2) under the same salinity conditions. These authors explain their results in terms of fluid shifts (i.e., plasma water diffusing into white muscle, which contains increased lactate concentrations) and other exchanges at the gills.

Hypercapnia

Hypercapnic (high-dissolved CO_2) stress is becoming more common in cultured fish, as culturists increase the densities of fish in rearing tanks. The use of O_2 injection systems to hyperoxygenate water flowing into the rearing tanks allows dissolved O_2 levels to remain high at higher fish densities. However, the larger respiring biomass of more densely cultured fish produces more excreted CO_2, resulting in the hypercapnic conditions. The CO_2 is hydrated in the water and makes carbonic acid, which partially dissociates to H^+ (and HCO_3^-) ions, lowering water pH. The H^+ ions diffuse into the fish, lowering their internal pH. Fish respond by conserving their metabolically produced HCO_3^- ions (to buffer the H^+) and increasing their excretion of H^+ ions to restore pH towards the levels that exist under normocapnic (low, normal CO_2 levels) conditions. The H^+ efflux occurs mostly at the gills, but some of the acid excretion [e.g., 16% in rainbow trout (72)] can be accounted for in the urine. Interestingly, conserving HCO_3^- ions decreases HCO_3^- efflux, decreasing Cl^- influx, according to the ionic exchanges outlined earlier (Fig. 3). Thus, plasma Cl^- decreases in freshwater fish such as Arctic grayling [*Thymallus arcticus* (73)], rainbow trout (74), and carp [*Cyprinus carpio* (75)]. The magnitude of this response varies among species. Channel catfish (*Ictalurus punctatus*) show progressive plasma Cl^- decreases as environmental hypercapnia intensifies (76), whereas freshwater American eels show very slight Cl^- decreases with hypercapnic exposure (77). Simultaneous with the decreases in plasma Cl^-, increases in H^+ efflux also increase Na^+ influx, according to the previously discussed ionic exchanges (Fig. 3), and result in an increase of plasma Na^+. These increases vary with species, but they return to normal levels faster in a seawater, at least in rainbow trout (78). The result of these plasma Cl^- decreases and Na^+ increases is a widening difference between these two important plasma ions' concentrations, under hypercapnia. The longer-term physiological implications of these compromises between the regulation of ionic concentrations vs. acid-base status associated with hypercapnia in cultured fish have yet to be resolved (79). However, long-term (weeks to months) exposure to hypercapnia is known to decrease the growth rate of juvenile white sturgeon [*Acipenser transmontanus* (80)] and produce nephrocalcinosis in rainbow trout (81).

Hyperoxia

Dissolved oxygen levels greater than air saturation (hyperoxia) may be observed in some fish culture systems. For example, lightly loaded ponds with a significant plant biomass may show higher-than-atmospheric levels of dissolved oxygen during daylight hours, due to photosynthetic production by the plants (algae or macrophytes). Interestingly, these systems may show significant hypoxia during nighttime periods (especially, predawn) when all of the animal and plant biomass is using oxygen for respiration, and the lack of light prevents photosynthetic oxygen production. Another example is hyperoxia that results from improperly controlled oxygen injection systems (over-injection) in high-density culture systems. Because fish (at least, rainbow trout) gill ventilation is controlled by oxygen receptors in the gills (82,83), the high environmental PO_2s limit adequate excretion of CO_2 from the gills. Consequently, hyperoxia produces an internal hypercapnia (84), and the hyperoxic responses parallel those to hypercapnic conditions. For example, Wilkes et al. (85) showed that white sucker (*Catostomus commersoni*) exposed to 2.2–3.5 times atmospheric oxygen levels showed a significant (ca. 15 meq/L) decrease in plasma Cl^- with no significant change in plasma Na^+, apparently due to a reduction in branchial Cl^-–HCO_3^- exchange. Goss and Wood (86) also showed significant branchial efflux of Cl^- in rainbow trout during hyperoxic exposure (3.6–4.3 times atmospheric oxygen levels).

Dietary Deficiencies

Deficiencies in dietary nutrients, either in a total sense, via food deprivation, or in a partial sense, via inadequate nutrient concentrations, may affect the concentration of plasma ions. Food-deprived tilapia (*O. mossambicus*), when acclimated to seawater, show increases in plasma Cl^- concentration (87). Rainbow trout that are fed five diets with various fatty acid compositions showed similar plasma Na^+ and Cl^- concentration responses to exposure to 33 ppt. seawater at both 2° and 17 °C (36 and 63 °F) (88). Finally, juvenile carp [ca. 100 g (3.3 oz)] showed low plasma Mg^{2+} concentrations when fed a low-magnesium diet for 17 weeks (89).

ACKNOWLEDGMENTS

I thank Dr. Christopher Myrick for his insights and assistance in preparing this manuscript. I was partially supported by a University of California Davis Agricultural Experiment Station grant (3455-H).

BIBLIOGRAPHY

1. C.M. Wood and T.J. Shuttleworth, eds., *Cellular Approaches to Fish Ionic Regulation*, Academic Press, San Diego, CA, 1995.
2. G.A. Wedemeyer, *Physiology of Fish in Intensive Culture Systems*, Chapman and Hall, New York, NY, 1996.
3. K.J. Karnaky, Jr., in D.H. Evans, ed., *The Physiology of Fishes, 2nd edition*, CRC Press, Boca Raton, FL, 1998, pp. 157–176.
4. W.T.W. Potts and P.P. Rudy, *J. Exp. Biol.* **56**, 703–715 (1972).
5. M. McEnroe and J.J. Cech, Jr., *Env. Biol. Fish.* **14**, 23–40 (1985).
6. I. Altinok, S.M. Galli, and F.A. Chapman, *Comp. Biochem. Physiol.* **120A**, 609–619 (1998).
7. R.H. Catlett and D.R. Millich, *Comp. Biochem. Physiol.* **55A**, 261–269 (1976).

8. M. Pisam, A. Caroff, and A. Rambourg, *Amer. J. Anat.* **179**, 40–50 (1987).
9. D. Kultz, R. Bastrop, K. Jurss, and D. Siebers, *Comp. Biochem. Physiol.* **102B**, 293–301 (1992).
10. P. Tytler and J. Ireland, *J. Therm. Biol.* **20**, 1–14 (1995).
11. K.J. Karnaky, Jr., *Amer. Zool.* **26**, 209–224 (1986).
12. P. Silva, R. Solomon, K. Spokes, and F.H. Epstein, *J. Exp. Zool.* **199**, 419–426 (1977).
13. S. Utida and T. Hirano, in W. Chavin, ed., *Responses of Fish to Environmental Changes*, Chas. C. Thomas, Springfield, IL, 1973, pp. 240–278.
14. M.N. Perrot, C.E. Grierson, N. Hazon, and R.J. Balment, *Fish Physiol. Biochem.* **10**, 161–168 (1992).
15. J. Fuentes, J.C. McGeer, and F.B. Eddy, *Fish Physiol and Biochem.* **15**, 65–69 (1996).
16. J. Fuentes and F.B. Eddy, *Physiol. Zool.* **69**, 1555–1569 (1996).
17. D.H. Evans, *Annu. Rev. Physiol.* **52**, 43–60 (1990).
18. D.H. Evans and Y. Takei, *News Physiol. Sci.* **7**, 15–18 (1992).
19. K.R. Olson and D.W. Duff, *J. Comp. Physiol. B* **162**, 408–415 (1992).
20. C.A. Loretz, *Amer. Zool.* **35**, 490–502 (1995).
21. W.H. Sawyer, J.R. Blair-West, P.A. Simpson, and M.K. Sawyer, *Amer. J. Physiol.* **231**, 593–602 (1976).
22. J.M. Grizzle, I.I. Mauldin, C. Alfred, D. Young, and E. Henderson, *Aquaculture* **46**, 167–171 (1985).
23. W.T.W. Potts and W.R. Fleming, *K. Exp. Biol.* **53**, 317–327 (1970).
24. S.E. Wendelaar Bonga, C.J.M. Löwik, and J.C.A. Van Der Meij, *Gen. Comp. Endocr.* **52**, 222–231 (1983).
25. G. Flik, J.C. Fenwick, and S.E. Wendelaar Bonga, *Am. J. Physiol.* **257**, R74–R79 (1989).
26. A. Krogh, *Osmotic Regulation in Aquatic Animals*, Cambridge University Press, London, 1939.
27. J. Maetz and F.G. Romeu, *J. Gen. Physiol.* **50**, 391–422 (1964).
28. P. Laurent, H. Hobe, and S. Dunel-Erb, *Cell Tissue Res.* **240**, 675–692 (1985).
29. F.B. Eddy, *Comp. Biochem. Physiol.* **73B**, 125–141 (1982).
30. D.H. Evans, *J. Exp. Biol.* **70**, 213–220 (1977).
31. H. Leivestad and I.P. Muniz, *Nature* **259**, 391–392 (1976).
32. R.W. Brocksen and R.E. Cole, *J. Fish. Res. Bd. Canada* **29**, 399–405 (1972).
33. D.E. Wohlschlag and J.M. Wakeman, *Cont. Mar. Sci.* **21**, 173–185 (1978).
34. C. Swanson, *J. Exp. Biol.* **201**, 3355–3366 (1998).
35. J.D. Morgan and G.K. Iwama, *Can. J. Fish. Aquat. Sci.* **48**, 2083–2094 (1991).
36. R. Febry and P. Lutz, *J. Exp. Biol.* **128**, 63–85 (1987).
37. J.N. Forrest, Jr., W.C. Mackay, B. Gallagher, and F.H. Epstein, *Am. J. Physiol.* **224**, 714–717 (1973).
38. D.H. Evans, in W.S. Hoar and D.J. Randall, eds., *Fish Physiology*, Academic Press, New York, NY, 1984, pp. 239–283.
39. J.K. Foskett, C.D. Logsdon, T. Turner, T.E. Machen, and H.A. Bern, *J. Exp. Biol.* **93**, 209–224 (1981).
40. S.D. McCormick, *Am. J. Physiol.* **259**, R857–R863 (1990).
41. A.J. Matty, *Fish Endocrinology*, Timber Press, Portland, OR, 1985.
42. W.S. Hoar, *J. Fish. Res. Bd. Canada* **33**, 1234–1252 (1976).
43. E.G. Grau, W.W. Dickhoff, R.S. Nishioka, H.A. Bern, and L.C. Folmar, *Science* **211**, 607–609 (1981).
44. J.L. Specker and C.B. Schreck, *Gen. Comp. Endocr.* **46**, 53–58 (1982).
45. C.A. Loretz, N.L. Collie, N.H. Richmann III, and H.A. Bern, *Aquaculture* **28**, 67–74 (1982).
46. B.A. Barrett and B.A. McKeown, *Can. J. Zool.* **66**, 853–855 (1988).
47. S.S. Madsen, *Gen. Comp. Endocr.* **79**, 1–11 (1990).
48. S.D. McCormick, T. Sakamoto, S. Hasegawa, and T. Hirano, *J. Endocr.* **130**, 87–92 (1991).
49. J. Maetz, W.H. Sawyer, G.E. Dickford, and N. Mayer, *Gen. Comp. Endocr.* **8**, 163–176 (1967).
50. T. Hirano, in C.L. Ralph, ed., *Comparative Endocrinology: Developments and Directions*, Alan R. Liso, Inc., New York, NY, 1986, pp. 53–74.
51. S.E. Wendelaar Bonga, G. Flik, C.J.M. Löwik, and G.J.J.M. Van Eyes, *Gen. Comp. Endocr.* **57**, 352–359 (1985).
52. W.T.W. Potts and W.R. Fleming, *J. Exp. Biol.* **54**, 63–75 (1971).
53. A.W. Cuthbert and J. Maetz, *J. Physiol.* **221**, 633–643 (1972).
54. D.T. Gundersen and L.R. Curtis, *Can. J. Fish. Aquat. Sci.* **52**, 2583–2593 (1995).
55. B.A. Barton and G.K. Iwama, *Ann. Rev. Fish Diseases* **1**, 3–26 (1991).
56. S.D. Reid, T.W. Moon, and S.F. Perry, *J. Exp. Biol.* **158**, 199–216 (1992).
57. J.H. Booth, *J. Exp. Biol.* **83**, 31–39 (1979).
58. A.P. Farrell and D.R. Jones, in W.S. Hoar, D.J. Randall, and A.P. Farrell, eds., *Fish Physiology Vol. 12A*, Academic Press, New York, NY, 1992, pp. 1–88.
59. D.J. Randall and S.F. Perry, in W.S. Hoar, D.J. Randall, and A.P. Farrell, eds., *Fish Physiology Vol. 12B*, Academic Press, San Diego, CA, 1992, pp. 255–300.
60. S.D. Reid and S.F. Perry, *J. Exp. Biol.* **158**, 217–240 (1991).
61. P. Pic, N. Mayer-Gostant, and J. Maetz, *Am. J. Physiol.* **226**, 698–702 (1974).
62. M. Mazeaud, R. Mazeaud, and E.M. Donaldson, *Trans. Amer. Fish. Soc.* **106**, 201–212 (1977).
63. G.J. Farmer and F.W.H. Beamish, *J. Fish. Res. Bd. Canada* **26**, 2807–2821 (1969).
64. R.J. Gonzales and D.G. McDonald, *J. Exp. Biol.* **190**, 95–108 (1994).
65. P.M. Mazik, B.A. Simco, and N.C. Parker, *Trans. Amer. Fish. Soc.* **120**, 121–126 (1991).
66. J.J. Cech, Jr., S.D. Bartholow, P.S. Young, and T.E. Hopkins, *Trans. Amer. Fish. Soc.* **125**, 308–320 (1996).
67. P.S. Young and J.J. Cech, Jr., *Can. J. Fish. Aquat. Sci.* **50**, 2094–2099 (1993).
68. G.J. Carmichael, J.R. Tomasso, B.A. Simco, and K.B. Davis, *Trans. Amer. Fish. Soc.* **113**, 778–785 (1984).
69. C. Swanson, R.C. Mager, S.I. Doroshov, and J.J. Cech, Jr., *Trans. Amer. Fish. Soc.* **125**, 326–329 (1996).
70. S. Thomas and G.M. Hughes, *Resp. Phys.* **49**, 371–382 (1982).
71. G. Claireaux and J.-D. Dutil, *J. Exp. Biol.* **163**, 97–118 (1992).
72. S.F. Perry, S. Malone, and D. Ewing, *Can. J. Zool.* **65**, 896–902 (1987a).
73. J.N. Cameron, *J. Exp. Biol.* **64**, 711–725 (1976).
74. S.F. Perry, S. Malone, and D. Ewing, *Can. J. Zool.* **65**, 888–895 (1987b).
75. J.B. Claiborne and N. Heisler, *J. Exp. Biol.* **126**, 41–61 (1986).

76. J.N. Cameron and G.K. Iwama, *J. Exp. Biol.* **133**, 187–197 (1987).
77. D.A. Hyde and S.F. Perry, *Physiol. Zool.* **62**, 1164–1186 (1989).
78. G.K. Iwama and N. Heisler, *J. Exp. Biol.* **158**, 1–18 (1991).
79. J.N. Cameron and G.K. Iwama, *Can. J. Zool.* **67**, 3078–3084 (1989).
80. C.E. Crocker and J.J. Cech, Jr., *Aquaculture* **147**, 293–299 (1996).
81. G.R. Smart, D. Knox, J.G. Harrison, J.A. Ralph, R.H. Richards, and C.B. Cowey, *J. Fish Diseases* **2**, 279–289 (1977).
82. C. Daxboeck and G. Holeton, *Can. J. Zool.* **56**, 1254–1259 (1978).
83. F.M. Smith and D.R. Jones, *Can. J. Zool.* **56**, 1260–1265 (1978).
84. D.J. Randall and D.R. Jones, *Resp. Phys.* **17**, 291–301 (1973).
85. P.R.H. Wilkes, R.L. Walker, D.G. MacDonald, and C.M. Wood, *J. Exp. Biol.* **91**, 239–254 (1981).
86. G.G. Goss and C.M. Wood, *J. Exp. Biol.* **152**, 521–548 (1990).
87. M.M. Vijayan, J.D. Morgan, E.G. Grau, and G.K. Iwama, *J. Exp. Biol.* **199**, 2467–2475 (1996).
88. B. Finstad and M.S. Thomassen, *Comp. Biochem. Physiol.* **99A**, 463–471 (1991).
89. J.A. Van der Velden, G. Flik, F.A.T. Spaning, T.G. Verburg, Z.I. Kolar, and S.E. Wendelaar Bonga, *J. Exp. Zool.* **264**, 237–244 (1992).
90. K. Schmidt-Nielsen, *Animal Physiology: Adaptation and Environment*, Cambridge University Press, London, 1975.
91. M. Morisawa, T. Hirano, and K. Suzuki, *Comp. Biochem. Physiol.* **64A**, 325–329 (1979).
92. C.E. Johnston and J.C. Cheverie, *Can. J. Fish. Aquat. Sci.* **42**, 1994–2003 (1985).
93. R. Fänge, U. Lidman, and A. Larsson, *J. Fish Biol.* **8**, 441–448 (1976).
94. F. Gaumet, G. Boeuf, A. Severe, A. Le Roux, and N. Mayer-Gostant, *J. Fish Biol.* **47**, 865–876 (1995).
95. A. Oikari, *Ann. Zool. Fennici* **15**, 84–88 (1978).
96. A.H. Houston and T.F. Koss, *J. Exp. Biol.* **97**, 427–440 (1982).
97. A.H. Houston and J.A. Madden, *Nature* **217**, 969–970 (1968).
98. S. Thomas, B. Fievet, G. Claireaux, and R. Motais, *Resp. Phys.* **64**, 77–90 (1988).
99. C.M. Wood and W.S. Marshall, *Estuaries* **17**, 34–52 (1994).

See also SALINITY.

OZONE

JOHN COLT
Northwest Fisheries Science Center
Seattle, Washington

EDWIN CRYER
Montgomery Watson
Boise, Idaho

OUTLINE

Ozone has been widely used in the disinfection of culture and rearing water. Ozone can achieve a higher level of disinfection than can ultraviolet radiation or the addition of chlorine and hypochlorite. Under ambient conditions, ozone gas or ozone gas dissolved in water is unstable and will rapidly decay back to oxygen. In freshwater, the half-life of ozone is typically measured in minutes, and the lack of a long-term residual is an important design advantage of ozone in aquatic systems. The ozonation of seawater can result in oxidation of the bromide ion into hypobromous acid, hypobromite ion, and bromate. These compounds (residual oxidants) are stable, and their toxicity must be considered in the design and operation of seawater ozone systems. The literature on aquatic uses of ozone is large, and general information on the uses of ozone in water treatment can be found in (1); the comprehensive bibliography prepared by Rosenthal and Wilson (2) is also very useful.

A general schematic of an ozone disinfection system is presented in Figure 1. Key components are (a) an air-preparation or liquid-oxygen supply, (b) two or more ozone generators, (c) an ozone contact system, (d) an ozone destruct system for off-gas, and (e) an ozone destruct system for removing ozone from the process water. Not all of these components may be present in a given system.

PHYSICAL AND CHEMICAL PROPERTIES OF OZONE

Physical Properties of Ozone

The ozone molecule contains three atoms of oxygen (O_3) with a molecular weight of 48 g/mole, and has a density

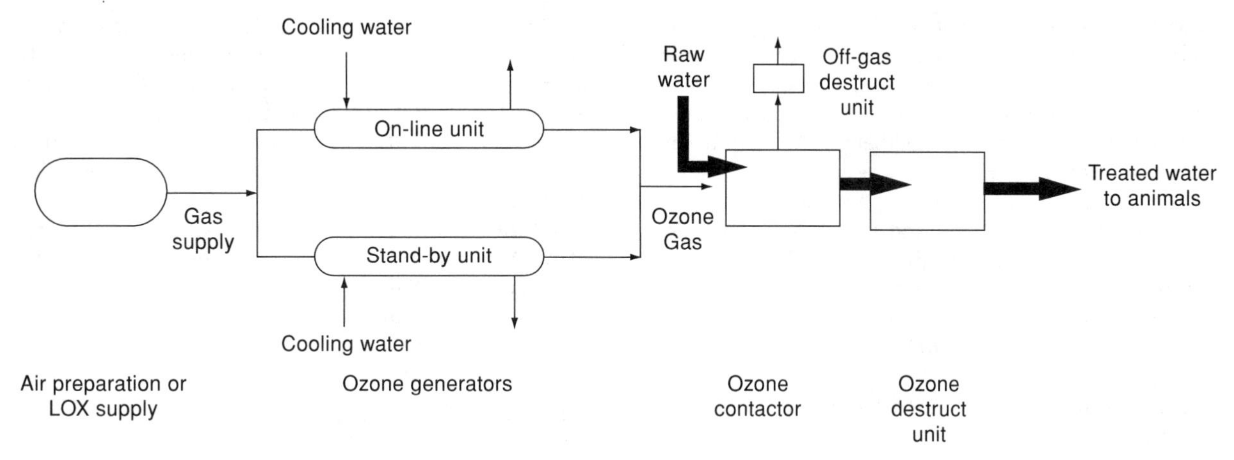

Figure 1. Typical process schematic diagram for an ozone system.

of 2.144 g/L [0 °C, (32 °F), 1 atm]. It is colorless and has a distinctive, pungent odor. Ozone gas is unstable and must be generated on-site. Ozone will react with a number of organic and inorganic molecules and will decay to the dioxygen molecule (O_2). The solubility of ozone at 0 °C (32 °F) is approximately 1,400 mg/L for pure gas or 70 mg/L for a 5% gas concentration (3).

Reactions of Ozone in Pure Water

The stability of ozone in water is affected by pH, temperature, exposure to ultraviolet radiation, and ozone concentration (3). The oxidization potential of ozone in an acidic solution is −2.07 v, compared to the −2.8 v for the hydroxyl radical. The hydroxyl radical is one of the most reactive unstable molecules known. Because of the reactivity of the hydroxyl radical, the effectiveness of ozone in a specific application may be related to the hydroxyl radical concentration rather that the concentration of dissolved ozone.

Reactions of Ozone with Inorganic and Organic Compounds

The reactions of ozone with inorganic compounds generally follow a first-order kinetic law with respect to both ozone and the oxidizable compound (1). The rate constant may depend on the pH and concentration of other chemical constituents.

Oxidation of Ammonia. Ammonia can be oxidized by the following reaction (3):

$$4O_3 + NH_3 \longrightarrow NO_3^- + 4O_2 + H_3O^+ \qquad (1)$$

The ammonia oxidation rate is pH-dependent and is slow for pH less than 9 (4). In typical freshwater applications, only 5–15% of the ammonia will be oxidized to nitrate by ozone. In seawater, the oxidation of ammonia by ozone is increased by the presence of the bromide ion (5). Unlike the direct ozonation of ammonia (Eq. 1), this reaction is affected very little by changes in pH.

Oxidation of Nitrite. Nitrite is quickly oxidized by ozone to nitrate (4):

$$NO_2^- + O_3 + \longrightarrow NO_3^- + O_2 \qquad (2)$$

Oxidation of Iron and Manganese. Reduced iron and manganese (Fe^{2+} and Mn^{2+}) are common in groundwaters and at the bottom of surface-water reservoirs. Oxidization of these compounds to their insoluble forms (Fe^{3+} and Mn^{4+}) allows their removal by sedimentation or filtration. Manganese is often more difficult to remove than iron, especially when organics and humic acids are present (3).

Oxidation of Aquatic Humic Substances. Humic substances are an extremely complex and diverse group of organic materials that are produced by decomposition of natural plant and animal matter. The build-up of humic acid substances is responsible for the yellow-brown to black color of natural waters. Ozone is the treatment of choice for decreasing the color of natural water and increasing the biodegradability of humic acids. The use of ozone in large marine aquariums is often for color control as well as for disinfection and removal of organic compounds.

Oxidation of Bromide. In seawater, ozone can react with the bromide ion and ammonia (6,7) to form the hypobromite ion, bromate ion, hypobromous acid, and bromamine. Hypobromite ion and hypobromous acid are unstable oxidants with half-lives ranging from 1 to 12 hours (8). Bromate ions and bromamines are stable compounds. In seawater systems that use marine water sources, residual bromide compounds may become critical concerns when attempts are made to culture sensitive species.

Reaction of Ozone with Microorganisms. The reactions of ozone with cell constituents of microorganisms are not well understood, even *in vitro*. It is thought that the disinfection reaction of ozone occurs between proteins that are contained in the cytoplasmic membrane. Effective inactivation of protozoa, bacteria, and virus has been demonstrated in a variety of culture applications.

Ozone Demand and Decay Rates. When ozone is added to water, there is commonly an immediate ozone demand. This demand is related to the chemical composition of the water, its solids content, and its temperature. This chemical demand must be satisfied before the disinfecting chemical oxidation reactions can proceed. Measurement of the ozone demand involves dosing a water sample with a known amount of ozone gas and measuring the ozone concentration in solution (residual). The decay rate of ozone is also quite specific to water. The test for decay rate involves adding ozone until a given residual is achieved, and measuring the ozone residual as it decays.

Kinetics of Disinfection. In general, it has been observed that for a given concentration of disinfectant, the longer the contact time, the greater the kill (9):

$$\log \frac{N_t}{N_0} = -k't \tag{3}$$

Here

N_t = Number of organisms at time t

N_0 = Number of organisms at $t = 0$

t = time (minute)

k' = constant (1/minute).

Because microorganism concentrations are sometime quite large, it is common to express these concentrations in terms of log units. Therefore, a concentration of 1,000,000 bacteria/mL would be equal to 6 log units. It takes the same contact time to reduce the concentration from 6 to 3 log units (1,000,000 to 1,000/mL) as it takes to reduce the concentration from 2 to −1 log units (100 to 0.1/mL). An infinite contact time is required to reduce the number of microorganisms to zero (sterilization).

Impact of Contact Time and Concentration. Within limits, it has been observed that concentration and contact time are related by the following relationship (9):

$$\text{Ct} = \text{Constant} \tag{4}$$

Here

C = concentration of disinfectant

t = time required to achieve a given kill.

To achieve a given level of disinfection, the product of Ct must equal the given constant. Therefore, the size of the contact basin can be made smaller if the concentration is increased, and vice versa.

GENERATION OF OZONE

The corona discharge method (3) is the most widely used technique for aquatic systems. Ozone is formed in a corona discharge caused by the interaction between an electrical discharge and oxygen molecules. An alternating current is applied across the two electrodes. One of the electrodes is coated with glass or ceramic dielectric material to evenly distribute the electrical discharge over the surface. An ozone generator can contain two to hundreds of separate tube assemblies, depending on the required output capacity.

Ozone generators are typically defined by their operating frequency. Generally, the higher the frequency, the more concentrated the output gas. Typically, the energy requirements of an ozone generator are approximately 3 kWh/kg (6–7 kWh/lb) of ozone.

MEASUREMENT OF OZONE

Ozone is difficult to measure accurately, especially in the liquid phase. In most facilities, it is necessary to measure ozone concentrations in both the gas and the liquid phases.

Gas Phase

The most commonly used, and most reliable, method for determination of continuous gas-phase ozone levels is the UV absorption technique. This method relies on the absorption of ozone by UV light at 260 nm, according to the Beer–Lambert law.

Liquid Phase

Numerous analytical methods have been developed for the determination of liquid-phase ozone concentrations. Because of the rapid decay of ozone in water, no holding time is allowed; samples must analyzed immediately. Two methods of liquid-phase analysis are recommended for aquaculture applications where low-concentration determination is needed. One is the indigo trisulfonate wet chemistry method for standardization; the other is the potentiometric method for on-line continuous monitoring.

Indigo Trisulfonate. This method is based on the rapid stoichiometric decolorization of indigo trisulfonate in acidic solution. The change in absorbance, measured at 600 nm, is proportional to the concentration of ozone. This method is not subject to interferences from ozone oxidation and decomposition products.

The Potentiometric Method. The dissolved ozone probe's (DOP) sensor consists of a working electrode, a reference electrode, and a buffered salt electrolyte solution. As ozone diffuses through the membrane, it is reduced at the cathode. The resulting flow of electrons is proportional to the ozone concentration.

Monitoring and Control of Ozone Systems

A monitoring and control program is necessary to safely and accurately control ozone dosage and to ensure that the residual ozone concentration in the water supply for the rearing units is safe for the culture species. Manual monitoring and control systems may be adequate for some applications.

For critical applications or where influent water quality changes significantly over the day, an on-line system

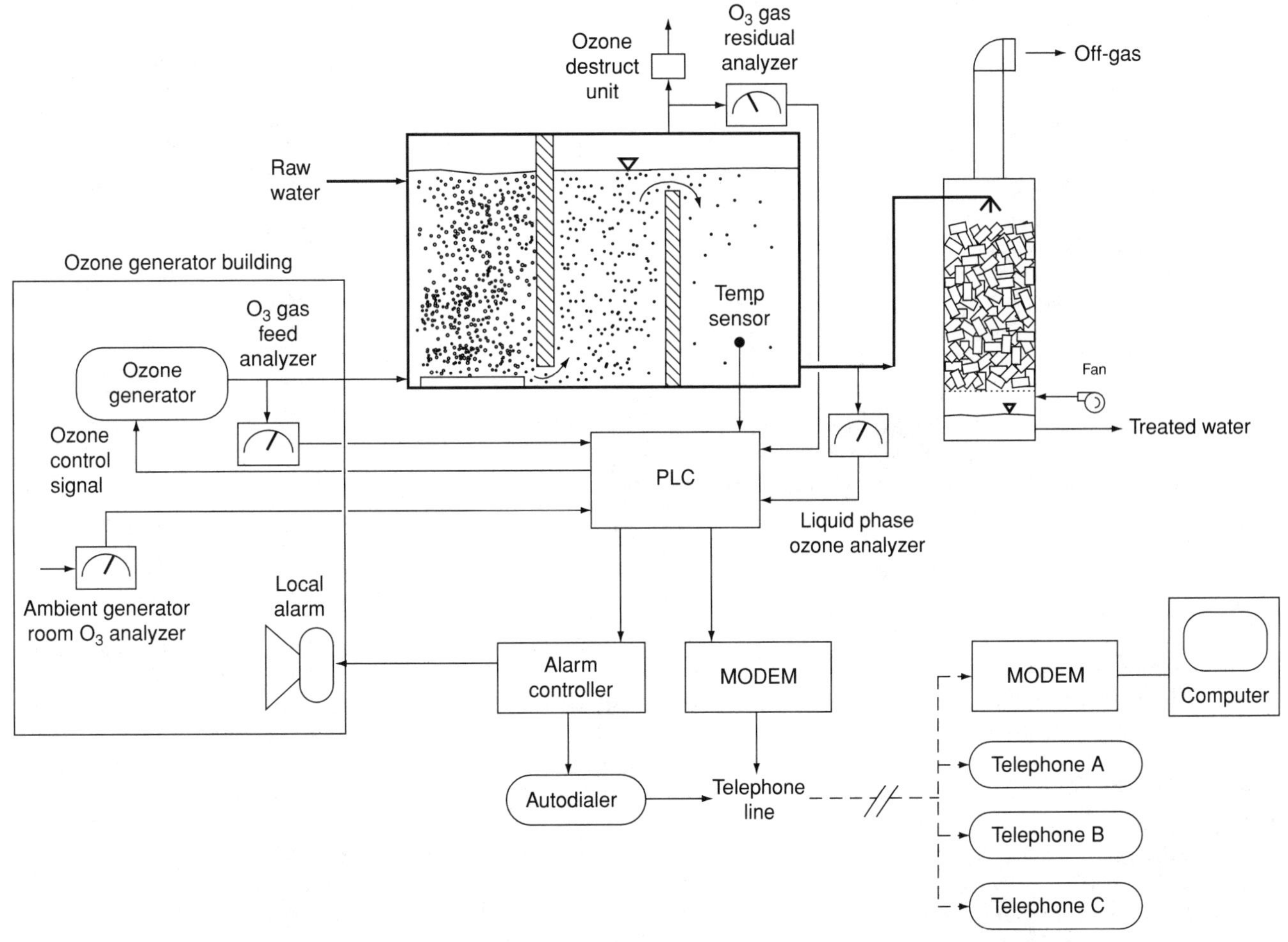

Figure 2. Typical schematic of an on-line ozone monitoring and control system (the PLC, or programmable logic controller, is a fault-tolerance industrial computer that is commonly used for process control).

may be necessary. An on-line monitoring and control system will consist of on-line gas-phase monitoring equipment, on-line liquid-phase monitoring equipment, and a programmable logic controller. Figure 2 provides a representative schematic of a typical on-line ozone monitoring and control system.

TOXICITY OF OZONE

Personnel

Ozone gas is toxic to humans, and in the United States (10), exposures are limited as follows:

0.1 ppm by volume (0.2 mg/m^3 NTP) determined as a time-weighted average over a full working day (8-hour maximum),

0.2 ppm by volume (0.4 mg/m^3 NTP) as maximum 10-minute exposure.

At low concentrations, ozone has a very distinctive smell, and it can be detected well below the daily safety criteria limit.

Pathogenic Organisms

Information on the impact of ozone on pathogenic organisms can be obtained through inoculation of a water sample with the test organism, followed by ozonation of the sample and measurement of the number of surviving organisms with time. A curve is then fitted through the experimental data. The required concentration, detention time, or Ct (concentration × time) is typically predicted for a 1-log (90%), 2-log (99%), 3-log (99.9%), or 4-log (99.99%) reduction in organism concentration. Detailed information on the impact of ozone on a number of freshwater and marine pathogens is presented in Table 1. While literature on effective inactivation rates is useful, pilot-scale work is generally necessary to determine the required dose or Ct.

Aquatic Animals

Lethal levels can be as low as 9.3 μg/L parts per trillion and sublethal effects have been observed below 5 μg/L parts per trillion (11). Under production conditions, ozone concentrations in the range of 10–20 μg/L parts per trillion (measured at the discharge from the contactor) have been found to have no impact on salmonid eggs and fry (12) because the chemical ozone demand from uneaten feed and waste products will quickly reduce the ozone residual to

Table 1. Values of Ct for a 99 and 99.9 Percent Inactivation for Various Pathogenic Organisms

Pathogen	Temperature (°C/°F)	Salinity (g/kg) [ppt]	Ct (mg/L·min)			Reference
			99% Kill	99.9% Kill	Design Value	
Enterococcus seriolicida	25/77	34.4	0.123	0.186	—	18
Vibrio anguillarum	25/77	34.4	0.056	0.084	—	18
Pasteurella piscicida	25/77	34.4	0.081	0.123	—	18
Mixed bacteria population	25/77	34.4	0.200	0.621	—	18
Infectious hematopoietic necrosis virus	4-16/39-61	0	—	—	2.0	19
Ceratomyxa shasta	12-17/54-62	0	—	—	1.0	12

nonsignificant levels. A general criteria of 10 μg/L parts per trillion should be protective of most species and life-stages; this criterion will be over protective for many culture animals that are reared under production conditions, and needs to be evaluated on a case-by-case basis.

OZONE CONTACTOR SYSTEMS

The most common ozone contact systems (a combination of gas mixing and detention time for inactivation) are the deep-tank contact basin and the injector system.

Deep-Tank Contact Basin

The deep-tank contact basin is conventional technology that has been used successfully in the drinking-water field for at least 40 years. The water depth is in the range of 5–7 m (15–21 ft). Commonly, the contact basin is divided by baffles into 3 or 5 separate chambers. The ozone-gas mixture is introduced into the first chamber by porous diffusers mounted on the bottom of the chamber.

Injector Systems

The most common type of injector system uses a number of venturi eductors or static mixers placed in parallel. Ozone gas is supplied to each eductor. The eductors require 170 to 200 kPa (25 to 30 psi) of pressure for operation, so a booster pump system is commonly needed. Static mixers operate best at lower water pressures

OZONE REMOVAL AND DESTRUCTION

Ozone removal and destruction is needed for both undissolved ozone gas from the contact system and for the residual ozone in the process water prior to use.

Gas Phase

For contact systems that do not result in 100% transfer into the water, the ozone gas in the off-gas must be collected and the ozone converted back into oxygen and vented to the atmosphere. This conversion and venting process is generally accomplished by passing the ozone-gas mixture through either a thermal ozone destruct unit or a catalytical destruct unit. Since 100% transfer does not always occur, a decision to leave out the destruction unit should be carefully considered.

Liquid Phase

Effective control of ozone residuals in the process water is necessary for protection of the culture animals. If effective control is not provided, the applied dose must be limited to levels that may not be effective for disinfection or for oxidation of the targeted organic.

Detention Time. The simplest process for ozone removal in the liquid phase is to provide enough detention time to allow the ozone to naturally decay back to oxygen gas. For large production flows, the required detention storage volume is significant; such storage offers the potential for reintroduction of pathogens from outside sources.

Air Stripping. Dissolved oxygen gas can be removed using diffused aeration, a cascade aerator, or an air-stripping column. This process will not remove ozone-produced bromide compounds. Air stripping is generally the most cost-effective way to remove residual ozone in a temperate climate. Careful design is important to prevent entrainment of air below the stripping column.

Granular Activated Carbon. In seawater applications, granular-activated carbon (GAC) can be used to remove both residual ozone, ozone-produced bromide compounds, and other organics. Carbon will be consumed by the reaction and spent carbon will need to be replaced or regenerated.

Ultraviolet Radiation. Ultraviolet radiation has been used to reduce residual ozone in the process water prior to its use in the rearing units. The primary application of ultraviolet radiation for ozone destruction has been in cold climates where air stripping would result in excessive heating costs. Ultraviolet radiation in the range of 250 to 260 nm catalyzes the decomposition of ozone to oxygen. An average dose of 90,000–150,000 μW·s/cm^2 is required to reduce an ozone residual of 0.3 to 1.0 mg/L (ppm) to a non-detectable concentration. Higher doses are required at lower temperatures.

Chemical Addition. Sodium thiosulfate will react with ozone-produced oxidants in seawater (13). The amount of ozone-produced oxidants removed per milligram of sodium thiosulfate ranges from 1.13 mg (1 hours of aeration) to 0.16 mg (48 hours of aeration).

GENERAL DESIGN CONSIDERATION

General design considerations that apply to all ozone systems are discussed in the following section.

Determination of System Capacity

Ozone output requirements depend on analytical testing to establish ozone demand [applied dose in mg/L (ppm)] and a knowledge of the flow requirements of the system. Ozone output is typically given in kg or lb/day of production. For example, a flow of 19 million liters (5 million gallons) per day, requiring 2.5 mg/L (ppm) of applied dose, would require 47 kg/d (104 lb/d). Assuming that the system would be operated at only 80% of capacity, a single ozone generator of 60 kg/d (132 lb/d) would be needed. Two units would be needed for critical applications where it was necessary to provide 100% standby duty.

System Control

Ozonators are controlled in several ways. The most common is to vary the power to the dielectrics. This can be accomplished either manually or by using an external signal (flow, ozone off-gas, or ozone concentration in the contactor). By using a programmable logic controller, the system can be automated and operated with constant attention if conditions vary.

Material Considerations

The extreme oxidation potential of ozone requires careful selection of construction materials to prevent corrosion and deterioration. Generally, the most serious problems occur with a mixture of ozone gas and water vapor. Fewer material problems are reported with dry or dissolved ozone gas.

Piping for dry ozone service should be stainless steel, AISI 316 or 316L, or better, and connection should be made with TIG welding and Teflon-filled gaskets. Valves should be 316 or 316L stainless steel or common steel with a Kynar body liner, 316 stainless steel disc, and Teflon seat and seals or 316 stainless steel, disc, and shaft, with TFE-filled seat and seal (14). Threaded pipe connections tend to leak with time and should be welded. However, flanges are necessary to remove fittings or connections. For small systems, Teflon tubing and fittings can be used. PVC should be used only after careful examination of the conditions and service computability. For example, in seawater applications, PVC may be preferable to 316 stainless steel due to chloride corrosion concerns.

Reinforced concrete has proven resistant to both dissolved ozone gas and ozone–water mixture in the head-space of contact chambers. The gas space concrete should be of Type 2 or Type 5 Portland cement using a low water/cement ratio (14). All fittings and seals should be stainless steel or Teflon.

Some fiberglass resins are resistant to dissolved ozone gas in the concentrations found in aquatic systems applications; the resistance of fiberglass to wet gas conditions depends very strongly on the specific resin used; fiberglass should be avoided for long-term critical applications.

DESIGN OF OZONE FACILITIES FOR AQUATIC SYSTEMS

Disinfection

Ozone has been shown to be effective at inactivating bacterial, viral, and protozoan agents at low dosages [0.5 to 1.5 mg/L (ppm) applied dose] in comparison to other chemical disinfectants. As discussed, the Ct for an ozone system is calculated based on the actual ozone residual [mg/L (ppm)] following the contact period, multiplied by the detention time of that mixing/contact unit. This approach provides a conservative method to establish the required ozone generator capacity.

General Improvements in Water Quality

Ozone has attracted a great amount of interest recently for the purpose of enhancing water quality in reuse systems for rearing shrimp and fish (15,16). These systems have generally not included effective ozone removal, so the applied ozone doses have been limited to prevent ozone toxicity problems. These systems have not been designed to provide a given level of disinfection, although some disinfection will occur (17). Improvements in water quality and solids removal have resulted from applied doses in the range of 0.025 to 0.045 kg ozone/kg of feed (16); but the difference between the required dose for water quality improvements and toxicity problems is very narrow. While this application may have the potential to improve general water-quality conditions in reuse systems, more focused research is necessary to rationally develop the design parameters and operational strategies for its use.

BIBLIOGRAPHY

1. B. Langlais, D.A. Reckhow, and D.R. Brink, eds., *Ozone in Water Treatment—Application and Engineering*, Lewis Publishers, Chelsea, MI, 1991.
2. H. Rosenthal and J.S. Wilson, *Can. Tech. Rep. Fish. Aquat. Sci.* No. 1542, 1987.
3. G. Bablon et al., in B. Langlais, D.A. Reckhow, and D.R. Brink, eds., *Ozone in Water Treatment—Application and Engineering*, Lewis Publishers, Chelsea, MI, 1991, pp. 11–132.
4. P.J. Colberg and A.J. Lingg, *J. Fish. Res. Board Can.* **35**, 1290–1296 (1978).
5. W.R. Haag, J. Hoigné, and H. Bader, *Water Res.* **18**, 1125–1128 (1984).
6. E.A. Crecelius, *J. Fish. Res. Board Can.* **36**, 1006–1008 (1979).
7. G. Grguric, J.H. Trefry, and J.J. Keaffaber, *Water Res.* **28**, 1087–1094 (1994).
8. L.B. Richardson, D.T. Burton, G.R. Helz, and J.C. Rhoderick, *Water Res.* **15**, 1067–1074 (1981).
9. Metcalf & Eddy, *Wastewater Engineering: Treatment, Disposal, Reuse*, McGraw-Hill, New York, 1979.
10. F. Damez, B. Langlais, K.L. Rakness, and C.M. Robson, in B. Langlais, D.A. Reckhow, and D.R. Brink, eds., *Ozone in Water Treatment—Application and Engineering*, Lewis Publishers, Chelsea, MI, 1991, pp. 469–490.
11. G.A. Wedemeyer, N.C. Nelson, and W.T. Yasutake, *J. Fish. Res. Bd. Can.* **36**, 605–614 (1979).
12. J.M. Tipping, *Prog. Fish-Cult.* **50**, 202–210 (1988).

13. J.F. Hemdal, *Prog. Fish-Cult.* **54**, 54–56 (1992).
14. W.D. Bellamy et al., in B. Langlais, D.A. Reckhow, and D.R. Brink, eds., *Ozone in Water Treatment—Application and Engineering*, Lewis Publishers, Chelsea, MI, 1991, pp. 317–468.
15. B. Reid and C.R. Arnold, *Prog. Fish-Cult.* **56**, 47–50 (1994).
16. S.T. Summerfelt and J.N. Hochheimer, *Prog. Fish-Cult.* **59**, 94–105 (1997).
17. G.L. Bullock, S.T. Summerfelt, A.C. Noble, A.L. Webster, M.D. Durant, and J.A. Hankins, *Aquaculture* **158**, 43–55 (1997).
18. H. Sugita et al., *App. Env. Microbiol.* **58**, 4072–4075 (1992).
19. D.E. Owsley, *Am. Fish. Soc. Symp. 10*, Bestheda, MD, 1991, pp. 417–420.

See also DISINFECTION AND STERILIZATION.

P

PERFORMANCE ENGINEERING

Douglas H. Ernst
Oregon State University
Corvallis, Oregon

OUTLINE

Fish performance engineering consists of the application of environmental physiology and bioenergetics to fish production in aquaculture systems. Fish performance engineering is a primary consideration in aquacultural engineering, in which fish metabolism, feeding, and growth represent the primary processes for the benefit of which fish culture systems are designed and managed. The objective of this entry is to provide an overview of the methods and models used in fish performance engineering, with a focus on analytical tools suitable for practical application. The scope of the term 'fish performance' in this context includes (*1*) fish response to environmental variables and (*2*) quantitative aspects of fish metabolism, feeding, growth, and survival. The methods presented here are applicable to aquaculture facility design and management for the production of both finfish and crustaceans (together referred to hereafter as "fish") in all types of aquaculture systems.

FISH PERFORMANCE PRINCIPLES

The fundamental principles of fish performance engineering are based upon physiological energetics (1–3). Physiological energetics, commonly referred to as bioenergetics, concerns the rates of energy losses and gains and the efficiencies of energy transformation for a whole organism or group of organisms; this mid-level focus distinguishes it from cellular energetics (4) and ecological energetics (5). In this respect, fish and fish foods can be expressed in terms of their material composition or in terms of the energy equivalent of these materials. Fish gain energy through food ingestion (I), store energy as growth (G), lose energy through excretion (E), and expend energy through metabolism (M). To apply physiological energetics to fish performance engineering, these energy sources and sinks are combined into a bioenergetic budget: $I = G + E + M$. In response to analytical objectives, this budget can be re-arranged to calculate energetic capacities (or scopes) for fish activity (6,7), growth (8–10), or feeding (11–13).

A flow chart of the sources, sinks, and pathways of fish bioenergetics is presented in Figure 1. For an individual fish, energy input begins with 'ingested energy' (I), and food conversion efficiency is the proportion of ingested energy incorporated as growth. In practical

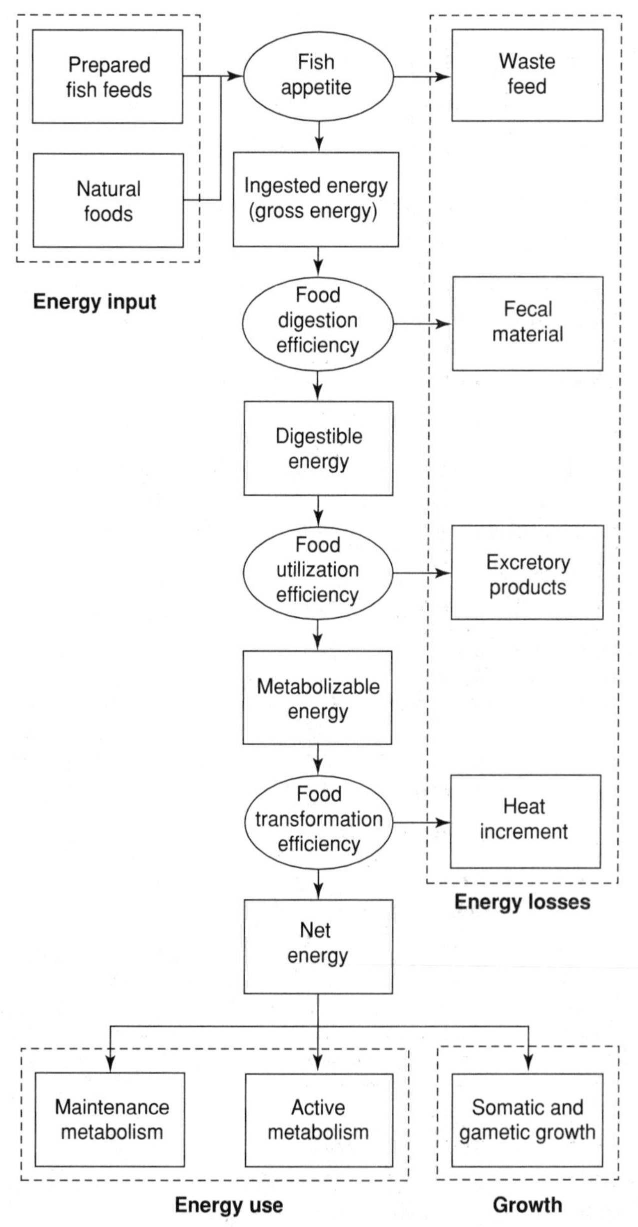

Figure 1. Flow diagram of fish bioenergetics, showing energy sources and sinks, controlling processes, and distribution paths.

applications of bioenergetics, however, fish are managed and analyzed as populations (fish production lots), and the quantities of feed applied, feed wasted, and fish mortality are also included in food conversion efficiency values. As shown, metabolic energy demand consists of several components. Maintenance metabolism includes (*1*) standard metabolism, defined as the rate of energy use by a fasting fish at rest, and (*2*) additional demands such as those due to responses to environmental stress. Active metabolism consists of the energy demands of swimming, aggression, and feeding (locomotion). Heat increment (or specific dynamic action) represents the energy costs of processing and assimilating food. Metabolic energy demands are supported through the catabolism of food substrates, in which oxygen is consumed and metabolites (or catabolites) are produced. In practical applications of bioenergetics, components of fish metabolism are combined into total oxygen consumption and metabolite excretion rates, usually expressed as *g (compounds)/kg (fish)/day* (grams of compound per kilogram of fish per day) or *mg (compound)/kg (fish)/hr*.

Rates of fish metabolism, feeding, and growth vary diurnally, seasonally, and over the life of a fish, in response to changes in fish and in environmental factors (14). Fish performance is a graded response to temporally and spatially graded environmental conditions, and the effect of environmental variables on fish activity, appetite, and growth is mediated through metabolism. The variables driving fish performance include (*1*) fish rearing-unit shape and dimensions, hydraulic characteristics, and water velocity, (*2*) water quality, (*3*) length of day and exposure of fish to direct light, (*4*) disturbances due to human activity and to fish-handling practices (e.g., fish grading), (*5*) fish density, as it affects fish behavior and access to food resources, (*6*) fish development state and size, (*7*) the availability and quality of natural food resources, and (*8*) the quality, application rates, and application methods of prepared feeds. The response variables of fish performance include (*1*) disease resistance, stress accumulation and compensation, and mortality, (*2*) swimming capacity, behavioral aggression, and competition for limited food resources, (*3*) appetite levels, food ingestion rates, and food assimilation efficiencies, (*4*) sexual maturation and reproduction, and (*5*) somatic growth.

FISH PERFORMANCE MODELING

For practical application of bioenergetics to fish performance engineering, bioenergetic budgets are formulated into quantitative models of fish metabolism, feeding, and growth. Development of these models relies on the use of simplifying assumptions about the underlying mechanistic processes and on the use of empirical data for model calibration. Fish performance models consist of one or more equations and rules of application, defined equation inputs and outputs (independent and dependent variables), and equation coefficients and exponential terms (parameters). Fish performance models may be used to analyze historical production data (through regression) or to predict future fish production (through simulation).

Values are established for model parameters by using procedures of model calibration and validation (15). Calibration is accomplished by using regression procedures on empirical datasets to generate parameter values. Validation is accomplished by using additional, independent datasets to test predictive accuracy. Aquaculture production trials are used to generate these datasets, in which data are collected at regular time intervals for fish length and weight samples, feed application rates, water quality, and other fish and environmental variables. Calibrated models are specific to a given fish species or stock, and closely related species share similar parameter values. When independent variables that significantly affect fish performance are not adequately accounted for in the model, calibrated models may be specific to an aquaculture system type or site location. While these site-specific models are not generally applicable, they are often useful for the site at which they were developed, and they can take advantage of simplified methods and reduced requirements for input data. After formulation, calibration, and validation, model application requires that values be provided for independent (input) variables. For analysis of historical fish performance, input variables can be specified using historical, empirical data. To simulate future fish performance, input variables can be based on past, empirical data, using daily, weekly, or monthly mean values, or they can be predicted by the use of system-level simulation models (16).

Models of fish metabolism, feeding, and growth intended for practical application are found throughout the aquaculture literature. These models utilize simplifying assumptions and aggregation of processes, in terms of the specific driving and response variables considered and of the functions used to represent bioenergetic processes. All of the potential driving variables of fish performance (listed earlier) may be considered in the design and management of aquaculture facilities, but fish size, water quality, food quantity, and food quality normally receive the most attention. The response variables of primary interest are mortality, somatic growth, feeding, and total metabolic rates. Models intended for practical use generally calculate the effects of environmental factors on fish growth and feeding rates directly, then calculate metabolic rates as a function of feeding rates. Feeding and growth rates are related by the use of food conversion efficiencies or ratios, and the metabolic rates of concern are limited to oxygen consumption and metabolite excretion.

ENVIRONMENTAL CRITERIA AND RESPONSE

Within the stated scope of this discussion, the environmental criteria of interest are water quality and related fish and feed loading variables. For this purpose, the criteria values listed in Table 1 are intended as approximate guidelines. Criteria values are dependent on the nutrition status, stress level, life stage, body size, and environmental acclimation of the fish. In addition, criteria for a specific variable may depend on levels of other variables (e.g., minimum criteria for dissolved oxygen increase as carbon dioxide levels increase). Environmental criteria can be used as parameters in fish performance models, as design

Table 1. Fish Environmental Criteria, as Defined by Low and High Values for Tolerance ($C_{min}t$ and $C_{max}t$) and Optimum ($C_{min}o$ and $C_{max}o$) Ranges, with Respect to Fish Growth[a]

Physical Variables	Criteria				
Temperature (C)	Fish Species	$C_{min}t$	$C_{min}o$	$C_{max}o$	$C_{max}t$
	Brook trout	3	8	13	19
	Pacific salmon	3	9	14	22
	Rainbow trout	3	11	17	23
	Atlantic salmon	3	12	18	22
	Yellow perch	8	18	22	27
Basis: primary controlling factor of fish performance	White sturgeon	6	19	22	26
	Striped bass	7	20	25	31
	Marine shrimp	10	20	30	36
	Centrarchids	12	20	28	33
	Freshwater prawns	10	22	28	33
	Hybrid striped bass	10	23	28	33
	Common carp	5	24	30	35
	Channel catfish	12	25	30	34
	Tilapia	15	28	33	37
Salinity (ppt)	• *Basis*: primary physiological criterion, and highly dependent on fish species and life stage. • *Values*: narrow ranges for stenohaline species, wider ranges for euryhaline species; criteria range from near zero to >35.0.				
pH	• *Basis*: primary controlling factor of fish performance. Exerts major impact on other water quality variables (e.g., un-ionized ammonia of total ammonia) and other facility processes (e.g., biofilter nitrifying bacteria). • *Values*: approximate $C_{min}t$–$C_{max}t$ range is 6.5–8.0 for trout, 6.0–9.5 for catfish and tilapia.				
Alkalinity (mg $CaCO_3$/L)	• *Basis*: closely related to pH and water hardness criteria; used mainly as a measure of buffering capacity against decreases in water pH. • *Values*: approximate ranges are ≥20 for flow through systems, ≥50 for phytoplankton systems, and ≥100 for recirculation systems, but ≤500 mg/L for all systems.				
Hardness (mg $CaCO_3$/L)	• *Basis*: calcium and magnesium are required nutrients for body composition and osmoregulation. • *Values*: approximate $C_{min}t$–$C_{max}t$ range is 50–350				

Fish Biomass Loading	Criteria
Biomass density (kg/m^3)	• *Basis*: biomass support and fish behavior. Values range widely and depend on fish species, feeding rate, and type of culture system. Used for systems with known biomass capacities. • *Values*: expressed as biomass density index (kg/m^3/cm-length). Approximate $C_{max}t$ are 0.05 for extensive production, 1.5–3.0 for salmon hatcheries, and 3.0–20.0 for intensive production.
Biomass loading (kg/m^3/d)	• *Basis*: biomass support. Values range widely and depend on fish species, influent water quality, and production intensity. Used for systems with known biomass loading capacities. • *Values*: expressed as biomass loading index (kg/m^3/d/cm-length). Approximate $C_{max}t$ are for 0.1 for intensive trout production and 0.3 for intensive tilapia production.
Feed loading (kg feed/m^3/d)	• *Basis*: system capacity to digest feed; related to oxygen demand and metabolite production. Highly dependent on fish species and culture system. Used for systems with known feed loading capacities. • *Values*: approximate $C_{max}t$ are for 0.1 for intensive trout production and 0.3 for intensive tilapia production; these example values are equivalent to biomass loading criteria for a 33-cm fish fed 3.0% of bw/day.
Water-exchange rate (number/day)	• *Basis*: biomass support and water velocity, and other considerations similar to biomass loading. Used for systems with known water-exchange rate requirements. • *Values*: values range from zero, for makeup of water loss only, to three exchanges per hour (72/day) for intensive, flowing water culture.
Cumulative oxygen consumption (COC, mg O_2/L)	• *Basis*: metabolite stoichiometry and criteria in relation to oxygen consumption. Used for systems with known relationships between COC values, pH values, and metabolite concentrations. • *Values*: for a pH range of 6.0–9.0, approximate $C_{max}t$ values range from 14 to 25 for carbon dioxide constraints and from 0.5 to 100 for un-ionized ammonia constraints.

(*continued*)

Table 1. *Continued*

Dissolved Gases	Criteria				
Dissolved oxygen (mg O_2/L)	Fish Species	$C_{min}t$	$C_{min}o$	$C_{max}o$	$C_{max}t$
Basis: primary limiting factor of fish performance.	Salmonids	5	7	(Use percent saturation)	(Use percent saturation)
	Striped/hybrid bass	4	6		
	Catfish	3	6		
	Carp	2	6		
	Tilapia	2	5		
Dissolved oxygen (% saturation)	All species	30	50–70	200	300
Dissolved carbon dioxide (mg CO_2/L)	• *Basis*: primary limiting factor of fish performance; minimum criteria are zero in value, maximum criteria decrease as dissolved oxygen levels decrease below lower optimum levels. • *Values*: approximate $C_{max}t$ range is 20.0–50.0, but values $\geq$100.0 are reported to be tolerated for some species.				
Total gas pressure (TGP; % saturation, or mm-Hg pressure difference between TGP and local barometric pressure)	• *Basis*: gas supersaturation can cause gas-bubble formation in fish blood and tissues (gas bubble disease). • *Values*: approximate $C_{max}t$ range is 105–110% sat. or 38–76 mm Hg. Approximate $C_{max}t$ is 102–103% sat. for sensitive species and life stages (e.g., eggs).				

Metabolites	Criteria				
Un-ionized ammonia (mg NH_3-N/L)	Fish Species	$C_{min}t$	$C_{min}o$	$C_{max}o$	$C_{max}t$
Basis: primary limiting factor of fish performance and concentration is a function of total ammonia, pH, temperature, and salinity.	Salmonids	0.0	0.0	0.01	0.2
	Striped/hybrid bass	0.0	0.0	0.02	0.3
	Carp	0.0	0.0	0.02	0.5
	Crustaceans	0.0	0.0	0.05	0.7
	Catfish	0.0	0.0	0.05	0.9
	Tilapia	0.0	0.0	0.10	1.0
Nitrite (mg NO_2-N/L)	• *Basis*: primary limiting factor of fish performance; minimum criteria are zero, and criteria are highly dependent on species, life stage, water salinity, hardness, and dissolved oxygen. • *Values*: approximate $C_{max}t$ is 0.1 for trout, 2.0 for catfish, and 4.0 for tilapia. $C_{max}t$ may range from 10–20 for tolerant species when sufficient chloride is present as natural or added salts.				
Nitrate (mg NO_3-N/L)	• *Basis*: effect on fish is likely limited to impaired osmotic regulation at very high concentrations; this effect may occur in recirculating fish culture systems with low system exchange rates and no denitrification processes. • *Values*: approximate $C_{max}t$ range is $\leq$50–300 for all species.				
Particulate solids (mg dry wt./L)	• *Basis*: impaired fish ventilation and gill abrasion; tolerance to solids is highly dependent on fish species. Includes inorganic, organic, and phytoplankton particulate solids. • *Values*: approximate $C_{max}t$ is 10–80 for solids of feed and fecal origin—for example, $C_{max}t$ is 25 for salmonids, a species with low tolerance to particulate solids.				

[a]Values represent an approximate consensus of reported values in the aquaculture literature. Use of the terms "controlling factor" and "limiting factor" conforms to their standard use in bioenergetics (1).

criteria for fish culture systems, and as management criteria for assessing results of environmental monitoring. For modeling applications, a fish growth and feed model generally includes water temperature as an independent variable or makes clear that its use is restricted to a given temperature range. Water quality variables in addition to temperature need to be considered if their levels fall significantly outside of optimum ranges. Fish biomass and feed loading variables can be used as indicators of water quality, but only when relationships between the loading levels and the associated concentrations of dissolved oxygen and metabolites have been determined. Relationships between loading and water-quality variables are highly dependent on the specific aquaculture system.

Up to four criteria values are used for each environmental variable: the minimum and maximum values for the tolerance and optimum ranges. In the optimum range, maximum growth rates and food conversion efficiencies are supported with respect to the given variable. The width of the optimum range is related to the capacity for environmental acclimation by the fish. The tolerance range exists outside of the optimum range. Conditions here cause a reduction in growth rates and accumulation of stress, depending on the elapsed time and extent of deviation from optimum limits. The lethal range exists outside of the tolerance range. Conditions here do not support fish growth and, when sustained, may result in mortality. Specification of only the lower or upper criteria may

Table 2. Typical Functional Forms Used to Calculate Performance Scalars as a Function of Given Water-Quality Criteria and Variable Values

Parameters and Variables	Definition
$C_{min}t$	Minimum tolerance criterion
$C_{min}o$	Minimum optimum criterion
$C_{max}o$	Maximum optimum criterion
$C_{max}t$	Maximum tolerance criterion
V	Water-quality variable (same units as associated criteria)
S	Individual scalar value (0–1)
CS	Combined scalar value (0–1) = minimum of individual separable scalars or product of interactive scalars

Scalar	Condition	Function
All	$V \leq C_{min}t$ or $V \geq C_{max}t$	$S = 0$
Linear	$C_{min}t < V < C_{min}o$	$S = (V - C_{min}t)/(C_{min}o - C_{min}t)$
	$C_{min}o \leq V \leq C_{max}o$	$S = 1.0$
	$C_{max}o < V < C_{max}t$	$S = (C_{max}t - V)/(C_{max}t - C_{max}o)$
Exponential	$C_{min}t < V < C_{min}o$	$S = \exp\{-k1 * [(C_{min}o - V)/(C_{min}o - C_{min}t)]^{k2}\}$
	$C_{min}o \leq V \leq C_{max}o$	$S = 1.0$
	$C_{max}o < V < C_{max}t$	$S = \exp\{-k1 * [(V - C_{max}o)/(C_{max}t - C_{max}o)]^{k2}\}$
Polynomial	$C_{min}t < V \leq C_{max}o$	$S = a + bV + cV^2$
(second order and higher)	$C_{min}o \leq V < C_{max}t$	$S = d + eV + fV^2$

be sufficient, depending on the specific aquaculture practices in use and the environmental variable in question. For example, water temperatures and dissolved oxygen levels often vary below maximum optimum levels, and only maximum criteria need to be considered for fish metabolites.

Fish growth and feeding rates are scaled from zero to maximum rates on the basis of environmental conditions relative to specified criteria. Scalar values are calculated for each environmental variable considered, and the overall scalar is estimated as the minimum individual scalar value or scalar product. (The latter is used for interactive criteria such as oxygen and carbon dioxide.)

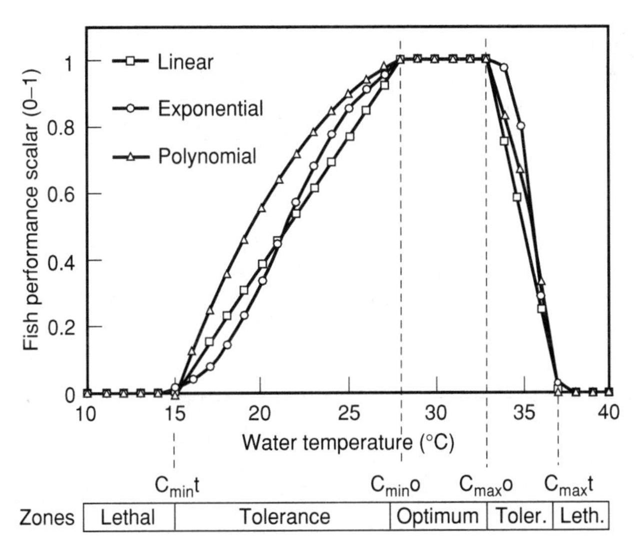

Figure 2. Linear, exponential, and polynomial scalar functions used for modeling fish response to environmental conditions, applied to the response of tilapia to water temperature.

Normally, depth-averaged water quality is used in design and management, but, if the cultured organism is a bottom dweller (e.g., shrimp) and water quality is vertically stratified, then benthic water quality is used. Various equation forms can be used to calculate scalar values (Fig. 2; Table 2), and paired functions are often required to express fish response adequately over the full range of a given variable (12,17). This functional discontinuity complicates both the regression and use of these functions, but, for a given environmental variable, consideration of values both above and below the optimum range often is not needed under practical conditions. Criteria values used in scalar functions can be taken from the literature or can be determined by regression using fish production datasets, in which scalar functions are substituted into fish growth and feeding models.

EXPRESSIONS FOR FISH SIZE, GROWTH, AND FEEDING

The following expressions for fish size, growth rate, and feeding rate are used in fish growth and feeding models. All fish and feed weights used in this discussion are expressed in terms of wet weight (moisture content included), unless use of dry weights is noted. Specific rates and index values can be expressed as factors (e.g., 0.1) or as percents (e.g., 10%), and factors are often used to simplify equations. Mathematical operators used in this discussion include the following abbreviations: log (base-10 log); ln (base-e log); and exp (e to given power). For the sake of readability, conversions from metric will be left to the reader.[a]

[a] 1 g = 0.0353 oz, 1 mm = 0.0394 in., 1 cm = 0.394 in., 1 kg = 2.205 lbs.

Table 3. Reported Values for Fish Condition Factor, where Le = 3.0 (18–20)

Fish Species	Fish Condition Factor (g/cm^3)
Northern pike	0.005013
Channel catfish	0.007964
Chinook salmon	0.008191
Steelhead	0.009425
Coho salmon	0.01034
Rainbow, brook, and brown trout	0.01122
Largemouth bass	0.01275
Blue tilapia	0.0233

Fish total length (L; cm) and body weight (W; g) are related by a length exponent (Le; range 2.5–3.5, typical value 3.0) and a fish condition factor (FCF; g/cm^3; Table 3). FCF varies with fish species, size, nutritional state, and water content (19,21). Parameters Le and FCF are established by linear regression, with log transformation (base 10 by convention) of geometric mean values (21):

$$W = FCF \times (L^{Le})$$

or

$$L = (W/FCF)^{(1.0/Le)}$$

or

$$FCF = W/(L^{Le})$$

where

$$\log(W) = \log(FCF) + [Le \times \log(L)]$$

Length growth rate (LGR; mm/day) is based on the change in fish length (L_o to L_t; cm) over a given time interval (t; days):

$$LGR = (10\ mm/cm) \times (L_t - L_o)/t$$

Absolute fish growth rate (GR; g increase/day) and specific growth rate (SGR; g increase/g fish/day or 1/day) are based on the change in fish weight (W_o to W_t; g) over a given time interval (t; days):

$$GR = (W_t - W_o)/t$$

and

$$SGR = [\ln(W_t) - \ln(W_o)]/t$$

Fish growth index (FGI; range 0–1) is based on the target or actual fish growth rate (GR) relative to the maximum growth rate obtained under satiation feeding (GRm) for a given growth interval:

$$FGI = GR/GRm$$

or

$$FGI = SGR/SGRm$$

For simulations of fish production over extended periods, in which it is desired to achieve a specified date and weight target (by control of feeding rates) but the varying of temperatures or other environmental scalars makes it difficult to predetermine GRm, FGI values can be adjusted over the course of iterative simulations by

$$FGInew = FGIprior \times [(W_{tt} - W_o)/tt]/[(W_t - W_o)/t]$$

where

FGInew = FGI to be used in an iterative simulation;
FGIprior = FGI that was used in the prior simulation;
W_{tt} = target fish weight (g) to be achieved at target time tt (days);
W_t = fish weight achieved in prior simulation (g) at time t (days);
W_o = initial fish weight (g).

Specific feeding rate (SFR; kg feed/kg fish/day or g feed/g fish/day) is based on the quantity of feed fed (FD; kg) over a given time interval (t; days) relative to total fish biomass (FB; kg) at the beginning or midpoint of the time interval:

$$SFR = FD/(FB \times t)$$

SFR may also be expressed as percent body weight per day (% bw/d). Fish feeding index (FFI; range 0–1) is based on the target or actual specific feeding rate (SFR) relative to the maximum (appetite satiation) specific feeding rate (SFRm) for a given growth interval:

$$FFI = SFR/SFRm$$

Food conversion efficiency (FCE; %) is based on the total feed applied (FD; kg) over a given time interval relative to the change in fish biomass (FB; kg) for this time interval (FB_o subtracted from FB_t; kg). The reciprocal value, food conversion ratio (FCR), is also commonly used. FCE and FCR values are typically calculated for fish populations, rather than for individual fish, so as to include losses due to applied feed that is not consumed and to fish mortality:

$$FCE = 100.0 \times (FB_t - FB_o)/FD;$$

$$FCR = FD/(FB_t - FB_o).$$

Interconversion of wet weight (ww) and dry weight (dw) values is based on the moisture contents of fish and feeds (FSmst and FDmst; range 0–1):

$$SFRdw = SFRww \times (1.0 - FDmst)/(1.0 - FSmst)$$

$$FCRdw = FCRww \times (1.0 - FDmst)/(1.0 - FSmst)$$

and

$$FCEdw = FCEww \times (1.0 - FSmst)/(1.0 - FDmst)$$

Typical moisture contents are 0.75 for fish, 0.10 for dry prepared feeds, 0.40 for moist prepared feeds, and 0.65–0.95 for natural foods.

FISH SURVIVAL

Fish survival is dependent upon a complex interaction among water quality variables, fish physiological status, and presence of pathogens. While aquaculture systems are designed and managed to avoid fish stress, disease, and losses beyond natural attrition levels, prediction of fish survival is necessarily an approximation. Nonetheless, some accounting of fish population numbers is required in order to generate fish biomass schedules based on mean fish weights. Assuming that fish mortality losses are proportional to population levels (i.e., exponential decay), predicted fish numbers over a culture period can be based on the expected fish number at the end of the culture period by

$$POP_t = POP_o \times \exp(SMR \times t)$$

and

$$SMR = \ln(POP_{tp}/POP_o)/tp$$

where

POP_t = fish number at elapsed time t (days),
POP_o = initial fish number at time zero,
SMR = specific mortality rate (1/day),

and

POP_{tp} = expected fish number after total period tp (days).

FISH GROWTH

The fish growth models presented in the aquaculture literature vary in their consideration of fish size, of the anabolic and catabolic components of fish metabolism, of food quality and quantity, and of water quality (21). Consequently, these models vary in their consideration of exponential, linear, and asymptotic fish growth stanzas (Fig. 3) and of profiles of predicted growth trajectories over time (Fig. 4). The transition from exponential to asymptotic fish growth is due to limits imposed by environmental variables and/or maximum (maturation) fish-size constraints. Reported growth models include (*1*) constant absolute weight growth rate (CAGR), (*2*) constant specific weight growth rate (CSGR), (*3*) length growth rate (LNGR; 18,22), (*4*) double-logarithmic specific growth rate (DSGR; 2,23), (*5*) von Bertalanffy growth function (VBGF; 24,25), and (*6*) anabolic-catabolic bioenergetic function (BIOE; 1,26,27). In reverse order, these models can be derived through successive levels of simplification of the fundamental bioenergetic equation of fish growth. A particular model is selected for use by choosing the model with the weight-time profile most similar to a plot of given weight-time data or by choosing the simplest model that adequately describes the data as indicated by regression analyses.

The LNGR and DSGR models are further described and applied here, because of their wide applicability and use in aquaculture and the ease of parameter calibration for them by linear regression. The LNGR and DSGR models are appropriate for exponential and

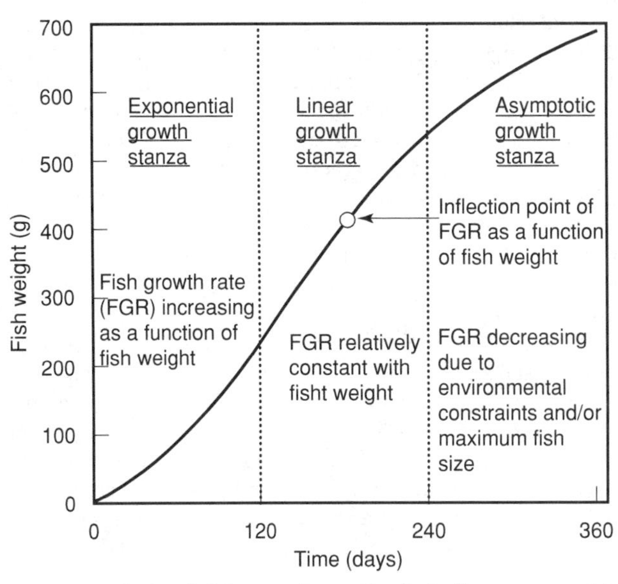

Figure 3. Idealized fish growth profile, including exponential, linear, and asymptotic growth stanzas.

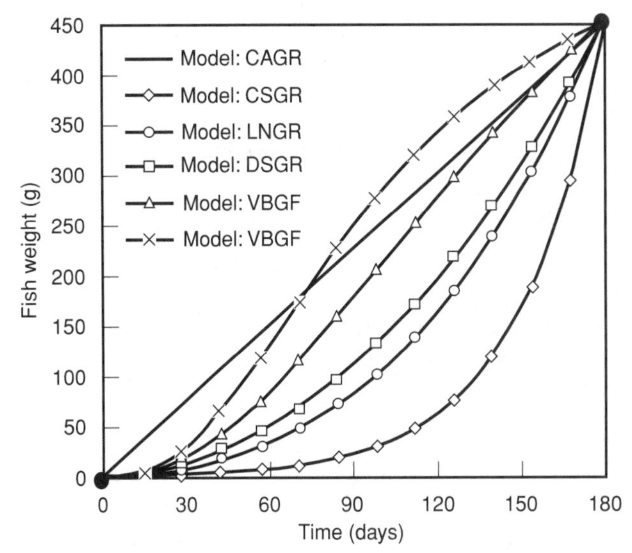

Figure 4. Representative fish growth profiles for each of the fish growth models identified in the text, using equivalent starting and ending weights, growth periods, and environmental conditions for each profile.

combined exponential-linear growth stanzas. With the inclusion of environmental scalar terms, these models can also consider growth that is asymptotic because it is limited by environmental constraints. Environment-limited asymptotic growth, when found in aquaculture systems, is most commonly due to deteriorating water quality or food availability, as fish and/or feed loading rates achieve maximum system capacities. The LNGR and DSGR models do not consider size-based asymptotic growth, but, under typical aquaculture conditions, fish weights are normally two-thirds or less of maximum maturation weights and are not significantly constrained by maximum fish size. For example, the maximum size of Nile tilapia (*Oreochromis niloticus*) is about 2.5 kg (5.5 lb) (28), but common harvest sizes are no more than 1.25 kg (2.8 lb), 50% of maximum size. For many other culture species, this differential between market size and maximum size is even greater (e.g., for channel catfish, rainbow trout, and sturgeon). When asymptotic growth may be due to maximum size, then the VBGF or BIOE model should be used. (Maximum fish-size scalars can be applied to the LNGR and DSGR models, but this approach complicates their use and is not covered in the current discussion.)

The LNGR model is based on the simplifying assumptions that length growth rate is constant with respect to fish length, length growth rate varies with water temperature, and water temperatures do not rise so high that growth is reduced. The LNGR model is well suited to fish growth in the exponential stanza and becomes increasingly less suitable as the linear stanza occupies an increasing proportion of the total growth profile. The LNGR growth model is

$$L_t = L_o + (k \times t)/10.0 \text{ mm/cm}$$

where

L_t = fish length (cm, total) at time t (days),
L_o = original fish length (at time = 0) (cm, total),

and

k = fish length growth rate constant (mm/day),

where

$k = FGI \times CS \times [La + (Lb \times T)]$
La = temperature function y-intercept (mm/day),
Lb = temperature function slope (mm/day/C),
T = temperature (C),
CS = combined environmental scalar (in addition to temperature; 0–1),

and

FGI = fish growth index (0–1).

To convert fish lengths and weights, the fish condition factor function given earlier is used. Combining the equations above into one equation, and assuming that CS and FGI both equal 1.0 (or are comparable between the model calibration and the application studies), length growth rate is calculated by the following equation. Reported values for La and Lb are provided in Table 4. This equation can also be used as a regression equation for estimating La and Lb from fish growth and temperature data:

$$L_t = L_o + [(La + (Lb \times T)) \times t]/10.0 \text{ mm/cm}$$

The DSGR model is well suited to both exponential and combined exponential-linear growth stanzas, as demonstrated in Figure 5. As demonstrated in Figure 6, the DSGR model shows poor predictive ability when it is applied without scalar terms to fish growth profiles having asymptotic components. The SGR exponent (SGRe) is apparently based on intrinsic physiological characteristics common to all fish (2). Reported values for SGRe tend to be about 0.33 for the exponential growth stanza and to increase to about 0.45 as greater proportions of the linear growth stanza are included. If SGRe is equivalent to the reciprocal of Le in the fish length–weight expression (e.g., Le = 3.0 and SGRe = 0.33), then the DSGR and LNGR models are equivalent. In contrast, the SGR coefficient

Table 4. Reported Parameters for the Length Growth Rate Model LNGR (22)[a]

Fish Species	Temperature Range (C)	La (mm/day)	Lb (mm/day/C)
Brook trout	5–12	−0.348	0.0944
Brook trout	4–19	+0.155	0.0355
Brook trout	7–19	+0.006	0.0455
Brook trout	7–16	−0.068	0.0578
Rainbow trout	4–19	−0.040	0.0505
Rainbow trout	7–19	+0.043	0.0450
Rainbow trout	7–16	−0.167	0.0660
Lake trout	4–16	+0.176	0.0426
Lake trout	4–13	+0.0622	0.0588
Steelhead	4–19	+0.0329	0.0294
Steelhead	7–16	−0.0407	0.0386
Atlantic salmon	4–19	+0.0043	0.0306
Atlantic salmon	7–16	−0.0429	0.0371
Channel catfish	24–30	+0.612	0.0298
Channel catfish	24–28	+0.195	0.0463
Tiger muskellunge (3–4 cm)	14–24	−0.0548	0.0912
Tiger muskellunge (12–13 cm)	18–24	+0.394	0.0471
Blue tilapia	20–30	−0.853	0.0480

[a]Emphasis on salmonid species reflects availability of values in the literature.

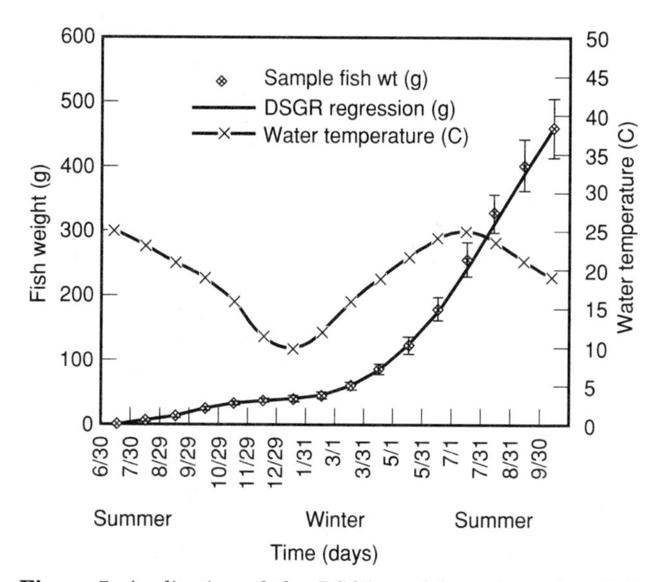

Figure 5. Application of the DSGR model to channel catfish growout in ponds, including over-wintering of fingerlings; a constant fish growth index is used, and water temperatures are accounted for by using a polynomial function within the DSGR model.

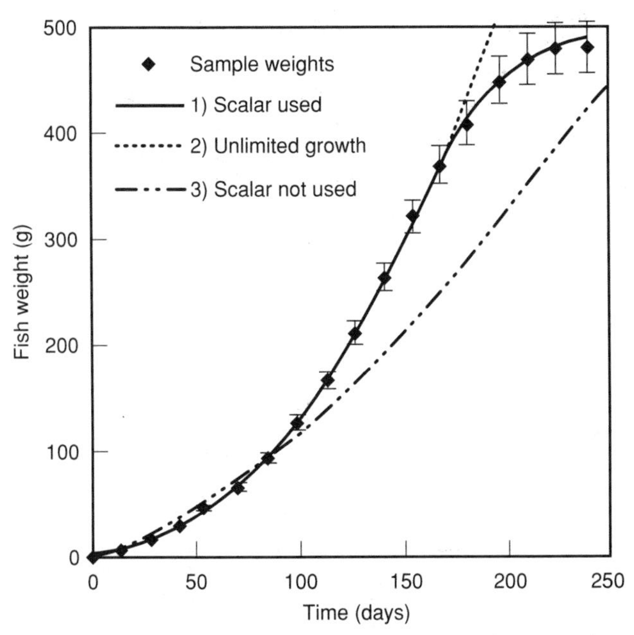

Figure 6. Application of the DSGR model to tilapia growout under intensive culture; growth is limited at fish weights over 400 g, as a result of limited feeding rates that are due to maximum system constraints. Curves derived by regression used are as follows: (*1*) feeding rate scalar; (*2*) data to day 150 only; and (*3*) all data but no scalar terms.

(SGRc) is a function of species, of environment variables, and of food availability, for which reported values range from 0.02 to 0.2. The DSGR growth model is

$$\mathrm{SGR} = \mathrm{SGRc} \times \mathrm{W}^{-\mathrm{SGRe}}$$

or

$$\mathrm{GR} = \mathrm{SGRc} \times \mathrm{W}^{1.0-\mathrm{SGRe}}$$

and, by integration

$$\mathrm{W_t} = [\mathrm{W_o}^{\mathrm{SGRe}} + (\mathrm{SGRc} \times \mathrm{SGRe} \times \mathrm{t})]^{(1.0/\mathrm{SGRe})}$$

where

SGR = specific growth rate (1/day),
SGRc = specific growth rate coefficient (1/day, constant or function),
SGRe = specific growth rate exponent,
GR = fish growth rate (g/day),
W_t = fish weight (g) at time t (days),
W_o = fish weight a time 0 (g).

Reported values for SGRe and SGRc are provided in Table 5. SGRe and SGRc can be determined by log-transform linear regression, when datasets are available. The simplest approach to this exercise occurs when water quality and feeding rate conditions are relatively uniform over time and are comparable between the model calibration and the application studies. In such cases, SGRc is a constant value. Alternatively, maximum growth rate data for fish under satiation feeding and optimum environmental conditions can be used to determine parameter values, and then the feeding and environmental scalar functions can be applied when the growth function is used. For a more rigorous approach, data for feeding rate and significant environmental variables must be included in the DSGR regression. For the latter two cases, SGRc is represented as a function of FGI, temperature, and additional water quality variables, as needed.

For all approaches, geometric-mean fish weight (Wgm, g) and SGR values are calculated for each growth interval of fish weight samples (W_i and W_{i+1}) and their natural

Table 5. Reported Parameters for the Weight Growth Model DSGR[a]

Fish Species	Temperature Range (C)	SGRc	SGRe	Reference
Sockeye salmon	15.0	NA	0.45	(14)
Pink salmon	15.0	NA	0.45	(14)
Coho salmon	15.5	NA	0.34	(29)
Salmonids	5.0–16.0	0.00303 × T	0.33	(30)
Rainbow trout	12.0	0.079	0.338	(31)
Rainbow trout	17.0	0.0686	0.323	(32)
Brook trout	11.0	NA	0.333	(33)
Brown trout	7.0–13.0	0.138 × (−0.3474 + 0.1053 × T)	0.325	(34)
Brown trout	12.8	0.042	0.325	(35)
Arctic char	7.0–13.0	0.126 × (−0.0815 + 0.0917 × T)	0.325	(34)
Arctic char	4.0–14.0	0.075 × (0.0219 + 0.0727 × T)	0.325	(36)
Channel catfish	NA	0.086	0.326	(2)
Common carp	NA	0.176	0.340	(2)
Florida red tilapia	26.0–30.0	0.176	0.428	(37)[b]
Mixed tilapia	28.0–32.0	0.175	0.444	(38)[c]
Nile tilapia	NA	0.836	0.713	(39)[d]

[a]SGRc and SGRe values are expressed for calculation of SGR in units of 1/day. SGRc may be expressed as a function of water temperature (T; C). emphasis on salmonid species reflects availability of values in the literature.
[b]Florida red tilapia: *Oreochromis mossambicus* × *O. hornorum* (seawater culture).
[c]Mixed species: *O. niloticus* and *O. niloticus* × *O. aureus*; DSGR parameters estimated from given data.
[d]The marked differences from the values typical in other entries reflect use of growth profiles consisting of linear and asymptotic growth stanzas under limited feed availability.

logs are used:

$$\ln(\mathrm{SGR}) = \ln(\mathrm{SGRc}) - [\mathrm{SGRe} \times \ln(\mathrm{Wgm})]$$

where

$$\mathrm{Wgm} = \exp\{[\ln(\mathrm{W_i}) + \ln(\mathrm{W_{i+1}})]/2.0\}$$

If SGRc is a function of environment variables and food availability, for example, water temperature (T; using polynomial scalar function) and feeding rate (expressed as FGI), then

$$\ln(\mathrm{SGR}) = \ln[\mathrm{FGI} \times (\mathrm{k1} + \mathrm{k2} \times \mathrm{T} + \mathrm{k3} \times \mathrm{T}^2)] - [\mathrm{SGRe} \times \ln(\mathrm{Wgm})]$$

where

$$\mathrm{SGRc} = \mathrm{FGI} \times (\mathrm{k1} + \mathrm{k2} \times \mathrm{T} + \mathrm{k3} \times \mathrm{T}^2)$$

As used in the LNGR model, a combined environmental scalar (CS) can be applied to SGRc, in addition to FGI and temperature.

NATURAL FISH PRODUCTIVITY

Food resources used by fish may consist of (*1*) natural (endogenous) food resources only, (*2*) natural foods plus supplemental prepared feeds, or (*3*) prepared feeds only. Possible natural food resources include bacterial-detrital aggregate, phytoplankton, zooplankton, invertebrates, and fish prey (polyculture). The contribution of natural foods, and the resulting level of natural fish productivity (NFP, kg fish/ha/day), is highly dependent on species (feeding habits), fish size (when feeding habits vary ontogenetically), biological productivity variables, and the type of aquaculture system. For herbivores and detritivores [such as carps, tilapia, and crayfish under extensive aquaculture production (500–5000 kg fish/ha)], natural foods can be a primary or sole food source. Natural foods may also be significant for fry and fingerling production of carnivorous species such as striped/hybrid bass, sea bass, and red drum, for which zooplankton food resources are utilized. Endogenous food resources can be indirectly managed by control of fish densities and by maintenance of nutrient levels for primary productivity.

Mechanistic models used to quantify natural food resources are relatively complex and are used mainly as research tools (17). For practical purposes, NFP can be approximated by empirically determined relationships between NFP and primary productivity (40) and between NFP and fish biomass density (FBD; kg/ha) (2). This method utilizes critical standing crop (FBDcsc) and carrying capacity (FBDcc) fish density parameters with respect to food availability. At fish densities less than FBDcsc, the availability of natural food resources exceeds maximum consumption rates by fish and does not limit fish growth rates. As fish density increases above FBDcsc (because of growth), natural food resources are utilized by fish beyond their sustainable yield and hence depleted; the result is a decline in natural fish productivity. When fish density achieves FBDcc, natural food resources are

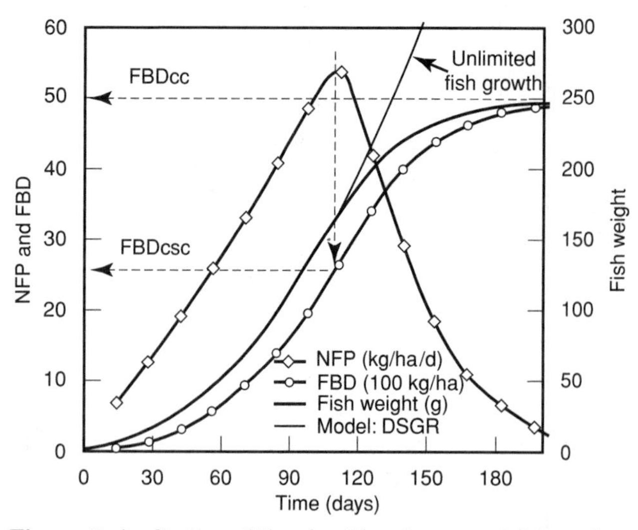

Figure 7. Application of the algorithm for natural fish productivity described in the text, in which supplemental feeding is not used, fish over utilize and deplete their natural food resources, and asymptotic fish growth results.

depleted to such a level that net fish growth is no longer supported, and natural fish productivity is reduced to zero. The ratio of FBDcc to FBDcsc typically ranges from 1.5 to 3.0. Use of natural fish productivity without supplemental feeding yields sigmoidal fish growth curves, as natural food resources are initially nonlimiting, then overwhelmed, then finally exhausted.

The calculation of fish feeding and growth rate equivalents of natural fish productivity by this simplified, empirical approach is outlined here and is illustrated in Figure 7. For projections over time, this procedure is repeated at each simulation time step, so that fish weights and related values are updated at each simulation time step.

1. For single-day calculations, or for the initial step of a simulation, the current mean fish weight is a given. Otherwise, a new fish weight is calculated from the fish weight and the growth rate in the prior simulation step. Current FBD and NFP values are updated accordingly, using the mean fish weight, the growth rate, and the number of fish.
2. Maximum potential NFP (NFPm) either is calculated as a function of primary productivity or is specified as a constant value (e.g., 50 kg/ha/day). When the current NFP exceeds NFPm, FBDcsc is set equal to the current FBD, and FBDcc is calculated as a multiple of FBDcsc. These FBDcsc and FBDcc values are fixed and are not recalculated in subsequent steps.
3. Maximum fish growth rate (FGRm) is calculated by using the selected growth model, for the existing environmental conditions and for an unlimited feeding rate. If FBD $\leq$ FBDcsc, then fish growth is equal to the maximum fish growth rate. If FBD $\geq$ FBDcc, then fish growth rate is zero, and this procedure is terminated. Otherwise, for the case FBDcsc $<$ FBD $<$ FBDcc, the fish growth index

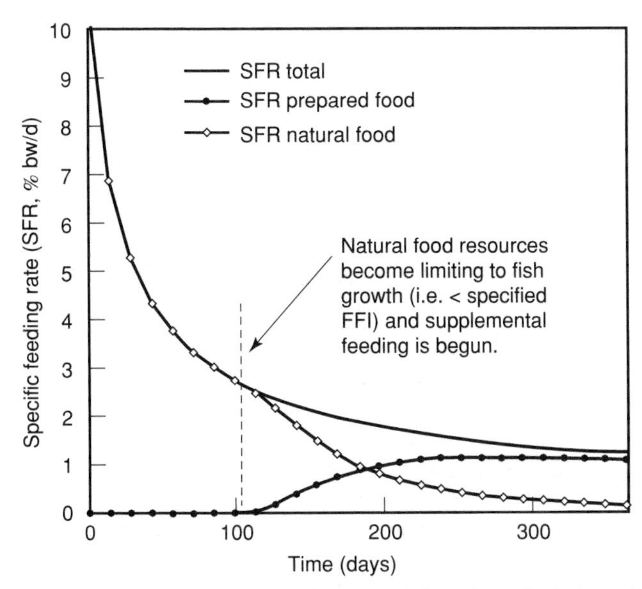

Figure 8. Application of the DSFR model to the calculation of feeding rates, where natural fish productivity is considered and a supplemental feeding schedule is determined that maintains a specified, constant fish feeding index (FFI).

(FGIn) and the growth rate (FGRn) supported by NFP are calculated by

$$\text{FGIn} = (\text{FBDcc} - \text{FBD})/(\text{FBDcc} - \text{FBDcsc})$$

and

$$\text{FGRn} = \text{FGIn} \times \text{FGRm}$$

4. If supplemental feeds are used, then the fish feeding index equivalent of natural fish productivity (FFIn) is calculated from FGIn, by use of the FFI–FGI function (given later). To achieve the target feeding index, as based on the target growth index, the FFI to be supplied by prepared feeds is equal to the target FFI minus FFIn (Fig. 8).

FISH FEEDING

Various calculation methods and management strategies for predicting and scheduling fish feeding rates are used in aquaculture. Feeding rates can be determined by the use of tables (which are indexed by water temperature and fish weight) or by the use of functions. For any method, the maximum feeding rates calculated for given environmental conditions must be corrected to actual feed application rates by the use of the fish feeding index (FFI) defined earlier, where FFI is based on target growth rates. The double-logarithmic specific feeding rate model (DSFR) is similar to the DSGR growth model in its functional form and in its parameter-estimation procedures, and these two models often share a calibration dataset. The DSFR feed model is

$$\text{SFR} = \text{SFRc} \times \text{W}^{-\text{SFRe}}$$

or

$$\text{FR} = \text{SFRc} \times \text{W}^{1.0-\text{SFRe}}$$

where

SFR = specific feeding rate (1/day),
SFRc = specific feeding rate coefficient (1/day, constant or function),
SFRe = specific feeding rate exponent,
FR = fish feeding rate (g/fish/day).

For parameter estimation by regression when SFRc is a constant, SFRc and SFRe are determined by this equation:

$$\ln(\text{SFR}) = \ln(\text{SFRc}) - [\text{SFRe} \times \ln(\text{Wgm})]$$

where Wgm is defined under the DSGR model.

If SFRc is a function of environment variables and of food availability—for example, water temperature (T; using polynomial scalar function) and feeding rate (expressed as FFI)—then

$$\ln(\text{SFR}) = \ln[\text{FFI} \times (\text{k1} + \text{k2} \times \text{T} + \text{k3} \times \text{T}^2)] - [\text{SFRe} \times \ln(\text{Wgm})]$$

and

$$\text{SFRc} = \text{FFI} \times (\text{k1} + \text{k2} \times \text{T} + \text{k3} \times \text{T}^2)$$

Alternatively, feeding rates can be determined from the fish growth rate and food conversion efficiency (FCE), where FCE can be determined by the use of either tables or functions. For calculating FCE, the DSGR and DSFR models can be mathematically combined to give FCE as a function of fish weight (Fig. 9). In this derivation, if the DSGR and DSFR models use the same environmental scalar criteria and functions, then these scalars cancel, and the calculated FCE is not responsive to water quality. While this approach may be a suitable simplification for practical aquaculture, FCE normally shows a concave-down profile over the full tolerance range of a given

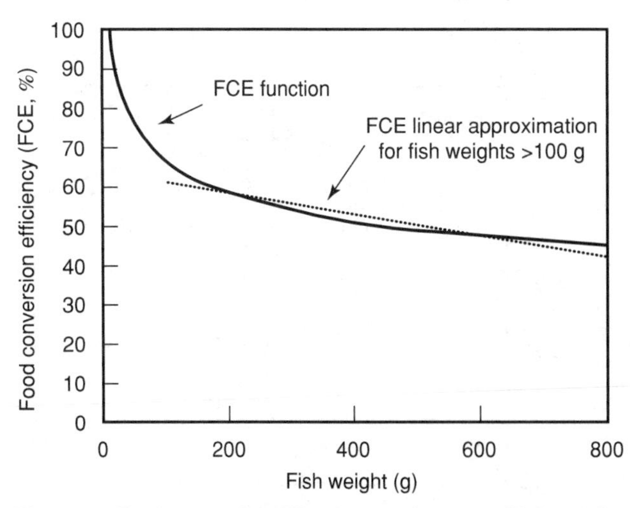

Figure 9. Food conversion efficiency as a function of fish weight, calculated by the methods given in the text. For larger fish sizes, it may be sufficient to represent this relationship with a straight line (i.e., FCE = a − b × W).

water quality variable, with a maximum achieved under optimum conditions. This example demonstrates that scalars for fish growth need to be somewhat more stringent than scalars for fish feeding, as controlled by the scalar criteria used. For the case where water scalars do cancel, FCE can be calculated as a function of fish weight (w; g) by

$$\mathrm{FCE} = 100.0 \times \mathrm{FCEc} \times \mathrm{W}^{\mathrm{FCEe}}$$

where

$$\mathrm{FCEc} = \mathrm{SGRc}/\mathrm{SFRc}$$
(note that $\mathrm{FCEc} < 100.0 \times [(1.0 - \mathrm{FDmst})/(1.0 - \mathrm{FSmst})]$)

and

$$\mathrm{FCEe} = \mathrm{SFRe} - \mathrm{SGRe}$$
(note that $\mathrm{FCEe} < 0.0$).

(See DSGR and DSFR models for additional definitions.)

FCE also varies with fish feeding rate for a given fish size (1,8,14). FCE first increases with increasing feeding rate above a maintenance ration (FFImnt, e.g., 10% or 0.1), then reaches a maximum efficiency at an optimum ration (FFIopt, e.g., 50% or 0.5), and finally declines as feeding rate is increased further to a maximum ration (FFImax: 100% or 1.0). The bioenergetic mechanisms underlying the response of FCE to feeding rate are complex. The shape of the FCE–FFI curve varies to some degree with fish species and with environmental conditions, and various functional approaches have been reported for expressing the FCE–FFI relationship. A somewhat simplified expression for the FCE–FFI relationship is given below and is illustrated in Figure 10. FFI is calculated as a function of FGI, and then FCE is calculated using this derived FFI value:

$$\mathrm{FGI} = 1.0 - [(\mathrm{FFImax} - \mathrm{FFI})/(\mathrm{FFImax} - \mathrm{FFImnt})]^{k}$$

or

$$\mathrm{FGI} = 1.0 - [(1.0 - \mathrm{FFI})/(1.0 - \mathrm{FFImnt})]^{k}$$

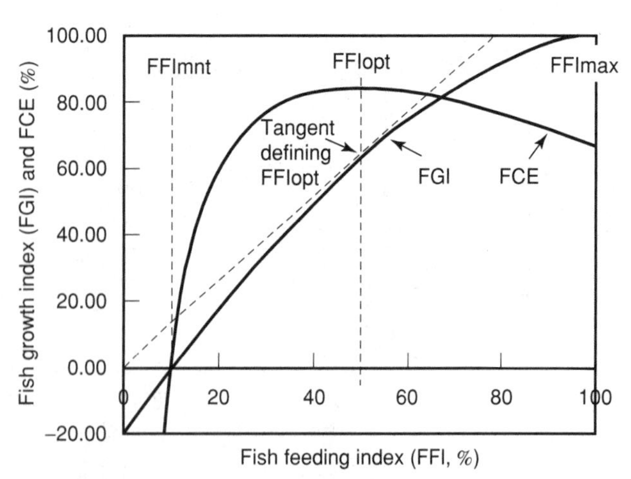

Figure 10. Fish growth rate (as FGI) and food conversion efficiency (FCE), as a function of fish feeding rate (as FFI), calculated by the methods given in the text.

and

$$\mathrm{FFI} = 1.0 - [(1.0 - \mathrm{FGI})^{1/k} \times (1.0 - \mathrm{FFImnt})]$$

and

$$\mathrm{FCE} = 100.0 \times (\mathrm{SGRm} \times \mathrm{FGI})/(\mathrm{SFRm} \times \mathrm{FFI})$$

where

k = exponent parameter used to adjust shape of FGI–FFI curve and region of optimum FCE (typical range 1.5 to 2.0; a value of 1.7 is used in Fig. 10),

SGRm = maximum SGR (1/day),

and

SFRm = maximum SFR (1/day).

The optimum feeding rate with respect to FCE is found at the highest point on the FCE–FFI curve; on the FGI–FFI curve, it is found where a tangent drawn from the origin intersects the curve. Feeding rates can also be optimized over the rearing period as a whole (41). However, feeding rates at which FCE is maximized can be two-thirds or less of maximum feeding rates (e.g., 50% FFI; 14) and can result in significantly slower growth rates and hence in longer growout periods. Especially for intensive aquaculture, the additional economic considerations of facility capacity utilization, fish biomass support, and production throughput for marketing objectives often overwhelm those of FCE optimization, and feeding rates well above FFIopt are normally used. As is evident in Figure 10, responses of FGI and FCE to feeding rates greater than FFIopt may be adequately represented with simple linear functions:

$$\mathrm{FGI} = a + (b \times \mathrm{FFI})$$

and

$$\mathrm{FCE} = c - (d \times \mathrm{FFI})$$

If feed quality is comparable between the model calibration and application studies, then feed quality can be ignored in calculating feeding rates and conversion efficiencies. If feed quality is a consideration, then feed quality scalars can be applied in a manner similar to that in which water quality scalars are. To accomplish this task, one assigns the prepared fish feed a nutritional quality scalar based on feed protein content (% dw), total gross energy (kcal/g), total metabolizable energy (kcal/g), or protein-energy ratio (mg/kcal) relative to nutritional criteria. This scalar is applied to FCE, so higher-quality feeds result in higher FCE values and hence lower SFR values. Corrections for differences in feed moisture content between the calibration and application studies may also be required, because the calculations presented here are in terms of wet weights.

FISH FEED DIGESTION AND METABOLISM

Fish oxygen consumption, metabolite excretion, and fecal egestion that occur in conjunction with food digestion and catabolism are of primary interest in aquaculture design and management. Fish oxygen demand, and the production of dissolved and particulate compounds [particularly carbon dioxide, nitrogen and phosphorous compounds, organic solids, and BOD (biochemical oxygen demand)], are fundamental design variables of fish culture systems and of aquaculture waste treatment. Excreted metabolites, including carbon dioxide and dissolved nitrogen and phosphorous compounds, are end products of ingested, digested, and catabolized food. Fish excrete nitrogen mainly as ammonia (generally $\geq$75%), with the remainder usually as urea, which breaks downs to ammonia and carbon dioxide. Rates of fish oxygen consumption and of metabolite excretion (g/kg fish/day or mg/kg fish/hr) are a function of feeding rate, feed quality, water quality, and fish metabolic demands (1–3,14). Egested particulate solids result from undigested foods, and their production rates are a function of feed ingestion and digestion rates. Applied feeds that remain uneaten (waste feed) may be an additional source of particulate solids in rearing units. The composition of fish fecal material is dependent on feed composition and digestion efficiency and can include nitrogen, phosphorous, and BOD. Generally, ash content is increased and organic and caloric contents are decreased relative to feed composition (1). The BOD content of fish fecal material is approximately 1.0 kg ultimate BOD per kg dry weight solids (43).

Fish oxygen consumption and metabolite/fecal excretion rates represent terms in the mass balance equations used to design and manage fish culture systems. Two general cases are found in their practical application. For extensive fish culture practices with no or low application rates of prepared feeds (gross fish yields $\leq$3000 kg/ha), the impact of fish metabolism on water quality is normally overwhelmed by other processes, for example, by the phytoplankton and bacterial processes of solar-algae ponds. Under these conditions, fish may exert an indirect impact on water quality, by representing significant sources of nutrients for primary productivity, but this impact normally is a consideration only when fish are fed. In contrast, for semi-intensive practices (gross fish yields 3000–20,000 kg/ha) and intensive practices (gross fish yields >20,000 kg/ha), which rely predominantly or wholly on prepared feeds, fish metabolic processes are a dominant component of the combined physical, chemical, and biological processes in rearing units and have direct impact on water quality.

In addition to mass-balance modeling for the design of fish culture systems, mass-balance equations can be used to determine actual fish metabolic rates, by measuring changes in water quality (e.g., dissolved oxygen) from the influent to the effluent of a rearing unit. This approach is particularly applicable to rearing units with high water exchange rates and a minimum of processes in addition to fish processes; it represents a logical extension of the single-fish respirometer chambers historically used in bioenergetic research. Measured metabolic rates can be used to parameterize fish metabolism models and to estimate FCE values based on feed application rates. Elevated oxygen-food consumption ratios can be used as indicators of fish stress and excessive activity.

Oxygen consumption rate and metabolite/fecal excretion rates can be calculated directly as a function of fish weight and temperature (2,14), but, more typically, those rates are based on fish feeding rates and on the stoichiometry of food catabolism. Ratios of oxygen consumption and of metabolite/fecal excretion to a unit of food consumption (g compound/g feed) are available in the literature and are often based on empirical observations. Alternatively, such ratios can be calculated as functions of food composition and of feeding rates. The first task of this procedure is to calculate ratios for complete feed oxidation as a function of food composition and catabolic parameters. These ratios are then corrected for food digestion and conversion efficiencies, to account for the undigested and stored feed fractions that are not catabolized. It is generally accepted that food digestion efficiency declines with increasing feed ingestion rate, but the magnitude of this response apparently depends on fish species and environmental conditions (2,43,44). Finally, daily mean oxygen consumption rates and metabolite/fecal excretion rates (g compound/kg fish/day) are calculated by multiplying compound-to-food ratios by the predicted (or actual) daily feed application rate (g feed/kg fish/day).

The correction of compound-to-food ratios from feed composition to food catabolism illustrates an important concept regarding feed quality and the utilization of feeds. Metabolizable energy contents of feeds (kcal/g) depend on the feed composition and on species-specific digestion capabilities (1). Because energy needs for maintenance and activity must be satisfied before growth can occur, dietry protein will be used for energy when the diet is deficient in non-protein calories (lipids and carbohydrates). As the use of protein for energy increases, nitrogen excretion and the exothermic energy losses of amino acid deamination (heat increment) also increase. Accordingly, nitrogen loading by fish on their culture system is increased, and the net energy content of the feed is reduced. To minimize the use of protein for energy and to maximize its use for growth, fish feeds can be formulated to achieve protein-to-energy ratios that spare protein as an energy source. Such formulations normally are achieved by increasing the proportion of lipids.

An example calculation of food catabolism is summarized in Tables 6 and 7. The results of this exercise demonstrate the pronounced impact of fish feeding rates on compound-to-food ratios (Table 7.2), an effect due to the dependency of both food conversion and digestion efficiencies on feeding rate. These results are highly dependent on the values used for food digestion efficiencies, conversion efficiencies, and energy and protein contents. For example, higher digestion efficiencies would *increase* oxygen-to-food ratios, and higher conversion efficiencies would *decrease* oxygen-to-food ratios. The metabolizable energy content of fish feeds varies from 2.0 to 3.5 kcal/g or more, compared to the 3.3 kcal/g used in the example, with a directly proportional impact on compound-to-food ratios for oxygen and carbon dioxide. In addition, the protein content of fish feeds varies from about 25 to 55% of dry weight (dw), and fish protein contents typically range from about 50 to 70% of dw (42). To simplify the example exercise, it was assumed that the protein content of both fish and

Table 6. Example Budget for Deriving Oxygen Demand and Carbon Dioxide Production Values for Fish Foods, Based on Given Food Composition and Mass-Energy Conversion Terms and Assuming Complete Oxidation of Feed (Column Letters Are Used to Identify Columns and Associated Footnotes)

Food Component[a]	Component Fraction (g cp/g fd)[b]	Caloric Content (kcal/g cp)[c]	Caloric Content (kcal/g fd)[d]	Oxycaloric Equivalent (g O_2/kcal)[e]	Oxygen Demand (g O_2/g fd)[f]	RQ (mol CO_2/mol O_2)[g]	Carbon Dioxide Production (g CO_2/g fd)[h]
Crude protein	0.45	4.0	1.800	0.313	0.563	0.9	0.696
Crude lipid	0.15	8.0	1.200	0.305	0.366	0.7	0.352
NFE	0.15	2.0	0.300	0.283	0.085	1.0	0.117
Crude fiber	0.04	0.0	0.000	0.000	0.000	1.0	0.000
Ash	0.11	0.0	0.000	0.000	0.000	0.0	0.000
Moisture	0.10	0.0	0.000	0.000	0.000	0.0	0.000
Total	1.00	—	3.300	—	1.013	—	1.165

[a]Abbreviations: food (fd) and food component (cp).
[b]Values represent a typical trout growout feed; nitrogen free extract (NFE; carbohydrates) is calculated from given values, phosphorus is contained in given components (e.g., 1.7%), and protein is assumed to contain 16% nitrogen.
[c]Caloric content values (metabolizable energy, kcal) of food components vary among fish species; the values here represent approximate averages of literature values (1–3,19).
[d]Value = b × c.
[e]Literature reference: Number 1.
[f]Value = d × e.
[g]Respiratory quotient (RQ); literature reference: Number 1.
[h]Value = f × g × (44 g CO_2/mol)/(32 g O_2/mol).

Table 7.1. Example Budget for Deriving Food Conversion Efficiencies (FCE) and Fractional Uses of Ingested Feed as a Function of Fish Feeding Rate (as FFI), where FCE Varies from 0.0 to 90% (wet wt. basis) and Digestion Efficiency Varies from 90 to 60% (dry wt. basis) as Feeding Rate is Increased from Maintenance (10% FFI) to Maximum (100% FFI) Levels

Feeding Rate (as FFI, %)	Food Conversion Efficiency (FCE, %)		Fractional Uses of Ingested Food (g use/g feed)		
FFI	FCE (wet wt.)	FCE (dry wt.)	Undi-gested[a]	Growth[b]	Catabo-lized[c]
10.00	0.00	0.00	0.10	0.00	0.90
20.00	60.00	20.00	0.13	0.20	0.67
40.00	75.00	25.00	0.20	0.25	0.55
60.00	90.00	30.00	0.27	0.30	0.43
80.00	75.00	25.00	0.33	0.25	0.42
100.00	60.00	20.00	0.40	0.20	0.40

[a]Undigested fraction—based on linear response of digestion efficiency to feeding rate.
[b]Growth fraction—equivalent to dry-weight food conversion efficiency, expressed as a fraction.
[c]Catabolized fraction = 1.0 − undigested fraction − growth fraction = digested fraction − growth fraction.

feed was the same, i.e. 50% of dw, and therefore, nitrogen-to-food ratios were a product of the catabolized fractions and the nitrogen content of the feed. In practical aquaculture, however, the protein level of feed is typically less than that of the fish, and feeds are normally formulated to allow protein sparing for growth. The resulting decrease in nitrogen-to-food catabolic ratios can be calculated by considering this protein differential in the nitrogen mass-balance of the fish. For example, if fish protein content is increased to 60% of dw in the example exercise, then the nitrogen-to-food ratio at the 80% FFI feeding level is 0.027 g N/g food (compared to 0.030). If feed protein content is reduced to 40% of dw in the example exercise, then the nitrogen-to-food ratio at the 80% FFI feeding level is 0.021 g N/g food (compared to 0.030), reflecting both the reduction in the nitrogen content of feed and the preferential nitrogen uptake by the fish. This concept of compound incorporation as a function of availability and body composition also applies to phosphorus and other feed minerals (ash content).

In the application of daily mean metabolic rates (from daily feeding rates) to facility design, it is critical to consider that over a 24-hour diurnal cycle, the hourly metabolic rates of a fish may vary by a factor of three or more; this variation is due to feed application events during daylight hours, diurnal temperature oscillations, and diurnal fish activity levels (1). Accordingly, peak biomass support demands on fish culture systems may exceed daily mean demands by a factor of two or more. To take into account peak metabolic demands, metabolic models can be expanded to include diurnal variations. More simply, daily mean metabolic rates can be multiplied by empirically based peak-to-mean ratios, ranging from 1.2 to 1.4 or higher (45). Diurnal variations can be significantly reduced by increasing the number of feedings per day and by lengthening the daily feeding period. Required peak-to-mean ratios for facility design also depend on management tolerance for short-term, suboptimal water quality. Depending on aquaculture management intensity, strategies used to address diurnal changes in fish metabolic loading differ on how closely hourly (say) levels of fish biomass support capacity are matched to demand. If biomass support capacity is maintained at a constant rate that satisfies peak requirements over a whole day, then biomass support capacity is not fully utilized during periods of low demand—but facility management is simplified.

Table 7.2. Example Budget for Deriving Compound-to-Food Ratios of Food Catabolism, Combining Feed Composition Values from Table 6 and Catabolic Fractions from Table 7.1 and Assuming that Fish and Feed Material Compositions Are Comparable (Other than in Moisture Content)

Feeding rate	Food Catabolism (g compound/g food)[a]					Total Metabolism[b]
FFI	Oxygen (O_2)	Carbon Dioxide (CO_2)	Total Ammonia (N)	Phosphorus (P)	Particulate Solids (dry wt.)	mg O_2/kg fish/hr
10.00	0.912	1.049	0.065	0.0153	0.100	114
20.00	0.675	0.777	0.048	0.0113	0.133	169
40.00	0.557	0.641	0.040	0.0094	0.200	279
60.00	0.439	0.505	0.031	0.0074	0.267	329
80.00	0.422	0.485	0.030	0.0071	0.333	422
100.00	0.405	0.466	0.029	0.0068	0.400	507
Reported values[c]	0.20–0.60	0.25–0.70	0.025–0.040	0.005–0.050	0.30–0.65	50–500+

[a]Expressed as g compound/g food, for wet-weight food at 10% moisture content.
[b]Calculated at a feeding rate of 30 g feed/kg fish/day (equivalent to 3% body weight per day).
[c]Ranges are based on a wide review of aquaculture literature sources.

APPLICATIONS TO AQUACULTURE DESIGN

The fish growth, feeding, and metabolism models presented earlier in this entry may be used singly or in combination, for a variety of aquaculture design and management tasks. Analyses performed for one point in time may be accomplished with a calculator. Projections over time (simulations) require many calculations and are best performed with computer spreadsheets or programs. Simulations are used extensively in fish performance engineering to generate schedules for the following: fish weights, numbers, and biomass; feed application rates; biomass and metabolic loading; and water-flow and -quality management. On the basis of this information, rearing volume requirements and the water transport and treatment systems required to maintain water quality criteria can be determined.

A typical spreadsheet format has columns of given and calculated values. From left to right, a typical column order is the following: (*1*) time (e.g., days); (*2*) given environmental variables (e.g., water temperature); (*3*) calculated fish number, body weight, growth rate, total biomass, and biomass density; (*4*) feed application rate, feed conversion rate, and metabolic rates; and (*5*) required fish rearing-unit volume and water flow-rate of given quality. With time increasing down the spreadsheet, each row represents an increment of time equivalent to the simulation time step (e.g., one day). Cumulative values in each row (e.g., fish weight) are based on values in the prior row; the first row holds initial values.

The aquaculture design procedure outlined below is used to generate fish management schedules and culture system design requirements. This procedure normally requires some iteration in order to match production objectives with available facility resources. At any point in the procedure, usage levels of one or more resources may be found to be under- or over-utilized; such a condition requires the return to a prior design step and the adjustment of production objectives or culture system specifications. This procedure consists of multiple stages, organized by increasing levels of analytical scope and resolution. Under this approach, the basic feasibility of rearing a given species and the target biomass of fish under expected environmental conditions is determined before the specific culture system, resource, and economic requirements necessary to provide this fish culture environment are developed.

1. Establish design objectives, specifications, and analytical tools, including the following:
   - 1.1. fish production targets: stocking and target release/harvest dates, weights, and numbers;
   - 1.2. fish environmental criteria: see Table 1;
   - 1.3. facility environmental regimes: historical (empirical) or predicted (simulated);
   - 1.4. facility specifications: tentative system design and management strategies;
   - 1.5. fish performance models: selection and calibration.
2. Assess the feasibility of production-trajectory objectives and generate associated schedules:
   - 2.1. Calculate the period mean FGI necessary to achieve the target fish weight, from the target growth rate relative to the maximum growth rate achieved under maximum feeding rate.
   - 2.2. Generate fish weight schedules on the basis of environmental conditions and FGI or on the basis of NFP (if prepared feeds are not used).
   - 2.3. Generate fish feeding schedules on the basis of environmental conditions, of contributions from NFP, and of the required FFI calculated as a function of FGI.
   - 2.4. Determine compound-to-food metabolic ratios from food quality and feeding rate.
3. Combine production trajectories and fish numbers in order to assess the feasibility of production-scale objectives, on the basis of rearing volume and water flow-rate requirements:
   - 3.1. Generate fish number schedules from initial numbers and survival estimates.
   - 3.2. Generate fish biomass schedules from fish number and weight schedules.

3.3. Generate use schedules for rearing units on the basis of biomass schedules (kg), of desired biomass density levels (kg/m^3), and of resulting rearing-volume requirements (m^3).
3.4. Generate rearing-unit water flow-rate schedules from desired water-exchange rates (no./day) or biomass loading rates (kg fish/m^3/day).
3.5. Generate feed application schedules for rearing units (kg feed/RU/day), on the basis of feeding rate (kg feed/kg fish/day) and biomass schedules.
3.6. Check on whether system capacity limits have been exceeded for water flow-rates and feed application rates.

4. Depending on the required rigor of design analyses, mass-balance analyses are applied to rearing units to better quantify biomass support requirements. These analyses are used to determine peak and mean capacity requirements for diurnal and seasonal periods. Objectives include the operational requirements of influent and effluent water systems and the scheduling of rearing-unit water flow-rates and of treatment processes performed within the fish rearing units (e.g., aeration).
4.1. Combine fish biomass density (kg/m^3) with metabolite/fecal production rates (g compound/kg fish/day) to get mass sources and sinks due to fish processes (g compound/m^3/day). Include any additional significant mass-transfer processes of the rearing unit, including (*1*) influent and effluent water flow rates and (*2*) passive and managed physical, chemical, and biological processes. Heat transfer between a rearing unit and its environment may also be an important consideration, especially for static ponds and at lower water exchange rates, where rearing unit temperatures may diverge significantly from initial or influent temperatures.
4.2. If the fish culture system is pre-defined in respect to rearing-unit water flow-rates, influent water quality, and in-pond treatment processes, then estimate fish impacts on water quality and determine whether water-quality criteria are maintained. If the fish culture system is to be designed, then determine the water flow-rates, the influent water quality, and/or in-pond water-treatment processes required to maintain water-quality criteria.
4.3. Assess management risk by determining the elapsed time between system failure and fish death (or excessive stress) due to deterioration in water quality past fish tolerance extremes. System failure includes loss of power or of critical components (e.g., water pumps and aerators) and is simulated by terminating all water flow through and treatment processes in the rearing unit.

BIBLIOGRAPHY

1. J.R. Brett and T.D.D. Groves, in W.S. Hoar, D.J. Randall, and J.R. Brett, eds., *Fish Physiology, Vol. 8, Bioenergetics and Growth*, Academic Press, New York, 1979, pp. 279–352.
2. B. Hepher, *Nutrition of Pond Fishes*, Cambridge University Press, 1988.
3. M. Jobling, *Fish Bioenergetics*, Chapman and Hall, Fish and Fisheries Series, London, 1994.
4. A.L. Lehninger, *Bioenergetics, The Molecular Basis of Biological Energy Transformations*, Benjamin, NY, 1965.
5. E.P. Odum, *Fundamentals of Ecology*, 3rd ed., Saunders, Philadelphia, PA, 1971.
6. F.E.J. Fry, *Univ. Toronto Stud. Bio. Sr.* **55**, 1–62 (1947).
7. F.E.J. Fry, in W.S. Hoar and D.J. Randall, eds., *Fish Physiology, Vol. 6, Environmental Relations and Behavior*, Academic Press, New York, 1971, pp. 1–98.
8. J.E. Paloheimo and L.M. Dickie, *J. Fish. Res. Board Can.* **23**, 1209–1248 (1966).
9. C.E. Warren and G.E. Davis, in S.D. Gerking, ed., *The Biological Basis of Freshwater Fish Production*, Blackwell Scientific, Oxford, 1967, pp. 175–241.
10. C.E. Warren, *Biology and Water Pollution Control*, Saunders, Philadelphia, PA, 1971.
11. M.L. Cuenco, R.R. Stickney, and W.E. Grant, *Ecol. Modelling* **27**, 169–190 (1985a).
12. M.L. Cuenco, R.R. Stickney, and W.E. Grant, *Ecol. Modelling* **27**, 191–206 (1985b).
13. M.L. Cuenco, R.R. Stickney, and W.E. Grant, *Ecol. Modelling* **28**, 73–95 (1985c).
14. J.R. Brett, in W.S. Hoar, D.J. Randall, and J.R. Brett, eds., *Fish Physiology, Vol. 8, Bioenergetics and Growth*, Academic Press, New York, 1979, pp. 599–675.
15. M.L. Cuenco, *Aquaculture Systems Modeling: An Introduction with Emphasis on Warmwater Aquaculture*, ICLARM Contribution No. 549, 1989.
16. D.H. Ernst, J.P. Bolte, and S.S. Nath, *Aquacult. Eng.* (in press).
17. Y.M. Svirezhev, V.P. Krysanova, and A.A. Voinov, *Ecol. Modelling* **21**, 315–337 (1984).
18. D.C. Haskell, *N.Y. Fish and Game J.* **6**, 204–237 (1959).
19. R.G. Piper, I.B. McElwain, L.E. Orme, J.P. McCraren, L.G. Fowler, and J.R. Leonard, *Fish Hatchery Management*, US Dept. of Interior, Fish and Wildlife Service, Washington, DC, 1986.
20. R.W. Soderberg, *The Progressive Fish Culturist* **52**, 155–157 (1990).
21. W.E. Ricker, in W.S. Hoar, D.J. Randall, and J.R. Brett, eds., *Fish Physiology, Vol. 8, Bioenergetics and Growth*, Academic Press, New York, 1979, pp. 677–743.
22. R.W. Soderberg, *The Progressive Fish Culturist* **54**, 255–258 (1992).
23. R.R. Parker and P.A. Larkin, *J. Fish. Res. Bd. Canada* **16**, 721–745 (1959).
24. W.E. Ricker, *Computation and Interpretation of Biological Statistics of Fish Populations*, Bulletin of the Fisheries Research Board of Canada, 191, 1975.
25. K.D. Hopkins, *J. World Aquacult. Soc.* **23**, 173–179 (1992).
26. E. Ursin, *J. Fish. Res. Bd. Can.* **24**, 2355–2453 (1967).
27. E. Ursin, *Symp. Zool. Soc. Lond.* **44**, 63–87 (1979).

28. J.D. Balarin and J.P. Hatton, *Tilapia: A Guide to Their Biology and Culture in Africa*, University of Stirling, Stirling, Scotland, 1979.
29. G.D. Stauffer, *A Growth Model for Salmonids Reared in Hatchery Environments*, Ph.D. Thesis, University of Washington, Seattle, WA, 1973.
30. G.K. Iwama and A.F. Tautz, *Can. J. Fish. Aquat. Sci.* **38**, 649–656 (1981).
31. A.H. Weatherley and H.S. Gill, *J. Fish Biol.* **23**, 653–674 (1983).
32. M. Jobling, *J. Fish Bio.* **22**, 471–475 (1983).
33. D.C. Haskell, *N.Y. Fish and Game J.* **6**(2), 204–237 (1959).
34. J.W. Jensen, *Aquaculture* **48**, 223–231 (1985).
35. J.M. Elliott, *J. Anim. Ecol.* **44**, 805–821 (1975).
36. M. Jobling, *J. Fish. Biol.* **22**, 471–475 (1983).
37. D.H. Ernst, L.J. Ellingson, B.L. Olla, R.I. Wicklund, W.O. Watanabe, and J.J. Grover, *Aquaculture* **80**, 247–260, Erratum (1989).
38. T.M. Losordo, in B.A. Costa-Pierce and J.E. Rakocy, eds., *Tilapia Aquaculture in the Americas, Vol. 1*, World Aquaculture Society, Baton Rouge, LA, 1997, pp. 185–211.
39. J.S. Diana, in H.S. Egna and C.E. Boyd, eds., *Dynamics of Pond Aquaculture*, CRC Press, Boca Raton, FL, 1997, pp. 245–262.
40. G. Almazan and C.E. Boyd, *Aquaculture* **15**, 75–77 (1978).
41. P.D. Corey and M.J. English, *Simulation* **44**, 81–93 (1985).
42. A.H. Weatherley and H.S. Gill, *The Biology of Fish Growth*, Academic Press, New York, 1987.
43. R.E. Speece, *Trans. American Fish. Soc.* **2**, (1973).
44. K.H. Meyer-Burgdorff, M.F. Osman, and K.D. Gunther, *Aquaculture* **79**, 283–291 (1989).
45. J. Colt and K. Orwicz, *Aquacultural Engineering* **10**, 1–29 (1991).

See also RECIRCULATING WATER SYSTEMS; RECIRCULATION SYSTEMS: PROCESS ENGINEERING.

PESTICIDES

ROBERT R. STICKNEY
Texas Sea Grant College Program
Bryan, Texas

OUTLINE

Types, Uses, and Threats
Bibliography

Certain pesticides have found limited use in aquaculture for controlling predatory insects and certain types of pests. The major interest of aquaculturists with regard to pesticides, however, is with respect to accidental exposure, as most aquatic animals are highly susceptible to these toxicants.

TYPES, USES, AND THREATS

Two general categories of pesticides have dominated the market for the past few decades: chlorinated hydrocarbons and organophosphates. Chlorinated hydrocarbons are persistent in nature. They or their breakdown products can remain active and lethal for many years. The most widely used organophosphates, on the other hand, become inactive within a few days after application. DDT was taken off the market in the United States in the 1970s, and many other chlorinated hydrocarbons were later banned for public use, though some can still be applied by persons who are licensed to employ them. Yet, DDT and related compounds not generally available in the United States are being produced and employed throughout much of the world.

While herbicides are used to control aquatic vegetation (see the entry "Aquatic vegetation control"), there is little need for pesticide applications in conjunction with aquaculture operations, although pesticides have been used to control sea lice in salmon net pens (1).

Pesticides have sometimes been used to control burrowing crawfish and other pests between crops of the target aquaculture species. It is possible to control aquatic insects with pesticides, but, such use is uncommon, because of potential toxicity to the aquaculture species. Aquatic insect populations are often controlled by pouring sufficient diesel fuel over a pond to cover the surface in a thin film, which will cause the insects to suffocate when they surface to obtain oxygen.

The use of any type of chemical in aquaculture facilities is closely controlled in the United States, and few biocides of any type have been approved for use by foodfish (including shellfish) producers. However, this is not the case in many countries, particularly those in the developing world, where chemical use is largely or totally uncontrolled. That having been said, there have been few, if any, reports of pesticide-contaminated fish or shellfish from the aquacultured products that reach the markets of the world.

Since many aquaculture facilities are located in agricultural areas, it is not at all uncommon for pesticide applications to occur in the immediate vicinity of fish or invertebrate culture facilities. In many cases, pesticides are applied, at least in the United States, through aerial application. Misjudgment on the part of pilots or wind drift of the chemicals can lead to disastrous results. Ponds and other outdoor rearing chambers are particularly susceptible, though contamination of incoming water must also be considered. In addition to the possibility of direct contamination of facilities or their water supply, there is the chance that runoff of active pesticide during and after rain events will contaminate water being used for aquaculture.

As a part of site selection (see the entry "Site selection"), the aquaculturist should evaluate the potential of pesticide-contaminated water and soils and have sample tests performed if there is any indication of a potential contamination problem. Once a facility has been established, aquaculturists should avoid causing pesticide contamination from their own activities and ensure to the greatest extent possible that neighbors do not spray pesticides on their crops on windy days.

Much of what is currently known about pesticide toxicity to fish and shellfish comes from toxicity studies, which have determined such things as LD_{50} concentrations. The LD_{50} (LD = lethal dose) is the lowest concentration of a chemical that will kill 50% of the animals exposed to the chemical in a given period. The periods used are usually 72 or 96 hours. While such acute toxicity information is useful, it is the chronic effects of pesticide contamination that are most worrisome for aquaculturists: If a pond happens to receive overspray or wind drift of pesticide from an aerial application, that fact will be apparent within several hours as fish begin to float up in affected ponds. Much less obvious is the cause of low-level, continuous mortalities over long periods, due to chronic exposure low levels of pesticide. Soil and waters tests should be virtually negative with respect to pesticides if the aquaculturist is to be assured that no chronic toxicity problem exists.

BIBLIOGRAPHY

1. A. Ross, *Marine Pollution Bulletin* **20**, 372–374 (1989).

pH

Gary A. Wedemeyer
Western Fisheries Research Center
Seattle, Washington

OUTLINE

The pH of the aquatic environment indicates the degree to which the water is acidic or basic (alkaline). This information is important to aquaculture, because plankton, fish, and other aquatic life can survive and grow only within a relatively narrow range of acidity. Knowledge of the pH of an environment is important to many other aspects of aquaculture management as well. For example, the pH is important in alkalinity and carbon dioxide measurements and affects such diverse factors as chlorine disinfection and ammonia toxicity. Thus, pH is one of the most frequently employed water chemistry tests in aquaculture.

DEFINITION OF pH

Acidic substances liberate hydrogen ions (H^+) into the water, while basic (alkaline) substances can accept H^+. The more acidic the water, the greater the hydrogen-ion concentration and the lower the pH. On the other hand, the more basic the solution, the lower the H^+ concentration and the higher the pH. Strictly speaking, acidity is the capacity of a water supply to neutralize alkaline substances (hydroxyl ions, OH^-). It is measured by titration and is expressed as mg/L of equivalent $CaCO_3$. Thus, like alkalinity (see the entry "Alkalinity"), acidity is a capacity factor. In biological work, however, it is more relevant to express acidity in terms of its intensity (i.e., the hydrogen-ion concentration itself), rather than as a capacity factor. To do so, the concept of pH, defined as the negative logarithm of the hydrogen-ion concentration (in moles/L), was developed. The "p" stands for puissance (power), and "H" is the symbol for hydrogen. By convention, the pH range extends from 0 to 14. A pH of 7.0 is the neutral point, a pH greater than 7 indicates alkaline conditions, and a pH less than 7 is acidic. By way of illustration, freshwater supplies used for aquaculture usually have pH of 5 to 9. Soft waters normally have a slightly acidic pH, while hard waters usually are slightly basic, because they contain bicarbonate and carbonate minerals. Seawater is strongly buffered at about pH 8.2. Sodium bicarbonate added to ponds to correct low alkalinity problems also adjusts the pH of the water to about 8.2. The gastric acid of salmonid fishes is about pH 3; their urine is pH 7.2, and their blood is pH 7.8. For some purposes, it is important to take into account that the neutral point of water is temperature dependent and is pH 7.0 only at 25 °C (77 °F); it ranges from pH 7.5 at 0 °C (32 °F) to pH 6.5 at 60 °C (140 °F) (1). Thus, the normal blood pH of fish and other poikilothermic animals is not fixed, but varies inversely with the water temperature.

MEASURING AND REPORTING pH

The pH of a solution is most accurately measured with a glass electrode system that produces an electrical potential, which is converted from millivolts into a pH value by an electronic circuit. Such pH meters are calibrated with standardized buffer solutions and are usually accurate to at least 0.1 pH unit. For routine monitoring work, portable water chemistry test kits using pH sensitive paper strips or color standards may be entirely adequate.

When reporting pH monitoring results, it is important to remember that the average pH value cannot be determined by simply adding the individual numbers and dividing by the number of samples. The individual pH values must first be converted back to hydrogen-ion concentrations in moles/L and then averaged. The negative logarithm of this number gives the average pH. Alternatively, it may be just as meaningful to report the pH range and explain any unusual deviations (2).

SOURCES AND EFFECTS OF UNFAVORABLE pH IN AQUACULTURE SYSTEMS

Acidity in freshwater used for aquaculture is generally due to carbon dioxide (CO_2) dissolved from the atmosphere or produced by fish metabolism. However, mineral acids from pollution (e.g., acid precipitation and acid drainage from coal mines), naturally occurring organic acids from humus deposits, and the hydrolysis of salts leached into water supplies from mineral deposits can also cause acidity.

Mine drainage water becomes acidic because iron bacteria oxidize pyrites (FeS_2) and other sulfide impurities in the ore to sulfuric acid (H_2SO_4). In hatcheries supplied with hard, alkaline water (greater than 200 mg/L as $CaCO_3$), introduced acidic compounds such as acid mine waste drainage may be partly or completely neutralized by the natural carbonate content of the water. However, the dissolved CO_2 that is then produced can be as deleterious to fish health as the low pH of the acid mine waste itself. Carbonate rock can also be placed in the mine waste drainage stream to neutralize H_2SO_4, but doing so may result in the production of semigelatinous hydrated iron oxide [$Fe(OH)_3$], which may physically damage gills (3).

The CO_2 produced by fish respiration itself can substantially lower the pH in hatcheries that use water low in total hardness (less than 50 mg/L as $CaCO_3$). This condition is rarely of concern in raceways, but can be a problem in fish transport operations, where CO_2 concentrations of 20–30 mg/L can accumulate in the tanks, lowering the pH to 6.0 or less within 30 minutes of loading. In heavily stocked ponds, the pH can fluctuate by one or two units, even in well-buffered water (i.e., alkalinity greater than 100 mg/L), because of the CO_2 produced by fish respiration.

The photosynthesis and respiration of phytoplankton can also strongly affect water pH. Phytoplankton photosynthesizing in intense sunlight may remove CO_2 faster than it can be replaced, causing the pH to rise. At night, plant respiration adds dissolved CO_2, decreasing the pH. The resulting pH fluctuations can be quite severe, depending on the size of the phytoplankton community and the buffering capacity of the pond water. A large phytoplankton community, typical of fed-fish ponds in waters of low alkalinity (i.e., less than 50 mg/L as $CaCO_3$), may cause the pH to fluctuate from pH 5 at night to pH 10 or above during daylight hours (3).

Proper management of water pH is fundamental to protecting the health of aquatic animals in aquaculture systems. If the pH is too low (i.e., pH less than 5), ion transport at the gills of salmonid fishes is affected, leading to osmoregulatory failure and death. Ammonia toxicity is also strongly affected by pH, because the equilibrium between nontoxic NH_4^+ and toxic NH_3 shifts toward the toxic NH_3 form as the pH rises. Thus, water containing ammonia from uneaten food or from fish excretion may be completely safe for salmonid fish at pH 6, but the same ammonia concentration at pH 8.0 may be toxic. Another well-known deleterious effect of pH is that toxic heavy metals in soils and bottom sediments can be solubilized and leached into the water, where they may kill fish and invertebrates at sensitive life stages. Aluminum toxicity is an important example of this problem. At pH 6.6, 30-day aluminum exposures at concentrations up to 57 μg/L (ppb) are safe for the egg and fry stages of brook trout. At pH 5.6, however, the safe concentration decreases to only 29 ppb (4). Low pH levels also interfere with the normal development of smoltification in juvenile anadromous salmonids, even in the absence of metal toxicity. For example, the parr–smolt transformation of juvenile Atlantic salmon is normal at pH 6.4–6.7, but seriously impaired at pH 4.2–4.7 (5). Spikes of acidity can occur during the spring runoff of freshwater streams when acid precipitation has accumulated in the snow pack. Finally, water pH strongly affects the metabolic activity of the nitrifying bacteria used in recirculating aquaculture systems to remove ammonia. Inhibition begins if the pH falls much below 7.0 and the alkalinity drops below about 80 mg/L (6).

MANAGEMENT RECOMMENDATIONS

The upper and lower limits for pH values that are safe for fish and invertebrates in aquaculture systems are not fixed values, but vary somewhat, depending on other environmental factors, such as temperature, metal and ammonia concentrations, and acclimation pH. Fish populations in natural waters can tolerate the pH extremes of 5–9 (without the effects of metal toxicity), but a more prudent range to protect the health of freshwater fish in aquaculture facilities is pH 6.5–9.0 (3). Fish in mariculture facilities, such as net pens, normally do not experience pH fluctuations, because seawater is strongly buffered at a pH of about 8.2.

BIBLIOGRAPHY

1. *Standard Methods for the Examination of Water and Wastewater*, 17th ed., American Public Health Association, Washington, DC, 1989.
2. E.C. Kinney, *Prog. Fish-Cult.* **35**, 93 (1973).
3. G. Wedemeyer, *Physiology of Fish in Intensive Culture Systems*, Chapman & Hall, New York, 1996.
4. L. Cleveland, E.E. Little, R.H. Wiedmeyer, and D.B. Buckler, in T.E. Lewis, ed., *Environmental Chemistry and Toxicity of Aluminum*, Lewis Publishers, Chelsea, MI, 1989.
5. R.L. Saunders, E.B. Henderson, P.R. Harmon, C.E. Johnson, and J.G. Eales, *Can. J. Fish. Aq. Sci.* **40**, 1203–1211 (1983).
6. J.C. Loyless and R.F. Malone, *Prog. Fish-Cult.* **59**, 198–205 (1997).

See also Carbon dioxide.

PLAICE CULTURE

Robert R. Stickney
Texas Sea Grant College Program
Bryan, Texas

OUTLINE

Plaice, *Pleuronectes platessa*, was one of the first flatfish to be developed as an aquaculture species. Culture and stocking of plaice in Norway goes back to the turn of the century, though the practice was nearly abandoned and not much interest existed until James Shelbourne with the Whitefish Authority in Great Britain began making significant breakthroughs in plaice larval culture in the 1950s and 1960s (1,2). That work paved the way for development of culture techniques for such other flatfish species as Atlantic halibut; sole; turbot; Japanese flounder; summer flounder; southern flounder; and two species closely related to plaice: winter flounder (*P. americanus*) and yellowtail flounder (*P. ferrugineus*). The culture of most of those species is described in detail in other sections of this encyclopedia.

GENERAL INFORMATION ON PLAICE CULTURE

One of the first breakthroughs in plaice larval culture was the finding by Shelbourne that heavy losses in the hatchery occurred as a result of the growth of bacteria in hatchery tanks. The problem was controlled by adding a mixture of penicillin and streptomycin to the water. Since that time, techniques for sterilizing eggs prior to introducing them into hatchery tanks have been developed. For example, gluteraldehyde is an effective chemical to sterilize plaice eggs. In addition, advances in the technology associated with water filtration and ozonation, improvements in ultraviolet sterilization units, and the development of hatching chambers that allow constant water exchange without damaging eggs have reduced or eliminated the need to use antibiotics in many instances.

The technique for plaice culture that was developed by the early 1970s involved holding adults in ponds, where they were allowed to spawn naturally. The floating eggs were collected and placed in the hatchery. Incubation tanks were supplied with recirculated seawater that was exposed to ultraviolet light for sterilization.

Hatching and rearing troughs were $4.9 \times 1.2 \times 1.2$ m deep ($12.4 \times 3 \times 3$ ft). Each trough was stocked with 30,000 to 40,000 eggs, which were hatched after three weeks when the water temperature was maintained at 6 °C (43 °F).

Brine shrimp nauplii were found to be a suitable first food for plaice larvae. It was also discovered that brine shrimp from different sources were slightly different in size, the difference being sufficient enough that nauplii from an inappropriate source were too large to be consumed by first-feeding larvae. Since the 1960s, many marine fish culturists have been producing rotifers, in particular *Brachionus plicatilis*, for first-feeding larvae. Rotifers are smaller than brine shrimp nauplii and have, in general, better nutritional quality. Larval fish quickly become too large to efficiently capture rotifers, so brine shrimp continue to be a food source for part of the larval development period.

Plaice, like other flatfish, are sight feeders. Research has shown that dark-colored, opaque tanks facilitate prey capture, as the brine shrimp nauplii could be clearly seen against the light that entered from the surface of the tank. Plaice larvae have been shown to feed successfully over a fairly wide range of illumination, unlike some other flatfish species.

After metamorphosis, plaice were, during the early years, fed chopped mussels or fish. Since then, prepared feeds have been developed. Live foods are still used, and a weaning period is required during which the fish are gradually adapted to prepared feeds by giving the feeds in combination with live or fresh food material initially and then gradually reducing the live or fresh food until it is eliminated.

Early nutritional work demonstrated that flatfish appear to have a high protein requirement. One study placed this requirement at 70% of a dry diet, but later research has suggested a lower apparent requirement of 57% (3). The protein requirement can best be met with animal protein sources, usually some type of fishmeal.

There are few other data available on the nutritional requirements of plaice. In addition to the aforementioned work on protein requirements, research has examined the vitamin C requirement of plaice

One problem with plaice and other flatfish has to do with pigment abnormalities, in particular the development of black pigment on the side that is down when the fish is swimming horizontally or lying on the substrate. Abnormal pigmentation reduces the value of fish and may even make them unmarketable. Early nutrition may play a role in the development of abnormal pigmentation. Incidence of the anomaly has been reduced by feeding enriched brine shrimp nauplii during first feeding. (Enriched nauplii are those that were fed a source of high-molecular-weight fatty acids prior to being offered to the plaice larvae.)

The initial work with plaice by the Whitefish Authority was aimed at producing fingerlings for enhancement stocking. Stocking of fjords has been ongoing for many years in the United Kingdom as a means of augmenting wild stocks. Once the techniques for larval culture had been worked out, interest in plaice culture within the commercial sector quickly developed, though that interest has since waned. Instead, other species have come into favor, for one reason or another. Much of the early and subsequent work with plaice has been of great importance to persons interested in other flatfish, which is the primary reason that a summary of the work with plaice is included here.

BIBLIOGRAPHY

1. J.E. Shelbourne, in W.J. McNeil, ed., *Marine Aquaculture*, Oregon State University Press, Corvallis, OR, 1968, pp. 15–36.
2. J.E. Bardach, J.E. Ryther, and W.O. McLarney, *Aquaculture*, Wiley-Interscience, New York, 1973.
3. J. Guillaume, M.-F. Coustans, R. Métailler, J. Person-Le Ruyet, and J. Robin, in R.P. Wilson, ed., *Handbook of Nutrient Requirements of Finfish*, CRC Press, Boca Raton, FL, 1991, pp. 77–82.

See also Brine shrimp culture; Flounder culture, Japanese; Halibut culture; Larval feeding—fish; Sole culture; Summer flounder culture; Winter flounder culture.

POLLUTION

Granvil D. Treece
Texas Sea Grant College Program
Bryan, Texas

OUTLINE

INTRODUCTION

The aquaculture industry is considered the fastest growing segment of agriculture. However, the industry has been experiencing growing pains as a result of water quality deterioration, and poor water quality has in turn contributed to disease outbreaks. For example, the shrimp aquaculture industry has experienced numerous disease outbreaks worldwide and production has decreased as a result. Because of these water pollution problems, production in the aquaculture industry has been slowed.

Pollution is at the heart of the seafood industry's dilemma. There is growing concern over the safety of ingesting shrimp, catfish, and other aquatic life taken from polluted waters. The U.S. Centers for Disease Control (CDC) reported that because of continuous pollution of the coastal wetlands and bays, shellfish caught in those waters are 18,000 times more likely to cause illness than shellfish grown elsewhere. This entry discusses water pollution and its relation to aquaculture.

EFFECTS OF POLLUTION

According to the U.S. National Marine Fisheries Service (NMFS), over one-third of all commercial and recreational fish species have dramatically declined in population or have completely disappeared in the past 15 years. This is mostly due to overfishing but is partly due to toxic pollutants and urban development of the coastlines. The NMFS has reported the status of U.S. fisheries to the U.S. Congress for the second time, and that report can be accessed through their Web site, http://kingfish.ssp.nmfs.gov/sfa/98stat.

In Canada, waters in and around Quebec and Montreal have been contaminated for over 20 years, due primarily to pesticides, sewer runoff, and pollutants from the St. Lawrence River. The Provincial Environmental Department monitors water quality and pollution levels regularly and issues guidelines on fish consumption to keep the public informed about how often they can eat various types of fish caught in those waters.

In 1990, approximately 1,400 beaches in the United States were closed due to pollution and toxic waste hazards, with another 1,700 threatened, as reported by the CDC. Toxic medical supplies discarded from Mexico were found washed up on the shorelines of Corpus Christi, Texas, resulting in the closing of some beaches in Texas. In 1997, Gulf of Mexico king mackerel were found to have high levels of mercury, and the Texas Department of Health issued guidelines for public consumption.

Off and on in the past, alarmed scientists have been concerned about 7,000 square miles (18,130 km^2) in the Gulf of Mexico, called the "dead zone," because hypoxia occurs in this area. It does not occur every year, but sometimes seasonally, especially during drought years, these areas are rendered almost lifeless by a lethal combination of agricultural fertilizer, sewage runoff, and diminished oxygen. Most of this comes from the Mississippi River. The process kills the food chain from the bottom up, rendering the area virtually lifeless. Some scientists are concerned that if the ecology of the affected ocean becomes stressed over long periods of time, it might become permanently damaged. When this major change takes place, forms of bacteria that thrive under anoxic conditions replace fish, shellfish, and crustaceans that need oxygen. However, these dead zones seem to come and go and are not always predictable. According to the U.S. National Oceanic and Atmospheric Administration (NOAA, Department of Commerce), 53% of U.S. estuaries experience hypoxia part of the year. The Gulf of Mexico is not the only water that suffers from dead zones. There are as many as a dozen dead zones in different areas of the world, all caused by some combination of pollution.

In certain cases, harmful algal blooms have resulted in fish kills, the deaths of numerous endangered West Indian manatees, beach and shellfish bed closures, threats to public health and safety, and concern among the public about the safety of seafood. According to some scientists, the factors causing or contributing to harmful algal blooms may include excessive nutrients in coastal waters, and other forms of pollution. There is a need to identify more workable and effective actions to reduce nutrient loading to coastal waters.

Since 1991, the Alabama Department of Environmental Management has been testing fish from major water bodies around the state for contaminants. In the latest round of testing, lab results showed that at some sites, polychlorinated biphenyls (PCBs) and mercury continue to be present in amounts exceeding the U.S. Food and Drug Administration (FDA) guidelines. As a result, consumption advisories for those areas have been issued. Even though the contaminant level may be very low in the water source itself, PCBs and mercury have a tendency to bioaccumulate. When a predator eats prey, the contaminants are deposited in the tissue of the predator. Over time, the contaminant can build up to unsafe levels.

As a result of growing health hazards, the Marriott Corporation (USA) placed a permanent ban, in all of its national hotel restaurants on the use of any shellfish caught in the Gulf of Mexico. Similarly, many more restaurants across the country make disclaimers warning of the possibility of food poisoning from the seafood they serve their customers. These problems are not isolated to the United States; in fact, they are often worse in other countries where laws are not as strict. One of the most prolific producers of cultured shrimp, Malaysia, has been plagued with a serious state of shrimp toxicity in polluted waters, which threatens their entire industry. China, the world's top producer in 1991 of pond-raised shrimp, was almost completely out of the shrimp business in 1995 due to pollution-induced diseases and the yellow head and white spot viruses. Indonesia and India have suffered similar fates of rapid aquaculture growth and near collapse as a result of pollution and water quality deterioration.

GREATER AWARENESS OF WATER POLLUTION

United States television broadcasts, Web sites, and published papers on public concerns about capture fishery policies and aquaculture development have increased the awareness of pollution problems. For example, CNN aired the program *Earth Matters* and included a segment titled, "Troubled Waters: The State of World Oceans." Worldwatch distributed a paper titled, "Rocking the Boat: Conserving Fisheries and Protecting Jobs." For more details on the press release, and ordering information, visit the Web site: http://www.worldwatch.org.

ACTIONS TAKEN TO COMBAT POLLUTION

Many countries with coastal aquaculture industries have enacted laws to address water pollution problems. However, legislative mandates can accomplish only so much toward a solution. The Food and Agricultural Organization of the United Nations (FAO) has published a set of guidelines for sustainable aquaculture, and most countries around the world are trying to adhere to them as they pass new laws to regulate their individual industries. More information on FAO and sustainable aquaculture can be obtained from their Web site http://www.fao.org/waicent/waicente.htm. An even greater alarm is the problem of radioactive leakage from the former USSR's decimated nuclear-powered naval fleet. Radioactive contamination of our oceans is going to continue to be a pollution problem in the future.

Aquaculture is important to the United States and to the world's fisheries. Both import and export markets for aquaculture products will expand and increase as research begins to remove physiological and other animal husbandry barriers. Overfishing of wild stocks will necessitate harvest regulations and replenishment through aquaculture. Future aquaculture development programs require an integrated public health approach to ensure that aquaculture does not cause unacceptable risks to public or environmental health and that aquaculture's potential economic and nutritional benefits are not harmed.

AQUACULTURE AND POLLUTION

Some aquaculture activities are considered pollution, but there are two aspects of water pollution in aquaculture. The first is the effect of pollution on aquaculture and the second is the pollution of the environment by aquaculture. Although aquaculture can pollute the environment, there are many other forms of pollution. Aquaculture requires high-quality water, and often the aquaculture industry suffers from other users' pollution.

Some of the pollution from aquaculture includes the misuse of therapeutic drugs, chemicals, fertilizers, and natural fishery habitat areas. As in agriculture, the use and misuse of antibiotics to control diseases is global and will probably increase as culturists move toward more intensive animal-rearing techniques and higher stocking densities. The illegal use of chloramphenicol in shrimp culture to control diseases may result in high levels in the harvested product, and this has been shown to cause liver damage in humans. Similarly, the improper or illegal use of chemicals (e.g., tributyl tin) to control pond pests such as snails can also result in human health hazards. The misuse of raw chicken manure as pond fertilizer may result in the transmission of *Salmonella* from manure to the cultured product.

Much attention has been given to the net-pen culture of salmon and to how that industry adds pollution to water in the form of feces, ammonia, and uneaten feed. It has been reported that one large salmon net-pen operation (exact size not given) can produce pollutants equivalent to untreated wastes from 10,000 people per day. Stickney (1) listed issues concerning net-pen culture. Other aquaculture industries have raised public concerns over pollution. Some of those have been the trout industry on the Snake River (western United States), and shrimp farms around the world.

The tilapia aquaculture industry is also facing a pollution-related dilemma. Some producers have found the bacteria *Streptococcus* in their fish, and this bacteria can be transmitted to humans. There has been at least one highly publicized case in Canada where this bacterium was transmitted from fish to human, but for the most part these occurrences are rare. Diseases associated with the handling of fish occur more often than diseases associated with human consumption of fish.

One of the most common pollutants from aquaculture activities is solids (clay and other soil particles that may be suspended as a result of moving water, especially during drain-harvesting). Suspended solids may cause low light attenuation and negatively affect seagrass growth or even smother seagrasses as the solids settle below the discharge area. Suspended solids may affect the overall aesthetics of the area. Such loading, with the addition of other organics, uneaten feed, and feces, may result in low oxygen and/or elevated ammonia levels in effluent waters. For the farmer, controlling pollution often becomes a balancing act between effluent dissolved oxygen (DO), biological oxygen demand (BOD), and ammonia.

Ammonia is another one of the common forms of pollution from aquaculture, but it is a natural element, and bacteria break it down in the environment. Nitrogen is one of several elements essential to the survival of aquatic organisms. Bacteria consume nitrogen in specific compounds that they can use and later excrete, frequently in a different, less harmful compound. Nitrogen cycles through the ecosystem by being absorbed and excreted by bacteria, plankton, and other aquatic life. In a normal cycle, bacteria transform nitrogen in ammonia ions into nitrite and later, ions. Denitrifying bacteria turn the nitrogen to gas that is released into the atmosphere from the soil or the water. Coastal waters have the ability to assimilate the pollutants that are released into them, as long as the capacity of those waters is not exceeded.

Treatment of pollution is the subject of much technology and may involve increased water use in a flow-through system, increased recirculation of water in a closed or semiclosed system, concentration or settling of solids, aeration, addition of beneficial bacteria, or a combination of these techniques. Treatment of solids is often done by routing the effluent into a settling pond or created wetland. Most of the solids can be settled by using baffles in the discharge canal and only allowing the discharge water to move slowly through the canal. Another technique is to hold water in a settling pond for a minimum of 6 hours (see settling time of solids in Fig. 1). Prevention of soil suspension is also another method used to control pollution and is done through soil erosion control techniques.

Permitting approaches often involve concentration of solids and monitoring the loading of solids and ammonia. Ambient conditions such as total suspended solids (TSS),

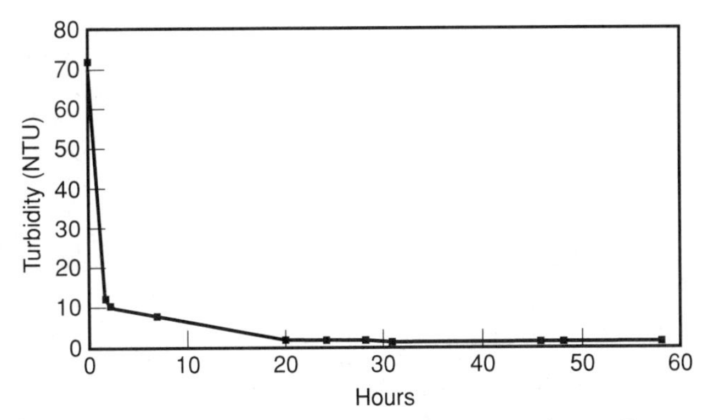

Figure 1. Turbidity (NTU) versus settling time, in hours. *Source*: David R. Teichert-Coddington, Auburn University, Alabama.

volatile suspended solids (VSS), inorganic suspended solids (ISS), ammonia, and, in most areas, biological oxygen demand (BOD) and dissolved oxygen (DO), are usually recorded, and effluent waters may not exceed set threshold criteria and limits over ambient conditions. VSS refers to those solids that can be incinerated, such as algae. ISS are those solids such as clay particles that are inorganic in nature and will not burn away in an oven.

The most common additives to aquaculture ponds are feeds, fertilizer, lime, and sometimes Zeolite (a natural mineral used for ammonia removal or blue-green algae control and is not toxic to the environment).

REGULATIONS TO CONTROL POLLUTION

The U.S. Environmental Protection Agency (EPA) and local state agencies regulate aquaculture effluents in the United States For those persons following aquaculture and National Pollutant Discharge Elimination System (NPDES) permit and effluent issues, a fact sheet and draft permit can be downloaded from the EPA homepage. The Internet sites are http://www.epa.gov/ and http:www.epa.gov/r10earth/offices/water/ow.htm. In 1977, the EPA considered developing effluent guidelines for aquaculture facilities and recommended issuance of Best Practicable control Technology (BPT) limitations, but regulations were not implemented. The EPA had until recently left effluent issues and regulation up to the states. If the state's regulations are not as strict as the EPA's, then the state must follow EPA guidelines. If state regulations are more strict, then they are followed.

Aquaculture operations in the United States are now controlled by the EPA's authority to regulate under the Clean Water Act, as amended (33 U.S.C. 1251 *et seq.*). Section 402 of the Act requires that an NPDES permit be issued by the EPA prior to discharge. In May, 1998 the EPA published their effluent guidelines plan. See the entry "Regulation and permitting" for more details on aquaculture permitting. Also see the EPA criteria referred to throughout this entry.

Some countries have developed Best Management Practices (BMPs) and effluent limitations and requirements to establish a minimum standard of operating in order to reduce pollution. Most monitoring involves sampling the discharge for flow solids (total and volatile), ammonia, and dissolved oxygen. Monitoring pH is sometimes done following the use of lime. Total residual chlorine monitoring may also be required following the use of chlorine.

ADDITIONAL SOLUTIONS TO POLLUTION

Control treatment technologies, which have been adopted by numerous industries, do reduce the pollutants of greatest concern at designated locations and can serve as a basis for BMPs. Improved feeding management practices, feed manufacturing, and diet formulations have contributed to gains in feed conversion efficiencies and reductions of solids and nutrient levels in discharges. New facility engineering designs and innovations have also refined solids waste reduction and removal for proper disposal.

POLLUTION AND TOXICOLOGY OF AQUATIC LIFE

The majority of available reports regarding pollution and invertebrate toxicology deal with penaeid shrimp, e.g., Overstreet (2). There have also been numerous studies by the EPA concerning the effects of pollution on commercially valuable penaeid shrimp of the United States, in the Atlantic States and Gulf Coast. Therefore, most of the information presented here is related to pollution and the following three species of penaeids: *Farfantepenaeus duorarum* (pink shrimp), *Farfantepenaeus aztecus* (brown shrimp), and *Litopenaeus setiferus* (white shrimp), all Atlantic and Gulf of Mexico species. References to other species of penaeids, nonpenaeid crustacea, and finfish are also made.

A brief discussion of the following pollutant categories follows: PCBs, organic chemicals other than petroleum and related compounds, petroleum, heavy metals, and other pollutants. Under each of these divisions some of the known toxicity levels are reviewed.

POLYCHLORINATED BIPHENYLS

PCBs are industrial pollutants that can influence aquatic ecology (3). They are a group of chlorinated hydrocarbons developed for commercial use as electrical transformer insulation fluids, hydraulic fluids, fire retardants, plasticizers, and extreme pressure oils and greases. For many years PCBs have been present in the aquatic environment as a result of waste effluents, disposal of dielectric fluids, and other industrial sources. It is a well-established fact that certain fresh and marine bodies of water are contaminated with various compounds of PCB. It is not as large a problem as it once was since laws controlling the discharge and burning of these waste products on offshore ships have been severely strengthened; however, PCBs are extremely stable and inert compounds. As a result, they have accumulated in native fish and wildlife in many parts of the world. Biphenyls may have anywhere from one to ten attached chlorine atoms, making possible more that 200 compounds. The 1998 FDA Hazard Analysis Critical Control Point (HACCP) (4) action level for PCBs in fish is 2 parts per million (ppm, Table 1). Other important information about FDA and seafood can be obtained at the Web site http://vm.cfsan.fda.gov/~mow/intro.html, and more information about the NMFS HACCP Program can be obtained at http://www.nmfs.gov/iss/manual.html or http://www-seafood.ucdavis.edu/haccp/Plans.htm.

According to the EPA, concentrations of PCBs should not be above 0.002 parts per billion (ppb) in water, and residue in aquatic animals should not be above 0.5 ppb. The minimum concentration causing mortality in invertebrates was 0.9 ppb. Penaeid shrimp suffered the greatest mortality when exposed during premolt (just before molting) and during molt. Most exposed shrimp and fish became lethargic and stopped feeding, and shrimp did not dig into the substrate (digging is a normal activity for penaeids). Dramatic chromatophore changes in the cuticle of exposed shrimp and the skin of fish were more frequent and obvious than in unexposed control animals.

ORGANIC AND OTHER PESTICIDE CHEMICALS

If the aquaculture area or site is suspected of having pollution such as organic chemicals, an analysis of organic chemicals (pesticides and their derivatives) should be performed for the following: endrin; lindane; methoxychlor; toxaphane; 2,4-D; 2,4,S-TP; and DDT; plus any other pesticides known to be used in the area (3). A composite analysis for all chlorinated hydrocarbons should also be performed. The HACCP action level for 2,4-D is 1.0 ppm for all fish (4).

In the past 50 years, many kinds of chemical pesticides have been released into the environment and are considered pollutants. The world has many insects and it is understandable that these chemicals are being used to successfully carry on agricultural activities. If the agricultural activities in an area require pesticides, then pesticides do pose a possible pollutant source, most likely during the rainy season when runoff from farming may occur. Aquatic life is exposed to these compounds because the rivers, estuaries, bays, and oceans often behave as a "sink" or receptacle for these compounds. Some pesticides, such as certain organochlorines or their metabolites, are slow to break down and thus tend to accumulate in various compartments of the aquatic environment. Shrimp and fish have been found to accumulate certain pesticide compounds in the laboratory, and feral or wild shrimp and fish have possessed detectable levels of compounds when taken directly from contaminated waters. The EPA has found that over several years of testing, penaeid shrimp generally are far more sensitive to toxic and ecological effects of most pesticides than are fish or molluscs because of the shrimp's close relation to insects. Vogt (5) reported further work with monitoring pesticides in aquaculture. The new United States HACCP Program (4) deals with natural toxins, environmental chemical contaminants and pesticides, methyl mercury, aquaculture drugs, toxin formations, and metal inclusion in seafood. Table 1 gives the environmental chemical

Table 1. Environmental Chemical Contaminant and Pesticide Tolerances, Action Levels, and FDA Guidance Levels[a]

Deleterious Substance	Level	Food Commodity	Reference
Aldrin/dieldrin[b]	0.3 ppm	All fish	Compliance Policy Guide, sec. 575.100
Benzene hexachloride	0.3 ppm	Frog legs	Compliance Policy Guide, sec. 575.100
Chlordane	0.3 ppm	All fish	Compliance Policy Guide, sec. 575.100
Chlordecone[c]	0.3 ppm	All fish	Compliance Policy Guide, sec. 575.100
	0.4 ppm	Crabmeat	
DDT, TDE, DDE[d]	5.0 ppm	All fish	Compliance Policy Guide, sec. 575.100
Diquat[e]	0.1 ppm	All fish	40 CFR 180.226
Fluridone[e]	0.5 ppm	Fish and crayfish	40 CFR 180.420
Glyphosate[e]	0.25 ppm	Fin fish	40 CFR 180.364
	3.0 ppm	Shellfish	
Toxic elements			
Arsenic	76 ppm	Crustacea	FDA Guidance Document
	86 ppm	Bivalves	FDA Guidance Document
Cadmium	3 ppm	Crustacea	FDA Guidance Document
	4 ppm	Bivalves	FDA Guidance Document
Chromium	12 ppm	Crustacea	FDA Guidance Document
	13 ppm	Bivalves	FDA Guidance Document
Lead	1.5 ppm	Crustacea	FDA Guidance Document
	1.7 ppm	Bivalves	FDA Guidance Document
Nickel	70 ppm	Crustacea	FDA Guidance Document
	80 ppm	Bivalves	FDA Guidance Document
Methyl mercury	1 ppm	All fish	Compliance Policy Guide, sec. 540.600
Heptachlor/heptachlor epoxide[f]	0.3 ppm	All fish	Compliance Policy Guide, sec. 575.100
Mirex	0.1 ppm	All fish	Compliance Policy Guide, sec. 575.100
Polychlorinated biphenyls (PCB's)[e]	2.0 ppm	All fish	21 CFR 109.30
Simazine[e]	12 ppm	Fin fish	40 CFR 180.213a
2,4-D[e]	1.0 ppm	All fish	40 CFR 180.142

[a]*Source*: Reference 4.
[b]The action level for aldrin and dieldrin are for residues of the pesticides individually or in combination. However, in adding amounts of aldrin and dieldrin, do not count aldrin or dieldrin found below 0.1 ppm.
[c]Previously listed as Kepone, the trade name of chlordecone.
[d]The action level for DDT, TDE, and DDE are for residues of the pesticides individually or in combination. However, in adding amounts of DDT, TDE, and DDE, do not count any of the three found below 0.2 ppm.
[e]The levels published in 21 CFR and 40 CFR represent tolerances rather than guidance levels or action levels.
[f]The action level for heptachlor and heptachlor epoxide is for the pesticides individually or in combination. However, in adding amounts of heptachlor and heptachlor epoxide, do not count heptachlor or heptachlor epoxide found below 0.1 ppm.

contaminant and pesticide tolerances, action levels, and guidance levels suggested in the HACCP program. As can be seen, the deleterious substances of concern are aldrin/dieldrin, benzene hexachloride, chlordane, chlordecone, DDT, TDE, DDE, diquat, fluridone, and glyphosphate. Among the elements considered toxic and of concern are arsenic, cadmium, chromium, lead, nickel, methyl mercury, heptachlor/heptachlor epoxide, mirex, PCBs, simazine, and 2,4-D. Amazingly, some of the action levels are very high. Arsenic, for example, has an action level of 76 ppm for crustaceans and 86 ppm for molluscs. Nickel has an action level of 70 ppm for crustaceans and 80 ppm for molluscs. All deleterious substances and toxic elements have action levels in the ppm (not ppb); therefore, tolerance levels appear to be very high.

Herbicides and pesticides fall under the larger heading and category of insecticides. Fortunately, most of the herbicides and pesticides either are not water-soluble or have a short half-life. The organophosphate Malathion has a half-life of days in an estuarine system rather than months or years. Unfortunately, some of the new insecticides used are much more deadly and have such complicated structures that bacteria are not able to break them down as readily. Therefore, they do what they were designed to do more effectively, and that is to kill. Banana farms are using fungicides to stop black spot, and these are suspected to cause problems for aquatic life in the estuaries when runoff occurs. Areas of insecticide and fungicide use should be avoided during site selection of aquaculture facilities. One must see to it that there is no widespread agricultural activity in the area which would be a potential source of pesticides, fungicides, and herbicides. If this is not done, the effects of the chemicals would indeed most likely show up during the wet season and would negatively affect production.

ORGANOCHLORINES

The following organochlorine pesticides have been studied relative to their ecological effects on fish and shrimp: chlordane, DDT, dieldrin, endrin, heptachlor, hexachlorobenzene, lindane, mirex, and toxaphene (3). For example, the

HACCP action level for heptachlor in all fish is 0.3 ppm, 0.3 ppm for benzene hexachloride in frog legs, and 0.3 ppm for aldrin/dieldrin in all fish (4).

White shrimp that died as a result of DDT exposure (banned in the United States) accumulated up to 40 ppb DDT and DDE (a decomposition product) in the hepatopancreas after 18 days of exposure to 0.20 ppb in flowing seawater. Exposure to a DDT concentration greater than 0.01 ppb was lethal to pink shrimp in 28 days. Sodium and potassium concentrations in shrimp exposed to 0.05 ppb DDT for 20 days were lower than in those not exposed. Magnesium, however, was not significantly lowered. The significance of reduced cation in the hepatopancreas of shrimp for pathophysiological behavior is not known, but a loss of ATPase activity in ion transport may be indicated. Blood protein levels have also been found to drop in shrimp exposed to DDT. In acute, high-concentration laboratory exposures, shrimp and fish showed tumors, hyperkinetic behavior, and paralysis. With DDT poisoning, shrimp did not become paralyzed but sank into lethargy, refused food, and then died (3). The HACCP action level for DDT in all fish is 5.0 ppm. Major problems have occurred mainly when inland ponds were located near row crops such as cotton. Ponds should not be located on land where DDT was used unless soil tests indicate that DDT is no longer there.

Some juvenile aquatic organisms died after laboratory exposure to low concentrations of mirex. However, all survivors from the test died after four days in mirex-free seawater, demonstrating a delayed toxic effect of this pollutant. Mirex poisoning in aquatic organisms produces loss of coordination and equilibrium, and finally, signs of lethargy and paralysis. The HACCP action level for mirex in all fish is 0.1 ppm.

The EPA conducted a detailed ecological monitoring study of the distribution of mirex in oysters, crabs, shrimp, and fishes from estuarine waters, from the Gulf of Mexico to Delaware Bay. Only shrimp in the Savannah, Georgia area had detectable concentrations of mirex (0.007 ppb in tissues). The Savannah, Georgia area has a long history of mirex usage. Mirex does not appear to be as widespread in estuarine regions as are PCBs and DDT. Mirex, usually applied as a particle-bait poison, would not be as directly available to many marine organisms as are broadcast liquids or powder formulations of other pesticides.

The EPA found that 25% of the pink shrimp tested died in the laboratory after seven days of exposure to only 1.0 ppb mirex. The EPA (3) also found that toxaphene affects the early metamorphic stages of pink shrimp and larval fish. The mysis stage of pink shrimp is most susceptible to toxaphene, particularly under various temperature and dissolved oxygen conditions.

ORGANOPHOSPHATES AND CARBAMATES

The organophosphate compounds tested have been shown to be approximately 1000 times more toxic to shrimp than most other pesticides tested, with penaeid shrimp showing greater sensitivity than fishes or molluscs (3).

Among compositions tested, baytex was toxic to penaeid shrimp in the laboratory. Naled (1,2-dibromo-2, 2-dichlorethyl phosphate) had little effect in field tests on shrimp. Malathion, at 14 ppb, caused hyperactivity, paralysis, and death in penaeids. Parathion's lethal concentration for 48 hours in pink shrimp was 0.2 ppb.

In the 1970s it was found that malathion, aerially applied to flooded marshes in Texas, caused from 14 to 80% mortality in brown and white shrimp held in cages. It was recommended that malathion not be applied to flooded marshes that maintained shrimp. It was also found that caged pink shrimp were killed when they received malathion via thermal fogging for mosquito control in salt marshes. In the 1980s Ralph Gouldy's hatchery in the Florida Keys had problems with aerial mosquito spraying of malathion and had to operate the facility as a closed system. The hatchery now operates under the name GMSB—Shrimp Culture, Inc., under the same conditions, producing postlarvae for a large Honduras farm. Additionally, the state of Florida will not allow them to draw directly from the bay system or to discharge directly into the bay system. They draw their water from wells, use a reverse osmosis system to eliminate hydrogen sulfide, and discharge into a series of settling ponds. The system is considered closed, and outside sources of pollution can therefore be avoided to some extent.

Both organophosphates and carbamates are reported to be potent growth inhibitors in the invertebrates and vertebrates. Inhibition as high as 75% was found in moribund shrimp experimentally exposed to malathion. Carbamate pesticides (e.g., sevin) were found to be quite toxic to aquatic animals in laboratory tests. Brown shrimp have a 96-hour LC_{50} (the concentration at which 50% of the organisms were killed) of 2.5 ppb, the lowest concentration of sevin reported to kill any crustacean tested.

Some naturally occurring insecticide chemicals such as pyrethrum, nicotine, rotenone, hellebore, ryania, and sabadilla could also be potential pollutants. Rotenone is a crystalline ketone insecticide extracted from roots of certain plant species in the bean family and is commonly used in ponds to kill fish (by blocking oxygen transport). Rotenone is a restricted use pesticide in the United States and could be considered a pollutant if not used properly because it may linger in cooler water for up to one month, it is commonly dissolved in diesel fuel for application, and this combination adds another pollutant to the environment.

PETROLEUM AS A POLLUTION THREAT TO AQUACULTURE

Annual spillage of oil and oil products into the world's oceans is over 4 million tons (not including the occurrence of occasional megaspills such as the Alaskan spill by Exxon). Most of these spills occur in estuaries or near coastal regions of significant biological value. Of particular importance is the fact that many coastal areas affected by oil spills (including potential spill areas) are in fish- and shrimp-producing regions and are prime sites for aquaculture activities. Detailed studies of ecological and physiological effects of oil on many aquatic animals have been published (3).

Almost all boats, including many sailing vessels with either inboard or outboard motors, use petroleum products and are potential pollution hazards. There are usually one or more deep ports in every country with a coastline that handles petroleum products. Barge traffic and other shipping occurs off the coasts of all countries. Even small boats can cause problems with water pollution. It has been found that when fish and penaeid shrimp are experimentally exposed to oil-contaminated seawater, they accumulate hydrocarbons in their tissues. Brown shrimp (*Farfantepenaeus aztecus*), exposed to oil in seawater and then depurated in clean seawater, released accumulated hydrocarbons more rapidly than clams (*Rangia cuneata*) and oysters (*Crassostrea virginica*). Shrimp can metabolize the hydrocarbons whereas molluscs have a limited, if any, capability to do so.

Couch (3) presented results of tests using the water-soluble fractions and oil-in-water dispersions of four oils (South Louisiana crude, Kuwait crude, refined No. 2 fuel, and Bunker C residual), against brown shrimp postlarvae, for toxicity. The tests showed that fractions of refined oils were generally more toxic to penaeid shrimp and to other aquatic species than were the fractions of crude oils. The crustacean species tested, including brown shrimp postlarvae, were more sensitive to oil fractions than fish species tested (*Cyprinodon* sp., *Menidia* sp., and *Fundulus* sp.). The 24-hour median tolerance limit (the concentration at which the median number of shrimp survived) of juvenile brown shrimp exposed to components of No. 2 fuel oil (naphthalenes, methyl naphthalenes, and dimethyl naphthalenes), ranged from 0.77 to 2.51 ppm. The naphthalenes were the most toxic water-soluble components of fuel oils to shrimp.

Mills and Culley (6) determined acute toxicity of four oils and two oil-spill dispersants on shrimp and fish and found that the dispersants were actually more toxic than the oil, with the 48-hour LC_{50} for the four oils being 1–40 ppt and 40-hour LC_{50} for dispersants being 2.5–5,000 ppm.

HEAVY METALS IN WATER CAUSING POLLUTION

A variety of heavy metals are found naturally in freshwater and seawater environments. These metals may exist in several oxidation states, with different reaction potentials, depending on their specific properties. Certain heavy metals are pollutants generated by industry, and some may be acted upon by microbes to produce other compounds accumulated by organisms, and which are potent toxicants.

Heavy metals occur naturally in the aquatic environment as a result of weathering and land drainage. Additionally, the use of various pesticides and fungicides, which contain metals, has added large quantities of heavy metals to the aquatic environment. Excessive additions of heavy metals to the aquatic environment could have an adverse effect both on animals and on humans who use these animals as food. There are a number of reports on the toxicity of heavy metals to aquatic animals (7–13). The biological effects of metals are complicated by their interactions with other metals (14) and by metal speciation (15). For example, zinc influences the toxicity of cadmium (16), and chelating agents such as tris, NTA, and EDTA reduce the toxicity of metals by sequestering reactive species (17–20). EDTA has been used for a variety of purposes (21), including as a chelator of heavy metals in aquaculture hatcheries to avoid toxic effects of metals to crustaceans during early stages of life.

Many studies have been carried out to determine the bioaccumulation and toxicity of heavy metals to aquatic organisms (e.g., 10,11,18,22,23). Chen and Liu (24) found that heavy metal concentration in *Artemia* nauplii increased linearly with an increase in the heavy metals in the water.

The heavy metals that can be methylated, such as mercury, tin, palladium, platinum, gold, and thallium, pose special threats as environmental pollutants. Other metals, such as cadmium, lead, and zinc, do not form stable alkyl-metals in aqueous solutions, but may have different modes of toxic action. The following is a description of various toxic metals that pose risks to aquaculture.

Cadmium

Biologically, cadmium (Cd) is generally recognized as a nonessential, nonbeneficial element with high toxicity potential. It is deposited and accumulated in various body tissues. Cadmium occurs chiefly as a sulfide salt, frequently in association with zinc and lead. Then EPA criteria are 0.01 ppm for domestic water supply, 0.04 ppm for freshwater aquatic life, and 0.05 ppm for marine aquatic life. This metal is a pollutant from several industrial effluents into aquatic systems. The HACCP action level for Cd is 3.0 ppm in crustaceans and 4.0 ppm in molluscs. Castille and Lawrence (19) found that Cd in a concentration of 0.2 ppm was lethal to *Litopenaeus stylirostris* nauplii.

The Cd concentration in seawater, at 35 ppt salinity, with a standardized nutrient level, is 1.1×10^4 ppm (Cd^{2+}) (25,26). Fujimura (27) reported hatchery water samples from Sabah, Malaysia with Cd levels of 0.05 ppm. Colt and Huguenin (28) recommend a Cd level of <0.005 ppm for aquaculture hatchery water. The Marine Products Export Development Agency (MPEDA) of India states that the 96-hour LC_{50} Cd level for postlarval shrimp is 0.42–0.8 ppm.

Iron

Dissolved iron (Fe) is not toxic to shrimp, but when it precipitates it can cause problems. Water treatment should be done by oxidizing the iron or by aeration of the water and filtering the precipitate before exposing animals.

Useful information concerning iron and other heavy metals in seawater can be found in Chen et al. (29).

Fe levels should be below 0.01 ppm to be in the optimum range for shrimp culture and below 1.0 ppm for any kind of shrimp production to occur. Fe combines with oxygen to give a rusty or orange color. This process uses up oxygen, and if not properly aerated, Fe forms a precipitate that can clog the gills of cultured organisms (Fe^{2+} is converted to Fe^{3+} which precipitates as $FeOH_3$). The 96-hour median tolerance limit (TL_m 96) of Fe to aquatic insects, mayflies, stoneflies, and caddisflies is 0.32 ppm.

In another study, Fujimura (27) reported hatchery water samples to contain 0.23 ppm Fe from Sabah, Malaysia and 0.3 ppm from Guam water samples. Natural seawater has been reported to have 0.02 ppm trace metal Fe. Colt and Huguenin (28) recommended Fe levels of <0.3 ppm for hatchery water. Care must be taken when taking the water sample for analysis. When sediment is in the water sample, Fe can and will leach into the sample before it is processed; therefore, the results obtained may not be a true indication of Fe levels.

Copper

Couch (3) reported that a copper (Cu) concentration of 0.5 ppb was lethal to nauplii, protozoea, and mysis of *Farfantepenaeus aztecus* and *Farfantepenaeus duorarum* that were exposed in a seawater–brine mixture similar to that derived from desalination plants. The same larval stages were able to grow normally in seawater (35 ppt) containing 0.025 ppm Cu. Kumaraguru et al. (23) reported a 96-hour LC_{50} of 0.57 ppm Cu to *Meretix casta*. Cu is toxic to juvenile fish and crustaceans. Criteria (upper limits) are 1.0 ppm for domestic water supply, and 0.1 times the 96-hour LC_{50} for freshwater and marine aquatic life.

Normal seawater contains 0.04–0.1 ppm Cu. Fujimura (27) reported 0.03 ppm in hatchery water from Sabah, Malaysia and Yuan et al. (30) reported 0.3–2.1 ppm levels of Cu from Qingdao, China water. Colt and Huguenin (28) recommend a Cu level of <0.003 ppm for hatchery water. Cu occurs as a natural or native metal in various mineral forms. Oxides and sulfates of Cu are used for pesticides, algaecides, and fungicides. Cu is present in seawater at a concentration of approximately 3 ppb (26,31). Cu is toxic to oysters at concentrations above 100 ppb, and to clams at levels of 20 ppb. The minimum reported concentration of Cu that begins to exhibit toxicity to some agricultural vegetation is 100 ppb. The toxicity of Cu to aquatic life is dependent on the alkalinity of the water, as the Cu ion is complexed by anions present. At lower alkalinity levels, Cu is generally more toxic to aquatic life. Other factors affecting toxicity are pH and the presence of organic compounds.

Mercury

Mercury (Hg) is widely distributed in the environment, but it is generally considered to be a nonessential and nonbeneficial element to organisms. The EPA's criteria are 2.0 ppb for domestic water supply, 0.05 ppb for freshwater aquatic life and wildlife, and 0.1 ppb for marine aquatic life.

Hg itself (as a metal) does not appear to have toxic effects on organisms; however, mercuric salts and methylated Hg are extremely toxic, with both short- and long-term chronic effects (3). These mercuric products were partly responsible for the 1988 closing of the area to fishermen around the Alcoa bauxite plant on Lavaca Bay, Texas. The area eventually became an EPA superfund cleanup site which cost the U.S. government millions of dollars. The HACCP action level for methyl Hg in fish is 1.0 ppm (4).

Zinc

Zinc (Zn) in the water column has been shown to cause a decrease in the rate of oxygen consumption of the freshwater shrimp (32). Correa (10) reported a differential reduction in respiration and ammonia excretion in shrimp, with static 96-hour LC_{50} values for Zn at 0.2 mg/L (ppm). Normal seawater has a level of 0–0.1 ppm Zn (8). As a comparison, Fujimura (27) reported Zn levels in hatchery water from Sabah, Malaysia at 0.06 ppm and at 0 levels in Hawaii and Guam, and Yuan et al. (30) reported 0.9–3.2 ppm in Qingdao, China water samples. Colt and Huguenin (28) reported that Zn should be at <0.05 ppm in shrimp hatchery water.

Lead

In addition to its natural occurrence, lead (Pb) and its compounds may enter and contaminate the global environment at any stage during mining, smelting, or industrial use. Pb enters the aquatic environment through precipitation, fallout, erosion, leaching, and industrial waste discharge. Criteria are 50 ppm for domestic water supply, and 0.01 times the 96-hour LC_{50} values for aquatic species.

Forstner and Wittman (8) reported the Pb level in normal seawater was between 0.005 and 0.01 ppm. The HACCP action level for Pb is 1.5 ppm in crustaceans and 1.7 ppm in molluscs (4).

Magnesium

Riley and Chester (33), as well as other sources, have reported magnesium levels in normal seawater (35 ppt) between 1,090 and 1,350 ppm. The level of Mg is dependent upon the salinity level. Generally, the higher the salinity the higher the Mg. A low salinity site of 8 ppt showed a Mg level of 285 ppm, and another with salinity of 15 ppt had a Mg level of 607 ppm. However, Mg is rarely found at high enough levels to be considered a pollutant.

Aluminum

Spotte (26) reported that 0.01 ppm aluminum (Al) is found as a trace metal in normal seawater. Water in an area with much industrial development might have elevated Al levels and should be checked. Although high levels of Al may be toxic to animals, little is known regarding its toxicity to aquatic species.

Manganese

Manganese (Mn) levels of less than 0.05 ppm are recommended by Colt and Huguenin (28) for aquaculture hatcheries. This metal is a required nutrient for aquatic animals, but it can be toxic to aquatic animals at relatively higher concentrations.

Nickel

Colt and Huguenin (28) recommended that hatchery water contain <0.05 ppm nickel (Ni). However, the HACCP action level for Ni is 70 ppm in crustaceans and 80 ppm in molluscs (4).

Chromium

Chromium (Cr) is a relatively abundant element in aquatic ecosystems. Fish are relatively tolerant to Cr, but some aquatic invertebrates are quite sensitive. The EPA's criteria are 50 ppm for domestic water supply and 100 ppm for aquatic life. Colt and Huguenin (28) recommended hatchery seawater levels at <0.025 ppm Cr. The HACCP action level for Cr is 12 ppm in crustaceans and 13 ppm in molluscs (4).

Silver

Biologically, silver (Ag) is a nonessential, nonbeneficial element recognized as causing localized skin discoloration in humans and as being systemically toxic to aquatic life. Criteria are 50 ppm for domestic oyster supplies, with a 96-hour LC_{50} of 0.01 ppm for many aquatic species.

Arsenic

Arsenic (Ar) has many diversified industrial uses such as hardening of Cu and Pb alloys, pigmentation in paints and fireworks, and the manufacture of glass, cloth, and electrical semiconductors. Arsenicals are also used in the formulation of herbicides. The EPA's criteria are 0.05 ppm for domestic water supply and 0.1 ppm for irrigation of crops. The HACCP action level is 76 ppm in crustaceans and 86 ppm in molluscs (4).

Barium

Barium (Ba) is a yellowish-white metal of the alkaline earth group and many of its salts are soluble in water or acid. It has various applications in the metallurgic, paint, and electronic industries. The EPA's criterion is 1 ppm for domestic drinking water. Experimental data show that soluble Ba concentrations in marine water would have to exceed 50 ppm before toxicity as opposed to nonlife would be expected.

Selenium

Biologically, selenium (Se) is an essential, beneficial element recognized as a metabolic requirement in trace amounts for animals, but toxic to them in amounts ranging from 0.1 to 10 ppm in food. The EPA's criteria are 10 ppb for domestic water supply, with a 96-hour LC_{50} 0.01 for sensitive resident aquatic species.

OTHER POLLUTANTS

Among metals, cyanides can be toxic depending upon the temperature, pH, and oxygen level in the water. Free cyanide can occur as hydrogen cyanide and is very toxic to fish if above 0.01 ppm in saltwater. According to the EPA, any cyanide level above 0.005 ppm can cause problems in freshwater.

Other forms of aquatic pollution that may affect aquaculture include urban usage of pesticides and petroleum products, commonly from storm water runoff, industrial pollution, and thermal pollution from power plants (34). Additionally, detergents containing phosphates can cause algal blooms that result in low dissolved oxygen.

Phosphorus appears to be more of a pollutant problem in freshwater than in saltwater. Phosphorus is the nutrient limiting algal growth and eutrophication in natural systems, and it is sometimes regulated in the freshwater aquaculture industry.

Further reading on the toxicology of aquatic pollution, especially the impact on fish biology of nitrite and oestrogenic substances, can be found in Taylor (35). Topics discussed are as follows: (*1*) water chemistry at the gill surface of fish and the uptake of xenobiotics; (*2*) bioaccumulation of waterborne 1,2,4,5-tetrachlorobenzene in tissues of rainbow trout; (*3*) dietary exposure to toxic metals in fish; (*4*) the physiology and toxicology of zinc in fish; (*5*) lethal and sublethal effects of copper upon fish: a role for ammonia; (*6*) the physiology and status of brown trout exposed to aluminum in acidic soft waters; (*7*) physiological and metabolic costs of acclimation to chronic sublethal acid and aluminum; (*8*) physiological effects of nitrite in teleosts and crustaceans; (*9*) metallothioneins in fish; (*10*) oestrogenic substances in the aquatic environment and their potential impact on animals, particularly fish; (*11*) the effect of genetic toxicants in aquatic organisms; (*12*) *in vitro* toxicology of aquatic pollutants: use of cultured fish cells; and (*13*) principles governing the use of cytochrome P-450 measurement as a pollution monitoring tool in the aquatic environment.

For additional information on aquaculture pollution, Ziemann et al. (36) characterized aquaculture effluents in Hawaii, Hastings and Heinle (37) discussed the effects of aquaculture in estuarine environments, and Piedrahita (38) and SEAFDEC (39) discussed managing environmental impacts in aquaculture. For additional information on pesticides, see (40).

BIBLIOGRAPHY

1. R.R. Stickney, *Aquaculture*, John Wiley, New York, 1994.
2. R.M. Overstreet, *Aquatic Tox.* **11**, 213–239 (1988).
3. J.A. Couch, in L.W. Hart and S.H. Fuller, eds., *Pollution Ecology of Estuarine Invertebrates*, Academic Press, New York, 1979.
4. Guidance on Seafood HACCP, *Fish & Fishery Products Hazards and Controls Guide*, 2nd ed., 1998, Ch. 9, p. 99, Food and Drug Administration, Washington, DC. Also available on the Internet at http://www.verity.fda.gov/
5. G. Vogt, *Aquaculture* **67**, 157–164 (1987).
6. E.R. Mills and D.D. Culley, *Proc. SE Assoc. of Game and Fish Comm.* **25**, 642–650 (1971).
7. D.W. McLeese, *J. Fish Res. Board Can.* **31**, 19–49 (1974).
8. U. Forstner and G.T.W. Wittmann, *Metal Pollution in the Aquatic Environment*, Springer-Verlag, Berlin, 1979.
9. D.W. McLeese and S. Ray, *Bull. Environ. Contam. Toxicol.* **36**, 749–755 (1988).
10. M. Correa, *Environ. Pollution* **45**, 149–155 (1987).
11. J. Del Ramo, J. Diaz-Mayans, A. Torreblanca, and A. Nunez, *Environ. Contam. Tox.* **38**, 736–741 (1987).

12. D.A. Holwerda, J. Hemelraad, P.R. Veenhof, and D.I. Zandee, *Bull. Environ. Contam. Tox.* **40**, 373–380 (1988).
13. D. Nugegoda and P.S. Rainbow, *J. Mar. Biol. Assoc. U.K.* **68**, 25–40 (1988).
14. P.L. Foster and F.M.M. Morel, *Limmol. Oceanogr.* **27**, 745–752 (1982).
15. C.D. Zamuda and W.G. Sunda, *Mar. Biol.* **66**, 77–82 (1982).
16. S. Dunlop and G. Chapman, *Environ. Res.* **24**, 264–274 (1981).
17. D.W. Engel and W.G. Sunda, *Mar. Biol.* **50**, 121–126 (1979).
18. S. Muramoto, *Bull. Environ. Contam. Toxicol.* **25**, 828–831 (1980).
19. F.L. Castille and A.L. Lawrence, *Compar. Biol. Chem. Physiol.* **70A**, 525–528 (1981).
20. A.L. Lawrence, J. Fox, and F.L. Castille, *J. World Maricult. Soc.* **12**(1), 271–280 (1981).
21. H.F. Mark, J.J. McKetta, and D.F. Othmer, *Kirk–Othmer Encyclopedia of Chemical Technology*, Wiley-Interscience, New York, 1965.
22. M.L. Evans, *Bull. Environ. Contam. Tox.* **24**, 916–920 (1980).
23. A.E. Kumaraguru, D. Selvi, and V.K. Venugopalan, *Bull. Environ. Contam. Tox.* **24**, 853–857 (1980).
24. J.C. Chen and P.C. Liu, *J. World Maricult. Soc.* **18**, 84–93 (1987).
25. M.S. Quinby-Hunt and K.K. Turekian, *Oceanogr. Rep. Eos, Trans. Am. Geophys. Union.* **64**, 130–131 (1983).
26. S.H. Spotte, *Water Management in Closed Systems*, Wiley-Interscience, New York, 1970, pp. 90–91.
27. T. Fujimura, in D. Akiyama, ed., *Proc. SE Asia Shrimp Farm Management Workshop*, American Soybean Assn., 1989, pp. 22–41.
28. J. Colt and J. Huguenin, in A. Fast and J. Lester, eds., *Marine Shrimp: Principles and Practices*, Elsevier, New York, 1990.
29. J.C. Chen, Y.Y. Ting, H. Lin, and T.C. Lian, *J. World Aquacult. Soc.* **16**, 316–332 (1985).
30. Y.X. Yuan, C.N. Gao, and D.X. Zhang, *J. World Aquacult. Soc.* **23**, 205–210 (1992).
31. K.W. Bruland, *Earth Planet. Sci. Lett.* **47**, 176–198 (1980).
32. B. Chinnayra, *Indian J. Exp. Biol.* **9**, 277–278 (1971).
33. J.P. Riley and R. Chester, *Introduction to Marine Chemistry*, Academic Press, London, 1971.
34. E.A. Laws, *Aquatic Pollution*, John Wiley, New York, 1981.
35. E.W. Taylor, ed., *Toxicology of Aquatic Pollution*, Cambridge University Press, Cambridge, U.K., 1996.
36. D.A. Ziemann, W.A. Walsh, E.G. Saphore, and K. Fulton-Bennett, *J. World Aquaculture Soc.* **23**(3), 180–191 (1992).
37. R.W. Hastings and D.R. Heinle, *Estuaries* **18**(1A), 1–20 (1995).
38. R.H. Piedrahita, *Bull. Natl. Res. Inst. Aquacult.* Suppl. **1**, 13–20 (1994).
39. SEAFDEC, *Aqua Farm News*, Vol. XIII (No. 2), Southeast Asian Fisheries Development Center, Ilollo City, Philippines, 1995.
40. Pesticide Analytical Manual (1968 and revisions), Vol. I (3rd ed., 1994) and II (1971), Food and Drug Administration, Washington, DC (available from National Technical Information Service, Springfield, VA 22161).

See also FERTILIZATION OF FISH PONDS; PESTICIDES.

POLYCULTURE

ROBERT R. STICKNEY
Texas Sea Grant College Program
Bryan, Texas

OUTLINE

Polyculture is the rearing of two or more species in the same culture system. In most cases, it involves two or more species of aquatic animals, but it could also mean rearing aquatic animals in conjunction with terrestrial or aquatic plants.

HISTORICAL PERSPECTIVE

Polyculture is not a new concept. The traditional method of rearing carp in China involves the culture of several species. (See the entry "Carp culture.") If the proper mixture of fish species is stocked, all of the available natural food in a pond will be preyed upon by a fish that has value as human food. In China, polyculture has always involved carp species almost exclusively, at least until recently, when such fish as tilapias (see the entry "Tilapia culture") were introduced into polyculture ponds. The practice goes back millennia and continues to exist today. Several species of carp are stocked, each of which has different food habits. Mud carp or common carp are benthos feeders, silver carp feed primarily on phytoplankton, bighead carp feed on zooplankton, and grass carp feed on higher aquatic plants (e.g., macrophytes) or on plant material supplied by the fish farmers (e.g., agricultural wastes). Ponds are fertilized with livestock manure and night soil. Aside from the previously mentioned agricultural wastes, no supplemental feed is provided.

In recent years, polyculture has been a concept that has been extended to various aquatic species. In most cases, it involves rearing compatible species that do not compete for food and that have commercial value. It may, however, also involve rearing of a noncommercial species that provides a service to the fish farmer. For example, predators may be stocked in tilapia ponds to consume fry and reduce overcrowding. (See the entry "Tilapia culture.") In Norway, wrasses have been stocked to control sea lice on Atlantic salmon (1).

The practice of using culture systems for the culture of fish or invertebrates in conjunction with hydroponically grown vegetables is another form of polyculture. Rice–fish culture, the rearing of fish in rice paddies, has been practiced at least for decades in some countries and can also be considered a form of polyculture.

ANIMAL POLYCULTURE

Multispecies carp culture in China continues to be practiced, and while that approach has not been widely adopted elsewhere, some research has been conducted that involved polyculture of two or more species used routinely by the Chinese. In addition, aquatic animal polyculture of various species, sometimes in combination with carp, has been adopted in a variety of situations. Grass carp (*Ctenopharyngodon idella*), one of the species used in Chinese polyculture, have found their way into many other aquaculture situations. For example, since aquatic weeds are a nuisance and can lead to oxygen depletions in ponds (see the entry "Pond culture"), their control is often considered not only desirable, but also necessary. (See the entry "Aquatic vegetation control.") Grass carp provide one means of controlling aquatic plants in ponds. Catfish farmers often stock a few grass carp in each of their production ponds, as do the culturists of other freshwater fish species.

There have been many studies on polyculture, but how widely the results of those studies have been adopted by producers is not clear. Among the species that have reportedly been stocked in various types of polyculture situations are the following:

- carp and Australian red claw crayfish;
- channel catfish and fathead minnows in separate ponds with water exchange;
- channel catfish and tilapias;
- channel catfish with Asian clams;
- common carp and *Colossoma* sp.;
- common carp and walking catfish;
- grass carp and *Colossoma* sp.;
- Indian carp and various other carp species;
- mullets and carp;
- salmon and scallops in net pens;
- salmon and sea urchins in net pens;
- tilapias and various species of carp (carp singly or in combinations);
- tilapias, common carp, and walking catfish;
- tilapias, common carp, grass carp, and pacu;
- tilapias and walking catfish;
- tilapias and mullets;
- tilapias and *Colossoma* sp.;
- tilapias and snakehead;
- tilapias, common carp, and freshwater shrimp;
- tilapias and freshwater shrimp;
- tilapias and Australian red claw crayfish;
- crayfish and snails;
- freshwater shrimp, grass carp, and *Colossoma* sp.;
- freshwater shrimp and bighead carp;
- freshwater shrimp and various species of carp (including Indian carp species);
- freshwater shrimp with giant gouramis;
- freshwater shrimp and mullets;
- marine shrimp and mullets;
- marine shrimp and milkfish;
- marine shrimp and rabbitfish;
- marine shrimp of various species;
- marine shrimp, abalone, and mussels;
- marine shrimp and clams;
- marine shrimp and scallops;
- marine shrimp, oysters, and clams;
- scallops and sea urchins;
- oysters and sea squirts.

ANIMAL/PLANT POLYCULTURE

There are significantly fewer examples of aquatic animals being polycultured with seaweeds than of aquatic animals being cultured with other aquatic animals. The following are a few examples of the former:

- marine fish with bivalves and seaweed;
- marine shrimp and seaweed;
- scallops, sea urchins, and seaweed.

More common is the culture of aquatic animals with higher plants and the use of plants for water treatment in aquatic animal culture systems.

One of the most widely used polyculture systems involving higher plants and aquatic animals is rice–fish culture. The practice of culturing rice and fish together is not new; it has been practiced for decades, if not centuries, in various parts of the world. Rice is a terrestrial plant that is routinely grown in flooded fields (paddies). Fish can also be reared in rice paddies. The most common species reared in rice–fish culture are in the group of species commonly referred to as tilapias (see the entry "Tilapia culture"), though other species, including various types of carp, also lend themselves to the practice.

In recent years, there has been a good deal of interest in the development of fish or aquatic invertebrate culture in conjunction with hydroponic fruit and vegetable production. Lettuce, tomatoes, cucumbers, and other fruits and vegetables have been reared in water that is also used for aquatic animal production. Tilapias are commonly used as the aquatic animals in such systems (2).

The waste from aquatic animals is not very rich in nutrients in comparison with organic wastes obtained from birds and mammals, so some augmentation of nutrients is often necessary. Water from aquaculture systems can also be used to irrigate and provide partial fertilization for row crops and ornamental plants, but that approach stretches the definition of polyculture. Sea lettuce (a type of marine algae), water hyacinths, and marsh grasses are among the plants that have been used to reduce the nutrient loads in aquaculture effluents before the water is released into natural environments. Some plants can be used as livestock food, though because of their high water content, they should be fed at a location immediately adjacent to the aquaculture facility, as hauling them any distance can be inordinately expensive.

BIBLIOGRAPHY

1. M.D.J. Sayer, J.W. Treasurer, and M.J. Costello, eds., *Wrasse Biology and Use in Aquaculture*, 1996.
2. J.E. Rakocy, in B.A. Costa-Pierce and J.E. Rakocy, eds., *Tilapias Culture in the Americas*, Vol. 1, World Aquaculture Society, Baton Rouge, LA, 1997, pp. 163–184.

See also CARP CULTURE.

POMPANO CULTURE

STEVEN R. CRAIG
Texas A&M University
College Station, Texas

OUTLINE

Recent interest in culturing Florida pompano (*Trachinotus carolinus*) commercially has resulted in renewed efforts to predictably spawn the species in captivity. Although pompano were the subject of many studies beginning in the early 1960s, several failed attempts at culture on a commercial level were experienced in the 1970s. Several factors are behind the renewed interest in the species. First, there has been a recent desire to culture more highly valued marine fish. Pompano is considered one of the most desirable and valuable table fish from tropical U.S. waters and sells in the United States for $2.00–3.50/kg ($4.00–8.00/lb) in the round, among the highest prices accorded any marine fish (1,2). Additionally, the U.S. pompano fishery has declined dramatically in recent years, the total catch in 1988 being just one tenth that of 1976 (3). Finally, pompano are hardy animals, able to tolerate a wide range of salinities, low dissolved oxygen concentrations, and high turbidity (1). These characteristics make them appealing candidates for the high-density recirculating systems that are becoming increasingly popular. Reliable and predictable spawning of Florida pompano has not yet been attained, although renewed efforts are underway by several major universities as well as enterprises in the private sector.

LIFE HISTORY

A member of the carangid, or jack, family, the pompano is found in an area ranging from North Carolina to Brazil. It is a small fish, with maximum length and weight of approximately 63.5 cm (25 in.) and 3.5 kg (7.7 lb), respectively (4). Pompano are found in the surf zone along beaches on the Gulf coast of the United States, and the appearance of juvenile fish throughout spring, summer, and fall indicates a prolonged spawning season (5). Additionally, spawning is believed to occur offshore, and in Florida, mature fish move to the shore in mid-April to mid-May. "Waves" of larvae appear about one month later and continue to show up through late October (4,6). Pompano generally spawn from April to October, with a peak in April and May (7), although spawning seasons vary according to geographic location. Shorter spawning periods are observed in the northern parts of the range, with potentially year-round spawning in the tropical parts of the range (8). Based on those observations, it appears as if pompano are continuous spawners, as is the case with many subtemperate and tropical marine species. Continuous-spawning fish are desirable in aquaculture, because they can potentially produce fry for stocking growout systems year-round (9).

Juvenile pompano are easily caught on the beaches of the Gulf coast of Mexico from late April until October, with fish staying in the surf zone until they reach approximately 60–80 mm (2–3 in.) in length (6), after which they migrate into deeper, offshore waters (4). Juvenile pompano are thought to be opportunistic feeders. As adults, they become more selective, consuming small fish, crustaceans, and clams, especially coquina clams (5).

Growth rates of pompano in the wild have been difficult to estimate. Several studies have investigated length-to-weight relationships (5,6,10). For example, Fields (6) and Finucane (5) estimated growth rates in wild fish at approximately 22 mm (0.8 in.) per month, while Bellinger and Avault (10) observed somewhat higher rates ranging from 27 to 42 mm (1.0 to 1.7 in.) per month, with a mean of 36 mm (1.4 in.).

Environmental Requirements

Pompano are tolerant of a wide range of environmental conditions. They thrive in salinities ranging from 0 to 35 g/L (ppt) (1) and adjust well to lower salinities if acclimated properly (4). Higher salinities are necessary in a hatchery application for the release, buoyancy, and survival of their eggs.

Temperature is a major constraint on pompano culture, as the species is relatively cold intolerant (1,4,5,11), with thermal stress and mortalities occurring in the range of 10–12 °C (50.0–53.6 °F) (5,11). This factor restricts the potential for outdoor culture to areas with winter temperatures higher than 12 °C (53.6 °F), but cultured pompano have recovered from temperatures as low as 9.7 °C (49.5 °F) (4). Growth and survival appear to be maximum at temperatures ranging from 26–30 °C (78.8–86.0 °F), although juveniles have been observed to thrive at a temperature of 34 °C (93.2 °F) (11).

In terms of their dissolved oxygen and pH requirements, pompano appear to be similar to other marine species currently being cultured. They can survive dissolved oxygen levels as low as 3.0 mg/L (ppm) for short periods of time, although levels at or below 2.5 mg/L (ppm) are lethal. Pompano can survive in water with pH ranging from 5–10 (4,11) and would likely survive any pH commonly encountered under culture conditions. Pompano also are known for their ability to survive in waters of high turbidity. These environmental characteristics make pompano an attractive candidate for commercial culture. Their capability to thrive in a wide range of environmental conditions also makes them a promising candidate for high-density recirculating systems. However, the major constraint on the development of such a system is the large capital outlay necessary to build and equip a commercial facility of this type. However, the ability to produce a product that commands such a traditionally high market price as pompano would be of great value to the aquaculture industry. Additionally, the use of indoor, high-density systems to culture pompano commercially would negate their relatively cold-intolerant nature as a factor limiting the development of an aquaculture industry focused on the species.

SPAWNING IN CAPTIVITY

Despite the attractiveness of pompano from a marketing and culturing standpoint, as of this writing, there has been little or no success in producing these fish economically via commercial aquaculture, a situation that is due primarily to the limited availability of young fish and to reduced growth efficiency of fish at advanced sizes (1). There were several commercial attempts at rearing pompano under aquacultural conditions in the 1960s and early 1970s in Florida and in the Dominican Republic (1). Those attempts generally met with failure, although McMaster (3) reported success with the culture of pompano by Oceanography Mariculture Industries in the Dominican Republic. That business reportedly developed reliable hatchery techniques, using gonadotropin injections and strip-spawning (i.e., manually expressing eggs from the female and fertilizing the eggs with milt from the male), but the growout stage had problems related to the design of the culture system and drastic declines in feed conversion as fish approached a size of 200 g (0.4 lb) (2). None of the techniques used by that venture have been published.

Numerous reports of spawning with the aid of human chorionic gonadotropin (HCG) injections by researchers in the 1970s (7,8,12) helped maintain interest in pompano culture. To date, however, there have been no published reports of successful spawning and culturing of pompano on a commercial level, and there have been no reports of a spawning regime for pompano that does not include hormone injections or implants. Several research institutions associated with universities, as well as a few enterprises in the private sector, have renewed attempts to close the life cycle of pompano. At the time of this writing, several laboratories in Texas, Louisiana, and Florida have pompano undergoing manipulation of photoperiod and temperature, in an effort to naturally spawn the fish, as is routinely done with red drum (*Sciaenops ocellatus*) (13).

Broodstock and Hatchery Conditions

Broodstock are collected by hook and line from the wild. Adult pompano are very active in the surf zones in the spring and summer. However, the fish are very nervous, and great care must be taken to transfer them to the hatchery without harm. Normal procedures for the hauling of any marine broodstock should be followed, including the use of oxygen to maintain dissolved oxygen levels above saturation and an anesthetic such as tricaine methane sulfonate (MS-222) at a level of 1–2 mg/L (ppm) to aid in calming the captured fish. By using these procedures, pompano can be transported for extremely long distances with little stress and minimal mortality. The fish are relatively hardy once relocated to spawning tanks and adapt well to the hatchery environment. They are voracious eaters and will begin to consume frozen squid or penaeid shrimp within two weeks of being brought into the hatchery. Typically, circular tanks 2.5 to 3.6 m (8 to 12 ft) in diameter are best for the hatchery system. Very little is known about proper male-to-female sex ratios or even how many broodfish to include in tanks of various diameters. Presently, 10–16 adult pompano in a tank 3.0 m (10 ft) in diameter appears to be an adequate stocking rate. The fecundity of pompano has been estimated to be between 400,000 to 650,000 eggs per female fish (12). A 1 : 1 or 2 : 3 male : female sex ratio should be sufficient to obtain adequate amounts of fertilized eggs. If significant numbers of eggs are observed as being unfertilized, the number of males in the hatchery tank should be increased.

Environmental Parameters

The hatchery system should include adequate biological filtration for the removal of nitrogenous wastes. Although no specific research on this issue has been conducted, hatchery systems designed for red drum have worked well with pompano. Levels of total ammonia and nitrite nitrogen of more than 0.5 mg/L (ppm) at pH 7.8–8.2 should not be sustained for prolonged periods of time. With proper sizing of the biological filter, and strict maintenance of pH through the addition of sodium bicarbonate, water quality should not pose a problem for the hatchery system. Although pompano are able to live and thrive in waters ranging widely in salinity, the preferred salinity for the hatchery system is 32–35 g/L (ppt), to ensure adequate floatation of the fertilized eggs for collection out of the spawning tank. Typically, water temperature is controlled using a heater and chiller combination, or, more effectively, by ambient-air control. The water temperature must be controlled within $\pm 0.5\,^{\circ}C$. Illumination should be provided by three separate banks of fluorescent lights, each connected to an individual digital timer, enabling the simulation of dawn, full daylight, and dusk, as well as precise control of photoperiod regimes.

Temperature and Photoperiod Spawning Regime for Pompano

Cycling of pompano in a hatchery may be achieved by techniques previously developed for the successful spawning of red drum. Because the beginning of the natural spawning season for pompano occurs in early spring, the cycle used for pompano should begin in summer and end the following spring. Spring-spawning fish typically respond to increases in temperature and photoperiod by beginning final oocyte maturation (9). The spawning cycle described next is presently being evaluated at Texas A&M University.

The cycle begins with conditions simulating summer and progresses through fall and winter, ending in conditions simulating spring. The environmental conditions for summer are 26–30 °C (78.8–86.0 °F) and 15 hours of daylight. The temperature is adjusted daily through the 30-day simulation of summer to temperatures resembling fall [20–24 °C (68.0–75.2 °F), 12 hours of daylight], winter [17–19 °C (62.6–66.2 °F), 9 hours of daylight], and, finally, spring conditions [21–25 °C (69.8–77.0 °F), 12 hours of daylight]. Due to the precise nature of the environmental controls, the changes occur on a daily basis, with each "season" represented by approximately 30 days. Once spring conditions are reached, the temperature and photoperiod are held constant until spawning begins. Minor alterations in temperature and photoperiod are enacted to induce spawning during the final 30-day period.

Other research laboratories are currently exposing groups of adult pompano to a natural year-long cycle, and as of this writing, they also have failed to produce successful spawning. Still other research laboratories are experimenting with hormonal implants. Implants typically are inserted into the broodfish, and the hormone is slowly released over time. Unfortunately, the method is relatively new in the field of aquaculture and has met with limited success. However, it is highly desirable over older manual injection techniques, due to the simplicity of the single event of inserting the hormone implant. Further, the technique greatly reduces stress on broodstock, and there are claims that it promotes more reliable and predictable spawning behavior. Still, there are substantial costs and labor involved with inserting the implant.

Successful Hormone-Induced Spawning

Early attempts to spawn pompano in captivity have met with varied degrees of success, with most methods involving the use of hormonal (gonadotropin) injections, followed by strip-spawning (3,7,8,12). Several successful attempts at spawning pompano occurred in Florida in the 1970s (7,8,12), using HCG injections ranging from 0.176 IU to 0.55 IU/g body weight. Higher dosages appeared to inhibit oocyte growth in response to the injections. The studies involved pompano that had been conditioned with photoperiod and temperature regimes, as well as unconditioned fish. Several successes were reported for techniques using both strip-spawning and seminatural spawning (i.e., allowing the injected female to return to the spawning tank to release eggs naturally). The use of hormone injections to spawn many different species of fish is fairly common, although natural spawning is far less labor intensive, less expensive, and, of course, far less stressful to the broodstock. However, problems associated with the methods used in the aforementioned studies are evident. For example, stress associated with the capture and injection of the broodstock is inevitable, and death of broodstock after strip-spawning is frequent and unavoidable. Clearly, it is desirable to spawn marine fish naturally, but so far that approach has met with no success with pompano. The future success of pompano culture is dependent upon reliable and predictable spawning procedures by any means. Until such procedures can be determined, pompano culture is destined to fail, as it did in the 1960s and 1970s.

CULTURE IN CAPTIVITY

Almost all research involving pompano has been carried out with wild-caught fingerlings, due to the unreliability and unpredictability of spawning adult pompano in captivity. Young juveniles have been easily caught with seine nets along the beaches of the Gulf coast of the southeastern United States (11). In the early 1970s, juvenile pompano were reared in 1-m^3 (35.3-ft^3) aluminum cages in Florida with some success (15). That effort concluded that juvenile pompano could be successfully reared to market size [approximately 454 g (1 lb)] in 47 to 51 weeks, starting with 7-g (0.02-lb) animals. Concrete tanks and wooden cages were used in Venezuela in the late 1970s, with mixed results. The cages were in an area of high salinity and temperature, which resulted in disease outbreaks and high mortality. Both growout studies concluded that several major constraints remained to be overcome before pompano culture could succeed on a commercial level. Included among the constraints were lack of a reliable source of fry and juveniles, lack of a formulated feed that would provide maximum growth, and lack of adequate disease control (1,15).

Diets for the culture of juvenile pompano have been problematic. Attempts in the 1960s to culture pompano involved primitive diets, mostly trout and salmon feed and fresh fish by-products (14). Those early attempts at growout studies, as well as commercial production, all documented a decrease in the feed conversion ratio as the fish approached market size. In some cases, feed conversion ratios went from 2:1 to over 6:1 in growout trials (1). The failure of previous commercial attempts to raise pompano was due primarily to the use of inferior diets that were not specifically formulated for pompano (1).

Pompano are relatively hardy fish and well suited for culture, even in high-density systems (1); however, disease outbreaks in early studies were prevalent. Protozoan parasites such as *Trichodina, Scyphidia* sp., and *Oodinium* sp. are the major external parasites found on wild pompano, while *Vibrio anguillarum* is the primary causative agent of bacterial disease outbreaks involving fin rot and skin ulcerations (16). Under good culture conditions, the most prevalent disease problem encountered in pompano culture is *Oodinium* sp. That gill parasite is exceedingly deadly and will rapidly spread throughout the culture system if the outbreak not controlled (17). *Oodinium* sp. can be easily diagnosed through observation of the fish and microscopic examination of gill filaments. When infected with *Oodinium* sp., pompano "flash" noticeably, by swimming on their sides near the bottom of the culture tank. Gill filaments of infected fish have noticeable brownish–blackish ovoid spots when observed at relatively low (400×) magnification. Although not approved for use as a chemotheraputic for food fish by the U.S. Food and Drug Administration (FDA), copper sulfate and chelated copper have been successfully used to treat pompano stocks by laboratories, which have found these chemicals to be extremely effective for control of *Oodinium* sp. outbreaks. Treatment levels for juvenile and adult pompano should not exceed 0.25 mg free copper/L (ppm). Higher doses irritate the gills of pompano, as evidenced by exaggerated opening and closing of the pompano's mouths and flaring of their gill opercula.

The second most prevalent disease problem with pompano is secondary bacterial infections resulting from wounds obtained in the culture tank. Pompano are very skittish fish, and sudden movements or noises lead to rapid and frenzied activity in tanks, often resulting in bruised or cut heads and other incidental scrapes. Under hatchery conditions with excellent water quality, such wounds normally heal in two to three weeks. However, under more intensive culture conditions, there is a paramount need to prevent secondary infection by bacteria present in the environment. Pompano broodstock have been successfully treated for bacterial infections with oral application of oxytetracycline (OTC). This course of treatment involves dissolving OTC in a gelatin solution (4–5 g OTC/mL of gelatin solution) and injecting this mixture into shrimp, which are then fed to the broodstock. Treatments continuing for four to five days have successfully controlled bacterial infections. If the infection persists after one week, the treatment may be repeated. Experimental treatment of juvenile pompano with OTC at a concentration of 55 mg/kg (ppm) body weight has been successfully achieved by coating commercial feed with OTC in gelatin and feeding it to the fish for one week. The use of drugs in aquaculture is an exceedingly important issue, and FDA approval (in the United States) of safe, therapeutic treatments to aid in the culture of fish is necessary for the future of the industry.

CONCLUSIONS

The three main constraints on pompano culture in the past currently remain: (*1*) The predictable and reliable spawning of broodstock in captivity has not been documented; (*2*) the quantitative nutritional requirements of the species have not been determined; and (*3*) the problems associated with disease have not been suitably addressed. With the current renewed interest in the species, it should not be long until the life cycle is truly closed on pompano and the predictable spawning of pompano becomes a reality. The well-documented nutritional problems, especially those relating to the increase in feed conversion rates, should not be a problem, given the steady advances in fish nutrition over the years. Diets formulated specifically for pompano would not only result in increased production efficiencies, but also aid in resistance to disease and reduction of stress. Additionally, the use or ultraviolet irradiation and, more recently, ozone for the oxidation of known pathogens in culture systems is increasing in the aquaculture community; these procedures may aid in the prevention and control of the disease outbreaks and mortalities that were so devastating to commercial ventures in the past. The price for pompano has historically been one of the highest for any marine fish, and that trend appears to be continuing. As long as the price for the species remains high and the public desires to consume this delectable fish, there will be a strong interest in commercial pompano culture.

BIBLIOGRAPHY

1. D. Jory, E.S. Iverson, and R.H. Lewis, *J. World Maricult. Soc.* **16**, 87–94 (1985).
2. M. McMaster, *Aquacult. Mag.* **14**(3), 28, 30–34 (1988).
3. W.O. Watanabe, in K.L. Main and C. Rosenfield, eds., *Culture of High-Value Marine Fishes in Asia and the United States*, The Oceanic Institute, Honolulu, HI, 1994, pp. 185–205.
4. C. Gilbert, *Biol. Rep.* **82**(11.42), (1986).
5. J.H. Finucane, *Trans. Amer. Fish. Soc.* **98**(3), 478–486 (1969).
6. H.M. Fields, *U.S. Fish Wild. Serv. Fish. Bull.* **207**, 62, 189–222 (1962).
7. F.H. Hoff, C. Rowell, and T. Pulver, *Proc. World Maricult. Soc.* **3**, 53–64 (1972).
8. F.H. Hoff, J. Mountain, T. Frakes, and K. Halcott, *Proc. World Maricult. Soc.* **9**, 279–297 (1978).
9. C.R. Arnold, in C.J. Sindermann, ed., *Proceedings of the Seventh U.S.–Japan meeting on Aquaculture: Marine Finfish Culture*, NOAA Tech. Rep. NMFS 10, Tokyo, Japan, October 3–4, 1978, pp. 25–27.
10. J.W. Bellinger and J.A. Avault, Jr., *Trans. Amer. Fish. Soc.* **2**, 353–358 (1970).
11. J.E. Bardach, J.H. Ryther, and W.D. McLarney, *Aquaculture*, New York Interscience, New York, 1972.
12. F.H. Hoff, T. Pulver, and J. Mountain, *Proc. World Maricult. Soc.* **9**, 299–309 (1978).
13. C.R. Arnold, W.H. Bailey, T.D. Williams, A. Johnson, and J.L. Lasswell, *Proc. Annu. Conf. Southeast. Assoc. Fish Wildl. Agencies* **31**, 437–440 (1977).
14. W.M. Tatum, *Proc. World Maricult. Soc.* **3**, 65–74 (1972).
15. T.J. Smith, University of Miami Sea Grant Technical Bulletin, No. 26, University of Miami, Miami, FL, January 1973.
16. D.W. Coombs, M.S. Thesis, Texas A&M University, College Station, TX, 1974.
17. M.S. Schwarz and S.A. Smith, *Commercial Fish and Shellfish Technology*, Virginia Cooperative Extension, Publication #600-200, 1998.

See also Brine shrimp culture; Larval feeding—fish.

POND CULTURE

Robert R. Stickney
Texas Sea Grant College Program
Bryan, Texas

OUTLINE

Ponds were the first type of culture system developed and continue to be much more widely used than any other type of culture unit (Fig. 1). A primary reason for the global popularity of ponds for culture is that under most circumstances, ponds are the least expensive way to produce aquatic animals.

Ponds can produce surprisingly high levels of aquatic animals if the ponds are properly managed. Management usually includes increasing the amount of food that is naturally available to the culture species, which can be done by fertilizing the pond to promote the growth of algae, zooplankton, and benthic organisms. Prepared feeds may also be provided, with or without fertilization.

Most aquaculture ponds are fairly shallow, although they vary widely in size and shape. A well-designed pond will not have side slopes that are so steep that they will easily erode and make access difficult or that have such a low angle that large amounts of shallow water, which promotes aquatic vegetation growth, are present. Ponds should have a reliable water supply, be fitted with a drain, and have a bottom that slopes toward the drain. They should be constructed in soils that have a high clay content, to prevent excessive seepage. Various types of liners may be used when facilities are constructed in porous soils. Ponds need to be carefully managed to avoid stressing the animals contained within. Additional details may be found in (1–4).

Figure 1. A typical fish culture pond, showing an inflow pipe to the left of the drain structure.

PONDS AS CULTURE CHAMBERS

Ponds are relatively small, man-made water bodies that are employed to water livestock, control flooding, and produce aquatic animals. In the United States, it is common practice for farmers to employ ponds for multiple uses, by using them for both livestock watering and recreational fishing, and to have ponds provide an alternative source of food for livestock.

The first aquaculture production facilities in China consisted of ponds, and ponds continue to be the predominant culture chamber employed throughout the world, particularly in conjunction with freshwater aquatic animal husbandry. It has been estimated, for example, that over 75% of the freshwater aquaculture production in China comes from ponds. Carp production in China currently exceeds seven million tons annually. Most tilapia culture is conducted in ponds. Catfish farming in the United States is conducted primarily in ponds, and throughout the world, shrimp farming is dominated by pond culture. Trout, which we commonly think of as being reared almost exclusively in raceways (see the entry "Tank and raceway culture"), have been satisfactorily reared in ponds as well. This brief list could be expanded to include many additional species.

The steps taken by the fish culturist to enhance the production of species being reared range widely, but, in general, involve the provision of more food for the species than would be available under natural conditions. If we take a farm pond as a baseline, we might expect that the annual production of stocked fish is around 100 kg/ha (100 lb/acre). If the pond is fertilized, either with chemical or organic fertilizer (manures), productivity can be increased severalfold. Sometimes, natural productivity is increased without direct assistance by humans. For example, fertilizers applied to cropland can run off into ponds and lead to increased productivity. Similarly, farm ponds in pastures may receive organic fertilizer inputs from the livestock in the pasture. If the pasture is fertilized, the ponds may also receive chemical fertilizers from runoff.

Options for growing aquatic animals in ponds include the following, where the production figures (in parentheses) are the range typically observed for channel catfish, in kg/ha (lb/acre) (4):

1. Stocking only (50 to 100);
2. Stocking and fertilizing (200 to 300);
3. Stocking, fertilizing, and providing supplemental feed (not a technique used with catfish);
4. Stocking and feeding (1,500 to 2,500);
5. Stocking, feeding, and providing supplemental management (4,000 to 15,000).

In the last option, supplemental management may include providing aeration when necessary, exchanging water, or both. If only aeration is provided in addition to stocking and feeding, the lower production level (4,000 kg/ha) may be achieved. Production can be improved somewhat with continuous aeration and can jump to the higher level (15,000 kg/ha) when high rates of water exchange are also employed.

To control the amount of nutrients that enter a pond, the aquaculturist must apply fertilizer in known amounts directly in the water. Fertilizers enhance plant growth, which, in most cases, means increased phytoplankton production. For some species, such as milkfish, fertilization techniques aimed at encouraging the growth of benthic algae have been developed. In most cases, promotion of the growth of rooted aquatic vegetation is not desirable. Phytoplankton provide food for zooplanktonic organisms, aquatic insects (in freshwater), and benthic organisms, all of which, in turn, may be directly fed upon by the species being cultured. In some cases, algae may be consumed directly by the culture species (for example, molluscs, such as clams and oysters, and some larval animals).

To further enhance productivity of the species being cultured, prepared feed may be provided, in the presence or absence of fertilization. Supplemental feed provides additional protein, fat, and energy for the culture species, but may not provide all of the required nutrients. The most sophisticated form of pond culture involves providing the culture animals with complete feeds. Complete feeds are feeds whose formulation is based on the known nutritional requirements of the animals. Complete feeds supply the proper levels of protein, fat, energy, vitamins, and minerals.

It has been said that no two ponds are alike. This statement does not refer to the general size or shape of a pond, as it is relatively easy to construct two ponds that share all of the same physical dimensions. What sets one pond apart from another relates to biology and chemistry. Even when ponds are stocked identically (i.e., with the same numbers of the same-size animals of the same species) and managed in the same fashion (e.g., the same amount of feed is given at the same time each day, the ponds share the same fertilization schedule and rate, etc.), there will be differences in the amount and type of biota present and in the water chemistry. Different species and concentrations of algae frequently grow, and differences in the zooplankton and benthos communities develop naturally in any pond. In many instances, bird's-eye examination of a group of ponds of the same size and shape will reveal distinct differences in water color that underscore the inherent variation that can occur. Because the biota vary, so will nonconservative water quality parameters, such as pH and levels of dissolved oxygen, carbon dioxide, ammonia, nitrite, and nitrate. As a result, management of individual ponds may vary considerably on any given aquaculture facility. Other culture systems, such as tanks and raceways, tend to be more uniform with respect to both their biota and water chemistry, because high water-turnover rates make physical factors more important than biological factors in controlling the culture environment.

SITING AND CONFIGURATION

Three types of ponds are recognized by aquaculturists and vary depending on hydrology (4). Levee, or embankment, ponds are the most common type of pond. They are made by constructing levees around the perimeter of the area that is to impounded. The original ground level becomes the pond bottom. Since the tops of the embankments are above the original ground level, water must be pumped into levee ponds. The source of water can be well water or surface water (such as lakes, streams, estuaries, or the ocean).

Excavated ponds are the second type of pond used for aquaculture. Earth from the pond-bottom area is used to construct the levees. In many instances, the tops of the levees are above the original ground level, and it may even be necessary to haul off excess earth (which may be used to construct levee ponds). If the bottom of an excavated pond is below the water table, it may be partially filled by seepage. Water may have to be pumped into excavated ponds to completely fill them, though rainfall sometimes is sufficient to maintain the water level. If the levee tops are at the same level as the elevation of the original land, there may also be some runoff to help keep the ponds full.

Both levee and excavated ponds can be of any shape. The shape of ponds may vary, because of the shape of the property on which they are constructed, but in most cases, rectangular ponds seem to be the most common type constructed.

The third type of aquaculture pond is the watershed pond. Watershed ponds are formed by damming a watercourse to impound runoff water. The ponds tend to be of irregular shape, though it is possible to construct rectangular ponds below a dam and distribute runoff water to them through canals or pipelines. Many catfish grown commercially in Alabama are raised in watershed ponds.

The location of an aquaculture facility is often based on the general locale, opportunity, or land availability, rather than suitability of the area for construction of ponds. Ideally, the land should be level to gradually sloping and have soil that has a high percentage of clay (a minimum of 25% clay is recommended). The soil should be free of toxic chemicals. Competition for space with other user groups, land costs, existing ownership by the prospective aquaculturist, proximity to markets, availability of suitable volumes and quality of water, and other factors may influence the choice of location (see the entry "Site selection").

Many coastal regions have sandy soils, which do not hold water well. In fact, some coastal sediments may be almost entirely composed of sand and shell material. However, in order to construct a facility in proximity to saltwater, it may be necessary to construct ponds in such soils. Inland sites may also have soils that do not hold water. There are various methods that can be used to prevent seepage, but all of them will add to the cost of construction. For example, bentonite, a clay mineral, can be mixed into the soil. When wet, bentonite expands greatly and is thought to help seal pond bottoms. However, many aquaculturists who have tried bentonite have experienced more frustration than success, so impermeable membrane liners have become more popular. Plastic and fabric lining materials, available from various sources, can be spread over the sides and bottoms of ponds to provide an impermeable layer. Liners vary greatly in cost, depending on the material they are made of and their thickness.

Ponds can be of virtually any size, though they typically range from less than 0.2 ha (0.5 acre) to 8 ha (20 acres). Sixteen-ha (40 acre) ponds were common in the catfish industry several years ago, but they were difficult to manage, and if a disease or water quality problem led to high levels of mortality, the economic consequences were disastrous. Most of the large ponds have now been divided into two or more smaller ponds.

Small ponds are often used for holding broodfish, for spawning, and for fingerling production. Large ponds tend to be used for growout. Large ponds also tend to be more common in areas where mechanized harvesting (e.g., seines pulled by trucks or tractors) is conducted. Where human power is used to harvest, smaller ponds are more practical. Also, when the culture species is fed from the bank by hand (as opposed to using boats or feed blowers towed behind tractors or trucks), it is easier to distribute food evenly from the bank of a small pond than a large one.

Prior to construction, surface vegetation, including grass, should be removed. Topsoil can be set aside for use on the completed levees in areas where grass will be encouraged to grow. It is extremely important to remove stumps, roots, and any other type of organic debris from the site and not to place any such material in the levees: As organic matter decays, it will create voids in the levees that water will eventually find. Breaches in levees and other catastrophic failures can result.

Levees should be constructed with slopes of 1 : 2 or 1 : 3 [i.e., 1 m (ft) high for every 2 or 3 m (ft) wide] (Fig. 2). Ponds have also been constructed with much steeper slopes; 1 : 1 is not uncommon, and even vertical walls — typically of wood or concrete — are sometimes used (Fig. 3). However, steep slopes make entry and exit from ponds by personnel difficult, unless stairways are provided. Also, the shallower the slope, the more shallow water that is available. This condition limits total water volume in the pond, but, more importantly, provides increased area where undesirable rooted aquatic vegetation can become established. Pond

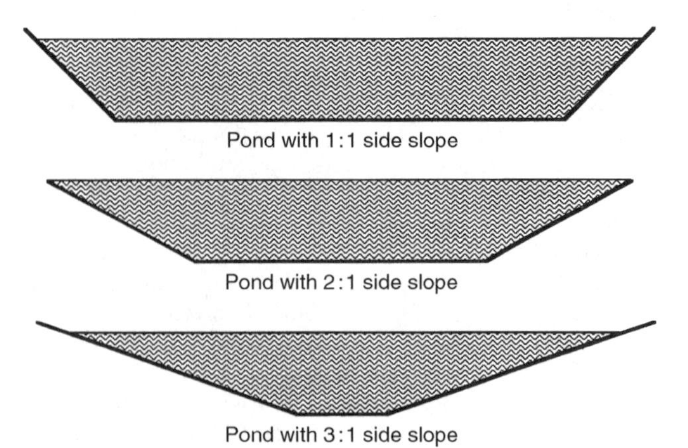

Figure 2. Cross-sectional diagrams of ponds with different side slopes.

Figure 3. Ponds with vertical sides constructed of wood. This approach allows for the construction of narrow between pond levees, but will add to the cost of construction.

bottoms should gradually slope toward the deep end. A 1 : 100 slope is sufficient for the bottom.

Most aquaculture ponds have a maximum depth of no more than about 2 m (4 ft). Shallow ponds are less expensive to construct than deep ponds, since less earth must be moved. Also, shallow ponds utilize less water, which may be important economically. In addition, it is much easier to seine a shallow pond than a deep one. Seines are used to harvest various species of fishes, and, at least in relatively small ponds, seines are often pulled manually. People who pull seines do not want to have to swim while pulling the seine; they should be able to walk along the banks without experiencing water depths much over waist high.

Deep ponds have greater volumes per unit of surface area than shallow ponds and tend to stratify during warm weather. If you were to wade into a pond out to a chest-deep level in a temperate region during summer, you might find that your upper body is in warm water while, from your waist down, the water becomes progressively and noticeably colder. The change in temperature, called a *thermocline*, can be several degrees in a typical pond. In shallow ponds, such as used in aquaculture, the thermocline goes to the bottom. In a deep pond, another zone, called the *hypolimnion*, will form below the thermocline. Water in the upper layer (the *epilimnion*) is mixed by the wind, but the thermocline forms a barrier that prevents mixing in the hypolimnion. Over a period of time, due to the respiration of living organisms and the decay of organic matter, the dissolved-oxygen level in the hypolimnion can be reduced to lethal levels. Thus, the aquaculture species will avoid the deep water, if possible. (Immobile species are not able to move to shallower, more oxygen-rich water). As a result, the effective volume of the pond is reduced to that of a shallower pond that does not have a hypolimnion. Therefore, there is no advantage to building a deep pond rather than a pond that, at its deepest, is about 2 m (6 ft).

Levees should be wide enough to allow for easy mowing. If they are too narrow, the chance that erosion will occur increases. One levee, usually the one at the drain end of the pond, should be wide enough to drive on and should be

Figure 4. An antiseep collar (this one fabricated from metal) being placed around a drain line that will be installed through the levee of a pond under construction.

Figure 5. Water flowing by gravity into a culture pond in Jamaica.

topped with gravel, to provide all-weather access. It will be necessary to drive along at least one side of each pond for harvesting and, in most cases, for feeding. The levees should be planted with grass, to prevent erosion.

Drain lines should be fitted with antiseep collars imbedded in the middle of the levee (Fig. 4). An antiseep collar is a device that prevents water from working through the levee along the drain line over time and eventually leaking from the pond. When water reaches an antiseep collar, it will be distributed laterally instead of being able to follow the pipe. Antiseep collars can be constructed of metal, concrete, or any other sturdy and nondegradable material.

Orientation of ponds with respect to prevailing winds is a sometimes a consideration, particularly in small ponds. Wind mixing will provide aeration and limit the time during which a thermocline can form, so the long axis of small ponds should line up with the direction of the prevailing wind if the culturist wishes to encourage mixing. At the same time, if the prevailing winds tend to be strong, it may be better to orient ponds with their narrowest aspect to the prevailing winds. This configuration will reduce the size of waves that form and thereby reduce bank erosion. In large ponds, wind mixing tends to be acceptable no matter what the orientation, so erosion prevention becomes the major consideration in pond orientation. Orientation is also often dictated by the configuration of the land on which the ponds are constructed.

WATER INFLOW AND DRAINAGE

Each pond should have plumbing to provide for inflow and drainage. While this requirement seems intuitive, there have been instances in which moveable pipe has had to be maneuvered into place each time a pond was to be filled. Inflow may also be diverted surface runoff, as mentioned previously for the case of watershed ponds. When water is diverted from a stream, reservoir, or irrigation canal, it may be routed into culture ponds by gravity through surface channels or through pipes (Fig. 5). Pressurized systems require pipelines and valves to each pond, to control inflow locations and volumes. When water under pressure is available, spraying the water into ponds may help improve the culture conditions during periods when levels of dissolved oxygen are low.

Pipelines should be sufficiently large to allow a pond to be filled within several days. Small ponds can often be filled in no more than three days, while it may take one to two weeks to fill large ponds. The smallest practical pipes and valves should be used to keep costs down. Splash blocks (concrete pads) are useful at inflow locations to avoid eroding pond bottoms during the early stages of filling.

Inflow pipes can be located anywhere, though they tend to be at either the upper (shallow) or lower (drain) end, or there may be one inflow line at both ends. If only one inflow line is used, the culturist needs to decide whether to place it at the upper end, where attempts to flush water through the entire pond will be facilitated, or at the drain end, where new, high-quality water can be added during periods when the pond is being drained for harvesting and the fish are concentrated in a small volume of water near the drain. Clearly, having water available at both ends of the pond is optimal.

Once a pond has been filled, the water level needs to be maintained. During dry periods, evaporation may require that water be added intermittently to maintain the level of the pond. A good management practice is to keep the water level somewhat below the top of the level or overflow pipe. A few centimeters (a little over an inch) of freeboard will allow additional free water to be obtained and retained when it rains.

The simplest and least expensive drain is a simple standpipe that is elbowed into a drain line. The water level can be controlled by the elevation of the standpipe above the pond bottom, since the standpipe swivels on the elbow. The standpipe can be located inside the pond or, if a drainage ditch is available, outside the pond (Fig. 6). Standpipes and drain lines of this nature can be made from metal pipe, which is recommended particularly for large ponds, because of its strength. Polyvinyl chloride (PVC) pipe is light in weight, nontoxic, will not corrode, and is much easier to cut and plumb than metal pipe. PVC

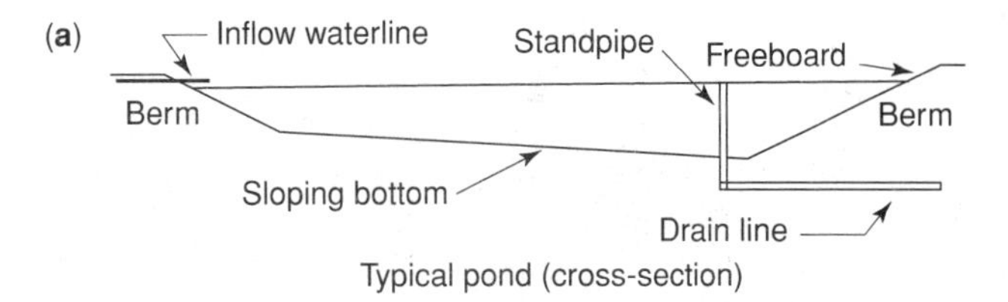

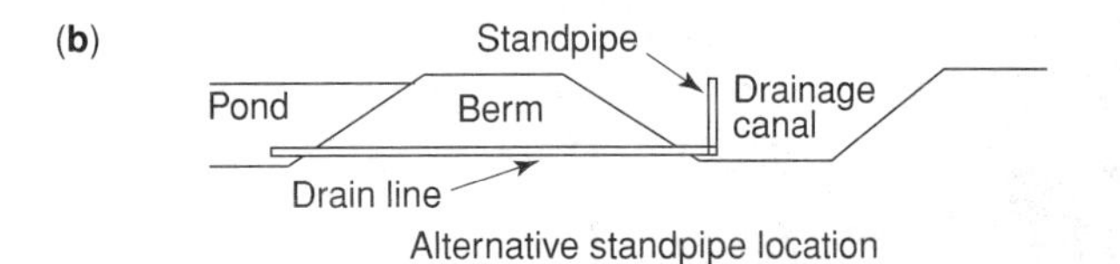

Figure 6. Cross-sectional diagrams of ponds with internal standpipe (**a**) and external standpipe (**b**) locations.

pipe is commonly used for both inflow and drain lines in small ponds.

When funding is not limiting, more elaborate drain structures are a better choice (Fig. 7). Such structures may have stairways, to provide easy access for personnel. They are usually constructed of concrete and are called *kettles*, or *monks* (Fig. 8). A typical kettle has a back and two sides made of concrete, with a concrete floor. Kettles are constructed in front of or within the levee at the deep end of the pond, with the top of the concrete walls extending above the maximum water level. A screen (or preferably a pair of screens) is fitted into slots at the front of the kettle

Figure 7. A concrete kettle with a catch basin in front. Note the stairway to the left, the valve that can be opened to completely drain the pond, and a moveable (up and down in this case) standpipe to control the water level when the pond is in use or being drained. The pipe on the right side of the kettle is an inflow line. The screens have been removed in this photograph.

Figure 8. A simple kettle design has two sets of boards set in slots to control the water level. Dirt is packed between the two sets of boards to deter leakage. Valves are not required in conjunction with this type of kettle, but a good deal of labor is needed to put the dirt in place and then remove it. The third pair of slots show the location of the screen.

so that water may enter, but fish are screened out. The dual-screen configuration (one screen behind the other) allows the culturist to remove one screen for cleaning if it gets fouled with debris, without allowing the culture animals to enter the kettle.

Concrete harvest basins are often constructed in front of kettles. Harvest basins of this type are particularly useful when harvesting freshwater or marine shrimp. Featuring concrete bottoms and side walls, harvest basins may be 30.5 cm (1 ft) or less in depth and of varying dimensions, though usually not more than a few meters (yards) on a side. Harvest basins provide a collection point for the animals being harvested to congregate when pond draining is nearly complete. By flowing new water into the harvest basin at the time of harvest, the level of dissolved oxygen can be maintained and excessively high temperatures due to radiational heating can be avoided.

POND PREPARATION AND STOCKING

It may be necessary to make some adjustments in water quality prior to stocking. Iron, carbon dioxide, and hydrogen sulfide may be present in the incoming water, particularly if it is from wells (see the entry "Water sources"). All of those undesirable substances can be removed, or chemically altered to reduce their toxicity, through aeration. If the soils in the pond bottom are acidic, it may be necessary to apply lime or crushed limestone to control the pH. Limestone ($CaCO_3$) can also be used to increase hardness and alkalinity if the water is deficient in one of these conditions (see the entries "Water hardness" and "Alkalinity"). The dissolved-oxygen level and temperature should be checked prior to stocking, to ensure that they are within the optimum range for the animals that will be introduced. If well water is used, the temperature of the incoming water may be different from that in ponds that have been allowed to come to ambient temperature, which may require several days.

Salinity, if too low, can be adjusted by adding salt, though this technique is not frequently practiced in ponds, because it is expensive. More commonly, the salinity is too high (or become too high as evaporation concentrates salts in the water), and refilling the pond with additional saltwater will not reduce the salinity to its original level. The solution for maintaining the desired salinity in marine ponds is to dilute the pond with freshwater.

Once a pond is filled, it may be fertilized to induce a plankton bloom or to encourage the growth of benthic algae, depending upon the species being cultured (see the entry "Fertilization of fish ponds" and specific information in contributions on individual species). Fry and fingerling ponds, as well as ponds in which filter feeders, milkfish, and tilapias are reared, are commonly fertilized to provide supplemental feed or the sole source of food. A well-established plankton bloom will shade out many of the rooted aquatic plants that might otherwise become established.

If an aquaculture facility runs its own hatchery, the amount of time between the capture of fish or invertebrates from the hatchery or early rearing facilities to the placement of the animals in growout ponds can be very short. On the other hand, if young animals are purchased from a distant hatchery, the amount of time between collection and stocking may be several hours. Further, the water temperature in hauling tanks will increase or decrease more rapidly than that in a pond. If the animals being stocked have been hauled for a period of at least several hours, the temperature may have changed significantly from what it was when the animals were put in the hauling tank. Also, the water source in which the animals are hauled may be of a temperature somewhat different from that of the water in the receiving pond. If the temperature of the hauling water is more than a few degrees different from that in the pond, it will be necessary to temper the animals. This can be done by pumping pond water into the hauling tank at a rate that will reduce the temperature by a few degrees per hour, until the water in the tank is the same temperature as that in the pond. When the temperatures have been equalized, the animals may be safely stocked.

MANAGEMENT OF PONDS

Day-to-day management of fish ponds involves feeding, evaluating and managing water quality, observing the culture animals for signs of disease and treating the animals if necessary, and keeping good records. The last item is sometimes neglected, but should be a primary management practice. Having accurate records for each pond is important to management decision making and for tracking performance from one pond to another. In many instances, information from pond records can be used to anticipate problems.

Feeding is a daily activity for aquaculturists. Some fish farmers feed six days a week, and some have even made statements to the effect that the fish will perform better if they go one day a week without feed. There is no validity in that opinion, but there is probably a lot of truth in the notion that the farmer wants to take off at least one day a week from performing farm duties. When prepared feeds are provided, the animals are typically fed once or twice daily, though when automatic or demand feeders are used (exclusively for fish), the animals may feed at their own volition. The number of daily feedings will vary to some extent, depending on the species being cultured, the water temperature, and other water quality conditions.

When sinking feeds are used, it may be necessary to subsample the animals periodically to determine growth rates and any changes that need to be made in the amount of feed offered. Floating feeds, which are appropriate for species that surface to obtain feed, provide a means by which the culturist can monitor the amount of feed being consumed and increase or decrease the daily ration, based on feeding activity. The use of floating feeds also provides the culturist with the opportunity to observe the animals and perhaps detect the onset of diseases before they become significant problems.

Water quality should be monitored in each pond, because of the intrinsic differences that exist between ponds. Feeding rates may be varied as a function of temperature. The amount of feed and supplemental aeration should be provided based on the level of dissolved oxygen, which should be measured at about dawn each day, particularly during warm, cloudy weather (see the entry "Dissolved oxygen"). The temperature need not be measured daily, though periodic measurements should be made, as is also the case for pH. Hardness and alkalinity measurements should be taken at least annually. Levels of ammonia, nitrite, hydrogen sulfide, and carbon dioxide may be measured when conditions warrant. An analysis of the water and soils for trace metals, pesticides, and herbicides may be desirable, or necessary prior to construction or after incidents that may have caused contamination. An example of such an incident might be suspected spray drift from pesticide application upwind of aquaculture ponds.

The water quality dynamics in a pond can be complex. As the rate of fertilization and/or use of prepared feeds increases with increased intensity of culture, there may be concomitant responses in terms of which water quality variable becomes first limiting. The first-limiting factor in ponds that are heavily fertilized and fed is usually the level of dissolved oxygen. Careful daily monitoring of oxygen is important, with measurement at dawn, as mentioned previously, being of particular importance for detection of dangerously low levels. Controlling the amount of oxygen through water pumping or aeration may allow other limiting factors to become dominant, including ammonia, nitrite, and, perhaps, carbon dioxide.

Ameliorating those factors when they become limiting is more difficult than simply providing aeration to deal with oxygen depletions.

Stress associated with low levels of dissolved oxygen, high levels of ammonia, rapid changes in temperature, handling, and various other factors can lead to the onset of disease in aquatic animals. The culturist should be particularly observant after any stressful incident. Epizootics typically occur after 24 hours, about three days, or two weeks following a stress event. Cessation of feeding is one of the first signs of a pending disease outbreak, so careful observation of the animals at feeding time is important, as mentioned previously.

In addition to maintaining grassy pond banks to avoid erosion, it may be necessary to perform intermittent routine maintenance on the portions of ponds that are underwater when the ponds are full. Uneaten feed, feces, and decaying organic matter from various sources can accumulate on pond bottoms, creating anaerobic areas that will reduce the water volume suitable for the culture species and may lead to overall oxygen depletion in the pond. Unavoidable erosion of banks, sometimes accelerated by the activities of culture species, such as rooting in pond banks by carp and digging into the banks by crawfish, may mean that repairs will become necessary.

After draining, ponds can be allowed to dry for a period of several weeks. The bottom may be disked to accelerate decomposition of organic matter, or pond bottoms contaminated with disease organisms or resting stages of parasites may be limed (such as with CaO) and disked to sterilize them. If slopes need to be repaired and reshaped, or if the bottom sediments need to be removed, it will be necessarily to employ tractors, bulldozers, backhoes, or draglines. In most cases, ponds can be in production for several years before needing extensive renovation.

Bird predation is a problem for many pond culturists. Noisemakers will work temporarily to ward off birds, but the birds will eventually ignore them. Netting can be installed over ponds to keep birds away from the water, but that solution is economically feasible only for small ponds. Patrols by dogs or humans will help keep birds away, but such patrols are not effective on large fish farms unless a large number of dogs or people are used, which may not be economical. Lethal means of dealing with birds are often employed as well.

In most instances, the animals stocked in ponds are allowed to roam freely within the pond, though there are cases in which cages of one species are used in conjunction with a second, free-roaming species. This type of polyculture (see the entry "Polyculture") may be employed if the two species are not compatible. In most cases, polyculture involves two or more compatible species that are allow to interact freely in an open pond.

HARVESTING

Ponds may be harvested one or more times a year to remove an entire crop, or they may be partially harvested periodically and restocked, a technique known as the "continuous harvest" method. Shrimp ponds will produce one crop a year in low temperate regions, such as South Carolina and Texas, and may produce two to three crops a year in tropical regions. The standard practice is to completely drain ponds during harvesting. The same technique is used for various other species as well.

The channel catfish industry once employed annual pond draining and harvesting almost exclusively, but has now gone to a continuous harvest approach. Fingerlings are stocked and fed for several months, then the pond is seined with a large mesh net, and marketable fish are removed, while submarketable fish escape through the seine. Periodically—perhaps as often as monthly—the pond is seined again. Restocking with additional fingerlings to replace market-size fish that were removed may follow each seining event or be done less frequently, depending on the availability of small fish. Some catfish ponds have been undergoing the continuous harvest regimen for over a decade without having been drained completely. One result of this technique is that water is conserved, and environmental damage associated with effluents from aquaculture ponds can be avoided.

Eventually, there may be an accumulation of stunted fish in ponds subject to continuous harvesting. Such fish consume feed, but do not reach market size, and end up being an expense, rather than a source of profit. Also, over time, pond banks may erode, and organic material will accumulate in the pond sediments and can impair water quality and negatively affect fish growth. Therefore, it may be wise to completely drain and harvest all ponds at least once every several years.

Continuous harvesting is desirable in that seine crews can spread their activity not only throughout the year, but also on a rotational basis, so that only one or a few ponds on a fish farm will be harvested on any given day. Continuous harvesting also ensures that there is a dependable year-round supply of fish for the market.

The technique of continuous harvesting may not always be the most economical way to raise fish. The problem of stunting has already been mentioned as a negative economic factor. Also some fish may reach market size, but escape the seine for extended periods of time. These fish will continue to grow and may reach sizes that are not particularly desirable, thereby bringing reduced prices at the processing plant. By harvesting an entire pond at one time, the farmer can, to some extent, time harvesting to take advantage of periods when prices are high, though timing is critical and may not always fit even the best planned growout schedule. Continuous harvesting spreads out and lessens the risk of bad timing, to some extent. For the catfish industry, in which off-flavors have been a major problem (see the entry "Channel catfish culture"), having several ponds containing harvestable fish virtually all of the time also helps ensure that at least some fish of suitable quality can be sold periodically.

A pond may be allowed to remain full, or it may be partially drained at the time of harvest. Seines should be of the proper mesh size to retain marketable animals, and they should be sufficiently deep to hold the bottom with the lead line while the float line reaches the surface, without stretching the seine tightly in the vertical direction. Having ponds no deeper than about 2 m (6 ft) reduces

Figure 9. Tilapia culturists seine a pond in the Philippines.

the necessary amount of netting, and thus the expense, of seines of any given length. The length of a seine should be $1\frac{1}{3}$ as long as the pond is wide. Having ponds of uniform size is desirable in that the number of seines required can be small relative to what would be needed if ponds of various sizes are used.

For small ponds, seines can be pulled by hand (Fig. 9). For large ponds, trucks and tractors are routinely used to pull seines. Seines usually have a bag in the middle where the harvested animals are concentrated. The fish or invertebrates may be herded from the seine bag into a live car (basically a net pen staked to the pond bottom), from which they are netted and moved to the trucks that will take them to the processing plant.

When ponds are totally drained during harvest, it is common practice to pull a seine through each full or partially drained pond, lower the pond level, seine again, and so forth in three or four steps until the majority of the animals have been removed and the water level has been reduced by perhaps 75 to 80%. Thereafter, the pond can be completely drained and the remaining animals captured from the harvest basin and the bare pond bottom.

There are many variations on the theme just presented. If harvest basins are sufficiently large, the pond may be drained all at once or after only one or two passes through it with a seine. If the drain line from a pond flows into an open drainage canal, it may be possible to put a net over the end of the line, to collect animals as they are flushed from the pond (in that case, the screens over the kettle would be removed). Shrimp are sometimes harvested in this manner.

It is important to avoid having the pond water become too warm during harvesting. For that reason, harvesting activities often take place as early in the morning as possible. A pond may be partially drained overnight, and seining will commence when there is sufficient daylight and be completed before the hottest part of the day. During the last stages of harvesting, when there is very little water in the pond and the animals are concentrated in one area, the water temperature can rise precipitously. Having a plentiful supply of new, cool water to add will greatly reduce stress and subsequent mortality, which are important considerations, even when the animals are on their way to market. For example, many fish are expected to reach the processor alive and in good condition. Shrimp and other invertebrates may be removed from ponds and placed directly on ice, but they, too, need to be in good condition when harvested, as they rapidly spoil after death when exposed to high temperatures.

BIBLIOGRAPHY

1. R.R. Stickney, *Principles of Aquaculture*, John Wiley & Sons, New York, 1994.
2. J.W. Avault, Jr., *Fundamentals of Aquaculture*, AVA Publishing Co., Baton Rouge, LA, 1996.
3. A.M. Kelly and C.C. Kohler, in H.S. Egna and C.E. Boyd, eds., *Dynamics of Pond Aquaculture*, CRC Press, Boca Raton, FL, 1997, pp. 109–133.
4. C.E. Boyd and C.S. Tucker, *Aquaculture Pond Water Quality Management*, Kluwer Academic Publishers, Dordrecht, The Netherlands, 1996.

See also Carp culture; Channel catfish culture; Fertilization of fish ponds; Harvesting; Shrimp culture.

PREDATORS AND PESTS

Michael P. Masser
Texas A&M University
College Station, Texas

OUTLINE

Predators on aquaculture facilities are at best a nuisance and at worst an economic disaster. Many predatory species are in small enough numbers or prey upon so few culture animals that, while being a nuisance, they cause little economic loss. On the other hand, some insects, birds, and a few other predators can decimate an entire crop, causing economic hardship for the aquaculturist.

Included in this discussion are animals that, while not predators, are pests on aquaculture facilities. Pests include burrowing mammals such as the muskrat, nutria, and beaver.

LOCATION AND DESIGN OF FACILITIES

Location and design of aquaculture facilities can significantly influence some predation problems. Aquaculture facilities located near rivers, major creeks, bays, marshes, or other wetlands, or known predatory bird roosting areas will increase the likelihood of significant predator interactions and problems. However, factors such as soil type and water availability critical to the production of the cultured species may override the problems associated with locating the facility within major predator corridors.

Facility design can also discourage predators or enhance predator control methods. For example, if a site is suitable for raceway culture, then it is more easily covered and protected from most predators than ponds. Small ponds are more easily protected than large ponds. Design of the facility to place fry, fingerlings, or other highly vulnerable life stages in areas nearest to human activities and where they can be easily observed is advantageous. Ponds and raceways at least 1 m (3 ft) deep with steep sides or band slopes and control of vegetation around facilities can discourage some types of predators. Finally, designing ponds or raceways that are covered or fenced, without disrupting normal production activities, can discourage many vertebrate predators.

INVERTEBRATES

Invertebrate predators include insects, crustaceans, molluscs, starfish, and leeches.

Insects

Certain aquatic insects prey on small fish (larvae and fry) and, in some cases, cause significant losses. Predatory insects include members of the following orders: Coleoptera, Hemiptera, and Odonata. The coleopterans and the hemipterans breathe air and must return to the surface periodically to replenish their supply, while odonates have gills. All of these insects colonize ponds as adults that fly in from other aquatic sites. Then they mate and lay eggs. Usually only after successful reproduction do they achieve numbers sufficient to cause serious damage.

The coleopterans are beetles and include several species of diving beetles (*Cybister*, *Dytiscus*, *Hydrophilus*, and *Dineutus* spp.) that prey on fish as larvae and adults. Diving beetle larvae resemble large segmented worms with large pincers or fangs. The fangs are modified mouth parts that pierce and hold prey, then suck out the body fluids. Adult diving beetles can reach 2 to 3 in. (5–8 cm) in length; they capture and kill prey with mouth parts that crush and tear the prey into pieces (1).

Hemipterans are true bugs and include predacious aquatic insects like the giant water bugs, water scorpions, water boatmen, and backswimmers. Hemipterans carry air bubbles trapped under their wings when they dive underwater and have mouth parts that are modified into beaks that pierce the prey. They inject their prey with digestive enzymes and then suck out the fluids. Typically it takes hemipterans 6 to 9 weeks to develop from eggs to adults and achieve numbers sufficient to cause major losses.

The giant water bug is among the largest insects, attaining a length of up to 4 in. (10 cm). The giant water bug swims by using its flattened hind legs, captures prey with modified front legs, then kills its prey and sucks the body fluids out using its piercing beak. Water bugs are seldom numerous enough to cause severe losses in aquaculture ponds.

Water scorpions are sticklike insects 1 to 2 in. long (2.5–5 cm). They are called water scorpions because the shape of their front legs resembles those of true scorpions. Prey is captured with the front legs, then pierced and consumed using the beak. Water scorpions are poor swimmers and usually cling to vegetation at the surface where they can breathe through their tubelike abdomen. Like water bugs, water scorpions are usually not numerous enough to cause severe losses of fish larvae and fry.

Water boatmen (Corixidae) are small bugs (less than 0.5 in.) that are not considered an important predator on fish larvae or fry, but compete with them for available natural food.

Backswimmers (*Notonectidae*) are small bugs (less than 0.5 in.) that swim upside-down by using their hind legs as oars. Backswimmers, although small, can inhabit ponds in such large numbers that they can severely impact larvae and fry. Backswimmers prey heavily on fish larvae, fish fry, and invertebrates.

Odonates include the damselflies and dragonflies. Large dragonflies (up to 5 in.) can prey on fish that they catch at the surface. The naiads (the aquatic larval stage) of dragonflies and damselflies have gills and do not need to come to the surface to breathe. The naiad stages of the dragonfly and damselflies are serious predators on fish larvae, fry, and also invertebrates like crayfish.

There are several methods to prevent and/or control aquatic insects. The most common method is to fill ponds immediately before stocking larvae or fry. This helps suppress the numbers of predatory insects by limiting the time for them to colonize, reproduce, and develop in the pond. However, immediately filling a pond before stocking may not be acceptable with certain fertilization and plankton development regimes (2).

Another common method is to treat ponds with chemicals to kill aquatic insects. In the United States, the following insecticides have been used to treat nursery ponds prior to stocking: trichlorfon, fenthion, and methyl parathion. However, insecticides may suppress zooplankton populations that are essential to the natural food chain of the larvae or fry being cultured. Insecticides are regulated by government agencies (e.g., in the United States, the Environmental Protection Agency or the Food and Drug Administration) and some "special local needs" (SLNs) exceptions have been granted. Current regulations should be checked before using any insecticides and the label followed explicitly (3). In most cases, insecticides cannot be used in ponds where foodfish will be cultured.

Probably the most common method for chemical control of insects is the use of diesel or kerosene. Vegetable oils (e.g., cottonseed oil) will also work but are more expensive than petroleum products. These products will form an oil slick on the surface of a pond. As air-breathing insects (hemipterans and coleopterans) come to the water's surface to get air, the petroleum coats or plugs their air passages, causing them to suffocate. Usually application rates are between 1 and 4 gal/ac (9.5 and 38 L/ha). Often motor oil is mixed with diesel at a ratio of 1 part oil to 3 parts diesel. These should be applied upwind on the pond and a slick will need to be maintained for at least an hour. Aquatic vegetation and high winds will reduce the effectiveness of these treatments.

Controlling aquatic vegetation will also reduce insect populations by denying the insects good habitat and breeding areas.

Crustaceans

Crayfish and crabs may prey on slow-swimming fish larvae and small fry that they can catch. They can also severely impact ponds where fish are spawned by preying upon the eggs (e.g., minnows) and often cause muddiness in ponds through their foraging activities. Muddiness reduces the effectiveness of fertilization regimes and can thereby diminish the natural food chain in the pond. Generally, crayfish are controlled by draining and drying out the pond. Burrows can be treated with calcium hypochlorite (HTH) at 200 ppm (4).

Crabs can prey heavily on molluscs. Control for crabs usually involves screening intake water to remove planktonic larvae, baited traps to capture adults, or placing molluscs in cages or baskets so crabs cannot get to them.

Molluscs

Oyster drills are marine conches (gastropods) that prey upon oysters and other molluscan shellfish. They can cause serious losses, in some cases as much as 85–90% (5,6). Some oyster drills are reported to secrete a toxin that paralyzes their prey. As the name suggests, the oyster drill drills a hole through the shell of its prey by using its radula (toothed tongue). After penetrating the shell, they extend their proboscis (tubular mouth) into the shell, cut up the soft tissues with the radula, and suck out the flesh. Conches have pelagic larvae that allow them to disperse and colonize over a wide area. This pelagic dispersal makes control of oyster drills difficult.

Oyster drills have been controlled physically by divers with vacuum devices and with underwater plows that bury them before seeding the shellfish. Exposing oysters to air and treating their surfaces with a strong salt solution has been effective in controlling oyster drills, as has moving the oysters to less saline water.

Snails can be a pest in aquaculture because they are an intermediate host for several species of parasitic flukes. They can also decimate beneficial aquatic vegetation. Drying pond bottoms and disking in hydrated lime at 1,500 to 2,000 lb/ac (1680–2240 kg/ha) can eliminate snails. Also, copper sulfate can be used to kill snails in ponds. However, the concentration of copper sulfate will depend on total alkalinity of the water (4). Eliminating rooted and filamentous aquatic vegetation can also reduce snail populations.

Starfish

Starfish are predators of oysters, clams, scallops, and mussels. Like conchs, starfish have pelagic larvae that can settle directly on spat (young molluscs) and start preying upon them (7).

Starfish have been controlled using the same methods as those for oysters drills. Starfish have also been controlled physically with "starfish mops." Mops are made of rope-yarn bundles attached to iron bars, which entangle starfish when they are dug over shellfish beds. Applying hydrated lime to shellfish beds at a rate of 270 lb/ac (300 kg/ha) also has been effective (2).

Leeches

Leeches can be simple external parasites or true predators. Different species of leeches can inhabit freshwater and marine ecosystems and vary in size from less than 1/2 to 8 in. (1–20 cm) (8). Predatory leeches feed only on invertebrates. Fish leeches are external parasites that suck blood from the host fish. In aquaculture, leeches seldom cause major problems, but can harm fish if their populations are high and can cosmetically damage the product (e.g., fillet), making it unacceptable to the consumer.

Control of freshwater leeches has been accomplished by using salt baths of 1–3% for up to 30 minutes. Masoten has also been used in ponds at 1–2 ppm, depending on temperature. As this chemical may not be legal in some countries, check the label.

FISH

Unwanted or wild species of fish compete with or prey upon cultured species. Wild fish can also introduce diseases into the culture system and increase harvest costs because of the need to separate them from the cultured species.

Preventing wild fish from entering ponds is the first step in controlling them. Prevention methods include filtering surface water that is used to fill ponds, constructing overflow spillways with at least a 3 ft (1 m) vertical drop, constructing levees and drains so that floodwaters cannot enter ponds, and not transferring wild fish into culture ponds during seining and stocking operations. Draining and drying ponds between crops and filling ponds just before stocking can also help control wild fish. Completely draining the pond and treating any puddles (and holes in levees that could contain puddles) will eliminate many species that can survive by keeping their gills moist for long periods of time. Filter boxes or socks can be used to screen out wild fish and their eggs (2). Screening drains during draw-downs can keep wild fish from entering the ponds through drains from drainage ditches. Seines that are not thoroughly dry can transfer unwanted fish eggs from one pond to another. Carefully checking shipments of fish for unwanted species is always prudent.

Once unwanted species of fish have contaminated a pond, they may be controlled biologically or chemically. Biological control usually involves stocking a predator fish in the pond. The predator must be smaller than the cultured species so that it cannot prey on them. This practice has worked in catfish and tilapia culture with the use of a variety of predatory species (9,10).

The most commonly used chemical to remove or kill fish is rotenone. Rotenone is a naturally occurring pesticide derived from the roots of certain plants (*Derris* and *Lonchoncarpus* spp.). Rotenone kills fish by blocking cellular respiration, but is not very toxic to mammals. Rotenone is best applied at temperatures above 60 °F (15 °C) and is more effective in acidic to neutral waters with low hardness (11). Rotenone comes in liquid and powdered formulations, usually with 5% as active ingredient. Rotenone is applied at 0.5 to 5 ppm, depending on water temperature and target species. Many times, after a pond is drained following harvest, rotenone is applied to the remaining water. Rotenone is detoxified rapidly by warm temperatures and intense sunlight. Usually rotenone is completely degraded in 3 to 14 days during warm weather. Rotenone is a restricted-use pesticide in the United States and many other countries.

Antimycin, chlorine, cyanide, tea seed cake, oil cake, and anhydrous ammonia have also been used to kill unwanted fish (12).

AMPHIBIANS

Frogs, particularly bullfrogs, are predators of larvae, fry, small fish, and crustaceans. Tadpoles compete for food and space with larvae and fry of some cultured species. The control of frogs and tadpoles includes drying ponds between crops and filling them just before stocking. Keeping pond levees and sides closely mowed and aquatic vegetation controlled will reduce the habitat that frogs prefer. Removing the frogs' floating egg masses can also help reduce their numbers. Encircling ponds with small meshed netting or solid fencing (e.g., sheet metal) to a height of 2 ft can keep out frogs. Finally, frogs can be physically removed by hunting them.

REPTILES

Snakes, turtles, alligators, and crocodiles can all prey on aquacultured species. Snakes can be discouraged by keeping pond levees and sides closely mowed and aquatic vegetation controlled. Bird netting and solid fencing around ponds can also discourage snakes and turtles. Turtles can be live trapped from ponds by using several different trap designs (13). Snakes and turtles are often hunted around aquaculture facilities.

Alligator and crocodiles (crocodilians) can be problematic. Crocodilians are usually more nuisance than major predator, although they have been known to prey on valuable brood fish. These animals migrate from surrounding areas, so a 6-ft (1.5-m) high chain-link fencing that is partially buried or has galvanized sheet metal buried to 18 in. (45 cm) deep can keep them out of most facilities. All crocodilians are protected by international treaty and special permits must be obtained before they can be captured or killed. Many states have animal damage personnel that will trap or remove nuisance crocodilians.

BIRDS

Undoubtedly, the most problematic predators around most aquaculture facilities are birds, as they can cause significant economic losses. In North America alone, more than 60 species of birds have been described as predators or pests on aquaculture facilities (14). Open water and a concentration of cultured aquatic animals attract predatory birds. Accurate identification and determination of the damage is critical to understanding potential impacts of a particular bird species (15). However, many species are attracted to aquaculture facilities without interfering with, or impacting, aquaculture production and profitability.

Predatory birds are highly mobile, efficient, and adaptable predators of aquatic organisms. Aside from preying directly on cultured organisms, birds also injure them (often leading to secondary diseases), spread diseases and parasites, and disrupt feeding and breeding activities. Common types of predatory birds that cause most problems on aquaculture facilities include cormorants, anhingas, pelicans, herons, kingfishers, gulls, terns, certain ducks, and particular hawks and eagles.

Cormorants, anhingas, and pelicans all belong to the same order (Pelecaniformes), which are characterized by having a pouched throat area, devoid of feathers, and highly webbed feet.

Cormorants (family Phalacrocoracidae) are major predators at aquaculture facilities around the world. Worldwide there are more than 30 species of cormorants (16). In North America, the double-crested cormorant is the most problematic predator on the majority of pond aquaculture facilities. Cormorants have long, flexible necks, and a straight, hook-tipped beak that efficiently captures slippery prey. They typically land and swim on the water's surface, then dive and swim through the water rapidly by using their feet for propulsion. Cormorants consume between 0.5 and 1 lb (0.25 and 0.5 kg) of fish per day (15). Cormorants do not have water-repellent feathers and, therefore, are often seen perched near the water with wings outspread, drying themselves after diving for food.

Anhingas (family Anhingidae), often called snake birds or water turkeys, are sometimes mistaken for cormorants. Anhingas also do not have water-repellent feathers and often swim with just their head out of the water. Like cormorants, they dive and swim to capture prey and perch with outspread wings while drying after diving. However, anhingas have a straight bill, not a hooked one, with which they spear prey. Anhingas are thought to consume about 0.5 lb (0.23 kg) of fish per day.

Pelicans (family Pelecanidae) capture fish near the surface with their scooplike pouch. All but one of the seven species of pelicans are predominately white and capture prey while swimming. The American and Eurasian white pelicans often fish cooperatively by swimming in lines while splashing the water with their wings and driving fish into the shallows, where the birds encircle them and capture them in their pouches. White pelicans are among the largest birds in the world and consume approximately 1 lb (0.5 kg) of fish per day. The brown pelican of North America can plunge-dive into the water from a soaring position and capture fish. Brown pelicans are typically only in marine habitats, while white pelicans inhabit both marine and freshwater environments.

Herons, egrets, bitterns, ibises, and storks (order Ciconiiformes) all prey to some extent on fish and shellfish. Members of this order are long necked and relatively long legged. They stalk their prey, often by wading through shallow water, then striking with a rapid thrust of the bill. Large herons and egrets (e.g., great blue heron and great egret) consume between 0.3 and 0.75 lb (0.14–0.35 kg) of fish per day, while smaller varieties consume from 0.15 to 0.3 lb (0.07–0.14 kg) per day. Some members of this group, like the great blue heron and the night herons, feed at dusk or at night, especially when harassed during the day. Ibises are common predators of crustaceans such as crayfish (17). Storks tend to prey on invertebrates and sick or weakened fish. Wading birds often collect around aquaculture facilities in large numbers and can cause heavy losses (18).

Kingfishers (family Alcedinidae) are small birds that are territorial and solitary daytime hunters that plunge-dive to catch prey. Worldwide there are more than 80 species of kingfishers. Because of their territorial nature, kingfishers seldom cause major fish losses, but can be problematic around ornamental and tropical farms consuming from 0.05 to 0.15 lb (0.02–0.07 kg) of fish per day (15).

Worldwide there are more than 80 species of gulls and terns (family Laridae). Gulls have hooked beaks, while terns have straight and pointed ones. Terns are solitary daytime hunters that feed by hovering above the water, then dive headfirst into the water to capture prey. Gulls feed in a similar manner, but also capture prey while swimming and can scoop prey from the water without plunging. Gulls usually feed in flocks, scavenge readily, and will consume aquaculture feeds off the surface of ponds or raceways. Gulls will consume between 0.15 and 0.3 lb (0.07 and 0.14 kg) per day, while terns consume only about 0.1 lb (0.05 kg) per day. Large feeding aggregations of these birds can cause serious economic losses of fish, crustaceans, or feed.

Certain species of ducks (family Anatidae), loons, and grebes will prey on fish and shellfish. Loons, grebes, mergansers, and other diving ducks capture prey by diving and swimming underwater. Many diving ducks feed extensively on molluscs and can cause heavy losses in cultured species of oysters, clams, and mussels. Mergansers and other diving ducks do prey on small fish and crustaceans and can become problematic for farms if in large flocks. Grebes are not usually considered a problem, except in crustacean culture (19).

The osprey and some eagles are fish predators and can catch large fish. Normally these solitary hunters are not a major concern for fish farms, but have been know to cause problems by preying upon brood fish.

Prevention and Control Methods

Most birds that cause problems on aquaculture facilities are considered migratory and are protected by either federal and state laws or international treaties. In some countries, as well as some states in the United States, permits can be obtained for lethal control of depredating birds, but these are relatively rare. In the United States permits to kill depredating birds can be issued by the U.S. Fish and Wildlife Service (USFWS) after assessments of the damage created by the birds have been confirmed by the U.S. Department of Agriculture—Animal and Plant Health Inspection Service—Animal Damage Control (USDA—APHIS—ADC). Lethal permits are never issued until nonlethal methods have been used correctly and damages were verified.

Methods to discourage or prevent bird depredation have included completely enclosing the facility, perimeter fencing, overhead wires or lines, automatic cannons or exploders, pyrotechnics, scarecrows, alarm or distress calls, water sprays, lights, ultralight or radio-controlled model aircraft, and adapting new management strategies. In some cases, harassment of roosting areas has been effective in moving entire colonies away from aquaculture facilities. The methods utilized depend on factors such as type and size of the facility, extent of the problem, the bird species involved, cost of the prevention method, and management implications (20–22).

Complete enclosure is the only method to entirely prevent bird depredation, but is usually limited to raceways and ponds smaller than 5 ac (2 ha). All other methods that have been employed have had mixed results, depending on the species targeted and the tenacity of the manager. A combination of methods has generally worked best.

Birds quickly adapt or habituate to scaring devices. Any device that is automatic or stationary, such as automatic gas cannons, scarecrows, electronic noise makers, distress calls, and strobe or flashing lights, must be moved every few days and, if possible, set so that they go off at random intervals and varying intensities.

Table 1 summarizes the techniques commonly used to harass and discourage predatory birds around aquaculture facilities.

MAMMALS

A few mammals, not excluding humans, can be serious predators on aquaculture facilities, including otters, seals, and sea lions. Sea otters, seals, and sea lions are protected in most countries and cannot be killed without special permits. They are usually problems only around sea cages. Generally, exclusion nets surrounding individual netpens or the entire sea cage complex has been used to control these predators. Harassment techniques for these species have involved underwater sonic devices, but their effectiveness has been questionable.

River otters can be serious predators of freshwater aquaculture facilities, especially brood fish. These otters are usually most problematic during the winter months when natural food becomes scarce. River otters can be controlled by trapping. Minks also predate fish but do not usually cause severe losses.

Many mammals around aquaculture facilities are more pest than predator. These include beaver, muskrats, nutria, rats, and mice (2). Beaver can dam up water

Table 1. Harassment Methods Frequently Used on Various Bird Predators[a]

Bird Species	Control Method[b]								
	AC	CE	DC	L	OW	PF	P	WS	SC[c]
Cormorants	X	X			X		X		X
Anhingas	X	X			X		X		X
Pelicans	X	X			X		X		
Herons, egrets, etc.[d]	X	X	X[e]	X[f]	X[g]	X	X		X
Kingfishers	X	X					X	X	
Gulls and terns	X	X	X[h]		X		X	X	X
Ducks	X	X					X		X
Hawks and eagles	X	X			X		X	X	

[a]Effectiveness is highly variable and a combination of methods are often utilized.
[b]Control methods are automatic cannons (AC), complete enclosure (CE), distress or alarm calls (DC), lights (L), overhead wires (OW), perimeter fencing (PF), pyrotechnics (P), water sprays (WS), and scarecrows (SC).
[c]Scarecrows can include human effigies, raptor models (e.g., owls), and reflective strips. Moving or animated scarecrows work better than stationary ones.
[d]This group includes herons, egrets, bitterns, ibises, and storks.
[e]Distress calls work only with some species (e.g., black-crowned night heron).
[f]Lights are only effective with night-feeding species (e.g., black-crowned night heron and great blue heron).
[g]Overhead wires are only somewhat effective on larger species (e.g., great blue heron and great egret) that fly in to land in shallow water.
[h]Distress calls are somewhat effective on gulls, but not on terns.

supplies, especially creeks. The burrowing activities of beaver, muskrats, and nutria can severely damage pond levees, causing ponds to leak and making vehicular access dangerous. Rats and mice can consume and contaminate aquacultural feeds. Prevention measures include proper storage of feeds and elimination of access points (i.e., "rat proofing") and alternative food sources (e.g., garbage). Control of all these pests usually involves trapping and/or poisoning.

Aquaculture, with its intensive confinement of aquatic organisms, is attractive to predators. Losses are often difficult to quantify. Hoy et al. estimated that wading birds could cause losses of up to $10,000 per week on some bait fish farms during fall migration (18). Stickley and Andrews estimated catfish losses in 1988 of $3.3 million due to double-crested cormorants (23). They also estimated that in the same year, bird depredation control efforts were costing the catfish industry $2.1 million. It is prudent that the aquaculturist work with regulatory agencies, be aggressive with prevention and harassment techniques, keep good records, and work with other local producers to combat predators.

BIBLIOGRAPHY

1. R.L. Usinger, *Aquatic Insects of California*, University of California Press, Berkeley, CA, 1963.
2. J.W. Avault, Jr., *Fundamentals of Aquaculture*, AVA Publishing, Baton Rouge, LA, 1996.
3. Joint Subcommittee on Aquaculture, *Guide to Drug, Vaccine, and Pesticide Use*, Texas Agricultural Extension Service, *http://ag.ansc.purdue.edu/aquanic/publicat/govagen/usda/gdvp.htm*, 1994.
4. M.P. Masser and J.W. Jensen, *Calculating Treatments for Ponds and Tanks*, Southern Regional Aquaculture Center, No. 410, 1991.
5. E.B. May and D.G. Bland, *Proc. Southeastern Assoc. Game Fish Comm.* **23**, 519–521 (1969).
6. J.B. Glude and K.K. Chew, *Proc. N. Pacific Aquaculture Symp.* University of Alaska, 1980.
7. C.G. Hurlburt and S.W. Hurlburt, in *Mussel Culture and Harvest: A North American Perspective*, Elsevier, New York, 1980.
8. R.D. Barnes, *Invertebrate Zoology*, W.B. Saunders, Philadelphia, 1966.
9. H.S. Swingle, *Trans. Amer. Fish. Soc.* **89**(2), 142–148 (1960).
10. D. Popper and T. Lichatowich, *Aquaculture* **5**(2), 213–214 (1975).
11. R.A. Schnick, *U.S. Fish and Wildlife Serv*, Rep. No. FWS-LR-74-15 (1974).
12. C.E. Boyd, *Water Quality in Ponds for Aquaculture* Alabama Agricultural Experiment Station, Auburn University, 1990.
13. D. Moss, *Alabama Dept. Conser.* **24**(6), 9 (1953).
14. W.P. Gorenzel, F.S. Conte, and T.P. Salmon, in *Prevention and Control of Wildlife Damage*, 1994.
15. A.R. Stickley, *Avian Predators on Southern Aquaculture*, Southern Regional Aquaculture Center, No. 400, 1990.
16. L. Line, K.L. Garrett, and K. Kaufman, *The Audubon Society Book of Water Birds*, Harry N. Abrams, Inc., New York, 1987.
17. J.V. Huner and G.R. Abraham, *Observations of Wading Bird Activity and Feeding Habits in and around Crawfish Ponds in South Louisiana With management Recommendations*, Mimeograph, 1981.
18. M.D. Hoy, J.W. Jones, and A.E. Bivings, *Eastern Wildl. Damage Control Conf.* **4**, 109–112 (1989).
19. J.L. Beynon, D.L. Hutchins, A.J. Rubino, and A.L. Lawrence, *J. World Maricult. Soc.* **12**(2), 63–70 (1981).
20. G.A. Littauer, *Avian Predators*, Southern Regional Aquaculture Center, No. 401, 1990.
21. G.A. Littauer, *Control of Bird Predation at Aquaculture Facilities*, Southern Regional Aquaculture Center, No. 402, 1990.
22. J.V. Huner, ed., *Management of Fish-Eating Birds on Fish Farms: A Symposium*, 1993.
23. A.R. Stickley and K.J. Andrews, *Eastern Wildl. Damage Control Conf.* **4**, 105–108 (1989).

PROBIOTICS AND IMMUNOSTIMULANTS

WENDY M. SEALEY
Texas A&M University
College Station, Texas

OUTLINE

Immunostimulants and probiotics are two promising options to aid in disease control. Interest in the use of immunostimulants and probiotics in aquaculture to improve aquatic animal health and prevent disease has increased rapidly in the last 10 years, due, in part, to limitations associated with the treatment of disease outbreaks in foodfish. Immunostimulants increase disease resistance by causing up-regulation of host defense mechanisms against pathogenic microorganisms. These substances are often grouped by either function or origin and consist of a heterogeneous group that includes vitamins and minerals; animal-, bacterial-, and fungal-derived compounds; and several synthetic compounds. Probiotics decrease the frequency and abundance of pathogenic or opportunistic pathogenic organisms in the environment. Probiotics are live microorganisms that are often introduced into the food chain to shift the microbial balance from disease-causing microorganisms to beneficial microorganisms.

IMMUNOSTIMULANTS

The immune system of fish and other cold-blooded vertebrates is similar to that of mammals in that it is divided into both specific and nonspecific branches. The immune system of invertebrates differs from that of all three of the former in that the immune system of invertebrates lacks a specific branch. An immunostimulant may stimulate the specific immune system when given along with an antigen, or it may stimulate the nonspecific immune system when given alone. The immunostimulatory effects of certain vitamins and minerals in aquatic species have been reviewed previously and will not be addressed here (1,2).

Immunostimulation of Specific Host Defense Mechanisms

A specific immune response is dependent upon the host having prior exposure to an antigen and requires recognition and subsequent activation of the immune response through a coordinated effort of antigen-presenting cells, B lymphocytes, and T cells. Vaccination is the best known method of specific immunostimulation. (See the entry "Vaccines.") Immunostimulatory products other than vaccines have also been effective at improving specific immune responses, by acting as adjuvants. Adjuvants are compounds that are used in conjunction with a specific antigen to increase intensity and duration of the specific response against that antigen. Classically, adjuvants were injected with vaccines to intensify antibody production by holding the antigen in tissues for slow release; most oil-based adjuvants intensify antibody production in this manner. Other products, such as dextran sulfate, act as adjuvants by protecting the antigen from rapid clearance, which could lead to a decreased response to the antigen.

Oil-based products, such as Freund's complete adjuvant (CFA) and various modifications of CFA, were initially used in fish and have been shown to increase antibody production in multiple species. When salmonids are injected with CFA or a modified version of incomplete Freund's adjuvant (MIFA) containing bacterins (whole-killed bacteria), an increase in antibody production is generally observed (3). Oil-based adjuvants also have been shown to increase protection against bacterial infections in fish. For example, injections of oil-based adjuvants combined with *Aeromonas salmonicida* bacterins increased protection of rainbow trout (*Oncorhynchus mykiss*), coho salmon (*O. kisutch*), and brook trout (*Salvelinus fontinalis*) against furunculosis (4,5).

Adjuvants also can increase the action of cell types involved in the specific immune response by attracting them to the injection site, thus allowing greater numbers of specific cell types to be exposed to the antigen. Compounds with mycobacterial components, such as CFA, bacille calmette guerin (BCG), muramyl dipeptide (MDP), and the synthetic drugs levamisole and FSK-565, function in this way by stimulating the attraction and subsequent increased production of T cells in terrestrial mammals. These compounds have also been used in aquatic species, with similar results.

Due to the damage of tissues surrounding the injection site often encountered with many of the classical compounds, newly developed adjuvant products focus on increasing cell activation. Recently, studies have examined the ability of β-glucans, polysaccharide derivatives from yeast and fungi, to increase antibody responses in fish and have turned up varied results (6–8). Glucans are believed to function as nonclassical adjuvants by increasing macrophage activation, thus, theoretically, increasing antigen presentation to T cells, resulting in a more vigorous antibody response. A positive aspect of glucans as adjuvants is their ability to be administered not only by injection, but also orally or by immersion; the latter two methods do not cause substantial tissue damage.

Immunostimulation of Nonspecific Host Defense Mechanisms

Most immunostimulatory compounds examined in fish have been shown to have immunoenhancing potential through improvement of nonspecific immune responses of fish. The nonspecific branch of the immune response, in contrast to the specific branch, does not require prior exposure to an antigen and consists of barriers, such as skin, scales, lytic enzymes, and phagocytic cells. The nonspecific immune response is also considered to be the first line of defense against invading organisms. It has been hypothesized that fish and invertebrates are more reliant on this branch of the immune response. For these reasons, a large portion of the research on immunostimulation has focused on up-regulating the nonspecific immune response. Mononuclear phagocytes and granulocytes are the key components of the cellular nonspecific response of fish. Hemocytes and the prophenoloxidase system are the primary defense mechanisms of shrimp. The activation states of these cells and enzyme systems are often used as measures of nonspecific immunostimulation. Other measures employed for this purpose include cell migration, phagocytosis, and bactericidal activity, as well as changes in numbers of leukocytes and the activation potential of cells upon stimulation, as measured by oxidative radicals and enzymes.

Common Immunostimulants Used in Aquaculture

Immunostimulants that have been examined most frequently for their ability to increase the nonspecific immune responses of aquatic animals include glucans (9–14) and the synthetic drug levamisole (15–18). Additionally, animal-derived products, such as chitin (19,20) and abalone extract (21); bacterial-derived products, such as MDP; and plant-derived alginates, such as k-carrageenan (22,23) and spirulina (24), have been examined.

Glucans appear to show the most promise of all immunostimulants thus far examined in fish and shrimp. β-glucans are insoluble polysaccharides consisting of repeating glucose units that can be joined through β1–3 and β1–6 linkages when derived from yeast and mycelia fungi (10), and β1–3 and β1–4 linkages when derived from barley (25). The source and extraction process by which these glucans are produced can greatly affect their immunostimulatory capacity. Engstad et al. (26) have suggested that the mechanism through which these agents induced protection in Atlantic salmon (*Salmo salar*) was by increasing lysozyme and complement activation; results observed in other fish species seem to support these results (27,28). Increased oxidative capacity of phagocytic cells also has been suggested as the mechanism through which β-glucans enhance nonspecific resistance (29,30).

In vitro studies have attempted to further examine the potential of glucans to stimulate the phagocytic cells of aquatic animals and the mechanisms by which the glucans exert their effects (31–33). When macrophages were cultured in the presence of glucans, they underwent spreading and membrane ruffling, which are indicative of activation. In addition, these cells demonstrated increased pinocytosis, increased superoxide anion, and elevated acid phosphatase levels. However, studies indicate a suppression of response at the highest glucan levels examined.

The immunostimulatory potential of levamisole in fish is of considerable interest in the United States and elsewhere, because it has approval by the U.S. Food and Drug Administration for treatment of helminth infections in ruminants (3). Levamisole, which is a synthetic phenylimidazothiazole (21), has been shown to have the ability to up-regulate nonspecific immune responses in carp (Cyprinidae) (20), rainbow trout (3), and gilthead seabream (*Sparus aurata* sp.) (21).

Certain animal products and extracts also have nonspecific immunostimulatory potential. For example, injections of abalone extract and chitin increase phagocytic response and natural killer cell activity in fish (23–25). Increased protection against *A. salmonicida* has been observed when brook trout were injected with chitin (34). Products derived from various algal species, such as carrageenans, linear sulfated poly-d-galactans composed of repeating disaccharide units (26), and the blue-green algae *Spirulina platensis*, (28) also show nonspecific immunostimulatory potential. Unfortunately, this up-regulation of the nonspecific response does not always correlate with resistance to bacterial infections.

The foregoing information provides a brief coverage of common immunostimulants used in aquaculture and their effects on the nonspecific immune response of fish. Many other immunostimulants have been tested for the ability to increase immune function in aquatic species (3,35,36).

Research on Immunostimulants in Disease Resistance of Aquatic Animals

Although the scientific literature on immunostimulants seems to indicate that they have a positive effect on nonspecific immune responses, their effect on disease resistance is less clear. Immunostimulation of disease resistance appears to be dependent on the type of aquatic species and disease agent. β-glucans have been the most studied immunostimulant in terms of disease resistance (3,35,36). β-glucans have been reported to increase protection against *Vibrio anguillarum*, *A. salmonicida*, and *A. hydrophila* in a variety of salmonid species, as well as in carp, gilthead seabream, milkfish (*Chanos chanos*), tilapia (*Oreochromis niloticus*), and turbot (*Scophthalmus maximus*). β-glucans also have been shown to increase protection against *Edwardsiella tarda* in carp, milkfish, and tilapia, as well as against *Streptococcus* sp. in yellowtail (*Seriola quinqueradiata*). However, β-glucans have not been shown to be effective at increasing protection against *Edwardsiella ictaluri* in channel catfish (*Ictalurus punctatus*) *Pasteurella* sp. in yellowtail (*S. quinqueradiata*) or *Yersinia ruckerii* in Atlantic salmon.

β-glucans have also received attention in recent years for their ability to increase disease resistance in shrimp, largely because the limitations of the shrimp immune response and the nature of the disease agents make the development of effective vaccines impractical. β-glucans have been shown to increase resistance of both kuruma (*Marsupenaeus japonicus*) and tiger shrimp (*Penaeus monodon*) to *Vibrio* infection (37).

Other Considerations

Certain types of immunostimulants appear to have potential for incorporation into disease prevention and control regimes—in particular, the β-glucans and levamisole. However, some concerns about immunostimulants must still be addressed. Many factors contribute to the success or failure of an immunostimulation regime, including the type of immunostimulant, the dosage (level, route, and application), the fish species, and the disease organism(s) against which protection is sought.

The type of immunostimulant chosen to be incorporated into an immunostimulation protocol is of obvious importance, due to the large variation of responses. The proper dosage of the immunostimulant also is critical, because failure to administer a sufficient amount could result in no enhancement of response, while excessive quantities can lead to immunosuppression. In addition, considerable variations in immunostimulant effectiveness have been noted with different routes of administration. Injection protocols clearly have produced the best results. However, injections are problematic, due to the considerable effort necessary for their administration, the excessive stress they cause on the fish, and the fact that they are often cost prohibitive. Bath immersion has also been shown to be somewhat effective, but is generally not practical for larger size fish. Oral application of immunostimulants therefore appears to be the route of choice. Unfortunately, oral applications also have the most varied success rates. In addition, to date, much of the research performed has focused on using immunostimulants to prevent disease by applying the compound prior to infection. Further research addressing dosage and timing of prophylactic and therapeutic oral treatments, and further investigation of the adjuvant activity of many of these compounds, seem warranted.

PROBIOTICS

Probiotics may be defined as live organisms that beneficially affect the microbial balance of the host (38). A variety of mechanisms have been proposed to explain the actions of probiotics, including their production of antimicrobial substances, competition for adhesion receptors, and the provision of nutrients and direct immunostimulatory effects (39).

Administration Strategy

The strategy often followed for administering probiotics includes isolating bacteria from mature fish and including these favorable bacteria in the feed of juvenile fish of the same species (40). Lactic acid bacteria, such as *Carnobacterium* sp., which produce bactericins, are often used in this manner (41). Lactic acid bacteria have been isolated from the intestine of Atlantic salmon and can inhibit the growth of pathogenic bacteria, such as *V. anguillarum*, *A. salmonicida* (42), and *A. hydrophila* (43). Challenge experiments have shown that feeding larval turbot with rotifers enriched in lactic acid bacteria (44) and giving cod (Gadidae) fry dry feed containing lactic acid bacteria (39) improved their resistance to mortality due to *Vibrio* infection. However, no effect on the protection of salmon fry against *A. salmonicida* was noted when lactic acid bacteria were added to dry feed (42).

Mechanisms

The mechanism(s) through which probiotics infer protection remain speculative. Gildberg et al. (39) were unable to demonstrate growth inhibition of *V. anguillarum* in an *in vitro* mixed culture of *V. anguillarum* and *Carnobacterium divergens*, even though the lactic acid bacteria had elicited disease protection when included in the diet. This observation seems to rule out the production of antimicrobial agents as the mechanism behind increased protection in Atlantic cod (*Gadus morhua*).

Considerations

One concern of including probiotics into the food chain or environment of larval fish is that the probiotics's colonization and proliferation within the host must remain under control. Gatesoupe (44) observed a negative correlation between high concentrations of lactic acid bacteria and survival in turbot larvae fed greater than 2×10^7 colony-forming units, indicating the need for careful determination of inclusion levels of lactic acid bacteria in order to obtain beneficial, not detrimental, effects of probiotics. In addition, these organisms have limited scope in their application to commercial aquaculture production. The successful isolation and cultivation of probiotic bacteria is a labor-intensive effort that seems best suited to larval fish culture when the environment or feed provided to mass quantities of fish may be treated with less probiotics and less labor.

CONCLUSIONS

Immunostimulants and probiotics appear to be useful tools to include in the arsenal of disease control and prevention. However, these compounds will not replace vaccines, proper nutrition, or good management techniques. The strength of these compounds appears to lie in their ability to enhance larval fish culture when the specific immune response has not yet developed and to improve nonspecific immune function against a variety of pathogens when a disease-causing agent has not been identified. Through such functions, the compounds may significantly aid the aquaculture industry, although some caution in their use is urged until dosage rates are optimized. Additional research is needed to define the specific dosage rates and efficacy of various compounds for a variety of aquatic species and their pathogens, to decrease costs of the immunostimulant or probiotic used and negate losses that could be encountered with improper supplementation levels.

BIBLIOGRAPHY

1. M.L. Landolt, *Aquaculture* **79**, 193–206 (1989).
2. V.S. Blazer, *Ann Rev. Fish Dis.* **2**, 309–323 (1992).
3. D.P. Anderson, *Ann. Rev. Fish Dis.* **2**, 281–307 (1992).
4. R.C. Cipriano and S.W. Pyle, *Can. J. Fish. Aquat. Sci.* **42**, 1290–1295 (1985).
5. G. Olivier, T.P.T. Evelyn, and R. Lallier, *Dev. Comp. Immunol.* **9**, 419–432 (1985).
6. M.O. Baulny, C. Quentel, V. Fournier, F. Lamour, and R. Le Gouvello, *Dis. Aquat. Org.* **26**, 139–147 (1996).
7. G. Rorstad, P.M. Aasjord, and B. Robertsen, *Fish and Shellfish Immunology* **3**, 179–190 (1993).
8. L. Nikl, T.P.T. Evelyn, and L.J. Albright, *Dis. Aquat. Org.* **17**, 191–196 (1993).
9. T. Yano, R.E.P. Mangindaan, and H. Masuyama, *Nippon Suisan Gakkaishi* **55**, 1815–1819 (1989).

10. B. Robertsen, G. Rorstad, R. Engstad, and J. Raa, *Journal of Fish Diseases* **13**, 391–400 (1990).
11. L. Nikl, L.J. Albright, and T.P.T. Evelyn, *Dis. Aquat. Org.* **12**, 7–12 (1991).
12. D. Chen and A.J. Ainsworth, *J. Fish Diseases* **15**, 295–304 (1992).
13. A.J. Ainsworth, C.P. Mao, and C.R. Boyle, in J.S. Stolen and T.C. Fletcher, eds., *Modulators of Fish Immune Responses* SOS Publications, Fair Haven, NJ, 1994.
14. P.L. Duncan and P.H. Klesius, *Journal of Aquatic Animal Health* **8**, 241–248 (1996).
15. A.K. Siwicki, D.P. Anderson, and O.W. Dixon, *Dev. Comp. Immunol.* **14**, 231–237 (1990).
16. T. Baba, Y. Watase, and Y. Yoshinaga, *Nippon Suisan Gakkaishi* **59**, 301–307 (1993).
17. V. Mulero, M.A. Esteban, J. Munoz, and J. Meseguer, *Fish and Shellfish Immunology* **8**, 49–62 (1998).
18. A.K. Siwicki, *J. Fish Biology* **31**, 245–246 (1987).
19. M. Sakai, H. Kamiya, S. Ishii, S. Atsuta, and M. Kobayashi, *Diseases in Asian Aquaculture* **1**, 413–417 (1991).
20. A.J. Siwicki, D.P. Anderson, and G.L. Rumsey, *Veterinary Immunology and Immunopathology* **41**, 125–139 (1994).
21. M. Sakai, H. Kamiya, S. Atsuta, and M. Kobayashi, *J. Appl. Ichthyol.* **7**, 54–59 (1992).
22. K. Fujiki, D.H. Shin, M. Nakao, and T. Yano, *J. Fac. Agr., Kyshu Univ.* **42**, 113–119 (1997).
23. K. Fujiki, D.H. Shin, M. Nakao, and T. Yano, *Fisheries Science* **63**, 934–938 (1997).
24. P.L. Duncan and P.H. Klesius, *J. of Aquat. Animal Health* **8**, 308–313 (1996).
25. W.S. Wang and D.H. Wang, *Taiwan J. Vet. Med. Anim. Husb.* **66**, 83–91 (1996).
26. R.E. Engstad, B. Robertsen, and E. Frivold, *Fish and Shellfish Immunology* **2**, 287–297 (1992).
27. M. Santarem, B. Novoa, and A. Figueras, *Fish and Shellfish Immunology* **7**, 429–437 (1997).
28. H. Matsuyama, R.E.P. Mangindaan, and T. Yano, *Aquaculture* **101**, 197–203 (1992).
29. R.A. Dalmo, J. Bogwald, K. Ingebrigtsen, and R. Seljelid, *Journal of Fish Disease* **19**, 449–457 (1996).
30. D.P. Anderson and A.K. Siwicki, *Prog. Fish-Cult.* **56**, 258–261 (1994).
31. R.A. Dalmo and R. Seljelid, *Journal of Fish Diseases* **18**, 175–185 (1995).
32. J.B. Jorgensen, G.J. Sharp, C.J. Secombes, and B. Robertsen, *Fish and Shellfish Immunology* **3**, 267–277 (1993).
33. A. Figueras, M.M. Santarem, and B. Novoa, *Fish Pathology* **32**, 153–157 (1997).
34. Y. Kajita, M. Sakai, S. Atsuta, and M. Kobayashi, *Fish Pathology* **25**, 93–98 (1990).
35. J. Raa, G. Rorstad, R. Engstad, and B. Robertsen, *Diseases of Asian Aquaculture* **1**, 39–50 (1992).
36. O. Vadstein, *Aquaculture* **155**, 401–417 (1997).
37. Y.L. Song, J.J. Liu, L.C. Chan, and H.H. Sung, in R. Gudding, A. Lillehaug, P.J. Midtylng, and F. Brown, eds., *Fish Vaccinology* Basel, Karger, Austria, 1997.
38. R. Fuller, *J. Appl. Bacteriol.* **66**, 365–378 (1989).
39. A.J. Montes and D.G. Pugh, *Vet. Med.* **88**, 282–288 (1993).
40. A. Gildberg, H. Mikkel, E. Sandaker, and E. Ringo, *Hydrobiologia* **352**, 279–285 (1997).
41. J. Nousiainen and J. Setala, in S. Salminen and A. von Wright, eds., *Lactic Acid Bacteria* Marcel Dekker, New York, 1993.
42. A. Gildberg, A. Johansen, and J. Bogwald, *Aquaculture.* **138**, 23–34 (1995).
43. C.B. Lewus, A. Kaiser, and T.J. Montville, *Appl. Environ. Microbiol.* **57**, 1683–1688 (1991).
44. F.J. Gatesoupe, *Aquat. Living Resour.* **7**, 277–282 (1994).

PROCESSING

Russell Miget
Texas A&M University
Corpus Christi, Texas

OUTLINE

Aquaculture can be broadly divided into the categories of subsistence food production (i.e., food production that relies on aquatic plant production as well as low-density production of animals that feed near the bottom of the food web, such as carp, tilapia, and milkfish), and commercial production of relatively high valued food commodities (food production that relies on high density monoculture of animals that generally feed at a much higher trophic level on prepared feeds). This practice can also be thought of as the conversion of food of low economic value (i.e., grains and fishmeal) to food of high economic value such as shrimp, salmon, and sea bass. In some countries both subsistence and commercial production are utilized; many of the commercial products are often exported. Of the thousand-plus species of aquatic plants and animals consumed by humans, only about five percent are cultured, and of this five percent, less than one percent have their entire life cycle under control. This system results in a reliable supply of "seed stock" year round, similar to that used in agricultural production.

This entry focuses on the processing of the relatively high-value molluscs, crustaceans, and finfish that are raised in captivity for domestic consumption and/or international trade. This approach will not include processing associated with subsistence aquaculture. Likewise, many traditional processes associated with seafoods, such as salting, curing, and mincing in surimi production, are not addressed, since these processes have historically relied on a relatively inexpensive supply of fish and shellfish in order to be profitable. Such products will continue to be economically supplied by the wild harvest fishery for the foreseeable future.

SPECIES

Molluscs

Aquacultured species consist principally of bivalves—oysters, mussels, clams, and scallops. Unlike finfish or crustacean aquaculture, which is most often carried out in monoculture ponds, molluscs are generally hatchery-reared and then raised to a size at which they can be placed in open waters for growout. Gastropods, such as abalone, queen conch, and certain species of whelks, are harvested primarily from the wild; however, these animals are in high demand, and there are active research programs underway to develop culture techniques for them. Cephalopods, which include squid, cuttlefish, and octopus, have thus far proven to be too expensive to raise in captivity, except in the specialty medical-research and ornamental markets.

Market Forms. Live bivalve molluscs are consistently the most valuable market form, but they require special handling and shipping procedures, and have a limited shelf life. Oysters (*Crassostrea virginica*, *Crassostrea gigas*, *Ostrea edulis*) are commonly consumed raw and therefore must be grown in waters that are certified by state shellfish authorities as meeting requirements to be (statistically) free of introduced human pathogens. Still, naturally-occurring pathogens which are concentrated by these filter-feeding animals can pose a health risk to people who consume them raw. Immunocompromised individuals are at the greatest risk. Unlike oysters, mussels and clams are generally cooked before consumption; the cooking effectively kills any pathogens which may be present.

Bivalve molluscs can survive for several days out of water simply by closing their shells tightly. However, in order to survive in that state, their metabolism must be significantly reduced through cooling. Most bivalves can survive for up to two weeks without significant mortality when stored at about 4 °C (40 °F). The exceptions to this rule are certain clam species for which 4 °C (40 °F) is lethal. The animals are stored dry in a cool, moist room. If stored in water, they will open up and begin pumping, and will eventually foul their own water and deplete the oxygen. Animals that die during live storage will open slightly (gape) and should be discarded. One exception to dry storage is soft-shell clams (*Mya arenaria*). These clams are routinely placed in clean seawater for a few hours to allow the discharge of grit and sand from the intestine, after which they are placed back in dry, cool storage.

There has been some thought given to wet holding (not storage) of oysters in order to improve their flavor. Because oysters are osmoconformers, shell stock placed in higher salinity waters will, over several hours, conform to the salinity of the surrounding water and result in a product some consumers find more flavorful. However, if such a process is employed, rigorous water quality standards (similar to those required for oyster depuration) would have to be adhered to. This process involves continuous filtration and disinfection of the recirculating high-salinity water.

More commonly, however, processing bivalve molluscs for the live market consists simply of breaking up clumps, washing, sorting to size, and packaging in breathable material like onion sacks. Mussels (*Mytilus edulis*) destined for the live market usually have their byssal threads (beard) removed first. The preferred mussel size is about 5 centimeters (2 in.) long.

Cultured mussels supply the live market. These, and marinating mussels, are the only types that have brought profit in the western hemisphere. Increasingly large quantities of green-lipped mussels (*Perna canaliculus*) are being imported to the United States from New Zealand. Korea, another significant exporter of mussels, ships mostly canned product to the rest of the world.

The live bivalve market has a few marketing idiosyncrasies of which one should be aware. The halfshell oyster market prefers medium-sized animals 7 to 9 centimeters (3 to 3.5 in.), whereas small hard clams (*Mercenaria mercenaria*) about 4 to 7 centimeters (1.5 to 3 in.), sold live, are more expensive than the larger animals, which are all wild-harvested and are usually processed further into soups and chowders.

The second most valuable market form for bivalves is the fresh market. Unlike many crustaceans and finfish that freeze exceedingly well (i.e., that make it difficult for the consumer to discern the difference between a fresh and a frozen product once it is prepared), certain bivalves (oysters and clams in particular) tend to diminish significantly in flavor and texture once frozen. Thus, considerable effort is made to develop and maintain fresh markets for these products.

Oysters are mechanically or hand-shucked to remove the meats for the fresh market. The entire animal, not just the muscle holding the shell together, is consumed, hence the public health concern discussed earlier. Mechanical shucking is faster but does not yield as uniformly intact a product as hand shucking and is therefore often employed when the product is going to be processed further, for example, breaded. Hand-shucked oysters are sorted by size as they are shucked, and are packed in gallon (3.8 L) containers (in the U.S) according to the number of meats (animals) per gallon (3.8 L): very small, over 500 meats; small (standard), 301 to 500; medium (select), 211 to 300 meats; large (extra select), 160–210; and extra large, less than 160. However, standards of identity for size no longer apply, so the aforementioned sizes are now voluntary grades. Unlike the live market, which prefers select size animals, the price for shucked oysters increases with size (except for *C. gigas*). After shucking and sorting, oysters are cleansed of shell fragments and other debris from the outside of the shell through a process of "blowing." This process consists of literally blowing the oyster meats around inside a container, using jets of air that enter from the bottom. This allows grit and shell attached to the shucked meats to fall to the bottom. After 10 to 15 minutes the air is shut off and the meats are bathed in running water for another 20–30 minutes, after which they are removed and allowed to drain. The oysters are then packed "dry" (without additional water) in appropriate containers–increasingly, plastic tubs vs. traditional metal cans. Plastic is less expensive, is easily imprinted with company logos, recipes, and nutritional information, and allows the consumer to see the product through the translucent material. Although packed without water, a certain amount of liquid leaves the oysters after they are packed. This is referred to as "free liquor" and may amount to between 5 and 30% of the weight of the packed oyster meats, depending on the season and harvest location (there is generally more loss from summer-harvested oysters), and the condition (post-spawn oysters tend to be more "watery").

Cultured scallops (*Pecten maximus* in Europe, *Patinopecten yessoensis* in Japan) are hand-shucked for the fresh market. It is interesting to note that, in the U.S., only the adductor muscle of the scallop is eaten, whereas in Europe the entire shucked animal is consumed.

Certain molluscs maintain their original texture and flavor much better than others after being frozen and stored for a period of time. Product quality depends not only on species differences, but on the rate of freezing and packaging, and the temperature while in frozen storage.

The freezing of oysters may result in a loss of up to 30% of internal fluids (free liquor) upon thawing. Oysters also tend to darken once frozen; the degree of color change is proportional to how slowly the product is frozen. Oysters are rarely frozen in traditional metal cans, since the relatively slow freezing rate results in excess liquor loss as well as flesh darkening. However, oysters that are rapidly frozen in flattened plastic bags in which all air is evacuated, and that are then stored at −18 °C (0 °F) maintain excellent quality for 10 months, (1). However, clams that are packaged, frozen, and stored similarly have a storage life of only four to six months before they become noticeably tough and rancid. Thus, most clams are destined for either the live or further-processed (chowder) markets.

Crustaceans

Tropical saltwater shrimp of the family Penaeidae are cultured worldwide in subtropical and tropical climates. Various species are grown in shallow coastal ponds, and reach market size in four to five months. Although large shrimp continue to command higher prices than small shrimp, aquaculture producers have determined that a medium-size animal [30–40 tails per pound (per 0.45 kg)] generally brings the best investment return. Growing animals appreciably beyond this size results in a lower feed-conversion ratio, increased mortality, additional operating costs, and, in climates where shrimp can be grown year-round, fewer annual crops.

Aquacultured shrimp have a distinct advantage over their wild-captured counterparts in that they generally fall within one or two size counts at harvest. This is a distinct advantage for companies that supply restaurants that are concerned with providing customers a consistent year-round "plate coverage" (i.e., number of shrimp per serving) at a specified cost.

Count size refers to the number of shrimp tails per pound (per 0.45 kg), beginning with the largest, which are simply classed as under 10, then grouped as 10–15s, 16–20s, and so forth until the smaller shrimp become classed by tens, beginning with 51–60s, 61–70s and so forth. It is unusual to find cultured shrimp smaller than 51–60s, since only larger warmwater shrimp are valuable enough to make a shrimp farming operation profitable. Many of the wild-harvested small shrimp (both warm- and coldwater species) are automatically peeled and subsequently breaded (popcorn shrimp) or used for value-added products like gumbo, or canned shrimp.

It is common to treat wild-harvested shrimp with a chemical to retard the darkening of pigments in the skin directly beneath the shell, a condition that is known as blackspot. While this blackening is esthetic and does not necessarily reflect the quality of the shrimp, it is perceived as a defect by the consumer and should thus be avoided. Shrimp vessels must use some type of chemical to retard blackspot formation, since they often remain at sea for a month or more, giving the enzymes that are responsible for the reaction plenty of time to work. On the other hand, due to the short period of time between harvest and processing (a matter of hours) some shrimp culturists have opted to eliminate the use of anti-melanosis agents (generally sodium bisulfite) as part of a chemical-free marketing strategy. While this has apparently worked well for some producers, others have been disappointed to find their untreated shrimp with a severely limited fresh shelf life (relative to blackspot development), or more commonly, turning black within hours after being thawed (following several months of frozen storage).

Unlike that which exists for bivalve molluscs, only a minuscule market exists for live shrimp. Only extremely upscale restaurants and a few specialty seafood shops offer live animals. Shrimp are considerably more difficult to ship and to subsequently maintain in aquaria than are live bivalves, which need only to be refrigerated. Thus, the cost for live shrimp becomes prohibitive for most consumers.

Although the market for fresh shrimp demands a premium price over similarly sized frozen product, the price difference in most retail settings is not great. This is because shrimp freeze extremely well, making it difficult for all but the most discerning consumers to tell the difference between fresh and previously frozen product. Shrimp are sold either whole (European market), head-off or tails (United States and Asian markets), peeled tails (worldwide market), fantail (peeled but leaving the uropod attached), or pieces (less than five of the six tail segments). Fantail shrimp are used for cocktail or specialty breading products whereas pieces are usually incorporated into salads or further processed products where shape is less important.

Since most of the world's production of cultured shrimp takes place in tropical countries for export, it is generally graded and then frozen in 425-kg (5-lb) waxed cardboard boxes with plastic bag inserts containing enough water to form a protective glaze around the entire block.

Increasingly, however, the largest customers (restaurants and supermarkets) are demanding individually quick frozen (IQF) shrimp which can be portioned by the piece while still frozen, and can then be rapidly thawed for either restaurant use or retail display.

Cultured crawfish include the red swamp (*Procambarus clarkii*), which is native to the United States, as well as the three Australian species, the marron (*Cherax tenuimanus*), the yabbie (*Cherax albidus-destructor*), and the red claw (*Cherax quadricarinatus*) These freshwater crustaceans survive well out of water and therefore bring the best price when sold live. Crawfish placed in high-humidity cool storage (8–9 °C or 46–48 °F) will survive for five to seven days. However, if storage temperatures fall below 3 °C (38 °F) the gills will frost over and the animals will suffocate. Likewise, crawfish should be transported in well-ventilated bags like onion sacks, never in tightly-sealed ice chests or plastic bags, or the animals will suffocate. Also, bags or other containers should not be so large as to impede the circulation of air to animals in the center. Crawfish intended to be sold live should be treated as gently as possible; i.e., sacks should not be thrown on and off trucks.

Some crawfish producers routinely "purge" live animals immediately after harvest by allowing them to remain in a shallow trough continuously supplied with clean water for 24 hours. This allows the animals to eliminate most of the content of their digestive tracts, which may contain gritty materials. Purged crawfish are reported to survive handling and cold-storage conditions better than untreated animals, and they also bring a better wholesale price, particularly from restaurateurs.

One relatively new market form that is increasing in popularity is that of soft-shell crawfish. Like all crustaceans, crawfish must molt in order to grow. Through manipulation of the water temperature and feeding of captive animals, they can be induced to molt, after which they are harvested, packaged and frozen. Prior to consumption, the head, which contains "stones," is removed.

Crawfish for the fresh market are generally sold as peeled tail meat. All tail meat is produced through hand peeling and is packed with or without the fat (i.e., digestive gland/hepatopancreas) attached.

Crawfish meat freezes extremely well; excess production can be frozen and stored for several months. However, if long-term storage is anticipated, the fat-laden hepatopancreas should be removed during the peeling process or the product will likely become rancid and result in a bitter taste. Dipping in 1.25% citric acid solution will retard discoloration during frozen storage. Until recently, most excess production was frozen and subsequently used in value-added products like bisque and etouffee. However, there is a developing market in both the U.S. and Europe for whole, boiled, frozen crawfish. In cooking, the live animals are added to boiling water, which is then allowed to come back to a boil. The crawfish are then removed, and immediately cooled and packaged for frozen storage.

Freshwater shrimp (*Macrobrachium rosenbergii*) culture developed rapidly in Hawaii during the 1980s, primarily to supply the local ethnic Asian market with live product. Animals can be live hauled in cool (21 °C, 70 °F) water at up to metric (0.5 lb/gal) for up to 24 hours with minimal mortality. The technology was soon introduced to the mainland U.S., but the demand for live animals was not sufficient to generate the high prices realized in Hawaii. To counter this economic problem, producers expanded the size of their farms to reduce production costs. This expansion necessitated the freezing of excess production. Most producers have subsequently found that only by selling a head-on product will they have a profitable operation. Freshwater shrimp (*M. rosenbergii*) lose 60% of their whole-body weight when deheaded, whereas saltwater shrimp suffer only a 40% loss. Thus, competing in the generic "shrimp tail" market is difficult. However, with their extremely long, slender-clawed appendages, whole animals provide attractive and sufficient "plate coverage" to consumers seeking an exotic menu alternative.

In the past, freezing head-on shrimp in conventional freezers resulted in a mushy texture. Recent work, however, has shown that whole animals that are individually quick-frozen (IQF) in cold tunnels and then stored at −18 °C (0 °F) have a shelf life of several months. Extreme care, however, must be taken when packaging IQF animals for shipping, to prevent breakage of the delicate long claws.

Alligators

Hide Processing. Since, in the recent past, alligators (*Alligator mississippiensis*) were considered an endangered species, each state regulatory agency restricts their processing to approved sites in order to assure that the animals were farm raised. Skinning, scraping, and curing must be carried out carefully to assure a top-quality product. Hides that are cut, scratched, or stretched, particularly the belly scales, have a greatly reduced value. Chilling prior to skinning makes this process easier. Hides are then scraped to remove fat and meat, and are then washed repeatedly to remove the remaining meat and blood residue. Fine-grained mixing salt is used to preserve the hide. Salt is rubbed thoroughly into the skin, which is then covered with about an inch of salt and rolled tightly to begin the curing process. Skins are allowed to drain and dry in a well-ventilated, cool enclosure. Hides are checked and resulted after about five days.

Meat Production. Alligator meat is cut in chunks and placed in waxed cartons (some states regulate the maximum amount of meat per carton) and sold either fresh or frozen. Deboned alligator meat comprises about 35–40% of the live weight.

Finfish

The culture of freshwater finfish species preceded marine species culture by several hundred years. There are two explanations for this: First, inland freshwater ponds, lakes, and streams offered relatively sheltered and controlled conditions conducive to the controlled production of aquatic plants and animals. Second, freshwater finfish generally produce much larger (but fewer) eggs which subsequently hatch into relatively large fry, which experience a relatively high survival rate to adulthood (all compared with marine finfish species).

It has been generally observed that freshwater finfish yield a somewhat longer fresh shelf life than their marine counterparts. This is thought to be due to a relative lack of osmoregulators in their body fluids. Osmoregulators organic compounds of low molecular weight that help keep saltwater fish from losing too much body fluid to their salty external surroundings via osmosis. However, these compounds are easily metabolized by spoilage bacteria once a fish dies. Freshwater fish, of course, have just the opposite problem, because they have body fluids that are saltier than the surrounding water. They deal with this problem by producing fewer organic osmoregulators and by excreting copious amounts of dilute urine.

Another general observation for any aquacultured finfish species, fresh or saltwater, is that the dark-colored flesh (usually beneath the lateral line) is inordinately oily and will likely become rancid if the product is stored for several months, in which case the dark flesh should be removed during processing.

The most common market forms of finfish include whole (small fish only), gilled and gutted (gills and entrails removed), dressed (head, intestines, and fins removed), steaks (cross-sections of dressed fish), or fillets (the sides of the fish cut away from the backbone). Fillets can be either boneless (i.e., rib cage bones removed) or not. In the United States, the preference tends toward a boneless piece of meat, which can significantly reduce yield depending on the species cultured.

As stated earlier, many of the more traditional processing treatments for finfish (i.e., canning, mincing, etc.) are not applicable to aquacultured products, due to the relatively high cost of production. The majority of these products are marketed fresh or frozen in one or more of the forms described in the previous paragraph. Although butchering and freezing are common to all aquacultured finfish, a few "preprocess" procedures which are unique to certain species will also be described.

Catfish (*Ictalurus punctatus*), one of the most commonly cultured freshwater finfish species, may acquire a muddy flavor in spring and summer, making it unacceptable to the consumer. Initially thought to be due to ingestion of sediment while feeding, further investigations revealed that the off-flavor is actually due to the uptake (through their gills and/or epithelial tissue) of one or both of the organic compounds geosmin and 2-methylisoborneal. These compounds are produced by blue-green algae and filamentous bacteria, which thrive in warm waters enriched with nutrients from uneaten feeds and fish excreta. Fortunately, these off-flavors can be "purged" from fish in a matter of days if the fish are transferred to an unaffected pond or if the algae/bacteria bloom subsides in the culture pond.

Net-pen-raised salmon (principally *Salmo salar*, the Atlantic salmon, and *Oncorhynchus kisutch*, the coho salmon) should be taken off their feed a week prior to harvest. The fish should be carefully removed from the cages with a minimum of struggle. This reduces the production of lactic acid, which is responsible for subsequent softening of the flesh. Fish should then be tranquilized in 1000-liter (263 gal) vats of chilled seawater through which carbon dioxide is bubbled. Individuals are then bled by severing the gill arch or by direct incision into the aorta. After bleeding, the salmon are returned to chilled seawater (below 8 °C; 46 °F), graded by size, gutted, and boxed with ice for the fresh market or for further processing or freezing.

A small percentage of cultured catfish find their way into the live market, both in restaurants and increasingly in upscale retail grocery stores. In contrast, the majority of cultured tilapia (*Oreochromis* sp.) are sold live. Metropolitan areas with substantial ethnic Asian populations have been largely responsible for the growth of this aquaculture industry. Both of these species can be easily raised in high-density culture (often in indoor recirculating systems close to metropolitan areas), and live-hauled for up to 24 hours with minimal mortality.

Other species of fish which have been profitably raised include rainbow trout (*Oncorhynchus mykiss*), hybrid striped bass (*Morone saxatilis* × *M. chrysops*), red drum (*Sciaenops ocellatus*), gilthead bream (*Sparus aurata*), sea bass (*Dicentrarchus labrax*), sole (*Solea vulgaris*), and turbot (*Scophthalmus maximus*).

HANDLING FRESH AQUATIC PRODUCTS

When one considers the complex process of getting wild-harvested aquatic products, particularly seafood items, from harvest to the consumer, it is understandable that it is conservatively estimated that approximately 50% of the product's remaining shelf life is lost before the product even reaches the processor (with the exception of processing on-board factory trawlers). Compared with other meat items, which reach the processing plant live, one can see how the post-processing shelf life of seafoods is severely reduced (Fig. 1). The captive production of aquatic products has greatly increased their post-processing shelf life, since the trauma of capture is reduced and the lengthy refrigerated storage on-board a vessel is eliminated. Much like other muscle foods, aquacultured products can be harvested to meet market demands. However, unlike beef, which may be "aged" under controlled conditions to improve its texture, aquatic products are naturally

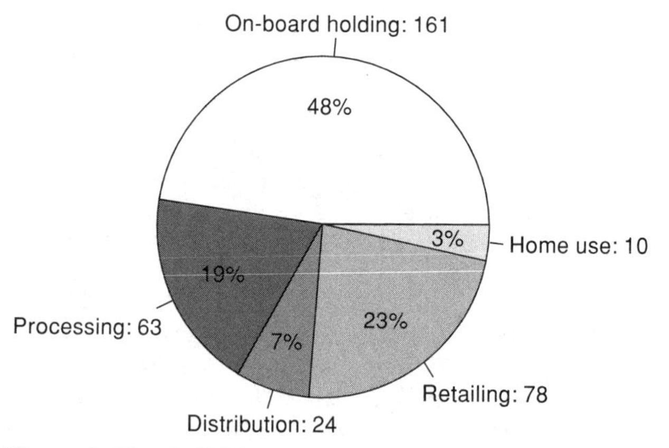

Figure 1. The shelf life timetable for wild-harvested, refrigerated seafood products. (in hours)

tender, and aging only serves to diminish their textural quality, due to catabolic enzymes which are active at lower temperatures than those in warm-blooded animals. Likewise, as previously mentioned, all aquatic animals (and seafoods in particular) contain compounds of low molecular weight that are readily metabolized by spoilage bacteria upon the death of the organism. Therefore, immediate chilling and subsequent maintenance of low temperature storage [near 0 °C (32 °F) or slightly cooler] of fresh product throughout processing and distribution is essential.

How long an aquacultured product can remain in prime freshness or acceptable quality under iced storage depends largely on its species. As a general rule, fatty fish[a] have an 8- to 10-day acceptable quality life, whereas lean species can be kept on ice for 14–18 days. Flatfish and tropical shrimp have been shown to maintain acceptable quality after more than 20 days.

Prime freshness and acceptable quality are not synonymous. Certain species of finfish which are highly sought after because of their extraordinary sensory quality (and are quite often consumed raw) are graded on their degree of freshness rather than degree of spoilage. Freshness can be numerically determined (and is referred to as the K-value) by measuring the chemicals released as a result of enzymatic (vs. bacterial) breakdown of cellular nucleotides in the first hours after death, well before bacterial spoilage takes place. Also, a fish in prerigor is considered fresh, whereas one which has resolved the stiffness of rigor mortis no longer displays the attributes of prime freshness that are demanded by raw fish connoisseurs. Following resolution of rigor, microbial inhibitors present in live animals are exhausted, allowing spoilage bacteria to outgrow on the abundant low-molecular-weight compounds that are naturally present in cellular fluids and resulting in the noticeable "off" odors that are associated with stale or spoiled fish.

For certain species of finfish, spiking the brain after chilling in ice water, or immediate slaughter (i.e., removing the brain), prolongs the onset of rigor by several hours. Aquacultured species have the potential to meet this market niche since, unlike for wild-captured animals, post-harvest operations (including shipping) can be carried out in a matter of hours rather than days.

The ideal cooling medium for fresh products is melting ice. Not only is the temperature of the meltwater 0 °C (32 °F), but the intimate contact between water and product results in efficient heat transfer. Ice storage prevents dehydration, while the washing action of meltwater has been shown to reduce the surface (skin) bacterial load. Flake ice is most often used to chill products because its large surface area leads to rapid delivery of meltwater over the product. Chunk and crushed ice tend to melt more slowly, and are more likely to damage the flesh during handling. Filleted or skinned products, however, should not be stored in melting ice, as the washing action tends to suck out soluble flavor components (freshwater has a lower osmotic pressure than body fluids), and may diminish the textural quality.

[a] Fatty fish are defined as those that have greater than 5% fat, whereas 2–5% is considered moderately fat, and less than 2%, lean. Most marine invertebrates are considered lean, because they have less than 2% fat. The remainder of the edible portion of aquatic animals is 18–20% protein and 75–78% water—along with trace amounts of carbohydrates, vitamins, and minerals.

A downside to the use of ice for shipping fresh product is its weight. Adequate icing can increase shipping charges by as much as 50%, a significant cost if products are being transported by air. The use of cryogenic chemicals (like those found in gel packs) to maintain low temperature during shipment offers some cost savings. Also, borrowing from the poultry industry, some shippers have begun blowing "snow" dry ice into master cartons of fresh seafood. Although this forms a frozen crust on products, if they are adequately packaged in moisture-proof materials the textural quality loss is negligible for most items.

Except for food safety and marketing concerns, the choice of packaging materials for the fresh market is not nearly as critical as it is for frozen products (which will be discussed later). Packaging, in general, should address three critical functions: those of protecting, selling, and ensuring convenience. Most fresh seafood items are simply overwrapped in styrofoam trays, using absorbent pads to soak up lost moisture. Such packaging for self-service cases is familiar to both meat/seafood store personnel as well as to the public, and has been readily accepted into the mix of fresh meat items. Grocery chains that have installed full service seafood display cases simply place unwrapped products on a cosmetic bed of ice and maintain a case air temperature between 2 and 4 °C (35 and 40 °F) to maximize shelf life. Shucked oysters packaged in sealed plastic tubs may be displayed in the seafood case, and/or in free standing ice only gondolas.

Modified atmosphere packaging (MAP) is a process in which air is evacuated from a gas-impermeable package and replaced with a gas mixture which inhibits growth of spoilage microorganisms, thereby increasing the shelf life of fresh products by as much as 8–10 days. The replacement gas is typically a mixture of nitrogen, carbon dioxide and/or oxygen. The goal is to find the appropriate gas mixture: one which will inhibit outgrowth of aerobic (oxygen-requiring) spoilage microorganisms while simultaneously preventing outgrowth of *Clostridium botulinum* microbes which require an anaerobic environment. Public health professionals are most concerned about elevated temperature abuse by the consumer, which could allow botulism toxin production without typical spoilage odors and could result in consumption of a highly toxic product. Vacuum packaging is a type of modified atmosphere (i.e., no atmosphere) and has been used successfully to display fresh seafood products, but usually only after they have been treated with a chemical such as potassium sorbate to inhibit outgrowth of *C. botulinum* in the event of temperature abuse.

FREEZING AQUATIC PRODUCTS

Freezing aquacultured products effectively eliminates bacterial spoilage, but it does not completely prevent lipid oxidation or other enzymatic activities which can also alter texture. However, proper packaging, freezing rate, and storage temperature can minimize quality losses associated with freezing. Initial freezing may destroy 50 to 90% of the bacterial flora on a product; the others remain viable but dormant: that is, slowly dying off as frozen storage time increases. However product is never held long enough to kill all spoilage microorganisms which, when thawed, will begin to spoil the product.

Another factor that influences finfish quality is related to rigor mortis. Basically, fish should not be allowed to go into rigor at warm temperatures (above 20 °C; 68 °F) because the stiffening will be excessive, the product will experience excessive drip loss when thawed, and the texture may toughen. Most wild-caught finfish go through rigor while iced, after which they are offloaded, processed and frozen. Aquacultured finfish, on the other hand, can easily be processed (i.e., filleted or dressed) and then frozen before rigor is resolved. In certain species of fish, a tough texture and a change in color from white to gray occurs if rigor is resolved during frozen storage. If the product is then cooked before frozen rigor is resolved it may contract and become tough and bland. In the industry, this is referred to as thaw rigor. Thus, if aquacultured finfish are going to be processed and then frozen, trials should be run to determine the best time (i.e., pre- or post-rigor) for product processing, prior to freezing. It may be advantageous to condition the fish prior to thawing and filleting by holding them at a relatively high frozen storage temp to induce and resolve rigor.

Species

In general, fatty fish have approximately half the frozen storage life of lean fish when all other conditions are similar (i.e., harvest method, rate of freezing, and storage temperature.) Thus, for example, small mackerel would remain at acceptable quality after three months at −18 °C (0 °F,) whereas flounder could be stored at the same temperature for up to six months. Moderately fat to fat species can be treated with approved antioxidants, such as sodium erythrobate, ascorbic acid, BHA, and BHT prior to freezing, to slow the rate of lipid oxidation. Species that contain a pronounced area of fatty tissue, such as that found beneath the lateral line of many finfish, should have this meat cut away in preparation for long term storage. Particularly bloody species should be carefully rinsed, because blood accelerates rancidity.

Frozen lean products have less of a rancidity problem, but tend to become tough and lose excessive water upon thawing due to protein denaturation during storage. Pigmented fish, such as salmon, should be stored at as low a temperature as economically possible, to prevent muscle color bleaching from red/pink to yellow/orange due to the oxidation of carotenoids.

Packaging

Gas-impermeable packaging is imperative for products that will be held in frozen storage. Nylon and polyvinylidine chloride (Saran) are excellent gas barrier materials, whereas polyethylene, polyvinyl chloride (PVC), and polypropylene allow gas transmission. Under prolonged frozen storage the oils in even moderately fat fish will enzymatically react with oxygen and become rancid. Therefore, wrapping product in materials which prevent oxygen transfer will improve storage life.

Another critical aspect of packaging for frozen storage is the reduction of freezer burn, a condition which leads to unacceptable visual and organoleptic (tough and dry) attributes. Freezer burn is the dehydration of tissues that occurs as moisture migrates from product (80% moisture) to a drier atmosphere (the storage cooler). This is particularly pronounced in freezers that cycle frequently, as the relatively warmer air in the cooler during the defrost cycle, which is attempting to reach a new humidity equilibrium, tends to draw moisture from the product.

Skintight moisture-proof packaging will not only slow the rate of freezer burn, it will also eliminate frost buildup on the inside of the package material. This visually unappealing condition occurs in loosely packaged products when moisture migrates (sublimates) from the product, condenses on the inside of the packaging, and then refreezes.

Lean fish fillets are often dipped in a 5% brine solution for 20–30 seconds prior to packaging. This treatment solubilizes surface proteins on the cut surfaces, forming a kind of skin to further help prevent moisture loss. Such a treatment should not be applied to fatty products because it will accelerate rancidity. Instead, various phosphate dips have been recommended to help retain moisture for these species.

Finally, "packaging" may consist of encasing the product in ice, which forms an effective skintight seal. This is one of the least expensive techniques and is employed with products which "freeze well." These products include shrimp, as well as many species of finfish. Three major drawbacks to this process have been as follows: (*1*) eventual sublimation of the ice coating, which results in exposure and freezerburn, (*2*) the brittle nature of ice, resulting in loss of portions of the glaze with moderate or greater handling, and (*3*) the slower rate at which the product can be frozen due to the addition of the water glaze.

Each of these problems has been addressed, the first by freezing the "block" inside a moisture impermeable plastic bag, which is usually contained in a waxed or plastic coated cardboard carton. The second is solved by layering products using dividers inside master cartons. The third is solved by glazing only products which "freeze well," and/or altering the freezing mechanism to increase the rate (i.e., cryogenic tunnel vs. blast air).

Rate of Freezing

The recommendation for rate of freezing is not dependent on species, size, or product form. It is always "as rapidly as economically possible." It is essential that muscle temperature pass through the "critical freezing zone" (CFZ) of 0 °C to −5 °C (32 °F to 23 °F) rapidly to reduce the size of ice crystal formation within the tissue. Slow freezing promotes the formation of relatively few, but large, crystals, which tend to disrupt cell tissue, and lead to excessive moisture loss upon thawing. Rapid freezing, on the other hand, results in numerous small crystals which are much less damaging. As water within the tissue begins to freeze, the remaining naturally salty fluid becomes even saltier (concentrating enzymes and catalysts) and, if left in this state for several hours, can denature proteins and result in a tough, fibrous texture and loss of water holding capacity.

There are three generally recognized freezing rates: *Slow*, in which the product remains in the CFZ for more than two hours (generally a freezer room or cabinet with little air circulation); *Fast*, where the product passes through the CFZ in less than two hours (blast or plate freezers brine immersion); and *Ultrarapid*, where products are frozen within minutes rather than hours (insulated tunnels where liquefied nitrogen, carbon dioxide, or Freon 12 are allowed to expand into their natural gas state creating very cold temperatures [−195 °C (−320 °F) nitrogen, −78 °C (−109 °F) carbon dioxide, and −30 °C (−22 °F) Freon]).

Frozen Storage Temperature

Storage temperature is critical, not so much for bacteria spoilage, but to reduce the rate of fat oxidation, other chemical reactions and certain enzymatic actions. These deteriorative processes require a certain amount of free water. At −1.1 °C (30 °F) only 32% of the water is frozen; at −7.8 °C (18 °F) 89% is frozen. Even at −18 °C (0 °F) and below a small amount of water still remains unfrozen. Maintaining seafood products at −18 °C (0 °F) is adequate for short term (1–2 months) storage, but for longer-term storage a temperature of −29 °C (−20 °F) is recommended. As a general rule, frozen storage at −20 °F (−28.9 °C) will typically increase the "high-quality shelf life" of product stored more than 2–3 months by two to three fold. Storage temperatures colder than −29 °C (−20 °F) generally have not been shown to result in quality maintenance noticeable enough to warrant the added expense. Maintaining uniform temperature during frozen storage minimizes moisture migration (freezer burn) and accretion (enlargement of ice crystals as they melt and slowly refreeze).

Thawing

Frozen products can be thawed by processors who wish to repackage or to produce value-added products. Although seafoods may be thawed in air or water, at the same temperature they will thaw faster in water due to its more efficient transfer of heat. As with freezing, as products are thawed they should be moved through the critical zone [−5 °C to 0 °C (23 °F to 32 °F)] as rapidly as possible, to prevent protein damage and subsequent excessive drip loss. When thawing in air or water, the temperature should be kept below 18 °C (65 °F). Thaw rate can be increased by keeping the air or water moving across the frozen surface. If the product is not sealed in moisture-proof packaging, the air should be humidified to prevent dehydration. Also, unwrapped products should not be thawed in water or they will become waterlogged and lose flavor through leaching of water-soluble components. Increasingly, both industry and consumers are employing microwave ovens to thaw products. Industrial use of microwaves is referred to as ohmic thawing.

HACCP

As of December 1997, all seafoods in the United States are required to be processed under a safety program entitled "Hazard Analysis Critical Control Point," or HACCP. In essence, this program requires that a hazard (biological/chemical/physical) analysis be completed for each processing operation and/or species, and any potential food safety concern(s) be addressed through establishment (and subsequent monitoring) of one or more critical control points in the operation(s). This program presupposes a continuation of existing Good Manufacturing Practices (GMP) which include such things as potable water, employee cleanliness and a work environment sanitation program.

PROCESSING TRENDS

Certain relatively inexpensive aquacultured products like catfish are now being incorporated into school lunch programs. Aquacultured products offer such programs a portion-controlled product of defined nutritional content at a forecasted price. Cultured products have also found increasing acceptance in fast-food restaurants for many of the same reasons.

Certain relatively bland products have successfully entered the meat mix through the incorporation of spices during processing. Enrobing is a process in which skinless fillets are coated with an oil or oil/water mixture containing spices such as lemon-pepper, cajun, or blackened flavorings. Phosphates may be injected to aid in the retention of moisture as well as to allow a means to carry flavors and spices into the core of the fillet. The process of vacuum marination, in which fillets are gently tumbled in a drum containing spices and flavors in an oil or oil/water base, also allows for good penetration of added ingredients.

A deboning machine can be used to remove meat that adheres to fish frames after filleting. This minced product can be reformed into patties and breaded for portion-control markets such as school lunches and military contracts. Incorporating boneless fillets of cultured species into the gourmet frozen dinner market as well as supplying specialty processors, such as airline catering services, have been suggested as means to increase the value of cultured products, which, unlike most wild-caught competitors, have a defined quality, shelf life, and year round supply.

Many molluscan products, as well as finfish and crustaceans, are battered and/or breaded as part of a value-added process. The United States is the world's largest consumer of breaded seafood products. Products are breaded to improve appearance and taste, increase portion size with relatively low-cost ingredients, and improve moisture retention while cooking. Batters are liquid mixtures of water, flour, starch, and seasonings into which the seafood item is immersed. The product is then cooked to set the batter, resulting in a batter-fry product. Alternatively frozen products can be coated with a dry breading mixture and cooked. Prior to battering, seafood products are usually given a light dusting with flour or dry batter mix to increase the uniformity of batter adhesion.

In order to retain optimal texture and flavor of high value seafood products, a process in which products are processed minimally was developed. Referred to as "sous vide," a French term which translates to "under vacuum," products are portioned, individually vacuum-packaged in plastic pouches or rigid containers which are impermeable to oxygen and moisture, and cooked in a water bath or high humidity oven. Cooking time and temperature are generally lower than those required for pasteurization. Extreme care must be taken to insure that the products are kept under constant refrigeration following processing, as the heat treatment is not sufficient to kill *C. botulinum*, which may subsequently produce toxins if temperature is abused.

BIBLIOGRAPHY

1. R.E. Martin and G.J. Flick, *The Seafood Industry*. Van Nostrand Reinhold, NY, 1990.

See also HARVESTING.

PROTEIN SOURCES FOR FEEDS

MENG H. LI
EDWIN H. ROBINSON
Mississippi State University
Stoneville, Mississippi

RONALD W. HARDY
Hagerman Fish Culture Experiment Station
Hagerman, Idaho

OUTLINE

Although nearly all feedstuffs contain protein, only those that contain 20% or more crude protein are considered to be protein feedstuffs in animal diets. However, in diets for carnivorous fish species (e.g., salmon and trout), only ingredients containing at least 35% crude protein are considered protein feedstuffs. Protein feedstuffs generally are classified based on their origin as either animal or plant proteins. Animal proteins that are used in fish diets are marine products or by-products, by-products from livestock or poultry processing, or dairy by-products. The primary plant protein feedstuffs used in fish diets are products of oilseeds, such as soybean meal and cottonseed meal. Animal proteins (e.g., fish meal) are generally considered to be of higher quality than plant proteins, primarily because of their superior profiles of essential amino acids. Compared to animal proteins, most plant proteins are deficient in lysine and methionine, the two limiting amino acids in fish diets. Also, certain plant proteins contain toxins and antinutritional factors that may or may not be inactivated during processing of the meal. In this contribution, an overview of protein feedstuffs used in fish diets is presented.

PROTEIN FEEDSTUFFS

Animal Proteins

Marine Products or By-Products

Fish Meal. Fish meal is prepared from dried, ground tissues of whole marine fish, such as menhaden, anchovy, and capelin, or from fish-processing waste. Fish meal contains 55 to 75% protein, depending on the species of fish used. Fish meal protein is of excellent quality, both in terms of amino acid profile and apparent digestibility, and is highly palatable to most fish. It contains 5 to 10% oil, making it rich in energy and essential fatty acids, and it also contains bones and other sources of essential minerals. Due to their high ash content, fish meals made from fish-processing waste and residues of canning plants are of lower quality than fish meals prepared from whole fish. High levels of fish meal are used in starter diets for most cultured fish and in growout diets for carnivorous species, such as eels, salmon, and trout. However, because of its high cost relative to that of plant protein feedstuffs, fish meal is used sparingly in diets for omnivorous fish species.

Fish Solubles, Condensed or Dried. Condensed fish solubles, which contain a minimum of 30% crude protein, are semisolid (50% solids) by-products obtained by evaporating water from "press water" produced during the processing of cooked fish in the manufacture of fish meal. Dried fish solubles are composed of the same material as condensed fish solubles, only then are dried to powder, and contain about 60% crude protein. Fish solubles are a highly palatable protein feedstuff for use in fish diets.

Shrimp Meal and Crab Meal. Shrimp meal is produced from the waste of shrimp processing and includes the head, shell, and/or whole shrimp. The exoskeleton is primary chitin and has limited nutritional value. Chitin may account for 10 to 15% of the total nitrogen in the meal. Shrimp meal contains approximately 32% crude protein and 18% ash and is a good source of n-3 fatty acids, cholesterol (essential for crustaceans), and astaxanthin (a red pigment). Astaxanthin is desirable in salmonid diets, because it gives the flesh the pink color favored by consumers; it also is used in the diets of some tropical fish to enhance their color. Astaxanthin is highly palatable and may serve as an attractant in diets for some species.

Crab meal is the by-product of the crab-processing industry and includes the shell, viscera, and flesh. It contains about 30% crude protein and 31% ash. Its high ash content limits its use in fish diets.

Fish Silage. Fish silage is prepared by grinding whole fish or fish-processing waste and then adding an acid, usually formic acid or a combination of sulfuric acid and formic acid, to prevent microbial spoilage. Well-prepared fish silage can be stored for years without spoilage. Good-quality silage made from fresh fish contains about 18% crude protein and 74% moisture. Because of their high content of free amino acids and short-chain peptides, which are absorbed and metabolized too quickly following a meal, fish silages do not appear to be as effective as whole-fish meals.

Rendered By-Products

Meat and Bone Meal. Meat and bone meal is the rendered product from beef or pork tissues and should not contain blood, hair, hoof, horn, hide trimmings, manure, or stomach and rumen contents, except in amounts as may be unavoidable during processing. Meat and bone meal contains approximately 45 to 50% crude protein, the quality of which is inferior to that of whole-fish meal, because meat and bone meal contains less lysine. In addition, protein quality may vary considerably among products. Meat and bone meal is a good source of minerals, but high ash content limits its use in fish diets, because of the possibility that a mineral imbalance may occur in the diet and because its phosphorus content is high, making it difficult to include in diets designed to have limited environmental impact. Research with channel catfish (*Ictalurus punctatus*) has shown that meat and bone meal can completely replace fish meal in growout diets (1,2). A level of 18% meat and bone meal can be used to partially replace fish meal in the diets of Japanese flounder (*Paralichthys oliverus*) (3). Meat and bone meal can replace 40% of fish meal in practical diets for sea bream (*Sparus aurata*) (4). However, including more than 20% meat and bone meal in place of fish meal lowers growth of rainbow trout (*Oncorhynchus mykiss*).

Meat Meal. Meat meal is similar to meat and bone meal, except that there is no added bone. It contains approximately 50 to 55% crude protein, and its ash content is lower than that of meat and bone meal. Meat meal can be used as a substitute for fish meal in channel catfish feeds.

Blood Meal. Blood meal is prepared from clean, fresh animal blood, excluding hair, stomach belchings, and urine, except in trace quantities that are unavoidable. Blood meal contains about 80 to 85% crude protein and is an excellent source of lysine, but is deficient in methionine. Research with palmetto bass (*Morone saxatilis* × *M. chrysops*) has shown that blood meal could replace 25% of the protein supplied by fish meal (5). A level of 10% blood meal can be used in diets for Nile tilapia (*Oreochromis niloticus*) (6). Trout diets typically contain 5 to 8% blood meal.

Blend of Meat and Bone Meal and Blood Meal. Special products are available for use in fish diets that consist of a mixture of meat and bone meal and blood meal. The two feedstuffs are mixed to mimic the nutritional profile of menhaden fish meal and to provide 60 to 65% protein. Blends of meat and bone meal and blood meal can completely replace fish meal in channel catfish (*I. punctatus*) diets (1,2), but apparently not in diets for species requiring higher levels of dietary protein.

Poultry Wastes

Poultry By-Product Meal. Poultry by-product meal is made of ground, rendered, or clean parts of the carcass of slaughtered poultry. It contains heads, feet, underdeveloped eggs, and visceral organs, but does not contain feathers. The product contains approximately 58% crude protein and 16% ash. Poultry by-product meal can replace 100% of fish meal in the diet of India major carp (*Labeo rohita*) (7), 25% of the protein supplied by fish meal in the diet of silver seabream (*Rhabdosargus sarba*) (8), and 50% of fish meal in the diet of Australian snapper (*Pagrus auratus*) (9). Poultry by-product meal is often used in limited amounts in rainbow trout diets.

Poultry Feather Meal, Hydrolyzed. Hydrolyzed poultry feather meal is prepared by the high-pressure treatment of clean, undecomposed feathers from slaughtered poultry. At least 75% of the protein should be digestible, as measured by pepsin digestion. It is high in protein (85%), but the quality of the protein is not as good as that of other animal protein feedstuffs. As a partial replacement of fish meal, a level of 25% can be used in the diet of Japanese flounder (10) and 20% in the diet of Indian major carp (11). Apparent protein digestibility coefficients for feather meal range from 75 to 86% for rainbow trout, depending on the source of the feather meal (12).

Plant Proteins

Oilseed Meals

Soybean Meal. Soybean meal is prepared by grinding the flakes that result after removal of the oil from soybeans by solvent extraction or by the expeller process. There are three types of soybean meal that can be used in fish diets: dehulled and solvent extracted, solvent extracted, and expeller processed. These types of soybean meal contain 48, 44, and 42% protein and 1, 0.5, and 3.5% oil, respectively. Soybean meal is the major protein feedstuff used in aquaculture diets. It has the best amino acid profile of all common plant proteins. Based on the requirement values published by National Research Council (13), dehulled, solvent-extracted soybean meal has a sufficient amount of all essential amino acids for, and highly palatable to, channel catfish. Antinutritional factors in soybeans, mainly trypsin inhibitor, are destroyed or reduced to insignificant levels by heating during oil extraction and are further inactivated by heating during extrusion of channel catfish diets. Additional heat treatment is beneficial for soy products used in salmon and trout diets.

Soybean meal is well utilized by many warmwater fish and is typically included in their diets at levels up to 60% or so. For example, soybean meal can be used at levels of about 50% in channel catfish diets (2,14), 55 to 60% in tilapia diets (15,16), 45% in common carp (*Cyprinus carpio*) diets (17), 44% in palmetto bass diets (18), 21% in silver seabream diets (8), and up to 66% in the diet of red drum (*Sciaenops ocellatus*), as long as 10% of the protein is supplied by fish meal (19). Soybean meal has been extensively studied as a trout diet component, and the consensus is that diets for postjuvenile trout can contain 25% soybean meal without reducing growth performance in carefully formulated diets. Juvenile trout are less tolerant of dietary soybean meal, and feed intake is reduced at levels of 10 to 15% soybean meal in the diet, depending on the other constituents of the diet.

Heated, Full-Fat Soybean Meal. Heated, full-fat soybean meal is prepared by grinding heated, full-fat soybeans. The meal contains 39% protein and 18% fat. It is rarely used in channel catfish diets, because of its high fat content, but a limited amount can be used as long as the total fat level in the diet does not exceed 6%. Nile tilapia can use up to 58% heated, full-fat soybean in the diet; however, the body fat of the fish increases when they are fed the full-fat meal (20). Growout-stage chinook salmon (*Oncorhynchus tshawytscha*) grow normally when up to 30% of dietary protein is supplied by full-fat soybean meal (21).

Cottonseed Meal. Cottonseed meal is obtained by grinding the cake remaining after the oil has been removed from cottonseeds, either hydraulically, by screw-press extraction, prepress solvent extraction, direct solvent extraction, or expander solvent extraction. The products generally contain 41% protein, but must not contain less than 36% protein. They are highly palatable to channel catfish, salmonids, and tilapia, but are deficient in lysine. Cottonseed meal contains free gossypol and cylcopropenoic acids, which can be toxic to monogastric animals. The amount of free gossypol in cottonseed meal depends on the processing method. The free gossypol contents of five types of cottonseed meal are as follows: screw press, 0.02 to 0.05%; hydraulic, 0.04 to 0.10%; prepress solvent, 0.02 to 0.07%; expander solvent, 0.06 to 0.21%; direct solvent, 0.10 to 0.50% (22). Currently, the expander solvent method is the method of choice for processing cottonseed into meal. Channel catfish can tolerate a dietary free-gossypol concentration of 900 mg/kg (ppm) (23). The levels of cottonseed meal should not exceed 30% of the channel catfish diet, unless supplemental lysine is used (2). A level of 35% cottonseed meal [free-gossypol concentration of 300 mg/kg (ppm)] can be used in the diet of Mozambique tilapias (*Oreochromis mossambicus*) (24).

Peanut Meal. Peanut meal is obtained by shelling peanuts, removing the oil, either mechanically or by solvent extraction, and then grinding the peanuts. Solvent-extracted peanut meal contains 48% protein, and the mechanically extracted product contains 45% protein. Peanut meal is highly palatable to fish and contains no known antinutritional factors; however, it is deficient in lysine. Without lysine supplementation, the levels of peanut meal used in channel catfish diets are restricted to 15 to 20%.

Sunflower Meal. Sunflower meal is prepared by grinding the residue remaining after mechanical or solvent extraction of the oil from sunflower seeds. Dehulled sunflower meal is prepared from sunflower seeds after the hulls are removed. Solvent-extracted, dehulled sunflower meal contains about 44% protein. As the hulls are not easily removed, the meal contains around 13% fiber. In fact, higher levels of fiber are found in meals that are not dehulled. Consequently, its low lysine content and high level of fiber limit its usefulness in fish diets. A level of 15 to 20% sunflower meal without lysine supplementation is acceptable for channel catfish diets (25). A level of up to 70% sunflower meal can be used in the diet of Mozambique tilapias (*O. mossambicus*) (24). Growth reduction was observed in rainbow trout fed diets in which 25% of the dietary protein was supplied by sunflower meal.

Rapeseed Meal and Canola Meal. Rapeseed meal is prepared by removing the oil from rapeseeds, using the solvent extraction method, and then grinding the remaining residue. Rapeseed meal contains glucosinolates (antithyroid factor) and erucic acid, which may be detrimental to fish growth. It can be included at a level of 28% in the diet of common carp without causing adverse effects (26). A level of 15% rapeseed meal can be used in the diet of Mozambique tilapias (with an initial size of 0.3 g/fish) (27). The level of rapeseed meal can be increased to 41% in the diet of larger Mozambique tilapias (with an initial size of 13 g/fish) without negatively affecting fish performance (24).

Canola meal is prepared from a selected variety of rapeseeds by solvent extraction to remove the oil. Compared to typical rapeseed meal, canola meal is low in glucosinolates and erucic acid. Canola meal contains about 38% protein and is relatively low in lysine as compared with soybean meal, but is higher in lysine than are other oilseed meals. Canola meal is palatable to fish and can be used at levels of up to 36% in channel catfish diets (29–31) and 32% in the diet of hybrid tilapia (*Tilapia mossambica* × *T. aurea*) (32).

Other Plant Proteins

Distillers' Dried Grains with Solubles. Distillers' dried grains with solubles are the primary residues from the fermentation of yeast in cereal grains and after the removal, by distillation, of the alcohol in the grains. The product contains approximately 27% protein and is highly palatable to fish. Levels up to 35% can be used in channel catfish diets (33). If higher levels are used, supplemental lysine may be needed.

Brewers' Dried Grains. Brewer's dried grains are residues obtained during the brewing of beers and ales after the removal of the starches and sugars of the grains, such as malted barley, corn, rice grit, and hops. This product contains about 28% crude protein and 12% fiber. It can be used in fish diets, but is deficient in lysine.

Corn Gluten Meal and Corn Gluten Feed. Corn gluten meal is the dried residue from corn after the removal of most of the starch and germ and after the separation of the bran by the process of wet milling of corn starch and corn syrup, or by enzymatic treatment of endosperm. There are two types of corn gluten meal, which contain 41% and 60% protein, and 4% and 2.5% fiber, respectively. Corn gluten meal is a good source of methionine, but contains high levels [200 to 350 mg/kg (ppm)] of the yellow pigment xanthophyll, which imparts an undesirable yellow color in the flesh of trout and catfish. This effect limits the use of corn gluten meal in rainbow trout and channel catfish diets. However, a dietary xanthophyll level of 11 mg/kg (ppm) can be used in channel catfish diets without accumulation of the yellow pigment (34), and supplemental astaxanthin can mask the yellow color in rainbow trout (35). A level of 20% corn gluten meal can be used in the diet of yellowtails without amino acid supplementation and without negatively affecting growth (36).

Corn gluten feed is the part of the corn that remains after most of the starch and gluten have been extracted by the process of wet milling of corn starch and corn syrup. Corn gluten feed contains about 21% crude protein and 10% fiber, as well as a low level of xanthophyll [approximately 11 mg/kg (ppm)]. A level of about 20% corn gluten feed can be used in the diet of Nile tilapia raised in cages without leading to a reduction in growth (37).

Alfalfa Meal. Alfalfa meal is prepared by grinding dried alfalfa plants. It contains 15 to 20% crude protein and 20 to 30% fiber. It is deficient in lysine. A level of 15 to 20% can be used in grass carp diets (38,39). A level of 5% in the diet reduced growth in blue tilapias (*T. aurea*) (40).

Miscellaneous Plant Protein Feedstuffs. Several other plant protein feedstuffs can be used in fish diets and may be important proteins in certain parts of the world. These include coconut or copra meal (about 25% protein) (24,41), groundnut meal (about 60% protein) (24,41), leucaena meal (about 30% protein) (24,41), linseed meal (about 38% protein) (41,57), mustard meal (about 35%) (41,42,57), salicornia meal (about 35% protein) (43), and sesame meal (about 50% protein) (41,57). These protein feedstuffs are generally low in lysine; some contain antinutritional factors, and some are high in fiber, which may limit their use in fish diets.

Plant Protein Concentrates. Plant protein concentrates are prepared by various methods that extract protein from plant feedstuffs. The concentrates contain high levels of protein, and several of the products have been tested in fish diets. Alfalfa leaf protein concentrates can be used at levels up to 35% of the dietary protein in Mozambique tilapias (44). Rapeseed protein concentrate can be used to replace 39% of the fish meal in the diet of gilthead seabream (*S. aurata*) (45) and all of the fish meal in rainbow trout diets if palatability is enhanced by adding betaine (28). Cowpea protein concentrate can replace up to 30% fish meal in the diet of Nile tilapia fry (46). A level of 20% Bermuda grass protein isolate can be used in channel catfish diets (47). However, these products are not generally available and may be too costly to be economical for use in diets of cultured fish.

Soybean protein concentrate (SPC) is commercially available, albeit at a relatively high cost compared with that of fish meal. SPC protein is highly digestible (95%) and highly palatable to rainbow trout. While some researchers have reported no reduction in growth at 100% fish meal replacement with SPC, optimum replacement levels are 45 to 50%, as measured by growth responses and feed efficiencies. Concentrates from other oilseeds, including sunflower, have been evaluated in trout diets and generally can replace 25% of dietary fish meal without reducing fish growth (48).

NUTRITIVE VALUE OF PROTEIN FEEDSTUFFS

The nutritive value of proteins, or protein quality, is based on the amino acid composition of the protein, particularly the concentration of essential amino acids and the biological availability of the amino acids. Several criteria can be used to determine the nutritive value of protein feedstuffs, including protein digestibility, amino acid availability, protein efficiency ratio, net protein utilization, percentage protein deposited, and essential amino acid index. These criteria are described in more detail elsewhere in this encyclopedia.

Protein Digestibility

The apparent protein digestibility coefficients of several commonly used protein feedstuffs by various species of fish are listed in Table 1. The apparent protein digestibility coefficients are defined as the percentage of protein consumed by the fish that is not excreted in the feces. They can be determined by direct or indirect methods (see other contributions for details). Some variations exist in the reported values, due to differences in fish species, experimental methodology, fecal collection methods, and diet composition. Processing of dietary ingredients also affects protein digestibility. For example, overheating reduces protein digestibility of soybean meal and fish meal. Protein digestibility coefficients can be used to estimate the value of a particular feedstuff or of a finished feed, and may also be used to estimate the availability of amino acids when these data are unknown. However, the availability of individual amino acids is variable and does not always correlate to the protein digestibility coefficient.

Amino Acid Availability

Values of the apparent amino acid availability of several commonly used protein feedstuffs by channel catfish, common carp, rainbow trout, and yellowtail are listed in Tables 2–4. Apparent amino acid availability is defined as the percentage of a specific amino acid consumed by the fish that is not excreted in the feces. It can be determined by direct or indirect methods (see other contributions for details). Factors affecting the protein digestibility coefficients also affect the availability of amino acids of the protein sources.

Protein Efficiency Ratio

Protein efficiency ratio (PER), which is defined as the grams of wet-weight gain per gram of protein consumed, is widely used as a measure of feed quality, because it is rather simple to calculate and does not require chemical analyses: PER = g wet weight gain/g crude protein fed. The major criticism of the assay is that it assumes that all of the protein is used for growth and makes no allowance for maintenance. Diet composition, fish size, and other factors that affect growth may also have an impact on the results. The protein efficiency ratio should be determined

Table 1. Apparent Protein Digestibility Coefficients

		Apparent Protein Digestibility Coefficient (%)					
Feedstuff	International Feed Number	Channel Catfish[a]	Common Carp[b]	Niie Tilapia[c]	Palmetto Bass[d]	Rainbow Trout[e]	Red Drum[f]
Alfalfa meal	1–00–023	13	—	66	—	—	—
Blood meal	5–00–381	74	—	—	86	69	100
Brewers' grain	5–02–141	—	—	63	—	—	—
Corn gluten meal	5–04–900	92	—	—	—	97	—
Cottonseed meal (mechanically extracted)	5–01–617	—	—	—	84	—	76–85
Cottonseed meal (direct solvent extracted)	5–01–621	81–83	—	—	—	76	—
Fish, anchovy meal	5–01–985	90	—	—	—	94	—
Fish, herring meal	5–02–000	—	89	—	—	95	—
Fish, menhaden meal	5–02–009	70–87	—	85	88	90	77–96
Fish, menhaden meal (select)	5–01–977	—	—	—	—	—	88
Fish, sardine meal	5–02–015	—	—	86	—	—	—
Mustard meal		—	—	85	—	—	—
Meat and bone meal	5–00–388	61–82	—	78	73	—	74–79
Meat meal (deboned)		—	—	—	—	97	—
Linseed meal	5–02–045	—	86	—	—	—	—
Peanut meal	5–03–640	74–86	—	—	—	—	—
Poultry by-product meal	5–03–798	65	—	74	—	96	49
Poultry feather meal	5–03–795	74	—	—	—	86	—
Sesame meal	5–04–220	—	79	—	—	—	—
Soybean meal (44%)	5–04–604	77	—	91–94	80	—	80
Soybean meal (48%)	5–04–612	84–97	—	—	—	90	86
Reference		(54–56)	(57)	(58,59)	(60)	(13,61)	(62,63)

[a] *Ictalurus punctatus*.
[b] *Cyprinus carpio*.
[c] *Oreochromis niloticus*.
[d] *Morone saxatilis* × *M. chrysops*.
[e] *Oncorhynchus mykiss*.
[f] *Sciaenops ocellatus*.

Table 2. Apparent Amino Acid Availabilities[a] for Various Feedstuffs, Determined for Channel Catfish (*Ictalurus punctatus*)

Amino Acid	Cottonseed Meal	Meat and Bone Meal	Menhaden Fish Meal	Peanut Meal	Soybean Meal
Alanine	70.4	70.9	87.3	88.9	79.0
Arginine	89.6	86.1	89.2	96.6	95.4
Aspartic acid	79.3	57.3	74.1	88.0	79.3
Glutamic acid	84.1	72.6	82.6	90.3	81.9
Glycine	73.5	65.6	83.1	78.4	71.9
Histidine	77.2	74.8	79.3	83.0	83.6
Isoleucine	68.9	77.0	84.8	89.7	77.6
Leucine	73.5	79.4	86.2	91.9	81.0
Lysine	66.2	81.6	82.5	85.9	90.9
Methionine	72.5	76.4	80.8	84.8	80.4
Phenylalanine	81.4	82.2	84.1	93.2	81.3
Proline	73.4	76.1	80.0	88.0	77.1
Serine	77.4	63.7	80.7	87.3	85.0
Threonine	71.8	69.9	83.3	86.6	77.5
Tyrosine	69.2	77.6	84.8	91.4	78.7
Valine	73.2	77.5	84.0	89.6	75.5
Average	75.1	74.3	82.9	88.4	81.0

[a] Expressed as a percentage; from (64).

by using single-protein diets that contain a suboptimal level of protein. If a high level of protein is fed, the amino acid deficiency may be masked. Protein efficiency ratio values for selected protein sources, determined by using channel catfish, are shown in Table 5.

Essential Amino Acid Index

The essential amino acid index is based on comparison of the essential amino acid content of the protein with the essential amino acid requirement values of the fish. The

Table 3. Apparent Amino Acid Availabilities[a] for Various Feedstuffs, Determined for Common Carp (*C. carpio*)

Amino Acid	Fish Meal (Herring)	Linseed Meal	Mustard Meal	Sesame Meal
Alanine	91.2	86.1	86.4	83.5
Arginine	83.3	84.5	84.6	81.1
Aspartic acid	90.4	86.7	85.4	83.5
Cystine	90.4	84.0	85.6	85.8
Glutamic acid	91.8	86.3	86.4	86.5
Glycine	90.1	86.2	86.6	80.2
Histidine	—	87.1	86.6	84.1
Isoleucine	90.4	85.6	86.1	82.4
Leucine	91.8	85.8	85.9	82.9
Lysine	91.2	83.7	84.8	80.5
Methionine	94.6	85.0	84.9	82.5
Phenylalanine	89.0	85.5	86.0	81.0
Proline	90.2	85.8	85.4	79.8
Serine	90.6	85.8	86.2	82.5
Threonine	90.1	85.5	85.9	82.0
Tyrosine	91.5	85.1	85.5	80.5
Valine	90.7	85.9	85.9	81.5
Average	90.5	85.6	85.8	82.4

[a]Expressed as a percentage; from (57).

Table 4. Apparent Amino Acid Availabilities[a] for Various Feedstuffs, Determined for Yellowtail (*Seriola lalandi*)

Amino Acid	Brown Fish Meal	Corn Gluten Meal	Meat Meal	Soy Protein Concentrate	Full-Fat Soybean Meal
Alanine	89.7	47.3	86.0	84.9	77.3
Arginine	92.5	47.6	82.2	89.9	85.4
Aspartic acid	89.3	44.1	79.2	90.3	82.0
Cystine	90.3	47.1	43.8	87.2	85.0
Glutamic acid	91.9	48.8	81.6	92.0	86.4
Glycine	92.0	42.4	89.8	81.8	74.8
Histidine	93.0	50.8	86.0	92.5	53.0
Isoleucine	90.2	45.0	75.9	87.9	79.3
Leucine	90.7	46.5	77.5	86.9	78.0
Lysine	93.1	47.6	85.0	91.2	83.4
Methionine	92.2	50.2	83.8	86.8	76.0
Phenylalanine	88.8	47.1	78.4	88.9	79.4
Proline	69.9	51.1	87.0	88.9	82.6
Serine	89.6	46.0	73.8	86.1	79.0
Threonine	88.9	43.4	73.8	83.0	74.8
Tyrosine	90.1	50.6	76.3	89.1	82.0
Valine	85.7	40.0	72.3	79.7	69.1
Average	89.3	46.8	78.4	87.5	78.1
APD[b]	88.7	49.7	80.3	87.3	83.2

[a]Expressed as a percentage; from (65).
[b]APD = apparent protein digestibility.

use of the index in predicting limiting amino acids (the amino acids present in the protein at a level below that required by the fish) is illustrated in Table 6. Soybean meal contains adequate amounts of each of the essential amino acids to meet the requirements of channel catfish. Cottonseed meal and peanut meal are deficient in lysine. This method, however, does not consider the availability of amino acids in the diet.

Table 5. Protein Efficiency Ratio Values for Selected Protein Sources, Determined with Channel Catfish

Protein Source	International Feed Number	PER[a] Reference (66)	PER[a] Reference (67)
Fish meal, menhaden	5–02–009	2.48	2.41
Fish meal, anchovy	5–01–985	2.44	—
Catfish waste, dry		2.08	2.01
Soy-catfish scrap[b]		—	2.47
Soy-liquid fish[c]		—	2.40
Soybean meal (44% protein)	5–04–604	1.70	2.04
Soybean meal (48% protein)	5–04–612	1.80	—
Meat and bone meal	5–00–388	1.64	—
Poultry feather meal	5–03–795	0.97	—

[a]PER = amount of weight gain/amount of protein fed.
[b]Mixture of 42% defatted soyflakes; 28% dehulled, full-fat soybeans; and 30% catfish scrap (offal).
[c]Mixture of 43% defatted soyflakes; 30% dehulled, full-fat soybeans; and 27% liquid fish (catfish offal hydrolyzed by an enzymatic process).

Table 6. Essential Amino Acid Contents of Selected Proteins[a]

Amino Acid	Requirement for Channel Catfish[b]	Cottonseed Meal	Peanut Meal	Soybean Meal
Arginine	4.3	11.2	11.10	9.27
Histidine	1.5	2.68	2.32	3.17
Isoleucine	2.6	3.24	4.29	6.34
Leucine	3.5	5.85	9.02	9.27
Lysine	5.0–5.1	<u>4.17</u>[c]	<u>4.32</u>	7.80
Methionine + cystine	2.3	2.83	2.80	3.63
Phenylalanine + tyrosine	5.0	7.36	9.63	10.22
Threonine	2.0	3.22	2.83	4.88
Tryptophan	0.5	1.15	1.22	1.71
Valine	3.0	4.59	4.59	6.59

[a]Expressed as a percentage of protein; from (68).
[b]From (13).
[c]Underlined values indicate deficiency.

ALL-PLANT PROTEIN DIETS

Feedstuffs of animal origin, especially marine fish meals, are generally considered to be of higher quality than feedstuffs of plant origin. However, because animal protein is expensive and its availability is often limited, efforts are being made to reduce or eliminate animal protein in diets for some species. Diets for coldwater fish such as salmon and trout traditionally contain relatively high levels of animal protein, especially fish meal, making it more difficult to completely replace fish meal with plant proteins. Warmwater carnivorous fish, such as striped bass (*M. saxatilis*) and largemouth bass (*Micropterus salmoides*), also require high levels of animal protein in their diets. Animal protein appears to be required for normal growth and survival in the diets of omnivorous fish, such as channel catfish, carp, and tilapia, during their early life stages, and research with channel catfish showed that fry and fingerlings require fish meal for maximum growth (49). However, under typical culture conditions, all-plant protein diets composed of soybean meal, cottonseed meal, corn, and wheat middlings are satisfactory for normal growth of channel catfish from fingerling to marketable size in ponds (2,14,50–52). A laboratory study with fingerling blue catfish (*Ictalurus furcatus*) indicated that fish meal cannot be totally replaced by soybean meal in the diet (33), but can be totally replaced by soybean meal plus supplemental methionine (53). There is also evidence that tilapias (15,16,37) can be fed a soybean-meal-based, all-plant diet without reduction in weight gain. Common carp (*C. carpio*), reared in plastic tanks, cages, or ponds and fed soybean-meal-based, all-plant diets with supplemental lipid and lysine grew as well as fish fed diets containing fish meal (17). When they can be used, all-plant protein diets are generally more economical and are also generally lower in phosphorus than are diets containing animal proteins. The lower level of phosphorus inherent in soybean-meal-based diets may be beneficial in that those diets can be supplemented with a highly digestible source of phosphorus, thus potentially reducing phosphorus loads in rearing waters.

SUMMARY

Although many protein feedstuffs have the potential to be used, relatively few of them are used in the commercial diets of various fish species, essentially for three reasons: Not many feedstuffs are available at a reasonable cost; not many feedstuffs contain the essential nutrients sufficient for optimum fish growth; and antinutritional factors are absent. Generally, proteins of animal origin are of higher quality, but more expensive than those of plant origin. Soybean meal has been the major protein feedstuff used in aquaculture diets worldwide, mainly because it has the best amino acid profile among all plant feedstuffs. Fish meal is the major protein feedstuff for coldwater and marine species, because the replacement of fish meal with plant proteins has not been successful, even though small amount of plant proteins can be used in diets for these species. Research efforts should continue to explore the use of less expensive alternative feedstuffs in aquaculture diets.

BIBLIOGRAPHY

1. E.H. Robinson, *Catfish J.* **6**(12), 12 and 21 (1992).
2. E.H. Robinson and M.H. Li, *J. World Aquacult. Soc.* **25**, 271–2276 (1994).
3. K. Kikuchi, T. Sato, T. Furuta, I. Sakaguchi, and Y. Deguchi, *Fish. Sci.* **63**, 29–32 (1997).
4. S.J. Davies, I. Nengas, M. Alexis, in S.J. Kaushik and P. Luquet, eds. *Fish Nutrition in Practice*, Institut National De La Recherche Agronomique, Paris, 1993, pp. 907–911.
5. M.L. Gallagher and M. LaDouceur, *J. Appl. Aquacult.* **5**, 57–66 (1995).
6. S.O. Otubusin, *Aquaculture* **65**, 263–266 (1987).
7. M.R. Hasan and P.M. Das, in S.J. Kaushik and P. Luquet, eds. *Fish Nutrition in Practice*, Institut National De La Recherche Agronomique, Paris, 1993, pp. 793–801.
8. A.F.M. El-Sayed, *Aquaculture* **127**, 169–176 (1994).
9. N. Quartararo, G.L. Allan, and J.D. Bell, *Aquaculture* **166**, 279–295 (1998).
10. K. Kikuchi, T. Furuta, and H. Honda, *Fish. Sci.* **60**, 203–206 (1994).
11. M.R. Hasan, M.S. Haq, P.M. Das, and G. Mowlah, *Aquaculture* **151**, 47–54 (1997).
12. D.P. Bureau, A.M. Harris, and C.Y. Cho, "Abstract," in *Proc. VIth Int. Symp. Feeding and Nutrition in Fish*, College Station, TX, 1996.
13. National Research Council, *Nutrient Requirement of Fish*, National Academy Press, Washington, DC, 1993.
14. E.H. Robinson and M.H. Li, *Catfish J.* **8**(10), 4 (1994).
15. S. Viola, Y. Arieli, and G. Zohar, *Aquaculture* **75**, 115–125 (1988).
16. S.G. Hughes and T.S. Handwerker, *Feed Management* **44**(5), 54–58 (1993).
17. S. Viola, S. Mokady, U. Rappaport, and Y. Arieli, *Aquaculture* **26**, 223–236 (1982).
18. M.L. Gallagher, *Aquaculture* **126**, 119–127 (1994).
19. B.B. McGoogan and D.M. Gatlin, III, *J. World Aquacult. Soc.* **28**, 374–385 (1997).
20. K.L. Wee and S.W. Shu, *Aquaculture* **81**, 303–314 (1989).
21. T.R. Wilson, Ph.D. Dissertation, University of Washington, Seattle, WA, 1992.
22. K. Eng, *Feedstuff* **64**, 62–63 (1990).
23. W.J. Dorsa, H.R. Robinette, E.H. Robinson, and W.E. Poe, *Trans. Am. Fish. Soc.* **111**, 651–655 (1982).
24. A.J. Jackson, B.S. Capper, and A.J. Matty, *Aquaculture* **27**, 97–109 (1982).
25. D.O. Balogu, H.F. Phillips, A. Gannam, O.A. Porter, and J. Handcock, *Arkansas Farm Research* **42**(3), 11 (1993).
26. K. Dabrowski and H. Kozlowska, in *Proc. World Symp. Aquaculture in Heated Effluents and Recirculating Systems*, Vol. 2, Berlin, 1981, pp. 263–274.
27. S.J. Davies, S. McConnel, and R.I. Bateson, *Aquaculture* **87**, 145–154 (1990).
28. D.A. Higgs, B.S. Dosanjh, A.F. Prendergast, R.M. Beames, R.W. Hardy, W. Riley, and G. Deacan, in C.E. Lim and D.J. Sessa, eds., *Nutrition and Utilization Technology in Aquaculture*, AOCS Press, Champaign, IL, 1995, pp. 130–156.
29. M.H. Li and E.H. Robinson, *Catfish J.* **9**(2), 14 (1994).
30. C.D. Webster, L.G. Tiu, J.H. Tidwell, and J.M. Grizzle, *Aquaculture* **150**, 103–112 (1997).
31. C. Lim, P.H. Klesius, and D.A. Higgs, *J. World Aquacult. Soc.* **29**, 161–168 (1998).
32. D.A. Higgs, B.S. Dosanjh, M. Little, R.J.J. Roy, and J.R. McBride, in M. Takeda and T. Watanabe, eds., *Proc. Third Int. Symp. Feeding and Nutrition in Fish*, Toba, Japan, 1989, pp. 1–25.
33. C.D. Webster, J.H. Tidwell, L.S. Goodgame, D.H. Yancey, and L. Mackey, *Aquaculture* **106**, 301–309 (1992).

34. T. Lovell, *Nutrition and Feeding of Fish*, Van Nostrand Reinhold, New York, 1989.
35. D.I. Skonberg, R.W. Hardy, F.T. Barrows, and F.M. Dong, *Aquaculture* **166**, 269–277 (1998).
36. S. Shimeno, T. Mima, T. Imanaga, and K. Tomaru, *Bull. Jpn. Soc. Sci. Fish.* **59**, 1889–1895 (1993).
37. Y.V. Wu, R. Rosati, D.J. Sessa, and P. Brown, *Prog. Fish-Cult.* **57**, 305–309 (1995).
38. B.O. Mgbenka and R.T. Lovell, *Prog. Fish-Cult.* **48**, 238–241 (1986).
39. T.J. Pfeiffer and R.T. Lovell, *Prog. Fish-Cult.* **52**, 213–217 (1990).
40. O.M. Yousif, G.A. Alhadhramim, and M. Pessarakli, *Aquaculture* **126**, 341–347 (1994).
41. M.R. Hasan, D.J. Macintosh, and K. Jauncey, *Aquaculture* **151**, 55–70 (1997).
42. B.S. Capper, J.F. Wood, and A.J. Jackson, *Aquaculture* **29**, 373–377 (1982).
43. I.E.H. Belal and M. Al-Dosari, *J. World Aquacult. Soc.* **30**, 285–289 (1999).
44. M.A. Olvera-Novoa, G.S. Campos, G.M. Sabido, and C.A. Martinez-Palacios, *Aquaculture* **90**, 291–302 (1990).
45. G.W. Kissil, I. Lupatsch, D.A. Higgs, and R.W. Hardy, *Israeli J. Aquacult.* **49**, 135–143 (1997).
46. M.A. Olvera-Novoa, F. Pereira-Pacheco, L. Olivera-Castillo, V. Perez-Flores, L. Navarro, and J.C. Samano, *Aquaculture* **158**, 107–116 (1997).
47. J.A. Buentello, D.M. Gatlin, III, and B.E. Dale, *J. World Aquacult. Soc.* **28**, 52–61 (1997).
48. R.R. Stickney, R.W. Hardy, K. Koch, R. Harrold, D. Seawright, and K.C. Massee, *J. World Aquacult. Soc.* **27**, 57–63 (1996).
49. A.A. Mohsen and R.T. Lovell, *Aquaculture* **90**, 303–311 (1990).
50. H.E. Leibovitz, M.S. Thesis, Auburn University, Auburn, AL, 1981.
51. E.H. Robinson and M.H. Li, *J. World Aquacult. Soc.* **29**, 273–280 (1998).
52. E.H. Robinson and M.H. Li, *J. World Aquacult. Soc.* **30**, 147–153 (1999).
53. C.D. Webster, D.H. Yamcey, and J.H. Tidwell, *Aquaculture* **103**, 141–152 (1995).
54. E.M. Cruz, Ph.D. Dissertation, Auburn University, Auburn, AL, 1975.
55. R.P. Wilson and W.E. Poe, *Prog. Fish-Cult.* **47**, 154–158 (1985).
56. P.B. Brown, R.J. Strange, and K.R. Robbins, *Prog. Fish-Cult.* **47**, 94–97 (1985).
57. M.A. Hossain and K. Jauncey, *Aquaculture* **83**, 59–72 (1989).
58. T.J. Popma, Ph.D. Dissertation, Auburn University, Auburn, AL, 1982.
59. F. Hanley, *Aquaculture* **66**, 163–179 (1987).
60. J.A. Sullivan and R.C. Reigh, *Aquaculture* **138**, 313–322 (1995).
61. S.H. Sugiura, F.M. Dong, C.K. Rathbone, and R.W. Hardy, *Aquaculture* **159**, 177–202 (1998).
62. T.G. Gaylord and D.M. Gatlin, III, *Aquaculture* **139**, 303–314 (1996).
63. B.B. McGoogan and R.C. Reigh, *Aquaculture* **141**, 233–244 (1996).
64. R.P. Wilson, E.H. Robinson, and W.E. Poe, *J. Nutr.* **111**, 923–929 (1981).
65. T. Masumoto, T. Ruchimat, Y. Ito, H. Hosokawa, and S. Shimeno, *Aquaculture* **146**, 109–119 (1996).
66. R.T. Lovell, Alabama Agricultural Experiment Station Bulletin 512, Auburn University, Auburn, AL, 1980.
67. E.H. Robinson, J.K. Miller, V.M. Vergara, and G.A. Ducharme, *Prog. Fish-Cult.* **48**, 233–237 (1985).
68. N. Dale, *Feedstuffs Reference Issue* **69**(30), 24–29 (1997).

See also ENVIRONMENTALLY FRIENDLY FEEDS; FEED MANUFACTURING TECHNOLOGY; NUTRIENT REQUIREMENTS.

PROTOZOANS AS DISEASE AGENTS

S.K. JOHNSON
Texas Veterinary Medical Diagnostic Laboratory
College Station, Texas

OUTLINE

Many kinds of protozoa are harmful to aquaculture animals. In some aquacultures, parasitic protozoa are the most important disease agents. This contribution emphasizes important protozoan associates of fishes, crustaceans, and molluscs. It aims to give a sufficient description of protozoans without using complex terminology and taxonomy that are unfamiliar to most readers.

PROTOZOA

Protozoa are unicellular organisms that live individually or in small groups. The membrane, or pellicle, that encompasses the nuclear material and cytoplasm is more or less flexible, and in some species, it secretes a hard covering. Individual size is usually less than 1 mm (0.039 in.), but giant forms up to 5 mm (0.195 in.) are known. The intake of nutriment at any or some specialized surface location is by permeation of the pellicle or by active engulfment of liquid or solid portions. A specialized cytostome functions in some protozoa for the consumption of small food particles, or, in the case of gulpers, swallowing of large prey. Reproduction is commonly accomplished by fission, either binary or multiple, and some groups produce new individuals by external or internal budding. Fertilization and sexuality are known to occur in many protozoa. Cyst formation occurs in some free-living types, and some cysts can resist harsh environments.

Structurally distinct stages of development are typical of entirely parasitic groups. They usually have complex developmental cycles, and a few involve alternate hosts. Some produce spores that are able to live for long periods in the water environment.

Protozoologists commonly include certain unicellular algae among protozoa, especially those capable of both animal-like and plantlike nutrition. Thus, parasitic algae are often considered protozoa. One encounters a similar inclusiveness with regard to certain microscopic organisms with undetermined taxonomic status. The myxosporans produce multicellular spores and other multicell stages in their developmental cycle, but for the sake of convenience and convention they are commonly discussed as protozoa.

PROTOZOA AS PARASITES OR HARMFUL ASSOCIATES

Many protozoan species are obligatory parasites, requiring a host for completion of the life cycle. Others, as facultative parasites, do not require this essential host connection, but have the capability for a parasitic life in relationship with suitable hosts. Some protozoan parasites gain nourishment by absorption of tissue and cellular fluids, while others use excretions or digestive content. Some feed directly on host tissue. Protozoan parasites inflict a broad array of structural damage, toxicity, and functional disturbance during their life on body surfaces, within tissues, or within cells. The potential to quickly increase in numbers upon a general deterioration of the host is common to many protozoan parasites. Inherent system characteristics, such as a stage of host development, reduction of water exchange, and host density, may also permit a rapid increase of a particular parasite. The intensity of protozoan parasitism usually determines its importance in aquaculture. The mere presence of a limited parasitism, which may be common, does not equate to a state of disease.

Certain free-living or otherwise benign protozoan associates have the ability to transform into facultative parasites. These protozoa may begin to invade and feed on tissues if susceptible hosts become vulnerable due to inadequate nutrition, exposure to overly stressful conditions, or internal access brought by abrasions and lesions. Some otherwise free-living forms are noninvasive, but colonize surfaces of weak hosts, causing irritation, host dysfunctions, and general burdens to the host.

Terms useful in distinguishing protozoan relationships are endoparasite (the protozoa exist in body lumen, tissues, and blood), ectoparasite (the protozoa exist on body surfaces), endocommensal (the protozoa consume the host's diet, but not the host's tissue), and ectocommensal (the protozoa affix to the host's body surfaces, but do not consume the host's tissue).

AMOEBAE

Amoebae use protoplasmic flow to produce body extensions useful in movement. Depending on the kind of amoeba, the retractable extensions are threadlike or blunt, but in some cases are branching. These simple, naked cells feed on bacteria and reproduce asexually by binary or multiple fission. Amoebae produce cyst stages, and some of the important aquatic animal associates also produce a flagellate stage. Though some species have a solid external cover or an internal skeleton, most amoebae that parasitize aquatic animals lack such hard parts (see Fig. 1).

Amoebae parasitize gill and skin surfaces, the intestine, and, occasionally, tissues of many aquaculture animals. Amoebae that invade internal tissues directly from the aquatic medium or the digestive tract are thought to be opportunistic pathogens. We know little about the biology of amoebae that exist within tissues of aquatic animals, but some harm results from the accumulation of large numbers of amoebae in body spaces and the structural disruption that occurs as amoebae move between cells. Infections are mostly noticed in tissue squashes and smears and in histological preparations. Plates of nonnutrient 1.5% agar that support "lawns" of bacteria are useful for isolation of amoebae from tissues and the amoebae's subsequent identification. Histology is essential in the determination of pathology.

A low prevalence of amoebae is sometimes found in normal fish populations. The occurrence of heavy infections in fish is sporadic and possibly relates to the presence of eutrophic water, abundant bacteria, and higher temperature. Aside from the benefits of maintaining sanitary systems and general animal health, we know little about specific interventions for the control of amoeba infections in aquatic animals. For chemotherapy of external infestations of gills and skin, formalin is commonly applied as a bath treatment of 150 to 200 ppm of water for 0.5 to 1 hour.

Amoebae have not shown great importance in molluscan or crustacean aquaculture. Blue crabs (*Callinectes sapidus*) are known to host *Paramoeba perniciosa* along the east coast of the United States. The intensity of infection becomes greater in warmer months, and more deaths occur at that time. Other *Paramoeba* spp. (possibly all *Paramoeba pemaquidensis*) are known to colonize and damage gills of salmon (Salmonidae), rainbow trout (*Oncorhynchus mykiss*), and turbot (*Scophthalmus maximus*).

Members of several genera of free-living amoebae opportunistically invade tissues or colonize surfaces of freshwater fish species. Not surprisingly, these genera are the same ones that externally invade tissues and cause eye infection of terrestrial animals, including humans. *Naegleria* spp. and *Vahlkampfia* spp. colonize gills and skin surfaces. *Acanthamoeba* spp. are also found on gills and surfaces, but with some regularity invade body tissues.

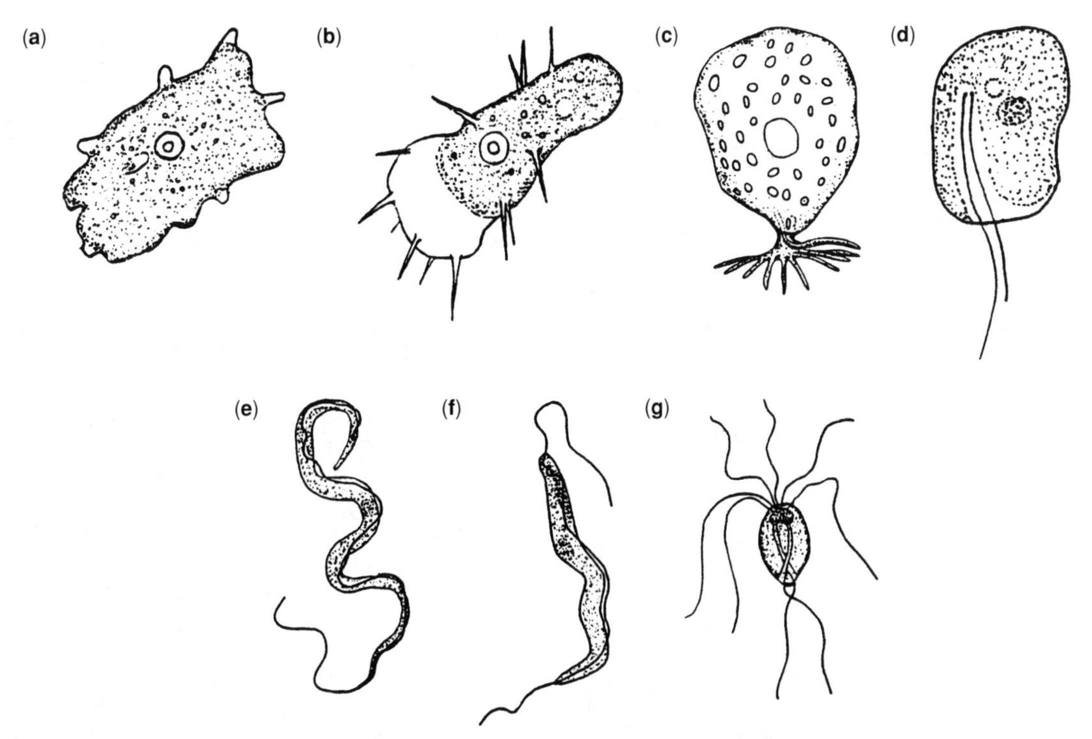

Figure 1. Examples of amoebae and flagellates. (**a**) *Paramoeba*; (**b**) *Acanthamoeba*; (**c**) *Amyloodinium*; (**d**) *Ichthyobodo*; (**e**) *Trypanosoma*; (**f**) *Cryptobia*; (**g**) *Hexamita*.

Relative to other fish species, tilapia (Cichlidae) have a greater history of *Acanthamoeba* infections. *Hartmanella* spp. are less commonly found in tissues of fish. Several fish, including the grass carp (*Ctenopharyngodon idella*), are known to host *Entamoeba* spp. within the intestine, but the importance of these amoebae to aquaculture is not understood.

FLAGELLATES

Possession of one to several locomotory flagella distinguishes the flagellates. Both plant and animal flagellates parasitize aquatic animals (see Fig. 1). Binary fission is the most common means of reproduction, and usually there is encystment.

Plant Flagellates

Among the plant flagellates, the dinoflagellates and euglenids have parasitic members. Much documentation exists about their parasitism of copepods, wherein they locate in gut, hemocoel, eggs, or general tissues and then influence the condition of the host (1). Such organisms have not become a nuisance in invertebrate aquaculture, but weak and dying larval animals often succumb to the invasion of opportunistic flagellates. In natural waters, *Hematodinium* spp. cause disease in lobsters, crabs, and pandalid shrimp during their molting season.

Parasitic plant flagellates of fishes are dinoflagellates, most of which are ectoparasitic. More important to aquaculture is *Amyloodinium ocellatus*, an ectoparasite of a variety of marine fish, and *Piscinoodinium* spp., parasites of freshwater aquarium fish. Unlike other dinoflagellates, *Amyloodinium* has lost its chloroplast, making it an obligate parasite. It is perhaps the greatest threat to the successful hatchery and nursery aquaculture of coastal fish species. At the optimal temperature of around 25 °C (77 °F), asexual multiplication is rapid, forming 256 infective dinospores in three days. The dinospores, whose structural features are responsible for taxonomic placement in the flagellates, must quickly find a host to continue the cycle. From there, the dinospore transforms into the harmful feeding stage. A penetrating and flexing organelle called the *stomopode* and the attachment base are important in inflicting pathology to epithelial cells that underlie the affixed parasite. This feeding stage, which is 80 to 100 μm (0.0031 to 0.0039 in.) in length, later detaches and transforms into the stage responsible for production of infective dinospores, which are about 8×12 μm in length. Under optimal conditions for the parasite, a cycle takes about a week to 10 days.

Controlling *Amyloodinium* is difficult. Chemical control with copper compounds (0.15 mg/L of water as free copper) has some advantage in the destruction of dinospores, but has the disadvantage of accumulating copper in systems. Other chemical approaches, such as chloroquine baths, have practical applications only in small systems. This is true also of rapid water exchange. Sudden exposures to freshwater results in some dislodgement, but the parasite is tolerant of salinities down to 1 ppt. Oestmann found that in small systems, a degree of dinospore reduction results from consumption by brine shrimp (2). Methods to enhance immunity have potential, but their validation requires much future work.

Certain dinoflagellates and other algae produce toxins that are harmful to aquatic species, especially fish. Toxic effects are occasionally seen in closed and recirculating systems, but the dynamics of these events are largely unknown. The exclusion of water entry is a normal practice when a farm's natural surface-water supply experiences a nuisance bloom. In marine cage culture, fish are moved to new areas when such conditions arise.

Two groups with members that parasitize molluscs and crustaceans are subjects of an occasional report, but do not have established importance in aquaculture: *Ellobiopsidae* and *Thraustochytridae*. Ellobiopsids are mostly seen with the unaided eye as dangling protrusions on the surfaces of pelagic crustaceans. They affix themselves on hosts through structures that penetrate and spread out like roots in the host's tissue. They produce biflagellate spores that resemble dinoflagellates. Ellobiopsids usually receive taxonomic treatment as a distinct group.

Perhaps because members of the thraustochytrids consume and parasitize plant cells, thraustochytrids receive more consideration as slime fungi than as protozoa. They are usually given a separate status when considered among the protozoa. In the past, some were grouped with amoebae. *Labyrinthula*-like thraustochytrids and *Labyrinthuloides* invade abalone, cephalopods, and probably other molluscs and may destroy tissue and cause death in only a few days. A free mobile stage with flagella and a parasitic amoeboid stage are part of the development cycle. After entering a host cell, the parasitic stage divides, forming a network. Apparently not so confined as in the cell walls of plants, the network expands throughout tissues of animal hosts. Parasite invasions into animals apparently favor young and poorly resistant individuals, but there is little understanding of the host-to-parasite relationships for this group, and the economic importance of such relationships to aquaculture is basically unknown.

Animal Flagellates

Ichthyobodo spp. (possibly all *Ichthyobodo necator*) are ectoparasitic on a wide variety of freshwater and marine fishes. The swimming form, which is 5 to 18 μm in length, has two flagella and is round to oval in shape. It attaches to skin and gill epithelium by its anterior end and assumes a tear shape. Several may attach to a single cell in heavy infections. The epithelial mucous cells produce in excess, and qualitative changes in skin cell type accompany infection. Some surface tissue is lost as fish rub on solid surfaces in response to irritation. Swelling of gill tissues constricts blood flow, and cell death is common.

The prevalence of *Ichthyobodo* infection of fish in aquaculture, particularly in salmonids, is relatively high. Young and malnourished fish are more vulnerable to heavy infestations. Damage to surface tissues affects the performance of salmon during transition to seawater. Application of a formalin bath (150–250 ppm for 0.5–1 hr) is probably the most common method of chemical control. Bithionol may have promise for tank application (3).

Hexamita and *Spironucleus* species are found in the digestive tract of freshwater and marine fish that are present in many climates. The round to tubular cells of less than 20 μm in length possess four anterior flagella. The flagellates feed on bacteria and particulate matter, and their cysts pass into the water medium with feces. A primary living site, depending on species, usually exists somewhere from the rectum to anterior intestine. Smaller and weaker fish are often the host to larger numbers of these parasites. Fish with many parasites show a deterioration in condition and become vulnerable to infections by opportunistic microbes. On occasion, the flagellates pass through the protective lining of the gut and invade various body tissues. *Hexamita* and *Spironucleus* commonly infect salmonids, cyprinids, cichlids, and a variety of aquarium fishes. *Hexamita* spp. are also known to exist in the intestines of oysters in aquaculture and as in fish, can cause disease when they bypass intestinal barriers of weak hosts and generally invade their tissues. Methods for controlling *Hexamita* and *Spironucleus* infections aim primarily at disinfection and removal of predisposing conditions. Therapy with metronidazole is used with aquarium fish.

Cryptobia spp. (including *Trypanoplasma*) parasitize a variety of aquatic animals, including freshwater and marine fish. Those that parasitize fish are typically ectoparasites or act in combination as ectoparasites and blood parasites. Intestinal forms of these parasites are also known. Depending on species, transmission is direct from fish to fish or involves a blood-sucking leech vector wherein the parasite multiplies between blood meals.

Cryptobia spp. are elongated protozoa with length-to-width ratios of 2 : 1 to 10 : 1. The anterior end is rounded, and the posterior end tapers. Two flagella extend from the anterior end; one projects forward and the other toward the back. The latter is attached for most of the protozoan's length until it extends freely beyond the posterior end. Along the attached portion, the pellicle stretches to accommodate flagellar movement, giving the false appearance of an undulating membrane.

A variety of clinical signs and pathogeneses accompany *Cryptobia* infections. For the blood forms, the sheer accumulation of large numbers of parasites due to multiplication promotes a wide array of pathology and dysfunction, not to mention a drain on immune potential and the general condition of the host. Infected and anemic fish generally do not respond well to environmental challenges. *Cryptobia* that infest gills also cause disease and mortality when present in excessive numbers.

Chemical control of ectoparasitic *Cryptobia* is possible in some circumstances. Leeches are not easily controlled in aquaculture water, especially when fish are present. Dehydration of systems to kill cocoons has shown promise. Vaccines have produced variable results, and selection of resistant fish strains shows promise.

Trypanosoma infect the blood of all vertebrate groups. In fish, parasite transmission is by blood-sucking leech vector. *Trypanosoma* species are elongated protozoa, typically with a length-to-width ratio of 10 : 1 and lengths of 20 to 50 μm. The cell tapers at both ends. The single flagellum typically originates near the posterior end and from there extends anteriorly affixed first to an undulating membrane and then becomes free. In contrast to what was previously thought about trypanosomas, there may be less host specificity and, due to structural variation, fewer species.

Though a degree of pathology is expressed in cyprinid infections by *Trypanosoma carassii* = syn. *T. danilewskyi* (4), overall, the trypanosomas have relatively little importance in aquaculture. Aside from protection of fish from leeches, there is little known about means for controlling *Trypanosoma*.

APICOMPLEXANS

Apicomplexans represent a taxonomic assemblage of parasitic protozoans whose kinship is based on a set of organelles located on the anterior end of the protozoan cell, called the *apical complex* (see Ref. 4 for illustrations). This complex is visible with electron microscopy and is important to penetration of the host by the infective stage. Important to aquaculture are the perkinseans, which infect bivalves and abalone, and the sporozoans, which infect fish, molluscs and crustaceans (see Fig. 2).

Perkinseans

Perkinseans are very important parasites of molluscs. Several species of *Perkinsus* are able to harm oysters, clams, scallops, and abalone. Densely set oysters rapidly succumb to infections of *P. marinus* in conditions of warm water temperature and high salinity. The parasite spreads and infects easily at salinities above 10 ppt in the warm season. Efforts to raise oysters in ponds and other confinements of warm regions have been largely unsuccessful, due to *P. marinus*. Areas with cooler water have less of a problem with the oyster parasite than do areas with warmer water, such as the Gulf of Mexico.

The mollusc host ingests the swimming infective stage, and parasitic development begins. Several multiplications increase the number of parasites. The parasite spreads throughout the body, inflicting severe damage to tissues as it locates within or outside of the host cells. When the parasite damages adductor muscle cells, the mollusc's shell opens (gapes) somewhat. Sick or dead oysters release cells, which divide and release new infective stages. Both swimming and multiplying stages are infective. *P. marinus* is generally endemic to growing areas, and disease occurs when conditions favor its expression.

Except for possession of a rudimentary apical complex, perkinseans do not have much in common with other apicomplexans. Perkinseans have a flagellated infective stage, and, information about their sexual reproduction is unknown. The round, mature "spore" is characteristic of infections. It is 5 to 20 (most likely 10) μm in diameter, and because of the large transparent vacuole and the smaller refractive cell content therein, it appears as a ring with a small inverted crown.

The vulnerability of molluscs in open aquaculture beds or suspensions presents a challenge for control of parasites. The growth period usually lasts two years in an unpredictable climate. The use of sites selected for stable and favorable conditions and the use of genetically resistant organisms have promise.

The work toward a consensus over the true taxonomic home for perkinseanlike organisms still continues today. Nowadays, some perkinseanlike organisms are considered as either apicomplexans or dinoflagellates. For many years, perkinseanlike organisms were discussed as fungi, mostly as *Dermocystidium* spp. Reports of work on this sort of parasite in fish continue to refer to it as a *Dermocystidium* species or as a *Dermocystidium*-like species. Parasitologists attribute importance to these parasitisms in the freshwater or marine aquacultures of salmon, eels, trout, and carp.

Sporozoans

Sporozoans are represented primarily by the coccidia, the piroplasmia, and the gregarinia. Coccidians of fishes are of two general types: (*1*) those that typically infect and produce oocysts in the intestinal lining—the common

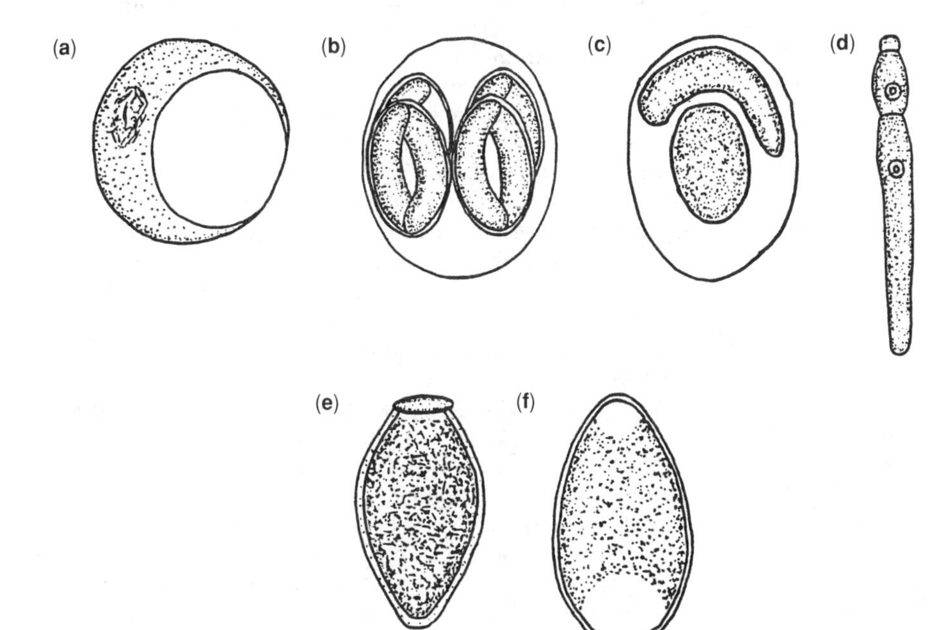

Figure 2. Examples of apicomplexans and microsporan. (**a**) *Perkinsus* (spore); (**b**) *Goussia* (oocyst containing infective stages, the sporozoites); (**c**) a blood-infecting coccidian within a red blood cell, with the parasite to the periphery of the cell nucleus; (**d**) *Nematopsis*; (**e**) *Haplosporidium* (spore); (**f**) microsporan (spore).

coccidia, and (*2*) those that typically infect fish blood cells with formation of oocysts in a second host, usually a blood-sucking leech. Piroplasmians also involve a leech and infect fish blood cells in the life cycle.

The dynamics of the fish–parasite relationship and the economic importance of parasitism by piroplasmians and blood-infecting coccidia are poorly known in natural environments and aquaculture systems.

The importance of the common coccidia is better known. In fish aquaculture, coccidian disease occurs in carp, goldfish, eels, red drum, (*Sciaenops ocellatus*) and tilapias. Most disease is caused by *Goussia* and *Eimeria* species. Infective stages enter and develop within the epithelial cells that line the intestine, eventually causing cell death. Many such cellular infections thus lead to disease. Oocysts pass with the feces and transmission normally occurs when the infective stages they contain are eaten by a new host. Upon release from the host, the oocyst of some species contains stages that are not yet infective. Depending on the species, the parasite must undergo further maturation in the oocyst after its release or after some transformation in the gut of a small organism that has eaten the oocyst.

Important coccidian infections of fishes occur in culture ponds. The traditional method for control is a disinfection that combines dehydration with a broadcast application of quicklime or chlorinated lime. Application of coccidiostat and other drugs to fish food has had some success, but the practical benefit of such chemotherapy is poorly understood. Coccidian infections of mollusc kidneys are apparently common and are possibly important to molluscan aquaculture.

Human infection by a coccidian, *Cryptosporidium*, via drinking water is a subject of current public-health interest. Attempts to transmit *Cryptosporidium* and comparable fish-infecting species, *Piscicryptosporidium* spp., between fish and people have been unsuccessful.

Most gregarinians are single-host parasites, but the ones important to crustacean aquaculture have alternate invertebrate hosts. Gregarinians locate in the gut of commercially important crabs, lobsters, and shrimp. There, a slender infective stage of approximately 30 μm in length emerges from the spore and initially attaches to a cell's surface, rather than entering the cells, as in the case of coccidians. The parasite then forms into a cigar-shaped structure that contains distinct components and expands greatly in size to 0.1–1.0 mm as it absorbs nutrients (see Fig. 2). Most parasites remain attached, but some glide about the intestinal lumen either as detachments or as parasites that did not make an initial attachment. Eventually, the parasite moves to the rectum, where it attaches and forms an encystment by coiling up and secreting a wall. Following multiplication therein, stages release into the water that are capable of being taken up and infecting a second host. In that host, the gregarinian apparently parasitizes a single cell and, without multiplication, transforms into a spore that contains the infective stage. This spore is consumed by a crustacean that eats the second host or some of its parasite-containing releases.

Crustacea show little pathology from gregarinians, except when very heavy parasitisms result in epithelial damage to the gut, a physical blockage of the gut, or, possibly, predisposing tissues for bacterial infection. *Nematopsis* spp. are common in crustaceans and occur widely in bivalves, which are often alternate hosts. In bivalves, the parasites have little or no reputation as important pathogens. Control efforts in aquaculture focus on elimination of the alternate hosts during system dryouts.

MICROSPORANS

Microsporans commonly parasitize insects and crustaceans. Other invertebrates and members of all vertebrate classes also host microsporans. Some species are important in aquaculture when exceptionally heavy infections cause disease or when obvious masses in skin and musculature degrade the market-ready product. Diseases of reproductive tissue affect the reproductive potential of some hosts.

The foremost defining and obvious structural feature of this group is a small (usually less than 10 μm in length) spore containing an extrusion apparatus (see Fig. 2). The extrusion apparatus appears as a coiled filament packed to the periphery of the spore. Also within the spore is a small infective agent. Transmission occurs upon consumption the infective agent within the spore. In some cases, infection is possible only if the spore first passes through the gut of another appropriate animal. Certain insect microsporans are known to involve a distinctive cycle in alternative, totally unrelated hosts. Perhaps future study will show such life cycles to be characteristic also of some aquatic animal microsporans.

Once a microsporan is inside the gut of its host, the filament bursts out from within the spore. The extruded filament is stiff enough to penetrate a cell. The infective unit passes through its hollow core and gains entry to the host's tissues. Once inside the host's tissue, it somehow associates with susceptible tissues and begins to grow and divide. Larger ameboid cells containing multiple divisions may grow to around 50 μm in length before their smaller internal units release to infect additional cells. As parasite cells continue to develop, some cells multiply to produce spores. These spores are eventually released in various ways, including release from a dead and disintegrating host.

Microsporan species are more or less specific in the selection of tissues to infect. Infection of musculature is common in crustaceans, and, due to transparency of the cuticle, many with infections give hosts a whitish appearance. During heavy parasitism of a fish's intestinal linings, the fish's gut appears completely displaced by opaque masses of the parasite. Skin infections may show discoloration, lumps, or nodules. *Enterocytozoon salmonis* infects the blood cell nuclei of chinook salmon which manifest a leukemic condition (5), but the chinook disease called plasmacytoid leukemia may have a viral cause (6). Some microsporans develop in great numbers within a host cell, causing the cell to enlarge greatly. In some species, these cystlike xenomas expand to the millimeter or centimeter range, with the infected cell membrane still intact.

Control of microsporans in most aquaculture systems is difficult. The effectiveness of oral medicines remains unclear, and the practice of their application has not been adopted by aquaculturists. Spores may survive for more than a year, and the efficiency of disinfectants is not sure. Facilities with troublesome infections may benefit most from species changes and general sanitary practice.

HAPLOSPORANS

Haplosporans parasitize invertebrates. Much of their importance in aquaculture comes from parasitism of oysters. The genus *Haplosporidium* is most important. *Haplosporidium nelsoni* has been responsible for many epidemics in eastern oysters along the eastern coast of the United States. Other *Haplosporidium* spp. cause disease to oysters and clams worldwide, but species assignments are seldom made, because of the rarity of spore stages in many infected hosts. Infected individuals grow poorly, show emaciation, and may die.

Spores are easily noticeable in infected tissues through use of using light microscopy, but except for their size of 5 to 7 µm in length, there is little to distinguish them as haplosprans (see Fig. 2). *Haplosporidium* spp. have a flat "cap" on one pole of their oval spore, and other genera may have peculiar surface structures. Electron microscopy is needed to determine a cellular part known as a haplolsporosome, the presence of which suggests the inclusion of otherwise ambiguous spores among the haplosporans.

The haplosporan infective stage enters a host oyster at the gills and develops into larger cells that contain multiple divisions. These cells eventually infect their hosts throughout the body, some associating with tissues next to the digestive tract and gonads. This process may be important to subsequent production of spores in the intestine, especially of young animals. A portion of the parasites that multiply in the gills release into the mantle cavity and from there are swept to the outside. This process may be important in the rapid development of the parasite in oyster beds. Details of the sexual reproduction of these oyster parasites are unknown in spite of long study of the subject. This suggests that a second host may be involved in the developmental cycle.

H. nelsoni lives best in salinities of 10 to 30 ppt. Culturing oysters below 10 ppt may have an advantage for controlling the parasites. The use of resistant strains in recent years has also shown promise.

The taxonomy of haplosporans is somewhat confusing. Many authors continue to write about haplosporans as part of a group known as the ascetosporans. Others treat haplosporans as a distinct and separate group, as is done in this text. Paramyxans of importance to aquaculture are generally considered among the haplosporans, but differ in the manner of spore formation. *Bonamia* spp. usually receive separate treatment as a taxon, but they have many haplosporan features.

Paramyxans

Paramyxans of *Martelia* spp. cause disease to young European flat oysters, mussels, scallops, and a variety of other molluscs. These parasites infect the stomach and move to the digestive gland. The number of parasites increase in the tissues of the digestive organs, where dividing stages of 6 to 30 µm in length may seen in histological preparations. A disruption of the digestive gland tissues during the formation and release of spore stages is probably important in the development of disease. The diameter of a spore is around 3 µm. Heavily infected populations of molluscs may become greatly reduced by *Martelia* infections. The best known paramyxan is *Martelia refringens*, a parasite of the flat oyster and mussels.

Bonamia and Mikrocytos

Bonamia ostrea is a small (2 to 3 µm in length) parasite found worldwide in populations of the flat oyster, *Ostrea edulis*. It is of major importance to aquaculture of this species. *B. ostrea* infects blood cells. On an occasion of disease, it is unclear whether oyster loss is due directly to the parasite or to its predisposing effect for bacterial infection. The disease occurs most easily at temperatures between 12 and 20 °C (54 and 68 °F). Newly exposed stocks are particularly vulnerable to mortality. *Bonamia* spp. infect several oyster species in a number of growing regions.

Mikrocytos spp. are small (2 to 3 µm in length) organisms that cause a winter disease of oysters. More is known about *Mikrocytos mackini*, which causes disease in Pacific oysters (*Crassostrea gigas*). Various superficial tissues and adductor muscles show greenish lesions in infected oysters. These parasites apparently do not develop when temparatures are above 15 °C (59 °F).

CILIATES

Ciliates possess rows of cilia that are useful in mobility and also in directing food bacteria toward their cytostomes. Structurally, cilia and flagella are basically identical. Cilia are, of course, more numerous and proportionally smaller in respect to the individual protozoan. Ciliate motion is gliding, whereas flagellates dart and wiggle. Ciliates range in size from 10 µm to around 3000 µm in length, but the ones that occur on and in fish are typically 30 to 100 µm in length (see Fig. 3). An obvious exception is the large feeding stage of *Ichthyophthirius*, which may grow to 1000 µm. Asexual reproduction occurs by simple or multiple fission and by budding. Some ciliates produce cysts.

Most internal and external associations of ciliates with hosts are not harmful. When conditions stressful to the host or system favor excessive ciliate proliferation, otherwise harmless ciliates may cause tissue irritation, asphyxiation if on gills, and, in larval animals, immobility. Only a few ciliates feed on tissue or have the ability to invade tissue.

Ectoparasitic and Ectocommensal Ciliates

Mantle cavities of molluscs harbor a number of commensal and parasitic ciliates. The genera *Ancistrocoma*,

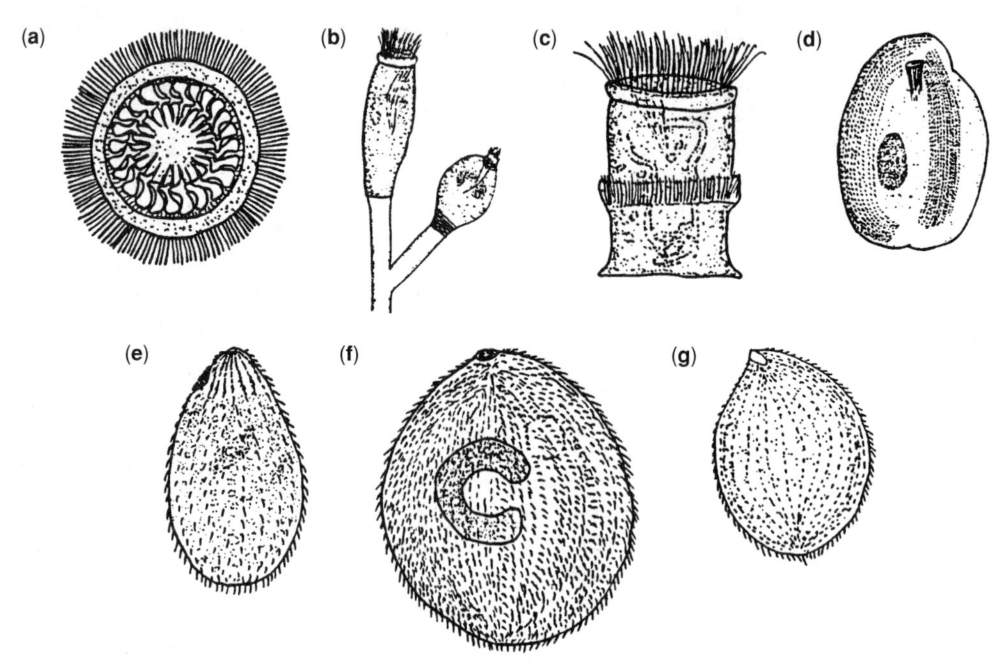

Figure 3. Examples of ciliates. (**a**) *Trichodina*; (**b**) *Epistylis*; (**c**) *Ambiphrya*; (**d**) *Chilodonella*; (**e**) *Tetrahymena*; (**f**) *Ichthyophthirius*; (**g**) *Cryptocaryon*.

Sphenophrya, and *Trichodina* are common, but cause little harm to cultured molluscs.

The crustacean cuticle, or exoskeleton, is the habitat for apostome ciliates. There, apostomes are found within encystments or as newly excysted feeding stages. After feeding occurs, internal division produces more infective stages, which settle on a new host. Some of the parasites feed on nutriment associated with the exoskeleton after it is shed at molting. Other types merely consume the unshed exoskeleton or excyst when there is a break in it and consume internal tissues. Apostomes have not become important in crustacean aquaculture.

Crustaceans and fish host a variety of sessile ciliates on their gill and body surfaces. Many associate with inanimate objects as well as the aquatic hosts. Others favor direct associations with aquatic animals, and some form even more specific associations with a host group or a species. A variety of structural types occur on crustacean hosts. The basic feeding cell of some is borne on the terminus of a long stalk or its branches. While this type dangles freely into the water, other kinds attach directly and snugly to the surface of the host. Suctoreans, with their obvious needlelike tentacles, make up part of this array of ciliates, though they do not normally possess obvious cilia during host attachment. Certain bacteria and algae also attach among and upon the ciliates. All organisms together contribute to a general burdening of the animal. Sessile ciliates fare better during periods of poor host condition and higher system eutrophication.

Sessile ciliates become important in crustacean aquaculture when heavy infestations occur. Species of *Zoothamnium*, *Ephelota*, and *Epistylis* are able to debilitate mobility of very small early larval stages in hatchery tanks. Excessive infestations by these and other sessile ciliates reduce marketability, because they mar the appearance of the whole product. Heavy infestations of the gills may impair respiration and promote disease during conditions of low levels of dissolved oxygen.

Ambiphrya, *Apiosoma*, *Capriniana*, and *Epistylis* species attach to gill and body surfaces of a great variety of marine and freshwater fish. Heavy infestations may contribute to host asphyxiation. *Epistylis* spp. are often the direct or indirect cause of lesions at attachment sites.

Mobile parasites of gill and skin surfaces are more harmful than sessile species. *Chilodonella pisciola* and *Chilodonella hexasticha* infest a great variety of freshwater fish. *Brooklynella* spp. infest a variety of marine aquarium fishes, but do not match *Chilodonella* spp. in terms of importance. Cells of these species vary around dimensions of 60×40 μm and have shapes similar to a fingerless human palm. Aside from ciliary rows covering the pellicle, there is a conspicuous structure called the pharyngeal basket that is associated with the cytostome. Heavy infestations of most *Chilodonella* cause parasitic disease to fish in aquaculture systems. *Chilodonella* is common on cyprinids and often infests all fish in mixed populations where cyprinids are grown as forage species.

Trichodinid ciliates are the most common mobile ectoparasites of marine and freshwater fish. Heavy trichodinid infestations, especially in hatchery fry, are responsible for fish mortalities in aquaculture. Large numbers of trichodinids can damage the gill epithelium and subsequently feed on cellular debris. This damage may derange functions associated with gill excretion and provide a portal of entry for infective bacteria. Some trichodinid species are found in urinary bladders. *Trichodina* is by far the most common trichodinid genus.

Bowl-shaped trichodinids are generally around 25 to 75 μm in length. Obvious cilia fringe the cell and also coil toward the cytostome. Characteristic denticles, which appear as jagged skeletal needles blunted to the periphery and pointed toward the cell center, overlap as a circular fan around the cell's relatively flat ventral surface. The height of the bowl-like cell differs according to species. Some are short like a frisbee, and others are tall like a bell jar.

Most methods to control ciliates that infest the surfaces of aquaculture animals usually rely on adjusting environmental conditions or adding mild toxicants that are safe for the aquaculture animal but are destructive to the protozoans. Adding sodium chloride to provide a concentration of 0.3% for indefinite periods is effective in ridding freshwater fish of ciliates. Dips for several minutes in concentrations of 1 to 3% salt in water are also effective against ciliates and are tolerable by fish of all but the smallest sizes. Applications of formalin, copper sulfate, potassium permanganate, and other chemicals are also effective at levels that are nontoxic to fish when used as baths or static treatments of indefinite duration. Water replacement to restore a lower level of organic content and microbial flora to the culture medium is effective in reducing ciliate burdens on the cuticular surfaces of crustaceans.

Invasive Ciliates

Fish and crustaceans of poor health in aquaculture systems succumb to invasions by certain free-living or commensal ciliates that possess an invasive potential. Better-known examples of this type of invasiveness are *Tetrahymena corlissi* of poecilid fishes, *Mesanophrys maggi* of Dungeness crabs, *Anophryoides hemophilia* of lobsters, and *Uronema marinum* of marine fish. These ciliates are generally oval shaped and measure near 50×30 μm.

The ciliate of greatest importance to aquaculture is *Ichthyophthirius multifiliis*, an obligate parasite of freshwater fish. Known as "Ich," this parasite goes through several development stages. The infective stage contacts a host's skin or gill and burrows into the epithelium, where it begins a stage of growth while feeding on the host's tissue. Upon gaining a considerable size that approaches 0.5 mm (even up to 1.5 mm in cold water) in length, it abandons the host, secretes a fragile cyst wall, and divides therein into several hundred to more than a thousand individuals. Upon the last division, the individuals are released into the water as infective stages. The complete life cycle of the parasite is as short as three to four days at 25 to 28 °C (77 to 82 °F). Seasonal influence encourages polycyclic infection. Ich disease in pond fishes typically follows the arrival of warmer weather, when the duration of the ciliate's life cycle decreases from weeks to a few days. Water temperatures of below 5 °C (41 °F) slow the cycle to as long as three months. Above 30 °C (86 °F) development does not take place.

These ciliates invade only superficial, epithelial tissues of skin and gills. Much mechanical damage to tissue is the result of parasite movement. Epithelial thickening occurs at first, and later the epithelium becomes detached, exposing the dermal layer of the skin to the water. Heavily infected fish with respiratory and osomoregulatory problems become listless and, when stocked in ponds, begin to accumulate at edges or leeward shores.

A low prevalence of subclinical infection must be normal for natural systems, but epidemics in streams during times of low flow and fish crowding have been observed. Populations of scaleless fish possessing a thick epithelium may be more susceptible to Ich. Infectivity obviously benefits from higher fish densities. The presence of the parasite in unnatural systems, such as fingerling ponds, during periods when the temperature is optimal for the parasite is a sure set for an epidemic. Acquired immunity has an influence on infection dynamics. Exposure of naive fish to infected, partially immune fish usually results in quick and severe infection of the former. Fish in live markets often acquire severe infections in this way, due to continual introduction of fish.

Ich epidemics often occur soon after fish holding or transport. Such handling crowds the fish, and if some are infected, a window of opportunity is given the parasite to massively infect. Fish are lost in large numbers two or three days later, usually at some new destination. Keepers of tropical ornamental fish often attribute outbreaks to a sudden lowering of temperature. In such cases, successful infections probably result from new conditions that are optimal for Ich and too cool for the fishes.

Approaches for avoidance of Ich disease include not consolidating stocks, using ground water rather than infected surface water for system supply, and using seed stock that is known to be Ich free. Periodic checks of fingerling stocks for infection are helpful for early detection. Control of the parasite in fish stocks is difficult. Chemical treatments must be directed to the postemergence stages and therefore require multiple applications to culture water. Success with chemical treatments in pond systems occurs only during transitional seasons, when temperatures favor a short parasite cycle. If conditions permit, such as in small growing and handling systems, continual exposure of fish to a salt concentration of three ppt of water results in control of new infections and eradication of the parasite within a few days to weeks, depending on water temperature.

Cryptocaryon irritans, a parasite of marine fish, produces a parasitism very similar to that of *I. multifiliis*. The developmental stages and host damage are essentially the same. The parasite lives better at warmer temperatures and higher levels of salinity. The preventive approaches previously mentioned are suitable for *Cryptocaryon*, and for active infections, a drop in salinity to levels below 20 ppt effectively controls the parasite.

Endoparasitic ciliates of fish are found in the digestive tract, urinary tract, and oviduct. The pathological importance of these ciliates is probably small in aquaculture, though a report on grass carp (7) describes a detrimental invasion of intestinal tissues by *Balantidium ctenopharyngodonis*.

MYXOSPORANS

Myxosporans are parasites of fish, amphibians, reptiles, and some invertebrates. They occur in fresh, brackish,

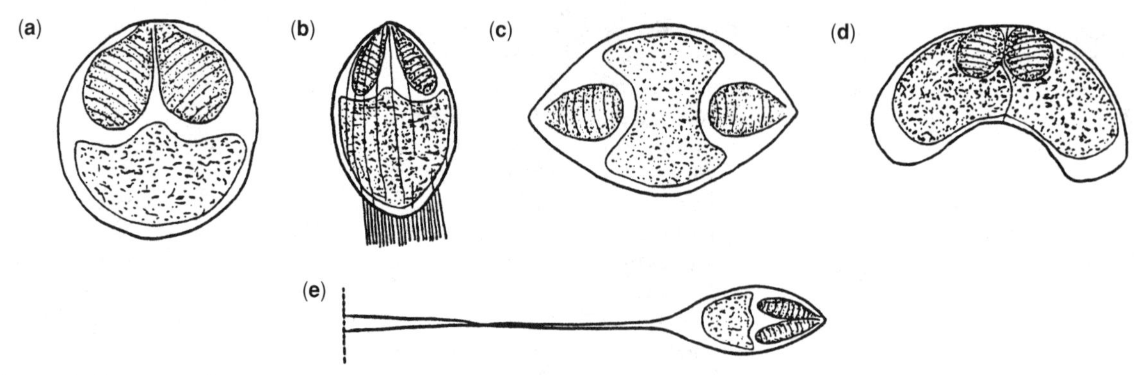

Figure 4. Examples of myxosporans (spores). (**a**) *Myxobolus*; (**b**) *Hoferellus*; (**c**) *Myxidium*; (**d**) *Ceratomyxa*; (**e**) *Henneguya*.

and marine water systems. Many are found in gall bladders, but species more important to aquaculture parasitize muscle, skeletal, gill, intestinal, kidney, and skin tissues. Some species are very important in aquaculture. Myxosporans may be the most important cause of fish disease in natural populations.

Myxosporans produce unique and characteristic multicellular spores, with dimensions usually between 10 and 20 µm in length (see Fig. 4). The spore's rigid shell and easily recognizable sporoplasm and polar capsules are representative of different cells. The shell formation results in portions known as valves. Spores and their valves have certain variability of size and shape, but the structure is distinctive enough to determine genus and species. The polar capsules contain filaments that extrude in appropriate conditions and act as holdfasts. Concurrent with filament extrusion is a release of the sporoplasm through open valves.

The means of entry into a new host differs according to species (e.g., via gut or body surface), but once entry is made, the sporoplasm migrates to the susceptible body site. Following further development, and depending on parasite species, the now small or large multinuclear and multicellular plasmodium of a few µm in length to more than 2 µm in length undergoes production of spores. Depending on species, one, two, or many spores will form in a plasmodium. The large plasmodia of some species contain many fully developed spores and have a cystlike appearance within gill, epithelial, or other tissue. The small plasmodia (pseudoplasmodia) of other species produce only one or two spores and are not visually apparent, unless perhaps in an assembly.

As previously mentioned, spores occur in various tissues of their hosts. They are released from the host at death, through excretions, ruptures of cysts, and after transport by blood to release sites. A second host is involved in some myxosporan development cycles. Transmission and infectivity studies show that some myxosporan species that infect aquatic worms (oligochaetes) also infect fish. In fish, the worm's myxosporan parasite transforms its appearance to that characteristic of a fish myxosporan. In other words, what was previously thought to be two distinct and structurally dissimilar spore-forming species is really one species with a development cycle that uses second hosts.

Many myxosporans do not cause disease when present in aquaculture. Those that do produce a varying degree of pathology, whether in the kidney tubules, digestive tract, gills, skin, or other tissue. Sometimes, damage is due merely to the intensity of the parasitism and the inability of the host to heal. Nevertheless, under certain aquaculture conditions, myxosporans have great potential for harm.

Several myxosporan diseases are notable in aquaculture. *Myxobolus cerebralis* causes "whirling disease" in young salmonids. The cartilage of fry is infected, and spores form a number of weeks later, depending on the water temperature. Erosion of cartilage by the parasite affects the nervous system, causing a whirling, or tail-chasing, behavior. This process and structural abnormalities may eventually lead to death. *Sphaerospora* and *Hoferellus* spp. have very small pseudoplasmodia that multiply several times prior to formation of spores. This proliferation of parasites results in an enhancement of infection. As numerous parasites locate in tissues, swelling and other tissue reactions become manifest, and disease occurs. Such parasites cause gill and kidney diseases in carp, catfish, salmonids, and other fish. Examples are proliferative gill disease of channel catfish and proliferative kidney disease of salmonids. *Ceratomyxa shasta* causes generalized infections in salmonids following parasitism of the intestinal epithelium. After causing serious lesions in the intestine, the parasite moves further into the host, affecting a variety of tissues.

Some important aquaculture diseases may result from harm done to "wrong" hosts by prespore stages of yet-unidentifiable myxosporans. According to this hypothesis, massive numbers of infective stages of myxosporans somehow become present in the culture system, and subsequent invasion of the abnormal fish host produces a pathology sufficient to cause disease.

Muscle-invading myxosporans produce milky-white, streaky, wormy-looking, or spotted flesh that is unacceptable in the market. Flesh heavily infected by developing stages of some myxosporans becomes soft, jellylike, or liquefied a few hours after death at ambient temperatures, due to postmortem release of proteolytic enzymes by the parasites. Skin-infecting myxosporans, such as some members of *Myxobolus*, *Henneguya*, and *Myxidium*, may alter fish appearance to a degree that the fish are unmarketable.

The application of sterilants to pond sediments and dehydration are the favored approaches for controlling myxosporans in aquaculture. The use of preventative drugs for myxosporan infection has some history of experimentation, but has not been adopted in aquaculture practice. The development of control methodologies for myxosporans is largely lacking.

Recent reviews of protozoan parasites of aquatic animals are helpful for their study. Important references that discuss fish (4,8,9) and molluscs and crustaceans (1) are available.

BIBLIOGRAPHY

1. P.C. Bradbury, in J.P. Kreier, ed., *Parasitic Protozoa*, Volume 8, 2nd ed., Academic Press, San Diego, CAS 1994, pp. 139–164.
2. D.J. Oestmann, D.H. Lewis, and B.A. Zettler, *J. Aquatic Animal Health* **7**, 257–261 (1995).
3. J.L. Tojo, M.T. Santamarina, J. Leiro, F.M. Ubeira, and M.L. Sanmartin, *J. Fish Dis.* **17**, 135–143 (1994).
4. J. Lom and I. Dykova, *Protozoan Parasites of Fishes*, Elsevier, Amsterdam, 1992.
5. R.P. Hedrick, J.M. Groff, and T.S. McDowell, *Dis. Aquat. Org.* **8**, 189–197 (1990).
6. M.L. Kent and S.C. Dawe, *Dis. Aquat. Org.* **15**, 115–121 (1993)
7. K. Molnar and M. Reihardt, *J. Fish Dis.* **1**, 151–156 (1976)
8. P.K.T. Woo, ed., *Fish Diseases and Disorders*, Volume 1, CAB International, Wallingford, United Kingdom, 1995.
9. J.P. Kreier, ed., *Parasitic Protozoa*, Volume 8, 2nd ed., Academic Press, San Diego, CA, 1994.

See also DISEASE TREATMENTS.

PURE OXYGEN SYSTEMS

JOHN COLT
Northwest Fisheries Science Center
Seattle, Washington

OUTLINE

The use of pure oxygen can increase the carrying capacity of an aquatic culture system when dissolved oxygen is the most limiting factor. The actual increase in carrying capacity will depend primarily on temperature, pH, and alkalinity. Pure oxygen systems can economically saturate or supersaturate water with dissolved oxygen. Supersaturated dissolved oxygen in the absorber effluent significantly reduces the volume of water that must be treated to satisfy a given oxygen demand. Unlike air contact systems, pure oxygen systems have the capability of reducing dissolved nitrogen to or below saturation for the purposes of controlling gas bubble disease.

SOLUBILITY OF OXYGEN IN PURE OXYGEN SYSTEMS

The solubility of oxygen depends on its temperature, salinity, gas composition, and total pressure (1):

$$C^* = (1{,}000K\beta)\chi\left(\frac{P - P_{H_2O}}{760}\right) \quad (1)$$

where K = ratio of molecular weight to molecular volume (1.42903 mg/mL); β = bunsen coefficient L/(L•Atm); χ = mole fraction of oxygen in dry gas (%/100) (dimensionless); P = total system pressure (mm Hg); and P_{H_2O} = vapor pressure of water (mm Hg).

This equation can be used to compute the solubility of a gas of arbitrary gas composition and pressures. Tabular values of β and P_{H_2O} can be found in standard references (1,2). At 15 °C (59 °F) and 1 atm pressure, the air solubility of oxygen ($\chi = 0.20946$) is equal to 10.07 mg/L (ppm), while the solubility of pure oxygen ($\chi = 1.0$) is equal to 48.09 mg/L (ppm). Oxygen solubility may be further increased by increasing the total system pressure, P in Equation 1. Compared to air contact systems, increased solubility in the pure oxygen systems increases both the rate of oxygen transfer and the potential maximum effluent dissolved oxygen concentration.

SOURCES OF PURE OXYGEN

Standard conditions for the reporting of volume are 20 °C (68 °F) and 1 atm for gas and the normal boiling point for liquids. Useful physical data for oxygen gas and liquid are presented in Table 1.

Enriched oxygen may be obtained from three sources: high-pressure oxygen gas, liquid oxygen (LOX), and on-site oxygen generators. In many facilities, at least two sources of oxygen are used.

Table 1. Physical Properties of Oxygen Gas and Liquid

Parameter	SI Units	English Units
	Gas Phase	
Density	1.331 kg/m^3	0.08309 lb/ft^3
Specific volume	0.7513 m^3/kg	12.04 ft^3/lb
Selling units	1.331 kg/m^3	8.31 lb/100 ft^3
	Liquid Phase	
Density	1.141 kg/L	9.52 lb/gal
Specific volume	0.877 L/kg	0.105 gal/lb
	Gas–Liquid Relations	
Volume of gas/volume of liquid	0.857 m^3 gas/L liquid	115 ft^3 gas/gal liquid

High-Pressure Oxygen Gas

High-pressure oxygen gas (98–99%) can be obtained in cylinders containing 3–7 m^3 (100–250 ft^3) at 17.6 MPa (2550 psi). A number of cylinders can be connected together using standard manifold assemblies to increase the total capacity. The use of large numbers of gas cylinders is expensive and, therefore, commonly restricted to small transport or backup systems.

Liquid Oxygen

Liquid oxygen (98–99%) is produced on a large-scale basis by distilling liquefied air. Liquid oxygen (LOX) is generally trucked to the use site in bulk, then pumped into the storage tank. At 1 atm pressure, liquid oxygen boils at −182.96 °C (−297.3 °F). To reduce the amount of oxygen converted to gas, liquid oxygen must be stored in an insulated container. Liquid oxygen containers range from 100 L (30 gal) to greater than 40,000 L (10,000 gal). A liquid oxygen supply system will consist of a storage tank, evaporator, filters, and pressure regulator (Fig. 1). The maximum gas pressure in these containers is generally in the range of 1000–1400 kPa (150 to 200 psi). If the pressure exceeds these values, gas is vented to the atmosphere through a safety valve. Approximately 0.25% of the liquid will be lost per day, even if no oxygen is used. With the exception of direct impact from bullets and vehicles, there are few things that can go wrong with a LOX supply. It is common to lease the LOX tank and evaporator from the company supplying the LOX.

Pressure Swing Adsorption Oxygen Generators

Manufacturers have adapted industrial pressure-swing adsorption (PSA) systems for aquaculture applications. Single units are available to produce 0.50 to 10.0 m^3 oxygen/hour (15 to 400 ft^3 gas/hour) and may be connected in parallel to produce larger quantities. These units require a source of 600–1000 kPa (90–150 psi) filtered air and generate enriched oxygen gas of 85 to 95% oxygen. Oxygen generators work on a demand basis, so oxygen is produced only when needed. When used as a primary source of oxygen, a standby electrical generator is generally needed.

Production Oxygen Systems

For production systems, more than one source of oxygen may be required. Selection of the oxygen source(s) will depend on a number of site-specific conditions and the biases of designers and hatchery managers. Isolated production facilities may find PSA systems to be more economic due to high LOX transport costs. Sites near LOX production plants may find LOX more attractive due to reduced capital and maintenance costs. A LOX supply is much more reliable than a PSA.

Steel pipe and copper tubing are recommended for oxygen lines. Special care must be used in construction and installation of pure oxygen piping and fittings to remove grease and oil which may present a fire hazard when in contact with pure oxygen. It is possible to order specially cleaned and packaged valves and fittings for pure oxygen applications.

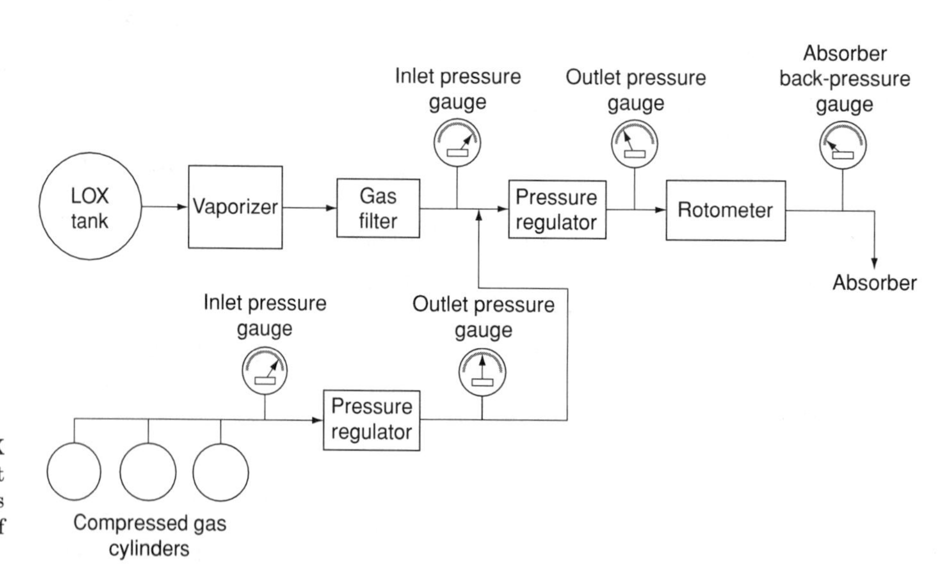

Figure 1. Major components of a LOX supply and regulation system that incorporate auxiliary compressed gas cylinders as an emergency source of oxygen.

TYPES OF AERATORS USED FOR PURE OXYGEN SYSTEMS

When compared to air, pure oxygen is relatively expensive, and only systems with high absorption efficiencies are generally economical. At least seven types of oxygen absorption equipment are being used.

U-Tubes

The u-tube aerator consists of a vertical shaft 10–45 m (30 to 150 ft) deep, either partitioned into two sections or consisting of two concentric pipes (3). Oxygen is sparged at the top of the down-leg of the u-tube and transferred into the water as the gas–liquid mixture is carried through the contact loop. Some commercially available u-tubes may include proprietary modifications to increase mixing and oxygen transfer. If 2 to 3 m of head are available, the system can be operated without pumping (Fig. 2a). To increase the oxygen absorption efficiency, the oxygen-rich off-gas may be collected and recycled (Fig. 2b) (4).

Packed Columns

A pure-oxygen packed column consists of a sealed column packed with a high specific-surface area medium (Fig. 3). Water is distributed uniformly over the upper surface of the medium by a perforated plate or spray bar. Pure-oxygen packed columns are efficient nitrogen strippers, and most production columns are designed to both increase dissolved oxygen and strip out nitrogen (5,6).

"Michigan" Spray Columns

A spray column is similar to a packed column except it does not contain media, and the size of the discharge is reduced (Fig. 4). This type of column may be required in reuse systems due to problems with media fouling. The reduced discharge produces a vacuum of 50–120 mm Hg (7) and results in excellent degassing of nitrogen.

Low-Head Oxygenator

The low-head oxygenator (LHO) is a proprietary oxygen transfer unit that can be readily retrofitted into serial

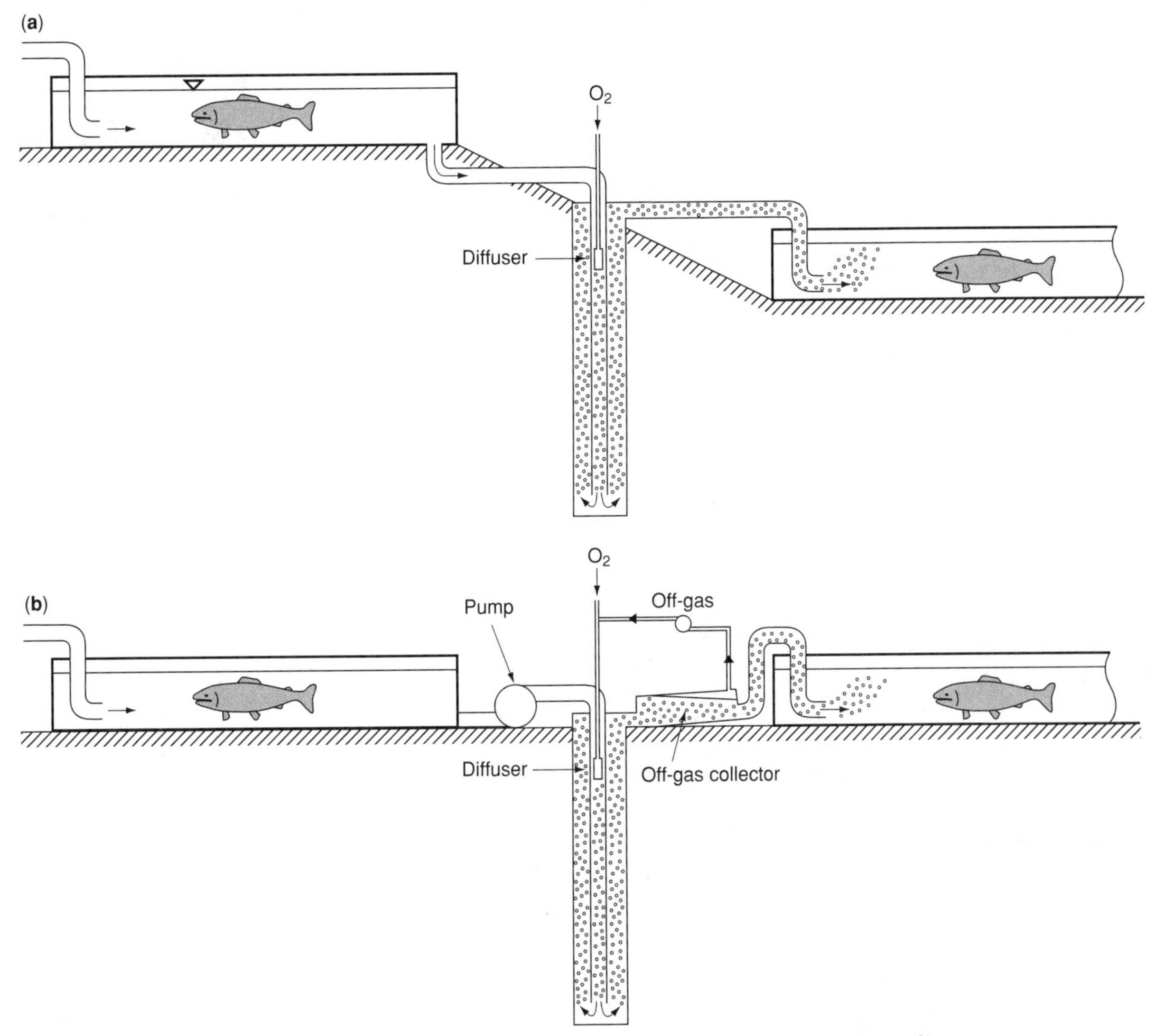

Figure 2. U-tube (**a**) gravity system and (**b**) pumped system with off-gas recycling.

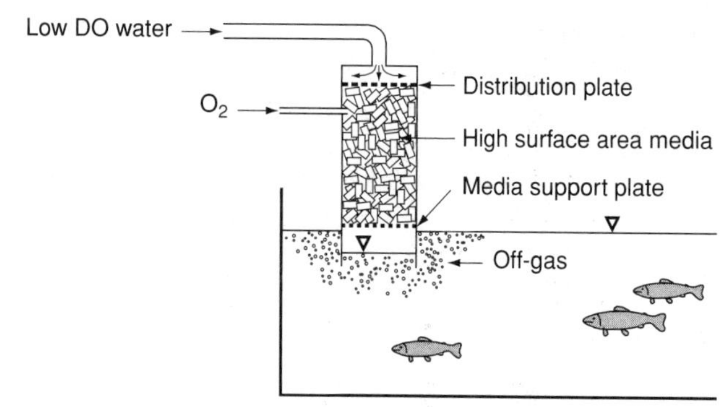

Figure 3. Packed column mounted in rearing unit.

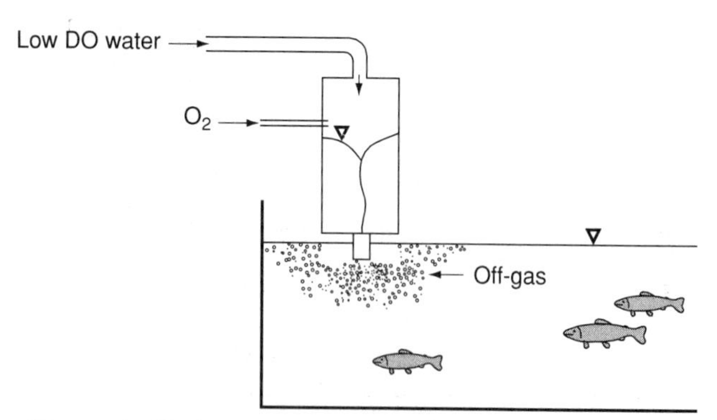

Figure 4. "Michigan" spray column mounted in rearing unit.

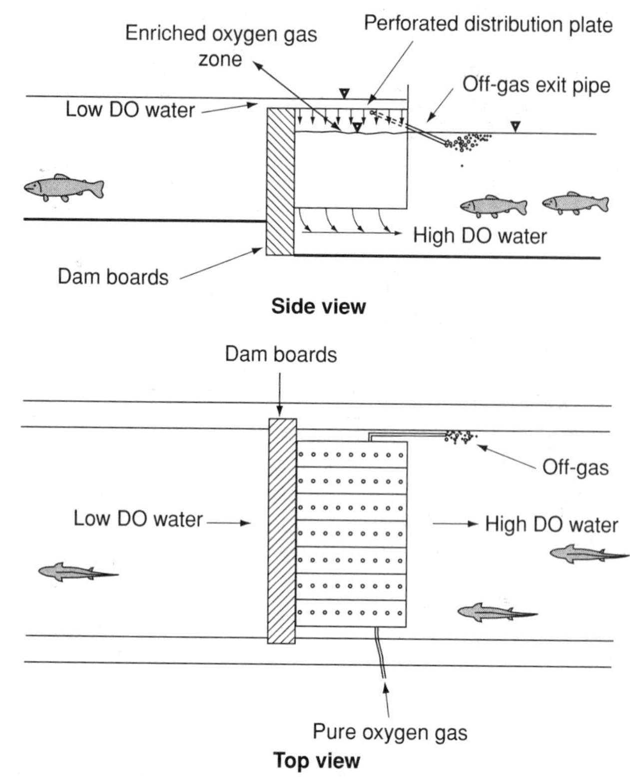

Figure 5. Low-head oxygenator (LHO) system. Water falls vertically through the perforated distribution plate into the oxygen-enriched gas zone and then flows out the bottom and down the raceway. The pure oxygen gas is introduced on the left side and exits on the right side (top view). The unit is divided into seven compartments, so there is minimal backmixing of the gas as it moves from the first compartment to last compartment. Within each compartment, the gas phase is well mixed.

raceway systems. The unit consists of a distribution plate positioned over seven rectangular chambers (Fig. 5). Water flows over the dam boards at the end of a raceway, through the distribution plate and then falls through the rectangular chambers. All of the pure oxygen is introduced into the outer rectangular chamber, flows through the series of individual chambers, and is finally vented to the atmosphere. Each of the rectangular chambers is gas tight, and the orifices between the chambers are constructed to reduce back-mixing between chambers (8). The LHO is most efficient when the required change in dissolved oxygen is in the range of 3–5 mg/L (ppm) (9).

Aeration Cones

The aeration cone, bicone, or downflow bubble-contact aerator (DBCA) consists of a cone-shaped cylinder with a high turbulence zone at the top of the cone (Fig. 6). Oxygen is injected into the high-turbulence zone, resulting in efficient gas transfer. As the bubbles are carried down toward the large end of the cone, the downward velocity of the water is reduced until it equals the upward buoyant velocity of the bubble and the bubbles are trapped inside the cone. Some of the gas may be vented for nitrogen gas control (10). These units are commercially produced and have been widely used in Europe.

Oxygen Injection

The most widely used form of oxygen injection injects oxygen through a venturi on the discharge side of a

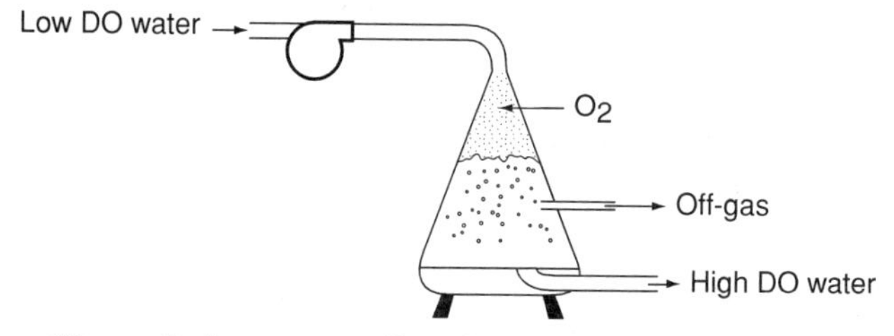

Figure 6. Aerator cone (downflow bubble contact aerator).

high-pressure pump (Fig. 7a) or into a high-pressure supply line. Relatively high pressures and contact times are required to achieve satisfactory absorption (11). This method can achieve 100% absorption of oxygen, but total gas pressure will also be elevated. A number of proprietary contact systems are available (11,12). These systems generally treat only a portion of the total flow and require a pressurized water supply.

Diffused Aeration

Pure oxygen diffused aeration generally has a low-absorption efficiency and has been limited to emergency use in production systems or for transport systems. Some recently developed fine-bubble diffusers have absorption

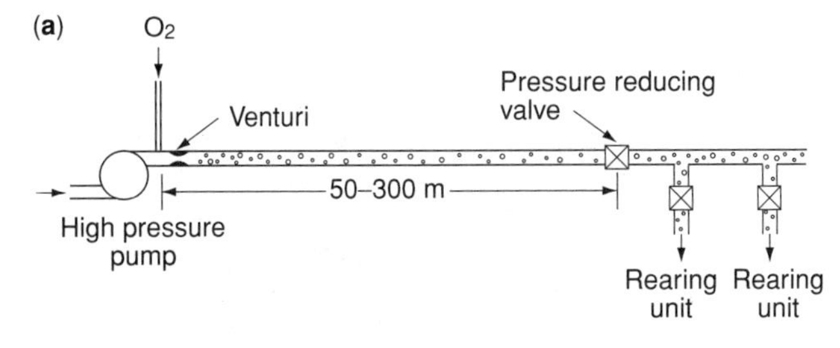

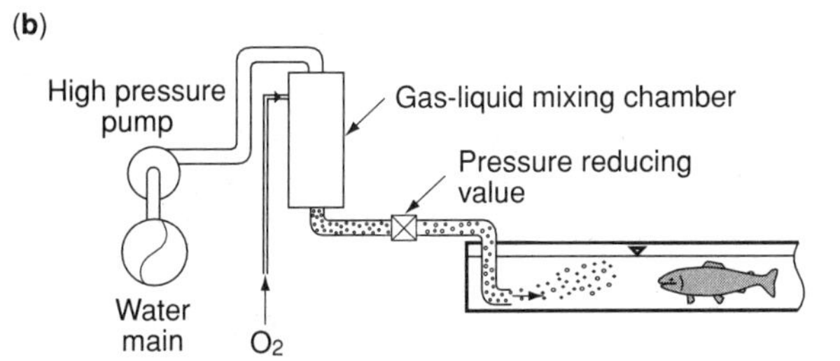

Figure 7. High-pressure oxygen injection. (**a**) High-pressure absorber; (**b**) pressurized side-stream unit.

efficiencies in the range of 40 to 60% when submerged at 1–2 m or more.

MODELING OF GAS TRANSFER IN PURE OXYGEN SYSTEMS

The rate at which a slightly soluble gas such as oxygen is transferred into water is proportional to the area of the gas-liquid interface and the gradient between the saturation and existing concentration of the gas in the water as in the following model (13):

$$\frac{\mathrm{dC}}{\mathrm{dt}} = (K_L)(a)(\mathrm{C}^* - \mathrm{C}) \tag{2}$$

where

$\frac{\mathrm{dC}}{\mathrm{dt}}$ = Rate of mass transfer (kg/hr)

K_L = Overall liquid phase mass-transfer coefficient (m/hr)

a = Area of interfacial contact between gas and liquid (m^2)

C^* = Saturation dissolved gas concentration at a given temperature, pressure, and mole fraction (mg/L)

C = Dissolved oxygen concentration (mg/L).

Substitution of Equation 1 into Equation 2 results in the following:

$$\frac{\mathrm{dC}}{\mathrm{dt}} \propto (K_L a)\left\{(1000K\beta)\chi\left(\frac{P - P_{\mathrm{H_2O}}}{760}\right)\right\} \tag{3}$$

For a given system and temperature, $K_L a$, $1000K\beta$, and $P_{\mathrm{H_2O}}$ are constants. The rate of oxygen transfer can be increased by increasing the mole fraction of oxygen (χ) and the system pressure (P). The pressure may be increased by (*1*) pressurizing the entire aeration unit or (*2*) increasing the depth of the contact system. The value of χ can be increased by increasing the gas-flow rate (or gas-to-liquid ratio), but at increased gas costs.

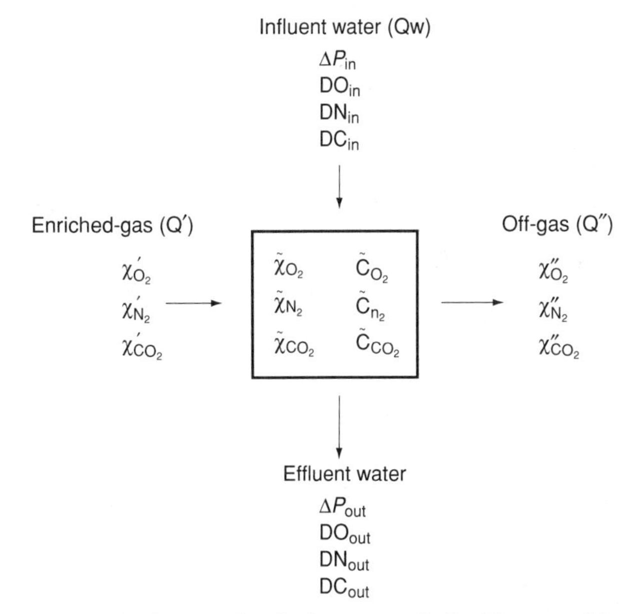

Figure 8. Definition sketch for gas and liquid composition relationships in pure oxygen systems.

A generalized sketch for the gas and water composition relationship in pure oxygen systems is presented in Figure 8. To model the performance of a general pure oxygen system, it is necessary to write the simultaneous mass transfer equations for oxygen, nitrogen + argon, and carbon dioxide. At any time and place, these three transfer equations are coupled by the following relationship:

$$\chi_{\mathrm{O_2}} + \chi_{\mathrm{N_2}} + \chi_{\mathrm{CO_2}} + \chi_{\mathrm{H_2O}} = 1 \tag{4}$$

In well-mixed systems, a single value of both C* and C can be determined. For other systems, both C* and C may vary both temporally and spatially within the aeration device. These variables are written with a tilde—$\tilde{\chi}_{\mathrm{O_2}}$ in Figure 8.

In general, the modeling of pure oxygen systems has involved an iterative solution. Published mass-transfer models are available for the u-tube (4), jet aerator (14), and packed column (15). A noniterative model has been developed for the packed column aerator (8).

PERFORMANCE OF PURE OXYGEN SYSTEMS

Important measures of performance for a pure oxygen system include the following:

- ΔDO Change in dissolved oxygen through the contact system
- ΔP_{out} Effluent ΔP from the contact system
- $\mathrm{DC_{out}}$ Effluent dissolved carbon dioxide concentration from the contact system
- COC Cumulative oxygen consumption
- AE Absorption efficiency (% of oxygen transferred into water)

Each of these parameters will be discussed shortly. It is important to note that important trade-offs may be required among certain parameters. For example, it may not be possible to simultaneously produce a high ΔDO, a high-absorption efficiency, and a low-variable cost.

Based on the transfer model developed for the packed column aerator, the performance of a typical pure-oxygen absorber system is presented in the following sections. Specific parameters used in this model are water temperature = 15 °C (59 °F), barometric pressure = 760 mm Hg, pressure inside column = 760 mm Hg, media type = 2.54-cm (1 in.) pall rings, $G = 2.0$, oxygen purity = 100%, DO_{in} = 10.07 mg/L (saturation), DN_{in} = 17.18 mg/L (105% of saturation), and a well-mixed gas phase inside the column. This model is used only to illustrate many of the complex interactions between the different operating parameters. Other types of pure oxygen systems may have significantly different characteristics and constraints.

ΔDO (Change in Dissolved Oxygen)

One of the major advantages of pure oxygen systems is that the effluent DO concentration can be varied by changing the G/L ratio (Fig. 9). In contrast, the effluent DO concentration from typical surface or gravity aerators used in aquaculture cannot exceed the air saturation concentration. Adjustment of the gas-flow (and G/L ratio) can be used to vary the capacity as a function of the (*1*) time of day, (*2*) production cycle, or (*3*) operational requirements. While it is possible to adjust the ΔDO value, it does affect absorption efficiency.

ΔP_{out} (Effluent ΔP)

The allowable ΔDO from the absorber is also a function of the criteria for ΔP. For a raceway, the allowable ΔDO is computed from

$$\Delta DO = \frac{\Delta P_{out} - \Delta P_{in}}{\left| F_{O_2} + \dfrac{F_{N_2}}{\Delta DO/\Delta DN} \right|} \tag{5}$$

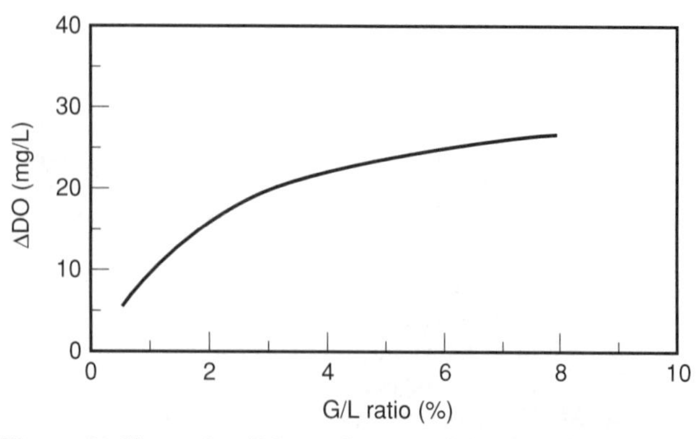

Figure 9. Change in ΔDO as a function of the gas-to-liquid ratio.

where

ΔDO = Allowable change in dissolved oxygen through the contact systems (mg/L)
ΔP_{out} = Maximum allowable ΔP in the raceway (mm Hg)
ΔP_{in} = Influent ΔP in the contactor (mm Hg)
F_{O_2} = Conversion factor between gas concentration and pressure for oxygen (mm Hg)/(mg/L)
F_{N_2} = Conversion factor between gas concentration and pressure for oxygen (mm Hg)/(mg/L)
$\Delta DO/\Delta DN$ = Change in dissolved oxygen/change in dissolved nitrogen + argon (mg/mg).

The allowable ΔDO is controlled by the oxygen:nitrogen stripping ratio, influent ΔP, and effluent ΔP. The F values are a constant at a given temperature.

The allowable ΔDO as a function of ΔDO/ΔDN for a plug-flow reactor is presented in Figure 10a for the following conditions: temperature = 15 °C, (59 °F) ΔP_{out} =

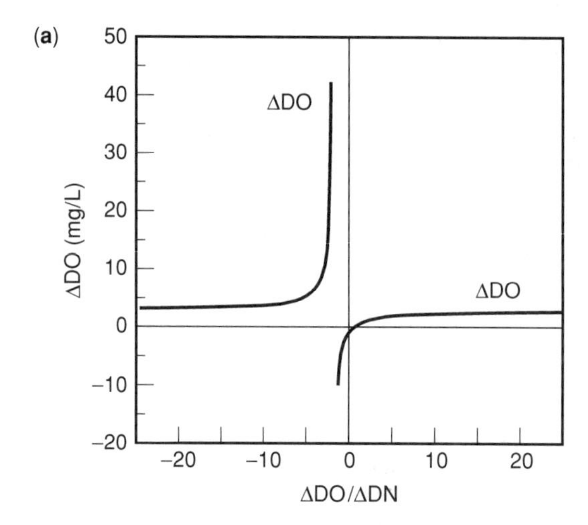

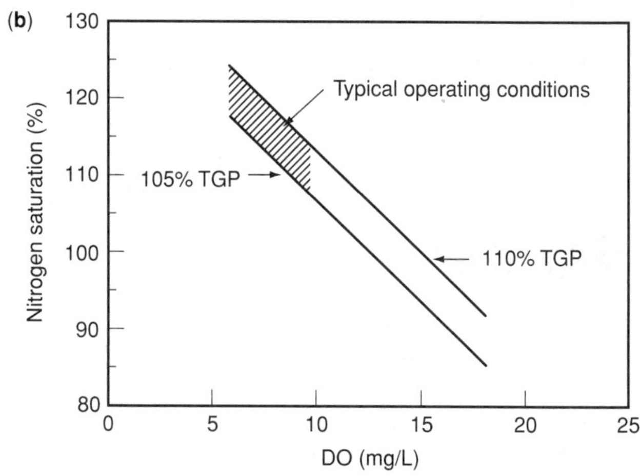

Figure 10. Limitation in maximum dissolved oxygen due to gas supersaturation considerations. (**a**) Allowable ΔDO for raceway and (**b**) conditions where nitrogen stripping is needed in a circular tank.

76 mm Hg $\Delta P_{in} = 29.5$ mm, influent DO = saturation, and influent DN = 105% of saturation).

In a pure oxygen aeration application, the operating conditions should be in the second quadrant ($+\Delta DO$, $-\Delta DN$, $-\Delta DO/\Delta DN$). At large positive and negative values of $\Delta DO/\Delta DN$, the ΔDO approaches a horizontal asymptote equal to 2.99 mg/L (ppm), which depend only on $\Delta P_{out} - \Delta P_{in}$. The value of $\Delta DO/\Delta DN$ also approaches a vertical asymptote at $-\{F_{N_2}/F_{O_2}\}$. The maximum positive ΔDO (second quadrant) is limited to 42.3 mg/L (ppm) due to conservation of mass considerations for nitrogen gas. At this point, all of the nitrogen gas has been stripped out and replaced with oxygen gas. For maximum ΔDOs, the absorber should be operated with $\Delta DO/\Delta DN$ slightly less than $-\{F_{N_2}/F_{O_2}\}$. If the value of $\Delta DO/\Delta DN > -\{F_{N_2}/F_{O_2}\}$, the value of ΔDO must be reduced to prevent problems with gas-bubble trauma. The value of $\Delta DO/\Delta DN$ can be adjusted by changing the system pressure and gas composition (8).

The allowable ΔDO across the absorber for a circular tank can be several times larger than for the raceway due to the rapid mixing of absorber effluent with process water in the rearing unit. The maximum allowable DO in the circular tank as a function of total gas pressure (TGP) (%) and nitrogen saturation (%) are presented in Figure 10b. Over typical operating conditions, stripping is required when the influent nitrogen saturation exceeds 106 to 126%. Therefore, if the influent water is degassed below these limits, no nitrogen stripping is needed in the oxygen absorber units.

DC_{out} (Effluent Dissolved Carbon Dioxide Concentration)

Due to the high solubility of carbon dioxide, little dissolved carbon dioxide gas will be removed in the off-gas from the absorber unit. Other types of aeration equipment will be needed to remove carbon dioxide gas.

COC (Cumulative Oxygen Consumption)

The utilization of oxygen produces both carbon dioxide and ammonia. The depletion of oxygen may not always be the most limiting parameter; and when ammonia or carbon dioxide is more limiting, aeration will have little effort on carrying capacity (16). Maximum cumulative oxygen consumption ($DO_{in} - DO_{out}$) based on pH, carbon dioxide, and un-ionized ammonia limitations are presented in Figure 11 for water quality criteria typical of salmon and trout culture.

AE (Absorption Efficiency)

The absorption efficiency is an important operating parameter and has a significant impact on operating costs. Over the range of 25–82% absorption efficiency, the G/L ratio decreases from 8 to 0.5% (Fig. 12a). The highest absorption efficiencies occur at low values of the G/L ratio. Over the range of absorption efficiencies modeled, the ΔDO decreases from 26.6 to 5.4 mg/L (ppm) (Fig. 12b). Therefore, the system can produce either high ΔDOs at low absorption efficiencies or low ΔDOs at high absorption efficiencies. Over the range of absorption efficiencies modeled, the ΔP_{out} decreases slightly from 21.0

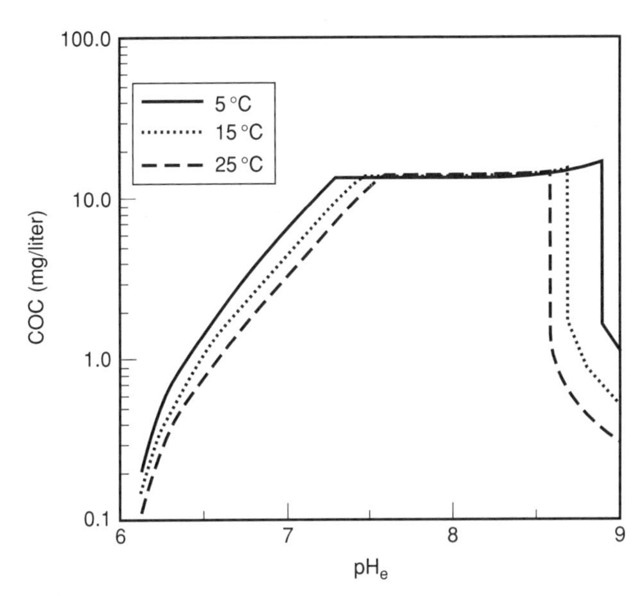

Figure 11. Maximum cumulative oxygen consumption (mg/L) as a function of equilibrium pH_e (pH of a solution in equilibrium with the atmosphere). At low pH_es COC is limited by the pH criteria, at intermediate pH_es by the carbon dioxide criteria, and at high pH_es by the un-ionized ammonia criteria (8).

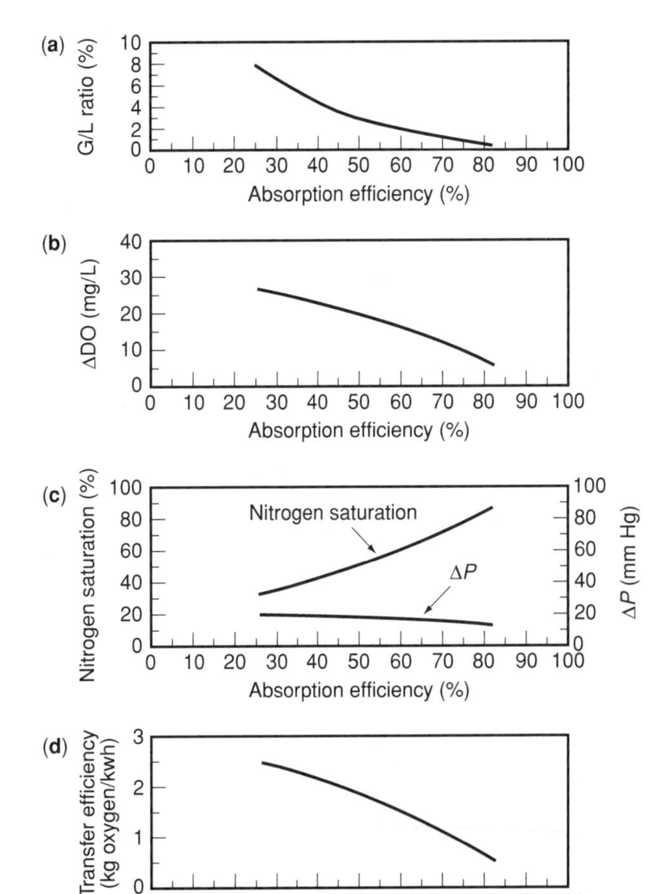

Figure 12. Operating points for (**a**) gas–liquid ratio vs. absorption efficiency, (**b**) ΔDO vs. absorption efficiency, (**c**) effluent nitrogen (%) and ΔP vs. absorption efficiency, and (**d**) transfer efficiency vs. absorption efficiency.

Table 2. Operational Advantages and Disadvantages of Different Pure Oxygen Systems

Type	Advantages	Disadvantages
U-tube	May be operated with no external power input if adequate head is available	Excavation may be expensive under some conditions; not an off-the-shelf device
Packed column	Simple to build or retrofit to existing hatcheries; good nitrogen stripping characteristics; can be ordered off the shelf	Fouling may be a significant problem in reuse applications
Spray column	Very resistant to fouling; good nitrogen stripping characteristics when operated at a vacuum (Michigan columns)	Moderate transfer efficiency
Pressurized packed column	Can be designed for 100% oxygen transfer efficiency	Poor nitrogen stripping characteristics
Low-head oxygenator (LHO)	Operate with 12–24 inches of drop; can be retrofitted into existing raceways	Poor nitrogen stripping characteristics
Aeration cones		Will require special basin; difficult to retrofit existing facility; limited commercial availablity in some parts of the world
Oxygen injection	Available from several vendors; simple to install in new or existing facility	Moderate transfer efficiency
Diffused aeration	Very simple to install; transfer efficiency has improved in the fine-bubble diffusers	High head losses; clogging can be a problems in fine-bubble diffusers

to 13.5 mm Hg, while the effluent N_2 increases from 32.9 to 87.8% (Fig. 12c). If the column was operated at 100% absorption efficiency, the ΔP_{out} would be equal to 473 mm Hg at a G/L ratio equal to 2%. Therefore, the absorption efficiency must be limited to allow stripping of nitrogen gas to prevent gas supersaturation problems. Increasing the absorption efficiency from 25 to 82% decreases the transfer efficiency from 2.49 to 0.51 kg O_2/kwh (4.09 to 0.84 lb O_2/hph) (Fig. 12d). Therefore, at high absorption efficiencies, both the transfer efficiencies and ΔDOs are low (Fig. 12a).

SELECTION OF PURE OXYGEN SYSTEMS

Some common advantages and disadvantages of different types of pure oxygen systems are presented in Table 2. Due to differences in site conditions, operational modes, and specific absorber units, it is impossible to suggest a single design procedure (17).

BIBLIOGRAPHY

1. J. Colt, *Computation of Dissolved Gas Concentrations in Water as Functions of Temperature, Salinity, and Pressure*, Spec. Pub. No. 14, Am. Fish. Soc., Bethesda, MD, 1984.
2. A.E. Greenberg, L.S. Clesceri, and A.D. Eaton, eds., *Standard Methods for the Examination of Water and Wastewater*, American Public Health Association, Washington, DC, 1992.
3. R.E. Speece and R. Orosco, *J. San. Eng. Div., ASCE* **96**, 715–725 (1970).
4. B.J. Watten and L.T. Beck, *Aquacult. Eng.* **4**, 271–297 (1985).
5. B.J. Watten, *Aquacult. Eng.* **9**, 305–328 (1990).
6. W.P. Dwyer, J. Colt, and D.E. Owsley, *Prog. Fish-Cult.* **53**, 72–80 (1991).
7. J. Colt, J. Sheahan, and G.R. Bouck, *Aquacult. Eng.* **12**, 141–154 (1993).
8. B. Watten and C.E. Boyd, *Aquacult. Eng.* **9**, 33–59 (1990).
9. E.J. Wagner, T. Bosakowski, and S.A. Miller, *Aquacult. Eng.* **14**, 49–57 (1995).
10. R.E. Speece, M. Madrid, and K. Needham, *J. San. Eng. Div., ASCE* **97**, 433–441 (1971).
11. B. Ludwig and G. Gale, *Fisheries Bioengineering Symposium*, Am. Fish. Soc. Sym. 10, American Fisheries Society, Bethesda, MD, 1991, pp. 437–444.
12. A.R. Schutte, *Prog. Fish-Cult.* **50**, 243–245 (1988).
13. W.K. Lewis and W.C. Whitman, *J. Ind. Eng.* **16**, 1215–1220 (1924).
14. J. Colt and H. Westers, *Trans. Am. Fish. Soc.* **111**, 342–360 (1982).
15. B.J. Watten, J. Colt, and C.E. Boyd, *Fisheries Bioengineering Symposium*, Am. Fish. Soc. Sym. 10, American Fisheries Society, Bethesda, MD, 1991, pp. 474–481.
16. J. Colt and K. Orwicz, *Aquacult. Eng.* **10**, 1–29 (1991).
17. J. Colt and B. Watten, *Aquacult. Eng.* **7**, 397–441 (1988).

See also DISSOLVED OXYGEN.

R

RABBITFISH CULTURE

Robert R. Stickney
Texas Sea Grant College Program
Bryan, Texas

OUTLINE

Rabbitfish (family Siganidae) are Indo-Pacific fish that are economically important in capture fisheries, particularly of a subsistence nature. Rabbitfish can also be found in the Red Sea and the Mediterranean Sea. Aquaculturists have expressed some interest in rabbitfish culture since the 1970s. Various species have reportedly been cultured in Tanzania, the Middle East, the Philippines, Indonesia, Guam, and elsewhere, but total production is limited, so rabbitfish are currently considered to be a minor contributor to total world aquaculture production. There are presently 27 species of rabbitfishes recognized by fish taxonomists.

APPROACHES TO CULTURE

In addition to being a popular foodfish in the regions of the world where they dwell, rabbitfish can tolerate the high densities required of fish under culture. They also efficiently utilize a variety of natural and prepared feeds and tolerate handling stress, low levels of dissolved oxygen, a wide range of salinity, and water temperatures up to 34 °C (93 °F) (1). However, there have been reports of ciguatera poisoning associated with the consumption of rabbitfish in some parts of the world, and the spines of rabbitfish are sharp and venomous (2), making handling somewhat dangerous for humans.

Rabbitfish can be spawned naturally in ponds (3) or in the laboratory (4) through injections of human chorionic gonadotropin and other hormones. (See the entry "Reproduction, fertilization, and selection.") Natural spawning typically occurs during the new moon, from about midnight to dawn (2). The eggs of most species are demersal and slightly adhesive, though at least one species, *Siganus argenteus*, lays pelagic eggs (4).

The eggs tend to be smaller than 0.7 mm (0.03 in.) in diameter and will hatch between 18 and 35 hours after being deposited (2). The time taken from fertilization to hatching varies as a function of such factors as water temperature and fish species.

Because of difficulties associated with captive spawning and rearing of the very small larvae produced by rabbitfish, the capture of juveniles near beaches provides an alternative source of fish for stocking. Rabbitfish have been reared in captivity from eggs through adulthood, but mass production of juveniles in hatcheries continues to be a major problem. Thus, capturing wild juveniles and placing them in growout ponds or cages is an attractive approach. This is particularly the case in developing nations, where suitable hatchery facilities and the associated technology need to spawn adults in captivity are lacking. (Capturing wild juveniles for growout has been practiced with various other aquaculture species—for example, milkfish in southeast Asia and shrimp in Ecuador. However, as an aquaculture industry develops, the sustainability of natural populations can be affected by the removal of large numbers of fry or juveniles, so the practice may become severely regulated or even be banned.)

Strictly herbivorous fish are the exception rather than the rule, but rabbitfish consume primarily filamentous algae as adults, though they have been observed to be omnivorous in culture (2). Algal diets are not suitable for young rabbitfish. Larval and early juveniles may, like such other herbivores as grass carp (see the entry "Carp culture"), require animal protein in their diet for rapid growth and as a source of essential amino acids.

A critical stage in the life cycle of rabbitfish, as is the case with many other marine fish species, is associated with the time of first feeding. At that time, the fish must make the transition from obtaining their nutrients from the yolk sac to feeding on exogenous sources of energy and nutrition. Suitable prey must be available to the larval rabbitfish at the proper time and in the proper density. Starvation can occur very quickly if food is not available. The food needs to be present in sufficient quantity, so that the larvae do not have to swim far to capture prey. Larvae are such weak swimmers that they can exhaust their energy reserves quickly if they have to exert themselves in seeking food. Food particles need to be of the proper size so that the fish can engulf them, and the food items must be recognizable; that is, they must have a shape, color, and behavior (movement in the water column) that allows the larval fish to identify the particles as food. Texture may also be an important factor. If a food item does not have the proper texture, it may be ignored. Odor can be an important factor associated with food recognition, though in larval fishes that sense may not be well developed at first feeding. For many species, first feeding is often facilitated by a well-developed visual sense, which may be the case with rabbitfish. In such cases, the fish must be able to see the food particles in order to know to approach and consume them. Thus, light, water clarity, and background color may be important factors in helping the larvae proceed through the critical first feeding stage.

First feeding rabbitfish will accept brine shrimp (*Artemia* sp.) nauplii, copepods, rotifers, and other natural foods. Larval rabbitfish have also been successfully fed with prepared feeds under laboratory conditions.

Older rabbitfish, including adults, will also accept prepared feeds, and while there is little information available on the nutritional requirements of rabbitfish at any stage of life, their food habits indicate that they might grow as well on a diet formulated for tilapias (many species of which are herbivorous or omnivorous and are able to grow well on diets high in plant proteins; See the entry "Tilapia culture") as on one formulated for trout or salmon. In the limited amount of nutritional research that has been conducted, the protein requirement, the protein-energy requirement, and the effect of dietary lipids on growth and egg quality have been investigated, though definitive information has yet to be developed on some of those subjects, as well as on many other topics of interest to fish nutritionists. There have also been a number of feeding trials conducted in various types of culture systems. At least one review paper has been published on the subject of rabbitfish nutrition (5). The ability of rabbitfish to digest various types of carbohydrates has been confirmed through examination of enzyme activity (6).

A limited amount of information exists on the diseases of rabbitfish. One survey of metazoan parasites in wild rabbitfish off Kenya (7) found incidences of monogenetic and digenetic trematodes, copepods, isopods, acanthocephalans, and nemaodes, but the impacts of parasites on cultured rabbitfish have yet to be determined. There appears to be very little information available on other diseases that affect rabbitfish.

BIBLIOGRAPHY

1. A.F.M. El-Sayed, K.A. Mostafa, J.S. Al-Mohammadi, A.Z. El-Dehaimi, and M. Kayid, *J. World Aquacult. Soc.* **26**, 212–226 (1995).
2. T.J. Lam, *Aquaculture* **3**, 325–354 (1974).
3. D. Popper and N. Gundermann, *Aquaculture* **7**, 291–292 (1976).
4. D. Popper, R. Pitt, and Y. Zohar, *Aquaculture* **16**, 177–181 (1979).
5. M. Boonyaratpalin, *Aquaculture* **151**, 283–313 (1997).
6. U. Sabapathy and L.H. Teo, *J. Fish Biol.* **42**, 595–602 (1993).
7. E. Martens and J. Moens, *Afr. J. Ecol.* **33**, 405–416 (1995).

See also Brine shrimp culture; Larval feeding—fish.

RAFT, POLE, AND STRING CULTURE

Robert R. Stickney
Texas Sea Grant College Program
Bryan, Texas

OUTLINE

Molluscs, such as oysters, mussels, scallops, and clams, are often cultured on sediments in appropriate locations. To take advantage of the entire water column for increased shell growth and to reduce predation, many culturists have elected to rear molluscs, and, in some cases, algae above the sediments. When suspended culture systems are employed, it is necessary for the aquaculturist to have control over the area in which the animals are being grown. Boat traffic, for example, could be highly disruptive. The culture animals are reared by suspending them on ropes or in cages from rafts, on long lines strung between floats, or attached to poles driven into the sediments. Some species of macroalgae can be grown on nets. This contribution introduces the concept of this alternative means of culture, but is not meant to be a comprehensive treatment of mollusc or algae culture.

RAFT CULTURE

Mussels and oysters can be cultured on ropes suspended from floating rafts. Spain pioneered the development of raft culture for mussels and continues to be a leading producer of raft-cultured molluscs. Oyster raft culture was developed in Japan and continues to be widely used in that country and, to a lesser extent, elsewhere.

Both oysters and mussels attach to substrates in the larval stage, initially planktonic, when they settle to the bottom. The mechanism of attachment is different for the two types of molluscs.

Aquaculturists can provide a suitable substrate (cultch material) for the settling oyster larvae (spat). Suitable cultch materials on which to capture spat include oyster and scallop shells. The shells can be placed in the natural environment or, if the molluscs are being produced in a hatchery, in the tank or raceway into which the spat are introduced. Once the oyster spat have attached to the cultch material, the shells are strung on ropes that are suspended below rafts in a suitable environment.

For mussels, the technique is somewhat different. Juveniles are collected from natural beds and are tied in clumps around ropes 12 to 25 mm (0.5 to 1 in.) in diameter. A wooden peg is inserted into the ropes at intervals of 30 to 45 cm (12 to 18 in.) to keep the clumps of mussels from sliding down the ropes (1).

The ropes may be quite long, though they should be located in the photic zone (above the 1% light level), where phytoplankton live. The depth of the photic zone will vary as a function of turbidity caused by suspended inorganic particles and organic materials, including the phytoplankton upon which the molluscs feed. The length of the ropes used will depend on water clarity and available depth. The ropes, sometimes called strings, should remain above the bottom at all tidal stages, to prevent benthic predators such as starfish and oyster drills from attacking the crop.

Rafts should be located in an appropriate environment. Requirements include proper water quality conditions and a good supply of phytoplankton. The environment should also be free from sewage and other types of pollutants that could concentrate in the molluscs and make them unfit for human consumption.

Figure 1. A net bag containing small scallops and attached to a long line.

LONG-LINE CULTURE

A modification of the raft culture technique is called *long-line culture*. Instead of hanging from surface rafts, the ropes are strung horizontally between floats. The long lines may be many hundreds of meters long. Oysters can be reared attached to cultch material and suspended on a secondary rope that hangs vertically from the main string. Mussels can be attached to secondary ropes in the manner described previously for raft culture.

Long-line culture of scallops, which is well developed in Japan, involves rearing young scallops in net baskets suspended from the main line, since scallops do not attach to substrates (see Fig. 1). When the scallops become sufficiently large, they can be hung individually. At that time, a hole is drilled in one of the so-called wings of each scallop, and a monofilament line is strung through the hole. Individual scallops are then hung from the lines until they reach harvest size.

POLE CULTURE

Poles have been used to suspend oysters above the bottom of ponds and other water bodies in the Philippines and have been used for mussel culture in the Philippines and France. The French grow mussels on poles in a system known as *bouchet culture* (1). In this system, ropes are placed in locations where natural settling of mussels occurs. Once the ropes are covered with young mussels, they are taken to culture locations and are wrapped around oak poles. The poles are then driven into intertidal sediments. Plastic is placed around the base of each pole to prevent benthic predators from climbing up to attack the mussels.

Poles are also used in conjunction with algae culture in Japan. Thousands of poles may be employed to anchor nets, upon which such seaweeds as nori (*Porphyra* sp.) are grown (see Fig. 2). Spores obtained from reproducing nori are collected on nets in indoor tanks (see Fig. 3). When the nets are covered with spores, they are placed into the environment and attached to the poles. The nets float at the surface and move up and down the poles with the tide. When the plants reach the appropriate size [about 15 cm (6 in.) long], they are harvested by being cut off. The nets are allowed to remain in place after the first harvest, and the plants then produce a second crop. Half of the colonized nets are kept in cold storage until the second

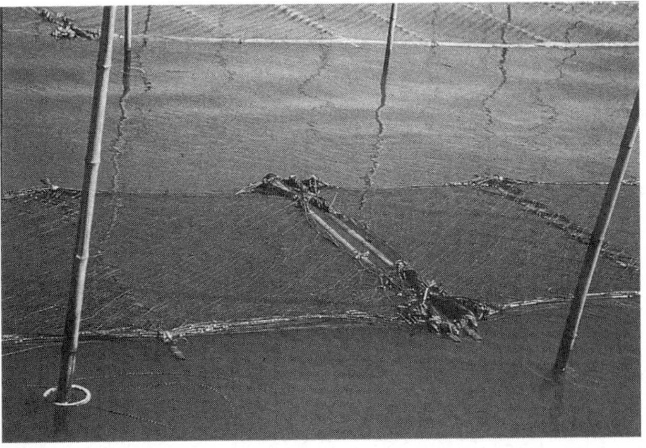

Figure 2. Koroshima Bay in Japan is devoted to pole culture of nori, a seaweed.

Figure 3. Indoor nori settling tanks.

cutting occurs, after which they are used to replace the first batch of nets. The result is four crops of nori per year.

BIBLIOGRAPHY

1. J.E. Bardach, J.H. Ryther, and W.O. McLarney, *Aquaculture*, Wiley Interscience, New York, 1972.

See also MOLLUSCAN CULTURE.

RAINBOW TROUT CULTURE

RONALD W. HARDY
Hagerman Fish Culture Experiment Station
Hagerman, Idaho

GARY C.G. FORNSHELL
University of Idaho Cooperative Extension System
Twin Falls, Idaho

ERNEST L. BRANNON
University of Idaho
Moscow, Idaho

OUTLINE

Rainbow trout are members of the genus *Oncorhynchus*, which also includes Pacific salmon, and members of the family Salmonidae (e.g., Atlantic salmon, trout, char, graylings, whitefish, and several other groups). Rainbow trout are native to cold-water environments in the north temperate zones and are distributed from southern California through Alaska, the Aleutians, and the western Pacific areas of the Kamchatka Peninsula and Okhotska Sea drainages. Rainbow trout are thought of as freshwater fish, but in the eastern Pacific, seawater forms called steelhead trout exhibit an anadromous life history, meaning that they spend a part of their life in the ocean, but return to lakes and rivers to spawn. Rainbow trout have been widely transplanted around the world and are established in South America, Japan, China, Europe, parts of Africa, Australia, and New Zealand.

The Kamchatka rainbow trout was originally named *Salmo mykiss* by Walbalm in 1792 (1) and was later named *S. gairdneri* by Richardson, regarding fish taken from the Columbia River at Fort Vancouver in 1836 (2). Rainbow trout appear to have survived glaciation in two refuges: the Pacific coast, south of ice-age glaciers; and the Bering area, north of the Alaskan peninsula. From there, they became reestablished in western North America as glaciers receded. For many years the rainbow trout native to North America (*S. gairdneri*) was thought to be a different species than the trout native to Kamchatka (*S. mykiss*), but they are now considered to be the same species. Prior to 1989, rainbow trout were considered to be part of the trout genus, having the scientific name *S. gairdneri*, but they are now classified as *O. mykiss* (3).

Figure 1. Two rows of serial raceways used for trout production in Idaho.

Rainbow trout is by far the most widely farmed trout in the world, mainly because it is a prized food fish and because it is relatively easy to culture (see Fig. 1). Rainbow trout can survive a variety of environmental conditions, such as water temperatures from 0 to 28 °C. They spawn successfully in water temperatures from 2 to 15 °C and grow in water up to 25 °C. Depending upon their diet, rainbow trout can have pigmented (red) or nonpigmented (white) flesh. Once they have reached a certain size, they can be farmed in freshwater or in seawater. In North America, Britain, Denmark, France, and Italy, most trout farming occurs in freshwater facilities, using flow-through water-supply systems. In Chile and Scandinavian countries, rainbow trout are initially grown in freshwater farms followed by growout to harvest in marine cages. Total production of rainbow trout was 358,456 metric tons in 1995, making it the second largest production segment of the salmonids, behind Atlantic salmon (4). Until 1994, global production of rainbow trout exceeded that of other salmonid species. Top trout producing countries are France, Chile, Denmark, and Italy, which accounted for 48% of global production in 1995. The United States accounts for slightly over 7% of global trout production and in 1997 produced 25,777 metric tons, 74% of that in the state of Idaho. Other top trout producing states were North Carolina, California, and Pennsylvania.

LIFE HISTORY AND FARMING SYSTEMS

Rainbow Trout Life History

Rainbow trout live in lakes, streams, and rivers. They consume zooplankton as fry, followed by insects, crustaceans, and other fish as they grow. Spawning occurs in spring, with rising water temperatures, although considerable variability is found, with coastal rainbow trout spawning in late December. Females deposit anywhere from 500 to 2,500 large eggs (50–150 mg per egg) in nests dug in gravel, while the males fertilize the eggs as they are deposited. The time required for fertilized eggs to develop and hatch depends upon water temperature. At 4.5 °C (40 °F), rainbow trout eggs require 80 days to hatch; at 10 °C (50 °F), 31 days to hatch; and at 15 °C (60 °F), 19 days to hatch (5). Eggs are extremely sensitive to handling and shock from 2 days postfertilization until the blastophore is completely closed, 9 days at 10 °C. Once the eggs become pigmented (about 16 days at 10 °C), the period of sensitivity is over, and the eggs can be handled up until just before hatching. At hatching, the fry still have large egg yolks attached to them and resemble small fish on beach balls. These fry are called yolk-sac fry, or alevins, and they burrow into spaces within the gravel, where they continue to develop and grow, utilizing their yolk sac for energy supply and nutrients. When the yolk sac is nearly gone and has been surrounded by skin on the ventral side of the fish, they are said to be "buttoned-up" fry. The time needed for alevins to reach this stage depends on water temperature, but is approximately 20 days at 10 °C and 10 days or less at 15 °C from hatching. The fry are then ready to feed and emerge from the gravel to seek food on the water's surface. At this point they are said to be "swim-up" fry. The entire sequence from spawning to emergence from the gravel is timed such that the fish emerge when natural food is abundant in spring. Since streams differ in water temperature and food abundance throughout the geographical range of rainbow trout, local populations are adapted to local conditions, and spawning and fry emergence are timed appropriately.

Rainbow trout growth rates depend on water temperature and food abundance, and wild fish generally reach maturity at three to four years of age. Most spawning trout are first spawners, but a small proportion of spawners, mainly females, survive to spawn again. Growth and maturation in rainbow trout is indeterminate, meaning that there is no set rate or age. Rather, environmental factors determine growth and maturation, with fish in cold, harsh environments generally living longer than those in warmer, more benign environments. Maximum size is variable, with 17–23 kg rainbow trout sometimes captured in Kooteney Lake, British Columbia. These fish would be five to six years old (1). However, rainbow trout in streams typically weigh 100 g at one year of age and 300–450 g after three years.

Rainbow Trout Farming

Rainbow trout farming in the United States was described in 1872 in a paper presented to the American Fish Culturists' Association by Livingston Stone (6). In the paper, Stone calculated the economics of trout farming, figuring four years to market, and a sales price of $0.50 to $1.25 per lb. In 1879, a trout spawning station was established on the McCloud River in California, and after various problems, the station produced 179,000 eggs in 1881. In subsequent years, the station shipped eggs throughout the country to state and federal agencies and private individuals interested in undertaking rainbow trout rearing (6). Rainbow trout culture became a farming business in the early 1900s, with a third of farms being fee-fishing operations, at least until the early 1950s. In Idaho, the first commercial trout farm was started in 1909, near Twin Falls. This area contains many suitable trout farm sites supplied with abundant, constant temperature (14.5 °C) springwater from the Eastern Snake River Aquifer, over 23 m^3/sec in flow (7). Trout farming expanded greatly in the early 1950s, supported in part by the development of pelleted feeds, which reduced the cost of production by eliminating the need for farms to prepare their own feed. Production was about 500 MT in the early 1950s, but increased sixfold by 1960, doubling by 1970, and doubling again to 12,000 MT by 1975. Current production in Idaho is about 19,000 MT, or 74% of U.S. production (see Fig. 2).

Several characteristics of rainbow trout contributed to the expansion of its culture. First, rainbow trout are easy to spawn, and their fry are large at first feeding compared with most other freshwater fish. Thus, they can thrive on prepared feed from first feeding. Second, they grow rapidly and are in demand as a food fish. Third, as mentioned earlier, they tolerate a wide range of water temperatures, and there are numerous water sources in temperate regions in which they can be grown. A final characteristic of rainbow trout that contributed to their success as a farmed fish was that their spawning time could be manipulated by selection and by photoperiod adjustment to make eggs available year-round. This allowed farms to supply rainbow trout to markets throughout the year, hence increasing sales.

Farming systems for rainbow trout are similar everywhere in the world. Fish are raised in flowing water in earthen or concrete raceways, with rearing densities depending upon water flow and water quality (e.g., temperature and dissolved-oxygen content). In Idaho, springwater flows by gravity through a series of raceways, each usually no more than 30–40 meters long, with five to seven raceways receiving water in series. Rearing density varies with water quality, which is higher in upper raceways than in lower raceways. Typical rearing densities are 1.8 kg per liter/min water flow (15 lb/gpm) in raceways receiving first-use water, and up to 9.6 kg/lpm (80 lb/gpm) when all raceways in a series are combined (7; see also

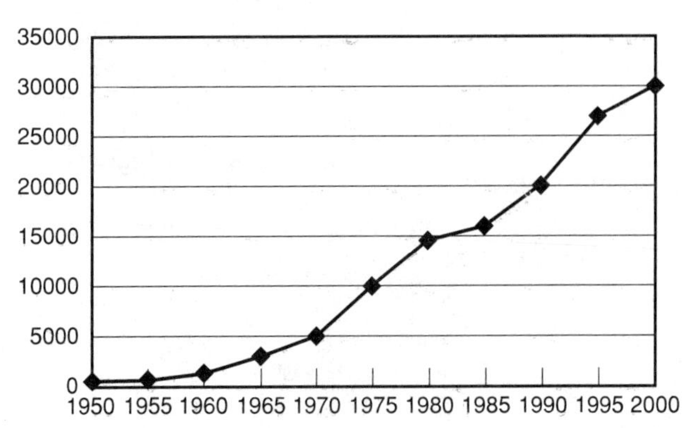

Figure 2. Annual production (metric tons) of rainbow trout in the United States.

Fig. 1). In Italy, long earthen raceways, sometimes more than 1,000 meters long and resembling wide, shallow streams, are used to raise trout.

Fish are hatched indoors, usually in hatching jars with upwelling water sufficient to cause the eggs to slightly roll (see Fig. 3). Hatching success is typically 95%. When fish hatch, they float out of the hatching jars into troughs, where they remain through yolk absorption and first feeding. Feed is provided nearly continuously during the first 7–10 days. Fish are moved to larger troughs or circular tanks as they grow, and are stocked into outdoor raceways when they reach 3–4 inches in length. They remain in raceways until harvest, and often are moved as they grow to raceways receiving third to last-use water (see Fig. 5).

The first limiting factor for rainbow trout in raceways and ponds is usually the oxygen content of the water. Rainbow trout can survive in water containing as little as 3 ppm oxygen, but 5 ppm is considered the minimum oxygen content for fish to thrive. The oxygen saturation level of water depends upon water temperature and elevation. At sea level, for example, dissolved oxygen at saturation is 11.3 ppm in 10 °C water and 10.0 in 15 °C water. At 910 m (3000 ft) elevation (the approximate elevation of most Idaho trout farms), the oxygen levels at saturation are 10.1 ppm and 8.9 ppm for 10 and 15 °C water, respectively. Activity associated with feeding, cleaning, or harvesting lowers the dissolved-oxygen content of trout raceways, and raceways are stocked and managed accordingly. Some oxygen is added by splashing water from one raceway to another, but this rarely restores dissolved-oxygen levels to saturation levels.

Figure 3. Typical upwelling incubator used to hatch rainbow trout eggs.

Other dissolved gases can be a problem for rainbow trout, particularly nitrogen under conditions of supersaturation, such as occurs downstream from large dams. If air is sucked into a pipe supplying a hatchery, and if the head (vertical distance between the source of the water and the farm) is sufficient, nitrogen supersaturation can occur, resulting in gas-bubble disease. This condition results from nitrogen gas coming out of solution in the blood of fish. Subcutaneous bubbles are visible in fins and around the eyes of affected fish. Other dissolved gases, including hydrogen sulfide (H_2S) and hydrogen cyanide (HCN), are toxic to trout, but farms avoid water sources containing such gases (8).

The second limiting factor for rainbow trout is ammonia, the excretory product of protein catabolism. In flow-through rearing systems, dissolved ammonia rarely reaches toxic levels (0.0125 ppm as un-ionized ammonia, NH_3), but in farms using recycled water, ammonia buildup must be monitored, and nitrification through the use of biological filters or ion-exchange must be employed.

In the United States, most rainbow trout are harvested at 450–600 g, a weight achieved after 10–15 months of rearing, depending upon water temperature (see Fig. 4). A significant proportion of U.S. trout production provides fish for recreational fishing, and these fish are delivered

Figure 4. Farm-raised trout ready for market.

Figure 5. Typical earthen ponds used in Idaho. Note the bird netting to keep birds out.

at nearly all sizes up to harvest size. In Europe, rainbow trout are harvested at 1–2 kg, while in Chile, Norway, Sweden, and Finland, rainbow trout are harvested at 3–5 kg after being grown in marine cages. Trout in the United States are expected to have white meat, so they are fed diets lacking the carotenoid pigments that give trout and salmon fillets their typically red color. In nature, these pigments are naturally present in their food. Trout raised in Europe and Chile to larger sizes are expected to have pigmented meat, so the feed for these fish is supplemented with astaxanthin, the carotenoid pigment found in wild fish. In trout feeds, astaxanthin can be supplied by natural sources, such as krill, yeast, or algae or by astaxanthin produced by chemical synthesis.

RAINBOW TROUT FEEDS AND FEEDING

Rainbow trout are carnivorous fish, consuming zooplankton, insects, crustaceans, and other fish. They have a relatively short digestive tract, consisting of an acid stomach, a pyloric cecae (where digestive enzymes are released), a small intestine (where most nutrient absorption occurs), and a large intestine for water and electrolyte reabsorption. The total length of the digestive tract, from mouth to anus, is barely the length of a trout, in contrast to humans (three to four times body length). Embody and Gordon (1924) conducted the first survey of the natural food of trout, finding that it averaged 45% protein, 16–17% fat, and 12% ash. Trout feeds were similar to that until the development of high-energy feeds within the last several years. In general, feeds for fry and fingerlings are higher in protein than feeds for growout or broodstock trout. The requirements of trout for essential dietary nutrients have been estimated, and trout are known to require about 40 nutrients, similar to other animals. They can obtain most minerals directly from water if rearing water levels are sufficient. Trout feeds contain adequate amounts of most essential minerals to meet their needs directly, regardless of rearing water supplies.

The first trout feeds were prepared by hatchery staff for each hatchery, and contained materials that could be obtained in the vicinity of the hatchery. The materials used in these feeds included slaughterhouse, fish, dairy, and other by-products. These were ground and mixed together to form a wet mixture that was hand-fed to fish in ponds. These feeds dissolved in water quite rapidly, lowering water quality, limiting stocking densities, and contributing to disease. Wet–dry mixtures were the next step in trout feed production. Wet products were combined with flour or other milling by-products to extend the wet products and to produce a mixture that formed a pellet. Pelleting technology was crude, often consisting of a meat grinder or noodle maker. In the mid-1950s, compressed, steam pellets were first produced by Clark Co. for trout; other feed companies soon followed suit. These feeds were based upon formulations developed by the Cortland Fish Nutrition Laboratory in New York State and contained 36% protein and 5% fat. By 1961, all trout farms used compressed pellets, a development that was critical to growth of the industry. In the mid-1990s, cooking–extrusion was introduced as a pelleting method in trout feed production. Pellets made by cooking–extrusion are harder, but less dense. They absorb high amounts of added fish oil and permit the production of high-energy trout feeds (i.e., those over 16–17% fat). Dietary protein levels in trout feeds have increased from 35% to 45% over the last 35 years, and dietary fat levels now exceed 22% in high-energy feeds. These feeds are more nutrient dense, and thus support more growth per unit feed than earlier feeds. In the 1960s, feed conversion ratios (amount of feed/fish weight gain) were about 2.0, meaning that for every 2 lb of feed fed to trout, 1 lb of gain resulted. Today, the best commercial, high-energy feeds yield feed conversion ratios of 1.2 : 1 to as low as 0.8 : 1, when their use is combined with good feeding practices.

Feed formulations for rainbow trout utilize fish meal, fish oil, grains, and various by-products derived from production of food oils and other food products, (e.g., meats and poultry) (Table 1). Efforts have been made to lower the percentage of fish meal used in rainbow trout feeds, substituting other protein sources such as soybean meal, poultry by-product meal, and small amounts of blood meal and feather meal. In the past decade, the amount of fish meal used in trout feeds has decreased by about 50% as a result of the use of alternate protein ingredients.

Rainbow trout fry are fed frequently, as often as every 15–30 min. Feeding frequency decreases as fish grow and their ability to consume more feed at each feeding increases. Feed can be delivered by hand, by demand feeders, or by mechanical feeders. Hand feeding simply involves the broadcast of feed over the surface of a raceway or pond. Automatic and demand feeders are devices that deliver feed to troughs, tanks, and raceways. Each contains a feed hopper and a delivery system, the difference being that automatic feeders operate according to a program containing the number of feedings and the duration of feed delivery at each feeding, while demand feeders only deliver feed when the fish activate the feeder. Automatic feeders are electrically, mechanically, or pneumatically driven, using vibration, a motorized belt, a chamber or

Table 1. Generalized Feed Formulations Used for Rainbow Trout Reared in Seawater and in Freshwater

Feed Ingredient	Seawater Feed (g/kg)	Freshwater Feed (g/kg)
Fish meal	550	300
Poultry by-product meal		100
Blood meal		0–50
Soybean meal	50	50–100
Wheat grain and by-products	144	170–300
Vitamin premix	10	10
Trace mineral premix	1	1
Choline chloride (60%)	4	4
Ascorbic acid	1	1
Fish oil	240	120–210
Proximate composition		
Moisture	8%	8%
Crude protein	43%	45%
Crude fat	28%	18–26%

solenoid system, or air pressure to deliver feed to the fish. Automatic feeders are very useful for feeding fry, and some farms use them throughout the production cycle. Demand feeders are simply tapered feed hoppers, open at the bottom, with a rod extending from the hopper through the bottom into the water. An adjustable platform is fitted on the rod within the tapered bottom, and can be moved up or down the rod, thus changing the clearance between the outside edges of the platform and the tapered sides of the feeder. Fish bump the rod, causing the platform to move and feed to fall off into the water. The amount of feed falling off with each bump depends on the clearance between the platform and the sides of the feeder. The advantage of demand feeders in rainbow trout farming is that feed is only delivered when the fish want to feed. This eliminates feeding when fish are not interested, ensuring that fish, not the water, are being fed. Trout using demand feeders tend to feed intensively in the morning and evening, but also feed intermittently nearly all day (see Fig. 6). Fish move in and out of the feeding area. Reductions in dissolved-oxygen content associated with broadcast feeding and the frantic feeding that accompanies it do not occur. The disadvantage of demand feeders is that the clearance must be properly set to avoid feed delivery associated with wind or water movement.

The appropriate quantity of feed for a given size and number of fish is determined by using feed tables that contain recommended amounts of feed for different water temperatures and fish weights. Generally, the daily feed ration is divided into several feedings, with the aim being to feed between 0.75% and 1.0% of total biomass in a tank or raceway per feeding. The amount is chosen to minimize the range in size within a group of fish. If a lower amount is fed, aggressive fish will consume most of the feed, thereby increasing size variation within a group of fish. The idea is to feed enough at a feeding to satiate aggressive fish, allowing less aggressive fish the opportunity to eat. Many farmers fill demand feeders with a programmed amount of feed per day, determined from feed tables or farm records, in an effort to feed the amount needed to support rapid, but economical growth. Maximum feed efficiency (lowest feed conversion ratio) occurs at about 75% of maximum feed intake in trout, but maximum growth occurs at or near maximum feed intake. Trout farmers strike a balance between maximum growth and maximum feed efficiency by feeding between these levels at the growout stage, when most feed is fed during the production cycle.

Figure 6. Outdoor raceways. Note the demand feeders.

RAINBOW TROUT HEALTH MANAGEMENT

Disease is one of the most important problems facing aquaculture, including rainbow trout farming. Disease organisms are frequently present in water or in fish, and when fish are stressed or exposed to suboptimal environmental conditions, disease outbreaks occur. Crowding, low dissolved-oxygen content, high particulate levels in water, and stress associated with handling are examples of conditions that can result in a disease outbreak in rainbow trout. Fish health management in rainbow trout farming is based upon prevention; once a disease outbreak occurs, it is very difficult to treat or control. Elements of prevention include sanitation (clean raceways), high-quality feed, prevention of overcrowding, elimination of disease vectors, and vaccination. Vaccination has been extremely effective in preventing some important diseases in rainbow trout. Treatment of disease in rainbow trout with antibiotics delivered via the feed is relatively uncommon. There are only two antibiotics approved for use in the United States to treat fish, and they are not effective against many trout disease strains. There are strict regulations about using antibiotics in trout farming that, in combination with the cost of adding antibiotics to feeds, limits their use. Health management by prevention is a more economical approach in rainbow trout farming. Birds are a major disease vector in trout farms because they move from farm to farm and eat diseased fish. Disease organisms pass through the gut of birds and remain pathogenic. Most farms in Idaho use netting to restrict bird access to trout raceways (see Fig. 5). Vehicles, equipment, and boots should be disinfected when moved between farms. Buildup of uneaten feed and feces in raceways should be avoided, and dead fish should be promptly removed.

ENVIRONMENTAL CONSIDERATIONS IN TROUT FARMING

Most forms of trout farming are not a consumptive use of water, but water from the farms is enriched with nutrients that in turn, enrich lakes and rivers receiving farm water flows. These nutrients reduce water quality and increase growth of algae and aquatic plants. Thus, rainbow trout farms are subject to regulations limiting the levels of solids and nutrients in hatchery water effluents. Phosphorus is one nutrient of concern in hatchery effluents. Unassimilated nutrients originate in uneaten feed, feed dust, feces, and metabolic excretions (urine and gill wastes). Many farms create quiescent zones at the ends of raceways where fish are excluded and particles settle without being disturbed. Settled material is regularly removed and applied to fields, sometimes after composting. Phosphorus in hatchery effluents is present in two forms: solid particles (e.g., bones and other insoluble forms) and soluble phosphorus excreted by fish in urine. Solid phosphorus can be collected and removed, but soluble phosphorus cannot be removed economically because it is present in very low concentrations in very large quantities of water. Thus, limiting the amount of digestible phosphorus in feeds to the amount needed by the fish is the approach used in rainbow trout feed formulation. By the use of this approach, the amount of soluble phosphorus excreted by fish has been reduced to very low levels. Reducing the amount of insoluble phosphorus in rainbow trout feeds requires two approaches: (*1*) using low-phosphorus feed ingredients and (*2*) increasing the bioavailability of phosphorus in feed ingredients. This is an area of active research.

RAINBOW TROUT PROCESSING AND PRODUCTS

Rainbow trout are processed within minutes or hours after harvesting and are supplied to markets either as fresh or frozen products. Kept on ice, their shelf life is 10–14 days or more. As is the case with all fish, storage of fresh fish at higher temperatures shortens shelf life rapidly. Trout are marketed as gutted fish, as fillets (often boneless), or as value-added products, such as smoked trout. Trout fillets typically contain 19% protein and 5–7% fat, with the remainder being water (~70%) and ash (2%). The omega-3 fatty acid content of trout fillets depends upon the feed, but averages 22% of fillet fat. In the United States, 60–70% of food-size trout production is sold as head-on, gutted fish, with the remainder sold as portion-sized fillets, while in Europe, most trout are sold as whole fillets from larger fish, sufficient to feed several people (see Fig. 7).

RAINBOW TROUT FARMING INDUSTRY STRUCTURE

The U.S. trout farming industry was dominated by individual entrepreneurs during its growth phase in the 1950s and 1960s, but industry consolidation and vertical integration have characterized the trout farming industry over the past decade. The industry was composed of trout egg producers, growers, fish processors, distributors and brokers, and feed manufacturers. Businesses that combine trout farming, processing, and sales are most common. Egg production has remained a specialized business, as has feed manufacturing, although businesses that combine

Figure 7. Hand filleting rainbow trout. Note the butterfly fillet in the foreground.

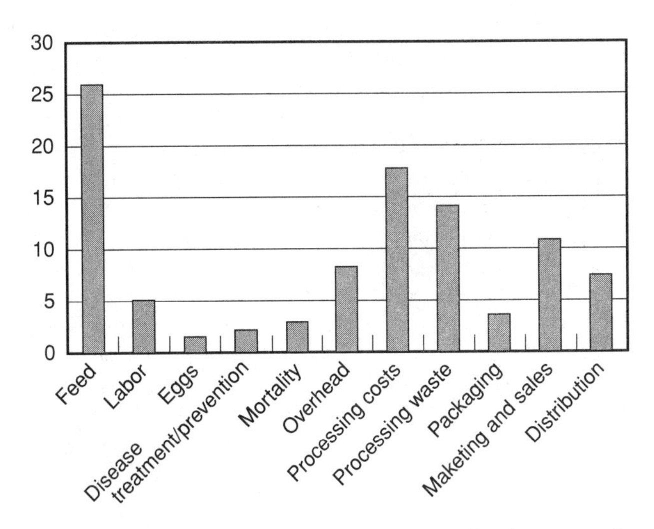

Figure 8. Percentage of rainbow trout production costs by category (7).

these sectors with production do exist. Some companies engage in contract rearing by small farmers, whereby the companies provide fingerlings, feed, and management advice to small growers who produce trout for the company, thereby expanding their production. The general impression of the U.S. trout farming industry is of a stable, maturing industry that has concentrated on lowering the cost of production and markets fish products that are less expensive than most other seafood products. The cost of producing trout varies from company to company, but in general, feed constitutes the largest single operating expense (Fig. 8). Increasingly, regulatory compliance (e.g., water testing and treatment) and HACCP compliance, constitute a rising share of operating expenses.

Complete utilization of available groundwater supplies in many trout rearing areas of the United States, particularly in Idaho, limits expansion of trout farming as it is currently practiced. New approaches and techniques will be necessary for trout production to increase. As mentioned earlier, oxygen concentration is the first limiting water-quality parameter for trout. Increasing the oxygen content of raceway water will increase the carrying capacity of raceways, thus increasing production from a given quantity of water. Simple filtration and even partial recirculation of water within farms offers the potential of increasing the productivity of existing trout farms, assuming that the economics of such approaches merit their use. Other innovative approaches to trout farming include the use of cages or pens to rear trout within large rivers and the use of irrigation water before it is applied to fields. Large fingerlings stocked into pens in the Columbia River as water warms in spring reach harvest size of 3–5 kg within 12–18 months, which is similar to the growth rates of rainbow trout reared in marine cages off the coast of Maine or in Chile.

FUTURE OUTLOOK

Worldwide, farmed trout production is increasing, but at a lower rate than Atlantic salmon production (4).

Trout are an efficient fish with respect to converting feed ingredients not consumed by humans into human food. Currently, the conversion of trout feed into trout weight gain is about 1:1, meaning that for each ton of feed, a ton of trout is produced. Feeds are composed of by-products of edible oil production (soybean meal), fish meal, grain by-products, and recovered protein from poultry and meat production. Trout yield more than 50% of edible product after processing, and this product is high in essential fatty acids and protein and low in saturated fats relative to animal proteins. Trout production is predicted to increase by 5% per year for the foreseeable future and will likely maintain its place in the top 15 finfish and crustacean aquaculture species produced in the world, as well as remain in the top 10 species, with respect to total value.

BIBLIOGRAPHY

1. R.J. Behnke, *Native Trout of Western North America*. American Fisheries Society Monograph 6, 1992.
2. J.D. McPhail and C.C. Lindsey, *Fisheries Research Board of Canada*, Bulletin 173. Ottawa, 1970.
3. G.R. Smith and R.F. Stearley, *Fisheries* **14**(1), 4–10 (1989).
4. A.G.J. Tacon, *International Aqua Feed Directory and Buyer's Guide 1997/98*, Turret Rai PLC., Middlesex, U.K., 1998.
5. E. Leitritz and R.C. Lewis, *Trout and Salmon Culture*. Fish Bulletin 164, State of California, Department of Fish and Game, 1976.
6. R.R. Stickney, *Aquaculture in the United States, A Historical Survey*. John Wiley & Sons, New York, 1996.
7. E.L. Brannon and G. Klontz, *Northwest Environ. J.* **5**, 23–35 (1989).
8. R.G. Piper, I.B. McElwain, L.E. Orme, J.P. McCraren, L.G. Fowler, and J.R. Leonard, *Fish Hatchery Management*, U.S. Dept. Interior, Fish and Wildlife Service, Washington, DC, 1982.
9. G.C. Embody and M. Gordon, *Trans. Amer. Fish. Soc.* **54**, 185–200 (1924).

See also TANK AND RACEWAY CULTURE.

RECIRCULATING WATER SYSTEMS

ROBERT R. STICKNEY
Texas Sea Grant College Program
Bryan, Texas

OUTLINE

INTRODUCTION

Recirculating or closed water systems continue to be used primarily for experimental work and for the rearing of larval organisms in commercial and research facilities, though increasingly, profitable commercial applications are being demonstrated. They are also used in public aquaria. The typical home aquarium is one type of a closed system. The popular press and aquaculture trade magazines have published many articles about systems that, their developers claim, work effectively and profitably. Such systems often do work. One can produce fish that demonstrate good growth rates and feed conversion ratios (wet weight increase divided by dry weight of feed offered) and that are acceptable to consumers. However, entirely closed commercial systems producing food-size aquatic animals in the United States are the exception rather than the rule. The exceptions tend to be associated with the availability of free heat (geothermally or industrially heated effluent used to provide increased temperature to the aquaculture facility through heat exchangers) or with the production of early life stages or high-priced ornamental species.

Given increasing land prices, water shortages that are occurring and are projected to occur in the future, and the increasing governmental regulation on the effluents from aquaculture facilities, there are many who feel that closed water systems will be needed if aquaculture is to continue growing in developed countries. Closed systems for marine species are currently being advocated as a means of avoiding the opposition that exists for conducting culture activities in coastal waters. Such systems may also be used to produce fingerlings for stocking in marine enhancement programs and may, at least for now, be more economical than net-pen or cage systems placed far offshore where they are exposed to storms.

Many innovations have occurred as a result of research and development, and ultimately, increasing numbers economically viable systems will be put into operation. In the meantime, claims of economic success should be looked upon with some skepticism until all the data are made available for close scrutiny.

The standard closed-water system is typically housed in a building to help maintain a constant environment, though outdoor systems have been designed and operated. Recently, shrimp farmers in Texas have converted from pond systems that relied upon enormous supplies of continuously flowing water to recirculation (achieved by setting aside some pond area as settling basins). When classical closed systems using tanks or raceways are placed outdoors, operators tend to experience problems with the growth of unwanted algae throughout the system. Covers over the various components of the system will reduce or eliminate the problem by drastically reducing the level of incident light. Such problems may be avoided through increasing the scale of the operation as indicated by the Texas shrimp culture example. Indoors, light levels are usually insufficient to support significant amounts of algae growth, although plants may be purposefully grown in hydroponic chambers for tertiary treatment (described in the next section).

BASIC COMPONENTS

Some of the components of recirculating water systems are unique to closed and semiclosed systems (water systems where some percentage of total water volume is replaced on a continuous basis), while others have broad application in aquaculture. Recirculating systems are generally comprised of three basic types of components: one or more settling chambers, a biological filter (or biofilter), and the culture chambers (1). The appearance of the various components and auxiliary types of equipment that have been added to recirculating water systems vary considerably, but most systems operate in basically the same manner. As discussed by Lucchetti and Gray (2), the functions of such systems include ammonia removal, disease and temperature control, aeration, and particulate removal.

Water leaving the culture chambers usually, but not always, is allowed to enter a primary settling chamber where solids are removed. The water is then passed into a biofilter, where bacteria detoxify ammonia and nitrite. The biofilter is typically followed by a secondary settling chamber, where bacteria that sloughs from the biological filter is removed. After secondary settling, the treated water is returned to the culture chambers. Some systems use mechanical filters in place of, or in addition to, one or both settling chambers.

Well-designed systems pump the water only once and utilize gravity as much as possible. If water is pumped more than once, it is necessary to balance the pumping rates precisely to keep portions of the system from being either pumped dry or caused to overflow.

Biofiltration is a form of secondary waste treatment, where the settling of solids represents primary treatment. Tertiary treatment, which removes nutrients like nitrates and may reduce the levels of trace elements and dissolved organics, has been added to some recirculating systems. Supplemental aeration is generally provided in closed systems. Sterilization with ozone or ultraviolet radiation, foam removal, and other auxiliary components have also been used. Backup systems for all motors and pumps are critical.

The various components may be quite large or rather small, depending on the purpose for which the system has been designed. A closed system designed to produce 0.5 kg (1.1 lb) for rainbow trout (*Oncorhynchus mykiss*) or channel catfish (*Ictalurus punctatus*), for example, would in all likelihood have larger components than one designed for the rearing of shrimp larvae. In the same vein, research facilities are typically much smaller than commercial growout operations.

Components may be physically separated or, at least some, may be combined (e.g., a settling chamber and biofilter could be placed within a single unit). A single biofilter and associated settling tanks may service a number of culture chambers, or each culture chamber may be served by its individual biofilter and settling chambers. Various configurations have been designed for both freshwater and marine use. Some designs, such as those employed for research, have employed materials as readily available as plastic pails and garbage cans, while others, including most commercial facilities, have been fabricated from concrete or fiberglass. The use to which a system is to be put, the space into which the system is to be placed, and the resources available play roles in the materials chosen.

Materials used for construction of not only the culture chambers, but also the other components of closed systems and other types of water systems should be nontoxic to the culture animals. The same applies to paint used on any of the components. Metal should be avoided as much as possible because of potential toxicity. In saltwater systems, corrosion of metal can be a significant problem. Pumps with plastic impellers are recommended, although if operated continuously, metal impellers will hold up in salt water, but can still cause elevated levels of trace metals in the water.

Plumbing in modern aquaculture systems usually comprises plastic pipe, typically polyvinyl chloride (PVC). PVC pipe comes in various sizes and strengths. Pipe designed for drains should not be used in pressurized parts of the system. Pipe labeled "Schedule 40" or "Schedule 80" can be placed under pressure and can carry hot water. PVC made for drains should not be used outdoors in climates where freezing temperatures occur, unless it is buried. Drain PVC will shatter when water standing in it freezes.

Culture Chambers

Closed, semiclosed, and open culture systems typically employ relatively small culture chambers compared with ponds. Typical culture chambers are circular tanks (also called circular raceways), linear raceways, and silos (tall circular raceways). The material used for construction of culture chambers can vary widely, depending on such factors as the preference of the culturist, the availability of materials, the cost, and in some cases the species to be cultured. The most commonly used materials are concrete, wood, fiberglass, various types of molded plastic, and sheet metal (usually aluminum or stainless steel, and usually used only in freshwater systems because of the corrosion problems associated with salt water). Stainless steel is expensive and galvanized metal should be avoided because of zinc toxicity. Glass aquaria have been popular with some researchers and represent yet another type of construction material that can be employed.

Culture-chamber dimensions are quite variable, although most circular and rectangular tanks used in conjunction with closed systems are less than 10 m (30 ft) in diameter and seldom exceed 1 m (3 ft) deep. (The sides may be somewhat higher to provide freeboard above the waterline.) Commercial tanks are generally larger than those used in research.

Circular tanks have an advantage over rectangular culture chambers. When water inflow and drainage are properly designed, circular tanks are characterized by uniform water flow throughout. Raceways tend to have areas of static water in the corners where waste accumulates; however, linear raceways with rounded ends have been designed to help reduce the problem. Circular tanks usually have center drains and water is typically introduced tangentially to the surface, causing it to move clockwise or counterclockwise in the tanks. The motion concentrates waste feed and fecal material around the drain, where it can be automatically removed if the proper type of drain is used. Some self-cleaning will occur in linear raceways that are equipped with venturi drains, which are very efficient in removing particulate matter from circular tanks. A venturi drain involves an inner standpipe that controls water level, with a larger standpipe surrounding the internal one. Holes at the bottom of the larger standpipe pull water in at the tank bottom from where it rises to exit at the top of the smaller internal standpipe. Waste material tends to collect between the two standpipes. Some of the material will flow out through the overflow and down the drain, and the remainder can be removed periodically by pulling the inner standpipe and allowing sufficient water to exit to remove the accumulated waste. The self-cleaning action is facilitated, particularly as the diameter of the tank increases, if the bottom of the tank slopes toward the drain. Venturi drains can also be placed on the outside of culture tanks through some modifications in plumbing design.

Circular tanks usually drain either into pipelines or into a channel imbedded in the floor of the building containing the system or ground on which an outdoor system is constructed. Pipes are typically used when gravity is depended on to carry water to the next component in a recirculating system. Either pipes or floor drain channels can lead to a sump in which a pump moves the water to the next component.

In linear raceways, water typically enters at one end and is discharged through a standpipe at the other. A venturi drain can be used in conjunction with a linear raceway and will help in waste removal. As previously indicated rounded corners can be used in conjunction with linear raceways to avoid dead spaces, but solids will tend to settle throughout much of the long axis of such chambers unless the flow rate is quite high. Screens are often placed in front of the standpipe in linear raceways to keep fish from being lost down standpipes that are not protected with an outside venturi pipe.

In tanks and raceways that have standpipes associated with the drains, water level is maintained in the event of a loss of inflow. Some designs employ bottom or siphon drains with balanced inflow. Depending on the design, such culture chambers may lose all their water in the event of a pump or power failure.

The exchange rate and manner in which water is introduced into raceways influences current velocity. Maintenance of some current not only is important in helping flush waste products toward the drain; it may also benefit the culture species by accommodating normal behavior in the case of species that tend to orient themselves with respect to a current, such as trout. There are some who feel that the exercise associated with constant swimming into a current is beneficial in terms of final product quality and may stimulate more active feeding and growth, although scientific evidence is scanty. Too much current can exhaust the culture species, or in the case of larval and early juveniles, may push them into screens and lead to heavy mortality. Employing a screen in front of the standpipe in a linear raceway will prevent the loss of animals down the drain. In circular tanks, large areas associated with the outer standpipe can be screened, which will avoid concentrating the effluent into small areas through which accelerated flow will occur. The size of the screen mesh in linear raceways and in association with outside standpipes used in circular tanks should be sufficiently fine to prevent escapement of the animals, but as large as possible to reduce clogging and the restriction of flow.

Culture tanks and raceways that employ venturi drains should be small enough to allow those drains to function properly. While the diameter of the standpipes can be increased as the size of the culture chamber is increased, in practice, outside standpipes larger than about 30 cm (1 ft) and inside standpipes larger than 10 cm (4 in.) in diameter are not normally used. Circular tanks larger than about 20 m (60 ft) in diameter may be impractical. Most are no larger than 10 m (30 ft) in diameter.

Raceways with screens placed in front of the standpipe area to prevent escapement can be quite large and can be fitted with two or more standpipes at the drain end if necessary to accommodate the flow. Indoor raceways used in hatcheries and research facilities tend to be relatively small. They are usually no more than 3 to 5 m (9 to 15 ft) wide and 25 to 50 m (75 to 150 ft) long. Many are less than 1 m (3 ft) wide and only 2 to 3 m (6 to 9 ft) long.

It should be possible to drain and refill culture tanks rapidly during harvesting and stocking. These processes are facilitated when oversized inflow and drain lines are used. During normal operation, inflow lines can be fitted with flow-regulation devices to control the amount of water entering. When the raceways are being filled, the flow regulators can be removed or bypassed to allow much higher flow rates. Similarly, while a large internal standpipe and drain lines may not carry anything like their design capacity during normal operation of the raceway system, when culture chambers are being drained, the job can be accomplished quite rapidly by having oversized plumbing. Also, it may be possible to drain several raceways simultaneously without exceeding the capacity of the drains to handle the increased water flow.

For large outdoor raceways, flow is generally controlled with valves alone. In the laboratory and in the hatchery, it may be necessary, or useful, to maintain constant, limited, and equal flow rates in a number of raceways. Homemade flow regulators have been produced for that purpose, but economical commercially produced flow regulators are also readily available through plumbing and irrigation supply houses. A flow regulator can be made by drilling a hole of the desired size through a PVC cap that is then glued onto the end of an inflow pipe. The flow rate through this type of flow regulator will vary if water pressure fluctuates.

Commercial flow regulators are designed to provide constant flow as long as reasonable water pressure is maintained. They adjust to changes in water pressure above some specified minimum. Such changes occur when the amount of water being used is changed, which occurs very commonly when raceways are being filled or cleaned, or when water is diverted to other uses such as filling hauling tanks or washing down the floor.

Some of the most effective commercial flow regulators are made from stainless steel and plastic. The metal ones can be screwed onto the end of a threaded PVC pipe, while the plastic ones are disc-shaped and can be placed into a fitting that will screw onto a faucet (also available in PVC). When stainless-steel flow regulators are used, it is helpful to place a valve in the water line upstream from each flow regulator so that water supply to the individual raceways can be independently turned on and off. Either type of flow regulator can be easily removed when higher volumes of water are needed to fill the raceways.

Flow regulators come in a variety of sizes. Flow regulators that will deliver less than 1 L/min (0.26 gal/min) can be used with small raceways [e.g., 20 L (5 gal) aquaria], while larger units may require flow regulators that deliver several liters per minute. The flow rate through such regulators is controlled by the size of the opening through which the water passes.

Water exiting the types of flow regulators previously described produces a thin stream under pressure. When that stream is allowed to run directly into the raceway, the incoming water is actually injected into the water within the culture chamber, causing turbulence and consequent aeration. A saturated level of dissolved oxygen (DO) can often be maintained in the presence of high fish biomass when such flow control devices are employed. In large tanks it may be necessary to have several water injection streams. If the culturist wishes to reduce the turbulence, as might be necessary when larvae or sensitive juveniles are being cultured, a length of pipe or plastic tubing can be placed downstream of the flow regulator to dampen the flow. While that may also reduce the oxygenation rate, in larval tanks, density of animals and low and maintenance of sufficient levels of DO are not usually a problem. Supplemental aeration can also be provided as necessary in all cases.

Each recirculating water system may employ one or more culture chambers in conjunction with each of the other main components of the system. The effluent from several raceways may be collected in a common drain system and flowed into the settling chambers and biofilter before reentering the tanks, or each culture chamber may have its own set of other components (which will add significantly to the expense of construction and operation). In systems wherein a number of culture chambers employ a single set of settling chambers and a commercial biofilter, the latter components need to be properly sized to accommodate the total volumes of water and waste that the system will be required to handle. Advantages to building large settling chambers and biofilters include economical use of space, especially indoors; limitation on the number of backup components required; and reduction in the need to duplicate plumbing. In general, the cost of constructing a few large units is less than that incurred in conjunction with building several smaller ones with the same combined capacity.

When the effluent from several culture chambers is pooled for treatment, an equipment malfunction can result in complete mortality of the animals in the system. Total power failures can lead to serious consequences in any closed water system; however, the failure of a single pump or aerator would be less crucial to the producer if each unit had its own mechanical devices than if several culture chambers shared the same pump and aerator. Duplicating primary and backup systems is expensive, but so is the loss of one or more tanks of animals.

A disease outbreak in a recirculating water system with more than one culture chamber can rapidly spread. Further, if the proper precautions are not taken, diseases can be spread by the culturist from one closed system to another. Disease treatment is often difficult in closed systems because exposing the biofilter to chemicals can kill the beneficial bacteria that reside therein. Isolating the culture chambers for treatment is not generally feasible, unless there is a sufficient volume of suitable quality water to put the culture chambers in a flow-through mode during treatment.

Settling Chambers

In most closed aquaculture systems, effluent water from the culture chambers passes through one settling chamber before entering the biofilter and another settling chamber after leaving the biofilter. In closed systems, mechanical filtration through sand or other types of filters has also been frequently used to remove particulate matter from culture tank and biofilter effluents. Removal of particulates using a hydrocyclone device prior to biofiltration has also been found effective (3).

A settling chamber is merely a tank in which the exchange rate is sufficiently slow that particulates will settle. In many such chambers, the water is made to flow in and out at the top so that the material that has already settled does not become resuspended. A valve should be located at the bottom of the settling chamber to provide a means by which settled material can be removed periodically. Sloping the bottom of the chamber to the drain will make removal of solids more complete and require less water usage during the sludge-draining process. In some designs, the primary settling chamber is ignored or is incorporated into the biofilter.

Solids removal with a primary settling chamber is important because it cuts down on the loading of the biofilter. The material that is removed from the settling chamber is nutrient-rich and can be used as fertilizer. The material is composed largely of feces (and pseudofeces when certain invertebrates are under culture), unconsumed feed, and bacterial floc.

Water leaving a biofilter will contain bacterial floc that sloughs off the biofilter medium. While bacteria will certainly grow on the walls of pipes and on the walls of culture chambers and settling basins, the bulk of the bacteria in the system is contained within the biofilter, and most of the sloughing occurs in that part of the system. Removal of sloughed material in a secondary settling chamber helps keep the water in the culture chambers from becoming turbid.

Proteins dissolved in the water from various sources will form masses of bubbles when the water is sprayed from one of the units in the system to another. Protein skimmers have been designed to remove the bubbles, thereby also reducing loading on the system. A skimmer may simply be a vertical arm that confines the bubbles and causes them to spill over the sides of the settling chamber, or it may be more elaborate, such as a moving belt with paddles on it that pass over the surface of the settling chamber and push the bubbles over the side or into a collection area leading to a drain. Some commercial biofilters also have foam removal components incorporated into them.

Biofilter Media and Function

Mechanical filters have been used on both experimental and commercial recirculating water systems in place of biofilters, but clogging and the resultant channeling and poor filtration efficiency have produced serious water-quality problems. Sand filters tend to clog rapidly when subjected to heavy loading. Gravel filters are less subject to clogging, but channeling is not uncommon if they are not frequently backwashed. When a mechanical filter becomes clogged, organic material, including bacteria and other organisms that have colonized the medium, will die and begin to decay. The filter can quickly become anaerobic. Not only does it lose the biological filtration activity that it may have once had; it will also begin to emit hydrogen sulfide, ammonia, and other toxic substances, as well as release water that is either very low in or devoid of oxygen. Thus, the filter will begin doing just the opposite of what it was designed to do.

Many functioning biofilters have been designed by using media that will not easily clog. Various types of plastics have been used, for example. For small-scale systems, pieces of plastic pipe cut into lengths of a few centimeters have worked well. Closed systems have employed Teflon rings, fiberglass, Styrofoam beads, and all sorts of plastics as media. Anything that bacteria will colonize can serve as a suitable medium for a biofilter. Commercial biofilter media are available as well. Those media are typically honeycombed sheets or cylinders of plastic with holes or slots cut in the sides and with protrusions of various kinds. In all cases, the void space is very large relative to the surface area of the material. Surface area for bacterial growth is enormous in sand or gravel filters compared with that of biofilters containing plastics, but the efficiency of the former tends to be very poor because of the clogging and channeling problem previously mentioned.

In recent years, advancements have been made in the utilization of more tightly packed media in biofilters. Ion exchange media (4,5) and the application of fluidized-bed technology with activated charcoal (6) have been successfully employed.

The major function of a biofilter is the nitrification of ammonia. Ammonia is excreted into the water by way of the gills of fishes and other organisms and tends to be one of the most important natural toxic chemicals produced in water systems of all types. The nitrification of ammonia occurs in two steps, each undertaken by a different genus of bacteria. *Nitrosomonas* converts ammonia (NH_3) to nitrite (NO_2^-), and *Nitrobacter* converts nitrite to nitrate (NO_3^-). The two genera of bacteria are aerobic and will only live and perform their function when there is oxygen in the water. When deprived of oxygen, even for a brief period, the bacteria will succumb and the biofilter will begin producing high levels of ammonia and nitrite.

Aerobic conditions can be maintained by bubbling air, compressed oxygen, or liquid oxygen into the biofilter chamber. Splashing water into the biofilter or exposing the bacteria to oxygen periodically (See the section on "Rotating biodisc filters") are also effective at maintaining aerobic conditions.

When a new recirculating system is put into use or when a system is restarted after having been harvested, drained, and cleaned, time must be allowed for colonization of the biofilter with the appropriate types of bacteria. At startup of a commercial system, the biomass of aquatic animals carried in the system may be sufficiently small that water quality deterioration will not occur before the bacteria become active. However, in most instances, it is best to ensure that the system is operating effectively before the aquatic organisms are added. This can be done by seeding the system with ammonia (e.g., in the form of ammonium salts), putting in a source of organic material that will deteriorate and form ammonia (e.g., prepared feed), stocking fish or other animals at low biomass, or some combination of these approaches.

The desirable bacteria are cosmopolitan and will soon colonize the biofilter without being inoculated. Commercial solutions containing bacteria are available. Reviews as to the effectiveness of those commercial preparations have been mixed.

Typically, *Nitrosomonas* will become active from several hours to several days before *Nitrobacter*; thus, spikes in nitrite production can be expected before the biofilter becomes fully functional. The time required for colonization by biofilter bacteria will vary as a function not only of loading, but also temperature and salinity. More rapid colonization generally occurs in warm as compared with cold water. Marine culture systems require a much longer colonization period than freshwater systems. Typical colonization rates are from a few to several days in freshwater to in excess of a month in saltwater. More rapid colonization has occurred in saltwater, but the colonization rate is generally slower than in freshwater.

The sizing of biofilters relates to anticipated ammonia production and the efficiency of the system in converting the ammonia to nitrate (2). At neutral pH (7.0), the activity of the biofilter is a function of temperature and the available surface area (7,8). Speece (7) produced a series of graphs that can help aquaculturists determine the amount of biofilter volume needed over a temperature range of 5 to 16 °C (41 to 61 °F) when the food conversion ratio of the aquaculture animals and the surface area of the biofilter medium are known.

Biofilters can be installed either indoors or outdoors, but the internal portion should be protected from exposure to sunlight and bright artificial lights to prevent the growth of undesirable algae. Algal growth can lead to clogging of the biological filter, and if blue-green algae become established, there is the potential of off-flavors or even direct toxicity from certain metabolites produced by those organisms. The biofilter should also be protected from large temperature fluctuations, so outdoor filters should be insulated.

Four basic biofilter designs have received the attention of aquaculturists. They are trickling, submerged, updraft, and rotating biodisc filters. In addition, fluidized beds can act as biofilters.

Trickling Biofilters. Water enters trickling filters from the top and is allowed to pass by gravity through the filter at a rate that does not allow the medium to become submerged, although all internal portions of the filter are continuously wetted. Municipal sewage treatment plants often employ trickling filters with rock as the medium. Those units are much larger than the ones used in most aquaculture operations because the volume of waste from even small cities is dramatically higher than from most aquaculture facilities.

Submerged Biofilters. The design of submerged biofilters is often similar to that of settling chambers, except that the submerged filter contains a medium on which bacteria become established. Water enters one end of the filter, flows through the medium, and exits from the opposite end in most designs.

Submerged biofilters can be operated by gravity flow or, with the incorporation of a watertight cover, water can be pushed through them under pressure. If pressure is used, inflow and outflow pipes can be at any desired height without danger of losing water because of overflow.

Since the filter medium is constantly underwater in a submerged filter, it is necessary to ensure that sufficient aeration is provided to keep the filter aerobic. If the filter becomes anaerobic, ammonia will begin to be produced instead of eliminated.

Updraft Biofilters. An updraft filter is essentially the same as a submerged biofilter in that the medium is continuously submerged. The difference is that instead of the water moving horizontally through the filter, it enters from the bottom and exits near the top. A sedimentation chamber (primary settling chamber) can be incorporated into an updraft filter, thereby allowing the settling of solids below the influent pipe elevation. A drain valve at the base of the settling chamber allows solids to be removed as necessary. As is the case with submerged filters, those of the updraft variety need to be well oxygenated to maintain aerobic conditions. Alternatively, as described by Paller and Lewis (9), an updraft biofilter can be "dewatered" periodically to place the bacteria in contact with atmospheric oxygen.

Rotating Biodisc Filters. A rotating biodisc filter or rotating biological contactor (RBC) utilizes a concept somewhat different from that of the filter types previously described. In this case, the medium is moved through the water, rather than the water being moved past the medium. Rotating biodisc media are composed of numerous circular plates placed on an axle and set in a trough with half of each disc submerged and half exposed to the atmosphere. Fiberglass is commonly used as disc material, though Styrofoam and various types of plastic sheets have also been employed. The discs are rotated slowly (only a few revolutions per minute), usually powered by an electric motor and appropriate reduction gears. Bacteria colonize the plates as in other types of biofilters. Alternating exposure to the metabolite-laden water in the trough and to the atmosphere provide the bacteria with a continuous supply of nutrients and oxygen. Various systems have been established (both for research and on a commercial scale) employing homemade and commercially available rotating biodisc filter units.

Fluidized Beds. As previously indicated, fluidized beds can act as biofilters. A fluidized bed is typically composed of material the size (and usually composition) of sand that is placed in suspension by flowing water through it. Both aerobic and anaerobic fluidized-bed biofilters have been used in aquaculture. Anaerobic fluidized beds can remove nitrate (10), which, while normally not a problem, can reach undesirable levels in some instances. When nitrate levels reach several hundred parts per million in closed systems (such as has happened in large marine aquarium systems), remedial action may be required.

Given sufficient energy input, many solids can be made to behave as liquids. Fluidized beds have commonly employed sand or ion exchange resins, but they may also employ activated plastic beads, charcoal, limestone, or crushed oyster shell. Depending on the type of medium used, fluidized beds may act as biofilters, mechanical filters, or as a means of adding chemicals to the water.

Biofilter Size. The size of the biofilter required for a particular culture system depends on many factors. Reliable formulas for calculating the relative size required as a function of the number, kind, and biomass of animals in the culture chambers, total water volume in the system, and flow rate remain to be thoroughly developed.

Early in the growing season, the size requirements for biofilters may be small, since the biomass in the system is low and the amount of waste to be treated is not great. As the culture animals increase in size, the efficiency of the biofilter may increase because proper loading has been reached and then decline as the quantity of waste exceeds the capacity of the biofilter. Additional biofiltration capacity is required when efficiency begins to drop off because of overloading. Routine water quality monitoring is required so that the aquaculturist can determine when and whether to alter the biofiltration capacity of the system. If a new biofilter is placed on line to augment an existing system, sufficient time for colonization of the medium in the new filter should be provided.

Maintenance of pH

The accumulation of dissolved chemicals in recirculating water systems leads to depression of the pH, unless the system is buffered. As the water becomes more acid, stress is placed on the culture organisms, and if the pH becomes too low, death will eventually occur. Microorganisms colonizing the biofilter may also be adversely affected by low pH. Organic acids and carbon dioxide are the primary causes of increased acidity. Wetzel (11) noted that ammonia is strongly sorbed to particulate matter at high pH — another compelling reason for preventing the water from becoming acidic. Since bacteria are also associated with surfaces, the nitrification process may be enhanced when adsorbed ammonia and microorganisms are placed in close proximity on a waste particle or on the biofilter medium.

For most freshwater aquaculture systems, the pH should be in the vicinity of 7.0 (with an acceptable range of 6.5 to 8.5), while saltwater systems should be maintained at a pH in excess of 8.0 (reflecting the differences in normal pH between natural fresh and saline waters). To accomplish pH control, calcium carbonate is often used as a buffering agent. This material may be in the form of crushed limestone, or more commonly, whole or crushed oyster shell. As hydrogen ions are produced in the system, calcium carbonate slowly dissolves. The carbonate ions remove hydrogen ions from solution to form bicarbonate.

Limestone and oyster shell are relatively inexpensive and require little or no attention once incorporated into the water system. Bacterial mats and other types of particulate matter can build up on the surface of the buffering material and may interfere to some extent with dissolution of the calcium carbonate. Cleaning may be necessary at appropriate intervals to ensure that the buffering capacity of the system is maintained.

The amount of buffer material present in the biofilter is not particularly critical. Most culturists utilize a few kilograms per cubic meter (pounds per cubic yard) of filter capacity. The buffering agent can be located nearly anywhere, but is commonly placed either in conjunction with the biofilter influent or effluent.

Moving Water

As the number mechanical devices in a recirculating system is increased, so are the chances of a failure. One high-quality continuous-duty water pump is often all that is required to move the water within such a system. The pump can be placed between any two components, except when an updraft filter is used, in which case it is desirable to place the pump on the influent side of the biofilter. In all types of systems, units downstream from the pump can obtain water through gravity flow, assuming they are placed at appropriate elevations relative to one another. When more than one pump is utilized, the design should ensure that water flow is balanced among the components in the system and that there is no potential for a component to be pumped dry.

Pumps can be expected to run intermittently or continuously for up to several years if they are properly maintained. Submersible pumps that run while submerged in water are handy for some applications, and pumps with impellers that are not corroded by salt water are available for use in mariculture facilities. Pumps of various types were discussed by Wheaton (12). Metal impellers will function well in salt water if the pump is allowed to run continuously, but trace metals may be released into the water. When present, even at low concentrations, some such metals can be highly toxic, particularly to larvae and, when used in closed systems where their concentration will continuously increase until the system is drained and refilled with new water.

The amount of water to be moved through a water system is related to the size of the system and the optimum flow rate for the culture organisms and the biofilter. Optimum flow rates for a given species may change, depending on the life stage of the organism. For some species, few data are available and flows may have to be determined by trial and error. Turnover times in culture chambers may exceed 24 hr for larvae and may be several times an hour for juveniles.

Water can be moved with airlifts, such as those used in conjunction with submerged filters in home aquaria. Simple airlifts can be constructed from PVC pipe mounted vertically in the water column. These devices can be used to move water within a culture chamber, as well as to move water between culture chambers. In addition to moving water, airlifts provide aeration. They can operate at relatively low pressures and are quite economical in certain situations.

Tertiary Treatment

There are three phases involved in complete water treatment. The first, which involves the settling of solids and their removal from the waste stream, is known as primary treatment. Secondary treatment involves transformation of toxic chemicals into less toxic forms and is accomplished by biofilters such as those described earlier. The removal of dissolved chemicals, such as nitrates and phosphates, from water is known as tertiary treatment. Tertiary treatment is rare in sewage treatment plants, which are typically designed for secondary treatment at best, though some tertiary treatment has been employed in aquaculture systems.

Nitrates, phosphates, and micronutrients accumulate in closed water systems even if a highly efficient biofilter is in operation. Although both nitrates and phosphates are required nutrients for plant growth and are not directly toxic to aquaculture animals when present at normal environmental levels, nitrate is known to be toxic at extremely high concentrations (13) and, for some life stages of certain species may be toxic at relatively low concentrations.

Nitrate can be removed from aquaculture systems through denitrification to elemental nitrogen. Effective removal of both nitrates and phosphates can be achieved by incorporating an additional culture chamber in the system for the production of plants. The plants, either terrestrial or aquatic, might best be located immediately following either the biofilter or secondary settling chamber.

Aquatic plant candidates for tertiary treatment include water hyacinths (*Eichhornia crassipes*) and Chinese water chestnuts (*Eleocharis dulcis*). Both grow rapidly, and each is efficient in removing dissolved nutrients from the water. Water chestnuts have economic value as human food. Water hyacinths, on the other hand, are generally considered a nuisance when they invade natural waters, but they do hold some potential as livestock feed, as do a large number of other aquatic plant species (14,15). Since aquatic plants have very high water content, they can only be economically used as livestock feed when they do not have to be transported from the site of harvest. Drying the plants before shipment is not generally considered to be a practical or economical alternative.

An often more attractive type of tertiary treatment with plants involves hydroponics with terrestrial plants. The plants are grown either in water or in an inert medium that is continuously or intermittently wetted with nutrient-rich water. On an experimental level, various types of vegetables (including tomatoes, lettuce, and cucumbers) have been grown in association with the recirculated water of aquaculture systems in which several types of fish have been produced. Greenhouses have been used in conjunction with many such systems to provide the appropriate environmental conditions for the plants. In all cases where plants are utilized, sufficient natural or artificial light must be provided to maintain active photosynthesis.

Maintaining the proper nutrient concentration for plants raised in conjunction with a recirculating water system is not a simple matter. The amounts of nutrients being produced and taken out of the system are not in equilibrium, since both the aquaculture animals and the plants are continuously growing and will be harvested from time to time (but often not at the same time). In many instances it may actually be necessary to supplement the system with nutrients to help maintain plant growth. Macronutrients like phosphates and nitrates, and various micronutrients such as trace metals may require supplementation.

Backup Systems and Auxiliary Apparatus

In theory, a closed system can be operated with only the components outlined in the above sections. In reality, redundancy needs to be built in so that disaster can be avoided if, for example, there is a pump failure. In addition, various additional features can be added to help the aquaculturist avoid disease problems and increase the efficiency of the system.

Backup Systems. Each mechanical component associated with an intensive aquaculture system of whatever type should have a backup so that the integrity of the system can be maintained in the event of an equipment failure. Aeration is an auxiliary component featured in many water systems, but some aquaculturists depend on flow regulators to provide aeration. A backup pump should definitely be available, and most systems have a backup aeration system such as an air blower or air compressor. Backup systems need not be identical to the primary system. An air compressor might, for example, take over for a failed blower; bottled gas, liquid oxygen, or agitators could also serve in backup roles. Many backup systems require electricity, so both primary and secondary systems can be lost during a power failure. A gasoline or diesel generator should be available to provide power during electrical outages.

Modern technology has dramatically improved the effectiveness of backup systems. State-of-the-art generators remain continuously warmed up and may even start-up periodically for a few minutes to run self-diagnostics and ensure that they are operating properly. Autodialers can be used that will dial one or even several telephone numbers to alert aquaculturists of a power failure during nights, holidays, and weekends when the culture system may be unmanned. The more automated a backup system is, the more expensive it will be, but the cost will be more than offset if the backup components prevent a crop loss.

Auxiliary Apparatus. Backup devices are essential for the operation of intensive aquaculture facilities. In addition there are various types of auxiliary components that can be considered optional, though desirable. Auxiliary apparatus may be used in pretreatment, on-line operation, or posttreatment of the water used in a recirculating system.

Air compressors provide high-pressure air, which means that airlifts can be effectively operated when a considerable amount of head is present. Since air-compressor motors cycle frequently (depending on how rapidly the air is bled from the pressure tank), the starters and motors have been known to fail with unacceptable frequency. Air blowers, on the other hand, seem capable of operating continuously and reliably for extended periods of time. They produce large volumes of low pressure air at economical prices. Because the air is delivered under low pressure, such systems cannot operate against significant head pressures, although they are effective in tanks of standard depth.

Liquid oxygen has become popular with many aquaculturists. It is readily available in most areas, including those that one would generally consider to be quite remote. Large volumes of oxygen can be stored in a fairly small amount of space when liquefied. Until the 1980s, bottled oxygen and compressed air were often used as sources of emergency aeration. Liquid oxygen can serve that purpose and, in some operations, is used routinely as a means of maintaining oxygen at or near saturation in aquaculture systems.

Some means of pretreating incoming water may be required, particularly when the water is from either a municipal or surface source. Chlorine will have to be removed from municipal water, unless only small quantities are being used to replace splashout and evaporation. Filtration may be used in conjunction with surface water to eliminate unwanted organisms and turbidity.

Clinoptilolite, a natural zeolite with an affinity for ammonia, has been effectively used in aquaculture systems (2). As indicated by Slone et al. (1981), by removing ammonia, clinoptilolite indirectly controls nitrite levels as well (16). The material can be regenerated by backwashing with a sodium chloride solution when the absorption sites become filled (2). There are ion exchange resins that will remove ammonia from water efficiently, but they are much more expensive than clinoptilolite.

Air stripping, which involves agitation and aeration, can also be used to remove ammonia from water if the pH is greater than 10.0 (17,18). Removal of ammonia in either of these ways can reduce the volume of biofilter medium required, or eliminate the need for the biofilter altogether in small systems.

Ultraviolet Treatment. As previously mentioned, disease control is difficult in closed water systems. One of the methods that has been used to keep the level of circulating pathogens below detectable levels in such systems is continuous ultraviolet (UV) irradiation of the water. The effectiveness of UV light is dependent on the size and species of the pathogenic organism and water clarity. Levels of UV irradiation required to kill various pathogens have been determined (2,19).

Fluorescent bulbs are used in both commercial and homemade UV sterilization systems. The bulbs are kept from contact with the water by housing them in quartz tubes around which the water flows, or they may be placed around a transparent quartz or plastic pipe through which the treated water passes. The UV lights should be in as close proximity as possible to the water. In many cases the water is exposed to the light in a thin film or passes in front of the light in confined channels that are a few millimeters to a few centimeters thick. Some systems have been designed that allow thin films of water to pass under UV lights suspended from above.

The output of UV bulbs declines with time. According to Lucchetti and Gray (2), UV output decreases by 10% during the first 100 hr of operation and drops to 70% of the initial level during the first six months of use. The bulbs should be replaced every several months or sooner. Particulate matter tends to sediment out on the quartz and plastic tubes that separate the UV bulbs from the water, thereby further diminishing the efficiency of the sterilization system. Prefiltering the water through a sand filter before exposing it to the UV light will help reduce the problem considerably, but it may still be necessary to clean the system periodically and even to replace the protective quartz or plastic on occasion.

Ozonation. Ozone generators have been used in aquaculture systems to both oxidize organic matter and kill pathogenic organisms. (See the entry "Ozone.") Other beneficial effects of ozonation are reduction in biological oxygen demand, ammonia, and nitrite. Ozone (O_3) itself is extremely toxic. While little information is available on aquaculture species, the fathead minnow (*Pimephales promelas*) is killed at ozone concentrations of 0.2 to 0.3 mg/L ppm (20). Sublethal pathology has been observed in rainbow trout exposed to low levels of ozone, and oyster larvae can be killed by only trace amounts of ozone (21,22).

At relatively low concentrations, ozone will lead to improved water quality as was demonstrated with respect to Atlantic salmon (*Salmo salar*) (23). If used for pathogen control, higher concentrations may be required, and sufficient time must be allowed for the gas to convert back to molecular oxygen (O_2) before the water is exposed to either the culture chamber or the biofilter after being ozonated. Alternatively, active deozonation may be applied. The half-life of ozone is about 15 min (24), so it can take a considerable amount of time for safe levels to occur. Aeration assists greatly with the conversion of ozone to molecular oxygen and can reduce the time required for keeping ozonated water away from the culture animals and biofilter. Activated charcoal can also be used to strip ozone from water.

Ozone seems to be much more effective in controlling pathogens in seawater than in freshwater. While the chemistry is apparently not entirely understood, it is believed that ozone reacts with bromine in seawater to produce lethal bromides. The reaction apparently occurs very rapidly, and the efficacy relative to pathogen control is high.

Some systems have been designed in which a portion of the total water flow is ozonated during each pass. When the treated side stream is returned to the main flow, the ozone is diluted to a sublethal concentration. With continuous exposure of portions of the flow to ozone, both water-quality improvement and pathogen control can be achieved.

Careful monitoring of ozone is necessary and adjustments to the amount of ozone being added may be required at intervals. While sensors to monitor ozone in water have been developed, they have not been perfected and most monitoring involves sensing ozone in the atmosphere above the deozonating component of the system. Equipment for that purpose is quite effective. Automatic alarms can be set for a desired range of atmospheric ozone in conjunction with the deozonation chamber. If the level moves outside of the desired range, the system will alert the culturist by sounding an alarm. Such systems can also be connected to autodialers that will telephone the culturist to indicate that a problem exists during periods when the facility is not being physically monitored.

Adjusting Water Temperature. Temperature control in closed water systems can be achieved to a degree by placing such systems in a well-insulated building that can provide proper ventilation during hot weather and protection from the cold during winter. Supplemental heating or cooling of the water may be required depending on the species being cultured and on ambient water temperature. Adjusting the temperature of the water is very expensive. Chillers and various types of water-heating devices are available, but the energy costs associated with most of them are prohibitively high. Solar heating has been advocated by some, though when needed the most, the method may not be very efficient due to extensive periods of cloud cover. The initial investment for solar systems of the capacity required for an aquaculture system may also be too high to be practical.

Heat can be removed from water with heat pumps, as long as the water temperature is above the freezing point. Depending on the needs of the culturist, the water that is cooled may be used in the culture system or discarded. The heat removed from the water by a heat pump can be dissipated into the air or used to warm water that will be used in the culture system. Thus, by employing heat pumps, the culturist can produce quantities of both chilled and heated water and use one or both in conjunction with the aquaculture operation. The application of heat pumps to aquaculture has been described (25,26) earlier.

ACKNOWLEDGMENTS

This entry is adapted from Stickney (1). Permission has been obtained for its use.

BIBLIOGRAPHY

1. R.R. Stickney, *Principles of Aquaculture*, John Wiley & Sons, New York, 1994.
2. G.L. Lucchetti and G.A. Gray, *Prog. Fish-Cult.* **50**, 1–6 (1988).
3. K.R. Scott and L. Allard, *Prog. Fish-Cult.* **46**, 254–261 (1984).
4. C.M. Horsch and C.M., *Salmonid* **7**, 12–15 (1984).
5. S.E. Hoergensen, *Vatten* **41**, 110–114 (1985).
6. M.H. Paller and W.M. Lewis, *Prog. Fish-Cult.* **50**, 141–147 (1988).
7. R. Speece, *Trans. Amer. Fish. Soc.* **102**, 323–334 (1973).
8. W.J. Hess, in L.J. Allen and E.C. Kinney, eds., *Proceedings of the Bio-Engineering Symposium for Fish Culture, Fish Culture Section Publication No. 1*, American Fisheries Society, Washington, DC, 1981, pp. 63–70.
9. M.H. Paller and W.M. Lewis, *Aquaculture Eng.* **1**, 139–151 (1982).
10. J. van Rign and G. Rivera, *Aquaculture Eng.* **9**, 217–234 (1990).
11. R.G. Wetzel, *Limnology*, W.B. Saunders, Philadelphia, 1975.
12. F.W. Wheaton, *Aquaculture Engineering*, Wiley-Interscience, New York, 1977.
13. J.E. Colt and D.A. Armstrong, in L.J. Allen and E.C. Kinney, eds., *Proceedings of the Bio-Engineering Symposium for Fish Culture, Fish Culture Section Publication No. 1*, American Fisheries Society, Washington, DC, 1981, pp. 34–47.
14. C.E. Boyd, *Hyacinth Control J.* **7**, 26–27 (1968).
15. C.E. Boyd, *Econ. Botony* **22**, 359–368 (1968).
16. W.J. Slone, D.B. Jester, and P.R. Turner, in L.J. Allen and E.C. Kinney, eds., *Proceedings of the Bio-Engineering Symposium for Fish Culture, Fish Culture Section Publication No. 1*, American Fisheries Society, Washington, DC, 1981, pp. 104–115.
17. T.P. O'Farrell, F.P. Frauson, A.F. Cassel, and D.F. Bishop, *J. Water Pollution Control Fed.* **44**, 1527–1535 (1972).
18. J.G. Reeves, *J. Water Pollution Control Fed.* **44**, 1895–1908 (1972).
19. J.B. Gratzek, J.P. Gilbert, A.L. Lohr, E.B. Shotts, Jr., and J. Brown, *J. Fish Dis.* **6**, 145–153 (1983).
20. J.W. Arthur and D.I. Mount, *Symposium for Fish Culture*, American Fisheries Society, Fish Culture Section, Bethesda, MD, 1975.
21. G.A. Wedemeyer, N.C. Nelson, and W.T. Yasutake, *Ozone Sci. Eng.* **1**, 295–318 (1979).
22. J.M. DeManche, P.L. Donaghay, W.P. Breese, and L.F. Small, *Sea Grant College Program Publication ORESUT-003*, Oregon State University, Corvallis, 1975.
23. M. Sutterlin, C.Y. Couturier, and T. Devereaux, *Prog. Fish-Cult.* **46**, 239–244 (1984).
24. R.F. Layton, R.F., in F.L. Evans, ed., *Ozone in Water and Wastewater Treatment*. Ann Arbor Science Publishers, Ann Arbor, MI, 1972, pp. 15–18.
25. T.J. Fuss, J.T., *Prog. Fish-Cult.* **45**, 121–123 (1983).
26. R.R. Stickney and N.K. Person, *Prog. Fish-Cult.* **47**, 71–73 (1985).

See also ENHANCEMENT; RECIRCULATION SYSTEMS: PROCESS ENGINEERING; TANK AND RACEWAY CULTURE.

RECIRCULATION SYSTEMS: PROCESS ENGINEERING

MICHAEL B. RUST
Northwest Fisheries Science Center
Seattle, Washington

OUTLINE

Recirculation systems can be thought of as a series of unit processes. However, how those unit processes fit and work together can be the difference between an easy-to-manage well-functioning system and failure. Puting unit processes together is the subject of process engineering. Unit processes such as aeration, filtration, sedimentation, ozonation, chlorination, and disinfection are covered elsewhere in this encyclopedia. Fish performance engineering, which is related to this topic as well, is also discussed as its own topic. This entry is intended as a qualitative overview of unit processes and configurations that are used in intensive systems of aquaculture. More detailed information can be found on each unit process, water chemistry, and other aspects of intensive aquaculture elsewhere in this volume. For further information the reader is also referred to classic texts by Spotte (1,2) and Wheaton (3), and recent texts by Spotte (4), Timmons and Losordo (5), and Lawson (6).

FACTORS THAT AFFECT DESIGN

Objectives of Water Treatment

Objectives differ depending on the facility's needs for water treatment. These needs can be broken down into four areas: pretreatment, reuse treatment, recirculation treatment, and posttreatment.

Pretreatment encompasses all the unit processes needed to treat water that is being brought into a facility. It can include a wide range of unit processes and is dictated by the water source being tapped. For more information on pretreatments that might be appropriate for differing water sources, see the entry on "Water sources".

Reuse and recirculation are similar. Reuse treatment occurs between culture units in facilities where the total water through the culture units is equal to the total new water introduced to the system and the total effluent. In other words, a reuse facility is similar to a flow-through facility, except that water is treated after the first culture unit(s) in a series, so it can be reused by different culture units downstream. In a reuse facility water is not returned to the point where new water is added. It is just treated between culture units. An example would be treatments between raceways arranged in series, where water is constantly supplied to the first raceway from a water source and the water from the last raceway in a series is discharged from the facility. Treatment in reuse systems occurs between rearing units and within rearing units.

Recirculation occurs when water discharging from the culture units is treated and then returned to the same culture units. In recirculating systems, water must be pumped. In fact, new water makes up only a certain percentage of the total flow to the culture units, as most of the water is used over and over. Treatment in recirculation systems can occur between rearing units, within rearing units, and at a centralized location.

Posttreatment relates to effluent discharges and is often dictated by law. Effluents may need to be screened to prevent fish from escaping into the environment or may require disinfection to keep pathogens from getting introduced into the receiving water. Guidelines on upper levels of temperature, salinity, dissolved nutrients, and solids may also dictate treatment to protect the receiving waters. For more information about this topic, see the entry on "Effluents: dissolved compounds".

Treatments needed for recirculation are often more comprehensive than for the other three operational objectives. The discussion from here on relates to all four objectives, but is focused specifically on recirculation systems. Typical water quality variables that are commonly monitored and controlled in aquaculture systems and the implications of some key processes are given in Table 1.

Choice of Organism(s)

Perhaps no other choice influences the function and selection of unit treatment processes more than the organism(s) to be cultured. The environmental requirements of the species under cultivation will determine how acceptable water quality is defined. Salinity may be very important to a stenohaline organism, but not for a euryhaline organism. Minimum acceptable oxygen levels for trout might be 6 mg/L (ppm), while for tilapia it might be 3 mg/L (ppm). Multispecies systems may require different water-quality criteria for each species. An example would be serial culture units, where striped bass are reared in the first culture unit, followed by sturgeon or catfish in the second serial culture unit, and then finally, tilapia are introduced in the third culture unit, with minimal or no water treatment between serial culture units. The environmental requirements for each species can be met as water goes from unit to unit. Acceptable water quality also changes from unit to unit, depending on species or unit process.

Since environmental criteria are usually determined by the culture organism, unit processes need to be chosen and sized to provide those environmental conditions. For example, a biofilter for a tilapia system will experience near optimal temperatures for the nitrifying bacteria, but could be low on dissolved oxygen, while the same biofilter used for salmon at sub optimal temperatures would have adequate dissolved oxygen. The specific surface area of media (area/volume of media used) that is needed to provide nitrification for the same biomass in each system would vary significantly because of the impact of temperature and oxygen on nitrification.

Feed

The feed used for different species will also influence the choice of unit processes. Ammonia comes primarily from the deamination of amino acids, the building blocks of proteins. The more a species uses protein for energy, the more ammonia will be produced. Carp and tilapia are typically fed low-protein feed that is made primarily of plant proteins. Salmon, on the other hand, are fed high-protein feed made from animal by-products. Animal proteins are typically more digestible and have a better amino acid balance than plant proteins. Well-balanced protein will be more efficiently used and will produce less ammonia than poorly balanced protein. Also, plant products often contain higher amounts of indigestible material than animal products, resulting in more solids. Feed choice will influence the design of nitrification and solids removal systems. For more information see the many feed-related topics in this encyclopedia.

Solids

In terms of the effect on other treatment unit processes, solids removal may be the most significant. Solids in the water increase biochemical oxygen demand (BOD), lower dissolved oxygen; clog biofilters; increase heterotrophic bacteria (which can limit nitrifying bacteria), decrease the effectiveness of ultraviolet radiation (UV) and aeration; and increase the demand for ozone or chlorine for oxidation or sterilization. Because of these wide ranging, mostly negative effects, early and efficient solids removal is very important.

Hardness, Alkalinity, and Salinity

Hardness and alkalinity affect pH, nitrification, foam fractionation, and fish stress tolerance. Salinity affects all of those and others. Species requirements and water ionic composition should be considered when sizing and choosing unit treatment processes. Materials such as stainless steel, which might be fine for a heat exchanger in a freshwater system, are not recommended for saltwater. In general, only glass, plastic, or titanium should contact saltwater in a recirculation system.

Water Clarity

Water clarity can directly influence the effectiveness of UV sterilization. Indirectly, high-water turbidity or off-colored water can be indicative of high solids or dissolved organic chemicals. Dissolved organics can lead to slightly

Table 1. Typical Water-Quality Variables that are Commonly Monitored and Controlled in Recirculation Systems and the Effects of Some Common Processes Used in Aquaculture

Process	Dissolved Oxygen	Carbon Dioxide	TAN	Nitrite and Nitrate	pH	Alkalinity and Buffer	Salinity	Solids	BOD	Temperature
Water Transport:										
Mechanical Pumps								Can break up solids making, them harder to remove		Can add heat to water
Airlifts	Increases	Decreases			Can reduce CO_2, increasing pH		Can increase evaporation, raising salinity	Same as above, but to a lesser extent	Increased oxygen satisfies some demand	Can shift temperature toward air temperature
Sump/head Tanks							Same as above	May be used to settle solids		Heat gain/loss if not covered
Physical Treatments:										
Chilling/heating	Changes the saturation concentra-tion of the water when temperature is changed	Changes the saturation concentra-tion of the water when temperature is changed	Rate of biological and chemical process are affected by temperature	Rate of biological and chemical process are affected by temperature			Heating increases evaporation, raising salinity		Rate of biological process are affected by temperature	Controlled by heating and chilling
Gas transfer	Directly affected	Directly affected	Oxygen needed for nitrification	Anaerobic conditions needed for denitrification	Can reduce CO_2, increasing pH				Increased oxygen satisfies some demand	Can shift temperature toward air temperature
Mechanical Filtration								Removes solids	Reduces BOD	
Sedimentation								Removes solids	Reduces BOD	
Chemical Treatments:										
Carbon	Can remove oxygen from water when carbon is new				Can lower pH when carbon is new					
Clinoptilolite			Removes ammonia	Lowers by reducing ammonia			Salt may be used to recharge the resin. Will not work in saltwater			

(continued)

Table 1. *Continued*

Process	Dissolved Oxygen	Carbon Dioxide	TAN	Nitrite and Nitrate	pH	Alkalinity and Buffer	Salinity	Solids	BOD	Temperature
Chlorination/ dechlorination										
Ozone	Can increase DO especially if source gas is pure oxygen		Not very effective at oxidation of ammonia except at high pH	Very effective at removal			Can form hypobromous acid, bromate, and bromamines, which can be toxic. Especially a problem with algae and invertebrates	Reduces dissolved solids by improving sedimentation	Reduces BOD	
Ultraviolet light								Does not function well in water with high solids load		
Buffers and salts	Higher salinity can reduce gas saturation concentration	Higher salinity can reduce gas saturation concentration	Affects nitrification	Affects denitrification	Directly affects	Directly affects	Directly affects			
Biological Treatments:										
Nitrification	Consumes oxygen	Produces carbon dioxide	Oxidizes ammonia	Produces nitrite and nitrate	Produces acids, lowering pH	Reduces alkalinity	Efficiency changes with salinity	Is inhibited by high solids loads	Part of the BOD	Efficiency changes with temperature
Denitrification	Anaerobic process		Can increase ammonia	Reduces nitrite and nitrate			Efficiency changes with salinity			Efficiency changes with temperature
Hetrotrophic processes	Consumes oxygen	Produces carbon dioxide	Produces ammonia		Produces acids, lowering pH	Reduces alkalinity	Efficiency changes with salinity		Part of the BOD	Efficiency changes with temperature
Culture Units:										
Algae and Higher Plants	Adds oxygen when growing (with light)	Consumes carbon dioxide when growing (with light)	Consumes ammonia when growing (with light)	Consumes nitrite and nitrate when growing (with light)	Raises pH			Microalgae may be inhibited by suspended solids	Reduces BOD	
Animals	Consume oxygen	Produce carbon dioxide	Produces ammonia	Produces nitrite and nitrate via ammonia and urea	Lowers pH			Produce dissolved and suspended solids	Increases BOD	

increased BOD, increased heterotrophic bacteria, and increased demand for ozone. In some cases, it may be more cost-effective to operate the system with water the color of dark tea, rather than try to maintain crystal-clear water. Tilapia, for example, often grow quite well under these conditions, and as long as there is no off flavor associated with the dark-colored water, there may be no reason to attempt to make the water clear. In this case, UV would be a poor choice as a method of disinfection. In other cases, such as in aquaria, where fish are to be on display, water clarity is critically important.

Heating

Maintaining an acceptable environmental temperature for the culture organism cost-effectively is the primary advantage of recirculation systems. Water cost and availability is often the second most important factor. The system temperature, set by culture species and production schedule, will affect biological and chemical processes. Heating and cooling can have a significant effect on the partial pressure of dissolved gases. When water is heated in an enclosed vessel (such as a pipe), the effluent water may be supersaturated with dissolved gases. This can lead to fish stress and gas bubble disease. This is why it is a good idea to place aeration after enclosed heating processes. On the other hand, it might be advantageous to inject pure oxygen into the coldest water in the system and, under pressure, to maximize transfer efficiency.

Oxygen

Oxygen is needed to maintain the health of organisms and to facilitate chemical oxidative processes. Low oxygen levels can stress aquatic animals and reduce biofilter function. In general, aquaculturists seek to keep oxygen near or above saturation. For more information, see this topic and the topic on fish "Performance engineering" elsewhere in this encyclopedia.

Pumping

There are a great variety of pumps on the market that can deliver water at almost any flow and pressure. Most unit processes, including pumps, have a range of maximum efficiency. When designing a recirculation system, it is important to determine flow and pressure requirements for each culture and treatment unit. Once this is done, a pump that can supply the proper flow and pressure can be chosen. Head-pressure losses through each unit treatment should be added to head-pressure losses from the plumbing and elevation changes in the system to determine the total head pressure to be overcome by the pump. Due to the wide variety of options with pumps, it is relatively easy to find a pump that will supply the required flow and pressure at high efficiency.

THE ORDER OF THINGS

Parallel or Series

Several factors need to be considered in developing a treatment process train, including upstream treatments needed, pressure requirements, flow needed, down stream processes, and the need for a break in head. (Break in head refers to where water in an enclosed vessel under pressure is exposed to atmospheric pressure in an open vessel. At this point, water will flow downhill or must be pumped out. Examples include fish culture tanks, drum filters, packed columns, and head tanks.) The need for a break in head may require repumping or elevation of some components to allow gravity flow through the remaining treatment train. Once this information is determined, it will aid in selection and sizing of unit processes and the location of processes within the process train.

Unit processes can be arranged in either a parallel or series arrangement (Fig. 1). Often, treatment unit processes are arranged in series, while culture units are arranged in parallel. There are exceptions. For example, tanks or raceways can be put in series that use minor treatment (usually aeration) between culture units. In this scenario, water can undergo more treatment following the final culture unit before being returned to the first culture unit in the series (recirculation), or the old water can be discharged as new water is added to the first culture unit (reuse). In general, serial facilities allow for staged processing, while parallel facilities allow modular management, when servicing or responding to loading rates, and thus spread the risk of a unit process failure.

Treatment unit processes are often put together in series to take advantage of the positive effects of upstream processes on downstream processes. For example, early removal of solids improves the efficiency of biofiltration, disinfection, and aeration processes downstream. In highly loaded systems, aeration may be needed both prior to and after biofiltration to provide oxygen for nitrification and for the culture organism. Efficiencies can be gained by staging biological bacterial processes with carbonaceous oxidation by heterotropic bacteria occurring before nitrification.

Conversely, one unit process may negatively affect a downstream process and this should be avoided. For example, turbulent processes such as pumping can break solids into finer particles. This makes removal more difficult and increases rates of decay and associated oxygen consumption and ammonia production. One approach to the early removal of solids is the use of two drains in the culture units. One drain is to carry the majority of

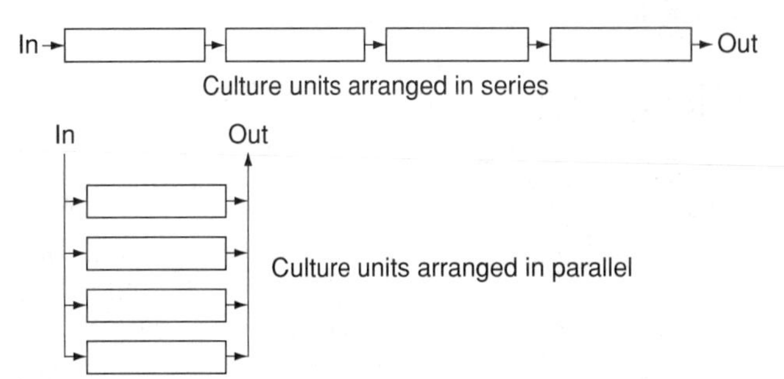

Figure 1. Arrangement of treatments and culture units in series or parallel.

(**a**) Single treatment loop with serial culture units.

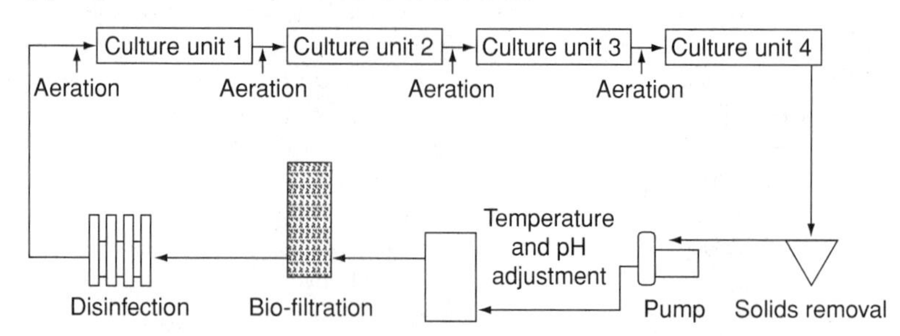

(**b**) Multi-loop/side-loop treatment system

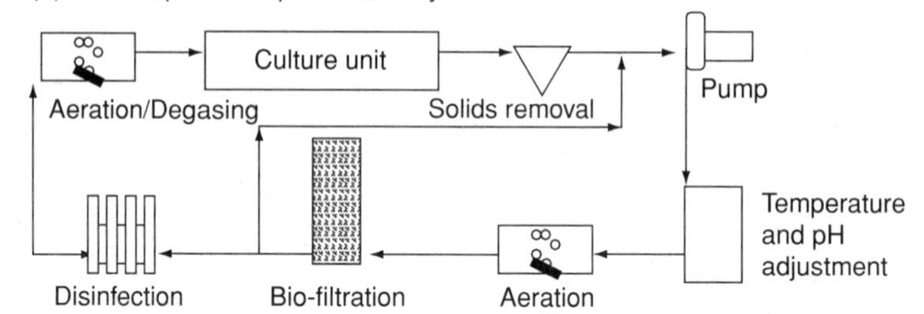

Figure 2. Examples of two different treatment trains.

the flow, and is placed to take water from an area in the culture unit away from were solids settle. The second smaller drain is located where solids collect. The smaller stream of concentrated solids can then be treated more effectively. For more information, see the entry on "Tank and raceway culture".

The ordering of treatment processes into a treatment system (or train) will depend on the acceptable water-quality requirements needed and the set-point conditions desired. Examples of different treatment trains are given in Figure 2.

Complexity

The creation of systems with a high degree of complexity should be avoided if possible. The more unit processes in a system, the more chance there is that one or more of them will malfunction. In complex systems, some redundancy may be desirable for key unit processes. Competing unit processes should also be evaluated based upon complexity and the ability of staff to maintain and manage each process. It may be desirable to choose a less complex process over a more efficient one if the chance for malfunction or maintenance requirements are high. These types of considerations require an understanding of the needs and skills of the end users of a recirculation system and its intended use.

BIBLIOGRAPHY

1. S. Spotte, *Fish and Invertebrate Culture*, John Wiley & Sons, New York, 1979.
2. S. Spotte, *Seawater Aquariums*, John Wiley & Sons, New York, 1979.
3. F. Wheaton, *Aquacultural Engineering*, John Wiley & Sons, New York, 1977.
4. S. Spotte, *Captive Seawater Fishes: Science and Technology*, John Wiley & Sons, New York, 1992.
5. M. Timmons and T. Losordo, (ed.) *Aquaculture Water Reuse Systems: Engineering Design and Management*, Elsevier, Amsterdam, 1994.
6. T. Lawson, *Fundamentals of Aquacultural Engineering*, Chapman & Hall, New York, 1995.

See also RECIRCULATING WATER SYSTEMS.

RED DRUM CULTURE

DELBERT M. GATLIN, III
Texas A&M University
College Station, Texas

OUTLINE

INTRODUCTION

The red drum (*Sciaenops ocellatus*), also known as redfish or channel bass, is a euryhaline sciaenid that has comprised important commercial and recreational fisheries in the Gulf of Mexico and the Atlantic Ocean for many decades. Overfishing in the Gulf of Mexico resulted in the state of Texas prohibiting the sale of native red drum in September 1981 (1). Harvest restrictions imposed by various state and federal regulatory agencies in the early 1980s prompted heightened research efforts in the culture of this species for stock enhancement and food production.

NATURAL HISTORY

The natural range of the red drum has been reported to be from the Gulf of Mexico off the coast of Tuxpan, Mexico, to the Atlantic Ocean off the coast of Massachusetts (2,3). Red drum have a rather complex life cycle that begins when adult fish move from the open ocean to gulf–bay passes to spawn in the fall. Spawning typically occurs from August through January, with peak activity in September or October (1). During courtship before spawning, the male produces a drumming noise, from which the fish's common name is derived. Spawning fish generally produce several hundred thousand eggs that measure less than 1 mm (0.04 in.) in diameter. Surface currents carry the buoyant eggs and larvae from the spawning areas into brackish estuarine nurseries. The eggs hatch approximately 24 hours after being spawned, and the larvae derive nourishment from their yolk sac for at least another 48 to 72 hours. After they begin exogenous feeding, the larvae consume primarily zooplankton such as rotifers and copepods until reaching a size of approximately 50 mm (2 in.) (4). Red drum juveniles of 60 to 100 mm (2.4 to 4 in.) consume small bottom-dwelling invertebrates, for which their subterminal mouth is particularly well suited, as well as small fish. Juveniles, along with subadults, remain within the estuary for approximately three to four years, until they reach sexual maturity. Shrimp, crabs, and fish constitute the major food items for red drum larger than 100 mm (4 in.). As red drum grow, they consume more fish relative to crustaceans (5). Sexually mature adult red drum permanently emigrate from estuaries to areas that range from 16 to 113 km (10 to 70 mi) from shore (1).

CAPTIVE BREEDING

Techniques for controlled spawning of red drum in captivity were developed in the 1970s and have allowed for the production of large quantities of eggs and larvae, with limited effort on the part of the aquaculturist (6,7). A common feature of state hatcheries and commercial facilities consists of circular fiberglass tanks ranging in size from 4 to 6 m (12 to 20 ft) in diameter. The tanks typically contain natural seawater or freshwater amended with synthetic sea salts to achieve a salinity of approximately 32%. The water quality is maintained within acceptable levels by biological and mechanical filtration. The water also may be treated with ozone or ultraviolet light for sterilization.

Sexually mature broodstock for the hatchery are initially captured from the wild, most typically from gulf–bay passes during the fall. These fish may range in weight from 10 to 25 kg (22 to 55 lb). A variety of collection methods can be used, although hook-and-line capture seems to be the least stressful. State regulatory agencies should be contacted to determine regulations governing the collection of broodstock from public waters.

Once potential broodfish have been collected, they are typically stocked in tanks in groups of four to six fish, with equal numbers of males and females. Broodfish are typically fed combinations of shrimp, squid, and fish at the rate of approximately 2.5% of body weight per day, every other day. The fish are conditioned to spawn by subjecting them to seasonal variations in temperature and photoperiod that are commonly compressed into cycles of 120 or 150 days. The water temperature may be controlled directly, with the aid of heating and chilling units, or indirectly, by controlling the temperature of the ambient air. After passing through simulated winter, spring, and summer seasons, the fish typically spawn when the water temperature is about 23–25 °C (73.4–77 °F), with a photoperiod of 11 hours of light and 13 hours of dark. Once the fish begin spawning, the environmental conditions can be adjusted slightly so that spawning is continuous for several months. The fish typically produce over one million eggs per spawning session. The eggs, which are buoyant, are commonly removed from the brood tank with the aid of a skimming device at the water surface and are then concentrated in a bag constructed of 500-μm (0.02-in.) netting. The eggs are subsequently removed to another tank, where they hatch in approximately 24 hours.

REARING OF LARVAE

Red drum larvae obtain endogenous nutrients from their yolk sac for approximately three days after hatching. After that time, an exogenous source of nourishment is required and typically provided in the form of zooplankton. The type of zooplankton provided depends on the larval culture system employed. If larval red drum are to be reared under controlled conditions in tanks, the marine rotifer *Brachionus plicatilis* is a preferred first food (8,9). Stock monocultures of unicellular algae such as *Tetraselmis chuii* and *Isochrysis* sp. are usually used to support the growth of rotifers (10). Algae and rotifers have been successfully propagated separately in continuous culture, as well as together in batch culture. Rotifers also have been grown with yeast and fish oil emulsion, with or without algae (11). Typically, rotifers are provided to larval red drum cultured in tanks at a density of 3 to 5 rotifers/mL (89 to 148 rotifers/oz) to ensure that there is an adequate number of prey (12). This density of rotifers is generally maintained for 7 to 10 days, after which the larval red drum should be large enough to consume newly hatched *Artemia* nauplii, which are provided at a density of 0.2 to 2 organisms/mL (6 to 60 organisms/oz) (13). Procedures involved in hatching *Artemia* cysts and preparing the nauplii for feeding to larval fish are well established (14). Enriching rotifers and *Artemia* nauplii with highly unsaturated fatty acids by culturing them with fish oil emulsions has proven beneficial in the rearing of red drum larvae (15). During the time larvae are fed rotifers and *Artemia* nauplii, inert microparticulate diets also may be provided. However, the commercial larval diets that are currently available have not been able to replace live food organisms for larval red drum cultured intensively.

As the red drum larvae make the transition to juveniles, they may be gradually weaned to an artificial dry diet. After 10 to 14 days of being fed *Artemia*, a high-protein starter crumble for salmonids can be introduced to the fish. Fish will typically convert from the live food to the artificial diet in 3 to 4 days. The feeding of *Artemia* to the fish should continue until all fish are weaned onto the artificial diet. Cannibalism may be a problem in rearing red drum in intensive tank culture and can be minimized by maintaining uniformity in the size distribution of the fish.

An alternative way of providing zooplankton to larval red drum is to stock the red drum into fertilized, brackish-water ponds when the water temperature is approximately 22 °C (72 °F) or higher. This approach is generally less labor intensive and is commonly employed at commercial production facilities and state hatcheries. Various combinations of organic and inorganic fertilizers have been used to promote the growth of phytoplankton and zooplankton. Inorganic nutrients may include nitrogen in the form of urea and phosphorus as phosphoric acid. One of the most common organic fertilizers used in ponds for red drum larvae is cottonseed meal. A variety of fertilization regimes, which are initiated approximately 10 to 21 days before stocking of the red drum larvae, have been successfully used in growing the fish to juvenile size in ponds (16). The preferred zooplankton for larval red drum in brackish-water ponds include rotifers and copepods. Approximately two weeks after the larval red drum are stocked, a high-protein, finely groundfish starter diet may be distributed in the pond, so that the fish will gradually become accustomed to eating the prepared diet. Any uneaten feed also may serve as organic fertilizer.

Under normal conditions, red drum larvae stocked in fertilized ponds will grow to 37.5 mm (1.5 in.) in 30 days. It is desirable for each fingerling pond to be equipped with a concrete harvest basin into which the juvenile fish can be concentrated as the water is drained from the pond. Care must be taken to harvest the fish with soft nylon nets, in order to avoid injuring the fish.

ENVIRONMENTAL REQUIREMENTS

As previously mentioned, red drum are euryhaline fish that can tolerate substantial variation in salinity. Although red drum in nature normally spawn in full-strength seawater ranging from salinity levels of 32 to 40%, levels that allow their eggs to remain buoyant, larvae are generally exposed to much lower salinity levels in their estuarine nurseries (17). Some aquaculture facilities have reported successful rearing of larval red drum to juvenile size in ponds with salinities as low as 4 ppt. The most critical step in this process is the gradual acclimation of larvae to low salinity levels prior to introduction into the pond.

Once red drum reach juvenile size, they can tolerate rapid transitions in salinity. The fish can grow in water with less than 1 ppt salinity if the hardness of the water is greater than 100 mg/L (ppm) as $CaCO_3$ and chloride ions are at a concentration greater than 150 mg/L (ppm) (17). Chloride ions appear to be particularly important, as a level of 500 mg/L (ppm) resulted in better growth, feed conversion, and survival of juvenile red drum compared to when chloride ions were at a level of 150 mg/L (ppm) (18).

Another important water quality characteristic, especially in intensive fish production, is the level of dissolved oxygen. Red drum can tolerate levels of dissolved oxygen below 2 mg/L (ppm), but levels maintained that low for any length of time will serve as a stressor and reduce the growth of the fish (19). A dissolved-oxygen level above 5 mg/L (ppm) is generally recommended for red drum production.

Ammonia and its oxidation products are the major metabolites in aquaculture that may have adverse effects on red drum. Unionized ammonia concentrations as low as 0.3 mg/L (ppm) have been reported to adversely affect survival of three-week-old red drum; however, older fish are generally more tolerant (17). When the water has some salinity, red drum juveniles can tolerate rather high levels of nitrite without showing adverse health effects (20,21).

Another environmental factor that significantly influences aquaculture of red drum is temperature. This species is relatively tolerant of elevated temperatures, as it can survive in estuaries at 33 °C (91.4 °F). As with many other warmwater species, the temperature range for optimum growth of red drum is approximately 24 to 30 °C (75.2 to 86 °F) (17). However, this fish is relatively intolerant of cold temperature and has been reported to die between 8 and 10 °C (46.4 and 50 °F), which makes its overwintering in outdoor ponds, even in the southern portion of the Gulf coast states, a considerable risk. The absolute lower lethal temperature of red drum may be influenced by a number of factors, including the previous thermal history of the fish, as well as water salinity and hardness (22,23).

The red drum's relative intolerance to cold temperatures may constrain production of the species in earthen ponds in the southern United States. A variety of research efforts have addressed this constraint. For instance, advancements have been made in altering cold tolerance of red drum through dietary lipid manipulation. In a series of experiments, red drum fed diets containing saturated lipids had lethal temperatures up to 4.5 °C (8.1 °F) higher than fish fed diets containing menhaden oil, which contains highly unsaturated fatty acids of the n-3 series (24). The increased tolerance to lower water temperature appeared to be mediated by a progressive increase in unsaturation of tissue phospholipids.

In addition to dietary manipulation, advancements also have been made in overwintering red drum in ponds with the development of thermal refuges. These devices, which have included a variety of designs, have been evaluated as to their ability to retain heated water in a portion of the pond such that it attracts the fish when they sense that water in other parts of the pond has become unacceptably cold (25). Commercial facilities along the Texas coast have successfully overwintered red drum in outdoor ponds with the aid of thermal refugia. However, several factors, including the energy costs associated with heating water for the refuge, the efficiency of the refuge design in retaining heat, and the duration and severity of cold weather conditions all may influence the economics and ultimate success of overwintering red drum in outdoor ponds. Culturing these fish indoors in areas where cold winter temperatures may threaten their survival is generally the safest approach but may increase production costs.

CULTURE SYSTEMS FOR FOODFISH PRODUCTION

A variety of culture systems have been used to grow juvenile red drum to a market size of 1–2 kg (2.2–4.4 lb). The systems include earthen ponds, recirculating raceways, cages, and net pens. Red drum can be grown from a weight of 1 g (0.002 lb) to approximately 1 kg (2.2 lb) in 8 to 12 months if environmental conditions, especially temperature, are favorable (26). At the time of this writing, earthen ponds appear to be the most cost-effective production system for red drum, although overwintering the fish in such ponds may be problematic in certain locations, as previously discussed.

Recirculating culture systems of several different designs have been used to produce red drum. The species appears to be relatively tolerant of the high densities and marginal water quality that are often encountered in the systems. However, the costs of operating such systems under intensive conditions may limit their profitability (27). Nonetheless, intensive production of fish in indoor, recirculating systems generally has been considered to offer such advantages over outdoor production systems as greater environmental control, flexibility in facility location, and reduced environmental and resource-use impacts.

The production of red drum in cages and net pens has been practiced commercially in some areas. In most cases, the bodies of water in which these units have been placed could not be drained and harvested by traditional means. In general, red drum are tolerant of such systems as long as critical water quality characteristics remain within acceptable ranges.

NUTRITION AND FEEDING

Nutrition is a critical factor in intensive aquaculture, because of its influence on fish growth and health, as well as on production costs. Development of nutritious and cost-effective diets is dependent upon knowing a species' nutritional requirements and meeting those requirements with balanced feed formulations and appropriate feeding practices. Studies to determine the red drum's requirements for specific nutrients were initiated in the early 1980s (28).

Satisfying the minimum dietary requirement for protein or a balanced mixture of amino acids is of primary concern because satisfying this requirement is necessary to ensure adequate growth and health of the fish. However, providing excessive levels is generally uneconomical, as protein is the most expensive dietary component. Several studies have been conducted with young, rapidly growing red drum to determine their minimum dietary protein requirement for maximum weight gain. The required values have generally ranged from 35 to 45% of the diet (29,30). Most recently, the metabolic protein requirements of red drum for maintenance and maximum growth were established (31). In addition to supplying amino acids for protein synthesis, dietary protein also may be catabolized for energy. Carnivorous fish species, in particular, appear to be very proficient at using dietary protein for energy, due to the efficient way in which ammonia from deaminated protein is excreted via the gills with limited energy expenditure (32). An available energy level of approximately 15 kJ/g diet (1600 kcal/lb diet), or 35 to 45 kJ energy/g protein (3800 to 4800 kcal/lb protein), was determined to be adequate for maximum weight gain and desirable body composition of the red drum (29,30).

Dietary requirements of red drum for some of the indispensable (essential) amino acids have been determined (33–36). Requirements for total sulfur amino acids (methionine plus cystine) and lysine are typically the most critical to quantify because the levels of these amino acids in feedstuffs are usually most limiting relative to the amounts required by fish. The total sulfur amino acid requirement of red drum was determined to be 3.0% of dietary protein (34). This sulfur amino acid requirement appeared to be more limiting than the lysine requirement, which was quantified to be approximately 4.4% of dietary protein (33,35). The threonine requirement of red drum was quantified at 2.3% of dietary protein (36). Established relationships between patterns of indispensable amino acids in muscle tissue and levels required in the diet may allow all amino acid requirements of a species to be estimated after quantifying the requirement for only one or two of the most limiting amino acids (34,37).

Fish do not have a specific dietary requirement for carbohydrates, but the presence of these compounds in diets may provide an inexpensive source of energy. However, the ability of fish to use dietary carbohydrate for energy varies considerably, with most carnivorous species having more limited ability than omnivorous and herbivorous species (32). Although the red drum is a carnivorous fish in nature, it is not adversely affected by high levels of soluble carbohydrate in the diet, although it preferentially uses lipid more effectively than carbohydrate (30,38).

Fibrous carbohydrate, primarily in the form of cellulose, is essentially indigestible by fish and does not make a positive contribution to their nutrition. In fact, the level of crude fiber in fish feeds is typically restricted to less than 7% of the diet, in order to limit the amount of undigested material entering the culture system.

Lipids are important components of fish diets, because they provide a concentrated source of energy that is typically well used. In addition, dietary lipids supply essential fatty acids that cannot be synthesized by the fish.

The red drum, like most other carnivorous species, has been shown to efficiently use dietary lipid for energy (30,38,39). Between 7 and 11% menhaden oil in diets containing 40% crude protein have produced maximum weight gain in red drum. Marine oils containing highly unsaturated fatty acids (HUFAs) of the linolenic acid (n-3) family are needed to satisfy the essential fatty acid requirements of red drum (40). Juvenile red drum were determined to require eicosapentaenoic acid (20 : 5n-3) and docosahexaenoic acid (22 : 6n-3) at approximately 10% of the level of dietary lipid (40,41). The red drum also appears to have a very limited ability to elongate and desaturate shorter-chain fatty acids (40).

It has been established that fish generally require the same minerals as terrestrial animals for tissue formation and other metabolic functions, including osmoregulation (32). However, dissolved minerals in the aquatic environment may contribute to the satisfaction of the metabolic requirements of fish and may interact with dietary requirements.

Supplementation of phosphorus in fish diets is usually most critical, because the presence of phosphorus in the water and its use by fish are limited. The dietary phosphorus requirement of red drum has been determined to be 0.86% of their diet (42). The availability of phosphorus from feedstuffs also may vary considerably (43); thus, supplementing diets on the basis of available phosphorus is important. In addition to phosphorus supplementation, inclusion of 2% sodium chloride and/or 2% potassium chloride to practical diets has been shown to have positive effects on growth of red drum in freshwater and brackish (6 ppt salinity) water, but no positive effects were observed in full-strength artificial seawater (44). The beneficial effects of dietary salt supplementation for red drum in dilute waters appeared to be due to provision of ions that were scarce in these hypotonic environments.

Of the microminerals, selenium and zinc have been demonstrated in some fish species to be most important to supplement in diets, due to their low levels in practical feedstuffs and/or interactions with other dietary components that may reduce bioavailability (32). At this time, only the minimum dietary zinc requirement of red drum has been determined and is 20–25 mg Zn/kg diet (45), although higher levels are generally supplemented. Supplementation of practical diets with other microminerals has not been shown to be necessary in most instances. However, an inexpensive trace-mineral premix is typically added to most nutritionally complete diets for fish to ensure adequacy of the diets (32).

Fifteen vitamins have been established as essential for terrestrial animals as well as for the fish species that have been examined to date. Currently, there is specific information only on the quantitative requirement for choline by red drum (46). However, a typical vitamin premix for other warmwater species, such as channel catfish, is added to nutritionally complete diets to provide adequate levels of each vitamin, independent of levels in ingredients, and thus allow a margin of safety for vitamin losses associated with processing and storage.

Feed costs generally constitute the largest variable cost in intensive fish production; therefore, formulation of cost-effective feeds can significantly influence the profitability of fish production. The accuracy of diet formulation can be improved if information about the digestibility or availability of various nutrients from feed ingredients is known. This information has been obtained for red drum for most of the commonly used feedstuffs (43,47). Feedstuffs of marine origin, including various fish meals, have been most effective in diet formulations for carnivorous fish species, such as the red drum, because they are generally quite palatable and high in protein, lipid, and energy. However, these feedstuffs are usually quite expensive and substantially increase the cost of diet formulations. Other feedstuffs from animal by-products have been used to replace fish meal, with some success; however, their quality can be variable (48). Several protein feedstuffs of plant origin, such as soybean meal, cottonseed meal, and canola meal, are less expensive but have had more limited use in diet formulations because they may be deficient in at least one indispensable amino acid and are usually less palatable to many carnivorous species, including the red drum (49,50). A recent study indicated that relatively high dietary levels of soybean meal could be included without reducing growth or feed intake of red drum if a minimum of 10% of dietary protein was provided by fish meal (51).

A variety of grain by-products from corn, wheat, and rice have been used in fish diets to supply available carbohydrate for energy and to improve pellet stability. Relatively high levels of these feedstuffs are included in red drum diets that are manufactured by extrusion processing. Pellets produced by extrusion processing have high water stability and low density such that they float on the water or slowly sink. This characteristic may assist the aquaculturist in monitoring the feed intake of fish, especially in large culture systems. For this reason, diets prepared by extrusion processing are generally preferred for red drum production. However, use of floating pellets in ponds may increase the potential for interactions between birds and the fish feeding at the water's surface.

Appropriate feeding schedules and practices must be employed in aquaculture to ensure that maximum benefit can be derived from prepared diets. It is generally desirable for fish to consume as much of the diet as they desire on a regular basis. However, excessive feeding should normally be avoided because it not only wastes expensive diet, but also may deteriorate water quality.

Feeding schedules for red drum, like those for other species, are influenced by the size of the fish and water temperature. In general, smaller fish consume more feed, expressed as a percentage of body weight, than do larger fish. Smaller fish also should be fed more frequently than larger fish due to their higher metabolic rates. In addition, water temperature may significantly influence feed intake, with reduced consumption occurring above and below the optimal temperature. Specific feeding schedules for red drum have been empirically derived. The means by which diet is administered to red drum is largely dictated by the design and size of the culture systems.

PARASITES AND DISEASES

Red drum subjected to aquacultural production are susceptible to various disease-causing organisms, including some bacteria, viruses, and protozoan parasites. The most common of these pathogens is *Amyloodinium ocellatus*, a parasitic dinoflagellate (52). This organism attacks the gills, where it causes hyperplasia and fusion of the gill lamellae resulting in reduced oxygen uptake by the fish. Infestations of *Amyloodinium* are typically most serious in larval or juvenile red drum that are being cultured intensively and can result in high mortality rates in less than 24 hours if not properly diagnosed and treated. Chelated copper compounds, such as Cutrine®, have proved quite effective as a treatment at doses of 2 to 6 mg/L (ppm) (52).

Salinity appears to influence the occurrence of *Amyloodinium* infestations with fish generally being less susceptible at a salinity of 6 ppt or less.

One viral disease that has been reported to occur infrequently in red drum is lymphocystis (52). The virus associated with this disease invades the cells of the skin and causes an increase in cell size. This disease typically is not fatal, but may negatively affect the marketability of the fish. As with other viral diseases of fish, no treatment is known for lymphocystis.

Bacteria in the genera *Vibrio* and *Aeromonas* have been isolated from red drum. The red drum is generally resistant to many bacterial pathogens, except when predisposed by stressful conditions, such as crowding, handling, or deteriorated water quality. These diseases are most easily treated by administering an antibiotic to the fish through its feed. Currently in the United States, no antibiotics are registered for use with red drum, although extralabel use of certain antibiotics approved for other aquaculture species, such as Terramyacin® and Romet-30®, has occurred on occasion. Due to the extremely limited treatment options for bacterial diseases, it is much better to prevent the diseases by avoiding conditions that may predispose the fish.

HARVESTING AND PROCESSING

Harvesting methods employed in red drum aquaculture are generally dependent upon the design of the culture system and size of the fish. Larval and juvenile red drum are rather delicate and should be handled with extreme care. Use of soft knotless nylon netting is recommended for these fish. Red drum that are advanced in size are rather resistant to the stress associated with handling, although care should be taken whenever the fish are handled. Fish cultured in earthen ponds are generally harvested by seine once they attain a weight of 1–2 kg (2.2–4.4 lb). Red drum that are confined in more limited areas, such as raceways, have been removed from these systems with fish pumps that were originally designed for salmonids.

At the time of this writing, the market for red drum is limited primarily to states along the Gulf coast of the United States. Most cultured red drum are sold to wholesale seafood markets in the round. Some red drum also have been sold as processed filets in grocery stores. An increase in the demand for red drum and expansion of markets for this fish will likely determine whether commercial production of this fish will continue to expand.

BIBLIOGRAPHY

1. G.C. Matlock, in G.W. Chamberlain, R.J. Miget, and M.G. Haby, eds., *Red Drum Aquaculture*, Texas A&M University Sea Grant College Program, Galveston, TX, 1990, pp. 1–21.
2. S.F. Hildebrand and W.C. Schroeder, *Bull. U.S. Bur. Fish.* **XLIII**, 1–388 (1928).
3. H.B. Bigelow and W.C. Schroeder, *U.S. Fish Wildl. Serv. Fish. Bull.* **53**, 1–577 (1954).
4. R.J. Bass and J.W. Avault, Jr., *Trans. Am. Fish. Soc.* **104**, 35–45 (1975).
5. R.M. Overstreet and R.W. Heard, *Gulf Res. Rep.* **6**, 131–135 (1978).
6. C.R. Arnold, J.D. Williams, A. Johnson, W.H. Bailey, and J.L. Lasswell, *Proc. Annu. Conf. S.E. Assoc. Fish Wildl. Agencies* **31**, 437–440 (1976).
7. C.R. Arnold, in C.R. Arnold, G.J. Holt, and P. Thomas, eds., *Red Drum Aquaculture: Proceedings of a Symposium on the Culture of Red Drum and Other Warm Water Fishes*, Contributions in Marine Science, Supplement to Volume 30, 1988, pp. 65–70.
8. E. Lubzens, *Hydrobiologia* **147**, 245–255 (1987).
9. P. Trotta, *Aquacultural Engin.* **2**, 93–100 (1983).
10. P. Trotta, *Aquaculture* **22**, 383–387 (1981).
11. N.S. Wohlschlag, L. Maotang, and C.R. Arnold, in G.W. Chamberlain, R.J. Miget, and M.G. Haby, eds., *Red Drum Aquaculture*, Texas A&M University Sea Grant College Program, Galveston, TX, 1990, pp. 66–70.
12. J. Holt, R. Godbout, and C.R. Arnold, *Fishery Bull.* **79**, (1981).
13. G.D. Treece and N. Wohlschlag, in G.W. Chamberlain, R.J. Miget, and M.G. Haby, eds., *Red Drum Aquaculture*, Texas A&M University Sea Grant College Program, Galveston, TX, 1990, pp. 71–77.
14. P. Sorgeloos and G. Personne, *Aquaculture* **6**, 303–317 (1975).
15. S.R. Craig, C.R. Arnold, and G.J. Holt, *J. World Aquacult. Soc.* **25**, 424–431 (1994).
16. C.W. Porter and A.F. Maciorowski, *J. World Maricult. Soc.* **15**, 222–232 (1984).
17. W.H. Neill, in C.R. Arnold, G.J. Holt, and P. Thomas, eds., *Red Drum Aquaculture: Proceedings of a Symposium on the Culture of Red Drum and Other Warm Water Fishes*, Contributions in Marine Science, Supplement to Volume 30, 1988, pp. 105–108.
18. M.G. Pursley and W.R. Wolters, *J. World Aquacult. Soc.* **25**, 448–453 (1994).
19. C.J. Stahl, D.M. Gatlin, and W.H. Neill, *World Aquaculture '97 Book of Abstracts*, p. 438.
20. D.J. Wise, C.R. Weirich, and J.T. Tomasso, *J. World Aquacult. Soc.* **20**, 188–192 (1989).
21. J.R. Tomasso, *Rev. Fish. Sci.* **2**, 291–314 (1994).
22. L.S. Procarione, M.S. Thesis, Texas A&M University, College Station, TX, 1986.
23. R.S. Boren, M.S. Thesis, Texas A&M University, College Station, TX, 1995.
24. S.R. Craig, W.H. Neill, and D.M. Gatlin, III, *Fish Physiol. Biochem.* **14**, 49–61 (1995).
25. P.W. Dorsett, M.S. Thesis, Texas A&M University, College Station, TX, 1994.
26. P.A. Sandifer, J.S. Hopkins, A.D. Stokes, and R.D. Smiley, *Aquaculture* **118**, 217–228 (1993).
27. S.G. Thacker and W.L. Griffin, *J. World Aquacult. Soc.* **25**, 86–100 (1994).
28. E.H. Robinson, in C.R. Arnold, G.J. Holt, and P. Thomas, eds., *Red Drum Aquaculture: Proceedings of a Symposium on the Culture of Red Drum and Other Warm Water Fishes*, Contributions in Marine Science, Supplement to Volume 30, 1988, pp. 11–20.
29. W.H. Daniels and E.H. Robinson, *Aquaculture* **53**, 243 (1986).
30. J.A. Serrano, G.R. Nematipour, and D.M. Gatlin, III, *Aquaculture* **101**, 283–291 (1992).
31. B.B. McGoogan and D.M. Gatlin, III, *J. Nutr.* **128**, 123–129 (1998).

32. *Nutrient Requirements of Fish*, National Academy Press, Washington, DC, 1993.
33. P.B. Brown, D.A. Davis, and E.H. Robinson, *J. World Aquacult. Soc.* **19**, 109 (1988).
34. H.Y. Moon and D.M. Gatlin, III, *Aquaculture* **95**, 97–106 (1991).
35. S.R. Craig and D.M. Gatlin, III, *J. World Aquacult. Soc.* **23**, 133–137 (1992).
36. R.S. Boren and D.M. Gatlin, III, *J. World Aquacult. Soc.* **26**, 279–283 (1995).
37. R.P. Wilson and W.E. Poe, *Comp. Biochem. Physiol.* **80B**, 385–388 (1985).
38. S.C. Ellis and R.C. Reigh, *Aquaculture* **97**, 383–394 (1991).
39. C.D. Williams and E.H. Robinson, *Aquaculture* **70**, 107–120 (1988).
40. R.T. Lochmann and D.M. Gatlin, III, *Aquaculture* **114**, 113–130 (1993).
41. R.T. Lochmann and D.M. Gatlin, III, *Fish Physiol. Biochem.* **12**, 221–235 (1993).
42. D.A. Davis and E.H. Robinson, *J. World Aquacult. Soc.* **18**, 129–136 (1987).
43. T.G. Gaylord and D.M. Gatlin, III, *Aquaculture* **139**, 303–314 (1996).
44. D.M. Gatlin, D.S. MacKenzie, S.R. Craig, and W.H. Neill, *Prog. Fish-Cult.* **54**, 220–227 (1992).
45. D.M. Gatlin, III, J.P. O'Connell, and J. Scarpa, *Aquaculture* **92**, 259–265 (1991).
46. S.R. Craig and D.M. Gatlin, III, *J. Nutr.* **126**, 1696–1700 (1996).
47. B.B. McGoogan and R.C. Reigh, *Aquaculture* **141**, 233–244 (1996).
48. H.Y.L. Moon and D.M. Gatlin, III, *Aquaculture* **120**, 327–340 (1994).
49. R.C. Reigh and S.C. Ellis, *Aquaculture* **104**, 279–292 (1992).
50. D.A. Davis, D. Jirsa, and C.R. Arnold, *J. World Aquacult. Soc.* **26**, 48–58 (1995).
51. B.B. McGoogan and D.M. Gatlin, III, *J. World Aquacult. Soc.* **28**, 374–385 (1998).
52. S.K. Johnson, in C.R. Arnold, G.J. Holt, and P. Thomas, eds., *Red Drum Aquaculture: Proceedings of a Symposium on the Culture of Red Drum and Other Warm Water Fishes*, Contributions in Marine Science, Supplement to Volume 30, 1988, pp. 113–130.

See also LARVAL FEEDING—FISH; ZOOPLANKTON CULTURE.

RED SEA BREAM CULTURE

ROBERT R. STICKNEY
Texas Sea Grant College Program
Bryan, Texas

OUTLINE

The red sea bream (*Pagrus major*) is a member of the porgy family (*Sparidae*). It is a marine fish that is valued for its pink flesh, which can be enhanced in aquaculture by feeding the fish carotenoids, including astaxanthin derived from crayfish (1). It is known as the king of fishes in Japan where it is consumed at weddings, graduations, and other ceremonies (2,3). Red sea bream are widely distributed in the Far East and are actively cultured in Japan where production is second only to yellowtail, (*Seriola quinqueradiata*), Hong Kong, Taiwan, and Korea. Studies examining the potential of red sea bream as a culture candidate were being conducted in Japan as early as the late nineteenth century (3). Captive rearing is typically conducted in net pens. Major areas of production include the Nagasaki, Mie, and Kumamoto Prefectures (3). The fish has been the subject of enhancement stocking in Japan since 1962. Kagoshima Bay in southern Japan has been a major site for releases of cultured red sea bream into the environment (2).

During 1995, about 30 million juveniles were produced in Japan. Some 22 million were used in enhancement stocking programs, with the remainder being reared on commercial farms (4).

REPRODUCTION AND LARVAL CULTURE

Fingerlings used for stocking net pens or for release in enhancement efforts can be readily produced in hatcheries. Red sea bream are multiple spawners that reproduce during the fall and winter in Taiwan (5) and from late February until early June in southern Japan (4). One six-year-old captive female in Taiwan was observed to spawn 18 times over a one-month period (late February to late March). Average numbers of eggs spawned during each episode was 898,000. Fertilization rate ranged from over 98% to 100%, while hatching rate ranged from 66.5% to 96.0% (5). Hormones can be employed to induce gonadal development during the off-season, thereby making it possible to undertake year-round reproductive activities in the hatchery (6). Egg quality can be improved by feeding broodstock feed supplemented with cuttlefish meal or raw krill for about one month prior to the onset of spawning (7).

Natural spawning in southern Japan occurs when the water temperature is between 15° and 23 °C (59° and 73 °F) (4). At a temperature of 20 °C (68 °F), hatching requires from 60 to 64 hours (8). At first feeding (about three days of age), red sea bream will accept such small items as oyster larvae and rotifers (*Brachionus* spp.). They later consume brine shrimp (*Artemia* sp.) nauplii. Microbound prepared diets have been employed as a substitute for live feeds (9,10). The dietary value of brine shrimp nauplii to red sea bream can be improved through enrichment with certain types of lipids (11,12), as has been shown for various other marine fish larvae. Larvae transform to the juvenile stage at 26 to 33 days of age (13) and can be weaned to prepared feeds or minced fish. The fish are about 10 mm (0.4 in.) long when the transition from brine shrimp begins (2).

Broodfish are maintained in spawning tanks at male:female ratios of 1 : 1 at the Kagoshima Prefecture

Mariculture Center (2), a production facility for red sea bream used to stock Kagoshima Bay. The adults are fed fresh fish, squid, and small shrimp. Spawning at the facility is from January to June with the peak coming between March 15 and April 15. The fish spawn naturally, and eggs are collected from the tanks with fine-meshed nets. The Center attempts to produce 3.5 million fingerlings annually—2.5 million for enhancement stocking, and the remainder for net pen growout.

NUTRITION AND FEEDING

For commercial rearing, both dry and wet prepared feeds can be used (Fig. 1). Wet feeds are often prepared at the growout site by grinding trash fish and, sometimes, mixing in dry ingredients such as plant meals and vitamins. Extrusion can increase the nutritional quality of some ingredients by gelatinizing starch (14), but wet diets have been shown to be very effective. Adding various types of algae, including *Spirulina* spp. (15), *Ascophyllum* spp. (15–17), *Porphyra yezoensis* (17), and *Ulva* spp. (17–19) to prepared diets has resulted in improved fingerling performance.

A significant amount of research has been conducted on the nutritional requirements of red sea bream. For example, the requirements for various amino acids have been established (20–22). Diets containing 52% crude protein result in better growth and food conversion efficiency than diets containing 42% protein (23). The requirements of red sea bream for eicosapentaenoic acid (EPA) and docosahexaenoic acid (DHA) have been evaluated (24,25). For juveniles, EPA and DHA levels should be 1% and 0.5% of the diet. Excessive levels of n-3 highly unsaturated fatty acids can lead to reduced growth (26).

Adding carnitine to the diet does not seem to promote growth, but it has been shown to both spare lysine (27) and promote the utilization of long-chain fatty acids (28). The dietary requirements of the vitamins choline, pantothenic acid, and vitamin C have all been established (29).

Figure 1. Freshly ground wet feed being loaded into boat for use in feeding fish in net pens in Japan.

PRODUCTION

Between the time young red sea bream leave the hatchery and are stocked, they are maintained in land-based tanks (Fig. 2). Net pens used for the culture of red sea bream (Fig. 3) are the same as those used for yellowtail (*S. quinqueradiata*). Fish as small as 30 mm (1.2 in.) can be stocked in net pens (with the appropriate sized mesh). Fish of that size are approximately two-months old.

Red sea bream are susceptible to various diseases, though high incidences of mortality do not appear to be very common under normal culture conditions. One interesting phenomenon is the occurrence of black lines

Figure 2. Tanks used to rear young red sea bream to the size used for stocking net pens.

Figure 3. Net pen used for rearing red sea bream, yellowtail, and other fishes in Japan.

that can occur in the muscle of both wild and cultured red sea bream. Providing shading from the sun for at least 17 days has been shown to be 90% effective in reducing the black line syndrome (30). Providing shade in net pens, produces a fish with high consumer appeal.

Enhancement stocking, also called red sea bream ranching in Japan, was developed as a result of declining catches that began in 1967 (2). Results of ranching nationally in Japan are not clear, though success has been documented in the Kagoshima area of southern Japan (Kyushu Island) where net pen culture of red sea bream is also occurring.

Prior to release, juvenile red sea bream, initially 30 mm (1.2 in.) long, are stocked in net pens where they are fed minced fish and prepared feeds for two months until they reach 70 mm (2.8 in.) in length. At that time they are released into Kagoshima Bay.

Released fish enter the capture fishery after about a year, though if they avoid capture they will survive for up to 11 years. Maximum growth occurs during the first three years (2), when the fish can reach some 37 cm (14.6 in.) and weigh 0.77 kg (1.7 lb). Maximum recaptures occur during the third year after release.

BIBLIOGRAPHY

1. J.W. Avault, Jr., *Fundamentals of Aquaculture*, AVA Publishing, Baton Rouge, LA, 1996.
2. J.R. Ungson, Y. Matsuda, and H. Hirata, *World Aquaculture* **26**(1), 6–12 (1995).
3. E.E. Brown, *World Fish Farming: Cultivation and Economics*, AVI Publishing, Westport, Connecticut, 1983.
4. A. Mihelakakis and T. Yoshimatsu, *Aquaculture International* **6**, 171–177 (1998).
5. K.-J. LIN, J.-Y. Twu, and C.-H. Chen, *J. Taiwan Fish. Res.* **1**, 35–42 (1993).
6. M. Matsuyama, H. Takeuchi, M. Kashiwagi, and K. Hirose, *Fish. Sci.* **61**, 372–477 (1995).
7. T. Watanabe, T. Koizumi, H. Suzuki, S. Satoh, T. Takeuchi, N. Yoshida, T. Kitada, and Y. Tsukashimas, *Bull. Japanese Soc. Sci. Fish.* **51**, 1511–1521 (1985).
8. Y. Zhang, C. Chen, Y. Xie, and J. Lin, *J. Xiamen Fish. College* **16**, 16–27 (1994).
9. M. Hayashi, *Bamidgeh* **47**, 119–128 (1995).
10. A. Kanazawa, S. Koshio, and S.-I. Teshima, *J. World Aquacult. Soc.* **20**, 31–37 (1989).
11. T. Takeuchi, M. Toyota, and T. Watanabe, *Bull. Japanese Soc. Sci. Fish.* **58**, 283–289 (1992).
12. O. Fukuhara, *Aquaculture* **95**, 117–124 (1991).
13. M. Hayashi, K. Toda, Yoneji, O. Sato, and S. Kitaoka, *Bull. Japanese Soc. Sci. Fish.* **59**, 1051–1058 (1993).
14. K.-S. Jeong, T. Takeuchi, and T. Watanabe, *Bull. Japanese Soc. Sci. Fish.* **57**, 1543–1549 (1991).
15. M.G. Mustafa, T.-A. Takeda, T. Umino, S. Wakamatsu, and H. Nakagawa, *Applied Biol. Sci.* **33**, 125–132 (1994).
16. H. Nakagawa, T. Umino, and Y. Tasaka, *Aquaculture* **151**, 275–281 (1997).
17. M.G. Mustafa, S. Wakamatsu, T.-A. Takeda, T. Umino, and H. Nakagawa, *Fish. Sci.* **61**, 25–28 (1995).
18. H. Nakagawa and S. Kasahara, *Bull. Japanese Soc. Sci. Fish.* **52**, 1887–1893 (1986).
19. K. Satoh, H. Nakagawa, and S. Kasahara, *Bull. Japanese Soc. Sci. Fish.* **53**, 1115–1120 (1987).
20. J. Lopez-Alvarado and A. Kanazawa, *Fish. Sci.* **60**, 435–439 (1994).
21. X. Zhao, N. Bi, and H. Liu, *J. Dalian Fish. College* **10**, 13–18 (1995).
22. I. Forster and H.Y. Ogata, *Aquaculture* **161**, 131–142 (1998).
23. T. Takeuchi, Y. Shima, and T. Watanabe, *Bull. Japanese Soc. Sci. Fish.* **57**, 293–299 (1991).
24. H. Furuita, T. Takeuchi, M. Toyota, and T. Watanabe, *Fish. Sci.* **62**, 246–251 (1996).
25. T. Takeuchi, M. Toyota, S. Satoh, and T. Watanabe, *Bull. Japanese Soc. Sci. Fish.* **56**, 1263–1269 (1990).
26. T. Takeuchi, Y. Shima, and T. Watanabe, *Bull. Japanese Soc. Sci. Fish.* **58**, 509–514 (1992).
27. S. Chatzifotis, T. Takeuchi, and T. Seikai, *Aquaculture* **147**, 235–248 (1996).
28. S. Chatzifotis, T. Takeuchi, and T. Seikai, *Fish. Sci.* **61**, 1004–1008 (1995).
29. T. Yano, M. Nakao, M. Furuichi, and Y. Yone, *Bull. Japanese Soc. Sci. Fish.* **54**, 141–144 (1988).
30. S. Matsui, T. Tanabe, M. Furuichi, T. Yoshimatsu, and C. Kitayama, *Bull. Japanese Soc. of Sci. Fish.* **58**, 1459–1464 (1992).

See also LARVAL FEEDING—FISH; NET PEN CULTURE; ZOOPLANKTON CULTURE.

REGULATION AND PERMITTING

M. RICHARD DEVOE
South Carolina Sea Grant Consortium
Charleston, South Carolina

OUTLINE

INTRODUCTION

Institutional, legal, and regulatory constraints have been identified as critical reasons for the slow growth of aquaculture in many regions (1–8). The National Research Council (9) concluded that consistent growth in new business starts in aquaculture tends to be primarily inhibited by issues political and administrative rather than scientific and technological. Indeed, the Marine Board of the National Academy of Sciences conducted a review in 1992 of the issues constraining aquaculture development in the United States and found that those constraints remain (10).

Aquaculture represents a fairly new use of coastal and inland resources and must compete with existing, established uses. It becomes a "chicken-and-egg" situation: the development of an institutional and legal system for aquaculture will evolve as significant strides in the industry are made. However, the industry may not be able to get off the ground until a favorable regulatory structure is in place. If governments are decidedly interested in promoting the development and economic success of aquaculture, they must consider the formulation of comprehensive and clear policies and regulatory programs as early as possible. Such regulatory programs can succeed in balancing the needs of the aquaculturist with those of other users of public resources and may actually enhance aquaculture development.

The development of comprehensive regulatory programs for aquaculture in many parts of the world has not occurred. Some nations have not felt the need to formulate such programs, which they may feel will actually limit growth of the aquaculture industry in their countries. In other countries, existing regulatory policies that do not incorporate the needs of aquaculture may be difficult to change, with the same result. However, countries without clear policies for aquaculture development may subject the industry to future risks and conflicts.

REQUIREMENTS OF THE AQUACULTURE INDUSTRY

The development and growth of the aquaculture industry is dependent upon the attainment of five basic requirements (11): (*1*) government commitment, (*2*) high water quality locations, (*3*) access to the aquaculture site, (*4*) assertion of exclusive fishing and culturing rights, and (*5*) financial investment.

Commitment of Government to the Development of the Industry

The development of aquaculture can be expected to accelerate only with governmental support. The nature of this support sets the tone for how the industry is regulated. Policies adopted by governments to promote aquaculture: (*1*) formally define what is meant by the term "aquaculture," (*2*) provide supporting policy statements (e.g., "aquaculture can help supplement existing capture fisheries in providing food fish to meet increasing demand"), (*3*) offer a number of incentives to underscore its commitment, and (*4*) define and streamline regulatory and legal requirements. Without endorsement of the government, foreign investors and local citizens alike find growth in the aquaculture industry difficult.

High Water Quality Locations

The availability and maintenance of high water quality environments are critical needs for aquaculture. The most favorable water locations are those that are free of pollution, possess suitable temperature (and salinity, if applicable) regimes, and are in areas environmentally suitable to the culture of aquatic organisms. A culturist must be assured that existing and future uses of the adjacent aquatic environment do not affect water quality conditions in areas where species are being cultured. The operator must also be assured of a dependable supply of high-quality water for direct use. Suitable areas for aquaculture, especially in public waters, are, as a result, limited.

Access to the Aquaculture Site

A critical issue for aquaculturists, concerned citizens, and local planning officials is the availability of adequate and accessible sites. The problem of locating suitable sites is compounded by the need to site an aquaculture project where it is permissible.

In choosing a site, a culturist considers an array of environmental, operational, and logistical factors. A reliable supply of electricity and access to transportation networks are two examples. More importantly, aquaculture usually requires both an aquatic environment and an adjacent on-land base of operation. A culturist may have to either obtain permission, rent, lease, or purchase the waters and/or upland sites to assure access.

Assertion of Exclusive Fishing and Culturing Rights

In many countries, common law provides their citizenry with rights to use public waters for navigation, recreation, fishing, and other activities. However, the nature of existing and emerging aquaculture technologies may require exclusive use of public water areas. Multiple use conflicts resulting from such allocations are possible.

The requirement of exclusive or semiexclusive use of submerged bottoms and/or superjacent waters is addressed in some aquaculture regulatory programs. Such use can be conveyed to culturists through implementation of a lease program, for example. Provisions of leases (and other property conveyance mechanisms) usually indicate the level of exclusivity provided.

The same must be said for the species being cultured; that is, ownership of organisms in the possession of the culturist remains exclusively with the culturist in some regulatory regimes.

Financial Investment

Many aquaculture ventures require significant financial investment. However, most investors see the aquaculture industry as a risky business for many reasons, including the ambiguity or absence of laws to protect culture operations from theft, vandalism, takeover, and other threats. Some governments have established a comprehensive regulatory program, accompanied by industry promotion through the provision of low-cost aquaculture loans and other forms of financial assistance, to address these issues. Such commitment from government enhances its ability to attract domestic and foreign investment.

REGULATORY CONSIDERATIONS FOR AQUACULTURE

Programs established to regulate the aquaculture industry reflect its complexity and multiple requirements. The diverse nature of the industry; conflicts with other, more traditional uses of inland, coastal and ocean waters; environmental concerns; and the existing legal and policy climate are key factors that shape the regulatory and permitting programs for aquaculture.

Nature of the Industry

A number of finfish, shellfish, and crustacean species are cultivated in the United States, including catfish, trout, salmon, striped and hybrid bass, tilapia, hard clams, oysters, mussels, crawfish, and penaeid shrimp. The industry is technologically diverse, with ponds, raceways, silos, circular pools, closed (water reuse) systems, cages and net pens, sea ranches, rafts, and long lines used according to the species cultured (12). Aquaculture remains a relatively young scientific discipline that is developing rapidly, with incorporation of a variety of modern technologies, most not yet fully adapted for widespread use (13). Indeed, there has been a trend toward intensification in both traditional and contemporary culture systems.

Aquaculture practices range from extensive, with few inputs and modest output, to intensive, with high inputs and output. On an annual yield per hectare of water basis, increased intensification requires greater resource use, ranging from simple pond culture to intensive tank and closed system aquaculture (14). These varying technologies are what make aquaculture the diverse industry it is, but they have wide-ranging resource needs, produce differing environmental impacts, and require a suite of technological and management responses.

Use Conflicts

While not yet a major problem for culturists with privately owned farm ponds, use conflicts represent one of the primary issues aquaculturists must face, and are likely to become more pronounced and frequent in the future (15). DeVoe et al. (16) found through a survey of the marine aquaculture industry and state regulatory agencies that the competing use of the coastal zone by recreational users, commercial fishermen, and developers was frequently encountered. The escalating costs of acquiring access to coastal land and waters in the country exacerbated the problem.

In 1992, the National Research Council of the National Academy of Sciences predicted that, due to increasing pressures along the coastal zone, the best opportunities for future commercial aquaculture development are in recirculating (closed) systems on land and in confinement systems in the open ocean (10).

Aquaculture and the Environment

Much has been published since 1980 on the environmental impacts of aquaculture (e.g., 17–22; also see Estuaries, Vol 18 : 1A, 1995). However, ecological concerns had been raised by a number of authors in the 1970s (23,24). One of the major challenges to the aquaculture industry around the world will be how it responds to these environmental sustainability issues (15).

Aquaculture practices can generate environmental impacts as a function of (*1*) the applied technique, (*2*) site location, (*3*) size of the production, and (*4*) capacity of the receiving body of water (17). These can include impacts on water quality, the benthic layer, the native gene pool, and the ecosystem as a whole, and impacts from nonnative species, disease, and chemicals.

The state of knowledge regarding the environmental impacts of aquaculture is rapidly improving. Whereas two decades ago very few research data were available, there has been a surge in the number and scope of research and monitoring programs seeking to document these effects. Much work worldwide has focused on the effects of netpen culture on the environment, with the International Council for the Exploration of the Sea (ICES) leading the way. In the United States, early research efforts dealt with fish hatchery effluents and catfish ponds. As the domestic industry diversified, so did environmental research, with major federal studies examining the impacts of marine shrimp pond culture and salmon netpen culture, and the issues regarding species introductions, the use of chemicals in aquaculture, and effluent discharges.

Legal and Regulatory Structures

The current regulatory environment for aquaculture in the United States is a major constraint to its development (9,10,25). No formal federal framework exists to govern the leasing and development of private commercial aquaculture activities in public waters (10).

In a 1981 study commissioned by the Joint Subcommittee on Aquaculture, the Aspen Corporation examined the federal and state regulatory framework for aquaculture (26). As many as 11 federal agencies are directly involved in regulating aquaculture and another 10 are indirectly involved. However, only a limited number of permitting and licensing requirements are directly imposed by federal agencies. More characteristic are federal agency programs that indirectly regulate fish farmers (e.g., restrictions on drug use and federal laws administered by states).

Some 50 federal statutes (with accompanying regulations) were found to have a direct impact on the aquaculture industry, although the actual number of statutes that affect an individual operation vary depending on its size, location, the species being cultured, and other factors. In total, over 120 statutory programs of the federal government were found to significantly affect aquaculture development. Slightly over one-half require direct compliance from the fish farmer.

Seven federal agencies have regulatory programs that directly affect the marine aquaculture industry: the U.S. Army Corps of Engineers, the U.S. Environmental Protection Agency, the U.S. Fish and Wildlife Service, the U.S. Food and Drug Administration, the U.S. Department of Agriculture, the NOAA National Marine Fisheries Service, and the U.S. Coast Guard. Federal oversight of the marine aquaculture industry is fragmented; there is no overall federal framework to address aquaculture development in inland areas, the coastal zone, or in offshore waters. Further, while recent evaluations of marine aquaculture suggest that offshore locations may represent a viable alternative (10), no formal policies have been developed to manage aquaculture development in the U.S. Exclusive Economic Zone. As a result, existing federal policies vary from one agency to another (and may even differ among divisions within the same agency), and the permitting process can be time-consuming, complex, and costly.

The majority of laws and regulations that specifically authorize, permit, or control aquaculture are usually found at the state level. The Aspen Corp. study examined 32 state regulatory programs and discovered that over 1,200 state laws have some significant bearing on aquaculture operations. Policies and regulations were found to affect aquaculture in eight major areas: aquaculture species use, water quality, water use, land use, facility and hatchery management, processing, financial assistance, and occupational safety and health.

Major aquaculture problems that arise from state laws and regulations are caused by the lack of uniformity of laws among the states, the sheer number of permits, licenses, and certifications that must be obtained, and the difficulty in obtaining them (9,10). Each state has its own unique legal, political, and economic climate for aquaculture, and culturists must navigate the regulatory environment differently in each. Only a few states have developed the information management capability to present the applicant with a comprehensive list of all the legal requirements that must be met. State regulatory programs can be and are usually more restrictive than federal guidelines and regulations dictate. The result is that state agencies vary greatly as to what standards they apply to aquaculture (27), and some still apply laws designed for other applications such as those for public fisheries management (9,10).

Federal agencies that establish the ground rules that most state agencies must follow have adopted vague, confusing, and poorly conceived regulations, or none at all (27). This translates into inconsistencies in the development and application of laws and regulations at the state level (28). Few states have a comprehensive regulatory plan that satisfactorily balances economic development and environmental protection. As a result, regulations governing aquaculture are scattered throughout state statutes and do not necessarily fit aquaculture (29). Complicating matters is the fact that existing permit programs do not have provisions for determining the capacity of the coastal and estuarine system for aquaculture, land-based or in situ (28).

The complexity that results from the involvement of many federal, state, and local agencies responsible for all aspects (including advocacy, promotion, conduct, and regulation) of aquaculture leads to an array of planning acts, policies, and regulations (10). Federal laws are applied differently in various geographic regions of the country (9), and the industry remains concerned about the lack of coordination among agencies regulating aquaculture (25). Unfortunately, the federal government has yet to make any significant headway in reducing regulatory constraints (27).

Another limitation to the current regulatory regime for aquaculture in the United States is the lack of long-range and whole-systems planning (28). Aquaculture policy appears to be made by granting permits on a case-by-case basis (30), and the requirements are often determined using regulations and technical standards not originally developed or intended for aquaculture (31). Each permit is considered individually by the issuing agency, usually with no provision for examining cumulative impacts (28).

The problem is not just with the sheer volume of regulations to be complied with, or the difficulty in obtaining the required permits. It also includes the fact that only a few states have themselves developed the information management capability to present the applicant with a comprehensive list of all that will be required of the prospective culturist in starting up an operation. This is not an easy task, however, because each aquaculture operation is different.

THE AQUACULTURE REGULATORY ENVIRONMENT

Policies and regulations affect aquaculture in seven major areas: (*1*) siting an operation; (*2*) environmental quality controls; (*3*) aquaculture species; (*4*) facility/hatchery management; (*5*) processing and sale of aquaculture products; (*6*) commercial and financial assistance; and (*7*) occupational safety and health.

Siting of an Aquaculture Operation

Aquaculture requires the exclusive use of an aquatic environment, or portion thereof, and an adjacent on-land base of operation, which can take many forms. In the instance where water is brought in to fill a farm pond for the culture of trout, for example, the culturist must own the land or receive permission from the landowner to use it. On the other hand, a mariculturist growing mussels or oysters on strings or suspended from rafts must receive some form of right to use that saltwater area. In the former, land use rights are paramount; in the latter, water column use is the critical variable. Thus, before any investment in aquaculture is undertaken, issues of land tenure and water rights must be resolved, especially in public areas where use conflicts may exist. If not, the aquaculturist faces an unstable and risky environment that could undermine the viability of the operation (32).

Three major land use issues face aquaculture:

1. Siting approvals are regulated at the city, county, regional and state levels. In many cases, strict application of the regulations or ordinances may constrain aquaculture development.
2. Aquaculture is a water-dependent industry. Locating culture facilities on property adjacent to growing areas is becoming very difficult due to escalating costs of waterfront property.
3. Very few local (municipal) land use planning and zoning ordinances acknowledge aquaculture as a legitimate land use. Those municipalities that do are not consistent with their classifications. Some classify it as industrial, others as commercial, still others as agricultural.

Land Use and Zoning. Unregulated aquaculture operations have the potential to be damaging to the environment (33). The need for a properly developed site and an operational plan is important in minimizing these impacts. If such a plan is developed and implemented, aquaculture can be compatible with many other uses.

Conte and Manus (33) suggested that a major issue facing the aquaculture industry is the location of adequate sites. Although many potentially attractive sites exist, the question of how culturists obtain access to those sites must be addressed. One important consideration is the "newcomer" status of aquaculture as a legitimate use of coastal and inland water areas. Most land use regulations do not acknowledge the existence of aquaculture; those jurisdictions with oversight usually do not have experience with aquaculture operations (33).

In countries where aquaculture has not been established, it is generally because (*1*) there is no tradition of eating fish and therefore demand is low, (*2*) fish harvested in coastal waters or rivers adequately supply the existing markets, and/or (*3*) there is a strong focus on agriculture and authorities have decided to use public lands (that might be suitable for fish farming) for further development of agriculture (32).

For instance, culturists who have (or plan to have) operations in the Caribbean have discovered that lands are not readily available to the culturist because of the perceived need to keep them productive for agriculture. Indeed, farmers continue to outnumber fishermen in the eastern Caribbean, and it is the farmers who hold the important political offices in the islands (34). Further, some say that "fish farming generally requires a higher level of management than conventional agriculture, in the sense that the technology as yet lies mainly in the realm of art rather than science" (35). This may, in part, be the reason that *Macrobrachium* sp. culture was unsuccessful in the 1970s in Jamaica; many problems were encountered with the leasing of suitable lands from the government (36).

The availability of appropriate land sites for use by culturists is another issue. For instance, land policies in Puerto Rico favor the use of good agricultural lands for agriculture. It has been suggested (34) that mariculture activities on prime agricultural fields may result in salt contamination of those lands, reducing their value for agriculture. However, in many cases, culturists must locate their operations near freshwater or brackish water areas in order to control salinity, areas that are found near river/ocean boundaries. Yet, because of regulations that prohibit building in areas subject to 100-year floods, the potential of these lands may never be realized. Furthermore, the construction of levees or dikes to reduce the likelihood of flooding is prohibited because farm and other lands upstream may be subject to flooding.

Another reason for the difficulty lies with the costs of obtaining land. Where does a small agrarian farmer get US$10,000 needed to purchase 0.4 ha (1 acre) of land in Jamaica? And in Ecuador, land acquisitions were only available in 10-year concessions and were not available for outright purchase. Ordinary businessmen may have to wait from one to three years to obtain permission, unless they expedite the process through "a series of unofficial payments given to members of the various government agencies," where a US$10,000 payment is not unusual for a 100-ha (250-acre) concession (37).

There are a number of programs that address concerns regarding land use for aquaculture. In the United States, for example, a number of state governments have developed policies that recognize aquaculture as a form of water "farming," or agriculture. This provides an expeditious route to get aquaculture recognized as an appropriate use of land and water resources, and has also provided aquaculture with a number of economic and incentive program benefits that agriculture already enjoys. It is instant legitimacy.

Detailed site surveys to determine the extent of available acreage for aquaculture (and other uses) are encouraged; the selection of sites through an early demarcation program has been suggested to preserve optimum sites for aquaculture (38). These efforts should be followed up with more detailed analyses which incorporate and integrate all users (actual and potential) in a land-use system. With a site plan in place, it is easier to identify and locate the types of aquaculture operations most suitable for particular areas.

Obtaining the necessary permissions to use identified lands for aquaculture operations must be achieved. In most cases, lands available are those for which no other primary uses are sought. Those areas are usually owned by the government or a community, and it is not uncommon for a culturist to gain access to them through long-term leases or permanent ownership (3).

Countries interested in promoting aquaculture might use land use planning (zoning) to identify areas available for siting such operations. In the United States, zoning is a state police power, usually delegated to local or municipal authorities (9), used to allocate lands for specified classifications of development. Localities develop "master plans" which provide guidelines for establishing zoning designations. But, because aquaculture represents a relatively new land use, most existing master plans do not recognize it. The adoption of new or the modification of existing zoning regulations can be critical to the development of aquaculture as it competes with pre-existing and intended uses of valuable upland sites.

Use of Public Waters. The success of an aquaculture operation also depends on the ability of the culturist to exert control over the culture area through some means of property right. For operations that require the use of public waters, regulations that balance the needs of aquaculture with those of other uses must be followed. The major use conflicts associated with aquaculture are navigation, fishing, recreation (including tourism and aesthetics), and water quality.

Shipping and Navigation. Conflicts between aquaculture and navigation occur primarily in the territorial seas and major harbors and rivers of coastal states. Under United States law, the public has a right to use the navigable waters of a state for navigation, a right that is protected by the federal government (39); it is also a basic principle of international law. Although the right of navigation is subject to regulation and restriction, aquaculture activities are usually sited away from major navigation channels to avoid conflicts.

Fishing. Citizens of many countries have a common-law right to fish in public waters. Many countries have large recreational and commercial fisheries; the conflict between aquaculture and fishing can become a very exhausting battle for a number of reasons. Fishermen want to protect their prime fishing spots. They are also concerned with having to compete with cultured products at the marketplace. Aquaculture operations proposed for areas where major conflicts with fishing interests might occur are usually discouraged by government regulators. By designating prime aquaculture sites, states and countries alike could avoid conflicts of interest between these two sectors.

Recreation. Recreational fishing, bathing, and boating in navigable waters are traditional rights provided by many governments to their citizens. However, these rights are not absolute; states have the authority to regulate recreational activities. Because of increases in population and in leisure time of visitors and tourists, recreational activities are increasing and placing a greater demand on land and water resources. Inevitably, conflicts between recreation and aquaculture will occur. Governments have to decide which of the two activities will have higher priority, although some forms of recreation, such as swimming and diving, are highly compatible with aquaculture (1).

Water Quality. A viable aquaculture operation depends on the availability of clean water, whether the species being cultured are held under natural conditions or artificially in tanks or ponds. Low-quality waters can negatively impact the cultured species, affecting growth, survival, and quality of marketable product. The location of any aquaculture operation must include a high-quality environment.

Aquaculture Leases. The complexity of acquiring exclusive use of lands and waters depends on the type of waters to be used (i.e., fresh, brackish, or saltwater areas) and the geographic location of the proposed facilities. Agrarian aquaculture conducted in farm ponds on private lands is usually lightly regulated and therefore, requires few, if any, permits, while more intensive operations which involve the use of public lands and/or water resources may require a number of permits and approvals.

Unless aquaculturists have a property interest in the lands they intend to use for their operations, they cannot expect to be protected by the law. For instance, Jamaica lacks any laws that govern the use of seawater or the sea bottom (36). This was also the case in the United States in 1989, where only 12 of 23 coastal states had established regulatory programs for siting "contemporary" aquaculture operations (11).

Lease arrangements are one form of property interest frequently granted by states to confer certain use rights. Conditions of the lease arrangement determine the degree of protection afforded to an aquaculturist as well as the associated costs of such protection (40). Leases specify the types of areas that can be put into exclusive or semiexclusive use. Governments offering submerged bottoms and superjacent waters for lease will be able to accommodate a wide range of aquaculture technologies.

Existing land lease laws might provide governments with the authority to lease submerged lands for aquaculture. For example, submerged lands could be leased by the Department of Lands and Surveys of the Ministry of Agriculture, Fisheries, and Local Government under land lease laws established by the government of the Bahamas (41).

Ideally, an aquaculture lease program seeks to balance the needs of the aquaculturist with the rights of other resource users. In doing so, the leasing authority should recognize that the prospective lessee will want to know the following (42):

1. How large an area can be leased?
2. What type of area can be leased?
3. How are parcels or tracts allocated (e.g., through first application or competitive bid)?
4. How much total acreage may one individual hold and for how long?
5. Is the lease renewable and, if so, how often, and under what conditions can the lessee lose it?
6. What degree of exclusivity is granted (e.g., does the culturist acquire all fishery rights)?
7. What protection does the lessee have from competition from other marine uses (e.g., recreation, navigation, fishing)?
8. What agency or agencies grant and manage these leases?

Governments that have established an aquaculture lease program usually have designated a lead agency responsible for making submerged lands and superjacent waters available for aquaculture development. Leasing programs typically convey the necessary degree of exclusivity to the culturist to minimize risks caused by pollution, vandalism, theft, and other forms of encroachment while protecting the rights of the public. These protections are critical to the long-term economic potential of aquaculture operations, without which financial investment will be difficult to obtain.

Leases usually set minimum and maximum limits on leased acreage: optimal conditions for the culture systems to be used and the overall amount of lands and waters available for lease are considered. Leases of five years or more, with renewal options, appear to offer the culturist enough time to become financially stable; leases of more than ten years may not provide the government, as steward of public lands and waters, with enough flexibility to manage its resources.

Aquaculture leasing programs may also include performance criteria that outline production, use, and resource protection within the lease. Sites not managed for optimum production by culturists may be more valuable to the government and its citizenry in their natural state or for other uses. To ensure adequate use, some leases include provisions, including the execution of bonds, to guarantee maximum productivity.

Finally, leasing programs may include other terms and conditions; for instance, user fees, royalty payments, assignability, and termination of lease agreements might be included in lease agreements.

The establishment of a program to provide land and water column leasing for aquaculture (or any other use) is usually within the context of overall land use planning. The creation of "zones" inland and along the coast where special protection could be offered to aquaculture is one approach some governments have taken. This amounts to a public (or private) decision to permanently protect those areas from incompatible uses and creeping water quality deterioration. For instance, the government of the Philippines forbids the sale of public lands that are suitable for aquaculture ponds and requires that they be leased for culture only (41). Hong Kong offers similar protections for its waters, where the Director of Aquaculture may designate such areas as "fish culture zones" (41). The Bahamas has also designated certain areas of the country to be used to attract aquaculture development.

Conversely, a government could prohibit or exclude aquaculture from particular areas, deciding instead to foster other uses at the expense of a location's potential for aquaculture.

Protection of Navigation and Water Resources. Leasing programs provide the aquaculturist with rights, guarantees, and certain responsibilities. On the other hand, the interests of the government, its citizens, and other users of public land and water resources must be protected as well. In many situations, a government may provide public protection for navigation and the use of water resources through the regulation of aquaculture operations.

The regulatory programs of the federal and state governments in the United States provide examples. At both levels, activities proposed in navigable waters must be conducted so as to prevent their unnecessary alteration or obstruction and to "protect and maintain the quality of... water resources" (40). Navigable waters can include rivers, creeks, lakes, and wetlands.

At the federal level, the Rivers and Harbors Act of 1899 regulates that those who wish to place structures or dredge in navigable waters must first obtain a permit. If the activity involves the discharge of dredge and fill materials into navigable waters, a permit is required under the amended Federal Water Pollution Control Act of 1972. These are described in more detail below.

Environmental Quality Controls

Little has yet been developed in the nature of environmental quality controls that suggests potential constraints upon aquaculture; nevertheless, protections afforded to and from aquaculture operations range from adequate to nil. The perception of the nature of aquaculture is that it is a "clean" industry, that it requires clean water for success. However, when water quality is considered, two points emerge: pollution affects aquaculture, and pollution is a by-product of aquaculture.

Aquaculture requires high-quality water locations and protection from external pollution discharges. It, thus, must rely on proper enforcement of existing pollution laws. Present-day concerns center primarily on non-point source (NPS) pollution. NPS pollution remains a serious problem because it creates a higher risk for investment. Pollution as a by-product of aquaculture has in recent years become a serious concern for the aquaculture industry. The two major pollutants emanating from culture facilities — organics and chemicals — are the subject of extensive research by federal and state entities. Relatively little information exists on the ecological impacts caused by these discharges; therefore, U.S. Environmental Protection Agency regulations, which provide for exemptions from NPDES requirements under certain conditions, are interpreted and enforced in varying degrees by the states.

Issues regarding water-quality impacts on aquaculture have already been covered. It bears repeating that aquaculture requires high water quality locations and protection from pollutant discharges. This is reflected, for example, in Puerto Rico's 1987 Aquaculture Task Force recommendations that call for the establishment of "protective measures for areas particularly promising for aquaculture to prevent environmental and water resource degradation which could hamper the development of aquaculture..." (43). Culturists must rely on proper enforcement of water quality laws. In Hong Kong, for example, any authorized person may arrest anyone polluting waters in the fish culture zone (41). However, these laws do not cover all sources of potential impact; pollution from non-point sources is an example. Adjacent uses, such as pesticide spraying on farmlands near culture sites, may not be compatible. Integrated land and water use policies and zoning is encouraged (38). Unless water-quality impacts can be anticipated and ameliorated, higher risks for investment and success will be created.

Water-Quality Impacts. Aquaculture practices can generate environmental impacts as well, as a function of (*1*) the applied technique, (*2*) site location, (*3*) size of the production, and (*4*) capacity of the receiving body of water (17). Water exchange has traditionally been required in all forms of aquaculture to prevent self-pollution from organic wastes and resulting oxygen depletion. Of most concern are the high concentrations of nutrients, nitrogenous wastes, and biochemical oxygen demand (BOD) that can be produced (especially from high-density pond or tank culture systems) and discharged in effluent waters (10). However, information on the effect of marine fish farms on water quality in and around culture facilities has been insufficient to allow a detailed evaluation (19). Released phosphorus is responsible for the greatest effects in inland waters; nitrogen is usually of most importance in coastal areas. In addition, low-quality feeds can present a special problem, readily releasing their nutrients into the water and their fiber into the effluent.

Benthic Layer Impacts. All forms of aquaculture produce organic-rich particulate wastes. Oysters grown in rafts can produce tons of fecal and pseudofecal material (19). The impacts of uneaten food and feces falling on benthic communities beneath salmon cage operations is a worldwide issue (10). Finfish operations not only generate fecal waste, but also feeds not ingested by the culture species add to the particulate load in those operations where feed is provided. Rosenthal et al. (19) noted the following physical and chemical changes in the substrate: (*1*) increased organic carbon, (*2*) increased sediment oxygen consumption rates, (*3*) decreased sediment redox potentials, (*4*) generation of hydrogen sulfide and methane, (*5*) increased organic and inorganic nitrogen content, (*6*) increased phosphorus (*7*) increased silicon, and (*8*) increased sodium, copper, and zinc.

Water-Quality Protections. In the United States, the U.S. Environmental Protection Agency regulates discharges from aquaculture facilities ("concentrated aquatic animal production facilities") that are considered point sources. According to the regulations, a hatchery, fish farm, or other aquaculture facility is subject to a permit if it contains, grows, or holds aquatic animals in either of the following categories:

- Coldwater fish species and other coldwater animals in ponds, raceways, or other similar structures which discharge at least 30 days per year or produce more than 9,090 kg (20,000 lb) of aquatic animals per year, or are fed more than 2,272 kg (5,000 lb) of food during the calendar month of maximum feeding
- Warmwater fish species and other warmwater animals in pounds, raceways, or other similar structures which discharge at least 30 days per year or produce more than 45,000 kg (100,000 lb) of aquatic animals per year

This permit contains water-quality monitoring requirements for contaminants of concern. Conditions and stipulations are usually added to such permits requiring regular monitoring and site inspections. In addition, states may employ more stringent requirements than those set out in federal regulations.

Whether the regulation of discharges from culture facilities is necessary depends on the nature of the aquaculture operation and the condition and quality of the receiving water body. Extensive, low-technology operations with no supplemental feeding would probably not require such oversight. On the other hand, large, intensive operations using significant amounts of water and supplemental feeding may. Where the line is drawn is not yet that clear; thus, the collection of much more basic information on the types and quality of effluents discharged from culture facilities is necessary. Currently, most decisions are made on a case-by-case basis.

Aquaculture Species

The regulatory framework for each species used in aquaculture is widely divergent, with catfish the least restricted and marine species the most restricted. Regulatory requirements vary according to the choice of species being cultured, the degree to which the species cultured require the use of private versus public lands and waters, and the nature of containment and control measures. Historically, catfish culture has a long "track record" and was established before the environmental movement took hold in the early 1970s. There is also the perception that catfish culture is truly an agricultural activity.

Because aquaculture is a relatively new practice in North America, regulatory frameworks designed specifically for cultivated species have not been developed (38). Instead, many governments attempt to regulate culture species under traditional fishery laws. In many instances, regulations promulgated to manage natural (wildstock) fisheries are applied in aquaculture situations, due to the lack of formalized aquaculture policy frameworks and the need by the industry for agency response. Shellfish regulations offer an excellent example. Most coastal states administer size, season, and harvest regulations to manage natural shellfisheries; in many cases, these regulations have been applied to aquaculture operations.

Species regulations typically cover all aspects of "fish and fishing." Generally, governments place restrictions on methods of harvest, sizes, and seasons for freshwater and marine species, and may set limits through quotas and other restrictions on the amounts that can be taken. Permits, licenses, and certifications may be required for fishing, harvesting and equipment use, possession of, and packaging, selling, and transporting the animals. It is not surprising that fishery regulations are at times inappropriately applied to cultured organisms. However, more serious are the instances where protections and safeguards are not applied at all (40). This latter point is particularly true when exotic species importation and disease transmittal are considered.

The selection by a culturist of a candidate species will depend on a number of factors. Bardach et al. (35) stated that there are five desirable characteristics in a cultured animal:

1. Be responsive to techniques to induce reproduction in a captive environment
2. Produce eggs and larvae that can withstand the methods used by culturists in the hatchery
3. Have feeding requirements that are easy to satisfy
4. Can be grown in highly concentrated densities and maintain good rates of growth
5. Be resistant to pollution and disease

However, there are not many species worldwide that can meet all five criteria. Those species that are highly desirable by culturists must usually be brought in from other countries or jurisdictions. The potential for the introduction of disease, infestation by pests, and competition with indigenous species is thus great. Most countries are aware of these concerns and are looking to maintain strict control over the introduction of exotics.

Importation of Exotic Species. In the United States, much of the production from agriculture is based on introduced species (20). Many times in aquaculture, exotic species exhibit more highly desirable characteristics (e.g., growth rates, hardiness, disease resistance) than native populations. However, introductions require careful thought and screening and must consider the political, environmental, and cultural implications, as well as production values (10). Nonnative species can be introduced via (*1*) translocation beyond their natural range by water traffic, (*2*) deliberate transplantation of organisms into new areas, (*3*) accidental introductions in connection with the transfer of other species, and (*4*) escape of organisms transferred for purposes other than deliberate introduction (13). Concerns with nonnative species introductions include the potential for "genetic pollution," ecosystem disturbances (e.g., competition with native populations), and introduction of disease. See Mann (44) and DeVoe (20) for a discussion of these issues in greater detail.

Disease Impacts. Disease remains a major concern for the culturist and can also become a problem in the surrounding environment. Many states have some form of disease testing or certification program for animals being imported across state lines; however, the established programs are limited primarily to freshwater species (10). Salmon egg and smolt importations are highly regulated and, in some states, a quarantine period exists prior to introduction. For marine species, the necessary expertise to conduct effective inspections is lacking (10). Also, routine shipments of live oysters, clams, and crabs, intended for direct sale to consumers, are seldom ever checked for diseases, parasites, and competitors, nor are most shipments of bait organisms. Diseases within an aquaculture operation also represent a potential problem. Outbreaks can occur with little or no warning and spread rapidly throughout the often highly dense culture population. Water-borne diseases can be transferred out of the production unit via the normal water exchange protocol used by many culturists. Internal pathogens can be transferred with accidental (or intentional) release of organisms into the natural environment.

Genetic Impacts. The contamination of wild stocks through the escape or release of mariculture organisms can be problematic. Potential impacts are believed to be severe (10,45); however, very little documentation exists. These impacts can be grouped into two categories. First, the potential exists for overwhelming the "wild" gene pool with the more restricted gene pool of a hatchery stock through repeated and massive intentional stock enhancement efforts (e.g., salmon and striped bass) (10). Second, there exists the possibility for weakening the "wild" gene pool as a result of interbreeding among native wild stocks and accidentally released nonnative culture species (45).

Aquaculture Species Management. Some government regulations established to manage and control exotic species introductions for aquaculture use the following criteria (46, modified by Sandifer, in 34):

1. The degree of need for the importation of exotics
2. Possible species competition with valuable native stocks
3. Their ability to live and grow, but not reproduce, in the natural environment
4. That "enemies, parasites, and diseases" which could attack native stocks not be brought in along with the animals
5. Their susceptibility to pathogens or parasites in the introduced environment
6. Their ability to "live and reproduce in equilibrium with its culture environment"

Many authors (including Sandifer, in Refs. 34 and 36) strongly suggest that governments should encourage the development of culturable indigenous species first and, if exotics are to be imported, ensure that adequate safeguards are put into place.

Ownership of Cultured Organisms. As Pillay (3) pointed out, the practice of aquaculture requires holding and ownership of the stocks being cultivated. Problems with ownership of the cultured animals exist in private farm pond operations as well as with aquaculture in public waters. Fisheries have become part of the commons; resources that are available to all. Without policies that clearly provide vested rights to the organisms being cultured, the culturist or the government may have difficulty exercising "ownership rights for aquaculture stocks" (3).

Wildsmith (4) provided a detailed review of the property rights culturists have in the cultured animals. It appears from his review that possession of animals has much to do with capture. Further, there is little in common law that protects rights to cultured organisms once they are released from the culturist's possession. With respect to cultured plants, it appears that ownership in the lands (fee simple, lease) gives rise to property rights in the plants. It should be noted that Wildsmith's analysis deals primarily with the legalities of cultured organism ownership, as decided by the courts of the United States and Canada.

The practical question that requires attention is what protections are or will be afforded to the culturist against theft or removal of his stock? In Jamaica, for example, only about 10% of oyster production is suitable for market, in part, because of poaching (47). The development of several types of aquaculture is constrained in the Bahamas because of the difficulty of preventing theft of cultured species (41). Government regulations in some cases include clearly stated provisions over property rights to the organism, alone or incorporated into leasing agreements with the government. Regulatory programs that lack such protections increase the risks in the industry to a point where few will attempt to enter.

Aquaculture Facility and Hatchery Management

There are several cases where the regulation of aquaculture facilities and hatchery management may be necessary to ensure the production of high-quality, disease-free product and maintain healthy environmental quality conditions. A government may require permits or licenses to operate a fish or shellfish hatchery. Some jurisdictions require fish-breeding licenses and permits for acquiring wildstock for spawning, depending on its availability. And the importation of eggs, larvae, or fish may require certification of freedom from disease or parasites, which is sometimes difficult to acquire due to a lack of agency facilities and diagnosticians.

Drugs, Chemicals, Vaccines, and Pesticides. In the United States, any drugs or chemicals used on food fish must be registered and cleared. It is a costly and time-consuming process; however, these safeguards are necessary to protect the health of seafood consumers. Any drugs or chemicals used on food fish must be registered and cleared with the U.S. Food and Drug Administration. The problem is that only a few drugs or chemicals have been approved for use. Since the prospective market is small, few drug manufacturers have pursued registration.

Vaccines are regulated and must be certified separately for each species cultured. The use of vaccines is regulated by the U.S. Department of Agriculture. Few are registered (again, due to time, cost, and the lack of markets). Each vaccine must be separately certified for each species grown in culture.

Pesticides, herbicides, and other chemicals used for predator control are regulated by the U.S. Environmental Protection Agency. Permits may be required in the United States before any chemicals are applied.

Whether the regulation of drugs and chemicals is necessary will depend on the nature of the culture industry in any particular locale. Many low-technology aquaculture operations will not require the use of these substances and therefore, these considerations may not be necessary. However, in large-scale, intensive systems, the use of drugs and/or chemicals may be necessary and will require approvals.

Water Use. Aquaculture may require significant amounts of water for the operation of hatchery, nursery, and growout facilities. Generally, species exhibit improved health, growth, and survival with adequate water exchange. The use of fresh surface water, brackish and marine surface waters, or groundwater might be necessary, depending on the type and location of the operation. If water resources are limited, a government may have to consider allocating water use rights among users.

This could take several forms. All operations might be asked to report water use on a regular basis if single-day or seasonal maximums established by the government are exceeded. In South Carolina, "capacity use areas" are designated for locations where significant groundwater use threatens the underlying aquifer. Activities needing groundwater and proposed in a capacity use area require a permit. Alternatively, a government may establish maximum water use levels and allocate water use in a specified area among existing and proposed users.

Regardless of the approach, significant water use by aquaculture operations may require government monitoring to ensure that all public and private needs remain satisfied.

Water issues that culturists may face include groundwater availability and allocation and interbasin transfers. However, a number of secondary, but significant, issues emerge, including: (a) water rights and riparian ownership law; (b) proscriptions on the use of public waters; (c) competition from other water uses (multiple use conflicts), and (d) federal, state, and local water management programs. Direct contact with state permitting officials should be made to clarify regulatory requirements.

Processing and Sale of Aquaculture Products

One of the advantages of aquaculture is that with control over the organism, the culturist can plan to harvest and sell product at almost any time of the year. The market may also demand seafood products that do not meet minimum sizes or weights as established for the capture fisheries. Some governments have adopted policies that exempt the aquaculture industry from minimum size and weight requirements and closed season laws, which provide marketing advantages to the aquaculture industry.

Other regulations that aquaculturists may face include licensing, operational, and labeling requirements meant to inform and protect the consuming public. These requirements are administered by the U.S. Food and Drug Administration at the federal level and by food, drug or health agencies at the state level. Health officials may require processing facilities to be licensed and certified as clean and safe. Problems regarding fish contamination and depuration of fish products may exist. Procedures for processing seafood products may be established, and site inspections frequently made. Minimum standards may also be set for package labeling. While many of these regulations are not unique to aquaculture, they may actually be useful in product marketing in that product differentiation between domestic and imported, or harvested versus cultured, products may be desired.

Sanitation and human health concerns regarding aquaculture products are usually addressed through regulatory programs. In some jurisdictions (e.g., Jamaica) no laws exist that cover these concerns (36).

It is important that aquaculture products be of high and consistent quality regardless of the species. Uniform standards throughout the industry are desirable. In some cases, industry and government have joined together to set, advertise, and maintain quality standards for cultured products. A federal seafood inspection program using the Hazard Analysis Critical Control Point (HACCP) methodology has been implemented for seafood and aquaculture operations in the United States.

Commercial and Financial Programs

Rules and regulations must also be followed by culturists when obtaining loans and financing. Those of the Internal Revenue Service (IRS) and the Securities and Exchange Commission (SEC), in particular, can have a detrimental effect on small business investment in the United States. There are regulations also in place for investment, financing, taxation, marketing, and insurance.

Labor and Occupational Safety

Labor and safety concerns in the United States are covered under Occupational Safety and Health Administration (OSHA) regulations, primarily. The unique aspects of aquaculture are not acknowledged.

FEDERAL PERMITTING PROCEDURES AND REQUIREMENTS IN THE UNITED STATES

Introduction

Although most of the permits required to operate an aquaculture facility are issued on the state level, several federal agencies are also involved in the permitting process. The major permits issued by federal agencies include the Section 10 and Section 404 permits issued jointly by the U.S. Army Corps of Engineers (ACOE) and the U.S. Environmental Protection Agency (EPA). These permits regulate the placement of structures and the discharge of dredge and fill material in the nation's waterways, respectively. The EPA also issues National Pollution Discharge Elimination System (NPDES) Permits that control the amount of waste and effluents released from aquaculture and other industrial facilities. This permit can be issued by states as long as the relevant state program has been designated with the authority by the EPA.

Several other federal agencies are also involved in regulating the aquaculture industry, including the U.S. Fish and Wildlife Service (FWS) and the U.S. Food and Drug Administration (FDA). These agencies regulate species selection and drug, feed, and pesticide use by aquaculture operations, respectively. The EPA and the ACOE are also required to solicit and consider comments from these and other agencies before issuing the Section 10, Section 404, and NPDES permits. Comments are also solicited from the U.S. Coast Guard and the NOAA National Marine Fisheries Service during the permitting process.

Many of the permits presented below require similar information for consideration by the agencies. Culturists should always begin the permitting process by contacting the permitting officials of each of these agencies to determine beforehand the information required for the permit application. In some instances, the information needed is quite technical in nature which may require the culturist to solicit the aid of a consultant to assemble the information necessary to begin the permitting procedure. Some states have set up aquaculture permitting assistance offices to help culturists identify the permits that would be required to begin an operation. Although these offices do not operate as permit processing centers, they can be extremely helpful in identifying the state and federal laws and regulations that may apply to a particular operation and the permits that may need to be obtained.

Section 10 and Section 404 Permits

Introduction. The U.S. Army Corps of Engineers (ACOE) has been regulating activities in the nation's waterways since 1890. However, the focus of their regulatory activities has shifted from protecting navigation to the consideration of the full public interest for the protection and utilization of water resources. The ACOE has regulatory jurisdiction over the obstruction or alteration of navigable waters under Section 10 of the Rivers and Harbors Act. The ACOE has also been given administrative responsibility for Section 404 of the Clean Water Act by the U.S. Environmental Protection Agency (EPA). Section 404 regulates the discharge of dredge and fill material into the waters of the United States. These permits can be obtained through a joint application issued by the ACOE.

Section 10 of the Rivers and Harbors Act of 1899. The ACOE is the lead federal agency for permits that involve protection and utilization of the water resources of the United States. These activities are covered under Section 10 of the Rivers and Harbors Act of 1899 (33 U.S.C. 403). Activities regulated under this authority include the placement of structures within navigable waters and the obstruction or alteration of navigable waters of the United States. The purpose of this permit is to ensure the free passage of ships and water-based traffic in public waters.

Section 10 of the Rivers and Harbors Act prohibits "the creation of any obstruction... to the navigable capacity of any of the waters of the United States." This law also makes it unlawful "to build or commence the building of any... structures in any... water of the United States... except on plans recommended by the Chief of Engineers and authorized by the Secretary of War." Section 10 also makes it unlawful to "excavate or fill, or in any manner to alter or modify the course, location, condition, or capacity of... any navigable water of the United States, unless the work has been recommended by the Chief of Engineers and authorized by the Secretary of War prior to beginning the same." This Act created the Section 10 regulatory program.

The Section 10 permit issued by the ACOE is required for all prospective culturists whose operations involve locating a structure in navigable waters. All water-based aquaculture operations will require a Section 10 permit. Examples of structures that would require a Section 10 permit include piers, intake and discharge pipes, floating docks, netpens, and any open water growout or depuration facilities (48).

Section 404 of the Clean Water Act. The EPA has statutory authority on all permits that fall under Section 404 of the Clean Water Act (33 U.S.C. 1344); however, administrative responsibility has been given to the ACOE on behalf of the EPA. Any activity that involves the discharge of dredge or fill materials into navigable waters requires a Section 404 permit. Some examples of aquaculture activities that would require a Section 404 permit include the construction or alteration of impoundments, bulkheads, road fills, and the dredging of canals or channels.

Section 404 of the Clean Water Act authorizes the Secretary of the Army, acting through the Chief of Engineers, to "issue permits, after notice and opportunity for public hearings, for the discharge of dredged or fill material into the navigable waters at specified disposal sites." All public notices must be made within 15 days of the date that all application materials are received. The Administrator of the EPA is allowed to prohibit the specification of an area as a disposal site. The administrator is also allowed to deny or restrict the use of an area for specification as a disposal site as long as an explanation for the decision is made public.

Section 404 also allows the ACOE to issue general permits on a state, regional, or nationwide basis for the discharge of dredge or fill material. The activities covered under a general permit must be "similar in nature, will cause only minimal adverse environmental effects when performed separately, and will have only minimal cumulative adverse effects on the environment." General Permits can only be issued for a period of up to five years. If, after opportunity for public hearings, the ACOE finds that a general permit has significant environmental impacts, it may revoke or modify the permit.

Section 404 requires permits for the following activities: (*1*) the discharge of dredged or fill material into navigable waters that brings the waters into a use that it was not previously subject, (*2*) the discharge of dredged or fill material that impairs the flow water in navigable waters, or (*3*) the discharge of dredged or fill material that reduces the reach of navigable water. However, normal farming activities, the maintenance or emergency repair of structures such as dikes, dams, and bridges, the construction of farm or stock ponds and temporary basins, and the construction of farming and logging roads that use best management practices and that require the discharge of dredged or fill material are exempt from the regulations.

The EPA retains two major functions under Section 404. The first is to develop and provide environmental standards to the ACOE for its use when evaluating a permit that involves the discharge of dredge or fill material into waters and wetlands of the United States. The EPA is also authorized to veto or restrict an ACOE permit that allows the discharge of dredge or fill material in a wetland. The EPA reviews all applications to determine if a proposed discharge will have a significant impact on municipal water supplies, shellfish beds, recreational areas, and wildlife and fishing areas (49).

Permitting Process. Obtaining a permit usually takes two to three months; therefore, culturists are encouraged to begin the application process as early as possible. It is important that culturists begin the permitting process with a preapplication interview. During this initial meeting, the ACOE can advise the applicants on what types of permits the project will require. Applicants must be flexible to meet the requests of the ACOE and other federal agencies that must be consulted during the public comment period. Generally, most permits are issued if the permittee is willing to make the necessary design changes.

The Engineer Form 4345, *Application for a Department of Army Permit*, is used to apply for both Section 10 and Section 404 permits. These applications are available from ACOE district regulatory offices. Some states may use a slightly modified form to facilitate joint processing with a state agency, such as a coastal management agency. The information required by these joint applications is similar to the standard ACOE application. Some projects may be previously authorized under nation-wide or region-wide general permits. The permitting process may also be abbreviated for other projects and only require a letter of permission from the ACOE. Most projects, however, require the full permitting procedure, including public notice and consultation with other federal agencies. The ACOE must balance the need and expected benefits of the proposal against the probable impacts of the project when reviewing an application. The ACOE is also required to take into consideration all public comments and other relevant factors in a process known as the public interest review.

The ACOE uses the following general criteria in its evaluation of permit applications:

1. The relative extent of the public and private need for the proposed activity
2. The practicability of using reasonable alternative locations and methods to accomplish the objective of the proposed activity
3. The extent and permanence of the beneficial and/or detrimental effects which the proposed activity is likely to have on the public and private uses to which the area is suited

The ACOE also identifies the possible impacts that the proposed activity may have on several other items when deciding whether to grant or deny a permit during the public interest review. These items include the following: conservation, economics, aesthetics, general environmental concerns, wetlands, cultural values, fish and wildlife values, flood hazards, floodplain values, food and fiber production, navigation, shore erosion and accretion, recreation, water supply and conservation, water quality, energy needs, safety, needs and welfare of the people, and considerations of private ownership.

The ACOE must also evaluate a permit application using the criteria established under Section 404(b)(1) of the Clean Water Act if the proposed activity will involve the discharge of dredged or fill material. These guidelines restrict discharges into aquatic areas where less environmentally damaging alternatives exist. However, final veto authority for Section 404 permits belongs to the EPA. They may override any decision reached by the ACOE regarding the discharge of dredge and fill material.

The ACOE is also required by federal laws and executive orders to solicit comments on proposals from certain federal agencies during the public review period. The following agencies are required to review applications under both Section 10 and Section 404 permits (adapted from 48):

Agency	Law/Regulation
1. U.S. Environmental Protection Agency	Fish and Wildlife Coordination Act,
2. U.S. Fish and Wildlife Service	16 U.S.C. 661 et seq. and Executive
3. National Marine Fisheries Service	Order 11990 (for nos. 1–3)
4. Federal Emergency Management Agency	Executive Order 11988, Floodplain Management
5. U.S. Coast Guard	Rivers and Harbors Act of 1899, Section 10

Evaluation factors for Section 404 permits are based upon sequential criteria, including the following: (*1*) Can the proposed project be avoided or moved elsewhere? (*2*) Have environmental impacts been minimized? (*3*) Can wetland losses be minimized?

During review for a Section 404 permit, the first criterion must be satisfied before the second is considered, and the second satisfied before the third is considered. The landowner must also comply with all other local, state, and federal requirements before a final permit is issued. The permit application must also be distributed to the federal agencies already mentioned as well as any applicable state agencies, for review. It is possible that mitigation measures might be required if an aquaculture facility is constructed in a wetland (30).

Permitting decisions are usually made within 60 days of receipt of the application. However, complexities of a project, incomplete applications, changes to the project, and public hearings, if required, can all increase the amount of time it takes to reach a decision. Fees are only assessed when a permit is issued. The fee for noncommercial activities is $10, for commercial activities $100.

The ACOE has prepared a document that outlines all of the requirements for the permitting process. It is available at all regional and district offices.

National Pollution Discharge Elimination System Permit

Introduction. Aquaculture operations have the potential to discharge significant quantities of dissolved and particulate wastes in large volumes of effluents into surface waters. The effluent stream also represents the most direct route for accidental escape of nonnative species; chemical, drug, and pesticide residues; and disease organisms. As a result, the aquaculture industry has faced heavy scrutiny from federal, state, and local resource officials who are concerned about the impacts of aquaculture on the aquatic environment. Therefore, the industry is subject to the rules and regulations of the National Pollutant Discharge Elimination System (NPDES) program. The NPDES program is designed to (*1*) limit discharge according to federal "technology-based" discharge standards or state water quality standards, (*2*) provide schedules for compliance, and (*3*) require monitoring and reporting for effluents (30).

Federal Policies. Effluent discharges into waters of the United States are regulated by the EPA to maintain and improve potability, aesthetics, and recreational quality of the receiving waters, under provisions of the Clean Water Act (CWA), including the Federal Water Pollution Control Act as amended in 1972 (PL 92–500; U.S. Congress 1972), the Clean Water Act of 1977 (PL 95–217; U.S. Congress 1977), and the Water Quality Act of 1987 (PL 100-4; 50). The 1972 amendments created the NPDES program, which requires that anyone discharging wastewater from a point source to a "water of the United States" apply for a discharge permit. Under NPDES regulations (40 CFR Part 122), which are generally administered at the state level, a "concentrated aquatic animal production facility" is defined as "a hatchery, fish farm or other facility which meets the criteria in Appendix C" (outlined below), or "any such facility which the Director determines is a significant contributor of pollution to the waters of the U.S. based on a non-site inspection of the facility" (50).

Under Appendix C, a hatchery, fish farm, or other facility is a concentrated aquatic animal production facility if it contains, grows, or holds aquatic animals in the following categories:

1. Coldwater fish species or other coldwater aquatic animals (including the Salmonidae family of fish, e.g., trout and salmon) in ponds, raceways, or other similar structures which discharge at least 30 days per year, but does not include (a) facilities that produce less than 9,090 harvest weight kilograms (20,000 lb) of aquatic animals per year, and (b) facilities that feed less than 2,272 kilograms (5,000 lb) of food during the calendar month of maximum feeding.
2. Warmwater fish species or other warmwater aquatic animals (including the Ameiuride, Centrarchidae, and Cyprinidae families of fish, e.g., catfish, sunfish, and minnows, respectively) in ponds, raceways, or similar structures that discharge at least 30 days per year, but does not include (a) closed ponds that discharge only during periods of excess runoff, or (b) facilities that produce less than 45,454 harvest weight kilograms (100,000 lb) of aquatic animals per year.

The CWA defines a pollutant as "dredged spoil, solid waste, incinerator residue, sewage, garbage, sewage sludge, munitions, chemical wastes, biological materials, radioactive materials, heat, wrecked or discarded equipment, rock, sand, cellar dirt, and industrial, municipal, or agricultural wastes discharged into water." The wastes produced by most aquaculture facilities meet these criteria and, therefore, are subject to the conditions of the NPDES permitting program. Likewise, aquaculture facilities located within a "defined managed area" of U.S. waters determined by the EPA to be ineligible for an initial exemption, or a continued exemption and discharge into that area for maintenance of production of harvestable fresh water, estuarine, or marine plants or animals, are also subject to the NPDES permit program.

Permitting Procedures. The process for obtaining the NPDES permit varies from state to state because permitting authority can be delegated by the EPA to state agencies. Generally, prospective culturists are encouraged to contact the regional office of the EPA to determine what applicable state agency is charged with issuing NPDES permits. To determine if an NPDES permit is necessary, the applicant sends a letter of determination to the applicable agency that includes a description of the facility, operation plans, preliminary or conceptual designs, and information on anticipated wastewater discharges. The permitting agency reviews the information and informs the applicant of any required permits for the specific operation.

States given the authority by the EPA to handle NPDES permits oversee the entire permitting process, including the assurance that all federal standards are met. When a state is not delegated to administer the NPDES program, culturists are required to comply with both federal discharge standards and any state discharge water quality standards (51). Currently, all but the following states issue their own permits under the NPDES program: Arkansas, Idaho, Illinois, Maine, Massachusetts, New Hampshire, New Mexico, Oklahoma, and South Dakota (49).

When issuing a NPDES permit, the permitting agency must apply the following criteria: (*1*) federal (EPA) technology-based standards (industry-specific process or end-of-pipe discharge standards or effluent criteria required for any plant in that industry), and (*2*) more stringent state water quality-based standards if the discharge is likely to affect the water quality-based objectives for the receiving waters. Under the CWA, states designate uses for bodies of water (such as swimming, fishing, and drinking), and may establish stricter standards than those of the federal to maintain designated uses and prevent water quality degradation (30).

NPDES permit program regulations promulgated in 1973 revised the classification of fish hatcheries from "critical industry" status to that of an agricultural facility, effectively reducing the need for the EPA to develop maximum technology-based discharge standards for the aquaculture industry. Although proposed regulations were published in the Federal Register in 1974 (40 CFR Part 115), the EPA has not issued effluent guidelines and minimum levels of treatments for aquacultural discharges. However, in 1998, the EPA published a Notice of Proposed Effluent Guidelines Plan (Fed. Reg. 1998) that sought public comments on its intent to develop new and revised effluent guidelines; fish hatcheries and farms are included as possible candidates for such attention. Currently, permit requirements for aquaculture are established on a case-by-case basis, taking into consideration related guidance that has been issued and any specific water-quality standards applicable to the receiving waters (50). Additionally, states are authorized to place additional requirements on these discharges, and, in some cases, effluent monitoring in aquaculture facilities is required even if the production capacity is less than the limits defined in the NPDES protocol (10).

The lack of a properly prepared EPA guidance document for effluent discharges from aquaculture operations has resulted in inconsistencies in regulating such activities. This is not a recent phenomenon; EPA's regional offices and the states administering the NPDES permit program were using different criteria for aquaculture discharge permits during the 1970s. For example, it became extremely difficult in Hawaii to discharge aquaculture effluents due to a lack of knowledge and communication in the following areas: (*1*) characteristics of aquaculture species and technology, (*2*) economic feasibility of conventional wastewater treatment alternatives before discharge, (*3*) environmental impacts, both positive and negative, of nutrient-rich effluent on the nearshore clean environment, (*4*) time and cost involved in completing the permit applications (e.g., consultant costs for reports and environmental assessments), and (*5*) inexperience and uncertainty in granting and administering the permit (52).

Other Regulations

Introduction. Although the Section 10, Section 404, and NPDES permits previously described represent the bulk of federal permits required before a culturist can begin operation, several other agencies are also involved with regulating the aquaculture industry. Federal regulations that affect aquaculture focus on the coastal zone, species selection, pesticide approval, and drug and vaccine limitations. The agencies that are involved in these regulations include the NOAA Office of Ocean and Coastal Resource Management, U.S. Fish and Wildlife Service, U.S. Food and Drug Administration, and the U.S. Environmental Protection Agency.

NOAA Office of Ocean and Coastal Resources Management. If an aquaculture operation is located within the "coastal zone" of the United States, it is subject to the federal Coastal Zone Management Act of 1972 (CZMA; PL104–150). The CZMA is administered by the Office of Ocean and Coastal Resources Management within the National Ocean Service of the National Oceanic and Atmospheric Administration in the U.S. Department of Commerce. Through the CZMA, funds are provided to states and territories to develop and implement coastal management programs (CMPs). A CMP is a comprehensive state plan designed to identify and protect coastal resources and minimize any environmental impacts associated with activities proposed within the coastal zone of the state (51). Many CMPs rely on a variety of regulatory and permitting programs conducted by the state under its statutory authority. An individual or entity proposing a significant activity or development within the coastal zone must obtain the requisite coastal management approvals.

In 1996, the CZMA was amended. Included in the revisions was new authorization for states to use a portion of their CZMA funding for the adoption of procedures and policies to evaluate and facilitate the siting of public and private aquaculture facilities in the coastal zone. This revision enables states to develop, administer, and implement strategic plans for marine aquaculture. However, only a handful of states have taken advantage of this opportunity.

In some states, such as South Carolina, the coastal permitting process has only a minimal impact on the freshwater aquaculture industry. However, for any aquaculture activity that is proposed within the state's eight coastal counties, the culturist will be required to obtain a "Critical Area" permit.

All coastal management regulatory programs are administered at the state or local level. Prospective aquaculturists are encouraged to contact the coastal management program office in the state or territory of interest before proceeding with their plans.

U.S. Fish and Wildlife Service. The U.S. Fish and Wildlife Service (FWS) is primarily responsible for the protection and management of fish, migratory birds, and wildlife and is the lead agency for several laws and permits that affect the aquaculture industry. Several programs that are administered by the FWS applicable to the aquaculture industry include review and comment responsibilities on proposed construction projects and the regulation of fish and wildlife imports and exports (48).

The Lacey Act and the Lacey Act amendments of 1981 (16 U.S.C. 3371) were enacted to protect indigenous species and prevent the trade of endangered or threatened wildlife (30). Its purpose is to restrict the importation of a species that might be injurious to human beings, agriculture, horticulture, forestry, or wildlife resources (51).

Aquaculture operations are also prohibited from raising endangered or threatened species because these species cannot be sold, offered for sale, imported, exported, taken, received, or shipped in interstate commerce. The federal government protects these species through the Endangered Species Act of 1973 (ESA) (16 U.S.C. 703–712). The regulation of imports and exports of fish and wildlife across both international and state boundaries is presented in Title 50 CFR, Parts 10–24. This jurisdiction of the FWS is based upon the above and the following laws: the Marine Mammals Protection Act (16 U.S.C. 1531–1543), the Migratory Bird Treaty Act (16 U.S.C. 3371), and the Injurious Wildlife Act (16 U.S.C. 152).

The FWS is also the lead agency in issuing the Fish and Wildlife Import/Export License. The license is required for any person who imports or exports animals or fish with a value exceeding $25,000 per year for purposes of propagation or sale. The FWS is allowed 60 days to process applications (90 days if endangered species are involved). The license fee is $125 per year and a fee of $25 is assessed for each import or export shipment. A completed "Declaration for Importation or Exportation of Fish and Wildlife" clearance form must also be completed and submitted to the FWS inspector at the port-of-entry for approval. This approval is required to obtain a shipment release from the U.S. Customs Service (48).

U.S. Food and Drug Administration. The U.S. Food and Drug Administration (FDA) is responsible for approval and regulation of drugs that can be used in aquaculture operations, based upon the Federal Food, Drug, and Cosmetic Act (FDCA; 21 U.S.C. 301). Drug regulation can include the use of drugs as additives in feeds as well as drugs for the treatment of diseases and parasitic infestations in aquatic animals sold for human consumption (48). The primary goal of the FDCA is to protect the health and safety of the public by preventing deleterious, adulterated, or misbranded articles from entering interstate commerce (51).

The FDA is charged with approving the use of drugs that culturists can use to fight the threat of disease. However, there are currently very few approved drugs available for culturists to use to fight diseases caused by bacteria, viruses, parasites, and fungi. The FDA uses a two-step process to approve drugs for use in aquaculture operations. First, the drug itself must be approved by the FDA based on the research conducted by the manufacturer of the drug. Next, the use and dosage of the drug must be approved for aquaculture applications. Culturists should be careful to follow FDA guidelines when using any type of drugs in their operations. Mishandling of these drugs can result in serious fines for the culturist, their products being declared unfit for human consumption, or their product being confiscated from the market (48).

Occasionally, the FDA grants Investigational New Animal Drug (INAD) exemptions to ease the approval process for "minor use" compounds in major agricultural industries. However, the FDA has recently tightened the requirements for INADs due to a concern over the effects on public health and lack of a drug residue monitoring program for the aquaculture industry (30). Additionally, the FDA has started working closely with state and federal agencies and the aquaculture industry to address drug and chemical use issues in aquaculture.

U.S. Environmental Protection Agency. The U.S. Environmental Protection Agency (EPA) regulates the use of pesticides under the Federal Insecticide, Fungicide, and Rodenticide Act (FIFRA) (7 U.S.C. 136). This legislation requires that all chemicals intended to kill pests must be registered with the EPA. The EPA requires the manufacturer of the chemical to show that the product performs as claimed, that the labeling is appropriate, that there is no unreasonable adverse effect on the environment, and that its use is safe. Although no formal permitting procedures are required by the EPA, culturists must be sure to adhere to the regulations set forth by the EPA when using these chemicals to avoid severe fines and penalties. The use of unregistered chemicals by a culturist could also result in penalties. In fact, some chemicals require application only by a professional who must be registered with the EPA (51).

ACKNOWLEDGMENTS

I offer my thanks to the editor, Bob Stickney, for the opportunity to contribute this article, and to Ross Nelson for providing assistance in collecting background references and information. I also wish to acknowledge the support of the S.C. Sea Grant Consortium and the State of South Carolina, without which the preparation of this article would not have been possible.

BIBLIOGRAPHY

1. J.M. Gates, G.C. Matthiessen, and C.A. Griscom, Marine Technical Report Series No. 18, University of Rhode Island Sea Grant Program, Kingston, RI, pp. 72 (1974).

2. H. Powles, Marine Science Center Manuscript Report No. 29, McGill University, Montreal, Canada, pp. 63 (1975).
3. T.V.R. Pillay, *Planning of Aquaculture Development: An Introductory Guide*, Food and Agriculture Organization of the United Nations. Fishing News Books, Farnham, Surrey, England, pp. 71 (1974).
4. B.H. Wildsmith, *Aquaculture: The Legal Framework*, Emond-Montgomery, Toronto, Canada, pp. 313 (1982).
5. Food and Agriculture Organization of the United Nations, FAO Report RLAC/84/9-PES-1, Trinidad (1984)
6. A.M. Wilson, *Oceanus* **30**(4), 33–41 (1987).
7. M.R. DeVoe, *Bull. Natl. Res. Inst. Aquacult.*, Suppl. **1**, 111–123 (1994).
8. M.R. DeVoe, in *Interactions Between Cultured Species and Naturally Occurring Species in the Environment*, Proceedings of the 24th U.S.–Japan Aquaculture Panel Symposium, Oct. 8–10, 1995, Texas A&M University Sea Grant College Program, 1997.
9. National Research Council (U.S.), *Aquaculture in the United States: Constraints and Opportunities*, National Academy Press, Washington, DC, pp. 1123 (1978).
10. National Research Council (U.S.), *Marine Aquaculture: Opportunities for Growth*, National Academy Press, Washington, DC, pp. 290 (1992).
11. M.R. DeVoe and A.S. Mount, *J. Shellfish Res.* **8**(1), 233–239 (1989).
12. Joint Subcommittee on Aquaculture, *National Aquaculture Development Plan: Volume I*, Washington, DC, pp. 67 (1983).
13. H. Rosenthal, *GeoJournal* **10**(3), 305–324 (1985).
14. J.F. Muir, *Endeavour, New Series* **9**(1), 52–55 (1985).
15. G. Chamberlain and H. Rosenthal, *World Aquacult.* **26**(1), 21–25 (1995).
16. M.R. DeVoe, R.S. Pomeroy, and A.W. Wypyszinski, *World Aquacult.* **23**(2), 24–25 (1992).
17. H. Ackefors and A. Sodergren, *Int. Counc. Explor. Sea. C.M.* 1985/E:40 (1985).
18. D.P. Weston, The environmental effects of floating mariculture in Puget Sound, Report 87-16 to Washington Dept. Fisheries and Ecology, pp. 148 (1986).
19. H. Rosenthal, D. Weston, R. Gower, and E. Black, *Int. Counc. Explor. Sea* 1988/No. 154., pp. 83 (1988).
20. M.R. DeVoe, ed., Proceedings of a Conference and Workshop on *Introductions and Transfers of Marine Species: Achieving a Balance Between Economic Development and Resource Protection*, S.C. Sea Grant Consortium, Charleston, pp. 201 (1992).
21. R. Goldburg and T. Triplett, *Murky Waters: Environmental Effects of Aquaculture in the United States*, Environmental Defense Fund, Washington, DC, pp. 196 (1997).
22. R.L. Naylor, R.J. Goldburg, H. Mooney, M. Beveridge, J. Clay, C. Folke, N. Kautsky, J. Lubchenko, J. Primavera, and M. Williams, *Science* **282**, 883–884 (1998).
23. W.E. Odum, *Environ. Conserv.* **1**(3), 225–230 (1974).
24. H. Ackefors and C.-G. Rosen, *Ambio* **8**(4), 132–143 (1979).
25. Joint Subcommittee on Aquaculture (JSA), *Aquaculture in the United States: Status Opportunities and Recommendations*, Report to the Federal Coordinating Council on Science, Engineering and Technology, pp. 21 (1993).
26. Aspen Corporation, *Aquaculture in the United States: Regulatory Constraints*, Final Report, Contract No. 14-16-009-79-095 to U.S. Fish and Wildlife Service, pp. 51 (1981).
27. H.D. McCoy II, *Aquaculture* (6), 39–46 (1989).
28. P.L. deFur and D.N. Rader, *Estuaries* **18**(1A), 2–9 (1995).
29. P.W. Breaux, *Comparative Study of State Aquaculture Regulation and Recommendations for Louisiana*, LCL 93, Louisiana Sea Grant Legal Program, Baton Rouge, LA, pp. 8 (1992).
30. M.C. Rubino and C.A. Wilson, *Issues in Aquaculture Regulation*, Bluewaters, Bethesda, MD (1993).
31. J.W. Ewart, J. Hankins, and D. Bullock, NRAC Bull. No. 300-1995, Northeastern Regional Aquaculture Center, North Dartmouth, MA, pp. 24 (1995).
32. A. Sfeir-Young, *Fishery Sector Policy Paper*, International Bank for Reconstruction and Development/The World Bank, Washington, DC, pp. 79 (1982).
33. F.S. Conte and A.T. Manus, *Aquaculture and Coastal Zone Planning*, Cooperative Extension Sea Grant Marine Advisory Program, University of California, pp. 21 (1980).
34. M. Goodwin, M. Orbach, P.A. Sandifer, and E. Towle, Fishery sector assessment for the Eastern Caribbean: Antigua/Barbuda, Dominica, Grenada, Montserrat, St. Christopher/Nevis, St. Lucia, St. Vincent & Grenadines, U.S. Agency for International Development, Regional Development Office/Caribbean, pp. 141 (1985).
35. J.E. Bardach, J.H. Ryther, and W.O. McLarney, *Aquaculture Farming and Husbandry of Freshwater and Marine Organisms*, Wiley-Interscience, New York, pp. 868 (1972).
36. C. Rogers, Report to the aquaculture task force of the U.S. Business Committee on Jamaica on the development of an aquaculture industry in Jamaica: Problems and opportunities, Fund for Multinational Management Education, New York, unpublished report (1982).
37. S.K. Meltzoff and E. LiPuma, *Coastal Zone Management J.* **14**(4), 349–380 (1986).
38. Food and Agriculture Organization of the United Nations, *Aquaculture Planning in Latin America*, Report of the regional workshop on aquaculture planning in Latin America, Caracas, Venezuela, November 24–December 10, Publication Number ADCP/REP/76/3, Rome, Italy, pp. 173 (1976).
39. T. Kane, *Aquaculture and the Law*, University of Florida Sea Grant Program, Miami (1970).
40. Joint Legislative Committee on Aquaculture, *Strategic Plan for Aquaculture Development in South Carolina. Volume I: Summary and Recommendations*. Columbia, SC, pp. 27 (1989).
41. C.P. Idyll and B.H. Wildsmith, *Aquaculture Legislation for the Commonwealth of the Bahamas*, A report to the Ministry of Agriculture and Fisheries, FAO, FI:DP/BHA/82/002, pp. 109 (1984).
42. C.C. Hanson and J.M. Collier (with J.P. Craven, G.W. Grimes, and G.M. Sheets), in *Open Sea Mariculture*, J.A. Hanson, ed., Oceanic Foundation, Hawaii, Dowden, Hutchinson and Ross, Stroudsburg, PA, pp. 410 (1974).
43. Aquaculture Task Force, Recommendations by the Aquaculture Task Force to President of the Senate Consumer Affairs Commission, Commonwealth of Puerto Rico, unpublished report (1987).
44. R. Mann, ed., *Exotic Species in Mariculture*, Proceedings from a symposium on exotic species in mariculture, Woods Hole Oceanographic Institution, Woods Hole, MA, MIT Press, Cambridge, MA, pp. 363 (1978).
45. L.J. Lester, in M.R. DeVoe, ed., Proceedings of a Conference on Introductions and Transfers of Marine Species, S.C. Sea Grant Consortium, Charleston, SC, pp. 201 (1992).
46. D. Dean, in R. Mann, ed., *Exotic Species in Mariculture*, MIT Press, Cambridge, MA (1979).
47. D.E. Jory and E.S. Iversen, *Mar. Fish. Rev.* **47**(4), 1–10 (1985).

48. M. Hightower, C. Branton, and G. Treece, *Governmental Permitting and Regulatory Requirements Affecting Texas Coastal Aquaculture Operations*, Texas A&M University Sea Grant College Program, Galveston, TX (1990).
49. Joint Subcommittee on Aquaculture (JSA), *State/Territory Permits and the Regulations Affecting Aquaculture*, U.S. Department of Agriculture (1995).
50. R.K. Bastian, *Water Farming J.* November 28, 1991, 7–10 (1991).
51. E.M. Peel and R.J. Rychlak, *Water Log* **11**(4), 10–13 (1991).
52. J.S. Corbin and L.G.L. Young, Proc. B.C. Conference (1988).

REPRODUCTION, FERTILIZATION, AND SELECTION

William K. Hershberger
National Center for Cool and Cold Water Aquaculture
Lectown, West Virginia

OUTLINE

Text
Bibliography

Reproduction in aquatic species, as in other animals, serves two important basic biological functions: (*1*) perpetuation of the species and (*2*) transmission and recombination of genetic material. From a more utilitarian perspective, the production of aquatic species (i.e., aquaculture) can only be successful as long as there is an adequate supply of 'seed' in the form of fertilized eggs or juvenile animals; this can only be assured through careful management and control of reproduction.

Many cultured aquatic species were initially propagated by collecting seed from the natural environment (1,2), and this is still practiced with some aquacultured species (e.g., mussel culture) (3). However, relying on the natural production of seed has the three major drawbacks. First, natural production of seed does not yield consistent supplies over time. Year to year variation in weather conditions and other environmental parameters can have major effects on the success of natural reproduction and on the survival of juvenile animals. Second, this practice does not allow for the genetic manipulation of stocks to increase production efficiency. As will be discussed later in this section, genetic selection is based on the ability to cross-breed specific animals with desirable traits; there is no assurance that those with the traits needed for improved aquaculture efficiency will mate in the natural environment. Finally, the natural productivity of a species can be negatively impacted when the demands for seed reach high levels. Thus, the ability to manage reproduction is not only a major factor that contributes to the success of aquaculture, it is also crucial to its sustainability.

Aquatic species exhibit one of the most diverse collection of reproductive modalities and strategies in the animal kingdom. However, the species currently raised in aquaculture (about 150) exhibit a fairly limited array of the reproductive strategies found in the natural environment.

Most aquaculture species are dioecious; that is, they have separate and distinct sexes. The genes necessary to define maleness or femaleness are transmitted to each individual at fertilization. In most higher vertebrates these genes are concentrated on morphologically distinctive chromosomes (e.g., "X" or female chromosome and "Y" or male chromosome in mammals), and some species of fish have been shown to contain similar types of sex-determining systems (4). However, few aquatic species exhibit physically discernible chromosomes (5), and the genetic determination of sex has been found to vary from a few genes to more complex combinations of genetic elements (6). The absence of chromosomes that are easily identified in one sex makes cytological determination of sex in these species practically impossible. Furthermore, the apparently diffuse nature of sex-determining genes may contribute to the relative ease with which phenotypic sex can be altered in some species (7) and also may play a role in the hermaphroditism (i.e., both sexes develop in a single animal) exhibited by some aquaculture species (8,9). Although most aquaculture species show inheritance of sex in a straight-forward and predictable manner, the mechanisms by which the genes are expressed as a sexual phenotype are not completely understood.

Depending on the outcome of the sex-determining process, each individual will begin the formation of either ova (oogenesis) or sperm (spermatogenesis) in the gonadal tissues. Despite the wide variation in reproductive strategies and modalities, the process by which gametes are produced is quite similar in a wide array of fish species—in fact, in all vertebrates (10). In both oogenesis and spermatogenesis there is an initial period of genial proliferation in which oogonia and spermatogonia proliferate by mitosis (cell division). This is the initial determinant of the number of gametes that can be produced. Both types of sex cells then go through a defined sequence of genetic and cell structure modifications that ultimately lead to functional oocytes in females and spermatozoa in males. For details in fish, see Refs. 11 and 12; for invertebrates see Ref. 13.

Although the basic processes for the development of mature ova and sperm are very similar, there are two points of divergence that are important to be aware of with respect to reproduction in aquaculture species. First is the timing of meiosis (division in sex cells to reduce chromosome numbers). In the female primary oocytes (oogonia that have become enveloped by somatic cells) begin meiotic division, but arrest during the first meiotic prophase (11). Meiosis is not begun again until the oocytes are ready to be ovulated, or released from the surrounding somatic cells. With invertebrates, meiotic division is not reinitiated until after activation by penetration of sperm. On the other hand, spermatocytes go through meiosis during development without being arrested and then go through final maturation stages. This sequence of events makes artificial manipulation of chromosome numbers possible in aquaculture species (14).

The second difference is the energetic requirements for achieving the final maturation of ova and sperm (15). Since most of the currently cultured species are oviparous (i.e., fertilization and early development occur externally) and the resulting embryo is, essentially, free living, it must be endowed with all the nutrients required for the completion of its development (16). This is accomplished by packaging in the egg all the nutrients (primarily protein and lipid) needed for metabolism and tissue formation (17). These materials are synthesized and placed in the egg during the process of vitellogenesis, which occurs prior to ovulation. On the other hand, sperm does not contain the quantity of materials that ova do, and, consequently, the energetic "cost" for production is lower.

The effect of these requirements is that the management of reproduction with aquaculture species should commence with the husbandry of the adult animals that will produce the gametes to create the next generation. Research with fish has shown that ration size, feeding rates, nutritional status, and stress have an impact on the size of eggs, number of eggs produced per female, and quality of eggs produced (18). For example, rainbow trout females fed a low ration (0.4% of body weight per day) for the first four months of the annual reproductive cycle had lower fecundities and a lower percentage of fish-producing mature gametes than those fed a high ration (1.0% of body weight per day) (19). Husbandry conditions that may affect sperm production and quality have received less attention, but research has shown that sperm quality can be altered by feeding regime, quality of feed, and the rearing temperatures experienced by males (20). Consequently, appropriate culture conditions and management procedures for adult fish used to produce the next generation are essential to the reliable production of seed for growout.

Exogenous (environmental) factors also play a major role in the control of gamete maturation, and release. Most aquaculture animals are seasonal breeders and changes in environmental factors (e.g., water temperature, water flow, or day length) provide the cues for initiation and adjustment of reproduction. The pathway by which this control is exercised in vertebrates is diagrammed in Figure 1. Environmental information received by the sensory system is interpreted internally by the brain. Based on the signals that are received, the hypothalamic region of the brain will secrete substances that activate the pituitary, a small organ at the base of the brain. The pituitary then releases hormones (gonadotropins) that are relayed to the gonadal tissue via the blood stream. This activates the processes of gamete recrudescence, maturation, and release in the gonads, which also produce hormones (steroids) that affect reproductive development. At various points in the pathway, both positive and negative feedback mechanisms can accelerate or halt the development of gonadal tissue (21). Consequently, this pathway is not a unidirectional, temporal flow, but has multiple inputs throughout the entire cycle that are coordinated by the stimuli received by the central nervous system (CNS). While the overall system of gamete

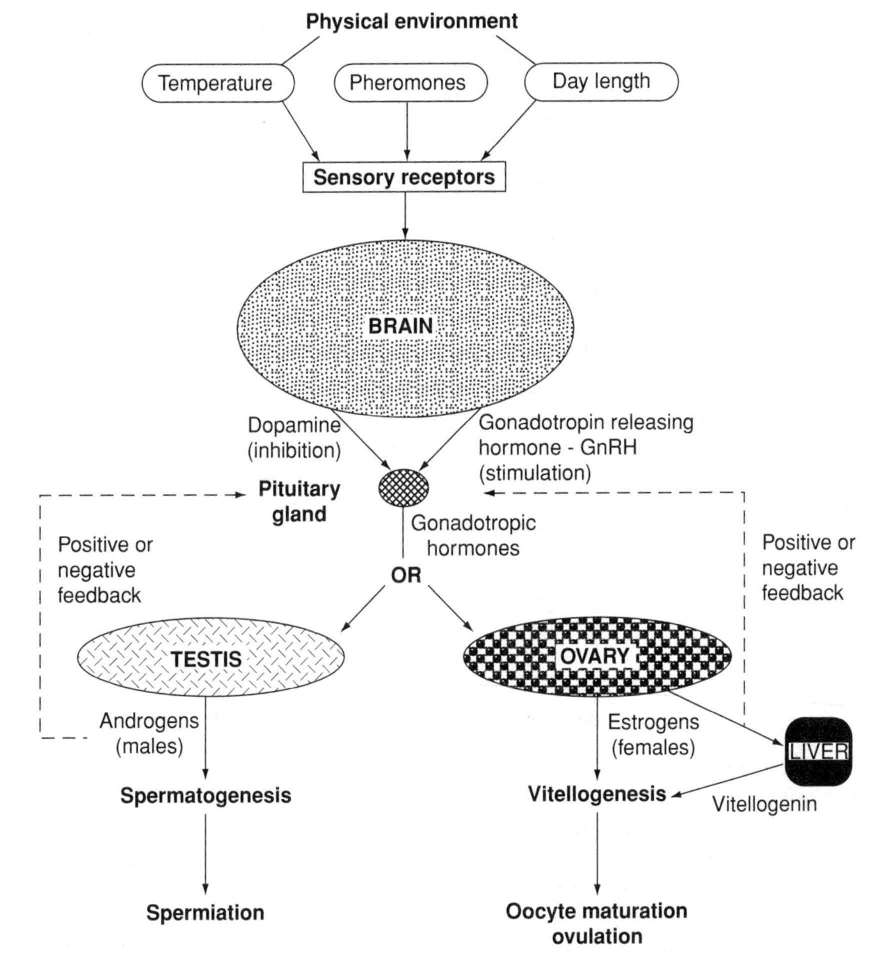

Figure 1. Diagram of the neuroendocrine axis, the pathway by which reproductive development is controlled in many vertebrate aquaculture species. Dashed lines indicate where feedback mechanisms operate to enhance or moderate development of gametes.

production, maturation, and release in invertebrates is conceptually similar to that outlined in Figure 1, there is little known about the specific controlling mechanisms or their methods of action.

There have been a number of procedures developed to utilize these inputs to manage reproduction in aquaculture operations (22). The most extensively employed procedures are manipulation of photoperiod (change in day length) and manipulation of temperature. For example, with rainbow trout, photoperiod alteration can advance or delay reproduction, depending on when during the reproductive cycle it is used and whether length of the day is increased or decreased (23). In fact, photoperiod control can be used to induce the normally annually spawning rainbow trout to mature again after only 6 months (24). Increases in water temperature are routinely used to induce and synchronize reproduction in molluscan shellfish species (25). Pacific oysters, for example, that have been conditioned by intensive feeding at high temperatures (20 °C) for a period of time can be induced to spawn by raising the water temperature from 20 °C to 28–30 °C (26). The use of environmental manipulation to change spawning time is particularly attractive for those species with very high fecundities (i.e., number of eggs per female), since only small numbers of broodstock need to be maintained under these artificial conditions.

In many circumstances, the physical environment is simply not appropriate to elicit the final stages of gamete maturation, release, and deposition. In these situations, intervention with hormone therapy can be used to overcome a missing control element. Furthermore, hormone treatment can be used to alter the natural time sequence of the events leading to final maturation and to synchronize final maturation of gametes. For example, with shrimp, the practice of eyestalk ablation is routinely employed to induce final maturation and can be considered an induced endogenous hormone manipulation (27). Other aquaculture species require exogenous treatment with hormones that are derived from fish and other animals (most commonly mammals) or are produced synthetically.

Initial work on the use of hormones to induce spawning concentrated on the pituitary gland and its hormonal products. In the 1930s, studies in Brazil laid (28) the foundation for this approach to spawing fish, and its practical application has revolutionized fish culture (22). Subsequent investigations have shown that some mammalian gonadotropins, as well as some steroids obtained from gonadal tissue, can be effectively used in fish (29–31). Recently, the synthesis and successful use of superactive analogs of gonadotropic hormones with fish has expanded the potential for use of hormone treatment in aquaculture species (29).

Hormone delivery is accomplished by adding it to the food or by injecting the material into either the muscle (i.m.) or the peritoneal cavity (i.p.). The major goal in these treatments is to introduce adequate amounts of material to increase the circulating hormone levels to a high enough concentration for an ample period of time to precipitate gamete maturation and ovulation or spermiation. If fed as a part of the diet, the hormone will be exposed to the digestive enzymes and hence must be resistant to degradation so that the levels that eventually get to the blood stream are high enough to be effective. Drawbacks to injection include the limited capacity of the muscle to accommodate introduced material and the possible damage to internal organs when using the i.p. route. Also, both of these methods may necessitate repeated handling for multiple injections, which increases stress and possibly affects endogenous endocrine functions associated with gamete maturation or ovulation/spermiation. Recently, techniques for administrating drugs or hormones via implants have been developed, which are as effective, but less stressful on the fish (32–34).

It should be noted that management of reproduction through the use of hormones is almost entirely limited to influencing the terminal stages of gametogenesis. Thus, for the procedures to be effective, the broodstock must have progressed to a point where they are responsive to the treatment. Consequently, the treatment schedule and dose administered to adult fish must be carefully controlled for the treatment to be successful; many of these procedures are detailed in Donaldson and Hunter (29).

In some aquaculture species, hormone treatment can lead to spontaneous spawning (35). However, in others the stage of readiness to release the gametes is not so apparent and, furthermore, the removal of gametes prior to complete maturity can adversely affect quality, especially that of ova (36). Thus, with some species, it is desirable to periodically check the status of gamete development to ensure collection at the optimum stage. For the most part, sperm maturation has not been as problematic as that of eggs, and thus, analyses have emphasized evaluation of the status of maturation of the ovum. In some species, such as trout and salmon, gentle abdominal pressure can be used to expel a few eggs for analysis (37). In others, collecting egg samples may involve catheterization (38) or surgical removal of a sample of eggs (39). There are a variety of indicators for determinating the status of maturation of fish ova (22), but the most definitive assessment is by evaluating the position of the germinal vesicle (GV), or nucleus. When maturation is complete, the GV migrates from the center of the ovum to the periphery and the nuclear membrane disintegrates, a process know as germinal vesicle breakdown (GVBD).

In many aquaculture operations, spawning and fertilization take place under seminatural conditions in the water column, and the fertilized eggs or young animals are collected subsequently (40–42). This minimizes handling and stress on the broodstock and ensures, as much as possible, that the culminating environmental cues before spawning (e.g., courtship and substrate conditions) lead to successful reproduction. However, the approach is not always suitable, and it may be desirable to have more control over the process of fertilization by manually removing the gametes.

Where seminatural spawning is not possible or desirable, the eggs and sperm can frequently be collected from the female and male animals, respectively, by manual stripping (37) or surgical removal (39), and fertilization is conducted outside of the water column in containers. Fish ova are usually stripped into a dry container, avoiding contamination with water until the milt (sperm + seminal fluid) has been added and mixed with the eggs. The change in ionic concentration resulting from the addition of water initiates rapid changes in the gametes (36,43), including activation of the sperm to penetrate the egg and start the process of fertilization.

Fertilization with most aquaculture species involves, essentially, two steps: A sperm enters the egg, either through a small hole in the egg (termed a micropyle) or via enzymatic action at a specific site on the surface of the egg (44). Then, this step activates the egg and initiates the final stages of meiosis to reduce the chromosome number to a haploid (1N) number. Subsequent to the completion of meiosis, the egg and sperm pronuclei join to form a diploid nucleus in the embryo and the process of embryonic development commences (16).

There are numerous factors that can affect the success of artificial fertilization. Premature addition of water can cause rapid changes in the gametes, including closing of the micropyle due to water absorption by ova (45) and premature activation of very short-lived sperm (46). Fertilization with species that have adhesive eggs presents another set of problems with respect to fertilization. These have been addressed by the use of physical coatings with organic or inorganic materials (39,47), treatment with chemicals (48), or the use of enzymes (49) to eliminate adhesiveness prior to, or during, fertilization. The amount of sperm per egg used can also have an impact of fertilization. The concentration of sperm in fish milt can range from 2×10^6 to 5.3×10^{10} cells per ml (50). Under normal circumstances, the minimal sperm: egg ratios for successful fertilization are about $3–6 \times 10^5 : 1$ (49). Problems can obviously arise when there are too few active sperm in situations where males are either not fully matured or have been excessively used (52) or where cyropreserved sperm are employed (53). In fact it has been recommended that the ratio of sperm:egg be increased 5- to 10-fold to ensure adequate fertilization rates with cryopreserved sperm (53). On the other hand, polyspermy (entry of multiple sperm) can be a problem in some species and this can be minimized by diluting the sperm (46). Saline solutions have been developed for this purpose and for extending sperm activity (54,55).

Control of reproduction in aquacultured species has led to two major improvements in management of seed production. First, aquaculturists can obtain specific information about the production potential of their operation through measurement of gamete yield, rates of fertilization and hatching, and survival during early life-history stages. Egg production in aquaculture species is generally high compared with other vertebrates (16,56), but fecundity varies with reproductive strategy. Enumeration of egg numbers can be accomplished by a variety of methods (45). Fertilization rate can be most accurately evaluated after gastrulation, since developmental mortality associated with gamete quality is highest during cleavage stages (16). Subsequent survival can be managed and enumerated in a well-designed incubation system in which appropriate environmental conditions are maintained (45). Use of these data provide the information for allocation of incubation, rearing, and growout facilities, and for accurate prediction of production potential.

The second improvement in management is the opportunity to genetically improve or domesticate desirable stocks. Controlled breeding is central to genetic selection for the improvement of stock, and artificial fertilization allows specific crossing schemes to be defined that will meet the needs of selection. Controlling fertilization makes possible a broad repertoire of methods by which gametes can be combined to meet the preferences, desirable specifications, or particular demands of aquaculture.

Whether in nature or in captivity, selection of animals with desirable traits and their subsequent propagation are integral parts of the "creation" of an organism. In natural situations, selection acts through the influence of the environment on the survival and reproduction of particular sets of traits (phenotypes). Not all phenotypes are equally fit to compete in a particular environment, and those phenotypes defined by genotypes (genetic composition) with low survival potential will be eliminated from the population (57). This process of natural selection is how a population of organisms becomes adapted to the environment in which it must exist. Organisms grown for food production are also subjected to artificial selection, which refers to a set of rules designed by humans to govern the probability that an individual survives and reproduces (58). Judicious selection of breeding stock combined with mating practices that maximize the probability of combining the most beneficial traits in the offspring comprise the process of selective breeding.

The major difficulty with selective breeding as just outlined is that it is based on the phenotypes of organisms, which result from both the expression of the genes in the organisms and the environmental influences on the traits. This can be briefly expressed by

$$\text{Phenotype} = \text{Genotype} + \text{Environment}$$

or

$$P = G + E$$

By way of exemplifying this concept, we can use a trait such as size at harvest measured as either weight (e.g., grams/fish) or length (e.g., cm/fish). The magnitude of such a measurement can be affected by the genes the animals contain (i.e., the genotype) and many other factors such as water temperature, quantity and quality of food, incidence of diseases, stress, water quality, etc. (i.e., environmental factors). Without the appropriate analyses, it is not possible to determine how much of the phenotype expressed in breeding animals is based on genetic differences that will be transmitted to future generations, or on environmental factors that may be rather transient and are not transmissible to the offspring of the breeders. Thus, to ensure that selective breeding is based on the genetic variation in the population and not on transient environmental factors, the relative influence of the genotype on the phenotype must be estimated.

The field of quantitative genetics has developed to provide these types of analyses and to predict the outcome of various selection and breeding approaches. Information on the techniques and methodology required to obtain genetic estimates for traits can be found in a number of publications (57,59,60). One of the most useful values that can be derived from these analyses is the heritability (designated h^2) for a trait. Heritability is broadly defined as the fraction of the total variability that is due to genetic differences; for example, $h^2 = 0.25$ means that 25% of the total observable variability in a particular trait is due to genetic differences.

Heritability can be used to develop a second concept important to selective breeding, estimation of the response of organisms to selection. The relationship between the selection of breeders and the response of their offspring is

$$R = h^2 S$$

where R = response, h^2 = heritability, and S = selection differential. Thus, by calculation of the selection differential (i.e., the difference between the mean value for a trait in the selected breeders and the mean value for the same trait in the source population) and the heritability of a trait, an animal breeder can estimate the gain/loss anticipated for that trait in the next generation. Expectations for the use of these values are diagrammatically shown in Figure 2.

While there are some crucial assumptions made in this diagram, several inferences can be derived about the components of this relationship and about the expectations from conducting a selection program. First, if a trait does not have some level of genetic determination, no response will be exhibited in the offspring. Next, the magnitude of change due to selection is a product of the selection differential applied and the amount of genetic variability. Thus, to some degree, a low amount of genetic variability can be compensated for by an increased selection differential. Finally, this relationship points out one of the values of planned selective breeding, the ability to predict results in future generations.

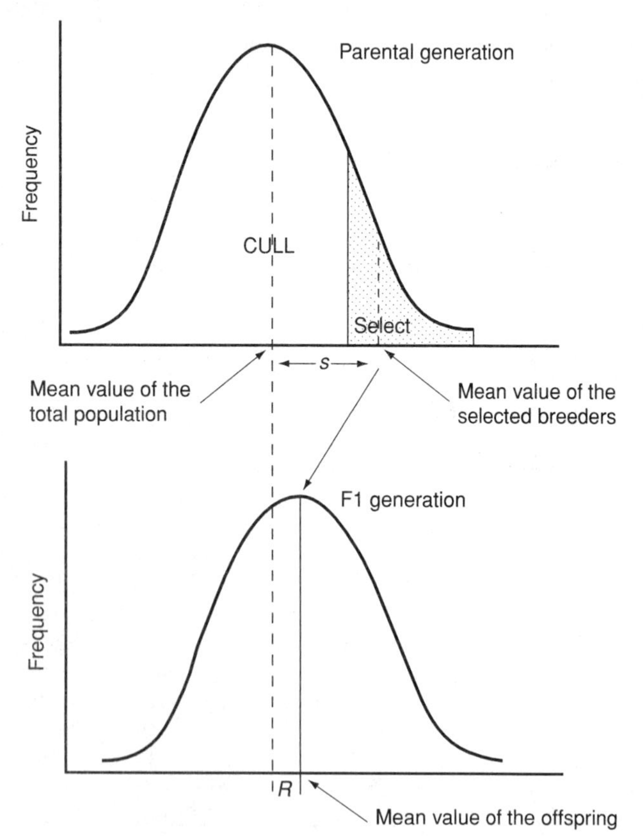

Figure 2. Schematic representation of the relationship between selection differential (S) and selection response (R) in a selection program for increasing a normally distributed trait.

The process of selective breeding for broodstock improvement for aquaculture is, basically, a two-step process. First, a prediction is made about the breeding value of the individuals that are to be used to reproduce the next generation. That is, the "worth" (i.e., breeding value) of an animal is estimated on the basis of a set of criteria that define the animal likely to produce offspring with desirable traits (e.g., rapid growth and good body conformation). Through the use of quantitative genetic analyses, statistical estimates can be utilized to provide more informed assessments of breeding values. After these estimates are obtained, the active, or second part of the process can begin; that is, breeding animals are chosen, crosses are made, and offspring are retained or discarded. There are many combinations of selection and breeding approaches that can be utilized (57), but the maximum response will be obtained only by giving strong consideration to the genetic information prior to making the selection and breeding decisions.

The next question that should be asked about selective breeding is "What is the suitability of aquacultural species with respect to such genetic manipulation?" One of the most important factors in the realization of genetic improvement is the amount of phenotypic variability in aquacultured species, and analyses have shown that these species have large quantities of variation relative to other species (61,62). While this may not directly equate to larger amounts of strictly genetic variability (62), research results indicate that "satisfactory" amounts are present to allow reasonable advancement in most traits. A summary of the h^2 estimates for a number of traits in aquaculture species can be found in Tave (63).

In addition to this, most aquaculture species exhibit two other biological characteristics that make their potential responsiveness to selective breeding programs more certain. First, a major proportion of the species used in aquaculture produce large numbers of offspring from a single cross (generally, in excess of 1,000). With such high rates of reproduction, a high selection intensity can be applied, which can, potentially, increase the magnitude of the response that can be achieved (57). Second, the oviparous reproductive nature of most aquaculture species allow the use of a wider range of breeding designs than can be utilized with many agricultural animals. Combining these characteristics with the large amount of variability exhibited by aquacultural species suggest that the potential for improvement via selective breeding is very good.

Although relatively few species have been subjected to a selective breeding program and the traits that have been investigated are relatively limited (64), the results of programs that have been conducted indicate that responses to selection can lead to major improvements in their performance (65–68). In most of these programs, the selection responses exceeded overall responses in similar traits reported for agricultural animals, although the programs with aquacultural species are still relatively short term.

Rather surprisingly, the promising results to date have not led to a major emphasis on selective breeding in the industry. There are several explanations for this (62), but increasing competitiveness in the future will require that adequate quantities of more efficient selectively bred stocks be available for production. This will undoubtedly be facilitated by modern molecular methods and genetic engineering (69,70). However, these approaches will not transcend the need for selective breeding in the production of improved stock for aquaculture. They will provide complementary techniques for changing the genome of aquaculture species in beneficial way, in much the same way that artificial fertilization provides a very necessary complement to selective breeding.

BIBLIOGRAPHY

1. J.E. Bardach, J.H. Ryther, and W.O. McLarney, *Aquaculture: The Farming and Husbandry of Freshwater and Marine Organisms*, John Wiley & Sons, New York, NY, 1972.
2. S.W. Ling, *Aquaculture in Southeast Asia*, Washington Sea Grant Publications, University of Washington, Seattle, WA, 1977.
3. D.B. Quayle and G.F. Newkirk, *Farming Bivalve Molluscs: Methods for Study and Development*, World Aquaculture Society, Baton Rouge, LA, 1989.
4. F. Yamazaki, *Aquaculture* **33**, 329–354 (1983).
5. J.R. Gold, in W.S. Hoar, D.J. Randall, and J.R. Brett, eds., *Fish Physiology*, Vol. 8, Academic Press, New York, 1979, pp. 353–405.
6. K.D. Kallman, in B.J. Turner, ed., *Evolutionary Genetics of Fishes*, Plenum Press, New York, 1984, pp. 95–171.
7. G.A. Hunter and E.M. Donaldson, in W.S. Hoar, D.J. Randall, and E.M. Donaldson, eds., *Fish Physiology*, Vol. 9B, Academic Press, New York, 1983, pp. 223–303.
8. S.T.H. Chan and W.S.B. Yeung, in W.S. Hoar, D.J. Randall, and E.M. Donaldson, eds., *Fish Physiology*, Vol. 9B, Academic Press, New York, 1983, pp. 171–222.
9. L.E. Haley, *J. Heredity* **68**, 114–115 (1977).
10. R.R. Tokarz, in R.E. Jones, ed., *The Vertebrate Ovary*, Plenum Press, New York, 1978, pp. 145–179.
11. R.A. Wallace and K. Selman, *Am. Zool.* **21**, 325–343 (1981).
12. Y. Nagahama, in W.S. Hoar, D.J. Randall, and E.M. Donaldson, eds., *Fish Physiology*, Vol. 9A, Academic Press, New York, 1983, pp. 223–275.
13. A.C. Giese and J.S. Pearse, eds., *Reproduction of Marine Invertebrates*. Academic Press, New York, 1979.
14. C.E. Purdom, *Aquaculture* **33**, 287–300 (1983).
15. R.J. Wootton, in P. Tytler and C. Calow, eds., *Fish Energetics: New Perspectives*, Croom Helm, New York, 1985, pp. 231–254.
16. J.H.S. Blaxter, *The Early Life History of Fish*, Springer-Verlag, New York, NY, 1974.
17. E. Kamler, *Early Life History of Fish: An Energetics Approach*, Chapman & Hall, New York, 1992.
18. N. Bromage, in N.R. Bromage and R.J. Roberts, eds., *Broodstock Management and Egg and Larval Quality*, Blackwell Science Ltd., London, 1995, pp. 1–24.
19. N. Bromage, J. Jones, C. Randall, M. Thrush, J. Springate, J. Duston, and G. Barker, *Aquaculture* **100**, 141–166 (1992).
20. R. Billard, J. Cosson, L.W. Crim, and M. Suquet, in N.R. Bromage and R.J. Roberts, eds., *Broodstock Management and Egg and Larval Quality*, Blackwell Science Ltd., London, 1995, pp. 25–52.
21. V.J. Bye, in G.W. Potts and R.J. Wootton, eds., *Fish Reproduction: Strategies and Tactics*, Academic Press, New York, 1984, pp. 187–205.
22. W.L. Shelton, *Rev. Aquat. Sci.* **1**, 497–535 (1989).
23. N. Bromage and J. Duston, *Rep. of the Inst. of Freshwater Research Drottningholm* **63**, 26–35 (1986).
24. N. Bromage, J.S. Elliot, J. Springate, and C. Whitehead, *Aquaculture* **43**, 213–223 (1984).
25. W.P. Breese and R.E. Malouf, *Hatchery Manual for the Pacific Oyster*, Oregon State Univ. Sea Grant Spec. Rep. No. 443, Corvallis, OR, 1975.
26. Y. Le Borgne, in G. Barnabe, ed., *Aquaculture Vol. 1*, Ellis Horwood, Chichester, GB, 1990, pp. 275–284.
27. C.L. Browdy and T.M. Samocha, *J. World Aquacult. Soc.* **16**, 236–249 (1985).
28. G.E. Pickford and J.W. Atz, *The Physiology of the Pituitary Gland of Fishes*, New York Zoological Soc., New York, 1957.
29. E.M. Donaldson and G.A. Hunter, in W.S. Hoar, D.J. Randall, and E.M. Donaldson, eds., *Fish Physiology*, Vol. 9B, Academic Press, New York, 1983, pp. 352–390.
30. R. Billard, *Reproduction, Nutrition and Development* **26**, 877–920 (1986).
31. T.J. Lam, in C.S. Lee and I.C. Liao, eds., *Reproduction and Culture of Milkfish*, Oceanic Institute, Waimanalo, HI, 1985, pp. 14–56.
32. L. Crim, A.M. Sutterlin, D.M. Evans, and C. Weil, *Aquaculture* **35**, 229–307 (1983).
33. Y. Zohar, G. Pagelson, Y. Gothilf, W.W. Dickhoff, P. Swanson, S. Duguay, W. Gombotz, J. Kost, and R. Langer, *Proc. Int. Symp. Controlled Release Bioact. Mater.* **17**, 8–9 (1990).
34. C.C. Mylonas, Y. Tabata, R. Langer, and Y. Zohar, *J. Controlled Release* **35**, 23–34 (1995).
35. R.S.V. Pullin and C.M. Kuo, in J.T. Manassah and E.J. Briskey, eds., *Advances in Food-Producing Systems for Arid and Semiarid Lands*, Academic Press, New York, 1981, pp. 899–978.
36. J. Stoss, in W.S. Hoar, D.J. Randall, and E.M. Donaldson, eds., *Fish Physiology*, Vol. 9B, Academic Press, New York, 1983, pp. 305–350.
37. E. Leitritz and R.C. Lewis, *Trout and Salmon Culture (Hatchery Methods)*, CA Fish Bull. No. 164, Univ. of CA, Berkeley, CA, 1980.
38. C. Markmann and S.I. Doroshov, *Aquaculture* **35**, 163–169 (1983).
39. S.I. Doroshov, W.H. Clark, P.B. Lutes, R.L. Swallow, K.E. Beer, A.B. McGuire, and M.D. Cochran, *Aquaculture* **32**, 93–104 (1983).
40. R.L. Busch, in C.S. Tucker, ed., *Channel Catfish Culture*, Elsevier, Amsterdam, 1985, pp. 13–84.
41. J.H. Kerby, in R.R. Stickney, ed., *Culture of Nonsalmonid Fishes*, CRC Press, Boca Raton, FL, 1993, pp. 251–306.
42. R.W. Rottmann and J.V. Shireman, *Aquaculture* **17**, 257–261 (1979).
43. R. Billard, R. Christen, M.P. Cosson, J.L. Gatty, L. Letellier, P. Renard, and A. Saad, *Fish Physiol. Biochem.* **2**, 115–120 (1986).

44. K.R. Foltz, J.S. Partin, and W.J. Lennarz, *Science* **259**, 1421–1425 (1993).
45. R.G. Piper, I.B. McElwain, L.E. Orme, J.P. McCraren, L.G. Fowler, and J.E. Leonard, *Fish Hatchery Management*, U.S. Dept. of the Interior, Fish and Wildlife Service, Washington, DC, 1982.
46. A.S. Ginzburg, *Fertilization in Fishes and the Problem of Polyspermy*, Acad. Sci, USSR (translated from Russian), National Technical Information Service, Springfield, VA, 1972.
47. J.G. Nickum and R.R. Stickney, in R.R. Stickney, ed., *Culture of Nonsalmonid Fishes*, CRC Press, Boca Raton, FL, 1986, pp. 231–249.
48. S. Rothbard, *Bamidgeh* **33**, 103–121 (1981).
49. W.F. Krise, L. Bulkowski-Cummings, A.D. Shellman, K.A. Kraus, and R.W. Gould, *Prog. Fish-Cult.* **48**, 95–100 (1986).
50. L.K.-P. Leung and B.G.M. Jamieson, in B.G.M. Jamieson, ed., *Fish Evolution and Systematics: Evidence from Spermatozoa*, Cambridge Univ. Press, Cambridge, 1991, pp. 245–295.
51. R. Billard, *Mar. Behaviour and Physiol.* **14**, 3–21 (1988).
52. G.H. Aas, T. Refstie, and B. Gjerde, *Aquaculture* **95**, 125–132 (1991).
53. F. Lahnsteiner, B. Berger, T. Weismann, and R. Patzner, *Prog. Fish-Cult.* **58**, 149–159 (1996).
54. R. Billard, J. Petit, B. Jalabert, and D. Szollosi, in J.H.S. Blaxter, ed., *The Early Life History of Fish*, Springer-Verlag, New York, 1974, pp. 715–723.
55. J.P. Rienets and J.L. Millard, *Prog. Fish-Cult.* **49**, 117–119 (1987).
56. T.B. Bagenal, in S.D. Gerking, ed., *Ecology of Freshwater Fish Production*, Blackwell Scientific, Oxford, 1978, pp. 75–101.
57. D.S. Falconer, *An Introduction to Quantitative Genetics*, John Wiley & Sons, Inc., New York, 1989.
58. L.D. Van Vleck, E.J. Pollack, and E.A.B. Oltenau, *Genetics for the Animal Sciences*, W.T. Freeman and Co., New York, 1987.
59. W.A. Becker, *Manual of Quantitative Genetics*, Academic Enterprises, Pullman, WA, 1984.
60. M. Lynch and B. Walsh, *Genetics and Analysis of Quantitative Traits*, Sinauer Assoc., Sunderland, MA, 1997.
61. T. Gjedrem, *Aquaculture* **6**, 23–29 (1975).
62. F. Allendorf, N. Ryman, and F. Utter, in N. Ryman and F. Utter, eds., *Population Genetics and Fishery Management*, Univ. of WA Press, Seattle, 1987, pp. 1–19.
63. D. Tave, *Genetics for Fish Hatchery Managers*, Van Nostrand Reinhold, New York, 1993.
64. W.K. Hershberger, *Food Rev. Int.* **6**, 359–372 (1990).
65. T. Gjedrem, B. Gjerde, and T. Refstie, in B.S. Weir, E.J. Eisen, M.M. Goodman, and G. Namkoong, eds., *Proc. Second International Conference on Quantitative Genetics*, Sinauer Assoc., Sunderland, MA, 1987, pp. 527–535.
66. W.K. Hershberger, J.M. Myers, R.N. Iwamoto, W.C. McCauley, and A.M. Saxton, *Aquaculture* **85**, 187–197 (1990).
67. L. Siitonen and G.A.E. Gall, *Aquaculture* **78**, 153–161 (1989).
68. R.A. Dunham and R.O. Smitherman, *Aquaculture* **33**, 89–96 (1983).
69. E.M. Hallerman and J.S. Beckman, *Can. J. Aquat. Sci.* **45**, 1075–1087 (1988).
70. R. Lewis, *Bioscience* **38**, 225–227 (1988).

S

SALINITY

Wade O. Watanabe
The University of North Carolina at Wilmington
Wilmington, North Carolina

OUTLINE

Salinity has been traditionally defined as the number of grams of all solid material, including both the inorganic ions (e.g., sodium and chloride, organic phosphorus and nitrogen) and organic compounds (e.g., vitamins and plant pigments) dissolved in 1 kg (2.2 lb) of seawater with all the carbonate converted to oxide, bromine and iodine replaced by chlorine, and the organic matter completely oxidized (1). The symbol for salinity is S and is measured in parts per thousand (ppt). The salinity of freshwater is less than 0.5 ppt, while the open ocean averages about 35 ppt and ranges from 33 to 37 ppt. Salinity can vary widely in bays and estuaries, which are affected by tidal flow, freshwater runoff, and evaporation. Aquatic organisms are adapted to survive and to grow best under salinities that characterize their natural habitats. Depending on species, effective salinity ranges for reproduction and growth may be narrow or broad, can vary with life stage, and may be modified by other environmental factors, particularly temperature. Knowledge of the combined effects of salinity and temperature on survival, growth, and reproduction of an organism allows the aquaculturist to develop environmental management strategies for optimizing hatchery production of seedstock and for growout to marketable sizes.

MEASURING SALINITY

Since salinity is difficult to measure by direct chemical methods, it has been expressed in terms of chlorinity (Cl o/oo), which is defined as the total number of grams of chlorine, bromine, and iodine contained in 1 kg (2.2 lb) of seawater, assuming that bromine and iodine have been replaced by chlorine (2). The relationship between these variables is as follows: Salinity $= 0.30 + 1.805 \times$ chlorinity. To determine chlorinity, a sample of seawater is titrated with silver nitrate in the presence of an indicator (1). The halides precipitate out, and the amount of silver nitrate can be converted into equivalent amounts of chlorinity.

Chlorinity, however, is seldom measured by aquaculturists who prefer simpler, less time-consuming methods of determining salinity. These include measurements of density, conductivity, and refractive index, each of which can be converted into salinity. The least expensive method of measuring salinity involves use of a glass hydrometer to determine a water sample's density, which can be converted to salinity by using tables that correct for sample temperature. Glass hydrometers also provide a high level of accuracy (± 0.1 ppt). Mechanical hydrometers are now available that permit direct reading in ppt and do not require temperature corrections. While less accurate (± 1.5 ppt), they are inexpensive and useful for spot checking. Conductivity meters are relatively expensive and can be used to measure salinity with a high degree of accuracy (± 0.1 ppt) and are therefore used mainly by researchers. They measure electrical conductivity through leads in a probe, which often break with repeated use. The simplest and most rapid means of measuring salinity is refractometry, which measures the refractive index of light passing through a sample solution. Handheld refractometers are moderately priced, compact, durable, require only a drop of sample, are temperature compensated, read directly in ppt, and provide a level of accuracy (± 0.5 ppt) suitable for most aquaculture applications.

OSMOREGULATION

Osmoregulation is the maintenance of a relatively constant internal salt concentration against external salinities that are higher, lower, or fluctuating. Internal salt concentration is generally expressed as osmotic concentration, which depends on the number of undissociated molecules and ions per unit volume or weight of solvent. For example, one gram molecule (mole) per liter [0.25 g/L/(=2.2 lb)] (kg) of water of a substance (e.g., glucose) that does not dissociate has an osmolality of 1 osmole per kg (Osm/kg) or 1,000 mosmoles per kg (mOsm/kg) (3). On the other hand, one mole of sodium chloride per kg of water has an osmolality close to 2 Osm/kg, because this compound dissociates nearly completely in solution. Freshwater has a salinity of 0 g/L and an osmolality of 0 mOsm/kg. Seawater, with a salinity of 34 g/L, has an osmolality of 1,000 mOsm/kg.

Fish, in general, do not have an internal salt concentration that closely match the water in which they swim. In freshwater fish, internal salt concentration tends to be in the range of 8.5 to 10.4 ppt (248–305 mOsm/kg), which is hypertonic to (higher than) the surrounding medium (0 ppt) (3). Fish living in marine waters maintain internal salt concentrations in the range of 12–15 ppt (352–440 mOsm/kg), which is hypotonic to (lower than) the external medium (33–37 ppt). Fish living under estuarine conditions are exposed to variable external salt concentrations, but tend to osmoregulate at around 9–15 ppt (263–440 mOsm/kg). Most marine invertebrates have internal salt concentrations nearly isotonic to (the same as) the surrounding water (4,5). Estuarine crustaceans, such as the juvenile stages of some penaeid shrimp species, are good osmoregulators and can maintain internal salt concentrations around 18–30 ppt (540–870 mOsm/kg) (6–8).

Through a process called osmosis, water moves through a semipermeable membrane, such as the gill epithelium of a fish, from the region of lower salt concentration to that of higher concentration until the salt concentration on both sides of the membrane is the same. Hence, freshwater fishes, which are hypertonic to their surrounding medium, continually gain water from the environment and lose salts in the urine and through diffusion at the gills. Freshwater fish must therefore eliminate excess water from their tissues and replace salts. This osmoregulatory problem is solved by having surface membranes of reduced permeability to inhibit water entry and kidneys that produce copious amounts of very dilute urine. Freshwater fish depend upon the salt intake in their food and active uptake through specialized "chloride cells" in the gills.

In contrast, marine fishes, which are hypotonic to the surrounding medium, continually lose water to the surrounding sea and gain salts by diffusion. To replace lost fluids, these fish must drink large quantities of seawater and then eliminate the excess salt. Kidneys of marine fish produce small volumes of urine containing relatively high concentrations of salt so that the urine is nearly isotonic (i.e., the same concentration) or slightly hypertonic to blood. The gills of marine fish also actively excrete ions that are absorbed.

SALINITY TOLERANCE

Fish are categorized according to their tolerance salinity. Those that tolerate a wide range of salinity at some phase in their life cycle are called euryhaline species. Anadromous euryhaline species commence life in freshwater, migrate to the sea as juveniles, but return to freshwater for breeding. Cultured anadromous species include Atlantic salmon *Salmo salar*, Pacific salmon, *Oncorhynchus* spp., striped bass, *Morone saxatilis*, and the white sturgeon, *Acipenser transmontanus*. Euryhaline marine fish species include red drum, *Sciaenops ocellatus* (9), which is grown in the United States and striped mullet, *Mugil cephalus*, (10) and milkfish, *Chanos chanos*, (11) which are grown in Asia and the Pacific. All are cultured in ponds over a broad range of salinity. The Asian sea bass, *Lates calcarifer*, is a highly euryhaline species that requires saline water (28–32 ppt) for spawning, but lives in brackishwater estuaries and in freshwaters (12). A remarkable euryhaline freshwater fish species is the mossambique tilapia, *Oreochromis mossambicus*, which grows faster in brackish water and seawater than in freshwater (13–15) and is able to adapt to salinities as high as 120 ppt (16).

Euryhaline invertebrate species of aquaculture interest include the American oyster, *Crassostrea virginica*, which can adapt to a salinity range of 3 to 35 ppt (17), although growth and flavor are adversely affected at low salinities. The hardshell clam, *Mercenaria mercenaria*, has an optimal salinity range of 25 to 35 ppt, with salinities above 38 ppt or below 20 ppt producing reduced growth (18). Most cultured penaeid shrimp are euryhaline, although salinity tolerance varies from species to species. *Litopenaeus japonicus*, grown in Japan and Taiwan, has a broad salinity tolerance range of 10–55, but cannot tolerate low salinities and is recommended for culture within a range of 30 to 40 ppt (19,20). The tiger shrimp, *L. monodon*, grown in Asia, can tolerate 0.2 to 40 ppt, with optimal growth between 10 and 25 ppt (12). *L. vannamei*, which can withstand salinities ranging from 0 to 50 ppt, is the species of preference for pond farming in the western hemisphere, including Central and South America, where very low salinities are characteristic of the annual rainy season (12).

Fish that can tolerate only a narrow range of salinity are known as stenohaline species. Marine fish species such as gilthead sea bream, *Sparus aurata*, and sea bass, *Dicentrarchus labrax*, grown in Europe and the Middle East, and the red sea bream, *Chrysophrys major*, and yellowtail, *Seriola quinqueradiata*, grown in Japan, are considered stenohaline (2). Other stenohaline marine fish species of culture interest include the dolphinfish or mahimahi, *Coryphaena hippurus* (21), Atlantic halibut, *Hippoglossus hippoglossus*, and Pacific halibut, *H. stenolepis*, (22,23). While it is assumed that stenohaline marine species are unable to tolerate salinities much lower than full strength seawater, experimental data on lower lethal salinity tolerance limits and effective ranges of salinity for growth are lacking for most species.

Cultured stenohaline freshwater fish include common carp, *Cyprinus carpio*, grass carp, *Ctenopharyngodon idella*, silver carp, *Hypopthalmichthys molitrix*, and channel catfish, *Ictalurus punctatus*. These species have upper salinity tolerance limits of approximately 15 ppt (24), 14 ppt (25,26), 10 ppt (27,28), and 14 ppt (29–31), respectively.

THE EFFECT OF SALINITY ON REPRODUCTION

Aside from anadromous species, salinity exerts marked effects on the reproductive performance of many cultured species. For example, some euryhaline marine finfish, such as red drum, striped mullet, and milkfish, may be grown under brackishwater conditions, but require high salinities for reproduction. This is evidenced by annual migrations from estuarine to offshore waters for spawning. Hatchery installations for such species must be located near the sea or use synthetic seawater. For euryhaline freshwater species, choice of salinity for hatchery operation can have a significant effect on the economics of fry production when freshwater resources are in scarce supply. For example, in the euryhaline tilapia, *O. mossambicus*, fry production was three times higher at salinities of 8.9–15.2 ppt than in freshwater (32). Hence, production of fry in brackishwater can reduce freshwater requirements for maintaining broodstock and for early rearing of fry (32–35).

In some cultured freshwater crustaceans, brackishwater is essential for reproduction. The freshwater shrimp, *Macrobrachium rosenbergii*, migrates from freshwater to estuarine areas to spawn and requires about 12 ppt salinity for proper larval development. The larvae move upstream into fresh water as they grow. Hence, hatchery protocols for this species must provide brackishwater rearing facilities for broodstock and for early rearing of larvae. In penaeid shrimp culture, the need for high salinity water for hatchery operation may also require that hatchery installations be located near the sea, rather than in the brackishwater areas where the growout ponds may be located (12).

ONTOGENETIC VARIATION IN SALINITY TOLERANCE

Smoltification is a physiological process by which anadromous salmonids adapt from fresh to saltwater (36). As transformation from dark fingerling to silvery smolt occurs, the fish become increasingly tolerant of salt water and will, in nature, begin their migration to sea. The ability to survive in seawater and hypoosmoregulatory ability, or the ability to regulate internal salt concentrations below those of the external medium, develop during the early freshwater phase of the life cycle, and preadapt them to subsequent seawater existence (36–38). These ontogenetic changes in salinity tolerance are known to be closely related to size, which is therefore an important criterion for determining the optimum time for release of hatchery-reared juveniles or for their transfer to seawater pens (39). Failure to attain a critical size before transfer results in mortality and stunting in seawater.

Ontogenetic alteration in salinity tolerance may be a general phenomenon among teleosts, including both steno- and euryhaline freshwater and marine species. Upper salinity tolerance in the stenohaline freshwater catfish, *I. punctatus*, increased from 8 ppt at hatching to 12.5 ppt at 6 months (40). Osmoregulatory capabilities in the euryhaline marine mullet, *M. cephalus*, improved with growth and reached a definitive state in juveniles at 7.5–8.5 months of age (41). In a number of freshwater euryhaline tilapias, *O. mossambicus*, *O. niloticus*, *O. aureus*, hybrid *O. mossambicus* × *O. niloticus*, and red hybrid tilapia, age (or size) at time of seawater transfer influences survival, with newly hatched fry being less tolerant than older individuals (42–45). Premature acclimation to seawater impairs survival in these fish and selection of proper transfer time, based on knowledge of ontogenetic variation in salinity tolerance, improves survival.

Improved salinity tolerance with age in fish is related to maturation events, such as a decreasing body-surface-to-volume relationship, with growth that reduces osmotic stress (46), ontogenetic changes from juvenile to adult hemoglobins, which are adaptive to more saline environments (47,48); and functional development of the hypoosmoregulatory system, including the salt-secreting chloride cells of the gills (49,50) and the endocrine system (51).

ACCLIMATION TO SEAWATER: OPTIMUM RATE

While euryhaline species are generally unable to tolerate abrupt transfer from low to high salinity, gradual acclimation can assure very high survival. For example, whereas none of the commercially important tilapia species tolerate direct transfer from freshwater (0–2 ppt) to full-strength seawater (≥32 ppt) (42–44,52), a gradual, step-wise increase in salinity until full-strength seawater is reached has been successfully used to acclimate a number species. Required acclimation periods range from 4 days (2 days at 18 ppt, 2 days at 27 ppt) *O. aureus*, *O. mossambicus*, and *O. spilurus* to 8 days (4 days at 18 ppt, 4 days at 27 ppt) for *O. niloticus* and *O. niloticus* × *O. aureus* hybrids, depending on the degree of euryhalinity (57).

Salinity acclimation allows physiological changes to take place that enable the fish to regulate internal osmotic concentrations within permissible limits under changing external osmotic conditions. Following direct transfer of *O. mossambicus* from freshwater to 30 ppt seawater, a rapid elevation of plasma osmotic concentration to excessive levels occurs within 1 h (osmotic shock) leading to death within 6 h (53). On the other hand, transfer to a lower, sublethal salinity allows osmoregulatory mechanisms to adapt to, and gradually reduce, the rising plasma osmotic concentration to a new equilibrium level within 44–96 h (52,54). This seawater adaptation process (55) involves the functional activation of the salt-secreting chloride cells of the gills (56–58) and elevates Na-K ATPase (59), which enables subsequent transfer to higher salinities. For *O. mossambicus*, activation of chloride cells occurs within 12 to 24 h after acclimation to 20 ppt seawater (53,58). Seawater adaptation can be simplified through single-step acclimation, where pre-acclimation at one intermediate salinity (e.g., 12 ppt for 48 h or 20 ppt for 24 h) preceeds direct transfer to 36 ppt seawater (58,60). While not as effective as gradual salinity adaptation, pre-transfer feeding of a high salt diet (7–10% NaCl) in freshwater has been shown to stimulate the seawater adaptation process in *O. mossambicus*, *O. spilurus* and *O. aureus* × *O. niloticus* hybrids (61) and in chinook salmon, *Oncorhynchus tshawytscha* (62). Starvation reduces Na, K-ATPase activity and survival during seawater acclimation (14).

ACCLIMATION TO SEAWATER: THE INFLUENCE OF SALINITY EXPOSURE DURING EARLY DEVELOPMENT

Early salinity exposure through spawning and hatching under elevated salinities enhances salinity tolerance of young tilapia fry and may facilitate acclimation to seawater (63). In *O. niloticus*, 96-hour median lethal salinity increased from 19.2 ppt for broods spawned in freshwater to less than 32 ppt for broods spawned at 15 ppt (63). In red hybrid tilapia, fish spawned and reared through early ontogenetic development at 18 ppt had better growth and feed conversion at 18 ppt and

36 ppt than those spawned in 4 ppt (64). Formation of the perivitelline fluid after fertilization (65) modifies the environment in which the embryo develops and may induce adjustments that persist through later development, an irreversible "nongenetic adaptation" (66) to environmental salinity.

EFFECTS OF SALINITY ON GROWTH

Because osmoregulation is a metabolic process requiring active transport of ions to maintain internal salt concentration, the more salt the body of an aquatic animal must take in or excrete, the more energy is required and growth is restricted. Stenohaline freshwater and marine fish species have become physiologically adapted to salinities different from their internal salt concentrations. These species will generally show optimal growth in water with salinities similar to those in which the fish occur in nature. Euryhaline species may grow well over a broad range of salinity (2).

In euryhaline tilapia, growth response to salinity may be quite variable among species and related to degree of euryhalinity. Some species (e.g., *O. niloticus* and *O. aureus*) may be acclimated to full strength seawater, but growth is impaired under these conditions (35). In other species (*O. mossambicus* (13–15), *O. mossambicus* × *O. hornorum* hybrid (67), *O. spilurus* (68), and red hybrid tilapia (69)), growth is higher in brackish and seawater than in freshwater, indicating an advantage of farming these tilapias in brackish- or seawater (35,70). In red hybrid tilapia, this was attributed to increased food consumption (appetite) and declining feed conversion ratios with increasing salinity.

The effects of salinity on growth in fish are related not only to total concentration of dissolved solids (i.e., salinity per se), but are also influenced by the concentrations of divalent ions (Ca^{2+} and Mg^{2+}) due to their effects on membrane permeability and osmoregulation (71,72). High concentrations of environmental calcium help reduce salt loss through the gills and body surfaces in freshwater environments, so less demand is placed on the kidneys to maintain stable concentrations of blood salts, allowing survival of some marine species in freshwater. While red drum eggs and larvae appear to require salinities above 25 ppt, juveniles can be reared in freshwater if hardness is at least 100 mg/L (9,71,73).

Food can be an important source of salt for some euryhaline fish. In the euryhaline marine red drum, *S. ocellatus*, growth limitation related to salt deficiency in hypotonic media can be overcome by adding salt (e.g., NaCl) to the diet (74). In Atlantic salmon, *S. salar*, smolts, however, dietary salt supplementation had negligible effects on growth and feed efficiency of fish reared in freshwater (75).

Relative growth at different salinities may not necessarily reflect the cost of osmoregulation at these salinities, depending on the magnitude of nonosmoregulatory effects on metabolism. In *O. mossambicus* × *O. urolepis hornorum* hybrids, salinity-related differences in total metabolic rates could not be solely attributed to changes in osmoregulation costs, indicating that other nonosmoregulatory (e.g., behavioral) factors must also affect metabolic rate (76). Territorial aggression can modify growth response to salinity (77,78) in tilapias. The variability in reported salinity-growth responses in fish suggest that uniformity of physiological state is often lacking (79).

TEMPERATURE INTERACTION WITH SALINITY

Temperature tolerance and optima for growth in fish may be modified by salinity (66,80–84) due to their interactive effects on osmoregulation (81,85). In red hybrid tilapia, feed consumption and growth at 0 ppt reached a maximum at 27 °C (80 °F), while at 18 and 36 ppt, consumption and growth were highest at 32 °C (90 °F) (87). This suggested that in freshwater, heating water to temperatures above 27 °C (80 °F) would not be justifiable, whereas in brackish- and seawater, heating water to 32 °C (90 °F) can increase growth rates.

In *Marsupenaeus japonicus*, lower salinity tolerance was 5.4 ppt at 25 °C (77 °F), but only 19.3 ppt at 10 °C (50 °F). Under low temperatures, lowest mortalities occurred when the salinity of the water was isosmotic with the hemolymph of shrimp (88).

OVERWINTERING

Cold tolerance in some species is influenced by the salinity. For example, in red hybrid tilapia, maximum growth efficiency under rearing temperatures of 22, 28, and 32 °C (72, 82, and 90 °F) was higher at 18 ppt than at 0 or 36 ppt, although these differences were pronounced at 22 °C (72 °F) (87). This suggested an advantage of brackishwater rearing to improve growth efficiency of this red hybrid tilapia strain under suboptimum temperatures. Studies with a number of tilapia species, including *O. aureus*, *Sarotherodon melanotheron*, *O. mossambicus*, and red hybrid tilapias, show that fish have better cold tolerance when maintained in low-salinity brackishwater (5–12 ppt) than in freshwater or full-strength seawater (85,89–91), presumably because osmoregulatory stress is minimized at near isosmotic salinities. Hence, osmoregulatory failure is prevented at temperatures that would be lethal in hypo- or hyperosmotic media (91). In red drum, cold tolerance is also improved by rearing the fish in water of 5 to 10 ppt salinity (14).

SALINITY FOR DISEASE CONTROL

Salinity can often be used as a form of disease control when the salinity tolerance limits of the cultured fish exceed those of its parasite. For example, in the channel catfish, *I. punctatus*, the freshwater parasitic dinoflagellate *Ichthyophthirius multifiliis* can be controlled in water containing 2 ppt salt (92). Euryhaline freshwater tilapias cultured in full-strength seawater are susceptible to infection by the marine monogenetic trematode *Neobenedenia melleni*. While seawater-cultured tilapias tolerate a reduction in salinity to 15 ppt, the ectoparasite is unable to survive these hyposaline conditions for more than 5 days, which makes this an effective therapeutic procedure (93,94).

CONTROL OF SALINITY IN AQUACULTURE SYSTEMS

Marine and brackishwater cage or net-pen aquaculture operations in embayments and estuaries encounter unstable salinities that may vary well below full-strength seawater due to freshwater runoff. The use of euryhaline species that can tolerate such changes may be important in some locations, since control of salinity is generally difficult under these conditions. While still in the early stages of research and commercial development, open-ocean mariculture, which reduces competition for space and water resources in the coastal zone, is becoming an increasingly attractive alternative. Stenohaline marine species would be grown in such systems were salinities are stable year round.

Land-based brackishwater and marine aquaculture systems, primarily for culture of fish and shrimp, have traditionally employed earthen ponds supplied by tidal flow. Alternatively, water may be pumped to ponds or tanks from a near shore location. A growing awareness of the need to conserve land and water resources, to preserve the environment, and to decrease reliance on natural water resources that may vary in quality, will require that aquaculture rely increasingly on recirculation systems or fish production systems that reuse water by removing wastes from the water and providing oxygen to the fish (95). Such systems utilize only a fraction of the water required by traditional fish production techniques and provide for the potential of a higher degree of control of the environment of the fish being grown, including salinity. In such systems, water of required salinities may be prepared from freshwater and artificial sea salts. Since there is little net loss of salt associated with the removal of wastes, required salinities are maintained by diluting with freshwater to replace water lost through evaporation. Net consumption of both water and salts are thereby minimized. Such systems can potentially allow marine fish farms to operate inland to avoid high-priced coastal land and may reduce the environmental impacts of waste disposal.

BIBLIOGRAPHY

1. F.W. Wheaton, *Aquacultural Engineering*, Robert E. Krieger Publishing Company, Malabar, FL, 1987.
2. R.R. Stickney, *Principles of Aquaculture*, John Wiley and Sons, New York, New York, 1994.
3. C.E. Bond, *Biology of Fishes*, W.B. Saunders Company, Philadelphia, 1979.
4. G.F. Warner, *The Biology of Crabs*, Van Nostrand Reinhold Co., New York, 1977.
5. L.H. Mantel and L.L. Farmer, in L.H. Mantel, ed., *The Biology of Crustacea*, Vol. 5, Academic Press, New York, 1983, pp. 54–161.
6. F.L. Castille, Jr. and A.L. Lawrence, *Comparative Biochem. Physiol.* **68A**, 75–80 (1981).
7. R.P. Ferraris, F.D. Parado–Estepa, J.M. Ladja, and E.G. De Jesus, *Comparative Biochem. Physiol.* **83A**, 701–708 (1986).
8. J.-C. Chen and J.-L Lin, *Aquaculture* **125**, 167–174 (1994).
9. W.H. Neill, in G.W. Chamberlain, R.J. Miget, and M.G. Haby, eds., *Manual on Red Drum Aquaculture*, Texas Agricultural Extension Service and Sea Grant College Program, Texas A&M University, College Station, 1990, pp. 105–108.
10. I.-C. Liao, in O.H. Oren, ed., *Aquaculture of Grey Mullets*, Cambridge University Press London, 1981, pp. 361.
11. I.-C. Liao and T.I. Chen, in C.-S. Lee, M.S. Gordon, and W.O. Watanabe, eds., *Aquaculture of Milkfish (Chanos chanos): State of the Art*, The Oceanic Institute, Waimanalo, Hawaii, 1986, pp. 209–238.
12. T.V.R. Pillay, *Aquaculture Principles and Practices*, Blackwell Scientific Publications, Inc., Cambridge, MA, 1993.
13. P. Canagaratnam, *Bulletin of the Fisheries Research Station, Ceylon* **19**, 47–50 (1966).
14. K. Jurss, T. Bittorf, T. Vokler, and R. Wacke, *Aquaculture* **40**, 171–182 (1984).
15. T.T. Kuwaye, D.K. Okimoto, S.K. Shimoda, R.D. Howerton, H.-R. Lin, P.K.T. Pang, and E.G. Grau, *Aquaculture* **113**, 137–152 (1993).
16. A.K. Whitfield and S.J.M. Blaber, *Environ. Biol. Fishes* **4**, 77–81 (1979).
17. A.S. Pearse and G. Gunter, in J.W. Hedgpeth, ed., *Treatise on Marine Ecology and Paleoecology*, Vol. 1, Memoir 69, Geological Society of America, New York, 1957.
18. D. Vaughan, L. Creswell, and M. Pardee, *Aquaculture Report Series: A Manual for Farming Hard Shell Clam in Florida*, Florida Department of Agriculture and Consumer Services, Tallahassee, FL, 1988.
19. G.J. Dalla Via, *Aquaculture* **55**, 297–305 (1986).
20. G.J. Dalla Via, *Aquaculture* **55**, 307–316 (1986).
21. J. Szyper, in J.P. McVey, ed., *Handbook of Mariculture Vol. II. Finfish Aquaculture*, CRC Press, Inc., Boca Raton, FL, 1991, pp. 228–240.
22. H.W. Liu, R.R. Stickney, and S.D. Smith, *Prog. Fish-Cult.* **53**, 189–192 (1991).
23. R.R. Stickney, H.W. Liu, and S.D. Smith, in R.S. Svrjcek, ed., *Marine Ranching, Proceedings of the 17th U.S. — Japan Meeting on Aquaculture; Ise, Mie Prefecture, Japan, NOAA Tech. Rep. NMFS 102*, U.S. Department of Commerce, Washington, DC, 1991, pp. 9–13.
24. A.I. Payne, in L. Fishelson and Z. Yaron, ed., *Proceeding of the International Symposium on Tilapia in Aquaculture, Nazareth, Israel*, Tel Aviv University, Tel Aviv, Israel, 1983, pp. 534–543.
25. M.J. Maceina and J.V. Shireman, *Prog. Fish-Cult.* **41**, 69–73 (1979).
26. R.V. Kilambi and Z. Zdinak, *J. Fish Biol.* **16**, 171–175 (1980).
27. J. Chervinski, *Aquaculture* **11**, 179–182 (1977).
28. J.-A. Oertzen, *Aquaculture* **44**, 321–332 (1985).
29. W.G. Perry, Jr., *Proceedings Annual Conference Southeast Association Game Fish Commisioners* **21**, 436–444 (1967).
30. K.O. Allen and J.W. Avault, Jr., *Proceedings Annual Conference Southeast Association Game Fish. Commisioners* **23**, 319–331 (1969).
31. W.G. Perry, Jr. and J.W. Avault, Jr., *Proceedings Annual Conference Southeast Association Game Fish Commisioners* **23**, 592–605 (1969).
32. R.N. Uchida and J.E. King, *U.S. Fish and Wildlife Service Fishery Bulletin* **199**(62), 21–47 (1962).
33. W.O. Watanabe, K.M. Burnett, B.L. Olla, and R.I. Wicklund, *J. World Aqua. Soc.* **20**, 223–229 (1989).

34. D.H. Ernst, W.O. Watanabe, L.J. Ellingson, R.I. Wicklund, and B.L. Olla, *J. World Aqua. Soc.* **22**, 36–44 (1991).
35. W.O. Watanabe, R.I. Wicklund, B.L. Olla, and W.D. Head, in B.A. Costa-Pierce and J.E. Rakocy, eds., *Tilapia Aquaculture in the Americas, Vol. 1*, World Aquaculture Society, Baton Rouge, LA 1997, pp. 54–141.
36. L.C. Folmar and W.W. Dickhoff, *Aquaculture* **21**, 1–37 (1980).
37. W.S. Hoar, *J. Fish. Res. Board Canada* **33**, 1234–1252 (1976).
38. G.A. Wedemeyer, R.L. Saunders, and W.C. Clarke, *Marine Fish. Rev.* **42**, 1–14 (1980).
39. C. Mahnken, E. Prentice, W. Waknitz, G. Monan, C. Sims, and J. Williams, *Aquaculture* **28**, 251–268 (1981).
40. K.O. Allen, Ph.D., Dissertation, Louisiana State University and Agricultural and Mechanical College, Baton Rouge, LA, 1971.
41. F.G. Nordlie, W.A. Szelistowski, and W.C. Nordlie, *J. Fish Biol.* **20**, 79–86 (1982).
42. W.O. Watanabe, C.-M. Kuo, and M.-C. Huang, *Aquaculture* **47**, 353–367 (1985).
43. P.W. Perschbacher and R.B. McGeachin, in R.S.V. Pullin, T. Bhukaswan, K. Tonguthai, and J.L. Maclean, eds., *The Second International Symposium on Tilapia in Aquaculture, ICLARM Conference Proceedings 15*, Department of Fisheries, Bangkok, Thailand and International Center for Living Aquatic Resources Management, Manila, Philippines, 1988, pp. 415–420.
44. C.T. Villegas, *Aquaculture* **85**, 281–292 (1990).
45. W.O. Watanabe, L.J. Ellingson, B.L. Olla, D.H. Ernst, and R.I. Wicklund, *Aquaculture* **87**, 311–321 (1990).
46. G. Parry, *J. Experi. Biol.* **37**, 425–434 (1960).
47. H.J.A. Koch, *Aquaculture* **18**, 231–240 (1982).
48. J.E. Perez and N. Maclean, *J. Fish. Biol.* **9**, 447–455 (1976).
49. R.D. Ewing, H.J. Pribble, S.L. Johnson, C.A. Fustish, J. Diamond, and J.A. Lichatowich, *Canadian J. Fish. Aquatic Sci.* **37**, 600–605 (1980).
50. W.C. Clarke, *Aquaculture* **18**, 177–183 (1982).
51. L.M.H. Helms, E.G. Grau, S.K. Shimoda, R.S. Nishioka, and H.A. Bern, *General and Comparative Endocrinology* **65**, 48–55 (1987).
52. M.M. Al-Amoudi, *Aquaculture* **65**, 333–342 (1987a).
53. P.P. Hwang, C.M. Sun, and S.M. Wu, *Marine Biol.* **100**, 295–299 (1989).
54. H. Assem and W. Hanke, *Comp. Biochem. Physiol.* **64**(A), 17–23 (1979).
55. P. Prunet and M. Bornancin, *Aquatic Living Resources* **2**, 91–97 (1989).
56. M. Dharmamba, M. Bornancin, and J. Maetz, *J. Physiol.* **70**, 627–636 (1975).
57. L. Fishelson, *Environ. Biol. Fishes* **5**, 161–165 (1980).
58. P.P. Hwang, *Marine Biol.* **94**, 643–649 (1987).
59. A.D. Dange, *Marine Biol.* **87**, 101–107 (1985).
60. J.K. Foskett, C.D. Logsdon, T. Turner, T.E. Machen, and H.A. Bern, *J. Exper. Mar. Biol.* **43**, 209–224 (1981).
61. M.M. Al-Amoudi, *Aquaculture* **64**, 333–338 (1987).
62. W.S. Zaugg, D.D. Roley, E.F. Prentice, K.X. Gores, and F.W. Waknitz, *Aquaculture* **32**, 183–188 (1983).
63. W.O. Watanabe, C.-M. Kuo, and M.-C. Huang, *Aquaculture* **48**, 159–176 (1985).
64. W.O. Watanabe, K.E. French, D.H. Ernst, B.L. Olla, and R.I. Wicklund, *J. World Aqua. Soc.* **20**, 134–142 (1989).
65. H.M. Peters, *ICLARM Translations 2*, ICLARM, Manila, Phillippines, (1983).
66. O. Kinne, *Comparative Biochem. Physiol.* **5**, 265–282 (1962).
67. T. Garcia and K. Sedjro, *Revista de Investigaciones Marinas* **8**, 61–65 (1987).
68. T.S. Osborne, *Field Report White Fish Authority*, Fishery Development Project Saudi Arabia **48**, (1979).
69. W.O. Watanabe, L.J. Ellingson, R.I. Wicklund, and B.L. Olla, in R.S.V. Pullin, T. Bhukaswan, K. Tonguthai, and J.L. Maclean, eds., *The Second International Symposium on Tilapia in Aquaculture, ICLARM Conference Proceedings 15*, Department of Fisheries, Bangkok, Thailand and International Center for Living Aquatic Resources Management, Manila, Philippines, 1988, pp. 515–523.
70. R.R. Stickney, *Prog. Fish-Cult.* **48**, 161–167 (1986).
71. W.A. Wurts and R.R. Stickney, *Aquaculture* **76**, 21–35 (1989).
72. J.A. Forsberg and W.H. Neill, *Environ. Biol. Fishes* **49**, 119–128 (1997).
73. M.G. Pursley and W.R. Wolters, *J. World Aquacult. Soc.* **25**, 448–453.
74. D.M. Gatlin, III, D.S. MacKenzie, S.R. Craig, and W.H. Neill, *Prog. Fish-Cult.* **54**, 220–227 (1992).
75. H.M. Shaw, R.L. Saunders, H.C. Hall, and E.B. Henderson, *J. Fish. Res. Board of Canada* **322**, 1813–1819 (1975).
76. R. Febry and P. Lutz, *J. Exp. Biol.* **128**, 63–85 (1987).
77. I.-C. Liao and S.L. Chang, in L. Fishelson and Z. Yaron, eds., *Proceedings of the International Symposium on Tilapia in Aquaculture, Nazareth, Israel*, Tel Aviv University, Tel Aviv, 1983, pp. 524–533.
78. J.R. Brett and T.D.D. Groves, in W.S. Hoar, D.J. Randall, and J.R. Brett, eds., *Fish Physiology. Vol. 8*, Academic Press, New York, NY, 1979, pp. 279–352.
79. J.R. Brett, in W.S. Hoar, D.J. Randall, and J.R. Brett, eds., *Fish Physiology, Vol. 8*, Academic Press, London, England, 1979, pp. 599–577.
80. O. Kinne, *Physiological Zoology* **33**, 288–317 (1960).
81. F.W.H. Beamish, *J. Fish. Res. Board Canada* **27**, 1209–1214 (1970).
82. D.S. Peters and M.T. Boyd, *J. Exp. Mar. Biol.* **7**, 201–207 (1972).
83. A.K. Whitfield and S.J.M. Blaber, *J. Fish Biol.* **9**, 99–104 (1976).
84. J.R. Stauffer, Jr., *Water Resources Bull.* **22**, 205–208 (1986).
85. B.R. Allanson, R.A. Bok, and N.I. van Wyk, *J. Fish Biol.* **3**, 181–185 (1971).
86. R.I. Tilney and C.H. Hocutt, *Environ. Biol. Fishes* **19**, 35–44 (1987).
87. W.O. Watanabe, D.H. Ernst, M.P. Chasar, R.I. Wicklund, and B.L. Olla, *Aquaculture* **112**, 309–320 (1993).
88. M. Charmentier–Daures, P. Thuet, G. Charmentier, and J.-P. Trilles, *Ressources Vivantes Aquatiques* **1**, 267–276 (1988).
89. J.R. Stauffer, Jr., D.K. Vann, and C.H. Hocutt, *Water Resources Bull., Amer. Water Resources Assoc.* **20**, 5 (1984).
90. Y.-Y. Ting, M.-H. Chang, S.-H. Chen, Y.-S. Wang, and W.-H. Cherng, *Bulletin of Taiwan Fisheries Research Institute* **37**, 101–115 (1984).
91. A.V. Zale and R.W. Gregory, *Trans. Amer. Fish. Soc.* **118**, 718–720 (1989).
92. S.K. Johnson, in *Proceedings of the 1976 Fish Farming Conference and Annual Convention of the Catfish Farmers of Texas*, Texas A&M University, College Station, 1976, pp. 91–96.

93. K.W. Mueller, W.O. Watanabe, and W.D. Head, *J. World Aquacult. Soc.* **23**, 199–204 (1992).
94. E.P. Ellis and W.O. Watanabe, *Aquaculture* **117**, 15–27 (1993).
95. T.M. Losordo and P.W. Westerman, *J. World Aquacult. Soc.* **25**, 193–203 (1994).

See also OSMOREGULATION IN BONY FISHES.

SALMON CULTURE

RONALD J. ROBERTS
RONALD W. HARDY
Hagerman Fish Culture Experiment Station
Hagerman, Idaho

OUTLINE

Current Farming Practices
Nutritional Requirements
Feed Formulation
Feeds and Feeding
Health Management
Genetics and Reproduction
Industry Trends
Bibliography

Salmon farming has grown from research/demonstration projects to a global industry over the past 25 years. In fact, farmed fish production now exceeds salmon production from capture fisheries (Fig. 1). In the marketplace, farmed salmon account for approximately 70% of all salmon and over 90% of fresh salmon served in restaurants or purchased for home consumption (Fig. 2). Salmon farming is a significant enterprise in Norway, Chile, Scotland, Canada, the Faroe Islands, Iceland, Tasmania, Ireland, and a number of other countries (Table 1). The total value of farmed salmon is over US$2.2 billion. Atlantic salmon (*Salmo salar*) is the most widely farmed salmon species with global production exceeding 640,000 mt in 1998. Chinook salmon (*Oncorhynchus tshawytscha*) and coho salmon (*O. kisutch*) are the other species of farmed salmon and are produced mainly in Chile (coho) and Canada (chinook).

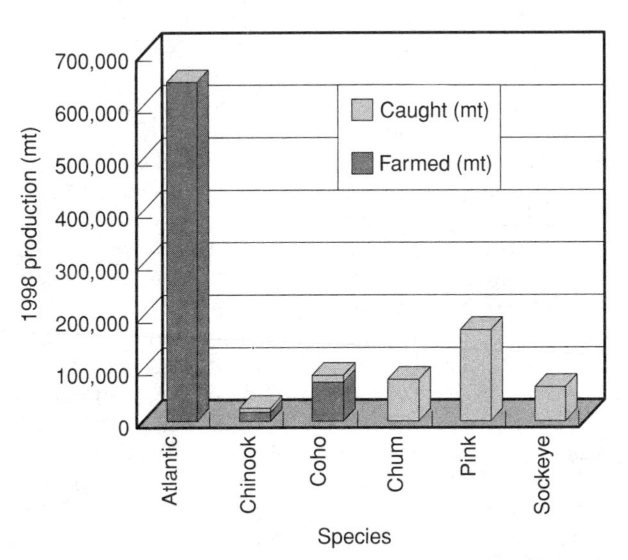

Figure 1. Salmon production from farming and capture fisheries, 1998.

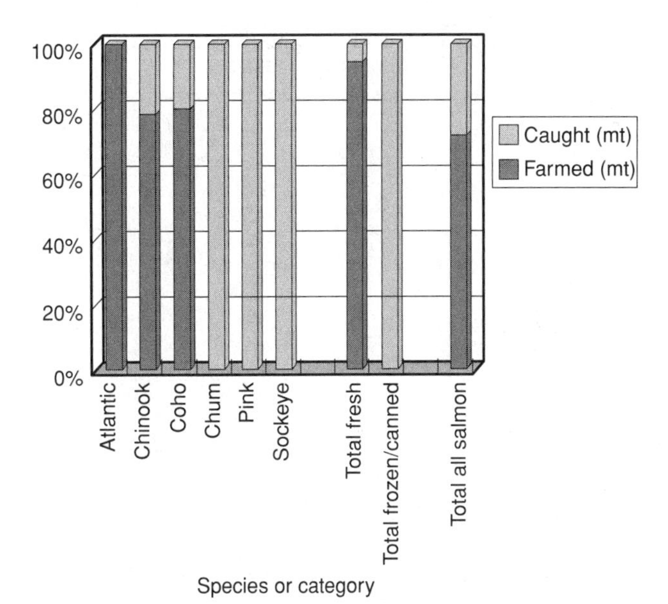

Figure 2. Percentage of salmon species produced from farming or from capture fisheries, 1998.

Table 1. Global Production and Value of Farmed Salmonids in 1998 (Preliminary Estimate, Scottish Salmon Board)

Salmonid Species	Annual Production (mt)
Atlantic salmon	639,200
Salmon trout[a]	87,900
Coho salmon	66,090
Chinook salmon	16,000

[a]Rainbow trout raised in sea cages.

The first commercial salmon farms were started in Scotland and Norway in the late 1960s and in Washington state (USA) in the early 1970s (1,2). In Washington state, early farms produced "pan-sized" coho salmon from marine net-pens, while in Scotland and Norway, larger (2–3 kg; 4.4–6.6 lb) Atlantic salmon were produced. Production levels were modest, and the combination of losses to disease, slow growth due to lack of appropriate feeds, poor site selection, and high costs of production limited expansion of salmon farming for a decade. Technical advances made in the areas of feed formulation and production, stock selection, vaccines, sex reversal in the case of chinook, and general husbandry practices resulted in rapid expansion of salmon farming and lower costs of production in the 1980s (3).

In Norway, individual farms were limited in size, in part to encourage individual rather than corporate ownership of farms; and this policy encouraged Norwegian farmers to

Table 2. 1998 Production of Atlantic Salmon by Country (Preliminary Estimate From Scottish Salmon Board)

Country	1997 Production (mt)
Norway	327,600
UK	105,000
Chile	107,000
Canada	63,000
Faroes	30,000
Ireland	13,650
Tasmania	10,000
USA	4,600
Iceland	6,180
Spain	1,900
France	1,000
Sweden	600

expand production elsewhere, notably Canada and Chile. Backed by generous government support of research and guarantees on investment loans, Norwegians invested heavily in overseas salmon farming, bringing with them equipment and husbandry techniques that soon dominated the salmon farming industry.

The Chilean salmon farming industry was started by Chilean entrepreneurs in the latter half of the 1980s. Being in the southern hemisphere, Chileans were positioned to supply fresh Pacific salmon to North American markets during winter and early spring, six months out of phase with North American wild harvests.

In North America and Chile, Pacific salmon farming was soon supplemented with Atlantic salmon farming as Scottish and Norwegian investments were made and technology was introduced. At present, Atlantic salmon production leads that of Pacific salmon (Table 2). Atlantic salmon have the advantage of being somewhat less susceptible to early male maturation in a portion of a given year and class than Pacific salmon, increasing the proportion of high-quality fish that can be harvested. Unlike farmed coho salmon, which must be harvested before they mature as two-year-old fish, Atlantic salmon mature first as three-year-old fish, giving farmers an additional year of rearing. In addition, production of out-of-season smolts for stocking farms has made it possible to supply markets with Atlantic salmon more or less year-round.

CURRENT FARMING PRACTICES

Briefly, salmon farming is divided into two segments: the freshwater segment during which first-feeding fry (<0.5 g; 0.02-oz) are reared to the smolt stage and the sea cage-rearing segment, during which the fish are reared from the smolt stage (ca. 50–80 g; 1.7–2.7 oz) to harvest (3–6 kg; 6.6–13.2 lb). In nature, coho and Atlantic salmon typically migrate to the sea in their second year of freshwater life, while chinook salmon migrate in their first year (ocean type) or second year (stream type), depending upon the stock. Ten years ago only a small proportion of Atlantic salmon could be brought to smoltification in one year (the so-called S1 or one-summer smolts). Now, as a result of improved husbandry and feeds, and selection for S1 smolts (S2s are normally culled), most farms produce a high proportion of S1 smolts. In addition, a proportion of the most rapidly growing fish are available for even earlier out-of-season transfer to seawater through photoperiod manipulation of smolting time. For reasons of economy, in areas where complete winter freezing of freshwater does not occur, fingerling salmon are placed in freshwater cages after the first-feeding stage. This is extremely cost effective in capital terms, but is coming under increasing pressure because of its eutrophication effect on the highly oligotrophic waters being used. As a result, tighter controls on phosphorus discharge from farms are being applied by environmental protection agencies. Thus, great pressures have developed for smolt cage farms to reduce stocking levels and thereby reduce discharges. This is forcing feed manufacturers to provide low-phosphorus feeds and will probably lead to almost all smolts ultimately being produced from flow-through or recirculation tank systems. The same improvements that have shortened the period of freshwater rearing for Atlantic salmon have had the same effect on Pacific salmon. Most smolts are stocked into sea cages in their first year of life.

Salmon growout to harvest is conducted in marine cages. A decade ago, the growout period in sea cages required at least two years; today fish often reach harvest size in 10–15 months after transfer to sea cages. Many factors have contributed to this improvement, including changes in feed formulation, feed pelleting technology, the introduction of effective vaccines, and the selection and domestication of farmed salmon stocks. The use of larger cages has also improved the growth rates of salmon, as has better site selection (higher water flow rates, deeper water).

NUTRITIONAL REQUIREMENTS

A great deal of research has been conducted on the nutritional requirements of Pacific salmon, but nearly all has focussed on studies with fry and juveniles in freshwater (4). In contrast to the situation with Pacific salmon, the scientific literature on the nutritional requirements of Atlantic salmon is far from complete, and what data do exist are often limited to fry or fingerlings, not growout fish (5). For both Pacific and Atlantic salmon, this is due to practical considerations. It is much easier and less expensive to conduct nutritional research on fry and fingerlings in freshwater research facilities than at sea. In sea cages, confounding factors such as variable water temperature, disease, natural prey (food), difficulties in quantifying feed intake, and so on, make it difficult to perform scientifically meaningful feeding trials. Nevertheless, limited research information on Atlantic salmon nutritional needs has been supplemented with information from Pacific salmon and rainbow trout studies and practical experience to permit the Atlantic salmon industry to reach its current level of production and efficiency. At present, feed costs account for 40–50% of the operating costs in a salmon farm (6).

Salmon require the same 10 essential amino acids as other vertebrates, but the quantitative dietary requirements for essential amino acids have only been determined for Pacific salmon. The nearly exclusive use of fish meal as a protein source for Atlantic salmon feeds has made determining the essential amino acid requirements of Atlantic salmon a relatively low priority, since fish meals contain those amino acids at levels above presumed dietary requirements. Protein levels in salmon fry and fingerling diets are higher than in growout diets (>45% and 42–44%, respectively), and dietary energy levels increase with fish size.

Atlantic salmon also presumably have similar dietary vitamin and mineral requirements to Pacific salmon or rainbow trout (*Oncorhynchus mykiss*), but relatively few dietary requirements for Atlantic salmon have been determined (4,7). For the few micronutrients for which quantitative requirements have been determined for Atlantic salmon, they have been found to be nearly the same as those of Pacific salmon. For example, the required dietary level of vitamins B_6, C, and E are similar for Atlantic salmon and rainbow trout, as are published values for phosphorus, manganese, copper, and iron (Table 3). In practical salmon diets, vitamins and trace minerals are supplemented at levels above known requirements to give a margin of safety in the event that vitamin levels are reduced by pelleting and feed storage. The cost of this extra supplementation is not a significant proportion of feed costs, so at present there is little incentive to precisely estimate the requirements of growout salmon for micronutrients. This may become a higher priority in the future, especially if plant protein sources replace a portion of fish meal in salmon feeds, thereby lowering the levels of trace minerals.

Atlantic salmon have the distinction of being the first salmonid species for which the carotenoid pigment astaxanthin has been demonstrated to be essential, albeit as a vertically transmitted micronutrient. Christiansen et al. (8) demonstrated that fry from female Atlantic salmon deprived of astaxanthin require it in their feed, while fry from females fed diets containing astaxanthin do not.

Salmon require a dietary source of n-3 fatty acids, between 1 and 2% of the diet. These fatty acids are supplied by fish oil. The only plant oil containing significant amounts of an n-3 fatty acid is linseed (flaxseed) oil. Other fat sources can be used in salmon feeds, providing that the diet contains at least 1–2% n-3 fatty acids. Soybean oil and beef tallow have been used to replace fish oil in Atlantic and Pacific salmon feeds; and, as expected, the fatty acid profile of fish tissues reflects that of the dietary fat source (9,10). The n-3 fatty acid levels fillets made from farmed salmon are equivalent to levels found in wild Pacific salmon when the feeds of farmed fish contain fish oils rich in n-3 fatty acids.

FEED FORMULATION

Formulating feeds for salmon is relatively simple. Feeds are not formulated by establishing minimum levels of essential dietary nutrients, selecting from a range of feed ingredients, establishing minimum and maximum levels for various feed ingredients, or commanding a computer to find the least-cost formulation. Rather, formulations are generally limited to just a few feed ingredients, with limits placed solely on dietary protein and fat levels and total phosphorus levels. Formulations for salmon have thus been developed empirically, with limits on total phosphorus and feed conversion ratios being the principal factors determining formulations (Table 4). Vitamin and mineral premixes provide these essential nutrients at appropriate levels. Carotenoid pigmentation for coloration of the final product is supplied by Carophyll Pink™, astaxanthin produced by chemical synthesis, or, less commonly, from natural sources such as krill, yeast, or algae. Because it is chemically synthesized, Carophyll Pink™ contains equal amounts of R and S isomers,

Table 3. Dietary Micronutrient Requirements and Their Deficiency Signs for Atlantic Salmon

Micronutrient	Requirement[a]	Deficiency Sign
Vitamin E	35 mg/kg diet	Muscle degeneration, anemia, reduced carcass protein, increased carcass moisture and fat
Pyridoxine	15–20 mg/kg diet	Nervous disorders, anorexia, reduced alanine transferase activity
Vitamin C	50 mg/kg diet	Scoliosis, lordosis, anemia, mortality
Vitamin K	Required	Not determined
Riboflavin	Required	Not determined
Pantothenic acid	Required	Not determined
Phosphorus	0.7% of diet	Low bone ash, calcium, phosphorus and magnesium, bone abnormalities, poor growth
Manganese	20 mg/kg diet	Reduced hematocrit and vetebral manganese
Copper	6 mg/kg diet	Reduced serum copper and liver cytochrome C oxidase activity
Iron	73 mg/kg diet	Reduced hematocrit, red blood cell count and tissue iron level
Selenium	Required	Muscular dystrophylike signs
Iodine	Required	Not determined
Astaxanthin	5 mg/kg	Poor growth, mortality in fry from females deprived of dietary astaxanthin

[a] Dietary requirements determined in one or two studies only, and generally with fry or fingerlings. Adapted from Hellend, Storebakken, and Grisdale–Helland, in R.P. Wilson, ed., *Handbook of Nutrient Requirements of Finfish*, CRC Press, Boca Raton, 1991, pp. 13–22.

Table 4. Generalized Formulations Used for Grower and Fingerling Atlantic Salmon Feeds

Feed Ingredient	Grower Feed (g/kg)	Fingerling Feed (g/kg)
Fish meal	550	630
Soybean meal	50	0
Wheat by-products	0	230
Ground whole wheat	104	0
Vitamin premix	10	10
Trace mineral premix	1	1
Choline chloride (60%)	4	4
Ascorbic acid	1	1
Fish oil	280	124

whereas the S isomer of astaxanthin predominates in natural sources (e.g., krill). Forensic analysis of astaxanthin isomers in tissue samples can be used to determine whether fish are of wild or farmed origin. Protein sources constitute the largest proportion of the ingredient costs of a typical Atlantic salmon feed, followed by carotenoid pigmentation, fat, and vitamin and trace mineral premixes (Fig. 1).

FEEDS AND FEEDING

When salmon farming began, fish were fed moist pellets originally developed for juvenile Pacific salmon raised in enhancement hatcheries. These moist feeds were composed of dry mixtures combined with fish hydrolysates (11). Dry pellets made by compression pelleting were soon available to farmers. Feed conversion ratios with these feeds were close to 2 : 1, partly because of the quality of ingredients and formulations used at that time and partly because compressed pellets are dense, which causes them to fall through cages before the fish could eat them all. In the 1980s, several events changed salmon feed formulations and pelleting methods. First, fish meal manufacturers began producing high-quality fish meal specifically for use in salmon feeds (12). These fish meals were produced from fresh raw material using low-temperature drying. Second, the introduction of cooking-extrusion pelleting for salmon feeds enabled feed manufacturers to produce less dense pellets that slowly sank in the marine cages, reducing the proportion of uneaten feed. Extruded pellets can be sprayed (top dressed) with fish oil to achieve much higher levels of fat than can be achieved with compressed pellets, making it possible to produce high-energy feeds. Third, limits were established in some countries on total phosphorus and feed conversion ratios to reduce the impact of fish farming on the aquatic environment. Limits on total phosphorus in the feed and on feed conversion ratios eliminated many feed ingredients from salmon feeds, especially those high in ash and indigestible fiber. High-ash fish meal contains large amounts of calcium and phosphorus from fish bone, which in addition to contributing to phosphorus pollution, was a cause of cataracts in young salmon because it reduced the bioavailability of dietary zinc (13–15); The overall result of these changes has been to lower feed conversion ratios to 1.0 : 1 or less, meaning that for every kg (or lb) of feed that is fed, the fish gain one kg (or lb) of weight, and to increase fish growth rates.

HEALTH MANAGEMENT

Given the high densities at which Atlantic salmon are reared, it is to be expected that maintenance of health can be difficult, as is the case with any farmed species. The aquatic medium and the ectothermic physiology of fishes means that, with the treatment of compared higher animals, there are many differences in approach to health management and disease control. Also, the principal infectious agents of salmonid fishes often have no direct counterparts in higher animals, so unique therapies must be used.

The principal diseases of cultured Atlantic salmon vary depending on the geographical location. In Norway and Scotland, for example, the three most serious conditions are infectious pancreatic necrosis (IPN), furunculosis, and salmon louse infection; whereas in Chile, only IPN and rickettsial infection are significant problems. In Australia, a different pathogen is prominent in salmon farming, namely a paramecium that invades the gills of fish during periods of high water temperatures.

A major development in Atlantic salmon health management has been the introduction in 1992 of oil-adjuvant antibacterial vaccines. These vaccines, originally monovalent against *Aeromonas salmonicida* (the cause of furunculosis), can now incorporate antigens to provide protection against a number of *Vibrio* strains and even an element of protection against IPN. Prior to the arrival of successful vaccination, the only means of control for microbial infections was the use of antibiotics in the feed and resting (fallowing) of sites for prolonged periods to break the infection cycle (16).

New diseases are continually being identified and addressed in the context of control and husbandry adjustment. Viral diseases are still a major concern because there are no treatments to prevent disease and because viral diseases can be transferred to new locations via egg exports. Currently, viruses of particular concern to salmon farmers are IPNV and infectious salmon anemia virus (ISAV). Egg producers have extensive monitoring and certification procedures in place to assure customers (and regulatory agencies) that salmon eggs are free from specific pathogens.

GENETICS AND REPRODUCTION

The closing of the reproductive cycle for farmed Atlantic salmon in the 1960s allowed the true domestication of the species for the first time. Extensive mass selection programs, particularly in Norway (17), allowed significant improvement in a number of heritable factors, including adaptability to confinement, proportion of S1 smolt production, growth rate, and feed conversion ratio. The exact contribution attributable to genetic selection as opposed to improvement in fish husbandry and nutritional quality of feeds is unquantifiable, but it is nevertheless likely to be significant (18).

The development of molecular genetics, and in particular its application to the analysis of the Atlantic salmon genome, has allowed microsatellite DNA probes to be defined leading to pedigree analysis at the family level (19). Subsequent work has allowed genetic fingerprinting to be applied to commercial aquaculture as a routine production tool.

Thus, the genetic fingerprint of highest performing production fish under commercial rearing conditions can be determined, and siblings among broodstock populations can be identified and their performance assessed. Best crosses can then be identified from among the broodstock based upon performance under commercial conditions, allowing for very rapid pedigree enhancement.

This technology, using standard breeding procedures rather than the controversial genetic engineering technology, is expected to contribute even further when specific genetic markers for particular desirable traits are identified. Such developments are expensive and technologically sophisticated and are, therefore, likely to lead to lead to further subdivision of the salmon farming industry into egg and smolt producers, large scale rearers, and downstream processors, as has happened in the swine and poultry industries.

Broodstock developments have not been confined to the genetic sector. Requirements for higher quality eggs, free from specific diseases and available across a long spawning period, have led to a variety of manipulations of the broodstock cycle. Diets have also improved over this period, although there is still a great deal to be learned about the nutritional requirements of salmon in the critical maturation period before fish cease feeding prior to spawning.

More important has been the increased understanding of the environmental cues that trigger the various stages of maturation and spawning (20). This has led to commercial exploitation of modified photoperiod, earlier freshwater exposure, and temperature control to allow normal egg and milt availability over a much longer timescale. Eggs can thus be produced and fertilized over a four-month period instead of the two-month window for natural spawning. Similarly, commercial exploitation of cryotechnology for both short-term and long-term cryopreservation of milt (though not of ova) has allowed genetically superior males to be used across wider ranges of maturing females and even across spawning generations (21).

The ultimate aim of such programs is to concentrate on the production cycle such that two crops of smolts per year can be produced for on growing in the sea on a 12-month time frame to harvest. Progress to this end is already considerable, but the final goal may be some years away.

INDUSTRY TRENDS

Atlantic salmon feeds have been overly dependent on fish meal as a source of dietary protein, but this is likely to change in the next few years. Currently, efforts are underway to evaluate alternate protein sources for Atlantic salmon feeds. Protein sources of plant origin under study include wheat gluten, corn gluten, soy protein concentrates, and soybean meal blends. In Chile, rendered products (e.g., poultry by-product meal, feather meal, and blood meal) are being used in limited amounts to supply a portion of the protein in Atlantic salmon feeds. In Canada, rapeseed protein concentrate is a possible protein source under evaluation in addition to the ingredients just listed. Information is beginning to accumulate on apparent digestibility coefficients of protein in these and other feed ingredients for Atlantic salmon.

Another future possibility in Atlantic salmon feeds is the "square diet," so-called because it is formulated to contain 35% protein and 35% fat. This diet would be fed to fish in sea cages, not juveniles in freshwater. The point of feeding such a diet would be to increase the percentage of dietary protein used by the fish to synthesize fish protein, as opposed to the fish using dietary protein as a source of metabolic energy. Whatever the future holds with respect to dietary fat level, it is critical that information be developed concerning the availability of amino acids in various protein sources used in Atlantic salmon feeds and that the dietary requirements of Atlantic salmon for essential amino acids be determined, thus allowing feed formulators to explore various possible alternate protein sources without unintentionally formulating a diet that lowers the growth rate of the fish.

Great emphasis is currently being placed on year-round smolt supplies and reduction of time to market from sea cages. This is leading to year-round availability of standard product at prices that compete well with other fish and white-meat products in the market. The added value of processed specialty products is also being widely addressed. Such developments push the product inexorably toward the high-volume commodity sector, as opposed to the high-value niche market, to which it originally was aimed.

Pressure on rearing sites for both freshwater and marine stages will tend to limit production in Europe and North America, unless more exposed, offshore locations can be utilized. However, the capital cost of large, offshore structures is high; and there is still considerable capacity for expansion in South America at lower cost.

Recirculation or, at least, water rehabilitation by removal of organic material and phosphates is slowly increasing in usage in the freshwater stage of salmon farming. The technologies are still developing and capital costs are high. Freshwater availability is likely to be the limiting factor for all aquaculture development in the long term, so replacement of lake-based cage systems with recirculation (water recycling) systems or engineered flow-through systems for high value, genetically improved salmon smolt production seem likely to develop.

Atlantic salmon are only 10 generations from the wild and already the domestication traits that have been expressed are distinctive and have contributed greatly to the productivity of the industry. Newer technologies involving family as opposed to mass selection and the use of sophisticated techniques such as DNA pedigree analysis, are being adopted by leading producers of eggs and smolts. Genetic engineering has been demonstrated to have great potential; but given the concern of consumers about genetic engineering, it is unlikely that this approach will be utilized by the salmon industry to improve farming productivity.

BIBLIOGRAPHY

1. C.V.M. Mahnken, *Proc. Sixth Annual Meeting*, World Mariculture Society, Louisiana State University, Baton Rouge, 1975, pp. 285–298.
2. A.J. Novotny, *Marine Fish. Rev.* **37**(1), 36–47 (1975).
3. O.J. Torrissen, J.C. Holm, G. Naevdal, and T. Hansen, *World Aqua.* **26**(3), 12–20 (1995).
4. J.E. Halver, in J.E. Halver, ed., *Fish Nutrition*, 2nd ed., Academic Press, NY, 1989, pp. 31–109.
5. R.W. Hardy and R.J. Roberts, *International Aqua. Feed*, July–August, (1998).
6. D.A. Higgs, B.S. Dosanjh, A.F. Prendergast, R.M. Beames, R.W. Hardy, W. Riley, and G. Deacon, in C.E. Lim and D.J. Sessa, eds., *Nutrition and Utilization Technology in Aquaculture*, AOCS Press, Champaign, IL, 1995, pp. 130–156.
7. National Research Council (NRC), *Nutrient Requirements of Fish*, National Academy Press, Washington, DC, 1993.
8. R. Christiansen, O. Lie, and O.J. Torrissen, *Aqua. Nutr.* **1**, 189–198 (1995).
9. R.W. Hardy, T.M. Scott, and L.W. Harrell, *Aquaculture* **62**, 267–277 (1987).
10. R.W. Hardy, T. Masumoto, W.T. Fairgrieve, and R.R. Stickney, *Proc. Third Int. Symp. on Feeding and Nutri. in Fish, Aug. 28–Sept. 1*, Toba, Japan, 1990, pp. 347–355.
11. R.W. Hardy, in J.E. Halver, ed., *Fish Nutrition*, 2nd ed., Academic Press, NY, 1989, pp. 473–546.
12. R.W. Hardy and T. Masumoto, in S. Keller, ed., *Proc. International Conference on Fish By-Product*, Alaska Sea Grant Program, Fairbanks, 1990, pp. 109–120.
13. H.G. Ketola, *J. Nutr.* **109**, 965–969 (1979).
14. R.W. Hardy and K.D. Shearer, *Can. J. Fish. Aquat. Sci.* **42**, 181–184 (1985).
15. N.A. Richardson, D.A. Higgs, R.M. Beames, and J.R. McBride, *J. Nutr.* **115**, 553–567 (1985).
16. A.L.S. Monro and T. Hastings, in V. Inglis, R.J. Roberts, and N.R. Bromage, eds., *Bacterial Diseases of Fish*, Blackwell Science, Oxford, 1993, pp. 122–142.
17. T. Gjedrem, *Aquaculture* **6**, 23–29 (1975).
18. T. Refstie, *Aquaculture* **85**, 163–169 (1990).
19. J.B. Taggart, P.A. Prodahl, and A. Ferguson, *Animal Genetics* **26**(1), 13–20 (1995).
20. N.R. Bromage, in N.R. Bromage and R.J. Roberts, eds., *Broodstock Management and Egg and Larval Quality*, Blackwell Science, Oxford, 1995, pp. 1–24.
21. K.J. Rana, N.R. Bromage, and R.J. Roberts, eds., *Broodstock Management and Egg and Larval Quality*, Blackwell Science, Oxford, 1995, pp. 53–75.
22. S. Hellend, T. Storebakken, and B. Grisdale–Helland., in R.P. Wilson, ed., *Handbook of Nutrient Requirements of Finfish*, CRC Press, Boca Raton, 1991, pp. 13–22.

SEA BASS CULTURE

George Wm. Kissil
Amos Tandler
Angelo Colorni
Abigail Elizur
National Center for Mariculture
Elat, Israel

OUTLINE

The European sea bass, a member of the recently revised family Moronidae, is found from the Black Sea westward across the Mediterranean Sea, out into the eastern Atlantic Ocean north to Ireland and the Baltic and North Seas, and south to Morocco and Senegal (1). The genus *Dicentrarchus* contains two species: *D. labrax* and *D. punctatus*, which are very similar in appearance except for the presence of black spots on the back and sides of the latter. Sea bass are highly prized throughout most of their range and, like the gilthead sea bream, are heavily fished and thus were early candidates for domestication. In 1996, the reported catch of sea bass reached close to 15,000 metric tons (mt) (2) (16,500 short tons) (st), while aquaculture production exceeded 26,500 mt (3) (29,150 st). Sea bass are farmed similarly to gilthead sea bream in seawater ponds and lagoons, with the bulk of production occurring in sea cages. Greece, Turkey, and Italy are the three biggest producers of sea bass, yielding over 70% of sea farm production. Sea bass culture has paralleled sea bream development, rising from 270 mt (297 st) in 1985 (4,5) to 26,500 mt in 1996 (3), with the 1998 production expected to reach 34,100 mt (6) (37,500 st).

LIFE HISTORY

Sea bass frequent coastal inshore waters, occurring in estuaries and brackish water lagoons, and sometimes venture upstream into fresh water (1). Their euryhaline nature allows them to grow in completely fresh water without reproducing. As a result sea bass have been used as natural predators to control the proliferation of unwanted fish in freshwater culture systems and water carriers (7). Spawning of sea bass takes place from November to March in the Mediterranean basin (8,9) with the spawning period being delayed in the northern parts of its distribution (1,8). Adults spawn in estuaries and inshore areas where the salinity of the water is close to that of seawater (30 ppt) (3.0%) (8).

Sea bass are round fish with elongated bodies and wide mouths, which they use to catch small fish and a variety of invertebrates, including shrimp, crabs, and squid and other cephalopods (10). Their bodies are grey to greenish black along the back, with silvery sides and white bellies. Both *D. labrax* and *D. punctatus* have dark spots on their back and sides as young fish. These spots disappear in *D. labrax* as they mature (7). Sea bass can reach maximum lengths of 1 m (40 in.) and weights of 15 kg (33 lb), although fish of 0.5 m (16.5 in.) and 4–6 kg (8.8–13.2 lb) are more commonly caught (11).

REPRODUCTION AND LARVAL REARING

Sea bass have been the subject of many studies, which have helped in understanding reproduction in marine fish (9,13–18). Sea bass were historically cultured along with other euryhaline species—gilthead sea bream and grey mullet—in coastal lagoons and tidal reservoirs for many years before a race to develop mass production of juveniles started in the late 1960s (5). Fish culture was initially associated with the production of salt in coastal evaporation pans and marshes. The salt was harvested during the high evaporation seasons of summer and autumn, and fish were cultured during winter and spring. The supply for this culture came from trapping schools of fish that lived in these estuarine areas and fed on the rich fauna flourishing in the nutrient-rich waters.

During late 1960s, research groups in both France and Italy competed to develop reliable mass-production techniques for juvenile sea bass. The fish were considered a "safe bet" for developing these techniques due to their large larvae, which are three times the size of sea bream at the yolk sac stage. By 1976–77, mass-production techniques were well enough developed to provide hundreds of thousands of larvae in France and Italy (5). This production reached 4.5 million by 1985, with Yugoslavia leading France and Italy in numbers produced (4). Some 147 million juveniles were reportedly produced during 1997 (12), some of which may have been from capture fisheries.

Sea bass are a dioecious species. Fish develop as either males or females, which, unless artificially induced, do not change sex during their lifetime. Males can produce sperm at one year, but females do not reach sexual maturity until the age of three. Photo and thermal cues control reproduction in sea bass. The fish respond to the shortened length of the day and to a reduction in water temperature, the latter being critical for egg quality.

Optimal reproduction can be obtained if brood stock are kept throughout the year at densities no higher than 5 kg/m^3 (11 lb/264 gal), in tanks that allow them to swim freely. Maintaining the temperature below 16 °C (61 °F) outside of the spawning season helps in obtaining high-quality eggs, as does feeding the fish squid a few times each week, starting two months prior to spawning. Dietary lipids have been found to be important for sea bass reproduction, because long-term deficiencies in *n*-3 highly unsaturated fatty acids (*n*-3 HUFA) can induce early gonadal atresia, lower fecundity, and subsequent reduction in egg survival (19). At the onset of the reproductive season, the fish can be transferred to smaller tanks (1 m^3 = 264 gal) at densities of up to 2.5 kg/m^3 (5.5 lb/264 gal). During this time, the maintenance of a low water temperature ($\leq$13 °C = 55.4 °F) is important for spontaneous spawning.

In the eastern Mediterranean, captive populations start spawning in November, about a month before the shortest day of the year, and will continue until March, provided that the water is maintained below 15 °C (59 °F). As the spawning season approaches, the female's oocytes undergo vitellogenesis. The process stops just before maturation, when the oocytes reach 600–750 μm (2.36–2.95×10^{-2} in.). The oocytes remain unchanged for over a month, at which point they undergo atresia if the female is not induced to spawn. Sea bass have an annual reproductive cycle with group synchronous ovarian development and are multiple spawners (20). In captivity, up to six induced spawnings per season have been observed, with the number of eggs produced by a female per spawn reaching 5–20% of her body weight, or a total exceeding 1 million eggs per season.

If spontaneous reproduction does not occur in sea bass, then hormonal treatment with luteinizing hormone releasing hormone (LHRH) analogs either by injection, controlled-release implants, or microcapsules can be used to trigger maturation, ovulation, and spawning in 54–68 hr. During the time from hormonal induction to actual spawning, germinal vesicle migration (GVM) and breakdown (GVBD), hydration, and ovulation can be observed in a proportion of the oocytes (>50%). After the fish have spawned these matured eggs, another batch of oocytes is recruited and respond to the hormonal treatment in the same way—undergoing GVM, ovulation, and spawning (20). This process is induced six times over a three-month period (National Center for Mariculture, unpublished data).

In addition, injections of gonadotropin-releasing hormone (GnRH) analogs have been consistently used to induce spawning in sea bass. Two injections, 24 hours apart, of 15 and 10 μg/kg (6 and 4×10^{-7} oz/2.2 lb) of fish, or a single injection of 10–15 μg/kg (4–6×10^{-7} oz/2.2 lb), have been effective in inducing spawning. The time of injection has an effect on the rate of GVBD and ovulation, both being faster if hormonal treatment is administered in the morning rather than at night (21). Males usually have good-quality sperm without any treatment, but when levels of milt appear low, a 5 μg/kg (2×10^{-7} oz/2.2 lb) injection of GnRH analog can be highly effective in increasing short-term spermiation. Sustained-release GnRH analog increases milt volume and prolongs spermiation while maintaining high-quality sperm (22).

The strong influence of photoperiod on sea bass reproduction makes this fish a candidate for extending its reproductive season throughout the year. Providing groups of parent stock sea bass with artificial lighting that mimics changes in the length of the day and that shifts the shortest day, coupled with reduced water temperatures, allows year-round spawning (23). Advancing the time of spawning can also be achieved, in this case, by maintaining a constant short photoperiod of 9 hr of light and 15 hr of darkness (24).

Larval rearing of sea bass is based on years of extensive research into environmental and biotic factors affecting growth and survival (25,26). The results of this research have been the development of successful mass rearing techniques that are presently used in hatcheries around the Mediterranean region. Techniques vary among the hatcheries, depending upon local experience, but all are based on the extensive research efforts carried out in the region since the 1970s.

Sea bass eggs are hatched and the larvae reared for 43 days at the National Center for Mariculture in Israel using the following procedure: Fertilized eggs are stocked at 100/L (380/gal) in either 400–1,700-L (104–442-gal) cylindroconical fiberglass tanks or 5,000–10,000-L (1,300–2,600-gal) flat-bottom tanks. Hatching success is estimated from aliquot samples. Rearing starts at seawater temperatures of $16.5 \pm 0.5\,^{\circ}\mathrm{C}$ ($61.7 \pm 0.9\,^{\circ}\mathrm{F}$) with the temperature gradually rising to $24.5 \pm 0.5\,^{\circ}\mathrm{C}$ ($76.1 \pm 0.9\,^{\circ}\mathrm{F}$) by the last 9 days of the 43-day rearing period. Filtered seawater ($10\ \mu\mathrm{m} = 3.9 \times 10^{-4}$ in.) is continuously supplied to these tanks at a rate of one to four exchanges each day, depending upon the age of the larvae. The larvae are continuously supplied with freshly enriched rotifers (*Brachionus plicatilis* at a concentration of 10/mL = 296/oz) as their first food followed by newly hatched or enriched Artemia nauplii (*Artemia salina* at a concentration of 1/mL = 29.6/oz) as the second. In addition, the tanks are continuously supplied with single-celled green algae (*Nannochloropsis* sp.) through a special delivery system designed to maintain the algal concentration at 5×10^5 cells/mL (148×10^5/oz) during the 15 hr of light the larvae are given. Larval survival under these rearing conditions will be 35% by day 43, when the sea bass reach an average weight of 80 mg (2.8×10^{-3} oz).

An easier rearing procedure, which eliminates the use of rotifers by going directly to Artemia nauplii at first feeding, is also in use (26). This technique was developed to reduce the dependence of hatcheries on mass culture of microalgae and rotifers. It exploits the internal energy reserves of newly hatched sea bass larvae and has been reported to provide a 60% survival rate. (26). Eggs and newly hatched larvae are reared in complete darkness up to the stage of eye pigmentation and opening of the mouth. This is believed to save energy by reducing the spontaneous activity of the larvae. In addition, the salinity is lowered to 25 ppt (2.5%) in their tanks, which reduces the energy necessary for osmoregulation by the fish larvae.

Rearing procedures for sea bass have been improved by developing enrichment techniques to modify the n-3 HUFA composition of the live rotifers and Artemia the larvae eat. Studies have shown that the growth and survival of sea bass, in contrast to larvae of gilthead sea bream, are independent of the dietary level of n-3 HUFA within the range of 7.4–41.6 mg/g of dry Artemia ($2.6–14.6 \times 10^{-4}$ oz/3.52×10^{-2} oz dry Artemia) (25). Although at present, sea bass larval nutrition is based mainly on the enrichment of live food, the development of microdiets to replace it has been the goal of many research groups in the region (27).

Once sea bass pass the critical first 43 days after hatching, special care is given to minimize mortalities due to cannibalism. This is done by mechanically grading the fish into 2–3 size groups which are then reared separately. The more equal the sea bass are in size, the more cannibalism is reduced and survival improved to the point where the fish are stocked into growout systems.

COMMERCIAL CULTURE

The European sea bass, like its main competitor in Mediterranean aquaculture, the gilthead sea bream, is grown mainly in sea cages, although more land-based systems are being developed as competition for cage sites increases. This competition is the result of increased urbanization, tourism, industry, and environmental regulation. The growout density of sea bass in these systems varies from approximately 2 kg/m^3 (1.7×10^{-2} lb/gal) in extensive ponds to 20–30 kg/m^3 (0.17–0.25 lb/gal) in intensive raceways with high water exchange to 20 kg/m^3 (0.17 lb/gal) in cages (11).

The sea bass is a slower growing fish than the sea bream, requiring up to two years to mature to a commercial size of 250–350 g (8.8–12.3 oz) in the Mediterranean region (11). Improvements in feeds, better management practices, and improved quality of juveniles have all contributed to a reduction in the time needed to reach commercial sizes. Today, sea bass reach market sizes in approximately 16 months in sea cages in the Mediterranean region. Commercial sizes for cultured sea bass vary from 300–1,000 g (10.6–35.2 oz), while much larger fish of 2–3 kg (4.4–6.6 lb) usually come from fishing.

NUTRITIONAL REQUIREMENTS

Since the start of R & D to develop sea bass for mariculture, over the past three decades there have been intensive efforts to identify the nutritional needs of the various life stages of the fish. The development of this information has not been completed, but enough information is available to allow the formulation of practical commercial feeds that are in use today wherever sea bass are being cultured. Table 1 summarizes the available information on the nutritional requirements of this fish at this time.

DISEASE PROBLEMS

The sea bass is a euryhaline and eurythermal fish that lives in marine, brackish and, occasionally, fresh water. Although a sturdy species, sea bass are subject to a wide range of diseases under culture conditions. These outbreaks have had devastating effects on commercial production and have prevented the expansion of the industry in some Mediterranean countries.

VIRAL DISEASES

Encephalitis Viruses

Bass viral encephalitis is characterized by whirling movements, a loss of balance, and hyperexcitability in

Table 1. Summary of Known Nutritional Needs of the European Sea Bass[a]

	Percent of Diet Unless Otherwise Indicated
Protein	
Total dietary level	
Larval/juvenile	50–60[28,29;cf]
Growout	45–50[29;cf]
Indispensable amino acids	
Arginine	3.9[30] (% of dietary protein)
Lysine	4.8[31] (% of dietary protein)
Methionine + cysteine	4.4[32] (% of dietary protein)
Tryptophan	0.5[33] (% of dietary protein)
Estimates of remaining (IAA)[34b]	
Histidine	1.6[34] (% of dietary protein)
Isoleucine	2.6[34] (% of dietary protein)
Leucine	4.3[34] (% of dietary protein)
Valine	2.9[34] (% of dietary protein)
Phenylalanine + tyrosine	2.6[34] (% of dietary protein)
Threonine	2.7[34] (% of dietary protein)
Lipid	
Total dietary level	15–20[29c]
n-3 HUFA = EPA + DHA	
Larvae (43–75 days)	1.0[35]
Growout to commercial size	1.5–2.7[cfd]
Broodstock	Essential[36e]
Energy	
Maintenance requirement[f]	40.6 kJ/kg fish$^{0.8}$ [g]
Growth requirement	20–23 MJ/kg feed[rv]
Vitamins	NRC recommendations for rainbow trout[h]
Minerals	Commercial premixes[rv]
P : E ratio	Only partial data available[37,38]

[a]Numerical superscripts on values listed in table indicate research data from published literature listed in references; "rv" = recommended values; "cf" = values found in commercial feeds for sea bass.
[b]Estimates of IAA based on whole-body IAA to total IAA ratios.
[c]High-energy diets may reach up to 26%.
[d]Levels in high-energy commercial diets.
[e]EPA + DHA known to be essential for good egg quality, but quantities and DHA – EPA ratios are unknown.
[f]Energy requirement at zero growth.
[g]I. Lupatsch and G.Wm. Kissil, unpublished data.
[h]NRC 1993 recommendations for rainbow trout have been shown to be sufficient for *D. labrax* when used in practical diets (39).

response to noise and light. In the brain and other nervous tissue, including spinal cord and retina, extensive vacuolization is typically observed. Mortality often occurs within one week from the onset of the first symptoms. The syndrome was described in European sea bass cultured in Martinique, in the French Caribbean Islands (40,41). Possibly, the picornalike virus described in the same host in France was responsible for the disease (42). Recently, these viruses were found to be more closely related to the Nodaviridae family (43). No effective therapy is known.

BACTERIAL DISEASES

Pasteurellosis

The bacterium responsible for this disease, *Pasteurella piscicida*, is a halophilic member of the Vibrionaceae family. A new name for the bacterium, *Photobacterium damsela* subsp. *piscicida*, has been proposed (44). The disease is widespread in the United States, in Japan, and all around the Mediterranean basin. The infection develops rapidly into an acute septicemic condition characterized by an enlarged spleen with typical foci of bacterial microcolonies. In advanced cases, these lesions appear on the surface of the spleen as whitish spots and patches. Typically, *P. damsela piscicida* is a rather unreactive, gram-negative, nonmotile, oval-shaped, 0.5×1.5-μm ($1.97 \times 10^{-5} \times 5.9 \times 10^{-5}$-in.) rod, seen through the microscope with pronounced bipolar staining.

Commercial vaccines against *P. damsela piscicida* are available, but so far their effectiveness has been limited, in particular because the vaccine is given during the most vulnerable stages of fish development (postlarval and juvenile), in which the fish have an underdeveloped immunological system. Early detection and the use of medicated feed, before the fish lose their appetite, offers the best chance of saving infected fish. A characteristic of this bacterium is the speed with which it develops resistance to a broad range of antibiotics. An antibiogram should be performed before any treatment, to evaluate the effectiveness of the drug to be used. *P. damsela piscicida* is highly contagious, and strict sanitary measures should be adopted to limit its spread.

Streptococcosis

The common symptoms of streptococcosis are exophthalmia, hemorrhages, and distension of the abdominal area. Serosanguinous fluid accumulates in the peritoneal cavity and intestine. Diagnosis of this disease can only be done by culturing the pathogen, which grows best when blood is added to the culture medium. Streptococci are non-motile, spherical, or ovoid bacteria that stain gram positive and form long chains in liquid culture. In Israel, the strain isolated from *D. labrax* was identified as *Streptococcus iniae*.

Mycobacteriosis

The etiological agents of mycobacteriosis, *Mycobacterium* spp., are slender, acid-fast rods that cause systemic, chronic, and subacute infections in fish and, on occasion, skin ulcers in humans. These bacteria require special media (Löwenstein–Jensen or Middlebrook) for culturing, which is usually a slow process, taking two to three weeks for the first colonies to become visible. Fish mycobacteriosis is widespread throughout the world in both marine and freshwater environments. *M. marinum*, the bacterium causing this disease, has been isolated from sea bass in Italy, Belgium, Denmark, Greece, and Israel (45,46). Superficial ulcers and exophthalmia are the only external signs of the disease, which can remain asymptomatic for a long time. The spleen and kidney, however, are severely affected and appear enlarged and covered with whitish nodules. In advanced cases, these lesions spread to all other visceral organs. Antibiotic treatment is generally unsuccessful.

PROTISTAN PARASITES

Amyloodiniosis

Amyloodinium ocellatus is a dinoflagellate that is highly adapted to parasitism. This organism is not selective in its host choice and is responsible for one of the most devastating parasitic diseases in temperate (47,48) and tropical mariculture (49–52). Its life cycle involves three main phases: trophont (the parasitic, feeding stage), tomont (the encysted, reproductive stage), and dinospore (the free-swimming, infective stage). The trophonts feed on the fish's epithelia by means of rhizoids, then drop from the fish, encyst on the bottom, and start dividing. The reproductive process finishes in two to three days at $24 \pm 2\,^{\circ}C$ ($75 \pm 3.6\,^{\circ}F$) with the release of 64–128 infective, highly motile dinospores from each tomont that begin the life cycle of the parasite again (52). Dinospores remain infective for several days.

Copper compounds are effective against *A. ocellatum*, but care must be taken when using them due to the toxic effects of copper on fish. The concentrations used to kill the pathogen are close to the levels at which copper acts as a poison to membranes and can harm the fish's gills, liver, kidney, and nervous system. In addition, its immunosuppressive nature is a major drawback to the use of copper in treating fish (53). Despite these drawbacks, the treatment of infected fish is carried out over a 12–14-day period by maintaining a concentration of 0.75 mg/L (2.39×10^{-3} oz/qt) of $CuSO_4$ in the water (51), using a slow drip of a concentrated solution and closely monitoring its concentration in the water.

Dipping infected fish in fresh water for a few minutes will dislodge most of the parasites, but will not kill them. This treatment causes the parasites to turn into tomonts, which start to divide as soon as the seawater flow is restored.

Cryptocaryonosis

Cryptocaryonosis is caused by the ciliate *Cryptocaryon irritans*, of the class Colpodea (54), which is known to have intraspecific variants (55–57). Although typical of tropical seas, this parasite has a worldwide distribution that extends well into temperate environments. *C. irritans* invades the skin, eyes, and gills of its host and impairs the functioning of these organs. External symptoms of the disease are the appearance of pinhead-sized whitish "blisters" on the skin and increased production of mucus. The diagnosis of cryptocaryonosis is made by microscopic examination of gill, fin, or skin tissue to determine the presence of the large, revolving ciliate. The ciliate's life cycle involves four phases, the first is parasitic (trophont) during which it feeds on the fish's epithelia. After three to seven days of growth, it leaves its host, loses its cilia (protomont), encysts, and starts dividing (tomont), eventually producing up to 200 free-swimming infective organisms (theronts). The number and size of theronts produced vary with geographic location, the species of the fish infected, and water temperature (58). Theronts have a life span of 24 hr, but their ability to infect a host decreases rapidly after 6–8 hr (59,60).

Myxosporean Infections

Myxosporeans are polynucleate, usually immobile parasites that can either reside in visceral cavities, such as the gallbladder, swim bladder, and urinary bladder (celozoic species), or settle as inter- or intracellular parasites in blood, muscle, or connective tissue (histozoic species). They produce spores that are subspherical in shape and several μm ($= 4 \times 10^{-5}$ in.) in size, with characteristic polar capsules and coiled filaments. *Sphaerospora testicularis* and *S. dicentrarchi* were detected in sea bass in Spain (61), and *S. dicentrarchi* was found in sea bass in Greece. Both species cause relatively inconspicuous infections in various organs of the fish.

METAZOAN PARASITES

Monogenean Flukes

Most monogenean flukes display a narrow host specificity and, consequently, a relatively benign relationship with the host they parasitize. Diplectanids are common on farmed Serranidae. These parasites infest gills and feed on mucus, but can cause severe hyperplasia and hemorrhages where the hooks of the parasite are inserted into the fish (62). The bodies of monogenean flukes are elongated and characterized by a large, flat attachment organ (opisthaptor) with squamodiscs at the posterior end. European sea bass cultured in the Mediterranean and Red Seas are infested with *Diplectanum aequans* and *D. laubieri* (63,64).

Sea Lice

Caged fish are particularly vulnerable to parasitic copepods and isopods, which can be a mild nuisance or cause debilitating infestations. Sea lice appear to be attracted by the fluid secretions from skin wounds, so that fish injured by handling are easy targets during the first few days after being stocked into cages. The life cycle of these crustaceans comprises several free-living and infective stages and is typically faster at higher temperatures.

Caligus minimus is a copepod known to parasitize only sea bass, selecting the roof of the fish's mouth and its gill rackers as preferred infestation sites (65). The copepods' second antennae and maxillipeds are claw shaped, so adhesion to the host is assured through pressure from the parasite's shield-like cephalothorax acting as a sucker. Active feeding may cause deep ulcers.

Pranizae (larval stages) of gnathiid isopods feed on blood. Most of them are highly specialized to specific hosts and stay continuously attached to them, while others such as *Gnathia piscivora* and *Elaphognathia* sp., are indiscriminate in their host selection and, once engorged with blood, abandon the fish. In the Red Sea, they feed and molt three times before maturation to nonparasitic adults (66).

BIBLIOGRAPHY

1. E. Tortonese, in P.J.P. Whitehead, M.L. Bauchot, J.C. Hureau, J. Nielsen, and E. Tortonese, eds., *Fishes of The North-Eastern Atlantic and The Mediterranean*, Unesco, Paris, 1984, pp. 793–795.
2. FAO, *Yearbook of Fishery Statistics*, capture production, vol. 82, 1996.
3. FAO, *Aquaculture Production Statistics*, FIDI/C815 (Rev. 10), 1987–1996.
4. D. Popper and Y. Zohar, *Proc. Aquaculture Int. Cong. Expo.*, 1988, p. 319.
5. R. Kirk, *A History of Marine Fish Culture in Europe and North America*, Fishing News Books Ltd., England, 1987.
6. FEAP, http://www.feap.org/basses.html
7. J. Chervinski, *Aquaculture* **6**, 249–256 (1975).
8. M. Carrillo, S. Zanuy, F. Prat, J. Cerda, J. Ramos, E. Mananos, and N. Bromage in N.R. Bromage and R.J. Roberts, eds., *Broodstock Management and Egg and Larval Quality*, Blackwell Science, UK, 1995, pp. 138–168.
9. S. Hassin, Z. Yaron, and Y. Zohar, *Proc. 4th Int. Symp. Reprod. Physiol. Fish.*, Norwich, UK, 1991, pp. 100.
10. G. Bini, in W. Fischer, ed., *FAO Species Identification Sheets — Mediterranean and Black Sea*, Vol. I, 1973.
11. D. Coves, G. Dewavrin, G. Breuil, and N. Devauchelle, in J.P. McVey, ed., *CRC Handbook of Mariculture*, Vol. II, CRC Press, Boca Raton, 1991, pp. 3–20.
12. FEAP, http://www.feap.org/juvenil.html
13. Y. Zohar, R. Billard, and C. Weil, in G. Barnabe and R. Billard, eds., *L'Aquaculture du Bar et des Sparids*, INRA Publ., Paris, 1984, pp. 3–24.
14. O. Kah, S. Zanuy, E. Mananos, I. Anglade, and M. Carrillo, *Cell Tissue Res.* **266**, 129–136 (1991).
15. J.M. Navas, E. Mananos, M. Thrush, J. Ramos, S. Zanuy, M. Carrillo, Y. Zohar, and N. Bromage, *Proc. Fifth Int. Symp. Reprod. Physiol. Fish*, Austin, TX, 1995, pp. 131.
16. J.M. Navas, M. Thrush, J. Ramos, M. Bruce, M. Carrillo, S. Zanuy, and N. Bromage, *Proc. Fifth Int. Symp. Reprod. Physiol. Fish*, Austin, TX, 1995, pp. 108–110.
17. L.A. Sorbera, C.C. Mylonas, S. Zanuy, M. Carrillo, and Y. Zohar, *Proc. Fifth Int. Symp. Reprod. Physiol. Fish*, Austin, TX, 1995, pp. 150.
18. E.L. Mananos, N. Rodriguez, F. Le, Menn, S. Zanuy, and M. Carrillo, *Reprod. Nut. Dev.* **37**, 1–11 (1997).
19. J. Cerda, S. Zanuy, M. Carrillo, J. Ramos, and R. Serrano, *Comp. Biochem. Physiol.* **111C**(1), 83–91 (1995).
20. J.M.R. Alvarino, M. Carrillo, S. Zanuy, F. Prat, and E. Mananos, *J. Fish Biol.* **41**, 965–970 (1992).
21. J.M.R. Alvarino, S. Zanuy, F. Prat, M. Carrillo, and E. Mananos, *Aquaculture* **102**, 177–186 (1992).
22. L.A. Sorbera, C.C. Mylonas, S. Zanuy, M. Carrillo, and Y. Zohar, *J. Exper. Zool.* **276**, 361–368 (1996).
23. S. Zanuy, M. Carrillo, and F. Ruiz, *Fish Physiol. Biochem.* **2**, 53–63 (1986).
24. S. Zanuy, F. Prat, M. Carrillo, and N. Bromage, *Aquat. Living Resour.* **8**, 147–152 (1995).
25. A. Tandler and O. Dvir, *Annual Report of the National Center for Mariculture*, E17/91, 1990.
26. B. Chatain, *Hydrobiologia* **358**(1–3), 7–11 (1997).
27. J.P. Le-Ruyet, J.C. Alexandre, L. Thebaud, and C. Mugnier, *J. World Aquacult. Soc.* **24**(2), 211–224 (1993).
28. F. Hidalgo, E. Alliot, and H. Thebault, *Aquaculture* **64**, 199–207 (1987).
29. P. Morris, *Fish Farmer — International File*, December: 6–9 (1997).
30. E. Tibaldi, F. Tulli, and D. Lanari, *Aquaculture* **127**, 207–218 (1994).
31. E. Tibaldi and D. Lanari, *Aquaculture* **95**, 297–304 (1991).
32. F. Hidalgo, E. Alliot, and H. Thebault, *Aquaculture* **64**, 209–217 (1987).
33. E. Tibaldi, F. Tulli, and M. Pinosa, *Eur. Aquacult. Soc. Spec. Publ.* **19**, 482 (1993).
34. S.J. Kaushik, *Aquat. Living Resour.* **11**(5), 355–358 (1998).
35. P. Coutteau, G. Van Stappen, and P. Sorgeloos, *Arch. Anim. Nutr.* **49**, 49–59 (1996).
36. J.M. Navas, M. Bruce, M. Thrush, B.M. Farndale, N. Bromage, S. Zanuy, M. Carrillo, J.G. Bell, and J. Ramos, *J. Fish Biol.* **51**, 760–773 (1997).
37. E. Tibaldi, F. Tulli, R. Ballestrazzi, and D. Lanari, *Zool. Nutr. Anim.* **17**, 313–320 (1991).
38. A. García–Alcázar, E. Abellán, M.R.L. Dehesa, M. Arizcun, J. Delgado, and A. Ortega, *Bol. Inst. Esp. Oceanogr.* **10**(2), 191–201 (1994).
39. S.J. Kaushik, M.F. Gouillou–Coustans, and C.Y. Cho, *Aquaculture* **161**, 463–474 (1998).
40. R. Bellance and D. Gallet de Saint Aurin, *Caraibes Medical*, 105–114 (1988).
41. D. Gallet de Saint Aurin, J.C. Raymond, and V. Vianas, *Actes de Colloque, Colloques sur L'Aquaculture* **9**, 143–160 (1990).
42. G. Breuil, J.R. Bonami, J.F. Pepin, and Y. Pichot, *Aquaculture* **97**, 109–116 (1990).
43. M. Comps, J.F. Pépin, and J.R. Bonami, *Aquaculture* **123**, 1–10 (1994).
44. G. Gauthier, B. Lafay, R. Ruimy, V. Breittmayer, J.L. Nicolas, M. Gauthier, and R. Christen, *J. Sys. Bac.* **45**, 139–144 (1995).
45. A. Colorni, *Isr. J. Aquacult. — Bamidgeh* **44**, 75–81 (1992).
46. A. Colorni, M. Ucko, and W. Knibb, *Sea bass and Seabream Culture: Problems and Prospects (Workshop)*, 1996, pp. 259–261.
47. A.R. Lawer, in C.J. Sindermann, ed., *Disease Diagnosis and Control in North American Marine Aquaculture*, Elsevier, Amsterdam, 1977, pp. 257–264.
48. A. Barbaro and A. Francescon, *Oebalia*, XI–2, N.S., 745–752 (1985).
49. M.C.L. Baticados and G.F. Quinitio, *Helgoländer Meeresuntersuchungen* **37**, 595–601 (1984).
50. I. Paperna, *J. Fish Dis.* **3**, 363–372 (1980).
51. I. Paperna, *Aquaculture* **38**, 1–18 (1984).
52. I. Paperna, *Annales de Parasitologie Humaine et Comparée* **59**, 7–30 (1984).
53. P. Cheng, in M.K. Stoskopf, ed., *Fish Medicine*, W.B. Saunders, Philadelphia, PA, 1993, pp. 646–658.
54. B.K. Diggles and R.D. Adlard, *Diseases of Aquatic Organisms* **22**, 39–43 (1995).
55. A. Diamant, G. Issar, A. Colorni, and I. Paperna, *Bulletin of the European Association of Fish Pathologists* **11**, 122–124 (1991).
56. A. Colorni and A. Diamant, *Eur. J. Protist.* **29**, 425–434 (1993).
57. B.K. Diggles and J.G. Lester, *J. Parasitology* **82**, 384–388 (1996).

58. A. Colorni and P. Burgess, *Aquarium Sciences and Conservation* **1**, 217–238 (1997).
59. T. Yoshinaga and H.W. Dickerson, *J. Aquatic Animal Health* **6**, 197–201 (1994).
60. B.K. Diggles and J.G. Lester, *J. Parasitology* **82**, 45–51 (1996).
61. A. Sitjà–Bobadilla and P. Alvarez–Pellitero, *Eur. J. Prot.* **29**, 219–229 (1993).
62. G. Oliver, *Zeit. Parasitenkunde* **53**, 7–11 (1977).
63. I. Paperna and F. Martinez–Gonzalez, *Stud. Rev. Gen. Fish Counc. Mediter.* **57**, 11–27 (1980).
64. R. Giavenni, *Rivista Italiana di Piscicoltura e Ittiopatologia* **18**, 167–176 (1983).
65. I. Paperna, *Annales de Parasitologie* **55**, 687–706 (1980).
66. I. Paperna and F.D. Por, *Rapp. P.-V. Réun. Comm. int. Mer Méditerr.* **24**, 195–197 (1977).

See also LARVAL FEEDING—FISH.

SEA TURTLE CULTURE: GENERAL CONSIDERATIONS

ROBERT R. STICKNEY
Texas Sea Grant College Program
Bryan, Texas

OUTLINE

Green sea turtle (*Chelonia mydas*) steaks were once a readily available and highly popular human food in Florida. As the population of wild turtles became depleted, the notion that commercial mariculture might be lucrative led to development of a commercial farming operation in the Cayman Islands (1,2). However, the Endangered Species Act placed the green sea turtle and other turtles under protection in 1973 and made it illegal to catch, sell, import, or own any part of the listed species. Sea turtles also gained protection under the Convention on International Trade in Endangered Species of Wild Fauna and Flora (CITES), which went into force in 1975 and has 125 nations as signatory members.

Green sea turtles continue to be reared in the Cayman Islands, partly for local consumption, but also for stocking in order to assist in recovery efforts. Sea turtle culture has been undertaken over the past century, for various reasons and with various species, in Australia, Barbados, Bermuda, Burma, Cambodia, the Cook Islands, Cuba, Cyprus, Fiji, French Polynesia, Grenada, Indonesia, Japan, Kiribati, Malaysia, Mauritius, Mexico, Palau, the Philippines, the Seychelles, the Solomon Islands, Suriname, Thailand, Tonga, the United States, Vietnam, and Western Samoa (2).

Most culture activities have been associated with enhancement efforts. Included are head-starting activities conducted by government agencies (see the entry "Sea turtle culture: kemp's ridley and loggerhead turtles"). Efforts on behalf of sea turtle conservation have provided a large amount of valuable information on the biology of these unique marine reptiles.

Through the efforts of many scientists and countless numbers of volunteers, some progress has been made to reverse the decline in the populations of at least some species of sea turtles. Nesting sea turtles have actually returned to at least a few beaches where they had not been observed in many years. Enhancement efforts have been augmented by reductions in the number of turtles taken incidental to trawling shrimp, through imposition by the National Marine Fisheries Service in the United States of a requirement that shrimp nets must be fitted with turtle excluder devices (TEDs). The U.S. government also bans the import of wild-caught shrimp from foreign nations if the shrimp were caught in nets that were not equipped with TEDs.

GENERAL BIOLOGY

There are several species and possibly a few subspecies of sea turtles in the oceans of the world (3,4). The recognized species are the green genus and species shown in first paragraph of entry, hawksbill, (*Eretmochelys imbricata*), leatherback (*Dermochelys coriacea*), loggerhead (*Caretta caretta*), flatback (*Natator depressus*), olive ridley (*Lepidochelys olivacea*), and Kemp's ridley (*L. kempi*). The species can most easily be distinguished by the pattern of plates (scutes) on the dorsal shell (carapace). The maximum size ranges widely between species, with Kemp's ridley, for example, reaching a few tens of pounds and leatherbacks, among others, reaching more than 1000 lb (>454 kg).

Many of the specific details on the general biology of sea turtles provided in the remainder of this section are drawn from information on the green sea turtle (3), but the overall patterns described can be generally applied to other species as well, unless otherwise indicated.

Sea turtles attain sexual maturity at various ages. Maturity may not occur in wild green turtles less than 30 years of age, though in captivity they may mature in less than 10 years. Males spend their entire lives in the ocean after leaving the beach as hatchlings. Females leave the sea only to lay their eggs.

Adults may travel distances of thousands of miles (1 mile = 1.67 km) from their feeding grounds to the nesting beaches. Each mated female hauls herself from the water at night and moves up the beach to a location above the high-tide mark. She then scoops out a body pit and digs a nest hole in the sand with her rear flippers. She deposits one hundred or more eggs in the nest pit, after which she covers both the nest and the body pits with sand. The female then returns to the ocean. She may come back to the beach at intervals of about two weeks to deposit subsequent batches of eggs.

The egg development rate is dependent on temperature and may typically be from 50 to 60 days (Fig. 1). Researchers have learned that the incubation temperature controls the sex of the developing embryos: Warmer temperatures produce females, while cooler temperatures produce males. Thus, in hatcheries, where temperatures

Figure 1. Newly hatched sea turtle and an unhatched egg. Photo courtesy of Texas Sea Grant College Program.

can be carefully controlled, it is possible to manipulate the sex of the turtles produced.

Various predators will dig up turtle eggs. Known predators of green sea turtle eggs include crabs, insects, reptiles, mammals, and birds. At one time, it was common practice for people to dig up turtle nests and harass the females on the beaches, but with the development of a strong conservation ethic and, more recently, active attempts to protect sea turtles and their eggs on the nesting beaches, that practice is now rare. More common is a female sea turtle disrupting an existing nest by digging a new nest pit in the same location.

All of the eggs that hatch in a nest do so over a brief time span. Hatchlings climb to the surface in a synchronous fashion over a period of days, using one another as "stepping stones" to move upward, in a virtual ball of baby turtles. Once they reach the surface, they scamper down the beach to the ocean (Fig. 2). Predators can take a heavy toll when the hatchlings are on the beach, but the vulnerability of the turtles to predation certainly does not end when they enter the water. Unable to dive until they have grown significantly, the young turtles are vulnerable to attacks from both above and below. Hatchlings have reportedly been preyed upon by crabs, fish, reptiles, mammals, and birds. Juveniles and adults have fewer predators, but there have been reports of attacks by a few species of fish, including sharks; a species of crocodile; and at least one type of marine mammal.

Figure 2. Hatchling heading for the water. Photo courtesy of Texas Sea Grant College Program.

Most sea turtles are carnivores, although the green sea turtle is an herbivore that consumes sea grasses and algae of many varieties. Being a vegetarian, the green turtle does not have a strong flavor and for that reason became the most desirable species for human consumption. When the Americas were being colonized and European explorers were charting the oceans of the world from sailing ships, it was common practice to capture green sea turtles and strap them, upside down, to the decks of the ships. The turtles would live for long periods and provide fresh meat to sailors, who would otherwise have to exist on salted meat and fish.

Turtles tend to be true to their nesting beaches. Females will faithfully nest on the same stretch of beach where they were hatched. However, many beaches that once were used by nesting turtles have been abandoned, because human activity has either prevented access or scared the turtles away. Overfishing and destruction of nests by humans and other predators have also contributed to the demise of some nesting beaches. Recovery efforts often require moving eggs from the beach on which they were laid to another beach, where final incubation occurs, often in a hatchery situation. The hatchlings will, through a process known as imprinting, have a "memory" of the beach implanted in their brains that will cause them to return to the beach where they entered the ocean as hatchlings, not where they were deposited as eggs. Thus, in theory, new nesting beaches can be established and abandoned ones put back into use.

MARICULTURE

Culture of turtles has been for the purpose of assisting in the recovery of threatened and endangered species and for food production. Early recovery efforts involved collection of eggs laid in nature, incubation of those eggs in hatcheries, and immediate release of hatchlings into the ocean. Because of high mortality rates, however, an additional step, called head-starting, is sometimes employed. This technique involves maintaining the young turtles in captivity for several months, until they are able to dive and have attained sufficient sizes and swimming speeds to allow them to avoid many of the common predators (see the entry "Sea turtle culture: kemp's ridley and loggerhead turtles" for more detail). In addition to head-starting Kemp's ridleys and loggerheads, similar activities have been undertaken with green and hawksbill turtles (1,2).

Mariculture, Ltd. was the first sea turtle farm. Established in 1968 on Grand Cayman Island in the British West Indies, it was acquired by Cayman Turtle Farm, Ltd., in 1975 and by the Government of the Cayman Islands in 1983 (1,2). It remains in business and is the largest and longest running sea turtle mariculture operation in the world. A great deal has been learned about green sea turtle biology and culture over the three decades that the farm has been operating. In addition, the farm has developed a captive breeding program with Kemp's ridley turtles.

Taking its initial stock of green turtle eggs and adults from the wild, the Cayman Island farm was able to develop a captive breeding program, though in the 1970s it continued to obtain eggs from nesting beaches to supplement its production. In the 1970s, annual production ranged from 12,000 to 15,000 turtles. Virtually every part of the turtles produced were marketed. Included were the meat, oil, fatty tissue (known as calipee), leather from the skin, and shell. Yet, the farm was not a profitable operation. The situation become worse when changes in CITES regulations and the listing of green sea turtles under the Endangered Species Act in the United States, banned the importation of turtle products and their transshipment through Miami.

Once the Government of the Cayman Islands took over the farm in 1983, it was developed into a tourist attraction as well as a production operation. The farm is located on 6.4 ha (16 acres) of land. It features, in addition to the tourist center, a breeding pond, a hatchery, rearing tanks, research laboratories, an administration building, and a processing plant.

By 1988, the farm was showing a profit, largely based on fees paid by tourists. It was also moving closer to being able to export some of its products under CITES criteria, which included a provision under which second-generation captively produced turtles were exempt from the distribution ban.

In 1973, five years after the farm was established, adult turtles that had been held in captivity since the farm was established finally produced viable eggs. The first generation of captively produced females (hatched in 1973 and 1974) reached maturity and laid eggs first in 1989. It is the offspring of the 1989 adults that are the second-generation turtles that meet the CITES criteria. By 1992, the captive breeding population on the farm had reached a total of 280 turtles, with a ratio of three females for each male being maintained.

The main emphasis of the farm in the past several years has been on research and head-starting, with only limited production for sale. Tourism and local meat sales are primary income sources. About 4,000 turtles are marketed annually at 3.5 years of age, when they weigh about 24 kg (48 lb). The farm's head-start program began in 1980. Tourists and residents become involved in the head-start program by paying a fee for the privilege of releasing the turtles. By 1992, more than 22,000 head started turtles had been released (5).

Sea turtles require clean water, or else they become susceptible to a variety of diseases, including skin lesions. At high densities, daily tank cleaning is required unless flow-through water is used in the culture system.

Unlike finfish and invertebrate culture species, each species of sea turtle seems to have its own personality. Green sea turtles are docile and do not mind being touched. Kemp's ridley and loggerhead turtles tend to be more aggressive than green turtles, but the public enjoys watching all sea turtle species when they are on display. While not as popular as marine mammals, sea turtles do have a large following among those who visit public aquaria and various tourist attractions that feature aquatic animals.

The major remaining Kemp's ridley sea turtle nesting beach in the world is at Rancho Nuevo, Mexico. A few green turtles and loggerheads also nest at Rancho Nuevo. Collaboration between U.S. and Mexican government agencies and scientists to protect Kemp's ridley turtles has been ongoing for several years. During the nesting season, people from both nations gather at Rancho Nuevo to count the number of ridley nests and in some cases gather eggs for incubation in corrals located elsewhere on the beach. Most of the turtles hatch naturally or are released at Rancho Nuevo, but some have been used to reestablish nesting beaches in the United States as well. As a result of these activities, aided in no small part by conservation efforts, the number of nesting females at Rancho Nuevo has increased considerably. Kemp's ridley head-starting efforts also seem to be paying off, as at least a few tagged individuals known to have been head-started have been observed on nesting beaches at Padre Island and Mustang Island in south Texas.

Many sea turtle populations continue to be threatened or endangered around the world, but their plight is not quite so perilous as it was only a few years ago.

BIBLIOGRAPHY

1. P. Fosdick and S. Fodick, *Last Chance Lost? Can and Should Farming Save the Green Sea Turtle? The Story of Mariculture, Ltd., Cayman Turtle Farm*, Irvin S. Naylor, York, PA, 1994.
2. M. Donnelly, *Sea Turtle Mariculture*, The Center for Marine Conservation, Washington, DC, 1994.
3. H.F. Hirth, *Synopsis of the Biological Data on the Green Turtle* Chelonia mydas *(Linnaeus 1758)*, U.S. Fish and Wildlife Service, Washington, DC, 1997.
4. P. Lutz and J. Musick, eds., *The Biology of Sea Turtles*, CRC Press, Boca Raton, FL, 1997.
5. T.A. Walker, *Aquaculture Magazine* March/April: 47–55 (1992).

See also SEA TURTLE CULTURE: KEMP'S RIDLEY AND LOGGERHEAD TURTLES.

SEA TURTLE CULTURE: KEMP'S RIDLEY AND LOGGERHEAD TURTLES

CHARLES W. CAILLOUET, JR.
National Marine Fisheries Service
Galveston, Texas

OUTLINE

On a worldwide basis, sea turtle aquaculture is conducted for commerce and conservation (1). The U.S. Departments of Commerce (DOC) and the Interior (DOI) are responsible for protecting sea turtles under the Endangered Species Act, thus limiting sea turtle aquaculture in the United States to research and conservation. In 1977, the National Marine Fisheries Service (NMFS) laboratory in Galveston, TX, initiated sea turtle aquaculture research related to conservation, first with loggerhead turtles (*Caretta caretta*), a threatened species, and then with endangered Kemp's ridley turtles (*Lepidochelys kempi*). This research requires scientific permits and threatened- and endangered-species permits from the Texas Parks and Wildlife Department (TPWD), the Florida Department of Environmental Protection (FDEP), the U.S. Fish and Wildlife Service (FWS), and Mexico. When eggs, hatchlings, or larger turtles are received from other countries, Convention on International Trade in Endangered Species of Wild Flora and Fauna (CITES) export and import permits also are required.

RATIONALE

At a meeting in Austin, TX, in January 1977, representatives of Mexico's Departamento de Pesca, the U.S. Department of the Interior [specifically, the National Park Service (NPS) and FWS], the U.S. Department of Commerce (specifically, NMFS), and TPWD established a Kemp's ridley recovery program (2–5). At that time, the annual number of Kemp's ridley nesters was less than 800 and declining at the species' primary nesting site near Rancho Nuevo, Tamaulipas, Mexico (6) (see Fig. 1). Program objectives were to increase protection of nesters, eggs, and hatchlings at Rancho Nuevo; reduce incidental capture of sea turtles in shrimp trawls; and encourage research leading to improved management and population recovery. Prior to the interagency meeting, NPS and FWS regional representatives had been discussing a potential feasibility study (later called the "Kemp's ridley headstart experiment") aimed at establishing a nesting colony of Kemp's ridleys at Padre Island National Seashore (PINS), near Corpus Christi, TX (3–5). The experiment involved collecting and incubating eggs, imprinting hatchlings, and then captive rearing, tagging, and releasing the turtles into the Gulf of Mexico within their first year of life. It protected them from predators and other causes of mortality associated with those early life stages. Mexican agencies, FWS contractors (initially, the Florida Audubon Society, Maitland, FL; then Gladys Porter Zoo, Brownsville, TX) and volunteers carried out the Rancho Nuevo operations (6,7); NPS conducted incubation and imprinting at PINS (5,8); and the NMFS Galveston Laboratory captive

Figure 1. Locations of the Galveston Laboratory, Padre Island National Seashore, and Rancho Nuevo.

reared, tagged, and released the turtles into the Gulf of Mexico (9–11).

Regulations requiring turtle excluder devices (TEDs) in shrimp trawls were implemented during the late 1980s and early 1990s (12). The regulations led to the development of procedures to test TEDs before they were certified for use in shrimp fisheries. Certification required that a sample of sea turtles, passed one at a time through a TED installed in a shrimp trawl, rapidly escaped through the TED. The Galveston Laboratory was and continues to be the source of turtles used in TED certification procedures. Both Kemp's ridleys and loggerheads have been used in TED certification procedures, but two-year-old loggerheads have become the standard.

In addition to their use in the headstart experiment and TED certification procedures sea turtles at the Galveston laboratory have been the subjects of research on physiology (reproductive, respiratory, metabolic, fitness, anesthetic, and submergence), chemical imprinting, tagging, sex determination, temperature–sex relationship, and diseases. The laboratory also attempts to rehabilitate sick and injured sea turtles obtained from the wild and then releases those that do not have incurable illnesses or debilitating handicaps.

LIFE HISTORY PATTERN

Sea turtle species include leatherback (*Dermochelys coriacea*), green (*Chelonia mydas*), flatback (*Natator depressus*), loggerhead, olive ridley (*Lepidochelys olivacea*), Kemp's ridley, and hawksbill (*Eretmochelys imbricata*) (13). These species all have similar life history patterns (14), with the greatest part of the multidecade life span spent at sea, where capture, observation, and tracking are difficult and costly. In contrast, much has been learned through studies of sea turtles in captivity, at nesting sites, and by examining carcasses found stranded on barrier beaches. Specific life history characteristics of Kemp's ridley (15,16) and loggerhead (17,18) turtles have been described in detail.

Adult female sea turtles ascend beaches to nest during spring and summer, sometimes more than once during a nesting season, but not always each year (19,20). Females dig nests with their rear flippers, deposit clutches containing around 100 eggs each (ranging from tens to fewer than 250 eggs per clutch, and varying with species), and then cover the eggs with sand before returning to the water. The whole process takes about 0.8–2.5 hours, depending on the species. The eggs hatch after incubating 6–13 weeks, depending on the species and the temperature (20). Hatchlings emerge from the nest and crawl down the beach to the water. Recent studies suggest that a hatchling's biological compass is set during this crawl down the beach, providing a mechanism for possible magnetic navigation back to the nesting site after the turtle matures (21).

Eggs and young sea turtles are especially vulnerable to environmental factors, predation, and human activities (22–24). Eggs are eaten by crabs, small to medium-sized mammals, reptiles, birds, and humans, but they also sometimes succumb to inundation by high tides, heavy rain, or flood waters and invasions by insects, microbes, fungi, and plant roots. Hatchlings are exposed to avian, mammalian, and crustacean predators on their way to the water, where they are then faced with added threats from fish, molluscs, other marine animals, and sea birds. They swim out to sea, drifting with ocean currents and foraging for food near the surface for a year to more than a decade, depending on the species (24,25). Little is known about the whereabouts and habits of turtles at this young life stage, which precedes the larger immature and adult stages that are capable of benthic feeding. Turtles at the larger stages are vulnerable to both deliberate and incidental capture by various fishing gears and methods (12,23,24,26). Diet and feeding habits, from omnivory to carnivory, vary considerably among species (27).

Estimates of age at sexual maturity vary within species and range from one to five decades among species (24,28). Copulation takes place on migratory routes to nesting beaches, as well as near nesting beaches (20; David Owens, personal communication, February 1998). Males have large, curved claws on their foreflippers; their plastrons soften and depress during the mating season; and they have long, thick tails and large, curved penises. All of these factors aid the males in grasping females during copulation, which can last for hours (20,28). Females can store sperm to fertilize multiple clutches of eggs during a nesting season (20,29).

AQUACULTURE OF KEMP'S RIDLEY AND LOGGERHEAD SEA TURTLES

The Galveston Laboratory first reared loggerhead hatchlings, provided by the Florida Department of Natural Resources, in Jensen Beach, FL (30), to gain experience before shifting to headstarting Kemp's ridleys (4). After 1977, loggerhead hatchlings reared at the Galveston Laboratory were received from Clearwater Marine Science Center, Clearwater, FL, or Mote Marine Laboratory, Sarasota, FL (31). One clutch of loggerhead eggs was obtained from a nest on Bolivar Peninsula, near Bolivar, TX, and incubated at PINS, and its hatchlings were provided to the Galveston Laboratory.

Egg Collection and Incubation

Kemp's ridley eggs for the headstart experiment were collected annually from 1978 to 1992 by FWS contractors, Mexican turtle camp personnel, and volunteers. The eggs were exposed to the Rancho Nuevo beach sand or to PINS sand transferred to Rancho Nuevo in polystyrene foam incubation boxes (5,7,8,32), under the working hypothesis that chemical imprinting took place in the egg or hatchling stage (21,33). Eggs to be placed in PINS sand were caught in plastic bags as they were laid, so that they would not come in contact with Rancho Nuevo sand (2,7), and the people who handled the eggs wore sterile gloves. Eggs from a clutch were stacked in an incubation box containing a layer of PINS sand; next, additional sand was packed around and on top of the eggs to prevent them from coming in contact with the box. The eggs were then transported by aircraft, vehicle, or both to PINS for incubation (5,7,8,34). Hatchlings for headstarting also were transported by aircraft from Rancho Nuevo (32) and from the Kemp's ridley breeding experiment at Cayman Turtle Farm, Grand Cayman, British West Indies (10).

The mainstay of the Kemp's ridley recovery program—protection of eggs and hatchlings at the Rancho Nuevo and contiguous nesting sites (2,15,16,35)—is a form of artificial culture. Typically, most of the eggs laid are collected and transplanted into artificial nests within fenced corrals on those beaches, so that they can be monitored during incubation and protected from predators and human exploitation. Hatchlings from the corrals are then protected during their crawl to the water, but must fend for themselves thereafter.

Sex Determination

Sex in sea turtles is determined by the temperature to which eggs are exposed during incubation. More females are produced at temperatures above the pivotal temperature (i.e., the temperature that produces a ratio of one male to one female), and more males are produced at temperatures below the pivotal temperature (36). As Kemp's ridley eggs were stacked in an incubation box, a temperature sensing probe was placed in the center of the clutch, with a wire leading outside the box for connection to a temperature recorder (5,7). The pivotal temperature for Kemp's ridleys was not known until 1985, and from that year through 1988, the NPS controlled incubation temperatures at PINS so as to produce mostly females (34,36,37). Prior to 1985, mostly males were produced (37,38). The Galveston Laboratory had no control over incubation of loggerhead eggs: All loggerheads reared at the Galveston Laboratory were obtained as hatchlings, with most coming from Florida and one clutch from PINS.

Imprinting

When the headstart experiment began, chemical imprinting was hypothesized to take place in the egg or hatchling stage (4,33). Kemp's ridley hatchlings that emerged from eggs incubated at PINS or Rancho Nuevo were allowed to crawl to the water before being netted and transferred to the Galveston Laboratory for captive rearing (5). If Kemp's ridleys return to their natal beach as adults, using magnetic navigation (5,21), then the procedures used in their crawl to the water as hatchlings should have sufficed to imprint them to their natal beach. The Padre Island incubation and imprinting phases of the headstart experiment were terminated in 1988, to focus on protecting, documenting, and monitoring Kemp's ridley nests at PINS (34).

Rearing Facilities and Seawater Management

The building housing the turtle rearing tanks is constructed of steel, and its concrete floor is slightly sloped toward drain troughs covered with fiberglass grates (see Figs. 2–4). Incoming seawater flows through PVC pipes laid in the same troughs that carry waste seawater from the turtle barn (see Figs. 2–4). Each fiberglass rearing tank is equipped with a seawater supply pipe and an elevated standpipe that maintains seawater level (see Fig. 4).

The sea turtles are reared in isolation from each other, to prevent incidents of biting, injury, and infections that arise when they are reared in groups (9,30) (see Figs. 4 and 5). Kemp's ridleys are especially aggressive, even as hatchlings, so isolation rearing increases their chance of survival. As hatchlings, the turtles are isolated in small plastic pots placed in groups of four within a plastic crate (see Fig. 5). After 60 days, the pots are removed, and hatchlings are redistributed, one per crate (see Fig. 5). The crates are bolted together with nylon fasteners into groups of 10 to facilitate handling (see Figs. 5 and 6). Each crate is lined on its four inside walls with high-impact styrene sheeting to prevent both dispersion of food and contact between turtles in adjacent crates. The bottom of each crate is fitted with vinyl-coated wire mesh to allow turtle excrement and uneaten food to sink to the bottom of the rearing tank. The crates are supported by a rack constructed of PVC pipe (see Figs. 3, 5, and 6). When the Kemp's ridleys are 9–11 months old, they are tagged and released into the Gulf of Mexico or adjacent bays (9–11,39,40).

Loggerheads approaching two years of age are required for TED certification. Since they are reared for a longer time than Kemp's ridleys in captivity, they outgrow the plastic crates and must be redistributed into larger vinyl-coated wire-mesh cages within the rearing tanks around one year of age (see Figs. 4–6). The walls of each cage are lined with high-impact styrene sheeting, and the bottom is made of vinyl-coated wire mesh (see Figs. 5 and 6). Cages, in groups of two, are supported by galvanized wire hangers or plastic cable ties attached to wooden poles laid across the rearing tank (see Figs. 4–6).

While isolation rearing is somewhat confining, the turtles exhibit high survival rates and good health. To accustom loggerheads to life in the wild prior to their use in TED certification and eventual release, the turtles are semiwild conditioned outside in seawater ponds or pens for about one month, while being closely monitored so that corrective measures can be initiated if aggressive biting occurs. They are fed squid (thawed after being purchased frozen) during that time and may also eat natural foods that they find within the ponds or pens. Except for Kemp's ridleys used in TED certification in the past, and turtles reared for more than one year in captivity (in some cases, to sexual maturity), captive-reared Kemp's ridleys have not been semiwild conditioned before release into the wild (4,9,10). However, an experimental exercise regimen was shown to improve swimming performance in captive-reared Kemp's ridleys (41).

Seawater is filtered through well points buried in the sand off the beach at Galveston as it is pumped from the Gulf of Mexico (9). It flows into a concrete-lined sump, where particulates are allowed to settle, and then is pumped into insulated fiberglass reservoirs (see Fig. 7). It receives no further filtration or chemical treatment, but is heated during winter with thermostat-controlled, electric immersion heaters placed in some of the reservoirs. Heated and unheated seawater are mixed to the appropriate temperature before being pumped into the rearing tanks, and the temperature is maintained thereafter by controlling the air temperature within the rearing building, using forced-air heaters during winter and exhaust fans during summer (see Fig. 2). The temperature of the seawater in the rearing tanks ranges annually from 23 to 31 °C (73 to 88 °F), with a mean near 27 °C (81 °F), and the salinity ranges 20–39 ppt, with a mean near 31 ppt. Three times per week, rearing tanks are drained to remove uneaten food and turtle wastes and are then refilled with clean seawater. Once per month, each tank is scrubbed with a high-pressure sprayer to remove algae and other materials adhering to the sides and bottom.

Keeping seawater clean and warm is of major importance to successful rearing and disease prevention in sea turtles. Various fungal infections, both external and internal, that eventually lead to death occur if the turtles are reared at temperatures below 20 °C (68 °F) (9,30). Whether this situation results from detrimental effects of cooler temperatures on the immune system, encouragement of growth of fungi by lower temperatures, or other factors is not known. Additional diseases and treatments have been described for Kemp's ridleys and loggerheads reared at the Galveston Laboratory (30,31). Sick or injured turtles are removed from the rearing building and treated in a separate "hospital," which is also used for quarantine, to avoid exposing healthy turtles to diseases.

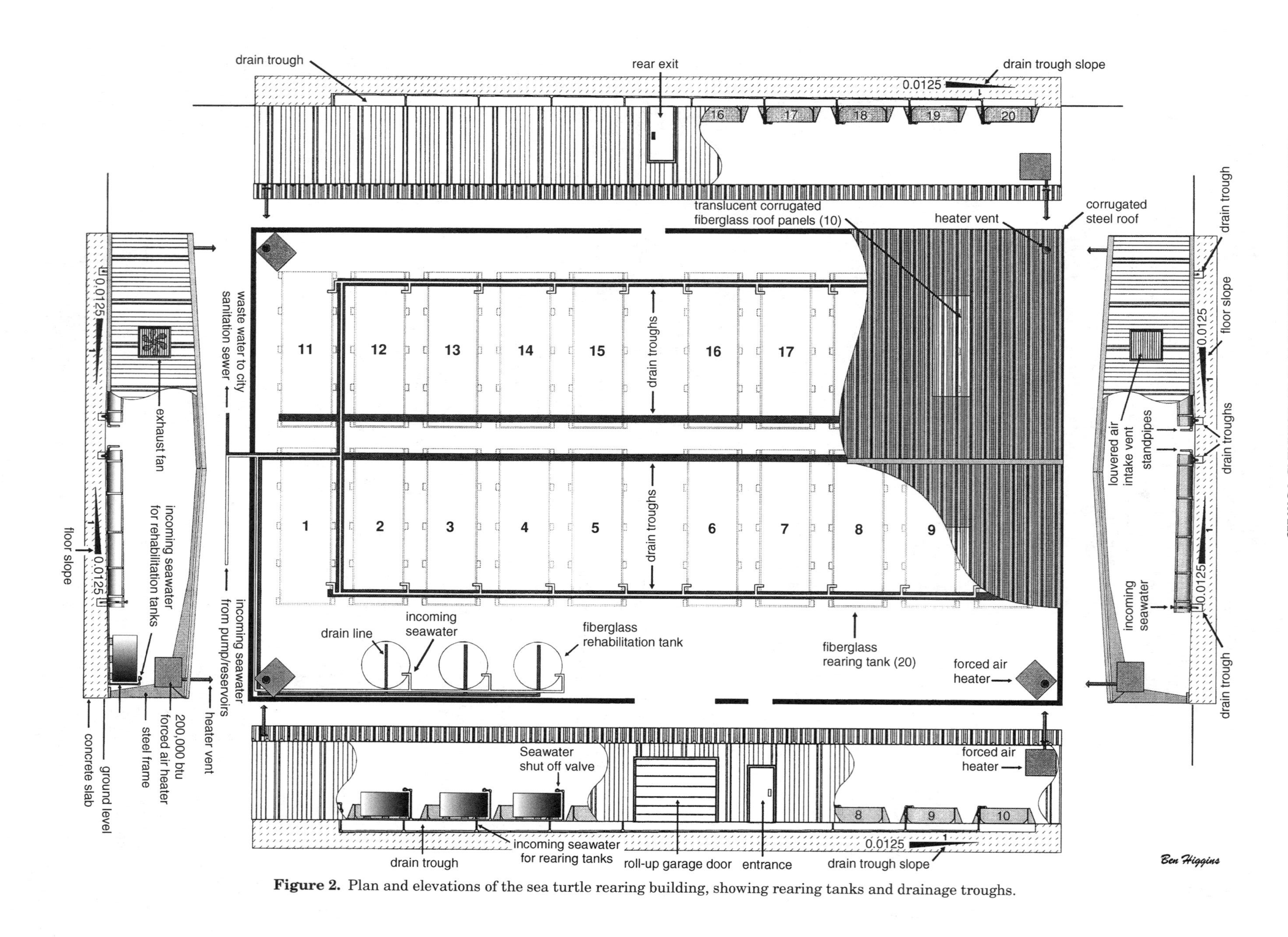

Figure 2. Plan and elevations of the sea turtle rearing building, showing rearing tanks and drainage troughs.

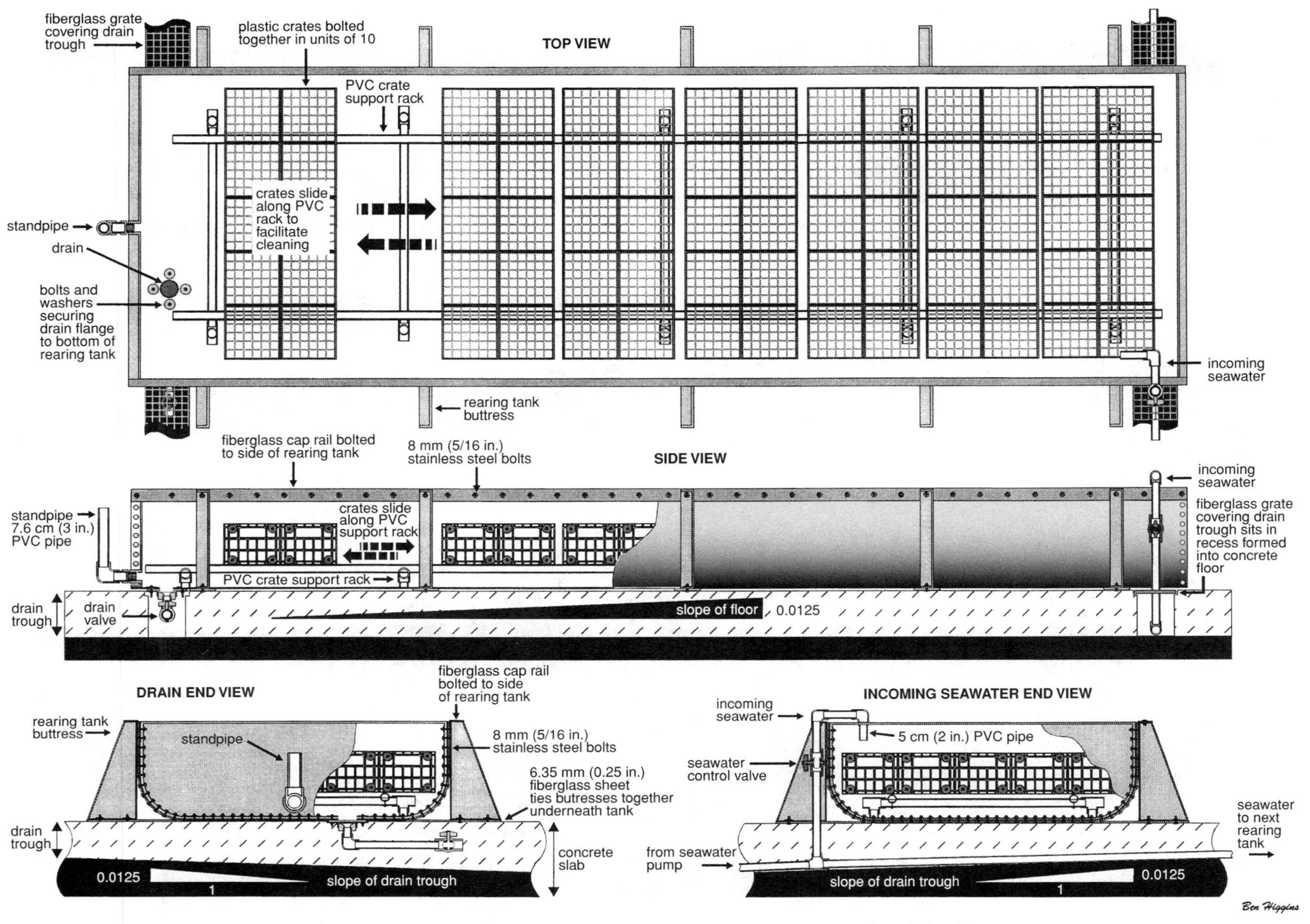

Figure 3. Rearing tanks, showing plastic crates and the PVC support rack and plumbing.

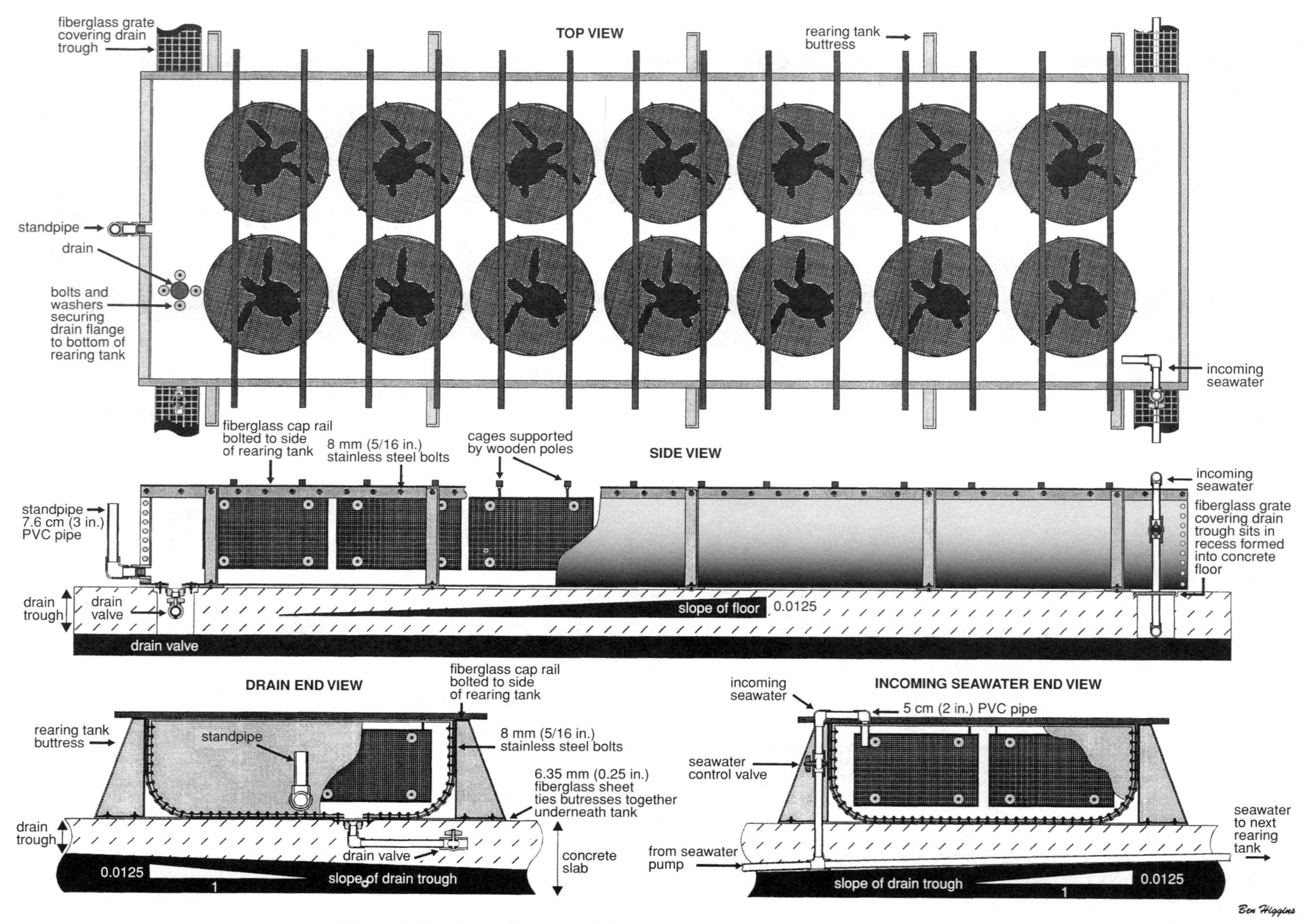

Figure 4. Rearing tanks, suspended cages, and the PVC support rack and plumbing.

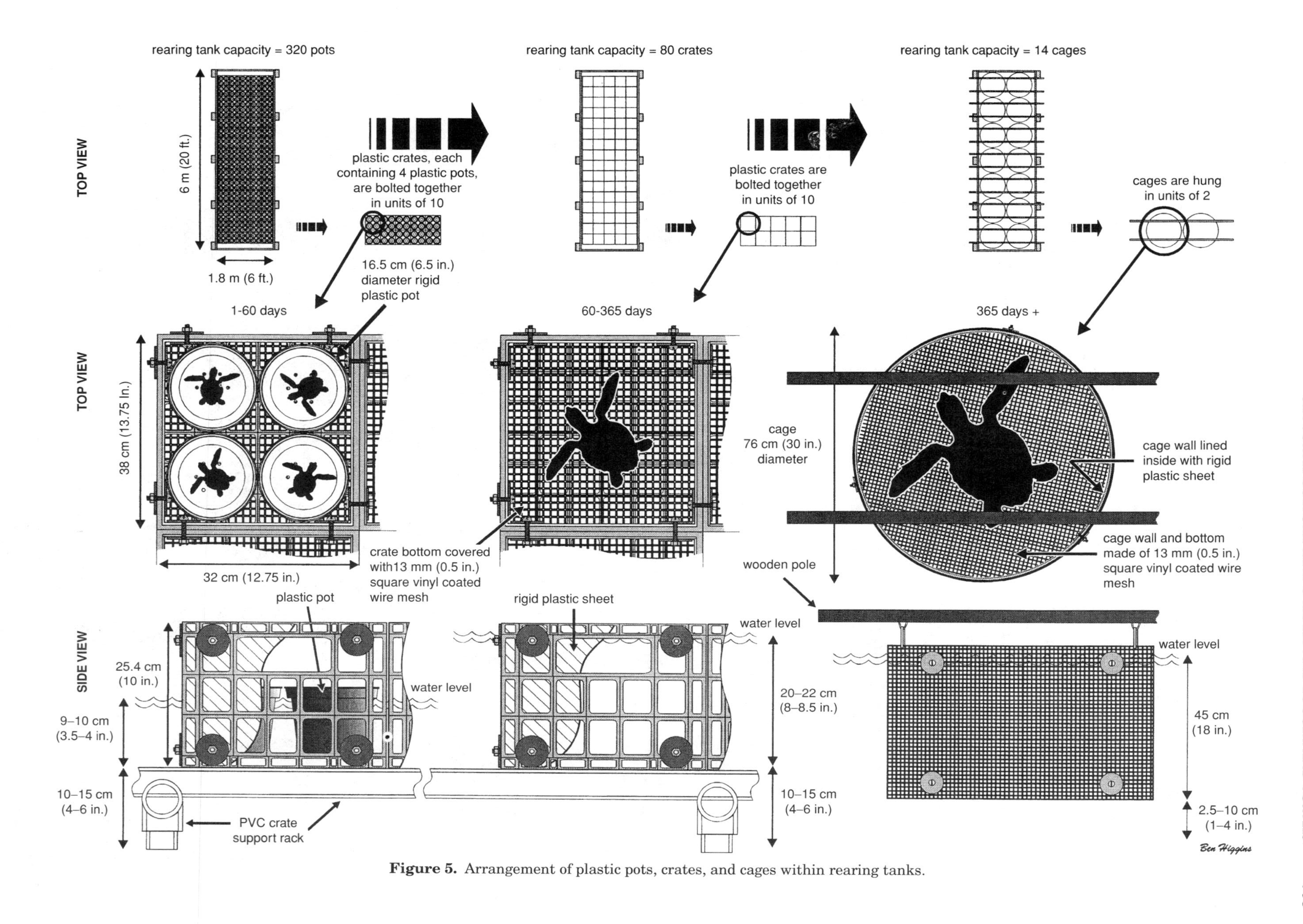

Figure 5. Arrangement of plastic pots, crates, and cages within rearing tanks.

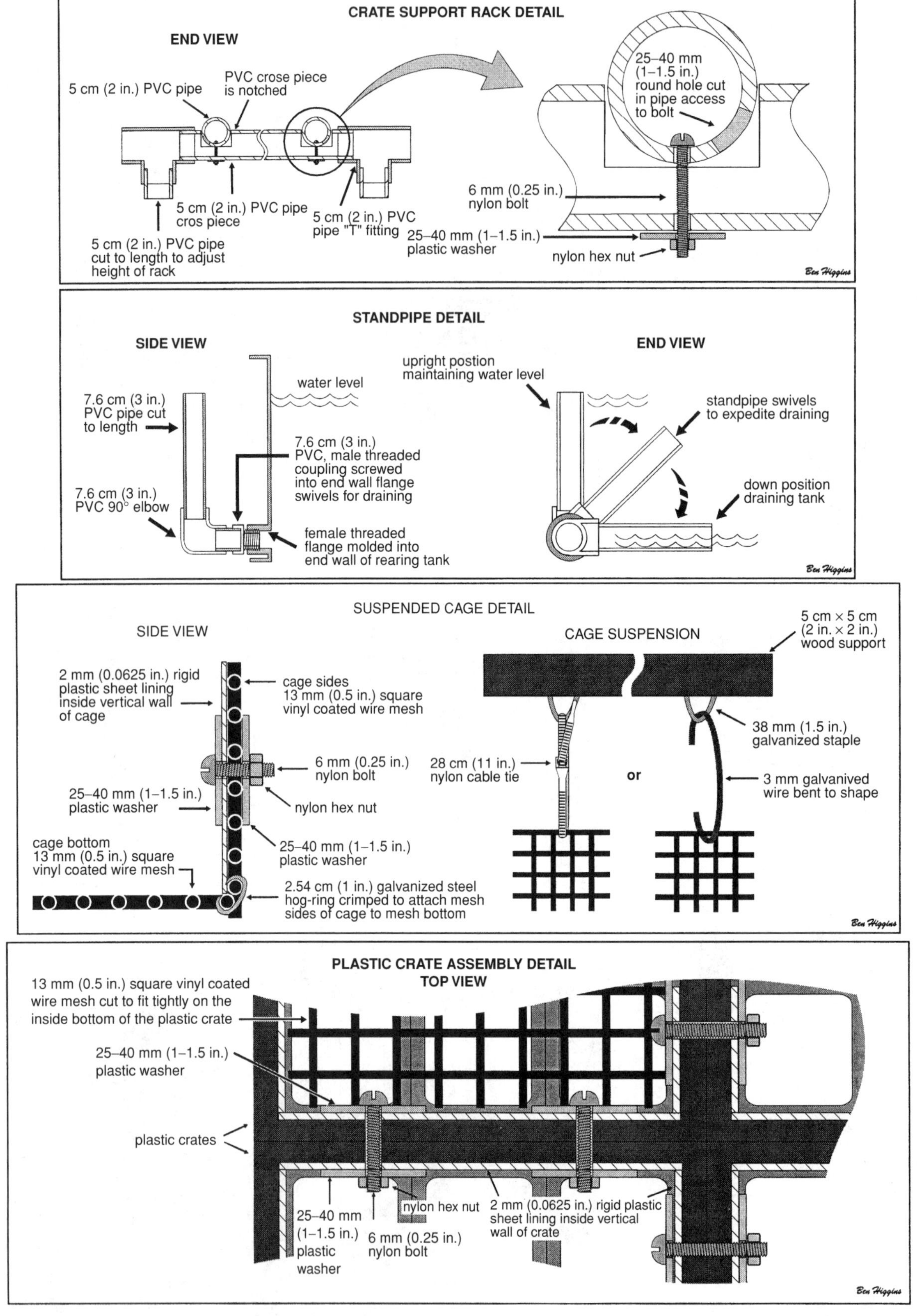

Figure 6. Details of the support rack, standpipe-drain plumbing, plastic crate assembly, and a suspended cage.

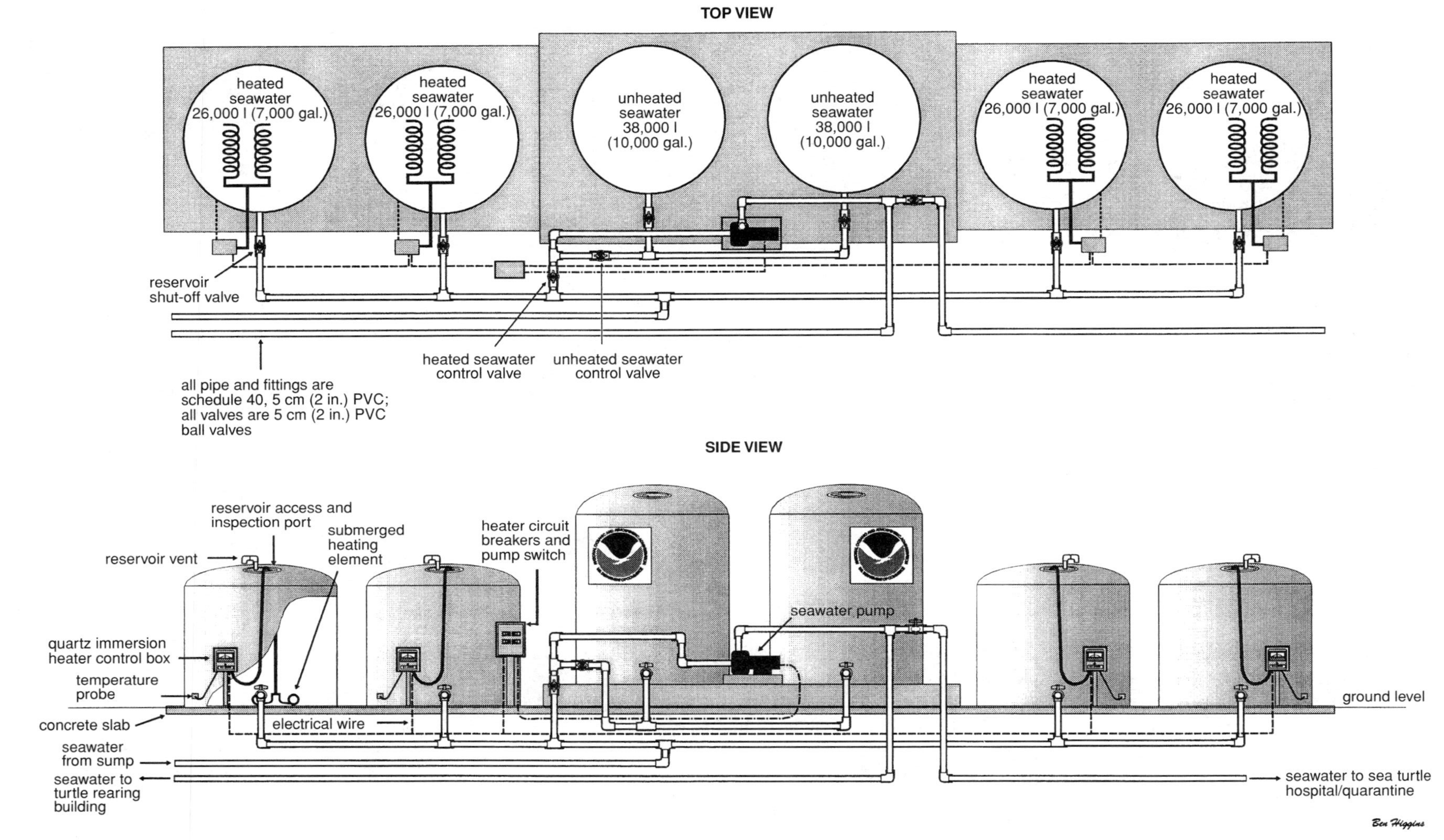

Figure 7. Seawater reservoirs, heaters, and PVC plumbing.

Food, Feeding, Growth, and Survival

The turtles are fed floating feed pellets manufactured by Purina Mills, Inc. (9,42). Each turtle receives a daily ration based on a percentage of the average body weight of a sample of turtles of the same age. Feeding of newly hatched sea turtles is postponed for up to nine days, to allow time for yolk sac absorption (43). The daily ration for hatchlings is about 2% of the average body weight per turtle and is reduced gradually to about 1% by the time that the turtles are one year old. The percentage can be decreased or increased to control growth rate. Sea turtles fed *ad libitum* or oily fish quickly become obese, which eventually leads to fatty degeneration of the liver (44). Each turtle is individually fed a daily ration divided into two equal portions, one fed in the morning and the remainder in the afternoon.

Kemp's ridleys and loggerheads readily eat squid, blue crab, and shrimp, if given the opportunity. They quickly adapt to eating natural foods after long periods of eating pelletized diets (45). However, once they encounter natural foods, they resist eating pellets unless they are extremely hungry.

Under the controlled conditions of captive rearing described herein, Kemp's ridleys can grow to an average weight of 1.26 kg (2.78 lb) and an average straight-carapace length of 19.5 cm (7.7 in.) in a year (11). The first-year survival rate in captivity for the 1978–1992 year-classes combined was 87% (11), and in recent years it has exceeded 95%. Under similar conditions, loggerhead growth and survival during the first year in captivity are comparable to that of Kemp's ridleys (Clark Fontaine, personal communication, February 1998).

Captive Breeding

Responding to recommendations by prominent sea turtle scientists (46,47), the government of Mexico and the Galveston Laboratory distributed Kemp's ridleys to cooperating marine aquaria (see Table 1) in the late 1970s to mid-1980s, to be reared to maturity and held as a breeding stock in case other conservation efforts failed (2,3,15,28,48). The turtles were dispersed among numerous marine aquaria, in order to distribute the costs of their maintenance and to avoid catastrophic losses due to disease outbreak or other causes of mortality (47). Cayman Turtle Farm, which received most of the turtles, was the first to breed Kemp's ridleys successfully, producing viable hatchlings from seven-year-old captive reared animals in 1986 (49). Under an agreement with the government of Mexico, Cayman Turtle Farm still maintains about 400 Kemp's ridleys. Included are survivors from the original stock received from Mexico, as well as offspring produced by captive breeding survivors from the stocks received from both Mexico and the United States (Rene Márquez–Milan, personal communication, November 1997). The Kemp's ridleys at Cayman Turtle Farm provided opportunities for studying reproductive behavior and physiology (28,50–52). Other marine aquaria also gained valuable experience with captive rearing (see Table 1). The experimental captive breeding program in the United States was terminated in 1988, and most of the surviving turtles in the dispersed brood stock have since been released into the wild. For example, survivors in the group of turtles originally obtained as yearlings by Cayman Turtle Farm from the United States were returned as adults by FWS to the Galveston Laboratory and released into Galveston Bay, TX, in 1992.

Table 1. Marine Aquaria that Received Kemp's Ridleys from the Galveston Laboratory for Purposes of Developing a Captive Brood Stock

Audubon Park Zoo, New Orleans, LA
Bass Pro Shops, Springfield, MO
Cayman Turtle Farm, Grand Cayman, British West Indies
Clearwater Marine Science Center, Clearwater, FL
Dallas Aquarium, Dallas, TX
Gulfarium, Fort Walton Beach, FL
Key West Aquarium, Key West, FL
Marine Life Park, Inc., Gulfport, MS
Marineland of Florida, Inc., St. Augustine, FL
Miami Seaquarium, Miami, FL
New England Aquarium, Boston, MA
North Carolina Marine Resources Center, Kure Beach, NC
University of Texas–Pan American, Coastal Studies Laboratory, South Padre Island, TX
San Antonio Zoological Gardens and Aquarium, San Antonio, TX
Sea-Arama Marineworld, Galveston, TX
Sea Turtles, Inc., South Padre Island, TX
Sea World of Florida, Orlando, FL
Sea World of Texas, San Antonio, TX
Theater of the Sea, Islamorado, FL
Turtle Kraals, Key West, FL

EPILOGUE

The Galveston Laboratory has reared, tagged, and released more than 23,000 Kemp's ridleys (40) and hundreds of loggerheads since 1978. Six nestings of headstarted Kemp's ridleys were documented from 1996 through 1998 at the Padre Island National Seashore, the beach to which they had been imprinted as hatchlings (53,54). Despite the Galveston Laboratory's successes in experimental captive rearing and reintroduction of Kemp's ridleys and loggerheads into the wild, the usefulness of such an approach in the conservation and management of sea turtle stocks remains in doubt (55). A listing of Galveston Laboratory reports and publications on sea turtle research (56) can be obtained by contacting the laboratory at 4700 Avenue U, Galveston, TX 77551 USA (Phone 409-766-3500, Fax 409-766-3508), or the National Technical Information Service (NTIS), 5285 Port Royal Road, Springfield, VA 22161 USA (Accession No. PB97-167415, Paper copy US$19.50 Microfiche US$10.00). A recent assessment of the status of Kemp's ridley and loggerhead stocks (57) can be obtained by contacting the NMFS Miami Laboratory, Sea Turtle Program, 75 Virginia Beach Drive, Miami, FL 33149 USA, or by contacting NTIS.

ACKNOWLEDGMENTS

Special thanks are due to Benjamin Higgins, who prepared the illustrations, and to Bradley Robertson, who provided data on rearing conditions. Clark Fontaine, William

Jackson, Maurice Renaud, Dickie Revera, Bradley Robertson, Wayne Witzell, and Roger Zimmerman reviewed the manuscript and provided many helpful suggestions.

This work was conducted under FWS Threatened and Endangered Species Permit PRT 676379, TPWD Permit SPR-0390-038, FDEP Permit TP #015, and permits from various Mexican government agencies (most recently, the Secretaria de Medio Ambiente Recursos Naturales y Pesca), as well as CITES.

BIBLIOGRAPHY

1. M. Donnelly, *Sea Turtle Mariculture: A Review of Relevant Information for Conservation and Commerce*, The Center for Marine Conservation, Washington, DC, 1994.
2. J.B. Woody, in A.S. Eno, R.L. Di Silvestro, and W.J. Chandler, eds., *Audubon Wildlife Report 1986*, The National Audubon Society, New York, NY, 1986, pp. 919–931.
3. J.B. Woody, in C.W. Caillouet, Jr., and A.M. Landry, Jr., eds., *Proceedings of the First International Symposium on Kemp's Ridley Sea Turtle Biology, Conservation and Management*, TAMU-SG-89-105, Texas A&M University, Sea Grant College Program, College Station, TX, 1989, pp. 1–3.
4. E.F. Klima and J.P. McVey, in K.A. Bjorndal, ed., *Biology and Conservation of Sea Turtles: Proceedings of the World Conference on Sea Turtle Conservation*, Smithsonian Institution Press, Washington, DC, 1982, pp. 481–487.
5. M.R. Fletcher, in C.W. Caillouet, Jr., and A.M. Landry, Jr., eds., *Proceedings of the First International Symposium on Kemp's Ridley Sea Turtle Biology, Conservation and Management*, TAMU-SG-89-105, Texas A&M University, Sea Grant College Program, College Station, TX, 1989, pp. 7–9.
6. R. Márquez M., A. Villanueva O., and P.M. Burchfield, in C.W. Caillouet, Jr., and A.M. Landry, Jr., eds., *Proceedings of the First International Symposium on Kemp's Ridley Sea Turtle Biology, Conservation and Management*, TAMU-SG-89-105, Texas A&M University, Sea Grant College Program, College Station, TX, 1989, pp. 16–19.
7. P.M. Burchfield and F.J. Foley, in C.W. Caillouet, Jr., and A.M. Landry, Jr., eds., *Proceedings of the First International Symposium on Kemp's Ridley Sea Turtle Biology, Conservation and Management*, TAMU-SG-89-105, Texas A&M University, Sea Grant College Program, College Station, TX, 1989, pp. 67–70.
8. D.J. Shaver, in S.A. Eckert, K.L. Eckert, and T.H. Richardson, compilers. *Proceedings of the Ninth Annual Workshop on Sea Turtle Conservation and Biology*, NOAA Technical Memorandum NMFS-SEFC-232, National Marine Fisheries Service, Miami, FL, 1989, pp. 163–165.
9. C.T. Fontaine, T.D. Williams, S.A. Manzella, and C.W. Caillouet, Jr., in C.W. Caillouet, Jr., and A.M. Landry, Jr., eds., *Proceedings of the First International Symposium on Kemp's Ridley Sea Turtle Biology, Conservation and Management*, TAMU-SG-89-105, Texas A&M University, Sea Grant College Program, College Station, TX, 1989, pp. 96–110.
10. C.W. Caillouet, Jr., C.T. Fontaine, S.A. Manzella–Tirpak, and D.J. Shaver, *Chelonian Conservation and Biology* **1**, 285–292 (1995).
11. C.W. Caillouet, Jr., C.T. Fontaine, T.D. Williams, and S.A. Manzella–Tirpak, *Gulf Research Reports* **9**, 239–246 (1997).
12. M.E. Lutcavage, P. Plotkin, B. Witherington, and P.L. Lutz, in P.L. Lutz and J.A. Musick, eds., *The Biology of Sea Turtles*, CRC Press, Boca Raton, FL, 1997, pp. 387–409.
13. B.W. Bowen, W.S. Nelson, and J.C. Avise, *Proceedings of the National Academy of Sciences* **90**, 5574–5577 (1993).
14. J.R. Hendrickson, *American Zoologist* **20**, 597–608 (1980).
15. R. Márquez M., compiler. *Synopsis of Biological Data on the Kemp's Ridley Turtle,* Lepidochelys kempi *(Garman, 1880)*, NOAA Technical Memorandum NMFS-SEFSC-343, National Marine Fisheries Service, Miami, FL, 1994.
16. National Marine Fisheries Service and U.S. Fish and Wildlife Service, *Recovery Plan for the Kemp's Ridley Sea Turtle* Lepidochelys kempi, National Marine Fisheries Service, Washington, DC, 1992.
17. C.K. Dodd, Jr., *Synopsis of the Biological Data on the Loggerhead Sea Turtle* Caretta caretta *(Linnaeus, 1758)*, Biological Report 88(14) and FAO Synopsis NMFS-149, U.S. Fish and Wildlife Service, Washington, DC, 1988.
18. National Marine Fisheries Service and U.S. Fish and Wildlife Service, *Recovery Plan for U.S. Population of Loggerhead Turtle* (Caretta caretta), National Marine Fisheries Service, Washington, DC, 1991.
19. H.F. Hirth, *American Zoologist* **20**, 507–523 (1980).
20. J.D. Miller, in P.L. Lutz and J.A. Musick, eds., *The Biology of Sea Turtles*, CRC Press, Boca Raton, FL, 1997, pp. 51–81.
21. K.J. Lohmann and C.M.F. Lohmann, *Journal of Navigation* **51**, 10–22 (1998).
22. S.E. Stancyk, in K.A. Bjorndal, ed., *Biology and Conservation of Sea Turtles: Proceedings of the World Conference on Sea Turtle Conservation*, Smithsonian Institution Press, Washington, DC, 1982, pp. 139–152.
23. National Research Council Committee on Sea Turtle Conservation, *Decline of the Sea Turtles: Causes and Prevention*, National Academy Press, Washington, DC, 1990.
24. A. Carr, *Conservation Biology* **1**, 103–121 (1987).
25. S.B. Collard and L.H. Ogren, *Bulletin of Marine Science* **47**, 233–243 (1990).
26. W.N. Witzell, *Marine Fisheries Review* **56**, 8–23 (1994).
27. J.A. Mortimer, in K.A. Bjorndal, ed., *Biology and Conservation of Sea Turtles: Proceedings of the World Conference on Sea Turtle Conservation*, Smithsonian Institution Press, Washington, DC, 1982, pp. 103–109.
28. D.W. Owens, in P.L. Lutz and J.A. Musick, eds., *The Biology of Sea Turtles*, CRC Press, Boca Raton, FL, 1997, pp. 315–342.
29. D.H. Gist and J.M. Jones, *Journal of Morphology* **19**, 379–384 (1989).
30. J.K. Leong, D.L. Smith, D.B. Revera, J.C. Clary, III, D.H. Lewis, J.L. Scott, and A.R. DiNuzzo, in C.W. Caillouet, Jr., and A.M. Landry, Jr., eds., *Proceedings of the First International Symposium on Kemp's Ridley Sea Turtle Biology, Conservation and Management*, TAMU-SG-89-105, Texas A&M University, Sea Grant College Program, College Station, TX, 1989, pp. 178–201.
31. B.A. Robertson and A.C. Cannon, *Texas Journal of Science* **49**, 331–334 (1997).
32. C.W. Caillouet, Jr., *Marine Turtle Newsletter* **68**, 13–15 (1995).
33. M.A. Grassman, D.W. Owens, J.P. McVey, and R. Márquez M., *Science* **224**, 83–84 (1984).
34. D.J. Shaver, *National Park Service Park Science* **10**, 12–13 (1990).
35. S.S. Heppell, *Marine Turtle Newsletter* **76**, 6–8 (1997).
36. D.J. Shaver, D.W. Owens, A.H. Chaney, C.W. Caillouet, Jr., P. Burchfield, and R. Márquez M., in B.A. Schroeder, compiler. *Proceedings of the Eight Annual Workshop on Sea*

Turtle Conservation and Biology, NOAA Technical Memorandum NMFS-SEFC-214, National Marine Fisheries Service, Miami, FL, 1988, pp. 103–108.
37. C.W. Caillouet, Jr., *Marine Turtle Newsletter* **69**, 11–14 (1995).
38. T.R. Wibbels, Y.A. Morris, D.W. Owens, G.A. Dienberg, J. Noell, J.K. Leong, R.E. King, and R. Márquez M., in C.W. Caillouet, Jr., and A.M. Landry, Jr., eds., *Proceedings of the First International Symposium on Kemp's Ridley Sea Turtle Biology, Conservation and Management*, TAMU-SG-89-105, Texas A&M University, Sea Grant College Program, College Station, TX, 1989, pp. 77–81.
39. C.T. Fontaine, D.B. Revera, T.D. Williams, and C.W. Caillouet, Jr., *Detection, Verification and Decoding of Tags and Marks in Head Started Kemp's Ridley Sea Turtles*, Lepidochelys kempi, NOAA Technical Memorandum NMFS-SEFC-334, National Marine Fisheries Service, Galveston, TX, 1993.
40. C.W. Caillouet, Jr., B.A. Robertson, C.T. Fontaine, T.D. Williams, B.J. Higgins, and D.B. Revera, *Marine Turtle Newsletter* **77**, 1–6 (1997).
41. E.K. Stabenau, A.M. Landry, Jr., and C.W. Caillouet, Jr., *Journal of Experimental Marine Biology and Ecology* **161**, 213–222 (1992).
42. C.W. Caillouet, Jr., S.A. Manzella, C.T. Fontaine, T.D. Williams, M.G. Tyree, and D.B. Koi, in C.W. Caillouet, Jr., and A.M. Landry, Jr., eds., *Proceedings of the First International Symposium on Kemp's Ridley Sea Turtle Biology, Conservation and Management*, TAMU-SG-89-105, Texas A&M University, Sea Grant College Program, College Station, TX, 1989, pp. 165–177.
43. C.T. Fontaine and T.D. Williams, *Chelonian Conservation and Biology* **2**, 573–576 (1997).
44. S.E. Solomon and R. Lippett, *Animal Technology* **42**, 77–81 (1991).
45. C.T. Fontaine, K.T. Marvin, W.J. Browning, R.M. Harris, K.L.W. Indelicato, G.A. Shattuck, and R.A. Sadler, *The Husbandry of Hatchling to Yearling Kemp's Ridley Sea Turtles*, NOAA Technical Memorandum NMFS-SEFC-158, National Marine Fisheries Service, Galveston Laboratory, Galveston, TX, 1985.
46. L.D. Brongersma, P.C.H. Pritchard, L. Ehrhart, N. Mrosovsky, J. Mittag, R. Márquez M., G.H. Hughes, R. Witham, J.R. Hendrickson, J.R. Wood, and H. Mittag, *Marine Turtle Newsletter* **12**, 2–3 (1979).
47. G.H. Balazs, *Marine Turtle Newsletter* **12**, 3–4 (1979).
48. C.T. Fontaine, T.D. Williams, and D.B. Revera, *Care and Maintenance Standards for Kemp's Ridley Sea Turtles* (Lepidochelys kempi) *Held in Captivity*, NOAA Technical Memorandum NMFS-SEFC-202, National Marine Fisheries Service, Galveston, TX, 1988.
49. J.R. Wood and F.E. Wood, *Herpetological Journal* **1**, 247–249 (1988).
50. D.C. Rostal, D.W. Owens, J.S. Grumbles, D.S. MacKenzie, and M.S. Amoss, Jr., *General and Comparative Endocrinology* **109**, 232–243 (1998).
51. D.C. Rostal, T.R. Robeck, D.W. Owens, and D.C. Kraemer, *Journal of Zoo and Wildlife Medicine* **21**, 27–35 (1990).
52. J.H. Heck, D.S. MacKenzie, D. Rostal, K. Medler, and D. Owens, *General and Comparative Endocrinology* **107**, 280–288 (1997).
53. D.J. Shaver, *Marine Turtle Newsletter* **74**, 5–7 and **75**, 25 (1996).
54. D.J. Shaver, and C.W. Caillouet, Jr., *Marine Turtle Newsletter* **82**, 1–5 (1998).
55. S.S. Heppell, L.B. Crowder, and D.T. Crouse, *Ecological Applications* **6**, 556–565 (1996).
56. C.W. Caillouet, Jr., *Publications and Reports on Sea Turtle Research by the National Marine Fisheries Service, Southeast Fisheries Science Center, Galveston Laboratory, 1979–1996*, NOAA Technical Memorandum NMFS-SEFSC-397, National Marine Fisheries Service, Galveston, TX, 1997.
57. Turtle Expert Working Group, *An Assessment of the Kemp's Ridley* (Lepidochelys kempi) *and Loggerhead* (Caretta caretta) *Sea Turtle Populations in the Western North Atlantic*, NOAA Technical Memorandum NMFS-SEFSC-397, National Marine Fisheries Service, Miami, FL, 1998.

See also SEA TURTLE CULTURE: GENERAL CONSIDERATIONS.

SHRIMP CULTURE

GRANVIL D. TREECE
Texas Sea Grant College Program
Bryan, Texas

OUTLINE

INTRODUCTION

Shrimp aquaculture, the production of saltwater and freshwater shrimp in impoundments and ponds, traces its origins to Southeast Asia, where, for centuries, farmers raised incidental crops of wild shrimp in tidal fish ponds (Fig. 1). The shrimp were not considered of great value. Time has changed this perspective, and shrimp culture has grown into one of the largest and most important aquaculture crops worldwide. Shrimp of all kinds (coldwater and warmwater) are highly desirable now in a world market. Most coastal countries have a harvest industry for shrimp, and about one hundred of those catch enough to export. Shrimp production has been increasing since 1975, when shrimp farming accounted for only 2% of the world market for shrimp. Of the shrimp-producing countries, over 50 practice shrimp aquaculture. Shrimp culture increased 300% from 1975 to 1985, 250% from 1985 to 1995, and if it increases 200% between 1995 and the year 2005, world shrimp culture production will be at 2.1 million metric tons (hereafter abbreviated as MT = 1.1 standard tons = 2,204.6 lb, or 1,000 kg). Present world shrimp culture production is 737 thousand MT annually, or 24.5% of the 3 million MT world market for shrimp (1). This entry discusses the major aspects of shrimp culture, its growth, and some of the problems encountered by the industry while producing shrimp for a world market. The major aspects (Fig. 2) of shrimp aquaculture are sourcing or obtaining brood for hatchery production, maturation and reproduction of broodstock, genetics, egg and nauplii production, larval rearing, postlarval holding and sales, growout in ponds and raceways, production of bait or edible shrimp, harvesting, processing, and sales to a world market.

Figure 1. Shrimp aquaculture.

HISTORICAL NOTES ON PENAEID SHRIMP AND SHRIMP CULTURE

The earliest record of Penaeidae is in Chinese history and traces back to the 8th century BC. Japanese literature referred to penaeids in 730 AD. The first scientific record of a penaeid was in 1759, when Seba of Amsterdam named and drew a North American penaeid. In 1815, Rafinesque recognized that penaeids were a distinct group within Decapoda and named them Penedia (corrected to Penaeidae by the International Commission on Zoological Nomenclature in 1955). Revision of these important decapods and the renaming continues today. Recently, Isabel Perez Farfante and Dr. Brian Kensley proposed some changes in the way scientists refer to the popular farmed shrimp species (2). Except for three species (*Penaeus monodon*, *P. esculentus*, and *P. semisulcatus*), the genus names were changed on other

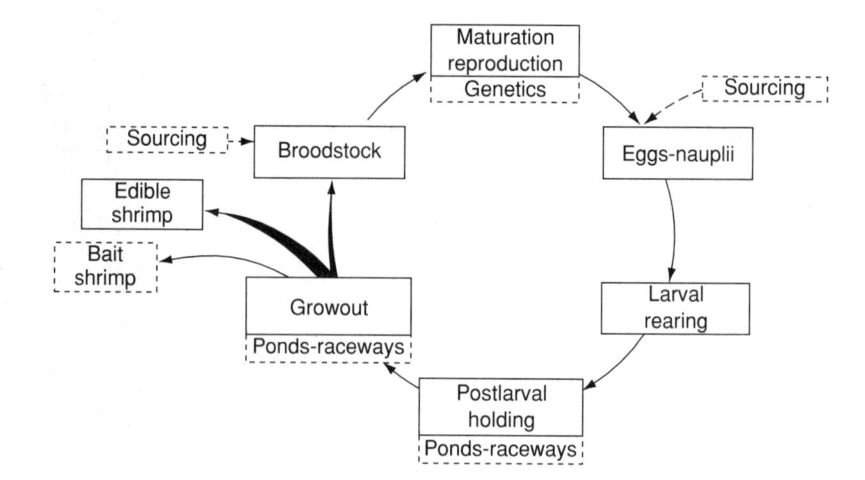

Figure 2. Major aspects of shrimp aquaculture.

penaeids to *Litopenaeus*, *Fenneropenaeus*, *Marsupenaeus*, or *Farfantepenaeus*. Although these proposed changes and the addition of new genera have not been officially accepted by the International Commission on Zoological Nomenclature, the majority of the scientific community accepted the change, and this article will refer to the new names.

For hundreds of years, shrimp farming had only been considered a secondary crop in traditional fish-farming practices in many Asian countries. Modern shrimp culture actually began in the 1930s when the Japanese succeeded in closing the life cycle of the kuruma shrimp, *Marsupenaeus japonicus*. The Japanese cultured larvae in the laboratory and succeeded in mass producing them on a commercial scale. The work of M. Fujinaga in 1933 [also found in the literature as Hudinaga (1935)] (3) with *Penaeus japonicus* (*M. japonicus*) opened the way to modern shrimp farming. Fujinaga published (4) and spread the technology from 1935 to 1967 (5), and his work contributed largely to getting the industry started. In the early 1960s, the first commercial farms were built in the Seto Inland Sea of Japan. In 1935, J.C. Pearson described the eggs of some penaeid shrimp from the western hemisphere, and in 1939 he described the life histories of some American penaeids. A familiarity with the penaeid shrimp life cycle (Fig. 3) emerged and was an important step in understanding what was required to obtain desired results in hatchery and growout procedures. Since adult shrimp migrate offshore to the more stable environment (better salinities and temperatures) of the ocean, where they mature and reproduce, commercial hatcheries found that they had to mimic natural conditions. Hatcheries worked better with higher salinities and cleaner water, whereas growout worked best in the back bays and estuaries with lower salinities.

Commercial shrimp growout attempts were made in Ecuador in the 1960s and in the U.S. starting in the late 1960s and early 1970s. The Ecuadorian industry was based upon *Litopenaeus vannamei* and *Litopenaeus stylirostris* and was started by accident when a broken dike on a banana farm allowed shrimp to enter. By the time the farmer repaired the dike, a crop of shrimp had been produced. Expansion of the Ecuadorian industry was made possible by an abundance of wild postlarval shrimp. Large groups of harvesters (Fig. 4) provide postlarvae (hereafter called PL) to the farm owners, by catching the seasonally appearing PL in scissor nets (Fig. 5). The PL prices generally fluctuated from US$2 for 1000 to US$20 for 1000, depending on their availability. After the industry matured and could not rely entirely upon the wild-caught PL for all of its needs, brood collection stations developed

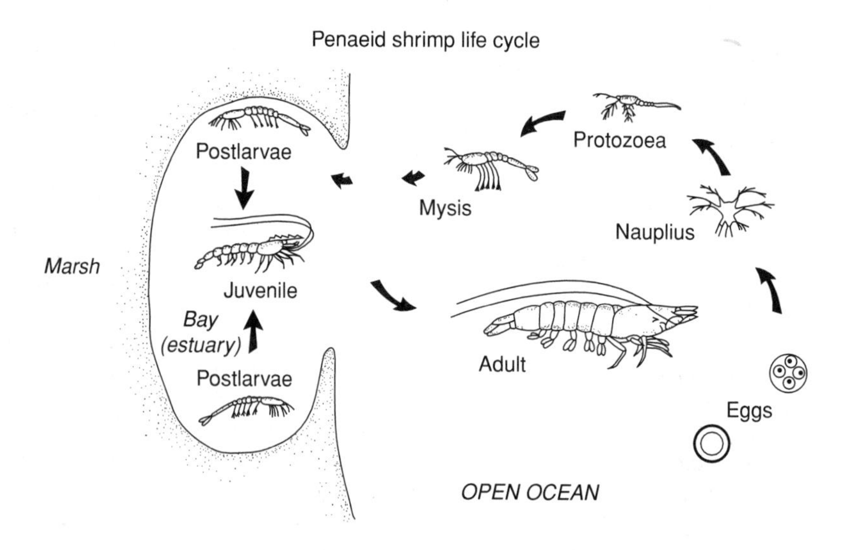

Figure 3. Penaeid shrimp life cycle.

Figure 4. Larval harvesters on the beach in Ecuador.

Figure 6. Twenty count (22.5 g), pond-harvested *L. vannamei*.

Figure 5. Scissor net used to harvest wild postlarval (PL) shrimp from surf in Ecuador.

Figure 7. Small, family-owned (called backyard) hatchery in Thailand.

along the coast and captured and spawned wild-mated females and provided an important nauplii source to meet the growing hatchery demand. The Ecuadorian industry has become more dependent on hatcheries as it has grown, and larvae from hatcheries have become stronger since new hatchery techniques and feeding combination diets have been developed. The initial U.S. industry attempts followed that lead, but were based upon native species of white, brown, and pink shrimp. When researchers grew exotic shrimp from the Pacific coast of Central and South America in the U.S., they proved to be easier to culture and more productive in the ponds. Gradually commercial producers in the U.S. concentrated on exotics, such as *L. vannamei*, now the most popular farm-raised shrimp in the western hemisphere (Fig. 6).

Once shrimp hatcheries (Figs. 7 and 8) began supplying large quantities of shrimp to farmers, the production of farm-raised shrimp expanded rapidly. The explosion of the industry continued into the early 1990s (Figs. 9–12), until problems began with disease outbreaks and water quality, which slowed worldwide production for a few years. In recent years, production has been on the increase again, as ways of controlling diseases have been found and water recirculation and reuse technologies are being more widely practiced (Fig. 13).

Figure 8. Large shrimp hatchery in India.

Many groups from different countries have continued to work on shrimp culture research and development. SEAFDEC (Philippines), TASPAC/MEPEDA (India), AQUACOP (France and Tahiti), NOAA/Sea Grant and USDA (USA), and United Nations/FAO (Rome, Italy) are just a few groups that have made progress through research.

Figure 9. Extensive pond culture in Indonesia.

Figure 12. Intensive culture, round pond farm in Indonesia.

Figure 10. Aquastar—intensive pond culture in Thailand.

Figure 13. Recirculating water practiced on farm in Thailand.

Figure 11. San Migael, semi-intensive and intensive culture farm on Negros, Philippines.

WORLD SHRIMP CULTURE PRODUCTION

Rosenberry (1) is one of the best sources to keep track of current world shrimp production and affairs. The total world production of cultured shrimp grew from about 30,000 MT in 1975 (about 2.3% of the world supply) to over 600,000 MT in 1993 (about 28% of world supply). World shrimp production declined slightly in 1993 for the first time in 11 years, but production still remained at 609,000 MT and rebounded quickly in 1994 with record breaking numbers of 733,000 MT. Most of this impressive growth in production has occurred in Asia, where traditionally about 75–80% of cultured shrimp are produced. The remaining 20–25% is largely produced in Central and South America. From its peak in 1994, until 1997, world production of cultured shrimp declined slightly every year (1995 = 712,000 MT, 1996 = 693,000 MT, and 1997 = 660,000 MT). In 1998, the world shrimp production reached its highest point to date at 737,000 MT (Table 1). Thailand produced 210,000 MT, despite the white-spot syndrome virus (WSSV), the yellow-head virus (YHV), and other diseases. Thailand's ability to adjust and adapt procedures to limit the spread of diseases helped keep production up. Figure 13 shows how one farm (Aquastar) took three to four ponds out of production and circulated water through them to supply one production pond (in the lower right of the figure). Ecuador's production has remained relatively stable (around 130,000 MT/yr) and has adjusted to losses from the Taura syndrome virus (TSV).

The acceleration in shrimp culture activity in the early 1990s was influenced by a combination of factors. Economic conditions encouraged a rapid increase in demand

Table 1. World Shrimp Aquaculture Production, 1986–1987 and 1989–1998

	1986	1987	1989	1990	1991	1992	1993	1994	1995	1996	1997	1998
Eastern Hemisphere	291	421	520	532	567	589	477	585	558	521	462	530
China	83	153	175	150	145	140	50	35	70	80	80	na[a]
Indonesia	41	52	97	120	140	130	80	100	80	90	80	50
Thailand	18	30	94	100	119	150	155	225	220	160	150	210
Philippines	30	35	48	35	30	25	25	30	15	25	10	35
India	19	22	25	32	35	45	80	70	60	70	40	70
Bangladesh	13	15	18	25	25	25	110	35	30	35	34	na[a]
Malaysia	2	3	4	5	5	4	5	4	8	4	6	8
Taiwan	70	90	32	30	30	30	25	25	15	na	14	na[a]
Vietnam	10	15	22	30	30	35	40	50	50	30	30	na[a]
Australia											1.6	2.2
Other	5	6	5	5	8	5	0	11	10	27	16.4	28.3
Western Hemisphere	46	88	63	101	123	123	132	148	154	172	198	207
Ecuador	36	70	40	73	100	95	90	100	100	120	130	130
U.S.	2	5	7	9	9	9	3	2	1	1	1.2	2
Mexico											16	17
Honduras											12	12
Other	4	9	11	13	9	9	39	46	53	16	39	46
Total	341	560	594	633	690	721	609	733	712	693	660	737

Sources: FAO, 1988–1992; Resenberry, 1989–1998; World Bank Report, 1992, U.S. Dept. of Commerce, 1992; and Asia Shrimp Culture Council Newsletter, 1994. *Note*: 1988 was not available in this format.

[a]na = Not available specifically, but it is lumped into "other."

for shrimp, while at the same time simultaneous new innovative culture technologies were being developed. Among the most important factors influencing the demand for shrimp were (*1*) rapid income growth in areas where shrimp was a popular premium food (primarily Asia and some parts of Europe), (*2*) a shift in U.S. and European consumer preference away from traditional protein sources (e.g., red meat) to seafood, and (*3*) adjustments in currency relative to the U.S. dollar that encouraged shrimp consumption in Japan, China, and Europe. The most important technological breakthroughs for countries producing farm-raised shrimp were (*1*) the commercialization of shrimp-hatchery technology, (*2*) the development of high-performance feeds, and (*3*) improvements in overall pond-management practices. Through the present, the industry has continued to adjust and improve technologies, including disease control and water recirculation.

Although cultured shrimp production now contributes over 24.5% of the total world production of shrimp. (It represents a much larger portion of export volume.) This is largely because the developing source countries have an economic incentive to export. Cultured shrimp are generally superior in quality to their trawled counterparts because cultured shrimp get to the processor sooner, without the use of chemicals. The shrimp farmer can usually schedule production and minimize delays between harvesting and processing of crops. Because of these factors, it is generally believed that any further increases in demand for shrimp, especially shrimp of export quality, will most likely be met by increases in cultured shrimp production. Shrimp diseases and poor water quality were the chief reasons that world shrimp production declined slightly from 1995 through 1997, but there appears to be evidence that the industry is rebounding from these problems and is now producing more shrimp than ever before (737,000 MT in 1998). However, the WSSV problem continues to plague the industry on a worldwide basis.

In India, Ecuador, Indonesia, and Thailand, industry revenues range from US$300 million per year to over US$1 billion per year. During the late 1980s Taiwan, Bangladesh, and the Philippines had developed large shrimp culture industries, but these markets have leveled off in recent years. China's shrimp-farming industry boomed until the early 1990s, but then crashed. India and Indonesia boomed until the mid-1990s and then leveled off or declined when problems related to overexpansion took place. More recently, Vietnam and Nicaragua have newly expanded shrimp culture industries. Honduras, Mexico, and Colombia all had substantial industries at one time, but production declined with diseases, water quality, and other problems. Smaller industries exist in Panama, Peru, Guatemala, Brazil, and Venezuela. Ecuador and Thailand have been leaders in the shrimp culture industry. These two countries have continued to adjust to problems in the rapidly changing industry and have continued to expand their industries despite troubles. Additionally, disease-resistant species and steps taken toward sustainable, environmentally friendly operations are assisting the industry recover. The most notable recovery took place in Mexico in the late 1990s, where the IHHN-resistant blue shrimp (*L. stylirostris*), also referred to as "Super Shrimp," is assisting the industry rebound. The disease-resistant strain was developed in Venezuela and is doing well there, as well as on the Pacific Coast of Mexico and a few other isolated areas, such as west Texas and the Dominican Republic.

Areas of the world that consume the majority of the shrimp produced are the United States, Japan, and western Europe. Also, more recently, China has begun to purchase shrimp from Thailand and other countries.

SHRIMP PRICES AND QUANTITIES CONSUMED IN THE UNITED STATES

The current prices for shrimp in the United States can be obtained on the internet free of charge at *http://www.st.nmfs.gov/st1/market_news/doc45.txt*, with quantities harvested by the shrimping industry. According to USDA's Aquaculture Outlook (6) (now available on the internet at the following web site address: *http://usda.mannlib.cornell.edu/emailinfo.html*), total shrimp imports in 1998 reached US$3.1 billion, an increase of 5% from 1997 and 27% from 1996. The increase was due to a 7% increase in volume, as the average price of all imported shrimp products declined 2% to US$9.86/kg (US$4.48/lb). The total trade deficit for the United States in 1999 averaged US$18.9 billion. Shrimp made up one-sixth of that deficit, and all seafood made up approximately one-third of it.

Imports of shrimp products totaled 313 million kg (695 million lb), with frozen products accounting for 86%, fresh shrimp for 1%, and prepared products (breaded, canned, precooked, etc.) for 13%. Shrimp imports are expected to continue to increase into the year 2000, as a strong domestic economy should increase both restaurant sales and home usage, and a strong U.S. dollar will encourage imports from major suppliers, such as Thailand, Ecuador, Mexico, and Indonesia. Although frozen products had dominated shrimp imports, a growing portion of imported shrimp is now being shipped as prepared products. In 1998, prepared-shrimp imports totaled 40 million kg (89 million lb), having a value of US$452 million, up 29% from 1997. Asian producers, notably Thailand, India, and Indonesia, were the major suppliers of prepared-shrimp products, accounting for 88% of total shipments in 1998. In 1999, shipments of prepared shrimp are again outpacing increases in fresh and frozen products. Higher away-from-home food consumption and the growth of prepared-meal sales at food stores drive the increases in prepared-shrimp imports. Imports from Thailand have been the fastest growing among the major shrimp suppliers. In 1998, imports from Thailand totaled 91 million kg (203 million lb) and were valued at US$1.1 billion (7). Thailand is estimated to be the largest shrimp-farming country, with total production in 1998 estimated at 210,000 MT on a head-on basis. Favorable exchange rates and a desire to gain foreign exchange earnings have bolstered Thai exports.

Imports of frozen-shrimp products reached 270 million kg (599 million lb) in 1998, up 5% from the previous year. Shipments of frozen unshelled shrimp are reported in nine different size categories and are grouped by count. The count sizes range from the largest shrimp, less than 33/kg (15/lb), to the smallest sized shrimp, with more than 155/kg (70/lb). Each country's role as a shrimp supplier to the United States varies with shrimp sizes. Mexico, India, and Bangladesh are major suppliers of large shrimp. Ecuador and Thailand dominate imports of middle-sized farmed shrimp. A number of Central American countries dominate imports of the smallest sized shrimp (6).

SALTWATER SHRIMP CULTURE IN OPEN POND SYSTEMS

Marine shrimp culture continues to generate a large amount of interest worldwide. Despite problems in an industry struggling to attain sustainability, average yields from most farms steadily increased in the early 1990s, partly due to improved techniques and these yields stimulated a higher level of interest among potential investors, entrepreneurs, and businesses wanting to diversify their holdings. In recent years, viruses such as TSV, WSSV, and YHV have taken their toll on production rates at shrimp farms. Even areas that practice High Health Genetically Improved (HHGI) technologies have been plagued with viruses, often resulting in 10–30% survivals in comparison to the normal 50–74% survivals during the four- to five-month culture period. Many have found that shrimp culture requires a large investment with hidden costs and that the returns can be low. But if planned and managed properly, it can also be a very profitable and rewarding business. A financial analysis spreadsheet is available for shrimp farming that assists in determining the economic feasibility of a proposed project over a 12-year horizon and calculates the internal rate of return (8). Another software program exists that assists shrimp farmers control parameters on the farm and forecast the production parameters, as well as look at the present financial status of a shrimp farm. This program is called AP/1 and is available in Ecuador.

Much progress has been made to fully domesticate shrimp, as the swine and poultry industries have done, but more technology development is needed in genetics, selection of disease-resistant strains, probiotics, and nutrition.

SALTWATER SHRIMP CULTURE IN SEMICLOSED OR RECIRCULATING SYSTEMS

With new environmental restrictions and regulations occurring in the shrimp culture industry, many businesses are further restrained and are finding it more difficult and expensive to conduct business under the new environmental constraints. Additionally, new viruses are discovered yearly. It is presently not as obvious in developing tropical countries, but in the United States, public sentiment falls on the side of sustainability and environmental soundness, and the laws reflect this. Agriculture and the seafood production industries (both wild harvest and aquaculture) are experiencing challenging and often difficult times. The wild harvest of seafood has reached its maximum sustainable yield and in some cases has declined. According to the Food and Agricultural Organization of the United Nations (FAO) (7), world commercial shrimp fishing production increased steadily from 1992 to 1995, but then dropped in 1996 and 1997. The number of live shrimp produced (in MT) in 1992 to 1997 were 2.22, 2.31, 2.38, 2.48, 2.40, and 2.35, respectively. Note that the 1997 figures dropped below the 1994 figures. According to this data, the world shrimp fishery began to decline in 1996 and has continued this trend. Data from 1998 and preliminary data from 1999 data are also indicative of this trend (7).

The recirculation technology is being used more now, because it is more cost-effective. In the late 1960s and in the 1970s, coastal property in Ecuador could be purchased for US$100 per ha (US$45 per ac) and a 250-ha (617.5-ac) farm cost approximately US$500 to US$1000 per ha (US$225 to US$450 per ac) to construct, including a pump station. Today, it costs approximately US$12,000 to US$15,000 per ha (US$5,400 to US$6,750 per ac) to develop new farms in Ecuador and in other areas of Central and South America. Additionally, most of the functional hatcheries have gone to recirculated operations, because they can control all phases of the hatchery cycle. This reduces losses by reducing the opportunities for opportunistic pathogens. During the past five years, the aquaculture industry has begun to modify average pond size by moving to smaller ponds. They have implemented and improved water treatment methods (both before and after use), have begun to implement biosecurity along with disease identification techniques, and treatment and prevention have now become top priorities. The degradation of the environment and more stringent regulations have encouraged the industry to move in this direction. At pond outlets, which are often someone else's intake, the industry has changed its attitude and is now doing what it can to discharge cleaner, higher quality water than it took in. A similar example from another industry is found in chicken production. That industry moved from barnyard production to 100% environmentally controlled growout units. Fifty years ago, it took approximately six months to produce a marketable chicken. Now, it generally takes six to seven weeks from hatchling to market. This is the direction where the aquaculture industry is headed. Traditional pond culture, like the barnyard, out of necessity, is being used less as time passes.

From 1990 through 1998, the Texas shrimp aquaculture industry produced 10.9 million kg (24 million lb) of shrimp with a farm-gate value of US$75 million, generating a US$225 million economic impact on the state's economy. This helps the United States offset some of its enormous seafood trade deficit. The shrimp farming industry generates more than 150 jobs in the state on a full-time basis, and the number increases during harvest season. The industry has positive affects on other areas, such as feed production (fishmeal, corn, rice bran, and other farm-raised products), processing, transportation, and sales. The same trend that has occurred worldwide has occurred in Texas. Production increased and expansion of the industry occurred until around 1995, when production took a serious dip because of diseases. The shrimp aquaculture production in Texas from 1987 to 1998 can be seen in Figure 14 and from 1990 to 1998 in more detail in Table 2.

Like the aquaculture industry, harvest fleets are being regulated more with limited entry, limited seasons, and limited gear. World markets are pressuring shrimp fleets and shrimp harvest countries to catch the product using certain standards, such as using fish-excluder devices to limit by-catch and avoid catching turtles.

Considering only natural products, next to oil, seafood is one of the greatest trade deficits in the United States. When all products are considered, seafood falls to fourth, after oil, automobiles, and electronics. Increased populations along the U.S. coasts have placed additional demands for seafood and have opened up new markets, but at the same time have placed additional burdens on the coastal environment. Seafood safety has become an issue, since environmental degradation continues. The United States officially implemented the Hazard Analysis Critical Control Point program (HACCP) in December, 1997, which placed additional controls upon the seafood industry. With limited entries, by-catch controversies, turtle- and dolphin-free industry requirements, the wild-caught seafood industry has no real area for expansion, and with new environmental regulations and restraints on our coasts, aquaculture is being restrained. The U.S. Food and Drug Administration (FDA), the U.S. Environmental Protection Agency (EPA), the National Marine Fisheries Service (NMFS), and others played roles in putting together the HACCP guidelines, which were designed to

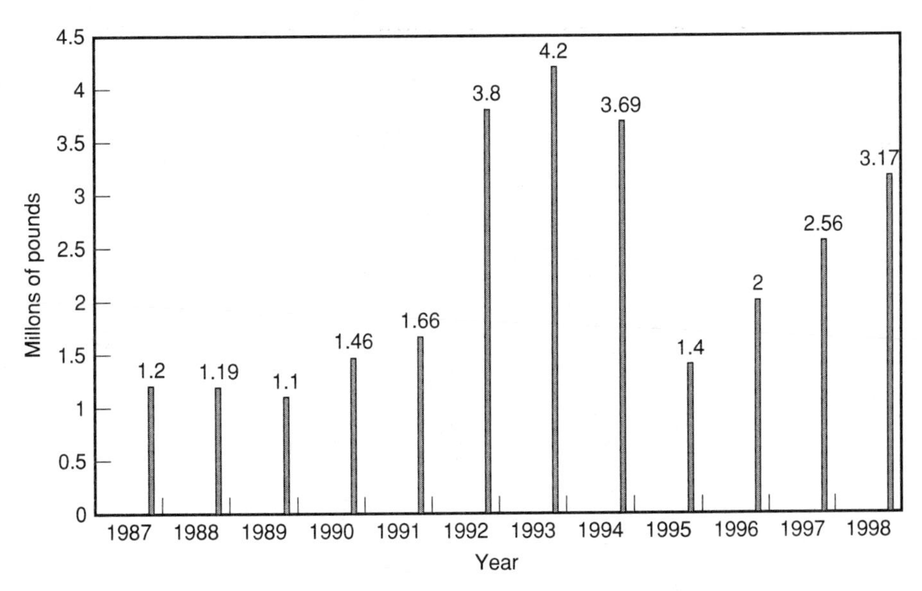

Figure 14. Texas farm-raised marine shrimp, 1987–1998.

Table 2. Texas Shrimp Aquaculture Production for the 1990s[a]

	Acreage Stocked									Heads on Production Pounds								
Shrimp Farms	1990	1991	1992	1993	1994	1995	1996	1997	1998	1990	1991	1992	1993	1994	1995	1996	1997	1998
Arroyo Aquaculture Assoc.,	190	230	180	365	395	395	210	210	345	500,000	300,000	1,104,000	985,500	1,680,000	550,000	600,000	600,000	1,402,356
Austwell Aqua Farms							20	60	60							20,000	42,000	42,000
Bowers' Shrimp Farm	37.8	37.8	37.8	111	200+	325	345	345	265	128,520	113,400	238,140	721,500	1,200,000	550,000	700,000	700,000	414,191
Harlingen Shrimp Farm, Inc.	120	180	450	450	370	450	200	200	337	320,000	550,000	1,300,000	1,100,000	960,000	150,000	183,000	269,000	595,000
M & M (Olivia, TX)	57.5	57.5				57.5	57.5	57.5	0	332,000	300,000					190,000	100,000	0
Regal (West Texas)						30	30	43	43						60,000	120,000	172,000	172,000
Port Lavaca (R & G)	28	28			28	28	70	70	56	94,000	0	0	0	152,000	150,000	50,000	50,000	84,000
St. Martin's Seafood					0	60	120	120	120			0				100,000	260,000	144,000
Southern Star		230	470	470	5	0	0	60	90		395,000	1,200,000	1,389,632	15,000	0	0	277,000	320,000
Triton (West Texas)			6	10	40				0			1,140	25,000	80,000				0
Totals	433 ac 175 ha	763 ac 309 ha	1144 ac 463 ha	1406 ac 569 ha	1038 ac 418 ha	1346 ac 545 ha	1063 ac 430 ha	1165 471 ha	1316 ac 533 ha	1.46 M lb	1.66 M lb	3.8 M lb	4.2 M lb	3.69 M lb	1.4 M lb	2 M lb	2.56 M lb	3.17 M lb
Estimated feed use in Texas (assuming av. FCR 2 : 1)										664 mt	1506 mt	3446 mt	3810 mt	3344 mt	1268 mt	1814 mt	2322 mt	2876 mt
Estimated farm-gate value of crop (in millions of U.S. dollars)[b]										$4 M	$4.7 M	$11 M	$11.2 M	$12.8 M	$3.5 M	$6 M	$11.49 M	$9.68 M

[a]M = million; lb = pounds; mt = metric tons; (1 mt = 2,205 lb); ac-acres; ha = hectares; (1 ha = 2.47 ac). Taura syndrome virus (TSV) did not appear in Texas Valley area in 1997–1998, but did occur on farms further north. Harlingen stocked 200 acres of *Litopenaeus setiferus* at low stocking densities as a precaution against TSV in 1996 & 1997. Bart Reid (Regal Farm) reported that at least 2 ponds produced 6,000 lb/ac with others producing from 4,000–5,000 lb/ac, using *L. stylirostris* (1996). Robert Smiley (Southern Star) reported figures for 1998. *L. stylirostris* can now be grown again in West Texas after being stopped by TPWD in 1997. Average state production from 1990–1998 was 2.66 million lb/year, and the average crop value over the same period was $8.2 M/year.

[b]Average farm gate, heads-on price in 1998 was down, ranging from 2.30–2.90/lb in Texas Valley. Total crop value estimated as follows: 595,000 lb at $2.30 = $1,368,500; 1.4 M lb at $2.90 av. = $4,066,832; 414,191 lb at $4.50 = $1,863,859; 172,000 lb at $4.00 = $688,00; 84,000 lb at $4.00 = $336,000; 144,000 lb at $3.00 = $432,000; and 320,000 lb at $2.90 = $928,000; totaled = $9,683,191.

check food safety within the fisheries and aquaculture industries. The HACCP guidelines, entitled *Fish and Fisheries Products Hazards and Controls Guide, January 1998*, contains 20 chapters in 267 pages. Chapters include the following: "Steps in Developing Your HACCP Plan," "Pathogens From the Harvest Area," "Parasites," "Natural Toxins," "Scombrotoxin (Histamine) Formation," "Other Decomposition-Related Hazards," "Environmental Chemical Contaminants and Pesticides," "Methyl Mercury," "Aquaculture Drugs," "Pathogen Growth and Toxin Formation (Other than *Clostridium botulinum*)," "*C. botulinum* Toxin Formation," "Pathogen Growth and Toxin Formation as a Result of Inadequate Drying," "*Staphylococcus aureus* Toxin Formation in Hydrated Batter Mixes," "Pathogen Survival through Cooking," "Pathogen Survival through Pasteurization," "Introduction of Pathogens After Pasteurization," "Food and Color Additives," "Metal Inclusion," as well as seven appendices, including FDA and EPA guidance levels.

Stricter environmental regulations and the desire to control the spread of shrimp diseases in the United States have forced some of the farms to recirculate water. Farms have learned to produce shrimp using far less water than ever thought possible.

SUSTAINABLE, ENVIRONMENTALLY FRIENDLY SHRIMP FARMING

In the mid-1990s Thailand led the way and started practicing the reuse, or recirculation, of water. In 1998, one Texas farm (Arroyo Aquaculture Association), produced over 637,435 kg (1.4 million lb) of shrimp on 139 ha (345 ac), or about 4,000 kg/ha (4,000 lb/ac) in a semiclosed system (Figs. 15 and 16). In 1999 the farm produced 816,000 kg (1.8 million lb) on 155 ha (385 ac). The management team is leading the way to sustainable, environmentally friendly shrimp farming by recirculating water (Fig. 17) and cutting its water use over time from 37,620 L/kg (4,500 gal/lb) of shrimp produced in 1994 to 2,508 L/kg (300 gal/lb) of shrimp produced in 1998 (Fig. 18). Most of the 2,508 L/kg (300 gal/lb) of water was used to fill the ponds and offset evaporation. The farm cut its stocking density from 50 shrimp/m^2 (50/10.7 ft^2) to 36 shrimp/m^2 (36/10.7 ft^2) and used more aeration, increasing from 8 to 10 hp/ac (Fig. 19). Additionally, the farm widened and deepened it discharge canal and placed aeration in it (Fig. 20). The results of these additions have dropped the total suspended solids (TSS) discharged per unit weight of shrimp produced. In 1994, the farm generated 1.6 kg (3.6 lb) of TSS/0.45 kg (1 lb) of shrimp, and in 1998 it only generated 0.45 kg (1 lb) of TSS for every 9 kg (20 lb) of shrimp produced. In 1994, it discharged almost 0.225 kg (0.05 lb) of ammonia (NH_3) for every pound of shrimp, and in 1998, it only discharged 0.45 kg (1 lb) of NH_3 for every 1,125 kg (2500 lb) of shrimp (Fig. 21). In 1994–1995, the farm discharged between 0.045 and 0.079 kg (0.1 and 0.17 lb) of carbonaceous biochemical oxygen demand (CBOD) for every 0.45 kg (1 lb) of shrimp, and by 1998 it only discharged 0.45 kg (1 lb) CBOD for every 45 kg (100 lb) of shrimp produced (Fig. 22). The amazing thing is that the farm surpassed its 1994 weight per pond produced in 1997 and 1998 (Fig. 23), and it is now producing in excess of 4,000 kg/ha (4,000 lb/ac). This is leading the way to future, sustainable shrimp farming in the coastal areas. See (9) for more detail on the treatment of harvest discharge from intensive shrimp ponds by settling.

Figure 15. Aerial photo of Arroyo Aquaculture Association (top) and Southern Star (bottom) farms in Texas.

Figure 16. Aerial photo of intake at Arroyo Aquaculture Association farm in Texas. (Photo by Louis Hamper.)

Figure 17. Recirculation water pump at Arroyo Aquaculture Association farm in Texas. (Photo by Louis Hamper.)

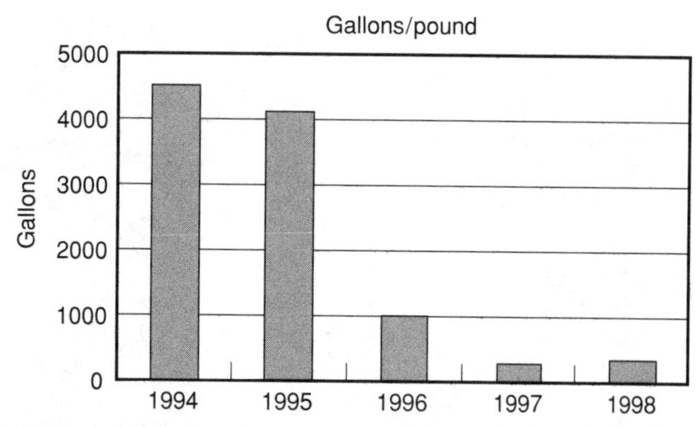

Figure 18. Decrease in water use at Texas shrimp farm. (Data from Louis Hamper.)

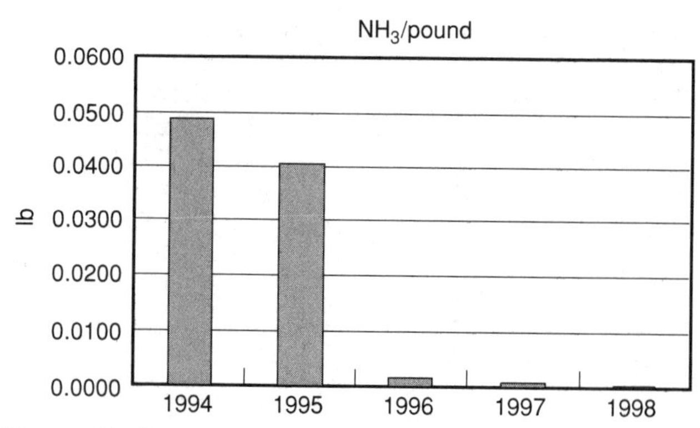

Figure 21. Decrease in ammonia at Texas shrimp farm. (Data from Louis Hamper.)

Figure 19. Paddlewheel aerators on 5-acre pond at Arroyo Aquaculture Association farm. (Photo by Louis Hamper.)

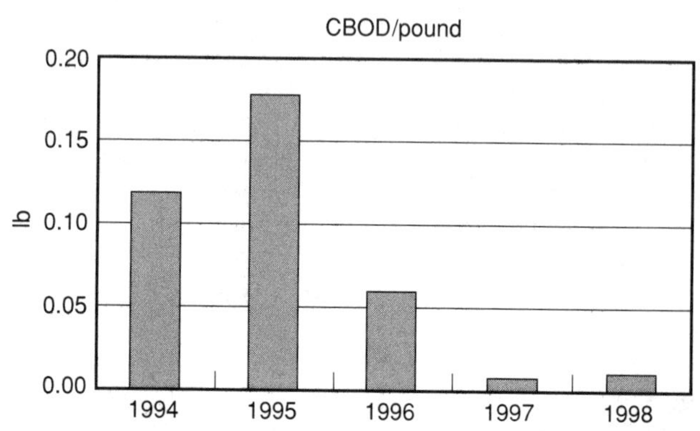

Figure 22. Decrease in CBOD at Texas shrimp farm. (Data from Louis Hamper.)

Figure 20. Discharge canal, baffles, aeration, and nets at Arroyo Aquaculture Association farm. (Photo by Louis Hamper.)

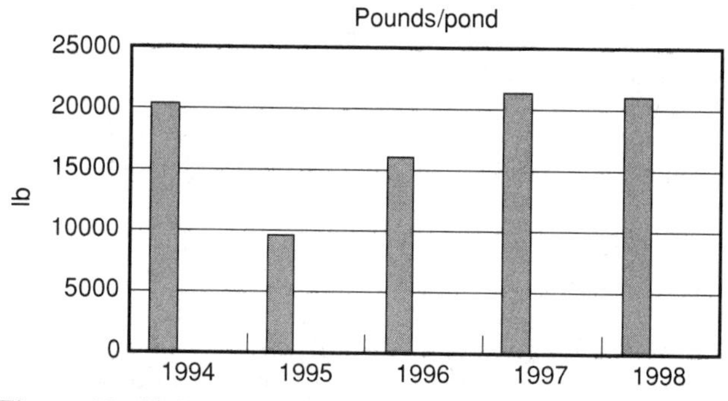

Figure 23. Shrimp production per 5-acre pond at Texas farm. (Data from Louis Hamper.)

BACKGROUND ON U.S. SHRIMP CULTURE RESEARCH

Many early identifications of penaeid shrimp appeared in the literature before shrimp culture had been considered in the U.S. An example is when A.B. Williams published the "Identification of Juvenile Shrimp in North Carolina" (10). National Marine Fisheries Laboratory (NMFS) researchers in Galveston, Texas, made some of the more significant U.S. government contributions toward shrimp culture. Research on the culture of larval shrimp started there in 1959 as part of an investigation into the life history of native shrimp in the Gulf of Mexico. Even though it had nothing to do with shrimp aquaculture, it did lay the foundation for later work on shrimp culture. For example, Harry Cook published a generic key to the protozoea, mysis, and postlarval (PL) stages of the littoral Penaeidae of the NW Gulf of Mexico in 1965 (11). Sindermann (12) provided a more detailed list of publications from this period.

Other groups also worked on larval rearing of penaeids in the United States, mainly in the state of Texas. The

Texas Parks and Wildlife Department and some of the universities published works on the subject very early. One example is Ewald (13). A significant aquaculture research and development effort continues through the U.S. Department of Commerce/NOAA/Sea Grant and the U.S. Department of Agriculture.

The significant early contributions from private industry in the United States came from Ralston Purina in Crystal River, Florida, Marifarms in Florida, and Dow Chemical in Texas. Texas now produces more farm-raised shrimp than any other state in the nation, and has the second largest shrimp hatchery in the U.S. Florida has the largest hatchery in the United States, and possibly in the western hemisphere. The hatchery is capable of producing 180 million PL/mo, but sends them to the parent company's farm in Honduras for growout. Additional information on this hatchery is available at *http://www.seafarmsgroup.com/hatchery.htm*. Texas is followed in the amount of shrimp produced in the United States by Hawaii, South Carolina, and Florida. Florida had a shrimp culture industry in the late 1970s and early 1980s, but the last of the early companies, Marifarms, Inc., moved their operation to Ecuador when a hurricane destroyed their operation in Florida. They stocked hatchery-raised, native shrimp into a bay and were harvesting when the storm destroyed the net that held the shrimp in the bay. Florida now has nine permitted operations growing shrimp in high-chloride water at inland locations. Problems have been experienced at these farms with mortality after handling of the stock, but once the problems are solved, there is potential for growth in the industry.

LIFE CYCLE OF PENAEID SHRIMP

Juveniles and adults migrate offshore, and in the stable environment of the ocean, they mature, mate, and spawn eggs in nearshore or offshore waters (Fig. 3). All but one species within the family Penaeidae follow this life cycle sequence, although the sequences vary highly among species. Most tropical shrimp eggs are 220 micrometers = microns (0.00003937 in.) in diameter (Fig. 24) and sink, but hatch within 14 hours at 28 °C (82.4 °F). The nauplius is the first larval stage and is attracted to light. In a natural setting, the shrimp PL are carried by currents in the ocean to the protection of the estuary. PL that are carried to the ocean beach generally die in the surf or die from predation or lack of food. In the bays and estuaries, larval shrimp are provided with a diet rich in various sources of nutrition and remain there until the late juvenile or early adult stage. PL are primarily benthic in orientation.

The growout phase in bays and ponds generally takes 4–5 months (16–20 weeks). In bays and estuaries and in culture ponds this length of time depends upon the environmental conditions, and shrimp species. Also, the length of time in bays depends on when shrimp migrate in mass to offshore areas. They generally commence migration from the estuary to restart the life cycle in the ocean during the full moon at night with an outgoing tide.

REPRODUCTIVE CHARACTERISTICS OF PENAEID SHRIMP

Grooved shrimp (those having grooves on both sides of the last abdominal segment) mate differently than do nongrooved shrimp. Female grooved shrimp have a closed thelycum [i.e., *P. monodon* (Fig. 25) and *Farfantepenaeus aztecus*]. The female molts and then mates while her shell is soft. The eggs then develop within the ovaries and are spawned. Grooved shrimp can spawn several times on one mating or on one sperm packet, or until the female molts and loses the sperm packet. The sperm packet is held inside the body and part of it is deposited during spawning.

Nongrooved shrimp, (i.e., *L. vannamei*, *L. setiferus*, and *L. stylirostris*) first develop eggs and then mate and spawn. They generally spawn within a few hours after the spermatophore is placed externally during mating. Once the eggs are spawned, they may not all develop in the same manner. Some may develop abnormally, ending up nonfertile.

Sexual maturity in male and female shrimp occurs as early as 34 g (1.2 oz) for *L. vannamei* and 60 g (2.1 oz) for *P. monodon*. Females may spawn numerous times during their mature lifetime; however, some may not spawn at all. Smaller shrimp live approximately 1.5 years, and larger ones may live as long as 3 or more years.

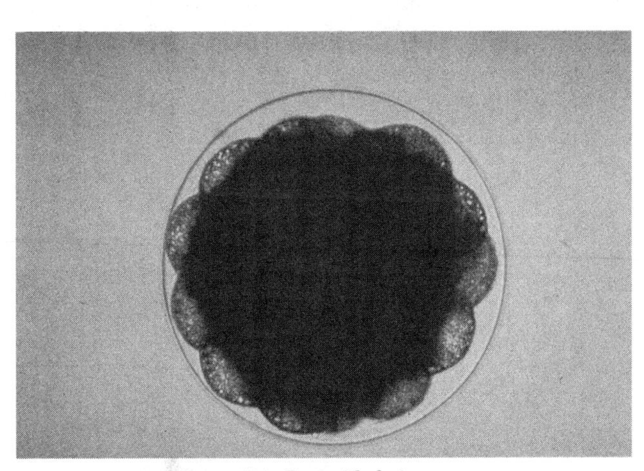

Figure 24. Penaeid shrimp egg.

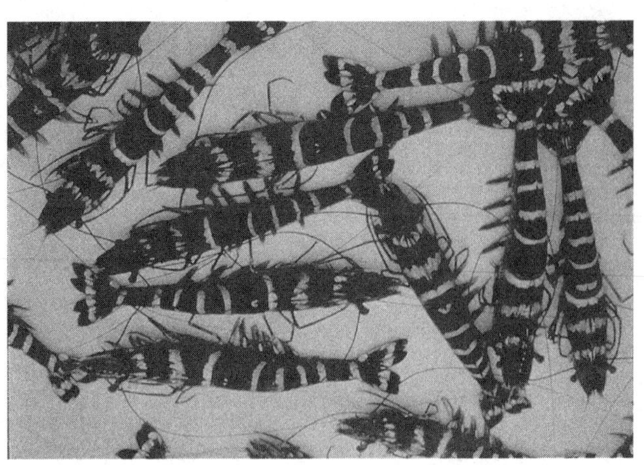

Figure 25. *P. monodon* broodstock.

DESCRIPTION OF CURRENT TECHNOLOGIES USED IN SHRIMP CULTURE

A female with ripe, egg-laden ovaries is said to be "gravid" (Fig. 26, top 3 shrimp; bottom is male). Once a gravid female is ready to spawn, she releases the eggs into the water, fertilizing the eggs by simultaneous rupturing of the spermatophore. The eggs exit the ovipositors, located at the base of the third pair of walking legs, and sink. In the nongrooved shrimp, the eggs brush back against the spermatophore (Fig. 27), as the female is continuously swimming. If the female stops swimming or if her swimming is interrupted, the eggs may fall straight down and are not likely to become fertilized.

Most cultured adult shrimp typically produce around 22,000 to 200,000 eggs per spawn, depending on the size of the female. The larger species, such as *P. monodon*, can produce 700,000 to over 1 million eggs during each spawn. For example, a 290-g (10.2-oz) female *P. monodon* might spawn 700,000 eggs, whereas a 454-g (1-lb) female might spawn 1.4 million eggs each spawn. The eggs hatch into the first larval stage, called the nauplius, which is only one of several larval stages. The microscopic larvae are planktonic and feed on their inner yolk sac for 48 hours. This is the best time to ship or transport young larvae. Starting at about 36 hours after hatching, the hatchery introduces or feeds microscopic single-celled algae and later other minute forms of zooplanktonic microcrustaceans (usually freshly hatched brine shrimp, *Artemia* nauplii). Nine to 11 days after hatching (at 28 °C, or 82.4 °F, and fewer days if at 30–32 °C, or 86–89.6 °F), the larvae change into a form more closely resembling a typical shrimp. They are then called postlarvae. Some hatcheries shorten the larval time in the hatchery by raising the temperature, but care must be taken because some bacteria grow faster at the higher temperatures. The object is to stay ahead of bacteria and other diseases through preventive measures.

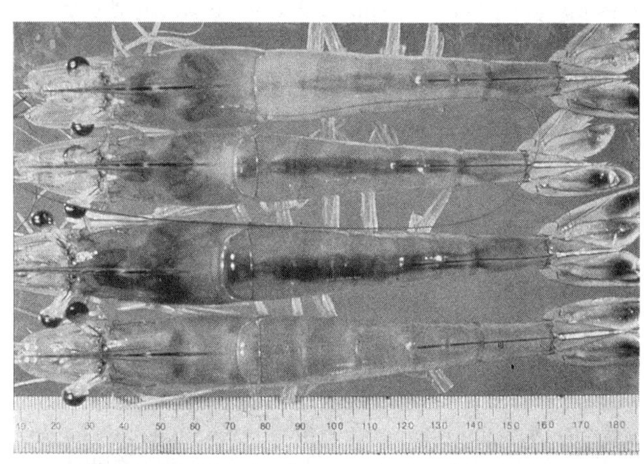

Figure 26. Egg development in the white shrimp.

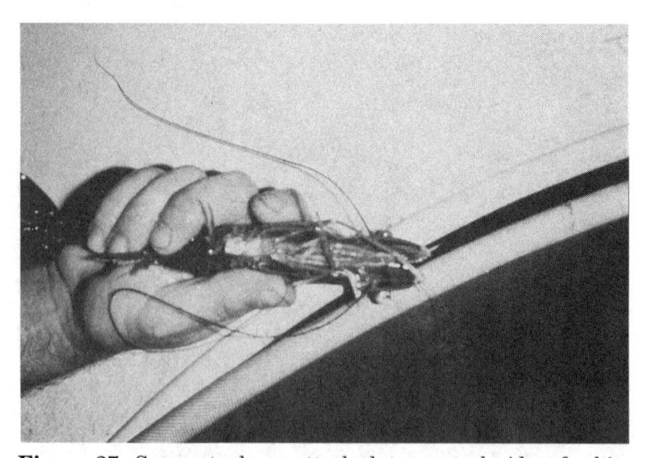

Figure 27. Spermatophore attached to ventral side of white shrimp.

It is estimated that nature is capable of yielding shrimp survival levels of around 2% through the whole life cycle. The commercial shrimp hatchery and subsequent growout phase are capable of achieving much higher survival rates. Nursery ponds are smaller to ponds and can serve as an intermediate phase for the elimination of substandard juveniles used in stocking growout ponds. The growout ponds are used for the production of marketable shrimp. Not all farms use the nursery phase. Many farms stock PL, either from the wild or the hatchery, directly to the growout pond. The market-sized product is generally sold to the processing plant, and the plant, in turn, sells the product in its various forms to the world market.

The shrimp hatchery is very important to the industry. One of the most important aspects with respect to both location and functionality of the shrimp hatchery is water quality. Almost all hatcheries require that oceanic quality water be available on a 24-hour basis. Salinity is the most important water parameter impacting the production of shrimp in the hatchery, and salinity levels must be maintained in a narrow range, between 27 and 36 ppt for best results with most of the species of Penaeidae. Some species of shrimp have very restricted salinity and temperature requirements and preferences.

One such species is the Argentine red shrimp, *Pleoticus muelleri*. It prefers salinities between 33.27 ppt and 33.94 ppt. It is a cold-water shrimp, preferring waters between 9 and 23 °C (48 and 73 °F), and it is found in relatively deep water (25 to 100 m, or 82 to 325 ft). Aquaculture of this species has been overlooked, because it has rather restricted environmental requirements and does not grow as rapidly as some of the tropical species. It takes approximately six months to one year to grow to market size, depending on the source of information, whereas most tropical shrimp can be grown in four months. It costs more to grow them because of the time factor. *L. stylirostris*, can grow to 28 g (1 oz) in 120 days at 26 °C (78.8 °F). *P. muelleri* can grow to up to 7 in. (17.7 cm), but most are closer to 4 in. (10 cm) long. It has an excellent taste, but the flesh can be soft, and handling and processing must be done with great care. If the shrimp are not processed rapidly, they spoil quickly, as is typical of many cold-water species. Some research has been conducted on the aquaculture of *P. muelleri* and other marine species in Argentina and nearby countries since the 1970s. *P. muelleri* does indeed have future potential for commercial aquaculture development, but more research and development are required before it is commercially viable. The Argentine red shrimp is found in the southwest

Atlantic from southern Brazil south along most of the coast of Argentina. Its natural distribution ranges between 20 °S and 48 °S, or from approximately Coratina and Vitória, Brazil to Deseado and Bahia Laura in Argentina, with most being found in the narrow band between 41° and 44 °S, or offshore from Carmen de Patagones to Cabo Raso in Argentina. This narrow band is from the upper reaches of the Gulf of San Matias to Cabo Raso. A marine shrimp, it likes salinities a little less than full ocean salinity, it prefers colder temperatures, and it prefers deeper water. Although it is generally thought of as a deepwater shrimp, some are caught in water shallower than 25 m (82 ft). These parameters are those noted from its natural distribution, but known tolerances outside these ranges are not well understood, which has contributed to its slow evolution as an aquaculture candidate species.

An alternative species suggested by experienced researchers offers additional future potential for commercial aquaculture. Silvio Peixoto, in southern Brazil, states that the species of choice for aquaculture in that area is *Farfantepenaeus paulensis*. This temperate species ranges from Ilheus, Brazil, to Mar del Plata, Argentina, or 150 °S to 38 °S. *F. paulensis* represents a sizable fishery resource in southern Brazil. The species is cold tolerant, and the capture of wild broodstock in adjacent waters has allowed for closing the life cycle in captivity. This is an important feature of *F. paulensis*.

Numerous groups must successfully accomplish maturation, larval rearing, and PL production in order to support an aquaculture industry. Researchers conducting work on these species can be reached through E-mail at *lovrich@satlink.com*, *rivelli@compunort.com.ar*, and *jfenucci@mdp.edu.ar*. Additional literature on the species can be obtained from the Web at *http://tierradelfuego.ml.org/alca*. A beautiful Brazilian hatchery can be seen at *http://www.aqualider.com.br*.

In the shrimp hatchery, seawater is typically checked for pesticides, trace metals, and dissolved organic content on a routine basis. Furthermore, water used in hatcheries is typically filtered to a particle exclusion level of 0.5–1.0 microns (0.000019–0.000039 in.), ozonated, or UV sterilized prior to entering the facility. These treatments help to reduce introduction of bacterial and viral pathogens, as well as organic contamination. Some *Vibrio* spp. are resistant to UV sterilization.

Shrimp hatcheries (Figs. 7 and 8) require relatively small tracts of land and are operated in a labor-intensive manner. Here, broodstock or spawner shrimp from the ocean are brought either into a special facility for subsequent sexual maturation and reproduction or, as in the case of mated females, are allowed to spawn in a nauplii production facility. Some hatcheries prefer to control all production inputs and, thus, source both male and female broodstock shrimp from the ocean. The shrimp are first quarantined to determine extent of possible infection with shrimp pathogens and then placed at densities of five to seven shrimp/m^2 (five to seven shrimp 10.7 ft^2) in large maturation tanks 13 ft, or 4 m, in diameter.

The most important parameters for successful maturation of penaeid shrimp are a constant temperature, acceptable levels of salinity, pH, and light, and good nutrition (Table 3). The object is to reproduce the near constant conditions found in the deeper oceans. Constancy is a must. Clear, pristine, oceanic quality seawater is the key to successful maturation. The following paragraphs describe the important parameters and give examples of how to maintain acceptable levels of these parameters.

Table 3. Parameters for Tropical Shrimp Maturation and Allowable Ranges/24 hr

Salinity	Temperature	pH	Light	D.O.
27–36 ppt +/–0.5	27–29 °C +/–0.2 (80.5–84.2 °F)	7.8 +/–0.2	14 L, 10 D	5 ppm +

1. Temperature for tropical shrimp: Optimum temperatures are 27–29 °C (80.5–84.2 °F) for most warmwater species. The author has found in a commercial setting that a minimum of 28 °C (82.4 °F) is required for *L. vannamei*, for best results, but 26 °C (78.8 °F) is sufficient for *L. stylirostris*. *P. monodon* will do best at 28 °C (82.4 °F). A 0.5° temperature fluctuation over a 24-hour period is acceptable. As much as ±2 °C has been experienced under commercial conditions by the author for *P. monodon* with no serious effects. The coldwater shrimp *M. japonicus*, has an optimum temperature range of 18–28 °C (64.4–82.4 °F) and a minimum spawning temperature of >20 °C (68 °F).

2. Salinity: Optimum salinity level is oceanic (32–35 ppt). Although maturation may occur at a lower or higher salinity (27–36 ppt), normal oceanic salinities are considered to be 35 ppt. Constancy is important. If the salinity is low in the hatchery, it is generally not a good idea to add salt and trace metals if the difference is more than 5 ppt. If artificial sea salts are used, consider using one of the best, which is made by Hawaiian Marine. One can save costs by using Morton's Salt (table salt quality) and buy trace metals separately. Rock salt is cheaper, but has many impurities and should not be used. Instant Ocean, Fritz "Supersalts" and others also sell trace metals. Salinities above 40 ppt should be avoided, because freshwater dilution may also interfere with trace-metal balances essential for maturation and high animal health. Carbon filters may also act to remove some trace metals from seawater.

3. pH: Optimum level is 8.0, but 7.8–8.2 is acceptable, compared with 6.5–10 for growout. Normal seawater, or average seawater, pH is considered to be 8.0. This can vary ±/–0.2 or more, depending on the location. If the pH is lower than 7.8 or higher than 8.2, the site should be avoided. Any addition of buffers or acids may interfere with the balance of trace metals and chemical reactions in seawater and should be avoided unless absolutely necessary. Some adjustments have been reported to maintain pH/alkalinity levels by using sodium hydroxide or calcium carbonate. Low pH affects blood affinity for oxygen. Growth is negatively impacted by pH levels below 5.0. Shrimp can tolerate high levels of pH for a short time. Phytoplankton often cause the pH in the pond to rise to 9 or 10, even sometimes higher, during the day, especially when there is a heavy algal bloom in

the pond. A high pH allows high levels of toxic, unionized ammonia to form.

4. Light: Dim light, with approximately 14 hours of light and 10 hours of darkness is sufficient. The light cycle should be longer than the dark. Ablated animals reproduce in a wide variety of light regimes, but dim light helps keep them docile.

5. Nutrition: One of the most important aspects in sustaining a maturation program is nutrition. A combination diet works best. There is strong evidence that bloodworms (*Glyceria dibrachiata*) are essential for commercial maturation of *L. vannamei*. There is no question that the animals prefer this food to others and become very excited when worms are placed in the tank. The source of the food needs to be carefully considered. Worms can be purchased in Panama or the United States. The United States, however, has strict entry requirements for worms from Panama, so check the rules before ordering. Worms can be purchased live or frozen.

Squid is most often used as the major food for penaeid maturation programs. Pathologists have found that Gulf of Mexico squid may carry Rickettsia (a microbe with similarity to both viruses and bacteria) and can infect the shrimp if used for feed. Therefore, it is recommended to purchase squid from a different area of the country and from a different climate to avoid contamination. Squid is relatively easy to obtain and inexpensive in comparison to bloodworms. Both food sources are high in protein, and squid is high in sterols, whereas bloodworms are high in long-chain, highly unsaturated fatty acids (HUFA).

Other parameters to consider in the maturation of penaeids are the following:

1. Nitrogen levels (especially ammonia and nitrites) should be very low to nonexistent. Average seawater has

0.02–0.04 mg/L (ppm) NH_4-N = ammonium ion (total ammonia nitrogen),

0.01–0.02 mg/L (ppm) NO_2-N (nitrite),

0.1–0.2 mg/L (ppm) NO_3-N (nitrate).

Any nitrogen levels above normal oceanic water levels should be taken seriously and dealt with, or maturation could be jeopardized. Chen and Chin (14) found that 0.1 mg/L (ppm) nitrite or above can affect reproduction.

2. Water should be provided on a flow-through basis or in conjunction with a good recirculated system to keep metabolic wastes and food by-products from building up in the tanks. Most sustainable hatcheries now recirculate at least 80% of the water in order to maintain better control of the environment and diseases, but none, to the author's knowledge, are able to utilize 100% water recirculation, and still maintain a commercial production output.

3. Total suspended solids, organics, brown or red tide, bacteria, and other debris in the incoming water should be filtered. Most suspended solids and organics should be removed during settling and slow sand filtration. All organics should be removed whether in flow-through systems or in closed or semiclosed systems. Food by-products, feces, and eggs provide substrata for bacteria, fungi, and protozoans. Brown tide or red tide can also become an occasional problem for the hatchery. If it is not economically practical to use carbon filters to remove these unwanted dinoflagulates, then chlorination may become necessary in a reservoir before water is brought into the hatchery. Depending on the organic load, normally 2–8 ppm chlorine treatment overnight is sufficient to kill dinoflagulates and bacteria in seawater. The treated seawater should be vigorously aerated to neutralize the chloramines. Extreme caution should be exercised if chlorine is used. Chlorine can form chloramines and other by-products in seawater as a result of chemical reactions with trace metals. Chloramines can cause stress to the animals and stop maturation. Chlorine can also be neutralized by sodium thiosulfate in a 1:3 to 1:6 ratio (one part thiosulfate for every three to six parts chlorine). Thiosulfate has been found to cause deformities in shrimp larvae, and it would be reasonable to assume that maturation and mating could be affected by a trace metal imbalance caused by the addition of yet another chemical, such as thiosulfate. Thiosulfate additions have caused deaths in broodstock of *P. monodon* in India at levels slightly above the suggested treatment ratio.

4. Tank color should be black, because animals see it better and seem to be more comfortable and at rest in a darker environment.

5. Noise levels should be kept low. All machinery, large air bubbles, or any human activity that would stress or disturb the shrimp should be avoided.

6. Obstructions in the maturation tank should be kept to a minimum (e.g., stand pipes, hoses, tubes, water inlets, air lines, nets, etc.). These interfere with swimming and mating, as well as with capturing the mated shrimp with a net. Obstructions outside the tank (e.g., pipes on the floor) should also be avoided for the safety of the workers.

7. Nets should be soft with small mesh, so as not to damage the animals during handling.

8. A large swimming pool vacuum head should be used with 3.8 to 5 cm (1.5 to 2 in.), and a flexible hose should be used when vacuuming the tank. Most vacuum heads ride on rollers a set distance from the bottom and have a strong vacuum capability to clean the tank rapidly without injuring the animals. Large pieces of uneaten squid, molts, and other debris should be removed during the cleaning routine. Tank sides and bottoms should be brushed once a week to clean them of algae, fungi, and other fouling.

9. Some hatcheries use bird bands or rubber tubing, placed over the shrimp eye, on the eye stalks, to mark the shrimp with eye tags. Nostril expanders can be used to stretch tubing like rubber bands. Other hatcheries simply cut a portion of the shrimp's uropod, or notch it, to mark the shrimp, when tagging is needed.

Females are induced to mature and spawn by a process known as unilateral eyestalk ablation, i.e., the removal, cauterizing, cutting, or ligation of the eyestalk (Fig. 28). See the entry on "Eyestalk ablation" for more detail. Within three to five days after ablation, and under the physical and nutritional conditions previously described,

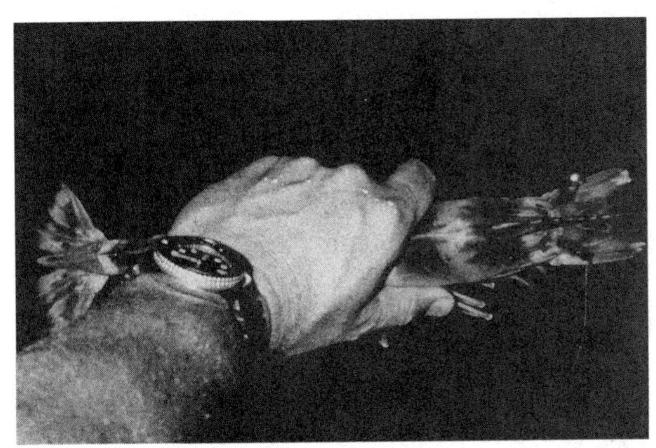

Figure 28. One-pound (0.45 kg, or 450 gram) female *P. monodon* brood, with right eye ablated.

Figure 29. Typical algae room arrangement in shrimp hatchery to feed larval shrimp.

the females should begin to produce eggs. The first spawn usually occurs within one week after ablation, and the ablated animals should be in full production in three weeks. Most females will mature and spawn on a continuous basis, about once every five days to two weeks. The hatchery usually receives approximately three months of continuous production from one set of brood before it becomes necessary to bring in new animals.

Hybridization was attempted with *L. setiferus* and *L. stylirostris*, as well as other species, at Texas A&M University in the 1980s, but their offspring were sterile. Cryopreservation of eggs has been attempted with limited success, both at Texas A&M University and at the University of California at Davis. Reproduction of penaeid species is detailed in (17).

On a daily basis, hatchery technicians evaluate female shrimp in maturation tanks, showing advanced ovarian maturation. In the case of *Litopenaeus* spp. the female is also evaluated for placement of a spermataphore on the exterior and should show signs of being full of eggs (Fig. 26). Those females possessing a spermataphore (Fig. 27) are carefully removed and placed into spawning tanks, taking care not to dislodge the spermatophore. For hatchery operations that only source with trawlers for wild-caught, gravid females, this is the point where the shrimp enter the hatchery. Wild-caught shrimp are typically disinfected prior to stocking into the spawning tanks to prevent diseases and entry of unwanted parasites from feral populations. Female shrimp in hatcheries typically spawn between 1800 and 2300 hours and always spawn before daylight. Once hatched, the young larvae or nauplii are disinfected and evaluated for physical quality attributes. Those possessing suitable quality are transferred to larval rearing tanks and stocked at densities ranging from 100 to 150 nauplii/L (379–568 nauplii/gal) in an intensive culture hatchery.

During the larval cycle, shrimp are planktonic and are generally fed live microalgae and planktonic microcrustaceans. Often this diet is supplemented with artificial feeds, especially when larvae reach the PL stage. With the intensive hatchery method, the production of live feeds (e.g., microalgae; Fig. 29) is undertaken in separate facilities within the hatchery with staff dedicated to this purpose (18–20). Once the young shrimp near the end of the larval stage, the amount of live feed they are offered is reduced, and the amount of artificial feed is increased. At the PL stage, shrimp are often transferred from the larval-rearing tank to a PL-rearing/holding tank. There they are stocked at densities of around 20,000–40,000 shrimp/m^2 (or the same number of shrimp per 10.7 ft^2).

Once PL have reached the PL8–18 stage (8- to 18-day-old postlarvae), they are typically sold to production farms. The species from the western hemisphere are generally sold at an earlier stage, such as PL8–10, whereas *P. monodon* is sold at PL18. Many hatcheries require transport by the farmer to the growout farm. This is an attempt on the part of the hatchery to reduce its liability with respect to shipment survival of PL. Often, the hatchery and the client farmer cooperate in harvesting and counting the PL in the hatchery, reducing the potential for disagreement. Most hatcheries provide customers with reports regarding the performance of the PL during their stay in the hatchery, as well as routine disease analysis reports. If requested by the client, some hatcheries will lower the salinity in the shipment boxes to assist with the acclimation process. Some farmers perform stress tests on the larvae and have them checked for diseases before purchasing them from the hatchery. However, for biosecurity reasons, most hatcheries have strict protocols for farmers visiting the hatchery.

PL are transported to production farms by a variety of means, largely dependent on their distance from the hatchery. In most cases, they are placed in 8–10 L (2–2.6 gal) of water, Styrofoam ice chests at densities between 625 and 2,000 shrimp/L (0.26 gal) (Fig. 30). The ice chests used in shipping contain two transparent plastic bags (one inside the other) that are filled with about 10 L (2.6 gal) of filtered hatchery water (Fig. 30). Once the shrimp have been placed in the water, the water temperature is slowly dropped to about 19 °C (66 °F). Small packets of granular-activated charcoal are added, and the bags are supersaturated with oxygen and sealed. The box is then closed, sealed, and all pertinent shipping information attached. Boxes are consolidated into one shipment and either placed on the transport truck for farm delivery or deposited with a freight expediter at the local

Figure 30. Placing shrimp larvae into styrofoam boxes and plastic bags for shipment.

airport (for long-distance shipping). Shipment of PL can be undertaken fairly efficiently if the total shipping time is less than 24 hours. Most competitive hatcheries allow 5% additional shrimp to reduce loss due to inevitable shipping mortality and have some form of transportation insurance.

PL can be transported in almost any container that holds water and that can be aerated. An example of PL transport in 1,000 L (263 gal) fiberglass tank via small truck can be seen in Figure 31. Upon arrival at the pond, a 12-volt submersible water pump brings water from the pond to the tank, where shrimp are gradually acclimated to pond conditions before being released.

SHRIMP GROWOUT CULTURE STRATEGIES (LEVELS OF INTENSITY)

Shrimp culture practices vary widely throughout the world, but have tended to evolve in line with each country's respective resource endowment. For example, in countries that have abundant supplies of wild PL in brackishwater tidal creeks of large river deltas (e.g., Ecuador and Bangladesh), extensive culture techniques evolved using wild stocking of mixed species of shrimp and other crustacean and fish species. Water enters the simple tidal ponds via gravity tidal water flow through crude water control gates. An extensive culture farm in Indonesia can be seen in Figure 9.

Figure 31. Postlarval (PL) transport from hatchery to pond via truck and 1,000 L fiberglass conical tank.

Extensive culture requires little water exchange and little or no feeding. Instead, tidal exchange is adequate to maintain water quality, and there is sufficient natural food available in the ponds for densities less than 2 shrimp/m^2 (2 shrimp/10.7 ft^2) being cultured. The key characteristic of extensive culture is the reliance on the natural food available in the pond. Yields are low from extensive ponds, typically averaging less than 200 kg/ha/crop (approximately 200 lb/ac/crop) each of shrimp and fish. Because costs for inputs and labor are low and minimal capital investment is required for extensive culture, production costs are generally the lowest in the industry.

As extensive farmers gain more experience, they usually upgrade their ponds to accommodate higher stocking densities, called "semi-intensive" culture (see upper portion of Fig. 11). Peru is one example of a country that has, over time, intensified culture methods (Fig. 32). The ponds located near Tumbes, Peru, are right on the beach in very sandy soils. To prevent erosion from wind and waves, the ponds are long and narrow, with the short dimension perpendicular and the long side running parallel to the shore. Typical stocking for semiintensive culture is 30,000 to 150,000 PL/hectare (or per 2.47 ac). In extensive shrimp culture systems, the shrimp feed upon the food organisms that grow naturally in the pond, and with semi-intensive culture, the farmer must add supplemental feeds, since the higher standing stock of

Figure 32. Shrimp ponds on the beach near Tumbes, Peru.

shrimp exceeds the natural carrying capacity of the pond. At the lower end of the stocking density range, the shrimp continue to derive a significant portion of their nutrition from natural production, and the supplemental feed can be quite simple. As the farmer increases stocking densities, increasingly more complete rations in larger quantities must be applied. Even a well-fertilized and well-managed pond will only support up to 300 kg/ha (300 lb/ac) without supplemental feeding.

With higher feeding rates, the farmer must also apply more management effort to maintain water quality, including pumping to exchange water (Figs. 33 and 34). At the higher end of the semi-intensive range, aeration and water mixing are also necessary. Farmers must also improve their water control gates to exclude predator and competitor species and to allow for increased water exchange. Yields from semi-intensive culture systems range from 500 to 4,000 kg/ha/crop (approximately 500 to 4,000 lb/ac/crop), and production costs typically range from US$3.30–8.80/kg (US$1.50–4.00/lb) of shrimp produced. Peru has generally adopted semi-intensive culture, with the additional management technique of using feeding trays. This technique has improved the water quality, but has added to the labor costs. However, the food conversion ratio (FCR) improved, and feed savings were realized, thus offsetting the increase in labor.

Figure 33. Large shrimp farm intake and reservoir in the Salinas, Ecuador.

Figure 34. One of 20 large intake pumps at farm in Salinas, Ecuador.

The most technologically advanced culture systems are intensive and were developed in such countries as Japan, Taiwan, and the United States, where wild PL are not readily available and where land and labor are expensive. To justify the high input costs and to maximize returns, high yields per unit area are required. Yields from intensive ponds can range from 3,000 kg/ha/crop to over 10,000 kg/ha/crop (approximately 3,000–10,000 lb/ac/crop). Some producers in Peru are bordering on intensive culture, but since they harvest shrimp at 12 g (0.4 oz), aeration is not needed, because very heavy biomasses or standing crops in the ponds are not reached, as they are during longer growing periods when larger animals are cultured.

At the high stocking densities typical in intensive culture systems, the nutritional contribution from natural food organisms in the pond is minimal. The farmer must provide the shrimp with a nutritionally complete ration. These feeds are very expensive compared with the supplemental feeds used in semi-intensive culture. Farmers routinely pay US$1.00 per kg (or per 2.2 lb) or more for intensive culture feeds, representing 60–70% of the cost of production. The feeding-tray method is becoming more popular in both semi-intensive and intensive farms, and as a result, much better FCR values are now being obtained (with a ratio often under 1 : 1, with the best at 0.6 : 1). The semi-intensive culture industry in Peru has proven that it is possible to produce up to 2,100 kg/ha (2,100 lb/ac) with each crop over a sustained number of years. Three crops per year are possible in a warm climate, although most farms average 2.6 crops on a year-round basis. Production levels during the cooler months will not be as high, and time is needed to treat pond bottoms between crops. Texas shrimp farms average 3,500–6,400 kg/ha/crop (3,500–6,400 lb/ac/crop) of heads-on shrimp, which generally fall into the 31–35 to 26–30 count tail sizes, with one crop per year. These are considered to be intensive farms and require aeration.

In addition to the added expense of extra feed, intensive farmers incur additional costs to control the degradation of water quality and the fouling of the pond bottom caused by the heavy organic load from the high feeding rates. These increased costs are related to the capital and operating expenses to build smaller, more manageable ponds (often with concrete walls), install pumps and wells to allow for high rates of water exchange or recirculation, and use mechanical devices to circulate and aerate the water (Figs. 11 and 19).

Taiwan, Japan, Thailand, and the United States are leaders in the development of intensive culture technology and have all averaged over 3,750 kg/ha (3,750 lb/ac) in the past, with some production exceeding 10,000 kg/ha (10,000 lb/ac). The cost to produce shrimp generally rises with increased culture intensity, due to increased stocking densities, feeding rates, and water-quality management efforts. The most cost-effective production strategy for any particular farmer depends on the size of the initial capital investment, the cost of available inputs (e.g., feed, PL, labor, fuel, and power), and the potential cost savings realized from economies of scale relative to the total area under culture. A typical intensive farm layout can be seen in Figure 35, with pump station, primary distribution canal, secondary distribution canal, inflow gates, harvest gates, drainage canal, and sedimentation pond. The primary distribution canal and reservoir also act as a sedimentation pond, where most solids settle within 6 hours. Most of the farms that have been built in the last 10 years have been either semiintensive or intensive, but the majority of the production still comes from extensive culture farms. India, Vietnam, Bangladesh, the Philippines, and Indonesia are good examples of countries with large numbers of extensive farms.

Shrimp farms utilize a two- or three-phase production cycle. With the three-phase cycle, there is a hatchery, nursery ponds, and growout ponds. In the two-phase cycle, the nursery is skipped, and PL are stocked directly into growout ponds. Experience in Texas has shown little difference in survival and growth of shrimp farmed with just two phases. The nursery system allows the farm to stockpile larvae and confine them in a smaller space, saving overhead. Juveniles must be moved to growout, and there is usually mortality associated with any movement of shrimp.

Some countries practice superintensive culture (even more intensive culture than previously described). This method of culture is sometimes done in a greenhouse or in a building. Culture is often in raceways or small, lined ponds. Only a small percentage of the industry practices this level of intensity, and most superintensive operations have not been sustainable.

WORLD CONTRIBUTIONS TO THE ADVANCEMENT OF SHRIMP AQUACULTURE

One of the most detailed semiintensive marine shrimp-growout procedure publications was produced was by Villalon (21). Tseng (22) described the intensive method of growout in Asia, and Wyban and Sweeney (23) described the intensive culture method practiced in the United States. Other major contributions to the advancement of shrimp aquaculture came from around the world from countless groups (24–99). Methods of culture and optimum parameters for cold-water shrimp culture can be found in Main and Fulks (100).

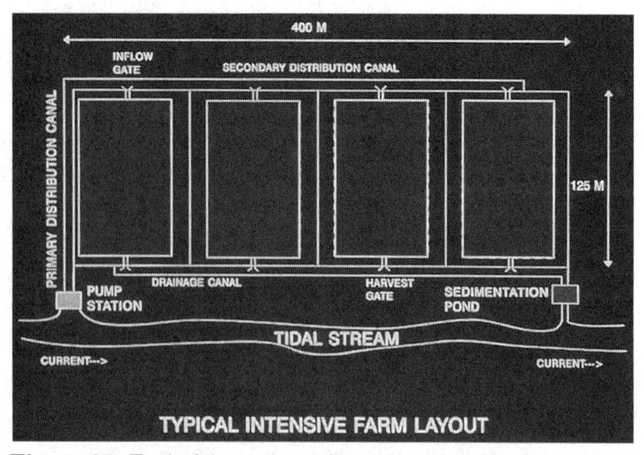

Figure 35. Typical intensive culture farm layout. (Drawing by Joe M. Fox.)

CULTURED SPECIES

The majority of the cultured shrimp in the past have fallen under the genus *Penaeus*. But recently taxonomists changed the genus names of the most commonly cultured shrimp, except *P. monodon*, *P. esculentus*, and *P. semisulcatus*. The shrimp *L. vannamei*, *L. stylirostris*, *P. schmitti*, *L. setiferus*, and *P. occidentalis* are now placed in the genus *Litopenaeus*. The pink and the brown shrimp previously referred to as *Farfantepenaeus duorarum* and *F. aztecus*, respectively, are now referred to as *F. duorarum* and *F. aztecus*, as is *F. brasiliensis*, *F. californiensis*, *F. notialis*, *F. subtilis*, and *F. paulensis*. *Fenneropenaeus chinensis*, is the new name for *P. orientalis*, as well as the new genus for *Fenneropenaeus penicillatus*, *Fenneropenaeus merguiensis*, and *Fenneropenaeus indicus*. *M. japonicus* replaces *P. japonicus*.

The giant tiger shrimp (*P. monodon*; Fig. 25) is named for its size and banded tail, and it dominates production everywhere in Asia, except Japan and China. Reaching a maximum length of 35 cm (13.7 in.) and 454 g (1 lb) in weight (Fig. 28), *P. monodon* is the largest and fastest growing of the farm-raised shrimp. Figure 36 presents growth curves of 11 penaeid shrimp under cultivation (33). It tolerates a wide range of salinities, as do other cultured species. There are often shortages of wild broodstock, and captive breeding can be difficult. Hatchery survivals are generally low (20–30%), and the shrimp tend to burrow into pond bottoms and require special harvesting techniques to retrieve them. The industry in Taiwan developed a specialized harvest technique with electrically charged 12-volt wires or chains that are used to “tickle” the shrimp out of the mud and then catch them in the net (Fig. 37).

The Pacific white shrimp, or the western white shrimp, (*L. vannamei*; Fig. 6) is the leading species produced in the western hemisphere. It can be stocked at small sizes and has a uniform growth rate, reaching a maximum length of 23 cm (9 in.). Its protein requirement (20–25%) is lower than that of *P. monodon*. Hatchery survival rates are high, ranging from 50–90%. During growout, *L. vannamei* has a reputation for being more forgiving or tougher than most penaeids. Harvesting is generally done with gravity draining into nets, but some of the intensive farms use an automated fish pump to remove the shrimp. The Magic Valley Heliarc pump, seen in Figure 38, is an example of the technology developed for the moving of fish and has been adapted for shrimp. Water is pumped into a large bin with a screen, and then shrimp and water are separated. The water is discharged, and the shrimp fall into a container of ice. Markets include the United States (70%), which takes raw, frozen, unshelled tails, called green-headless and value-added products, and Europe (30%), which takes whole frozen animals. Europeans like to see the animal with the head on, so that they can judge the freshness more easily. Japan is the newest market for western white shrimp, and there have been some efforts to grow the western white shrimp in the eastern hemisphere.

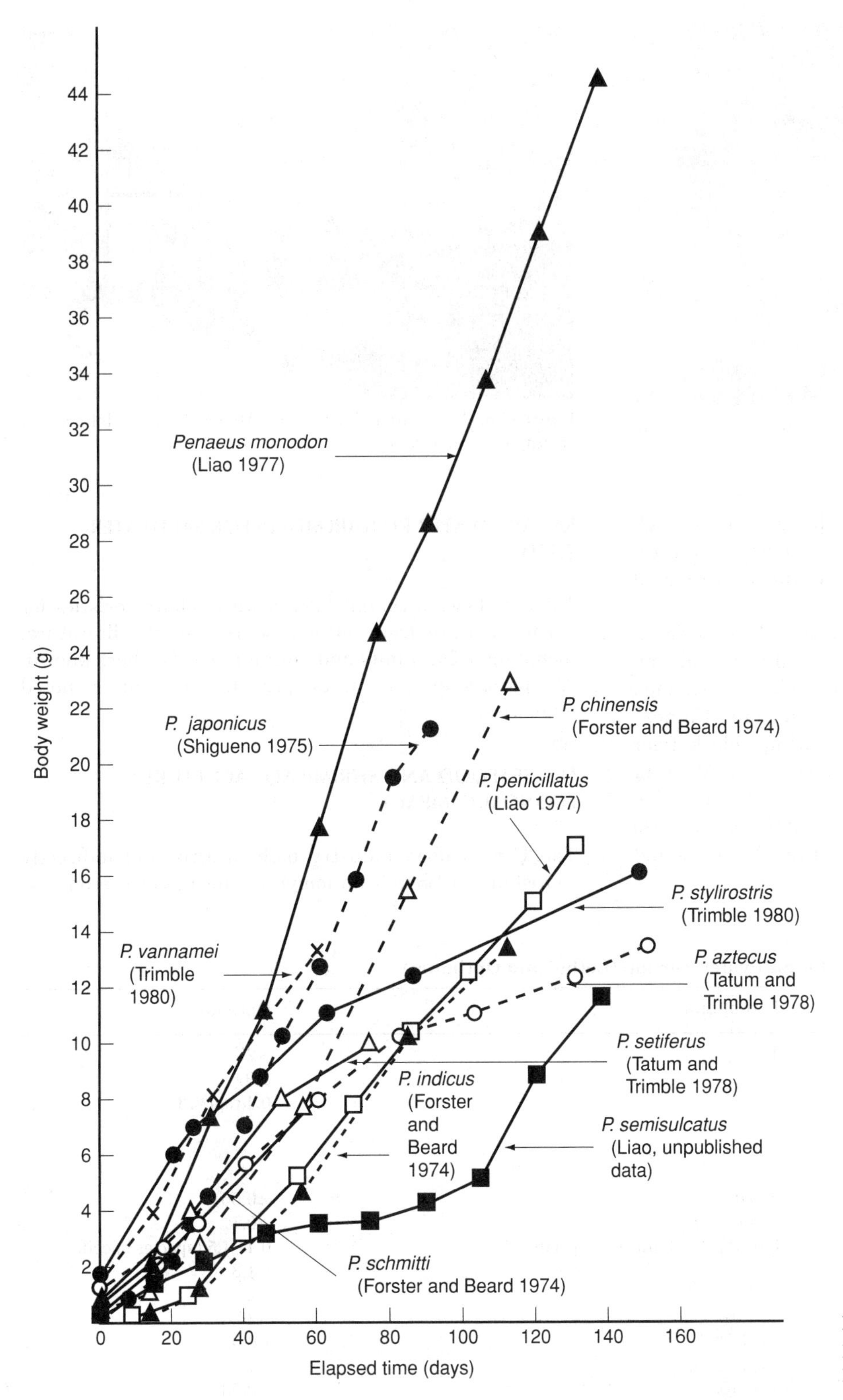

Figure 36. Growth curves of 11 penaeid prawns under cultivation (33) (with permission from CRC Press).

Other Important Species: *L. stylirostris* (the Pacific blue shrimp) (Fig. 1) occurs naturally from the Pacific coast of Mexico to Peru. The 'Super Shrimp,' is also a member of this species and was developed in Venezuela as an IHHN virus-resistant strain. *F. chinensis*, previously known as *P. orientalis*, (the Chinese white shrimp) is found in northern China and Korea. *M. japonicus* (the kuruma shrimp) lives in Australia, Japan, China, and Taiwan; *F. penicillatus* (Taiwan and China) and *F. merguiensis* and *F. indicus* are grown on extensive farms throughout southeast Asia and more intensively in South Africa. *F. merguiensis* and *F. indicus* have become more popular with growers in South Africa. As a result of droughts, wild populations of *P. monodon* became hard to find in certain locations. The two more popular species tolerate low water quality better

Figure 37. Specialized technique for harvesting *P. monodon* in Taiwan. (Photo by Henry Branstetter.)

Figure 38. Automated harvester (Magic Valley Heliarc) at shrimp farm in Texas.

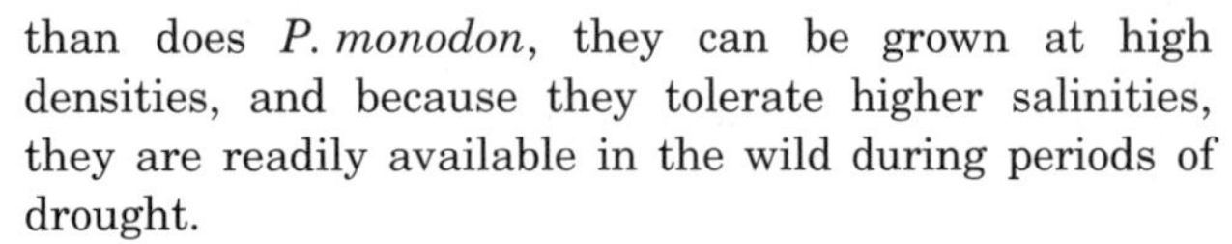

than does *P. monodon*, they can be grown at high densities, and because they tolerate higher salinities, they are readily available in the wild during periods of drought.

Researchers and farmers work with at least a dozen other species. Some common names of saltwater shrimp are banana shrimp, kuruma shrimp, yellow-leg shrimp, western white shrimp, greasy back shrimp, Chinese white shrimp, brown shrimp, pink shrimp, black tiger shrimp, Pacific white shrimp, Pacific blue shrimp, Mexican white shrimp, and the Argentine red shrimp. The freshwater prawn, *Macrobrachium rosenbergii*, is also farmed throughout Asia and in part of the Americas and is discussed next.

KNOWN WATER REQUIREMENTS FOR SALTWATER SHRIMP

Table 4 shows a compilation of water characteristics for shrimp culture from various sources in the literature, including lethal limits and optimum levels where known. All parameters in parts per million, unless noted otherwise.

U.S. SEAFOOD AND SHRIMP AQUACULTURE'S ECONOMIC IMPACT

The U.S. seafood industry both directly and indirectly contributes US$49.5 billion/yr to the nation's economy;

Table 4. Known Water Requirements for Saltwater Shrimp (Lethal and Optimum)

Parameter	Lethal Limit	Optimum
Bicarbonate alkalinity ($CaCO_3$)	10	>20
Boron (B) normal at 35 ppt = 4.5		3.9
Cadmium (Cd)		0.1 (opt <3)
Calcium (Ca)[a]		
Carbon dioxide (CO_2)[b]		
Chloride (Cl)[c]		
Chromium (Cr)[d]	2–20	<0.05
Copper (Cu)	0.300–1.0	<0.1
Dissolved oxygen (DO):	Growth, 2–3 (min) wt specific	0.1–1.5 (species specific)
Fluoride (F)		<1.3
Free CO_2	>20	<2
Hardness as $CaCO_3$[a,e]		>20
Hydrogen sulfide	>0.1	<0.02
Iron (Fe)[g]		<1
Lead (Pb)	1.0–40	<0.03
Magnesium (Mg)[a]		
Manganese (Mn)[m]	5.0	0.08
Mercury (Hg)	0.01–0.04	<0.01
Total ammonia nitrogen (NH_4–N)[h,i]		0.40
Un-ionized form (NH_3)	>0.1	>0.1
Nitrite[h–j]		For growth <1.0
Nickel[k]		0
pH	(For growth: 7–7.5 min, 10–11 max)[h,l]	
Phosphate (PO_4)		ponds 0.15

Table 4. ***Continued***

Parameter	Lethal Limit	Optimum
Salinity (ppt)	(unknown)	10–25 ppt (species specific)
Salinity	(for growth, 0–10 low, 30–40 high)[h]	
Sodium (Na)		Normal 10,000 ppm at 35 ppt
Sulfate (SO_4)[c]		
Suspended solids		<1
Temperature (°C) (tropical shrimp)	(12–15 low, 38 high)	26–29
Temperature	(for growth, 23–25 min and 33–34 max)	
Total dissolved solids (TDS)[f]		
Turbidity	(For growth Secchi disc of 8–10 in. or 20–25 cm)	
Zinc (Zn^{2+})[k]		1.0–10

[a]Hardness is the measurement of calcium and magnesium ions. Hardness should be above 20 ppm for water to be considered favorable for fish culture. Total hardness (calcium carbonate or CCO_3) should be 50–150 ppm for best results according to Boyd (101), between 20 and 200 ppm according to various other sources, and be above 20 ppm according to Johnson (102).

[b]Carbon dioxide (CO_2) is uncommon >60 ppm in well water. CO_2 is thought to be harmful to fish culture if levels are >20 ppm. The optimum for fish culture is thought to be <2.0 ppm.

[c]For best results in fish culture, chlorides and sulfates combined should be at levels of less than 50 ppm, but may be kept at levels ranging from 1 to 2,000+ ppm in wells considered to be freshwater. Common salt in water appears as ionized sodium and chloride (Na^+ and CL^-). In freshwater, the most common ionic constituents are calcium (Ca^{2+}), magnesium (Mg^{2+}), carbonate ($CO_3=$), and bicarbonate (HCO_3-). The chloride level in normal seawater should be 19,000 at 35 ppt salinity and 21,939 ppm at 41.6 ppt salinity.

Corrective actions should be taken when problems arise. Aeration of water at the well or other water source is usually adequate to correct for excess carbon dioxide, iron, sulfide, and DO deficiency. Low pH resulting from excess carbon dioxide may be corrected by aeration, but acidity caused by other dissolved constituents will be little affected by aeration. Adding chemicals can be helpful in some circumstances if the chemical, such as limestone, is inexpensive. When limestone dissolves, its hardness and alkalinity increase. Water already alkaline because of sodium carbonate content will be difficult to adjust for hardness because the solubility of the liming compound is lowered (102).

[d]Chromium concentrations of 0.1 ppm is U.S. coastal water quality criterion, and the detection limit is 0.08 ppm. Concentration of chromium may be higher in summer than in winter.

[e]Alkalinity is a measure of the water's ability to neutralize acid. Alkalinity is contributed by carbonate and bicarbonate ions. Normal seawater alkalinity is around 200 ppm at pH 8.3 and 35 ppt salinity. There is a linear relationship between the salinity level and alkalinity. For example, water with a salinity of 22 ppt should have an alkalinity level of 122 ppm, whereas seawater with a salinity of 12 ppt is expected to have an alkalinity of 67 ppm (assuming that the seawater has been diluted with distilled water). The actual alkalinity will be influenced by the alkalinity of the water with which the seawater is diluted. For example, the alkalinity of estuarine water can be lower if the seawater is diluted with runoff water high in tannic acid. It can be higher if the seawater is diluted with river water high in alkalinity. Both hardness and alkalinity should be above 20 ppm for water to be considered favorable for fish culture. Well water with a low hardness (<20 ppm) and high alkalinity (>200 ppm) is not desirable for fish culture.

[f]A total dissolved solids (TDS) reading of 1,000 ppm and above is considered saline, and a reading of less than 1000 ppm is considered freshwater. Drinking water usually has less than 500 ppm TDS. Livestock can drink water containing up to 2,000 ppm TDS, depending on their acclimation. Freshwater fish can endure 10,000 ppm TDS. Research concerning the effects of TDS on fish has been reported in salinity values. Salinity is a measure of total dissolved ions and closely approximates the numerical value of TDS for a water sample. Examples of maximum salinity readings are the following; Golden shiner fry and goldfish fry is 2,000 ppm; Channel catfish is 11,000 ppm, Grass carp is 12,000 ppm, common carp is 9,000 ppm, and tilapia is 20,000 to 30,000 ppm.

[g]The 96-hour median tolerance limit (TLm 96) to iron by aquatic insects, mayflies, stoneflies, and caddisflies is 0.32 ppm. Most iron appears as divalent (Fe^{2+}) in reducing conditions, but it is readily oxidized to the trivalent state (Fe^{3+}) as a colloid, which may also block the gills of shrimp.

[h]Many of these factors are directly or indirectly interrelated. For example, there is often a direct correlation between DO and Secchi disc readings/turbidity. The pH has a direct effect on the percentage of ionized and un-ionized ammonia in the water. The un-ionized form of ammonia is by far the most toxic, and the percentage of total ammonia that exists in the un-ionized state is a function of pH, temperature, and salinity. NH_4–N is safe in ionized form because the + charge prevents it from passing over gill cell membranes.

[i]NH_4–N = ammonium ion (total ammonia nitrogen): 96-hour LC_{50} (see below) for PL6 is 11.51 mg/L (ppm). In the hatchery, NH_4–N levels reach 0.808 mg/L (ppm) routinely, and the NO_2–N (nitrite) levels can reach 0.118 mg/L (ppm) routinely. Un-ionized ammonia (NH_3–N) is the most dangerous and can be lethal, depending upon the temperature, salinity, pH, DO, and age of the animal. The allowed nitrogen levels increase with the age of the shrimp. The 96-hour LC_{50} values for nitrite (NO_2–N) are as follows: nauplii = 5 ppm, zoea = 13.2 ppm, mysis = 20.6 ppm, and PL = 61.9 ppm (15).

Note: 96-hour LC_{50} is the (lethal) concentration that will kill 50% of the animals exposed to it in 96 hours. The safe level estimates for nitrite are as follows: nauplii = 0.11 ppm and PL = 1.36 ppm (15). Chin and Chen (16) found with different parameters present, the 96-hour LC_{50} for nitrite and PL6 *P. monodon* is 13.5 ppm. The authors also reported that a safe nitrite level for growout was 4.5 ppm.

Ten percent of the 96-hour LC_{50} is usually considered a safe nitrite concentration. This often leads to a rather conservative figure that may be difficult or unreasonable to attain. Freshwater fish farmers have so many problems with nitrites because of low calcium and chloride levels in the water. In brackishwater shrimp culture, low chloride levels tend to be less of a problem, and the higher levels of chlorides serve as a buffer. There are so many factors that influence nitrite toxicity that it is virtually impossible to make recommendations on lethal concentrations or safe concentrations of nitrite in aquaculture. These include chloride concentration, pH, animal size, previous exposure, nutritional status, infection, and DO concentration.

[j]Normal seawater levels are as follows: Ammonium ion, or NH_4–N = 0.02–0.04 ppm; Nitrite, or NO_4–N = 0.01–0.02 ppm; Nitrate, or NO_3–N, is not as toxic and may be found at 1 ppm in normal seawater or as high as 200 ppm in some fish culture systems without causing problems.

[k]Japanese fisheries' criterion for zinc and nickel is <0.1 ppm. Zinc has been reported to accumulate in marine animals in levels ranging from 6 to 1500 ppm (U.S. National Technical Advisory Committee, 1968). It has also been found that shellfish concentrated zinc from shrimp feed when grown in shrimp farm effluent waters.

[l]pH and salinity are usually lower in coastal waters than in the ocean. Groundwater pH is usually in the ranges of 5.5 to 8.0. The most desirable range for the culture of fish is 6.5 to 9.0 and for shrimp is 7.0 to 11. Shrimp can grow at pH levels above 7.8 no matter what the alkalinity level is, but a rule to follow is, if pH reading in the pond is below 7.8 at midafternoon, then the alkalinity reading should be above 100 ppm (see also the note on alkalinity for more detail). Normal seawater pH is 8.0 ± 0.4. If the culture water is out of the 7.6 to 8.4 range (before being treated in any manner, such as through fertilization, liming etc.), care should be taken to watch all of the parameters thought to be important during the culture cycle to be sure that the normal ranges are not deviated from or that unanticipated chemical reactions are not occurring.

[m]One ppm and above manganese concentration is considered inappropriate for fisheries in Japan. Some underground seawater (seawater from wells) has been found to have 10–100 times the manganese concentrations of coastal seawaters. Concentrations of iron, zinc, nickel, cadmium, lead, chromium, tin, and cobalt are other important metals to observe. For example, high levels of tin in well water were found to be the limiting factor in raising shrimp nauplii in a Roswell, New Mexico facility.

this total is projected to reach US$62.9 billion by the turn of the century. Of the US$49.5 billion currently contributed, the food service and processing sector accounts for the largest percentage (77%). Harvesting injects 16%, with distribution and retail stores contributing only 7% (Fig. 39).

In addition, aquaculture products produced in the United States generate a US$4.15 billion economic impact. This industry employees 140,000 full-time workers for a value of US$1.7 billion annually. Shrimp aquaculture plays a role in impacting the economy, generating 2 million kg (4.6 million lb) in 1994. However, due to diseases, it generated under 1.35 million kg (3 million lb) in 1995, 1996, and 1997. In 1998, production again reached 2 million kg (4.6 million lb) of shrimp (Fig. 40).

Texas is the largest producer of farm-raised shrimp in the United States. It produced an average of 1.2 million kg/yr (2.66 million lb/yr) from 1990 through 1998, with an average crop value of US$8.2 million over the same period, having a US$24.6 million/year impact on the state's economy. As can be seen in Table 2 and Figure 14, diseases lowered the crop amount in 1995, but production has been increasing steadily since then.

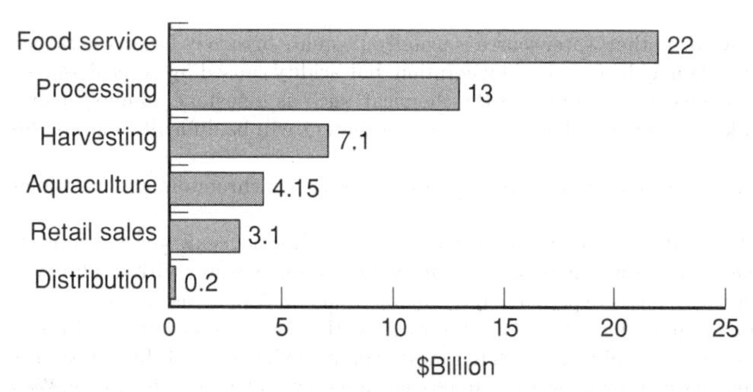

Figure 39. U.S. seafood industry economic impact. (From National Fisheries Education and Research Foundation.)

PROBLEMS IN THE SHRIMP CULTURE INDUSTRY

Open Systems

Sustainability and environmental degradation, coupled with numerous viruses and other diseases, are probably the biggest problems facing open system aquaculture. The costs to producers generated by new environmental regulations are of great concern. In the United States, the economies of scale are nonexistent for the shrimp culture industry, as they are for the catfish industry, and shrimp producers say that less expensive feeds and other products would certainly make their industry more appealing and profitable.

Turbidity is generally an indication of the amount and type of phytoplankton bloom in the pond and is maintained with pumping and fertilizing procedures. Turbidity can also be affected by the amount of suspended clay and silt in the water. If there is an abundance of suspended particles in the incoming water, settling may be necessary before the water goes into the ponds. Usually a reservoir serves this purpose, and most solids settle within 6 hours (9). To keep track of the total volatile solids in the pond or solids that can be burned off in an oven (excluding clay and silt), the farmer generally takes a reading with a Secchi disc and then maintains a reading of 20–25 cm (8–10 in.). This means that the disc is lowered into the

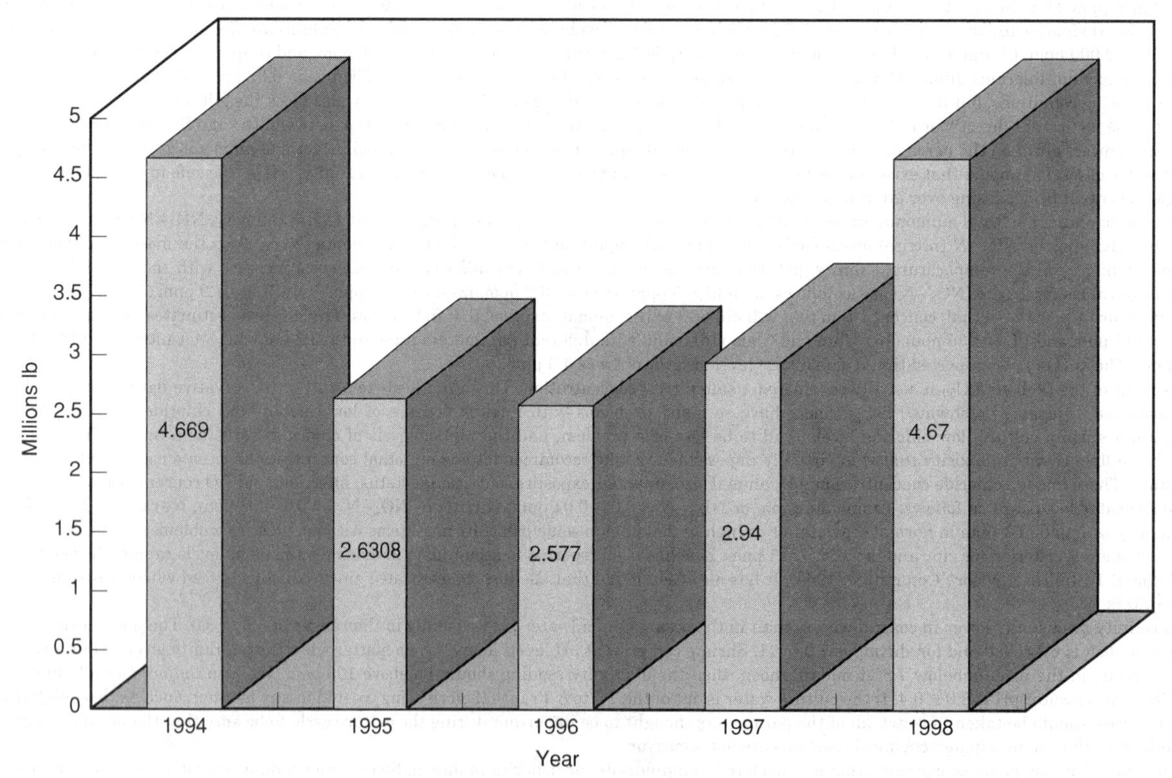

Figure 40. U.S. shrimp aquaculture production.

water between 20–25 cm (8–10 in.) before disappearing from sight. By either flushing with new water or fertilizing, the optimum reading obtained from the Secchi disc can be maintained, and, subsequently, the proper algal bloom can be maintained in the pond. This same turbidity can become a problem when released as effluent. Regulatory agencies generally start by controlling turbidity levels at the discharge. Total suspended solids (TSS), total volatile solids (TVS), biological oxygen demand (BOD), and carbomase biological oxygen demand (CBOD) generally have upper limits placed on them, but dissolved oxygen (DO) has a minimum placed on it.

Excretory products can cause problems. The pond must be designed and managed so that excretory products or metabolic wastes do not build up. In ponds, most excretory products will break down rapidly, with the help of bacteria. In intensive systems, excretory products that do not break down must be removed. Soluble metabolic by-products, such as un-ionized ammonia, and by-products made up of organic materials breaking down to nitrites can be a problem. Tolerance levels in growout are not well known, but high levels have been recorded (0.75 ppm to 2 ppm and higher at pH 8.3) without mortality. Some gill damage may occur when the level of un-ionized ammonia goes above 0.5 ppm or when other stresses are present (e.g., low DO or handling). Growth is also reduced with high ammonia levels present. The by-products can also cause problems in discharge waters and are generally controlled by regulatory agencies.

SEMICLOSED AND CLOSED SYSTEMS

Interests in closed systems extend beyond culturists in higher latitudes, where climatic conditions limit the profitable growing season. Many tropical producers are either retrofitting farms or seriously evaluating closed-pond production techniques as plausible alternatives to traditional flow-through practices that increasingly are accused of being unsustainable.

Design and management of recirculating systems requires greater attention to water chemistry than design and management of flow-through systems. The latter system may export potential water quality problems to the local environment simply by flushing the pond, especially during harvest. When raising animals in closed systems, any such problems must be dealt with effectively within the culture environment itself. This often requires confronting water quality treatment issues at a more technical level. Most operations do not have that capability.

Since there are so few closed-system aquaculture companies, they receive a good deal of publicity, especially if they fail. Usually a company is started with private investment and later moves to public offering. If the company fails while under private ownership, often little is known about it. On the other hand, public offerings of stock further publicize the industry and the company offering the stock. A number of failures have been apparent in the industry, and the publicly owned companies are of higher profile. Developers out to make a quick profit often make projections that can not be reached. These problems occur in all industries, but shrimp culture in the United States is small, and these occurrences are usually of higher profile than the other closely related agricultural enterprises. These failures cause the risk factor to be very high with shrimp culture ventures, and most lending agencies will not loan money unless there is a proven track record.

Recirculation, or water reuse, is a rapidly growing segment of the aquaculture industry, but for the most part is still widely unused or unproven. Some people in the industry have found that they can reuse far more water than ever thought possible and still obtain respectable production.

The areas in the United States where expansion of the aquaculture industry is possible are offshore or inland (or within the coastal plain, but with recirculating or closed systems). The offshore industry will be severely limited by costs, weather factors, and the U.S. Department of Interior's Minerals Management Service laws. The most logical area for the aquaculture industry to grow is on the coast with closed, or partially recirculating, systems and inland, in environmentally friendly, closed, recirculating systems, close to the markets or in brackishwater aquifer areas, such as West Texas, Arizona, and New Mexico. Much interest is being generated in these new growth areas. There is an increased use of closed systems in the aquaculture industry. For example, the largest shrimp hatchery in the western hemisphere is located in the Florida Keys, and due to environmental requirements, regulations, and cost savings, it recirculates water numerous times before discharging into a settling pond. The second largest shrimp hatchery in the United States is located in Texas, and it also recirculates as much water as possible (up to 90%) to save on heating and water filtration costs. Other species, such as tilapia and hybrid striped bass, are being grown successfully on a commercial scale in closed systems, and this is stimulating increased interest in closed systems for shrimp. Unfortunately, there have been far more commercial venture attempts with shrimp culture in fully closed systems that have failed than have succeeded. There is still much to be learned about this method of culture before it is commercially sustainable.

SHRIMP VIRUSES ARE A MAJOR PROBLEM PLAGUING THE INDUSTRY

Shrimp viruses are natural occurrences and can to some extent be controlled. Like viruses in man, some shrimp viruses are more severe than others. One of the first shrimp viruses identified was the baculovirus (BP), found in the waters of the Gulf of Mexico, and described by John Couch more than 20 years ago. Rolland Laramore was one of the first researchers to make an electron micrograph of an occlusion body of the baculovirus. Once researchers developed the techniques to recognize the virus and began to look for it, it was found to be fairly common in the pink shrimp (*F. duorarum*), but was also found in other species from the Gulf of Mexico.

BP is widely distributed in cultured and wild penaeids in the Americas, ranging from the Northern Gulf of Mexico south through the Caribbean and reaching at least as far as the state of Bahia in Central Brazil. On the Pacific coast, BP ranges from Peru to Mexico, and it has been observed in wild penaeid shrimp in Hawaii (96,103, and 104). BP affected shrimp aquaculture in the United States and other countries in the late 1970s and early 1980s, and, in some cases, shrimp larvae were pumped out on the ground and destroyed on Texas farms as a result of disease checks and directives from state regulatory agencies. The virus was controlled largely through the control measures implemented in shrimp hatcheries to recognize the virus in wild-caught brood animals. Thus, BP was no longer considered a major problem in either wild caught shrimp or hatchery animals once more was found out about its control.

The following is a compilation of the information known about baculovirus. Most of the information comes from Lightner (96,103, and 104) and other important sources cited. The known penaeid baculoviruses infect the epithelial cells of the hepatopancreas from protozoea through adult and the mid-gut epithelium of larvae and postlarvae. Baculovirus infections may result in disease in cultured penaeids that is accompanied by high mortality rates, but, for the most part, survival is usually above 30%. In hatcheries worldwide, the baculoviruses *Baculovirus penaei* (BP) in the western hemisphere, monodon baculovirus (MBV) (infecting *P. monodon*), and baculoviral mid-gut gland necrosis (BMN) (infecting *F. japonicus*) have been the cause of serious epizootic disease outbreaks in the larval and early PL stages of their principal host species. BP may cause disease and mortalities in juvenile and subadult animals, but BP has not been a major problem in recent years, because most hatcheries check their brood for it, and if they find it, the brood shrimp are destroyed. This is the most common way to control BP. The geographic distribution of these baculoviruses in cultured penaeid shrimp suggests that they are problems to shrimp culturists mainly in those areas where the virus is enzootic in local wild populations. This appears to be the case of BMN, which has thus far been observed in *F. japonicus* in hatcheries in Japan. However, MBV and BP have been documented as having been introduced into new geographic regions by the transfer of infected PL or broodstock to areas outside the normal range of the host species. According to Lightner (103), BP is widely distributed in cultured and wild penaeids in the Americas. In Brazil and Ecuador, BP infects larvae and PL of at least six penaeid species. It is significant that the imported Asian species *P. monodon* and *F. penicillatus* may also be infected. BP was found in Mexico in cultured larval and PL *L. stylirostris* at a facility near Guaymas, Sonora, in 1988 (104). BP and MBV infections may be readily diagnosed by identifying their characteristic occlusion bodies in either wet mounts or histological preparations of the hepatopancreas and midgut (104). BP occlusions are distinctive tetrahedral bodies easily detected by bright field or phase microscopy in unstained wet-mounts of tissue squashes (105), while MBV occlusions are spherical and difficult to distinguish from lipid droplets. The use of a stain-like 0.1 percentage aqueous malachite green in preparing wet mounts for MBV diagnosis aids in identifying the occlusions. The protein makeup the occlusion absorbs the stain more rapidly than do most of the host's tissue components, making them distinctive within a few minutes (104).

Unlike BP and MBV, which are Type-A baculoviruses because they produce occlusion bodies, BMN is a Type-C baculovirus that does not produce an occlusion body. Hence, the diagnosis of BMN infections is dependent on the clinical signs of the disease, histopathology, and transmission electron microscopy (TEM) (104). Sano et al. (106) developed a rapid fluorescent antibody test for BMN that reportedly simplifies the diagnosis of BMN.

A baculovirus (*plebejus baculovirus*) was reported in *Penaeus plebejus* (107).

Today, BP is rarely found in larvae coming from a good hatchery source. The best prevention is to catch it at the hatchery. If the farm is diagnosed with BP, if it has kept good records about where the shrimp came from, and if it did not mix sources, it can narrow the BP-infected shrimp down to one larval source. BP is generally spread from mother shrimp to larvae and is passed on in the egg. It does not spread further after larvae are placed in the ponds like the more serious viruses do. Therefore, the source of BP on a farm is most likely the hatchery or the wild seed purchased.

Today, more than 20 known viruses infect shrimp, and 4 of these continue to pose a major threat to the industry (infectious hypodermal and hematopoietic necrosis virus (IHHN), Taura syndrome virus (TSV), yellow head virus (YHV), and white spot syndrome virus (WSSV)). In the western hemisphere, 9 of the 20 viruses have killed shrimp, and 5 are considered serious pathogens. In the eastern hemisphere, 12 viruses have been found with 5 causing mass mortality.

TSV was found to cause a problem in the Taura River basin in Ecuador in 1992. Through histological records, it was shown to be present in some samples taken in Colombia as early as 1990, and in 1999, Rolland Laramore found histological evidence that TSV existed in Panama in 1987. This evidence came as a result of looking at stained slides of sick animals that had been kept for future reference. However, the preserved animals had been destroyed after five years and were not available for further confirmation. It is now thought that the reasons Panamanian shrimp (*L. vannamei*) appear to be more resistant to TSV than shrimp from other areas is that they have been exposed longer and that wild stocks have had time to build up resistance.

TSV spread north into Central America and into Texas in 1995. It arrived on the farms in May, shortly after the first migrating birds appeared from the south, but the exact method of transmission was never determined. Shrimp-processing plants were also suspected sources of bringing in the virus with imported frozen shrimp. The shrimp hatchery in Texas was checked for diseases and found to be clear, but was contaminated by TSV from the farm at a later date and closed down for disinfection. In 1996, the Texas hatchery stocked only *L. setiferus* from their hatchery on the farm to protect against TSV infection, and the hatchery continued to produce *L. vannamei* and *L. setiferus* PL for the other farms to stock. They attempted to do the same in 1997, but *L. setiferus* PL production levels were low. This approach apparently worked, because the hatchery remained disease-free, even through the 1999 season.

Ecuador's industries (both wild fishery and farmed shrimp) have continued even with the presence of TSV. There are some areas in Ecuador that have never been affected by TSV. The overall levels of farm-raised shrimp from Ecuador remained above 100,000 MT per year, mostly because of new and smaller farms starting up and taking the place of many of the larger farms. There are now 160,000 ha (395,000 ac) in production on 1,600 farms, supplied by 350 hatcheries in Ecuador (1). In 1997 and 1998, shrimp aquaculture production in Ecuador increased to 130,000 MT (1). The export value of the crop was worth US$871,723,000 in 1997 (7). Apparently resistant stocks from the wild and from selected stocks in the aquaculture industry have allowed both the aquaculture and the harvest industries to survive.

The threat of TSV appears to be subsiding in Ecuador, Panama, and other South American and Central American countries, but WSSV has now been recorded in all of the preceding countries and is causing serious mortality. TSV is also diminishing in Texas. In 1997 and 1998, the farms in south Texas did not have an epidemic of TSV and experienced a very good crop with 70% average survival, which was comparable to survival in 1994, before TSV hit. In 1997, Texas farms produced approximately 1,160,000 kg (2.56 million lb) of shrimp, and in 1998, they produced 1,440,000 kg (3.17 million lb) on 533 ha (1,316 ac). The Texas hatchery sent the same PL to the upper Texas coast farms in Palacios that had given the Texas Rio Grande Valley farms such a good crop in 1997–1999. The Palacios area has been hit with TSV every year since 1995 (including 1999), but survival in ponds increased to 31% in 1998, which was higher than in the previous two years (0–10% and 19%, respectively). In 1998, production in the Rio Grande Valley (without TSV) averaged 4,000 kg/ha (4,000 lb/ac), but in the Palacios area (with TSV), production averaged 1,200–1,500 kg/ha (1,200–1,500 lb/ac). In 1999 the Palacios area still had TSV, but production was up to 3,000–3,600 kg/ha (3,000–3,600 lb/ac). The virus was somewhat different in 1998 in that its presence was not obvious until the shrimp were tested at harvest time. There had been no obvious die-off or bird problems in relation to the disease as had been noticed in previous years. There is still much to be learned about shrimp diseases and their prevention.

Shrimp may become infected from many sources. A major potential exposure pathway to shrimp in the United States includes wastes from shrimp processing plants. Foreign shrimp boat harvesters may catch shrimp with diseases and ship them to the United States as a frozen product. Likewise, some foreign aquaculture operations harvest their ponds immediately upon finding disease and export the infected shrimp. Infected shrimp are now routinely found in U.S. retail markets. Shrimp viruses do not infect humans. The United States imported 118 million kg (262 million lb) of shrimp from the six largest exporting countries in Asia (China, India, Thailand, Vietnam, Indonesia, and the Philippines) in 1995 (7). Frozen shrimp, either unshelled or peeled, make up 97% of all shrimp imported into the United States from Asia. Imports, in the United States and other countries, are typically repackaged at processing plants that are located along coastal waters, because the plants were initially constructed to process shrimp from nearby coastal fisheries. The solid waste and effluent from these plants were disposed of, often untreated, into the local waters and landfills. In addition, frozen imported shrimp from countries in which viral diseases are epizootic, were often used as bait shrimp by sport fishermen. This possible method of transmitting the viruses affects both wild and cultured shrimp and is a US$2 billion industry. The U.S. shrimp-processing industry employs over 11,000 people in 182 companies. Any new requirements that may be necessary to reduce disease risks will increase costs to producers and processors, and ultimately to consumers, but serious consideration of new regulations to control these virus pathways should be undertaken by the appropriate agencies that control this industry. Some voluntary measures have been taken by processors at the encouragement of U.S. regulatory agencies.

Estimated losses due to YHV and WSSV, according to Tim Flegel of Thailand, are sizable. Thailand's production dropped from 225,000 MT in 1994 to 220,000 MT in 1995, mostly due to YHV. At approximately US$8/kg, this represented a shortfall of about US$40 million. However, given that Thai production had been rising by 20,000 to 30,000 MT/year, the actual lost production may have been in the order of 30,000 MT, or over US$240 million.

WSSV was first reported in Japan in farmed *M. japonicus* and spread (96). In Asia, WSSV overshadowed the other diseases. In Thailand, production dropped from 220,000 MT in 1995 to 160,000 MT in 1996, a difference (lost production) of 60,000 MT, over US$500 million worth of shrimp. Using figures from some Thai sources, the yearly totals are different, but the difference (i.e., the estimated production lost) is about the same. The mid-1997 figures looked like there would be a further drop of 10,000 MT, giving a total lost production of around US$600 million. Those figures are for Thailand alone.

For China, the story is also similar. Their production dropped from 150,000 MT in 1992 to 35,000 MT in 1993, probably due to YHV and WSSV. They have since recovered somewhat, but are still well below the 1992 peak (1).

The total losses for all of Asia have been estimated between US$1 billion to US$3 billion per year. C.G. Lundin (World Bank), addressing the second Asia–Pacific Marine Biotechnology Conference, National Center for Genetic Engineering and Biotechnology, Bangkok, stated, "If we calculate an average price of shrimp at about US$5 per kg (or per 2.2 lb) for shrimp still with heads and total disease-related losses at 540,000 MT, the total annual loss based on 1994 data would be US$3 billion. This indicates a very significant problem that has implications for the well-being of millions of people in developing countries." His estimates of 1994 included the Americas.

In an attempt to combat viruses the USDA Shrimp Farming Consortium in Hawaii conducts a selective breeding program for shrimp in state of Hawaii. They hold selected shrimp stocks in quarantine. Offspring from those stocks are challenged with various viruses in other locations away from the quarantine area. Results of the work are then transferred back to the central quarantine. This program is definitely making a difference in the United States, and is helping the industry control TSV and other diseases.

Until they were found in Texas in 1995, WSSV and YHV had mostly been found in the eastern hemisphere. There had been no reason to look for them to that point in the western hemisphere, and researchers had not fully developed dependable testing devices for detecting these viruses.

WSSV, or a "white spotlike virus," was diagnosed in native white shrimp *L. setiferus*, being pond cultured in south Texas in 1995, but did not appear anywhere else in Texas that year after the ponds were drained. In 1997, the same virus was found in a Texas research facility in wild-caught *L. setiferus*. In that facility, the virus did not infect other shrimp (*L. vannamei* and *L. stylirostris*), and all indications suggested that the virus had come in with the wild-caught animals. Apparently, three gene probes were then used in an attempt to identify the virus. The two probes developed in Asia did not show it to be the same as the WSSV that had been found on that continent. The probe developed by Don Lightner in Arizona did show it as WSSV. WSSV was found in Honduras, Nicaragua, and Ecuador in 1999. Histological records show that it existed in Ecuador back at least to 1995.

TSV hit South Carolina in 1996, but did not cause problems in 1997. Of the 20 farms in South Carolina, 7 were affected with an outbreak, and the state experienced a 62% reduction in production. Just as in Texas, some growers experienced losses as high as 90% of their crop. The outbreak started in May 1996, and drought conditions in South Carolina and higher than normal temperatures caused undo stress to farmed shrimp. This increased the susceptibility of the shrimp to disease, the transfer of the disease among the shrimp, and the mortality due to the disease. In 1995, before TSV, average shrimp survival was 60.4%, and average shrimp size was 18.72 g (0.04 lb), or 24 count. South Carolina did not have an epidemic of TSV in 1997.

WSSV was a problem in South Carolina in 1997 and 1998, and the white spotlike virus was detected in both cultured and wild stocks in South Carolina. Reports surfaced of WSSV having been found in crawfish that were being fed at an aquarium on the east coast of the United States. It is known that WSSV affects other crustaceans and has been found in crawfish, freshwater, and saltwater shrimp, as well as other crustaceans. WSSV was also found in shrimp from Texas bait stands in 1998. Paul Frelier, then a professor and researcher with Texas A&M University, College of Veterinary Medicine, and his graduate student, found the virus, but were not able to subsequently find it in four attempts. Lightner at the University of Arizona confirmed the virus diagnosis after 100 bait shrimp were held in quarantine for three weeks before the disease appeared.

We do not know for sure how long these viruses have been in Gulf of Mexico or how they arrived there. Thanks to technology developed by the shrimp aquaculture industry, techniques were developed to detect these viruses. The Texas Parks and Wildlife Department (TPWD) has only recently implemented a disease assessment of wild shrimp in Texas using the techniques developed by university and aquaculture industry researchers. This is not the first time that diseases have been looked for in wild populations in Texas waters, but the techniques for detecting WSSV and YHV were not used during earlier sampling. TPWD has found WSSV to be fairly widespread in the Gulf of Mexico, along the coast Texas, and it has been found in crustaceans, other than shrimp. The source of the virus is unknown, and the effects on wild populations appear minimal, because crustacean harvesting has been average or above in recent years.

Clinical signs of infected shrimp may vary from hemisphere to hemisphere or from strain to strain of WSSV. WSSV from the eastern hemisphere causes juveniles and subadults to exhibit white spots or patches about 0.5–2.0 mm (0.019–0.078 in.) in diameter; these spots begin in the cuticle of the carapace and on the fifth and sixth abdominal segments and then spread to the whole body. The cuticle can also become loose. The white spots are abnormal deposits of calcium salts and are most apparent on the inside surface of the carapace. Infected animals are lethargic, stop feeding, swim slowly on the pond surface, and eventually sink to the bottom and die. Moribund animals display a pink to reddish-brown coloration due to an expansion of cuticular chromatophores. Populations of shrimp with these symptoms typically have high mortality rates that can reach 100% within 3–10 days after the onset of clinical signs. However, some shrimp with WSSV in the western hemisphere have not shown the signs of having white spot or do not show the presence of white spots on them (96). A number of diagnostic methods are available (see below).

Transmission of exotic pathogens can occur through a variety of means, including migration with humans, birds, and other animals, and through the shipment of infected frozen food products.

Possible methods for pathogen introduction of WSSV and YHV into Texas after the 1995 occurrence have been examined and Nunan et al. (108) concluded that the "introduction and spread of WSSV and YHV may be through the importation of frozen shrimp product." From frozen imported shrimp sampled in 12 grocery stores in Arizona, California, and Texas, it was found that 5 of the 12 samples of shrimp had either WSSV or YHV. Freezing does not damage the virus. In fact, this is the way that most pathologists preserve the viruses. Frozen imported shrimp were then fed to live shrimp and infection and mortality occurred. Polymerase Chain Reaction (PCR) tests, described in the next section, showed that the actual cause of the mortalities of the Pacific blue shrimp used was YHV.

NEW TECHNOLOGIES IN DIAGNOSING SHRIMP DISEASES

Diagnostic and detection methods for shrimp viruses include direct bright field, light microscopy, phase contrast, dark field light microscopy, histopathology, enhanced histopathology, bioassay histopathology, transmission electron microscopy, fluorescent antibody, DNA probes, and PCR.

The majority of shrimp diseases in the past have been diagnosed using histology techniques and reviewing slides stained with H & E (stain used for most viruses), giemsa (stain for Rickettsia), or acid fast stain (for Mycobacteria) and require that the shrimp be iced, frozen, or preserved properly and sent to a certified diagnostic laboratory. Now, the rapid diagnostic techniques for shrimp viruses are based on genetic hybridization with probes. A series (for WSSV, TSV, IHHNV, NHP, HPV, MBV, BP, and YHV) of gene probe diagnostic kits are available (*http://www.diagxotics.com*). The shrimp specimens used for these kits are either hemolymph or unstained histology slides (in situ). The instructions are well written and straightforward and can be used by most shrimp-farming operations. This new advancement in technology allows the farmer to check for diseases on the farm, without sending the shrimp to a pathology laboratory. The DiagXotics "Spot-On, Rapid Field Test," for example, has four easy steps to follow and will give results in 15 min.

POLYMERASE CHAIN REACTION (PCR)

PCR is a widely used test to verify a shrimp virus and has been used by researchers to verify the results of other tests. The PCR test for WSSV and YHV was first developed in Asia. These tests seem to be dependable, but not all PCR tests are perfected enough to be considered dependable for commercial use. Some methods remain as research verification tests only. One such PCR test that was slow in development was the one for TSV. TSV accumulates in the lymphoid organ (small gland in front of the hepatopancreas or digestive gland of shrimp) and not in the blood, so it is much harder to detect without killing the animal. When a PCR test is given, the animal may not necessarily show the presence of TSV, because the virus is not in the blood. Paul Frelier (while at Texas A&M University, Veterinary Medical Center) said that the only effective way to look for TSV was through a bioassay, but Jeff Lotz (of Mississippi's Gulf Coast Research Laboratory) stated that, even then, the animals had to be destroyed to look at the lymphoid organ, and it was only a short-term test. After the 1995 epidemic hit, offspring from TSV survivors from Harlingen Shrimp Farm were held in Caldwell, Texas (an inland facility). The adults were sent to Lotz in Mississippi, where it was learned that the disease was not transmitted to the offspring, that the TSV survivors concentrated the TSV in the lymphoid organ, and that there was no sign of the TSV in the blood. Researchers now think that the animals that concentrate the TSV in the lymphoid organ may not pass on the TSV. More research is needed in this area. Other researchers have also stated that just because certain penaeid strains are resistant to a virus (IHHN, TSV, and WSSV), this does not mean that they do not carry the diseases. For example, even though *L. stylirostris*, or the Super Shrimp, is resistant to IHHN, it has been found to be a carrier of TSV.

A list of basic PCR equipment required for anyone wishing to do gene probing to detect WSSV or YHV can be obtained from Aquafauna Biomarine. It is best to seek the assistance and use of equipment from a government institution, hospital, or university, since most will have access to such equipment. PCR equipment is very expensive. Once you obtain the equipment or obtain access to its use, then you need to train a technician to run it. Then you need primers.

From the farmers' stand point, even if PCR tests are done at the hatchery and PL are shown to be clear, there is no assurance that WSSV won't come into the farm. A number of crustaceans and other arthropod species have been found to act as reservoirs for the virus. Detailed histological studies were required to confirm that suspected carriers had active viral infections. Three common crustacean residents of shrimp culture ponds (the sand crab *Portunus pelagicus*, the mud crab *Scylla serrata*, and krill *Acetes* sp.) have been known to become infected with WSSV. Normal histology, electron microscopy, and in situ DNA hybridization determined infection. Crab species and krill are considered viral reservoirs, since they are able to carry the infection and may persist for significant periods in the shrimp-farming environment (96). *M. rosenbergii* has been shown to exhibit WSSV. WSSV has also been found to exist in wild-caught shrimp (*P. monodon*, *M. japonicus*, *P. semisulcatus*, *F. penicillatus*, *L. setiferus*, *F. aztecus*) and crabs (*Charybdis feriatus*, *P. sanguinolentus*, and *Callinectes sapidus*) collected from the natural environment in coastal waters of Asia and the western hemisphere (Texas and South Carolina). Detection of WSSV in non-cultured arthropods collected from WSSV-affected shrimp farms revealed that copepods, the pest crab *Helice tridens*, small pest Palaemonidae (grass shrimp) and the larvae of an Ephydridae insect were also reservoir hosts of WSSV. Most hatcheries perform routine tests on their brood, and some perform PCR, but not on each batch of PL sent out as this would not be practical.

There are basic virus-avoidance procedures farms can take. Frelier wrote one such set of recommended procedures. The procedures are somewhat elaborate; therefore, the farmer has to decide which ones should be adopted. Chlorinating all incoming waters, for example, may not be economically practical or desirable, because it would kill even beneficial diatoms and zooplankton. Treating intake waters with an insecticide might be a better option, as discussed later in this article.

ADDITIONAL WAYS TO LOOK FOR SHRIMP VIRUSES

Virological recognition techniques include light microscopy, histological changes in host, electron microscopy with negative staining, and cell (tissue) culture.

Virological techniques for shrimp larvae are similar to those for larger shrimp. Routine examination of tissues for occlusion viruses, however, applies squash of a whole animal's mass. In larger larvae (PL), it is advantageous to remove a portion of the tail before making the wet-mount squash. This can be done by severing the tail at midpoint of the first tail segment with a scalpel while viewing through a dissecting microscope. This procedure will cause the digestive gland and anterior intestine to wash free of other body portions. Certain stains (e.g., 0.1% aqueous malachite green) are used in tissue squashes.

SELECTED STEPS TAKEN BY SOME IN THE INDUSTRY TO AVOID SHRIMP VIRUSES

When used correctly, Dipterex™ (or other insecticides) can eliminate crustacean carriers in farm water prior to stocking shrimp. In Thailand, 0.5–0.8 ppm (6.5 kg/ha or 5.8 lb/ac) Dipterex™ has been used to treat intake waters. The water is used safely for the stocking of shrimp after a period of 7–10 days. Most importantly, insecticide-treated water *is not* discharged into the environment. This step only represents a portion of the overall biosecurity program that is being used by farmers to fight WSSV. The shrimp farmers are moving toward mimicking the traditional animal-husbandry systems where biosecurity and disease prevention are essential parts of the overall management scheme. The animal-husbandry business has many powerful tools, such as genetically improved stock with overall better hardiness, as well as vaccines, whereas, the shrimp aquaculture industry presently only has the basic biosecurity program and not much more. There is no doubt that other viruses (such as TSV), which are transmitted through similar vectors, can also be effectively fought. Besides controlling the vectors and reservoirs that harbor these viruses and implementing a strong biosecurity program, there is little that today's farmers can do against these very powerful diseases. The successful use of an insecticide in one area does not necessarily mean that it will be effective in another area. Dipterex™, which is an organophosphate, does not kill the crabs in Nicaragua; therefore, another insecticide may be more effective there. Other names for Dipterex™ are 3-trichlorfon, Dylon, and Neguron trichlorfon dimethylphosphonate. Dipterex™ is similar to other organophosphates, such as benthiocarb, diaoxathon, diazionon (Spectracide), ethyl-parathion, fenthion (Baytex, Entex, Tiguvon), malathion (cythion), methyl-parathion, and mevinphos (phosdrin). Most organophosphate insecticides, including Dipterex™, break down very rapidly, especially in waters with a high pH or high alkalinity. Sunlight also breaks them down rapidly, just as it causes chlorine to break down. Most households in the United States routinely use organophosphates around the outside of the house to control insects. In Thailand, the Dipterex™/trichlorfon used is a 95% active ingredient and is applied as supplied. In Thailand, it kills all the crustaceans in approximately 24 hours, and farmers stock shrimp after normal pond fertilization and bloom development. There do not appear to be any residual effects on the shrimp, as growth and survival rates are similar to those found in untreated ponds. As examples of research studies found in the literature on organophosphates, malathion took 14 ppb and 48 hours to cause mortality with penaeids, and diazionon took 28 ppb in 96 hours to cause 50% mortality in *F. aztecus* (brown shrimp). The sizes of the shrimp were not given.

A pyrethroid insecticide known as fenvalerato, or fenvalerate, has been effectively used in Central America as an aerial spray on one farm. When the farm was drained and dried, the spray was used to eliminate the possible vectors and pathways that spread TSV. The farm did not have further problems with TSV after using this, among other steps implemented to contain disease. Fenvalerato was sprayed initially; then, after subsequent sprays the following year, the farm was sprayed with the basic insecticide used on citrus fruits and for mosquito control.

Another pyrethroid used in Central America to accomplish vector control is cypermetrina 25C™. This cypermethrin is Alfa-ciano-3 Fenoxibencil (+) cistrans 3-(2.2 Diclorovinil) −2.2 Dimetil ciclopropano 25% and 75% inert ingredients.

The two pyrethroids have in common Alfa-ciano-3 Fenoxibencil. Apparently, the pyrethroids are completely different in chemical makeup from the organophosphates and also work differently to control insects. Cypermetrina is a pyrethroid that acts upon the insect's nervous system. It closes the nerve channel, and as the nerves fire and continue to fire, the insect becomes hyperactive and dies. If the chemical is applied on dry ground that remains dry, the pyrethroids will break down in 10 days, at the most, in full sunlight. In river water, pyrethroids have a half-life of 5 days, and will hydrolyze very rapidly. They will bind rapidly with organic matter and be absorbed by clay. According to Larry Keeley at Texas A&M University (personal communication, 1999), the pyrethroids hold up a little better than the organophosphates, but they all break down rapidly compared with the organochlorides. (The following are the insecticides that should be avoided. Aldrin, DDT, Dieldrin, Endrin, Heptachlor, Lindane and Methoxychlor.)

Cypermethrin was tested, and the results are shown in Table 5.

Fenvalerate was also tested, and the results are shown in Table 6.

One can obtain profiles on insecticides at *http://ace.orst.edu/info/extoxnet/pips/ghindex.html*.

Pesticides should be only a small portion of a farm's overall biosecurity and ecosystem management program. If a farm reuses old water that it knows to be free of viruses because of previous good harvests, there is no need for treatment. It is used as long as possible, because there is fear that new water will introduce new diseases. After 60 or more days of very low to zero water exchange (in aerated ponds), the water is pumped through a series of recirculating ponds and used again. Farmers developed this method because the only water that can be trusted is the water that holds live shrimp. When used in a closed, recirculating system, Dipterex™ and other insecticides are environmentally benign. The fact that PL thrive in water that a week earlier killed older, tougher crustaceans implies that there is little residue.

Before these practices become widely adopted (and they will if they work, at least in countries without strong environmental controls), we need to know the following: (*1*) How persistent are pesticides in water, and what are the breakdown products? (*2*) Are there any residues in shrimp grown under these conditions?

Table 5. Results of Exposure to Cypermethrin on Various Crustaceans

Test Organism	Order	LC_{50} in ppb	Reference
Mysidopsis bahia	Mysidacea	0.019 (96)	Cripe et al., 1989
		0.005 (96)	Hill, 1985
		0.056 (96)	Clark et al., 1989
Homarus americanus	Decapoda	0.01 (96)	Schimmel et al., 1983
Daphnia magna	Cladocera	1.0–5.0 (24)	Day, 1989
Crangon septemspinosa	Decapoda	0.01 (96)	McLeese et al., 1980
Palaemonetes pugio	Decapoda	0.016 (96)	Clark et al., 1989
Farfantepenaeus duorarum	Decapoda	0.036 (96)	Clark et al., 1989
Uca pugilator	Decapoda	0.2 (96)	Clark et al., 1989

Table 6. Results of Exposure to Fenvalerate on Various Crustaceans

Test Organism	Order	LC_{50} in ppb	Reference
H. americanus	Decapoda	0.14 (96)	McLeese et al., 1980
D. magna	Cladocera	0.3 (24)	Day, 1989
		0.8–2.5 (48)	Day, 1989
Daphnia g.m.	Cladocera	0.16–0.29 (48)	Day, 1989
Ceriodaphnia locustris	Cladocera	0.21 (48)	Day, 1989
Deaptomus oregonensis	Cladocera	0.12 (24)	Day, 1989
M. bahia	Mysidacia	0.0008 (96)	Schimmel et al., 1983
F. duorarum	Decapoda	0.84 (96)	Schimmel et al., 1983
Nitocra spinipes	(Unknown)	0.38 (96)	Clark et al., 1989
P. pugio	Decapoda	0.003 (96)	Clark et al., 1989

The biosecurity programs generally consist of many distinct parts, with each part addressing a particular mode of transmission. Together, they help farmers eliminate or reduce the possibility of contamination. The theory holds true no matter the size of the farm or the size of the pond. Each farm needs to develop its program to fit local conditions.

The farm must start with clean animals and water and take steps to keep disease transmission down by cutting vector transmissions. The farm must eliminate the use of untreated water and use either no, low, or recirculating exchange systems to grow the shrimp. Additionally, the farm must keep the walking, swimming, and flying members of the animal kingdom (such as crabs, birds, and people) away from the shrimp and make sure that ponds on the same farm do not cross-contaminate. Swift and definitive action must be taken when the farm does identify diseased ponds. In the end, every little bit helps to better the odds, no matter what disease may come to the farm.

Health management on shrimp farms has become an important issue. At present, there are no cures or cost-effective treatments for emerging viruses. Management strategies have been developed to prevent the spread of viruses and improve the economics of disease-related problems.

One way of managing potential disease-related problems is to institute a health-management program to establish accurate, ongoing data collection and record keeping. Important parameters and information on shrimp health, diagnosis, and disease-control techniques must be decided and then routinely obtained, monitored, and stored. Farms have recently developed health-management programs incorporating (*1*) preventive strategies, (*2*) health monitoring, (*3*) diagnosis procedures, and (*4*) management techniques.

The following is a general protocol developed for limiting the introduction of diseases. If viral, bacterial, or fungal infections develop in production ponds, this disease-prevention practice can be set in place and modified to the particular needs of the farm.

- Implement a pond-bottom management procedure involving the removal of sediment, drying, and turning the soil when possible and applying a disinfectant.
- Disinfect the entire farm by spraying an insecticide that is not harmful to humans, birds, or large animals.
- Destroy all life in the reservoir supplying the growout farm. This is to inhibit any potential intermediate host from entering the water system and infecting the shrimp.
- Limit access to the facility, this includes disinfecting all PL containers, boxes, feed trucks, machinery, etc., before they are allowed on the premises, including the processing plant.
- Keep good records of all movement of shrimp, including information on shipping and transporting, so if a problem occurs, the specific source can be traced.
- At the time of stocking, put a subsample of PL in a survival cage to determine the survival rate 24 and 48 hours after stocking. Any moralities should be examined to determine the cause.

- Minimize the amount of water exchange as much as possible.
- Limit the escape of shrimp at harvest, and never release any shrimp into the wild for any reason.

Broodstock

Adult *L. vannamei* used for captive reproduction is either wild caught or grown out in pond areas a fair distance from the natural range. Generally speaking, wild broodstock collected from the ocean are preferred. They tend to produce larger quantities of nauplii that are stronger. However, experiments to improve *L. vannamei* captive production through pathogen control and selective breeding are showing potential, and disease resistance is being developed in a number of demonstration programs. In terms of pathogens, both wild-caught and pond-reared shrimp may have similar dispositions to associated microbial agents. Metazoan and some protozoan parasites are more likely to be present in wild-caught broodstock, whereas pond-raised shrimp may be more susceptible to emerging viruses.

Nauplii

The quality of nauplii is also an important concern. Historical data on nauplii used in hatcheries supplying stock to the farm should be collected and recorded. High-quality *L. vannamei* or *L. stylirostris* nauplii will be vigorous swimmers and have strong phototactic responses, be free of appendage deformities, and display high survival through the protozoa I stage. If particular females are producing inferior nauplii, reviewing the production records should reveal a pattern of reduced nauplii quality from the spawns of these females.

PL Stock

The need to stock virus-free PL on the farm from areas free of known viruses is imperative. Although hatcheries from infected areas may be free of viruses at the time of maturation and spawning. Viruses may be transmitted transovariantly (vertically) from an infected female into an egg or embryo. Here they may remain in eclipsed form until conditions are suitable for replication. There are limited ways to determine certain viruses, such as TSV, in the early PL stages as of yet with total accuracy. There is also the possibility of increasing the virulence of a virus when it is serially passed from one animal to another. There is some merit to holding the PL for a month to 6 weeks to check for the existence of TSV before releasing them into large growout ponds, but PL holding and rearing raceways would need to be designed into the farm for this procedure to become effective.

Stocking Densities

Certain management schemes have been developed to improve the economics of disease problems. Some people in the industry choose to stock at lower densities to limit the number of virus particles or bacterial infections being released into the environment, increasing the number of survivors. If the farm plans to utilize wild PL, it should stock at higher densities with quality animals, use feeding trays, and maintain high water quality to ensure the greatest possible survival rates. Many of the wild larvae may be other species that do not do well in pond culture, so overstocking will compensate for a mixture. Table 7 reflects some possible economic results of disease-infested populations and different stocking densities. The following assumptions were made in order to prepare the table: The average stocking rate is 15.5 PL/m^2 (10.76 ft^2), the average production is 1,961 kg/ha (1,961 lb/ac) [heads on] or 1,255 kg/ha (1,255 lb/ac) [tails] per crop, the survival rate is 67%, giving a harvest of 10.4 pieces/m^2 (42,088/0.45 ha or 42,088/ac), the average weight per tail is 0.01 kg (0.02982 lb), the price average is US$10/kg (US$4.55/lb) of tails, PL cost is US$8/1,000, growth of surviving shrimp will not always be the same each production cycle, and there are generally seasonal differences in temperature, water quality, and salinity that all interact to affect production.

From Table 7, if the survival rate is reduced to 40% by TSV, it is necessary to stock at a rate of 26 PL/m^2 (26 PL/10.76 ft^2) to get the same revenue per 0.45 ha (1 ac), as was obtained with a stocking rate of 15.5 PL/m^2 (15.5 PL/10.76 ft^2). The added cost of PL would be only US$340/0.45 ha (US$340/1 ac). *Note*: If TSV lowers survival rates to 30%, stocking at 26/m^2 (26 PL/10.76 ft^2) will not provide as much revenue per hectare or acre as was obtained earlier.

The table and stocking density scheme are subject to change from production cycle to production cycle, depending on the species being stocked and the degree of infestation on the farm.

Table 7. Possible Economic Results of Disease-Infected Populations and Different Stocking Densities

Stocking (PL/m^2)	Survival (%)	Pieces Harvested (per m^2)	Harvest Weight (lb tail/acre)	Revenue (per acre US$)	PL Cost (US$)
15.5	67	10.4	1,255	5,710	502
15.5	40	6.2	748	3,404	502
22	40	8.8	1,062	4,832	712
24	40	9.6	1,159	5,271	777
26	40	10.4	1,255	5,710	842
26	30	7.8	941	4,282	842

Hatchery Visit or Assessment of Wild-Caught Larvae

Hatchery and wild-caught PL examination protocol is designed to establish a knowledge of the farm's PL, encouraging the best quality PL purchased from the most reputable sources. The protocol is as follows:

- Historical background listing the location and source (wild vs captive) of broodstock and nauplii. It is good to know the progeny of captive broodstock to determine the genetic strain. Excessive inbreeding can produce genetically weak, or inferior offspring.
- Hatchery design, procedures, and operations are observed and documented to assure quality techniques.
- Clinical examinations of PL are conducted "on-site" in the hatchery prior to shipping.
- There are different criteria for judging PL quality; the most important is developmental stage along with size, age, stress tests, and physical condition.

Examination protocols for a subset of larval rearing tanks should be followed. A clinical index for each tank, or sample set, is established. Examine each PL and record the condition based on the following scale: 1 = poor, lowest, least; 2 = fair, mid-range, average; 3 = excellent, highest, most.

To remove crab larvae from wild-caught postlarval shrimp boxes and containers, place small pieces of white Styrofoam in the container for a short time. The crab larvae (stage dependent) are attracted to the Styrofoam and can be removed.

Use the following points to help identify healthy PL:

Tail

A PL tail should be noticeably open. The preferred age of PL at stocking for *L. vannamei* or *L. stylirostris* is 7–10 days old, and 18 days old for *P. monodon*.

Color

Good-quality PL have transparent bodies with star-like brown or dark brown pigmentation. Those with pink or red coloration indicate stress related to rearing or handling.

Activity

PL should be visibly strong and active, swimming from side to side. Active PL will swim against the current when the water is agitated, or whirled. While in a small container, PL may not be moving, but will react when the container is tapped, or if there is movement in the water.

Size

PL size should be uniform, and any significant variation is indicative of different age levels. Generally speaking, age variations are acceptable as long as size differences are minimal.

Appearance

PL are examined for signs of infestations (debris adhering to body, swimmerets, etc.) and deformities (broken rostrum, crooked body, undeveloped gills, etc.). Healthy PL appear clean and have no physical abnormalities.

Feeding

A dissecting microscope is needed to determine gut fullness. A healthy PL will have a full digestive tract, except after a long time in shipment. Empty guts in larval rearing tanks can be the result of underfeeding, disease, or stress.

Clean Shell

A clean shell represents frequent molts, indicating fast, consistent growth. Slow growth is indicated by the presence of protozoans, other dirt, and necrosis (black spots or brown lesions) on the shell.

Good Muscle Development

With the aid of a good microscope (better seen with three-dimensional phase microscopy) examine the sixth tail segment for good tail muscle development. The muscle should completely fill the shell from the gut down. When the PL is stressed the muscle has a grainy effect much like the grain in wood. The muscle will appear grayish or brown in color. Healthy PL have a clear, thick, smooth muscle.

Prestocking/Acclimation Examination of PL

Routine assessment of PL during acclimation consists of (*1*) microscopic examination of randomly select PL at the moment of arrival, or before being stocked into acclimation tanks and (*2*) routine assessment of PL throughout acclimation.

Microscopic Assessment

Microscopic assessment of PL should be done on PL upon arrival at the farm site to assess the effects of transportation on PL.

- Immediately upon arrival, collect twenty PL from one of the transport containers or bags while the acclimation supervisor examines the parameters of the water in the transport containers and bags.
- This evaluation should be done immediately to determine any stress resulting from transporting and not from being held in the acclimation container.
- Examine each PL sampled, and record the condition based on the clinical index mentioned earlier.
- Record all results on a PL State of Health Examination form.
- Collect 1 L (0.26 gal) of water from the acclimation tank (sampling from the bottom upwards), and carefully examine, with the aid of a dissecting microscope for more precise observation, PL for the following characteristics:

1. Index of gut fullness. Since water temperature should be maintained at around 22 °C (68 °F) during long transport, it is unlikely that the index of gut

fullness (formula given in next section, point 5) will be higher than two (average). However, a percentage estimate of gut fullness upon arrival can serve as a guideline for evaluating the intensity of feeding behavior during acclimation. Larvae collected from estuaries will probably contain higher gut fullness because the trip is shorter and the temperature is probably higher.

2. Mucus and debris on setae. Accumulation of mucus and debris on the setae, antennae, and appendages is a strong indicator of stress. Fine focus the microscope to get a three-dimensional effect, and examine the spaces between the setae and antenna hairs for any build-up of debris.
3. Opaqueness of swimmerets and tail muscle. An obvious sign of stress is change in opaqueness of the tail muscle. Normally, the tail muscle will be transparent with a few pigmentation spots. However, when stressed, the tail muscle and swimmerets may become opaque or, in extreme cases, completely white.
4. Morphological deformities. Although deformities are not directly related to stress, a large number of deformed PL within a population can be indicative of chronic diseases, such as IHHNV. Deformities can also be used as an indicator for future survival estimates in the growout pond. Morphological examination should be focused on the following:

- complete, well-developed, unbent rostrum
- no curvature or cramped tail
- well-formed eyes and eye stalks
- well-formed complete swimmerets
- overall physical appearance

Routine (Acclimation) Assessment

In addition to the microscopic examination, a routine assessment of PL should be done every hour during the acclimation period. Constant observations provide subjective assessments of the PL general condition during fluctuations of water quality inherent in acclimation. If indications of stress are observed, steps can be taken to slow down the acclimation process and decrease the stress. This routine assessment is performed by sampling a 1-L (0.3-gal) volume of water in the same manner as previously described and carefully examining the PL swimming in the container and the acclimation tank for the following:

1. Level of swimming activity. If PL accumulate on the surface, agitating aggressively, and if there DO level is low, inject pure oxygen into the water. If PL are lethargic and swimming activity is diminished, reduce the rate of water exchange, allowing PL more time to adjust physiologically.
2. Erratic swimming behavior. A periodic rhythm of cramping up and relaxing signifies an attempt to molt. A large percentage of the acclimation tank's population molting may indicate stress. If this occurs, reduce the rate of water exchange, increase feed, and reduce the temperature to less than 23 °C (less than 73 °F) to suppress cannibalistic activity by the nonmolting PL.
3. Opaqueness of tail muscle. As already mentioned, this sign is an indicator of physiological stress. If noted at a high frequency, slow down the acclimation schedule to allow for physiological adjustments.
4. Presence of molts. Floating exoskeletons indicate molting. Take the steps just mentioned to discourage cannibalism. A slow acclimation water exchange is recommended to prevent molting. Do not mistake exoskeletons for dying or dead PL. If a major molt occurs, check gill movement and heart activity. If no organs appear to be active, mass mortality may be occurring.
5. Index of gut fullness. Active feeding and high index of fullness are positive signs, indicating that there is little stress. Generally, PL will not feed if they are under stress. A low index of gut fullness and excessive feed present may indicate stress. If this occurs, suspend feeding and reduce water exchange. The index of gut fullness should be represented as a percentage. The length of the intestinal track is classified as 100%, and an estimated length of the full gut is made. Use the following scale to grade gut fullness: 1 = poorest, least (<30%); 2 = fair, mid-range (30–60%); and 3 = excellent, high (60–100%).
6. Presence of mortalities. Under normal conditions, a mortality rate greater than 3% should put the acclimation supervisor on alert. Check water quality, reduce rate of water exchange, and if guts are full, reduce the temperature to slow the PL's metabolic rate. These procedures may provide sufficient time for a thorough investigation into the cause of the mortalities.
7. Frequency of cannibalism. Generally, cannibalistic behavior indicates that there are insufficient quantities of feed or a high level of mortality. An increase in feed quantity and frequency should minimize cannibalism.

Survival/Holding Cages

Survival cages and a holding tank with subsampled PL may be used to determine the survival rate and to monitor mortalities as follows:

- Before stocking, collect 200 PL from several different tanks and place them into different survival cages.
- Place the cages near the edge of a pond at a minimum depth of 50 cm (20 in.). The cages should be firmly anchored to the mud, with a minimum of 15 cm (6 in.) water depth in the cage.
- Do not feed PL in the cages, and leave them undisturbed for 48 hours.
- After 48 hours, remove all PL from the cages and count them individually. Live PL should be counted separately from dead ones.
- Divide the number of live PL by the number of PL originally placed in the cages to calculate survival

rate. Multiply that number by 100 to obtain the percentage of PL that survived.

- After counting the PL, examine seams and corners for trapped or squeezed PL and for holes where they may have escaped.
- The survival count is a useful indicator of the acclimation process. A survial rate of greater than 85% indicates a successful stocking, one of 65–85% may indicate poor-quality PL or excessive stress due to acclimation, and a rate of less than 65% strongly suggests problems.

Data Collection

Parameters recorded to monitor the health of shrimp in production ponds should be routinely collected, easily obtained, and accurate. The farm should monitor physical, chemical, physiological, and environmental parameters on a daily, weekly, monthly, seasonal, and historical basis. The parameters and time of collection are as follows:

Daily

Water Quality. Temperature DO, salinity, pH, and turbidity.

Weekly

Water Quality. Alkalinity, ammonia, nitrite, phosphorus, calcium, iron, and silicon.

Natural Productivity. Evaluation of algae communities and population numbers.

Bottom Sampling. Evaluation of benthic flora, fauna, and the presence of hydrogen sulfide.

Shrimp Sampling. Estimation of survival, growth rates, feeding rates. Clinical examination of shrimp for all or for a subset of the ponds on the farm for lethargy (inactivity), anorexia (empty intestinal tract), soft shell, fouled gills, or black areas on the shell, red tail, antennae, and appendages.

Monthly

Environmental. Phytoplankton communities, bacteria, and soil sediment.

Seasonal

Trends. In growth rates, feeding rates, nutrients, phytoplankton communities, survival rates.

Historical

Background Information On. Farm, individual pond construction, hatcheries, other farms in the area, and past production records.

Keeping good records of all parameters and physiological condition of the shrimp population can provide information for a retrospective study to determine the cause and onset of any problems that may occur.

Routine Health Monitoring

Sampling Populations. Shrimp should be randomly caught from various locations around the pond. The locations should be clearly marked on the sample sheet. For consistency, one individual should be designated to do the clinical examination.

- Sampling should be done weekly, or daily if a problem occurs, on 50–100 randomly caught shrimp.
- Percentage of shrimp demonstrating clinical signs of disease should be determined (e.g., shells, gills, tails, spots, deformities, bacterial necrosis).
- The percentage of sick animals can be followed to help determine when a specific problem first begins.

Clinical Examinations. If problems or deformities are noted during routine weekly sampling, clinical examinations should be done pond-side with live animals. The clinical exam determines the percentage of shrimp in the sample that are clinically ill.

- A clinical disease index for each pond during the production season should be established.
- A clinically ill animal in the sample may demonstrate the following signs: lethargy (inactivity), anorexia (empty intestinal tract), soft shell, fouled gills, or black areas on the shell, red tail, antennae, or appendages.
- Clinical examinations of shrimp are done from 100% of the ponds on a weekly or daily basis during heavy disease period. The samples ideally should consist of 50–100 shrimp per pond. However, during periods of outbreaks and low survival rates, it may not be possible to collect this many samples, and a smaller sample set will have to suffice.

Define Problem and Occurrence. When a problem or disease occurs, it must be defined and the onset determined.

- Using data from weekly clinical examinations, determine when a specific problem first began. Additional resources can then be used to define the problem, point of entry, and possible causes.
- Record a general statement of the overall health of the pond. Include a subjective evaluation by the pond manager regarding feed consumption as indicated by feed trays, bird activity, or the presence of dead or moribund shrimp.
- Define the problem in terms of survival, growth rate, feeding rate, extent of problem on the farm, historical data on the farm, and in the general area. A disease outbreak may be the result of certain management procedures that could be changed to inhibit the occurrence of the disease.

If survival rates are low, examine pond bottoms weekly to determine when and where animals are dying. Dive a predetermined section of the pond bottom along a path of 16–32 m (50–100 ft) in front of the harvest and outlet, while counting the number of dead shrimp. Mark the path with a rope or stake along the path and at each end to ensure that the same pond area is being examined during the entire growout season. If there is a prevailing wind creating directional currents in the pond, check the opposite shores for debris and moribund shrimp.

Accurate Diagnosis

If a problem is disease related, it should be described clinically, bacteriologically, and, if applicable, morphologically, including a histological examination. An accurate diagnosis is required for a proper treatment. There are laboratories that will process shrimp samples bacteriologically, histologically, or using an in situ hybridization test. It may be necessary to preserve samples and send them to the lab for testing, depending on what the lab recommends. Eventually, it would be advantageous for the farm to develop these capabilities within their own laboratory and at least methods for preliminary testing. This would drastically reduce the cost of outside testing.

DNA probes-samples are first examined by conventional histopathology, and if a diagnosis is possible, then the more costly in situ hybridization test with gene probes is not necessary. To keep costs down and ensure higher probabilities of accurate diagnosis, collect nonrandom samples with apparent problems or disease.

Specimen Processing and Preservation Procedures. It is extremely important that proper sampling and preservative procedures be followed, so that histopathological analysis will not be jeopardized. The following procedures (96,103–105) should be followed explicitly to ensure proper preservation:

Collection

1. Collect shrimp by whatever means are available with a minimum of handling stress. For the study of presumably diseased shrimp, select those that are moribund, discolored, displaying abnormal behavior, or are otherwise abnormal (except in the case of intentional random sampling for estimation of disease prevalence). Shrimp sampled for normal histology should not be abnormal in appearance or behavior. Do not collect shrimp that are dead for any sample, unless it can be positively determined that they have died within the last few minutes. If recently dead shrimp must be sampled, be sure to make note of this condition and estimate the time since their death.
2. Transport the shrimp to the laboratory via a bucket with pond water. Supply adequate aeration to the bucket if the shrimp are to be left for a short period before actual fixation.

Fixation or Preservation

1. Have on hand an adequate supply of fixative. A rule of thumb is that a minimum of approximately 10 times the volume of fixative should be used for each specimen. [For example, a shrimp of 10 mL (0.3 fl. oz) volume would require 100 mL (3 fl. oz) of fixative.]
2. Davidson's fixative should be made as follows (1 mL = 0.03 fl. oz):
   a. 330 mL (9.9 fl. oz) 95% ethyl alcohol
   b. 220 mL (6.6 fl. oz) 100% formalin (saturated aqueous solution of formaldehyde gas, 37–39% solution
   c. 115 mL (3.45 fl. oz) glacial acetic acid
   d. 335 mL (10.05 fl. oz) distilled water
   e. store at room temperature
3. Inject the fixative 0.1 to 10 mL (0.0034 fl. oz to 0.3 fl. oz), depending on the size of shrimp), via needle and syringe (needle gauge also depends on the shrimp size) into the living shrimp. The site of injection should be laterally in the hepatopancreas proper, in the region anterior to the hepatopancreas, in the posterior abdominal region, and in the anterior abdominal region. Precautions should be taken to avoid skin and eye contact with the fixative (wear surgical gloves and eye protection). The fixative should be divided between the different regions, with the cephalothoracic region, specifically the hepatopancreas, receiving a larger share than the abdominal region. A good rule of thumb is to inject an equivalent of 5–10% of the shrimp's body weight; all signs of life should cease.
4. Immediately following injection, slit the cuticle with dissecting scissors from the sixth abdominal segment to the base of the rostrum, paying particular attention not to cut deeply into the underlying tissue. The incision in the cephalothoracic region should be just lateral of the dorsal midline, while that in the abdominal region should be approximately mid-lateral.
5. Shrimp larger than 12 g (1/2 oz) should then be transversely slit once at the junction of the abdomen and cephalothorax or again mid-abdominally.
6. Following injection, incisions, and bisection/trisection, immerse the specimen in the remainder of the fixative.
7. Allow the shrimp to remain in the fixative at room temperature for 24 to 72 hours, depending on the size of the shrimp.
8. Following proper fixation, the specimen should be transferred to 50% ethyl alcohol, where it can be stored for an indefinite period.
9. Record a complete history of the specimen at the time of collection: gross observations on the condition of the shrimp, species, age, weight, source (pond, tank or raceway identifying number), source of parent stock, and any other pertinent historical information that may at a later time provide clues to the source and cause of the problem. Use soft-lead pencil on paper, waterproof paper if possible.

Transportation or Shipment for Processing

1. Remove the specimens from the 50% ethyl alcohol.
2. Wrap with paper towels to completely cover the specimen.
3. Place towel-wrapped specimen in a sealable plastic bag and saturate with 50% ethyl alcohol.
4. Include the history, as previously explained with the shipment.

5. Place the bag within a second sealable bag.
6. Multiple small sealable bags can again be placed within a large sealable bag.

Interpretation of Histological Results. An analysis of the samples will include descriptions of the findings in terms of severity and morphology. When reading the results, refer to Table 8, which is a generalized scheme for assigning a numerical qualitative value to assess the severity of infections, surface infestations, and disease syndrome severity. The University of Arizona uses this assessment method.

MANAGEMENT SCHEMES

While several different pond-management and disease-treatment schemes are being implemented throughout the world, the most important is pond preparation and disinfecting prior to production. There are several different types of vaccines and immunostimulant products on the market used in the prevention of diseases. Most are still experimental and not economical. Presently, there are no policies or procedures for using organic or inorganic compounds for prevention or treatment. If the farm production ponds are affected by an epizootic infestation, such as TSV, or bacterial infections, such as *Vibrio* spp., it would be advisable to establish a prevention or treatment program utilizing some of the aforementioned products.

A typical farm scenario with a disease problem is as follows: The pond manager would notice that shrimp had a loss of appetite, observe cannibalism, would probably notice a change in the water color in the pond, and might see shrimp slowly turning a reddish color, followed by mortality. At first, there might be a few mortalities noticed each day. Then, after 3–4 days, the manager would notice 100 or more dead daily. Other nearby farms might be experiencing the same problems, so it is a good idea to communicate with neighbors. If the manager were to wait until harvest and not act to prevent further loss, survival could be very poor.

Without the benefit of a proper histopathological examination, the farm manager is operating on instinct, not certain knowledge. To properly manage a disease outbreak, the manager should collect a sample of moribund shrimp from one of the affected ponds, and split it into two subsamples. Collect hemolymph from one of the subsamples, and make plate cultures on TSA and TCBS media. If a high number of colonies (CFU) are detected, the manager should conduct an antibiogram of the isolate from one of the green colonies using the most common and effective antibiotics offered by the feed manufacturer. He should also check the plates at night for luminescence. The other sample should be preserved in Davidson's fixative (or less ideal, 10% buffered formalin), and the manager should submit all to the pathology lab. If the shrimp are *P. monodon*, *L. vannamei*, or *L. stylirostris*, the lab should check for WSSV, YHV, IHHN, LOC, BP, NHP, and signs of vibriosis. Corrective measures for the ponds should only be made based on the results of the health examination. By testing shrimp in a pathology lab, the manager can determine if the farm has a transmittable disease (and not a water quality problem). An easy and unsophisticated way to confirm a transmittable disease without sending samples off to a lab would be to collect moribund shrimp and feed macerated portions of them in an aquarium to healthy shrimp collected from another pond or tank. Look for development of identical disease signs in the experimental group. This will only tell the manager that there is a transmittable disease and will not identify the disease agent. One of the better references on shrimp diseases that emphasizes information that can be applied to understanding problems encountered in shrimp culture operations (especially *L. vannamei*) is Brock and Main (105).

PHILOSOPHIES OF DISEASES AVOIDANCE AND CONTROL

Ecuador, Mexico, Panama, the United States, and others have taken precautionary steps in preventing possible contamination from other areas of the globe by banning imports of nauplii, worms, crabs, etc., that could serve as possible vectors or pathways for diseases. The rationale of using lower densities to control stress is a factor being considered and tried in most countries. Asian shrimp farm production comes mainly from cooperative efforts, whereas

Table 8. Generalized Scheme for Assigning a Numerical Qualitative Value to Severity Grade of Infections

Severity Grade	Clinical Findings
0	No signs of infection by pathogen, parasite, or epicommensal present. No lesions characteristic of syndrome present.
1	Pathogen parasite, or epicommensal present, but in numbers or amounts just above diagnostic procedure minimum detection limits. Lesions characteristic of syndrome present, but disease not significant. Prognosis is for insignificant effect, except in developing infections by highly virulent pathogens.
2	Low to moderate numbers of pathogen, parasite, or epicommensal present. Light to moderate lesions characteristic of syndrome present. Prognosis is for possible production losses or slight increases in mortality if no treatment is applied.
3	Moderate numbers of pathogen, parasite, or epicommensal present. Moderate to severe lesions characteristic of syndrome present. Potentially lethal prognosis if no treatment is applied.
4	High numbers of pathogen, parasite, or epicommensal present. Severe lesions characteristic of syndrome present. Lethal prognosis.

Ecuador has multinational companies well structured with formidable power of investment that aggressively sought technologies and implementation of quarantines. Educational campaigns from the Chamber of Aquaculture in Ecuador were commonplace. The technical support and scientific community are strong in Eucador.

Multinational companies are generally willing to pay for solutions. The abundant stocks of larvae that were present during the 1998 season assisted the country in stocking most of its ponds. Environmentally speaking, Ecuador is a leader of organic shrimp production methods and bioremediation. Many natural and environmentally friendly policies have been implemented there because of the lack of working capital, and the cost of purchasing and importing chemical products is virtually impossible. Companies like GreenAqua, Inc. have established a set of operation standards or best management practices (BMPs) for ethical, social, and environmental issues governing the shrimp culture industry. The philosophy there is that most shrimp are healthy, but once the carrying capacity in ponds is exceeded, problems arise and become unmanageable. GreenAqua, Inc. produces Bioregulator®, which is a product that contains citric acid and garlic; Bioregulator® has been used successfully to treat TSV, NHP, BVP, IHHN and "colita roja" (red tail) outbreaks in Ecuador. Other ingredients in Bioregulator® that may assist in fighting diseases are Selenium and Germanium. The same company produces a 100% natural (nonnitrogen) fertilizer-like, pond additive made from Leonardita, a mineral/carbon less than 300 million years old. They market this product, called Biorganic®, as a rich mixture of humic and fulvic acids, which act to quick start primary and secondary productivity in ponds. The company has data and 15 years of experience to back their claims that these products promote rich soils, lower operating costs, promote higher survival rates, give better food conversion ratios, cleaner ponds, and good harvest yields. Apparently, WSSV was present in Ecuador as early as 1995, according to histological slide evidence, but techniques had not been developed to recognize it. It has also been suggested that the abuse of sodium metabisulfates in that country and others may have contributed to stress and eventually lead to infection and transmission of the viruses. The heavy use of Tilt as a fungicide by the banana industry to prevent black spots on bananas is also suspected of causing additional stresses in the estuarine waters of Ecuador.

The control of temperature or any sudden change in environment is important in the control of stress and WSSV.

OTHER MANAGEMENT PRACTICES

Saponin, or Tea Seed Cake, as it is commonly called, is made from a tea seed or from the lilly or camellia in the eastern hemisphere. It is used as an aquatic pesticide and is lethal to crabs, fish, and insects that would prey upon shrimp at relatively low levels (10–15 ppb); at higher levels it is lethal to shrimp. It is used to eliminate predator fish and crabs from ponds or in wild-caught shrimp transport containers. Saponin is used in terrestrial farming as an organic insecticide. It is effective against snails and slugs in rice fields and in vegetable and fruit farms.

Tea seed cake (or powder) is a residue remaining after the oil has been extracted from the seeds of certain plants in the camellia family. It is compressed into a cake shape (or powder form) and contains saponin (a toxin reacting in the blood that is suitable for many applications). Saponin in plants (as soapwort or soapbark) makes a characteristic soapy lather. Hydroscopic saponin mixtures are used as foaming and emulsifying agents and detergents. Saponin detoxifies quickly in water and is not injurious to cattle or people who may use the water, and it leaves no cumulative adverse residues.

According to C.P. Shrimp News, tea seed cake from the eastern hemisphere can be applied to ponds at 20–30 ppm for 6 hours, after which water is exchanged to take care of shrimp carapaces that are not smooth, fuzzy, and are dirty (caused by *Zoothamnium* fouling, algal fouling, or colloidal fouling). Saponin can be used to control *Zoothamnium*, and *Epistylis* on PL shrimp (20,103). A 10% saponin solution diluted and added at 5 ppm concentration can be used to treat larvae.

Saponin is also made in the western hemisphere, mainly in the United States. However, unlike saponin from the eastern hemisphere, U.S.-made saponin comes from the yucca plant and is not toxic to crustaceans and fish even at levels above 10–15 ppb. In fact, recent research by a private company (DPI) in the United States has shown that the yucca derivative saponin is actually beneficial to shrimp when placed in the diet. It changes the flora and fauna of the gut by attacking gram-negative bacteria. It has been shown that young shrimp actually grow faster with saponin added to the diet. Saponin does not effect the growth of plants, and compared with other chemicals, it is safe, inexpensive, and easy to use. The foaming agent in the soft drink root beer is from saponin; therefore, used in small amounts, it is not harmful to humans.

Saponin is generally not a pleasant substance to handle, and most chemical companies do not carry it or sell it within the United States. However, importers can obtain container loads relatively easy.

Some farms have used chlorine to disinfect their incoming waters before use in recirculating systems. Chlorine will eliminate all living things in the water. Treatment with chlorine is neither necessary nor desirable, because it leaves water with no algae and other beneficial organisms that are important for maintaining a healthy ecosystem. Insecticides and pesticides generally target certain organisms and have been found to be more effective and useful in controlling disease vectors.

ONE METHOD OF POND BOTTOM PREPARATION

Pond bottoms, especially in low areas, are tilled so that blocks of soil are broken and will oxidize by exposure to air. The UV light from the sun also helps bake or sterilize the soil. The internal canals should be kept free of sediment and dried. The depth of tilling in the central area should be restricted to 1–2 cm (2–3 in.), because this is the reactive depth between pond soil and water. Liming rates for soil

Table 9. Liming Rates for Soil pH[a]

pH	Dolomite (lb/acre) (kg/ha)	Burnt Lime (lb/acre) (kg/ha)
<5	2000	600
5–6	1000	400
6–7.5	500	200

[a] 1 lb = 0.45 kg and 1 ac = 0.4047 ha.

pH are shown in Table 9. Soils should be maintained at or near neutral (pH7).

After a pond is prepared for the next crop, it is ready to fill. One method of filling a pond starts by flushing or flooding approximately 60% of the pond bottom, ensuring that the interior canals are covered with water. The water is drained to flush unwanted hydrogen sulfide, organics, and possibly heavy metals that may have leached from the dikes or pond bottom. Harvest structures and effluent gates are re-sealed, and the pond is flooded again. No fertilizer is added until flushing is completed. Filling ponds for stocking is to be completed within 10 days after they are made ready for filling. The water level by stocking day will be 30 cm (12 in.) below the full mark. The water level from 8 to 15 days after stocking will be raised to 20 cm (8 in.) below full. The water level from 16 to 22 days after stocking will be raised to 10 cm (4 in.) below full. The water level from 23 to 30 days after stocking will be raised gradually to the full level. Microscreens (0.5 mm) are also removed from the inlets at this time. They are then cleaned, dried, labeled, and stored for the next crop. Once water exchange has begun, harvest and outlet screens should be thoroughly cleaned at the bottom to ensure that poor quality bottom water is flushed from the pond. Harvest and outlet screens should be checked regularly for damage. A microscreen bag at the discharge end of the pipe in the harvest catch basin should be set up to determine if any PL are escaping during water exchange. Mosquito screens (7 mm mesh) are removed from harvest basins and outlets once the shrimp population reaches an average weight of 5 g (0.01 lb).

FERTILIZATION LEVELS AND COSTS

One of the most important factors to keep in mind on fertilizers is the ratio of nitrogen to phosphorus (N : P) used. To promote beneficial diatoms and algae (and not blue greens, especially with higher salinities in the dry season) you need to keep this ratio at least a 5 : 1. A 10 : 1 ratio is better, and best is 20 : 1 or higher, according to most successful pond managers. This ratio is the quantity placed in the pond, not the %N versus %P. The following example is from Nicaragua: Nutrilake™–10 kg: TSP–3.3 kg/ha (22 lb : 7.26 lb/2.47 ac). This is a 3 : 1 (in quantity) ratio and is not high enough. Urea–6 kg/ha: TSP–2 kg/ha (13.2 lb/2.47 ac : 4.4 lb/2.47 ac) also is a 3 : 1 ratio and too low. The source of nitrogen is important, but not as important as the quantity ratio with Phosphorus. If Nutrilake™ works better than urea, then use it. The cost of fertilizer is far below the cost of PL and feed. The manager should spend time trying to economize with feed and PL more than with fertilizers. Use what works best and shrimp growth rates should make up the difference in any added costs of nitrogen.

ONE METHOD OF INITIAL FERTILIZATION

Some of the best fertilizers to use are DI-ammonium phosphate 17-46-0 (DAP) in pellet form, triple superphosphate 0-46-0 (TSP), ammonium nitrate 33.5-0-0 (AM), urea 46-0-0, sodium nitrate 16-0-0, and molasses. The ratio of urea to DAP/TSP used in pond fertilization may vary from time to time from 5 : 1 to 10 : 1 (or higher), depending on pond nutrient level and the quality of the incoming reservoir water. The manager should decide which ratio is in effect at any given time. Either method (1) or (2) may be used for initial fertilization of ponds, depending on the requirements of the water.

1. Before filling ponds, apply 22.5 kg/0.4 ha (50 lb/ac) of diammonium phosphate (DAP) or triple superphosphate (TSP) and 22.5 kg/0.4 ha (50 lb/ac) of urea over dry pond bottoms. Spread fertilizer as uniformly as possible. To increase substrate area in some ponds, 225–450 kg/0.4 ha (500–1000 lb/ac) of rice processing water, or 113 kg/0.4 ha (250 lb/ac) bagasse may be applied to the pond bottom. The bagasse needs to be soaked for a few days in pond water while the pond is filling before being spread over the pond surface. While the pond is filling, add 15–19 L/0.4 ha (4–5 gal/ac) of molasses at the inlet. Fill the pond to 60–70% bottom cover, and hold this level for a couple days or until a bloom develops. Continue filling the pond to stocking level, 30 cm (12 in.) below full, applying routine fertilization daily or as required.
2. Initial fertilization begins the day after step 2 of pond filling begins. Fill the pond to cover at least 60% of the bottom. While the pond is being filled, fertilize with 9 kg (20 lb) urea and 0.9 kg or 1.8 kg (2 lb or 4 lb) DAP/TSP per hectare (2.4 ac) of total pond surface area at the inlets, so that water can gradually dissolve the fertilizer. When the water is deep enough, spread urea by hand directly from a boat. DAP/TSP is dissolved in a container and slowly poured into the water behind an outboard engine on the moving boat. This mixes the solution with pond water by the propeller action. After one day, begin bringing the level up to 50% pond volume. While filling, add 13.6 kg (30 lb) urea and 1.3 kg or 2.7 kg (3 lb or 6 lb) DAP/TSP per ha (2.47 ac), ensuring adequate distribution over the pond surface area. Skip another day. Continue to fill the pond to 5 cm (2 in.) below full mark (maximum operative level). While filling, add 22.6 kg (50 lb) urea and 2.2 kg or 4.5 kg (5 lb or 10 lb) DAP/TSP per ha. Additional fertilizer may be applied at the discretion of the manager if a bloom has not developed before the stocking date. The ammonia level must be checked daily preceding stocking to ensure that it is at an acceptable level for PL. If the ammonia level is unacceptably high, it can be reduced through pond

flushing. Fertilizer may be applied at reduced levels immediately before and after stocking to prevent excessive ammonia levels.

- Whenever the morning secchi disc reading is greater than 35 cm (14 in.), routine fertilization will be applied a minimum of two times per week.
- In the first few weeks of pond culture, it may be necessary to initially fertilize on consecutive days and after that every other day, and eventually biweekly routine fertilization should be enough to maintain turbidities of 35–40 cm (14–17 in.).
- When the turbidity reading is less than 30 cm (12 in.), suspend fertilization for that pond until transparency increases to 35 cm (14 in.) or greater.
- If turbidity readings are less than 20 cm (8 in.), suspend routine fertilization, as well as daily feed ration, and increase the water exchange rate by 20%, flushing until turbidity readings are greater than 35 cm (14 in.).
- If a phytoplankton crash causes low DO levels, emergency fertilization should be implemented by adding 6.8 kg (15 lb) urea and 0.67 kg or 1.36 kg (1.5 lb or 3 lb). DAP/TSP per hectare (2.47 ac) of pond surface area.
- TSP and DAP in solid form are never spread directly into the water. Always dissolve a container or secure the bags in the inlets. Villalon (21) gives a detailed account of Ecuadorian fertilization procedures. Table 10 comes from a farm in Belize that follows similar procedures to the ones described in Villalon (21).

IMPORTANCE OF NATURAL PRODUCTIVITY, SUCH AS PLANKTON, IN PONDS

One can produce good-quality shrimp without natural productivity, but it may cost more (in fertilizers and higher feed inputs), and the growout cycle duration might be longer than in a pond that has high natural productivity. Plankton is very important in extensive systems, relatively important in semi-intensive systems, and much less important in intensive systems. Plankton provide the essential micronutrients missing in many commercial shrimp feeds. In the absence of dietary input from natural productivity (e.g., indoor, clear water culture systems), the commercial shrimp feed used must be complete or growth will be poor. Newly stocked larvae will generally favor natural foods over crumbled feeds. If the newly stocked pond is deficient in plankton, the larval survival is at risk. Phytoplankton and meiofauna constitute the food sources for secondary productivity. If ponds are deficient in algae and bacteria, there will be very little zooplankton production, which may impact shrimp growth. Algae are natural biofilters and are very effective removers of soluble nitrogenous waste products, such as ammonia. Phytoplankton and suspended solids shade the water column, assist in hiding the shrimp from birds and other predators, and create a more favorable environment for the shrimp, which generally dislike high light intensities. The most economical way to aerate or oxygenate pond water is through algal photosynthesis. In ponds built in acid sulfate soil conditions, algae will assist in elevating water pH. Algae and bacteria will help in assimilating and neutralizing any residual antibiotic or pesticide that may be present. The type of plankton is also important. Diatoms are preferred for their superior nutritional content. Some algae possess natural antibacterial properties. Blue-greens and dinoflagellates are considered undesirable because they may cause unstable water chemistry and health problems (hemocytic enteritis, growth inhibition). One such algae is *Gleosystus major* and is suspected of causing problems in ponds in Texas. It has an envelope that carries bacteria and is thought to infect shrimp, but more research is needed to find the exact mechanism of infection.

Good pond managers try to identify and quantify the plankton in their ponds. There is some debate as to whether the resulting information justifies the cost and manpower requirement, but most farmers like to know what they have in their pond water.

Table 10. Routine Fertilization (1 lb = 0.45 kg; 1 gal = 3.79 L)[a]

Pond	Urea (lb)	DAP (lb)	Molasses (gal)	Pond	Urea (lb)	DAP (lb)
1	50	5 (or 10)	15	14	40	4 (or 8)
2	50	5 (or 10)	15	15	40	4 (or 8)
3	50	5 (or 10)	15	16	40	4 (or 8)
4	20	2 (or 4)	5	17	40	4 (or 8)
5	40	4 (or 8)	10	18	40	4 (or 8)
6	40	4 (or 8)	10	19	40	4 (or 8)
7	40	4 (or 8)	10	20	40	4 (or 8)
8	40	4 (or 8)	10	21	40	4 (or 8)
9	40	4 (or 8)	10	22	40	4 (or 8)
10	40	4 (or 8)	10	23	40	4 (or 8)
11	40	4 (or 8)	10	24	40	4 (or 8)
12	40	4 (or 8)	10	25	40	4 (or 8)
13	40	4 (or 8)	10	26	10	1 (or 2)

[a] Multiply urea lb by 1.5 if substituting with ammonium nitrate and by 3 if substituting with sodium nitrate.

ECONOMICS OF PRODUCING SHRIMP IN PONDS

Shrimp are produced in ponds, raceways, and tanks. Production costs vary from US$5.00/0.45 kg (US$2.50/lb) to US$10.00/0.45 kg (US$5.00/lb) because of the varied output costs associated with each form of production. Feed, processing, and larvae are the three highest costs. It generally takes 0.91 kg (2 lb) of feed to produce 0.45 kg (1 lb) of shrimp. The principal economic problems with culturing shrimp in the United States are availability of low-cost, high-quality feed; short growing season; one crop only in some areas because of temperatures; high cost of land, labor, high operating costs (e.g., power); foreign competition; and price fluctuations. Because of the cold climate, outdoor culture in the continental United States is limited to nine months in extreme southern regions. Roughly 40% of the body weight of shrimp is in the head. Shrimp producers generally contract a processing plant to process the shrimp. An average of US$1.38/kg (US$0.63/lb) is charged for processing (icing, deheading, grading, packing, freezing in plate freezer, and one month in cold storage). Production in ponds often ranges from 2,000–8,000 kg/ha per crop (2,000–8,000 lb/ac per crop) in the United States with average U.S. pond production at 3,500 kg/ha per crop (3,500 lb/ac per crop). The relationship between total length and weight of shrimp is given in Table 11. A breakdown of the cost analysis of shrimp tails can be seen in Table 12.

A 20% profit margin is considered to be excellent in Ecuador. The real price of Ecuadorian shrimp has declined sharply since 1980. Profit margins have been narrowing because of rising feed, labor, and fuel costs, and the exchange rates have hurt the farms. Government exchange rates have actually acted like high taxes to the farmers, and most people want to see a free exchange rate.

Table 11. Length and Weight Relationships

Total Length (mm)	Weight of Whole Shrimp (g)
135	19.10
140	21.25
145	23.56
150	26.02
155	28.65
160	31.46
165	34.43
170	37.59
175	40.93
180	44.46
185	48.19
190	52.12
195	56.25
200	60.59
205	65.15
210	69.93
215	74.94
220	80.17
225	85.65
230	91.36
235	98.74
240	105.91
245	113.60
250	121.85
255	130.70
260	140.19
265	150.38
270	161.30
275	173.01
280	185.58
285	199.05
290	213.51
295	229.01
300	245.65
305	263.49
310	282.62
315	303.15
320	323.15
325	343
330	363
335	383
340	400
345	420
350	440
355	450

Table 12. Cost Analysis per Pound of Shrimp Tails

Operating Expense Items	Cost/lb (0.45 kg)	%
1. Feed	1.011	28.5
2. Processing and harvesting	0.630	17.8
3. Postlarvae	0.549	15.5
4. Interest	0.393	11.1
5. Salaries and wages	0.361	10.2
6. Pumping costs	0.170	4.8
7. Aerator utilities	0.118	3.3
8. Management consultant	0.105	3.0
9. Supplies, miscellaneous	0.091	2.5
10. Maintenance	0.078	2.2
11. Land lease	0.039	1.1
Total	US$3.54	100.0

Present (1999) costs for a pilot shrimp farm in south Texas, consisting of four ponds, each 2 ha (5 ac), with a settling basin attached to each pond and one common 6 ha (14.8 ac) constructed wetland can be seen in Table 13. The facility is now in operation and was designed to treat water on site and to discharge water only during harvest. The farm takes in water only to fill ponds and offset evaporation and other water loss, and the facility is capable of producing 36 MT of shrimp per year (approximately 4,000 lb/ac). The average construction cost for the 14.1-ha (35-ac) facility is US$5,315/ha (US$13,130/ac).

The capital and operating costs of three different sized, flow-through system, hypothethical shrimp farms can be seen in Tables 14 and 15, respectively. The 20.2 ha, 40.4 ha, and 76.8 ha (50 ac, 100 ac, and 190 ac) facilities are itemized in detail. Not included are the extra costs of recirculation design, equipment, and wetland water treatment systems given in Table 13. As can be seen in Table 14, the average estimated construction cost is US$24,803/ha for a 20.2-ha farm, US$15,657/ha for a 40.4-ha farm, and US$15,882–17,126/ha for a 76.8-ha facility (US$10,042/ac for the 50-ac farm, US$6,339/ac for the 100-ac farm and US$6,430–6,934 for the 190-ac facility).

SHRIMP NUTRITION AND FEEDS

Feed prices vary, but current price for 35% protein feed in Texas, if purchased in bulk, ranges between US$.60/kg and US$0.68/kg (US$0.27 to US$0.31/lb). Standard quality feeds generally fall apart as soon has they hit water. Most of the feed companies offer a lower quality feed, but promote the higher grade feeds by saying that the FCR will be better, which is usually true. Many feed companies in the eastern hemisphere promise FCRs of around 1.8 kg feed: 1 kg shrimp (1.8 lb feed: 1 lb shrimp) produced, and an average growth of 3 g/10 days (0.09 oz/10 days) for *P. monodon*, with their top line feed and FCRs of 2.5 : 1 with their lower grade feed.

Table 13. List of 1999 Costs for a Pilot Shrimp Farm in Texas (Price in US$)

Contractual	
Construction Management, Equipment Operator, and Rental	20,045
Earth Moving	72,982
Electricity Establishment (Includes Electricity to Aerators)	25,000
Fencing 1,500 Meters or (5000 ft) Installed	10,815
Insurance, Repairs and Maintenance, Dues, Water Analysis, Miscellaneous	6,000
Legal Fees (Permitting, etc.)	50,000
Supplies	
Wetland Vegetation, Truck Fuel, Grass Seed, Tools, Miscellaneous	18,400
Pipe, Lumber, Hardware	20,073
Equipment	
Land 20 at $3,750/ha or (50 ac at US$1500/ac)	75,000
Pumps	5,000
Feed Equipment: Pond Feeder, Bulk Bin (8 ton)	10,770
Aerators, Controllers and Wire, 60 at 2 hp Each (US$476 ea)	28,565
Emergency Aerator	4,449
Tractor (Used, 140 hp), Truck (Used, 3/4 ton, 4wd)	25,000
Electrical Generator (Pto Driven, 50 kva)	5,000
Drains, Harvest Basins	7,953
Scraper Blade, Mower	4,000
Screens, Nets, Pl Acclimation Equipment	3,500
Trailer and Furniture (Office, Storage and Occasional Housing)	11,000
Water Quality Lab Equipment	8,500
Repairs, Contingencies, Miscellaneous	11,000
Personnel	
On-Farm Labor, Consultants	36,500
Total Costs	US$459,552

Source: Dr. Ronald Rosati, Texas A&M University in Kingsville, Texas.

Feed manufacturers have come a long way to improve the stability of shrimp feed. Formaldehyde-based binders were discarded, and vitamin-C stabilizers were added, to name a few improvements. Cruz (109) and SEAFDEC (110) covered shrimp feeding principles and practices. Some of the more useful feed and nutrition references are Piedad-Pascual (111), American Soybean Association (112), and New, Saram, and Singh (113). A good live-feed reference is Dhert and Sorgeloos (114).

Some of the protein requirements reported in the literature for various species of shrimp are the following: *F. aztecus* at 23–31%, *F. californiensis* at 35%, *F. duorarum* at 28–32%, *F. indicus* at 43%, *M. japonicus* at greater than 60%, *F. merguiensis* at 34–42%, *P. monodon* at 35–50%, *F. chinensis* at 40%, *F. penicillatus* at 22–27%, and *L. setiferus* at 25–28%. A nutritional review of *L. stylirostris* was done by Cuzon (115), and one for *L. vannamei* by Pedrazzoli et al. (116). The World Aquaculture Society published an excellent reference on crustacean nutrition in 1997 (117). A number of shrimp feed formulas can be seen in New (118).

SHRIMP GROWTH RATES

Growth of shrimp varies greatly with species, stocking density, and food supply (Fig. 36). Other conditions, such as water quality and temperature, are also major factors. Under ideal conditions, *L. vannamei* can reach 20 g (0.6 oz) in 120 days, whereas *P. monodon* can attain 35 g (1.05 oz) and *L. stylirostris* can attain 28 g (1 oz) in the same period. The normal weight in which *L. vannamei* are harvested is 16–18 g (0.48–0.54 oz). Generally, 62% of that weight is in the tail meat. Growth rate comparisons of other species can be seen in Figure 36. An example of a shrimp feeding regime can be seen in Table 16. The management of ponds stocked with blue shrimp (*L. stylirostris*) was described by Clifford (119). Optimum water parameters for shrimp farming can be seen in Table 17, and seawater analyses from around the world have been compiled in Tables 18a and 18b. Soil types that are conducive to shrimp culture in ponds are listed on Table 19 (120), and the optimum soil parameters, with normal, highs, lows, and ranges, are listed in Table 20 (120).

SPECIFIC ADVANTAGES OF WORKING WITH SALTWATER SHRIMP

Saltwater shrimp taste good to most people. The texture is firm, and it keeps well when frozen, because the shell protects the meat. It can be thawed and refrozen with minimal damage to the product. Market demand for such shrimp is generally high. Price is generally high in comparison with other seafood, and the product has nutritional benefits. Commercial feeds are available for shrimp growout. Processors, exporters, and importers often prefer cultured shrimp, because it is not as seasonal as wild-caught shrimp and is considered more reliable, more uniform in size, a higher quality, and a fresher product. Cultured shrimp availability can be predicted and planned for. This allows the industry to adapt to consumer demand regarding species and size. Shrimp can be provided live or near live to restaurants that serve seafood. Curricula are now available for teachers wishing to cover the foregoing aspects of shrimp culture in the classroom [see Treece (121).] Dall et al. (122) is a must-have text for the shrimp culture library.

TASTE DIFFERENCES IN SHRIMP

Shrimp reared at high salinity are more flavorful, while the shrimp grown in freshwater usually have a bland taste. Shrimp osmoregulate to adjust to osmotic pressure differences between fresh and saltwater. As salinity goes up, the shrimp increase free amino acids in the muscle, and some free amino acids are associated with taste.

FRESHWATER SHRIMP CULTURE

Introduction

The Malaysian prawn (*M. rosenbergii*, seen in Fig. 41) has been grown for centuries in Asia, but it has only been the subject of research and commercial enterprise

Table 14. Capital Costs for 50, 100, and 190 acre (20, 40, and 77 ha) Hypothetical Shrimp Farm Built in 3 Phases

Number of ponds	10	20	38
Water acres per pond	5	5	5
Number of acres	50	100	190
Water delivery system			
Criteria			
Maximum exchange rate (%)	20	20	20
Mean depth (ft)	4	4	4
Hours pumping per day	18	18	18
Total dynamic head (ft)	18	18	18
GPM per 24″ pump	15,000	15,000	15,000
Cost per 24″ pump and motor	15,000	15,000	15,000
Cost per cubic yard	0.6	0.6	0.6
Pump station			
Pumps and motors			
Pond volume/acre	1,303,489	1,303,489	1,303,489
Total pond volume	65,174,472	130,348,944	247,662,994
GPM required	12,069	24,139	45,864
Number of 24″ pumps	1	2	3
Cost	15,000	30,000	45,000
Dredging (200′L, 30′W, 6′D with 3 : 1 slopes)			
Estimated cubic yards	2,133		2,133
Hydraulic dredging cost	20,000		20,000
GLO cost for sand and marl	533		533
Total dredging cost	20,533		20,533
Pump housing			
Bulkhead and foundation	20,000	5,000	25,000
Electrical controls	10,000	2,500	5,000
Pump house	2,500	5,000	7,500
Cost of pump housing	32,500	12,500	37,500
Subtotal	68,033	42,500	10,3033
Distribution canal			
Primary canal			
Distance (pump station to secondary canal)	1,500		
Mean elevation	8		
Required elevation at levee top	16		
Crown width	12		
Side slope	3		
Calculated height	8		
Cross-sectional area	288		
Cubic yards	32,000	0	0
Earthmoving cost	19,200		
Secondary Canal			
Distance (primary canal to end sec. canal)	5,500	7,600	7,600
Mean elevation	9	9	9
Required elevation at levee top	16	16	16
Crown width	12	12	12
Side slope	3	3	3
Calculated height	7	7	7
Cross-sectional area	231	231	231
Cubic yards	94,111	0	130,044
Earthmoving cost	56,467	0	78,027
Total cubic yards	126,111	0	130,044
Earthmoving cost	75,667	0	78,027
Cost for water delivery system	143,700	42,500	181,060
Per acre cost for water system	2874	425	953
Cumulative per acre cost	2874	—	1,080
Pond construction			
Desired length : width ratio	2	2	2
Height of levee	4.5	4.5	4.5
Maximum depth (′)	6	6	6
Crown width at deep end	18	18	18
Crown width at shallow end	16	16	16
Side slope	3	3	3
Desired freeboard	1.5	1.5	1.5

(continued)

Table 14. *Continued*

Calculated dimensions			
Freeboard compensation	4.5	4.5	4.5
Crown compensation			
Deep end	9	9	9
Shallow end	8	8	8
Water area per pond	217,800	217,800	217,800
Width of water area	330	330	330
Length of water area	660	660	660
Levee centerline dimensions			
Length	686	686	686
Width	355	355	355
Water area per pond (%)	89.4	89.4	89.4
Calculation of perimeter distance			
Number of levees			
Shallow width	10	20	38
Deep width	10	20	38
Length	20	40	76
Number of shared levees			
Shallow width	1	0	0
Deep width	0	0	0
Length	12	16	36
Number of distribution canal borders			
Shallow width	12	16	36
Deep width	0	0	0
Length	1	1	1
Total number of lengths	7	23	39
Total number of shallow widths	3	4	2
Total number of deep widths	10	20	38
Total levee perimeter distance	7,287	24,298	40,954
Calculation of levee C-S area			
Levee length cs	133	133	133
Shallow width cs	133	133	133
Deep width cs	142	142	142
Calculation of cubic yards			
Lengths	23,610	77,575	131,541
Shallow widths	5,236	6,982	3,491
Deep widths	18,638	37,275	70,823
Total levee earthmoving	37,011	121,832	205,854
Cost for levee earthmoving	22,207	73,099	123,512
Per acre cost for earthmoving	444	731	650
Gravel surfacing (4″)			
Desired gravel thickness (″)	4	4	4
Width of gravelled area (′)	10	10	10
Levee crown area			
Lengths	48,020	252,448	428,064
Widths			
Shallow	10,650	22,720	11,360
Deep	35,500	127,800	242,820
Distribution canal			
Primary	15,000	0	0
Secondary	55,000	91,200	91,200
Total crown area	142,870	494,168	773,444
Cubic yards of gravel needed	2,293	7,931	12,413
Price per cubic yard	8.5	8.5	8.5
Total cost of gravel	19,490	67,414	105,513
Per acre cost of gravel	390	674	555
Drain and harvest basin			
Plastic culvert pipe (30″)			
Depth of culvert below grade(′)	5	5	5
Length	56.5	41.5	41.5
Price/ft	16.5	16.5	16.5
Price/pond	932.25	684.75	684.75
Harvest basin	1,000	1,000	1,000

Table 14. *Continued*

Fiberglass channel board slots			
Number needed/gate	8	8	8
Length of each channel (ft)	8	8	8
Total length needed	64	64	64
Price/ft (4″ channel)	2.75	2.75	2.75
Price/pond	176	176	176
Installation	422	930	930
Price per pond	2,530	2,874	2,874
Price per phase	25,299	57,478	109,207
Price per acre	506	575	575
Additional drainage (15″)			
15″ Corrigated HDPE pipe			
Distance	165	165	165
Price/ft	5.51	5.51	5.51
Subtotal	909	909	909
Installation	182	182	182
Price per pond	1,091	1,091	1,091
Price per phase	10,910	21,820	41,457
Price per acre	218	218	218
Water inlet			
Concrete structure	500	500	500
Length of 18″ inlet pipe	25	25	25
Price/ft for pipe	8.95	8.95	8.95
Price/pond for pipe	224	224	224
Materials cost	724	724	724
Installation	145	145	145
Total price/pond	869	869	869
Total price/acre	174	174	174
Total price/phase	8,685	17,370	33,003
Riprapping water streams			
Pond water inlet			
Number of inlets	10	20	38
Cost/inlet	300	300	300
Subtotal	3,000	6,000	11,400
Pond water outlet			
Number of outlets	10	20	38
Cost/outlet	500	500	500
Subtotal	5,000	10,000	19,000
Total price	8,000	16,000	30,400
Total price/acre	160	160	160
Drainage ditch culverts			
Number of crossings	5	3	5
Number of 36″ culverts/crossing	2	2	2
Length of culverts	20	20	20
Cost/ft for concrete culvert	16.5	16.5	16.5
Materials cost	3,300	1,980	3,300
Installation cost	1,650	990	1,650
Subtotal	4,950	2,970	4,950
Stabilizing slopes against erosion			
Acres of levee slopes			
Primary distribution canal	1	0	0
Secondary d. canal	3	4	4
Pond levees	2	8	13
Total acres of side slopes	6	11	16
Cost/acre for topsoil dressing	807	807	807
Cost/acre (seed and fertilizer)	20	20	20
Cost/acre for planting	50	50	50
Total cost	5,029	9,814	14,339
Electrical distribution (3 ph 440 V)			
Wiring distance			
Along deep ends	3,550	7,100	13,490
Branching to lengths	4,116	5,488	12,348
Total length	7,666	12,588	25,838

(*continued*)

Table 14. *Continued*

Cost/ft for wire	0.8	0.8	0.8
Cost of wire	6,132.8	10,070.4	20,670.4
Components of switch box			
Plastic box housing			
Breaker			
Plugs (2)			
Timers (2)			
Treated 4 × 6″ post			
Size required (ft)	8	8	8
Cost/ft	1.29	1.29	1.29
Cost of post	10	10	10
PVC conduit (2″ DWV)			
Length needed	24	24	24
Price/ft	0.71	0.71	0.71
Cost of conduit	17	17	17
Total cost/box	600	600	600
Number of boxes	18	33	60
Cost of boxes	10,800	19,800	36,000
Cost of materials	16,933	29,870	56,670
Installation	3,387	5,974	11,334
Total cost/phase	20,319	35,844	68,004
Per acre cost	406	358	358
Major pond equipment			
Bulk feed storage bin			
Number of 25-ton bins	2	2	3
Cost/bin	9,000	9,000	9,000
Total cost	18,000	18,000	27,000
Harvest machine			
Number needed	1	1	1
Cost per unit	15,000	15,000	15,000
Total cost	15,000	15,000	15,000
Stocking and harvest tank/trailer			
Capacity (tons)	5	5	5
Number needed/pond harvest	0	4	4
Total number needed	1	8	4
Cost/unit	4,000	4,000	4,000
Total cost	4,000	32,000	16,000
Tractor/90 HP/used			
Number needed	1	1	2
Cost/unit	10,000	10,000	10,000
Total cost	10,000	10,000	20,000
Pond aerators (5 HP)			
Number needed/pond	4	4	4
Cost/unit	2,000	2,000	2,000
Total cost	80,000	160,000	304,000
PTO aerator			
Number needed	1	1	2
Cost/unit	3,000	3,000	3,000
Total cost	3,000	3,000	6,000
Feed blower			
Number needed	1	1	2
Cost/unit	5,000	5,000	5,000
Total cost	5,000	5,000	10,000
Truck/gooseneck/4 wd			
Number needed	1	0	1
Cost/unit	12,000	12,000	12,000
Total cost	12,000	0	12,000
Gooseneck trailer			
Number needed	1	0	1
Cost/unit	3,000	3,000	3,000
Total cost	3,000	0	3,000
All-terrain vehicle			
Number needed	1	1	2
Cost/unit	2,400	2,400	2,400
Total cost	2,400	2,400	4,800

Table 14. ***Continued***

Fuel storage tanks			
Number needed			
Diesel (500 gal cap)	10	20	38
Daily fuel usage	2	2	2
Feedings/day	15	15	15
Minutes/pond feeding	4	4	4
Fuel efficiency (gal/hr)	20	40	76
Total	1	0	1
Number of tanks needed	1	0	0
Gas (500 gal cap)	600	600	600
Cost/unit	1,200	0	600
Total cost			
Boat/motor/trailer			
Jon boat 14′, 38″ bottom	1,000	0	0
Boat motor, 18HP	2,000	0	0
Boat trailer (capacity 500+ lb)	750	0	0
Total cost	3,750	0	0
Implements			
Disc (14′ used)	2000	0	0
Drag	1200	0	0
Side mower (6′)	2000	0	0
Miscellaneous	1500	0	0
Total	6700	0	0
Total major pond equipment cost	164,050	245,400	418,400
Miscellaneous equipment and supplies			
Disposable pond materials			
Filter bags for inlet pipes			
Bags/pond	2	2	2
Sq yds/bag (24″ dia, 10′ long)	7	7	7
Price/sq yd (Tetko HC7-500)	10	10	10
Cost/pond	140	140	140
Total cost	1,396	2,791	5,303
Inlet filter screens			
Screens/pond	2	2	2
Components of screen			
3 × 4′ Frame (2 × 6″ wood)			
Linear ft needed	14	14	14
Price/ft	0.75	0.75	0.75
Subtotal	11	11	11
3 × 4′ vinyl wire (1/2 × 2″ 14 ga)			
Square ft needed	12	12	12
Price/sq ft	0.54	0.54	0.54
Subtotal	6	6	6
Tetko HC-500 filter cloth			
Square yd needed	1.3	1.3	1.3
Price/sq yd	10	10	10
Subtotal	13	13	13
Labor	25	25	25
Subtotal	55	55	55
Price/pond	111	111	111
Total cost	1,106	2,213	4,204
Drain screen			
Vinyl-coated wire screen			
Square feet (3′ dia × 6′ ht)	57	57	57
Price/sqft (1/2″ × 1″, 14 ga)	0.5	0.5	0.5
Subtotal	28	28	28
Tetco filter cloth			
Square yd (3′ dia × 6′ ht)	6	6	6
Price/sq yd (Tetko HC7-500)	10	10	10
Subtotal	63	63	63
Total materials cost	91	91	91
Installation	18	18	18
Price per pond	109	109	109
Price per phase	1,093	2,185	4,152
Price per acre	22	22	22

(continued)

Table 14. *Continued*

Harvest basin filter screens			
Screens/pond (stack two 3 × 4′ screens to form 6 × 4′)	2	2	2
Components of screen			
3 × 4′ frame (2″ × 6″ wood)			
Linear feet needed	14	14	14
Price/ft	0.75	0.75	0.75
Subtotal	11	11	11
3 × 4′ vinyl wire (1/2 × 2″ 14 ga)			
Square feet needed	12	12	12
Price/sq ft	0.54	0.54	0.54
Subtotal	6	6	6
Tetko HC-500 filter cloth			
Square yields needed	1.3	1.3	1.3
Price/sq yd	10	10	10
Subtotal	13	13	13
Labor	25	25	25
Subtotal	55	55	55
Price/pond	111	111	111
Total cost	1,106	2,213	4,204
Harvest basin boards (2″ × 6″ × 6′)			
Sets needed	4	4	4
Number/set	14	14	14
Board ft/pond	336	336	336
Cost/board foot	0.75	0.75	0.75
Cost/pond	252	252	252
Total cost	2,520	5,040	9,576
Water inlet boards (2″ × 6″ × 3′)			
Sets needed	2	2	2
Number/set	8	8	8
Board ft/pond	48	96	96
Cost/board foot	0.75	0.75	0.75
Cost/pond	36	36	36
Total Cost	648	648	648
Total cost of pond supplies	7,869	15,090	28,087
Shop equipment	Quantity	$/Unit	Total
Phase I			
Vise	1	60	60
Come-along winch	1	40	40
Battery charger	1	50	50
Voltmeter	1	150	150
Jack 12 ton	1	400	400
Wheelbarrow	1	90	90
Hand tools	1	3,000	3,000
Shovels			
Posthole diggers			
Axe			
Sledge hammer			
Wrenches			
Generator supplies	1	1,000	1,000
Generator (small)	1	1,000	1,000
Hand drill 3/8	1	120	120
Drill 1/2	1	350	350
Jigsaw	1	100	100
Circular saw	2	150	300
Ladder	1	50	50
Miscellaneous			671
Subtotal			7,381
Phase III			
Grinder	1	250	250
Welder	1	300	300
Acetylene torch	1	300	300
Air compressor	1	650	650
Drill press	1	500	500
Table saw	1	400	400
Miscellaneous			240
Subtotal			2,640

Table 14. *Continued*

Office equipment			
Phase I			
Desk and chair	3	750	2,250
Blackboard	1	100	100
Bookshelves	5	150	750
Filing cabinet	10	150	1,500
IBM computer system	1	10,000	10,000
Typewriter	1	1,000	1,000
Telephone with recorder	1	250	250
Xerox machine	1	2,500	2,500
Calculator	1	75	75
Air conditioner	1	2,000	2,000
Miscellaneous			1,843
Subtotal (phase I)			22,268
Phase II			
Desk and chair	1	750	750
Bookshelves	2	150	300
Filing cabinet	2	150	300
Subtotal (phase II)			1,350
Phase III			
Desk and chair	3	750	2,250
Bookshelves	3	150	450
Filing cabinet	6	150	900
IBM computer system	1	10,000	10,000
Subtotal (phase III)			3,600
Monitoring equipment			
Water quality			
Phase I			
Dissolved oxygen meters	2	1,100	2,200
Refractometer	2	500	1,000
Secchi Discs	2	15	30
Balances	2	90	180
Microscopes	1	2,000	2,000
Hemacytometer	1	70	70
Test kits	5	30	150
Refrigerator	1	1,000	1,000
PH meter	2	300	600
Miscellaneous			723
Subtotal (phase I)			7,953
Phase II			
Dissolved-oxygen meters	1	1,100	1,100
Refractometer	1	500	500
Secchi discs	1	15	15
Microscopes	1	2,000	2,000
Test kits	5	30	150
Subtotal (phase II)			3,765
Phase III			
Dissolved-oxygen meters	2	1,100	2,200
Refractometer	2	500	1000
Secchi discs	2	15	30
Microscopes	1	2,000	2,000
Spectrophotometer	1	1,000	1,000
Test kits	5	30	150
Subtotal (phase III)			11,038
Sampling gear			
Seines	2	250	500
Cast nets	10	50	500
Waders	8	60	480
Subtotal (repeat for ea phase)			1,480
Total monitoring cost/phase	9,433	5,245	12,518
Buildings			
Shop/storage shed (phase I)	750	30	22,500
Expansion (phase II)	750	30	22,500
Expansion (phase III)	1,500	30	45,000

(continued)

Table 14. *Continued*

Annual Total	524,589	656,393	1,266,691
Number of ponds	10	20	38
Water acres per pond	5	5	5
Number of acres	50	100	190
Summary of capital costs			
Water delivery system			
Pump station	68,033	42,500	103,033
Primary canal	19,200	0	0
Secondary canal	56,467	0	78,027
Subtotal	143,700	42,500	181,060
Pond construction			
Earthmoving	22,207	73,099	123,512
Gravel surfacing	19,490	67,414	105,513
Drain and harvest basin	25,299	57,478	109,207
Additional drainage	10,910	21,820	41,457
Water inlet	8,685	17,370	33,003
Riprapping	8,000	16,000	30,400
Drainage ditch culverts	4,950	2,970	4,950
Levee stabilizing	5,029	9,814	14,339
Electrical distribution	20,319	35,844	68,004
Subtotal	124,889	301,809	530,386
Major pond equipment	164,050	245,400	418,400
Miscellaneous equipment and supplies			
Disposable pond materials	7,869	15,090	28,087
Building	22,500	22,500	45,000
Shop equipment	7,381	0	2,640
Office equipment	22,268	1,350	3,600
Monitoring equipment	9,433	5,245	12,518
Subtotal	69,450	44,185	91,845
Total cost	502,089	633,893	1,221,691
Cost/acre	10,042	6,339	6,430

Table 15. Operating Costs of Hypothetical Shrimp Farm

	Phase		
Item	I	II	III
Farm size			
Number of ponds	10	20	38
Water acres per pond	5	5	5
Number of acres	50	100	190
Key biological assumptions			
Species utilized			
L. vannamei			
Percentage of crop	90	75	75
Stocking density (000s/acre)	100	100	100
Size at stocking (g)	0.001	0.001	0.001
Thousands of PLs needed	4,500	7,500	14,250
Price/thousand postlarvae	10	10	10
Cost of *L. vannamei* PLs	45,000	75,000	142,500
P. monodon			
Percentage of crop	10	25	25
Stocking density (000s/acre)	100	100	100
Size at stocking (g)	0.001	0.001	0.001
Thousands of PLs needed	500	2500	4750
Price/thousand postlarvae	20	18	15
Cost of *P. monodon* PLs	10,000	45,000	71,250
Total cost of postlarvae	55,000	120,000	213,750
Survival rate at harvest (%)			
L. vannamei	70	70	75
P. monodon	70	70	75

Table 15. *Continued*

Item	Phase		
	I	II	III
Growth data			
Length of growing season (days)	165	165	165
L. vannamei			
Wt after month 1 (g)	1	1	1
Grams/wk after month 1	1	1	1
Size at harvest (g)	20.3	20.3	20.3
P. monodon			
Wt after month 1 (g)	1	1	1
Grams/wk after month 1	1.5	1.5	1.5
Size at harvest (g)	29.9	29.9	29.9
Feed utilization			
L. vannamei			
Food conversion ratio	2.5	2.3	2.2
Feed requirements			
Per acre	7,810	7,185	7,364
Per pond	39,050	35,926	36,819
Per phase	351,450	538,890	1,049,329
Feed cost			
Per pound	0.27	0.27	0.27
Per acre	2,109	1,940	1,988
Per pond	10,544	9,700	9,941
Per phase	94,892	145,500	283,319
P. monodon			
Food conversion ratio	2.5	2.5	2.2
Feed requirements			
Per acre	11,523	11,523	10,864
Per pond	57,613	57,613	54,320
Per phase	57,613	288,063	516,043
Feed cost			
Per pound	0.27	0.27	0.27
Per acre	3,111	3,111	2,933
Per pond	15,555	15,555	14,666
Per phase	15,555	77,777	139,332
Total feed cost	110,447	223,277	422,651
Production results			
L. vannamei			
Head-on yield			
Per acre	3,124	3,124	3,347
Per pond	15,620	15,620	16,736
Per phase	140,580	234,300	476,968
% Sold head-on to processor	100	100	100
Calculation of head-on farm-gate price			
Percentage tail wt	60	60	60
Average tail count (#/lb)	37	37	37
Wholesale value of tails/lb	0	0	0
Processing expenses (per lb)			
Ice	0.04	0.04	0.04
Transportation	0.06	0.06	0.06
Heading	0.25	0.25	0.25
5-lb carton	0.01	0.01	0.01
Grading and freezing	0.22	0.22	0.22
Processor's profit	0.17	0.17	0.17
Cold storage (6 mos)	0.05	0.05	0.05
Broker's fee	0.05	0.05	0.05
Subtotal for IQF add \$0.23/lb; for peeled, −17% of GH wt	0.80	0.80	0.80
Calculation of farm-gate value (head-on)			
Per pound	−0.48[a]	−0.48[a]	−0.48[a]
Per acre	−1,507[a]	−1,507[a]	−1,615[a]
Per pond	−7,535[a]	−7,535[a]	−8,073[a]
Per phase	−67,816[a]	−113,026[a]	−230,089[a]

(continued)

Table 15. *Continued*

Item	Phase		
	I	II	III
P. monodon			
Head-on yield			
Per acre	4,609	4,609	4,938
Per pond	23,045	23,045	24,691
Per phase	23,045	115,225	234,565
% sold head-on to processor	100	100	100
Calculation of head-on farm-gate price			
Percentage tail wt.	60	60	60
Average tail count (#/lb)	25	25	25
Wholesale value of tails/lb	0	0	0
Processing expenses (per lb)	0.80	0.80	0.80
Calculation farm-gate value (head-on)			
Per pound	−0.48[a]	−0.48[a]	−0.48[a]
Per acre	−2,223[a]	−2,223[a]	−2,382[a]
Per pond	−11,117[a]	−11,117[a]	−11,911[a]
Per phase	−11,117[a]	−55,585[a]	−113,154[a]
Average head-on yield of *L. vannamei* and *P. monodon*			
Per acre	3,273	3,495	3,745
Per pond	16,363	17,476	18,725
Total yield of *L. vannamei* and *P. monodon*	163,625	349,525	711,533
Average value of *L. vannamei* and *P. monodon*			
Per acre	−1,579[a]	−1,686[a]	−1,807[a]
Per pond	−7,893[a]	−8,431[a]	−9,033[a]
Total value of *L. vannamei* and *P. monodon*	−78,933[a]	−168,611[a]	−343,244[a]
Pond Fertilization			
Initial treatment (gal/acre)	1	1	1
Monthly Treatments (gal/acre)	0.5	0.5	0.5
Total gallons required			
Per acre	3.5	3.5	3.5
Per pond	17.5	17.5	17.5
Per phase	175	350	665
Cost per gallon	1.00	1.00	1.00
Total cost			
Per acre	4	4	4
Per pond	18	18	18
Per phase	175	350	665
Electricity			
Cost/KWH	0.073	0.073	0.073
Pumping cost			
Average daily exchange rate (%)	12.17	12.17	12.17
Pond volume (gal/acre)	2,097,089	2,097,089	2,097,089
Average daily GPMs/acre	255,216	255,216	255,216
Horsepower/pump	115	115	115
Estimate GPM flow/pump	16,000	16,000	16,000
Pumping hours/acre/day	0.27	0.27	0.27
KW requirements/HP	0.9,960,662	0.9,960,662	0.9,960,662
KWHs of pumping/acre/day	30	30	30
KWHs/acre/growing season	5,025	5,025	5,025
Cost/Growing Season			
Per acre	367	367	367
Per pond	1,834	1,834	1,834
Per phase	18,340	36,680	69,692
Aeration costs			
Average HP operating/acre/day	1.4	1.4	1.4
Average Hours operating/acre/day	17.7	17.7	17.7
Average HP hours/acre/day	25	25	25
Average KWHs/acre/day	25	25	25
KWHs/acre/crop	4,073	4,073	4,073
Cost/growing season			
Per acre	297	297	297
Per pond	1,487	1,487	1,487
Per phase	14,865	29,730	56,487

Table 15. *Continued*

Item	Phase		
	I	II	III
Cost of pumping and aeration	33,205	66,410	126,179
Production overhead			
Insurance	7,500	7,500	15,000
Maintenance	25,000	25,000	50,000
Depreciation	35,000	35,000	70,000
Subtotal	67,500	67,500	135,000
Labor requirements			
Management			
General manager			
Responsibilities			
Major production decisions			
Supervise management staff			
Interact with investors			
Legal and accounting activity			
Approve purchases			
Negotiate contracts			
PLs			
Feed			
Labor			
Processing and marketing			
Number needed	1	1	1
Annual salary/individual	40,000	45,000	50,000
Subtotal	40,000	45,000	50,000
Pond manager			
Responsibilities			
Oversee status of ponds			
Supervise assistant pond managers			
Interact with general manager			
Summarize production data			
Water quality			
Growth rates			
Survival estimates			
Anticipate supply needs			
Personnel recommendations			
Number needed	1	1	1
Annual salary/individual	30,000	30,000	35,000
Subtotal	30,000	30,000	35,000
Assistant pond manager			
Responsibilities			
Day-to-day pond decisions			
Feeding rates			
Fertilization			
Water flow			
Aerator usage			
Adjust feeding rates			
Supervise technicians			
Interact with pond manager			
Number needed	0	0	1
Annual salary/individual	25,000	25,000	25,000
Subtotal	0	0	25,000
Management subtotal	70,000	75,000	110,000
Direct labor			
Biologists			
Responsibilities			
Day-to-day pond management			
Measure water quality	10	10	10
Operate pump station	5	5	5
Adjust pond water flow	10	10	10
Operate aerators	5	5	5
Cast net sampling	10	10	10

(continued)

Table 15. *Continued*

Item	Phase		
	I	II	III
Supervise laborers	2	2	2
Interact with assistant pond manager	2	2	2
Total hours/week	51	103	195
Hours expected/individual	50	50	50
Number of biologists needed	0	0	3
Annual salary/individual	18,000	18,000	18,000
Subtotal	0	0	54,000
Technicians			
Responsibilities			
Day-to-day activities			
Clean screens	10	10	10
Check feeding trays	10	10	10
Distribute feed	30	30	30
Mow Grass	2	2	2
Interact with biologists	2	2	2
Subtotal: hours/day	9	18	34.2
Weekly activities			
Maintain vehicles	2	2	4
Clean office	2	2	4
Subtotal: hours/week	4	4	8
Total hours/week	67	130	247
Hours expected/individual	50	50	50
Number/growing season	2	3	5
Seasonal activities			
Assistant with Pond harvest	2	2	4
Total man-months required	14	20	34
Monthly salary/individual	700	700	700
Subtotal	9,800	14,000	23,800
Direct labor subtotal	9,800	14,000	77,800
Clerical labor			
Secretary/receptionist			
Responsibilities			
Answer phone			
Type correspondence			
Input data			
Conduct tours			
Number needed	0	0	1
Annual salary/individual	9,000	9,000	9,000
Subtotal	0	0	9,000
Bookkeeper			
Responsibilities			
Establish accounts			
Pay bills			
Maintain budgets			
Prepare financial reports			
Number needed	1	1	1
Annual salary/indiv.	18,000	18,000	18,000
Subtotal	18,000	18,000	18,000
Subtotal for clerical labor	18,000	18,000	27,000
Total annual labor cost	97,800	107,000	214,800
General and administrative expenses			
Office supplies	1,000	1,000	2,000
Telephone	1,750	1,750	2,400
Legal	2,400	2,400	3,600
Accounting	2,400	2,400	3,600
Fringe benefits	8,000	8,000	12,000
Subtotal	15,550	15,550	23,600

Table 15. *Continued*

Item	Phase I	Phase II	Phase III
Summary of revenue and operating expenses			
Total value of *L. vannamei* and *P. monodon*	−78,933[a]	−168,611[a]	−343,244[a]
Total cost of postlarvae	55,000	120,000	213,750
Total feed cost	110,447	223,277	422,651
Fertilizer cost	175	350	665
Cost of pumping and aeration	33,205	66,410	126,179
Production overhead	67,500	67,500	135,000
Total labor cost	97,800	107,000	214,800
General and administrative costs	15,550	15,550	23,600
Total operating expenses	379,677	600,087	1,136,645
Net income	−458,610[a]	−768,698[a]	−1,479,888[a]
Corporate taxes (35%)[b]	−160,513[a]	−269,044[a]	−517,961[a]
Net Income after taxes	−298,096[a]	−499,654[a]	−961,927[a]
Net margin after taxes	3.78	2.96	2.80
Breakeven price (US$/lb or 0.45 kg)	2.32	1.72	1.60

[a]Negative values the first year of the 3 phases are typical, but net profits would be shown during later years of operation after capital investment is recovered.
[b]Very high tax rate.

Table 16. Example of Shrimp Feeding Regime[a]

Production Week	Feed Size (g)	Body Weight per Shrimp (1.5×10^6)	Number Surviving	Percentage Survival	Percent Body Weight Fed Daily		Feed (g) Each Shrimp /Day	Total lb Feed/Day	lb Feed Each Feeding
Week 1	Fry 2	0.005	1,500,000	100	20	0.001	1,500	3.3	1 lb
Week 2	Fry 2	0.05	1,425,000	95	20	0.01	14,250 (or less)	31 (or less)	10
Week 3	Fry 2	0.5	1,350,000	90	18	0.09	121,500 (or less)	267 (or less)	89
Week 4	Fry 2	0.75	1,275,000	85	16	0.12	153,000 (or less)	337 (or less)	112
Week 5	Fry 2	1.0 (g)	1,200,000	80	15	0.15	180,000	396	132
Week 6	Pellet (1 × 1.5 mm)	2 g	1,170,000	78	13.5	0.27	315,900	695.5	231
Week 7	Pellet (1.5 × 2.5 mm)	3 g	1,140,000	76	11.5	0.34	393,300	866	288
Week 8	Pellet (1.5 × 2.5 mm)	4 g	1,110,000	74	9.8	0.392	435,120	958	319
Week 9	Pellet (1.5 × 2.5 mm)	6 g	1,110,000	72	8.1	0.486	524,880	1156	385
Week 10	Pellet (2.5 × 8–10 mm)	8 g	1,050,000	70	6.4	0.512	537,600	1184	394
Week 11	Pellet (2.5 × 8–10 mm)	10 g	1,035,000	69	5.5	0.55	569,250	1253	417
Week 12	Pellet (2.5 × 8–10 mm)	12 g	1,020,000	68	4.7	0.564	575,280	1267	422
Week 13	Pellet (3.2 × 13–16 mm)	14 g	1,005,000	67	4.2	0.588	590,940	1301	433
Week 14	Pellet (3.2 × 13–16 mm)	16 g	990,000	66	3.8	0.608	601,920	1325	441
Week 15	Pellet (3.2 × 13–16 mm)	18 g	975,000	65	3.5	0.63	614,250	1352	450
Week 16	Pellet (3.2 × 13–16 mm)	20 g	975,000	65	3.2	0.64	624,000	1374	458

[a]Author's recommended feeding regime for *L. vannamei* (Pacific White Shrimp), 1.5 million stocked and held in small nursery for 1 month and moved to growout 113 days, 65% sur., 20 g animals, 975,000 × 20 g = 19,500,000 g ÷ 454 g = 42,951 lb ÷ 7 ac = 6,135 lb/ac production (actually tested at Plantation Seafoods, Post Lavaca, Texas.)

Table 17. Optimum Water Parameters for Penaeid Shrimp Culture: How Your Analyses Compare (all ions in ppm)

	Source								Ranges Found in Actual Samples by Author	How Your Sample Compares
	Liu, 1989	MPEDA (India)	Chien, 1992	Boyd, 1992	Chiang et al., 1989	Clifford, 1985	New, 1990 (optimum)	New, 1990 (range)		
Ammonia (Total)		<1			<3	1–40[a]	0.1	1.0	1–40	
Arsenic as AsO_4H_5										
Bicarbonate (HCO_3)										
Bicarbonate alkalinity ($CaCO_3$)									19.7–200	
Biological oxygen demand (BOD_5)									1.5–8.5	
Boron (B)										
Bromide (Br^-)										
Cadmium (Cd)	<3 ppb	<.015					<0.15		0.15–3 ppb = 3,000 ppm	
Calcium (Ca)										
Carbonate (CO_3)										
Chloride (Cl)										
Chromium (Cr)	<0.05								<0.05	
Cobalt (Co)										
Copper (Cu)	<0.1	<0.1					<0.1		<0.006–<0.1	
Depth of water growout	23–59″ 60–150 cm								23–59″	
Dissolved oxygen	>5	3–12	4		7.5	2–3 growth	4–7	3–12	2–12	
Dissolved solids: Residue on evaporation at 180 °C calculated									7,405	
Fluoride (F)									0.69	
Free CO_2	<2.0								<2.0	
Hydrogen sulfide	<0.02	<0.25					0	0.25	0.02–0.25[b]	
Iron (Fe)	<1.0	<0.01					<0.01		<0.01–<1.0	
Lead (Pb)	<0.03								<0.03	
Magnesium (Mg)									>300	
Manganese (Mn)										
Mercury (Hg)	<0.05	<0.0025					<0.0025		<0.0025–<0.05	
Nitrate (NO_3)									2.0	
Nitrite	<0.25	<0.25			<0.25		0	0.25	0.25	
Nitrogen (N)	<110%									
Noncarbonate hardness as $CaCO_3$										
pH	8–8.5	7.5–8.7	7.5–8.5		7.5–9	7–11	8–8.5	7.5–8.7	7–11	
Phosphate (PO_4)									0.05	
Potassium (K)										

Salinity (ppt) (growout)	10–25	15–35	10–25		25 (PL12)	0–40	15–25	10–35	0–47
Salinity (ppt) (hatchery)	32–35				20–25 (PL20)				
Secchi disk						8–10″	30–40 cm	25–60 cm	25–60 cm
Selenium (Se)									
Silica (SiO_2)									
Sodium (Na)									
Specific conductance (micromhes at 25 °C)									14,810
Strontium (Sr)									
Sulfate (SO_4)									700
Sulfide					<0.002				<0.002
Temperature (C°)	28–33	26–33				23–34	29–30	26–33	17–34
Total hardness as $CaCO_3$	20–200			50–150					20–200
Total solid content									7,405
Turbidity, cm	20–70	25–60	30–40 (summer) 20–30 (winter)				20–30		20–70
Undissociated ammonia	<0.1	<0.25			<0.1		0	0.25	0.1–0.25
Unionized H_2S			0.005						0.005
Zinc (Zn^{2+})	<0.25	<0.25					<0.25		<0.25 (soft water)

[a]Depends upon pH, temperature, and salinity.
[b]Evidence of odor from decaying organic matter, which could result in the production of hydrogen sulfide, should be noted and further investigated.

Table 18a. Seawater Analyses Compiled from Around the World for Comparison (all ions in ppm unless noted)

	Hawaii[a]	Guam[a]	Sabah Malaysia[a]	Normal Seawater[b]	Normal Surface Seawater[c]	Qingdao China[d]	Normal Seawater[e]	Fumba Zanaibar Africa[f]	Bumbwini Zanzibar Africa[g]	Godavari Estuary India[h]	Borewell Water Bhaira-Valanka India[h]	Ponnada India Hatchery[h]	Saveru Island India Growout[h]	Godavari River India[h]	Elements in Seawater[i]	Demak Indo-nesia[j]	Demak Indo-nesia[k]	Lake Kollura Area India[l]	Estuary on Bay of Bengal, India
Ammonia as NH_3											48.6*	0.15	0.45						
Ammonia as NH_4											51.4*	0.16	0.47						
Ammoniacal nitrogen			0.02																8.6*
Arsenic (As)									1*		<0.005		<0.005		0.003			<0.005	
Bicarbonate (HCO_3)	162	160					142												488*
Bicarbonate alkalinity as $CaCO_3$	133	117	109								516*	120/0	144						400*
BOD5 (ppm)																0.41	0.42		13.1*
Boron							4.5								4.6				
Bromide (Br^-)							67.4								65				
Cadmium (Cd)			0.05		0.01	0.01–0.13		0.01	0.01						0.00011	0.03	0.048	<0.006	0.004
Calcium (Ca)	390	450		400			412			2,960*	224		960 272		400			663	192
Calcium hardness as $CaCO_3$ mg/L			20								560	1,040	680						1,920
Carbonate (CO_3)													1,485/ 158						0
Chloride (Cl)	19,000	17,460	16,300	18,978			19,344			20,561	18,293	18,805/ 19,994	17,515 12,305	19,852	19,000			19,700	6,720*
Chromium (Cr^{6+})			0.03								<0.001		0.03		0.00005			0	—
Cobalt (Co)			0.24												0.00027				
Copper (Cu)			0.03		0.04–0.1	0.3–2.1*		0.08	0.09		0.03		0.04		0.003			0.02	0.02
Cyanide as Cn^-																		0	
Dissolved oxygen (ppm)			4.3					5.0	4.9			2.1							
Dissolved solids			33,400					17,000	39,000		31,114	31,504 35,692	22,468					42,900	9,702*
Fluoride (F)	1	0.7					1.3				0.3	0.3	0.2		1.3			1.15	0.3
Carbon dioxide (CO_2)											4.7	7.0	18.8*						
Hardness as $CaCO_3$	6,120	5,720	5,300							8,600	4,000	6,800 5,800	6,640 3,920	6,200				4,368	1,920*
Hydrogen sulfide											0.01	0.002	0.003			0.19*	0.29*		0
Iron (Fe) (Total)		0.3*	0.23*	0.02				0.25–0.26*	0.30*	0.1	0.03	0	0.02	0.001	0.01			0.01	0.03
Iron–ferrous (Fe^{2+})										0.1	<0.005	0	<0.005	0.1	0.01	0.15	.12		
Lead (Pb)			0.14		0.005– 0.01	0.1–2.0*		0.08–0.11	0.08	0.16	<0.03	0.12	<0.03 0.09	0.13	0.00003	0.26*	0.44*	0.07	0.23*
Magnesium (Mg)	1,250	1,175		1,272			1,294	837–879	900–940		0.77		0.12		1,350			659	351
Manganese (Mn)		0.06	0.04	0.01						576	1,000	1,459	980 1,530	1,440	0.002			0.1	0.8
Mercury (Hg)										0	<0.001	0	<0.001 0	0	0.000008			<0.001	0

Nitrate (NO_3)		3.5	0.23							13.3	0.9	0/3.6	13.3 0	5.0				9.8	17.7*
Nitrite			0.01								0	0.004	0.003						0.01
Noncarbonate hardness as $CaCO_3$	5,990	73,000*								4,330		6,068	2,960	1,320					
pH[m]	6.6*	7.7	7.4	7.4				8.0	7.3	7.86	7.3	7.8–7.96	7.7	7.99					7.84
Phosphate (PO_4)											<0.1	0.1	<0.1		0.07	1.5	3.7		1.2
Potassium (K)	418	350		380			399			400	220	450	300 220	350	380				82*
Residue on evaporation at 180 °C calculated	34,400		34,200																
Salinity (ppt)			30	35				34	33	9.27/9.02	32.2	33.2/9	21.9 7.65	9.95					8.51
Selenium (Se)											<0.005		<0.005		0.00009			<0.001	
Silica (SiO2)	14*	1.5	Trace								16*		7		3				6.5
Sodium (Na)	10,500	9,400		10,556			10,773			27,600	7,500	21,000	26,000 14,600	16,000	10,500				2,400*
Specific conductance (micromhes at 25 °C)	51,000	73,000						69,300	62,700	65,700	63,300	71,300/ 65,700	45,000						
Strontium							7.9								8.1				
Sulfate (SO_4)	2,780	2,510	1,600	2,648			2,712	2,822	2,769	2,075	1709	2,517/ 1,940	158	1,775	885			270	615
Suspended solids			1,150									35,692							472
Total coliform (MPN/100 mL)																		7	
Total solid content			34,500																
Turbidity, NTU			8.5					50 cm	40 cm	20	564	64/ 10/25	198/0			15		56	1,940*
Zinc (Zn)			0.06		0.1	0.9–3.2				0.18	<0.005	0.16	<0.005 0.18	0.08	0.01			<0.005	0.4

[a]Fujimura, 1989.
[b]Sverdrup, et al., 1970.
[c]Förstner and Wittman, 1979.
[d]Yuan, et al., 1991.
[e]Riley and Chester, 1979.
[f]Treece and Yates, 1992 (unpublished) open ocean site.
[g]Treece and Yates, 1992 (unpublished) mangroves swamp-lowtide.
[h]Treece, 1992 (unpublished) feasibility study on sites in India. Note: "/" indicates separation of two samples.
[i]Spotte, 1970.
[j]Seawater from Demak, Indonesia (Sample 1).
[k]Seawater from Demak, Indonesia (Sample 2).
[l]Underground water at *Macrobrachium* hatchery.
[m]Most sources state that normal pH of seawater is 8.0 at 35 ppt (but experience indicates that a range of 7.6 to 8.2 is not uncommon).
*Indicates that the reading is out of the normal range.

Table 18b. Seawater Analyses Compiled from Around the World for Comparison (all ions in ppm unless noted)

	San Blas, Mexico (Filtered Seawater)	San Blas, Mexico (Unfiltered Seawater)	Port Mansfield, Texas, USA Seawater	Port Mansfield, Texas, USA Well Water	Laguna Agua Dulce (Pacific Coast of Mexico)	Estero El Ermita (Pacific Coast of Mexico)[g]	Fontera, TABASCO, Mexico (Gulf Coast)	Placentia Lagoon, Belize, C.A.	Parrita, Costa Rica, C.A. Farm (Intake)	Parrita, Costa Rica, C.A. Farm (Estuary)	"Normal" Seawater (Average and Lethal Limits)[a–d]
Ammonia as NH_3											
Ammonia as NH_4								0.02			0.02–0.04
Total nitrogen											
Arsenic (As)	0	0									
Bicarbonate (HCO_3)	116	123	136	77							142
Bicarbonate alkalinity as $CaCO_3$[f]							140	19.7–pond 98.1–lagoon			200[i]
BOD_5[h]							30	1.1–2.1			1.5 normal estuary 8.5 normal intensive pond
Boron	19.3*	3.8	3.8	9.7*							4.5
Bromide (BR^-)											67.4
Cadmium (Cd)	0	0			<0.005	<0.005		<0.005			0.01 (0.08–0.4 lethal)
Calcium (Ca)	536	515	500	525	197–206	77	92	111*–485	44	71	400–412
Coliform (MNP/100 ml) total/fecal							30/22				
Carbonate (CO_3)	3	3	19								
Chloride (Cl)	20,945	18,815	16,700	1,250			6,062				18,978–19,344
Chromium (Cr)											<0.05–<0.1 (2–2.0 lethal)
Cobalt (Co)											
Copper (Cu)	0	0			0.03	<0.03		<0.03–0.04			0.04–0.1 (0.3–1.0 lethal)
Cyanide as Cn^-											
Dissolved oxygen (ppm)							6.0	6.5–8.1			>5
Dissolved solids					20,816	6,298–6,333	26,000				

Fluoride (F)	0.8	0.8	0.8	0.7					<1.3
Carbon dioxide (CO_2)									<2.0 opt >20.0 harmful
Total hardness as $CaCO_3$[e]	196.5 gr/gal	178.4 gr/gal	144.6 gr/gal				4.5687*		
Hydrogen sulfide									<0.02–0.1 and low pH (lethal)
Total iron (Fe)	0	0	—	—	0.12	0.32		0.44–27*	<0.01
Iron–ferrous (Fe^{2+})									0.02 (>0.32 problems)
Lead (Pb)	0	0	—	—	<0.05	<0.05		<0.05	0.005–0.1 (1–40 lethal)
Magnesium (Mg)	1,717	1,542	1,200	125	589–592	189	26	285–1,100	1,272–1,294
Manganese (Mn)	0	0	—	—					0.01 (toxic above 1 ppm)
Mercury (Hg)									<00025 0.01–0.04 lethal
Nitrate (NO_3)	0	0	0	0			0		Not toxic 1–200
Nitrite							0		Max. 1 (ionized) 0.1 (un-ionized) 0.25
Noncarbonate hardness as $CaCO_3$							10		
pH	8.0	8.0	8.4	7.7	8.6	8.3–8.4	8.0	7.1–7.3*	7.4–8.2
Phosphate (PO_4)								<0.10–0.89	
Potassium (K)	522	483	324	28					380–399
Residue on evaporation at 180 °C, calculated									

(continued)

Table 18b. *Continued*

	San Blas, Mexico (Filtered Seawater)	San Blas, Mexico (Unfiltered Seawater)	Port Mansfield, Texas, USA Seawater)	Port Mansfield, Texas, USA Well Water	Laguna Agua Dulce (Pacific Coast of Mexico)	Estero El Ermita (Pacific Coast of Mexico)[g]	Fontera, TABASCO, Mexico (Gulf Coast)	Placentia Lagoon, Belize, C.A.	Parrita, Costa Rica, C.A. Farm (Intake)	Parrita, Costa Rica, C.A. Farm (Estuary)	"Normal" Seawater (Average and Lethal Limits)[a–d]
Salinity (ppt)											
open	35	35						32			32–36
ocean/bay								15–17			10–27
Selenium (Se)											1.5
Silica (SiO_2)							2.0				
Sodium (Na)	13,725	12,924	10,900	2150	4890–4902	1485		2288–9357			10,556–10,773
Specific conductance (micromhes at 25 °C)	71,000	65,000	57,000	13,400	38.6	12.5–12.6	44,000				
Strontium											7.9
Sulfate (SO_4)	4,207	3,740	3,498	5,301*	1,348	391–434	140	10.5–29.7			2,648–2,712
Suspended solids							28				
Temperature (°C)	28	28			28	28	30	28–30			26–33 for tropical shrimp
Total solid content							26,000				
Transparency (cm)							28				30–40 cm opt. 20–70 cm range
Turbidity, NTU					4.8	11.0					
Zinc (Zn)											

[a]Sverdrup et al., 1970.
[b]Förstner and Wittman, 1979.
[c]Riley and Chester, 1979.
[d]Miscellaneous sources.
[e]Hardness is the measurement of calcium (Ca) and magnesium (Mg) ions. Add these two and the total should be above 20 ppm for water to be considered favorable for fish culture. Total hardness (calcium carbonate or $CaCO_3$) should be between 50–150 ppm for best results according to Boyd, 1992; between 20–200 according to isc. sources and 20 ppm according to Johnson (SRAC pub. #A0901).
[f]Salinity dependent.
[g]Estuary with river inflow.
[h]BODS–1.5 ppm is normal for estuary and 8.5 is normal for intensive culture pond.
[i]Alkalinity (should be no lower than 20 ppm. Optimum range is 50–150 ppm in seawater at 35 ppt and pH 8.3).
*Indicates that the sample reading is out of the ordinary or not normal.

Table 19. Soil Types that are Conducive to Shrimp Culture in Ponds[a]

pH Acidity	Total Nitrogen	Phosphorus	Potassium	Calcium	Magnesium	Zinc	Iron	Manganese	Copper	Sodium	Sulfur
<4–>9	<0.15–>0.5%	<20–>400 ppm	<100–>1700 ppm	<1,000–>8,000 ppm	<700–>4,000 ppm	<2–>14 ppm	<60–>1200 ppm	<10–>350 ppm	<1–>11 ppm	<2500–>25,000	<0.05–>1.5
Average of five actual samples		Average of five actual samples	Average of five actual samples	Average of five actual samples	Average of five actual samples	Average of five actual samples			Average of five actual samples	Average of five actual samples	
5.74		51	695	4,921	2,860	11			6	9,464	
Comments:		Comments:	Comments:	Comments:	Comments:	Comments:			Comments:	Comments:	
Low pH[b]		Average	Average	Average	Average	Average			Average	Average	

[a]Range of 346 Brackishwater Aquaculture Pond Soil Test Analyses (Boyd et al., 1994). Average of five actual soil samples and comments about each. Results in either ppm or % if indicated.
[b]Lime can be used to adjust the soil pH to neutral (7.52). Refer to literature for suggested amounts per hectare at a certain pH.

Table 20. Optimum Soil Parameters for Saltwater Shrimp Culture: How Your Soil Analyses Compare (All Ions in ppm unless Stated Otherwise)

Parameter	Soil Analyses Ranges Found in the Literature						Ranges Found in Actual Samples Taken by Author and Others	How Your Sample Compares
	Normal Readings	High Readings	Boyd, 1994 av. of 346	Range of Boyd, 1994	Optimum Boyd, 1994	Low Reading		
Iron (Fe)		20–50	626	+/− 113		<3	60–6,768	
Manganese (Mn)			157	+/− 27			10–350	
Zinc (Zn)			7.9	+/− 1.05			2–14	
Copper (Cu)			7.6	+/− 7.16			1–11	
Aluminum (Al)			371	+/− 43			100–600	
Barium			1.55	+/− 0.12			0.5–3.5	
Magnesium (Mg)			2,258	+/− 142			700–4,000	
Calcium (Ca)			3,450	+/− 363			1,000–8,000	
Silica (Si)			316	+/− 46			30–750	
Potassium (K)			822	+/− 61	30–60		100–1,700	
Sodium (Na)			15,153	+/− 2,980			2,500–25,000	
Boron (B)			31	+/− 21			4–24	
Cobalt			2.35	+/− 0.42			0.5–3.5	
Molybdenum			0.94	+/− 0.33			0.3–1.2	
Chromium (Cr)			3.35	+/− 0.35			1–7	
Lead (Pb)			5.38	+/− 3.71			2–9	
Sulfur (Su)	<10 ppm		0.49%	+/− 0.06%	[a]		0.05–1.5%	%
pH			6.5	+/− 1.49	6.5–7.5		4–9	
Organic Carbon (%)		<2.5%	1.79%	+/− 0.16%	0.5–2.5%		0.5–4%	%
Nitrogen (%)			0.30%	+/− 0.01%			0.15–0.5%	
Pyrite		<0.48%						
Available nitrogen					250–750			
Phosphorus							20–400	
Arsenic	<1	>1					0–2	

[a] Sulfur in soils above 0.75% indicates potential acid sulfate soil; Average of 346 ponds was 0.49%. Potential acid-sulfate soils often have 1–5% total sulfur.

Figure 41. *Macrobrachium rosenbergii* male (note long pinchers, claws, or chela).

in the United States during the past 40 years. The information in this section, except where noted, was taken directly from or summarized from D'Abramo and Brunson (123). Other species of *Macrobrachium* are indigenous to the United States, but none are as suitable for aquaculture as the Malaysian prawn. The other species (*M. acanthurus*, *M. carcinus*, *M. ohione*, and *M. olfersii*) do not reach sizes that are considered desirable for aquaculture (124). Basic production techniques were developed in the late 1950s in Malaysia and refined in Hawaii and Israel during the past 30 years. Mistakidis (124) published an excellent biological account of the freshwater shrimp, with line drawings of eggs and larval stages.

A major breakthrough occurred when Fujimura and Okamoto at the University of Hawaii made the mass production of PL possible in 1970 (125). Once this bottleneck was removed, freshwater shrimp culture began to spread to areas such as Mauritius, French Polynesia (126), Israel, and the state of Florida. Weyerhaeuser, in Florida, started an R&D program in 1974; soon after, other countries started similar programs (Puerto Rico in 1975, Martinique, French West Indies in 1977, Jamaica and the Dominican Republic in 1978, Central America in 1979 and Brazil in 1981). Bardach et al. wrote a classic paper on the species in 1972 (127).

At the same time, Weyerhaeuser, in Florida, and companies in other countries were developing R&D

programs. The states of South Carolina, Texas, and Louisiana conducted research into basic production techniques, as well as marketing, processing, and hatchery procedures. In 1974, Sun Oil Company established a pilot freshwater shrimp-farming company (Aquaprawns, Inc.) near Brownsville, Texas. The firm developed several new techniques to cultivate freshwater and saltwater shrimp, including the use of a harvest pump (128). In 1977, Hanson and Goodwin (129) reviewed the culture practices developed for *M. rosenbergii*, and S.K. Johnson described the diseases found in the species (130). In 1978, Sun Oil Co. closed its non-petroleum-related subsidiaries, and a new company (CSCI), was formed. In 1980, CSCI built a 27.2-ha (68-ac) freshwater shrimp farm near Los Fresnos, Texas. The operation was located several miles inland, but used saline groundwater to operate the hatchery.

Even though it is called a freshwater shrimp, a certain part of its life cycle is spent in saltwater. The natural life cycle involves the adult shrimp migrating down rivers toward estuaries to have their young, and juveniles return to the rivers to complete the cycle. Freshwater shrimp require brackish water (12–15 ppt) for larval development and can tolerate up to 5 ppt during growout. They are tropical in temperature requirements and do not do well in water temperature below 10 °C (50 °F). In 1981, population profile development and morphotypic differentiation in the species was described (131), and a fact sheet was published on the culture of the species (132).

In 1983, Aquaculture Enterprises, Inc. acquired an unsuccessful prawn farm in Puerto Rico (Shrimps Unltd., Inc.), and John Glude restarted the farm at Sabana Grande on the southwest coast. It experienced a large debt service and construction delays for five years before it became what was considered an economically viable size in 1988 (58 ha or 143.3 ac). The Weyerhaeuser technology was inappropriate for the environment in Puerto Rico, and production failed to achieve projected levels. A change in production strategy, termed the "Modified Batch System" was developed and tested in 1989 and 1990, and a production rate of 3,000 kg/ha (3,000 lb/ac) per year was achieved. A disease called the "white PL disease," caused by Rickettsia, hit the company hard, while a recession in the United States caused a drop in demand for the product, and the company was forced to put production on hold in 1992 (133). Additionally, by 1992, inexpensive Taiwanese frozen shrimp had appeared in the world market at US$10/kg (US$4.54/lb) and created fierce competition in the industry. Many producers' costs were higher than the shrimp were bringing on the market.

In Texas, CSCI produced large amounts of shrimp, but closed in 1985, unable to find a large, high-value market for the product. At least three companies in Texas produced and sold freshwater shrimp. In south Texas, Sweet Water Aqua-Farms, Inc. reopened the CSCI farm in 1989, raised freshwater shrimp, and distributed nationwide. Sweet Water had been marketing the Malaysian prawn for a number of years from Brooklyn, New York, and decided to move into production. They sold mainly to "white table cloth" restaurants, had a toll-free telephone number, and had an agreement with one of its investors (Federal Express) for overnight delivery of the product. In 1990, Sweet Water Aqua-Farms, Inc. produced 544 kg (1,200 lb) of *Cherax* (Australian red claw) and 9,979 kg (22,000 lb) of freshwater shrimp. Although the numbers are not large when compared with the Texas marine shrimp production, the product was marketed as a specialty item for a niche market. The freshwater shrimp were shipped fresh (killed and heads-on at 1–3 °C or 35–38 °F). The farm suffered the loss of 9,979 kg (22,000 lb) of shrimp (valued at US$254,540) due to an Arctic cold front in November, 1991. The owner was able to save the broodstock and still made a little money, despite the freeze loss. The farm relocated to Puerto Rico, where year-round production was possible without the threat of cold weather.

Freshwater shrimp PL are quite expensive relative to saltwater shrimp. Even in the 1980s, freshwater shrimp PL cost between US$25–50/1000. Now they are generally selling for US$50–65/1000. By comparison, saltwater shrimp PL sell for US$8–10/1000 if purchased in the United States and US$6–7/1000 purchased from Central or South America or in large quantities. Part of the reason for the price difference is the larval cycle is longer for freshwater shrimp (25–45 days), whereas the saltwater shrimp hatchery cycle generally takes 18 days, or even less if the temperature is raised above 28 °C (82.4 °F).

Cannibalism has been a major problem in the freshwater shrimp industry. As is the case in most crustaceans, the larger shrimp prey upon the smaller ones. Producers of freshwater shrimp provide habitat or hiding places, vegetation in the pond, and harvest the larger animals routinely to minimize the problem. Another problem faced by the freshwater shrimp industry is that two-thirds of the animal is head and one-third tail muscle. Most producers are forced to sell the product fresh and with the head on. Digestive enzymes in the cephalothorax cause of the muscles to deteriorate if not properly handled after harvest, and producers say that the animal cannot be held on ice very long because the shells becomes soft.

During recent years, new management practices have dramatically increased the potential for economic success of freshwater shrimp culture in the southern United States. Research efforts have been complemented by demonstration projects designed to evaluate methods under large-scale commercial conditions. Freshwater prawns, like all crustaceans, have an exoskeleton or shell that must be shed regularly in order for growth to occur. As crustaceans grow, they shed the shell, or molt, and weight and size increase occur soon after each molt. When crustaceans molt, they have approximately 12% more water in their bodies, and they are soft, lethargic, and subject to attack by others. Because of these periodic molts, growth occurs in distinct increments, rather than on a continuous basis.

Females generally become reproductively mature before six months of age. Mating can occur only between hard-shelled males and freshly molted females. The male deposits sperm into a gelatinous mass that is held on the ventral side of the female, between the fourth pair of walking legs (123).

Eggs are laid within a few hours after mating and are fertilized by the sperm contained in the gelatinous mass. The female then transfers the fertilized eggs to the ventral tail, into a brood chamber, where they are kept aerated and cleaned by movement of the abdominal swimming appendages (pleopods). The eggs remain attached to the abdomen until they hatch. As in saltwater shrimp, the number of eggs produced in each spawn is directly proportional to the size of the female. As long as water temperature exceeds 21 °C (70 °F), multiple spawns per female can occur annually, and eyestalk ablation (see the entry "Eyestalk ablation") is not necessary with the freshwater shrimp, as it is for commercial production of saltwater shrimp. Females carrying eggs, or berried females, are easily recognized by the bright yellow to orange color of newly spawned eggs, which gradually change to orange, then brown, and finally gray a few days before hatching. At 28 °C (82 °F), the eggs hatch approximately 20 days after spawning.

After hatching, larvae are released and swim upside down, tail first, like the mysis stage of saltwater shrimp. The larvae cannot survive in freshwater more than 48 hours and survive best in brackish water of salinity 9–19 ppt. As larvae grow, they become aggressive sight feeders and feed almost continuously, primarily on small zooplankton, worms, and larval stages of other aquatic invertebrates. Daniels et al. (134) found that larval feeding habits could be modified or improved with light manipulations and by keeping bacterial counts low. That work can be summarized as follows: In a 30-day larval cycle, 60 larvae/L (or per 0.26 gal) stock, 80% survival after 30 days; allow one week for wash down, dry out, and disinfecting; building 9.1 m × 27.4 m (29.8 ft × 89.9 ft); 6.3 million larvae for two runs, 3.15 million each run; large center drain for cleaning, disinfecting; 3.1 m × 3.1 m (10 ft × 10 ft) required to house blowers; six, 11,000 L (2,860 gal), conical tanks, not to exceed 1% slope; 6% volume of rearing tanks should be the biofilter's size 3,960/L (1,029 gal); five tanks used for conditioning water, saltwater storage, conditioning of filter media; broodstock held at 4 per m^2 (4 per 10.76 ft^2) in intense light; from juveniles on, growout in freshwater or below 5 ppt salinity is best; ponds should be stocked at 39,520/ha (16,000/ac), a higher stocking, level will stunt growth; larvae salinity range 7–15 ppt, 12–15 ppt is acceptable, but 7–10 ppt is best; larvae stocked at 50 to 60/L (0.26 gal), up to 80–90/L (0.26 gal), should result in 90% survival.

Most larvae are fed *Artemia* nauplii throughout the hatchery phase, up to 45 days. Fuller et al. (135) looked at the economics of operating a closed, recirculating "clearwater" hatchery for the commercial production of PL.

During the hatchery period, larvae undergo 11 molts, each representing a different stage of metamorphosis. Following the last molt, larvae transform into PL. Transformation from newly hatched larvae to PL requires 15 to 40 days, depending upon food quantity and quality, temperature, and a variety of other water quality variables. Optimum temperatures for growth are 28–31 °C (82.4–87.8 °F).

After metamorphosis to PL, the shrimp resemble miniature adults, about 7 to 10 mm (0.3 to 0.4 in.) long and weigh 6 to 9 mg (50,000 to 76,000/lb). PL change from planktonic to benthic crawling individuals. When they do swim, they move like adults with the dorsal side up and swim or crawl in a forward direction. PL can tolerate a range of salinities and migrate to freshwater upon transformation. In addition to the types of food they consume as larvae, larger pieces of animal and plant materials are ingested. The diet includes larval and adult insects, algae, molluscs, worms, fish, and feces of fish and other animals. At high densities, or under conditions of limited food, freshwater shrimp become cannibalistic. PL are translucent and may have a light orange or pink head. As they change to the juvenile stage, they take on the bluish to brownish color of the adult stage. PL are juveniles, but through common usage, the term juvenile is reserved for the stage between PL and adult; however, no standard definition for the juvenile stage exists.

Older juveniles and adults usually have a distinctive blue-green color, although sometimes they may take on a brownish hue. Color is usually the result of the quality and type of diet. Adult males (Fig. 41) are larger than the females (Fig. 42), and the sexes are easily distinguishable. The claws (chela) and the head region of males are larger than those of the females. The base of the fifth or last pair of walking legs (periopods) of males is expanded inward to form a flap or clear bubble that covers the opening (gonopore) through which sperm are released. The walking legs of males are set close together in nearly parallel lines, with little open space between them, which helps distinguish immature males from females. A wide gap exists between the last pair of walking legs in females, and they have a genital opening at the base of the third pair of walking legs. Three types of males have been identified, based upon external characteristics. Blue claw males are easily distinguishable and are characterized by long, spiny blue claws (Fig. 41). Eventually, the male will either die or molt and return to a growth phase and later regain its blue claw status. Two other classes of non-blue claw males exist, orange-claw and strong orange claw males (123).

There are three phases of culture of freshwater shrimp: hatchery, nursery, and growout. For detailed information on the pond growout phase, refer to (136). Those contemplating starting a freshwater shrimp production enterprise should forego the hatchery phase at least initially and possibly the nursery phase by purchasing

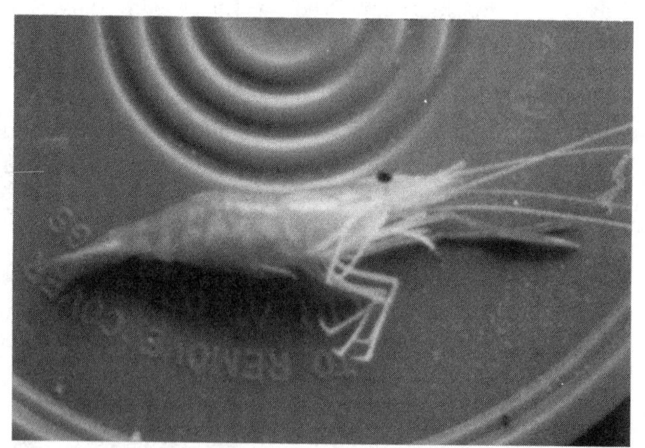

Figure 42. *M. rosenbergii* female (note eggs on underside of tail).

juveniles from a supplier (137). As production increases through successful pond growout, plans can be made to develop a nursery and possibly a hatchery. There is a limited number of juvenile shrimp suppliers, but increased demands will lead to a need for more enterprises that deal exclusively in the production and sale of PL.

Ponds should have a minimum depth of 0.6 to 0.9 m (2 to 3 ft) at the shallow end and a maximum depth of 1 to 1.5 m (3.5 to 5 ft) at the deep end. The slope of the bottom should allow for rapid draining. Publications that provide additional information on pond design and construction are available on the Internet at http://www.msstate.edu/dept/srac/fslist.htm.

A soil sample should be collected from the pond bottom to determine whether lime is needed. Take soil samples from about six different places in the pond, and mix them together to make a composite sample that is then air dried. Send the soil sample to a soil testing laboratory and request a lime requirement test. There may be a small charge for this service. If the pH of the soil is less than 6.5, you must add agricultural limestone to increase the pH to a minimum of 6.5, and preferably 6.8.

The final phase of freshwater shrimp production is growout of juveniles to adults for market as a food product. Research in Mississippi, Kentucky, and other southern states in the United States has demonstrated that this can be a profitable enterprise (136–138). Unless you have a hatchery/nursery, you must purchase juveniles for the pond growout phase.

Shipping costs can be minimized if the hatcheries are located within a one-day driving distance of the growout facility. Otherwise, it is best to have the shrimp shipped via air or express courier, but this significantly increases the cost.

Ponds used for raising freshwater shrimp should have many of the same basic features of ponds used for the culture of channel catfish (136). A good supply of freshwater is important, and the soil must have excellent water retention qualities. Well water of acceptable quality is the preferred water source for raising freshwater shrimp. Surface runoff water from rivers, streams, and reservoirs can be used, but quality and quantity can be highly variable and subject to uncontrollable change. The quality of the water source should be evaluated before any site is selected (see the entry "Site selection").

The surface area of growout ponds ideally should range from 0.4 to 2 ha (1 to 5 ac), but larger ponds have been successfully used. The pond should be rectangular to facilitate distribution of feed. The bottom of the pond should be smooth and free of obstructions to seining (136). After filling the pond, fertilize it to provide an abundance of natural food organisms for the shrimp and to shade unwanted aquatic weeds (see the entry "Fertilization of fish ponds"). If a water source other than well water is used, it is critically important to prevent fish, particularly members of the sunfish family (e.g., bass, bluegills, and green sunfish) from getting into the pond when it is filled. Screening or filtering the incoming water is advised if it is not from a well. The effects of predation by these kinds of fish can be devastating. Birds, especially cormorants and hingas can also be a problem. (See the entry "Predators and pests.") If there are fish in the pond, remove them before stocking shrimp, using 0.95/L (0.25 gal or 1 qt) of 5% liquid emulsifiable rotenone per acre-foot or 1,233 cu m (325,900 gal) of water when water temperatures exceed 21 °C (70 °F). Rotenone is a restricted use pesticide, and either a commercial or private pesticide applicator license is required to purchase and apply this material in the United States.

Juvenile freshwater shrimp must be gradually acclimated to conditions in the growout pond to prevent temperature shock or other types of stress. The water in which they will be stocked should gradually replace water in which PL and juveniles are transported. This acclimation procedure should not be attempted until the temperature difference between the transport and culture water is less than 2 to 4 degrees. The temperature of the pond water at stocking should be consistently at least 20 °C (68 °F) to avoid stress because of low temperatures. Juvenile freshwater shrimp are more susceptible than are adults to low water temperature exposure. Juveniles, preferably derived from populations that have been size graded, ranging in weight from 0.1–0.3 g (0.0002–0.0006 lb), should be stocked at densities of 29,640–39,520 shrimp/ha (12,000–16,000 shrimp/ac). The size grading results in more uniform growth and helps to reduce the percentage of smaller, nonmarketable individuals at harvest. Lower stocking densities will yield larger shrimp, but lower total harvested weight. If the market demands whole, live or fresh ice-packed shrimp, stocking at lower densities will result in larger, more marketable individuals. The duration of growout depends on the water temperature, and the time generally is 120 to 180 days in the southern United States. Freshwater shrimp could be grown year-round if a water source is found that provides a sufficiently warm temperature for growth (136).

Juveniles stocked into growout ponds are able to initially obtain sufficient nutrition from natural pond organisms. At the stocking densities recommended by D'Abramo and Brunson (123), begin feeding when the average weight is 5.0 g (0.01 lb) or greater. See Table 21 for feeding rates for semi-intensive pond growout. Commercially available, sinking channel catfish feed (28–32% crude protein) is an effective and economical feed at the recommended stocking densities. The feeding rate is based on the mean weight of the population. A feeding schedule can be developed based on three factors: (*1*) a feed-conversion ratio of 2.5, (*2*) 1% mortality

Table 21. Weight-Dependent Feeding Rates for Semi-intensive Pond Growout of the Freshwater Shrimp (*M. rosenbergii*) (123)

Mean Wet Weight (g)	Daily Feeding Rate (% of body weight)[a]
<5	0
5 to 15	7
15 to 25	5
>25	3

[a]As-fed weight of diet/wet biomass of freshwater shrimp × 100.

in the population per week, and (*3*) mean individual weight determined from samples obtained every 3 weeks. At the end of the growout season, survival may range from 60 to 85%, if you have practiced good water quality maintenance. Yields typically range from 600 to 1,200 kg/ha (600–1,200 lb/ac). Weights of shrimp range from 35 to 45 g (10–13 shrimp/lb or 22–28 shrimp/kg). These yields and average sizes will be significantly influenced by initial stocking density.

Water quality is important in raising freshwater shrimp, as it is in raising saltwater shrimp, catfish, or any other aquatic species. Dissolved oxygen is particularly important and must be monitored several times daily, especially in the early morning hours.

Selective harvest of large shrimp during a period of 4–6 weeks before final harvest is recommended to increase total production in the pond. Selective harvesting usually is performed with a 2.54-cm to 5-cm (1 in. to 2 in.) bar-mesh seine, allowing animals that pass through the seine to remain in the pond and to continue to grow, while the larger shrimp are removed. Selective harvest may also be accomplished with properly designed traps. Shrimp can be trapped using an array of traditionally designed crawfish traps. Selective harvest can help extend the duration of the availability of the fresh or live product to the market. However, there is a lack of research to show whether selective harvesting or complete bulk harvesting is the most economical approach. Regardless of the harvest method employed, some shrimp will remain in the pond and will have to be manually picked up. Rapid draining or careful seining can minimize this residual crop. Harvested shrimp should be quickly chilled to preserve the integrity of the muscle tissue, thus maintaining a firm, high quality texture. The product may be marketed fresh on ice, processed and frozen, or frozen whole for storage and shipment (123).

Culture of freshwater shrimp in combination with fingerling catfish has been successfully demonstrated under small-scale experimental conditions and appears possible under commercial conditions. Before introducing catfish fry, D'Abramo and Brunson (136) recommends stocking juvenile shrimp at a rate of 7,410–12,350 per ha (3,000–5,000 per ac) and recommends stocking catfish fry at a density to insure that they will pass through a 2.54-cm (1-in.) mesh seine used to harvest the shrimp at the end of the growing season. Soft water (<7 ppm total hardness) can be expected to cause a softening of the shell. Hard water (>300 ppm) has been implicated in reduced growth and lime encrustations on freshwater shrimp.

Polyculture of channel catfish and freshwater shrimp may be best achieved through cage culture of the fish. A scheme for intercropping of freshwater shrimp and red swamp crawfish was developed and evaluated in the United States. Intercropping is the culture of two species that are stocked at different times of the year with little, if any, overlap of their growth and harvest seasons. Intercropping provides for a number of benefits that include: (*1*) minimizing competition for resources, (*2*) avoiding potential problems of species separation during or after harvest, and (*3*) spreading fixed costs of a production unit (pond) throughout the calendar year. Adult mature crawfish are stocked at a rate of 8,892 per ha (3,600 per ac) in summer (late June or early July). Juvenile shrimp are stocked at a density of 39,520 per ha (16,000 per ac) in late May and harvested from August through early October. In late February, seine harvest of the crawfish begins and continues through late June before stocking of new adult crawfish. Freshwater shrimp are small enough to pass through the mesh of the seine used to harvest crawfish during the May–June overlap period. Other intercropping scenarios involving such species as bait minnows, tilapia, and other fish species may be possible, but to date no research has been conducted in the United States (123).

Nitrites at concentrations of 1.8 ppm have caused problems in hatcheries, but there is no definitive information as to the toxicity of nitrite to shrimp in ponds. High nitrite concentrations in ponds would not be expected given the anticipated biomass of shrimp at harvest. Levels of un-ionized ammonia above 0.1 ppm in fish ponds can be detrimental. Concentrations of un-ionized ammonia as low as 0.26 ppm at a pH of 6.83 have been reported to kill 50% of the shrimp in a population in 144 hours. Therefore, every effort should be made to prevent concentrations of 0.1 ppm or higher un-ionized ammonia.

High pH can cause mortality directly through pH toxicity and indirectly because a higher percentage of the total ammonia in the water exists in the toxic, un-ionized form. Although freshwater shrimp have been raised in ponds with a pH range of 6.0 to 10.5 with no apparent short-term adverse effects, it is best to avoid a pH below 6.5 or above 9.5, if possible. Constant high pH stresses the shrimp and reduces growth rates. High pH values usually occur in waters with total alkalinity of 50 ppm or higher and when a dense algae bloom is present. Liming ponds that are built in acid soils can help minimize severe pH fluctuations. Another way to avoid problems with high pH is to reduce the quantity of algae in the pond by periodic flushing the top 30 cm (12 in.) of water. Alternatively, organic matter, such as corn, grain or rice bran, can be distributed over the surface area of the pond. This procedure must be accompanied by careful monitoring of oxygen levels, which may dramatically decrease due to the decay processes.

Other than the "white PL disease," caused by Rickettsia, discussed earlier, diseases do not appear to be a significant problem in the production of freshwater shrimp, but as densities are increased to improve production, disease problems are certain to become more prevalent. We know that white spot syndrome virus (WSSV) is spreading worldwide and affects many crustaceans.

Production levels and harvesting practices should match marketing strategies. Without this approach, financial loss due to lack of adequate storage (holding) facilities or price change is inevitable. Marketing studies strongly suggest that a "heads off" product should be avoided and that a specific market niche for whole freshwater shrimp needs to be identified and carefully developed. To establish year-round distribution of this seasonal product, freezing, preferably individually quick frozen (IQF), is an attractive form of processing. Block frozen is an alternative method of processing. Adult freshwater shrimp can be successfully live-hauled for at least 24 hours at a density of 0.22 kg/3.8 L (0.5 lb/gal) with little mortality and no observed effect on exterior quality of the product. Transport under these conditions requires good aeration. Distributing shrimp on shelves stacked vertically within the water column assists in avoiding mortality due to crowding and localized poor water quality. Using holding water with a comparatively cool temperature 20–22 °C or (68–72 °F) minimizes the incidence of water quality problems and injury by reducing the activity level of the freshwater shrimp (136).

ECONOMICS OF RAISING FRESHWATER SHRIMP

Based on an average feed cost of US$250 to US$300/907 kg (or per 2,000 lb), a cost of US$65/1,000 juveniles, a 2.5 : 1 FCR, expected mean yields of 1,000 kg/ha (1,000 lb/ac), and a pond-bank selling price of US$9.35/kg (US$4.25/lb), the expected return can be as high as US$4,940 to US$6,175/ha (US$2,000 to US$2,500 per ac). Revenue and ultimate profitability depend on the type of market that is used. This estimated return does not include labor costs or other variable costs that differ greatly from operation to operation. Some thorough economic evaluations that incorporate annual ownership and operating costs under different scenarios for a synthesized firm of 17.4 ha (43 ac), having 4 ha (10.25 ac) of water surface in production, are provided in (137).

NUTRITION

The nutritional requirements of *M. rosenbergii* were summarized and compared with species of penaeid shrimp by D'Abramo (139). Other important contributions toward our knowledge of the nutrition requirements and other aspects of these animals have been published (140–149). According to D'Abramo (139), the quantitative amino acid requirements for *M. rosenbergii* remain undefined, a situation generally attributed to the common lack of success in using crystalline sources of amino acids in shrimp diets to supplement protein sources deficient in one or more essential amino acids. In contrast, crystalline amino acids have been successfully used in investigations of amino acid requirements of fish. Farmanfarmaian and Lauterio (150) showed evidence of growth enhancement achieved with a 1% supplementation of either arginine, phenylalanine, leucine, or isoleucine to a commercial diet. Analysis of the free amino acid content of whole body and tail muscle tissue of juvenile shrimp revealed that arginine is the predominant amino acid (151). The quantitative dietary protein requirement for juveniles has generally fallen within the range of 30–40% (dry weight), but lower values have been reported (152). With use of soybean meal, fishmeal, and shrimp meal, the optimum dietary protein levels are between 35 and 40% (153 and 154). All other dietary requirements were detailed by D'Abramo (139). Distinct dietary differences exist between *M. rosenbergii* and other species of *Macrobrachium*.

A list of citations on the freshwater shrimp since 1972 may be obtained from reference (155).

ACKNOWLEDGMENTS

The author would like to thank the following people and organizations for sharing information used in this entry: Bob Phillips and Nova Companies (Belize) Ltd., Louis Hamper and Arroyo Aquaculture Assoc., Joe Fox, Texas A&M University-Corpus Christi, and Ron Rosati, TAMU-Kingsville.

BIBLIOGRAPHY

1. B. Rosenberry, ed., *World Shrimp Farming 1998*, Shrimp News International, San Diego, CA, 1998.
2. I.P. Farfante and B. Kensley, *Penaeoid and Sergestoid Shrimps and Shrimp of the World*, Museum National D'Histoire Naturelle, Public, 1997.
3. M. Hudinaga, *Rep. Hayatomo Fish. Inst.* **1**(1), 1–51 (1935).
4. M. Hudinaga, *Nat. Res. Council of Jpn.* **10**(2), 305–393 (1942).
5. M. Hudinaga and J. Kittaka, *Bull. Plankt. Soc.* (Japan) 35–67 (1967).
6. Aquaculture Outlook, Economic Research Service, U.S. Dept. of Agriculture, LDP-AQS-9, Washington, DC, 1999. Order electronically from USDA web site: *http://usda.mannlib.cornell.edu/emailinfo.html*.
7. Food and Agriculture Organization of the United Nations, Rome, Italy, internet home page, *http://www.fao.org/* and specific database located at *http://www.fao.org/WAICENT/FAOINFO/FISHERY/fishbase/fishbase.html*.
8. W.L. Griffin and G.D. Treece, *A Guide to the Financial Analysis of Shrimp Farming* (a spreadsheet for microcomputers), Texas A&M University, Sea Grant College Program, pub. #TAMU-SG-99-502, 1999.
9. D.R. Teichert–Coddington, D.B. Rouse, A. Potts, and C.E. Boyd, *Aquacult. Eng.* **19**, 147–161 (1999).
10. A.B. Williams, *J. Elisha Mitchell Sci. Soc.* **69**, 156–160 (1953).
11. H. Cook, *Fish. Bull.* **65**(2), 437–447 (1965).
12. C.J. Sindermann, ed., *Reproduction, Maturation and Seed Production of Cultured Species*, NOAA Technical Report NMFS 47, 1987. (1st published in Proc. of the 12th U.S.–Jpn. meeting on Aquacult. Baton Rouge, LA, 1983).
13. J.J. Ewald, *Bull. Mar. Sc. Gulf and Caribb.* **15**(2), 436–449 (1965).
14. J.C. Chen and T.S. Chin, *Aquaculture* **69**, 253–262 (1988).
15. J.C. Chen and T.S. Chin, *J. World Aquacult. Soc.* **19**(3), 143–148 (1988).
16. T.S. Chin and J.C. Chen, *Aquaculture* **66**, 247–253 (1987).
17. W.A. Bray and A.L. Lawrence, in A.W. Fast and L. Lester, eds., *Marine Shrimp Culture: Principles and Practices*, Elsevier Sci. Pub. Amsterdam, The Netherlands, 1992, pp. 93–170.
18. L. Smith, J.M. Fox, and G.D. Treece, in J.P. McVey, ed., *CRC Handbook of Mariculture*, 2nd ed., CRC Press, Inc., Boca Raton, FL, 1993, pp. 3–13.
19. L. Smith, J.M. Fox, and G.D. Treece, in J.P. McVey, ed., *CRC Handbook of Mariculture*, 2nd ed., CRC Press, Inc., Boca Raton, FL, 1993, pp. 153–172.
20. G.D. Treece and J.M. Fox, *Design, Operation and Training Manual for an Intensive Culture Shrimp Hatchery, with Emphasis on P. Monodon and L. Vannamei*, Texas A&M University, Sea Grant College Program, Pub. #93–505, College Station, TX, 1993.

21. J.R. Villalon, *Practical Manual For Semiintensive Commercial Production of Marine Shrimp*, Texas A&M University, Sea Grant College Program, Pub. #91–501, College Station, TX, 1991.
22. W. Tseng, *Shrimp Mariculture, A Practical Manual*, Chien Cheng Publisher, Republic of China, 1987.
23. J.A. Wyban and J.N. Sweeney, *Intensive Shrimp Production Technology, The Oceanic Institute Shrimp Manual*, The Oceanic Institute, HI, 1991.
24. J.C. Chen, T.S. Chin, and C.K. Lee, in Maclean, Dizon, and Hosillos, eds., *The First Asian Fisheries Forum*, The Asian Fisheries Society, Manila, Philippines, 1986, pp. 657–662.
25. S.N. Chen and G.H. Kkou, *J. of Fish Disease* **12**, 733–776 (1989).
26. K.H. Mohamed, P.V. Rao, and M.J. George, *FAO Fish. Rep.* **57**(2), 489–504 (1970).
27. H. Loesch and H. Avila, *Clavespara identificación de camarones peneidos de interés comercial en el Ecuador* (Identification Keys for Commercial Ecuadorian Penaeid Shrimp), Instituto Nacional de Pesca del Ecuador, Boletín científico y técnico., Guayaquil, Ecuador, 1965.
28. I. Perez Farfante, *U.S. Dept. Comm. NOAA Rep.* **559**, 1–26 (1970).
29. T.U. Bagarinao, N.B. Solis, and W.R. Villaver, *Important Fish and Shrimp Fry in Philippines' Coastal Waters: Identification, Collection and Handling*, Aquacult. Ext. Manual No. 10, SEAFDEC, Philippines, 1986.
30. R.G. Adiyodi and T. Subramoniam, in K.G. and R.G. Adiyodi, eds., *Arthropoda-crustacea, Reproductive Biology of Invertebrates*, John Wiley, New York, 1983, pp. 443–495.
31. L.M. Bielsa, W.H. Murdich, and R.F. Labisky, *U.S. Fish & Wildl. Ser.* **82**, 11–17 (1983).
32. J.E. King, *Biol. Bull. Wood's Hole* **94**(3), 244–262 (1948).
33. J.P. McVey, eds., *CRC Handbook of Mariculture*, 1st ed., CRC Press, Inc., Boca Raton, FL, 1983. 2nd ed., 1993.
34. A.B. Al-Haji, C.M. James, S. Al-Ablani, and A.S. Farmer, *Kuwait Institute Sci. Research, Bull. of Mar. Sc.* **6**, 143–154 (1985).
35. Aquacop, *Actes de Colloques du CNEXO* **4**, 179–191 (1977).
36. Y. Taki, J.H. Primavera, and J.A. Llobrera, eds., *Proceedings of the First International Conference on the Culture of Penaeid Prawns/Shrimps*, SEAFDEC, Philippines, 1985.
37. J.M. Biedenbach, L.L. Smith, T.K. Thomsen, and A.L. Lawrence, *J. World Aquacult. Soc.* **20**(2), 61–71 (1989).
38. H.L. Cook and M.A. Murphy, *Fishery Bull.* **69**(1), 223–239 (1971).
39. H.L. Cook and M.A. Murphy, *Trans. Am. Fish. Soc.* **98**(4), 751–754 (1969).
40. H.L. Cook, *FAO Fish Rep.* **57**(3), (1969).
41. H.L. Cook and M.A. Murphy, *Proc. Conf. SE Assoc. Game Comm.* **19**, (1966).
42. S. Dobkin, *Fish. Bull.* **61**(190), 321–349 (1961).
43. W.D. Emmerson and B. Andrews, *Aquaculture* **23**, 45–47 (1981).
44. W.D. Emmerson, *Mar. Bio.* **58**, 65–73 (1980).
45. I. Furukawa, K. Hidaka, and K. Hirano, *Bull. Fac. Agr. Miyazaki Univ.* **20**(1), 93–110 (1973).
46. C.T. Fontaine, D.B. Revera, H.M. Morales, and C.R. Mock, *Spec. Publ. NMFS, NOAA*, Galveston, Texas, 1981.
47. C.T. Fontaine and D.B. Revera, *Proc. World Maricult. Soc.* **11**, 211–218 (1980).
48. D.R. Fielder, J.G. Greenwood, and J.C. Ryall, *Aust. J. Mar. Freshwat. Res.* **26**, 155–175 (1975).
49. K. Gopalakrishnan, *Aquaculture* **9**, 145–154 (1976).
50. A.K. Hameed, S.N. Dwivedi, and K.H. Alikunki, *Bull. Central Inst. Fish. Edu.* (India) **9**, 2–3 (1982).
51. J.M. Heinen, *Proc. World Maricult. Soc.* **7**, 333–343 (1976).
52. H. Hirata, Y. Mori, and M. Watanabe, *Mar. Bio.* **29**, 9–13 (1975).
53. S. Imamura and T. Sugita, *Culture Tech. Res.* **1**(2), 35–46 (1972).
54. D.A. Jones, K. Kurmaly, and A. Arshard, *J. World Aquacult. Soc.* **18**(1), (1987).
55. D.A. Jones, A. Kanazawa, and K. Ono, *Mar. Biol.* **54**, 261–267 (1979).
56. A. Kanazawa, S. Teshima, H. Sasada, and S.A. Rahman, *Bull. of Jpn. Soc. Sci. Fish.* **48**, 195–199 (1982).
57. F.D. Kuban, J.S. Wilkenfeld, and A.L. Lawrence, *Aquaculture* **47**, 151–162 (1985).
58. P. Kungvankij, *The design and operation of shrimp hatcheries in Thailand*, In Working party on small-scale shrimp/prawn hatcheries in South East Asia, Semarang, Central Java, Indonesia, II Tech. Report, SCSFD & C.P., Manila, Philippines, 1982.
59. N. Kureha and T. Nakenishi, *Culture Tech. Res.* **1**(1), 41–46 (1972).
60. P. Leger and P. Sorgeloos, *J. World Aquacult. Soc.* **18**(1), 17A (1987).
61. P. Leger, *Oceanogr. Mar. Biol. Ann. Rep.* **24**, 521–623, Aberdeen Univ. Press (1986).
62. G. LeMoullac, R. Leguedes, G. Cuzon, and J. Goguenheim, *J. World Aquacult. Soc.* **18**(1), 18A (1987).
63. I.C. Liao, ed., *Manual on Propagation and Cultivation of Grass Prawn, P. Monodon*, Tungkang Mar. Lab., Taiwan, 1977.
64. I.C. Liao and T.L. Huang, in V.R. Pilag, ed., *Coastal Aquaculture in the Indo-Pacific Region*, Fishing News (Books), Ltd., 1972, pp. 328–354.
65. I.C. Liao, T.L. Huang, and K. Katsutani, *J.C.R.R. Fish. Ser.* **8**, 67–71 (1969).
66. Y. Maeda, *Fish Culture* **5**(2), 64–70 (1968).
67. O. Millamena and E.J. Aujero, SEAFDEC, Quarterly Report **2**, 15–16 (1978).
68. C.R. Mock, D.B. Revera, and C.T. Fontaine, *Proc. World Aquacult. Soc.* **11**, 102–117 (1980).
69. C.R. Mock and M.A. Murphy, *Proc. World Maricult. Soc.* **1**, 143–156 (1970).
70. H. Motoh, *Bull. Jpn. Soc. Sci. Fish.* **45**(10), 1201–1216 (1979).
71. L.G. Pinto and J.J. Edward, *Bol. Cent. Invest. Biol.* Univ. Zulia, (12), 1–61 (1974).
72. R. Platon, *Design, Operation and Economics of a Small-Scale Hatchery for Larval Rearing of Sugpo, P. Monodon (Frabricius)*, Extension Manual No. 1, SEAFDEC, Philippines, 1978.
73. N.G. Primavera, ed., *Prawn Hatchery Design and Operation*, Aquaculture Extension Manual No. 9, 2nd ed., SEAFDEC, Philippines, 1985.
74. J.M. San Felici, F. Muñoz, and M. Alcaraz, *Stud. Rev.* GFCM **52**, 105–121 (1973).
75. SEAFDEC, *Crustacean Hatcheries*, SEAFDEC Aquacult. Dept., Ann. Rep., 1981, pp. 13–17.

76. K. Shigueno, *Shrimp Culture in Japan*, Assoc. for Int. Tech. Promotion, Tokyo, Japan, 1975.
77. K. Shigueno, *Problems of Prawn Culture in Japan*, Overseas Tech. Coop. Agency Publ., 1972.
78. C.M. Simon, *J. World Maricult. Soc.* **12**(2), 322–334 (1981).
79. S. Skokita, *Biol. Mag. Okinawa* **6**, 34–36 (1970).
80. P. Sorgeloos, *Manual for the Culture and Use of Brine Shrimp* Artemia *in Aquaculture*, State Univ. of Ghent, *Artemia* Reference Center, Belgium, 1986.
81. P. Sorgeloos, E. Bossuyt, P. Lavens, P. Leger, P. Yanhaecke, and D. Versichele, in J.P. McVey, ed., *CRC Handbook of Mariculture*, 1st ed., CRC Press, Inc., Boca Raton, FL, 1983, pp. 71–96.
82. P. Sorgeloos, in G. Persoone, P. Sorgeloos, O. Roels, and E. Jaspers, *The Brine Shrimp* Artemia, Universa Press, Wetteren, Belgium, 1980, 3rd Vol.
83. R. Sudhakaro, *CMFRI Bull.* **28**, 60–64 (1978).
84. D.C. Tabb, W.T. Yang, Y. Hirono, and J. Heinen, *A Manual for the Culture of Pink Shrimp,* Farfantepenaeus duorarum *from Eggs to Postlarvae Suitable for Stocking*, U.S. Dept. Of Commerce, NOAA, Florida Sea Grant College Program, Spec. Bull. No. 7, Univ. of Miami, Florida, 1972.
85. A.G.J. Tacon, FAO Report ADCP/MR/86/23, Rome, Italy, 1986.
86. H.C. Tang, *Chin. Fish. Mon.* **290**, 2–7 (1977).
87. S. Teshima and A. Kanazawa, *Bull. Jpn. Soc. Sci. Fish*, 1983, pp. 1893–1896.
88. Y.Y. Ting, T.T. Lu, and M.N. Lin, *Chin. Fish. Mon.* **292**, 22–28 (1977).
89. G.D. Treece and M.E. Yates, *Laboratory Manual for the Culture of Penaeid Shrimp Larvae*, Texas A&M University, Sea Grant College Program, pub. #TAMU-SG-88-202, Texas A&M University, College Station, Texas, 1988. 2nd ed., 1990, 3rd ed., 1993. (Available in 4 languages, including English, Indonesian, Spanish, and Chinese; English and Spanish versions are available from the publisher that is listed.)
90. G.D. Treece and J.M. Fox, *J. World Aquacult. Soc.* **18**(1), 20A (1987).
91. G.D. Treece, in G.W. Chamberlain, M.G. Haby, and R.J. Miget, eds., *Texas Shrimp Farming Manual*, Texas Agricultural Extension Service, Corpus Christi, TX, 1985, pp. 43–64.
92. W.T. Yang, *A Manual for Large-Tank Culture of Penaeid Shrimp to the Postlarval Stages*, U.S. Dept. of Commerce, NOAA, Florida Sea Grant College Program, Technical Bulletin 31, 1975.
93. D.K. Villaluz, A. Villauz, B. Ladera, M. Sheik, and A. Gonzaga, *Philippines J. Sci.* **98**, 205–236 (1972).
94. *An Evaluation of Potential Shrimp Virus Impacts on Cultured Shrimp and Wild Shrimp Populations in the Gulf of Mexico and Southeastern U.S. Atlantic Coastal Waters*, A Report to the Joint Subcommittee on Aquaculture, Dept. of Commerce, NOAA/NMFS, USDA, Animal and Health Inspection Service, National Center for Environmental Assessment, US EPA, US Fish and Wildlife Service, US Dept. of Interior, June, 1997.
95. S.K. Johnson, *Handbook of Shrimp Diseases*, Texas A&M Univ. Sea Grant College Program, Tamu-SG-90-601 (R), Bryan, TX, 1995.
96. D.V. Lightner, (in press), *J. World Aquacult. Soc.*, The Penaeid Shrimp Viruses: Current Status, Available Diagnostic Methods, and Management Strategies, 1999.
97. V. Alday de Graindorge and T.W. Flegel, Interactive CD entitled "Diagnosis of Shrimp Diseases, with Emphasis on the Black Tiger Shrimp (*Penaeus monodon*)," FAO, Rome, Italy and Multimedia Asia Co., Ltd., Bangkok, Thailand Tel. (662) 298-0646-49, (1999).
98. I. Dore and C. Frimodt, *An Illustrated Guide to Shrimp of the World*, ISBN 0-943738-20-2, Osprey Books, Huntington, New York, 1987.
99. R. Mallikarjuna and V. Gopalkrishnan. *IPFC/C68/Tech.* **13**, 1–4 (1968).
100. K.L. Main and W. Fulks, eds., *The Culture of Cold-Tolerant Shrimp: Proceedings of an Asian-U.S. Workshop on Shrimp Culture*, Oceanic Institute, HI, 1990.
101. C.E. Boyd, *Water Quality in Warmwater Fish Ponds*, Auburn University, Auburn, Alabama, 1979. (2nd ed., 1990.)
102. S.K. Johnson, Southern Regional Aquaculture Center (SRAC) pub. #A0901, 1986.
103. D.V. Lightner, in J.P. McVey, ed., *CRC Handbook of Mariculture*, CRC Press, Boca Raton, FL, 1983, pp. 289–320.
104. D.V. Lightner, R.M. Redman, and E.A. Almada Ruiz, *J. Invert. Path.* **51**, 137–139 (1988).
105. J.A. Brock and K.L. Main, *A Guide to the Common Problems and Diseases of Cultured* Litopenaeus Vannamei, The Oceanic Institute, HI, and the University of Hawaii Sea Grant College Program, UNIHI-SEAGRANT-CR-95-01, 1995.
106. T.T. Sano, H. Nishimura, H. Fukuda, T. Hayashida, and K. Momoyama, *Helgolander Meeresunters* **37**, 255–264 (1984).
107. R.J.G. Lester, A. Doubrovsky, J.L. Paynter, S.K. Sambhi, and J.G. Atherton, *Diseases of Aquatic Organisms* **3**, 159–165 (1987).
108. L.M. Nunan, B.T. Poulos, and D.V. Lightner (in press), *Aquaculture* **160**, 19–30 (1998).
109. P.S. Cruz, *Shrimp Feeding Management: Principles and Practices*, Kabukiran Enterprises, Inc., Philippines, 1991.
110. SEAFDEC, *Biology and Culture of Penaeus monodon*, Brackishwater Aquaculture Information System, Aquaculture Department, SEAFDEC, Iloilo, Philippines, 1988.
111. F. Piedad–Pascual, *SEAFDEC Asian Aquacult.* **12**(2), 5–8 (1990).
112. A.L. Lawrence, D.M. Akiyama, and W.G. Dominy, eds., *Penaeid Shrimp Nutrition for the Commercial Feed Industry*, revised, American Soybean Association, St. Louis, MO, 1994.
113. M.B. New, H. Saram, and T. Singh, eds., *Technical and Economic Aspects of Shrimp Farming*, Infofish, Kuyala Lumpur, Malaysia, 1990.
114. P. Dhert and P. Sorgeloos, *Infofish International* **2**, 31–38 (1995).
115. G. Cuzon, *Rev. Fish. Sci.* **6**(1,2), 129–141 (1998).
116. A. Pedrazszoli, C. Molina, N. Montoya, S. Townsend, A. Leon–Hing, Y. Parades, and J. Calderon, *Rev. Fish. Sci.* **6**(1,2) 143–151 (1998).
117. L.R. D'Abramo, D.E. Conklin, and D.M. Akiyama, eds., *Crustacean Nutrition*, World Aquaculture Society, Baton Rouge, LA, 1997.
118. M.B. New, *Feed and Feeding of Fish and Shrimp, A Manual on the Preparation and Presentation of Compound Feeds for Shrimp and Fish in Aquaculture*, ADCP/REP/87/26, United Nations Dev. Prog. FAO, Rome, Italy, 1987.
119. H.C. Clifford, *Proc. of the 1st Latin American Shrimp Culture Congress*, Panama, 1998.
120. C.E. Boyd, M.E. Tanner, M. Madkour, and K. Masuda, *J. World Aquacult. Soc.* **25**(4), 517–534, (1994).

121. G.D. Treece, *Aquaculture Curriculum Guide, Species Specific Module, Saltwater Shrimp*, USDA Cooperative State Research Service and The National Council for Agricultural Education, Alexandria, VA, 1993. (Once US$50 each, the curriculum guides can now be downloaded free of charge at *http://ag.ansc.purdue.edu/aquanic/publicat/govagen/ncae/saltshp.pdf.*)
122. W. Dall, B.J. Hill, P.C. Rothlisberg, and D.J. Sharples, in J.H.S. Blaxter and A.J. Southward, eds., *Advances in Marine Biology, Vol. 27, The Biology of the Penaeidae*, Academic Press, New York, 1990.
123. L.R. D'Abramo and M.W. Brunson, Biology and Life History of Freshwater Prawns, Southern Regional Aquaculture Center (SRAC) Pub. 483, Miss. State Univ., MS, 1996. (SRAC publications are available on the internet at *http://www.msstate.edu/dept/srac/fslist.htm.*)
124. M.N. Mistakidis, ed., *FAO Fisheries Rep.* **5**(3), (1969).
125. T. Fujimura and H. Okamoto, *Indo. Pac. Fish Counc.*, 14th session, Bangkok, Thailand, 1970.
126. Aquacop, *J. World Maricult. Soc.* **8**, 311–319 (1977).
127. J.E. Bardach, J.H. Ryther, and W.O. McLarney, *Aquaculture: the Farming and Husbandry of Freshwater and Marine Organisms*, John Wiley and Sons, Inc., New York, 1972.
128. G.D. Treece, *Texas Aquaculture, History and Growth Potential for the 1990s*, Texas A&M University, Sea Grant College Program, pub. # GT103, 1993.
129. J.A. Hanson and H.L. Goodwin, *Shrimp and Prawn Farming in the Western Hemisphere*, Dowden, Hutchinson and Toss, Inc., Stroudsburg, PA, 1977.
130. S.K. Johnson, *Crawfish and Freshwater Shrimp Diseases*, Texas A&M University, Sea Grant College Program, pub. #TAMU-SG-77-605, 1977.
131. D. Cohen, Z. Ranan, and A. Barnes, *J. World Maricult. Soc.* **12**, 231–243 (1981).
132. R.W. Brick and J.T. Davis, *Farming Freshwater Shrimp*, Texas A&M University, Texas Agricultural Extension Service, Fact Sheet 3M-9-81, 1981.
133. Special Report, *J. World Aquacult. Soc.* **25**(1), 5–17 (1994).
134. W.H. Daniels, L.R. D'Abramo, and L. De Parseval, *J. Shell. Res.* **11**(1), 65–73 (1992).
135. M.J. Fuller, R.A. Kelly, and A.P. Smith, *J. Shell. Res.* **11**(1), 75–80 (1992).
136. L.R. D'Abramo and M.W. Brunson, *Production of Freshwater Prawns in Ponds*, Southern Regional Aquaculture Center (SRAC) Pub. 484, Miss. State Univ., MS, 1996. (SRAC publications are available on the internet at *http://www.msstate.edu/dept/srac/fslist.htm.*)
137. Mississippi Agriculture and Forestry Experiment Station, *Economics of freshwater shrimp production in Mississippi*, Mississippi Agriculture and Forestry Experiment Station Bulletin #985, Department of Agricultural Economics, Miss. State Univ., MS, 1996.
138. C.D. Webster and J.H. Tidwell, *Aquacult. Mag.*, Nov./Dec. 47–60 (1995).
139. L.R. D'Abramo, *Rev. Fish. Sci.* **6**(1,2), 153–163 (1998).
140. L.R. D'Abramo and S.S. Sheen, *Rev. Fish. Sci.* **2**(1), 1–21 (1994).
141. L.R. D'Abramo, S.R. Malecha, M.J. Fuller, W.H. Daniels, and J.M. Heinen, in P.A. Sandifer, ed., *Shrimp Culture in North America and the Caribbean*, World Aquacult. Soc., Baton Rouge, LA, 1991, pp. 96–123.
142. P. Bartlett and E. Enkerlin, *Aquaculture* **30**, 353–356 (1983).
143. M.R.P. Briggs, K. Jauncey, and J.H. Brown, *Aquacult.* **70**, 121–129 (1988).
144. L.R. D'Abramo, J.M. Heinen, H.R. Robinette, and J.S. Collins, *J. World Aquacult. Soc.* **20**, 81–89 (1989).
145. L.R. D'Abramo and S.S. Sheen, *Aquaculture* **115**, 63–86 (1993).
146. R.C. Reigh, *Fatty Acid Metabolism in the Freshwater Shrimp* Macrobrachium rosenbergii, Ph.D. dissertation, Texas A&M University, College Station, Texas, 1985.
147. S.S. Sheen and L.R. D'Abramo, *Aquaculture* **93**, 121–134 (1991).
148. J.H. Tidwell, C.D. Webster, J.A. Clark, and L.R. D'Abramo, *J. World Aquacult. Soc.* **24**, 66–70 (1993).
149. J.H. Tidwell, C.D. Webster, D.H. Yancey, and L.R. D'Abramo, *Aquaculture* **118**, 119–130 (1993).
150. A. Farmanfarmaian and T. Lauterio, *Proc. World Maricult. Soc.* **10**, 674–688 (1979).
151. L. Reed and L.R. D'Abramo, *J. World Aquacult. Soc.* **20**, 107–113 (1989).
152. G. Gomez Diaz, H. Nakagawa, and S. Kasahara, *Bull. Jpn. Xoc. Sci. Fish.* **54**, 1401–1407 (1988).
153. G.H. Balazs and E. Ross, *Aquaculture* **7**, 200–213 (1976).
154. M.R. Millikin, A.R. Fortner, P.H. Fair, and L.V. Sick, *Proc. World Maricult. Soc.* **11**, 382–391 (1980).
155. Quick Bibliography Series, USDA, National Agricultural Library, Beltsville, MA, 1987. (Web page at *http://www.nalusda.gov/.*)

See also BRINE SHRIMP CULTURE.

SILVER PERCH CULTURE

ROBERT R. STICKNEY
Texas Sea Grant College Program
Bryan, Texas

GEOFF ALLAN
Port Stephens Research Centre
Taylors Beach, Australia

OUTLINE

Culture Requirements
Bibliography

The silver perch (*Bidyanus bidyanus*) belongs to the family Teraparidae, (Fig. 1) which includes freshwater grunters, or perches (1). They are endemic to Australia, occurring through most of the Murray–Darling drainage system from southern Queensland in the north to Victoria in the south (1). They are now uncommon in the wild. Silver perch are omnivorous, and key features, such as rapid growth rates and high survival rates when cultured at high stocking densities in earthen ponds, tolerance to relatively poor water quality, and excellent eating attributes, have made silver perch a popular fish for culture. Research conducted by New South Wales Fisheries in Australia has shown that this species has the potential to form a large industry (greater than 10,000 tons/yr) based on high-value

Figure 1. Silver perch.

and low-cost production (2). Official production was about 135 tons in 1996–1997 (3).

The authors thank S.J. Rowland for providing some of the information used in this entry.

CULTURE REQUIREMENTS

Silver perch are temperate species that can tolerate water temperatures from 2 to 38 °C (36 to 100 °F) (2), but with optimal temperatures between about 23 and 28 °C (73 to 82 °F). In the wild, silver perch spawn after migrating to upstream areas behind the peaks of floods (1). In ponds, silver perch mature at about two years of age for males and three for females, but will not usually spawn. Females are fecund (approximately 150,000 eggs/kg) and in hatcheries are induced to spawn using hormones such as HCG (200 iu/kg HCG; Rowland, 1984).

The hatchery phase [broodstock to 30 mm (1.2 in.) fry] usually takes about six to eight weeks, commonly over the spring or summer months. Fry are then usually reared in fertilized earthen nursery ponds at stocking densities up to about 100,000 fish/ha (40,000/acre) for 3 to 4 months over summer and autumn (2). The growout phase usually occurs in drainable, aerated earthen ponds. Stocking densities of up to 21,000 fish/ha (8,400/acre) have been shown not to reduce fish performance (4), and fish can grow from 50-g (1.7-oz) fingerlings to 500-g (16.7-oz) adults over 5 to 7 months if summer growth is included as a part of the rearing period. If fingerlings are stocked in autumn, after the nursery phase, the growout phase will take up to 12 months (2).

In common with many species reared in freshwater, silver perch can acquire off-flavors associated with blue-green algae and some types of pond bacteria. Farmers ameliorate off-flavors by purging live fish for one to three weeks in tanks supplied with water that does not contain the offensive material. Fish are not fed during this process. Purged silver perch have qualities excellent for eating; they are firm, white-fleshed fish with an excellent fillet yield of about 40%). Silver perch are an excellent source of n-3 fatty acids, which are thought to provide health benefits for humans.

Silver perch have been bred for many years for stocking into farm ponds. Their diseases are well known (5,6). Major ectoparasites include white spot, chilodonelliasis, ichthyobodiasis, and trichodiniasis. Fungal diseases include epizootic ulcerative disease and fungus or cotton wool disease. Bacterial diseases include tail rot, columnaris, and goldfish ulcer disease (2,5,6).

Silver perch are omnivores with relatively low protein requirements. Recent research has established that silver perch can be farmed on diets without fishmeal that are based on agricultural ingredients, such as meat meal, poultry offal meal, grain legumes, such as lupins and peas, and wheat (7). Diets that contain 30 to 35% protein can yield growth rates of around 3 g/day (0.1 oz/day) and apparent food conversion ratios of 1.5–2.0 : 1 (8). Current research on further ingredient evaluation, nutritional requirements, and feeding strategies should result in a further decrease in the cost of feeding silver perch.

BIBLIOGRAPHY

1. R.M. McDowall, ed., *Freshwater Fishes of Southeastern Australia*, Reed, Sydney, NSW Australia, 1980.
2. S.J. Rowland, in K.W. Hyde, ed., *The New Rural Industries—A Handbook for Farmers and Investors*, Rural Industries Research and Development Corporation, Canberra, Australia, 1998, pp. 134–139.
3. G. Allan, *World Aquaculture* **30**(1), 39ff (1999).
4. S.J. Rowland, G.L. Allan, M. Hollis, and T. Pontifix, *Aquaculture* **130**, 317–328 (1995).
5. R.B. Callinan and S.J. Rowland, in *Proceedings of Silver Perch Aquaculture Workshops, Grafton and Narrandera, April 1994*. Austasia Aquaculture for NSW Fisheries, 1995, pp. 67–75.
6. S.J. Rowland and B.A. Ingram, *Bachelor of Arts, Fisheries Bulletin 4*, NSW Fisheries, Sydney, 1991.
7. G.L. Allan, D.A.J. Stone, and M.A. Booth, *Book of Abstracts, World Aquaculture '99*, World Aquaculture Society, Baton Rouge, LA, 1999, p. 18.
8. G.L. Allan, S.J. Rowland, M.A. Booth, and D.A.J. Stone, *Book of Abstracts, World Aquaculture '99*, World Aquaculture Society, Baton Rouge, LA, 1999, p. 19.

SITE SELECTION

GRANVIL D. TREECE
Texas Sea Grant College Program
Bryan, Texas

OUTLINE

The selection of a suitable site is one of the first and most important steps toward making an aquaculture project a success. The selection of the site could determine the success or failure of the project and should not be taken lightly. The site selected depends upon a number of important criteria, which are discussed in this entry. Site selection is crucial, because it can determine the tank, raceway, or pond design; farm layout; supportive infrastructure; production methodology; management strategy; and hatchery location.

INITIAL CONSIDERATIONS PRIOR TO SITE VISIT

Before the potential aquaculture site is selected and visited, the overall objectives of the company should be determined. The market for the product to be grown at the site should be assessed and tested, if possible. Conducting a market survey and analysis is recommended. Unless the project is totally integrated and the company will have its own processing operation, the processing plant in the area is usually the market. Most large farms are integrated and sell to a world market, where competition is fierce. Know your market channels and realize what it takes to get the product there in good shape.

Once the market is researched and tested, determine how much of the product your company wants to produce or what production level is desired. Determine the level of technology that is best for your company and the site; the desired and feasible number of crops per year; and the required land area, which is a function of the desired production level, the number of crops per year, and the production strategy. The species to be cultured should be chosen, and the overall productivity of the farm should be predicted as accurately as possible. A farmer can select a site to suit the species or a species that will be best suited for a particular site.

The financial returns of the project should be assessed and the preset return thresholds met. Because of many failures in the industry, most investors and lending agencies consider aquaculture to be a high-risk operation. Most lending institutions will not consider funding a project unless it has already been proven to work on a small scale at the particular site and the management team has proven experience and plans to work full time at the site. In addition, lending institutions will not be interested in a project unless it meets a given internal-rate-of-return threshold. Generally, the projected financial returns are much higher than the actual returns, because there are many unexpected costs; therefore, the financial assessment should be done very conservatively. In my opinion, if the internal rate of return is projected to be above 20%, and preferably in the 30–50% range over a 12-year horizon, then the site selection procedure should be continued. Depending on the company's goals, if the projected returns fall below 20% per annum, then the project should be dropped before any further money is spent. Generally speaking, to be competitive, entrepreneurs entering the business for the first time must have added benefits (such as already owning the land), to assist the overall financial return.

OTHER IMPORTANT ASPECTS TO CONSIDER IN THE SITE SELECTION PROCESS

The political stability in the area should be reviewed; areas with political instability should be avoided. It is also important that aquaculture sites be located away from any significant sources of pollutants, including agricultural runoff (e.g., from crop spraying), sewage outfalls (both municipal and industrial), storm runoff canals, harbors, and refineries. Water samples should be collected and tested on a routine basis, and the results should be placed in a database, to monitor future pollution, help enforce existing regulations, assist in designing cleanup operations (if needed) in severely impacted areas, and help identify suitable sites for aquaculture development in the future. It is also important that the site be located away from wetlands, mangrove areas, or any other environmentally sensitive location. Most countries now have laws governing the use of these areas, although it may still be possible to negotiate the use of these areas through mitigation.

SOURCES OF INFORMATION ON SITES

If the financial return of the aquaculture operation looks good, then all possible sources of information concerning the site should be looked into before it is visited. These sources include, but are not limited to, the Internet; fisheries reports; regional development plans; zoning regulations; local, state, and national government aquaculture regulations; economic development plans; remote satellite sensing; aerial photographic surveys; oceanographic surveys; hydrological surveys; land-use surveys; soil-type surveys; and engineering and topographical surveys in the area.

SITE ASSESSMENT

The next step is to visit the site and gather as much information as possible, through personal reconnaissance and by reviewing topographical, hydrological, meteorological, and biological data. A site and zone description or general description of the entire area should be obtained, as well as more specific information about the site, such as specific location; types of vegetation, water, roads; and drainage potential. Obtain the following information while on the site-assessment trip:

General Information on the Site

This includes a map, details on how to get access, to the site, and information on the location; wave action; water salinity at high and low tide; water pH; elevation above seawater source; land topography; benthic topography or description; distance of seawater inlets from the site; type of inlet prescribed; water depth at the inlet site; pollution potential; organics in water; turbidity; nearest river; source, price, and conditions of broodstock; electrical power in the area; source of freshwater; aquaculture distributors in area; source of trained workers, and potential for a well.

Detailed Information on the Site and Details About the Land

This includes information on the total amount of land available, expansion potential, current or previous use of the land, owner of the land, cost of the land, available lease terms (if any), and topography.

Soil Details for Construction Purposes

This category includes soil samples taken by location, a map of the area from which the samples were drawn, the depth at which the samples were taken, and the site stability determined by soil type.

Details on Seawater Quality at the Site

This category includes information on the potential source of seawater, distance of the seawater from the site, color of the seawater, potential for storms, appearance of the beach, erosion, accretion, freshwater plume, proximity of rivers to the site, tidal flux, type of tide (diurnal or semidiurnal), navigation traffic, industrial activity on or near the source of seawater, and a description of the nearest port.

Details on the Freshwater at the Site

This includes information on the nearest river; size of the river; other sources of freshwater; rainy- and dry-season river depth and width; estimated hydraulic flow during seasons; potential for flooding; last major flood damage; distance of the source of freshwater from the site; potential for building well, a pipeline, or canal; potential sources of effluent; and agricultural pollution.

Meterological Data at the Site

This includes data on the duration of dry and rainy seasons and temperature variance in seasons.

Flora and Fauna at the Site

This includes a description of the marine and freshwater fisheries, including all relevant details, and information on the distance to broodstock and fishing grounds, typical fishing vessels, and the nearest fishing port.

Utilities and Resources at the Site

This category includes information on the electrical supply, nearest transformer, type of current, capacity, cost of electricity, potential for power shortages, availability of petroleum and other lubricants, building materials, laboratory equipment, heavy construction equipment, and available computers.

Socioeconomic Data at the Site

This includes data on the nearest city, industries, labor force, construction contractors, engineers, aquaculturists, university and research backup, business schools in the area, nearest village or town, local industries, local government, degree of political peace and tranquility or unrest, other aquaculture projects (such as hatcheries and farms) in the area, and extent of government assistance.

OTHER INFORMATION NEEDED AT MARICULTURE SITES

A description of the climate in the region should be obtained, including air and water temperature, wind, rainfall, solar radiation, and evaporation. It is pertinent to know the air temperature, because it can change the temperature of the water. Wind and solar radiation also influence water temperature. Data on air temperature (in °C or °F) should be recorded if local data are unavailable.

Knowing the water temperature is important, because it determines which species will grow best at the site or which species should be grown during a particular season. Table 1 serves as an example of preexisting water-temperature data found during a site assessment for a tropical shrimp farm.

Table 1. Water Temperatures at a Site

Month	AM Temperature °C (°F)	PM Temperature °C (°F)
December	22–23 (71.6–73.4)	26–28 (78.8–82.4)
January	22 (71.6)	24 (75.2)
February	19–20 (66.2–68)	21–23 (69.8–73.4)
March	25 (77)	28 (82.4)
April	21–25 (69.8–77)	29–32 (84.2–89.6)
May	26–27 (78.8–80.6)	31 (87.8)
June	27–30 (80.6–86)	31–33 (87.8–91.4)
July	No data	No data
August	29 (84.2)	33–35 (91.4–95)
September	No data	No data
October	22–24 (71.5–75.2)	25–32 (77–89.6)
November	19–27 (66.2–80.6)	28–30 (82.4–86)

From these preexisting records, it appears that the cool months are November through February (four months out of the year) and that the hotter afternoon temperatures of concern fall in the month of August. These data can be useful in assisting management decisions on species to culture during the different seasons.

The prevailing or dominant wind direction should also be obtained. This information can assist in pond design and layout. Wind has three major effects on water in culture ponds. First, it circulates the water, thus mixing the layers of differing density that tend to form in ponds (a phenomenon called "stratification"). The circulation helps oxygen- and phytoplankton-rich surface waters to reach the organism being grown. It also moves water containing less oxygen, with higher concentrations of waste (metabolites), to the surface, thus encouraging reoxygenation and stimulating the biodegradation of metabolites. Second, wind tends to lower the water temperature, by increasing evaporative cooling. Third, it generates waves in a pond, causing erosion of pond banks.

Hurricanes, typhoons, and large storms may be a threat to a site. Storm winds can rip out pond liners if water is not in the pond and can destroy tanks and raceways if they are not weighted with water. A storm surge can cause flooding and loss of a crop in tanks, raceways, and ponds and may cause erosion and dike collapse in ponds. For example, Hurricane Mitch caused much damage to the aquaculture industry in Central America in 1998.

Information on the average yearly precipitation for the site should be obtained. Rainfall influences aquaculture in several ways: (*1*) It dilutes saltwater, thus lowering salinity in brackish areas; (*2*) it erodes the pond banks, roads, etc.; (*3*) it promotes the growth of erosion-preventative grasses on the pond dikes; (*4*) it lowers the water temperature.

In the tropics, rains are intense and seasonal. The result is that the heavy rains may cause damage to dikes and roads, while the long period of dry weather discourages the growth of protective vegetation. On the other hand, monsoon rains usually tend to lower the water temperatures of ponds during the hottest time of the year, when they need to be somewhat lower. Data on rainfall are generally available from government sources, such as a national weather service.

The main concern regarding rainfall is potential for flooding. One should try to obtain the 20-, 100-, or even 200-year flood-plain data for the area, if available. One should also try to obtain data on the maximum intensity of rainfall recorded, the duration and frequency of rainfall, yearly precipitation, and the watershed characteristics. Data on rainfall, soil properties, types of vegetative cover, topography, and the extent of the watershed that drains into the area determine the potential for flooding. The outer-dike or perimeter-dike elevation is generally based on the highest known flood and its height.

There is usually a definite season of low water temperatures at any given site, even in the tropics. Solar radiation influences water temperature and growth of phytoplankton in ponds. Phytoplankton rely on sunlight to convert carbon dioxide into carbon and oxygen. Carbon is essential for growth, and oxygen is released into the environment. Phytoplankton are fundamental to pond growout technology. Not only do they provide the primary production in the natural food chain, but they are also the major source of dissolved oxygen in ponds. In addition, phytoplankton bind or convert certain toxins, such as metabolites, thus purifying the water. However, in the absence of sunlight, phytoplankton become a net consumer of oxygen. Thus, during prolonged periods of overcast weather, water may be stripped of oxygen, which may stress aquatic animals, inhibiting their growth, and, in severe cases, lead to mortality. Solar radiation may also cause evaporation, which causes a rise in salinity in brackish environments. Evaporation during the dry season may also contribute to evaporative cooling at night, such that water temperatures in the pond drop considerably and may even slow growth or stress the animals being cultured.

GENERAL TIDAL CHARACTERISTICS AT MARICULTURE SITES

Tidal amplitude is important at the site, because it can determine pond-bottom elevation, dike elevations, pumping costs, and the drainage pattern. Tidal frequency determines the pumping schedule, fresh- and saltwater mixing, and the drainage schedule. The tidal fluctuation, or flux, at the site may either be small or be large and dynamic. A large tide allows for good flushing action, whereas special arrangements may be necessary to discharge water without a tide. Sites wherein tides are dynamic should have minimal problems with regard to the removal of discharge water at low tide, but might experience problems at high tide. Another problem might occur at low tide, when there is no intake water. Accordingly, farms in many areas oversize the pump station to compensate for pumping at high tide only and have very large reservoirs to supply water during low tide.

SALTWATER AND FRESHWATER RESOURCES

Large quantities of good-quality water with the proper chemical composition are generally required for aquatic farming. In a brackish environment, the salinity at the site is an important consideration: During the production phase, the species selected for culture needs to achieve maximum growth rates, and salinity levels influence growth rates. For example, most tropical shrimp grow best in salinities ranging from 5 to 25 ppt. Although they can survive at salinities outside of that range (to 0.5 ppt or less and 50 ppt or higher), growth rates would be depressed and, at the extremes, cease altogether. If shrimp culture is planned at the site, it is essential to source brackish water in the range of 5 to 25 ppt or to dilute full-strength seawater (30–35 ppt) with freshwater or brackish water. Catfish and other freshwater species may not tolerate or grow well at salinities above 5–10 ppt. See Table 2 for an example of analysis of selected water samples from Costa Rica.

Salinity is not the only factor that needs to be considered in evaluating potential saltwater sources. Other water quality parameters that need to be assessed include pH,

Table 2. Analysis of Selected Water Samples from Costa Rica[a]

Location	Temperature °C (°F)	Salinity (ppt)	pH	DO (ppm)	NO^3 (ppm)	NH_3 (ppm)	PO_4 (ppm)
Guanacaste	28 (82.4)	32	7.9	>6.0	0.45	0.67	13
North Punta	29 (84.2)	28	7.7				
South Punta	28 (82.4)	30	7.7				
Isla Negritos	28 (82.4)	30	8.1				
Rio Viscaya	26 (78.8)	10	7.6	6.0	0.50	0.31	0.25
Rio Estrella	24 (75.2)	0	7.2				
Estero Negro	24 (75.2)	21	7.7				

[a]Data from FAO.

alkalinity, hardness, sulfate, iron, nitrate, phosphate, biochemical oxygen demand, bacteria, and pollutants (such as pesticides).

The hydrogen-ion concentration (pH) is a measure of acidity. A pH of 7 is neutral (i.e., neither acid nor alkaline). Extremes in pH have detrimental effects on organisms and can also facilitate adverse chemical reactions in the water and soil of ponds. The pH of seawater at a chosen site should be expected to be in the neutral range throughout most of the year, ranging between 7 and 8.3. During the rainy season, water pH readings can be expected to be slightly lower. Low pH values usually correlate with lower salinities during and after rains and are attributable to dilution by rainwater runoff. Normal ocean water is generally in the pH range of 7.6–8.2. The pH of freshwater ranges widely, but is generally between 6.0 and 9.0. Alkalinity, on the other hand, as the term is used in natural water chemistry, measures the degree to which a solution is able to resist pH change. Note that this definition of "alkalinity" is different than the more familiar definition used in chemistry, where "alkaline" is synonymous with "basic." Speaking in terms of chemistry, solutions with a pH greater than neutral are thus "alkaline," and "alkalinity" refers to the concentration of hydroxide ions (OH^-).

In pond culture, the pH is usually lowest at or near dawn and highest at midafternoon. In most finfish culture, the desired pH range is 6.5–9.0. The point at which acidity becomes deadly occurs at a pH of approximately 4, and the alkalinity becomes deadly at approximately pH 11. When the pH is outside the desirable range, fish growth is slowed, reproduction reduced, and susceptibility to disease increased.

Most references state that normal ocean alkalinity is in the 100–200 ppm range. Sometimes, alkalinity is measured in meq/L; a normal reading at pH 8.3 would be an alkalinity of 2.3 meq/L. I have measured bicarbonate alkalinity ($CaCO_3$) at different sites and found it to be between 19 ppm and 200 ppm. Alkalinities 20 ppm are considered to be detrimental to aquaculture. If the alkalinity at a site is this low, many chemical reactions occur, and it is no doubt difficult to maintain phytoplankton in ponds at the site. Alkalinity has a straight-line relationship with salinity. If the alkalinity in a body of water is low compared with the salinity, the alkalinity in the water may be being consumed by acid entering the water from the soil (e.g., through seepage or rain runoff from land). To combat this problem, white lime [calcium carbonate ($CaCO_3$)] can be added to the water after each heavy rain. Generally, ponds with acidic soil conditions and soft water are not good finfish production units.

Normally, the water source for the site should be within 1 km (0.6 mi) of the site, to keep construction and pumping costs reasonable. It should be free of pollution and from possible future sources of pollution. The physical parameters of the water source should be relatively stable on a year-round basis, to keep from stressing the cultured animals.

The water source should be matched with the requirements of the species to be cultured. Sometimes, a seawater source is mixed with freshwater in mixing ponds to bring the salinity to the desired level.

Hybrid striped bass can be grown in freshwater or seawater up to 45 ppt salinity, but grow best in freshwater. Usually, inland site permitting is less stringent than at coastal sites; therefore, an inland freshwater site should be considered for hybrid striped bass. A penaeid shrimp hatchery is most suitable at full-strength seawater (32–35 ppt salinity), whereas penaeid shrimp growout is best at 10–25 ppt in brackish water. Check the salinity of the water source by sending water samples to a qualified and reputable laboratory. It is best to check the water source over a one-year period, through rainy and dry seasons, to assess any seasonal changes. Some experts prefer to take their testing programs even further and test the water daily for one lunar cycle. If the site is in a tidal area, take water samples at high and low tides. Samples should be taken from both the top and the bottom of the water column, because salinities and other parameters vary with depth. A refractometer can be used to check salinity. If many samples are to be taken, one may want to use a portable water test kit and then verify the result with a few laboratory tests.

IMPORTANT COMPOUNDS, METALS, AND WATER PARAMETERS

In well water, the amount of dissolved gases and chemicals are important. Iron (Fe), oxygen (O_2), carbon dioxide (CO_2), hydrogen sulfide (H_2S), and ammonia (NH_3) may or may not be present. Most fish need at least 5 ppm of oxygen to remain healthy. Carbon-dioxide levels greater than 20 ppm are potentially dangerous to fish; some wells produce water at more than 60 ppm CO_2. H_2S is common in well water and can be toxic to fish at 0.1 ppm. Aeration is generally used as a practical correction for O_2, CO_2, and H_2S problems. The hardness and alkalinity of well water should be at least 20 ppm. The total amount of dissolved solids should be less than 500 mg/L (ppm).

The phosphate (PO_4–P) level in the water source is of importance, since algae production in ponds will most likely require the adding of triple phosphate to culture water. Normally, phytoplankton removes phosphorous in ponds, but if acid sulfate is present at the site, most of the phosphate removal will be due to iron and aluminum binding with phosphate and silicate in the water source and in the sediments. The normal solution to this problem is to eliminate the acid in the soils by liming and to fertilize the algae blooms with an amount of fertilizer that will promote the growth of diatoms or beneficial phytoplankton. Ten parts or greater nitrogen (urea) to one part triple phosphate (10 : 1), up to a 46 : 1 ratio, is recommended. The ratio of the fertilizer components is important to promote the correct type of bloom. As a general rule, the criterion for dosage application is to ensure a water-source nutrient concentration of 1.3 ppm nitrogen and 0.15 ppm phosphorous in production ponds.

The sulfate (SO_4) level in the water is another important parameter to consider. Normal seawater at 35 ppt salinity has 2,648–2,712 ppm SO_4. The amount of SO_4 in average brackish water is 995 ppm, and the average level of SO_4 in freshwater is 16 ppm.

The total hardness of seawater (the calcium-ion level plus the magnesium-ion level) is another important parameter to consider in the water source. The typical magnesium-ion level in full-strength seawater is 1,350 ppm and in brackish water is 125 ppm. The amount of calcium (Ca) in normal brackish water is 308 ppm. The normal Ca-ion level in freshwater is 42 ppm. Ca hardness levels found in the literature range from 340 to 11,560 ppm. The optimum total hardness for most brackish water aquaculture is 850 to 2,550 ppm, with 340 ppm as a minimum. Freshwater hardness is addressed in more detail later.

Fecal coliform bacteria is another parameter that should be looked at in the water source. The total coliform bacteria level in water is normally around 1,330 MPN (most probable number)/100 mL (3.38 fl. oz or 13/mL), and the normal fecal coliform bacteria count is around 950 MPN/100 mL (3.38 fl. oz or 9.5/mL). As a comparison, the median fecal coliform bacterial concentration in swimming water should not exceed 20,000/100 mL, and for shellfish, harvesting should not exceed 14 MPN/100 mL. Waters used in processing (e.g., for washing, freezing) should meet standards for drinking water.

Biochemical oxygen demand (BOD) is a standard test for organic material. It is defined as the oxygen used in meeting the metabolic needs of aerobic microorganisms in water rich in organic matter (such as water polluted with sewage). It is sometimes referred to as biological oxygen demand as well. The BOD of the water source should be determined. BOD is usually reported in ppm, pph, or parts per total test time and is usually measured over a five-day period. On a per-hour basis, 0.5 ppm is considered rich and 0.05 ppm light. A per-total-test-time BOD of 1.5 ppm is considered to be average for an estuary, and 8.5 ppm is considered to be average for an intensive-culture shrimp pond. Often, watershed and river runoff bring organic debris with them, which can add to the BOD. A settling canal can be added to take care of most of the suspended solids if they become a problem at the site. This action should eliminate high BOD levels. Fish communities on pond bottoms and organic buildup on pond bottoms from metabolic wastes and uneaten food can add to the BOD as well. Careful pond management (e.g., feeding, water exchange, etc.) is the key to controlling BOD levels.

Water tests for heavy metals should be conducted. A variety of heavy metals are found in almost any water source. These metals may exist in several oxidation states with different reaction potentials, depending on their specific chemistries. Some metals are considered essential for growth, but may be toxic at high levels. Some heavy metals are generated at high levels by industry and become significant pollutants. Estuarine microbes may produce alkyl-metallic compounds that can be accumulated by estuarine species and may become potent toxicants.

Although heavy metals occur naturally in the aquatic environment as a result of weathering and land drainage, in recent years, the use of various metals in pesticides and fungicides has added large quantities of heavy metals to the aquatic environment. Excessive additions of heavy metals to the aquatic environment can have an adverse effect both on animals and on people who eat the animals as food. There are a number of reports on the toxicity of heavy metals to aquatic animals (see the entry "Pollution"). Additional results of water analyses for aquaculture sites can be found in (see the entry "Shrimp culture").

Trace metals in water are mainly associated with chlorides and other inorganic complexes. If one is ever faced with the problem of heavy-metal contamination, there are a number of ways to address it. For example, if elevated iron levels begin to influence algae blooms in ponds, aeration of the water will assist in neutralizing the effects, by oxidizing iron and forming a precipitate of $Fe(OH)_3$ that will settle out of solution.

GENERAL COMMENTS ON FRESHWATER SOURCES

Freshwater is mixed with seawater to achieve proper salinity, is the key medium for freshwater aquaculture, and, among other uses, serves as drinking water. Freshwater is generally required during construction and by workers during and after construction. The freshwater source at the chosen aquaculture site can become especially important during the dry season. It may be necessary to water grasses on pond dikes to prevent erosion. Any farm site will need a reliable source of freshwater, whether it is a freshwater aquaculture facility or a mariculture facility. The freshwater source is evaluated by calculating the required rate of flow (i.e., required pond depth, prescribed filling time, total losses due to evaporation, and percolation); the available rate of flow must be equal to or greater than the required rate of flow. Freshwater wells, especially shallow wells, can be unpredictable, of low capacity, and can deplete local drinking water supplies. Deep wells require exploratory drilling and water analysis and can be expensive. Most sites require that a hydrological survey or study be conducted, to accurately assess the ground water.

Some of the most common measurements taken in testing freshwater determine the amount or level of the following factors: pH, salinity, specific conductance (or conductivity), total dissolved solids, suspended solids (or turbidity), trace elements, dissolved gases, organic material, total organic carbon, chemical oxygen demand (COD), biochemical oxygen demand (BOD), chlorophyll *a*, alkalinity, hardness, oil and grease, phenol, cyanide, total ammonia nitrogen (TAN), nitrite, and pesticides. COD is a speedy and reliable estimate of organic load; it is reported in ppm. A normal measurement is less than 10 ppm, and a high measurement is 60 ppm. Chlorophyll *a* gives an estimate of the amount of plant life in the water; unfertilized water might return a value up to 20 parts per billion (ppb), whereas fertilized water might range from 20 to 150 ppb. Phenol causes flavor problems in fish from polluted waters. Phenol levels in normal water are less than 1 ppm. Cyanide will kill fish at less than 1 ppm, and the total organic carbon level in normal surface water is expected to contain 10 ppm.

SITE VEGETATION

The types of vegetation at the site usually vary according to elevation and soil type. A high density of vegetation yields increased site development costs and should be avoided if possible. Mangrove or other swamps should be avoided. Table 3 describes some factors linked to the presence of certain species of vegetation at sites with brackish water.

Avoid sites that have nypa palm or other palms, because the palms are generally found in very sandy soils. Palms that stand straight up usually indicate that there is some clay in the soil, while palms that lean indicate that there is more sand in the soil.

SOIL CRITERIA RELEVANT TO SITE SELECTION

The physical factors of soils at the site should be considered. Will the soil type hold water? If not, seepage might be a major problem. Water that seeps through soil can carry heavy metals, which can then influence the algae blooms in ponds and cause health problems in the cultured

Table 3. Examples of Vegetation Indicators in Brackish-Water Areas

Physical/Chemical	Species Indicator
Elevated areas	*Avicennia* (mangrove)
Low-lying areas	*Rhizophora*, *Melaneura*, *Phoenix*
Clay soils	*Avicennia*
Sandy soils	Nypa palm, grasses
Peaty soils	Nypa palm, *Melaneura*
High organic content	*Rhizophora*
Potentially acidic	Nypa, *Rhizophora*, *Melaneura*
Less acidic	*Avicennia*

animals. The load-bearing capacity of soils should also be considered, as well as permeability.

The production methodology to be used depends on the soil criteria at the site. Use of lower levels of technology (such as in extensive culture) requires higher percentages of clay in the soils. Higher levels of technology allow for the use of artificial substrates or liners, due to the capability of high levels of technology to produce more than lower levels. Certain soil types provide for economical dike construction, and the larger the pond, the less soil that must be moved to build the pond. Smaller ponds require more soil to be moved, because of more and generally higher dikes or deeper ponds. Good soils help support primary productivity in a pond, while poor soils can work against primary productivity.

Soil texture at the site is also an important consideration. Soils may be classified by their texture, which is determined by the size of the particles that make up the soil. Sand, clay, and silt are the most common soil texture classifications, and a combination of the three types is found most often in the field. See (1) for ratings of soil types and water limitations in pond aquaculture.

According to engineers, a clay content of 30–50% is not always optimum for dike construction and pond bottoms, it was once thought to be, and Boyd (2) has stated that the old myth that good soil must be 30–50% clay is not true. Even soil that is 5–10% clay is fine, provided that the particle-size distribution is suitable. For example, heavy clay soils have many problems when used in aquaculture that are not associated with lighter, loamy soils. One should avoid soils with very high percentages of sand or clay, if possible. Although beach sites with high percentages of sand are used successfully for aquaculture in Peru, the construction costs are increased, because more land is required, and thicker, larger dikes with less of a slope are needed to avoid seepage and erosion. It is possible to import clay to use as the top layer of ponds and dikes, but construction costs are increased and are generally considered cost prohibitive for this method.

The best agricultural soils are loamy soils, made up of a mixture of sand and clay of different particle sizes (3). A sandy soil will not retain much water and has only a small capacity for adsorbing and holding nutrients. On the other hand, a soil with a majority of clay-sized particles binds water and nutrients tightly and is often sticky and difficult to till. A loamy soil is intermediate between the two extremes. A general rule is that soils that are considered desirable for terrestrial agriculture, that occur in suitable locations for pond sites, and that contain enough clay to provide a barrier to seepage can be used for aquaculture ponds (3).

SOIL SAMPLING HINTS AND SOIL CHEMISTRY ANALYSIS

Observe the entire site. Develop a sample matrix in which all areas are observed. Sample anomalies or abnormal areas as well as normal areas. Sample from different depths. Take at least one soil sample for every 16 ha (40 ac) (40). Check with the laboratory doing the soil analyses to see how much soil they require for each test. A reputable laboratory should look at the following soil parameters:

wet and air-dried pH, total potential acidity, total sulfide, exchangeable aluminum, total ferric hydroxide (iron), total organic carbon, and soil texture.

ACID SULFATE SOIL

Many factors may contribute to low productivity and low profitability of an aquaculture site, such as poor engineering, poor site selection, poor production methodology, poor management, and acid sulfate soil conditions. Acid sulfate soil can be identified in the field and confirmed by laboratory analysis. It has a bulk density of 1.0–1.4 g/cm^3 (0.036–0.050 lb/in.^{3}), is typically clay in texture, and has a pH range of 3 to 6.5, an organic carbon range of 1.5 to 18%, a total potential acidity level of 10 meq/100 g, and a total sulfur content range of 0.1 to 0.75%. Acid sulfate soils usually occur in low-lying areas or swamps. This type of soil is responsible for poor pond productivity, due to iron and aluminum binding with phosphate and silicate. Development of acid sulfate conditions occurs when sufficient amounts of sulfate and iron are present, along with a high content of organic matter. Inability of the tidal exchange to buffer carbonates, as well as limited aeration, will assist in the formation of acid sulfate soil. Pond productivity is generally low under acid sulfate conditions, because of a poor fertilization response. For example, benthic diatoms need silicate for the formation of their tests or shells. If the silicate is all bound up, then it is not available to the diatoms (which provide a primary food source in the pond).

To recognize acid sulfate soil in the field, look for the presence of pale yellow mottles, called jarosite, or reddish, brown ferric hydroxide. Ponds with acid sulfate soil may exhibit low water pH, and there is generally poor growth of vegetation on the dikes.

Potential acid sulfate soil is more difficult to identify. It is generally associated with sulfate-reducing root systems (*Rhizophora*) and a high organic content. Potential acid sulfate soil has gray to black specks and mottles. More information on acid sulfate soils can be obtained from (4).

Characteristics of optimum soil for shrimp culture are addressed in the entry "Shrimp culture", where examples of soil analyses can also be found.

HATCHERY SITE SELECTION

The most important parameters of hatchery site selection are water supply, water quality, proximity to broodstock, fishing industry in the area, physical access, pollution, pesticides, flood and storm potential, prevalence of disease in the area, proximity to growout ponds, culture methodology, degree of integration, competition, amount of training required for local workforce, technical support in the area, availability of building materials, availability and cost of utilities, freshwater source, and personnel comfort.

A hatchery site should be located near a disease-free broodstock supply. It should have a continuous supply of good-quality water. Shrimp hatcheries and stenohaline marine fish hatcheries should have year-round constant salinity. A hatchery site should be integrated with growout or the farms that it will be supplying. The hatchery should be located within the temperature range at which the species of choice grows best. The turbidity should be less than 100 ppm, so that expensive water treatment is avoided. The hatchery should be located in an area with low levels of dissolved organics (less than 10 ppm). There should be no hydrocarbon pollution in the area, and there should be no agricultural pesticides used in the immediate area.

BIOLOGICAL FACTORS

For hatcheries growing shrimp, the availability of healthy, wild postlarvae in the region for stocking ponds is a luxury. Milkfish are also produced primarily from wild fry. Studies to detect concentrations of shrimp postlarvae or fish fry in the region should be made. Note that their availability is not always predictable in numbers and in season; in other words, the postlarvae or fish fry are generally seasonal and are not always available when needed. As a result, the site may, at times, depend on hatchery-produced shrimp postlarvae or finfish fry for stocking. Other hatcheries in the area should be assessed. When rearing finfish, determine whether brood are available in the area, or if fingerlings can or must be purchased from a hatchery instead.

Other biological observations should also be made. For example, is the ground covered with a well-established growth of plants? If the site is in a coastal area, are the plants salt resistant? Does the terrestrial animal community include a large variety of birds and insects? If not, there may be a source of toxicity in the area. If the site is to be developed to grow fish or shrimp, are there healthy fish or shrimp already growing in the area?

Estuaries contain minerals and organic matter that have been leached from the soils of their extensive watershed and carried by runoff to the coastal area. Estuaries are where the freshwater mixes with seawater to form brackish water. Benthic algae should be found in the resulting nutrient-rich brackish water, which should provide the base for an extensive aquatic animal community. This faunal community should include a variety of fish and crustaceans that can be captured by local fishermen. Turbidity may be high in an estuary during the rainy season and may limit some of the aquatic growth. Oysters should be found naturally in the area, but may not be present if there are toxins or if the oysters have been overharvested.

The appearance of plant and animal life in an estuary has several relevant implications. First, carnivorous fish will be a direct threat to the production of small fish or shrimp larvae if the fish are allowed to enter the ponds. Second, noncarnivorous fish can also influence yields, by competing for available food and by increasing the metabolic load and BOD in the ponds. However, the presence of a variety of potential predators and competitors in the area should not be viewed in a negative manner. Their variety and appearance are a result of a high level of natural productivity at the site. This is a very favorable indication for the potential productivity, and the natural carrying capacity (i.e., food organisms that the cultured crop can eat) of the ponds can be expected to be relatively high. Indeed, this is why the semiintensive culture system, rather than a more intensive one, is usually suggested, to take advantage of natural productivity. A less natural condition (e.g., intensive culture) requires considerably greater addition of fertilizer and higher quality feed. The appearance of aquatic life, particularly postlarvae, juvenile shrimp and finfish, is an obvious indication that the physical and chemical characteristics of the water at the site are well suited to the survival and growth of aquatic organisms.

INFRASTRUCTURE

Road and water access to the site should be assessed. Materials and heavy equipment will need to be transported to the site for construction purposes and deliveries. If adequate roads and water access to the site to support such efforts do not exist, then they may have to be constructed. Boats and barges should also be considered as viable methods of transportation to and from the site.

The power supply to the area should be assessed. If a utility supply does not exist, then generators can supply electric power until utilities can be brought in from the nearest supplier. All pumps can be powered by diesel engines, though the cost of pumping will be greater than with electric motors, or else electric motors can be used to run the pumps, powered by generators. Generators can also be used for lighting, laboratory use, and domestic needs. Most utility suppliers charge the user for the expense of bringing in power (e.g., poles, wires), but in some cases, the materials are provided as part of the service connection, while everything on the inside of the site boundary is the responsibility of the owner. In some countries, such as India and Nicaragua, the government generates and supplies the power at reduced rates. Arrangements must be made with proper authorities to ensure that an ample supply of power can be provided once the expense of providing the infrastructure is completed. It may be more desirable to use the low-cost government utilities and use generators as backup. In some areas, however, such as the U.S. Virgin Islands, electrical prices from the Water and Power Authority are in excess of US$0.22/kwh, and thus it may be cheaper for the company to produce its own power. Determine whether 440-volt, three-phase electrical power is available. Although it can be more dangerous, this power supply is generally cost efficient, with less energy loss in line transmittal.

Find out if aquaculture feed mills are available. Can they provide a constant supply of feed in the needed quantities? If not, make sure to add the extra costs of importing feed.

Communications at the site should also be assessed. A telephone system will be used to link the site with the world and is a necessity at any aquaculture site. A two-way radio will provide additional communication. With the cellular telephone systems available today, communications should not be a problem to obtain, even at the most remote site.

Other goods and services near the site should be assessed as well. The site should be relatively close to sources of building materials, equipment, and contractors. To minimize construction costs, the site should be within close proximity to heavy equipment. If the goods or services are not available in the area, they will have to be obtained from elsewhere, and some may require importation. The process of procuring supplies and equipment may need to be initiated well in advance, especially if they are to be imported and if customs must be cleared, duties must be paid, and other paperwork must be processed. This procedure calls for standardized systems and rigorous management. The development of nearby hatcheries and existing processing plants should greatly assist in providing stock and making marketing-channel connections.

SOCIAL AND CULTURAL CONSIDERATIONS

Social considerations at the site should be determined and will vary considerably from developed countries to developing countries. Neighbors are a very important factor, and the characteristics of the villages and communities near the site should be taken into account. The inhabitants can be employed at the site, but the potential effects of the site's development should be studied carefully before the decision to select the site is made. To counter any opposition by the inhabitants, the benefits that will result from the development should be pointed out to them, and any grievances should be addressed fairly and quickly. Local residents should be given priority in hiring at the site or for any other positions for which they might qualify. This will help to build good will in the community. A policy of promoting a good relationship with the community can often be the most effective means of maintaining security at the site. Wherever possible, trainees for technical positions should also be drawn from the surrounding villages. A security fence is usually a requirement to protect the crop from poaching, though fences may also disrupt the normal routine or travel routes of villagers. The Chilka Lake aquaculture project in India is one example for which a government declined a project development permit as a result of protests from fishermen; the potential displacement of the fishermen outweighed the project's benefits.

The labor profile in the area should be obtained. The labor profile is generally a compromise between skilled and unskilled workers, and most skilled labor is usually brought in. Also, the local customs and traditions of the employees should be considered. For example, in Panama and other Central American countries, a "13th month" is given from before Christmas to the first of the year as vacation to employees; this vacation must be budgeted into the project.

Although efforts should be made to employ as many local residents as possible, it is anticipated that a portion of the personnel for the operation of the site must be recruited from outside of the area. Graduates from technical colleges and universities are good candidates for on-the-job training for positions as technicians and technical supervisors. Technical aides should have had some secondary-school education. Mechanics, electricians, and personnel with office and management skills are often required to operate an aquaculture site. Security guards should be hired from outside the area, so that they are not familiar with or related to the locals. If you plan to raise finfish, ensure that you have available a biologist or ichthyologist who is competent to make immediate diagnoses and proceed with proper chemical treatment of diseases, parasites, low levels of dissolved oxygen, and related problems.

The availability of potential housing for employees should also be assessed. To attract the highly skilled staff that is required for the site's development and management, it is important that pleasant living conditions exist or will be made available. Some of the technical staff must be available at all times at the production site and should be housed there. The majority, however, will live off site. Bailey et al. take a more detailed look into the social dimensions of aquaculture development (5).

OTHER IMPORTANT POINTS TO CONSIDER

When conducting the site survey, keep in mind that no site is perfect, that no two sites are the same, and that the survey should determine both technical and nontechnical parameters. It should include agricultural and economic data from the area. An expert who has selected sites for other aquaculture operations and has followed up on the results should conduct the site assessment or survey; a learner should not be in charge of site selection. The site assessor should use as much local assistance, extension services, and experience as possible, and the assessor should refer only to data that are documented and reputable.

SITE SURVEY EQUIPMENT

Site survey equipment might include a shovel, handheld auger, or backhoe for taking soil samples; soil and water pH meters; a handheld salinometer or refractometer; and a portable water test kit. Other equipment might include radios, a compass, plastic or glass water sample bottles, plastic soil sample bags, binoculars, a camera, maps, insect repellant, drinking water, and personal items. If soil samples are to be collected outside of the United States and shipped back to the United States for analysis, they must be mailed or sent via courier service; the United States Department of Agriculture (USDA) will not allow the personal transport of soils. A permit from the soil analysis laboratory is required and must be placed in the shipment. This permit ensures the USDA that the soil will be disposed of properly after it is analyzed. Some countries require the export of soil to be permitted by the Department of Mines.

AVAILABILITY OF CONSTRUCTION EQUIPMENT AND MATERIALS

Heavy equipment may not always be found or used in an area. Importation of equipment will add to the construction cost. The load-bearing capacity of the soil at the site must also be taken into consideration, as it will determine the type of equipment that can be used. If low-lying areas are to be developed, a drag line, Hymac, or other special equipment may be needed. A drag line can build a road or dike in front of itself by removing soil on either side. Hymacs can knock brush or trees over and use them for a base while excavating. The availability of cheap raw materials will influence construction costs. The local price of sand, for example, can make a big difference in the cost of a hatchery.

OTHER LAND AND TOPOGRAPHICAL CRITERIA

The distance of the site from the water source can make a big difference with regard to costs and selection of pumps and piping. The slope towards the ocean or other drainage area should be considered. For best results, the area should have a minimum slope, or drop, of 2/1000 (i.e., for every 1000 units traveled horizontally, the elevation drops 2 units). Undulations will require fill or excavation and will add to the construction cost. The location of ground water is also important, as it will determine whether the site has sufficient elevation for proper pond drainage. A minimum of 30 cm (1 ft) is needed between the lowest part of the pond and the water table.

QUICK METHOD FOR EVALUATING A SITE

A quick method to objectively evaluate a site is to use a weighted ranking system. This system allows you to score the criterion from 1 to 5, where 5 is best and 1 is worst, and to assign a relative weight of importance to each criterion, on a scale of 1 to 3, where 3 is most important and 1 is least important. Multiply the score by the weight to obtain the weighted score. Add the weighted scores and compare them to the totaled ideal scores to obtain an objective evaluation. Tables 4 and 5 exemplify this process.

Table 4. Weighted Ranking System

Criterion	Your Score	×	Weight =	Weighted Score vs.	Ideal
Soil quality	3	×	3	9	15
Water quality	3	×	3	9	15
Social and economic factors	4	×	3	12	15
Topography	5	×	2	10	10
Elevation	5	×	2	10	10
Available area	5	×	2	10	10
Potential flooding	2	×	2	4	10
Ownership	5	×	2	10	10
Accessibility	5	×	1	5	5
Vegetation	4	×	1	4	5
Mechanization	3	×	1	3	5
Weighted score total (added from above) =				86 vs.	110

Table 5. Site Evaluation and Scoring

Range of Scores (%)	Evaluation
100–80	Excellent site
79–60	Good to fair site
59–40	Marginal site
<40	Avoid site

In this site evaluation example, a score of 86 was obtained in Table 4, and thus the location was considered to be an excellent site (Table 5).

ENGINEERING SURVEY

Once the site is selected, to finalize the precise elevation and positioning of the ponds and water supply and drainage systems, it is necessary to have more detailed knowledge of the site's shape, topography, relationship to the surface-water source or estuary (if applicable), lateral profiles, natural drainage, and future locations of structures such as supply canals, dikes, drainage, and infrastructure. An engineering survey is generally required to be able to plan the project adequately. Physical references and benchmarks should be established that correspond to plotted positions on the site plans. The chosen site must be surveyed, because the engineers will require exact distances, directions and other detailed knowledge of the area.

See References 6–8 for additional information on aquaculture site selection.

BIBLIOGRAPHY

1. B.F. Hajek and C. Boyd, *Aquacultural Engineering* **13**, 115–128 (1994).
2. C.E. Boyd, Personal communication, 1999.
3. C.E. Boyd and J.R. Bowman, in H.S. Egna and C.E. Boyd, eds., *Dynamics of Pond Aquaculture*, CRC Press, Boca Raton, FL, and New York, 1997, pp. 135–162.
4. H.J. Simpson and M. Pedini, *Brackishwater Aquaculture in the Tropics: The Problem of Acid Sulfate Soils*, Food and Agricultural Organization of the United Nations, Rome, Italy, 1985.
5. C. Bailey, S. Jentoft, and P. Sinclair, eds., *Aquacultural Development: Social Dimensions of an Emerging Industry*, Westview Press, Boulder, CO, Cumnor Hill, Oxford, 1996.
6. J.W. Avault, *Fundamentals of Aquaculture*, AVA Publishing Co., Inc., Baton Rouge, LA, 1996.
7. H.H. Webber, *Proc. Gulf and Caribbean Fish. Inst.* **24**, 117–125 (1972).
8. T.L. Welborn, *Site Selection of Levee-type Fish Production Ponds*, Southern Regional Aquaculture Center Publication No. 2409, 1990.

SMOLTING

W. Craig Clarke
Pacific Biological Station
Nanaimo, Canada

OUTLINE

INTRODUCTION

Many salmonid fishes used in commercial aquaculture have an anadromous life cycle in which the adults migrate into rivers from the ocean to spawn. The juveniles reside in freshwater for a period of time and then migrate to the sea, where they grow into adults. Smolting is the process of development of juvenile salmonids from the parr stage, which lives in freshwater, to the smolt stage, which migrates out to sea and is capable of living in saltwater. This process is also known as the parr–smolt transformation, or smoltification.

There are two types of salmon aquaculture that involve the production of smolts. The first is salmon farming, an intensive form of aquaculture involving two distinct phases: a freshwater hatchery phase for the incubation of eggs and the rearing of juveniles to the smolt stage, followed by a seawater phase in which salmon are grown out to market size, usually in floating net pens. As a result of the expansion of salmon aquaculture around the world, many millions of smolts are produced for transfer to saltwater. In Norway alone, more than 100 million smolts are transferred into seawater net pens each year.

Salmon ranching is an alternative form of extensive aquaculture that involves the release of smolts from hatcheries and the harvest of the adults returning from the ocean. Several billion smolts are released annually in many countries, usually by public agencies or cooperatives in support of commercial or recreational fisheries. However, low rates of marine survival, concerns about the impacts of privately released salmon on wild stocks, and socioeconomic conflicts have constrained the development of commercial salmon ranching (1).

Currently, the value of salmon produced from intensive aquaculture exceeds that from the traditional capture fisheries. The main species used for salmon farming are Atlantic salmon (*Salmo salar*), chinook salmon (*Oncorhynchus tshawytscha*), coho salmon (*O. kisutch*), and rainbow trout (*O. mykiss*). The most important species raised in ocean ranching and enhancement programs are pink salmon (*O. gorbuscha*), chum salmon (*O. keta*), sockeye salmon (*O. nerka*), chinook salmon, coho salmon, Atlantic salmon, masu salmon (*O. masou*), and steelhead trout (*O. mykiss*).

The smolts produced must be of consistently high quality in order to achieve predictable growth and survival during the seawater phase of salmon aquaculture. The freshwater stage of the production system must be managed so that smolting is completed successfully. Hatchery managers must have a thorough understanding of the biology of the parr–smolt transformation and its control in order to produce high-quality smolts precisely at the times required by the production schedule. The transfer of smolts to saltwater before smolting is complete will cause osmotic stress, resulting in poor growth and excessive losses of stock due to death from osmotic shock or from outbreaks of disease.

The parr–smolt transformation is synchronized by environmental factors, mainly the annual cycle of changes in the length of the day (the photoperiod) and changes in temperature. Smolting involves a variety of morphological and physiological changes that prepare the juvenile salmon for life at sea. From the perspective of salmon farming, the most important characteristics are the development of the ability to adapt to saltwater and the change in growth pattern from the seasonally variable juvenile pattern to the more sustained growth observed in adults. Smolts produced for salmon ranching must be able to adapt to saltwater at a time when there is an adequate supply of prey available in order to allow them to grow and survive in the ocean.

BIOLOGY OF SMOLTING

Smolting involves a coordinated set of morphological, physiological, and behavioral changes that develop gradually under natural conditions over a period of several months in the spring. Salmon parr are dark, cryptically colored, and deeper bodied, while smolts are silvery, more slender, and have dark fin margins. The growth rate of parr in freshwater is highly seasonal, rapid during spring and summer and slow during the winter. Under natural conditions slow growth during winter is an adaptation to changes in food availability and is triggered by short-day photoperiod. In hatcheries, periods of slow growth can be minimized or avoided entirely by use of photoperiod manipulation to accelerate development.

A major change that occurs during smolting is the development of the capacity for regulation of salt and water balance in high salinity. Salmonids maintain the osmotic concentration of body fluids within a narrow range that is equivalent to one-third of seawater whether they are in freshwater or saltwater. In freshwater, the body fluids of the juvenile salmon are more concentrated than the fish's surroundings are, so water tends to diffuse inward and salts diffuse outward along the concentration gradient. Specialized cells termed chloride cells in the gills actively take up ions from the water, and excess water is excreted in the kidney. The urinary bladder reabsorbs sodium and chloride ions so that a dilute urine is produced. Once the fish is in seawater, the osmotic challenge is reversed: The external environment is more concentrated than the body fluids, so there is a tendency for salts to enter across body surfaces and for water to leave, causing dehydration. During smolting, the gill chloride cells develop the ability to actively excrete salts. Once smolting is complete, juvenile salmon can withstand direct transfer from freshwater to saltwater and can rapidly adjust their regulatory organs to maintain an appropriate salt and water balance. After entering salt water, smolts actively drink water, absorb the salts from the gut, and excrete them through the gill chloride cells. The kidney retains water, so urine volume is reduced and the urinary bladder no longer reabsorbs ions (2).

Life History and Anatomical Changes

On the one hand, the life cycle of most trout and charr species is completed in freshwater; on the other hand, juvenile Atlantic salmon, as well as Pacific salmon and steelhead trout, typically become smolts and enter the sea to grow for at least one year, until they reach adult size and migrate back to freshwater for spawning. The duration of the freshwater growth stage varies from a few weeks or months up to one or two years, depending on the species. As a result, the size at which smolting occurs also varies considerably among species (Table 1). Changes in body shape and color are more definite in species, such as the Atlantic salmon, coho salmon, and steelhead trout, that develop into smolts after one or more years of growth in freshwater. Of the many morphological changes taking place, the most noticeable is the increase in silvering of the body surfaces and darkening of the fin margins (3). Chinook salmon have two juvenile life histories, termed ocean type and stream type. Ocean-type chinook salmon enter the sea in their first year, whereas stream-type chinook smolts are at least one year old. Most stocks of chinook salmon used for commercial aquaculture have the ocean-type life history.

Domesticated rainbow trout are descended from wild freshwater trout and are sufficiently closely related to the migratory steelhead trout to be considered the same species. Although rainbow trout do not undergo a typical smolting, several strains are used for seawater aquaculture.

The condition factor, which is a measure of weight per unit length (calculated as weight $\times$ 100 $\times$ length^{-3}, where weight is expressed in grams and length is measured in centimeters) declines in smolts relative to parr. This change in body shape of smolts results from a faster

Table 1. Approximate Age and Size of Smolts Under Natural Conditions

Species	Age	Weight
Atlantic salmon	2–3 yr	30–40 g (0.066–0.088 lb)
Coho salmon	1–2 yr	7–10 g (0.015–0.022 lb)
Chinook salmon (ocean type)	3 mo	4–6 g (0.0009–0.013 lb)
Chinook salmon (stream type)	1–2 yr	8–10 g (0.018–0.022 lb)
Masu salmon	1–2 yr	12–15 g (0.026–0.033 lb)
Sockeye salmon	4 mo–2 yr	2–5 g (0.004–0.011 lb)
Pink salmon	<1 mo	0.13–0.26 g (0.0003–0.001 lb)
Chum salmon	<1 mo	0.3–1 g (0.001–0.002 lb)
Steelhead trout	2–3 yr	40–50 g (0.088–0.11 lb)

growth in length than in weight. The amplitude of the change in condition factor is influenced by temperature and feeding rate. Condition factor is increased in response to a high feeding rate and in warm water. Therefore, under hatchery conditions, smolts that are growing very rapidly may exhibit only a slight decrease in condition factor.

Environmental Control

The photoperiod is the main environmental signal that synchronizes the seasonal cycles of growth, smolting, and reproduction in salmonids. Water temperature also has a significant role in the smolting process. Species differ with respect to their sensitivity to these two factors (Table 2). A rule of thumb is that species that live more than a year in freshwater before smolting are influenced strongly by the photoperiod, while those that enter the sea during their first year are not cued by the photoperiod. Growth of juveniles of ocean-type chinook, chum, and pink salmon, for example, is controlled mainly by water temperature and food supply; smolting is not cued by photoperiod. In contrast, Atlantic salmon fry emerge from the spawning gravel under conditions of a long spring photoperiod and then experience a sequence of long days during summer, followed by short winter days. The fry develop into smolts during the second or subsequent year as the photoperiod increases. Under intensive culture in a hatchery setting, care must be taken to provide a sequence of short photoperiods followed by long ones in order to synchronize the growth and smolting of species that are photoperiod controlled. If smolts are needed at the usual time in spring, it is sufficient to simulate the seasonal photoperiod cycle. However, if smolts are required for transfer to the sea at other times of the year, the hatchery manager must provide a coordinated regulation of both photoperiod and temperature conditions to allow growth and development to be completed successfully.

Water temperature has an important influence on smolting. First, it affects the rate of growth of fry and thus can control the attainment of the necessary body size for smolting. Temperature also affects the process whereby the fish adapts to seawater. Tolerance of salinity is greatest at moderate temperatures of 8–14 °C (46–57 °F), but is reduced considerably at temperatures above 17 °C (63 °F) or below 5 °C (41 °F). Daily and seasonal temperature cycles also influence smolting and initiation of seaward migration.

Table 2. Classification of Important Anadromous Salmonid Species According to Photoperiod Control of Smolting

Photoperiod Controlled Smolting	Photoperiod Independent Smolting
Atlantic salmon	Ocean-type chinook salmon
Coho salmon	Chum salmon
Stream-type chinook salmon	Pink salmon
Masu salmon	
Sockeye salmon	
Steelhead trout	

Hormonal Changes

Numerous changes in hormonal activity control physiological and morphological development during smolting. Although research is usually conducted only on one or a few hormones at a time, the various developmental changes that constitute smolting are controlled by a number of hormones acting in a coordinated manner. In smolts, an increase in thyroid hormone secretion is associated with silvering, increased growth, and olfactory imprinting. Administering the thyroid hormone triiodothyronine in the diet can induce the development of several characteristics of smolts, but tolerance of salinity is usually not affected (4,5). Cortisol secretion from the interrenal tissue is also increased at smolting. Cortisol promotes differentiation of the chloride cells in the gills and increases the concentration of the ion transport enzyme sodium, potassium-adenosine triphosphatase (Na^+, K^+-ATPase) in preparation for entry of the fish into saltwater. Other effects of cortisol include suppression of the immune system and a decrease in body lipid content. Growth hormone secreted by the pituitary gland promotes growth and facilitates adaptation to saltwater (6–8).

INTENSIVE SMOLT PRODUCTION SYSTEMS

Hatchery Tanks

Hatchery tanks traditionally fall into two categories: the flow-through or raceway type and the circulating tank. Raceways were commonly used in hatcheries that produced smolts for release to the sea. Circulating tanks are typically used in modern commercial facilities that produce smolts for stocking cages. Circular tanks provide better control over water velocity and fish distribution. Where feasible, hatcheries may use groundwater to avoid the introduction of infectious agents and maintain more moderate temperatures. Control of the photoperiod is also more readily achieved in these systems.

Net Pens

Often, net pens in freshwater lakes or ponds are used for rearing fish from the fry to the smolt stage. A major advantage of net pens is that costs for their construction and for operation are greatly reduced relative to costs of land-based hatcheries. Also, an ample water supply allows the production of greater biomass. However, a disadvantage is that control over environmental conditions is reduced, because water temperatures often vary widely and it is not feasible to exclude light to provide a short-day photoperiod stimulus. In addition, the stock may be exposed to pathogens from other fishes in the lake.

As an alternative to lakes, net pens may be suspended in brackish water in estuaries (9). The salinity should not exceed 20 ppt until the parr–smolt transformation is completed. The advantages of brackish water rearing are that the sea offers greater thermal stability than most lakes and the juveniles are able to acclimate to higher salinity more gradually. It is important that vaccination be completed before transfer of the presmolts into brackish water; the common marine bacterium *Vibrio anguillarum* can cause severe mortality of juvenile salmonids that lack resistance to it.

Floating Enclosures

An improvement over netpens is the use of floating enclosures (10). Water is pumped through the enclosure in a manner analogous to the way a circulating tank works in a conventional land-based hatchery. In comparison with net pens, floating enclosures provide more control over water quality, because the intake for the pump can be placed at variable depths. The enclosures can be used in either fresh or saltwater. For use in the sea, the salinity of the incoming water can be controlled by a venturi device to mix fresh and saltwater (11). Thus, floating enclosures are intermediate between netpens and land-based hatchery systems, in that they have the portability of netpens together with the hydraulic characteristics of circulating tanks. Trials with a commercial version of the floating enclosure have demonstrated that it is possible to obtain more rapid growth and reduced mortality at considerably higher stocking densities than are possible in conventional netpens (12).

Stress

Extra care must be taken to minimize stress in smolt production facilities. The scales are more easily removed in smolts than in parr. Descaling makes the fish more susceptible to disease and impairs osmoregulation in saltwater. Smolts are also more susceptible to stress compared with parr (13), and their immune system is suppressed by hormonal changes, making them more sensitive to infection (14). Handling and transportation are two common sources of stress when smolts are transferred from hatchery facilities to growout facilities in seawater. The reduction of stress due to handling can greatly diminish mortality associated with entry of the fish to saltwater (15).

ANALYZING SMOLT READINESS AND QUALITY

Frequently, large and apparently healthy hatchery juveniles, though silvery in appearance, are not fully functional smolts (16). This is particularly true under intensive rearing conditions in which the synchrony of the various developmental events constituting smolting can break down. For that reason, it is necessary to have an objective method for assessing smolt quality and readiness for transfer to saltwater.

Seawater Challenge Test

The seawater challenge hypo-osmoregulatory test measures the ability of a fish to control plasma ion levels quickly after abrupt transfer to saltwater. The test was developed originally to measure the readiness of accelerated coho salmon smolts for transfer into seawater. The seawater challenge test exhibited a regular developmental pattern even at the increased temperatures used to accelerate the animal's growth (17). Its other advantages are that it is cheaper and simpler to conduct than are assays for gill Na^+, K^+-ATPase or plasma thyroid hormone.

To conduct the seawater challenge test, a sample of smolts is transferred to saltwater at the acclimation temperature. Salinity of the saltwater used for the test should be in the range of 28–32 ppt. If it is necessary to make artificial saltwater, a balanced salt mixture must be used because single ion solutions are toxic. After 24 hours, the smolts are anesthetized and a blood sample is collected. Plasma is separated from the blood cells by centrifugation and can be stored for up to 24 hours in a refrigerator or for longer periods in a freezer prior to analysis. Regulatory ability can be assessed by measuring either sodium, chloride, or osmolality of the plasma (18–20). Expected plasma sodium levels in smolts after 24 hours are in the range of 160–165 mmol/L. The corresponding level for osmolality is 330–340 mosmol/L, and for plasma chloride it is 140–150 mmol/L.

If mortality is observed during the test, the smolts have very poor hypo-osmoregulatory capacity, possibly due to gill damage or other health problems. Other factors that can influence the results obtained from the seawater challenge test include the temperature and density of the water during the test, descaling of the smolts, pollutants such as copper, and contamination of the blood samples with saltwater.

A single test is not sufficient to determine whether the fish's performance is increasing or decreasing. Therefore, it is advisable to perform the test regularly during the last two to three months before the planned date for transferring the fish to saltwater for growout.

Salinity Tolerance

The salinity tolerance test involves counting the number of dead fish 96 hours after they are transferred to hypersaline saltwater. Salinities of 35–40 ppt are necessary for the test because many large parr will tolerate 30 ppt indefinitely, despite the fact that they are unable to grow. Salinity must be measured precisely because mortality rises quickly with increasing salinity. The tolerance test is conducted in temperature-controlled containers so that the fish can be held at their acclimation temperature. The smolts are made to fast for 24 hours and are not fed during the test, to avoid fouling of the containers. Smolts that survive for 96 hours in high salinity are capable of direct transfer to typical salinities of 30–32 ppt for growout. More information can be obtained if a range of salinities is used simultaneously; this strategy allows probit analysis to compute the 96-hour median lethal salinity with confidence limits (21).

The main advantage of the tolerance test is its simplicity. However, its shortcomings relative to more complex tests are that a large number of fish are required for statistical comparisons and it causes more distress and mortality to the smolts.

Gill Na^+, K^+-ATPase

Biologists have shown that the level of Na^+, K^+-ATPase in gill tissue increases in smolts and have used it as an indicator of when smolting is complete, particularly in research. Na^+, K^+-ATPase is an enzyme that is concentrated in the chloride cells and is involved in ion transport across the gills. Several methods are used for measuring gill Na^+, K^+-ATPase activity, including microassays, which can be used for nonlethal biopsies (22–24). Essentially, the assay involves the determination of enzyme activity (the breakdown of adenosine triphosphate) in a microsomal preparation from gill tissue. The amount of protein in the preparation is also measured, and the activity is expressed per unit wet weight of protein. The values obtained are dependent upon the conditions used for the assay. Therefore, it is important to maintain uniform conditions so that the results can be compared among assays.

Hormone Assays

As noted earlier, a number of hormonal changes occur at the time of the parr–smolt transformation. One of the best studied is that of the thyroid hormone thyroxine, which rises to a peak and then decreases (25). Folmar and Dickhoff (26) proposed using the proportion of the peak in plasma thyroxine levels that had been completed as an indicator of the time of transfer of smolts into saltwater.

OFF-SEASON TRANSFERS

Salmon respond to the length of the day in the same way that many plants and animals do: by measuring the length of the dark period. When they are exposed to light during a period of high sensitivity, a photoperiodic response is induced (27). The length of the period of light that must be exceeded in order to produce a photoperiodic response is termed the critical daylength and is usually between 10 and 14 hours (27). For juvenile coho salmon, the critical daylength is between 11.5 and 12 hours (28). The practical significance of this fact is that simple light-control systems can be used to provide the necessary stimuli, and changes in the photoperiod can be imposed abruptly. It is possible to produce short-day conditions by using lightproof covers that are placed on and removed from the rearing tanks according to a regular schedule each day.

Accidental exposure of juvenile salmon to artificial light during the night must be avoided in order to prevent disruption of the smolting process. This is particularly important during the short-day portion of the photoperiod cycle. Extensive research has shown that the parr–smolt transformation can be induced out of season by a sequence of a short-day photoperiod followed by a long-day photoperiod (29). However, there are species-related differences in the size at which the photoperiod treatment can be initiated. On the one hand, coho and stream-type chinook salmon fry can be given eight weeks of short-day exposure from the time of first feeding, followed by a long-day photoperiod, provided that water temperatures are sufficient to permit rapid growth to smolt size. The minimum growing temperature when one starts with the fry stage is 8 °C (46 °F), but larger and more uniform smolts are obtained at temperatures of 11–14 °C (52–57 °F) (30). On the other hand, Atlantic salmon must reach approximately 10 cm (4 in.) in length before they become competent to respond to the short-day treatment (31). It is customary to expose Atlantic salmon to a long-day photoperiod from the first feeding and then impose a short-day photoperiod about four months prior to the desired date for transferring the fish to saltwater. The smolts are ready for transfer to saltwater about 400 degree-days (°C) (720 degree-days, °F) after initiation of the long-day treatment, or approximately six to eight weeks, depending on the temperature (32).

FURTHER READING

For additional information on smolts, consult (33–36).

BIBLIOGRAPHY

1. W.R. Heard, in W. Pennell and B.A. Barton, eds., *Principles of Salmonid Culture*, Elsevier, Amsterdam, 1996, pp. 833–869.
2. S.D. McCormick and R.L. Saunders, in M.J. Dadswell, R.J. Klauda, C.M. Moffiit, R.L. Saunders, R.A. Rulifson, and J.E. Cooper, eds., *Common Strategies of Anadromous and Catadromous Fishes*, American Fisheries Society, Bethesda, MD, 1987, pp. 211–229.
3. A. Gorbman, W.W. Dickhoff, J.L. Mighell, E.F. Prentice, and F.W. Waknitz, *Aquaculture* **28**, 1–19 (1982).
4. G. Boeuf, A.M. Marc, P. Prunet, P.Y. Le Bail, and J. Smal, *Aquaculture* **121**, 195–208 (1994).
5. J.E. Shelbourn, W.C. Clarke, J.R. McBride, U.H.M. Fagerlund, and E.M. Donaldson, *Aquaculture* **103**, 85–99 (1992).
6. B.T. Björnsson, *Fish Physiol. Biochem.* **17**, 9–24 (1997).
7. G. Nonnote and G. Boeuf, *J. Fish Biol.* **46**, 563–577 (1995).
8. B.T. Björnsson, S.O. Stefansson, and T. Hansen, *Gen. Comp. Endocrinol.* **100**, 73–82 (1995).
9. A.C. Wertheimer, W.R. Heard, and R.M. Martin, *Aquaculture* **32**, 373–381 (1983).
10. R.M. Martin and W.R. Heard, *Aquaculture* **61**, 295–302 (1987).
11. W.R. Heard and F.H. Salter, *Prog. Fish-Cult.* **40**, 101–103 (1978).
12. H. Kreiberg and V. Brenton, *Aquaculture Update 79, 81*, Pacific Biological Station, Nanaimo, BC, 1997, 1998.
13. B.A. Barton, C.B. Schreck, R.D. Ewing, A.R. Hemmingsen, and R. Patiño, *Gen. Comp. Endocrinol.* **59**, 468–471 (1985).
14. A.G. Maule, C.B. Schreck, and S.L. Kaattari, *Can. J. Fish. Aquat. Sci.* **44**, 161–166 (1987).
15. T.A. Flagg and L.W. Harrell, *Prog. Fish-Cult.* **52**, 127–129 (1990).
16. G.A. Wedemeyer, R.L. Saunders, and W.C. Clarke, *U.S. Natl. Mar. Fish. Serv. Mar. Fish. Rev.* **42**(6), 1–14 (1980).
17. W.C. Clarke, *Aquaculture* **28**, 177–183 (1982).
18. J. Blackburn and W.C. Clarke, *Can. Tech. Rep. Fish. Aquat. Sci.* **1515**, 1–35 (1987).
19. E.G. Grau, A.W. Fast, R.S. Nishioka, H.A. Bern, D.K. Barclay, and S.A. Katase, *Aquaculture* **45**, 121–132 (1985).
20. T. Sigholt, T. Järvi, and R. Lofthus, *Aquaculture* **82**, 127–136 (1989).
21. J. Johnsson and W.C. Clarke, *Aquaculture* **71**, 247–263 (1988).
22. W.S. Zaugg, *Can. J. Fish. Aquat. Sci.* **39**, 215–217 (1982).
23. S.D. McCormick, *Can. J. Fish. Aquat. Sci.* **50**, 656–658 (1993).
24. R.M. Schrock, J.W. Beeman, D.W. Rondorf, and P.V. Haner, *Trans. Am. Fish. Soc.* **123**, 223–229 (1994).

25. W.W. Dickhoff, L.C. Folmar, and A. Gorbman, *Gen. Comp. Endocrinol.* **36**, 229–232 (1978).
26. L.C. Folmar and W.W. Dickhoff, *Aquaculture* **23**, 309–324 (1981).
27. E. Bünning, *The Physiological Clock Circadian Rhythmns and Biological Chronometry*, 3rd ed., Springer-Verlag, New York, 1973.
28. W.C. Clarke, *Canadian Technical Report of Fisheries and Aquatic Sciences* **1831**, 133–139 (1991).
29. W.C. Clarke, *World Aquaculture* **23**(4), 40–42 (1992).
30. W.C. Clarke and J.E. Shelbourn, in N. De Pauw, E. Jaspers, H. Ackefors, and N. Wilkins, eds., *Aquaculture—A Biotechnology in Progress*, European Aquaculture Society, Bredene, Belgium, 1989, pp. 813–819.
31. J.B. Kristinsson, R.L. Saunders, and A.J. Wiggs, *Aquaculture* **45**, 1–20 (1985).
32. T. Sigholt, T. Åsgård, and M. Staurnes, *Aquaculture* **160**, 129–144 (1998).
33. G. Boeuf, in J.C. Rankin and F.B. Jensen, eds., *Fish Ecophysiology*, Chapman and Hall, London, 1993, pp. 105–135.
34. W.C. Clarke and T. Hirano, in C. Groot, L. Margolis, and W.C. Clarke, eds., *Physiological Ecology of Pacific Salmon*, UBC Press, Vancouver, 1995, pp. 317–377.
35. W.C. Clarke, R.L. Saunders, and S. McCormick, in W. Pennell and B.A. Barton, eds., *Principles of Salmonid Culture*, Elsevier, Amsterdam, 1996, pp. 517–567.
36. W.S. Hoar, in W.S. Hoar and D.J. Randall, eds., *Fish Physiology*, Volume 11B, Academic Press, New York, 1988, pp. 275–343.

See also OSMOREGULATION IN BONY FISHES; SALMON CULTURE.

SNAPPER (FAMILY LUTJANIDAE) CULTURE

D.A. DAVIS
K.L. BOOTES
C.R. ARNOLD
Marine Science Institute
Port Aransas, Texas

OUTLINE

INTRODUCTION

The snapper family contains a large number of species that are found throughout the world in tropical and subtropical areas (1). They are primarily bottom-oriented predators, occurring from shallow inshore areas to deep offshore waters. The snapper family represents an important fisheries resource in almost all areas in which they are found (2). As with many of our world fish stocks, most snapper fisheries are being harvested at or beyond their maximum sustainable yield. Because of their wide acceptance as an excellent food fish, high market price, and limited harvests from wild stocks, there is considerable interest in culturing a variety of snapper species. Despite the economic importance of this family, information pertaining to culture techniques is quite limited.

Although literature is scarce, techniques for spawning, larval rearing, and growout are available for several species of snapper such as mangrove red snapper (*Lutjanus argentimaculatus*) (3), John's snapper (*L. johni*) (4), mutton snapper (*L. analis*) (5), and yellowtail snapper (*Ocyurus chrysurus*) (6). Several other species, such as grey or mangrove snapper (*L. griseus*) and red snapper (*L. campechanus*), have been cultured in the laboratory through the larval stages (7,8). When considering raising species of this family, one must recognize that although eggs and larvae may be obtained using a variety of methods, in general the larvae are relatively small (1.5–2.7 mm total length) and difficult to raise. Considerable effort has been invested in the development of larval rearing techniques for snappers. Success has been limited and techniques are considerably behind those developed for other marine species. Because of their high market value and wide acceptance as an excellent food fish, small scale commercial growout of wild juveniles has developed in a variety of countries; however, if snapper are to become established species for aquaculture, techniques for controlled spawning and larval production must be developed to supply juveniles to the commercial industry.

BROODSTOCK MANAGEMENT

Fish, like many other animals, exhibit rhythmic physiological and behavioral patterns generally known as biorhythms. Fish reproduction (maturation, mating, and spawning) is often rhythmic and strongly correlated with the interrelated seasonal cycles of light, water temperature, and food supply. This naturally leads to a period of the year which can be defined as the spawning season. Based on natural history data for over 40 species, Grimes (9) reported that reproduction falls into two general patterns: (*1*) continental-shelf populations, which demonstrate restricted seasonal spawning, centered around summer, and (*2*) populations and species associated with small oceanic islands, which mostly reproduce year-round, with pulses in the spring and fall. Spawning frequently occurs at night, with distinctive courtship behavior culminating in an upward spiral accompanied by the release of gametes into the water column.

Under hatchery conditions, many species of snapper will spawn when suitable seasonal changes occur and food is not limiting. If natural spawning is to be encouraged, the fish must be maintained in a culture system that will ensure adequate water quality parameters and give sufficient room for mating behavior. Broodstock can either be obtained from the wild as adults or cultured from juveniles. If broodstock are to be collected from the wild, it is preferable to capture them by hook and line. This method offers the least amount of stress and a high rate of success. Abdominal distension, owing to overinflation of the air bladder, is a common problem when obtaining fish from deep water. If this occurs, the air can be removed by puncturing the abdomen with a sterile hypodermic needle. Anesthetics, such as quinaldine sulfate or tricaine methanesulfonate, should be used when handling and transporting the fish.

Most snappers acclimatize quickly to captivity and begin to feed within a short period of time. Initially, fresh food items such as squid, shrimp, and fish should be offered on a regular basis, and the fish treated for any disease or parasite infections. Once the fish have been acclimated to captive conditions, they should be transferred to the hatchery or growout facilities. Juveniles may be grown under typical growout conditions by using commercially prepared feeds; however, once the fish approach sexual maturity they should be switched to a high-quality maturation diet.

The nutritional status of broodstock is one of the primary determinants of egg quality and consequently should be considered as a critical component of the production cycle. Although the overall nutritional quality of the diet is important, a variety of trace nutrients [e.g., polyunsaturated fatty acids (PUFA), particularly of the n-3 series, vitamins C and E, as well as carotenoids] are considered very important (10). The requirement for trace nutrients is generally satisfied by providing a mix of high-quality food items such as fresh squid, cuttlefish, shrimp, krill, and fish. If desired, both marine lipid and vitamin supplements may be included in the diet. Although semimoist and dry compounded diets are sometimes used as maturation diets, they are generally accompanied by a fresh component. Snapper have been spawned using both indoor [*L. campechanus* (11), *O. chrysurus* (12), *L. stellatus* (13)] and outdoor culture systems [*L. argentimaculatus* (14)]. In the laboratory, temperature and photoperiod can be adjusted to mimic seasonal changes (11). A typical indoor maturation and spawning system will consist of culture chamber, biological filter, particulate removal system, supplemental aeration, circulation pump, and temperature and photoperiod control. The annual cycle utilized to induce maturation and spawning can vary in length from a full year to a four-month condensed cycle. An example of a subtropical artificial cycle is presented in Table 1. As the fish pass though the artificial seasons, changing temperature and day lengths will result in the natural development of maturation and spawning. *Lutjanus campechanus*, and *O. chrysurus* are examples of summer spawners which, under summer conditions, will spawn, releasing thousands of small floating eggs. Once spawning has initiated, it will continue as long as the fish are held under these conditions and nutritional reserves are adequate for ovulation. With the advent of controlled spawning techniques, a fish can be spawned any time of the year and held in spawning condition for an extended period of time. Additionally, the same fish can be utilized as broodstock year after year. It should be noted that under artificial conditions the spawning season can be extended for extremely long periods of time. For some species, extended spawning seasons and/or inadequate overwintering to allow gonadal reabsorption will often result in a decrease in egg and larval quality. Signs of overspawning vary between species, but often include reduced spawning frequency and egg production, changes in egg and oil globule diameter, and increases in larval sensitivity to stresses and/or difficulties in larval rearing.

Table 1. Manipulation of the Temperature (°C) and Photoperiod (hr) for Summer Spawning Broodstock Captured in the Summer and Acclimated to the Laboratory for 2–4 Months

Time (months)	Light : Dark	Temperature
Acclimation 2–4 months	15:9	27
2	12:12	22
2	9:15	18
2	12:12	22
1–4[a]	15:9	27

[a]Spawning should begin.

If maturation and spawning does not occur using environmental manipulations or laboratory facilities for the manipulation of temperature and photoperiod are not available, spawning can be induced by using hormone treatments. Hormone treatments have been shown to be reliable methods of inducing spawning of various species of snapper, including *L. johni* (4), *L. argentimaculatus* (14), *L. campechanus* (15), *L. synagris* (8), *O. chrysurus* (8), and *L. analis* (5). The primary substances used for hormone-induced spawning are (*1*) pituitary extracts and purified gonadotropins (e.g., human chorionic gonadotropin, HCG), which are used to stimulate the ovaries and testes, or (*2*) luteinizing hormone-releasing hormone analogs (LHRHa) to stimulate the pituitary. Hormone treatments may be administered to fish with well-developed oocytes either by direct injection, utilization of an implant designed to slowly release the hormone, or oral administration (16). Monitoring of ovarian development is a critical component of hormone therapy, and only females with eggs at the yolk globule stage should be induced to spawn. Ovarian development can be determined from eggs sampled by catheterization, i.e., removal of intraovarian eggs by using a polyethylene cannule [e.g., 2.5 mm (0.1 in.) tubing attached to a suction device such as a syringe]. Single or multiple injections of HCG (500–1500 IU/kg body weight) as well as single injections of LHRHa (100 μg/kg body weight) have been successfully utilized to induce spawning in mature snapper. Induced spawning generally occurs within 24–38 hr after hormonal stimuli to prematured eggs. Both *L. johni* and *L. argentimaculatus* females (4–6 kg) are reported to release 1 to 2 million eggs per female, and they may spawn for four consecutive nights.

In general, spawning occurs at night and is highly synchronized. The fertilized planktonic eggs can be collected either automatically during the night or by hand in the morning. Most egg collection systems rely on a simple screening system, which includes (*1*) inflow of water containing eggs, (*2*) 500–750 μm screen or collection bag designed to retain the eggs, and (*3*) return of culture water to the tank or into an effluent drain. Once the eggs are collected, they should be counted by volumetric methods (e.g., 1,000 eggs/mL for *O. chrysurus*), treated to reduce the transfer of pathogens (e.g., 1 hr bath in 10 ppm formalin), and then transferred to hatching or rearing tanks. Care should be taken to make sure that the rearing tanks are at the same temperature and salinity as the spawning tanks. Although time to hatch is temperature dependent, tropical and subtropical species will generally hatch within 24 hr of fertilization.

LARVICULTURE

Snappers are members of the group of fishes known as pelagic spawners. The reproductive strategy employed by these fishes involves the production of large numbers of very small, buoyant eggs which, after being released into the water and fertilized, drift freely in surface currents with no further parental attention. Newly hatched larvae typically carry an elliptical yolk sac containing an oil globule. Energy stores in the yolks of these eggs tend to be minimal, resulting in small larvae that have both a small mouth gape at first feeding and little ability to survive on food of suboptimal nutritional value. Providing prey items of appropriate size and nutritional characteristics is a major concern in the culture of these larvae. Snappers produce particularly small eggs and larvae (1.5–2.7 mm total length), and defining efficient techniques for their rearing has been the primary constraint on the expansion of the commercial culture of snapper species.

Techniques that have been utilized for the rearing of larval snapper range from high-density, also called intensive, systems utilizing monocultures of phytoplankton and zooplankton to low-density, or extensive, systems that rely on wild plankton. One of the first reports of successful laboratory rearing of lutjanid larvae through metamorphosis to a juvenile was conducted by Richards and Saksena (7), who reared *L. griseus* larvae on a diet of size-sorted, wild zooplankton. Although only a few larvae reached metamorphosis, this marked the initiation of successful laboratory-scale larval rearing trials. Since this initial work, a variety of species have been raised in the laboratory and in pilot-scale operations.

Lutjanus analis, *L. synagris*, *O. chrysurus*, and a hybrid (female *L. synagris* × male *O. chrysurus*) larvae were cultured by Clarke et al. (8). They used cultured rotifers, algae, and size-sorted wild zooplankton for the first 13 days, followed by the feeding of *Artemia* nauplii through the end of the culture period. A similar technique was used by Riley et al. (12) in rearing *O. chrysurus*. Those initial successes have led to pilot-scale evaluations with *L. analis* and *O. chrysurus*.

Work by Watanabe et al. (5) described initial success of mass production of *L. analis* juveniles in a 30,000 L (7,926 gal) outdoor larval rearing tank. The culture tank was prepared by filling with unfiltered seawater, fertilizing with inorganic fertilizer and inoculating the culture water with *Nannochloropsis oculata* followed by ss-type rotifers, *Brachionus plicatilis*. Eggs were then stocked into the system at a density of 10.5 eggs/L (39.7 eggs/gal). Both rotifers and algae were supplemented to the culture tanks from batch cultures. Newly hatched *Artemia* nauplii, as well as lipid-enriched nauplii, were fed at 1/L (3.8/gal) from day 7 to 35 posthatch with artificial feeds introduced at day 24 posthatch. On day 38 posthatch, the fish averaged 0.3 g and had a survival from day two posthatch of 14.3%.

Similarly, ongoing research with *O. chrysurus* at the University of Texas at Austin, Marine Science Institute, Fisheries and Mariculture Laboratory, has resulted in closure of the life cycle for that species. Utilizing a modified version of the intensive larval rearing techniques developed for *Sciaenops ocellatus*, several thousand juvenile *O. chrysurus* have been reared from F1 generation broodstock. The procedure currently being evaluated includes the stocking of 9.4 eggs/L (35.4 eggs/gal) in 1,600 L (422 gal) culture tanks containing presterilized seawater. After hatching, the tanks are inoculated with algae (*N. oculata* or *Chlorella minutissima*) and rotifers from batch culture tanks. During the first four days of feeding, the rotifers are maintained at 10 to 20/mL after which, the rotifers are flushed from the tank each night and replaced with enriched rotifers on a daily basis. The feeding of microparticulate larval feeds is initiated at day 7 posthatch, and enriched *Artemia* nauplii are introduced at day 14 and maintained through day 30 posthatch. Initial trials have resulted in a survival rate of approximately 4% from egg to juvenile.

The commercial importance of the red snapper (*L. campechanus*) in the coastal fisheries of the southeastern United States has led to substantial interest in developing techniques for the culture of this species. Arnold et al. (11) first described the successful induction of spawning of captive broodstock using temperature and photoperiod manipulation. However, rearing of larvae to the juvenile stage was not successful (17). Repeated attempts at rearing red snapper larvae by this and other groups, [e.g., Minton et al. (18)], using the same controlled rearing techniques, were also unsuccessful and resulted in 100% mortality by day 21 at the latest. In this rearing approach, larvae are typically hatched and maintained in relatively sterile recirculating systems, and food organisms are collected and transferred from other environments (e.g., coastal waters, aquacultural growout ponds, and algae-based rotifer cultures).

A less-controlled mesocosm approach in which fertilized eggs were allowed to hatch in selectively managed wild zooplankton cultures in outdoor tanks resulted in modest survival of red snapper larvae (19). Pond-filtered zooplankton, collected during the day and consisting predominantly of rotifers, was supplemented to the culture water from day 4 through day 15 posthatch. Additions of night-collected copepods began at day 4 and continued through day 17 posthatch. The feeding of fatty acid-enriched second-instar *Artemia* nauplii began at day 16 posthatch, and juvenile snapper were offered chopped, dried krill starting at day 27. These cultures contained a variety of potential prey organisms, particularly copepod nauplii and various protozoan species. Both copepods and protozoans are significant components of marine zooplankton (20) and may be critical for proper nutrition of the larvae. Survival of larvae occurred in 4 of 11 culture tanks and averaged 3%.

The rearing attempts with red snapper (19) were based on outdoor techniques utilized in Asia for the culture of the mangrove red snapper, *L. argentimaculatus* (14), and the John's snapper, *L. johni* (4). As noted by Duray et al. (21), rearing trials with *L. argentimaculatus* have been carried out for a number of years, using a variety of techniques, with limited success. Extensive larval rearing techniques were initially found to produce the best results. In this production system, eggs were hatched and larvae reared in 190 m^3 (50,198 gal) outdoor tanks in which populations of various prey organisms had been established. Copepods, collected from earthen ponds, were cultured in separate tanks and transferred to the larval rearing tank from day 3 to day 8 posthatch. Rotifers cultured with baker's yeast and algae were offered from day 6 to 12 posthatch, and *Artemia* nauplii from 10 days to the end of the rearing period at 20 days. Copepod nauplii were the only organisms observed in the guts of larvae through day 11 posthatch. Rotifers were first observed at day 12. Up to 18,000 larvae, or nearly 100 larvae/m^3 (380 larvae/1000 gal), at three weeks posthatch were produced.

The initial success of extensive larval rearing techniques for *L. argentimaculatus* has led to the development of a semi-intensive larval rearing technique using small rotifers (21). In this production protocol, newly hatched larvae were reared indoors in 3 m^3 (792 gal) tanks at an initial stocking density of 30 larvae/L (113 larvae/gal) and water temperature of 25.5 to 28.7 °C (77.9 to 83.7 °F). *Chlorella* was added daily at $1-3 \times 10^5$ cells per mL as a food source for rotifers and as a water conditioner. The best results were obtained when harvested rotifers were screened to <90 µm during the first 14 days of feeding. Rotifers were maintained at 20/mL through day 20 and 10/mL from day 20 to day 30. Newly hatched *Artemia* nauplii were offered from day 21 to 25 and enriched instar II *Artemia* offered from day 26 to 50 at 1–3 *Artemia*/mL. Average survival at day 24 was 27%. During the second phase of rearing (day 25–55), increasing sizes of *Artemia* were fed, and starting at day 40 minced fish was provided. The mean survival during this phase of production was reported to be 10.6%.

Although results with several snapper species are promising, protocols for the reliable production of juveniles have not been established, and techniques are far behind those established for other species. There appears to be a consistent trend that snapper larvae perform best in large culture systems. This trend may be due to both muting of environmental changes (e.g., daily temperature shifts) as well as more stable food webs. Also, the larvae are very small and have limited endogenous nutritional reserves, so the first feeding and the nutritional quality of the feed is critical. Several authors have noted that snapper larvae are often sensitive to environmental changes and handling. This is most likely due to poor-quality spawns and/or inadequacies of the larval foods. With respect to larval feeds, there are several nutrient deficiencies that could explain the observed problems. Essential fatty acids (EFA) have been demonstrated to be critical for larval development, and deficiency symptoms include reductions in the resistance to stress (22). Although the EFA requirement for snapper larvae is unknown, a deficiency in EFAs and/or other nutrients would explain some of the problems encountered with larval survival. Consequently, it is recommended that the polyunsaturated fatty acid content of live food organisms should be optimized (23), and cofeeding of high-quality artificial feeds in conjunction with live prey items should be incorporated into production techniques whenever possible. As our understanding of snapper larvae nutrition expands and species-specific culture techniques are developed and/or adapted from other species, reliable techniques for larval rearing should be available in the near future.

GROWOUT

Many species of snapper may not exhibit growth rates and market prices that will justify commercial production, but several species are commercially cultured, and a number of species have characteristics that make them good candidates for culture. Unlike the larvae, which are generally difficult to raise, juveniles are quite hardy and adapt very quickly to culture conditions. Even juveniles captured from the wild will readily adapt to culture conditions and prepared feeds. Quite often they will initiate feeding on the same day that they are captured and transported into captivity.

Current literature indicates that only two species of snapper, *L. argentimaculatus* and *L. johni*, are considered economically important culture species (4,14). However, because of wide acceptance and marketability of snapper, a number of species are being evaluated for their culture potential. Despite the economic importance of snapper, data on the growth rates in the wild or under culture conditions are quite limited. Manooch (24) summarized the growth rates of a number of snapper species in the wild. Natural history data show that many species of snapper exhibit relatively slow growth rates, requiring several years to reach a marketable size. Four species that are currently being evaluated for their culture potential in the United States are *O. chrysurus*, *L. campechanus*, *L. griseus*, and *L. analis*. Based on predictive growth equations reported for *O. chrysurus* (25,26) and *L. griseus* (27) those species will only reach about 200 g after two years in the wild. By utilizing semiclosed recirculating systems, *O. chrysurus* has been raised to 489 g in 767 days posthatch (6). These growth rates are similar to the results obtained by Thouard et al. (28) using wild juveniles cultured in cages in Tahiti. Similarly, wild juvenile *L. griseus* (3 g initial weight) raised under controlled conditions reached about 500 g after two years of culture (6). Based on growth estimates of *L. analis* (29), this species will reach about 116 g after one year or 362 g after two years in the wild. Watanabe et al. (5) reported final weights of *L. analis* from a pilot-scale growout. In this growth trial, a mean body weight of 140 g was achieved at 265 days posthatch. These preliminary trials show that it is clear that growth rates can be achieved above those in the wild. Although it can be assumed that growth rates observed in these preliminary trials will improve as species-specific culture techniques and feeds are developed, a two- to three-year production period will most likely be required for these species.

The most extensive data on culture conditions for snapper are from Thailand and Singapore, where largely wild juvenile *L. argentimaculatus* and *L. johni* are used to supply small-scale net cage production. These species are primarily raised as an alternative to sea bass, *Lates calcarifer*, and groupers (e.g., *Epinephelus tauvina*, *E. salmoides*, *E. malabaricus*, and *E. bleekeri*) (4,30). In this production situation, wild juveniles are cultured in relatively small (e.g., 3 m × 2 m) cages and fed low-valued fish. Good growth rates and survival have been reported when juveniles are stocked at 60–90 fish/m^3. When acceptable culture conditions are maintained, growth rates, feed efficiency (wet weight gain/wet weight of feed × 100), and survival are very good, with both species reaching market size within 10 months. This type of culture is exemplified by the report of Chaitanawisuti and Piyatiratitivorakul (30). In those studies, juvenile *L. argentimaculatus* having a mean initial weight of 20 g were stocked at an initial density of 90 fish/m^2 in 3 m diameter floating cages and offered chopped carangids, *Selaroides* spp., to satiation twice daily. Mean environment conditions for water temperature, salinity, and dissolved oxygen were reported to be 28.9 °C, 31 ppt, 7.8 mg/L (ppm). At the conclusion of the 10-month growth trial, an average final weight of 806 g was achieved, with a feed efficiency of 15.6% and survival of 83%. Based on the results of this study, it was concluded that the results presented were in agreement with those reported by other authors and that the growth rates for *L. argentimaculatus* are similar to those obtained with sea bass and several groupers.

One of the primary restrictions for commercial culture of snapper is a reliance on wild stock, which is limited in supply and availability (4). Another problem, albeit one far less severe, is limited information on the nutritional requirements and the acceptance of prepared feeds by snapper species. However, it would appear that snappers readily accept commercially prepared feeds and that high-quality rations designed for other warmwater marine species, such as *S. ocellatus* or *L. calcarifer*, should be adequate.

SUMMARY

A number of snapper species have been spawned either by using hormone induction or by environmental manipulations. The literature indicates considerable variation in fertilization and hatch rates. This variable egg quality may be due to the use of hormone therapy and the presence of underdeveloped oocytes, inadequate broodstock nutrition, or natural variation in spawn quality. Despite some problems with egg quality, a number of species have been spawned in captivity and a good supply of eggs and newly hatched larvae can be routinely obtained. Currently, the development of mass-rearing techniques for lutjanid larvae fall far behind those which have been established for other marine species, such as *L. calcarifer* or *S. ocellatus*. Reports of larval rearing experiments indicate that larvae often have numerous prey items and prepared feeds in their digestive systems. Yet there is a general trend of high variability of larval survival and stage-specific larval mortalities. This may indicate a nutrient deficiency (e.g., essential fatty acids or vitamins). If a reliable supply of juveniles is to be developed, it is crucial that larval research be given high priority, particularly with respect to ontological development, nutritional requirements, feeding protocols, and larval rearing techniques. Once larval rearing techniques are developed and a steady supply of juveniles is available, commercial culture of a number of snappers will develop rapidly. Juveniles are quite hardy, readily accept commercial feeds, and adapt to intensive culture conditions. Given the high market price and consumer acceptability, commercial culture of snapper should be a viable business in a number of countries.

BIBLIOGRAPHY

1. W.D. Anderson, Jr., in J.J. Polovina and S. Ralston, eds., *Tropical Snappers and Groupers: Biology and Fisheries Management*, Westview Press, Boulder, Colorado, 1987, pp. 1–32.
2. J.J. Polovina and S. Ralston, eds., *Tropical Snappers and Groupers: Biology and Fisheries Management*, Westview Press, Boulder, Colorado, 1987.
3. T. Singhagraiwan and M. Doi, *Thai Mar. Fish. Res. Bull.* **4**, 45–57 (1993).
4. L.C. Lim, L. Cheong, H.B. Lee, and H.H. Heng, *Singapore J. Pri. Ind.* **13**, 70–83 (1985).
5. W.O. Watanabe, E.P. Ellis, S.C. Ellis, J. Chaves, and C. Manfredi, *J. World Aquacult. Soc.* **29**, 176–187 (1998).
6. D.A. Davis, C.R. Arnold, and G.J. Holt, *Proc. Third Annual Int. Conf., Open Ocean Aquaculture 1998*, Corpus Christi, TX, May 10–15, 1998.
7. W.J. Richards and V.P. Saksena, *Bull. Mar. Sci.* **30**(2), 515–521 (1980).
8. M.E. Clarke, M.L. Domeier, and W.A. Laroche, *Bull. Mar. Sci.* **61**(3), 511–537 (1997).
9. C.B. Grimes, in J.J. Polovina and S. Ralston, eds., *Tropical Snappers and Groupers: Biology and Fisheries Management*, Westview Press, Boulder, Colorado, 1987, pp. 239–294.
10. N. Bromage, in N.R. Bromage and R.J. Roberts, eds., *Broodstock Management and Egg and Larval Quality*, Blackwell Science Ltd., Cambridge, MA, 1995, pp. 1–24.
11. C.R. Arnold, J.M. Wakeman, T.D. Williams, and G.D. Treece, *Aquaculture* **15**, 301–302 (1978).
12. C.M. Riley, G.J. Holt, and C.R. Arnold, *Fish. Bull.* **93**, 179–185 (1995).
13. S. Hamamoto, S. Kumagai, K. Nosaka, S. Manabe, A. Kasuga, and Y. Iwatsuki, *Jpn. J. Ichthyology* **39**, 219–228 (1992).
14. M. Doi and T. Singhagraiwan, *Biology and Culture of The Red Snapper*, Lutjanus Argentimaculatus, The Eastern Marine Fisheries Development Center (EMDEC), Department of Fisheries, Ministry of Agriculture and Cooperatives, Thailand, 1993.
15. A.C. Emata, B. Eullaran, and T.U. Bagarinao, *Aquaculture* **121**, 381–387 (1994).
16. P. Thomas, C.R. Arnold, and G.J. Holt, in N.R. Bromage and R.J. Roberts, eds., *Broodstock Management and Egg and Larval Quality*, Blackwell Science Ltd, Cambridge, MA, 1995, pp. 118–137.
17. N.N. Rabalais, S.C. Rabalais, and C.R. Arnold, *Copeia* **1980**(4), 704–708 (1980).

18. V.P. Minton, J.P. Hawke, and W.M. Tatum, *Aquaculture* **30**, 363–368 (1983).
19. K.L. Bootes, *Culture and Description of Larval Red Snapper, Lutjanus campechanus*, Master's thesis, Department of Fisheries and Allied Aquacultures, Auburn University, Auburn, AL, 1998.
20. C.M. Lalli and T.R. Parsons, *Biological Oceanography*, Pergamon Press, New York, 1993.
21. M.N. Duray, L.G. Alpasan, and C.B. Estudillo, *Isr. J. Aquaculture-Bamidgeh* **48**, 123–132 (1996).
22. J.R. Rainuzo, K.I. Retan, and Y. Olsen, *Aquaculture* **155**, 103–115 (1997).
23. J.R. Sargent, L.A. McEvoy, and J.G. Bell, *Aquaculture* **155**, 117–127 (1997).
24. C.S. Manooch, III, in J.J. Polovina and S. Ralston, eds., *Tropical Snappers and Groupers: Biology and Fisheries Management*, Westview Press, Inc., Boulder, CO, 1987, pp. 329–374.
25. C.S. Manooch, III and C.L. Drennon, *Fish. Res.* **6**, 53–68 (1987).
26. A.G. Johnson, *Trans. Amer. Fish. Soc.* **112**, 173–177 (1983).
27. M. Baez Hildago, L. Alvarez-Lajonchere, and B. Pedrosa Tabio, *Rev. Invest. Mar.* **1**, 135–150 (1980).
28. E. Thouard, P. Soletchnik, and J.P. Marion, in Advances in *Tropical Aquaculture, Selection of Finfish Species for Aquaculture Development in Martinique (F.W.I.)*, AQUACOP. IFREMER, Actes de Collogue 9, 1990, pp. 499–510.
29. D.L. Mason and C.S. Manooch, III, *Fish. Res.* **3**, 93–104 (1985).
30. N. Chaitanawisuti and S. Piyatiratitivorakul, *J. Aqua. Trop.* **9**, 269–278 (1994).

See also LARVAL FEEDING—FISH.

SOLE CULTURE

B.R. HOWELL
Centre for Environment, Fisheries and Aquaculture Research
Weymouth, United Kingdom

OUTLINE

Sole are not yet cultured commercially on a significant scale, but in Europe, two species (*Solea solea* and *S. senegalensis*) have long been considered to be prime candidates for farming, because of their widespread popularity and consistent high value. In the United Kingdom, for example, the Dover, or common, sole (*S. solea*) is among the most valuable fish landed, attracting a price comparable to that of halibut (*Hippoglossus hippoglossus*) and turbot (*Scophthalmus maximus*), both of which are now being farmed commercially.

Research on sole culture has a long history, dating back to the turn of the century, when the French biologists Fabre–Domerque and Biétrix (1) claimed limited success in feeding early sole larvae. Interest at that time was focused more on supplementing natural recruitment by releasing eggs and yolk sac stages into the sea than on developing rearing procedures for later stages, and hence little further progress was made. Significant progress towards the development of techniques for rearing the larvae to and beyond metamorphosis was not made until about 60 years later, when Shelbourne began his pioneering work in the United Kingdom on rearing marine flatfish larvae (2). Shelbourne's immense contribution to fish culture exploited the earlier observation by the Norwegian biologist Rollefsen (3) that plaice (*Pleuronectes platessa*) larvae could readily be fed on the naupliar stages of brine shrimp (*Artemia salina*). This live food organism remains the basis of hatchery production of commercial fish and shrimp farming.

Despite the demonstration that, by using brine shrimp as live food, sole juveniles could be produced readily in hatcheries, a concerted effort to develop farming techniques, primarily in the United Kingdom and France during the 1970s and 1980s, failed to achieve commercial viability, because of difficulties in feeding the subsequent juvenile stages. Problems initially centered on the attractiveness of formulated feeds, but even when this difficulty was alleviated, through the identification of chemical feeding stimulants (4), growth and survival remained unacceptably low. Recent advances in feed technology, combined with the desire of the now well-established marine fish farming industry in Europe to diversify its activities, have rekindled interest in the farming of sole.

THE SOLEIDAE

Soleidae is a family of flatfish that are typically found in relatively warm water, with most European species reaching the northern limit of their distribution around the British Isles (5). The family comprises a large number of species that generally inhabit sandy or muddy grounds in the relatively shallow waters of the continental shelf. However, the common sole, *S. solea*, and, to a somewhat lesser extent, the Senegal sole, *S. senegalensis*, are the two European species that have attracted the greatest attention in a farming context. *S. solea* has a more northern distribution and extends further into the Mediterranean Sea than the latter, but although differences between the two species are evident, particularly in relation to their response to temperature (reflecting their slightly contrasting geographic distribution), the two species have been found to present similar problems in culture. *S. solea*, however, has been the subject of more extensive study than its more southerly counterpart, and thus the information presented in this contribution has largely been derived from work on that species.

General Biology

Sole are batch spawners, with each female ovulating and releasing batches of eggs every few days over a period of several weeks. The testes of the males are small, a feature that is indicative of the need for intimate courtship behavior to ensure that the gametes are simultaneously released in close proximity (6). Spawning is initiated when the water temperature reaches about 10 °C (50 °F) and thus occurs earlier in the year in the southern part of the fish's distribution than in its northern range (7).

The eggs, which range in diameter from 1.0 to 1.6 mm (0.04 to 0.06 in.), are buoyant and hatch after 5 days at a temperature of 12 °C (54 °F). The emergent larvae are without a functional mouth or pigmented eyes, features that develop about 3 days later, when the larvae begin to feed. The larvae metamorphose and settle to the bottom after 25–30 days at 14–16 °C (57–61 °F), when they are just over 1 cm (0.4 in.) long. Whereas the larvae are visual feeders, the demersal stages are nocturnal, becoming active at dusk and burying themselves in the substrate during the day (8).

The juveniles feed on benthic organisms, primarily polychaetes, molluscs, and small crustacea. Sole have a poorly developed stomach, in contrast to that of more predatory flatfish, such as turbot and halibut (9). An important consequence of this is that they need to feed relatively frequently, being unable to meet their dietary energy requirements with relatively large, infrequent meals.

CURRENT STATUS OF CULTURE METHODS

Broodstock Management

A reliable supply of eggs of consistently high quality is readily obtainable from sole, a species that spawns spontaneously in captivity, without the assistance of administered hormones. Spawning has been obtained under a wide range of environmental conditions (10), with fish held in tanks of 0.15–1 m (0.5–3.3 ft) depth and 0.15–25 m^3 (5.3–882.2 ft^3) volume, at densities of 1–6 fish/m^2 (0.1–0.6 fish/ft^2), and at sex ratios ranging from 0.5 to 3 males for each female. Illumination has been provided by natural light and artificial light with intensities of 25–1,500 lux. A variety of temperature regimes have been used, but there is evidence that a cyclic temperature appears to be essential for successful egg production, with relatively high winter temperatures appearing to reduce both the quantity and quality of the output. Out-of-phase spawning has been obtained by the synchronous manipulation of photoperiod and temperature, or even by manipulating temperature alone. The diet for spawning has consisted of a variety of polychaetes, molluscs, and, occasionally, crustacea, fed between three and six times a week. A diet consisting only of frozen molluscs does not appear to support good egg production. Mixed diets containing a live component invariably meet with greater success.

Reported values for egg production have ranged from 11 to 140 eggs/g (312 to 3,976 eggs/oz) of live weight of females (10). Since it is clear that many of the females within a stock—perhaps as many as 70–80%,—may not contribute to egg production in any one year, the relative fecundity of individual females may be much higher. Fertilization rates are also highly variable, with an average of around 60%. Assuming that an average relative fecundity of 100 eggs/g (2,840 eggs/oz) is achieved, a 10-m^3 (353-ft^3) tank stocked with 15 kg (33 lb) of females and about half that weight of males should, on average, yield about one million eggs during a single spawning season.

Egg Incubation and Larvae Rearing

Eggs are generally stocked at a density of about 300 eggs/l (1,350 eggs/gal) and hatch after five days at 12 °C (54 °F). Survival rates of fertilized eggs are usually in excess of 80%, probably reflecting, to a large extent, the high quality of eggs obtained from natural spawning. After two to three days of hatching, the yolk sac larvae are transferred to rearing tanks, about one day before the eyes have pigmented, the jaw has become functional, and the larvae are ready to begin feeding.

Sole is among the easiest of all marine fish to rear through the larval stages. Although some success has been obtained with formulated feeds (11), live food remains an essential requirement. The larvae can be reared on the rotifer *Brachionus plicatilis* (12). However, that organism is rarely used, because *Artemia* nauplii are more convenient and are acceptable from the outset of feeding as well. In laboratory-scale tanks stocked at an initial density of about 100 larvae/L, survival rates to metamorphosis have generally been in excess of 70% (12–14). Such survival rates have been achieved without prefeeding the nauplii on "booster" diets to enhance their content of highly unsaturated fatty acids (HUFAs), though recent unpublished work indicates that the hardiness of juveniles is improved if HUFA-rich diets, are provided. Although marine fish larvae are considered to have an essential dietary requirement for HUFAs (15), particularly eicosopentaenoic acid (EPA) and docosahexaenoic acid (DHA), the requirement of sole seems to be limited to the former (16). This almost certainly contributes appreciably to the relatively high survival rates obtained with cultured sole, since *Artemia* nauplii, depending on their origin, contain significant quantities of EPA, but negligible quantities of DHA.

As with other flatfish, reared sole may display various forms of pigment abnormalities. The cause of this phenomenon remains poorly understood, but the incidence of such fish among reared sole is generally less than 20%. Other defects, such as skeletal abnormalities, are rare as well.

Weaning

The transfer of metamorphosed sole to a formulated feed was the major obstacle confronting biologists in the 1970s. A marked reluctance to accept fish-based diets appeared to be related to the importance of olfaction in the feeding behavior of this nocturnal, invertebrate-feeding species. The problem was partially overcome by inclusion in diets of either invertebrate tissue (17,18) or chemical attractants (4,19), but despite those measures, growth rates were generally low and survival rates highly variable. Studies focused on the attractiveness of diets, but an experiment in which survival was positively related to the level of hydrolyzed fish protein concentrate (20) suggested that the ability of the fish to use the diets may also be an important consideration. More recent trials were conducted with a larval feed produced by a process of agglomeration (SSF, Fyllingsdalen, Norway). That feed was based on high-quality protein with a relatively high soluble fraction and was supplemented with chemical taste attractants (21). Sole 2–3 cm (0.8–1.2 in.) in length accepted the feed, and mortality rates were negligible, while growth rates similar to those reported for live feeds were sustained. These advances now allow fully weaned sole to be produced on a routine basis with a high survival rate.

Growout

Growout is the most important phase with regard to economic viability and the least well-developed phase for sole. In intensive culture systems, profitability depends on the rate of production, which is a function of both the mean growth rate of the fish and the stocking density.

There is little documented information on the potential growth rate of sole to market size under optimum conditions and on the extent to which growout to market size may be achieved on commercially acceptable formulated feeds. Experimental evaluations of the dependence of growth rate on temperature in which the fish were fed live food were used by Howell (21) to provide an indication of potential for growth. Results of the study indicated that fish 5 cm (2 in.) in length would reach a minimum market size of 125 g (4.4 oz) in less than 300 days at temperatures close to the optimum of 20 °C (68 °F). Day et al. (20) grew fish to that size on a formulated feed in just under 600 days, but in that trial, temperatures were subject to seasonal variation and were consistently well below the optimum. Although this result is encouraging, there clearly is scope for considerable improvement.

With regard to the effects of stocking density, there is some indication that sole may be less suited to crowded conditions than other species. Their browsing mode of feeding may result in high levels of interaction between individuals, leading to reduced growth rates and high variation in size. Howell (21) reported that, over a 12-week period, the growth rate of 1.5-g (0.05-oz) fish stocked at 500 fish/m^2 (46 fish/ft^2) was about 15% less than that of those stocked at 17 fish/m^2 (1.6 fish/ft^2). The difference was attributable to social factors, as the experimental design eliminated the potentially suppressive effects of water quality variables. Ways in which these effects may be reduced need to be developed, which may involve regular grading to reduce size variation, and the adoption of feeding strategies that minimize or eliminate opportunities for individuals to dominate the food supply.

PROSPECTS FOR COMMERCIAL EXPLOITATION

For most marine fish, rearing through the early developmental stages presents the greatest obstacle to the realization of industrial-scale farming operations. In this regard, however, sole have a particular advantage, in that a regular supply of good-quality eggs can readily be obtained from captive broodstocks, and survival rates through the egg and larval stages are consistently high. The weaning of small juveniles to a formulated feed can also be readily accomplished with minimal losses. The current uncertainty is whether sufficiently high growth rates can be achieved on formulated feeds at commercially realistic stocking densities. In this regard, the inherent feeding behavior of sole puts them at a distinct disadvantage as compared with some other species. In contrast to turbot, for example, sole need to take their food from the bottom, rather than from the water column, and this spacial requirement may impose an unacceptably low limit on the stocking densities that can be achieved. It remains to be seen whether these difficulties can be overcome by imaginative system design and the adoption of appropriate husbandry practices.

An alternative opportunity may be to exploit the relative ease with which the early stages of sole can be reared to develop a more extensive form of cultivation. The relatively high growth rates achieved on live food suggest that the species may have potential for extensive or semiextensive cultivation in regions that have a suitable topography and climate.

BIBLIOGRAPHY

1. P. Fabre–Domerque and E. Biétrix, *Travail du Laboratoire de Zoologie Maritime de Concarneau*, Vuibert et Nony, Paris, 1905.
2. J.E. Shelbourne, *Advances in Marine Biology* **2**, 1–83, (1964).
3. G. Rollefsen, *Rapp. P.-V. Réun. Cons. Perm. Int. Explor. Mer.* **109**, 133, (1939).
4. A.M. Mackie, J.W. Adron, and P.T. Grant, *J. Fish Biol.* **16**, 701–708, (1980).
5. A. Wheeler, *The Fishes of the British Isles and North West Europe*, Michigan State University Press, East Lansing, MI, 1969.
6. S.M. Baynes, B.R. Howell, T.W. Beard, and J.D. Hallam, *Neth. J. Sea Res.* **32**, 271–275 (1994).
7. J.W. Horwood, *Advances in Marine Biology* **29**, 215–367 (1993).
8. H. Kruuk, *Neth. J. Sea. Res.* **1**, 1–28 (1963).
9. S.J. de Groot, *J. Cons. Int. Explor. Mer.* **32**, 385–394 (1969).
10. S.M. Baynes, B.R. Howell, and T.W. Beard, *Aquaculture and Fisheries Management* **24**, 171–180 (1993).
11. S. Applebaum, *Aquaculture* **49**, 209–221 (1985).
12. B.R. Howell, *J. Cons. Int. Explor. Mer.* **35**, 1–6 (1973).
13. J.E. Shelbourne, *Ministry of Agriculture, Fisheries and Food: Fishery Investigations*, Series II, 27, no. 9, Her Majesty's Stationary Office, London, 1975.
14. J. Fuchs, *Aquaculture* **26**, 321–337 (1982).
15. T. Watanabe, F. Oowa, C. Kitajima, and S. Jujita, *Bull. Jpn. Soc. Sci. Fish.* **44**, 1115–1121 (1978).

16. B.R. Howell and T.S. Tzoumas, in P. Lavens, P. Sorgeloos, E. Jaspers, and F. Ollevier, eds., *Larvi '91: Fish and Crustacean Larviculture Symposium*, European Aquaculture Society, Special Publication No. 15, 1991, pp. 63–65.
17. P.J. Bromley, *Aquaculture* **12**, 337–347 (1977).
18. R. Métailler, B. Menu, and P. Morinière, *J. World Maricult. Soc.* **12**, 111–116 (1981).
19. R. Métailler, M. Cadena–Roa, and J. Person–Le Ruyet, *J. World Maricult. Soc.* **14**, 679–684 (1983).
20. O.J. Day, B.R. Howell, and D.A. Jones, *Aquaculture Research* **28**, 911–921 (1997).
21. B.R. Howell, *Aquaculture* **155**, 355–365 (1997).

See also BRINE SHRIMP CULTURE; FLOUNDER CULTURE, JAPANESE; HALIBUT CULTURE; LARVAL FEEDING—FISH; PLAICE CULTURE; SUMMER FLOUNDER CULTURE; WINTER FLOUNDER CULTURE.

STRESS

BRUCE A. BARTON
University of South Dakota
Vermillion, South Dakota

OUTLINE

During the past decade aquaculturists have become increasingly interested in stress in fish and many now recognize that managing for stress is part of normal operations along with nutrition, disease, and genetics management. While a fish's response to stress is considered adaptive, stress is a concern in aquaculture because of its possible detrimental effects on important fish performance features such as metabolism and growth, disease resistance, and reproductive capacity. Despite this general acceptance of the importance of considering stress in aquaculture and fisheries management, the phenomenon of stress is still not completely understood.

Stress has been defined as the response of an organism to any demand placed on it such that it causes an extension of a physiological state beyond its normal resting state to the point that the chances of survival may be reduced (1,2). This is a useful working definition for aquaculture, as it incorporates both the notion of a physiological change occurring within the organism in response to a stimulus and the idea that, as a result, some aspect of fish performance may be compromised. This is only one view and a precise definition of stress still eludes scientists despite the many years of research dedicated to this subject. The concept of stress from a physiological or medical perspective, which implies a threat to the maintenance of an organism's homeostasis (3,4) is somewhat different from the view of environmental stress that somehow impairs fitness in a Darwinian or evolutionary sense (5). Nevertheless, a common theme exists among widely ranging perceptions about stress; that is, there is a biological response to a stimulus at some level of organization.

Hans Selye, whose contribution to this branch of science was recognized recently by a major proceedings publication (6), developed the nucleus of the present underlying concept of stress more than a half-century ago. Stress in fish has been studied for about three decades and readers are encouraged to refer to a number of collected reviews in books (7–9), review papers in other publications (10–15), and other reviews cited herein for a more thorough background on this subject. The purpose of this entry is to describe what the scientific community currently knows about stress in fish and why an understanding of stress is important in aquaculture. Confusion still exists among scientists with respect to appropriate terminology used to describe stress. For purposes of this entry, the term "stressor" (or "stress factor") means the stimulus that inflicts stress on the fish. "Stress" (or "stressed state" or "stressful experience") refers to the altered state of the fish, whereas "stress response" is those physiological or behavioral manifestations that can be measured to indicate the degree of stress experienced.

CONCEPTS AND BACKGROUND

A misconception among biologists and practicing aquaculturists is that stress, in itself, is detrimental to the fish. The acute response to stress is an adaptive mechanism that provides the fish with a means to cope with the stressor in order to maintain its normal or homeostatic state. If the stressor is overly severe or long lasting, the continued response can become detrimental to the fish's health and well-being, or become maladaptive, a state often associated by many with the term "distress." This view is consistent with the original general adaptation syndrome (GAS) paradigm of Selye (16), which considered that an organism passes through three stages in response to stress: (a) an alarm phase consisting of the organism's perception of the stimulus and recognition of it as a threat to homeostasis, (b) a stage of resistance during which the organism mobilizes its resources to adjust to the disturbance and maintain homeostasis, and (c) a stage of exhaustion that follows if the organism is incapable of coping with the disturbance. The final stage is the maladaptive phase normally associated with the development of a pathological condition or mortality in fish culture.

Stressors in aquaculture are typically physical disturbances such as those caused by handling, grading, or transporting that invoke acute stress or those that are chronically stressful—for example, poor water quality and overcrowding. Responses to these types of stressors have been grouped as (a) primary, which include the initial neuroendocrine responses, particularly both stimulation of the hypothalamic-pituitary-interrenal (HPI) axis, culminating in the release of corticosteroid hormones, and the direct release of catecholamines; (b) secondary, which include changes in blood glucose and lactate, electrolytes and osmolality, and hematological features that relate to physiological functions such as metabolism and hydromineral balance; and (c) tertiary, which refer to aspects of whole-animal performance such as changes in growth, overall resistance to disease, metabolic scope for activity, behavior, and ultimately survival (11) (Fig. 1). This grouping is convenient, but simplistic. However, stress affects fish at all levels of organization, from molecular and biochemical to population and community (8). Moreover, responses to stress at different levels of organization are not only interrelated functionally to each other, but often interregulated as well. Thus, to appreciate how fish can be affected by stress in aquacultural operations, it is useful to consider their responses in an integrated or holistic sense, rather than simply observe isolated physiological phenomena. The stressor may be a real threat to homeostasis (e.g., altered water quality, acute physical disturbance) or one simply perceived by the fish as a threat (e.g., predator presence, human presence near tank), but, in either case, the typical GAS-type of response appears to require some form of sensory input such as fright, pain, or discomfort (17). In that context, Mason (18) argued that an organism's response to stress is as much behavioral as physiological. How fish behave when stimulated by a stressor provides an outward whole-animal manifestation of the complex neurological and physiological changes that occur, thus allowing interpretation of the response at an ecological level (19).

Much of our present knowledge about stress in fish has been gained from studying the primary responses of the HPI axis and subsequent or secondary effects on metabolism, reproduction, and the immune system (2,12). The investigation of heat-shock proteins in fish as a general cellular response to various stressors is a recent and rapidly emerging field (20). Although most research in this area is descriptive at this point, the elucidation of possible functional relationships between the cellular responses to stress and neuroendocrine, immune and other physiological systems may provide useful applications for aquaculture in the future.

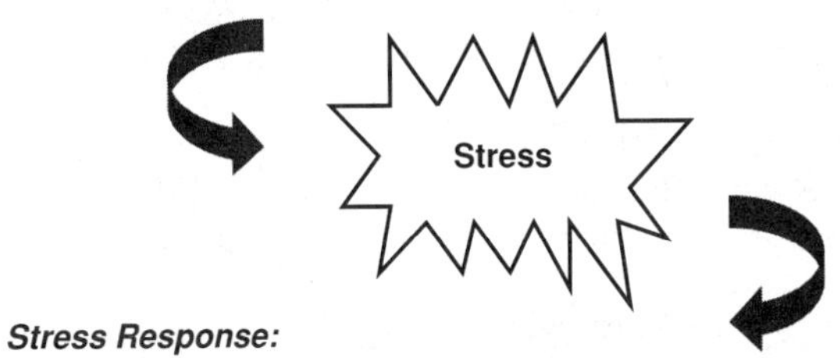

Figure 1. Typical stressors encountered in aquaculture evoke stress responses that can be classed as primary, secondary, and tertiary, which can be used as indicators to evaluate the degree of stress experienced by the fish.

When fish are exposed to a stressor, the stress response is initiated by the perception of a real or perceived threat by the central nervous system (CNS). The response of the HPI axis begins with the release of corticotropin-releasing hormone (CRH), or factor (CRF), chiefly from the hypothalamus in the brain, which stimulates the corticotrophic cells of the anterior pituitary, or adenohypophysis, to secrete adrenocorticotropin (ACTH). Circulating ACTH, in turn, stimulates the interrenal tissue (adrenal cortex homologue) located in the kidney to synthesize and release corticosteroids, mainly cortisol, into circulation for distribution to target tissues. Control of cortisol release is through negative feedback of the hormone to all levels of the HPI axis (21,22). Regulation of the HPI axis is far more complicated than this description implies, however. For additional details, Wendelaar Bonga (14) and Sumpter (23) recently provided more complete descriptions of the endocrine stress axis in fish, and Chouros (24) presented a thorough synthesis of the neuroendocrinology of stress in higher vertebrates, including a review of the multiple roles of CRH in the organism's response to stress.

Concurrent with the elevation of circulating corticosteroids during stress is the release of catecholamines, mainly epinephrine (adrenaline) and norepinephrine (noradrenaline), following sympathetic stimulation of the chromaffin tissue (adrenal medulla homologue) in the kidney (23). Other hormones, including thyroxine, somatolactin, gonadotropins, and reproductive steroids in circulation and serotonin and its derivatives in the brain can become either elevated or suppressed during stress (2).

The primary response, the release of corticosteroids and catecholamines, affects the secondary stress responses involved in metabolism (e.g., blood glucose and lactate, liver and muscle glycogen), hydromineral balance (e.g., blood ions and osmolality), and those related to hematology (e.g., circulating erythrocytes and leukocytes, differential leukocyte ratios, hemoglobin). Moreover, the corticosteroids and catecholamines can directly or indirectly affect aspects of fish performance of particular concern to aquaculturists, including disease resistance, scope for growth, feeding and avoidance behavior, and reproductive capacity.

One of the most rapidly appearing manifestations of acute stress is a change in behavior, which can occur within seconds after perception of the stressor (19). This is understandable as the natural response of the fish presumably would be to escape or avoid the noxious stimulus it perceives as an immediate threat to its well-being. However, fish already experiencing stress display a lowered ability for avoidance response behavior when threatened compared with unstressed fish (25). The initial perception of the disturbance and the concomitant behavior associated with it are followed by the neuroendocrine responses, which take seconds

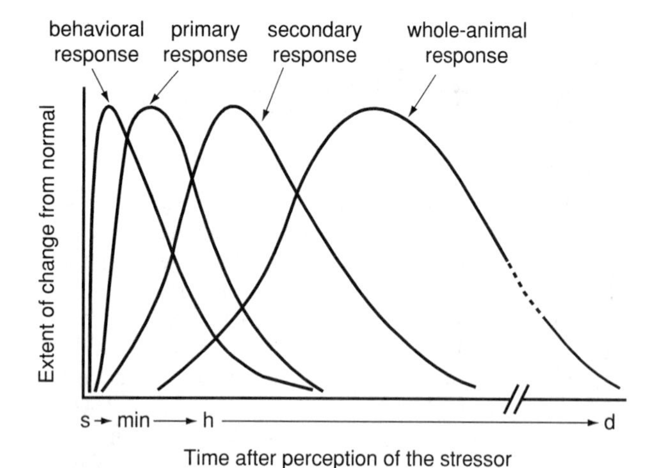

Figure 2. Stress responses such as behavioral alterations can occur within seconds to minutes following perception of the stressor, whereas primary and secondary physiological responses can take from minutes to hours to manifest themselves as measurable changes. Changes in whole-animal performance characteristics may take much longer to occur and also recover from stress, but the timing of such changes is variable depending on the nature of the particular performance of interest.

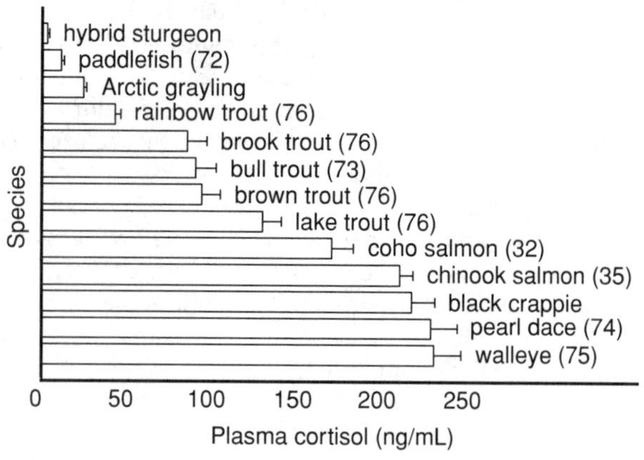

Figure 3. Fish species subjected to the same stressor may show considerable variation in their primary response to stress. Examples of peak poststress levels of plasma cortisol (+SE) in various fishes measured 1 hour after being subjected to the same 30-s aerial emersion stressor in a dip net. (Species without reference citation are from the author's unpublished data.)

(e.g., epinephrine) to minutes (e.g., cortisol) to become measurably elevated in circulation (Fig. 2). The secondary physiological changes that occur tend to take longer to manifest themselves in circulation, from minutes to hours (e.g., glucose, lactate, chloride), but often remain altered for more extended periods (Fig. 2). Timing of changes in whole-animal performance characteristics may be variable; for example, swimming stamina may be affected relatively quickly, whereas alterations in the immune system or reproductive function may not appear for hours or days or even weeks. Nevertheless, most research tends to support the notion that the magnitude and duration of the response reflect the severity and duration of the stressor. Thus, many of the primary and secondary stress responses documented in fish have become well established as useful monitoring tools to assess the degree of stress experienced by cultured fishes. Indeed, much of the following discussion is based on stress-related changes in plasma cortisol because of its proven utility as a sensitive acute-stress indicator (11,12).

FACTORS INFLUENCING STRESS RESPONSES

The normal responses of fish to stressful encounters are influenced by a suite of nonstress factors affecting both the magnitude of the response and recovery from it. These factors are genetic, developmental, or environmental. Fishes exhibit a wide variation in their responses to stress, particularly endocrine responses (12), ranging over as much as two orders of magnitude (Fig. 3). The cause for differences among major taxonomic groups is likely attributed to genetic factors. At our current level of understanding, however, it is unknown whether fishes that display relatively low stress responses are actually "less stressed" than others, or are as "stressed," but have a different capacity to respond to stress. Consistent response differences are not only evident among fish species, but also among strains or stocks within the same species (26,27) and even within the same population (28), a trend that appears to be at least partially heritable (29). The potential for selective breeding programs designed to produce fish with an attenuated stress response and, thus, show improved disease resistance is presently an attractive possibility for finfish aquaculture (30).

The developmental stage of the fish will affect its responsiveness to stress. A fish's ability to respond to a disturbance develops very early in life, for example, as early as two weeks after hatching in salmonids (31). Little evidence exists to suggest that fish show a consistent increase in stress responses as they develop, but they do appear to have heightened responses during periods of metamorphosis. For example, juvenile anadromous salmonids are particularly sensitive to stress during the period of parr-smolt transformation (32,33). As fish mature, primary stress responses may actually decrease in magnitude. Pottinger et al. (34) suggested that the threshold or "set point" for regulatory feedback may be lowered with the onset of maturity, thereby resulting in reduced responses of both cortisol and ACTH following stress.

Almost all environmental nonstress factors examined to date can influence the degree to which fish respond to stress. These include, but are not limited to, acclimation temperature, external salinity, nutritional state, water quality, time of day, overhead light, fish density, and even background color of the tank (2). An awareness of the extent to which these nonstress factors modify responses is important to both researchers and aquaculturists wishing to interpret experimental results and compare them with published values. For example, the fish's acclimation temperature or nutritional state is likely to have an appreciable effect on the magnitude of poststress elevations of cortisol and glucose, particularly the latter (35–38). In certain instances, stress-modifying factors that are themselves chronically stressful, such as poor water quality, can actually attenuate the cortisol response to an acute second stressor (39,40).

Repeated acute stressors can have a cumulative response effect in fish (25,41,42). The implication of this phenomenon for aquaculture is that the fish's ability to recover from individual stressful events could be prevented if sufficient intervals are not allowed between separate acute disturbances such as tank transfer, grading, or hauling. Over a long term, however, fish can habituate or become accustomed to repeated disturbances such that responses to stress are lessened (43,44). Habituation to repeated stressful stimuli by using a positive conditioning protocol may have potential as an approach for improving the survivorship of hatchery fish following their transport and release (19,44).

MEASURING AND INTERPRETING STRESS RESPONSES

As judged by the prevalence in the literature and inferred earlier, the most popular approaches for evaluating stress in fish in aquaculture-related situations are measurements of plasma cortisol, glucose, lactate, chloride (and other ions), and osmolality and hematological features. Typical resting and stress-elevated values for those and other features are listed in Table 1. However, readers should note that these are approximate values to serve as a guideline only and have limited diagnostic value, as stress responses are highly variable, depending on genetic makeup, early life history, and the fish's environment. Extensive data indicating the point at which certain physiological features may actually indicate a life-threatening situation are not available, but plasma chloride and osmolality concentrations less than 90 meq/L and 200 mOsm/kg, respectively, have been suggested as indicative of compromised osmoregulatory ability in salmonids (45).

Methods of stress assessment in fish have been described (11,46) and include simple test kits (e.g., glucose, lactate) and easy-to-use meters (e.g., chloride, osmolality) for many of the physiological features of general interest. Measuring hormones like cortisol is more complicated, however, and usually involves a radioimmunoassay or enzyme-linked immunosorbant assay technique. Inexpensive, readily available, portable meters, such as those used clinically for glucose and other features, have been tested for their efficacy in fish stress monitoring (46) and show promise as a future useful tool for the aquaculture industry.

Table 1. Approximate Typical Resting and Stress-Elevated Values of Primary and Secondary Physiological Parameters Used as Indicators of Stress in Fish[a]

Physiological Parameter	Resting	Poststress
Plasma epinephrine (nmoles/L)	<3	20–70
Plasma cortisol (ng/mL)	<10–50	30–300+
Plasma glucose (mg/dL)	50–150	100–250+
Plasma lactate (mg/dL)	20–30	40–80
Plasma chloride (meq/L)	100–130	≈10% ↑ or ↓[b]
Plasma sodium (meq/L)	150–170	≈10% ↑ or ↓[b]
Plasma osmolality (mOsm/kg)	290–320	≈10% ↑ or ↓[b]
Plasma hemoglobin (g/dL)	5–9	<4
Hematocrit (% packed cell volume)	25–40	40–50+

[a]Compiled from references 11,12,48, and author's unpublished data. However, considerable variation among these values and exceptions outside of these ranges exist depending on genetic background, rearing history and environmental conditions. (See text.).

[b]Blood ions and other features related to hydromineral status will fluctuate upward or downward, depending on whether fish is marine or freshwater species, respectively.

More problematic than actual measurement of stress in fish is the interpretation of results for three major reasons. First is the modifying effects alluded to earlier, that the various genetic, ontogenetic, and environmental factors have on the magnitude and duration of the response. Without knowing the extent to which other nonstress factors may have altered the response, it is difficult to interpret the biological significance of that response in a relative context. Second is the variation and apparent inconsistency among fishes in the responses of different blood-chemistry characteristics. For example, a species that shows the greatest endocrine response increase (e.g., plasma cortisol) compared with others may not be the same species that elicits the greatest increase in a secondary response, such as glucose or lactate, when subjected to the identical stressor (12). Thus, a species or group that appears "most stressed" by one feature may not necessarily be so if measured by another. Such discrepancies among different physiological indicators emphasize the importance of not relying on a single indicator and the need for appropriate controls in stress assessment. Third is the nature of the stress response itself. The response to stress is a dynamic process and physiological measurements taken during a time course are only representative instantaneous "snap-shots" of that process. A time lag, ranging from minutes to hours or more, exists that complicates the interpretation of results. A significant delay, depending on the level and type of response, occurs from the initial perception of the stressor by the CNS to the time when the physiological feature of interest reaches a peak level of response (Fig. 2). Thus, the measurement of a particular stress indicator may not necessarily reflect the degree of stress experienced by the fish at that instant of time, but more likely earlier.

IMPLICATIONS OF STRESS FOR FISH PERFORMANCE

Stress affects the performance of fish in four areas that have direct implications for aquaculture: metabolism, immunocompetence, reproduction, and behavior.

Metabolism

Among the most important factors for successful aquaculture are acceptable growth rates and food conversions of the target species. The speculated effect of stress on metabolism is based on the premise that coping with stress is a process that requires energy within the fish's overall scope for activity, which is then unavailable for other performances such as growth (12,17). Certainly when fish are stressed, energy reserves are mobilized, with characteristic increases in plasma glucose and lactate and decreases in tissue glycogen occurring as a result (47,48). In the short term, the acute metabolic cost associated with stress is reflected by an increase in oxygen consumption, which is a direct measure of metabolic rate (49,50). However, the long-term effects of stress on metabolic functions, such as growth, and the involvement of the various endocrine axes are not yet clear (47). Moreover, it is difficult to attribute observed, apparently stress-mediated suppressions in growth rate to single isolated stressors because of possible confounding effects of individual factors encountered in fish husbandry, such as density, water quality, feeding regime, and social interaction (12), as well as the involvement of multiple endocrine axes (47).

Immunocompetence

Stress has a profound effect on the immune system of fish and increases their susceptibility to infectious diseases (51–53). Evidence is increasing that demonstrates that mediation of the immune system is through endocrine pathways, but precise mechanisms in fish are not yet well known, although the HPI axis is certainly involved (14,54,55). Peripheral hormones of the HPI axis affect a number of immunological features in fish, including a reduction in circulating lymphocytes, antibody production, and macrophage respiratory-burst activity (2,12,55).

Recent findings in higher vertebrates support the view that immune and endocrine systems are connected reciprocally and interact directly to affect health and disease outbreaks in stressed fish (14,24,55). In particular, the cytokine proteins involved in inflammation have directly stimulatory effects on the hypothalamic-pituitary-adrenal (HPA) axis, notably on stimulation of CRH and ACTH, whereas activation of the HPA axis has an appreciable inhibitory influence on the inflammatory immune response (24,56). However, the direct influence of the HPA axis on immune function is far more complicated than previously thought, as effects of corticosteroids can be both stimulatory and inhibitory, depending on timing, other immune factors, and host status (56). Detailed studies are still needed to clarify the existence of such stimulatory factors and their immune-endocrine interactions in fish (54,55), although many similarities exist between fish and mammals (14). Nevertheless, most experts agree that managing fish culture to reduce or minimize stress should decrease the likelihood of costly fish losses due to poor health and presence of disease, even if exact mechanisms of action are not understood.

Reproduction

Stress can have a significant inhibitory effect on the reproductive capacity of fish, particularly by suppression of the gonadal steroids. Substantial evidence indicates that this effect is mediated in large part by the HPI axis, particularly by cortisol (2,12,14,47). For example, a number of studies have documented a general suppressive effect of both stress and cortisol on plasma levels of testosterone, 11-ketotestosterone and estradiol (2,12,14). However, all evidence collectively is equivocal and investigations showing neutral or even stimulatory effects complicate interpretation of existing literature (47).

Stress may also affect reproduction negatively by acting directly or indirectly on other hormonal pathways and on vitellogenesis (23,47). Regardless of mechanisms, important concerns for aquaculture are whether stress affects gametogenesis, gamete quality, and survivorship. Although limited evidence suggests that gamete quality and progeny survival are reduced by culture-related stressors (57,58), this is an area of aquaculture research that remains largely unexplored (47). In contrast, the effects of chronic environmental stressors, such as chemical pollutants on embryonic development, gamete quality, hatching success, and larval quality are reasonably well established (59), although it may be difficult in many instances to ascertain whether the documented impairment occurred as a result of a generalized stress response or a specific action of the chemical stressor in question.

Behavior

Behavior patterns in fish important to aquaculture include feeding, avoidance, and aggressive behaviors, all of which can be affected by stress (19,60,61). As with other stress responses, initial behavioral reactions to a stressful event, such as avoidance of the stressor, may be to maximize chances of survival and, thus, are adaptive. However, in the face of continued or additional stress, behavioral responses may deviate sufficiently from normal to become deleterious, for example, decreased avoidance response ability or loss of appetite. Unlike other physiological responses that incur a time lag before they are detectable, such as blood-chemistry alterations, some stress-induced changes in behavior mediated directly through the CNS can occur and be observed in the fish almost instantaneously (19).

Major stressors encountered in aquaculture that can alter fish behavior include handling, transport, crowding, and poor water quality (19,60). For example, simply handling fish increases their susceptibility to being captured as prey (42,62–64) and reduces their ability to avoid other noxious stimuli (25). Normal social interactions such as aggressive or dominance behavior in captive fish populations can also be disrupted by stress (19,65).

A variety of adverse water-quality characteristics are known to affect fish behavior (19,60); while most of those documented are associated with environmental pollution, those with relevance to aquaculture are elevated ammonia, reduced dissolved oxygen, suboptimal pH, and altered thermal regime. Feeding behavior, in particular, is affected by altered water quality, as well as by physical stressors, such as handling, with the obvious responses being a cessation or reduction in feeding activity or loss of appetite (19,60). Endocrine mechanisms underlying stress-induced behavioral alterations in fish are not well elucidated, but at the level of the CNS, appear to involve monoamine neurotransmitters, specifically dopamine, norepinephrine, and serotonin (19,66). Notably, increased brain serotonergic activity is linked with subordinate behavior and reduced food intake in fish (67), suggesting one possible mechanism for reduced growth under stressful conditions in aquaculture.

Regardless of the mechanism, fish that are not feeding properly and subsequently not growing well in culture systems, or those hatchery fish incapable of effectively avoiding predators after being stocked, represent appreciable economic losses to the program.

APPROACHES FOR REDUCING EFFECTS OF STRESS

Over the long term, knowing that clear genetic differences exist in fish in their responses to stress has led to the possibility of selecting stress-tolerant strains for aquaculture (30). Although preliminary evidence is suggestive, research results to date have not definitively demonstrated a distinct consistent relationship between a reduced stress response and improved performance in growth or disease resistance. The potential also exists to improve fish performance and increase their tolerance to stress through the use of dietary supplements (68,69), but this is also a new research field.

In the short term, practical approaches for mitigating the detrimental effects of stress are relatively straightforward, employ common sense, and can be implemented at any time in the program. Guidelines and complete details are available elsewhere (45,70,71) and include, for example, (a) maintaining optimum water quality for the target species, (b) adhering to recommended loading densities, and (c) using established methods for disease prevention, including vaccination, water treatment, and facility and equipment disinfection. Healthy fish are more capable of tolerating stress than those whose health may be compromised because of poor husbandry practices.

Similarly, fish transportation methods should incorporate approaches designed to reduce stress and ensure that fish arrive at their destination in the best condition possible. Such methods employ the following: (a) proper oxygenation or aeration to provide adequate dissolved oxygen levels, (b) proper tank venting to prevent excessive carbon dioxide buildup, (c) appropriate in-tank water circulation and maintenance of water quality to reduce accumulation of toxic un-ionized ammonia, (d) prevention of overcrowding or overloading hauling tanks, (e) minimal use of out-of-water handling during loading and unloading, (f) addition of mineral salts to minimize osmoregulatory disturbance, and (g) possible use of anesthetics and other drugs to sedate fish in transit (45,70,71).

BIBLIOGRAPHY

1. J.R. Brett, in P.A. Larkin, ed., *The Investigation of Fish-Power Problems*, H.R. MacMillan Lectures in Fisheries, University of British Columbia, Vancouver, BC, 1958, pp. 69–83.
2. B.A. Barton, in G.K. Iwama, A.D. Pickering, J.P. Sumpter, and C.B. Schreck, eds., *Fish Stress and Health in Aquaculture*, Soc. Exp. Biol. Ser. 62, Cambridge University Press, Cambridge, UK, 1997, pp. 1–33.
3. H. Weiner, *Perturbing the Organism*, University of Chicago Press, Chicago, IL, 1992.
4. F. Toates, *Stress: Conceptual and Biological Aspects*, John Wiley & Sons, West Sussex, UK, 1995.
5. R. Bijlsma and V. Loeschcke, *Environmental Stress, Adaptation and Evolution*, Birkhäuser Verlag, Basel, Switzerland, 1997.
6. P. Csermely, ed., *Ann. N.Y. Acad. Sci.* **851**, 1–547 (1998).
7. A.D. Pickering, ed., *Stress and Fish*, Academic Press, London, 1981.
8. S.M. Adams, ed., *Am. Fish. Soc. Symp. Ser.* **8**, 1–191 (1990).
9. G.K. Iwama, A.D. Pickering, J.P. Sumpter, and C.B. Schreck, eds., *Fish Stress and Health in Aquaculture*, Soc. Exp. Biol. Sem. Ser. 62, Cambridge University Press, Cambridge, UK, 1997.
10. M.M. Mazeaud, F. Mazeaud, and E.M. Donaldson, *Trans. Am. Fish. Soc.* **106**, 201–212 (1977).
11. G.A. Wedemeyer, B.A. Barton, and D.J. McLeay, in C.B. Schreck and P.B. Moyle, eds., *Methods for Fish Biology*, American Fisheries Society, Bethesda, MD, 1990, pp. 451–489.
12. B.A. Barton and G.K. Iwama, *Ann. Rev. Fish Dis.* **1**, 3–26 (1991).
13. U.H.M. Fagerlund, J.R. McBride, and I.V. Williams, in C. Groot, L. Margolis and W.C. Clarke, eds., *Physiological Ecology of Pacific Salmon*, University of British Columbia Press, Vancouver, BC, 1995, pp. 461–503.
14. S.E. Wendelaar Bonga, *Physiol. Rev.* **77**, 591–625 (1997).
15. A.D. Pickering, in K.D. Black and A.D. Pickering, eds., *Biology of Farmed Fish*, Sheffield Academic Press, Sheffield, UK, 1998, pp. 222–255.
16. H. Selye, *Br. Med. J.* **1**(4667), 1383–1392 (1950).
17. C.B. Schreck, in A.D. Pickering, ed., *Stress and Fish*, Academic Press, London, 1981, pp. 295–321.
18. J.W. Mason, *J. Psychiat. Res.* **8**, 323–333 (1971).
19. C.B. Schreck, B.L. Olla, and M.W. Davis, in G.K. Iwama, A.D. Pickering, J.P. Sumpter, and C.B. Schreck, eds., *Fish Stress and Health in Aquaculture*, Soc. Exp. Biol. Sem. Ser. 62, Cambridge University Press, Cambridge, UK, 1997, pp. 145–170.
20. G.K. Iwama, P.T. Thomas, R.B. Forsyth, and M.M. Vijayan, *Rev. Fish Biol. Fish.* **8**, 35–56 (1998).
21. J.N. Fryer and R.E. Peter, *Gen. Comp. Endocrinol.* **33**, 215–225 (1977).
22. C.S. Bradford, M.S. Fitzpatrick, and C.B. Schreck, *Gen. Comp. Endocrinol.* **87**, 292–299 (1992).
23. J.P. Sumpter, in G.K. Iwama, A.D. Pickering, J.P. Sumpter, and C.B. Schreck, eds., *Fish Stress and Health in Aquaculture*, Soc. Exp. Biol. Sem. Ser. 62, Cambridge University Press, Cambridge, UK, 1997, pp. 95–118.
24. G.P. Chouros, *Ann. N.Y. Acad. Sci.* **851**, 311–335 (1998).
25. L.A. Sigismondi and L.J. Weber, *Trans. Am. Fish. Soc.* **117**, 196–201 (1988).
26. C.C. Woodward and R.J. Strange, *Trans. Am. Fish. Soc.* **116**, 574–579 (1987).
27. G.K. Iwama, J.C. McGeer, and N.J. Bernier, *ICES Mar. Sci. Symp.* **194**, 67–83 (1992).
28. T.G. Pottinger, A.D. Pickering, and M.A. Hurley, *Aquaculture* **103**, 275–289 (1992).
29. T.G. Pottinger, T.A. Moran, and J.A.W. Morgan, *J. Fish Biol.* **44**, 149–163 (1994).
30. T.G. Pottinger and A.D. Pickering, in G.K. Iwama, A.D. Pickering, J.P. Sumpter, and C.B. Schreck, eds., *Fish Stress and Health in Aquaculture*, Soc. Exp. Biol. Sem. Ser. 62, Cambridge University Press, Cambridge, UK, 1997, pp. 171–193.

31. T.P. Barry, J.A. Malison, J.A. Held, and J.J. Parrish, *Gen. Comp. Endocrinol.* **97**, 57–65 (1995).
32. B.A. Barton, C.B. Schreck, R.D. Ewing, A.R. Hemmingsen, and R. Patiño, *Gen. Comp. Endocrinol.* **59**, 468–471 (1985).
33. A.G. Maule, C.B. Schreck, and S.L. Kaattari, *Can. J. Fish. Aquat. Sci.* **44**, 161–166 (1987).
34. T.G. Pottinger, P.H.M. Balm, and A.D. Pickering, *Gen. Comp. Endocrinol.* **98**, 311–320 (1995).
35. B.A. Barton and C.B. Schreck, *Aquaculture* **62**, 299–310 (1987).
36. K.B. Davis, M.A. Suttle, and N.C. Parker, *Trans. Am. Fish. Soc.* **113**, 414–421 (1984).
37. B.A. Barton, C.B. Schreck, and L.G. Fowler, *Prog. Fish-Cult.* **50**, 16–22 (1988).
38. K.B. Davis and N.C. Parker, *Aquaculture* **91**, 349–358 (1990).
39. A.D. Pickering and T.G. Pottinger, *J. Fish Biol.* **30**, 363–374 (1987).
40. A. Hontela, *Rev. Toxicol.* **1**, 1–46 (1997).
41. B.A. Barton, C.B. Schreck, and L.A. Sigismondi, *Trans. Am. Fish. Soc.* **115**, 245–251 (1986).
42. M.G. Mesa, *Trans. Am. Fish. Soc.* **123**, 786–793 (1994).
43. B.A. Barton, C.B. Schreck, and L.D. Barton, *Dis. Aquat. Org.* **2**, 173–185 (1987).
44. C.B. Schreck, L. Jonsson, G. Feist, and P. Reno, *Aquaculture* **135**, 99–110 (1995).
45. G.A. Wedemeyer, *Physiology of Fish in Intensive Culture Systems*, Chapman & Hall, New York, NY, 1996.
46. G.K. Iwama, J.D. Morgan, and B.A. Barton, *Aquacult. Res.* **26**, 273–282 (1995).
47. N.W. Pankhurst and G. Van Der Kraak, in G.K. Iwama, A.D. Pickering, J.P. Sumpter, and C.B. Schreck, eds., *Fish Stress and Health in Aquaculture*, Soc. Exp. Biol. Sem. Ser. 62, Cambridge University Press, Cambridge, UK, 1997, pp. 73–93.
48. G.K. Iwama, *Ann. N.Y. Acad. Sci.* **851**, 304–310 (1998).
49. B.A. Barton and C.B. Schreck, *Trans. Am. Fish. Soc.* **116**, 257–263 (1987).
50. L.E. Davis and C.B. Schreck, *Trans. Am. Fish. Soc.* **126**, 248–258 (1997).
51. S.F. Snieszko, *J. Fish Biol.* **6**, 197–208 (1974).
52. A.E. Ellis, in A.D. Pickering, ed., *Stress and Fish*, Academic Press, London, 1981, pp. 147–169.
53. A.D. Pickering, in *Fish Diseases, A Threat to the International Fish Farming Industry*, AquaNor 87, Conference 3, Norske Fiskeoppdretteres Forening, Trondheim, Norway, 1987, pp. 35–49.
54. J.E. Bly, S.M.-A. Quiniou, and L.W. Clem, *Dev. Biol. Stand.* **90**, 30–43 (1997).
55. P.H.M. Balm, in G.K. Iwama, A.D. Pickering, J.P. Sumpter, and C.B. Schreck, eds., *Fish Stress and Health in Aquaculture*, Soc. Exp. Biol. Sem. Ser. 62, Cambridge University Press, Cambridge, UK, 1997, pp. 195–221.
56. B.S. McEwen, C.A. Biron, K.W. Brunson, K. Bulloch, W.H. Chambers, F.S. Dhabhar, R.H. Goldfarb, R.P. Kitson, A.H. Miller, R.L. Spencer, and J.M. Weiss, *Brain Res. Rev.* **23**, 79–133 (1997).
57. P.M. Campbell, T.G. Pottinger, and J.P. Sumpter, *Biol. Reprod.* **47**, 1140–1150 (1992).
58. P.M. Campbell, T.G. Pottinger, and J.P. Sumpter, *Aquaculture* **120**, 151–169 (1994).
59. E.M. Donaldson, *Am. Fish. Soc. Symp.* **8**, 109–122 (1990).
60. T.L. Beitinger, *J. Great Lakes Res.* **6**, 495–528 (1990).
61. J.E. Thorpe and F.A. Huntingford, eds., *The Importance of Feeding Behavior for the Efficient Culture of Salmonid Fishes*, World Aquaculture Workshops No. 2, The World Aquaculture Society, Baton Rouge, LA, 1992.
62. B.L. Olla and M.W. Davis, *Aquaculture* **76**, 209–214 (1989).
63. B.L. Olla, M.W. Davis, and C.B. Schreck, *Trans. Am. Fish. Soc.* **121**, 544–547 (1992).
64. B.L. Olla, M.W. Davis, and C.B. Schreck, *Aquacult. Res.* **26**, 393–398 (1995).
65. C. Ejike and C.B. Schreck, *Trans. Am. Fish. Soc.* **109**, 423–426 (1980).
66. S. Winberg and G.E. Nilsson, *Comp. Biochem. Physiol.* **106C**, 597–614 (1993).
67. S. Winberg, C.G. Carter, I.D. McCarthy, Z.-Y. He, G.E. Nilsson, and D.F. Houlihan, *J. Exp. Biol.* **179**, 197–211 (1993).
68. V.S. Blazer, *Ann. Rev. Fish Dis.* **2**, 309–323 (1992).
69. T.C. Fletcher, in G.K. Iwama, A.D. Pickering, J.P. Sumpter, and C.B. Schreck, eds., *Fish Stress and Health in Aquaculture*, Soc. Exp. Biol. Sem. Ser. 62, Cambridge University Press, Cambridge, UK, 1997, pp. 223–246.
70. G.A. Wedemeyer, in W. Pennell and B.A. Barton, eds., *Principles of Salmonid Culture*, Dev. Aquacult. Fish. Sci. Vol. 29, Elsevier Science BV, The Netherlands, 1996, pp. 727–758.
71. G.A. Wedemeyer, in G.K. Iwama, A.D. Pickering, J.P. Sumpter, and C.B. Schreck, eds., *Fish Stress and Health in Aquaculture*, Soc. Exp. Biol. Sem. Ser. 62, Cambridge University Press, Cambridge, UK, 1997, pp. 35–71.
72. B.A. Barton, A.B. Rahn, G. Feist, H. Bollig, and C.B. Schreck, *Comp. Biochem. Physiol.* **120A**, 355–363 (1998).
73. B.A. Barton and W.P. Dwyer, *J. Fish Biol.* **51**, 998–1008 (1997).
74. B.G. Rehnberg, R.J.F. Smith, and B.D. Sloley, *Can. J. Zool.* **65**, 2916–2921 (1987).
75. B.A. Barton and R.E. Zitzow, *Prog. Fish-Cult.* **57**, 267–276 (1995).
76. B.A. Barton, *N. Am. J. Aquacult.* **62**, in press (2000).

See also BACTERIAL DISEASE AGENTS; DISSOLVED OXYGEN; GAS BUBBLE DISEASE; OSMOREGULATION IN BONY FISHES.

STRIPED BASS AND HYBRID STRIPED BASS CULTURE

CHRISTOPHER C. KOHLER
Southern Illinois University
Carbondale, Illinois

OUTLINE

The anadromous striped bass (*Morone saxatilis*) is one of four *Morone* species. The others are white bass (*M. chrysops*), yellow bass (*M. mississippiensis*), and white perch (*M. americana*). *Morone* spp. were originally placed in the family Percichthyidae, however, Johnson (1) placed them in their own family, Moronidae. The Names of Fishes Committee of the American Fisheries Society recently adopted the nomenclature suggested by Johnson. The striped bass is a major sport and commercial species native to the Atlantic and Gulf Coast of the United States, but stockings have expanded its range through much of North America as well as to other continents. When a reproducing population of striped bass was recognized in landlocked Santee Cooper Reservoir, South Carolina (2,3), fisheries biologists became interested in stocking striped bass in reservoirs to serve as a recreational sportfish and as a predator to control underutilized forage species. *Morone* hybridization programs were initiated in the 1960s. Efforts were aimed at combining the trophy fish potential of striped bass with the higher adaptability to landlocked, freshwater systems of its congenerics (4,5). Of the various crosses, backcrosses, and outcrosses made, only the striped bass X white bass cross gained wide acceptance (4,6–8). Striped bass and white bass hybrids were first made experimentally by Robert Stevens (South Carolina Wildlife Resources Department) in 1965 (9). The first cross, of the striped bass female with the white bass male, was initially called the original cross-hybrid striped bass, but is now referred to as the palmetto bass. The second, or reciprocal cross (of the white bass female with the striped bass male), is called sunshine bass (Fig. 1). Hybrid striped bass are still being used for recreational fish stockings, but interest in foodfish production emerged beginning in the late 1970s. Striped bass/hybrid striped bass is considered to be the fastest growing segment of the U.S. aquaculture industry (10).

Figure 1. The sunshine bass is a cross between a female white bass and a male striped bass. It is also referred to as the reciprocal cross-hybrid striped bass. The original cross, or palmetto bass, is similar in appearance to sunshine bass and is produced by crossing a female striped bass with a male white bass. (Photo by J. Rudacille.)

Production of food-size hybrid striped bass in the United States has risen from about 455,000 kg (1 million lb) in 1990, to over 4.5 million kg (10 million lb) in 1996 (11). Unless otherwise noted, the information that follows pertains to both striped bass and hybrid striped bass culture.

HISTORICAL PERSPECTIVE

The first hatchery for striped bass was built on the Roanoke River at Weldon, North Carolina in the 1880s (12). Ovulating females and running ripe males were collected on their spawning run, and then eggs and milt were stripped and mixed with water. The developing embryos were incubated in MacDonald jars much as they are to this day. It was not until the 1960s that aquaculturists were able to utilize nonovulating female striped bass. South Carolina Wildlife Resources Department personnel, led by Robert Stevens, developed procedures to induce ovulation in female striped bass that had undergone final gonad maturation (13–15). The department also initiated the production of hybrid striped bass by crossing striped bass with white bass (9) at the Monck's Corner Hatchery, South Carolina, in 1965 (9,16). Both crosses produced viable eggs, but poor survival was experienced with sunshine bass fry (7,9). From 1966 through 1973, production of hybrid striped bass was limited to palmetto bass. In 1973, Florida researchers began successfully culturing sunshine bass at the Richloam Fish Hatchery (17). Intensive rearing of one-day-old striped bass began in the mid-1970s at Southern Illinois University at Carbondale (18). Pond production of hybrid striped bass as a foodfish began in North and South Carolina around 1980 (19). Today, striped bass and both hybrid crosses are reared in ponds and in intensive systems. Rearing of larval sunshine bass is still problematic due to the smaller size of white bass eggs (20).

BIOLOGY AND ECOLOGY

Geographic Range

The anadromous striped bass has been stocked throughout much of North America. They were introduced to the coastal waters of California in 1879 (21) and were subsequently distributed along much of the Pacific coast. Stockings have been directed to freshwater impoundments since the 1960s (22), although marine restoration and enhancement programs have also taken place along the east coast (23).

White bass are also native to North America, and their range extends from the St. Lawrence River west through the Great Lakes (excluding Lake Superior) to South Dakota and the Mississippi and Ohio River drainages south to the Gulf of Mexico (24–26). Like striped bass, the range of white bass has been greatly expanded by stocking and now includes the Gulf and south Atlantic states, as well as New Mexico, Utah, Colorado (27), California (28), and Nevada (29).

Although natural hybridization between striped and white bass is conceivable (30), there are no known natural populations of hybrids in existence (31). Hybrid striped bass are artificially produced and stocked in freshwater impoundments to enhance recreational fisheries and to control forage fish populations (22). Hybrids are believed to have most of the positive recreational attributes of striped bass, while having the adaptability of white bass to varying environments (4,5). They are cultured for food in numerous states throughout the U.S., as well as in Israel, Taiwan, and the People's Republic of China.

Morphological Characteristics

Striped bass, white bass, and their hybrids have horizontal stripes laterally and an overall silver color with upper sides of olive gray shading to white on head and belly (24–26). The striped bass has an elongated body, which is laterally compressed and with the deepest part below the posterior portion of the spinous dorsal fin. The white bass body is robust, deep, and strongly compressed laterally and has similar coloring. The spinous dorsal and soft dorsal fins are entirely separate. Hybrids are intermediate of parentals in physical appearance and most meristic characters:

	Striped Bass	White Bass	Hybrids
• Lateral line scales	50 to 72	52 to 58	Intermediate
• Scales above lateral line	9 to 13	7 to 9 (usually 8)	Intermediate
• Soft anal rays	9 to 11	12 to 13	Intermediate
• Soft dorsal rays	12	12 to 13	12 to 13
• Teeth on tongue	2 patches	1 patch	Intermediate
• Parr marks	Present	Absent	Present

Fingerling white bass are easy to distinguish from fingerling striped bass, which have parr marks (7) and typically two tooth patches (32). Conversely, it is sometimes difficult to distinguish the young of white bass and hybrids of striped bass, including F_1, F_2 and backcrosses (33).

Age and Growth

Landlocked striped bass tend to initially grow faster than coastal stocks (21), but the latter grow considerably larger. Scott and Crossman (26) reported a striped bass caught in North Carolina in 1891 that weighed over 50 kg (110 lb). Striped bass are known to live more than 30 years (31), with females growing to the larger sizes (34,35). White bass grow considerably slower than striped bass, rarely exceed 2 kg (4.4 lb), and generally live no more than 9 years (20). Hybrid striped bass generally have superior growth rate to parentals (weight not length) in the first 2 years of life (8), but striped bass ultimately get larger (Table 1).

Food Habits

Newly hatched striped bass feed in schools on mobile zooplankton (31). Cladocerans and copepods are the primary dietary constituent during the early summer (49–51). Insects, other crustaceans and larval clupeids become more prevalent in late summer (49). Juvenile stripers usually switch to a piscine diet in autumn with soft-rayed clupeids being preferred (31,52).

Newly hatched white bass initially feed on rotifers and similar size organisms for two or three weeks when they reach a size that allows them to feed on microcrustaceans. Copepods and cladocerans comprise the bulk of the natural diet, at least through midsummer (41,53). However, insects were found to predominate in young white bass during spring in Beaver Reservoir, Arkansas (54). By midsummer, young white bass will switch to a largely piscine diet, provided forage fish are available in suitable size and abundance. Alternatively, the fish may continue to consume invertebrates (55). Adults are primarily piscivorous.

Hybrid striped bass food habits are initially very similar to their maternal parent (7). The larger newly hatched palmetto bass feed on cladocerans and copepods, while the initial food of sunshine bass consists mainly of rotifers. As adults, hybrid striped bass food habits are essentially equivalent to those of pure striped bass (56–59).

Natural Spawning

Female striped bass in the wild usually mature at age 4 years or older, while males usually mature at age 2 (60,31). Age at maturity is a function of size so those in warm waters mature earlier (31). Fecundity is a function of age, length, and weight (61). Most females produce between 110,000 to 220,000 ova/kg (50,000 to 100,000 ova/lb) of wet body weight (31). The spawning temperature for striped bass is typically between 13.9 to 21.1 °C (57 to 70 °F) (31). Striped bass are polygamous broadcast spawners in freshwater rivers (62). Ripe eggs range from 1.0 to 1.6 mm (0.03 to 0.06 in.) in diameter (63–65), are semibuoyant, and are fertilized when released into a current (26). Spawning times are highly variable (31). Hatching time is depends upon the

Table 1. Calculated Total Length in Centimeters (inches in parentheses) of Striped Bass, White Bass, and Palmetto Bass (SB female × WB male) in Various Landlocked Populations

Taxon	Location	Age 1	Age 2	Age 3	Age 4	Reference
Striped bass	South Carolina	21.6 (8.5)	39.9 (15.7)	50.3 (19.8)	58.3 (23.0)	3
	Oklahoma	25.8 (10.2)	45.5 (17.9)	54.1 (21.3)	60.6 (23.9)	36
	Kentucky	25.1 (9.9)	40.4 (15.9)	55.9 (22.0)	65.3 (25.7)	37
	Virginia/North Carolina	12.9 (5.1)	28.0 (11.0)	41.5 (16.3)	56.1 (22.1)	38
	Virginia	21.4 (8.4)	38.7 (15.2)	51.9 (20.4)	58.3 (23.0)	39
White bass	Lake Erie	11.9 (4.7)	20.8 (8.2)	27.7 (10.9)	31.5 (12.4)	40
	South Dakota	10.9 (4.3)	24.4 (9.6)	30.2 (11.9)	32.8 (12.9)	41
	Nebraska	11.7 (4.6)	25.1 (9.9)	32.0 (12.6)	35.8 (14.1)	42
	New York	13.5 (5.3)	26.2 (10.3)	31.4 (12.4)	33.8 (13.3)	43
	Virgina	14.8 (5.8)	26.8 (10.6)	33.4 (13.1)	38.7 (15.2)	39
Palmetto bass	Alabama	23.4 (9.2)	35.8 (14.1)	47.2 (18.6)	55.4 (21.8)	44
	Georgia	27.9 (11.0)	42.9 (16.9)	49.0 (19.3)	53.6 (21.1)	45
	Kentucky	25.1 (9.9)	42.7 (16.8)	50.8 (20.0)	56.6 (22.3)	46
	Ohio	17.0 (6.7)	35.3 (13.9)	48.8 (19.2)	— —	47
	Illinois[a]	26.9 (10.6)	44.7 (17.6)	55.9 (22.0)	65.0 (25.6)	48

[a]Heated power cooling pond.

temperature, ranging from 29 h at 22 °C (71.6 °F) to 80 h at 11 °C (51.8 °F) (31). At hatching, larval striped bass average 3.1 mm (0.1 in.) TL in the wild (31), but are as large as 5.0 to 7.0 mm (0.2 to 0.3 in.) TL from laboratory-reared fish (66,67).

Female white bass in the wild usually mature when reaching 3 years old, while males generally mature a year earlier (41). The spawning temperature for white bass is typically between 14.4 to 18.3 °C (58.0 to 65.0 °F) (41). White bass migrate up tributaries when available and spawn in shallow waters on firm gravel or sand (24). They will spawn on any suitable shoreline structure in the absence of tributaries. Spawning occurs during both day and night, but spawning fish are most active crespuscularly (55). Eggs are adhesive, are demersal, and increase little in diameter when hardened by water (68). No nest construction or care is provided to the eggs that stick to gravel and vegetation. Mature white bass females can each produce several hundred thousand eggs, ranging in diameter between 0.61 and 0.68 mm (0.02 to 0.03 in.) at ovulation (4). Hatching occurs in about two days, with fry being approximately 3.0 mm (0.1 in.) TL (41).

Morone hybrids are not sterile and have bred with each other in the wild (69,70), as well as having been outcrossed with striped bass (71). No reports of natural hybrid outcrossing with white bass were found in the literature.

HATCHERY PHASE

Collection and Transportation of Broodstock

Striped bass broodstock are usually collected during spawning migrations in river headwaters above and below dams (72). Electrofishing, gillnets, and pound nets have been successfully employed (72,73). Electrofishing has been shown to be less stressful than gillnetting (74). The smallest males spermiating and females with secondary oocytes are preferred if the fish are to be maintained as broodstock. Striped bass in excess of 5 kg (11 lb) obtained from the wild do not adapt well to captivity (73). North Carolina researchers (73) recommend allowing females to resorb their eggs as an energy reserve to carry them through adaptation to dry feed. Males will generally spermiate the subsequent spawning season, whereas females may require two years before they produce good quality eggs (73). State hatcheries usually obtain all their striped bass broodstock from the wild and spawn them immediately.

Adult white bass can readily be caught by hook-and-line (75) or trap nets (72). Electrofishing can be used for capturing white bass, but with varying degrees of success. Many producers purchase their white bass broodstock from commercial fishermen on Lake Erie (20). White bass are usually hardier than striped bass, and are often spawned immediately after collection.

Ideally, hauling tanks should be filled with water from where the broodfish are collected. Salt (NaCl or synthetic sea salts) should be added to the water to raise the salinity to 8–12 ppt for striped bass (72) and 5 ppt for white bass (20). Depending on water temperature where the fish are collected, hauling tank water temperature should be 18.3 °C (65 °F) or less. Ice can be added if needed. The hauling tank should be equipped with pure oxygen regulated to maintain dissolved oxygen concentrations at a minimum of 7 ppm (76). Depending on fish condition and hauling duration, approved therapeutic treatments to control disease should be considered. The use of the anesthetic MS-222 during hauling does not appear to be necessary, but can be used at appropriate doses based on experience of haulers. Due to possible shifts in water pH when using MS-222, always buffer this anesthetic with sodium bicarbonate (500 ppm $NaHCO_3$: 100 ppm MS-222).

REARING FACILITIES

Training to Formulated Feed

Kohler et al. (75) demonstrated that adult white bass can be trained to accept and then be maintained on dry feed for over two years. They stressed that newly collected white bass should not be fed until all therapeutic treatments are completed. This delay assures that the fish will be hungry when feed is first offered and prevents habituating fish to ignore feed by presenting it to them when stress or disease agents impede their appetites. The fish are initially trained to formulated feed by hand-feeding moist pellets, which are prepared by mixing equivalent amounts of commercial dry trout feed [broodstock diet; 40 : 11 (% protein : % fat)] with raw gizzard shad, *Dorosoma cepedianum*, and a vitamin premix (coolwater fish; U.S. Biochemical, Cleveland, Ohio). Healthy white bass readily take this feed. By the second day of active feeding, a small proportion of the dry trout feed (floating) is mixed with the moist pellets. The proportion of dry feed is slowly increased in the diet until the fish accept 100% dry feed, a process usually completed in two weeks or less. This procedure should also work for striped bass. However, it may be necessary to feed striped bass pieces of fresh fish or even live fish if formulated feed is rejected after several days in captivity. Both white bass and striped bass tend to rapidly switch to dry feed once a few fish begin to accept the feed. Broodfish already trained to formulated feed can be placed with newly captured fish to accelerate formulated feed acceptance.

At the hatchery, striped bass and white bass broodstock should be placed for several hours in stress-recovery tanks containing salinities similar to those used for hauling. Salinity can be slowly reduced to 5 ppt by flushing. Both species seem to perform best in circular tanks, but can be held in rectangular tanks if necessary.

Domestication of Broodstock

A few efforts have been made to develop domesticated broodstocks of striped bass (77–80) and white bass (20,75,81). The difficulty with striped bass is the long time between generations, particularly for females. The shorter generation time of white bass makes this parental species a simpler candidate to domesticate (Fig. 2). Sullivan et al. (73) suggested that unless a producer is prepared to maintain and propagate domestic broodstocks of both parental lines, the simplest solution may be to use captive striped bass males crossed with captive white bass females to produce sunshine bass. Consequently, the producer would not be burdened with producing and maintaining mature striped bass females. However, without the domestication of both sexes of striped bass and white bass broodstock, there will be no significant advancements made in terms of genetic improvement of heritable traits of the parental lines used to make the hybrids for the industry. Such domestication efforts are being carried out at various research centers and well-established hybrid striped bass commercial enterprises.

Figure 2. White bass broodfish are highly amenable to captivity and will readily undergo sexual maturation if held under proper photothermal regimes. (Photo by D. Russell.)

Controlled Spawning

Striped bass and white bass females are normally injected with human chorionic gonadotropin (HCG) to induce ovulation. Chorulon® is the recommended brand, because it was recently approved by the U.S. Food and Drug Administration for fish use. Recommended dosages for female striped bass are between 275 and 330 IU/kg (125–150 IU/lb) wet-body weight (4,5,8,14). Bonn et al. (5) stated that white bass females can be successfully induced to spawn at dosages between 1,000 and 2,000 IU/kg (500 and 1,000/lb). This was not a recommendation, but has been taken to be one. Recently, Kohler et al. (81) showed that HCG dosages as low as 50 IU/kg (110 IU/lb) are more efficacious. Accordingly, it appears that white bass should be injected at dosages similar to those commonly used for striped bass [i.e., 275–330 IU/kg (125–150 IU/lb) for females and 110–165 IU/kg (50–75 IU/lb) for males].

Only females in spawning condition — based on plumpness and/or visual examination of oocytes collected by a 3.0 mm (0.1 in.) outside diameter plastic catheter (68) for striped bass; or a 1.5 mm (0.005 in.) outside diameter plastic catheter for white bass (83) — should be injected with HCG. Reinjection of females with HCG usually results in abortion of eggs (68). However, multiple injections, using lower HCG dosages, have not been well examined. White bass males have been shown to continuously spermiate for several months when held at spawning temperature of about 16 °C (68 °F) (75). A monthly dosage of 100 IU/kg (45 IU/lb) is recommended if males are to be reused as broodstock over an extended period of time (20).

Before injection with HCG, the fish should be anesthetized with 50–100 ppm MS-222 with five times that amount of sodium bicarbonate to serve as a buffer. The fish can then be weighed, and the proper dosages administered intramuscularly just ventral to the first dorsal fin above the lateral line.

Production of palmetto bass can be problematic due to difficulties sometimes encountered in obtaining female striped bass in the late stages of ovarian maturation (84). The stress on broodstock associated with repetitive handling to check for maturation can impair reproductive performance and cause mortalities. To address this issue, researchers have demonstrated that a synthetic analogue of mammalian gonadotropin releasing hormone (mGnRHa) implanted in striped bass can reliably induce spawning (73,85,86). Commercial application of this treatment awaits necessary studies to obtain regulatory approval.

Female striped bass and white bass should be checked for ovulation every 2 hours, starting at 16 hours post-HCG injection by lightly exerting abdominal pressure to extrude a small number of eggs. Oocytes can be staged for both species by using similar procedures described for striped bass (8,68). Ovulation generally occurs between 24–36 hours post-injection at 16–18 °C (60.8–64.4 °F), but a window of only 1–2 hours is available to obtain properly ripened eggs. In general, ovulation is indicated by the occurrence of clear, free-flowing, uniform-shaped, yellowish-tinged eggs with fully intact inner chorion surfaces. Ovulated females should be anesthetized with MS-222, as previously described. Ripe males usually do not need to be anesthetized because of the relative ease with which their gametes can be expressed.

Before manually removing the gametes, fish should be dried with a paper towel to avoid water contamination that might prematurely activate sperm. Eggs are removed from females by firmly exerting abdominal pressure starting just above and slightly posterior to the pelvic fins and progressing posteriorly and vertically toward the genital opening (Fig. 3). Semen can be expressed in a similar fashion. To avoid any contamination with urine, semen can be collected by inserting a Pasteur pipette in the urogenital opening and applying suction (75). It is advisable to collect semen from two or more males for fertilizing each egg batch to improve genetic heterozygosity. This also decreases the possibility of fertilizing eggs with low-quality semen (e.g., semen of low motility).

Figure 3. *Morone* are usually spawned by manually stripping gametes. Here, semen from a male striped bass is being stripped into a metal pan containing striped bass eggs. (Photo by G. Brown.)

The "wet" method for fertilization, in which semen is added to a mixture of eggs and water, is often used when males are stripped because urine often prematurely activates spermatozoa. Some hatcheries use a modified "wet" fertilization method, in which semen and water are added simultaneously to the eggs (33). The "dry" method, in which semen is mixed with eggs followed by water, should probably be limited to when semen is collected by a pipette. Regardless of the method employed, at least one minute of gentle stirring in a Teflon® pan (white bass eggs are adhesive) should be allowed for fertilization to take place (20).

Incubation Techniques

Embryos can be incubated in aquaria (5), Heath trays (75), or in MacDonald-type jars. However, in the case of white bass eggs, they cannot be incubated in jars, unless their adhesiveness is neutralized. Fuller's earth, silt, clay, starch, sodium sulfite, and tannic acid have all been used with varying degrees of success (33). Rottmann et al. (87) provided a detailed protocol (20) for eliminating adhesiveness of white bass eggs. Key elements of the protocol are the use of sodium chloride and urea to clear the embryos, so that developmental events can be seen, and tannic acid to reduce adhesiveness. Formalin-F can be used at 50 ppm to reduce fungal infections if hatchery water is being recycled through the MacDonald jars. Otherwise, concentrated bath treatments of 286–429 ppm can be used by injecting 2 to 3 ml (0.07 to 0.1 oz) of full strength formalin into the top of the shad tubes to provide a rapid flush treatment (33). Dead embryos usually turn opaque after 30 minutes and, when practical, should be removed.

Development is temperature dependent. At 16–18 °C (60.8–64.4 °F), the embryos will begin to hatch 36–48 hours post-fertilization and will usually be complete within another 24 hours (75). Depending on temperature, an additional 72–96 hours are required for the larvae to absorb their yolk sacs. Rees and Harrell (68) provided characteristics of various stages of development for striped bass, while white bass developmental stages were described by Yellayi and Kilambi (88) and Bayless (4).

POND PRODUCTION

Morone culture is divided into four phases: the hatchery phase as previously described; phase I in which larval fish are reared for 30–45 days; phase II, where fish harvested from phase I ponds are reared through the first growing season (6–9 months); and phase III, where fish harvested from phase II ponds are reared to market size in the second growing season. For a detailed description of these procedures see Harrel et al. (89), Harrell (90), and Hodson (19). Pond production procedures are similar for all *Morone* spp., except during phase I.

Phase I

The first foods of *Morone* are zooplankton with the smaller white bass and sunshine bass starting on rotifers, while striped bass and palmetto bass start on cladocerans and copepods. Accordingly, successful phase I production requires a proper fertilization scheme to ensure that the right zooplankton communities are present in terms of quality and quantity. Prior to stocking, ponds should be drained, refilled, and fertilized with a mixture of organic fertilizers (cottonseed meal, alfalfa hay, and animal manure) and inorganic fertilizers (ammonium nitrate, 52%; and phosphoric acid, 32%). Organic fertilizers may be applied at 200 to 500 kg/ha (200 to 500 lb/acre), while inorganic fertilizers should be applied at 2.5 kg per ha (2.5 lb per acre). Pond filling and fertilization should take place 2 to 3 days to one week prior to stocking white bass and sunshine bass and two weeks prior to stocking striped bass and palmetto bass. This timing scheme ensures that rotifers dominate the zooplankton community when the former are stocked and that cladocerans and copepods prevail when the latter are stocked. In either case, inorganic fertilizers should be added weekly at the same rate as for organic fertilizers through at least 3 weeks poststocking. Additional organic fertilizer should also be applied on the third week poststocking at about 25 kg/ha (25 lb/acre). These are general recommendations, and the timing and quantities may need to vary depending on site characteristics.

The stocking rate is about 250,000 to 500,000 of 4-day posthatch fry/ha (100,000 to 200,000 /acre) for striped bass/palmetto bass and white bass/sunshine bass, respectively. Salmon starter feeds are usually provided at day 21 at 5 to 10 kg/ha (5 to 10 lb/acre) per day (fed three times per day). Progressively larger feed sizes and higher ration sizes should be used as fish grow. It is advisable to continue feeding some starter feed throughout phase I to ensure most surviving fish switch to prepared feed. Fish should readily be consuming no. 1 crumbles by the end of phase I. It is essential to maintain good water quality throughout this phase. Phase I generally takes 30–45 days when fish attain lengths of 2.5 to 5.0 cm (1.0 to 2.0 in.) TL and weigh about 1.0 g (0.03 oz). Survival in excess of 15% is considered good for white bass and sunshine bass, while striped bass and palmetto bass should exceed 45% (19).

Phase II

Harvested phase I fish should be graded to separate out runts (fish <2.5 cm or 1.0 in. TL). This is usually accomplished by holding fingerlings in raceways or tanks. Further feed training also takes place at this time. Larger fish should also be graded out, since they are likely to become cannibals. Stocking rate ranges from 20,000 to 30,000 fingerlings per ha (8,000 to 12,000/acre). Fish are initially fed no. 1 or no. 2 crumbles (40% crude protein) at 15–25% body weight daily in three equal feedings for the first month. Fish can be fed in two feedings thereafter at 15% body weight. The feed rate is gradually reduced to 3% body weight by fall. Near the end of phase II, fish should be able to consume no. 4 crumbles. Survival at the end of phase II should exceed 85%, and fish should weigh approximately 100 g (0.22 lb) or more. Phase II fish can be harvested and restocked for phase III in the first fall, or alternatively, the fish can be overwintered and then harvested and restocked in the following spring. Runts should be graded out prior to restocking.

Phase III

Phase III fish are stocked at 7,500 to 10,000 fish/ha (3,000 to 4,000/acre). Fish are fed floating feeds of 36–40% crude protein at 3% body weight per day (usually in two feedings). Ponds should not be provided with more than 250 kg/ha (250 lb/acre) on a daily basis, even with supplemental aeration (91). Fish are harvested at the end of the growing season in late fall and should weigh 0.7 kg (1.5 lb) or more. The harvested fish can be stunned in super-chilled water and then packed on ice for delivery to the buyer.

INTENSIVE PRODUCTION

Striped bass and hybrid striped bass can be cultured in indoor recirculating systems (92). Researchers at Southern Illinois University (18) were the first to succesfully rear striped bass in such systems. That research was the basis for a striped bass hatchery on the Hudson River built by Commonwealth Edison Utility Company. Subsequently, commercial producers have used intensive systems to raise both striped bass (93) and hybrid striped bass (94). In fact, approximately half of all food-size hybrid striped bass grown in the U.S. are produced in intensive systems. The reader is referred to Hochheimer and Wheaton (92) and other contributions in this encyclopedia for more details on the use of such systems for *Morone* culture.

WATER QUALITY

Morone can be cultured in a wide range of water-quality variables (19). Oxygen requirements are slightly higher than that required for channel catfish, but are not as high as are needed for trout. Recommended levels for the major water quality variables are as follows:

Variable	Desirable
Dissolved oxygen	>4.0 ppm
Temperature	15–17 °C (77–86 °F) optimal
	16–32 °C (60–90 °F) growth occurs
	<10 °C (50 °F); >33 °C (93 °F) no growth
Alkalinity/hardness	>100 ppm $CaCO_3$
pH	7–8.5 optimum
	6–9.5 growth occurs
Ammonia	<1.0 ppm

The reader is referred to Boyd (90) for details on water-quality management.

NUTRITION

A review of micro- and macronutrient requirements of striped bass and hybrid striped bass appears in

Gatlin (91). The general recommendation for diet formulation from juvenile to market-size fish is as follows:

Composition	As fed basis
Protein	40%
Lipid	8%
Fiber	1.7%
Estimated available energy	3.3 kcal/g feed
Energy and protein ratio	8.25 kcal/g protein

Producers have successfully produced phase III hybrid striped bass with diets of lower quality. It is not advisable to use feeds with less than 36% crude protein. Broodstock diets should be high quality; salmonid diets in excess of 40% crude protein and 10% crude fat are sufficient.

DISEASES AND TREATMENTS

Morone are susceptible to most of the disease-causing organisms associated with aquaculture (97). Viruses have not been problematic. On the other hand, bacterial diseases are common, particularly columnaris and motile aeromonas septicemia (*Aeromonas* and *Pseudomonas*). Clinical signs may include exophthalmia (pop eye), hemorrhage in skin and fins, frayed fins, distended abdomen, bloody fluid in body cavity, and anemia (pale gills). Columnaris (*Flexibacter columnaris*) is also common, but is usually confined to skin, fins, or gills. Lesions on skin and gills, as well as frayed fins are the prevailing clinical signs. *Streptococcus* septicemia has recently been found in some striped bass operations (98). Stress from handling and movement seems to trigger this disease where *Streptococcus* is endemic (97). Clinical signs include darker than normal color, erratic swimming (spiraling), and curvature of the body. *Morone* reared in saltwater are susceptible to vibriosis (*Vibrio* spp.). The disease is usually stress related, and the clinical signs are similar to motile aeromonas septicemia. *Morone* reared in intensive systems are susceptible to mycobacteriosis (usually *Mycobacterium marinum*). Clinical signs include dark coloration, hemorrhaging in the skin, pale and granulanatous livers, as well as granulomas in the spleen, heart, kidney, and mesenteries.

Fish with any of the aforementioned diseases may also be afflicted with fungal diseases. White, cottony growths on their bodies is the classic sign of fungal disease.

Flagellated and ciliated protozoan ectoparasites can infect all *Morone* species and hybrids. Ichthyophthiriasis ("Ich") is the most problematic of these parasites. Clinical signs include small white spots on skin and gills, lethargy, gasping at the water surface, and excessive production of mucus. All protozoan parasites tend to cause discomfort and lethargy, and fish may go off feed.

Morone are also subject to being parasitized by gillworms, grubs, nematodes, crustaceans, and so forth. These are usually not life threatening, although grubs can render the fish unusable to human consumers.

Standard therapeutic treatments used in other fish are effective for *Morone*. However, in the United States there are no therapeutics approved by the U.S. Food and Drug Administration for *Morone*. Salt and Formalin F are commonly used to treat parasites. The producer is referred to veterinarians for all other treatments.

ECONOMICS AND MARKETING

Operating Costs

Once the production system and major equipment are in place, operating costs will primarily consist of cost for stocking (or hatchery production), feed, labor, utilities, and water-quality monitoring. Fry can be purchased for about US$0.01 each, while feed-trained fingerlings cost from US$0.15 to US$0.25, depending on quantity. Feed costs will range from US$0.44 to US$0.66/kg (US$0.20 to US$0.30/lb). An enterprise budget for a new 24 ha (60 acre) pond operation for hybrid striped bass is presented in Hodson (19), which is based on a US$1,000,000 beginning balance (US$250,000 equity and US$750,000 bank loan financed at 11% for 20 years). Once the first harvest is sold, monies from fish sales meet all fixed and variable costs and a positive balance occurs by the third year.

Marketing

Hybrid striped bass and striped bass appear to be equally acceptable in the marketplace. Even with the recent rebounding of the wild-striped bass fishery, the wholesale price of aquacultured fish has only slightly declined. Prices, on average, range from US$5.50 to US$7.70/kg (US$2.50 to US$3.50/lb) for whole fish on ice. Producers have not reported problems with respect to selling their product.

CONCLUSION

The commercial aquaculture of *Morone* has greatly advanced over the past decade. Hybrid striped bass are currently raised in many parts of the United States and also in many other areas of the world. The hybrid, in particular, is poised to become a global seafood delicacy in the twenty-first century.

BIBLIOGRAPHY

1. G.D. Johnson, in H.G. Moser, W.J. Richards, D.M. Cohen, M.P. Fahay, A.W. Kendall, Jr., and S.L. Richardson, eds., *Ontogeny and Systematics of Fishes*, American Society of Ichthyologists and Herpetologists, Special Publication 1, 1984, pp. 464–498.
2. G.D. Scruggs, Jr. and J.C. Guller, Jr., *Proc. Ann. Conf. Southeast. Assoc. Game Fish Commrs.* **8**, 64–69 (1954).
3. R.E. Stevens, *Proc. Ann. Conf. Southeast. Assoc. Game Fish Commrs.* **11**, 253–264 (1958).
4. J.D. Bayless, *Artificial Propagation and Hybridization of Striped Bass,* Morone saxatilis *Walbaum*, South Carolina Wildlife and Marine Resources Department, Columbia, 1972.
5. E.W. Bonn, W.M. Bailey, J.D. Bayless, K.E. Erickson, and R.E. Stevens, eds., *Guidelines for Striped Bass Culture*, Striped Bass Committee, American Fisheries Society, Bethesda, MD, 1976.
6. W.B. Smith, W.R. Bonner, and B.L. Tatum, *Proc. Ann. Conf. Southeast. Assoc. Game Fish Commrs.* **20**, 324–330 (1967).

7. J.D. Bayless, *Proc. Ann. Conf. Southeast. Assoc. Game Fish Commrs.* **21**, 233–244 (1968).
8. J.H. Kerby, in R.R. Stickney, ed., *Culture of Nonsalmonid Freshwater Fishes*, CRC Press, Boca Raton, FL, 1986, pp. 127–147.
9. R.D. Bishop, *Proc. Ann. Conf. Southeast. Assoc. Game Fish Commrs.* **21**, 245–254 (1968).
10. USDA, *Aquaculture Situation and Outlook Report*, United States Department of Agriculture, Washington, DC, 1992.
11. R.M. Harrell and D.W. Webster, in R.M. Harrell, ed., *Striped Bass and Other Morone Culture*, Elservier, Amsterdam, 1997.
12. S.G. Worth, *Bull. U.S. Fish. Comm.* **4**(15), 225–230 (1884).
13. R.E. Stevens, O.D. May, Jr., and H.J. Logan, *Proc. Ann. Conf. Southeast. Assoc. Game Fish Commrs.* **17**, 226–237 (1965).
14. R.E. Stevens, *Prog. Fish-Cult.* **28**, 19–28 (1966).
15. R.E. Stevens, *Proc. Ann. Conf. Southeast. Assoc. Game Fish Commrs.* **18**, 525–538 (1967).
16. H.J. Logan, *Proc. Ann. Conf. Southeast. Assoc. Game Fish Commrs.* **21**, 260–263 (1968).
17. F.J. Ware, *Proc. Ann. Conf. Southeast. Assoc. Game Fish Commrs.* **28**, 48–54 (1975).
18. W.M. Lewis, R.C. Heidinger, and B.L. Tetzlaff, *Tank Culture of Striped Bass Production Manual*, Fisheries Research Laboratory, Southern Illinois University, Carbondale, 1981.
19. R.G. Hodson, *Farming a New Fish: Hybrid Striped Bass*, North Carolina Sea Grant, Raleigh, 1995.
20. C.C. Kohler, in R.M. Harrell, ed., *Striped Bass and Other Morone Culture*, Elsevier, Amsterdam, 1977.
21. J.A. Goodson, in A. Calhoun, ed., *Inland Fisheries Management*, California Department of Fish and Game, Sacramento, 1966, pp. 407–412.
22. J.R. Axon and D.K. Whitehurst, *Trans. Amer. Fish. Soc.* **114**, 8–11 (1985).
23. L.C. Woods, III, *Fisheries* **12**, 2–8 (1987).
24. G.C. Becker, *Fishes of Wisconsin*, University of Wisconsin Press, Madison, 1983.
25. D.S. Lee, C.E. Gilbert, C.H. Hocutt, R.E. Jenkins, D.E. McAllister, and J.R. Stauffer, eds., *Atlas of North American Freshwater Fishes*, North Carolina Biological Survey, Raleigh, 1980.
26. W.B. Scott and E.J. Crossman, *Freshwater Fishes of Canada*, Fisheries Research Board of Canada, Ottawa, 1973.
27. R.M. Jenkins, in N.C. Benson, ed., *A Century of Fisheries in North America*, American Fisheries Society, Bethesda, MD, 1970, pp. 173–182.
28. C. von Geldern, *Calif. Fish Game* **52**, 303 (1966).
29. T.R. Trelease, *Nevada Outdoors Widl. Rev.* **4**, 5–10 (1970).
30. C.L. Woods, III, C.C. Kohler, R.J. Sheehan, and C.V. Sullivan, *Trans. Amer. Fish. Soc.* **124**, 628–632 (1995).
31. E.M. Setler, W.R. Boynton, K.V. Wood, H.H. Zion, L. Lubbers, N.K. Mountford, P. Frere, L. Tucker, and J.A. Mihursky, *Synopsis of Biological Data on Striped Bass,* Morone saxatilis Nat. Mar. Fish, Circ., Washington, DC, 1980.
32. H.M. Williams, *Proc. Ann. Conf. Southeast. Assoc. Fish. Wildl. Agencies* **29**, 168–172 (1976).
33. J.H. Kerby and R.V. Minton, eds., *Culture and Propagation of Striped Bass and Its Hybrids*, American Fisheries Society, Bethesda, MD, 1990, pp. 159–190.
34. W.L. Plieger, *The Fishes of Missouri*, Missouri Department of Conservation, Jefferson City, 1975.
35. P.W. Smith, *The Fishes of Illinois*, University of Illinois Press, Urbana, 1975.
36. G.C. Mensinger, *Proc. Ann. Conf. Southeast. Assoc. Game Fish Commrs.* **24**, 447–463 (1971).
37. J.R. Axon, *Walleye Stocking Evaluation at Rough River Lake*, Kentucky Department of Fish and Wildlife, Frankfort, 1979.
38. R.J. Domrose, *Warmwater Fisheries Management Investigations—Striped Bass Study*, Virginia Commission Game and Inland Fisheries, Richmond, 1963.
39. C.C. Kohler, *Trophic Ecology of an Introduced, Land-Locked Alewife (Alosa psuedoharengus) Population and Assessment of Alewife Impact on Resident Sportfish and Crustacean Zooplankton Communities in Claytor Lake, Virginia*, Ph.D. Dissertation, Virginia Polytechnic Institute and State University, Blacksburg, 1980.
40. J. Van Oosten, *Michigan Acad. Sci. Arts, Letters* **27**, 307–334 (1941).
41. R. Ruelle, *Trans. Amer. Fish. Soc. Spec. Pub.* **8**, 411–423 (1971).
42. D.B. McCarraher, M.L. Madsen, and R.E. Thomas, *Trans. Amer. Fish Soc. Spec. Pub.* **8**, 299–311 (1971).
43. J.L. Forney and C.P. Taylor, *New York Fish Game J.* **10**, 194–200 (1963).
44. J.L. Moss and C.S. Lawson, *Proc. Ann. Conf. Southeast. Fish Wildl. Agencies* **36**, 33–41 (1982).
45. J.F. Germann and Z.E. Bunch, *Proc. Ann. Conf. Southeast. Assoc. Fish Wildl. Agencies* **37**, 267–275 (1983).
46. B.T. Kinman, *Evaluation of Hybrid Striped Bass Introductions in Herrington Lake*, Kentucky Department Fish and Wildlife Resources, Frankfort, 1987.
47. M.R. Austin and S.T. Hurley, *Evaluation of Striped Bass X White Bass Hybrid Introduction in East Fork Lake, Ohio*, Ohio Department of Natural Resources, Columbus, 1987.
48. R.C. Heidinger, K.C. Clodfelter, and W.M. Lewis, *Sport Fishing Potential of Power Plant Cooling Reservoirs*, Commonwealth Edison Company, Chicago, IL, 1988.
49. R. Gomez, *Proc. Okla. Acad. Sci.* **50**, 79–83 (1970).
50. J.L. Harper and R. Jarman, *Proc. Ann. Conf. Southeast. Assoc. Game Fish Commrs.* **25**, 501–512 (1972).
51. E.T. Humphries and K.B. Cummings, *Proc. Ann. Conf. Southeast. Assoc. Game Fish Commrs.* **25**, 522–536 (1972).
52. F.J. Ware, *Proc. Ann. Southeast. Assoc. Game Fish Commrs.* **24**, 439–447 (1971).
53. R.R. Priegel, *Trans. Amer. Fish Soc.* **99**, 440–443 (1970).
54. L.L. Olmsted and R.V. Kilambi, *Amer. Fish. Soc. Spec. Pub.* **8**, 397–409 (1971).
55. C.W. Voightlander and T.E. Wissing, *Trans. Amer. Fish. Soc.* **103**, 25–31 (1984).
56. J.F. Germann, *Proc. Ann. Conf. Southeast. Assoc. Fish Wildl. Agencies* **36**, 53–61 (1982).
57. B.A. Saul and J.L. Wilson, *Proc. Ann. Conf. Southeast. Assoc. Fish Wildl. Agencies* **35**, 311–316 (1981).
58. W.K. Borkowski and L.E. Snyder, *Proc. Ann. Conf. Southeast. Assoc. Fish. Wildl. Agencies* **36**, 74–82 (1982).
59. L.A. Jahn, D.R. Douglas, M.J. Terhaar, and G.W. Kruse, *N. Amer. J. Fish. Manage.* **7**, 522–530 (1987).
60. B.A. Rogers, D.T. Westin, and S.B. Saila, *Development of Techniques and Methodology for the Laboratory Culture of Striped Bass,* Morone saxatilis *(Walbaum)*, U.S. Environmental Protection Agency, Narragansett, RI, 1978.
61. D.T. Westin and B.A. Rogers, University Rhode Island Marine Technical Report 67, 1978.
62. T.J. Hassler, *Species Profiles: Life Histories and Environmental Requirements of Coastal Fishes and Invertebrates (Pacific*

Southwest)—Striped Bass, U.S. Fish and Wildlife Service, Washington, DC, 1988.
63. C. Woodhull, *Calif. Fish Game* **33**, 97–102 (1947).
64. E.C. Raney, *Bull. Bingham Oceanogr. Collect., Yale Univ.* **14**(1), 5–97 (1952).
65. R.M. Lewis, *Trans. Amer. Fish. Soc.* **91**, 279–282 (1962).
66. R.J. Mansueti and A.J. Mansueti, *Maryland Tidewater News* **12**(7), 1–3 (1955).
67. B.A. Rogers, D.T. Westin, and S.B. Saila, *Life Stages Duration Studies on Hudson River Striped Bass*, Rhode Island Sea Grant Technical Report 31, 1977.
68. R.A. Rees and R.M. Harrell, in R.M. Harrell, J.H. Kerby, and R.V. Minton, eds., *Culture and Propogation of Striped Bass and Its Hybrids*, American Fisheries Society, Bethesda, MD, 1990, pp. 43–72.
69. J.C. Advise and M.J. Van Den Avyle, *Trans Amer. Fish. Soc.* **113**, 563–570 (1984).
70. A.A. Forshage, W.D. Harvey, K.E. Kulzer, and L.T. Fries, *Proc. Ann. Conf. Southeast. Assoc. Fish. Wildl. Agencies* **40**, 53–61 (1986).
71. R.M. Harrell, X.L. Xu, and B. Ely, *Mol. Mar. Bio. Biotech.* **2**, 291–299 (1993).
72. D.M. Yeager, J.E. Van Tassel, and C.M. Wooley, in R.M. Harrell, J.H. Kerby, and R.V. Minton, eds., *Culture and Propagation of Striped Bass and Its Hybrids*, American Fisheries Society, Bethesda, MD, 1990, pp. 43–72.
73. C.V. Sullivan, D.L. Berlinsky, and R.G. Hodson, in R.M. Harrell, ed., *Striped Bass and Other Morone Culture*, Elsevier, Amsterdam, 1997, pp. 11–73.
74. R.M. Harrell and M.A. Moline, *J. World Aqua. Soc.* **23**, 58–63 (1992).
75. C.C. Kohler, R.J. Sheehan, C. Habicht, J.A. Malison, and T.B. Kayes, *Trans. Amer. Fish. Soc.* **123**, 964–974 (1994).
76. R.G. Piper, I.B. McElwain, L.E. Orme, J.P. MCraren, L.G. Fowler, and J.R. Leonard, *Fish Hatchery Management*, U.S. Fish and Wildlife Service, Washington, DC, 1982.
77. T.I.J. Smith and W.E. Jenkins, *Proc. Ann. Conf. Southeast. Assoc. Fish. Wildl. Agencies* **40**, 152–162 (1988).
78. L.C. Woods, III, J.G. Woiwode, M.A. McCarthy, D.D. Theisen, and R.O. Bennett, *Prog. Fish-Cult.* **52**, 201–202 (1990).
79. L.C. Woods, III, R.O. Bennett, and C.V. Sullivan. *Prog. Fish-Cult.* **54**, 184–188 (1992).
80. L.C. Woods, III, R.M. Harrell, and B. Ely, *Aquaculture* **137**, 41–44 (1995).
81. C.C. Kohler, R.J. Sheehan, V. Sanchez, and A. Suresh, *Aquaculture '94*, abstract, World Aquaculture Society, New Orleans, LA, 1994.
82. T.I.J. Smith, W.E. Jenkins, and L.D. Heyward, *Prog. Fish-Cult.* **58**, 85–91 (1996).
83. T.I.J. Smith, in R.S. Svrjcek, ed., *Genetics in Aquaculture: Proceedings of the Sixteenth U.S.–Japan Meeting on Aquaculture*, National Marine Fisheries Service, Washington, DC, 1990, pp. 53–61.
84. L.C. Woods, III and C.V. Sullivan, *J. Aquacult. & Fish Mgmt.* **24**, 213–224 (1993).
85. Y. Zohar, A. Elizur, N.M. Sherwood, J.F. Rivier, and N. Zmora, *Gen. Comp. Endocrinology* **97**, 289–299 (1995).
86. C.C. Mylonas, Y. Tabata, R. Langer, and Y. Zohar, *J. Controlled Release* **35**, 23–34 (1995).
87. R.W. Rottmann, J.V. Shireman, C.C. Starling, and W.H. Revels, *Prog. Fish-Cult.* **50**, 55–57 (1988).
88. R.R. Yellayi and R.V. Kilambi, *Proc. Ann. Conf. Southeast. Assoc. Fish. Game Commrs.* **23**, 261–265 (1970).
89. R.M. Harrell, J.H. Kerby, and R.V. Minton, eds., *Culture and Propagation of Striped Bass and Its Hybrids*, American Fisheries Society, Bethesda, MD, 1990.
90. R.M. Harrell, ed., *Striped Bass and Other Morone Culture*, Elsevier, Amsterdam, 1997.
91. R.M. Harrell, in R.M. Harrell, ed., *Striped Bass and Other Morone Culture*, Elsevier, Amsterdam, 1997, pp. 75–97.
92. J.N. Hochheimer and F.W. Wheaton, in R.M. Harrell, ed., *Striped Bass and Other Morone Culture*, Elsevier, Amsterdam, 1997, pp. 127–168.
93. J.M. Carlberg, J.C. Van Olst, M.J. Massingil, and T.A. Hovanec, in J.P. McCraren, ed., *The Aquaculture of Striped Bass: Proceedings*, Maryland Sea Grant, University of Maryland, College Park, MD, 1984.
94. J.C. Van Olst and J.M. Carlberg, *Aquac. Mag.* January/February, 49–59 (1990).
95. C.E. Boyd, *Water Quality in Warmwater Fish Ponds*, Auburn University, Auburn, AL, 1979.
96. D.M. Gatlin, III, in R.M. Harrell, ed., *Striped Bass and Other Morone Culture*, Elsevier, Amsterdam, 1997, pp. 235–251.
97. J.A. Plumb, in R.M. Harrell, ed., *Striped Bass and Other Morone Culture*, Elsevier, Amsterdam, 1997, 271–313.
98. T. Kitao, in V. Inglis, R.J. Roberts, and N.R. Bromage, eds. Bacterial Diseases of Fish, Blackwell Scientific Press, Oxford, 1993, pp. 196–210.

SUMMER FLOUNDER CULTURE

DAVID A. BENGTSON
University of Rhode Island
Kingston, Rhode Island

GEORGE NARDI
GreatBay Aquafarms
Portsmouth, New Hampshire

OUTLINE

The summer flounder, or fluke, (*Paralichthys dentatus*) is a "left-handed" flounder (both eyes are on the left side) that ranges from Maine to Florida along the east coast of the United States. Commercial landings of this species have been in decline since the 1950s. The U.S. government has been funding research and development for aquaculture of summer flounder since 1990, and commercial hatchery production began in 1996. The biologically similar turbot (*Scophthalmus maximus*) is commercially cultured in Europe, and the

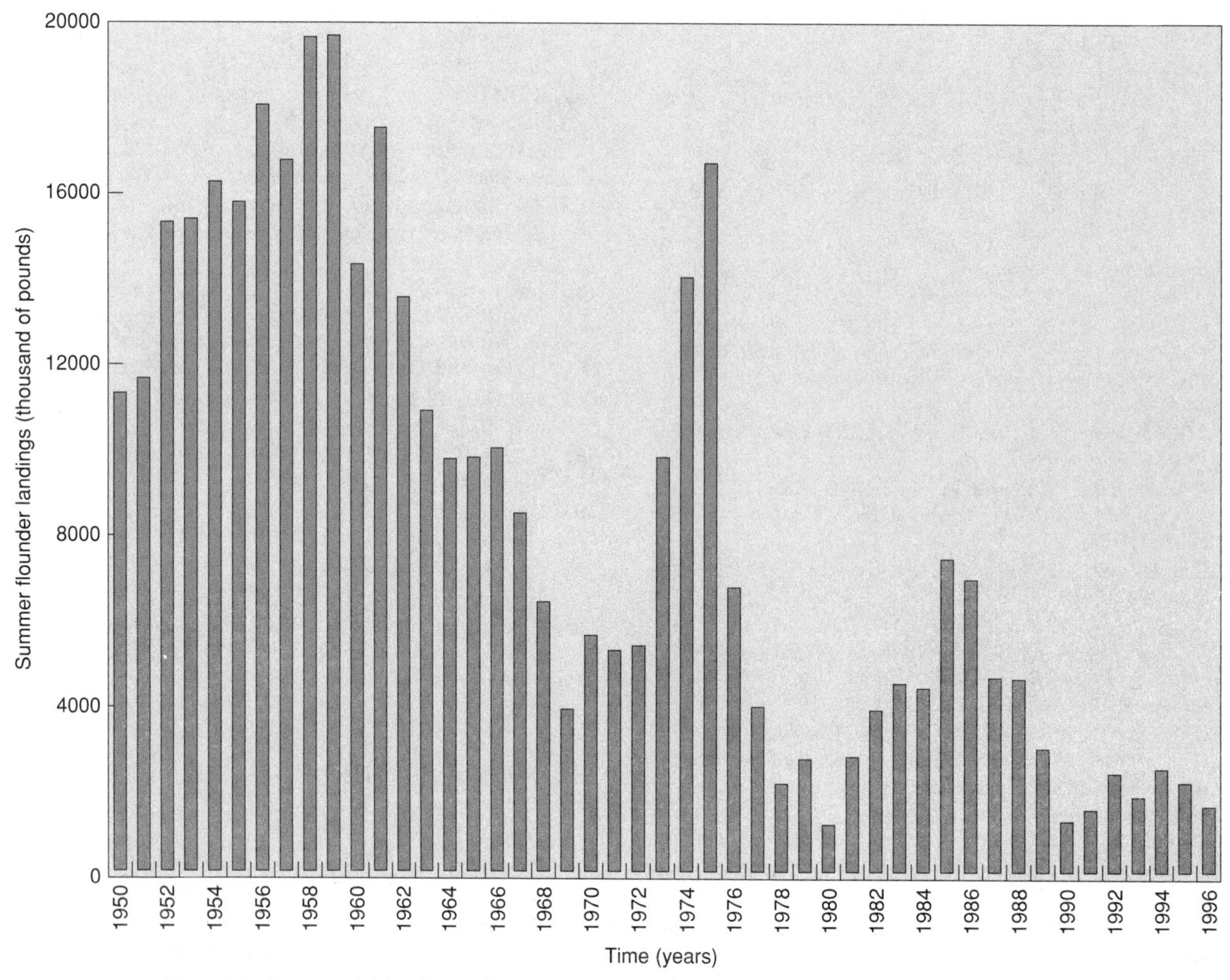

Figure 1. Commercial landings of summer flounder on the East Coast of the United States, 1950–1996 (National Marine Fisheries Service, 1998).

congeneric Japanese (or olive) flounder (*Paralichthys olivaceus*) is commercially cultured in Asia, extensively in public hatcheries for stock enhancement in Japan. Thus, successful models already existed for commercial culture of left-handed flounders and some techniques for rearing those species were readily adaptable to the culture of summer flounder.

The first attempts to spawn and raise summer flounder for both research and as a potential aquaculture candidate occurred in the 1970s (1,2). However, landings in the mid-1970s were close to the 1950s highs, so the need for cultured flounder was not yet apparent. In addition, rearing of larval marine fish was in its infancy at that time and the considerable advances that took place, primarily in Japan and Europe, in the 1970s to 1990s were necessary before the culture of summer flounder could take off. The U.S. National Marine Fisheries Service, through its Saltonstall–Kennedy grants program, funded the initial research in the early to mid-1990s that led to GreatBay Aquafarms establishing itself as the first commercial summer flounder hatchery and Mariculture Technologies establishing itself as the first commercial summer flounder growout facility. The U.S. Department of Commerce, through its Sea Grant program, and the Department of Agriculture, through its Northeastern Regional Aquaculture Center, also began funding important summer flounder aquaculture research in the mid-1990s. At the moment, two hatcheries are in operation, along with four growout facilities. Thus, both the research and the production techniques described herein are in their nascent stages, compared with those of more established species like turbot and Japanese flounder. The research and development of the industry has recently been reviewed elsewhere (3).

BROODSTOCK

To this point, only wild-caught broodstock fish have been used for both research and commercial production. The hatcheries are in the process of domesticating broodstock and beginning selection programs. When captured broodstock are brought in to the hatchery, they require a few weeks to adapt to the tanks and begin feeding on nonliving food (either frozen fish or squid or pelleted diets). It seems best to allow them several months in the tank before spawning induction is attempted. Because

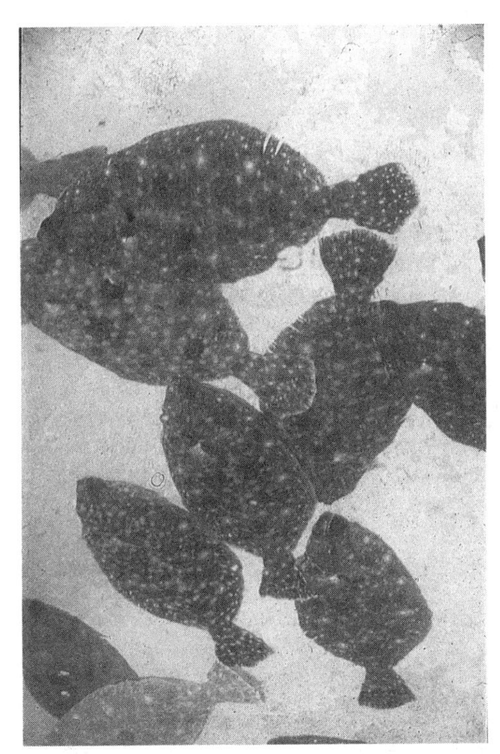

Figure 2. Broodstock summer flounder used for commercial production.

the species spawns during autumn, the environmental cues required for spawning are temperatures less than 18 °C (65 °F) and photoperiod less than 12 hours of light. It has been possible to induce spawning of lab-held broodstock fish in all months of the year by manipulation of environmental cues in association with hormonal manipulations. Once the female fish have gone through hormonal manipulation in the hatchery, they sometimes ripen in response to the environmental cues alone. A recent experiment (4) indicated that females will not ripen in response to environmental cues alone even after they have been in the hatchery for one year. On the other hand, males can be brought into ripe condition by manipulation of environmental cues only after one year in the hatchery.

Hormonal manipulation with injections of carp pituitary extract, originally described by Smigielski (1), has continued to be the most effective inducer of female spawning (5,6). However, the requirement for multiple injections (usually 2 mg/kg/d or 3.2×10^{-5} oz/lb/d) places great stress on the fish. Slow-release implants of GnRHa have therefore been used and are effective as long as the oocytes are greater than about 0.5–0.6 mm ($2–2.4 \times 10^{-2}$ in.) in diameter at the time of implantation (5).

Current spawning methods involve hand-stripping of the ripe fish, which is stressful to them and does not necessarily yield the highest quality eggs. Therefore, development of methods for volitional spawning are a top priority for research. Based on work with Japanese flounder, it is believed that volitional spawning requires stocking broodfish at very low densities; therefore, it is important to know that the fish chosen have the potential to produce quality eggs and sperm. Another area of priority research is broodstock nutrition, about which we know nothing. The tendency is still to feed broodstock on a diet of frozen fish and squid. Anecdotal information suggests that egg and larval quality may be related to dietary changes.

HATCHERY TECHNIQUES

Fertilization and Incubation of Embryos

Milt stripped from males is usually not very abundant and is often collected by pipette so that contact of the sperm with seawater is avoided and the sperm are not activated (sperm are active for only about two minutes after seawater activation). A small amount is examined by microscope to verify sperm motility, and the full quantity is maintained dry (i.e., without addition of seawater) in a beaker on ice until the eggs are obtained. Small samples of eggs are also examined to determine their quality. Good-quality eggs are spherical and float; poor quality eggs may be misshapen and sink. Eggs judged worthy of fertilization are stripped from females into a dry basin or beaker. A female releases tens to hundreds of thousands of eggs. After a sufficient quantity of eggs has been obtained, seawater is finally added to the milt, which is then added to the eggs. The gametes are thoroughly mixed and allowed to settle for several minutes, during which time fertilization takes place. The embryos are then suspended in 34 (ppt) seawater in a cylindrical or cylindro-conical container in order to separate floating (good) eggs from sinking (bad) eggs. Frequent removal of the bad eggs will reduce the substrate for bacterial buildup and decomposition. Alternatively, one can transfer the good eggs to new vessels with clean seawater at frequent intervals. In any case, the embryo incubation period is relatively short and the larvae will hatch in about 60 hours at an incubation temperature of about 20 °C (68 °F). Although hatching is delayed by up to 12 hours, there is anecdotal evidence that larvae hatch more uniformly and are more vigorous when incubated at 16–17 °C (61–63 °F).

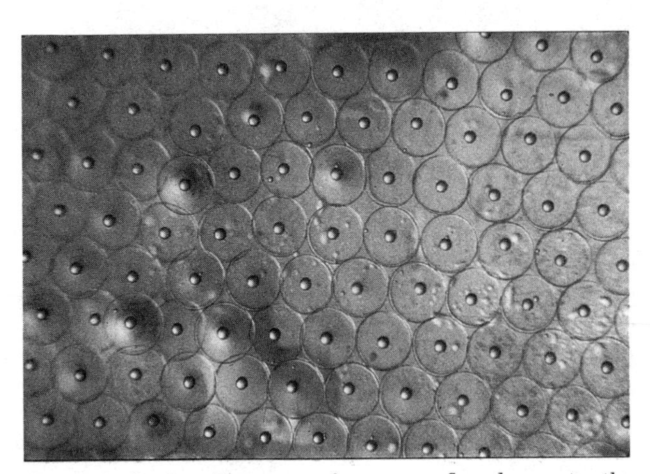

Figure 3. Good-quality eggs of summer flounder; note the spherical shape and the oil droplet.

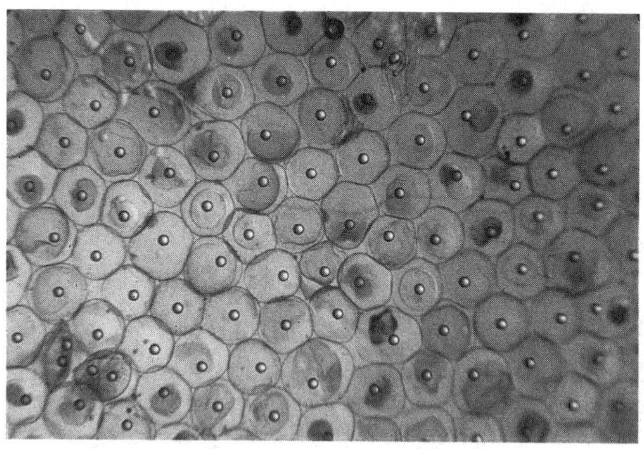

Figure 4. Poor-quality eggs of summer flounder.

Figure 5. Summer flounder embryo.

Figure 6. Tanks for incubation of summer flounder embryos.

Larvae

Larvae are very rudimentary at hatching, with neither fully developed eyes nor digestive tracts. They exhibit very little swimming activity during the first two days and are normally observed hanging head down in the water column. Larvae can be readily transferred from their incubation vessels to rearing tanks up to 2 days after hatch (DAH). Optimal stocking density appears to be in the range of 20–30 larvae/L (75–115 larvae/gal) (6,7). Larvae survive better in conditions of "green-water" rearing (i.e., with algae added to the tanks) than in conditions of "clear-water" rearing (no algae) (8). The eyes and digestive tract are sufficiently developed by 3 DAH that the larvae can begin feeding on rotifers (*Brachionus plicatilis*) (9). The larvae reach the "point of no return" by 6 DAH (at 21 °C, 70 °F) to 11–12 DAH (at 12 °C, 54 °F) if they do not receive food (10). Particularly at high densities, the larvae may suffer severe mortality during the first 10 DAH, but that mortality is due to rearing conditions, rather than inability to initiate feeding. Alves et al. (11) demonstrated that replicate batches of larvae showed high variability in mortality among rearing chambers, but that all were feeding well up to the time of mortality. The tank microbiological environment most likely changes as feed types are changed. This change may introduce or provide the opportunity for opportunistic bacteria that stress or contribute to tank mortality. An increase in mortality is often seen during and following the transition of the fish from rotifers to brine shrimp nauplii (*Artemia* sp.) and then to a dry diet. The change in food is accompanied by a change in the bacteria associated with that food, thus exposing the fish to new microbes during a time of nutritional and physiological stress. Current probiotic research is oriented toward influencing the microbial fauna associated with the feeds, particularly the rotifers and *Artemia* sp., in a way that benefits the fish. Larvae can be reared at a wide range of salinities, although their tolerance to rapid changes in salinity varies greatly with age; and larvae grow better at 14 (ppt) than at 38 (ppt) (12). At the age of about 15–20 DAH, larvae can begin to ingest brine shrimp nauplii (*Artemia* sp.), so a combination of rotifers and brine shrimp is provided to them for a period of about one week. Feeding rates have been shown to increase from about 50 rotifers/larva/day at 6 DAH to about 300 rotifers/larva/day at 13 DAH and from about 50 brine shrimp/larva/day at 23 DAH to about 400 brine shrimp/larva/day at 47 DAH (13). Because those food consumption estimates were obtained in experiments with a 12L : 12D light regime and we assume that the larvae are capable of visually feeding only in the light, it would be wise to double the estimates of food consumption for larvae reared in constant light, as is the practice at GreatBay Aquafarms. As the larvae grow, they change from thin, elongated rudimentary larvae to deeper-bodied, bilaterally compressed larvae with more complex organ systems. About 4–5 weeks after hatch, they undergo the process of metamorphosis that ends the larval stage.

Metamorphosis

Metamorphosis in flatfish is one of the most spectacular examples of body reorganization among the vertebrates. In the case of summer flounder, the right eye migrates across the midline of the body to its juvenile position on the left side of the head, and the fish is transformed from a bilaterally symmetrical swimming larva to an asymmetrical settled juvenile. Meanwhile, the fins attain their juvenile configuration, and the digestive tract completes its development with the formation of a stomach. The process of metamorphosis, divided

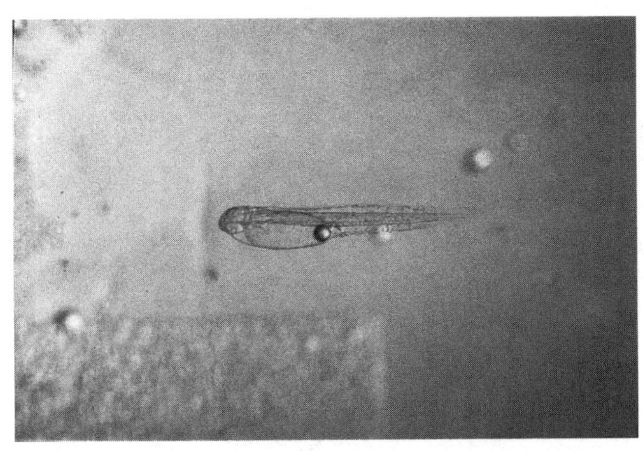

Figure 7. Summer flounder larva with yolk sac.

Figure 8. Summer flounder larva nearing metamorphosis.

into stages of premetamorphosis, prometamorphosis, and metamorphic climax, is primarily under the control of the thyroid hormone and is highly correlated with growth rate (14); thus, fast-growing larvae metamorphose and settle to the bottom earlier than do slow-growing larvae. The result is that settlement of the juveniles occurs from about 35 DAH to about 65 DAH, and the wide range of sizes is conducive to cannibalism if grading is not performed. Francis (15) found that juveniles of about 55 mm ($2\frac{1}{4}$ in.) length could consume siblings up to about 40% of their own size. At both laboratory and commercial hatchery scale, one method of grading is simply to remove settled fish from the larval rearing tanks on a weekly basis and to place them in juvenile rearing tanks. Removal by siphoning is an effective technique, but a bit tedious and not 100% successful. One modification to the technique is to drain the tank into a larval collector or remove the swimming larvae with a bucket after a few days of thinning the tank by siphoning.

The development of a stomach during metamorphosis allows the fish to be weaned from live feed to commercially available formulated diets. Prior to metamorphosis, the larvae are unable to survive on formulated diets (16). Treatment of larvae with exogenous thyroid hormone accelerates development of the stomach (17), but larvae treated this way are still unable to survive any better on formulated diets than are untreated larvae (16). Weaning appears to be more successful the longer the start of weaning can be delayed and the longer the duration of the weaning period (8). Nevertheless, weaning is still a period of unacceptably high mortality, and there is plenty of opportunity for research on both diet formulation and behavioral strategies to induce fish to consume the formulated diets. King et al. (18) recently found that 50% or 55% protein diets yielded better survival and growth during weaning than did a 45% protein diet.

NURSERY AND TRANSITION TO GROWOUT

Once the juvenile flounders have been weaned onto an artificial diet and reach a size of approximately 2 g (7.1×10^{-2} oz), they can be easily netted and graded into larger nursery tanks. At this stage the fish grow rapidly and care must be taken to manage the fish through the system, being sure to have available empty tanks into which the fish can be graded and moved as the densities increase. A simple box grader that floats in the tank works well with fish from about 2–5 g (7.1×10^{-2}–1.8×10^{-1} oz). After this size, a grading table may be more efficient. Commercial automatic graders for flatfish have recently debuted. Fish may have to be graded up to 3 times between the size of 2 and 10 grams (7.1×10^{-2} oz and 3.6×10^{-1} oz).

The nursery system may be either recirculating or flow-through with the choice being determined by the attributes of the site, regulatory climate, and economics. Either system will work well. Tanks may range from 3–6 m^3 (105–210 ft^3) and be of varying design: round, square, or raceways with rounded corners (D-ended). The tanks should be cleaned regularly and uneaten feed and feces removed as quickly as possible. Use of a self-cleaning tank design and drain system is advisable. Flounder, by virtue of their swimming on the bottom, help to move solids toward the drain. Oxygen levels in excess of 5 ppm should be maintained, preferably as close to 100% saturation as possible. Un-ionized ammonia should also be kept <0.015 ppm and nitrite <5 ppm. As a rule, saltwater fish are much more tolerant of nitrite than freshwater fish are. Salinity may range from 15 ppt to full strength at

Figure 9. Juvenile summer flounder in nursery tank.

35 ppt. Research is ongoing to evaluate whether these fish can be cultured in salinities less than 15 ppt, possibly less than 10 ppt. Other environmental criteria, such as alkalinity, turbidity, pH, CO_2, or heavy metals will be based on system technology employed, i.e., recirculating vs. flow-through and site-specific criteria. Some mode of culture water disinfection should also be employed, such as ultraviolet light.

As with other species of marine flatfish, such as turbot or Japanese flounder, the young fish are fed more frequently than older or larger fish. The 2 g (7.1×10^{-2} oz) fish may be fed to satiation up to 6 times per day, while a 10 g (3.6×10^{-1} oz) fish may be fed 4 times per day. Protein content of the feed should be 50% or better, while fat content should be <20%. Common marine fish diets range from 50 to 62% protein and from 11 to 19% fat. The protein is best if derived from fish and be >94% digestible. Weaning or starter diets may start at 300 microns (1.2×10^{-4} in.) in size and progress up through to a 2- to 3-mm (1/8 in.) pellet when the fish reach 5 g (1.8×10^{-1} oz). Summer flounder have a large mouth and prefer large pellets. By efficient grading, a fish culturist can maximize feeding efficiency by feeding the largest possible pellet acceptable to the population of the tank. At 20 °C (68 °F), one can expect to produce a 5 g (1.8×10^{-1} oz) fingerling in five months, at which point they could be transferred to a growout operation.

GROWOUT

The growout of summer flounder is in its infancy and optimum systems and procedures have not yet been established. Both recirculation systems and net pens are currently being investigated by different companies, some of which have received funding from the U.S. government for these demonstration projects. The fish seem to survive and grow well in both systems, but the demonstrations are ongoing and comparative data on biological and economic performance are not yet available. Comparisons of interest include (a) the growth performance and product quality of fish reared in more or less constant conditions in a land-based recirculation facility vs. those of fish reared in the highly variable temperature in a net pen and (b) the qualitative and quantitative differences in costs of production for the different systems. Researchers funded by the Northeastern Regional Aquaculture Center (NRAC) in the United States are currently assessing the survival, growth, behavior, and health of summer flounder reared in net pens with those of fish reared in recirculation systems. As of this writing, fish have not been maintained throughout the winter in net pens.

Specific environmental requirements are not known, but it seems likely that optimal growth will be obtained at temperatures around 20 °C (68 °F), which are characteristic of coastal waters in the northeastern U.S. during summer months. Growth at 10, 20, or 30 ppt salinity seems to be equivalent, at least for early juveniles (19). Since no specific summer flounder diets exist, fish in growout facilities are being fed diets similar in composition to those for turbot (50–55% protein, 12–20% fat). NRAC-funded research is also under way to identify nutritional requirements of summer flounder and food conversion ratios (dry weight of feed offered/wet weight gain) of about 1.4 are being measured in those studies. Growth data on laboratory-reared summer flounder in a variety of laboratory-scale and production-scale systems over the years indicate that summer flounder exhibit growth rates very similar to those of turbot (19,20) (GreatBay Aquafarms, unpublished data; R. Link, Mariculture Technologies, personal communication). Thus, the average time for growth to market size is expected to be 24–28 months, at least prior to selective breeding programs, which may reduce that time. As this is being written, no cultured summer flounder have yet been brought to market. Fish currently being grown are targeted at the live and freshly killed markets for Asian and Asian-American consumers.

Figure 10. Recirculation system growout tank for summer flounder.

SUMMARY AND FUTURE EXPECTATIONS

The U.S. government funded both research and demonstration projects for the development of a summer flounder aquaculture industry during the 1990s. That investment has resulted in several fledgling companies engaged in summer flounder production, so far employing relatively few people. Knowledge gained from the turbot and Japanese flounder industries has been applicable to summer flounder production and has aided the rapid development of the industry. It is noteworthy that most of the companies involved in production include personnel who are making the transition from commercial fishing to fish farming. That trend may very well continue as the industry grows and increasingly strict limitations are placed on capture fisheries.

Hatchery production is reasonably stable, although continued research is necessary to reduce mortality and to bring down the costs of production. Problems of weaning and cannibalism at the nursery stage remain to be overcome and will also require more research. Growout production has not yet been well studied and the greatest amount of future research should be concentrated on this production phase, particularly in the areas of selective breeding and health management (e.g., breeding for rapid growth and disease resistance, and development of vaccines and diagnostic techniques).

The economics of production are still a question mark. The current costs of production require sale of product to the very expensive live and freshly killed markets. The size of that market and the acceptance of cultured products by the market have only been estimated at this point (21). The number of companies presently producing summer flounder or building production capabilities may be able to supply that market if they all reach projected production capacity. In the long term, production costs need to be brought down to the point where summer flounder can be produced for the U.S. retail filet market and for restaurants. The industry should establish a marketing strategy to identify cultured summer flounder as a high quality, safe, fresh seafood product that is continuously available at a reasonable price, not subject to seasonal variation in price and quality. With the continuing crisis in the world's fisheries and consumer demand for a well-accepted product like flounder, the future for the summer flounder industry should be bright.

ACKNOWLEDGMENTS

We thank Tessa Simlick for the preparation of Figure 1.

BIBLIOGRAPHY

1. A.S. Smigielski, *Prog. Fish-Cult.* **37**, 3–8 (1975).
2. R.R. Stickney and D.B. White, *Trans. Am. Fish. Soc.* **104**, 158–160 (1975).
3. D.A. Bengtson, *Aquaculture* **176**, 39–49 (1999).
4. N. King, M. Huber, W.H. Howell, and D.A. Bengtson, Comparison of summer flounder *Paralichthys dentatus* maturation, spawning, and egg quality by hormal induction vs. environmental cues. Written for *J. World Aquacult. Soc.*
5. D.L. Berlinsky, W. King, R.G. Hodson, and C.V. Sullivan, *J. World Aquacult. Soc.* **28**, 79–86 (1997).
6. N. King, W.H. Howell, M. Huber, and D.A. Bengtson, Effects of larval stocking density on laboratory-scale and commercial-scale production of summer flounder, *Paralichthys dentatus*. Submitted to *J. World Aquacult. Soc.*
7. G. Klein–MacPhee, *P.-v. Reun. Cons. int. Explor. Mer.* **178**, 505–506 (1981).
8. D.A. Bengtson, L. Lydon, and J.D. Ainley, *N. Amer. J. Aquacult.* **61**, 239–242 (1999).
9. G.A. Bisbal and D.A. Bengtson, *J. Fish Biol.* **47**, 277–291 (1995).
10. G.A. Bisbal and D.A. Bengtson, *Mar. Ecol. Prog. Ser.* **121**, 301–306 (1995).
11. D.A Alves, J.L. Specker, and D.A. Bengtson, *Aquaculture* **176**, 155–172 (1999).
12. J.L. Specker, A.M. Schreiber, M.E. McArdle, A. Poholek, J. Henderson, and D.A. Bengtson, *Aquaculture* (in press).
13. D.A. Bengtson, M.A. Hossain, and T.R. Gleason, *N. Amer. J. Aquacult.* **61**, 243–245 (1999).
14. A.M. Schreiber and J.L. Specker, *Gen. Comp. Endocrinol.* **111**, 156–166 (1998).
15. A.W. Francis, Jr., *Cannibalism in larval and juvenile summer flounder,* Paralichthys dentatus, Masters thesis, University of Rhode Island, 1996.
16. D.A. Bengtson, T.L. Simlick, A.M. Schreiber, E.W. Binette, R.R. Lovett, IV, D. Alves, and J.L. Specker, Written for *Aquaculture Nutrition* (in press).
17. L. Huang, A.M. Schreiber, B. Soffientino, D.A. Bengtson, and J.L. Specker, *J. Exp. Zool.* **280**, 413–420 (1998).
18. N. King, W.H. Howell, D.A. Bengtson, R.A. Cooper, and M. Subramanyam, *J. World Aquacult. Soc.* (submitted).
19. G. Klein–MacPhee, *Growth, activity and metabolism studies of summer flounder* Paralichthys dentatus *(L.) under laboratory conditions*, Ph.D. dissertation, University of Rhode Island, 1979.
20. D.A. Bengtson, G. Bisbal, H. Iken, and R.P. Athanas, *J. Shellfish Res.* **13**, 312 (1994).
21. D.A. Zucker, *Economic analysis of flounder aquaculture*, Ph.D. dissertation, University of Rhode Island, 1998.

See also BRINE SHRIMP CULTURE; FLOUNDER CULTURE, JAPANESE; HALIBUT CULTURE; LARVAL FEEDING—FISH; PLAICE CULTURE; SOLE CULTURE; WINTER FLOUNDER CULTURE.

SUNFISH CULTURE

MARTIN W. BRUNSON
H. RANDALL ROBINETTE
Mississippi State University
Mississippi State, Mississippi

OUTLINE

INTRODUCTION

The Sunfish Family (Centrarchidae) includes 30 species (1) and is limited exclusively to North America. This family comprises the black bass (*Micropterus* spp.), crappie (*Pomoxis* spp.), and the bream (*Lepomis* spp.), as well as the genera *Amboplites*, *Elassoma*, *Enneacanthus*, *Centrarchus*, *Archoplites*, and *Acantharcus*. This family, especially the black bass, crappie, and bream, represents one of the most popular and widely known game fish groups in North America. Sunfish are widely sought by anglers because of their fierce tenacity when caught by hook and line; their firm, white flesh; and their favored status as "bread-and-butter fish" (2). In addition to their utility as sport and food fish, the lepomid sunfish species are generally stocked as forage fish for predators such as bass in impounded waters. Few North American anglers are not familiar with these fish, and thus, their appeal and recognition are broad based. Bass, crappie, and bream have been cultured for the past 50 years primarily to provide fingerlings for stocking recreational ponds and lakes or for use in research. Recently, however, there is an increasing interest in culturing these fishes for human consumption and for use in fee-fishing operations. This entry addresses only the culture of species from the genus *Lepomis*, commonly referred to as bream, sunfish, sun perch, or simply panfish (see the entries "Black bass/largemouth bass culture" and "Crappie culture" for information on other members of the sunfish family).

SPECIES PROFILES

There are 11 species in the genus *Lepomis*, but only the bluegill (*L. macrochirus*), redear sunfish (*L. microlophus*), warmouth (*L. gulosus*), green sunfish (*L. cyanellus*), and their hybrids have been widely cultured as sportfishes. The centrarchids are perciform fish and thus, characteristically, have a spiny dorsal fin (with 6–13 spines), followed by a soft dorsal fin. The anal fin has at least 3 spiny rays at its origin, followed by numerous soft rays. The pelvic fins are located immediately beneath the pectoral fins and contain one spine and 5 soft rays. The caudal fin has 17 principal rays. Proper identification is critical to successful production, since the genus *Lepomis* readily hybridizes (3). The species profiles that follow are adapted from Etnier and Starnes (1).

Bluegill

The bluegill may be the most well known of all the sunfishes and is certainly the most popular with anglers and consumers alike. Originally distributed from the Great Lakes south to the Gulf of Mexico, it has been stocked throughout North America as a game fish. It is equally at home in lakes or streams, but is most abundant in shallow, eutrophic lakes and ponds. The bluegill is identified by its deep, laterally compressed head and body, and small mouth. The opercular flap is black, and individuals longer than 51 mm (2 in.) exhibit a dark blotch at the posterior base of the dorsal fin. The sides usually exhibit 8–10 sets of double-vertical bars that are chainlike in appearance. Body colors range from olivaceous to purple, with a white to orange belly.

Redear

The redear, like the bluegill, has been widely introduced throughout much of the United States and, perhaps, North America as a game fish and a companion to the bluegill in managed systems. Known by several common names such as "shellcracker" and "chinquapin," the redear prefers sluggish waters. It is identified by its olive-yellow to straw-yellow body with gray or dusky spots. The breast is bright yellow to orange, and the opercular flap is short with a distinct scarlet outside margin or spot.

Warmouth

The warmouth, also commonly called "goggle-eye" is an inhabitant of sluggish waters, typically preferring weedy structure or other debris such as logs, stumps, and brush piles. This species has not been widely used for stocking recreational waters, but is a common component of the panfish angler's creel in areas where it is abundant. Its primary importance to aquaculture has been in production of hybrids with the other primary species. It exhibits a dark olivaceous to brown body, with dark splotches on the sides and fins. The cheeks and opercula have 3–4 dark bars radiating posteriorly from the eye, and the eye is often reddish. The mouth is larger than that of the bluegill and the redear, and the body is not as dramatically laterally compressed as these species.

Green Sunfish

The green sunfish may be one of the most adaptable, abundant, and environmentally tolerant of the sunfishes, being found in a wide variety of habitats ranging from ponds and lakes to river systems. Characteristic colors of the green sunfish include a blue-green to dusky dorsum and sides, with a yellow to white belly. A dark basal spot is usually present on the posterior base of the dorsal fin. All fins are yellow to orange tinted, with occasional bright orange areas and white margins. Green sunfish are known to hybridize readily with other *Lepomis* species (3,4).

SUNFISH REPRODUCTIVE HABITS

An understanding of the reproductive habits of the lepomid sunfishes is vital to the successful production of fingerlings destined either for stocking recreational waters or for culture to adult stages. Breder (5) was perhaps the first scientist to observe and report extensively on the courtship and reproductive behavior of these fishes. Since that time, it has been clearly confirmed that sunfish are, in general, colonial nestbuilders and multiple spawners. Spawning begins in early to midspring and continues, to some degree, until early fall, depending upon the species and the geographic location (1,2,5–7). The general courtship ritual and nesting behavior are similar for all lepomids, with slight variations by species.

When the temperature first reaches the preferred range for the respective species, the male typically prepares a nest in shallow water. The depth at which these nests are constructed depends upon water temperature, which in turn depends greatly upon the time of year. In general, nesting will occur first in the warm shallows in early spring, and nesting activity will gradually migrate toward deeper waters (up to 2 m, 6 ft) as the shallows become progressively warmer. Nest building is done primarily by sweeping or fanning the caudal peduncle across a sandy or gravelly substrate to form a saucerlike depression, ranging from 51 mm (2 in.) to 156 mm (6 in.) deep and 101 mm (4 in.) to 303 mm (12 in.) in diameter, depending upon the size of the male. These sunfishes are colonial nesters (8) and a single colony, often called a "bed" by anglers, may contain as many as five dozen nests and cover several dozen square meters (1 square m = 10.8 square ft).

Once nests are constructed, the courtship ritual begins. The males, which are usually the more brightly colored gender, initiate a complex circling behavior around the perimeter of the nest, usually accompanied by a series of gruntlike calls (1,8–10). Females are attracted to the nest by the male and, after a short response ritual, deposit their eggs and depart. The males fertilize the eggs by releasing milt across the egg mass and then vigorously and aggressively defend the nest from invading predators, periodically fanning the nest to aerate the eggs. Egg incubation takes 1–6 days at temperatures above 21 °C (70 °F) (11,12).

BROODSTOCK

Broodstock selection is a critical step in the production of sunfish and their hybrids. Improper identification of the broodfish can lead to contaminated stocks of offspring, since these fish readily hybridize (3,13). Proper identification of both sexes is vital to successfully producing the desired offspring. Bluegill and green sunfish males in breeding condition are easily distinguished from females, but redears and warmouth sexes are more difficult to differentiate. Usually, experienced culturists can visually distinguish males from females based solely upon external characteristics such as color, body shape or size, distension of the abdomen due to enlarged ovaries, or on the size and shape of the urogenital sinus (14). Additionally, eggs or milt can usually be freely expressed from ripe broodstock. However, broodstock often are not ripe at the time of stocking and identification, so other methods have proven useful. Dupree and Huner (13) described a simple method of probing for eggs by using a 5–10 cm (2–4 in.) long capillary tube that is 1.1–1.2 mm (0.04–0.05 in.) in diameter. The tube is inserted into the urogenital sinus, then angled slightly back toward the tail and slightly to one side. With the application of light pressure, the tube passes through the oviduct into the ovary. A finger is then placed on the open end of the tube to seal the tube before it is removed for inspection. If no eggs are retrieved, the fish should be rejected as a broodfish. This is a simple procedure, but must be done with care to avoid damage to the fish.

Handling fish can be stressful to them, and every attempt should be made to reduce handling stress. In general, broodstock should be handled in cool water, which is conveniently available during the stocking season (late winter–early spring). Fish should be handled as little as possible, sedated when practical, and transported in well-oxygenated tanks. Thermal shock should be avoided at all costs. Ideally, sexes should be kept separate until the time for stocking into spawning ponds (13). This approach allows broodstock to be sexed in advance and held for a period of time in conditioning ponds to prepare them for spawning, eliminating or at least minimizing stressful handling, sexing, and sorting immediately prior to the spawning season.

Bluegills and redears should be at least two years old and 110 to 225 g (0.25 to 0.5 lb) for maximum productivity (15), while warmouth and green sunfish will spawn at much smaller sizes. Smaller bluegills and redears will spawn, but stocking rates should be increased to compensate for reduced fecundity and greater variability of spawn size, consistency; and success (15). Most culturists agree that broodstock should be stocked in the winter or at least by early spring (15,16) at the rate of 50–100 pairs/ha (20–40 pairs/ac). Sex ratios of broodfish in spawning ponds are typically 1:1 (14,17). Supplemental feeding of broodstock with a floating pellet is desirable as soon as the water temperature reaches a level where feeding activity is stimulated. Feeding rate should be 2–3% of body weight/day (14,15).

POND PREPARATION

Sunfish can be produced in many types of facilities and are often spawned in laboratory settings for research purposes (14,18,19). Most culture, however, is practiced in ponds of varying sizes, utilizing either open ponds or cages. In general, spawning ponds should be 0.6 to 1.5 m (2–5 ft) deep, with a smooth, uniformly sloped bottom to facilitate harvest of fingerlings. Pond size should be less than 1.2 ha (3 ac), although larger ponds have been successfully used. Drainpipes allow manipulation of water levels, and a water source (preferably ground water) is necessary to fill ponds and replace evaporative losses. Higginbotham (15) recommended that ponds be filled with well water at least 2–4 weeks prior to initiation of spawning activity. Ponds should be completely free of other fish species. Failure to completely remove resident fish populations is a common mistake that spells disaster for hatchery production. The presence of other sunfish can pose the risks of both hybridization and depredation, while the presence of other piscivorous species poses risk of depredation of the fingerlings and loss of a potential crop of fish. A plankton bloom should be established prior to spawning activity, using either inorganic or organic fertilizers, or some combination of the two (18).

Once water temperatures reach appropriate levels, spawning will be initiated. Warmouth and green sunfish are the first to commence spawning, with activity beginning at 21 °C (70 °F). Redear begin spawning when water temperatures reach 24 °C (75 °F), while bluegills do not spawn until temperatures reach 26–27 °C (78–80 °F) (4). Nest-building and territorial behavior for all species will commence at temperatures several degrees lower than optimum spawning temperature.

SPAWNING TECHNIQUES AND FINGERLING PRODUCTION

Except for limited manual stripping of gametes and laboratory fertilization, sunfish spawning in culture systems is done almost exclusively in ponds or other extensive facilities. Childers and Bennet (21) manually stripped gametes into petri dishes, mixing milt and eggs with water to accomplish fertilization. Fertilized eggs were placed into clean petri dishes containing aged tap water for water hardening. Fertilized eggs were then placed into aerated aquaria. When larvae became free-swimming fry, they were transferred to rearing ponds. Toetz (22) conducted detailed studies on larval rearing of bluegill, reporting up to 79% hatch from a procedure modified from that of Childers and Bennett (21). Several investigators have successfully induced spawning in the laboratory or under simulated field conditions by manipulating temperature and photoperiod (14,21,23–26).

In pond spawning systems, broodstock are allowed to free spawn in ponds. Production of up to 375,000 fry/100 broodfish can be obtained, with an average of about 247,000 fry/ha (100,000 fry/a) (14,15). Once optimum conditions of temperature have been reached, and with good water quality, broodfish should produce offspring almost immediately. Fry should be observable soon after hatching, and a feeding program should be initiated. A fry powder or mash should be offered initially; then, as fry grow, the size of feed particles can be matched to size of the fish. With a feeding program, initial growth of up to about 2.5 cm (1 in.) per month can be obtained, with fingerlings reaching stocker size of 51–76 mm (2–3 in.) in 60–100 days (15).

HARVEST AND MARKETING

Harvest of small (<2.5 cm; 1 in.) sunfish is stressful to the fish, and thus most producers do not attempt harvest until the fish reach an average of 50 mm (2 in.). This is the most commonly stocked size for bluegills, redears, and sunfish hybrids. Transport for stocking purposes is usually delayed until the fall months when prevailing low water temperatures help minimize stress that may rise during handling and shipping. This also corresponds with the recommended time frame for stocking sunfish into small impoundments for recreational fishing (27–29).

Traditionally, sunfish have been cultured almost exclusively to produce fingerlings for stocking natural and manmade waters to enhance recreational fishing (14,16). Bluegills and redear sunfish have been the mainstay forage species stocked in combination with largemouth bass in small impoundments throughout much of the United State (30,31), providing not only forage, but also excellent sportfishing as well. Thus, most culture systems have concentrated upon fingerling production. These markets continue to expand, as the interest in recreational fisheries management, on both public and private waters, increases. Until recently, through various state and federal fish stocking programs (31), most state and federal hatcheries produced large numbers of sunfish for public distribution. However, increasing numbers of private hatcheries have developed which specialize in the production of game fish for stocking purposes. Simultaneously, many public hatcheries and state agencies have reduced or eliminated their programs for producing and distributing game fish to the public.

Recently, aquaculture interest in these fishes has expanded to include potential use as foodfish for human consumption (32) and for use in fee-fishing operations (33). Production of sunfish and their hybrids for human consumption will require culture of these fishes to adult sizes suitable for marketing to consumers, as well as conformity to several marketing criteria such as appearance, texture, flavor, and consumer recognition (34). Bluegill and redear, as well as at least a couple of sunfish hybrids, have a favorable combination of flavor, texture, firm, flaky, low-fat flesh (16) with good storage qualities (2), and strong consumer recognition and acceptance. Other considerations for sunfish as a marketable foodfish include biological criteria such as temperature and water quality requirements (35), growth rates (36–38), trainability to manufactured feeds (39–41), and fecundity and ease of spawning in captivity (17,23). The lepomids, in general, all meet these criteria. Specifically, bluegill and bluegill hybrids appear to best meet considerations for culture to adult sizes. The most common hybrid is the male bluegill X female green sunfish, which has a desirably high male : female sex ratio in the F1 generation. Other hybrids may hold potential despite variable sex ratios if techniques such as ploidy manipulation (42) can be perfected.

The interest in sunfishes and their hybrids as aquaculture foodfish candidates or for use in fee-fishing operations as an alternative to channel catfish has stimulated development of another production and marketing scenario. In this scenario, fingerlings would presumably be stocked at high densities, fed a manufactured feed, and grown to adult sizes suitable for marketing as foodfish (225–340+ g or 0.5–0.75+ lb) or to a minimum acceptable catch size of 110 g (0.25 lb) (43). Limited work has been conducted to define the potential that these fishes possess for these markets, but preliminary results appear promising. More emphasis has been placed on the culture of hybrids for this purpose (33,37,40) than for pure species populations, but it is anticipated that increasing consumer interest, especially in bluegills, will stimulate additional research into the propagation of these fishes for nontraditional markets. Research is needed in larval nutrition and in development of appropriate, economical growout procedures.

BIBLIOGRAPHY

1. D.A. Etnier and W.C. Starnes, *The Fishes of Tennessee*, The University of Tennessee Press, Nashville, 1993.
2. G. Becker, in *Fishes of Wisconsin*, University of Wisconsin Press, Madison, WI, 1983, pp. 844–851.
3. W. Childers, *Illinois Nat. Hist. Surv. Bull.* **29**(3), 159–214 (1967).
4. K. Carlander, *Handbook of Freshwater Fishery Biology* **2**, 44–45 (1977).
5. C. Breder, Jr., *Zoologica* **21**, 1–48 (1936).
6. C. Manooch, *Fisherman's Guide to the Fishes of the Southeastern United States*, N.C. State Museum of Natural History, Raleigh, NC, 1988.
7. D. Lee, C. Gilbert, C. Hocutt, R. Jenkins, D. McAllister, and J. Stauffer, Jr., *Atlas of North American Freshwater Fishes*, N.C. State Museum of Natural History, Raleigh, NC, 1990.
8. M. Gross and A. MacMillan, *Beh. Ecol. Sociobiol.* **8**, 163–174.
9. J. Gerald, *Evolution* **25**, 75–87 (1971).
10. V. Avila, *Am. Midland Naturalist.* **96**, 195–207 (1981).
11. M. Huet, *Textbook on Fish Culture: Breeding and Cultivation of Fish*, Fishing News (Books), Ltd., London, 1970.
12. C.C. Mischke and J.E. Morris, *Prog. Fish-Cult.* **59**(4), 297–302 (1997).

13. H.K. Dupree and J.V. Huner, *Third Report to the Fish Farmers*, United States Fish and Wildlife Service, 1984.
14. T. McComish, *Prog. Fish-Cult.* **30**, 28 (1968).
15. B. Higginbotham, *Forage species: Production techniques*, SRAC Pub. 141, Southern Regional Aquaculture Center, Stoneville, MS, 1992.
16. W. McClarney, *The Freshwater Aquaculture Book*, Hartley & Marks, Point Roberts, WA, 1987, pp. 485–508.
17. B. Simco, H. Williamson, G. Carmichael, and J. Tomasso, *Culture of Non-Salmonid Freshwater Fishes*, CRC Press, Boca Raton, FL, 1985, pp. 73–89.
18. M. Bryan, J. Morris, and G. Atchison, *Prog. Fish-Cult.* **56**, 217–221 (1994).
19. C.C. Mischke and J.E. Morris, *Prog. Fish-Cult.* **60**(3), 206–213 (1998).
20. J. Steeby and M. Brunson, *Fry-Pond Preparation for Rearing Channel Catfish*, Mississippi Cooperative Extension Service, Mississippi State, Information Sheet 1553, 1996.
21. W. Childers and G. Bennett, *Illinois Natural History Survey Biological Notes, No. 46*, Carbondale, IL, 1961.
22. D. Toetz, *Investigations of Indiana Lakes and Streams* **7**, 115–146 (1966).
23. A. Banner and M. Hyatt, *Prog. Fish-Cult.* **34**, 173–180 (1972).
24. R. Smitherman and F. Hester, *Trans. Am. Fish. Society* **91**, 333–341 (1962).
25. J. Merriner, *Trans. Am. Fish. Society* **100**, 611–613 (1971).
26. W. Smith, *Prog. Fish-Cult.* **37**, 227–229 (1975).
27. M. Brunson, *Managing Mississippi Farm Ponds and Small Lakes*, Mississippi Cooperative Extension Service, Mississippi State, Pub. 1429, 1992.
28. M. Masser, *Managing Recreational Fish Ponds in Alabama*, Circular ANR-577, Alabama Cooperative Extension System, Alabama A&M and Auburn Universities, Auburn, AL, 1996.
29. M. Austin, H. Devine, L. Goedde, M. Greenlee, T. Hall, L. Johnson, and P. Moser, Ohio Pond Management Handbook, Division of Wildlife, Ohio Department of Natural Resources, 1996.
30. H.S. Swingle, *Trans. Am. Fish. Soc.* **76**, 46–62 (1946).
31. H. Regier, *Prog. Fish-Cult.* **24**(2), 99–111 (1962).
32. J. Tidwell, C. Webster, and S. Coyle, *Aquaculture* **145**, 213–223 (1996).
33. J. Tidwell, C. Webster, J. Clark, and M. Brunson, *Aquaculture* **26**, 305–313 (1994).
34. H. Webber and P. Riordan, *Aquaculture* **7**, 107–123 (1976).
35. R.C. Heidinger, *Trans. Am. Fish. Soc.* **104**, 333–334 (1975).
36. L. Krumholz, *Trans. Am. Fish. Soc.* **76**, 190–203 (1946).
37. M.W. Brunson and H.R. Robinette, *North Am. J. Fish. Mgt.* **6**(2), 156–167 (1986).
38. J. Breck, *Trans. Am. Fish. Soc.* **122**, 467–480 (1993).
39. T. Ehlinger, *Animal Behav.* **38**, 643–658 (1989).
40. J. Tidwell, C. Webster, and J. Clark, *Prog. Fish-Cult.* **54**(4), 234–239 (1992).
41. M.W. Brunson and H.R. Robinette, *Proc. Ann. Conf. S.E. Assoc. Fish and Wildl. Agencies* **36**(1982), 157–161 (1984).
42. P.S. Wills, J.M. Paret, and R.J. Sheehan, *J. World Aqua. Soc.* **24**(4), 507–511 (1994).
43. D. Ellison and R. Heidinger, *Proc. Annual Conf. SE Assoc. Fish and Wildlife Agencies* **104**, 333–334 (1978).

See also BLACK BASS/LARGEMOUTH BASS CULTURE.

SUSTAINABLE AQUACULTURE

ROBERT R. STICKNEY
Texas Sea Grant College Program
Bryan, Texas

OUTLINE

The term "sustainability" is widely utilized with respect to the management and conservation of natural resources. One frequently hears the term used in conjunction with agriculture, fisheries, and, increasingly, aquaculture. Sustainability involves the establishment of production systems that can exist, at least theoretically, in perpetuity. There are various perceptions as to how sustainability can be achieved, but in most instances the concept is associated with limited inputs of nonrenewable resources.

THE SUSTAINABILITY CONCEPT

Sustainability is one of the buzz words of the 1990s. The concept is visualized in different ways by its proponents, with one of the most thoughtful definitions being put forward in a 1991 meeting (1) of the Food and Agriculture Organization (FAO) of the United Nations. The FAO definition of "sustainable development" involves the management and conservation of the natural resources of the world with an "orientation of technological and institution change in such a manner as to ensure the attainment and continued satisfaction of human needs for present and future generations." FAO goes on to indicate that such development will conserve "land, water, plant and animal genetic resources, is environmentally non-degrading, technically appropriate, economically viable and socially acceptable."

The concept of sustainability has been widely discussed with respect to agriculture, including forestry. The U.S. Department of Commerce's National Marine Fisheries Service has as one of its primary objectives building sustainable fisheries. This is in response to the collapse or declines in fish standing stocks due in large part to overfishing. There is increasing interest in how aquaculture can be developed in a more sustainable manner, and a book on the subject has been published (2). One aquaculturist defined "sustainable aquaculture" as that which leads to production of a crop with no net loss of natural resources (3).

AQUACULTURE SUSTAINABILITY

Total conservation of natural resources may be achievable in low intensity aquaculture operations. For example, subsistence culture, which provides a relatively safe place, such as a small pond, for the culture species to exist meets the sustainability objectives if it does not include the use of inorganic fertilizers or prepared feeds. Subsistence aquaculturists often use organic fertilizers (manures, including night soil) and may feed agricultural waste products to supplement nutrition. Production levels are modest, ponds tend to be small, and water is supplied by runoff, springs, or surface water diversion. There is no pumping, aeration, or other use of electricity or fossil fuels.

Somewhat in parallel are extensive mollusc culture systems, such as those used for rearing oysters, which use natural reproduction to provide the young oysters and natural food (phytoplankton in the water column) to provide nutrition. The culturist may move young oysters from the location where spat settling occurred to another location for growout and may employ some type of predator control, but supplemental feeds are not used and water quality is not managed.

Most other forms of aquaculture require modest to significant inputs of natural resources, some of which are not renewable, such as the energy obtained from petroleum. The more intensive the system, the more inputs of resources it requires. For example, a pond system stocked at moderate density and receiving supplemental feed and sufficient inputs of water (which may or may not be pumped) to replace evaporation and seepage losses (see the entry "Pond culture") is clearly operating in a more sustainable manner than a closed recirculating water system that requires constant inputs of energy and the provision of complete feeds. (See the entry "Recirculating water systems.")

Developing culture systems that require significant infusions of natural resources may actually result in a net utilization of those resources. Rearing fish in a closed system within a metropolitan area will obviate the need for hauling the product long distances to market. Fuel, icing down or freezing a product, and refrigerating the trucks used to haul it represent net losses in natural resources that may offset, at least partially, added utilization of natural resources operating closed culture systems located close to the market. With increasingly heavy demands on land and water, along with the need to maximize production per unit volume of water, pressure to further intensify production from aquaculture facilities, particularly in developed nations is growing. Increased intensity is typically not compatible with increased sustainability because of socioeconomics. Balancing the goal of increasing sustainability with the need to generate a reasonable profit will continue to challenge aquaculturists.

It is often argued that aquaculturists should not employ high quality protein sources such as fish meal to produce other fish because the fish meal could be used more directly as human food. Some species that are currently used solely as sources of fish meal could be processed into surimi analogs (e.g., artificial crab), but the demand is not currently sufficient to warrant a major change in the approach to processing.

During the 1970s, the processing of low-priced fish (such as those used in fish meal) into fish protein concentrate (FPC), which was a tasteless white protein-rich powder grew. Adding one to two ounces (30 to 60 g) of FPC to a person's diet daily quickly overcame problems associated with malnutrition. FPC was promoted by the United States in its international development programs, but was abandoned when use of the product was banned in the United States (because it was made from whole fish). Developing countries became suspicious about a product that was illegal in the nation of origin.

Fish meal (and other animal proteins such as meat and bone meal) contain essential amino acid levels that cannot be matched by most plant proteins; therefore, many aquatic species animal proteins or plant proteins supplemented with purified amino acids are necessary in feed formulations. Research continues to find alternative plant proteins or feed formulations that will meet the nutritional requirements of various aquatic species while reducing or eliminating the need for animal protein. One example of a success utilizing that approach is the feed currently employed for the culture of channel catfish. A few decades ago much of the protein in catfish growout feeds came from fish meal. By the 1980s, fish meal content had been reduced to about 15%. Today, catfish in Mississippi are typically fed rations containing no more than 4% fish meal once the fish reach the advanced fingerling stage. Fry and small fingerlings have higher protein and amino acid requirements that can only be met with animal proteins.

Marine enhancement programs and ocean ranching move away from totally captive rearing to at least partial production that depends on resources available in nature. Reproduction and early rearing occur in hatcheries; but once the fish reach a size at which they have a reasonable chance for survival on their own, they are released into the natural environment for later capture. Input of some natural resources is greatly reduced as the bulk of the life cycle occurs in the wild, not in a culture facility.

Animals released for enhancement purposes will consume higher weights of food to reach the same size as their counterparts reared entirely in captivity. Food conversion ratios of 1.5 to 2.0 (one or two unit weights of feed will be required to produce one unit weight of growth) are commonly seen in aquaculture facilities, while it may take 10 units of natural food to produce 1 unit of weight gain in nature. Prepared feeds are specifically designed to meet the animal's nutritional requirements and typically contain only a few percent water, whereas the percentage of water in natural foods can exceed 90% and a combination of foods may need to be ingested to meet all the nutritional requirements. Aquatic animals living in the wild can be expected to utilize more food energy for foraging activities than their captive counterparts.

The major savings in natural resources come from the decreased energy associated with captive rearing systems (water pumping, aeration, and perhaps heating or cooling), growing or capturing the ingredients, and manufacturing the prepared feeds employed to rear many varieties of captive culture animals. Additional energy expenditures are required to construct the larger facilities needed to rear animals to market size and to operate the equipment and vehicles associated with construction and operation of such facilities. Those outlays are somewhat offset by the energy requirements for capturing enhanced species from nature, rather than from a pond, raceway, or net pen.

Aquaculturists are becoming increasingly aware of the need to limit the amounts of natural resources that are consumed in conjunction with their activities and the potential environmental problems that can result from many forms of aquaculture. Nutrient and suspended solids loading into receiving waters from effluents are significant problems in some types of aquaculture. Various methods to control the quality of effluents, including the use of filtration systems, settling basins, and constructed wetlands are being employed by some culturists and are increasingly mandated by policymakers in some regions. Aquaculturists can also be more environmentally responsible by modifying feed formulations to reduce nitrogen and phosphorus losses to the environment (more complete utilization by the culture animals). The use of exotic species in aquaculture has been criticized by individuals and groups fearing that the nonnative species will escape and negatively impact native populations. Aquaculturists are taking increased care to ensure that exotics do not escape, particularly from facilities that have effluents which enter natural waters (as compared with those being recycled, used for irrigation, or that are put into sanitary sewers).

Most aquaculturists will steadfastly stand behind the view that they are conservationists and, as a consequence, are interested in the maintenance of environmental quality. While mitigating against environmental damage from aquaculture can be expensive, most aquaculturists recognize that it is to their long-range benefit to take appropriate steps. Many culturists in Japan learned that lesson the hard way when self-pollution caused both economic loss and stricter controls on the numbers and sizes of facilities that can be established within sensitive coastal areas.

Aquaculture may never achieve complete sustainability, but efforts are being made in many circles to approach that goal. With the growing human population around the world, the pressures to move toward sustainability and provide additional aquatic foods through aquaculture production will increase dramatically. How aquaculturists and policymakers respond to the challenge remains to be seen.

BIBLIOGRAPHY

1. J.E. Bardach, in J.E. Bardach, ed., *Sustainable Aquaculture*, John Wiley & Sons, New York, 1997, pp. 1–14.
2. J.E. Bardach, ed., *Sustainable Aquaculture*, John Wiley & Sons, New York, 1997.
3. J.W. Avault, Jr., *Fundamentals of Aquaculture*, AVA Publishing Co., Baton Rouge, LA, 1996.

See also ENHANCEMENT.

T

TANK AND RACEWAY CULTURE

STEVEN T. SUMMERFELT
The Conservation Fund's Freshwater Institute
Shepherdstown, West Virginia

MICHAEL B. TIMMONS
Cornell University
Ithaca, New York

BARNABY J. WATTEN
U.S. Geological Survey
Kearneysville, West Virginia

OUTLINE

INTRODUCTION

Tank-based fish farming methods vary widely, and external pressures are constantly forcing improvements. Tank-based fish farms in the United States are especially challenged by lower farm-gate prices and by the implementation of new state and federal regulations that govern water use, the total maximum daily load (TMDL) of wastes, and/or the concentration of wastes discharged from fish farms. Therefore, better fish culture strategies and technologies are being adopted to reduce fish production costs, water requirements, and waste discharges. Culture tank design is especially important because it influences both physical and biological variables in fish culture (Fig. 1). The culture tanks are also the asset in which much of a farm's fixed and variable costs are invested, and they are the point of waste production and water use. A large portion of fish farm capital can go toward the purchase of culture tanks, of their water distribution and collection components, of support equipment (e.g., fish feeders, oxygen probes, and flow or level switches), and of the floor space that they require. Also, a large portion of fish farm operating costs can go into the labor required to manage culture tanks and control their waste discharge. Therefore, a large financial incentive exists to select the best culture tank, scale, and operating strategy to optimize fish farm profitability. This article discusses the design rationale, use, and cost implications of circular culture tanks, raceway tanks, and a new mixed-cell raceway-type tank. Several of the features, advantages, and disadvantages of each culture-tank design are summarized. First, however, this article briefly reviews how issues of culture-tank scale, carrying capacity, and stock management influence fish production and how rapid solids removal from the culture tank can affect waste management.

Scale Issues

The number and size of culture tanks is an important factor to consider during the design of the fish farm and of its stock-management plan. It is now becoming more common practice for fish farms to use few, but relatively large culture tanks to meet volume requirements. For example, Karlsen (1) described how more recent land-based salmon smolt farms in Norway use only 6 to 8 production tanks, many fewer than earlier farms used. No matter where the fish farm is located or what species the farm is producing, it often becomes readily apparent that fewer (maybe 6 to 10), but relatively larger, culture tanks can provide culture volume much more cost-effectively than can many (maybe 30 to 100), but correspondingly smaller, tanks. Also, the costs of miscellaneous equipment and labor decrease when a given culture volume can be achieved with a few large culture tanks rather than with many small tanks. Use of fewer but larger tanks reduces the purchase and maintenance cost of feeders, dissolved oxygen probes, level switches, flow meters/switches, flow-control valves, and effluent standpipe structures (2). Use of fewer but larger tanks also reduces the time required to analyze water quality, distribute feed, and perform cleaning chores (because the times are about the same for a large tank as for a small tank) (2). Also, the time and logistics of fish management in a large number of tanks can become quite costly.

However, the advantages achieved through the use of larger tanks must be balanced against the risk of larger economic loss if a tank fails for mechanical or biological reasons. There are also difficulties that could arise in larger culture tanks when removing mortalities, grading and harvesting fish, and controlling flow hydraulics (e.g., water velocities, tank mixing, dead spaces, and settling zones). Therefore, large culture tanks must be designed properly to allow for fish management and for control over flow hydraulics, as will be discussed.

Carrying Capacity Issues

Production in culture tanks can be boosted by increasing the culture tank's carrying capacity, which (in simplistic terms) is the maximum fish biomass that can be supported at a selected feeding rate. Dissolved oxygen is usually the first water quality parameter to limit culture-tank carrying capacity. The amount of dissolved oxygen available in fish culture tanks is dependent upon the water flow rate multiplied by the concentration of available dissolved oxygen (i.e., the inlet dissolved-oxygen concentration minus the minimum allowable dissolved-oxygen concentration). If there is no in-tank aeration

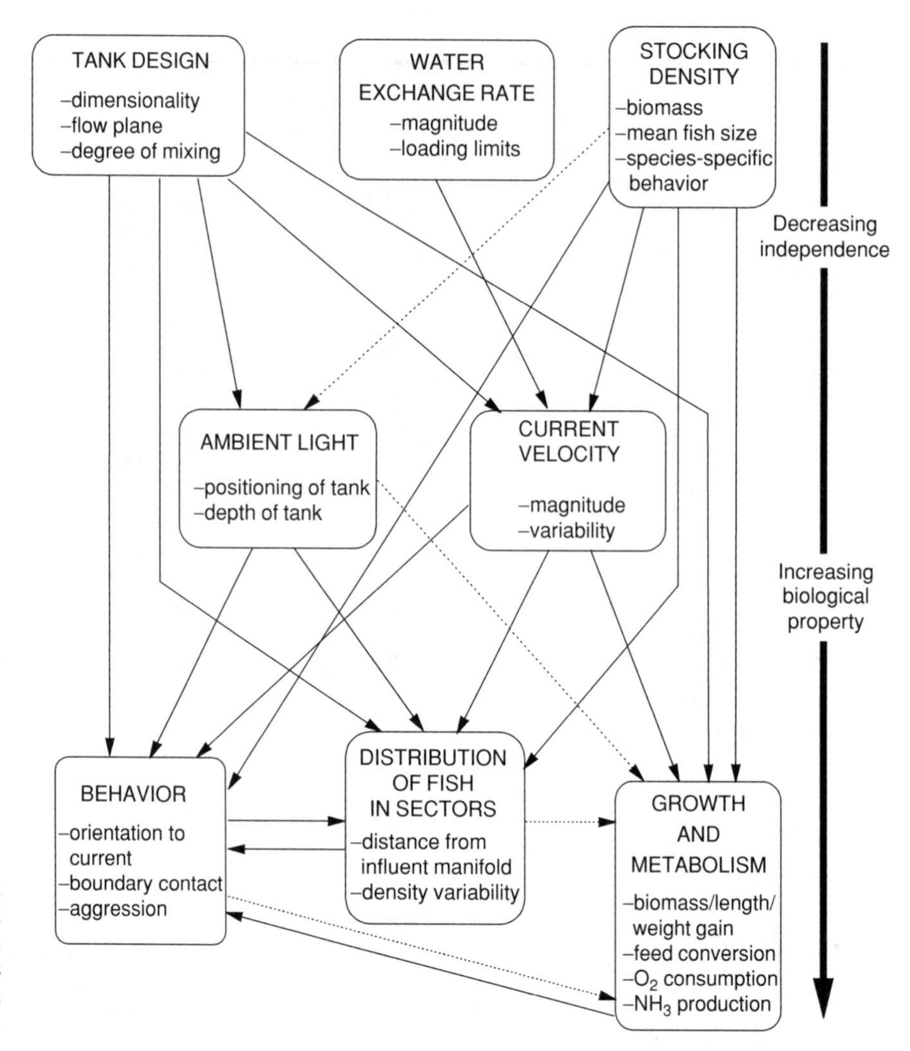

Figure 1. Rearing unit design is often dependent upon physical and biological variables, as has been illustrated by Ross and Watten (42). Solid arrows depict relationships demonstrated by research; dashed arrows depict hypothesized effects.

or significant photosynthesis, doubling the water flow through a culture tank will double the carrying capacity of the tank; however, moving more water through the culture volume is not always possible and often requires larger pumps and pipes or higher water-pressure requirements that can increase the farm's fixed and variable costs. Alternatively, supersaturating water with dissolved oxygen prior to use is popular and is often a more cost-effective method of improving the profitability of tank-based fish farms (3,4). For example, if we assume that a minimum allowable outlet dissolved oxygen concentration is 7 mg/L (ppm), increasing the dissolved oxygen concentration entering a culture tank from 10 to 16 mg/L (ppm) would triple the culture tank's available oxygen and thus triple its carrying capacity (if there are no other limitations). Supersaturating water with dissolved oxygen can be achieved cost-effectively with many different oxygen transfer devices, even in low-head applications (3,5).

Other fish metabolites, such as dissolved carbon dioxide, ammonia, and suspended solids, can limit culture tank carrying capacity when the level of dissolved oxygen no longer does so. Fish produce approximately 28 to 32 g (1 oz) of total ammonia nitrogen, 300 to 400 g (10–14 oz) of carbon dioxide, and 250 to 400 g (9–14 oz) of total suspended solids for every 1.0 kg (2.2 lb) of feed consumed. In terms of dissolved oxygen consumption, fed fish produce roughly 1.0–1.4 mg/L (ppm) of total ammonia nitrogen, 13–14 mg/L (ppm) of dissolved carbon dioxide, and 10–20 mg/L (ppm) of total suspended solids (TSS) for every 10 mg/L (ppm) of dissolved oxygen that they consume. Dissolved carbon dioxide and un-ionized ammonia concentrations can rapidly accumulate to toxic levels when fish consume large concentrations of dissolved oxygen without some form of ammonia or carbon dioxide control. Colt et al. (4) developed a method for estimating the carrying capacity of water when dissolved oxygen is not limiting. The method uses mass balances and chemical equilibrium relationships to predict the amount of dissolved oxygen concentration that fish can consume before the dissolved carbon-dioxide or ammonia concentrations limit further oxygen consumption. Accordingly, intensive fish farms can use the water flow without concern about ammonia or carbon dioxide limitations (in the absence of biofiltration and air stripping) up to a cumulative dissolved oxygen consumption of about 10 to 22 mg/L (ppm), depending upon pH, alkalinity, temperature, fish species, and life stage (4). After reaching this cumulative oxygen demand, the water flow cannot be used again until the dissolved carbon dioxide and ammonia concentrations are reduced.

Stock Management Issues

Fish farm production can also be increased (approximately doubled) through the use of a continuous production strategy rather than of a batch-production strategy (6–9). The maximum economic productivity of the culture system can be obtained by year-round fish stocking and harvesting, because continuous production maintains the culture system at or just below its carrying capacity. Also, harvesting at a given size increases product value by regularly providing more uniform-sized fish for the market. In a full-scale production experiment, Heinen et al. (10) showed that rainbow trout (*Oncorhynchus mykiss*) stocked every eight weeks and harvested weekly could achieve a ratio of steady-state annual production (kg/yr) to maximum system biomass (kg), P:B, of 4.65:1/yr. This research demonstrated that continuous-production techniques can provide exceptionally high P:B rates. In practice, however, most commercial farms cannot operate at such high P:B ratios, and P:B ratios of 3:1 per year (or lower) are more common. Incorporation of stock management strategies that can increase the P:B ratio on a commercial farm would have a large and beneficial effect on production costs.

Continuous stocking and harvesting strategies also require the culturing of several size groups at the same time, frequent handling of the fish, and improvement in inventory accounting techniques, which can stress the fish (and increase labor costs if automated equipment is not used). To keep the cost of handling and grading fish (and the associated stress on the fish) to a minimum, convenient mechanisms for sorting the fish by size, counting them, and moving them to other locations should be incorporated into design of the culture tank and facility. Simply netting the fish out of the tank or using a net to crowd the fish for harvest or grading is an obvious solution. More sophisticated crowding and grading can also be achieved by using crowder and grader frames or gates that move down the length of a raceway or pivot around the center of a circular culture tank (1,11,12). Fish of small size can swim through the grader bars, while the larger fish are retained behind the gate. Use of crowder and grader gates is thought to be less stressful on the fish, because their use does not require handling the fish or moving them out of the water. Once crowded, fish can also be induced to swim through channels, pipes, or raceways to another location with relatively little stress. Crowded fish can also be moved rapidly to other areas by using more aggressive fish pumps or brail nets and cages, but these techniques should be used with care to avoid damaging fish.

Hand grading devices, such as box graders, are common at many hatcheries and have proven to work well on small farms, where the cost of automatic grading and counting equipment cannot be justified. At larger farms, however, labor and fish stress can be reduced by the wise use of automated grading and inventory tracking equipment. Commonly used automatic graders include mechanically driven belt graders and roller graders. These mechanical graders usually require removing the fish from water for a brief period as they pass through the sizing mechanism. Although mechanical graders can produce some stress and trauma, many of the established commercial mechanical grading machines are considered safe, reliable, and fast methods to sort large numbers of fish by size and count them. As with all new technologies, however, it is best to check with other fish farmers who have used the equipment before purchasing it.

There is a great deal of interest in affordable inventory tracking equipment because sampling fish with dip nets is labor-intensive and not extremely accurate. Exciting new technologies that use ultrasonics, infrared light, and video systems are now commercially available to estimate fish size distributions within culture tanks and cages. This type of inventory tracking equipment may sometimes also be used to track fish growth and feed conversion (if feed input and fish numbers are known) and to estimate the fraction of fish reaching harvest size.

Waste Management Issues

The concentration of waste discharged from most tank-based coldwater fish farms is relatively low under normal operating conditions; however, the large flow rates involved can make the cumulative waste load (i.e., total maximum daily load) discharged from fish farms significant (13–15). Consistently meeting strict discharge standards can also be difficult because cleaning routines for pipes, channels, and tanks can produce fluctuations in discharge flowrates and in the consistencies and concentrations of waste material. The distribution of the nutrients and organic matter between the dissolved, suspended, and settleable fractions affects the choice of method used and the difficulty of effluent treatment. The filterable or settleable solids contain most of the phosphorus discharged from tanks (50–85%) but relatively little of the total effluent nitrogen (about 15%) (13,16). Most of the effluent nitrogen released (75–80%) is in the form of dissolved ammonia (or nitrate, when nitrification is promoted). The variability in the nutrient and organic material fractionation between dissolved and particulate matter is largely dependent on feed formulation and on the opportunity for particulate matter to break apart, because the production of smaller particles increases the rate of nutrient and organic matter dissolution. Fecal matter, uneaten feed, and feed fines can be broken rapidly into much finer and more readily soluble particles by water turbulence, fish motion, scouring, and pumping. It is much more difficult to remove dissolved and fine particulate matter than larger particles. Therefore, culture tank designs, and operating strategies that remove solids rapidly and with little turbulence, mechanical shear, and opportunity for microbiological degradation are important in helping the fish farm meet discharge limits (to be discussed in more detail later in this entry).

CIRCULAR CULTURE TANKS

Circular tanks have been used widely in land-based trout and salmon smolt farms (Fig. 2) in Norway, Scotland, and

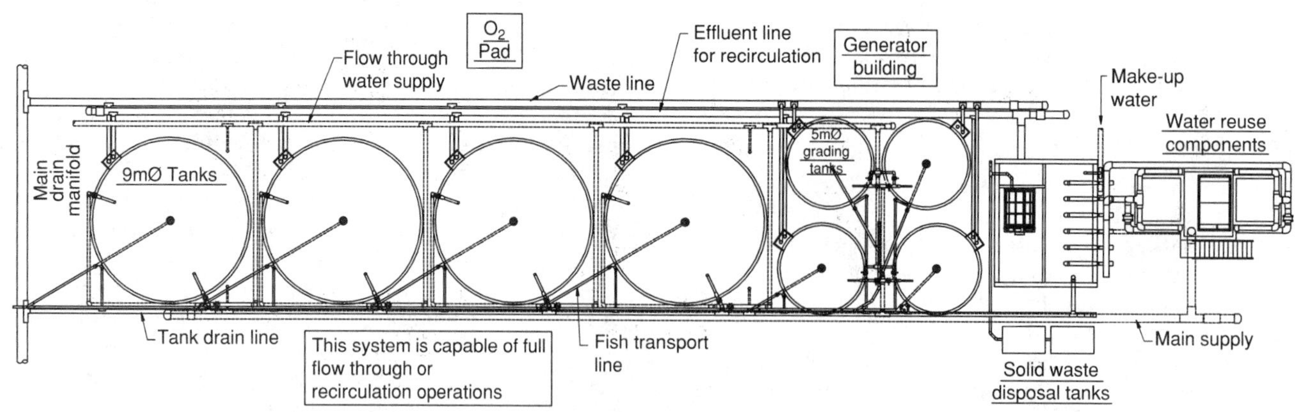

Figure 2. Circular culture tanks in one of the salmon-smolt production units at Target Marine Hatcheries in British Columbia. (Drawing courtesy of PRAqua Technologies, Ltd., Nanaimo, British Columbia, Canada.)

Iceland (1,17,18), as well as in North America and other parts of the world (11,12). Circular tanks used for salmonid production are generally large — usually between 12 and 42 m (40 and 140 ft) in diameter (although smaller tanks are used in hatcheries and on smaller farms). Diameter-to-depth ratios typically range from 3 : 1 to 10 : 1 (1). Circular tanks are also used for raising many species other than salmonids, including hybrid striped bass (*morone* sp.), tilapia (*Oreochromis* sp.), yellow perch (*Perca flavescens*), sturgeon (*Acipenser* sp.), walleye (*Stizostedion vitreum vitreum*), and red drum (*Sciaenops ocellatus*).

Circular tanks have several advantages that make them particularly attractive (2,19–21): they can provide a uniform culture environment; they can be operated under a wide range of rotational velocities to optimize fish health and condition; they can be used to concentrate and remove settleable solids rapidly; they allow for good feed and fish distribution; and they can permit designs that allow for visual or automated sensing and regulation of sinking feeds (22).

Relatively complete mixing of the water in circular culture tanks is necessary to prevent flow from short-circuiting along the tank bottom and to produce uniform water quality throughout the tank. The water exchange rate can then be set to provide the fish with good water quality throughout the entire culture tank, even when it is operating up to its maximum carrying capacity. The velocity of the water rotating in the culture tank must be swift enough to make the tank self-cleaning, but not faster than the desired fish swimming speed. The tank becomes self-cleaning at water rotational velocities that exceed 15 to 30 cm/s (6–12 in./s), which are adequate to create a secondary radial flow strong enough to move settleable solids (e.g., fish feed and fecal matter) along the tank's bottom to its center drain (20–24). To maintain fish health, muscle tone, and respiration, water velocities should be 0.5–2.0 times fish body length per second (25). For salmonids, the following equation can be used to predict safe nonfatiguing water velocities (26):

$$V_{safe} < 5.25/(L)^{0.37}$$

Here L is the fish body length in cm (in times 2.54) and V_{safe} is the maximum design velocity (about 50% of the critical swimming speed) in fish lengths per second. In circular tanks, water velocities are somewhat lower in a toroid region about the tank center, a circumstance that allows fish to select a variety of water velocities.

The self-cleaning effectiveness of the circular tank is also affected by the overall rate of the flow leaving the bottom center drain and how much the swimming motion of the fish resuspends the settled materials. Only about 5 to 20% of the total flow through the tank is required to flush settleable solids from the tank's bottom center drain, depending upon water exchange rate, because the water rotational velocity and the swimming motion of the fish control the transport of settleable solids to the tank's bottom center drain. This is the principle behind the use of dual-drain tanks to concentrate settleable solids (2). Therefore, the flow through the culture tank does not have to be increased beyond that required to support a selected carrying capacity, if the water inlet structure is properly designed.

The water inlet structure must be designed correctly to obtain uniform water quality, to achieve specific water rotational velocities, and provide for the rapid flushing of solids. According to recent studies (20,21) from the SINTEF Norwegian Hydrotechnical Laboratory, the tank rotational velocity is roughly proportional to the velocity through the openings in the water inlet structure at a given water exchange rate through the tank. The impulse force created by injecting the flow into the tank controls the rotational velocity in the tank and can be regulated by adjusting the inlet flow rate and the size, number, and orientation of the inlet openings (19). Injecting flow through an open-ended pipe creates poor mixing in the central toroid zone, much higher velocity profiles along the tank wall than in the central toroid region, resuspension of solids to all tank depths, and poor flushing of solids from the bottom (20,21). In contrast, distributing the inlet flow by using a combination of both vertical and horizontal branches can achieve uniform mixing in the culture tank, prevent short-circuiting of flow along the bottom, produce more uniform velocities throughout the tank, and more effectively transport waste solids along the tank bottom to the center drain.

Circular fish culture tanks can be managed as "swirl settlers" when the bulk flow passing through the tank is discharged from a location distant from the settleable solids concentrated at the bottom and center of the tank. Ideally, the bulk of settleable solids will be transported out of the tank's bottom center drain by using only 5–20% of the total flow. Here, the majority of flow is withdrawn (relatively free from settleable solids) from an elevated drain. Dual drains have been used to help remove settleable solids from fish culture tanks since 1930 (24,27–32). Patents covering specific features of dual-drain designs have been awarded (33,34). A nonproprietary design, the "Cornell-type" dual-drain tank (Fig. 3), is a circular culture tank with a center drain on the tank bottom and an elevated drain part-way up the tank side wall (2). The separation of the two drains so that one is part-way up the tank side wall and the other is in the tank center makes the Cornell-type dual-drain tank easy to install, even as a retrofit on existing circular culture tanks. No matter which dual-drain design is selected, removing settleable solids from the bulk flow at the culture tank can have large economic implications — a reduction in capital cost as well as in space and water-head requirements (2).

The concentrating of settleable solids in the discharge leaving the bottom center drain in a dual-drain culture tank will depend largely on how rapidly fish fecal matter and waste feed settle. The settling properties of fecal matter and waste feed are influenced by several factors, including diet formulation. Feed pellets have been reported to settle at velocities ranging from 15.2 to 17.9 cm/s (6–7 in./s) (35); intact fecal matter can settle at velocities ranging from 2 to 5 cm/s (1–2 in./s) (36), but finer and/or less dense particles can be produced and may settle at only 0.01 cm/s (0.004 in./s) (14), a speed that would not allow solids to concentrate effectively at the bottom center of dual-drain tanks. Research at the Freshwater Institute (Shepherdstown, WV) on solids removal within a Cornell-type dual-drain tank has shown that the mean total suspended solids (TSS) concentrations discharged through the elevated side-wall drain averaged only 1.5 mg/L (ppm), while the bottom drain discharge contained an average of 20 mg/L (ppm) TSS (S.T. Summerfelt, Freshwater Institute, unpublished data). This study was performed to determine how solids removal from the culture tank was influenced by hydraulic retention time (one and two culture-volume exchanges per hour), by diameter : depth ratio (12 : 1, 6 : 1, and 3 : 1), and by the percentage of flow discharged through the bottom drain (5, 10, and 20%). In that study, the tank contained rainbow trout at a density of 60 kg/m^3 that were fed 1% body weight per day. Others (32) have reported concentrating 91% of the fecal matter and 98% of uneaten feed within the bottom flow leaving dual-drain culture tanks.

A simple, fast, and reliable method that can be used to remove the occasional dead fish from the bottom-center drain should also be built in during the tank design, in order to decrease labor costs, reduce the spread of fish disease, and maintain water level in the culture tank (2). Dead fish are usually netted from the bottom drain, but automated mortality removal methods are sometimes used in relatively large and deep culture tanks.

RACEWAY TANKS

Raceways are the most common rearing-tank design prevailing in locations where aquaculture has tapped into huge groundwater resources. In these instances, the lay of the land allows for gravity flow from the water source

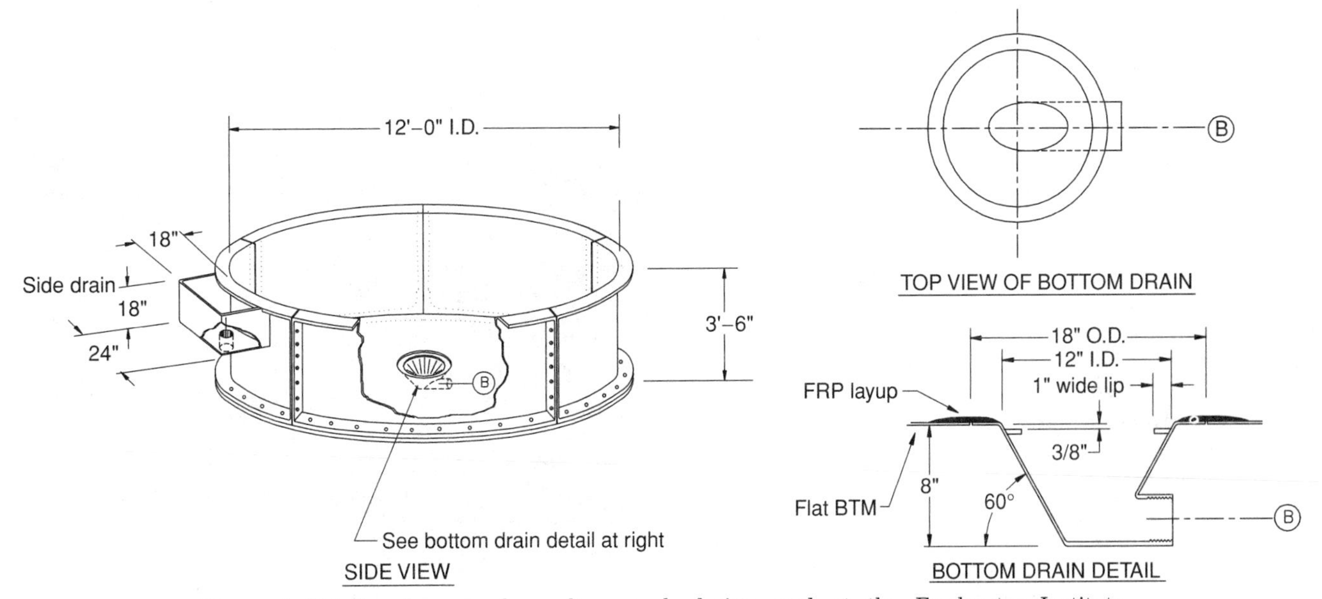

Figure 3. Details of a circular culture tank design used at the Freshwater Institute (Shepherdstown, WV) to concentrate settleable solids at its bottom center drain while discharging the majority of water flow (i.e., 80–95% of the total flow) through its "Cornell-type" side-wall drain. (Drawing courtesy of Red Ewald, Inc., Karnes City, Texas.)

through the raceways. Such is the case in Idaho, where some of the world's largest producers of rainbow trout are located (37). In Idaho, raceways are typically around 3–5.5 m (10–17 ft) wide, 24–46 m (80–155 ft) long, and 0.8–1.1 m (32–44 in.) deep (14). Raceways usually have a length-to-width ratio of 1:10, a depth <1.0 m (3 ft), and a requirement for a high water-exchange rate (e.g., one tank-volume exchange every 10 to 15 minutes) (38). Raceways used for warmwater species are typically much shorter, often only 7–13 m (23–43 ft) long.

Water enters the raceway at one end and then flows through the raceway in a plug-flow manner, with minimal back-mixing. The plug flow produces a concentration gradient along the axis in dissolved oxygen and in such metabolites as ammonia and carbon dioxide. The water quality is best at the head of the tank, where the water enters, and then deteriorates along the axis of the raceway toward the outlet. Oxygen is often the limiting criterion, so fish may congregate at the head of the raceway and cause an unequal distribution of fish density throughout the raceway. It is also more difficult to distribute feed throughout raceways than within circular tanks.

The velocity of water through the raceway is generally 2–4 cm/s (1–2 in./s), so a substantial amount of solids settles in the rearing area; however, these solids are slowly moved downstream through the rearing area by the swimming activity of larger fish (14,38). A series of baffles spaced at intervals equaling the width of the raceway and placed perpendicular to the flow can be used to create high water velocities (20–30 cm/s—8–12 in./s) between the bottom of the raceway and the bottom edge of the baffle. The baffles allow solids to be swept continuously from the raceway (38); however, the Idaho Division of Environmental Quality reports (14) that baffles can be troublesome, because they must be moved to work the fish and can provide a substrate for biosolids growth in the summer.

In practice, raceways are managed on the basis of their oxygen requirements rather than their cleaning requirements (26). The velocity required to flush solids from unbaffled raceways is much greater than the velocity required to supply the oxygen needs of the fish. In a practical sense, raceways are incapable of producing the optimum water velocities recommended for fish health, muscle tone, and respiration.

Raceways are designed to minimize cross-sectional area and thus maintain as high a water velocity as possible, typically at least 2–4 cm/s (1–2 in./s). For this reason, many raceway systems are operated in series, with the discharge of the upstream raceway serving as the inflow water of the next one downstream. Hydraulic drops between raceways in series provide some reaeration. Long and narrow raceways are very convenient culture tanks for managing fish during crowding or grading. Crowders or graders can be placed in the raceway at one end and slowly worked down the axis of the raceway. Raceways can be constructed side by side, with common walls, to maximize the utilization of floor space and to reduce construction costs (Fig. 4). When constructed without common walls, raceways require 1.5 to 2.0 times as much wall length as circular tanks because of their large aspect ratio (L:W).

Figure 4. Raceways are often constructed side by side, with common walls. In some instances, the raceway units may be stair-stepped down a hillside, as shown at Leo Ray's fish farm (Buhl, ID), in order to provide some water aeration before serially reusing the flow in downstream raceways.

Circular tanks can also better handle the weight of the confined water structurally and can thus use thinner walls than rectangular tanks.

A quiescent zone devoid of fish is usually placed at the end of a raceway tank to collect the settleable solids that are swept out of the fish-rearing area (14,38). These solids-collection zones are the primary means for solids removal aimed at meeting discharge permit requirements at many large trout farms (14). The overflow rate recommended for capturing solids in the quiescent zone is <1 cm/s (0.4 in./s) (e.g., 0.01 m^3/s flow per square meter of surface area). Settled solids should be removed from these quiescent zones as frequently as possible; settling zones are cleaned at least twice per month, and occasionally as often as daily (14). Prolonged storage allows for some nutrient leaching, solids degradation, and solids resuspension (due to denitrification and fermentation of the organic matter). Suction through a vacuum pump is the most common method for removing solids from the quiescent zone (14)

Figure 5. Some of the solids swept from a raceway's fish-rearing area can settle in quiescent zones, but the captured solids must be manually removed. Shown here is a quiescent zone being cleaned at the Pennsylvania Fish and Boat Commission's Big Spring Fish Culture Station (Newville, PA), where it takes about one day for a single person to clean all 80 of the hatchery's quiescent zones.

and sometimes also from the fish-rearing areas. Quiescent zones are also cleaned through a central drain after flow through the zone is temporarily stopped (Fig. 5). Even with efficient techniques, operating labor for solids removal has been reported to exceed 25% of the total farm labor (14).

MIXED-CELL RACEWAY TANKS

We have noted that circular tanks offer the advantages of elevated water velocities, uniform water quality, and good solids removal characteristics, whereas linear raceways make better use of floor space and facilitate harvesting, grading, and flushing operations. The crossflow tank is a recent hybrid design that incorporates the desirable characteristics of both circular tanks and linear raceways (39,40). Water is distributed uniformly along one side of a cross-flow tank (via a submerged manifold), and is collected in a submerged perforated drain line running the length of the opposite side. The influent is jetted perpendicular to the water surface, with sufficient force to establish a rotary circulation about the longitudinal direction. Comparative production trials with hybrid striped bass (39), tilapia (40), rainbow trout (41), and lake trout (42) have been positive, but application has been hampered by the need for the small-diameter, fixed, and submerged inlet jets and drain ports, as well as costs associated with rounding the lower side areas to streamline flow. The rectangular mixed-cell tank (43) avoids these problems while achieving the same overall objective: a hybrid tank design. Here, a standard raceway section is modified to create horizontal, counter-rotating mixed cells with cell length equal to vessel width (Fig. 6). Cells receive water from vertical pipe sections extending to the tank floor and positioned in the corners of the cells. Vertical pipe sections incorporate jet ports that direct water into the cell tangentially to establish rotary circulation. The pipe sections can be swung up and out of the water during fish crowding or grading operations. Water exits each cell through a centrally located floor drain. Hydraulic characteristics of the tank have been established and indicate that tank performance approximates that of a circular tank (mixed-flow reactor), both with and without fish present. Water velocities averaged 15, 12, and 12 cm/s (6, 5, and 5 in./s) for tank-surface, mid-depth, and bottom regions and were sufficient to scour and purge fecal solids. Cell interaction was significant with cell-to-cell exchange rates

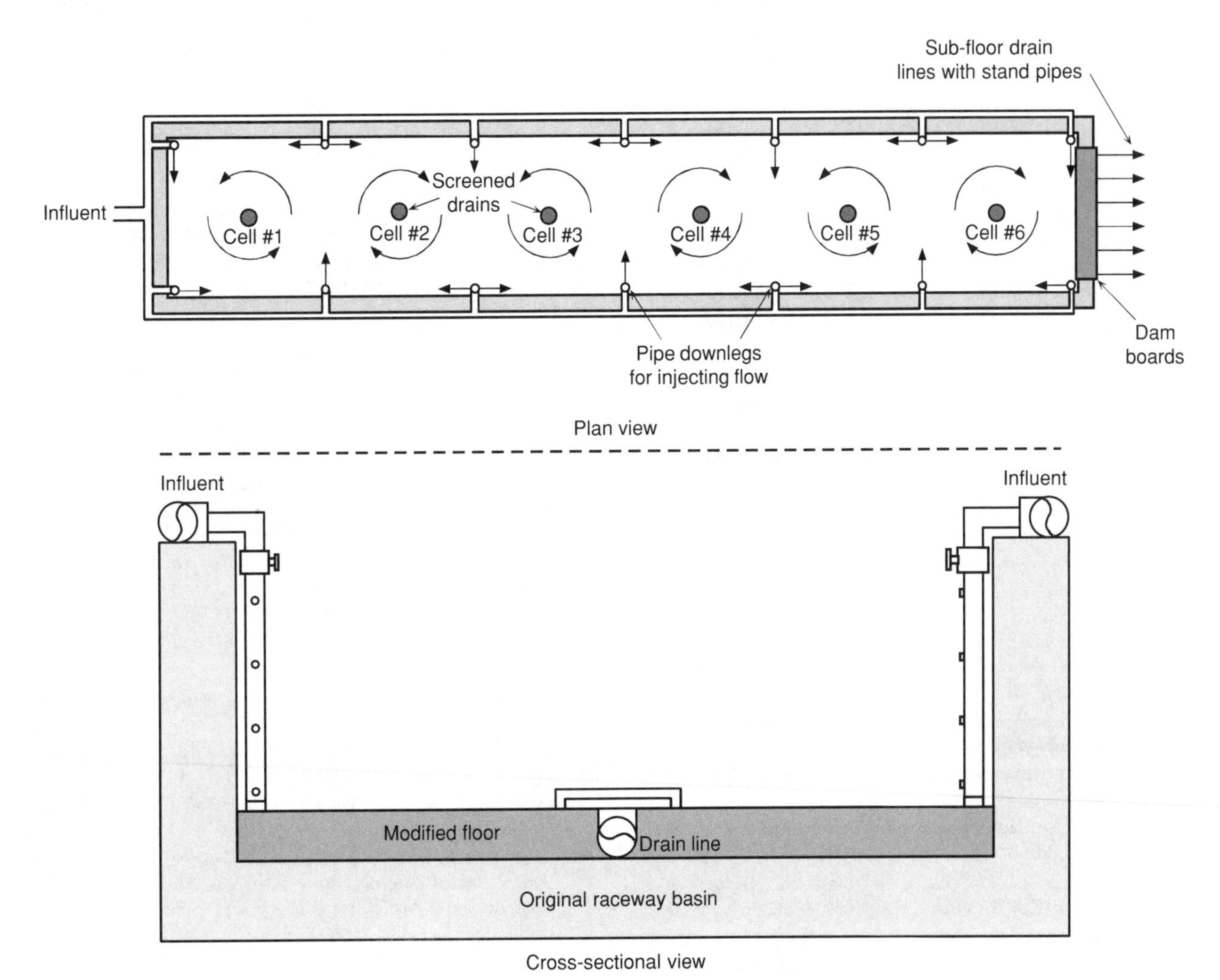

Figure 6. Illustration of a mixed-cell raceway tank.

representing about three to four times the tank inflow rate. This characteristic contributed to the observed uniform distribution of fish throughout the vessel. Further, the energy requirement of the design was kept low [just 1.32 m (52 in.) of water head] through use of a large number of low-velocity inlet jets. Given that the tank's drain is similar to that of a circular tank, application of dual-drain solids concentration is feasible and desirable.

BIBLIOGRAPHY

1. L. Karlsen, in K. Heen, R.L. Monahan, and F. Utter. *Salmon Aquaculture*, Fishing News Books, Oxford, England, 1993, pp. 59–82.
2. M.B. Timmons, S.T. Summerfelt, and B.J. Vinci, *Aquacultural Eng.* **18**, 51–69 (1998).
3. J.E. Colt and B.J. Watten, *Aquacultural Eng.* **7**, 397–441 (1988).
4. J.E. Colt, K. Orwicz, and G. Bouck, in J.E. Colt and R.J. White, eds., *Fisheries Bioengineering Symposium 10*, American Fisheries Society, Bethesda, MD, 1991, pp. 372–385.
5. C.E. Boyd and B.J. Watten, *CRC Rev. in Aquatic Sci.* **1**, 425–473 (1989).
6. B.J. Watten, *Aquacultural Eng.* **11**, 33–46 (1992).
7. S.T. Summerfelt, J.A. Hankins, S.R. Summerfelt, and J.M. Heinen, in J.-K. Wang, ed., *Techniques for Modern Aquaculture*, American Society of Agricultural Engineers, St. Joseph, MI, 1993, pp. 581–593.
8. J.A. Hankins, S.T. Summerfelt, and M.D. Durant, in M.B. Timmons, ed., *Aquacultural Engineering and Waste Management*, Northeast Regional Agricultural Engineering Service, Ithaca, NY, 1995, pp. 70–86.
9. S.T. Summerfelt and R.C. Summerfelt, in R.C. Summerfelt, ed., *Walleye Culture Manual*. NCRAC Culture Series 101, North Central Regional Aquaculture Center Publications Office, Iowa State University, Ames, IA, 1996, pp. 215–230.
10. J.M. Heinen, J.A. Hankins, A.L. Weber, and B.J. Watten, *Prog. Fish Cult.* **58**, 11–22 (1996a).
11. R.G. Piper, I.B. McElwain, L.E. Orme, J.P. McCraren, L.G. Fowler, and J.R. Leonard, *Fish Hatchery Management*, U.S. Fish and Wildlife Service, Washington, DC, 1982.
12. E.L. Brannon, in R.R. Stickney, ed., *Culture of Salmonid Fishes*, CRC Press, Boca Raton, FL, 1991, pp. 22–55.
13. B. Braaten, in N. De Pauw and J.N. Joyce, eds., *Aquaculture and the Environment*, European Aquaculture Society Special Publication 16, Gent, Belgium, 1991, pp. 79–101.
14. Idaho Division of Environmental Quality (IDEQ), *Idaho Waste Management Guidelines for Aquaculture Operations*, Idaho Department of Health and Welfare, Division of Environmental Quality, Twin Falls, ID, 1998.
15. S.T. Summerfelt, in E.H. Bartali, F. Wheaton, and S. Singh, eds., *CIGR Handbook of Agricultural Engineering, Volume II: Animal Production and Aquacultural Engineering*, American Society of Agricultural Engineers, St. Joseph, MI, 1999, pp. 309–350.
16. J.M. Heinen, J.A. Hankins, and P.R. Adler, *Aquaculture Res.* **27**, 699–710 (1996b).
17. M. Ingram, in L.M. Laird and T. Needham, eds., *Salmon and Trout Farming*, Ellis Horwood, New York, 1988, pp. 155–189.
18. T. Needham, in L.M. Laird and T. Needham, eds., *Salmon and Trout Farming*, Ellis Horwood, New York, 1988, pp. 87–116.
19. K. Tvinnereim and S. Skybakmoen, in N. De Pauw, E. Jaspers, H. Ackefors, and N. Wilkens, eds., *Aquaculture: A Biotechnology in Progress*, European Aquaculture Society, Bredena, Belgium, 1989, pp. 1041–1047.
20. S. Skybakmoen, *AquaNor Conference 3—Water Treatment and Quality*, AquaNor, Trondheim, Norway, 1989, pp. 17–21.
21. S. Skybakmoen, *Fish Rearing Tanks—Aquaculture Series Brochure*, AGA AB, Lidingö, Sweden, 1993.
22. S.T. Summerfelt, K.H. Holland, J.A. Hankins, and M.D. Durant, *Water Sci. and Tech.* **31**(10), 123–129 (1995).
23. R. Burrows and H. Chenoweth, *Prog. Fish Cult.* **32**, 67–80 (1970).
24. T. Mäkinen, S. Lindgren, and P. Eskelinen, *Aquacultural Eng.* **7**, 367–377 (1988).
25. T.M. Losordo and H. Westers, in M.B. Timmons and T.M. Losordo, eds., *Aquaculture Water Systems: Engineering Design and Management*, Elsevier, New York, 1994, pp. 9–60.
26. M.B. Timmons and W.D. Youngs, in P. Giovannini, ed., *Aquaculture Systems Engineering*, American Society of Agricultural Engineers, St. Joseph, MI, 1991, pp. 34–45.
27. W.W. Cobb and J.W. Titcomb, *Trans. Am. Fish. Soc.* **60**, 121–123 (1930).
28. E.W. Surber, *Prog. Fish-Cult.* **21**, 1–14 (1936).
29. T. MacVane, *Fish Culture Tank*, U.S. Patent No. 4,141,318, 1979.
30. W.J. Slone, D.B. Jester, and P.R. Turner, in L.J. Allen and E.C. Kinney, eds., *Fisheries Bioengineering Symposium 1*, American Fisheries Society, Bethesda, MD, 1981, pp. 104–115.
31. B. Eikebrokk and Y. Ulgenes, in H. Keinertsen, L.A. Dahle, L. Jorgensen, and K. Tvinnereim, eds., *Fish Farming Technology*, Balkema, Rotterdam, 1993, pp. 361–366.
32. T. Lunde and S. Skybakmoen, in H. Keinertsen, L.A. Dahle, L. Jorgensen, and K. Tvinnereim, eds., *Fish Farming Technology*, Balkema, Rotterdam, 1993, pp. 465–467.
33. T. Lunde, S. Skybakmoen, and I. Schei, *Particle trap*, U.S. Patent No. 5,636,595, 1997.
34. W. Van Toever, *Water Treatment System Particularly for Use in Aquaculture*, U.S. Patent No. 5,593,574, 1997.
35. J.E. Juell, *Aquacultural Eng.* **10**, 207–217 (1991).
36. I. Warrer-Hansen, in J. Alabaster, ed., *Report of the EIFAC Workshop on Fish Farm Effluents* (EIFAC Technical Paper No. 41.), 1982, FAO, Rome, 1982, pp. 113–121.
37. R. MacMillan, in *National Livestock, Poultry, and Aquaculture Waste Management*, American Society of Agricultural Engineers, St. Joseph, MI, 1992, pp. 185–190.
38. H. Westers, in C.B. Cowey and C.Y. Cho, eds., *Nutritional Strategies and Aquaculture Waste*, University of Guelph, Guelph, ON, Canada, 1991, pp. 231–238.
39. B.J. Watten and L.T. Beck, *Aquacultural Eng.* **6**, 127–140 (1987).
40. B.J. Watten and R.P. Johnson, *Aquacultural Eng.* **9**, 245–266 (1990).
41. R.M. Ross, B.J. Watten, W.F. Krise, and M.N. DiLauro, *Aquacultural Eng.* **14**, 29–47 (1995).
42. R.M. Ross and B.J. Watten, *Aquacultural Eng.* **19**, 41–56 (1998).
43. B.J. Watten and D.C. Honeyfield, in M.B. Timmons, ed., *Aquacultural Engineering and Waste Management*, Northeast Regional Agricultural Engineering Service, Ithaca, NY, 1995, pp. 112–126.

See also RECIRCULATING WATER SYSTEMS.

TEMPERATURE

Robert R. Stickney
Texas Sea Grant College Program
Bryan, Texas

OUTLINE

Temperature is one of the water-quality parameters most critical to the aquaculturist. Animals reared under aquaculture conditions are poikilothermic, so their metabolism is closely tied to the temperature of the water. Each species has a thermal optimum—the temperature (or temperature range) at which growth is optimized, and a thermal tolerance range—the range of temperature within which the species will survive. Aquatic animals tend to fall into one of three categories: warmwater, coolwater, and coldwater species. In general, the thermal optima for the three groups tend to be around 30, 25, and 20 °C (86, 78, and 68 °F), respectively. This entry is adapted from a previously published book (1) and is used with permission.

THERMAL REQUIREMENTS

Selection of an aquaculture species usually takes temperature into consideration, particularly when the site has been selected and the aquaculturist knows the type of water that will be used in the facility. Water temperature may fluctuate significantly in temperature on a seasonal basis, depending on the source of the water being used. Some culture species can survive a broad range of temperatures; others cannot. As indicated above, each species has a relatively narrow temperature range within which growth is optimum. Catfish, tilapia, striped bass, and trout provide good examples.

The original range of occurrence of the channel catfish (*Ictalurus punctatus*), for example, was from the Great Lakes region and the prairie provinces of Canada to the Gulf states of the USA (2). The species can thus survive water temperatures approaching freezing as well as those that rise above 30 °C (86 °F). Growth rate differs significantly from the North to the South; the optimum temperature generally accepted lies within the range from 26 to 30 °C (from 79 to 86 °F) (3,4), though a broader optimum (21 to 30 °C; 70 to 86 °F) has also been reported (5). Temperatures in that band may never be reached at the northern end of the fish's range, and, in high latitudes, several years might be required for a fish to reach marketable size (about 450 g; 1 lb). In the southern United States, market size is generally reached in 18 months. If the temperature can be maintained within the optimum range at all times, the length of time from egg to market can be reduced by another 8 to 10 months. On the basis of its optimum temperature, the channel catfish is considered to be a warmwater fish (Table 1).

Table 1. Classification of Selected Aquaculture Species as a Function of Temperature Required for Optimum Growth

Optimum Temperature	Name
Coldwater invertebrate	Pacific oyster (*Crassostrea gigas*)
Coldwater fishes	Chum salmon (*Oncorhynchus keta*)
	Coho salmon (*Oncorhynchus kisutch*)
	Pink salmon (*Oncorhynchus gorbuscha*)
	Rainbow trout (*Oncorhynchus mykiss*)
	Chinook salmon (*Oncorhynchus tshawytscha*)
	Plaice (*Pleuronectes platessa*)
	Atlantic salmon (*Salmo salar*)
	Brown trout (*Salmo trutta*)
	Sole (*Solea solea*)
Coolwater fishes	Northern pike (*Esox lucius*)
	Muskellunge (*Esox masquinongy*)
	Striped bass (*Morone saxatilis*)
	Yellow perch (*Perca flavescens*)
	Walleye (*Stizostedion vitreum vitreum*)
Warmwater invertebrates	American oyster (*Crassostrea virginica*)
	Freshwater shrimp (*Macrobrachium rosenbergii*)
	Northern quahog (*Mercenaria mercenaria*)
	Southern quahog (*Mercenaria campechiensis*)
	Blue mussel (*Mytilus edulis*)
	Kuruma shrimp (*Marsupenaeus japonicus*)
	Tiger shrimp (*Penaeus monodon*)
	Blue shrimp (*Litopenaeus stylirostris*)
Warmwater fishes	Bighead carp (*Aristichthys nobilis*)
	Goldfish (*Carassius auratus*)
	Milkfish (*Chanos chanos*)
	Mud carp (*Cirrhina molitorella*)
	Walking catfish (*Clarias batrachus*)
	Grass carp (*Ctenopharyngodon idella*)
	Common carp (*Cyprinus carpio*)
	Sea bass (*Dicentrarchus labrax*)
	Silver carp (*Hypopthalmichthys molitrix*)
	Bigmouth buffalo (*Ictiobus bubalus*)
	Blue catfish (*Ictalurus furcatus*)
	Channel catfish (*I. punctatus*)
	Red drum (*Sciaenops ocellatus*)
	Yellowtail (*Seriola quinqueradiata*)
	Rabbitfish (*Siganus* spp.)
	Gilthead sea bream (*Sparus aurata*)
	Blue tilapia (*Oreochromis aureus*)
	Mossambique tilapia (*Oreochromis mossambicus*)
	Nile tilapia (*Oreochromis niloticus*)

Tilapia (most species of aquaculture being in the genus *Oreochromis*) are tropical fishes that can tolerate temperatures above those at which many warmwater species succumb. Tilapia are, however, not tolerant of cold water. Death generally occurs when the water falls below 10 to 12 °C (50 to 54 °F) (6–8). Growth is generally poor below about 20 °C (68 °F), and disease epizootics, extremely rare when the water is around 30 °C (86 °F), become very common when the temperature approaches the lower lethal range.

Rainbow trout (*O. mykiss*) represent a coldwater species that has a temperature tolerance range of from about 1 °C (34 °F) to nearly 26 °C (79 °F) (5). While rainbow trout and other salmonids can survive in relatively warm water, the optimum temperature range for rainbows is about 10 to 16 °C (50 to 61 °F). Growth is retarded at higher and lower temperatures.

Striped bass (*M. saxatilis*) can be considered a coolwater (also called midrange) species with respect to temperature tolerance. Striped bass survive a temperature range of from about 2 to 32 °C (36 to 90 °F) (5), but have an optimum range of from 13 to 24 °C (55 to 75 °F). Other species that are often considered to be midrange in terms of their temperature optima, such as northern pike (*E. lucius*), muskellunge (*E. masquinongy*), and walleye (*S. vitreum vitreum*), may actually be better classified as coldwater species, according to their optimum ranges (5).

Temperature-tolerant coldwater, coolwater, and warmwater species may exist in water bodies that lie in close proximity; two or three types may even coexist in the same water body. For example, the deep waters of reservoirs that feature primarily warmwater species may have sufficiently cold water at depth that coldwater species can survive through the summer under the thermocline or in proximity to cold springs. Coolwater species that are fairly adaptable to a wide range in temperature can also thrive.

Latitude, altitude, and water source are all factors that affect the suitability of a given area for various species of aquaculture interest. The relatively low latitudes in which the southern United States occur are generally conducive for the production of warmwater fishes, yet at the higher elevations in many southern states it is possible to produce coolwater and even coldwater fishes. Most southern states have trout hatcheries in them. Northern climates where coolwater and coldwater species predominate will have native populations of some warmwater fishes [e.g., channel catfish and largemouth bass (*Micropterus salmoides*)], and aquaculture of warmwater fishes may be possible in geothermal water or the effluent of power plants nearly anywhere.

Virtually all aquaculture candidates are poikilothermic. For reasons that are largely unknown but that undoubtedly entail genetic differences in enzyme systems among the various species, temperature tolerances and optima vary greatly. As a general rule, coolwater and coldwater fishes grow much more slowly than warmwater fishes, though that does not mean that warmwater fishes tend to be larger on average. There are many small warmwater fishes, just as there are large coolwater and warmwater ones.

There has been interest expressed in recent years in genetically engineering fish to tolerate, or even grow more rapidly in, water temperatures outside of the range in which those fishes are normally found now. For example, it might be possible, through the insertion of appropriate genes, to engineer a trout or salmon that can tolerate and grow well at 30 °C (86 °F) or a tilapia that will not succumb to disease and possibly experience mortality as the water temperature falls below 15 °C (59 °F). The latter would involve insertion of a so-called antifreeze gene. Certain polypeptides are present in the blood plasma of fishes that exhibit antifreeze protection. Presence of the antifreeze gene sounds impressive, and it is certainly of importance to tilapia culturists in certain regions where slightly more tolerance to winter minimum temperatures could mean the difference between success and devastating mortality, but presence of the polypeptides that allow fish to live at unusually cold temperatures can fail to impart as much as a one-degree advantage. Adult cod (*Gadus morhua*), for example, freeze at about −1.2 °C (29.9 °F), as compared with juveniles, which can tolerate −1.55 °C (29.2 °F), and as compared with fish like halibut and salmon, which lack the antifreeze polypeptide and generally freeze at −0.7 to −0.9 °C (30.8 to 30.4 °F) (9), though Pacific halibut juveniles have been found to survive temperatures below −1.0 °C (30 °F) (10).

Some success has been obtained not only in the transfer of a winter flounder (*Pseudopleuronectes americanus*) antifreeze gene into Atlantic salmon (*S. salar*), but also in the expression of that gene in a cross between a transgenic male and a normal female salmon (11). Better tolerance to low temperature could come from the transfer to Atlantic salmon of genes from other species that have higher concentrations of the antifreeze polypeptides (12). These kinds of alterations may soon become routine, but are they desirable?

One of the certainties associated with the practice of aquaculture is that, unless strict quarantine restrictions are imposed, culture animals will eventually escape into the natural environment. Restrictions have already been imposed on genetically engineered animals that ensure, to the extent possible, that the fish or their progeny do not escape into the wild. So far, transgenic animals of aquaculture interest have been developed and evaluated only in a research environment. If they are released for general aquaculture use, it will be very difficult to maintain the same stringent controls that are possible at research institutions.

It is difficult to know what the impact on native populations of warmwater trout or coldwater tilapia would be, but studies have shown that tilapia would compete with other species for nesting sites and would also compete with various species for food. At present, the threat of tilapia to native fish populations is low, because all but a very few locations in the United States are too cold in winter for tilapia to survive. While the aquaculturist might benefit from a cold-tolerant tilapia, native populations might suffer. Consideration of the welfare of native populations should supersede the interest of the aquaculturist in instances where the consequences could alter the natural ecological balance of a region.

THE INFLUENCE OF GENETICS AND AGE ON RESPONSE TO TEMPERATURE

While the tendency is to indicate that a species will respond to temperature stress in a predictable manner, the facts are that genetics can play a significant role and that different stocks of animals may respond differently to the same stressor. For example, coho salmon (*O. kisutch*) from six hatcheries responded differently to thermal increases of 1 °C/hr, as confirmed by measurements of changes in the critical thermal maximum tolerated by the fish (13). Similarly, it has been determined that a southern population of chinook salmon was better able to survive a challenge by high temperature than was a more northern population. (The southern stock would experience higher summer maximum water temperatures than would the northern stock (14).) The response was affected by fish size: Larger (and presumably older) fish survived the temperature challenge better than smaller fish.

The ability in Black Sea golden gray mullet (*Liza aurata*) to tolerate low temperatures increases with age until the fish reach sexual maturity, after which it goes back down (15). The relationship between age and temperature tolerance has not been well documented for most aquaculture species.

Genetic influences on the cold tolerance of tilapia have also been demonstrated. For example, *O. niloticus* from strains originating in Egypt, the Ivory Coast, and Ghana showed different lower lethal temperatures, ones ranging from 10 °C (50 °F) for the Egyptian strain to 14 °C (57 °F) for fish from Ghana (16). The Ivory Coast strain was intermediate (12 °C; 54 °F). The responses to low temperature were correlated with the normal ranges of temperatures at the geographic origins of the three strains.

RESPONSE TO SEASONAL PATTERNS IN WATER TEMPERATURE

The temperature of a surface water body typically fluctuates to one extent or another on a seasonal basis. The range of annual fluctuation tends to be maximized in temperate regions and minimized in Arctic and tropical ones. There are locations in temperate regions where temperatures approach the optimum for various culture species throughout much of the year. Surface waters in southern Florida and extreme south Texas, for example, can support good growth of warmwater species throughout most of the year. Temperatures are typically sufficiently warm to support tilapia survival during all but exceptionally cold winters. The climate in Hawaii is tropical and will support year-round growth of warmwater species. By contrast, the waters of Puget Sound in Washington and those off upper New England are sufficiently cold to support year-round salmon production. Parts of Alaska are subarctic and will also support year-round salmon growth.

There are instances wherein water that is unusually cold or atypically warm for a given region occurs. There are also places where water temperature does not fluctuate to any extent seasonally. Cold springs may reduce the extent of spring and summer warming in some surface waters; geothermal water may be used to produce warmwater species throughout the year in regions where even summer water temperatures might otherwise be too low to allow for growth, or in some cases, survival of a particular warmwater species. Facilities producing tilapia and channel catfish in parts of Idaho where geothermal water is available are good examples. Similar, and also in Idaho, is the production of rainbow trout in the Hagerman Valley, where underground rivers of virtually constant temperature provide ideal year-round growing conditions. The water flows through raceways that have a short residence time shortly after it erupts from the ground, so there is little temperature change at any time of year.

Water that has had its temperature altered by the activities of humans can sometimes be used for aquaculture also. The best example is the warm water associated with fossil-fuel and nuclear power plants, but many other industries produce heated effluents that may be suitable for use in aquaculture, either directly or through heat exchange with water in the culture chambers.

Most species that are being successfully produced by commercial aquaculturists reach the market in less than two years; one year or less for growout is even more desirable. Tilapia can be grown to market size (about 500 g, 1.1 lb) in under a year in tropical regions or when geothermal or artificially heated water is available. Channel catfish can also be reared to the same size within a year if water temperature is constantly maintained within the optimum range for the species. In most parts of the United States where catfish are reared in waters that are subject to seasonal temperature fluctuations, growout requires about 18 months. In tropical regions it is easy to obtain at least two crops of penaeid shrimp annually; in temperate regions, however, one crop per year is typical.

Some species of commercial importance require two or more years to reach market size. Pacific oysters and Atlantic salmon are examples, with three-year production cycles from egg to market being typical, though selective breeding is leading to a reduction in the time required to reach market size in Atlantic salmon. Atlantic halibut (*Hippoglossus hippoglossus*) appear to require about three years for growout to a size of 5 kg (11 lb), which is considered to be market size for that species. One of the reasons for slow growth in halibut is related to their requirement for cold water.

At least modest reductions in the time required to produce marketable species in commercial culture and those of suitable stocking size in the recreational fish production arena may be possible as a result of improved nutrition, genetic manipulation, and aggressive culture system management. Research has already been initiated with respect to the introduction of growth hormone into fish. Yet, given the small initial size of the larvae and fry of aquatic species of aquaculture interest, there will be limits to how fast marketable animals can be produced. We are not likely to see fish or oysters reach the market in a few weeks as broiler chickens do, but we can anticipate some reduction in the growout period as technology advances. Maintenance of optimum temperature will be one key to reducing the period required for growout.

For most species, the maximum temperature experienced during the year is not within the temperature range optimum for most rapid growth. At temperatures both above and below the optimum range, growth rate decreases. Metabolic rate is reduced when the water temperature falls below the optimum range and increases when the temperature rises above that range. Low metabolism means reduced feed intake and slower growth. At temperatures above the optimum range, it usually happens that feed consumption rate increases to accommodate the higher metabolic rate, yet growth does not increase—and there is an economic impact associated with meeting the energy demand by providing more food, for no additional increase in rate of weight gain. As temperature approaches the upper tolerance point for a given species, metabolic rate begins to fall as various systems fail. Feeding activity declines, and, ultimately, death occurs.

Given a range of temperature, most species will select one that is within the optimum for growth, unless some mitigating factor is operating. For example, a fish that is approaching sexual maturity may select a temperature that is more suited to egg survival and development than one that is optimum for growth. In nature, aquatic animals can often make such selections because water bodies tend not to have uniform temperatures throughout. Thermal stratification and inflowing warm or cold springs provide opportunities for aquatic organisms to be somewhat selective in terms of temperature. In water systems in which temperature is controlled in some way, conditions may be more uniform.

Aquaculture ponds tend to have zones of temperature. Even a pond that is less than 2 m (6 ft) deep may partially stratify, and the shallow water around the edges of culture ponds is often considerably warmer during the day than is water in the middle, particularly on sunny days during summer.

SYNERGISMS WITH TEMPERATURE

The influence of temperature on the growth, metabolism, disease resistance, and survival of aquaculture animals may be influenced by various other parameters. While many of these synergisms have not been evaluated in any detail, and none has been examined across a broad range of aquaculture species, there is some information documenting at least a few of the factors that can interact with temperature and influence performance of aquaculture species.

The cold tolerance of *Tilapia aurea* at salinities ranging from 0 to 35 ppt has been examined (17), and it was determined that fish that were isosmotic with the medium survived lower temperatures than those for which the external medium was at a salinity higher or lower than that of the tissues of the fish. One implication of the relationship between temperature tolerance and salinity is that the species could be expected to have an extended range of distribution into estuarine waters as compared with into fresh water or marine waters (see the entry "Salinity" for further explanation).

Various other chemical parameters associated with the water can influence the temperature tolerance of aquaculture species. The ability of channel catfish to tolerate high temperature is reduced not only in the presence of ammonia (the un-ionized percentage of which increases with increasing temperature) but also when the fish are exposed to elevated nitrite levels.

EFFECTS OF TEMPERATURE ON DEVELOPMENT

Egg and embryo development are heavily influenced by temperature. A minimum temperature, below which larval development does not occur, has been identified for most species of aquaculture interest and importance. As the temperature increases above the minimum, the development rate also increases. However, once the temperature to which eggs or developing larvae are exposed rises above some level, which varies from species to species, abnormal development occurs, and the rate of mortality increases until it becomes total as the upper temperature limit is reached. The upper limit for proper development of eggs and larvae may not be as high as the temperature maximum that can be tolerated by later stages in the life history of the animals. A series of abnormalities can occur when temperature becomes too high during egg and larvae development. Some of the common ones in fishes are unusual numbers of vertebrae and tail abnormalities. A special condition related to development, temperature, and other variables in anadromous salmonids is smoltification, a physiological process that allows the fish to make the transition from fresh water to the ocean.

THERMAL SHOCK AND TEMPERING

Thermal shock results when aquatic animals are exposed to rapid changes in ambient water temperature. The stress associated with such shock may weaken the immune system and thereby reduce the resistance of the animals to disease. If severe enough, temperature shock can lead to death. The daily changes in temperature that occur as a result of daytime warming and nighttime cooling of the water within a culture chamber are usually small enough that no thermal shock occurs. An exception might be a small static tank or pond that is only partially filled; in such a case, daytime warming could be significant.

Thermal shock is most commonly associated with harvesting, live hauling, and transfer of aquatic animals from one water body to another. When aquatic animals are harvested at times of the year at which air and water temperatures are disparate, there can be a considerable amount of thermal stress that occurs during the harvesting process. When a pond is being harvested, the degree of temperature change can be significantly enhanced if the water level is greatly reduced. Total pond harvesting often involves seining, followed by reduction in water volume, followed by re-seining, and so on. Rather small volumes of water associated with harvest basins (or the immediate vicinity of the drain in ponds without harvest basins) may contain surprisingly large numbers of animals that avoided the seine. While those animals are being collected, the water can warm or cool considerably, depending on season of the year. Heating of small volumes of water has been a more significant problem than cooling. In warm climates during summer or fall, the water temperature can rise several degrees in only a few minutes. The resulting stress may be acceptable if the animals are being hauled a short distance for processing, but it can be devastating when the affected fish or invertebrates are being moved for restocking. This type of thermal stress can lead directly to mortality or reduce the disease resistance of affected animals.

In addition to thermal stress associated with the final stages of pond harvest, there is often significant reduction in dissolved oxygen (DO). Low DO occurs because the water can carry less oxygen at saturation when it is heated and because the metabolism in the animals that are being harvested increases. The problem can be alleviated by adding copious amounts of new, oxygen-rich pond water to the drain area during the latter stages of harvesting.

Water in hauling trucks should be at approximately the same temperature as the water from which the aquatic animals are collected. In many instances, water from the culture chamber being harvested is used to fill the hauling tanks, thereby assuring a minimal initial change in temperature in most cases.

Whether a hauling tank or some other type of container is used for transporting fish, the temperature of the water is subject to change with time. Insulated boxes and hauling tanks are helpful in maintaining temperature, but it may be necessary to add ice to keep water within acceptable limits during summer if the boxes or hauling tanks are exposed to high temperature and direct sunlight.

Once live hauling has been completed, and before the animals are stocked, the temperature in the transportation tank and receiving water should be compared. If the difference in temperature is more than about 2–3 °C (4–5 °F), though the conversion is not linear, the animals should be gradually tempered. For animals transported in tanks, common practice is to introduce water from the receiving site into the tank slowly, until the temperature within the tank equilibrates with that of the receiving water body. Tempering should not exceed a rate of more than a few degrees per hour, yet, in all instances, the process should be completed within 10 to 12 hr. Care should be taken to maintain a high level of DO at all times during the tempering process.

When fish are hauled in small containers such as plastic pails, plastic bags, or others that readily conduct heat, the containers can be floated in the receiving water body until the temperature of the water within the container equilibrates. This technique should not be used if a dramatic difference in initial temperature exists, because the rate of tempering may be too rapid and, therefore, stressful.

Rapid changes in temperature and exposure of aquatic animals to inappropriate temperatures sometimes cannot be avoided. It appears that such changes are less stressful to aquatic animals when the temperature is being changed toward, rather than away from, the thermal optimum of the species. Temperatures above and below the optimum can lead to increased rates of metabolism and to other physiological changes that can be detrimental. Whether tempering can be conducted more rapidly when the temperature change is toward the thermal optimum has not been well researched, but there is at least some circumstantial evidence to support that approach. When fish are exposed to heat stress in an almost completely drained pond during summer, rapid reduction in the temperature by adding cool water is often an effective means of avoiding subsequent mortality, even though the change in temperature may occur very rapidly.

Aquaculture animals can generally be handled without damage or severe stress in cool weather, but great care should be taken when handling them in the summer, particularly in temperate and tropical latitudes. Handling often leads to injury, and the high metabolic rate of aquatic animals in the summer may make them more active when caught, a condition leading to an increase in the incidence of self-inflicted injury. Also, bacterial activity is higher in warm as compared with cold water; thus, wounds are more likely to become infected during the summer than in the winter. An exception occurs in tilapia, which are quite disease resistant during summer but are extremely sensitive to various types of infections when exposed to cool water.

Animals that have been exposed to water temperatures above their normal optimum may be more vulnerable to parasitic and bacterial epizootics than those that have not, especially when exposure to unusually high temperatures is coupled with handling. It is generally a good idea to avoid handling aquatic animals during the summer insofar as possible. When it becomes necessary to handle animals during hot weather, they should be caught early in the morning, when the water has reached its coolest temperature of the day. Care should be taken to ensure that a high level of DO is maintained at all times. The animals should be handled gently and returned to the water as quickly as possible. If the animals are to be weighed, the operation should be accomplished in pre-weighed, water-filled containers; to avoid damage, animal density in the weighing containers should be as low as practicable.

OVERWINTERING

Most aquaculture is conducted in systems designed to operate under ambient water-temperature conditions. Outdoor aquaculture systems in temperate climates can vary in temperature from 4 °C to over 30 °C (from 39 °F to over 97 °F), though there have been a few attempts to alter the normal pattern; among them are the use of geothermal water and the use of heated effluents. Heating or chilling production units in the absence of free or very inexpensive sources of heat energy has typically been unviable economically in commercial growout systems, though both heating and cooling have been used in conjunction with research laboratories, in hatcheries, and when animals are being held over the winter.

Maintenance of certain minimum temperatures through artificial heating during the winter is often necessary for tropical species being reared in temperate climates. Various species of tilapia and certain species of penaeid and freshwater shrimp require supplemental heat during winter. Overwintering can be accomplished by maintenance of proper air temperatures in buildings used to house such species in static or recirculating systems. In flow-through systems, water heating can lead to significant and even staggering energy costs.

Overwintering is typically practiced in conjunction with the maintenance of broodstock, though young animals may also be held during winter for subsequent stocking and growout the following spring and summer. It is generally not necessary (or even desirable) to maintain the overwintering temperature at the optimum for the species involved, because the goal is to hold the animals, not have them grow. Thus, overwintering temperatures should be sufficiently low to slow the metabolic rate of the animals without stressing them and reducing their resistance to disease. For species such as tilapia, an overwintering temperature of around 20 °C (68 °F) may be appropriate. Fish at that temperature can be expected to eat a sufficient amount of food to maintain their body weight and can be expected to survive well.

The number of overwintered broodstock of species with high fecundity that spawn during spring need not be very high, so modest facilities can be used to house sufficient adults for fairly large growout operations. Species vary in their ability to accept crowding, but, for most species being reared commercially today, density limitations are usually not a factor as long as suitable water quality is maintained.

When a source of inexpensive heated water is not available, heated buildings, including greenhouses, are typically used for overwintering. Most commonly, the aquaculture animals are maintained in circular or linear raceways, though ponds covered by greenhouses may be practical in low-temperate regions where soil temperatures remain fairly high throughout the winter.

Stickney and Person (18) examined various ways of heating water before designing a heat exchanger used in conjunction with a recirculating water system. Such systems remove thermal energy from water that is of any initial temperature above freezing, thereby producing both a heat source that can be used to warm water and a source of cold water (the incoming water from which heat has been removed). Utilizing such a system could provide water suitable for the maintenance of both warmwater and coldwater species, if the temperature of the incoming water falls between the optima of the two types of animals.

BIBLIOGRAPHY

1. R.R. Stickney, *Principles of Aquaculture*, John Wiley & Sons, New York, 1994.
2. T.L. Wellborn and C.S. Tucker, in C.S. Tucker, ed., *Channel Catfish Culture*, Elsevier, New York, 1985, pp. 1–12.
3. R.W. Kilambi, J. Noble, and C.E. Hoffman, *Proceedings of the Southeastern Association of Game and Fish Commissioners* **24**, 519–531 (1992).
4. J.W. Andrews and R.R. Stickney, *Trans. Amer. Fish. Soc.* **101**, 94–99 (1972).
5. R.G. Piper, I.B. McElwain, L.E. Orme, J.P. McCraren, L.G. Fowler, and J.R. Leonard, *Fish Hatchery Management*, U.S. Fish and Wildlife Service, Washington, DC, 1982.
6. P. Chimits, *FAO Fish. Bull.* **10**, 1–24 (1957).
7. L.G. McBay, *Proceedings of the Southeastern Association and Game and Fish Commissioners* **15**, 208–218 (1961).
8. J.W. Avault, Jr. and E.W. Shell, *FAO Fish. Rep.* **44**, 237–242 (1968).
9. M.J. King, M.H. Kao, J.A. Brown, and G.L. Fletcher, *Bull. Aquaculture Assoc. Canada* **89**(3), 47–49 (1989).
10. G.P. Goff, D.A. Methven, and J.A. Brown, *Bull. Aquaculture Assoc. Canada* **89**(3), 53–55 (1989).
11. G.L. Fletcher, S.-J. Du, M.A. Shears, C.L. Hew, and P.L. Davies, *Bull. Aquaculture Assoc. Canada* **90**(1), 70–71 (1990).
12. K.V. Ewart and G.L. Fletcher, *Bull. Aquaculture Assoc. Canada* **89**(3), 25–27 (1989).
13. J.C. McGeer, L. Baranyi, and G.K. Iwama, *Canadian J. Fish. Aquaculture Sci.* **48**, 1761–1771 (1991).
14. T.D. Beacham and R.E. Withler, *Aquaculture Fish. Manage.* **22**, 125–133 (1991).
15. P.B. Shekk, N.I. Kulikova, and V.I. Rudenko, *J. Ichthyology* **30**, 132–147 (1990).
16. A.A. Khater and R.O. Smitherman, in R.S.V. Pullin, T. Bhukaswan, K. Tonguthai, and J.L. Maclean, eds., *The Second International Symposium on Tilapia in Aquaculture, ICLARM Conference Proceedings No. 15*, International Center for Living Aquatic Resources Management, Manila, Philippines, 1988, pp. 215–218.
17. A.V. Zale and R.W. Gregory, *Trans. Amer. Fish. Soc.* **118**, 718–720 (1989).
18. R.R. Stickney and N.K. Person, *Prog. Fish-Cult.* **47**, 71–73 (1985).

See also RECIRCULATING WATER SYSTEMS; SITE SELECTION.

TILAPIA CULTURE

ROBERT R. STICKNEY
Texas Sea Grant College Program
Bryan, Texas

OUTLINE

Tilapia are a group of fish species that are among the most popular fishes being cultured around the world today (Fig. 1). Native to Africa and the Middle East, tilapia have been introduced throughout the tropical world and are also cultured in temperate regions, where protection against winterkill from cold temperatures must be provided. Tilapia exhibit various characteristics that make them

Figure 1. A blue tilapia (*Oreochromis aureus*) adult and small fingerling.

a desirable species for culture, such as rapid growth, ease of reproduction, and tolerance for both crowding and relatively poor water quality conditions. Also, tilapia can be reared on somewhat inexpensive prepared feeds and are resistant to diseases, unless stressed.

DISTRIBUTION AND TAXONOMY

Tilapia are freshwater fishes in the family Cichlidae that are said to have evolved from a marine progenitor. Much of Africa is dominated by cichlids. Various genera can be found in virtually every ecological niche; tilapia are known primarily as a group that has species which are herbivorous or omnivorous. The fact that tilapia feed relatively low on the food chain is among their attributes as culture species.

As previously mentioned, in addition to being native to Africa, tilapia are found in the Middle East. The fish in the biblical story of Jesus feeding a mass of people on only several loaves and fish is often believed to have been tilapia. Various species of tilapia are native to Lake Kinneret (the Sea of Galilee) and are considered highly desirable foodfish in Israel.

The introduction of tilapia into parts of Asia apparently began in the 1930s when *Oreochromis mossambicus* were imported to Java as an aquarium fish (1). During World War II, the Japanese distributed tilapia more broadly throughout much of southeast Asia, and today tilapia are considered native in many nations in Asia. Tilapia were first introduced into Caribbean nations in the 1940s and subsequently found their way throughout much of Latin America and into the United States. By the late 1950s, tilapia had become the main subject of aquaculture research at Auburn University.

Tilapia are not only produced as foodfish. In California, one species, *Tilapia zillii*, an herbivore, was stocked in irrigation canals to control aquatic vegetation. Tilapia have also been introduced into sewage lagoons to consume vegetation. Some of the more colorful species of tilapia have been marketed as aquarium fishes by producers of ornamental fishes. Many other genera of cichlids are also popular aquarium fishes.

Some tilapia species and hybrids can tolerate moderate to very high salinities. In recent years, interest has developed in the rearing of salt-tolerant tilapia in coastal ponds and marine cages. Interest for rearing tilapia in saltwater continues to be strong in the Bahamas and some Caribbean nations.

The tilapia of interest to aquaculturists were once all classified as members of the genus *Tilapia*. However, that changed during the 1970s when a taxonomist examined the situation and determined that most species of interest to the fish culture community should actually be placed in the genus *Sarotherodon* (2). Most of the world accepted the change, but the American Fisheries Society, which publishes a list of accepted names of fish that can be found in North America, did not. By the early 1980s, the taxonomy was reconsidered once again, and it was concluded that many of the species that had been placed in the genus *Sarotherodon*—again, including the most popular aquaculture species—should actually be placed in the genus *Oreochromis* (3). Once more, the American Fisheries Society experts initially failed to go along with the change, but have recently accepted it. However, the common name "tilapia" continues to be applied, regardless of the taxonomic status of these fish. Much of the world quickly accepted the changes in taxonomy, resulting in a mixture of scientific names during the years when some organizations accepted the changes and others retained the earlier taxonomy.

The most commonly reared species of tilapia are the blue tilapia, Nile tilapia, and Mozambique tilapia. Their scientific names appear in the literature as follows:

Blue tilapia	*Tilapia aurea*
	Sarotherodon aureus
	O. aureus (Fig. 1)
Nile tilapia	*Tilapia nilotica*
	Sarotherodon niloticus
	Oreochromis niloticus
Mozambique Tilapia	*Tilapia mossambica*
	Sarotherodon mossambicus
	O. mossambicus

Since the genus *Oreochromis* is now widely accepted around the world for tilapia, it is used here when scientific names are provided.

Various hybrid red tilapia have been produced and go by such common names as the Taiwanese red hybrid, Florida red hybrid, and Israeli red hybrid tilapia. The hybrids are composed of two- or three-way crosses, which may or may not include *O. mossambicus*. Crosses that include *O. mossambicus* tend to be more salt tolerant than crosses that do not include a contribution from that species.

Many other species of tilapia exist, some of which have been evaluated or employed as aquaculture species. However, the major emphasis continues to be on the species mentioned in this contribution. The Mozambique tilapia was once a widely cultured species, particularly in Asia, since it was the first tilapia introduced to that continent. However, many culturists have turned to blue tilapia and Nile tilapia as the primary species of interest, due to their better dressout percentages, later maturity, and more desirable color (Mozambique tilapia have a large amount of dark pigment on their scales and a black peritoneum, which is seen as being undesirable). Mozambique tilapia continue to be reared extensively in Asia, though other species are gaining in popularity as well. Salt-tolerant red hybrids are cultured in coastal areas where they can be sold in markets competitively with red or silvery marine species.

Attributes of tilapia culture are as follows (4):

- tolerance to handling
- tolerance to crowding
- tolerance of poor water quality
- resistance to disease
- efficient conversion of natural and prepared feeds
- controllable reproduction
- good marketability
- rapid growth to marketable size.

CULTURE PRACTICES

Water Systems

In tropical regions, tilapia are most often reared commercially in earthen ponds (see the entry "Pond culture"). Ponds (Fig. 2) vary considerably in surface area, but most are no more than 2 m (6 ft) in depth and have slopes of 2 : 1 or 3 : 1. A well-designed tilapia pond should be equipped with a drain, to facilitate harvesting. Seines are also used for harvesting, but large numbers of tilapia are able to avoid capture by burrowing into the mud. After several passes of a seine through a partially drained pond, it is common practice to drain the pond completely and pick up the remaining fish by hand. This process can be protracted, however, since the fish may work their way to the surface and be subject to capture only after they have been buried for several hours. The fact that a fish which has been buried in the mud of a pond bottom for several hours will often recover with no apparent lasting effects after being placed in back in good-quality water is testimony to the heartiness of tilapia.

Tilapia can be reared in temperate ponds, but they should be harvested and marketed prior to the time that the water falls below about 20 °C (68 °F), when growth generally ceases. Broodstock need to be overwintered at temperatures well above about 12 °C (54 °F), or else they may die. When maintained at cool, but nonlethal, temperatures, tilapia become more susceptible to disease than when they are maintained at their optimum temperature, which is about 30 °C (86 °F). Pond culture in temperate areas can result in one crop a year, while two to three crops a year can be obtained in tropical regions.

Although there has been some intensive tank and raceway culture of tilapia in tropical regions (Fig. 3), such systems are much more common in temperate regions. Intensive culture (Figs. 3 and 4) systems may be the flow through or recirculating type (see the entry "Recirculating water systems"). If a source of heated water is available,

Figure 2. A tilapia pond in Jamaica, with the inflow line supplied by pumping water from an irrigation canal.

Figure 3. Flow-through tank culture of tilapia in Idaho. The water source is geothermal.

Figure 4. Flow-through raceway culture of tilapia in Idaho. The water source is geothermal.

Figure 5. Concrete-lined saltwater pond culture of red tilapia at an experimental facility in the Bahamas.

tilapia can be reared year round (Fig. 5). Depending on the temperature of the available water and the ability of the culturist to alter the temperature to maintain the system within a narrow range of fluctuation, the tanks may be placed outdoors, or they can be housed within a greenhouse or another structure.

In the southernmost parts of Texas and in much of Florida, tilapia will survive many, but not all, winters without supplemental heating of the water. In more northern regions, however, winter temperatures are lethal annually. Yet, tilapia have been reared in such unlikely places as the high desert of Idaho and the prairie of North Dakota, where winter temperatures routinely reach −34 °C (−30 °F).

Heating water can be a very expensive proposition, though aquaculturists have, in some instances, overwintered broodstock or even tried to grow out fingerlings in systems heated with oil, natural gas, or electric heating elements. The economics of using fossil fuels or electricity to heat water are highly unfavorable. Some inexpensive or, preferably, free source of heat is needed for economically successful commercial tilapia production in temperate regions.

In some regions, geothermal water is available from wells. While many geothermal wells are high in toxic chemicals, there are some hot aquifers containing very pure water that can be flowed directly into raceways or tanks containing tilapia. During portions or all of the year, depending on the temperature of the well water, it may be necessary to allow the water to stand in a pond for some period prior to exposing it to the fish, or to mix the incoming water with colder water to reduce the temperature. On the other hand, if the ambient geothermal water is at or near the optimum temperature for tilapia when it comes out of the ground, it may be necessary to provide supplemental heat during cold periods.

Artesian geothermal wells provide the best option, because no pumping is required when flow rates are sufficiently high. Some producers have to pump all of their geothermal water when there is no artesian flow, while others who have an artesian flow, but insufficient volume to supply their needs, may employ pumps to augment the flow.

A second option is to use warm wastewater from a power plant or industrial source. Coal, oil, natural gas, and nuclear power plants all produce large quantities of heated water as a result of electrical production. Many industries also produce copious amounts of hot water. In many cases, the warm water cannot be used directly, because it contains supersaturated gas levels (see the entry "Gas bubble disease") or toxins that have been added to prevent fouling within the facility producing the heated water. For example, power plants often flush condenser pipes with chlorine to remove fouling organisms. However, exposing fish to chlorinated water can result in total mortality.

Hot water from nearly any source, including geothermal water that may not be of a quality suitable for direct exposure to fish, can be passed through heat exchangers, which transmit the heat to the water in culture tanks. In some cases, power plants or other industries with large volumes of available warm water have been willing to work with aquaculturists in providing free or low-cost access to that water.

Reproduction and Hatching

While tilapia have a number of positive attributes that, altogether, are not shared by many aquaculture species (4), there are problems associated with their culture. One of those problems is early maturity, accompanied by slow growth of females and overpopulation of culture systems with submarketable fish. Mozambique tilapia, for example, mature as early as three months after hatching. Nile and blue tilapia typically do not mature until about six months of age. In both cases, maturity occurs well in advance of a size that can be sold in most markets. [Exceptions to the problem of marketing small, mature fish occur in nations where fish of 100 g (0.22 lb) or less can be sold].

The most common species of tilapia currently being cultured around the world are mouthbrooders. The males construct nests in pond bottoms (Fig. 6) and then go through courtship display behavior to attract a female. As the female extrudes her eggs into the nest, the males fertilizes them. The female then picks the eggs up in her mouth and retires, allowing the male to seek out additional mates.

Figure 6. Tilapia nests in a drained spawning pond.

Incubation of the eggs takes place in the mouth of the female and requires about one week. The fry remain in the mouth of the female through yolk sac absorption. When they begin foraging for food, the fry remain in a school near the female and will move back into her mouth if they sense danger. Thus, for two weeks or more, the female does not eat, since her mouth and buccal cavity are filled with eggs or fry.

A typical female can spawn as often as every 30 days, with the spawning season being virtually year round in the tropics, though studies have shown that a female usually does not spawn more than about eight times a year (5). Because of the high frequency of spawning, which requires that a considerable amount of the energy consumed by females go into egg development, and due to the fact that the female does not eat for two weeks a month during the protracted spawning season, growth of females is much slower than that of males.

Tilapia produce relatively small numbers of eggs — usually not more than a few hundred — during each spawning episode. Because of the high frequency of spawning, a brood pond stocked with a few hundred females per ha (2.4 ac) will, during the course of a spawning season, produce hundreds of thousands of fry and fingerlings. If the fry or small fingerlings are not removed periodically, competition for food can lead to stunting, even though, as the spawning season progresses, the number of fry observed decreases, apparently due to cannibalism by fingerlings. Because it is virtually impossible to collect all of the young fish without draining the pond, significant numbers of fingerlings are often found at the end of the growout period, even when efforts are made to remove fry or fingerlings periodically.

Tilapia fry will school for several days after leaving the protection of the female. The schools of fry tend to remain near the edge of the pond, at least in temperate areas, apparently because the water is warmest along the pond's margin. Schools can be dipped out with a fine-mesh seine for transfer to rearing ponds. Alternatively, entire brood ponds can be seined periodically to remove fingerlings.

Tilapia are not always reproduced in open ponds, even in developing countries. In some countries, small, very fine-mesh cages are used for spawning and fry rearing. These cages, called "hapas," allow the fish farmer easily to capture fry, since they are confined within a small volume of water (Fig. 7).

Tilapia can be stripped of their eggs and milt for egg incubation in hatching jars, or the eggs can be removed from the mouth of the female and hatched. Some hatcheries use one or both techniques, but natural spawning is the most commonly used method.

Several methods have been developed to avoid the problem of early spawning and overpopulation of stunted fish in growout ponds (5). One method involves hand-sexing, wherein trained individuals examine each fish individually, retain the males for stocking, and discard the females. The fish need to be about 60 mm (2.5 in.) long and near maturity before differences in the genital region of the two sexes can be detected visually. Mistakes are bound to occur, either in making the sex determination or in placing a given fish into the wrong container following sexing. This method is also very labor intensive and time consuming.

Figure 7. Hapas are often used in the Philippines and other countries to spawn tilapia and for early rearing of fry.

Various species of predatory fish have been stocked in tilapia ponds to consume fry, so that the ponds do not become over populated. The predator fish should not become large enough to consume the tilapia that are intended for growout, of course. While this method can be fairly efficient if the proper number of predators is stocked, it does not eliminate the stunting of females' growth, however, since the females are still allowed to spawn and mouthbrood.

A third method of controlling reproduction has been to rear tilapia in cages placed in ponds. The theory behind this technique is that when the females spawn, the eggs will fall through the bottom of the cages, thereby eliminating the two or more weeks of nonfeeding that would occur if the females were mouthbrooding, as well as effectively eliminating fry and fingerling production. The method apparently works to some extent, but some successful spawning has occurred when females are able to gather fertilized eggs before the eggs can fall from the cage.

Some hybrid crosses are reported to result in all-male or nearly all-male offspring. There are some problems associated with using hybridization to produce all-male populations, however. First, since many species of tilapia freely hybridize and produce fertile offspring, a large percentage of the fish in culture today are undoubtedly cross-bred to some extent. While they may be called *O. niloticus* or *O. aureus*, they may, in reality, have some contributions from other species in their genomes. Thus, fish that have been identified as belonging to both species *A* and *B* might produce all-male populations on one farm when hybridized, but fish thought to be of the same two species on another farm may not produce a ratio of males to females substantially different from 1:1. Secondly, it is virtually impossible to avoid mixing the two species on a fish farm. No matter how much care is taken, fish will inevitably be placed in the wrong pond or other holding facility. Extensive genetic testing to retain pure species is possible, but is prohibitively expensive.

Sex reversal has become a widely used method of producing all-male tilapia populations. Various hormones have been used by fish culturists to produce sex-reversed tilapia, with some form of methyl or ethyl testosterone being the most common. The standard technique is to dissolve the hormone (commonly used hormones have been synthesized and are readily available) in alcohol. The alcohol is then poured over prepared feed and allowed to evaporate, resulting in the feed being coated with small amounts of the hormone. [Methyl testosterone works well with 30 to 60 mg/kg of feed (30 to 60 ppm).] Immersion in hormone solutions has also been used for sex reversal (6). First-feeding fry are provided with the treated feed, usually for about three weeks. If the process is done properly, 95 to 100% male populations can be produced.

The most recent development involves the production of YY males—that is, males with two Y chromosomes (diploid males are normally XY). The process involves exposing normal male tilapia fry to estrogens (female hormones) to produce females that are XY instead of XX (3). When feminized males are bred with normal XY males, the offspring produced will have the following genotypes: 1 XX female: 2 XY males: 1 YY male. Mating YY males with normal XY females will result in all-XY-male offspring. Producing YY males requires extreme care in identification of the genotype of each fish produced from the pairing of feminized males with normal males.

Water Quality Requirements

Tilapia are generally resistant to degraded water quality. That resistance is one of the characteristics that make tilapia popular aquaculture fish. The basic water quality requirements of tilapia have been known for at least a few decades (7).

Temperature is a critical water quality variable for all aquatic animals. As previously indicated, tilapia growth ceases when the water temperature drops below about 20 °C (68 °F), and most species die at temperatures between 8 and 12 °C (46 and 54 °F). Optimum growth occurs at about 30 °C (86 °F), and the upper lethal limit may exceed 40 °C (104 °F) for at least some species. As the upper temperature limit is approached, feeding activity will decline and eventually cease. In most pond culture, maximum temperatures, at least in deep water, where a thermocline may form, will not exceed about 34 °C (93 °F), and tilapia will continue to feed and grow.

Tilapia are unusually tolerant of low levels of dissolved oxygen. When the supply of dissolved oxygen becomes depleted, tilapia are able to surface and skim the water surface, to run the upper microlayer of water, which is saturated with oxygen by diffusion, across their gills. However, long periods of exposure to low levels of dissolved oxygen, while possibly not lethal, will stress tilapia, and the fish may not feed during periods when dissolved-oxygen levels are very low. In many instances, oxygen depletions occur near dawn, as respiration causes the level of dissolved oxygen to drop throughout the night. Once photosynthesis begins in the morning, oxygen levels will rise and may even reach supersaturation by afternoon.

Most tilapia hatcheries, even those that produce fish for rearing in saline waters, use freshwater or water of very low salinity, although *O. mossambicus* has been known to spawn over a wide range of salinities (8), including those somewhat higher than full-strength seawater (35 ppt). Fry may not be tolerant of saltwater, but fingerlings of many species can be either directly transferred or acclimated to saltwater, and some can even tolerate hypersaline conditions. Some strains of red tilapia are salt tolerant and may even reproduce in saltwater (9).

Tilapia have a high tolerance for turbidity, grow well over hardness and alkalinity ranges commonly observed in ponds, and can tolerate relatively high levels of ammonia.

Nutrition and Feeding

Most species of tilapia are omnivores. First-feeding stages typically consume zooplankton, but may eventually convert to plant-only diets or mixtures of plant and animal diets. Tilapia have the ability to digest carbohydrates, including starches, so they can be fed diets low in, or even devoid of, animal protein and still grow rapidly. Tilapia will accept prepared feeds from first feeding, provided that the size of the particles is appropriate. In areas where the ingredients are plentiful and can be used economically, tilapia are often fed prepared feeds containing some animal protein (e.g., fishmeal), along with such items as corn meal, wheat meal, soybean meal, and cottonseed meal. Vitamin and mineral supplements may also be provided. Because of the ability of tilapia to use vegetable protein efficiently, prepared feeds tend to be considerably less expensive than the feeds used for carnivorous fishes such as salmonids.

In developing countries, a variety of alternative ingredients have been evaluated, some of which are in use. A partial list of nontraditional tilapia feed ingredients is provided in Table 1. In developing countries, prepared feeds routinely contain rice bran, which has a low protein content and is high in indigestible fiber.

Since tilapia have a general tolerance for poor water quality conditions, they are often produced in ponds that receive heavy inputs of organic fertilizers from such terrestrial animal sources as ducks, chickens, swine, and cattle. The technique of fertilization as a means of promoting natural algae and zooplankton production is widely practiced

Table 1. Nontraditional Dietary Ingredients that are Used or have been Evaluated for Use in Prepared Tilapia Feeds[a]

Alfalfa	Algae
Azolla meal	Brewery waste
Cassava	Cocoa cake
Coffee pulp	Copra meal
Cowpeas	Cracked rice
Duckweed	Groundnut
Leucaena leaf meal	Lettuce
Linseed meal	Macadamia presscake
Mango seed	Mustard oil cake
Palm kernal meal	Rapeseed (canola) meal, cake, or oil
Salt bush	Sesame meal
Sesbania	Silage grass clippings
Soldier fly larvae	Sugarcane bagasse
Sweet lupin	Sweet-potato meal
Sunflower seed meal	Tapioca

[a]From (10).

in developing countries where the ingredients required to produce good-quality formulated feeds may be unavailable or are prohibitively expensive.

Specific nutritional information has been developed for various species of tilapia; most requirements are similar among the species investigated thus far. A summary of the general nutritional requirements of tilapia is provided in Table 2. The actual lipid and carbohydrate requirements of tilapia have not been determined in detail. Research indicates that at least some species seem to require fatty acids in the linoleic (n-6) and linolenic (n-3) acid families (see the entry "Lipids and fatty acids"). Tilapia seem highly capable of digesting simple sugars and starches.

As with other fish, fry tilapia are fed to excess, perhaps as much as 60% of their body weight daily. The feed should be evenly distributed over the culture chamber, to ensure that the fish will not have to swim long distances in search of food particles. Juvenile tilapia are typically fed 3 to 4% of their body weight daily. The number of daily feedings employed by tilapia farmers is highly variable, ranging from one to several times daily. The most common feeding schedule involves two feedings daily. Demand feeders may also be used, which allow the fish to feed *ad libitum*.

Table 2. General Nutritional Requirements of Tilapia[a]

Nutritional Factor	Requirement
Crude Protein Requirements (% of dry diet)	
Fry	35–40
Fingerlings	25–40
Amino Acid Requirements (% of dry diet)	
Arginine	1.13–1.18
Histidine	0.43–0.48
Isoleucine	0.80–0.87
Leucine	0.95–1.35
Lysine	1.43–1.51
Methionine	0.40–0.75[b]
Phenylalanine	1.00–1.05[b]
Threonine	1.05–1.17
Trypophan	0.28
Valine	0.78–0.88
Vitamin Requirements (mg/kg[c] of diet, unless otherwise indicated)	
Choline	None[a]
Pantothenic acid	6–10
Riboflavin	5–6
Vitamin B_{12}	None[d]
Vitamin C	40–125[e]
Vitamin D	375 I.U.[f]
Vitamin E	25–100
Mineral Requirements (% of dry diet)	
Calcium	0.70
Manganese	0.90
Phosphorus	0.45–0.50
Zinc	0.20–0.30

[a]From (10).
[b]The requirement is variable, depending on the level of cystine in the diet.
[c]mg/kg = part per million (ppm).
[d]Research has been conducted, but no requirement level could be established.
[e]The range may relate to the form in which the vitamin was incorporated into the diet, as well as to the method of diet manufacture.
[f]I.U. = international units.

DISEASES

In many culture situations, tilapia are remarkably disease free, as compared with various other aquaculture species. However, A notable exception occurs when tilapia are stressed. In temperate ponds, reduced water temperature is the most common stressor. Tilapia populations that may have shown no signs of disease for months will develop a variety of problems when the water temperature drops in the fall. The high-density conditions that exist in closed systems, which may also involve long periods of exposure of the fish to somewhat degraded water quality, have also resulted in disease outbreaks.

No viral diseases have been reported from tilapia reared in the Americas (11), but various other diseases have occurred. In other parts of the world, there have been a few reports of viral diseases associated with tilapia (12). Viruses observed to date include *Lymphocystis* sp., (a marine virus), a birna virus, and a rickettsial virus.

Numerous occurrences of diseases, due to fungi and crustacean parasites have been reported in cultured tilapia around the world (11,12). The fungus *Saprolegnia* sp., has been known to attack the skin and fins of tilapia that have been damaged during handling. That fungus is restricted to freshwater. Marine counterparts to *Saprolegnia* exist, but have not been reported to attack tilapia reared in saltwater. Other fungi reported include *Achya* sp., *Aphanomyces* sp., *Aspergillus* sp., and *Dictyuchus* sp. Crustacean parasites that have infected tilapia include *Argulus* sp., *Caligus* sp., *Dolops ranarum, Ergasilus* sp., *Lamproglena* sp., and *Lernaea* sp.

Bacterial infections of tilapia include *Aeromonas* sp., *Pseudomonas* sp., *Vibrio* sp., *Edwardsiella tarda*, *Flavobacterium columnare* (previously known as *Flexibacter columnaris*), *Mycobacterium* sp., and *Streptococcus* sp. Protozoan parasites that attack the gills and skin of tilapia include *Amyloodinium* sp., *Chilodonella* sp., *Epistylis* sp., *Ithyobodo* sp., *Ichthyophthirius multifiliis*, *Trichodina* sp., and *Tripartiella* sp. Internal protozoan infestations include *Eimeria* sp., *Myxobolus* sp., *Myxosoma* sp., and *Trypanosoma* sp.

The monogenetic trematodes *Cichlidogyrus* sp., *Dactylogyrus* sp., *Enterogyus cichlidonum*, and *Gyrodactylus* sp., have also been reported in cultured tilapia. In addition to those freshwater species, a marine monogenean, *Neobenedenia melleni*, has been reported in tilapia held in marine cages. At least one digenetic trematode ectoparasite, *Transversotrema* sp., has been reported in tilapia as well. Monogenetic and digenetic endoparasites have also been reported.

Treatment chemicals that have been effective with regard to the control of tilapia diseases include the following (12): acriflavin, to control bacterial infections of eggs; formalin, to control parasites and external monogenetic parasites; furanace, to control bacteria; malachite green, as a fungicide on eggs and a protozoan parasiticide when mixed with formalin; masoten, as a parasiticide; nitrofuran, oxolinic acid, and potassium permanganate, as bacteriocides; and salt (sodium chloride), to control bacteria, fungi, and parasites. Either each chemical is added to the culture water or the fish are exposed by dip or bath treatment. The antibiotics oxytetracycline and sulfamerazine have been incorporated in tilapia feed to control bacterial infections. While clearance for some of the aforementioned drugs and chemicals is still being sought, there are currently no therapeutic agents approved for use on tilapia in the United States.

MARKETING

Depending on the particular market, tilapia may be sold live, in the round, gutted, headed and gutted, or filleted. In tropical areas, tilapia tend to be sold in the round. Shoppers in developing nations, who often do not have refrigerators, will purchase and consume the fish the same day. Live tilapia and tilapia in the round on ice are sold in various ethnic markets in the United States and Canada. Live tilapia are, for example, often seen in Asian restaurants, where customers are able to select from an aquarium display the fish that they want to be prepared for their meal.

Headed and gutted tilapia for home consumption are popular in some markets. They may be sold fresh on ice or frozen. In the United States, shoppers often prefer filleted fish, so that they do not have to deal with scales, skin, and bones. However, the dressout percentage for tilapia fillets is not particularly high (approximately 33% for 500-g (1.1-lb) fish, slightly more for larger fish and fish that are carefully hand filleted, and perhaps no more than 25% for smaller fish). Fillets are also considerably more expensive than other forms in which tilapia are marketed, and fillets of other species may be less expensive, so marketing fillets can be a problem.

Tilapia are marketed as tilapia, St. Peter's fish, and under various other names. The name "tilapia" is becoming increasingly familiar to the buying public in the developed world. In the United States, the modest level of tilapia production that currently exists [in excess of 7,700 metric tons (7,700 short tons) in 1997] is augmented with high levels of imports (28,000 tons in 1997). The highest percentage of imports is from Taiwan. Other nations from which significant amounts of tilapia are imported are Costa Rica, Ecuador, Indonesia, Columbia, Jamaica, and Thailand (13).

REGULATIONS

In the United States and some Latin American nations, the exotic status of tilapia continues to be recognized and has led to regulations aimed at controlling the spread of the fish into the wild. Importation by and movement of tilapia within various nations are controlled through the regulatory process. Some species are allowed to be imported or moved, while others are restricted. Tilapia have been reared in most, if not all, of the states of the United States, as well as in Canada.

In states where the growing season is insufficient to produce a crop before winterkill occurs, production of tilapia is confined to greenhouses and other types of buildings where intensive culture is practiced and supplemental heat is available. Intensive culture, often in recirculating systems, is also practiced in areas where regulations prohibit culture in systems where the fish can escape (or be carried from one place to another by birds or other animals) and thereby invade public waters.

The following states have some type of restriction on tilapia culture: Arizona, California, Colorado, Florida, Hawaii, Illinois, Louisiana, Missouri, Nevada, and Texas. In those states, a permit may be required to culture tilapia, or tilapia can be reared only if the species of interest appears on a list of approved fishes. Before obtaining tilapia, prospective tilapia culturists should check with the appropriate government agencies to determine which, if any, permits are required.

BIBLIOGRAPHY

1. W.R. Courtenay, Jr., in B.A. Costa-Pierce and J.E. Rakocy, eds., *Tilapia Aquaculture in the Americas*, World Aquaculture Society, Baton Rouge, LA, 1997, pp. 18–33.
2. E. Trewavas, in R.S.V. Pullin and R.H. Lowe-McConnell, eds., *The Biology and Culture of Tilapia*, International Center for Living Aquatic Resources Management, Manila, 1982, pp. 3–13.
3. L. Fishelson and Z. Yaron, eds., *International Symposium on Tilapia in Aquaculture*, Tel Aviv University, Tel Aviv, 1983.
4. H.S. Egna and C.E. Boyd, eds., *Dynamics of Pond Aquaculture*, CRC Press, Boca Raton, FL, 1997.
5. R.R. Stickney, *Principles of Aquaculture*, John Wiley & Sons, New York, 1994.
6. B.W. Green, K.L. Veverica, and M.S. Fitzpatrick, in H.S. Egna and C.E. Boyd, eds., *Dynamics of Pond Aquaculture*, CRC Press, Boca Raton, FL, 1997, pp. 215–243.
7. D.D. Balarin and J.P. Hatton, *Tilapia: A Guide to their Biology and Culture in Africa*, University of Stirling, Scotland, 1979.
8. R.R. Stickney, *Prog. Fish-Cult.* **48**, 161–167 (1986).
9. W.O. Watanabe, B.L. Olla, R.I. Wicklund, and W.D. Head, in B.A. Costa-Pierce and J.E. Rakocy, eds., *Tilapia Aquaculture in the Americas*, World Aquaculture Society, Baton Rouge, LA, 1997, pp. 55–141.
10. R.R. Stickney, in B.A. Costa-Pierce and J.E. Rakocy, eds., *Tilapia Aquaculture in the Americas*, World Aquaculture Society, Baton Rouge, LA, 1997, pp. 34–54.
11. J.A. Plumb, in B.A. Costa-Pierce and J.E. Rakocy, eds., *Tilapia Aquaculture in the Americas*, World Aquaculture Society, Baton Rouge, LA, 1997, pp. 212–228.
12. K. Tonguthai and S. Chinabut, in H.S. Egna and C.E. Boyd, eds., *Dynamics of Pond Aquaculture*, CRC Press, Boca Raton, FL, 1997, pp. 263–287.
13. Anonymous, *Aquaculture Magazine*, September/October: 6–12 (1997).

See also EXOTIC INTRODUCTIONS.

TURBOT CULTURE

Joe McElwee
Galway, Ireland

OUTLINE

Fertilization
Diseases/Health
Future Prospects

Turbot are large flatfish, circular in shape, with both eyes on the upper side of the head, and a large fan shaped flattened tail. They are dispersed throughout most parts of the seas that surround Europe, including the Caspian and Black seas. They are normally dark brown to black on the dorsal side and white on the underside. Their flesh is white and succulent in taste. They also have small lumps on the dorsal side, termed tubercules, which are believed to act as sensory organs, when they are buried in the sand. Turbot tend to habitate in up to 30 m (90 ft) of water and tend to meander between that depth and closer to the shore, depending on migratory and food patterns.

FERTILIZATION

The sexually mature fish tend to come into the shore to spawn around August and September as water temperature increases to circa 14–16 °C (57–61 °F). It is not uncommon for the dorsal side to change in color from dark to light brown, depending on the surrounding bottom habitat.

The females can produce up to three million eggs over a spawning period of 5–8 weeks and are, for the most part, larger in size than their male counterparts, with both male and female reaching sexual maturity in about four to five years, weighing from 4 kg (8.8 lb) upwards and averaging 3–4 kg (6.6–8.8 lb) in size at spawning. A female may spawn several times during this period, with up to 500,000 eggs per spawn. Fertilization is external, and, accordingly, the rate of success is variable, depending on prevailing tides and water currents. The fertilized eggs are suspended in the water column, and after consuming their yolk sac in the first two days, will start to seek out their food, which consists of *Artemia* and rotifers. Juvenile turbot are born bilaterally symmetrical with eyes on either side of the head, similar to salmonids, and will start to metamorphose circa day 25–30, to the flattened shape, with the left eye migrating to the upper side. Turbot will also start to descend to the sea-floor to assume their new existence and adjust to their new surroundings, in the sandy gravel benthos. As with any small living entity, the juveniles are prone to predation by larger demersal fish until they metamorphose properly, and can outmanouver predators in the open deeper waters. There may also be increased mortality levels, due to starvation and swimbladder deformities.

Since turbot feed on the bottom, they have adapted mouths quite large in proportion to the head size. In the wild, they can grow in size to 30 kg (66 lb) or more. However, due to consistent overfishing, it's rare to catch one of that size. In relation to their body size, their gut cavity is relatively small and is located on the dorsal side, to the right of the head, as one would look at the fish from head on.

DISEASES/HEALTH

Like any animal living in natural or cultured conditions, turbot are susceptible to parasites (both internal and external) and diseases caused by both bacteria and viruses. There are a number of bacteria and viruses naturally present in the water, which, if the health of the fish is compromised, will take a quick and often deadly hold. Some bacterial problems will naturally dissipate through time, or if the fish's own immune is strong enough to fight it off. These will tend to occur through the lifespan of the fish as it forages for food or sustains flesh damage from skirmishes with potential predators, or in-house fighting. The turbot's own immune system is strong enough to generally fight off most bacterial problems. In intensive culture systems, the constant risk of infection is heightened, due to the numbers being stocked in such close proximity. The following are some of the bacterial affiliations common to turbot:

Vibrio's spp.
Pseduotuberculosis sp.
Furunculosis sp.
Streptococcus septicimic disease
Alteromonas sp.
Aeromonas salmonicida
Flexibacter sp.

Common treatments for bacterial problems in cultured regimes are the use of varying antibiotics in feed and immersion or bath treatments, again with an antibiotic chemical.

Viruses are naturally harder to fend off and can result in dramatic health problems and rapid mortalities in both wild and farmed stocks. While after time, fish in the wild may develop antibodies and a resistance. In intensive systems, vaccines are actively employed as a preventative measure. These are in constant development, due to the nature of viruses. Transmission of these viruses can be from direct contact with already infected stock or from other fish species. The following are some of the more prevalent viruses:

(1) Viral Heamoraghic Septicemia (VHS)
(2) Iridovirus/Rodovirus
(3) Rickettsial-Epitheliocystsis
(4) Paramyxovirus

There are a few others, but these are the major dangerous viruses that turbot contend with at present, and vaccines are being constantly being adapted to combat newer strands of viruses.

Turbot may also be afflicted by a number of other health problems that can be more attributed to environmental and water quality problems and thus may arise commonly in farmed systems.

External parasites on the gills and skin can cause not only unslightly sores, but also give rise to bacterial and viral problems relatively quickly if not treated. Some of the main parasites that cause problems for turbot are the following:

Protozoa:
- Ciliates
  - *Trichodina* spp.
  - *Cryptocaryon* spp.
  - *Scyphydia* spp.

Flagellates:
- *Ichthyobodo necator*
- *Costia* sp.

Microsporidia:
- *Tetramiera brevifilum*

Trichodina, a ciliate parasite, can settle in the gill lamellae or the skin surfaces. It can create sores if left untreated and respiratory problems within the gills. A variety of myxobacteria can also cause problems, and *Flexibacter* can lead to vibriosis and other serious ailments. These, for the most part, are easily treated and show evident physical damage prior to serious stages. Most are treated with tetracycline or quinolones (oxolinic acid and flumequine) and nitrofuranes and sulfonamides, by treatment in one-hour baths or inclusion in the feed.

Future Prospects

Turbot is a high priced delicacy in Europe at present and enjoys a niche market. It is cultured in a number of European countries in land-based pump-ashore systems, many using expensive filtration and heating facilities. Over the past few years, technology and biological advances have enabled hatcheries to successfully produce large numbers of juveniles, which are subsequently transported to grow out facilities. There is extensive detailed work being done with regard to genetic's and subsequent follow-up work with the sibling generations being reared.

There is a future for turbot farming, and as the market requires and demands more diverse and nutritionally healthy products, the rearing of different species should occur. As the fish species in the world are constantly being overfished, there is a huge opportunity in this area of the aquaculture industry.

VACCINES

Larry A. Hanson
Mississippi State University
Mississippi State, Mississippi

OUTLINE

Vaccines are useful tools for increasing efficiency of fish production. The basic purpose of a vaccine is to enhance the immune system's ability to recognize a disease-causing infectious agent and therefore reduce the severity of a disease. Optimally, this results in better growth, reduced use of therapeutics, and decreased losses. One must consider the costs of administering the vaccine, its effectiveness, its potential side effects, and regulatory issues before the vaccine is developed for an industry and made commercially available. The aquaculturist must consider the same issues, since they relate to the unique characteristics of the specific production facility. For some diseases, the aquaculturist has several options as to the vaccine or vaccination methodology. There are five groups of vaccines: killed, subunit, live, genetic, and inducers of nonspecific defenses. Vaccines can be administered by injection, immersion, or incorporation into the feed. The focus of vaccine development is to maximize the immune response to the pathogen, while minimizing cost and negative side effects such as handling stress. Factors that influence vaccine effectiveness include the immune status of the fish; the quantity, form, and persistence of the antigen; and the route of antigen expression/uptake in the fish. Because of potential benefits and the complex characteristics of vaccines used in aquaculture, there have been a large research focus and several review articles and books written on fish vaccinology (1–5).

KILLED VACCINES

Killed vaccines are crude antigen preparations derived from the disease-causing agent. They are processed in such a way as to kill or inactivate the agent. This is the most common type of vaccine in aquaculture. This type of vaccine contains most of the antigens produced by a pathogen, and, at least for bacterial pathogens, is relatively inexpensive to produce. The pathogens are generally inactivated by using heat or a chemical, such as formalin. Because the agent is killed, storage of the vaccine and safety of the vaccine are generally of less concern than they are for live vaccines. Safety concerns for killed vaccines usually involve residual activity of toxins produced by a bacterial agent. The susceptibility of fish to bacterial toxins varies with species and age of the fish. In some killed vaccines, the protective antigen may not persist in the fish at a high enough level to induce a strong response. This may result in shorter duration or lower levels of immunity than those provided by live vaccines. This lower-level response can be enhanced by providing multiple exposures or by incorporating adjuvants into the vaccine. Adjuvants are substances that enhance the immune response of the fish to the antigens in the vaccine. These substances can change the physical characteristics of the vaccine, or influence the immune system directly, to improve the ability of the fish to respond to the vaccine (6). Another concern in the use of killed vaccines is the potential for the culture methods or the methods used to inactivate the agent to alter the form or quantity of a critical antigen, and therefore reduce vaccine effectiveness. Also, if the agent is an intracellular pathogen (like a virus), the killed vaccine will not induce as strong a cell-mediated immune response as would a live agent. Successful commercially available killed vaccines include vaccines for enteric red mouth disease for salmonids (7), furunculosis for salmon (8), and mono- and multivalent *Vibrio* vaccines for salmon and other marine fish (9). An inactivated vaccine for spring viremia of carp has been commercially produced (10) but is no longer available (11). Research has demonstrated the effectiveness of killed vaccines for enteric septicemia of catfish (ESC). Two ESC vaccines have been evaluated for commercial production: one conditionally licensed, the other fully licensed by the U.S. Department of Agriculture. Neither product is commercially available at this time (12). *Photobacterium damselae* bacterins for marine fish show promising results (13).

SUBUNIT VACCINES

Subunit vaccines are purified antigen preparations that consist of a single protein or portion of a protein that can induce a protective immune response in fish. Subunit vaccines have the advantage of focusing the immune response of the fish toward the specific antigen. They also allow the mass production of the specific antigen by chemical synthesis or biological expression systems. These characteristics are especially important in the establishment of practical vaccination systems against infectious organisms that are expensive to produce, such as viruses and parasites. Also, because the fish is not exposed to the entire array of antigens that the pathogen expresses, the serologic characteristic of a fish exposed to a subunit vaccine would be different from that of a fish exposed to the pathogen. This allows for the differentiation of vaccinates from potential carrier fish in serologic assays. Subunit vaccines have the same constraints of killed vaccines; i.e., dosage, duration, and inefficiency in inducing cell-mediated immune responses. No subunit vaccines are commercially available for aquaculture at this time, but promising research indicates that subunit vaccines against several viruses will be effective. The A segment encoding VP2, VP3, and NS from infectious pancreatic necrosis (IPN) virus was expressed in the bacterium *Escherichia coli*, and lysates from this expression system that were used in immersion exposures also protected against IPN (14). Recent field trials have been performed on Atlantic salmon (*Salmo salar*) using injection vaccines containing *E. coli* expressed VP2 with encouraging results (15). Subunit infectious hematopoietic necrosis (IHN) vaccines have been produced from the viral glycoprotein expressed in *E. coli* that induce protection against IHN after immersion vaccination (16,17). This subunit vaccine has shown promising results in field trials (1). Also, maternal transfer of antibodies was demonstrated from broodfish that were given injection vaccinations of a similar subunit vaccine; these results indicate some protection of fry up to 25 days posthatch (18). Injection-administered subunit vaccines based on the glycoprotein of viral hemorrhagic septicemia (VHS) virus have been effective against VHS in laboratory trials (19).

LIVE VACCINES

Live vaccines are generally mutated strains of an infectious agent that have a reduced ability to cause disease and are referred to as attenuated. During vaccination, fish are infected with these attenuated agents and develop protective immunity that will provide protection when the fish are exposed to the virulent disease-causing agent. These vaccine agents can be developed by inducing mutations in the disease-causing agent by growing it in an unnatural host, exposing it to a mutagen, or using genetic engineering to inactivate certain genes that are involved in inducing disease. In some cases, a naturally occurring nonvirulent organism related to the disease-causing agent can be identified which will infect the fish and induce a protective response to the disease-causing pathogen. The effects of live vaccines are based on the ability of the agent to actively infect and replicate in the fish. This ability provides the persistence and dosage needed to induce a strong immune response even when using inefficient application methods. Additionally, the mode of infection is the same as in the disease-causing agent, allowing the induction of focused local immunity at critical locations and the induction of cell-mediated specific immunity in the case of intracellular pathogens. Disadvantages of live vaccines are related to storage and safety. Because the vaccine agent must be kept alive, storage conditions must be critically controlled. Improper storage, preparation, or application can result in vaccine failure. Safety concerns are related to the effect of the infectious vaccine agent on the physiology of the fish. Attenuated agents occasionally cause severe disease in an immunosuppressed host, and they can cause mild disease and temporary immune suppression during the initial stages of the infection. This may lead to reduced feeding activity, reduced growth rate, and a predisposition to other diseases. Because of potential side effects from live vaccines, timing can be critical for the most effective and safest administration of these vaccines. Another factor that must be considered in the administration of attenuted agents that are derived from uncharacterized mutations is the ability of some attenuated agents to mutate back (revert) to a more virulent form in the host. Genetically engineered, live attenuated agents are generally designed so that reversion cannot occur. Another type of vaccine that shows promise for aquaculture is the vaccine vector. In this system, an attenuated, infectious agent is genetically manipulated so that it expresses protective antigens from other pathogens. In this type of system, one application of vaccine could provide protection against several pathogens. No attenuated agents are currently licensed for use in aquaculture, but laboratory studies have demonstrated effective vaccinations with attenuated live vaccines for furunculosis in salmonids (20), ESC (21,22), IHN (23), VHS (24), spring viremia of carp (10), and channel catfish virus disease (25,26). Vaccine vector potential has been demonstrated for attenuated *Aeromonas salmonicida* (27) and channel catfish virus (28).

GENETIC VACCINES

In 1990, it was shown that genes in purified DNA could be expressed when injected into the muscle of mammals (29) and that this expression could be used to induce an immune response to the gene products (30). Since then, research has been performed on a variety of fish and has shown that this methodology can be used in fish as well (31–33). The DNA injection method of vaccination induces an immune response to a specific protein or proteins encoded by the injected DNA and can induce a cell-mediated response. At present, vaccines of this type are not used in aquaculture, but they are used as tools for vaccine and immunology research.

INDUCERS OF NONSPECIFIC DEFENSES

When a fish is exposed to a pathogen, the focus of its physiology changes from maintenance and growth to defending itself from disease. Certain infectious agents or components of pathogens strongly induce this response, and these agents can be used to heighten the nonspecific defenses of fish at critical times. Cell wall components of fungus, gram negative bacteria, and acid fast bacteria have all been shown to induce nonspecific defenses and have been used to strengthen fish immunity or immune responses to vaccines in the form of adjuvants (6). Additionally, challenging a fish with an infectious agent that has a low ability to cause disease can induce a heightened nonspecific immunity. The nonpathogenic chum salmon reovirus has been shown to induce a response in rainbow trout (*Oncorhynchus mykiss*) that provides protection against IHN (34). This induction of nonspecific defenses may be responsible for some short-term cross protection (observed after application of a vaccine) and can obscure true induction of specific acquired immunity in experimental vaccine trials. In the short term, this protective response can be useful, especially at early life stages, when the acquired immune system is not fully functional.

ROUTE OF ADMINISTRATION

Route of administration is largely determined by the type of vaccine available, the characteristics of the fish, and the severity of the problem. Most vaccines for fish are applied using an extensive application technique such as oral or immersion administration. This is generally the case because of the high numbers of fish that must be vaccinated, the relatively low value of the individual fish, and the stress-induced losses that are encountered when are uses injected vaccines.

Oral administration is accomplished through the incorporation of the vaccine in the feed. This method of vaccination eliminates stress-induced side effects, because the fish are not handled. However, oral administration is generally considered the least effective method of vaccinating the fish per application. Orally administered vaccines result in disproportionate application of the vaccine, due to disproportionate consumption of feed by the individuals in the population. Also, many antigens are not taken up by the gut or are poorly recognized by the immune system when absorbed via the gut. The low recognition of the immunogens is at least partially counteracted by the relative ease of application, allowing several applications sequentially over several days and thus providing antigen persistence.

Immersion vaccination is better for the equal distribution of the antigen within a population. Generally, the fish are netted or crowded and placed in a concentrated suspension of antigen for a short period of time and then released. All of the fish are exposed to the same amount of antigen. When killed or subunit antigen preparations are used, the fish absorb the antigen through the gills, skin, and gut. The persistence, concentration, and structure of the antigen influence the effectiveness of the vaccination on the fish.

In live vaccines, the agent actively infects the host, and replicates to provide the persistence and dosage needed to induce a strong response. Therefore, extensive methods of vaccination are generally used. Many of the live agents infect through the gut, making oral application of the agent feasible. However, more reliable dosing, ease in preparation, and maintenance of viability makes immersion exposure the most common method of application for live vaccine studies.

Vaccines administered by injection can provide the highest level of antigen in a single application. The formulation used can be an emulsion that slowly releases the vaccine, allowing for persistence. Practical routes of injection are intramuscular or intraperitoneal. Generally, intraperitoneal injections are used for food fish, since damage that could occur to the musculature from an intramuscularly administered vaccine would reduce the value of the flesh. Also, a larger volume of vaccine can be injected into the peritoneal cavity and retained after a single injection than by the intramuscular route. The negative aspects of injection vaccination are the manpower required and the stress induced by handling. Efficiency can be increased substantially by using an automated injection device or by setting up an assembly line with repeatable injectors. Nevertheless, individual handling and injection is practical only for fish that have high value and that are of a size that can be easily handled. In addition, extreme care must be taken when handling fish for injected vaccines. The fish must be anesthetized and processed quickly with minimal damage to the skin. Handling injury and induced stress is substantially higher when injected vaccines are given.

IMMUNE STATUS

The immune status of the fish is a critical parameter that determines the effectiveness of a vaccination. Vaccines work best in healthy, well-nourished fish that have a fully developed immune system and have not been environmentally or metabolically stressed. These conditions are often not convenient or practical to aquaculture operations. Often times, vaccination is desired at a very early stage in fish development. In fish species that have been evaluated, the acquired immune response is not fully functional until well after the fish have begun feeding. Also, the developmental stage at which effective vaccination can be done in fish is dependent on the type of antigen that elicits a protective response, and therefore will vary with the vaccine as well as the species vaccinated (35,36). In a broad sense, T-dependent antigens, like most proteins, are responded to later in developmental than are T-independent antigens (such as the complex sugar subunits of the lipopolysaccharides in gram-negative bacteria). If fish are vaccinated before they are developmentally mature enough to recognize the antigens, not only do they not respond, but they can develop an inability to respond to the antigen at a later date. This phenomenon is known as antigenic tolerance (a similar phenomenon known as anergy can occur if the immune system is overloaded with antigens and it is difficult to differentiate the two effects in developing fish). This phenomenon could result in a vaccinated fish population having higher losses than nonvaccinated fish during an outbreak of the disease that the vaccine was designed to prevent. Other metabolic factors that will affect vaccination efficacy are hormonally induced immune suppression due to smoltification in anadromous species and spawning and temperature-induced immune suppression. The timing of immunizations are critical for the desired effect. Generally, fish are thought to have a less sophisticated form of acquired immunity than mammals and birds. The boosting effect seen from giving fish multiple exposures to a vaccine is not as fast, as strong, or as long lasting as that seen in mammals. Also, this boosting should occur while there is a low persistent effect from the previous exposure. For best effect, boosting or primary exposure should be done within one to three months of when the first outbreaks of the disease are expected to occur. If the disease usually occurs during periods of immune suppression, such as smoltification or spawning, the fish should be vaccinated three to four weeks before the immune suppressive event occurs so that residual humoral immunity can help protect the fish.

Finally, because the immune system is most effective in healthy nonstressed fish, the use of vaccination does not reduce the need for constant environmental management or allow the aquaculturist to push the productive capacity of the system. Reduced water quality due to overstocking and overfeeding will result in stressful conditions. If severe enough, the long-term stress response induced by the stress hormones will cause suppression of the immune system and seriously reduce the effectiveness of the vaccine.

NEGATIVE EFFECTS

Negative side effects of vaccinating fish can include reduced growth, reduced feeding, handling induced injuries, and stress-induced metabolic changes. Many of these side effects are simply the effect of the normal stress response due to the handling of the fish. These effects can be minimized by keeping physical handling to a minimum, keeping all surfaces that the fish contacts wet and smooth (preferably transported in a cushion of water), using anesthetics as soon as possible, avoiding temperature changes and maintaining optimal water quality throughout the operation. Supplementing water in recovery tanks with salt, magnesium, and calcium for freshwater fish during recovery will reduce osmotic imbalances induced by handling.

Other side effects are due to the makeup of the vaccine. Attenuated live vaccines are infectious agents, and certain components of killed vaccines induce the mediators of the immune system. The release of high amounts of these mediators into the bloodstream can temporarily switch the homeostatic focus of the fish from growth to induction of specific and nonspecific immunity. These mediators are the factors that induce fever and malaise in humans. In fish the effects of these mediators may result in a temporary reduction in feeding activity and a short term loss in feed conversion.

Another potentially negative result of using vaccines is related to the regulatory surveillance of pathogens. The induction of specific antibodies may result in the misidentification of populations of fish as endemic for a regulated pathogen when they are just vaccinated and not carrying the pathogen. The use of antibody specific assays such as serum neutralization and ELISAs for screening previously exposed fish can result in this false positive identification, which can in certain instances affect the marketability and transport of the fish. Alternatively, the use of modified live vaccines and subunit vaccines that have a different antigenic profile from the wild-type pathogen and antigen-based assays can be designed that can differentiate vaccinated fish from those exposed to the disease-causing agent.

BENEFITS

Benefits of implementing an effective vaccination program against a disease in endemic areas include not only the short-term effects of increased survival, reduced expenses on therapeutics, and increased growth, but also the longer-term effects of reduced medication induced antibiotic resistance of bacterial pathogens in the system and the potential for eliminating a pathogen from the population. The use of a vaccine can not only reduce disease induced deaths in a population but also it can reduce the amount of subclinical disease. Generally, when it infects a healthy host, a pathogen does not kill the host. Disease outbreaks that result in detectable mortality in a population are a small fraction of the total number of diseases that pass through the population. In aquaculture these diseases would generally go unnoticed or would be recognized as a temporary reduction in feeding behavior. These diseases can last for one to several weeks in the individual fish resulting in reduced feeding and weight loss. In more severe cases damage to vital organs can cause poorer performance and a predisposition to other pathogens in the environment. It has been shown that when fish in endemic areas were effectively vaccinated against enteric red mouth disease, or vibrio, they demonstrated significantly higher growth rates as well as better survival. The reduced need for antibiotics, in the case of effective vaccines against bacterial diseases, reduces the selective pressure on all of the bacterial pathogens in the system to develop antibiotic resistance. Therefore, over time, the aquaculture facility that effectively uses vaccination programs to reduce disease incidence is less likely to encounter problems from antibiotic resistance in the common bacterial pathogens in the system. Therefore, when an important bacterial disease outbreak occurs, the available antibiotics can be effectively used. The ultimate goal of a vaccination program is to reduce the incidence of the disease-causing pathogen in the population. This can be accomplished when the vaccination provides protective immunity that results in a substantially reduced ability of the pathogen to infect and to be shed from the host. Although vaccinations cannot induce protective immunity in all of the individuals in a population, if a large enough percentage of the population has protective immunity to the pathogen, the pathogen cannot maintain itself in the population (37). Thus, through a stringent program of vaccination and pathogen avoidance it is possible to cure an endemic population of the pathogen.

The decision to use a vaccine on an aquaculture operation is often not easy. The cost, management changes, risks, and benefits in using a vaccine are difficult to assess and are different for each operation. Often, the benefits are not clear cut; some vaccines reduce the incidence of a disease, but do not eliminate the disease. Also, user experience and skill and environmental effects can often confound the issue. Many managers will try a vaccine on a portion of the operation. This allows them to learn how to best incorporate the management tool into their production system and evaluate the vaccine's effectiveness before applying it to the entire system.

BIBLIOGRAPHY

1. J.C. Leong and J.L. Fryer, *Ann. Rev. Fish Dis.* **3**, 225–240 (1993).
2. S.G. Newman, *Ann. Rev. Fish Dis.* **3**, 145–185 (1993).
3. R. Gudding, A. Lillehuaug, P.J. Midtlyng, and F. Brown, eds., *Developments in Biological Standardization: Fish Vaccinology*, Vol. 90, Karger, Basel, 1997.
4. International Association of Biological Standardization, *Developments in Biological Standardization: Fish Biologics: Serodiagnostics and Vaccines*, Vol. 49, Karger, Basel, 1981.
5. A.E. Ellis, ed., *Fish Vaccination*, Academic Press, London, 1988.
6. D.P. Anderson, *Ann. Rev. Fish Dis.* **2**, 281–307 (1992).
7. R.M.W. Stevenson, in R. Gudding, A. Lillehuaug, P.J. Midtlyng, and F. Brown, eds., *Developments in Biological Standardization*, Vol. 90, Karger, Basel, 1997, pp. 117–124.
8. A.E. Ellis, in R. Gudding, A. Lillehuaug, P.J. Midtlyng, and F. Brown, eds., *Developments in Biological Standardization*, Vol. 90, Karger, Basel, 1997, pp. 107–116.
9. A.E. Toranzo, Y. Santos, and J.L. Barja, in R. Gudding, A. Lillehuaug, P.J. Midtlyng, and F. Brown, eds., *Developments in Biological Standardization*, Vol. 90, Karger, Basel, 1997, pp. 93–105.
10. N. Fijan, in A.E. Ellis, ed., *Fish Vaccination*, Academic Press, 1988, pp. 204–215.
11. P. Dixon, in R. Gudding, A. Lillehuaug, P.J. Midtlyng, and F. Brown, eds., *Developments in Biological Standardization*, Vol. 90, Karger, Basel, 1997, pp. 221–232.
12. R.L. Thune, J.P. Hawke, D.H. Fernandez, M.L. Lawrence, and M.M. Moore, in R. Gudding, A. Lillehuaug, P.J. Midtlyng, and F. Brown, eds., *Developments in Biological Standardization*, Vol. 90, Karger, Basel, 1997, pp. 117–124.
13. J.L. Romalde and B. Magarinos, in R. Gudding, A. Lillehuaug, P.J. Midtlyng, and F. Brown, eds., *Developments in Biological Standardization*, Vol. 90, Karger, Basel, 1997, pp. 167–177.
14. D.S. Manning and J.C. Leong, *Virology* **179**, 16–25 (1990).
15. K.E. Christie, in R. Gudding, A. Lillehuaug, P.J. Midtlyng, and F. Brown, eds., *Developments in Biological Standardization*, Vol. 90, Karger, Basel, 1997, pp. 191–199.
16. R.D.J. Gilmore, H.M. Engelking, D.S. Manning, and J.C. Leong, *Bio/Technology* **6**, 295–300 (1988).
17. L. Xu, D.V. Mourich, H.M. Engelking, S. Ristow, J. Arnzen, and J.C. Leong, *J. Virol.* **65**, 1611–1615 (1991).
18. S. Oshima, J. Hata, C. Segawa, and S. Yamashita, *J. Gen. Virol.* **77**, 2441–2445 (1996).
19. F. Lecocq-Xhonneux, M. Thiry, I. Dheur, M. Rossius, N. Vanderheijden, J. Martial, and P. de Kinkelin, *J. Gen. Virol.* **75**, 1579–1587 (1994).
20. L.M. Vaughan, P.R. Smith, and T.J. Foster, *Infect. Immun.* **61**, 2172–2181 (1993).
21. R.K.I. Cooper, E.B.J. Shotts, and L.K. Nolan, *J. Aquat. Anim. Health* **8**, 319–324 (1996).
22. M.L. Lawrence, R.K. Cooper, and R.L. Thune, *Infect. Immun.* **65**, 4642–4651 (1997).
23. J.R. Winton, in R. Gudding, A. Lillehuaug, P.J. Midtlyng, and F. Brown, eds., *Developments in Biological Standardization*, Vol. 90, Karger, Basel, 1997, pp. 211–220.
24. A. Benmansour and P. de Kinkelin, in R. Gudding, A. Lillehuaug, P.J. Midtlyng, and F. Brown, eds., *Developments in Biological Standardization*, Vol. 90, Karger, Basel, 1997, pp. 279–289.
25. E.J. Noga and J.X. Hartmann, *Can. J. Fish. Aquat. Sci.* **38**, 925–929 (1981).
26. H.G. Zhang and L.A. Hanson, *Virology* **209**, 658–663 (1995).
27. B. Noonan, P.J. Enzmann, and T.J. Trust, *Appl. Environ. Microbiol.* **61**, 3586–3591 (1995).
28. H.G. Zhang and L.A. Hanson, *J. Fish Dis.* **19**, 121–128 (1996).
29. J.A. Wolff, R.W. Malone, P. Williams, W. Chong, G. Acsadi, A. Jani, and P.L. Felgner, *Science* **247**, 1465–1468 (1990).
30. D.C. Tang, M. De Vit, and S.A. Johnston, *Nature* **356**, 152–154 (1992).
31. E. Hansen, K. Fernandes, G. Goldspink, P. Butterworth, P.K. Umenda, and K.-C. Chang, *Fed. Euro. Biochem. Soc.* **290**, 73–76 (1991).
32. P.H. Russell, T. Kanellos, I.D. Sylvester, K.C. Chang, and C.R.H. Howard, *Fish Shellfish Immun.* **8**, 121–128 (1998).
33. E.D. Anderson, D.V. Mourich, S. Fahrenkrug, S. LaPatra, J. Shepard, and J.C. Leong, *Mol. Marine Biol. Biotech.* **5**, 114–122 (1996).
34. S.E. LaPatra, K.A. Lauda, and G.R. Jones, *Vet. Res.* **26**, 455–459 (1995).
35. A.E. Ellis, in A.E. Ellis, ed., *Fish Vaccination*, Academic Press, 1988, pp. 20–31.
36. A.G. Zapata, T. Torroba, A. Varas, and E. Jimenez, in R. Gudding, A. Lillehuaug, P.J. Midtlyng, and F. Brown, eds., *Developments in Biological Standardization*, Vol. 90, Karger, Basel, 1997, pp. 23–32.
37. R.M. Anderson and R.M. May, *Nature* **318**, 323–329 (1985).

See also DISEASE TREATMENTS.

VIRAL DISEASES OF FISH AND SHELLFISH

YOLANDA J. BRADY
Auburn University
Auburn, Alabama

OUTLINE

INTRODUCTION

The effects of viral diseases of finfish and shellfish may range from little effect to catastrophic. Viruses may be present without detrimental effects, but when the environment deteriorates and fish succumb to environmental stressors, disease outbreak is usually the result. New viruses are being continually discovered and their effects on the host organism may not be fully known. Often viruses are detected in cell culture but detection is limited to the availability of compatible cell culture systems. Viruses have also been detected visually by electron microscopy and molecular techniques. As more interest in husbandry of new species arises, it is likely that more new viruses will continue to be discovered. Currently there are no effective treatments for viral diseases in fish and shellfish. Management practices must be used to prevent the introduction of viruses into a culture system or, once present, managed by manipulation of water temperature, good nutrition, and best management practices to keep viral disease outbreaks to a minimum. The following is a summary of descriptions of the major viral diseases in fish and shellfish.

ADENOVIRIDAE

White Sturgeon Adenovirus

White sturgeon adenovirus (WSA) was first identified in juvenile white sturgeon in the mid-1980s at a farm in northern California. Fish appeared lethargic, were anorexic and emaciated, and had pale livers and no food in the intestine. Mortality may reach 50% with this virus, however, no severe disease problems have been reported since 1986 (1,2).

BACULOVIRIDAE

Baculoviral Midgut Gland Necrosis

Baculoviral midgut gland necrosis (BMN) is classified as a type-C baculovirus and causes serious peracute epizootics in hatchery reared *Marsupenaeus japonicus* in southern Japan and in *Penaeus monodon* in Australia. There are no specific clinical signs except that the disease has a sudden onset with a high mortality rate of up to 98% (3,4).

Baculovirus Penaei

Baculovirus penaei (BP) is a type-a occluded baculvorus. It is also known as BP virus disease and nuclear polyhedrosis disease. Several strains are likely to exist based on different morphological characteristics of the virion, especially size. Distribution is widespread in cultured and wild penaeids in the Americas from northern Gulf of Mexico through the Caribbean to central Brazil. BP can cause serious epizootics in larval and postlarval juvenile stages, with an acute onset and high mortality rates. Epizootics range from chronic to acute with high mortality, however, presence of the virus may not always indicate disease. Severely affected mysis stage larvae and post larvae may exhibit a white midgut line through the abdomen. The principal clinical sign is the presence of tetrahedral occlusion bodies in the hepatopancreas and midgut epithelial cells. Hosts include *Farfantepenaeus aztecus*, *Litopenaeus vannamei*, *Farfantepenaeus duorarum*, *Litopenaeus setiferus*, and *Litopenaeus stylirostris* (5).

Cherax Baculvirus

Also known as baculovirus of blue crayfish and hepatopancreatic baculovirus of *Cherax* this was the first reported baculovirus in a decapod crustacean, described in *Cherax quadricarinatus* in Australia. First seen in Australia, then in second-generation crayfish cultured in California, the impact on the host is uncertain, since up to 52% of crayfish examined in Australia were infected but no distinct clinical signs were present. In California, infected animals grew poorly but mortality was not significantly higher than animals not infected (5,6). The virus was detected in hepatopancreatic tubular epithelial cells. Animals were in poor condition but were not experiencing unusually high mortalities. The disease occurs in juvenile, subadult, and adult crayfish (3).

Monodon Baculovirus

Monodon baculovirus (MBV) is classified as a type-A occluded baculovirus. It is worldwide in distribution, including the southeastern United States. The primary clinical sign is the presence of a single or multiple spherical inclusion body in the hepatopancreas and midgut epithelial cells. The virus causes moderate to heavy infections in the hepatopancreas and anterior midgut. Mortalities may exceed 90%, with clinical signs of reduced feeding and growth rates and increased surface and gill fouling due to epicommensals. Also present are lethargy, anorexia, and dark pigmentation. Severely affected larvae and postlarvae may exhibit a white midgut line through the abdomen. Species infected include *P. monodon*, *Fenneropenaeus merguiensis*, *P. semisulcatus*, *P. kerathurus*, *L. vannamei*, *P. esculentus*, *P. penicillatus*, *P. plebejus*, and *Metapenaeus ensis* (3,4).

Tau Virus of Crabs: Baculoviruses A and B, Subgroup C Nonoccluded Baculovirus

Tau virus causes hepatopancreatic lesions and destruction of epithelial cells. Decreased aggressive behavior is evident, followed by lethargy, anorexia, and death (3). Species affected are *Callinectes sapidus*, *Carcinus maenas*, *C. mediterraneus*, and *Macropipus depurator* (4).

White Spot Syndrome Baculovirus Complex

At least three viruses have been described in the white spot syndrome baculovirus (WSBV) complex and have been reported in China, Japan, Korea, Thailand, Indonesia, Taiwan, Vietnam, Malaysia, India, and Texas. Natural infections were observed in *P. monodon*, *M. japonicus*, *Fenneropenaeus chinensis*, *Fenneropenaeus indicus*, *F. merguiensis*, and *L. setiferus*. The onset of disease is usually acute showing rapid reduction of feeding, lethargy, and loose cuticle with white spots, which are mostly apparent on the inner surface of the carapace. White spots represent abnormal deposits of calcium salts by the cuticular epidermis. In many cases, moribund shrimp display a pink to reddish brown coloration due to the expansion of the cuticular chromatophores. Mortalities may reach 100% within 3 to 10 days of the onset of clinical signs (3).

Yellow Head Virus (YHV) Disease, Yellow Head Baculovirus

Initially described in Thailand and probably widespread in cultured *P. monodon* in India and southeast Asia, YHV, along with mixed infection of TSV and WSBV, were observed in pond-reared juvenile *L. setiferus* from south Texas. Shrimp packing plants are thought to be the cause. Behavioral clinical signs include abrupt increase in feeding rate for several days, then total cessation of feeding followed by moribund shrimp swimming at the surface pond edge. The primary clinical sign is a bright yellow cephalothorax. Massive mortality begins by day three after feeding cessation. Also evident are white or pale yellow to brown gills (3,4).

BIRNAVIRIDAE

Eel Virus European

Eel Virus European (EVE) was isolated from European eels cultured in Japan and Japanese eels and tilapia in Taiwan. Clinical signs include rigidity with muscle spasms and congested anal fins, hyperplasia of lamellar epithelium, clubbed filaments, ventral petechiation, enlarged kidney, and ascites. EVE is found in cultured young European and Japanese eels. Disease occurs at 8 to 14 °C. Mortalities may reach 60% (2,7).

Infectious Pancreatic Necrosis

A subacute disease of salmonid fry and fingerlings, infectious pancreatic necrosis (IPN) or IPN-like viruses have been isolated from marnie and freshwater fish and invertebrates. Combined into the aquatic birnavirus group, IPN was the first proven viral disease of fish. IPN is found in 22 countries including North and South America, Europe, and Asia. Brook and rainbow trout are most susceptible, but virtually all salmonids are susceptible. Larger fry and fingerlings are the first to develop clinical signs. Externally infected trout have dark pigmentation, distended abdomen, exopthalmia, hemorrhages on ventral surface and fins, and pale gills. Internally, there is general hemorrhage with obvious petechiae throughout the viscera, pyloric caeca, and adipose tissue. The spleen, heart, liver, and kidneys are pale. The body cavity is filled with a pale yellow fluid, and the posterior stomach contains gelatinous mucoid plug that is pathogomonic. IPN infections range from subclinical to acute with up to 100% mortality in trout populations. Optimum temperature is 10 to 15 °C. Survivors become carriers and shedding virus is seasonal (2,7–10).

FISH TUMOR VIRUSES

Over 50 types of fish tumors or tumorlike proliferations have been linked to viruses. Many of these have not been isolated or characterized and have only been visualized with electron microscopy (7,11).

HERPESVIRIDAE

Anguillid Herpesvirus

Also known as anguillid herpesvirus (HAV) 1, eel herpesvirus, and *Herpesvirus anguillidae*, HAV has been found only in limited populations of eels in Japan. Eels showed varying degrees of erythema on skin and gills. Optimum temperature for the disease is 20 to 25 °C (12).

Bifacies Virus of Crabs

Bifacies virus of crabs (BFV), also known as herpeslike virus of blue crabs (HLV), has been seen in *C. sapidus* and others on the east coast of the United States and possibly worldwide. The hemolymph fails to clot and is chalky due to viral lysis of hemocytes (3).

Channel Catfish Virus Disease

Channel catfish virus disease (CCVD) is an acute viral disease infecting juvenile channel catfish, *Ictalurus punctatus*, in most Southern states and worldwide where channel catfish are cultured. CCVD was first discovered in 1968 experimentally in fingerling blue catfish and channel × blue hybrids by injection. Brown and yellow bullheads are not susceptible. European catfish are also resistant. Clinical signs include distended abdomen, expothalmia, pale-to-hemorrhagic gills, and hemorrhage at the base of fins, throughout the skin, and on the ventral surface. Infected fish swim erratically, sink to the bottom, and respire weakly before death. Internally there is a clear yellow-colored fluid in the peritoneal cavity, general hyperemia and pale liver and kidneys, dark red enlarged spleen, no food in the stomach or intestine, and the presence of a mucoid secretion. CCVD usually occurs during the summer when temperatures are above 25 °C (2,7).

Epizootic Epitheliotropic Disease

Acute mortalitiy in juvenile lake trout in the Great Lakes Basin was caused by epizootic epitheliotropic disease (EED). The virus effected lake trout × brook trout hybrids. Clinical signs include lethargy, erratic swimming, and hemorrhage in the eyes and mouth. Epithelial hyperplastic lesions and gray-to-white mucoid blotches develop on the jaw, inside the mouth, and on the body and fins. Secondary fungal infections develop in the eyes, fins, and body. Internally, infected fish have a swollen spleen. HAV is a significant juvenile disease in lake trout populations (13).

Fish Pox, Herpesvirus of Carp, Carp Herpesvirus, *Herpesvirus cyprini*

Carp pox has been reported in Europe, China, Japan, Korea, Israel, Malaysia, and the United States Common carp and koi carp are the primary hosts. Fish pox are benign, hyperplastic papillomatous growths on the epithelium of carp. Tumors are white to gray, elevated to 1 to 3 mm above the skin, and are commonly found on the head, fins, and body. Tumors are generally small, but may increase in size and then regress and disappear. Fish show no specific behavioral signs and usually no morbidity. Juvenile carp can suffer high mortality. Clinical signs include anorexia, distended abdomen, expothalmia, dark pigmentation, and hemorrhages on the operculum and abdomen. Young fish exposed to the virus may not show clinical signs for up to one year (2,7).

Goldfish Herpesvirus

Also known as herpesviral hematopoietic necrosis (HVHN), this virus was the cause of a severe epizootic in Japan with no apparent clinical signs. Not much is known about the virus (14).

Herpeslike Virus of Crabs

This virus causes extensive destruction of the bladder and antennal gland. No gross clinical signs have been reported. Species affected are *Paralithodes platypus*, *P. camtschatica*, and *Lithodes aequispina* in Alaskan waters over a large geographical range (3).

Herpes-Type Virus Disease of Oysters

A herpes-type virus was found in the eastern oyster, *Crassostrea virginica*, in the Piscataqua River, Maine. The virus was also seen in the Pacific oyster, *C. gigas*, in hatcheries in New Zealand and France. Clinical signs are pale digestive glands, which may be caused by poor conditions, elevated temperatures and crowding (3).

Salmonid Herpesvirus Type 1

Salmonid herpesvirus type 1 (SHV-1) comprises 10 herpesviruses isolated from trout and salmon. SHV-1 causes mild disease of rainbow trout. SHV-1 was first isolated in Washington State and later in California. Steelhead and rainbow trout are natural hosts, while chum salmon fry and chinook salmon are experimentally susceptible. Atlantic salmon and brook and brown trout are refractive. Clinical signs include dark pigmentation, experimental exposure, anorexia, lethargy, erratic swimming, exopthalmia, hemorrhage, pale gills, distended abdomen, and mucoid fecal casts. Internal signs include bloody gelatinous ascites, flaccid intestine, hemorrhagic liver, mottled or friable liver, and pale kidneys. SHV-1 is of minimal importance and low pathogenicity (15,16).

Salmonid Herpesvirus 2, *Oncorhynchus masu* virus (OMV)

Salmonid herpesvirus 2 (SHV-2) causes mortality in juvenile masu and coho salmon and tumors on adult fish in Japan. In juveniles, the acute disease kills 30- to 150-day-old fish. Clinical signs include anorexic, exopthalmic, petechial hemorrhage on the body. Internally, infected fish have a mottled white liver and swollen spleen. Neoplastic tumors develop in survivors, primarily around the mouth but also sometimes on the operculum, body, and cornea. SHV-2 is a major problem in land-locked salmon populations and cage-cultured coho in northern Japan (17).

White Sturgeon Herpesvirus

Two types of herpesviruses have been isolated from white sturgeon and were designated white sturgeon herpesvirus-1 (WSHV-1) and white sturgeon herpesvirus-2 (WSHV-2). WSHV-1 was first reported in California sturgeon while WSHV-2 was detected in sturgeon from commercial farms in California and wild sturgeon in Oregon. No specific clinical signs for either viral disease have been described. Fish appear normal and continue to feed until death (2,18,19).

IRIDOVIRIDAE

Catfish Iridovirus

Catfish iridovirus (CIV) was first isolated from cultured black bullheads in France. Clinical signs include body and muscle edema, petechial hemorrhage, pale gills, and ascites. Disease occurs from 15 to 25 °C. Mortality may reach 100%. The full impact of this disease on cultured bullheads is uncertain (20).

Eel Iridovirus

An eel iridovirus (EV-102) was isolated from cultured Japanese eels in Japan. This virus was specific for Japanese eels since it had no effect on European eels also cultured in Japan. Clinical signs include depigmentation of skin and increased mucus. Mortality may reach up to 70% in three to five days at temperatures of 15 to 20 °C (7,21).

Epizootic Hematopoietic Necrosis Virus

Epizootic hematopoietic necrosis virus (EHNV) is thought to be the first viral disease of fish in Australia. The virus has been isolated from yellow perch and rainbow trout. Clinical signs include a slow spiraling behavior, dark pigmentation, anorexia, petechial hemorrhage of skin, fins, head, and internal organs, abdominal distension, swollen kidney and spleen, and mottled liver (22,23).

Gill Disease of Portuguese Oysters

Also known as gill necrosis viral disease, gill disease of portuguese oysters is caused by an iridovirus. This virus has been found in cultured *C. angulata* and *C. gigas* in Portugal, France, Spain, and Great Britain. Clinical signs include extreme gill erosion and yellow spots on gills that progress to brown discoloration indicating necrosis of the gill. Yellow or green pustules may occur on the mantle or adductor muscle (3).

Hemolytic Infection Virus Disease

Hemolytic infection virus (HIV) disease had been described in oysters, *C. angulata* and *C. gigas*, in France and Spain. There are no distinctive clinical signs, but atrophy and weakness of the adductor muscle has been observed (3).

Largemouth Bass Iridovirus

This is the first debilitating systemic virus reported from centrarchids. Moribund fish lose equilibrium and float at the surface. There are no external lesions but internally the swim bladder is greatly enlarged. The virus occurs around 30 °C and its significance is not yet known (24).

Lymphocystis Virus, Cellular Hypertrophic Disease

Oldest and best known of all fish viruses, the lymphocystis virus was first recognized as a disease in 1874, proposed to be of viral origin in 1920, and finally confirmed in the 1960s. The most widely distributed virus lymphocystis affects fish in North America, South America, Europe, Africa, Australia, and Asia. Lymphocystis has been reported in 11 orders, 45 families, and 141 fish species. Lymphocystis cells are large gray or whitish cells that appear singularly or grouped together in grape-like clusters occurring primarily on the fins and body. Lesions are most prevalent on the fins, head, and lateral body surface. Lymphocystis is a chronic, benign condition that rarely causes morbidity or mortality and occurs in all sizes and ages of fish. Lesions depend on temperature and appear 5 to 12 days at 20 to 25 °C or in up to 6 weeks at 10 to 15 °C (7,25,26).

Oyster Velar Virus Disease, Blister Disease

Oyster velar virus disease (OVVD) infects *C. gigas*. It is found in Washington State, however, it is believed to be more widespread in distribution where *C. gigas* is found. The virus causes infection of the velum and may cause up to 100% mortality in hatcheries. OVVD usually occurs from March to May, but can also appear in the summer and is seen in larvae more than 150 μm in shell length and more than 10 days old. Gross signs of the disease are blisters on the velum and a lack of cilia (3).

Sheathfish Iridovirus

Sheathfish iridovirus (SHV) was first isolated from cultured European sheathfish (catfish) in a recirculating system in Germany. Sheathfish fry show spiral swimming and general hemorrhage on the body and internal organs. Mortality may reach 100% (27).

Viral Erythrocytic Necrosis

Viral erythrocytic necrosis (VEN) is also known as piscine erythrocytic necrosis. VEN affects erythrocytes and a wide variety of marine and anadromous fish. Reported in a variety of fish, primarily marine, its clinical signs include severe anemia, pale gills, colorless blood, and discolored liver. Hematocrits may be as low as 2 to 10% (28,29).

White Sturgeon Iridovirus

White sturgeon iridovirus was isolated in juvenile-cultured white sturgeon in northern California. Clinical signs include weakness, anorexia, and pale, necrotic gills. Internally there is no body fat, a pale liver, and an intestine void of food (2,30,31).

NODAVIRIDAE

Viral Nervous Necrosis Virus

Viral nervous necrosis virus (NNV) is also referred to as striped jack nervous necrosis virus (SJNNV). The virus was first seen in Japan in cultured striped jack but has also been seen in Greece, Martinique, and France in larval striped jack. A virus thought to be SJNNV was also reported from barramundi, groupers, other jacks, flounders, puffers, and a variety of marine fish in several geographic locations. Clinical signs include emaciation, enlarged swim bladder, spinal deformities, dark pigmentation, and exopthalmia (32–34).

ORTHOMYXOVIRIDAE

Infectious Salmon Anemia Virus

Occuring in Atlantic Salmon in Norway, the clinical signs for this virus are pale gills, exopthalmia, ascites, enlarged liver and spleen, petechial hemorrhage of the intestine, pale internal organs, and high mortality. Classification of the virus is not complete; however, it has tentatively been identified as an orthomyxoviridae (35,36).

PAPOVAVIRIDAE

Papillomaviruslike Papovavirus

This virus has shown up primarily in *C. virginica*, with unconfirmed reports in other species areas affected include Atlantic Canada, the eastern United States, and similar lesions in the western United States, Korea, and Japan. Clinical signs are mass hypertrophy of gametes and gametogenic epithelium. Host response is negligible, and there is generally low infection with no associated mortality (3).

PARVOVIRIDAE

Hepatopancreatic Parvovirus

Hepatopancreatic parvovirus (HPV) is seen in wild and cultured penaeids in Australia and in cultured penaeids in the yellow Sea area of China, in Korea, Taiwan, Philippines, Indonesia, Malaysia, Singapore, Kenya, Kuwait, and Israel. It is also seen in North and South America in *L. vannamei* and *L. stylirostris*. Clinical signs are not specific but include whitish and atrophied hepatopancreas, poor growth rate, anorexia, increased gill fouling due to epicommensal parasites, occasional opacity of abdominal muscles, and secondary opportunistic bacteria such as *Vibrio* spp. (3,4).

Infectious Hepatopoietic and Hypodermal Necrosis Virus

Infectious hepatopoietic and hypodermal necrosis virus (IHHN) is also known as runt deformity syndrome (RDS). IHHN is widely distributed in culture facilities in North and South America and Asia and is also found in wild penaeids. Natural infections have been observed in *L. stylirostris*, *L. vannamei*, *L. occidentalis*, *F. californiensis*, *P. monodon*, *P. semisulcatus* and *M. japonicus*. *L. setiferus*, *F. duorarum*, and *F. aztecus* have been experimentally infected. IHHN causes acute mortality in juvenile *L. stylirostris*. Gross signs are not specific, but reduction of feeding and changes in behavior and appearance are noted. This disease is typically chronic in *L. vannamei* and linked to runt-deformity syndrome. Clinical signs in juvenile shrimp include bent or deformed rostrums, wrinkled antennal flagella, and rough cuticle (3,4).

Lymphoidal Parvolike Virus

Lymphoidal parvolike virus (LPV) has only been observed in Australia, although it is likely to be in the Indo-Pacific area or southeast Asia. LPV has been seen in *P. monodon*, *F. merguiensis*, and *P. esculentus*. No consistent clinical signs have been reported (4).

Parvolike Virus of Crayfish

Parvolike virus in freshwater crayfish, *Cherax destructor*, was the first systemic virus in freshwater crayfish and is indistinguishable from the virus that infects freshwater prawn, *Macrobrachium rosenbergii*. Crayfish are extremely moribund. Clinical signs include opaque musculature on the ventral surface of the abdomen (37).

PICORNAVIRIDAE

Chesapeake Bay Virus

Chesapeake bay virus (CBV) is also known as picornalike virus of crabs. This disease affects the blue crab, *C. sapidus*, and other crabs on the east coast of the United States (3). The CBV infection causes destruction of bladder epithelium, epidermis, gills, and neurosecretory cells of the central nervous system. Clinical signs include abnormal behavior and blindness (3).

Taura Syndrome Virus

Taura syndrome virus (TSV) is geographically limited to the Americas and was first recognized near the Taura River, near the Guyaquil Equador in June, 1992. Natural infections occur in *L. vannamei*, *L. stylirostris*, and *L. setiferus*. TSV causes high mortalities at PL 12 stage and older and is best known as a disease of the nursery phase. *L. vannamei* are affected within 14 to 40 days of stocking as postlarvae. Typically affected are small juveniles less that 5 g. The disease has two distinct phases: the peracute/acute phase and the chronic/recovery phase. In the peracute/acute phase, gross signs in moribund shrimp include expansion of the red chromatophores, giving an overall reddish coloration, with the uropod and pleopods distinctly red. These animals usually die during ecdysis. The exoskeleton is soft and the gut is empty. In the chronic/recovery phase, shrimp show multifocal melanized cuticular lesions similar to bacterial shell disease. Shrimp may or may not have soft cuticles and red coloration and may behave and feed normally. These shrimp are survivors of the peracute/acute phase. Losses may reach 95% in affected ponds. Survivors typically display survival rates of 60% to harvest size (4).

REOVIRIDAE

Catfish Reovirus

First isolated from healthy channel catfish during a CCVD survey in California, catfish reovirus (CRV) is considered insignificant because no significant pathology has occurred in experimentally infected channel catfish (38,39).

Chum Salmon Virus

Chum salmon virus (CSV) is a reovirus pathogenic to chum salmon, resulting in mortality and moderate to severe liver necrosis (40).

Golden Shiner Virus

The aquareovirus genus was formed to include reoviruses of fish and shellfish. Golden shiner virus (GSV) is the type species. In general, aquareoviruses exhibit little or no pathogenicity. GSV was first isolated from cultured golden shiners, *Notemigonus crysoleucas*. GSV appears to be confined to the southeastern United States and California. The virus has also been isolated from grass carp. Clinical signs include lethargic swimming and erythema on the back and head as a result of hemorrhages in the muscle and skin. Infected fish may hold fins close to the body and may be on the bottom of the tank. GSV can cause significant loss and may reach 75% when fish are stressed in holding tanks. The disease occurs in summer when temperatures are 25 °C or higher (41).

Grass Carp Hemorrhagic Virus

Grass carp hemorrhagic virus (GCHV) is a reovirus first reported in grass carp in China and is distinct from grass carp reovirus. Carp develop hemorrhage at the base of fins, operculum, and mouth, are exopthalmic, and have pale gills. Internally there is severe hemorrhage of the intestine. The disease primarily infects fry and fingerlings with up to 80% mortality and occurs during the summer when temperatures range from 25 to 30 °C (2,42).

Reolike Virus of Crabs

This virus occurs in a complex with a rhabdolike virus A. Death of crabs is caused by nerve and hemocyte dysfunction. Clinical signs include lethargy followed by paralysis. The hemolymph fails to completely clot (3).

Reolike Virus of Penaeids

At least two types of reolike viruses are known in penaeids. They are designated REO III and REO IV and are classified in the genus aquareovirus (REO I and REO II are found in other decapod crustacea). REO III is found in cultured *M. japonicus*, *P. monodon*, and *L. vannamei*. REO IV is found in cultured and wild *F. chinensis* in Asia. Clinical signs include lethargy, poor resistance to stress, eroded and melanized appendage tips, shell disease lesions, and black gills. The hepatopancreas is necrotic, pale, and atrophied. *M. japonicus* signs include poor growth rate, anorexia, lethargy, gill and surface fouling, and occasionally opaque abdominal muscle (3,4).

RETROVIRIDAE

Epidermal Hyperplasia

Also known as walleye epidermal hyperplasia and discrete epidermal hyperplasia, epidermal hyperplasia is found in adult walleye in Saskatchewan and Manitoba, Canada, and Lake Oneida, New York. Lesions are raised, which are clear translucent mucoid patches that appear on the body and fins. The virus seems to have little effect on the fish (7).

Walleye Dermal Sarcoma

Walleye dermal sarcoma (WDSV) was first described in 1947 from walleye in Lakes Oneida and Champlain in New York State. The virus produces spherical lesions that can occur anywhere on the body and fins of adult walleye. The lesions may be pink to white in color. Optimum temperature for tumor production is 10 °C. The disease usually does not cause mortality, but fish are rejected commercially (7,41,42).

RHABDOVIRIDAE

Eel Rhabdovirus

Several rhabdoviruses have been isolated from American eel and European eel. Clinicals signs are limited and the effects of these viruses are unclear (43,44).

Grass Carp Rhabdovirus

Grass carp rhabdovirus was first isolated in Hungary from healthy two-year-old grass carp. The role of this disease is still uncertain (2,45).

Infectious Hematopoietic Necrosis Virus

Infectious hematopoietic necrosis virus (IHNV) was first detected in fish from California to Washington State in the 1940s to 1950s. Originally thought confined to tributaries that flowed into the Pacific Ocean from California to Alaska, IHNV was later discovered in Minnesota, South

Dakota, Montana, West Virginia, New York, and Idaho. It was also reported in China, France, Germany, Italy, Korea, and Taiwan. Species susceptible are trout, chinook, sockeye, and chum salmon. Brook and brown trout are susceptible to a lesser degree than rainbow trout. Clinical signs in fry and fingerlings are lethargy, avoiding currents, moving to the edge of ponds or raceways, weak respiration, and swimming in circles. They also become extremely dark and have exopthalmic swollen abdomen, pale gills, and hemorrhage at the base of their fins, mouth, and body. Chevron-shaped hemorrhages are seen in the musculature. An opaque mucoid fecal cast trails from the vent. Internally, organs are pale with petechaial hemorrhage in the mesenteries, peritoneum, air bladder, liver, and kidney and pale yellow fluid is in the body cavity. Survivors may show malformed heads, scoliosis, or lordosis. Mortality can range from low to nearly 100% depending on species, age, size, environment, and virus strain. IHNV normally does not occur in water above 15 °C (47,48).

Pike Fry Rhabdovirus

Also known as red disease and head disease, pike fry rhabdovirus is an acute, highly contagious disease of cultured fry and fingerling northern pike and is found in the Netherlands, Germany, and Hungary, but not in the United States. Species affected include grass carp, tench, white bream, European catfish, brown trout, roach, and gudgeon. Two forms of the disease exist. Head disease affects swimming fry. Clinical signs include hydrocephalus, exopthalmia, and poor growth. Red disease occurs in larger fish. Clinical signs include hemorrhages along lateral trunk musculature, pale gills exopthalmia, abdominal distension, and ascites. Infected fish lose schooling behavior and become lethargic. Up to 100% mortality can occur at the optimum temperature of 10 °C (49,50).

Spring Viremia of Carp

Spring viremia of carp (SVC) *Rhabdovirus carpio* occurs primarily in Europe and has not been reported elsewhere even though carp are cultured worldwide. Species infected are common carp, goldfish, bighead carp, silver carp, and grass carp. External clinical signs include dark pigmentation, distended abdomen, exopthalmia, prolapsed inflamed anus, and pale gills. Internally, signs are general hyperemia with hemorrhages in the kidney, liver, and air bladder. Optimum temperature for the disease is 16 to 17 °C but the disease may occur at 12 to 22 °C. Spring viremia is an extremely important disease of cultured carp in Europe and may cause up to 70% mortality in yearling fish (2,7,51).

Stomatopapiloma Virus

Stomatopapiloma virus is also known as Cauliflower disease. Cauliflower disease is one of the earliest diseases thought to be of viral origin and was first recognized in 1910. Only European eels in northern Europe and Great Britain are affected (7,52,53).

Viral Hemorrhagic Septicemia

Viral hemorrhagic septicemia (VHS) is also known as Egtved disease. This is an acute, highly infectious disease of cultured salmonids in Europe. Cultured rainbow trout are most susceptible; brown trout are also a natural host. VHS has been considered a European virus until isolated from Washington State in 1988. VHS in salmonids occurs in three forms based on pathological changes and mortality patterns: acute, chronic, and nervous. Acute is the most serious form and is associated with high mortalities. Behavior of infected fish includes erratic and spiral swimming, exophthalmia, distended abdomen, dark pigmentation, and pale gills with petechial hemorrhage on the gills. Internally, signs include hemorrhages in the skeletal muscle, swim bladder, and swollen kidney and liver. VHS has subacute to chronic lower mortality. Internally, the liver is pale. There is a nervous stage with low mortality, but fish exhibit poor balance in swimming and are anemic. The optimum temperature for VHS disease to occur is between 8 and 12 °C and typically does not occur above 14 °C. Younger fish are more susceptible to disease, but as fish become larger, they gain greater resistance (54).

TOGAVIRIDAE

Erythrocytic Inclusion Body Syndrome

Erythrocytic inclusion body syndrome causes anemia in cultured Coho and Chinook salmon in tributaries of the Columbia River in Oregon and Washington. It is also reported in Canada, Chile, Ireland, Japan, and Norway. Chum and masu salmon can also become susceptible, but have mild infections. Clinical signs include primarily severe anemia and pale gills, and internal organs. No external or behavioral clinical signs have been reported. Mortality occurs due to severe anemia; mortality is due to secondary bacterial and fungal infections (54,55).

Lymphoid Organ Vacuolization Virus Disease

Lymphoid organ vacuolization virus disease (LOVV) is found in cultured *L. vannamei* and *L. stylirostris* in the Americas and Hawaii. There are no recognized gross clinical signs, and prevalence and pathogenicity of the virus is not known (3).

BIBLIOGRAPHY

1. R.P. Hedrick, J. Spears, M.L. Kent, and T. McDowell, *Can. J. Fish. Aqua. Sci.* **42**, 1321–1325 (1985).
2. J.A. Plumb, *Health Maintenance and Principal Microbial Diseases of Cultured Fish*, Iowa State University Press, Ames, IA, 1999.
3. S.M. Bower, S.E. McGladdery, and I.M. Price, *Ann. Rev. Fish Dis.* **4**, 1–199 (1994).
4. D.V. Lightner, *A Handbook of Shrimp Pathology and Diagnostic Procedures of Cultured Penaeid Shrimp*, World Aquaculture Society, Baton, Rouge, LA, 1996.
5. J.M. Goff, T. McDowell, C.S. Friedman, and R.P. Hedrick, *J. Aquat. Anim. Health* **5**, 275–279 (1993).

6. I.G. Anderson and H.C. Prior, *J. Invert. Pathol.* **60**, 265–273 (1992).
7. K. Wolf, *Fish Viruses and Fish Viral Diseases*, Cornell University Press, Ithaca, NY, 1988.
8. E.M. Wood, S.F. Snieszko, and W.T. Yasutake, *Am. Med. Assoc. Arch. Path.* **60**, 26–28 (1955).
9. K. Wolf, S.F. Snieszko, C.E. Dunbar, and E. Pyle, *Soc. Exp. Bio and Med.* **104**, 105–108 (1960).
10. B.J. Hill and K. Way, *Ann. Rev. Fish Dis.* **5**, 55–77 (1995).
11. K. Anders and M. Yoshimizu, *Dis. Aqua. Org.* **19**, 215–232 (1994).
12. M. Sano, H. Fukuda, and T. Sano, in F.O. Perkins and C.T. Cheng, eds., *Pathology in Marine Science*, Academic Press, New York, 1976, pp. 15–31.
13. P.E. McAllister and R.L. Herman, *Dis. Aqua. Org.* **4**, 101–104 (1989).
14. S.G. Jung and T. Miyazaki, *J. Fish Dis.* **18**, 211–220 (1995).
15. M. Tanaka, M. Yoshimizu, and T. Kimura in T. Kimura, ed., *Salmonid Fish Diseases*, Hokkaido University Press, Sapporo, Japan, 1992, pp. 111–117.
16. K. Wolf and W.G. Taylor, *Fish Health News* **4**, 3–4 (1975).
17. T. Kimura, M. Yoshimizu, and M. Tanaka, *Fish Path.* **15**, 149–153 (1981).
18. R.P. Hedrick, T.S. McDowell, J.M. Groff, S. Yun, and W.H. Wingfield, *Dis. Aqua. Org.* **11**, 49–56 (1991).
19. H.M. Engelking and J. Kaufman, *Am. Fish. Soc./FHS News* **24**, 4–5 (1996).
20. F. Pozet, M. Morand, A. Moussa, C. Torhy, and P. deKinkelin, *Dis. Aqua. Org.* **14**, 35–42 (1992).
21. M. Sorimachi, *Bull. Nat. Res. Inst. Aqua.* **6**, 71–75 (1984).
22. J.S. Langdon, *J. Fish Dis.* **12**, 295–310 (1989).
23. J.S. Langdon, J.D. Humphrey, L.M. Williams, A.D. Hyatt, and H.A. Westbury, *J. Fish Dis.* **9**, 263–268 (1986).
24. J.A. Plumb, J.M. Grizzle, H.E. Young, A.D. Noyes, and S. Lamprecht, *J. Aquat. Anim. Health* **8**, 265–270 (1996).
25. K. Wolf, *Virology* **18**, 249–256 (1962).
26. K. Wolf, M. Gravell, and R.G. Malsberger, *Science* **151**, 1004–1005 (1966).
27. W. Ahne, M. Ogawa, and J.J. Schlotfeldt, *J. Vet. Med.* **37**, 187–190 (1990).
28. R. Walker, *Am. Zoologist* **11**, 707 (1971).
29. R. Walker and S.W. Sherburne, *J. Fish Res. Bd. Can* **34**, 1188–1195 (1977).
30. R.P. Hedrick, T.S. McDowell, J.M. Groff, S. Yun, and W.H. Wingfield, *Dis Aqua. Org.* **12**, 75–81 (1992).
31. R.P. Hedrick, J.M. Groff, T.S. McDowell, and W.H. Wingfield, *Dis Aqua. Org.* **8**, 39–44 (1990).
32. T. Nakai, K. Mori, T. Nishizawa, and K. Muroga, *Asian Fish. Soc.* Special Publication **10**, 147–152 (1995).
33. K. Muroga, *Fish Path.* **30**, 71–85 (1995).
34. A. Le Breton, L. Girsez, J. Sweetman, and F. Olevier, *J. Fish Dis.* **20**, 145–151 (1997).
35. K. Thorud and H.O. Djupvic, *Bull. Eur. Assoc. Fish Path.* **8**, 109–111 (1988).
36. C.W.R. Koren and A. Nylund, *Dis. Aqua. Org.* **29**, 99–109 (1997).
37. B. Edgerton, R. Webb, and M. Wingfield, *Dis. Aqua. Org.* **29**, 73–78 (1997).
38. D.F. Amend, T. McDowell, and R.P. Hedrick, *Can. J. Aqua. Sci.* **43**, 807–811 (1984).
39. R.P. Hedrick, R. Rosemark, D. Aronstein, J.R. Winton, T. McDowell, and D.F. Amend, *J. Gen. Virol.* **65**, 1527–1534 (1984).
40. S.E. LaPatra and R.P. Hedrick, *Am Fish. Soc/FHS News* **23**, 3–4 (1995).
41. J.A. Plumb, P.R. Bowser, J.M. Grizzle, and A.J. Mitchell, *J. Fish. Res. Rd. Can.* **36**, 1390–1394 (1979).
42. Y. Chen and Y. Jiang, *Kexue-Tongboa* **29**, 832–835 (1984) [Chinese with English abstract].
43. R. Walker, *Anal. Rec.* **99**, 559–560 (1947).
44. P.R. Bowser, G.A. Wooster, S.L. Quackenbush, R.N. Casey, and J.W. Casey, *J. Aquat. Anim. Health* **8**, 78–81 (1996).
45. T. Sano, *Fish Path.* **19**, 221–226 (1976).
46. J. Castric, D. Rasschaert, and J. Bernard, *Ann. Virol* **135**, 35–55 (1984).
47. D.F. Amend, W.T. Yasutake, and R.W. Mead, *Trans. Am. Fish. Soc.* **98**, 796–804 (1969).
48. F.M. Hetrick, J.L. Fryer, and M.D. Knittel, *J. Fish Dis.* **2**, 253–257 (1979)
49. R. Bootsma, *J. Fish Biol.* **3**, 417–419 (1991).
50. P. deKinkelin, B. Galimard, and R. Bootsma, *Nature* **241**, 465–467 (1973).
51. I.S. Shchelkunov and T.I. Shchelkunova, in W. Ahne, ed., Viruses in Lower Vertebrates, Springer-Verlag, Berlin, 1989, pp. 333–348.
52. J. Delves-Broughton, J.K. Fawell, and D. Woods, *J. Fish Dis.* **3**, 255–256 (1980).
53. G. Peters, *J. Fish Biol.* **7**, 415–422 (1975).
54. R. Holt and J. Rohovec, *Am. Fish. Soc./FHS News* **12**, 4 (1984).
55. S.L. Leek, *Can. J. Fish Aqua. Sci.* **44**, 685–688. (1987).

See also DISEASE TREATMENTS; VACCINES.

VITAMIN REQUIREMENTS

J. GABAUDAN
Research Centre for Animal Nutrition and Health
Saint-Louis Cedex, France

RONALD W. HARDY
Hagerman Fish Culture Experiment Station
Hagerman, Idaho

OUTLINE

Vitamins are defined as organic compounds which are required in small quantities and not used as energy sources, but are essential for growth and maintenance and must be supplied in the diet. Vitamins were first discovered through efforts to find cures for human and animal diseases. It has been known for millenia that some human conditions, such as night-blindness, could be cured by eating certain foods. Vitamin C deficiency signs, known as scurvy, were a major problem for sailors, whose diet on sailing ships was limited. The French explorer, Cartier, learned of a cure for scurvy from North American Indians, who extracted vitamin C from spruce needles. Lind, in 1753, wrote a book on scurvy for the British navy, in which he concluded that it could be cured by eating citrus fruits. This is how British sailors came to be know as 'Limeys'. Although certain foods were known to prevent or cure diseases caused by vitamin deficiencies, it was only through development and use of purified diets in the twentieth century that identification of vitamins and the establishment of dietary requirements was accomplished.

Although fish have been raised for thousands of years, vitamin deficiencies became apparent only when fish were raised exclusively on prepared feeds. In the 1920s, McCay and Dilley (1) attempted to identify the compound(s) in fresh meat that prevented anemia in trout. Twenty years later, folic acid and vitamin B12 were found to be the essential vitamins needed to prevent anemia in trout (2). Establishment of the essential vitamin requirements of fish could not be made until a purified diet was developed for fish. Through the research efforts of McLaren et al. (3), Wolf (4), and Halver (5), who continued to refine and develop purified diets for trout and salmon, the qualitative and eventually the quantitative vitamin requirements of fish were established. Purified diet formulations for trout and salmon continue to evolve as new information is developed on specific nutrient requirements of these fish and as purified diets for salmonids are adapted to other species of fish.

Table 1. Vitamin Specifications for Salmonid Feeds (mg/kg)

Vitamin	NRC (1993) Recommendations[a]	Typical Added Levels in Feeds
Vitamin A (IU)	2500 (1500)	6000
Vitamin D_3 (IU)	2400 (200)	2000
Vitamin E	50 (10)	300–500
Vitamin K_3	R[b] (0.53)	10
Thiamin	1 (1.8)	15
Riboflavin	4–7 (3.6)	25
Pyridoxine	3–6 (3.5)	15
Pantothenic acid	20 (10)	50
Niacin	10 (35)	180
Biotin	0.15 (0.15)	0.6
Folic acid	1 (0.55)	8
Vitamin B_{12}	0.01 (0.01)	0.03
Inositol	300	130
Choline	1,000 (1,300)	1,000
Ascorbic acid	50	150[c]

[a]Values in parentheses are NRC (1994) recommendations for broilers.
[b]Required, but not quantitatively determined.
[c]When a stable and bioavailable source of vitamin C is used.

Table 2. Primary Functions of Vitamins in Fish

Vitamin	Primary Function
Fat-soluble vitamins	
Vitamin A	Normal vision
Vitamin D	Calcium metabolism, bone formation
Vitamin E	Cell membrane maintenance
Vitamin K	Blood clotting
Water-soluble vitamins	
Thiamin	Carbohydrate metabolism
Pyridoxine	Amino acid metabolism
Riboflavin, niacin	Hydrogen transport, amino acid metabolism
Pantothenic acid	Energy metabolism
Biotin	Energy metabolism, integrity of scales
Folic acid, vitamin B_{12}	Synthesis of nucleotides
Ascorbic acid	Collagen synthesis, intracellular antioxidant
Choline, inositol	Component of phospholipids

VITAMIN REQUIREMENTS OF FISH

All species of fish examined so far require the same 15 vitamins that are required by birds and mammals, including ascorbic acid, which is not required in the diet of most birds and mammals because they can synthesize ascorbic acid in their tissues. Based upon early studies of vitamins, in which they were extracted from foods, vitamins are classified as being fat soluble or water soluble. Originally, it was thought that there were two vitamins, one fat-soluble and the other water-soluble. Further research revealed that there was more than one active compound within each extract; eventually four separate vitamins were discovered within the fat-soluble fraction and 11 or more within the water-soluble fraction. The fat-soluble vitamins (A, D, E, and K) can be stored in the body, while the water-soluble vitamins supposedly cannot. Fish do have some capacity to store water-soluble vitamins, but the amount that is stored differs among the vitamins.

Vitamins have the same metabolic roles in fish as they do in other animals, and, in general, the required dietary level of those vitamins for fish is not substantially different from that for animals (Table 1). The fat-soluble vitamins are associated with vision (A), calcium metabolism (D), protection against free-radical lipid oxidation (E), and blood clotting (K). Water-soluble vitamins fall into two categories: the B-vitamins, which act primarily as co-factors for metabolic enzymes; and the other water-soluble vitamins (e.g., choline, inositol, and ascorbic acid), which have more complicated functions (Table 2).

VITAMIN DEFICIENCY SIGNS IN FISH

Vitamin deficiency signs in fish are related to the function of the vitamin for some vitamins but seemingly unrelated to function for others (Table 3). Deficiencies of all but two vitamins (B_{12} and K_3) cause loss of appetite, usually as the first overt sign of deficiency. Loss of appetite (anorexia) is thus a sign of vitamin deficiency (not an exclusive sign), but one not specific to an individual vitamin deficiency. Because fat-soluble vitamins can be stored in the body,

Table 3. Primary Vitamin-Deficiency Signs in Fish

Vitamin	Anorexia	Primary Deficiency Signs
Vitamin A	Yes	Vision problems
Vitamin D	Yes	Impaired bone calcification
Vitamin E	Yes	Anemia, ascites, membrane fragility
Vitamin K	No	Anemia, prolonged prothrombin time
Thiamin	Yes	Hyper-irritability, convulsions
Riboflavin	Yes	Lens cataracts
Pyridoxine	Yes	Convulsions, erratic swimming
Pantothenic acid	Yes	Clubbed gills
Niacin	Yes	Skin lesions
Biotin	Yes	Muscle atrophy, skin depigmentation
Folic acid	Yes	Macrocytic anemia
Vitamin B_{12}	No	Anemia
Inositol	Yes	Distended abdomen, edema
Choline	Yes	Hemorrhagic kidney and intestine
Ascorbic acid	Yes	Lordosis, scoliosis, hemorrhages

long periods of feeding deficient diets are needed to cause deficiency for these vitamins, particularly in large fish. Fry and fingerlings have lower body reserves of vitamins, and this scarcity, coupled with their rapid growth rate, makes it easier to create vitamin deficiencies in them than in large fish. As an example, deficiency signs of pantothenic acid (clubbed gills) are evident after about 8–12 weeks of feeding a diet devoid of pantothenic acid in fry and fingerling salmonids (2), but 28–30 weeks of feeding are needed in larger fish (6).

Specific vitamin-deficiency signs are known for some vitamins, but not for others. Deficiencies of thiamin and pyridoxine cause neurological problems, manifested as hyperirritability in trout deficient in thiamin and tetany in salmonids deficient in pyridoxine. Pantothenic-acid deficiency causes a characteristic clubbing of the gills. Riboflavin deficiency is characterized by the presence of cataracts. Vitamin B_{12} and folic-acid deficiency cause anaemia, which can be distinguished by examining blood smears (2). Niacin deficiency causes skin lesions in rainbow trout exposed to ultraviolet radiation. Ascorbic acid deficiency is characterized by broken-back syndrome in channel catfish (*Ictalurus punctatus*) and rainbow trout (*Oncorhynchus mykiss*). Salmonids exhibit deformities of the gill cover as an early sign of ascorbic acid deficiency.

For vitamins such as choline, inositol, biotin, and niacin, specific clinical deficiency signs are observed only in cases of severe and prolonged depletion. In all cases of vitamin deficiency in fish, by the time clinical signs of deficiency are evident, the fish will not eat, and so it is nearly impossible to reverse the deficiency by diet supplementation. Thus, tests to detect subclinical vitamin deficiencies are important. Such tests typically measure the activity of an enzyme for which the vitamin is required as a cofactor. When enzyme activity decreases, the fish are in the early stages of vitamin deficiency, but the condition can usually be reversed by restoring or increasing the dietary vitamin level.

For a few vitamins, deficiencies can be induced when fish are fed semipurified diets lacking the vitamin, but, in practical diets and under controlled conditions, deficiencies cannot be caused, even when there is no supplementation of certain vitamins to the diets. For example, the clinical signs of biotin deficiency in rainbow trout are well described from studies in which semipurified diets were fed (2), but deficiency cannot be induced in rainbow trout fed practical feeds (7). Similar findings are reported in channel catfish fed practical diets (8). Intestinal synthesis of inositol has been demonstrated in channel catfish, making dietary supplementation with this vitamin unnecessary (9). Vitamin B_{12} is another vitamin for which intestinal synthesis is sufficient to supply the needs of some fish [e.g., Nile tilapia (*Oreochromis niloticus*) (10)]. Nevertheless, it is a good practice, under intensive farming conditions, to supplement the feed with a complete vitamin premix, in order to compensate for possible vitamin loss during feed processing and impaired intestinal synthesis of vitamins caused by the use of medicated feed additives in case of disease outbreak.

HYPERVITAMINOSIS IN FISH

Because fat-soluble vitamins are stored in tissues, excessive intake of these vitamins can cause health problems in fish. The occurrence of hypervitaminosis requires massive dietary vitamin levels. The maximum tolerable level of vitamin A in rainbow trout is 900,000 IU/kg, which represents over 130 times the average practical supplementation level for this species (11). Vitamin A hypervitaminosis symptoms in trout include growth depression, necrotic fins, pale liver, and mortalities. Similar levels of vitamin A (from *Artemia* by enriched 920,000–1,000,000 IU/kg) cause compression of vertebrae in larval flounder fed the brine shrimp at the time of notochord segmentation (12). Such extreme levels are unlikely to be encountered in production feeds. Furthermore, vitamin A is sensitive to heat and relatively unstable through feed processing and storage. The probability of observing cases of vitamin A hypervitaminosis in fish farming is therefore marginal. Massive vitamin D_2 supplementation (3.75 million IU/kg) increases circulating calcium in brook trout and reduces growth (13). In juvenile channel catfish, 50,000 IU D_3/kg impairs growth, although it does not alter calcium content in vertebrae (14). Although it is not well documented, it is suggested that excessive doses of vitamin K may lead to hypervitaminosis symptoms in fish. This danger does not seem to be the case for vitamin E, perhaps because it is widely distributed throughout all the lipids in the body.

ESTIMATING DIETARY VITAMIN REQUIREMENTS OF FISH

In the early years of vitamin-requirement studies in fish, requirements were estimated by feeding semipurified diets to groups of fish. The diets contained all known (at the time) essential nutrients at levels in excess of dietary requirements, except for the vitamin being studied; that one would be supplemented to diets at increasing levels. Researchers would feed these diets to fish, then measure growth (weight gain), feed conversion ratios, survival, and, in some cases, vitamin levels in the liver or in blood smears. Absence of deficiency signs, both those clinically evident and those identifiable through histology, was also a response variable in these studies. From these data, the dietary requirement would be estimated. All early studies were conducted using small fish over relatively short experimental periods.

The early studies resulted in estimated dietary vitamin requirements that were in some cases much higher than those established for farmed animals or poultry. Part of this was the result of using liver vitamin levels as a response variable, but part was associated simply with the necessity for being safe. The rationale for this was that for hatchery salmon, at least, the cost of having a dietary vitamin deficiency was much higher in terms of lost production than was the cost of overfortifying the diets with vitamins.

Dietary vitamin-requirement estimates for fish became more accurate in the 1980s as a result of new, sensitive measures of vitamin status, most of which were based on the activities of vitamin-dependent enzymes (15). In addition, the expansion of fish farming, which made it the driving force in fish nutrition research, brought economic considerations, feed cost being a major one, to the forefront of fish nutrition research. Finally, improved semipurified diets made it possible to approximate more closely the growth rates seen in production settings (early semipurified feeds did not support such growth rates). The culmination of these advances is that vitamin requirements are now estimated more accurately, and at values more closely in line with those of other animals.

Recently, the attention of researchers has been focused on the relationship between the intake of several vitamins and disease resistance in fish. Extensive work in mammals has shown that vitamin nutrition plays a significant role in the development and function of the immune system. Vitamin C and vitamin E are potent antioxidants. They scavenge oxygen-rich free radicals and thus protect cells against oxidative damage. Studies show that an elevated vitamin C dietary level (1000 mg/kg) enhances cellular immune response (macrophage and lymphocyte functions) in rainbow trout (16,17). The response of the cells is correlated to their ascorbate content, which is related to the vitamin C intake. Enhanced phagocytosis and humoral immunity upon increased vitamin C intake is also observed in channel catfish (18). Infection trials further demonstrate the positive effect of high vitamin C intake on the resistance of channel catfish and Atlantic salmon (*Salmo salar*) to bacterial infections (18,19) and on resistance of rainbow trout to bacterial, viral, and parasitic infections (20–22).

Vitamin E prevents oxidation of polyunsaturated phospholipids in cellular and subcellular membranes. It also improves both humoral and cellular defenses in fish. Rainbow trout fed a diet containing vitamin E at 450 mg/kg (ppm) exhibit a significantly higher antibody response to vaccination and lymphocyte proliferation than do fish fed a diet supplemented with 45 mg/kg (ppm) (23). Rainbow trout experimentally infected with *Yersinia ruckeri* exhibit significantly less mortality when fed a diet containing 806 mg/g (ppm) compared to fish receiving a diet with 86 mg/kg (ppm) (24).

Vitamin C and vitamin E requirements for optimal immune response are higher than the levels necessary for normal growth and prevention of deficiency symptoms. An appropriate dietary supplementation of these two vitamins as a prophylactic tool contributes to improving the health status of fish and to reducing losses due to infectious diseases.

FACTORS THAT INFLUENCE DIETARY VITAMIN RECOMMENDATIONS

Vitamin requirements of fish are influenced by several factors. First, there are species differences, as noted earlier. Second, there are differences associated with age, size, rearing conditions, and maturation. Third, dietary components can increase the physiological need for some vitamins—For examples, the presence of oxidizing fats in the feed increases the need for tocopherol (vitamin E) and possibly vitamin C; and egg white contains avidin, a compound that binds biotin, making it unavailable. Fourth, any leaching of vitamins from feeds that remains in the water for a period before being eaten by fish or shrimp increases the amount that must be added to feeds. In addition, any manufacturing condition that destroys vitamins must be taken into account, and the feeds must be fortified accordingly. Fifth, for pond-reared fish and shrimp, natural food supplies a portion of dietary vitamin needs. As the supply of natural food diminishes or the stocking density of ponds increases, vitamin fortification of feeds becomes more critical. Lastly, and most importantly, exposure to stressful rearing conditions, including exposure to pathogens, increases the requirement for some vitamins.

SUMMARY

Vitamins are essential nutrients in the diets of fish. Vitamins have the same metabolic roles in fish as in mammals and birds, and the dietary requirements of fish are similar to those of other animals. Vitamin needs increase when feeds contain oxidizing lipids, when fish are at an early stage of development or subjected to diseases or stressful rearing conditions, and possibly when fish mature.

BIBLIOGRAPHY

1. C.M. McCay and W.E. Dilley, *Trans. Am. Fish Soc.* **57**, 250 (1927).
2. J.E. Halver, in J.E. Halver, ed., *Fish Nutrition*, 2d ed., Academic Press, NY, 1989, pp. 31–109.
3. B.A. McLaren, E. Keller, D.J. O'Donnell, and C.A. Elvehjem, *Arch. Biochem. Biophys.* **15**, 169 (1947).
4. L.E. Wolf, *Prog. Fish-Cult.* **13**, 17 (1951).
5. J.E. Halver, *Trans. Am. Fish. Soc.* **83**, 254 (1953).
6. T. Masumoto, R.W. Hardy, and R.R. Stickney, *J. Nutr.* **124**, 430–435 (1994).

7. A.J. Castledine, C.Y. Cho, S.J. Slinger, B. Hicks, and H.S. Bayley, *J. Nutr.* **108**, 698–711 (1978).
8. R.T. Lovell and J.C. Buston, *J. Nutr.* **114**, 1092–1096 (1984).
9. G.J. Burtle and R.T. Lovell, *Can. J. Fish. Aquat. Sci.* **46**, 218–222 (1989).
10. R.T. Lovell and T. Limsuwan, *Trans. Am. Fish. Soc.* **111**, 485–490 (1982).
11. J.W. Hilton, *J. Nutr.* **113**, 1737–1745 (1983).
12. J. Dedi, T. Takeushi, T. Seikai, T. Watanabe, and K. Hosoya, *Fisheries Science* **63**, 466–473 (1997).
13. H.A. Poston, *Fisheries Res. Bull.* **32**, 44–47 (1969).
14. C.A. Launer, O.W. Tiemeier, and C.W. Deyoe, *Prog. Fish-Culturist* **40**, 16–20 (1978).
15. National Research Council, *Nutrient Requirements of Fish*, National Academy Press, Washington, DC, 1993.
16. V. Verlhac, A. Obach, J. Gabaudan, W. Schüep, and R. Hole, *Fish and Shellfish Immunology* **8**, 409–424 (1998).
17. V. Verlhac and J. Gabaudan, *Aquaculture and Fisheries Management* **25**, 21–36 (1996).
18. Y. Li and R.T. Lovell, *J. Nutr.* **115**, 123–131 (1985).
19. L.J. Hardie, T.C. Fletcher, and C.J. Secombes, *Aquaculture* **95**, 201–214 (1991).
20. O. Navarre and J.E. Halver, *Aquaculture* **79**, 207–221 (1989).
21. Y. Suzuki and T. Ai, *Bull. Shizuoka Pref. Fish. Exp. Stn.* **24**, 25–29 (1989).
22. T. Whali, R. Frischknecht, M. Schmitt, J. Gabaudan, V. Verlhac, and W. Meier, *J. Fish Dis.* **1995**, 347–355 (1995).
23. V. Verlhac, A. N'Doye, J. Gabaudan, D. Troutaud, and P. Deschaux, in *Les Colloques* **61**, 167–177 (1991).
24. M.D. Furones, D.J. Alderman, D. Buske, T.C. Fletcher, D. Knox, and A. White, *J. Fish Biol.* **41**, 1037–1041 (1992).

See also FEED MANUFACTURING TECHNOLOGY; NUTRIENT REQUIREMENTS; VITAMINS SOURCES FOR FISH FEEDS.

VITAMINS SOURCES FOR FISH FEEDS

J. GABAUDAN
Research Centre for Animal Nutrition and Health
Saint-Louis Cedex, France

RONALD W. HARDY
Hagerman Fish Culture Experiment Station
Hagerman, Idaho

OUTLINE

Vitamin supplements for fish feeds must meet the dietary requirements of fish. Fish feeds are a challenging environment for vitamins, in part because of the conditions of manufacture and in part because of the fact that the pellets must remain intact in water for up to several hours before fish or shrimp consume them. Thus, vitamins must be provided to fish feeds in forms that can tolerate high heat and pressure during feed processing and conditions encountered during drying. Leaching of vitamins from pellets exposed to water is a major problem in raising shrimp or fish that consume feed slowly. The use of fish oil, which is rich in polyunsaturated fatty acids that are susceptible to oxidation, as the primary lipid source in fish feeds also must be considered when the chemical forms of certain vitamins are being chosen. Oxidizing lipids can destroy some vitamins unless they are protected. Most vitamins used to supplement fish and animal feeds are produced in forms that disperse well in feeds and resist destruction during pelleting and storage. However, few of the vitamins are completely stable in fish feeds. Thus, levels of supplementation must be chosen to account for the inevitable losses that occur in feed preparation and storage.

SOURCES, FORMS, AND ACTIVITY OF VITAMINS ADDED TO FISH FEEDS

Sources

Most vitamins added to fish feeds are produced by industrial processes, generally either chemical synthesis, fermentation, or a combination of both, rather than by extraction from natural sources, which is expensive and gives low yields. In addition, vitamins are manufactured in forms that are designed to flow evenly, mix well, and resist degradation during feed processing and storage. This is not a simple process, since optimal stability and full bioavailability are not necessarily compatible characteristics. Fat-soluble vitamins formerly were only available from natural sources, such as cod or shark liver oil. These vitamins generally occur in several active forms in nature, and the biological activities of these forms differ (Table 1). Today, vitamins A, D, E, and K are produced by chemical synthesis. Of the water soluble vitamins, most that are used in feeds are produced by chemical synthesis or by fermentation and are presented as vitamin salts or isomers. Often crystalline forms used in pharmaceutical products are produced by synthesis, while dilution products used in feeds are produced by fermentation.

Forms

Fat-soluble vitamins. Vitamin A is a chemically synthesized product that is enclosed in a beadlet to protect

Table 1. Relative Biological Activity of Different Forms of Vitamin E (1)

Vitamin E Form	Relative Vitamin E Activity (%)
Alpha-tocopherol	100
Beta-tocopherol	30
Gamma-tocopherol	15
Delta-tocopherol	Almost inactive
Alpha-tocotrienol	20
Beta-tocotrienol	5

against oxidation. Indeed, vitamin A is one of the vitamins most affected by the aggression of mill machinery. Beadlets typically contain a matrix such as cross-linked gelatin, with vitamin A dispersed throughout the matrix. Within the matrix the vitamin can be additionally protected by an antioxidant. Beadlets are then coated with a protective layer, such as corn starch, which improves handling. The form of vitamin A in such beadlets is vitamin A acetate. Several manufacturers produce beadlets containing both vitamin A and vitamin D. Stability of vitamin A beadlets through extrusion and three months of room temperature storage is approximately 80% at best and 40% at worst for different vitamin A beadlet products, the latter having no cross-linked matrix.

Vitamin D is added to feeds as a beadlet, enclosing cholecalciferol (D_3), or as a spray-dried product. The other form of vitamin D, ergocalciferol (D_2), is not used as a vitamin D supplement in animal or fish feeds. Antioxidant addition plus encapsulation within a beadlet provide protection of vitamin D_3 against oxidation. Typical stability ranges between 75–100% after extrusion and three months of room temperature storage.

Vitamin E is supplied to feeds as *dl*-alpha-tocopheryl acetate, which is an acetate ester of alpha-tocopherol. The acetate moiety is attached at the active site on the tocopherol molecule, thus preventing any other reactions from occurring that might result in loss of tocopherol activity. The most worrisome reaction in feeds is associated with oxidizing lipids, in which tocopherol donates a hydrogen atom, thus becoming a sacrificial antioxidant. The presence of the acetate moiety prevents oxidation, but also renders *dl*-alpha-tocopheryl acetate inactive with respect to antioxidant function in feeds. Once in the gut, the acetate moiety is enzymatically removed, restoring the antioxidant property to the tocopherol molecule. Vitamin E is relatively stable in extruded feeds when supplemented in the protected form, with no more than 10% loss after pelleting and extrusion.

Vitamin K is supplied as a menadione (K_3) salt. There are four forms of menadione salts used in feeds: menadione sodium bisulfite (MSB, 50% active K_3), menadium nicotinamide bisulfite (MNB, 43% active K_3), menadione sodium bisulfite complex (MSBC, 33% active K_3), and menadione dimethlypyrimidinol bisulfite (MPB, 45.4% active K_3). All are affected by heat, moisture, and the presence of trace minerals. After extrusion pelleting and three months of room temperature storage, between 20 and 50% of vitamin K activity remains.

Water-soluble vitamins. Thiamin (vitamin B_1) is commercially available as crystalline mononitrate or hydrochloride salts. Thiamin mononitrate (1 g thiamine = 1.088 g thiamine mononitrate) is typically used in animal feeds, while thiamin hydrochloride is typically used in liquid parenteral or oral vitamin products because it is more water soluble. Between 60% and 80% retention of thiamin activity is typically observed after feeds are extruded and stored at room temperature for three months.

Riboflavin (vitamin B_2) is produced as a crystalline compound or as a product of fermentation. The crystalline product is electrostatic and hygroscopic and does not distribute well when blended into a feed mixture. Its handling properties are significantly improved when it is formulated into a spray-dried powder. Riboflavin is relatively unaffected by extrusion pelleting and storage, with no more than 10% loss occurring after three months of storage of extruded pellets.

Pyridoxine (vitamin B_6) is typically added to feeds as crystalline pyridoxine hydrochloride, which is 82.3% active. Pyridoxine is relatively unstable, especially in premixes exposed to moisture (e.g., high humidity, and containing trace minerals). Up to 50% of pyridoxine activity in premixes exposed to abusive conditions can be lost after three months of storage. The stability of pyridoxine partly depends on the size of its crystal particles. Therefore, fine granular crystals have improved stability during feed processing compared to very fine crystals. Properly formulated pyridoxine is relatively stable during pelleting, with typical extrusion and storage losses of 10–20%.

Pantothenic acid is normally added to feeds as calcium *d*-pantothenate, which contains 92% of *d*-pantothenic acid. Calcium *dl*-pantothenate also exists, but has half of this activity because the *l* forms of pantothenate are not biologically active. Calcium *d*-pantothenate is relatively stable during pelleting and feed storage, with losses after extrusion and storage of no more than 20%.

Niacin is added to feeds as niacinamide and nicotinic acid, both having the same biological activity. Both forms are quite stable during extrusion, pelleting and storage, with losses generally 10% or less.

Biotin is added to feeds as *d*-biotin, the biologically active form. The isomer *l*-biotin has no biological activity. The activity of *d*-biotin products is 2% on a weight basis, and the stability of biotin during extrusion pelleting and room temperature storage for three months ranges from 70 to 90%.

Folic acid is synthesized and added to vitamin premixes as a dry dilution, either as a crystalline form or a spray-dried form. The crystalline form is electrostatic and tends to adhere to the machinery, while the spray-dried form does not and therefore contributes to a higher recovery of the vitamin in the feed. Stability of folic acid after extrusion and feed storage is relatively low, ranging from 50 to 65%.

Vitamin B_{12} is produced by fermentation and used in feeds as a dry dilution having 1% activity, on a weight basis. Vitamin B_{12} stability in extruded feeds after three months of room temperature storage ranges from 40 to 80%.

Choline is produced for feed use as a chloride salt, which is available as a dry dilution product having either 25, 50, or 60% activity on a weight basis or as a liquid having 70% activity. On a molecular weight basis, choline chloride is 86.8% choline. Choline is completely stable during feed pelleting and storage, but it is a hygroscopic substance and a strong base. Its presence reduces the activity of other vitamins, such as vitamin E and vitamin K when it is included in vitamin premixes. Thus, it should be added separately to feed mixtures.

Inositol is a hexahydric cyclic alcohol with several isomeric forms. The only isomer with biological activity is myoinositol.

Vitamin C, or ascorbic acid, was once the most problematic of all the essential vitamins with respect to stability in fish feeds. Crystalline ascorbic acid (100% active on a weight basis) is extremely susceptible to oxidation, and early tests with the Oregon Moist Pellet® indicated that within three days without frozen storage, all vitamin C activity was gone. Thus, feed was kept frozen until use, and coated forms (fat-coated, ethylcellulose-coated) more resistant to oxidation were used. In dry, pelleted fish feeds, approximately 20% of vitamin C activity remained after steam pelleting and storage, so feed formulators added five times more crystalline or coated ascorbic acid to ensure that enough remained to meet the dietary requirements of fish at the time of feeding. The use of extrusion pelleting in fish feed production introduced additional heat and pressure to the pelleting process, sufficient to melt the fat coating and otherwise accelerate the loss of ascorbic acid activity in fish feeds. Thus, conjugates were developed that added a functional group to the second carbon position of ascorbic acid, thus protecting ascorbic acid from oxidation. The first such product was ascorbate-2-sulfate, which was very stable, but had low biological activity to salmon (ca. 30%) and catfish (ca. 10%). The second such product was ascorbate-2-polyphosphate, which had full biological activity, but relatively low activity on a molecular weight basis, due to the relative weight of the polyphosphate moiety.

More recently, ascorbate-2-monophosphate has been developed, increasing the ascorbic acid content to ca. 49% on a molecular weight basis. This product, a Na/Ca salt of ascorbate-2-phosphate consists of ascorbate-2-monophosphate (equivalent to 33% ascorbic acid activity) and small amounts of ascorbate-2-polyphosphate (equivalent to an additional 2% ascorbic acid activity). It is in wide use in the fish feed industry today and exhibits less than 15% loss of activity in extrusion pelleting and three months of room temperature storage, compared to 70–90% loss of activity for ethylcellulose-coated or fat-coated ascorbic acid. For steam pelleted feeds, losses of crystalline ascorbic acid range from 30 to 70%, depending upon pelleting, drying, and storage conditions, compared with less than 10% loss of ascorbate-2-phosphate.

ADDITIONAL FACTORS THAT INFLUENCE VITAMINS IN FISH FEEDS

Vitamins are affected by many factors associated with feed ingredients and feeds, but not all vitamins are affected equally. As mentioned earlier, vitamins are produced in forms that resist deterioration and loss of activity. Several processes are used in the manufacture of feed-grade vitamin products. These include encapsulation (beadlets), spray drying, coating, adsorption, compaction, and high-shear granulation. The formulation process affects the stability, handling, miscibility, and bioavailability of the vitamins (2). However well a vitamin is formulated, losses still may occur. The main factors that lower vitamin levels in premixes or feeds are moisture, oxidation, reduction, trace minerals, heat, light, and pH (low or high) (Table 2). Conditions of pelleting (temperature, moisture), length of storage, and the composition of the feed, particularly the presence of oxidizing polyunsaturated fatty acids affect vitamin stability (Table 3). In the case of vitamin premixes, the inclusion in the premix of trace minerals accelerates the destruction of vitamins, as many trace mineral ions act as catalysts of oxidation reactions, while oxides and hydroxides of metals increase the pH. The hydroscopic property of choline chloride increases moisture levels in vitamin premixes, thereby affecting stability of certain vitamins, notably vitamin K. The importance of these two factors is summarized in Table 4. Finally, some compounds added to fish feeds or present in ingredients can seriously inhibit or destroy vitamins. For example, thiaminases present in certain fish degrade thiamine, lindane (a common insecticide) is an antagonist of inositol, and sulfaquinoxaline inhibits vitamin K.

Because vitamins are affected by different factors, it is not possible to define a set of practical conditions favorable to all of them. Nevertheless, following a few simple rules will contribute to maximize vitamin stability and bioavailability in feeds: (*1*) select good

Table 2. Relative Influence of Various Factors on Vitamin Stability[a]

Vitamin	Moisture	Oxidation	Trace Minerals	Heat	Light	pH (low)
Vitamin A (beadlet)	S	S	S	MS	MS	S
Vitamin D (beadlet)	S	S	S	MS	MS	S
Vitamin E (acetate)	R	R	MS	R	R	MS
Vitamin K (MSBC)	VS	R	VS	MS	S	MS
Thiamin mononitrate	R	MS	MS	MS	R	R
Riboflavin	R	R	R	R	MS	R
Pyridoxine-HCl	R	R	MS	R	S	R
Ca-pantothenate	S	R	R	MS	R	S
Niacin	R	R	R	R	R	R
d-Biotin	R	R	R	S	R	MS
Folic acid	R	MS	S	MS	MS	S
Vitamin B_{12}	R	MS	MS	MS	S	MS
Choline chloride	VS	R	R	R	R	R
Ascorbate-2-phosphate	R	R	R	R	R	R

[a]Code: R: resistant, S: susceptible, MS: moderately susceptible, VS: very susceptible.

Table 3. Stability of Vitamins in Premixes, Steam Pellets, and Extruded Pellets after Three Months of Room-Temperature Storage[a]

Vitamin	Premixes (%)	Steam Pellets (%)	Extruded Pellets (%)
Vitamin A (beadlet, cross-linked)	70–90	85–95	70–90
Vitamin D (beadlet, cross-linked)	80–100	90–100	75–100
Vitamin E (acetate)	90–100	90–100	90–100
Vitamin K (MNB)	65–85	70–90	40–70
Thiamin	70–80	85–100	60–80
Riboflavin	90–100	90–100	90–100
Pyridoxine	80–90	90–100	80–90
Pantothenic acid	80–100	90–100	80–100
Niacin	90–100	90–100	90–100
Biotin	80–100	90–100	70–90
Folic acid	50–70	70–90	50–65
Vitamin B_{12}	50–80	60–90	40–80
Inositol	100	100	100
Choline	Not added	100	100
Ascorbic acid, crystalline	30–70	30–70	10–30
Ascorbate-2-phosphate	90	90	90

[a]*Source*: Reference 5.

Table 4. Retention (%) of Vitamins in Premixes With and Without Choline and Trace Minerals after Three and Six Months of Storage

	Vitamins Alone		With Choline and Trace Minerals	
Vitamin	3 months	6 months	3 months	6 months
Vitamin A (beadlet)	97	94	74	58
Vitamin D (beadlet)	98	96	77	65
Vitamin E (acetate)	98	96	88	82
Vitamin K (MSBC)	96	90	21	0
Thiamin mononitrate	99	94	72	52
Riboflavin	99	98	78	59
Pyridoxine	98	97	76	56
Pantothenic acid	99	98	79	58
Niacin	99	98	79	58
Biotin	99	98	77	57
Folic acid	99	97	63	43
Vitamin B_{12}	99	98	95	89
Choline	Not added		997	91
Ascorbic acid, crystalline	78	65	22	0
Ascorbate-2-phosphate	98	96	77	65

quality, protected vitamin products; (*2*) store vitamin premixes in a cool, dark, and dry room and reseal bags; (*3*) use good quality feed ingredients, particularly oils; (*4*) add vitamin premix to the mix after grinding; (*5*) use dry steam and lowest processing temperatures compatible with pellet quality; and (*6*) pack and store feeds with care.

MEASURING VITAMIN LEVELS IN FEEDS

Historically, most vitamins were measured by microbiological assay, both qualitatively and quantitatively. The development of high-performance liquid chromatography (HPLC) has reduced the use of microbiological and wet chemistry methods for vitamin analysis. Today, vitamins A, D_3, E, K_3, thiamin, riboflavin, niacin and ascorbic acid are determined by HPLC. Pyridoxine, B_{12}, pantothenic acid, biotin, and folic acid contents in feeds are evaluated with microbiological assays (3,4).

SUMMARY

The manufacturing conditions used to pellet fish feeds are abusive and result in destruction of many vitamins, unless they are supplemented in protected forms with good handling and miscibility properties. The use of these forms protects the vitamins against loss, but generally not completely. Thus, care must be taken to add sufficient amounts of vitamins to fish feeds to counter the inevitable losses that occur during manufacture and storage. Vitamins are added to feeds mainly as a vitamin premix, with the exception of choline chloride and ascorbic acid, which are added separately to avoid combining these compounds with other vitamins.

BIBLIOGRAPHY

1. J. Le Grusse and B. Watier, in J. Le Grusse and B. Watier, eds., *Les Vitamines, Données Biochimiques, Nutritionnelles et Cliniques*, CEIV Produits Roche, Neuilly-sur-Seine, 1993, pp. 81–100.
2. M. Putnam and A. Taylor, *Feed Tech.* **1**, 39–43 (1997).
3. J. Augustin, B.P. Klein, D. Becker, and P.B. Venugopal, eds., *Methods of Vitamin Assay*, 4th ed., John Wiley & Sons, NY, 1985.
4. H.E. Keller, ed., *Analytical Methods for Vitamins and Carotenoids in Feed*, F. Hoffmann La Roche Ltd, Basel, 1988.
5. F. Hoffmann-LaRoche *Technical Document No. 51098*, F. Hoffmann-LaRoche Ltd., Basel, 1998.

See also VITAMIN REQUIREMENTS.

WALKING CATFISH CULTURE

ROBERT R. STICKNEY
Texas Sea Grant College Program
Bryan, Texas

OUTLINE

Walking-catfish (family Clariidae) have long been studied by zoologists because of their ability to survive for long periods out of water by breathing air directly. Walking-catfish were named for their ability to use their modified pectoral fins to essentially "elbow" their way across the land, often from pond to pond. The physiological and morphological adaptations allow the fish to either aestivate by burying themselves in mud like lungfish or to leave ponds that are becoming desiccated in order to find new bodies of water. Walking-catfish are also popular foodfish in many countries, not only the tropical regions to which they are native, but also in some European nations, where they are reared in recirculating water systems. The Netherlands has been actively involved in research and development. Walking-catfish have been reared in the United States primarily as aquarium fishes, but fears that they could escape and walk across land areas to invade various bodies of water—where they might displace more desirable species—prompted the prohibition of walking-catfish in the United States. There have been stories about walking-catfish eating dogs in Florida, though such rumors have never been substantiated and can probably be classified as urban legends.

DISTRIBUTION AND GENERAL CHARACTERISTICS

Walking-catfish are distributed throughout southeast Asia, including the Indian subcontinent, Africa, and parts of the Near East (1). In Asia, Thailand has been a leader in research associated with walking-catfish culture, as well as a leading producer nation. A considerable amount of research has also been conducted in India and the Philippines. Important species in Asia are *Clarias batrachus* (Fig. 1) and *C. macrocephalus*. There is also some culture of *C. gariepinus* and hybrids in Asia.

In Africa, walking-catfish inhabit tropical swamps, lakes, and rivers. Most of the interest by African culturists has been associated with the sharptooth catfish, *C. gariepinus* (other species have been recognized in Africa, but there is a view among some taxonomists that *C. gariepinus* is the only actual African species). *C. gariepinus* is found from the Orange River in South Africa to the Nile River. The species can also be found in parts of Turkey and is the species of interest to culturists in Europe. Most of the activity associated with walking-catfish culture in Europe is centered in the Netherlands (2). Markets have been established in Germany, Italy, and the United Kingdom, in addition to the Netherlands.

A few other species of walking-catfishes have also received the attention of aquaculturists. Among them are *C. fuscus*, *C. anguillaris*, *C. lazera*, and *C. leather*, reared in China. Research has also been conducted on hybrids between *C. gariepinus* and another clariid fish, *Heterobranchus longifilis*. However, there is not much information available on these hybrids and minor species, and they do not appear to be produced in large numbers as compared with the three dominant culture species.

Walking-catfish are tropical species. Growth occurs when water temperature is above 20 °C (68 °F), with best growth occurring in the range of 25 to 30 °C (77 to 86 °F). Because of the presence of an accessory air-breathing organ, walking-catfish are able to exist for hours out of water; they are often sold live on the streets in Asia from buckets containing fish with no water. Of more interest to fish farmers is the fact that walking-catfish can tolerate oxygen-depleted water and even survive in moist mud. Their propensity for leaving ponds and foraging on land, particularly at night, means that fences must be placed around ponds to keep the fish from escaping and taking up residence in the wild or at a competitor's culture facility. Vertical pond walls of earth or concrete are also useful in helping to constrain walking-catfish (Fig. 2). Walking-catfish are bottom feeders. They grow relatively rapidly and reach maturity after about one year at weights of 200 to 400 g (0.44 to 0.88 lb). As is true of other families of catfishes, the clariids produce a relatively small number of large eggs. Mature females typically produce from 2,000

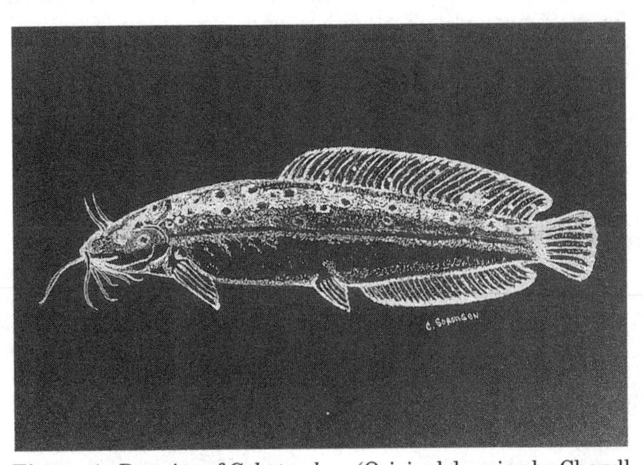

Figure 1. Drawing of *C. batrachus*. (Original drawing by Cheryll Sorensen.)

Figure 2. Walking-catfish pond in the Philippines with vertical walls and fencing.

to 5,000 eggs that are round, yellowish brown in color, and range from 1.3 to 1.6 mm (0.05 to 0.06 in.) in diameter (2). Hatching occurs within 24 hours (1) when temperature ranges from 25 to 32 °C (77 to 90 °F).

Walking-catfish are not a major aquaculture commodity. In 1995, global production was about 89,000 tons (3).

COLLECTION AND CAPTIVE SPAWNING OF FRY

The simplest method of obtaining young fish for stocking is to allow them to spawn naturally and collect the resulting fry. Depending on species, walking-catfish spawn in nests constructed during the rainy season in the bottoms of ponds and rice fields or in holes in the banks of a body of water. Plant material may be placed in the nests to facilitate egg collection (4). Spawning generally occurs in 20 to 50 cm (8 to 20 in.) of water (1). The fry can then be removed from the nests with nets. This method of spawning and collection of fry is simple and does not require indoor facilities, use of hormones, or other technology that may be unavailable or prohibitively expensive in some regions.

Fish farmers in Thailand, prompted by a shortage of walking-catfish fry from wild spawns, began in the 1950s to employ methods similar to those used in the United States for spawning channel catfish. (See the entry "Channel catfish culture.") Instead of placing spawning containers in the ponds, Thai fish farmers dug suitable-sized horizontal holes in their pond banks, stocked pairs of *C. batrachus* near the holes, and were rewarded with a high degree of successful spawns (1). Fry could be collected in the usual manner from the nests, which were 20 to 35 cm (8 to 14 in.) in diameter and constructed about 1 m (3 ft) apart.

Pituitary hormone injections were found to be effective for inducing spawning in *C. macrocephalus* by Thai researchers who were employing the technique by the early 1970s. By utilizing the proper level of hormone, the fish could be induced to spawn well over 50% of the time within 14 to 16 hours following injection (1). Various other hormones, including human chorionic gonadotropin and leuteinizing hormone-releasing hormone or LHRH (see the entries "Hormones in finfish aquaculture" and "Reproduction, fertilization, and selection"), have in more recent years been shown to be effective at inducing spawning in walking-catfish (5). Pituitary extracts that are effective can be obtained from fish, such as carp, and from other animals, including chickens, toads, and frogs. To reduce the need for indoor broodstock spawning facilities, hormone injections and subsequent spawning activities can be undertaken in small cages placed in ponds.

After fertilization, eggs can be collected and hatched in troughs or jars. At present, Thai fish culturists produce *C. batrachus*, *C. macrocephalus*, and a hybrid composed of a cross between *C. gariepinus* and *C. macrocephalus*. Hybrid vigor is one positive result of such crosses. Other species of less interest to fish culturists have also been induced to spawn with hormones.

GROWOUT SYSTEMS

When fertilized eggs are incubated in hatcheries, the resulting larvae are often maintained through absorption of the yolk sac for about five days, then stocked in small earthen ponds (1). Troughs or circular raceways should be used for early rearing of fry. Sufficient oxygen levels should be maintained until the fish begin air breathing if, as some advocate, the fry are retained in the hatchery for several days after hatching (6). Fry have traditionally been maintained in ponds at high densities until they reach fingerling size, at which time they are restocked into production ponds or cages (1).

Production ponds are of various sizes and have depths that are typical of aquaculture ponds throughout the temperate and tropical world. (See the entry "Pond culture.") The difference is that the walls of the ponds may be more vertical than is the case with the standard aquaculture pond, and fences of about 50 cm (20 in.) may be constructed around each pond to prevent the fish from escaping (Fig. 2). While most production continues to be in ponds, flow-through raceways have also been employed in some places, including developing countries. Walking-catfish are most commonly reared in monoculture, though they have been polycultured with tilapia and are sometimes grown in rice-fish culture. (See the entry "Polyculture.")

Live and prepared feeds are commonly used in conjunction with walking-catfish farming, though many farmers depend upon fertilization alone. Organic fertilizers, including night soil, are commonly used, often in lieu of feeding with prepared rations. Organic fertilizers stimulate the production of natural food organisms. Manure may be carried to the ponds intermittently, or it may be continuously provided by rearing livestock in facilities constructed over, or immediately adjacent to, the fishponds. Apparently, there are no studies showing that manure ingested directly provides walking-catfish with some nutritional benefit, but there is evidence that dried poultry waste can be used in feeds to substitute for other ingredients in walking-catfish diets. Organic fertilizers commonly used in conjunction with the rearing of walking-catfish include manure from poultry, swine, water buffalo, and cattle. Ducks are a popular source of organic fertilizers for ponds in various Asian countries and are also undoubtedly used to fertilize walking-catfish ponds. It is important in many parts of Asia and Africa to produce fish in facilities that involve little capital investment and only limited inputs, such as prepared feeds, which require significant expense. Subsistence culture is common, as are small family-operated commercial farms. Family farms can be operated in the absence of weekly or monthly payrolls. Ponds can be dug by hand or with the aid of livestock-drawn implements, natural spawning and fry collection eliminate the need for hatcheries, and the provision of organic fertilizers from other farm animals eliminates the need for purchasing feed or fertilizers. Relatively small fish are accepted by consumers in many countries, so walking-catfish can often be sold at sizes of less than 200 g (0.45 lb). If the fish can be sold at relatively small sizes, it is possible to produce three crops per year in the tropics.

WATER-QUALITY CONSIDERATIONS

As mentioned earlier, one of the characteristics of walking-catfish that make them desirable as aquaculture species is their tolerance of degraded water quality. They can be grown in ponds, which, at least during part of the time, have dissolved oxygen levels approaching or reaching zero parts per million. In the marketplace, one often sees live walking-catfish that have been maintained for several hours out of water. Even after lying out of water in the sun for several hours, the fish can be seen actively moving about.

Walking-catfish, like other freshwater catfishes, have limited tolerance to salinity. Salinities much above one-third strength seawater are not tolerated.

Because of the requirement of the tropical walking-catfishes for warm water, aquaculture activities in the Netherlands are centered around production in recirculating water systems. (See the entry "Recirculating water systems.") Among the water-quality variables that can cause toxicity in such systems is the buildup of ammonia and nitrite (NO_2^-) in the water. Walking-catfish do not appear to be more tolerant of nitrite than are channel catfish (family Ictaluridae) and are less tolerant than various other species of fishes and marine shrimp (5). Ammonia, particularly in the un-ionized form (NH_3), is highly toxic to fish, yet *C. batrachus*, which is intolerant of high nitrite, is quite tolerant of un-ionized ammonia (5) compared with channel catfish. The nitrification of ammonia by bacteria results in the production of toxic nitrite, which is then converted to low-toxicity nitrate (NO_3^-).

FEEDS, FEEDING, AND NUTRITION

Walking-catfish are omnivorous (7). They are aggressive and known to be cannibalistic. As first-feeding fry, they are normally provided with live zooplankton, though yeast has been used as a sole first feed for at least one species (5). Brine shrimp nauplii (*Artemia* sp.) are often used for several days, followed by cladocerans mixed with brine shrimp nauplii, and then cladocerans only. Monocultures of the mentioned zooplanktonic species can be grown, though it is much less difficult and requires little technology to rear natural zooplankton communities in fertilized tanks or small ponds. After a few days of providing live food, culturists can successfully convert fry to wet or dry prepared feeds.

In places where trash fish are readily available — for example, parts of Thailand — a mixture of 90% ground fish and 10% rice bran has been found to be readily utilized as food by walking-catfish fingerlings (8). Dry diets employing fishmeal, silkworm pupae, and various vegetable proteins have been used in feeding walking-catfish. Rice bran, while low in overall nutritional value, is widely used at some level in prepared feeds, largely perhaps because it is readily available and inexpensive. Rice and groundnut cake have also been mentioned as dietary ingredients. The protein requirement of walking-catfish is currently placed at about 40% (7), though there have been studies that placed the level 10% lower (3). Growth is better facilitated by animal protein than by plant protein. When the dietary crude protein level is 40%, *C. batrachus* have an optimum lipid requirement of between 7 and 9% (9). Little is known about the vitamin requirement of walking-catfish, though the ascorbic acid requirement of *C. batrachus* fry is about 70 mg/kg (70 parts per million) of their diet (10). The use of high levels of carbohydrate in conjunction with walking-catfish feeds remains questionable, as the ability of those fishes to utilize various types of carbohydrates has yet to be determined in detail.

DISEASES

Walking-catfish are not only extremely hearty and tolerant of degraded water quality; they seem also to have some resistance to aquatic animal diseases. Among the bacteria that have been reported in conjunction with walking-catfish under culture have been *Aeromonas hydrophila* and *Pseudomonas* sp. In the case of *A. hydrophila*, very high bacterial levels must be present in the musculature before lesions develop (11). Parasites of various kinds have been reported from collections of walking-catfish from natural water bodies.

BIBLIOGRAPHY

1. J.E. Bardach, J.H. Ryther, and W.O. McLarney, *Aquaculture*, Wiley-Interscience, New York, 1972.
2. R.R. Stickney, *World Aquaculture Mag.* **22**(2), 44–54.
3. FAO, *Fisheries Circular No. 815, Rev. 9*, Food and Agriculture Organization of the United Nations, Rome, 1997.
4. C.F. Knud-Hansen, T.R. Batterson, C.D. McNabb, Y. Hadiroseyani, D. Dana, and H.M. Eidman, *Aquaculture* **89**, 9–19 (1990).
5. R.R. Stickney, *Principles of Aquaculture*, John Wiley & Sons, New York, 1994.
6. G.S. Haylor, *Aquacult. Fish. Manage.* **24**, 245–252 (1993).
7. Van Weerd, *Aquat. Living Resour.* **8**, 395–401 (1995).
8. Brown, *World Fish Farming: Cultivation and Economics*, AVI Publishing Co., Westport, CT, 1983.

9. M.F. Anwar and A.K. Jafri, *J. Appl. Aquacult.* **5**, 61–71 (1995).
10. S. Mishra and P.K. Mukhopadhyay, *Indian J. Fish.* **43**, 157–162.
11. G.D. Lio-Po, L.J. Albright, and E.M. Leano, *J. Aquat. An. Health* **8**, 340–343 (1996).

See also EXOTIC INTRODUCTIONS.

WALLEYE CULTURE

ROBERT C. SUMMERFELT
Iowa State University
Ames, Iowa

OUTLINE

Walleye (*Stizostedion vitreum*) is a member of the perch family (Percidae), which, in Canada and the United States, includes sauger (*S. canadense*), yellow perch (*Perca flavescens*) and about 138 species of darters and logperch (1). Blue pike (*S. vitreum glacum*), presumed to be extinct, is a subspecies of walleye that once thrived in Lake Erie. Walleye presently occur in 32% of the freshwater habitat in the United States and Canada (2). Walleye are targeted as a sport fish in 34 states, seven provinces, and one territory (3). In the United States in 1996, walleye and sauger were fished by about 11% the 35.2 million U.S. residents 16 years old and older that fished in freshwater other than the Great Lakes and 36% of anglers in the Great Lakes (4). In 1990, walleye represented 16.3% of the total freshwater fishes captured by anglers in Canada (3), and walleye are said to be "the most economically valuable species in Canada's inland waters" (5).

Wild-caught walleye, blue pike, and sauger were once a substantial part of the commercial fisheries of the Great Lakes, especially from Lake Erie, the Mississippi River, and many of the numerous glacial lakes of the United States and Canada. In Lake Erie, in the interval 1879–1959, the three congeners — walleye, sauger, and blue pike — contributed 13.7 to 18.3% of the total commercial harvest of fish. The blue pike harvest was 49.5 to 79% of harvest of the three taxa until they abruptly declined in the 1960s (6) (Fig. 1). Continued commercial exploitation of walleye from Lake Erie may have contributed to the extinction of the blue pike (7). Currently, an overwhelming majority of the commercial harvest of walleye in North America is from the Canadian shore of Lake Erie and many isolated lakes of western Ontario and the Canadian Prairie Provinces. In the United States, a small harvest is made on the Great Lakes by tribal fishers for subsistence purposes, and a few tribes support a small commercial market.

There has been interest in walleye culture for over 100 years; the 1900 *Manual of Fish-Culture* published by the U.S. Commission of Fish and Fisheries included a chapter on propagation of "The Pike Perch or Wall-Eyed Pike" (8). That *Manual* describes spawning and spawn taking, use of "swamp muck" to prevent adhesion of eggs, egg incubation, transportation of eggs, description of cannibalism (including some excellent photographs of cannibalism), and prey selectivity by first feeding fry when lake water containing zooplankton was used as the water supply. By 1948, public hatcheries of 44 States and the U.S. Fish and Wildlife Service distributed 596.4 million "wall-eyed" pike, 79.6 million "yellow" pike perch fry, and 485.4 million "unclassified" pike perch (9). If we presume that all of these categories are walleye, the total was 1.16 billion fry. That level of production seems to have persisted for forty years. A survey for production years 1983–1984, indicated a similar number of fry were stocked annually by state, federal, and provincial agencies in the United States and Canada (10).

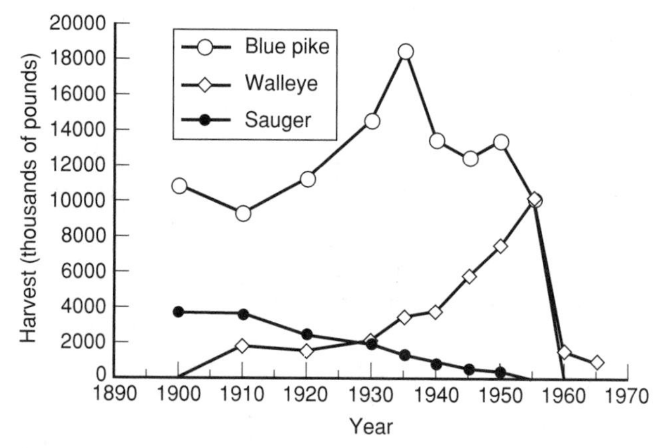

Figure 1. Changes in harvest of walleye, blue pike (subspecies of walleye), and sauger from Lake Erie, 1900–1969 (data from Hartman 1973: Table 3).

AQUACULTURE POTENTIAL

Walleye is recognized as a species with substantial aquaculture potential (11). There is a market for all life stages from egg to adult. Presently, the objective of most walleye aquaculture, both public and private, is to produce fry and fingerlings for stocking. At this time, there are few growers of food-size walleye. Encouraging market factors that may lead to development of walleye food-fish culture include excellent reputation for the food-quality of walleye, its name recognition by consumers, a high retail price (>US$16.48–21.89/kg, $7.49–9.95/lb), and a small and shrinking supply of competitive sources of wild-caught walleye for the food-fish market.

Biological characteristics that are important for a decision on fish suitability for aquaculture include desirable reproductive traits, lack of cannibalism, suitable growth rate, acceptance of prepared food, tolerance of crowding and other hatchery conditions, disease resistance, and palatability (12).

Reproductive Traits

Public hatcheries have spawned, incubated, and hatched walleye for more than 100 years. Females have been striped of as many as 300,000 eggs, although the number varies with size of the female and the fecundity ranges from 55,000 to 88,000/kg (25,000 to 40,000/lb) (13). Walleye eggs are easily incubated in conventional hatchery containers. The "McDonald jars" are glass or plastic cylinders with parabolic bottoms. Large batteries of the jars are commonplace at state and federal hatcheries (Fig. 2). The number of days for incubation varies with the water temperature used at individual hatcheries, but it typically ranges from 12–21 days to the start of hatch and 3–7 days more for all eggs to hatch from a given hatchery container. The incubation interval can be extended to 42 days by manipulating the incubation temperature. First, the eggs are incubated at a minimum of 7.8 °C (46 °F) for 5 days for initial embryo development; next, the water temperature is lowered until the eggs are close to hatching; and finally, the temperature is increased to a minimum of 13.3 °C (54 °F) for hatching so the fry will be vigorous enough to shed the egg shell (14). The newly hatched larval walleye (prolarva) is about 7.4 mm (0.3 in.) in total length and has a large yolk sac, hence, the name yolk-sac fry (Fig. 3).

The annual reproductive cycle of gonadal and hormonal changes in walleye have been described (15); hormonal treatments (LHRHa and hCG) to induce oocyte maturation have been evaluated (16); and environmental manipulation (light and temperature) have been combined with hormonal treatment human chorionic gonndotropin) to induce ovulation in walleye about 10 weeks prior to natural spawning. These methods have been used by the Iowa Department of Natural Resources to advance spawning of walleye from normal interval in April to January or February (17). The early spawn extends the growing season in order to produce larger fall fingerlings than otherwise possible.

Figure 2. Large-scale, walleye incubation facility with many hatching jars. On hatch, the larvae swim to the top of the jar and go with the flow to a holding (catch) tank.

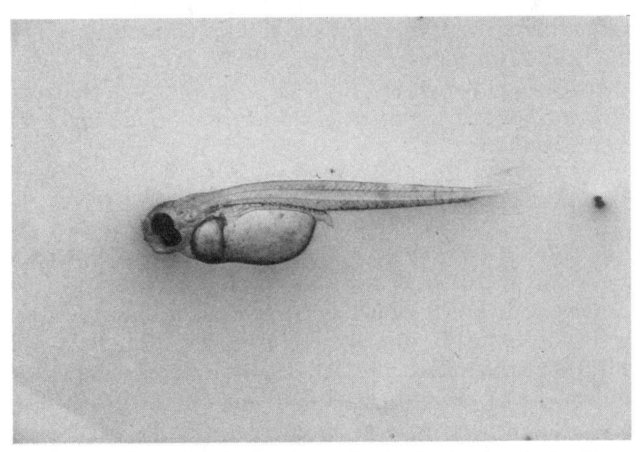

Figure 3. Newly hatched, yolk-sac fry (prolarva stage) of walleye.

Cannibalism

Walleye is a cannibalistic species; indeed, their first meal may be a sibling of about equal size (cohort cannibalism) (18). Cannibalism has been considered a serious problem in tank and pond culture of fry to fingerlings and during the 30-day interval of habituating pond-raised fingerlings (phase I fingerlings) to formulated feed (19,20). Nonetheless, cannibalistic tendencies vary with stocks (20). It may be reduced by genetic selection (21); and in intensive fry culture, it can be managed to insignificance by use of turbid water (22,23) and by providing adequate numbers of live brine shrimp (24). Frequent feeding and an adequate feeding rate are critical to prevent cannibalism. Nearly all walleye used by both public and private aquaculturists are progeny of wild-caught stock. For commercialization of walleye as a food fish, domestication, which seems to reduce the cannibalistic character of walleye, will be required, (25).

Growth

In nature, the major limits on growth are the abundance of prey and environmental temperatures. Assuming a minimum weight for a food-size fish is about 540 g (1.2 lbs) in nature, walleye of this size are not typically seen until their fourth summer of life (26). In cage culture of walleye, where food is not limiting but environmental temperature strongly influences growth rate, walleye reach food-size in the middle or latter part of their third year of life (27), one summer earlier than in nature. In intensive culture where food is abundant and temperature nearly optimum, walleye can reach food-size in 16 months from hatch (28).

Acceptance of Formulated Feed

Techniques were developed almost 30 years ago to habituate pond-raised fingerlings to formulated feed in tank culture (29,30). When environmental conditions are right and low light intensity (31) or in-tank lighting (32) is used, small (35–70 mm, 1.4–2.7 in.) pond-reared fingerlings can be habituated to manufactured feed and raised to about 150 mm (6 in.), with about 90% survival of the initial stock for the intensive phase in practical culture conditions (32). First feeding fry readily feed on brine shrimp (*Artemia* sp.) nauplii (24,33) and they can be habituated to commercially available formulated feed (e.g., the Fry Feed Kyowa [FFK], Biokyowa, Inc., Chesterfield, Missouri) when requisite environmental conditions are provided (34).

Tolerance to Hatchery Conditions

With the exception of a single domesticated hatchery stock developed and maintained at the London Fish Hatchery, London, Ohio (25), nearly all walleye used in aquaculture today are progeny of wild stock. It is not surprising that offspring of feral walleye raised in hatchery raceways and tanks seem skittish and easily disturbed by overhead movement and hatchery activities (e.g., tank cleaning) (35–37). However, when low (<20 lux) overhead light (31) or in-tank lighting (35,37) is used, walleye of all ages are more tolerant of hatchery tanks. Subadult walleye seem to tolerate high-density culture; and, for short-term exposure, they are surprisingly tolerant of low oxygen and high temperature (28). Domesticated stock seem to show less response to fish hatchery conditions (32). Hybrid walleye, which are a cross between female walleye and male sauger, are quite docile and exhibit faster growth than walleye, at least as juveniles (35–37).

Disease

Walleye are susceptible, but not more susceptible than other cultured fish, to commonplace external protozoan parasites such as *Ichthyophthirius multifiliis* and *Trichodina* spp. (38). Columnaris disease (*Cytophaga columnaris*) is the most serious and commonplace bacterial disease of walleye (39,40). Columnaris infections often result from mechanical injuries when fingerling fish are harvested or handled at temperatures favorable to the bacterium (39). Environmental stress is certainly a significant factor contributing to an epizootic in walleye (40). To date, no viral infections have been found to cause epizootic mortality in walleye, but adult walleye are susceptible to several viral diseases of the integument: lymphocystis, epidermal hyperplasia, and dermal sarcoma (41–44).

Palatability

Cookbooks that include walleye recipes invariably provide very positive endorsements; for example, walleye "...one of the best eating of all freshwater fishes..." (45), and "...the walleye is one of the most delicious of freshwater fishes. Its snow-white flesh is both delicately and distinctively flavored, ..." (46). Walleye also have favorable name recognition in small, upscale, white-tablecloth restaurants, and have been on the menu of some national franchise chains. Given these facts and limited commercial supply, skin-on walleye fillets sell for more than cultured catfish, salmon, or trout. Surveys in 1990 and 1992 of retail, wholesale, and other firms that comprise the traditional marketing channel for fish and seafood products within the Midwest indicated that walleye had high marketing potential as foodfish (47,48). The scales and skin of walleye are 10.8 to 13.0%. Therefore, to avoid substantial economic loss at the retail level and for better presention in restaurants, walleye are generally sold scaled with their skin on (49). As a percentage of their body weight, female walleyes have significantly larger heads than males (12.9 and 9.8%, respectively). Thus, fillet yield of males is higher than females. The range in yield of two cohorts for scaled, skin-on fillets of male, Spirit Lake, Iowa walleyes was 43.2 to 44.5% for male and 39.3 to 41.0% for female walleye (49). Using an unweighted mean of 42% for Spirit Lake walleye (47), the fillet yield of walleye is similar to that of channel catfish (40–45%) and 10% higher than tilapia (32%) (50). A 540 g (1.2 lb) walleye is sufficient to obtain two 113.5 g (4 oz) skin-on fillets, and a 811 g (1.8 lb) walleye would provide two 170 g (6-oz) fillets (49).

STRATEGIES FOR WALLEYE CULTURE

Several types of culture systems are used for walleyes, such as

- Pond culture
- Tandem pond to tank culture
- Pond to tank to pond culture
- Cage culture
- Intensive culture
  - — Fry to fingerlings
  - — Fingerlings to food fish

Some topics have been more thoroughly researched than others; e.g., there is much more information on pond culture in drainable ponds than undrainable ponds because most state and federal agencies use drainable ponds, although the largest commercial production of fingerlings in the United States takes place in undrainable ponds in Minnesota. Also, nearly all research has been conducted or sponsored by natural resource agencies whose interests are in production of fingerlings for enhancement stocking, and only limited information has been published on the culture of walleye as foodfish.

POND CULTURE

Walleye harvested from ponds 30–50 days after fry stocking (typically between mid-June to early July) are called "summer fingerlings" or "phase I fingerlings" and have a total length of 32–76 mm (1.25 to 3.0 in.). Juvenile walleye raised to the end of the growing season (September and October) are called "fall fingerlings" or "phase II fingerlings"; they range between 100 to 200 mm (4.0 to 7.9 in.) total length. Pond culture of fingerling walleyes dates from at least the early 1920s (50). In 1948, pond production generated more than four million fingerlings (9). By 1983/84, more than nine million pond cultured fingerlings were produced annually by public agencies (10).

Pond culture of fingerling walleye is carried out in drainable and undrainable ponds. Drainable ponds are constructed with a levee on all four sides. Undrainable ponds are a diverse assortment of water bodies, including farm and ranch ponds, shallow natural lakes, marshes, borrow pit ponds, and dug ponds. Most fingerling walleyes produced in Minnesota are raised in natural prairie pothole and shallow lakes that do not sustain a fish population because of winterkill (51–54). Methodology for walleye culture in undrainable (54) and drainable (55) ponds was recently reviewed. Those references should be consulted for details of culture technology. The following is a summary of major issues concerning pond culture of walleye, with an emphasis on drainable ponds.

Variability

It has been said that pond culture of walleye fingerlings (i.e., survival and yield) is "plagued by extreme variability of results" (56), probably in reference to pond-to-pond variability within the same year. For example, in Ontario, within-year, pond-to-pond variation in survival ranged from 0.7 to 73% and yield ranged from 1.8 to 26.1 g/m^3 (0.00012 to 0.0016 lb/ft^3) (57). In a state facility in Ohio, 55% of walleye ponds within the same year had $<10\%$ survival, a problem reduced by carefully controlling N : P ratios with frequent addition of inorganic fertilizers (58,59).

Far less information has been reported on long-term variability at the same hatchery. In a 19-year interval at a state operated pond culture facility in western Nebraska, year-to-year variation in survival and yield (kg/ha) was greater than within-year, pond-to-pond variability (60). At an Illinois public hatchery with plastic-lined ponds, survival ranged from 55% to 91% from one year to another (61).

Pond Management

Management of drainable ponds includes pond preparation, scheduling pond filling in relation to the anticipated hatching date of walleye, stocking density and date after pond filling, fertilization (kinds, amounts, and application schedule), zooplankton inoculation, control of problem organisms (aquatic insects, clam shrimp and vegetation), water quality management (aeration etc.), and harvest methods (drain to catch-basin, drain and seine, or partial harvest with lights) (56).

Pond Preparation. If the time between draining and next production cycle allows it, the pond bottom may be seeded with an annual rye grass (62). If seeding is not done after the last harvest of the season, the ponds may be dried and disked; if needed, agricultural lime ($CaCO_3$) may be added to increase alkalinity. Caustic (hydrated) lime, $Ca(OH)_2$, may be applied after pond draining and harvest to kill parasites or problem organisms.

Pond Filling. Water supply for ponds may be from groundwater or surface sources. If ponds are filled from surface water sources, the water must be filtered with nylon screens or bags to exclude fish and fish eggs, tadpoles, and other problem animals. When ponds have been left empty over winter, filling is usually timed to the expected date that walleye eggs hatch and in relationship to strategies for zooplankton development. The time between pond filling and fry stocking varies from 1 to 28 days or "just before fry are ready to be stocked" (63), i.e., 9 days (64), 10 days (51), 2 weeks (65), and 4 weeks (66) after pond filling.

Once pond filling commences, most hatcheries fill ponds as quickly as the water supply will allow, which is usually 1 to 3 days. A staged- or gradual-filling process is a purposeful procedure to concentrate newly stocked fry and zooplankton, and it provides multiple hatches of zooplankton because new zooplankton hatches are stimulated by each increment in pond filling. Also, when zooplankton are abundant in the water supply, slowly filling the ponds will continually add zooplankton. Enhancing zooplankton by continuous addition of water has not been the subject of research, but it is inadvertently done when a continuous addition of water is used to replace water loss from evaporation and/or seepage. In one report, the addition of water was equivalent to the total volume of the pond every 10 days. The inflow water was from a surface source that contained about 500 zooplankton/L (1900/gal). The inflow represented 15% of the average density (360/L; 1368/gal) of zooplankton in the pond (60).

Stocking Density. Because they are more intensively managed, stocking densities in drainable ponds are substantially greater than reported for undrainable ponds. Fry stocking density in undrainable ponds by private producers in Minnesota typically ranged from 6,178 to 24,710/ha (2,500 to 10,000/acre), but 49,420 to 74,120 fry/ha (20,000 to 30,000 fry/acre) if the pond is aerated or known to be exceptionally productive (67). The Minnesota DNR typically stocks 7,410 to 24,710 fry/ha (3,000 to 10,000 fry/acre), and the Michigan DNR stocks 74,130 to 185,325 fry/ha (30,000 to 75,000 fry/acre) (68).

There is a substantial range reported in stocking density in drainable, production ponds at state, federal, and provincial hatcheries: 250,000/ha (101,736/acre) for plastic-line production ponds in Illinois (61); 250,000 to 375,000/ha (101,736 to 151,760/acre) in earthen ponds in Nebraska (60); 335,938/ha (135,952/acre) average for three hatcheries in Ohio (63); 410,000 to 494,000 fry/ha (166,00 to 200,000/acre) at a federal hatchery in North Dakota (64); and 600,000/ha (242,817/acre) at a provincial hatchery in Ontario (69). Generally, state and federal hatcheries that use drainable ponds have higher stocking densities, more intensive management, and a smaller size at harvest than private hatcheries using undrainable ponds.

At a Nebraska state fish hatchery, an increase in stocking density from 250,000/ha to 375,000/ha in unfertilized ponds increased the numerical harvest of fingerling walleye from 153,509/ha to 228,647/ha without reducing survival (61.8% vs. 61.3%), mean fish size [36 mm (1.4 in.) vs. 33.5 mm (1.3 in.)], or yield (60). Yield was 52.0 kg/ha (46 lb/acre) at a stocking density of 250,000 and 57.2 kg/ha (50 lb/acre) at 375,000/ha (150,000/acre). Reduced yield and variability in returns of pond-reared walleye may occur when low fish densities allow zooplankton to overgraze algae, causing declines in both algae and zooplankton (59).

Commercially, the value of fingerling walleye varies by size and number; thus, an increase in number of fish harvested per unit area may be more important than an increase in yield (kg/ha). In most studies, fish size at harvest is inversely related to stocking density, survival rate, and number of fish harvested/unit area. When several ponds are stocked on the same date, but harvested over a 10 day interval, an inverse relationship in fish survival and length of the culture interval is evidenced, while there is a positive relationship between length of the culture interval and fish length at harvest (60).

Fertilization. The producer of phase I fingerlings may rely on natural fertility without supplemental fertilization, such as is typical of undrainable ponds. Generally, in drainable ponds, efforts are made to enhance survival and production by adding organic matter or inorganic fertilizers, or a combination of the two. Traditionally, organic fertilizers were preferred in walleye culture, especially in northern latitudes. A justification for this preference may have been based on results from early studies on pond fertilization for rearing walleyes. Research in Minnesota over 50 years ago did not indicate a positive correlation between total phosphorus and total nitrogen and yield of fingerlings in 66 walleye culture ponds (70); they concluded that "...basic fertility is not converted into a crustacean crop early enough in the spring to be available when most needed by the small yellow pike perch (i.e., walleye)." Thus, from that study and personal experience of many culturists, it has been widely held that algal populations are too slow to develop at low spring time temperatures when walleye fry are first stocked. With a short interval between pond filling and fry stocking, zooplankton foods of walleye develop more rapidly with organic fertilization. This view has been changing as noted below in the review of strategies for use of inorganic fertilizers.

Organic fertilizers that have been used for fertilization of walleye culture ponds include flooded rye grass, animal manures, ground hay, alfalfa pellets, alfalfa meal, soybean meal, and Torula yeast (55). The type and particle size of organic fertilizers is important because they affect both the rate of decomposition and the ability to stimulate development of microbial populations (71). Organic matter is used to nourish a microbial-detrital (heterotrophic) pathway to fish production. Associated with the detritus is a complex community of decomposer organisms (bacteria, fungi, and many kinds of protozoa, but particularly the ciliates) and a diverse assemblage of small benthic metazoans that consume detritus meiofauna (nematodes, copepods, turbellarians, gastrotrichs, small annelids, hydrozoans, and larger invertebrates). The pathway from organic fertilizers to fish begins with decomposer microorganisms (bacteria and fungi) that are consumed by detritivores (e.g., infusoria, other detritus-feeding organisms), and their predators. Zooplankton and aquatic insect larvae that are eaten by walleye are either detritivores (i.e., heterotrophic organisms that consume organic matter) or predators on detritivores. Although organic matter also functions like a slow release fertilizer, releasing inorganic nutrients (N and P) that eventually stimulate algal production, it is seldom applied on the basis of its nutrient content.

A diverse complex of factors affect the abundance and composition of zooplankton communities in ways that are still unpredictable. Needed are experimental studies (74) to determine specific types and particle sizes of organic fertilizers and application schedules to optimize the abundance of particulate organic matter, its associated microbes, and zooplankton numbers and composition. Alfalfa, for example, is used as ground hay, pellets, or meal, and vary in particle size from large to small; smaller particles decompose faster and are food for zooplankton. A succession of experiments have been carried out at the White Lake Fish Culture Station, Ontario, to evaluate stocking density and fertilization regimens using organic and inorganic fertilizers (65,69,72).

Inorganic fertilizers stimulate an autotrophic food chain from algae to zooplankton. Autotrophic means "self-feeding," because algal cells produce their own carbohydrates ("food"). Synthetic inorganic fertilizers, which have a high concentration of nitrogen (N) and available phosphorus (P), are used to stimulate crops of small algae (diatoms, coccoid greens, and flagellates) that are eaten by zooplankton. Culver (58,59,63) described methods for measurement of N and P and the calculations needed to determine the precise amounts and mixtures of fertilizers to achieve a desirable N:P ratio. After pond filling, the inorganic concentration of N and P are raised to 600 mg [0.005 lb/gal] N/L and 30 mg [0.0002 lb/gal] PO_4-P/L and are maintained at those levels by weekly applications of liquid inorganic fertilizers. This procedure has the potential for improved management of algal abundance and algal community structure, as well as lower costs (exclusive of the costs for equipment, glassware, and reagents for determining N and P) for fertilizers.

A few experimental comparisons of organic and inorganic fertilization strategies for production of phase I walleye fingerlings have been reported. Jahn et al. (61) found that organic fertilizers were superior to liquid inorganic fertilizer in nearly every category of production of walleye in plastic-lined ponds; also, the inorganic fertilizers were said to promote poor water quality (e.g., high pH, low-dissolved oxygen). Soderberg et al. (56) compared walleye fingerling production in ponds fertilized with periodic doses of alfalfa meal and Torula yeast with ponds fertilized with liquid fertilizer to maintain 600 μg/L of N and 30 μg/L of P. Differences in survival and yield between fertilizer treatment were not statistically significant, but the authors recommended the use of inorganic fertilizer treatment because it was less expensive and easier to apply. Tice et al. (73) carried out a similar experiment at the same production site in Pennsylvania. They also found statistically similar results with both treatments. Yield was 27% greater (47 kg/ha; 42 lb/acre) from ponds fertilized with organic fertilizers than from ponds fertilized with inorganic fertilizer (37 kg/ha; 32.9 lb/acre). Also, pond-to-pond variation (standard error) in survival and yield within treatments were greater in ponds treated with inorganic fertilizers. Fox et al. (74) compared production of fingerling walleye in ponds fertilized with a soybean meal slurry with production in ponds given the same dosage of soybean meal and supplemented with weekly applications of 8-32-16 (N-P-K) inorganic fertilizer. Although the difference was not statistically significant, walleye grew faster and biomass harvest was 42% higher in the ponds treated with inorganic fertilizers. These studies do not provide unequivocal conclusions regarding fertilizer strategies.

Bacterial decomposition of organic matter consumes oxygen, and oxygen depletion may cause fish kills; but oxygen depletion also occurs in ponds fertilized with inorganic nutrients (74). To prevent critically low oxygen from causing mortality, some production sites add aerated water (44,66,74) or employ oxygen diffusers (66) or a portable paddlewheel aerator (55) to increase oxygen concentrations.

Natural Foods. The first food of walleye may be diatoms (75), rotifers, and copepod nauplii (61,70); cyclopoid copepods (78); or small soft-bodied cladocerans (79). As they grow, walleye progressively switch to from copepod nauplii to small then larger cladocerans, especially *Daphnia* (69,76,77,79,80), and within three weeks, they are consuming immature aquatic insects, typically chironomid larvae and pupae (69,77,80). A sequence from copepod nauplii, to copepods, to small then large cladoceran species, and finally to chironomid larvae and pupae has been observed in gut contents of walleye fingerlings in production ponds in North Dakota and Nebraska (55). In that study, walleye did not consume diatoms or rotifers, even though rotifers were the most abundant zooplankton. Smaller copepods and cladocerans were preferred early in the culture season, but food preference shifted to large cladocerans (up to 2.0 mm; 0.08 in.) and then chironomids in the latter stages of the culture period for phase I fingerlings.

Zooplankton Inoculation. The purposeful addition of zooplankton (zooplankton inoculation, or zooplankton "seeding") to culture ponds is a potential, but largely unstudied, strategy to initiate a rapid increase in desirable zooplankton numbers. Zooplankton inoculation has been recommended for pond production of phase I fingerling striped bass (81). Zooplankton inoculation has special appeal for hatcheries in northern climates where the interval between pond filling and stocking is limited, water temperatures are low, and there is little time to optimize zooplankton populations by other means (71). Ponds filled with water from surface water sources (even filtered water) invariably contain zooplankton which serve as an initial inoculum for the newly filled pond. The kind, size, and abundance of zooplankton in a water supply will vary temporally. Quantification of zooplankton in a water supply was reported for a walleye hatchery in western Nebraska at 490 to 506 zooplankton/L (1862 to 1923/gal); and based on pond volume, the addition was 54 zooplankton/L (205/gal) in the ponds, or about 9.9 to 15% of the average zooplankton density in the ponds (60).

Daily (53) reported that the Minnesota Department of Natural Resources stocked amphipods (0.56 kg/ha; 0.49 lb/acre) into undrainable walleye culture ponds when amphipods were not present because "they provide important forage for the fish." Based on typical size of amphipods, they would not be suitable food for small walleye. An inoculum of cladoceran species too large for consumption by first feeding larval walleye may actually reduce walleye survival (77). Zooplankton stocking is limited by the ability to obtain large quantities of zooplankton of the right kind and size when needed. Skrzypczak et al. (82) described a system for larval culture of Eurasian perch (*Perca fluviatilis*) and pikeperch (*Stizostedion lucioperca*) that involved attracting zooplankton with lights into a fine-mesh cage and then pumping them to a fry culture cage. Certainly, zooplankton inoculation for walleye culture ponds requires considerably more research than has been reported to date.

Monitoring Growth and Survival. Walleye may be seined or trapped to monitor growth and to obtain an index of abundance (survival). Walleye have a strong phototactic response to light at 32-mm (1.3 in.) total length (83) and gradually become photonegative by 40 mm (1.6 in.). The attraction of walleye to light allows them to be captured in a light trap (55); the catch per unit effort (CPUE) in the light trap can provide a database that may be used to determine the status of the populations. Fish may be monitored to determine growth and presence of food in the gut, which can indicate the adequacy of the food supply and determine the need for additional fertilization and timing of harvest. It is also useful to examine fish for gas bladder inflation and incidence of parasitism. After the attraction of fingerlings to light wanes, nighttime seining is usually effective for sampling.

Problem Organisms. Predacious, aquatic hemipteran insects, such as backswimmers (Family Notonectidae) and giant water bugs (Family Belostomatidae), reduce survival of stocked fry; and the stout beak of notonectids is capable of a painful sting to personnel harvesting fish from seines. It has been common practice to control these insects in ponds by applying oil to the water surface (52).

Abundant clam shrimp are a nuisance in ponds used for production of fingerling fish because they clog screens during draining and they have been reported to reduce the production of fingerling northern pike (84,85).

Clam shrimp are branchiopod crustaceans; their chitnous exoskeleton resembles the valves of large ostracods or the shell of a fingernail clam (family Sphaeriidae, class Bivalvia). Clam shrimp densities as high as 4,488 m^2 (3,740/yd^2) have been reported in walleye culture ponds (85). Clam shrimp are not predators on fish — they consume detritus, diatoms, and green algae — but they compete for the same food resources used by copepods and cladocerans (85). Strategies for control of clam shrimp include: keeping ponds full of water over winter and not dried before stocking; flushing newly hatched clam shrimp nauplii with a fill-drain-and-refill strategy; sterilizing ponds with slaked lime (calcium hydroxide) at the rate of 2,000 kg/ha (1,760 lb/acre) to the moist pond bottom after fish harvest; and if regulations permit, applying insecticide trichlorfon (85).

Harvest

Fingerling walleyes may be harvested from undrainable ponds with traps or seines. The same gear may be use in drainable ponds, but they are usually harvested by draining water into a catch basin located inside or outside of the pond. Although drainable ponds may be partially harvested with seines or fyke nets, or seined during draining, most state and federal production hatcheries drain the ponds to a catch basin and harvest the entire pond in one day (55,64). If desired, walleye may be partially harvested over several days by using traps, seines, night harvest with lights, or a drain-and-seine procedure.

A night-harvest technique described for yellow perch (86) has been modified and used for harvesting walleye during the photopositive phase (56), which lasts until they are about 32 mm (1.3 in.) (83). A special advantage of the night harvesting technique is that fingerlings may be captured without the usual abundance of tadpoles and salamander larvae, which were a nuisance with daylight pond seining.

TANDEM POND-TANK CULTURE

Because it is difficult to raise many walleye to sizes >100 mm (3.9 in.) in ponds without the addition of forage fish, many public hatcheries use the tandem pond-tank culture method (i.e., extensive–intensive) to raise phase II fingerlings. Pond-reared fingerlings are transferred to indoor culture tanks where they are habituated to formulated feed. Thereafter, they are typically raised to 125–200 mm (5–8 inches). A variety of cultural factors have been studied, including stocking density, temperature, light, diet, and feeding frequency. There are two periods of high mortality in tandem pond-tank culture. The first occurs in the pond-culture interval, as is obvious by survival data from pond cultured fish. The second period of high mortality occurs in the intensive culture system during the first 21 days when fish are habituated to formulated feed (31). In earlier studies, survival through the feed-training interval varied from 20 to 60% (29,30,88,92), and 6.5 (89) to 84.3% in cages (93). Survival rates of 65 to 90% have been achieved in tanks in other studies (31,32).

A key element to successfully habituating pond-reared walleye to formulated feed includes starting with healthy fingerlings that have a good condition factor. Feed training should begin immediately after removal from ponds. Poor survival during the training interval will occur when fingerling fish have been held in a holding tank for several days before the feed-training interval begins. These fish usually refuse to feed and most starve. Also, rough handling when harvesting or during transport, especially at temperatures greater than 20 °C (68 °F), often results in high mortality from columnaris disease (39).

Cultural practices for successfully feed-training pond-reared walleye include: temperature (20 to 25 °C; 68 to 77 °F); low intensity (20 lux) overhead light or intank lighting; suitable feed quality; high initial feeding rate (10% body weight per day); frequent feeding (every 5 minutes, 18 hours/day); and initial densities of less than 3 kg/m^3 (0.19 lb/ft^3) of tank volume and maintenance of suitable water quality (31,32).

POND TO TANK TO PONDS

Phase I, fingerling walleyes harvested from ponds are often transferred to raceways and habituated to formulated feed. Nagel (88) returned feed-trained fingerlings to ponds where they were reared for several years to obtain captive broodstock.

A pond-to-cage-to-pond strategy was described by Coyle et al. (90) in which pond-raised, phase II fingerlings were over-wintered and, in early spring, habituated to formulated feed in cages in small (0.04 ha; 0.16 acre) ponds. In a 47-day interval (March 24–May 11), a yield of 47% "usable" survivors was obtained. Thereafter, the feed-trained fingerlings were released to ponds and fed once per day. Survival was 67% after 6 months (May 11–November 9) of culture. These findings suggested that if overwintered another year, a food-size fish may be produced in the third summer. Unfortunately, considerable mortality occurs in each of the several culture steps: between fry stocking and harvest of a fall fingerling; during overwintering fingerlings in ponds; training pond-reared fingerlings to formulated feed in cages; through the second summer; over another overwintering; and until the fish reach market size in the third summer. A serial multiplication of their survival data ($0.83 \times 0.45 \times 0.673$) — 83% survival for overwintering phase I fingerlings, a feed-training success in cages of 45%, and survival to the end of the second summer of 67.3% — implies a cost of \$2.28 to \$3.50 for fingerlings at the end of their second summer. Raisanen (51) points out that for the private sector, the cost of a fall fingerling raised in undrainable winterkill ponds in Minnesota is likely to be \$0.28 to 0.57 each based entirely on cost of fry, not including any production costs. Variable production costs include input for feed, feeders, cages, seines, labor, energy, and capital costs for land, ponds, and buildings. Unless the survival rates through each of the production phases can be increased, this cultural procedure is uneconomical.

Figure 4. Cage culture system for walleye tethered to raft in gravel quarry lake (89).

CAGE CULTURE

Cage culture of walleye has been in water-filled gravel and rock quarries, natural and artificial lakes, and farm ponds (93). Cages may be attached to rafts (Fig. 4), piers, or docks. Cage culture has been used to raise walleye fry to fingerlings (94), to raise a phase I pond-reared fingerlings to phase II fall fingerlings (150–200 mm; 5.9–7.9 in.), for enhancement stocking (95–97), and for culture of a food-size walleye (27,95).

Stevens (27) was the first to growout walleye to food size in cages, and Bushman (95) overwintered fingerlings in aerated cages for the purpose of growout to food size. Coyle et al. (89) overwintered fingerlings in a pond in north central Kentucky, then feed-trained the yearling walleye in cages. Starting with young fingerlings in midsummer, survival to fall is more successful when feed-trained fingerlings are used rather than training pond-raised fingerlings in cages to accept commercial feed (97).

When walleye are overwintered in cages in northern climates, aeration apparatus is required to keep the area around the cages free of ice (27,95). Overwinter survival of walleye fingerlings in cages was 40 to 60% in southern Iowa (27) to 93 to 98% in northeast Iowa (95).

Because fish growth rates are a function of water temperature, seasonal variation in water temperature in ponds and lakes will rarely be optimal for growth of walleye in cages or in pond culture; thus, it has not been proven to be an economical method to raise walleye to food size.

INTENSIVE CULTURE

High density, finfish culture in flowing water systems using single pass (one use), serial reuse (stair-step raceway system) or recycle systems are collectively referred to as intensive culture. High density (expressed as kg/m^3 or lbs/ft^3) is achieved by using a high exchange rate of water (single pass, reused, or recycled water) to supply oxygen and remove dissolved wastes (i.e., ammonia) from the culture tank. Recycle systems are designed to remove suspended solids and to nitrify ammonia to nitrates. Most intensive culture of walleye has been used to habituate phase I fingerings to formulated feed with the purpose of growing them to a fall fingerling. Providing growth rates are sufficient and production costs reasonable, feed-trained fingerlings may be raised to a size for the food fish market in intensive culture.

One advantage of recycle culture is that it "closes" the production cycle. All aspects of culture, from holding broodstock, spawning, and culture to food size, can be done indoors under optimum conditions for growth. Also, intensive culture is an alternative for hatcheries that lack pond facilities and sites for cage culture and where space or water supply is a constraint. Advancing the spawning season 90 days to late January requires an indoor environment to hold and spawn broodfish, as well as for egg incubation and fry culture. Obviously, ice-covered ponds cannot be stocked with fry derived from an early-season spawn. Also, training fingerlings or fry to formulated rations is best accomplished in controlled conditions characteristic of intensive culture.

FRY TO FINGERLINGS

Twenty years ago the prospects for intensive culture of walleye fry seemed remote: "It is extremely speculative to suggest a culture method for walleye fry at this time, ..." (98). Major problems encountered in intensive fry culture were nonfeeding, noninflation of the gas bladder (NGB), clinging behavior, and interrelationships between these problem areas. Intensive fry culture is a now an effective production technology for large-scale production walleye fingerlings (98–100).

Advantages of intensive culture include control of temperature to enhance growth rates and extend the growing season. In intensive culture, growth rates can be accelerated or decreased by temperature manipulation to meet production schedules for fish of different size. Intensive culture is the only technology that can be used to raise fry produced by out-of-season (early) spawning. In nature, the growing season (i.e., the number of days when water temperatures are adequate for growth) is usually too short to raise a phase II fingerling to a target size of 150 to 200 mm (5.9 to 7.9 in.) by fall in ponds or cages. Fry produced by spawning in winter months cannot be stocked in ponds in the Midwest. Pond stocking begins in mid-April in Ohio (59) and western Nebraska (68), late April to early May in central Michigan (101), early May to early June in North Dakota (55,77), the third week of May in northern Michigan (104), and late May to early June in Ontario (65). Intensive culture of walleye fry in recycle systems has been used for intensive culture at sites as far north as near Gunton, Manitoba (98,99).

Methodology

Intensive culture of walleye from hatch to small fingerling requires careful attention to system design, environmental factors, feed and feeding, tank hygiene, and water quality, including the following factors:

- Culture tanks, such as size, shape, and color.
- Screens

- Surface sprays
- Aeration and pumping
- Light and temperature
- Stocking
- Feeds, feeders and feeding
- Tank hygiene
- Water quantity and quality

Culture Tanks. Many tank shapes and drain systems have been evaluated for intensive culture of walleye fry. Barrows et al. (102) proposed a tank-within-a-tank design to maximize screen size and to develop up-flow pattern to keep fry in suspension. Kindschi (103) described a similar apparatus that has a stainless steel mesh basket insert for a circular, cone-shaped tank. To maintain a high feed particle density, a cuboidal-shaped tank with an upwelling water circulation pattern was designed to resuspend formulated feed (34). In an experimental comparison, fry survival and gas bladder inflation for larval walleyes were higher in cylindrical than cuboidal tanks (107). Fry survival in a round-bottomed trough was 47% compared to 13% in a round tank with a hemisherical bottom (98). Moodie and Mathias (99) described a production-scale, trough-shaped tank with an upwelling current for intensive production of walleye fry. Conventional rectangular raceways (33) and cylindrical (circular) tanks (100–107) that are typically used in hatchery production of other fishes are effective for production-scale culture of walleye fry on brine shrimp (*Artemia* sp.) and formulated diets (Fig. 5).

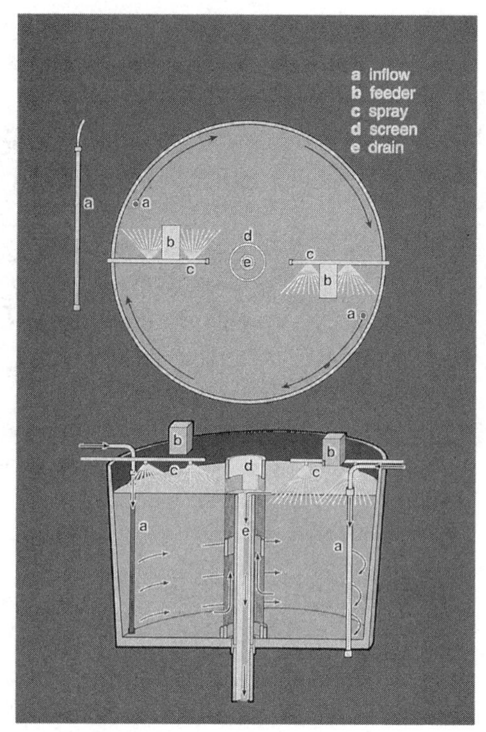

Figure 5. Design features of production-scale walleye fry-culture tank (123). The fiberglass tank measured 123 cm across the top, 109 cm at the bottom, and 76 cm deep. In use, the water depth is 67 cm and the rearing volume 680 L. The tank is supplied with four surface sprays to enhance gas bladder inflation (123). The sprays are located about 20 cm above the water surface and directed 90° from the water surface.

Larval walleye are strongly phototactic (86). In tank culture, they are attracted to direct or reflected light, and they cling to the tank wall, drain screens, or any other surface that reflects light (Fig. 6). Their clinging behavior affects initial feeding, growth, and survival (22,23). Tank size, tank color, light intensity, and turbidity influence fry clinging behavior. A greater percentage of the fry will cling to the sides of smaller than larger tanks because the surface area per unit of tank volume of smaller tanks is larger than that of larger tanks. Corazza and Nickum (109) observed greater larval dispersal in tanks with gray walls than in tanks with white, yellow, or green walls. Most investigators now use black over lighter colors for the interior of culture tanks. Clinging behavior has been reduced by culturing fry in turbid water (22,23) or by using high-intensity (680 lux) overhead light (105).

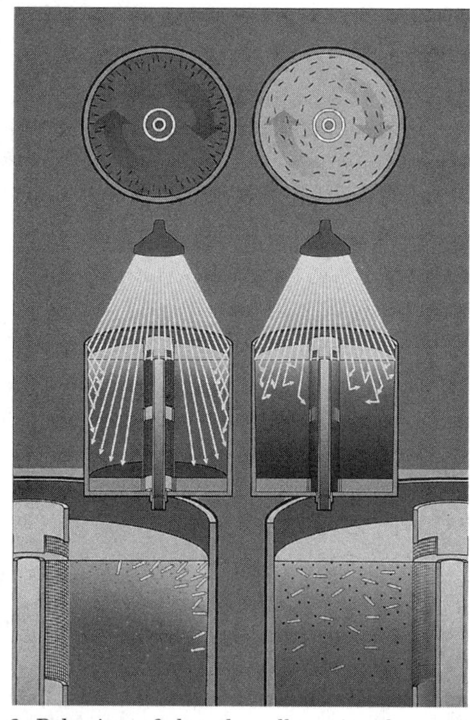

Figure 6. Behavior of larval walleye in clear-water (left) and turbid-water culture (right). Larval walleye are strongly phototactic (83). In clear water, their attraction to light results in clinging of the larvae to the tank walls; but in turbid water, they disperse across the tank and orient to the water current (22,23).

Screens. The drain must be equipped with a screen with a mesh small enough to retain the fry. A screen not larger than 710 μm with no more than 53% open area is said to retain 3- to 5-day old larvae (110). Walleye fry produced out-of-season and fry of hybrid walleye, however, are small enough to pass through a 0.704 mm (0.03 in.) mesh, which implies that a somewhat smaller screen size is needed for these smaller size larvae. After 21 days posthatch, however, effluent flow and tank hygiene can be improved by switching to 1 mm (0.04 in.) mesh with 58% open area. In intensive culture, walleye typically grow about 0.67 mm/d (0.026 in./d) in the first 30 days, from 7.5 mm (0.3 in.) at hatch to 27 mm (1.0 in.) in 30 days (22,23).

Obviously, food must be smaller than the gap of the mouth, thus brine shrimp nauplii or particles of formulated feed will pass through mesh small enough to retain larval walleye. The 400 μm (0.016 in.) Fry Feed Kyowa (BioKyowa, Chesterfield, Missouri), which has a particle range 240 to 675 μm (0.009 to 0.026 in.) (105) and brine shrimp nauplii, in the range 200 to 250 μm (0.008 to 0.010 in.), will pass through the 700 μm (0.028) mesh that will retain fry. In an effort to reduce loss of feed particles, a 500 μm (0.02 in.) mesh has been used to retain FFK B-400 feed (34), and screens with openings of 200 μm (0.008 in.) have been used to prevent loss of brine shrimp nauplii (105).

Surface Spray. Larval walleye, in common with most fish, must penetrate the water surface to gulp air to fill their gas bladder (Fig. 7). Gas bladder inflation (GBI) was typically poor until it was discovered that a spray of water to the surface would enhance gas bladder inflation (106,107). Walleye typically inflate their gas bladder between 6 and 12 days posthatch. The gas bladder is first round but elongates quickly (Fig. 8). Oil film on the water surface of the culture tank causes noninflation of the gas bladder because larvae cannot penetrate the surface to gulp air for first filling of the gas bladder (108). Maximum thickness of the oil film on the water surface of fry culture tanks occurs when fish are 7 to 18 days posthatch. In the critical period when gas bladder inflation occurs, about 65% of the total oil thickness is derived from fish and 35% from the feed (109). A water spray to the tank surface removes the oil film and cleans the surface of feed and debris (106,107). In circular tanks with a circular flow pattern, the water passes under the spray head with each revolution of the water mass. It is important that the spray impacts the water surface with enough force to produce a slight depression in the water under the spray. About one spray per 5,000-cm^2 (775-in.^{2}) of tank surface seems to be adequate.

Aeration and Pumping. Degassing and aeration of the water supply should be done before the water is delivered to the culture tanks. Compressors should not be used to aerate water destined for use in intensive culture of fry because they often contaminate the air with oil, which leaves an oil film on the tank surface and interferes with gas bladder inflation. In some hatcheries, an air line is placed around the center standpipe to keep fry from being impinged on the screen. However, this may cause undesirable turbulence, and fast rising air bubbles will even throw fry out of the water where they will stick to the side walls above the water line.

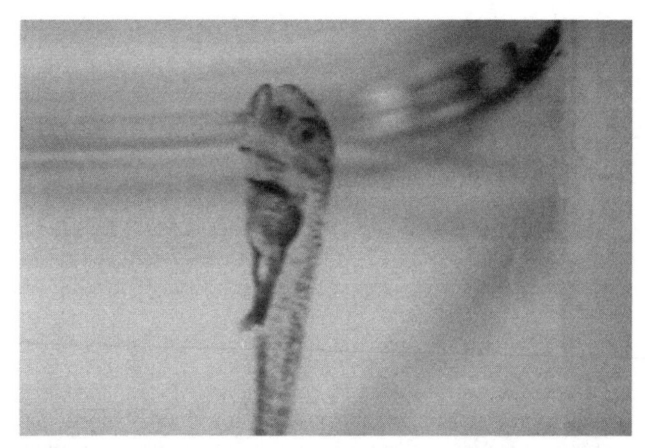

Figure 7. Larval walleye (postlarval I) penetrating the water surface for first filling of the gas bladder (124). Gas bladder inflation is prevented by an oil film on the water surface that prevents surface pentration by the larva. (Photo courtesy of Phillip Rieger.)

Figure 8. Sixteen-day-old larval walleye showing fully inflated gas bladder. This 17-mm juvenile was cultured at a mean temperature of 18 °C (64.4 °F), 289 daily temperature units. (Units of scale are 1 mm.)

Light and Temperature. Fluorescent lights, flood lamps, and natural light are all acceptable, but a diffuse light source may deter fry from clinging to the sides of the rearing tank. Light intensities of 100 to 700 lx have been used. A light of 680 lx at the water surface is said to aid in achieving fry dispersal and attracting fry to the surface (33). Moore (100) recommended 100 to 100 lux (lx) for intensive culture of larval walleye with formulated feed.

Feed acceptance and survival is greater at 18.4 °C (65 °F) than at 12.8 °C (56 °F), with an optimum temperature 18.4 °C (65 °F) (99). A minimum water temperature of 12.8 °C (56 °F) is required to start walleye on brine shrimp (33). A constant temperature of 20 °C (68 °F) throughout a 30 day fry culture interval has also been used (100). To prevent water quality problems with a system having low water exchange rates, a constant temperature of 16 °C (61 °F) was recommended for the first 23 days when fry are fed with brine shrimp (24). A temperature protocol was proscribed that began with an initial temperature of 14 °C (57 °F), a sudden increase of 2.5 °C (5 °F) in the fifth day posthatch when feeding was started, and a gradual increase to 20 °C (68 °F) at 30-days posthatch (108).

Stocking Density. To reduce cannibalism, fry for stocking the same culture tank should be of similar age; that is, hatched within a short time interval, preferably within 12 hours, but not more than 24 hours. To reduce variance in age, the tanks that receive fry coming from the hatching jars ("catch tanks") should be purged of all fish every day.

In large-scale production of walleye fry in intensive culture, stocking has been reported to be 15 (98), 19 (100), 20 (98), 21 (33), 40 (24,99), 56 (100) and 60 fry/L (57, 72, 76, 80, 152, 212 and 227 fry/gal) (24). Survival of fry through 23 days of culture at a stocking density of 40 fry/L (152 fry/gal) was 36.1% compared with 33.7% survival at a stocking density of 60 fry/L (228 fry/gal) when feeding brine shrimp (24). Using formulated feed, survival to 30 days was 47% at an initial stocking density of 19 fry/L (72/gal) and 23% at 56 fry/L (213/gal) (100). Moore (99) reported survival from 35 to 70% at 25 days posthatch with a stocking density of 40 fry/L (152/gal). Problems with cannibalism, poor water quality, and disease (bacterial gill disease and columnaris disease) seem to be directly proportional to increased density. Generally, however, even though survival rates decline with high density, yield of fingerlings can be much higher, thus increasing return on investment in tanks and hatchery space.

Feeds, Feeders, and Feeding. Larval feeding is not required until the yolk sac has been absorbed. Fry may be fed brine shrimp, then weaned to dry diets (24,33) or, as is more commonly the practice, started directly with formulated feed (22,23,98–100,102–104,106–108). Feed size, color, and texture have been considered the most important factors affecting acceptability. Failure of the fish to feed or digest the feed has been cited frequency as a major factor in the failure of walleye culture on formulated feeds. First-feeding walleye (postlarva I) have a mouth width of 0.7 mm (0.3 in.) and a gape of 1.5 mm (0.06 in.) (111). The mouth is large enough to cannibalize similar-sized siblings (cohort cannibalism) or to consume the 0.4-mm (0.016 in.) size common starter feed.

Nearly all research on formulated feed for intensive culture of larval walleye has been with the closed-formula diet Fry Feed Kyowa (FFK) B-series (B-400 and B-700) feeds manufactured in Japan and sold in the United States by Biokyowa Inc., Chesterfield, Missouri. The 400 and 700 represent approximate median particle sizes of 400 μm (0.4 mm) and 700 μm (0.7 mm) respectively (34). The manufacturer describes the B-series feeds as "krill-based," and they are more expensive than the FFK C-series, which the company calls a "minced-fish" feed. To reduce costs, a "phase" feeding strategy was proposed (106), which is starting first-feeding fry on FFK-B series (B-400), followed by transition to C-700 rather than B-700. Once on feed, feed sizes are gradually increased, from C-1000 (1.0 mm) to C-1500 (1.5 mm), then shifted to the 2 mm walleye grower diet, WG-9206 (112). The latter has been manufactured by Nelson and Sons, Inc., Murray, Utah. Colesante (33) feeds brine shrimp nauplii for 30 days, then starts feeding formulated feed ("New York State" diet) along with brine shrimp (period of "mixed" feeding) through 44 days posthatch.

Precise feeding of small quantities of fry feed is needed to prevent food deprivation. Excessive feeding that leads to water quality deterioration and higher incidence of bacterial gill disease. A variety of feeders have been developed. A hand-made scraper feeder using a clock mechanism (115) is commonplace in many research laboratories. A commercial vibrator feeder is also used to dispense fry feed (34). In experimental studies in tanks with small capacity (circa 150 L; 39 gal), vibratory feeders cannot be regulated with accuracy or consistency (33,98). For experimental purposes, custom made feeders have been developed that deliver small amounts of feed more precisely than other types (100). A custom-made auger feeder has been used for practical, large-scale intensive culture of larval walleye in 679 L (170 gal) tanks (108).

Feeding rates for larval walleye with formulated feed have been based on fish size (99,108) or amounts needed to achieve a particle density of 100 feed particles/L; (380/gal) (98,100). Live brine shrimp have been fed at 800 to 1,000 nauplii per fish per day for the first 30 days then 1,300 nauplii/fish/day from 30 through 44 days posthatch when fish are habituated to formulated diets (33). Typically, feeding frequency for first-feeding fry with formulated feed is at 3 to 5 min intervals at least 22 hours/day, stopping only to clean the tanks (99,108). Survival has been poor when feeding was stopped for more than an 6 hours per day.

Tank Hygiene. Careful attention to tank hygiene is an essential component to successful fry culture (100,108). Cleaning regimens vary with each hatchery, but in feeding formulated feed, which is fed in excess to habituate the fry to formulated feed, the tank must be cleaned each day. A tank cleaning protocol may involve siphoning the bottom to remove waste food and dead fish, wiping biofilm from the tank walls, and removing the standpipe for pressure washing (108).

Water Flow. The prolarvae (yolk sac fry) are poor swimmers and do not feed; therefore, at this stage a low-water exchange rate may be used, being careful to maintain sufficient oxygen and avoid exceeding conventional standard for unionized ammonia [0.02 mg/L (ppm)]. Once feeding begins, however, the exchange rate should be increased. Exchange rates have not been evaluated and rates from the literature are not often given. In a research facility, exchange rates start with 0.5 per hour before feeding commences, and increase to 1.0 exchange by 21 days (108). In a production facility where brine shrimp are used as the only feed for 30 days, calculations of exchange rates based on tank size and inflow rates ranged from 0.44 to 0.81 per hour (33). Peterson et al. (24) reported an exchange rate of 0.3/hour for 1,200 (315 gal) tanks and a stocking density ranging from 40 to 60 fry/L (152 to 228/gal) using brine shrimp as the starter feed. Whatever the exchange rate, current velocities should not overtax the swimming ability of larvae.

Fingerling to Food Fish

Both extensive, intensive, and a tandem system of extensive-intensive culture systems can be used to raise food-size walleye: pond-to-tanks, pond-to-tank-to-cage and pond-to-cage, pond-to-cage-to-pond, and intensive culture of fry to food size. Substantial differences in production costs occur among these methods.

Pond to Tanks. Fry stocked and reared in ponds to midsummer and harvested as phase I fingerlings (32–64 mm; 1.3–2.6 in.) can be transferred to tanks to habituate (i.e., train) them to formulated feed. This culture strategy is common practice by public fisheries agencies to produce advanced fingerlings (phase II fingerlings) for fall stocking. Similar procedures may be used by commercial fish producers, but instead of stocking fish in the fall, the feed-trained fingerlings would be reared to food-size fish.

Pond to Tank to Cage and Pond to Cage. Phase I fingerlings (32–64 mm; 1.3–2.6 in.) may be transferred to tanks, habituated to formulated feed, then stocked in cages for growout to food size. Or, phase I fingerlings may be trained to accept formulated feed in cages and rearing continued in cages until they reach a food size. There are several constraints to cage culture. The major constraint is the slow growth of caged fish at ambient temperature. In Iowa, food-size walleye cannot be produced in cages until the middle to latter part of the third summer. The culture interval needed to reach food size is even longer in the colder climates. In Minnesota, a portion of cage-cultured walleye reached marketable size in three growing seasons, with the first year in ponds and two additional years in cages (120).

Pond to Cage to Pond. In this protocol, phase II fingerlings were overwintered in ponds and in spring were stocked in cages where they were habituated to formulated feed. After they were feed-trained, they were released to unconfined culture in ponds for further rearing to fall (90). In that study, at a site in Kentucky, the size of fish at the end of the second summer indicated they would not reach a food size until the third summer. The major impediment to this procedure for production of a food fish is the slow growth at ambient pond temperatures, whether at large or in cages in a pond. Also, considerable mortality occurs during the feed-training interval, two overwinter periods, and nearly three summers of culture. Stevens (27) reported similar growth rates in a pond-to-cage culture regimen in southern Iowa.

Intensive Culture: Fry to Food-Size. Culture systems for intensive fry culture have been described. Intensive fry culture allows production of a fingerling that is trained to formulated feed in 21 to 60 days. After fingerlings are habituated to formulated feed, they may be raised to food size in several types of single pass, reuse, or recycle systems. Several experimental trials have been conducted on intensive culture of walleye to food-size (116–118).

Intensive culture systems for rearing food-size walleye include single pass, serial reuse, and recycle aquaculture systems. The major constraint to single pass and serial reuse culture of walleye is the requirement for an abundant supply of water in a desirable temperature range for good growth (20 to 25 °C; 68 to 77 °F). Growth of walleye in intensive culture at temperatures less than 15 to 17 °C (59 to 63 °F) are economically impractical. Thus, the major constraint to general use of flow-through culture is availability of sufficient water sources with desirable water temperatures.

A detailed description of components of recycle culture system for walleye culture has been described (119). The advantages of recycle culture are the same for walleye as any other species: controlled water temperature and water quality (pH, DO); a 12-month growing season; low water requirements relative to production capabilities; a small volume of concentrated waste; and the ability to locate a facility close to major markets. In recycle culture, fish are stocked at high densities and raised on pelleted feeds, and there is an intentional effort to minimize the use of new water to $\leq 5\%$ or less of total system volume per day. The effluent from the culture tanks must undergo several treatment processes before it is returned to the culture tank: clarification to remove solids; nitrification (biofiltration) to convert ammonia to nitrate; reaeration or reoxygenation; and disinfection by passing water through tubes with UV lamps or by ozone injection.

A survey of fish producers in the North Central Region indicated that 89% used ponds, but a surprising 14% used some form of recycle system to produce a variety of fishes (47,48). There are many recycle culture facilities in the Midwest and elsewhere that raise food-size tilapia, and others that raise hybrid striped bass. Moode and Mathias (98,100) described a recycle culture system for intensive culture of larval walleye on formulated feed at a site in Manitoba, Canada. A sensory evaluation of fillets from walleye raised in a recycle system was compared with walleye purchased from a market culture (122). In that study, the organoleptic qualities—those characteristics (aroma, flavor, and texture) that are evaluated by one's senses—of cultured walleye were equal to those of commercial, wild-caught walleye.

CHARACTERISTICS OF FOOD-SIZE WALLEYE

The most common market form of walleye is a scaled, skin-on fillet, but some consumers prefer skinless fillets because of the rubbery nature of the skin and taste problems that develop in frozen products from oxidative rancidity in the fatty layer under the skin. Dressed yield is essential for cost/benefit or break-even analysis for walleye food-fish production. A highly processed form of the fish (skinned fillets) results in more waste, higher processing costs per unit weight, and a more expensive final product.

Dressed yield is the percent of the live weight obtained for a specific processed product. The live weight to dressed weight relationships for different forms of a food-fish product strongly influences which form of the dressed product will be marketed as well as the overall economic feasibility for commercial food-fish production of that species. Obviously, yield of a dressed fish product affects the live weight of the fish needed to yield fillets of a commercial size.

Although a processing yield as high as 50% for walleye has been assumed (87), laboratory data by Flickinger (117) indicate a filleted yield of 42% (range 39 to 46%) for cultured walleye weighing 377 g (0.83 lb), and 45% (range of 39 to 46%) for wild-caught walleye of similar size. Yager et al. (122) reported variation in processing yield of skin-on fillets from different year classes, but the yield range was relatively small, 39.3 to 44.5% of whole body weight.

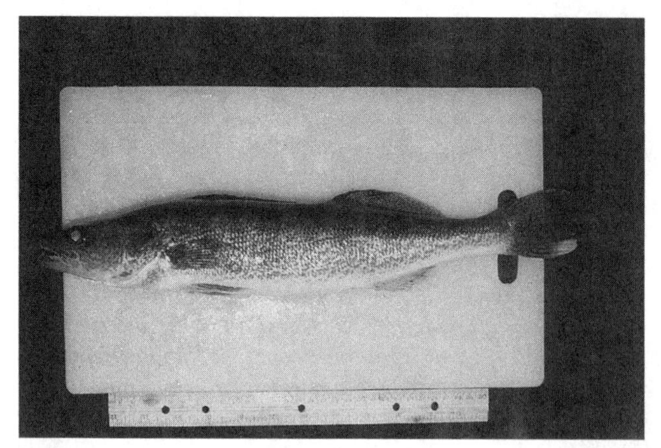

Figure 9. Intensively reared, 16-month posthatch, food-size walleye (28).

They found that males had consistently higher yield of skin-on fillets than females, as much as 5.3% greater in one year-class.

Assuming an average yield of skin-on fillets is about 42%, the live weight of a food-size walleye would be about 540 g (1.2 lbs) to obtain two, 114 g (4 oz) skin-on fillets (Fig. 9). An 810-g (1.78-lb) walleye would be needed to obtain two, 170.1-g (8-oz) fillets. Because body mass of walleye, as for most fish, increases in proportion to the cube of the length, it is reasonable to assume that the fillet yield as a percent of total body weight will increase with size. However, regression analysis of fillet yield on body weight for several cohorts, stocks within cohorts, and for as many as six age groups within the same cohort did not support the hypothesis (122). Body weight did not account for more than 6.6% of the variability of yield. Thus, within the range of fish size examined in that study, there was not a trend for the older (i.e., larger) fish to have higher processed yields.

In the Midwest, traditional food-size walleye fillets (scaled, but skin on) have been in three basic size groups: 114 g (4 oz), 170 g (6 oz), and 227 g (8 oz). Usually, a single 114 g (4 oz) fillet may be served for lunch, and one 170 g (8 oz) or two, 114 g (4 oz) fillets for dinner. Fillet size is of considerable importance in defining the size of fish raised for the food market. Obviously, if the market can be persuaded to embrace a 114 g (4 oz) fillet, perhaps labeled "pan-sized walleye," or "petite walleye" it will increase turnover time of the stock in the culture system and enhance profitability. With a 45% dressout yield, a fish that will yield two 114 g (4 oz) fillets would be about 504 g (1.1 lbs); a dress-out yield of 40% would require a fish of 568 g (1.25 lbs).

BIBLIOGRAPHY

1. C.R. Robins, R.M. Bailey, C.E. Bond, J.R. Brooker, E.A. Lachner, R.N. Lea, and W.B. Scott, *Common and Scientific Names of Fishes from the United States and Canada*, 5th ed., Spec. Pub. 20, American Fisheries Society, 1991.
2. K.D. Carlander, J.S. Campbell, and R.J. Muncy, *Amer. Fish. Soc. Spec. Pub.* **11**, 100–108 (1979).
3. R. Fenton, J.A. Mathias, and G.E.E. Moodie, *Fisheries* **21**(1), 6–12 (1996).
4. U.S. Department of the Interior, Fish and Wildlife Service and U.S. Department of Commerce, Bureau of the Census. *1996 National Survey of Fishing, Hunting, and Wildlife-associated Recreation*, U.S. Department of the Interior, Fish and Wildlife Service, Washington, DC, 1997.
5. W.B. Scott and E.J. Crossman, *Freshwater Fishes of Canada*, Bulletin 184, Fisheries Research Board of Canada, Ottawa, 1973.
6. W.L. Hartman, *Technical Report 22*, Great Lakes Fishery Commission, Ann Arbor, MI, 1973.
7. W.M. Zarbock, *Fisheries* **2**(2), 2–4; 26–33 (1977).
8. U.S. Commission of Fish and Fisheries, *A Manual of Fish-Culture*, rev. ed., Washington, DC, 1900, pp. 165–179.
9. A.V. Tunison, S.M. Mullin, and O.L. Meehean, *Prog. Fish-Cult.* **11**, 253–262 (1949).
10. M.C. Conover, in R.H. Stroud, ed., *Fish Culture in Fisheries Management*, American Fisheries Society, Bethesda, MD, 1986, pp. 31–39.
11. Joint Subcommittee on Aquaculture, *National Aquaculture Development Plan*, vol. 1 and 2, Washington, DC, 1983.
12. R.C. Summerfelt, in D.C. Beitz, ed., *Animal Products in Human Nutrition*, Nutrition Foundation Monograph Series, Academic Press, NY, 1982.
13. J.E. Harvey and S.E. Hood, in R.C. Summerfelt, ed., *Walleye Culture Manual*, NCRAC Culture Series 101, North Central Regional Aquaculture Center, Iowa State Univ., Ames, 1996, pp. 25–28.
14. D. Paddock, in R.C. Summerfelt, ed., *Walleye Culture Manual*, NCRAC Culture Series 101, North Central Regional Aquaculture Center, Iowa State Univ., Ames, 1996, pp. 29–30.
15. J.A. Malison, L.S. Procarione, T.B. Kayes, J.F. Hansen, and J.A. Held, *Aquaculture* **163**, 151–161 (1998).
16. T.P. Barry, J.A. Malison, A.F. Lapp, and L.S. Procarione, *Aquaculture* **138**(1–4), 331–347 (1995).
17. Alan A. Moore, Personal Communication, Rathbun Fish Hatchery, Iowa Department of Natural Resources, Moravia, Iowa.
18. N.L. Loadman, G.E. Moodie, and J.A. Mathias, *Can. J. Fish. Aq. Sci.* **43**, 613–618 (1986).
19. R.G. Howey, G.L. Theis, and P.B. Haines, U.S. Fish Wildl. Serv., *Lamar Fish Cultural Develop. Ctr. Information Leaflet 80-05*, Lamar, PA, 1980.
20. J. Hey, E. Farrar, B.T. Bristow, C. Stettner, and R.C. Summerfelt, *J. World Aquacult. Soc.* **27**, 40–51 (1996).
21. A.R. Kapuscinski, M. Hove, W. Senanan, and L.M. Miller, in R.C. Summerfelt, ed., *Walleye Culture Manual*, NCRAC Culture Series 101, North Central Regional Aquaculture Center, Iowa State Univ., Ames, 1996, pp. 331–338.
22. B.T. Bristow, R.C. Summerfelt, and R.D. Clayton, *Prog. Fish-Cult.* **58**, 1–10 (1996).
23. B.T. Bristow and R.C. Summerfelt, *J. World Aquacult. Soc.* **25**, 454–464 (1994).
24. D.L. Peterson, R.F. Carline, T.A. Wilson, and M.L. Hendricks, *Prog. Fish-Cult.* **59**, 14–19 (1997).
25. T.O. Nagel, in R.C. Summerfelt, ed., *Walleye Culture Manual*, NCRAC Culture Series 101, North Central Regional Aquaculture Center, Iowa State Univ., Ames, 1996, pp. 75–76.
26. K.D. Carlander, *Handbook of Freshwater Fishery Biology, Vol. 3*, Iowa State University Press, Ames, 1997.
27. C.G. Stevens, in R.C. Summerfelt, ed., *Walleye Culture Manual*, NCRAC Culture Series 101, North Central Regional

Aquaculture Center, Iowa State Univ., Ames, 1996, pp. 273–274.

28. S.T. Summerfelt and R.C. Summerfelt, in R.C. Summerfelt, ed., *Walleye Culture Manual*, NCRAC Culture Series 101, North Central Regional Aquaculture Center, Iowa State Univ., Ames, 1996, pp. 215–230.
29. W.F. Cheshire and K.L. Steele, *Prog Fish-Cult.* **34**, 96–99 (1972).
30. G.B. Byerele, *Prog Fish-Cult.* **27**, 103–105 (1975).
31. K.L. Kuipers and R.C. Summerfelt, *J. App. Aqua.* **42**(2), 31–57 (1994).
32. T.O. Nagel, in R.C. Summerfelt, ed., *Walleye Culture Manual*, NCRAC Culture Series 101, North Central Regional Aquaculture Center, Iowa State Univ., Ames, 1996, pp. 205–207.
33. R.T. Colesante, in R.C. Summerfelt, ed., *Walleye Culture Manual*, NCRAC Culture Series #101, North Central Regional Aquaculture Center, Iowa State Univ., Ames, 1996, pp. 191–194.
34. N.L. Loadman, G.E.E. Moodie, and J.A. Mathias, *Prog. Fish-Cult.* **51**, 1–9 (1989).
35. G.L. Siegwarth and R.C. Summerfelt, *Prog. Fish-Cult.* **52**, 100–104 (1990).
36. G.L. Siegwarth and R.C. Summerfelt, *Prog. Fish-Cult.* **55**, 49–53 (1992).
37. J.A. Malison, T.B. Kayes, J.A. Held, and C.H. Amundson, *Prog. Fish-Cult.* 73–82 (1990).
38. J. Marcino, in R.C. Summerfelt, ed., *Walleye Culture Manual*, NCRAC Culture Series #101, North Central Regional Aquaculture Center, Iowa State Univ., Ames, 1996, pp. 369–370.
39. M. Hussain and R.C. Summerfelt, *Iowa Academy of Science* **98**, 93–98 (1991).
40. R. Horner, in R.C. Summerfelt, ed., *Walleye Culture Manual*, NCRAC Culture Series 101, North Central Regional Aquaculture Center, Iowa State Univ., Ames, 1996, pp. 339–341.
41. P.E. McAllister, in R.C. Summerfelt, ed., *Walleye Culture Manual*, NCRAC Culture Series 101, North Central Regional Aquaculture Center, Iowa State Univ., Ames, 1996, pp. 355–367.
42. P.R. Bowser, G.A. Wooster, and K. Earnest-Koons, *J. Aq. An. Health* **9**, 274–278 (1997).
43. R.G. Getchell, J.W. Casey, and P.R. Bowser, *J. Aq. An. Health* **10**, 191–201 (1998).
44. L.A. Lapierre, D.L. Holzschu, G.A. Wooster, P.R. Bowser, and J.W. Casey, *J. Virology* **72**(4), 3484–3490 (1998).
45. G. Carmichael, M. Ring, and J. McCraren, *Sea Fare Cookbook*, American Fisheries Society, Bethesda, MD, 1992.
46. A. Cameron and J. Jones, *The L. L. Bean, Game And Fish Cookbook*, Random House, NY, 1983.
47. L.J. Hushak, C.F. Cole, and D.P. Gleckler, *North Central Region, Technical Bulletin Series #104*, North Central Regional Aquaculture Center, Iowa State Univ., Ames, 1992.
48. L.J. Hushak, *North Central Regional Aquaculture Industry Situation and Outlook Report, Volume 1*, North Central Regional Aquaculture Center, Iowa State Univ., Ames, 1993.
49. R.C. Summerfelt, R.D. Clayton, T.K. Yager, and K.L. Kuipers, in R.C. Summerfelt, ed., *Walleye Culture Manual*, NCRAC Culture Series #101, North Central Regional Aquaculture Center, Iowa State Univ., Ames, 1996, pp. 241–250.
50. E.W. Cobb, *Trans. Am. Fish. Soc.* **53**, 95–105 (1923).
51. G. Raisanen, in R.C. Summerfelt, ed., *Walleye Culture Manual*, NCRAC Culture Series 101, North Central Regional Aquaculture Center, Iowa State Univ., Ames, 1996, pp. 131–134.
52. R.E. Kinnunen, in R.C. Summerfelt, ed., *Walleye Culture Manual*, NCRAC Culture Series 101, North Central Regional Aquaculture Center, Iowa State Univ., Ames, 1996, pp. 135–145.
53. J.B. Daily, in R.C. Summerfelt, ed., *Walleye Culture Manual*, NCRAC Culture Series 101, North Central Regional Aquaculture Center, Iowa State Univ., Ames, 1996, pp. 147–150.
54. J.D. Lilienthal, in R.C. Summerfelt, ed., *Walleye Culture Manual*, NCRAC Culture Series 101, North Central Regional Aquaculture Center, Iowa State Univ., Ames, 1996, pp. 85–87.
55. R.C. Summerfelt, C.P. Clouse, L. Harding, and J. Luzier, in R.C. Summerfelt, ed., *Walleye Culture Manual*, NCRAC Culture Series 101, North Central Regional Aquaculture Center, Iowa State Univ., Ames, 1996, pp. 89–108.
56. R.W. Soderberg, J.M. Kirby, D. Lunger, and M.T. Marcinko, *J. App. Aqua.* **7**(2), 23–29 (1997).
57. M.G. Fox, J.A. Keast, and R.J. Swainson, *Environ. Biol. Fishes* **26**, 129–142 (1989).
58. J. Qin and D.A. Culver, *Aquaculture* **108**, 247–276 (1992).
59. D.A. Culver, S.P. Madon, and J. Qin, *J. App. Aquac* **2**(3/4), 9–31 (1993).
60. L.M. Harding and R.C. Summerfelt, *J. of Applied Aquaculture* **2**(3/4), 59–79 (1993).
61. L.A. Jahn, L.M. O'Flaherty, G.M. Quartucci, J.H. Kim, and X. Mao, *Completion. Rept., Federal Aid Proj. F-50-R*, Illinois Dept. Conser., Springfield, 1989.
62. J. Dobie, *Prog Fish-Cult.* **17**, 51–57 (1956).
63. D.A. Culver, in R.C. Summerfelt, ed., *Walleye Culture Manual*, NCRAC Culture Series 101, North Central Regional Aquaculture Center, Iowa State Univ., Ames, 1996, pp. 115–122.
64. J.E. Call, in R.C. Summerfelt, ed., *Walleye Culture Manual*, NCRAC Culture Series 101, North Central Regional Aquaculture Center, Iowa State Univ., Ames, 1996, pp. 109–110.
65. D.D. Flowers, in R.C. Summerfelt, ed., *Walleye Culture Manual*, NCRAC Culture Series 101, North Central Regional Aquaculture Center, Iowa State Univ., Ames, 1996, pp. 123–128.
66. L.J. Wawronowicz and W.G. Allen, in R.C. Summerfelt, ed., *Walleye Culture Manual*, NCRAC Culture Series 101, North Central Regional Aquaculture Center, Iowa State Univ., Ames, 1996, pp. 111–113.
67. J. Gunderson, in R.C. Summerfelt, ed., *Walleye Culture Manual*, NCRAC Culture Series 101, North Central Regional Aquaculture Center, Iowa State Univ., Ames, 1996, pp. 157–160.
68. C. Gustafson, in R.C. Summerfelt, ed., *Walleye Culture Manual*, NCRAC Culture Series #101, North Central Regional Aquaculture Center, Iowa State Univ., Ames, 1996, pp. 153–155.
69. M.G. Fox and D.D. Flowers, *Trans. Am. Fish. Soc.* **119**, 112–121 (1990).
70. L.L. Smith and J.B. Moyle, *Trans. Am. Fish. Soc.* **73**, 243–261 (1943).
71. A. Barkoh and C.F. Rabeni, *Prog Fish-Cult.* **52**, 19–25 (1990).

72. P.D. Richard and J. Hynes, *Walleye Culture Manual*, Ontario Min. Nat. Res., Fish Cult. Sect., Toronto, Ontario, Canada, 1986.
73. B.J. Tice, R.W. Soderberg, J.M. Kirby, and M.T. Marcinko, *Prog. Fish-Culturist* **58**, 135–139 (1996).
74. M.G. Fox, D.D. Flowers, and C. Waters, *Aquaculture* **106**, 27–40 (1992).
75. M.H. Hohn, *Ohio J. Sci.* **66**(2), 193–197 (1966).
76. G.A. Raisanen and R.L. Applegate, *Prog Fish-Cult.* **45**, 209–214 (1983).
77. C.P. Clouse, *Evaluation of Zooplankton Inoculation and Organic Fertilization For Pond-Rearing Walleye Fry to Fingerlings*, Master's thesis, Iowa State Univ., Ames, 1991.
78. M.G. Fox and D.D. Flowers, *Trans. Amer. Fish. Soc.* **119**, 112–121 (1990).
79. G.A. Raisanen and R.L. Applegate, *Prog. Fish-Cult.* **45**, 209–214 (1983).
80. R.C. Summerfelt, C.P. Clouse, and L.M. Harding, *J. Appl. Aqua.* **2**(3/4), 33–58 (1993).
81. J.G. Geiger, *Aquaculture* **35**, 331–351 (1983).
82. A. Skrzypczak, A. Mamcarz, D. Kucharczyk, and R. Kujaw, *Prog. Fish-Cult.* **60**, 239–241 (1998).
83. L. Bulkowski and J.W. Meade, *Trans. Am. Fish. Soc.* **112**, 445–447 (1983).
84. R.W. Dexter and D.B. McCarraher, *Prog Fish-Cult.* **29**, 105–107 (1967).
85. J.M. Luzier and R.C. Summerfelt, *J. Appl. Aqua.* **6**(4), 25–38 (1996).
86. W.E. Manci, J.A. Malison, T.B. Kayes, and T.E. Kuczynski, *Aquaculture* **34**(1–2), 157–164 (1983).
87. J.A. Malison and J.A. Held, in R.C. Summerfelt, ed., *Walleye Culture Manual*, NCRAC Culture Series #101, North Central Regional Aquaculture Center, Iowa State Univ., Ames, 1996, pp. 199–204.
88. T.O. Nagel, *Prog. Fish-Cult.* **38**, 90–91 (1976).
89. T. Harder and R.C. Summerfelt, in R.C. Summerfelt, ed., *Walleye Culture Manual*, NCRAC Culture Series #101, North Central Regional Aquaculture Center, Iowa State Univ., Ames, 1996, pp. 267–271.
90. S.D. Coyle, J.H. Tidwell, and F.T. Barrows, *Prog. Fish-Cult.* **59**, 249–252 (1997).
91. T.O. Nagel, *Prog. Fish-Cult.* **47**, 121–122 (1985).
92. J.G. Nickum, *Am. Fish. Soc. Spec. Pub.* **11**, 187–194 (1978).
93. D.L. Bergerhouse, in R.C. Summerfelt, ed., *Walleye Culture Manual*, NCRAC Culture Series #101, North Central Regional Aquaculture Center, Iowa State Univ., Ames, 1996, pp. 251–266.
94. G.T. Brugge and D.J. McQueen, *Prog. Fish-Cult.* **53**, 91–94 (1991).
95. R.P. Bushman, in R.C. Summerfelt, ed., *Walleye Culture Manual*, NCRAC Culture Series #101, North Central Regional Aquaculture Center, Iowa State Univ., Ames, 1996, pp. 261–266
96. K. Blazek, in R.C. Summerfelt, ed., *Walleye Culture Manual*, NCRAC Culture Series #101, North Central Regional Aquaculture Center, Iowa State Univ., Ames, 1996, pp. 275–276.
97. T. Harder and R.C. Summerfelt, in R.C. Summerfelt, ed., *Walleye Culture Manual*, NCRAC Culture Series #101, North Central Regional Aquaculture Center, Iowa State Univ., Ames, 1996, pp. 267–271.
98. G.E. Moodie, J.A. Mathias, and N.L. Loadman, *Aquacultural Engineering* **11**, 171–182 (1992).
99. A. Moore, in R.C. Summerfelt, ed., *Walleye Culture Manual*, NCRAC Culture Series #101, North Central Regional Aquaculture Center, Iowa State Univ., Ames, 1996, pp. 195–197.
100. G.E. Moodie and J.A. Mathias, in R.C. Summerfelt, ed., *Walleye Culture Manual*, NCRAC Culture Series #101, North Central Regional Aquaculture Center, Iowa State Univ., Ames, 1996, pp. 187–190.
101. G.M. Wright, in R.C. Summerfelt, ed., *Walleye Culture Manual*, NCRAC Culture Series #101, North Central Regional Aquaculture Center, Iowa State Univ., Ames, 1996, pp. 129–130.
102. F.T. Barrows, W.A. Lellis, and J.G. Nickum, *Prog. Fish-Cult.* **50**, 160–166 (1988).
103. G.A. Kindschi and E. MacConnell, *Prog Fish-Cult.* **51**, 220–226 (1989).
104. G.A. Kindschi and F.T. Barrows, *Prog. Fish-Cult.* **53**, 180–183 (1991).
105. R.T. Colesante, N.B. Youmans, and B. Ziolkoski, *Prog. Fish-Cult.* **48**, 33–37 (1986).
106. F.T. Barrows, R.E. Zitzow, and G.A. Kindschi, *Prog Fish-Cult.* **55**, 224–228 (1993).
107. A. Moore, M.A. Prange, R.C. Summerfelt, and R.P. Bushman, *Prog Fish-Cult.* **56**, 100–110 (1994).
108. R.C. Summerfelt, in R.C. Summerfelt, ed., *Walleye Culture Manual*, NCRAC Culture Series #101, North Central Regional Aquaculture Center, Iowa State Univ., Ames, 1996, pp. 161–185.
109. C.T. Boggs and R.C. Summerfelt, *Northeast Regional Agricultural Engineering Service*, NRAES-98, volume 2, 1996, pp. 623–630.
110. R. Zitzow, *Proceedings of the Coolwater Fish Culture Workshop*, Missouri Department of Conservation, Mt. Vernon, MO, 1991, pp. 11–16.
111. S. Li and J.A. Mathias, *Trans. Am. Fish. Soc.* **111**, 710–721 (1982).
112. F.T. Barrows and W.A. Lellis, in R.C. Summerfelt, ed., *Walleye Culture Manual*, NCRAC Culture Series 101, North Central Regional Aquaculture Center, Iowa State Univ., Ames, 1996, pp. 315–321.
113. L. Corazza and J.G. Nickum, in L.J. Allen and E.C. Kennedy, ed., *Bio-Engineering Symposium for Fish Culture*, Fish Culture Sect. Amer. Fish. Soc. Pub. No. 1, 1981, pp. 48–52.
114. G.A. Kindschi and F.T. Barrows, *Prog. Fish-Cult.* **53**, 53–55 (1991).
115. R.W. McCauley, *Prog Fish-Cult.* **32**, 42 (1970).
116. C.R. Stettner, R.C. Summerfelt, and K.L. Kuipers, *Proceedings of the Annual Conference Southeastern Association of Fish and Wildlife Agencies* **46**, 402–412 (1992).
117. S.A. Flickinger, in R.C. Summerfelt, ed., *Walleye Culture Manual*, NCRAC Culture Series #101, North Central Regional Aquaculture Center, Iowa State Univ., Ames, 1996, pp. 233–235.
118. G.L. Siegwarth and R.C. Summerfelt, *Prog. Fish-Cult.* **55**, 229–235 (1993).
119. S.T. Summerfelt, in R.C. Summerfelt, ed., *Walleye Culture Manual*, NCRAC Culture Series #101, North Central Regional Aquaculture Center, Iowa State Univ., Ames, 1996, pp. 277–309.
120. J. Mittlemark and A. Kapuscinski, *Walleye Culture in Minnesota*, Minnesota Sea Grant Publication, St. Paul, 1993.
121. J.A. Held and J.A. Malison, in R.C. Summerfelt, ed., *Walleye Culture Manual*, NCRAC Culture Series #101, North

Central Regional Aquaculture Center, Iowa State Univ., Ames, 1996, pp. 231–232.

122. T. Yager and R.C. Summerfelt, in R.C. Summerfelt, ed., *Walleye Culture Manual*, NCRAC Culture Series #101, North Central Regional Aquaculture Center, Iowa State Univ., Ames, 1996, pp. 237–240.
123. A. Moore, M.A. Prange, B.T. Bristow, and R.C. Summerfelt, *Prog. Fish-Cult.* **56**, 194–201 (1994).
124. P.W. Rieger and R.C. Summerfelt, *J. Fish Biol.* **53**, 93–99 (1998).

See also LARVAL FEEDING—FISH.

WATER HARDNESS

GARY A. WEDEMEYER
Western Fisheries Research Center
Seattle, Washington

OUTLINE

Determinations of Hardness
Importance in Aquaculture
Management Recommendations
Bibliography

Hardness is the term used to describe the collective concentration of divalent ions that results when mineral salts are leached into water from limestone and other minerals in soils or rock formations. Hardness is expressed in units of mg/L or ppm (parts per million) of $CaCO_3$ equivalents and is due primarily to the carbonate and bicarbonate salts of calcium and magnesium. However, silica and dissolved metals such as iron and manganese may also contribute to a minor degree. Because aquatic animals and plants require calcium, magnesium, and carbonates (HCO_3^-, CO_3^{-2}) for normal growth and development, both hardness and alkalinity (see the entry "Alkalinity") have long been employed in aquaculture as important measures of water type and potential for biological productivity. Water low in hardness (soft) is usually also acidic, whereas harder water tends to be alkaline (basic).

In water from limestone aquifers, the total hardness and alkalinity values will often be similar, and both may be useful as a measure of buffering capacity. However, ground water can also be quite hard, but have little or no alkalinity. Natural waters can usefully be classified in terms of total hardness as follows (1): (a) soft, 0–75 mg/L (as $CaCO_3$); (b) moderate, 75–150 mg/L; (c) hard, 150–300 mg/L; and (d) very hard, 300 mg/L and above.

DETERMINATIONS OF HARDNESS

Historically, hardness was defined as the capacity of water to precipitate soap to form a scum and was measured by shaking standardized soap solutions with the water sample in question. However, this concept is not useful in aquaculture, so that, for biological purposes, total hardness is best calculated from the results of separate calcium and magnesium determinations (2):

$$\text{Total Hardness (mg/L } CaCO_3) = 2.497\ [Ca] + 4.118\ [Mg]$$

The simpler ethylenediaminetetraacetic acid (EDTA) titration method may also be used. An indicator dye is added, and the water sample is titrated dropwise with EDTA. When all the Mg^{+2} and Ca^{+2} ions present have been complexed by the EDTA, a color change occurs that marks the end point. The calcium hardness is sometimes also measured and expressed as mg/L (ppm) $CaCO_3$. For routine monitoring purposes, commercially available water chemistry test kits provide adequate accuracy and precision for both total hardness and calcium hardness determinations.

For aquaculture work, total hardness is most usefully expressed as mg/L (or ppm) $CaCO_3$ equivalents. However, a variety of other units may also be encountered. The calcium hardness is often expressed as mg/L Ca^{+2} instead of in $CaCO_3$ equivalents (1 mg/L $CaCO_3$ = 0.4 mg/L Ca^{+2}). In the aquarium industry, hardness is often reported as degrees of hardness (dH); either degrees of carbonate hardness (KH, from the German *karbonate*), or as degrees of general hardness (GH, calcium + magnesium). One unit (degree) of hardness (dH) = 17.9 mg/L $CaCO_3$. In the water softening industry, water hardness is often expressed as grains per gallon (1 gpg = 17 mg/L $CaCO_3$).

As evident from its name, the KH is a measure of the amount of carbonate, CO_3^{-2}, and bicarbonate, HCO_3^-, present and is only indirectly related to water hardness itself. Unfortunately, KH and alkalinity are often used interchangeably even though hydroxides, borates, silicates, and phosphates may also contribute to alkalinity. In seawater rearing systems, however, nearly all of the alkalinity is due to carbonate, and the values for alkalinity and carbonate hardness will be essentially equivalent.

IMPORTANCE IN AQUACULTURE

The total hardness (expressed as mg/L $CaCO_3$) of a water supply is an important environmental factor of which aquaculturists should be aware. As mentioned, aquatic plants and animals reared in aquaculture systems require calcium, magnesium, and carbonates for normal growth and development. Soft water is low in the calcium and magnesium needed for fish health, but can be tolerated by most species if dietary intake is sufficient. For example, both juvenile and adult salmonids and channel catfish (*Ictalurus punctatus*) can absorb the required amounts of Ca^{+2} and Mg^{+2} either from the water or from their diet. However, a minimum of 5 mg/L of calcium hardness is recommended to assure normal egg hatching and fry development (3). In addition, harder water reduces the physiological stress of fish culture procedures, such as handling, transportation, and disease treatments. On the negative side, hard water causes scale deposits that can clog pipes, air diffusers, and other submerged equipment.

Another important effect of water hardness on fish and invertebrate health is the protection it affords against the toxicity of heavy metals, such as zinc and copper. In hard or alkaline water (>150 mg/L, pH >7), these metals form insoluble salts and precipitate out of solution rendering them less biologically available. For this reason, there is also less risk of fish toxicity when copper containing disease therapeutants or algicides are used. The criteria originally developed by Alabaster and Lloyd (4) for levels of copper safe for rainbow trout (*Oncorhynchus mykiss*) as a function of water hardness still serve as a useful guideline.

Water Hardness (mg/L as $CaCO_3$)	Safe Level (mg/L)
10	0.001
50	0.006
100	0.01
300	0.28

Salmonid fish may be slightly less susceptible in hard water to disease such as infectious pancreatic necrosis (IPN) and bacterial kidney disease (BKD), although the incidence may actually be inversely proportional to the water's ionic composition, rather than to hardness *per se* (5). Epizootic ulcerative syndrome (EUS), a disease affecting milkfish (*Chanos chanos*), and other warmwater fishes cultured in Southeast Asia, appears to be most severe in waters of low total hardness (6).

If necessary, water hardness (GH but not the KH) can be increased in ponds or aquaria by adding calcium sulfate. Calcium carbonate additions will increase the KH as well, but the proces is likely to be slow because of solubility problems. Calcium sulfate (gypsum) is relatively inexpensive, moderately water soluble, and does not cause areas of locally high pH if applied unevenly. Calcium sulfate in the form of gypsum, which is mined, rather than the calcium sulfate produced as a byproduct of phosphoric acid manufacture, should always be used because the latter may contain residual acid. To increase the hardness of the smaller volumes of water used in egg hatching or fry rearing, the required amount of calcium chloride can be inexpensively added to the incoming water supply using a drip system or metering pump.

MANAGEMENT RECOMMENDATIONS

As a guideline, water hardness in the range of 50–200 mg/L with a pH of 6.5–9 and alkalinity of 100–200 mg/L (as $CaCO_3$) is widely considered desirable for the intensive culture of both cold- and warmwater fishes and invertebrates (7).

BIBLIOGRAPHY

1. Environmental Protection Agency, *Quality Criteria for Water*, U.S. Environmental Protection Agency, Washington, DC, 1986.
2. American Public Health Association (APHA), *Standard Methods for the Examination of Water and Wastewater*, 17th edition, APHA, Washington, DC, 1989.
3. C.S. Tucker and E.H. Robinson, *Channel Catfish Farming Handbook*, Van Nostrand Reinhold, New York, 1990.
4. J.S. Alabaster and R. Lloyd, *Water Quality for Freshwater Fish*, Butterworth, Sydney, 1980.
5. J.L. Fryer and C.N. Lannan, *Fisheries Research* **17**, 15–33 (1993).
6. M.K. Das and R.K. Das, *Environmental Ecology* **11**, 134–145 (1993).
7. G. Wedemeyer, *Physiology of Fish in Intensive Culture Systems*, Chapman & Hall, New York, 1996.

WATER MANAGEMENT: HATCHERY WATER AND WASTEWATER TREATMENT SYSTEMS

David E. Owsley
Dworshak Fisheries Complex
Ahsahka, Idaho

OUTLINE

INTRODUCTION

It has often been said that oxygen and ammonia, in that order, are the limiting factors in aquaculture production. Today, we have ways to treat these limiting factors. The biggest challenge to the aquaculturist is not the level of oxygen or ammonia, but how to treat hatchery effluents. Unless you are in the domestic wastewater treatment business, it is hard to justify the costs of secondary and tertiary treatments. For hatcheries to coexist in this world, aquaculturists must look to the future and apply these technologies. Therefore, aquaculturists must become skilled in the water and wastewater treatment processes. In this regard, managers and administrators should now be looking for ways to improve and use an important resource: water. For example, large cities use municipal wastewater to irrigate golf courses and parks. Water is a valuable resource, whether in pure form or in the form of wastewater.

WATER REQUIREMENTS

The demand for aquaculture products is rapidly increasing in the North American continent. This demand comes from people who fish for sport and from a viable market for aquaculture products. While the demand in North America may never reach that of Europe, it is steadily increasing. However, the oceans have limitations and cannot sustain the current need alone. Therefore, fish culturists must assume a role in meeting the demand. Europeans use every available source of water. For example, Germans rear eels in distillery wastewater. The Japanese use domestic wastewater to rear carp. Infact, every country in the world is engaged in some type of aquaculture.

Unlike some other continents, the North American continent has an abundant supply of water. For example, the Hagerman Valley, in southern Idaho, produces a large percentage of the trout in the United States. The spring waters in that area are rich in nutrients, ideal in temperature, and gravity fed to the hatcheries. However, all of the water is currently being used, and production is at an upper limit. What happens when the demand increases for this product? There are very few "Hagerman Valleys," and most of the good water sources have been already used. What does this situation mean to the present and future fish culturist? It means that they must produce more and be more efficient with available resources.

A statement by J.W. Atz (1) puts this thought into perspective and has a message for those who are concerned about the future of aquaculture:

> For more than 4,000 years, men have kept fish alive in ponds or tanks for food and pleasure. Fish culture has been practiced since the times of the ancient Romans and Chinese, and some of the methods used today are virtually the same as those devised centuries ago. Fish culturists have been much slower to take advantage of the benefits of science and technology than have their counterparts, the farmer and animal husbandman. This has been especially true of the maintenance and control of the water in which captive fishes must live. So poor has been the fish culturist's understanding of the complicated interactions between the fish and its water element that he has failed to recognize bad water as the underlying cause of most of his failures. Man's lack of understanding of the fish's way of life stems primarily from the fact he's a terrestrial animal, while the fish, still bathed in the fluid in which life itself arose, maintains the most intimate relationship with its surrounding. Evolution has provided fishes with limited homeostatic mechanisms, which are all too often overtaxed by the conditions they find ponds and hatchery troughs alike.
>
> Nevertheless, by locating his hatcheries and fish ponds near unlimited supplies of water, or by dealing only with species preadapted to polluted environments, the fish culturist has managed to be successful enough in the past, in spite of his ignorance. But water is fast becoming a scarce commodity, and in both Europe and America fish are now being reared in recirculated water. Moreover, the exigencies of mariculture and the mass culturing of new and more delicate kinds of fishes demands a degree of water control that the old-time fish culturists would never dreamed about.

If we look closely at our counterpart, animal husbandry, we can see the achievements that have been made in that field. Feed lots reproduce more animals in less space and at less cost. Waste is being recycled back to feed. The key word that separates aquaculture from animal husbandry is "environment." The person engaged in animal husbandry lives in the same environment in which he or she rears the animals. It is therefore much easier for that individual to understand and deal with a species with which he or she shares a common surrounding. The fish culturist, however, is faced with rearing a product in an environment that is foreign to him or her, an environment that can change its physical and chemical characteristics without the culturist's knowledge.

With today's rising energy costs, it is becoming more difficult to operate within the restricted environment known as water. Furthermore, there are certain options that remain, such as the many aspects of water treatment that are used in the domestic water and wastewater treatment industry.

Aquaculturists must not separate themselves from other disciplines. They must use the knowledge and technology that are found in the water and wastewater treatment industry. Then, and only then, can the aquaculturist progress and move into the next century.

According to the works of Burrows and Combs (2), there are five major aspects of any potential hatchery water supply: (*1*) quantity, (*2*) quality, (*3*) temperature, (*4*) environment (pathogenic or pathogen free), and (*5*) location.

Quantity

The quantity of water available must meet all of the hatchery's water requirements. Wheaton (3) lists four major needs that must be met for an adequate water supply: (*1*) evaporation, (*2*) seepage, (*3*) oxygen, and (*4*) waste disposal. In addition to the water supply for fish culture needs, a certain amount of water is needed for fire protection, irrigation, and use in residences.

Quality

The quality of water must match the requirements of the propagated species. Water can be "good" or "bad," depending on its use. Good-quality water is defined as water capable of supporting the desired species and of maintaining the sanitary conditions necessary to allow the harvested species to be used as intended. Under this definition, it is possible for raw municipal sewage to be considered good-quality water if algae are grown in it, processed, sterilized, and fed to animals. But this same water could be considered bad for human consumption, even after undergoing tertiary treatment.

Temperature

The water temperature must be within the optimum growth range for the desired species. Temperature is the most critical factor affecting respiration. Due to intensified fish culture techniques, fish are, at times, reared in less than optimum temperature ranges. The final result is a subquality product, or, worse, a product that experiences high losses throughout a production cycle. For hatchery operations, the water temperature must be within a range in which the species being cultured can survive and grow.

Environment

Disease prevention and pathogen control are critical in any type of hatchery operation. If a water supply meets all the requirements for a species, except for the ability to prevent and/or control disease outbreaks, then the water in a reuse system should be used only with adequate treatment. While it is generally economically unfeasible to treat all of the water in a single-pass hatchery, it may be possible to treat a portion of the total flow, such as in a reuse system, if all other requirements are met.

Location

The location of a hatchery is important in order to ensure that water is available for present and future needs. The site should be large enough to accommodate housing, water treatment facilities, and future expansion, as well as to meet fish-rearing needs. Accessibility to visitors and the security needs of the facilities during nonworking hours are also important considerations. Hatcheries are often located in very remote, as well as urban, areas.

Natural springs that meet all, or even just two or three, of the aforementioned aspects for hatchery sites are almost nonexistent in this day and age. If a body of water does not meet even one of these major aspects, it is not feasible for an open system. On the other hand, a body of water can lack two or more of the major aspects and still be feasible for a closed system. An open system is one in which water is continuously discarded and replenished from a natural source. An example of an open system is a single-pass hatchery. A closed system is one in which water is recycled and used again. A closed system can be completely closed, or it can be semiclosed. An example of a completely closed system is an aquarium. A semiclosed system is one in which some new water is continuously added to the recycled water and some of the water that has been recycled is discarded. A reuse hatchery, which is an example of a semiclosed system has been described by Owsley (4).

WATER TREATMENT

Filtering Systems

The importance of the aeration process cannot be overemphasized with respect to the design and operation of fish hatcheries. Oxygen is the controlling factor in pond loadings and is important to all aspects of hatchery operation. Piper (5) recommends for salmonids that, for every part of oxygen below saturation, the pond loadings are decreased by 5%. A good aeration system should provide a minimum level of 90% oxygen saturation.

One of the most important parameters for the aquaculturist to remember with respect to aeration is that air is 78% nitrogen and 21% oxygen. (Pure oxygen can be 98% or more.) Oxygen is critical to the survival and growth of fish and is just as critical to nitrification of bacteria in the wastewater treatment process.

Filtration can be classified into three general categories: (*1*) mechanical, (*2*) biological, and (*3*) chemical. Mechanical filtration is the physical separation of suspended particulate matter from the water. It is accomplished by passing the water solution through a suitable medium that traps the particles. Examples of mechanical filtration include sand filters (pressure and gravity flow), stationary screens, rotary screens, micro strainers, and diatomaceous earth filters. Mechanical filters usually exhibit a high head loss and require more energy to operate than biological filters.

Biological filtration is defined as the mineralization of organic nitrogenous compounds by bacterial action. It can be either aerobic or anaerobic. Aerobic biological filtration is called *nitrification*, and anaerobic biological filtration is called *denitrification*. Of the biological treatment methods studied for the removal of ammonia, nitrification appears to be the most efficient. The reasons for this are as follows: (*1*) high potential removal efficiency, (*2*) process stability and reliability, (*3*) easy process control, (*4*) small land-area requirement, and (*5*) moderate costs.

Nitrification is the oxidation of ammonia to nitrite and thence to nitrate by autotrophic bacteria, namely, *Nitrosomonas* and *Nitrobacter*. The two steps of the nitrification transformation are as follows:

$$\text{Step 1} \quad 2NF_3{-}N + 3O_2 \xrightarrow{\textit{Nitrosomonas}} 2NO_2^{-}{-}N + 2H_2 + 2H^{-}$$

$$\text{Step 2} \quad 2NO_2^{-}{-}N + O_2 \xrightarrow{\textit{Nitrobacter}} 2NO_3^{-}{-}N$$

The nitrosomonas reaction is the rate-controlling step. The requirements for nitrification are the presence of ammonia, oxygen, nutrients, a low level of organic carbon, and a substrate for the bacterial to attach to. This substrate is called *medium* and is available in various forms. Some of the more commonly used media are oyster shell and rock, expanded shale, plastic rings, plastic beads, plastic saddles, porcelain saddles, styrofoam beads, and dolomite rock.

Chemical filtration is the removal of substances from a solution on a molecular level by adsorption on a porous substrate or by direct chemical oxidation. Examples of chemical filtration include ion exchange wherein a natural zeolite, such as clinoptilolite, is used as the resin for the adsorption of ammonia. Clinoptilolite is abundant in the southwestern areas of the United States and is common throughout the world. Carbon adsorption is another example of chemical filtration, as is foam fractionation. Air stripping, breakpoint chlorination, and ozonation are examples of chemical oxidation. Clinoptilolite may be the best filtration method available to the fish culturist.

Another type of filtration to be considered in hatcheries is hydroponics, which is the use of plants to filter nutrients from the water. This type of filtration can actually produce a crop from nutrients contained in wastewater.

Disinfection Systems

A number of disinfection systems are available to the fish culturist. Some work well, others hold promise, and still others have failed.

Ozone. Owsley (6) has reported on the use of ozone at Dworshak National Fish Hatchery (NFH). Ozone is a three-atom allotrope of oxygen. It is a colorless gas and can be readily detected by its odor at very low concentrations. It is unstable and the strongest oxidizing agent commercially available. Ozone is produced by passing air or oxygen through a high-frequency electric field, which must be generated at the point of application.

Commercial ozone generators are available in many designs. Most use a high-voltage corona discharge system consisting of two surfaces separated by a space. High voltage is impressed across the space, and air or oxygen is passed between the surfaces, where the oxygen molecules are excited sufficiently to form ozone. Ozone can also be produced by ultraviolet (UV) radiation. However, UV radiation does not produce the quantity of ozone achieved by an ozone generator.

Although ozone is toxic to aquatic organisms and to humans, it is highly effective as a disinfectant, having about twice the oxidizing capabilities as chlorine and killing both bacteria and viruses on contact with equal effectiveness and speed. Ozone reacts very quickly as compared with compounds such as chlorine and is much less affected by pH and temperature than is chlorine. Unlike UV radiation, ozone is not limited by low-turbidity water, nor does it appear to leave harmful residues in water, such as chloramines, which are produced by chlorine.

Both organic and inorganic materials exhibit a demand for ozone. Ozone is used for removal of color, odor, and turbidity. In order to achieve the same disinfection level, water containing organic matter must be treated with a higher concentration of ozone than similar water without organic matter. Inorganic materials such as iron and manganese can be oxidized to the insoluble oxide forms by ozone, which decomposes back to oxygen. The decomposition rate is temperature dependent, rapidly increasing with increased temperature. Thus, an ozone destruction unit is basically a heater.

For many years, ozone has been widely used in Europe for water disinfection. Europe does not have the luxury of the abundant clean water that exists in the United States and Canada. However, even though the use of chlorine and UV radiation was more economical in obtaining clean water, we are finding that there are certain diseases that chlorine and UV radiation cannot effectively combat. One example is giardiasis, which is a disease that affects humans, and is caused by a protozoan, *Giardia lambia*, commonly found in high-mountain streams and lakes. Ozone will effectively kill giardia and its cysts, while chlorine is ineffective. An example of water treatment in an aquaculture facility is the treatment for a virus called *infectious hematopoietic necrosis virus* (IHN virus). Infectious hematopoietic necrosis is an epizootic virus disease that causes high rates of mortality in most salmonids. Fish that survive may become carriers of the disease. Adult fish may become infected by disease causing organism, or may shed the virus during spawning. Natural transmission occurs primarily through water (7). External symptoms of fish with IHN include hemorrhaging under the skin, exophthalmia (i.e., protruding eyes), swollen abdomens, lethargy, darkening of skin color, and hemorrhaging at the base of the fins. Internally, the liver, spleen, and kidneys are usually pale. The stomach and intestine may be filled with fluid. The work at the Seattle laboratory of Wedemeyer et al. (8) indicates that ozone destroys the IHN virus at low dosages and low contact times (see Table 1).

Other aquaculture studies have shown positive results when ozone was used to treat water supplies. Ceratomyxosis, caused by the myxosporean parasite *Ceratomyxa shasta*, was controlled at the Cowlitz Hatchery in Washington (9). The Coleman National Fish Hatchery in California has had success using ozone for controlling whirling disease, caused by another myxosporean parasite, *Myxobolus cerebralis* (10). Ozone was found to be superior to chlorine for inactivating the bacterial fish pathogens *Aeromonas salmonicida* (furunculosis) and *Yersinia ruckeri* (enteric redmouth) and pathogenic viruses, including infectious pancreatic necrosis virus (IPNV), the causative agent of infectious pancreatic necrosis (11).

A study by Oakes et al. (12) at the Dworshak NFH showed that nitrite could be virtually eliminated by using ozone in reuse water. Williams et al. (13) determined the oxidation rates for ammonium, nitrate, and biological oxygen demand (BOD), separately and in combination, by using an improved contact-chamber design. The upflow design, which included an improved foam removal system, demonstrated good results in the study.

Foam can be a problem in an ozone system. Foam is made up of protein, suspended and dissolved solids, and organic and inorganic compounds. It is mainly found in the contact chamber, and most systems are designed with some type of foam removal system, such as a protein skimmer. Foam in incubation, especially during nursery rearing, can be a hindrance to fish culture operations. In nursery tanks, the foam prevents the starter feed from reaching the fish. A foam remover, such as Dow Corning antifoam FG–10, can be used to clear the water.

The feasibility of treating lesser quality water, as opposed to developing good-quality supplies, was studied at Coleman National Fish Hatchery in northern California (14). Four water supply systems were evaluated: wells, chlorination and dechlorination, UV sterilization, and ozonation. Table 2 summarizes the findings of that study, using UV sterilization as the base. At this facility,

Table 1. Ozone Dosages and Contact Times for Treatment of IHN Virus for Different Water Sources

Ozone Level (mg/L)	Contact Time (min)	Water Source at 10 °C
0.001	0.5 or 1.0	Phosphate-buffered distilled water
0.01	10.0	Soft lake water
0.01	10.0	Hard lake water

Table 2. Cost Comparison of Development of Good-Quality vs. Treating of Lesser Quality Water Supplies[a]

Water Supply	Construction Costs	Annual Operation and Maintenance
Wells	366	45
Chlorination and dechlorination	70	61
UV sterilization	100	100
Ozonation	94	36

[a]Use of UV sterilization is cost basis. All numbers are percentage of cost base.

ozonation was the logical choice to pursue, based upon cost-effectiveness.

As new systems are developed, ozonation will become more cost-effective. One new development, the aquatector (15), designed to efficiently add oxygen into water, could greatly alleviate, or even eliminate, the need for ozone contact basins. The aquatector uses microbubble technology that could improve the efficiency of adding ozone to water. This improvement would, in turn, reduce the size of equipment needed, as well as reduce costs. On-site oxygen-generating equipment can double the production of ozone and reduce costs accordingly.

As mentioned previously, ozone is a very toxic compound. Its toxicity is a function of time and concentration. The maximum allowable concentration for humans in an eight-hour day is 0.1 mg/L. Ozone can normally be detected in the air by the human nose in the range of 0.05 mg/L. It is very important that work areas be free of ozone. Ozone can be converted back to oxygen by using a heat system prior to discharging the ozone into the atmosphere.

Ozone is also very toxic to fish and must be removed from the water supply. Wedemeyer et al. (16) determined that the permissible safe exposure level of ozone for salmonids was 0.0002 mg/L. Several other studies have verified this level, while others have reported much higher exposure levels. The discrepancy between Wedemeyer's work and other studies is probably due to sampling accuracy and production vs. laboratory conditions. The conventional method of removing ozone is through the use of detention chambers. Since ozone has a short half-life, it can be removed by allowing the water to be held in a chamber for a period of time.

A faster, more economical way to remove ozone is to strip it out of the water by using packed columns (17). Removal efficiency of packed columns varies from 70% to 95%. Complete removal can be accomplished by using stripping towers (18). A stripping tower is a packed column that uses a countercurrent air flow. However, stripping towers raise the energy costs of removing ozone, due to the additional height and blower requirements. Carbon filters will also remove ozone very effectively, but will accommodate only small systems and are not as economical as packed columns or stripping towers.

Ultraviolet Light. Ultraviolet (UV) light, or energy in a wavelength of approximately 254 nm, has been shown to be effective in killing both bacterial and viral organisms. Light in the UVB and UVC spectra are responsible for the majority of the disinfection and sterilization attributed to this type of system. High-quality bacteria-free water (99.9% removal) has reportedly been produced by using UV disinfection for low-turbidity supply sources.

Since UV light must effectively penetrate a water source in order to ensure that the shortwave energy is imparted to the biological organism, it is critical to ascertain that the water source entering the UV contractor is very low in turbidity and suspended solids and that turbulent conditions are maintained for adequate mixing. In order to achieve the necessary influent water quality, prefiltration is often recommended, to remove particulate matter capable of blocking light penetration and fouling the UV tubes.

The U.S. Environmental Protection Agency (EPA) has not recognized UV treatment as an allowable method of disinfecting potable water without another treatment used in conjunction with it, since it is impossible to develop a corollary concentration and detention time relationship to chemical oxidation. However, this does not mean that UV disinfection is not effective, only that it has not been widely used for treating water supplies destined for human consumption.

The use of UV energy to deactivate waterborne microorganisms in the aquaculture industry is a well-known and well-understood technology. UV systems have become an integral part of many aquaculture operations that provide disinfected water to areas of the hatchery and rearing operation where there is an established need to maintain control of fish pathogens (i.e., bacteria, viruses, molds, fungi, and protozoans). According to Caufield (19) the following are the advantage of UV radiation: (*1*) nontoxic, (*2*) adds nothing to the water, thereby preventing formation of toxic chemical residuals, (*3*) does not affect water chemistry, (*4*) is effective against a wide range of organisms, (*5*) can be designed to suit any flow rate, and (*6*) can be supplied with an automatic duty standby capability.

It is generally accepted that the germicidal nature of UV energy is the result of disruption of the microorganisms' DNA molecules by irradiation with wavelengths of 220–290 nm. With most organisms, this effect is maximized at 260 nm. It has also been shown that the survival ratio after UV treatment is related to the UV dose applied (normally reported in milliwatt seconds or millijoules per square centimeter — $mWsec/cm^2$ or mJ/cm^2, respectively).

The UV dose received by an organism is dependent on (*1*) the energy output of the UV source, (*2*) the flow rates of the water and the organism's residence time under the influence of the UV source, (*3*) the ability of the fluid to transmit the germicidal wavelengths, often referred to as *UV transmittancy*, and (*4*) the geometry of the radiation chamber.

In many cases, UV systems have been designed to provide a high kill rate for organisms that are easy to kill with UV radiation; such as *Escherichia coli*. *Escherichia coli* is a common bacteria found in wastewater treatment plants. A theoretical UV dose of 3.2 mJ/cm^2 will provide a 90% kill, or one log reduction. (The actual UV dose

Table 3. Dosages (in mJ/cm²) of UV Required to Reduce Fish Pathogens[a]

Pathogen	90% Reduction	99.9% Reduction
Viral hemorrhagic septicaemia virus (VHS virus)	10	30
Saprolegnia (fungal disease)	13	39
Ichthyophthirius (white spot, or "ich")	40	120
Infectious pancreatic necrosis virus (IPN virus)	60	190

[a]*Source*: Blake, Ref. 20.

applied depends on factors such as UV transmittancy and TSS.) Table 3 shows the dosage of UV radiation needed to achieve a 90% and a 99.9% reduction of certain fish pathogens (20).

In many aquaculture applications, seasonal changes in water quality, due to a high concentration of undesirable organisms and high suspended solids or turbidity levels, affect UV treatment. These facts, together with a lack of accurate, continuous monitoring or validation of the UV system operation, have contributed to the poor performance of UV equipment, which was originally designed to treat consistently clear water. Over the years, several hatcheries including Dworshak National Fish Hatchery, have installed large-scale UV systems. However, as for as can be determined, Alaska's Trail Lake Hatchery, which used UV for posthatchery disinfection, is the only major facility in the northwest United States still using UV on a routine basis. The Trail Lake system (2,000 gpm) is working satisfactorily.

Chlorine. According to a report by Montgomery Engineers (18), chlorine is not a viable disinfectant alternative for most fish production facilities. Chlorination is the most common type of disinfection used for water treatment systems. Use of chlorine gas, HTH (hypochlorite), a chloramine, or other materials to produce a solution containing the oxidant is practiced by the majority of domestic water purveyors throughout the United States. Chlorine disinfection has been used on several aquaculture projects within the northwest United States, but has several inherent drawbacks. The most serious is the necessity to remove the residual chlorine prior to introducing the disinfected water into the fish hatchery in order to avoid the possibility of creating a toxic environment that would affect fish production. However, Wedemeyer et al. (6) and Bedell (21) found chlorine effective in inactivating both *C. shasta* and the IHN virus. The only large-scale facility in the northwest United States using chlorine disinfection in salmonids, Ore-Aqua, relies on careful control and constant monitoring, in combination with sufficiently large postchlorination storage to provide the extended detention necessary for dissipation of any harmful chlorine residual. Chloramines (ammoniated chlorine) are not suitable for aquaculture systems due to their known toxicity to aquatic organisms and are not considered as a feasible method of disinfecting hatcheries.

Iodine. Iodine is a nonmetallic element with an atomic weight of 126.92. It is the heaviest of the members of the halogen group (i.e., chlorine, bromine, iodine, and fluorine). The halogens are a group of elements that form with metal compounds, such as common salt—sodium chloride. Iodine is the only halogen that is solid at room temperature. It is a shining blackish brown crystal solid with a specific gravity of 4.93 and peculiar chlorinelike odor. Iodine is always found combined and can be prepared from kelp or crude chile saltpeter. It is only slightly soluble in water.

Iodine and its compounds have been used in medicine since the early 1800s. The first use of iodine in water was in World War I for sterilization of water for troops. Iodine as a disinfectant for water supplies has been recognized for a long time, but it has never been feasible to use iodine in the same capacity as chlorine is used. However, iodine does have some advantages over chlorine as a water disinfectant. In the late 1970s, the Idaho Department of Health and Welfare recommended that the domestic water-supply treatment method at Dworshak National Fish Hatchery be changed from chlorine to iodine. This change was to combat a parasite, *G. lambia*, which is chlorine resistant.

Iodine is widely used in fish culture practices for egg disinfection. The term *iodophores* includes commercial forms of iodine, the two most common being Wescodyne and Betadine. Wescodyne is an iodine solution with a detergent base. Betadine is an iodine solution that contains povidone as the organic base. Both iodophores contain iodine as the active ingredient, but have different organic bases. Both are considered to be effective bactericides and virucides.

Since 1982, due to a severe infectious hematopoietic necrosis virus (IHN virus) problem at Dworshak, steelhead eggs have been water hardened in an iodophor solution each year to prevent the virus from erupting and infecting other fish. Based upon personal communications with Dr. Bob Busch, Director of Rangens Laboratory in Buhl, Idaho, iodine was selected to be tested using a continuous-drip method on a small group of fish from Brood Year 1989 steelhead egg production. Dr. Busch has been successful in reducing losses of rainbow trout due to IHN virus by applying a continuous dosage of iodine, between 0.3 to 0.5 mg/L, to rainbow trout fry and fingerling. Dr. Busch has also stated that iodine is an effective treatment against bacteria gill disease. This report was further substantiated by Jim Winton of the Service's National Fisheries Research Center in Seattle, WA. Dr. Winton was able to effectively kill the IHN virus in a laboratory setting at levels from 0.3 to 0.5 mg/L, which were the exact levels that Dr. Busch had reported. With these data, a pilot study was set up for the Dworshak hatchery (22).

Results of the test showed that iodine was not successful in destroying IHNV in the water, even with a continuous level of iodine present. Ozone-treated and control groups both showed very little signs of the virus. These results are based upon the literature and direct observation of fish.

Dworshak's water supply has a pH value that can range from 6.3 to 7.3 over a complete rearing cycle. Iodine is not a good viricide with this low-to-neutral pH range; however, it is a good bactericide. Second, the low level of iodine on a continuous exposure could have created a stress on the fish that may have initiated an IHN virus outbreak.

Further evaluation of iodine needs to be conducted at various levels of pH and water quality. Iodine does have the potential to help alleviate a serious disease problem in fish culture, such as the IHN virus, as long as the water is of good quality and has an adequate PH level.

Hydrogen Peroxide. Hydrogen peroxide is clear, colorless, and waterlike in appearance and has a characteristic pungent odor. A nonflammable substance, it is miscible with water in all proportions and is sold as an aqueous solution. The amount of hydrogen peroxide in commercial solutions is expressed as a percentage of the solutions weight. Thus, a 35% solution contains 35% hydrogen peroxide and 65% water by weight. Most industrial applications call for 35, 50, or 70% concentrations.

Hydrogen peroxide is not a particularly hazardous substance. It is considerably safer to handle and store than chlorine gas, which is widely used in wastewater treatment. It does not have the highly corrosive and dangerously toxic characteristics of chlorine. In fact, when properly handled and contained, hydrogen peroxide has been safely stored on street corners and the center strips of residential streets, where it is injected into sewer mains. At the same time, a basic understanding of the properties of hydrogen peroxide is necessary for proper handling.

The U.S. Department of Transportation classifies solutions of hydrogen peroxide as "oxidizers" (DOT yellow label). In addition, the chemical is a strong oxidizing agent, liberating oxygen and heat when it decomposes. In dilute solutions, the heat is readily absorbed by the water, but in more concentrated solutions, the heat raises the temperature of the solution and accelerates the decomposition rate. Hydrogen peroxide itself will not burn, but its decomposition liberates oxygen, which supports combustion. For water treatment, hydrogen peroxide holds much promise and as a by-product, it reverts back to oxygen. However more research needs to be done to evaluate the use of hydrogen peroxide in aquaculture.

Photozone. The simultaneous application of UV radiation with ozone can accelerate otherwise sluggish reaction rates significantly. Acceleration is brought about by catalytic formation of hydroxyl free radicals, which are stronger oxidizing agents than ozone itself. Ozone and UV radiation complement each other, requiring lower doses of ozone and better operation and maintenance of the UV radiation system.

Perozone. The simultaneous application of hydrogen peroxide with ozone can accelerate otherwise sluggish reaction rates significantly. Acceleration is brought about by catalytic formation of hydroxyl free radicals, which are stronger oxidizing agents than ozone itself.

Microfiltration. Microfiltration effectively cleans a water supply, but does not kill viruses or bacteria. Its use is limited to small flows of water, and it is generally used with a disinfectant.

Heat. Heat will destroy all bacteria and viruses, but it is not practical or economical.

Pure Oxygen. Although pure oxygen is not considered as a disinfectant, high levels of pure oxygen can relate to a clean environment and reduce the toxic effects of some bacteria and viruses.

SOLIDS REMOVAL

There are various methods of removing solids from the rearing space. They include, but are not limited to the following systems: (*1*) vacuum systems: These involve transfer of solids using a siphon device. These systems are very efficient when used in conjunction with baffles and screens. (*2*) mechanical systems: These range from swimming-pool cleaners to floor sweepers and are usually complex and difficult to use in a hatchery environment; (*3*) hydraulic systems: These drain the space while brushing and use hydraulic energy to remove the solids.

There are a number of systems and designs that are used to enhance removal of solids in a hatchery, including (*1*) baffles, as used in Michigan to promote the removal of solids from the upper end of a raceway to the tail end via a velocity concept, and (*2*) circular or square tank or pond designs for better solids removal with decreased flow.

The most important thing to remember with respect to the removal of solids is not to reduce the particle size of the solids, if at all possible. However, this requirement complicates the treatment process.

WASTEWATER TREATMENT

Effluent Treatments

The degree of treatment depends upon the receiving body of water. A recreational lake or small stream will require a higher degree of treatment than a large free-flowing river. The large river has a high rate of turnover, and the large volume of water can accept a higher nutrient load.

There are three types of treatment for wastewater: primary, secondary, and tertiary. Primary treatment includes sedimentation and clarification and can usually be expected to remove 50 to 60% of the suspended solids and 25 to 35% of the biological oxygen demand (BOD). Examples of primary treatment include settling ponds, lagoons, and clarifiers. Settling ponds and lagoons operate from detention time and require large volumes of space. These systems are effective for removing settleable solids. Insects and odors can be problems associated with ponds and lagoons. The efficiency of ponds and lagoons can be increased by aeration, which increases the settling efficiency of solids and reduces BOD by bacterial action. Aerated ponds and lagoons require less space than nonaerated ponds and lagoons, and aeration reduces odor and insect problems. A clarifier is a controlled settling pond where the flow and solids removed are regulated in the design.

Secondary treatment uses conventional biological processes. Secondary treatment can remove up to 90% of the suspended solids and 75 to 90% of BOD. Examples of secondary treatment include trickling filters, biological filters, activated digesters, separators, and, ofcourse, filtration.

Tertiary treatment refers to methods and processes that remove more nutrients and contaminants from wastewater than are usually taken out by secondary treatment. Examples of tertiary treatment include microscreens, fine filtration, chemical filtration, air stripping, ion exchange, and hydroponics.

Other aspects of a hatchery water supply include fire and maintenance, domestic, and irrigation. Water is a scarce commodity in some areas and a valuable commodity in all areas. It should be used conservatively and wisely. Properly treated wastewater can be used on golf courses and parks. If hatcheries are to survive, fish culturists must be willing to make changes. They must start thinking about water and wastewater treatment as well as rearing fish. The time has come to merge water and wastewater treatment technologies with the technology of hatcheries.

OPERATION AND MAINTENANCE

Operation and maintenance go hand in hand. It is estimated that for every 100 man-hours of plant operation, at least 40 man-hours must be spent to maintain equipment in a water or wastewater treatment plant. A hatchery could be a combination of a water and wastewater treatment plant. The actual time spent depends upon the complexity of the plant and the age and type of equipment. Regardless of size or complexity, every hatchery must have a maintenance program.

A maintenance program may be as complex as necessary, but the maintenance requirements must be reduced to a fixed schedule. A gear box may require annual maintenance, whereas a pump may require daily checks and service. There must be specific tasks to do at specific times or else maintenance will be disorganized and cause equipment failures, which hatcheries cannot afford.

A basic maintenance management system may include equipment records, planning and scheduling, inventory control, maintenance personnel, maintenance guidelines, and a budget. Equipment records are vital for a successful maintenance program. Such records are used to keep track of the maintenance of equipment. The records will tell when equipment needs major overhaul or replacement. Records can also be used to keep track of costs and labor for equipment. The record system can take many forms, ranging from a log book to a card to a computerized system. Moreover, there are many available computer systems for water and wastewater treatment plants that are applicable for hatcheries.

Planning and scheduling preventive maintenance is the critical part of a good maintenance management system. It is based on the size and capabilities of the maintenance staff, quantity of work, and the time required to complete the work. A work schedule listing job priorities, work assignments, personnel, and timing is vital. The schedule may be divided into any convenient period of time: daily, weekly, monthly, quarterly, semiannually, or annually. Scheduling should take advantage of weather, seasons, low production, and flow periods. The biggest error is to perform preventive maintenance work as time permits. This practice is the direct opposite of scheduling and will cause problems. The manufacturers maintenance manuals are a good basis for preventive maintenance instructions and scheduling.

Critical and hard-to-get spare parts should be maintained. The manufacture's recommendations and work experience should help in determining which spare parts should be stored. Spare parts should be replaced immediately when they are used.

Only properly trained personnel may be expected to perform satisfactory inspections, repairs, and preventive maintenance. Even with properly trained maintenance personnel, some work will be beyond staff capabilities. Consultants or factory representatives may be called in to perform certain maintenance work. Some work must be sent out, such as an electric motor that needs to be rewound. Contractors may be used to perform infrequent and labor-intensive work. Trained personnel should have a knowledge of the functions and operations of the equipment as well as the maintenance procedure for the service to be performed. Specific skills are required in the operation and maintenance of mechanical and electrical equipment. As there is danger when dealing with electrical equipment, particularly when high voltage is involved, only a qualified electrician should be hired to maintain such equipment.

Maintenance guidelines can be found in the literature supplied by the manufacturer. Some may be known from operational experience. All information should be kept on record in a central location. Drawings and blueprints should also be kept at this location. A manual describing the preventive maintenance of the facility is a necessity.

All maintenance programs should have a budget for normal operations. Scheduled repairs and replacement of equipment and parts should be included. These costs should be part of the records for preventive maintenance.

BIBLIOGRAPHY

1. J.W. Atz, *Some Principles and Practices of Water Management for Marine Aquariums*, U.S. Department of the Interior, Bureau of Sports Fisheries and Wildlife, Research Report 63. 192, 1964.
2. R.E. Burrows and B.D. Combs, *Prog. Fish-Cult.* **30**, 123–136 (1968).
3. F.W. Wheaton, *Aquaculture Engineering*, Wiley Interscience, 1977, pp. 229–247.
4. D.E. Owsley, Reuse Systems, *Proceedings of the Ontario Trout Farmers Association Annual Convention*, Toronto, Ontario, 1983.
5. R.G. Piper, Know the Proper Carrying Capacities of Your Farm, *American Fishes U.S. Trout News*, 1970, pp. 4–6.
6. D.E. Owsley, Operation of the Ozone Pilot System at Dworshak National Fish Hatchery, *Proceedings of the Northwest Fish Culture Conference*, Tacoma, Washington, 1984.

7. D.E. Owsley, Reuse Systems, *Proceedings of the Ontario Trout Farmers Association Annual Convention*, Toronto, Ontario, 1983.
8. G.A. Wedemeyer, N.C. Nelson, and C.A. Smith, *J. Fish Res. Board Can.* **35**, 875–879 (1978).
9. J.M. Tipping and K.B. Kral, *Evaluation of a Pilot Ozone System to Control* Ceratomyxa shasta *at the Cowlitz Trout Hatchery*, Washington State Game Department, Bulletin 85–18, Olympia, WA, 1985.
10. B. Baker, Ozonation of Hatchery Water Supply at Coleman National Fish Hatchery, *Proceedings of the Northwest Fish Culture Conference*, Eugene, OR, 1986.
11. G.A. Wedemeyer, N.C. Nelson, and C.A. Smith, *J. Fish. Res. Board Can.* **34**, 429–432 (1978).
12. D.P. Oakes, Cooley, and L. Edwards, *Ozone Disinfection of Make-Up and Recycle Water at Dworshak National Fish Hatchery*, Chemical Engineering Department, University of Idaho, Moscow, ID, 1978.
13. R.C. Williams, S.G. Hughes, and G.L. Rumsey, *Prog. Fish-Cult.* **44**, 102–105 (1982).
14. Sverdrup and Parcel Engineering, *An Evaluation of Pilot Ozone System to Control* Ceratomyxa shasta *at the Cowlitz Trout Hatchery*, Washington State Game Department Bulletin 85–18, Olympia, WA, 1986.
15. A.R. Schutte, *Evaluation of the Aquatector, an Aeration System for Intensive Fish Culture*, U.S. Fish and Wildlife Service Information Leaflet 42, Bozeman, MT, 1986.
16. G.A. Wedemeyer, N.C. Nelson, and W.T. Yasutake, *Ozone: Sci. Eng.* **1**, 295–318 (1979).
17. D.E. Owsley, in L.J. Allen and E.C. Kinney, eds., *Proceedings of the Bio Engineering Symposium for Fish Culture*, American Fisheries Society, Bethesda, MD, 1979, pp. 71–82.
18. Montgomery Engineers, Cold Lake Fish Hatchery Water Treatment Modifications Conceptual Design Report, 1991.
19. J.D. Caufield, in J. Colt and R.J. White, eds., *Proceedings of the Bioengineering Symposium for Fish Culture*, American Fisheries Society, Bethesda, MD, 1988.
20. M. Blake, *Aquionics Aquaculture*, Technical Report, 1988.
21. J. Bedell, *Prog. Fish Cult.* **33**, 51–54 (1971).
22. D.E. Owsley and J.M. Streufert, The Effects of Iodinated and Ozonated Water Supplies on Steelhead Trout Susceptibility to the IHN Virus, *Proceedings of the Northwest Fish Culture Conference*, Gleneden Beach, OR, 1989.

See also SITE SELECTION.

WATER SOURCES

MICHAEL B. RUST
JOHN COLT
Northwest Fisheries Science Center
Seattle, Washington

OUTLINE

Water has been dubbed the drink of the gods and the basis of life. It is little wonder that the source of water for an aquaculture facility is so important. Water sources can be from underground (groundwater), on the ground (surface waters), or above the ground (rain). Each source has unique water-quality parameters associated with it. A summary of the sources of water and associated parameters relevant to aquaculture is presented in Table 1. Many other topics in this encyclopedia also discuss various aspects of water quality and quantity. This entry is intended to provide an overview of water sources and a framework from which to judge their suitability for aquaculture.

GROUNDWATER

Groundwater refers to water that is contained in subsurface geological formations (1). It is brought to the surface either by pumping from a well or flowing naturally from a spring or artesian well. Water has a mass of about 1000–1028 kg/m^3 (62–64 lb/ft^3) depending on salinity and temperature. To move a large mass of water needed for fish culture any great height or distance requires energy. Water sources that are located deep in the earth or at great distance from the farm will always be more costly than water that is free flowing or shallow (2). The impact that this will have on the economic viability of the farm depends on many factors related to species and site selection.

Because groundwater is contained far from the surface and can remain underground for millions of years, the water takes on characteristics that are related to the rock formations that the water is in contact with. Alkalinity, hardness, pH, dissolved minerals, and dissolved gasses (each of these topics are covered elsewhere in this volume) in groundwater can all be at very different levels from water that is in contact with the atmosphere. Oxygen and biological processes are limited underground, so ground water typically contains few pathogens and little oxygen, although carbon dioxide and argon gasses can be supersaturated.

Groundwater temperatures below 10 m (33 ft) are quite stable in a given location relative to surface waters. Temperature is moderated by the thermal mass of the earth, so groundwater tends to have a relatively minor seasonal temperature change. Groundwater varies with latitude, being warmer near the equator and colder away from the equator. For example, groundwater temperatures in the United States range from about 26 °C (79 °F) in southern Florida to 3 °C (37 °F) in northern Minnesota (1). In general, the average annual temperature of groundwater is a degree or two (°C) higher than the mean annual air temperature (3). In addition, groundwater temperature increases 1–5 °C (average about 2.5 °C) per 100 m depth (4). Higher increases may be due to local geothermal activity. If the groundwater is in proximity to volcanic activity, then it may be naturally

Table 1. Water Sources and Some Associated Water-Quality Parameters

Source	Quantity	Pathogens and Predators	Salinity	Temperature	Suspended Solids	Dissolved Oxygen	Other Dissolved Gasses	Total Gas Pressure	Metals	Oxidizers	Buffer (pH)
Groundwater	Generally stable, but may vary with season or year	Not usually a problem	Fresh to full strength seawater	Stable over the short term, can vary seasonally	Low	Low	Carbon dioxide, and argon can be high depending on geology	Can be above saturation variable	Iron and manganese may be a problem in water with low DO.	None	Depends on geology
Rivers, streams, and lakes	Streams and rivers variable, lakes generally stable, but can vary with seasons	Common and should be expected	Fresh	Variable short term and seasonally, varies more than ground-water	Varies, can be high during runoff events	Low to high variable	Generally low, but carbon dioxide can be high and variable if there is a lot of respiration	Same as above	Depends on industrial and domestic discharges in proximity to inflow lines	None	Depends on geology and source of water
Oceans, seas, and bays	Stable, although elevation can vary with tides	Common and should be expected	Brackish to seawater	Variable	Varies, can be high during rujoff events and storms	Low to high, variable	Same as above	Generally below saturation at depth	Same as above	None	Generally well buffered, locally may reflect buffer capacity of rivers flowing into it.
Municipal water sources	Limited and expensive	Reduced, but not eliminated from source	Fresh	Variable	Low	Low to high	Low	Variable. Can be high in summer	Should not be a problem	Chlorine and/or ozone is often used	Depends on source and treatment
Rain	Highly variable	Low pro-bability	Fresh	Variable	Very low	High	Low	Saturated	Depends on industrial air pollution upwind of catchment area.	None	Poorly buffered, can be acidic
Recirculation systems	Can intensify use of existing source.	Requires treatment, difficult to eliminate pathogens completely	Fresh to full strength seawater	Set by operator, stable	Depends on treatment	Depends on treatment	Carbon dioxide can be high	Generally not a problem, unless pump has a suction leak	May build up from feed	Ozone is often used to maintain ORP and kill pathogens	Highly buffered and controlled

heated by a geothermal source. Much of the trout industry in Idaho, USA benefits from geothermally heated water.

Groundwater is usually fresh, but saltwater aquifers also exist. Saltwater aquifers are common near the coasts, but can also extend for hundreds or thousands of miles inland. The salinity of the groundwater is variable. Discharge into freshwater canals or streams may be problematic if the site is located very far inland or in agricultural areas.

SURFACE WATER

Surface waters include streams, rivers, canals, ponds, lakes, seas, and oceans. Because they are exposed to the atmosphere and typically support diverse and abundant biological ecosystems, surface waters have different characteristics from groundwater. Surface waters can be fed either by rain or groundwater or both. Water quality parameters such as alkalinity, hardness, pH, dissolved minerals, and other factors will be somewhat dependent on the source. In addition, biological processes tend to change water quality and add competing organisms, pathogens, and predators. Biological processes tend to add acids to water, there by lowering pH and depleting alkalinity. They also tend to reduce dissolved compounds in water and lower the oxidative-reductive potential (ORP) of the water. Water temperature follows seasonal and local weather patterns, therefore surface water is more variable than groundwater.

RAIN

Rain is not usually a reliable source of water for aquaculture; however, in certain regions it can be. For example, many catfish (*Ictalurus punctatus*) raised in Alabama, are from ponds supplied primarily by rainwater (4). Rainwater may also be a useable supply for makeup in recirculation systems or other systems where a continuous high volume supply is not needed. Since rainwater contains almost no buffer and can be affected by airborne pollution, careful consideration must be given to its quality and dependability as a primary source of supply. In some cases, it may help reduce demand on other sources.

MUNICIPAL WATER SOURCES

Municipal water can originate from groundwater, surface water, or rainwater sources. While it may be convenient to simply hook up to a municipal-water source, there are two drawbacks. First of all, municipal water is treated to make it safe for human consumption. Typically, this includes treatment with strong oxidizers, such as chlorine or ozone, to kill human pathogens. These compounds are lethal to fish and need to be removed or treated to make the water safe for fish (see the entries "Ozone" and "Chlorination/dechlorination"). Second, municipal water is expensive and a limited resource in most areas. The use of municipal water for aquaculture is largely limited to small-scale experimental systems, holding for live retail sales, or as emergency makeup for recirculation systems.

NATURAL OR REHYDRATED SEAWATER

Small amounts of seawater can be made from salt mixtures and freshwater (6–8). This might be sufficient for inland holding facilities, aquariums, or small research recirculation systems, but due to the expense and labor needed to prepare seawater, high-flow production aquaculture will likely need a large body of natural saltwater nearby. The advantage of making up seawater from salts is that very clean water can be made. This might be important for algae, zooplankton, egg, and larval culture, no matter where the facility is located. Also, if a seawater recirculation system is used with ozone as an oxidizer, then a salt mixture lacking bromine can be used to minimize hypobromide production. Hypobromides are toxic to many types of aquatic organisms (see the entry "Ozone").

RECIRCULATION AS A SOURCE

Just as recycling of paper and plastic can reduce the use of trees and petroleum resources, water recirculation technologies can reduce the demand on the water supply. Water recirculation can increase the flow to the culture units by 10–1000 times the flow of new water (9). The downside is the high capital cost of the treatment components.

Energy requirements may be greater or lesser than a single-pass system, depending on the amount of water that needs to be pumped in each system and the need for temperature adjustment. Temperature and other water-quality parameters can be adjusted in a recirculation system to suit the species and life stage of interest. This is a major advantage and may or may not offset high capital and energy costs and may offer higher growth rates. This advantage is particularly important with photoperiod and temperature-manipulated broodstock, egg incubation, and larval rearing. (For more information, see the many related topics about recirculating systems in this Encyclopedia.)

MATCHING WATER SOURCES TO AQUACULTURE PRODUCTS

What comes first, the water or the fish? Often, aquaculture projects come about from one of two situations. Either someone is knowledgeable about a species that they are hoping to culture and wants to find a site with a source of water to fit the species or has a site with an available source of water and is looking for a suitable species to rear. Water source and species selection are closely linked. If large quantities of water are involved, such as is the case with pond or flow-through culture, then it is very difficult and expensive to treat water beyond using simple screens and aeration. Pumping from deep wells, heating, or chilling large volumes of water is energy intensive, expensive, and unlikely to be economically or environmentally sustainable over the long haul. For all these reasons, water quality from the source should be well matched to the environmental optima of the species to be cultured if the project is to be a success.

Temperature is often one of the most important factors when matching a water source and a culture species. Except for small water volumes used for egg or broodstock holding or for recirculation systems, attempting to significantly change the temperature of the culture water by heating or cooling is costly. Exceptions to this rule are geothermal or waste heat from a power plant to heat water, or possibly the use of deep, cold marine water for cooling warm water (such as is done at the Natural Energy Laboratory of Hawaii Authority site).

The amount of water needed for a fish production system will depend on the intensity of culture. In extensive systems, water requirements may be based upon management concerns, such as the time to fill a pond, or the amount water needed to makeup for evaporation or leakage. Larson (10) recommended that enough water be available for aquaculture ponds to fill in two weeks. Amounts needed for evaporation or leakage will depend on the area where the ponds are located.

For intensive systems, water requirements will be dictated by the first limiting water-quality parameter. Typically, oxygen is first limiting, followed by carbon dioxide and ammonia. A mass-balance approach can be used to determine water requirements for a given species and level of production. Conversely, this approach can also be used to determine the potential production levels of a given species for a known water source. The mass balance approach is very powerful and can be used to examine and identify which water-quality parameter is limiting for a given set of circumstances, and which water treatments will have the greatest impact on production (11).

The mass balance approach accounts for all inputs/outputs and production/consumption of a compound of interest in a defined system. Compounds can get into or out of the system either predissolved in the water flow or they can transfer into the water once it is already in the system (form air, feed, light, and so on). Flow (Q, in volume/time, m^3/hr) in a system times concentration (C, in mass/volume, g/m^3) equals the rate of mass flowing (g/hr) into or out of a system predissolved in the water. Transfers (T), production (P), and consumption (R) are also rates in mass/time (g/hr). Transfers of mass into or out of the system come from or go outside of the system. For example, oxygen may be transferred into a system and carbon dioxide out of a system due to aeration. Compounds can also be taken up or released by the fish, bacteria, or some other internally generated activity of the system, to or from the water. Typically this relates to the biomass (B) of organisms contained in the system. For example, the consumption of oxygen by fish depends on the size and number of fish. Accounting for all the ins and outs and conversions that occur can be expressed mathematically. Mass is conserved, so the mass of a given compound into and out of a system is in balance, and this can be expressed in the equation below. The same approach can be used for energy, which is also conserved.

$$Q_{in}C_{in} + T + (P - R)B = Q_{out}C_{out}$$

where Q_{in} is the influent water flow (e.g., Lpm), C_{in} is the influent concentration of the compound in water (e.g., mg/L), T = transfer of the compound into (or out of, if negative) the tank (e.g., mg/hr), P = production of the compound in the tank (e.g., mg/hr/kg of organism), R = consumption of the compound in the tank (e.g., mg/hr/kg of organism), B = biomass of the organisms in the tank (e.g., Kg), Q_{out} = effluent water flow (e.g., Lpm), and C_{out} = effluent concentration of the compound in water (e.g., mg/L).

For example, if we have a spring that produces 100 m^3/hr of water at 15 °C (59 °F), and it contains 10 mg/L of oxygen, and we want to know the biomass of rainbow trout (*Oncorhynchus mykiss*) that can be maintained with this water, we can use the above equation to figure it out. We need to know what the trout will do to the oxygen in the water and what the minimum level of oxygen should be to keep our trout healthy. For this example, we will assume that the trout removes oxygen at the rate of 200 mg O_2/kg of fish/hr and requires a minimum of 5 mg/L to stay healthy (1). The concentration to keep the fish healthy will be equal to the outflow concentration since when it gets to that level, we want to get rid of it.

For our example the equation has the following values:

Flow ($Q_{\text{in and out}}$) = 100 m^3/hr or 100,000 L/hr both in and out

Transfer of oxygen (T) = 0 mg O_2/hr (we have no aeration in this example)

Production of oxygen (P) = 0 mg O_2/kg of biomass/hr (if there were plants or algae in the system then this might be a nonzero number)

Consumption of oxygen (R) = 200 mg O_2/kg of fish/hr

Concentration of oxygen in the inflow (C_{in}) = 10 mg/L (this is saturation at 15 °C)

Concentration of oxygen in the outflow (C_{out}) = 5 mg/L (set as the minimum acceptable level)

Plugging in the values:

$$Q_{in}C_{in} + T + (P - R)B = Q_{out}C_{out}$$

$$(100{,}000 \text{ L/hr} \cdot 10 \text{ mg/L}) + 0 \text{ mg/L/hr} + (200 \text{ mg/kg hr} \cdot B \text{ kg}) = (100{,}000 \text{ L/hr} \cdot 5 \text{ mg/L})$$

Solving for B, (kg of trout)

$$B = 2500 \text{ kg (5500 lbs)}$$

The same exercise, repeated with each water-quality factor, can be used to determine the first limiting water-quality parameter. This approach can also be used to determine which water treatments will be necessary to increase production. In the foregoing example, the effect of aeration can be addressed by adding in a value other than zero for the production term. The calculations become more difficult when recirculation and multiple treatments are considered.

Surface water sources, especially those from established ecosystems, will contain pathogens and potential predators. Pollution may also be a concern for rain, surface, and groundwater sources that are in proximity to a pollution source. Various screens and sterilizers are available to reduce predators and pathogens, but treatments for pollution will depend on the nature of the pollution, and may not be treatable.

Regulations for the removal of water from either a groundwater or surface-water source exist at multiple levels in almost all countries. This is true for marine or freshwater sources. Additional regulations govern discharge of water from aquaculture facilities. Property owners should check with local authorities and government agencies regarding rights and permits for development of the water source.

BIBLIOGRAPHY

1. R.W. Soderburg, *Flowing Water Fish Culture*, Lewis Publishers, Boca Raton, FL, 1994.
2. R.G. Piper, I.B. McElwain, L.E. Orme, J.P. McCraren, L.G. Fowler and J.R. Leonard, *Fish Hatchery Management*. U.S. Fish and Wildlife Service, Washington DC, 1982.
3. J.D. Meisner, J.S. Rosenfeld, and H.A. Regier, *Fisheries* **13**, 2–8 (1988).
4. H. Bouwer, *Groundwater Hydrology*, McGraw-Hill, New York, 1978.
5. R. Stickney, *Principles of Aquaculture*, John Wiley & Sons, New York, 1994.
6. S. Spotte, *Fish and Invertebrate Culture*, John Wiley & Sons, New York, 1979.
7. S. Spotte, *Seawater Aquariums*, John Wiley & Sons, New York, 1979.
8. S. Spotte, *Captive Seawater Fishes: Science and Technology*, John Wiley & Sons, New York, 1992.
9. M. Timmons and T. Losordo, eds., *Aquaculture Water Reuse systems: Engineering Design and Management*, Elsevier, Amsterdam, 1994.
10. T. Lawson, *Fundamentals of Aquacultural Engineering*, Chapman & Hall, New York, 1995.
11. R. Piedrahita, *Aquaculture Systems Engineering*, American Society of Aqricultureal Engineers, St. Joseph, MI, 1991.

See also SITE SELECTION.

WINTER FLOUNDER CULTURE

W. HUNTTING HOWELL
University of New Hampshire
Durham, New Hampshire

MATTHEW K. LITVAK
University of New Brunswick
St. John, Canada

OUTLINE

The winter flounder, *Pseudopleuronectes americanus*, is a right-eyed flounder (family Pleuronectidae) found along the east coast of North America from Labrador, Canada to Georgia, USA. The species is typically found on mud and sand substrates, at depths ranging from the shallow subtidal to 37 m (1). Maximum size is usually 2.25 kg and 65 cm, although they occasionally grow to over 3 kg (2). Among the flatfish species found along the coast of New England, winter flounder are the heaviest per unit length (3). The species has been exploited both commercially and recreationally for well over a century. Their stocks have been in decline over the past 20 years, and the species is currently considered overexploited (4).

The first propagation of winter flounder occurred in the late 1880s, at three government hatcheries located in Massachusetts and Maine, and operated by the U.S. Fish and Fisheries Commission. The work was undertaken in an effort to rebuild declining wild populations, and tens of millions of early larvae were released before the last of these hatcheries closed in the early 1950s (5,6). Although the success of those efforts was probably minimal due to the small size of the larvae released, some of the basic culture techniques developed through those efforts are still in use today.

Recent declines in the fishery, combined with a demand for high quality flatfish, have once again stimulated interest in the culture of various flounder species, including winter flounder (7). Although a pilot attempt to rear winter flounder in New Hampshire had only minimal success (8), advances since then, by a number of researchers, have improved the ability to culture this species. Researchers and culturists have also relied heavily on the wealth of information available on production techniques for other flounder species, including turbot (9,10), Japanese flounder (11), and summer flounder (12).

The purpose of this entry is to review what is known about winter flounder culture. Regrettably there are some topics for which little information is available (e.g., broodstock management, growout systems). In some cases, techniques come from our own recent studies, which are not yet published. Nevertheless, we are confident that the techniques work, and since we are anxious to give the reader the most recent information, we have included some of these results and observations.

COLLECTION AND MAINTENANCE OF BROODSTOCK

Currently broodstock are being kept by Huntsman Marine Science Centre in Saint Andrews, NB, Canada, and by Sambro Fisheries, Sambro, NS, Canada. Save for a few fish kept at Sambro that were grown by Litvak's laboratory in 1994–1997, these broodstock are wild caught. For the most part, researchers have relied on wild-caught adults to produce the early life-history stages needed for their studies (13–25). Brood fish are collected by trawl net, gillnet, fyke net, or by divers in the weeks preceding the natural spawning season, and returned to the laboratory. Ripe females are identified by their swollen ovaries, which indicate hydration, and spermiating males are identified as those which release milt upon slight abdominal pressure. Maintenance of captive adults for short periods of time is relatively easy. Although females fast during the spawning season, males readily consume clams, squid, chopped menhaden, silversides, clam worms, earthworms, and chopped capelin during the spawning season (16,26–29).

Wild winter flounder adults caught prior to spawning will usually undergo gametogenesis when provided with photoperiods and temperatures that are "normal" to their place and time. Spawning times vary from December to June, beginning earlier in the southern part of the fish's range and later in the northern parts of the range (30,31). Final maturation and spawning are commonly induced through hormonal manipulation. Smigielski (32) experimented with a variety of hormones, including human chorionic gonadotropin (HCG), pregnant mare serum gonadotropin (PMSG), deoxycorticosterone (DOCA), oxytocin, and freeze-dried carp pituitary extract. Smigielski's best success, in every case, was with freeze-dried carp pituitary extract, at doses of 0.5 or 5 mg/454 g female body weight. The extract was mixed in an isotonic (to fish blood) solution of sodium chloride as a carrier, and injected intramuscularly on a daily basis until spawning. Fish receiving the higher dose spawned after only three injections, while those receiving the lower dose required six injections. All but one female (which was later determined to be sexually immature) hydrated and spawned, and all resultant eggs and larvae were normal. Water temperature appeared to be a critical factor in producing ovulation, as the majority of fish did not hydrate at temperatures above 6 °C, even under hormonal treatment.

More recently, Harmin and Crim (33,34) conducted additional research on hormonal induction of maturation and spawning. Intraperitoneal injections of gonadotropic releasing hormone analog (GnRH-A), 20 μg per kg body weight, three times per week, resulted in a few females ovulating at temperatures as low as 0 °C, and accelerated ovulation and increased spawning reliability in prespawning flounders maintained at 5 °C. Harmin and Crim were able to induce spawning in February, which was three months prior to normal spawning season in their area (Newfoundland). Furthermore, egg and larval quality (as indicated by rates of fertilization, hatching, and larval survival) was good after this accelerated spawning. The researchers also observed that when using intramuscular implants of GnRH-A (100–120 μg slow release, or 40 μg fast release) there were rapid and predictable ovulatory responses from the fish.

Harmin and Crim (35) have also used GnRH-A injections to treat prespawning males. Maturing fish treated during winter (December/January), with a single injection of either 20 or 200 μg/kg body weight, showed increased levels of testosterone and 11 ketotestosterone within 12 hours and these levels remained high for several days. Single injections advanced spermiation in some individuals, but only small amounts (<50 μL) of milt were produced. By March, following GnRH-A treatment, all males were spermiating. In fish that were injected twice, there was a significant increase in sperm production and milt volume.

Although wild fish are easily collected and spawned in the laboratory, the development of a winter-flounder aquaculture industry will require the establishment of captive adult populations (broodstock) to produce a reliable supply of high-quality eggs. There are several disadvantages associated with the use of wild brood fish. Perry et al. (22) have shown, for example, in brood fish collected from areas that differed in habitat quality, that habitat affected the viability and health of resultant embryos and larvae. Fish from anthropogenically contaminated areas produced eggs with high incidence of chromosome damage and mitotic abnormalities. Similarly, brood fish collected from different spawning locations in Long Island Sound, CT, and Narragansett Bay, RI, spanning about 200 km, produced larvae with different mean sizes at hatching and different biochemical content (20). This was important, because there was a direct correlation between these variables and survival through the first month. In a related study (21), it was found that spawning time and female size affected the composition and viability of the eggs and larvae. Larger females produced larger eggs, and mean egg weight from the population decreased as the spawning season progressed. Therefore, large, early spawning fish produced the largest eggs, and these in turn had the highest viability.

COLLECTION AND INCUBATION OF EGGS

Collection

The fecundity of winter flounder has been described by a number of authors for a number of areas (2,36–38) and generally ranges from 100,000 to 3.3 million eggs per female, depending on female size. Ripe winter-flounder eggs are spherical, range in diameter from about 0.71 to 0.96 mm diameter (39), have a specific gravity of 1.085–1.095 (40), and are adhesive (15). Thus, eggs form demersal clumps when extruded. Artificial fertilization is typically done using the "dry" technique, in which the ripe eggs are stripped from the female into a dry container. Typically all the eggs from a female are extruded by a single stripping. A mixture of milt and clean seawater is then added to the eggs, and the mixture is swirled for several minutes to ensure fertilization. Fertilization percentages are generally quite high and range from 78 to 93% (20,21). Because high egg mortalities have been attributed to clumping (15), the fertilized eggs are often treated with a diatomaceous earth solution to prevent them from adhering to one another. Newly fertilized eggs are spread into a polyethylene pan, then covered by a dense slurry of diatomaceous earth solution (50 g diatomaceous earth in 1 L of sterile seawater). The mixture is swirled for five minutes and then rinsed to remove excess earth (15). Litvak's lab has switched to Pyrex trays instead of polyethylene pans because plastic is prone to scratches, making them harder to disinfect after use and increasing the risk of bacterial outbreak. Alternatives to this procedure include the "plating out" of eggs into single layers on panels of either glass or fine plastic screening, which allows each egg more contact with the seawater (Klein-MacPhee, unpublished).

Very ripe females spawn volitionally when placed together with males in small (100 L) spawning tanks supplied with ambient (6 °C, 30 ppt) flowing seawater (25). No attempt is made to control photoperiod or temperature, but the amount of ambient light entering the spawning tanks is reduced by covering them with black window screens. Stocking ratio of males to females is 3:1. Depending on ripeness, the fish spawn within 1 to 10 days, and spawning always occurs at night. Although winter flounder are reputably batch spawners (30), our experience is that each female releases most of her eggs in a single spawning. Occasionally, however, smaller groups of eggs are released on subsequent days. After spawning is completed, the adults are removed from the tank. Eggs are allowed to clump and are then moved to 100 L tanks supplied with flowing seawater. This technique of volitional spawning usually resulted in very high (>95%) fertilization rates (Howell, unpublished). Although spawning behavior was not observed by King and Howell (25), the spawning behavior of captive winter flounder has been described (41).

Incubation Systems

A variety of embryo incubation systems have been used successfully. Smigielski and Arnold (15) used simple "incubation baskets," which were 15 L rectangular plastic containers into which windows, covered with 505 μm plastic screening, had been cut in the sides and bottom. Baskets of fertilized eggs were then incubated in flowing seawater troughs. Similar systems have been used by other researchers (16,20,22). Six-liter acrylic hatching jars, supplied with filtered seawater at ambient temperatures and salinities (4–6 °C; 31–33 ppt) have also been used successfully (42–44). Static methods, in which embryos are incubated in small, temperature-controlled containers, and in which a portion of the water (25–50%) is periodically (every 1 to 3 days) replaced with filtered, ultraviolet treated seawater, have also been used extensively (24,45,46). Stocking densities of embryos in static systems have ranged from 40–1250 per liter (16,19,43). Litvak's lab incubates eggs at a density as high as 5000 per liter in Pyrex trays. These relatively low densities prevent ammonia buildup and oxygen depletion. With containers placed in a flowing seawater trough, Smigielski and Arnold (15) stocked eggs at 17,000–34,000 per liter. King and Howell (25) allowed entire egg masses to incubate in 100 L, flowing seawater tanks. Stocking density of eggs is estimated as 10,000–30,000 per liter.

Incubation Conditions

Winter flounder embryos are relatively eurythermal, but survival to hatching is generally higher at temperatures less than 10 °C. Williams (47) incubated eggs over a range of water temperatures from −1.8 to 18 °C and found that viability was high over a wide temperature range. Percentage of eggs surviving to hatch at low temperature (less than −1 °C) was variable, ranging from 0 to 79%. Survival to hatching was consistently higher than 75% for embryos incubated between 0 and 10 °C and consistently less than 75% for those incubated at temperatures between 10 and 15 °C. It was noted, however, that egg mortality may have resulted from microbial infection rather than directly from temperature. The upper lethal thermal limit was given as 15 °C. Buckley et al. (19) reported hatching percentages of 70–85% at 4, 7, and 10 °C, and noted that percentage was significantly higher at the lower temperatures. They also noted that hatching percentage was unaffected by the acclimation temperatures of the adults, which were 2 and 7 °C. Rogers (16) experimentally investigated various combinations of incubation temperature (3–14 °C) and salinity (0.5–45 ppt). She found that viable hatching was highest at 3 °C, lowest at 14 °C, and was similar at 5, 7, and 12 °C. Moreover, she found that temperature and salinity interacted, such that the highest viable hatch (78%) occurred at 3 °C over a salinity range of 15–35 ppt, while at temperatures above 3 °C, the optimal salinity range decreased to 15–25 ppt.

Embryonic development has been described (14) and is typical of other flatfish species. Rate of embryonic development is temperature dependent (16,47). Rogers (16) reported that time from fertilization to 50% hatching can last from 17 to 31 days (mean of 25 days) at 3 °C and from 5–10 days (mean of 7 days) at 14 °C. Near the end of the embryonic period, larvae move within the egg capsules, and this movement assists with rupturing the capsule at hatching (13).

Solutions of penicillin G and streptomycin (0.02 mg/mL of each) have been used prophylactically to disinfect embryos (24,45,48,51). These antibiotic solutions were added to the culture water 24 hours after fertilization, or periodically, as water in the cultures was changed.

LARVICULTURE

Development and Size

Good descriptions of larval development, from hatching to the end of the second month, are available (13,49). For a thorough review of the older literature on early life-history stages, the reader is referred to Martin and Drewry (50). Newly hatched winter flounder larvae range in total length from about 3.5 to 3.8 mm (13,45,46). Acclimation temperature of the adults, incubation temperature of the embryos, and geographic origin of the broodstock can affect both length and weight at hatching, as well as the biochemical composition of the larvae (17,18). Size at hatching affects survival potential, with larger larvae having a higher probability of surviving through the first month of life (19,20).

Time to yolk sac absorption is dependent on temperature, and generally occurs at 7 days posthatching (dph) at 12 °C (48), 9–10 dph at 5 °C, 12–14 dph at 4 °C, and 14 dph at 2 °C (17,19).

Larval Rearing Systems

Winter flounder larvae have been reared in static, flow-through and *in situ* systems. In static systems, which are used in research, but not aquaculture, the larvae are reared in containers with no flowing water. To prevent the buildup of nitrogenous waste products and to ensure adequate amounts of dissolved oxygen, a portion of the water is replaced either daily, or every second or third day. Generally, one-third to one-half of the volume is changed on each occasion. Because larvae that have been raised in static systems have been used for experiments, container sizes are relatively small, ranging from 0.4 to 40 L. The containers are typically dark walled and made of plastic or glass. Water used is typically filtered (0.45 μm) and sterilized by ultraviolet light. Antibiotics (an equal mixture of penicillin G and streptomycin) have been added (25 ppm) to static larval culture systems (42–44). Light aeration is provided, and overhead lights are used for illumination, which facilitates the visual feeding of the larvae. The advantage of static systems is that temperature is easily controlled by placing the container in a constant temperature room or water bath. An additional advantage is that prey organisms are not washed out of the system. Static systems have been used extensively (15–25, 42–45).

Flow-through systems, in which larvae are reared in 100 L circular tanks, supplied with 5 μm filtered, ultraviolet light-treated seawater at ambient temperatures and salinities (5–10 °C, 28–33 ppt) have also been used successfully (25). Flow rate to the tanks is 1–2 L (0.25–0.50 gallon) per minute. Loss of larvae is prevented by a high surface area, screened (80 μm) outflow that is located in the middle of the water column. The advantage of the flow-through system is its larger volume and improved water quality. Maintenance of desired prey levels is achieved by feeding the fish multiple times per day. Litvak's laboratory also uses a flow-through system, except they use an upwelling cylindroconical tank (48). Both experimental-sized (100 L) and production-sized cylindroconical tanks (1,000 L) have been used to successfully produce metamorphosed fish.

An *in situ* system to monitor the growth and survival of winter flounder larvae has also been used (52). It was a large (11.5 m^3) open-mesh enclosure, suspended from a surface floatation collar, deployed in a subestuary of Narragansett Bay, RI. Mesh size (505 μm) was small enough to prevent the escape of the larvae and large enough to allow their natural food to enter. The system was stocked with 1,000 four-week-old, laboratory-reared larvae, and these were held in the enclosure for two weeks in April and May. Results of the experiment with this system were very encouraging. Physical conditions (temperature and salinity) within the enclosure were optimal, and prey (copepods, rotifers, polychaete larvae, barnacle nauplii, and cladocerans) concentrations were high (10–8700 per liter). Of the fish placed in the enclosure, 76.8% survived and all metamorphosed. Daily specific growth rate was 10.7% dry weight and 1.9% standard length. The advantage of this system is the ability to raise large numbers of larvae at relatively low cost. Perceived disadvantages are the lack of control of the rearing conditions (e.g., temperature, prey availability) and the danger of losing the system in storm conditions.

Photoperiod and Light Intensity

Downing and Litvak (unpublished data) found that there was no effect of light intensity (5 vs. 100 lux) on larval winter flounder growth and survival. Continuous light, however, significantly improved both growth and survival of winter flounder larvae (48). Larvae raised under continuous light showed a five-fold increase (50 vs. 10%) in survival to metamorphosis, compared with those raised under ambient photoperiod (48). The continuous light treatment also reduced the time to metamorphosis by five days. Both our laboratories now use continuous light in rearing winter flounder larvae (25,48).

Larval Stocking Densities

Because static systems have been used for experimental studies, the number of larvae stocked per liter has varied with experimental design. Numbers range from 2 larvae/L (44) to nearly 300 larvae/L (23). More typically, 10–40 larvae/L are stocked (19,25,42,43). In flow-through systems, stocking density is as high as 100/L (Howell, unpublished).

Green Water

The addition of cultured microalgae to larval rearing tanks ("green water" treatment) has been widely accepted as a technique for commercial marine finfish production. Cultured microalgae has occasionally been added to winter flounder larval culture systems as a "water conditioner." Species of microalgae used have included *Nannochloropsis* sp., *Isochrysis galbana*, *Dunaliella tertiolecta*, and *Tetraselmis souscii* (20,21,23,25,42,43). Buckley et al. (20,21) inoculated larval tanks prior to rotifer introduction with dense cultures of *Tetraselmis* sp. at a rate of 1 L per 36 L of aquarium water. King and Howell (25), working in static systems, determined that when

3 L (200,000 cells per mL) of *I. galbana* were added to 20 L larval cultures every third day, the larvae in green water grew significantly faster than those in clear water. The importance of microalgae to winter flounder larvae has been further documented in a recently completed experiment in which microalgal species (*Nannochloropsis*, *Dunaliella*, or *Tetraselmis*) were provided at 3 L of dense (200,000 cells/mL) microalgae per 36 L tank per day, for varying lengths of time (one to four weeks) following yolk absorption (Bidwell, unpublished). Results of the study indicated that larvae cultured in green water for at least two weeks had significantly higher growth rates than those provided either no microalgae (control) or those given microalgae for only one week. Moreover, the larvae provided with microalgae for at least two weeks were more completely metamorphosed after four weeks than those which received either no microalgae or microalgae for only two weeks. We strongly recommend the use of microalgae during larval culture.

Larval Feeding

Hatchlings spend the first several days feeding endogenously as they absorb nutrients from their yolk sac. First feeding occurs within one day of absorption of the yolk sac (19,53). Total lengths at first feeding range from about 4 to 4.4 mm, depending on adult acclimation temperature and larval incubation temperature (19). Winter flounder larvae are continuous, visual, daylight feeders (18). As with virtually all small marine fish larvae, first feeding winter flounder larvae require small (<200 μm) live food items. This need has been met by feeding the larvae the cultured rotifer (*Brachionus* sp.), field collected zooplankton, or a mix of rotifers and wild zooplankton. Feeding rates of rotifers range from 2000 to 5000/L/day (24,25,51).

A 1 : 1 or 1 : 2 mixture (wild plankton:cultured rotifers) has been used by Buckley et al. (20,21) and Klein-MacPhee (42,43), respectively. Feeding rates of these mixtures ranged from 1000 to 3000 prey per liter. The exclusive use of small (48–200 μm) wild zooplankton (principally copepod nauplii) has also been used successfully (17,18,25). Initial feeding rates ranged from 2000 to 2100 prey per liter. Larger sized (up to 500 μm) wild plankton are used as the larvae grow (17,25,45). King and Howell (25) found no difference in growth or survival between larval winter flounder fed wild zooplankton and cultured rotifers that had been enriched with the microalgae *I. galbana*.

Rotifers or wild zooplankton are typically given for the first four to five weeks of feeding, but the larvae need larger food particles as they continue to grow. In most instances, this need has been met by feeding nauplii of brine shrimp (*Artemia salina*) to the late stage larvae. *Artemia* feeding is typically started at four to six weeks after hatching (22,23,41–44,51); however, this is dependent on larval size and *Artemia* can be fed to the larvae as early as 21 dph (48). Klein-MacPhee et al. (42,43) tested commercially available brine shrimp from different geographical locations and found that the geographic origin of the brine shrimp affected the growth and survival of late stage (42–71 dph) winter flounder. They suggested that the poor performance of some larvae may have resulted from the presence of various pesticides, which were relatively high in some of the brine shrimp strains tested. The differences may have also have been due to differences in the nutritional value of the different strains, particularly in the amount of long chain polyunsaturated fatty acids (e.g., 20 : 5n-3 series) that were present. In a related study Shauer and Simpson (44) found that *Artemia* are able to bioconvert short-chain fatty acids (18 : 2n-6, 18 : 3n-3) to longer chain forms (20 : 5n-3, 22 : 6n-3). Further, winter flounder juveniles accumulated the long-chain fatty acids, and were able to convert 18 : 3n-3 gotten from the *Artemia* to 20 : 5n-3 and to lesser extent, 22 : 6n-3.

Because cultured rotifers and brine shrimp may lack the essential fatty acids required for optimal growth and survival, many fish culturists, including those of winter flounder, enrich these live prey organisms. This is usually done through the use of commercially available emulsions (e.g., Selco™ products, Inve Aquaculture) or by feeding the rotifers and brine shrimp microalgal species that have the desired fatty acids.

The amount of food required by winter flounder larvae held at 8 °C, from yolk absorption to metamorphosis, has been documented (18). It was found that (*1*) all larvae fed less than 100 prey/L died within two weeks; (*2*) the number of prey consumed increased curvilinearly with prey density, particularly in fish that were five to seven weeks old; (*3*) daily specific growth in dry weight was similar at prey concentrations ranging from 500 to 3000/L. Mean daily specific growth in dry weight was 5.72% at 500 prey/L, 7.68% at 1000 prey/L, and 8.62% at 3000 prey/L; (*4*) mortality rates decreased as prey density increased; and (*5*) complete digestion of gut contents occurred within 5–8 hours. In the same paper, Laurence developed a bioenergetic model that simulated the effects of a number of variables, including temperature, prey density, and larval size on the ability of larvae to obtain the food energy needed to meet the needs of experimentally determined growth and metabolism. Results of the model simulations indicate that (*1*) the amount of time feeding must change as the larvae develop, and this in turn is related to prey density. Depending on concentration, larvae need to feed from about 3 to 18 hours per day; (*2*) wild larvae need a minimum of 300–800 prey/L to meet their feeding needs within the 12 hour of light that is normally available to them; and (*3*) the theoretical number of nauplii or older stage copepods needed to be eaten per day increased with the dry weight of larvae (from about 25 to 200 nauplii, or from 2 to 16 older-stage copepods) over the size from first feeding to metamorphosis.

Larval Growth Rates

The growth of winter flounder, from hatching to metamorphosis has been extensively studied, and can be influenced by temperature (17–19), prey concentration (18), culture conditions (25), and nutritional quality of prey (42,43). Laurence (17) studied the growth of winter flounder larvae from yolk absorption through metamorphosis at 2, 5, and 8 °C. Larvae were fed wild zooplankton (principally copepod nauplii) at a rate of 2000/L/day. Mean daily specific growth (dry weight) was 10.1, 5.8, and 2.6% at 8, 5, and 2 °C, respectively. He also provided regression equations that related larval dry weight to weeks after yolk absorption. Growth in mass was found to be exponential. Growth rates using different units of measurement are also available. Chambers and Leggett (45) reported that the average daily growth rate, from hatching to metamorphosis was 0.068 mm/day at 6.9 °C. King and Howell (25) reported specific growth rates (length increases per week) of larvae reared using different combinations of green and clear water and wild zooplankton and rotifers. Rates were 15.4% (green water/wild zooplankton), 14.2% (green water/rotifers), 12.2% (clear water/rotifers), and 9.6% (clear water/wild zooplankton). Bertram et al. (24) found that there was some variation in growth rates between individual larvae and that growth in length was nonlinear. Increase in length was rapid up to 30 days posthatch, then slowed, or even decreased at metamorphosis. In a related finding, Laurence (17) found that oxygen consumption increased as the larvae grew, decreased at metamorphosis, and then increased again. Jerald et al. (23) described the development of daily growth increments of otoliths and provided a growth equation (length) for laboratory-reared fish up to 50 days posthatch. The best fit of the growth data was achieved using a Gompertz-type curve. The length-weight relationship for laboratory-reared larval winter flounder has been reported (54).

Age and Size at Metamorphosis

Mean age at metamorphosis, which is functionally defined as the migration of the left eye to the right side of the head and loss of pigmentation on the blind side (45), has been reported by a number of investigators, and ranges from 49 to 64 days posthatch at incubation temperatures ranging from about 7–10 °C (17,18,23,24,45,46,51). At a higher mean incubation temperature of 15 °C, metamorphosis occurred from 26 to 33 days posthatch (55), while at a lower incubation temperature of 5 °C, metamorphosis was delayed until 80 days posthatch (17). Mean length at metamorphosis has also been reported by a number of investigators, and ranges from 6.1 to 10.1 mm total length (23,24,45,46,51). Laurence (17) found that all larvae maintained at 2 °C died before reaching metamorphosis. Length at metamorphosis is less variable than age at metamorphosis and that larvae that metamorphose at a later age do so at a larger size (45,46).

GROWOUT

Weaning of Juveniles

Lee and Litvak (55) used wild young-of-the-year winter flounder juveniles to develop a weaning protocol. They were able to wean wild juvenile flounder onto dry feed (BP Nutrition™, Aquaculture Research Centre, Stavanger, Norway) by cofeeding live *Artemia* over one week. In a study using recently metamorphosed laboratory-reared winter flounder juveniles, they further examined the weaning protocol and also tested two different diets: nonsalmonid and salmonid starter feed (56). The study found that the locally produced inexpensive salmonid pellet (Hi-Pro™, Corey Feed Mills, Fredericton, NB, Canada) performed as well as the specialty marine dry pellet (Nippai SFI-3, CATVIS, Hertogenbosch, The Netherlands). However, they did see a slight decrease in growth rates immediately after the switch to dry diets, suggesting that there is room for improvement in their weaning protocol during this critical stage.

Juvenile Rations and Diets

Little information is available on how much food juvenile winter flounder need, and for this reason, fish in captivity are normally fed *ad libitum*. Juveniles fed a diet of chopped bivalve (*Mya arenaria*) siphons *ad libitum*, and held at 20 °C (4 °C above their normal seasonal limit) consumed between 193 and 973 mg food/day but lost weight because of the temperature stress (57). Fish at cooler temperatures (12° and 16 °C) consumed more food per day (1,118–2,088 mg/day) and gained weight. Maintenance ration was calculated to be about 1.5% wet-body weight per day at 12–16 °C, and gross caloric conversion efficiencies ranges from 13.9 to 36.8% (57).

Little work has been completed on juvenile diets. The only published work that we are aware of is that of Hoornbeek et al. (8), who found that wild-caught juveniles would feed on a bound mixture of frozen shrimp, herring meal and oil, and a vitamin premix. That diet was 38% protein. Research is currently being conducted at the Department of Fisheries and Oceans (Cheryl Hebb and John Castell, DFO Saint Andrews Biological Station, NB, Canada) on diet formulation for on-growing juvenile winter flounder. Winter flounder do not seem to have the capability to spare protein and require at least 50% protein in their diets. However, there have been promising results suggesting that winter flounder protein requirements may be partially satisfied with either a portion of soy meal or canola protein (John Castell, personal, communication). Clearly, the capability to digest and utilize plant protein would be a boost to winter flounder's potential for aquaculture.

Juvenile Growth Rates

Winter flounder juvenile growth rate compares well with other commercially grown flatfish (56). Daily specific growth rates of recently metamorphosed fish have reached 3.11% weight gain per day (Table 1). This high growth rate was maintained through their first year of growth (Casey and Litvak, unpublished data) in which the specific growth rate exceeded 2.8% per day. From this research, it is clear that the culture of winter flounder juveniles will have to be conducted in warm or heated water to be profitable. In a further study of photoperiod manipulation, Casey and Litvak (unpublished data) found that winter flounder juveniles, like larvae, grow fastest under continuous light.

Producing Adults

To our knowledge, winter flounder have yet to be grown from egg to market size in captivity. Growth to market size, which we assume would be about 30 cm total length, takes between two and four years in the wild, depending on latitude (30). Presumably this time could be shortened in an aquaculture setting, where fish would be provided with optimal diets and warmer year-round temperatures (8).

Table 1. Daily Specific Growth Rates for Juvenile Winter Flounder (*Pseudopleuronectes americanus*) Reared at Different Temperatures

Reference	Age (years)	Source[a]	SGR[b] (%lgt/d)	SGR (%wgt/d)	Temperature (°C)
Frame (57)	1	W	—	0.27	12
	1	W	—	0.47	16
	1	W	—	−0.23	20
Hoornbeek	0–1	H	0.09	0.21	1–15
et al. (8)		W	0.24	0.71	10–15
Lee and Litvak (55)	0	W	0.12	0.53	5.3
Lee and Litvak (56)	0	H	1.36	3.11	15.0
Casey and Litvak	1	H	0.67	2.30	10.0
(unpublished data)	1	H	0.82	2.81	15.0
	1	H	0.79	2.71	20.0

[a]H = hatchery; W = wild-caught tested in the laboratory.
[b]Specific growth rate = [ln(final weight) − ln(initial weight)]/time × 100%.

Adult winter flounders held in captivity and provided with *ad libitum* amounts of clams and cubed beef liver daily consumed 2% of their body weight per day. Gross caloric conversion efficiency ranged from 0.10 to 0.22 (27).

Growout systems, although not developed, will probably be similar to those used for other flatfish species, including land-based tanks or raceways and net pens. There is no reason to suspect that winter flounder could not be raised in recirculating systems. Litvak (unpublished observation) has conducted preliminary experiments on cage grow out of winter flounder. A plastic coated, wire mesh cage, with a flat bottom, was suspended from a 10 m octagonal collar. Winter flounder were placed in the cage in the fall and their weights taken in December. The greatest weight gain was 40% in this period. Fish grown in the cage were 1.8 times heavier than were wild-caught fish of similar size, suggesting that the yield (fillet weight) could be higher from cultured fish than from wild-caught winter flounder.

DISEASE AND PARASITES

Diseases and parasites of wild winter flounder have been reviewed by Klein-MacPhee (30). Because winter flounder spawn and hatch during a cold part of the year, disease problems do not appear to be as great as in some warmwater species. Prophylactic antibiotics are often administered to egg and larval cultures as a precautionary measure (42,43). Fin erosion of captive juveniles has been observed (56), and smaller individuals appeared to be more susceptible to this disease. Most mortalities of wild caught juveniles held in the laboratory for extended periods of time resulted from systemic bacterial infections and parasites (8). *Vibrio anguillarum* was isolated during the periods of heaviest mortality. Symptoms included the cessation of feeding, fin rot, and hemorrhagic areas on the ventral surface. The most effective treatment was mixing furazolidone into the feed, provided at a rate of 12 g/100 g of fish per day, for 10 days. Two myxosporidian parasites were also identified from these fish. In Howell's laboratory, cultured juveniles have occasionally developed an unidentified fungal infection. This has been effectively treated by placing the fish in a dilute (250 ppm) hydrogen peroxide solution for 20 minutes. Treatment is repeated three times over 10 days.

ECONOMIC VALUE AND PRODUCT POTENTIAL

Winter flounder market value varies seasonally, depending on quantities being harvested and on the size, catch location, and name given by the broker. Current price (May 1998) for whole winter flounder in New York City is between US$5.50 and US$6.00 per pound, which is more than $2.00 higher than either yellowtail flounder and cod. Winter flounder is also sold fresh dressed and filleted. Considering that this fish is eurythermal, euryhaline, and extremely hardy, it is an excellent candidate for the live fish markets of Asia and the developing live fish markets in North America. Another favorable attribute of the species is that it is very cold tolerant. Because they produce a set of antifreeze polypeptides (58), they survive even in ice-laden seawater at temperatures below −1 °C. For this reason, they are one of the few candidates for flow-through, or net pen aquaculture at high latitudes, particularly in Canadian waters. Attempts have also been made to transfer the antifreeze producing genes from the winter flounder to the genome of the Atlantic salmon, thereby making the salmon more cold tolerant (59).

BIBLIOGRAPHY

1. F. McCracken, *J. Fish. Res. Board Can.* **20**, 551–586 (1963).
2. H. Bigelow and W. Schroeder, *Fish. Bull. U.S.* **53**, 1–577 (1953).
3. F. Lux, *Fish. Bull. U.S.* **71**, 505–512 (1973).
4. Anonymous, *NOAA Tech. Memo.* NMFS-F/NEC-101, (1993).
5. W. Richards and R. Edwards, in R. Stroud, ed., *Fish Culture in Fisheries Management*, American Fisheries Society, Bethesda, MD, 1986, 75–80.
6. R. Stickney, *Bull. Natl. Res. Inst. Aquacult.* Suppl. **3**, 135–140 (1997)
7. E. Waters, Univ. of North Carolina Sea Grant College Program, Raleigh, N.C., Pub. No. UNC-SG-96-14, 1996.

8. F. Hoornbeek, P. Sawyer, and E. Sawyer. *Aquaculture* **28**, 363–373 (1982).
9. J. Person-LeRuyet and et al., in J. McVey, ed., *Handbook of Mariculture, Vol. II, Finfish Aquaculture*, CRC Press, Boca Raton, FL, 1991, pp. 21–41.
10. P. Dhert, P. Lavens, M. Dehasque, and P. Sorgeloos. *Special Publ. No. 22, European Aquaculture Soc.* (1994).
11. K. Fukusho, *Intern. Symp. on Sea Ranching of Cod and Other Species*, Arendal, Norway, 1993.
12. G. Nardi, Final Report of Saltonstall-Kennedy Grant No. NA26FD0042-01, (1996).
13. W. Sullivan, *Trans. Amer. Fish. Soc.* **44**, 125–136 (1915).
14. C. Breder, *Bull. U.S. Bur. Fish* **38**, 311–315 (1924).
15. A. Smigielski and C. Arnold, *Prog. Fish Cult.* **34**, 113 (1972).
16. C. Rogers, *Fish. Bull. U.S.* **74**, 52–58 (1976).
17. G. Laurence, *Mar. Biol.* **32**, 223–229 (1975).
18. G. Laurence, *Fish. Bull. U.S.* **75**, 529–546 (1977).
19. L. Buckley, A. Smigielski, T. Halavik, and G. Laurence, *Fish. Bull. U.S.* **88**, 419–428 (1990).
20. L. Buckley, A. Smigielski, T. Halavik, E. Caldarone, B. Burns, and G. Laurence, *Mar. Ecol. Prog. Ser.* **74**, 117–124 (1991).
21. L. Buckley, A. Smigielski, T. Halavik, E. Caldarone, B. Burns, and G. Laurence, *Mar. Ecol. Prog. Ser.* **74**, 125–135 (1991).
22. D. Perry, J. Hughes, and A. Hebert, *Estuaries* **14**, 306–317 (1991).
23. A. Jerald, S. Sass, and M. Davis. *Fish. Bull. U.S.* **91**, 65–75 (1993).
24. D. Bertram, T. Miller, and W. Leggett, *Fish. Bull. U.S.* **95**, 1–10 (1997).
25. N. King and W. Howell, *Symposium on Marine Finfish and Shellfish Aquaculture, Marine Stock Enhancement, and Open Ocean Engineering*, Univ. of New Hampshire, Durham, NH, p. 12, (1997).
26. A. Tyler and R. Dunn, *J. Fish. Res. Board Can.* **33**, 63–75 (1976).
27. M. Burton and D. Idler, *J. Fish Biol.* **30**, 643–650 (1987).
28. M. Burton, *J. Fish Biol.* **39**, 909–910 (1991).
29. M. Burton, *J. Zool. Lond.* **233**, 405–415 (1994).
30. G. Klein-MacPhee, *NOAA Tech. Rept. NMFS Circular* **414**, 43 (1978).
31. W. Scott and M. Scott, *Can. Bull. Fish. Aquat. Sci.* **219**, 731 (1988).
32. A. Smigielski, *Fish. Bull. U.S.* **73**, 431–438 (1975).
33. S. Harmin and L. Crim, *Proc. 1989 Ann. Mtg. Aqua. Assoc. of Canada* **89**(3), 46 (1989).
34. S. Harmin and L. Crim, *Aquaculture* **104**, 375–390 (1992).
35. S. Harmin and L. Crim, *Fish Physiol. Biochem.* **10**, 399–407 (1993).
36. S. Saila, *Proc. Gulf Caribb. Fish. Inst. 14th Ann. Sess.* **1961**, 95–109 (1962).
37. R. Topp, *J. Fish. Res. Bd. Can.* **25**, 1299–1302 (1967).
38. V. Kennedy and D. Steele, *J. Fish. Res. Board Can.* **28**, 1153–1165 (1971).
39. M. Fahay, *J. Northw. Atl. Fish. Sci.* **4**, 423p. (1983).
40. W. Pearcy, *Bull. Bingham Oceanogr. Collect.* **18**, 39–64 (1962).
41. C. Breder, *Copeia* **102**, 3–4 (1922).
42. G. Klein-MacPhee, W. Howell, and A. Beck, in: G. Personne, P. Sorgeloos, O. Roels, and E. Jaspers, eds., *The Brine Shrimp, Artemia: Ecology, Culturing, Use in Aquaculture.* Universa Press, Wetteren, Belgium, 1980, pp. 305–312.
43. G. Klein-MacPhee, W. Howell, and A. Beck. *Aquaculture* **29**, 279–284 (1982).
44. P. Shauer and K. Simpson, *Can. J. Fish. Aquat. Sci.* **42**, 1430–1438 (1985).
45. R. Chambers and W. Leggett, *Can. J. Fish. Aquat. Sci.* **44**, 1936–1947 (1987).
46. R. Chambers, W. Leggett, and J. Brown, *Mar. Ecol. Prog. Ser.* **47**, 1–15 (1988).
47. G. Williams, *Mar. Biol.* **33**, 71–74 (1975).
48. M. Litvak, *Bull. Aquacult. Assoc. Can.* **91**, 4–8 (1994).
49. W. LaRoche, *Fish. Bull. U.S.* **78**, 897–910 (1980).
50. F. Martin and G. Drewry, *Development of Fishes of the Mid-Atlantic Bight*, Vol. 6, Center for Environmental and Estuarine Studies, Univ. of Maryland Contrib. no. 788, (1978).
51. D. Bertram, R. Chambers, and W. Leggett, *Mar. Ecol. Prog. Ser.* **96**, 209–215 (1993).
52. G. Laurence, T. Halavik, B. Burns, and A. Smigielski, *Trans. Amer. Fish. Soc.* **108**, 197–203 (1979).
53. L. Buckley, *Mar. Ecol. Prog. Ser.* **8**, 181–186 (1982).
54. G. Laurence, *Fish. Bull. U.S.* **76**, 890–895 (1979).
55. G. Lee and M. Litvak, *J. World Aqua. Soc.* **27**, 30–38 (1996).
56. G. Lee and M. Litvak, *Aquaculture* **144**, 251–263 (1996).
57. D. Frame, *Trans. Amer. Fish. Soc.* **3**, 614–617 (1973).
58. G. Fletcher, K. Haya, M. King, and H. Reisman, *Mar. Ecol. Prog. Ser.* **21**, 205–212 (1985).
59. G. Fletcher, M. Shears, and M. King, *Can. J. Fish. Aquat. Sci.* **45**, 352–357 (1988).

See also BRINE SHRIMP CULTURE; FLOUNDER CULTURE, JAPANESE; HALIBUT CULTURE; LARVAL FEEDING—FISH; PLAICE CULTURE; SOLE CULTURE; SUMMER FLOUNDER CULTURE.

YELLOWTAIL AND RELATED SPECIES CULTURE

Makoto Nakada
Nisshin Feed Co.
Tokyo, Japan

OUTLINE

INTRODUCTION

Since publication of a yellowtail (*Seriola quinqueradiata*) aquaculture review paper in the 1980s (1), yellowtail farmers have had a difficult time due to drastic declines in Japanese sardine resources and the stagnant economy in Japan. Farmers were able to maintain annual production of 140,000 to 160,000 metric tons until 1998 (Fig. 1). At the same time, more valuable species than yellowtail, such as amberjack (*Seriola dumerili*), goldstriped amberjack (*Seriola lalandi* or *Seriola aureovittata*), and striped jack (*Caranx delicatissimus*), which are members of the same family as yellowtail, were becoming more attractive to fish farmers, and the number of yellowtail juveniles stocked for aquaculture started to decline beginning in 1995 (Fig. 2).

Sudden decreases in the number of available juvenile yellowtail, called Mojako in Japan (Fig. 3), and significant drops in the price of cultured yellowtail in 1987 forced the fish farmers to increase production of red sea bream (*Pagrus major*) (Fig. 4). Although production costs increased with the rising costs of feed due to drastic declines in the volume of sardines caught in the waters around Japan and a poor supply of Mojako, the market price of cultured yellowtail was sluggish because of the economic depression. Thus, the number of yellowtail farms decreased to 1,815 by 1996 from 3,991 in 1977. Yellowtail culture, which used to be highly profitable (1), is facing difficulty today. Farmers are trying to introduce new species such as those previously cited to make fish farming more profitable.

High-density culture is becoming common practice to compensate for falling profits, which in turn stimulates pollution of the culture areas. It is recognized that under overcrowded conditions fish eat less, resulting in poor growth and increased susceptibility to diseases. Since the fish farmers have to use more expensive formulated feeds instead of raw fish, effective utilization of feed is essential and is a more environmentally friendly practice. In order to overcome these problems, the fish farmers are becoming aware of the importance of maintaining good records on stocking density. Many of them have begun to use personal computers to assist in farm management (2).

Use of formulated feeds with balanced nutrients made possible the production of high-quality fish with firm flesh and no fishy odor. The practice also creates a new demand for culture products for fillets used in various dishes in addition to sashimi. As consumption of cultured fishes increases, the public has become pickier about quality.

The wholesale prices of cultured yellowtail, amberjack, and red sea bream in Tokyo markets are shown in Table 1. The species bringing the highest price is amberjack, the meat of which maintains its brilliant color and firm texture longer than does yellowtail (3). Because of its high quality, amberjack usually brings a much higher price than cultured red sea bream and yellowtail at wholesale markets.

Another related species, goldstriped amberjack, is especially popular in the northern Kyushu area. Compared with yellowtail, it has less dark muscle. It gets the highest evaluation as a sashimi material, especially during the summer, because of low fat content (4). In 1997, 2.5 million large juvenile goldstriped amberjack, called Hiramasa in Japanese, were caught in the waters around the Goto Islands and cultured. The products are now being shipped to the Kanto area.

Techniques for artificial propagation of species related to yellowtail are almost completed at the Fisheries Experiment Station of Nagasaki and Mie Prefectures and the Japan Sea-Farming Association (5). Reproduction using cultured broodstock has been developed so that artificial propagation of the species on a commercial scale will soon be feasible (6).

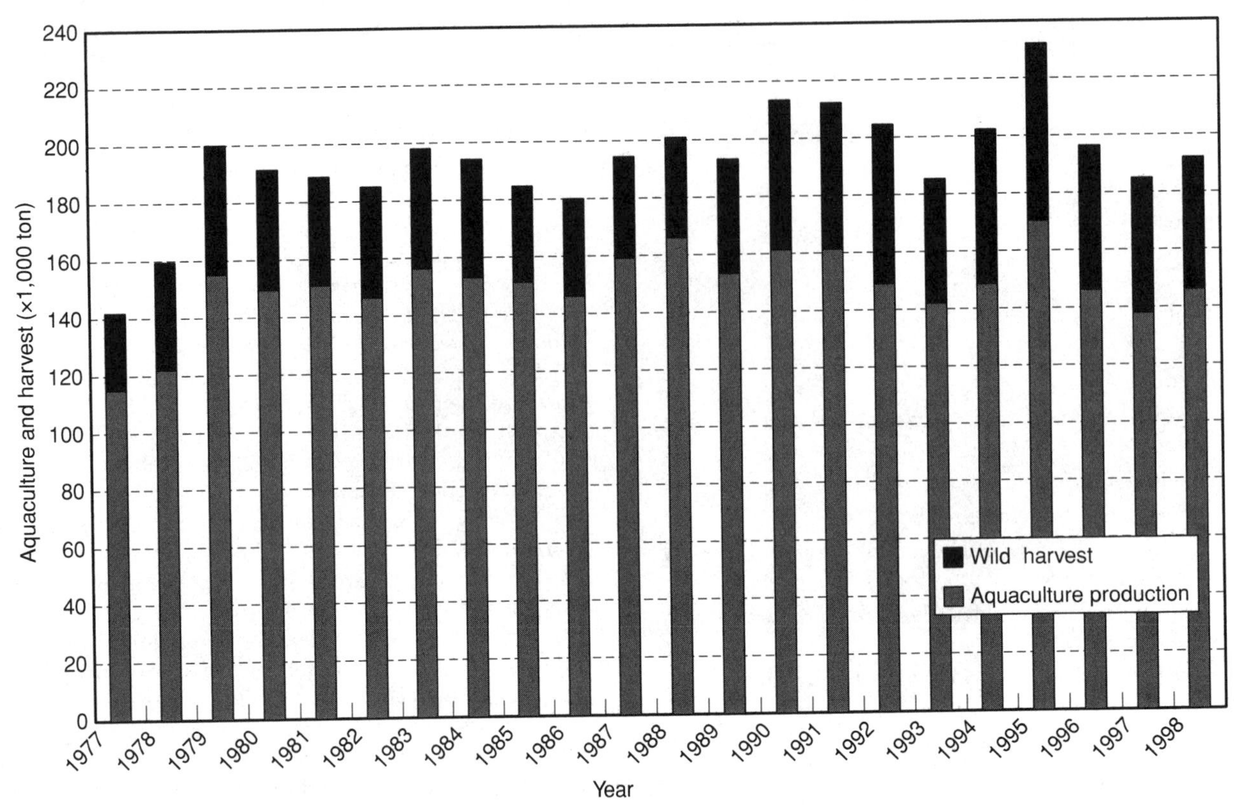

Figure 1. Aquaculture production and wild harvest of yellowtail in Japan. (Statistics and Information Department, Ministry of Agriculture, Forestry and Fisheries.)

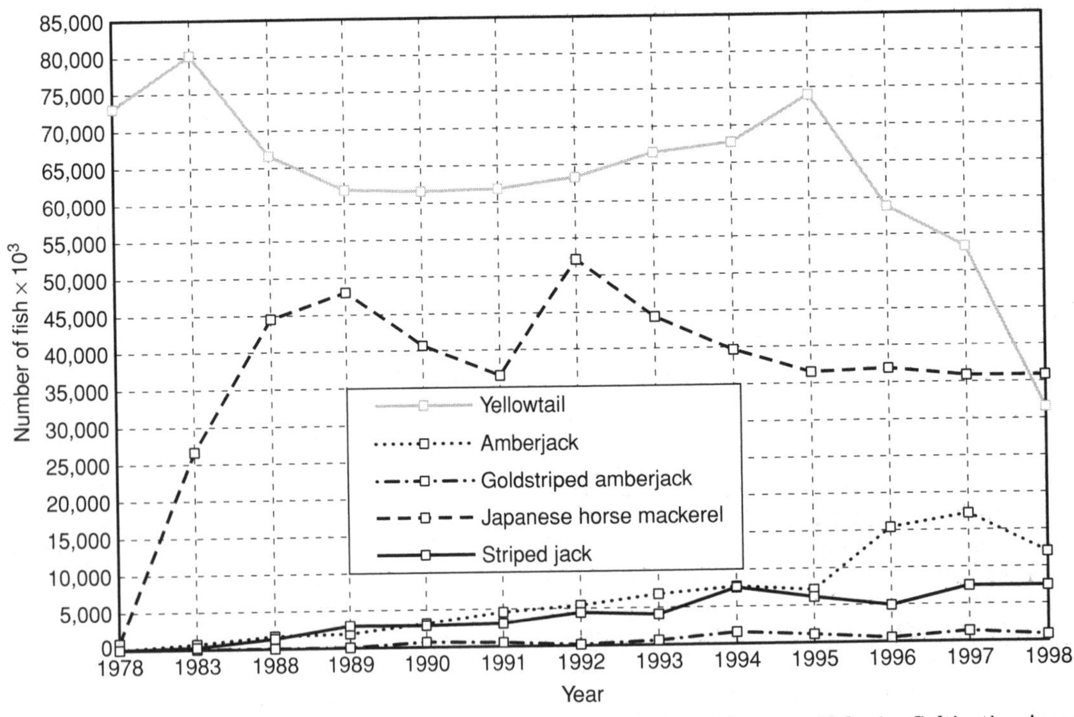

Figure 2. Total number of yellowtail and related species culture (data from Japan Seawater Fisheries Cultivation Association).

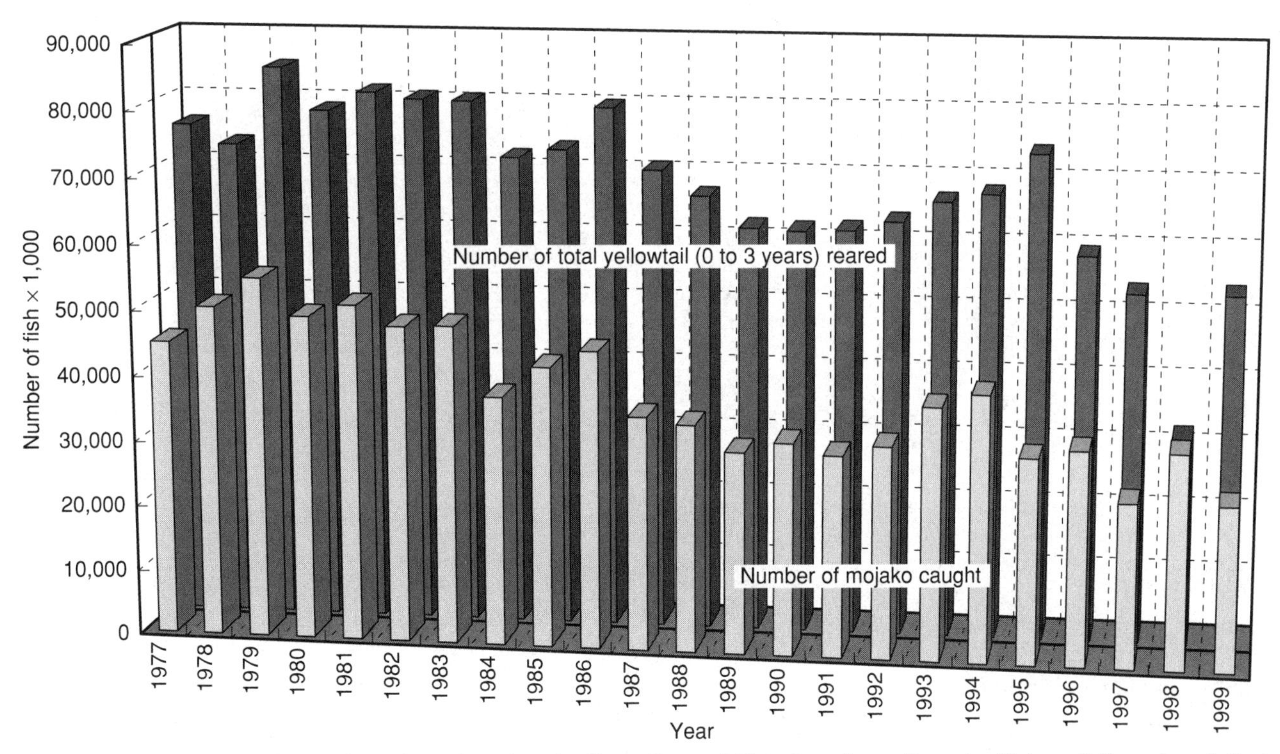

Figure 3. Number of Majako caught and number of total yellowtail reared (data from Japan Seawater Fishery Culture Association).

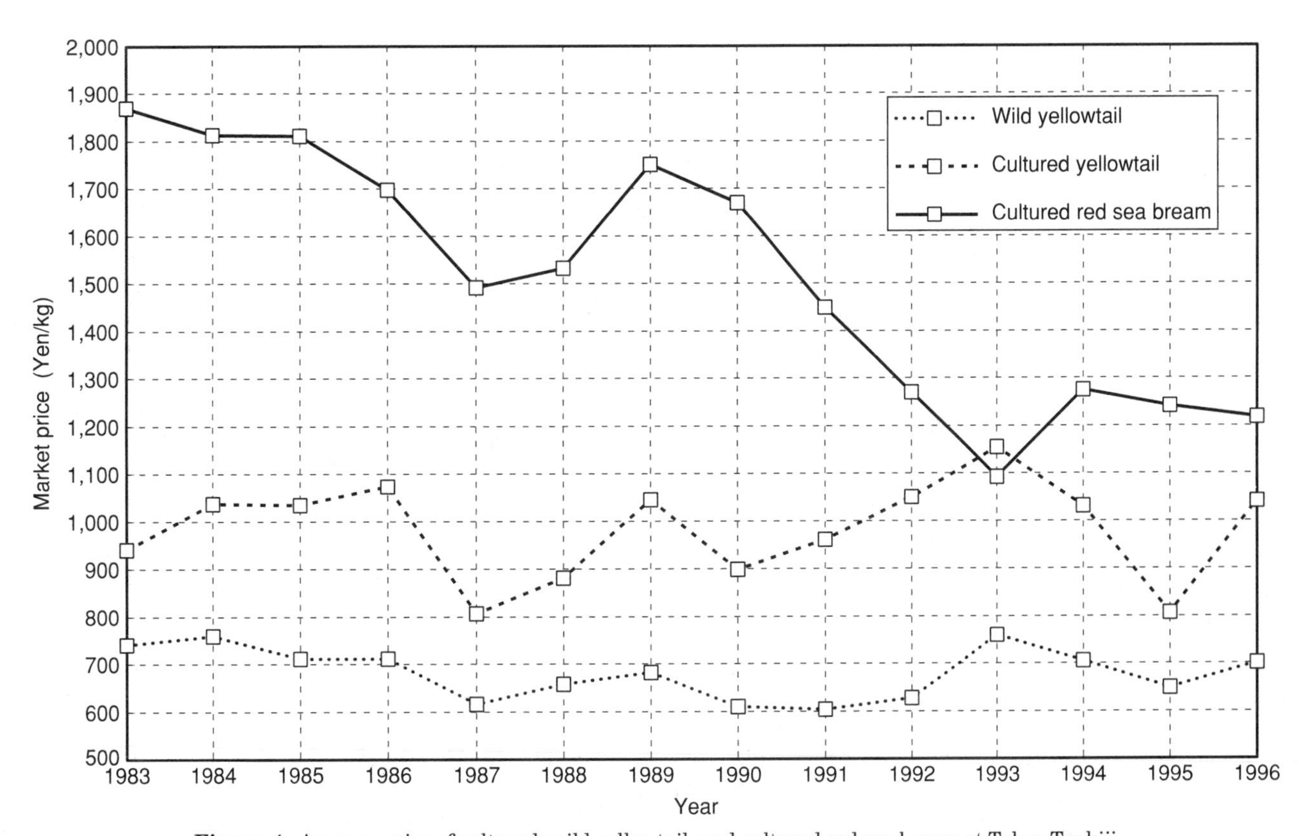

Figure 4. Average price of cultured, wild yellowtail, and cultured red sea bream at Tokyo Tsukiji market (statistics and Information Department, Ministry of Agriculture, Forestry and Fisheries).

Table 1. Volume and Price of Cultured Red Sea Bream, Yellowtail, and Amberjack Wholesaled at Tokyo Tsukiji Market[a]

	Cultured Red Sea Bream			Cultured Yellowtail			Cultured Amberjack			Total			Ratio of Amberjack	
Year	Tons	Yen/kg	Million Yen	Tons	Yen/kg	Million Yen	Tons	Yen/kg	Million Yen	Tons	Yen/kg	Million Yen	Tons %	Million Yen %
1994	2,594	1,203	3,120	11,587	1,006	11,655	2,622	1,392	3,650	16,803	1,097	18,425	15.6	19.8
1995	2,679	1,185	3,173	16,528	699	11,554	3,104	1,459	4,528	22,311	863	19,256	13.9	23.5
1996	2,891	1,156	3,341	12,290	1,007	12,373	2,943	1,680	4,944	18,124	1,140	20,657	16.2	23.9
1997	2,933	1,156	3,390	9,574	1,187	11,368	3,823	1,503	5,747	16,330	1,256	20,505	23.4	28.0
1998	3,910	880	3,441	11,652	922	10,743	5,020	1,258	6,315	20,582	996	20,499	24.4	30.8

[a] Data from Flesh Fish Department of Chuo Gyorui Co., Ltd.

CLASSIFICATION AND ECOLOGY OF YELLOWTAIL AND RELATED SPECIES

Even though they are all in the family Carangidae (7), each of the species of jacks being cultured in Japan has different characteristics and requires different culture methods (Table 2). Carangids are only a few of the many fish species being cultured in Japan (Table 3).

Yellowtail (*Seriola quinqueradiata*)

Yellowtail spawn offshore from southern Kyushu to the Chugoku area of the Sea of Japan. They migrate north to near Hokkaido, feeding for three to five years until reaching sexual maturity. Then they migrate south for spawning (8). From season to season, various sizes of yellowtail can be caught in different parts of Japan; therefore, special names are given to them in the different regions (9). Differences in migratory populations include growth rate and nutritional status (10). All juveniles weighing less than 50 g (2 oz) are called Mojako. Cultured yellowtail weighing less than 5 kg (11 lb) are called Hamachi, and those heavier than 5 kg (11 lb) are called cultured Buri (to be distinguished from wild Buri) (11).

Amberjack (*Seriola dumerili*)

Aquaculture of amberjack has been growing rapidly, and the species has become a rival of yellowtail. Amberjack are distributed throughout the world. In our research (12), we have found amberjack can grow faster with better feed efficiency than yellowtail if the water temperature is higher than 17 °C (63 °F). Amberjack have higher fat content and better flavor than salmon (13,14).

Goldstriped Amberjack (*Seriola lalandi*, or *Seriola aureovittata*)

The goldstriped amberjack is a new rival of yellowtail. Both goldstriped amberjack and amberjack, especially those weighing over 5 kg (11 lb), are known to sometimes cause ciguatera poisoning. If they are raised in a net pen and fed formulated feeds, they may not accumulate the poison. Before the culture of this species is pursued, it is important to make sure that cultured fish will not contain ciguatoxin (15,16).

Goldstriped amberjack is called Hiramasa in Japan and carries the scientific name *S. aureovittata*. That species may in fact be a synonym of *S. lalandi* (17). The total annual catch of goldstriped amberjack is less than either yellowtail or amberjack.

Horse Mackerel or Jack Mackerel (*Tracurus japonicus*)

Horse mackerel have an important position in Japanese aquaculture (18). Aquaculture of horse mackerel began in 1970 when a large number of wild fish were caught and stocked alive in a net pen (19). Horse mackerel raised solely on sardines contains 20 to 40% fat in their muscle tissue (20,21). Vitamin deficiencies and an improper protein : energy ratio caused poor survival. Based on a feeding experiment with horse mackerel conducted in Mie in 1974, a formulated feed was developed, and it is now possible to raise healthy horse mackerel having meat quality as good as the wild fish (Fig. 5).

Striped Jack (*Caranx delicatissimus*)

Many Japanese select striped jack as the best fish for sashimi. Striped jack culture started about 1963. At that time, experimental operations were initiated at the Fisheries Experimental Station and Fisheries Cooperative Association in Tokyo and Nagasaki and at Kinki University (22–24). Production of juveniles was attempted because the quantity of wild fish had declined. In 1984, Harada et al. (25) succeeded in their attempts to induce spawning with injection of an artificial hormone. After that, Takamatsu et al. (26) succeeded in natural spawning and hatching through thermal stimulation. Mass production of juveniles was not possible due to unstable spawning and poor fertilization rates. Murai et al. (27) thought that the quality of the eggs was influenced by stress experienced during attempts at artificial spawning. Subsequently, their research resulted in the development of techniques for stable juvenile production. In addition, they found that injuries to the body surface caused by rough seas and handling make the fish susceptible to infectious diseases and high mortality. Selective breeding has produced higher quality striped jack broodstock (Table 4).

In the 1990s, the juvenile production of striped jack in the Kyushu area was disrupted by iridovirus infections (VNN). Thus, there are still more problems to be overcome before stable production of striped jack can occur. Imaizumi et al. (28) from the Japan Sea-Farming Association are trying to maintain fingerlings that are free of virus.

Table 2. Comparison of Yellowtail and Related Species Cultured in Japan

English Name	Yellowtail	Amberjack or Great Amberjack	Goldstriped Amberjack or Yellowtail	Horse Mackerel or Jack Mackerel	Striped Jack
Japanese name:	Mojako: under 200 g, Hamachi: 200 to 5,000 g, Buri: over 5,000 g	Kanpachi, (Akahara)	Hiramasa	Maaji	Shimaaji
Scientific name:	*S. quinqueradiata*	*S. dumerili*	*S. lalandi* or *S. aureovittata*	*T. japonicus*	*Pseudocaranx dentex* (*C. delicatissimus*)
Market size:	Up to 6 kg for fillet, 3.5–4.5 kg for sashimi	3.5–5.5 kg for sashimi	Up to 4.0 kg for fillet and sashimi	80–200 g	0.8–2.5 kg for sashimi
Price (yen/kg)[a]:	600–900: Producer, 1,200–2,500: Consumer	800–1,300: Producer, 1,500–3,000: Consumer	700–1,200: Producer, 1,500–3,000: Consumer	400–700: Producer, 1,200–2,000: Consumer	1,000–2,000: Producer, 2,000–4,000: Consumer
Maximum size:	Up to 15 kg	Up to 70 kg	Up to 50 kg	500 g	Up to 20 kg
Rearing number in 1997:	$53,303 \times 10^3$	$17,200 \times 10^3$	$2,500 \times 10^3$	$35,901 \times 10^3$	$7,443 \times 10^3$
Juvenile supply:	Wild juvenile called Mojako	Wild juvenile mainly and prefeeding juveniles imported from China and Vietnam	Large juvenile of about 700 g caught in waters around the Goto Islands	Wild juvenile	Propagated juvenile

Source: Data from Nakada et al., Nisshin Feed Co., Ltd.
[a]Estimated by Hamazaki, Nisshin Feed Co., Ltd.

Table 3. Number of Major Fishes Cultured in Japan (1998)

Japanese Name	Scientific Name	English Name	×1000 Fish
Buri	*S. quinqueradiata*	Yellowtail	48,700
Kanpachi	*S. dumerili*	Amberjack	16,180
Hiramasa	*S. aureovittata*	Goldstriped amberjack	800
Maaji	*Trachurus japonicus*	Japanese horse mackerel	31,038
Shimaaji	*P. dentex* (*C. delicatissimus*)	Striped jack	10,170
Madai	*P. major*	Red sea bream	160,376
Ishidai	*Oplegnathus fasciatus*	Japanese striped knifejaw	201
Kurodai	*Acanthopagrus schlegeli*	Black sea bream	538
Chidai	*Evynnis japonicus*	Crimson sea bream	968
Hirame	*Paralichthys olivaceus*	Japanese flounder	6,064
Torafugu	*Takifugu rubripes*	Ocellate puffer	18,107
Suzuki	*Lateolabrax japonicus*	Japanese sea bass	6,131
Isaki	*Parapristipoma trilineatum*	Three-line grunt	3,331
Kurosoi	*Sebastes schlegeli*	Jacopever	256
Maguro	*Thunnus thynnus*, *Thunnus maccoyi*	Bluefin tuna and Southern bluefin tuna	21
Mejina	*Girella punctata*	Rudder fish	435
Kasago and Mebaru	*Sebastiscus marmoratus*, *Sebastes inermis*	Scorpion fish and Japanese stingfish	4,193
Ohonibe	*Nibea mitsukurii*	Nibe croaker	213
Hata-family	Grouper family		55

Source: Data from Japan Sea Water Fisheries Cultivation Association.

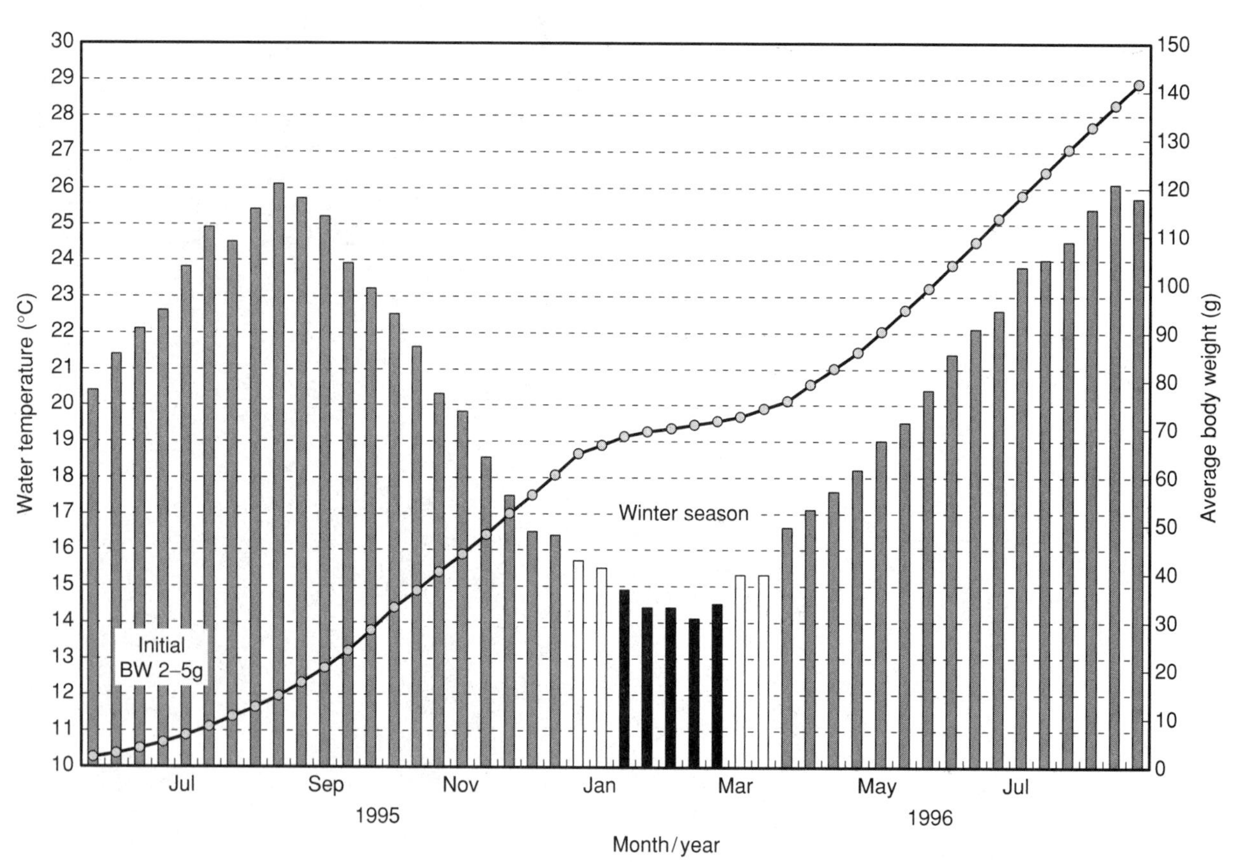

Figure 5. Standard growth of horse mackeral and water temperature in Shizuoka prefecture (data from Senkaisangyo in Suruga-Bay, calculated by Matsumoto Nisshin Feed Co., Ltd.).

Table 4. Effect of Different Feeding Patterns on Performance of Striped Jack

Factor	Normal Feeding	Skipped Feeding
Total fed days	510	424
Total rear days	926	926
Feed operation ratio (%)	55.1	45.8
Initial body weight (g)	10	10
Number of fish stocked	20,000	20,000
Initial total weight (kg)	200	200
Final body weight (g)	1,181	1,256
Number of fish harvested	15,587	16,960
Survival (%)	77.9	84.8
Final total weight (kg)	18,408	21,302
Total feed used	208,145	184,360
MP10 (feed powder 10: raw fish 90)	94,380	83,670
MP20 (feed powder 20: raw fish 80)	99,120	87,790
MP30 (feed powder 30: raw fish 70)	14,645	12,900
Total weight increase (kg)	18,208	21,102
Feed conversion rate	11.4	8.74
Profit Analysis, × Thousand Yen (% of Gross Earnings)		
Gross earning	27,618	31,958
Cost of juvenile	320 (1.2)	320 (1.0)
Cost of feed	18,697 (67.7)	16,558 (518)
Cost of labor	6,200 (22.4)	5,150 (16.1)
Cost of fuel	1,200 (4.3)	1,000 (3.1)
Sundry expenses	430 (1.6)	430 (1.3)
Miscellaneous	170 (0.6)	170 (0.5)
Gross profit	601 (2.8)	8,330 (26.0)

Source: Data from Watanabe Suisan in Ohoita, June 1987–December 1989, calculated by Chaki and Nakada, Nisshin Feed Co., Ltd.

GROWTH

Performance of marine fish raised in floating net pens varies considerably. Exact weights of individual Mojako are not usually determined before stocking. At most, the total weight of the fish being stocked is determined and then divided by the estimated total number of fish to determine the average weight. The number of Mojako in each cage is estimated further by subtracting mortalities as they occur. Using those data, the daily amount of feed is calculated by the farmers. In my 30-year career, I have often encountered farmers who do not record the amount of feed offered in their net pens. This is why there are many cases where farmers cannot calculate a reasonable feed conversion ratio.

Effects of Water Temperature

Water temperatures vary in the areas where yellowtail culture is carried out (Fig. 6). In each region, yellowtail farmers developed a characteristic way of rearing by taking water temperature changes into consideration.

Representative growth curves with average harvest sizes, a summary of the environmental conditions required for yellowtail culture, and stomach evacuation times for yellowtail on various types of feed are shown in Figures 7–9. Information on the recommended feeding frequency for yellowtail is presented in Table 5. Depending on water temperature, Mojako can usually be stocked from April through July. Harvest size varies to some extent depending on mean annual water temperature.

In subtropical regions such as Okinawa and Kagoshima, the average water temperature range is from 20 to 24 °C (68 to 75 °F). Temperature remains within the optimum range for yellowtail for more than 75% of the year. By stocking young Mojako in these areas, it is possible to obtain more than 6 kg (3.2 lb) yellowtail within two years.

The range of average annual water temperature in the Kyushu area, which includes Kumamoto and Nagasaki, is 17 to 19 °C (63 to 66 °F). Temperature is optimum for yellowtail culture for about 50% of the year. Because of the shorter period when temperature is optimal, over 70% of the yellowtail reared in the Kyushu region are three years old at harvest.

The region where yellowtail can be grown but temperature is less optimal than the regions previously mentioned is the Honshu area, which includes Shizuoka and Yamaguchi. Average annual water temperature is 18 to 19 °C (64 to 66 °F), and temperature is within the optimum range only 50 to 60% of the year. More than three years are required to produce 6 kg (13.2 lb) yellowtail. A specific feature of this region is its short autumn, which provides the fish with insufficient time to prepare for winter. If the fish are pushed to grow rapidly during

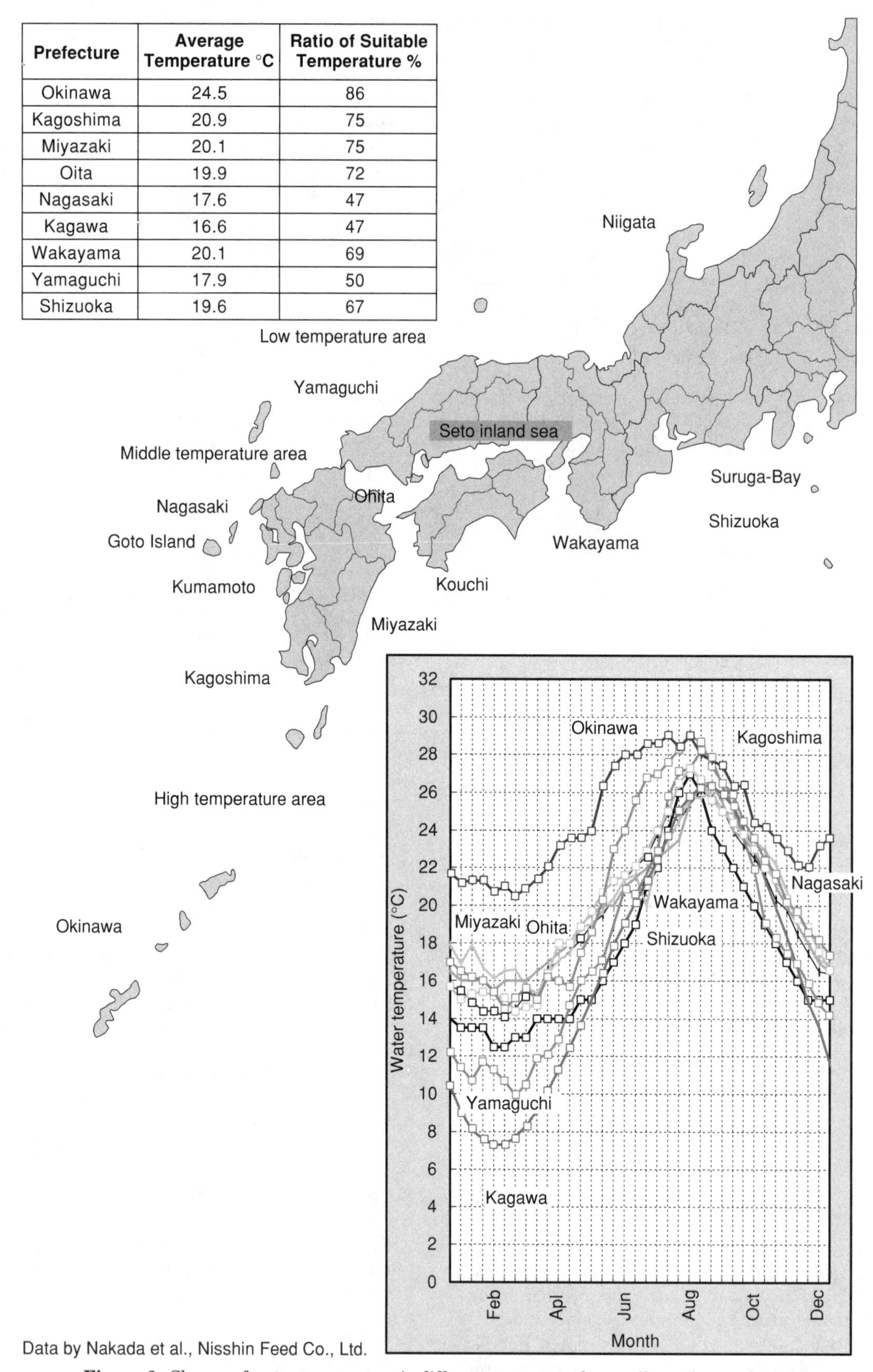

Prefecture	Average Temperature °C	Ratio of Suitable Temperature %
Okinawa	24.5	86
Kagoshima	20.9	75
Miyazaki	20.1	75
Oita	19.9	72
Nagasaki	17.6	47
Kagawa	16.6	47
Wakayama	20.1	69
Yamaguchi	17.9	50
Shizuoka	19.6	67

Data by Nakada et al., Nisshin Feed Co., Ltd.

Figure 6. Change of water temperature in different sea areas where yellowtail are cultured.

Table 5. Recommended Feeding Frequency for Yellowtail Culture (times per day)

Average Body Weight (g)	Feeding Stage	Seawater Temperature (°C)	Feeding Frequency (times/day)
Mojako			
0.5–10	Introducing		5
10–50	Domestication	18–28	3
50–200	Selection		2
Hamachi			
(200–2,000)	Start rearing	22–27	2
	Optimum temperature	28–30	1
	High temperature	31–34	Once/two days
	Descending temperature	33–23	1
	Low temperature	22–18	Once/two days
	Wintering	17–11	Once/three days
Hamachi and Buri			
(2,000–12,000)	Spring start	9–11	Once/three days
	Ascending temperature	15–19	Once/two days
	Optimum temperature	20–26	1
	High temperature	27–30	Once/two days
	Unusual temperature	31–34	Once/three days
	Desending temperature	33–21	1
	Low temperature	20–15	Once/two days
	Wintering	14–9	Once/three days

Source: Data from Nakada et al., Nisshin Feed Co., Ltd.

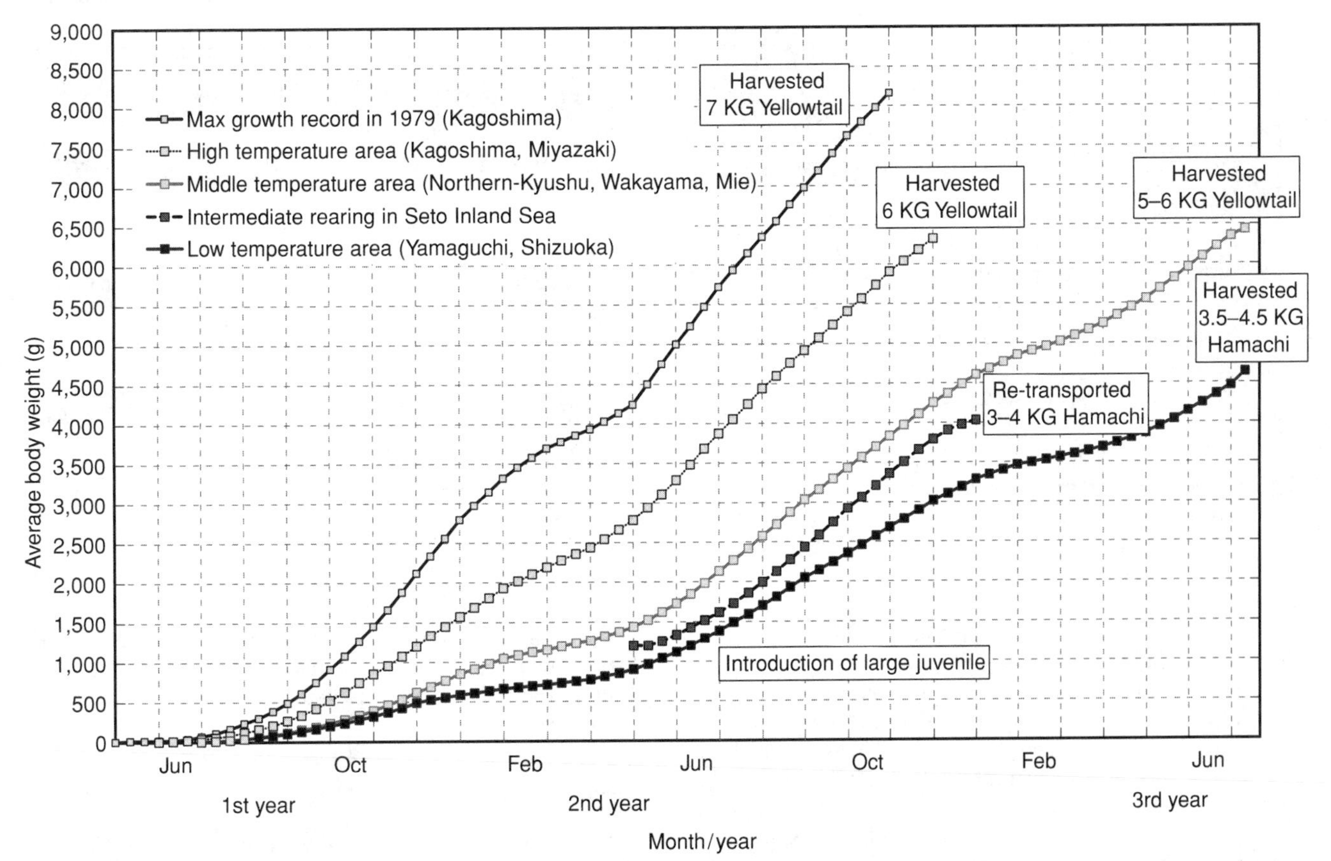

Figure 7. Typical growth performance of yellowtail in different sea areas in Japan (data from Nakada et al., Nisshin Feed Co., Ltd.).

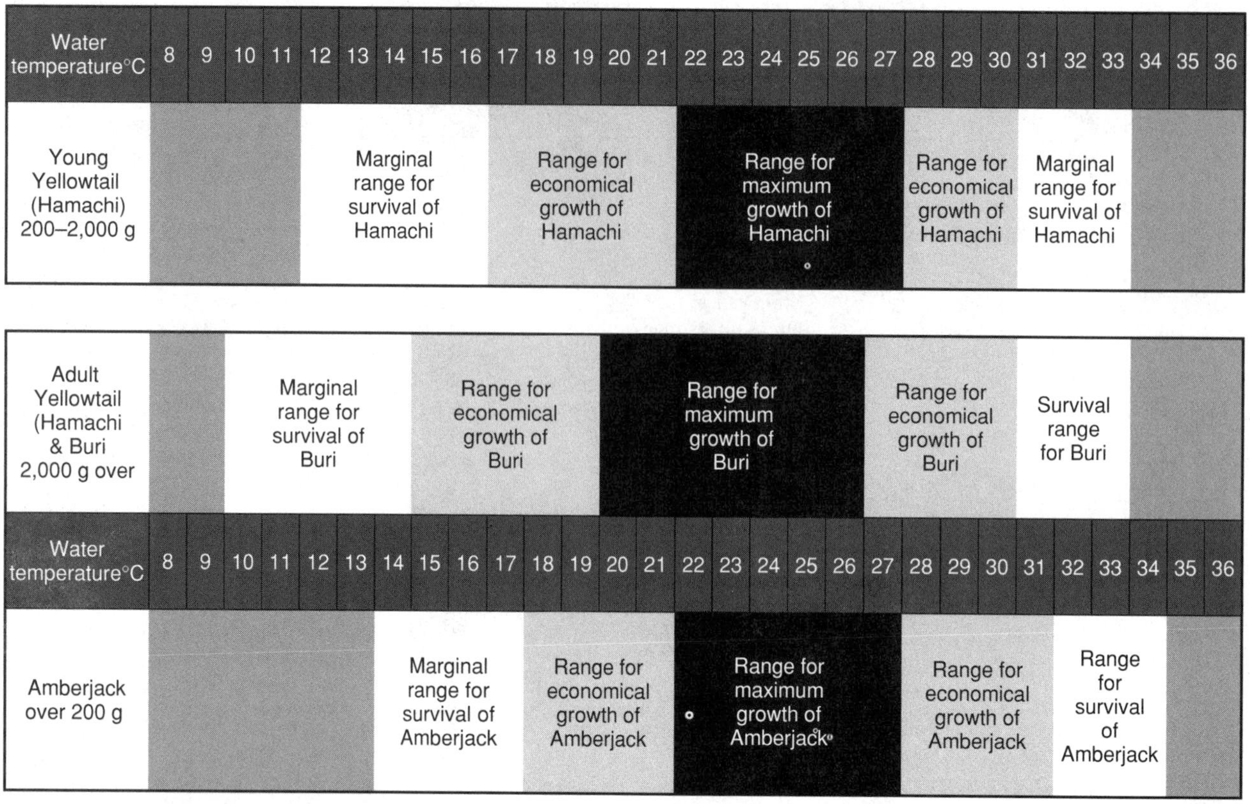

Water temperature°C	8	9	10	11	12	13	14	15	16	17	18	19	20	21	22	23	24	25	26	27	28	29	30	31	32	33	34	35	36
Young Yellowtail (Hamachi) 200–2,000 g					Marginal range for survival of Hamachi					Range for economical growth of Hamachi					Range for maximum growth of Hamachi						Range for economical growth of Hamachi			Marginal range for survival of Hamachi					
Adult Yellowtail (Hamachi & Buri 2,000 g over			Marginal range for survival of Buri					Range for economical growth of Buri					Range for maximum growth of Buri							Range for economical growth of Buri				Survival range for Buri					
Water temperature°C	8	9	10	11	12	13	14	15	16	17	18	19	20	21	22	23	24	25	26	27	28	29	30	31	32	33	34	35	36
Amberjack over 200 g							Marginal range for survival of Amberjack				Range for economical growth of Amberjack				Range for maximum growth of Amberjack						Range for economical growth of Amberjack				Range for survival of Amberjack				

Salinity concentration
Range of optimum salinity for Yellowtail
29.8~36.3%
Specific gravity (1.022–1.027; cl:16.50–20.12%)*
Suboptimum range (decreased feeding rate)
Under 27.1%
Specific gravity (under 1.020, cl:15.1%)*

Dissolved oxygen concentration		
Condition of fish	mg/L	Satulation %
Active feeding	5.7 over	70 over
Decreased feeding	4.3~5.7	50~70
Unusual activity	2.9~4.3	40~50
Respiration difficulty	1.4~2.9	20~40
Suffocate to death	under 1.4	under 20

*Harada et al., Kinki University of Aquaculture, Report No. 1; 1–275, 1966

Figure 8. Optimum environmental conditions for yellowtails culture (data from Nakada et al., Nisshin Feed Co., Ltd.).

fall, high winter and early spring mortalities may occur. Therefore, yellowtail weighing from 3.5 to 4.5 kg (7.7 to 9.9 lb), a size range popular for sashimi, is produced.

In the Seto Inland Sea, the average annual water temperature is lower than 17 °C (63 °F), with less than 50% of the year being conducive to yellowtail growth. Temperature falls below 10 °C (50 °F) during the last two months of winter, at which time yellowtail may experience mass mortalities. To avoid the mortality problem, fish can be transferred to warmer areas such as Kochi and Miyazaki for overwintering. When the water temperature rises again in spring the fish may be returned to the Seto Inland Sea and reared to the size appropriate for use in sashimi. Another widely used approach is to stock large juveniles from other districts in the spring. It is then possible to produce fish suitable for sashimi within a growing season.

Juvenile Resources

The number of Mojako caught annually along with the number of yellowtail farmers is shown in Table 6. In 1966, the Fisheries Agency imposed regulations limiting the number of Mojako that can be caught annually for purposes of aquaculture to about 40 million in order to conserve the resource. Allocations are made to

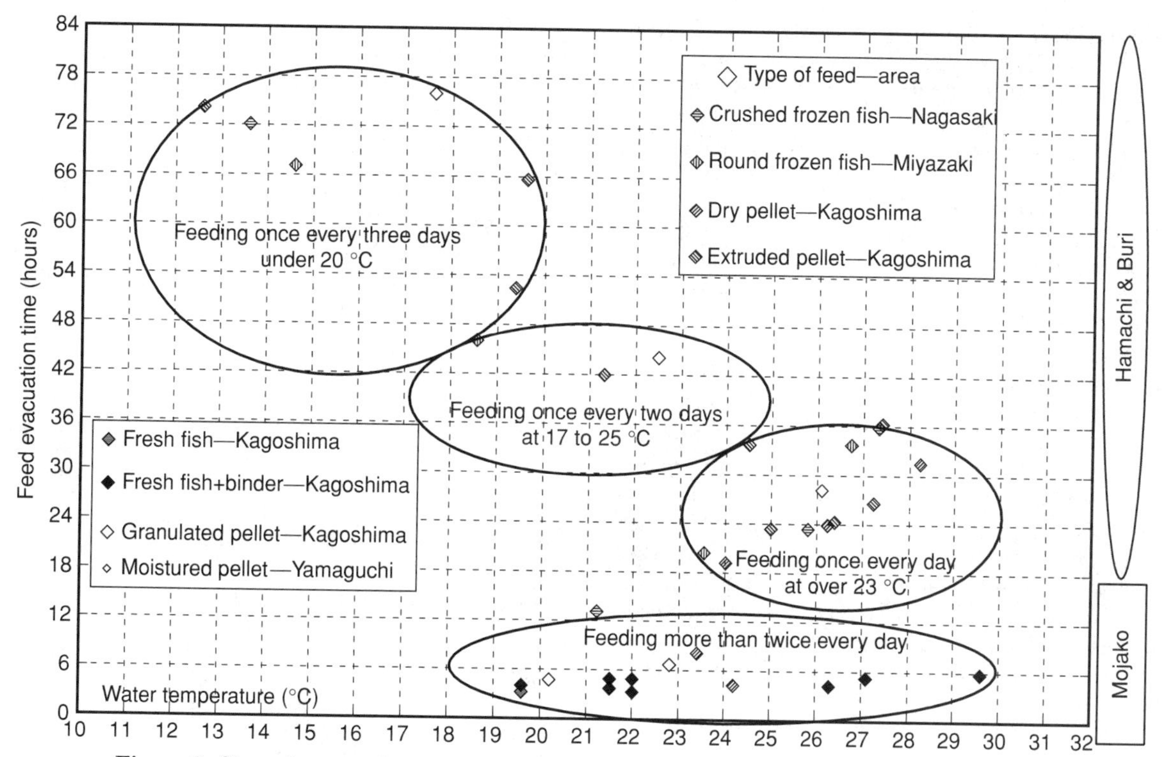

Figure 9. Stomach evacuation time of yellowtail fed various types of feed in different temperature areas (data from Nakada et al., Nisshin Feed Co., Ltd.).

Table 6. Number of Mojako Caught and Number of Yellowtail and Amberjack Culture Farmers by Prefecture (per 1,000 fish)[a]

Yellowtail Rearing Number in Net Pen 1st January, 1999					
Prefecture	Enterprise	1st Year	2nd Year	3rd Year	Total
Chiba	5	26	26	14	66
Sizuoka	27	141	133	35	309
Ishikawa	0	0	0	0	0
Fukui	6	6	1	0	7
Mie	57	454	135	0	589
Kyoto	10	0	53	2	55
Wakayama	5	302	21	3	326
Hyogo	7	115	52	1	168
Tottori	1	80	90	10	180
Hiroshima	10	0	270	150	420
Shimane	4	231	33	15	279
Yamaguchi	15	121	62	20	203
Tokushima	15	587	38	0	625
Kagawa	0	0	0	0	0
Ehime	270	7357	3490	212	11059
Kouchi	58	2126	457	0	2583
Fukuoka	6	11	4	0	15
Saga	16	38	64	58	160
Nagasaki	145	2693	1397	257	4347
Kumamoto	51	785	619	0	1404
Oita	68	2507	1276	100	3883
Miyazaki	25	693	101	6	800
Kagoshima	280	3358	711	20	4089
Total	1081	21631	9033	903	31567
September 98	1308	23479	18783	3041	45303
1999/1998	82.6	92.1	48.1	29.7	69.7

Amberjack Rearing Number in Net Pen 1st January, 1999					
Prefecture	Enterprise	1st Year	2nd Year	3rd Year	Total
Chiba	0	0	0	0	0
Sizuoka	0	0	0	0	0
Ishikawa	0	0	0	0	0
Fukui	0	0	0	0	0
Mie	0	0	0	0	0
Kyoto	0	0	0	0	0
Wakayama	4	4	2	2	8
Hyogo	0	0	0	0	0
Tottori	0	0	0	0	0
Hiroshima		No searching			
Shimane	0	0	0	0	0
Yamaguchi	0	0	0	0	0
Tokushima	5	2	4	1	7
Kagawa	0	0	0	0	0
Ehime	18	348	371	34	753
Kouchi	80	803	1073	0	1876
Fukuoka	0	0	0	0	0
Saga	0	0	0	0	0
Nagasaki	9	6	12	5	23
Kumamoto	13	104	348	0	452
Oita	7	104	348	0	452
Miyazaki	49	1588	645	2	2235
Kagoshima	185	5285	851	10	6146
Total	370	8244	3654	54	11952
September 98	470	9071	7093	35	16199
1999/1998	78.7	90.9	51.5	154.3	73.8

Source: Data from Japan Sea Water Fisheries Cultivation Association.

[a]Wakayama, Nagasaki, and Oita had not completed the searching data from some areas.

each prefecture by the Japan Seawater Fishery Culture Association (29). Each prefectural government decides on the allowable period for catching Mojako and allots the number of fish allowed to be caught to the individual Federation of Fisheries Cooperatives in the prefecture. In 1977, the number of Mojako actually caught was about 45 million. The number has fluctuated between 30 and 50 million for about 20 years, but dropped to 25 million in 1997. Fish farmers were, however, able to maintain a total production level of about 150,000 t despite the decrease in available Mojako. The production in 1995 was the highest at 170,000 t.

Nisshin Flour Milling Co., Ltd., in response to the development of prepared feeds such as Umisachi and Otohime, produced the increased production related to high survival of Mojako. By using the proper prepared feed, it is possible to raise healthy Mojako that initially weigh less than 2 g (0.08 oz), which was not possible when raw minced fish was fed. Another reason for increased production of cultured yellowtail was the availability of imported Mojako from other countries to compensate for the decreased domestic supply. For example, over eight million Mojako were imported from Korea during the 1980s.

Recently, the domestic supply of Mojako showed a significant decrease, and a few million were once again imported from Korea. Juvenile amberjack are usually caught with Mojako and at one time the two species were cultured together. However, amberjack are vulnerable to the parasitic worm, *Benedinia*, and the parasites will spread to yellowtail. To avoid the problem, some farmers separated amberjack juveniles from Mojako and raised them separately.

The price of amberjack juveniles is 500 and 1,500 yen for fish weighing 50 and 600 g (2 and 24 oz). The high price has made possible commercial production of propagated juveniles. Japan, via Hong Kong, has imported wild juveniles caught in China and Vietnam since 1986.

Selection of Juveniles

Fry of yellowtail and related species inhabit seaweed that breaks away from the bottom. They feed on microorganisms and small fishes while drifting north with the seaweed. During the day they swim around the seaweed, and hide inside it at night. Small Mojako of 4 to 5 cm (1.6 to 2.3 in.) stay under or inside of floating seaweed. Larger fish 0.5 to 2 m (1.7 to 6.5 ft) swim below the surface. After reaching a size of 10 to 14 cm (4 to 5.5 in.), they disperse from the floating seaweed and swim toward the shore where they are caught in set nets (30,31).

Mojako feed actively at sunrise and sunset when swarms of zooplankton also can be seen. During daytime they feed on small fishes (32,33).

Propagation of Juveniles

The Japan Sea-Farming Association and several prefectural experimental stations have already established the techniques for artificial production of about 60 marine species (34). Significant quantities of juvenile yellowtail, amberjack, goldstriped amberjack, and striped jack have been developed by aquaculturists (35–38). Viable eggs are being obtained from both wild fish and broodstock cultured on high-quality formulated feeds. Hormone injections are necessary to stimulate maturation in many instances (39,40). By using mass-produced food organisms, such as rotifers and brine shrimp nauplii fortified with n-3 highly unsaturated fatty acids (HUFA) and formulated feeds, production of healthy fry is possible (41–43).

Domestication and Rearing of Mojako

Wild juveniles are weaned to prepared feed after capture and weak individuals are eliminated. They are then sold to producers who put them in net pens. Small juvenile yellowtail and related species are sensitive to feed deprivation. If a fishing boat catches Mojako far away from port, the fish will cannibalize one another in holding tanks (44). If fasted for more than three days, Mojako fail to adapt to prepared feed (45,46). It is well known that a prolonged fasting period before first feeding in net pens has a significant negative effect on later growth rate. If a good quality prepared feed is accepted while the fish are on the collecting boat, the problem can be overcome (47).

It is very important to obtain a good quality fish for stocking, whether Mojako or large-size juveniles. Careful observation to determine that the fish look and behave normally and careful record keeping with respect to the use of medications is important to document the health of the fish to buyers. Condition information for yellowtail of various sizes is presented in Table 7.

The improved technology both in domestication and transportation made it possible to import high quality juvenile amberjack at the size of 8 to 10 cm (3.1 to 3.9 in.) from China and Vietnam (48). Amberjack culture has been started around Tainan and Penhu-dao in Taiwan. In the near future, large-scale production there is scheduled to begin in floating net pens.

Problems associated with producing yellowtail and related species in warm waters include muscle parasites

Table 7A. Condition Factor in Various Stages of Yellowtail[a]

Growth Stage of Yellowtail	Average Body Weight (g)	Abnormally Fat	Fat	Normal	Thin	Abnormally Thin
Mojako	~200	20.0 over	16.0–19.0	13.0–16.0	10.0–13.0	Under 10.0
Hamachi	200 ~ 2000	21.0 over	17.0–20.0	14.0–17.0	11.0–14.0	Under 11.0
Hamachi and Buri	2000~	22.0 over	18.0–21.0	15.0–18.0	12.0–15.0	Under 12.0

Source: Data from Nakada et al., Nisshin Feed Co., Ltd.

[a] Condition factor = [Body weight/(fork length)^3] × 1000.

Table 7B. Change in Condition Factor of Yellowtail During its Growth

Fork Length (cm)	Average Body Weight (g)	Condition Score (CS)
12.2	25	13.8
19.7	110	14.4
25.5	250	15.1
32.0	500	15.3
36.5	775	15.9
40.5	1,100	16.6
43.0	1,350	17.0
44.0	1,500	17.6
45.0	1,650	18.1
46.2	1,800	18.3
48.5	2,000	17.5
50.5	2,250	17.5
53.0	2,600	17.5
55.9	3,000	17.2
57.9	3,400	17.5
60.1	3,800	17.5
62.0	4,300	18.0
64.3	4,800	18.1
65.7	5,400	19.0
66.2	5,800	20.0

and ciguatera. In the waters south of Kagoshima, aquaculture of these species is not feasible because of parasitism with kudoa in the muscles and the internal organ (49). In some cases, cultured juvenile yellowtail, amberjack, and striped jack have been killed due to infection of iridovirus, which was originally introduced with wild juveniles imported from tropical areas (50). Unlike wild fishes, cultured fishes are vulnerable to diseases because they are always under a certain amount of stress.

WATER QUALITY

As the price of dissolved oxygen (DO) meters and salinometers has fallen in recent years, it became possible to measure such parameters conveniently at low cost. However, few farmers record the change of DO and salinity, probably due to the confidence in their intuition fostered by long experience (51). Annual differences in water quality (Fig. 10) over periods of several years provide some insight as to what farmers can expect, but spatial and interannual variations can be difficult to predict.

There can also be large differences in temperature from the surface to the bottom of the water column as shown in Figure 11. Tateishi, a colleague who stresses that measurement of the water temperature at the depth of 10 cm (4 in.) water each day at 0800 is not sufficient. He determined diurnal changes in vertical water temperature at a Nagasaki aquaculture site in August, 1984 (Fig. 12). The data clearly indicate that it is not a good practice to judge water quality from scanty data.

In order to develop an automatic feeder that responds to fish feeding activity, we devised a system that automatically inputs data into a personal computer. The daily change in vertical water temperature was continuously recorded, using our system at a yellowtail culture site in the Goto Islands, Nagasaki, from June 16

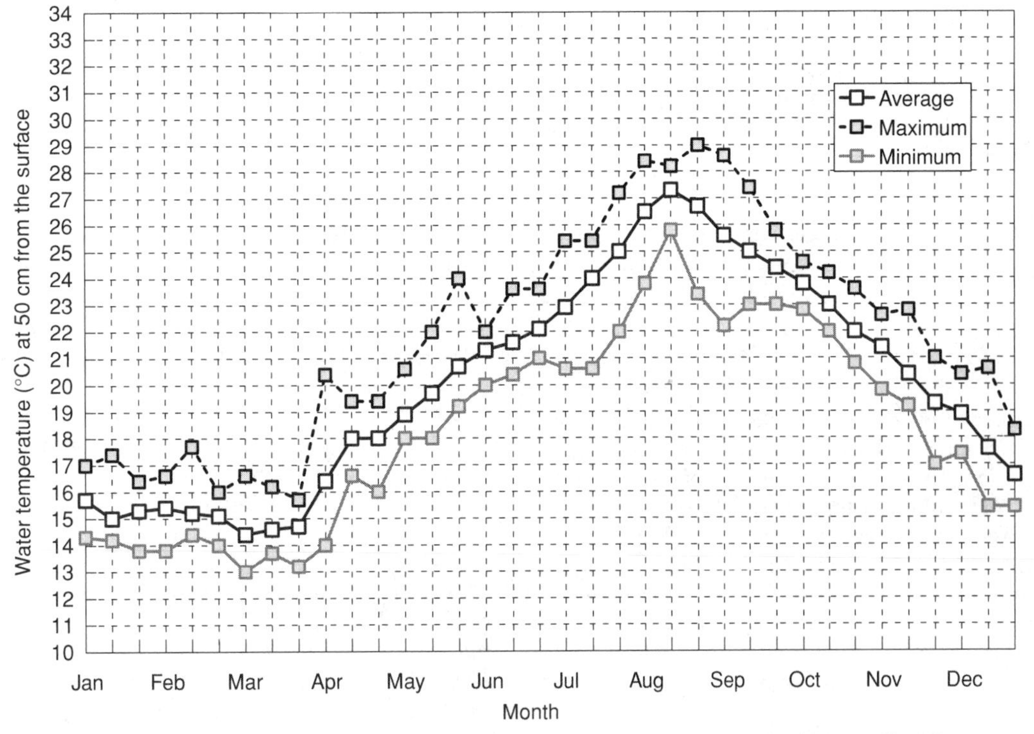

Figure 10. Annual difference of water temperature in the past 10 years in Wakayama (data from Fukaya et al., Nisshin Feed Co., Ltd., 1967–1976).

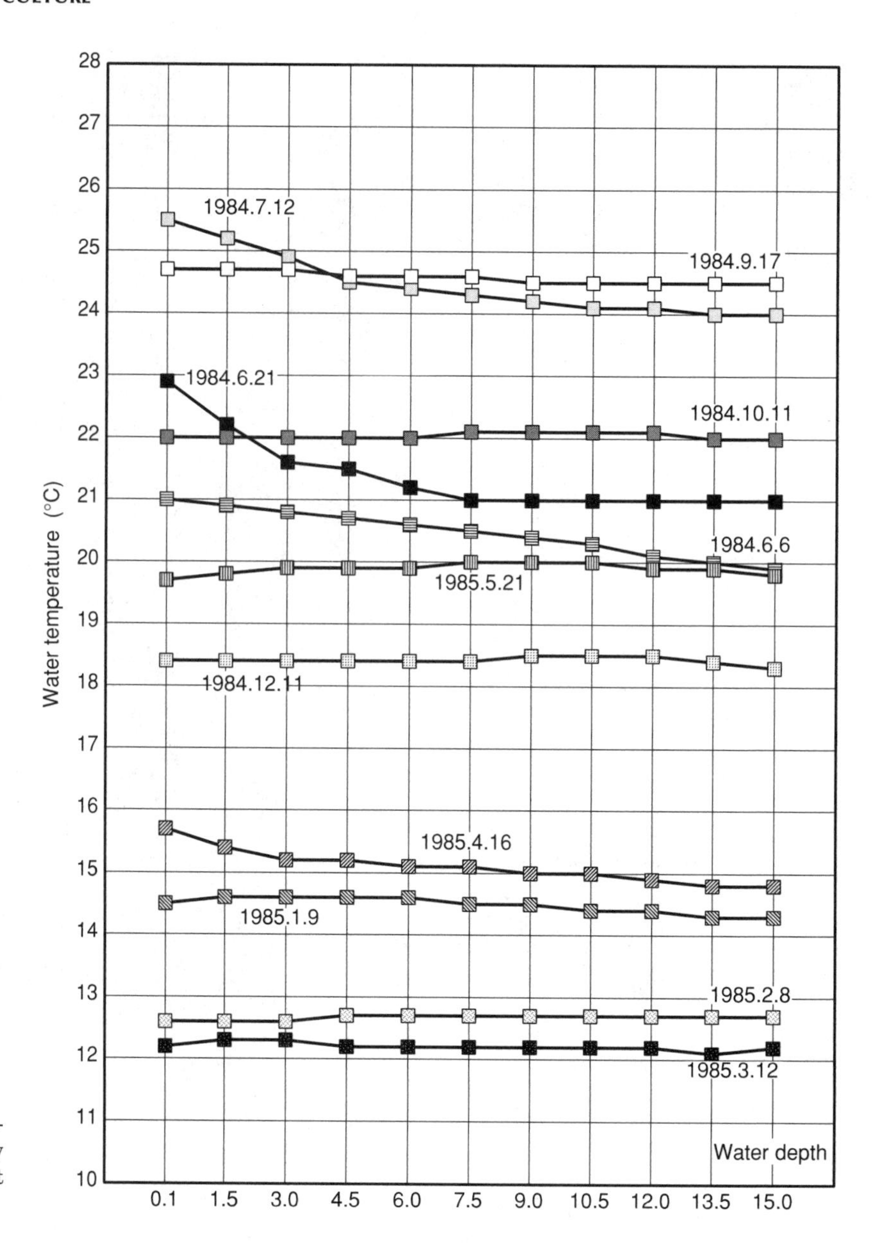

Figure 11. Seasonal change in the water temperature at different depths (measured by Tateishi; Nisshin Feed Co., Ltd., 1984–1985 at Floating Net Pen in Nagasaki).

to 20, 1992 (Fig. 13). At the same time, we determined diurnal changes of the depths at which yellowtail were swimming in a floating net pen (Fig. 14).

AQUACULTURE FACILITIES

Yellowtail culture started in 1927 when juvenile marine fishes like jacks, mackerel, red sea bream, black porgy, and young yellowtail caught in large set nets were released into embankment-type enclosures in AdoIke in Kagawa Prefecture. Fish culture in embankment-type enclosures and net enclosures depended on tidal flow through a few sluice gates for water exchange. Red tide episodes and oxygen depletions associated with poor water exchange, along with waste feed and fecal accumulations, caused heavy mortalities in such facilities (52).

Harada et al. developed net pens (8 × 8 × 8 m, 26 × 26 × 26 ft), which were less expensive than the previously described systems (53). This method has advantages such as a high water exchange rate, lower maintenance costs, and easier fish harvesting. Thus, net pen systems were adopted quickly and successively, and almost all the farmers use this system with various sizes of structures at present.

Net-pen culture requires frequent net exchanges because of biofouling that restricts water exchange. The problem was overcome through the use of tri-butyl tin (TBT) which was ultimately banned because of toxicity problems. New chemicals are now being sought as replacements.

As net pen culture developed, increasingly larger pens were used. Pens of 15 × 15 × 15 m (49 × 49 × 49 ft) are most commonly used, and the frameworks have changed from wood to metal and reinforced plastics. Even larger pens, up to 50 × 50 × 50 m (164 × 164 × 164 ft), are in use. For large fish like yellowtail, enough space for exercise helps them build firm muscle. Recommended

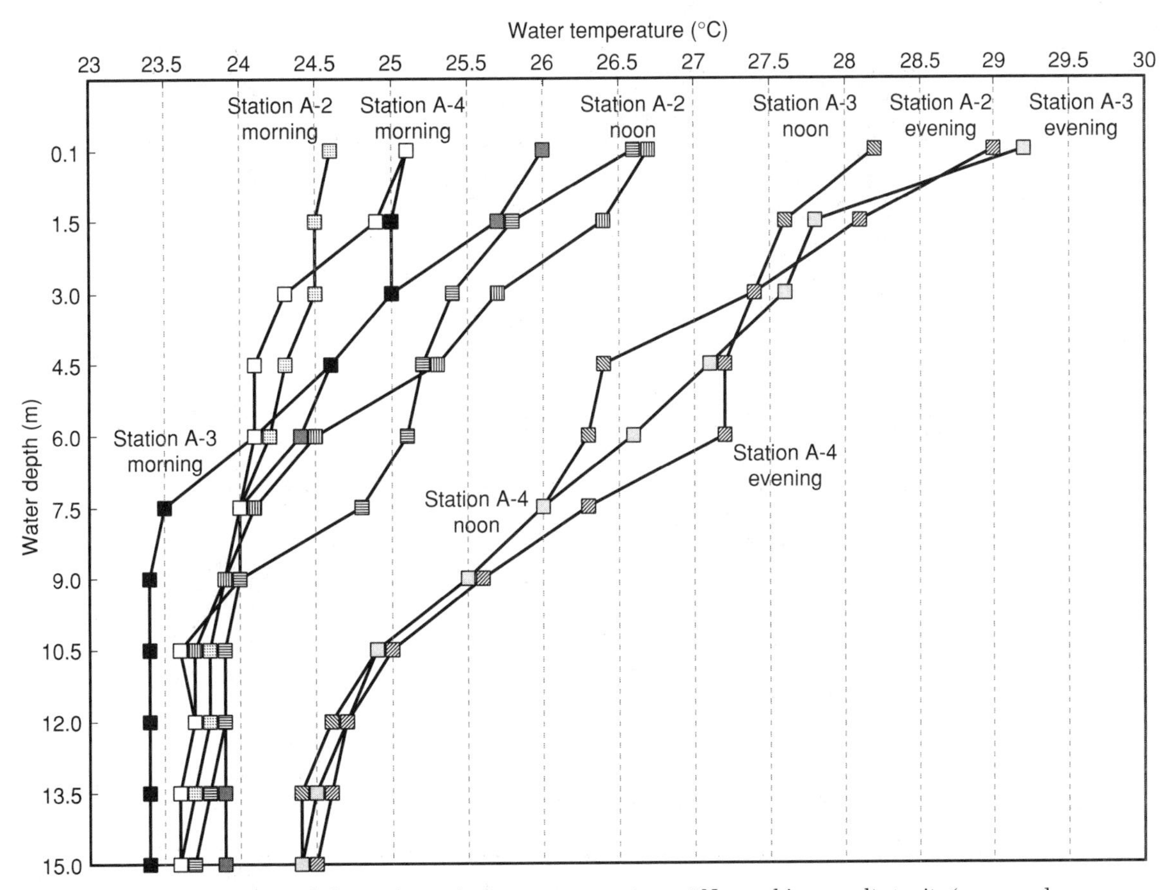

Figure 12. Diurnal change in vertical water temperature at Nagasaki aquaculture site (measured by Tateishi; Nisshin Feed Co., Ltd., on August 10, 1984, at Floating Net Pen in Nagasaki).

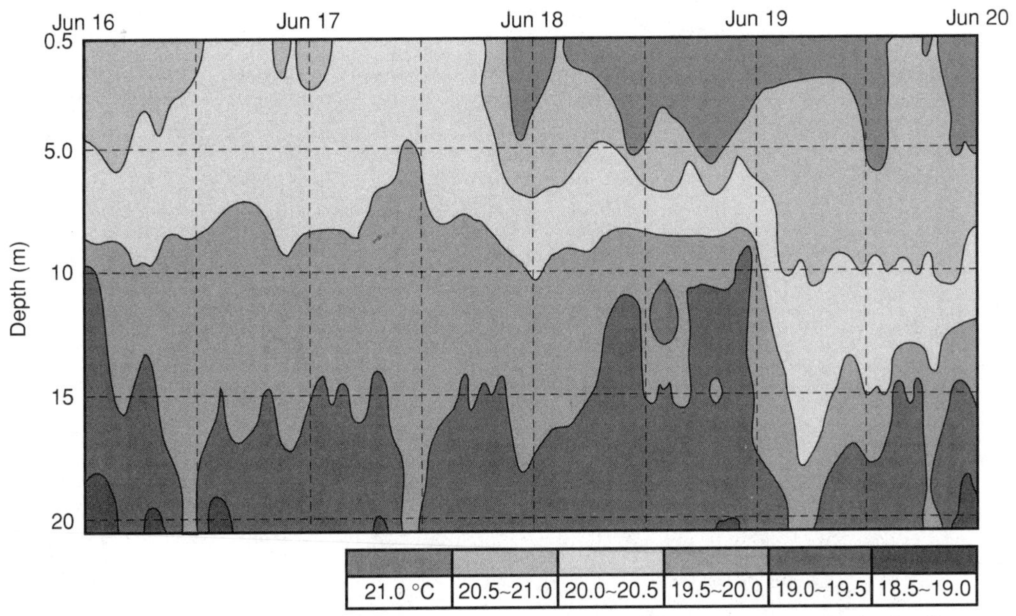

Figure 13. Diurnal change in vertical water temperature at yellowtail culture site in the Goto Island, Nagasaki (data from Komatsu, Hirota, and Teramachi; Nisshin Technical Research Laboratory; and from Nakada, Ohkubo, Kinpara, and Ohike, Nisshin Nasu Feed Research Laboratory).

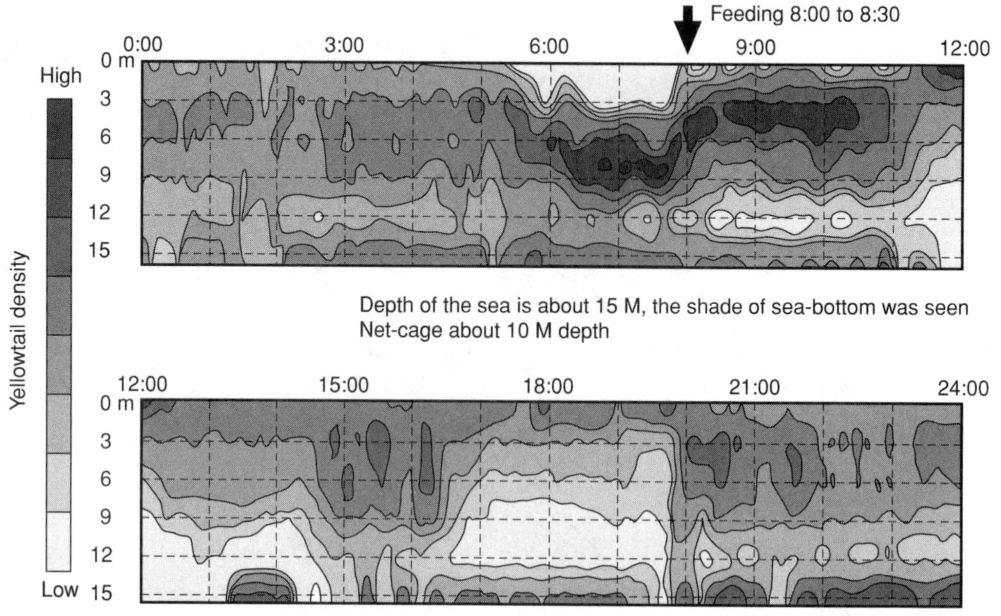

Figure 14. Diurnal change of the swimming layer of yellowtail in floating net pen (date of June 17, 1992, at Goto Tsuruta-Suisan in Nagasaki by Nisshin Technical Research Laboratory; Komatsu, Hirota, Teramachi, Uehara, and by Nisshin Nasu Feed Laboratory; Nakada, Ohkubo, Shiratori, Kinpara, Ohike).

Table 8. Recommended Stocking Density of Yellowtail in Floating Net Pen

Classification Period	Pen Size	Rearing Type	Average Body Weight (g)		Number of Fish Stocked			Total Body Weight (kg)		Stocking Density (kg/m³)	
			Initial	Final	Initial	Survival (%)	Final	Initial	Final	Initial	Final
Mojako	5 × 5 × 5 125 m³	M-1	0.5	20	30,000	90	27,000	15	540	0.12	4.3
		M-2	2.0	50	20,000	90	18,000	40	900	0.32	7.2
		M-3	5.0	100	15,000	90	13,500	75	1,350	0.60	10.8
		M-4	10.0	200	10,000	90	9,000	100	1,800	0.80	14.4
Young Hamachi	8 × 8 × 8 512 m³	H-1	20	300	20,000	95	19,000	400	5,700	0.78	11.1
		H-2	20	400	20,000	95	19,000	400	7,600	0.78	14.8
		H-3	50	500	18,000	95	17,100	900	8,550	1.76	16.7
		H-4	50	600	18,000	95	17,100	900	10,260	1.76	20.0
		H-5	100	700	16,000	95	15,200	1,600	10,640	3.13	20.8
		H-6	100	800	16,000	95	15,200	1,600	12,160	3.13	23.8
		H-7	200	900	14,000	95	13,300	2,800	11,970	5.47	23.4
		H-8	200	1,000	14,000	95	13,300	2,800	13,300	5.47	26.0
Buri	10 × 10 × 10 1000 m³	B-1	600	3,500	7,000	97	6,790	4,200	23,765	8.20	23.8
		B-2	800	4,500	6,000	97	5,820	4,800	26,190	9.38	26.2
		B-3	1,000	5,000	6,000	97	5,820	6,000	29,100	11.72	29.1
		B-4	1,200	6,000	5,000	97	4,850	6,000	29,100	11.72	29.1
		B-5	1,200	5,500	5,000	97	4,850	6,000	26,675	11.72	26.7
		B-6	1,400	6,500	4,000	97	3,880	5,600	25,220	10.94	25.2
Three years Buri	10 × 10 × 10 1,000 m³	B-7	1,200	3,500	5,000	97	4,850	6,000	16,975	11.72	17.0
		B-8	1,200	4,000	4,500	97	4,365	5,400	17,460	10.55	17.5
		B-9	1,500	3,500	4,000	97	3,880	6,000	13,580	11.72	13.6
		B-10	1,500	4,000	3,500	97	3,395	5,250	13,580	10.25	13.6

Source: Data from Nakada et al., Nisshin Feed Co., Ltd.

stocking density of yellowtail in floating net pens is shown in Table 8. Modifications must be made based on the environmental conditions at each aquaculture site. The use of large net pens makes for high-quality meat with proper fat level (54).

Selection of the Culture Area

A good area for young yellowtail and amberjack culture should have water temperatures higher than 22 °C (72 °F) as indicated in Figure 8 (55). In nature, yellowtail migrate north and south along the coast of Japan and their body composition changes with growth and season. Fish cultured in net pens show similar tendencies to some extent. It is interesting to note that in winter feeding, activity and growth are depressed in the southeast area like Amami and Okinawa where water temperature remained above 20 °C (68 °F) (56,57). DO affects not only digestion and absorption of feeds but also fish health, with the effects increasing with density. Without sufficient DO, the fish become more susceptible to infectious diseases (58). Under the net pens, oxygen depletions may occur due to decomposition of accumulated waste materials. During autumn the oxygen-depleted layer may rise due to convection and cause oxygen depletions and associated mortalities in net pens. Eutrophication in culture areas can lead to the development of red tide phytoplankton blooms that have caused mortality at levels of thousands of tons of cultured fish annually.

FEEDS AND NUTRITION

Early yellowtail culture depended upon the locally available trash fishes as feed. In early times, only a few farmers started yellowtail culture depending solely on locally available trash fishes, and they enjoyed high profit without occurrence of severe disease and high mortality problems. As the culture technology was disseminated to various areas, demand for trash fishes exceeded production; and commercially available sardines, which were abundant and cheap, become a main feed. Supported by Government funds, freezing equipment became available in each area; and frozen sardines supported quick development of yellowtail culture thereafter. Formerly, minced frozen sardines were fed. Considerable amounts of food remained uneaten, resulting in deterioration of culture grounds. By using frozen instead of thawed fish, the deterioration problem has been alleviated considerably. However, it became apparent that use of sardines as the sole feed for yellowtail led to nutritional disorders because of unsuitable protein and energy levels. Miyazaki (59) found that the use of frozen fish did not cause any ill effects and was better than feeding thawed sardines. The use of frozen fish alleviated deterioration of the feed and reduced environmental pollution. Body components of sardine, especially fat content, changes drastically along with seasons and harvest areas as shown in Tables 9 and 10. Fat content also differs among the Pacific Ocean stock and the Japan Sea stock as illustrated in Figure 15 (60–71).

While a good system for distributing sardines was developed that made them readily available to fish farmers, there was no control over their fat content. Sato et al. (Nisshin Nasu Feed Research Laboratory, unpublished data) found a high linear correlation between moisture and fat content of sardines landed at Kyushu, Sanin, and Kushiro. The amount of water and fat in the bodies of sardines is highly negatively correlated (water and fat equation in sardine; $y = -1.138x + 90.121$, x: moisture 57–74%, y: crude fat 25.3–59%, $n = 35$, $\gamma = 0.9702$). Thus, fat content could be estimated roughly by measuring moisture content more simply than by direct fat measurement. A more simple method is available to small-scale farmers. In Table 11, a simple estimation table is provided.

In 1979, the Fisheries Agency started a large project to develop a moist pellet diet adequate for yellowtail culture to prevent pollution coming from marine

Table 9. Seasonal Change in Nutritional Components of Spotlined Sardine Caught in the Waters Off Kushiro[a]

Classification[b]	Month	Moisture (%)	Crude Fat (%)	Crude Protein (%)	Crude Ash (%)	Total Calorie kcal/kg (kJ/kg)	Calorie/ Protein Ratio (C/P)	Ca (mg%)	Na (mg%)	P (mg%)	Fe (mg%)
Large	July	56.0	26.1	13.9	4.0	2714 (11355)	195	507	764	406	3.18
	August	53.4	30.6	13.7	2.3	3065 (12824)	224	578	179	409	3.80
	September	51.3	33.0	12.7	1.3	3212 (13439)	253	557	396	347	3.09
	October	46.8	39.0	12.2	2.0	3669 (15351)	301	501	141	356	3.30
	Average	51.9	32.2	13.1	2.4	3165 (13242)	241	536	370	380	3.34
Middle	July	58.2	23.8	14.5	3.5	2557 (10698)	176	541	594	446	3.11
	August	56.8	26.5	14.3	2.4	2764 (11565)	193	580	214	458	3.48
	September	52.3	32.3	12.7	2.7	3156 (13209)	248	620	287	420	2.98
	October	45.5	39.9	12.7	1.9	3764 (15749)	296	630	130	403	2.76
	Average	53.2	30.6	13.6	2.6	3060 (12803)	226	593	306	432	3.08
Small	July	59.8	22.7	14.3	3.2	2460 (10293)	172	577	423	431	3.21
	August	56.4	27.9	13.1	2.6	2822 (11807)	215	553	339	364	3.16
	September	56.7	28.0	13.2	2.1	2834 (11857)	215	518	230	379	3.30
	October	57.8	25.4	14.4	2.5	2680 (11213)	186	464	304	362	3.62
	Average	57.7	26.0	13.8	2.6	2699 (11293)	196	528	324	384	3.32

Source: Data from Japan Aquatic Oil Association.

[a]Moisture and fat equation in sardines: $y = -1.138x + 90.121$, where x is moisture 57–74%, y is crude fat 25.3–5.9, n = 35, $\gamma = -0.9702$. From Sato et al., Nisshin Nasu Feed Laboratory.

[b]For large size, fork length is 20.0–22.0 cm, average fork length is 21.0 cm, body weight is 149–170 g, and average body weight is 162 g. For middle size, fork length is 19.5–21.0 cm, average is 20.0 cm, body weight is 116–130 g, and average is 122 g. For small size, fork length is 18.0–20.0 g, average fork length is 19.0 g, body weight is 84–122 g, and average body weight is 102 g.

Table 10. Seasonal Change in Nutritional Contents of Spotlined Sardines Caught in the Waters Off West Kyushu

Classification[a]	Month (Harvest)	Moisture (%)	Crude Fat (%)	Crude Protein (%)	Crude Ash (%)	Total Calorie kcal/kg (kJ/kg)	Calorie/ Protein Ratio (C/P)	Ca (mg%)	Na (mg%)	P (mg%)	Fe (mg%)
Large	January (inital)	67.7	13.6	15.6	2.8	1,759 (7,360)	113	629	81	485	4.55
	March (inital)	75.7	5.3	15.5	3.4	1,091 (4,565)	70	909	127	608	4.06
	March (final)	77.0	5.6	14.1	3.2	1,054 (4,410)	75	660	439.0	464.0	4.1
	April (inital)	75.6	4.9	15.2	4.2	1,046 (4,376)	69	689	660	523	4.00
	June (middle)	70.0	11.0	16.2	2.7	1,577 (6,598)	97	648	116	472	3.34
	Average	73.2	8.0	15.7	3.5	1,311 (5,485)	84	669	388	498	3.67
Middle	May middle	72.1	8.7	16.2	2.9	1,393 (5,828)	86	553	226	433	3.26
	June middle	70.0	11.2	16.3	2.4	1,597 (6,682)	98	600	122	449	3.16
	Average	71.1	10.0	16.3	2.7	1,495 (6,255)	92	577	174	441	3.21

Source: Data from Japan Aquatic Oil Association.

[a]For large size, the fork length is 19.5–25.0 cm, average fork length is 22.0 cm, body weight is 78–171 g, and average body weight is 114 g. For middle size, fork length is 17.0–20.5 cm, average fork length is 18.7 cm, body weight is 58–107 g, and average body weight is 74 g.

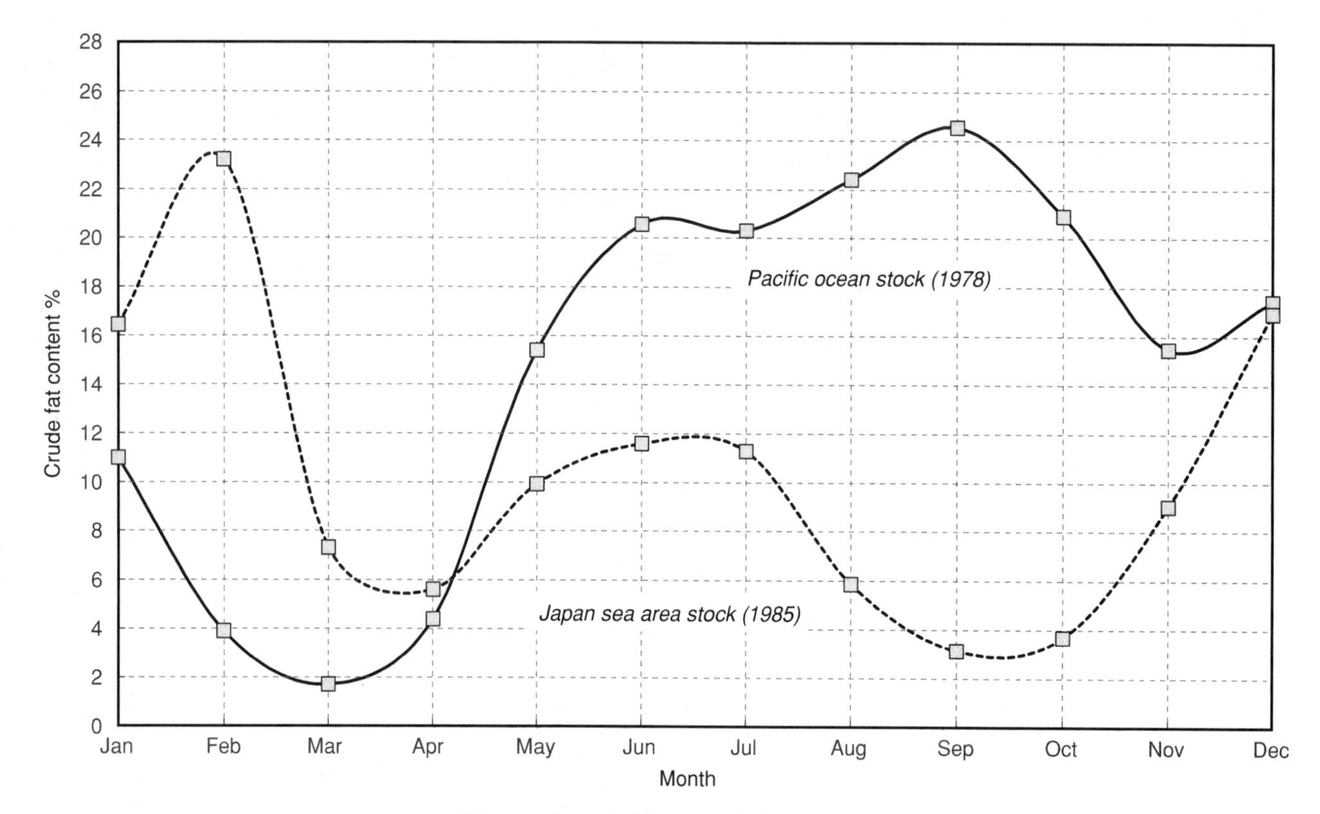

Difference in crude fat content by harvest area

Sea Area	Range of Fat Content %	Average %
East Hokkaido	19.9 to 39.9	26.8
Boso to Joban	8.8 to 22.5	14.4
Sanin	1.4 to 22.3	13
Kyushu	4.9 to 13.6	8.6

Figure 15. Seasonal change in fat content of spotlined sardine caught in different areas (data from Japan Aquatic Oil Association).

aquaculture. However, introduction of the moist pellet to yellowtail culture failed until the early 1990s, mostly due to an abundant and extremely cheap supply of domestic sardines. At that time, fish farmers finally became conscious of the severely damaged environmental conditions around their aquaculture grounds. In 1988, a new type of dry pellet diet produced by extrusion was found to be useable in yellowtail culture.

Table 11. Simple Estimation Methods for Fat Content in Sardines

Classification of Sardines	Subcuta-neous Fat	Fat in Abdominal Cavity	Dark Muscle Side Fat	Estimated Crude Fat Content (%)
Small	None	None	None	2.0
(~39 g)	None	None	Yes	3.9
	None	Yes	None	5.5
	None	Yes	Yes	7.3
	Yes	None	None	9.0
	Yes	None	Yes	12.0
	Yes	Yes	None	15.0
	Yes	Yes	Yes	18.0
Middle	None	None	None	6.5
(40 ~ 79 g)	None	None	Yes	9.8
	None	Yes	None	13.0
	None	Yes	Yes	15.3
	Yes	None	None	17.5
	Yes	None	Yes	20.3
	Yes	Yes	None	23.0
	Yes	Yes	Yes	26.0
Large	None	None	None	7.5
(80~ g)	None	None	Yes	12.0
	None	Yes	None	16.5
	None	Yes	Yes	19.5
	Yes	None	None	22.5
	Yes	None	Yes	26.3
	Yes	Yes	None	30.0
	Yes	Yes	Yes	34.0

Source: Data from Nakada, Nisshin Feed Co., Ltd.

Improvement in fish feeds has been important to the development of aquaculture (72). World production of fish meal and fish oil has fluctuated drastically so the supply of those ingredients, which are important to aquaculture, is very unstable (Fig. 16). Thus, further improvement in fish feeds, including use of alternative protein sources, has been undertaken.

The quantity of formulated feed used for yellowtail culture has increased almost linearly (Table 12). Yet more research is needed to develop dry pelleted feed and appropriate feeding techniques and to evaluate the use of inexpensive feed materials (73–75). The Fisheries Agency of Japan has been promoting development of high-quality formulated fish feeds. The Japan Fish Feed Association is involved in that effort with the help of researchers from universities and fisheries experimental stations (76). Various substitutes for fish meal can be used to decrease the amount of fish meal and fish oil in yellowtail feeds by half without adverse effects (77–79).

If fish are fed only Japanese anchovy for a long period of time, feeding activity decreases and mortality results due to vitamin B_1 deficiency. The problem can be avoided if a vitamin mix is added to the diet. Not only is vitamin B_1 added, but also vitamins C and E to prevent oxidation and fat deterioration.

Feed efficiency from minced raw fish can be improved almost twofold by adding certain binding agents; and daily feeding rate can be reduced by 20% to 30%, accompanied by a better feed conversion ratio and reduced water pollution (Table 13). The use of the right feed in the appropriate

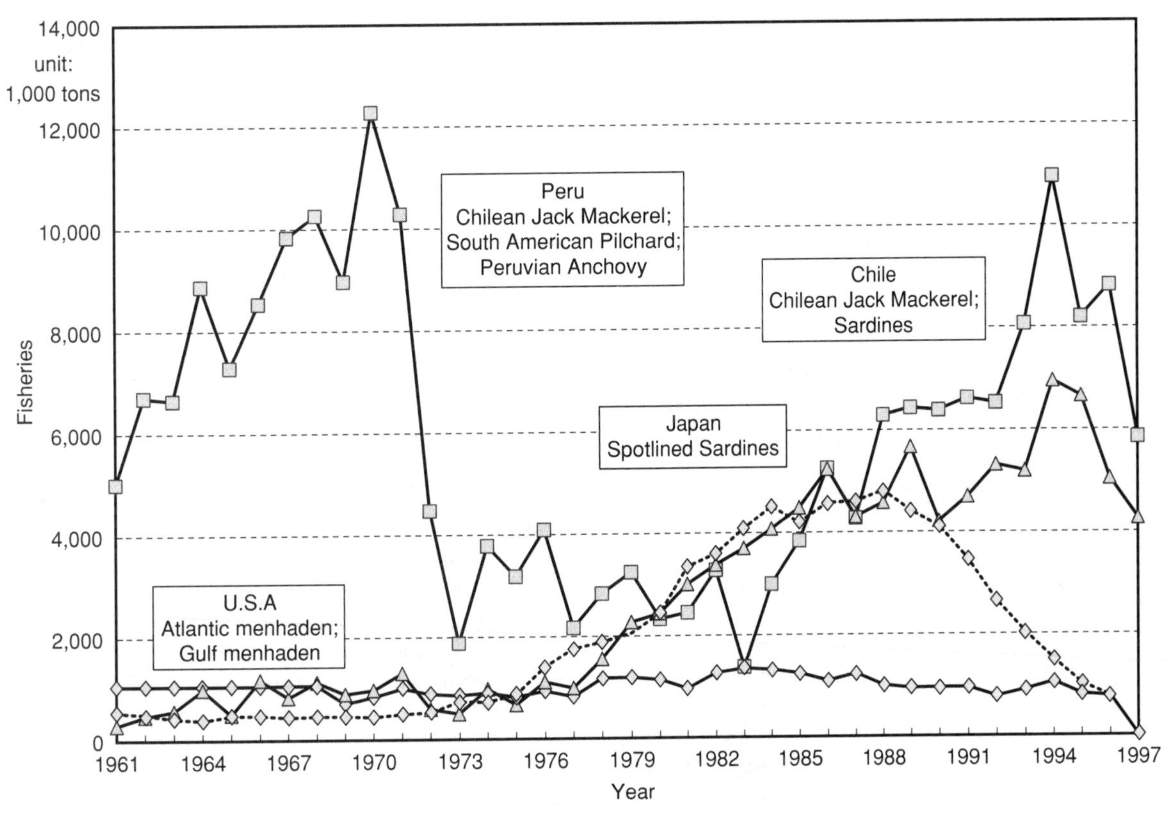

Figure 16. Main fisheries for fish oil and fish meal productions in the world (statistics fisheries data by FAO and the National Fisheries Statistics Reports).

Table 12. Production of Formulated Feed for Major Species Culture in Japan (tons)

Fish Species	1963	1973	1983	1993	1995	1996	1997	1998
Rainbow trout	6,993	29,738	26,033	23,527	23,217	20,066	21,424	19,555
Carp	3,167	51,597	32,406	24,752	26,198	20,308	20,735	18,025
Ayu-fish	211	9,006	19,950	17,192	15,548	13,058	13,215	14,043
Eel								
Powder		37,998	52,660	38,812	37,780	31,907	27,135	23,464
Solid				2,909	2,706	1,890	1,393	1,015
Total	0	37,998	52,660	41,721	40,486	33,797	28,528	24,479
Yellowtail								
Powder	2	5,448	2,385	44,210	86,449	59,273	60,889	61,527
Solid				13,789	54,282	48,977	60,063	65,207
Total	2	5,448	2,385	57,999	140,731	108,250	120,952	126,734
Red sea bream								
Powder			27,944	79,473	78,719	67,371	73,280	73,658
Solid				36,772	66,502	77,450	102,280	105,193
Total	0	0	27,944	116,245	145,221	144,821	175,560	178,851
Silver salmon				8,033	6,515	8,104	8,298	10,943
Tilapia				5,619	2,290	1,634	1,627	1,395
Shrimp				3,925	4,132	5,450	4,630	4,441
Others	80	11,029	11,225	26,036	21,777	23,892	26,975	26,909
Total	10,453	144,816	172,603	325,049	426,115	379,380	421,944	425,375

Source: Data from Japan Fish Feeds Association.

Table 13. Feeding Intake Efficiency and Feed Conversion Rate by Type of Feed

Type of Feeds for Yellowtail	Feed Intake Efficiency (%)	Feed Conversion Rate (dry base)	Growth Stage of Yellowtail
Minced raw fish	20–30	15.0–20.0 (6.1)	Mojako
	20–30	10.0–15.0 (4.4)	Hamachi
MRF + binder	40–50	7.0–10.0 (3.2)	Hamachi
Round raw fish	40–60	8.0–12.0 (3.5)	Hamachi
	40–60	6.0–9.0 (2.6)	Buri
Moist pellet 0% all fish	40–60	7.0–15.0 (3.9)	Hamachi and Buri
MP 30% artificial feed	50–70	5.0–12.0 (4.4)	
MP 50% artificial feed	60–80	4.0–8.0 (3.8)	
MP 100% artificial feed	70–90	3.5–8.5 (3.9)	
High-fat dry pellet	60–80	3.0–4.5 (3.6)	Majako
	70–80	4.5–6.5 (5.0)	Hamachi
Extrude pellet	80–90	0.8–1.2 (0.9)	Majako
	70–80	1.0–2.0 (1.4)	Hamachi
	60–70	1.8–3.5 (2.4)	Buri

Source: Data from Nakada et al., Nisshin Feed Co., Ltd.

amount is a very important factor for efficient and sustainable production. We have also recognized that this practice can improve cultured fish quality. Comparison of feed costs for yellowtail culture on various feeds can be seen in Table 14.

When raw fishes were used as the primary feed material, it is difficult to predict fish growth precisely because nutritional composition of the feed varies significantly. As information on the protein and vitamin requirements of yellowtail are developed (80), production of various types of moist pellets and formulated feeds becomes possible. Now that the production cost for yellowtail fed moist pellets or formulated feeds is less than that with raw fish, many feed makers and pharmaceutical companies have started to produce fish feeds.

If defatted and dried fish meal are used in aquaculture, fish feeds may have an insufficient lipid content in the absence of added fat (81). During the 1980s we recognized that if the ratio of formulated feed to moist pellets offered to yellowtail was increased, growth of young yellowtail deteriorated due to a decrease in protein : calorie ratio in the moist pellets. During the same period, it was common practice to provide supplemental oil in feeds used for the rearing of freshwater fishes; but the same oils could not be used for the marine species since there are significant differences in the fatty acid requirement

Table 14. Comparison of Feed Cost for Yellowtail Culture by Type of Feed

Type of Feed for Yellowtail	Growth Stage and Average Body Weight (g)	Unit Cost of Feed (yen/kg)[a,b]	Feed Conversion Rate	Production Cost (yen/kg)	Remarks
Minced raw fish sand eel	Mojako: 5–50	125	8.4	1,050	High stable quality, but high leaching
Minced raw fish spotlined sardine	Mojako: 5–200	60	16.3	978	Unstable quality and high leaching
Minced raw fish spotlined sardine	Hamachi: 50–2,000	65	12.1	787	Unstable quality and high leaching
Minced raw fish + binder	Hamachi: 50–2,000	70	6.8	476	Decreased leaching problem
Round raw fish + supplement	Hamachi: 500–2,000	60	7.6	456	Economical, but small harvest
Round raw fish + supplement	Buri: 1,000–8,000	60	8.1	486	Economical, but small harvest
Moist pellet; MP					
MP50 (50[a]: 50[b])	Mojako: 5–200	105	2.6	273	Technical difficulty in feed production
MP30 (30: 70)	Hamachi: 50–2,000	88	4.1	361	Economical and good for fish healthy
MP0 (0: 100)	Hamachi: 500–2,000	64	6.3	403	Low feed cost, but unstable quality
MP100 (100: 0)	All season for medication	140 + medicine	2.4	336 + α	Used especially for medical treatment
High fat dry pellet (C-fat more than 12%)	Mojako: 5–200	170	1.8	306	Highest performance until the end of August
High fat dry pellet (C-fat more than 12%)	Hamachi: 50–2,000	160	2.7	432	Suitable feed for least feeding rate during winter season
Extruded pellet	Mojako: 5–200	230	1.1	253	Highest growth and economical performance
	Hamachi: 50–2,000	180	1.6	288	Low feed conversion rate during winter season
	Buri: 1,000–8,000	170	2.7	459	Low feeding efficiency for large size yellowtail

Source: Data from Nakada et al., Nisshin Feed Co., Ltd.
[a]Formulated powder; 135 yen/kg.
[b]Frozen sardines; 60 yen/kg, feed oil; 170 yen/kg.

between freshwater and marine species. We developed a feed oil suitable for marine species that was tested in commercial production trials with yellowtail and produced fish similar in lipid composition to wild fish. However, the quality of oil containing high levels of HUFA vary, so the Society of Aquaculture Feed Oil Investigation has set up standards of feed oils recommended for aquaculture diets (Table 15). In the future, fish meal, which is a major protein source in formulated feeds, will be replaced by soybean meal or poultry meal and a certain amount of fish oil will be added along with soybean oil or coconut oil.

Extruded pellets formulated to contain more than 20% fat are efficiently utilized by yellowtail. By using extruded pellets, farmers have achieved feed conversion ratios (dry weight of feed offered/wet weight of body weight increase) as low as 1.2 during the production of one-year-old fish. Satisfactory growth has also been achieved during the second year using the same type of feed if temperatures

Table 15. Standard Feed Oil Recommended for Aquaculture Feed[a]

Factors	Specifications
Components:	Main fatty acids and glycerin ester
Color:	Yellow and yellowish brown
Condition:	Liquid in normal temperature
Smell:	Fishy odor, but not putrefaction smell
Acid value:	Max. 2.0
Iodine value:	130–160
Peroxide value:	Max. 10 meq/kg
Unsaponifiable matter:	Max. 5.0%
Water and other:	Max. 1.0%
Vitamin A:	~500 IU/g
Vitamin D:	~100 IU/g
Vitamin E:	~30 mg/100 g

Source: Data from Riken Products Technical Bulletin.
[a]Specification standard of the Society of Aquaculture Feed-Oil Investigation, Riken Vitamin Co., Ltd.

are optimal. After the water temperature falls in winter, feeding activity diminishes and the fish do not consume sufficient calories for growth.

Yellowtail larger than 3 kg (6.6 lb) prefer raw fish to extruded pellets; and it is difficult to attain daily feeding rates of 2% on extruded pellets, especially during winter. The author thinks development of an extruded diet that contains more than 25% fat will be required for the economical production of yellowtail larger than 3 kg (6.6 lb), especially during periods of low water temperature.

DISEASES

If it turns out to be profitable, people will copy those who have been successful. This adage is true in fish culture so that the same kind of fish is often overproduced in a given area, which can lead to outbreaks of diseases and pollution (82). The importation of wild fry, fingerling, or juvenile fish is also a source of disease (83). In Japan, 13.9 billion yen were spent on medications for use in fish culture during 1997. Kagoshima spent 2.9 billion, with lesser amounts spent in Ehime, Kochi, and Kagawa with the amount spent falling proportional to the fish farming activity in those areas (84). The types of medications purchased annually changes based on the situation and in response to the occurrence of drug resistance (85).

Disease is usually not a problem during the initial phases of rearing a particular aquaculture species. It becomes a problem when several people try to raise the same fish in floating net pens within proximity to one another. The fish will become more susceptible to disease if fed too much and maintained in overcrowded conditions. Moreover, deterioration of the environment and nutritionally deficient feeds will aggravate the situation (86).

Nishioka et al. (87) compiled information on the occurrence of disease during juvenile production of marine fishes from 1989 to 1994. Incidence was as follows: viruses, 24%; bacteria, 23.7%; mycotic granulomatosis, 14.6%; parasites, 2.4%; and unknown, 35.2%. A large portion of the unknown category can be explained by the fact that target organisms were extremely small and the techniques for juvenile production remain to be perfected.

In order to prevent fish disease, it is necessary to maintain a healthy environment and, based on experience, predict when and what kind of diseases might be anticipated under existing conditions (88). If conditions indicate the potential for a disease epizootic, stocking density can be reduced and feeding can be curtailed. We have observed increases in the level of *Enterococcus seriolicida* in the waters around fish farms before epizootics.

Recently observed mass mortalities can be attributed to several factors (89). Environmental and nutritional stress increased the need for supplemental vitamin C and vitamin E, which can be supplemented in the diet. The most serious problem in recent years has been Iridovirus infection introduced from Southeast Asia. It causes mass mortalities in yellowtail (90) and striped jack (91,92).

Records of the amount of feed consumed and numbers of mortalities should be kept to help determine when an epizootic is occurring. Books are available to assist farmers with diagnosis of specific diseases (93–95).

The most common disease in yellowtail, *E. seriolicida*, is diagnosed by simply identifying Gram-positive bacteria by using STAN agar (96). For diagnosis of Pasteurellosis, a specific iodine solution is dropped on the spleen obtained from a sick fish. Infection with Iridovirus is confirmed by presence of abnormally hypertrophic cells from the spleen.

Mortality in cultured yellowtail can be divided into four major categories: (*1*) physical damage from handling and transportation and from contact with netting during storms and strong tides, (*2*) turbidity and pollutants, (*3*) nutritional deficiencies and the feeding of oxidized raw fish and nutritionally inadequate feeds, and (*4*) diseases. Disease often occurs in response to one or more of the other four factors. An important publication on controlling fish disease is being published by Sano (97).

For treatment of bacillus diseases, the Ministry of Agriculture Forestry and Fisheries approves 25 kinds of drugs for aquatic species. Japan has a system in which any person or officials of prefectural government may be trained and obtain a license to diagnose and treat fish diseases. Several trade publications have published special issues on the subject of disease treatment, which are used by fish farmers (98). Hara described the research needed to establish a program in marine medicine (48).

The first important step for disease treatment is to remove the causative factor. In order to prevent recurrence of the same problem in the future and to identify the actual causes, detailed records should be kept when mass mortality takes place. Also, daily management records and treatments used during epizootics should be kept.

Removal of dead fish is the first step in the prevention of further spread of disease. Both sick and dead fish should be removed from affected net pens. The amount of feed consumed in net pens where disease has occurred should be kept. Sick fish will not feed as well as healthy fish, and it is usually necessary to reduce the feeding rate to 60 to 70% of normal.

Methods for mixing medication into feed depend on feed type. For dry pellets, the medication can be dispersed in water or oil in advance, and then the solution is poured over the feed. For moist pellets, the medicine should be well mixed with the other feed material and then extruded. In both cases, it is necessary to make sure the proper amount of medication is uniformly included in the final feed. Drying the surface of pellets or coating them with oil will decrease leaching of the medication. The medicated feed should be offered quickly to the entire net pen. If fed slowly, only active fish will consume the feed.

MANAGEMENT TECHNIQUE

Proper management is critical to any fish culture operation. Overfeeding results in poor quality meat and unhealthy fish. Also, overfeeding pollutes the growout area and can promote the occurrence of red tide outbreaks that can result in mass mortality. In order to escape this vicious circle, raising fish at the proper stocking density and providing them with the proper amount of feeding are critical.

Best management practices include maintaining daily records of fish health, feeding activity, and environmental quality (Table 16). The determination of water quality can be made with the assistance of the latest analytical equipment, but the final judgment has to be made by humans. On large farms, managers may be assigned to particular sectors. For instance, a manager for the feeding sector has the responsibility for determining the right amount of feed to provide in each net pen using his intuition. The overall manager has to confirm by checking the daily records that the feed manager is properly performing the feeding. The same would be true of other sectors, such as disease and water quality.

Proper feeding level relates to sea condition. Only people who watch the sea every day can properly evaluate the environmental conditions. The water quality depends on weather, tide, and wind velocity.

With the use of an underwater camera, the swimming activity of fish in a net pen can be documented and unusual behavior of individual fish observed. An accurate judgment as to whether a few diseased fish and mortalities represent the whole group of fish in a net pen is critical. If the wrong conclusion is drawn, the result could be either mass mortality or a waste of money on medication.

Table 16. Daily Record for Management, for Estimation of Fish Health, Feeding Activity, and Environmental Conditions

Period	Start	10th Day	20th Day	End of Month
Average body weight (g)	5	7.8	12.4	20.1
Number of fish	6000	5990	5980	5970
Total weight (kg)	30	46.6	74.3	119.7
Rearing dencity (kg/M^3)	0.12	0.19	0.30	0.49
Total feed used (kg)	30	50	82	162
Provisional FCR	1.8	1.8	1.8	1.8
Estimated increase (kg)	16.7	27.8	45.6	90
Number of fish lost	10	10	10	30
Weight loss	0.05	0.08	0.12	0.25
Corrected increase (kg)	16.6	27.7	45.4	89.7

1st 10 days	1	2	3	4	5	6	7	8	9	10	Total
Sea condition	○	○	○	○	○	○	○	○	○	○	○
Fish health	△	△	△	△	△	△	○	○	○	△	△
Feeding activity	–	–	△	△	○	△	△	○	○	○	△
Total check	–	–	△	△	△	△	○	○	○	○	○
Special treatment	Release		Start feed								
Feed planned to use (kg)	0	0	2	2.5	3.5	4	4	4	5	5	30
Actually fed (kg)	0	0	2	2.5	3.5	4	4	4	5	5	30
Number of fish lost	3	2	1	1	0	0	1	1	0	1	10

2nd 10 days	11	12	13	14	15	16	17	18	19	20	Total
Sea condition	○	○	○	○	△	○	△	×	×	×	△
Fish health	△	△	○	○	△	○	△	○	×	×	△
Feeding activity	△	○	○	○	○	△	△	△	×	×	△
Total check	△	○	○	○	○	△	△	×	×	×	△
Special treatment											
Feed planned to use (kg)	5	5	5	5	5	6	6	6	6	6	55
Actually fed (kg)	5	5	5	6	6	7	4	2	5	5	50
Number of fish lost	0	0	3	1	1	0	2	0	0	3	10

3rd 10 days	21	22	23	24	25	26	27	28	29	30	31	Total
Sea condition	○	○	○	○	○	○	○	○	○	○	○	○
Fish health	△	△	△	△	△	△	△	△	△	△	△	△
Feeding activity	×	×	×	×	△	△	○	○	○	○	○	×
Total check	×	×	×	×	×	○	△	△	○	○	○	×
Special treatment		Stop feed										
Feed planned to use (kg)	7	7	8	8	8	9	9	9	10	10	10	95
Actually fed (kg)	6	0	3	5	6	8	10	10	12	12	10	82
Number of fish lost	3	1	1	2	2	0	0	1	0	0	0	10

Source: Data from Nakada, Nisshin Feed Co., Ltd.

Another important practice is to check for fish that are not healthy and may be swimming around the corners of the net pen. This can be done with an underwater mirror before and after each feeding. Observations of swimming speed of individual fish while feeding, swimming activity of the group, and fish color are also important.

The feeding activity of the fish is the clearest indicator of their health. In order to determine what portion of the fish in the net pen are participating in feeding and how vigorously they feed, daily measurement of the time spent feeding a particular net pen can be used. Thus, not only total amount of feed being used but also the time spent in offering the feed to each pen should be recorded. Table 17 shows a management record of the intermediate rearing of yellowtail as an example of feeding. For estimation of the final total weight, feed conversion ratio plays a vital role.

The major reason why the growth potential of properly fed fish is not fully realized is that stocking density has exceeded the carrying capacity of the system. Technical keys to proper rearing are shown in Figure 17. An optimum density and proper feeding rate are musts for economical production. If rearing records are accumulated at a particular site for at least three years, the optimum stocking density and feeding rate for maximum growth and feed efficiency relative to season and fish size can be estimated. Table 18 shows the proper daily feeding rate calculated from the results of actual Mojako rearing data from 1986 to 1993. Figure 18 shows recommended daily feeding rate by size for yellowtail with the use of extruded pellets. Recommended calorie : protein ratio by season and size of yellowtail is shown in Table 19.

SHIPMENT OF THE PRODUCT

The strongest competitor for cultured yellowtail is not pork or beef; it is wild small Buri, 50 to 60 cm (20 to 23 in.) in body length, which are caught in set nets. If a large quantity of young Buri are landed at one time, their market price drops as low as 200 to 300 yen/kg (91 to 136 yen/lb), while 800 yen/kg (364 yen/lb) or more is a lowest price for cultured yellowtail. Moreover, amberjack, goldstriped amberjack, striped jack, and jack mackerel are becoming competitors in recent years.

The market for cultured fish can be divided largely into that for high-class Japanese restaurants that deal mainly with live fish, wholesale stores and supermarkets dealing with fresh and frozen fishes, and direct delivery of fillets processed to the individual restaurant and home

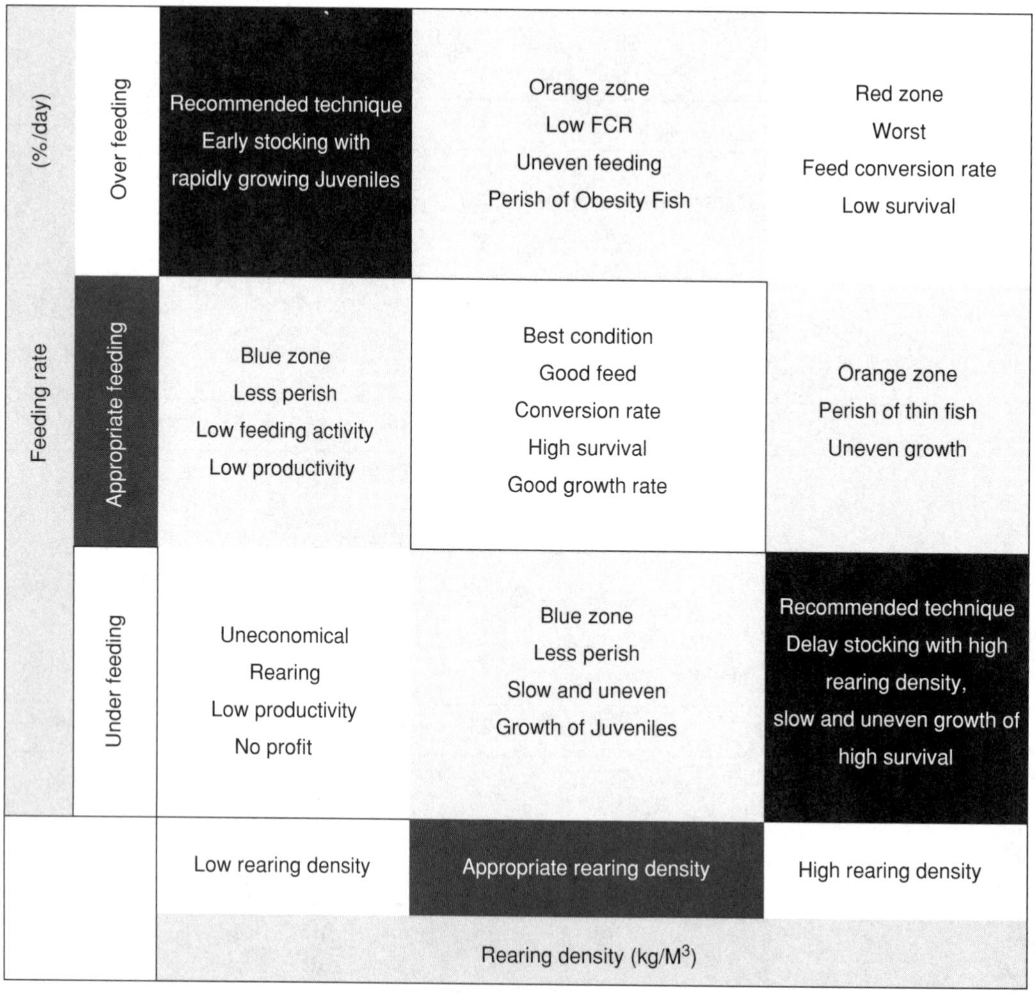

Feeding rate (%/day)	Low rearing density	Appropriate rearing density	High rearing density
Over feeding	Recommended technique Early stocking with rapidly growing Juveniles	Orange zone Low FCR Uneven feeding Perish of Obesity Fish	Red zone Worst Feed conversion rate Low survival
Appropriate feeding	Blue zone Less perish Low feeding activity Low productivity	Best condition Good feed Conversion rate High survival Good growth rate	Orange zone Perish of thin fish Uneven growth
Under feeding	Uneconomical Rearing Low productivity No profit	Blue zone Less perish Slow and uneven Growth of Juveniles	Recommended technique Delay stocking with high rearing density, slow and uneven growth of high survival
	Rearing density (kg/M^3)		

Figure 17. Technical keys for a proper rearing (data from Nakada, Nisshin Feed Co., Ltd.).

Table 17. Managemental Record of Intermediate Rearing of Yellowtail in Kagoshima, 1984

Month	Rearing Days	Feeding Days	Average Water Temperature (°C)	Average Body Weight (g)	Rearing Number of Fish in Pen	Total Weight of Fish (kg)	Average Daily Feeding Rate (%)	Amounts of Feed Used Daily (kg)	Feed Amount for 10 Days (kg)	Feed Conversion Rate	Increase in 10 Days (kg)	Number of Fish Lost	Weight Loss (kg)	True Gain (kg)
May	31	24	20.0	0.3	10,000	3	24.7	3	66	0.9	73	293	0	73
June	30	29	23.0	7.8	9,707	76	7.3	9	263	1.0	263	284	2	261
July	31	24	26.0	36	9,423	337	4.6	21	500	1.0	500	276	10	490
August	31	24	28.3	90	9,147	827	3.0	28	661	1.1	601	268	24	577
September	30	23	26.0	158	8,879	1,403	2.8	39	887	1.1	806	260	41	765
October	31	20	23.0	252	8,619	2,168	2.6	59	1,173	1.2	977	252	64	914
November	30	17	20.0	368	8,367	3,082	2.5	85	1,445	1.2	1,204	245	90	1,114
December	31	13	17.3	517	8,122	4,196	2.4	113	1,468	1.3	1,129	238	123	1,006
January	31	12	16.0	660	7,884	5,202	2.3	136	1,627	1.3	1,252	231	152	1,099
February	28	12	14.0	823	7,653	6,302	2.2	153	1,830	1.4	1,307	224	185	1,123
March	31	12	14.7	999	7,429	7,425	2.2	166	1,992	1.4	1,423	218	217	1,205
April	30	15	17.0	1,197	7,211	8,630	2.1	187	2,811	1.5	1,874	211	253	1,621
Final	365	225	245	1,464	7,000	10,251			14,722		11,409	3,000	1,161	10,248

Source: Data from Mouri and Ishizaki, Nisshin Feed Co., Ltd. Nisshin's Umisachi (extruded pellet) raising yellowtail at Shimozu Suisan Kagoshima, May, 1997 to April, 1998.

Table 18. Optimum Daily Feeding Rate Calculated from the Actual Practice of Majako Rearing (1986–1993)[a,b]

Average Body Weight (g)	Increasing Body Weight Rate		Feeding Rate (%/day)
	(In 10 days)	(%/day)	
1	1.60	6.0	7.8
2	1.59	5.9	7.7
3	1.58	5.8	7.5
4	1.57	5.7	7.4
5	1.56	5.6	6.7
6	1.55	5.5	6.6
8	1.53	5.3	6.4
10	1.51	5.1	6.1
15	1.49	4.9	5.9
20	1.46	4.6	5.5
25	1.44	4.4	5.3
30	1.41	4.1	4.9
35	1.39	3.9	4.7
40	1.37	3.7	4.4
45	1.35	3.5	3.9
50	1.34	3.8	4.2
55	1.33	3.7	4.0
60	1.31	3.4	3.8
65	1.30	3.3	3.7
70	1.29	3.2	3.5
75	1.28	3.1	3.4
80	1.27	3.0	3.3
85	1.26	2.9	3.2
90	1.25	2.8	3.1
95	1.25	2.8	3.1
100	1.24	2.7	2.9
110	1.23	2.6	2.8
120	1.22	2.4	2.7
130	1.22	2.4	2.7
140	1.21	2.3	2.6
150	1.20	2.2	2.4
160	1.20	2.2	2.4
170	1.19	2.1	2.3
180	1.19	2.1	2.3
190	1.19	2.1	2.3
200	1.18	2.0	2.2

Source: Data from Nakada et al., Nisshin Feed Co., Ltd.
[a]Feed frequency 10–9–8 days/10 days.
[b]Feed conversion rate 1.3–1.2–1.0.

(99). Fish farmers are having difficulty making a profit due to the stagnant economy and excessive competition among themselves.

Recently, direct delivery from the producer to the consumer has begun. The Internet, mail, or fax can be used to order the merchandise and reliably collect the money. Producers maintain lists of reliable customers and can estimate future demands. The consumer has recognized the difference in quality of the product, so this type of delivery system has a promising future and could help stabilize fish farming.

In order to maintain high product quality, the fish should be fasted before harvesting. The main purpose of fasting is evacuation of the ingested feeds as they may contribute to rapid deterioration in fish quality as well as lead to increased oxygen consumption and water pollution due to vomiting feed during transportation.

In order to keep product freshness as long as possible, fish should be killed immediately after being taken from the water by severing the medulla oblongata. They should be bled by cutting the caudal artery with a knife. If it is impossible to treat the fish individually they should be dumped into a tank with a large amount of chipped ice. If the moribund state is prolonged or the fish are shipped without enough chilling, rigor mortis will start earlier and reduce product quality (100).

PRODUCT QUALITY

Yellowtail were once sold strictly by weight, but consumers have become pickier about product quality, so the farmers have started to produce higher quality fish. Currently, a special brand of cultured yellowtail will fetch a higher price than the ordinary products. Having stable quality of product by discarding the second grade fish and paying special attention to maintaining freshness has become highly valued by buyers. At supermarkets and retail fish stores, sales have been expanded through the marketing of special brands produced by such organizations as the Kagawa and Kagoshima Federation of Fisheries Cooperatives.

The quality of fish deteriorates much faster than that of land animals. Therefore, it is vital when dealing with fish to get the product to consumers quickly after harvest. The fish meat can be served as sashimi for about three days in cold storage, depending on rearing conditions and treatment after harvest. Rapid killing, bleeding, filleting, and proper packaging and refrigeration can lead to excellent yellowtail (101). Amberjack and goldstriped

Table 19. Proper Calorie/Protein Ratio of Feed by Season and Size of Yellowtail

Growth Stage	Average Body Weight (g)	Season	Month	Recommended C/P Ratio, kcal (kJ)[a]	Nutrient, kcal/g (kJ/g)
Mojako	0.2–10	Catching and acclimation	April–June	60–80 (251–335)	Crude protein, 4.5 (18.8)
	10–200	Acclimation and rearing	April–August	80–90 (335–377)	Crude fat, 8.0 (33.5)
Hamachi	100–800	High temperature period	May–September	90–100 (377–418)	Carbohydrate, 2.3 (9.6)
	500–2,500	Low temperature period	September–March	100–120 (418–502)	
Yellowtail	800–5,000	Rising temperature period	March–September	120–140 (502–586)	
	3,000–	Descending temperature period	September–	140–160 (586–669)	

Source: Data from Nakada, Nisshin Feed Co., Ltd.
[a]Calorie/protein ratio should be changed by checking following items of fish being reared: condition factor; increase or decrease in abdominal cavity fat; and activity of feeding.

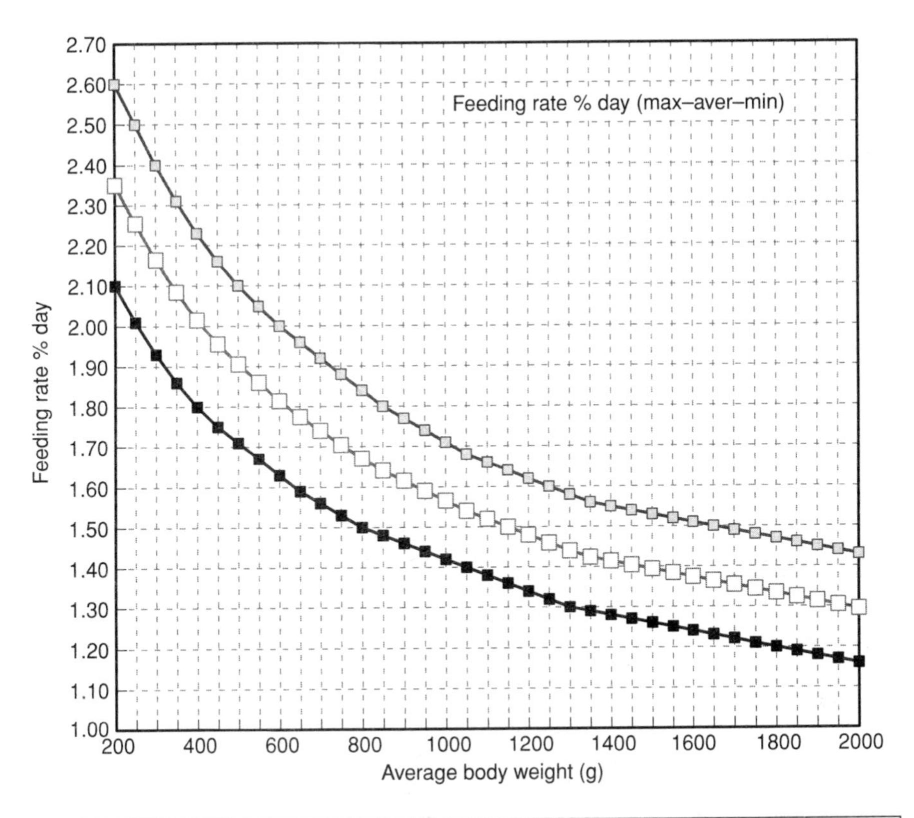

Umisachi[a] Series	Feed Size	Approximate Composition of Feed			Fish Size	
		C-Protein	C-Fat	D-Energy	Fork Length	Body Weight
No.	mm	%.	%	Kcal/kg	cm	g
1	0.62–0.92	48.0	10	3,408	under 4.5	under 2.0
2	0.92–1.41	48.0	10	3,408	4.5–5.5	under 2.0
3	1.7	48.0	13	3,732	5.5–7.0	2–5
4	2.3	48.0	13	3,732	7–10	5–13
5	3.3	47.0	16	3,871	10–16	13–55
6	4.7	47.0	16	3,871	16–23	55–15
7	6.7	47.0	16	3,871	23–28	150–300
8	9.6	44.0	20	4,028	28–35	300–800
9	13.0	44.0	20	4,028	35–44	800–1,500
10	16.5	40.0	24	4,168	44–53	1,500–2,600
11	20.0	40.0	24	4,168	53–60	2,600–3,800
12	25.0	40.0	24	4,168	60 over	3,800 over

[a]Manufactured by Nisshin Flour Milling Co., Ltd.

Figure 18. Daily feeding rate for yellowtail using extruded pellet (Umisachi; Nisshin Flour Milling Co., Ltd.) (data estimated from Field Hamach Rearing Data 1986–1993 by Nakada et al., Nisshin Feed Co., Ltd.).

amberjack are more popular than yellowtail because they can be kept for more than three days under refrigeration without losing their flavor, color, and firmness. Currently, demand for them exceeds the supply.

ECONOMICS

As the cost analysis in Figure 19 shows, the gross profit for the past 10 years has not been good compared to previous years because the total income declined due to low prices. In addition, the proportion of total expenditures dedicated to the cost of juveniles has increased. Feed cost is also increasing because of a drastic decline in the sardine resource around Japan. It is not clearly shown in the figure, but total production has changed little even though the number of fish farmers is declining. This indicates that some farmers have expanded their production.

FUTURE NEEDS

The production of better quality juveniles that show good growth rates and less vulnerability to diseases through selective breeding is an urgent need of the yellowtail culture industry. To accomplish that task, different strains

10,000 Yen	Total Income	Feed Cost	Juvenile Cost	Labor Cost	Depreciation	Miscellaneous	Gross Profit
1992	9,321	4,008	1,585	280	373	1,305	1,864
1993	9,641	3,542	1,957	186	373	1,119	2,144
1994	9,061	4,754	1,957	186	373	1,119	839
1995	8,748	5,861	1,487	262	437	1,225	(524)
1996	11,363	5,113	2,386	227	455	1,449	1,733
1997	13,500	5,670	2,700	135	405	1,485	2,835

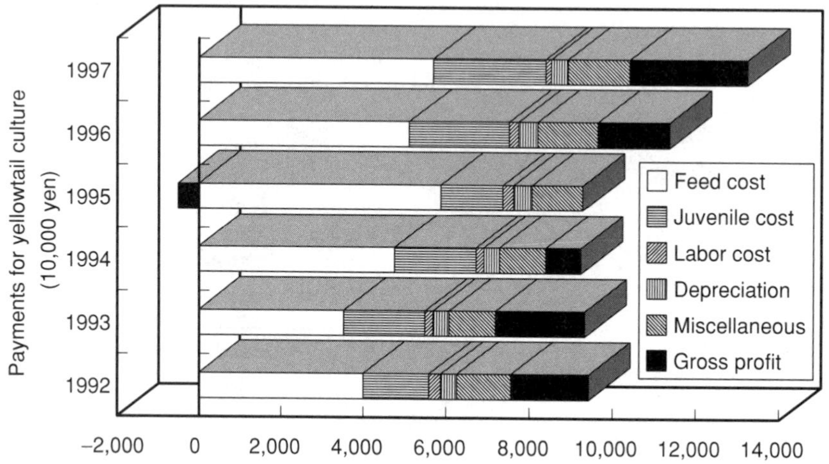

Figure 19. Cost analysis of an actual yellowtail Culturist (Statistics and Information Department, Ministry of Agriculture, Forestry and Fisheries).

of yellowtail have to be collected to select the best strain for developing the required brood stocks.

In the past, no one thought about culturing marine species on land because of the high initial cost for facilities, but it now may be a feasible approach. First of all, the fish can be raised in water, the quality of which can be controlled by humans. Under well-controlled conditions, the fish may have fewer disease problems and exposure to pollutants can be avoided. Artificial seawater systems that perform better than natural seawater systems for larval production have been developed (102). Techniques for closed systems and automatic feeding systems are improving day by day (103). Moreover, it can be predicted that people will more readily consume cultured fish if they know that the fish have not been medicated.

Finding suitable heat sources to control water temperature remains a problem. If culture of yellowtail and related species becomes possible on land without polluting coastal areas, it will be a welcome approach for producing high-quality protein to supply an ever-increasing human population (104).

The author developed a moist pellet for yellowtail 10 years ago and has also developed a formulated feed. However, we do not have proper countermeasures for declining productivity of the fish in the growout regions or for controlling disease in intensively cultured fish. For economical and sustainable fish culture, it is indispensable to maintain an optimum stocking density based on carrying capacity. It is also important to prevent the spread of infectious diseases. Hirata and his colleague's (105) proposed developing a distribution graph of DO concentrations around culture areas to aid in proper management. Recently, real time information on the dissolved oxygen and water temperature of particular areas became available through the respective Fisheries Experimental Station and Fisheries Cooperative Association.

In order to alleviate the environmental problems associated with marine fish farming, various measures, such as dredging accumulated sediment from the bottom of the sea, using chemicals to stimulate decomposition of organic materials, prohibiting the use of minced raw fish, and prohibiting the culture of large yellowtail in favor of culturing smaller, less polluting fish, are appropriate. Moreover, increasing the propagation of lugworms, which consume organic material in the mud, and cultivating algae in close proximity to fish pens should be emphasized. The algae absorb dissolved nutrients last from feed and excreted by fish. The comprehensive utilization of natural productivity may be a direction of aquaculture in the future (106).

It is time to think about a comprehensive culture approach that utilizes the natural purification ability of the environment. Such an approach may involve polyculture not only of several species of fishes but also of crustaceans and algae (107).

ACKNOWLEDGMENTS

I would like to thank the many people who are in the Aquaculture division of Nisshin Flour Milling Co., Ltd., for their support and guidance. Also, I would like to thank Dr. Takeshi Murai, Director of Coastal Fisheries and Aquaculture, Seikai Fisheries Research Institute; Dr. Toshio Takeuchi, Professor of the Tokyo University of Fisheries; Dr. Takeshi Nose, Executive Adviser of Japan Sea Water Fisheries Cultivation Association; and Dr. Robert R. Stickney of the Texas Sea Grant College Program, for taking their time revising this manuscript. and last, I would like to thank Kristi Nguyen for her help with the translation into English.

BIBLIOGRAPHY

1. M. Nakada and T. Murai, *Handbook of Mariculture*, Volume II, CRC Press, Boston, 1991.
2. Y. Izumi, *Youshoku* **34**(5), 65–67 (1997).
3. T. Ohoshima, *Meat-Color and Fat Oxidation of Yellowtail and Amberjack*, Tokyo Univ. Fisheries (unpublished data), 1998.
4. K. Itoho, *Sakananomekiki (Estimation of Fresh Fish)*, Tokuma-shoten, Tokyo, 1985.
5. K. Tachihara, M.K. El-Zibdeh, A. Ishimatsu, and M. Tagawa, *J. World Aquacult. Soc.* **28**(1), 33–44 (1997).
6. E. Kumai, *Youshoku* **32**(2), 45–47 (1997).
7. T. Tomiyama and K. Hibiya, *Fisheries in Japan Horse Mackerel and Yellowtail*, All Nippon Fisheries Photographs Data Aggregate Corporation, Tokyo, 1977.
8. T. Abe and A. Homma, *Modern Fish Dictionary*, NTS Inc. Tokyo, 1997.
9. Y. Suehiro and T. Abe, *Illustrated Book of Fishes and Shellfish*, Shogakukan, Tokyo, 1994.
10. T. Abe, *Illustrated Fishes of the World*, Hokuryu-kan, Tokyo, 1987.
11. T. Abe, *Fish Data—I. Fish Name and Illustrate*, 75th Anniversary Establishment of Nihon Suisan Co., Ltd., Tokyo, 1986.
12. T. Abe, *Fish Food Material Illustrated Data*, Tsuribito-shya, Inc. Tokyo, 1995.
13. G.R. Allen and D.R. Robertson, *Fishes of the Tropical Eastern Pacific*, University of Hawaii Press, Honolulu, 1992.
14. C. Richard Robins and G. Carleton Ray, *A Field Guide to Atlantic Coast Fishes North America*, Houghton Mifflin, Boston, 1994.
15. H.T. Boschung, Jr., J.D. Williams, D.W. Gotshall, D.K. Caldwell, M.C. Caldwell, C. Nehring, and J. Verner, *The Audubon Society Field Guide to North American Fishes, Whales, and Dolphins*, Alfred A. Knopf, New York, 1988.
16. M. Deguchi, *Sakana Arekore Jiten*, Seito-Sha, Tokyo, 1986.
17. H. Masuda, K. Amaoka, T. Araga, T. Ueno, and T. Yoshino, *Great Picture Book of Japan Fishes*, Tokai University Printed Association, Shizuoka, 1988.
18. E. Kumai, *Katsugyo-Daizen (Yellowtail)*, Fuji-Techno. Co., Ltd., Tokyo, 1990.
19. Y. Suzuki, *Youshoku* **32**(2), 202–204 (1997).
20. E. Kumai, *Katsugyo-Daizen (Hosemackerel)*, Fuji-Techno. Co., Ltd., Tokyo, 1990.
21. N. Kunisaki, *Youshoku* **24**(1), 106–109 (1987).
22. N. Tanimoto, Y. Ohoshima, S. Hanaoka, S. Inomata, and T. Sudo, *Shallow-Sea Mariculture 60-Fishes*, Taiseishuppan, Inc., Tokyo, 1965.
23. N. Tanimoto and Y. Ohoshima, *Handbook of Aquaculture*, Suisanshya, Tokyo, 1969.
24. T. Harada, *Youshoku* **23**(8), 48–51 (1986).
25. T. Harada, O. Murata, and S. Miyashita, *Kinki Daigaku Suisan Kenkyusho Hokoku* **2**, 151–158 (1984).
26. S. Takamatsu, *Youshoku* **23**(8), 52–55 (1986).
27. M. Murai, *Publication of The Metropolitan Fisheries Experiment Station*, Tokyo, No. 365, 1992.
28. K. Mushiake, T. Nakai, K. Muruga, S. Sekiya, and I. Furusawa, *Suisan-Zoshoku* **41**(3), 327–332 (1993).
29. M. Inagaki, *Suisan-Zoshoku* **38**(3), 300–314 (1990).
30. T. Senda, O. Sano, K. Ikehara, A. Nishimura, M. Ohono, H. Ida, K. Yamamoto, M. Tashiro, T. Senda, F. Hanaoka, and S. Kozima, *Kaiyoukagaku Symposim* **197**, 681–732 (1986).
31. K. Ikehara, *Nihonkaiku Regional Fisheries Research Laboratory Report*, 1984, pp. 221–232.
32. Y. Sakakura and K. Tsukamoto, *J. Fish Biol.* **48**, 16–29 (1996).
33. M. Anraku and M. Azeta, *Seikai Regional Fisheries Research Laboratory Report* **214**(35), 41–50 (1967).
34. K. Fukusho, *Suisan-Zoshoku* **44**(4), 539–546 (1996).
35. K. Kawabe, K. Kato, J. Kimura, Y. Okamura, K. Ando, M. Saito, and K. Yoshida, *Suisan-Zoshoku* **44**(2), 151–157 (1996).
36. T. Arakawa, M. Takaya, T. Kitazima, N. Yoshida, K. Yamashita, H. Yamamoto, M.S. Izquierdo, and T. Watanabe, *Bull. of Nagasaki Prefectural Institute of Fisheries* **13**, 31–37 (1987).
37. K. Tachihara, R. Ebisu, and Y. Tukashima, *Nippon Suisan Gakkaishi* **59**(9), 1479–1488 (1993).
38. K. Kawanabe, K. Kato, J. Kimura, Y. Okamura, T. Takenouchi, and K. Yoshida, *Suisan-Zoshoku* **45**(2), 201–206 (1997).
39. K. Mushiake, S. Arai, A. Matsumoto, H. Shimma, and I. Hasegawa, *Bull. Japan. Soc. Sci. Fish* **59**(10), 1721–1726 (1993).
40. Nagasaki Prefecture Fisheries Experimental Station, *Monthly Kansui* **403**, 44–44 (1998).
41. V. Verakunpiriya, K. Mushiake, K. Kawano, and T. Watanabe, *Fish. Sci.* **63**(5), 816–823 (1997).
42. V. Verakunpiriya, K. Watanabe, K. Mushiake, K. Kawano, T. Kobayashi, I. Hasegawa, V. Kiron, S. Satoho, and T. Watanabe, *Fish. Sci.* **63**(3), 433–439 (1997).
43. O. Fukuhara, T. Nakagawa, and T. Fukunaga, *Bull. Japan. Soc. Sci. Fish* **52**(12), 2091–2098 (1986).
44. K. Kawabe, M. Murai, and F. Takashima, *Suisan-Zoshoku* **44**(3), 279–283 (1996).
45. K. Mushiake and S. Sekiya, *Suisan-Zoshoku* **41**(2), 155–160 (1993).
46. K. Mushiake, H. Fujimoto, and H. Shimma, *Suisan-Zoshoku* **41**(3), 339–344 (1993).
47. K. Matsuzato, *Katsugyo-Daizen (Hamachi)*, Fuji-Techno System, Tokyo, 1990, pp. 484–489.
48. T. Ohono, *Youshoku* **32**(2), 205–209 (1997).
49. A. Sugiyama, *Youshoku* **34**(5), 77–80 (1997).
50. T. Hara, *Aquaculture and Medicines for Fish, JVPA Digest* **12**, 3–32 (1997).
51. M. Kasahara, *Youshoku* **29**(8), 212–225 (1992).

52. S. Hamamoto *Suisan-Zoshoku* **38**(3), 301–302 (1990).
53. T. Harada, *Suisan-Zoshoku* **38**(3), 304–305 (1990).
54. M. Amano, *Youshoku* **29**(8), 156–165 (1997).
55. K. Watanabe, H. Aoki, Y. Hara, Y. Ikeda, Y. Yamagata, V. Kiron, S. Satoh, and T. Watanabe, *Fish. Sci.* **65**(5), 744–752 (1998).
56. K. Date and Y. Yamamoto, *Nippon Suisan Gakkaishi* **54**(6), 1041–1047 (1988).
57. Y. Yamamoto and K. Date, *Youshoku* **37**(1), 105–107 (1989).
58. S. Kadowaki, *Suisan-Zoshoku* **37**(1), 27–33 (1989).
59. T. Miyazaki, *Youshoku* **23**(9), 101–103 (1986).
60. Japan Aquatic Oil Association, *Annual Report for Nutritional Analysis Data of Spotlined Sardine in Offshore Kushiro of the Pacific Ocean*, 1979.
61. Japan Aquatic Oil Association, *Annual Report for Nutritional Analysis Data of Spotlined Sardine in Offshore Sanin of the Japan Sea*, 1980.
62. Japan Aquatic Oil Association, *Annual Report for Nutritional Analysis Data of Spotlined Sardine in Western Sea Area of Kuushu*, 1981.
63. Japan Aquatic Oil Association, *Annual Report for Nutritional Analysis Data of Spotlined Sardine in Offshore of Choushi, Chiba-Prefecture*, 1982.
64. Japan Aquatic Oil Association, *Annual Report for Nutritional Analysis Data of Spotlined Sardine in the Pacific Ocean No. 1*, 1983.
65. Japan Aquatic Oil Association, *Annual Report for Nutritional Analysis Data of Spotlined Sardine in the Pacific Ocean No. 2*, 1984.
66. Japan Aquatic Oil Association, *Annual Report for Nutritional Analysis Data of Spotlined Sardine in the Japan Sea No. 1*, 1985.
67. Japan Aquatic Oil Association, *Annual Report for Nutritional Analysis Data of Spotlined Sardine in the Japan Sea No. 2*, 1986.
68. Japan Aquatic Oil Association, *Annual Report for Nutritional Analysis Data of Spotlined Sardine in the Japan Sea No. 1*, 1987.
69. Japan Aquatic Oil Association, *Annual Report for Nutritional Analysis Data of Spotlined Sardine in the Japan Sea No. 2*, 1988.
70. Japan Aquatic Oil Association, *Annual Report for Nutritional Analysis Data of Spotlined Sardine in the Japan Sea No. 3*, 1989.
71. Japan Aquatic Oil Association, *Annual Report for Nutritional Analysis Data of Spotlined Sardine in the Japan Sea No. 4*, 1990.
72. T. Nose, *Youshoku* **22**(3), 44–49 (1975).
73. M. Nakada, *Youshoku* **34**(5), 60–64 (1997).
74. S. Shimeno, M. Masaya, and M. Ukawa, *Nippon Suisan Gakkaishi* **63**(6), 971–976 (1997).
75. H. Nakayama, *Youshoku* **34**(5), 21–25 (1997).
76. S. Matsumoto, *Youshoku* **34**(5), 49–51 (1997).
77. T. Watanabe, *Suisan-Zoshoku* **44**(2), 227–229 (1996).
78. S. Shimeno, *Youshoku* **34**(5), 52–55 (1997).
79. S. Shimeno, M. Takeda, K. Takii, and T. Ono, *Nippon Suisan Gakkaishi* **59**(3), 507–513 (1993).
80. T. Takeuchi, Y. Shiina, T. Watanabe, S. Sekiya, and K. Imaizumi, *Nippon Suisan Gakkaishi* **58**(7), 1333–1339 (1992).
81. M. Nakada, *Youshoku* **29**(8), 38–60 (1992).
82. Y. Mizuno, *Youshoku* **31**(5), 67–71 (1994).
83. I. Zintou, *Youshoku* **31**(5), 72–74 (1994).
84. T. Kayaki, *Estimation of Market Scale of Veterinary Pharmaceuticals (1996)*, Crecon Report Veterinary Pharmaceuticals, Crecon Research & Consulting, Inc., Tokyo, 1997.
85. S. Ohoshima, *The Present Situation and the Future of Aquaculture*, Japan Veterinary Pharmaceutical Association Digest, 1996.
86. Y. Takahashi, *Youshoku* **31**(2), 44–48 (1994).
87. T. Nishioka, T. Furusawa, and Y. Mizuta, *Suisan-Zoshoku* **45**(2), 285–290 (1997).
88. Y. Mizuno, *Youshoku* **31**(2), 67–71 (1994).
89. H. Endo, K. Fujisaki, Y. Ohokubo, T. Hayashi, and E. Watanabe, *Fish. Sci.* **62**(2), 235–239 (1996).
90. M. Matsuoka, *Youshoku-Extra Issue (Fish Disease)*, Midorishobo Tokyo, 1994.
91. Y. Fukuda, *Youshoku-Extra Issue (Fish Disease)*, Midorishobo Tokyo, 1994.
92. S. Jung, T. Miyazaki, M. Miyata, Y. Danayadol, and S. Tanaka, *Fish. Sci.* **63**(5), 735–740 (1997).
93. T. Miyazaki, *Visual Diagnosis for Yellowtail Diseases*, Fujisawa Chemical Industry, Inc., Ohosaka, 1987.
94. A. Ochiai, *Atlas of Fish Anatomy*, Midorishobo, Tokyo, 1977.
95. K. Hatai, K. Ogawa, and K. Hirose, *The Picture Book for Fish Diseases*, Midorishobo, Tokyo, 1977.
96. T. Sakata and T. Kawazu, *Mem. Fac. Fish Kagoshima Univ.* **39**, 151–157 (1990).
97. T. Sano, *Apride Ichiology*, 1997.
98. S. Ikeda, *Youshoku-Extra Issue (Fish Disease)*, Midorishobo, Tokyo, 1994.
99. R. Satoho and A. Homma, *Handling of Live Fish*, Fuji Technosystem, Inc., Tokyo, 1990.
100. H. Yamanaka and K. Hirayama, *Price Formation and Quality Control of Mariculture Products, Suisangaku Series 78*, Kouseisha Koseikaku, Tokyo, 1990.
101. S. Ishida, *Youshoku* **34**(11), 52–56 (1997).
102. B. Baskerville-Bridges and L.J. Kling, *Bull. Aquacult. Assoc. Can.* **96**(3), 27–28 (1996).
103. M. Nakada, *Auto Feeding System and Formulated Fish Feed-Study of Auto Feeding Marino Forum 21*, 1997, pp. 168–172.
104. K. Kikuchi, *New Land Aquaculture System*, Brochure of Technical Bulletin of Hitachi Metals, Tokyo: 1–4, Jan. 1998.
105. H. Hirata, S. Kadowaki, and S. Ishida, *Bull. Natl. Res. Inst. Aquaculture Suppl.* **1**, 61–65 (1994).
106. H. Tsutsumi and S. Montani, *Nippon Suisan Gakkaishi* **59**(8), 1343–1347 (1993).
107. K. Hamauzu and M. Yamanaka, *Suisan-Zoshoku* **45**(3), 357–363 (1997).

See also LARVAL FEEDING—FISH.

Z

ZOOPLANKTON CULTURE

GRANVIL D. TREECE
Texas Sea Grant College Program
Bryan, Texas

OUTLINE

Plankton makes up the primary and secondary food chains in most bodies of water and is generally passively floating, or weakly swimming, minute animal or plant life. Zooplankton is considered the animal portion of plankton, whereas phytoplankton is considered the plant portion of plankton. Zooplankton generally feed upon phytoplankton. Thus, phytoplankton is referred to as the base of the food chain. Phytoplankton are considered autotrophic, because they can produce their own food from a carbon source, a simple nitrogen source, and sunlight. Heterotrophic organisms require more complex organic compounds of carbon and nitrogen to sustain life and often feed upon the autotrophes. Heterotrophic production of microorganisms, such as bacteria and protozoans, helps to provide feed for zooplankton. Zooplankton are also considered heterotrophic, and they feed upon autotrophes and other heterotrophs. Zooplankton, in turn, provides an important food source for larval fish and shrimp in natural waters and in aquaculture ponds. Zooplankton populations can be increased through fertilization management and can be harvested with plankton nets or by other means.

Zooplankton can be harvested and fed to fry, or, if fry or postlarval shrimp are stocked directly in the pond, they will prey upon the zooplankton, which thus provide an important natural food source. Although bacteria and protozoans also play important roles in the diet of larval fish and crustaceans, zooplankton is naturally their major food source until the larvae reach a certain size. The dominant zooplankton groups in ponds include Rotifera (rotifers) and Copepoda (copepods), a subclass of Crustacea; these groups remain the preferred prey for shrimp and fish. The most common zooplankters used in aquaculture are rotifers, cladocerans (water fleas), copepods, tintinnid ciliates, and, most importantly, *Artemia* (brine shrimp).

ROTIFERS AS LIVE FEED

The rotifer *Brachionus* (see Fig. 1) is one of the most important zooplankters and food organisms for the mass cultivation of larval fish in hatcheries around the world. *Brachionus plicatilis* is the most common species used. *B. plicatilis* is a euryhaline species and is extensively used as a first food for larval fish, because of its size, mobility, and nutritional value. The culture and use of *B. plicatilis* as food for larval fish were first developed and studied in Japan in the 1950s. Rotifers were investigated first because their blooms caused problems with oxygen levels in eel production ponds. Later in the 1950s, researchers in Japan discovered the use of rotifers as food for marine fish larvae, and the technology spread rapidly. In 1965, rotifers were first used to feed the commercially important Red Sea bream. In the late 1960s, rotifers were cultured on baker's yeast, but the commercial use of yeast as a culture medium did not spread until the 1970s. Since then, many different methods for culturing the rotifer have been developed and used. Presently, more than 60 species of marine finfish are cultured worldwide using the rotifer as live food. The developmental phases of rotifer technology are as follows:

- Introduction as feed
- Development of mass culture
- Evaluation and improvement of nutritional value
- Development of biological and genetic information
- Nutritional requirements
- Environmental control
- Automation

B. plicatilis is well suited to mass culture because of its life cycle. It is a planktonic filter feeder which feeds on organic particles that are brought to its mouth by the movements initiated by the corona, which is a ciliated organ on the head region that characterizes rotifers and serves as a means of locomotion.

B. plicatilis varies in size, depending on strain and culture conditions. Adult sizes range roughly from 100 to 300 μm (0.0039 to 0.01 in.) in length, including the egg mass on females. Strain selection is important, because reproduction rate, size, and optimum culture conditions (temperature and salinity) can all vary with strain. Two of the best known strains or morphotypes of *B. plicatilis* are the large (L-) and small (S-) types. The mean dry weight of the L-type is 0.33 μg/rotifer (0.000000009 oz/rotifer) and of the S-type is 0.22 μg/rotifer (0.000000006 oz/rotifer). The size range of the S-type is usually 126–172 μm (0.005–0.006 in.) in length, and the size range of the L-type is 183–233 μm (0.007–0.009 in.). Rotifers may tolerate 0–60 ppt salinity, but most production facilities indicate that 1–20 ppt is the optimum salinity range for the best growth and reproduction rates.

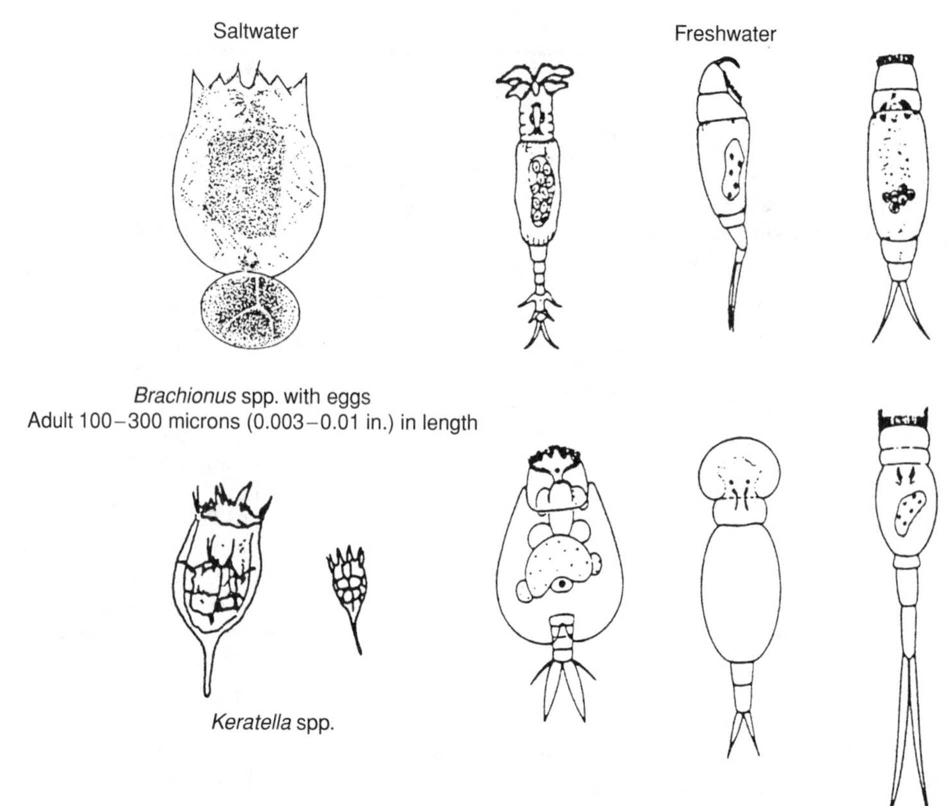

Figure 1. Rotifers. [Figure modified from (1) and (16), with permission from W.H. Freeman & Co., New York.]

There are many freshwater rotifers. The S-type rotifer is most commonly cultured outdoors. Temperature, salinity, and feed concentration all affect the rate of growth of rotifers, but temperature is the most critical factor affecting growth. The most suitable temperature range for the strains described previously is 28–32 °C (82.4–89.6 °F). Above 28 °C (82.4 °F), the salinity and size of the strain are not very critical, but the density of feed is very important. Below 28 °C (82.4 °F), the bigger strains grow faster than the smaller ones. Decreasing the salinity to 10–20 ppt increases the growth rate of both strains.

Stock rotifer cultures are used as starter cultures to initiate production in larger culture containers. The cultures should be maintained in a separate area from mass culture tanks, to prevent contamination. Stock rotifer cultures can be maintained in 1- to 2-L (0.2641 to 0.5283-gal) flasks, fed algae at 24–25 °C (75–77 °F), and placed within a light cycle of 12 hours of light followed by 12 hours of dark. The cultures are maintained in a slightly cooler environment to slow down growth and development during the stock culture period, which allows for better water quality and less maintenance. The culture should be restarted periodically (at least every month, and more often if environmental factors are poor).

Rotifers have broad nutritional requirements. They ingest many types of feed, including bacteria, as long as the size of the particle is appropriate. Rotifers also require vitamin B_{12} and vitamin A. However, the nutritional requirements of larval fish are more specific. Watanabe et al. (2) determined that highly unsaturated fatty acids (n-3 HUFAs), especially 20 : 5n-3, are essential for survival and growth of marine finfish larvae. While some species can synthesize long-chain HUFAs from short-chain HUFAs, many marine fish cannot. This is why certain feeds containing HUFAs can be as valuable as rotifer feed. Depending upon the food source, the proximate composition of rotifers consists of 52–59% protein, up to 13% fat, and 3.1% n-3 HUFAs.

There are many recognized culture methods for rotifers, both extensive (low-density culture) and intensive (high-density culture). An early method involved daily tank transfers of rotifers to fresh tanks of the same size after most of the algae was consumed. Following this procedure, batch, semicontinuous, continuous, and feedback culture techniques evolved. Each system has advantages and disadvantages. Batch culture is the most reliable method, but the least efficient. Semicontinuous culture is less reliable than batch culture, but more efficient; however, the former allows buildup of wastes, causing contamination. Continuous cultures are the most efficient and consistent, but are maintained under strictly defined conditions and are almost always "closed" and kept indoors, limiting the size and increasing the cost of operation. Another technique is the "Galveston method," in which rotifers are cultured in an open shed with unfiltered seawater and toruluse yeast as food. Rotifers are also grown in tanks that are 0.5-m (20-in.) deep and are harvested with skimmers. The feedback system, developed in Japan, uses wastes from rotifer culture (treated by bacteria and the nutrients retrieved) as fertilizer for algae cultured in a separate tank. The Japanese consider this method to be the most efficient and reliable technique.

The highest fecundity of rotifers (21 offspring/female every week) has been reported to occur when the rotifers were raised on a pure diet of *Isochrysis galbana* (Tahiti strain) at a temperature of 20–21 °C (68–69.8 °F). Lubzens (3) and Arnold and Holt (4) have described methods of culturing rotifers, including techniques using baker's yeast and emulsified oils; algae (*I. galbana*), yeast, and emulsified oil; algae as a sole nutrient source; algae and rotifers together; and outdoor culture of rotifers. Rotifers are excellent feed for larval fish because they are small, slow swimming, can be grown in high densities, and have a high rate of reproduction. But their nutritional value can vary widely, greatly affecting larval survival and growth.

Rotifers can be enriched with fatty acids and antibiotics and are used to transfer those substances to fish larvae. For example, larval red drum (*Sciaenops ocellatus*) begin feeding on the third day after hatching (earlier if temperatures are high, later if temperatures are low), with the development of their mouth parts. Rotifers are fed to fish at this time at a rate of 3 to 5 rotifers/mL (0.034 fl. oz) until the larval fish can consume larger foods. The optimum food density varies between fish species and larval stocking densities. In comparison, when they first begin to feed, larval mullets (*Mugil* spp.) require a food density of 10 rotifers/mL (0.034 fl. oz) for densities of 25 to 50 larvae/L (33.8 fl. oz). Because the nutritional value of rotifers decreases when the rotifers are held for over six hours, it is best to feed them to fish at least two times per day, or whenever the rotifer density drops below a designated number per mL (0.034 fl. oz). For example, in red drum larval culture, feeding occurs when the rotifer density drops below 3 rotifers/mL (0.034 fl. oz).

Variations in feeding techniques occur between species of larval finfish, but generally, rotifers are fed to fish larvae as soon as the larvae have developed mouth parts [for example, 2 days after hatching for the striped mullet (*Mugil cephalus*) and, as described earlier, 3 days after hatching for the red drum]. Since one fish larva can eat as many as 1,900 rotifers/day, from 13,300 to 57,000 rotifers are needed to feed one fish larva through this period (depending on the species of fish and the size of the rotifers). Most producers estimate that three times this amount of rotifers are actually eaten ($1,900 \times 3 = 5,700$ rotifers/day are fed per larva). Therefore, as many as 39,900 rotifers (for a 7-day period), or as many as 171,000 rotifers (for a 30-day period), may be required to feed one fish larva. However, too many rotifers present in the tank can cause the fish to ingest so much that assimilation becomes a problem.

In the past, red drum larvae were generally fed rotifers from day 3 posthatch to day 10 and then were fed *Artemia* (another zooplankter) nauplii from day 11 to 15. Weaning larvae from rotifers to *Artemia* was commonly practiced in the industry for better larval survival. More recently, red drum larvae have been weaned to artificial diets earlier in their life by cofeeding them rotifers and artificial diets (5). The protocols for feeding red drum larvae change with time and are moving away from the dependence of feeding live feeds.

If striped mullet (*M. cephalus*) larvae are fed rotifers, the quantities to use are 5 to 20/mL (0.034 fl. oz) starting on day 2 posthatch and continuing until day 40. For most marine finfish species that are reared indoors, weaning of fish from live rotifers and *Artemia* to dry food should begin a few days before transformation and finish by the time the fish are juveniles. The timing of this transition might be done in three days or take as long as two weeks, but should be done gradually. The size of food particles should be the largest that can easily be swallowed by the fish (one fourth to one half of the fish's mouth width). Starter feeds should contain 50–60% high-quality protein. Thus, the rotifer is an excellent starter feed, as it is high in good-quality protein.

Rotifers were fed to larval shrimp during the early developmental stages of shrimp aquaculture, but, for the most part, their feeding has been discontinued in modern intensive hatcheries. Larval shrimp are fed algae, and then *Artemia* and artificial diets in saltwater culture, whereas freshwater shrimp are started on *Artemia*. Most shrimp hatcheries have done away with the added step of rotifer culture.

Keratella is another genus of rotifer found in salt water (see Fig. 1), but it is not used indoors as commonly as *Brachionus*. Other freshwater rotifers can be seen in Figure 1 as well.

COPEPODS, CLADOCERANS, AND TINTINNID CILIATES AS LIVE FEED

Copepods are common zooplankton in both freshwater and brackish water and are a natural food for many finfish and crustacean larvae and juveniles (see Fig. 2). In the field, most marine larvae feed on copepod eggs and nauplii during the first few weeks of life (6). Only a few copepods, such as *Tigriopus japonicus*, have been mass cultured successfully, and even the technique for their culture employs the combination of rotifer culture and the use of baker's yeast or omega yeast as feed. The amount of yeast used to produce the copepod and rotifer combination outdoors is fairly high. At Florida State University, Nancy Marcus has cultured copepods in the laboratory under conditions designed to induce diapause egg production. In the last two decades, Marcus and others have shown that many coastal planktonic copepods spend a portion of their life in the sea bed as resting eggs (7). The eggs have been collected, incubated, and hatched to obtain nauplii to feed larval fish and crustaceans, but the technique for their culture has not been developed for commercial use.

Dr. G. Joan Holt at the University of Texas Marine Science Institute's Mariculture Research Center has been researching the possibility of using copepods as a feed source for marine ornamental fish and of using copepods to make a better larval diet. Results thus far are promising, but there is still much work to be done before copepods can be used for more than supplemental feeds. Culture and harvest techniques still need to be developed so that the copepods can actually substitute for currently used feed methods and be considered commercially viable. Marine organisms such as the pygmy angelfish (*Centropyge argi*), spotfin hogfish (*Bodianus puchellus*), yellowtail snapper (*Ocyurus chrysurus*), cubbyu (*Pareques umbrosus*), and cleaner shrimp (*Lysmata ambionensis*) have all been spawned in the laboratory by Dr. Holt, but larval rearing

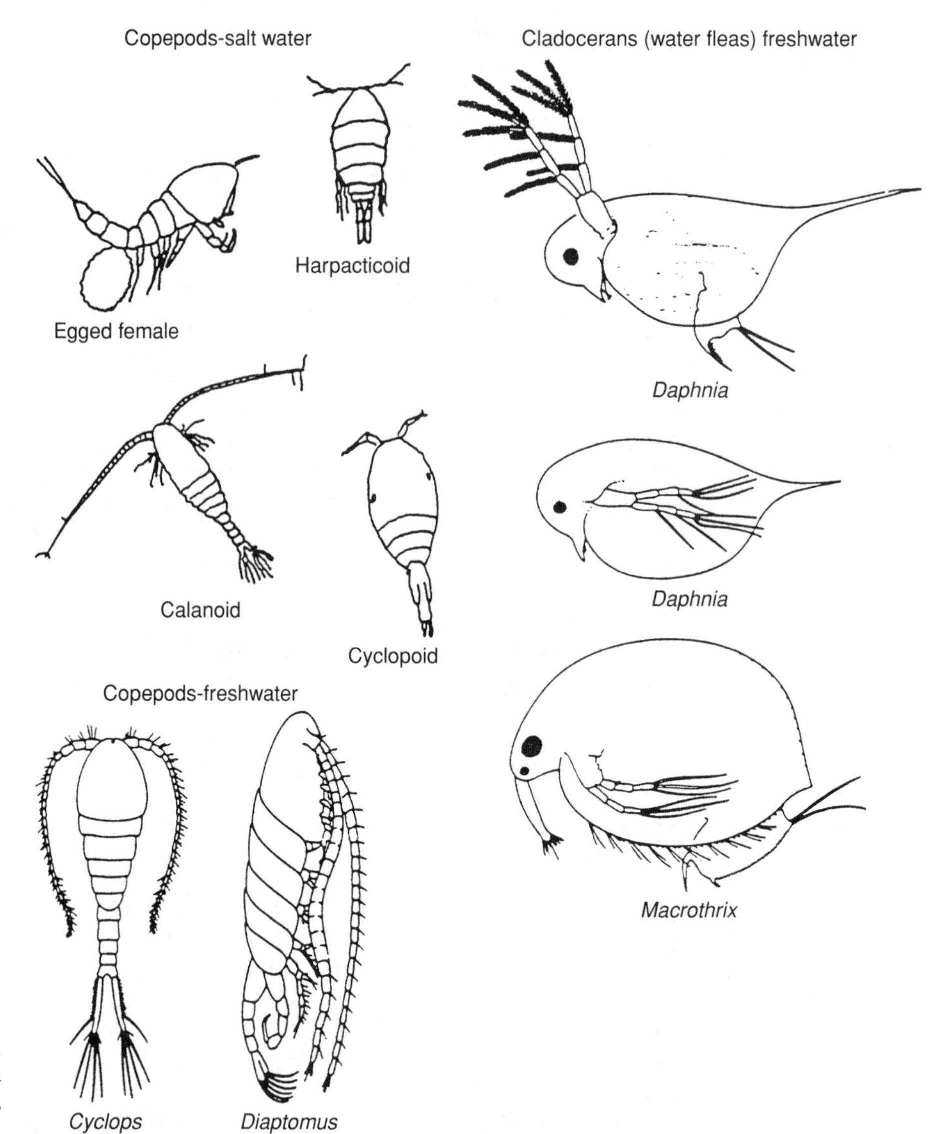

Figure 2. Copepods and cladocerans. [Figure modified from (1) and (16), with permission from W.H. Freeman & Co., New York.]

has been the bottleneck. According to Dr. Holt, easily cultured zooplankton such as rotifers and brine shrimp, along with artificial diets, have been unacceptable as the first feed for those organisms. However, copepods have been a suitable first feed, but using wild-caught zooplankton is time consuming, and species composition is not easy to reproduce.

A major hindrance to the cultivation of marine tropicals has been the inability to rear the larvae of most species of reef fish. The greatest success has been with substrate spawners, such as damselfish [*Dascyllus albisella* (Gill) and *D. aruanus* (L.)] (8). These fish have relatively large eggs and receive parental care during embryonic development, unlike most marine ornamentals, which produce tiny planktonic eggs and provide no parental care.

Mass culture of copepods has not yet been adopted for full commercial use, but it does offer future potential, because copepod eggs can be collected in large numbers and stored for months, like *Artemia* and rotifer cysts. It has been shown that photoperiod and temperature determine, in large measures, the production of copepod resting eggs and that laboratory production of those eggs is possible, but has not yet proven to be economically feasible. It is hoped that copepods, as a food source, can improve the culture of species such as red drum, by reducing the size variability and the mortality rate. Copepods have not been used extensively in aquaculture because wild zooplankton are not a reliable source, and the species that have been cultured require continuous attention. Copepods also have a reputation in aquaculture as being difficult to maintain on a continuous basis. Most of the studies conducted on copepods as potential food sources for finfish have relied on wild net-collected plankton (9). This approach has generally resulted in good growth and survival of the fish and has shown that when offered mixed-plankton diets, young turbot (*Scophthalmus maximus*) larvae consumed more copepod nauplii than rotifers and preferred copepod nauplii, due to the differences in size and swimming patterns of the two prey types.

Stottrup and Norsker (10) reported on the production and use of copepods, especially harpacticoids, in marine fish culture. They also mentioned the potential of using resting eggs of copepods, similar to the method currently practiced with *Artemia*. Other studies have relied on laboratory-cultured copepods for use as live feed. Klein Breteler (11) reared copepods in vessels as large as 100 L (26 gal) for experimental purposes and suggested that the copepods could be useful food items for mariculture. Stottrup et al. (12) described a 450 L (118 gal) system for rearing *Acartia tonsa* and reported that the system would provide 250,000 nauplii per day for fish larvae cultures. According to Watanabe et al. (2), the Japanese have routinely cultured the copepods *Tigriopus* and *Acartia* for rearing fish larvae approximately 7 mm (0.28 in.) in length. Kraul et al. (13,14) compared the growth and biochemical composition of mahimahi (*Coryphaena hippurus*) larvae that were cultured in 700 L (185 gal) tanks and fed brine shrimp, rotifers, and the copepod *Euterpina acutifrons*. Larvae fed copepods survived better under stressful conditions. More recently, Sun and Fleeger (15) have described a system for the mass culture of a benthic marine harpacticoid copepod, which, they suggested, would be useful for aquaculture.

Herbivorous copepods are primarily filter feeders and typically feed on very small particles. But they have the ability to feed upon larger particles, which gives them an advantage over rotifers. Copepods can also eat detritus, called "marine snow" (16). Copepods differ from *Artemia* (brine shrimp) and rotifers in that they lack the ability to reproduce asexually. Copepods mate sexually after maturing, and the female produces between 250 to 750 fertilized eggs, depending on the species and the size of the female. Unlike rotifers, which have a small brood size of 15–25 per female and exhibit rapid development, with life spans of 5 to 12 days, the life span of a copepod can range from 40 to 50 days and has a longer generation time (1 to 3 days for rotifers vs. 7 to 12 days for copepods).

A cylindrical shape characterizes copepods, with a trunk composed of 10 segments, consisting of head, thorax, and abdomen. Adult copepods have body sizes ranging from 0.5 to 5.0 mm (0.01 to 0.19 in.) (16). The larval stages consist of six naupliar and 6 copepodite stages. The main suborders of copepods found in brackish water ponds are calanoids (*Acartia*, *Calanus*, and *Pseudocalanus* spp.), harpacticoids (*Tisbe* and *Tigriopus* spp.), and cyclopoids (see Fig. 2 for shape differences between these copepods).

Other copepods considered to be promising species for mass culture are as follows: *Acartia clausi*, *A. longiremis*, *Eurytemora pacifica*, *E. acutifrons*, *Oithona brevicornis*, *O. similis*, *Pseudodiaptomus inopinus*, *P. marinus*, *Microsetella norvegica*, and *Sinocalanus tenellus*.

Cladocerans, or water fleas (see Fig. 2), such as *Daphnia magna*, have been cultured as live food. Pennak (17) has discussed the life history of cladocerans. Other cladocerans considered to be promising species for mass culture are *Evandne tergestina*, *Penilia avirostris*, and *Podon polyphemoides*. The cladoceran *Moina macrocopa* has been used in Southeast Asia as feed for sea bass fry immediately after weaning from *Artemia* sp., and prior to feeding minced fish flesh. During this period, sea bass, being a catadromous species (i.e., a species that moves into freshwater for a portion of its life cycle), may be reared at lower salinities, to allow feeding of freshwater zooplankton; but this practice is not commonly used and has not proven to be viable on a commercial scale. Many laboratories use *Daphnia* as the invertebrate of choice for conducting toxicity tests, because it is easy to culture and maintain in the laboratory. Cladocerans are mainly freshwater zooplankters, do not tolerate salinities higher than 3 ppt, and are generally not found in brackish water ponds (16).

Tintinnid ciliates, which are consumed by larval fish and crustaceans in the wild, have also been considered to be promising candidates for mass production. However, since the technology for mass production of rotifers is well established and microparticulated diets are being cofed with rotifers or have been developed to partially substitute for live food, the role of copepods, cladocerans, and tintinnid ciliates in aquaculture is not as important as that of rotifers.

ARTEMIA AS LIVE FEED

Artemia (brine shrimp) are probably the most important and most widely used zooplankter in aquaculture. Brine shrimp eggs, or cysts, are easily purchased and hatch readily when placed in seawater overnight. Before the eggs are placed in seawater, they are generally exposed to active household bleach for a specified period of time. This process is called *decapsulation*. Decapsulation disinfects and softens the shell and makes it easier for the nauplii to emerge from the eggs. Once brine shrimp nauplii have hatched, they can easily be separated from the shells and other debris and are then fed to the fish or crustacean larvae being cultured. The ease of feeding *Artemia* nauplii to cultured animals and the superior nutritional value of *Artemia* ensure that brine shrimp will be used in hatcheries for many years to come. However, artificial diets are slowly being developed that may substitute for *Artemia* in the future.

MAINTAINING ZOOPLANKTON IN TANKS AND PONDS

Zooplankton blooms can be managed indoors and outdoors in tanks and ponds by fertilizing the water with organic products, such as manure, cottonseed cake or meal, fatted soybean meal, and so on. It is important not to remove the fat in the soybean meal, in order to have proper results with zooplankton blooms in ponds. The fertilizer acts to stimulate phytoplankton growth and density, which are in turn fed upon by zooplankton. Zooplankton also feed upon the organic matter of the fertilizer itself. Survival and adequate growth of cultured finfish and crustaceans in tanks or ponds are strongly dependent on the availability of suitable live food organisms of the proper size and in sufficient numbers. The presence of zooplankton ensures the initiation of feeding by larvae or fry and subsequent growth and survival during the rearing period of the juvenile crustacean or fingerling stage of finfish.

In ponds, live zooplankton can be caught with a plankton net (18), pulled from the bank or towed from a boat

(19) or harvested by using a low-volume impeller pump that concentrates the zooplankton in a net (20). All of these methods are inefficient and labor intensive, and the impeller pump mutilates some zooplankton (21). Although large-scale harvest of zooplankton from marine environments has been successful (22), the bulky harvesting apparatus may not be suitable for use in small hatchery ponds. Graves and Morrow (21) modified a Fresh-Flo propeller-lift pump (model MD, 1/20 hp, Fresh-Flo Corp., Cascade, WI) originally designed for aeration to make a zooplankton harvester (see Fig. 3). After modifications, the pumping capacity was 178 L/min (47 gal/min). Pumped water was discharged into a 0.91 m × 0.45 m × 0.45 m (36 in. × 18 in. × 18 in.) floating Saran basket with 0.5-mm (0.02 in.) mess apertures. Harvesting zooplankton at night with a light was the most effective collection method (21), and only the size of the catch basket appeared to limit the amount harvested. Harvest periods longer than six hours resulted in clogged basket meshes and losses of zooplankton due to overflows from the catch basket. With one catch basket, up to 2.7 kg (6 lbs) of live zooplankton can be harvested at night from fertilized ponds. A larger mesh of 2 mm (0.08 in.) placed in front of the smaller meshed nets allowed passage of most zooplankton, but retained undesirable aquatic insects. The system provides zooplankton for feeding and for inoculation.

A zooplankton tube sampler (see Fig. 4) (23) or a plankton net can be used to assess zooplankton populations in ponds. The tube sampler is used by rapidly lowering the vertical tube to the desired depth. The check valve is self-operating, opening when the sampler is lowered and closing immediately when the sampler is raised, trapping a known amount of water. The sample is released by lifting the check valve with a finger. A sample can be concentrated by straining it through netting of the appropriate mesh size.

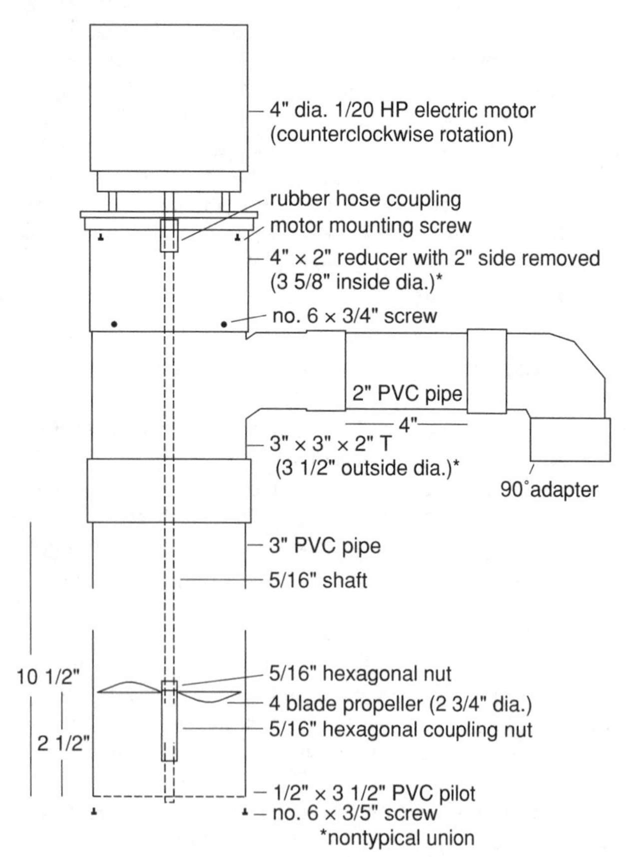

Figure 3. Diagram of modified propeller-lift pump for harvesting zooplankton. From (21), with permission from *The Progressive Fish-Culturist*. All measurements are in inches.

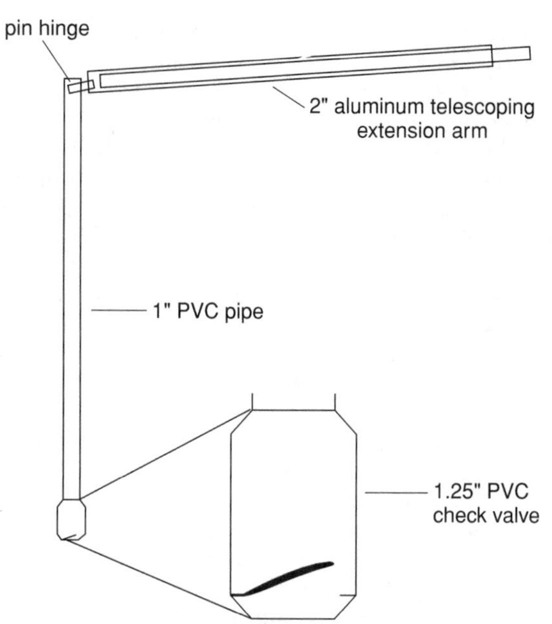

Figure 4. Zooplankton tube sampler. From (23), with permission from *The Progressive Fish-Culturist*. All measurements are in inches.

Zooplankton may be sampled from a pond with a Wisconsin-style plankton net, which may be obtained from Wildco Supply Company, Saginaw, MI. This sampler is available with either 80 μm (0.0031 in.) or 153 μm (0.006 in.) nitex netting. The smaller mesh is recommended for retaining rotifers and copepods. Plankton net-sampling methods have been standardized by the American Public Health Association (APHA), and oblique tows, in which the net is lowered to some predetermined depth in the pond and raised at a constant speed for a known distance, allow the sampler to estimate the amount of zooplankton in a pond.

As in microalgae production, indoor zooplankton production costs are high. The ratio of rotifer biomass to target species is generally 3 : 1. The estimated cost of rotifer mass production using large-scale batch methods is US$4.50/1 million rotifers. Of these production costs, 72% is for feed (50% for live algae, 22% for yeast). Continuous cultures using chemostats offer future promise for improving indoor rotifer mass culture economics, but at the present time, these systems are still very expensive to operate, remain in the research and development phase, and are not commercially viable. As with most indoor-versus outdoor comparisons, the outdoor production of zooplankton is less expensive, but, for reasons associated with climate, may not always be possible.

POND MANAGEMENT FOR ZOOPLANKTON

Management of ponds for zooplankton may require fertilization, liming, chemical treatment, and inoculation. Most healthy water sources have an abundance of zooplankton, but the ability of water to support the zooplankton depends on the availability of nutrients for the food chain. The water source generally has a variety of zooplankters. The types of zooplankton in a pond can be controlled with the use of chemicals. For example, low concentrations of organic phosphorus acid esters kill cladocerans, but do not affect rotifer populations (24). Most ponds are fertilized with either organic or inorganic fertilizers or a combination of both. Good organic fertilizers are alfalfa, cottonseed meal, fatted soybean meal, and manure. Combining the two types of fertilizers (i.e., organic and inorganic) promotes a diverse autotrophic and heterotrophic microbial community, necessary to improve zooplankton abundance. Phytoplankton, the base of the food chain, require inorganic nutrients, carbon dioxide, water, and sunlight to produce their own food. The key nutrient in regulating phytoplankton or autotrophic production in freshwater is phosphorus, whereas nitrogen is generally considered the limiting nutrient in brackish water. Inorganic fertilizers promote phytoplankton growth, and organic fertilizers and their decomposition promote the heterotrophic production of microorganisms, such as bacteria and protozoans, that help feed the zooplankton.

There are numerous fertilization regimes that have been used by pond managers for years, but the most successful ones generally provide a way for the fertilizer to solubilize quickly, disburse rapidly and evenly, and generally have a low carbon-to-nitrogen ratio or, in other words, a 10–20 : 1 N : P ratio. Other information on fertilization and zooplankton can be found in (25–36). Additional information is also available on pond management (37,38), fertilization (39), liming (40), chemical treatments (41), inoculation (42), and zooplankton identification (43,44).

BIBLIOGRAPHY

1. P. Abramoff and R.G. Thomson, *Laboratory Outlines in Biology IV*, formerly *Separate No. 846, Light Microscopy*, W.H. Freeman and Co., New York, 1986, p. 15.
2. T. Watanabe, C. Kitajima, and S. Fujita, *Aquaculture* **34**, 115–143 (1983).
3. Lubzens, *Hydrobiologia* **147**, 245–255 (1987).
4. C.R. Arnold and G.J. Holt, in W. Fulks and K.L. Main, eds., *Rotifer and Microalgae Culture Systems*, The Oceanic Institute, Honolulu, HI, 1991, pp. 119–124.
5. G.J. Holt, *J. World Aquacult. Soc.* **24**, 225–230 (1993).
6. E.D. Houde, *Bull. Mar. Sci.* **28**, 395–411 (1978).
7. N.H. Marcus, *Hydrobiologia* **320**, 141–152 (1996).
8. B.S. Danilowicz and C.L. Brown, *Aquaculture* **106**, 141–149 (1992).
9. T. Naess, M. Germain-Henry, and K.E. Naas, *Aquaculture* **130**, 235–250 (1995).
10. G. Stottrup and N.H. Norsker, *Aquaculture* **155**, 231–247 (1997).
11. W.C.M. Klein Breteler, *Mar. Ecol. Prog. Ser.* **2**, 229–233 (1980).
12. J.G. Stottrup, K. Richardson, E. Kirkegaard, and N.J. Pihl, *Aquaculture* **52**, 87–96 (1986).
13. S. Kraul, A. Nelson, K. Brittain, H. Ako, and A. Ogasawara, *J. World Aquacult. Soc.* **23**, 299–306 (1992).
14. S. Kraul, K. Brittain, R. Cantrell, T. Nagao, H. Ako, A. Ogasawara, and H. Kitagawa, *J. World Aquacult. Soc.* **24**, 186–193 (1993).
15. B. Sun and J.W. Fleeger, *Aquaculture* **136**, 313–321 (1995).
16. L.N. Sturmer, in G.W. Chamberlain, R.J. Miget, and M.G. Haby, eds., *Red Drum Aquaculture*, Texas A&M University, Sea Grant College Program, College Station, TX, publication TAMU-SG-90-603, 1990, pp. 80–90.
17. R.W. Pennak, *Freshwater Invertebrates of the United States*, The Ronald Press Company, New York, 1953.
18. W. Gale, *Prog. Fish-Cult.* **45**, 102 (1983).
19. K.E. Norman, J.B. Blakely, and K.K. Chew, *Proc. World Maricult. Soc.* **10**, 116–122 (1979).
20. B.W. Farquhar and J.G. Geiger, *Prog. Fish-Cult.* **46**, 209–211 (1984).
21. K.G. Graves and J.C. Morrow, *Prog. Fish-Cult.* **50**, 184–186 (1988).
22. P.L. Shafland, J.W. Davis, and D.H. Kuel, *Prog. Fish-Cult.* **41**, 204–205 (1979).
23. K.G. Graves and J.C. Morrow, *Prog. Fish-Cult.* **50**, 182–183 (1988).
24. Tamas, in E. Styczynska-Jurewicz, T. Backiel, E. Jaspers, and G. Persoone, eds., *Cultivation of Fish Fry and Its Live Food*, European Mariculture Society, publication No. 4, 1979, pp. 199–209.
25. R.D. Barnes, *Invertebrate Zoology*, W.B. Saunders Co., Philadelphia, PA, 1968.
26. G.E. Hutchinson, *A Treatise on Limnology: Vol. I: Geography, Physics, and Chemistry*, John Wiley & Sons, New York, 1957.
27. G.E. Newell and R.C. Newell, *Marine Phytoplankton: A Practical Guide*, Hutchinson & Co., London, 1977.
28. R.W. Pennak, *Freshwater Invertebrates of the United States*, John Wiley & Sons, New York, 1978.
29. J.G. Geiger, *Aquaculture* **35**, 331–351 (1983).
30. J.G. Geiger, *Aquaculture* **35**, 353–369 (1983).
31. W.T. Edmondson, *Ecol. Monogr.* **35**, 61–111 (1965).
32. K. Walker, *Hydrobiologia* **81**, 159–167 (1981).
33. S.M. Marshall and A.P. Orr, *The Biology of a Marine Copepod*, Calanus finmarchicus (*Gunnerus*), Oliver & Boyd, Edinburgh, Scotland, 1955.
34. J.H. Wickstead, *Proc. Zool. Soc. London* **139**, 545–555 (1962).
35. D.T. Gauld, in H. Barnes, ed., *Some Contemporary Studies in Marine Science*, Allen & Unwin, London, 1966, pp. 313–334.
36. W.T. Edmondson, H.B. Ward, and G.C. Whipple, eds., *Freshwater Biology*, John Wiley & Sons, New York, 1959.
37. C.E. Boyd, *Water Quality in Warmwater Fish Ponds*, Auburn University Agricultural Experiment Station, Auburn, AL, 1979.
38. H.S. Egna and C.E. Boyd, *Dynamics of Pond Aquaculture*, CRC Press, Boca Raton, FL, 1997.
39. R.L. Colura, in G.W. Chamberlain, R.J. Miget, and M.G. Haby, eds., *Red Drum Aquaculture*, Texas A&M University, Sea Grant College Program, College Station, TX, publication TAMU-SG-90-603, 1990, pp. 78–79.

40. J.R. Villalon, *Practical Manual for the Semi-intensive Commercial Production of Marine Shrimp*, Texas A&M University, Sea Grant College Program, College Station, TX, publication TAMU-SG-91-501, 1991.
41. C.S. Tucker and C.E. Boyd, *Trans. Am. Fish. Soc.* **107**(2), 316–320 (1978).
42. S.G. Lawrence, ed., *Manual for the Culture of Selected Freshwater Invertebrates*, Can. Spec. Publ. Fish. Aquat. Sci., No. 54, 1981.
43. L.H. Hyman, *The Invertebrates: Acanthociphala, Aschelminthes and Ectoprocts*, Vol. III, McGraw-Hill, New York, 1951.
44. C. Davis, *The Marine and Freshwater Plankton*, Michigan State University Press, East Lansing, MI, 1955.

See also BRINE SHRIMP CULTURE; LARVAL FEEDING—FISH; SHRIMP CULTURE.

INDEX

Page references in **bold** type indicate main articles. Page references followed by italic *t* indicate material in tables. If a main article appears as a subheading, there may be additional index terms at the main heading location of that topic. Countless disease agents and diseases are discussed in the context of specific cultured species. A complete list of every species covered in a main article is contained in the index entry "Aquaculture (of specific organisms)".